THE VNR CONCISE ENCYCLOPEDIA OF MATHEMATICS

SECOND EDITION

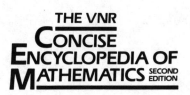

THE VNR CONCISE ENCYCLOPEDIA OF MATHEMATICS

SECOND EDITION

W. Gellert · S. Gottwald
M. Hellwich · H. Kästner · H. Küstner
Editors

K. A. Hirsch · H. Reichardt
Scientific Advisors

VNR VAN NOSTRAND REINHOLD
New York

Mathematics at a Glance
First American Edition 1977
Second American Edition 1989

Library of Congress Catalog Card Number 88-26992
ISBN 0-442-20590-2

Made in the German Democratic Republic.

Published by Van Nostrand Reinhold
115 Fifth Avenue
New York, New York 10003

Van Nostrand Reinhold International Company Limited
11 New Fetter Lane
London EC4P 4EE, England

Van Nostrand Reinhold
480 La Trobe Street
Melbourne, Victoria 3000, Australia

Macmillan of Canada
Division of Canada Publishing Corporation
164 Commander Boulevard
Agincourt, Ontario MIS 3C7, Canada

16 15 14 13 12 11 10 9 8 7 6 5 4 3 2 1

Library of Congress Cataloging-in-Publication Data

Main entry under title:

The VNR concise encyclopedia of mathematics.

First published under title: Mathematics at a glance.
Includes index.
I. Mathematics-Handbooks, manuals, etc. I. Gottwald, S.
II. Van Nostrand Reinhold Company.

QA40.VI8 1989 510–dc19 88-26992
ISBN 0-442-20590-2

Contents

Preface

It is commonplace that in our time science and technology cannot be mastered without the tools of mathematics; but the same applies to an ever growing extent to many domains of everyday life, not least owing to the spread of cybernetic methods and arguments. As a consequence, there is a wide demand for a survey of the results of mathematics, for an unconventional approach that would also make it possible to fill gaps in one's knowledge. We do not think that a mere juxtaposition of theorems or a collection of formulae would be suitable for this purpose, because this would overemphasize the symbolic language of signs and letters rather than the mathematical idea, the only thing that really matters. Our task was to describe mathematical interrelations as briefly and precisely as possible. In view of the overwhelming amount of material it goes without saying that we did not just compile details from the numerous text-books for individual branches: what we were aiming at is to smooth out the access to the specialist literature for as many readers as possible. Since well over 700000 copies of the German edition of this book have been sold, we hope to have achieved our difficult goal.

Colours are used extensively to help the reader. Important definitions and groups of formulae are on a *yellow* background, examples on *blue*, and theorems on *red*. The course of more complicated calculations is indicated by red arrows. Also, in the illustrations in the text colours show up the essential features. Ample examples help to make general statements understandable. Frequently the numerical calculations have been arranged separately so that a problem can be read as an explanatory text, without reference to calculations, while the latter can be regarded as worked examples with explicit details. Physical units, which occur in some examples, are given in the SI-system, which is coming more and more into legal and practical use. Everyday examples are given in everyday units, both metric and others.

A systematic subdivision of the material, many brief section headings, and tables are meant to provide the reader with quick and reliable orientation. The detailed index to the book gives an easy access to specific questions.

In the plates at the end numerous photographs and colour plates help to make the material more vivid and provide interesting glimpses of the history of mathematics.

We thank the authors of the various chapters, specially to acceding to our request for generally understandable diction even at the risk of deviating from the usual terminology. Above all in the brief reports on special topics many an author has found it difficult to be content with mere indications about a topic in which he is an expert.

Our particular thanks are due to our advisors, Professor K. A. Hirsch, Queen Mary College, University of London, and Professor H. Reichardt, Section for Mathematics, Humboldt University of Berlin. They have worked untiringly for the improvement of the book and have helped to create a work which is a reliable source of information for every user and should convince everyone that mathematics is essentially a simple and learnable discipline.

The Editors and the Publishers

Plates

Index of mathematicians

Introduction

The great achievements of technology in all its forms, which deeply influence the life of every human being, have led to a widespread recognition of the importance of mathematics: everybody knows, or at least believes, that without mathematics these achievements in their entirety could not have come about. Interest in mathematics has therefore grown steadily, and with it the need for information about this science.

Now in many respects mathematics is an exceptional science, in particular, as regards the presentation of its problems and results. While in medicine, zoology, botany, geography and geology, or in languages, history, astronomy, a scholar, fully equipped with the knowledge of his time, can explain to a layman the majority of his problems and results, perhaps even his methods or the fundamental principles of his special interests, in such a way that he succeeds in conveying an impression of the contents of this field, in present-day chemistry and physics this is far more difficult – and in mathematics well-nigh impossible. Not only has the volume of results grown phenomenally, but the problems are so difficult to treat and lie so deep that even mathematicians can have no more than a superficial view of the whole of mathematics.

One tries to counteract the fragmentation of mathematics into many *special branches* by extracting as far as possible from various domains common features, which sometimes do not lie at all close to the surface, and by creating from them a new and even more abstract theory: in just this way new links are forged between at first sight widely diverging directions. This process can be regarded as a repeated abstraction: whereas the basic disciplines such as algebra and geometry have their origin in abstractions from everyday experience, one arrives at such a unifying theory by further abstractions, for example, from algebra and geometry: and under certain circumstances such abstracting processes can be repeatedly piled on top of one another. Here '*abstract*' has to be understood in the literal meaning of the word as 'removing', as leaving aside everything inessential for the context in question or for a particular purpose; for example, ignoring colour in geometric figures, which may very well play a role in ornaments.

From all this it follows that it is quite impossible to give a layman even a glimpse of the whole of contemporary mathematics. Here a *layman* is not only one whose knowledge is limited to the normal contents of a school syllabus. Even a mathematician with a diploma or a B. Sc., even a teacher of mathematics, has to be regarded as a layman in many special branches. It is simply impossible to acquire specialized knowledge of all branches of mathematics in three or four years of study. Therefore this book cannot have the ambition of imparting knowledge in all special fields of mathematics – restriction is essential.

In its historical development mathematics first proceeded in quite a naive manner. It started out from the *numbers* 1, 2, 3, ... and from the intuitively obvious *figures* of geometry such as points, segments, lines, planes in space, angles, triangles, circles, etc.; gradually it ascended to more complex formations, with the realm of numbers and that of figures not developing as separate entities, but connected through the notion of *measuring*. It was in this development, progressing from the intuitively simple and obvious to more complicated problems, that mathematics was built up, for example, in Babylonia and Egypt; astonishing achievements were reached in astronomy, such as the prediction of lunar eclipses. But it was the Greeks who lifted mathematics to a completely new level of development when they felt compelled not always to forge ahead, but also to reflect: what is it that one does in pursuing mathematics? The result was that through them mathematics became a sience in the present-day sense. On the one hand, they recognized that a *proof* consists in reducing a mathematical proposition to other known facts by the simplest logical conclusions, supported and made convincing sufficiently often by evidence or experience. On the other hand, they realized that such a reduction process cannot go on indefinitely but only as far as certain simplest properties of numbers or figures, which appear secure by virtue of intuition or experience.

In this way they compiled for the first time consciously a system of *fundamental facts*, for example, that there is precisely one straight line passing through two points, and they created the foundation of *logic*. Together these two features lead to a systematic build-up of geometry, rising from the simple to the complex.

For a long time this Euclidean geometry, apart from a few minor supplements, remained the model of a science. However, no comparable attempt was made for about two thousand years to treat algebra and later analysis in the same manner. The basic properties of the *natural numbers* were something obvious for the Greeks, but questions of divisibility and problems concerning prime numbers were of interest to them. They knew how to manipulate *common fractions*, but they did not pursue the idea of introducing *negative numbers*. However, in connection with a right-angled isosceles triangle they stumbled on the fact that fractions are insufficient to describe the ratios of all quantities: they noticed that in such a triangle the ratio of side to base cannot be represented by a fraction. But from this they did not by any means draw the conclusion that the domain of fractions ought to be extended in such a way that this ratio, and as far as possible all other geometric ratios, could be described numerically in terms of the new numbers of the more extensive domain. They did precisely the opposite: they geometrized their algebra. True, this led to a theory that is equivalent to our theory of real numbers; but the geometrization gave rise to such complications that Greek mathematics ground to a halt.

Centuries later the practical needs of astronomers and mariners required urgently trigonometric calculations, which could only be mastered with the aid of tables of certain trigonometric functions. Since observational values could only be measured with limited accuracy, it was sufficient to give *approximately* the quantities to be calculated. This led gradually to the invention of terminating decimals, which proved much more suitable for practical computations than the common fractions. Most probably the conviction grew that the results would be the more exact the more decimal places used, and even that every preassigned accuracy can be achieved by using sufficiently many decimal places. In the last analysis this approach grasps the very essence of the *real numbers*; indeed, mathematicians no longer shied away from talking of decimal fractions with infinitely many places. If this theory had been developed consistently, the result could have been an exact theory of the real numbers.

An interesting example of fundamental significance shows how this notion in a slightly different form appears as early as in Archimedes' work, when he tries to calculate the area of certain parts of the plane with curvilinear boundaries. First of all, in his famous *exhaustion method* he succeeded in calculating the area bounded by part of a parabola and one of its chords. It turned out that a certain ratio of areas of paramount importance was 1/3. But Archimedes did not succeed in finding a correspondingly simple result for the area of a circle. To solve the problem he would have had to calculate the number π. As we now know, he could not succeed, being only in possession of fractions; he had to be satisfied with proving that the number π lies between two fractions, namely $3^1/_7$ and $3^{10}/_{71}$. For this purpose he calculated, by a repeated application of Pythagoras' theorem, the areas of the regular convex polygons with 96 sides inscribed in, and circumscribed to, the circle and gave approximate values for them. Clearly ARCHIMEDES was aware that by taking the number of sides and vertices sufficiently large he could include π within ever narrower limits and even calculate it with any prescribed accuracy. But this possibility of determining a number approximately with a prescribed accuracy by means of fractions is a characteristic feature of the real numbers.

This feeling of familiarity with the nature of the real numbers became firmly established in the course of time on diverse occasions, for example, – long before the foundation of the differential and integral calculus – in the composition of logarithmic tables, in Descartes' analytic geometry, where the points of a plane or of space are specified by coordinates, and then to a large degree in the development of the differential and integral calculus, which was started by LEIBNIZ and NEWTON and continued, as if intoxicated by the joy of discovery, by the BERNOULLIS, by EULER and FERMAT, by CAUCHY, GAUSS and others. No one imagined that the foundation of the theory of the real numbers would require a further intensive study.

However, questions of foundation played their part in two other branches, geometry and algebra. As already indicated, *Euclid's geometry* takes as its starting point a system of very simple geometric propositions from which further theorems of geometry can be derived. These simple propositions, called *axioms*, represented an extract of the geometric knowledge of the time and were intuitively so clear that no one felt the need to prove them. An exception was the *parallel axiom* (or *postulate*). This states that to a given line and a given point not on the line there is one and only one line passing through the given point without intersecting the given line. Was it perhaps possible to remove this statement from the system of axioms by deriving it from the remaining axioms? – For 2000 years mathematicians wrestled in vain with the problem, until GAUSS in Germany, LOBACHEVSKII in Russia, and BÓLYAI in Hungary succeeded in showing that the parallel axiom is independent of the other axioms. The significance of this result only becomes clear in connection with other developments.

In *algebra* the formula for the solution of quadratic equations can lead to the expression $\sqrt{-1}$, which at first sight is meaningless. But as long as one calculates with it just as with ordinary roots like $\sqrt{2}$, $\sqrt{3}$ or even $\sqrt{\pi}$, the results invariably makes sense. This strengthened the belief in the right of citizenship of this formation $\sqrt{-1}$, for which the notation i had meanwhile been accepted. Nearly 300 years elapsed before GAUSS and others showed that what one had done until then can be interpreted in a completely sensible manner as an *extension* of the *domain of real numbers* in which there exists a new number whose square is equal to -1.

Even GAUSS was so thoroughly familiar with real numbers that he had no scruples in using them without justification. Only when certain difficulties emerged in the process of clarifying the concept of *limit* at the hands of CAUCHY and other mathematicians of the time, did the real numbers become an object of serious thought. It was recognized that a theory of the real numbers can be founded, in fact in different ways, on a reduction to fractions. The latter, in turn, could be reduced to the natural numbers, and again it appeared that in the domain of natural numbers all their properties could be united in a few perfectly obvious fundamental facts, the *Peano axioms*.

With this reduction to the natural numbers a basis was given for the theory of the real and complex numbers, and also for the whole of real and complex *analysis* and beyond, even for geometry; for in analytic geometry it is shown how to master the basic objects of geometry, above all the points, by means of their coordinates, which are real numbers.

In this context another development should be mentioned, which started rather tentatively about 150 years ago. It was common knowledge that some rules for the multiplication of numbers and some for the addition show a strong formal similarity. Similarly, quite simple formal laws were observed in other *mathematical operations*, for example, in carrying out several motions in succession. But only very slowly did mathematicians proceed to the next logical step of extracting the common basic properties and of deriving from them new and ever deeper properties by purely logical processes. This field developed gradually to the present-day *theory of groups*, and again one sees, just as in Euclidean geometry, the emergence of an *axiom system* with all the subsequent developments.

Nowadays large parts of mathematics, above all algebra, but to an ever increasing extent analysis and geometry, are built up axiomatically. The procedure is roughly as follows: given is a collection, usually called a *set*, of mathematical objects, the *elements* of the set, together with some *system of axioms* that describes the basic properties of these objects. Now the following tasks arise: first of all, to draw the most far-reaching conclusions from the axioms, in other words, to carry the theory of such a structure as far as possible; next, to gain a survey of all specific ways of *realizing* the axiom system in question. It can happen that essentially there is only *one* possibility of realization, or *several*, or perhaps even *infinitely many*; it is also possible that *no* such realization can be found, for example, when the given axioms contradict each other. If there are several models, that is, ways of realizing the axioms, then one searches for *characteristic features* by which the various possibilities can effectively be distinguished in finitely many steps. For some structures these tasks have been solved completely, for others we are still far away from a solution. This indicates, incidentally, how closely interwoven axiomatics and mathematical logic are.

Even more imperative became the demand for an efficient mathematical logic when at the turn of the century contradictions arose in one of the new structural theories, the theory of sets. *Set theory* is the simplest *structural theory*, inasmuch as it is concerned with completely arbitrary collections whose elements are not subject to any axioms, such as points, numbers, motions, functions, figures, but equally well men, stars, chairs or what have you. Since no structural assumptions are made, two such sets are to be regarded as equivalent or *equipotent* if they have equally many elements. In the case of finite sets the meaning of this is immediately clear to everyone; but it was a magnificent achievement to define even for infinite sets something like the number of its elements, the so-called power or *cardinality*. True, this fails to have some of the properties with which we are familiar when the number of elements of a finite set is involved. For example, in this sense there are just as many natural numbers as there are fractions, but not as many fractions as real numbers, and the set of points on a line has the same cardinality as that of the points in the plane. All these are things which in spite of their apparent lack of intuitiveness are entirely unobjectionable from the point of view of total mathematical rigour. *Contradictions* appeared, however, in the unrestrained formation of sets; for example, the concept of 'set of all sets' is contradictory in itself. Nevertheless, this was not a *crisis in mathematics*, as the phenomenon was sometimes called; on the contrary, mathematicians took occasion to reflect more thoroughly on what is involved in defining mathematical concepts. Indeed, a systematic mathematical logic was developed, and today one knows precisely how to avoid such contradictions.

One might think that this utmost abstraction, in the form of axiomatization of the very general structural theories and of mathematical logic, could lead further and further away from down-to-earth applied mathematics. This is by no means the case; it was no accident that LEIBNIZ, who apart from his immediate creative mathematical work occupied himself with some fundamental questions of logic, has already constructed a workable *calculating machine*.

The appearance of factory-made calculating machines, operated by hand or by a motor, did not give rise to important discussions of principles. But this state of affairs changed radically with the creation of *electronic computing machines*, by which the speed of calculation was increased drastically. True, these machines work on a *simple black-white principle*, because in each of their components current does or does not flow. Nevertheless, they can cope with calculations that otherwise would be practically impossible: they perform huge numbers of the simplest operations with an unimaginable speed and so can go through a complicated and protracted *program* in an acceptable time. Naturally, the duration of such a calculation depends on the skill that goes into the making of the program.

After some preliminary work that had been done before the invention of electronic computing machines it soon turned out that in programming certain regularities are observed, which also play a role in mathematical logic, for example, in the *theory of algorithms*. This once more demonstrated the practical advantages of certain purely mathematical investigations which had been carried out merely for theoretical needs – a truly classical example of the close natural relationship between *pure* and *applied mathematics*, in this case computing techniques.

In this context it seems appropriate to draw attention to the difference between the *theoretical* and the *practical solubility* of a mathematical problem. Quite frequently in mathematics it is not individual, numerically given, problems that are discussed, but general problems depending on certain data, whose numerical values can be chosen in many, as a rule infinitely many, ways. A simple example: to determine the area of a triangle depending on the lengths of its three sides. There is a formula for this area that is valid for all triangles, although there are infinitely many possibilities for the length of each side.

Such a problem is regarded as solved when a formula, an *algorithm*, can be given by means of which the solution can be calculated in each individual case. Here one postulates that the formula or the procedure leads to the numerical result in finitely many steps. When this is the case, a pure mathematician considers the problem as solved. Nevertheless, in practice the problem can still be insoluble if the number of necessary steps is finite, but for reasons of time or economy is too large. This can lead to new and interesting problems of pure mathematics: to find more effective procedures – unless one is satisfied with approximate solutions or one builds faster computers.

An enormous step forward in this respect was the invention of the electronic computing machines. It had the consequence that new branches arose, above all in applied mathematics, branches which had not been developed previously, because it was clear from the outset that their main problems could not possibly be attacked and solved within a practically acceptable period. Two examples of problems *soluble in principle* are the games of 'nine men's morris' and chess. They are soluble in principle, because by the rules there are only finitely many possible games. Nine men's morris is also solved in practice, in that one can give to the first player exact instructions how to react to all possible moves of the opponent so as to win in every case. The same question, whether in chess the white player can always win, is still unsolved in spite of the finiteness of the problem; even if all electronic computers at present available in the whole world were used solely to solve the chess problem, a solution could not be reached: this would require computers working unimaginably faster than the present ones.

The development of mathematics, which has been roughly sketched here, led from the simplest fundamental concepts of number, operation, figure, and measure to its present-day thoroughly axiomatized form of a wealth of highly abstract structures and to the modern computing automata whose possibilities are far from being exhausted. A comparison of this development with the table of contents of this book indicates many direct and indirect relationships.

Thus, the material of the first part '*Elementary mathematics*' agrees to a large extent with mathematics as it was developed from antiquity through the Middle Ages and before the foundation of the differential and integral calculus. Only here arithmetic, the theory of numbers, and geometry are not set forth side by side, but one after the other. We begin with the natural numbers, together with the rules for the *elementary operations*, just as they present themselves as perfectly obvious to a naive person. But the *axiomatic build-up* follows immediately, starting from the natural numbers and leading up to the complex numbers.

Even for these simple concepts a notation is used that was unknown to the Greeks and whose absence was one reason for an extremely cumbersome and unwieldy presentation: the use of *letters* for numbers. Today it is taken for granted in schools. Here the notation is admirably suited to the basic mathematical concepts, but it is so easy to handle that sometimes there is the danger of thoughtless and mechanical manipulation of letters. This suggestive effect must be strongly opposed, especially in schools: the primary thing is the *mathematical idea*, and the computational working details are secondary – not the other way round. On this theme GAUSS wrote to SCHUMACHER in a letter of 1 September 1850: 'It is a characteristic of modern mathematics ... that in our language of symbols and names we possess a lever by which the most complicated arguments are reduced to a certain mechanism ... How often is this lever handled just mechanically, although in most cases the authority to do so implies certain tacit assumptions. I postulate that in every application of the calculus, in every use of concepts, one should always remain conscious of the original conditions and should never regard results produced by the mechanism as mathematical property beyond the clearly permitted limits.'

Many tasks require unknown quantities to be determined from given quantities. As a rule, the use of letters enables us to state such tasks simply and lucidly. It then happens frequently that problems which at first sight appear totally distinct have one and the same form in the resulting equations or systems of equations. This points again to the parallelism between the mathematical formulation of problems and the abstraction that consists in disregarding the meaning of the given and the required quantities and leaving only the mathematical nucleus.

A characteristic feature of modern mathematics is *functional thinking*. This means that one is concerned with functional relationships, such as the dependence of certain quantities on certain others, for example, the area or the angles of a triangle on the lengths of its sides. We shall become acquainted with other examples of this kind of thinking in the analysis of the notion of a function.

Elementary geometry deals with points, segments, angles, straight lines, triangles, quadrangles, circles, tetrahedra etc. in a plane or in space. An essential role is played here by the concept of number as developed previously, owing to the need for measuring the objects. Naturally, this must not lead to a neglect of pure geometrical thinking, especially in the solution of problems. One tries to solve geometric problems by purely geometric means, that is, by constructive drawings. How to treat problems in space by drawings in the plane is the topic of *descriptive geometry*. The most intimate fusion between geometry and calculation occurs in *analytic geometry*: by means of the concepts of coordinates geometric problems can be transformed into numerical problems: in this way geometry becomes accessible to the far-reaching methods of analysis.

The rudiments of *analysis* itself are treated in the second main part 'Steps towards higher mathematics'. Although the concept of limit is already used in elementary mathematics in an intuitive fashion, higher mathematics begins just with a rigourous *theory of limits*. This in turn is the basis, on the one hand, for the theory of *infinite series* of numbers and functions, on the other hand, for the notion of continuity of functions as well as for the *differential* and *integral calculus*, whose significance is fundamental not only for the entire framework of mathematics, but also for the applications in physics, technology etc. Many problems of geometry and physics present themselves in the form of *differential equations*, that is, in relations between a function and its derivatives. The theory, which has grown by now to a very large volume, can only be sketched here in its simplest parts. An attractive branch is *differential geometry*, an application of the differential and integral calculus to the theory of curves in a plane and in space and to surfaces in space.

As we remarked above, the theoretical solution of a problem is frequently far removed from an immediate application of specific cases, because the necessary numerical calculations become too extensive. It is the task of *graphical representations* and of *numerical methods* to transform theoretical solutions into directly applicable ones. *Probability theory* and *statistics* also play an important part in applications.

In the last main part 'Brief reports on selected topics' an attempt is made to give an insight into a number of research fields of contemporary mathematics. For the reasons stated at the beginning, a more detailed account of the individual problems is impossible, and domains that at present are still in a nascent stage or in the process of deep reorganization could not be included. The reader who wishes to acquaint himself more thoroughly with one branch or another would do well to refer to the specialized literature – and this applies equally well to the first two main parts.

Hans Reichardt

Authors and translators

The authors of the 'Kleine Enzyklopädie der Mathematik' are:

G. Berthold
Prof. O. Beyer
Prof. L. Bittner
Prof. H. Boseck
Dr. H. G. Bothe
Dr. G. Czichowski
J. Dähnn
Dr. C. Frischmuth
Dr. D. Göhde
W. Göhler
Prof. L. Görke
Dr. M. Hellwich
Dr. H. Herre
Prof. M. Herrmann
H. Kästner
G. Lisske
Dr. G. Lorenz
Dr. G. Maess
Dr. W. D. Müller
Dr. F. Neigenfind
Prof. F. Nožička
Dr. S. Oberländer
Prof. M. Peschel
Dr. G. Pietzsch
Dr. B. Renschuch
Prof. H. Sachs
Prof. H. Salié
H. Schlosser
Dr. E. Schröder
Dr. L. Stammler
A. Steger
Prof. R. Sulanke
Prof. H. Thiele
Dr. H. Thiele
Prof. W. Tutschke
Dr H. Vahle
Dr. L. Wagner
Prof. W. Walsch
Dr. V. Wünsch
Dr. G. Wussing
Prof. H. Wussing

The present English version of the 'Kleine Enzyklopädie der Mathematik' was prepared under the editorship of Professor K. A. Hirsch and with the collaboration of

Dr. O. Pretzel
Dr. E. J. F. Primrose
Professor G. E. H. Reuter
Dr. A. Stefan
Dr. A. M. Tropper
Dr. A. Walker

I. Elementary Mathematics

1. Fundamental operations on rational numbers

1.1. The natural numbers N

Numbers and digits

What are natural numbers? Two kinds of activity made our ancestors face the necessity of occupying themselves with numbers; this led to the development of *cardinal* and *ordinal* numbers.

Cardinal numbers. Man had to compare various *sets* of things, for example, flints, dogs, hunting companions, in order to ascertain which set contains more elements (constituents, members). Today one does this, as a rule, by counting and comparing the quantities so obtained; this presumes an ability to count, that is, a knowledge of the numbers. But there is an easier way: if one wishes to find out, for example, whether men and horses are present in equal numbers, one simply places a rider on every horse. In other words: one sets up a matching, a *correspondence*, between men and horses. This matching may tally – then there are just as many horses as there are men, and one says: *the sets are equipotent*, – or some of one kind are left over; then there are more of this kind (Fig.). In laying a table one arranges a correspondence among sets of cups, saucers, spoons, etc. All sets between which such a matching of pairs can be established therefore have the corresponding number as a common property (Fig.). This is the way in which even today our children gain their knowledge of the *cardinal* numbers.

1.1-1 Men and horses — without matching

1.1-2 Men and horses — with matching. One man is left over

1.1-3 Common number: three

Cardinal numbers are counting numbers	1, 2, 3, ...

Ordinal numbers are place numbers	1st, 2nd, 3rd, ...

Abstraction has not progressed this far in all stages of civilization. There are primitive tribes who use distinct numerals when they refer to distinct objects. *Two* women is then something other than *two* arrows; here the abstraction of number from the other properties of the sets has not yet been achieved.

Ordinal numbers. The second need consisted in creating *order* within one and the same set. For example, it had to be laid down according to some point of view – say, the height, the age or the bravery of the rider – who would ride first, second, ... at the hunt (Fig.). Something quite similar occurs when one counts through the elements of set; only, the order so obtained is, as a rule, without significance. In this way, there arise the *ordinal numbers.*

Natural numbers **N**	0, 1, 2, 3, ...

1.1-4 Set of four hunters, unordered, and ordered by height

Cardinal and ordinal numbers have developed in close interconnection and form the two aspects of the *natural numbers*; frequently the zero (or null) is, by convention, reckoned to belong to them.

Numerals and number symbols. For the purpose of oral and written communication and of memorizing cardinal and ordinal numbers *number words* (numerals) and *number symbols* are required, the latter particularly for abbreviation and ease of calculation (Fig.). The strong similarity between the words for corresponding cardinal and ordinal numbers in all languages or writings is a sign of their close connection. In English most ordinal numbers have the ending -th (four—the fourth; a hundred—the hundredth), in writing a stop is added (for example, NEWTON was born on 25. 12. 1642). In the United States the month is placed before the day: 12/25, 1642.

Because of the great similarity, in what follows it is sufficient to confine our attention to cardinal numbers; corresponding arguments apply to ordinal numbers.

number word: *nine*
number symbol: 𝍸IIII *or* IX *or* 9

1.1-5 Three symbols for the number word 'nine'

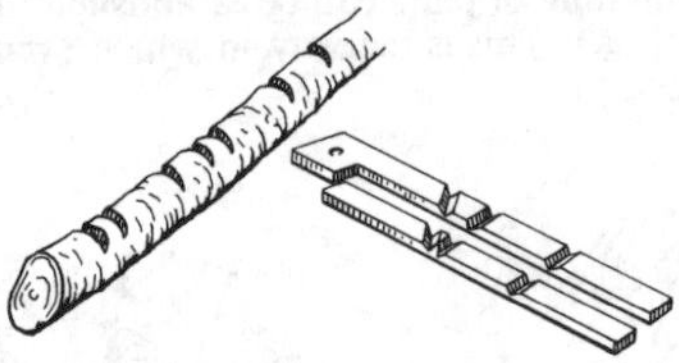

1.1-6 Tally sticks

basic symbols:	I 1	X 10	C 100	M 1000
auxiliary symbols:	V 5	L 50	D 500	
example:	MDCCLXVIII 1768			

1.1-7 Roman number symbols

Representations of numbers. The simplest representations of numbers occur in *tally sticks* (Fig.), pieces of wood scored across with notches to record the items. Frequently they were split into halves of which each party kept one. The method of *strokes,* by which Robinson CRUSOE counted days, is still in use, particularly for tedious countings. But very soon, when the numbers become larger, this representation loses its perspicuity: it can be restored by appropriate groupings. Something quite similar occurs when *new words* for numbers are formed or new symbols are invented: it would be most uneconomical to introduce a completely new word and a new symbol for every number. Instead one composes words and symbols for larger numbers from those of smaller ones, and these building bricks themselves have arisen by the combination of units or smaller groups. According to the method of this grouping and the arrangement of the symbols one distinguishes between *addition systems* and *position systems*.

Addition systems. The best known example for an addition system is the *Roman method* of writing numbers. Of the basic symbols ten each were combined to the next higher group; in between there are auxiliary symbols (Fig.). By the way, the origin of these symbols is not completely clear. Some of them, for example M (*mille*) for 1000, have been in use in this form only since the middle ages. The Romans wrote CIↃ for 1000. The essence of an addition system is that all number symbols are formed by juxtaposition of as few of these symbols as possible (in our case seven symbols, see Fig. 1.1-7). A rule prescribes that *the symbol for the larger number always stands to the left of that for the smaller number*. An exception to this rule is motivated by the endeavour to use as few symbols as possible. The number nine can be represented as VIIII $(5 + 4)$ or IX $(10 - 1)$. The latter writing is preferred. Therefore, if the symbol of a smaller number stands at the left, then the corresponding number has to be subtracted, not added. However, *it is not permitted to place several basic symbols or an auxiliary symbol in front*: MCMLIX for 1959; CML (not LM) for 950. An addition system has disadvantages: in general, the number symbols are very long and therefore lack in clarity; when the numbers grow (in the present case, beyond 10 000), one has to keep inventing new symbols to avoid representations of excessive length; written calculations in an addition system are exceedingly troublesome.

Position systems. Our present-day *position system* goes back to the Hindu from whom it came to us by way of the Near East (Arabic digits). In this perfection it is a fairly late achievement in the historical development of representations of numbers. In the system ten individuals (Units U) are combined to a new group, a Ten T and again ten of these to a Hundred H etc. However, no new symbols are introduced for these groups of higher rank (as in the Roman system), but they are distinguished by their position within the entire numerical symbol. In the Roman symbol XXX for thirty each of the three letters has the same numerical value 10, and since it is an addition system, the total number is obtained by adding the three individual values. In the symbol 444 for four-hundred-and-forty-four the three digits also have the same numerical value four; but within the total symbol they stand in different places and therefore have different positional values; the rightmost position indicates the units:

321 means: $3\,\mathrm{H} + 2\,\mathrm{T} + 1\,\mathrm{U}$,

CCCXXI means: $100 + 100 + 100 + 10 + 10 + 1$.

Since the gathering occurs in groups of ten each, one talks of a *decimal system* (Latin, *decem* 10) or a *decadic* positional system (Greek, *deka* 10). Accordingly, the Roman number system is a decimal addition system. The number ten is called the *base* of the system. The positional values are the *powers* of ten, some with their own names such as 1 *million* for $10^6 = 1\,000\,000$, 1 *milliard* for 10^9, 1 *billion* for 10^{12}, 1 *trillion* for 10^{18}. There follow 1 *quadrillion*, 1 *quintillion*, etc., each time with six more zeros. Formations such as 1 *billiard* for 10^{15} are rarely used; in the U.S.A. and U.S.S.R., 10^9 is called 1 billion, 10^{12} a trillion, and 10^{15} a quadrillion. It is probable, but not certain, that the choice of ten as a base is connected with the number ten of our fingers. In old measuring units (one *dozen*, one *gross*) one finds traces of a vanished duodecimal system with the base 12; the French word *quatre-vingt* for eighty points to a (non-positional) system with base 20, and the word *score* for a group of 20 objects is still in frequent use. Our time measures (1 h = 60 min, 1 min = 60 s), as well as the division of the full angle into 360° recall the *sexagesimal system* (base 60) of the *Babylonians*. This system already showed clearly some features of a positional system. But the complete development of such a system was hampered by lack of the consistent use of a symbol for *empty places*, a zero. The introduction of zero is one of the greatest achievements of the Hindu (around 800 A. D.).

Not only 10, 12, 20, or 60 are suitable as bases of a positional system. Every natural number $b > 1$ can serve as base, because then every natural number a has exactly one *b-adic* representation $a = a_n b^n + a_{n-1} b^{n-1} + \cdots + a_1 b + a_0$, in which the natural numbers a_i, $i = 0, \ldots, n$ satisfy $0 \leqslant a_i < b$. The a_i are called the *digits* of a. Every positional system requires exactly b distinct digits.

The binary system. Of particular technical importance is the *binary system*, which is also called *dyadic* or *dual* system. In it the position values are the powers of the base 2, that is, 1, 2, 4, 8, 16, 32, 64, 128, ... These position values are considerably closer to each other than those of the decimal system; therefore the number symbols become comparatively long. On the other hand, one only needs two digits: 0 and 1. For the binary unit the notation L is in frequent use:

$$7 = 1 \cdot 4 + 1 \cdot 2 + 1 \cdot 1 = 1 \cdot 2^2 + 1 \cdot 2^1 + 1 \cdot 2^0 = \mathrm{LLL},$$

$$9 = 1 \cdot 8 + 0 \cdot 4 + 0 \cdot 2 + 1 \cdot 1 = 1 \cdot 2^3 + 0 \cdot 2^2 + 0 \cdot 2^1 + 1 \cdot 2^0 = \mathrm{LOOL},$$

$$22 = 1 \cdot 16 + 0 \cdot 8 + 1 \cdot 4 + 1 \cdot 2 + 0 \cdot 1 = 1 \cdot 2^4 + 0 \cdot 2^3 + 1 \cdot 2^2 + 1 \cdot 2^1 + 0 \cdot 2^0 = \mathrm{LOLLO}.$$

This binary system is often used in digital computers.

Order of the natural numbers N. Every natural number has exactly one immediate successor; for example, 96 is successor of 95. This means that the sequence of natural numbers has no last member, it never breaks off. The number 0 is not a successor; every natural number other than 0 has exactly one immediate predecessor; this means that the sequence of natural numbers has a beginning in its first member 0.

For any two natural numbers n_1 and n_2 exactly one of the following relations holds: $n_1 < n_2$, that is, n_1 is smaller than n_2, for example, $3 < 7$, or $n_1 = n_2$, that is n_1 is equal to n_2, for example, $5 = 5$, or $n_1 > n_2$, that is, n_1 is greater than n_2, for example, $8 > 6$.

If one wishes to express that a number n_1 is at most as large as n_2, one writes $n_1 \leqslant n_2$, n_1 is less than or equal to n_2. Accordingly, $n_1 \geqslant n_2$, that is, n_1 is greater than or equal to n_2, means that n_1 is at least as large as n_2; therefore both $4 \leqslant 19$ and $11 \leqslant 11$ are correct statements.

These relations have a property called transitivity; for the relationship 'smaller' it takes the form: from $n_1 < n_2$ and $n_2 < n_3$ it follows that $n_1 < n_3$. The *relationship 'larger'*, respectively, *'smaller'*, orders the natural numbers 'linearly'. An illustration of this linear order is the *number ray* (Fig.). In it the natural numbers are represented by a set of isolated (*discrete*) points. The fact that n_1 is less than n_2 then means that the point on the number ray belonging to n_1 lies to the left of the point n_2.

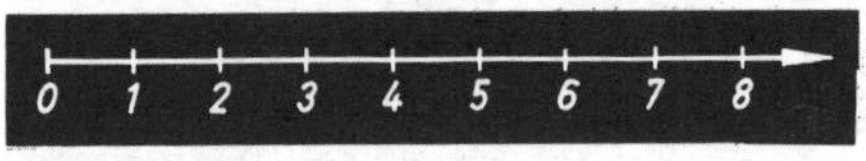

1.1-8 Number ray

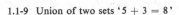

1.1-9 Union of two sets '5 + 3 = 8'

Calculations with natural numbers N

Addition and subtraction. Addition is the simplest operation on natural numbers, and subtraction is its inverse. They are the arithmetical operations of the *first* kind.

Addition. Addition reflects the joining, the *union* of two sets (Fig.). The operation symbol is + (read *plus*). The addition can also be regarded as an abbreviated counting forward: $5 + 3$ as $5 + 1 \rightarrow 6 + 1 \rightarrow 7 + 1 \rightarrow 8$. The two numbers to be added are called *summands*, the result is their *sum*.

Addition	summand plus summand equals sum 3 + 2 = 5

Commutative law of addition
$a + b = b + a$

The name sum is used in two meanings: 8 is the sum of 5 and 3; the expression $5 + 3$ is a sum. The addition of two natural numbers can always be carried out, that is, two natural numbers always determine a third, their sum. Several laws hold for the addition of natural numbers.

Commutative law. The order of the summands has no influence on the result; for example, $5 + 3 = 3 + 5 = 8$. Since this *commutativity* of the summands holds for all natural numbers, one writes briefly: $a + b = b + a$.

Here and in what follows, a and b are symbols for arbitrary natural numbers.

Associative law. In the first instance, addition is defined for only two summands. If three numbers are to be added, then two of them have to be added first, and a new addition of two summands can be formed from this sum and the third number. Here the order of combining the numbers has no influence on the result.

This *law of the* (*sequence of*) *combination* also holds for all natural numbers. It means that brackets may be omitted: $5 + 3 + 4 = 12$. Similarly, additions of more than three summands may be written without brackets.

Associative law of addition
$(a + b) + c = a + (b + c)$

Examples: 1. $5 + 3 + 4 = (5 + 3) + 4 = 8 + 4 = 12$,
2. $5 + 3 + 4 = 5 + (3 + 4) = 5 + 7 = 12$.

Monotonic law. The relationship 'smaller' between two natural numbers is preserved when the same number is added to the two numbers; for example, from $3 < 4$ it follows that $3 + 7 < 4 + 7$. This law, too, is valid for all natural numbers.

Monotonic law of addition	from $a < b$ it follows that $a + c < b + c$

Subtraction. The process opposite to adding, namely taking away or deducting, leads to this arithmetical operation. Its symbol is — (read *minus*). Subtraction can also be interpreted as an *abbreviated counting backwards*, for example, 7 — 3 as 7 — 1 → 6 — 1 → 5 — 1 → 4 (Fig.). From addition one comes to subtraction if for a given sum one asks for one of these summands, for example, 4 + x = 7; x = 7 — 4. Accordingly subtraction is the *inverse* of addition. The number from which the other number is to be subtracted is called the *minuend*; the number to be subtracted is called the *subtrahend* and the result is the *difference*.

Subtraction						
minuend	minus	subtrahend	equals	difference		
7	—	3	=	4		

1.1-10 7 — 3 = 4

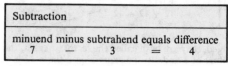

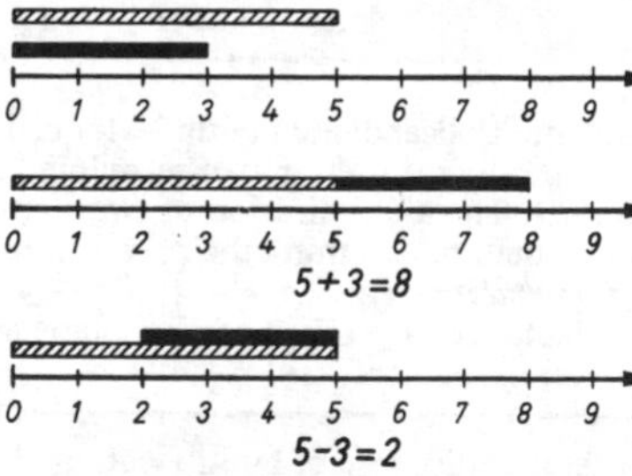

1.1-11 Operations on the number ray

Like the word 'sum', so also 'difference' is used in two meanings: the difference of 7 and 3 is 4; the expression 7 — 3 is a difference.

In contrast to addition, the subtraction of two natural numbers cannot always be carried out; for example, the problem 2 — 9 does not have a natural number as a solution. The condition for solubility is: the minuend must not be smaller than the subtrahend.

Operations of the first kind on the number ray. Addition and subtraction of natural numbers can be illustrated on the number ray as addition and subtraction of segments (Fig.).

Written addition. The summands are written one under the other, so that equal position values stand in the same column. The addition begins with the units, in any order on account of the commutative law, and then successively from right to left towards higher position values. If in one column the sum exceeds the next position unit, the corresponding amount is carried:

Example:

```
Th H T U              Th H T U
 7  3 6 2              7  3 6 2                  7 3 6 2
+1  6 8 4    or   +    1  6 8 4    written as  + 1 6 8 4
    1 4 6              1↖ 1↖                     -------
    -                  ---------                 9 0 4 6
 1  0                  9 10 14 6
 -
 9
```

Addition of more than two summands follows the same pattern.

```
 74627
− 3193
 71434
```

Written subtraction. Subtractions can be performed in two slightly different ways: (t) taking away: 7 *take* 3 *gives* 4; (s) supplementing: *from* 3 *to* 7 *is* 4.

Accordingly two methods of written subtraction are in common use, (t) and (s). In both methods the subtrahend is written below the minuend (units under units etc.) and a start is made with the units. The following example illustrates the difference in the case of the Tens.

Method (t): Take 9 from 2 does not go; one H is dissolved into ten T, and now: take 9 from 12 gives 3. Next, instead of 6 H there are only 5 H left and take 1 H from 5 H gives 4 H.

Method (s): From 9 to 2 does not go. From 9 to 12 is 3. The dissolved H is not taken away from the minuend but is added to the subtrahend. This gives the same result from 2 H to 6 H is 4 H.

Examples:

```
  6311
−  768
−  229
− 1046
  4268

 70003
−11628
 58375
```

The (s)-method is more lucid, for example, if the problem requires several dissolutions of position values because the minuend contains several consecutive zeros. Also it permits to carry out the subtraction of several subtrahends in a single step. In the example above the mental process at the units is: 6 + 9 + 8 = 23; from 23 to 31 is 8. The three dissolved Tens are added to the subtrahend; for the Tens the calculation is 3 + 4 + 2 + 6 = 15, from 15 to 21 is 6, etc.

Multiplication and division. Multiplication and division are the arithmetical operations of the second kind.

Multiplication. Multiplication can be arrived at in various ways, for example, by addition of several equal summands, $12 + 12 + 12 = 3 \times 12 = 36$ (Fig.). The operational symbol is · (read *times*) or a lying cross ×.

Multiplication						
multiplicator	times	multiplicand	equals	product		
3	·	12	=	36		
factor	times	factor	equals	product		

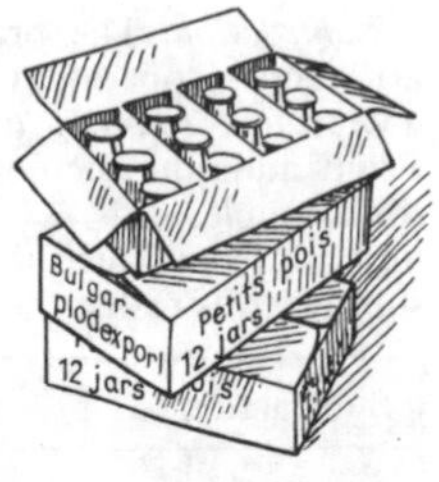

1.1-12
$12 + 12 + 12 = 3 \cdot 12 = 36$

Since multiplicand and multiplicator can be interchanged, they are both called the factors. Again, the word product is used in two meanings. 36 is the product of 3 and 12; the expression 3 times 12 is a product. The multiplication of two numbers *can always be carried out*, that is, two natural numbers always determine a third, their product. For all natural numbers a: $a \cdot 0 = 0 \cdot a = 0$ and $a \cdot 1 = 1 \cdot a = a$.

Commutative law. $3 \cdot 4 = 4 + 4 + 4 = 12$ and $4 \cdot 3 = 3 + 3 + 3 + 3 = 12$, hence $4 \cdot 3 = 3 \cdot 4$. The factors of a product can be interchanged without alterating the result. This is true for all natural numbers as factors:

Commutative law of multiplication
$a \cdot b = b \cdot a$

Associative law of multiplication
$(a \cdot b) \cdot c = a \cdot (b \cdot c)$

Associative law. If three numbers are to be multiplied, first two of them are multiplied and then the product is multiplied by the third. Here the order of combining the factors has no influence on the result. For example, $3 \cdot 4 \cdot 7 = (3 \cdot 4) \cdot 7 = 12 \cdot 7 = 84$; $3 \cdot 4 \cdot 7 = 3 \cdot (4 \cdot 7) = 3 \cdot 28 = 84$. This law also holds for all natural numbers. Therefore brackets may be omitted: $3 \cdot 4 \cdot 7 = 84$. For more than three factors one proceeds similarly.

Monotonic law. $3 < 4$ leads to $3 \cdot 8 < 4 \cdot 8$ but $3 \cdot 0 = 4 \cdot 0$. This is true for all natural numbers a, b, c:

Monotonic law of multiplication	from $a < b$ and $c > 0$ it follows that $a \cdot c < b \cdot c$

Division. Two distinct everyday problems lead to the other basic arithmetical operation of the second kind, division:

(*i*) Sharing: twelve pears are to be shared equally among four persons; each receives three pears (Fig.);

(*ii*) Being contained: how many times is 4 cm contained in 12 cm? – 3 times (Fig.).

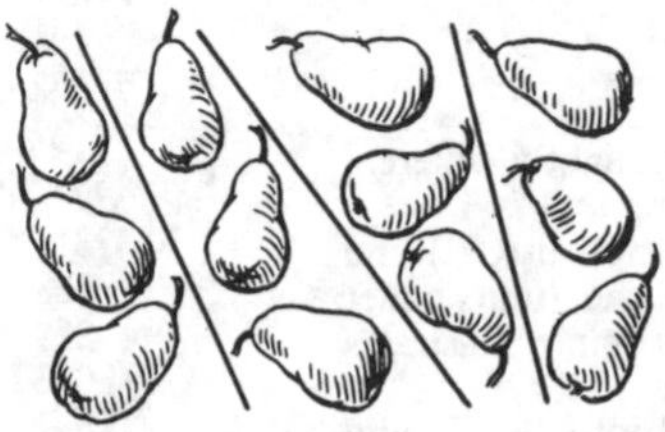

1.1-13 Sharing: 12 pears are divided into 4 equal parts

1.1-14 Being contained

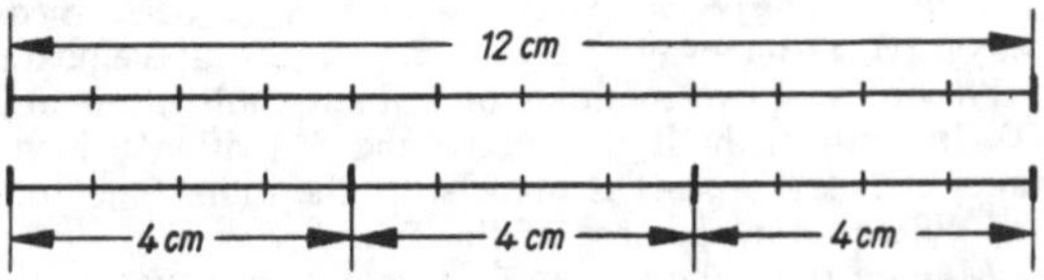

Mathematically, division is arrived at as the inverse operation to multiplication: a product and one factor are given and the other factor is required; both $3 \cdot \boxed{x} = 15$ and $\boxed{x} \cdot 3 = 15$ lead to $\boxed{x} = 15 : 3$.

Division	dividend by divisor equals quotient
	15 : 3 = 5

Since the factors can be interchanged, the two questions corresponding to sharing and to being contained lead to the same division problem. The operational symbol is the colon (read *divided by*). Again, the word quotient is used in two meanings: *the quotient of* 15 *by* 3 *is* 5; *the expression* 15 : 3 *is a quotient.*

Feasibility of division. The division of two natural numbers cannot always be performed within the domain of the natural numbers; for example, there is no natural number n for which $3n = 17$, for $3 \cdot 5 = 15$ and $3 \cdot 6 = 18$, hence 17 is not divisible by 3. The use of the equality sign in writing

$17:3 = 5$ remainder 2 is inexact; the expression $17 = 3 \cdot 5 + 2$ (dividend equals divisor times quotient plus remainder) is unobjectionable. Division by zero is impossible; for $5:0 = n$ would mean that $n \cdot 0 = 5$ but for every n the product is 0, never 5. Even for the dividend 0 the division by zero is impossible, because the problem does not have a unique result. One could claim that $0:0 = 17$ because $0 \cdot 17 = 0$, but also $0:0 = 193$ because $0 \cdot 193 = 0$.

Division by zero cannot be performed.

Sequence of arithmetical operations. If in a problem operations of different kind occur, the sequence of performing them can influence the result: $7 \cdot 5 + 3$ leads to $7 \cdot 8 = 56$ if addition is first carried out, but to $35 + 3 = 38$ if multiplication is first carried out. The situation is similar for subtractions or divisions. Therefore the sequence of the operations must be agreed upon:

The operation of higher kind is performed first.

If in a given case the operations are to be performed in a different sequence, then brackets have to be introduced. The contents of each bracket are treated first:

$$(12 + 96):3 - 8 \cdot (5 - 2) = 108:3 - 8 \cdot 3 = 36 - 24 = 12.$$

Distributive law. This law of distribution expresses a connection between arithmetical operations of different kinds; for example, $5 \cdot (4 + 3) = 5 \cdot 7 = 35$, but also $5 \cdot 4 + 5 \cdot 3 = 20 + 15 = 35$; therefore $5 \cdot (4 + 3) = 5 \cdot 4 + 5 \cdot 3$. This way in which for a multiplication of a sum the other factor is distributed over the summands is the same for all natural numbers a, b, c:

Distributive law
$a \cdot (b + c) = a \cdot b + a \cdot c$

From the distributive law one derives the relations $(a - b) \cdot c = a \cdot c - b \cdot c$; $(a + b):c = a:c + b:c$ for $c \neq 0$; $(a - b):c = a:c - b:c$ for $c \neq 0$. For natural numbers a, b, c these equations only have a meaning if the subtraction $a - b$ and the division $a:c$ and $b:c$ can be performed.

Written multiplication. Written multiplication utilizes the distributive law, so that a knowledge of the multiplication table up to $9 \cdot 9$ is sufficient. In principle the process is as follows:

$$\begin{aligned} 2356 \cdot 473 &= 2356 \cdot (400 + 70 + 3) \\ &= 2356 \cdot 4(00) + 2356 \cdot 7(0) + 2356 \cdot 3 \\ &= 9424(00) + 16492(0) + 7068 = 1114388. \end{aligned}$$

```
2356 · 473          2356 · 473
----------          ----------
 9424          or         7068
  16492                  16492
    7068                9424
----------          ----------
 1114388             1114388
```

In the second line the multiplication of the first factor 2356 is carried out by splitting this factor into its Units, Tens etc. The final addition of the partial products is done in writing. Instead of attaching zeros to the partial products belonging to higher position values of the multiplicand, one shifts the number appropriately.

Written division. In the written division process the dividend is split into Units, Tens, Hundreds; for example: $86:2 = (80 + 6):2 = 80:2 + 6:2 = 40 + 3 = 43$. Since division is the inverse process to multiplication, the product of quotient and divisor must yield the dividend, and this must hold equally for the partial quotients. This leads to the following division scheme:

```
      487
23) 11208
     92
     ---
     200
     184
     ---
      168
      161
      ---
        7
```

Or more briefly, by performing the individual subtractions mentally:

```
      487
23) 11208
     200
      168
        7
```

The quotient is 487 and the remainder is 7.

Elementary number theory

Divisibility. $12:4 = 3$, that is, the number 12 is *divisible* by 4; but 15 is not divisible by 4. Hence 4 is called a *divisor* of 12 or 4 *divides* 12 (in symbols $4 \mid 12$); 4 is not a divisor of 15 ($4 \nmid 15$). In general, a natural number a is said to be divisible by another b if there exists a natural number n such that $a = n \cdot b$; b as well as n are then called divisors of a. On the other hand, a is called a multiple of b and of n. The number 0 is divisible by all numbers $a \neq 0$ and is a multiple of every number. Every number $a \neq 0$ is divisible by 1 and by itself; these divisors are called *improper*.

Prime numbers. Prime numbers are numbers that have only improper divisors; for example, 5 is only divisible by 1 and 5, 13 only by 1 and 13; hence 5 and 13 are prime numbers. The number 1 itself is not counted among the prime numbers so that the sequence begins with 2.

Prime numbers	2, 3, 5, 7, 11, 13, 17, 19, 23, ...

$$120 = 4 \cdot 30$$
$$4 = 2 \cdot 2 \qquad 30 = 2 \cdot 15$$
$$15 = 3 \cdot 5$$
$$120 = 2 \cdot 2 \cdot 2 \cdot 3 \cdot 5.$$

Factorization. Every natural number is either itself a prime number or can be written as a product of prime numbers, can be split into prime factors:
The same decomposition is obtained by starting out from, say, $120 = 10 \cdot 12$. By means of *Euclid's algorithm* it can be proved that the factorization into primes is unique apart from the order, in other words, apart from the order there is only one way of splitting a natural number into prime factors (see the derivation at the end of this chapter). The statement of the theorem would be incorrect if 1 were counted among the primes. By the use of *powers* the prime factorization of natural numbers can be written down more conveniently, for example, $1008 = 2 \cdot 2 \cdot 2 \cdot 2 \cdot 3 \cdot 3 \cdot 7 = 2^4 \cdot 3^2 \cdot 7$.

Sieve of Eratosthenes. ERATOSTHENES of Kyrene (approx. 276–194 B. C.) indicated the following method of obtaining all the primes in a segment of the natural numbers: delete after 2 every second number (every number divisible by 2), then after 3 every third number (every number divisible by 3), then after 5 every fifth number (every number divisible by 5) etc. The remaining numbers of this segment are primes. As one can see in the following table, up to 100 only 4 deletions are required, the last one for all numbers divisible by 7. The reason is that $7 \cdot 7 = 49$ is less than 100, but already $11 \cdot 11 = 121 > 100$; and 11 is the first non-deleted number after 7, hence the next prime.

	2	3	4	5	6	7	8	9	10
11	12	13	14	15	16	17	18	19	20
21	22	23	24	25	26	27	28	29	30
31	32	33	34	35	36	37	38	39	40
41	42	43	44	45	46	47	48	49	50
51	52	53	54	55	56	57	58	59	60
61	62	63	64	65	66	67	68	69	70
71	72	73	74	75	76	77	78	79	80
81	82	83	84	85	86	87	88	89	90
91	92	93	94	95	96	97	98	99	100

1st 2nd 3rd 4th deletion

If one wishes to find out whether or not a given number, say 1303, is a prime, one need not carry the sieve method right up 1303. It is sufficient to check whether 1303 is divisible by prime numbers p for which $p^2 < 1303$. The reason is that if 1303 can be factored at all, $1303 = m \cdot n$, then the square of one factor is at most 1303, that of the other at least 1303. For 1303 the division has to be tried only for the primes $p = 2, 3, 5, \ldots, 31$ because 37^2 is $1369 > 1303$. In fact, it turns that 1303 is a prime.

Endlessness of the sequence of primes. Already EUCLID (approx. 300 B. C.) raised the question whether the sequence of prime numbers breaks off or whether there are infinitely many prime numbers. He proved indirectly that *there cannot be a largest prime.* Suppose that there is a largest prime P; then one forms the natural number $N = 2 \cdot 3 \cdot 5 \cdot 7 \cdot 11 \cdots P + 1$, the product of all primes up to and including P increased by 1. This number N is not divisible by any of the prime numbers up to P, because upon each division it leaves the remainder 1. Hence it is either itself a prime or it has prime divisors that do not occur in the sequence $2, 3, \ldots, P$. Both contradict the assumption that P is the largest prime – hence the sequence of primes is infinite. The Appendix contains a table of the primes between 1 and 1000 and of their natural logarithms. The largest prime number known at present is $2^{19937} - 1$; it has 6002 digits. An unsolved problem is whether there exist infinitely many prime twins, that is, whether or not the sequence of pairs of consecutive odd numbers that are both primes, like [5; 7], [59; 61], [641; 643] or [1451; 1453], breaks off.

Common divisors and multiples. *Greatest common divisor.* If t is a divisor of a, then the factorization of t can only contain primes occurring in the factorization of a, and at most to the exponent in the factorization of a; for example, $12 \mid 60$; $12 = 2^2 \cdot 3$; $60 = 2^2 \cdot 3 \cdot 5$. If t is a *common divisor* of a and b, then t can only contain prime factors occurring in a and b, and at most to the smaller of the powers in a or b; for example, 12 is a common divisor of 48 and 360; from the prime factorizations $12 = 2^2 \cdot 3$, $48 = 2^4 \cdot 3$, and $360 = 2^3 \cdot 3^2 \cdot 5$ one sees that 48 and 360 have several common divisors: 1, 2, 3, 4, 6, 8, 12, 24. Of these $24 = 2^3 \cdot 3$ is the greatest. One says, 24 is the greatest common divisor (gcd) of 48 and 360; gcd (a, b) is the greatest among all numbers dividing both a and b. Every common divisor of a and b divides the greatest common divisor of a and b; for this is the product of all prime factors occurring both in a and b, and exactly to the smaller of the relevant powers. This is the basis for a method of finding the gcd, which is equally applicable for several numbers, as the example shows.
If two numbers a and b have no common divisor (except 1), so that gcd $(a, b) = 1$, than a and b are called *coprime* or *relatively prime*.

Example:

$$\begin{aligned} 1260 &= 2^2 \cdot 3^2 \cdot 5 \cdot 7 \\ 3024 &= 2^4 \cdot 3^3 \cdot 7 \\ 5544 &= 2^3 \cdot 3^2 \cdot 7 \cdot 11 \\ \hline \text{gcd} \quad & 2^2 \cdot 3^2 \cdot 7 = 252 \end{aligned}$$

Euclid's algorithm. For larger numbers the decomposition into prime factors is frequently very tedious, because it has to be done by trial and error; for example, 23 613 864 709 is the product of the two primes 112 843 and 209 263. If one wishes to determine the greatest common divisor of such numbers, it is appropriate to use a method that avoids the prime factorization – *Euclid's algorithm.* Without proof it is illustrated in the example of the numbers 53 667 and 25 527:

$$\begin{aligned} 53\,667 &= 25\,527 \cdot 2 + 2\,613, \\ 25\,527 &= 2\,613 \cdot 9 + 2\,010, \\ 2\,613 &= 2\,010 \cdot 1 + 603, \\ 2\,010 &= 603 \cdot 3 + 201, \\ 603 &= 201 \cdot 3 + 0. \end{aligned}$$

Hence gcd (53 667, 25 527) = 201.

For the case of coprime numbers one obtains

$$\begin{aligned} 87 &= 41 \cdot 2 + 5, \\ 41 &= 5 \cdot 8 + 1, \\ 5 &= 1 \cdot 5 + 0. \end{aligned}$$

Hence gcd (87, 41) = 1, the numbers 87 and 41 are coprime.

If one wishes to determine the gcd of more than 2 numbers by Euclid's algorithm, say of a, b and c, one can proceed step by step: one determines first gcd $(a, b) = d$, and then gcd $(a, b, c) =$ gcd (d, c).

Least common multiple. 60 is a common multiple of 6 and of 15, because 60 is a multiple of 6 as well as of 15. There are other, in fact infinitely many, common multiples of 6 and 15. For if a number m is a multiple of a and b, then all multiples of m are also common multiples of a and b. The common multiples of 6 and 15 are 30, 60, 90, 120, ..., and among them 30 is the smallest; one says: 30 is the least common multiple (lcm) of 6 and 15; it divides every other common multiple. If $m =$ lcm $(a \cdot b)$, then m must contain all prime factors occurring in the decomposition of a or of b, and each to the highest occurring power. This fact makes it possible to determine the lcm according to the following scheme for three numbers:
For larger numbers this method is again unsuitable on account of the prime factorization. An expedient is first to determine the gcd by Euclid's algorithm and then to utilize the relation lcm $(a, b) \cdot$ gcd $(a, b) = a \cdot b$. But for more than two numbers there is no such simple relationship.

Example:

$$\begin{aligned} 40 &= 2^3 \cdot 5 \\ 36 &= 2^2 \cdot 3^2 \\ 126 &= 2 \cdot 3^2 \cdot 7 \\ \hline \text{lcm} \quad & 2^3 \cdot 3^2 \cdot 5 \cdot 7 = 2520 \end{aligned}$$

Rules of divisibility. The number 84 is divisible by 4 and by 3, therefore also divisible by $4 \cdot 3 = 12$. This conclusion is only permitted if the two divisors are relatively prime. In general:

If a is divisible by m and n and if gcd $(m, n) = 1$, *then a is also divisible by* $m \cdot n$.

The determination of divisors and, if possible, the immediate recognition of divisibility by certain numbers is advantageous not only for the prime factorization, but above all also for cancellation in fractions. The relevant rules utilize simple laws of decimal writing; for example, for divisibility by 2 or by 5, multiples of 10 need not be taken into account because 10 is divisible by 2 and by 5. Similarly for multiples of 100 in respect of divisibility by 4 and 25, finally for multiples of 1 000 in respect of divisibility by 8 and 125. All the powers of 10, that is, 10, 100, 1 000 etc., leave the remainder 1 on division by 3 or by 9. From the rules for the calculation with remainders it follows that, for example, $600 = 6 \cdot 100$ then leaves the remainder $6 \cdot 1 = 6$ and for $230 = 2 \cdot 100 + 3 \cdot 10$ the remainder is $2 \cdot 1 + 3 \cdot 1 = 5$. With respect to the divisors 3 or 9 the *cross sum* of every number has the same remainder as the number itself. Here the cross sum is defined as the sum of all the digits; 7309 has the cross sum $7 + 3 + 0 + 9 = 19$ and is not divisible by 3 or 9.

All the even powers of 10, that is, 100, 10 000, 1 000 000, etc., leave the remainder 1 on division by 11, and all the odd powers (10, 1000, 100 000 etc.) leave the remainder 10 or $10 - 11 = -1$.

Here the *alternating cross sum* has the same remainder as the number itself. How to form the alternating cross sum is illustrated by an example.

Example: 8 5 9 7 6

$8 + 9 + 6 = 23$
$5 + 7 = 12$ alternating cross sum
$23 - 12 = 11$

Therefore 85 976 is divisible by 11.

A number is divisible by
- 2 if the last digit is divisible by 2;
- 4 if the last two digits represent a number divisible by 4;
- 8 if the last three digits represent a number divisible by 8;
- 5 if the last digit is divisible by 5, that is, 5 or 0;
- 25 if the last two digits represent a number divisible by 25;
- 3 if its cross sum is divisible by 3;
- 9 if its cross sum is divisible by 9;
- 11 if its alternating cross sum is divisible by 11.

Tests for accuracy. *Calculations with remainders.* To express the fact that a and b leave the same remainder r on division by d one writes $a \equiv b \pmod{d}$ (read *a congruent to b modulo d*), for example, $17 \equiv 42 \pmod{5}$. For the remainder r one then has $a \equiv r \pmod{d}$ and $b \equiv r \pmod{d}$. Thus, $17 \equiv 2 \pmod{5}$ and $42 \equiv 2 \pmod{5}$. The fact that a is divisible by d can also be written as $a \equiv 0 \pmod{d}$. The following rules hold:

Suppose that $a_1 \equiv b_1 \pmod{d}$
and $a_2 \equiv b_2 \pmod{d}$
Then: $a_1 + a_2 \equiv b_1 + b_2 \pmod{d}$
$a_1 - a_2 \equiv b_1 - b_2 \pmod{d}$
$a_1 \cdot a_2 \equiv b_1 \cdot b_2 \pmod{d}$

Example:
$22 \equiv 4 \pmod{6}$
$15 \equiv 3 \pmod{6}$
$37 \equiv 7 \pmod{6} \equiv 1 \pmod{6}$
$7 \equiv 1 \pmod{6}$
$330 \equiv 12 \pmod{6} \equiv 0 \pmod{6}$

The examples also show how to *reduce* numbers to the *simple remainder system* $0, 1, 2, \ldots, d-1$ (here $0, 1, \ldots, 5$) by adding or subtracting d suitably often. The rules for the calculation with remainders are applied in checking calculations by replacing numbers by their remainders modulo d rather than repeating the calculations with the numbers themselves; since the remainders are so easily calculated, one usually chooses $d = 9$, frequently also $d = 11$. If an inconsistency shows up, one is certain that there has been a calculating error. But when the test gives a consistent result, there is no guarantee that the calculations are correct: the error may be a multiple of d; therefore a test with $d = 2$ is of little value.

The nine test. Every number on division by 9 leaves the same remainder as its cross sum. This is a generalization of the rule for divisibility by 9. This makes it easy to carry out the nine test for the basic arithmetical operations:

The remainder on division by nine of a sum (a difference, a product) is equal to the sum (the difference, the product) of the individual remainders.

Example 1: Problem	→ cross sum	→ remainder
412	7	7
+3 964	22	+4
+4 722	15	+6
9 098	$26 \equiv 8$	$17 \equiv 8$

Example 2: Problem	→ cross sum	→ remainder
7 428	21	3
−3 986	26	−8
3 442	$13 \equiv 4$	$-5 \equiv 4$

Example 3: Problem	→ cross sum	→ remainder
617	14	5
· 382	13	· 4
234 694	$28 \equiv 1$	$20 \equiv 2$

Addition: Reduce to the simple remainder system. The calculation can only be wrong by a multiple of 9, for example, owing to the interchange of two digits.

Subtraction: Here the reduction to the simple remainder system is important if the subtraction of the residues leads to a negative result.

Multiplication: Here the remainder of the product does not agree with the product of the remainders, hence the product cannot be correct; correct is 235 694.

The eleven test. Just as the cross sum yields the remainder of a number on division by 9, so the alternating cross sum leads to the remainder of division by 11; attention has to be paid to what

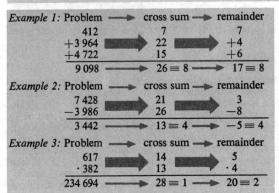

Example 4: Problem → alternating cross sum → remainder (mod 11)

2 468	$12 - 8 = 4$	4
+4 293	$5 - 13 = -8$	3
6 761	$8 - 12 \equiv -4 \equiv 7$	7

places contribute to the minuend **and** what to the subtrahend of the alternating cross sum. In the example the sum can only be wrong by a multiple of 11. If one uses the nine and the eleven test for the same problem, one obtains information on the correctness of the calculation to within multiples of 99.

1.2. The integers $\mathbb{Z}$

Foundations

Why integers? – There are situations in everyday life in which the natural numbers are insufficient to characterize certain quantities, because two opposing *tendencies*, two opposite *directions*, are possible for them; for example, the statement that a temperature is 23°C is incomplete; it has to be added whether it is measured above or below freezing point (Fig.). An amount of £ 100 is always the same. But if this sum is mentioned in connection with the property of a person, then it is important to know whether it is a credit in the savings account or a loan from the bank. For the elevation of a place it is essential whether it lies on a hill 895′ above or in a deep depression 895′ below sea level. To characterize these opposing tendencies the relevant numbers are provided with a sign, for example, +23°C and −23°C or +895′ or −895′, in chronology also −300 and +300 for events before and after the beginning of our era. A *point of reference* must exist here from which the measurements are taken. As a rule, it is laid down from a practical point of view, but in principle it is arbitrary; for example, there are scales of temperature (Fahrenheit) with a different zero-point. In cases when the variation of the quantities is in one direction only, then for a suitable point of reference the sign can be omitted (absolute temperature scale of KELVIN). Theoretically, elevations on the earth can be measured from the centre of the earth. The *positive numbers* +1, +2, +3, ... and the *negative numbers* −1, −2, −3, ..., which are obtained *when direction is taken into account*, together with zero (strictly speaking ±0), are called the *integers*.

Feasibility of subtraction. In mathematics the introduction of the integers is necessary so that subtraction, the inverse operation to addition, can always be carried out; for example, the subtraction $7 - 11$ has no solution in natural numbers. One says: the equation $x + 11 = 7$ is not soluble in the domain of the natural numbers.

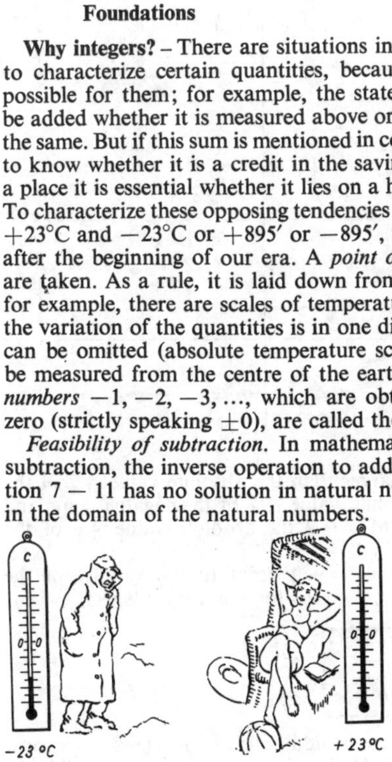

1.2-1 Temperature 23°C

The integers form a number domain in which every subtraction problem has a solution.

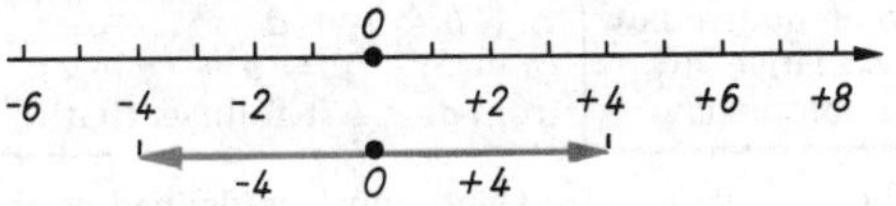

1.2-2 Number line and the numbers (+4) and (−4) opposite to one another

Opposite numbers. Just as the number ray illustrates the natural numbers, so the number line serves to illustrate the integers (Fig.). The non-negative integers correspond to the natural numbers. On the number line there is, for every integer other than 0, exactly one having the same distance from zero, but lying on the other side. Two such numbers which differ only by their sign are called *opposite*, for example, −4 and +4. The formation of the opposite number is expressed by a minus sign as prefix, so that $-(-4) = +4$ and $-(+4) = -4$. For the zero point one sets $-0 = 0$.

Absolute value. Two opposite numbers on the number line having the same distance from zero are said to have the same *absolute value*. The absolute value is defined as the non-negative one of the two numbers:

$$|a| = \text{for } a \geqslant 0 \quad \text{and} \quad |a| = -a \quad \text{for} \quad a < 0.$$

Order. Of two different integers the smaller one is that which lies further to the left on the number line. For any two integers n_1 and n_2 there always holds exactly one of the three relations $n_1 < n_2$, or $n_1 = n_2$, or $n_1 > n_2$, for example, $-3 < +2$, $+5 < +7$, $+8 = +8$, $-1 > -7$, $+3 > -5$. Every integer has exactly one immediate predecessor and exactly one immediate successor, that is, the sequence of integers contains neither a smallest nor a largest, neither a first nor a last, number.

Calculations with the integers Z

Arithmetical operations of the first kind. In order to distinguish between the *computational* or *operational* symbols + and − and the signs, which have the same outward appearance, one encloses the complete number symbols (with sign) in brackets (Fig.). The minus sign − has the additional function of indicating opposite numbers.

To define the addition of integers one is guided by the addition of natural numbers, that is, $(+3) + (+4) = +7$ because $3 + 4 = 7$.

If the two summands have the same sign, then the sum of the natural numbers corresponding to their absolute values is provided with the sign of the summands.

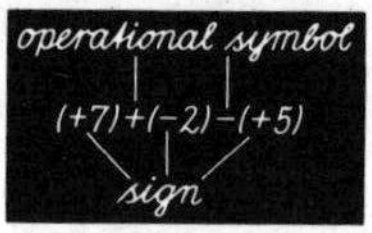

1.2-4 Sign and operational symbol

In $(-3) + (-2) = -5$ the sum $3 + 2 = 5$ receives the negative sign.

If the two summands have different signs, then the difference of the natural numbers corresponding to their absolute values is provided with the sign of the summand with the larger absolute value.

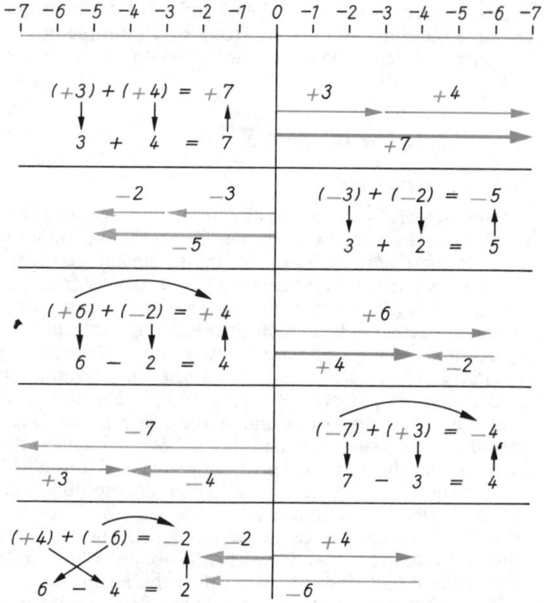

1.2-3 Addition on the number line

In $(+6) + (-2) = +4$ the positive summand $(+6)$ is larger than the absolute value $+2$ of the negative summand (-2); in $(-7) + (+3) = -4$ the absolute value $+7$ of the negative summand is larger than the positive summand, and in $(+4) + (-6) = -2$ the absolute value $+6$ of the negative summand is greater than the positive summand (Fig.).

Examples such as $(+4) + (-6) = -2$ and $(-6) + (+4) = -2$ point to the validity of the commutative law. The other *laws of addition* also hold and are stated for arbitrary integers.

Commutative law	$a + b = b + a$
Associative law	$(a + b) + c = a + (b + c)$
Monotonic law	from $a < b$ it follows that $a + c < b + c$

The *subtraction* of integers must be defined as the converse to addition; $(-7) - (+3) = x$ must have the same meaning as $x + (+3) = -7$. But in accordance with the definition of addition $(-10) + (+3) = -7$, consequently $x = -10$, that is, $(-7) - (+3) = -10$.

On the other hand, $(-7) + (-3) = -10$. Hence for every subtraction:

An integer is subtracted by adding the number opposite to it.

Therefore subtraction can be carried out without restriction in the domain of the integers, because this is so for addition, and every integer has its opposite number; for example:

$$(+28) - (-16) = (+28) + (+16) = +44.$$

Algebraic sums. Since among integers every subtraction can be replaced by a corresponding addition, one defines as algebraic sums expressions in which the terms are combined only by operations of the first kind. If they contain more than two summands, then the computation is performed conveniently as follows:

$$\begin{aligned} &(+15) - (+27) + (-11) - (-9) + (+31) && \text{transform} \\ &= (+15) + (-27) + (-11) + (+9) + (+31) && \text{rearrange} \\ &= (+15) + (+9) + (+31) + (-27) + (-11) && \text{combine} \\ &= (+55) + (-38) && \text{final addition} \\ &= +17. \end{aligned}$$

Arithmetical operations of the second kind. To define the multiplication of integers one is also guided by the multiplication of natural numbers and then proceeds step by step.

If both factors are positive, their product is the positive number corresponding to the product of the corresponding natural numbers.

The product $(+4) \cdot (-7)$ is interpreted by analogy to $4 \cdot 7 = 7 + 7 + 7 + 7$ as repeated addition of equal summands and is determined by $(+4) \cdot (-7) = (-7) + (-7) + (-7) + (-7) = -28$. If the multiplicator is negative, one agrees that the commutative law remains valid: $(-7) \cdot (+4) = (+4) \cdot (-7)$.

The product of two factors of opposite sign is negative, and its absolute value is the product of the absolute values of the factors.

$$(+3) \cdot (-7) = -21 \qquad (-5) \cdot (+8) = -40$$
$$3 \cdot 7 = 21 \qquad 5 \cdot 8 = 40$$

A comparison of the stipulations made so far shows that the sign of a product changes when that of one of the factors changes. Therefore one agrees to set: $(-4) \cdot (-7) = +28$.

The product of two negative factors is positive, and its absolute value equal to the product of the absolute values of the factors.

Summary of multiplication:

If two integers have equal signs, their product is positive, otherwise negative; the absolute value of the product is equal to the product of the absolute values.

Commutative law	$a \cdot b = b \cdot a$
Associative law	$(a \cdot b) \cdot c = a \cdot (b \cdot c)$

Examples:
1. $(-13) \cdot (+5) = -65$. *2.* $(-8) \cdot (-12) = +96$.
3. $(+3) \cdot (-4) \cdot (-9) = (+3) \cdot (+36) = +108$.
4. $(-3)^4 = (-3) \cdot (-3) \cdot (-3) \cdot (-3) = +81$.

Laws of multiplication. As is evident from the introduction of multiplication and the example of three factors, for the multiplication of integers a, b, c the commutative and associative law hold. For natural numbers ($c > 0$) the monotonic law holds: from $a < b$ it follows that $a \cdot c < b \cdot c$. This law does not hold for integers; for negative c it follows from $a < b$ that $ac > bc$, as is shown by the following example:

$$(+5) < (+7), \quad \text{but} \quad (+5) \cdot (-3) > (+7) \cdot (-3).$$

Division. Division is the inverse to multiplication. Therefore $(+12) : (-4) = x$ is equivalent to $(-4) \cdot x = +12$. But this holds for $x = -3$ only. Similarly one can derive the rules of division for all combinations of signs. In general:

If dividend and divisor have the same sign, the quotient is positive, otherwise negative; its absolute value is equal to the quotient of the absolute values.

Examples: $(+72) : (+6) = +12$, $(+119) : (-17) = -7$,
$(-75) : (+25) = -3$, $(-91) : (-7) = +13$.

Since the arithmetical operations for the *non-negative integers*, that is, the *positive integers* and *zero*, have been determined just as for natural numbers, the sign + of positive integers can be omitted. They are replaced by the corresponding natural numbers and lead to a simpler representation; for example, $(+9) + (-17) - (+6) + (+21) - (-2) = 9 - 17 - 6 + 21 + 2 = 9$ or $7 \cdot (-9) = -63$ or $(-56) : (-7) = 8$. Whether they are then natural numbers of counting character or positive integers will be clear from the appropriate context.

On the history of the integers. The negative numbers are not known in Greek mathematics, but first traces can be found in the writings of DIOPHANTOS (around 250 A. D.). In India (around 700 A. D.) the calculation with negative numbers was already completely developed by the Hindu. It is interesting that their names for *positive* and *negative* are derived from their words for credit and debit. The negative integers play an important role in the Hindu theory of equations.

In Europe the negative numbers gained a foothold comparatively late; the reason is probably that the *Arabs*, who formed the mathematical bridge between India and Europe, refused to accept the negative numbers. The break-through was made by Michael STIFEL in his *Arithmetica integra* (1544). The ultimate foundation of the integers within mathematics was not made until 1867, by Hermann HANKEL.

1.3. The rational numbers Q

Foundations

What are fractions? – If 6 apples have to be shared equally among 3 children, one calculates $6:3 = 2$ and then knows that each child receives 2 apples. But if only 2 apples are available for sharing out, one has to solve the division problem $2:3$. Within the natural numbers this problem cannot be solved. Nevertheless one accomplishes division by resulting to the knife (Fig.). In this case the share of each child is indicated by the fraction 2/3. All similar cases of sharing lead to fractions.

Explanations. Every fraction is of the form $\frac{p}{q}$. The numerator p indicates the number of the entities divided, the denominator q the number of parts. The fraction line runs horizontally. If no confusion is to be feared, a slanting fraction line (*solidus*) is allowed, for example 3/4, particularly in a running text. Fractions whose numerator is 1 are called *unit fractions*, for example 1/3; 1/8; 1/12. Fractions whose numerator is smaller than the denominator are called *proper*; for example, 2/3, 1/7, 5/9, 10/11. Fractions in which the numerator is greater than or equal to the denominator are called *improper*, for examples 3/2, 16/3, 9/8, 5/5.

Fractions arise when one or several whole entities are divided up.

The denominator 0 is always excluded.

1.3-1 2 apples for 3 children

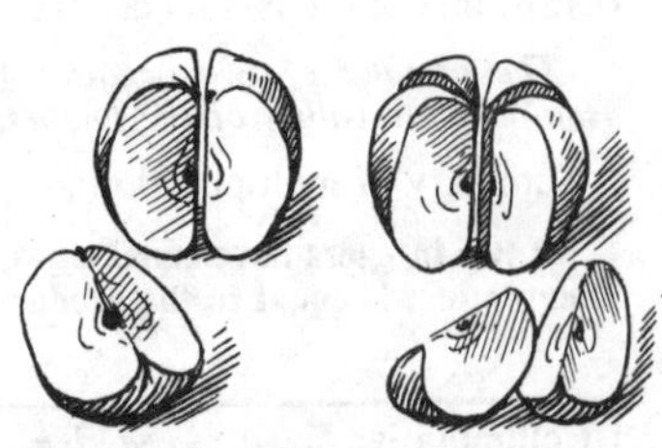

1.3-2 One third is the same as two sixths

If the numerator of one fraction is equal to the denominator of another and vice versa, the two fractions are called *reciprocal*; for example 3/5 and 5/3, 17/6 and 6/17.

Numerator or denominator of a fraction may be negative, for example, $-3/5$, $-2/-9$, $7/-4$. By the sign rules for the calculations with integers $-3/5 = 3/-5 = -\ 3/5$, $-3/-5 = 3/5$. Usually the signs are written before the fraction line or solidus rather than in numerator or denominator. With this convention the definitions above of proper and improper fractions also hold for negative fractions. The fact that in what follows positive fractions are predominantly used in examples is motivated only to simplify the presentation; with the appropriate modifications everything holds equally well for negative fractions.

Equivalent fractions. If one divides an apple into three parts (Fig.) and takes one part, one has the same quantity as when one takes two parts of a division into six parts, $1/3 = 2/6$. Similarly, for example, $2/5 = 4/10$, $5/3 = 20/12$, $2/3 = 4/6 = 6/9 = \cdots = 24/36 = \cdots$.

Extension. If two fractions are such that numerator and denominator of one fraction are equal multiples of numerator and denominator of the other fraction, for example, $8/9 = 40/45$, then the second fraction is said to arise from the first by *extension*:

Extension	$\frac{a}{b} = \frac{ac}{bc}$

To extend a fraction means to multiply numerator and denominator by the same number $c \neq 0$.

Cancellation. The inverse process is called *cancellation* of a fraction.

Cancellation	$\frac{ac}{bc} = \frac{a}{b}$

To cancel a fraction means to divide numerator and denominator by the same number $c \neq 0$.

Every fraction in which numerator and denominator have common factors can be cancelled. In $2/7 = (2 \cdot 3)/(7 \cdot 3) = 6/21$ the transition from left to right is extension, from right to left is cancellation. Since a cancellation diminishes numerator and denominator, it is usually advantageous to cancel fractions as far as possible. The fraction is then called *reduced*.

Rational numbers. All fractions that represent the same quantity, that is, can be carried into one another by extension or cancellation or both, for example, (3/4, 6/8, ..., 27/36, ...) are combined into a single number, a so-called *rational number*. The fractions 3/4, 6/8, 27/36 are merely different espressions for one and the same rational number. It is customary to write this number in the reduced form, in this case 3/4. Consequently, 3/4, like all reduced fractions, has a two-fold meaning: firstly it is a fraction, secondly it represents a *rational number* and stands for the totality of all fractions arising from extension, which are different expressions for one and the same number. In computations every expression of a rational number can be replaced by any other expression for the same number according to circumstances. Fractions with the denominator 1 and those that arise by extension from them, like $3/1 = 6/2 = \cdots = 18/6 = \cdots$ are subsumed among the rational numbers and are equivalent to integers, for example, $15/3 = 5/1 = 5$, $8/8 = 1/1 = 1$. Here, too, one expression can be replaced, if necessary, by another. The integer 0 is represented by all fractions having the numerator 0.

Order of the rational numbers. Just as for natural numbers and integers, so for any two rational numbers r_1 and r_2 one and only one of the relations $r_1 < r_2$, or $r_1 = r_2$, or $r_1 > r_2$ holds. For positive a, b, c and $a < b$ one always has $a/c < b/c$ and $c/a > c/b$. For equal denominators the fraction with the larger numerator represents the larger number, for example, $2/7 < 6/7$. For equal

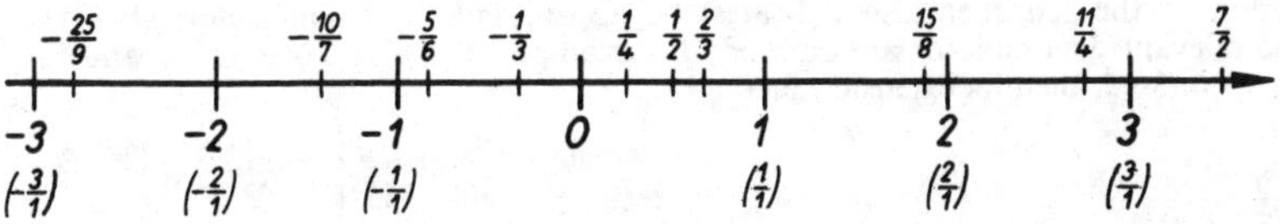

1.3-3 Number line

numerators the fraction with the smaller denominator represents the larger number, for example $5/9 < 5/6$. If for two positive rational numbers one wishes to determine which one is the larger, one finds representations for them with *equal denominators* and compares the numerators; from $7/12 = 35/60$ and $11/20 = 33/60$ it follows that $11/20 < 7/12$. Two such expressions with equal denominators can always be found for a/b and c/d; in any case the product $b \cdot d$ of the two denominators is a common denominator. The appropriate numerators are then $a \cdot d$ and $b \cdot c$.

$a/b < c/d$ if and only if $a \cdot d < b \cdot c$, provided that $b, d > 0$.

In the domain of rational numbers, too, there is *neither a smallest nor a largest number*. The *successor relation also fails to hold*: nò rational number has an immediate precessor nor an immediate successor.

Between any two distinct rational numbers r_1 and r_2, with $r_1 < r_2$, say, there always lie further, in fact infinitely many, rational numbers r satisfying $r_1 < r < r_2$.

On account of their order the rational numbers can be represented by points or arrows on the number line (Fig.). Here every point is given by one expression for the rational number. In illustrating all fractions the totality of expressions for a rational number would occupy the same place. On the number line the smaller rational number always stands to the left of the larger.

Calculations with common or vulgar fractions

Calculations with rational numbers are explained in the first instance by their expression as common or vulgar fractions. For example, one talks of the *addition of fractions with different denominators*; it would be more accurate, though more cumbrous, to talk of the *addition of rational numbers given by fractions with equal denominators*. Subsequent sections then treat the decimal fractions as a further expression of rational numbers for which the computations are somewhat different. The name *common* or *vulgar* fraction, which originally emphasized the distinction to the sexagesimal fraction, nowadays emphasizes the distinction to the decimal fraction and has nothing to do with vulgarity.

Addition and subtraction. The fractions concerned can have equal or unequal denominators.

Fractions with equal denominators are added or subtracted by adding or subtracting their numerators; the denominator remains unchanged.

$$\frac{a}{c} \pm \frac{b}{c} = \frac{a \pm b}{c}$$

Examples: 1. $3/7 + 5/7 = 8/7$. 2. $4/11 - 7/11 = -3/11$.
3. $5/17 + 9/17 - 18/17 + 13/17 - 2/17 = 7/17$

Consequently, every *improper fraction* can be split into two summands of which the first is an integer and the second a proper fraction: $8/7 = 7/7 + 1/7$, $22/5 = 20/5 + 2/5$. Improper fractions are therefore frequently written as *mixed numbers*, for example, $8/7 = 1^1/_7$; $22/5 = 4^2/_5$; an addition sign between the integer and the proper fraction has to be imagined.

$$\frac{a}{b} \pm \frac{c}{d} = \frac{ad}{bd} \pm \frac{bc}{bd} = \frac{ad \pm bc}{bd}.$$

Fractions with unequal denominators are added or subtraced by first writing them in forms with equal denominators and then adding or subtracting the numerators.

In the simplest case one denominator is a common multiple of all the others; for example, for 2/3 — 7/12 + 5/4 the number 12 is the least common denominator. Therefore one extends 2/3 and 5/4 by 4 or 3 so that they have the denominator 12; 2/3 — 7/12 + 5/4 = 8/12 — 7/12 + 15/12 = 4/3 = $1^1/_3$. As a rule, one simplifies by writing the second sum at once in the form of a fraction whose numerator is the sum of the individual numerators: 2/3 — 7/12 + 5/4 = (8 — 7 + 15)/12 = 4/3 = $1^1/_3$.

For 1/6 + 3/10 — 11/15 a common multiple of the individual denominators has to be found first. This can be done by guessing, say 60, or by taking the product. The fractions then have to be extended in turn by 10, 6 and 4. In order to keep the numerators as small as possible, one chooses for the least common denominator lcd the least common multiple lcm of the individual denominators. The lcm is found in the usual way, for example by decomposition into prime factors. The product representation of the lcd then also indicates the appropriate extension factor e.f. if the prime factors of the relevant denominator are omitted; for example, if the lcd is $2 \cdot 3 \cdot 5$, and the last denominator is $15 = 3 \cdot 5$, then its extension factor is 2.

Example:

	e.f.
$6 = 2 \cdot 3$	5
$10 = 2 \cdot 5$	3
$15 = 3 \cdot 5$	2
lcd $2 \cdot 3 \cdot 5 = 30$	

$1/6 + 3/10 - 11/15 = (5 + 9 - 22)/30 = -8/30 = -4/15.$

Example: $3^{17}/_{21} - 3/8 - 11/12 + 2 = 80/21 - 3/8 - 11/12 + 2$

	e.f.
$21 = 3 \cdot 7$	$2^3 = 8$
$8 = 2^3$	$3 \cdot 7 = 21$
$12 = 2^2 \cdot 3$	$2 \cdot 7 = 14$
lcd $2^3 \cdot 3 \cdot 7 = 168$	

In determining the lcd by prime factorization of the individual denominators, integers can be ignored, because they have the denominator 1. For example, having found the lcd 168 one obtains

$$\frac{80}{21} - \frac{3}{8} - \frac{11}{12} + 2 = \frac{640 - 63 - 154 + 336}{168} = \frac{759}{168} = \frac{253}{56} = 4\frac{29}{56}.$$

For 3/5 + 1/4 — 2/9 the denominators have no common factors, they are *coprime in pairs.* The lcd is therefore the product of the individual denominators; the result is 113/180.

Multiplication and division. The arithmetical operations of the second kind on common fractions are easier to carry out than those of the first kind, because the determination of the lcd is not necessary. On need not distinguish between operations on fractions with equal or unequal denominators.

$$\frac{a}{b} \cdot \frac{c}{d} = \frac{a \cdot c}{b \cdot d}$$

The product of fractions is a fraction; its numerator is the product of the numerators, its denominator the product of the denominators.

Integers are again interpreted as fractions with the denominator 1. In order to avoid large numbers, which can lead to errors, one cancels as far as possible before multiplying out.

Examples:

1. $\frac{2}{5} \cdot \frac{3}{8} = \frac{2 \cdot 3}{5 \cdot 8} = \frac{1 \cdot 3}{5 \cdot 4} = \frac{3}{20}$. *2.* $\frac{4}{7} \cdot 3\frac{9}{32} = \frac{4}{7} \cdot \frac{105}{32} = \frac{4 \cdot 105}{7 \cdot 32} = \frac{1 \cdot 15}{1 \cdot 8} = \frac{15}{8} = 1\frac{7}{8}$.

3. $7 \cdot \frac{9}{28} \cdot \frac{2}{3} = \frac{7 \cdot 9 \cdot 2}{1 \cdot 28 \cdot 3} = \frac{1 \cdot 3 \cdot 1}{1 \cdot 2 \cdot 1} = \frac{3}{2} = 1\frac{1}{2}$. *4.* $\frac{11}{17} \cdot \frac{17}{11} = \frac{11 \cdot 17}{17 \cdot 11} = \frac{1 \cdot 1}{1 \cdot 1} = 1$.

5. If a tug pulling barges has the velocity of $4^1/_2$ m.p.h., then in $2^3/_4$ hours it travels $9/2 \cdot 11/4$ miles = 99/8 miles = $12^3/_8$ miles.

Example 4 shows that the product of two reciprocal fractions is 1. This property can be utilized in a simple definition of reciprocity of rational numbers:

Two rational numbers are reciprocal to each other if and only if their product is 1.

Accordingly, for every rational number other than 0 there is a reciprocal, for example —3 and —1/3 are reciprocal, because $-3 \cdot (-1/3) = 1$.

$$\frac{a}{b}:\frac{c}{d}=\frac{a}{b}\cdot\frac{d}{c}$$ **The division by a fraction is carried out as multiplication by its reciprocal.**

The validity of this method of division is shown by the following argument: $2/3 : 3/4 = 2/3 \cdot 4/3 = 8/9$; since division is the inverse to multiplication, the product of quotient and divisor must give the dividend, and in fact, $8/9 \cdot 3/4 = 2/3$, or generally $\frac{ad}{bc}\cdot\frac{c}{d}=\frac{a}{b}$.

Examples: 1. $\frac{7}{12}:\frac{5}{8}=\frac{7\cdot 8}{12\cdot 5}=\frac{7\cdot 2}{3\cdot 5}=\frac{14}{15}$. *2.* $\frac{3}{5}:6=\frac{3\cdot 1}{5\cdot 6}=\frac{1}{10}$.

3. $5:\frac{6}{7}=\frac{5\cdot 7}{1\cdot 6}=\frac{35}{6}=5\frac{5}{6}$. *4.* $2\frac{4}{13}:\frac{16}{39}=\frac{30\cdot 39}{13\cdot 16}=\frac{15\cdot 3}{1\cdot 8}=\frac{45}{8}=5\frac{5}{8}$.

5. $\frac{11}{12}:\frac{11}{12}=\frac{11\cdot 12}{12\cdot 11}=1$.

6. A motor scooter covering $58^1/_2$ miles in two and a quarter hours has an average speed of 117/2 miles : 9/4 h = 117/2 · 4/9 m.p.h. = 2 · 13 m.p.h. = 26 miles per hour.

Division of rational numbers has been reduced to multiplication; therefore:

Among rational numbers division – except by zero – can always be carried out.

Double fractions. The division sign and the fraction line or solidus are generally interchangeable, for example, $2:3 = 2/1 : 3/1 = 2/1 \cdot 1/3 = 2/3$. Consequently every division of common fractions can be represented as a double fraction whose numerator and denominator are not integers but fractions. Conversely, every double fraction can be simplified by division. But an expression is uniquely determined only when the principal fraction line is clearly indicated, for example, by its length or by the position of a following equality sign.

Examples: 1. $\dfrac{3}{\frac{7}{5}}=3:\frac{7}{5}=3\cdot\frac{5}{7}=\frac{15}{7}$. *3.* $\dfrac{\frac{1}{3}}{9}=\frac{1}{3}:9=\frac{1}{3}\cdot\frac{1}{9}=\frac{1}{27}$.

2. $\dfrac{3}{\frac{7}{5}}=\frac{15}{7}$, but $\dfrac{\frac{3}{7}}{5}=\frac{3}{35}$. *4.* $\dfrac{\frac{2}{5}}{\frac{3}{8}}=\frac{2}{5}:\frac{3}{8}=\frac{2}{5}\cdot\frac{8}{3}=\frac{16}{15}=1\frac{1}{15}$.

The commutative and associative law of addition and of multiplication as well as the distributive law, which were previously stated for integers, also hold for computations with rational numbers.

Decimal fractions

Foundations. In a positional system the digits within a number symbol have a positional value apart from their numerical value; for example, in 3752 the 5 indicates by its position 5 Tens (T). In a *decimal positional system* every positional value is 1/10 of that to its left. As long as only natural numbers or integers are presented in this system, the units (U) must occupy the last position.

The system can be continued beyond the units to represent the rational numbers. After the units one has from left to right the positional values *one tenth* t, *one hundredth* h, *one thousandth* th etc. If previously the number of positions was bounded to the right by the units, it is now unbounded in both directions. Here a particular position must be distinguished, as point of reference, as it were. For this purpose a stop is placed between the units and the tenths.

This agrees with the treatment of decimal measures: 7.5 cm means 7 cm and 5 mm because a millimetre is a tenth of a centimetre and 3.75 m means 3 m and 75 cm because 1 m and 100 cm are the same length.

The places after the decimal point are read individually (for 2.31 read *two-point-three-one*) because otherwise ambiguities could occur; for example: which number is larger: three-point-eleven or three-point-nine? – The places after the point are called the *decimals*. The first decimal therefore represents tenths, the second hundredths etc. The number 4.81 has three places but only two decimals. Every number other than an integer, written down in the decimal system, for example, 0.375 or 17.8, is called a *decimal fraction*. This has to be distinguished from a vulgar fraction whose denominator is a power of 10, that is, 10, 100, 1000 etc., for example, 3/10, 17/100 000.

Transformations. *From vulgar fractions to decimal fractions.* Every fraction whose denominator is a power of 10 can immediately be written as a decimal fraction by putting its numerator into the position in the decimal system indicated by the denominator.

Examples: 1. $3/10 = 0.3$. *2.* $23/100 = (20 + 3)/100 = 2/10 + 3/100 = 0.23$.
3. $70\,{}^{105}/_{1000} = 70.105$.

Since $10 = 2 \cdot 5$, all powers of 10 contain only the prime factors 2 and 5. Therefore all fractions whose denominator contains no other prime factors can be extended so that the denominator is a power of 10 and can then be written as a decimal fraction: $7/20 = 35/100 = 0.35$; $1\,{}^{3}/_{8} = 1375/1000 = 1.375$. Such a transformation is impossible if the denominator of the common fraction contains in the reduced form prime factors other than 2 and 5. Such vulgar fractions cannot be written as decimal fractions in the previous form. Here the following arguments are helpful. Within the domain of natural numbers the division $2:7$ cannot be performed. In the domain of rational numbers there are two possibilities:

a) $2:7 = 2/1 : 7/1 = 2/1 \cdot 1/7 = 2/7$;
b) $2:7 = 0.285\,714\,28 \ldots$

```
20
 60
  40
   50
    10
     30
      20
       60
        ...
```

Mentally:

2 divided by 7, quotient 0, remainder $2 = 20/10$
20/10 divided by 7, quotient 2/10, remainder $6/10 = 60/100$
60/100 divided by 7, quotient 8/100, remainder $4/100 = 40/1000$
etc.

The rearrangement of the remainder to the next power of 10 corresponds to advancing by one decimal. The method goes like written division of natural numbers; on transition from units to tenths in the dividend the same has to be done in the quotient, that is, a point has to be placed.

The two results of the division $2:7$ are equated, $2/7 = 0.285\,714\,28 \ldots$ On division by 7 the only possible remainders are 1, 2, 3, 4, 5, 6. The remainder 0 is excluded because none of these numbers multiplied by 10 is divisible by 7. This means that the decimal fraction is *infinite*, that the sequence of its digits never breaks off. These digits must repeat as soon as a remainder occurs for the second time. The decimal fraction is *periodic*, and on division by 7 the *period* can have at most 6 digits.

If p/q is a reduced fraction and if q contains prime numbers other than 2 and 5, then the appropriate decimal fraction is periodic, and its period has at most $q - 1$ digits.

Periodicity is indicated by writing down the period once only and placing a bar on top:

$$1/3 = 0.33\ldots = 0.\overline{3}; \quad 34/99 = 0.\overline{34}; \quad 17/12 = 1.41\overline{6}; \quad 11/26 = 0.4\overline{230769}$$

(read *nought-point-three-four-period-three-four* or *one point four-one-six-period six*).

In the first two examples the decimal fractions are *purely periodic*: the period begins immediately after the decimal point. The last two examples have digits between the decimal point and the beginning of the period; such decimal fractions are called *mixed*; they arise always if the denominator contains among others the factors 2 or 5.

From a decimal fraction to a common fraction. The transformation of a finite decimal fraction into a common fraction follows from its definition, for example, $0.17 = 1/10 + 7/100 = (10 + 7)/100 = 17/100$; $6.05 = 605/100 = 121/20$. One places the digits of the decimal fraction, omitting the point and its "initial zeros" as the numerator and as the denominator the power of 10 that corresponds to the number of decimals. Also, every periodic decimal fraction can be transformed into a common fraction. For a purely periodic decimal fraction one places the digits of the period into the numerator, and for the denominator one takes the power of 10 corresponding to the length of the period, diminished by one; for example $0.\overline{3} = 3/9 = 1/3$; $0.\overline{27} = 27/99 = 3/11$; $0.\overline{253} = 253/999$.

Examples: 1.
$$\begin{aligned} p/q &= 0.\overline{369} \\ 1000p/q &= 369.\overline{369} \\ \hline 999p/q &= 369 \\ p/q &= 369/999 \\ p/q &= 41/111 \end{aligned}$$

2.
$$\begin{aligned} p/q &= 0.3\overline{58} \\ 100p/q &= 35.8\overline{58} \\ \hline 99p/q &= 35.5 = 355/10 \\ p/q &= (355/10)/99 = 355/990 \\ p/q &= 71/198 \end{aligned}$$

This method of transformation is based on the fact, so far unproved, that calculations with infinite periodic decimal fractions can be performed just as with finite ones. The application of the method, which is illustrated in the examples *on the left*, also assumes some knowledge of the working with equations. The application of the rule stated leads to the same result if one splits and transforms:

$$0.3\overline{58} = 0.3 + 0.0\overline{58} = 0.3 + 0.\overline{58} \cdot 0.1 = \frac{3}{10} + \frac{58}{99} \cdot \frac{1}{10} = \frac{297 + 58}{990} = \frac{355}{990} = \frac{71}{198}.$$

Every common fraction can be written as a finite or a periodic decimal fraction. Every finite and every periodic decimal fraction can be transformed into a common fraction. Common fractions on the one hand, and finite or periodic decimal fractions on the other hand, are two distinct ways of writing the same kind of number: the rational numbers.

In order to avoid the distinction of the two possibilities – finite or periodic decimal fraction – one can argue as follows: by the method indicated it can be shown that $0.\overline{9} = 1$. Therefore every finite decimal fraction and accordingly every integer can be transformed into a periodic decimal by diminishing the last non-zero digit by 1 and attaching the period 9:

$$0.84 = 0.83\overline{9}; \quad 3.156 = 3.155\overline{9}; \quad 17 = 16.\overline{9}.$$

Computations with decimal fractions

Only finite decimals are treated here; periodic decimal fractions have to be rounded off suitably before the computation; or one has to calculate with common fractions.

Addition and subtraction. In written addition and subtraction of decimal fractions one proceeds just as for natural numbers or integers: equal places are written one under the other, decimal point under decimal point, one proceeds column by column from right to left, taking account of the appropriate transfer, and on transition from tenths to units the decimal point is placed in the result. The simplicity of the arithmetical operations of the first kind is an essential advantage of the decimal fractions compared with common fractions.

Examples: 1.
```
  713.25
+   1.085
+  22.9
---------
  737.235
```
2.
```
  38.023
 - 9.13
 - 0.0258
---------
  28.8672
```

Example:
```
0.175 · 3.5
-----------
    525
     875
-----------
  0.6125
```

Multiplication. Every finite decimal fraction can be transformed to a common fraction whose denominator is a power of 10. The multiplication of such fractions can be carried out in the usual way, for example, $0.175 \cdot 3.5 = 175/1000 \cdot 35/10 = (175 \cdot 35)/(1000 \cdot 10) = 6125/10\,000$, without any cancellation. The denominator of the result is again a power of 10; the numerator is the product of the individual numerators, and when the result is changed into a decimal fraction, there are as many decimals as in the factors taken together. The calculation on the right gives the same result.

Two decimal fractions are multiplied by multiplying irrespective of the decimal point as for natural numbers and attaching to the result as many decimals as the factors have, taken together.

Multiplication by powers of ten is performed simply by changing the points by as many decimals to the right as the power has zeros: $7.136 \cdot 100 = 713.6$.

Division. A quotient remains unchanged when dividend and divisor are multiplied by the same number (see Extension of fractions), for example, $12:4 = 48:16 = 120:40 = 1.2:0.4 = 6:2 = 3$. In order to imitate the division of natural numbers, one rearranges dividend and divisor, making use of this fact, so that the *divisor becomes an integer*, preferably by multiplication by a power of 10, for example, $33:6.5 = 330:65$; $6.729:13.58 = 672.9:1358$.
Now the division can be performed just as with natural numbers, and on transition from units to tenths in the dividend the same transition is made in the quotient by placing a decimal point.

Examples: 1. 47.275 : 3.1
```
         15.25
   31) 472.75
       162
        77
        155
          0
```
2. $714.5:100 = 7.145$
3. $1.92:1000 = 0.00192$

If a number a is to be multiplied (divided) by a power b of 10, say $b = 10^k$, then the decimal point of a is shifted to the right (left) by as many places as b has zeros, namely k.

Abbreviated methods of calculation. Multiplication and division of decimal fractions, in general, yield results having more places than the initial numbers. If the initial numbers are not absolutely exact, but are *approximations,* rounded off or carrying measuring errors, then these places are inadmissible or, more precisely, meaningless, because they give a false impression of a non-existing calculating or measuring accuracy (see Chapter 28.).

In calculations of the first kind with approximate numbers the result must not show more *reliable decimal places* than the smallest number of decimal places among the original numbers. On multiplication and division the result has only as many *valid digits* (not decimals!) as the original number with the smallest number of valid digits.

Valid digits of a number are all its digits except the zeros before the first non-zero digit: for example 307.6 as well as 0.0002643 have four valid digits. To save the calculation of places going beyond the reliable number of places one uses abbreviated methods in which the result straightaway has only the required or the admissible number of places.

Abbreviated addition and subtraction. If the summands have the same number of reliable decimals, then addition and subtraction proceed in the usual way. If the decimal numbers are unequal, then one proceeds as follows: let k be the smallest occurring number of decimals, then all values given with greater accuracy are rounded off to $k + 1$ decimals and are then added or subtracted, where the last place is only taken into account at the transfer, so that the final result has k decimals. If for the sum or difference initially an accuracy up to the kth decimal is postulated, then one chooses the summands also, as far as possible, accurate up to the $(k + 1)$th decimal.

Example:

2.7362	→	2.736
+ 0.8749		0.875
+17.53		17.53
+ 8.665		8.665
		29.81

Example: 27.8673 · 49.23

27.8673 · 49.23	$27\dot{8}.\dot{6}\dot{7} \cdot 4.923$
1114692	111468
2508057	25080
557346	557
836019	83
1371.907179	1371.9 ≈ 1372

Abbreviated multiplication. What matters here is not the number of decimal places but of valid digits. If one factor has k valid digits and the other more, the latter is rounded off to $k + 1$ digits (1 extra place). It is advantageous to write the factor with $k + 1$ digits as multiplicator and, particularly for large k, to see to it by suitable transformations that the multiplicand only has units before the decimal point, for example, $27.8673 \cdot 49.23 = 278.673 \cdot 4.923$. The subsequent procedure is best illustrated by placing next to one another the normal and the abbreviated method of multiplication.

While in the first multiplication the entire multiplicator occurs, in the subsequent partial products place after place are omitted. To avoid errors one marks each time the digit that is only used for carrying to the next higher place, by setting a stop over it. For example, the third partial product digit 6 is marked, and one calculates: $2 \cdot \dot{6} = 12$ (carry 1), $2 \cdot 8 + 1 = \mathbf{17}$, $2 \cdot 7 + 1 = \mathbf{15}$, $2 \cdot 2 + 1 = \mathbf{5}$. In the final addition of the partial products the last place is only taken into account for carrying. If then the final result, as in the example above, still shows one place beyond the maximal number of valid digits, a further rounding off has to be done. If in the calculation of a product accuracy of k places is postulated, then one performs the calculation, as far as possible, with factors having each $k + 1$ valid digits, and then rounds off.

Abbreviated division. The number of valid digits to be shown in the quotient is also equal to the smallest number of valid places in the original numbers (dividend or divisor), so that one can round off from the outset leaving one extra place. If the quotient is required to k places, one chooses, if possible, $k + 1$ places in dividend and divisor; in the kth place of the quotient rounding off has to be observed. To illustrate the method the abbreviated division is again placed next to the normal division. The quotient 674.283 divided by 439.17 has to be calculated to three places.

Example:

1.535	1.54
43917) 67428.3	$43\dot{9}\dot{2}$) 6743
43917	4392
235113	2351
219585	2196
155280	155
131751	176
235290	−21

Instead of adding zero each time to the remainder, in the abbreviated division the divisor is shortened each time by one place. However, this place has to be taken into account for carrying in the formation of the intermediate product: $2351 : 439 = 5$; $5 \cdot \dot{2} = 10$ (carry 1); $5 \cdot 9 + 1 = \mathbf{46}$; $5 \cdot 3 + 4 = \mathbf{19}$; $5 \cdot 4 + 1 = \mathbf{21}$. At the next step: $155 : 44 = 4$, because $44 \cdot 4 = 176$ is nearer to 155 than $44 \cdot 3 = 132$.

Historical remarks. The theory of the common or vulgar fractions and the calculations with them, as they are conducted nowadays, is the achievement of the Hindu (BRAHMAGUPTA). From there fractions came to us by way of the Arabs and the Italian merchants. However, already the arithmetic book of AHMES (Papyrus Rhind, about 1700 B. C.) exhibits a remarkable well developed calculation with fractions. Apart from 2/3, only unit fractions are used, and all other fractions are transformed to them, for example, $5/6 = 1/2 + 1/3$. The transformations themselves are made less on the basis of definite rules than by compiled tables; therefore calculations with fractions were comparatively tedious. The *Babylonians* used sexagesimal fractions, derived from the division of time and angles. In a certain sense these fractions are predecessors of the decimal fractions, because they are built on the positional system with the basis 60, which however was not fully developed. Owing to the fact that no denominators were written down, the calculations became comparatively simple. The *Greeks* did not develop a system of fractions. That of the Romans is meagre, strictly speaking, it only knows fractions with a denominator 12, derived from the measure of weight 1 as = 12 ounces; other fractions were approximated by fractions with a denominator 12. In *Germany* vulgar fractions did not come into common use until the Middle Ages; but is was about 1700 when calculations with fractions were introduced into the school syllabus. Even then at first only the most necessary parts were offered, as a rule without foundation and in the form of mnemonics. Decimal fractions appeared comparatively late. The founder of the theory of decimal fractions was the merchant and engineer Simon STEVIN (1548–1620). In his book, which marked the breakthrough of decimal fractions appropriate to the decimal positional system, he postulated among other things the introduction of decimal monetary systems and weights and measures in all countries. But STEVIN had precursors; among them above all Johannes REGIOMONTANUS (1436–1476), VIETA (1540–1603) and Christoff RUDOLFF (born around 1500).

1.4. Proportionality and proportions

Direct proportionality. The heavier the suspended body, the greater is the extension of a helical spring (Fig.). In a certain spring a load of x units of weight caused an extension of y units of length:

x	50	100	125	175	240	300
y	10	20	25	35	48	60

From the numbers x for the load the numbers y for the appropriate extensions arise each time by multiplication by 0.2; that is, $y = 0.2x$ or $y/x = 0.2$.

In general, two quantities x and y are said to be directly proportional if 1. to every value of one quantity there corresponds exactly one value of the second quantity and if 2. from every measure of x the appropriate measure of y arises by multiplication by one and the same real number c.

Direct proportionality	$y = c \cdot x$ or $y/x = c$

If this connection is represented in a rectangular coordinate system, the points (x, y) lie on a line through the origin. The number c is called *proportionality factor*. It characterizes the prevalent practical situation. In the example the spring constant $c = 0.2$ is characteristic for the spring used.

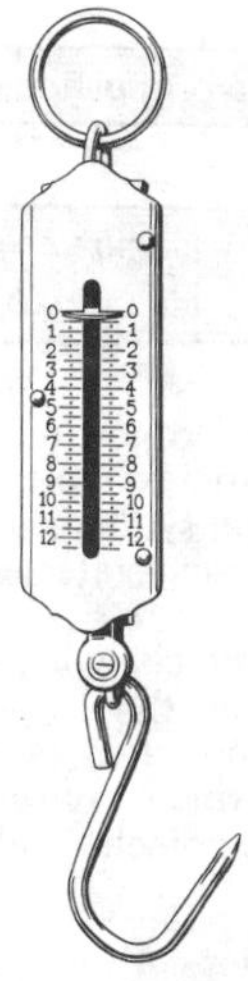

1.4-1 Spring balance

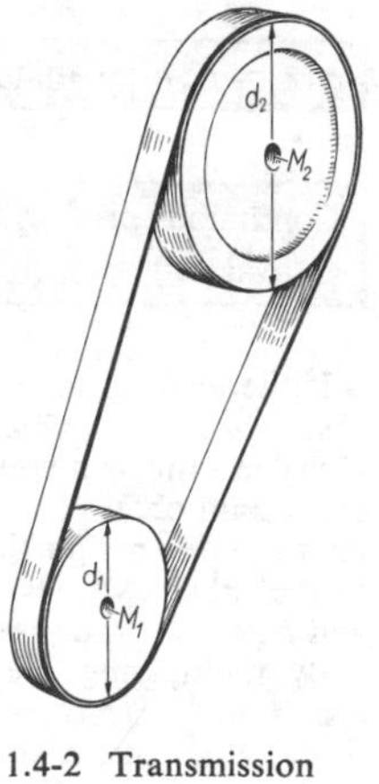

1.4-2 Transmission

Indirect or inverse proportionality. If in a transmission (Fig.) one of the two wheels has a diameter of 20 ins. and rotates once, the rotation number y of the other wheel is the larger, the smaller its diameter x in inches:

x	4	5	10	15	20	30
y	5	4	2	4/3	1	2/3

For corresponding values of x and y the relation $y \cdot x = 20$ or $y = 20/x$ always holds. The same relationship holds in the force between the two wheels transmitted by friction or by cogged teeth.

In general, two quantities x and y are said to be inversely proportional if 1. to every value of one quantity there corresponds exactly one value of the second and if 2. from every measure of x the corresponding measure of y is obtained when one and the same real number c is divided by the measure of x.

Indirect proportionality	$y = c/x$ or $y \cdot x = c$

If this relationship is represented in a rectangular coordinate system, the points lie on an equilateral hyperbola. Here the number c, on account of $y = c/x = c \cdot 1/x$, is also called *proportionality factor*.

Ratio. A train travels in one hour 80 miles, a plane 400 miles, that is, 320 miles more. The comparison becomes clearer if one says that the plane covers five times the distance of the train. In this form it is independent of the time interval. One obtains the number 5 as quotient $400:80 = 5:1 = 5$, and one says: *the distances covered in equal times are in the ratio* $5:1$.

The ratio of two quantities of the same kind is the quotient of their measures.

For numbers instead of quantities the ratio is defined accordingly; in both cases the ratio is a number.

One can also form the ratio of quantities of different kinds. If a man walking takes 4 hours to cover 11 miles one forms the ratio 11 m : 4 h = 11/4 m.p.h. (read *eleven over four miles per hour*). In this case the formation of the ratio leads to the new concept of velocity with the measuring unit m.p.h. or m h^{-1}.

In direct proportionality the associated values always have the same ratio, in an inverse proportionality they have the same product.

Since a quotient does not change when dividend and divisor are multiplied or divided by the same number $c \neq 0$, one and the same ratio can be given in various way, for example, $5:1 = 10:2 = 30/6 = 1:0.2 = 650:130$. One also speaks of extending or cancelling a ratio. As a rule, one chooses the expression with the smallest natural numbers, for example $5:1$.

Equality of ratios or proportion. An equality of two ratios is called proportion, for example, $2:3 = 1:1.5$ or $4/5 = 8/10$; this is read: *four is to five as eight is to ten*, or briefly, *four to five as eight to ten*. A proportion is true or valid if the same ratio, but differently expressed, does, in fact, stand on the two sides; $4:5 = 5:4$ is a false proportion.

If a valid proportion has the same inner terms, this quantity or number is called the *middle proportional* of the outer terms; for example, since $12:6 = 6:3$ the number 6 is the middle proportional to 12 and 3. By the product equation (see Theorems on proportions) the geometric mean $m_g = \sqrt{(a \cdot b)}$ of two positive numbers a and b is their middle proportional.

$a:c = c:b$	c middle proportional to a and b

Continuous proportion equivalent to	$a:b:c = d:e:f$ $a:d = b:e = c:f$

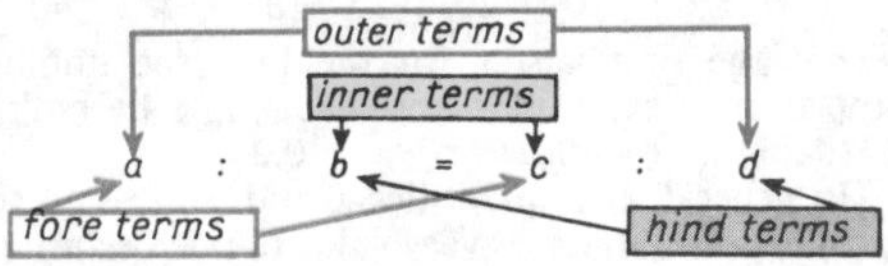

1.4-3 Terms of a proportion

If the hind terms of a proportion are equal to the fore terms of another, one frequently writes a *continuous proportion*, for example, for $2:5=4:10$ and $5:8=10:16$ one writes $2:5:8=4:10:16$. However, this is merely a symbolic notation, for if one were to interpret the two sides as quotients, one would obtain the wrong statement $1/20 = 1/40$. In general, $a:b:c = d:e:f$ is an abbreviation for the three proportions $a:b = d:e$, $b:c = e:f$ and $a:c = d:f$, of which each is a consequence of the other two. By interchanging the inner terms of the first two, say, one obtains $a:d = b:e$ and $b:e = c:f$, consequently $a:d = c:f$ or $a:c = d:f$, that is, the third proportion.

By multiplying the first two proportions by b/d or c/e one obtains the chain of equations $a/d = b/e = c/f$ from which, conversely, the proportions may be obtained: for example, the sine theorem of plane trigonometry can be put in the form $a/\sin\alpha = b/\sin\beta = c/\sin\gamma$ or $a:b:c = \sin\alpha : \sin\beta : \sin\gamma$.

Theorems on proportions. Every proportion may be transformed like an equation, for example, by interchanging the two sides. But there are also special rules leading from one valid proportion $a:b = c:d$ to another valid statement.

If one multiplies $a/b = c/d$ on both sides by bd, one obtains the product equation $a \cdot d = b \cdot c$.

Product equation. In every valid proportion the product of the inner terms is equal to the product of the outer terms.

Conversely, from the equality $a \cdot d = b \cdot c$ ($\neq 0$) one can obtain proportions. Division by $b \cdot d$ yields $a:b = c:d$, division by $a \cdot b$ yields $d:b = c:a$ etc. This leads to the *exchange theorems.*

Exchange theorems. In every valid proportion the interchange of the two outer terms, the two inner terms or the inner terms with the outer terms leads to another valid proportion.

If in $a : b = c : d$ or in $b : a = d : c$ one adds or subtracts 1 on both sides, then by *corresponding addition* one obtains $(a + b) : b = (c + d) : d$ and $(a + b) : a = (c + d) : c$, and by *corresponding subtraction* $(a - b) : b = (c - d) : d$.

Division of corresponding proportions for $a \neq b$, hence $c \neq d$, leads to $(a + b) : (a - b) = (c + d) : (c - d)$. These formulae are only the most important particular cases of the *general law of corresponding addition and subtraction.*

Corresponding addition and subtraction	From $a/b = c/d$ it follows that $(pa + qb)/(ra + sb) = (pc + qd)/(rc + sd)$ for arbitrary p, q, r, s with $ra + sb \neq 0$.

The validity of this statement can be seen by substituting in it the values $a = bk$ and $c = dk$ obtained from $a : b = c : d = k$ and then cancelling by b or d.

Proportionality and proportions. If direct proportionality holds between two quantities x and y, then $y_1/x_1 = y_2/x_2 = \cdots = y_n/x_n = c$ for any n associated values x_i, y_i. Repeated interchange of the inner terms leads to $y_1 : y_2 : y_3 : \cdots : y_n = x_1 : x_2 : x_3 : \cdots : x_n$. This means that in a direct proportionality the associated values y_i and x_i always have the same ratio and that any two values x_i and x_j have the same ratio as the associated values y_i and y_j.

In an indirect proportionality between the quantities x and y associated values x_i and y_i satisfy $x_1 \cdot y_1 = x_2 \cdot y_2 = \cdots = x_n \cdot y_n = c$. From all these product equations one can derive proportions such as $y_1 : y_2 = x_2 : x_1$ and from them finally $y_1 : y_2 : y_3 : \cdots : y_n = x_n : \cdots : x_3 : x_2 : x_1$. This means that in an indirect proportionality any two values x_i and x_j have the inverse ratio of the associated values y_i and y_j.

Solutions of proportions. The proportions $50 : 140 = 10 : x$ and $x : 2 = 50 : 80$ contain one variable each. To solve them means to find numbers giving equal ratios on substitution for x. One also talks of the task of *determining the fourth proportional.* In the first case one sees immediately that $x = 28$, in the second $x = 5/4$. The correctness can be checked by means of the product equation. In difficult cases one uses the product equation to find a solution

Example 1: $(8x - 7) : (4x - 1) = (6x - 5) : 3x$

By means of the product equation one obtains

$$3x(8x - 7) = (4x - 1)(6x - 5)$$
$$24x^2 - 21x = 24x^2 - 26x + 5$$
$$5x = 5$$
$$x = 1$$

Check:

left-hand side	$(8 \cdot 1 - 7) : (4 \cdot 1 - 1)$	$= 1 : 3$
right-hand side	$(6 \cdot 1 - 5) : 3 \cdot 1$	$= 1 : 3$
Comparison		$1 : 3 = 1 : 3$

Example 2: Of two bodies of equal volume the first has the density $\varrho_1 = 7.3$ oz./cub. in., $\varrho_2 = 2.7$ oz./cub. in. What is the mass of the second body if that of the first is 4.8 lbs.? –

$$4.8 : x = 7.3 : 2.7$$
$$x = 1.775.$$

The mass of the second body is 1.775 lbs.

Example 3: A wire of length $l_1 = 400$ m and diameter $d_1 = 4$ mm has the mass $m_1 = 36.7$ kg. How many meters of wire of the same material, but of diameter $d_2 = 6$ mm have the mass $m_2 = 90$ kg? – Since the wires consist of the same material, their masses are in the ratio of their volumes. Hence

$$m_1 : m_2 = (d_1^2/4)\,\pi l_1 : (d_2^2/4)\,\pi l_2,$$
$$m_1 : m_2 = d_1^2 l_1 : d_2^2 l_2.$$

By the product equation the solution is:

$$l_2 = \frac{m_2 d_1^2 l_1}{m_1 d_2^2},$$

in which corresponding quantities (l_1 and l_2; d_1 and d_2; m_1 and m_2) must be measured in the same units. In this numerical example one calculates the length of 436 m.

Example 4: A service station has enough fuel for 24 days if the daily sale is 1000 gallons. For how long does the fuel last if the daily sale is 1200 gallons? –
The fuel lasts for 20 days.

$$24 : x = 1200 : 1000$$
$$x = \frac{24 \cdot 1000}{1200}$$
$$x = 20.$$

Historical remarks. The theory of proportions occupied a central position in ancient mathematics, because the most diverse problems lead to proportions.

In Greek mathematics the fourth proportional was constructed geometrically, by the method of geometrical algebra. But the computational treatment of proportions and the calculus of the 'rule of three' were not developed in Europe until the 15th–17th century, in particular, in connection with commercial calculations. Such problems form one of the main constituents of the widely used arithmetic books and the principal teaching topic of the arithmeticians and 'cossists'. The best known among them is Adam RIES (1492–1559).

Proportions also played an important role in the visual arts of the Renaissance. To be esteemed as beautiful, buildings and representations of human beings (in paintings and sculptures) had to be built according to a specific 'canon', that is, parts of the total had to stand in definite ratios. For example, head to body length = 1 : 8; head : face = 5 : 4; trunk : thigh = thigh : leg; height of a building : width = 3 : 7 etc. An important role was also played by the Golden Section (see Chapter 7.). Even today the word 'well-proportioned' is used in the sense of 'satisfying the aesthetic sense'. During the Renaissance, above all Leonardo DA VINCI (1452–1519) and Albrecht DÜRER (1471–1528) occupied themselves with this branch of the visual arts (see Table 18).

1.5. Working with numerical variables

The fact known as the distributive law for rational numbers is expressed briefly in the form $a \cdot (b + c) = a \cdot b + a \cdot c$. Here a, b and c stand for arbitrary rational numbers, they are *variables* for rational numbers. In general, variables, which are usually represented by letters, represent an empty space into which an arbitrary element (or its symbol) from a fixed set can be substituted. The letters a, b and c above are numerical variables (occasionally called general numerical symbols) for which the set of rational numbers is their special *domain of variability*.

Variables are useful in two ways: they make it easy to state laws, and the solution of a problem expressed in terms of variables yields the result for arbitrarily many individual cases without new calculations, by mere substitution.

An *expression* is a combination of numerical symbols, numerical variables, and (meaningful) juxtapositions of such symbols with operational symbols and brackets, for example, $1/4$; $12 - 5$; $3 \cdot a$; $5 \cdot (17 + 60)$; $(2z - 13) : (5z + 10)$. However, the combinations $5 : 0$ or $7 + \cdot a($ are not expressions; and $7 + 8 = 15$ or $a - 3 < 3a$ are not expressions but *propositions*. In the expression $(2z - 13)/(5z + 10)$ the set of rational numbers may be the domain of variability of z. But the domain of definition of the expression differs from it, because the expression is not defined for $z = -2$ (see Chapter 4.).

In a *specification* of a variable in an expression a particular element (or its symbol) of the domain of variability is substitued at every place where the variable occurs. Distinct variables can be specified by means of distinct elements or by one and the same element; for example, for $a = b = -3/7$ the expression $2a + 5b$ yields the value -3.

Two *equivalent expressions* contain the same variables, and for every specification of these variables by equal elements the two expressions take the same value; for example, $3a + 7a$ and $10a$ are equivalent, and so are $a \cdot (b + c)$ and $a \cdot b + a \cdot c$.

Simple transformations. A proper calculation with variables, such as with integers, for example, a calculation of the sum $2a + 3b$, is impossible. Expressions with variables can only be transformed into equivalent expressions, for example, $3a + 7a$ into $10a$; here the same laws hold as for the calculation with numbers in the appropriate domain of variability.

Commutative law	$a + b = b + a$; $a \cdot b = b \cdot a$
Associative law	$a + (b + c) = (a + b) + c$; $a \cdot (b \cdot c) = (a \cdot b) \cdot c$
Distributive law	$a\,(b + c) = a \cdot b + a \cdot c$

Although for expressions only such transformations are possible, names such as sum or product are also used for them.

Under *addition* and *subtraction* only terms of the same kind with the same variables can be contracted, in accordance with the distributive law. For example, $5a - 2a = (5 - 2)\,a = 3a$, or by an additional exploitation of the commutative and associative law of addition $5a + 7c - 3b + 6c - 2a - 7b - 5c = 3a - 10b + 8c$. Also, under *multiplication* and *division* expressions with distinct variables, for example, $m \cdot n$ or $s : t$, are not calculated or contracted; for equal factors the notation of powers is used. In products it is customary to omit the multiplication sign between the variables and between numerical symbols and variables, for example, to write ab instead of $a \cdot b$ or $6(p + q)$ instead of $6 \cdot (p + q)$.

Examples: *1.* $4m \cdot 3n \cdot 15k = 180kmn$. *2.* $(-320pq) : (-80q) = 4p$.
3. $125c^2 \cdot (-3/7d) \cdot 14/75cd = -10c^3d^2$. *4.* $93s^2t^4 : 31st^2 = 3st^2$.

Algebraic sums. The notion of *opposite numbers* can also be transferred to variables and expressions. And then again every subtraction can be represented as an addition. Algebraic sums with variables are frequently called polynomials (Greek *poly*, many), but this name is also used in a different meaning. A *monomial* (Greek *mono*, single) is then an expression with one term, a *binomial* (Latin *bi*, double) contains two terms, a *trinomial* (Latin *tri*, triple) three terms.

Lexicographic order. Owing to the validity of the commutative laws the order of summands or factors is arbitrary, but it is customary for the sake of clarity to order the variables as far as possible according to the order in the alphabet or *lexicographically*, as this has been done in all the preceding examples; instead of $28b^3af^2d$ it is better to write $28ab^3df^2$, instead of $36vw + 2.5uv - 3.2uw$ better $2.5uv - 3.2uw + 36vw$. If the same variable occurs several times with distinct exponents, one orders as a rule by falling or occasionally by rising powers; for example, instead of $2s^2 - 3s^4 + s^5 - 8s$ better $s^5 - 3s^4 + 2s^2 - 8s$.

Working with algebraic sums

Addition and subtraction. Brackets can occur in the addition or subtraction of algebraic sums, for example, $(7a - 3b) + (5c - 3b - 6a) - (7b - 8a + 2c)$. Contractions and simplifications are not possible until the brackets are dissolved.

Dissolution of brackets. A numerical example for the written way of solving mentally an addition and subtraction problem illustrates the method:

$$\begin{aligned} &227 + 36 - 213 - 198 + 29 \\ =\ &227 + (30 + 6) - (200 + 13) - (200 - 2) + (30 - 1) \\ =\ &227 + 30 + 6 - 200 - 13 - 200 + 2 + 30 - 1. \end{aligned}$$

If a plus sign stands before a bracket, the bracket can be omitted. If a minus sign stands before it, on omission of the bracket all signs and operational symbols occurring in it have to be reversed.

For example, if $6a - (4a - b)$ is to be calculated, then the number to be subtracted from $6a$ is smaller by b than $4a$. Therefore, if one subtracts $4a$, one has subtracted b too much, so that b must be added, and one obtains $6a - (4a - b) = 6a - 4a + b = 2a + b$. Similar arguments hold for the other cases.

Example:

$$\begin{aligned} &8p - (15r - 7q + 6p) + (8q - p + 7r) \\ =\ &8p - 15r + 7q - 6p + 8q - p + 7r \\ =\ &p + 15q - 8r. \end{aligned}$$

Multiple brackets. If in an expression algebraic sums are again contracted, then one distinguishes the brackets suitably by different forms. In such cases it is frequently advantageous to begin with a dissolution of the inner brackets:

$$\begin{aligned} &17m + [6n - (3m + 4n)] - \{(8m - n) - [5m + (3n - 6m)]\} \\ =\ &17m + [6n - 3m - 4n] - \{8m - n - [5m + 3n - 6m]\} \\ =\ &17m + [-3m + 2n] - \{8m - n - [-m + 3n]\} \\ =\ &17m - 3m + 2n - \{8m - n + m - 3n\} \\ =\ &14m + 2n - \{9m - 4n\} = 14m + 2n - 9m + 4n \\ =\ &5m + 6n. \end{aligned}$$

If the outer brackets are dissolved first, one obtains the same result:

$$\begin{aligned} &17m + [6n - (3m + 4n)] - (8m - n) + [5m + (3n - 6m)] \\ =\ &17m + 6n - (3m + 4n) - (8m - n) + 5m + (3n - 6m) \\ =\ &22m + 6n - 3m - 4n - 8m + n + 3n - 6m \\ =\ &5m + 6n. \end{aligned}$$

Multiplication. Algebraic sums can be multiplied by a number, a *monomial*, or again by an algebraic sum.

Multiplication by a monomial. Here one is concerned with an application of the distributive law. As to the operational symbols, one only has to keep clearly in mind that every subtraction can be changed into an addition of the appropriate opposite number and vice versa.

Because $a(b + c) = ab + ac$ one has $a(b - c) = a[b + (-c)] = ab + a(-c) = ab + (-ac)$, therefore $a(b - c) = ab - ac$.

The sign rules known from calculation with integers have to be observed.

Example:

$$\begin{aligned} &6x + 7(3x - 2y) - 5x(3 - 6y) - 3y(10x + 9) \\ =\ &6x + (21x - 14y) - (15x - 30xy) - (30xy + 27y) \\ =\ &6x + 21x - 14y - 15x + 30xy - 30xy - 27y \\ =\ &12x - 41y. \end{aligned}$$

If one is sufficiently skilled in the use of the sign and operational rules, one can go from the first line straight to the third.

Multiplication of algebraic sums. The rules for the procedure of multiplication of several algebraic sums is obtained by repeated application of the distributive law, taking account of the sign rules (Fig.).

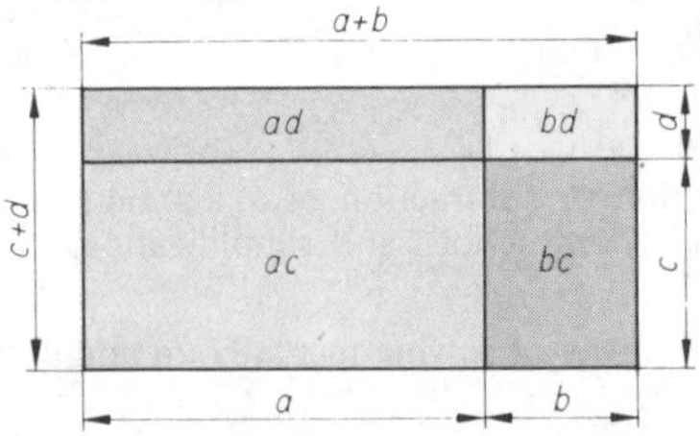

$$(a + b)(c + d) = a(c + d) + b(c + d)$$
$$= ac + ad + bc + bd$$

Algebraic sums are multiplied by multiplying every term of one sum into every term of the other and adding these products.

1.5-1 Illustration of the multiplication of two binomials if a, b, c, d are positive $(a + b)(c + d) = ac + ad + bc + bd$

In the subsequent examples sums of several terms occur; for more than two factors one proceeds step by step.

Examples: *1.* $(7u - 3v)(4u + 5v)$
$= 28u^2 + 35uv - 12uv - 15v^2$
$= 28u^2 + 23uv - 15v^2$.

2. $(2s - 3t)(5r - 7s + 2t)$
$= 10rs - 14s^2 + 4st - 15rt + 21st - 6t^2$
$= 10rs - 15rt - 14s^2 + 25st - 6t^2$.

3. $(u + 7v)(3u + v)(9u - 6v)(2u - v)$
$= (3u^2 + 22uv + 7v^2)(18u^2 - 21uv + 6v^2)$
$= 54u^4 + 333u^3v - 318u^2v^2 - 15uv^3 + 42v^4$.

Factorizations. The distributive law $a(b + c) = ab + ac$ can be used not only from left to right, but also in the reverse direction from the sum to the product. The procedure is called factorization. It is always possible when several summands have equal factors. This factor can be emphasized in an intermediate step. The transformation into a product of two algebraic sums usually takes place in several steps.

Examples:
1. $44p - 77q + 99r = 11 \cdot 4p - 11 \cdot 7q + 11 \cdot 9r = 11(4p - 7q + 9r)$.
2. $54a^3b^2c^3 + 18a^2b^3c^2 - 36a^2b^2c^2 = 18a^2b^2c^2(3ac + b - 2)$.
3. $18am - 24bm + 15an - 20bn = 6m(3a - 4b) + 5n(3a - 4b) = (3a - 4b)(6m + 5n)$.

Binomial formulae. A particularly important special case of the multiplication of algebraic sums is expressed by means of the binomial formulae. For example, $(a + b)(a + b) = a^2 + ab + ab + b^2 = a^2 + 2ab + b^2$.

Binomial formulae	$(a + b)^2 = a^2 + 2ab + b^2$ $(a - b)^2 = a^2 - 2ab + b^2$ $(a + b)(a - b) = a^2 - b^2$

The formula for $(a - b)^2$ is superfluous, strictly speaking, because it is sufficient to observe the sign rules in applying $(a + b)^2 = a^2 + 2ab + b^2$. It also results from substituting $-b$ for b in this formula (Fig.).

By applying these formulae the square of an algebraic sum can be written down and a sum can be factorized. In both directions one obtains aids for mental calculations.

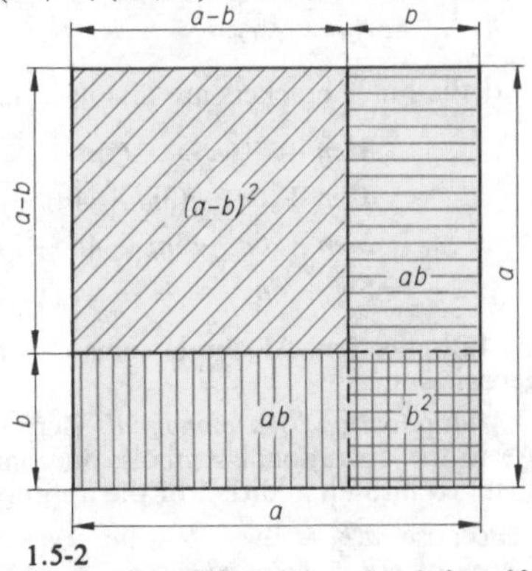

1.5-2
Illustration of $(a - b)^2 = a^2 - 2ab + b^2$

Examples:
1. $(7uv - 5vw)^2 = 49u^2v^2 - 70uv^2w + 25v^2w^2$.
2. $(5m + {}^1/_2n)(5m - {}^1/_2n) = 25m^2 - {}^1/_4n^2$.
3. $1.96r^2 + 1.4rs + 0.25s^2 = (1.4r + 0.5s)^2$.
4. $16a^2 - 56ab + 49b^2 - 64c^2 = (4a - 7b)^2 - 64c^2 = (4a - 7b + 8c)(4a - 7b - 8c)$.
5. $394 \cdot 406 = (400 - 6)(400 + 6) = 160000 - 36 = 159964$.
6. $204^2 = (200 + 4)^2 = 40000 + 1600 + 16 = 41616$.
7. $47^2 - 43^2 = (47 + 43)(47 - 43) = 90 \cdot 4 = 360$.

Higher powers. Just as for $(a+b)^2$, so also corresponding binomial formulae for higher exponents than 2 exist:

$$(a+b)^2 = a^2 + 2ab + b^2,$$
$$(a+b)^3 = a^3 + 3a^2b + 3ab^2 + b^3,$$
$$(a+b)^4 = a^4 + 4a^3b + 6a^2b^2 + 4ab^3 + b^4,$$
$$(a+b)^5 = a^5 + 5a^4b + 10a^3b^2 + 10a^2b^3 + 5ab^4 + b^5,$$
etc.

If a term of a binomial is replaced by its opposite, then all odd powers of this term have the negative sign, for example, $(a-b)^3 = a^3 - 3a^2b + 3ab^2 - b^3$. Clearly the exponent of a decreases from term to term, while that of b increases, and in $(a+b)^n$ the sum of the two exponents in every term is n. The preceding factors are called the binomial coefficients $\binom{n}{k}$ (read n over k). They stand for:

$$\binom{n}{k} = \frac{n(n-1)(n-2)\dots(n-k+1)}{1\cdot 2\cdot 3\cdots k} = \frac{n(n-1)(n-2)\dots(n-k+1)}{k!} = \frac{n!}{(n-k)!\cdot k!}.$$

If one puts $\binom{n}{0} = 1 = \binom{n}{n}$, then the binomial theorem can also be simplified by means of the summation symbol (see Chapter 18.).

Binomial theorem	$(a+b)^n = \sum_{k=0}^{n} \binom{n}{k} a^{n-k}b^k = \binom{n}{0} a^n + \binom{n}{1} a^{n-1}b + \binom{n}{2} a^{n-2}b^2 + \dots + \binom{n}{n-1} ab^{n-1} + \binom{n}{n} b^n$

Accordingly the value of the fifth term, that is, the fourth mixed term, of $(a+b)^6$ with $n = 6$, $k = 4$ is:

$$\binom{6}{4} a^{6-4}b^4 = \frac{6\cdot 5\cdot 4\cdot 3}{1\cdot 2\cdot 3\cdot 4} a^2b^4 = 15a^2b^4.$$

Pascal's triangle. This triangle (Fig.) makes it possible to determine the binomial coefficients even for someone who is not familar with the formation of $\binom{n}{k}$. It is obtained by writing the coef-

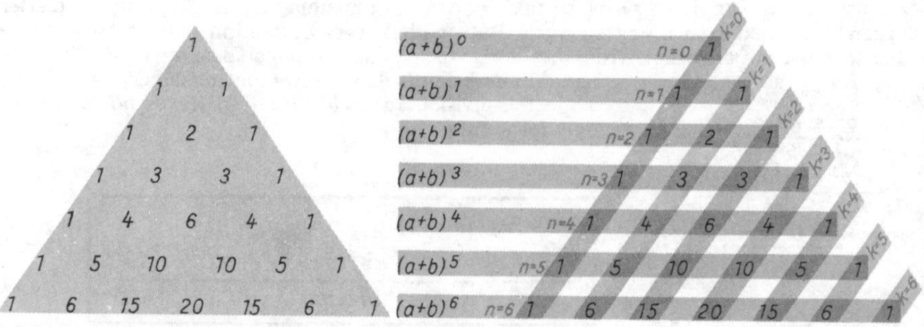

1.5-3 Pascal's triangle

ficients in triangular form under each other, beginning with $(a+b)^0 = 1$ and $(a+b)^1 = a+b$, or by using the equation $(a+b)^{n+1} = (a+b)^n(a+b)$. Here the numbers of each row arise by adding the two adjacent numbers in the row above, for example, $\binom{6}{4} = \binom{5}{3} + \binom{5}{4} = 10 + 5 = 15$.

In general, this relation between the binomial coefficients is $\binom{n}{k} + \binom{n}{k+1} = \binom{n+1}{k+1}$, on account of

$$\binom{n}{k} + \binom{n}{k+1} = \frac{n!}{(n-k)!\,k!} + \frac{n!}{(n-k-1)!\,(k+1)!}$$
$$= \frac{n!\,[(k+1)+(n-k)]}{(n-k)!\,(k+1)!} = \frac{(n+1)!}{(n-k)!\,(k+1)!} = \binom{n+1}{k+1}.$$

Division. Also in division one has to distinguish between division by a number, by a monomial, that is, a one-term expression, and by an algebraic sum.

Division by a monomial. Substituting $d = 1/c$ in the distributive law $(a + b)\,d = ad + bd$ and

$$\boxed{(a + b) : c = a : c + b : c}$$

taking into account that the fraction line can also be regarded as a symbol of division, one obtains the rule for the division of a sum by a number. Algebraic sums are divided term by term. Here the sign rules have to be taken into account.

Example: $(28m^2n - 63m^2n^2 + 84mn^2) : 7mn = 4m - 9mn + 12n.$

Division by an algebraic sum. Frequently in problems requiring the division by an algebraic sum a knowledge of factorizations or of the binomial formulae is sufficient. One splits the dividend suitably to factors, having rearranged it if necessary.

Example: $(0.54fg - 0.3eh - 0.45fh + 0.36eg) : (0.2e + 0.3f)$
$= (0.36eg - 0.3eh + 0.54fg - 0.45fh) : (0.2e + 0.3f)$
$= [0.2e(1.8g - 1.5h) + 0.3f(1.8g - 1.5h)] : (0.2e + 0.3f)$
$= [(1.8g - 1.5h)(0.2e + 0.3f)] : (0.2e + 0.3f) = 1.8g - 1.5h.$

If such a transformation of dividend cannot be achieved, which is the case, in particular, if the required division leads to a remainder, then one has to adopt the method of stepwise division.

Stepwise division. This method is merely a generalization of the ordinary written division, in which one proceeds in principle in exactly the same way, even if not every step is written down, as the example $286 : 22 = 13$ illustrates.

Example:

$$\begin{array}{l} 286 : 22 \\ = (200 + 80 + 6) : (20 + 2) = 10 + 3 \\ \underline{-\,(200 + 20)} \\ \quad 0 \quad 60 + 6 \\ \quad\ \ \underline{-\,(60 + 6)} \\ \qquad\quad 0 \end{array}$$

Example: $(13a^2x + 3x^3 - ax^2 + 10a^3) : (2a + 3x) =$

$$\begin{array}{rl} & 5a^2 - ax + x^2 \\ (2a + 3x)\big) & (10a^3 + 13a^2x - ax^2 + 3x^2) \\ & \underline{-(10a^3 + 15a^2x)} \\ & 0 \quad - 2a^2x - ax^2 + 3x^2 \\ & \underline{-(-\,2a^2x - 3ax^2)} \\ & 0 \quad 2ax^2 + 3x^2 \\ & \underline{-(2ax^2 + 3x^2)} \\ & 0 \end{array}$$

In the division of algebraic sums one proceeds similarly, taking into account that dividend and divisor are ordered in the same way. Just like writing down zeros, so taking down all remaining terms of the dividend after every step can be omitted to save written work. But one has to pay attention that no term is forgotten. Incidentally, not only the division $(a^3 - b^3) : (a - b)$ has no remainder; $(a^n - b^n)$ for every natural number n is divisible without remainder by $(a - b)$, while the division $(a^n - b^n) : (a + b)$ leaves no remainder only for even n.

Example: $(a^3 - b^3) : (a - b) =$

$$\begin{array}{rl} & a^2 + ab + b^2 \\ (a - b)\big) & (a^3 - b^3) \\ & \underline{-(a^3 - a^2b)} \\ & \quad a^2b \\ & \underline{-(a^2b - ab^2)} \\ & \quad ab^2 - b^3 \\ & \underline{-(ab^2 - b^3)} \\ & \quad 0 \end{array}$$

$$\boxed{(a^n - b^n) : (a - b) = \underbrace{a^{n-1} + a^{n-2}b + a^{n-3}b^2 + \cdots + ab^{n-2} + b^{n-1}}_{n \text{ terms}}.}$$

Division with remainder. Even when a division leaves a remainder, the method is unchanged; in writing one proceeds by analogy to $47 : 5 = 9 + 2/5$.

Example:

$$\begin{array}{rl} & x^2 - 4x - 2 \\ x^2 + 3x + 9\big) & x^4 - x^3 - 5x^2 - 40x + 7 \\ & \underline{-(x^4 + 3x^3 + 9x^2)} \\ & -\,4x^3 - 14x^2 \\ & \underline{-(-\,4x^3 - 12x^2 - 36x)} \\ & -\,2x^2 - 4x \\ & \underline{-(-2x^2 - 6x - 18)} \\ & 2x + 25 \end{array}$$

; quotient $x^2 - 4x - 2$, remainder $2x + 25$

Fractions with variables

Extension and cancellation. Extensions and cancellations are only transformations of the rational numbers represented by fractions. These transformations are also possible in fractions with variables.

Extensions. Numerator and denominator are multiplied by the same factor: $a/b = (a \cdot k)/(b \cdot k)$; similarly $5m/9n = (3 \cdot 5m)/(3 \cdot 9n) = 15m/27n$ and $7/(3a + 3b) = (7a - 7b)/(3a^2 - 3b^2)$ (extended by $a - b$).

Cancellations. Numerator and denominator of a fraction are divided by the same expression, for example, $6cd/22de = 3c/11e$. Here a sufficient skill in fractorization and in the application of binomial formulae is particularly important, because inside algebraic sums cancellation is not permitted:

$$\frac{15u^2 - 24uv}{12u^2} = \frac{3u(5u - 8v)}{12u^2} = \frac{5u - 8v}{4u}.$$

Of course, mental factorization of numerator and denominator and cancellation of common factors is permitted; for example, $(p^4 - 1)/(3p^2 + 3) = (p^2 - 1)/3$.

Addition and subtraction. Addition and subtraction of fractions with equal denominators present no problem in the calculation with variables; $a/c + b/c = (a + b)/c$.

Example: $$\frac{7i + 5k}{3k^2} - \frac{5i - 4k}{3k^2} = \frac{7i + 5k - (5i - 4k)}{3k^2}$$
$$= \frac{7i + 5k - 5i + 4k}{3k^2} = \frac{2i + 9k}{3k^2} \quad \text{(observe brackets!)}$$

Here, too, intermediate results can be omitted when sufficient skill has been acquired. One has to pay particular attention to the signs or operational symbols, because the second fraction is subtracted.

Fractions with unequal denominators. Such fractions have to be extended first so that the denominators become equal. One usually chooses the least common denominator, that is, the simplest denominator containing all the factors of the individual denominators. Frequently it can be obtained mentally without written work:

Examples: 1. $$\frac{2y}{3z} + \frac{5x}{6z} - \frac{y + 2x}{4z} = \frac{4 \cdot 2y + 2 \cdot 5x - 3(y + 2x)}{12z}$$
$$= \frac{8y + 10x - 3y - 6x}{12z} = \frac{4x + 5y}{12z}.$$

2. $$4 + \frac{3a}{a - b} - \frac{2b}{b - a} = 4 + \frac{3a}{a - b} + \frac{2b}{a - b} = \frac{4(a - b) + 3a + 2b}{a - b} = \frac{7a - 2b}{a - b}.$$

If the denominators are more complicated, it is advantageous to determine the lcd in writing.

Determination of the least common denominator lcd. In the problem

$$\frac{3}{12u - 18v} - \frac{2u - v}{36u^2 - 81v^2} + \frac{6u - 5v}{8u^2 + 24uv + 18v^2}$$

one determins the lcd as in an ordinary calculation with fractions, by decomposing the individual denominators into indecomposible factors:

		extension factors
$12u - 18v =$	$2 \cdot 3 \cdot (2u - 3v)$	$3 \cdot (2u + 3v)^2$
$36u^2 - 81v^2 =$	$3^2 \cdot (2u - 3v)(2u + 3v)$	$2 \cdot (2u + 3v)$
$8u^2 + 24uv + 18v^2 =$	$2 \cdot (2u + 3v)^2$	$3^2 \cdot (2u - 3v)$
lcd	$2 \cdot 3^2 \cdot (2u - 3v)(2u + 3v)^2$	

For example, in the preceding problem

$$\frac{3^2(2u + 3v)^2 - 2(2u + 3v)(2u - v) + 3^2(2u - 3v)(6u - 5v)}{18(2u - 3v)(2u + 3v)^2}.$$

Such a fraction is then simplified further by contracting in the numerator and possibly by cancellation.

Multiplication and division. Since in the second kind operations with fractions the determination of the lcd is unnecessary, the operations are simpler to carry out than those of the first kind. Both in multiplication and division attention has to be paid to the possibility of cancellations before the operations.

Multiplication. For the multiplication of fractions one has $a/b \cdot c/d = (a \cdot c)/(b \cdot d)$.

Examples: 1. $\dfrac{32r^2}{35q}\cdot\dfrac{25p}{24r}=\dfrac{32r^2\cdot 25p}{35q\cdot 24r}=\dfrac{4r\cdot 5p}{7q\cdot 3}=\dfrac{20pr}{21q}$.

2. $\dfrac{7p}{15m-25n}\cdot(6m-10n)=\dfrac{14p}{5}$ after cancelling by $(3m-5n)$.

Division. Division can always be carried out as multiplication by the reciprocal of the divisor, $a/b : c/d = (a\cdot d)/(b\cdot c)$.

Examples: 1. $\dfrac{14m}{9k^2}:\dfrac{7mn}{6k}=\dfrac{14m\cdot 6k}{9k^2\cdot 7mn}=\dfrac{4}{3kn}$. 2. $\dfrac{18s-18t}{u}:(12s^2-12t^2)=\dfrac{3}{2u(s+t)}$.

3. $95e^4f^3g^2:\dfrac{38e^2f^3g^4}{3h}=\dfrac{95e^4f^3g^2\cdot 3h}{38e^2f^3g^4}=\dfrac{15e^2h}{2g^2}$.

Double fractions. Double fractions occur when numerator or denominator of a fraction again contain fractions:

$$\frac{\dfrac{a}{7}}{3b};\quad \frac{x}{\dfrac{y}{x}+9};\quad \frac{\dfrac{1}{m^2}+\dfrac{2}{mn}+\dfrac{1}{n^2}}{\dfrac{3}{m}+\dfrac{3}{n}}.$$

The transformation is performed as in double fractions with numbers, by regarding the principal fraction line as symbol of division.

Example:
$$\frac{\dfrac{1}{m^2}+\dfrac{2}{mn}+\dfrac{1}{n^2}}{\dfrac{3}{m}+\dfrac{3}{n}}=\frac{\dfrac{n^2+2mn+m^2}{m^2n^2}}{\dfrac{3n+3m}{mn}}=\frac{(m^2+2mn+n^2)\,mn}{m^2n^2(3m+3n)}=\frac{m+n}{3mn}.$$

Uniqueness of the decomposition of a natural number into prime factors. As an example of how the use of variables can demonstrate the validity of mathematical arguments for all numbers within a domain, there follows a proof of the theorem of elementary number theory that the decomposition of a natural number into prime factors is unique apart from the order. The proof starts out from Euclid's algorithm. For 13013 and 390 or for two numbers a and b, one divides the larger one by the smaller one, then the smaller by the remainder r, and each remainder by the next one, etc. In the section Elementary number theory it was claimed that in the numerical example the number 13, and in general r_n, is the greatest common divisor gcd (13013, 390) or gcd (a, b). The remainders r_i are each time *smaller by at least* 1 than the divisor, therefore after finitely many steps a remainder r_{n+1} must be zero. When the equations are read from bottom to top, one sees that r_n is a divisor of r_{n-1}, hence of r_{n-2} etc. therefore also of b and of a, that is, r_n is a common divisor of a and b. From the equation $r_1 = a - bq_1$ it also follows that every common divisor of a and b divides r_1, hence r_2 etc. finally that r_n contains every common divisor of a and b and is therefore the greatest common divisor; one writes gcd $(a, b) = r_n$.

$$\begin{array}{rl}
13013 = 390\cdot 33 + 143, & a = b\cdot q_1 + r_1,\\
390 = 143\cdot\ 2 + 104, & b = r_1\cdot q_2 + r_2,\\
143 = 104\cdot\ 1 + \ 39, & r_1 = r_2\cdot q_3 + r_3,\\
104 = \ 39\cdot\ 2 + \ 26, & \dots\dots\dots\dots\dots\dots\\
39 = \ 26\cdot\ 1 + \ 13, & r_{n-2} = r_{n-1}\cdot q_n + r_n,\\
26 = \ 13\cdot\ 2 + \ \ 0, & r_{n-1} = r_n\cdot q_{n+1} + 0.
\end{array}$$

From the last equation but one gcd $(a, b) = r_n = r_{n-2} - r_{n-1}q_n$. Replacing in it r_{n-1} by $r_{n-3} - r_{n-2}q_{n-1}$ and r_{n-2} by $r_{n-4} - r_{n-3}q_{n-2}$ etc. one obtains two natural numbers x and y for which gcd $(a, b) = r_n = ax - by$. In particular, if the numbers a and b are coprime, then gcd $(a, b) = r_n = 1 = ax - by$. This leads to the theorem:

If two numbers a and b are coprime and b divides ac, then b divides c.

By assumption there is a natural number k such that $a = b\cdot k$ and because gcd $(a, b) = 1$ one has also $1 = ax - by$ or $c = acx - bcy$, hence $c = bkx - bcy = b(kx - cy)$ that is, b divides c.

Corollary: If the product ab is divisible by a prime number p, then at least one of the factors a or b is divisible by p.

Now the uniqueness of the decomposition of a natural number n into prime factors can be proved. For if $n = p_1\cdot p_2\cdot\dots\cdot p_r = q_1\cdot q_2\cdot\dots\cdot q_s$ are two decompositions, then p_1 divides the product $q_1\cdot q_2\cdot\dots\cdot q_s$, hence divides one of the prime factors q_i. But this is only possible when they are equal. By a suitable numbering of the q_i it may be assumed that $p_1 = q_1$. The same argument holds for $p_2\cdot p_3\cdot\dots\cdot p_r = q_2\cdot q_3\cdot\dots\cdot q_s$ and shows that $p_2 = q_2$ may be assumed. By a repeated application one finally obtains $r = s$, and $p_r = q_r$.

Historical remarks. In the initial steps of mathematical activity calculations, theorems and formulae were only expressed in words, not in symbols. Since this procedure led to complications and lack of clarity, abbreviations were used for objects of frequenty occurrence. For example, the Greeks denoted points, lines and surfaces by letters. DIOPHANTOS of Alexandria generally used a letter for unknown numbers; it ought to be observed that the Greeks always used letters for digits.

The development among the *Hindu* and *Arab* mathematicians concerned mainly the theory of equations. This is the reason why no account of it is given here (see Chapter 4.), although the name *algebra* apart from the theory of equations, is frequently used in elementary mathematics in the extended sense of working with numerical variables.

Letters and variables are used to a greater extent for the first time by LEONARDO of Pisa (1180–1228). He also used the fraction line; but symbols for operations were unknown to him. The proper originator of consistent working with variables is François VIÈTE (latinized VIETA, 1540–1603), who lived at the Royal court as a law officer.

René DESCARTES (latinized CARTESIUS, 1596–1650) also emphasized the importance of variables; the present notation for powers is due to him.

The operational symbols + and − occur for the first time in 1489 in an arithmetic book of Johannes WIDMANN of Eger; in 1631 William OUGHTRED introduced the symbol × for multiplication. The stop for multiplication and the colon for division were introduced by Gottfried Wilhelm LEIBNIZ (1646–1716), the sign for equality by Robert RECORDE (1557) (see Chapter 4.).

2. Higher arithmetical operations

Addition, subtraction, multiplication, and division are called the four basic arithmetical operations. Just as repeated addition of the same summand leads to a new arithmetical operation, namely multiplication, so repeated multiplication by the same factor leads to a new arithmetical operation: raising to a power or exponentiation. Like addition or multiplication, this operation can be inverted, but this time one obtains two distinct inverse operations: extracting roots and taking logarithms.

2.1. Calculations with powers and roots

Historical remarks. *Powers* were already known in antiquity, by their applications in geometrical calculations, or by their occurrence in quadratic and higher degree equations. The *Babylonians* had tables of squares and powers. They knew how to solve problems of compound interest by means of powers of 2. In the 'Elements' of EUCLID of Alexandria (4th century B.C.) one finds the formula for $(a + b)^2$, an astonishing achievement at that time. The concept of a power can be traced back to the Greek mathematician HIPPOCRATES of Chios (5th century B. C.). Subsequently it was used more frequently, for example by PLATO (427–347 B. C.). Originally only the second power was intended. BOMBELLI of Bologna (16th century) is believed to have been the first to use the word potenza (latin *potentia*, power, ability, faculty). He also used it to denote the square of the unknown; the present-day general meaning of the notion of power is of a later date. Our notation for powers essentially goes back to René DESCARTES (1596–1650). But he used it only for integral exponents greater than 2. He still wrote the square of a number a as $a \cdot a$. Powers with fractional exponents have also been known for a considerable time. Some theorems on calculations with fractional powers can already be found in the writings of Nicole ORESME (1323–1382).

Like the powers, so the *roots* were known in antiquity. Thus, the *Babylonians* had tables of rational square roots. The irrational square roots were calculated approximately by the method of the arithmetic-geometric mean. The formula used was $\sqrt{(a^2 + b)} \approx a + b/(2a)$. The *Greeks* knew that the square roots of the numbers 2, 3, ..., 17 other than 4, 9, and 16, are irrational. The proofs of the irrationality of these roots are attributed to HIPPASOS of Metapontum (c. 450 B. C.) or to THEODOROS of Cyrene (c. 430 B. C.). In the 'Elements' of EUCLID arithmetical operations of the second order are applied to roots.

In the Middle Ages calculations with roots were further developed steadily. As early as the 9th century the *Hindu* knew that the solution of a quadratic equation and the square root of a number are two-valued and also that the square root of a negative number cannot be real. They also could calculate square roots and cube roots approximately. Michael STIFEL (1487–1567) wrote about the numerical extraction of up to the seventh root. He extended the theory of irrationalities of the form $\sqrt{(a+\sqrt{b})}$ to expressions of the form $\sqrt[m]{(a+\sqrt[n]{b})}$. Gradually the root sign acquired its present-day form (derived from the letter *r* for *radix*, root), whereas Christoff RUDOLFF (16th century) used the following symbols: $\surd$ for $\sqrt[2]{\ }$, $\surd\surd\surd$ for $\sqrt[3]{\ }$ etc. It was also recognized that roots can be represented as powers (which were then familiar) with fractional exponents.

Powers

The concept of a power. It happens frequently that equal quantities have to be added: $3.7 + 3.7 + 3.7 + 3.7 + 3.7$. This sum of equal terms can be written as the product $5 \cdot 3.7$. Multiplication of equal quantities occurs just as often. Here, too, an abbreviated notation was introduced; for example, in geometry the area of a square of side a is calculated as side a times side a or $A = a \cdot a$ or briefly $A = a^2$ (read: second power of a, or a upper 2, or a squared). Correspondingly one obtains for the volume of a cube of side a: $V = a \cdot a \cdot a = a^3$. Generally, for positive integers n:

Power	$a \cdot a \cdots a = a^n$ n factors a; $n > 0$, integral

(read: nth power of a, or a to the nth, or a upper n).

Here a is the *basis* and n is the *exponent* of the power. Hence the nth power of a number is an abbreviated expression for a product of n numbers equal to the given one; in this sense one sets $a^1 = a$. Example: $2^5 = 2 \cdot 2 \cdot 2 \cdot 2 \cdot 2 = 32$. One also says that 2 *is to be raised to the fifth power*, and the operation is called *exponentiation*; it is a repeated multiplication by the same quantity. Since exponentiation is built up on the second kind arithmetical operation of multiplication, it is called a higher operation, of the third kind.

Since $0 \cdot 0 = 0$, one has generally: $0^n = 0$ for $n > 0$. Similarly, 1 is reproduced by exponentiation: $1^n = 1$ for $n > 0$. In exponentiation the basis and the exponent cannot be interchanged, as a rule; for example, $2^3 = 8 \neq 3^2 = 9$, in fact, $a^b = b^a$ holds, of course, for $a = b$, but when $a \neq b$, only for $2^4 = 4^2 = 16$.

One distinguishes between *even* and *odd* powers, according as the exponent is even (divisible by 2) or odd. Thus, 6^4, c^{16}, and generally a^{2n} are even powers, whereas 6^7, c^{13}, and generally a^{2n-1} are odd powers.

Examples: Powers occur in many formulae and laws of mathematics, science, and technology; for example, in geometry $4\pi r^3/3$ represents the volume of a sphere of radius r, $(s^2/4)\sqrt{3}$ the area of an equilateral triangle of side s; in physics $gt^2/2$ is the distance-time law of the free fall, and in the calculus of compound interest $b \cdot (r^n - 1)/(r - 1)$ is the formula for an annuity.

Of particular importance are the powers of 10. They are used (in rough estimates or slide rule calculations etc.) to obtain an idea of the *order of magnitude* of a number or to write very *large* or very *small* numbers in an abbreviated and perspicuous form. $100 = 10 \cdot 10 = 10^2$, $1000 = 10 \cdot 10 \cdot 10 = 10^3$, a *million* $= 10^6$, etc.; for instance, 1291000 can be written $1.291 \cdot 10^6$ or $1291 \cdot 10^3$. Also units of measurement are represented in the power notation, such as m^2 (square meter), cm^3 (cubic centimeter), m/s^2 (meter per second squared) etc.

Powers whose base lies between 0 and 1 decrease when the exponent increases: $(1/2)^2 > (1/2)^3 > (1/2)^4 \ldots$, but increase when the basis is greater than 1: $2^2 < 2^3 < 2^4 \ldots$ They grow very rapidly; the following problem is in the oldest arithmetic book, named after AHMES (1700 B. C.):

Each of 7 persons owns 7 cats, every cat eats 7 mice, every mouse eats 7 ears of barley, every ear of barley could yield 7 measures. How many measures is this? Solution: this is 7^5 or 16807 measures.

Sign of powers. Since negative numbers can be multiplied, the *basis* of a power may be *negative*; by the standard sign rules one obtains, for example, $(-3)^4 = (-3) \cdot (-3) \cdot (-3) \cdot (-3) = +81$ or $(-5)^3 = (-5) \cdot (-5) \cdot (-5) = -125$. It is immediately evident that the product of two negative factors is positive, that of three negative, that of four positive, and so on alternately. If the number of minus signs is even, the power has a positive value, if the number is odd, a negative value. The exponent indicates the number of (equal) factors.

A power with negative basis has a positive value for an even exponent and a negative value for an odd exponent.

To make the essence of this rule quite clear one chooses the basis (—1). Together with the obvious fact that for a positive basis the power is positive one obtains for every positive integer n:

$$(+1)^n = +1, \quad (-1)^{2n} = +1, \quad (-1)^{2n-1} = -1.$$

Multiplication and division of powers. Powers whose basis and exponent are distinct cannot be contracted on multiplication or division, for example, a^4c^3/x^7.

Powers with equal exponents. If one raises a product to a power, for example $(ab)^n$, one obtains n factors $a \cdot b$, altogether $2n$ factors, namely n factors a and n factors b alternately. Since the factors may be interchanged (commutative law), the product can be rearranged as a product of n factors a and n factors b.

Examples: 1. $(2xyz)^5 = 2^5x^5y^5z^5 = 32x^5y^5z^5$.
2. $(3a)^3 = 3a \cdot 3a \cdot 3a = 3 \cdot 3 \cdot 3 \cdot a \cdot a \cdot a = 3^3a^3 = 27a^3$.
3. $2^8 \cdot 5^7 = 2 \cdot 2^7 \cdot 5^7 = 2 \cdot (2 \cdot 5)^7 = 2 \cdot 10^7 = 20\,000\,000$.

$$(a \cdot b)^n = a^n \cdot b^n.$$

First law for powers. A product is raised to a power by raising every factor to the same power and multiplying the powers so obtained. Conversely, powers with the same exponent are multiplied by raising the product of the bases to the power given by the common exponent.

Similarly, a power $(a/b)^n$ whose basis is a fraction is obtained by multiplying n equal factors a/b, hence is a fraction whose numerator consists of n factors a und whose denominator of n factors b, that is, a^n/b^n.

Examples: 1. $\left(\frac{5}{6}\right)^3 = \frac{5}{6} \cdot \frac{5}{6} \cdot \frac{5}{6} = \frac{5 \cdot 5 \cdot 5}{6 \cdot 6 \cdot 6} = \frac{5^3}{6^3} \cdot \frac{125}{216}$.

2. $\left(\frac{5x}{2a}\right)^3 = \frac{5^3x^3}{2^3a^3} = \frac{125x^3}{8a^3}$.

3. $\frac{17^4}{34^5} = \frac{17^4}{34 \cdot 34^4} = \frac{1}{34} \cdot \left(\frac{17}{34}\right)^4 = \frac{1}{34} \cdot \left(\frac{1}{2}\right)^4 = \frac{1}{34 \cdot 2^4} = \frac{1}{544}$.

$$(a/b)^n = a^n/b^n.$$

A fraction (quotient) is raised to a power by raising the numerator (dividend) and denominator (divisor) individually to the same power and dividing the powers so obtained. Conversely, powers with the same exponent are divided by dividing their bases and raising the quotient so obtained to the power given by the common exponent.

Powers with equal basis. By the definition of the power the multiplication of two powers a^m and a^n with the same basis a means that m factors a are to be combined with another n factors a; one then has $m + n$ factors, that is, the $(m + n)$th power.

Examples: 1. $3^4 \cdot 3^2 = (3 \cdot 3 \cdot 3 \cdot 3) \cdot (3 \cdot 3) = 3 \cdot 3 \cdot 3 \cdot 3 \cdot 3 \cdot 3 = 3^{4+2} = 3^6$.
2. $56a^5b \cdot 98a^7b^5 \cdot 14a^2b^3 = 2^3 \cdot 7 \cdot a^5b \cdot 2 \cdot 7^2 \cdot a^7b^5 \cdot 2 \cdot 7 \cdot a^2b^3$
$= 2^{3+1+1} \cdot 7^{1+2+1}a^{5+7+2}b^{1+5+3} = 2^5 \cdot 7^4a^{14}b^9$.

$$a^m \cdot a^n = a^{m+n}.$$

Second law for powers: Powers with the same basis are multiplied by raising the basis to the power given by the sum of the exponents.

Division. Since the result of every division can be regarded as a fraction in which the dividend is the numerator and the divisor the denominator, one obtains on dividing the power a^m by the power a^n a fraction with m factors a in the numerator and n factors a in the denominator. If n is the smaller exponent, then after n cancellations the denominator becomes 1, and the numerator has n factors a fewer, that is, only $m - n$ factors, hence the value a^{m-n}. On the other hand, if m is the smaller exponent, then the numerator becomes 1 after cancellation, and there are $n - m$ factors left in the denominator; one obtains $1/a^{n-m}$. If the two exponents are equal, then both numerator and denominator become 1 after cancellation, and the division leads to the value 1 for every basis.

Examples: 1. $7^6 : 7^4 = \frac{7 \cdot 7 \cdot 7 \cdot 7 \cdot 7 \cdot 7}{7 \cdot 7 \cdot 7 \cdot 7} = 7^{6-4} = 7^2 = 49$.

2. $11^3 : 11^5 = \frac{11 \cdot 11 \cdot 11}{11 \cdot 11 \cdot 11 \cdot 11 \cdot 11} = \frac{1}{11^{5-3}} = \frac{1}{11^2} = \frac{1}{121}$.

$\frac{a^m}{a^n} = a^{m-n}$ when $m > n$
$\frac{a^m}{a^n} = \frac{1}{a^{n-m}}$ when $n > m$
$\frac{a^m}{a^n} = 1$ when $m = n$

Compared with the result for the multiplication of two powers with equal basis the result obtained for division is unsatisfactory: there the product had the *sum of the exponents*, here *one of the differences* $m - n$ or $n - m$ occurs as the exponent for the quotient, or even the *number* 1, which at first sight has nothing to do with powers. Since division is the inverse operation of multiplication, one should expect that the result is determined in every case by the difference $m - n$ of the exponent m of the numerator and n of the denominator; this would lead to the third law for powers.

Third law for powers: Powers with equal basis are divided by raising the basis to the exponent given by the difference of the exponents.

According to the so-called *principle of permanence*, which was formulated in 1867 by HANKEL, one tries to retain the validity of calculating rules, but to extend the concepts of the mathematical objects connected by them. The difference $m - n$ of the exponents, which occurs in the third law for powers, has a meaning in the first instance for $m > n$ only. If, in accordance with the principle of permanence, this law is to remain valid also for $m = n$ and for $m < n$, then the exponent 0 or negative exponents occur, which have no meaning under the definition adopted hitherto: 'a^n means n equal factors a'. Therefore one extends the notion of power by the following two definitions:

Extension of the notion of power	$a^0 = 1$ and $a^{-n} = 1/a^n$ for all $a \neq 0$

Then $a^m : a^n = a^{m-n}$ always holds without exception and in agreement with the previous results. For one has

1. for $m > n$ the original definition;
2. for $m = n$, however, $a^{m-n} = a^0 = 1$; and
3. for $m < n$ according to the new definition $a^{m-n} = a^{-(n-m)} = 1/a^{n-m}$.

Examples: 1. $a^3 : a^5 = a^{3-5} = a^{-2} = 1/a^2$.

2. $25 \cdot \left(\frac{1}{a}\right)^{-n} \cdot (2n)^0 \cdot 5^{-3} \cdot \left(\frac{a}{x}\right)^{-n} = 5^{2-3} \cdot a^n \cdot a^{-n} \cdot x^{-(-n)} = 5^{-1} \cdot a^0 \cdot x^n = x^n/5.$

3. $27a^4b^4 \cdot 56a^2b^{-3} \cdot 42a^{-2}b^3$
$= 3^3 \cdot a^4b^4 \cdot 7 \cdot 2^3 \cdot a^2b^{-3} \cdot 7 \cdot 3 \cdot 2 \cdot a^{-2}b^3 = 2^4 \cdot 3^4 \cdot 7^2a^4b^4 = (2^2 \cdot 3^2 \cdot 7a^2b^2)^2.$

4. What is the energy in kWh (1 kWh $= 3.6 \cdot 10^{13}$ gcm² s⁻²) corresponding to a mass defect of 2 mg? – $E = m \cdot c^2$ (E energy, m mass, c velocity of light $\approx 3 \cdot 10^{10}$ cm/s).

One obtains $\frac{2 \cdot 10^{-3} \cdot (3 \cdot 10^{10})^2}{3.6 \cdot 10^{13}}$ kWh $= \frac{2 \cdot 9 \cdot 10^{-3} \cdot 10^{20}}{3.6 \cdot 10^{13}}$ kWh $= 5 \cdot 10^{+4}$ kWh.

Consequently, if 2 mg of a substance are transformed completely into energy, the energy liberated is 50 000 kWh.

Powers with negative exponents are frequently used for *units of measurement* and for the basis 10; for example, ms^{-1} = m/s for the velocity in meters per second, gcm^{-3} = g/cm³ for the density in gram per cubic centimeter etc. *One uses powers of* 10 *with negative exponents* because they give a better picture of very small numbers, such as the elementary electric charge $e = 1.602 \cdot 10^{-19}$ C or the diameter of hydrogen atom $d = 1.06 \cdot 10^{-8}$ cm. To give an approximate idea in the latter case of the size of the diameter of the atom, it can be stated that the diameter of a hydrogen atom is to that of a football roughly as that of the football to that of the earth.

Exponentiating a power. To exponentiate a^m, that is, to calculate $(a^m)^n$, means according to the original definition: to form a product of n equal factors a^m each of which consists of m equal factors a. Hence altogether $m \cdot n$ factors a are to be multiplied. This argument also holds for m equal factors $1/a^n$ or n equal factors $1/a^m$, so that the integers m and n can also be negative.

$(a^m)^n = a^{m \cdot n}$	**Fourth law for powers. Powers are exponentiated by raising the basis to the exponent given by the product of the exponents.**

Since the order of the factors can be changed, so can that of the exponents. Therefore the exponent of a power may be factorized, and the order of the factors is immaterial: $a^{m \cdot n} = (a^m)^n = (a^n)^m$.

Examples: 1. $(2^2)^4 = (2^4)^2 = 16^2 = 256$.

2. $\frac{(-9a^2b^3)^5}{(-6a^2b)^4} = \frac{(-1)^5\,(3^2a^2b^3)^5}{(-1)^4\,(2 \cdot 3a^2b)^4} = -\frac{3^{10}a^{10}b^{15}}{2^4 \cdot 3^4a^8b^4} = -\frac{3^6a^2b^{11}}{2^4}.$

3. The largest integer that can be written by means of three digits, using addition, multiplication, and exponentiation only, is $9^{(9^9)}$. For $9 + 9 + 9 < 9 \cdot 9 \cdot 9 < 999 < 99^9 < (9^9)^9 < 9^{99} < 9^{(9^9)} = 9^{387\,420\,489}$. To write this number down in the decimal system one needs a strip of paper stretching nearly from London to Stockholm; or one could fill 33 books of 800 pages each with 14 000 digits per page.

Tables for squares and cubes

Looking up squares. In a table of squares one finds in the column headed 0 the squares of the numbers from 1.0 to 9.9 (which stand in the leftmost *entry column*); for example, $6.4^2 = 40.96$. The subsequent columns headed 1, 2, ..., 9 contain the squares of all three-digit numbers rounded off to four valid figures (Fig.), namely the squares of numbers ending in 1 in the column marked 1, those ending in 2 in the column marked 2 etc. Thus, the square of 6.44 stands at the intersection of the row 6.4 with the column 4: $6.44^2 = 41.47$, *rounded off to four figures*, whereas the true value is 41.4736. This gives at the same time the squares of the numbers 64.4, 644, 0.644, 0.0644 etc. with the same accuracy, because a rough estimate shows that, say, 64.4^2 must lie between $60^2 = 3600$ and $70^2 = 4900$, hence must be 4147 to four significant figures. Of course, today instead usually a pocket calculator or some type of computer is used. Nevertheless it is sometimes helpful to be able to use tables, e. g. in case one likes to get many significant figures.

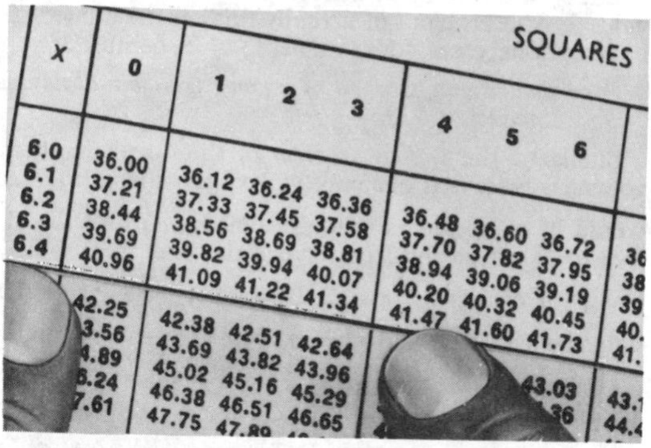

SQUARES

x	0	1	2	3	4	5	6
6.0	36.00	36.12	36.24	36.36	36.48	36.60	36.72
6.1	37.21	37.33	37.45	37.58	37.70	37.82	37.95
6.2	38.44	38.56	38.69	38.81	38.94	39.06	39.19
6.3	39.69	39.82	39.94	40.07	40.20	40.32	40.45
6.4	40.96	41.09	41.22	41.34	41.47	41.60	41.73
[illegible]	42.25	42.38	42.51	42.64	[illegible]	[illegible]	[illegible]
[illegible]	3.56	43.69	43.82	43.96	[illegible]	[illegible]	[illegible]
[illegible]	.89	45.02	45.16	45.29	[illegible]	[illegible]	[illegible]
[illegible]	.24	46.38	46.51	46.65	[illegible]	[illegible]	[illegible]
[illegible]	7.61	47.75	[illegible]	[illegible]	[illegible]	[illegible]	[illegible]

2.1-1 Looking up the square $6.44^2 = 41.47$

If the basis whose square is required has *four digits*, then the difference *d* of the squares of the neighbouring three-digit numbers (table difference) is divided evenly among the ten intervals between the possible fourth digits 0, 1, 2, 3, 4, 5, 6, 7, 8, 9, 10. For example, since according to the table 6.447^2 lies between $6.440^2 = 41.47$ and $6.450^2 = 41.60$, the difference of 0.13 (or 13 units in the 4th place) is divided among 10 intervals of which each is allotted $0.13 : 10 = 0.013$ (or 1.3 units); hence in 6.447^2 the required 7 units in the fourth place of the basis correspond to $7 \cdot 0.013 = 0.091$ (or $7 \cdot 1.3 = 9.1$ units). Generally, for the difference *d* and *z* units in the fourth place of the basis the share is $c = d \cdot z/10$ units. The correction *c* of the table value rounded off to four significant figures is 0.09; so one obtains $6.447^2 = 41.47 + 0.09 = 41.56$. Because of the even distribution of the table difference one speaks of *linear interpolation.* Whereas the curve of the squares is a parabola, one assumes a straight line, a chord of the parabola, between the points corresponding to the squares 6.44^2 and 6.45^2. The lesser the difference between neighbouring table values, the smaller is the discrepancy between the value found by interpolation and the true value. For 7.607 one obtains in succession $7.60^2 = 57.76$, $7.61^2 = 57.91$, $d = 15$, $c = d \cdot z/10 = 15 \cdot 7/10 = 10.5 \approx 10$, hence $7.607^2 = 57.86$. Tables of cubes are arranged similarly (Fig.).

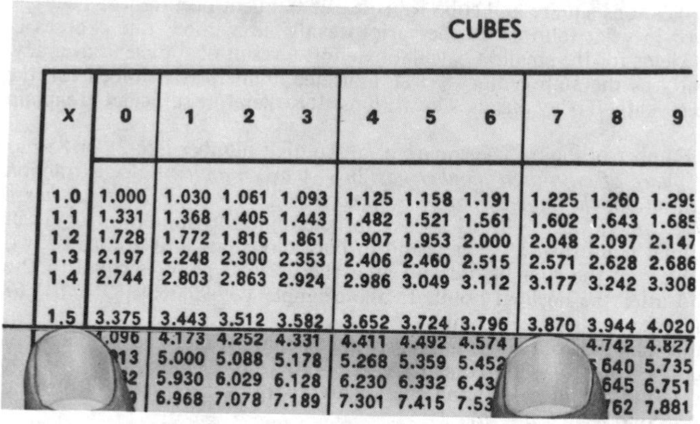

CUBES

x	0	1	2	3	4	5	6	7	8	9
1.0	1.000	1.030	1.061	1.093	1.125	1.158	1.191	1.225	1.260	1.29
1.1	1.331	1.368	1.405	1.443	1.482	1.521	1.561	1.602	1.643	1.68
1.2	1.728	1.772	1.816	1.861	1.907	1.953	2.000	2.048	2.097	2.147
1.3	2.197	2.248	2.300	2.353	2.406	2.460	2.515	2.571	2.628	2.686
1.4	2.744	2.803	2.863	2.924	2.986	3.049	3.112	3.177	3.242	3.308
1.5	3.375	3.443	3.512	3.582	3.652	3.724	3.796	3.870	3.944	4.020
[illegible]	.096	4.173	4.252	4.331	4.411	4.492	4.574	[illegible]	4.742	4.827
[illegible]	13	5.000	5.088	5.178	5.268	5.359	5.45	[illegible]	640	5.735
[illegible]	2	5.930	6.029	6.128	6.230	6.332	6.43	[illegible]	645	6.751
[illegible]	[illegible]	6.968	7.078	7.189	7.301	7.415	7.53	[illegible]	62	7.881

2.1-2 Looking up the cube $1.57^3 = 3.870$

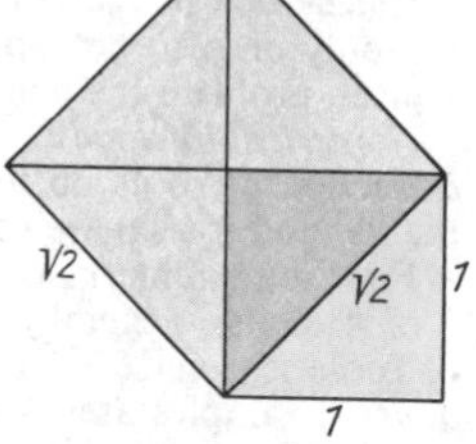

2.1-3 Square of twice a given area

Roots

The concept of a root. The Greeks were already familiar with the question of finding the length of the side of a square whose area is known. It is easy to state the length *x* of the side when the area x^2 is a square such as $4\ m^2$, $9\ m^2$, $16\ m^2$ etc.; from $x_1^2 = 3^2\ m^2$ or $x_2^2 = 0.5^2\ m^2$ it follows, of course, that $x_1 = 3$ m or $x_2 = 0.5$ m. For the general case, when the area is an arbitrary positive (real) number, no general solution was then known; in PLATO's dialogue 'Menon', SOCRATES explains by lengthy geometric discussions that the diagonal of a square of side 1 is itself the side of a square of area 2 (Fig.). Today Menon would summarize the geometric content of the dialogue in the statement that the side x of a square of area 2 has the value $x = \sqrt{2}$. Here the symbol $\sqrt{2}$ (read: *square root of* 2) denotes the number x which when multiplied by itself (squared) gives the value 2. The problem of actually finding the numerical value of this number x can be solved only in exceptional cases, for example, $\sqrt{9} = 3$, because $3^2 = 9$, or $\sqrt{0.0144} = 0.12$, because $0.12^2 = 0.0144$.

The square root $x = \sqrt{a}$ of a non-negative real number a is defined as the non-negative number x whose square is a: $x^2 = a$.

Similarly, the *Delian problem* of the Greek mathematicians: to find the side of a cube whose volume is twice that of a cube with edges of length 1, leads to a *third* or *cube root*. Today the problem would be expressed as follows: the edge e of a cube of volume 2 is the number $e = \sqrt[3]{2}$ (read: *cube root of* 2) whose third power has the value 2, $e^3 = 2$. To find this number is easy only in exceptional cases, for example, $\sqrt[3]{8} = 2$, because $2^3 = 8$, or $\sqrt[3]{0.125} = 0.5$, because $0.5^3 = 0.125$. Just as the statements $x = \sqrt{a}$ and $x^2 = a$, $x \geqslant 0$, or $e = \sqrt[3]{a}$ and $e^3 = a$ are equivalent, so for $b \geqslant 0$

$$a = \sqrt[n]{b} \quad \text{and} \quad a^n = b, \quad a \geqslant 0,$$

are defined to be equivalent.

Definition: The nth root $a = \sqrt[n]{b}$ of a non-negative real number b is that non-negative real number a whose nth power a^n has the value b: $a^n = b$.

The extraction of roots is to be regarded as an inversion of exponentiation. The number b from which the root is to be extracted is called the *radicand* and corresponds to the power; the value a corresponds to the basis of the power, and the term exponent is also used here for n.

Since $1^n = 1$ for every positive integer n, one has $\sqrt[n]{1} = 1$; also from $0^n = 0$ it follows that $\sqrt[n]{0} = 0$. For the sake of completeness one sets $\sqrt[1]{a} = a$. The equation $x^2 = 4$ has two solutions, $x_1 = +2$ and $x_2 = -2$, because $x_1^2 = x_2^2 = 4$. The root $\sqrt[n]{b} = x$ is uniquely determined. For even n the two solutions of the equation $x^n = b$ must therefore differ by their sign.

The equation $x^3 = -8$ has the solution $x = -2$. Since in the definition the radicand is assumed to be non-negative, one has to set $x = -\sqrt[3]{(-(-8))} = -\sqrt[3]{8}$. For odd n and $b < 0$ a solution of the equation $x^n = b$ is $x = -\sqrt[n]{(-b)}$. Frequently the notation $x = \sqrt[3]{-8}$ is used for the solution of $x^3 = -8$, and then it is tacitly agreed that for odd n the root of a negative number is the negative root of its absolute value. As long as one stays within the domain of *positive numbers*, extraction of roots and exponentiation are operations inverse to each other.

Calculation of roots. In applications square and cube roots are of frequent occurrence. They are the only ones to be considered in what follows. Of the various available methods one chooses in a given problem the one that yields for the smallest calculating effort a result of sufficient accuracy.

Numerical methods. As early as the 16th century STIFEL indicated numerical methods for the extraction of roots up to the seventh. Today one uses logarithms. It is therefore sufficient to explain the method for square roots.

First some remarks on the number of digits. The square of a two digit number like 21 or 85 has 3 or 4 digits. Generally, *the square of an n digit number has $2n - 1$ or $2n$ digits.* Since extraction of roots is inverse to exponentiation, *the square root of a number with $2n - 1$ or $2n$ digits has n digits.* Examples are $\sqrt{441} = 21$ and $\sqrt{7225} = 85$. It is easy to determine the number of digits of a root: starting from the decimal point divide the radicand in both directions into groups of two digits. Then the number of digits of the root before and after the decimal point is equal to the number of groups before and after the decimal point. In the example $\sqrt{39|90|06.98|89} = 631.67$ there are 3 places before and 2 after the decimal point.

Considering 441, one knows that the root must have two digits, that is, must be of the form $a + b$, where a is a multiple of 10. Consequently, $441 = (a + b)^2 = a^2 + 2ab + b^2$ or $441 = a^2 + (2a + b)\,b$. This is utilized in calculating the square root, by subtracting from the radicand first a^2 and then the second term $(2a + b)\,b$:

$$\begin{array}{rrl} & \sqrt{441} & = 20 + 1 = 21 \\ -a^2 & -400 & \quad a \quad\; b \quad\; a+b \\ \hline & 41 & \\ -(2a+b)\,b & -\;41 & \\ \hline & 0 & \end{array}$$

A root with more than two digits is computed analogously:

	$\sqrt{57\,45.64} = 70 + 5 + 0.8 = 75.8$ (with $70 = a$, $5 = b$, $0.8 = c$, $75.8 = a + b + c$)
$-a^2$	$-49\,00$
	$8\,45$
$-(2a + b)b$	$-7\,25$
	$1\,20.64$
$-(2a + 2b + c)c$	$-1\,20.64$
	0

Approximate formulae. Here approximation methods are mentioned that were known and used in antiquity. Some others are discussed in Chapter 21.

If a is very large compared with b, then in the expressions $(a+b/(2a))^2 = a^2+b+b^2/(4a^2)$ and $(a + b/(3a^2))^3 = a^3 + b + b^2/(3a^3) + b^3/(27a^6)$ the terms with powers of a in the denominator are small and negligible. This leads to approximate values for the square and cube root.

$$\left(a + \frac{b}{2a}\right)^2 \approx a^2 + b;\ \sqrt{(a^2 + b)} \approx a + \frac{b}{2a};\ \left(a + \frac{b}{3a^2}\right)^3 \approx a^3 + b;\ \sqrt[3]{(a^3 + b)} \approx a + \frac{b}{3a^2}.$$

Examples: 1. $\sqrt{35} = \sqrt{(36 - 1)} \approx 6 - 1/12 = 5.917$, the exact value being 5.91608...
2. $\sqrt[3]{730000} = \sqrt[3]{(729000 + 1000)} \approx 90 + 1000/(3 \cdot 90^2) = 90.041$, the exact value being 90.0411...

Extraction of roots by other means. The simplest way is to use a pocket calculator or some other type of computer. Also a slide rule or logarithmic table often has been used. Extraction of roots by logarithms instead of a slide rule has the advantage that roots with an arbitrary exponent can be calculated quickly and without much work.

To extract roots graphically one can utilize the graph of the power function $y^n = x$. When a sufficiently accurate drawing is available or the required accuracy is fairly low, then points lying

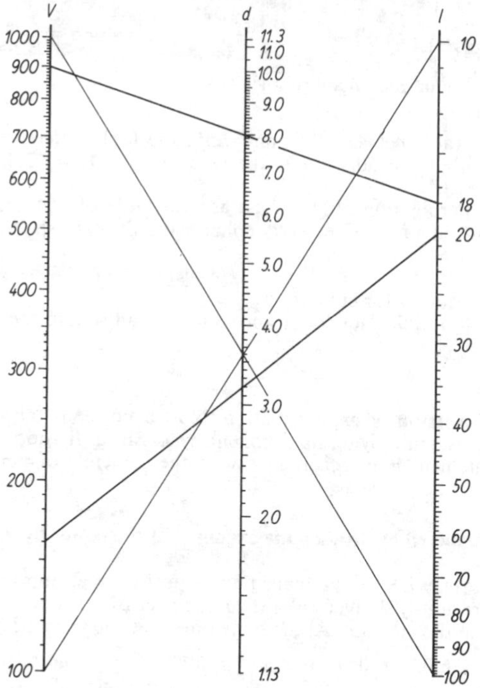

2.1-4 Nomogram for the computation of the diameter $d = \sqrt{[4V/(\pi l)]}$ of cylindrical bodies d, in cm, if V in cm³ and l in cm

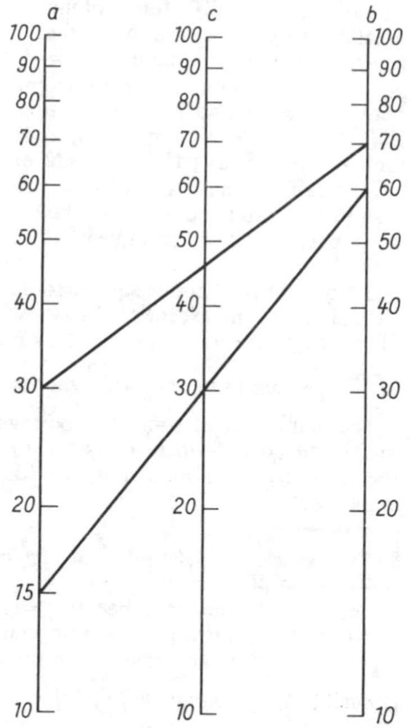

2.1-5 Square root of a product $c = \sqrt{(ab)}$

between those used in the construction of the curve can serve to determine for a given positive value x of the ordinate the appropriate value $y = \sqrt[n]{x}$ of the abscissa as the value of the nth root. *Nomograms* are also used very frequently. For example, one can represent the connection between the height l, the diameter d, and the volume V of a cylinder in a nomogram (Fig. 2.1-4).

Example: What is the diameter of a cylinder 20 cm long and of volume 160 cm^3? –
On the volume scale one marks the point 160 cm^3 and on the length scale the point 20 cm. One joins the points by a straight line which cuts the middle scale at the required diameter. In this case the diameter is 3.2 cm. Similarly: a cylinder of volume 900 cm^3 and diameter 8 cm has a length of 18 cm.

The *geometric mean* can also be represented graphically and then serves as a nomogram for the square root of a product (Fig. 2.1-5).

Extraction of roots by means of tables. A table of squares contains the squares, rounded off to four significant figures. By the definition of the square root they can be regarded as radicands. The sequence of digits of the root of a radicand results from the digits at the beginning of the row of the radicand and the digit at the head of the column in which the radicand stands; for example, $\sqrt{76.39} = 8.74$, $\sqrt{0.1136} = 0.337$, $\sqrt{2777} = 52.7$, or $\sqrt{2.403} = 1.55$ (Fig.).

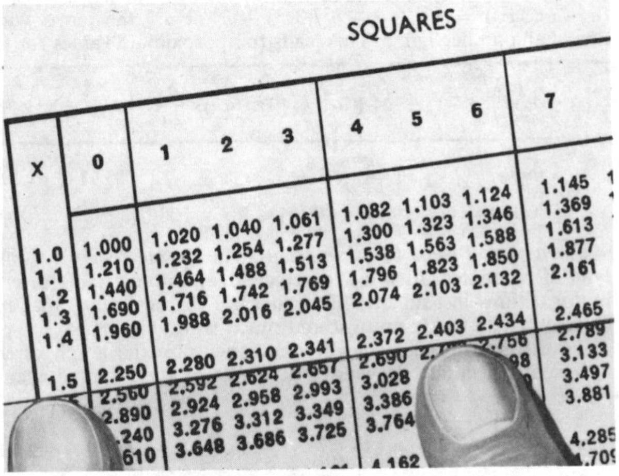

2.1-6 Looking up the square root $\sqrt{2.403} = 1.55$

If the radicand lies between two squares of the table, then apart from the table difference d the correction c is given, and from the formula $c = d \cdot z/10$ for looking up the squares one now finds the fourth significant figure $z = 10 \cdot c/d$; for example, $\sqrt{56.90}$ lies between $\sqrt{56.85}$ and $\sqrt{57.00}$ and therefore has a value between 7.54 and 7.55; the table difference is $d = 15$, the correction $c = 5$, and the fourth digit $z = 50/15 = 3$: the result is $\sqrt{56.90} = 7.543$ Similarly one finds $\sqrt{15.78} = 3.972$ or $\sqrt{6666} = 81.64$.

In a table of square roots the reading and interpolation proceed just as in a table of squares. It should be observed that now the radicand stands in a row of the entry column and the corresponding square root in the middle of the table.

Example: $\sqrt{55.3} = 7.437$.

Everything said about square roots holds analogously for cube roots.

Apart from the exceptional cases when the radicand is a square or an nth power, all square roots or nth roots are irrational numbers (see Chapter 3.).

Roots as powers with fractional exponents

The extraction of roots can be regarded as inversion of exponentiation. Now a power is taken to the nth power, where n is a positive integer, by multiplying its exponent by n. Since division is the inverse of multiplication, the following statement is in accordance with the principle of permanence:

$$a^{m/n} = \sqrt[n]{a^m}.$$

The nth root of a power is extracted by dividing the exponent of the power by n.

This defines a new number $a^{m/n}$, which must satisfy $a^{m/n} = \sqrt[n]{a^m}$ when a is a positive real number and m and n are two positive integers. By the previously derived rules the nth power of this number is $(a^{m/n})^n = a^m$, in agreement with the definition of the nth root. At the same time, this representation is *unique*: from $m/n = m'/n'$, for example, $4/6 = 6/9$, it follows that $\sqrt[n]{a^m} = \sqrt[n']{a^{m'}}$. For if one raises the root $\sqrt[n]{a^m}$ to the $(m' \cdot n)$th power and bears in mind that $mn' = m'n$, one obtains $(\sqrt[n]{a^m})^{m'n} = [(\sqrt[n]{a^m})^n]^{m'} = [a^m]^{m'} = a^{m'm}$. Similarly $(\sqrt[n']{a^{m'}})^{mn'} = [(\sqrt[n']{a^{m'}})^{n'}]^m = [a^{m'}]^m = a^{m'm}$, that is, the same value. This leads to a number of relations.

1. $\sqrt[n]{a}\cdot\sqrt[n]{b}=\sqrt[n]{(ab)}$, because $a^{1/n}\cdot b^{1/n}=(ab)^{1/n}$. 2. $\dfrac{\sqrt[n]{a}}{\sqrt[n]{b}}=\sqrt[n]{\dfrac{a}{b}}$, because $\dfrac{a^{1/n}}{b^{1/n}}=(a/b)^{1/n}$.

3. $\dfrac{1}{\sqrt[n]{b}}=\sqrt[n]{\dfrac{1}{b}}$, because $b^{-1/n}=(b^{-1})^{1/n}$. 4. $\sqrt[rq]{a^{sq}}=\sqrt[r]{a^{s}}$, because $a^{sq/rq}=a^{s/r}$.

Examples: 1. $\sqrt[3]{(24x^4)}=\sqrt[3]{(2^3\cdot 3\cdot x^3\cdot x)}=\sqrt[3]{2^3}\cdot\sqrt[3]{x^3}\cdot\sqrt[3]{(3x)}=2x\sqrt[3]{(3x)}$.

2. $\dfrac{\sqrt{12}}{\sqrt{27}}=\sqrt{\left(\dfrac{3\cdot 4}{3\cdot 9}\right)}=\sqrt{\dfrac{4}{9}}=\dfrac{2}{3}$. *3.* $\sqrt[3]{\dfrac{1}{64}}=\dfrac{1}{\sqrt[3]{64}}=\dfrac{1}{4}$. *4.* $\dfrac{1}{\sqrt{(x^2-a^2)}}=(x^2-a^2)^{-1/2}$.

5. $\sqrt[6]{9}=\sqrt[2\cdot 3]{3^2}=\sqrt[3]{3}$. *6.* $\sqrt[7]{a^5}=a^{5/7}$. *7.* $\dfrac{1}{\sqrt[7]{(14a)^3}}=(14a)^{-3/7}$.

8. $\sqrt[3]{(18a^2b)}\cdot\sqrt[3]{(12ab^2)}\cdot\sqrt[3]{(16ab)}=\sqrt[3]{(2\cdot 3^2\cdot 2^2\cdot 3\cdot 2^4\cdot a^4b^4)}$
$=\sqrt[3]{(2^6\cdot 2\cdot 3^3\cdot a^4b^4)}=2^2\cdot 3^1\cdot a^1\cdot b^1\cdot\sqrt[3]{(2ab)}=12ab\sqrt[3]{(2ab)}$.

9. $8\cdot\sqrt{\dfrac{13}{128}}=2^3\cdot\sqrt{\dfrac{13}{2^7}}=\dfrac{2^3}{2^3}\cdot\sqrt{\dfrac{13}{2}}=\sqrt{\dfrac{13}{2}}$.

10. $\sqrt{\left[\left(2+\dfrac{14}{25}\right)\dfrac{a^4}{p^2q^2}\right]}=\sqrt{\dfrac{64\cdot a^4}{25p^2q^2}}=\dfrac{8a^2}{5pq}$.

11. $12bc\sqrt{\dfrac{5a}{24b^2c}}=\sqrt{\dfrac{12^2b^2c^2\cdot 5a}{24b^2c}}=\sqrt{\dfrac{12c\cdot 5a}{2}}=\sqrt{(30ac)}$.

12. $(\sqrt[4]{16})^3=\sqrt[4]{(16)^3}=\sqrt[4]{(2^4)^3}=\sqrt[4]{2^{12}}=2^{12/4}=2^3=8$.

13. $[\sqrt[6]{(n^2v)}]^9=[(n^2v)^{1/6}]^9=n^{2\cdot 9/6}\cdot v^{9/6}=n^3\cdot v^{1+1/2}=n^3v\cdot v^{1/2}=n^3v\cdot\sqrt{v}$.

14. $\sqrt[n]{\left(a+\dfrac{a}{a^n-1}\right)}=\sqrt[n]{\dfrac{(a^{n+1}-a)+a}{a^n-1}}=\sqrt[n]{\dfrac{a^n\cdot a}{a^n-1}}=a\sqrt[n]{\dfrac{a}{a^n-1}}$;
for $a=3$, $n=2$ this means $\sqrt{[3+(3/8)]}=3\sqrt{(3/8)}$ and for
$a=2$, $n=5$ it means $\sqrt[5]{[2+(2/31)]}=2\sqrt[5]{(2/31)}$.

15. $\sqrt[5]{(\sqrt[2]{32})}=[(2^5)^{1/2}]^{1/5}=2^{5/10}=2^{1/2}=\sqrt{2}$.

16. $\sqrt[3]{(a^5\cdot\sqrt[4]{a^5})}=[a^5(a^5)^{1/4}]^{1/3}=a^{5/3}\cdot a^{5/12}=a^{5/3+5/12}=a^{25/12}=a^{2+1/12}=a^2\cdot\sqrt[12]{a}$.

17. $\sqrt{[3\sqrt{(3\sqrt{3})}]}=[3\cdot(3\cdot 3^{1/2})^{1/2}]^{1/2}=3^{1/2}\cdot 3^{1/4}\cdot 3^{1/8}=3^{1/2+1/4+1/8}=3^{7/8}=\sqrt[8]{3^7}$.

18. $\sqrt[3]{a^2}:(\sqrt{a})^3=a^{2/3}:a^{3/2}=a^{2/3-3/2}=a^{-5/6}=1/\sqrt[6]{a^5}$.

19. $\sqrt[5]{a^{2x}}\cdot\sqrt[6]{a^{5x}}=a^{(2/5)x}\cdot a^{(5/6)x}=a^{(2/5)x+(5/6)x}=a^{(37/30)x}=a^{(1+7/30)x}=a^x\cdot\sqrt[30]{a^{7x}}$.

20. a) $\sqrt[2]{\{\sqrt[5]{[\sqrt[2]{(\sqrt[5]{10})}]}\}}=\sqrt[10]{(\sqrt[10]{10})}=\sqrt[10^2]{10}=\sqrt[100]{10}=10^{0.01}$. b) $\sqrt[10^4]{10}=10^{0.0001}$.

Rationalizing a denominator. Roots of integers or rational numbers are, in general, irrational numbers and can be represented by infinite non-periodic decimal fractions. One therefore tries to avoid division by roots, that is, fractions whose denominator is a root of a rational number. In such cases one can always find a number by which the fraction can be extended so as to make the denominator rational. If the denominator is $\sqrt[n]{a^m}=a^{m/n}$, with a rational, then on multiplication by $a^m:a^{m/n}=a^{m(1-1/n)}=a^{m(n-1)/n}$ it assumes the value a^m and becomes rational. For example, the fraction $1/\sqrt[3]{p}$ must be extended by $p^{(3-1)/3}=p^{2/3}$.

Examples: 1. $\dfrac{1}{\sqrt[3]{p}}=\dfrac{1\cdot p^{2/3}}{p^{1/3}\cdot p^{2/3}}=\dfrac{p^{2/3}}{p}=\dfrac{1}{p}\cdot\sqrt[3]{p^2}$. 2. $\dfrac{1}{\sqrt{2}}=\dfrac{1\cdot\sqrt{2}}{\sqrt{2}\cdot\sqrt{2}}=\dfrac{1}{2}\sqrt{2}$.

A denominator of the form $\sqrt{a}-\sqrt{b}$ becomes rational on multiplication by $\sqrt{a}+\sqrt{b}$, because the result is $a-b$.

Example: $\dfrac{\sqrt{3}}{(\sqrt{3}-\sqrt{2})}=\dfrac{\sqrt{3}(\sqrt{3}+\sqrt{2})}{(\sqrt{3}-\sqrt{2})(\sqrt{3}+\sqrt{2})}=\dfrac{3+\sqrt{6}}{3-2}=3+\sqrt{6}$.

Powers with irrational exponents. Starting-out from the definition of the power as a product of an integral number of equal factors, its meaning has been extended in accordance with the principle of permanence first to negative and then to arbitrary rational exponents. It is plausible to go a step further and to admit irrational exponents. Let α be such an irrational positive exponent. The problem is to define b^α for a given base b. Now α can be represented as an *infinite non-periodic decimal* fraction, that is, in the form: $\alpha = a\,.\,a_1a_2a_3\,\dots\,a_i\,\dots$, which means

$$\alpha = a + a_1/10 + a_2/10^2 + a_3/10^3 + \cdots + a_i/10^i + \cdots.$$

Here a is a whole number, the a_i are digits between 0 and 9, a_1 indicates the number of tenths, a_2 that of hundredths etc. The digits of the decimal fraction are not all zero from a certain place onwards, nor do they repeat themselves regularly in a period. For example, for $a = \sqrt{2} = 1.41421356\dots$ one has $a = 1$, $a_1 = 4$, $a_2 = 1$, $a_3 = 4$, $a_4 = 2$, $a_5 = 1$, $a_6 = 3$, $a_7 = 5$, $a_8 = 6\,\dots$ If one breaks off the sum for α after the term $a_i/10^i$, one obtains an *approximation* α_i which differs from the true value α by less than $1/10^i$. Now it can be proved that as α_i comes closer and closer to the irrational number α, so for any positive basis b the numbers b^{α_i}, which are powers with a rational exponent come closer and closer to a certain number, which is then defined to be b^α.

For a negative exponent α the same arguments apply to the denominator of the fraction $1/b^{-\alpha}$ whose exponent is positive. It can be shown that with these definitions the power laws are valid for arbitrary real exponents.

2.2. Calculations with logarithms

150 years ago there were poets who regarded the logarithmic table as the very essence of mathematics. 'What the logarithms are in relation to mathematics, that is mathematics in relation to the sciences' (NOVALIS). Nowadays the logarithmic table has long since been dethroned in this sense and its place in numerical calculations has been taken by computers and pocket calculators (which however sometimes use logarithms in their hardware realizations).

Logarithmic laws and logarithmic systems

Multiplication by means of powers. If one compiles a list of the exponents l, the powers p, and their values n for a basis, say 2, one can easily carry out multiplication and division of the values n by means of their exponents l: for example, $4 \cdot 8 = 2^2 \cdot 2^3 = 2^{2+3} = 32$, or $16 : 64 = 2^4 : 2^6 = 2^{-2} = 1/4$. According to the power laws, if one uses the table on the left, one only has to calculate $2 + 3 = 5$ or $4 - 6 = -2$. Instead of exponentiating it is sufficient to multiply: $4^3 = (2^2)^3 = 2^6 = 64$; similarly, extraction of roots can be replaced by division, for example $\sqrt[4]{(1/16)} = 2^{-4/4} = 1/2$. *All arithmetical operations are reduced* to the next lower kind if instead of the *powers* occurring in a problem one calculates with the corresponding *exponents.* A disadvantage of the method is that for numbers lying between powers of 2 the appropriate exponents are not known. But once one has calculated, for example, $\sqrt[100]{2} = 2^{0.01} = 1.006956$, the powers of this number for all exponents between 0.01 and 1 yield the required values; for example, $2^{0.02} = (2^{0.01})^2 = 1.013960$ or $2^{0.1} = (2^{0.01})^{10} = 1.071773$. Intermediate values between the other powers can also be found. For example, $2^{1.01} = 2^{1+0.01} = 2 \cdot 2^{0.01} = 2.013912$ or $2^{3.1} = 2^{3+0.1} = 8 \cdot 1.071773 = 8.574184$. All these powers of 2 with exponents that are rational but not integral are irrational numbers; it is possible to find exponents α for which the power 2^α is an arbitrary given number.

l	p	n
-4	2^{-4}	1/16
-3	2^{-3}	1/8
-2	2^{-2}	1/4
-1	2^{-1}	1/2
0	2^0	1
1	2^1	2
2	2^2	4
3	2^3	8
4	2^4	16
5	2^5	32
6	2^6	64

It is easy to show that for a basis $b > 1$ and an arbitrary positive number x there *must be an exponent* l such that the power $b^l = x$. For sufficiently large positive exponents the powers of $b > 1$ exceed any given real number $x > 1$, and for sufficiently large negative exponents they are smaller than any given real number x with $0 \leqslant x < 1$. Hence there is an integer a such that $b^a \leqslant x < b^{a+1}$. If one divides the interval from a to $a + 1$ into *ten equal parts* of length 1/10, then one can find an integer a_1 between 0 and 9 such that $b^{a+a_1/10} \leqslant x < b^{a+a_1/10+1/10}$. If one continues to divide the interval between the last two exponents into ten equal parts, one finds for the smaller exponent a *decimal fraction* $\alpha = a\,.\,a_1a_2 \cdots a_i \cdots = a + a_1/10 + a_2/10^2 + \cdots + a_i/10^i + \cdots$. This may terminate, in which case $x = b^{\alpha_i}$, where $\alpha_i = a\,.\,a_1a_2\,\dots\,a_i$ is *rational*, or it comes *arbitrarily near* a real number α for which $b^\alpha = x$. Numerically these exponents $l = \alpha$ can be determined by a series expansion. They are called the logarithms to the basis b of the number x, and are written $l = \log_b x$. By the definition the basis b must be greater than 1 and the number x positive. All the logarithms to a fixed basis b form the *logarithmic system with basis b*.

The actual logarithms considered at the beginning were to the basis 2. The previous results can now be written as follows:

$\log_2 2 = 1$, for $2^1 = 2$; $\log_2 4 = 2$, for $2^2 = 4$;
$\log_2 32 = 5$, for $2^5 = 32$; $\log_2 (1/16) = -4$, for $2^{-4} = 1/2^4 = (1/16)$;
$\log_2 1.006\,956 = 0.01$, for $2^{0.01} = 1.006\,956$;
$\log_2 1.071\,773 = 0.1$, for $2^{0.1} = 1.071\,773$;
$\log_2 8.574\,184 = 3.1$, for $2^{3.1} = 8.574\,184$;
$\log_2 (4 \cdot 8) = \log_2 4 + \log_2 8 = 2 + 3 = 5 = \log_2 32$;
$\log_2 (16:64) = \log_2 16 - \log_2 64 = 4 - 6 = -2 = \log_2 1/4$;
$\log_2 4^3 = 3 \cdot \log_2 4 = 3 \cdot 2 = 6 = \log_2 64$;
$\log_2 \sqrt[4]{\frac{1}{16}} = \frac{1}{4} \log_2 \frac{1}{16} = \frac{1}{4}(-4) = -1 = \log_2 \frac{1}{2}$.

The underlying relations hold for any basis b, because they represent nothing but the power laws stated for exponents. The sequence of powers ..., $b^{-3}, b^{-2}, b^{-1}, 1, b^1, b^2, b^3$, ... gives rise to $\log_b b^2 = 2$, $\log_b b^3 = 3$, ..., $\log_b b^n = n$, $\log_b b = 1$, $\log_b 1 = 0$, $\log_b (1/b) = -1$, $\log_b (1/b^n) = -n$ that is, the logarithm of 1 is always 0, and further values are as in the table on the right. Because of their frequent application the rules for calculation with logarithms are stated as separate laws.

basis	numbers between	logarithms between
$b > 1$	$1 \dots b$	$0 \dots 1$
	$b \dots b^2$	$1 \dots 2$
	$b^n \dots b^{n+1}$	$n \dots n+1$
	$1 \cdots \frac{1}{b}$	$0 \cdots -1$
	$\frac{1}{b^{n-1}} \cdots \frac{1}{b^n}$	$-n+1 \cdots -n$

First law for logarithms: The logarithm of a product is equal to the sum of the logarithms of the factors.

$$\log_b (n_1 \cdot n_2) = \log_b n_1 + \log_b n_2$$

For from $l = \log_b (n_1 \cdot n_2)$, $l_1 = \log_b n_1$, $l_2 = \log_b n_2$ it follows that $b^l = n_1 \cdot n_2$, $b^{l_1} = n_1$, $b^{l_2} = n_2$ or $b^l = b^{l_1 + l_2}$, that is, $l = l_1 + l_2$.

Second law for logarithms: The logarithm of a quotient is equal to the difference of the logarithms of the dividend and the divisor.

$$\log_b (n_1/n_2) = \log_b n_1 - \log_n n_2$$

For from $l = \log_b (n_1/n_2)$, $l_1 = \log_b n_1$, $l_2 = \log_b n_2$ it follows that $b^l = n_1/n_2$, $b^{l_1} = n_1$, $b^{l_2} = n_2$ or $b^l = b^{l_1 - l_2}$, that is, $l = l_1 - l_2$.

Example: $\log_3 1/17 = \log_3 1 - \log_3 17 = -\log_3 17$.

$$\log_b (p^r) = r \log_b p$$

Third law for logarithms: The logarithm of a power is equal to the logarithm of the basis multiplied by the exponent.

For from $l = \log_b p^r$, $l_1 = \log_b p$ it follows that $b^l = p^r$, $b^{l_1} = p$ or $b^l = (b^{l_1})^r = b^{rl_1}$, that is, $l = rl_1$.

Examples: 1. $\log_b \frac{5^3 x^2}{6^4} = 3 \log_b 5 + 2 \log_b x - 4 \log_b 6$. *2.* $\log_b b^r = r \log_b b = r \cdot 1 = r$.

$$\log_b \sqrt[r]{w} = (1/r) \log_b w$$

Fourth law for logarithms: The logarithm of a root is equal to the logarithm of the radicand divided by the exponent.

For from $l = \log_b \sqrt[r]{w}$, $l_1 = \log_b w$ it follows that $b^l = \sqrt[r]{w}$, $b^{l_1} = w$ or $b^l = w^{1/r} = b^{(1/r)l_1}$, that is, $l = (1/r) \cdot l_1$.

Example: $\log_b \sqrt[3]{\frac{q^5}{s^2}} = {}^1/_3 (\log_b q^5 - \log_b s^2) = {}^5/_3 \log_b q - {}^2/_3 \log_b s$.

Logarithmic systems. Among all possible logarithmic systems (basis $b > 1$) only two are in common usage: the *natural* and the *decadic* logarithms. In higher mathematics the ones used almost exclusively are the *natural logarithms*. They are based on the transcendental number e, which is defined by the limit $\lim_{n\to\infty} (1 + 1/n)^n$ or by the infinite series:

e = 2.718 281 8...

$$e = \lim_{n\to\infty}\left(1+\frac{1}{n}\right)^n = 1 + \frac{1}{1!} + \frac{1}{2!} + \frac{1}{3!} + \cdots,$$

The powers of e with variable exponents form the exponential function e^x, which is suitable for the description of all events whose increase or decrease is proportional to the quantity present at any given moment, such as radioactive decay or the growth of a forest or a population. In fact, the first logarithms calculated by mathematicians of the 16th and 17th century belong to this system. Instead of $\log_e$ they are frequently denoted by ln: $\log_e x \equiv \ln x$. The series used to calculate logarithms yields the values of the natural logarithms. The *decadic* or *common logarithms* have the basis 10 and were first calculated by BRIGGS. In practical computations they are used almost exclusively and are then denoted just by lg instead of $\log_{10}$. If the reader comes across the symbol $\log x$ with no basis specified, he may assume that in elementary mathematics the intended basis is 10, in higher mathematics it is e. The notation $\lg x \equiv \log_{10} x$, used in this book, is recommended internationally and is gradually coming into use. The advantage of the common logarithms, which lies in the fact that their basis is the same as that of the number system, is clear from their integral values (see the table). This means that the logarithms to the basis 10 need only be calculated for the numbers between 1 and 10, or that for the calculation of logarithms only the sequence of digits of a number matters. For example, having found that lg 2.37 = 0.3747 one has at the same time the common logarithms of the numbers 23.7, 2 370, 0.237, 0.002 37 etc., that is, of every number that is a product of 2.37 and a power of 10 (see table).

number	logarithm
...	...
$1/10^3 = 10^{-3}$	−3
$1/100 = 10^{-2}$	−2
$1/10 = 10^{-1}$	−1
1	0
10	1
100	2
1000	3
...	...

Logarithms derived from lg 2.37 = 0.3747

number	conversion	logarithm	characteristic
$23.7 = 10 \cdot 2.37$	$\lg 10 + \lg 2.37$	$\lg 23.7 = 1.3747$	1
$2\,370 = 10^3 \cdot 2.37$	$\lg 10^3 + \lg 2.37$	$\lg 2370 = 3.3747$	3
$0.237 = 1/10 \cdot 2.37$	$\lg 1/10 + \lg 2.37$	$\lg 0.237 = 0.3747 - 1$	−1
$0.002\,37 = 1/10^3 \cdot 2.37$	$\lg 1/10^3 + \lg 2.37$	$\lg 0.002\,37 = 0.3747 - 3$	−3

The actual digits 3747 to be calculated for a logarithm are called its *mantissa*, and the integer before the decimal point its *characteristic*. This characteristic has the value 0 for a number greater than or equal to 1 and less than 10; the value 1 for numbers from 10 up to but excluding 100; and generally, the value $n - 1$ for numbers with n digits before the decimal point. If the number is a decimal fraction less than 1, then the characteristic is negative and indicates the number of places by which the decimal point has to be shifted to the right so as to stand behind the first non-zero digit of the number.

For logarithms to a basis b one need only calculate the values for numbers between 1 and b as mantissa; for numbers between b and b^2 the characteristic is 1, etc. For the decadic logarithms, whose basis is the same as that of the number system, the characteristic can be read off immediately.

Transition from one logarithmic system to another. The fact that the simplest expansions in series give natural logarithms, but that the common logarithms are needed for practical work makes it necessary to calculate the logarithm of a number n in a basis b from that of the same number in another basis a. Suppose that $l_a = \log_a n$ is known and that $l_b = \log_b n$ is required. Let $\bar{l}_a = \log_a b$ be the logarithm of the basis b of the new system referred to the basis a of the known system. In power notation the three equations are

$$a^{l_a} = n; \quad b^{l_b} = n; \quad a^{\bar{l}_a} = b.$$

Raising the third to the l_bth power one obtains

$$a^{\bar{l}_a \cdot l_b} = b^{l_b} = n = a^{l_a}, \quad \text{that is,} \quad a^{\bar{l}_a \cdot l_b} = a^{l_a} \quad \text{or} \quad \bar{l}_a \cdot l_b = l_a.$$

This relation is known as the *chain rule*.

Chain rule	$\log_a b \cdot \log_b n = \log_a n$

If the natural logarithms are taken as known, one has to set $a = e$ und $b = 10$ and obtains $\ln 10 \cdot \lg n = \ln n$. The common logarithms therefore arise from the natural logarithms after multiplication by the constant $1/\ln 10 = M_{10}$, which is called the *modulus of the logarithms to the basis* 10.

$\lg n = (1/\ln 10) \cdot \ln n = M_{10} \cdot \ln n$		$\ln n = \lg n/\lg e = \lg n/M_{10}$
$M_{10} = 0.434\,294\,5...$	$\lg M_{10} = 0.637\,7843 - 1$	$1/M_{10} = \ln 10 = 2.302\,5851...$

If, on the other hand, natural logarithms are to be calculated from a given table of common logarithms, then one has to set in the chain rule $a = 10$, $b = e$ and obtains $\lg e \cdot \ln n = \lg n$.

The fact that $\lg e = M_{10}$ can be seen immediately by setting $n = 10$: from $\ln n = \lg n/\lg e$ one has $\ln 10 = 1/\lg e$ or $\lg e = 1/\ln 10 = M_{10}$.

Inverse operations. Addition and multiplication have one inverse operation each, namely subtraction and division, for from $s_1 + s_2 = s$ it follows that $s_1 = s - s_2$ or $s_2 = s - s_1$. Similarly, from $f_1 \cdot f_2 = p$ it follows that $f_1 = p/f_2$ or $f_2 = p/f_1$. But if one wishes to calculate from a given power $r^q = p$ the basis r or the exponent q, then two distinct operations come into play: the *basis* arises as *root* $r = \sqrt[q]{p}$, the *exponent* as *logarithm* $q = \log_r p$. By substituting in the power $r^q = p$ the formal inversions one obtains either $(\sqrt[q]{p})^q = p$, that is, the definition of the root, or the equation $r^{\log_r p} = p$, that is, the definition of the logarithm. The *root* as inverse to exponentiation, however, assumes a *rational exponent*: when $q = t/s$, where t and s are coprime integers, then $r^{t/s}$ immediately leads to $r^t = p^s$, $r = \sqrt[t]{p^s}$. But if q has the irrational value α, then the root can only be interpreted as a power with a real exponent: $r = p^{1/\alpha}$. To calculate the power p or the root r for an irrational exponent α one utilizes logarithms. From $p = r^\alpha$ one obtains by taking logarithms on both sides, to the basis 10, say, $\lg p = \alpha \lg r$ and then $p = 10^{\alpha \lg r}$, or from $\lg r = (\lg p)/\alpha$ the value $r = 10^{(\lg p)/\alpha}$.

Logarithms of numbers other than powers of the basis b with rational exponents of the logarithmic system are *irrational*. For example, if $\log 2 = t/s$ were a rational number, with t and s coprime integers and $s > t$, then $10^{t/s} = 2$ or $10^t = 2^s$ or after cancelling $5^t = 2^{s-t}$, in contradiction to the theorem on the unique factorization of integers into prime factors. The argument may be generalized to any basis b and any integer n, which may be assumed to lie between 1 and b. The contradiction in the equation $b^t = n^s$ then results from the fact that b on account of $b > n$ must have a divisor other than n.

Application of logarithms. About the invention of logarithms LAPLACE said: 'The invention of logarithms shortens calculations extending over months to just a few days and thereby, as it were, doubles the life-span of the calculators.'

But the significance of logarithms is not exhausted by the immense simplification of computing. The concept of a logarithm is a working tool in many branches of higher mathematics, for example, in the differential and integral calculus, differential equations, in complex analysis, potential theory, and analytic number theory.

In *thermodynamics* the entropy S of a body or a system of bodies is directly proportional to the natural logarithm of the thermodynamic probability W, that is, $S = k \cdot \ln W$, where k is the Planck-Boltzmann constant ($k = 1.380 \cdot 10^{-16}$ erg/dg).

In *astronomy* the magnitude (brightness) m of a star is measured not by the energy I meeting the eye, but by its logarithm. Here $m - m_0 = -2.5 \lg (I/I_0)$, where I_0 is the radiation energy for the magnitude m_0 and I that for m.

Example: If the absolute magnitude of the sun is $M_0 = +4.7$ and that of the star Rigel in the constellation Orion is $M = -5.8$, one obtains $-5.8 - 4.7 = -2.5 \lg (I/I_0)$, $10.5/2.5 = 4.20 = \lg (I/I_0)$ or $I/I_0 \approx 16\,000$, so that the star Rigel radiates per second about 16 000 times as much energy as the sun.

The law can be regarded as a special case of the *Weber-Fechner law*, which states that the amount of a perception is proportional to the natural logarithm of the stimulus, in other words, not differences but quotients of stimuli are perceived as equal.

In the *barometric height formula* $h - h_0 = 60\,370(\lg b_0 - \lg b)$, h_0 and b_0 are the known height in feet and barometer reading in inches mercury of a place, and b is the barometer reading at another place whose height h is required.

Example: What is the height above ground of an *aeroplane* for which the pressure of the surrounding air is measured as 17.60 in. Hg, while a station on the ground reports $b_0 = 22.15$ in. Hg? – Since $\lg b_0 = 1.3454$ and $\lg b = 1.2455$, one obtains approximately $h = 60\,370 \cdot 0.1$ ft. $= 6037$ ft. as the height of the aeroplane above ground.

In *biology* the compound interest formula $b_n = b(1 + p/100)^n$ can be used, for example, to calculate the number of years needed for a certain increase in the volume of wood of a forest. Here b and b_n are the volumes at the beginning and the end of the period in question, and p is the annual rate of growth in percent. The formula holds generally for organic growth with a given annual percentage rate. For example, in the investigation of radioactive substances it was found that of n nuclei present at a given instant λn nuclei disintegrate, where λ is a number between 0 and 1. Hence the differential equation $dn/dt = -\lambda n$ holds, which can be integrated by separation of the variables: $dn/n = -\lambda\, dt$ or $\ln (n/n_0) = -\lambda t$, $n = n_0\, e^{-\lambda t}$, where the integration constant n_0 is the number of nuclei present at the time $t = 0$. The time T at which half the nuclei have disintegrated is called the *half-life*; it is given by $\ln (1/2) = -\lambda T$ or $T = (\ln 2)/\lambda$.

Practical logarithmic calculations

In numerical calculations the decadic or common logarithms have been of main importance. It is sufficient to discuss this system, because of the simple relations for the transitions from the logarithms of one system to those of any other. It was shown above that *logarithmic tables* need only contain the values for numbers between 1 and 10. All other numbers can be represented in the decimal system as products of one of these numbers and of a power of 10. The exponent of this power is an integer, which for numbers greater than 10 is positive, and for numbers less than 1 is negative; it is called the *characteristic* and is 0 for numbers between 1 and 10, whose logarithms are decimal fractions between 0 and 1. The sequence of their digits after the decimal point is called the *mantissa*.

Logarithmic tables. According to the number of digits to which the irrational values of the logarithms are rounded off one speaks of 4-, or 5-, or 7-place (or -figure) logarithms. Logarithms with more figures are rarely used and only for special purposes. The leftmost or *entry column* contains the first 2, or 3, or 4 digits of the number, that is, of the digits of numbers from 10 to 99 (4-figure). from 100 to 999 (5-figure), or from 1000 to 9999 (7-figure). Behind every number of the entry column there are the 10 mantissae of the numbers whose third, fourth, or fifth digits have the appropriate value between 0 and 9. In 5-figure tables many mantissae would have the same first two digits, for example 97 for the values lg 9.333 = 0.97002, ..., lg 9.340 = 0.97035, ..., lg 9.549 = 0.97996, altogether 217 mantissae! To avoid these unnecessary and unwieldy repetitions one places these first *two common digits in a separate column* before the column of the fourth digits headed 0, and only once at the beginning of the row in which *all* the mantissae start with these two digits; in the example (Fig.) to the right of 934. The parts 002 up to 030 of the preceding mantissae stand

930	848	853	858	862	867	872	876	881	886	890
931	895	900	904	909	914	918	923	928	932	937
932	942	946	951	956	960	965	970	974	979	984
933	988	993	997	*002	*007	*011	*016	*021	*025	*030
934	97 035	039	044	049	053	058	063	067	072	077
935	081	086	090	095	100	104	109	114	118	123
936	128	132	137	142	146	151	155	160	165	169
937	174	179	183	188	192	197	202	206	211	216

2.2-1 The rows 930 to 937 of the entry column of a 5-figure logarithmic table

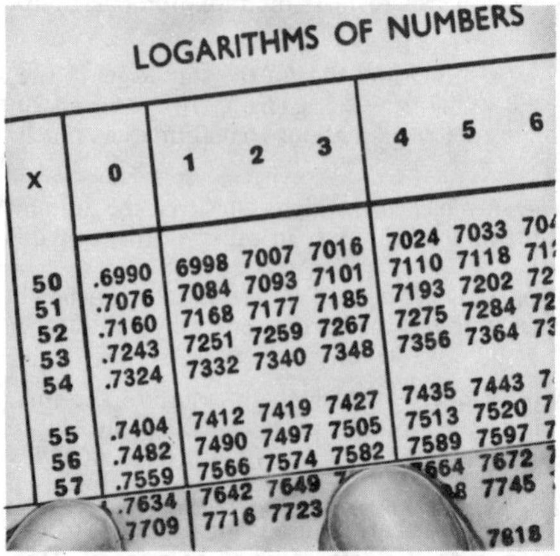

2.2-2 Looking up the mantissa of lg 5.728 = 0.7580

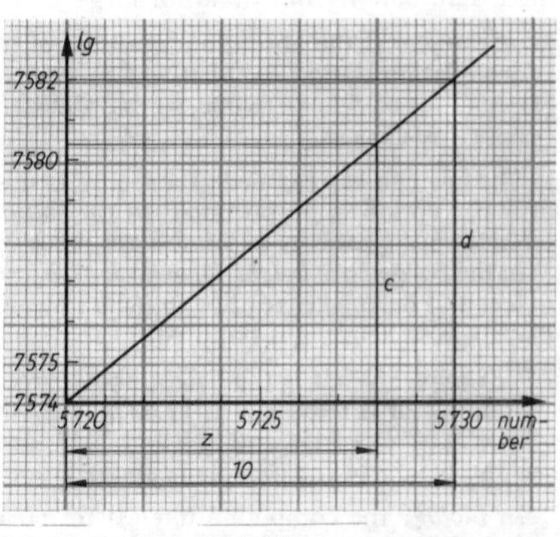

2.2-3 Graphical interpolation for the mantissa; lg 5.728 = 0.7580

therefore under the pair 96, but belong to 97. They are therefore marked with an asterisk: *002 up to *030 (Fig.) or by heavy type or in other manners. Similarly in 7-figure tables the first 3 digits of the mantissae are extracted.

The last digits z of the numbers are accounted for, as usual, by *linear interpolation* (see Squares). The *table difference* d of the neighbouring mantissae is divided into 10 equal parts $d/10$ whose z-fold $z \cdot d/10 = c$ gives the correction of the mantissa. For example, lg 5.728 lies between 0.7574 and 0.7582 (Fig.), the table difference d is $0.7582 - 0.7574 = 0.0008$, the next digit z of the number is 8, hence the correction to the mantissa is $c = z \cdot d/10 = 8 \cdot 8/10 = 6.4 \approx 6$ units of the fourth place after the decimal point: lg 5.728 = 0.7574 + 0.0006 = 0.7580 (Fig.). In a 4-figure table lg 5.7283 has the same value, but in a *5-figure* table it lies between lg 5.728 = 0.75 800 and lg 5.729 = 0.75 808, hence for $d = 8$, $z = 3$ and $c = 0.8 \cdot 3 = 2.4$ it has the value lg 5.7283 = 0.75 802. The same result is obtained for lg 5.728342, the digits 4 and 2 cannot be taken into account when working with a 5-figure table. But in a *7-figure table* (Fig.) one obtains lg 5.7283 = 0.7580258 and

28	758 0030	0106	0182	0258	0333	0409	0485	0561	0637	0712
29	0788	0864	0940	1016	1091	1167	1243	1319	1395	1470
30	1546	1622	1698	1774	1849	1925	2001	2077	2153	2228
5731	2304	2380	2456	2531	2607	2683	2759	2835	2910	2986
32	3062	3138	3213	3289	3365	3441	3516	3592	3668	3744

2.2-4 The rows 5728 to 5732 of the entry column of a 7-figure logarithmic table

lg 5.7284 = 0.758 0333; for the table difference $d = 75$ the table of proportional parts gives for the digit 4 the correction $c_4 = 30$, for 0.2 the correction $c_{0.2} = 1.5$, hence for 42 the correction $c = 31.5 \approx 32$, and so lg 5.728 342 = 0.7580290.

But if a logarithm is given and the appropriate number is required, then apart from the table difference d the correction c is known, and the next valid digit z is obtained from $z = 10 \cdot c/d$.

	75
1	7.5
2	15.0
3	22.5
4	30.0
5	37.5
6	45.0
7	52.5
8	60.0
9	67.5

Example 1: lg n_1 = 0.5412 lies between 0.5403 and 0.5416, hence n_1 between 3.47 and 3.48; here $d = 13$, $c = 9$, hence $z = 90/13 \approx 7$; consequently $n_1 = 3.477$.

Example 2: lg n_2 = 0.50000 lies between lg 3.162 = 0.49996 and lg 3.163 = 0.50010; here $d = 14$, $c = 4$ and $z = 40/14 \approx 3$, hence $n_2 = 3.1623$.

Example 3: lg n_3 = 0.240 9357 lies between lg 1.7415 = 0.240 9235 and lg 1.7416 = 0.240 9484. The table of proportional parts of $d = 249$ for $c = 122$ yields the sixth digit 4, because $4 \cdot 24.9 = 99.6$, and since $122 - 99.6 = 22.4$, the seventh digit has the value 9, so that n_3 = 1.741 549.

	249
1	24.9
2	49.8
3	74.7
4	99.6
5	124.5
6	149.4
7	174.3
8	199.2
9	224.1

Negative characteristics are often indicated by placing a bar (for the minus sign) on top, and not in front, of the number before the decimal point. For example, from lg 2 = 0.3010 one obtains lg 1/2 = lg 1 − lg 2 = −0.3010; this number can be written as 0.6990 − 1 or $\bar{1}$.6990, and the same result is obtained by starting from lg 5 = 0.6990, hence lg 0.5 = 0.6990 − 1 = $\bar{1}$.6990. Another method, which is sometimes used in tables of logarithms of trigonometric functions, avoids negative characteristics by imagining a characteristic of −10 placed behind the logarithm. In this notation lg 0.5 = 0.6990 − 1 = 9.6990 − 10, and the table then only gives the entry 9.6990.

Example 1: lg 0.723 = 0.8591 − 1 = $\bar{1}$.8591 = 9.8591 − 10, written as 9.8591.

Example 2: lg 0.00723 = 0.8591 − 3 = $\bar{3}$.8591 = 7.8591 − 10, written as 7.8591.

Example 3: To compute the value of $n = \dfrac{\lg 2}{\lg 1.03}$ a 4-figure table gives lg 2 = 0.3010 and lg 1.03 = 0.0128. To find the quotient $n = \dfrac{0.3010}{0.0128}$ one takes logarithms: lg n = lg0.3010 − lg0.0128 = (0.4786 − 1) − (0.1072 − 2) = 0.3714 + 1 = 1.3714. The number for this logarithm is $n = 23.52$. In the bar notation the calculation is: lg n = $\bar{1}$.4786 − $\bar{2}$.1072 = 1.3714.

Example 4: To compute the value of ln 2. – The formula for the transition from the common to the natural logarithms is:
ln 2 = (lg 2)/(lg e) = $0.3010/M_{10}$ = 0.3010/0.4343. By taking logarithms one obtains: lg(ln 2) = lg 0.3010 − lg 0.4343 = $\bar{1}$.4786 − $\bar{1}$.6378 = $\bar{1}$.8408. The number for this logarithm is ln 2 = 0.6931.

Graph of the function $y = \lg x$. The tables give the function values y for every positive real number x. By plotting these values in a Cartesian coordinate system one obtains the curve of the function $y = \lg x$ (Fig.). It meets the x-axis in the point $x_0 = 1$, $y_0 = \lg 1 = 0$. For decreasing x-values from 1 to 0 the curve falls rapidly and for $x \to 0$ it approaches asymptotically the negative y-axis.

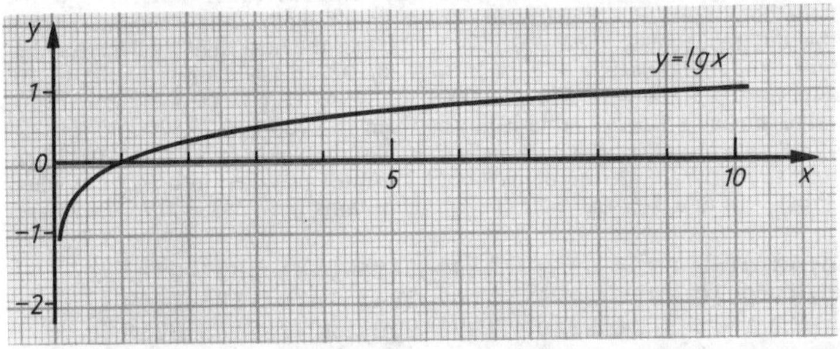

2.2-5 Graph of the function $y = \lg x$

For increasing x-values from 1 to 10 the function grows monotonically from 0 to 1. Since $dy/dx = 1/(x \ln 10) = M_{10}/x = 0.4343/x$, the angle α between a tangent to the curve and the x-axis is small, for $\tan \alpha = dy/dx$ decreases from 0.4343 for $x = 1$ to 0.043 43 for $x = 10$. Since the characteristic of the numbers from 10 to 100 is 1, the function grows in this interval of ten times the length by exactly the same amount as between 1 and 10, namely by $\Delta y = 1$. The same applies for the interval from $x = 1000$ to $x = 10000$. One sees by how little the curve differs from a straight line, which justifies linear interpolation in this case. The values given directly lie for a 4-figure table between 100 and 1000 and for a 7-figure table even between 10 000 and 100 000.

Worked examples. The numerical expressions which follow are built up in various ways. If they contain only higher arithmetical operations, they can be calculated *logarithmically throughout*, that is, the only number to be looked up from its logarithm is the final result. But if they contain sums or differences, then intermediate calculations are necessary to determine the individual summands and then their sums. Here it can happen that there occurs the logarithm of a number that is to be subtracted. Now *logarithms of negative numbers* do not exist in the real field. Therefore one often uses the device, especially in trigonometric calculations, of indicating by the symbol n (negative) or p (positive) the operation to be applied to a number. The usual sign rules for multiplication have to be modified accordingly: $(+) \cdot (-) = (-)$ now becomes $p + n = n$, and $(-) \cdot (-) = (+)$ becomes $n + n = p$ etc.

The *characteristic* and the *mantissa* are determined in different ways, but only when taken together, do they give the value of the logarithm. The operations to be performed on the logarithm are applied equally to the digits of the characteristic and of the mantissa. For example, if the logarithm of a number x has the value 5.6, $\lg x = 5.6$, then half the logarithm has the value 2.8 so that $^1/_2 \lg x = \lg \sqrt{x} = 2.8$, and similarly $3 \lg x = \lg x^3 = 16.8$.

Care must be taken in applying these obvious rules to *logarithms of positive real numbers* less than 1. These logarithms have *negative values*, and the notation differs somewhat from the common usage. On the numerical line, instead of going leftward from zero directly to the required number one goes in zig-zag fashion beyond it to a negative integer and then again to the right up to the number (Fig.); for example, $-0.35 = -1 + 0.65$ or $-1.62 = -2 + 0.38$. This procedure is motivated by the occurrence of powers of 10 in the decimal expansion of the number in question; it leads to *various representations of one and the same number*, for example, $-0.35 = -2 + 1.65 = -10 + 9.65$; $-1.62 = -5 + 3.38 = -20 + 18.38$ etc. The mantissae in these examples are 65 and 38; the characteristics consist of two integers: one positive or zero, and the other, usually in second place, negative. In the examples above they are $(0. \ldots -1) = (1. \ldots -2) = (9. \ldots -10)$ and $(0. \ldots -2) = (3. \ldots -5) = (18. \ldots -20)$. Multiplication, division, and exponentiation of numbers, that is, addition, subtraction, and multiplication of their logarithms, presents no difficulties if one bears in mind that in looking up a number one starts out from the next negative integer below it. In extracting roots, that is, dividing logarithms, one has to arrange that the characteristic

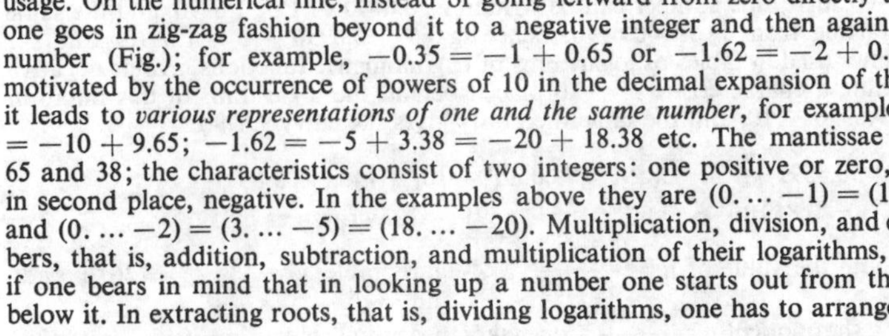

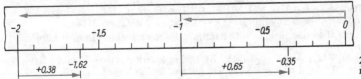

2.2-6 Logarithms with negative characteristics

is divisible by the exponent of the root so that after division it is again an integer. This can always be done.

Similar remarks apply to the other notation for negative values of logarithms: instead of 0. ... $-k$ one writes $\bar{k}$. ... and performs the arithmetical operations in the usual way, bearing in mind that $\bar{k}$ stands for $-k$.

Example 1: $x = \sqrt[3]{100}$,
$\lg x = {}^1/_3 \lg 100 = {}^1/_3 \cdot 2 = {}^2/_3 = 0.6667$,
$x = 4.642$.

Example 2: $x = \sqrt{0.2}$,
$\lg x = {}^1/_2 \lg 0.2 = {}^1/_2(0.3010 - 1)$
$= {}^1/_2(1.3010 - 2) = 0.6505 - 1$
or $\lg x = {}^1/_2(9.3010 - 10)$
$= {}^1/_2(19.3010 - 20)$
$= 9.6505 - 10$,
$x = 0.4472$.

Example 3: $x = \sqrt[3]{(1 - (0.927)^5)}$
$\lg 0.927 = \bar{1}.9671$,
$5 \lg 0.927 = \bar{1}.8355$,
$(0.927)^5 = 0.6847$
$1 - 0.6847 = 0.3153$,
$\lg 0.3153 = \bar{1}.4987$,
${}^1/_3 \lg 0.3153 = \bar{1}.8329$,
$x = \sqrt[3]{0.3153} = 0.6807$.

Example 4: $x = \boxed{160.6} \cdot \boxed{0.2856} \cdot \boxed{0.006\,998}$
$= \boxed{a} \cdot \boxed{b} \cdot \boxed{c}$,
$\lg x = \lg a + \lg b + \lg c$.

In the skeleton diagram the numbers N are separated from the logarithms lg by a vertical line:

N	lg	
$a = 160.6$ →	2.2057	+
$b = 0.2856$ →	0.4558 −1	
$c = 0.006\,998$ →	0.8450 −3	↓
$a \cdot b \cdot c$	3.5065 −4	
$x = 0.3210$ ←	0.5065 −1	

Example 5: $x = \boxed{0.075\,35} : \boxed{6.459}$
$= \boxed{a} : \boxed{b}$
$\lg x = \lg a - \lg b$

N	lg	
0.075 35 →	0.8771 −2	−
6.459 →	0.8101	↓
$x = 0.011\,67$ ←	0.0670 −2	

Example 6: $x = \boxed{56.07} : \boxed{992.6}$
$= \boxed{a} : \boxed{b}$
$\lg x = \lg a - \lg b$

N	lg	
56.07 →	3.7488 −2	−
992.6 →	2.9967	↓
$x = 0.056\,51$ ←	0.7521 −2	

Example 7: $x = \boxed{0.002\,934} : \boxed{0.000\,089\,98}$
$= \boxed{a} : \boxed{b}$
$\lg x = \lg a - \lg b$

N	lg	
0.002 934 →	1.4675 −4	−
0.000 089 98 →	0.9541 −5	↓
	0.5134 +1	
$x = 32.62$ ←	1.5134	

Example 8: $x = \boxed{0.07440}^{\,5}$
$= \boxed{a^5}$
$\lg x = 5 \cdot \lg a$

N	lg	
$a = 0.074\,40$ →	0.8716 −2	·5
a^5	4.3580 −10	↓
$x = 0.000\,002\,281$ ←	0.3580 −6	

Example 9: $x = \boxed{16.24}^{\,\pi}$
$= \boxed{a}^{\,\pi}$
$\lg x = \pi \lg a$
$\lg(\lg x) = \lg \pi + \lg(\lg a)$

N	lg	lg	
16.24 →	1.2106 →	0.0830	+
π →		0.4971	↓
$x = 6353$ ←	3.803 ←	0.5801	

Example 10: $x = \sqrt[5]{\boxed{0.009\,028}}$
$= \sqrt[5]{\boxed{a}} = \boxed{a}^{\,1/5}$
$\lg x = {}^1/_5 \lg a$

N	lg	
0.009 028 →	2.9556 −5	:5
$x = 0.3900$ ←	0.5911 −1	↓

Example 11: $x = \sqrt[6]{\left[\frac{\boxed{89.49}^{3.5}\cdot\sqrt{\boxed{0.006\,006}}}{\boxed{0.000\,010\,01}^{2}\cdot\boxed{3\,601\,000}^{4}}\right]} = \sqrt[6]{\left[\frac{\boxed{a}^{3.5}\cdot\sqrt{\boxed{b}}}{\boxed{c}^{2}\cdot\boxed{d}^{4}}\right]}$

$\lg x = {}^1/_6\,[3.5 \lg \boxed{a} + {}^1/_2 \lg \boxed{b} - 2 \lg \boxed{c} - 4 \lg \boxed{d}]$

$x = 0.017\,74.$

N	lg	operation	lg	
89.49 →	1.9517	· 3.5 →	6.8310	
0.006 006 →	1.7786 −4	: 2 →	0.8893 −2	
$a^{3.5}\sqrt{b}$			5.7203 →	17.7203 −12
0.000 010 01 →	0.0004 −5	· 2 →	0.0008 −10	
3 601 000 →	6.5564	· 4 →	26.2256	
$c^2 \cdot d^4$			16.2264 →	16.2264
0.017 74 ←	0.2490 −2 ←	: 6 ←		1.4939 −12

Example 12: $x = \frac{\boxed{\sqrt[3]{\boxed{2934}\cdot\boxed{1.843}^{2}} - \boxed{\boxed{11.55}\cdot\boxed{32.92}\cdot\boxed{0.8844}}}}{\boxed{\boxed{4.321}^{3}\cdot\sqrt{\boxed{0.9428}}\cdot\boxed{0.010\,36}}},$

$x = \frac{\boxed{\sqrt[3]{\boxed{a}\cdot\boxed{b}^{2}}} - \boxed{\boxed{c}\cdot\boxed{d}\cdot\boxed{e}}}{\boxed{\boxed{f}^{3}\cdot\sqrt{\boxed{g}}\cdot\boxed{h}}} = \frac{\boxed{A} - \boxed{B}}{\boxed{C}} = -354.8.$

N		N	lg	operation	lg	
	$\boxed{a}$	2934 →	3.4675	: 3 →	1.1558	
	$\boxed{b}$	1.843 →	0.2655	· 2 →	0.5310	
+ 48.62 ←	$\boxed{A}$ ←				1.6868	
	$\boxed{c}$	11.55 →	1.0626			
	$\boxed{d}$	32.94 →	1.5177			
	$\boxed{e}$	0.8844 →	0.9467 −1			
−336.50 ←	$\boxed{B}$ ←		2.5270			
−287.88	$\boxed{A}-\boxed{B}$ →		2.4592n →			2.4592n
	$\boxed{f}$	4.321 →	0.6356	· 3 →	1.9068	
	$\boxed{g}$	0.9428 →	1.9744 −2	: 2 →	0.9872 −1	
	$\boxed{h}$	0.01036 →	0.0153 −2	→	0.0153 −2	
	$\boxed{C}$				0.9093 −1 →	(0.9093 −1) p
	x	−354.8 ←	2.5499n ←			(1.5499 +1) n

Historical remarks. The origin of logarithmic calculations shows clearly the connections between the development of society and that of mathematics. The discovery of the sea route to India is closely linked with a flourishing period in astronomy, navigation, and trigonometry. Mathematics was an indispensible tool of the navigators. With the spreading of trade the commercial methods of calculation grew in importance, in this context mainly the *calculus of compound interest*. In both these spheres the demands made on professional calculators were extraordinary for the time; one should visualize the amount of computations which the astronomer Johannes KEPLER (1571–1630) needed to derive the laws named after him. The leading calculators were searching for simplifying methods, in particular, for a link between arithmetic and geometric numerical sequences which would replace multiplication by addition.

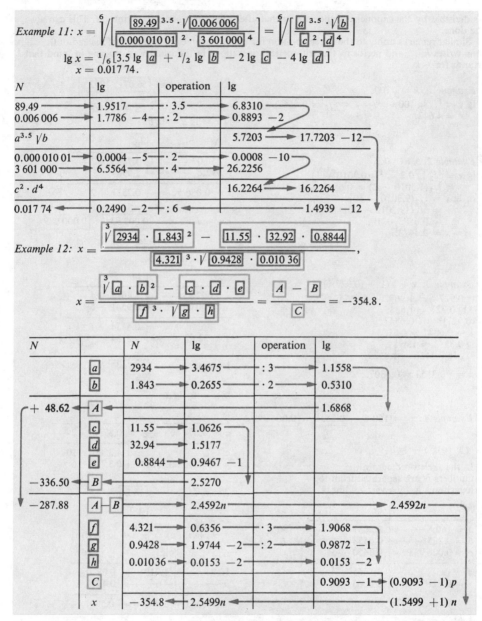

The mathematicians Paul WITTICH (1555–1587) and Christoph CLAVIUS (1537–1612) proposed in their book '*de Astrolabio*', which appeared in 1593, to reduce the multiplication of two positive numbers a and b less than 1 to an addition, by regarding them as the values of trigonometric functions, $a = \sin\alpha$ and $b = \cos\beta$. By the addition theorems (see Chapter 10.):

$$\sin(\alpha+\beta) = \sin\alpha\cos\beta + \sin\beta\cos\alpha$$
$$\sin(\alpha-\beta) = \sin\alpha\cos\beta - \sin\beta\cos\alpha$$
$$\tfrac{1}{2}[\sin(\alpha+\beta) + \sin(\alpha-\beta)] = \sin\alpha\cos\beta = a\cdot b.$$

For $a = 0.61566$ and $b = 0.93969$ they obtained $\alpha = 38°$ and $\beta = 20°$, hence $\alpha + \beta = 58°$ and $\alpha - \beta = 18°$, consequently $a \cdot b = \tfrac{1}{2}[0.84805 + 0.30902] = \tfrac{1}{2} \cdot 1.15707$ and $0.61566 \cdot 0.93969 = 0.57854$.

The mathematician Simon STEVIN (1548–1620), who at one time was Quartermaster General of the Dutch army, advocated the *Hindu-Arabic positional notation*, in particular, the *decimal notation for fractions*. He worked on tables for the calculation of compound interest, which were continued by Jost BÜRGI (1552–1632) and published by him in Prague in 1620 as '*Arithmetische und geometrische Progresstàbuln*'. He started out from the basis $1.0001 = 1 + 1/10\,000$, the powers of which are easy to compute because – in modern language – even a few terms of the binomial expansion give the required accuracy of 10 places postulated by BÜRGI. The 10000th power $(1 + 1/10000)^{10000}$ of his basis is 2.71814, close to the number e = 2.7182818..., which is defined as the limit $\lim_{n\to\infty}(1 + 1/n)^n$. The exponents divided by 10 000 are therefore *approximately equal to the natural logarithms* of the powers. The tables contain in the entry column the logarithms, and in place of the mantissae of present-day tables there are the numbers themselves (*antilogarithm tables*).

Even the distinguished mathematician Jon NAPIER (or NEPER), eighth laird of Merchiston (1550–1617), only partially achieved his aim in his work '*Mirifici logarithmorum canonis descriptio*', published in 1614. The function he chose was, in modern notation, $y = ge^{-x/g}$ with $g = 1000000$. For the sum $x = x_1 + x_2$ of two exponents this leads to $y = ge^{-x_1/g} \cdot e^{-x_2/g} = y_1 y_2/g$. Together with Henry BRIGGS (1556–1630), Professor of mathematics in London, who admired NAPIER, he decided in favour of the function $y = 10^x$. After NAPIER's death BRIGGS continued the calculations; his '*Arithmetica logarithmica*', which was published in 1624, contains the 14-figure logarithms of the numbers from 1 to 20 000 and from 90 000 to 100 000. The missing logarithms were calculated by the surveyors Ezechiel DE DECKER and Adrian VLACQ, whose first complete logarithmic table appeared in 1627.

To convert a table of antilogarithms into one of logarithms, BRIGGS made use of the relationship between an arithmetic progression and its associated geometric progression in which to the arithmetic mean $(a_1 + a_2)/2$ of two numbers a_1 and a_2 there corresponds the geometric mean $\sqrt{(g_1 g_2)}$ of the corresponding quantities g_1 and g_2. For example, in the sequence of powers to the basis 3

1	2	3	4	...
3	9	27	81	...

to the number $2.5 = (2 + 3)/2$ there corresponds the value $\sqrt{(9 \cdot 27)} = 9\sqrt{3}$, because $3^{(2+3)/2} = \sqrt{(3^2 \cdot 3^3)} = \sqrt{(9 \cdot 27)}$.

Choosing the basis 10 and denoting the terms of the arithmetic progression as logarithms by l_i and the values of the powers by n_i, one obtains in succession for the intervals $0 < l_i < 1$ and $1 < n_i < 10$, respectively:

$$\begin{aligned}
l_1 &= \tfrac{1}{2}(0 + 1) = 0.5 & n_1 &= \sqrt{10} = 3.162\,277\ldots\\
l_2 &= \tfrac{1}{2}(l_1 + 1) = 0.75 & n_2 &= \sqrt{(n_1 \cdot 10)} = 5.623\,413\ldots\\
l_3 &= \tfrac{1}{2}(l_1 + l_2) = 0.625 & n_3 &= \sqrt{(n_1 \cdot n_2)} = 4.216\,964\ldots\\
l_4 &= \tfrac{1}{2}(l_2 + l_3) = 0.6875 & n_4 &= \sqrt{(n_2 \cdot n_3)} = 4.869\,674\ldots\\
l_5 &= \tfrac{1}{2}(l_2 + l_4) = 0.718\,75 & n_5 &= \sqrt{(n_2 \cdot n_4)} = 5.232\,991\ldots\\
l_6 &= \tfrac{1}{2}(l_4 + l_5) = 0.703\,125 & n_6 &= \sqrt{(n_4 \cdot n_5)} = 5.048\,065\ldots\\
l_7 &= \tfrac{1}{2}(l_4 + l_6) = 0.695\,3125 & n_7 &= \sqrt{(n_4 \cdot n_6)} = 4.958\,067\ldots\\
l_8 &= \tfrac{1}{2}(l_6 + l_7) = 0.699\,218\,75 & n_8 &= \sqrt{(n_6 \cdot n_7)} = 5.002\,865\ldots
\end{aligned}$$

The eight steps listed yield $\lg 5.002865 = 0.69921875$, and the method shows how by a repeated extraction of square roots the value of lg 5 can finally be computed with an arbitrary accuracy.

The logarithmic slide rule

Historical remarks. As early as the second decade of the 17th century Edmund GUNTER (1561–1626) indicated the principle of logarithmic calculations along a graduated straight line. On his scales multiplication and division were performed as addition and subtraction of lengths by means of a pair of dividers. Shortly afterwards William OUGHTRED (1574–1660) used two of Gunter's lines sliding along each other. This made the use of dividers unnecessary. His lines were made both in straight and in circular form. Around the middle of the 17th century Edmund WINGATE (1593–1656) and Seth PARTRIDGE used a rule sliding between parts of a fixed stock, that is, an instrument similar to the present slide rule. It acquired its final shape in the course of the 19th century. Industrial mass production of slide rules began towards the end of the 19th century. Slide rules for special purposes, for example, for electricians and tradesmen, followed shortly afterwards.

The structure of a logarithmic slide rule. The following exposition refers to a system of slide rule in common use among scientists and engineers (Fig.). Other systems are arranged in slightly different ways and require some minor modifications in technique. As a rule, the length of the scales of a slide rule is 25 cm.

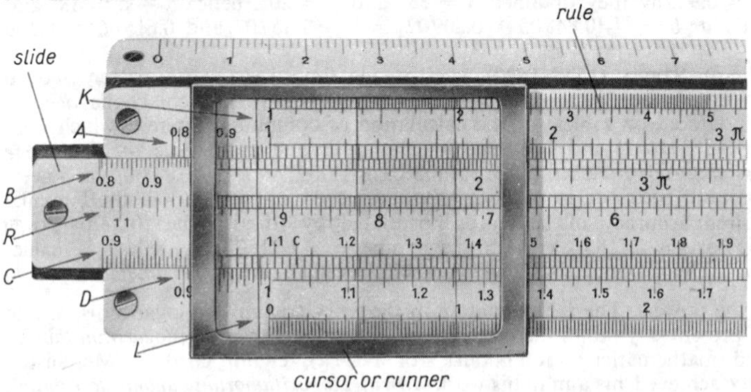

2.2-7 The logarithmic slide rule

A slide rule consists of three parts: the *rule*, the *slide*, and the *cursor* or *runner*. The rule contain s the scales marked A, D and K (and often also other scales). The slide moves in it in grooves. The front of the slide carries the scales marked C, R or CI, and B, and the back the scales of the trigonometric functions or others. The cursor with one or three vertical lines moves across rule and slide. As a rule, one uses the middle vertical, the others play a role in the calculation of circular areas and cylinder volumes.

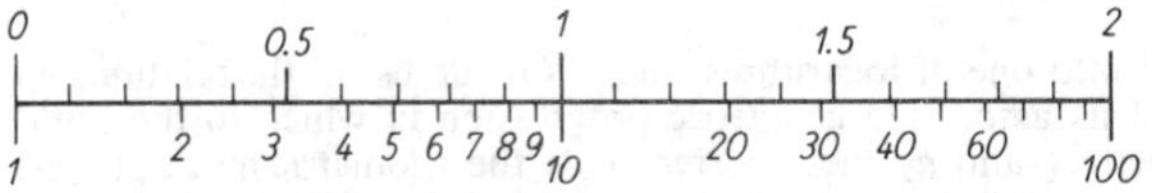

2.2-8 Logarithmic scale below and linear or ordinary scale above

The scales of a slide rule. The scales of a slide rule are related to the function $y = \lg x$. In these logarithmic scales numbers are placed at points whose distance from the origin is proportional to their logarithm. The *basic logarithmic scales* C and D refer to numbers from 1 to 10. The distance from the origin for a particular point on the scale is obtained by multiplying the length of the scale (250 mm) by the appropriate logarithm. For example, the distance from the initial point 1 to the point 2 is $\lg 2 \cdot 250\text{ mm} = 75.3\text{ mm}$. The uneven growth of the logarithms implies an uneven growth of the logarithmic scale (Fig.). One sees that as the numbers increase, their distances become smaller.

The *scales* A and B are also logarithmic. They consist of two parts of equal length. The distance between the numbers 1 and 10 is half as much as on the scales C and D. Hence a segment on the scale D of length $\lg n$ corresponds on the scale B to a segment of length $2 \lg n = \lg n^2$. This means that the numbers on the scales A and B are the squares of those on the scales C and D below them.

Similarly, the *scale* K, which consists of three parts of equal lengths, gives the cubes of the numbers on the scale D.

The *scale* L, often at the bottom of the rule or in the middle of the back of the slide, is not logarithmic and gives directly the mantissa of the common logarithm of the number standing above it.

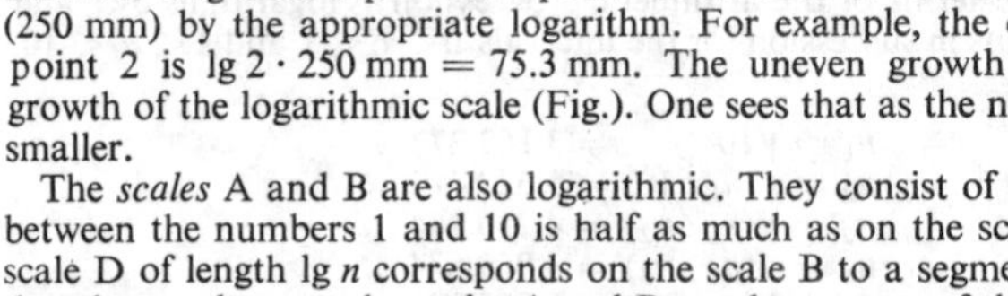

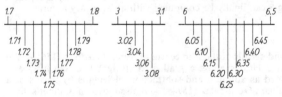

2.2-9 The various subdivisions of a slide rule

The *scale* CI or R contains the reciprocals of the numbers on the scale D. It has the same graduation, but in the reverse order from right to left and gives for every value x on the scale D the reciprocal $1/x$.

Reading and setting. One can read three digit numbers on a slide rule. The first two digits present hardly any difficulties, not more than reading numbers off a ruler. The third digit is a different matter. The uneven scales necessitate uneven divisions.

If one investigates the scale D, one recognizes *three sections*: the first section lies between 1.00 and 2.00. Between two neighbouring two-digit numbers there are 9 verticals; for example, 1.00 is followed by 1.01, 1.02, ... The subdivision is into *ten parts* each (Fig.).

In the second section, between 2 and 4, the division is into *five parts* each. Between two neighbouring verticals for two-digit numbers there are 4 further verticals whose distance corresponds to two units in the third place. For example, 2.00 is followed by 2.02, 2.04, ... Since the distance between numbers becomes smaller and smaller towards the right, in the third section, between 4 and 10, only a division into *two parts* is possible, and the distance between the verticals corresponds to five units of the third digit.

Reading and setting proceed with the help of the middle vertical on the *cursor*. This is placed exactly at the number needed for a calculation or just calculated. Further details can be found in the subsequent instructions for calculations with the slide rule. If one comes to a setting in between two vertical lines, one has to resort to an estimate.

Calculations with the slide rule. Just as every logarithmic computation involves the determination of the characteristic, so every slide rule computation requires a *rough estimate* or an approximate calculation in powers of 10, to determine the number of digits before the decimal point. A rough calculation is advisable, because it gives not only the right number of places, but also a certain control of the result. Calculations with the slide rule are geometric additions and subtractions of segments. By using the scales A, B and K one can double or treble segments, or divide them into two or three equal parts. Here are a few examples.

Multiplication. This is based on the law $\lg(a \cdot b) = \lg a + \lg b$. The addition of the two logarithms is carried out on the slide rule as addition of two segments of length $\lg a$ and $\lg b$. The initial point 1 of the slide scale C is placed at the point a of the rule scale D. Then the cursor is placed at the point b of the scale C and underneath the product $a \cdot b$ is read off on the scale D (Fig.).

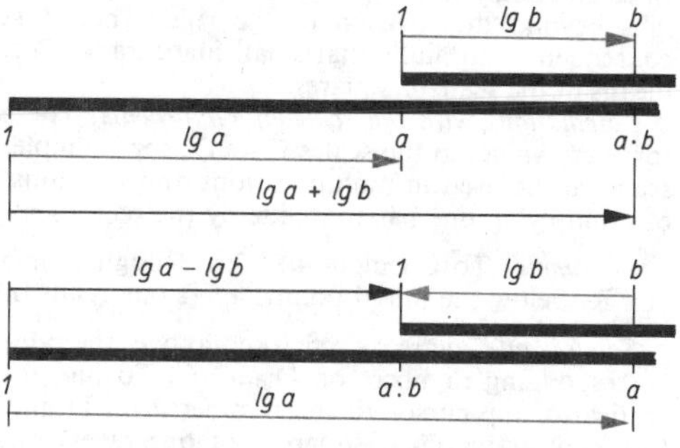

2.2-10 Illustration of multiplication and division

Example 1: To calculate $2 \cdot 1.5$. One places the 1 on the scale C above 2 on the scale D. Under 1.5 on C one reads off the product 3 on the scale D (Fig.).

Example 2: To calculate $2.84 \cdot 4.55$. Rough estimate: $3 \cdot 4 = 12$. This alone shows that the result can no longer be read off the scale D. One proceeds as follows: on the scale D one sets 2.84. Shifting the slide to the left one sets the end point 10 of the scale C above the point 2.84. One places the cursor at the point 4.55 of the scale C and underneath one reads off on the scale D the value of the product $2.84 \cdot 4.55 = 12.9$.

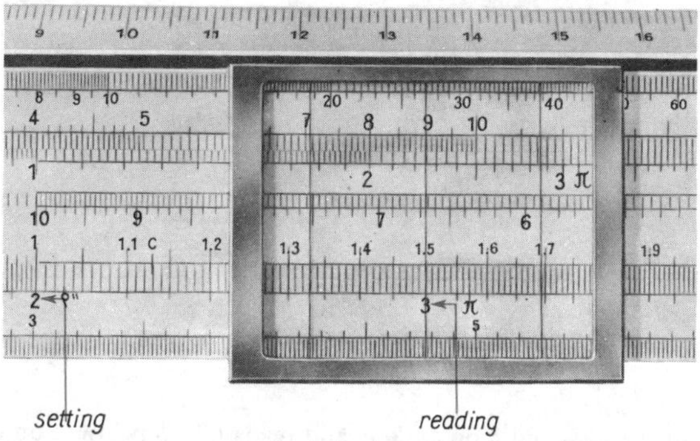

2.2-11 Example of multiplication by slide rule: $2 \cdot 1.5 = 3$

An explanation of this method of multiplication can be obtained by carrying out the same calculation on the two left sections of the scales A and B. The result then falls into the right section.

For a multiplication the 1 or 10 of the slide scale is placed at one of the factors and the result is read off at the other factor.

Division. The relevant law is $\lg(a/b) = \lg a - \lg b$. On the slide rule this means that the segment of length $\lg b$ has to be subtracted from the segment of length $\lg a$. Above the point a of the scale D on the rule one places the point b of the scale C on the slide. Then one reads off the quotient a/b on the scale D under the initial point or the end-point 10 of the scale C on the slide.

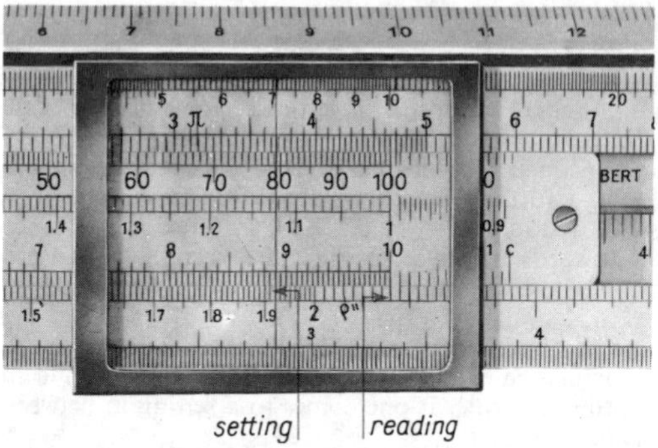

2.2-12 Division by slide rule: 19.2 : 89 = 0.216

Example 1: To calculate 88.5 : 0.515. Rough estimate: 90 : 0.5 = 180. Above the point 88.5 on D one places the point 0.515 of C. Under the initial point 1 of C one finds on D the quotient 172 = 88.5 : 0.515.

Example 2: To calculate 19.2 : 89. Rough estimate: 20 : 90 = 0.2. One proceeds just as before, but the quotient is read off under the end-point 10 of the scale C. Solution: 19.2 : 89 = 0.216 (Fig.).

For a division the divisor on the slide scale is placed at the dividend on the rule scale, and the result is read off at 1 or 10 of the slide scale.

To calculate expressions of the form $(a_1 \cdot a_2 \ldots)/(b_1 \cdot b_2 \ldots)$ one divides and multiplies alternately so as to have the least possible nubmer of shifts of the slide. In calculating a proportion of the form $y = a \cdot c/b$ one first divides a by b and then multiplies by c. This requires a single setting of the slide, whereas the multiplication $a \cdot c$ followed by division by b necessitates two settings of the slide. In calculating y according to the method proposed it can happen that the setting of b on the slide falls beyond the division of the rule. Then a second shift of the slide becomes necessary, and consequently an additional small inaccuracy. Therefore it is preferable to form such expressions by means of the scale of squares.

Calculations with the scale of reciprocals. The scale R or CI of reciprocals gives the reciprocal for every value on the scales C or D; for example, above 4 on C one finds the reciprocal 0.25. This scale can be used in multiplications and divisions. It is to be observed that $a \cdot b = a : b^{-1}$. Instead of multiplying one has to divide by the reciprocal.

Example: To calculate 4.8 · 3.6. Rough estimate: 5 · 3 = 15. Above 4.8 on D one places 3.6 on R. Below the initial point 1 of R one reads off on D the result 4.8 · 3.6 = 17.28.

Squares and roots. As remarked above, the numbers on the scales A and B are the squares of the corresponding numbers on D and C. To find the square of a number a one need only place the vertical of the cursor to the number a on D and read off above it the number a^2 on the scale A (Fig.). In extracting a square root one reverses the procedure: one sets the vertical of the cursor

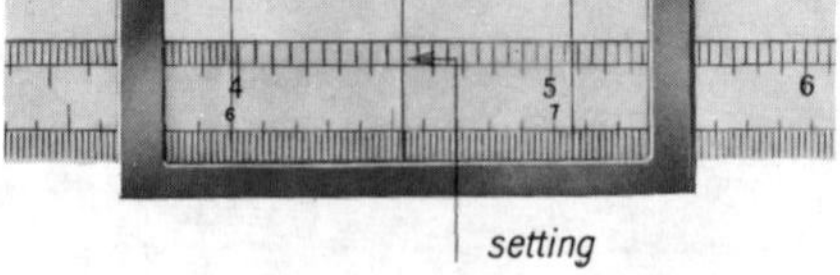

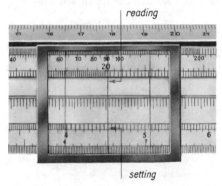

2.2-13 Example of squaring: $4.5^2 = 20.25$

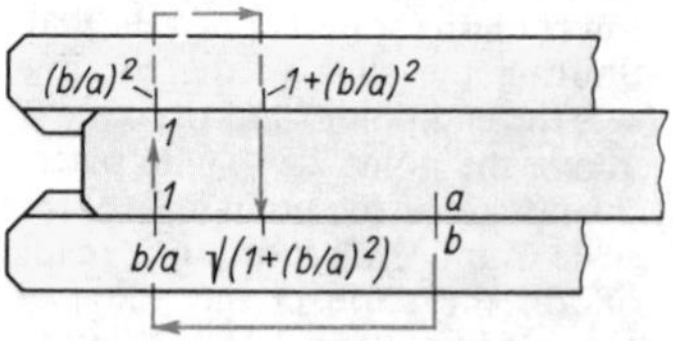

2.2-14 Position of slide rule for the calculation of $\sqrt{(1 + (b/a)^2)}$

at the radicand b on scale A and reads off $\sqrt{b}$ below it on the scale D. In extracting roots one has to observe the correct setting of b on the scale of squares:

$\sqrt{25} = 5$ Setting of 25 between 10 and 100.
$\sqrt{250} = 15.81$ Setting of 25 on the section between 1 and 10, because $\sqrt{250} = \sqrt{(2.5 \cdot 100)} = 10 \cdot \sqrt{2.5}$.

Radicands between 1 and 100 are set directly, other radicands are brought to a value between 1 and 100 after multiplication by a suitable power of 100.

Cubes and cube roots. On the scale K one finds the cubes of the numbers of the scale D. The calculation of cubes and cube roots follows the same pattern as that of squares and square roots.

Calculation of $c = \sqrt{(a^2 + b^2)}$. These expressions are of frequent occurrence. It may be assumed that $b > a$. The calculation proceeds according to the rule

$$c = a\sqrt{[1 + (b/a)^2]}.$$

One takes the following steps: calculation of b/a by means of scales C and D, reading off $(b/a)^2$ on the scale of squares, mental addition of 1 and setting of $1 + (b/a)^2$ on the scale of squares, reading off $\sqrt{[1 + (b/a)^2]}$ on D, and finally multiplication by a. The figure illustrates the procedure (Fig.).

Trigonometric, Pythagorean, and exponential scales. As a rule, the back of the slide contains further scales: a *sine and tangent scale* for small angles from 34.5′ to 6° for which the values of the sines and tangents are approximately equal: a *sine scale* from 5° 45′ to 90°; a *tangent scale*, which can be used for calculating tangents and cotangents. This scale goes from left to right for angles from 5° 45′ to 45° and in the reverse direction from 45° to 84° 15′. The *Pythagorean scale* records the values of $\sqrt{(1 - x^2)}$ and permits the calculation of the cosine values. Exponential scales require further explanations, which would take up too much space here. They are useful for a number of specialized computations. Only the basic principles of the slide rule have been described here. Modern refinements provide for more sophisticated numerical calculations.

Accuracy in calculations. The applications of the slide rule are limited only by the accuracy of the result that can be achieved. For a scale length of 25 cm the error of a careful reading can be assumed to be around 0.1%. For scales of length 12.5 cm the error in reading is doubled. Several settings in the course of a calculation increase the mean error according to the Gaussian law of propagation of errors. For four settings it amounts to about 0.2%. Even in simple calculations one should make several repetitions and take the average of the results obtained.

Calculating discs. In these instruments the carrier of the logarithmic scale is not a straight line but a circle. In *calculating wheels* these scales are attached to the faces of concentric discs which slide against one another. This special form of a slide rule has the advantage that even for an instrument of small dimensions the lengths of the scales are comparatively large and that shifting of the slide can be avoided because all parts of one scale are opposite parts of another. These advantages lead to an increased accuracy.

Since an increase in the lengths of the scales leads to greater accuracy, the scales of *calculating frames* are split into smaller parts and arranged parallel to each other in a plane. In *calculating cylinders* the parts of the scales are arranged along the surface of a cylinder. Calculating cylinders have been constructed by means of which one can achieve the same accuracy as in calculations with a 5-figure logarithmic table.

All those traditional calculating means have been overcome by the modern calculating devices, the computers. The slide rule almost completely is substituted by the pocket calculator. And for particular purposes, for example, for use in electronics, hydraulics, building in concrete, surveying, navigation, or optics specialized software or even specialized computers are available.

3. Development of the number system

There is no occupation in which a man would not have to perform simple calculations. The numbers used in such tasks are a means of grasping and mastering the world quantitatively. But statements of general validity can be based on arithmetical operations only when the laws underlying the calculations are known. The accumulated experience of millionfold applications in the historical development of mankind is laid down in computational rules that can be reduced to a few logical fundamental concepts and axioms. The foundations for this deductive build-up of the number system are the theory of sets and mathematical logic. On this basis first the natural numbers, and then the further domains of numbers, are constructed.

3.1. The natural numbers N

The natural numbers have their origin in a fusion of the *cardinal* or counting numbers with the *ordinal* or place numbers. But in what follows, no distinction will be made between cardinal and ordinal numbers.

The Peano axioms. As early as 1891 PEANO showed that all the properties of the natural numbers can be derived from five axioms, which now bear his name.

Peano's axiom system	**1. 0 is a number.** **2. Every number n has precisely one successor n'.** **3. 0 is not the successor of any number.** **4. Distinct numbers have distinct successors.** **5. If a set of natural numbers contains the number 0 and contains, together with any number, also its successor, then it is the set of all natural numbers.**

According to these axioms every natural number other than 0 can be described as a successor, or as a successor of ... of a successor, of 0. But instead of the notation $0, 0', 0'', \ldots$ one uses simply the decadic positional system and denotes the numbers by $0, 1, 2, \ldots$ In particular, Axiom 5 gives the justification for the method of *mathematical induction*.

Mathematical induction. This method of reasoning is used to prove that a proposition $P(n)$ concerning natural numbers n is valid for all natural numbers; to give but one example: the equality $1 + 2 + 4 + \cdots + 2^n = 2^{n+1} - 1$ holds for all natural numbers. In using mathematical induction as a method of proof one considers the set M of all natural numbers for which $P(n)$ is true. If one can show 1) that $0 \in M$, in words, that the proposition is valid for $n = 0$, (*basis of induction*), and 2) that from $n \in M$ it follows that $n' \in M$, in words, that the validity of the proposition for any natural number n (*inductive hypothesis*) implies its validity for the successor n' (*inductive step*), then by Peano's fifth axiom M must be the set of all natural numbers. In the example above, $P(0)$ is evidently true $(1 = 2 - 1)$, and if the proposition is assumed to be valid for a natural number n, then it follows that $1 + 2 + 4 + \cdots + 2^n + 2^{n+1} = 2^{n+1} - 1 + 2^{n+1} = 2^{n+1}(1 + 1) - 1 = 2^{n+2} - 1$, in other words, its validity for the natural number $n + 1$. Consequently the proposition is valid for all natural numbers.

Mathematical induction is used not only in proving theorems, as just now, but also in defining and constructing mathematical objects. Such definitions and constructions proceed *recursively* or step-by-step, and each step makes use of the previous steps. That this procedure cannot lead to logical complications was proved by DEDEKIND in his 'justification theorem'.

Calculations with natural numbers N. The *addition* is defined by $m + 0 = 0 + m = m$ and $(m + n') = (m + n)'$, the *multiplication* by $m \cdot 0 = 0 \cdot m = 0$ and $m \cdot n' = m \cdot n + m$. By these *recursive definitions* addition and multiplication are uniquely determined. The rules of calculation are then derived by mathematical induction.

Commutative law	$a + b = b + a$	$a \cdot b = b \cdot a$
Associative law	$(a + b) + c = a + (b + c)$	$(a \cdot b) \cdot c = a \cdot (b \cdot c)$
Distributive law	$a \cdot (b + c) = a \cdot b + a \cdot c$	

The proof, for the associative law of addition, for example, proceeds as follows: The statement is true for $c = 1$ (*basis of induction*); indeed, $(a + b) + 1 = (a + b) + 0' = [(a + b) + 0]' = (a + b)' = a + b' = a + (b + 1)$. Now assume that the statement is true for $c = n$ (*inductive hypothesis*): $(a + b) + n = a + (b + n)$. It is required to show that it is then also valid for $c = n + 1$ (*inductive step*). Now $(a + b) + n' = [(a + b) + n]' = a + (b + n)' = a + (b + n')$. By Peano's fifth axiom all natural numbers then have the property expressed in the associative law.

For more than two terms the operations are defined by mathematical induction, for example, $a_1 + a_2 + \cdots + a_n = (a_1 + a_2 + \cdots + a_{n-1}) + a_n$ and $a_1 \cdot a_2 \cdot \cdots \cdot a_n = (a_1 \cdot a_2 \cdot \cdots \cdot a_{n-1}) \cdot a_n$. In any sum or product brackets can then be inserted or removed arbitrarily, as can again be proved by mathematical induction.

Subtraction and *division* of two numbers a and b are defined as *inverse operations* to addition and multiplication: If for the given numbers a and b there exists a number x such that $a + x = b$, then $x = b - a$ is the *difference* of b and a; if it exists, it is uniquely determined. If for the given numbers $a \neq 0$ and b there exists a number y such that $a \cdot y = b$, then $y = b/a$ is the *quotient* of b and a; if it exists, it is also uniquely determined. Of course, as a rule, for given natural numbers a and b such an x or y need not exist, as the examples $3 + x = 2$ or $3 \cdot y = 2$ show.

Subtraction and division cannot always be carried out within the domain of natural numbers.

The wish to remove these restrictions gives rise to the formation of new number systems.

Exponentiation. For numbers $a \neq 0$ and n the *power* a^n is again introduced by a recursive definition: $a^0 = 1$ and $a^{n'} = a^n \cdot a$. The standard laws of powers (see Chapter 2.) hold for the multiplication and exponentiation of powers, as well as for the division, provided that the quotient exists. The operation of exponentiation is neither commutative nor associative, as the following examples show: $3^2 = 9$, but $2^3 = 8$; $(3^2)^3 = 9^3 = 729$, but $3^{(2^3)} = 3^8 = 6561$.

Roots and *logarithms*. If for given numbers a and n ($n \neq 0, 1$) there exists a number b such that $b^n = a$, then b is called the *nth root of a* and is denoted by $b = \sqrt[n]{a}$. If for the numbers $a \neq 0$ and b there exists a number n such that $a^n = b$, then n is called the *logarithm of b to the basis a* and is denoted by $n = \log_a b$; for example, $5 = \sqrt[3]{125}$, $6 = \log_2 64$.

Survey of the arithmetical operations
First kind: Addition with subtraction as inverse operation,
second kind: multiplication with division as inverse operation,
third kind: exponentiation with taking roots and logarithms as inverse operations.

An *order* $n' > n$ between neighbouring numbers is already provided by the successor relation. For arbitrary natural numbers a and b the relation $a > b$ or $b < a$ is defined to hold if there exists a natural number $c \neq 0$ such that $a = b + c$. It can be shown that the relation so defined is an *irreflexive order* (see Chapter 14.). It satisfies the *monotonic law of addition and multiplication* as well as the *Archimedean axiom*.

Monotonic law of addition and multiplication of natural numbers: If $a > b$, c arbitrary and $d \neq 0$, then $a + c > b + c$ and $a \cdot d > b \cdot d$.

Archimedean axiom: For arbitrary $a > 0$ and b there always exists a number n such that $a \cdot n > b$.

It can be shown that for any two distinct natural numbers a, b either $a > b$ or $b > a$. Therefore the relation $>$ on the domain of natural numbers is a *total* (or *linear*) and *Archimedean order*.

For calculations with natural numbers the usual rules hold: in the following examples it is assumed that all the operations occurring can be carried out:

$$a + (b - c) = (a + b) - c = a + b - c = a - c + b, \quad (a \cdot c) : (b \cdot c) = a : b,$$
$$a - (b + c) = (a - b) - c = a - b - c = a - c - b, \quad (b - c) : a = (b : a) - (c : a),$$
$$a - (b - c) = (a - b) + c = a - b + c = a + c - b.$$

From the properties of order it can be proved that for $a > 1$ the powers a^n form a sequence of steadily increasing numbers and that roots and logarithms, if they exist, are uniquely determined.

3.2. Absolute rational numbers. Fractions

The processes of sharing and of measuring require a system of numbers in which one can carry out not only the operations that can be performed on natural numbers, but in addition also unrestricted division. The postulate that in an extended system the laws of the old domain remain valid, as far as possible, is called the *principle of permanence*.

Construction of the extended system. One forms ordered pairs of natural numbers m and $n \neq 0$ and writes them like fractions m/n. Two such symbols m/n and p/q are said to be equal, written as $m/n = p/q$, if $mq = np$, for example, $3/1 = 6/2$ because $3 \cdot 2 = 6 \cdot 1$. This equality satisfies the conditions for an equivalence relation. All equal fractions can therefore be collected in a class such as $\{3/1, 6/2, 9/3, \ldots, 150/50, \ldots\}$, and every fraction belongs to one class. These classes are called *absolute rational numbers*. Each such rational number can be represented by any one fraction of its class; for example, $2/3$, $4/6$, $30/45$ are representatives of one and the same absolute rational number $\alpha = \{2/3, 4/6, 6/9, 8/12, \ldots, 18/27, \ldots, 30/45, \ldots\}$. The preferred representation is by a reduced fraction in which numerator and denominator have no common divisor greater than 1, in the last example one also writes $\alpha = \{2/3\}$. The curly brackets serve to distinguish between the class and its representative.

Arithmetical operations and order. In determining the arithmetical operations and the order for absolute rational numbers one is guided by the standard rules for fractions. The relevant definitions are chosen such that the extended number system has the properties that have proved sensible in centuries of calculating practice. The operations of the first and second kind for the numbers $\alpha = \{m/n\}$ and $\beta = \{p/q\}$ with $n \neq 0$ and $q \neq 0$ are defined as follows:

$$\alpha + \beta = \{(mq + pn)/nq\}, \text{ for example, } \{2/3\} + \{10/7\} = \{(2 \cdot 7 + 3 \cdot 10)/3 \cdot 7\} = \{44/21\},$$
$$\alpha - \beta = \{(mq - np)/nq\}, \text{ provided that } mq \geqslant np,$$
$$\alpha \cdot \beta = \{(mp)/(nq)\}, \text{ for example } \{3/5\} \cdot \{20/9\} = \{(3 \cdot 20/(5 \cdot 9)\} = \{60/45\} = \{4/3\},$$
$$\alpha/\beta = \{(mq)/(np)\}, \ (p \neq 0), \text{ for example, } \{3/5\}/\{7/1\} = \{(3 \cdot 1)/(5 \cdot 7)\} = \{3/35\}.$$

The last rule shows that division, except by $0 = \{0/q\}$, can be performed without restriction. Subtraction and division are the inverse operations to addition and multiplication. Subtraction cannot always be carried out. Operations on more than two numbers are defined correspondingly.

An *order* is defined by analogy to equality: $\alpha > \beta$ if $mq > np$; for example, $8/9 > 11/14$ because $8 \cdot 14 > 9 \cdot 11$.

All these definitions are meaningful only if they are independent of the choice of representatives for α and β. This will now be demonstrated for the order.

Assumption: $\{m/n\} > \{p/q\}$, that is, $mq > np$; let m_1/n_1 and p_1/q_1 be other fractions in the classes $\{m/n\}$ and $\{p/q\}$, respectively, that is, $m/n = m_1/n_1$ and $p/q = p_1/q_1$.

Assertion: Then $\{m_1/n_1\} > \{p_1/q_1\}$.

Proof: Multiplication of $mq > np$ by n_1q_1 shows that $mn_1qq_1 > nn_1pq_1$. Since, by assumption, $mn_1 = m_1n$ and $pq_1 = p_1q$, substitution gives $m_1nqq_1 > nn_1p_1q$. After cancelling the factor nq on both sides one obtains $m_1q_1 > n_1p_1$, as required.

The order relation on the domain of absolute rational numbers is total (*linear*) *and Archimedean.*

This domain can be mapped, with preservation of the order, point by point to a so-called number ray. The images lie *dense* on the ray, that is, between any two of them there is always another, for example, between the images of α and β that of $(\alpha + \beta)/2$.

The commutative, associative, and distributive laws for the operations and the other rules for natural numbers are also valid for the absolute rational numbers. Here is a proof for the associative law of multiplication: Let $\alpha = \{m/n\}$, $\beta = \{p/q\}$, $\gamma = \{r/s\}$. Then $[\alpha \cdot \beta] \cdot \gamma = \{(mp)/(nq)\} \cdot \{r/s\} = \{(mpr)/(nqs)\} = \{m/n\} \cdot \{(pr)/(qs)\} = \alpha \cdot [\beta \cdot \gamma]$.

Absolute rational and natural numbers. The extended number system contains as part a domain corresponding exactly to the natural numbers, namely that of the numbers $\{1/1\}$, $\{2/1\}$, $\{3/1\}$, ..., $\{k/1\}$, ... For if two numbers of this kind are added or multiplied, another such number arises: $\{k/1\} + \{l/1\} = \{(k+l)/1\}$ and $\{k/1\} \cdot \{l/1\} = \{k \cdot l/1\}$. Furthermore, $\{k/1\} > \{l/1\}$ holds if and only if $\{k > l\}$. If one now assigns to $\{1/1\}$ the natural number 1, to $\{2/1\}$ the natural number 2, and in general, to $\{k/1\}$ the natural number k, then it turns out that one calculates with the numbers $\{k/1\}$ just as with the natural numbers.

In the system of absolute rational numbers the subsystem consisting of the numbers of the form $\{k/1\}$ *is isomorphic to the system of natural numbers with respect to the operations and the order defined there.*

From now on the numbers $\{k/1\}$ can simply be written as k and can be treated as natural numbers, because they satisfy Peano's axioms. Also in the other absolute rational numbers the curly brackets can be omitted, by writing, for example, m/n instead of $\{m/n\}$. Confusion cannot arise as long as the rules for calculations with fractions are observed. True, there is a conceptual difference: the fractions 3/7 and 6/14 are not identical, because their numerators and denominators differ, but their values are equal. However, the absolute rational numbers 3/7 and 6/14 are simply identical. This simplification is the motivation for the definitions of the arithmetical operations and the order. The definitions are chosen just so that the isomorphism mentioned above holds. It corresponds to Hankel's *principle* (better: *postulate*) *of permanence* to define operations and order in the extended system in such a way that the laws which hold in the old system remain in force, as far as possible. In contrast to the principle of mathematical induction, which is a valid method of proof, Hankel's postulate only states what is desirable in an extension of a number system. This principle has no power of conviction. One can now state:

The system of absolute rational numbers is an extension of the system of natural numbers. It consists of the natural numbers and the fractions.

Operations of the third kind. The power $\alpha^\beta = \gamma$ in which $\beta = n$ is a natural number is defined in the obvious way. In accordance with this, for a given value of γ the numbers $\alpha = \sqrt[\beta]{\gamma}$ and $\beta = \log_\alpha \gamma$ are defined as quantities for which $\alpha^\beta = \gamma$, provided that they exist. But if β is an absolute rational number, $\beta = r/s$, then $\alpha^\beta = \gamma$ is to be interpreted as $\gamma = \sqrt[s]{\alpha^r}$. If such numbers exist in the system of absolute rational numbers, then they are uniquely determined by these stipulations.

Examples: 1. For $\alpha = 2/3$ and $\beta = 3$ one obtains $\gamma = \alpha^\beta = (2/3)^3 = (2/3) \cdot (2/3) \cdot (2/3) = 8/27$. When $\gamma = 8/27$ and $\beta = 3$ are given, one has $\alpha = \sqrt[\beta]{\gamma} = \sqrt[3]{(8/27)} = 2/3$, and for $\gamma = 8/27$ and $\alpha = 2/3$ one has $\beta = \log_\alpha \gamma = \log_{2/3}(8/27) = 3$.

2. For $\alpha = 9/16$ and $\beta = 1/2$ one obtains $\gamma = \alpha^\beta = (9/16)^{1/2} = \sqrt[2]{(9/16)} = 3/4$. When $\gamma = 3/4$ and $\beta = 1/2$ are given, one has $\alpha = \sqrt[\beta]{\gamma} = \sqrt[1/2]{(3/4)} = 9/16$, because $(9/16)^{1/2} = 3/4$; and for $\gamma = 3/4$ and $\alpha = 9/16$ one has $\beta = \log_\alpha \gamma = \log_{9/16}(3/4) = 1/2$.

3. For $\alpha = 2$ and $\beta = 1/2$ one should have $\gamma = \alpha^\beta = 2^{1/2} = \sqrt{2}$. But this is not a rational number, as will be shown later.

3.3. The rational numbers **Q**

In order to be able to subtract absolute rational numbers without restriction one introduces in the set of ordered pairs (see Chapter 14.) (α, β) of absolute rational numbers an equivalence relation, the so-called *equality of difference*: one writes $(\alpha, \beta) \sim (\alpha', \beta')$ if $\alpha + \beta' = \alpha' + \beta$. For example, $(12/5, 3/5) \sim (5/2, 7/10)$ because $12/5 + 7/10 = 5/2 + 3/5$. The equivalence class of the pair (m, n) is called a *rational number* and is denoted for the time being by $\{(m, n)\}$. For these new numbers the arithmetical operations are defined so that they are independent of the representatives and that subtraction can always be performed. The proofs of the laws of arithmetic (see Chapter 1.) are based on the fact that these laws hold for absolute rational numbers.

Signed numbers. To remove the clumsy notation $\{(m, n)\}$ one begins by observing that a number pair (m, n) for $m > n$ can be written in the form $(n + k, n)$, for $m < n$ in the form $(m, m + k)$, and for $m = n$ in the form (m, m); for example, $(2, 1/2) = (1/2 + 3/2, 1/2)$, $(3/4, 1) = (3/4, 3/4 + 1/4)$, $(2/5, 4/10) = (2/5, 2/5)$. One then introduces a new notation:

$$\alpha = \begin{cases} (+k) \text{ when } m = n + k \ (k > 0); & \text{for example, } \{(2, 1/2)\} = (+3/2) \\ (-k) \text{ when } n = m + k \ (k > 0); & \text{for example, } \{(3/4, 1)\} = (-1/4) \\ (0) \text{ when } m = n; & \text{for example, } \{(2/5, 4/10)\} = (0). \end{cases}$$

The fact that this notation is independent of the representative is a consequence of the equality of difference. The plus- and minus-sign are called the sign of the number. It must not be confused with the operational symbols of the same shape. The numbers $\alpha = (+k)$ are called *positive*, the numbers $\beta = (-k)$ *negative*, and k is called the *absolute value* of α or β, in symbols $k = |\alpha|$, $k = |\beta|$. Here k is an absolute rational number. An order for rational numbers is defined as follows:

$$\alpha > \beta \begin{cases} \text{when } \alpha \text{ is positive, } \beta \text{ positive or zero, and } |\alpha| > |\beta|, \text{ for example, } (+7/3) > (5/6); \\ \text{when } \alpha \text{ is positive and } \beta \text{ is negative, for example, } (+1/100) > (-10); \\ \text{when } \alpha \text{ is negative or zero, } \beta \text{ is negative, and } |\alpha| < |\beta|; \text{ for example, } (-2/3) > (-1). \end{cases}$$

This definition is also compatible with the equivalence relation and satisfies the condition of transitivity. The monotonic laws hold for addition, and also for multiplication by a positive factor. But the order relation is reversed when both sides are multiplied by the same negative numbers; for example,

$$(+5) > (+2) \text{ implies that } (+5) + (-1) > (+2) + (-1) \text{ and } (+5) \cdot (+3/4) > (+2) \cdot (+3/4),$$
$$\text{but } (+5) \cdot (-1) < (+2) \cdot (-1).$$

The Archimedean axiom, suitably modified, holds just as for absolute rational numbers.

Rational numbers and absolute rational numbers. The system of rational numbers contains as a proper part that of the positive rational numbers. By assigning to the number $(+k)$ its absolute value k this part is mapped one-to-one onto the system of absolute rational numbers. The arithmetical operations, as far as they can be performed, and the order are preserved, for example, $(+k) \cdot (+l) = (+k \cdot l)$.

The system of positive rational numbers is isomorphic to that of the absolute rational numbers with respect to the operations, as far as they can be performed, and to the order.

Since $(+k) + (-l) = (+k) - (+l)$, the notation can be further simplified by omitting the plus sign; for example, $(+3/4) + (-2/5) = (+3/4) - (+2/5) = 3/4 - 2/5$.

The power α^β for positive α is defined as on the left; α^β has a meaning only when $|\alpha|^{|\beta|}$ has a meaning in the system of absolute rational numbers; $\sqrt[n]{\alpha}$ and $\log_\beta \alpha$ are only defined for certain positive α and β $(\beta \neq 1)$ and then correspond to the definitions for absolute rational numbers.

$$\alpha^\beta = \begin{cases} |\alpha|^{|\beta|}, & \text{when } \beta > 0, \\ 1, & \text{when } \beta = 0, \\ 1/|\alpha|^{|\beta|}, & \text{when } \beta < 0. \end{cases}$$

The system of rational numbers is an extension of that of absolute rational numbers. In it the arithmetical operations of the first and second kind can be performed without restriction, but not those of the third kind.

3.4. The integers **Z**

Instead of constructing on the basis of the natural numbers first the system of absolute rational, and then that of rational numbers, as it has been done here following the practice in some schools, one could have followed another path (see Chapter 1.). There the domain of *integers* $0, \pm 1, \pm 2, \ldots$ is first constructed from ordered pairs of natural numbers with equal differences, and then the domain of rational numbers is obtained by means of pairs of integers with equal quotients. The domain of integers contains as part that of the positive integers, which is isomorphic to that of the natural numbers. The domain of rational numbers, in its turn, contains as part the domain of numbers $\pm k$, which is isomorphic to that of the integers. In both cases there are two extensions.

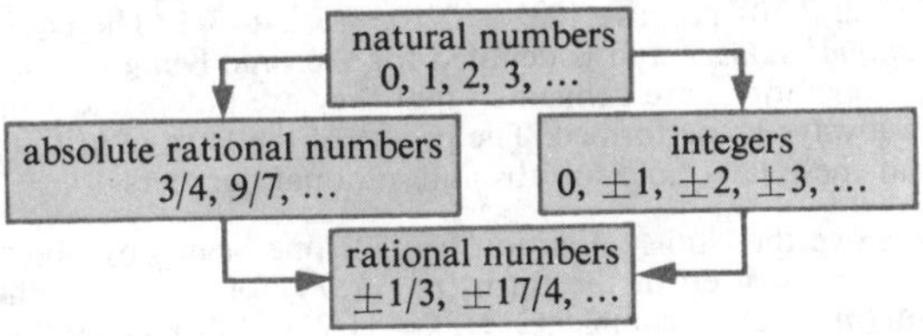

3.5. The real numbers R

Like the absolute numbers on the number ray, so the rational numbers are dense on the *number line*. But they do not fill the line without gaps, as the following arguments show. If to every segment s a positive number α is to be assigned as its length, then by means of a sequence of right-angled triangles segments s_1, s_2, s_3, s_4, ... can be constructed (Fig.) to which there belong the positive numbers $\alpha_1 = 1$, α_2, α_3, α_4, ..., say. By the theorem of Pythagoras $\alpha_2^2 = 2$, $\alpha_3^2 = 3$, $\alpha_4^2 = 4, \ldots$ One should therefore have $\alpha_2 = \sqrt{2}$, $\alpha_3 = \sqrt{3}$, $\alpha_4 = \sqrt{4}$, ... These numbers correspond on the number line to the points obtained by applying at the origin the segments s_1, s_2, s_3, s_4, ... But the numbers α_2 and α_3 are not rational; for example, if α_2 were a rational number r/s (with r and s coprime), then one should have $r^2/s^2 = 2$. However, this is impossible, because if r and s are coprime, then so are r^2 and s^2, and the fraction r^2/s^2 cannot be cancelled, hence cannot be an integer. The proof that α_3, α_5, etc. are not rational is quite the same, but for α_4, α_9, and generally for α_n, when n is a square, it leads to a fraction r/s with $s = 1$, that is, to an integer.

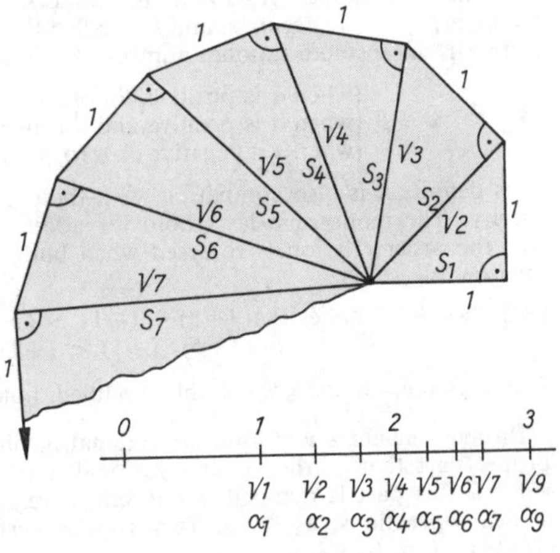

3.5-1 Construction of the segments s_1, s_2, ... and the points α_1, α_2, ... on the number line

Also, to the perimeter of a circle with rational diameter d there should correspond, by geometric theorems, the length $\pi \cdot d$, and already Johann Heinrich LAMBERT (1728–1777) has proved that this is not a rational number.

If every segment is to have a numerical measure as its length, then a new domain of numbers is needed, an extension of the domain of rational numbers. This new domain can no longer be constructed, as in the previous cases, by number pairs. But hints for its construction are provided by a theoretical analysis of the measuring process for segments. To measure a segment accurately in terms of a unit segment e, one uses in succession the segments e, $e/10$, $e/100$, ... Every subsequent measurement contributes a further decimal to the numerical values of the length. A decimal fraction represents a rational number only when it terminates or is periodic. In general, it neither terminates nor is periodic.

The infinite decimal fractions form the domain of real numbers.

The domain of real numbers comprises that of the rational numbers. For every terminating decimal fraction can also be written as an infinite decimal fraction with the digit 9 from a certain place onwards, for example, $7.58 = 7.57999\ldots$

Approximation by rational numbers. Every real number can be approximated with arbitrary accuracy by rational numbers. This is utilized in every measurement, which in practice breaks off after a definite number of decimal places; for example, if one third of a segment of length 16 m is to be given in the decimal system within an accuracy of 10^{-4} m, then four decimals are sufficient: $5^1/_3\ \text{m} \approx 5.3333\ \text{m}$.

A decimal fraction is regarded as completely determined if the integer before the decimal point is known and a rule by which the successive decimals can be formed. For example, to calculate the number α_2 with $\alpha_2^2 = 2$, one has the following scheme:

$$\begin{array}{llll} 1 < \alpha_2 < 2, & & 1^2 < 2 < 2^2, \\ 1.4 < \alpha_2 < 1.5, & \text{because} & 1.4^2 < 2 < 1.5^2, \\ 1.41 < \alpha_2 < 1.42, & & 1.41^2 < 2 < 1.42^2, \\ 1.414 < \alpha_2 < 1.415, & & 1.414^2 < 2 < 1.415^2, \text{ etc.} \end{array}$$

Therefore the number $\alpha_2 = \sqrt{2}$ lies within the infinitely many intervals with the rational end-points (1, 2), (1.4, 1.5), (1.41, 1.42), ..., which are nested within one another. A '*nest of intervals*' means that the left end-points of the intervals increase steadily, the right ones decrease, and the lengths of the intervals become arbitrarily small. By such a nest of intervals the infinite decimal fraction $\alpha_2 = \sqrt{2}$ is uniquely determined. All real numbers can be given correspondingly with arbitrary accuracy.

Order and arithmetical operations. An *order* for the real numbers can be laid down *lexicographically*: a positive decimal fraction $\alpha = a \,.\, a_1a_2a_3 \ldots$ is said to be greater than a positive decimal fraction $\beta = b \,.\, b_1b_2b_3 \ldots$ if $a > b$ or, in case $a = b$, if the first decimal a_i that differs from b_i is greater than b_i; for example, 3.78634... > 3.78629... If one orders the negative numbers by mirror image, the laws of transitivity, irreflexivity, and comparability are satisfied, so that one has a relationship of total or linear order. The real numbers can be mapped, with preservation of the order, onto the points of a line, the number line, which now is filled without gaps. The *arithmetical operations of the first and second kind* for real numbers given by nests of intervals are defined by means of the rational end-points of the intervals. A typical example is the division, and the other operations are treated similarly. To begin with, suppose that α/β is to be considered, with α and β positive. Let α and β be given by the rational nests of intervals (a_i, a_i') and (b_i, b_i'). Since $a_i < \alpha < a_i'$ and $1/b_i' < 1/\beta < 1/b_i$, one has $a_i/b_i' < \alpha/\beta < a_i'/b_i$. The intervals $\{(a_i/b_i', a_i'/b_i)\}$ form a nest of intervals for the quotient α/β: the lower end-points form a bounded increasing and the upper end-points a bounded decreasing sequence of rational numbers.

The length l_n of the nth interval is

$$l_n = \left| \frac{a_n'}{b_n} - \frac{a_n}{b_n'} \right| = \left| \frac{a_n'b_n' - a_nb_n}{b_nb_n'} \right| = \frac{|(a_n' - a_n)\, b_n' + a_n(b_n' - b_n)|}{|b_nb_n'|} .$$

The factors b_n' and a_n in the numerator are bounded. Since the expressions in brackets can be made arbitrarily small when n is sufficiently large, the same holds for the entire numerator. Since $\beta \neq 0$, there is a positive lower bound for all b_n and b_n' with sufficiently large n. The positive real number determined by this nest of intervals is denoted by α/β. If α or β or both are negative, the arguments suitably modified lead to the required result.

In order to define the power $\alpha^\beta (\alpha > 0)$ one forms the intervals $(a_i^{b_i}, a_i'^{b_i'})$. They represent again a nest of intervals; however, in general, the end-points of the intervals are no longer rational. But the following *completeness theorem* holds:

An arbitrary nest of intervals (ϱ_i, ϱ_i') with real end-points determines uniquely a real number ϱ with the property $\varrho_i \leqslant \varrho \leqslant \varrho_i'$ for all i.

The proof is omitted here. The nest of intervals $\{(a_i^{b_i}, a_i'^{b_i'})\}$ therefore determines a real number, which is denoted by α^β.

The arithmetical operations so defined satisfy the same laws as in the domain of rational numbers.

The domain of real numbers is an extension of the domain of rational numbers. It contains, in particular, all roots of positive numbers.

The construction of the real numbers could only be sketched here. It should be mentioned that in a formal treatment it is necessary to define an equivalence relation for nests of intervals and to form the corresponding classes. This method, carried through rigorously, is a genuine constructive way of obtaining the real numbers.

Other methods of defining the real numbers. EUDOXUS (about 408–355 B. C.) can be regarded as a harbinger for the development of a theory of the real numbers. His geometrically orientated ideas were taken up by Karl WEIERSTRASS (1815–1897) and Richard DEDEKIND (1831–1916) and were further developed, utilizing modern arithmetic and analytic methods. The method of *nests of intervals* goes back to WEIERSTRASS. DEDEKIND introduced the real numbers by *cuts* in the domain of rational numbers. Georg CANTOR (1845–1918) constructed them by means of *Cauchy fundamental sequences*.

The domains obtained by these methods can be mapped onto one another by order-preserving isomorphisms. Therefore, structurally there is only a single domain of real numbers.

3.6. Continued fractions

Continued fractions of order *n*. Let $b_0, b_1, b_2, \ldots, b_n$ be integers with $b_k > 0$ for $k > 0$. The *continued fraction* of order n with the denominators $b_1, b_2, \ldots, b_n$ and the initial term b_0 is defined by the following expression, which is also abbreviated as $[b_0; b_1, b_2, \ldots, b_n]$:

$$b_0 + {}^1/_{b_1} + {}^1/_{b_2} + \cdots + {}^1/_{b_n}.$$

Example: $n = 3$; $b_0 = 2$, $b_1 = 3$; $b_2 = 1$; $b_3 = 4$
$= [2; 3, 1, 4] = 43/19$
$2 + 1/[3 + 1/(1 + 1/4)]$.

Approximating fractions. Let α be a continued fraction of order n. By an *approximating fraction* for α of order k ($k < n$) one understands the continued fraction breaking off at the kth denominator. From the definition of a continued fraction of order k it is clear that it can also be written as an ordinary fraction. Then one obtains:

$$[b_0; b_1] = b_0 + 1/b_1 = (b_0 b_1 + 1)/b_1 = A_1/B_1,$$
$$[b_0; b_1, b_2] = b_0 + 1/(b_1 + 1/b_2) = (b_2(b_0 b_1 + 1) + b_0)/(b_1 b_2 + 1) = A_2/B_2,$$
$$[b_0; b_1, b_2, b_3] = b_0 + 1/[b_1 + 1/(b_2 + 1/b_3)]$$
$$= (b_3[b_2(b_0 b_1 + 1) + b_0] + b_0 b_1 + 1)/(b_3(b_1 b_2 + 1) + b_1) = A_3/B_3$$

Here all the A_i and B_i are integers; for example, for the continued fraction [2; 3, 1, 4, 2, 1, 2] the second approximating fraction is $[2; 3, 1] = 2 + 1/(3 + 1) = 9/4$, with $A_2 = 9$, $B_2 = 4$.

If one defines for the sake of completeness $A_0 = b_0$, $A_{-1} = 1$, $A_{-2} = 0$; $B_0 = 1$, $B_{-1} = 0$, $B_{-2} = 1$, then the following recursion formulae can be established by mathematical induction:

Recursion formulae	$A_k = b_k A_{k-1} + A_{k-2}$; $B_k = b_k B_{k-1} + B_{k-2}$

Example:										
	k	−2	−1	0	1	2	3	4	5	6	given quantities
	b_k	—	—	2	3	1	4	2	1	2	
by definition	A_k	0	1	2	7	9	43	95	138	371	calculated quantities.
	B_k	1	0	1	3	4	19	42	61	164	

The example [2; 3, 1, 4, 2, 1, 2] shows that in order to reach A_k (the case of B_k is analogous) one multiplies the b_k standing above A_k by the left neighbour A_{k-1} (which has already been computed) and adds its left neighbour A_{k-2}. At the two places marked by arrows in the diagram one calculates $0 + 1 \cdot 2 = 2$ and $7 + 9 \cdot 4 = 43$. This leads to the approximating fractions $A_1/B_1 = 7/3 = 2.33$; $A_2/B_2 = 9/4 = 2.25$; $A_3/B_3 = 43/19 = 2.2631$; $A_4/B_4 = 95/42 = 2.2619\ldots$; $A_5/B_5 = 138/61 = 2.2623\ldots$; $A_6/B_6 = 371/164 = 2.262195\ldots = [2; 3, 1, 4, 2, 1, 2]$.

A comparison of the final fraction with the approximating fractions indicates the justification for the name:

The approximating fractions come progressively closer to the final fraction, alternately from below and above, with increasing accuracy.

Every rational number can be expanded in a continued fraction.

Example: If 964/437 is to be expanded in a continued fraction, one obtains: $r = 964/437 = 2 + 90/437$; $r = b_0 + 1/r_1$; $b_0 = [r]$, where $[r]$ denotes the greatest integer not exceeding r, $r_1 = 437/90 = 4 + 77/90$; $r_1 = b_1 + 1/r_2$; $r_1 > 1$, if r_1 is not integral, $b_1 = [r_1]$, the greatest integer not exceeding r_1. The continuation of the method finally yields:

$$r = 964/437 = 2 + {}^1/_{4} + {}^1/_{1} + {}^1/_{5} + {}^1/_{1} + {}^1/_{12},$$
$$r = [2; 4, 1, 5, 1, 12].$$

The example $1/2 = 0 + 1/2 = 0 + 1/[1 + (1/1)]$, hence $[0; 2] = [0; 1, 1]$ shows that there is no uniqueness of the expansion in a continued fraction. This can be achieved by postulating that $b_n > 1$, which can always be satisfied because $[b_0; b_1, b_2, \ldots, b_n, 1] = [b_0; b_1, b_2, \ldots, b_n + 1]$. The expansion in a continued fraction makes it possible to approximate a rational number with an unwieldy numerator and denominator by others with smaller numerators and denominators, a fact that occasionally is of importance for technical problems.

Non-terminating continued fractions. Non-terminating continued fractions $[b_0; b_1, b_2, \ldots]$ are used to represent real numbers and can be treated correspondingly. The approximation is better than that by decimal fractions. Conversely, every real number can be represented by a (finite or infinite) continued fraction, in fact:

The expansion of a real number in a continued fraction terminates if and only if the number is rational.

Here, too, the approximating fractions come progressively closer to the real number, alternately from below and above, with increasing accuracy.

Example: $\alpha = \sqrt{2}$ is to be expanded in a continued fraction. The symbol $[\alpha]$ denoting the greatest integer less than, or equal to, α one obtains the sequence of denominators:

1. $\alpha = b_0 + (\alpha - [\alpha]), \quad b_0 = [\alpha], \quad \alpha - [\alpha] = 1/\alpha_1;$
2. $\alpha_1 = b_1 + (\alpha_1 - [\alpha_1]), \quad b_1 = [\alpha_1], \quad \alpha_1 - [\alpha_1] = 1/\alpha_2;$
3. $\alpha_2 = b_2 + (\alpha_2 - [\alpha_2]), \quad b_2 = [\alpha_2], \quad \alpha_2 - [\alpha_2] = 1/\alpha_3;$

....

Now one makes use of the inequality $1 < \sqrt{2} < 2$ and rearranges:

1. $\alpha = 1 + (\sqrt{2} - 1), \ b_0 = 1, \ \sqrt{2} - 1 = 1/\alpha_1, \ \alpha_1 = (\sqrt{2} + 1)/[(\sqrt{2} - 1)(\sqrt{2} + 1)] = \sqrt{2} + 1;$
2. $\alpha_1 = 2 + (\sqrt{2} - 1), \ b_1 = 2, \ \sqrt{2} - 1 = 1/\alpha_2, \ \alpha_2 = \sqrt{2} + 1;$
3. $\alpha_2 = 2 + (\sqrt{2} - 1), \ b_2 = 2, \ \sqrt{2} - 1 = 1/\alpha_3, \ \alpha_3 = \sqrt{2} + 1;$

....

By continuing the process one obtains the *periodic continued fraction* [1; 2, 2, ...] belonging to $\sqrt{2}$.

Approximating fractions can be found by means of the scheme for the A_k und B_k:

k	−2	−1	0	1	2	3	4	5	6
b_k	—	—	1	2	2	2	2	2	2
A_k	0	1	1	3	7	17	41	99	239
B_k	1	0	1	2	5	12	29	70	169

Hence, for example, $A_6/B_6 = 239/169 = 1.414\,201\ldots$, while $\sqrt{2} \approx 1.414\,214$.

The approximating fractions converge to $\sqrt{2}$.

The fact that the expansion of $\sqrt{2}$ in a continued fraction is periodic is no accident; periodic continued fractions always arise for the so-called *quadratic irrationalities*, that is, numbers of the form $(a + b\sqrt{D})/c$, where a, $b \neq 0$, $c \neq 0$, and D are integers, and $D > 1$ is square-free. LAGRANGE showed that the expansion of a quadratic irrationality in a continued fraction is periodic. The converse, that a periodic continued fraction represents a quadratic irrationality, had already been proved by EULER.

3.7. The complex numbers C

In the domain of real numbers the arithmetical operations of the first and second kind can be carried out without restriction. This is not the case for the operations of the third kind; for example, the power $a^{1/n} = \sqrt[n]{a}$ does not exist when a is negative and n is even: there is no real number $\sqrt[2]{-4}$. But square roots of negative real numbers are needed, for example, in the solution of a cubic equation by Cardano's formula (see Chapter 4.), in fact, just in the so-called casus irreducibilis, when there are three distinct real solutions. In order to remove this restriction the number system is extended once more.

Construction of the new numbers. One considers ordered pairs (a, b) of arbitrary real numbers a and b. This time the equivalence relation is the ordinary identity, that is, the pair (a, b) is called equivalent to (a', b') if and only if $a = a'$ and $b = b'$; every equivalence class consists of a single number pair.

Such a pair (a, b) is called a *complex number*. The arithmetical operations of the first and second kind are defined by:

$$z_1 \pm z_2 = (a_1 \pm a_2, b_1 \pm b_2), \qquad z_1 z_2 = (a_1 a_2 - b_1 b_2, a_1 b_2 + b_1 a_2),$$

$$z_1 : z_2 = \left(\frac{a_1 a_2 + b_1 b_2}{a_2^2 + b_2^2}, \frac{b_1 a_2 - b_2 a_1}{a_2^2 + b_2^2} \right), \qquad z_2 \neq (0, 0).$$

It is easy to verify that subtraction and division are the inverse operations to addition and multiplication. The commutative, associative, and distributive laws hold for addition and multiplication. For example, the distributive law is proved as follows: let $z_i = (a_i, b_i)$, $i = 1, 2, 3$, be three complex numbers. Then: $z_1[z_2 + z_3] = (a_1, b_1)(a_2 + a_3, b_2 + b_3)$
$= (a_1a_2 + a_1a_3 - b_1b_2 - b_1b_3,\ a_1b_2 + a_1b_3 + b_1a_2 + b_1a_3)$.
On the other hand, $z_1z_2 + z_1z_3 = (a_1a_2 - b_1b_2,\ a_1b_2 + b_1a_2) + (a_1a_3 - b_1b_3,\ a_1b_3 + b_1a_3)$ $= (a_1a_2 - b_1b_2 + a_1a_3 - b_1b_3, a_1b_2 + b_1a_2 + a_1b_3 + b_1a_3)$.
The two expressions are identical.

Furthermore, all the laws for these operations that hold in the domain of real numbers are also satisfied here, except those in which an order relation 'greater than' occurs. There are many ways of introducing a total order in the domain **C** of complex numbers, for example, first by absolute value and for equal absolute value by argument; but it can be proved that no relationship of total order in **C** is compatible with addition and multiplication.

Complex and real numbers. The domain of complex numbers contains as part that of the numbers $(a, 0)$, which is isomorphic to the domain of real numbers with respect to the permitted operations: $(a, 0) + (a', 0) = (a + a', 0)$ and $(a, 0) \cdot (a', 0) = (aa', 0)$. One can therefore treat such numbers by writing simply a instead of $(a, 0)$. The numbers $(0, b)$ are called *purely imaginary*. In particular, the complex number $(0, 1) = \mathrm{i}$ is called the *imaginary unit*. Brackets for the arithmetical operations can now be omitted, because errors need not be feared. One has $(0, b) = (b, 0) \cdot (0, 1) = b\mathrm{i}$ and $(a, b) = (a, 0) + (0, b) = a + b\mathrm{i}$.

Imaginary unit i	$\mathrm{i}^2 = -1$

Next, $\mathrm{i} \cdot \mathrm{i} = (0, 1) \cdot (0, 1) = (-1, 0) = -1$.

Every complex number can be represented as the sum of a real and a purely imaginary number: $z = a + b\mathrm{i}$. Here a is called the real part and b the imaginary part of z; a and b being real numbers.

Graphical representation of the complex numbers. In the plane one draws a Cartesian rectangular system of axes and marks on the x-axis the real numbers in the usual way, on the y-axis the imaginary numbers with i as unit. To the complex number $z = a + \mathrm{i}b$ one assigns the point z with the coordinates (a, b) or the vector z leading from the origin to this point. These correspondences are one to one. To the sum $z_1 + z_2$ there corresponds by vector addition (according to the parallelogram rule) the vector $z_1 + z_2$ (Fig.).

To give a geometric interpretation also for the product one represents $z = a + b\mathrm{i}$ in terms of the length r of z and the angle φ which this vector forms with the positive x-axis; r is called the *absolute value* or *modulus* and φ the *argument* or *amplitude* of z (Fig.). It should be observed that φ is determined only up to multiples of 2π and is measured anticlockwise.

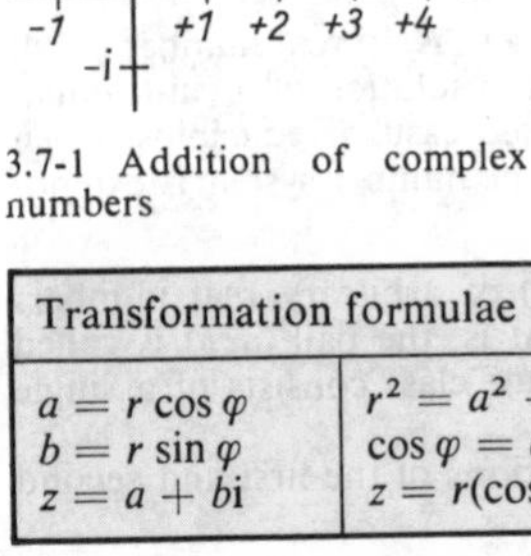

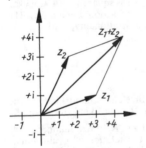

3.7-1 Addition of complex numbers

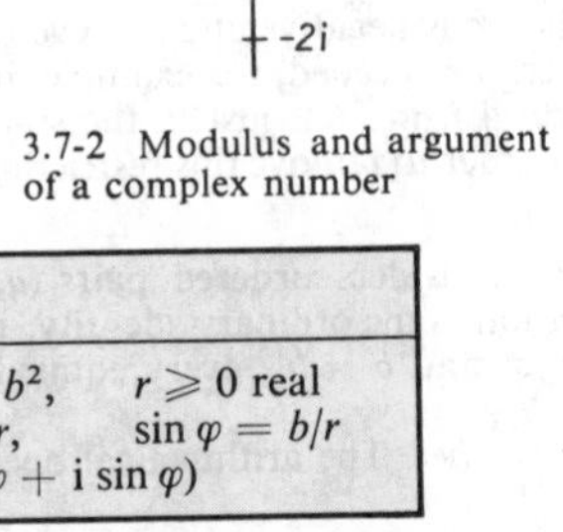

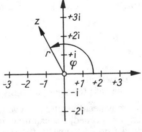

3.7-2 Modulus and argument of a complex number

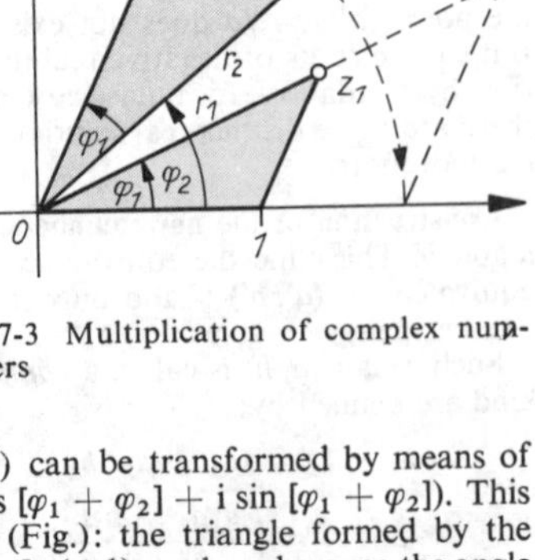

3.7-3 Multiplication of complex numbers

Transformation formulae	
$a = r\cos\varphi$	$r^2 = a^2 + b^2$, $r \geqslant 0$ real
$b = r\sin\varphi$	$\cos\varphi = a/r$, $\sin\varphi = b/r$
$z = a + b\mathrm{i}$	$z = r(\cos\varphi + \mathrm{i}\sin\varphi)$

The product $z_1z_2 = r_1(\cos\varphi_1 + \mathrm{i}\sin\varphi_1)\, r_2(\cos\varphi_2 + \mathrm{i}\sin\varphi_2)$ can be transformed by means of the addition theorems for sine and cosine into $z_1z_2 = r_1r_2(\cos[\varphi_1 + \varphi_2] + \mathrm{i}\sin[\varphi_1 + \varphi_2])$. This representation leads to the following geometric interpretation (Fig.): the triangle formed by the points O, z_2, and $z_1 \cdot z_2$ is similar to that formed by the points O, $(+1)$, and z_1, because the angle

φ_1 is common to the two triangles and the sides including it have the same ratio $r_1 r_2 : r_2 = r_1 : 1$. Thus, there is a simple geometric construction for the product.

Powers and roots. As usual, z^n for a natural number n is defined by $z^0 = 1$, $z^{n+1} = z^n \cdot z$. By means of the addition theorems and mathematical induction one derives the important formula of de Moivre:

de Moivre's formula	$z^n = r^n(\cos n\varphi + \mathrm{i} \sin n\varphi)$

For $n = -1$ one obtains $z^{-1} = r^{-1}(\cos(-\varphi) + \mathrm{i}\sin(-\varphi)) = r^{-1}(\cos\varphi - \mathrm{i}\sin\varphi)$ and by mathematical induction $z^{-n} = r^{-n}(\cos(n\varphi) - \mathrm{i}\sin(n\varphi))$.

By $\sqrt[n]{z}$ one means a complex number w whose nth power is equal to z, that is, a solution of the equation $w^n = z$. Let $w = \varrho(\cos\psi + \mathrm{i}\sin\psi)$. Then from $w^n = z = r(\cos\varphi + \mathrm{i}\sin\varphi)$ it follows by de Moivre's formula that:

$$\varrho^n = r, \quad \varrho = \sqrt[n]{r}, \quad \psi = \varphi/n + k \cdot 2\pi/n \quad \text{or} \quad w = \sqrt[n]{r}[\cos(\varphi/n + k \cdot 2\pi/n) + \mathrm{i}\sin(\varphi/n + k \cdot 2\pi/n)].$$

If $w \neq 0$, then n distinct values arise for $k = 0, 1, 2, \ldots, n-1$. In the domain of complex numbers the symbol $\sqrt[n]{z}$ is not restricted to a single value, but is many-valued. How this many-valuedness can be mastered is shown in the theory of Riemann surfaces (see Chapter 23.). In particular, when z is a positive real number, then the uniquely determined positive real nth root is called the *principal value*. The extraction of roots can be performed without restriction in the domain of complex numbers. Among the n values of $\sqrt[n]{z}$ is the number $r^{1/n}(\cos(\varphi/n) + \mathrm{i}\sin(\varphi/n))$.

Example: To obtain all values of $\sqrt[4]{(-1)}$, one sets:
$z = 1[\cos(180° + k \cdot 360°) + \mathrm{i}\sin(180° + k \cdot 360°)]$
$w = 1[\cos(180°/4 + k \cdot 90°) + \mathrm{i}\sin(180°/4 + k \cdot 90°)]$
$k = 0$ gives $w_0 = \cos 45° + \mathrm{i}\sin 45°$,
$k = 1$ gives $w_1 = \cos 135° + \mathrm{i}\sin 135°$,
$k = 2$ gives $w_2 = \cos 225° + \mathrm{i}\sin 225°$,
$k = 3$ gives $w_3 = \cos 315° + \mathrm{i}\sin 315°$.
For $k = 4$ one has $\psi_4 = 405° = 360° + 45°$, hence $w_4 = w_0$, $w_5 = w_1$, ... (Fig.).

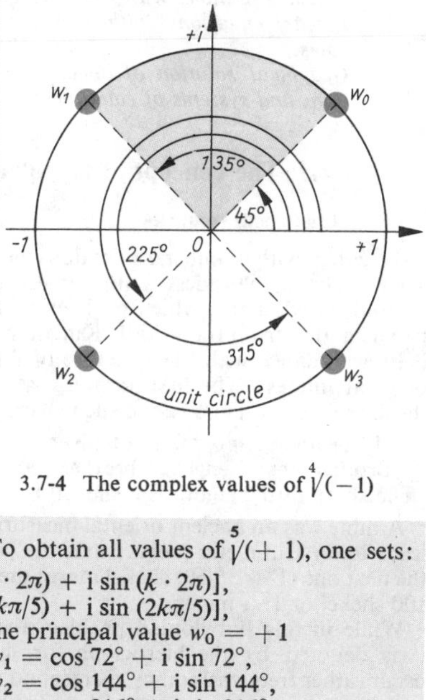

3.7-4 The complex values of $\sqrt[4]{(-1)}$

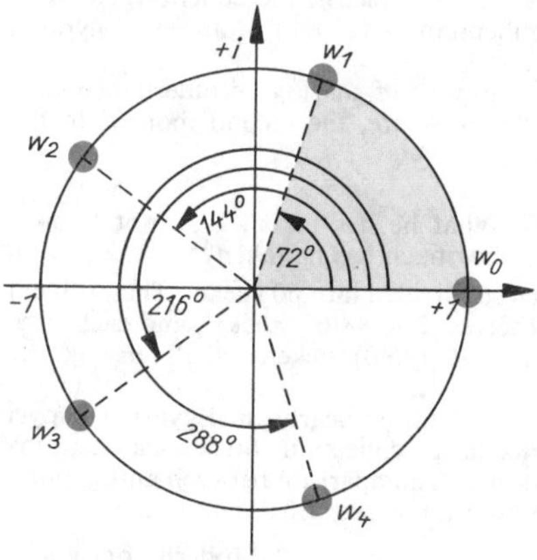

3.7-5 The complex values of $\sqrt[5]{(+1)}$ (fifth roots of unity)

Example: To obtain all values of $\sqrt[5]{(+1)}$, one sets:
$z = 1[\cos(k \cdot 2\pi) + \mathrm{i}\sin(k \cdot 2\pi)]$,
$w = 1[\cos(2k\pi/5) + \mathrm{i}\sin(2k\pi/5)]$,
$k = 0$ gives the principal value $w_0 = +1$
$k = 1$ gives $w_1 = \cos 72° + \mathrm{i}\sin 72°$,
$k = 2$ gives $w_2 = \cos 144° + \mathrm{i}\sin 144°$,
$k = 3$ gives $w_3 = \cos 216° + \mathrm{i}\sin 216°$,
$k = 4$ gives $w_4 = \cos 288° + \mathrm{i}\sin 288°$.
For $k = 5$ one has $\psi_5 = 360°$, hence $w_5 = w_0$, $w_6 = w_1$, ... (Fig.).

The nth *roots of unity* lie on the circumference of the unit circle and divide it into n equal parts.

For rational α the power z^α is defined in accordance with the corresponding definitions for absolute rational numbers and rational numbers. If $\alpha > 0$ and $\alpha = r/s$ with positive integers r and s, then z^α is to be interpreted as $\sqrt[s]{z^r}$; if $\alpha < 0$ and $z \neq 0$, then z^α is defined as $1/z^{-\alpha}$. Of great importance is the so-called fundamental theorem of classical algebra, which was first proved, and in more than one way, by Gauss.

The domain of complex numbers is algebraically closed; this means that every algebraic equation with complex coefficients has at least one solution in this domain.

Historical remarks on complex numbers. Roots of negative numbers were used since the middle of the 17th century and were since known as imaginary numbers. The mathematicians of the 17th century could rely on a book on algebra by Raffaele BOMBELLI, which dates from 1572 and contains a consistent theory of purely imaginary numbers. Later the theory of complex numbers was advanced by Johann BERNOULLI (1667–1748), Leonhard EULER (1707–1783), and above all by Carl Friedrich GAUSS (1777–1855). The representation of the complex numbers in the plane goes back to Caspard WESSEL (1745–1818) and Jean Robert ARGAND (1768–1822); it became a generally used form of representing complex numbers by the authority of GAUSS (the *Gaussian plane*). The complex numbers are the foundation of the theory of *functions of a complex variable*, or *complex analysis*.

4. Algebraic equations

4.1. The concept of an equation

Historical remarks

Together with numbers, equations belong to the first mathematical achievements of mankind. They occur in the oldest written mathematical documents, for example, in the cuneiform texts of the old Babylonians, which go back as far as the third millenium B. C., and in ancient Egyptian papyri dating from the Middle Kingdom, about 1800 B. C.

In accordance with the structure of Babylonian society questions of sharing an inheritance were of great interest. The first-born son always received the largest share, the second more than the third, and so on. Here is one such sharing problem:

'10 brothers; $1^2/_3$ mines of silver.
Brother has risen over brother (concerning his share). What he has taken, I do not know. The share of the eighth is 6 shekel. Brother after brother, how much has he taken?'

A mine was an ancient oriental measuring unit, which was subdivided into 60 shekel. The problem leads to an arithmetic progression; the youngest brother receives $(2 + 48/60)$ shekel, and each time the next one $(1 + 36/60)$ shekel more; the firstborn received $(17 + 12/60)$ shekel, and all ten together 100 shekel or $1^2/_3$ mines.

While in this Babylonian problem the unknown is described fairly clearly, in Egyptian papyri it is denoted by the hieroglyph for 'h', which represents heap, collection. Such h-calculations occur rather frequently; they correspond to our linear equations. A comparison between an Egyptian text from the Moscow papyrus (see Table 10) and modern notation makes this point clear:

Literal translation	Modern notation
Form of calculation of a heap, counted $1^1/_2$ times together with 4. He has reached 10. Now what is the heap's name?	$3x/2 + 4 = 10$
You calculate the magnitude of this 10 over this 4. What arises is 6.	$10 - 4 = 6$
You calculate with $1^1/_2$ to find 1. What arises is 2/3.	$1 : 3/2 = 2/3$
You calculate 2/3 of these 6. What arises is 4.	$6 \cdot 2/3 = 4$
Look: 4 is the name. You have calculated correctly.	$x = 4$

Before an algebraic symbolic language had been developed, equations had to be written in words. Even François VIÈTE (usually called VIETA, 1540–1603), who has great merits in the field of algebra, made do with the Latin verb *aequare* = to be equal. The equality sign = in common use nowadays was proposed by Robert RECORDE (1510–1558), Royal Court Physician, but it took a considerable time before it was generally accepted. He made this proposal in a textbook of algebra, written in dialogue form, with the title 'The Whetstone of Witte' (1557) and motivated it by saying (Fig.). 'I will sette as I doe often in woorke use, a paire of paralleles, of Gemowe (twin) lines of one lengthe, thus: ===, bicause noe .2. thynges can be moare equalle'.

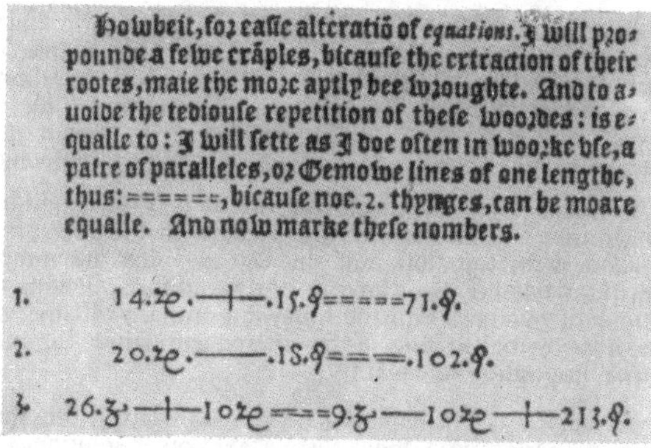
Howbeit, for easie alteratiō of equations. I will propounde a fewe exāples, bicause the extraction of their rootes, maie the more aptly bee wroughte. And to auoide the tediouse repetition of these woordes: is equalle to: I will sette as I doe often in woorke use, a paire of paralleles, or Gemowe lines of one lengthe, thus: =====, bicause noe .2. thynges, can be moare equalle. And now marke these nombers.

1. 14.ze.—+—.15.ʒ=====71.ʒ.
2. 20.ze.——.18.ʒ===.102.ʒ.
3. 26.ʒ—+—10ze===9.ʒ—10ze—+—213.ʒ.

4.1-1 From R. Recorde's 'Whetstone of Witte', 1557. First occurrence of the equality sign. Immediately above the formulae is the motivation for the choice of this symbol

Equations. Solution sets

One starts out from a definite set of numbers, the *fundamental domain*, and of *variables* for which elements of the fundamental domain or of a subset, the *domain of variability*, may be substituted. In specifying the fundamental domain and the domain of variability, **N** stands for the set of natural numbers, **Z** for that of the integers, **Q** for that of the rational, **R** for that of the real, and **C** for that of the complex numbers. In what follows, unless the contrary is stated, the fundamental domain is taken to be **R**. The concept of an equation can be explained by reference to the concept of an *expression*, which is defined inductively (see Chapter 15.).

Expressions. All numbers and all variables are expressions. Sums, differences, products, and quotients of two expressions are again expressions. Also exponentiation and the extraction of roots from expressions yield new expressions. Division by zero is excluded; for the time being, in exponentiation and extraction of roots the exponent is taken to be a positive integer and the radicand positive.

Examples of expressions: 5; 4/7; a; $4x$; $b + 7$; $5(a + b)$; $(4x + 3)/y$; $x^4/2$; $\sqrt[3]{a}$.

The concept of an expression can be extended to include, for example, $\sin x$, $\log_a x$, e^x.

An expression E_1 is said to be equivalent to an expression E_2 if they assume the same value for every substitution of the variables by the same numbers of the given domain of variability; for example, $4a + 5a$ and $9a$ are equivalent expressions with respect to the set **R** of real numbers, while the expressions $(x^2 + x)/x$ and $x + 1$ are not equivalent, because $(x^2 + x)/x$ is not defined for $x = 0$, whereas $x + 1$ for $x = 0$ assumes the value 1. These two expressions are equivalent for the set of all real numbers other than zero.

The following facts are evident:

1. Every expression E is equivalent to itself. 2. If E_1 is equivalent to E_2, then E_2 is equivalent to E_1. 3. If E_1 is equivalent to E_2, and E_2 to E_3, then E_1 is equivalent to E_3.

The *domain of definition of an expression* involving a variable is the set of all numbers of the domain of variability for which the expression goes over into a number of the domain of variability; for example, the domain of definition of the expression $(4a - 5)/3$ consists of all real numbers, while that of $x/(x - 3)$ contains all real numbers other than 3. The domain of definition for expressions with several variables is explained correspondingly.

Equations. If two expressions E_1 and E_2 are linked by the symbol of equality, an *equation* $E_1 = E_2$ arises. Here E_1 is called the *left-hand side*, and E_2 the *right-hand side*, of the equation.

The *domain of definition of an equation* is the intersection of the domains of definition of all the expressions with variables occurring in it (Fig.).

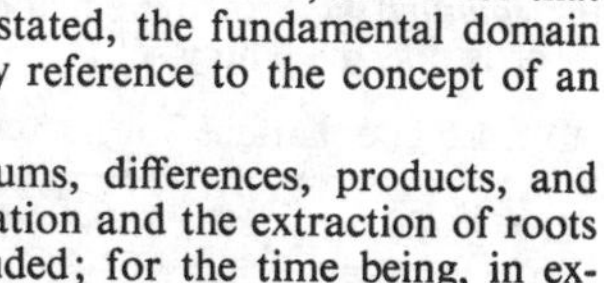

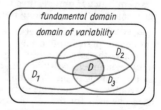

4.1-2 Domain of definition D of an equation with one variable as intersection of the domains of definition D_i of the expressions E_i with variables

An equation whose expressions do not contain variables is a *proposition* in the sense of mathematical logic, which can be true or false; for example, $3 + 2 = 5$ and $3 \cdot (5 + 2) = 20 + 1$ are true propositions, while $2 + 3 \cdot 4 = 15$ is a false proposition. But if the expressions contain variables, then the equation is a *predicate*, for example, the equations $3x = -12$, $4a + 3b = 1$ or $x^2 = (6x + 24)/3$. Only after numbers from the domain of definition of the equation are substituted for the variables, the predicate becomes a proposition, which may be true or false.

Solutions. Every number from the domain of definition of an equation with a single variable which after substitution for the variable makes the equation into a true proposition is called a *solution* of the equation, and one also says that the number *solves* or *satisfies* the equation. If an equation contains two, three, ..., or n variables, then a solution is an ordered pair, triple, ..., or n-tuple of numbers with the following property: if the variables are replaced with due regard to the order by the elements of the ordered pair, triple, ..., or n-tuple, then the equation goes over into a true proposition of equality.

Examples: 1. The equation $3x = -12$ is satisfied by the real number -4, for $3 \cdot (-4) = -12$ is a true proposition. Since there are no other solutions, -4 is *the* solution of the equation.

2. The equation $4a + 3b = 11$ is satisfied, for example, by the number pair (2, 1), for it is true that $4 \cdot 2 + 3 \cdot 1 = 11$. But there are further solutions, in fact, infinitely many, and (2, 1) is a solution of the equation.

3. The equation $x^2 = (6x + 24)/3$ has the numbers -2 and $+4$ as solutions, because both $(-2)^2 = [6 \cdot (-2) + 24]/3$ and $(+4)^2 = (6 \cdot 4 + 24)/3$ are true propositions. There are no other solutions.

4. If the domain of variability for the equation $x^2 = 2$ is taken to be the set $\mathbf{Q}$ of rational numbers, then the equation has *no* solution, because there is no rational number whose square is equal to 2.

Solution set. The set of *all* solutions of an equation relative to its domain of definition is called the *solution set* S of the equation. An equation is called *inconsistent* or *consistent* according as S is, or is not, the empty set $\emptyset$.

Consistent equations	Inconsistent equations
$7x = -28$ for $x \in \mathbf{Z}$; $S = \{-4\}$	$7x = -28$ for $x \in \mathbf{N}$; $S = \emptyset$
$x^2 = 9$ for $x \in \mathbf{R}$; $S = \{-3; +3\}$	$x^2 = -9$ for $x \in \mathbf{R}$; $S = \emptyset$
$4a^2 = 1$ for $a \in \mathbf{Q}$; $S = \{-1/2; +1/2\}$	$4a^2 = 1$ for $a \in \mathbf{Z}$; $S = \emptyset$
$2x + x = 3x$ for $x \in \mathbf{C}$; $S = \mathbf{C}$	$3x = 3x + 1$ for $x \in \mathbf{C}$; $S = \emptyset$

A consistent equation with one variable is called *universally valid* if all the elements of the domain of definition are solutions; for example, $2x + x = 3x$ is universally valid in the set of complex numbers. A consistent equation with n variables is called universally valid if every ordered n-tuple of numbers from the given domain of variability is a solution of the equation. For example, $(a + b)^2 = a^2 + 2ab + b^2$ for $a, b \in \mathbf{R}$ is a universally valid equation, because it is satisfied by every pair (a, b) of real numbers. Every transformation carrying one expression into an equivalent expression involves a chain of universally valid equations; for example, the transformation $(4a + 7a) \cdot 2 = 11a \cdot 2 = 22a$ is an equivalence relative to the set $\mathbf{R}$ of real numbers. But the transformation

$$\frac{a^2 - 16a + 64}{5a - 5} \cdot \frac{a - 1}{a^2 - 64} = \frac{(a - 8)^2}{5(a - 1)} \cdot \frac{(a - 1)}{(a + 8)(a - 8)} = \frac{a - 8}{5(a + 8)}$$

is an equivalence only relative to sets of real numbers not containing the numbers ± 8 or 1, because these numbers do not belong to the domain of definition of the expressions occurring.

Equations with parameters. An equation with several variables, say $2a + b = 5$ for $a, b \in \mathbf{R}$, can be interpreted in two ways:

Firstly, the two variables can be regarded as of equal standing and one can ask for all pairs (a, b) of numbers satisfying the equation. Then (2, 1), (1/2, 4), (−5, 15) are three of the infinitely many solutions of the equation.

Secondly, one can single out one of the variables and regard the others as auxiliary, as a *parameter*. Then one asks for the solutions of the equation in dependence on the parameter; a solution is an expression containing the parameter that satisfies the equation for every admissible value of the parameter. In the example above, if a is the variable and b the parameter, then $a = (5 - b)/2$, and $(5 - b)/2$ is the expression for the solution of the given equation, for $2 \cdot (5 - b)/2 + b = 5$ is a true proposition for all $b \in \mathbf{R}$. If b is the variable and a the parameter, then $b = 5 - 2a$, and $5 - 2a$ is the expression for the solution, because $2a + (5 - 2a) = 5$ is true for all $a \in \mathbf{R}$.

In an equation with several variables it must always be stated which are the true variables and which the parameters. For example, if in the equation $3x - 2y = 5a + 1$ the true variables are x and y, while a is to be regarded as a parameter, one speaks of an equation in x and y.

If in an equation with n variables there are no parameters and only true variables, then a solution of the equation is an ordered n-tuple of numbers from the respective domains of variability. If there are m true variables $(0 < m < n)$ and the remaining ones are parameters, then a solution is an ordered m-tuple of expressions in which, in general, the parameters enter.

Algebraic equations. In an *algebraic equation* the variables and the elements of the domain of variability are subject only to the so-called elementary algebraic or rational operations: addition, subtraction, multiplication, and division. Examples of algebraic equations in x are: $x^3 - 5x^2 - 8x + 12 = 0$; $4(x + a)^2 (x - b) = c/x$. An equation such as $9x - 7 = 4\sqrt{(5x - 31)}$ can also be subsumed under the name of algebraic equation.

Of course, here both the coefficients and the solutions may be transcendental numbers, as in the equation $\pi x^2 - 5 = 12$, which is algebraic in x. The equation $\sin^2 x - (1/2) \sin x - 1/2 = 0$ is not algebraic in x, but with a suitable extension of the concept can be regarded as algebraic in $\sin x$.

Algebraic equations			
in one variable		in several variables	
linear	non-linear	linear	non-linear
$a + 5 = 12$	$x^3 = 27$	$x + y + z = 4$	$(x + 4)^2 + y^2 = 16$
$3x - 4 = 27$	$x^2 + 3x - 4 = 0$	$4a + 3b - 6 = 0$	$x^2 + z^2 = y^3$

General form of an algebraic equation with one variable. The fundamental domain for the variable x is taken as large as possible, the set **C** of complex numbers. The a_i, $i = 1, 2, \ldots, n$, can be real or complex parameters; a_0 is called the absolute term. The exponent of the highest occurring power of the variable is called the *degree of the equation*. If $a_n \neq 0$, then the degree of the equation is n. If several variables occur in an equation, then one forms for every term the sum of the exponents of the variables and calls their maximum the degree of the equation. For example, the equation $(1/6)x^5 + 4x - 6 = 0$ is of degree 5, and here $a_5 = 1/6$, $a_4 = a_3 = a_2 = 0$, $a_1 = 4$, $a_0 = -6$; the equation $x^2y - xy + 3x = 1$ is of degree 3.

General form of an algebraic equation of degree n	$a_nx^n + a_{n-1}x^{n-1} + \ldots + a_1x + a_0 = 0,\ x \in \mathbf{C};\ a_i \in \mathbf{C},\ i = 0, 1, 2, \ldots, n;\ a_n \neq 0$

Normal form. An algebraic equation of degree n with one variable and with the highest coefficient $a_n = 1$ is said to be *monic* or in normal form. This can be obtained from the general form on division by $a_n \neq 0$.

Transcendental equations. All equations with variables that are not algebraic are called transcendental. Among them are exponential, logarithmic, and trigonometric equations. For their solution methods are required that transcend the means of algebra – *quod algebrae vires transcendit*, as EULER put it. Frequently graphical or approximation methods are used to solve them (see Chapter 10. – Trigonometric equations).

Equivalent equations

Two equations with variables are said to be *equivalent* if they have the same domains of definition and the same solution sets. Otherwise the equations are called *inequivalent*.

Examples: 1. The equations $4a + 2 = 10$ and $6x = 12$ are equivalent relative to the set **R** of real numbers, because the solution set of each consists of the number 2 only.

2. The equations $a^2 = 9$ and $x^3 = 27$ are inequivalent relative to the set **Z** of integers, because the solution set of the first equation consists of ± 3, that of the second of $+3$ only. However, relative to the set **N** of natural numbers these equations are equivalent, for then the solution set of each consists of the number 3 only.

Example 2 shows that the notion of 'equivalent equations' has a meaning only relative to given domains of variability or the resulting domains of definition, and the same is true of 'consistent', 'inconsistent', and 'universally valid'. Relative to equal domains of definition both universally valid equations and inconsistent equations are always equivalent.

Properties of equivalent equations

1. **Reflexivity: Every equation is equivalent to itself.**
2. **Symmetry: If one equation is equivalent to another, then the latter is equivalent to the first.**
3. **Transitivity: If one equation is equivalent to a second and the second to a third, then the first is equivalent to the third.**

Consequently, the equivalence of equations is an equivalence relation (see Chapter 14.).

Equivalent transformations. In transformations of equations with variables one distinguishes between equivalent and inequivalent transformations. If an equation (1) is transformed so that the resulting equation (2) is equivalent to (1), then one says that (2) arises from (1) by an *equivalent transformation*.

If S_1 and S_2 are the solution sets of the equations (1) and (2), then an equivalent transformation is therefore characterized by the fact that $S_1 = S_2$. In all other cases the transformation is said to be inequivalent. This is so, in particular, when $S_1 \subset S_2$, that is, when the transformation has led to additional solutions, or when $S_1 \supset S_2$, that is, when solutions have got lost in the transformation. In the case $S_1 \subset S_2$ those solutions of the equation (2) that are not solutions of the equation (1) can be sorted out by a *check* in (1).

Examples: 1. The transition from (1) $4x = 20;\ x \in \mathbf{N}$, to (2) $x = 5;\ x \in \mathbf{N}$, is an equivalent transformation, because $S_1 = S_2 = \{5\}$.

2. The transformation of the equation (1) $x = 6;\ x \in \mathbf{Z}$ into (2) $x(x+2) = 6(x+2);\ x \in \mathbf{Z}$, is inequivalent; in this case $S_1 = \{6\}$; $S_2 = \{-2; 6\}$; hence $S_1 \subset S_2$.

3. If one goes from (1) $x^3 = x^2 + 12x;\ x \in \mathbf{Z}$, to (2) $x^2 = x + 12;\ x \in \mathbf{Z}$, on division by x, one has $S_1 = \{-3, 0, 4\}$, $S_2 = \{-3, 4\}$; hence $S_1 \supset S_2$, that is, the transformation is inequivalent.

Transformations leading to a loss of solutions can occur, for example, on dividing an equation by an expression containing a variable or by extracting a root from the equation. If in the solution of the equation one performs inequivalent transformations, then additional investigations are required to determine the solutions that may have got lost or those that are not solutions of the original equation. Such complications can be avoided if only equivalent transformations are performed. Therefore it is very important to know what transformations of an equation are equivalent. The following theorems, in which the domain of definition is $\mathbf{R}$, give some relevant indications.

Proposition 1: An equation $E_1 = E_2$ is equivalent to an equation $E_1' = E_2'$ if the expressions E_1 and E_1' as well as E_2 and E_2' are equivalent.

According to this proposition one may, in particular, contract terms, divide fractions by numbers, or multiply brackets; for example, the equations $4x + 7 - 2x + 15 = 8x - 6x + 13 - 3x$ and $2x + 22 = -x + 13$ are equivalent because $S_1 = S_2 = \{-3\}$.

Proposition 2: An equation $E_1 = E_2$ is equivalent to $E_2 = E_1$, that is, by interchanging the sides an equation goes over into an equivalent one.

Proposition 3: By adding (or subtracting) to both sides of an equation $E_1 = E_2$ one and the same expression E_3, defined for the whole domain of definition of $E_1 = E_2$, then the equation $E_1 + E_3 = E_2 + E_3$ (or $E_1 - E_3 = E_2 - E_3$) is equivalent to the original equation.

Example:

(1) $8x - 29 = 4x + 31$	$\mid + (29 - 4x)$	$E_1 = E_2 \quad \mid + E_3$
$8x - 29 + (29 - 4x) = 4x + 31 + (29 - 4x)$		$E_1 + E_3 = E_2 + E_3$
(2) $4x = 60$		

According to Theorem 3 the equations $8x - 29 = 4x + 31$ and $4x = 60$ are equivalent. In fact, $S_1 = S_2 = \{15\}$.

Example which shows the necessity of restricting E_3:

(1) $x = 4 \quad \Big| + \dfrac{1}{x-4}$

(2) $x + \dfrac{1}{x-4} = 4 + \dfrac{1}{x-4}$.

While the solution set of (1) is $S_1 = \{4\}$, that of (2) is $S_2 = \emptyset$ because $1/(x-4)$ is not defined for the number 4. Consequently, since $S_1 \neq S_2$, the equations (1) and (2) are inequivalent.

Proposition 4: If one multiplies (or divides) both sides of an equation $E_1 = E_2$ by one and the same expression E_3, defined for the whole domain of definition of $E_1 = E_2$ and different from zero there, then the equation $E_1 \cdot E_3 = E_2 \cdot E_3$ (or $E_1/E_3 = E_2/E_3$) is equivalent to the original equation.

Example:

(1) $6a = -3 \quad \mid : 6$	$E_1 = E_2 \quad \mid : E_3$
$6a/6 = -3/6$	$E_1/E_3 = E_2/E_3$
(2) $a = -1/2$	

The equation $6a = -3$ and $a = -1/2$ are equivalent by proposition 4. In fact, $S_1 = S_2 = \{-1/2\}$. By way of contrast, the transition from

(1) $\dfrac{a}{a+4} = \dfrac{-4}{a+4} \quad \Big| \cdot (a+4)$

to (2) $a = -4$

is an inequivalent transformation, because the expression $(a + 4)$ takes the value 0 for $a = -4$. Since $S_1 = \emptyset$ and $S_2 = \{-4\}$, (1) and (2) are inequivalent equations.

The propositions above require proofs, which will, however, be omitted here. There is no such general equivalence theorem for raising to a power or extracting roots, because these operations can lead to inequivalent equations, as is shown by the following example:

Example:

$$\begin{aligned} (1)\ 1 + x &= \sqrt{(1 - x)} \quad | \text{ Squaring} \\ (1 + x)^2 &= (\sqrt{(1 - x)})^2 \\ 1 + 2x + x^2 &= 1 - x \\ (2)\ x^2 + 3x &= 0. \end{aligned}$$

With respect to the set **R** of real numbers the equations (1) and (2) are inequivalent, because here $S_1 = \{0\}$ and $S_2 = \{-3, 0\}$, that is $S_1 \neq S_2$. The number -3 is a solution of (2), but not of (1).

Solving equations. To solve an equation means to give *all* solutions relative to given domains of variability, in other words, to give the solution set whose elements can be numbers, number pairs, *n*-tuples of numbers, expressions with parameters, or *n*-tuples of such expressions.

In particular cases the task of solving an equation can be accomplished by systematic trial and error, in the case of the simplest equations by reading off the solutions directly, and in general by working through a method of solution or a *solution algorithm*. Such solution methods and algorithms consist in most cases of step-by-step equivalent transformations of the given equation, until finally an equation arises whose solutions can be read off.

Model solutions for a linear equation with one variable, by using the equivalence theorems. The aim of the transformations is to obtain an equation so simple that its solutions can be read off directly.

$$\begin{aligned} 7x - 2 - 5x &= -4x + 3 + 3 - 8 && \text{Proposition 1} \\ 2x - 2 &= -x - 5 && | + 2 + x \quad \text{Proposition 3} \\ 2x - 2 + 2 + x &= -x - 5 + 2 + x && \text{Proposition 1} \\ 3x &= -3 && | : 3 \quad \text{Proposition 4} \\ 3x/3 &= -3/3 && \text{Proposition 1} \\ x &= -1. \end{aligned}$$

This is a chain of equivalent equations. Since equivalence is transitive, the last equation $x = -1$ is equivalent to the original equation. The number -1 evidently is the only solution of the equation $x = -1$ and so the only solution of the original equation. Every equation in the chain has the solution set $S = \{-1\}$.

Check. When an equation with variables has been solved, it is necessary to check whether the solution set has been found correctly. If all the transformations are equivalent, the check has the purpose of spotting calculating errors and of verifying that the solutions belong to the domain of definition; if also *inequivalent* transformations have been used, then the check indicates whether additional solutions have arisen; it cannot tell whether solutions of the original equation have got lost.

The check has to be carried out in the original equation. In the first part of the check one replaces all variables by the numbers that have been found for them; for example, in the model above:

$$\begin{aligned} 7 \cdot (-1) - 2 - 5 \cdot (-1) &= -4 \cdot (-1) + 3 + 3 \cdot (-1) - 8 \\ -7 - 2 + 5 &= 4 + 3 - 3 - 8 \\ -4 &= -4. \end{aligned}$$

This is a true statement and confirms that the calculations have been correct. In the second part of the check it has to be verified that the numbers found belong to the domain of definition, in the present case to **R**. Since $-1 \in \mathbf{R}$ is a true statement, the solution set is, in fact, $S = \{-1\}$.

If another domain of variability is assumed, for example, $x \in \mathbf{N}$, then the first part of the check proceeds as above, but in the second part one obtains $-1 \notin \mathbf{N}$, so that the solution set is $S = \emptyset$.

Solution of problems

The problems in question concern either a mathematical situation expressed in natural language, or a practical situation from one of the domains of application, for example, the natural sciences, technology, or economics. In both cases the task is to translate the text into the formalized language of mathematics. This can result in equations with variables; for example, the text: 'If 7 is added to three times a natural number, the result is the same as subtracting this number from 13' leads to the equation: $3x + 7 = 13 - x$; $x \in \mathbf{N}$, by introducing the variable x in place of the required number.

Usually the 'translation' of a problem leads first to an equation between quantities and variables for quantities, and then from this to an equation with numbers and variables for numbers.

Example: A fir tree of height 9 yards breaks off 4 yd. above ground. How far from the foot of the tree does the tip of the tree hit the ground? – For the solution such verbal problems the following scheme is recommended:

1. *Fixing the variable, if possible by means of a sketch* (Fig.).
   The tip of the tree hits the ground x yd. from the foot of the tree.
2. *Setting up the equation(s) and determining the domains of variability.*
   Equation $(4 \text{ yd.})^2 + (x \text{ yd.})^2 = (5 \text{ yd.})^2$ with quantities or with numbers and variables for numbers $4^2 + x^2 = 5^2$ with $x \in \mathbf{R}$ and $x > 0$.
3. *Solving the equation(s).*
   $x^2 = 25 - 16 = 9$, hence $x_1 = 3$ and $x_2 = -3$.
4. *Check with reference to the meaning of the text.*
   From the text or the resulting domain of variability it is evident that only $x_1 = 3$ can be regarded, and in fact is, the solution of the problem expressed in words.
5. *Answer.*
   The tip of the tree hits the ground 3 yards from the foot of the tree.

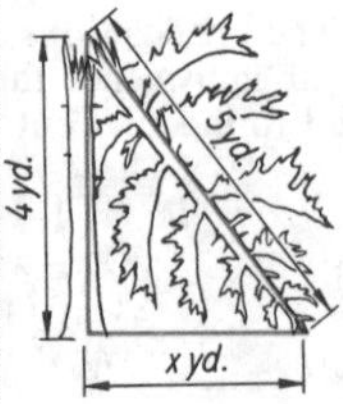

4.1-3 Broken fir tree

4.2. Linear equations

In a linear equation or equation of the first degree, all the variables occur only to the first power; for example, $5x - 2 = 8$, $3a + 2b = 4$, $4u + 5v + 3w - 1 = 0$ are linear equations with one, two, and three variables, respectively. The equation $(x + 4)(x + 3) = (x + 1)(x + 7)$ for $x \in \mathbf{R}$, although not linear, is equivalent to a linear equation, because it can be brought to the form $x - 5 = 0$ for $x \in \mathbf{R}$ by multiplying out and rearranging. However, the equation $(x + 4)(x + 3) = 6$ for $x \in \mathbf{R}$ is equivalent to the non-linear equation $x^2 + 7x + 6 = 0$; $x \in \mathbf{R}$. Also fractional equations and equations in root form can be equivalent to linear equations.

Linear equations with one variable

General form	$ax + b = 0$	$x \in \mathbf{R}$; $a, b \in \mathbf{R}$

In the general form x is the variable, a and b are real parameters. One calls ax the *linear term* and b the *absolute term*. The case $a = 0$ is included in the discussion, though in the following this case will be excluded because an equation $ax + b = 0$ in which $a = 0$ is not, strictly speaking, linear. By equivalent transformations every linear equation with one variable can be brought to this general form. In the solution of the equation $ax + b = 0$ three cases have to be distinguished.

Case distinction	I. $a \neq 0$, b arbitrary	II. $a = 0$, $b \neq 0$	III. $a = 0$, $b = 0$
solution	$ax + b = 0 \quad \vert -b$ $ax = -b \quad \vert :a$ $x = -b/a$	$0 \cdot x + b = 0 \quad \vert -b$ $0 \cdot x = -b$	$0 \cdot x + 0 = 0$ $0 \cdot x = 0$
number of solutions solution set	precisely one $S = \{-b/a\}$	none $S = \emptyset$	infinitely many $S = \mathbf{R}$
check	1. $a(-b/a) + b = 0$ $-b + b = 0$ $0 = 0$ true 2. $-b/a \in \mathbf{R}$, because $a, b \in \mathbf{R}$ and $a \neq 0$	since $0 \cdot x = 0$ for every $x \in \mathbf{R}$, but $b \neq 0$, no real number satisfies the equation	the product of every real number by zero is zero, hence every real number is a solution

When the domain of variability is altered, it is, of course, quite possible that $x = -b/a$ is not a solution or is a universally valid solution. For example, the equation $5x + 10 = 0$; $x \in \mathbf{R}$, has the solution set $S = \{-2\}$; but if the domain of variability is $\mathbf{N}$, then the solution set is empty: $S = \emptyset$, because $-2 \notin \mathbf{N}$. Finally, if the domain of variability is $\{-2\}$, then the only element of the domain of variability is also the only solution of the equation, which is therefore universally valid.

Example 1: Linear equations without a parameter.

$$4a/3 + 1/2 - a = -3/2 + 2a/3 + 5/2; \quad a \in \mathbf{Q}$$
$$a/3 + 1/2 = 2a/3 + 1 \quad \vert -1/2 - 2a/3;$$
$$-a/3 = 1/2 \quad \vert :(-1/3);$$
$$a = -3/2.$$

Check: 1. $(4/3)\cdot(-3/2)+1/2-(-3/2) = -3/2+(2/3)(-3/2)+5/2$ 2. $-3/2 \in \mathbf{Q}$ true
$-2+1/2+3/2 = -3/2-1+5/2$
$0 = 0$ true

The solution set is therefore: $S=\{-3/2\}$.

Example 2: Equation with the variable $x \in \mathbf{R}$ and the parameters $a, b \in \mathbf{R}$ that leads to a linear equation in x.

$$\begin{aligned}
(x+a)^2-(x-b)^2 &= 2a(a+b); \\
x^2+2ax+a^2-x^2+2bx-b^2 &= 2a^2+2ab \qquad \mid -a^2+b^2 \\
2ax+2bx &= a^2+2ab+b^2 \\
2x(a+b) &= (a+b)^2 \qquad \mid :2(a+b)
\end{aligned}$$

Here a case distinction is necessary:

First case: If $(a+b) \neq 0$, division leads to $x=(a+b)/2$, so that $S=\{(a+b)/2\}$.
Check: 1. $[(a+b)/2+a]^2-[(a+b)/2-b]^2 = 2a(a+b)$,
$[(3a+b)/2]^2-[(a-b)/2]^2 = 2a^2+2ab$,
$[9a^2+6ab+b^2-a^2+2ab-b^2]/4 = 2a^2+2ab$, $2a^2+2ab=2a^2+2ab$.
This is a true statement for all real numbers a and b.
2. $(a+b)/2 \in \mathbf{R}$, because $a, b \in \mathbf{R}$.

Second case: If $a+b=0$, that is, $b=-a$, the given equation is $(x+a)^2-(x+a)^2=2a(a-a)$. It is equivalent to $0\cdot x=0$ and has the solution set $S=\mathbf{R}$.
Check: $(x+a)^2-(x+a)^2=0$ is true for every real number and every parameter $a \in \mathbf{R}$.

In fractional equations at least one of the variables occurs at least once in the denominator of a fraction.

Example 3: $\dfrac{2}{x-2}+\dfrac{3}{x+2}=\dfrac{5}{x}$.

For all real numbers $x \neq \pm 2$ and $x \neq 0$ multiplication by the least common denominator $x(x-2)(x+2)$ is an equivalent transformation and leads to a linear equation.

$$\begin{aligned}
2x(x+2)+3x(x-2) &= 5(x-2)(x+2) \\
2x^2+4x+3x^2-6x &= 5x^2-20 \qquad \mid -5x^2 \\
-2x &= -20 \qquad \mid :(-2) \\
x &= 10
\end{aligned}$$

The check has to be made in the original equation.

1. $\dfrac{2}{10-2}+\dfrac{3}{10+2}=\dfrac{5}{10}$ 2. It is true that $10 \in \mathbf{R}$ and $10 \neq \pm 2$, $10 \neq 0$
$2/8+3/12=1/2$ true
$1/2=1/2$ true

The solution set is therefore $S=\{10\}$.

Example 4: The fractional equation contains the parameter a:

$$\frac{x+2a}{2a-x}+\frac{x-2a}{2a+x}=\frac{4a^2}{4a^2-x^2} \qquad \mid \cdot(2a-x)(2a+x) \quad x \neq 2a,\ x \neq -2a$$

$$\begin{aligned}
(x+2a)(x+2a)-(2a-x)(2a-x) &= 4a^2 \\
x^2+4ax+4a^2-4a^2+4ax-x^2 &= 4a^2 \\
8ax &= 4a^2.
\end{aligned}$$

First case: $a \neq 0$
$x=a/2$
$S=\{a/2\}$
The check shows that this is correct.

Second case: $a=0$
$x/(-x)+x/x = 0/(-x^2) \quad \mid \cdot(-x^2) \quad \mid x \neq 0$
$+x^2-x^2 = 0$
$0\cdot x^2 = 0$
All real numbers other than 0 are solutions of this equation.

In *equations with roots* at least one of the variables occurs at least once in the radicand of a root. In the simplest cases exponentiation eliminates the roots, but it has to be observed that this may be an inequivalent transformation leading to additional solutions.

Example 5: $\sqrt[3]{(x+2)}=3$ is a root equation equivalent to a linear equation.

$$\begin{aligned}
\sqrt[3]{(x+2)} &= 3 \qquad \mid \text{third power} \\
x+2 &= 27 \qquad \mid -2 \\
x &= 25 \\
S &= \{25\}
\end{aligned}$$

Check:
1. $\sqrt[3]{(25+2)}=3$
$3=3$ true
2. $25 \in \mathbf{R}$ true

Example 6: If several square roots occur, one of these can be isolated before exponentiation.

$$14 = \sqrt{(x-4)} + \sqrt{(x+24)}$$
$$\sqrt{(x-4)} = 14 - \sqrt{(x+24)} \quad | \text{ squaring}$$
$$x - 4 = 196 - 28\sqrt{(x+24)} + x + 24$$
$$28\sqrt{(x+24)} = 224$$
$$\sqrt{(x+24)} = 8$$
$$x + 24 = 64$$
$$x = 40$$
$$S = \{40\}.$$

Check:
1. $14 = \sqrt{(40-4)} + \sqrt{(40+24)}$
$14 = 6 + 8$
$14 = 14$ true
2. $40 \in \mathbf{R}$ true

Fractional and root equations reducing to quadratic equations can be found in the appropriate sections. The following examples of *applied problems*, which in every case lead to linear equations with one variable, are to be regarded as models of frequently occurring types.

Example 7: Mixing problem. In a Siemens-Martin furnace 20 t steel of 0.5% carbon content are melted together with 5 t pig iron of 5% carbon content. What is the percentage of carbon in the mixture? –

Let x% be the carbon content of the mixture, that is, 25 t of mixture contain $25 \cdot x/100$ t carbon. The 20 t steel contain $20 \cdot 0.5/100$ t, and the 5 t pig iron $5 \cdot 5/100$ t carbon. Since the sum of the carbon content of the parts must be equal to the total carbon content, one obtains the equation $25 \cdot 0.5/100 + 5 \cdot 5/100 = 25 \cdot x/100$, which shows that the carbon content of the mixture is 1.4%.

Example 8: Distribution problem. 3 excavators together move daily 31000 m³ earth. The second bulldozer moves 1000 m³ more than the third, and the first 4000 m³ less than twice the amount of the second. What is the amount of earth moved daily by each of the 3 bulldozers? – If the third bulldozer moves x m³ earth, then the second moves $(x + 1000)$ m³, and the first moves $[2(x + 1000) - 4000]$ m³ earth. The three bulldozers together, move $31000\,\text{m}^3 = \{x + (x + 1000) + [2(x + 1000) - 4000]\}\,\text{m}^3$. The calculation leads to $S = \{8000\}$; this means that the third bulldozer moves 8000 m³, the second 9000 m³, and the first 14000 m³, and all three together 31000 m³ as required.

Example 9: Simple problem of motion. A train of length 250 yd. passes through a tunnel of length 200 yd. at a speed of 50 miles per hour. How long does it take to pass through the tunnel? – Let x seconds be the time between the entrance of the locomotive into the tunnel and the exit of the last carriage. During this time the last carriage passes $\frac{50.1760}{60 \cdot 60} \cdot x$ yd., but this is the length of the tunnel plus the length of the train

$$200 + 250 = \frac{50.1760}{60 \cdot 60} \cdot x; \qquad S = \{18.4\}.$$

The passage takes 18.4 seconds.

Example 10: More complicated problem of motion. A barge going downstream reaches its destination in two hours. Going upstream for the same distance with the same machine power, it needs three hours. Its velocity in still water is 250 yd./min. What is the velocity of the moving water? – Let x yd./min be the velocity of the streaming water; then the steamer has the downstream velocity of $(250 + x)$ yd./min and takes 120 min; but upstream at the velocity of $(250 - x)$ yd./min it needs 180 min for the same distance. Therefore the equation is: $(250 + x) \cdot 120 = (250 - x) \cdot 180$. The velocity of the water is 50 yards per minute.

Linear equations with two variables

The *solution set* of a linear equation with two variables, for example, $4x + 3y - 10 = 0$ with $x \in \mathbf{R}$, $y \in \mathbf{R}$, consists of all ordered pairs of real numbers (x, y) which on substitution make the equation into a true statement; for example, $(1, 2)$ is a solution, because $4 \cdot 1 + 3 \cdot 2 - 10 = 0$ is a true statement. $(0, 10/3)$ and $(+3, 2/3)$ are likewise solutions. If one postulates that x and y are to be natural numbers, then solutions are pairs of natural numbers satisfying the equations; other domains of variability for x and y can also be prescribed.

Systems of two linear equations. Linear algebra provides methods of solving m linear equations with n variables. If it is required to satisfy simultaneously m equations with n variables, one speaks of a *system* of m equations with n variables. Every *solution* of such a system is an ordered n-tuple of numbers. Here only the case $m = n = 2$ will be treated in detail (for arbitrary m and n see Chapter 17.). Every *solution* of such a system of equations is an ordered pair (x, y) of numbers.

General form of a system of linear equations	(1) $a_1x + b_1y = c_1$ (2) $a_2x + b_2y = c_2$	variables $x, y \in \mathbf{R}$ parameters $a_1, a_2, b_1, b_2, c_1, c_2 \in \mathbf{R}$

Solving a system of two linear equations. To solve a system of two linear equations with two variables means to determine *all* ordered pairs (x, y) satisfying both the first and the second equation; in other words, one has to determine the intersection S of the solution sets S_1 and S_2 of the two equations. Here the only possible cases are the following three:

1. $S = S_1 \cap S_2 = \{(a, b)\}$. The system of equations has a *unique solution*;
2. $S = S_1 \cap S_2 = \emptyset$; the equations of the system are *inconsistent*, *incompatible*, or *contradictory*;
3. $S = S_1 \cap S_2 = S_1$ or S_2; the system of equations is not uniquely *soluble*; it then has infinitely many solutions.

This last case occurs if and only if the two equations are linearly dependent, that is, if one equation is a real multiple of the other equation. For the numerical solution of systems of two linear equations with two variables among the elementary methods available are the method of substitution, of equating, and of adding. They all aim at eliminating one of the variables, so that there are only two linear equations with one variable which each are to be solved. In the *substitution method* one equation is solved with respect to one of the variables, and the expression obtained is substituted in the other equation.

Example 1: (1) $x + y = -3$ → $x + (2x + 6) = -3$
(2) $-2x + y = 6$ → $y = 2x + 6$ → $3x + 9 = 0$
y is eliminated $x = -3$

$y = 2 \cdot (-3) + 6$
$y = 0$ y is calculated.

The *check* must be carried out for *both* initial equations:

1. (1 $-3 + 0 = -3$ true 2. $-3 \in \mathbf{R}$ true and $0 \in \mathbf{R}$ true
(2) $0 - 2 \cdot (-3) = 6$ true

The number pair $(-3, 0)$ is the only solution of this system of equations. The solution set is $S = \{(-3, 0)\}$.

In the *method of equating* the two equations are solved with respect to the same variable, and the expressions so obtained are equated; hence the method is based on the transitivity of the equivalence of expressions.

Example 2: (1) $x - 2y = 4$ → $x = 4 + 2y$
(2) $2x + 5y = 35$ → $x = (35 - 5y)/2$

$x - 2 \cdot 3 = 4$ $4 + 2y = (35 - 5y)/2$, here x is eliminated
$x = 10$ $y = 3$

The solution set is $S = \{(10, 3)\}$.

The addition method. By multiplying each equation by a suitable number it can always be achieved that the coefficients of one of the two variables in the two equations are opposite numbers. By adding the two equations one of the variables is eliminated.

Example 3:
(1) $12x - 8y = 4$ $| \cdot 3$ → (1′) $36x - 24y = 12$
(2) $18x - 15y = 3$ $| \cdot (-2)$ → (2′) $-36x + 30y = -6$

$12x - 8 \cdot 1 = 4$ $0 \cdot x + 6y = 6$
$x = 1$ $y = 1$

The solution set is $S = \{(1, 1)\}$.

It is a matter of experience to recognize which of the methods is the most suitable in an individual case.

To obtain a *survey of the solutions* of a system of two linear equations with two variables x and y on distinguishes between two principal cases: I. For $a_1 = a_2 = b_2 = b_1 = 0$ in the case $c_1 = c_2 = 0$ an arbitrary pair of real numbers is a solution of the system. If, however, one of the numbers c_1 or c_2 is different from zero, then there is no solution.

(1) $a_1 x + b_1 y = c_1$
(2) $a_2 x + b_2 y = c_2$ $x \in \mathbf{R}, y \in \mathbf{R}$

II. If at least one of the coefficients a_1, a_2, b_1, b_2 is different from zero, then there are three possible cases.

Equations	first case	second case	third case
	linearly *independent* and *consistent* for example: (1) $4x + y = 12$ (2) $x + 2y = 10$	linearly *dependent* for example: (1) $4x + y = 12$ (2) $8x + 2y = 24$	inconsistent for example: (1) $4x + y = 12$ (2) $4x + y = 10$
number of solutions	exactly one (2, 4)	infinitely many, for example: (1, 8), (2, 4), (3, 0), (4, −4), ...	none
solution set	$S = \{(2, 4)\}$	$S = \{(x, y) \mid 4x + y = 12\}$	$S = \emptyset$
graphical illustration	two intersecting straight lines; one point of intersection	two coincident straight lines; infinitely many points in common	two distinct parallel, straight lines no point in common

Example 4:

multiplying out and ordering

$$\begin{aligned}(1)\ & 4y(10x - 3) - 5x(8y + 7) + 165 = 0\\ (2)\ & 9x(4y - 7) + 3y(5 - 12x) = -114\end{aligned}$$

addition method

$$\begin{aligned}(1')\ & -35x - 12y = -165 \quad \cdot 5\\ (2')\ & -63x + 15y = -114 \quad \cdot 4\\ (1'')\ & -175x - 60y = -825 \quad +\\ (2'')\ & -252x + 60y = -456\\ & -427x = -1281 \quad | : (-427)\\ & x = 3\end{aligned}$$

$$-35 \cdot 3 - 12y = -165, \quad y = 5.$$

The check confirms that the calculation is correct. Here $S = \{(3, 5)\}$.

Example 5: The equations are given in fractional form. It is to be observed that the denominators must be different from zero. The solution set is $S = \{(3, 2)\}$.

$$(1)\ \frac{x + y + 1}{x + y - 1} = \frac{3}{2}; \quad (2)\ \frac{x - y + 1}{x + y + 1} = \frac{1}{3} \quad \longrightarrow \quad \begin{aligned}(1')\ & x + y = 5\\ (2')\ & 2x - 4y = -2\end{aligned}$$

Example 6: The variables are x and y, while a and b are real parameters. The solution set is $S = \{(a + b, a - b)\}$.

$$\begin{aligned}(1)\ & x + y = 2a\\ (2)\ & x - y = 2b\end{aligned}$$

$$+:\ 2x = 2a + 2b,\ x = a + b \qquad -:\ 2y = 2a - 2b,\ y = a - b$$

Example 7: The second equation is a multiple of the first; then every ordered pair satisfying the first equation satisfies also the second. There are infinitely many solutions.

$$\begin{aligned}(1)\ & v - 2n/3 = 1 & \longrightarrow\ (1')\ & 3v - 2n = 3\\ (2)\ & 6v - 6 = 4n & \longrightarrow\ (2')\ & 6v - 4n = 6\end{aligned}$$

Example 8: The equations are inconsistent. There is no solution. $S = \emptyset$.

$$\begin{aligned}(1)\ & 4a + 3b = 7 & \longrightarrow\ (1')\ & 4a + 3b = 7\\ (2)\ & 4(a - 2) = -3b & \longrightarrow\ (2')\ & 4a + 3b = 8\end{aligned}$$

Applied problems leading to a system of linear equations.

Example 9: Distribution problem. A water container can be filled from a hot and from a cold tap. If the hot tap is left on for three minutes and the cold tap for one minute, then 50 quarts have flown in. But if the hot tap is left on for one minute and the cold tap for two minutes, then 40 quarts have flown in. How many quarts of water flow through each tap in one minute? – To introduce the variables one assumes that the hot tap yields x quarts/min and the cold tap y quarts/min; so one obtains the system of equations to the right and then the solution: the warm tap yields 12 quarts per minute, and the cold tap 14 quarts per minute.

$$\begin{aligned}(1)\ & 3x + y = 50\\ (2)\ & x + 2y = 40\end{aligned}$$

Example 10: Mixing problem. To prevent the water in the cylinder block and cooling system of a car from freezing one adds at the beginning of winter an antifreeze of the density 1.135 referred to the density 1 of the water in the radiator. If the mixture has a density of 1.027, one obtains frost protection up to −10°C (14°F). How many quarts of antifreeze and how many quarts of water are to be mixed in order to obtain 100 quarts of the mixture? –

To introduce the variables one assumes that the amount of antifreeze is x quarts that of water is y quarts. Then one obtains the system of equations and from it the solution: one mixes 20 quarts of antifreeze with 80 quarts of water to obtain the desired mixture.

$$\begin{aligned} x + y &= 100, \\ 1.135x + y &= 1.027 \cdot 100 \end{aligned}$$

Graphical solution of linear equations and systems of equations

In solving equations graphically one sets up a one-to-one correspondence between the solutions and certain sets of points. By representing these point sets in a coordinate system one obtains approximate solutions for the equations. The coordinate system to be used here is rectangular Cartesian.

Graphical solution of one linear equation with one variable. To solve the equation $ax + b = 0$, $a \neq 0$, graphically one goes over to the function represented by the equation $y = ax + b$, $a \neq 0$. Its graph is a straight line (see Chapter 5.). The zero of the function, that is, the abscissa of the point of intersection of the line with the x-axis, is the solution of the given equation.

Example 1: From the equation $2x - 6 = 0$ one goes over to the function with the equation $y = 2x - 6$. Its graph is a straight line which cuts the x axis in the point $P = (3, 0)$. Hence 3 is the solution of the equation $2x - 6 = 0$ (Fig.).

Graphical solution of systems of two linear equations with two variables. The solution sets of the two equations are represented graphically, and their intersection is determined. For this purpose one interprets the given equations as functional equations and draws the graph of these functions. In general they are straight lines. The coordinates of all points of the first line, and only they, satisfy the first equation, and those of the second line, and only they, the second equation. From the drawing one determines *all* points lying both on the first and second line, that is, in their intersection. To the coordinates of each of these points there corresponds one-to-one a solution of the system of equations. Depending on the relative position of the lines one obtains exactly one, none, or infinitely many common points, that is, the system of equations is uniquely soluble, insoluble, or not uniquely soluble.

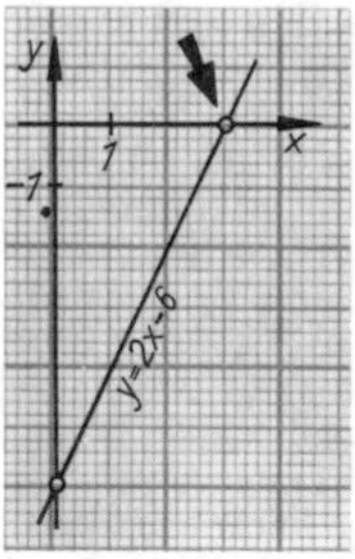

4.2-1 Graphical solution of the equation $2x - 6 = 0$

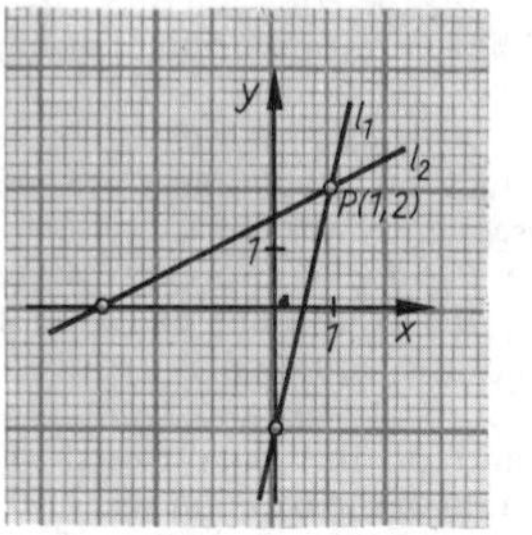

4.2-2 Graphical solution of the system of equations $4x - y = 2$, $x - 2y = -3$

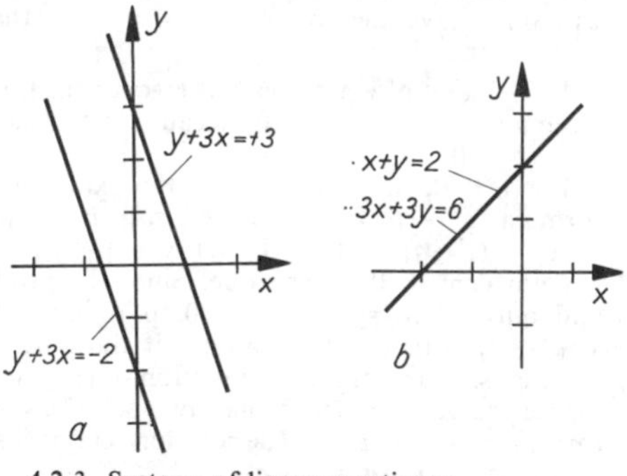

4.2-3 Systems of linear equations:
a) with no solutions,
b) with infinitely many solutions

Example 2: To solve the system of equations graphically one represents the functions with these equations graphically. The point of intersection $P = (1, 2)$ yields $(1, 2)$ as the only solution of the system of equations (Fig.).

$$\begin{aligned} &(1)\ 4x - y = 2 \\ &(2)\ x - 2y = -3 \end{aligned}$$

$$\begin{aligned} &(1)\ -x + y = 2 \\ &(2)\ -3x + 3y = 6 \end{aligned}$$

Example 3: In the graphical solution of the system of equations one is led to two coincident lines. Consequently, the coordinates of every point lying on the line given by $-x + y = 2$ is a solution of the system (Fig.).

The graphical representation of the equations (1) $y + 3x = +3$ and (2) $y + 3x = -2$ gives two parallel lines, that is, no point of intersection and therefore the solution set $S = \emptyset$.

4.3. Quadratic equations

In a quadratic equation or equation of the second degree with one variable this variable occurs at least once to the second power and not at all to a higher power; for example, $2x^2 + 5x = 16 - x$ is a quadratic equation in x and $a^2 = a^2/2 + 6$ is a quadratic equation in a. The fractional equation $3/(u-2) + 8/(u+3) = 2$, after multiplying both sides with the least common denominator $(u-2)(u+3)$ and rearranging, leads to a quadratic equation, namely $2u^2 - 9u - 5 = 0$. If several variables occur in an equation and if the sum of the exponents for at least one term is two, but never higher, the equation is also called quadratic; for example, $x^2 + y^2 = 4$ and $x \cdot y = \text{const}$ are quadratic equations with two variables.

The following treatment concerns quadratic equations with one variable.

General form of the quadratic equation in x	$Ax^2 + Bx + C = 0$	$x \in \mathbf{R}$; $A, B, C \in \mathbf{R}$; $A \neq 0$

Here x is the variable, and A, B, C are real parameters. The term Ax^2 is called the *quadratic term*, Bx the *linear term*, and C the *absolute term*. $A \neq 0$ is necessary, because otherwise the equation is linear.

If one divides the general form $Ax^2 + Bx + C = 0$ on both sides by $A \neq 0$ one obtains the equivalent equation $x^2 + (B/A)x + C/A = 0$. The abbreviation $B/A = p$ and $C/A = q$ leads to the *normal form*.

Normal form of the quadratic equation	$x^2 + px + q = 0$	$x \in \mathbf{R}$, $p, q \in \mathbf{R}$

It is characterized by the fact that the coefficient of the quadratic term is $+1$. If all the terms really occur, that is, $p \neq 0$ and $q \neq 0$, one speaks of the *mixed-quadratic equation* in normal form.

I. $p = 0$; $q = 0$	II. $p = 0$; $q \neq 0$	III. $p \neq 0$; $q = 0$	IV. $p \neq 0$; $q \neq 0$
$x^2 = 0$	$x^2 + q = 0$	$x^2 + px = 0$	$x^2 + px + q = 0$
pure quadratic equation without absolute term	pure quadratic equation	mixed quadratic equation without absolute term	mixed quadratic equation
special cases			

Numerical solution of quadratic equations

I. Solution of a pure quadratic equation without absolute term. The equation $x^2 = 0$ or $x \cdot x = 0$ can only have the solution $x_1 = x_2 = 0$, that is, the solution set $S = \{0\}$, since for $x \gtrless 0$ also $x^2 > 0$, and vice versa.

II. Solution of a pure quadratic equation. For $q > 0$ the equation $x^2 + q = 0$ with $q = \mathbf{R}$ cannot have a solution with $x \in \mathbf{R}$, because in this case the expression on the left-hand side always satisfies $x^2 + q > 0$.

For $q < 0$, hence $(-q) > 0$, the expression $x^2 + q$ can be written by means of the binomial formula $a^2 - b^2 = (a-b)(a+b)$ as a product of two linear expressions in x: $x^2 + q = x^2 - (\sqrt{-q})^2 = (x - \sqrt{-q})(x + \sqrt{-q})$. Consequently, the equation $(x - \sqrt{-q})(x + \sqrt{-q}) = 0$ is equivalent to the given one. Since the product $E_1 \cdot E_2$ of two expressions E_1 and E_2 is zero if and only if $E_1 = 0$ or $E_2 = 0$, it follows from the equivalent equation that $x - \sqrt{-q} = 0$ or $x + \sqrt{-q} = 0$. So the solution of the pure quadratic equation is reduced to that of two linear equations. From the first equation one obtains $x_1 = \sqrt{-q}$, and from the second $x_2 = -\sqrt{-q}$. Hence the given equation has two solutions x_1 and x_2; this is expressed in the combined *solution formula* $x_{1,2} = \pm\sqrt{-q}$. The solution set S is the union of the solution sets of the two linear equations, that is, $S = \{\sqrt{-q}, -\sqrt{-q}\}$.

Check. 1. $(\pm\sqrt{-q})^2 + q = 0$,
$-q + q = 0$,
$0 = 0$ true

2. $\pm\sqrt{-q}$ is real provided that $q < 0$.

But if one chooses for the domain of variability the set $\mathbf{C}$ of complex numbers, then for $q > 0$ there exist two imaginary solutions which differ by sign only and which can be obtained formally in the same way, by splitting the expression $x^2 + q$ into $(x + \sqrt{-q})(x - \sqrt{-q})$.

Pure quadratic equations	$x^2 + q = 0$, $q \in \mathbf{R}$	solution formula $x_{1,2} = \pm\sqrt{-q}$	$x \in \mathbf{R}$	$q \leq 0$; solutions real
				$q > 0$; no real solution
			$x \in \mathbf{C}$	$q > 0$; solutions imaginary

Examples:

1. $x^2 - 4 = 0$; $x \in \mathbf{R}$
$x_{1,2} = \pm\sqrt{4}$
$x_{1,2} = \pm 2$; $S = \{-2, +2\}$.

Check: 1. $(\pm 2)^2 - 4 = 0$
$4 - 4 = 0$ true
2. $+2 \in \mathbf{R}$ true
$-2 \in \mathbf{R}$ true

2. $x^2 + 144 = 0$
$x_{1,2} = \pm\sqrt{-144}$
For $x \in \mathbf{R}$ there is no solution, because $\sqrt{-144} \notin \mathbf{R}$; here $S = \emptyset$.
For $x \in \mathbf{C}$ one has $x_{1,2} = \pm 12\mathrm{i}$ and $S = \{-12\mathrm{i}, +12\mathrm{i}\}$.

Check: 1. $(\pm 12\mathrm{i})^2 + 144 = 0$
$-144 + 144 = 0$ true
2. $\pm 12\mathrm{i} \in \mathbf{C}$ true

III. Solution of a mixed quadratic equation without absolute term. Taking x before a bracket transforms the equation $x^2 + px = 0$ into the equivalent equation $x(x + p) = 0$. From this it follows that $x = 0$ or $x + p = 0$. The first of these two linear equations has the only solution $x_1 = 0$, the second $x_2 = -p$. Hence the mixed quadratic equation without absolute term always has two real solutions of which one is zero; the solution set is $S = \{0, -p\}$.

Check: $0^2 + p \cdot 0 = 0$ true for all $p \in \mathbf{R}$;
$(-p)^2 + p(-p) = 0$ true for all $p \in \mathbf{R}$.

Mixed quadratic equation without absolute term	$x^2 + px = 0$; $p \neq 0$	$x \in \mathbf{R}$ $p \in \mathbf{R}$	$S = \{0, -p\}$

Example: $7x^2 - 2x = 0$
$x(7x - 2) = 0$
$x(x - 2/7) = 0$
$S = \{0, 2/7\}$

Check:
for x_1: $7 \cdot 0 - 2 \cdot 0$ true and $0 \in \mathbf{R}$ true
for x_2: $7 \cdot (2/7)^2 \cdot -2 \cdot 2/7 = 0$
$4/7 - 4/7 = 0$ true and $2/7 \in \mathbf{R}$ true

IV. Solution of a mixed quadratic equation $x^2 + px + q = 0$. The idea of the solution is to make the expression $x^2 + px$ into a perfect square by adding a suitable term, and so to reduce the equation to a pure quadratic one; here the *quadratic supplement* is the square $(p/2)^2$ of half the coefficient of the linear term px in the normal form. To make an equivalent transformation of the given equation, one has to add $(p/2)^2 - (p/2)^2$. For example, for the equation $x^2 + 2x - 5 = 0$ one has $p = 2$ and $(p/2)^2 = 1$. By addition of $1 - 1$ this equation goes over into $x^2 + 2x + 1 - 5 - 1 = 0$ or $(x + 1)^2 - 6 = 0$, that is, into a pure quadratic equation in $(x + 1)$, and from its solutions those of the given equation can be obtained. From $(x + 1)_{1,2} = \pm\sqrt{6}$ it follows that $x_{1,2} = -1 \pm \sqrt{6}$. To achieve that in the following *solution method* the pure quadratic equation is always soluble, for a while the domain of variability is taken to be the set $\mathbf{C}$ of complex numbers.

Solution method: $x^2 + px + q = 0$, $\quad | + (p/2)^2 - (p/2)^2$
quadratic supplement: $x^2 + px + (p/2)^2 - (p/2)^2 + q = 0$,
pure quadratic equation: $(x + (p/2))^2 - [(p/2)^2 - q] = 0$
its solution: $(x + p/2)_{1,2} = \pm\sqrt{(p/2)^2 - q]}$,
solution formula: $x_{1,2} = -p/2 \pm \sqrt{[(p/2)^2 - q]}$.

Check: 1. $[-p/2 \mp \sqrt{[(p/2)^2 - q]}]^2 + p\{-p/2 \pm \sqrt{(p/2)^2 - q]}\} + q = 0$,
$(p/2)^2 \pm p\sqrt{[(p/2)^2 - q]} + (p/2)^2 - q - p^2/2 \pm p\sqrt{[(p/2)^2 - q]} + q = 0$,
$p^2/4 + p^2/4 - q - p^2/2 + q = 0$,
$0 = 0$ true

2. $-p/2 \pm \sqrt{[(p/2)^2 - q]} \in \mathbf{C}$ true

Mixed quadratic equation	$x^2 + px + q = 0$ $p, q \in \mathbf{R}$	$x \in \mathbf{C}$		solution formula $x_{1,2} = -p/2 \pm \sqrt{[(p/2)^2 - q]}$
Discriminant	$D = (p/2)^2 - q$	$x \in \mathbf{R}$	I $D > 0$	$x_{1,2} = -p/2 \pm \sqrt{D}$
			II $D = 0$	$x_1 = x_2 = -p/2$
			III $D < 0$	no real solution

The solution formula also contains the solutions for the special cases, as one can verify by substituting $p = 0$ or $q = 0$ or both. It is applicable whenever the equation is given in its normal form.

Discriminant. Evidently the nature of the solutions of the quadratic equation is determined by the radicand $D = (p/2)^2 - q$ of the root in the solution formula. It is called *discriminant*. If p and q

are real parameters and if one returns to the set **R** of real numbers as domain of variability, then three cases are to be distinguished: I with two distinct solutions, II with two equal solutions and III with no real solution.

If one chooses as domain of variability the set **C** of complex numbers, then two conjugate complex solutions occur in the case $D < 0$. Choosing as domain of variability a subset of the set of real numbers, the solution set may be different.

Example 1:

$$x^2 + 4x - 5 = 0; \quad x \in \mathbf{R}$$
$$x_{1,2} = -2 \pm \sqrt{(2^2 + 5)}$$
$$x_{1,2} = -2 \pm \sqrt{9}$$
$$x_{1,2} = -2 \pm 3$$
$$x_1 = 1$$
$$x_2 = -5$$
$$S = \{-5, 1\}$$

Check:
for x_1:
$1^2 + 4 \cdot 1 - 5 = 0$
$1 + 4 - 5 = 0$ true | $1 \in \mathbf{R}$ true
for x_2:
$(-5)^2 + 4(-5) - 5 = 0$
$25 - 20 - 5 = 0$ true | $-5 \in \mathbf{R}$ true

There are two distinct real solutions. But if one chooses, for example, $x \in \mathbf{N}$, then $S = \{1\}$, because $-5 \notin \mathbf{N}$.

Example 2:

$$2x^2 - 16x + 36 = 0; \quad x \in \mathbf{R}$$
$$x^2 - 8x + 18 = 0$$
$$x_{1,2} = 4 \pm \sqrt{(16 - 18)}$$
$$x_{1,2} = 4 \pm \sqrt{-2}$$

Check:
$2(4 \pm i\sqrt{2})^2 - 16(4 \pm i\sqrt{2}) + 36 = 0$
$2(16 \pm 8i\sqrt{2} - 2) - 64 \mp 16i\sqrt{2} + 36 = 0$
$32 \pm 16i\sqrt{2} - 4 - 64 \mp 16i\sqrt{2} + 36 = 0$
$0 = 0$ true

$S = \emptyset$; there are no real solutions, because $\sqrt{-2} \notin \mathbf{R}$. But if one chooses $x \in \mathbf{C}$, then there are two distinct complex solutions, $S = \{4 + i\sqrt{2}, 4 - i\sqrt{2}\}$.

Example 3: The equation $x^2 - 14x + 49 = 0$ with $x \in \mathbf{R}$ has two coincident real solutions; $S = \{7\}$, as the check shows.

Example 4: The fractional equation $\frac{x-1}{x+1} = \frac{4x-3}{5x-10} - \frac{7}{10}$ for $x \in \mathbf{R}$ is transformed equivalently after multiplication by the least common denominator $10 \cdot (x+1) \cdot (x-2)$ for $x \neq -1$ and $x \neq 2$ into the quadratic equation $9x^2 - 39x + 12 = 0$. The solution set of the fractional equation is $S = \{1/3, 4\}$.

Example 5:

$$\sqrt{[x + 2 + \sqrt{(2x + 7)}]} = 4 \quad | \text{ squaring}$$
$$x + 2 + \sqrt{(2x + 7)} = 16$$
$$\sqrt{(2x + 7)} = 14 - x \quad | \text{ squaring}$$
$$2x + 7 = 196 - 28x + x^2,$$
$$x^2 - 30x + 189 = 0,$$
$$x_1 = 21,$$
$$x_2 = 9.$$

Check:
for x_1: $\sqrt{(21 + 2 + \sqrt{(2 \cdot 21 + 7)})} = 4$,
$\sqrt{30} = 4$ false
for x_2: $\sqrt{9 + 2 + \sqrt{(2 \cdot 9 + 7)}} = 4$,
$4 = 4$ true.

Here squaring was an *inequivalent* transformation. As the check has shown, only $x_2 = 9$ is a solution of the initial equation; hence $S = \{9\}$.

Not every equation involving square roots leads to a quadratic equation. It is always possible to get rid of all roots with integral exponents; an example is the equation $\sqrt[3]{(x+7)^2} - \sqrt[3]{(x+7)} = 6$ which for $x \in \mathbf{C}$ has the solution set $S = \{-15, 20\}$.

Applied problems leading to quadratic equations.

Example 1: Echo soundings. To measure the depth of the sea bed one uses echo soundings. The source of sound is situated at A, the receiver at B (Fig.). The width of the ship is 52.5 ft. Sound is propagated in water with a velocity of 4956 ft/s. During the time measurement the ship is considered to be at rest. What is the depth of water for a time difference of 0.1 s?

Let x ft. be the depth of water. The distance passed by the sound to the sea bed is $(4956 \cdot 0.1)/2$ ft. By the theorem of Phythagoras one obtains

$$x^2 = (247.8)^2 - (52.5/2)^2,$$
$$x^2 = 60715.8,$$
$$x_{1,2} = \pm\sqrt{60715.8} \approx 246.3.$$

The depth of water is approximately 246 feet. The negative value has no physical significance.

Example 2: Depth of a well. To determine the depth of a well one can drop a stone into it and measure the time from the beginning of the fall to the moment when one hears the stone hitting the water in the well. Suppose that this time is 4 seconds (Fig.). The velocity of sound is taken to

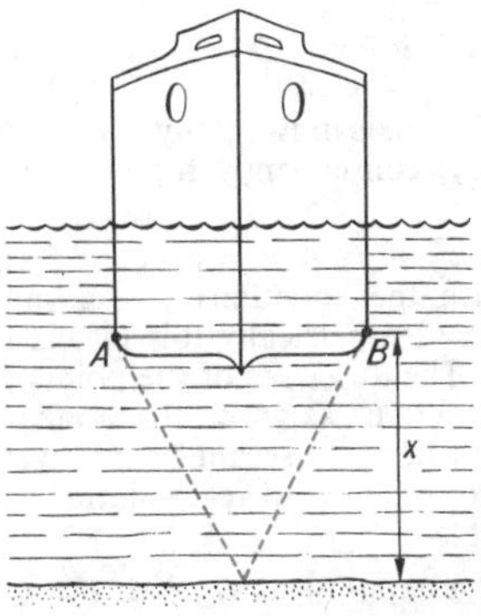

4.3-1 Echo soundings

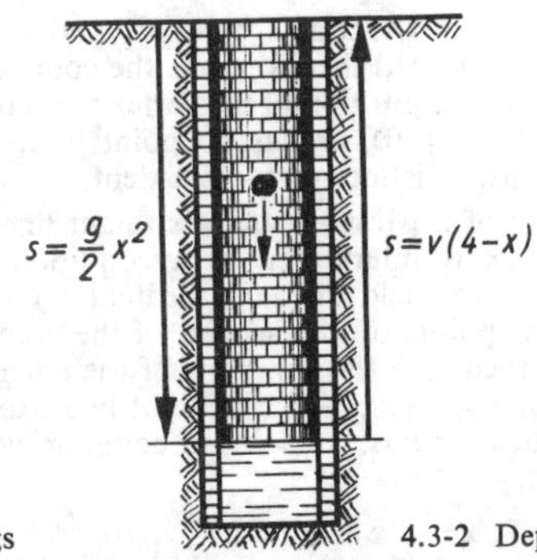

4.3-2 Depth of a well

be $v = 1092.9$ ft./s, and the acceleration due to gravity as $g = 32$ ft./s^2. What is the depth of the water level below the rim of the well? –

Let x s be the time up to the stone hitting the water; then it has covered a distance of $16x^2$ ft. For the return passage sound has taken $(4 - x)$ s, and during this time it has covered $(4 - x) \cdot 1092.9$ ft. Since the two distances are equal, one obtains the quadratic equation

$$16x^2 = (4 - x) \cdot 1092.9 \quad \text{or} \quad x^2 + 68.3x - 273.2 = 0.$$

One finds that the depth of the well is about 230 feet, because only the positive solution $x_1 \approx 3.79$ s of the quadratic equation has physical significance.

Example 3: Hardness testing. In testing the hardness of a material by the pressure method developed by BRINELL the impression h of a small steel ball of known diameter $d = 2r$ in the material to be tested is calculated from the diameter $\delta = 2\varrho$ of the circular impression (Fig.). What is the depth of impression h when the diameter of the sphere is $d = 2r = 10$ mm and the diameter of the spherical impression is $\delta = 2\varrho = 6$ mm? –

Let the depth of impression be h (in mm). By the theorem of Phythagoras one obtains $r^2 = (r - h)^2 + \varrho^2$ or $h^2 - 2rh + \varrho^2 = 0$. Of the two solutions $h_{1,2} = r \pm \sqrt{(r^2 - \varrho^2)}$ only $h_2 = r - \sqrt{(r^2 - \varrho^2)}$ can be used, because for depths $h > r$ the spherical impression always has the radius $\varrho = r$ and the method is not applicable. Substituting for r and ϱ the given quantities, one obtains $h = 1$ mm.

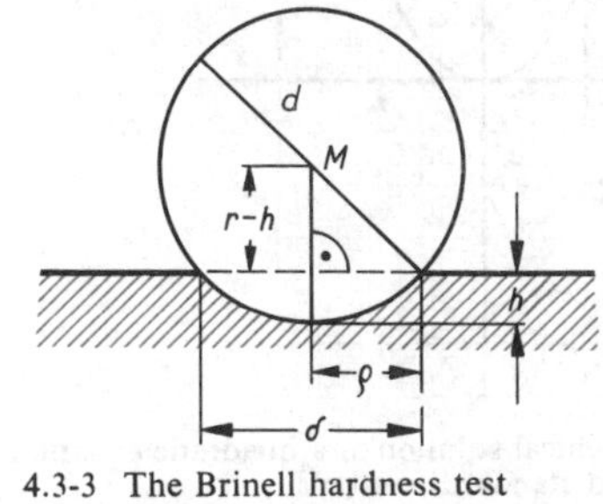

4.3-3 The Brinell hardness test

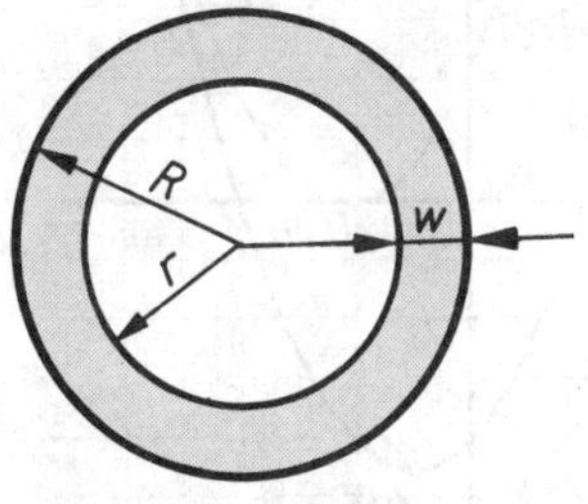

4.3-4 Section of a hollow sphere

Example 4: Stereometric problem. A hollow steel sphere has the mass $M = 160.72$ lb. The thickness of its wall is $w = 2.36$ in. (Fig.). What is its inner radius r and outer radius R if the density is $\varrho = 0.28$ pounds per cubic inch? –

If the inner radius r has the length x inches then the outer radius is $R = (x + w)$. The mass of the hollow sphere is $M = (4\pi/3)(R^3 - r^3)\varrho$, that is, in this case $M = (4\pi/3)[(x + w)^3 - x^3]\varrho$. This leads to the quadratic equation $x^2 + wx + w^2/3 - M/(4\pi w) = 0$ with the solutions $x_{1,2} = -w/2 \pm \sqrt{[M/(4\varrho w\pi) - w^2/12]}$.

Here only $x_1 = -w/2 + \sqrt{[M/(4\varrho\pi) - w^2/12]}$ is a solution of the problem. The required radii are $R = 5.52$ in. and $r = 3.16$ in.

Graphical solution of quadratic equations

Standard parabola in parallel displacement. The zeros of the quadratic function $y = x^2 + px + q$ or $y = (x + p/2)^2 + (q - p^2/4)$ yield the solutions of the quadratic equation $x^2 + px + q = 0$. The graph of this function is a standard parabola that has been subject to a parallel displacement in the direction of the x-axis by $-p/2$ and in the direction of the y-axis by $-D = +(q - p^2/4)$, whose vertex $V(x_v, y_v)$ therefore has the coordinates $x_v = -p/2$; $y_v = q - p^2/4$.

According to the position of the vertex the standard parabola cuts the x-axis in two points ($y_v < 0$), or touches it ($y_v = 0$), or has no point in common with it ($y_v > 0$); consequently the quadratic equation has two distinct, two coincident, or no real solutions.

Intersection of a parabola and a straight line. The given equation $x^2 + px + q = 0$ in the form $x^2 = -px - q$ is interpreted as a condition for the functions with the equations $y = x^2$ and $y = -px - q$ to yield the same ordinates y for certain abscissae x. Geometrically this means to determine the points of intersection of the graphs of these functions. The abscissae of the points of intersection then give the solutions of the equation. For $y = x^2$ one obtains as graph the standard parabola, for $y = -px - q$ a straight line. According as this straight line is a secant or a tangent to the parabola or has no point in common with it, one obtains two, one, or no real solutions of the equation.

Example 1: $x^2 - x - 2 = 0$.

One goes over to the function with the equation $y = x^2 - x - 2$. The graph is the displaced standard parabola with the vertex $V_1(^1/_2, -2^1/_4)$. Here $y_v < 0$. The parabola intersects the x-axis at $x_1 = -1$ and $x_2 = 2$ (Fig.).

One transforms the equation into $x^2 = x + 2$. The abscissae of the points of intersection of the standard parabola with the equation $y = x^2$ and the straight line with the equation $y = x + 2$ are $x_1 = -1$ and $x_2 = 2$ (Fig.).

The equation $x^2 - x - 2 = 0$ has two distinct real solutions. Its solution set is $S = \{-1, 2\}$.

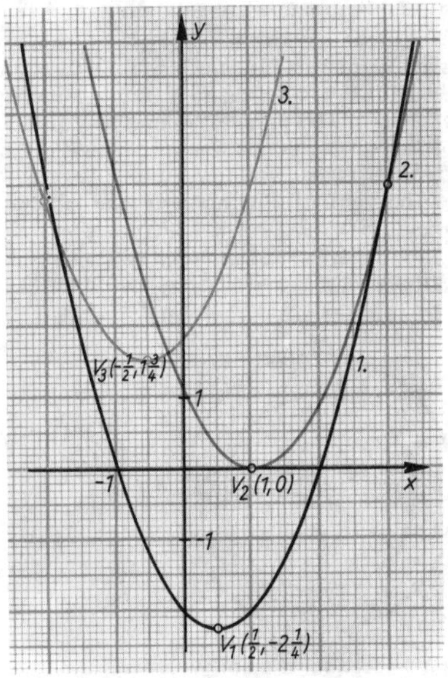

4.3-5 Roots of a quadratic equation as abscissae of the intersection of the x-axis with a standard parabola subject to parallel displacement

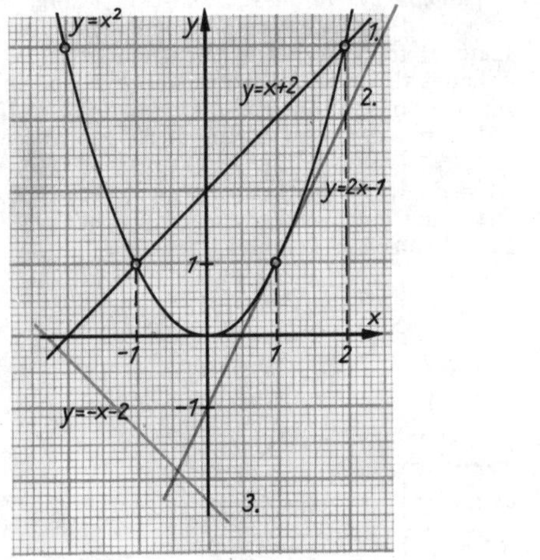

4.3-6 Graphical solution of a quadratic equation with a fixed standard parabola

Example 2: $x^2 - 2x + 1 = 0$.

One goes over to the function with the equation $y = x^2 - 2x + 1$ whose graph is the displaced standard parabola with the vertex $V_2(1, 0)$. Here $y_v = 0$. The parabola touches the x-axis at $x_1 = x_2 = 1$ (Fig.).

One transforms the equation into $x^2 = 2x - 1$. The standard parabola with the equation $y = x^2$ and the straight line with the equation $y = 2x - 1$ touch each other. The abscissa of the point of contact is $x_1 = x_2 = 1$ (Fig.).

The equation $x^2 - 2x + 1 = 0$ has two coincident real solutions. The solution set is $S = \{1\}$.

Example 3: $x^2 + x + 2 = 0$.

The function with the equation $y = x^2 + x + 2$ has as its graph the displaced standard parabola with the vertex $V_3(-^1/_2, 1^3/_4)$. Here $y_v > 0$. The parabola does not intersect the x-axis (Fig.).

The transformed equation is $x^2 = -x - 2$. The straight line with the equation $y = -x - 2$ does not intersect the standard parabola with the equation $y = x^2$ (Fig.).

The equation $x^2 + x + 2 = 0$ has no real solutions. The solution set is $S = \emptyset$.

Historical remarks

Practical needs, in particular, problems of mensuration (theorem of Pythagoras) led at an early stage to quadratic equations. Many such problems dating from *Babylonian* mathematics have come down to us in cuneiform tablets. Even systems of quadratic equations with several variables occur there. A problem dating from about 2000 B. C. in modern notation is: $x^2 - 29x + 210 = 0$ of slightly later date is, for example, the system $x^2 + y^2 = 1000$, $y = 2x/3 - 10$.

The *Greek* mathematicians treated algebraic problems in geometric form, that is, by construction. Since a square root can always be constructed by ruler and compass, the Greek mathematicians were in a position to treat all types of quadratic equations having real solutions. The classical account of these methods is in Book X of the 'Elements' of Euclid (about 300 B. C.), which in its contents goes back to THEAITETUS (410?–368 B. C.). The Hellenistic engineer and mathematician HERON of Alexandria (about 100 A. D.) took up the Babylonian and ancient Egyptian tradition of numerical treatment of quadratic equations, using approximate methods of extracting square roots. Traces of this approach can already be found in the writings of ARCHIMEDES (278?–212 B. C.). The discovery that roots occur in pairs is due to the *Hindu* mathematicians, above all BHASKARA (born 1114 A. D.). Their methods found their way into Europe, with scholars writing in Arabic as intermediaries, who themselves made further progress.

4.4. Equations of degree three and four

In general, the higher the degree of an algebraic equation, the more difficult is its solution. Therefore, quite a number of graphical and approximate methods of solution have been developed for the practical task of finding numerical solutions, which make it possible to calculate solutions to an arbitrary number of decimal places.

The cubic equation

General form of the cubic equation	$Ax^3+Bx^2+Cx+D=0$ $A \neq 0$	$x \in \mathbf{C}$ $A, B, C, D \in \mathbf{R}$

In the general form of the cubic equation or equation of the third degree x is the variable for which the set $\mathbf{C}$ of complex numbers is laid down as fundamental domain. A, B, C and D are real parameters. Ax^3 is called the *cubic*, Bx^2 the *quadratic*, Cx the *linear*, and D the *absolute term*. On dividing both sides by $A \neq 0$ and setting $B/A = r$, $C/A = s$, $D/A = t$ one obtains the equivalent normal form $x^3 + rx^2 + sx + t = 0$.

Normal form of the cubic equation	$x^3 + rx^2 + sx + t = 0$	$x \in \mathbf{C}$ $r, s, t \in \mathbf{R}$

Solubility. Special cases. In the domain of complex numbers every cubic equation has three solutions, some of which may be coincident. Since every polynomial of odd degree has at least one real zero, one of the solutions is always real. The other two are either also real or conjugate complex. If x_1 is a real root, then (see Chapter 5.) the cubic function can be split into a product of the linear factor $(x - x_1)$ and a polynomial of degree two. Since a product vanishes if and only if one of the factors does, the two other solutions of the cubic equations are the solutions of the resulting quadratic equation.

By the *theorem of Vieta* the product $x_1 \cdot x_2 \cdot x_3$ of the three solutions is equal to the negative of the absolute term $(-t)$. Therefore, if it is known that the given equation has integer solutions, then a real solution x_1 can be found as a factor of $(-t)$ by trial and error; for example, the cubic equation $x^3 - 5x^2 - 8x + 12 = 0$ has the solution $x_1 = +1$ and can be split into the product $(x - 1)(x^2 - 4x - 12) = 0$. Since the quadratic equation has the solution -2 and $+6$, the solution set of the given cubic equation is $S = \{-2, +1, +6\}$.

The cubic equation $x^3 + rx^2 + x = 0$ in which the absolute term t is zero, reduces by factorization to the equivalent equation $x(x^2 + rx + s) = 0$. Apart from the real solution $x_1 = 0$ the other two solutions of the given cubic equation are the solutions of the quadratic equation $x^2 + rx + s = 0$.

The pure cubic equation $x^3 + t = 0$ arises for $r = 0$, $s = 0$. It has the three solutions $x_1 = \sqrt[3]{-t}$, $x_2 = \omega_2 \sqrt[3]{-t}$ and $x_3 = \omega_3 \sqrt[3]{-t}$, where $\omega_2 = (-1 + \mathrm{i}\sqrt{3})/2$ and $\omega_3 = (-1 - \mathrm{i}\sqrt{3})/2$ are the complex cube roots of unity.

If in addition $t = 0$, that is, $x^3 = 0$, then only $x_1 = x_2 = x_3 = 0$ can be a solution, because for $x \neq 0$ one also has $x^3 \neq 0$, and vice versa.

Cardano's formula. This formula to calculate the roots of the cubic equation is obtained in two steps. First, the normal form $x^3 + rx^2 + sx + t = 0$ is brought by the substitution $x = y - (r/3)$ to the *reduced form* in which there is no quadratic term:

Reduced form of the cubic equation	$y^3 + py + q = 0$

Here the abbreviations $p = s - r^2/3$, $q = 2r^3/27 - sr/3 + t$ are used; for example, the reduction of $x^3 - 9x^2 + 33x - 65 = 0$ leads to $y^3 + 6y - 20 = 0$.

Next, the required solution y is put into two parts u and v, which will be determined separately. One sets tentatively $y = u + v$ and obtains $(u + v)^3 + p(u + v) + q = 0$ or $u^3 + v^3 + q + (u + v)(3uv + p) = 0$. One now has an equation in the two variables u and v. Therefore, one is free to use an additional condition on the connection between u and v. One chooses it so that the factor $3uv + p$, and hence the last summand, vanishes; $3uv + p = 0$. This yields a system of equations for the variables u and v.

$u^3 + v^3 = -q$	squaring →	$u^6 + 2u^3v^3 + v^6 = q^2$	+
$uv = -p/3$	four times the third power →	$4u^3v^3 = -4(p/3)^3$	−
		$(u^3 - v^3)^2 = q^2 + 4(p/3)^3$	
The system of equations obtained yields →		$u^3 - v^3 = \pm\sqrt{[q^2 + 4(p/3)^3]}$	
		$u^3 + v^3 = -q.$	

$$u^3 = -q/2 \pm \sqrt{[(q/2)^2 + (p/3)^3]} \text{ and } v^3 = -q/2 \mp \sqrt{[(q/2)^2 + (p/3)^3]}.$$

By interchanging the upper signs with the lower signs in the roots, u^3 goes over into v^3; but when u and v are interchanged, the equations $u^3 + v^3 + q = 0$ and $uv = -p/3$ remain unchanged. Therefore it is sufficient to consider only one of the pairs of signs, say the upper. Every cube root of a complex number has three values; apart from one solution x_1 there are the other solutions $\omega_2 x_1$ and $\omega_3 x_1$ in which ω_2 and ω_3 are the complex cube roots of unity. Consequently, for u and v one obtains the values

$$u_1 = \sqrt[3]{\{-q/2 + \sqrt{[(q/2)^2 + (p/3)^3]}\}}, \quad u_2 = u_1\omega_2, \quad u_3 = u_1\omega_3,$$
$$v_1 = \sqrt[3]{\{-q/2 - \sqrt{[(q/2)^2 + (p/3)^3]}\}}, \quad v_2 = v_1\omega_2, \quad v_3 = v_1\omega_3.$$

For $y = u_i + v_j$ one would obtain 9 solutions ($i = 1, 2, 3; j = 1, 2, 3$) of the cubic equation. But the number of solutions reduces to the following three; $y_1 = u_1 + v_1, y_2 = u_2 + v_3, y_3 = u_3 + v_2$, because the additional condition $u_i v_j = -p/3$ is satisfied only for u_1v_1, u_2v_3 and u_3v_2, since $\omega_2\omega_3 = (-1/2 + (i/2)\sqrt{3})(-1/2 - (i/2)\sqrt{3}) = 1/4 + 3/4 = 1$. Under the assumption that the radicand of the square root is non-negative, $(q/2)^2 + (p/3)^3 \geq 0$, the solution y_1 is real, while y_2 and y_3 are conjugate complex, as the calculation shows:

$$y_2 = u_1\omega_2 + v_1\omega_3 = -(u_1 + v_1)/2 + [(u_1 - v_1)/2] \cdot i\sqrt{3},$$
$$y_3 = u_1\omega_3 + v_1\omega_2 = -(u_1 + v_1)/2 - [(u_1 - v_1)/2] \cdot i\sqrt{3}.$$

Reduced form of the cubic equation $y^3 + py + q = 0$	Cardano's formula $y_1 = \sqrt[3]{\{-q/2 + \sqrt{[q^2/4 + p^3/27]}\}} + \sqrt[3]{\{-q/2 - \sqrt{[q^2/4 + p^3/27]}\}}$

Example: $y^3 - 15y - 126 = 0$. The equation is in reduced form. Here $p = -15$ and $q = -126$. Substitution into Cardano's formula gives

$$y_1 = \sqrt[3]{[63 + \sqrt{(63^2 - 5^3)}]} + \sqrt[3]{[63 - \sqrt{(63^2 - 5^3)}]} = \sqrt[3]{125} + \sqrt[3]{1} = 5 + 1 = 6,$$
$$y_2 = -[(5 + 1)/2] + [(5 - 1)/2] \cdot i\sqrt{3} = -3 + 2i\sqrt{3},$$
$$y_3 = -[(5 + 1)/2] - [(5 - 1)/2] \cdot i\sqrt{3} = -3 - 2i\sqrt{3}.$$

Hence $S = \{6, -3 + 2i\sqrt{3}, -3 - 2i\sqrt{3}\}$.

Casus irreducibilis, trigonometric solution. Apparently the solution of the cubic equation becomes particularly difficult when the radicand $(q/2)^2 + (p/3)^3$ of the square root is negative. Then one has to extract the cube root of complex numbers. On the other hand, a cubic equation always has at least one real solution. For a long time the mathematicians in the 15th and 16th centuries did not succeed in producing this real solution and called this case, which went beyond their means, as 'not reducible', as casus irreducibilis. VIETA succeeded in obtaining the solution, around 1600, by means of trigonometry. In fact, it turned out that in this apparently so complicated case all the *three solutions are real.*

Since $(q/2)^2 + (p/3)^3 < 0$, one must have $p < 0$; setting $p = -p'$ one has p' positive, and the reduced equation $y^3 + py + q = 0$ goes over into $y^3 - p'y + q = 0$, where $p'^3/27 - q^2/4 > 0$. The radicand of the cube root of u_1 or v_1 is then:

$$-q/2 \pm \sqrt{(-p'^3/27 + q^2/4)} = -q/2 \pm \sqrt{-(p'^3/27 - q^2/4)} = -q/2 \pm \mathrm{i}\sqrt{(p'^3/27 - q^2/4)}.$$

This complex value can be written in trigonometric form: $-q/2 \pm \mathrm{i}\sqrt{(p'^3/27 - q^2/4)} = r(\cos\varphi \pm \mathrm{i}\sin\varphi)$, where

$$r = \sqrt{(p'^3/27)}, \quad \cos\varphi = -q/2 : \sqrt{(p'^3/27)}, \quad \sin\varphi = \sqrt{(p'^3/27 - q^2/4)} : \sqrt{(p'^3/27)}.$$

By de Moivre's theorem one obtains for u_1 or v_1: $\sqrt[3]{r}(\cos\varphi/3 \pm \mathrm{i}\sin\varphi/3)$
and so: $y_1 = u_1 + v_1 = \sqrt[3]{r}[\cos\varphi/3 + \mathrm{i}\sin\varphi/3 + \cos\varphi/3 - \mathrm{i}\sin\varphi/3] = 2\sqrt[3]{r}\cdot\cos\varphi/3$.
Since on account of the periodicity of the cosine function the angle can also have the value $\varphi + 360°$ or $\varphi + 720°$, the other two solutions are:

$$y_2 = 2\sqrt[3]{r}\cdot\cos(\varphi/3 + 120°) \text{ and } y_3 = 2\sqrt[3]{r}\cdot\cos(\varphi/3 + 240°).$$

Cubic equation	$Ax^3 + Bx^2 + Cx + D = 0$	$x \in \mathbf{C};\ A, B, C, D \in \mathbf{R};\ A \neq 0$
normal form	$x^3 + rx^2 + sx + t = 0$	$r = B/A;\ s = C/A;\ t = D/A$
reduced form	$y^3 + py + q = 0$	$p = s - r^2/3$ $q = 2r^3/27 - rs/3 + t$
Cardano's formula	$(q/2)^2 + (p/3)^3 > 0$ one real solution and two conjugate complex solutions; for $(q/2)^2 + (p/3)^3 = 0$ three real solutions of which two coincide	$u_1 = \sqrt[3]{[-q/2 + \sqrt{(q^2/4 + p^3/27)}]}$ $v_1 = \sqrt[3]{[-q/2 - \sqrt{(q^2/4 + p^3/27)}]}$ $y_1 = u_1 + v_1$ $y_{2,3} = -(u_1 + v_1)/2$ $\pm [(u_1 - v_1)/2]\cdot\mathrm{i}\sqrt{3}$
casus irreducibilis	$(q/2)^2 + (p/3)^3 < 0$ three distinct real solutions	$r = \sqrt{(-p^3/27)}$ $\cos\varphi = -(q/2) : \sqrt{(-p^3/27)}$ $y_1 = 2\sqrt[3]{r}\cos\varphi/3$ $y_2 = 2\sqrt[3]{r}\cos(\varphi/3 + 120°)$ $y_3 = 2\sqrt[3]{r}\cos(\varphi/3 + 240°)$

Example 1: In the equation $y^3 - 981y - 11340 = 0$ the conditions of the casus irreducibilis are satisfied, and one obtains $r = \sqrt{327^3}$, $\cos\varphi = 5670/\sqrt{327^3}$. The logarithmic calculation yields the (approximate) value $\varphi = 16°\,30'$, hence $\varphi/3 = 5°\,30'$.

By logarithmic evaluation of the formulae for y_1, y_2, y_3 one finds $y_1 \approx 36$, $y_2 \approx -21$, $y_3 \approx -15$. The check shows that equality holds. Therefore $S = \{-21, -15, 36\}$.

Example 2: The axial cross-section of a normed glass funnel is an equilateral triangle (Fig.). What is the width d of the funnel if its volume is $V = 765\ \text{cm}^3$? – Since the width d is one side of the axial cross section, the height of the funnel is $h = (d/2)\sqrt{3}$ and the radius of the base $r = d/2$. From the volume formula for the cone $V = (1/3)\pi r^2 \cdot h$ it follows that

$765\ \text{cm}^3 = (1/3)\pi\cdot(d/2)^2\cdot(d/2)\sqrt{3}$ or $d^3 = 765\cdot 24/(\pi\sqrt{3})\ \text{cm}^3$.

Of the three values for d here only the real value is meaningful. One obtains for the width of the funnel $d \approx 15$ cm.

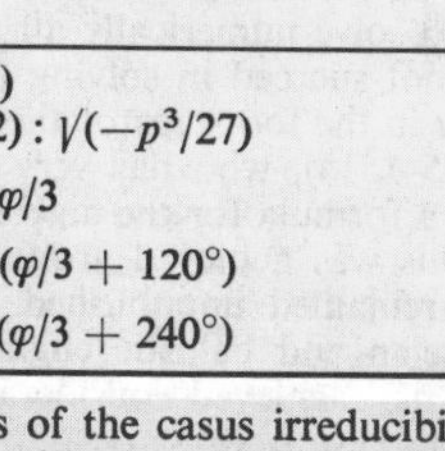

4.4-1 Normed glass funnel

Graphical solution of a cubic equation. From the cubic equation $Ax^3 + Bx^2 + Cx + D = 0$ one goes over to the function of the third degree with the equation $y = Ax^3 + Bx^2 + Cx + D$. The graph of this function intersects the x-axis in points whose abscissae give the solutions of the cubic equation. One obtains approximate solutions, which can be improved to any required degree, for example, by Newton's method. As a rule one is satisfied in finding one solution x_1 graphically, and then to divide the given cubic polynomial by the linear factor $(x - x_1)$, so that one obtains a quadratic equation which can easily be solved.

Example: From the equation $8x^3 - 20x^2 - 2x + 5 = 0$ one goes over to the function with the equation $y = 8x^3 - 20x^2 - 2x + 5$ (Fig.). Its graph can be drawn by means of the table

x ...	-2	-1	0	1	2	3	...
y ...	-135	-21	5	-9	-15	35	...

in such a good approximation that one would expect zeros to lie at $x_1 = -1/2$, $x_2 = +1/2$, $x_3 = +5/2$. Here the choice of different units on the y-axis and on the x-axis has no influence on the position of the zeros. Incidentally, the table indicates even before the graph is drawn, by the change of signs of the ordinates, that the zeros must lie between -1 and 0, between 0 and 1, and between 2 and 3.

By substituting the values into the equation one finds that the expected zeros are, in fact, solutions. But it would be sufficient to check for $x_2 = 1/2$, say, and after division of polynomials

$(8x^3 - 20x^2 - 2x + 5) : (x - 1/2) = 8x^2 - 16x - 10$

to solve the quadratic equation $8x^2 - 16x - 10 = 0$ whose solutions are $x_{1,3} = 1 \mp 3/2$. This is a verification of the solutions obtained graphically. Here $S = \{-1/2, +1/2, +5/2\}$.

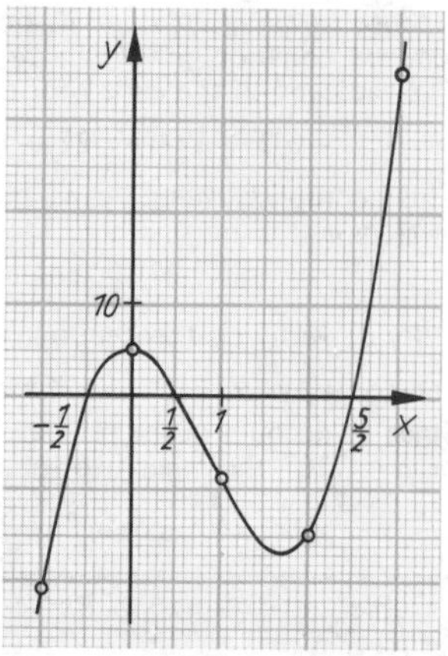

4.4-2 Graphical solution of the cubic equation $8x^3 - 20x^2 - 2x + 5 = 0$

Historical remarks. Simple cubic equations already occur in the ancient Greek, Hindu and Arabic mathematics. Since the Greek mathematicians treated algebraic problems by the methods of geometry, they were facing fundamental difficulties in the treatment of cubic equations.

The Hellenistic technologist and mathematician HERON of Alexandria (about 100 A. D.) took an important step forward in the treatment of cubic equations. By reverting to older Babylonian and Egyptian approximation methods for the numerical extraction of roots he succeeded in solving pure cubic equations. The real progress in the numerical treatment and the commencing algebraization of the computational steps is due to Hindu and above all to Arabic mathematicians. They could solve numerically all types of quadratic and the simplest types of cubic equations, but they did not succeed in solving the general equation. The European mathematicians followed immediately in the footsteps of the Arabs, also in the numerical treatment of equations. But Luca PACIOLI (1445–1514), who has very great merits in the development of algebra, thought it impossible to find a formula for the algebraic solution of the general equation of the third degree.

This was achieved around 1500 by Master Scipione del FERRO of Bologna (about 1465–1526), but remained unpublished. Quite independently, Niccolo TARTAGLIA (about 1500–1557), mathematician and ballistic engineer, had found the formula, which today is named after CARDANO, and had achieved considerable fame by applying it brilliantly in public problem solving contests which were customary at the time. The ambitious Professor Geronimo CARDANO (1501–1576) of Venice, who did not succeed in finding the formula for the solution, obtained it from TARTAGLIA in 1539 after years of intense pressure, but he had to swear a solemn oath to TARTAGLIA to treat it as a kind of professional secret. However, CARDANO broke his promise and included the result in his 'Artis magnae sive de regulis algebraicis' (that is, of the Great Art of the Rules of Algebra) of 1545. And since the formula appeared in print for the first time under CARDANO's name, it became known as Cardano's formula. Even Tartaglia's protest, which led to a violent quarrel, was of no avail. Incidentally, also the well-known method of universal suspension, for example, of a ship's compass, bears Cardano's name wrongly: it was in use long before him.

The quartic equations

General equation of the fourth degree	$Ax^4 + Bx^3 + Cx^2 + Dx + E = 0$; $A \neq 0$	$x \in \mathbf{C}$ $A, B, C, D, E \in \mathbf{R}$

There is also a general solution formula for the general equation of the fourth degree. However, this is much more complicated than that of the cubic equation and is therefore hardly used in the numerical determination of the solutions. Hence only a sketch of the method, without details of the calculation is given here. By the substitution $x = y - a/4$ one obtains from the *normal form* $x^4 + ax^3 + bx^2 + cx + d = 0$ the
reduced equation $y^4 + py^2 + qy + r = 0$ with new coefficients p, q, r.
Its four solutions y_1, y_2, y_3, y_4 with $2y_1 = \sqrt{z_1} + \sqrt{z_2} + \sqrt{z_3}$; $2y_2 = \sqrt{z_1} - \sqrt{z_2} - \sqrt{z_3}$;
$2y_3 = -\sqrt{z_1} + \sqrt{z_2} - \sqrt{z_3}$; $2y_4 = -\sqrt{z_1} - \sqrt{z_2} + \sqrt{z_3}$

can be obtained from the three solutions z_1, z_2, z_3 of the *cubic resolvant* of the given quartic equation:

$$z^3 + 2pz^2 + (p^2 - 4r)\,z - q^2 = 0.$$

An additional condition is that the product of the three solutions $z_1 z_2 z_3 = q^2$ must always be positive. For the solutions of the reduced quartic equation in the domain of complex numbers the following *three cases* can arise:

solutions z_1, z_2, z_3 of the cubic resolvant	solutions y_1, y_2, y_3, y_4 of the quartic equation
all real and positive	four real values
one positive, two negative	two pairs of conjugate complex values
one real, two conjugate complex	two real values and one pair of conjugate complex values

The biquadratic equation. A special case of the quartic equation occurs frequently and is easy to treat. This is the biquadratic equation $x^4 + px^2 + q = 0$. It is distinguished by the fact that the variable x only occurs to even powers. The equation can therefore be regarded as a quadratic in x^2: hence its name. For $y = x^2$ one obtains $y^2 + py + q = 0$. One solves the quadratic equation for y. By a subsequent solution of the equation $x^2 = y$ one obtains the solutions of the biquadratic equation.

Example: $x^4 - 29x^2 + 100 = 0.$

$y_1 = 25,\ y_2 = 4;\ x_1 = +5,\ x_2 = -5,\ x_3 = 2,\ x_4 = -2;$
$S = \{-5, -2, +2, +5\}.$

Historical remark. The formula for the solution of the general quartic equation was found by L. FERRARI (1522–1565), a pupil and collaborator of CARDANO. The formula was taken by CARDANO into his 'Ars Magna'.

4.5. General theorems

Fundamental theorem of algebra. The conjecture that in the domain of complex numbers an equation of degree n always has n solutions was made already by GIRARD (1595–1632). Attempts to prove this were made later by DESCARTES, Jean D'ALEMBERT and others, but it was only GAUSS who succeeded in 1799 in giving a rigourous proof without gaps, which formed the main topic of his dissertation; later GAUSS found other independent proofs for the fact that an algebraic equation always has a solution.

Fundamental theorem of algebra: Every equation of degree n

$$x^n + a_1 x^{n-1} + a_2 x^{n-2} + \dots + a_{n-1} x_n + a_n = 0,$$

in which the a_i ($i = 1, 2, \dots, n$) are real or complex numbers, has at least one solution in the domain of complex numbers.

Factorization. If a solution of the equation $x^n + a_1 x^{n-1} + a_2 x^{n-2} + \dots + a_{n-1} x + a_n = 0$, which is guaranteed by the fundamental theorem, is denoted by x_1 and if one subtracts the equation $x_1^n + a_1 x_1^{n-1} + a_2 x_1^{n-2} + \cdots + a_{n-1} x_1 + a_n = 0$, which is obtained from the given solution by substituting $x = x_1$, one obtains $(x^n - x_1^n) + a_1(x^{n-1} - x_1^{n-1}) + \cdots + a_{n-1}(x - x_1) = 0$. Every term contains the factor $(x - x_1)$, hence it follows by factorization that

$$(x - x_1) \cdot [x^{n-1} + \cdots + a_{n-1}] = 0.$$

The expression in the square bracket on the left-hand side is a polynomial of degree $n - 1$. By the fundamental theorem this also has a solution. If it is denoted by x_2, then the factor $(x - x_2)$ can be split off. One obtains $(x - x_1)(x - x_2)[x^{n-2} + \cdots + a_{n-2}] = 0$. If this method is continued, one obtains finally, the product representation.

Product representation of the equation of degree n	$x^n + a_1 x^{n-1} + \cdots + a_{n-1} x + a_n = (x - x_1)(x - x_2) \cdots (x - x_n) = 0$	$x \in \mathbf{C}$

Number of solutions. An important theorem follows immediately from the product representation: an equation of degree n in one variable always has exactly n solutions. These need not all be distinct from each other.

If a solution occurs 2, 3, …, k times, one speaks of a 2-, 3-, …, k-fold solution or *root*. If the coefficients of the equation are real and if the equation has a complex solution $a + ib$, then the conjugate complex number $a - ib$ is also a solution of the equation.

Vieta's root theorem. If one multiplies the right-hand side of the product representation and orders by equal powers of x, a comparison of coefficients leads to Vieta's root theorem.

Vieta's root theorem	$x_1 + x_2 + \cdots + x_n = -a_1$ $x_1x_2 + x_1x_3 + x_2x_3 + \cdots + x_{n-1}x_n = a_2$ $x_1x_2x_3 + x_1x_2x_4 + \cdots + x_{n-2}x_{n-1}x_n = -a_3$ $\vdots$ $x_1x_2x_3 \ldots x_n = (-1)^n a_n$

In addition the following theorem holds:

If an equation of degree n with integer coefficients in its normal form has an integral solution, this is a divisor of the absolute term.

For quadratic and cubic equations Vieta's root theorem takes the following form:

$x^2 + px + q = 0$	$x_1 + x_2 = -p$ $x_1x_2 = q$

$x^3 + rx^2 + sx + t = 0$	$x_1 + x_2 + x_3 = -r$ $x_1x_2 + x_2x_3 + x_3x_1 = s$ $x_1x_2x_3 = -t$

Solubility by radicals. The fundamental theorem of algebra guarantees the existence of the roots of the equation $x^n + a_1x^{n-1} + a_2x^{n-2} + \cdots + a_{n-1}x + a_n = 0$ *for all degrees*. For $n = 2, 3, 4$ a general formula for this solution can be written down. For $n = 3$ it consists in a succession of roots, one contained in the other; the solution is of the type $\sqrt[3]{(a + \sqrt{b})}$; for $n = 4$ it is of the type $\sqrt{\{a + \sqrt[3]{[b + \sqrt{(c + \sqrt{d})}]}\}}$. By a *radical* one understands an expression formed by superimposing roots whose exponents are positive integers. Using this notion one can say:

Algebraic equations of up to the fourth degree are soluble by radicals.

Evidently there is an unexhaustable variety of radicals, and one should think that somehow by a combination of superimposed roots the solutions of an equation of degree 5 could be obtained. But this is not the case. On the contrary: it is impossible for $n > 4$ to solve the general algebraic equation of degree n by radicals.

Historical remarks. After the solution formulae for the cubic and quartic equation had been found during the Renaissance, the mathematicians of the 17th and 18th centuries searched with great tenacity for corresponding solution formulae for equations of degree 5 and higher. In fact, some mathematicians, among them von TSCHIRNHAUS (1651–1708), believed that they had proved the possibility of a solution by radicals.

But gradually it was recognized that a solution of the general equation of higher degrees by radicals might be impossible; this was an opinion expressed by LAGRANGE and by GAUSS. After an attempted proof (1799) by RUFFINI, which contained gaps, Niels Henrik ABEL (1802–1829), a brilliant mathematical genius who died young from tuberculosis, succeeded in proving that the general equation of degree 5, and hence also equations of higher degrees, are not soluble by radicals. One of the reasons why it was so difficult to get an understanding of the solubility situations of equations of higher degree was the fact that special equations of higher degree can very well be soluble by radicals. A precise and complete survey of all equations of all degrees that are soluble by radicals is given in Galois' theory. Evariste GALOIS (1811–1832) started out from the results obtained by GAUSS on the problem of cyclotomy (division of the circle). He was a man of genius and an ardent republican. Like PUSHKIN in Tsarist Russia, he was mortally wounded in a duel in which his opponent was possibly an *agent provocateur* of the reactionary monarchist police.

In Galois theory a group is assigned to every equation; its structure gives information on whether the equation is soluble by radicals (see Chapter 16.).

4.6. Systems of non-linear equations

Certain types of systems of non-linear equations occur rather frequently, for example, in coordinate geometry or in connection with systems of ordinary differential equations. Here a few cases are selected from the multitude of systems of non-linear equations. A systematic treatment is not possible. In what follows the fundamental domain for all variables is $\mathbf{R}$.

One linear and one quadratic equation. By the method of substitution the system is easy to solve. It occurs, for example, when the points of intersection of a conic with a straight line are to be determined.

Example:

$$\begin{array}{l} x^2 + y^2 + 4x - 1 = 0 \\ x + y = -1 \end{array} \quad x = -y - 1 \quad (-y - 1)^2 + y^2 + 4(-y - 1) - 1 = 0 \quad y^2 - y - 2 = 0 \quad y_1 = 2;\ y_2 = -1 \quad x_1 = -3;\ x_2 = 0$$

The check confirms that these values are solutions. $S = \{(-3, 2), (0, -1)\}$.

Two quadratic equations. This problem occurs when two conics are intersected. If there are no mixed quadratic terms and if in the two equations the corresponding coefficients of the pure quadratic equations are equal apart from a constant factor $k \neq 0$, then after multiplication by $1/k$ and subtraction can be achieved the quadratic terms are absent and a linear equation results with the help of which one variable can be eliminated by substituting into one of the quadratic equations.

Example 1:

$$\begin{array}{l} x^2 + y^2 - 18x - 18y + 112 = 0 \\ x^2/2 + y^2/2 - 11x + 5y - 52 = 0 \quad \cdot 2 \end{array} \quad \begin{array}{l} x^2 + y^2 - 18x - 18y + 112 = 0 \\ x^2 + y^2 - 22x + 10y - 104 = 0 \\ \hline 4x - 28y + 216 = 0 \end{array}$$

$$x = 7y - 54 \qquad y^2 - 18y + 80 = 0 \qquad y_1 = 10;\ y_2 = 8 \qquad x_1 = 16;\ x_2 = 2$$

The check confirms that $S = \{(2, 8), (16, 10)\}$ is the solution set of the system.

Example 2:

$$\begin{array}{ll} (1) & x^2 + y^2 = a \\ (2) & x \cdot y = b \end{array}$$

Method I.

$$\begin{array}{ll} (x + y)^2 = a + 2b & (1) + 2 \cdot (2) \\ (x - y)^2 = a - 2b & (1) - 2 \cdot (2) \end{array}$$

$$x + y = \pm\sqrt{(a + 2b)}$$
$$x - y = \pm\sqrt{(a - 2b)}$$
$$y_{1,2} = \pm(1/2)\,[\sqrt{(a + 2b)} - \sqrt{(a - 2b)}]$$
$$x_{1,2} = \pm(1/2)\,[\sqrt{(a + 2b)} + \sqrt{(a - 2b)}]$$
$$y_{1,2}^2 = (1/4)\,[a + 2b + a - 2b \pm 2\sqrt{(a^2 - 4b^2)}]$$
$$x_{1,2}^2 = (1/4)\,[a + 2b + a - 2b \mp 2\sqrt{(a^2 - 4b^2)}]$$

Method II.

$$\begin{array}{ll} (1\,a) & b^2/y^2 + y^2 = a \\ (2\,a) & x = b/y \end{array}$$

$$y^4 - ay^2 + b^2 = 0$$

biquadratic equation

$$y_{1,2}^2 = a/2 \pm (1/2)\sqrt{(a^2 - 4b^2)}$$
$$x_{1,2}^2 = a/2 \mp (1/2)\sqrt{(a^2 - 4b^2)}$$

One sees that the two methods lead to the same result.

Three quadratic equations in three variables. A special system of equations of this kind arises by the problem in coordinate geometry of finding the equation of a circle through three points say $P_1 = (-8, 12)$, $P_2 = (-4, 4)$, $P_3 = (9, -5)$. Required are the coordinates of the centre $C(a, b)$ and the radius r of the circle. One obtains the system

$$\left| \begin{array}{l} (-8 - a)^2 + (12 - b)^2 = r^2, \\ (-4 - a)^2 + (4 - b)^2 = r^2, \\ (9 - a)^2 + (-5 - b)^2 = r^2 \end{array} \right|$$

for the variables a, b and r. Evaluating the squares and subtracting say, the second equation from the first, and the third from the first, two linear equations in the variables a and b are obtained, which lead to the values $a = 16$ and $b = 19$. Substituting the calculated values for a and b into one of the original equations one obtains a pure quadratic equation for r. Its positive solution is the required radius, in the present case $r = 25$.

4.7. Algebraic inequalities

The notion of an inequality, like that of an equation, is defined by means of the concept of an expression. If two expressions E_1 and E_2 are linked by one of the relation symbols $>$ 'greater than', $\geqslant$ 'greater than or equal to', $<$ 'less than', $\leqslant$ 'less than or equal to', or $\neq$ 'unequal to', then there arises one of the inequalities $E_1 > E_2$, $E_1 \geqslant E_2$, $E_1 < E_2$, $E_1 \leqslant E_2$, or $E_1 \neq E_2$; for example, $3x < 5$, $a^2 \geqslant 9$, $2 \leqslant 8$, $x + y > 6$, $1/2 \neq 1/3$ are inequalities. The only inequalities to be treated in what follows are of the forms $E_1 > E_2$ and $E_1 < E_2$.

Just as for equations, so one distinguishes among inequalities between those *without variables*, which are propositions on inequality that can be true or false, and those that are *predicates* on inequality; for example, $2 < 8$ and $1/2 > 1/3$ are propositions, while $a^2 < 9$ and $x + y > 6$ are predicates.

Solution set and solution of an inequality. Every number from the domain of definition which on substitution for the variable makes an inequality with one variable into a true proposition is called a *solution* of the inequality. Here the domain of definition of an inequality is defined by analogy to that of an equation. If the inequality contains two, three, ..., n variables, then a solution is an ordered pair, triple, ..., n-tuple of numbers. The *solution set* S is the set of *all* solutions of an inequality relative to its domain of definition. For example, the inequality $x < 4$ for the set $\mathbf{N}$ of natural numbers has the solutions 0, 1, 2, 3, that is, $S = \{0, 1, 2, 3\}$; but for $x \in \mathbf{Z}$ one has $S = \{..., -3, -2, -1, 0, 1, 2, 3\}$. For the inequality $x + y < 2$, $x \in \mathbf{N}$, $y \in \mathbf{N}$, the solution set is $S = \{(0, 0), (1, 0), (0, 1)\}$; for $x \in \mathbf{Z}$, $y \in \mathbf{Z}$ the solution set of this inequality consists of infinitely many solutions, namely all ordered pairs of integers for which $x + y < 2$, for example, $(-5, 1)$ or $(1, -4)$.

Consistent, inconsistent, and universally valid inequalities. One speaks of a consistent, respectively, inconsistent inequality according as the inequality with variables has, or does not have, solutions relative to its domain of definition.

Consistent inequalities	Inconsistent inequalities
$x < 0$ for $x \in \mathbf{Z}$: $S = \{...; -3, -2, -1\}$	$x < 0$ for $x \in \mathbf{N}$: $S = \emptyset$
$a^2 > 0$ for $a \in \mathbf{N}$: $S = \{1, 2, 3, 4, 5, ...\}$	$a^2 < 0$ for $a \in \mathbf{N}$: $S = \emptyset$
$2x > 3x$ for $x \in \mathbf{R}$: $S = \{x \mid x \in \mathbf{R}$ and $x < 0\}$	$2x > 3x$ for $x \in \mathbf{N}$: $S = \emptyset$
$y + 3 < y + 4$ for $y \in \mathbf{R}$: $S = \mathbf{R}$	$y + 3 < y + 3$ for $y \in \mathbf{R}$: $S = \emptyset$

Here the inequality $y + 3 < y + 4$ for $y \in \mathbf{R}$ is not only consistent, but even *universally valid*, because all $y \in \mathbf{R}$ are solutions. A consistent equation with n variables is called *universally valid* if all ordered n-tuples of numbers from the domain of definition are solutions of the inequality; for example, the so-called triangle inequality $|a + b| \leqslant |a| + |b|$ is satisfied for all pairs of real numbers, hence it is universally valid for $a \in \mathbf{R}$, $b \in \mathbf{R}$.

Equivalent inequalities. Two inequalities with variables are said to be equivalent if they have the same domain of definition and the same solution sets; otherwise the inequalities are called inequivalent; for example, $x + 4 < 7$ and $x < 3$ are equivalent relative to the set $\mathbf{N}$ of natural numbers, for the solution set of both inequalities is $S_1 = \{0, 1, 2\}$. Similarly $-2a > 4$ and $a < -2$ for $a \in \mathbf{Z}$ are equivalent inequalities, because they both have $S_1 = S_2 = \{..., -6, -5, -4, -3\} = S$. However, the inequalities $y > 0$ and $y > -2$ are equivalent over $\mathbf{N}$, but not over $\mathbf{Z}$. Transformations carrying an inequality into an equivalent one are called *equivalent transformations*. They are based on the fundamental laws of arithmetic, especially on the monotony properties of real numbers.

Propositions on equivalent transformations of inequalities with variables. The following inequalities are equivalent to $E_1 < E_2$:
1. $E_1' < E_2'$, where E_1 and E_1 as well as E_2' and E_2 are equivalent expressions;
2. $E_2 > E_1$;
3. $E_1 \pm E_3 < E_2 \pm E_3$, provided that the expression E_3 is defined in the entire fundamental domain of variability;
4. $E_1 \cdot E_3 < E_2 \cdot E_3$ and E_1/E_3, E_2/E_3, provided that the expression E_3 is defined and positive in the entire fundamental domain of variability;
5. $E_1 \cdot E_3 > E_2 \cdot E_3$ and $E_1/E_3 > E_2/E_3$, provided that the expression E_3 is defined and negative in the entire domain of variability.

Solving inequalities. The task of *solving* an inequality is that of determining *all* solutions relative to given fundamental domains of variability, in other words, of finding their solution sets. As for equations, so for solving inequalities the point is to carry out a sequence of suitable equivalent transformations and to arrive finally at an inequality so simple that its solution set can be read off. Frequently, especially in estimates, one uses the *transitivity of the relations* $<$ or $>$, which make it possible to deduce from $E_1 < E_2$ and $E_2 < E_3$ that $E_1 < E_3$. For a linear inequality with one variable there is a method of *solution* by utilizing the transformation theorems.

$2x + 2 + 3x < 3x - 8 + 4; \quad x \in \mathbf{R}$

Proposition 1

$5x + 2 < 3x - 4$

$| -3x - 2$ Proposition 3

$5x + 2 - 3x - 2 < 3x - 4 - 3x - 2$

Proposition 1

$2x < -6$

Proposition 4 $| :2$

$2x/2 < -6/2$

Proposition 1

$x < -3$

The solution set consists of all real numbers that are less than -3: $S = \{x \mid x \in \mathbf{R}$ and $x < -3\}$. This solution set can be illustrated graphically on the number line (Fig.).

Check: For inequalities, unlike for equations, it is not possible, in general, to check the correctness of the calculations by substituting all solutions for the variables. But it is advisable to make tests for individual elements of the solution set, in the example above, for $-5 \in \mathbf{R}$, say:

$$\begin{aligned} 2(-5) + 2 + 3(-5) &< 3(-5) - 8 + 4 \\ -10 + 2 - 15 &< -15 - 8 + 4 \\ -23 &< -19 \quad \text{true.} \end{aligned}$$

The test can also be carried out completely by writing all elements of S in the form $x = -3 - h\ (h > 0)$, substituting in the given inequality, and checking whether the resulting inequality proposition is true for all real $h > 0$.

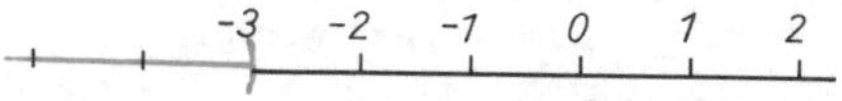

4.7-1 Graphical representation of the solution set $S = \{x \mid x \in \mathbf{R} \text{ and } x < -3\}$

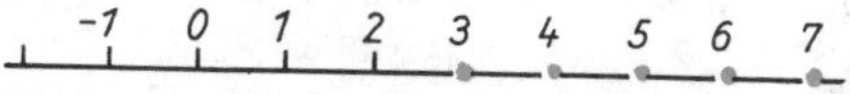

4.7-2 Graphical representation of the solution set $S = \{a \mid a \in \mathbf{N} \text{ and } a > 2\}$

Example 1:

$$\begin{aligned} 25 - 3a &< 22 - 2a;\ a \in \mathbf{N} && \mid + 2a - 25 \\ 25 - 3a + 2a - 25 &< 22 - 2a + 2a - 25 \\ -a &< -2 && \mid \cdot (-1) \\ a &> 2. \end{aligned}$$

The solution set consists of all natural numbers greater than 2: $S = \{a \mid a \in \mathbf{N} \text{ and } a > 2\} = \{3, 4, 5, \ldots\}$ (Fig.).

Example 2: $y + x < 4;\ y \in \mathbf{N},\ y \in \mathbf{N} \quad \mid -x$

$$y < -x + 4.$$

The solution set is $S = \{(0, 0), (0, 1), (0, 2), (0, 3), (1, 0), (1, 1), (1, 2), (2, 0), (2, 1), (3, 0)\}$ (Fig.).

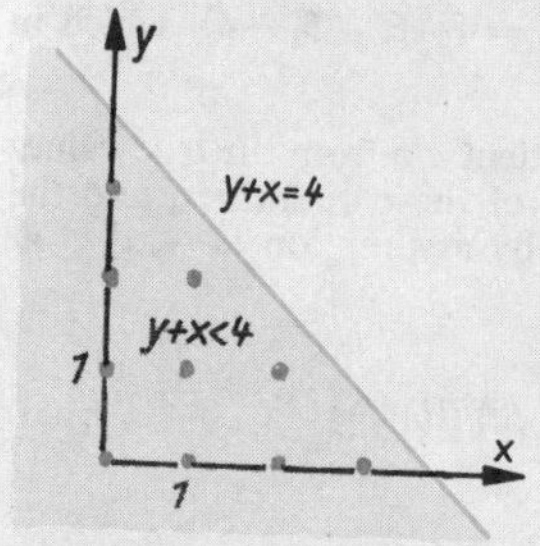

If the domain of variability for x and y is chosen to be $\mathbf{R}$, then the coordinate pairs of all points of the half-plane below the straight line with the equation $y = -x + 4$ are solutions of the inequality.

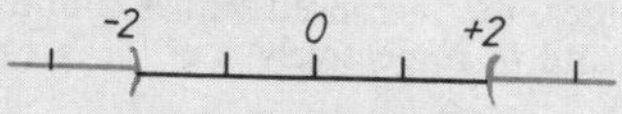

4.7-4 Graphical representation of the solution set of the inequality $x^2 - 4 > 0;\ x \in \mathbf{R}$

4.7-3 Graphical representation of the solution set of the inequality $x + y < 4$ for $x \in \mathbf{N},\ y \in \mathbf{N}$ and for $x \in \mathbf{R},\ y \in \mathbf{R}$

Example 3:

$$\begin{aligned} x^2 - 4 &> 0;\ x \in \mathbf{R}, \\ (x - 2)(x + 2) &> 0. \end{aligned}$$

A product is positive if and only if both factors have the same sign. This leads to two cases:

First case:

$$\begin{aligned} x - 2 &> 0 \quad \textit{and} \quad x + 2 > 0 \\ x &> 2 \quad \textit{and} \quad x > -2 \\ x &> 2 \end{aligned}$$

Second case:

$$\begin{aligned} x - 2 &< 0 \quad \textit{and} \quad x + 2 < 0 \\ x &< 2 \quad \textit{and} \quad x < -2 \\ x &< -2 \end{aligned}$$

The solution set therefore consists of all real numbers that are greater than 2 or less than -2 (Fig.).

Example 4: $x^2 - 4 < 0;\ x \in \mathbf{R}$,

$$(x - 2)(x + 2) < 0.$$

A product of two factors is negative if and only if the factors have opposite sign. This leads to two cases:

First case:

$$\begin{aligned} x - 2 &< 0 \quad \textit{and} \quad x + 2 > 0 \\ x &< 2 \quad \textit{and} \quad x > -2 \\ -2 &< x < 2 \end{aligned}$$

Second case:

$$\begin{aligned} x - 2 &> 0 \quad \textit{and} \quad x + 2 < 0 \\ x &> 2 \quad \textit{and} \quad x < -2 \quad \text{inconsistent} \end{aligned}$$

The solution set therefore consists of all real numbers in the interval between -2 and $+2$: $S = \{x \mid x \in \mathbf{R} \text{ and } -2 < x < 2\}$.

Example 5: $(x + 2)/(x - 1) > 4;\ x \in \mathbf{R}$. The domain of definition of the inequality is the set of all real numbers $x \neq 1$. In fractional inequalities one has to make case distinctions:

First case:

$(x+2)/(x-1) > 4$ *and* $x-1>0$
$x+2 > 4(x-1)$ *and* $x>1$
$x+2 > 4x-4$ *and* $x>1$
$6 > 3x$ *and* $x>1$
$x < 2$ *and* $x>1$

$S_1 = \{x \mid x \in \mathbf{R} \text{ and } 1 < x < 2\}$

Second case:

$(x+2)/(x-1) > 4$ *and* $x-1<0$
$x+2 < 4(x-1)$ *and* $x<1$
$x+2 < 4x-4$ *and* $x<1$
$6 < 3x$ *and* $x<1$
$x > 2$ *and* $x<1$

inconsistent $S_2 = \emptyset$.

The solution set of the given inequality is $S = S_1 \cup S_2 = S_1$, it consists of all real numbers x in the interval $1 < x < 2$.

Example 6: $|a+5| < 2$; $a \in \mathbf{Z}$. By definition of the absolute value: $|a+5| = a+5$ for $a+5 \geqslant 0$, or $|a+5| = -(a+5)$ for $a+5 \leqslant 0$. One has therefore to distinguish between two cases:

First case:

$a+5 < 2$ *and* $a+5 \geqslant 0$
$a < -3$ *and* $a \geqslant -5$
$S_1 = \{-5, -4\}$

Second case:

$-(a+5) < 2$ *and* $a+5 \leqslant 0$
$-a-5 < 2$ *and* $a \leqslant -5$
$-a < 7 \quad | \cdot(-1)$ *and* $a \leqslant -5$
$a > -7$ *and* $a \leqslant -5$
$S_2 = \{-6, -5\}$.

This leads to the solution set for the initial inequality $S = S_1 \cup S_2 = \{-6, -5, -4\}$, which in this case can easily be checked by substitution.

Example 7: It is required to compute the maximal error of the quotient a/b from the true values a and b of certain physical quantities, the measured values α and β of these quantities, and the measuring errors ε_1 and ε_2 for a and b, respectively. Let $|\beta| > \varepsilon_2$; by assumption $|a-\alpha| < \varepsilon_1$ and $|b-\beta| < \varepsilon_2$. Then:

$$a/b - \alpha/\beta = (a\beta - b\alpha)/b\beta = [\beta(a-\alpha) - \alpha(b-\beta)]/b\beta$$

$$|a/b - \alpha/\beta| = |[\beta(a-\alpha) - \alpha(b-\beta)]|/|b\beta| \leqslant [|\beta|\,|a-\alpha| + |\alpha|\,|b-\beta|]/(|b|\,|\beta|)$$
$$\leqslant [|\beta|\,\varepsilon_1 + |\alpha|\,\varepsilon_2]/(|b|\,|\beta|).$$

Since $|\beta| > \varepsilon_2$ is $|b| < \varepsilon_2 + |\beta|$, hence

$$|a/b - \alpha/\beta| \leqslant [|\beta|\,\varepsilon_1 + |\alpha|\,\varepsilon_2]\,/[|\beta|(|\beta| + \varepsilon_2)]$$

is the maximal error of the quotient.

Special inequalities. Throughout the fundamental domain of variability is the set of real numbers.

1. *Triangle inequality:* For all real numbers a and b one has $|a+b| \leqslant |a| + |b|$. By mathematical induction oneobtains: $|a_1 + a_2 + a_3 + \cdots + a_n| \leqslant |a_1| + |a_2| + |a_3| + \cdots + |a_n|$ for $n = 1, 2, 3, \ldots, a_1, a_2, \ldots, a_n \in \mathbf{R}$.
2. Similarly for all $a, b \in \mathbf{R}$ one has $||a| - |b|| \leqslant |a+b|$.
3. For all natural numbers n one always has $2^n > n$ (proof by mathematical induction).
4. For real numbers $a > 0$, $b > 0$ and $n = 1, 2, 3, \ldots$ one always has by the binomial theorem $a^n + b^n \leqslant (a+b)^n$.
5. *Bernoulli's inequality:* $(1+a)^n > 1 + na$ for natural numbers $n \geqslant 2$ and real $a \neq 0$ and $a > -1$.
6. For real numbers $a \geqslant 0$, $b \geqslant 0$ one always has $ab \leqslant [(a+b)/2]^2$ or $\sqrt{(ab)} \leqslant (a+b)/2$; in general, for $n \in \mathbf{N}$ and real numbers $a_1 \geqslant 0, \ldots, a_n \geqslant 0$, $\sqrt[n]{(a_1 a_2 \ldots a_n)} \leqslant (a_1 + a_2 + \cdots + a_n)/n$; in words: the *geometric mean* is always less than or equal to the *arithmetic mean*.
7. Between the *arithmetic mean* $A = (a_1 + a_2 + \cdots + a_n)/n$, the *geometric mean* $G = \sqrt[n]{(a_1 a_2 \ldots a_n)}$ and the *harmonic mean* $H = \dfrac{n}{1/a_1 + 1/a_2 + \cdots + 1/a_n}$ the following relation holds: $A \geqslant G \geqslant H$, where the a_i are non-negative real numbers and n is a natural number.
8. *Cauchy-Schwarz inequality:* for all real numbers $a_1, a_2, \ldots, a_n, b_1, b_2, \ldots, b_n$ one has $(a_1 b_1 + a_2 b_2 + \cdots + a_n b_n)^2 \leqslant (a_1^2 + a_2^2 + \cdots + a_n^2)(b_1^2 + b_2^2 + \cdots + b_n^2)$.

5. Functions

5.1. Basic concepts

Concept of a function

In accordance with a definition, which Euler had already given in 1749, a function is often explained as a *variable quantity that is dependent upon another variable quantity*. For many purposes such a definition of the concept of a function suffices. But in the course of the further development of mathematics it turned out to be necessary and useful to give a more general and abstract content to the concept of a function. The essence of the concept is not the dependence of quantities, by which one usually understands numbers that can be compared in a 'less than or greater than' relationship, but the fact of the *correspondence* itself, on the basis of which certain objects are regarded as being assigned to certain other objects. The concept of a function is reduced to set-theoretical definitions.

Correspondences. Every metal bar alters its length when heated. Suppose, for example, that a copper bar has a length of $l_0 = 200$ units u of length at 0°C, say centimetres or inches, then its length l at a temperature t°C is given by $l = 200(1 + 0.000\,016t)$. By this *formula* each value of t between 0°C and 100°C is made to correspond to a certain length l between $200u$ and $200.32u$. Similarly, to each quantity of a merchandise there corresponds a certain sum of money as its selling price, and to each page number in this book, a number stating how many letters occur on the page concerned.

Correspondences exist not only between numbers, but more generally between elements a in a set A and elements b in a set B; for example, each seat for a performance in a theatre corresponds to an entrance ticket or to a particular visitor. Thus, the correspondence is determined by a relation F defined on $A \cup B$ (see Chapter 14.) with domain of definition $D(F) \subseteq A$ and range $R(F) \subseteq B$. If with respect to this relation F one and only one element b of its range $R(F)$ corresponds to each element a of its domain $D(F)$, then the relation is said to be single-valued and one speaks of a *function* or *mapping* from the set A *into* the set B (Fig.). The element b of the range corresponding to the *original* element a of the domain is called the *image* of a. Consequently the function F is a set of *ordered pairs* (a, b) whose first element belongs to the domain of definition $D(F)$ and whose second element belongs to the range $R(F)$. For a mapping *of A into B* one has $D(F) = A$; that is, *every* element $a \in A$ occurs as

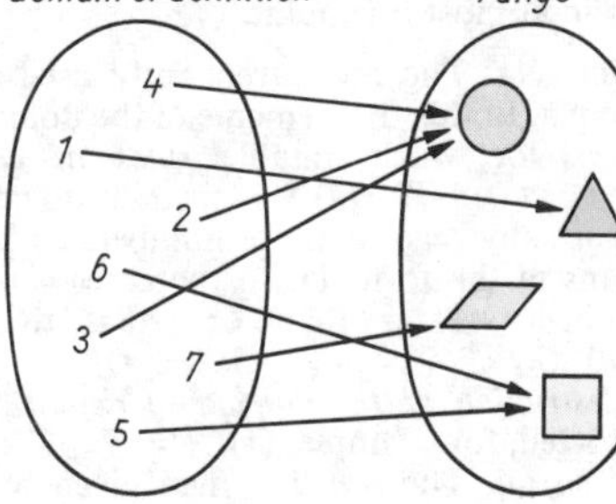

5.1-1 Graph of a function

an original element, and for a mapping *of A onto B*, in addition, every element $b \in B$ occurs as an image.

The element y that is assigned to the element x by the function f is often denoted by $f(x)$ and the correspondence is then written $x \to y = f(x)$, or more briefly $y = f(x)$. The element x is called the *argument* and the corresponding element y the *function value $f(x)$ at the point x*. The *domain of definition* (or just domain) of the function $x \to y = f(x)$ is denoted by X and the range by Y. If f is a function *from A into B*, then clearly $X \subseteq A$ and $Y \subseteq B$.

A function f is a mapping from a set A into a set B, that is, a non-empty set of ordered pairs $(x, y) \in f$ with $x \in X \subseteq A$, $y \in Y \subseteq B$ and with the property that to each $x \in X$ there corresponds *exactly one* $y \in Y$.

Representation of functions

To describe a function one must give its domain of definition and its range and the rule for the correspondence.

Graph. In the graph of a function the domain and the range are represented diagrammatically and the correspondence is indicated by arrows (Fig.). Only one directed line goes out from each element of the domain, but one or more of these lines may lead to any one element of the range.

domain of definition	1	2	3	4	5	6	7
range	△	○	○	○	□	□	▱

5.1-2 Table of values of a function

Table of values. The rule for the correspondence can also be set down in a table of values (Fig.) rather than by means of a graph. The elements of the domain are entered in the top line of the table and under each one is the corresponding element of the range. A table of values can give only finitely many ordered pairs; it is not sufficient for the complete description of an arbitrary function F.

Explanation in words. If the domain and the range of a function are not finite or are so extensive that it is no longer possible to represent the graph or the table of values on a sheet of paper, then it is sufficient to give an *exact description* of the domain and the range, together with a rule by which for every element of the domain the corresponding element of the range can be found. A function can be defined entirely without the use of mathematical symbols, by means of a sentence in everyday language; for example, a *function* is defined if to every first division match in the football league there corresponds the *quotient* of the number of *entrance tickets* sold and the number of *inhabitants* of the place where the match is played. This function can give a certain indication of the interest shown by the public in individual games. Many examples can be found of rules of correspondences that are formulated entirely or partly in words.

Example 1: To each real number x there corresponds either the value 0 or the value 1, according as x is irrational or rational. For example, $\sqrt{2} \to 0$; $(3/4) \to 1$.
Example 2: $g(x) = [x]$, where x denotes a real number and $[x]$ denotes the greatest integer that is less than, or equal to, x.

Diagram. A diagram likewise represents a function if one chooses a set of numbers of the horizontal axis as domain of definition and a set of numbers of the vertical axis as range, and assigns to the argument x of the domain precisely that value of y for which the point with the coordinates x, y is a point of the diagram. However, not every arbitrarily drawn curve in a coordinate system can be regarded as the representation of a function. The correspondence given by means of the curve must be *single-valued*. This is the case if the curve of the diagram is cut by each line parallel to the vertical axis in at most one point.

Formula. The most frequently used method of representing a function in mathematics is the formula. In this the elements of the domain and range are now only *numbers*, or at least *mathematical objects* for which suitable *rules of calculation* can be given; for example, (1) $y = 7x + 2$; (2) $y = \sqrt{(x - 4)}$; (3) $y = \sin x$. If no particular information is given about the domain of definition, one usually regards those numbers as belonging to it to which a definite value can be ascribed by means of the formula. In the cases (1) and (3) these are all real numbers, and in case (2) all real numbers greater than or equal to 4. The range is then given by: (1) $-\infty < y < +\infty$; (2) $0 \leqslant y < +\infty$; (3) $-1 \leqslant y \leqslant +1$.

Restriction of the domain of definition. The domain of definition can, however, be arbitrarily restricted, for example, (1)* $y = 7x + 2$ (for $-3 \leqslant x < 5$) or (1)** $y = 7x + 2$ (for $-8 < x < 0$), and so on. The *range* is then given by (1)* $-19 \leqslant y < 37$, and (1)** $-54 < y < 2$. Here it is essential that, according to the definition of the concept of a function, (1), (1)* and (1)** represent entirely different functions. Because two sets are equal precisely if they have the same elements, two functions f_1 and f_2 are likewise equal precisely if each pair of elements (x, y) that belongs to f_1, $(x, y) \in f_1$, also belongs to f_2, $(x, y) \in f_2$, and vice versa. This is not the case for the functions (1), (1)* and (1)**.

Example 1: If P is the sign for the price, p for the price of 100 gram of a certain merchandise and m the symbol for its mass in gram, then $P = p \cdot m/100$ is the connection between the mass and price of the material. Substituting 0.72 for p, and 100, 200, 300, ... for m in the formula, one obtains the values 0.72, 1.44, 2.16, ... for P.

Example 2: In the formula $l = l_0(1 + 0.000\,016t)$, for the *length of a copper bar when heated*, l and t are symbols with another meaning; l stands for numbers in the domain of lengths, t for numbers in the domain of temperatures. The formula is valid if t assumes values between 0°C and 100°C.

Example 3: The *calculation of the area* of a square is made according to the equation $A = a^2$. Here a is a symbol for the number of units of length of the side and A for the number of units of area.

Abstracting from the special content of individual examples, one arrives at the following state of affairs:

1. *Variables* are introduced for the elements of the domain of definition and of the range. In the examples above these are the symbols m, t, a and P, l, A. In mathematics one often uses the symbols x or y as variables in functions, and the symbols f, g, φ etc. to denote functions.

2. The rule for the correspondence is defined with the help of the variables by means of an *equation*. The element (y) of the range corresponding to an element (x) of the domain is obtained by first substituting for the variable x in the equation and then calculating y. For example, if the function is defined by the equation $y = -2x^2 + 4x - \sqrt{x}$, with domain of definition $0 \leqslant x < +\infty$, then the value corresponding to $x = 9$ is obtained by substitution: $y = -2 \cdot 9^2 + 4 \cdot 9 - \sqrt{9} = -129$. In this way the number -129 corresponds to the number 9 according to the given function. The value corresponding to every number of the domain of definition can be found in the same way.

The symbol for the elements of the domain of definition is called the *independent* variable and that for the elements of the range of a function is called the *dependent* variable. An equation by which the rule for the correspondence defined by a function is given is called the *equation of the function.*

Graphical representation. From the equation of the function one often arrives by means of a table of values at an intuitive representation of the function concerned. With the help of a plane coordinate system (see Chapter 13.) a point P of the plane is constructed to correspond to each number pair (x, y) and the totality of points P is called the graph of the function. According to the nature of the domain of definition and of the equation of the function, one obtains a sequence of isolated points, individual portions of curves or a connected *function curve.*

Example: If x is the independent variable in the domain of definition $-1 \leqslant x \leqslant +2$, then the function with the equation $y = x/2$ has the range $-1/2 \leqslant y \leqslant +1$. For individual values of x the accompanying table of values is obtained.

x	-1	$-1/2$	0	$+1/2$	$+1$	$+3/2$	$+2$
y	$-1/2$	$-1/4$	0	$+1/4$	$+1/2$	$+3/4$	$+1$

Individual points of the function curve in the domain $-1 \leqslant x \leqslant 2$ can first be drawn. If one calculates the values of the function for further arguments, one obtains an ever more dense sequence of points, which all lie on the same straight line (Fig.).

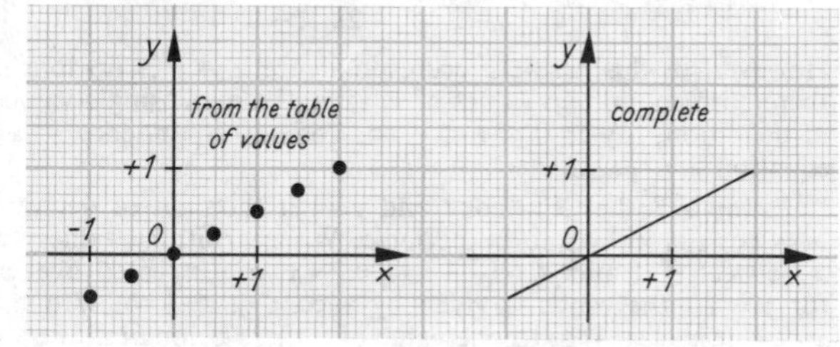

5.1-3 Graph of the function $y = x/2$ if $-1 \leq x \leq 2$

It is customary to display the values of the independent variable on the horizontal axis of a rectangular Cartesian coordinate system and those of the dependent variable on the vertical axis.

Explicit form. The form $y = A(x)$ of the equation of a function, in which $A(x)$ is an arbitrary expression that contains, besides the variable x, only numbers or elements of the basic number domain, is called an *explicit form.*

Implicit form. In contrast, an *implicit form* is characterized by the fact that both variables occur on at least one side of the equation, for example (1) $4x - 2y = 6$; (2) $xy = 1$; (3) $y = \sin x \cdot \sin y + x^2$; (4) $x^2 + y^2 = 16$; (5) $x^2 + xy + y^x = \sqrt{(xy)}$. If the equation of a function is presented in explicit form, then as a rule one regards the variable that is isolated on one side of the equation as dependent and the other as independent; it is immaterial whether they are denoted by x, y; u, v; s, t or in any other way. With an implicit form it is not always so obvious. When x and y are used, it is usual to regard y as the dependent variable, but it is often necessary to explain one's convention, especially if other variables are used. It is, however, also possible to regard both variables in an implicit equation as being of equal standing. It is important to note that an equation given in implicit form cannot always be rearranged in an explicit form. In examples (1) and (2) this is easily done; one obtains (1) $y = 2x - 3$ and (2) $y = 1/x$. Examples (3) and (5), however, defy all attempts to do this. In both examples neither y nor x can be isolated (see Chapter 4.). Another fact is shown clearly by the example (4). It is well known that $x^2 + y^2 = 16$ is the equation of the circle of radius 4 about the origin of the coordinate system. In this case for each value of x there are two values of y that satisfy the equation. Regarding y as the dependent variable, a correspondence is defined that is *not single-valued*. For this reason (4) is not the equation of a function. The explicit form $y = +\sqrt{(16 - x^2)}$, on the other hand, does represent a function. But its image consists of only the *upper* semicircle. The equation of the function belonging to the *lower* semicircle is $y = -\sqrt{(16 - x^2)}$. Sometimes both functions are combined in the form $y = \pm\sqrt{(16 - x^2)}$. It would be wrong, however, to regard this way of writing it as the equation of a function that is *many-valued*; functions are *single-valued* correspondences, by definition.

Parametric representation. This is concerned in the first instance with two explicit function equations, each of which determines a function. The domain of definition in both cases is the same. Thus, in general form one has $t \to x = f_1(t)$ and $t \to y = f_2(t)$. If one now assigns to each $x_0 = f_1(t_0)$ the value $y_0 = f_1(t_0)$, one obtains a mapping of the range of f_1 onto the range of f_2, which need not, of course, be single-valued.

Example 1: Let $x = 2t$ and $y = t/2$ with $-\infty < t < +\infty$. Then the table of values for both functions is:

t	−10	−8	−6	−4	−2	0	2	4	6	8	10
x	−20	−16	−12	−8	−4	0	4	8	12	16	20
y	−5	−4	−3	−2	−1	0	1	2	3	4	5

The first and second lines refer to the function $x = 2t$, and the first and third lines to the function $y = t/2$. The values of x and y belonging to the same values of t determine a new correspondence, which is displayed in the second and third lines of the table of values. This new correspondence is clearly *single-valued* and is described by the new function equation $y = x/4$, as one can see at once from the table of values. From $x = 2t$ it follows that $t = x/2$. Substituting the expression $x/2$ for t in $y = t/2$, one obtains $y = x/4$; the parameter t has been eliminated.

Example 2: Let $x = t^2$ and $y = t/2$ with domain of definition $-\infty < t < +\infty$. The table of values is:

t	−10	−8	−6	−4	−2	0	2	4	6	8	10
x	100	64	36	16	4	0	4	16	36	64	100
y	−5	−4	−3	−2	−1	0	1	2	3	4	5

However, in this case the correspondence $x \to y$ is no longer single-valued: to each value of x there correspond two values of y. It can be made single-valued by restricting the original domain of definition, say to $0 \leqslant t < +\infty$. The correspondence $x \to y$ is then again a function with the equation $y = \sqrt{x}/2$.

Example 3: Let $x = \cos t$ and $y = 2t$ (domain of definition $-\infty < t < +\infty$). The function $x = \cos t$ is known to be periodic. When arbitrary values are chosen for t, the same values for x between -1 and $+1$ $(-1 \leqslant x \leqslant +1)$ are repeated over and over again. On the other hand, for $y = 2t$ the range is given by $-\infty < y < +\infty$. If one now considers the correspondence $x \to y$, it is clear that to one value of x there belong infinitely many values of y. One special value suffices to make this clear. One obtains $x = 1$ for $t = 0, t = \pm 2\pi, t = \pm 4\pi$, etc. Thus, the values $y = 0, y = \pm 4\pi, y = \pm 8\pi$ etc. belong to $x = 1$. A single-valued correspondence is again achieved only by restricting the original domain of definition, say to $0 \leqslant t \leqslant \pi$. The function then defined has the equation $y = 2 \arccos x$ with domain of definition $-1 \leqslant x \leqslant 1$ and range $0 \leqslant y \leqslant 2\pi$.

If a function $x \to y = f(x)$ is represented by two separate functions of the form $x = f_1(t)$ and $y = f_2(t)$, the variable t is called a *parameter*. By means of such a parametric representation a given implicit relation between x and y can often be represented by two explicit functions; for example, $x^2 + y^2 = 1$ by $x = \cos t$, $y = \sin t$ with $0 \leqslant t < 2\pi$. To achieve uniqueness, the domain of definition for t must be suitably restricted.

Composite functions. If the element a corresponds under the mapping G to the element b, and under a further mapping F the element b corresponds to the element c, then by *successive application* of the two mappings F and G, one obtains a mapping under which the element a corresponds to the element c. The mapping defined in this way is called the *product* (or *compositum*) of the two mappings F and G; thus, $(a, c) \in F \cdot G$ if and only if there exists an element b, such that $(a, b) \in G$ and $(b, c) \in F$. Clearly the element b must belong both to the domain of definition X_F of F and to the range Y_G of G (Fig.). From this it follows that $F \cdot G$ can be formed only if $X_F \cap Y_G \neq \emptyset$. Furthermore, in carrying out the successive mappings the order is important, because, in general, $F \cdot G \neq G \cdot F$. If X_F, X_G, $X_{F \cdot G}$ denote the domains of definition and Y_F, Y_G, $Y_{F \cdot G}$ the ranges of F, G, $F \cdot G$, respectively, then $F \cdot G$ can be formed precisely when $X_F \cap Y_G \neq \emptyset$; $X_{F \cdot G} \subseteq X_G$; $Y_{F \cdot G} \subseteq Y_F$. Stated more precisely, $X_{F \cdot G}$ contains just those elements of X_G whose function values with respect to G lie in $X_F \cap Y_G$, and $Y_{F \cdot G}$ contains just those elements of Y_F whose arguments with respect to F lie in $X_F \cap Y_G$. The product $f \cdot g$ of two functions f and g with the function equations $y = f(x)$ and $y = g(x)$ is often written as $y = f[g(x)]$ and called the *compositum* of the two functions g and f, in this order. In this connection g is often called the *inner* function and f the *outer* function of the composite function $f \cdot g$.

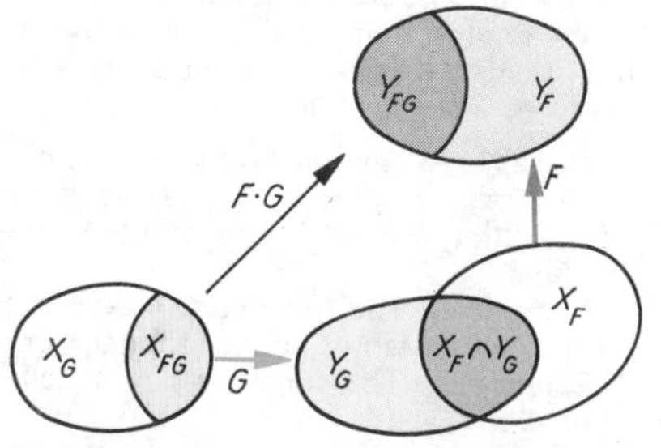

5.1-4 The composite $F \cdot G$ of the mappings G and F; the domain of definition X_{FG} (yellow) is the complete original with respect to G of the set $X_F \cap Y_G$ (green); the range Y_{FG} (grey) is the image of $X_F \cap Y_G$ with respect to F

Example: The domains and ranges of the functions $g(x) = x^2 - 2$ and $f(x) = \sqrt{x}$ are $X_g = (-\infty, +\infty)$, $Y_g = [-2, \infty)$ and $X_f = [0, \infty)$, $Y_f = [0, \infty)$. The composite function $f \cdot g$ has the equation $f[g(x)] = \sqrt{(x^2 - 2)}$ and its domain of definition $X_{f \cdot g}$ consists precisely of those elements of X_g whose function values with respect to g lie in $X_f \cap Y_g = [0, \infty)$. But these are all x with the property $x^2 \geqslant 2$, that is, the set of all real numbers with the exception of the interval from $-\sqrt{2}$ to $+\sqrt{2}$. The composite function $g \cdot f$ has the equation $g[f(x)] = (\sqrt{x})^2 - 2 = x - 2$ with domain of definition $X_{g \cdot f} = [0, \infty)$.

Special types of function

In what follows the only functions to be considered are those whose domain of definition and range are contained in the set of real numbers. They are usually called *real functions*. According to certain general properties special real functions are collected together in groups, for example, monotonic, bounded, even, odd, or periodic functions.

Monotonic functions. A function $x \to y = f(x)$ is said to be *monotonic increasing* in an interval $a < x < b$ if for the greater x_2 of two arbitrary values x_1 and x_2 in the interval the function value $f(x_2)$ also is always the greater; if $x_1 < x_2$, then $f(x_1) < f(x_2)$.

Example 1: The function $y = 2^x$ with domain of definition $-\infty < x < +\infty$ is a monotonic increasing function in the whole of its domain.

Example 2: The function $y = \sin x$ with domain of definition $-\infty < x < +\infty$ is monotonic increasing only in the intervals $-5\pi/2 < x < -3\pi/2$; $-\pi/2 < x < \pi/2$; $3\pi/2 < x < 5\pi/2$; and so on, but considered as a whole it does not represent a monotonic increasing function.

A function is said to be *monotonic decreasing* in an interval $a < x < b$ if $f(x_1) > f(x_2)$ whenever $a < x_1 < x_2 < b$.

Example 1: The function $y = 1/x$ decreases monotonically for $-\infty < x < 0$ and for $0 < x < +\infty$ and is not defined for the value $x = 0$.

Example 2: The function $y = x^2$ is monotonic decreasing for $-\infty < x \leqslant 0$. For $x \geqslant 0$ the function is monotonic increasing.

Example 3: The function $y = -3x + 5$ is monotonic decreasing in the whole of its domain of definition.

Sometimes a function is also called monotonic in an interval if $x_1 < x_2$ implies that always $f(x_1) \leqslant f(x_2)$ [or that always $f(x_1) \geqslant f(x_2)$]. More accurately such functions are called *non-de-*

creasing (or *non-increasing*), and in contrast the functions already considered are *strictly monotonic* (increasing or decreasing).

Bounded functions. A function $x \to y = f(x)$ is said to be *bounded in an* (open or closed) *interval* if there exists a number $B > 0$ with the property that $|f(x)| \leqslant B$ for every value of x in the interval. In particular, if $|f(x)| \leqslant B$ for every value of x in the domain of definition, then $x \to y = f(x)$ is said to be a *bounded function*.

Example 1: The function $y = x^3$ is bounded in every closed interval. For example, in the interval $0 \leqslant x \leqslant a$, $|f(x)| \leqslant B = a^3$. However, it is not a bounded function, because for the domain of definition $-\infty < x < +\infty$ no number B can be found that is not exceeded by any value of the function.

Example 2: The function $y = x^{-2}$ is bounded for every interval of the form $a \leqslant x < +\infty$, with $a > 0$. It is not bounded for $0 < x \leqslant b$.

Example 3: The function $y = \sqrt{(100 - x^2)}$ is bounded in the whole domain of definition $-10 \leqslant x \leqslant +10$, because $|\sqrt{(100 - x^2)}| \leqslant 10$ always holds (Fig.).

Example 4: The function $y = \dfrac{x^2 - 1}{x^2 + 1}$ is bounded in the whole domain of definition, as can be seen by writing it in the form $y = 1 - \dfrac{2}{x^2 + 1}$. Here $\left|1 - \dfrac{2}{x^2 + 1}\right| \leqslant 1$ for every value of x.

The graphical representation of a bounded function is characterized by the fact that two lines parallel to the x-axis can always be found so that the graph of the function lies entirely between them.

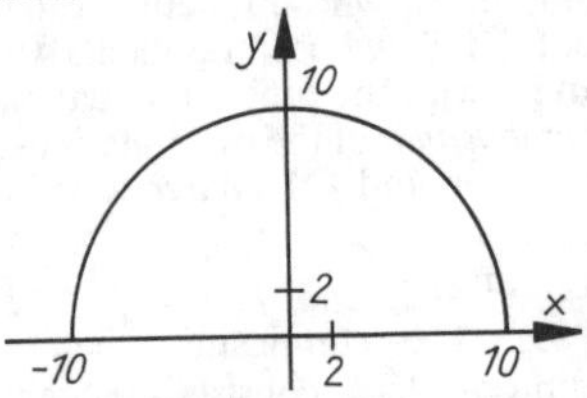

5.1-5 Graph of the function $y = \sqrt{(100 - x^2)}$

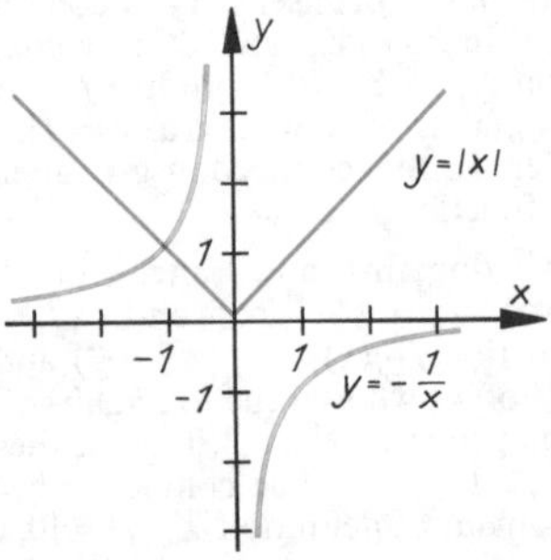

5.1-6 Graphical representation of the even function $y = |x|$ and the odd function $y = -1/x$

even functions	odd functions
$y = -x^2/2$	$y = x^3$
$y = \|x\|$	$y = -1/x$
$y = (x^2 - 1)/(x^2 + 1)$	$y = x/2$
$y = a \cdot x^{2n}$	$y = a \cdot x^{2n+1}$
$a \neq 0,\ n = 0, \pm 1, \pm 2, \ldots$	
$y = \cos x$	$y = \sin x$

Even and odd functions. A function $x \to y = f(x)$ is said to be *even* if $f(-x) = f(x)$ for every value of x in the domain of definition. A function $x \to y = f(x)$ is said to be *odd* if $f(-x) = -f(x)$ for every value of x in the domain of definition.

The graph of an *even* function is *symmetric about the y-axis*. The graph of an odd function is *symmetric about the origin* (0, 0). It goes into itself under a rotation through 180° about this point (Fig.).

Periodic functions. A non-constant function $x \to y = f(x)$ is said to be periodic if there exists a number $a > 0$ such that $f(x) = f(x + a)$ for every possible value of x. It then also follows that $f(x) = f(x + 2a)$ and $f(x) = f(x - a)$, in general, $f(x) = f(x + na)$ for every integer n, as long as the values $(x + na)$ belong to the domain of definition of the function. Each such number a is called a *period*, and the smallest positive number k for which $f(x) = f(x + k)$ is called the *primitive period* of the periodic function. The graphical representation of a periodic function is a graph that goes into itself when translated in the direction of the x-axis through a distance equal to an integral multiple of a period (Fig.). The best-known periodic functions are the *trigonometric functions*. From these further periodic functions can be constructed; for example, the functions $y = b \sin(ax)$ with $b \neq 0$ and $a \neq 0$ have the period $2\pi/a$. Combined functions such as $y = b_1 \sin(a_1 x) + b_2 \sin(a_2 x)$ are periodic, provided that the ratio of a_1 to a_2 is rational, that is, if $a_1/a_2 = m/n$, where m and n are relatively prime integers. The period of the first function is $2\pi/a_1$ and that of the second is $2\pi/a_2$, and their ratio is $(2\pi/a_1):(2\pi/a_2) = a_2/a_1 = n/m$. Thus, n periods of the

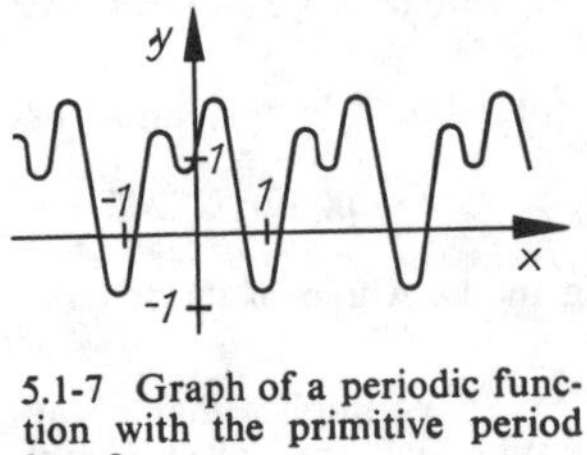

5.1-7 Graph of a periodic function with the primitive period $k = 2$

first function correspond exactly to m periods of the second function. Consequently the sum function has the period $m \cdot (2\pi/a_1) = n \cdot (2\pi/a_2)$.

Example: The periods of the individual functions of the sum function $y = \sin(2x) + 2\sin(3x/2)$ are π and $4\pi/3$ and their ratio is $\pi/(4\pi/3) = 3/4$. The given function therefore has the period 4π (Fig.).

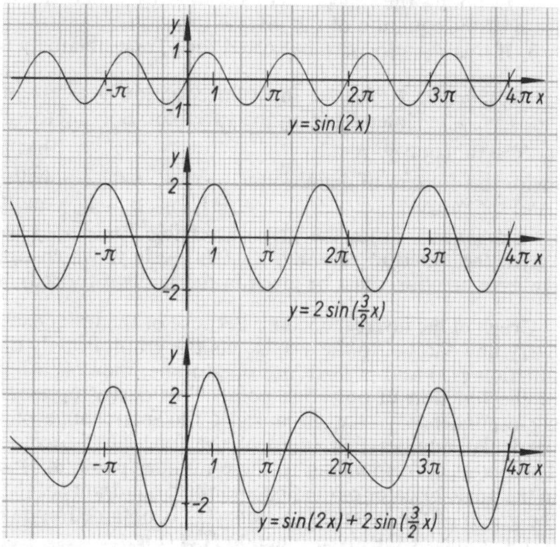

5.1-8 Graphs of the functions $y = \sin(2x)$, $y = 2\sin(3x/2)$ and $y = \sin(2x) + 2\sin(3x/2)$

Inverse of a function

Invertible functions. The single-valued correspondence determined by a function between the elements of the domain and the elements of the range, conversely also assigns to each element of the range one or more elements of the domain. Functions for which each element of the range occurs only once as the image of an element of the domain have a special significance, because the inverse of the correspondence is also single-valued. To each element r of the range there belongs only one element d of the domain. In this case the range of the given function f can be regarded as the domain of a new function φ. If the given function f determines the correspondence $d \to r = f(d)$, then for the new function φ one has $r \to d = \varphi(r)$. In other words, $(r, d) \in \varphi$ if and only if $(d, r) \in f$. Functions for which in this sense the correspondence between the domain X and the range Y can be inverted are called invertible functions (Fig.). These are one-to-one correspondences of X onto Y. Monotonic functions belong to the class of invertible functions: a monotonic function is always invertible. On the other hand, an invertible function need not necessarily be monotonic; for example, the domain and range may not be ordered sets, so that the concept of monotonicity is not defined. Again, a non-monotonic function can also be invertible, for instance if the domain and the range consist of only finitely many elements. An example of this is the function given by the following table of values:

x	1	2	3	4	5	6	7	8	9	10
y	0	2	4	6	8	1	3	5	7	9

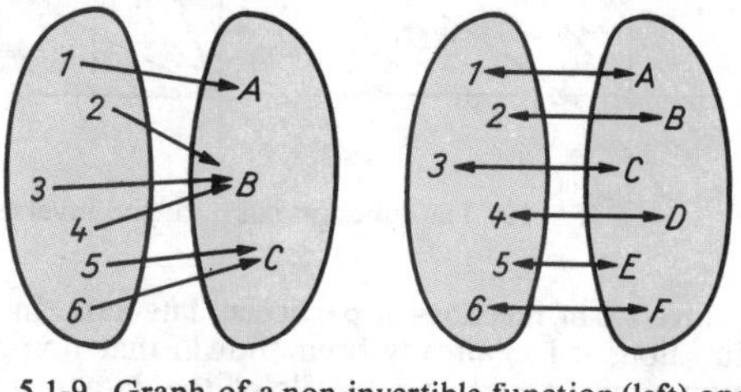

5.1-9 Graph of a non-invertible function (left) and an invertible function (right)

Inverse function. If one regards the range Y of an invertible function f as domain of definition of a new function φ, whose range is the domain X of f, and if one reverses the singlevalued correspondence between the sets X and Y given by the function f, then one obtains the *inverse function* φ of the given function f. The inverse function is itself an invertible. By considering $d \to r = f(d)$ and $r \to d = \varphi(r)$ it is easy to see that the inverse function of the inverse function of a given function f is f itself. Thus, one is justified in calling f and φ *mutually inverse* functions.

Example 1:

Function f

domain	1	2	3	4	5
range	a	b	c	d	e

Inverse function φ of f

domain	a	b	c	d	e
range	1	2	3	4	5

If $y = f(x)$ is the equation of an invertible function, then the same equation naturally also de scribes the inverse function, only y must then be the independent and x the dependent va riable It is agreed, however, that in a function equation of this form x shall always denote the independen and y the dependent variable and, whenever possible, the explicit form of the function equation shall be given. One therefore rearranges the equation as follows:

1. In the given function equation $y = f(x)$, y is regarded as the independent and x as the dependent variable.
2. Denoting the independent variable by x and the dependent one by y, $x = f(y)$ is an implicit form of the equation of the inverse function.
3. If this equation can be solved for y, one obtains $y = \varphi(x)$ as its explicit form.

Example 2: From the function equation $y = x/2$ of a given invertible function one obtains $x = y/2$ after interchanging the variables. Solving for y gives $y = 2x$.

The function $y = x/2$		has the inverse function $y = 2x$	
with the domain of definition	$-1 \leqslant x \leqslant 2$	with the domain of definition	$-1/2 \leqslant x \leqslant 1$
and the range	$-1/2 \leqslant y \leqslant 1$	and the range	$-1 \leqslant y \leqslant 2$.

Example 3: From the given invertible function $y = 3x + \sin x$, interchange of the variables gives the function equation $x = 3y + \sin y$, which cannot be solved explicitly for y. Thus, the inverse function must be given in the implicit form $3y + \sin y - x = 0$.

Graph of the inverse function. Because of the uniqueness of the mapping represented by a function, every line parallel to the y-axis cuts the graph in only one point. If the function $f(x)$ has an inverse function $\varphi(x)$ and is therefore one-to-one, then each line parallel to the x-axis also cuts the graph of the function in only one point. This curve represents both the correspondence $x \to y$ and the correspondence $y \to x$. Because of the interchange of the variables in the inverse function each particular number pair (a, b) of the function f becomes a number pair (b, a) of the function φ. The points corresponding to these number pairs (a, b) and (b, a) are mirror images of one another in the angle bisector of the first and third quadrants of the Cartesian coordinate system. Consequently the graph of the inverse function $\varphi(x)$ is obtained by taking the mirror image in this angle bisector of the graph of the given function $f(x)$ (Fig.).

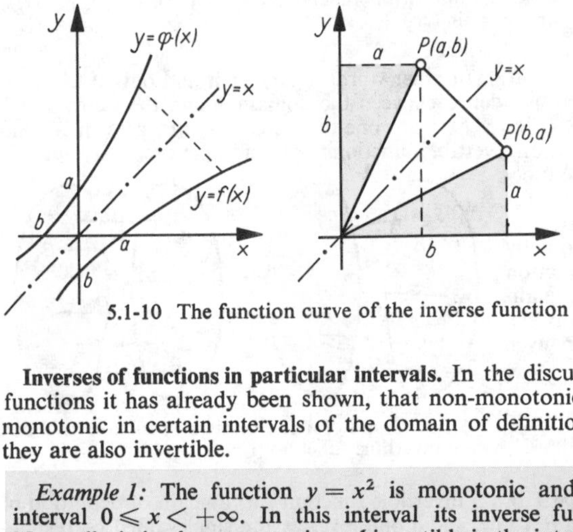

5.1-10 The function curve of the inverse function

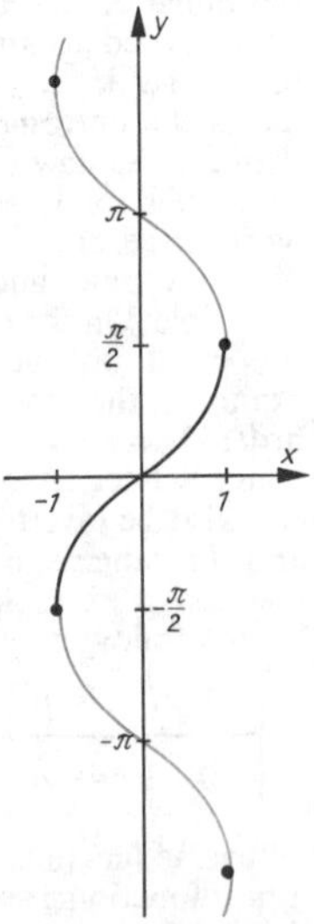

5.1-11 Graph of $y = \arcsin x$; principal value $y = \text{Arcsin}\, x$ is drawn in black

Inverses of functions in particular intervals. In the discussion of monotonic functions it has already been shown, that non-monotonic functions may be monotonic in certain intervals of the domain of definition. In these intervals they are also invertible.

Example 1: The function $y = x^2$ is monotonic and invertible in the interval $0 \leqslant x < +\infty$. In this interval its inverse function is $y = \sqrt{x}$. Naturally it is also monotonic and invertible in the interval $-\infty < x \leqslant 0$. Here the inverse function is $y = -\sqrt{x}$.

Example 2: The domain of definition of $y = \sin x$ can be split up into intervals in which the given function is monotonic. The inverse function is denoted by arcsin x, but in each case the range must be stated, because otherwise it is not clear in which interval of monotonicity the inverse has been formed. For example, if $y = \sin x$ is inverted in the interval $3\pi/2 \leqslant x \leqslant 5\pi/2$, then the inverse function should be denoted by $y = \arcsin x$ $(3\pi/2 \leqslant y \leqslant 5\pi/2)$. If the range is not specified, then arcsin x is always understood to mean the *principal value*, which lies in the interval $[-\pi/2, +\pi/2]$ and is denoted by Arcsin x (Fig.).

Example 3: Also for the other trigonometric functions intervals can be chosen in which they are monotonic, so that in them circular functions are defined as their inverses. The function $y = \cos x$, for example, decreases monotonically in the interval $0 \leqslant x \leqslant +\pi$ from $y = +1$ to $y = -1$ and in doing so assumes all values of its range exactly once. Hence in this interval an inverse function exists. It is denoted by $y = \arccos x$. Its domain of definition is $-1 \leqslant x \leqslant +1$ and its range is $\pi \geqslant y \geqslant 0$. If the function $y = \cos x$ is inverted in another interval in which it is monotonic, say in the interval $\pi \leqslant x \leqslant 2\pi$, then $y = \arccos x$ has the range $\pi \leqslant y \leqslant 2\pi$. In order to specify which inverse function is intended, the range must be given in each case. If it is not specified, then arccos x is to be understood as the *principal value*, which is characterized by $0 \leqslant \arccos x \leqslant \pi$ and is denoted by Arccos x.

A similar result holds for the function $y = \arctan x$ in the interval $-\pi/2 < \text{Arctan}\, x < +\pi/2$ and for $y = \text{arccot}\, x$ in the interval $0 < \text{Arccot}\, x < +\pi$ (see Chapter 10.).

5.2. Polynomial and rational functions

The concept of a rational function

An expression of the form $a_n x^n + a_{n-1} x^{n-1} + \cdots + a_1 x + a_0$, where n is a natural number, the *coefficients* a_r are arbitrary real numbers, and $a_n \neq 0$, is called a *polynomial of degree n*.

A *rational function* is a function of the form p/q, where p and q are polynomial functions and at least one coefficient of q is not 0.

Examples of rational functions. 1. $y = 8x - 3$.
2. $y = \dfrac{4x^2 + 1}{x(x^3 - 2)}$. *3.* $y = \sqrt{10} \cdot x^2 - \dfrac{\ln 5}{x}$. *4.* $y = 1/x^4$.

Examples of non-rational functions. 1. $y = \sqrt{x^3}$. *2.* $y = \cos^2 x$.
3. $y = x - \dfrac{x^3}{3!} + \dfrac{x^5}{5!} - \dfrac{x^7}{7!} + \cdots = \sum_{n=0}^{\infty} \dfrac{x^{2n+1}}{(2n+1)!}$.

For polynomial functions the domain **R** of all real numbers can be chosen as domain of definition. If no restriction is indicated on account of special conditions, **R** is always regarded as the domain of definition. The same holds for rational functions, except that those values for which the denominator vanishes must be excluded. It should also be pointed out that rational functions are continuous and differentiable arbitrarily often in the whole of their domain of definition.

In the following, first the polynomial functions and then the rational functions are considered. Before establishing general properties, certain special types of such functions that occur particularly often are examined first.

Linear functions

The functions $y = mx$. The *tables of values* of the functions $y = x$, $y = x/2$, and $y = -4x/3$ give number pairs (x, y) from which one obtains *points* of the graphs of these functions in a Cartesian coordinate system (Fig.).

x	...	-3	-2	-1	0	1	3	3	...
$y = x$	...	-3	-2	-1	0	1	2	3	...
$y = x/2$	...	$-3/2$	-1	$-1/2$	0	$1/2$	1	$3/2$	...
$y = -4x/3$	...	4	$8/3$	$4/3$	0	$-4/3$	$-8/3$	-4	...

Because the pair of values (0, 0) always occurs, the curve always passes through the origin of coordinates. The curves are straight lines, because from $y = mx$ the coordinates of arbitrarily chosen points $P_1, P_2, \ldots, P$ always satisfy $y_1/x_1 = y_2/x_2 = \cdots = y/x = m$, where for each func-

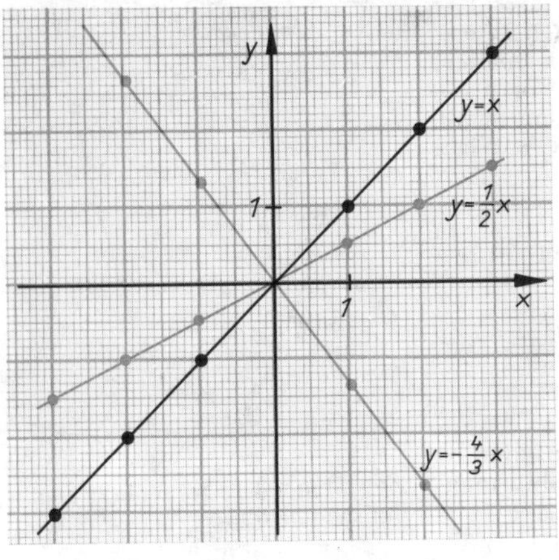

5.2-1 Graphs of the functions $y = x$, $y = x/2$, $y = -4x/3$; $m = 1$, $m = 1/2$, $m = -4/3$

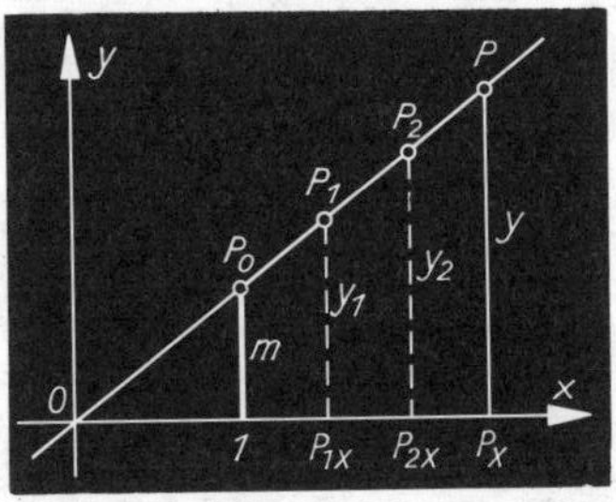

5.2-2 The graph of $y = mx$ is a straight line

tion m is a constant (Fig.). If $P_{1x}, P_{2x}, \ldots, P_x$ are the projections of the points $P_1, P_2, \ldots, P$ onto the x-axis, then the triangles $OP_1P_{1x}, OP_2P_{2x}, \ldots, OPP_x$ are similar. Since the points $P_{1x}, P_{2x}, \ldots, P_x$ lie on a straight line, the points $P_1, P_2, \ldots, P$ must also lie on a straight line. Because m is a constant, the corresponding coordinates of different points are proportional, $y_1/x_1 = y_2/x_2$. The magnitude of y is directly proportional to the magnitude of x; the constant m is the *factor of proportionality*. If the rate L for a job is proportional to the working time t in hours, the connection between the two is represented by the linear function $L = mt$. The constant of proportionality represents the rate per hour.

5.2-3 Sign of a steep hill

From the graph of the linear function $y = mx$ and the table of values of the special functions $y = x$, $y = x/2$ and $y = -4x/3$ it can be seen that the function is monotonic; for positive m it is monotonic increasing and for negative m it is monotonic decreasing. With reference to roads and railways the constant is called the *gradient* (Fig.). In mathematics the gradient is defined as the ratio of the *difference in height BC* to the *horizontal distance AB* (Fig.). It is given as a ratio or as a percentage. For example, 1 : 50, 3/150, 1/50, 2/100, 2% = 0.02 all have the same meaning.

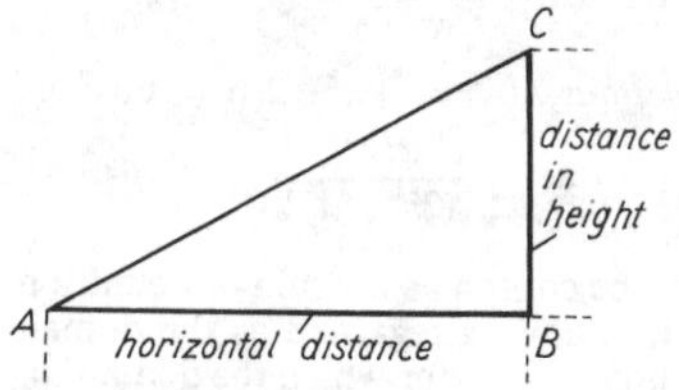

5.2-4 Gradient

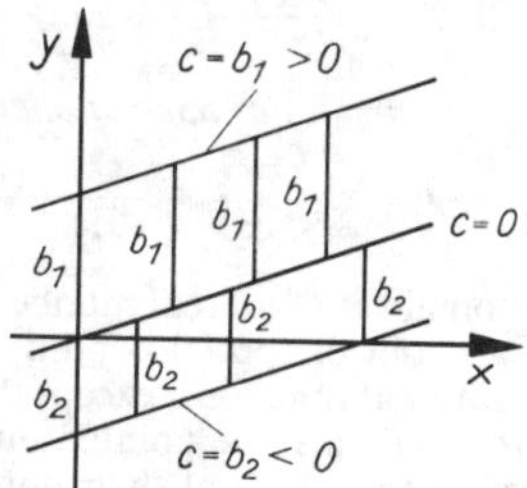

5.2-5 The function $y = mx + c$

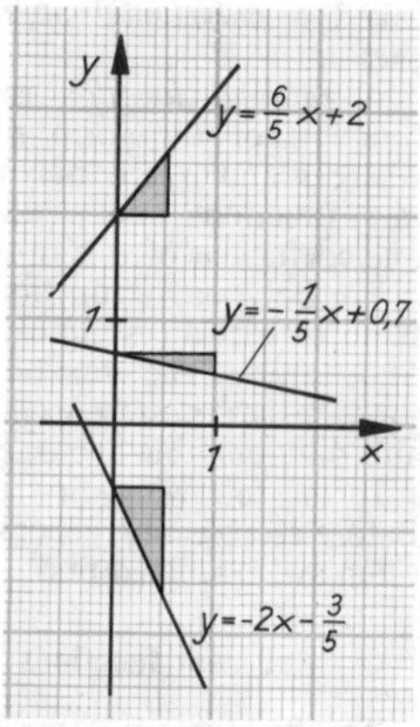

5.2-6 Graphs of further functions $y = mx + c$

The functions $y = mx + c$. If for every value of x on the straight line $y = mx$ a fixed quantity c is added to or subtracted from the ordinate y, this signifies a translation of the line $y = mx$, which one recognizes most easily as the intercept c on the y-axis (Fig.). Thus, the graph of the function $y = mx + c$ is a *straight line with gradient m and intercept c on the y-axis* (see Chapter 13. – Cartesian normal form).

In drawing the line one need not actually carry out the translation. The graph of the line is obtained by first marking off the intercept c on the y-axis, thinking of a parallel to the x-axis through its end-point and constructing the gradient with respect to it (Fig.).

Implicit representation of the linear function. The graphical representation of $Ax + By + C = 0$ in a Cartesian coordinate system is always a straight line (see Chapter 13.), provided that A and B are not both equal to zero. Moreover, $Ax + By + C = 0$ can be expressed as a linear function in explicit form only if $B \neq 0$. The rearrangement in explicit form then gives $y = -(A/B)\,x - (C/B)$ or $y = mx + c$, with $m = -(A/B)$ and $c = -(C/B)$. For $A = 0$ and $B \neq 0$ the result is a *constant function*, whose graph is a line parallel to the x-axis. For $A \neq 0$ and $B = 0$ the equation does not represent a function at all. The graphical representation of the equation $Ax + C = 0$ is a line parallel to the y-axis.

Quadratic functions

The function $y = x^2$. The function equation $y = x^2$ leads to a curve known as the *standard parabola* (Fig.).

Table of values for $y = x^2$

x	−3	−2	−1	0	1	2	3
y	9	4	1	0	1	4	9

Intermediate values to those of the table of values are given by a *table of squares*, which is nothing more than a skilfully arranged table of values of the function $y = x^2$.

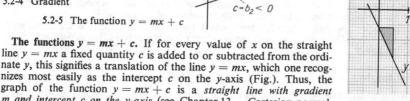

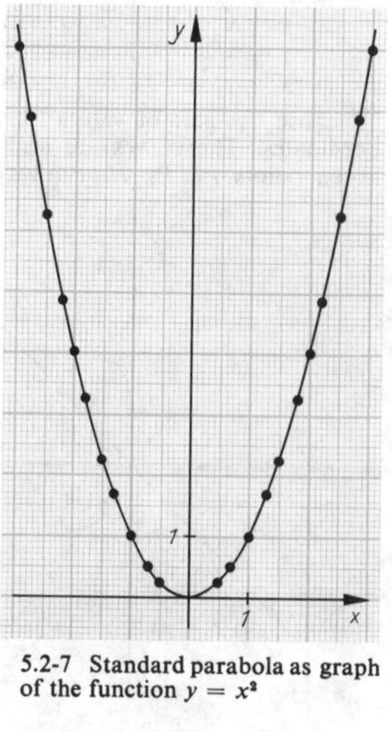

5.2-7 Standard parabola as graph of the function $y = x^2$

Properties. Because $x^2 \geqslant 0$ for every value of x, the curve always remains above the x-axis; thus, to the domain of definition $-\infty < x < +\infty$ there corresponds the range $0 \leqslant y < +\infty$. The standard parabola is symmetric about the y-axis (*axially symmetric*). The zero point, which is symmetric with itself, is called the *vertex*. The *curvature* of the standard parabola, in contrast to the straight line, shows up calculations by the fact that y changes by ever greater amounts as $|x|$ increases uniformly. In the table the difference sequences Δx and Δy are introduced, as well as the sequence of differences for Δy, which is denoted by $\Delta^2 y$. It shows that Δy increases for constant Δx and only the *second difference sequence* $\Delta^2 y$ is constant.

Δx	...		1		1		1		1		1		1		1		1		...
x	...	-2		-1		0		1		2		3		4		5		6	...
y	...	$+4$		$+1$		0		$+1$		$+4$		$+9$		$+16$		$+25$		$+36$	...
Δy			-3		-1		$+1$		$+3$		$+5$		$+7$		$+9$		$+11$		...
$\Delta^2 y$	...			2		2		2		2		2		2		2		...	

For an intuitive understanding of *curvature* one can imagine that a motor car is travelling along the curve in the direction of increasing values of x. If the steering wheel must be turned to the left in order to remain on the curve, then the curve is said to have a *positive curvature*, and if to the right, then it has a *negative curvature*. Thus, the standard parabola has positive curvature throughout.

The functions $y = x^2 + px + q$. By completing the square that is, by introducing the square of half the coefficient of the linear term px, the given function can be expressed in the form $y = (x - a)^2 + b$ by $y = x^2 + px + q = x^2 + px + (p/2)^2 - (p/2)^2 + q = (x + p/2)^2 + (q - p^2/4)$. Writing $a = -p/2$, $b = (q - p^2/4)$, one obtains, in fact, $y = (x - a)^2 + b$ or $(y - b) = (x - a)^2$, that is, $\eta = \xi^2$, where $y - b = \eta$ and $x - a = \xi$. This means that the graph of the function in the ξ, η-coordinate system is again the standard parabola $\eta = \xi^2$. But the ξ, η-system is transformed into the x, y-system by the linear transformation $x - a = \xi$, $y - b = \eta$, corresponding to a translation (Fig.). In the x, y-system the vertex V of the standard parabola $\eta = \xi^2$ has the coordinates (a, b);

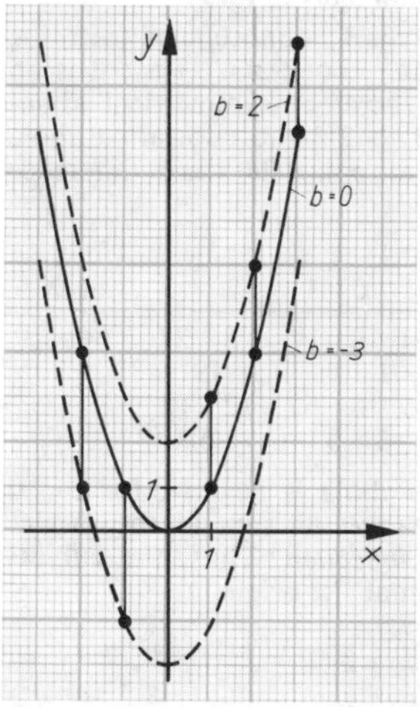

5.2-8
Graphs of the functions
$y = x^2 + b$

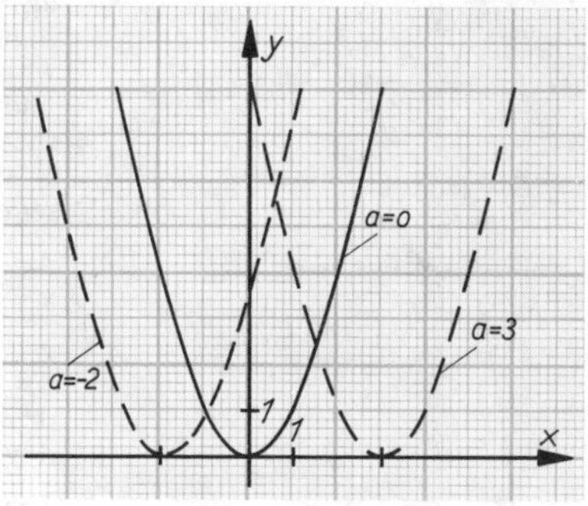

5.2-9 Graphs of the functions $y = (x - a)^2$

expressed in terms of the coefficients p and q of the given quadratic function $y = x^2 + px + q$, the coordinates of the vertex are $(-p/2, q - p^2/4)$ (Fig.).

Example: Completing the square in the equation of the function $y = x^2 + 6x + 11$ gives $y = (x^2 + 6x + 9) - 9 + 11$, or $y = (x + 3)^2 + 2$. Thus, one can read off at once that the graph is a translated standard parabola with vertex $V(-3, 2)$.

The general quadratic function $y = Ax^2 + Bx + C$. In this equation it is assumed that $A \neq 0$, otherwise the function is not quadratic at all. Thus, the factor A can be taken out: $y = A[x^2 + (B/A)\,x + (C/A)] = A \cdot Y$. The graph of the quadratic function

$$Y = x^2 + (B/A)\,x + (C/A) = x^2 + px + q,$$
$$Y = (x + p/2)^2 + (q - p^2/4)$$
$$= [x + B/(2A)]^2 + [C/A - B^2/(4A^2)],$$

where $p = B/A$, $q = C/A$, is known. The values of Y are given

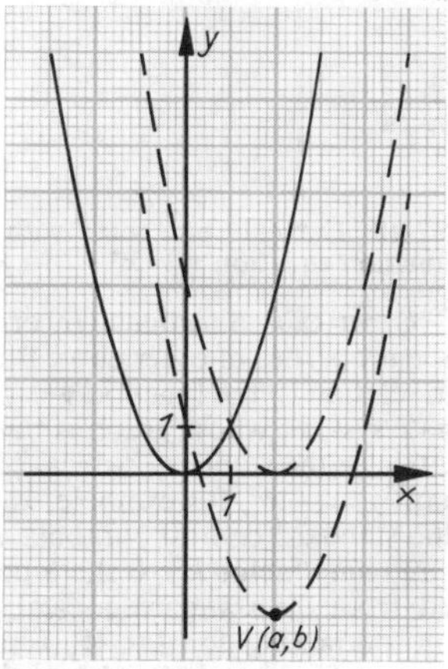

5.2-10 Graph of $y = (x - a)^2 + b$ obtained by translating the standard parabola

as the sum of the ordinate values $[x + B/(2A)]^2$ of the standard parabola, whose vertex has been translated by an amount $p/2 = -B/(2A)$ in the direction of the $+x$-axis, and the quantity $b = q - p^2/4 = [C/A - B^2/(4A^2)]$ of the translation in the direction of the $+y$-axis. But the equation $y = A \cdot Y$ states that each of these values of Y is to be multiplied by the number A. For $A > 1$, all the ordinates of the standard parabola and also the segment $[C/A - B^2/(4A^2)]$ are *stretched* in the ratio $A : 1$, while for A between 0 and 1 they are *contracted* in the same ratio (Fig.). If A takes negative values, this stretching ($|A| > 1$) or contraction ($|A| < 1$) is followed by a *reflection* in the x-axis.

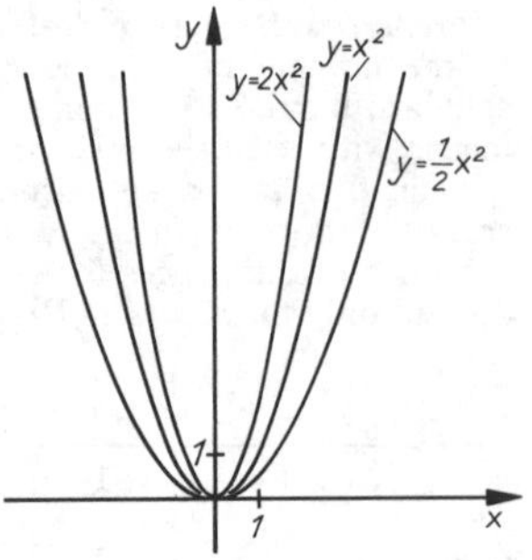

5.2-11 Graphs of the functions $y = x^2/2$ and $y = 2x^2$

Example 1: The graph of the function $y = -x^2$ is the mirror image in the x-axis of the standard parabola.

Example 2: The graph of the function $y = x^2/4$ is the standard parabola contracted in the ratio $(1/4) : 1 = 1 : 4$.

Example 3: The quadratic function $y = 3x^2 - 4x - 1/6$, by taking out the factor 3 and completing the square, can be written in the form:

$$y = 3[x^2 - (4/3)\,x + (2/3)^2 - (4/9 + 1/18)] = 3[(x - 2/3)^2 - 1/2].$$

Hence the graph of the function is a standard parabola stretched in the ratio 3 : 1, whose vertex V has the coordinates $(2/3, -3/2)$.

Cubic functions

The function $y = x^3$. In a table of cubes one has available an extensive and clear table of values of the function $y = x^3$; with its help one obtains the graph of the function, the *cubical parabola* or *parabola of degree three* (Fig.).

Properties. The function increases monotonically in the whole of its domain of definition $-\infty < x < +\infty$; it is an odd function, and its graph is therefore symmetrical about the origin of coordinates. For $|x| > 2/3$ the cubical parabola is steeper than the quadratic parabola; its third difference sequence is constant:

Δx	...		1		1		1		1		1		1		1		1	...
x	...	−4		−3		−2		−1		0		1		2		3		4 ...
y	...	−64		−27		−8		−1		0		1		8		27		64 ...
Δy	...		37		19		7		1		1		7		19		37	...
$\Delta^2 y$	...			−18		−12		−6		0		6		12		18		...
$\Delta^3 y$	...				6		6		6		6		6		6			...

The curvature of the cubical parabola is negative for $x < 0$ and positive for $x > 0$, and it changes its sign at the origin. Such points are called *points of inflection.* Thus, the cubical parabola has a point of inflection at the origin.

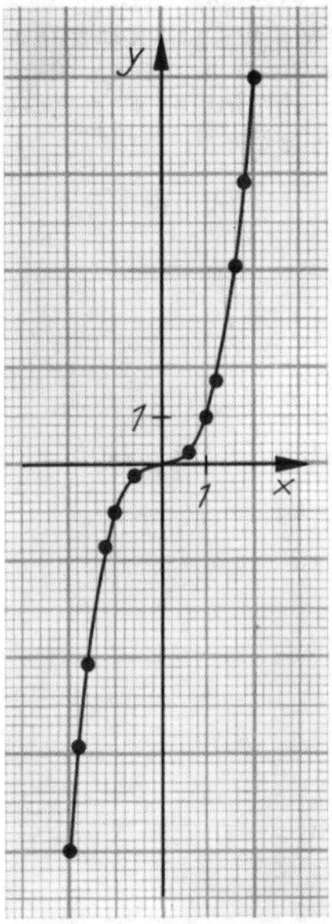

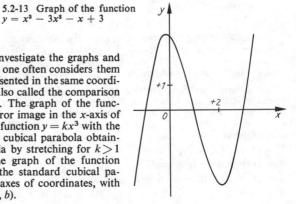

5.2-12 Cubical parabola with the function equation $y = x^3$

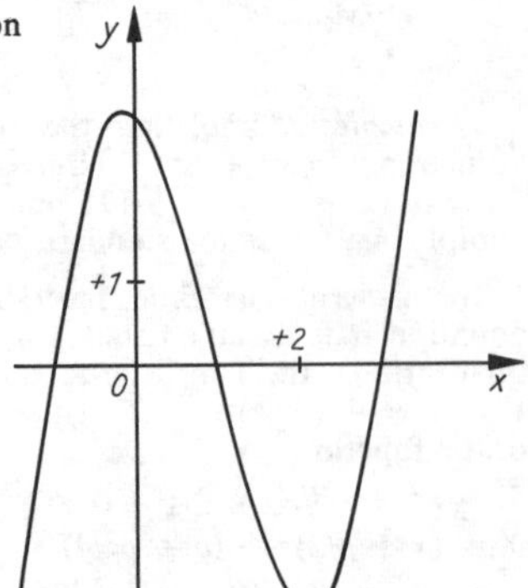

5.2-13 Graph of the function $y = x^3 - 3x^2 - x + 3$

Other cubic functions. In order to investigate the graphs and the properties of other cubic functions, one often considers them in relation to the function $y = x^3$ represented in the same coordinate system, whose graph is therefore also called the comparison cubic or the standard cubical parabola. The graph of the function $y = -x^3$, for example, is the mirror image in the x-axis of the standard cubical parabola. To the function $y = kx^3$ with the stretching factor $k > 0$ there belongs a cubical parabola obtained from the standard cubical parabola by stretching for $k > 1$ or contracting for $k < 1$. Finally, the graph of the function $y = (x - a)^3 + b$ is obtained from the standard cubical parabola by translation parallel to the axes of coordinates, with the new centre of symmetry $Z = (a, b)$.

The *general cubic function* $y = Ax^3 + Bx^2 + Cx + D$ always has *three zeros* of which, under certain conditions between the coefficients, two can be conjugate complex. In the differential calculus it is shown, in addition, that when it has three real zeros, the function has two *extrema*, one (local) maximum and one (local) minimum. The example shows that the graph of such a function cannot be obtained by simple transformations from the standard cubical parabola $y = x^3$ (Fig.).

Example: $y = x^3 - 3x^2 - x + 3$.

Table of values

x	-2	-1	-0.15	0	$+1$	2.15	$+3$
y	-15	0	$+3.08$	$+3$	0	-3.08	0

Power functions with positive exponents

Concept of a power function. A function $y = x^n$, in which n is an integer, is called a power function. If n is positive, the function is a polynomial, but if n is negative, $n = -\nu$ ($\nu > 0$, an integer), then the function can be expressed in the form $y = 1/x^\nu$ and is a rational function.

The polynomial functions $y = x^n$ are even if their exponent $n = 2m$ is even; they decrease monotonically for $-\infty < x \leqslant 0$ and increase monotonically for $0 \leqslant x < +\infty$. For odd exponents $n = 2m + 1$ the functions $y = x^n$ are odd; they increase monotonically everywhere.

Even polynomial power functions $y = x^{2m}$. The curves represented by these functions are symmetrical with respect to the y-axis, and their curvature is everywhere positive (Fig.). Each of them contains the origin (0, 0) and the points $Q(-1, +1)$ and $P(+1, +1)$. In the neighbourhood of the vertex (0, 0) the tangents are the flatter the larger m is, whereas in a particular neighbourhood of the points Q and P they are the steeper, the larger m is. To every point (x_1, y_1) on $y = x^{2m_1}$ a point (x_2, y_2) on $y = x^{2m_2} (m_2 > m_1)$ can be determined by means of the differential calculus, so that the tangents at the two points are parallel. These curves are called *parabolas of order* $2m$ (Fig.).

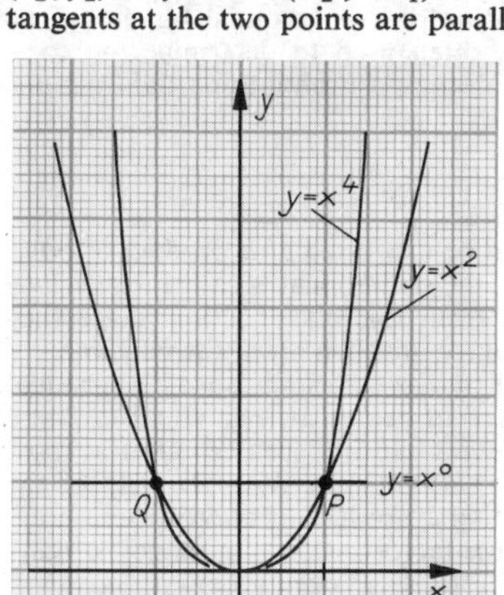

5.2-14 Graphs of the functions $y = x^{2m}$ for $m = 0, 1, 2, \ldots$; $y = x^0$ is not defined for $x = 0$

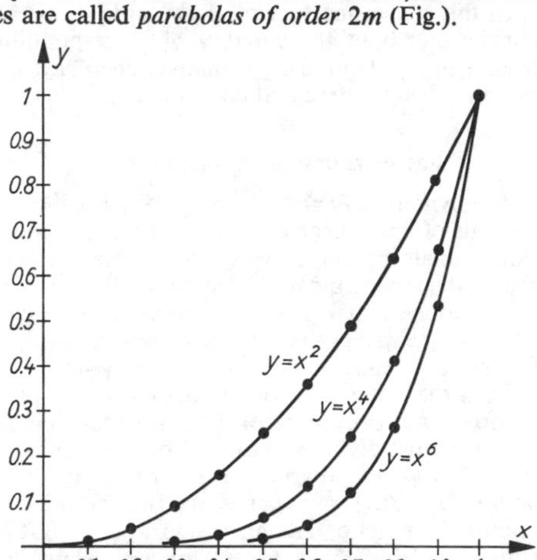

5.2-15 Portions of the curves of the functions $y = x^{2m}$ as parabolas of order $2m$

Odd polynomial power functions $y = x^{2m+1}$. The curves are centrally symmetric about the origin. Except for the angle bisector of the quadrants I and III ($y = x$) they have negative curvature for all negative values of their domain of definition ($-\infty < x < 0$) and positive curvature for positive values ($0 < x < +\infty$), and so they have a point of inflection at the origin. Each of these *parabolas of order* $2m + 1$ contains the points $(+1, +1)$ and $(-1, -1)$, and in the neighbourhood of these points their tangents are the steeper, the greater m is; but in the neighbourhood of their common point of inflection (0, 0) the tangents are the flatter, the greater m is (Fig.).

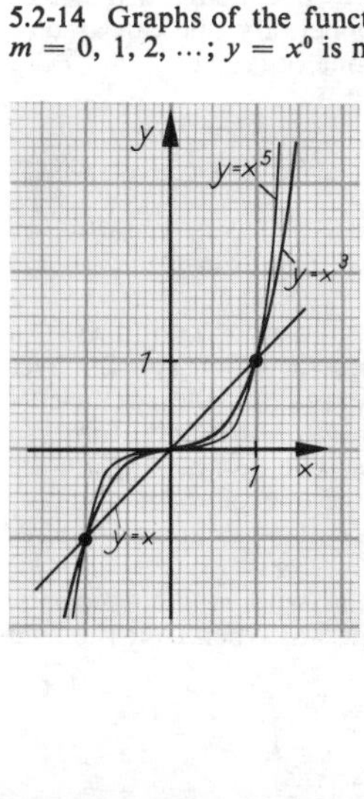

5.2-16 Graphs of the functions $y = x^{2m+1}$ for $m = 0, 1, 2$

Polynomial functions

The expression $a_n x^n + a_{n-1}x^{n-1} + \cdots + a_1 x + a_0$, where $a_n \neq 0$, has been defined above to be a polynomial of degree n. A polynomial is a special kind of rational function.

Example:

$$y = 2(x^2 - 1)^2 + (x + 2)(x^3 - 2) - 2x + x^2 - 1$$
$$= 2x^4 - 4x^2 + 2 + x^4 - 2x + 2x^3 - 4 - 2x + x^2 - 1, \text{ or } y = 3x^4 + 2x^3 - 3x^2 - 4x - 3$$

is a polynomial of degree 4 with the coefficients $a_4 = 3$, $a_3 = 2$, $a_2 = -3$, $a_1 = -4$, $a_0 = -3$.

Uniqueness of polynomial representation. The assumption that two different polynomials can represent the same rational function leads to a contradiction. Because the two polynomials

$$y_a = a_n x^n + a_{n-1} a^{n-1} + \cdots + a_1 x + a_0 \quad \text{and} \quad y_b = b_m x^m + b_{m-1} x^{m-1} + \cdots + b_1 x + b_0$$

are assumed to be different, it must be that either $n \neq m$, or if $n = m$, then $a_i \neq b_i$ for at least one pair of coefficients. Their difference

$$(a_n x^n + a_{n-1} x^{n-1} + \cdots + a_1 x + a_0) - (b_m x^m + b_{m-1} x^{m-1} + \cdots + b_1 x + b_0)$$

can be arranged according to powers of x; it is a polynomial with at least one non-zero coefficient, whose degree does not exceed the greater of the two numbers m and n. It represents a polynomial function, which has only finitely many zeros, not more than its degree. But since y_a and y_b are assumed to be the same function, their difference must be identically zero, that is, zero for all values of x. This contradiction leads to the conclusion that both polynomials have the same degree, $m = n$, and that their corresponding coefficients are equal, $a_i = b_i$, because only then is the difference of the two polynomials identically equal to zero.

In this sense one speaks of the uniqueness of the representation of a polynomial function. The conclusion about the equality of corresponding coefficients is often used to determine the coefficients of a polynomial by *equating coefficients*, for example, in decomposition into partial fractions and in solving differential equations.

Factorization of polynomials

A polynomial $P(x)$ of degree $n \geqslant 1$ is called *reducible* if it can be expressed as a product of polynomials of lower degree. If such a representation is not possible, the polynomial is called *irreducible*. Polynomials of degree zero are constants; they are excluded from this classification, since they are neither reducible nor irreducible. Polynomials of the first degree are then always irreducible.

If a polynomial $P(x)$ of degree n is reducible, that is, if it can be split up as a product $P(x) = p_1(x)\, p_2(x)$, then the polynomials $p_1(x)$ and $p_2(x)$ must be of degree at least equal to 1 and necessarily smaller than n. If $p_1(x)$ or $p_2(x)$ is reducible, the process can be repeated; after at most n steps the polynomial is factorized into a product $P(x) = g(x)\, h(x)\, k(x) \ldots$ With the help of the theorem (not proved here) that an irreducible polynomial dividing a product of two or more polynomials must divide *at least one* of them, it can be shown that the factorization of a reducible polynomial is *unique* apart from the order and up to constant factors. If $P(x) = g_1(x)\, h_1(x)\, k_1(x) \ldots$ and $P(x) = g(x)\, h(x)\, k(x) \ldots$ are two factorizations into irreducible factors, then $g_1(x)$ must divide one of the polynomials $g(x), h(x), k(x) \ldots$ But because these are themselves irreducible, it must be equal to one of them, except for a constant factor c_1. Without loss of generality one can assume that $g(x)$ is this polynomial. Then $g_1(x) = c_1 g(x)$. Dividing $P(x)$ by $g(x)$, one obtains:

$$c_1 h_1(x)\, k_1(x) \ldots = h(x)\, k(x) \ldots$$

By the same argument it follows that $h_1(x) = c_2 h(x)$, say, and $c_1 c_2 k_1(x) \ldots = k(x) \ldots$ Thus, except for constant numerical factors the two factorizations agree overall.

The question as to whether a polynomial is reducible depends essentially, of course, on which *number system* its coefficients and those of the irreducible factors belong to. For example, $x^2 - 1/4$ is irreducible over the domain of integers, but reducible over that of the rational numbers; $x^2 - 2$ is rationally irreducible, but factorizes into $(x - \sqrt{2})(x + \sqrt{2})$ over the real numbers; and $x^2 + 4$ is irreducible over the domain of real numbers, whereas it factorizes into $(x + 2i)(x - 2i)$ over the domain of complex numbers. If arbitrary complex numbers are allowed for the coefficients of the polynomial, then the fundamental theorem of algebra shows that each polynomial function of degree n can be factorized into n linear factors $(x - \alpha_k)$, $k = 1, 2, \ldots, n$. The numbers α_k are the zeros of the function. In the case of polynomials with real coefficients, if one of these numbers $\alpha = a + bi$ is complex, then the conjugate complex number $\bar{\alpha} = a - bi$ also occurs as a zero. The product of the corresponding linear factors is then $(x - \alpha)(x - \bar{\alpha}) = (x - a - bi)(x - a + bi) = (x - a)^2 + b^2 = x^2 - 2ax + (a^2 + b^2)$, that is, a *quadratic polynomial* with real coefficients. If all pairs of conjugate complex linear factors are collected together in this way, then over the field of real numbers the irreducible factors are either *real linear factors* or *quadratic polynomials* with real coefficients. When the coefficients are restricted to the field of real numbers, it follows that all irreducible polynomials are of degree at most 2. If it is further required that the coefficients of the given polynomial and its factors shall be rational, this theorem no longer holds. For example, the factorization (for real numbers) $x^4 - 5 = (x^2 + \sqrt{5})(x^2 - \sqrt{5})$ is no longer allowed.

Zeros

A number α is called a *zero* of a function $x \to y = f(x)$ if the number 0 is assigned to the number α by the function, that is, if $\alpha \to f(\alpha) = 0$. Thus, for a polynomial f, $f(\alpha) = a_n\alpha^n + a_{n-1}\alpha^{n-1} + \cdots + a_1\alpha + a_0 = 0$.

In the graph of a function a real zero appears as an intersection or point of contact of the curve with the x-axis.

If α is a zero of a polynomial $f(x)$, then $f(x)$ is divisible by $(x - \alpha)$; thus, there exists a polynomial $g(x)$ such that $f(x) = (x - \alpha)\,g(x)$.

In any case the polynomial $f(x)$ can be divided by $(x - \alpha)$, where α is arbitrary. By this division one obtains a quotient function $g(x)$ of lower degree than $f(x)$, and the remainder r must be of lower degree than $(x - \alpha)$, and must therefore be a constant: $f(x) = (x - \alpha)\,g(x) + r$. If α is a zero, then for $x = \alpha$ this equation becomes $0 = 0 \cdot g(\alpha) + r$. Thus, the remainder r *must be zero*, the function $f(x)$ is divisible by the linear factor $(x - \alpha)$ without remainder, and $f(x) = (x - \alpha)\,g(x)$. A generalization of this theorem can be proved by the method of induction:

If $\alpha_1, \alpha_2, \alpha_3, \ldots, \alpha_k$ are zeros of a polynomial $f(x)$, then the product $(x - \alpha_1)(x - \alpha_2) \cdots (x - \alpha_k)$ is a factor of $f(x)$, that is, $f(x)$ can be expressed in the form $f(x) = (x - \alpha_1)(x - \alpha_2) \cdots (x - \alpha_k)\,g(x)$.

Example: The polynomial $f(x) = x^3 - 5x^2 + 7x - 3$ has the zero $x = 3$. Division by $(x - 3)$ gives the quotient $x^2 - 2x + 1$, so that the polynomial can be expressed in the form $f(x) = (x - 3)(x^2 - 2x + 1) = (x - 3)(x - 1)^2$.

A polynomial $f(x) = a_nx^n + a_{n-1}x^{n-1} + \cdots + a_1x + a_0$ has at most n distinct zeros.

Proof by induction: 1. For $n = 1$, that is, for the polynomial $a_1x + a_0$ with $a_1 \neq 0$ (since otherwise the polynomial is not of the first degree), the theorem holds, because this polynomial has the single zero $x = -a_0/a_1$.

2. Let $f(x)$ be a polynomial of degree $n + 1$ and α a zero of this polynomial. By the previous theorem, $f(x)$ can be expressed in the form $f(x) = (x - \alpha)\,g(x)$, where $g(x)$ is only of degree n. The product $(x - \alpha)\,g(x)$ can be zero only if at least one factor is zero. The first factor is zero for $x = \alpha$, and by the inductive hypothesis the second factor is zero for at most n further values of x. Thus, the product, and hence $f(x)$, is equal to zero for at most $n + 1$ different values of x. The theorem is now proved.

Multiplicity of a zero. It can happen that a polynomial with a zero α is divisible not only by $(x - \alpha)$, but also by $(x - \alpha)^2$, $(x - \alpha)^3$ or a still higher power of $(x - \alpha)$. If $f(x)$ is divisible by $(x - \alpha)^k$ but not by $(x - \alpha)^{k+1}$, then α is called a *k-fold zero* or a *zero of multiplicity k* ($k \geqslant 1$, an integer) of $f(x)$.

Example: The polynomial $x^4 - 9x^3 + 27x^2 - 31x + 12$ has a double zero for $x = 1$, that is, it is divisible by $(x - 1)^2$ but not by $(x - 1)^3$.

$$f(x) = (x - 1)^2 (x^2 - 7x + 12) = (x - 1)^2 (x - 3)(x - 4).$$

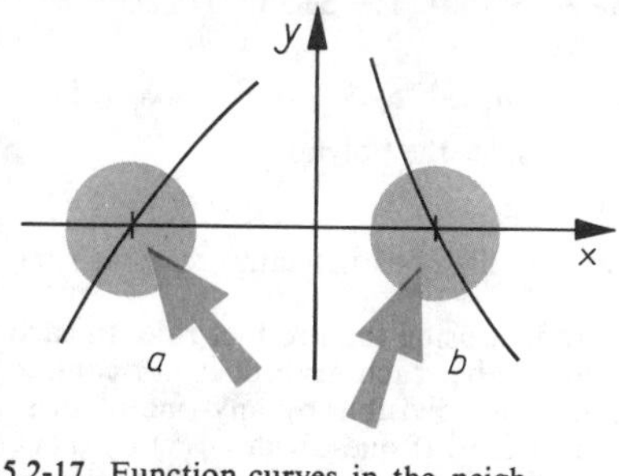

5.2-17 Function curves in the neighbourhood of a simple zero

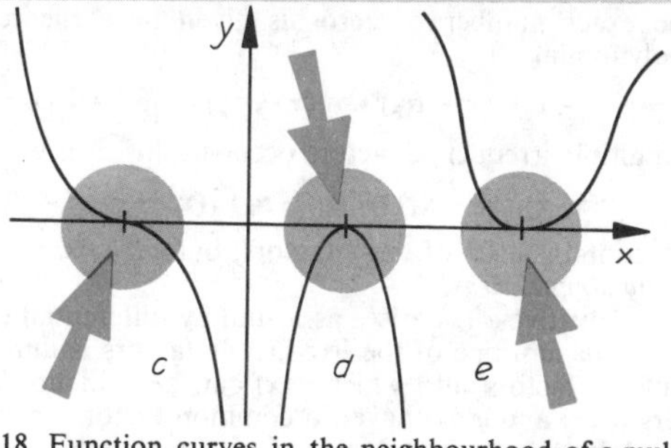

5.2-18 Function curves in the neighbourhood of a multiple zero, in c of odd order, in d and e of even order

Simple and multiple zeros. Different multiplicities of a zero give rise to different behaviour of the graph of a function in its neighbourhood. At a simple zero (of multiplicity 1) the curve always has a slope different from zero (positive or negative), while at a multiple zero the slope is zero, so that the tangent to the curve at such a point coincides with the x-axis (Fig.).

Zeros of even and odd order. The behaviour of the curve also differs according to whether the *multiplicity, or order, of a zero* is even or odd. Let α be a zero of $f(x)$ of multiplicity k. Then $f(x)$ can be factorized in the form $f(x) = (x - \alpha)^k g(x)$, where $g(x)$, because of its continuity, is different from zero in a neighbourhood of α, and consequently does not change sign in this neighbourhood. Thus, there exists an $\varepsilon > 0$, such that $g(x) \neq 0$ for each x with $|x - \alpha| < \varepsilon$. But the linear factor $(x - \alpha)$ changes its sign as x passes from $x < \alpha$ to $x > \alpha$ (Fig.). The polynomial $f(x) = (x - \alpha)^k g(x)$ therefore changes its sign in this passage if and only if k is odd (Fig.). For even k, $f(x)$ keeps the same sign.

Zeros and factorization. Every polynomial function can be expressed as a product of irreducible factors in the form

$$f(x) = c(x - \alpha_1)^{r_1} (x - \alpha_2)^{r_2} \dots (x - \alpha_k)^{r_k} (x^2 + a_1x + b_1)^{s_1} (x^2 + a_2x + b_2)^{s_2} \dots (x^2 + a_lx + b_l)^{s_l}.$$

In this expression c is a polynomial of degree zero, that is, a constant different from zero, α_i, a_j and b_j denote real numbers, and r_i and s_j are natural numbers. The exponents satisfy the condition $n = \sum_{i=1}^{k} r_i + 2 \sum_{j=1}^{l} s_j$. It is immediately obvious that the α are *zeros* of the function. Moreover, there are no other real zeros. For each value α of x other than $\alpha_1, \alpha_2, \dots, \alpha_l$, each linear factor $(x - \alpha_i)$, $i = 1, 2, \dots, k$, is different from zero. If one of the quadratic polynomials $x^2 + a_jx + b_j$ were equal to zero for $x = \alpha$, then it would be reducible, contradicting the assumption. An irreducible quadratic polynomial is different from zero for all real values of x, because it has only two conjugate complex zeros. As a consequence of these arguments one obtains the theorem:

The number of real zeros of a polynomial, counted according to their multiplicity, is even precisely when the degree of the polynomial is even, and odd when it is odd. In particular: A polynomial of odd degree has at least one real zero.

Sturm's theorem. By means of an approximation method, for example, that of NEWTON, every root of each polynomial function can be calculated with arbitrary accuracy if a value of x in the neighbourhood of the zero is known. Already DESCARTES, NEWTON and FOURIER, had tried to find criteria to decide whether a root of a polynomial lies in a given interval of the domain of definition. By suitable choice of the interval values of x in the neighbourhood of the zero can be found.

Descartes' rule of signs. DESCARTES considered the signs of the coefficients of the polynomial $f(x) = a_nx^n + a_{n-1}x^{n-1} + \dots + a_1x + a_0$. Here it can be assumed that neither a_n nor a_0 is zero. Other coefficients that are zero are not included in the sequence $a_n, a_{n-1}, \dots, a_1, a_0$. If two neighbouring coefficients have different signs, one then speaks of a sign change.

Descartes found that the number of positive zeros of a polynomial is equal to the number of its sign changes, or differs from it by a positive even number. The number of negative zeros is obtained similarly from the number of sign changes in the sequence of coefficients of the polynomial $f(-x)$.

Example: The polynomial $f(x) = x^5 - x^4 + 2x^3 + x^2 - 3x + 2$ has four, two, or no positive zeros, because four sign changes occur in the sequence of coefficients $1; -1; 2; 1; -3; 2$. If one forms $f(-x) = -x^5 - x^4 - 2x^3 + x^2 + 3x + 2$, one sees that $f(x)$ must have exactly one negative zero, because one sign change occurs in the sequence of coefficients of $f(-x)$.

The exact number of zeros is given by a theorem due to STURM. He begins by factorizing the polynomial

$$f(x) = c(x - \alpha_1)^{r_1}(x - \alpha_2)^{r_2} \cdots (x - \alpha_k)^{r_k} \cdot (x^2 + a_1x + b_1)^{s_1} (x^2 + a_2x + b_2)^{s_2} \cdots (x^2 + a_lx + b_l)^{s_l}.$$

If multiple irreducible factors occur in this, it is enough to consider the polynomial

$$\varphi(x) = (x - \alpha_1) \cdots (x - \alpha_k) \cdot (x^2 + a_1x + b_1) \cdots (x^2 + a_lx + b_l),$$

which contains each of these factors, but *each one only once*; $\varphi(x)$ then has the same zeros as $f(x)$, but only *simple zeros*.

The derivative $\varphi'(x)$ arises as a sum by differentiating the product using the product rule. In each term of the sum one of the irreducible factors is differentiated, so that each term does not contain one of the factors into which $\varphi(x)$ can be split up. The sum is not divisible by any one of these factors: $\varphi(x)$ and $\varphi'(x)$ have no common factor except for a constant. If one divides $\varphi(x)$ by $\varphi'(x)$, one obtains a polynomial $q_1(x)$ and a remainder $-\varphi_2(x)$, which is a polynomial of lower degree than $\varphi'(x)$; $\varphi(x) = q_1(x)\,\varphi'(x) - \varphi_2(x)$. Dividing $\varphi'(x)$ by $\varphi_2(x)$ therefore gives a new remainder term $-\varphi_3(x)$, for which $\varphi'(x) = q_2(x)\varphi_2(x) - \varphi_3(x)$. This procedure must terminate after finitely many steps; in a simplified notation one obtains the accompanying scheme. From the last equation, the preceding one, and step by step back to the first one, it is obvious that φ_r is a factor of φ_{r-1}, of φ_{r-2}, of φ_{r-3} and so on, and finally also of φ' and of φ. Because φ and φ' have no common factor, φ_r can only be a non-zero constant. The sequence of these functions $\varphi, \varphi', \varphi_2, \varphi_3, \dots, \varphi_r$ is called the *Sturm chain.* Substituting for x a particular value a in the polynomials of the Sturm chain gives a sequence of real numbers $\varphi(a), \varphi'(a), \varphi_2(a), \dots, \varphi_r(a)$. If two *neighbouring* numbers $\varphi_i(a)$ and $\varphi_{i+1}(a)$ in this sequence have different signs, one speaks of a *sign change.* $W(a)$ denotes the *numbers of sign changes* in the Sturm chain for the value $x = a$. Clearly the number $W(x)$ of sign changes can alter only if the argument x passes through a zero of one of the polynomials of the chain $\varphi, \varphi', \varphi_2, \dots, \varphi_{r-1}$. First let one of the polynomials $\varphi', \varphi_2, \dots, \varphi_{r-1}$ following φ be equal to zero for $x = \xi$. From the above scheme it can be read off at once that its two neighbouring terms are not zero and have different signs. Consequently the number $W(x)$ of sign changes cannot alter in passing through this value; this can only happen if x passes through a zero of φ itself. In this case, in fact, $W(x)$ decreases by 1 exactly, because φ has only simple zeros, so that φ changes sign in passing through the zero, whilst φ' is different from zero in some neighbourhood of this point and because of its continuity has a constant sign there. This proves the following theorem.

$$\begin{aligned}
\varphi &= q_1\varphi' - \varphi_2\\
\varphi' &= q_2\varphi_2 - \varphi_3\\
\varphi_2 &= q_3\varphi_3 - \varphi_4\\
\varphi_3 &= q_4\varphi_4 - \varphi_5\\
\varphi_4 &= q_5\varphi_5 - \varphi_6\\
\dots &= \dots\\
\varphi_{r-2} &= q_{r-1}\varphi_{r-1} - \varphi_r\\
\varphi_{r-1} &= q_r\varphi_r
\end{aligned}$$

Sturm's theorem. If *$\varphi(x)$ is a polynomial with only simple zeros, if $a < b$ and $\varphi(a) \neq 0$, $\varphi(b) \neq 0$, then $W(a) - W(b)$ is equal to the number of zeros of the polynomial $\varphi(x)$ in the closed interval $[a, b]$.*

In order to determine with the help of this theorem the exact number of all the zeros of the polynomial $\varphi(x)$, one chooses for x_1 and $x_2 > x_1$ numbers $-M$ and $+M$ whose absolute values are greater than the maximum of the absolute values of the zeros; thus, $M > \max(|\alpha_1|, |\alpha_2|, \dots, |\alpha_k|)$, where $\alpha_1, \alpha_2, \dots, \alpha_k$ are the zeros of $\varphi(x)$. For this purpose one has to determine M without knowing the zeros. This is possible, because the estimate

$$\max(|\alpha_1|, |\alpha_2|, \dots, |\alpha_k|) < 1 + |a_{n-1}| + |a_{n-2}| + \cdots + |a_1| + |a_0|$$

holds for the absolute values of the zeros of the polynomial $\varphi(x) = x^n + a_{n-1}x^{n-1} + \cdots + a_0$ in question. Each polynomial with $a_n \neq 1$ can be normalized on dividing by a_n. The zeros remain unchanged by this. One can therefore choose $M = 1 + |a_{n-1}| + \cdots + |a_1| + |a_0|$ and is then sure that all the zeros of $\varphi(x)$ lie in the interval $[-M, M]$. A proof of this fact would lead too far here. It will, however, be plausible if one thinks of the connection between the zeros x_1 and x_2 (assumed to be real) of $f(x) = x^2 + ax + b$ and the coefficients a and b. It is known that $x_1 + x_2 = -a$ and $x_1x_2 = b$. It is clear from this that $|x_1|$ and $|x_2|$ cannot be very large, and $|a|$ and $|b|$ very small at the same time. In other words: the absolute values of the zeros cannot exceed certain bounds that depend on the absolute values of the coefficients.

Example: In order to determine the number of real zeros of the polynomial $\varphi(x) = x^5 - 2x^4 - x + 2$, the Sturm chain must be calculated.

To simplify the calculations, the chain polynomials in the given example are multiplied by positive numbers; this clearly does not alter the number of sign changes. One recognizes that $\varphi(x)$ has only simple zeros (and that Sturm's theorem is therefore applicable) from the fact that the Sturm chain terminates with a polynomial of degree zero.

The following summary applies to the given polynomial:

Sturm chain	calculation	scheme	signs at the interval limits	
$\varphi(x) = x^5 - 2x^4 - x + 2$			$\varphi(-6) = -\ 10\,360$	$\varphi(+6) = +\ 5\,180$
$\varphi'(x) = 5x^4 - 8x^3 - 1$ →	$5\varphi : \varphi'$ →	$\varphi = q_1\varphi' - \varphi_2$	$\varphi'(-6) = +\ 8\,207$	$\varphi'(+6) = +\ 4\,751$
$\varphi_2(x) = 16x^3/5 + 4x - 48/5$ →	$4\varphi' : 5\varphi_2/4$ →	$\varphi' = q_2\varphi_2 - \varphi_3$	$\varphi_2(-6) = -\ 724^4/_5$	$\varphi_2(+6) = +\ 705^3/_5$
$\varphi_3(x) = 25x^2 - 100x + 100$ →	$5\varphi_2/4 : \varphi_3/25$ →	$\varphi_2 = q_3\varphi_3 - \varphi_4$	$\varphi_3(-6) = +\ 1\,600$	$\varphi_3(+6) = +\ 400$
$\varphi_4(x) = -53x + 76$ →	$53\varphi_3/25 : \varphi_4$ →	$\varphi_3 = q_4\varphi_4 - \varphi_5$	$\varphi_4(-6) = +\ 394$	$\varphi_4(+6) = -\ 242$
$\varphi_5(x) = -16^{52}/_{53}$			$\varphi_5(-6) = -\ 16^{52}/_{53}$	$\varphi_5(+6) = -\ 16^{52}/_{53}$

For the polynomial $\varphi(x) = x^5 - 2x^4 - x + 2$, M can be taken to be $1 + |-2| + |-1| + |+2| = 6$; thus, all the zeros of the polynomial lie in the interval $[-6, 6]$. In the summary the values for $\varphi(-6), \varphi'(-6), \varphi_2(-6), \varphi_3(-6), \varphi_4(-6), \varphi_5(-6)$ and for $\varphi(6), \varphi'(6), \varphi_2(6), \varphi_3(6), \varphi_4(6), \varphi_5(6)$ are tabulated. By counting sign changes one obtains $W(-6) = 4$, $W(6) = 1$; hence the polynomial has exactly $4 - 1 = 3$ real zeros.

Separation of the zeros. By *separation of the zeros* one understands the determination of intervals within which exactly one zero lies. By means of the above example it will be shown how this is possible with the help of Sturm's theorem.

Substitution $x = 0$ in the Sturm chain belonging to $\varphi(x) = x^5 - 2x^4 - x + 2$ shows that $W(-6) - W(0)$ gives the number of zeros in the interval $[-6, 0]$. Since $\varphi(0) = +2$, $\varphi'(0) = -1$, $\varphi_2(0) = -9^3/_5$, $\varphi_3(0) = +100$, $\varphi_4(0) = +76$, $\varphi_5(0) = -16^{52}/_{53}$, it follows that $W(0) = 3$. But $W(-6) = 4$, and so $W(-6) - W(0) = 1$; hence exactly one zero lies in the interval $[-6, 0]$. The other two must therefore lie in the interval $[0, 6]$. To separate them one can again halve this interval and examine the Sturm chain for $x = 3$. One finds that $W(3) = 1$. Because $W(0) - W(3) = 2$, both zeros must lie in the interval $[0, 3]$, and thus they are still not separated. Halving again yields $W(1.5) = 2$. Now $W(0) - W(1.5) = 1$ and $W(1.5) - W(3) = 1$, so that exactly one zero lies in each of the intervals $[0, 1.5]$ and $[1.5, 3]$. The three zeros are now separated.

Extension of Sturm's theorem. The assumption that no *multiple factors* occur in the factorization of $f(x)$ will now be considered. For a given polynomial it is not always possible to decide at once whether this condition is satisfied. In spite of this one can follow Sturm's method. If the greatest common divisor of $f(x)$ and $f'(x)$ is determined with the help of the *Euclidean algorithm*, then there are two possibilities:

a) if $f(x)$ satisfies the given condition, then the gcd is a constant different from zero, and Sturm's theorem is immediately applicable.

b) if $f(x)$ does not satisfy the given condition, then $f(x)$ has factors of the form $(x - \alpha_i)^{r_i}$ or $(x^2 + a_jx + b_j)^{s_j}$ with $r_i > 1$ or $s_j > 1$. Then $(x - \alpha_i)^{r_i - 1}$ or $(x^2 + a_jx + b_j)^{s_j - 1}$ is a factor of $f'(x)$, as can be seen from the product rule for differentiation. The product of these factors and possibly a constant c ($c \neq 0$, $c \neq 1$) then appears as the gcd of $f(x)$ and $f'(x)$. When $f(x)$ is divided by this gcd, the quotient satisfies the assumptions of Sturm's theorem, and the theorem can then be applied to it. The gcd of $f(x)$ and $f'(x)$ need not be further investigated for zeros, because it can only have zeros that are also zeros of the quotient polynomial.

The behaviour of polynomial functions at infinity

Besides the zeros, other special properties of polynomial functions are often of interest: extrema, points of inflection, gradient at zeros and points of inflection, among others. The corresponding investigations are made with the help of the infinitesimal calculus and will therefore not be taken further at this stage. It is, however, common to all these investigations that only an interval of the domain of definition bounded on both sides is considered. The question arises, how does the graph of a polynomial function look outside such an interval, what values can it take if $|x|$ is greater than the maximum of the absolute values of all the zeros, extrema, points of inflection, and so on, of the function. The answer to this question is usually called the *behaviour at infinity*. Taking the first term a_nx^n out of the expression $f(x) = a_nx^n + a_{n-1}x^{n-1} + \cdots + a_1x + a_0$ leads to

$$f(x) = a_nx^n\left(1 + \frac{a_{n-1}}{a_nx} + \frac{a_{n-2}}{a_nx^2} + \cdots + \frac{a_0}{a_nx^n}\right).$$

From this representation it can be seen that for unbounded increasing $|x|$, $|f(x)|$ also increases beyond all limits, because the expression in the brackets tends to 1 in this case, while $|a_nx^n|$ becomes arbitrarily large. This property is often expressed symbolically in the form $\lim\limits_{|x| \to \infty} |f(x)| = \infty$. The sign of the function $f(x)$ for $|x| \to \infty$ depends only on a_nx^n, because the expression in the bracket is certainly positive from a certain x_p onwards, for all $|x| > x_p$. There are only the possibilities collected together in the table.

a_n	n	$x \to$	$f(x) \to$
>0	even	$+\infty$	$+\infty$
		$-\infty$	$+\infty$
	odd	$+\infty$	$+\infty$
		$-\infty$	$-\infty$
<0	even	$+\infty$	$-\infty$
		$-\infty$	$-\infty$
	odd	$+\infty$	$-\infty$
		$-\infty$	$+\infty$

Example 1: The function $y = x^4 - x^3 - x^2 - x - 2$ has the real zeros $x = -1$ and $x = 2$. A minimum of the function occurs for $x \approx 1.3$, and points of inflection for $x \approx 0.73$ and $x \approx -0.23$. For the behaviour of the function at infinity, $\lim\limits_{x \to +\infty} f(x) = +\infty$ and $\lim\limits_{x \to -\infty} f(x) = +\infty$. By determining some values of the function the following table of values is obtained:

x	−2	−1.6	−1.3	−1	−0.7	−0.23	0	0.3	0.73	1	1.3	1.6	2	2.3	3
y	20	7.6	2.6	0	−1.2	−1.8	−2	−2.4	−3.4	−4	−4.3	−3.7	0	6.2	40

The function can now be represented graphically (Fig.).

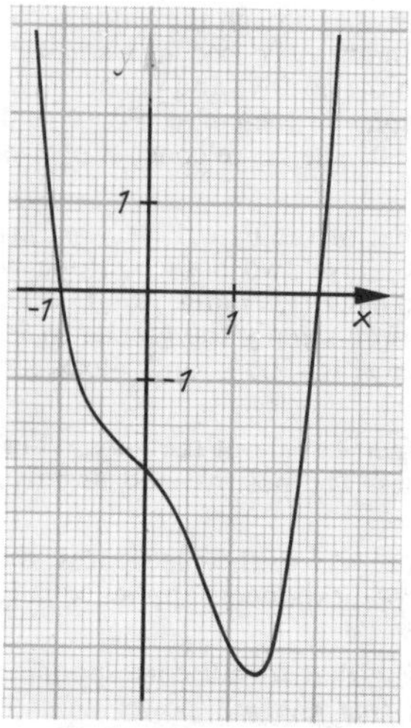

5.2-19 Graph of the function $y = x^4 - x^3 - x^2 - x - 2$

5.2-20 Graph of the function $y = 0.025x^5 + 0.05x^4 - 0.6x^3 - 0.55x^2 + 2.575x - 1.5$

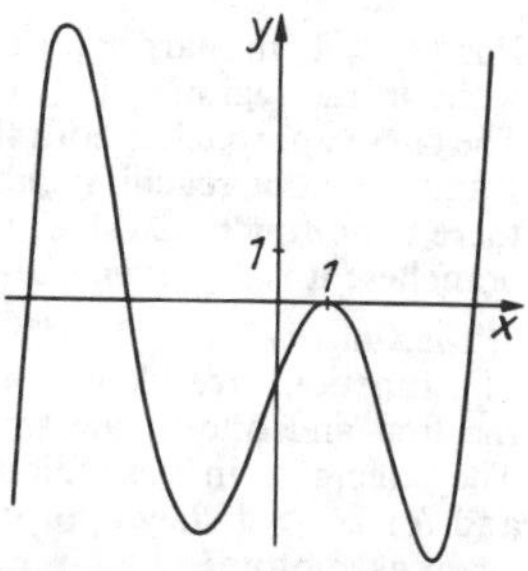

Example 2: The function $y = 0.025x^5 + 0.05x^4 - 0.6x^3 - 0.55x^2 + 2.575x - 1.5$ has simple zeros for $x = -5$, $x = -3$ and $x = 4$ and a double zero for $x = 1$. Local minima occur for $x = -1.53$ and $x = 3.17$ and local maxima for $x = -4.24$ and $x = 1$. Points of inflection occur for $x = -3.22$, $x = -0.29$ and $x = 2.32$. (These numbers are approximate values.) The behaviour of the function at infinity is characterized by $\lim_{x\to+\infty} f(x) = +\infty$ and $\lim_{x\to-\infty} f(x) = -\infty$. The following table of values is used to construct the graph of the function (Fig.).

x	−5.3	−5	−4.7	−4.23	−3.8	−3.22	−3	−2.3	−1.53	−1	−0.3
y	−6.4	0	3.6	5.3	4.3	1.25	0	−3.25	−4.5	−4	−2.3
x	0	0.5	1	1.5	2	2.32	3	3.16	3.7	4	4.3
y	−1.5	−0.42	0	−0.45	−1.73	−2.9	−4.8	−5	−3.2	0	5.6

Power functions with negative exponents

The simplest *rational* functions (other than polynomials) are those whose rule of correspondence can be expressed in the form $y = 1/x^n$ $(n = 1, 2, 3, \ldots)$. These are called *power functions with negative exponents*, because one can also write x^{-n} for $1/x^n$. They will be investigated first.

The function $y = 1/x$. This is clearly an odd function and its graph is therefore centrally symmetric with respect to the origin (Fig.).

Table of values for $y = 1/x$:

x	−4	−3	−2	−1	−1/2	−1/100	1/1000	1/50	1/5	1
y	−1/4	−1/3	−1/2	−1	−2	−100	1000	50	5	1

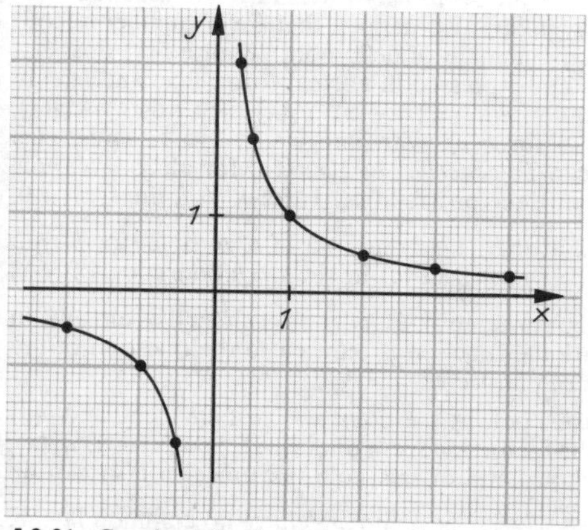

5.2-21 Graph of the function $y = 1/x$

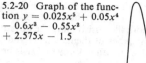

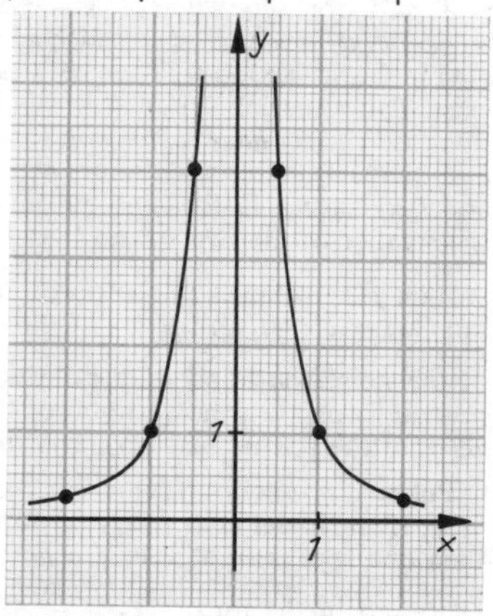

5.2-22 Graph of the function $y = 1/x^2$

For $|x| > 1$, the larger $|x|$ becomes, the nearer the ordinates of the curve approach the value zero, while in the region $-1 < x < +1$ the ordinates increase beyond all limits as $|x|$ becomes smaller. The curve approaches both the positive and the negative x-axis, and also the positive and the negative y-axis, without reaching either of them. The x- and y-axes are *asymptotes* of the curve. For $x = 0$ there is no function value: the function $y = 1/x$ is not defined for $x = 0$. The curve consists of two branches; it is a *rectangular hyperbola*.

The functions $y = 1/x^{2m+1}$. The shape of the curves is similar to that of the hyperbola $y = 1/x$. The functions are likewise odd. The curves are not defined for $x = 0$, have two branches, one in the first and one in the third quadrant, and all pass through the points $P(1, 1)$ and $R(-1, -1)$. They decrease in the region $-1 < x < 0$ and in $0 < x < +1$, the more steeply the greater m is, and for $|x| > 1$ they approach the x-axis, the more quickly the greater m is. The x- and y-axes are again asymptotes.

The function $y = 1/x^2$. It is an even function, whose graph is symmetrical about the y-axis (Fig. 5.2-22). For $x = 0$ it is not defined, and it therefore has two branches. The positive and negative x-axis and the positive y-axis are asymptotes.

The function $y = 1/x^{2m}$. These curves are similar to that of $y = 1/x^2$. As to their steepness, the remark made about the branches of the power functions with negative odd exponents applies equally to them. The points $P(1, 1)$ and $Q(-1, 1)$ are common to all of them.

Power functions and proportionality. Because it follows from $y = kx^n$ that for all corresponding values the ratio $y_1/x_1^n = y_2/x_2^n = \cdots = y/x^n = k = \text{const}$, one says that the nth power of x is proportional to y.

In the correspondence $y = k/x$, the smaller y is, the larger x becomes, and vice versa. Such a relationship is known as *inverse proportion* and defined by the property that the product of corresponding values is constant, $xy = k$. In both cases k is called the *constant of proportionality*.

In free fall the distance s fallen is proportional to the square of the time; the force of attraction F between two masses is inversely proportional to the square of their distance r apart. Hence the corresponding laws must have the form $s = kt^2$ and $F = m/r^2$, where the constant of proportionality can be calculated if one pair (s, t) or (r, F) is known.

General form of rational functions

As for the polynomial functions, so for rational functions there exists a representation that may be regarded as a *normal form*.

The correspondence rule for each rational function $f(x)$ can be expressed as the quotient of two polynomials $p(x)$ and $q(x)$ having no common factor, that is, $x \to f(x) = p(x)/q(x)$.

If the polynomial $q(x)$ in the denominator has degree 0, then it is a constant, and the special case of a polynomial function arises. In the following it will be assumed that the degree of $q(x)$ is at least one, so that the rational function in question is not a polynomial.

Zeros and poles of rational functions

Zeros. A rational function can take the value zero only for values of x for which the numerator $p(x)$ of the normal form $p(x)/q(x)$ is zero, and at the same time $q(x)$ is not zero. Thus, a number α is a *zero* precisely when $p(\alpha) = 0$ and $q(\alpha) \neq 0$. For an arbitrarily given rational function $f(x) = g(x)/h(x)$ in which $g(x)$ and $h(x)$ are polynomials, it can also happen that for a number α both $g(\alpha) = 0$ and $h(\alpha) = 0$. Then α does not belong to the domain of definition, because $f(\alpha)$ does not exist. This situation arises from the fact that $f(x)$ is not expressed in normal form: $g(x)$ and $h(x)$ in this case clearly *have a common factor*. A representation $g(x) = (x - \alpha)^k g_1(x)$ then exists for $g(x)$, and similarly a representation $h(x) = (x - \alpha)^l h_1(x)$ for $h(x)$, with integers $k, l \geqslant 1$. Thus, $g(x)$ and $h(x)$ have a factor $(x - \alpha)^m$ in common that can be *cancelled* in $g(x)/h(x)$ for all $x \neq \alpha$; m is the smaller of the two values k and l. There are then three possibilities: 1. for $k > l$, $\lim_{x\to\alpha} f(x) = 0$, so that $f(x)$ behaves in the neighbourhood of α as in the neighbourhood of a zero; 2. for $k = l$, $\lim_{x\to\alpha} f(x) = c \neq 0$; 3. for $k < l$, $\lim_{x\to\alpha} f(x) = \infty$, so that $f(x)$ behaves in the neighbourhood of α as in the neighbourhood of a pole (see the following paragraph). If $f(x) = p(x)/q(x)$ is already in normal form, then the question of the zeros of a rational function is reduced to the question of the zeros of the polynomial $p(x)$, and this has already been answered.

Poles. The function $f(x) = p(x)/q(x)$ is said to have a pole at the point $x = \alpha$ if $q(\alpha) = 0$ and $p(\alpha) \neq 0$. If the linear factor $(x - \alpha)$ occurs k times in the factorized form of $q(x)$, $q(x) = (x - \alpha)^k q_1(x)$, then one speaks of a *pole of order k*. In the neighbourhood of this pole the function $f(x)$ can be expressed in the form $f(x) = \dfrac{p(x)}{q(x)} = \dfrac{1}{(x-\alpha)^k} \cdot \dfrac{p(x)}{q_1(x)}$. If $p(x)$ and $q(x)$ have no common factor, then neither $p(x)$ nor $q_1(x)$ has a zero in a neighbourhood of $x = \alpha$, hence they do not change their signs, and their quotient therefore has a bounded positive or negative (non-zero) value. But the function $1/(x - \alpha)^k$ increases without limit as $x \to \alpha$. If one approaches the pole in the sense of increasing values of x ($x < \alpha$), then $(x - \alpha)$ is negative and $1/(x - \alpha)^k$ tends to $-\infty$ for odd values of k and to $+\infty$ for even values of k. If one approaches the pole in the sense of decreasing values of x ($x > \alpha$), then $(x - \alpha)$ is positive and so $1/(x - \alpha)^k$ always tends to $+\infty$. This property of the function $1/(x - \alpha)^k$ is changed by the factor $p(x)/q_1(x)$ only to the extent that for negative values of the factor the sign of the function $f(x)$ is reversed (Fig.). The line $x = \alpha$ is an asymptote of the function.

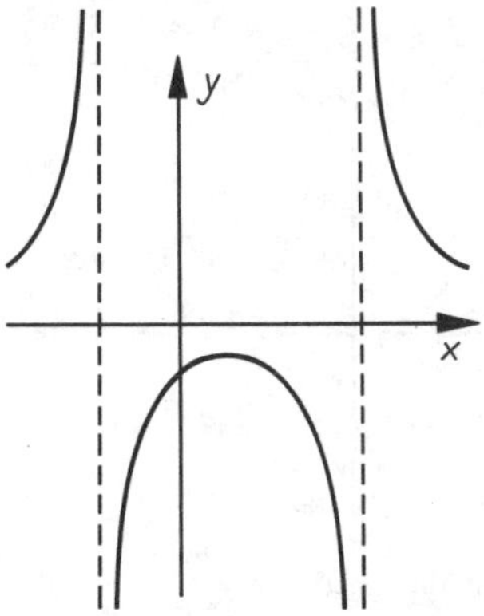

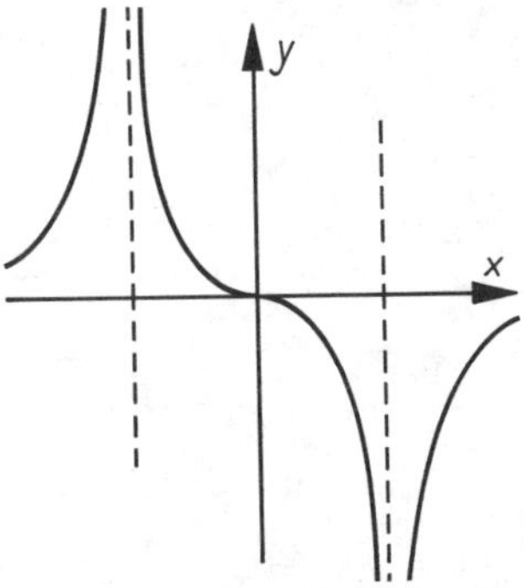

5.2-24 Graph of a function with poles of even order

5.2-23 Graph of a function with poles of odd order

The behaviour of rational functions at infinity

Beginning with the general form

$$f(x) = \frac{p(x)}{q(x)} = \frac{a_m x^m + a_{m-1}x^{m-1} + \cdots + a_1 x + a_0}{b_n x^n + b_{n-1}x^{n-1} + \cdots + b_1 x + b_0},$$

there are three possibilities to consider, namely $m < n$, $m = n$, $m > n$. If the degree of the numerator polynomial $p(x)$ is greater than or equal to the degree of the denominator polynomial $q(x)$ ($m \geqslant n$), the function $f(x)$ is called an *improper rational function*. Dividing the numerator by the denominator, a polynomial function $g(x)$ of degree $(m - n)$ can always be split off:

$$f(x) = p(x)/q(x) = g(x) + r(x).$$

In the case $m = n$, $g(x)$ is the constant a_m/b_n. The remainder $r(x)$ is always a *proper rational function*, that is, the degree of its numerator is less than that of its denominator. But the behaviour of a polynomial function at infinity is known; it only remains to investigate that of the proper rational function. Dividing the numerator and denominator by x^m ($m < n$), one obtains:

$$f(x) = \frac{a_m + a_{m-1}/x + \cdots + a_1/x^{m-1} + a_0/x^m}{b_n x^{n-m} + \cdots + b_1/x^{m-1} + b_0/x^m}.$$

As $|x| \to \infty$, the numerator tends to the value a_m, while at the same time the absolute value of the denominator can take arbitrarily large values; thus, $|f(x)| \to 0$ as $|x| \to \infty$. Consequently the x-axis is an *asymptote* of the function $f(x)$. According to the signs of a_m and b_n and the degree $n - m$, the graph of the function approaches the x-axis asymptotically from above or below, through positive or negative values. For example, if $a_m > 0$, $b_n > 0$ and $n - m$ is odd, then $f(x)$ is positive as $x \to +\infty$ and negative as $x \to -\infty$. The corresponding result holds for the remainder $r(x)$ after the polynomial function $g(x)$ is split off from the improper rational function $f(x)$. The function $f(x)$ approaches the function $g(x)$ asymptotically as $|x| \to \infty$, *from above* if $r(x)$ has small, but positive values, and *from below* if $r(x)$ tends to zero through negative values. The graph of the function $g(x)$ is called the *limiting curve*. In particular, if $m = n$ and $g(x) = a_m/b_n$, the *parallel* to the x-axis at a distance a_m/b_n from it is an asymptote of the function $f(x)$ for $|x| \to \infty$.

Example 1: The function $y = \dfrac{3x - 6}{x^3 - 3x + 2}$ has a zero for $x = 2$, a pole of the first order for $x = -2$ and a pole of the second order for $x = 1$. Two local extrema, both maxima, occur for $x \approx -0.74$ and $x \approx 2.74$. The behaviour of the function at infinity is characterized by $\lim\limits_{|x| \to \infty} y = 0$. Thus, the x-axis is an asymptote of the function curve. To consider the sign of the function values for the whole domain of definition one writes the equation of the function in the more useful form $y = \dfrac{3(x-2)}{(x-1)^2\,(x+2)}$. It is easily seen that y is positive for $-\infty < x < -2$, negative for $-2 < x < 1$ and $1 < x < 2$, and again positive for $x > 2$ (Fig.). For a more precise graphical representation a table of values is needed:

x	-5	-4	-3	-2.5	-2.2	-1.8	-1.5	-1	-0.74	-0.3	0
y	0.2	0.36	0.94	2.20	6.15	-7.27	-3.36	-2.25	-2.14	-2.4	-3

x	0.5	1.3	1.5	1.8	2	2.74	3
y	-7.2	-7.1	-1.72	-0.25	0	0.16	0.15

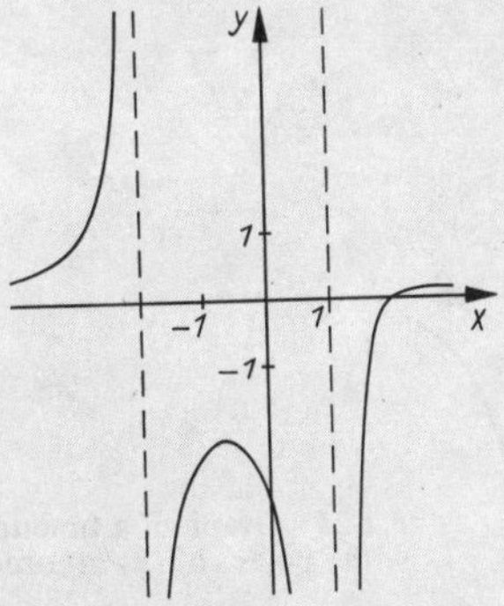

5.2-25 Graph of the function $y = \dfrac{3x-6}{(x-1)^2\,(x+2)}$

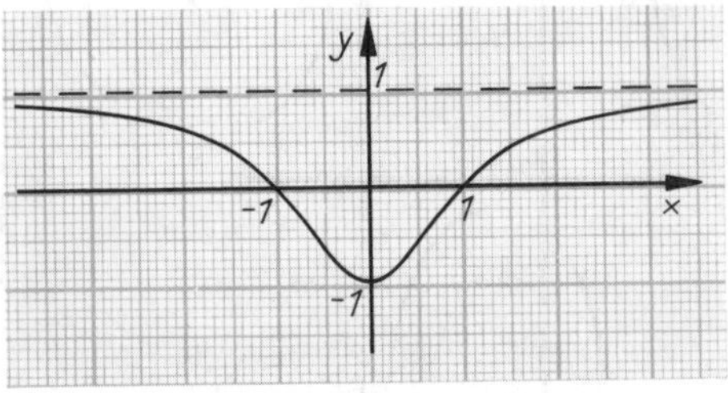

5.2-26 Graph of the function $y = \dfrac{x^2-1}{x^2+1}$

Example 2: The function $y = (x^2-1)/(x^2+1)$ has zeros at $x = -1$ and $x = 1$; it has no poles. An extremum (minimum) occurs for $x = 0$, and points of inflection for $x \approx -0.57$ and $x \approx 0.57$. By division one obtains $y = 1 - 2/(x^2+1)$. The line with the equation $y = 1$ is an asymptote of the function curve, because $\lim\limits_{|x|\to\infty} y = 1$. It is easy to see, in addition, that the graph of the function lies entirely below the asymptote (Fig.).

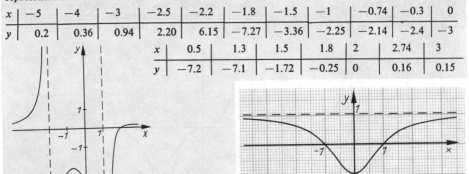

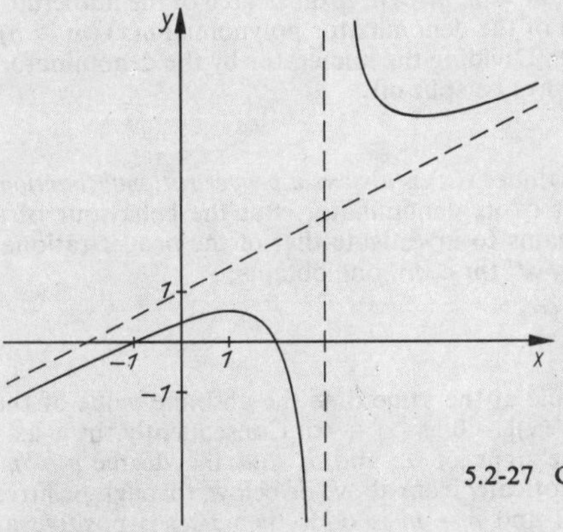

Table of values:

x	0	± 0.3	± 0.5	± 1	± 1.5	± 2	± 3	± 5
y	-1	-0.84	-0.6	0	0.38	0.6	0.8	0.92

Example 3: The function $y = (x^2 - x - 2)/(2x - 6)$ has zeros $x = -1$ and $x = 2$, a pole at $x = 3$, and local extrema for $x = 1$ (maximum) and $x = 5$ (minimum). Separation of the polynomial part yields the representation $y = x/2 + 1 + 4/(2x - 6)$. Thus, $y = x/2 + 1$ is an asymptote of the function curve. The curve approaches the asymptote from below as $x \to -\infty$ and from above as $x \to +\infty$ (Fig.).

5.2-27 Graph of the function $y = \dfrac{x^2 - x - 2}{2x - 6}$

Table of values:

x	-5	-3	-2	-1	0	1	1.5	2	2.5	2.8	3.5	4	5	6	7
y	-1.75	-0.83	-0.4	0	0.33	0.5	0.42	0	-1.75	-7.6	6.75	5	4.5	4.67	5

Example 4: For $y = f(x) = (x^3+2)/(2x)$ the curve $y = x^2/2$ is the limiting curve, because $(x^3+2)/(2x) = x^2/2 + 1/x$ and $\lim\limits_{|x|\to+\infty} [(x^3+1)/(2x) - x^2/2] = \lim\limits_{x\to+\infty} (1/x) = 0$.

Decomposition into partial fractions

To integrate a rational function $f(x)$ one has to express it as a sum of partial fractions. In its normal form $f(x) = p(x)/q(x)$ the numerator $p(x)$ and the denominator $q(x)$ have no common factor. If the degree of the numerator $p(x)$ is greater than or equal to the degree of the denominator, then a polynomial function $g(x)$ can be split off by division, $f(x) = g(x) + p_1(x)/q(x)$. The denominator $q(x)$, for its part, can be expressed as a product of linear factors, except for a constant factor that can be incorporated into the numerator:

$$q(x) = (x - \alpha_1)^{r_1} (x - \alpha_2)^{r_2} \dots (x - \alpha_k)^{r_k} (x - \beta_1)^{s_1} (x - \bar{\beta}_1)^{s_1} \dots (x - \beta_l)^{s_l} (x - \bar{\beta}_l)^{s_l},$$

in which the k real zeros α_i with multiplicities r_i, and the l pairs of conjugate complex zeros β_j and $\bar{\beta}_j$ with multiplicities s_j occur. The product of two conjugate complex linear factors is a real polynomial of the second degree; $(x - \beta)(x - \bar{\beta}) = x^2 - (\beta + \bar{\beta})x + \beta\bar{\beta} = x^2 + ax + b$. The substitutions $a = -(\beta + \bar{\beta})$ and $b = \beta\bar{\beta}$ have been made here. Then $q(x)$ can be expressed as the following product of polynomials that are irreducible in the domain of the real numbers:

$$q(x) = (x - \alpha_1)^{r_1} (x - \alpha_2)^{r_2} \cdots (x - \alpha_k)^{r_k} (x^2 + a_1 x + b_1)^{s_1} \cdots (x^2 + a_l x + b_l)^{s_l}.$$

Proper fractions whose denominator is a power of a linear factor or of an irreducible polynomial of the second degree are called partial fractions; in the first case the numerators are constants A, and in the second case linear polynomials $B + Cx$. The proper rational function $p_1(x)/q(x)$ can be expressed in the following form:

Partial fractions decomposition	$$\begin{aligned} \frac{p_1(x)}{q(x)} = {} & \frac{A_{11}}{x-\alpha_1} + \frac{A_{12}}{(x-\alpha_1)^2} + \dots + \frac{A_{1r_1}}{(x-\alpha_1)^{r_1}} \\ & + \frac{A_{21}}{x-\alpha_2} + \frac{A_{22}}{(x-\alpha_2)^2} + \dots + \frac{A_{2r_2}}{(x-\alpha_2)^{r_2}} \\ & + \dots\dots\dots \\ & + \frac{A_{k1}}{x-\alpha_k} + \frac{A_{k2}}{(x-\alpha_k)^2} + \dots + \frac{A_{kr_k}}{(x-\alpha_k)^{r_k}} \\ & + \frac{B_{11} + C_{11}x}{x^2 + a_1 x + b_1} + \frac{B_{12} + C_{12}x}{(x^2 + a_1 x + b_1)^2} + \dots + \frac{B_{1s_1} + C_{1s_1}x}{(x^2 + a_1 x + b_1)^{s_1}} \\ & + \frac{B_{21} + C_{21}x}{x^2 + a_2 x + b_2} + \frac{B_{22} + C_{22}x}{(x^2 + a_2 x + b_2)^2} + \dots + \frac{B_{2s_2} + C_{2s_2}x}{(x^2 + a_2 x + b_2)^{s_2}} \\ & + \dots\dots\dots \\ & + \frac{B_{l1} + C_{l1}x}{x^2 + a_l x + b_l} + \frac{B_{l2} + C_{l2}x}{(x^2 + a_l x + b_l)^2} + \dots + \frac{B_{ls_l} + C_{ls_l}x}{(x^2 + a_l x + b_l)^{s_l}} \end{aligned}$$

Here A_{ij}, B_{ij}, C_{ij} are real constants. That such a decomposition exists when the denominators are powers of linear factors can be seen as follows. Let α be a zero of order r of the denominator $q(x)$, so that $q(x) = (x - \alpha)^r q_1(x)$, where α is not a zero of $q_1(x)$. If the partial fraction $A/(x - \alpha)^r$ is separated from the given proper rational function $p_1(x)/q(x)$, one obtains:

$$\frac{p_1(x)}{(x-\alpha)^r q_1(x)} - \frac{A}{(x-\alpha)^r} = \frac{p_1(x) - Aq_1(x)}{(x-\alpha)^r q_1(x)} = \frac{\Phi(x)}{(x-\alpha)^r q_1(x)}.$$

Because neither $p_1(x)$ nor $q_1(x)$ is zero for $x = \alpha$, the undetermined constant A may be taken to be the number $p_1(\alpha)/q_1(\alpha)$. This ensures that the function $\Phi(x) = p_1(x) - Aq_1(x)$ has a *zero* for $x = \alpha$, and hence $\Phi(x) = (x - \alpha)\varphi(x)$. *Cancellation* gives

$$\frac{p_1(x)}{(x-\alpha)^r q_1(x)} - \frac{A}{(x-\alpha)^r} = \frac{\varphi(x)}{(x-\alpha)^{r-1} q_1(x)}.$$

This function is again a *proper rational function*, because the degree of $\varphi(x)$ is 1 smaller than that of $\Phi(x)$, whose degree is at most equal to that of $p_1(x)$ or of $q_1(x)$, and in any case less than that of $q(x) = (x - \alpha)^r q_1(x)$. By the same procedure a partial fraction $A_1/(x - \alpha)^{r-1}$ can again be split off from the function $\dfrac{\varphi(x)}{(x-\alpha)^{r-1} q_1(x)}$. A similar result holds for the other real zeros of the denominator $q(x)$.

If one admits *complex* numbers temporarily, then the same considerations hold also for zeros β and $\bar{\beta}$ of the denominator $q(x)$. One must remember that substitution of $\bar{\beta}$, the conjugate complex of β, in either of the functions $p_1(x)$ and $q_1(x)$ with *real coefficients*, gives rise to the conjugate complex values: thus, $A_1 = p_1(\bar{\beta})/q_1(\bar{\beta}) = \bar{p}_1(\beta)/\bar{q}_1(\beta) = \bar{A}_1$. Hence for each partial fraction $A/(x - \beta)^r$ the partial fraction $\bar{A}/(x - \bar{\beta})^r$ also occurs; their sum $\dfrac{A(x - \bar{\beta})^r + \bar{A}(x - \beta)^r}{(x^2 - [\beta + \bar{\beta}]x + \beta\bar{\beta})^r}$ is unchanged if each complex number is replaced by its conjugate, which means that it must be real and must have the form $h(x)/(x^2 + ax + b)^r$, where $h(x)$ is of degree at most r. If its degree is greater than 1, then $h(x)$ can be divided by $(x^2 + ax + b)$, giving

$$h(x) = h_1(x)(x^2 + ax + b) + (Bx + C) \quad \text{or}$$

$$\frac{h(x)}{(x^2 + ax + b)^r} = \frac{Bx + C}{(x^2 + ax + b)^r} + \frac{h_1(x)}{(x^2 + ax + b)^{r-1}}.$$

If the degree of $h_1(x)$ is again greater than 1, then it can again be divided by $(x^2 + ax + b)$. This decomposition is unique, as can be seen by multiplying by $(x - \alpha_\lambda)^{r_\lambda}$ and then equating coefficients.

Decomposition into partial fractions in practice. There are various possibilities for carrying out in practice the decomposition into partial fractions of a rational function. One can, for example, proceed step by step as in the proof given above. However, another method is usually more convenient, which will be explained by the following examples. It is the *method of undetermined coefficients.*

Example 1: The function $y = \dfrac{2x - 1}{(x + 2)^2 (x - 1)}$ is to be expressed as a sum of partial fractions. From the general theorem one knows that the decomposition must have the form:

$$\frac{2x - 1}{(x + 2)^2 (x - 1)} = \frac{A_1}{(x + 2)^2} + \frac{A_2}{x + 2} + \frac{A_3}{x - 1}.$$

If one multiplies this equation by the denominator $(x + 2)^2 (x - 1)$, it follows that $2x - 1 = A_1(x - 1) + A_2(x + 2)(x - 1) + A_3(x + 2)^2$. Removing the brackets in this identity and collecting together like powers of x leads to $2x - 1 = (A_2 + A_3)x^2 + (A_1 + A_2 + 4A_3)x + (-A_1 - 2A_2 + 4A_3)$. Equating coefficients one obtains the following system of equations for the determination of A_1, A_2 and A_3: I. $A_2 + A_3 = 0$; II. $A_1 + A_2 + 4A_3 = 2$; III. $-A_1 - 2A_2 + 4A_3 = -1$. It has the unique solution $A_1 = 5/3$, $A_2 = -1/9$, $A_3 = 1/9$. Hence the required decomposition into partial fractions is:

$$\frac{2x - 1}{(x + 2)^2 (x - 1)} = \frac{5}{3(x + 2)^2} - \frac{1}{9(x + 2)} + \frac{1}{9(x - 1)}.$$

Example 2: The decomposition into partial fractions of $y = p_1(x)/q(x)$ with $p_1(x) = x^2 + 5x$ and $q(x) = x^4 - 2x^3 + 2x^2 - 2x + 1 = (x - 1)^2 (x^2 + 1)$ starts out from $\dfrac{x^2 + 5x}{(x - 1)^2 (x^2 + 1)} = \dfrac{A_1}{(x - 1)^2} + \dfrac{A_2}{(x - 1)} + \dfrac{B + Cx}{(x^2 + 1)}$. Multiplying by the common denominator one obtains $x^2 + 5x = A_1(x^2 + 1) + A_2(x^2 + 1)(x - 1) + (B + Cx)(x - 1)^2$ and further $x^2 + 5x = (A_2 + C)x^3 + (A_1 - A_2 + B - 2C)x^2 + (A_2 - 2B + C)x + (A_1 - A_2 + B)$. The coefficients A_1, A_2, B and C must satisfy the following system of equations: I. $A_2 + C = 0$; II. $A_1 - A_2 + B - 2C = 1$; III. $A_2 - 2B + C = 5$; IV. $A_1 - A_2 + B = 0$. One finds that $A_1 = 3$, $A_2 = 1/2$, $B = -5/2$ and $C = -1/2$. Hence the required decomposition is:

$$\frac{x^2 + 5x}{x^4 - 2x^3 + 2x^2 - 2x + 1} = \frac{3}{(x - 1)^2} + \frac{1}{2(x - 1)} - \frac{5 + x}{2(x^2 + 1)}.$$

5.3. Non-rational functions

The definition of non-rational functions – also known as *irrational functions* – is given already by their name: they are functions that are not rational.

Root functions

The function $y = \sqrt{x}$. Corresponding to the definition of the square root, the function $y = \sqrt{x}$ is defined only for non-negative values of x; if the maximal domain $0 \leqslant x < +\infty$ is chosen as domain of definition, then the range is $0 \leqslant y < +\infty$. It follows further from the definition that $y = \sqrt{x}$ is the inverse function of $y = x^2$ in the interval $0 \leqslant x < +\infty$. The graph of the function $y = \sqrt{x}$ can be constructed with the help of a table of square roots or by taking the mirror image in the line determined by $y = x$ of the "half" parabola given by $y = x^2$ in the interval $0 \leqslant x < +\infty$ (Fig.). The inverse function of $y = x^2$ in the interval $-\infty < x < 0$ is $y = -\sqrt{x}$, with $0 < x < +\infty$.

The function $y = \sqrt[3]{x}$. Strictly speaking, in the first instance this function is also defined only for non-negative values of the argument. In the interval of definition $0 \leqslant x < +\infty$ it is the inverse of the function $y = x^3$ with $y \geqslant 0$ and hence $x \geqslant 0$. Its graph can again be obtained by taking the mirror image in the line $y = x$ of the "half" cubical parabola $y = x^3$ with $0 \leqslant x < +\infty$ (Fig.).

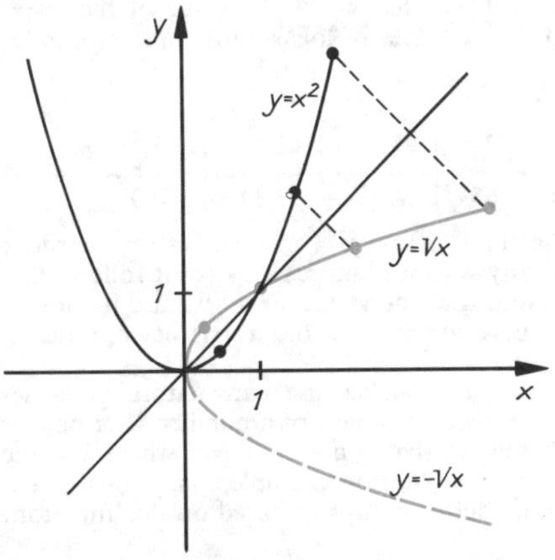

5.3-1 Graphs of the functions $y = \sqrt{x}$ and $y = x^2$ as mirror images

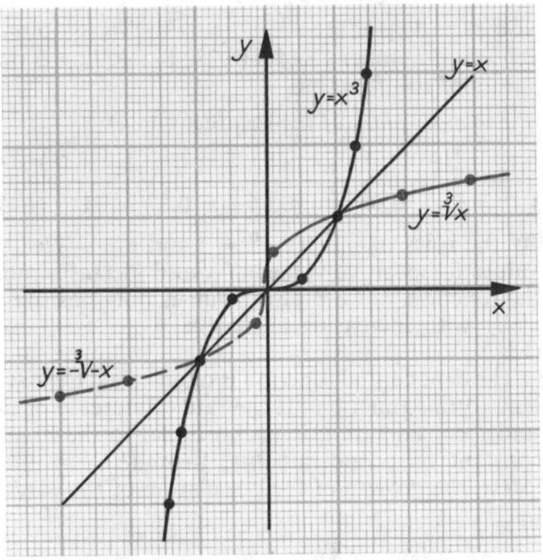

5.3-2 Graph of the inverse function of $y = x^3$

On the other hand, the inverse function of $y = x^3$ with $y < 0$, and hence $x < 0$, is described by the equation $y = -\sqrt[3]{(-x)}$ in the interval of definition $-\infty < x < 0$ (see the dotted curve in the figure). Consequently two equations are needed to describe explicitly the *complete inverse* function of $y = x^3$, which must exist because $y = x^3$ is one-to-one. But the two parts are usually subsumed under the single expression $y = \sqrt[n]{x}$ for all real values of x (see Chapter 3.).

The function $y = \sqrt[n]{x}$. According to the general definition of a root, n is assumed to be an integer greater than 1. The cases $n = 2$ and $n = 3$ have already been considered; the investigation of greater values of n yields nothing essentially new. For the domain of definition $0 \leqslant x < +\infty$ and range $0 \leqslant y < +\infty$, all the functions increase monotonically, but differ from one another in that for $n < m$, $\sqrt[n]{x} < \sqrt[m]{x}$ for $0 < x < 1$, and $\sqrt[n]{x} > \sqrt[m]{x}$ for $x > 1$. The functions $y = \sqrt[n]{x}$ are inverses of the power functions. If n is even, then $y = x^n$ is monotonic increasing in the interval $0 \leqslant x < +\infty$, and consequently is invertible there with the inverse function $y_1 = \sqrt[n]{x}$; in the interval $(-\infty, 0]$, the functions $y = x^n$ and $y = -\sqrt[n]{x}$ are inverses to one another (Fig.). For odd values of n, $y = x^n$ is monotonic increasing in the whole domain of definition $(-\infty, +\infty)$ and has the inverse function $x = y^n$, whose explicit form is $y = \sqrt[n]{x}$ for $0 \leqslant x < +\infty$ and $y = -\sqrt[n]{(-x)}$ for $-\infty < x \leqslant 0$ (Fig.) or $y = \sqrt[n]{x}$ for all real values of x.

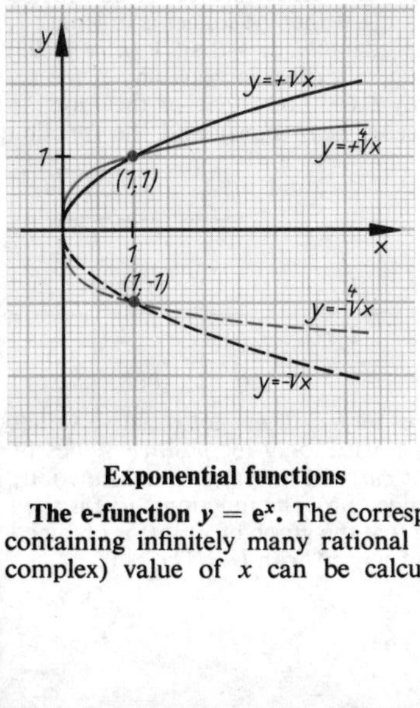

5.3-3 Graphs of the functions $y = +\sqrt{x}$, $y = -\sqrt{x}$, $y = \sqrt[4]{x}$ and $y = -\sqrt[4]{x}$

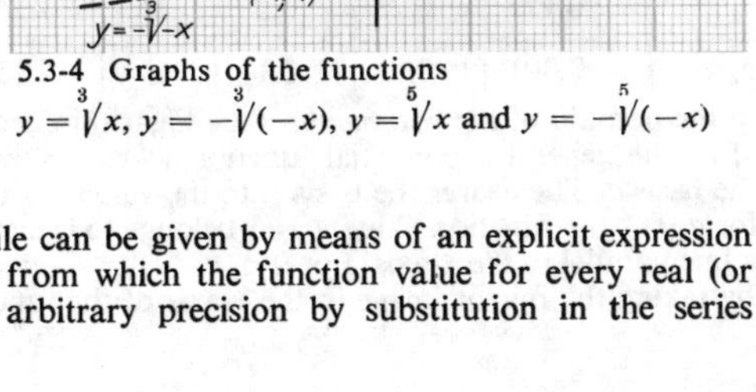

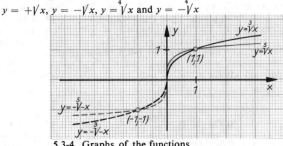

5.3-4 Graphs of the functions $y = \sqrt[3]{x}$, $y = -\sqrt[3]{(-x)}$, $y = \sqrt[5]{x}$ and $y = -\sqrt[5]{(-x)}$

Exponential functions

The e-function $y = e^x$. The correspondence rule can be given by means of an explicit expression containing infinitely many rational operations, from which the function value for every real (or complex) value of x can be calculated with arbitrary precision by substitution in the series

$y = e^x = 1 + x/1! + x^2/2! + x^3/3! + \cdots$. For the particular value $x = 1$ the value of the transcendental number $e = 2.718\,281\,828\,459 \ldots$ is obtained. Some tables of logarithms contain rounded values of this function.

Table of values of the function $y = e^x$

x	… −3	−2	−1	0	1/3	1/2	1	2	3	…
y	… 0.05	0.14	0.37	1	1.4	1.65	2.72	7.39	20.09	…

By the rules of indices, $e^0 = 1$ and $e^{-x} = 1/e^x$. Because the function $y = e^x$ assumes only positive values for positive values of x and increases monotonically without limit as $x \to \infty$, it follows that the function $y = e^{-x}$ also assumes only positive values, and that the values of y decrease monotonically with increasing argument x. As $x \to +\infty$, the curve approaches the x-axis asymptotically (Fig.).

The exponential function is often called the natural *growth function*, because many natural processes lead to this function. If the rate of increase or decrease $\pm dN/dt$ at time t of a number N of objects under consideration is proportional to the number N itself, so that $\pm dN/dt = Nk$, where k is the constant of proportionality, then $dN/N = \pm k\, dt$, or $e^{\pm kt} = N$. For example, the growth of a forest, the growth of the population of the earth, and radioactive decay are based on this function.

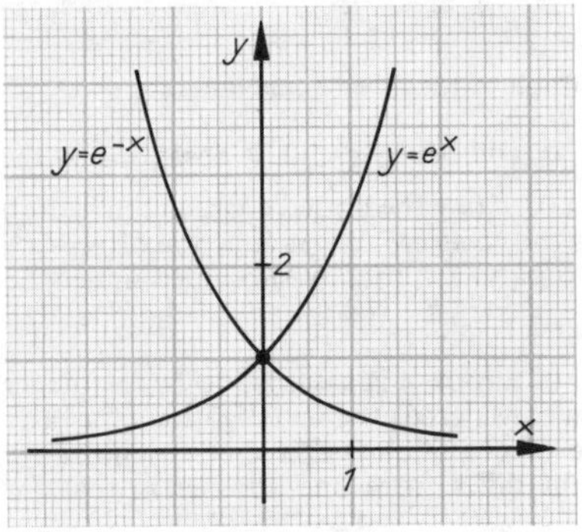

5.3-5
Graphs of the functions $y = e^x$ and $y = e^{-x}$

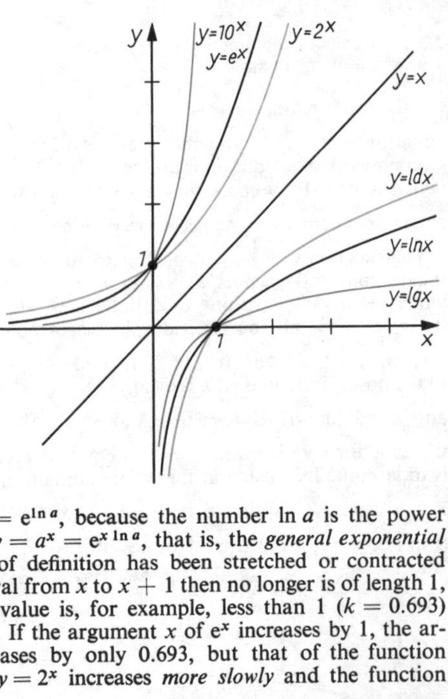

5.3-6
Graphs of the logarithmic and exponential functions

The function $y = a^x$. By the rules of indices, $a = e^{\ln a}$, because the number $\ln a$ is the power to which e must be raised to give a. One obtains $y = a^x = e^{x \ln a}$, that is, the *general exponential function* is an e-function $y = e^{kx}$, whose interval of definition has been stretched or contracted uniformly by the constant factor $k = \ln a$. The interval from x to $x + 1$ then no longer is of length 1, but $1 \cdot k = \ln a = \lg a/\lg e = 2.30259 \ldots \lg a$. This value is, for example, less than 1 ($k = 0.693$) for $a = 2$ and greater than 1 ($k = 2.30$) for $a = 10$. If the argument x of e^x increases by 1, the argument kx of the function $y = 2^x = e^{x \ln 2}$ increases by only 0.693, but that of the function $y = 10^x = e^{x \ln 10}$ increases by 2.30. The function $y = 2^x$ increases *more slowly* and the function $y = 10^x$ *more quickly* than the e-function e^x (Fig.).

In general, $y = a^x$ increases more slowly than $y = e^x$ for $1 < a < e$, and more quickly for $a > e$.

$$y = a^x = e^{x \ln a}$$

Table of values for the functions $y_1 = 2^x$ *and* $y_2 = 10^x$

x	… −3	−2	−1	0	1/3	1/2	1	2	3	…
y_1	… 0.125	0.25	0.5	1	1.26	1.41	2	4	8	…
y_2	… 0.001	0.01	0.1	1	2.15	3.16	10	100	1000	…

In the domain of the *real numbers* the logarithm is defined neither for $a = 0$ nor for negative values of a; the general exponential function $y = a^x = e^{x \ln a}$ therefore exists only for *positive* values of the basis a. The nearer the basis is to the value 1, the flatter is the *curve* of the exponential function; for $a = 1$ the function value $y = 1$ belongs to every arbitrary value of x. The function curve is then a line parallel to the x-axis. For $0 < a < 1$, $y = a^x = (1/a)^{-x}$, so that the graph of $y = a^x$ is obtained by taking the mirror image in the y-axis of the function curve of $y = b^x$ with $b = 1/a > 1$.

The functions $y = k \cdot a^x$. On account of the constant positive factor k, the ordinate values y of the function are *stretched* ($k > 1$) or *contracted* ($k < 1$) in the ratio $1 : k$. It can be shown that in this way the curve goes into itself up to a translation parallel to the x-axis. Because $k = e^{\ln k}$, it follows that $y = k \cdot a^x = e^{\ln k} \cdot e^{x \ln a} = e^{x \ln a + \ln k}$, so that there is a translation parallel to the x-axis by an amount $c = -\ln k$.

Logarithmic functions

The function $y = \log_a x$. This function is the inverse of the exponential function $y = a^x$, which is, of course, monotonic in its whole domain of definition. Because the range of the exponential function is $0 < y < +\infty$, the logarithmic function can be defined only for *positive* values of the argument, and thus has the domain of definition $0 < x < +\infty$. Special inverse functions are $y = \ln x$ for $y = e^x$, and $y = \lg x$ for $y = 10^x$. Consequently their *graphs* are obtained by taking the mirror images of the graphs of $y = e^x$ and $y = 10^x$ in the line $y = x$ (see Fig. 5.3-6). Information about $y = \lg x$ and about the slide rule are contained in Chapter 2.

The function $y = \log_a x^k$. Clearly $y = k \log_a x$, and the function values are therefore given by multiplying by the constant k. Its value can also be negative, because for $k = -\varkappa$ with $\varkappa > 0$ one obtains $y = \log_a x^{-\varkappa} = \log_a (1/x^\varkappa) = -\log_a x^\varkappa = -\varkappa \log_a x$. For $k = -1$ in particular, the graph of the function is the result of taking the mirror image in the $+x$-axis of the graph of the function $y = \log_a x$. The function $y = -\log_a x = \log_a x^{-1} = \log_a (1/x)$ is the inverse of the function $y = a^{-x}$.

The function $y = \log_a (kx)$. For positive values of the constant k, because $y = \log_a (kx) = \log_a k + \log_a x$, the graph of this function is obtained from that of $y_1 = \log_a x$ by translation parallel to the $+y$-axis through a distance $d = +\log_a k$. For negative values $\varkappa = -k (k > 0)$ the function is defined only for negative values of x, and then because $\varkappa x = |kx|$ the function values are the same as those of $y = \log_a (kx)$ with $0 < x < +\infty$.

Trigonometric and circular functions

The *trigonometric* or *angular functions* and the *circular* or *arc functions* are a very frequently occurring type of irrational function. They are investigated more thoroughly in Chapter 10.

Connections between trigonometric and circular functions. Because the circular functions are defined as *inverse functions* of the trigonometric functions, the following connections are immediate: $\sin(\arcsin x) = x$; $\cos(\arccos x) = x$ and so on. For positive x, if the principal value is taken on both sides, it is also true that $\text{Arccot}\, x = \text{Arctan}(1/x)$, because $\cot x = 1/\tan x$. For this reason the function $y = \text{arccot}\, x$ can be dispensed with for most investigations. The following further relations are of interest:

$$\sin(\text{Arccos}\, x) = \sqrt{(1 - x^2)},\quad \sin(\text{Arctan}\, x) = \frac{x}{\sqrt{(1 + x^2)}},\quad \cos(\text{Arcsin}\, x) = \sqrt{(1 - x^2)},$$
$$\cos(\text{Arctan}\, x) = \frac{1}{\sqrt{(1 + x^2)}},\quad \tan(\text{Arcsin}\, x) = \frac{x}{\sqrt{(1 - x^2)}},\quad \tan(\text{Arccos}\, x) = \frac{\sqrt{(1 - x^2)}}{x}$$

It is enough here to give the justification for the first relation, because the others can be obtained in exactly the same way. From $\sin^2 y + \cos^2 y = 1$ it follows that $\sin y = \pm\sqrt{(1 - \cos^2 y)}$, or $\sin(\text{Arccos}\, x) = \pm\sqrt{(1 - x^2)}$, for the principal value of $\arccos x$, $0 \leqslant \text{Arccos}\, x \leqslant \pi$. In this domain of definition the sine function has no negative values, $\sin(\text{Arccos}\, x) \geqslant 0$; the square root can therefore have only a positive sign, and $\sin(\text{Arccos}\, x) = +\sqrt{(1 - x^2)}$, as given in the table. Similarly for the principal value $-\pi/2 \leqslant \text{Arcsin}\, x \leqslant +\pi/2$ it follows that $\cos(\text{Arcsin}\, x) = +\sqrt{(1 - x^2)}$, because the cosine function takes no negative values in this interval. Using a connection between the trigonometric functions and the exponential function, first found by EULER, a relation between the *inverse trigonometric functions* and the *logarithmic* function can be obtained. These relations are valid in the *domain of complex numbers*.

Euler's formulae	$e^{i\varphi} = \cos\varphi + i\sin\varphi$ $e^{-i\varphi} = \cos\varphi - i\sin\varphi$
$\sin\varphi = \dfrac{e^{i\varphi} - e^{-i\varphi}}{2i}$	$\cos\varphi = \dfrac{e^{i\varphi} + e^{-i\varphi}}{2}$
$\tan\varphi = -i\,\dfrac{e^{i\varphi} - e^{-i\varphi}}{e^{i\varphi} + e^{-i\varphi}}$	$\cot\varphi = +i\,\dfrac{e^{i\varphi} + e^{-i\varphi}}{e^{i\varphi} - e^{-i\varphi}}$

From the relations derived above, for $\sin\varphi = x$ and the principal value $\text{Arcsin}\, x = \varphi$ the first Euler formula becomes $e^{i\varphi} = ix + \sqrt{(1 - x^2)}$. By taking the logarithm one obtains $i\varphi = i\,\text{Arcsin}\, x = \ln(ix + \sqrt{(1 - x^2)})$. By similar calculations for the other inverse functions the following relations are obtained:

$\operatorname{Arcsin} x = -\mathrm{i} \ln (x\mathrm{i} + \sqrt{(1 - x^2)})$	$\operatorname{Arccos} x = -\mathrm{i} \ln (x + \mathrm{i} \sqrt{(1 - x^2)})$
$\operatorname{Arctan} x = -\mathrm{i} \ln \sqrt{[(1 + x\mathrm{i})/(1 - x\mathrm{i})]}$	$\operatorname{Arccot} x = -\mathrm{i} \ln \sqrt{[(x\mathrm{i} - 1)/(x\mathrm{i} + 1)]}$

Hyperbolic functions

The correspondences defined by the following function equations are called *hyperbolic functions*:

1. *Hyperbolic sine:* $y = \sinh x = (\mathrm{e}^x - \mathrm{e}^{-x})/2$;
2. *Hyperbolic cosine:* $y = \cosh x = (\mathrm{e}^x + \mathrm{e}^{-x})/2$;
3. *Hyperbolic tangent:* $y = \tanh x = \dfrac{\mathrm{e}^x - \mathrm{e}^{-x}}{\mathrm{e}^x + \mathrm{e}^{-x}}$;
4. *Hyperbolic cotangent:* $y = \coth x = \dfrac{\mathrm{e}^x + \mathrm{e}^{-x}}{\mathrm{e}^x - \mathrm{e}^{-x}}$.

The function $y = \sinh x$. From the definition it follows that $y = (\mathrm{e}^x - \mathrm{e}^{-x})/2$ is defined for all values of x. The function has one zero at $x = 0$. If $x \to +\infty$, e^{-x} becomes arbitrarily small. Because at the same time e^x increases without limit, the values of the function become arbitrarily large. As $x \to -\infty$, e^{-x} becomes arbitrarily large and e^x approaches the value zero, so that the value of the function tends to $-\infty$. From the defining equation it follows further that $\sinh x = -\sinh(-x)$. Thus, the function is *odd* and its graph is *centrally symmetric* about the origin of coordinates; the *range* is $-\infty < y < +\infty$ (Fig.).

The function $y = \cosh x$. This function is also defined for all values of x and its range is given by $1 \leqslant y < +\infty$, as can be seen from the equation of the function $y = (\mathrm{e}^x + \mathrm{e}^{-x})/2$. The function is *even* and its graph is *symmetrical about the y-axis* (Fig.). The shape of a heavy chain or rope hanging under its own weight is given by the graph of the function $y = a \cosh (x/a)$, where a is a suitable positive constant depending on the material (see Fig. 19.5-11).

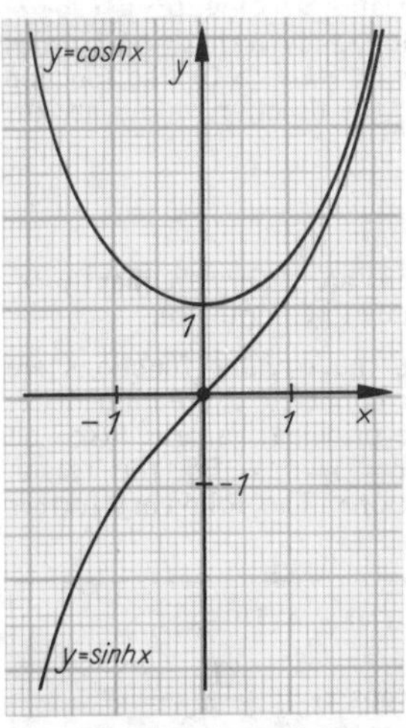

5.3-7 Graphs of $y = \sinh x$ and $y = \cosh x$

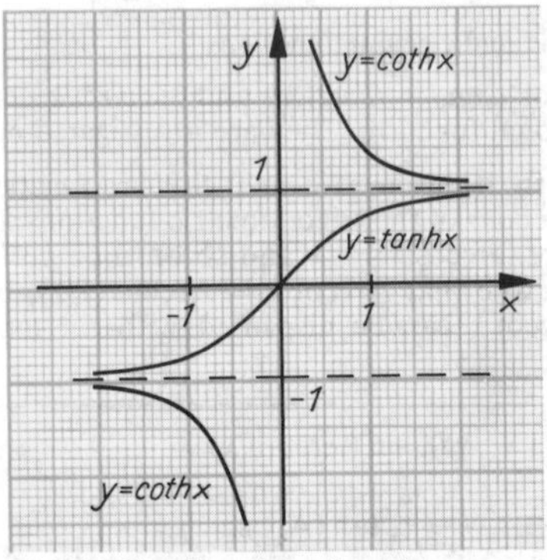

5.3-8 Graphs of the functions $y = \tanh x$ and $y = \coth x$

The functions $y = \tanh x$ and $y = \coth x$. The first of these two functions is defined for all values of x, but for the second the value $x = 0$ must be excluded. The range of $y = \tanh x$ is bounded: $-1 < y < +1$. By contrast, the range of $y = \coth x$ is by $-\infty < y < -1$ and $+1 < y < +\infty$. Both functions are *odd* (Fig.).

Relations between the hyperbolic functions. From the equations of the functions the following identities are immediate consequences:

$$\tanh x = \sinh x/\cosh x, \quad \coth x = 1/\tanh x, \quad \cosh^2 x - \sinh^2 x = 1$$

The close similarity of these relations to those between the trigonometric functions justifies the use of the terms hyperbolic sine, hyperbolic cosine and so on. The reason one speaks of hyperbolic functions is clear when one considers the third relation. Putting $\cosh x = X$ and $\sinh x = Y$, this relation becomes $X^2 - Y^2 = 1$. This is the equation of a hyperbola in the X, Y-plane, but of course only the right-hand branch is represented because $\cosh x \geqslant 1$.

The inverse functions of the hyperbolic functions

The hyperbolic functions are invertible. For $y = \sinh x$ and $y = \tanh x$ this can be seen from their graphs; for $y = \coth x$ this follows because the function is one-to-one. For $y = \cosh x$ an inverse function can be defined in each of the two intervals $-\infty < x \leqslant 0$ and $0 \leqslant x < +\infty$ in which the function is monotonic.

Inverse hyperbolic sine, $y = \sinh^{-1} x$. This is the inverse function of $y = \sinh x$. Solving the equation $y = (e^x - e^{-x})/2$ for x gives first $2y = e^x - e^{-x}$, or after multiplying by e^x, $2ye^x = e^{2x} - 1$, or $e^{2x} - 2ye^x - 1 = 0$. This is a quadratic equation for e^x. Only the solution $e^x = y + \sqrt{(y^2+1)}$ is relevant, because $y - \sqrt{(y^2+1)}$ is always negative, whereas e^x can take only positive values. Taking logarithms finally yields $x = \ln[y + \sqrt{(y^2+1)}]$. From this one obtains at the same time $y = \ln[x + \sqrt{(x^2+1)}]$ as the *explicit form of the equation of the inverse function*; it represents $y = \sinh^{-1} x$. The graph of the function $y = \sinh^{-1} x$ is the *mirror image* in the line $y = x$ of the curve with the equation $y = \sinh x$.

Inverse hyperbolic cosine, $y = \cosh^{-1} x$. To invert $y = \cosh x$ one proceeds as for $y = \sinh x$ by similar steps to the equation $e^{2x} - 2ye^x + 1 = 0$, leading to $e^x = y \pm \sqrt{(y^2-1)}$. Finally one obtains the inverse functions $y = \ln[x - \sqrt{(x^2-1)}]$ for the interval $-\infty < x \leqslant 0$, and $y = \ln[x + \sqrt{(x^2-1)}]$ for the interval $0 \leqslant x < +\infty$. For both the domain of definition is $1 \leqslant x < +\infty$, and the range is $-\infty < y \leqslant 0$ in the first case and $0 \leqslant y < +\infty$ in the second. The inverse function with the range $0 \leqslant y < +\infty$ is called the principal value of the inverse hyperbolic cosine, which therefore has the explicit form $y = \cosh^{-1} x = \ln[x + \sqrt{(x^2-1)}]$.

Inverse hyperbolic tangent, $y = \tanh^{-1} x$. From the equation of the function $y = \dfrac{e^x - e^{-x}}{e^x + e^{-x}}$, first multiplying numerator and denominator by e^x, one obtains $y = (e^{2x} - 1)/(e^{2x} + 1)$, so that $ye^{2x} + y = e^{2x} - 1$, or $ye^{2x} - e^{2x} = -y - 1$. Multiplying by (-1) and collecting together the terms in e^{2x} leads to $e^{2x}(1-y) = 1 + y$, or $e^{2x} = (1+y)/(1-y)$, $e^x = \sqrt{[(1+y)/(1-y)]}$. Taking logarithms and finally interchanging the variables gives the *explicit equation* $y = \ln\sqrt{[(1+x)/(1-x)]}$ or $y = (1/2)\ln[(1+x)/(1-x)]$ for the function $y = \tanh^{-1} x$. The domain of definition is limited to $-1 < x < +1$, but the range consists of all real numbers.

For completeness the equation of the *inverse hyperbolic cotangent*, $y = \coth^{-1} x$ is given here: $y = (1/2)\ln[(x+1)/(x-1)]$, with domain of definition $-\infty < x < -1$ and $1 < x < +\infty$.

Geometrical significance of the inverse functions. The inverse functions of the trigonometric functions can be regarded geometrically as the *arc lengths* for which the functions sine, cosine, tangent or cotangent have the given value x. If this arc length is taken as parameter t, then $x = \cos t$ and $y = \sin t$ represent a point on the unit circle in the x,y-plane, because of the relation $\cos^2 t + \sin^2 t = x^2 + y^2 = 1$. Similarly a parametric representation $x = \cosh t$ and $y = \sinh t$ can be introduced for the *hyperbolic functions*. Because $\cosh^2 t - \sinh^2 t = 1$, the point $P(x, y)$ lies on the rectangular hyperbola $x^2 - y^2 = 1$ (Fig.). It can be shown that in this case the parameter t represents twice the area between the segment $|OV|$ of the x-axis, the arc VP of the hyperbola up to the point $P(x_0, y_0)$ and the line joining P to the origin O. With the help of integral calculus this area is obtained as follows:

$$\text{Area } VAP = \int_1^{x_0} \sqrt{(x^2-1)}\,dx = {}^1/_2 x_0 \sqrt{(x_0^2-1)} - {}^1/_2 \ln|x_0 + \sqrt{(x_0^2-1)}|,$$

$$\text{Area } OVPB = \int_0^{y_0} \sqrt{(y^2+1)}\,dy = {}^1/_2 y_0 \sqrt{(y_0^2+1)} + {}^1/_2 \ln|y_0 + \sqrt{(y_0^2+1)}|.$$

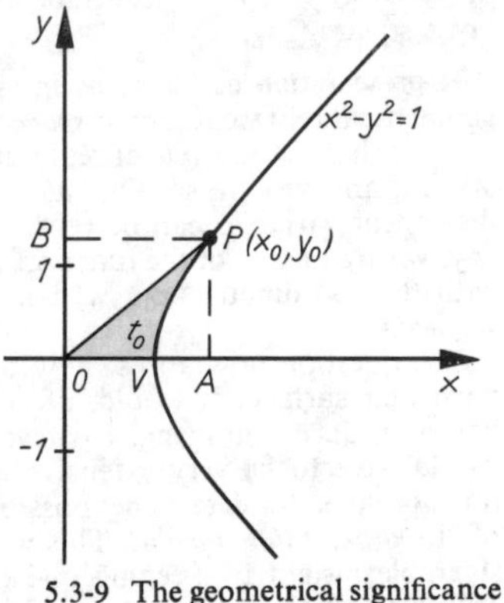

5.3-9 The geometrical significance of the inverse hyperbolic functions

From this the area $t_0/2 = OVP$ can be calculated in two ways.

$$1.\ OVP = OAP - VAP = {}^1/_2 x_0 y_0 - VAP$$
$$= {}^1/_2 x_0 \sqrt{(x_0^2-1)} - {}^1/_2 x_0 \sqrt{(x_0^2-1)} + {}^1/_2 \ln|x_0 + \sqrt{(x_0^2-1)}|$$
$$= {}^1/_2 \ln|x_0 + \sqrt{(x_0^2-1)}|.$$

$$2.\ OVP = OVPB - OPB$$
$$= {}^1/_2 y_0 \sqrt{(y_0^2+1)} + {}^1/_2 \ln|y_0 + \sqrt{(y_0^2+1)}| - {}^1/_2 y_0 \sqrt{(y_0^2+1)} = {}^1/_2 \ln|y_0 + \sqrt{(y_0^2+1)}|.$$

As was found in considering the inverse hyperbolic cosine function, ${}^1/_2 \ln|x_0 + \sqrt{(x_0^2-1)}| = {}^1/_2 \cosh^{-1} x_0$, so that $t_0 = \cosh^{-1} x_0$. On the other hand, in considering $\sinh^{-1} x$ it was shown that ${}^1/_2 \ln|y_0 + \sqrt{(y_0^2+1)}| = {}^1/_2 \sinh^{-1} y_0$, so that $t_0 = \sinh^{-1} y_0$.

5.4. Functions with more than one independent variable

General definition

If $M_1, M_2, \ldots, M_n$ are n non-empty sets, not necessarily distinct from one another, then one can select one element from each set, say x_1 from M_1, x_2 from $M_2, \ldots, x_n$ from M_n, having regard to the order of the sequence. The set $(x_1, x_2, \ldots, x_n)$ of all these elements is called an n-tuple. If exactly one element of a set N is assigned to each n-tuple ordered according to the given sequence, then one speaks of a function of n independent variables; in general this is written $y=f(x_1, x_2, \ldots, x_n)$.

Real functions with two independent variables

In the following functions the domain of definition consists of ordered pairs of real numbers, while the range is contained in the set of real numbers. In general form this is usually written as $z=f(x, y)$, where z is used for the dependent variable, and x and y for the independent variables.

Representation of the domain of definition in the plane. The domain of definition of a real function of two independent variables may have geometrical significance. Because each ordered pair of real numbers corresponds to a unique point of a plane provided with a coordinate system, and conversely, plane regions of the most diverse shape can serve as domains of definition; for example, the domain may be *connected*, or at the other extreme, it may consist of isolated points only.

Example 1: The domain of definition given by $-\infty < x < +\infty$ and $0 \leqslant y < +\infty$ corresponds to the *upper half-plane* of the x, y-plane, including the x-axis (Fig.).

Example 2: With the condition $x^2 + y^2 < 1$, the domain of definition is the interior of the unit circle (Fig.).

Example 3: The conditions $-\infty < x \leqslant -1$ or $1 \leqslant x < +\infty$ and $-\infty < y < +\infty$, together with $-1 < x < +1$ and $+1 \leqslant y < +\infty$ or $-\infty < y \leqslant -1$, determine as domain of definition the whole plane apart from the interior of a square (Fig.).

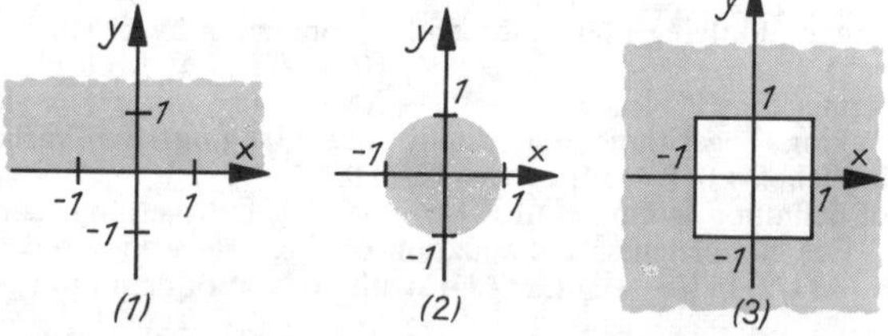

5.4-1 Geometrical representations of the regions given in the examples

Representation of functions in space. Because altogether three variables occur, one uses for the geometrical representation a *space coordinate system* with three axes, usually a rectangular system.

To each ordered *triple* of real numbers there corresponds exactly one point in the space coordinate system, and vice versa. On the basis of this unique correspondence all real functions of two independent variables can be represented geometrically. If the number z_0 is associated with the pair (x_0, y_0) by means of the function $z=f(x, y)$, then this corresponds geometrically to the point P_0 with the coordinates (x_0, y_0, z_0). The function is assumed to be such that its graph represents a *surface*.

The question now arises, how can one obtain in an individual case an idea of the nature of the particular surface. It would, of course, be possible in principle to draw up a *table of values* and hence produce a drawing. However, to arrive in this way at an intuitive picture, the table of values would have to be very extensive. In practice, therefore, one usually makes use of other methods; for example, to determine possible extrema, saddle points, and so on, one draws on the results of the *differential calculus*. This will not be gone into more deeply here. One certainly obtains considerable insight by keeping one of the three variables of the function constant. One selects, for example, all those pairs (x, y) from the domain of definition, whose x-value is equal to a fixed number c. From $z=f(x, y)$ one obtains the function equation $z=f(c, y)$, which now contains only one independent variable. Its graph is a curve, the *curve of intersection* of the surface determined by $z=f(x, y)$ with the fixed plane given by $x=c$. If one constructs these curves for different fixed values of x, one obtains a *family of curves* that can give an approximate picture of the surface. Naturally the same method can also be applied to the variable y. The matter is slightly different if the dependent variable z is kept constant. Each special value of z then leads to an equation with two variables: $z=f(x, y)$ leads to $f(x, y)=c$. The set of solution pairs (x, y) that satisfy this equation is interpreted geometrically as a point set in the plane given by $z=c$. Moreover, if one postulates that c belongs to the range of the function, this point set is not empty. In general, it forms certain curves. Because one usually thinks of the z-axis as being vertical, these curves are called *contour lines* or *level curves*. In principle they are the same as the contour lines in a *geographical map*. In both cases the points lying on them are those that are at the same height above or below some standard level; in geographical maps this is usually sea level and in the case in question it is the x, y-plane.

Example 1: The function $z = x + y$ is defined in the whole x, y-plane. Its range is clearly $-\infty < z < +\infty$. If x is kept constant, then to each special value of x there corresponds an equation of the form $z = y + c$. Geometrically this represents a family of straight lines. If one investigates the *contour lines* given by $x + y = c$, one obtains a family of parallel straight lines. The graph of the function $z = x + y$ is a plane (Fig.).

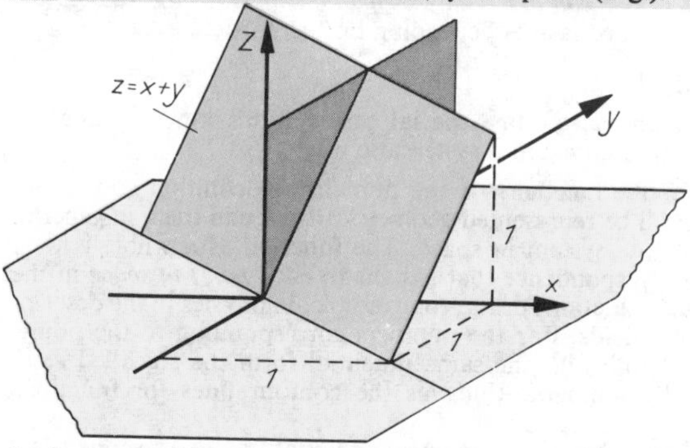

5.4-2 Geometrical representation of the function $z = x + y$

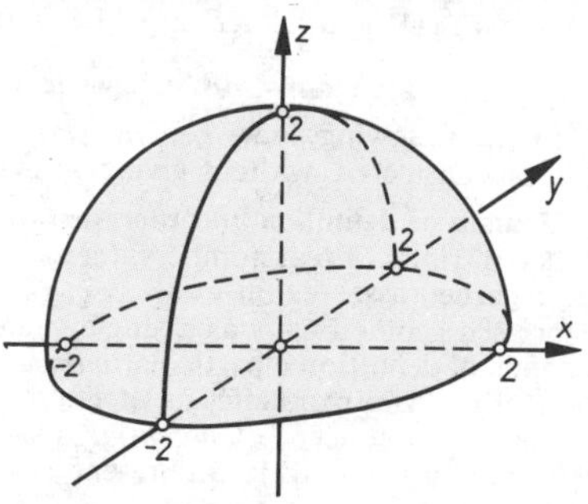

5.4-3 Geometrical representation of the function $z = \sqrt{(4 - x^2 - y^2)}$

Example 2: The function $z = \sqrt{(4 - x^2 - y^2)}$ is defined only in the region $x^2 + y^2 \leqslant 4$; thus, the *domain of definition* is a circle with centre at the origin and radius 2. Its range is bounded: $0 \leqslant z \leqslant 2$. Keeping x constant leads to an equation of the form $z = \sqrt{\{(4 - c^2) - y^2\}}$. This represents geometrically a semicircle. The result is the same if y is kept constant. The contour lines are given by $c = \sqrt{(4 - x^2 - y^2)}$, which can be rearranged in the form $x^2 + y^2 = 4 - c^2$. This represents a circle of radius $\sqrt{(4 - c^2)}$. Consequently the geometrical representation of $z = \sqrt{(4 - x^2 - y^2)}$ is a hemisphere (Fig.).

Example 3: The function $z = xy$ is defined in the whole x, y-plane. Its range is $-\infty < z < +\infty$. Keeping x constant, one obtains the functions $z = cy$ whose graphs are straight lines. Of course these are not parallel, as they were in Example 1. The same result is obtained when y is kept constant. In this case the *contour lines* are hyperbolas with the equation $xy = c$ $(c \neq 0)$. For $c = 0$ the x-axis and the y-axis are obtained as contour lines (Fig.).

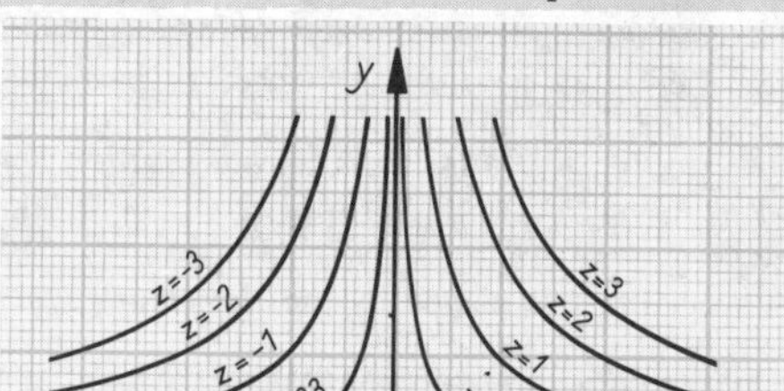

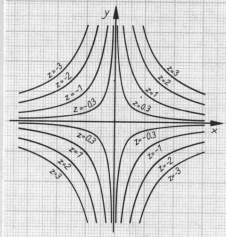

5.4-4 Contour lines of the function $z = xy$

If the surface is cut by a plane perpendicular to the x, y-plane, then the curves of intersection are parabolas, provided that the intersecting plane is not parallel to the x, z-plane or to the y, z-plane; in these excluded cases the intersections are the lines already found above.

This can be seen as follows: The intersecting planes all have equations of the form $Ax + By + C = 0$, which can be rearranged in the form $y = ax + b$. Substituting the expression $ax + b$ for y in $z = xy$, one obtains $z = ax^2 + bx$. But these are equations of parabolas. Their *vertices* lie in the planes $x = y$ or $x = -y$.

The function $z = xy$ represents geometrically a surface known as a *hyperbolic parabola*.

Occurrences in other fields. Real functions of two independent variables are used to express not only mathematical, but also physical and technological relationships, among others. Examples are: area formulae such as $A = ab$ and $A = gh/2$; volume formulae such as $V = \pi r^2 h$ and $V = a^2 h/3$; solution formulae for equations such as $x = -p/2 + \sqrt{(p^2/4 - q)}$; formulae for Ohm's law $I = E/R$ or for the connection $s = vt$ between distance, speed and time; formulae for the cutting speed of lathes $v = \pi\, d\, n/1000$, and so on. The function formulae can also occur in *implicit form*, in which it is not specified in advance which variable is to be regarded as dependent. An example of this is given by the *equation of state* for ideal gases $pV_m = RT$, giving the mutual dependence of pressure p, volume V_m (volume of a mole of gas) and absolute temperature T; R is the absolute gas constant. Each of these three variables can be regarded as dependent. One must just take into account that from physical considerations only positive values of the variables are applicable here. Curves resulting from keeping T constant are called *isotherms* and are in principle just like the contour lines of the function $z = xy$ for positive values of x and y.

A somewhat more complicated example is the *van der Waals's equation of state* for real gases, $(p + a/V^2)(V - b) = RT$, in which a and b are constants depending on the particular gas.

Real functions with independent variables

In the following some general properties and also some special cases of this kind of function will be considered, without giving an exhaustive survey in a systematic way.

Domain of definition and representation of the functions. If the domain of definition consists of ordered triples of real numbers, then it can still be represented geometrically. It can then, in general, be regarded as a region in an x, y, z-coordinate system in space. The function as a whole is then expressed geometrically as a single-valued correspondence that assigns to each *point in space* in the domain of definition a particular *number*. Such functions occur, for example, in physics in the description of electric, magnetic, or of gravitational fields. For the numbers corresponding to the points in space, the concept of *potential* is used. Points with the same potential form the so-called *equipotential surfaces*, which represent essentially the same thing as the contour lines for functions of two independent variables.

If one is concerned with functions having more than three independent variables, then a geometrical representation in the sense so far discussed – that is, a visual realization of the function – is no longer possible.

Symmetric functions. A real function of n independent variables is said to be symmetric if one can interchange the independent variables arbitrarily among themselves without thereby altering the function. The most important are polynomial and rational symmetric functions. A polynomial function $y = f(x_1, x_2, \dots, x_n)$ is said to be symmetric, if for every permutation $\begin{pmatrix} x_1 & x_2 & \dots & x_n \\ x_{i_1} & x_{i_2} & \dots & x_{i_n} \end{pmatrix}$ of the variables $x_1, x_2, \dots, x_n$, one has $f(x_1, x_2, \dots, x_n) = f(x_{i_1}, x_{i_2}, \dots, x_{i_n})$.

Examples of polynomial symmetric functions:

1. $y = x_1 + x_2 + \dots + x_n$, hence, in particular, the case of the function $z = x + y$ already examined.
2. $y = x_1 x_2 \dots x_n$, for example, the function $z = xy$ already considered.
3. $y = x_1 x_2 + x_1 x_3 + x_2 x_3$.
4. $y = x_1^2 + x_1 x_2 + x_2^2$.

A special role is played by the *elementary symmetric functions*. They are given here in full for the case $n = 4$:

Elementary symmetric functions	$\sigma_1(x_1, x_2, x_3, x_4) = x_1 + x_2 + x_3 + x_4$ $\sigma_2(x_1, x_2, x_3, x_4) = x_1x_2 + x_1x_3 + x_1x_4 + x_2x_3 + x_2x_4 + x_3x_4$ $\sigma_3(x_1, x_2, x_3, x_4) = x_1x_2x_3 + x_1x_2x_4 + x_1x_3x_4 + x_2x_3x_4$ $\sigma_4(x_1, x_2, x_3, x_4) = x_1x_2x_3x_4$

By Vieta's root theorem, these elementary symmetric functions give the coefficients, up to their signs, of the polynomial that has $x_1, x_2, \dots, x_n$ as its roots. For example, for $x^4 + ax^3 + bx^2 + cx + d = 0$ with roots x_1, x_2, x_3 and x_4:

$$a = -\sigma_1(x_1, x_2, x_3, x_4); \quad b = +\sigma_2(x_1, x_2, x_3, x_4);$$
$$c = -\sigma_3(x_1, x_2, x_3, x_4); \quad d = +\sigma_4(x_1, x_2, x_3, x_4).$$

The following theorem holds for the elementary symmetric functions:

Every polynomial symmetric function with n independent variables can be expressed as a polynomial in the elementary symmetric functions $\sigma_1, \sigma_2, \dots, \sigma_n$.

Instead of a proof, which would require fairly complicated explanations, here is a simple example, which will make the situation clear. The symmetric function $f(x_1, x_2) = x_1^2 - x_1x_2 + x_2^2$ can be expressed in the form $g(\sigma_1, \sigma_2) = \sigma_1^2 - 3\sigma_2$, as can be verified by substitution: $\sigma_1^2 = x_1^2 + 2x_1x_2 + x_2^2$ and $3\sigma_2 = 3x_1x_2$ in fact give $\sigma_1^2 - 3\sigma_2 = x_1^2 - x_1x_2 + x_2^2$.

A corresponding theorem holds for rational functions.

Every rational symmetric function can be expressed as the quotient of two polynomial symmetric functions.

Finally, it should be noted that in cases in which symmetric functions can be represented geometrically, they also exhibit properties of geometrical symmetry. For example, the representations of the functions $z = x + y$ and $z = xy$ already considered both have the plane $x = y$ as a plane of symmetry.

Homogeneous functions. A function with n independent variables is called *homogeneous of degree m* if the function value is multiplied by t^m when each individual independent variable is multiplied by t; this means that $f(tx_1, tx_2, ..., tx_n) = t^m f(x_1, x_2, ..., x_n)$. Homogeneous *polynomial* functions are of special interest. A homogeneous polynomial function of degree m is often called a *form* of degree m. For $m = 2$, one speaks of a *quadratic form*, for $m = 3$, a *cubic form*. In the case $m = 1$ such a polynomial is called a *linear form*.

Examples of homogeneous polynomials: 1. $f(x_1, x_2, x_3) = x_1^2 + x_2^2 + x_3^2$.
2. $f(x_1, x_2) = x_1^4 + x_1^3 x_2 + x_1^2 x_2^2$.
3. $f(x_1, x_2, x_3, x_4) = x_1 + x_2 - x_3 - x_4$.

The following theorem holds for homogeneous polynomials:

The product of non-zero homogeneous polynomials is again a homogeneous polynomial. Its degree is equal to the sum of the degrees of the individual polynomials.

The proof of this theorem follows at once from the multiplication rule for polynomials.

Homogeneous functions, in particular homogeneous polynomials, play a part in various branches of mathematics. For example, a determinant with n rows and n columns is a homogeneous function of degree n, with n^2 independent variables. Quadratic forms such as $F(x, y) = Ax^2 + Bxy + Cy^2$, that is, homogeneous functions of the second degree having two independent variables, occur in the theory of quadratic number fields. From another point of view certain quadratic forms will be investigated in analytic geometry.

6. Percentages, interest and annuities

6.1. Percentages

Percentage and actual value. In very many areas of everyday life one comes across the concept of *percentage*. It is stated, for example, by how many per cent in a certain time interval production has increased or net costs have been increased, or what percentage of the population is male or female. In all these statements a comparison is made. The *reference values*, for example, the production or net costs at a particular point in time or the total population, are taken to be equal to 100, and the numbers to be compared, called *actual values*, for example, the production or net costs at another point in time or the female population, are referred to 100. The numerators of these fractions with denominator 100 are called percentages and denoted by %. If p is the percentage, a the reference value and b the actual value, then $b/a = p/100$ or $p = 100 \cdot b/a$.

$$\frac{\text{percentage}}{100} = \frac{\text{actual value}}{\text{reference value}} \qquad \frac{p}{100} = \frac{b}{a}$$

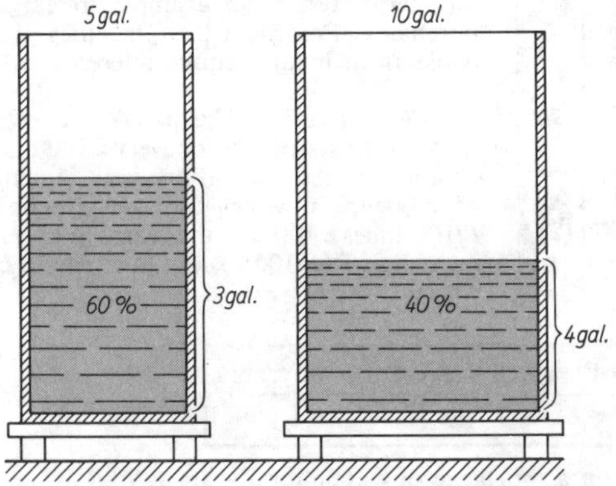

6.1-1 Utilization of capacity

If two vessels have capacities of 5 gal. and 10 gal., respectively, but contain 3 gal. and 4 gal. of liquid, then the 10 gal. vessel contains more liquid than the 5 gal. one, but it is less well utilized in proportion to its capacity, namely in the ratio 4 : 10, compared with 3 : 5 for the 5 gal. vessel. If one calculates relative to the reference value 100, then $4/10 = x_1/100$, or $x_1 = 4 \cdot 100/10 = 40$, while $3/5 = x_2/100$, or $x_2 = 3 \cdot 100/5 = 60$. Thus, 40% of the capacity of the 10 gal. vessel is utilized, and 60% of the 5 gal. vessel (Fig.).

Example 1: Among 1500 employees in a factory there are 300 women. Thus, among 100 employees there are on average $300/15 = 20$ women. Although it is obvious from this that 20% of the workers in this factory are women, the result can be obtained formally by substitution in the formula derived. From the reference value $a = 1\,500$ and the actual value $b = 300$ the percentage p is given by $p = 300 \cdot 100/1\,500 = 20$; that is, 20% of the employees are women (Fig.).

Example 2: How many pounds of titanium are contained in 275 lb of a steel alloy if it contains 4% of titanium? – Here the actual value b is required and the reference value $a = 275$ and the percentage $p = 4$ are given. Thus, the above formula is to be solved for the actual value. This gives $b = p \cdot a/100$ and for the numbers of the example $b = 4 \cdot 275/100 = 11$. The steel alloy contains 11 lb of titanium.

Example 3: The average annual milk yield of 2800 kg per cow is raised by 8% in the course of a year. From $b = p \cdot a/100$ one obtains $b = 8 \cdot 2800/100 = 224$. The milk yield is increased *by* 224 kg *to* 3024 kg per cow (Fig.).

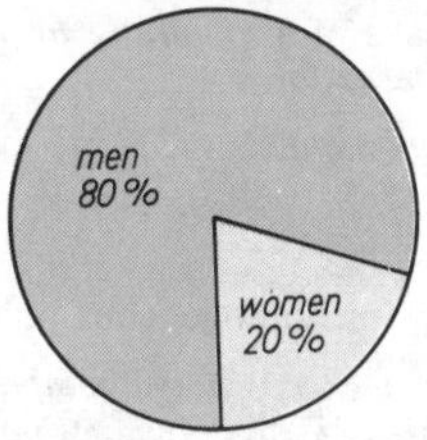

6.1-2 Composition of employees illustrated by areas

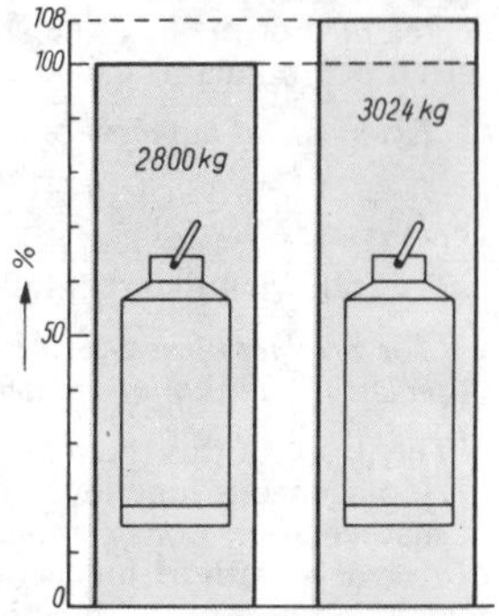

6.1-3 Raising of the milk yield

Example 4: By improved planning the cost of transport of bricks in a quarter can be reduced by £ 4800 or 12%. Here the reference value a is not known, but can be calculated from $a = b \cdot 100/p$. One finds that $a = 4800 \cdot 100/12 = 40000$. The cost of transport before was £ 40000; now it is £ 35200.

Example 5: In one year 3600 articles were manufactured. Production has increased to 120% compared with the previous year. How many articles were produced the previous year? – From the actual value $b = 3600$ and the percentage $p = 120$ the reference value a can be calculated: $a = 3600 \cdot 100/120 = 3000$. In the previous year 3000 items were produced.

6.2. Simple interest

In financial transactions it is customary to pay compensation for the loan of a sum of money; that depends on the amount of money borrowed and on the time and is called *interest*. For example, the National Savings Bank paid 1970 interest to the public at the rate of 3.5% of the amount of money deposited per annum (Fig.). These deposits, however, do not lie idle; on the contrary the banks use them to provide short-term and long-term credits, for example, for large purchases. For these loan facilities the banks in their turn require interest.

6.2-1 Extract from a Savings Bank book

Rate of interest. The *percentage rate of interest r* states that for every £ 100 one obtains interest of £ r in one year. A sum of money, or *principal*, £ P contains $P/100$ times £ 100, and thus earns in 1 year interest of £ $(P/100) \cdot r$ and in n years £ I, where $I = P \cdot r \cdot n/100$.

Interest formula (years)	$I = \dfrac{P \cdot r \cdot n}{100}$	$\text{interest} = \dfrac{\text{principal} \cdot \text{rate of interest} \cdot \text{no. of years}}{100}$

Example 1: How much interest is payable on a mortgage of £ 2000 for 5 years at 7%? – The principal is $P = 2000$, the rate of interest $r = 7\%$ and the number of years $n = 5$ are given. The

interest formula gives $I = (2000 \cdot 7 \cdot 5)/100 = 700$. The interest payable on the mortgage is £ 700 in 5 years.

Example 2: What is the rate of interest if a principal of £ 1200 earns interest of £ 576 in 6 years? – In this case the principal $P = 1200$, the interest $I = 576$ and the number of years $n = 6$ are given. The interest formula gives for the rate of interest $r = (100 \cdot I)/(P \cdot n)$, or with the values of this example, $r = (100 \cdot 576)/(1200 \cdot 6) = 8$. Thus, the rate of interest is 8%.

Interest formulae for	*n* years	*m* months	*d* days
interest	$I = \frac{P \cdot r \cdot n}{100}$	$I = \frac{P \cdot r \cdot m}{12 \cdot 100}$	$I = \frac{P \cdot r \cdot d}{365 \cdot 100}$
principal	$P = \frac{100 \cdot I}{r \cdot n}$	$P = \frac{1200 \cdot I}{r \cdot m}$	$P = \frac{36500 \cdot I}{r \cdot d}$
rate of interest	$r = \frac{100 \cdot I}{P \cdot n}$	$r = \frac{1200 \cdot I}{P \cdot m}$	$r = \frac{36500 \cdot I}{P \cdot d}$
time	$n = \frac{100 \cdot I}{P \cdot r}$	$m = \frac{1200 \cdot I}{P \cdot r}$	$d = \frac{36500 \cdot I}{P \cdot r}$

Example 3: What is the interest on a principal of £ 400 in 5 months at 9% and at 11%? – At 9% interest, $I = \frac{400 \cdot 9 \cdot 5}{12 \cdot 100} = 15.0$. At 11%, $I = \frac{400 \cdot 11 \cdot 5}{12 \cdot 100} = 18.33$. Hence in 5 months £ 400 earns interest of £ 15.0 at 9% and £ 18.33 at 11%.

6.3. Compound interest

Savings banks add the interest accruing at the end of a year to the principal; hence in the following year interest is also calculated on this interest. This method of calculation is called *compound interest*. Annuities and repayments of loans are also calculated with compound interest. Annuities are sums of money paid at fixed intervals of time, for example, yearly. In insurance the payment depends either on reaching a fixed point in time, for example, for an old-age pension, or on the occurrence of particular circumstances, for example, for sickness and accident benefit. A principal P_0 yields interest of $P_0 \cdot r/100$ in one year at r%.

From the beginning of the second year the principal $P_1 = P_0 + P_0 \cdot (r/100) = P_0(1 + r/100) = P_0 R$ with $R = (1 + r/100)$ will earn interest.

In one year this increases by $P_1 \cdot r/100$ to $P_2 = P_1 + P_1(r/100) = P_1 R = P_0 R^2$. The same argument holds for the 3rd, ..., nth year. At the end of the nth year the principal P_0 has grown by compound interest to the sum, called the *amount*, $P_n = P_0 R^n$, where $R = (1 + r/100)$ is the *growth factor*.

Compound interest	$P_n = P_0 R^n$

Example: At 6% a principal of £ 1500 grows to the amount £ 2007.33 in 5 years, and interest for the individual years is shown in the following table.

Year	Principal in £ at beginning of year	Interest in £ at end of year	Amount in £ at end of year
1	1500.00	90.00	1590.00
2	1590.00	95.40	1685.40
3	1685.40	101.12	1786.52
4	1786.52	107.19	1893.71
5	1893.71	113.62	2007.33

If the interest had been simple instead of compound, the principal would have grown to the amount £ 1950 after 5 years.

Figure 6.3-1 shows the growth of the principal $P_0 = 1$ at simple interest and at compound interest if $r = 6\%$. From the compound interest formula both the number of years n and the percentage rate of interest r can be calculated. Taking logarithms of both sides of the formula one finds $\lg P_n = \lg P_0 + n \lg R$ or $n = (\lg P_n - \lg P_0)/\lg R$; thus, one can calculate the number of years n. Taking the nth root of both sides one finds $R = \sqrt[n]{(P_n/P_0)} = 1 + r/100$, and from this

the percentage rate of interest is given by
$r = 100[\sqrt[n]{(P_n/P_0)} - 1]$.

Number of years	$n = (\lg P_n - \lg P_0)/\lg R$
Percentage rate of interest	$r = 100[\sqrt[n]{(P_n/P_0)} - 1]$

Example: In how many years does a principal of £ 500 double itself at 10% per annum compound interest? –
Besides the principal $P_0 = 500$ and the amount $P_n = 1000$, the growth factor $R = 1.10$ is given. The number of years n is therefore given by $n = (\lg 1000 - \lg 500)/\lg 1.10 = 7.27 \ldots$ After about 7 years the principal has doubled.

Discount. For a known amount P_n and growth factor R the initial principal P_0 can be calculated from the compound interest formula: $P_0 = P_n/R^n = P_n \cdot V^n$, where $V = 1/R$. This process is known as *discounting*, and V as the *discount factor*. One says that the amount P_n after n years is discounted to the *present value*.

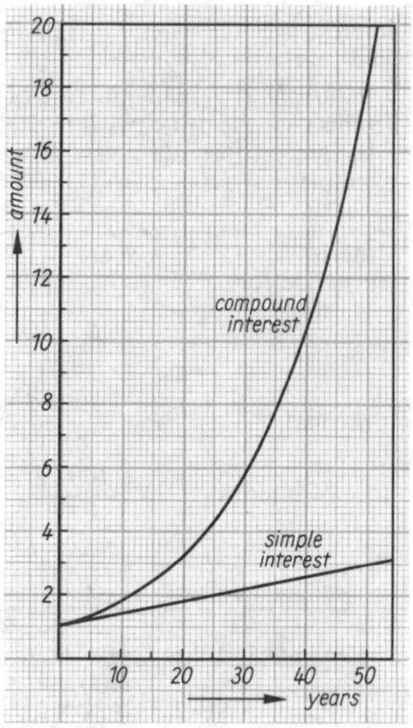

6.3-1 Growth of the principal $P_0 = 1$ at simple interest and at compound interest of 6% in 50 years

Example: A married couple intends to make a purchase to the value of £ 750 in 5 years' time. A sum of money is to be paid into a savings bank today (at 9% interest) in order to ensure that the full amount will be available after 5 years.
For the calculation of the required sum, the amount after 5 years $P_5 = 750$, the rate of interest $r = 9\%$ and the number of years $n = 5$ are known. From the table of discount factors one reads off for $n = 5$ and $r = 9$ the value $V^5 = 0.6499$ and calculates the present value $P_0 = 750 \cdot 0.6499 = 487.43$. At the present time £ 487.43 must be paid into the savings bank.

Interest factors

	Growth factor R^n			Discount factor V^n			
Years	Rate of interest			Rate of interest			Years
n	7%	9%	11%	7%	9%	11%	n
1	1.07	1.09	1.11	0.9346	0.9174	0.9009	1
2	1.1449	1.1881	1.2321	0.8734	0.8417	0.8116	2
3	1.2250	1.2950	1.3676	0.8163	0.7722	0.7312	3
4	1.3108	1.4116	1.5181	0.7629	0.7084	0.6587	4
5	1.4025	1.5386	1.6851	0.7130	0.6499	0.5935	5
6	1.5007	1.6771	1.8704	0.6663	0.5963	0.5346	6
7	1.6068	1.8280	2.0762	0.6227	0.5470	0.4817	7
8	1.7182	1.9926	2.3045	0.5820	0.5019	0.4339	8
9	1.8385	2.1719	2.5580	0.5439	0.4604	0.3909	9
10	1.9672	2.3674	2.8394	0.5083	0.4224	0.3522	10
11	2.1049	2.5804	3.1518	0.4751	0.3875	0.3173	11
12	2.2522	2.8127	3.4985	0.4440	0.3555	0.2858	12
13	2.4098	3.0658	3.8833	0.4150	0.3262	0.2575	13
14	2.5785	3.3417	4.3104	0.3878	0.2992	0.2320	14
15	2.7590	3.6425	4.7846	0.3624	0.2745	0.2090	15

Compound interest tables. For calculating the amount P_n or the present value P_0 banks use so-called *compound interest tables*, in which the powers R^n of the growth factor R and the powers V^n of the discount factor V are given for different rates of interest r and different numbers of years n. The growth and discount of a sum P_0 with growth factor R and discount factor V can be represented graphically with the help of a time line (Fig.).

Example: What is the rate of interest r if a sum of £ 400 grows to £ 787 in 10 years? –

The principal $P_0 = 400$, the amount $P_n = 787$, and the number of years $n = 10$ are known. The rate of interest can be calculated either with the help of compound interest tables, by calculating $R^{10} = 787/400 = 1.9675$ and reading off the rate of interest $r = 7$ in the table, or from the formula

$r = 100 \cdot [\sqrt[10]{(787/400)} - 1] = 7.$

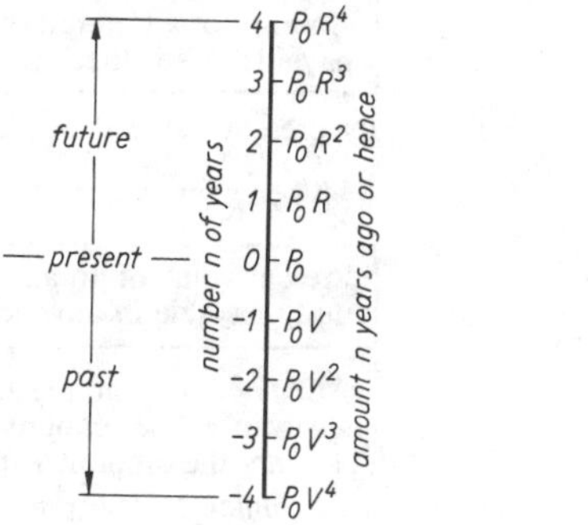

6.3-2 Representation of the growth and discount of a sum P_0 with the help of a time line

6.4. Annuities

The most important form of income from investment is the annuity, that is, a sequence of payments at previously determined points of time over a certain number of years. The individual payments of agreed amounts, called *instalments*, are usually paid at the end of each of the time intervals considered (in arrears) and seldom at the beginning (in advance). The *amount* of an annuity is the sum to which the instalments would accumulate at the end of the period if they were invested at $r\%$ compound interest at once on being received. The *present value* is the sum of money that has to be paid at the beginning of the period if the amount of the annuity is to be secured by a single payment.

Annuity payable in arrears. The instalments b payable at the end of each year accumulate at compound interest. After n years the first instalment has the value bR^{n-1}, the 2nd the value bR^{n-2}, and so on, and the last (b) has just been paid. Consequently the total amount s_n is the sum of a geometric series:

$$s_n = b + bR + \cdots + bR^{n-1}$$
$$= b \cdot \sum_{i=0}^{n-1} R^i = b\frac{R^n - 1}{R - 1}.$$

Amount of an annuity payable in arrears	$s_n = b \cdot \dfrac{R^n - 1}{R - 1}$
Present value of an annuity payable in arrears	$a = \dfrac{b}{R^n} \cdot \dfrac{R^n - 1}{R - 1}$

An annuity of £ 1500 payable in arrears has the amount after 5 years at 11% of

$$s_5 = £\,1500 \cdot (1.11^4 + 1.11^3 + 1.11^2 + 1.11^1 + 1) = £\,9341.7.$$

To calculate the present value a, the amount s_5 has to be discounted to the time zero; with the discount factor $V = 1/1.11 = 0.9009$, one obtanis $a = s_5V^5 = s_5 \cdot 0.5935 = 5544$. In general: $a = s_nV^n = [b(R^n - 1)]/[R^n(R - 1)]$.

Example 1: An annuity of £ 2000 is to be paid in arrears for 11 years. What is the amount of the annuity at 11% compound interest? – For the growth factor $R = 1.11$ one reads off $R^{11} = 3.1518$ from the table of interest factors. Thus, $s_{11} = 2000 \cdot (3.1518 - 1)/(1.11 - 1) = 39123.6$. The amount at the end of the 11th year is £ 39123.6.

Example 2: What sum of money will secure the annuity of Example 1? – With the help of the table of interest factors one obtains $a = s_{11} \cdot V^{11} = 39123.6 \cdot 0.3173 = 12414$. The present value of the annuity is £ 12414.

Example 3: After how many years does an annuity of £ 1000 at 11% compound interest, payable in arrears, have the amount £ 18000? – From the formula $s_n = b \cdot (R^n - 1)/(R - 1)$ one obtains $s_n(R - 1)/b = R^n - 1$ and $R^n = s_n(R - 1)/b + 1$, so that $n = \dfrac{\lg\,[s_n(R - 1)/b + 1]}{\lg R}$.

With the numerical values $s_n = 18000$, $b = 1000$ and $R = 1.11$ this gives $n = \lg 2.98/\lg 1.11 \approx 10.463$. The amount is reached after about 11 years.

Annuity payable in advance. For an annuity payable in advance each instalment b will be subject to interest for one year longer. From the compound interest formula the amount $\bar{s}_n$ of the annuity is therefore R times as much as the annuity payable in arrears: $Rb \cdot \dfrac{R^n - 1}{R - 1} = \bar{s}_n$.

Amount of an annuity payable in advance	$\bar{s}_n = Rb \cdot \dfrac{R^n - 1}{R - 1}$

Its present value $\bar{a}$ is again given by discounting the amount $\bar{s}_n$; one obtains $\bar{a} = \bar{s}_n V^n = Rb \cdot \dfrac{R^n - 1}{R - 1} \cdot \dfrac{1}{R^n} = b \cdot \dfrac{1 - V^n}{1 - V}$.

Present value of an annuity payable in advance	$\bar{a} = \dfrac{b}{R^{n-1}} \cdot \dfrac{R^n - 1}{R - 1}$

Example 1: For the data of Example 1 in the previous section, but for an annuity payable in advance, the amount is given by $\bar{s}_{11} = 2000 \cdot 1.11 \cdot (3.1518 - 1)/(1.11 - 1) = 43427.2$. After 11 years the amount is £ 43427.2.

Example 2: The present value of the annuity of Example 1, payable in advance, is given by $\bar{a} = 2000 \cdot 0.3522 \cdot (3.1518 - 1)/(1.11 - 1) = 13779$. Thus, the sum of £ 13779 will purchase the annuity.

Example 3: For the data of Example 3 above, but for an annuity payable in advance,

$$n = \frac{\lg [(18000 \cdot 0.11)/(1000 \cdot 1.11) + 1]}{\lg 1.11} = 9.81 .$$

The amount is reached after about 10 years.

Repayment of a loan. The repayment of credits and mortgages, for example, for investments or new buildings, is usually arranged in such a way that a yearly fixed sum, the *annual instalment*, is paid, which does not vary for the duration of the loan. This is composed of two parts, the *interest* and the *repayment of principal*. The course of the repayments for a credit of £ 10000 at 7% compound interest can be seen from the following *repayment scheme*, in which the annual instalment is fixed at £ 1000.

Year	Debt at beginning of year	Annual repayment	Interest payment	Repayment of principal	Debt at end of year
1	10000.00	1000.00	700.00	300.00	9700.00
2	9700.00	1000.00	679.00	321.00	9379.00
3	9379.00	1000.00	656.53	343.47	9035.53
4	9035.53	1000.00	632.49	367.51	8668.02
5	8668.02	1000.00	606.76	393.24	8274.78
.	...	...	...	...	...

It is seen that as the period of the loan progresses, the interest payment decreases and the repayment of principal increases.

This repayment scheme is clearly based on the following rules. At the end of the first year interest of $S \cdot (r/100)$ has to be paid on a loan S. Thus, out of the annual instalment A there remains a sum $T_1 = A - S(r/100)$ for repayment of principal. At the end of the second year interest of $S_1 \cdot (r/100)$ is due for the smaller loan $S_1 = S - T_1$; out of the annual instalment A there remains a sum $T_2 = A - S_1 \cdot (r/100) = A - S \cdot (r/100) + T_1 \cdot (r/100) = T_1 + T_1 \cdot (r/100) = T_1 R$ for repayment of principal. A sum $S_2 = S_1 - T_2$ is still to be repaid. At the end of the year the interest is $S_2 \cdot (r/100)$ and the repayment of principal $T_3 = A - S_2 \cdot (r/100) = A - S_1 \cdot (r/100) + T_2 \cdot (r/100) = T_2 R$. At the end of the nth year the repayment of principal is $T_n = T_{n-1} R = T_{n-2} R^2 = \dots = T_1 R^{n-1}$. In the example quoted the repayment of principal in the eleventh year is $T_{11} =$ £ $300 \cdot 1.07^{10}$ = £ 590.16.

The sum s_n of the repayments of principal of the first n years corresponds to the amount after n years of an annuity payable in arrears, $s_n = T_1 \cdot (R^n - 1)/(R - 1)$. In the example, after 11 years $s_{11} =$ £ $300 \cdot (1.07^{11} - 1)/(1.07 - 1) =$ £ 4734.90. The loan will have been paid off when the total repayment of principal is equal to the loan: $s_n = S$ or $T_1 \cdot (R^n - 1)/(R - 1) = S$.

Repayment formula	$S = T_1 \cdot \dfrac{R^n - 1}{R - 1}$

From this equation one can calculate the number of years n after which the loan will have been discharged: $n = \dfrac{\lg[(S/T_1)(R - 1) + 1]}{\lg R}$; in the example, $n = 17.79$, or 18 in round numbers. The debt is discharged after 18 years.

Life insurance. A further field of application of compound interest and annuity calculations consists of the different forms of life insurance. One distinguishes, among others, between *death and accident insurance, sickness and old-age insurance* and *contributory pensions.* In all these insurances the insurance companies enter into a contract, called the *insurance contract*, with the insured person. Between the payments of the person taking out the insurance and the commitments of the insurance company, which depend, among other things, on the type of insurance, an equivalence must exist, so that the insurance contract does not lead to a loss for any of those concerned. Naturally the equivalence does not hold for the individual insured, in which case the insurance would be superfluous, but only for the totality of all insured persons. Consequently, in the mathematical formulation of the equivalence principle, besides compound interest and annuity calculations, *demographic assumptions* play a role.

Mortality tables. The most important demographic aid consists of the mortality tables. These tables, which are drawn up partly on the basis of population census and partly by experience over many years of the insurance company, start from a definite (but arbitrarily chosen) number l_n of persons of the same age, namely n-year olds, and state how many of these reach their xth year. This number is denoted by l_x and is called the *number of survivors* to the age x. In addition the following relations are contained in the table: $l_x - l_{x+1} = d_x$ is the number of those dying at the age x, $p_x = l_{x+1}/l_x$ is the probability of surviving of the x-year old, $q_x = d_x/l_x$ is the probability of dying of the x-year old and $e_x = (1/l_x) \sum_{k=0}^{\infty} l_{x+k} - 1/2$ is the mean life expectancy.

Figure 6.4-1 is a section of a typical general mortality table. One reads in this that of 100000 males of the same age born say 50 years ago, 89104 are still alive, and that for each of these the mean life expectancy is 24.01 years.

6.4-1 General mortality table

Males	Males					Females				
completed years of age	of 100000 live-born at the same time		probability of dying	life expectancy in years		of 100000 live-born at the same time		probability of dying	life expectancy in years	
	still living	dead		of all those still living	of each one still living	still living	dead		of all those still living	of each one still living
x	l_x	d_x	q_x	$e_x l_x$	e_x	l_x	d_x	q_x	$e_x l_x$	e_x
46	90 859	332	0.003 653 149	2 499 889	27.51	93 612	241	0.002 579 387	2 926 801	31.27
47	90 527	345	0.003 812 489	2 409 196	26.61	93 370	248	0.002 660 978	2 883 310	30.35
48	90 182	447	0.004 954 396	2 318 841	25.71	93 122	342	0.003 672 863	2 740 064	29.42
49	89 736	632	0.007 041 835	2 228 882	24.84	92 780	468	0.005 044 604	2 647 114	28.53
50	89 104	801	0.008 988 733	2 139 462	24.01	92 312	543	0.005 883 410	2 554 568	27.67
51	88 303	849	0.009 609 744	2 050 759	23.22	91 768	547	0.005 959 474	2 462 528	26.83
52	87 454	807	0.009 226 817	1 962 881	22.45	91 222	525	0.005 753 766	2 371 033	25.99
53	86 647	795	0.009 180 458	1 875 830	21.65	90 697	522	0.005 758 713	2 280 074	25.14
54	85 852	870	0.010 138 656	1 789 581	20.85	90 174	546	0.006 059 439	2 189 638	24.28
55	84 981	990	0.011 647 731	1 704 164	20.05	89 628	586	0.006 536 162	2 099 737	23.43

7. Plane geometry

Plane geometry is that part of geometry (Greek, measuring the earth) which deals with two-dimensional figures. Although we live in a three-dimensional world, the study of plane geometry can deepen our insight into some of the properties of our surroundings.

Just as the notion of number was abstracted from the visible world, so also the concepts at the basis of geometry were gained by a process of abstraction extending over many centuries. By ignoring inessential differences, for example, of mass, colour, form or surface texture, and disregarding further irregularities of real objects one arrived at spatial forms in three dimensions: length, breadth and height. Accordingly we say that a solid body has three dimensions, but a surface only two, a line, for example, an edge of intersection of two surfaces, one dimension, and finally a point, regarded as the intersection of two lines, has the dimension zero.

In plane geometry a plane is always taken as given. Geometric investigations are, in general, carried out within this plane, but in individual cases it is advantageous to consider also Euclidean space (EUCLID, Greek mathematician, about 300 B. C.) as a basic geometric object containing the given plane.

7.1. Points, lines, rays and segments

Points and lines

Points and lines (more accurately, straight lines) are the basic concepts of elementary plane geometry. Intuitively, a *line* is often defined as the path of a point that moves in a plane in such a way that it always takes the *shortest route* between any two of its positions and does not change direction; even in a more rigorous approach no definition of lines and points is given. But in modern mathematics the relationships between these two kinds of geometric objects are fixed by *axioms* (see Chapter 40.).

Number of intersections of several lines. Two lines in a plane have at most *one point* in common unless they coincide (that is, have all points in common). Two lines in a plane that have no points in common are called *parallel*.

If three lines in a plane do not all contain a common point, and if not two of them are parallel or coincide, then there are exactly *three* points of intersection between pairs of these lines.

Four lines of which no two coincide or are parallel and of which no three have a point in common have exactly *six* points of intersection (see Complete quadrilateral).

If n lines in a plane are given such that no two coincide or are parallel and no three have a common point, then each line has $n - 1$ intersections with the others; since each intersection is counted twice, the total number of points of intersection is $n(n - 1)/2$.

Number of lines through several points. There is exactly one line through any two distinct points. If three points are not collinear (that is, do not all lie on the same straight line), then there are three lines, each containing two of them. These three points, or any two of the three lines, or one of the lines and the point not on it completely determine the *plane* in which they lie.

If n distinct points in a plane are given such that no three are collinear, then every point lies on a single line through each of the other $n - 1$. As each line is counted twice, there are $n(n - 1)/2$ possible lines; thus, if $n = 4$, there are six lines (see Complete quadrangle).

Pencils of lines. There are infinitely many lines through any point in the plane. The set of all lines in the plane going through a single given point is called a *pencil of lines*. The point that is common to all the lines is called the *carrier* of the pencil. An upper case letter in brackets such as (P) is used to denote the pencil with the carrier P. By analogy, the set of all lines in the plane parallel to a given line is called a *pencil of parallels*.

If the lines of a pencil (P) or of a pencil of parallels are intersected by two straight lines l_1 and l_2 that do not belong to the pencil, then the lines of the pencil induce a *perspective map* of all the points of l_1 onto those of l_2.

Rays and segments

Rays. A *ray* (or half-line) contains precisely those points of a line that lie on one side of a point O of that line; the point O is included in the ray. In other words, the ray contains those points of the line that can be reached by travelling along the line, starting at O, in a particular direction (without reversing!). The concept of a ray was obtained, like all mathematical concepts, by abstraction. It is intuitively easy to grasp if one thinks of a ray of light emitted by the sun, or a line of sight, which is straight and is bounded by the eye of the observer (one idealizes the sun or the eye to a single point).

Segments. A *segment* (or interval) AB contains precisely those points of the line through A and B that lie between A and B; A and B themselves are included in the segment. The segment is the *shortest path* in the plane connecting its two end-points. If it is important to emphasize direction, the symbol $\overrightarrow{AB}$ is used to signify that the segment is directed from A to B. To avoid the somewhat cumbersome use of arrows it is agreed that the segment is directed from the first named point to the second, that is, the first named point is the initial point, and the second the terminal point, of a directed segment. A further convention is that a minus sign indicates the reversal of the direction of the segment. Thus, one can replace $\overrightarrow{AB} = \overleftarrow{BA}$ by $\overrightarrow{AB} = -\overrightarrow{BA}$. The length $|AB|$ of the segment AB is the *distance* between A and B. It is measured by comparing it with another segment, the *unit segment*. The length of the unit segment serves as the unit of measure for lengths.

Units of length. The basic unit of length is the *metre*. It is now defined as 1650763.73 wavelengths of the orange line in the spectrum of the Krypton isotope 86, measured in vacuo.

Two of the admissible multiples of the metre, the *decametre* (1 dm = 10 metres) and the *hectometre* (1 hm = 100 metres) have not found general use.

Units of length derived from the metre

Unit	*Symbol*	*Relation*
terametre	Tm	$1\ \mathrm{Tm} = 10^{12}\ \mathrm{m}$
gigametre	Gm	$1\ \mathrm{Gm} = 10^{9}\ \mathrm{m}$
megametre	Mm	$1\ \mathrm{Mm} = 10^{6}\ \mathrm{m}$
kilometre	km	$1\ \mathrm{km} = 10^{3}\ \mathrm{m}$
decimetre	dm	$1\ \mathrm{dm} = 10^{-1}\ \mathrm{m}$
centimetre	cm	$1\ \mathrm{cm} = 10^{-2}\ \mathrm{m}$
millimetre	mm	$1\ \mathrm{mm} = 10^{-3}\ \mathrm{m}$
micrometre	µm	$1\ \mu\mathrm{m} = 10^{-6}\ \mathrm{m}$
nanometre	nm	$1\ \mathrm{nm} = 10^{-9}\ \mathrm{m}$
picometre	pm	$1\ \mathrm{pm} = 10^{-12}\ \mathrm{m}$
femtometre	fm	$1\ \mathrm{fm} = 10^{-15}\ \mathrm{m}$
attometre	am	$1\ \mathrm{am} = 10^{-18}\ \mathrm{m}$

Units in U.K. and U.S.A.

inch	1 in = 1″ = 0.0254 m
foot	1 ft = 1′ = 12″ = 0.3048 m
yard	1 yd = 3′ = 0.9144 m
mile	1 statute mile = 1 mi = 5280′ = 1609.344 m
	1 imperial nautical mile = 6080′ = 1853.181 m = 1 meridian second
	1 US nautical mile = 1852 m

1 m $\approx 6.215 \cdot 10^{-4}$ statute miles = 1.094 yd = 3.281′ = 39.37″

1 km $\approx$ 0.6215 statute miles

Parallel and orthogonal lines

Parallel lines. Two parallels have no point in common. If a straight line l is parallel to a line l' (notation: $l \| l'$), then l and l' have the same direction. One can obtain l' from l by moving all the points of l through a segment equal to $\overrightarrow{PP'}$ (such a transformation is called a *translation*). This is the basis of the construction of parallels by ruler and set square (Fig.). If one connects a point of a line l to all the points of a parallel line l' by straight line segments, then among those segments there is a *shortest* one. The length of this segment defines the distance d between the two parallel lines. It is *the same whichever point of l* is chosen, in other words, parallels do not come closer or move further apart.

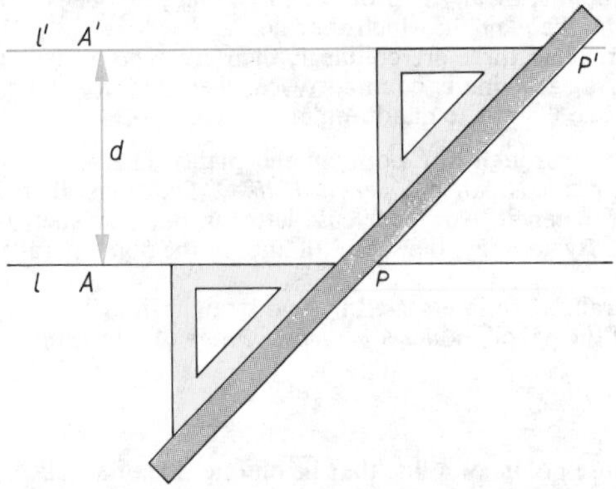

7.1-1 Construction of a parallel l' to l at a distance d by shifting (translation)

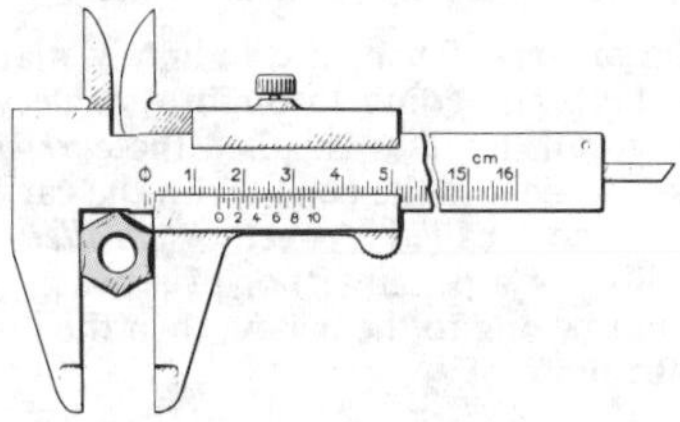

7.1-2 The distance between parallel surfaces of a work-piece (measurement of thickness)

The normal distance between the parallel rails of a railway line (the gauge) is 1.435 m ($56^1/_2$ ins.); in the Soviet Union it is 1.524 m, and in Spain and Portugal it is 1.670 m. The parallel jaws of a pair of *calipers* can be moved to measure the breadth of objects with parallel sides (Fig.).

Every line l has exactly *one parallel* l' on each side at any given distance d. The construction of l' with ruler and set-square is shown in the figure (see Fig. 7.1-1). If l is a line and P' a point not on l, then there is exactly one line l' through P' parallel to l; it lies in the plane through P' and l. This statement, the existence and uniqueness of a parallel through P' is the *parallel axiom* (or *postulate*) of Euclidean geometry.

Orthogonal lines. The distance between parallel lines l and l' is measured along a segment AA' that meets l in A and l' in A' in pairs of equal angles. These angles are called *right angles*. *Orthogonal* lines are lines that intersect at right angles; they are also called *perpendicular* to one another. Orthogonality or parallelism are properties of two lines relative to one another; the lines need not be vertical or horizontal.

7.2. Angles

Two rays a and b with the same initial point S can be made to coincide by a rotation about S, which determines the *angle* between a and b [notation $\sphericalangle(a, b)$ or (a, b)].

An *orientation* of the plane is given by fixing the direction of the rotation. In mathematics the positive orientation is chosen as the one corresponding to an anti-clockwise rotation; in surveying the clockwise rotation is taken to be positive. If the direction of rotation is important, one must distinguish between the angles $\sphericalangle(a, b)$ and $\sphericalangle(b, a)$: $\sphericalangle(a, b) = -\sphericalangle(b, a)$. If A is a point on a and B is a point on b, the angle $\sphericalangle(a, b)$ is usually written as $\sphericalangle ASB$. S is called the *vertex* of the angle, the rays are called its *arms*. Each arm defines a particular direction and the angle is a measure of the difference of these directions in the oriented plane (Fig.).

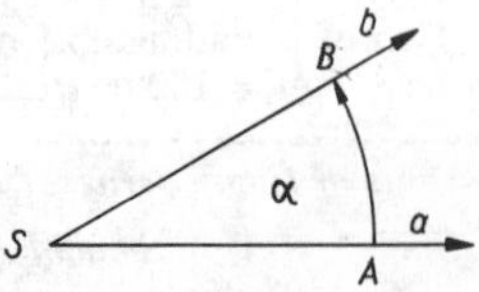

7.2-1 Definition of an angle

Classification of angles

Angles are classified by the amount of the difference of the directions of their arms. If the angle corresponds to a rotation through a quarter of the full circle, it is called a *right angle*; a *straight angle* corresponds to a rotation through half of a full circle. If the inclination of the arms is less than a right angle, the angle is called *acute*; if the inclination is greater than a right angle but less than a straight angle, the angle is called *obtuse*; if the inclination is greater than a straight angle, the angle is called *reflex*; a reflex angle whose arms coincide is a *full* angle.

7.2-2 Angles

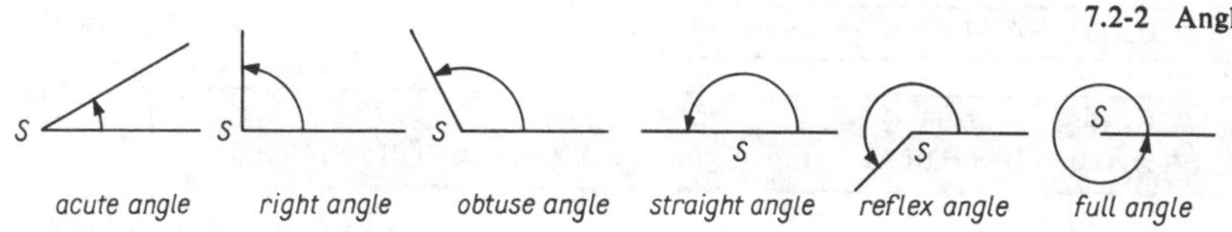

Units of measurement of angles

All methods of measuring angles are based on division of the circle (Fig.). There are two types of unit, based on measurement by degrees and by arc length, respectively.

Degrees. If a circle is divided into 360 equal parts by radii, the angle between two neighbouring radii is called a *degree* (notation: 1°). Thus, a degree is one 360th of a full angle or one 90th of a right angle. The degree is divided into sixty *minutes* (notation: 1′) and the minute into sixty *seconds* (notation: 1″). Although the subdivisions of a degree have the same names as those of the hour, it is important that the notation should distinguish between units of angle and units of time.

$1° = 60' = 3600''$; $1\ \text{h} = 60\ \text{min} = 3600\ \text{s}$

In geodesy there is also a system of *grades*, where the circle is divided into four hundred parts and the right angle has 100 parts. Each of these is one *gon* or one grade and is divided decimally (notation: 1 gon).

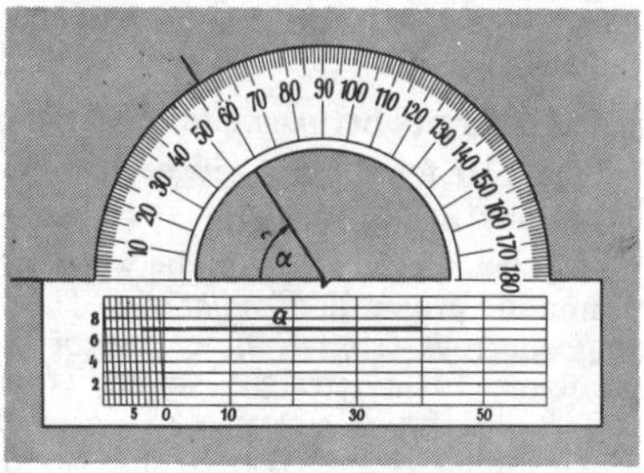

7.2-3 Protractor with linear scale; $\alpha = 57°$, $a = 43.7$ mm

For a right angle the following equations hold:

$$1 \text{ right angle} = 90° = 5400' = 324000'' = 100.0000 \text{ gon}.$$

Example 1: To transform 62° 48′ 15″ into gons.

$48' = 48°/60 = 0.8°$; $15'' = 15°/3600 = 0.004167°$,
$62°\,48'\,15'' = 62.804167 \cdot 100\ \text{gon}/90 = 69.7824\ \text{gon}$.

Example 2: To transform 135.4682 gon into degrees, minutes and seconds.

$135.4682\ \text{gon} = 100\ \text{gon} + 35.4682\ \text{gon} \triangleq 90° + 35.4681 \cdot 90°/100 = 121.92138°$,
$0.92138° = 0.92138 \cdot 60' = 55.2828'$;
$0.2828' = 0.2828' \cdot 60'' = 16.968''$,
$135.4682\ \text{gon} \approx 121°\,55'\,17''$.

Equivalents	
$90° \triangleq 100$ gon	$100\ \text{gon} \triangleq 90°$
$1° \triangleq 10/9\ \text{gon} = 1.111111\ldots$ gon	$1\ \text{gon} \triangleq 9°/10 = 54'$
$1' \triangleq 1/60 \cdot 10/9\ \text{gon} \approx 0.018519$ gon	$0.01\ \text{gon} \triangleq 54'/100 = 32.4''$
$1'' \triangleq 1/3600 \cdot 10/9\ \text{gon} \approx 0.000309$ gon	$0.0001\ \text{gon} \triangleq 324''/1000 = 0.324''$

Arc units. In a circle the length of an arc *a* between two radii is proportional to the angle between them to their length. The following ratios hold if the angle is measured in degrees.

circumference : arc = full angle : angle subtended at centre	$2\pi r : a = 360 : \alpha$.

From this it follows that ratio between the lengths of the arc and the radius depends only on the angle at the centre subtended by the arc

$$a : r = (\alpha/360) \cdot 2\pi = 2\pi\alpha/360 = (\pi/180) \cdot \alpha.$$

Thus, if the radius of a circle is known, the length of an arc on the circumference can be used to measure the corresponding angle at the centre. The *arc* of an angle is therefore defined as the quotient

$$a/r = (\pi/180)\,\alpha = \hat{\alpha} = \text{arc}\,\alpha; \qquad a = r \cdot \hat{\alpha} = (\pi/180)\,\alpha \cdot r.$$

The unit of this measurement is called a *radian* (Notation: 1 rad). 1 rad is the angle at the centre subtended by an arc of length equal to the radius of the circle:

$1\ \text{rad} \triangleq 57.29578° = 57°\,17'\,44.8'' \approx 57°\,17'\,45''$ (Fig.). $a : r = \hat{\alpha} = \text{arc}\,\alpha \qquad \alpha = \hat{\alpha} \cdot 180°/\pi$

1 rad ≙ 57° 17′ 44.8″ ≙ 63.6620 gon;

$\alpha = \hat{\alpha} \cdot 180/\pi$ in degrees, $\alpha_1 = \hat{\alpha} \cdot 200/\pi$ in gon; $1° \triangleq \pi/180$ rad $\approx 0.0174\,53$ rad, 1 gon $\triangleq \pi/200$ rad ≈ 0.0157076 rad

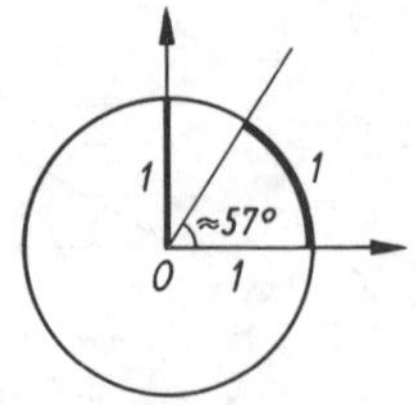

7.2-4 1 rad is the unit of radian measure

The unit circle is the circle with radius equal to one unit of length. Thus, for the unit circle the measure of an angle in radians is the length of the corresponding arc on the circumference.

Survey of angles and measures

	acute angles			right angle	obtuse angles	straight angle	reflex angles	full angle
degrees	30°	45°	57° 17′ 45″	90°		180°	270°	360°
gons	$33^1/_3$gon	50 gon	63.6620 gon	100 gon		200 gon	300 gon	400 gon
radians	$\pi/6$	$\pi/4$	1	$\pi/2$		π	$3\pi/2$	2π

Angles between intersecting lines

There are four angles between two intersecting lines of a plane. One distinguishes between adjacent angles and opposite angles.

Adjacent angles. Angles between two intersecting lines having their vertex S and one arm in common are called *adjacent*. The other arms are the two opposite rays from S along one of the lines, so that the angles combine to a straight angle; for example, in the figure $\alpha + \beta = \beta + \gamma = \gamma + \delta = \delta + \alpha = 180°$. Two adjacent angles (for example α and β) need not be equal. If they are, then each must be half of 180°, that is, a right angle. This fact was exploited in the definition of orthogonal lines and is the basis of the following definition.

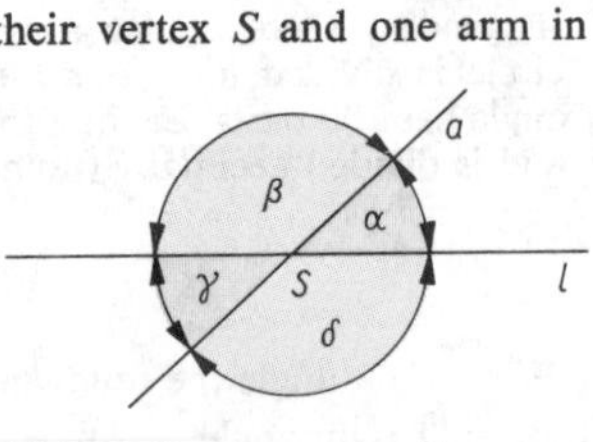

7.2-5 Adjacent angles, for example, α and β, and vertically opposite angles, for example, α and γ

A right angle is any one of the four angles between two lines that intersect so that adjacent angles between them are equal.

Any two angles whose sum is 180°, whether they are adjacent or not, are called *supplementary*. Angles whose sum is 90° are called *complementary*. Adjacent angles are supplementary, but supplementary angles are adjacent only if they have an arm in common.

Vertically opposite angles. Two angles between intersecting lines with a common vertex but no common arm are called *vertically opposite*, or simply opposite if no confusion is to be feared. Since two opposite angles have no arm in common, the directions of the arms of one must be opposite to those of the other. In the figure α and γ, and β and δ are opposite angles.

Opposite angles are equal, because the sum of each with a common adjacent angle is 180°.

Angles between two parallels and a line

If a pair of parallels is intersected by a third line (Fig.), they form eight angles, which fall into two sets of four equal angles each, for example, $\alpha = \gamma = \alpha' = \gamma'$ and $\beta = \delta = \beta' = \delta'$.

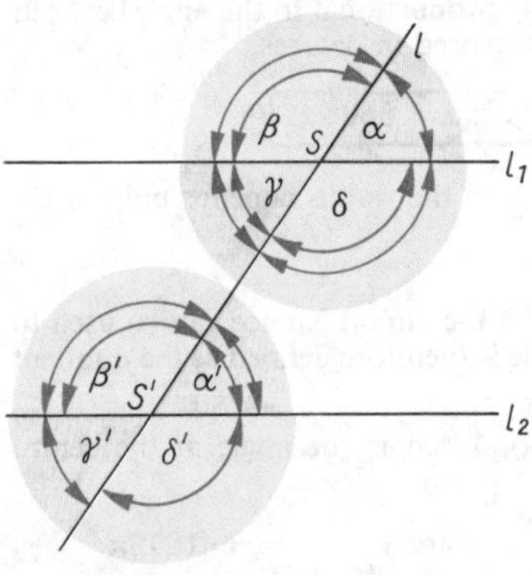

7.2-6 Angles between parallels and a straight line intersecting them

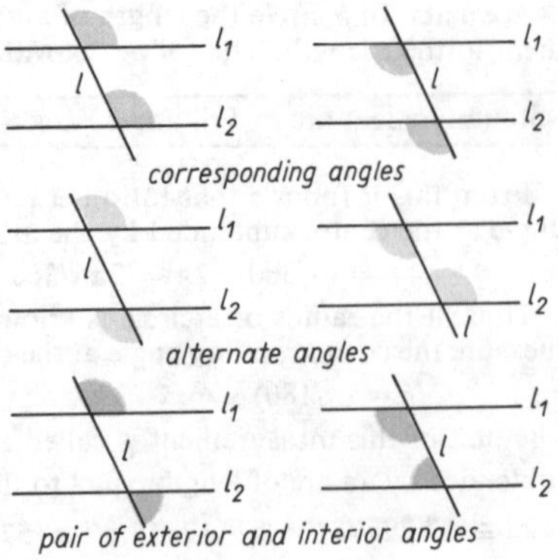

7.2-7 Examples of angles between parallels and a line intersecting them

Pairs of angles with a *common vertex* are opposite or adjacent. They are *opposite* if their arms are opposed, for example $\alpha = \gamma$, or $\beta' = \delta'$. They are *adjacent* if they have one arm in common and their other arms are opposed, for example, $\alpha + \beta = 180°$, or $\gamma + \delta = 180°$.

Pairs of angles with *distinct vertices* are classified as follows (Fig.):

1. *Corresponding angles:* the arms of one are directed in the same sense as those of the other, for example, $\alpha = \alpha'$, or $\gamma = \gamma'$.

2. *Alternate angles:* the arms of one are directed in the opposite sense to those of the other; for example, $\alpha = \gamma'$, or $\gamma = \alpha'$.

3. *Interior and exterior angles:* the arms on the parallels are directed in the same sense and those on the intersecting lines are opposite. If these arms contain the segment between the parallels, the angles are interior, otherwise exterior; for example, $\alpha + \delta' = 180°$, exterior angles; and $\gamma + \beta' = 180°$, interior angles.

If two parallel lines are intersected by a third line, then corresponding angles are equal, alternate angles are equal, and pairs of interior or exterior angles are supplementary.

These theorems have converses, which state that if two lines are intersected by a third and if two corresponding or alternate angles are equal, or two interior or exterior angles are supplementary, then the two lines are parallel.

The relative position of the three classes of angles can be described even without assuming that the two lines are parallel. Alternate angles are then defined as lying on opposite sides of the intersecting line, and either both between the two lines or both outside them. The other types lie on the same side of the intersecting line. Exterior angles are both outside the two lines, interior angles both between them, and of two corresponding angles one is inside and one is outside.

Constructions of angles

Set squares. The usual models of *set squares* have a right angle and angles of 45° or 60° and 30°. Other angles can be constructed by *adding* or *subtracting* these angles (Fig.). Bisecting with a ruler (or set square) and compass gives further angles, for example 22.5°; 15°; 7.5°; and the angles obtained by adding or subtracting these new angles. In adding or subtracting angles it is sometimes necessary to shift the position of a previously constructed angle. A method of accomplishing this is described in the following paragraph.

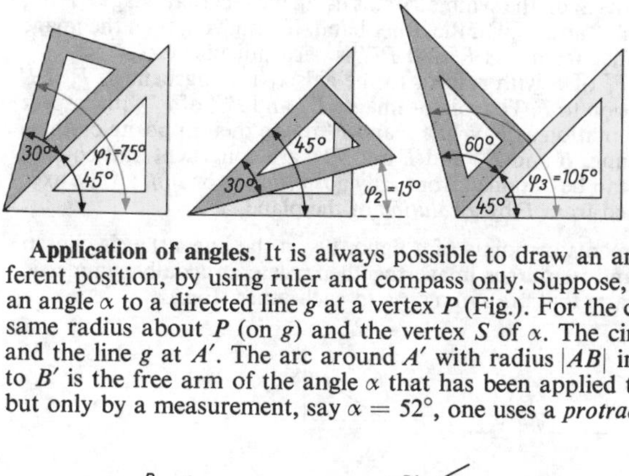

7.2-8 Construction of special angles φ with set-squares, $\varphi_1 = 75°$, $\varphi_2 = 15°$, $\varphi_3 = 105°$

Application of angles. It is always possible to draw an angle equal to a given angle, but in a different position, by using ruler and compass only. Suppose, for example, that it is required to apply an angle α to a directed line g at a vertex P (Fig.). For the construction one draws two circles of the same radius about P (on g) and the vertex S of α. The circles intersect the arms of α at A and B and the line g at A'. The arc around A' with radius $|AB|$ intersects the circle at B'. The ray from P to B' is the free arm of the angle α that has been applied to g at P. If α is not given by a drawing but only by a measurement, say $\alpha = 52°$, one uses a *protractor*.

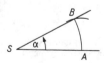

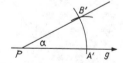

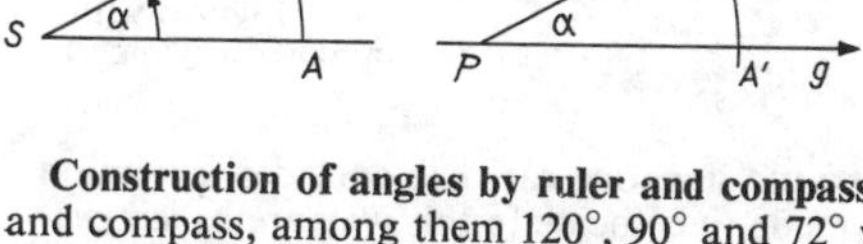

7.2-9 Application of an angle

Construction of angles by ruler and compass. Only exceptional angles can be constructed by ruler and compass, among them 120°, 90° and 72° which result from the ruler and compass construction of the equilateral triangle square, and regular pentagon (see Regular convex n-gons). Continued bisections of these lead to 60°, 30°, 15°, 45°, 36°, 18° and 9° (to mention only angles of a whole number of degrees). By adding 15° and 9° one obtains 24°, and hence 12°, 6° and 3°. Thus, all multiples of 3° can be constructed by ruler and compass. In fact, they are the only constructible angles of a whole number of degrees.

7.3. Symmetry

Axial symmetry

Every plane E is divided into two half-planes by any straight line s in E (Fig.). A rotation in space of 180° about s maps each of these half-planes onto the other. Every point S of the axis s is its own image, $S = S'$, the axis is a *fixed line*. Furthermore, any line AS forms with s the same angle as its image $A'S$, and AS and $A'S$ are equal in length. The line AA' connecting a point to its image is perpendicular to s and is bisected by s. Every figure F is congruent to its image F'.

Reflecting the half-plane in a mirror perpendicular to E and intersecting it in s has the same effect as the operation just discussed. F and its image F' are therefore called *mirror images* or *reflections* of each other, and the map can be called a reflection in a line. The technical term is, however, *axial symmetry*. It is a geometric map, or *transformation*, and is uniquely determined by any point P and its image P' (if distinct from P) or by the axis of symmetry s.

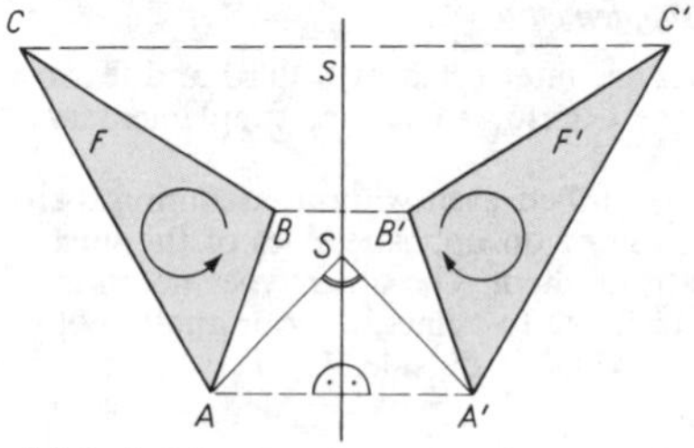

7.3-1 Axial symmetry

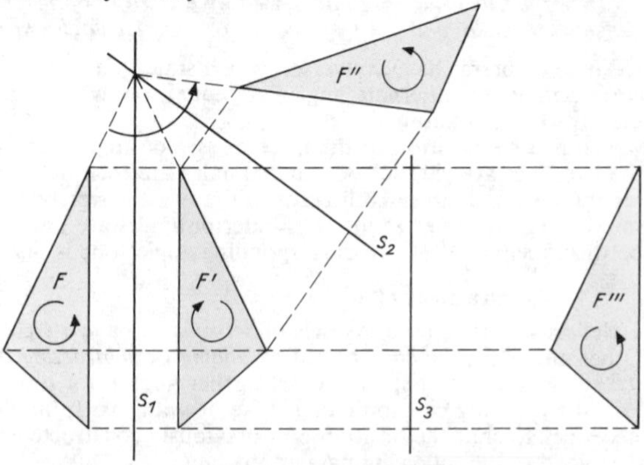

7.3-2 Two reflections can be replaced by a shift or a rotation

If P and P' are given, one can find the points S_i on the symmetry axis using the fact that $|PS_i| = |P'S_i|$, by intersecting circles of equal radius about P and P'. On the other hand, if s is given, then the image of an arbitrary point P has the same distance from s as P' and PP' is perpendicular to s.

In the figure, though the mirror image F' of F with respect to the axis s_1 is congruent to F, if E is oriented, then it has the opposite orientation to F. The mirror images F'' and F''' of F' with respect to further axes s_2 and s_3 have the same orientation as F. While F and F' retain their opposite orientation under any transformation of the plane, F and F'' and F and F''' are congruent and equally oriented. In particular, if $s_1 \parallel s_3$, then F''' can be obtained from F by a *translation* or *shift*; if the axes s_1 and s_2 intersect, then F''' can be obtained from F by a *rotation* of the plane.

Axially symmetric figures. If certain segments or points of a figure lie on the line s that is chosen as the axis of symmetry, then the figure and its mirror image together create an axially symmetric figure, that is, a figure consisting of two parts that are symmetric to each other (Fig.).

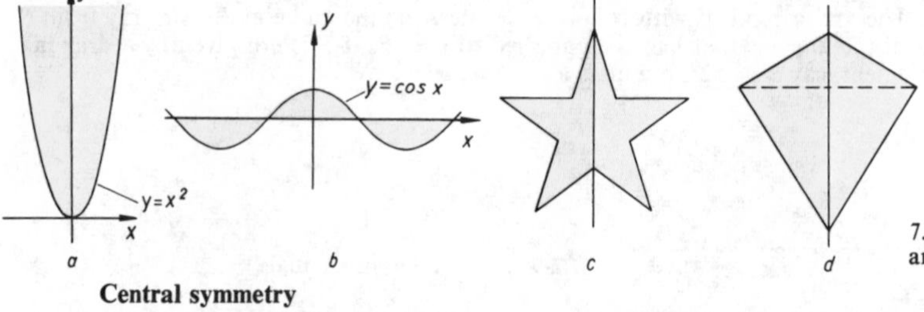

7.3-3 Figures with an axis of symmetry

Central symmetry

Apart from rotating a plane by 180° through space about a line, one can also consider a rotation of 180° in the plane about a point. Figures that can be made to coincide in this way are called *centrally symmetric* to each other, and the point is called the *centre of symmetry* (Fig.).

Under a central symmetry every point B of the plane is mapped to a point B' such that the segment between them has the centre of symmetry S as its mid-point. This transformation is also called *reflection in a point*. Like every other rotation, it is a *congruence transformation*, that is, the size and shape of figures are not changed by the transformation. In contrast to axial symmetry, a point

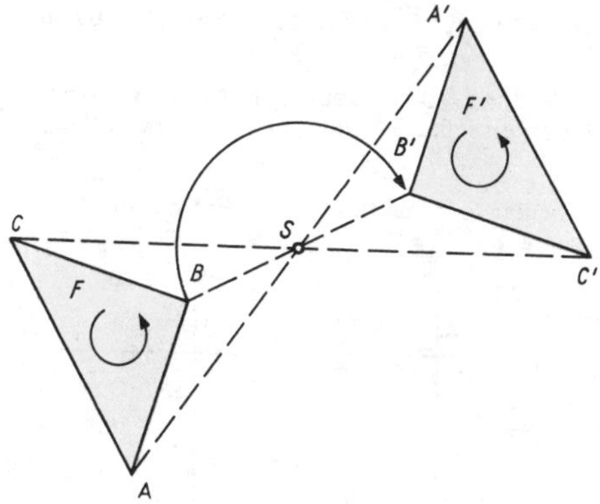

7.3-4 Central symmetry

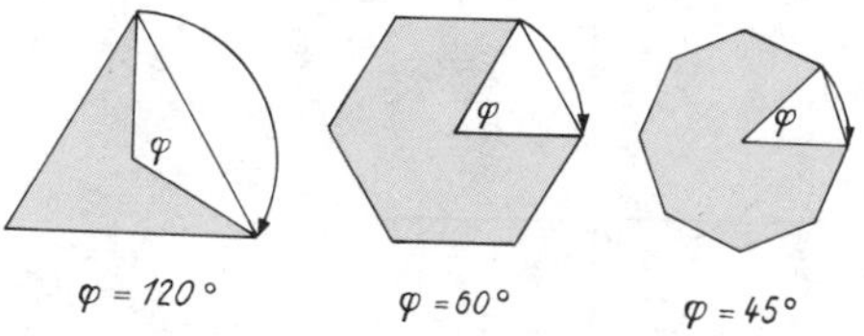

7.3-5 Radial symmetry of regular polygons

symmetry preserves the orientation of figures, so that a figure and its image under a point symmetry are *congruent and equally oriented*. The images under two consecutive reflections in the same point are coincident with the original figure. The only fixed point under a central symmetry is the centre of symmetry.

Figures that can be made to coincide with themselves by a rotation through an angle φ about a point P are called *radially symmetric*. Central symmetry is the special case of radial symmetry corresponding to $\varphi = 180°$. All regular polygons are radially symmetric (Fig.).

Basic constructions

Construction of the mid-point of a segment. This is effected by constructing the *axis of the symmetry* that carries the end point A of AB into B.

About each of the points A and B as centres draw circles of equal radius greater than half the length of AB. These intersect in points S_1 and S_2 on the symmetry axis. Therefore the line s through S_1 and S_2 intersects AB in its mid-point and is perpendicular to AB; it is called the perpendicular bisector of AB (Fig.).

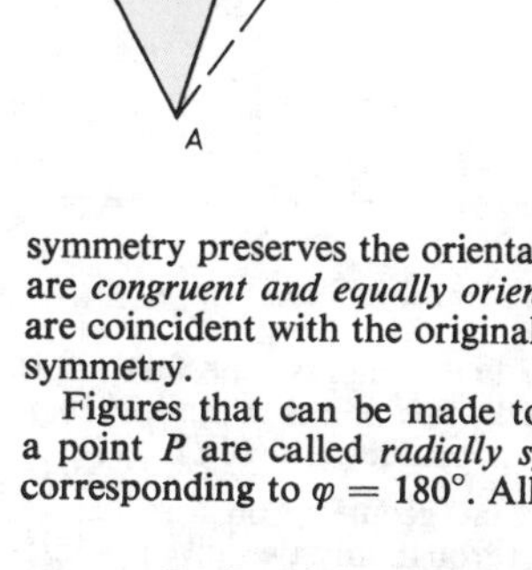

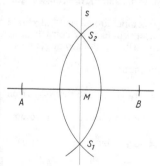

7.3-6 Bisecting a segment

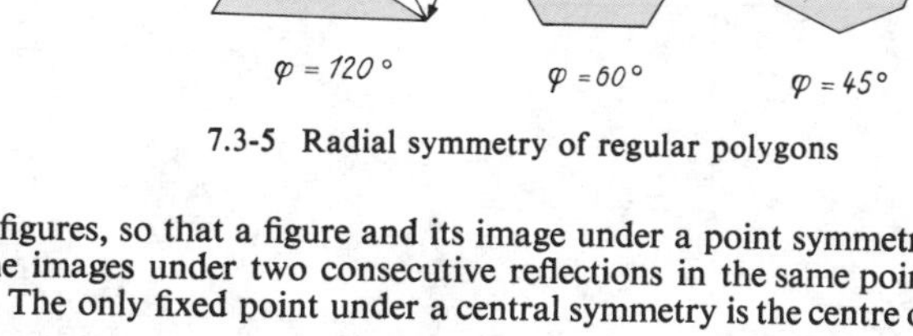

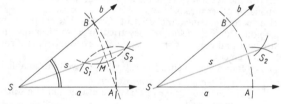

7.3-7 Bisecting an angle

Construction of the bisector of an angle. Angles are also bisected by using the properties of axially symmetric points. An arc of arbitrary radius is drawn about the vertex S of the angle $\sphericalangle(a, b)$. It intersects the arms of the angle in A and B. The *perpendicular bisector* of AB is the symmetry axis of the figure and bisects the angle. Since S is on this axis, it is only necessary to construct *one* further point on it. Such a point, for example S_2, is found by intersecting two arcs of equal radius about A and B (Fig.).

Unlike bisection, *trisection* of an angle with ruler and compass alone is possible only in exceptional cases.

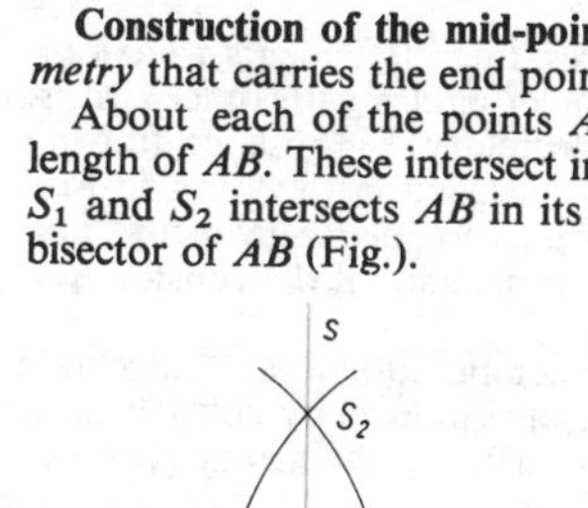

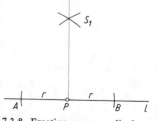

7.3-8 Erecting a perpendicular to a line in a point on it

Construction of perpendiculars. To erect the perpendicular on a line l in a point P on that line, draw a circle with P as centre of arbitrary radius r (Fig.). It intersects l in A and B. One draws circles about A and B with equal radii greater than r. One of their points of intersection, say S_1, is connected to P. The line S_1P is the required perpendicular.

If P is the mid-point of a segment AB, the perpendicular in P to AB can be constructed by the method given above for bisecting a segment.

To drop a perpendicular. To drop the perpendicular to a line l from a point P not on l one draws a circle with P as centre of sufficiently large radius. This intersects l in A and B. The perpendicular bisector of AB is the required line (Fig.).

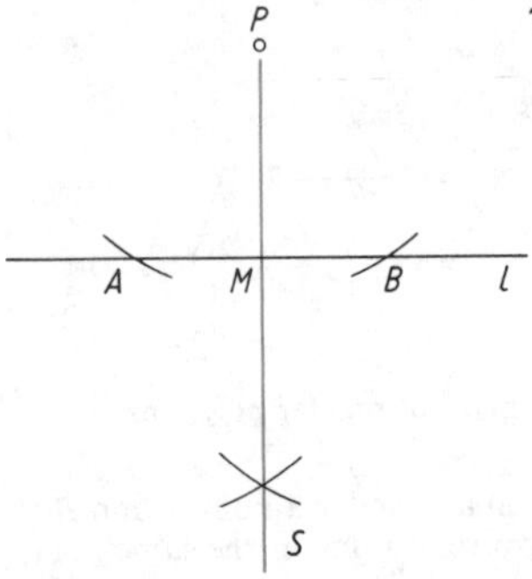

7.3-9 Dropping a perpendicular to a line through a point not on it

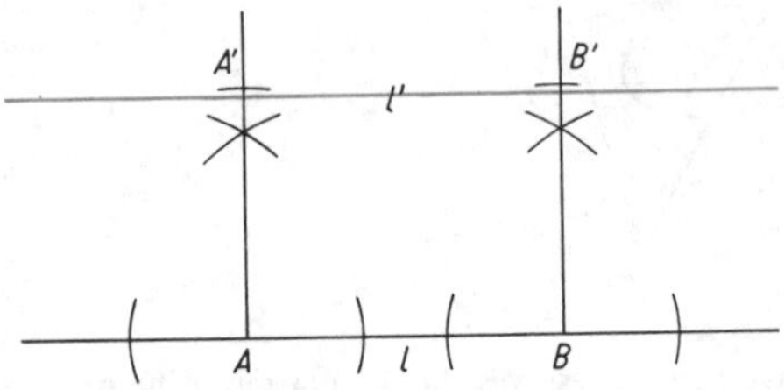

7.3-10 Drawing a parallel to a line

Construction of parallels. To draw a parallel to a line l with ruler and compass, one first erects the perpendiculars to l in two points A and B on l. On these one marks two points A' and B' *equidistant* from l and on the same side of l. The line through A' and B' is parallel to l (Fig.).

Ruler and compass constructions in general. Euclid founded his plane geometry on a system of axioms which ensure that it is always possible (1) to draw a line through any two given points, and (2) to draw a circle whose radius is the distance between two given points and whose centre is a given point.

Thus, his theorems were always proved by a technique that reduces their statements to axioms or basic theorems on the intersections of lines with lines or circles, or of circles with other circles. Consequently, the only means admitted for construction in Euclidean geometry are those for drawing straight lines (ruler), and circles (compass). The requirement for a construction to be made *with ruler and compass alone* is therefore connected with the choice of axioms for plane geometry, and not with the accuracy of the result. Indeed, the result is frequently more accurate if the construction is made by other means.

Furthermore, since the ancient Greeks had only rudimentary computational and algebraic techniques at their disposal, they tried to solve all major mathematical problems by constructions with ruler and compass. For instance, they found square roots by constructing the geometric mean of two segments.

Three famous problems, in particular, proved intractable by this method:

the trisection of an arbitrary angle,

the squaring of the circle: the construction of a square whose area is equal to that of a given circle, and

the doubling of the cube (the Delic problem): to construct a cube with double the volume of a given cube.

Modern methods have proved that all three problems cannot be solved by ruler and compass alone (see Chapter 16.2. – Galois theory).

7.4. Triangles

The parts of a triangle, classification of triangles

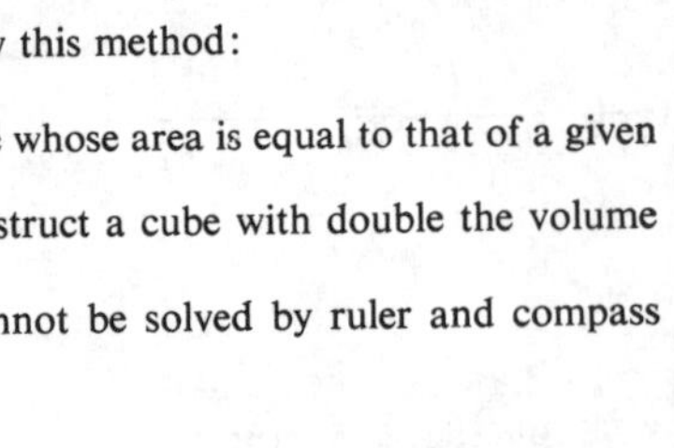

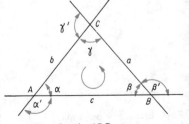

7.4-1 The triangle ABC

If three points in a plane, not on a single straight line, are given, then there are exactly three lines joining them. The closed figure formed by these lines (or rather the segments between the points) is called a *triangle*. The points are the three *vertices* of the triangle, the segments between them are its *sides*. A triangle is convex: this means that with any two points it also contains the segment between them. The vertices are usually labelled A, B and C, the sides with the lower case letters corresponding to the opposite vertex, that is, $|AB|$ by c, $|BC|$ by a, and $|CA|$ by b. Any two sides of a triangle form the arms of an interior angle of the triangle. The interior angles are labelled by a sequence of vertices, or by the lower case Greek letter corresponding to their vertex (Fig.). Thus,

$$\sphericalangle CAB = \alpha,\ \sphericalangle ABC = \beta,\ \sphericalangle BCA = \gamma.$$

If a side is extended, the angle its *extension* makes with the following side is called an exterior angle of the triangle. Thus, α' is the angle between CA produced beyond A and AB, β' is the angle between AB produced beyond B and BC, and γ' is the angle between BC produced beyond C and CA. The whole triangle is written $\triangle ABC$. The triangle is called *positively oriented* if the direction of rotation is $\overrightarrow{AB} \to \overrightarrow{BC} \to \overrightarrow{CA}$.

A triangle is called *isosceles* if two sides (a) are equal, the third side is called the *base* (c), the opposite vertex is called the *apex*. In an *equilateral* triangle all three sides are equal.

A triangle is called *acute* if all interior angles are acute, *right-angled* if one interior angle is a right angle, and *obtuse* if one is obtuse. In a right-angled triangle the side opposite the right angle is called the hypotenuse (Fig.).

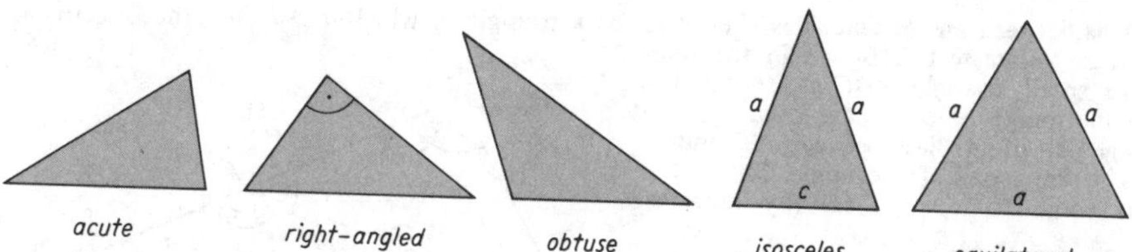

7.4-2 Types of triangles

Basic facts about triangles

Relations between the sides. From any vertex it is possible to reach any other vertex by traversing the sides in two different ways: either by going along the connecting side or along the other two, via the third vertex (for example, from A to B along c or along b and a via C). Since the straight line is the shortest distance between two points this implies that $c < a + b$, $b < a + c$, and $a < b + c$, from which six further inequalities can be deduced by subtraction: $c - a < b$ or $a - c < b$, $c - b < a$ or $b - c < a$, and $b - a < c$ or $a - b < c$. But only three of these make sense geometrically, since the difference between two segments must have non-negative length.

In a triangle the sum of any two sides is greater than the third side. In a triangle any side is greater than the difference between the other two sides.

For instance, it is possible to construct a triangle with sides of lengths 3, 4 and 5 units u; but there is no triangle with sides measuring $3u$, $4u$ and $8u$, because $3u + 4u < 8u$, and thus the sum of two sides would be less than the third.

Relations between the angles. If a parallel g to a side is drawn through the opposite vertex (for example, a parallel to $c = |AB|$ through C), then this gives a straight angle at C which is subdivided by two sides of the triangle. The two parallels g and c are intersected by a and b. Thus, the alternate angles are equal: $\delta = \alpha$ and $\varepsilon = \beta$. Now $\delta + \gamma + \varepsilon = 180°$ and so $\alpha + \beta + \gamma = 180°$ (Fig.).

The sum of the interior angles of a triangle is 180°.

Since *every exterior angle* is adjacent to an interior angle the following equations hold (Fig.):

$$\alpha + \alpha' = 180°,\quad \beta + \beta' = 180°,\quad \gamma + \gamma' = 180°.$$

Adding both sides of the equations one obtains

$$(\alpha + \beta + \gamma) + (\alpha' + \beta' + \gamma') = 540°.$$

Since $\alpha + \beta + \gamma = 180°$, it follows that $\alpha' + \beta' + \gamma' = 360°$.

The sum of the exterior angles of a triangle is 360°.

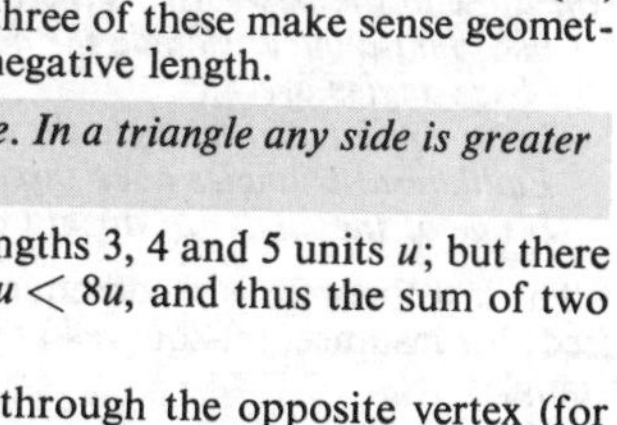

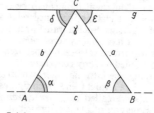

7.4-3 $\alpha + \beta + \gamma = 180°$

Now every exterior angle is *supplementary* to the adjacent interior angle. But it has been proved that the sum of the interior angles is also 180°. Hence:

Every exterior angle is equal to the sum of the two opposite interior angles: $\alpha' = \beta + \gamma$, $\beta' = \gamma + \alpha$, $\gamma' = \alpha + \beta$

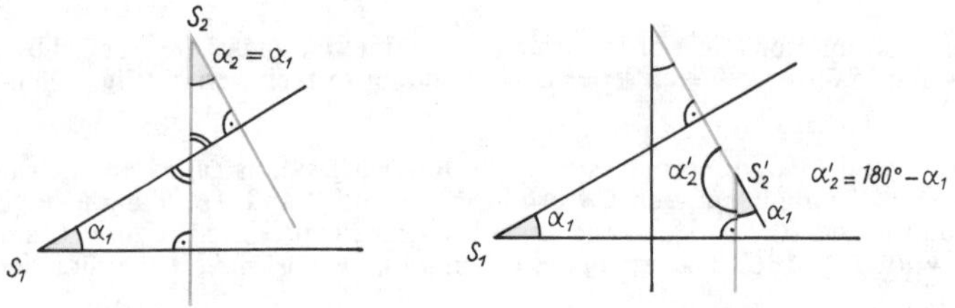

7.4-4 Angles with pairwise orthogonal arms

From these theorems it is easy to deduce the following statement, which is particularly useful in applications to physics (Fig.).

Angles with pairwise orthogonal arms are equal if the vertex of one does not lie inside the arms of the other, nor on one of them. If the vertex lies inside them or on one of them, then the angles are supplementary.

Relations between angles and sides. Let ABC be a triangle in which $a > b$. Let the bisector w_γ of the angle γ intersect $|AB| = c$ in the point D. If the small triangle ADC is reflected in w_γ then the image of b is part of a, so hat the image A' of A lies between B and C (Fig.). A' is the vertex of the angle $CA'D = \alpha'$, which is equal to $\alpha = \sphericalangle CAD$ (by reflecting in w_γ).

Now α' is an exterior angle of DBA' and so is equal to the sum of β and $\sphericalangle BDA'$, and, in particular α' is greater than β. Since $\alpha = \alpha'$ and hence $\alpha > \beta$, this proves that $a > b$ implies that $\alpha > \beta$. The converse can also be proved: if $\alpha > \beta$, then $a > b$.

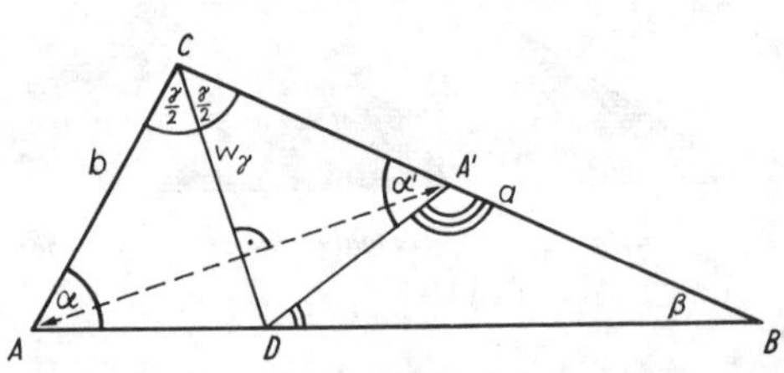

7.4-5 In the triangle ABC $a > b$ implies that $\alpha > \beta$

In a triangle the angle opposite the longer of two sides is greater than the angle opposite the other. The side opposite the greater of two angles is longer than the side opposite the other. Angles opposite equal sides are equal and vice versa. Every isosceles triangle is symmetrical. The perpendicular from the apex to the base bisects the base and the angle at the apex. The angles at the base are equal.

In a right-angled triangle the acute angles are complementary. In an isosceles right-angled triangle, the base angles are 45°.

In an equilaterial triangle the interior angles are all equal; each is 60°.

Equilateral triangles have three axes of symmetry.

If one of the angles of a right-angled triangle is 30°, then the opposite side is half the hypotenuse.

The last theorem is a consequence of the symmetries of the equilateral triangle and is frequently used; for instance, set-squares are usually right-angled triangles of this type or isosceles right-angled triangles.

Congruence of triangles

General remarks. Plane figures are called congruent if they are of the same *shape* and the same *size*. Congruent figures can be carried into each other by a transformation that moves points, but does not change incidence relations (between points and lines), angles between lines, and lengths of segments. Such a transformation also preserves areas and leaves parallel lines parallel. If congruent figures have the same *orientation* (with respect to some fixed orientation of the plane), they can be transformed into each other by a sequence of translations and rotations of the plane. Such figures are called *directly congruent*. If they do not have the same orientation, then a sequence can be found taking one into the other, which apart from successive translations and rotations has a single reflection in a straight line. Such figures are called *inversely congruent*. Translations, rotations, and reflections are called *congruence transformations* and can be used as criteria of congruences in the investigation of plane figures; but this by no means exhausts their usefulness as a tool of discovering new geometrical facts.

Four theorems on congruence of triangles. In the definition of congruence it is required that the figures agree in all aspects, in particular, that the lengths of corresponding sides and the angles between them are equal. The theorems of this section state that for triangles in certain cases it is sufficient to test three parts as a check for congruence – if they are equal for two triangles, then the triangles are congruent. Here are the theorems.

1. Two triangles are congruent if the length of a side of one is equal to the length of the corresponding side of the other and two angles of one are equal to the corresponding angles of the other (α, s, α).

2. Two triangles are congruent if the lengths of two sides of one are equal to the lengths of the corresponding sides of the other and the angles between these sides are equal (s, α, s).

3. Two triangles are congruent if the lengths of two sides of one are equal to the lengths of the corresponding sides of the other and the angles opposite the larger sides are equal (s, s, α).

4. Two triangles are congruent if the lengths of the three sides of one are equal to the lengths of the corresponding sides of the other (s, s, s).

If one tries to construct triangles with three given sides and angles, one sees that if these correspond to one of the theorems, and only then – the triangle is uniquely determined. On the other hand, if two sides and the angle opposite the smaller side are given, say $a = 3$ units, $c = 5$ units, and $\alpha = 20°$, then the figure shows that there are *two* possible triangles with these measurements. For if a line is drawn at 20° to the segment $|AB| = c$ and an arc is drawn about B with radius a, then this arc intersects the line in two points C' and C''. Both the triangles ABC' and ABC'' satisfy the requirements of the construction. Again, if $\alpha = 80°$, then the arc of radius a does not intersect the free arm of α at all, and there is no triangle that fits the requirements (Fig.).

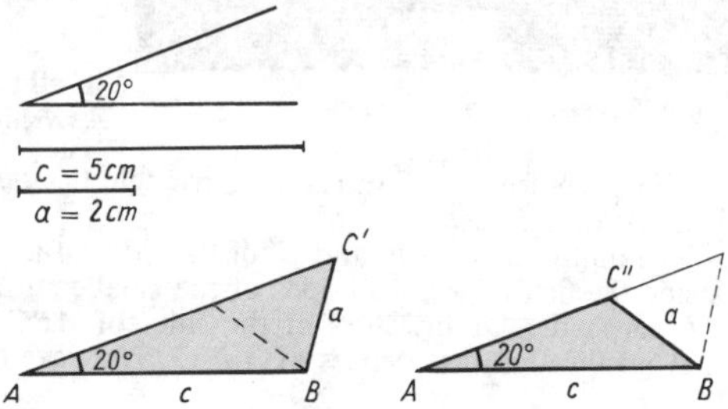

7.4-6 These triangles agree in three data, but are not congruent

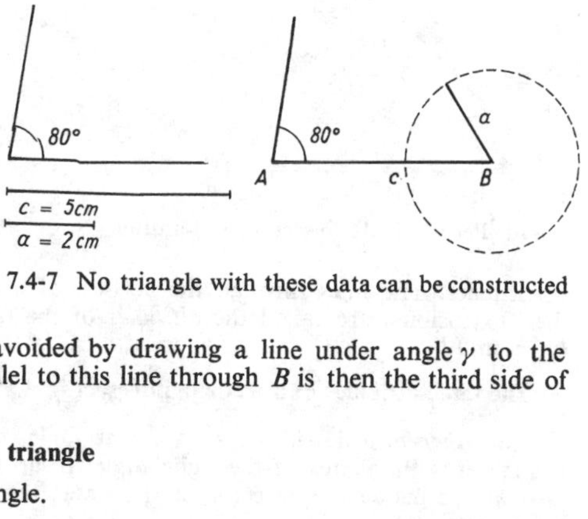

7.4-7 No triangle with these data can be constructed

In constructions using the first congruence theorem it is convenient first to obtain the two angles *adjoining* the given side. If one of them is not given in the data, it can be found by the theorem on the sum of the interior angles of a triangle. For instance, if c, α and γ are given, $\beta = 180° - \alpha - \gamma$ can be computed. Then the triangle can be constructed simply by drawing lines at the appropriate angles through the end-points of the given segment. Alternatively, the computation can be avoided by drawing a line under angle γ to the free arm of α in an arbitrary point C'. The parallel to this line through B is then the third side of the triangle ABC.

Transversals and distinguished points of a triangle

A *transversal* is any line that intersects the triangle.

Perpendicular bisectors. The perpendicular bisector of a side is the line that intersects the side in its mid-point and makes a right angle with the side.

The perpendicular bisectors of the sides of a triangle intersect in a single point M, the circumcentre of the triangle.

The points on the perpendicular bisector of a segment are the points that are *equidistant* from its end points. Hence the intersection of two perpendicular bisectors, say m_a and m_b, is *equidistant* from the end-points of the two sides BC and CA, that is, from the three vertices of the triangle. But then it must also lie on the third perpendicular bisector m_c. In fact, M is the centre of the circle through the vertices of the triangle. This circle is called the *circumscribed circle* or *circumcircle* of the triangle and M is called the *circumcentre* (Fig.). The radius of the circumcircle is the distance from M to any of the vertices of the triangle. In acute triangles the circumcentre lies *inside* the tri-

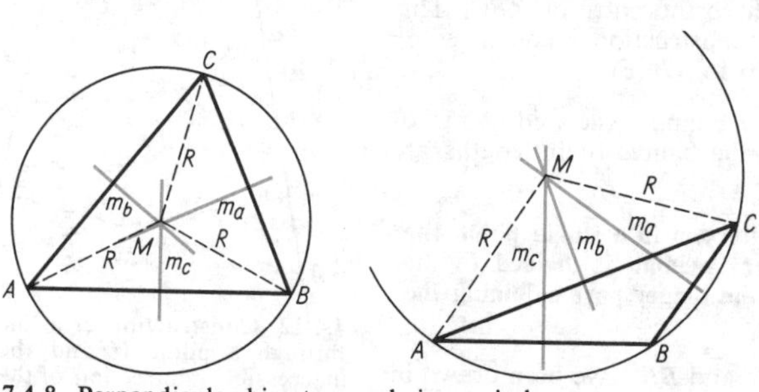

7.4-8 Perpendicular bisectors and circumcircle

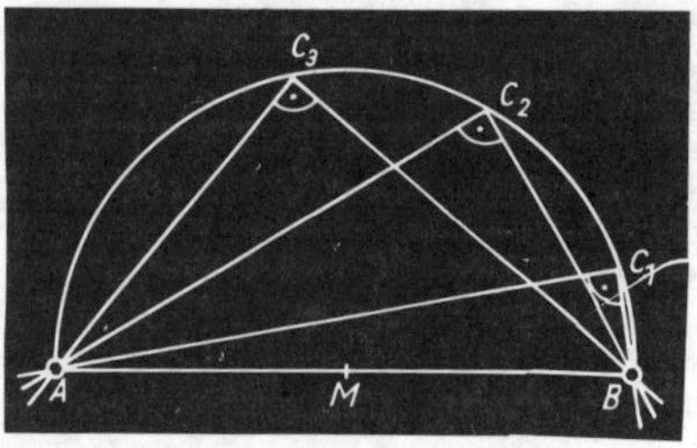

7.4-9 The theorem of Thales

angle the circumcentre of an obtuse triangle lies *outside* the triangle, and the circumcentre of a right-angled triangle lies on the hypotenuse. Thus, the hypotenuse is the diameter of the circumcircle, and the vertices C of all right-angled triangles with hypotenuse AB must lie on the circle with AB as diameter. The discovery of this fact is attributed to THALES of Miletus (about 624–547 B. C.), and the circle is sometimes called Thales' circle on his honour (Fig.).

Theorem of Thales. The locus of the vertices C_i of all right-angles whose arms go through to fixed points A and B is the circle with AB as diameter.

The circumcentre of an isosceles triangle lies on the axis of symmetry, which is the perpendicular bisector of the base.

The mid-points A', B' and C' of the sides of a triangle form the vertices of a *smaller* triangle lying inside the first (Fig.). The sides of the smaller triangle are parallel to the sides of the original. Thus, the perpendicular bisectors of the sides of ABC are also perpendicular to the sides of $A'B'C'$, and they go through the vertices of $A'B'C'$. They are the *altitudes* of $A'B'C'$.

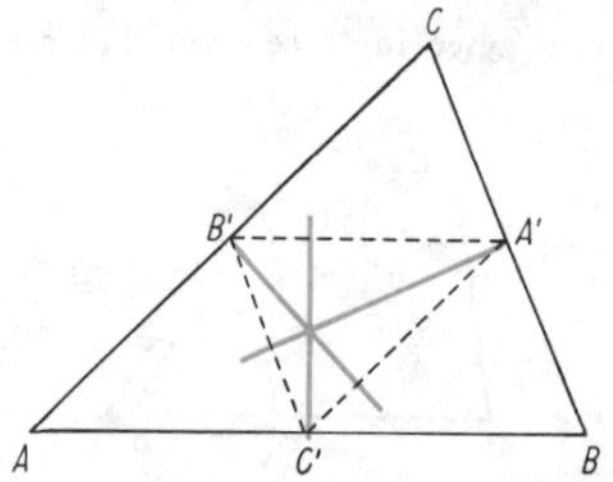

7.4-10 Perpendicular bisectors and altitudes

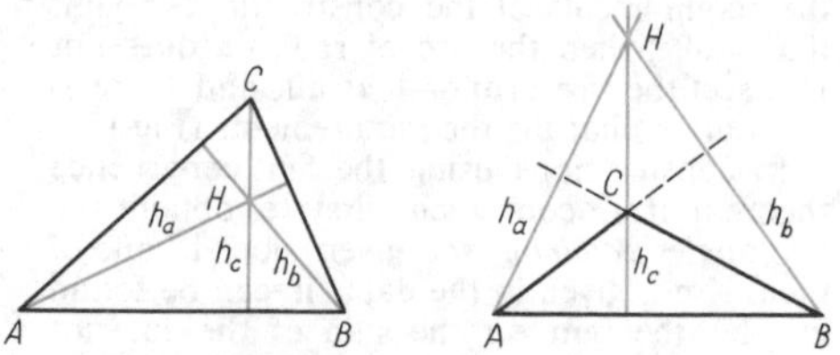

7.4-11 The altitudes of a triangle

Altitudes. The lines through the vertices of a triangle perpendicular to the opposite sides (or their extensions) are called the *altitudes* of the triangle. In the figures their lengths are marked by h_a, h_b and h_c.

The three altitudes of a triangle intersect in a single point, the orthocentre of the triangle.

The *orthocentre* lies inside an acute triangle, *outside* an obtuse triangle, and in a right-angled triangle it is the vertex of the right angle. In an isosceles triangle the altitude through the apex is also the perpendicular bisector of the base; the two transversals both coincide with the axis of symmetry.

The orthocentre can be used to solve the following problem. Using only a restricted area of the plane, say a sheet of paper, to find the line through a given point H and the intersection C of two non-parallel lines l_1 and l_2 that do *not intersect in the permitted area*. The method is to drop perpendiculars from H to l_1 and l_2; each of these intersects the other line, say in the points A and B. The perpendiculars are altitudes of the triangle ABC. Therefore, the third altitude must go through C and H, the orthocentre of ABC. This third altitude is the required line. The construction is completed by drawing the perpendicular c through H to AB (Fig.).

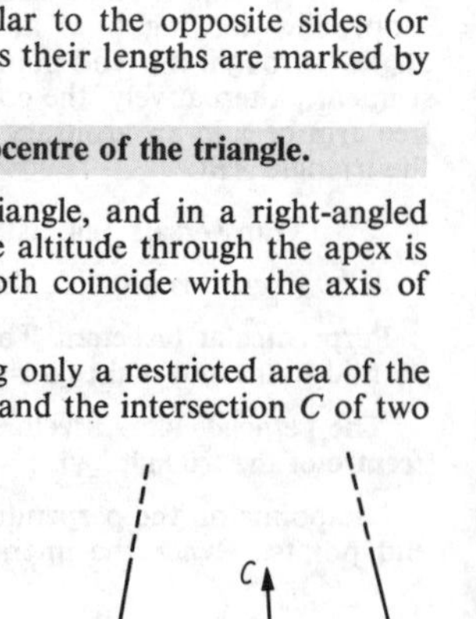

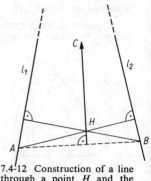

7.4-12 Construction of a line through a point H and the inaccessible intersection of the non-parallel lines l_1 and l_2

Medians. The *medians* of a triangle connect the mid-points of the sides to the opposite vertices. In the figures their lengths are marked by s_a, s_b and s_c.

The three medians of a triangle intersect in a single point, the centre of gravity of the triangle. Every median is divided in the ratio 2 : 1 by the centre of gravity, the longer part adjoining the vertex.

To prove this the two medians AD and BE have been drawn in the figure; their intersection is S. The lines ED and AB intersect

CA and CB and AD and EB. Since $|CB|:|CD| = |CA|:|CE| = 2:1$, it can be proved (using the intercept theorems, or rather their valid converses) that AB and ED are parallel and that the ratio of their lenghts is $2:1$. Therefore $|SA|:|SD| = |SB|:|SE| = |AB|:|ED| = 2:1$. The same observations can be made with a different pair of medians, say CF and BE, and their point of intersection must be S again, since there is only one point that divides $|BE|$ in a ratio $2:1$.

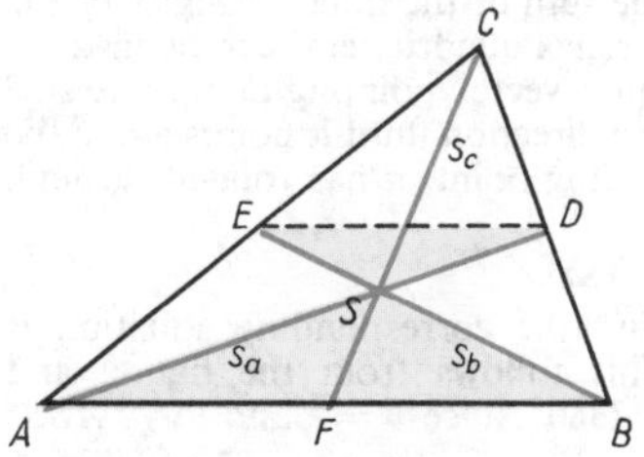

7.4-13 The three medians of a triangle intersect in a common point S

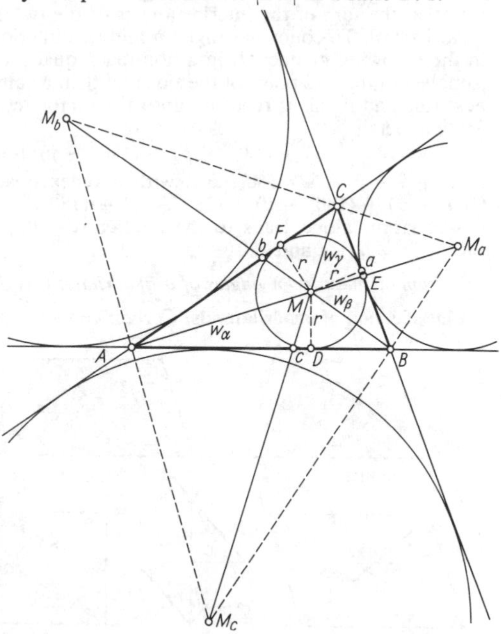

7.4-14 The three escribed circles of a triangle

Angle bisectors. The *bisectors* of the angles of a triangle are marked w_α, w_β, and w_γ in the figures.

The three angle bisectors of a triangle intersect in a single point M, the incentre of the triangle.

The intersection M of w_α and w_β is *equidistant* from b and c (since w_α bisects α) and from c and a (since w_β bisects β). So it is equidistant from a and b and hence must lie on w_γ. This distance from the sides is the radius of the *inscribed circle* or *incircle* of the triangle, which touches each side of the triangle in a single point. M is the *incentre* of the triangle. The points D, E and F in which the incircle touches the sides are the intersections of the perpendiculars from M to those sides (Fig.). The bisectors of the external angles (or their opposite angles) intersect in the three centres M_a, M_b, and M_c of the three *ecircles* (or *escribed circles*) of the triangle. An ecircle touches one side and the extensions of the other two sides in a single point each. In an isosceles triangle the bisector of the angle at the apex coincides with the median, the perpendicular bisector of the base, and its altitude.

7.5. Quadrilaterals

Generalities

Sum of the interior angles of a quadrilateral. In projective geometry four points in a plane no three of which are collinear define a *complete quadrangle*, which has 6 sides. But in ordinary plane geometry a quadrangle has only 4 sides which depend on the sequence in which the points are given.

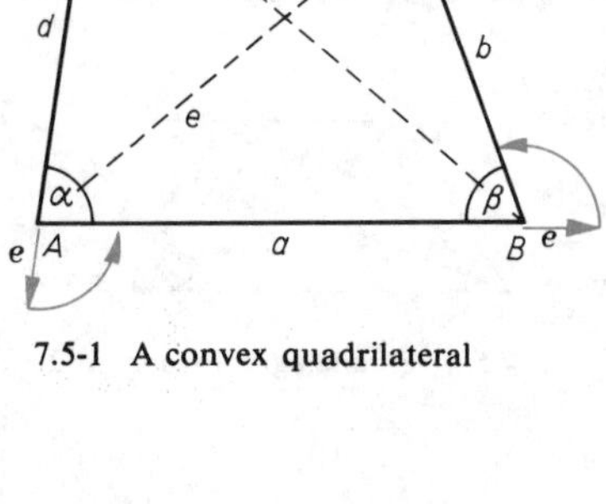

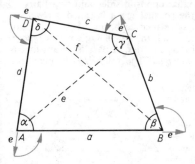

7.5-1 A convex quadrilateral

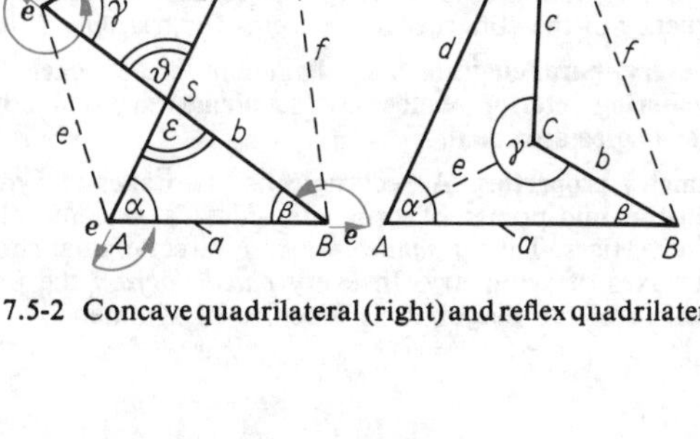

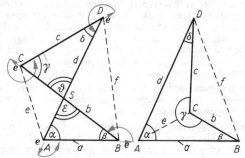

7.5-2 Concave quadrilateral (right) and reflex quadrilateral (left)

If the points are A, B, C and D, then the segments $|AB| = a$, $|BC| = b$, $|CD| = c$, and $|DA| = d$, are the sides of the quadrangle or quadrilateral, the other two connecting segments, $|AC| = e$, and $|BD| = f$, the diagonals, are not counted among the sides. Usually the sequence chosen is that which gives a *convex* quadrilateral, but it can be *concave* or even as an extreme case *reflex*.

If a quadrilateral is not reflex (Fig.), it has a diagonal that divides it into two triangles in such a way that the sum of the interior angles of the two triangles is the sum of the interior angles of the quadrilateral. To count the angles ϑ and ε as interior angles of a reflex quadrilateral can be justified in the following manner. If in a non-reflex quadrilateral a direction vector pointing along one side is moved to the end-point of the side and then rotated in a positive direction until it points along the next side and if this is repeated until the vector returns to its starting point, it has rotated through 360°; in detail:

$$(180° - \alpha) + (180° - \beta) + (180° - \gamma) + (180° - \delta) = 360°,$$

or $\alpha + \beta + \gamma + \delta = 360°$. But with a reflex quadrilateral (Fig.) the corresponding equation is $(180° - \alpha) + (180° - \beta) + (180° + \gamma) + (180° + \delta) = 720°$. This follows from the figure, and this equation alone leads to the relation $\alpha + \beta + \gamma + \varepsilon + \vartheta = 360°$ since $\vartheta = \varepsilon$ and in $\triangle ABS$: $180° - \alpha = \beta + \varepsilon$ and $180° - \beta = \alpha + \varepsilon$.

The sum of the interior angles of a quadrilateral is 360°.

Classification of quadrilaterals. Certain convex quadrilaterals have special names (Fig.).

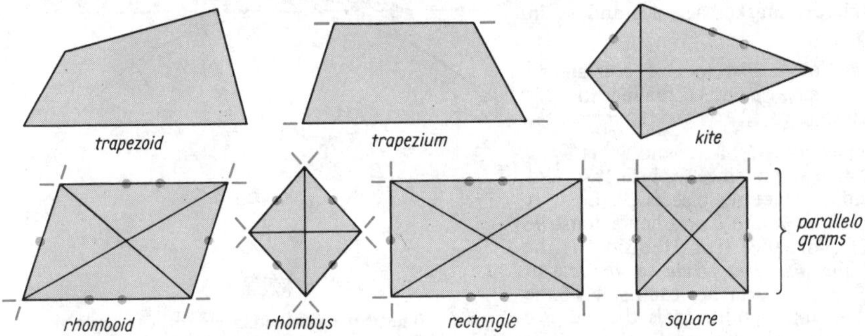

7.5-3 Special types of quadrilaterals

Classification by *lengths of sides*:

general quadrilateral	all four sides of different lengths,
kite	two pairs of equal adjacent sides,
parallelogram	two pairs of equal opposite sides,
rhombus	all four sides equal.

Classification by *relative positions of sides*:

general quadrilateral	no pairs of parallel sides,
trapezium	one pair of parallel sides,
parallelogram	two pairs of parallel sides.

Parallelograms

General remarks. Every *parallelogram* has two pairs of equal opposite sides and two pairs of equal opposite interior angles. If all four sides are equal, the parallelogram is called a *rhombus*. Parallelograms with four equal angles are called *rectangles*, since then each angle is a quarter of 360° or a right angle. If all four sides and all four angles are equal, the parallelogram is a *square*. A square is a rhombus with equal angles and a rectangle with equal sides. A parallelogram that is neither a rhombus nor a rectangle is sometimes called a *rhomboid*. The following theorem is a consequence of the congruence theorems for triangles (Fig.).

In every parallelogram the diagonals bisect each other. Neighbouring interior angles are supplementary and opposite interior angles are equal.

Symmetry properties. A rectangle has two axes of symmetry through the mid-points of opposite sides, a *rhombus* through opposite vertices. Thus, a *square*, which is a rectangular rhombus, has four axes of symmetry. In every *parallelogram* the point of intersection of the diagonals is a centre of symmetry.

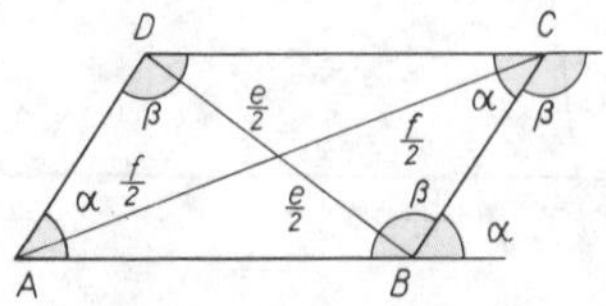

7.5-4 Parallelogram (rhomboid)

The diagonals of a parallelogram divide it into two pairs of congruent triangles.

The diagonals of a rectangle (and a square) have the same length. The diagonals of a rhombus (and a square) intersect at right angles and bisect the interior angles. They divide the rhombus into four congruent triangles.

Construction of parallelograms. Since each diagonal divides a *parallelogram* into two congruent triangles, three independent data are needed to determine it, just as for a triangle. For a general quadrilateral two further data are needed, that is, five altogether, because the two triangles are no longer congruent but have a side in common (the diagonal). For a *rectangle* or a *rhombus* only two data are needed, and for a *square* only one. As a rule, a square is given by the length of its side or diagonal, a rectangle by two different sides or a side and a diagonal. On the other hand, with the rhombus one of the data can be an angle, the other being a side or a diagonal. All constructions proceed by triangles in the figure, hence do not differ in principle from constructions of triangles.

Trapezia

A *trapezium* is a convex quadrilateral with (at least) one pair of parallel sides. If the non-parallel sides of a trapezium have the same length, it is called *isosceles*. A trapezium with two pairs of parallel sides is a parallelogram. The two parallel sides of a trapezium are called its *base lines* and the longer of the two is called its *base*. The segment connecting the mid-points E and F of the non-parallel sides is called the *mid-line* m of the trapezium (Fig.). The mid-line is parallel to the base lines $|AB| = a$ and to $|CD| = c$, for otherwise the parallel to AB through E would intersect BC in a point F'. By the intercept theorems (with S as the carrier of the pencil) the equation $|DE| : |EA| = |CF'| : |F'B|$ would then hold. But by the hypothesis, $|DE| : |EA| = |CF| : |FB| = 1$, and so $F = F'$, hence $EF \parallel AB$. If G is the intersection of the line through D and F with the extension of AB, the triangles BGF and CDF are congruent. Therefore, $|AG| = a + c$, and by considering the triangle AGD one obtains $m = (a + c)/2$.

The mid-line of a trapezium is parallel to the base lines and half as long as their sum; $m = (a + c)/2$ (*arithmetic mean*).

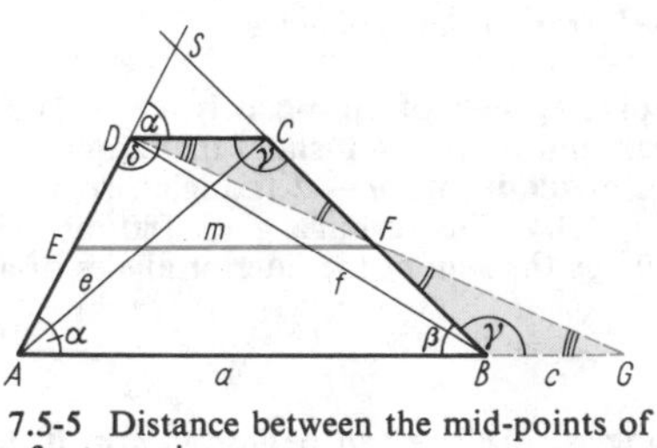

7.5-5 Distance between the mid-points of the diagonals of a trapezium

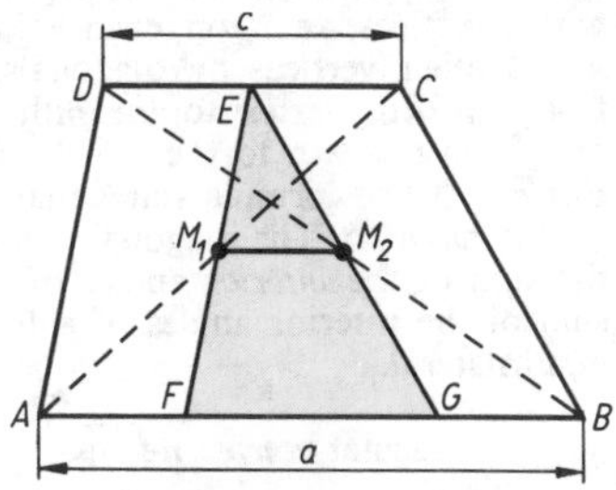

7.5-6 Trapezium

The mid-line goes through the *mid-points of the diagonals* of a trapezium.

The distance between the mid-points of the diagonals of a trapezium is half the difference of the lengths of the base lines; $m' = (a - c)/2$.

Let the base-lines of the trapezium be $|AB| = a$ and $|CD| = c$; and let M_1 and M_2 be the mid-points of the diagonals AC and BD. The parallels through each mid-point to the nearer side of the trapezium intersect AB in F and G and CD in a single point E, because $|DE| = c/2 = |CE|$ (Fig.). Since $|DE| = |AF| = c/2 = |CE| = |BG|$ it follows that $|FG| = a - c$. Now M_1M_2 bisects the sides of the triangle EFG and is therefore half as long as the base, so that $|M_1M_2| = m' = (a - c)/2$.

From the theorem on parallel lines it follows that the interior angles on each side of the trapezium are supplementary.

The sum of the interior angles of a trapezium on each of the two non-parallel sides is 180°.

To construct a trapezium it is sufficient to have four independent data, for an isosceles trapezium only three are needed.

In an *isosceles* trapezium the diagonals have the same length and the angles at the base are equal.

Kites and deltoids

A convex quadrilateral with two pairs of equal adjacent sides is called a *kite*.

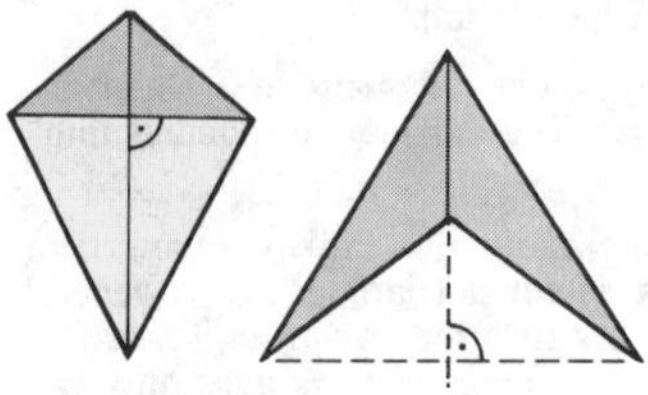

7.5-7 Kite (left) and deltoid (right)

The diagonals of a kite are perpendicular to one another; one of them is an axis of symmetry and divides the kite into two congruent triangles, the other divides it into two isosceles triangles (Fig.). A kite with four equal sides is a rhombus.

A non-convex quadrilateral with two pairs of adjacent equal sides is a *deltoid* (Fig.). Just as with a kite, the *diagonals* are perpendicular to one another and one is an *axis of symmetry* and divides the deltoid into two congruent triangles. The other diagonal lies outside the deltoid and is the base of two isosceles triangles whose other sides are the pairs of equal sides of the deltoid. The deltoid can be used to construct the mid-point of a segment lying near the edge of a restricted area of the plane, without going outside that area. Three independent data are needed to determine a kite or a deltoid.

7.6. Polygons

General polygons

Closed plane figures with straight edges are called polygons. They can be classified by the number of their vertices. Triangles and quadrilaterals are special types of polygons. If two non-adjacent sides of a polygon intersect, it is called *reflex*. If segments connecting any two points inside the polygon always lie inside that polygon, it is *convex*. In that case the interior angle at any vertex is less than 180°. Otherwise the polygon is called *concave* and has one or more inverted corners where the interior angle exceeds 180° (Fig.).

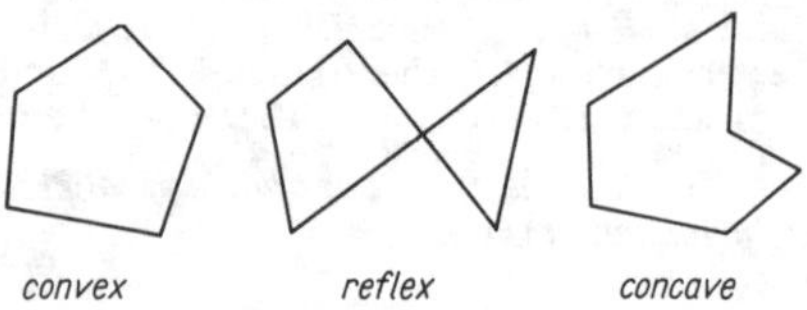

7.6-1 Various types of polygon

The segments connecting neighbouring vertices of a polygon are called its *sides*; the other segments connecting vertices are *diagonals*. The number of sides is equal to the number of vertices. In a polygon with n vertices, an n-gon, each vertex is connected to $n - 3$ other vertices by diagonals, but the diagonal from the kth vertex to the mth is the same as the one from the mth to the kth. It follows that the number of diagonals of an n-gon is $n(n - 3)/2$. For $n = 3$ the formula states that a triangle has no diagonals and for $n = 4$ it states that a quadrilateral has two. The diagonals from one vertex of an n-gon divide it into $n - 2$ triangles, so that the sum of the *interior* angles of a *convex* n-gon is $(n - 2) \cdot 180°$. This formula gives 180° as the sum of the interior angles of a triangle and $2 \cdot 180° = 360°$ as the sum of the interior angles of a quadrilateral.

Regular convex *n*-gons

A regular convex n-gon has n equal sides and n equal angles. A *circle* can be *inscribed* in it in such a way that the sides of the regular n-gon are tangents to the circle, and another circle can be *circumscribed* around the regular n-gon in such a way that the sides of the n-gon are chords of the circle. Since the sum of the interior angles of a convex n-gon is $(n - 2) \cdot 180°$ and the interior angles of a regular n-gon are equal, it follows that $\alpha = (n - 2) \cdot 180°/n = 180° - 360°/n$ for each interior angle. The radii r from the centre M of the circumscribed circle to the vertices of the regular n-gon divide it into n congruent isosceles triangles. The angle at the apex of each of these is $360°/n$, hence this angle and r together determine the regular n-gon completely.

The regular hexagon. In this case the six congruent triangles are *equilateral*, and the lengths of all their sides are r. If three vertices, no two of which are neighbours, are connected, the result is an equilateral triangle of side $r\sqrt{3}$.

Sequences of regular *n*-gons. Conversely, given an equilateral triangle inscribed in a circle of radius r, it is possible to find the vertices of the inscribed regular hexagon by intersecting the perpendicular bisectors of the sides of the triangle with the circle. This method works for any number of sides and makes it possible to construct the regular $2n$-gon from the regular n-gon. Thus, beginning

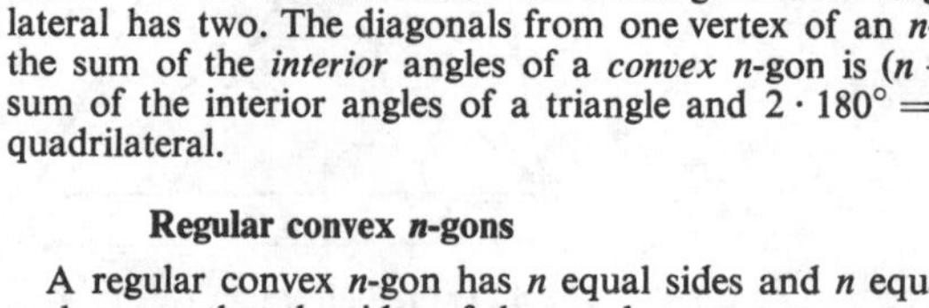

7.6-2 a) Regular hexagon, b) regular triangle, c) and regular tetragon

with an equilateral triangle one can construct regular 6-, 12-, 24-, ..., $3 \cdot 2^n$-gons.

The regular tetragon. The regular tetragon or 4-gon is a square. Its corners are the intersections with the circumference of two mutually perpendicular diameters of a circle (Fig.). The method of bisecting sides yields the regular 8-, 16-, ..., 2^n-gons.

The regular decagon (10-gon). If two neighbouring vertices P and Q of the regular decagon inscribed in a circle are connected to the centre, then the angle at the centre is 36° and hence for the base angles β of the resulting isosceles triangle one has $\beta = (180° - 36°)/2 = 72°$ (Fig.). The bisector QR of the angle $\sphericalangle PQM$ divides the triangle into two isosceles triangles, PQR and QMR. For these $|PQ| = |QR| = |MR| = s_{10}$, the side of the regular decagon. Further, $|RP| = r - s_{10}$ and the triangles PQR and QMP are equiangular and thus similar. Hence $r : s_{10} = s_{10} : (r - s_{10})$. The solution of the quadratic equation $s_{10}^2 + rs_{10} = r^2$ is $s_{10} = (r\sqrt{5} - r)/2$. A point X is said to divide a segment AB in the *golden section* if $|AB| : |AX| = |AX| : |XB|$.

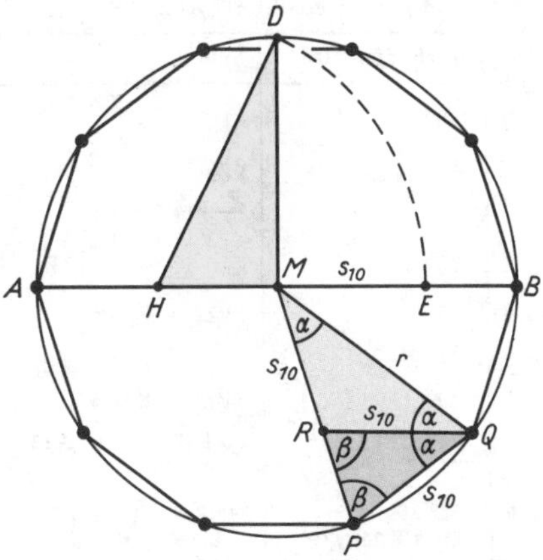

7.6-3 Regular decagon

If a regular decagon is inscribed in a circle of radius r, and r is divided in the golden section, then the larger of the two resulting segments is the side of the decagon.

The regular decagon can be constructed by the following method. One draws the diameter AB of the circle of radius r and constructs perpendicular to it the radius MD. Let H be the midpoint of AM. By Pythagoras' theorem $|HD|^2 = |DM|^2 + |HM|^2 = (r/2)^2 + r^2 = (5/4)\,r^2$. Thus, $|HD| = r\sqrt{5}/2$. The arc with radius $|HD|$ about H intersects MB in E. Then $|ME| = |HE| - |HM| = (r\sqrt{5} - r)/2 = s_{10}$. Taking alternative vertices of the regular decagon one obtains the *regular pentagon* (*5-gon*) and thus the whole sequence of regular 5-, 10-, 20-, ..., $5 \cdot 2^n$-gons.

The regular 17-gon. The constructions of the regular polygons discussed above and also of the regular 15-gon (whose central angle of 24° is constructed as $(60/4)° + (72/8)° = 15° + 9°$) were known to the mathematicians of ancient Greece. It was not until 1796 that any further constructions of regular polygons were discovered. In that year GAUSS, who had just turned nineteen, proved that it is possible to construct the regular 17-gon with ruler and compass. He writes in the introduction to his first scientific publication (1 June 1796):

"It is well known to every beginner in geometry that certain regular polygons, namely those with three, five, and fifteen sides and those that can be obtained from these by doubling the number of sides, may be constructed by geometrical means. Knowledge had advanced thus far in Euclid's time, and it seems that since then men have been of the conviction that the domain of elementary geometry extends no further; at least I know of no successful attempt to expand its boundaries in this direction. All the more worthy of notice, it seems to me, is the discovery that many other regular polygons, notably that of seventeen sides, may be constructed by geometrical means.

This discovery is merely a corollary to a far-reaching theory which is not yet quite complete and on completion will be presented to the public.— C. F. Gauss of Brunswick, student of mathematics at Göttingen."

The theory of which Gauss speaks is his theory of the *cyclotomic* (Greek: dividing the circle) *equation* $x^n - 1 = 0$ (n a natural number). The nth roots of unity, the roots of this equation, lie in regular intervals along the circumference of the unit circle of the Gaussian number plane, whose centre is the intersection of the real and imaginary axes and whose radius is 1. Gauss showed that this circle can be divided into n equal parts with ruler and compass alone if n is a prime number of the form

$$2^{2^k} + 1 \qquad (k = 0, 1, 2, 3, \ldots).$$

And indeed the prime numbers 3, 5 and 17 can be obtained for $k = 0$, 1 and 2.

GAUSS gave to his pupil GERLING the following real representation of $\cos\varphi$ for $\varphi = (360/17)°$, which contains only rational numbers and square roots:

$$\cos\varphi = -{}^1/_{16} + {}^1/_{16}\sqrt{17} + {}^1/_{16}\sqrt{(34 - 2\sqrt{17})}$$
$$+ {}^1/_8\sqrt{[17 + 3\sqrt{17} - \sqrt{(34 - 2\sqrt{17})} - 2\sqrt{(34 + 2\sqrt{17})}]}.$$

Survey of regular convex polygons (r = radius of circumscribed circle)

n	Angle at centre	Side	Circumference	Area
3	120°	$r\sqrt{3}$	$2r \cdot 2.59807621\ldots$	$\frac{3r^2}{4}\sqrt{3} \approx 1.29990\, r^2$
4	90°	$r\sqrt{2}$	$2r \cdot 2.82842712\ldots$	$2r^2$
5	72°	$\frac{r}{2}\sqrt{(10 - 2\sqrt{5})}$	$2r \cdot 2.93892626\ldots$	$\frac{5r^2}{8}\sqrt{(10 + 2\sqrt{5})} \approx 2.3776\, r^2$
6	60°	r	$2r \cdot 3$	$\frac{3r^2}{2}\sqrt{3} \approx 2.5981\, r^2$
8	45°	$r\sqrt{(2 - \sqrt{2})}$	$2r \cdot 3.06146746\ldots$	$2r^2\sqrt{2} \approx 2.8284\, r^2$
10	36°	$\frac{r}{2}(\sqrt{5} - 1)$	$2r \cdot 3.090\,169\,994\ldots$	$\frac{5r^2}{4}\sqrt{(10 - 2\sqrt{5})} \approx 2.9389\, r^2$
12	30°	$r\sqrt{(2 - \sqrt{3})}$	$2r \cdot 3.10582854\ldots$	$3r^2$
15	24°	$\frac{r}{2}\sqrt{[7 - \sqrt{5} - \sqrt{(30 - 6\sqrt{5})}]}$	$2r \cdot 3.11867536\ldots$	$\frac{15r^2}{8}\sqrt{[7 + \sqrt{5} - \sqrt{(30 + 6\sqrt{5})}]} \approx 3.0505\, r^2$
16	22° 30′	$r\sqrt{[2 - \sqrt{(2 + \sqrt{2})}]}$	$2r \cdot 3.12144515\ldots$	$4r^2\sqrt{(2 - \sqrt{2})} \approx 3.0615\, r^2$
17	21° 10′ 35$^5/_{17}$″	$\approx 0.36749904\, r$	$2r \cdot 3.12374180\ldots$	$\approx 3.0706 r^2$
20	18°	$r\sqrt{\{2 - \sqrt{[(5 + \sqrt{5})/2]}\}}$	$2r \cdot 3.12868930\ldots$	$\frac{5r^2}{2}\sqrt{(6 - 2\sqrt{5})} \approx 3.0902\, r^2$
24	15°	$r\sqrt{[2 - \sqrt{(2 + \sqrt{3})}]}$	$2r \cdot 3.\,132\,628\,61\ldots$	$6r^2\sqrt{(2 - \sqrt{3})} \approx 3.1058\, r^2$
	in general	$s_{2n} = \sqrt{[2r^2 - r\sqrt{(4r^2 - s_n^2)}]}$		

7.7. Mensuration of figures bounded by straight lines

Measurement of area

The basic unit of area is the *square metre* m^2. It is defined as the area of a square of side 1 metre. Further units derived from the square metre are:

square kilometre	1 km²	= 10^6 m²	
hectare	1 ha	= 100 a	= 10^4 m²
are	1 a	= 100 m²	
square decimetre	1 dm²	= 10^{-2} m²	
square centimetre	1 cm²	= 10^{-4} m²	
square millimetre	1 mm²	= 10^{-6} m²	

In the Anglo-Saxon countries areas are still based on the square yard, 1 sq. yd. $\approx 0.8361\ \mathrm{m}^2$ or $1\ \mathrm{m}^2 \approx 1.196$ sq. yd. From this one obtains:

1 square mile $\approx 2.59\ \mathrm{km}^2$ or $1\ \mathrm{km}^2 \approx 0.3861$ sq. miles $= 0.3861\ \mathrm{mi}^2$,
1 square foot $\approx 929\ \mathrm{cm}^2 \approx 0.0929\ \mathrm{m}^2$ or $1\ \mathrm{m}^2 \approx 10.76$ sq. ft. $= 10.76\ \mathrm{ft.}^2$,
1 square inch $\approx 6.452\ \mathrm{cm}^2$ or $1\ \mathrm{cm}^2 \approx 0.155$ sq. in. $= 0.155\ \mathrm{in}^2$.

The most common derived unit is the acre, one acre = 4840 sq. yd. $\approx 3\,377.844\ \mathrm{m}^2$.

Mensuration of simple figures

Squares. A square of side a can be completely covered by unit squares. One obtains a strips, each containing a unit squares, hence $a \cdot a = a^2$ unit squares (Fig.).

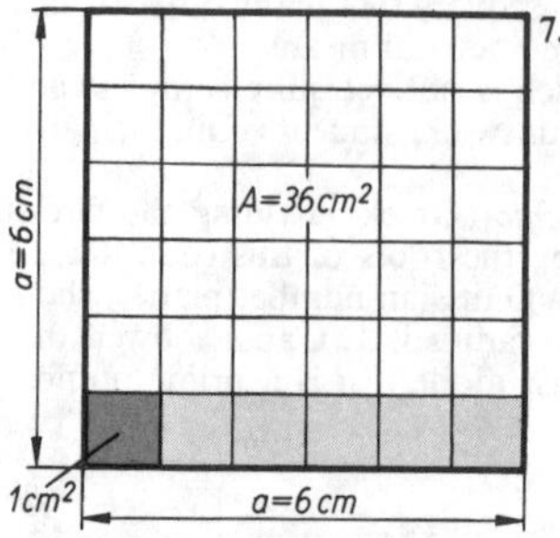

7.7-1 Area of a square

Area of a square of side a	$A = a^2$

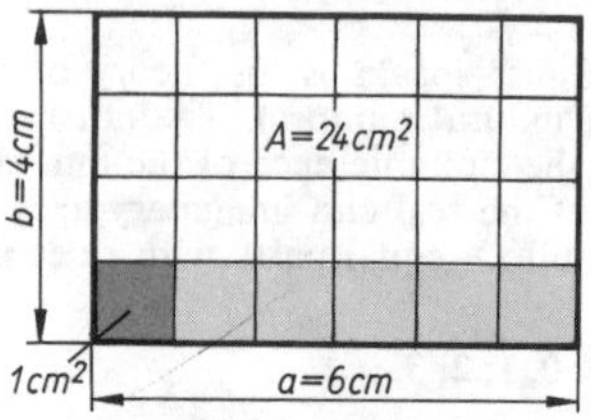

7.7-2 Area of a rectangle

Rectangles. A rectangle of sides a and b can be covered by a strips, each containing b unit squares (Fig.). Therefore its area is $a \cdot b$ units.

Area of a rectangle of sides a and b	$A = a \cdot b$

In both cases it is assumed that there exist unit squares whose sides are *commensurable* with the sides of the square or rectangle to be measured. But the formulae also hold if the side lengths are arbitrary real numbers. If a and b are rational multiples $a = (p_1/q_1) \cdot e$ and $b = (p_2/q_2) \cdot e$ of the unit length e, then the area can be covered by squares of side length e', where $e = [q_1, q_2] \cdot e'$ and $[q_1, q_2]$ is the least common multiple of q_1 and q_2.

In Chapter 3. it was shown that every real number can be approximated to any required accuracy by rational numbers. Computation of areas of figures whose boundaries can be described by continuous functions in a suitable coordinate system is possible by means of the integral calculus.

Parallelograms. A general parallelogram can be transformed into a rectangle of the same area by removing a right-angled triangle from one side and replacing it on the other (Fig.). If the height of the parallelogram is defined as the length of the perpendicular from one side to the opposite side, then the area is given by the product of the length of a side and the corresponding height: $A = a \cdot h_a = b \cdot h_b$.

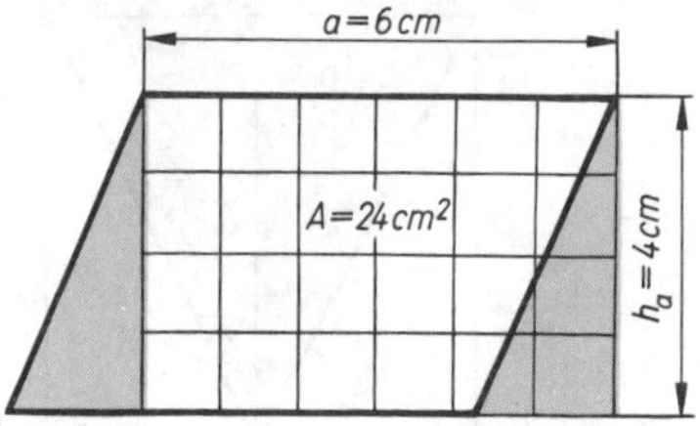

Parallelogram	$A = a \cdot h_a = b \cdot h_b$

The area of a parallelogram is the product of the length of one side and the corresponding height.

7.7-3 Area of a parallelogram

Triangles. Any triangle can be regarded as half parallelogram (Fig.). Therefore the area of a triangle is half the product of the base and the height: $A = c \cdot h_c/2$. If in a right-angled triangle the base is chosen as one of the two shorter sides, the height is the other. Thus, if a and b are the two shorter sides, the area is $a \cdot b/2$. The height of an equilateral triangle of side a is $a\sqrt{3}/2$ (this is a corollary to Pythagoras' theorem); $A = a^2\sqrt{3}/4$.

The area of a triangle is half the product of base and height.

Triangle
$A = a \cdot h_a/2 = b \cdot h_b/2 = c \cdot h_c/2$

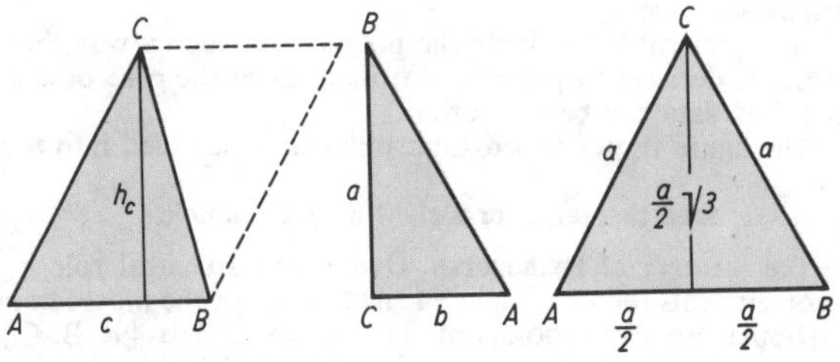

7.7-4 Area of a triangle

The formula of HERON gives the area of a triangle in terms of its sides alone:

$$A = \sqrt{[s(s-a)(s-b)(s-c)]}, \quad \text{where} \quad 2s = a + b + c.$$

Probably this formula was already known to ARCHIMEDES. A *Heron triangle* is one in which all the sides and the area can be expressed as rational numbers. For instance, if $a = 13$, $b = 14$ and $c = 15$, then $A = \sqrt{[21 \cdot (21-13)(21-14)(21-15)]} = \sqrt{(21 \cdot 8 \cdot 7 \cdot 6)} = \sqrt{(4^2 \cdot 3^2 \cdot 7^2)} = 84$ units of area. The area of a triangle is also dependent on its angles and the radii of the circumcircle or the incircle (see Chapter 11. – Further theorems and Applications).

Trapezia. To find the area of a trapezium $ABCD$ it is reflected in the mid-point of one of its (non-parallel) sides (or rotated 180° about that point, which is the same thing) (Fig.).

The resulting parallelogram $AD'A'D$ has twice the area of the trapezium. The length of one side is $a + c$, and the corresponding height h is the height of the trapezium. Therefore $A = (1/2) \cdot (a + c)\, h$.

Since the length of the mid-line of the trapezium is $m = (a + c)/2$, one can also write $A = mh$.

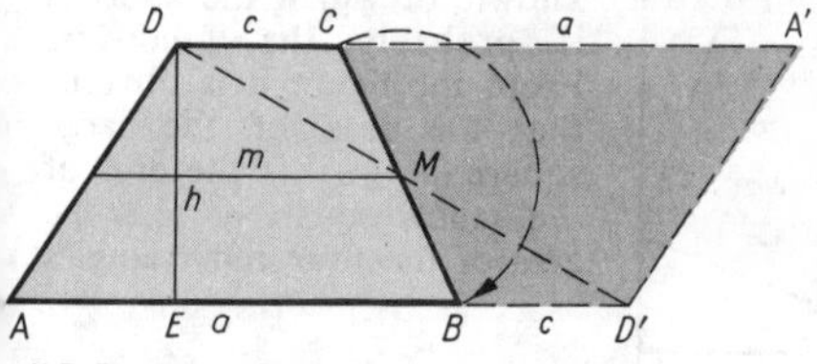

7.7-5 Area of a trapezium

Area of a trapezium of mid-line m and height h
$A = m \cdot h = (a + c)\, h/2$

The area of a trapezium is the product of its mid-line and its height.

If in a trapezium $a = c$, then it is a parallelogram and the formula reduces to $A = (a + a)\,h/2 = a \cdot h$, which is the formula for the area of a parallelogram. If $c = 0$, the trapezium degenerates to a triangle, and $A = (a + 0)\,h/2 = a \cdot h/2$ becomes the formula for the triangle.

Kites. Kites are divided by their diagonals e and f into four right-angled triangles (Fig.). This gives the formula $A = e \cdot f/2$, in which only the diagonals occur. The formula is also valid for a rhombus, which is a special case of a kite, and for a square with diagonal d it becomes $A = d^2/2$.

Area of a kite of diagonals e and f	$A = e \cdot f/2$

The area of a kite is half the product of the lengths of the diagonals.

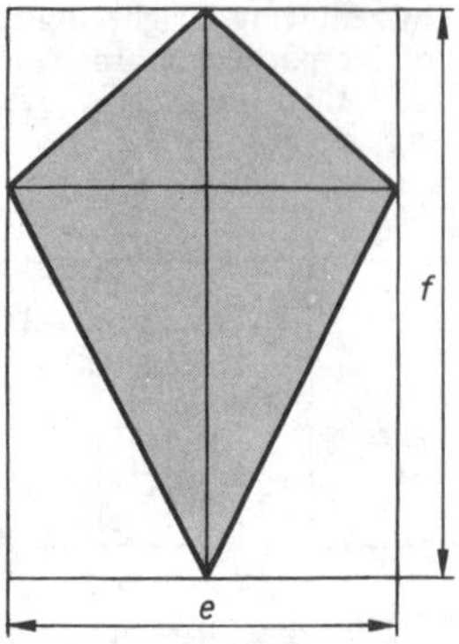

7.7-6 Area of a kite

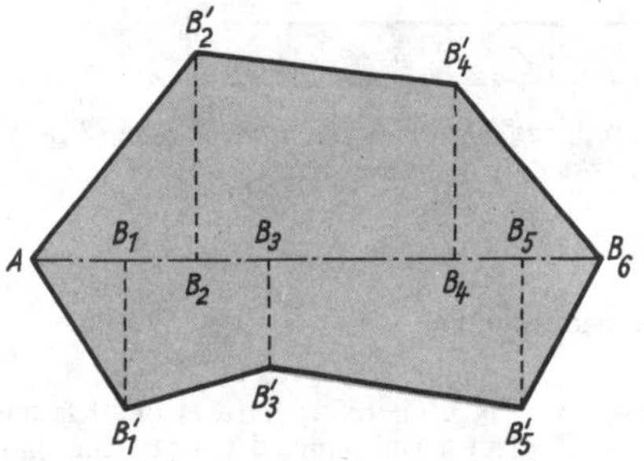

7.7-7 Area of a general multilateral

General polygons. It is not customary to set up formulae for the areas of general polygons except for regular polygons. The area can be calculated by subdividing the polygon into triangles or trapezia. Frequently it becomes necessary to calculate further auxiliary data for the heights and sides of the triangles.

It is preferable to divide the polygons in such a way that the formulae given above can be used. What is decisive in practice is not so much the ease of calculation as the accuracy with which the required data can be measured.

The figure shows an irregular polygon subdivided into triangles and trapezia.

Area theorems for a right-angled triangle

The theorem of Pythagoras. Owing to its central role in both calculations and proofs in plane geometry this theorem rightly ranks as one of the most famous of the subject. Its discovery is usually attributed to PYTHAGORAS of Samos (about 580–496 B. C.), but this is certainly not strictly true. Historically verifiable details of Pythagoras' life are very sparse. There are, however, many myths and legends about him and his life, and his theorem has even inspired poets.

Theorem of Pythagoras. In a right-angled triangle the square on the hypotenuse is equal to the sum of the squares on the other two sides (Fig.).

Theorem of Pythagoras
$a^2 + b^2 = c^2$, c hypotenuse, a, b the other two sides

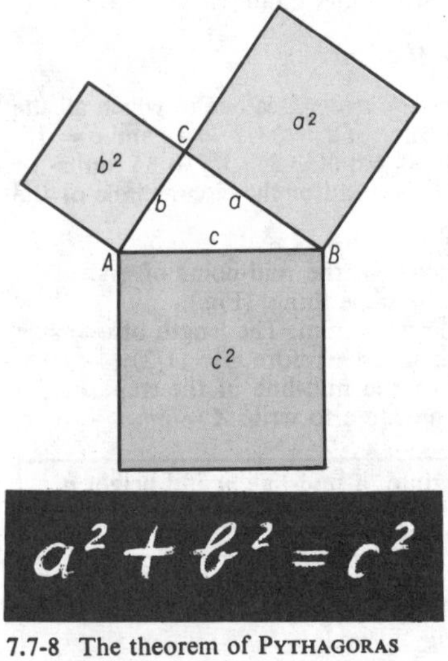

7.7-8 The theorem of PYTHAGORAS

More than 100 different proofs of this theorem are known, of which the shortest is probably the following. From the figure it is obvious that the area of the large square $(a + b)^2$ is the area of the yellow square c^2 plus the area of the four red triangles

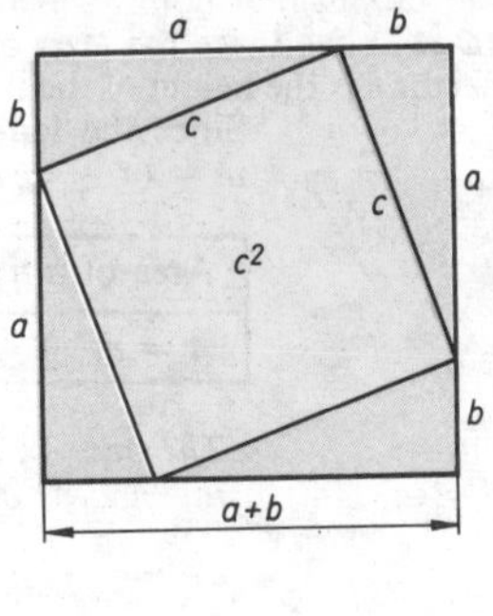

7.7-9 A proof by dissection

$4ab/2 = 2ab$. Thus,

$$(a + b)^2 = c^2 + 2ab \quad \text{or} \quad a^2 + 2ab + b^2 = c^2 + 2ab \quad \text{and hence} \quad a^2 + b^2 = c^2.$$

The theorem of Euclid. The classical proof of Pythagoras' theorem is based on the following *theorem of Euclid*:

In a right-angled triangle the square on one of the two shorter sides has the same area as the rectangle whose sides are the projection of that side onto the hypotenuse and the hypotenuse itself (Fig.).

$a^2 = p \cdot c$	$b^2 = q \cdot c$

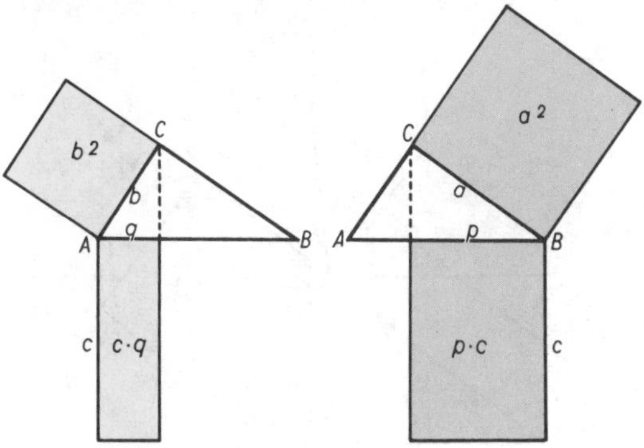

7.7-10 Equal areas by theorem of EUCLID

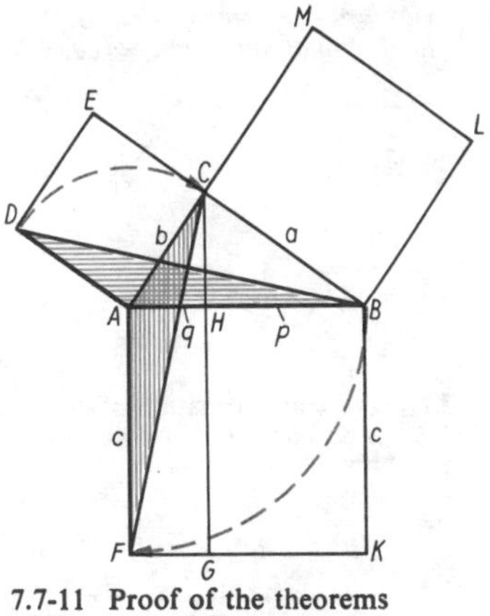

7.7-11 Proof of the theorems of EUCLID and PYTHAGORAS

The triangle ABD has the same base and the same height $|AC|$ as the square $ACED$ so that it has half the area of the square (Fig.). If it is rotated by 90° about A, then D becomes C and B becomes F. The new triangle AFC is congruent to ABD. It has the same base $|AF|$ and the same height $|AH|$ as the rectangle $AFGH$. Thus, it has half the area of the rectangle, and the areas of the rectangle and the square must be equal, in other words, $b^2 = q \cdot c$, and similarly $a^2 = p \cdot c$. If these two expressions are added, one obtains Pythagoras' theorem:

$$a^2 + b^2 = cp + cq = c(p + q) = c \cdot c = c^2.$$

Some of the many applications of Pythagoras' theorem are: the computation of the areas of regular polygons, of the distance between points in analytic geometry, of the height of an equilateral triangle, or of the height of a tetrahedron.

The altitude theorem. A further interesting theorem on right-angled triangles is the altitude theorem:

In a right-angled triangle the square of the altitude on the hypotenuse is equal in area to the rectangle on the two segments of the hypotenuse.

Altitude theorem	$h^2 = q \cdot p$

The proof uses the theorems on similarity (Fig.). Since the altitude $|CD|$ divides ABC into two equiangular, hence similar, triangles $\triangle ADC \sim \triangle CDB$, it follows that

$$q : h = h : p \quad \text{or} \quad h^2 = p \cdot q.$$

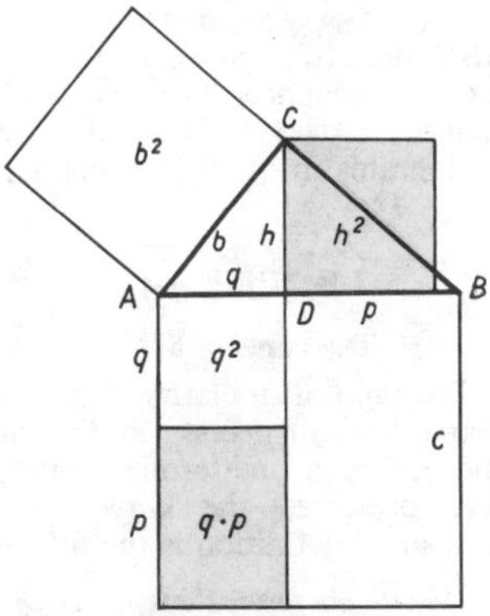

7.7-12 The altitude theorem

Transformation of areas

Every convex polygon can be transformed into a square of equal area. Since triangles with the same base and height have the same area, the vertex S_1 of the polygon can be moved parallel to the diagonal $d_n = |S_nS_2|$ to a point S_1' on the line $S_{n-1}S_n$. Now the area of $\triangle S_nS_1S_2$ is equal to that of $\triangle S_nS_1'S_2$, so that the polygon has been transformed into an $(n - 1)$-gon of equal area (Fig.). This process can be continued until the result is a triangle ABC (Fig.) of area equal to that of the original polygon. Its altitude is $h_D = |B'C|$. The rectangle $CDEF$ with the sides $|FC| = h_D/2$ and $|CD| = |AB|$ has the same area as the triangle. If one takes $|FC| = p$ and $|EF| = q$ as the segments of the hypotenuse (intersected by the altitude) of a right-angled triangle FGK, then the square $CIHK$ has the area $h^2 = p \cdot q$ (by the altitude theorem), and thus the area of the original polygon. To find the triangle FGK construct $|CG| = |CD|$ on the line through F and C and draw Thales' circle on the diameter $|FG|$.

Transformations of other areas can be obtained from the following theorem on supplementary parallelograms. Like the preceding theorems, this is also contained in the first book of EUCLID; it states:

If from an arbitrary point on the diagonal of a parallelogram the parallels to the sides of the parallelogram are drawn, then of the four resulting parallelograms those two that are not intersected by the diagonal have equal area. (They are called supplementary parallelograms.)

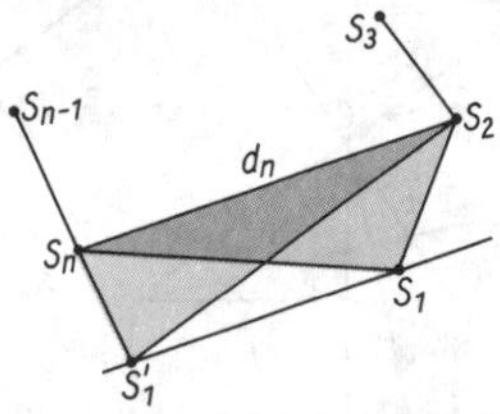

7.7-13a) Transformation of the n-gon $S_1S_2 \ldots S_n$ into a $(n-1)$-gon of equal area

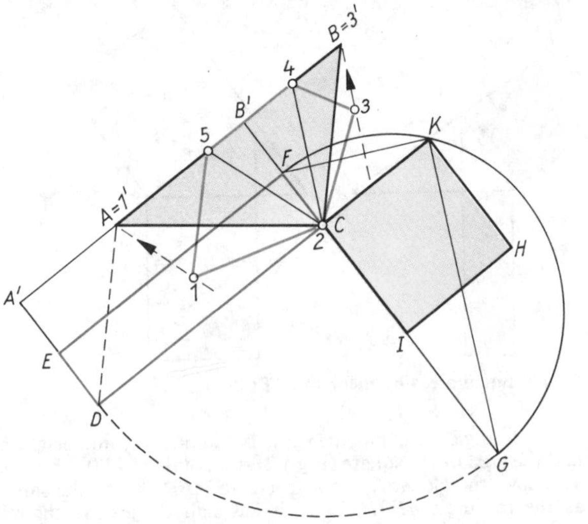

7.7-13b) Transformation of a pentagon into a square of equal area

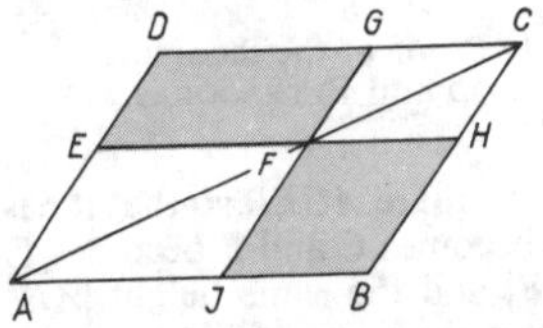

7.7-14 The theorem on supplementary parallelograms

The proof is as follows (Fig.). The diagonal AC divides the parallelogram into congruent triangles ABC and ACD. Since $JG \| AD$ and $AB \| EH$, it is clear that $|AJ| = |EF|$ and $|JB| = |AE|$. Therefore the triangles AJF and AEF are congruent. Similarly FCG and FHC are congruent. If these congruent triangles are taken away from the congruent triangles ABC and ACD, the two parallelograms in question remain. Hence they must have the same area.

7.8. Similarity

The concept of similarity

The ratio of similarity. Similarity is the geometric relationship between figures having the same shape, but not necessarily the same dimensions (Fig.). Similar figures can be transformed into one another by a one-to-one geometric transformation that preserves the *angles* of the figures. An equivalent definition is the following:

In similar figures segments of one are in a fixed ratio to the corresponding segments of the other.

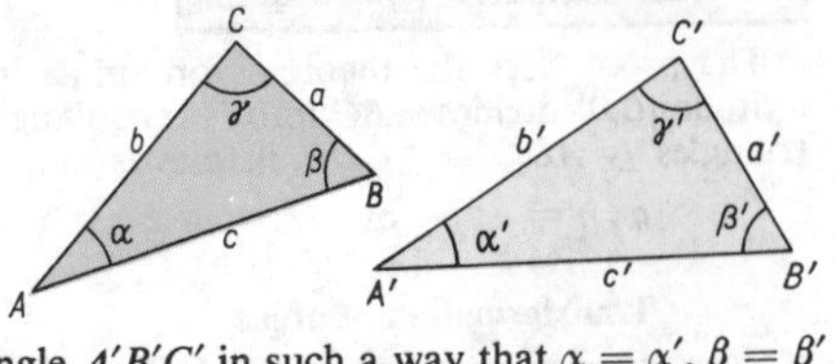

7.8-1 Similar triangles

For example, if a triangle ABC is mapped to a triangle $A'B'C'$ in such a way that $\alpha = \alpha'$, $\beta = \beta'$ and $\gamma = \gamma'$, then these triangles are similar (notation $\triangle ABC \sim \triangle A'B'C'$) (Fig.). The equality of the angles also implies that $a' : a = b' : b = c' : c = k$. This constant ratio k between the corresponding segments of similar figures is called the *ratio of similarity*. If $k > 1$, the image is larger than the original and the transformation may be called an *amplification*. If $k = 1$, the image is congruent to the original and the transformation is a congruence transformation. For $0 < k < 1$, the image is smaller than the original and the transformation is a *contraction*.

Perspective. Two similar figures can be moved by congruence transformations into special relative positions in which corresponding segments are parallel. In this position they are said to be *in perspective*. If two figures are in perspective, the transformation mapping one onto the other may be obtained from a pencil of (possibly parallel) lines. The carrier of the pencil is called a *centre of perspective*.

From the ratios given above for similar triangles it is easy to derive the following proportions: $a:b = a':b'$; $a:c = a':c'$; $b:c = b':c'$. These relations are often expressed by the continued proportion $a:b:c:\ldots:n = a':b':c':\ldots:n'$, from which one reads off that in similar figures the ratio between two segments of one is equal to the ratio between the corresponding segments of the other.

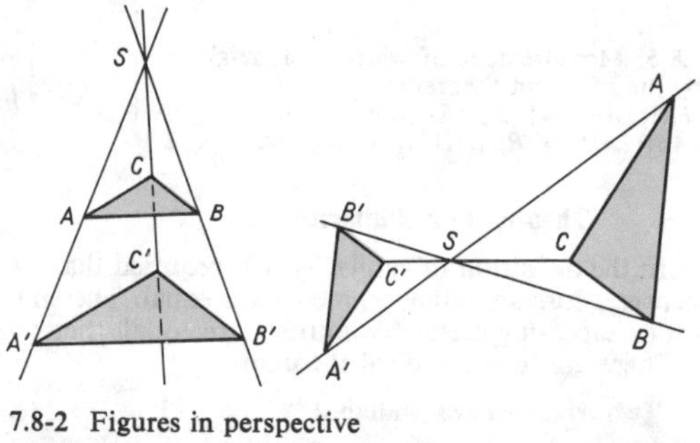

7.8-2 Figures in perspective

The intercept theorems

The following theorems are immediate consequences of the definition of similarity.

If the lines of a pencil are intersected by two parallel lines, then
a) the segments on any one line of the pencil are in the same ratio as the segments on any other;
b) the segments on the two parallel lines between any two fixed lines of the pencil are in the same ratio as the segments on any line of the pencil between each of the parallels and the carrier of the pencil; and
c) the segments on the two parallel lines between any two lines of the pencil are in the same ratio as the segments between any two other lines of the pencil.

Thus, in the figure the following proportions hold:

by *a)* $|SA|:|AB| = |SC|:|CD|$, $|SA|:|SB| = |SC|:|SD|$, $|SA'|:|SB'| = |SC'|:|SD'|$, $|SA'|:|A'B'| = |SC'|:|C'D'|$;

by *b)* $|AC|:|BD| = |SA|:|SB|$, $|A'C'|:|B'D'| = |SA'|:|SB'|$;

by *c)* $|A_1A_2|:|B_1B_2| = |A_2A_3|:|B_2B_3| = |A_3A_4|:|B_3B_4|$.

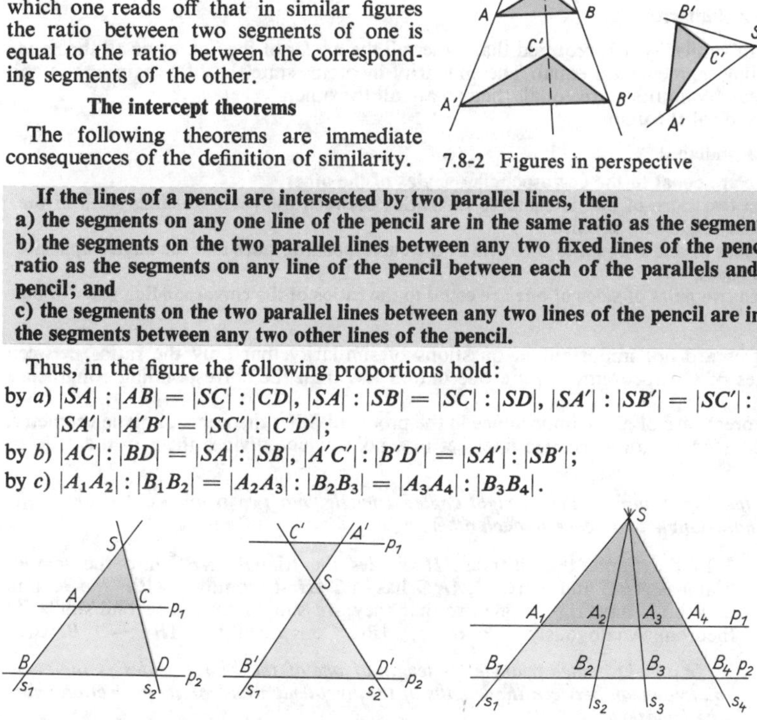

7.8-3 The intercept theorems

These theorems are concerned with the possibility of comparing the lengths of two segments and are closely related to the classical problem of *commensurability*. Two segments are commensurable if by choosing a suitable *unit* of measurement they can both be made to have rational lengths. The concept of commensurability is historically important because in classical mathematics only *rational*, not arbitrary real numbers were admitted for the computations. The theorems are widely used in many proofs, construction and measuring processes. An example is the triangular gauge. In the example of the figure, the theorems yield the following proportions:

case 1: $x:1 = 5.2:10$
or $x = 0.52$ cm,

case 2: $x:1 = 6.3:10$
or $x = 0.63$ cm.

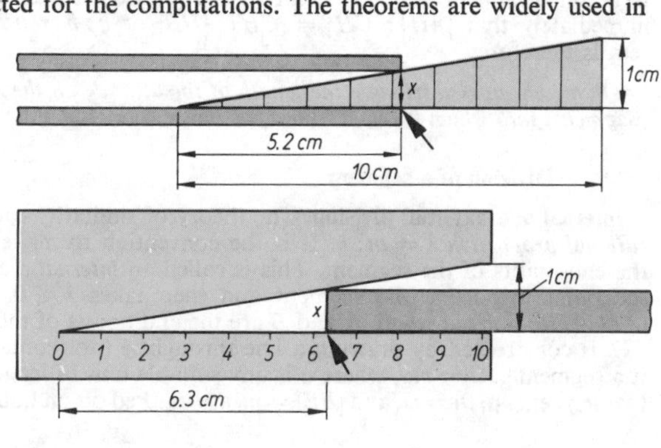

7.8-4 Wedge gauges

The theorems can also be used to measure heights (say of a tree) or distances (such as the width of a river) (Fig.).

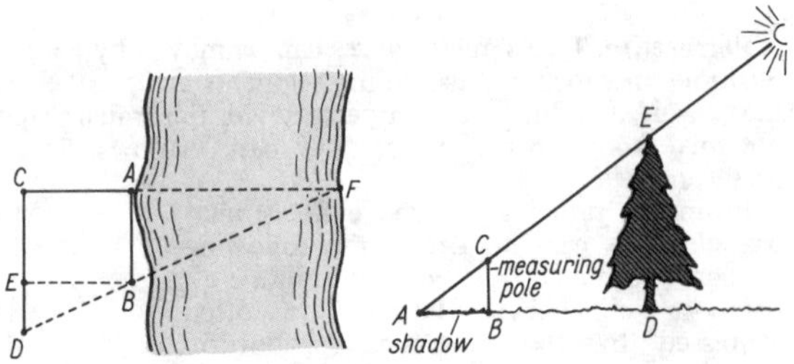

7.8-5 Measurement of width and height by the intercept theorems, $|FA|:|BE| = |AB|:|ED|$ or $|AB|:|AD| = |BC|:|DE|$, respectively

Theorems on similarity

In the definition of similarity it is required that either all the *angles* of the figures or all the *ratios* between corresponding segments are equal. The similarity theorems state that if certain angles and ratios satisfying certain conditions are equal, then so are all the others.

There are four principal theorems.

Two triangles are similar if

1. two angles of one are equal to the corresponding angles of the other,
2. the ratio between two sides of one is equal to the ratios between the corresponding sides of the other and the enclosed angles are equal,
3. the ratio of two sides of one is equal to the ratio of the corresponding sides and the angles opposite the larger sides are equal,
4. the ratios between two pairs of sides of one are equal to the ratios of the corresponding pairs of the other.

Since actual lengths are not important in questions of similarity, but only the ratios between them, the hypotheses of the theorems require one datum less than the corresponding congruence theorems.

The similarity theorems are of great importance in the proofs of other theorems of plane geometry, for instance, that the medians of a triangle intersect in a point which divides them in the ratio of 2 : 1.

The altitude on the hypotenuse divides a right-angled triangle into two triangles that are both similar to the original triangle and hence to each other.

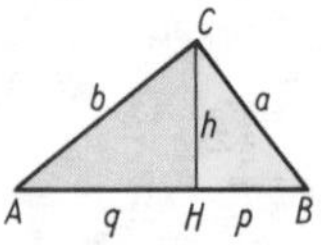

7.8-6 Theorems on right-angled triangles

In the figure the altitude AH divides the triangle ABC into the smaller triangles AHC and BHC. $\triangle AHC$ has $\sphericalangle CAH$ in common with $\triangle ABC$ and they both have right angles, so that they are similar by the second similarity theorem. Analogously $\triangle BCH \sim \triangle ABC$. Consequently $\triangle AHC \sim \triangle BHC$.

In a right-angled triangle the length of one of the shorter sides is the geometric mean between the lengths of the hypotenuse and of its projection onto the hypotenuse.

From $\triangle AHC \sim \triangle ABC$ it follows that $|AH| : |AC| = |AC| : |AB|$ and from $\triangle BHC \sim \triangle ABC$ it follows that $|BH| : |BC| = |BC| : |AB|$.

If this is rewritten as $q : b = b : c$ or $p : a = a : c$, and then transformed to $b^2 = qc$ and $a^2 = pc$, it becomes evident that this theorem is equivalent to the theorem of Euclid. Similarly it follows immediately that $|AH| : |CH| = |CH| : |HB|$ or $q : h = h : p$. The analogue to the altitude theorem is therefore:

In a right-angled triangle the length of the altitude on the hypotenuse is the geometric mean of the segments into which it divides the hypotenuse.

Division of a segment

Internal and external division. The theory of similarity can be used to divide any segment in any *rational proportion* $\lambda = m : n$. It is the convention to make $\lambda > 0$ if the dividing point is between the end-points of the segment. This is called an *internal division*. One speaks of *external division* if the point is outside the segment and then takes $\lambda < 0$. Thus, λ is always taken as the ratio $|AT| : |TB| = m : n$, where A and B are the end-points of the segment and T is the dividing point.

T is constructed by drawing a line through A (not containing the segment) and marking off on it a segment $|AC| = mu$, where u is any suitable unit of length. On the parallel to AC through B the two segments $|BD| = nu$ and $|BE| = nu$ are marked off on both sides of B using the same unit of length

u (Fig.). The intersections T_i and T_e of the lines CD and CE with AB divide the segment internally and externally in the ratio $m : n$. The points T_i and T_e are said to divide the segment harmonically, because the ratios of the partial segments are equal, and on an oriented line have opposite signs.

The ratio λ can also be an arbitrary real number if arbitrary, possibly incommensurable, segments $|AC|$ and $|BD| = |BE|$ are admitted. If m and n are integers, the construction can also be used to divide AB into $m + n$ equal intervals.

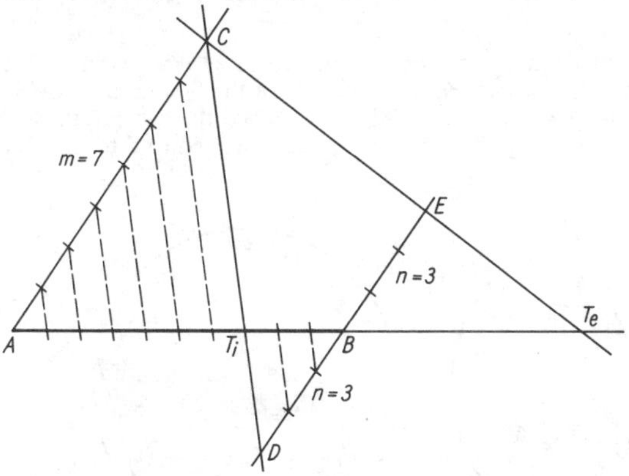

7.8-7 Division of a segment

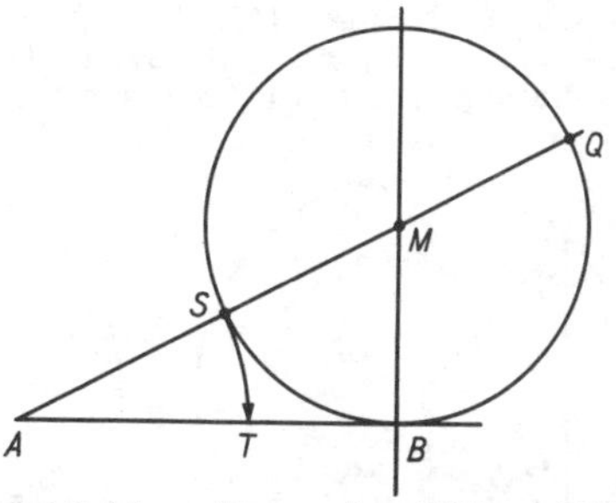

7.8-8 The golden section, $|BM| = |AB|/2$,

Golden section. It is also possible to divide a segment in such a way that the ratio of the whole segment to the larger part is equal to the ratio of the larger part to the smaller part, that is, $|AB| : |AT| = |AT| : |TB|$. The figure shows the construction. By the secant-tangent theorem $|AB|^2 = |AS| \cdot |AQ|$ or $|AB| : |AQ| = |AS| : |AB|$. Since $|AS| = |AT|$ and $|AB| = |SQ|$, subtraction of corresponding quantities gives $|AB| : (|AQ| - |SQ|) = |AS| : (|AB| - |AT|)$ or $|AB| : |AT| = |AT| : |TB|$. This division is called the *golden section*. It is of historical importance in aesthetics, because it has frequently been held that a condition for ideal beauty of figures (including the human form) is that the various parts should have the proportions of the golden section.

7.9. Circles

Notation

A *circle* is the set of all points in a plane whose distance from a given point is a fixed constant length. To distinguish between the *disc* bounded by the circle and the circle itself, this boundary is called the *circumference* of the circle. The point equidistant from all the points on the circle is called its *centre*. A straight-line segment (or *interval*) from the centre of a circle to a point on the circumference is called a *radius*. Any interval between two points on the circumference lies entirely inside the circle, so that the circle is a *convex figure*. A line through two points on the circumference is called a *secant* and the interval between them a *chord*. Chords containing the centre of the circle, the so-called *diameters*, are the longest chords in any circle. Lines that have only one point in common with a circle are called *tangents*. The segment of the circumference between two points on it is called an *arc*. Angles whose arms are secants and whose vertex lies on the circumference are said to be *subtended* by the arc (or chord) between their arms. Angles whose vertex is the centre and whose arms are radii are said to be *subtended at the centre* by the arc (or chord) between their arms. The portion of the disc between two radii is called a *sector*, and

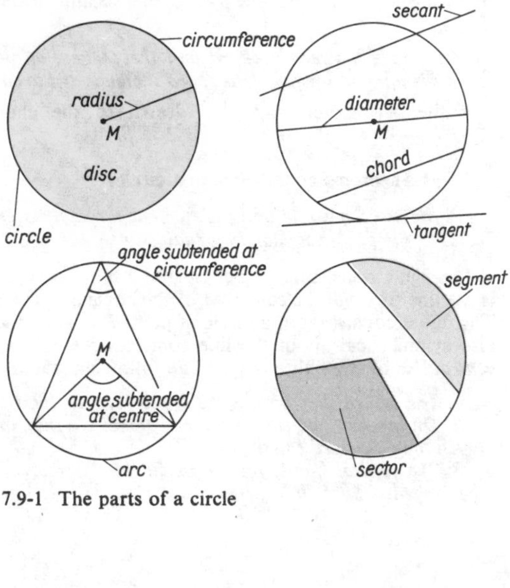

7.9-1 The parts of a circle

the portion of a sector between the chord connecting the two end-points of the radii and the circumference of the circle is called a *segment* of the circle. (To avoid confusion straight-line segments will be called intervals in this section.) (Fig.)

Theorems on angles in circles

Any angle subtended at the circumference of a circle is half the angle subtended at the centre by the same arc.

The proof distinguishes three cases (Fig.), depending on whether the centre M lies *on one* of the arms of the angle (left), *between* its arms (middle), or *outside* them (right). In the first case $\triangle AMS_1$ is isosceles, because $|AM| = |MS_1| = r$. Thus $\sphericalangle AS_1M = \sphericalangle MAS_1 = \beta$. Since the central angle α is an external angle of $\triangle AMS_1$, it follows that $\alpha = 2\beta$. The two other cases are reduced to the first by drawing the diameters S_2D and S_3E.

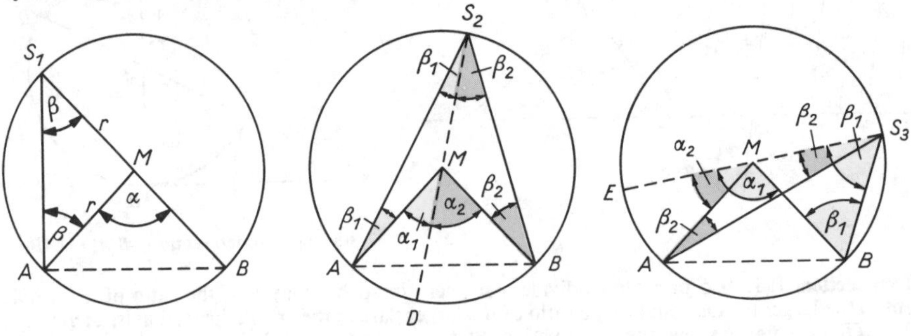

7.9-2 Angles at the circumference and the centre

Every central angle has a unique arc and vice versa. An angle at the circumference has a unique arc, but an arc subtends infinitely many angles at the circumference.

Angles at the circumference subtended by the same arc are equal. Angles subtended at the circumference by a semicircle (or diameter) are right-angles.

If the vertex of an angle at the circumference is moved from A to B, then as it approaches B, one of its arms approaches the tangent at B and the other the secant through A and B (Fig.).

The angle between a chord and the tangent at one of its end-points is equal to the angle the chord subtends at the circumference.

If the perpendicular ML is drawn to the chord AB, then $\sphericalangle ABT = 90° - [90° - \alpha/2] = \alpha/2 = \beta$.

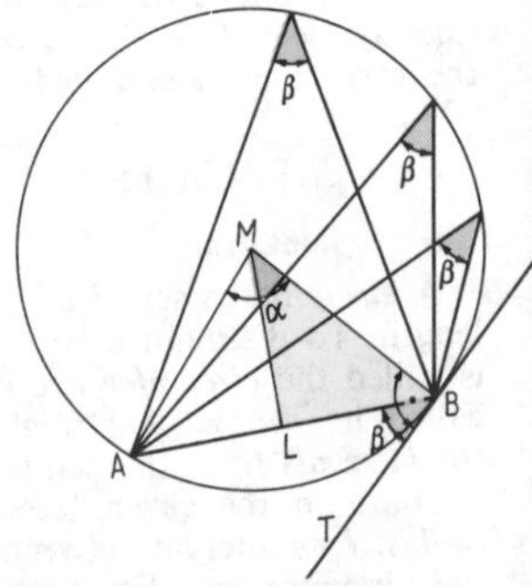

7.9-3 Angles at the circumference and the angle between a chord and a tangent

Theorems on tangents to a circle

A radius of a circle and the tangent through its end-point are perpendicular to each other. Conversely, the perpendicular to a radius in its end-point is a tangent.

The figure consisting of a circle and a tangent is symmetrical about an axis. The axis of symmetry is the line through the centre M of the circle and the point B at which the tangent touches the circle. The figure consisting of a circle, a point P outside the circle and the tangents from P to the circle is also symmetrical about the line connecting P to the centre M of the circle. This line is called the *central line* of the figure (Fig.). The following statements are consequences of this symmetry.

1. The central line bisects the angle between the two tangents from a point to a circle.
2. On the two tangents from P to the circle the intercepts between P and the points at which they touch the circle are equal.
3. The central line is the perpendicular bisector of the chord connecting the two points at which the tangents touch the circle.

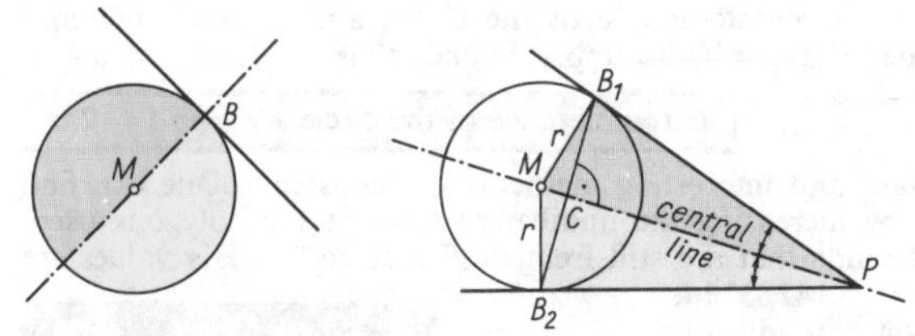

7.9-4 Tangents to a circle

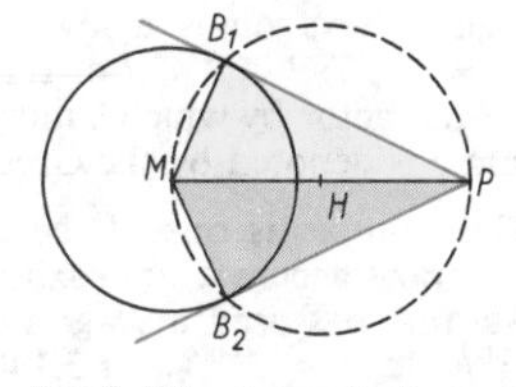

7.9-5 The tangents from P to a circle about M

Constructions. The above theorems on circles and tangents are the basis of all constructions involving tangents.

To construct a tangent from a point outside a circle to that circle, another circle of radius $|HP|$ is drawn with centre in the mid-point H of the central line $|MP|$. This intersects the circle in B_1 and B_2. Then PB_1 and PB_2 are the required tangents (*theorem* of Thales) (Fig.).

To construct a tangent at a point B on the circle. The radius BM is produced through B to C with $|BC| = |BM|$. With C and M as centres arcs of radius greater than $|MB|$ are drawn. The line connecting their intersections D_1 and D_2 is the required tangent (Fig.).

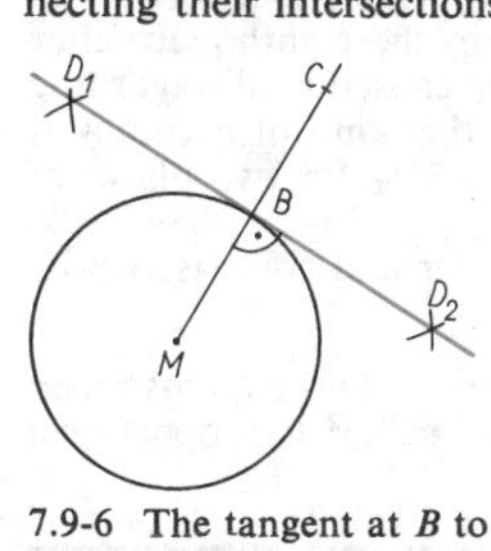

7.9-6 The tangent at B to a circle about M

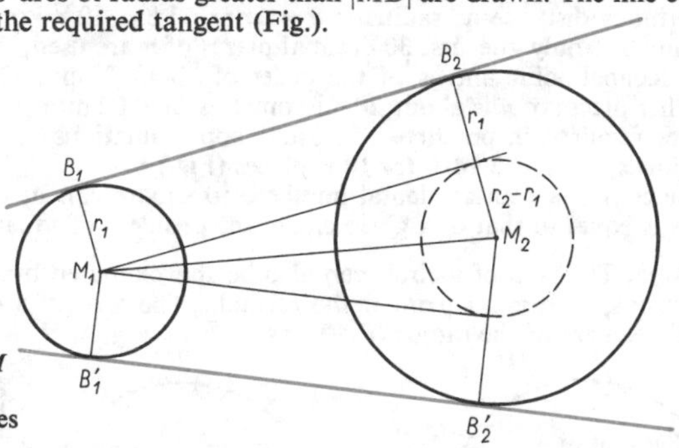

7.9-7 External tangents to two circles

To construct the external common tangents to two circles. Let M_1 and M_2 be the centres of the circles, and their radii r_1 and r_2 ($r_1 < r_2$). With M_2 as centre a circle of radius $r_2 - r_1$ and the tangents from M_1 to this circle are drawn. The parallels B_1B_2 and $B_1'B_2'$ to these tangents at a distance r_1 are the required tangents (Fig.).

These tangents give the shape of a belt drive around two wheels of radius r_1 and r_2, respectively.

To construct the internal common tangents to two circles. With M_2 as centre a circle of radius $r_1 + r_2$ and the tangents from M_1 to this circle are drawn. The parallels B_1B_2 and $B_1'B_2'$ to these tangents at a distance r_1 are the internal tangents (Fig.).

In this case the shape is that of a reversing belt drive.

7.9-8 Internal tangents to two circles

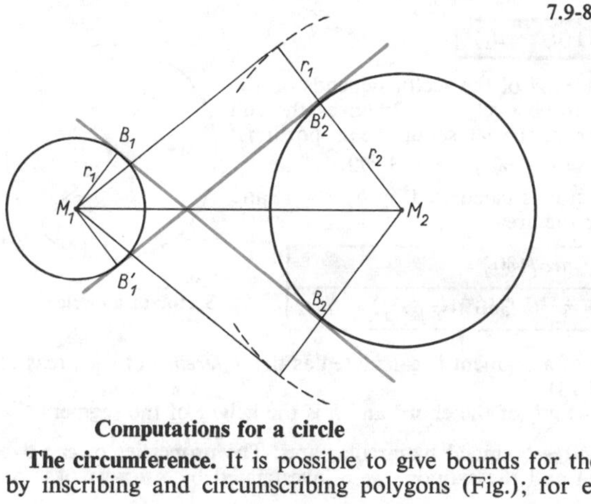

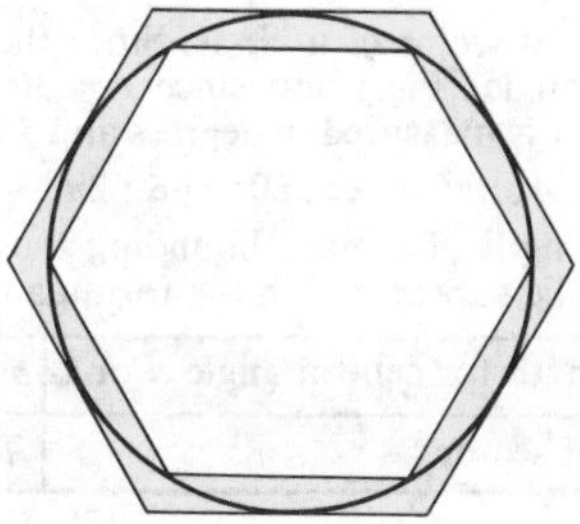

7.9-9 A circle with inscribed and circumscribed hexagons

Computations for a circle

The circumference. It is possible to give bounds for the circumference of a circle of diameter d by inscribing and circumscribing polygons (Fig.); for example, the circumference $c_1 = 3d$ of a regular hexagon is a lower bound for the circumference c of the circle, and the circumference $c_e = (2\sqrt{3})\, d < 3.47d$ of the circumscribed hexagon is an upper bound, that is: $3d < c < 3.47d$.

The factor by which d must be multiplied to obtain c is denoted by the Greek letter $\pi : c = \pi \cdot d$.

Circumference of the circle	$c = \pi d = 2\pi r$

This number is one of the most important and interesting mathematical constants. One can find arbitrary accurate approximations of π by increasing the number of sides of the polygons used. ARCHIMEDES used a 96-gon and found bounds that are still frequently used today. His values are $3^{10}/_{71} < \pi < 3^{10}/_{70}$ or $3.14084507 < \pi < 3.14285714$.

The first 40 places after the decimal point are given by

$\pi = 3.14159\ 26535\ 89793\ 23846\ 26433\ 83279\ 50288\ 41971\ \ldots$

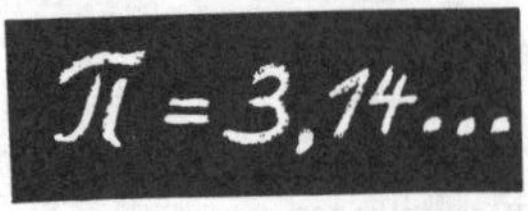

7.9-10 The number π to two places of decimals

The following rough calculation shows what an accuracy of "only" 30 places of decimals means. A system of stars that astronomers can just make visible by hour-long exposures on photographic plates using the most powerful telescope, emitted the light that is trapped by the plate about 2000 million years ago. Since light travels about $9.5 \cdot 10^{12}$ km per year, these stars are about $2 \cdot 10^9 \cdot 9.5 \cdot 10^{12}\,\text{km} = 1.9 \cdot 10^{22}$ km away from the earth. The circumference of a circle with this enormous distance as radius is $c = 2\pi r = 3.8\pi \cdot 10^{22}$ km. If in calculating this circumference in kilometres only the first 30 decimal places of π are used, the error occurs in the eighth place after the decimal point and is of the order of about 2 units. That is, the error caused by disregarding further places of π is about 20 micrometres or 0.02 mm. It is obvious that this kind of accuracy is never required in practice. The usual approximations are $\pi \approx 3.14$ or $\pi \approx 3^1/_7$ for two places of decimals, or $\pi \approx 3.1416$ for four places (Fig.).

Since π is a transcendental number, no square can be constructed by ruler and compass whose area is equal to that of a given circle (the problem of squaring the circle).

Area. The area of a circle can also be approximated by the areas of inscribed and circumscribed polygons, with π occurring in the formula. The area of a circle $A = \pi r^2 = \pi(d/2)^2$ is proportional to the square of the radius (Fig.).

Area of the circle	$A = \pi r^2 = \pi(d/2)^2$

Area of an annulus. The difference of the areas of two concentric circles of diameters d_1 and $d_2 > d_1$ is (Fig.):

$$A = \pi d_2^2/4 - \pi d_1^2/4 = (\pi/4)(d_2 + d_1)(d_2 - d_1).$$

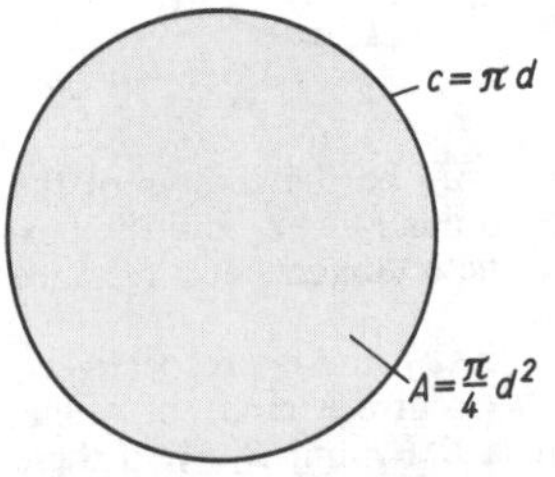

7.9-11 Circumference and area of a circle

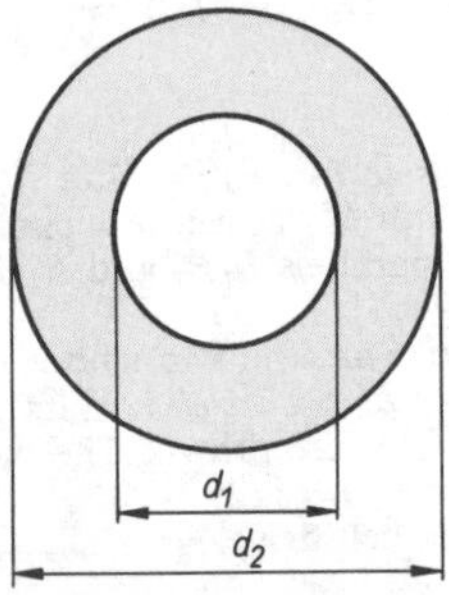

7.9-12 Annulus

Area of annulus	$A = (\pi/4)(d_2 + d_1)(d_2 - d_1)$

Area of a sector of a circle. Since the area A of the sector depends on the central angle (Fig.) and since $\alpha = 360°$ respectively $\hat{\alpha} = 2\pi$ gives the full circle, if α is measured in degrees and $\hat{\alpha}$ in rad, one can set up the proportions:

$$A : \pi r^2 = \alpha : 360° = \hat{\alpha} : 2\pi \rightarrow A = (\alpha/360°)\,\pi r^2 = \hat{\alpha} r^2/2.$$

The length of the arc bounding the sector is calculated by $b : \hat{\alpha} = r$, and this can be substituted in the formula for the area.

Arc length for central angle α or $\hat{\alpha}$	$b = \pi r\alpha/180° = r\hat{\alpha}$
Area of sector	$A = \pi r^2(\alpha/360°) = \hat{\alpha} r^2/2 = br/2$

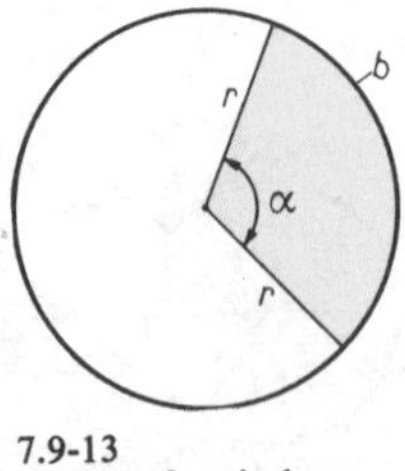

7.9-13 Sector of a circle

Area of a segment of a circle. The area of a segment is calculated as the *difference* of the areas of the sector and the *triangle AMB* (Fig. 7.9-14).

$A = br/2 - s(r - h)/2$, where s is the length of the chord and h is the height of the segment.

Area of an arbelos. There are many figures bounded by circular arcs. The properties of the two shown here, the arbelos (cobbler's knife) and the salinon, were investigated by ARCHIMEDES. An arbelos consists of two small semicircles inside a large one, that such the sum of the diameters of the small semicircles is the diameter of the large one (Fig.). If the common tangent DC of the two small semicircles is drawn, then the circle K on the semichord $|DC| = h$ as diameter has the same area as the arbelos. This semichord $|DC|$ can be taken as the altitude of a right-angled triangle ABC with $|AB| = d$ as the hypotenuse. Thus, by the altitude theorem, $h^2 = q(d - q)$. Hence, if A_h is the area of the circle of diameter h, the following equation holds: $A_h = \pi h^2/4 = \pi q(d - q)/4$. The area of the arbelos is

$$A_{\text{Arb}} = (1/2)\,(A_{AB} - A_{AD} - A_{DB}) = (\pi/8)\,(d^2 - q^2 - (d - q)^2) = \pi q(d - q)/4.$$

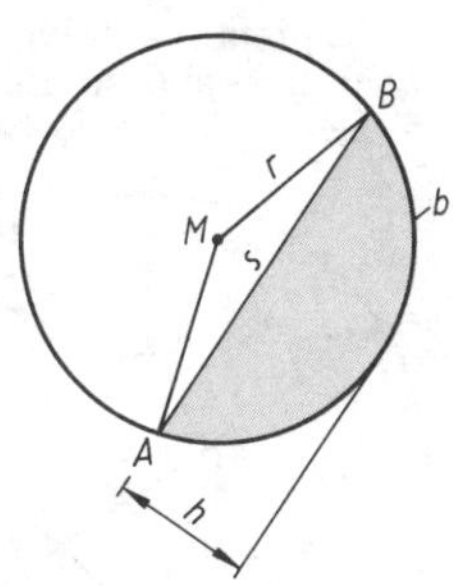

7.9-14 Segment of a circle

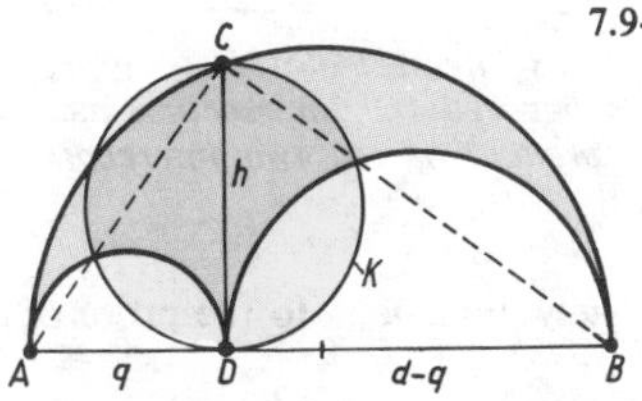

7.9-15 Arbelos

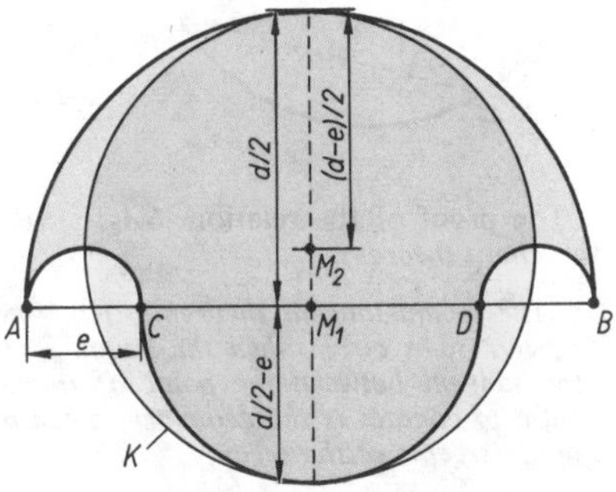

7.9-16 Salinon

Area of a salinon. A salinon consists of two small semicircles of equal diameter e inside a large semicircle of diameter d and a further semicircle of diameter $d - 2e$ between them on the opposite side of d (Fig.). The area of a salinon is equal to that of a circle of diameter $d - e$, which is the sum of the radii of the large semicircle and the semicircle on the other side of its diameter. By the figure, the area of this circle K and the salinon S are given by

$$A_K = \pi(d - e)^2/4;$$

$$A_S = (A_{AB} - 2A_{AC} + A_{CD})/2 = \pi[d^2 - 2e^2 + (d - 2e)^2]/8 = \pi(d^2 - 2de + e^2)/4 = \pi(d - e)^2/4.$$

The lunulae of Hippocrates. There are also some famous moonshaped figures. The best known of these are the *crescents* (or lunulae) *of Hippocrates*. By the theorem of Thales the *triangle ABC* in the left-hand figure is right-angled (Fig.); thus $c^2 = a^2 + b^2$. The semicircle on $|AB| = c$ has the area $A_{AB} = \pi c^2/8$; the sum of the areas of the semicircles on $|AC|$ and $|BC|$ is $A_{AC} + A_{BC} = \pi(b^2 + a^2)/8$ and is thus equal to A_{AB}. From this it follows that:

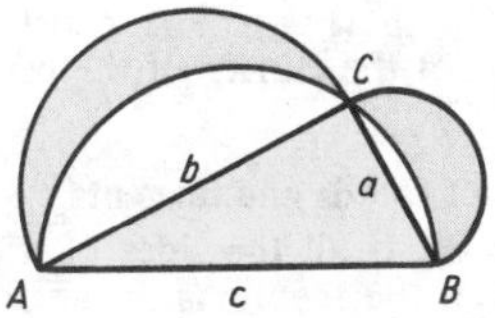

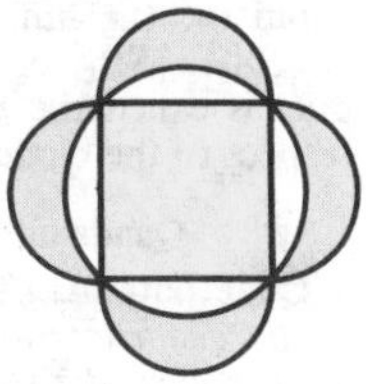

7.9-17 The lunulae of Hippocrates

The sum of the areas of the two crescents is the area of the triangle.

Similarly the sum of the areas of the crescents in the right-hand figure is the area of the square. Misled by the remarkable fact that π does not occur in these formulae many mathematicians of ancient times (including Hippocrates himself) continued the hopeless search for a method of sqaring the circle.

Theorems on chords, secants and tangents

By rotating a circle about its centre one sees immediately that:

Chords of equal length are equidistant from the centre; conversely, equidistant chords have the same length.

If in a circle a *chord* c_1 is *longer* than a *chord* c_2, it is *closer* to the centre than c_2, hence the diameters are the longest chords.

If the lines of a pencil intersect a circle, then the *products* of the *two intercepts* on each line between the carrier of the pencil and the circumference of the circle are equal to a fixed *constant*. This is a summary and generalization of the statements of the three following theorems.

The chord theorem: If two chords of a circle intersect, then the product of the intercepts on one chord is equal to the product of the intercepts on the other.

The proof is as follows (Fig.): the angles $\sphericalangle B_1A_1B_2$ and $\sphericalangle B_2A_2B_1$ are equal, because they subtend equal arcs, and so are $\sphericalangle A_1B_2A_2$ and $\sphericalangle A_2B_1A_1$. Hence $\triangle A_1SB_2 \sim \triangle A_2SB_1$ and $|SA_1| : |SB_2| = |SA_2| : |SB_1|$ or $|SA_1| \cdot |SB_1| = |SA_2| \cdot |SB_2|$.

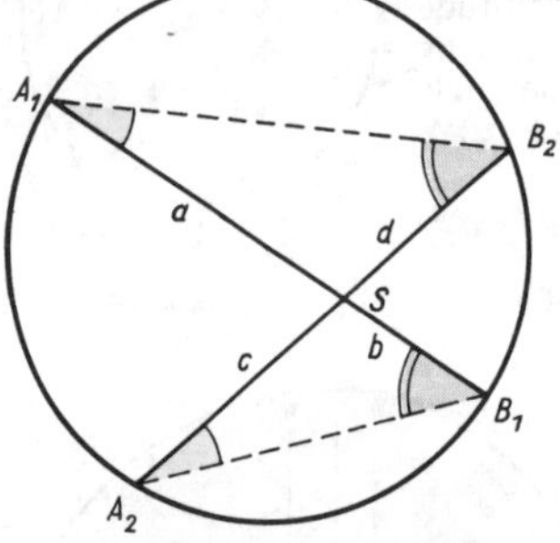

7.9-18 The chord theorem

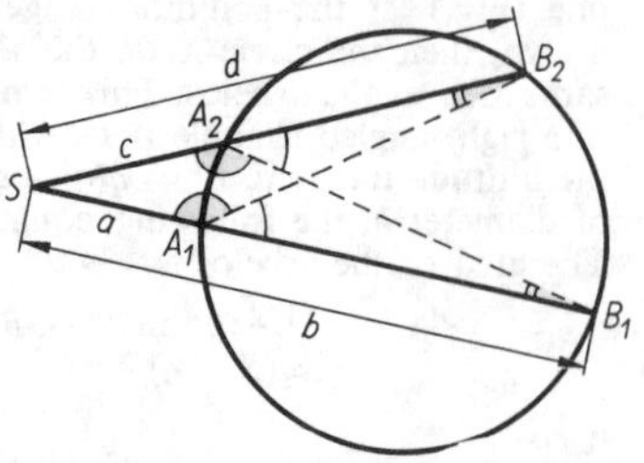

7.9-19 The secant theorem

The secant theorem: If two secants intersect outside a circle, then the product of the intercepts between the intersection and the circle on one is equal to the product of the two intercepts on the other (Fig.).

The proof of the relation $|SA_1| \cdot |SB_1| = |SA_2| \cdot |SB_2|$ is completely analogous to the proof of the *chord theorem.*

The secant-tangent theorem: If a secant intersects a tangent to a circle, then the length of the intercept on the tangent between the point of intersection and the point of contact is the geometric mean of the lengths of the intercepts of the secant.

The proof is obtained by taking this as the extreme case of the *secant theorem.* As one secant tends to a tangent, the lengths of the two intercepts on it become equal (in the limit) and the theorem reduces to $|SA|^2 = |SA_1| \cdot |SB_1|$ or $t^2 = a \cdot b$ (Fig.). The constant product is called the *power* of the carrier of the pencil with respect to the circle.

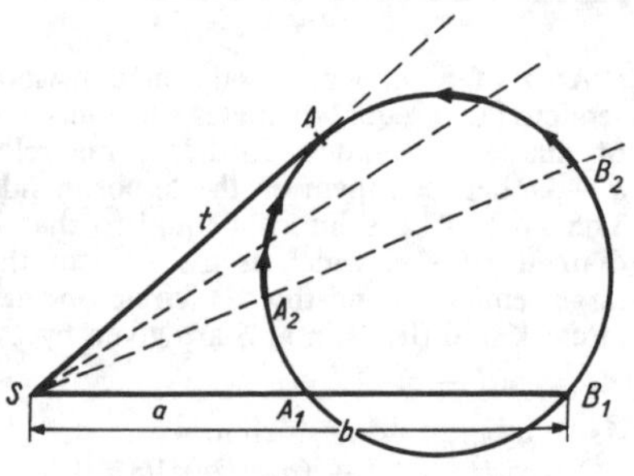

7.9-20 The secant-tangent theorem

Quadrilaterals of chords and tangents

Quadrilaterals of chords. If all the sides of a quadrilateral are chords of a circle, it is called a *cyclic quadrilateral.* Then the theorems on angles subtended by an arc at the centre and circumference yield the following result:

In a cyclic quadrilateral the sum of opposite interior angles is equal to 180° (Fig.).

Conversely, if the sum of opposite interior angles of a quadrilateral is 180°, then it is cyclic.

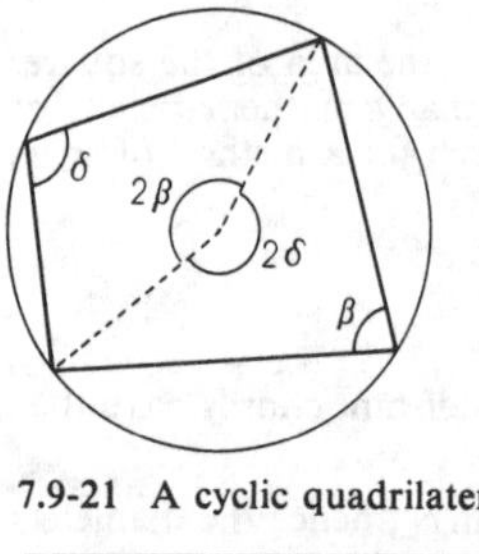

7.9-21 A cyclic quadrilateral

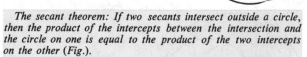

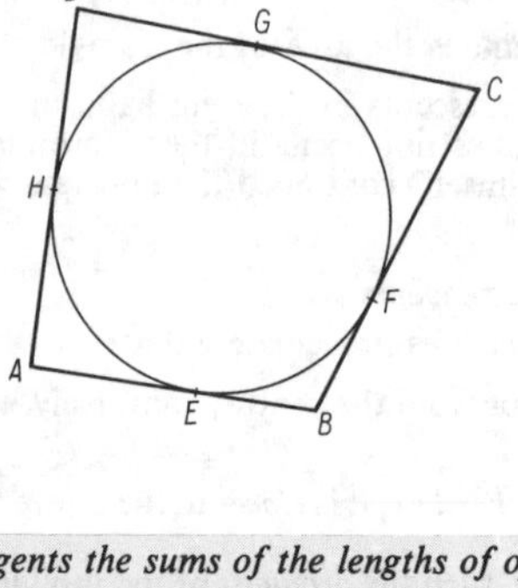

7.9-22 A quadrilateral of tangents

Quadrilaterals of tangents. If the sides of a quadrilateral touch a circle, then (Fig.) $|AE| = |AH|$, $|BE| = |BF|$, $|CF| = |CG|$ and $|DG| = |DH|$, because the intercepts of two tangents from a single point to a circle are equal. From this it follows that $|AE| + |EB| + |CG| + |GD| = |BF| + |FC| + |DH| + |HA|$, hence $|AB| + |CD| = |BC| + |DA|$ or $a + c = b + d$.

In a quadrilateral of tangents the sums of the lengths of opposite sides are equal.

The validity of the converse can also be seen from the figure.

These theorems have many corollaries; for instance, it follows that a square has an inscribed and a circumscribed circle, that a rectangle has a circumscribed but no inscribed circle and that a rhombus has an inscribed but no circumscribed circle. A general parallelogram has neither an inscribed nor a circumscribed circle.

7.10. Geometric loci

A locus is a *set of points* defined by a rule that makes it possible to decide for any given point whether or not it belongs to the set. The locus then contains exactly the points of that set and no others. Thus, the locus of all points in space at a fixed distance r from a given point M is the sphere of radius r with centre at M. Similarly, the locus of all points in space having constant distance from a line is the curved surface of a cylinder with the given line as axis.

In the plane, loci can be curves such as circles, parabolas or straight lines. Frequently, a point is determined as an intersection of two loci. For instance, the position of the point C whose distance from the end-points A and B of a segment c is a or b, respectively, is determined up to a reflection in the straight line AB by the condition that it must be in the locus of points of distance a from B and also in the locus of points of distance b from A (Fig.).

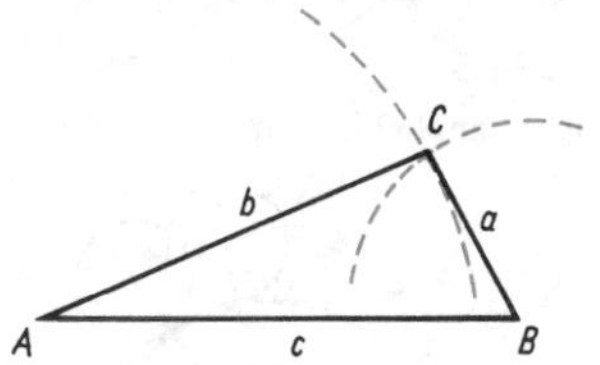

7.10-1 Geometric locus: position of the point C

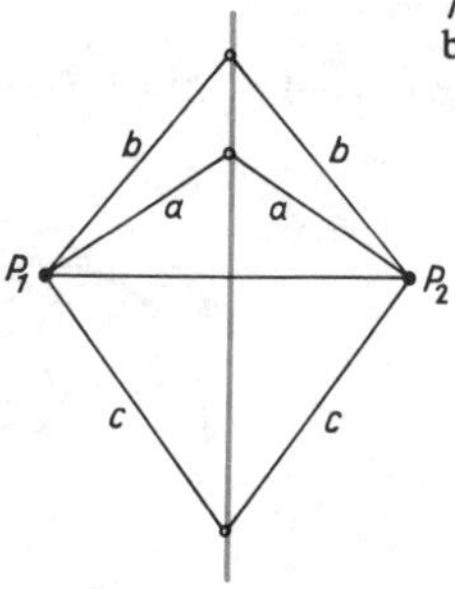

7.10-2 Geometric locus: perpendicular bisector

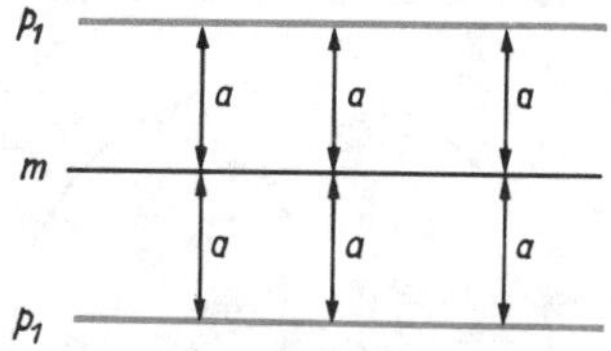

7.10-3 Geometric locus: pair of parallels and mid-parallel

Certain elementary loci.

1. The locus of points equidistant from two given points is the perpendicular bisector of the segment between those points (Fig.).

2. The locus of points at a fixed distance from a given line is the pair of parallels at that distance from the line (Fig.).

3. The locus of points equidistant from two parallels is the parallel midway between them (Fig.).

4. The locus of points equidistant from two intersecting lines is the pair of bisectors of the vertical angles at the intersection (Fig.).

5. The locus of all points at a fixed distance from a given point is the circle with that point as centre and the distance as radius.

6. The locus of all points at which a given segment subtends a fixed angle is a circular arc with the segment as a chord (Fig. 7.10-6).

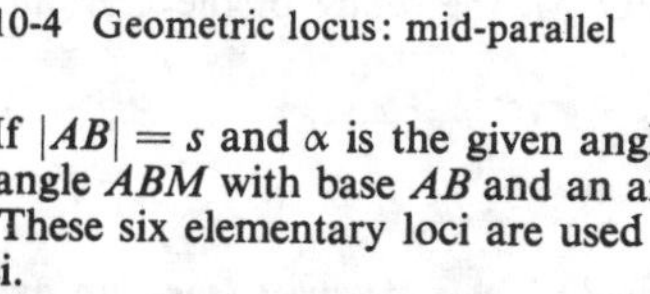

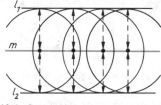

7.10-4 Geometric locus: mid-parallel

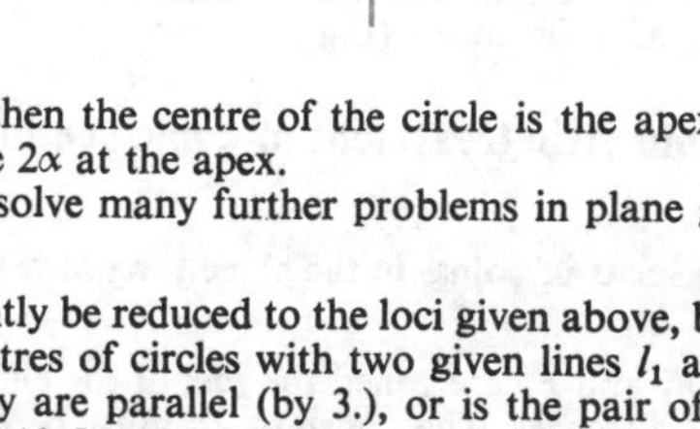

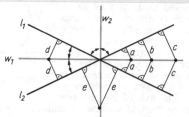

7.10-5 Geometric locus: angle bisector

If $|AB| = s$ and α is the given angle, then the centre of the circle is the apex M of an isosceles triangle ABM with base AB and an angle 2α at the apex.

These six elementary loci are used to solve many further problems in plane geometry, involving loci.

Applications. Constructions can frequently be reduced to the loci given above, but this may require several steps. For instance, the set of centres of circles with two given lines l_1 and l_2 as tangents is the parallel halfway between them if they are parallel (by 3.), or is the pair of angle bisectors w_1 and w_2 if they intersect (by 4.) (see Fig. 7.10-5 and 7.10-7).

New loci result if further conditions are added. For instance, the locus of the centres of all circles of radius r that touch a given straight line l at a point P is the perpendicular to l at P (Fig.), and the locus of the centres of all circles that touch a given circle of radius ϱ is the circle of radius $r + \varrho$

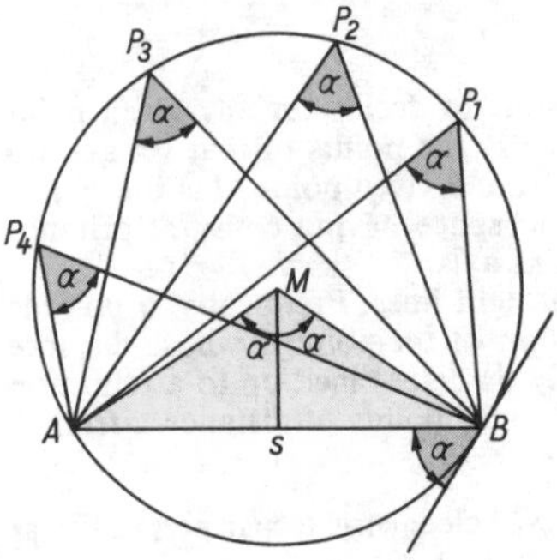

7.10-6 Geometric locus: circular arc on a chord

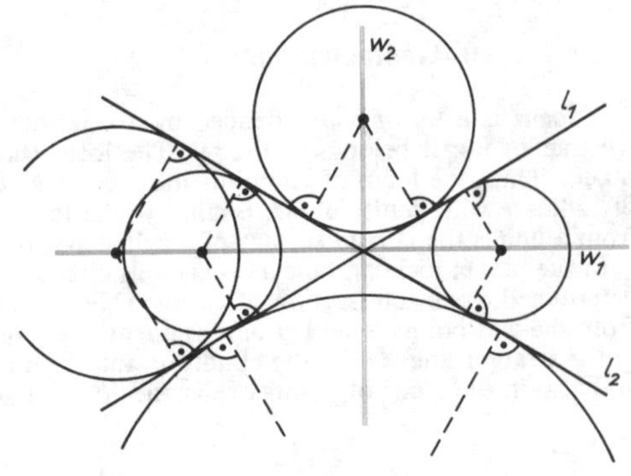

7.10-7 Geometric locus: angle bisector

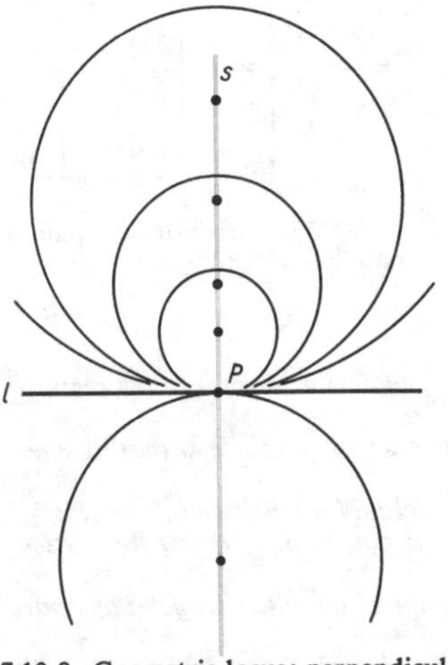

7.10-8 Geometric locus: perpendicular

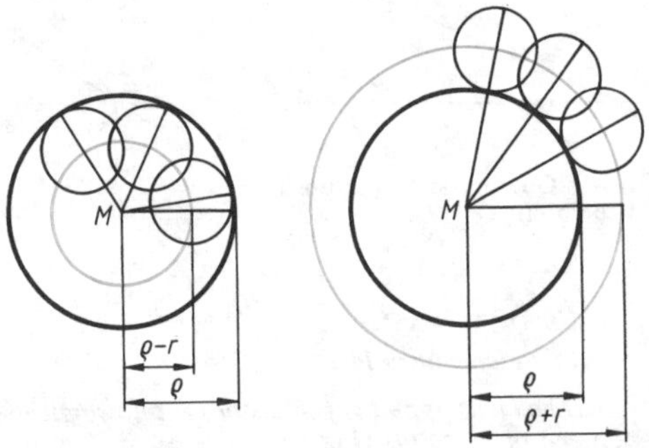

7.10-9 Geometric locus: concentric circles

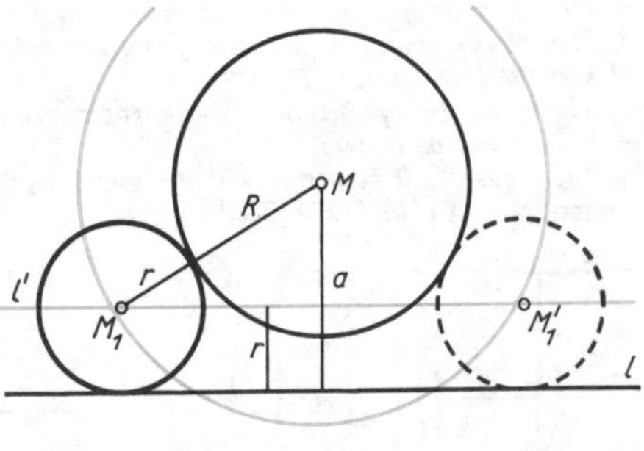

7.10-10 Geometric locus: straight line and circle

concentric to the given circle if they touch externally, and the concentric circle of radius $r - \varrho$ if they touch internally and $r < \varrho$ (Fig.).

Points are frequently constructed by intersecting two loci. For instance, a gear wheel of radius R can be made to move a rack l at distance $a > R$ from its centre by interposing a pinion of radius r. The axis of the pinion is obtained as the intersection of a circle of radius $R + r$ about M, the centre of the gear, and a parallel to the rack l at a distance r (Fig.).

7.11. Planimetric treatment of conic sections

The ellipse

The ellipse is the locus of points in the plane for which the sum of their distances from two fixed points is a constant $2a$.

The fixed points F_1 and F_2 are called the *foci* of the ellipse and the distances r_1 and r_2 of P from F_1 and F_2 its *radial distances*. The constant $2a$ must be *larger* than the distance between the foci. Then the points of intersection P_1 and P_2 of two circles of radius r_1 and r_2 about F_1 and F_2, respectively, where $r_1 + r_2 = 2a$, both lie on the ellipse. If the foci are inverted, the same process yields two further points P_3 and P_4 on the ellipse (Fig.). The points P_1 and P_2 and P_3 and P_4, and indeed the whole ellipse are *symmetrical* about the line through the foci. But they are also symmetrical about the perpendicular bisector of F_1F_2. The intersection C of the two axes of symmetry is a *centre of symmetry* of the ellipse and is called its *centre*. The distance of each focus from the centre of the ellipse is called the *linear eccentricity* of the ellipse, $|CF_1| = |CF_2| = e$. On any line through the centre of the ellipse the interval inside the ellipse is called a *diameter*. The largest diameter is called the *major axis* and its end-points, the *principal vertices* V_1 and V_2, have distance $a + e$ and $a - e$ from the focal points and a from the centre C. The smallest diameter is called the *minor axis* and its end-points the *subsidiary vertices* W_1 and W_2 have distance a from both foci. Their distance b from the centre of the ellipse can be computed from the semi-major axis a and the linear eccentricity e by the theorem of Pythagoras: $b^2 = a^2 - e^2$.

Ellipse	$r_1 + r_2 = 2a$ $a^2 - b^2 = e^2$

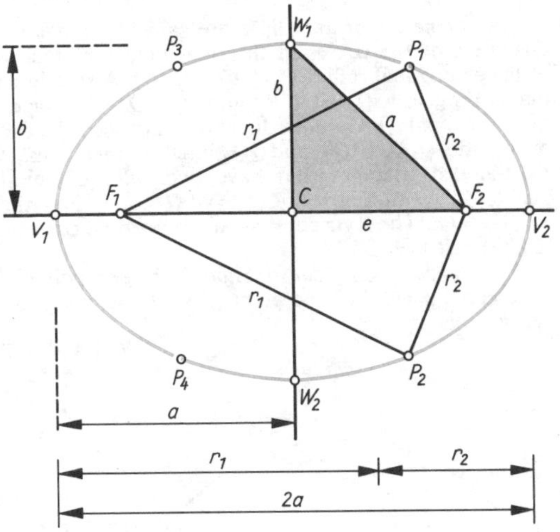

7.11-1 Ellipse

The *shape* of the *ellipse* is determined by any two of the quantities a, b and e, for instance, by the rectangle of sides $2a$ and $2b$. If b is small, the ellipse is very flat and elongated, in the limit it degenerates into the interval $|F_1F_2| = |V_1V_2|$ on which every point is counted twice. As b increases in relation to a, the ellipse becomes more and more like a circle, which is reached for $b = a$, when the rectangle is a square. The circle can thus be regarded as an ellipse of linear eccentricity 0 for which $a = b = r_1 = r_2 = r$.

Thread construction. (This construction is used for laying out elliptical flower beds.) Mark the foci with two drawing pins (or sticks) and attach a string of length $2a > 2e$ to them. If a pencil or a third stick is hooked into the string and moved in such a way that the string stays taut, then it describes an ellipse, for $r_1 + r_2 = 2a$ always holds (Fig.).

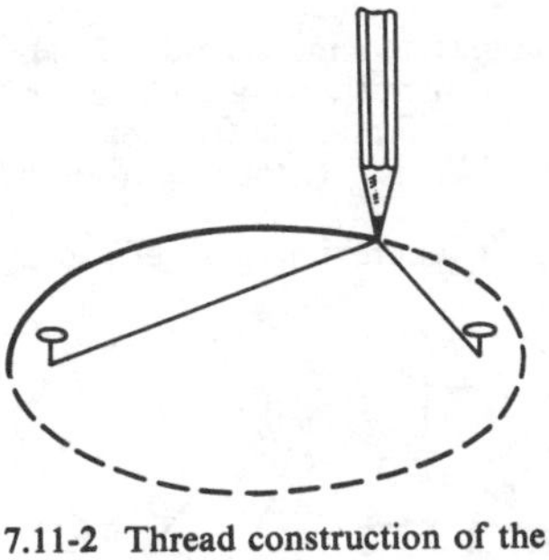
7.11-2 Thread construction of the ellipse

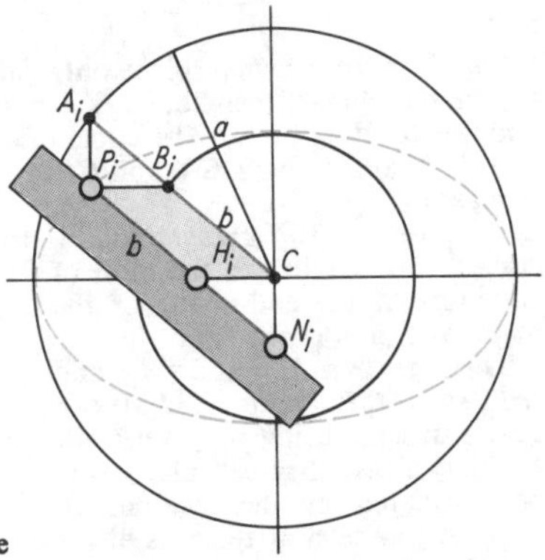

7.11-3 The principle of an ellipse-drawing device

The ellipse as a perspective-affine image of a circle. The *two circle construction* of descriptive geometry obtains a point P_i on the ellipse as the intersection of parallels to the axes through the two points A_i and B_i on a common radius of the concentric circles about C of radius a and b, respectively (Fig.). A parallel to CA_i through P_i intersects the axes in M_i and N_i. The parallelogram $CB_iP_iM_i$ shows that $|M_iP_i| = b$, and the parallelogram $CA_iP_iN_i$ shows that $|P_iN_i| = a$. Thus, one can construct an ellipse in the following fashion. On the edge of a piece of paper mark off P_i, M_i and N_i in such a way that $|M_iP_i| = b$ and $|N_iP_i| = a$. If M_i and N_i are kept on two mutually orthogonal lines and moved to and fro, then P_i describes an ellipse. This '*paper-strip*' *construction* illustrates the principle on which most devices for drawing ellipses are based: two points M_i and N_i of fixed distance move on mutually orthogonal lines.

Two *diameters* of an ellipse are said to be *conjugate* if they are the images of two orthogonal diameters of the circles in the two-circle construction. For instance, CP and CQ are the images of the orthogonal radii $CP_1 \perp CQ_1$ (Fig.). A rotation of 90° therefore takes $\triangle CPP_1$ into $\triangle CP^*Q_1$, that is, P_2 goes to Q_2, P to P^*, and P_1 to Q_1. The quadrilateral $QQ_2P^*Q_1$ is a rectangle. The extension of its diagonal QP^* in both directions intersects the axes of the ellipse in X and V, and the extensions of the sides Q_1Q, P^*Q_2, and QQ_2 intersect them in R, S and T, respectively. The figure now contains right-angled triangles that have two sides of equal length and equal angles. The congruence $\triangle XRQ \cong \triangle CSQ_2$ gives $|CQ_2| = |XQ| = b$, and the congruence $\triangle RCQ_1 \cong \triangle QTV$ gives $|CQ_1| = |QV| = a$. The Rytz construction relies on a consequence of these equalities, namely the fact that $|UX| = |UC| = |UV|$.

A second paper-strip construction. If the end-points X and V of the segment $a + b$ move along two orthogonal axes, then the point Q on the segment for which $|XQ| = b$ and $|QV| = a$ describes an ellipse.

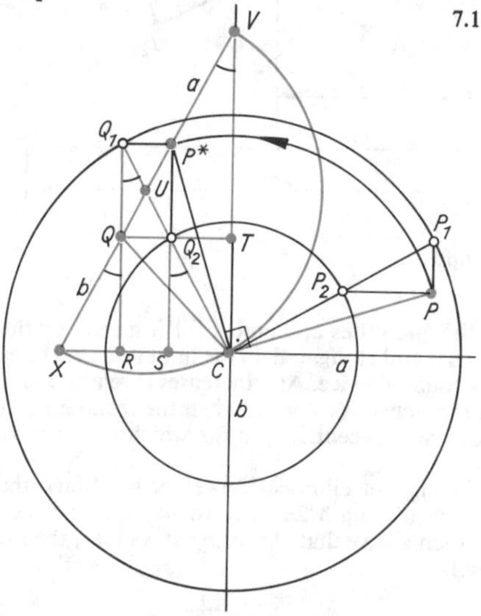

7.11-4 The Rytz construction

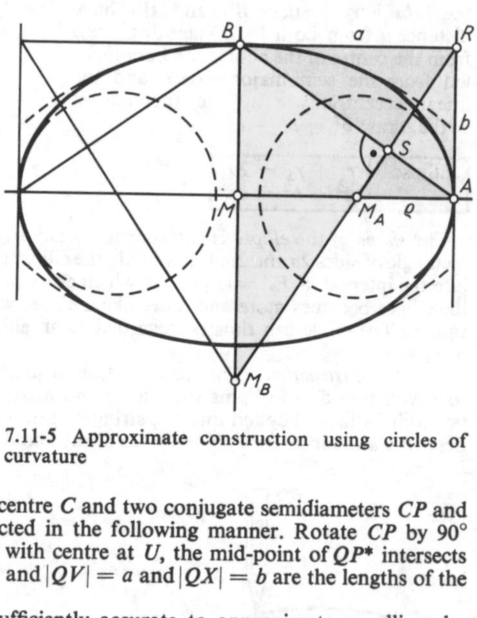

7.11-5 Approximate construction using circles of curvature

The Rytz construction. If the only data are the centre C and two conjugate semidiameters CP and CQ of the ellipse, then the axes can be constructed in the following manner. Rotate CP by 90° about C to obtain CP^*. The circle of radius $|UC|$ with centre at U, the mid-point of QP^* intersects QP^* in X and V. Now CX and CV are the axes and $|QV| = a$ and $|QX| = b$ are the lengths of the semiaxes.

Principal radii of curvature. It is frequently sufficiently accurate to approximate an ellipse by circles whose radii are the radii of curvature at the end-points of the major and minor axes (Fig.).

Their centres M_A and M_B and radii $r = |M_BB|$ and $\varrho = |M_AA|$ can be constructed using only very few auxiliary lines (they can also easily be calculated by the methods of analytic geometry if the axes of the coordinate system are chosen as the axes of the ellipse). Draw the rectangle with vertices $M(0, 0)$; $A(a, 0)$; $R(a, b)$; and $B(0, b)$ and its diagonal $AB(y = -(b/a)\,x + b)$. The perpendicular to this line through $R(y = (a/b)\,x - (a^2 - b^2)/b)$ intersects it in $S(x_S = a^3/(a^2 + b^2)$, $y_S = b^3/({}^2a + b^2))$; $|AS| = b^2/\sqrt{(a^2 + b^2)}$, $|SB| = a^2/\sqrt{(a^2 + b^2)}$ and the axes in the required centres M_A and M_B.

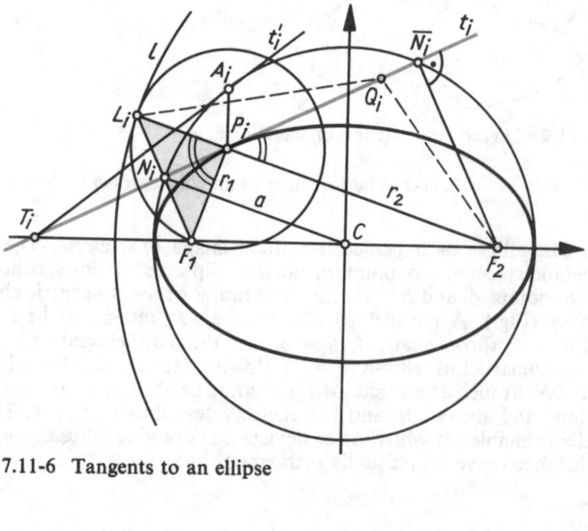

7.11-6 Tangents to an ellipse

The radii are calculated by similar triangles. The lines RSM_B and ASB are intersected by the parallels $RB \| AM$ and $BM_B \| RA$. Thus, $\varrho : a = |AS| : |SB| = b^2 : a^2$, or $\varrho = b^2/a$ and $r : b = |SB| : |SA| = a^2 : b^2$ or $r = a^2/b$.

Tangents to an ellipse. If P_i is an arbitrary point on an ellipse with foci F_1 and F_2 (Fig.), then the *continuation of the radius* $|F_2P_i| = r_2$ to L_i with $|P_iL_i| = |P_iF_1| = r_1$ gives a point L_i whose distance from F_2 is always equal to $r_1 + r_2 = 2a$. The circle of radius $2a$ about F_2 is sometimes called the *leading circle*. The *perpendicular bisector* P_iN_i of the segment F_1L_1 bisects the angle at the apex of the isosceles triangle $F_1P_iL_i$ and is the tangent t_i to the ellipse in P_i, since for every other point Q_i on this line the triangle $F_2Q_iL_i$ gives $|F_2Q_i| + |Q_iL_i| > |F_2L_i| = 2a$, and therefore Q_i is not on the ellipse. This also provides a new definition for the ellipse.

The ellipse is the locus of points in the plane that are centres of circles touching internally a given circle, the leading circle (of radius $2a$ and centre F_2), and passing through a fixed interior point F_1.

From the figure it can be seen that the radii r_1 and r_2 *intersect the tangent* t_i in *equal angles*. Hence, sound or light waves emitted from one focus of the ellipse are reflected to the other focus. In elliptical whispering galleries soft sounds made at F_1 can be clearly heard at F_2, but nowhere else.

If C is the centre of the ellipse, then N_iC bisects the sides F_1L_i and F_1F_2 of the triangle $F_1F_2L_i$ and is hence parallel to the third side and half as long. Hence all the points N_i that are the intersections of tangents with the perpendiculars from F_1 onto them lie on the circle of radius a about C.

The feet of perpendiculars from the foci of an ellipse to its tangents lie on the circumscribed circle whose radius is the semi-major axis a.

If one regards the ellipse as the affine image of the circumscribed circle (Fig. 7.11-6), then P_i is the image of the point A_i on that circle for which A_iP_i is perpendicular to the major axis. The major axis is called the axis of affinity, and the tangent t_i' to the circle at A_i intersects the tangent t_i to the ellipse in P_i in a point T_i on the extended major axis. This point and t_i' can be used to construct t_i.

The area of an ellipse. An affine map $x = x'$, $y = (b/a)\,y'$ transforms the circle of radius a into an ellipse with semi-major axis a and semi-minor axis b. The area formula πa^2 of the circle becomes πab.

Area of ellipse	$A = \pi ab$

The hyperbola

The hyperbola is the locus of points in the plane for which the difference between their distances from two fixed points is a constant $2a$.

The given points F_1 and F_2 are the *foci* of the hyperbola. The intervals r_1 and r_2 from a point on the hyperbola to the foci are the *radial distances* of that point (Fig.). The constant $2a$ must be smaller than the distance between the foci. If two arcs of radii r_1 and r_2 with $r_1 - r_2 = \pm 2a$ are drawn about F_1 and F_2, respectively, their intersections P_1 and P_2 are points of the hyperbola. Interchange of the radii gives two further points P_3 and P_4. The line through F_1 and F_2 is an *axis of symmetry* for P_1 and P_2 and also P_3 and P_4, and indeed for the whole hyperbola. The perpendicular bisector of F_1F_2 is also an *axis of symmetry*. The intersection C of the two axes is a *centre of symmetry* of the hyperbola and is called its *centre*. The distance of the foci from the centre is called the *focal distance* or *linear eccentricity* $e = |CF_1| = |CF_2|$. The intersections of the hyperbola with its principal axis F_1F_2 are called its *vertices* V_1 and V_2, their distance from the centre is a and from the foci $e - a$ and $e + a$, respectively.

Hyperbola	$r_1 - r_2 = 2a, \quad e^2 = a^2 + b^2$

It is shown in analytic geometry that the hyperbola has two *asymptotes*, which intersect the perpendiculars at V_1 and V_2 to the principal axis at a distance b from V_1 and V_2, respectively. Then b can be computed by the formula $b^2 = e^2 - a^2$.

The hyperbola lies completely between the arms of two vertically opposite angles of the asymptotes. For the limiting case $b = 0$, $e = a$, the two half-lines on the principal axis outside $|V_1V_2| = |F_1F_2|$

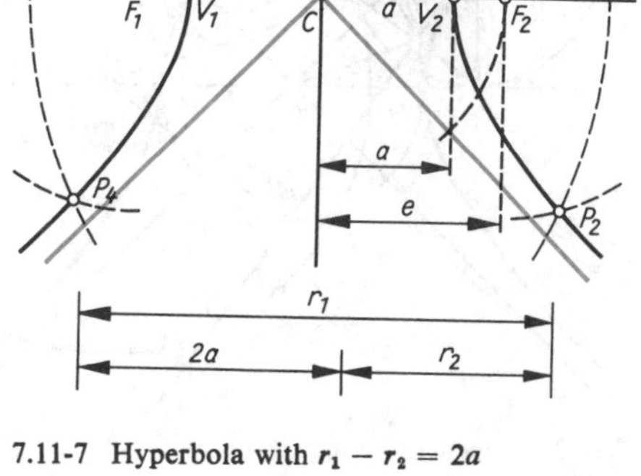

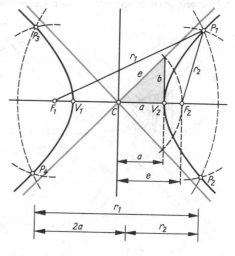

7.11-7 Hyperbola with $r_1 - r_2 = 2a$

each counted with multiplicity 2 form a *degenerate hyperbola*. On the other hand as b becomes large, the curvature of the hyperbola decreases; in the limit $b \to \infty$ the two perpendiculars to the principal axis in F_1 and F_2 are taken to be the degenerate hyperbola. If $a = b$, the asymptotes are perpendicular to one another and the hyperbola is called *equilateral*.

Thread construction. Suppose that the foci F_1 and F_2 of the hyperbola and the segment $2a$ are given. A ruler of length l is fixed at one end to F_1 in such a way that it is free to rotate. A thread of length $k = l - 2a$ is attached to the other end of the ruler and its free end is fastened at F_2. A pencil is hooked behind the thread and pressed against the ruler in such a way that the thread is taut. If the ruler is now rotated and the pencil moved so that it stays against the ruler and the thread stays taut, the pencil describes an arc of a hyperbola (Fig.). This follows from the relations

$$l_2 + l_3 = l, \quad l_1 + l_3 = k \quad \text{and}$$
$$l_2 - l_1 = l - k = 2a.$$

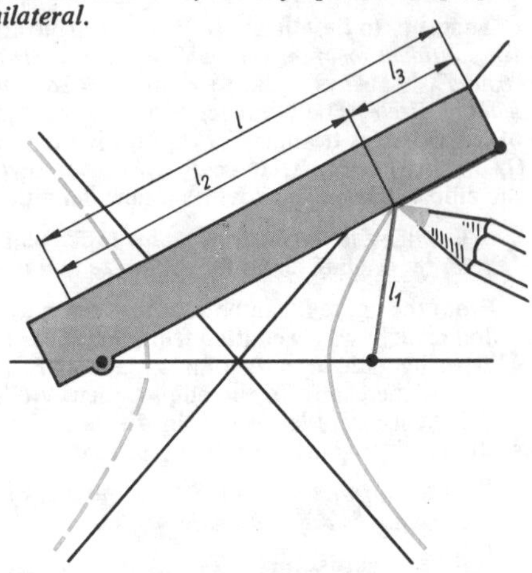

7.11-8 Thread construction of the hyperbola

Tangents to a hyperbola. If P_i is an arbitrary point on the hyperbola with foci F_1 and F_2 and the radius $|P_iF_2| = r_2$ is *shortened* by a segment equal to $|P_iF_1| = r_1$, then the point of division L_i has *constant distance* $|F_2L_i| = 2a$ from F_2. The *perpendicular bisector* P_iN_i of F_1L_i bisects the angle at the apex P_i of the isosceles triangle $F_1L_iP_i$ and is the tangent t_i to the hyperbola in P_i, because for every other point Q_i on the line an inspection of the triangle $F_2Q_iL_i$ shows that $|F_2Q_i| - |Q_iL_i|$ is less than $|F_2L_i| = 2a$, so that none of these points can lie on the hyperbola. This also provides a new definition for the hyperbola.

The hyperbola is the locus of the centres of all circles touching externally a given circle, the leading circle (about F_2 of radius $2a$), and passing through a given exterior point F_1.

From the figure it can be seen that:

A tangent to the hyperbola at a point P_i bisects the angle between the radii through that point.

If C is the *centre* of the hyperbola, then the line CN_i bisects the sides F_1F_2 and F_1L_i of the triangle $F_1F_2L_i$ and hence is *parallel* to the third side and *half as long*. Therefore the feet of the perpendiculars from a focus to the tangents of a hyperbola lie on a circle about C of radius a.

The feet of perpendiculars from the foci to the tangents of a hyperbola lie on a circle touching the hyperbola in its vertices and having the same centre C as the hyperbola.

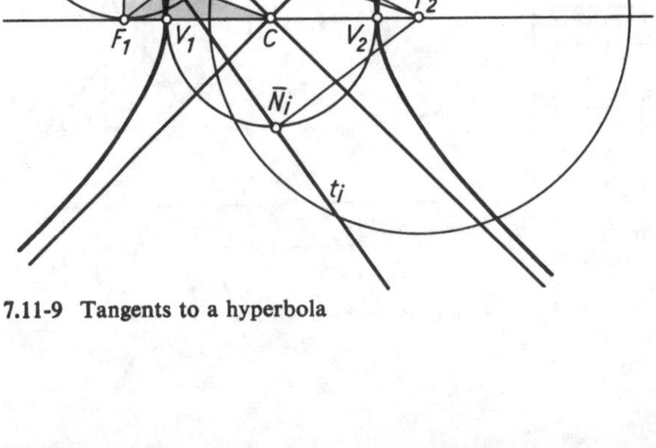

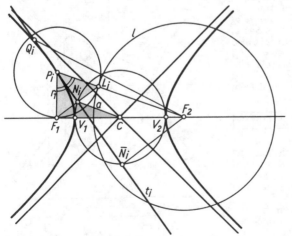

7.11-9 Tangents to a hyperbola

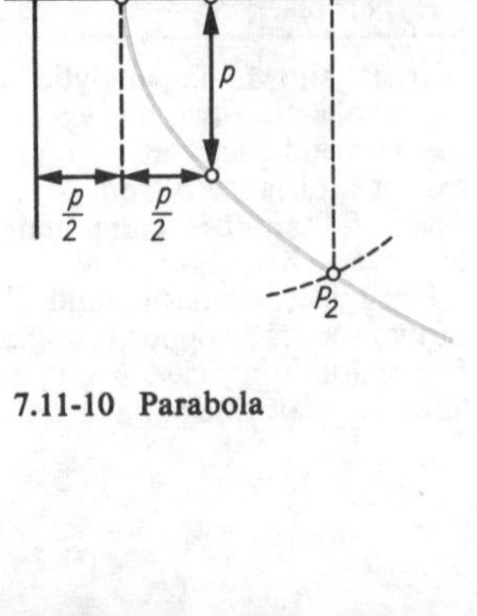

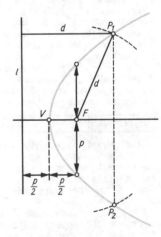

7.11-10 Parabola

The parabola

The parabola is the locus of points in the plane that are equidistant from a given point and a given line.

The point is called the *focus* F and the line is called the *directrix*. The distance from the focus to the directrix is the *semi-parameter* p of the parabola (Fig.). Every parallel to the directrix at a distance d greater than $p/2$ is intersected by a circle of radius d about F in two points P_1 and P_2 on the parabola. The points, and hence the parabola, are symmetrical about the perpendicular from the focus to the directrix. This line is called the *axis* of the parabola and intersects it in its *vertex* V, whose distance from the focus and the directrix is $p/2$. It is clear from the definition of the parabola that the chord through the focus parallel to the directrix has length $2p$.

Thread construction. The parabola also has a *construction using a thread.* A set square (or ruler) with its shortest side AB on the directrix is free to move up and down; a thread of length $|BC|$ is attached to C and its free end is fastened at the focus. If the thread is pushed against the set square with a pencil and held taut, and the set square is moved up and down, the pencil describes an arc of a parabola, because its distance from the directrix is always equal to its distance from the focus (Fig.).

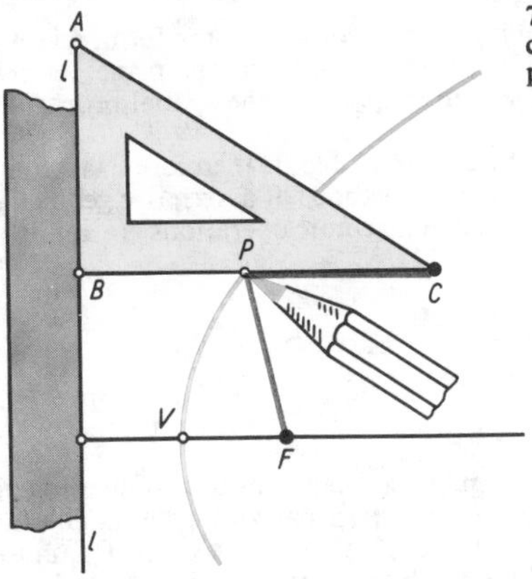

7.11-11 Thread construction of the parabola

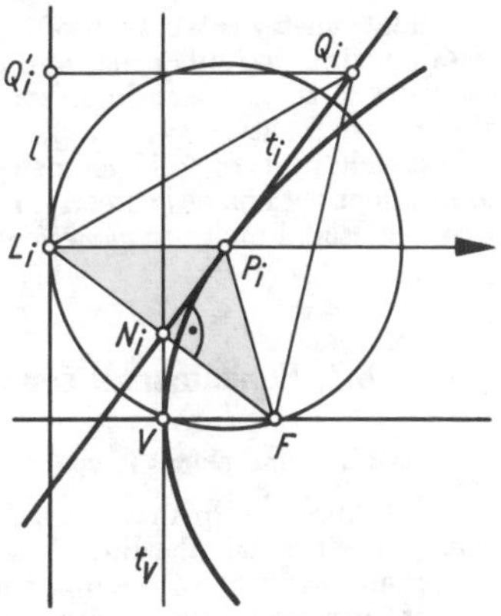

7.11-12 Tangents to a parabola

Tangents to a parabola. If P_i is a point on the parabola, its distance $|P_iL_i|$ from the *directrix* l is equal to $|P_iF|$. The triangle FL_iP_i is hence isosceles, and the *perpendicular bisector* P_iN_i of FL_i bisects the angle L_iP_iF and is the *tangent* t_i to the parabola in P_i. For if Q_i is any other point on t_i, the distance $|Q_iQ_i'|$ from Q_i to the directrix is less than $|Q_iL_i| = |Q_iF|$ (Fig.). This again provides a new definition for the parabola.

The parabola is the locus of the centres P_i of all circles having a given line l (the directrix) as tangent and passing through a given point F.

From the figure it can be seen that the angle between the tangent at P_i and the radius P_iF is the same as the angle between the tangent and the parallel to the axis through P_i. All rays from the focus of the parabola are reflected in such a way that they become parallel to the axis; vice versa, parallels to the axis are reflected into the focus.

Since the point N_i lies on the vertical tangent t_V through the vertex V and this is perpendicular to the axis, it follows that:

The parabola is the envelope of the free arms of all right angles whose vertices are on the vertical tangent and whose fixed arms go through the focus.

If a tangent t_j is perpendicular to t_i and they intersect in T, then FN_iTN_j is a rectangle. Its diagonal N_iN_j is a segment of the vertical tangent t_V and so it must be parallel to the directrix. The triangles N_iN_jF and N_iN_jT are congruent and therefore have the same altitude on the base N_iN_j. Therefore T has the distance $p/2$ from the vertical tangent and lies on the directrix (Fig.).

Pairs of orthogonal tangents to a parabola intersect on the directrix.

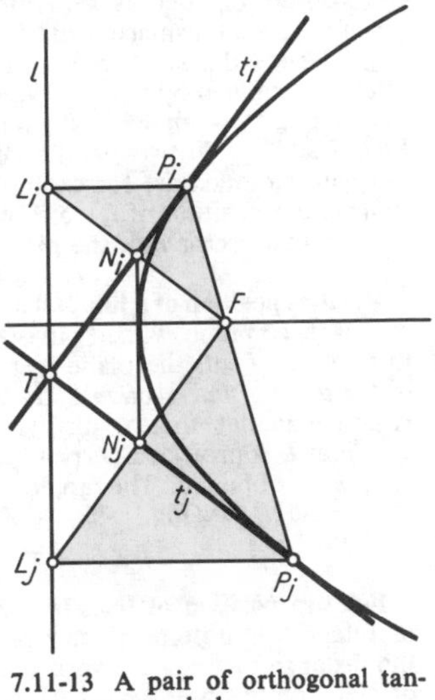

7.11-13 A pair of orthogonal tangents to a parabola

8. Solid geometry

Solid geometry is a branch of Euclidean geometry. Its subject matter consists of the form, relative position, size, and other metric properties of geometric figures that do not lie in one plane. As the geometry of three-dimensional space, solid geometry gives a deep insight into the spatial properties of objective reality.

In certain parts of solid geometry a restriction to a single plane is possible. It therefore has strong connections with plane geometry. Furthermore, in solid geometry the methods of descriptive geometry are often used. Finally, in *numerical solid geometry*, arithmetical and algebraic operations are applied.

8.1. Fundamental concepts

Lines and planes in space

Points, lines, and planes are the foundation stones of elementary geometry in three-dimensional space; in particular, the bounding surfaces of geometric *solids* are often parts of planes. The intuitive interpretation of the fundamental concepts of point and line, given in plane geometry, must be supplemented by the fundamental concept of plane and the relative position of lines and planes in space.

The plane. The family of lines through a fixed point A that intersect a line l_1 not passing through A, or are parallel to l_1, form a plane (Fig.). A plane in space can also be regarded as generated by a line l that is given a parallel displacement along a line l_1 intersecting l. Hence the *position of a plane* in space is uniquely determined by the following subsets: 1. a line l_1 and a point A not lying on l_1; 2. two intersecting lines l and l_1; 3. two parallel lines; 4. three points not lying on one line, for example, A and two points that fix the position of l_1; 5. a point A and a vector (the normal vector $\boldsymbol{n}$ of the plane).

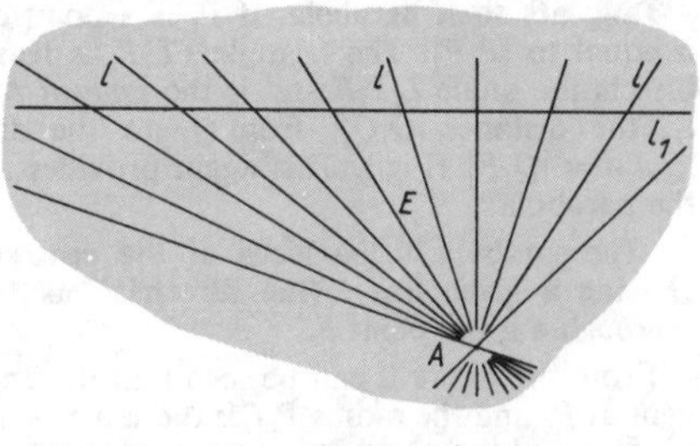

8.1-1 Formation of a plane in space

Relative position of a line and a plane in space. A line l lies entirely in a plane E if it has two points A and B in common with it; it is parallel to E if it has no point in common with E or lies entirely in E. A line l cuts the plane if it has exactly one point in common with it, the *point of intersection L*. It is perpendicular to the plane at L if it is perpendicular to two distinct lines l_1 and l_2 of E. If a line l that cuts E at L is projected perpendicular to E, the normal projection l' of l on E is obtained. The angle of inclination α of l to E is defined by $\alpha = \sphericalangle(l, l')$ (Fig.); if $l \perp E$, then $\alpha = 90°$ and if $g \parallel E$, then $\alpha = 0°$.

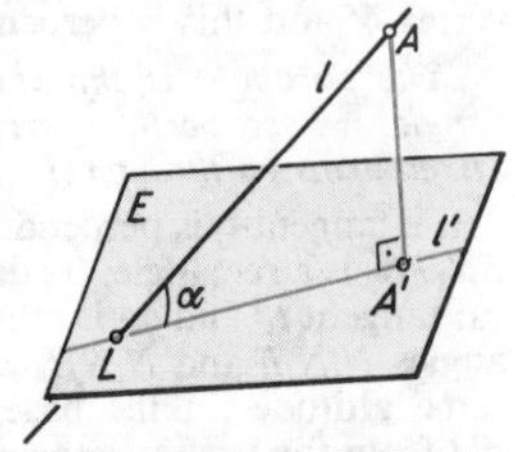

8.1-2 Angle of inclination of a line l to a plane E

Relative position of lines in space. If two lines l_1 and l_2 are parallel or intersect in a point, then a plane can be drawn through them, and the distance or angle between them can be determined by the methods of plane geometry.

Two *skew lines* l_1 and l_2 (Fig.) are not parallel and have no common point. Their angle of inclination α is defined as the angle between one of the lines and the line through one of its points parallel to the other, for example, $l_2' \parallel l_2$ through N on l_1. The line n_{12} perpendicular to l_1 and l_2' at N is perpendicular to l_2. The plane E spanned by l_1 and n_{12} cuts l_2 in D_2, and the line through D_2 parallel to n_{12} cuts l_1 at D_1. The line D_1D_2 is the *common perpendicular* of l_1 and l_2, and the length d of D_1D_2 is the *distance between the skew lines.* It is the shortest distance between any point of l_1 and any point of l_2.

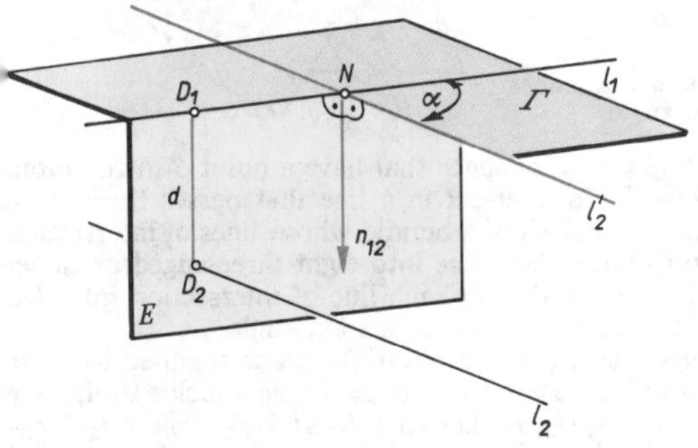

8.1-3 Distance between skew lines

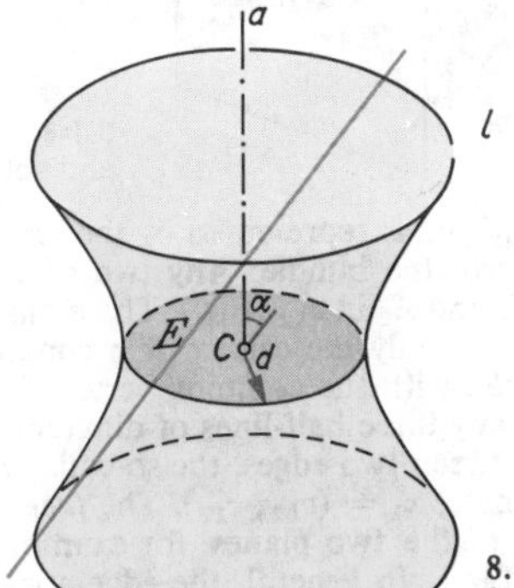

8.1-4 One-sheet hyperboloid of rotation

For any line a in space there are arbitrarily many lines l that make a given angle α with a and have a given distance d from a. If the choice is restricted to those lines l whose common perpendicular with a lies in a plane E perpendicular to a, two families of lines are obtained. They are the generators of a one-sheeted *hyperboloid of rotation* with a as axis of rotation (Fig.). The plane E cuts the hyperboloid in its smallest circle and the axis in the centre of this circle. Furthermore, each line of one system of generators cuts each line of the other system (except the one that is parallel to it), while any two lines of one system are skew. In particular, if $\alpha = 0$, then the surface is a *cylinder of rotation,* which has just one system of generators, parallel to one another.

The set of those lines l that cut a fixed line a at a point Z of a and at a fixed angle α generate a *cone of rotation.* Z is the vertex, a the axis, and the lines l are the generators of the cone.

The set of lines through a point P in space form a *bundle of lines.* An arbitrary plane through P cuts this bundle in a *pencil of lines.* If P is a point at infinity (improper point), then the corresponding systems of lines are a *parallel bundle of lines* in space and a *parallel pencil of lines* in a plane (see Chapter 25.).

Relative position of planes in space. Two planes in space have at most one line in common if they do not coincide. Two planes that have no common point or coincide are said to be *parallel.*

The set of planes that contain a line s form a *pencil of planes,* whose carrier is s (Fig.). A plane Π perpendicular to s cuts the planes of the pencil in lines. The angle of intersection of lines in Π is by definition equal to the angle of intersection of the corresponding planes, for example, $\beta = \sphericalangle(l_1, l_2) = \sphericalangle(E_1, E_2)$. Hence the angle of intersection of two planes is reduced to the angle of intersection of two lines.

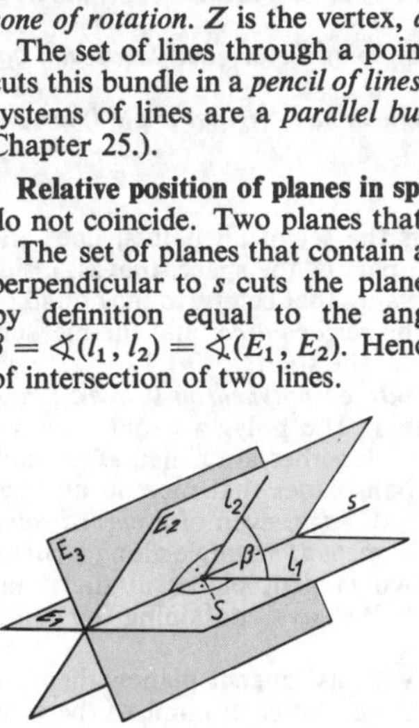

8.1-5 Pencil of planes with a line s as carrier

8.1-6 Parallelism of the lines of intersection c_{12}, c_{13}, c_{23} of three planes E_1, E_2, E_3

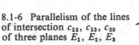

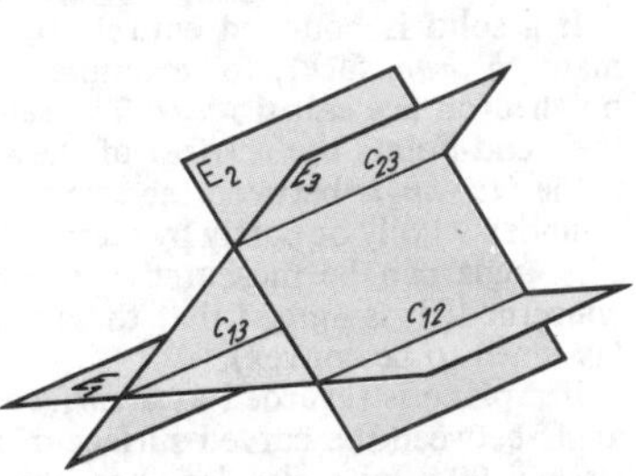

If $\beta = 90°$, the planes are perpendicular. A plane E_2 parallel to the line of intersection c_{13} of two planes E_1 and E_3 cuts these in parallel lines, $c_{12} \parallel c_{13} \parallel c_{23}$ (Fig.). A *parallel pencil of planes* consists of planes such that none of them has a point in common with any other (Fig.). A line perpendicular to one plane of a parallel pencil is perpendicular to each of the planes. The segments cut on the perpendicular line determine the distances of the corresponding planes from one another; for example, $|P_2P_3|$ is the distance between the planes E_2 and E_3.

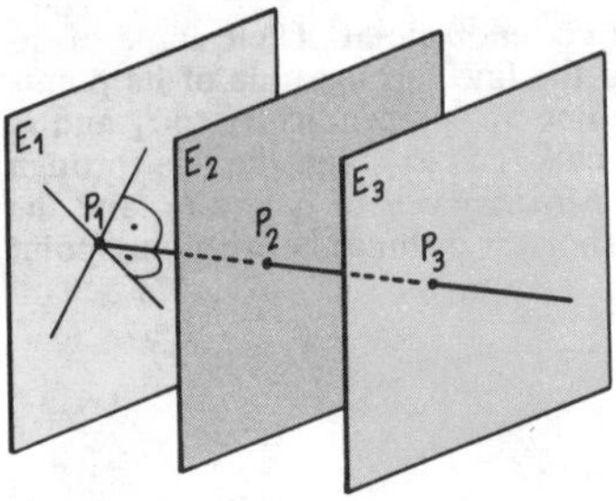

8.1-7 Pencil of parallel planes

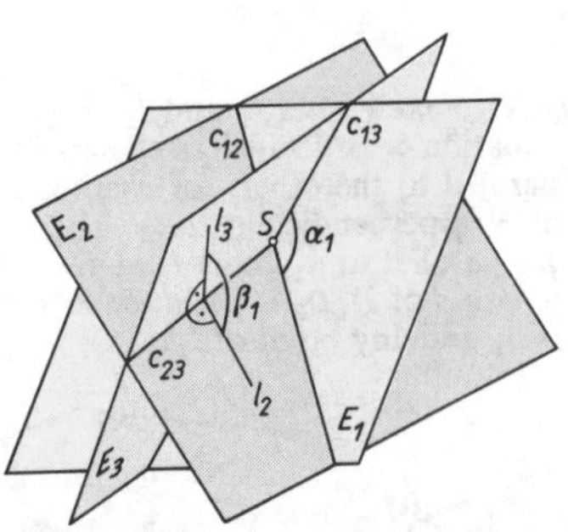

8.1-8 Bundle of planes and solid angle

A *bundle of planes* is represented by the set of all planes of space that have a point S in common. S is the carrier of the bundle. Any two of these planes intersect in a line that passes through S, for example, E_1 and E_2 in c_{12} (Fig.). Three planes E_1, E_2, E_3 of a bundle whose lines of intersection c_{12}, c_{13}, c_{23} have only the carrier S in common divide the space into eight three-edged or three-faced *solid angles* with the common vertex S. The vertex divides each line of intersection into two half-lines, and any three half-lines of different lines form the *edges* of a solid angle.

The angle between two edges, the so-called *edge-angle*, is measured in the plane spanned by these edges, for example, $\alpha_1 = (c_{12}, c_{13})$. The *face-angle* is measured in a plane perpendicular to the line of intersection of the two planes, for example, $\beta_1 = \sphericalangle(l_2, l_3)$ between E_2 and E_3, where $l_2 \perp c_{23}$ and $l_3 \perp c_{23}$ (Fig.). In general, the edge-angles α and the face-angles β are different. Three-edged solid angles whose edge-angles and face-angles are all right angles occur at the vertices of a cuboid. Solid angles of this kind occur in descriptive geometry: they are formed by the three planes of projection.

If perpendiculars are drawn to the faces of a solid angle from an interior point, another solid angle is formed with the point as vertex and the perpendiculars as edges. This new angle is called the *polar angle* of the original angle, which can, in turn, be regarded as the polar angle of the new angle (see Chapter 12.).

Every solid angle is the polar angle of its polar angle.

The following theorems hold for solid angles:

1. The sum of all the edge-angles of an n-edged solid angle is less than 360°.

2. The sum of all the face-angles of an n-edged solid angle is greater than $n \cdot 180° - 360°$ *and less than* $n \cdot 180°$.

3. An edge-angle of the polar angle and the corresponding face-angle of the original angle add up to 180°.

5. A face-angle of the polar angle and the corresponding edge-angle of the original angle likewise add up to 180°.

Solids

Fundamental concepts. A solid in the sense of solid geometry is the set of all points, lines and planes of three-dimensional space that lie inside a complete closed part of the space, that is, inside the bounding surfaces of the solid, including those points, lines and planes that belong to the bounding surfaces. The sum of the areas of the bounding surfaces is called the *surface area*, and the measure of that part of space completely enclosed by it is called the *volume* of the solid.

If a solid is bounded entirely by planes, it is called a *planar body* or *polyhedron* (Greek *polys*, many, *hedron*, face); for example, the cube, cuboid, prism, pyramid. The polygons that bound a polyhedron are called *faces*. The segments in which two faces come together are called *edges*, and their end-points the *vertices* of the solid. The angle between two half-planes that meet at an edge is the *face-angle* between the two faces. In a wider sense one speaks of *edges* even of a *curved solid*, bounded wholly or partly by curved surfaces, when two of its surfaces meet at an angle along a curve. The angle can be measured between the perpendiculars to the two tangent planes at the point concerned; it is agreed that these perpendiculars are to lie in the half-spaces containing the solids (assumed to be convex).

If a plane is regarded as a surface of zero curvature coinciding with its tangent planes, then the angle between the curved surface of a right circular cylinder and its base at each point of the base circle is 90°; for a circular cone, this is the angle of elevation. Examples of curved surfaces without edges are the sphere, ellipsoid, and torus.

The *surface area* of a solid can be determined in principle as the sum of the areas of the individual bounding surfaces. By deriving certain formulae, the process of summing the partial areas, which may be troublesome in practice, can sometimes be avoided.

The *volume* of a solid can be determined with the help of the following tables of cubic content and capacity.

Units of measurement

Cubic content	Symbol	Relations
cubic kilometre	km^3	$1\ km^3 = 10^9\ m^3 = 10^{12}\ dm^3$
cubic metre	m^3	$1\ m^3 = 1\ m^3 = 10^3\ dm^3$
cubic decimetre	dm^3	$1\ dm^3 = 10^{-3}\ m^3 = 1\ dm^3$
cubic centimetre	cm^3	$1\ cm^3 = 10^{-6}\ m^3 = 10^{-3}\ dm^3$
cubic millimetre	mm^3	$1\ mm^3 = 10^{-9}\ m^3 = 10^{-6}\ dm^3$

Cubic content. The cubic metre (denoted by m^3) is the volume of a cube of edge-length 1 m. Larger and smaller units of cubic content are derived from the cubic metre.

In *shipping* various units for measuring content are in use. Weight is measured in long tons of 2240 lb. The *displacement tonnage* is equal (by Archimedes' principle) to the weight of water displaced by a floating vessel. In cargo vessels the *lightweight displacement tonnage* corresponds to the weight of the hull, machinery, equipment, plus the weight of the crew and their effects; the *full displacement tonnage* takes account of the additional maximum weight of bunkers and cargo, and the difference between the two is the *deadweight tonnage*.

Harbour and canal dues are based on the *volumetric ton*, which measures the capacity of the enclosed space, with one ton equal to 100 cu. ft. or 2.83 m^3. Finally, for freight rates on cargo the *freight ton* corresponds to only 40 cu. ft. or 1.13 m^3.

Capacity. Whereas in scientific measurements capacity is assigned by the units of cubic content in common life it is assigned by *litres*. The litre, denoted by l, is defined by $1\ l = 1\ dm^3$. The multiple $1\ hl = 100\ l$ and the parts of one litre $1\ cl = 0.01\ l$ and $1\ ml = 0.001\ l$ are also in use.

In the Anglo-Saxon countries cubic contents and capacities are still based on the cubic yard (see table).

	1 cubic yard	$ft.^3$	$in.^3$	1 gallon (Imp.)	1 gallon (USA)	1 dm^3	1 m^3
1 cubic yard	1	27	46 656	168.2	140.17	764.553	0.765
1 cubic foot	0.037	1	1728	6.229	5.191	28.32	0.028
1 cubic inch	0.000 02	0.000 58	1	0.0036	0.004 329	0.016 4	0.000 16
1 gallon (Imperial)	0.059 45	0.160 5	277.42	1	6/5	4.546	0.004 55
1 gallon (USA)	0.049 54	0.133 7	231	5/6	1	3.788	0.003 79
$1\ l = 1\ dm^3$	0.001 31	0.035 3	61.02	0.22	0.183	1	0.001
1 m^3	1.308	35.314	61 020	220	183	1000	1

8.2. Cube and cuboid

Cubes and cuboids are polyhedra. The *cube* has eight rectangular solid angles, twelve edges of equal length and is bounded by six equal squares.

The *cuboid* has, like the cube, eight rectangular solid angles, and twelve edges, equal and parallel in fours. It is bounded by three pairs of congruent rectangles lying in parallel planes.

The cube is a special case of the cuboid (Fig.).

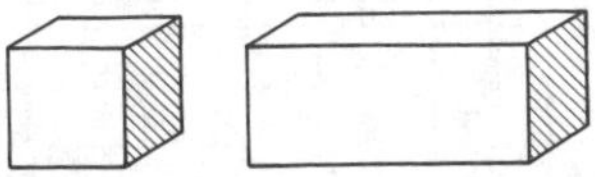

8.2-1 The cube is a special case of a cuboid

Surface area

If a model of the surface of a polyhedron is cut along sufficiently many edges, it can be placed in one plane to form a connected system of bounding surfaces. This is called a *net* of the polyhedron.

Conversely, the net of the polyhedron can be bent along certain edges and stuck together to form the model of the polyhedron.

The *net of the cube* consists of a connected system of six equal squares (Fig.). There are different ways of arranging the squares, one of which is shown in the figure.

The *net of the cuboid* consists of a connected system of three pairs of congruent rectangles. Here too there are

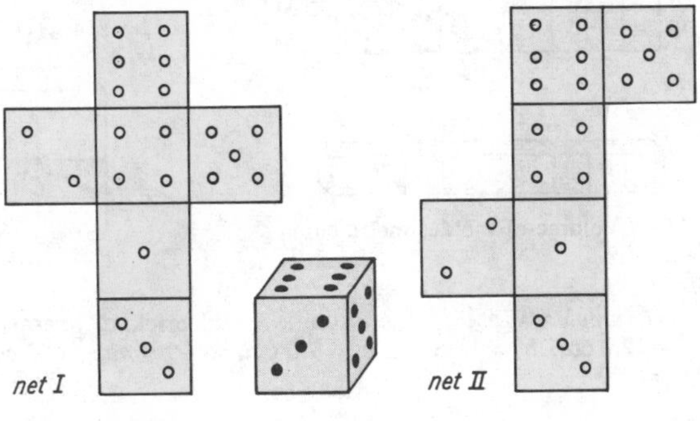

8.2-2 Two nets of a cube

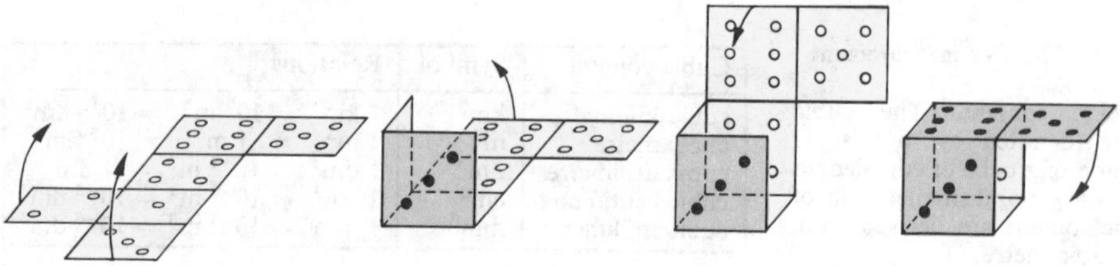

8.2-3 Building a model of a cube from the net

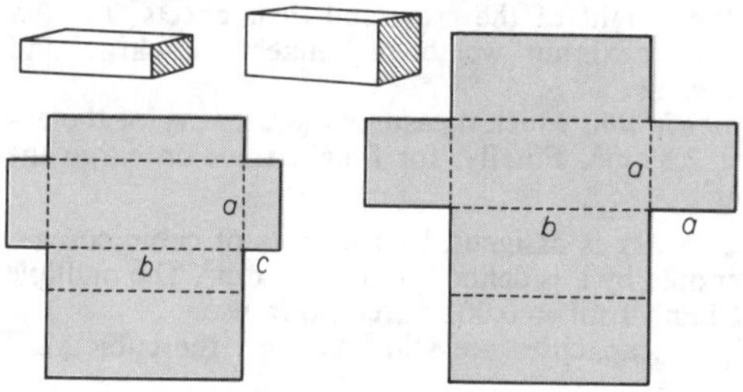

8.2-4 Two nets of cuboids

different ways of arranging them. The figure shows two cuboids; the right-hand one has one pair of square faces; the remaining four faces are then congruent.

If the lengths of the edges of the cuboid are a, b, c, then the areas of the three rectangles are ab, ac and bc, so that the surface area S is given by

$$S = 2ab + 2ac + 2bc = 2(ab + ac + bc).$$

Surface area of a cuboid	$S = 2(ab + ac + bc)$
Surface area of a cube	$S = 6a^2$

A cuboid with one pair of square faces ($a = c$) has surface area $S = 2a^2 + 4ab$. Finally, for a cube, where $a = b = c$, $S = 6a^2$.

Volume

In plane geometry the measure of the area of a figure, for example, a square or rectangle, is defined by covering the figure with unit squares. Similarly, the measure of a volume, for example, of a cube or a cuboid, can be defined by filling the space with unit cubes.

The *volume of a cube* with edge-length a (say 10 units) can be completely filled in this way (Fig.).

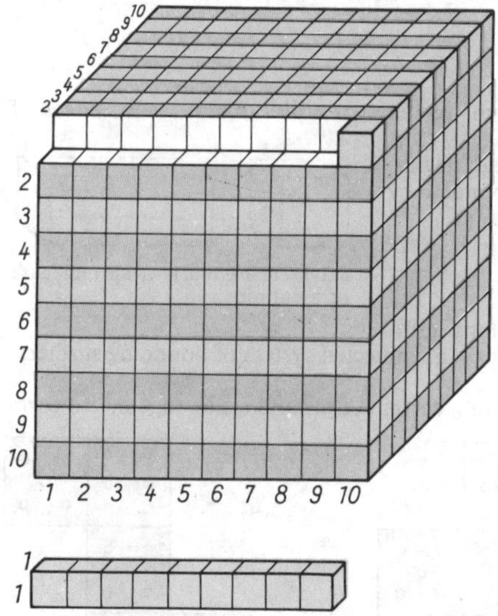

8.2-5 Volume of the decimetre cube

Volume of a cube	$V = a^3$

There are a (= 10) *layers*, each having a (= 10) *rows* of a (= 10) *unit cubes*, and so altogether $a \cdot a \cdot a = a^3$ ($10 \times 10 \times 10 = 1000$) units cubes.

Volume of a cuboid	$V = abc$

The *volume of a cuboid* with edge-lengths a, b, c can be filled by c layers each containing b rows of a unit cubes. The volume therefore amounts to $a \cdot b \cdot c$ unit cubes (Fig.).

For the special case of a cuboid with one pair of square faces, $V = a^2b$.

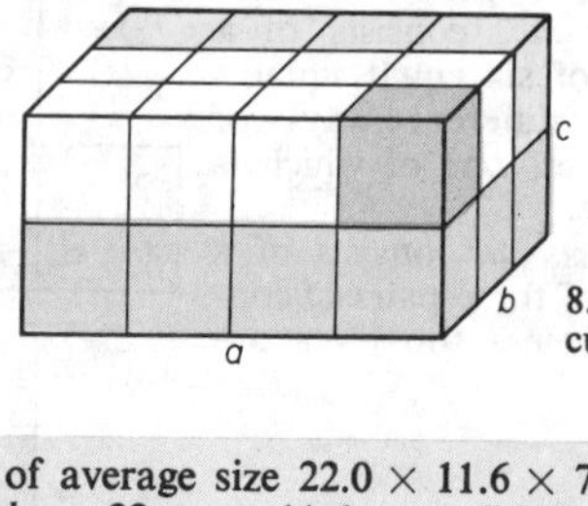

8.2-6 Volume of a cuboid

Example: What is the volume of a brick of average size $22.0 \times 11.6 \times 7.0$ cm? – Here $a = 22.0$ cm, $b = 11.6$ cm, $c = 7.0$ cm, so $V = abc = 22\text{ cm} \times 11.6\text{ cm} \times 7.0\text{ cm} = 1786.4\text{ cm}^3$.

These formulae also hold if the length of one or more edges is not an integral multiple of the edge-length e of the unit cube. If the edge-lengths a, b, c are rational multiples of e, for example, $a = (p_1/q_1)\,e$, $b = (p_2/q_2)\,e$, $c = (p_3/q_3)\,e$, the argument holds for a smaller unit cube whose edge-length e' is the kth part of e, where k is the least common multiple of q_1, q_2, q_3. An irrational multiple of e can be approximated arbitrarily closely by a sequence of rational numbers (see Chapter 3.). In general, the calculation of the volume of a solid whose surface can be defined mathematically is a topic of the integral calculus.

Special relations

Diagonals of a cuboid. One distinguishes between *face-diagonals* and *space-diagonals*, according as the two non-neighbouring vertices that are joined by the diagonal lie in a face of the solid or not. The cuboid has 12 face-diagonals, four each of equal length, and four space-diagonals all of equal length.

The lengths of all the diagonals can be calculated as the hypotenuses of right-angled triangles by means of the theorem of Pythagoras (Fig.). If a, b, c are the three edge-lengths, then the lengths f_1, f_2, f_3 of the face-diagonals are given by

$$f_1 = \sqrt{(a^2 + b^2)}, \quad f_2 = \sqrt{(a^2 + c^2)}, \quad f_3 = \sqrt{(b^2 + c^2)}.$$

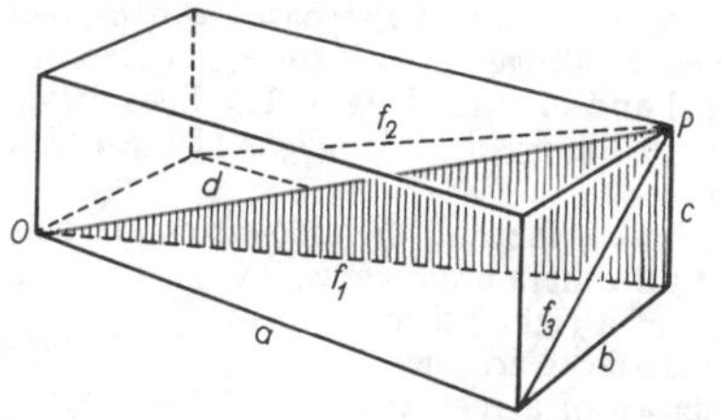

8.2-7 Length of the space diagonals of a cuboid

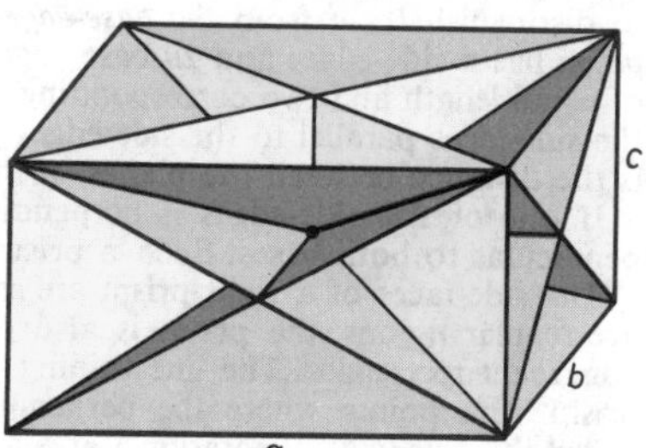

8.2-8 Three pairs of diagonal planes of a cuboid

The length d of the space-diagonals can be calculated as the hypotenuse of right-angled triangle whose sides are a face-diagonal and the third edge:

$$d = \sqrt{(f_1^2 + c^2)} = \sqrt{(f_2^2 + b^2)} = \sqrt{(f_3^2 + a^2)} = \sqrt{(a^2 + b^2 + c^2)}.$$

The four space-diagonals of the cuboid form six *diagonal planes*, which cut the cuboid in rectangles (Fig.). These rectangles are congruent in pairs and are bounded by the face-diagonals and edges of the cuboid. Their areas D_1, D_2, D_3 are given by:

$$D_1 = cf_1 = c\sqrt{(a^2 + b^2)}; \quad D_2 = bf_2 = b\sqrt{(a^2 + c^2)}; \quad D_3 = af_3 = a\sqrt{(b^2 + c^2)}.$$

Cuboid	face-diagonals	$f_1 = \sqrt{(a^2 + b^2)}, f_2 = \sqrt{(a^2 + c^2)}, f_3 = \sqrt{(b^2 + c^2)}$
	space-diagonals	$d = \sqrt{(a^2 + b^2 + c^2)}$

Diagonals of a cube. Since $a = b = c$, the face-diagonals, space-diagonals and area of the diagonal sections are given by $f_1 = f_2 = f_3 = f = \sqrt{(a^2 + a^2)} = a\sqrt{2}$; $d = \sqrt{(a^2 + a^2 + a^2)} = a\sqrt{3}$ and $D_1 = D_2 = D_3 = D = af = a\sqrt{(a^2 + a^2)} = a^2\sqrt{2}$.

The edge-length a, the face-diagonal f and the space-diagonal d are in the ratios

$$a : f : d = a : a\sqrt{2} : a\sqrt{3} = \sqrt{1} : \sqrt{2} : \sqrt{3}.$$

Cube	face-diagonals	$f = a\sqrt{2}$
	space-diagonals	$d = a\sqrt{3}$

Centre. The cube and cuboid both have the property that all the space-diagonals meet in a point C and are bisected there. C is called the *centre* of the solid and is also the *centre of gravity* of a uniform solid of the appropriate type. Finally, C is the common centre of a cube and the circumscribed and inscribed spheres. The radius of the circumscribed sphere is half the length of a space-diagonal, that is, $r = {}^1/_2 a\sqrt{3}$; the radius of the inscribed sphere is equal to half the length of an edge, that is $\varrho = a/2$.

Section through a cube. A plane section of a cube can be arranged so that it is a regular hexagon. Then the centre C of the cube coincides with the centre of the hexagon and the vertices of the hexagon with the midpoints of six edges of the cube, which can be described in one circuit and are such that no three lie in one plane (Fig.).

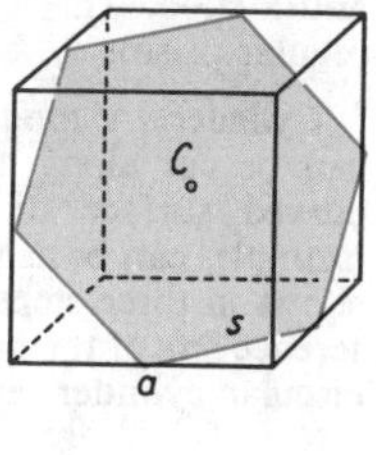

8.2-9 Regular hexagon as plane section of a cube

This regular hexagon consists of six equal equilateral triangles with side-lengths $s = {}^1/_2 a \sqrt{2}$ (half a face-diagonal of the cube) and area $A = {}^1/_4 s^2 \sqrt{3}$. The area S of the hexagon is given by $S = 6A = 6 \times {}^1/_4 s^2 \sqrt{3} = {}^3/_2 s^2 \sqrt{3} = {}^3/_4 a^2 \sqrt{3}$.

8.3. Prism and cylinder

General

Prism. If a line moves in space, without altering its direction, along the perimeter of a plane n-gon ($n = 3, 4, \ldots$), then it describes a *prismatic surface*; if it passes through a vertex of the n-gon, it is an *edge* of this surface.

An n-gon can be interpreted as the section of the prismatic surface by a plane that cuts all its edges. If a second plane, parallel to the first, cuts the prismatic surface, then the section is a second n-gon congruent to the first; the two sections and the prismatic surface completely enclose a part of space. This solid is called a prism (greek, sawn off), the two n-gons are its *bases*, or *base* and *top*, and the part of the prismatic surface belonging to the prism is called its *surface*. The segments of the edges of the prismatic surface that join corresponding vertices of the bases are called *side-edges*, to distinguish them from the *base-edges*, which correspond to the sides of the bases. An *n-sided prism* has n side-edges and $2n$ base-edges, and therefore $3n$ edges altogether. All the side-edges are of equal length and two corresponding base-edges are parallel and of equal length. The lines inside the side-faces parallel to the side-edges are called *generators* of the prism. The *height* of the prism is the distance between the planes of the base and the top.

If one of the side-edges is perpendicular to one of the bases, then all the side-edges are perpendicular to both bases. Such a prism is called *right*, and all others are *oblique*.

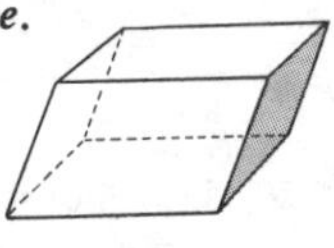

8.3-1 Oblique parallelepiped

The side-faces of a right prism are rectangles. If the bases of a right prism are regular n-gons, the prism is also called *regular*. The side-faces are then congruent rectangles. The line joining the centres of the bases of a regular prism (the points where the perpendicular bisectors of the sides meet) is called the *axis* (axis of rotation), and a section of the prism by a plane containing the axis is called an *axial section*. A skew 4-sided prism with a parallelogram as base (Fig.) is called a *parallelepiped*.

Cylinder. If a line, the *generator*, moves in space, without altering its direction, along a curve, the *guide curve*, then it describes a *cylindrical surface*. A *cylinder* is a solid bounded by a cylindrical surface with a closed guide curve and two planes that are parallel to each other but not to the generators. The segments of the generators of the cylindrical surface between the parallel planes are also called the *generators* of the cylinder, and they are of equal length. The part of the cylindrical surface between the parallel planes is called the *curved surface* of the cylinder. The *base* and the *top*, which are cut on the cylindrical surface by the parallel planes, are congruent. Their perpendicular distance is the *height* of the cylinder. Every cylinder has at least two edges in the wider sense, the boundary of the base and the top. If at each point of these edges the angle between the base or the top and the curved surface is 90°, the cylinder is called *right*, otherwise *oblique*.

According to the type of base there are different types of cylinder. In particular, if the base is a circle, one speaks of a *circular cylinder*. A right circular cylinder is also called a *cylinder of rotation*.

If one thinks of the cylinder as made of solid material, and a smaller cylinder as bored out in such a way that the bases of the cylinders are concentric circles, then the remainder is a *hollow cylinder*. Hollow cylinders are often used in technology, for example as gas-holders, boilers, petrol-tanks, tar-barrels, and so on. *Pipes* are very long hollow cylinders; they are used, for example, to transport gases (natural gas or steam) or liquids (water or petrol).

Surface area

Prism. From a model of the surface of a prism one can obtain its *net* (or conversely, the model can be built from the net), just as for a cuboid and a cube. The figure shows the net of a 6-sided regular prism.

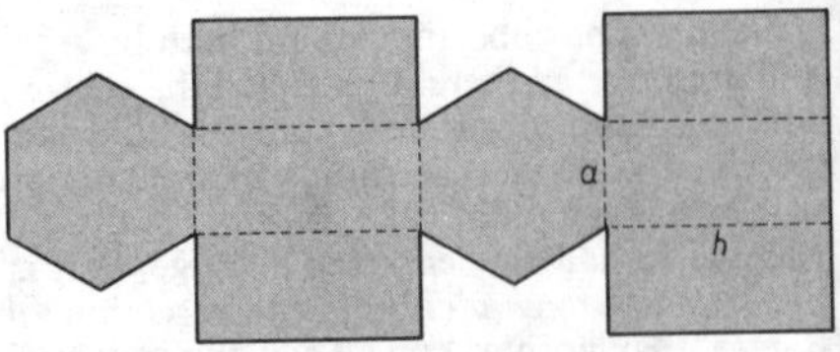

8.3-2 Net of a six-sided regular prism

Cylinder. A model of the surface of a cylinder can be cut along a generator and both bases. The curved surface of a right circular cylinder, for example, can be *developed* into a plane, as the figure shows in three steps. Its net consists of a rectangle with the generator s as height h and the circumference $2\pi r$ of the base as side, and the two circular bases (Fig.). The developed surface of an oblique circular cylinder cut by two parallel planes is bounded by two parallel lines and two sine curves

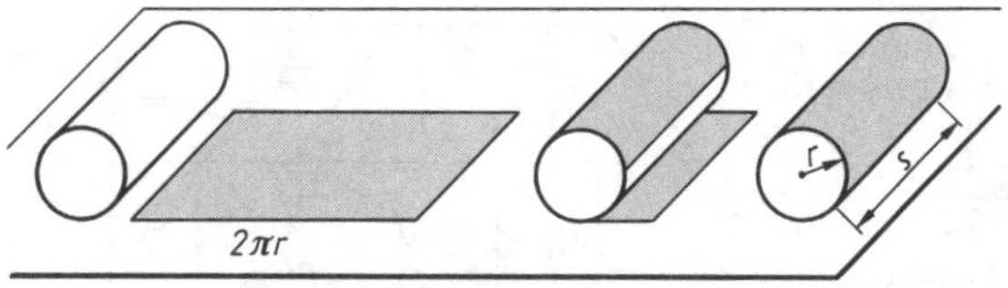

8.3-3 Curved surface of a right circular cylinder

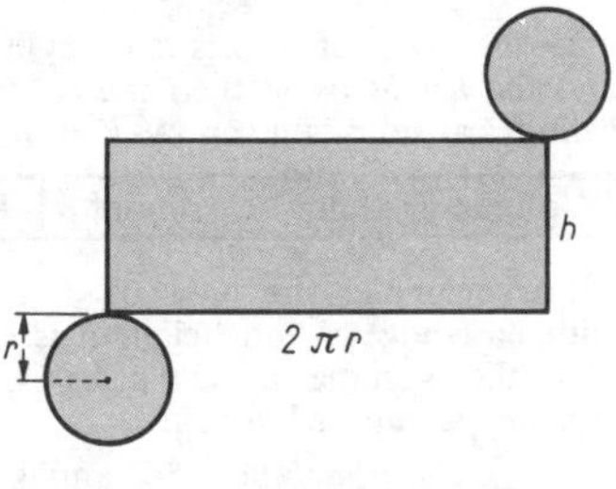

8.3-4 Net of a right circular cylinder

of the same phase and amplitude (Fig.). In principle, every cylindrical surface is developable. In practice, the development even of the right circular cylinder assumes the rectification of the circumference of a circle. An approximation to within 0.002% is given by the construction of A. KOCHANSKY (1685), which is easily carried out by ruler and compass; it gives $\pi \approx \sqrt{(13^1/_3 - 2\sqrt{3})}$ (Fig.).

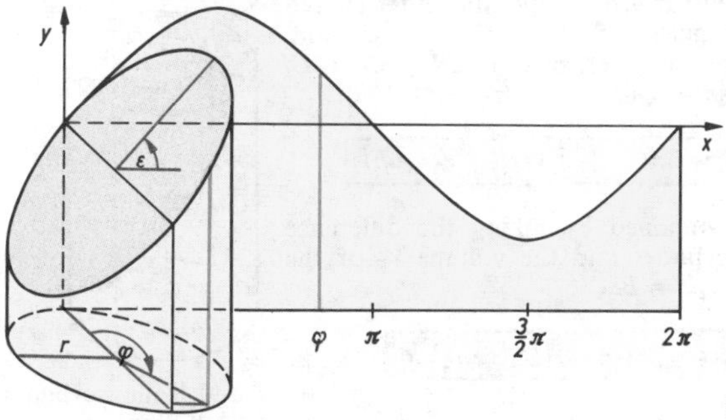

8.3-5 Oblique sections of a right circular cylinder

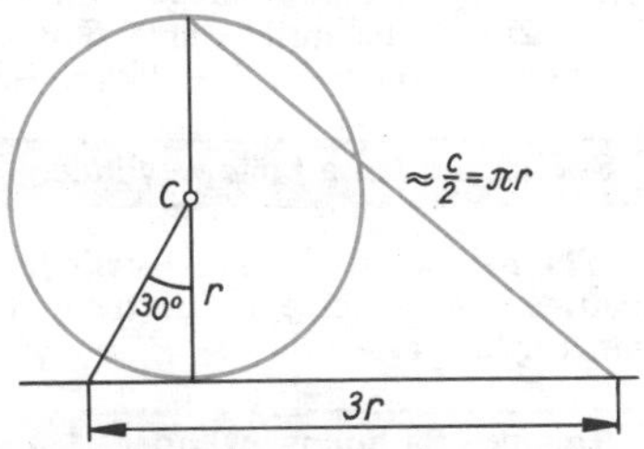

8.3-6 Kochansky's approximate construction for the circumference of a circle

The surface area S of any prism or cylinder can be obtained by adding the area C of the side-faces or curved surface and the areas (each B) of the bases. The formula can be specialized as required.

Surface area for prism and cylinder
$S = 2B + C$

Example 1: To find the surface area of a regular 6-sided prism with base-edge $a = 3u$ and height $h = 4u$, where u is the unit of length. –

$B = {}^3/_2 a^2 \sqrt{3}$ and $C = 6ah$, and so $S = 2 \times {}^3/_2 a^2 \sqrt{3} + 6ah = 3a^2 \sqrt{3} + 6ah = 3a(a\sqrt{3} + 2h)$. By substituting the given values one obtains $S \approx 119u^2$ (square units, say square inches).

Example 2: For the surface area of a steel bolt of circular cross-section with diameter $d = 50u$ and height $h = 60u$, one obtains $B = \pi d^2/4$ and $C = \pi dh$, and so $S = 2\pi d^2/4 + \pi dh = \pi d(d/2 + h)$ and, by substituting the given values, $S = 4250\pi u^2 \approx 13\,352u^2$.

Cavalieri's principle

To calculate the volume of a prism or cylinder one uses a principle published in 1629 by CAVALIERI. a pupil of GALILEI.

Cavalieri's principle: Solids with the same height and with cross-sections of equal area have the same volume; in particular, prisms or cylinders with equal bases and heights have the same volume.

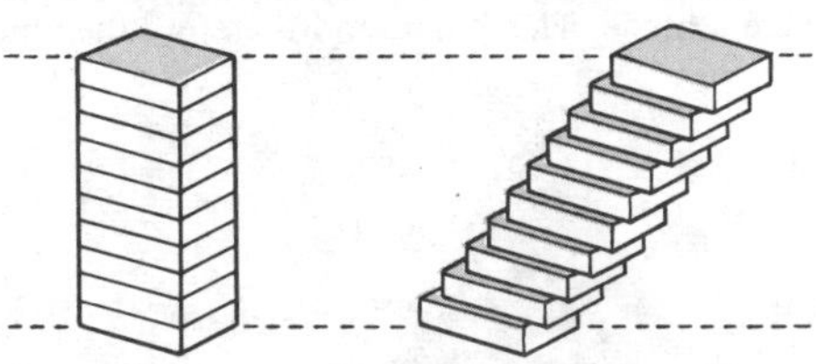

8.3-7 Illustration of Cavalieri's principle

The theorem can be made plausible by elementary methods, by constructing a solid from prismatic sheets of very small height and then, by displacing the sheets, giving it a different form with evidently the same volume (Fig.). The bases of the sheets are the cross-sections, and therefore area at the same height. The smaller the height of the sheets, the more nearly the surface of the 'staircase' approaches the form of a surface described by a continuous function; for example, a pile of equal circular sheets of very thin paper can represent an oblique circular cylinder very closely (Fig.). By means of limiting arguments the theorem can be proved by the methods of the integral calculus.

If the *base* of a prism of cylinder is denoted by B and the *height* by h, then the *volume* of any prismatic or cylindrical solid is given by $V = Bh$.

Volume of prism or cylinder	$V = Bh$

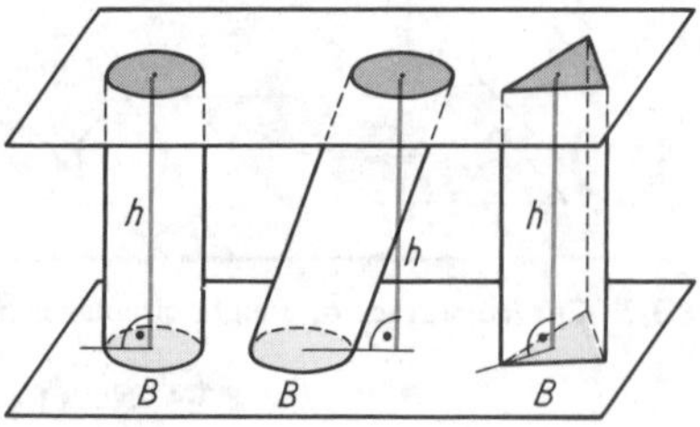

8.3-8 Calculation of volume by Cavalieri's principle

For example, the base of a *regular triangular prism* with base-edge a and height h is *given by* $B = {}^1/_4 a^2 \sqrt{3}$ and the volume is given by $V = {}^1/_4 a^2 h \sqrt{3}$. For a *regular pentagonal prism,*

$$B = (5a^2/4) \cot 36° \text{ and } V = (5a^2h/4) \cot 36°.$$

Volume of circular cylinder	$V = \pi r^2 h = \pi d^2 h/4$

The base and top of a hollow cylinder (Fig.) are congruent circular rings of area $B = \pi r_1^2 - \pi r_2^2$; the inner curved surface C_i and the outer curved surface C_0 are developed into rectangles of area $C_i = 2\pi r_2 h$ and $C_0 = 2\pi r_1 h$; the surface area S is therefore given by $S = 2B + C_0 + C_i = 2\pi(r_1 + r_2)(r_1 - r_2) + 2\pi r_1 h + 2\pi r_2 h = 2\pi(r_1 + r_2)(r_1 - r_2 + h)$.

Surface area of a hollow cylinder	$S = 2\pi(r_1 + r_2)(r_1 - r_2 + h)$

The *volume* of a hollow cylinder is obtained by taking the difference between the volume V_0 of the outer cylinder and the volume V_i of the inner cylinder: $V = V_0 - V_i = B_1 h - B_2 h = Bh$.

Volume of a hollow cylinder	$V = Bh = (r_1 + r_2)(r_1 - r_2) h$

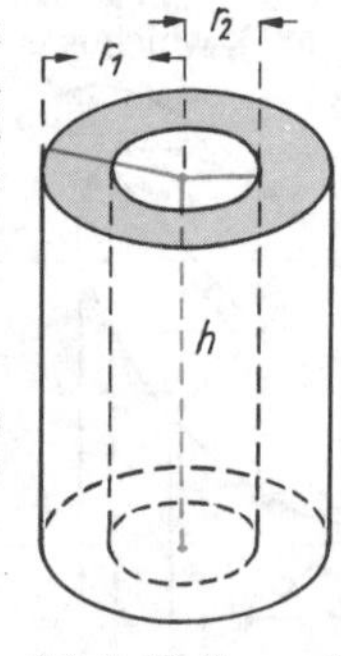

8.3-9 Hollow cylinder

8.4. Pyramid and cone

General

Pyramid. If a ray emanating from a fixed point Z of space moves round the perimeter of a plane n-gon ($n = 3, 4, \ldots$) whose plane does not pass through Z, then the ray describes a pyramidal surface. The rays through the vertices of the n-gon are the *edges* of the pyramidal surface. The n-gon and the part of the pyramidal surface between it and the point Z enclose a completely bounded space; this geometric solid is called a *pyramid* (Fig.).

The n-gon is called the *base*, the point Z the *vertex*, and the part of the pyramidal surface belonging to the solid is called the surface of the pyramid. The segments of the edges of the pyramidal surface that lie between the vertices of the base and the vertex Z are called the *side-edges* of the pyramid, to distinguish them from the *base-edges*, which correspond to the sides of the base. An *n-sided pyramid* has n side-edges and n base-edges and therefore $2n$ edges altogether, and n triangles as *side-faces*. The line-segments in the side-faces that join any point of a base-edge to the vertex are called *generators* of the pyramid.

The *height* of a pyramid is understood to be the distance from the vertex to the plane of the base, measured along the perpendicular. This perpendicular meets the plane of the base in the foot Z' of the altitude. This point, and therefore the altitude, may lie outside the base of the pyramid (Fig.)

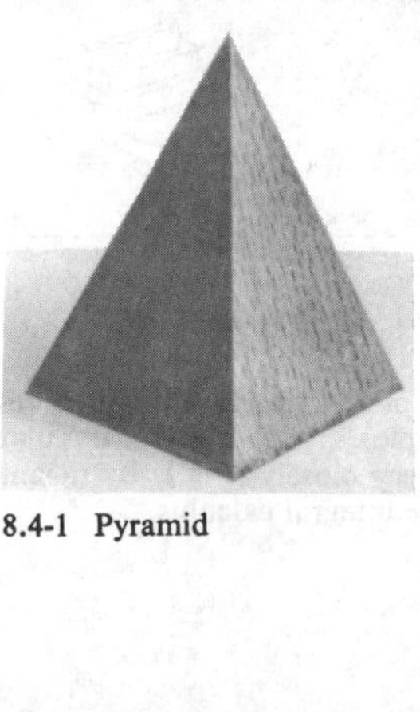

8.4-1 Pyramid

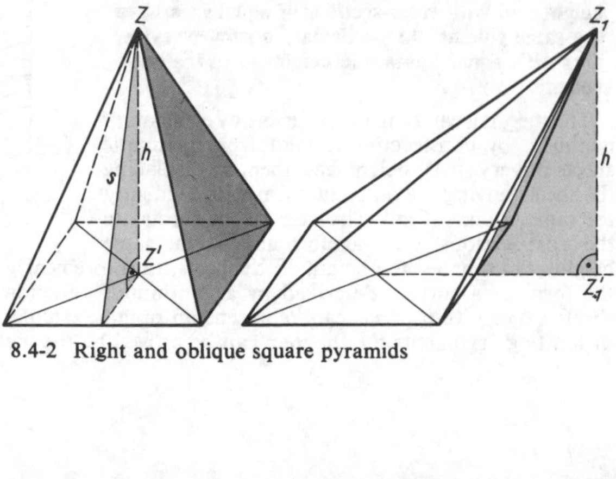

8.4-2 Right and oblique square pyramids

The base of a *regular pyramid* is a regular n-gon; if the foot of the altitude coincides with the centre of the base, the pyramid is called *right*, otherwise *oblique*. The side-faces of a right regular pyramid are congruent isosceles triangles. The height of a right pyramid is its *axis*, and any section by a plane containing the axis is an *axial section*. The (regular) *tetrahedron* is a pyramid whose base- and side-faces are equilateral triangles.

The famous tombs of the ancient Egyptian kings are right square pyramids; the best-known are the pyramids in the southern outskirts of Cairo near Giza. The largest pyramid has a base-edge of approximately 227 m and a height of approximately 137 m.

Cone. A line (*generator*) that passes through a fixed point Z of space and moves along a curve, the *guide-curve*, describes a *conical surface* (Fig.). A *cone* is a solid bounded by a conical surface with *closed guide-curve* and a plane that does not pass through Z. The conical surface cuts the plane in the *base* of the cone. The point Z is called the *vertex* of the cone and its distance from the base is the *height*. The *curved surface of the cone* is the part of the conical surface between the vertex Z and the base. The parts of the generators of the conical surface that belong to the cone are called the *generators* of the cone. One speaks of a *double cone* if a conical surface with closed guide-curve is cut by two parallel planes on opposite sides of Z. The two bases are similar, and the sum of the heights is the distance between the parallel planes. According to the type of base one distinguishes the *circular cone*, *elliptic cone* and other types of cone. If the base has a centre Z' and if the line ZZ' is perpendicular to the base, the cone is called *right*, otherwise *oblique*. A right cone for which all plane sections through ZZ' are congruent can be regarded as a solid of rotation, for example, the right circular cone by rotation of a right-angled triangle about one of the sides of the right angle (Fig.).

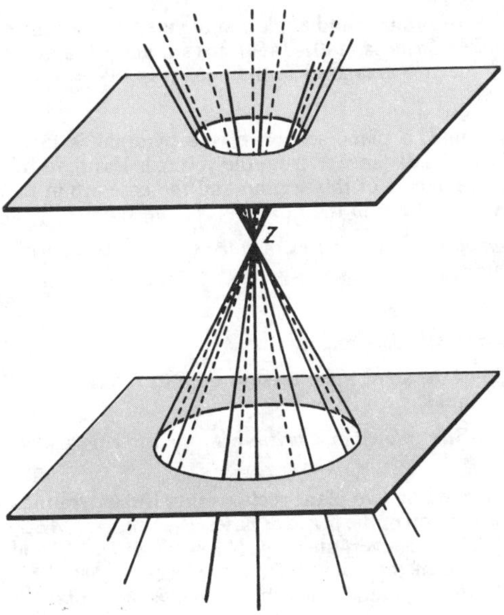

8.4-3 Conical surface and double cone

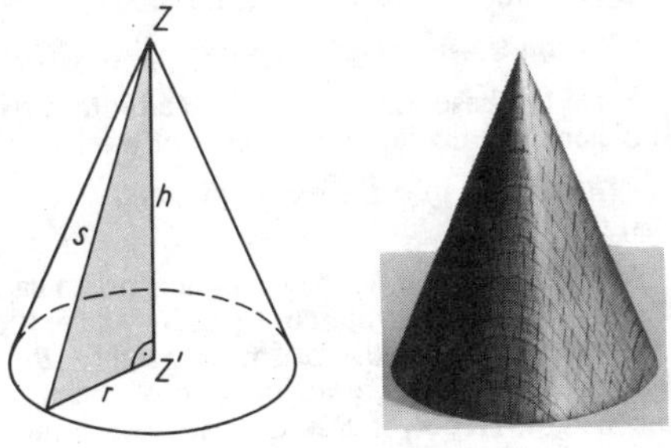

8.4-4 Right circular cone

Examples of *conical shapes in technology* are the roof of a tower, parts of containers (for example the lower part of a cement silo), parts of machinery, conical vawes and conical couplings in powered vehicles. However, conical shapes also occur in nature: many mountains of volcanic origin have the shape of a cone, as well as a heap formed by slowly running sand or soil.

Surface area

The *net of a pyramid* can be obtained from a model of its surface by cutting it along one side-edge and all but one of the base-edges or along all the side-edges and opening out the side-faces into the plane of the base. The figure shows the net of a square pyramid.

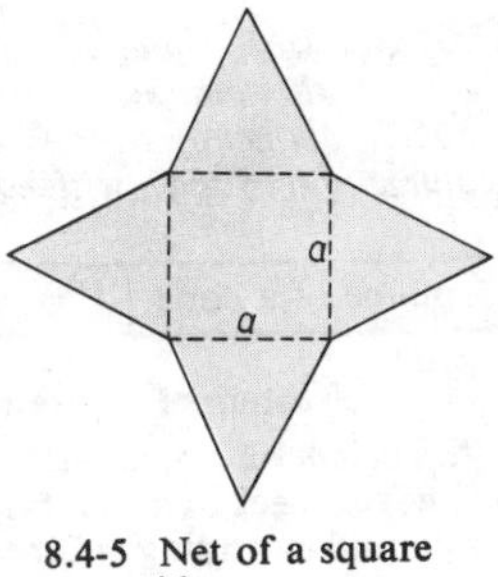

8.4-5 Net of a square pyramid

The model of the surface of a *cone* has to be cut along one generator and the base-edge. For example, the curved surface of a right circular cone can be *developed* into a plane as for a right circular cylinder. It consists of a circular sector (Fig.). The net of a right circular cone consists of the circular sector, whose radius ϱ is the length s of a generator and whose arc b is the length $2\pi r$ of the circumference of the base, and the circular base of the cone itself. If, C is the area of the curved surface, then $C : \pi\varrho^2 = b : 2\pi\varrho$, $C = (b\varrho^2)/(2\varrho) = (2\pi rs)/2 = \pi rs$. If h is the height of the cone, then $s = \sqrt{(r^2 + h^2)}$

Area of the curved surface of a circular cone	$C = \pi rs$

If the surface area of any pyramid or cone is denoted by S, that of the base by B, and that of the side-faces or curved surface by C, then $S = B + C$.

According to the type of solid, these formulae can be specialized. For the surface area of a regular tetrahedron with edge a, $C = 3B$, and since $B = {}^1/_4 a^2 \sqrt{3}$, $S = 4B = {}^1/_4 \cdot 4a^2 \sqrt{3} = a^2 \sqrt{3}$.

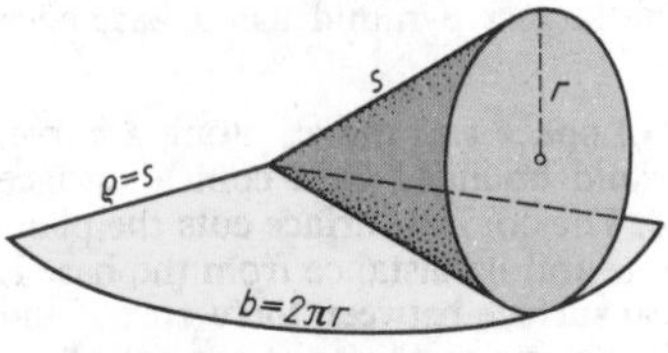

8.4-6 Developing the curved surface of a right circular cone

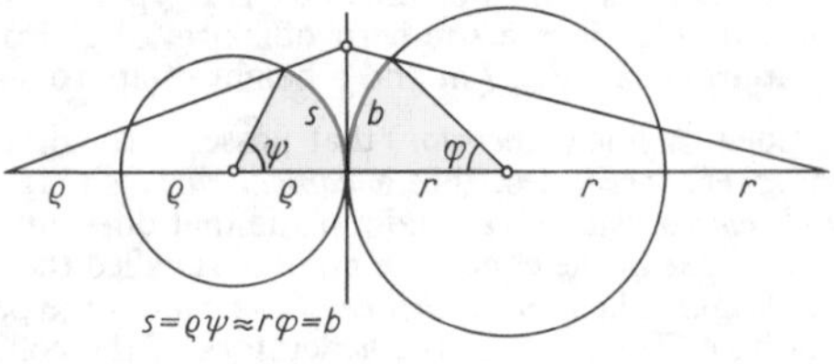

8.4-7 Cusanus' construction of corresponding circular arcs

For the problem of developing the arc s of a circle of radius r and angle φ subtended at the centre onto a circle of radius ϱ, the *construction of Nicolaus Cusanus* (1401–1464) gives a good graphical approximation if the angles φ and ψ subtended by the two arcs are less than 45° (Fig.).

Volume

In order to apply Cavalieri's principle to a pyramid, a plane section of the pyramid is taken parallel to the base. The section is similar to the base. Its distance h' from the vertex is less than the height h of the pyramid. All corresponding parallel segments of this section and the base are in the ratio of the heights, $s' : s = h' : h$, and the areas A' and A are in the ratio $A' : A = h'^2 : h^2$.

The areas of the base of a pyramid and any section parallel to the base are in the ratio of the squares of the corresponding distances from the vertex (heights).

By Cavalieri's principle this implies:

Pyramids with equal bases and equal heights have equal volumes.

Since the base can be transformed into a triangle of the same area, or split up into triangles, it is sufficient to calculate the volume of a triangular pyramid.

The volume of a triangular pyramid is one third of the volume of a prism with the same base and height.

As the figure shows, the triangular prism can be divided by two plane sections into three pyramids V_1, V_2, V_3 of equal volume. V_1 and V_2 have the base and top of the prism as bases, $\triangle DEF = \triangle ABC$, and the height of the prism as height, $|BE| = |CF|$. However, since $\triangle ACF = \triangle AFD$, V_2 and V_3 have equal bases, and for each of them the distance of the point B from the side-face $ACFD$ is the height. Hence, if B is the base and h the height of a pyramid, then the volume is $V = Bh/3$.

Volume of a pyramid	$V = Bh/3$

Cavalieri's principle is concerned only with the equality of area of parallel sections, and not with the form of these sections. A cone can be regarded as a special pyramid with base $B = \pi r^2$ and so:

Cones with equal bases and equal heights have equal volumes.
The volume of a cone is one third of the volume of a cylinder with equal base and height.

Volume of a cone	$V = \pi r^2 h/3 = \pi d^2 h/12$

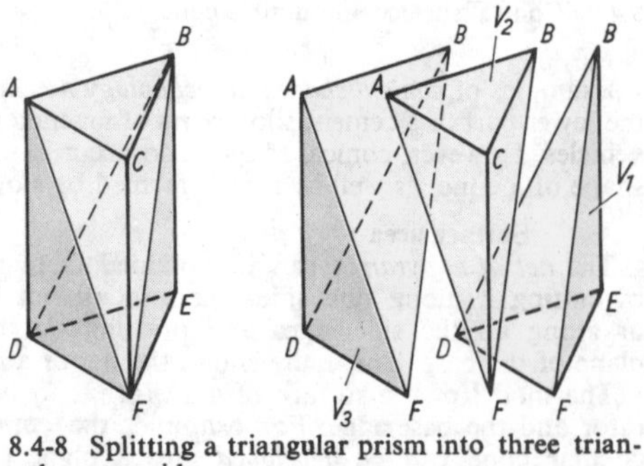

8.4-8 Splitting a triangular prism into three triangular pyramids

Frustum of a pyramid and cone

A *frustum of a pyramid* is a solid bounded by planes, whose base and top are parallel and whose side-edges meet in a point outside the solid (Fig.). A frustum of a pyramid can always be made into a pyramid by adding another pyramid on top.

Area of a frustum of a pyramid	$S = B_1 + B_2 + C$

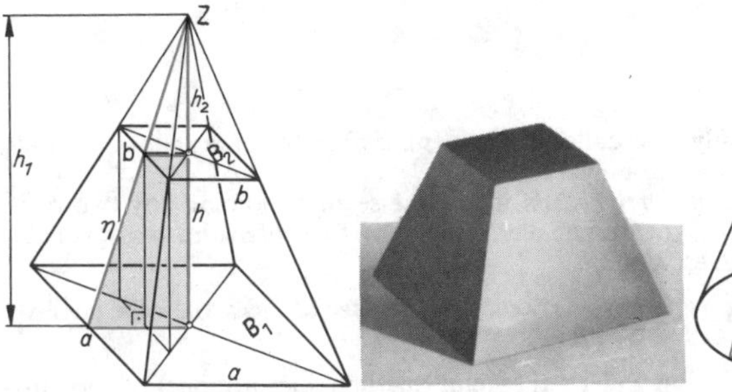

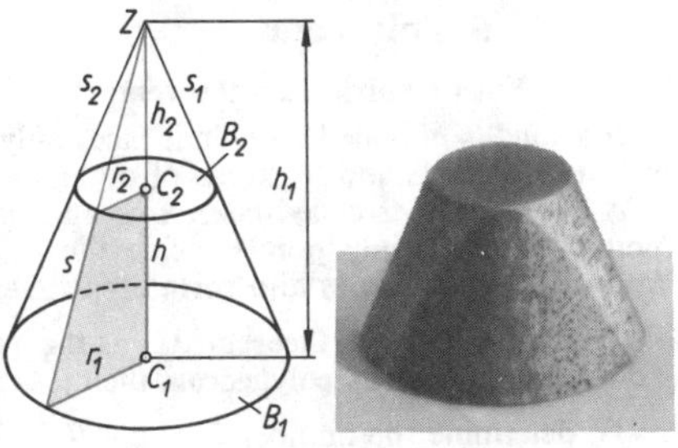

8.4-9 Frustum of a pyramid 8.4-10 Frustum of a cone

If B_1 is the base and h_1 the height of the completed pyramid, and B_2 the base and h_2 the height of the completing pyramid, then $h = h_1 - h_2$ is the height and B_1 and B_2 the areas of the base and top of the frustum. This is called *oblique*, *right*, or *regular* according as the completing pyramid is oblique, right, or regular. Its surface area S is composed of the bases B_1 and B_2 and the side-surface C, which consists of n trapezia, if B_1 is an n-gon. The frustum therefore has $2n$ base-edges and n side-edges. A *right square frustum* of a pyramid with base-edges a and b has isosceles trapezia as side-faces; its height is $\eta = \sqrt{[(h^2 + (a-b)^2/4]}$, and so its surface area is given by $S = 4[(a+b)/2]\,\eta = a^2 + b^2 + 2\eta(a+b)$.

In the same way a *frustum of a cone* can be obtained from a cone by a section parallel to the base (Fig.). Its curved surface C can be developed. If r_1 is the radius of the base-circle, h_1 the height and s_1 the length of generator of a right circular cone and r_2, h_2 and s_2 the corresponding lengths of the cone that is cut off, then for the frustum that is left the height is $h = h_1 - h_2$ and the length of a generator is $s = s_1 - s_2$. In an axial section, $r_1 : r_2 = s_1 : s_2$ and so $(r_1 - r_2) : r_1 = s : s_1$ and $(r_1 - r_2) : r_2 = s : s_2$. Therefore $s_1 = sr_1/(r_1 - r_2)$ and $s_2 = sr_2/(r_1 - r_2)$, hence $C = \pi s_1 r_1 - \pi s_2 r_2 = \pi s(r_1 + r_2)$ and $S = \pi r_1^2 + \pi r_2^2 + \pi s(r_1 + r_2)$ for the surface area of a right frustum of a cone.

Curved surface C and surface area S of a frustum of a right circular cone	$C = \pi s(r_1 + r_2)$ $S = \pi[(r_1^2 + r_2^2) + s(r_1 + r_2)]$ $= {}^1/_4\pi[d_1^2 + d_2^2 + 2s(d_1 + d_2)]$

Many articles in daily use are like a frustum of a pyramid, for example a laundry basket, a baking tin and a wheelbarrow; the form of a frustum of a cone is illustrated by a lampshade, a flower vase and a drinking glass.

Volumes. If B_1 and h_1 are the base and height of a completed pyramid and B_2 and h_2 the base and height of the completing pyramid, then the volume of the frustum of a pyramid is given by $V = (1/3)\,(B_1h_1 - B_2h_2)$. Since the areas of parallel sections are proportional to the squares of their distances from the vertex, $h_1 : h_2 = \sqrt{B_1} : \sqrt{B_2}$. Hence $h_1 = h\sqrt{B_1}/(\sqrt{B_1} - \sqrt{B_2})$ and $h_2 = h\sqrt{B_2}/(\sqrt{B_1} - \sqrt{B_2})$. The volume V of a frustum of a pyramid is therefore given by

$$V = \frac{h}{3} \cdot \frac{B_1\sqrt{B_1} - B_2\sqrt{B_2}}{\sqrt{B_1} - \sqrt{B_2}} = \frac{h}{3} \cdot \frac{B_1^2 - B_2\sqrt{(b_1B_2)} + B_1\sqrt{(B_1B_2)} - B_2^2}{B_1 - B_2}$$

$$= \frac{h}{3}(B_1 + \sqrt{(B_1B_2)} + B_2).$$

A corresponding relation holds for a frustum of a cone with $B_1 = \pi r_1^2$ and $B_2 = \pi r_2^2$.

	Volume	Approximate formula
Frustum of a pyramid	$V = \frac{h}{3}(B_1 + \sqrt{(B_1B_2)} + B_2)$	$V \approx \frac{B_1 + B_2}{2}h$
Frustum of a cone	$V = \frac{\pi h}{3}(r_1^2 + r_1r_2 + r_2^2)$	$V \approx \frac{\pi h}{2}(r_1^2 + r_2^2)$ or $V \approx \frac{\pi h}{4}(r_1 + r_2)^2$

In practice, sufficiently accurate results can often be obtained from the *approximate formulae*. These results are the more accurate the more the form of the frustum of a pyramid approximates to that of a prism ($B_1 \approx B_2$) or that of the frustum of a cone approximates to that of a cylinder ($r_1 \approx r_2$). The first two approximation formulae always give results that are too large, and the other too small.

8.5. Polyhedra

Euler's polyhedron theorem

If a solid is bounded by planar faces only, it is called a *planar solid* or *polyhedron*; cubes, cuboids, prisms, pyramids and frustums of pyramids are polyhedra.

A polyhedron is called *convex* or an *Euler polyhedron* if the line-segment joining any two of its points contains only points inside the polyhedron. Euler's theorem for polyhedra was probably known to ARCHIMEDES and certainly to DESCARTES.

Euler's polyhedron theorem: If v is the number of vertices, f the number of faces and e the number of edges of a convex polyhedron, then $v + f - e = 2$.

To determine the number $E = v + f - e$ in Euler's theorem, one imagines a model of the polyhedron covered with a rubber sheet of which one face has been cut out. The number φ of faces remaining is $f - 1$, so that $E = v + \varphi + 1 - e$. If the remaining surface is spread out into one plane, the edge-lengths and angles are altered, but not v, φ and e. In the *Schlegel diagram* of the polyhedron so obtained (Fig.) each of the φ faces can be divided into triangles by means of diagonals. Each diagonal increases e and φ by 1, that is, E remains constant. If then one edge belonging to exactly one triangle is removed from the boundary, then e and φ are decreased by 1, and again E remains constant. If one edge and one of its vertices, which no longer belong to any face, are removed, then v and e are decreased by 1 and E remains constant. By repeatedly applying these steps, one triangle is finally left, for which $v = 3$, $e = 3$ and $\varphi = 1$, so that $E = v + \varphi + 1 - e = 2$. Hence $v + f - e = 2$ is generally true.

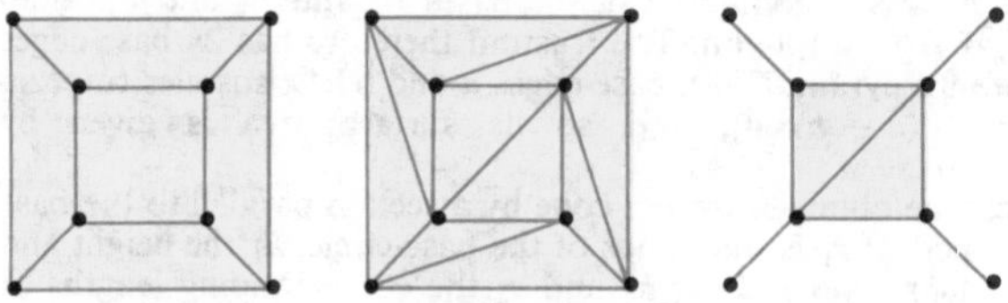

8.5-1 Plane net of a cube to prove Euler's polyhedron theorem

Regular polyhedra

The five regular polyhedra. A convex polyhedron is called *regular* if it is bounded by regular congruent polygons and the same number of edges meet at each vertex. The five solids of this kind (Fig.) are also called *Platonic*, after PLATO.

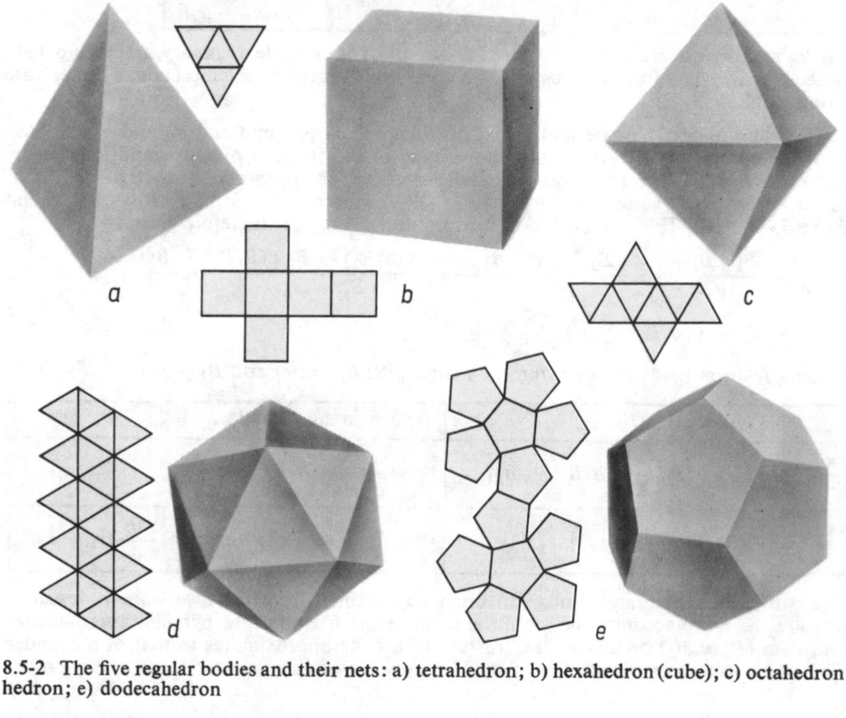

8.5-2 The five regular bodies and their nets: a) tetrahedron; b) hexahedron (cube); c) octahedron; d) icosahedron; e) dodecahedron

By the theorem that the sum of the edge-angles at a vertex is less than 360°, there can only be five regular solids. 1. If the polyhedron is bounded by *equilateral triangles*, then, since the edge-angles are 60°, only three, four or five faces can meet at a vertex; if there were six faces, the sum of the edge-angles would be $6 \cdot 60° = 360°$. 2. If the polyhedron is bounded by *squares* (each edge-angle 90°) or 3. by *regular pentagons* (each edge-angle 108°), then only three faces can meet at a vertex.

Regular hexagons (each edge-angle 120°) are impossible, since $3 \cdot 120°$ is not less than 360°.

These arguments lead to five possible types of regular solid, and the numbers v of vertices, f of faces and e of edges are summarized in the following table, n denoting the number of faces at a vertex:

Bounding faces	n	v	f	e	Regular solid
equilateral triangles	3	4	4	6	tetrahedron
equilateral triangles	4	6	8	12	octahedron
equilateral triangles	5	12	20	30	icosahedron
squares	3	8	6	12	hexahedron or cube
regular pentagons	3	20	12	30	dodecahedron

The *inscribed and circumscribed spheres* are an important characteristic of this class of solids, since the centre of a regular polyhedron is also the common centre of these spheres. The surface of the circumscribed sphere passes through the vertices of the polyhedron, and the surface of the inscribed sphere touches each face at its centre. Hence: the perpendiculars to the faces at their centres meet at the centre of the polyhedron.

If n is the number of sides of a face, m the number of edges that meet at a vertex, v the number of vertices of the polyhedron, f the number of faces and e the number of edges, then if a is the length of an edge, S the surface area and V the volume, the following results are obtained:

Regular solid	n	m	v	f	e	S	V
tetrahedron	3	3	4	4	6	$a^2\sqrt{3}$	$\frac{1}{12}a^3\sqrt{2}$
hexahedron	4	3	8	6	12	$6a^2$	a^3
octahedron	3	4	6	8	12	$2a^2\sqrt{3}$	$\frac{1}{3}a^3\sqrt{2}$
dodecahedron	5	3	20	12	30	$3a^2\sqrt{[5(5+2\sqrt{5})]}$	$\frac{1}{3}a^3(15+7\sqrt{5})$
icosahedron	3	5	12	20	30	$5a^2\sqrt{3}$	$\frac{5}{12}a^3(3+\sqrt{5})$

Duality. The crosses in the table mean that the solids connected in pairs are *dual*. The numbers of vertices and faces are interchanged; the figure shows the example of the cube and octahedron. The number of edges remains the same, by Euler's theorem: $v_1 + f_1 = f_2 + v_2 = e + 2$.

The tetrahedron is self-dual.

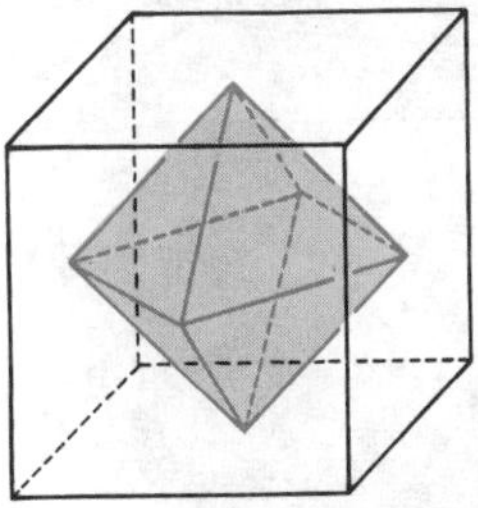

8.5-3 Duality between cube and octahedron

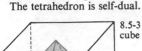

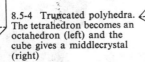

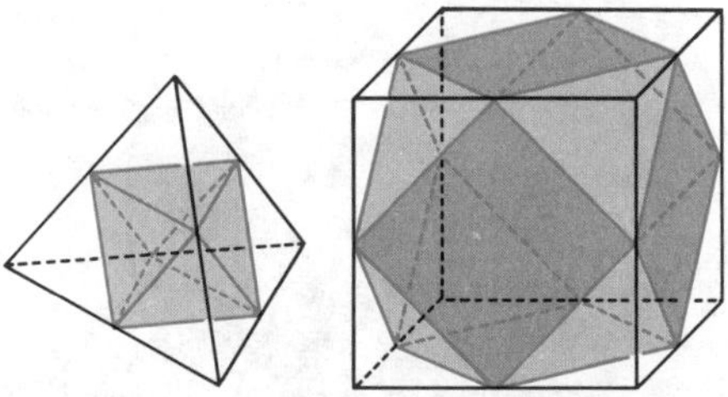

8.5-4 Truncated polyhedra. The tetrahedron becomes an octahedron (left) and the cube gives a middlecrystal (right)

Truncated polyhedra. If the vertices of a regular polyhedron are cut off in such a way that the plane sections are regular and congruent, then the remaining solid is again a regular polyhedron or a *semiregular* (Archimedean) solid, according as all the faces of the truncated solid are congruent, or regular n-gons with different numbers of vertices meet at each vertex of the solid (Fig.). The truncated cube in the figure is called a *middle crystal*, since it can be generated equally well from an octahedron by taking sections through the mid-points of the edges. However, a cube can be so truncated that a regular octagon is obtained on each face.

Crystals

While most solids that occur naturally are irregular, crystals arise directly as mathematical solids. Niels STENSEN (1638–1686) discovered the *law of constancy of angle*, according to which the angle

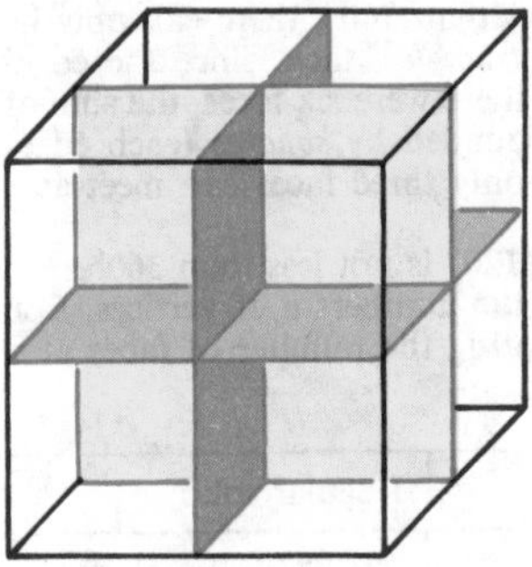

between corresponding faces of all crystals of the same material has the same value. To describe the symmetry properties of crystals one needs the ideas of *centre of symmetry*, *axis of symmetry*, and *plane of symmetry*. For the cube there are three axes of symmetry joining the centres of opposite faces, and nine planes of symmetry. The three *principal planes of symmetry* or briefly *principal planes* pass through the centre of the cube and are parallel to pairs of opposite faces. Perpendicular to each of these are two planes of symmetry, each of which contains two space-diagonals of the cube (Fig.).

8.5-5 Principal planes of the cube

8.6. Sphere

General

If a semicircle is rotated about its diameter, then its circumference describes a *spherical surface* or sphere. The part of space completely enclosed by a spherical surface is called a *sphere*. A spherical surface is the locus of all points of space that have a constant distance from a fixed point of space. The fixed point is the *centre* of the sphere. The spherical form plays an important part in everyday life; one can think of a ball bearing, a ball-and-socket joint, a toy ball, and the celestial sphere.

Relative position of a line or plane and a sphere. A line has 0, 1 or 2 points in common with a spherical surface.

Secants, chords. A secant cuts the sphere in two points. The chord is the segment of it that contains no points outside the sphere. The longest chord is a *diameter* of the sphere; it is bisected at the centre C of the sphere. Every segment from C to a point of the spherical surface is a *radius*.

Tangents, tangent planes. A tangent t to the sphere has exactly one point in common with the sphere, the point of contact B of t (Fig.). The pencil of planes through t cuts the sphere in circles, which all touch t at B. That plane which contains the centre cuts the sphere in a great circle. The plane of the pencil perpendicular to this is the *tangent plane* to the sphere with B as point of contact.

A plane that is not a tangent plane either misses the sphere or cuts it, in general, in a *small circle* (Fig.), but a *great circle* if the plane passes through C.

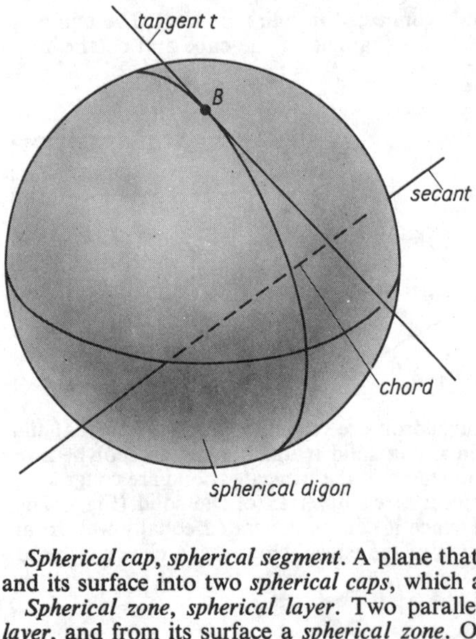

8.6-1 Sphere and line

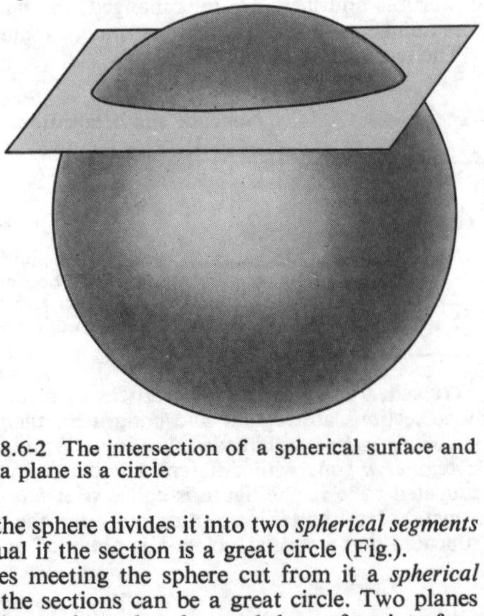

8.6-2 The intersection of a spherical surface and a plane is a circle

Spherical cap, spherical segment. A plane that cuts the sphere divides it into two *spherical segments* and its surface into two *spherical caps*, which are equal if the section is a great circle (Fig.).

Spherical zone, spherical layer. Two parallel planes meeting the sphere cut from it a *spherical layer*, and from its surface a *spherical zone*. One of the sections can be a great circle. Two planes having a diameter in common divide the sphere into four *spherical wedges* and the surface into four *spherical digons*. Two opposite wedges or digons are congruent.

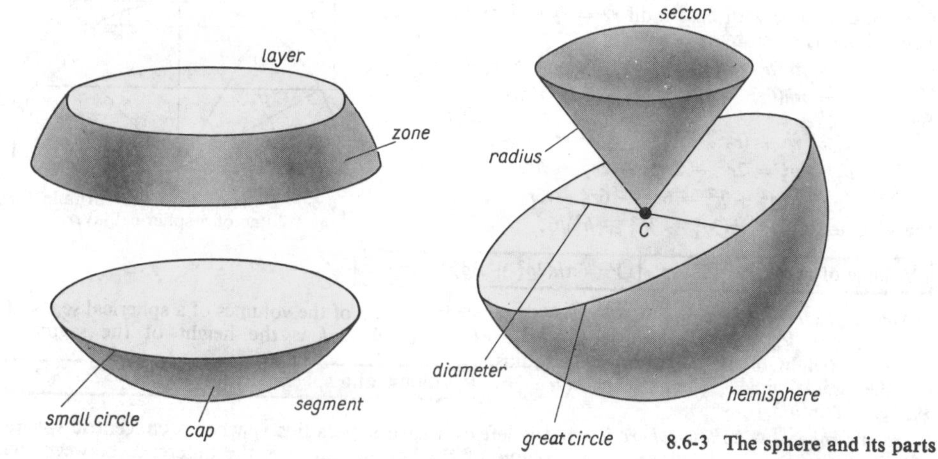

8.6-3 The sphere and its parts

Spherical sector. If a radius of the sphere moves along a small circle of the sphere as guide-curve, then it describes a conical surface and divides the sphere into two *spherical sectors*. If the guide-curve is a great circle, then the surface separating the two spherical sectors degenerates into the plane of the great circle, and the spherical sectors are hemispheres.

Volume

Volume of a sphere. By Cavalieri's principle a hemisphere of radius r has the same volume as a *right circular cylinder* of radius r and height r from which a *right circular cone* with the same base-radius r and the same height r has been bored (Fig.).

A plane at an arbitrary distance $r_1 (r_1 < r)$ from the base and parallel to it cuts the hemisphere in a circle of radius $\varrho_1 = \sqrt{(r^2 - r_1^2)}$ and the remaining solid in a *circular ring* with radii r and r_1. The *cross-sections* therefore have the areas $A_1 = \pi\varrho_1^2 = \pi(r^2 - r_1^2)$ and $A_2 = \pi r^2 - \pi r_1^2$, that is, the *same area*. This shows that the hemisphere has the volume $V_H = \pi r^3 - \pi r^3/3 = 2\pi r^3/3$ (Fig.).

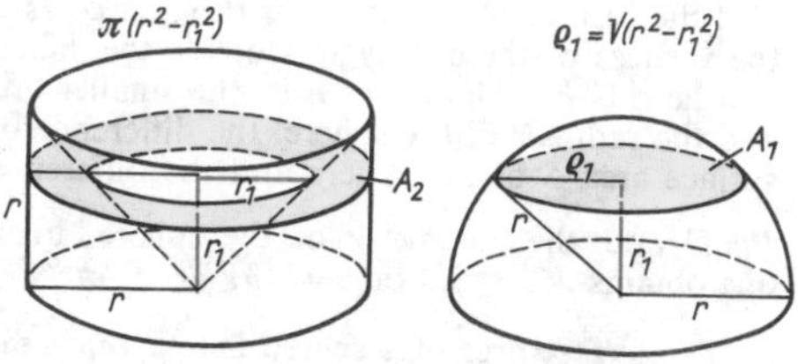

8.6-4 Derivation of the formula for the volume of a sphere

Volume of a sphere	$V = 4\pi r^3/3$

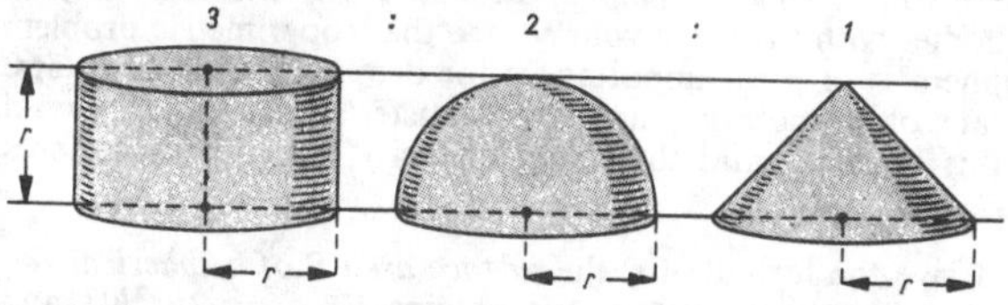

8.6-5 The volumes of these three bodies are in the ratios 3 : 2 : 1

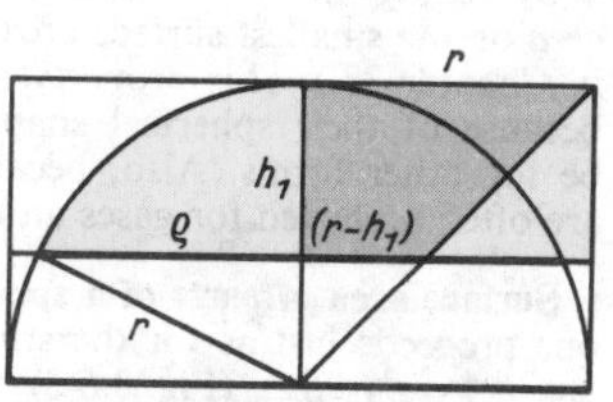

8.6-6 Derivation of the formula for the volume of a spherical segment

Volume of parts of a sphere. *Spherical segment*. The formula for the volume of a spherical segment (Fig.) is derived by the same principle (comparison of the hemisphere with the residue of the cylinder). Here, instead of the cone it is a frustum of a cone:
$V = \pi r^2 h_1 - {}^1/_3\pi h_1[r^2 + (r - h_1) r + (r - h_1)^2] = {}^1/_3\pi h_1(3rh_1 - h_1^2) = {}^1/_3\pi h_1^2(3r - h_1)$.
Since $\varrho^2 = r^2 - (r - h_1)^2 = 2rh_1 - h_1^2$ or $6rh_1 - 2h_1^2 = 3\varrho^2 + h_1^2$, $V = \pi h_1(3\varrho^2 + h_1^2)/6$, where r is the radius of the sphere, h_1 the height of the segment and ϱ the radius of the base-circle.

Volume of a spherical segment	$V = \pi h_1^2(3r - h_1)/3 = \pi h_1(3\varrho^2 + h_1^2)/6$

Spherical layer. If the spherical layer between two circular sections with radii ϱ_1 and ϱ_2 has height h, then by Cavalieri's principle its volume is the difference between that of a cylinder, $\pi r^2 h$, and a frustum of a cone with base-radii $r_1 = r_2 + h$ and r_2 (Fig.). The volume is given by

$$V = \pi r^2 h - \pi h[(r_2 + h)^2 + (r_2 + h) r_2 + r_2^2]/3$$
$$= \pi h(6r^2 - 6r_2^2 - 6r_2 h - 2h^2)/6.$$

Because

$$\varrho_1^2 = r^2 - (r_2 + h)^2, \quad \varrho_2^2 = r^2 - r_2^2,$$
$$\varrho_1^2 + \varrho_2^2 = 2r^2 - 2r_2 h - 2r_2^2 - h^2,$$
$$3\varrho_1^2 + 3\varrho_2^2 + h^2 = 6r^2 - 6r_2^2 - 6r_2 h - 2h^2,$$

the volume is $V = \pi h(3\varrho_1^2 + 3\varrho_2^2 + h^2)/6$.

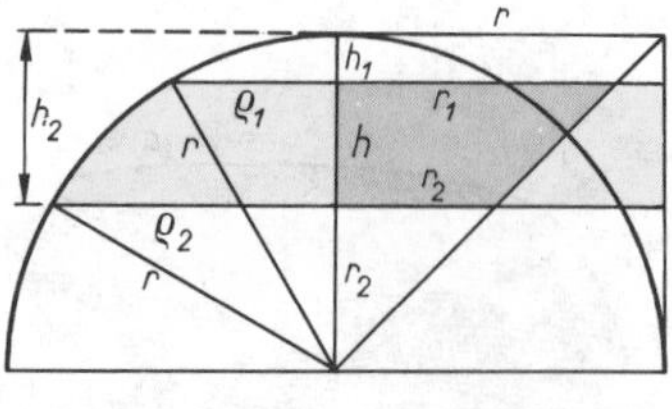

8.6-7 Derivation of the formula for the volume of a spherical layer

Volume of a spherical layer	$V = \pi h(3\varrho_1^2 + 3\varrho_2^2 + h^2)/6$

Spherical sector. The volume of a *spherical sector* is the sum of the volumes of a spherical segment and a cone: $V_{sector} = {}^1/_3\pi h^2(3r - h) + {}^1/_3\pi\varrho^2(r - h)$, where h is the height of the segment, $r - h$ the height of the cone and ϱ the radius, of the guide-circle. Since $\varrho^2 = h(2r - h)$, $V_{sector} = 2\pi r^2 h/3$.

Volume of a spherical sector	$V = 2\pi r^2 h/3$

Hollow sphere. A *hollow sphere* is what is left of a sphere of radius r_1 when a concentric sphere of radius $r_2 (r_1 > r_2)$ is removed. The volume of the hollow sphere is the difference between the volumes of the two spheres:

$V_{hollow\ sphere} = {}^4/_3\pi r_1^3 - {}^4/_3\pi r_2^3$.

Volume of a hollow sphere	$V = {}^4/_3\pi(r_1^3 - r_2^3)$

Surface area

Surface area of a sphere. In contrast to the curved surface of a cone or cylinder, the surface of a sphere cannot be developed into a plane. To derive the formula for the surface area of a sphere limiting arguments are necessary.

If the surface of the sphere is thought of as divided into n small *polygons*, then the radii through the vertices of these polygons divide the spherical space into n *pyramidal subspaces* with bases B_i and heights h_i. The larger n is, the smaller are the bases B_i, the difference between the heights h_i and the radius r of the sphere, the difference between the sum of the areas of the bases B_i and the surface area S of the sphere and the difference between the sum of the volumes of the subspaces $B_i h_i/3$ and the volume V of the sphere. From $\lim_{n\to\infty} \sum_{i=1}^{n} B_i = S$, $\lim_{n\to\infty} h_i = r$ and $\lim_{n\to\infty} \sum_{i=1}^{n} B_i h_i/3 = V$ one obtains $V = rS/3$ or $S = (3V)/r = 4\pi r^2$.

The surface area of a sphere is four times the area of a great circle of the sphere.

Surface area of a sphere	$S = 4\pi r^2$

By analogy with the circle, *the sphere has the greatest volume of all bodies with the same surface area* or the smallest surface area of all bodies with the same volume (see the Isoperimetric problem in Chapter 38.). This property of the sphere is of great importance; for drops of liquid and stars, because of their spherical shape, the rate of evaporation and heat transfer is less than it would be for other forms. Also, because of this property and their large capacity, spherical containers are often preferred for gases and liquids.

Surface area of parts of a sphere. To derive the formula for the *surface area S of a spherical cap* one proceeds just as for the surface area of a sphere, that is, one obtains $V_{sector} = 2\pi r^2 h/3$ and $2\pi r^2 h/3 = {}^1/_3 r S_{cap}$ (Fig. 8.6-3). Hence $S_{cap} = 2\pi rh$, where h is the height of the corresponding segment. The surface area of a *spherical zone* can be regarded as the difference between the areas of two spherical caps. If h is the height of the corresponding spherical layer, and h_1, h_2 the heights of the smaller and larger spherical caps, then $S_{zone} = S_{cap\,2} - S_{cap\,1} = 2\pi rh_2 - 2\pi rh_1 = 2\pi r(h_2 - h_1)$, and since $h_2 - h_1 = h$, $S_{zone} = 2\pi rh$. – Note the formal equality of the two formulae for S_{cap} and S_{zone}, though h has different meanings in the two cases.

The surface area of a *spherical sector* is the sum of the surface areas of a spherical cap and the curved surface of a cone: $S_{sector} = 2\pi rh + \pi\varrho r = \pi r(2h + \varrho)$, where h is the height of the corresponding segment, ϱ is the base-radius of the cone and r is the radius of the sphere and the length of a generator of the cone.

Surface areas					
Spherical cap	$S = 2\pi rh$	Spherical zone	$S = 2\pi rh$	Spherical sector	$S = \pi r(2h + \varrho)$

8.7. Further solids

Solids of rotation. Since the invention of the potter's wheel, *solids of rotation* have had many applications. Every plane through the axis of rotation cuts a *surface of rotation* in a *meridian* or *profile*, and every plane perpendicular to the axis cuts the surface in a *parallel circle*. Every surface of rotation can be covered by an orthogonal net of meridians and parallel circles. The surface normals ***n*** along a parallel circle consisting of regular points generally form a cone of rotation. For circles of maximum and minimum radius the cone degenerates to a plane, and for a circle along which a tangent plane touches the surface, the cone degenerates into a cylinder (Fig.).

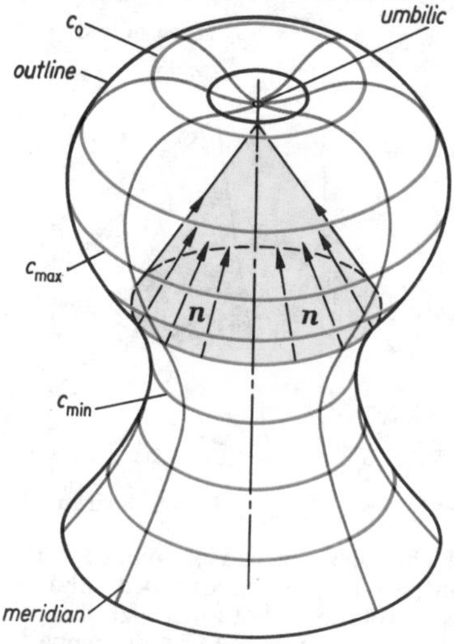

8.7-1 Solids of rotation; normal ***n***, the normal cone along a circular section is coloured yellow; along a circle in a tangent plane it becomes a circular cylinder, along a circle whose radius is a relative maximum or minimum it degenerates to a plane

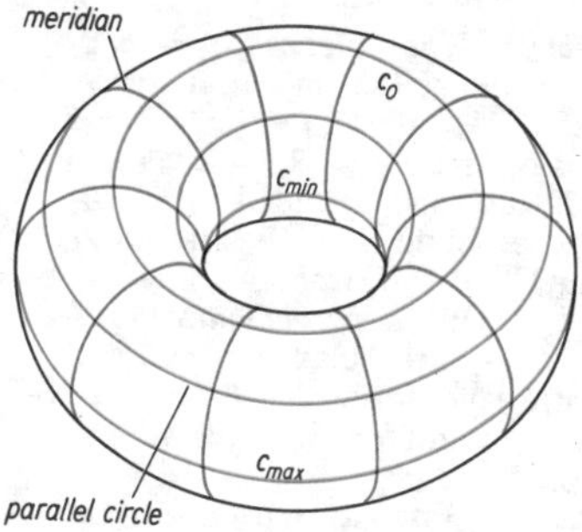

8.7-2 Torus with intermediate positions of the rotating circle

Well-known solids of rotation are the sphere, cone of rotation, cylinder of rotation, paraboloid of rotation, one-sheet hyperboloid of rotation, two-sheet hyperboloid of rotation, ellipsoid of rotation (see Chapter 24.), the torus, pseudosphere and catenoid.

A *torus* is obtained by rotating a circle of radius r about an axis in the plane of the circle at a distance $a \geqslant r$ from its centre (Fig.). The torus is a *tubular surface*.

The *pseudosphere* (see Table 56) is obtained by rotating the tractrix about its asymptote. If the x-axis of a Cartesian coordinate system is taken as the asymptote, and if a is the distance of the cusp A along the y-axis, then $V = 2\pi a^3/3$ is the volume of the pseudosphere. At each regular point, the surface has constant negative Gaussian curvature. Because of this property the pseudosphere serves as a model for non-Euclidean hyperbolic geometry, just as a sphere for non-Euclidean elliptic geometry.

The centres of curvature of the tractrix lie on a *catenary*, which is therefore the evolute of the tractrix (see Fig. 19.5-11). Rotation of the catenary about its directrix gives the *catenoid*, which is the only real minimal surface of rotation.

Pappus' rules. To calculate the volume and surface area of solids of rotation Pappus of Alexandria (end of the 3rd century A. D.) gave rules, which are derived nowadays by means of integral calculus.

Pappus' rule for surface area. If a plane curve C is rotated about a line l in its plane such that C lies on one side of l, then the area S of the resulting surface of rotation is equal to the product of the length of the generating curve C and the length of the path of the centre of gravity of C under the rotation.

Pappus' rule for volume. If part of a plane A is rotated about a line in the plane that has at most boundary points in common with A, then the volume V of the resulting solid of rotation is equal to the product of the area of A and the length of the path of the centre of gravity of A under the rotation.

Example 1: For a torus (Fig. 8.7-2) the surface area S is given by $S = 2\pi a \cdot 2\pi r = 4\pi^2 ar$ and the volume by $V = 2\pi a \cdot \pi r^2 = 2\pi^2 ar^2$.

Example 2: By rotating a semicircle about its diameter one obtains the well-known values for the surface area and volume of a sphere, and so the distance ϱ_C and ϱ_A of the centres of gravity of the semicircular arc and the semicircular disc from the axis can be calculated. From $S = 4\pi r^2 = s \cdot 2\pi\varrho_C$ with $s = \pi r$ one obtains $4\pi r^2 = 2\pi^2 r\varrho_C$ and so $\varrho_C = 2r/\pi$. From $V = 4\pi r^3/3 = 2\pi\varrho_A A$ with $A = \pi r^2/2$ one obtains $4\pi r^3/3 = \pi^2 r^2 \varrho_A$ and so $\varrho_A = 4r/(3\pi)$.

Kepler's rule, Simpson's rule. Certain approximate formulae for the volume of a solid are very useful in practice. In many special cases the formulae give the exact values.

In a large work on the solid geometry of a barrel, KEPLER (1571–1630) gave an approximate formula to determine the volume V of a barrel, where B_0, B_2, B_1 are the areas of the top and bottom surface and the section half way between them and h is the height of the barrel.

This formula gives the *exact value* for the frustum of a pyramid including a pyramid, sphere, elliptic paraboloid, hyperboloid of one sheet, ellipsoid and all layers of these bodies obtained by taking sections perpendicular to the axis.

Kepler's rule	$V = h(B_0 + 4B_1 + B_2)/6$

Example 1: The planes $z_0 = c$, $z_2 = -c$, $z_1 = 0$ cut the one-sheet *hyperboloid* $(x^2/a^2) + (y^2/b^2) - (z^2/c^2) = 1$ in sections of area $B_0 = B_2 = 2\pi ab$ and $B_1 = \pi ab$. The solid bounded by B_0, B_2 and the hyperboloid (Fig.) has height $2c$ and volume $V = (8/3)\,\pi abc$.

Example 2: On the *paraboloid of rotation* $z = x^2 + y^2$ the planes $z_0 = 1$ and $z_2 = 9$ cut a layer for which $B_0 = \pi$, $B_2 = 9\pi$, $B_1 = 5\pi$ and $h = 8$, so the volume is $V = 40\pi$.

Example 3: For a *tetrahedron* of edge-length a (Fig.), $B_0 = B_2 = 0$, $B_1 = a^2/4$ and $h^2 = h_a^2 - a^2/4$ and so $h = a/\sqrt{2}$. The volume V is given by Kepler's rule: $V = a^3\sqrt{2}/12$.

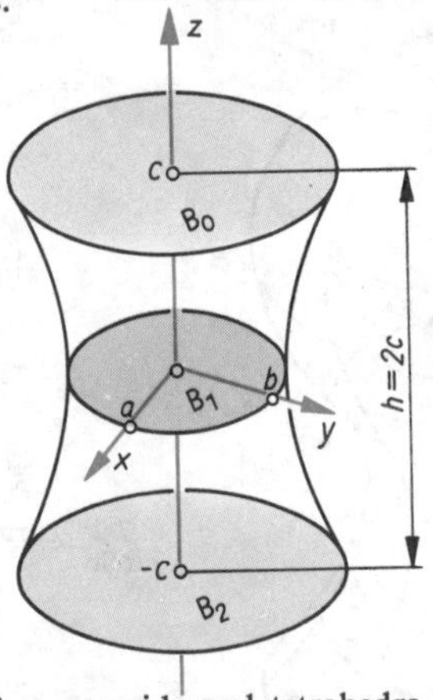

8.7-3 Hyperboloid

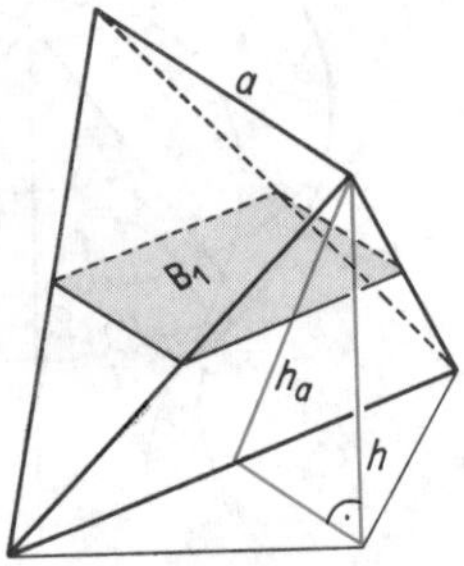

8.7-4 Tetrahedron

Since Kepler's rule gives exact results for pyramids and tetrahedra, it can be applied without error to the prismoids. These provide good approximations for barrels, cask-shaped solids and tree-trunks that are not too long. It breaks down for solids of rotation whose meridian curve has discontinuities in the direction of the tangent and for solids whose height is large in comparison with the mean diameter. In critical cases greater accuracy can be obtained at the cost of more measurement by dividing the height h into $n = 2k$ equal parts and applying Kepler's rule to the k arising pairs of layers. If B_i is the area of the ith section, the volume is obtained from a rule named after SIMPSON (1710–1761).

Simpson's rule	$V = h/(3n)\,\{B_0 + 4(B_1 + B_3 + \cdots + B_{n-1}) + 2(B_2 + B_4 + \cdots + B_{n-2}) + B_n\}$

Conoids. In technical applications conoids are of practical importance. In general, they are generated in the following way. Given a *guide-curve* c, a *guide-line* l and a *direction plane* not parallel to l, the conoid is formed from the set of lines that meet c and l and are parallel to the plane. If the guide-curve is a circle whose plane does not contain the guide-line, one speaks of a *circular conoid.* For a *right circular conoid* the guide-line is perpendicular to the direction plane and is cut at right angles by the axis of the circle outside the plane of the circle. A plane parallel to the plane of the circle cuts the conoid in an ellipse (Fig.). Kepler's rule, applied to a right circular conoid, gives the exact volume. If r is the radius of the base-circle and h the height, then $B_0 = \pi r^2$, $4B_1 = 2\pi r^2$, $B_2 = 0$ and $V = \pi r^2 h/2$.

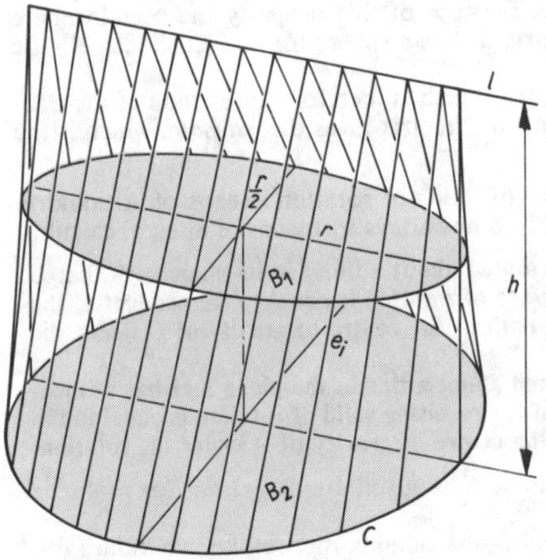

8.7-5 Conoid

Prismoids. A *prismoid* is a polyhedron with two parallel polygons as top and bottom faces and triangles or trapezia as side-faces. A prism, pyramid and frustum of a pyramid are special forms of a prismoid.

A further special form is the *wedge*, where the top surface has reduced to a line parallel to the base, called the knife edge. A plane parallel to the base of a right wedge cuts off another wedge, leaving a *pontoon* (Fig.). The side-trapezia of this pontoon are congruent in pairs. There is no similarity between the base and top face. If the height of a pontoon is very much greater than the sides of the base and top, the solid is called an *obelisk*.

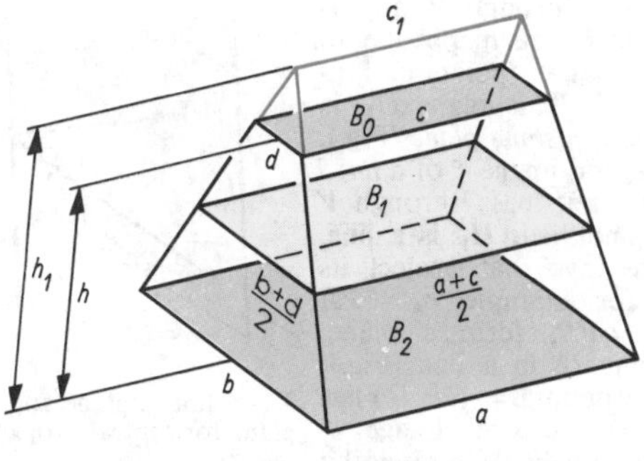

8.7-6 Prismoid, pontoon, wedge

The application of Kepler's rule to prismoids gives exact values. For a pontoon with a and b as sides of the base, c and d as sides of the top and h as height, $B_0 = ab$, $4B_1 = (a+c)(b+d)$, $B_2 = cd$, and so $V = h[2(ab+cd) + ad + bc]/6$. For $d = 0$ the solid is a wedge with edge-length $c = c_1$ and volume $V = h_1 b(2a + c_1)/6$. In technology the wedge acts as a splitting tool and machine element, for example as a stop. Pontoons are well known as floating elements of transportable bridges and floating docks. Obelisks occur as stone monuments and religious symbols. Milestones often take these forms. Finally, various forms of roofs can be comprised under the general title of prismoid.

9. Descriptive geometry

Descriptive geometry investigates and applies mappings of three-dimensional space onto a plane drawing-board. In order to carry over the constructive methods of plane geometry one gives preference to mappings that make lines in space correspond to lines on the drawing-board. In the choice of mappings two requirements have priority: perspicuity and preservation of measurements.

Clear images are given, for example, by central projection, because the image on the drawing-board imitates what is seen by the eye. The most usual way of preserving the measurements is to use normal projection. Since one dimension is lost in mapping spatial objects onto a plane drawing-board, preservation of proportions is possible only under certain restrictions. Norm-preserving axonometric images give a clear representation of a spatial object by reconstructing the proportions. Among these the frontal axonometric images are generally preferred because of their simplicity.

If technical drawings and constructions are to be an additional means of communication, side-by-side with speech and writing, they must be produced in accordance with certain conventions introduced in descriptive geometry. The working out of these conventions suitable for practical needs goes back essentially to Monge (1746–1818), who because of his famous work 'Géométrie Descriptive' and his teaching and research in the subject became the founder of descriptive geometry.

9.1. Mappings in descriptive geometry

Central projection

In central projection a mapping is carried out by means of a bundle of rays whose carrier, the *vertex of projection V*, lies outside the *plane of projection Π*. For an arbitrary point $P \neq V$ the *central image* or *perspective image* P^c is the point of intersection $P^c = r_P \cap \Pi$ of the ray $r_P = VP$ with the image plane Π. Under this projection all points of a plane Π_V that passes through V and is parallel to Π are mapped onto improper points (points at infinity) of Π. This plane Π_V is called the *vanishing plane* (Fig.).

The central image l^c of a *line* l that does not pass through V and does not lie in Π_V is a line, since the rays that project its points, for example $r_A = VA$ and $r_B = VB$, form a plane, which cuts Π in a line (Fig.). The trace point $L = (l \cap \Pi)$ lies on l^c. The central image l^c of l is uniquely determined by the central images A^c and B^c of two of its points A and B. The point of intersection L_V of l with Π_V is called the *vanishing point* of l; its image is the improper point of l^c. The image of the improper point of l is the point of intersection L_u^c of Π with the ray r_l parallel to l through V. This *vanishing point* L_u^c is the image of the common point at infinity of all lines parallel to l. The vanishing point of all lines perpendicular to Π is the foot of the perpendicular from V to Π and is called the *principal point* or *principal vanishing point* H. The length d of the segment VH is called the distance. The vanishing points of all lines that cut Π at an angle of 45° lie on a circle with centre at H and radius d, the distance circle.

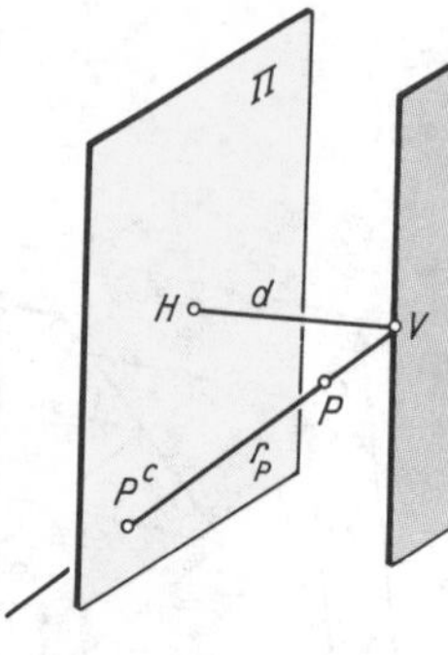

9.1-1 Image plane and vanishing plane for central projection

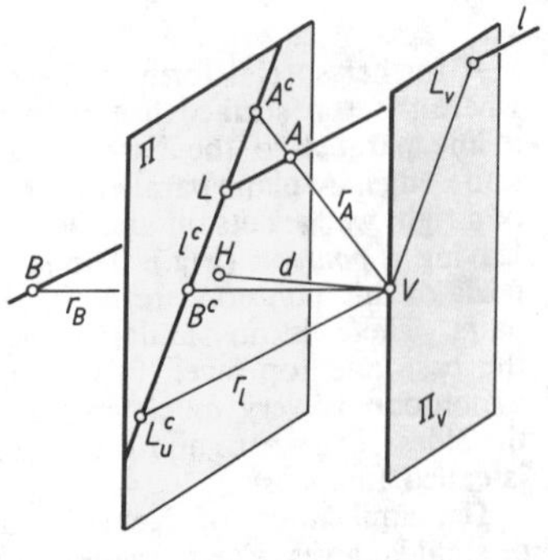

9.1-2 Mapping of a line by central projection

Parallel projection

If the vertex V of the bundle of rays giving the mapping is at infinity, the projection is a parallel projection of the points P of space onto the points P' of Π. Since the rays that project the points of a line p form a plane, the image p' of a line is a line, in general, and the images of two parallel lines p and q are parallel. Only when the given line l_0 is parallel to the projecting rays, then its image is a point $L_0 = (l_0 \cap \Pi)$. For lines that do not lie along a projecting ray the following theorems hold:

The ratio of division of three points on a line is invariant under this mapping, for example $|AB| : |BC| = |A'B'| : |B'C'|$ (Fig.).

The ratio of two segments that lie along parallel lines is invariant under parallel projection, for example, $|AB| : |DE| = |A'B'| : |D'E'|$.

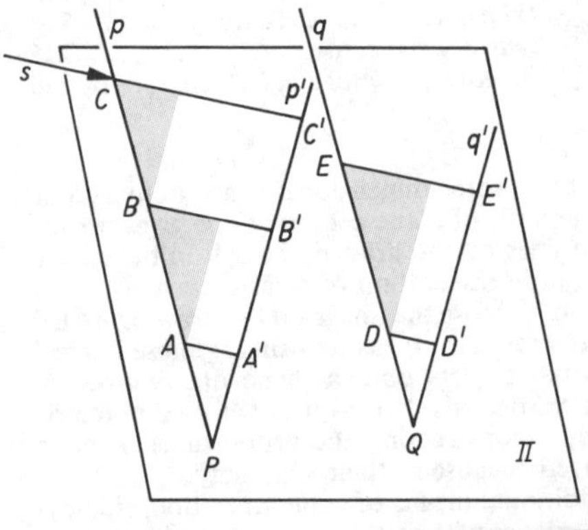

9.1-3 Invariance of ratio of division under parallel projection

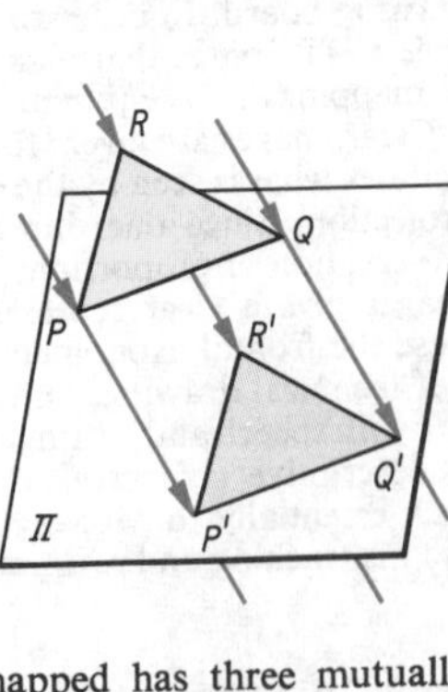

9.1-4 The image of a plane figure lying parallel to the image plane is congruent to the original under parallel projection

The image of a figure in a plane parallel to Π is congruent to the original figure, for example, $\triangle PQR \cong \triangle P'Q'R'$ (Fig.).

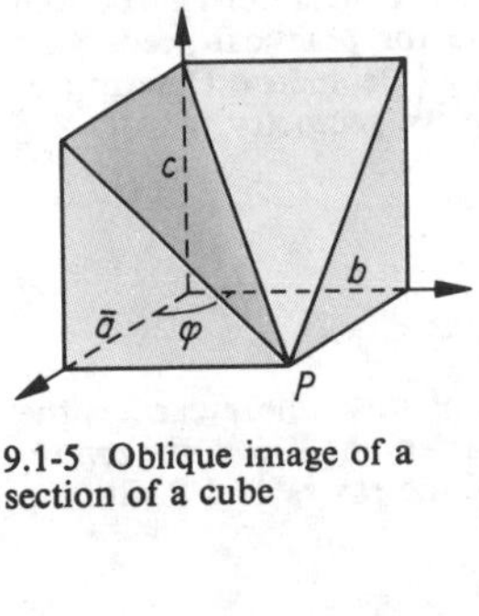

9.1-5 Oblique image of a section of a cube

Oblique projection. If the solid to be mapped has three mutually perpendicular edges or axes of symmetry a, b, c, then an oblique projection of the solid (skew parallel projection) can be obtained constructively. If the image plane Π is taken to be vertical, then the solid can be brought into the '*ordinary position*', where two of the three axes are parallel to Π, with one, say b, horizontal and the other, say c, vertical (Fig.). The third axis a and all lines parallel to it are then perpendicular to Π and are called *depth lines*. Their images are parallel and make with the image of b the *distortion angle* $\varphi = (\bar{a}, b)$, where $\bar{a}$ is the image of a. The ratio $\lambda = \bar{a} : a$ of the image segment to the original segment on the depth line is called the *distortion ratio*. From the distortion angle φ and the distortion ratio λ the measurements of the spatial figure can be reconstructed from the drawing. For ease of construction φ is chosen to be 30°, 45°, 60° or 120°, which is convenient to draw with set squares, and the distortion ratio λ is chosen to be a simple rational number such as 1, 1 : 2, 2 : 3, 1 : 3 or 3 : 4.

Normal projection or orthogonal projection. In this mapping the parallel projecting rays are perpendicular to the image plane Π. The much reduced clarity is compensated in two ways.

1) Chosen points or lines of the object are labelled by means of their heights above the horizontal reference plane. This leads to a one-plane projection with heights. It is applied particularly in the projection of earthworks and representation of landscape.

2) The normal image is compared with a second normal image in the same drawing. This is done in such a way that the directions of projection and the image planes giving the two normal images are perpendicular. The method of two-plane projection sketched here, or the corresponding normal images, are applied in machine construction and architecture.

9.2. The two-plane method

In the two-plane method the spatial object is mapped onto two perpendicular planes Π_1 and Π_2 by normal projection (Fig.). These planes divide space into four numbered quadrants (see Fig. 9.2-3). A spatial object is mapped onto $\underline{\Pi}_1$ by the first and onto Π_2 by the second projecting parallel line bundle. In this way there arise two normal projections of one object in two perpendicular planes, namely the *plan* in $\underline{\Pi}_1$ and the *elevation* in Π_2. The convention is that the plan is horizontal and the elevation vertical. The object to be represented is *preferably* drawn in the first quadrant. For the convenience of the draughtsman in working with these two images of one object, the elevation is placed in the drawing-board and the plan is rotated about the horizontal ground line x_{12} into the plane of the drawing-board. After this rotation the elevation lies above the ground line and the plan below it. The plan and elevation of a point P lie on a line perpendicular to the ground line, This is called an *order line*. One says that the plan P' and elevation P'' of a point P are in the Monge position. The distance of P'' from the ground line is called the *first distance* d_1 and that of P' from the ground line the *second distance* d_2 of P. Corresponding to the plan P' and the elevation P'' of P are the first projection ray r_1 and the second projection ray r_2, both passing through P.

If the object lies entirely in the first quadrant, then the elevation is always above and the plan always below the ground line. In the course of spatial constructions one may come across points from the other three quadrants. For example, the points A, B, C, D lie in the quadrants I, II, III, IV (Fig.). The distances d_i satisfy the following inequalities: for $A: d_1, d_2 > 0$, for $B: d_1 > 0, d_2 < 0$,

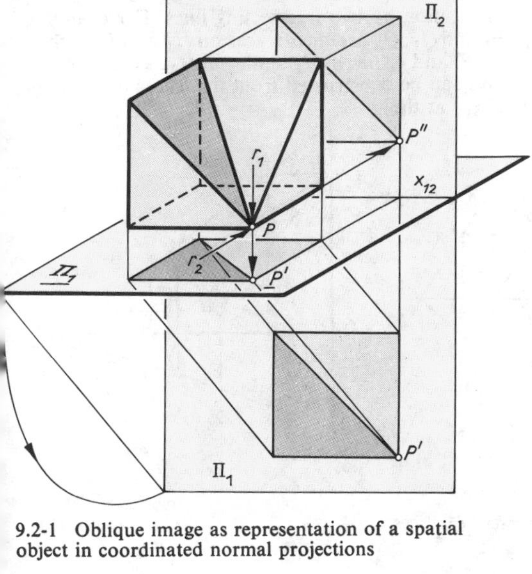

9.2-1 Oblique image as representation of a spatial object in coordinated normal projections

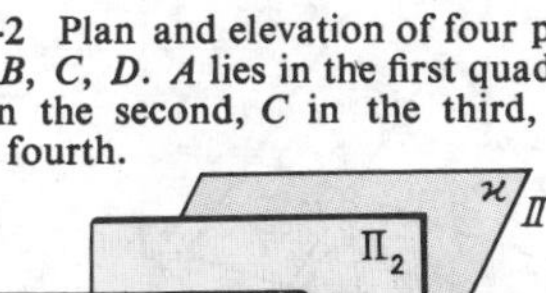

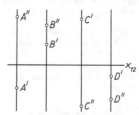

9.2-2 Plan and elevation of four points A, B, C, D. A lies in the first quadrant, B in the second, C in the third, D in the fourth.

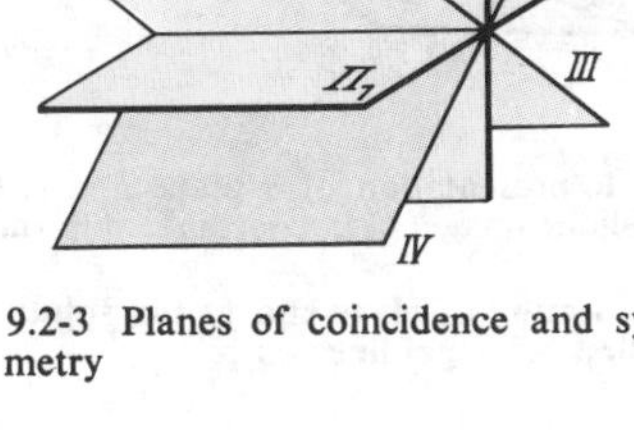

9.2-3 Planes of coincidence and symmetry

for $C: d_1, d_2 < 0$, for $D: d_1 < 0, d_2 > 0$. For points whose plan (elevation) lies on the ground line $d_2 = 0$ ($d_1 = 0$). All points of the plane of symmetry σ (coincidence plane $\varkappa$) satisfy the equation $d_1 = d_2$ ($d_1 = -d_2$) (Fig.). In the two-plane projection two image planes cover one another in the drawing-board. By a suitable combination of constructions of plane geometry, applied to the two images of an object, problems of spatial construction can be solved. The ground line x_{12} is therefore not to be regarded as a line of separation of the plan and elevation, but it illustrates the position of the line of intersection of the two image planes before the rotation into the drawing-board.

Representation of lines and planes

Representation of a line. The projections l' and l'' of a line l are uniquely determined by the projections of two of its points. For a *first principal line* h_1, parallel to Π_1, the elevation h_1'' is parallel to the ground line x_{12}; for a *second principal line* h_2, parallel to Π_2, the plan h_2' is parallel to x_{12}. The position of an arbitrary line l that does not cut the ground line and is not a principal line is determined by its *traces*, its points of intersection with Π_1 and Π_2. Here $L_1 = (l \cap \Pi_1)$ is called the first trace and $L_2 = (l \cap \Pi_2)$ the second trace (Fig.); L_1'' and L_2' lie on x_{12}. The point of intersection $K' = K''$ of the projections of the line characterizes the point of intersection K of the line l with the coincidence plane. For a *first projecting line* l perpendicular to Π_1 one has $l' = L_1$, and for a *second projecting line* l perpendicular to Π_2 one has $l'' = L_2$. Two lines p and q intersect in a point S of space if and only if the points of intersection $1' = (p' \cap q')$ and $2'' = (p'' \cap q'')$ are in the Monge position. Then $1' = S'$ and $2'' = S''$. Otherwise the lines p and q are skew (Fig.).

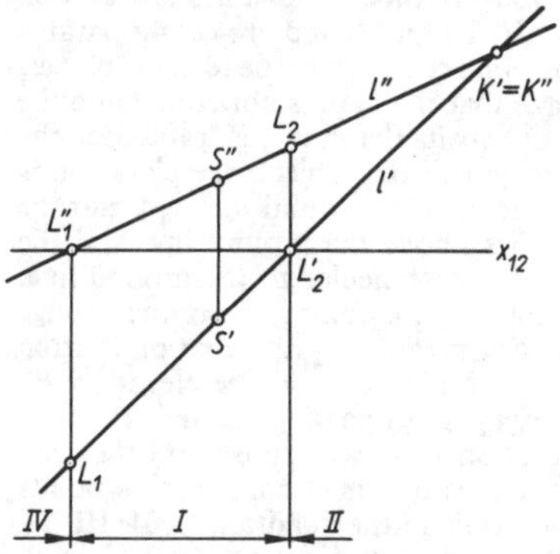

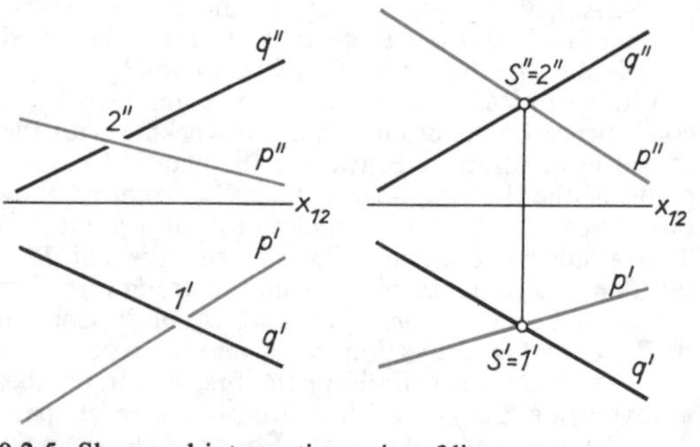

9.2-5 Skew and intersecting pairs of lines

9.2-4 Plan and elevation of a line l with its trace points L_1 and L_2; since $d_1 = -d_2$, $K' = K''$ lies in the coincidence plane

Representation of a plane. A plane E is fixed in position by two intersecting lines. For example, if the lines p and q intersect in P (Fig.) and P_1, P_2 and Q_1, Q_2 are their traces on Π_1 and Π_2, then the lines $e_1 = P_1Q_1$ and $e_2 = P_2Q_2$ lie in the plane E, and e_1 lies in Π_1 and e_2 in Π_2. These lines e_1 and e_2 are the traces of the plane in Π_1 and Π_2 and can be constructed from the traces of p and q. The traces e_1 and e_2 intersect on the ground line x_{12} at the *node* K of the plane.

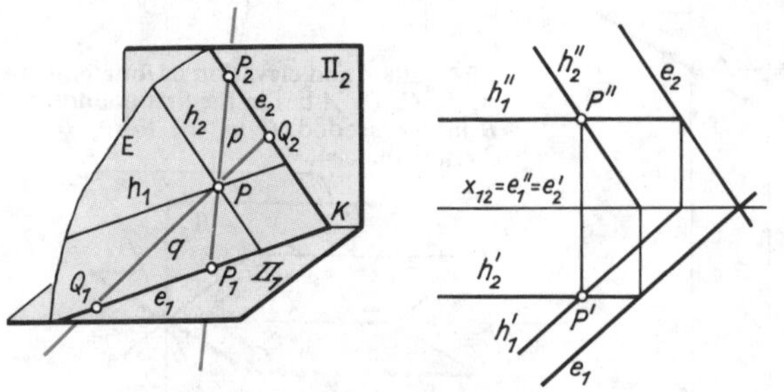

9.2-6 Representation of a plane and lattice of a point P; a) in the oblique image, b) in coordinated normal projections

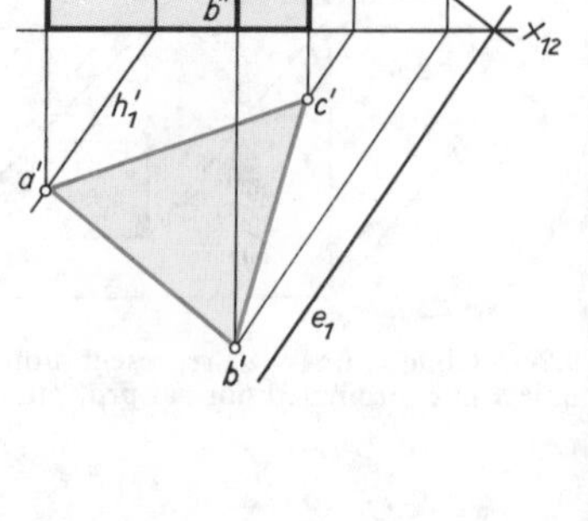

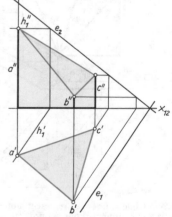

9.2-7 Section of plane and prism, solution by means of lattices with first principal lines

For constructions with in a given plane the *principal lines* or *trace parallels* are useful as an auxiliary tool. First principal lines, first trace parallels or *height lines* are parallel to e_1; second principal lines, second trace parallels or *front lines* are parallel to e_2. For example, it is easy to establish whether a point P determined by P' and P'' lies in a plane E determined by e_1 and e_2 or not. For if the elevation h_1'' of a first principal line h_1 of E is drawn through P'', then the plan h_1' of this principal line is uniquely determined. If P' lies on the line h_1' found in this way, then P is a point of E, otherwise P lies outside E. The model described here is called the *lattice* of a point P. This lattice is also possible for an arbitrary line in E. An application of lattices enables, for example, the construction of the intersection of a prism perpendicular to Π_1 with a plane. The plans of the points of intersection of a, b, c coincide with a', b', c'. The elevations of the points of intersection are found by means of first principal lines, using lattices (Fig.). *Fall lines* intersect the height lines at right angles and point in the direction of greatest slope of the plane with respect to a horizontal plane. Their plans are therefore perpendicular to the plans of the height lines.

Special positions of the plane E can be characterized by the position of the traces e_1 and e_2. For a ***first projecting plane*** perpendicular to Π_1 one has $e_2 \perp x_{12}$, for a ***second projecting plane*** perpendicular to Π_2 one has $e_1 \perp x_{12}$, and for a ***first and second projecting plane*** both traces are perpendicular to x_{12}. On such a plane one obtains a *side elevation* to supplement the plan and elevation of a solid. For a *desk plane* e_1 and e_2 are parallel to the ground line. If they coincide with it, then the plane contains the ground line and can only be represented by principal lines.

A plane is called *direct* if the plan and elevation of a triangle ABC in the plane have the same sense of rotation. If the senses of rotation of the two images of the triangle are opposite, the plane is called *alternating*.

Determination of true measurement from the corresponding normal projections. The distance between two points A and B is equal to that of the segment AB measured along the line $l = AB$. If l is a principal line with respect to the image plane, then normal projection of the segment gives its true value, for example, h_1 in Π_1 and h_2 in Π_2. If A and B lie on a line l in general position, then l can be rotated about a first or second projecting line as axis so as to lie in one of the above positions relative to the image plane. For example, l can be rotated about the first projecting line a through A into the position of a second principal line (Fig.). If C is the foot of the perpendicular from B to a, then the right-angled triangle ABC can be rotated about the side $|AC| = |A''C''|$. The other side is $|CB| = |A'B'|$. The point B moves in space on an arc of a circle parallel to Π_1 with centre at C and radius $|A'B'|$. In Π_2 the point B'' moves on the line through B'' parallel to the ground line x_{12} towards the point $B^{*''}$, which is determined by $|B^{*''}C''| = |A'B'|$ (Fig.). Apart from the true distance $|AB|$, this construction gives the first angle of inclination α_1 of the line $l = AB$ to the ground plane Π_1; $\alpha_1 = \sphericalangle AL_1L_2' = \sphericalangle ABC = \sphericalangle A''B^{*''}C''$. Similarly the second angle of inclination α_2 that l makes with Π_2 is obtained by rotation about a second projecting line perpendicular to Π_2.

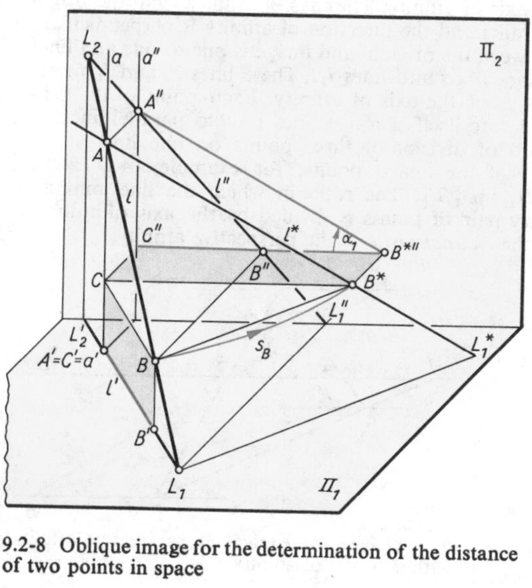

9.2-8 Oblique image for the determination of the distance of two points in space

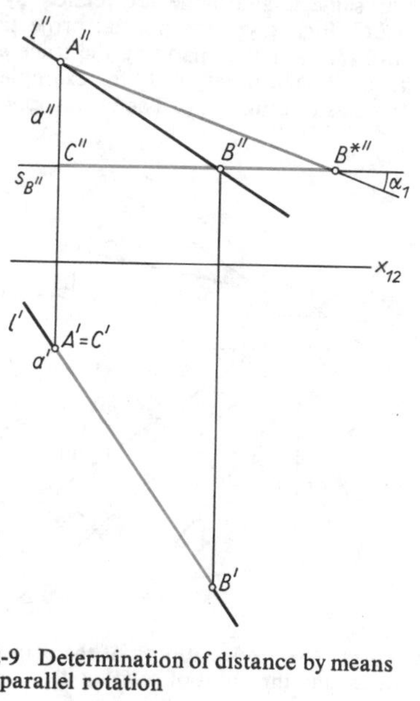

9.2-9 Determination of distance by means of parallel rotation

The *true shape of a plane figure* can be determined by rotation parallel to a plane of projection. It is natural to use a first or second principal line of the plane of the figure as axis of rotation.

Example: The true shape of a given triangle ABC can be determined from the plan and elevation by rotating it about the first principal line h_1 through A into a position parallel to the ground plane. Under the rotation A remains fixed, while B and C describe arcs of circles whose planes have the axis of rotation h_1 as common normal (Fig.). The radius r_C of the arc described by C is the hypotenuse of a right-angled triangle with $|C'M_C'|$ and h_C as sides, where M_C' is the foot of the perpendicular from C' to h_1' and the segment h_C is taken from the elevation. If this is drawn from M_C' along h_1' to its end-point S', then $|S'C'| = r_C$. By laying off this segment r_C from M_C' along the order line through C' perpendicular to h_1' the point C_1' is obtained, which is the result of rotating C about h_1. The point B is treated similarly.

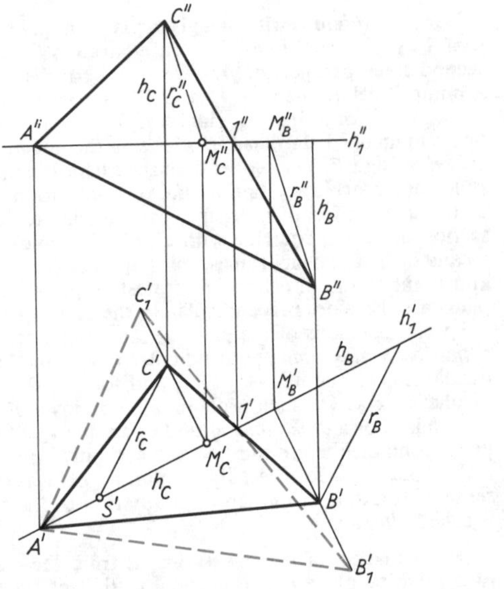

9.2-10 True form of a plane figure by application of the double compass method

Since in this method, in addition to the order line perpendicular to the principal line through the point to be rotated, two distances have to be laid off with compasses, it is called the *double compass method*. As a check on the drawing, one should verify that the lines $B'C'$ and $B_1'C_1'$ meet in a point on the axis of rotation h_1. The true shape of a plane figure also takes care of the angle of intersection of two intersecting lines in space.

Perspective affinity

If an arbitrary plane figure in space is rotated about one of its traces into the image plane corresponding to that trace, then the resulting figure and the normal projection of the original figure onto the same image plane are related by an *orthogonal perspective affinity*; for example, the triangle ABC (Fig.) gives the normal projection $A'B'C'$ and the triangle $A_1B_1C_1$ under rotation about e_1 into Π_1. For this mapping the trace e_1 is the axis of affinity. The rays of affinity from the original to the image point, $A'A_1$ for example, are parallel and the direction of affinity is perpendicular to the axis of affinity e_1. The correspondence between the original and image is one-to-one and linear, since lines l' go into lines l_1. These lines l' and l_1 meet at a point L of the axis of affinity. Each point of the axis is mapped onto itself. Parallel lines go into parallel lines, and the ratio of division of three points on one line is equal to that of the image points, for example, $|A'D'| : |D'B'| = |A_1D_1| : |D_1B_1|$. The ratio in which the line joining an arbitrary pair of points is divided by the axis of affinity is called the *characteristic* of the perspective affinity.

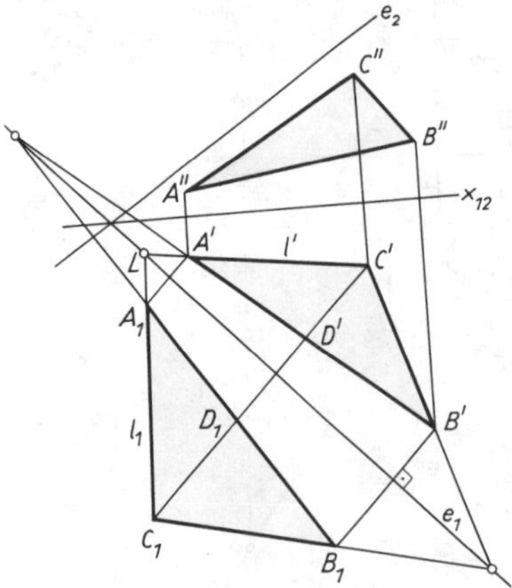

9.2-11 Perspective affinity of the plan of a plane figure and the result of rotating it

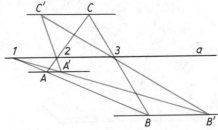

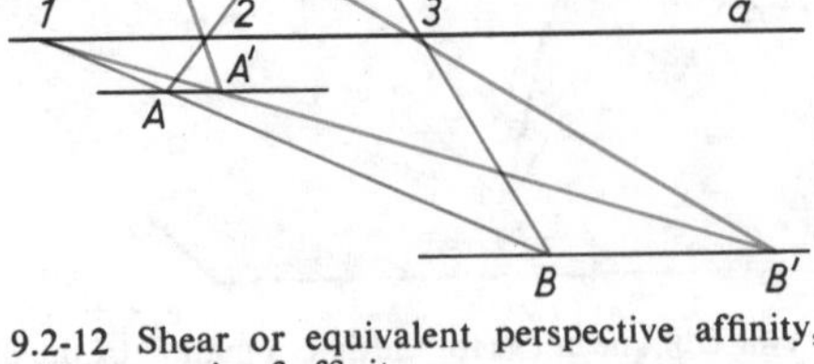

9.2-12 Shear or equivalent perspective affinity, with a as axis of affinity

Under a *skew* or *general* perspective affinity the direction of affinity can make any angle with the axis of affinity.

Under a *shear* or equivalent-perspective affinity the rays of affinity are parallel to the axis of affinity (Fig.).

A perspective affine mapping in the plane is uniquely determined by the axis of affinity and a pair of points that correspond under the mapping.

On the drawing-board there is a perspective affine relationship between the plan and elevation of a plane figure. The order lines of all points, $D'D''$ for example, give the direction of affinity, and the images l' and l'' of any line l of the plane figure intersect at a point of a line s in the drawing-board. This is the axis of affinity of the perspective affine mapping. The plan and elevation of the line of intersection $k = (\varkappa \cap E)$ of the coincidence plane $\varkappa$ and the plane E of the figure coincide with this, that is, $k' = k'' = s$ (Fig.).

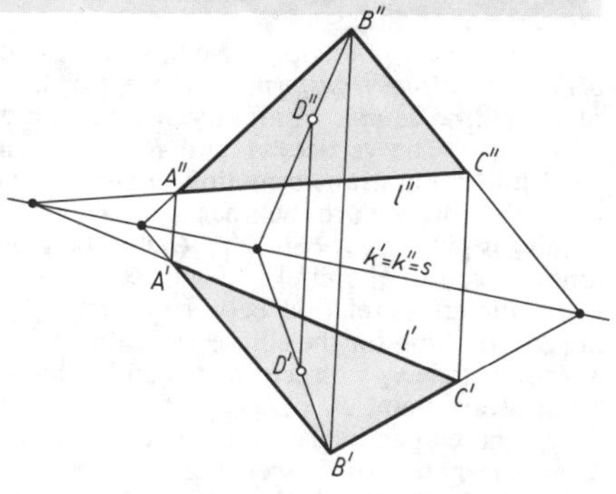

9.2-13 Perspective affinity of the plan and elevation of a plane figure

Between the plan and elevation of a plane figure there is a perspective affine point relation. The rays of affinity coincide with the order lines and the axis of affinity s coincides with the identical images of the coincidence line k in the drawing-board.

This geometric property can be used constructively, for example, to obtain the lattice of a point D lying in the plane.

The ellipse as perspective affine image of the circle. The perspective affine mapping of a circle with centre at C is determined by the axis of affinity s and the image C' of the centre C. The direction of affinity is CC'. Since any line cuts its image on the axis of affinity, the image of any point P can be constructed as the point of intersection of two lines, for example, the image P' of P is obtained from the two conditions $CC' \| PP'$ and $(CP \cap s) = (C'P' \cap s)$.

The images of perpendicular diameters of the circle are called conjugate diameters of the ellipse (see Chapters 7. and 25.), for example, $P'R'$ and $Q'S'$, the images of $PR \perp QS$. Since parallels go into parallels and the ratio of division remains unchanged, there are important relations between parallel chords and tangents.

A diameter of an ellipse bisects all chords parallel to its conjugate diameter, and the tangents at the end-points of a diameter of an ellipse are parallel to the conjugate diameter; for example, $P'R'$ bisects the chords parallel to $Q'S'$, and the tangents at P' and R' are parallel to $Q'S'$.

Among the pairs of conjugate diameters there is exactly one that forms an orthogonal pair of diameters. These are the *major* and *minor axes* of the ellipse. If K_1 and K_2 (Fig.) are their points of intersection with the axis of affinity s, then both C and C' lie on the circle with K_1K_2 as diameter. Its centre N is the point of intersection of the perpendicular bisector of the segment CC' with s; its radius is $|NC| = |NC'| = |NK_1| = |NK_2|$.

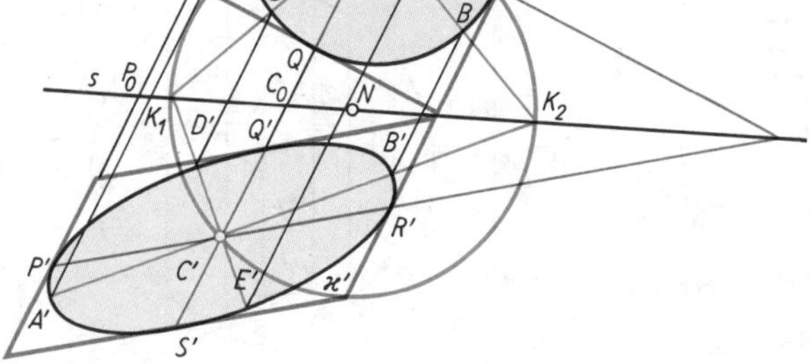

9.2-14 Ellipse as perspective affine image of a circle

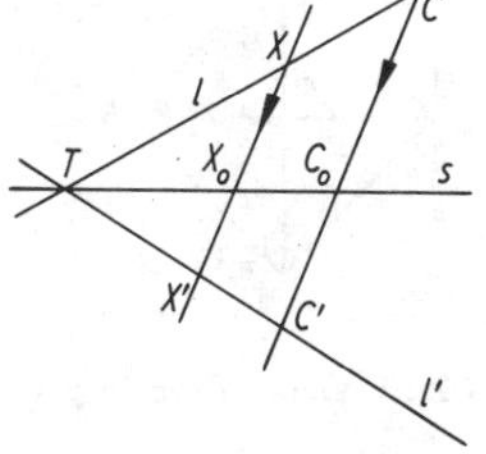

9.2-15 Ratios of division for perspective affine lines

The axis of affinity s and the point pair C, C' are sufficient to construct the affine transformation of the circle $\varkappa$ into the ellipse $\varkappa'$. The axis s cuts CC' in C_0 and PP' in P_0. For the construction of a further arbitrary pair of points X, X' the properties of the affine relation are used (Fig.): $CC' \parallel PP' \parallel XX'$ and $|PP_0| : |P'P_0| = |CC_0| : |C'C_0| = |XX_0| : |X'X_0| = k$, where k is the ratio of affinity.

To construct points of the ellipse with ruler and compass an *orthogonal affine transformation* is applied, with the major axis of the ellipse as axis of affinity and the minor axis as direction of af finity. The vertices A and B on the major axis remain fixed under this transformation, while the vertices D' and E' on the minor axis are the images of D and E (Fig.). The ratio of affinity is given by $k = |CD'| : |CD|$. The ellipse is therefore the affine image of the circle $\varkappa_1$ with centre at C and radius $|CA|$. From the affine relation between $\varkappa_1$ and the ellipse the following construction for the ellipse can easily be derived: One draws a second circle $\varkappa_2$ with centre at C and radius $|CD'|$. One chooses an arbitrary point P_1 on $\varkappa_1$. The line CP_1 cuts $\varkappa_2$ at P_2. One draws the perpendicular from P_1 to s and the line parallel to s through P_2. Then the perpendicular and parallel intersect in a point P of the ellipse.

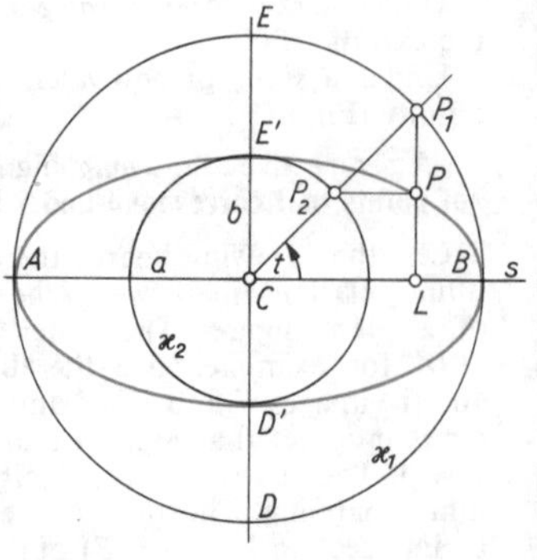

9.2-16 Two circle construction of the ellipse

Proof. By construction $|CD'| : |CD| = |CP_2| : |CP_1| = |LP| : |LP_1| = k$. Because the direction of affinity is perpendicular to s, P is the image of P_1. This construction for the ellipse is known as the *two circle construction*. There are other constructions for the ellipse (see Chapter 7.) and a parametric representation can be derived for it (see Chapter 13. – Equations of the ellipse).

Side elevations, rotations, and representations of solids

The preceding fundamental constructions, in terms of coordinated normal projections, are often applied to the representation of spatial objects, as the following examples show.

Example 1: From the plan and elevation of an octahedron in special position with respect to the image planes a representation in the general position can be obtained by means of two side elevations. For example, the point $1'''$ has the same distance from x_{23} as $1'$ from x_{12} and 1^{IV} has the same distance from x_{34} as $1''$ from x_{23} (Fig.). By suppressing the projecting lines and surfaces one obtains a natural representation of the solid. However, it is no longer possible to obtain the measurements of the solid directly from this projection. Apart from its usefulness in giving

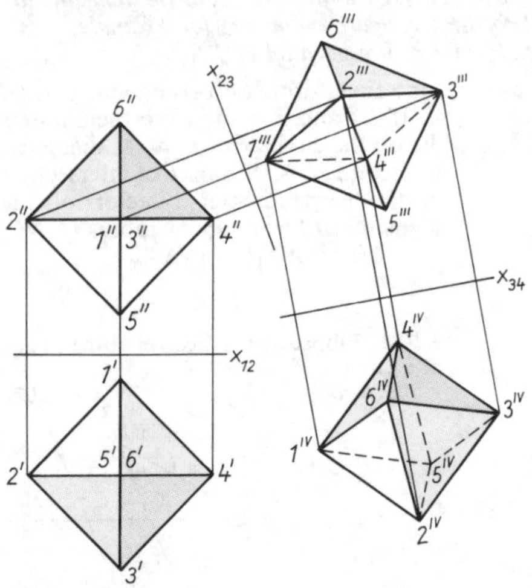

9.2-17 Octahedron in plan, elevation and side elevation

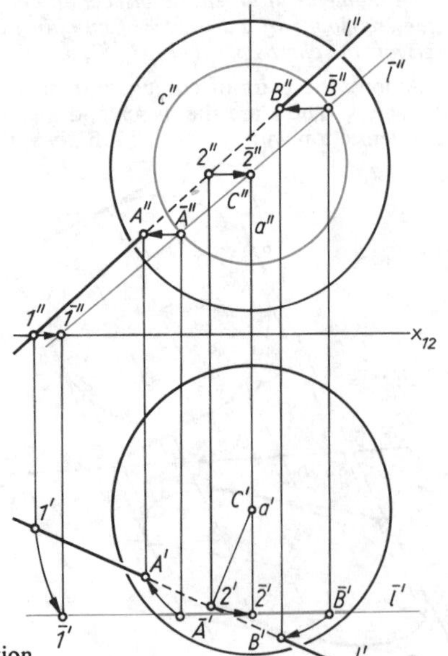

9.2-18 Intersection of line and sphere: application of rotation of a solid as a principle of construction

a sketch of the appearance of an object, the side elevation is applied as a principle of construction.

The transfer of points to a new side elevation is carried out according to the rule: the distances of points of the suppressed projection from the suppressed axis are transferred from the new ground line along the corresponding order lines to the new projection.

Example 2: The *intersection of a line with a sphere* can be constructed by means of rotation. The first projecting diameter a of the sphere is chosen as axis of rotation (Fig.). The line l is rotated about a so that its final position $\bar{l}$ is parallel to Π_2. Two suitably chosen points 1 and 2 of l go into $\bar{1}$ and $\bar{2}$, and the sphere goes into itself. The first projecting plane Γ through $\bar{1}$ cuts the sphere in a circle c, whose elevation coincides with its true form. Hence the points of intersection $\bar{A}''$ and $\bar{B}''$ of c'' with $\bar{l}''$ are the elevations of the points of intersection of $\bar{l}$ with the sphere. The lines through $\bar{A}''$ and $\bar{B}''$ parallel to the ground line cut l'' in A'' and B''. This has the effect of reversing the rotation in the elevation. The plans of the points of intersection A and B of l with the sphere are found by means of the order lines through A'' and B''.

Example 3: The *tangent plane to a sphere* at a given point P of the sphere can be constructed by means of the first and second principal lines. Since the tangent plane T is perpendicular to the radius r_P through P, the two principal lines h_1 and h_2 of T that intersect at P make right angles with r_P (Fig.). Hence $\sphericalangle h_1' r_P' = \sphericalangle h_2'' r_P'' = 90°$. Since h_1'' and h_2' are parallel to the ground line, the traces e_1 and e_2 of the tangent plane spanned by the principal lines h_1 and h_2 can be constructed.

Example 4: To find the intersections of a line l with a cone whose vertex Z and trace curve s in Π are known. The principle of solution can be seen from an oblique image (Fig.). The plane Γ through Z and l has the trace line c in Π. If c cuts the trace curve s in points P_1, P_2, P_3, P_4, then the lines ZP_i are generators of the cone lying in Γ. Their points of intersection 1, 2, 3, 4 with l are therefore the points of intersection of l with the cone.

Example 5: To construct a cone of rotation whose base circle lies in a plane E determined by its traces e_1 and e_2, where the height h of the cone, its base radius r and the plan C' of the centre C of the base circle are given (Fig.). The construction can be divided into steps. 1. By means

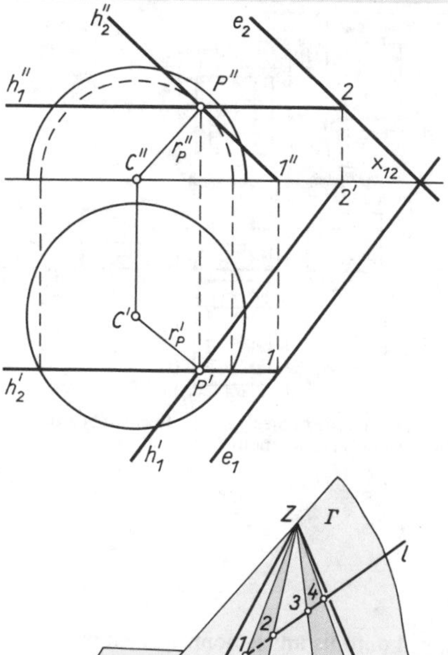

9.2-19 Tangent plane to a sphere

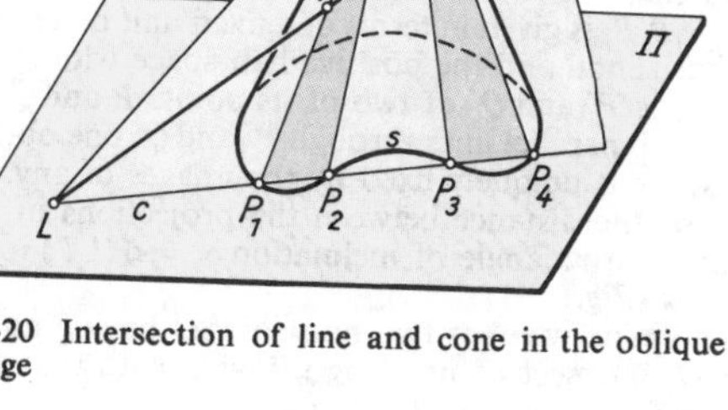

9.2-20 Intersection of line and cone in the oblique image

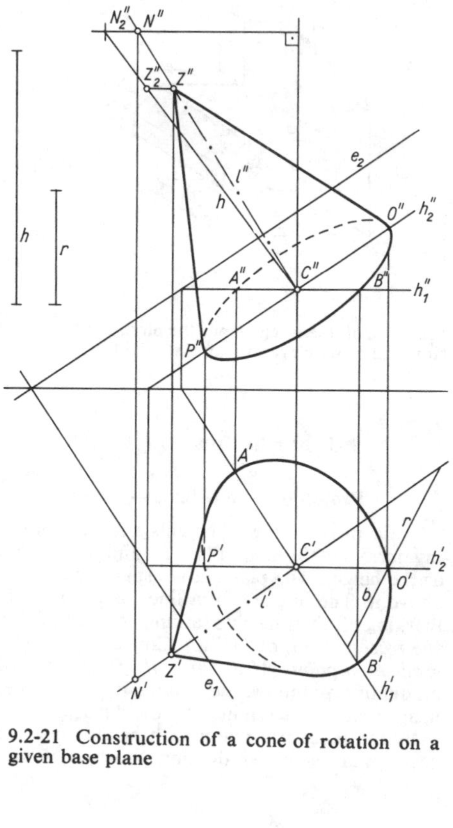

9.2-21 Construction of a cone of rotation on a given base plane

of the first and second principal lines the elevation C'' of C from C' has to be constructed. 2. The diameter of the base circle is mapped in the plan on h_1' and in the elevation on h_2'' by its true length. $|A'B'|$ and $|P''Q''|$ are the major axes of the two image ellipses. 3. By means of order lines, one finds the points A'', B'' on h_1'' and P', Q' on h_2'. 4. By the paper strip construction one finds the minor axes of the image ellipses and hence these ellipses themselves as images of the base circle of the cone. 5. On the perpendicular line l through C perpendicular to E with projections $l' \perp h_1'$ and $l'' \perp h_2''$ one takes an arbitrary point $N \neq C$ and rotates the segment CN, keeping C fixed, so that it is parallel to Π_2. 6. If CN_2 is the position of the perpendicular after rotation, a segment of length h is marked off along $C''N_2''$ starting from C'' to the point Z''. The horizontal line through Z_2'' cuts $C''N''$ in Z''. The order line through this elevation of the vertex of the cone gives the plan Z'. 7. Tangents from Z' and Z'' to the corresponding images of the base circle can be constructed by using a perspective affinity, for example. This determines the required projections of the cone of rotation.

The six principal projections. The side elevation whose plane Π_3 is perpendicular to Π_1 and Π_2 is called the *cross elevation* (Fig.). Their lines of intersection x_{12}, x_{23} and x_{13} are perpendicular in space. The figure shows that not every spatial object can be uniquely reconstructed from its plan and elevation.

Quite generally, an object can be projected at right angles onto 6 planes that form the surface of a cube (Fig.). One obtains the six *principal projections* of a spatial structure in the *European representation*. In the *American representation* the appearance of the object is treated in the opposite sense.

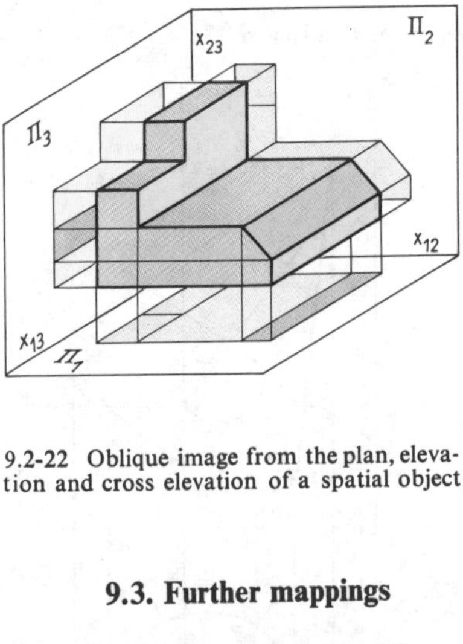

9.2-22 Oblique image from the plan, elevation and cross elevation of a spatial object

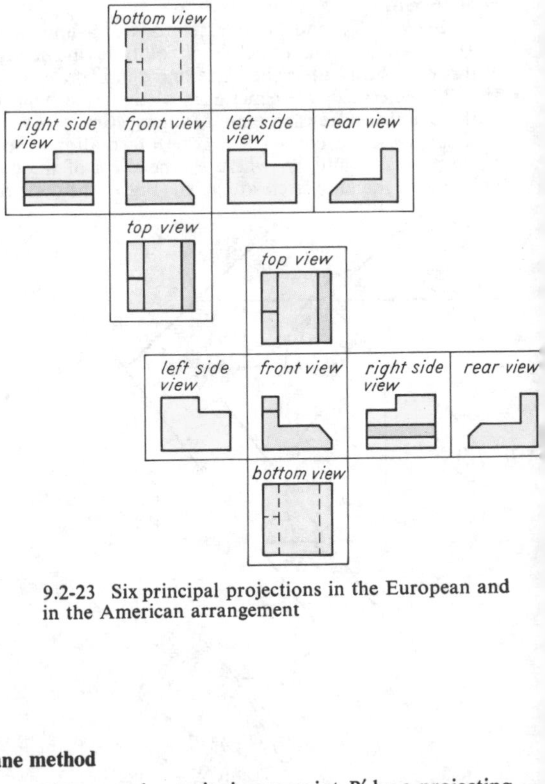

9.2-23 Six principal projections in the European and in the American arrangement

9.3. Further mappings

Projection with heights — the one-plane method

In the projection with heights a *point* P of space is mapped onto its image point P' by a projecting ray normal to the image plane Π, and its distance $k = |P'P|$ is given in terms of a fixed unit of length e as the height. The plane Π is usually taken to be horizontal and the positive half-space with $k > 0$ above it. The image l' of a line l is fixed by the images P' and Q' of two of its points P and Q. By marking off their heights (taking account of sign) on two parallel lines through P' and Q' one obtains the *trace point* L of l (Fig.). Conversely, a line in space is uniquely fixed by the images of any two *graduation points*. If its *interval* i is understood to be the distance between the projections of two graduation points whose heights differ by one unit, then its angle of inclination $\alpha = \sphericalangle(l, l')$ to the image plane is determined by the equation $i = e \cot \alpha$ (Fig.).

By means of projection with heights one can determine whether two non-parallel lines a and b, each given by two graduation points A, B and P, Q, intersect or not (Fig.). Points with the same

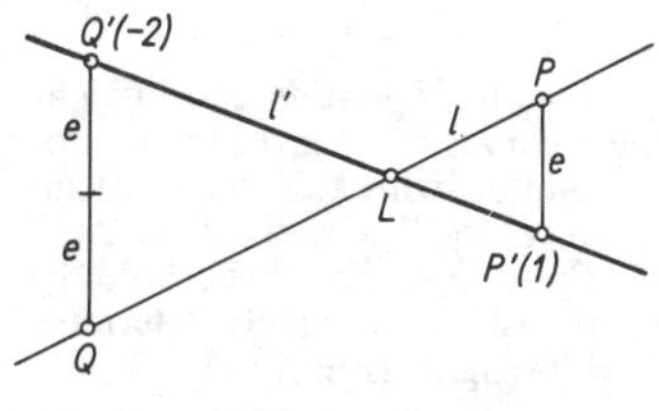

9.3-1 Representation of point and line in the projection with heights

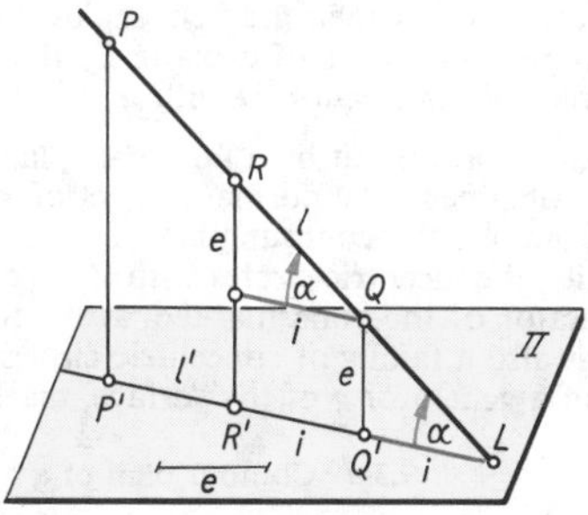

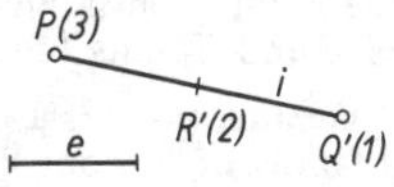

9.3-2 Interval and angle of inclination of a line; oblique image, projection with heights

height lie on a plane parallel to Π, for example, $P(2)$ and $B(2)$ on the plane $k = 2$. A plane through $P(2)$, $B(2)$ and $A(3)$ cuts the plane $k = 3$ in a line p parallel to $P(2)\,B(2)$, which contains the line a. The line b cuts a only if it lies in this plane, that is, if $Q(3)$ lies on p.

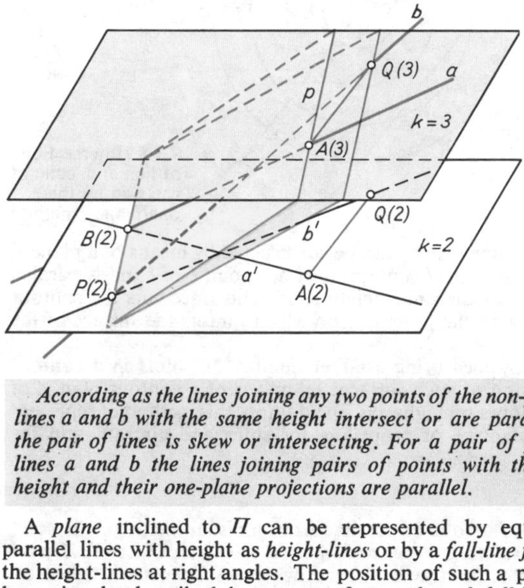

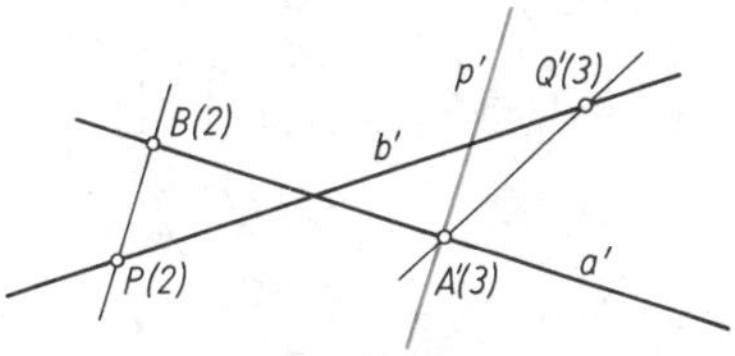

9.3-3 Projection with heights of two skew lines a and b;
a) oblique image, b) projection with heights

According as the lines joining any two points of the non-parallel lines a and b with the same height intersect or are parallel, so the pair of lines is skew or intersecting. For a pair of parallel lines a and b the lines joining pairs of points with the same height and their one-plane projections are parallel.

A *plane* inclined to Π can be represented by equidistant parallel lines with height as *height-lines* or by a *fall-line f*, cutting the height-lines at right angles. The position of such a plane can be uniquely described by means of a *graduated fall-line*. The height-line with height 0 is the *trace of the plane* (Fig.).

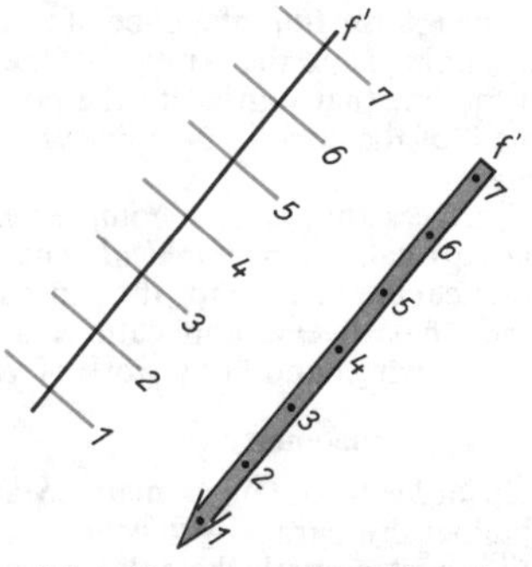

9.3-4 Representation of a plane by means of the fall scale

The intersection of two planes given by graduated fall-lines can be obtained by finding the intersections of height-lines with the same height (Fig.). This principle is applied in problems concerned with slopes and roofs.

9.3-5 Intersection of two planes in the one-plane method

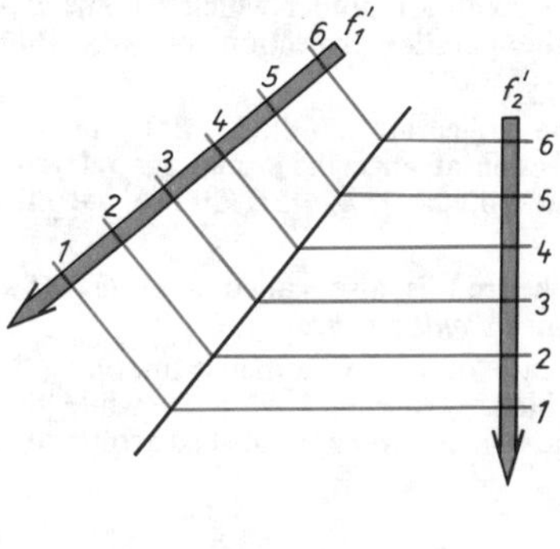

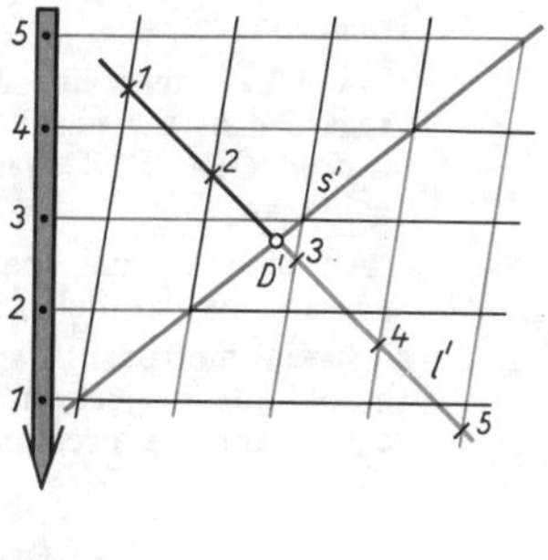

9.3-6 Intersection of a line and plane in the one-plane method

If a line l and a plane E are given by their gradient scales, then the family of parallel lines through the graduation points of l are height-lines of a plane E_1 that contains l. The line of intersection s of E and E_1 cuts l in its point of intersection D with E (Fig.).

Contour plan. If a general surface is cut by a family of planes parallel to Π with integral heights, a family of contour lines is obtained. The normal projection of these curves on Π gives the contour plan of the surface. For example, the contour plan of a cone of rotation with axis perpendicular to Π and vertex Z is a family of concentric circles with Z' as common centre. The radii of the circles can be taken from a generator of the cone marked with heights. For a one-sheet hyperboloid of rotation the contour plan is also a family of concentric circles, whose radii are uniquely determined by the normal projection of a generator g of the surface, marked with heights (Fig.).

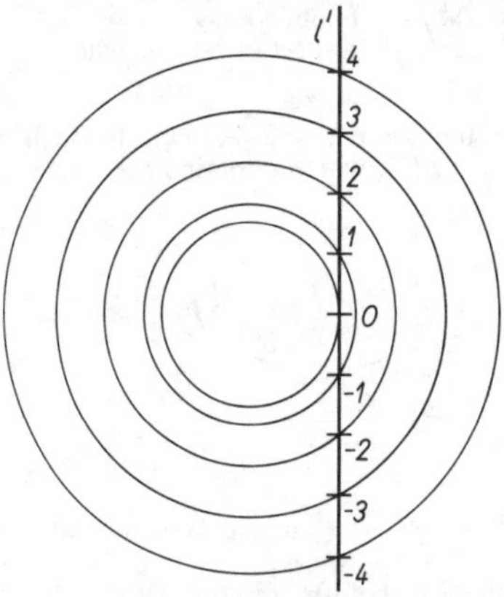

9.3-7 Contour plan of a one-sheet hyperboloid of rotation

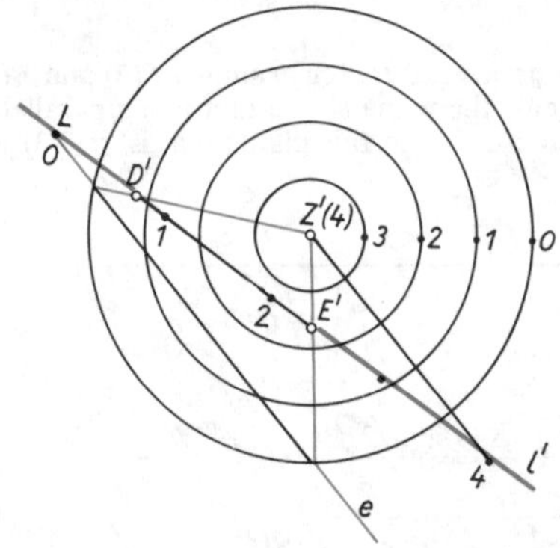

9.3-8 Intersection of line and cone of rotation by the one-plane method

The intersection of a cone of rotation with a line l (Fig.) can be constructed by means of a plane Γ containing l and the vertex Z of the cone. Its trace e in Π contains the trace point L of l and is parallel to the line that joins Z to the point of l that has the same height as Z. This trace cuts the contour line 0 of the cone in its point of intersection with the generator on which l meets the surface of the cone.

Surfaces that can be comprehended only by measuring a larger number of points in a contour plan are called *topographical*. The representation of such surfaces by means of a contour plan is of practical importance for the construction of slopes on highways. In addition, this method of representation has diverse applications in the manufacture of propellors for ships and aircraft, and for aircraft wings and body work of vehicles.

Axonometry

In order to obtain as many measurements of a solid as possible from a visual image of the solid obtained by parallel projection, the solid is referred to an *orthonormal trihedron* $O\ (X, Y, Z)$ and this, together with the solid, is projected onto the drawing-board, where its image is a trihedron $O^s\ (X^s, Y^s, Z^s)$. According to the direction of incidence of the projecting rays one distinguishes between *general*, or *skew*, *axonometry* and *orthogonal*, or *normal*, *axonometry*. Instead of the single unit segment $|OX| = |OY| = |OZ| = e$ in the original figure, there are three, $|O^sX^s| = e_x$, $|O^sY^s| = e_y$ and $|O^sZ^s| = e_z$ in the image, which can have different lengths and are obtained from the image $O^s(X^s, Y^s, Z^s)$ (Fig.).

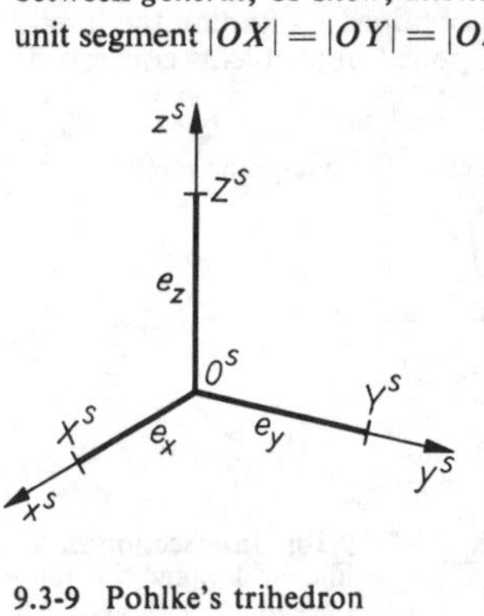

9.3-9 Pohlke's trihedron

Pohlke's theorem contains a condition under which a plane trihedron can be regarded as the parallel projection of a spatial trihedron.

Pohlke's theorem. A plane trihedron $O^s(X^s, Y^s, Z^s)$ can be regarded as the parallel projection of an orthonormal spatial trihedron $O(X, Y, Z)$ if the four points O^s, X^s, Y^s, Z^s are not all collinear.

An orthonormal spatial trihedron is also called a *vertex of a cube* and its parallel projection a *Pohlke trihedron.*

Special methods. A special case of an axonometric mapping is the oblique projection, for which $e_y = e_z = 1$, $y^s \perp z^s$, while the scale e_x and the direction of the x^s-axis can be prescribed arbitrarily.

This is a question of a *dimetric skew axonometry*, which is also known as *frontal axonometry*. The *cavalier perspective* is a special case of the oblique projection method; in this case, $e_x = e_y = e_z = 1$, $y^s \perp z^s$ and $\sphericalangle(x^s, y^s) = 135°$ (Fig.). The *military perspective* or *bird's-eye view* is characterized by $e_x = e_y = e_z = 1$, $x^s \perp y^s$ and z^s vertical (Fig.). Like the cavalier perspective, it represents an *isometric axonometry* and is used to give a more lifelike impression of a building than the plan does.

In practice one frequently uses isometric, dimetric, or trimetric representation, observing the rule of standardization demonstrated by the example of a section of a cube (Fig.).

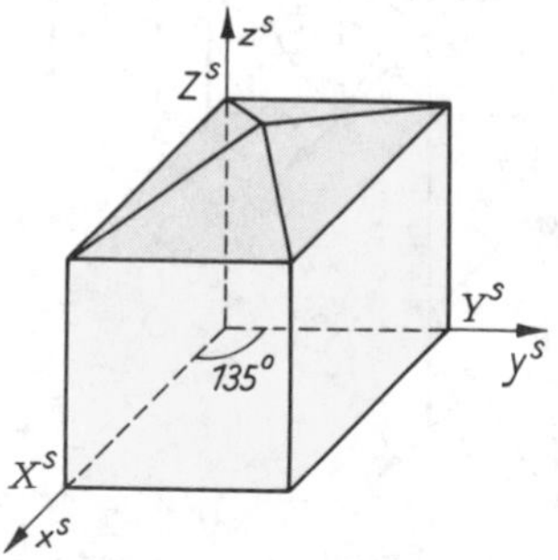

9.3-10 Model of a house in cavalier perspective

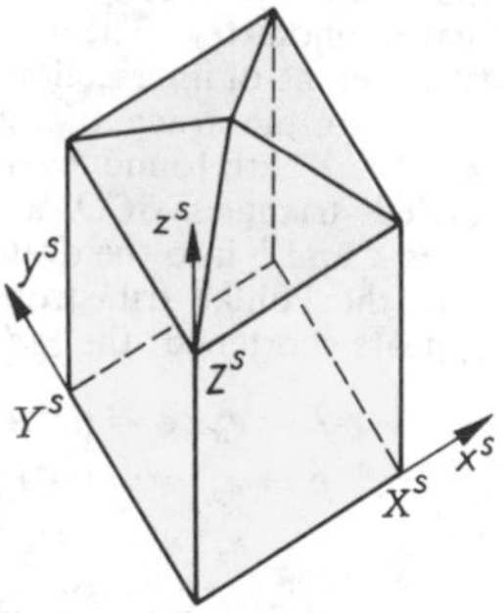

9.3-11 Model of a house in military perspective (bird's eye view)

Isometry	Dimetry	Trimetry
$e_x : e_y : e_z = 1 : 1 : 1$ $\alpha = 30°, \beta = 30°,$	$e_x : e_y : e_z = 0.5 : 1 : 1$ $\alpha = 42°, \beta = 7°$	$e_x : e_y : e_z = 0.5 : 0.9 : 1$ $\alpha = 18°, \beta = 5°$

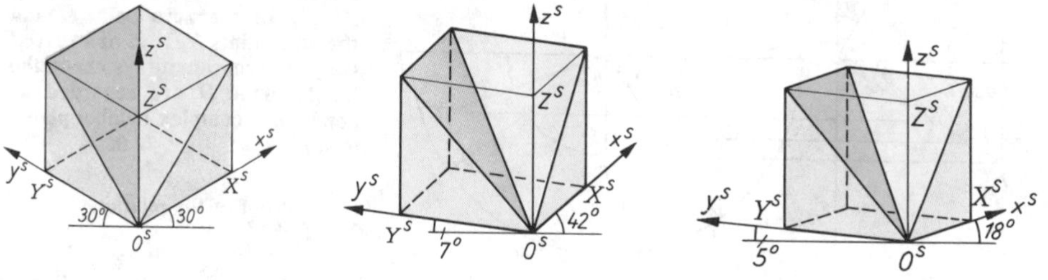

9.3-12 Images of a section of a cube: a) isometric, b) dimetric and c) trimetric

If a representation of a spatial object is given by means of coordinated normal projections, then an axonometric image can be obtained by the *indenting method* due to L. ECKHART. The two images are separated and placed arbitrarily on the drawing-board. For each projection an indenting direction is prescribed arbitrarily. The image points of the axonometric image lie at the points of intersection of the corresponding indenting rays (Fig.). The method requires some practice in the arrangement of projections and in the choice of indenting directions, to obtain an axonometric image that is not too much distorted.

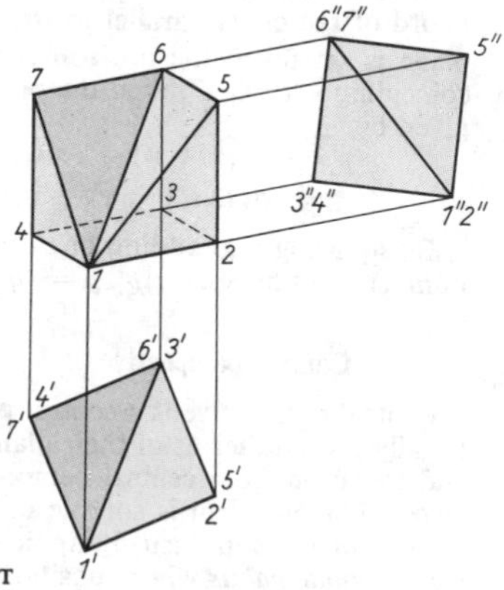

9.3-13 Axonometric image by the indentation method of L. ECKHART

Normal axonometry. To obtain clear images it is assumed that none of the axes of the orthonormal spatial trihedron O (X, Y, Z) is parallel to the image plane Π. Then its traces A, B, C are proper points. The trace triangle ABC with sides a, b, c is, from spatial considerations, acute angled and it uniquely determines the Pohlke trihedron O^n (X^n, Y^n, Z^n) by normal axonometry. The normal projection O^n is the point of intersection of the altitudes h_a, h_b, h_c of the trace triangle, and the points X^n, Y^n, Z^n are found by reflecting the right-angled triangles BCO and CAO about the sides a and b into the drawing-board (Fig.).

In the Pohlke trihedron the unit segment e appears shortened; the *reduction factors* are

$$\lambda = e_x : e = |O^nA| : |OA|,$$
$$\mu = e_y : e = |O^nB| : |OB|,$$
$$\nu = e_z : e = |O^nC| : |OC|.$$

These satisfy a relation which is obtained in the course of deriving Gauss's theorem.

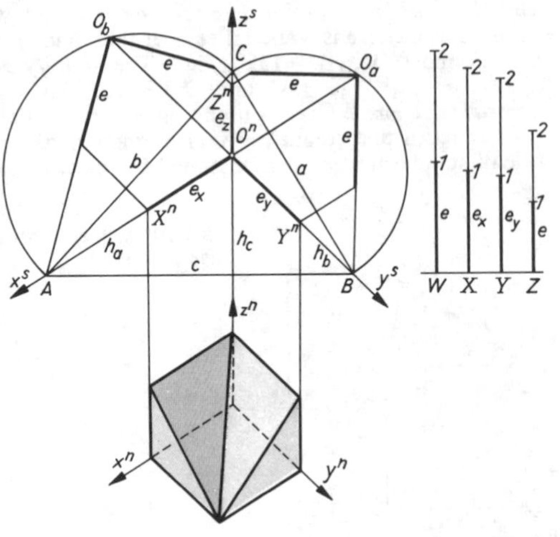

9.3-14 Normal axonometric Pohlke trihedron and its application to a section of a cube

Relation between the reduction factors
$\lambda^2 + \mu^2 + \nu^2 = 2$

Gauss's theorem. If the normal projections $O' = O$, $X' = p$, $Y' = q$, $Z' = r$ of the zero point O and the unit points X, Y, Z of a spatial Cartesian coordinate system on the image plane Π are regarded as points of a complex number plane, then $p^2 + q^2 + r^2 = 0$.

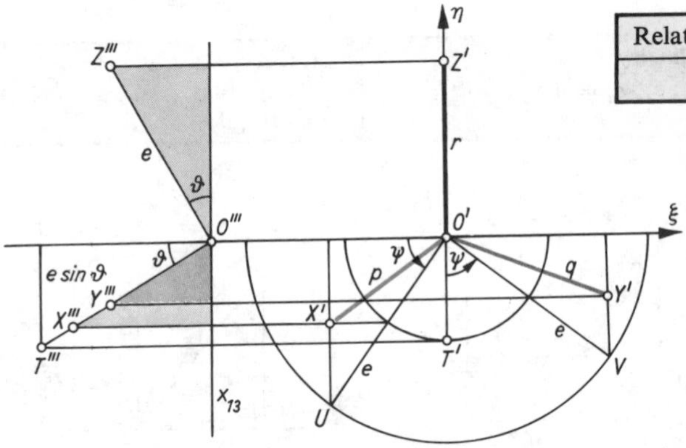

9.3-15 Proof of the relation $p^2 + q^2 + r^2 = 0$

To prove this, one represents the orthonormal trihedron $O(X, Y, Z)$ by means of coordinated normal projections (plan and side elevation), where the side elevation is taken parallel to the OZ-axis (Fig.). It follows that $|O'Z'| = |r| = e \cos\vartheta$. Furthermore, $O'X'$ and $O'Y'$ are conjugate semi-diameters of an ellipse, which is derived from a circle with centre O' and radius e by reducing each chord of the circle parallel to the ground line x_{13} in the ratio $\sin\vartheta : 1$. If a Gaussian ξ, η-number plane is set up in the horizontal plane of projection with O' as origin and the imaginary η-axis coinciding with $O'Z'$, then the points X', Y', Z' correspond to complex numbers p, q, r. These are given by

$$p = \cos(\pi + \psi) + \mathrm{i} \sin\vartheta \sin(\pi + \psi) = -\cos\psi - \mathrm{i} \sin\vartheta \sin\psi,$$
$$q = \cos(3\pi/2 + \psi) + \mathrm{i} \sin\vartheta \sin(3\pi/2 + \psi) = \sin\psi - \mathrm{i} \sin\vartheta \cos\psi, \qquad r = \mathrm{i} \cos\vartheta.$$

By squaring and adding one obtains $p^2 + q^2 + r^2 = 0$. If one takes the moduli of the complex numbers and puts $\lambda = |p|$, $\mu = |q|$, $\nu = |r|$, then it follows that $\lambda^2 + \mu^2 + \nu^2 = 2$.

Central perspective

Central perspective is a central projection and is used to construct clear images of spatial objects, usually given in terms of their plan and elevation. The converse problem of reconstructing the plan and elevation from central perspective images, usually photographs, is the domain of *photogrammetry*. This problem is soluble only if for an *oriented camera* its position relative to the object is known, or if, as in a cartographic survey by means of aerial photographs, the photographs contain certain *guide points* whose position is known.

To determine a central perspective mapping it is sufficient to know the centre of perspective or *point of sight* O, a horizontal *position plane* Γ not passing through O and a vertical *image plane* Π not passing through O. The plane Ω parallel to Γ through O cuts the image plane Π in the *horizon* h. The image plane and the position plane intersect in the *position line* l. The foot of the perpendicular from O to Π is the *principal point* H. It lies on the horizon h. The segment $d = |OH| = |O'H'|$ is the *eye distance*.

Intersection and vanishing point methods. If the plan and elevation of a model of a house are given, then the central perspective image of the model relative to a vertical image plane Π and point of sight O can be constructed point by point by the rules of the two-plane method. This is shown for the point P (Fig.). $P'O'$ cuts the position line l in $P^{c\prime}$, and $P''O''$ cuts the order line through $P^{c\prime}$ in $P^{c\prime\prime}$. In this way the image P^c of P can be found. It can be transferred to another drawing-board, which corresponds to a reflection of Π about l and contains the horizon, the principal point and the vanishing points. By the rules of central projection all parallel lines have the same vanishing point. For lines parallel to the position plan Γ these vanishing points lie on the horizon h. For the family of parallel lines determined by AB or DC it is F_1. Its plan is the intersection of l with the line through O' parallel to DC; similarly, F_2' is the plan of the vanishing point for all parallel lines characterized by DA or CB. The vanishing point of all lines perpendicular to the image plane is the principal point H. Furthermore, the points of intersection of the base edges with the position line are helpful for the construction. The points 1, 2, 3, 4 on l and F_1, F_2 on h, obtained in the present example, can be brought into the central projection by orientation on the principal point H. In this way the construction of the perspective plan is reduced to the joining of points and the intersection of lines. Now the heights of the first line and gutter line of the model of a house are taken from the elevation and laid off on the perpendiculars to l through 1, 2, 3 and 4 in Π. By using the vanishing points F_1 and F_2 the perspective image of the model can be completed.

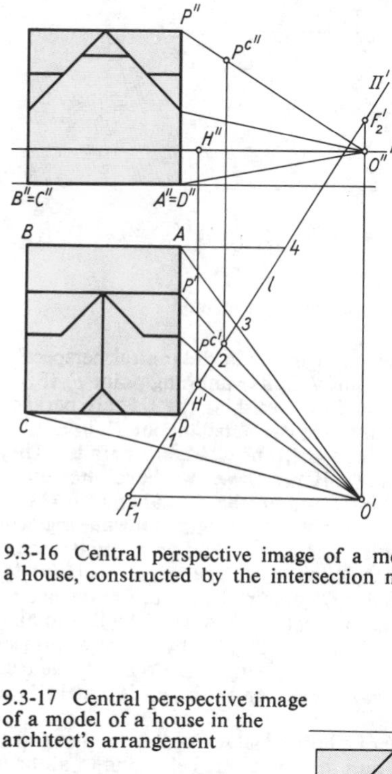

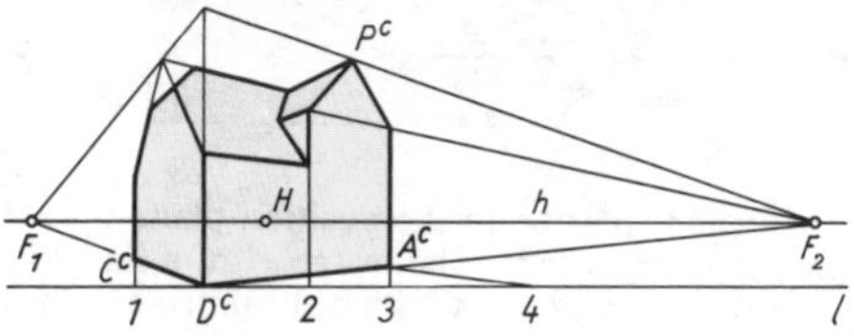

9.3-16 Central perspective image of a model of a house, constructed by the intersection method

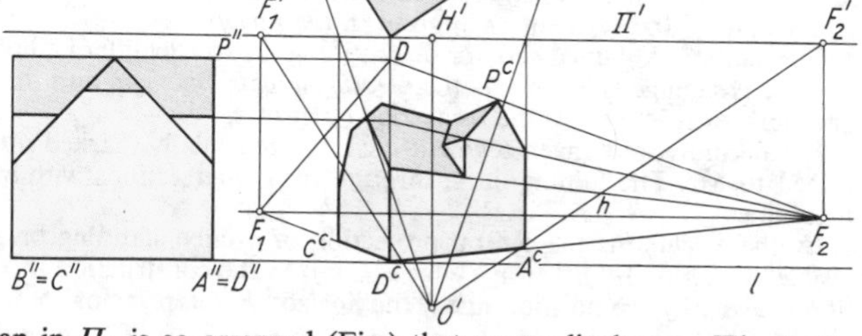

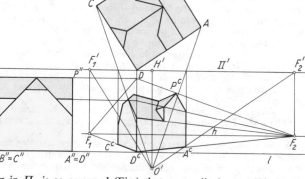

9.3-17 Central perspective image of a model of a house in the architect's arrangement

The *intersection method* was described by BRUNELLESCI (1377–1446). It has the disadvantage of being very space-consuming and requiring transfer of measurements, and is avoided in the *architect's arrangement*. The plan in Π_1 is so arranged (Fig.) that perpendiculars to Π_1' give the perpendicular edges of the house in the central perspective image. The heights of all the front edges of the house with respect to the point of sight are carried over from the elevation, and the apparently reduced heights in the image are constructed not in the elevation, but by means of vanishing lines.

Measurement problems of perspective. In the treatment of measurement problems of central perspective several tools are needed, such as the determination of the true length of a segment AB whose central projection A^cB^c is given. For segments perpendicular to the position plane Γ the solution is simple. For example, if the true length of a segment m perpendicular to Γ, given by $m^c = |A^cB^c|$, is to be found (Fig.), a point F_s is taken arbitrarily on h and joined to A^c and B^c. The line $s^c = A^cF_s$ lies in Γ and cuts l in S, a point of the image plane Π. Let t^c denote the line B^cF_s, it intersects the perpendicular to l at S in T. Now T also lies in Π, hence $|ST|$ is the true length of the perpendicular m given by a central projection.

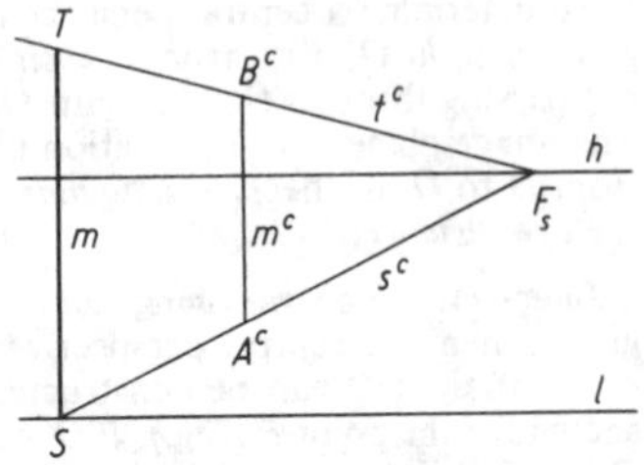

9.3-18 True length of a perpendicular to the position plane Γ

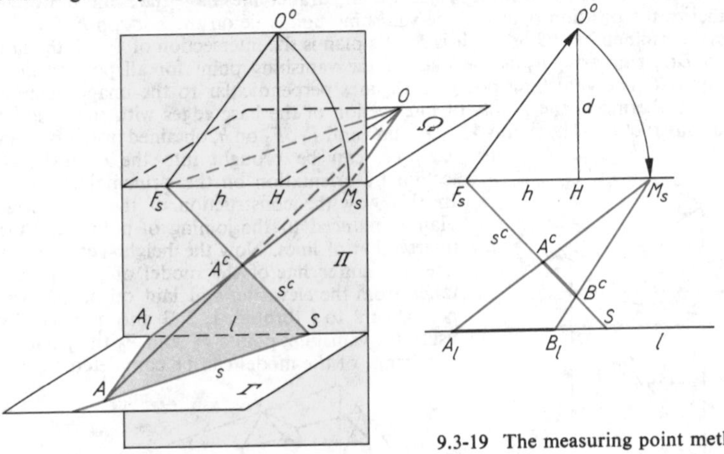

9.3-19 The measuring point method

For lines that lie in the position plane the laying off of given segments in the central perspective images is important. The image s^c of a line s in Γ cuts l in S and h in its vanishing point F_s (Fig.). On s there is a point A with central projection A^c. The line s and the vanishing ray OF_s are parallel. To derive the construction s is rotated in Γ about S to l and OF_s is rotated about F_s in Ω to h. Under the rotations, A goes into A_l and O into M_s. The lines AA_l and OM_s are parallel. They therefore span a plane which contains the lines OA and M_sA_l. Hence these two lines intersect in space. Since their common point must lie on the one hand in Π and on the other hand on OA, it can only be the image point A^c of A. Since the segment $|SA_l|$ gives the true length of the line segment $|SA|$ given by a central projection, the true length of any image segment laid off on s^c can be reconstructed with the help of the point M_s by means of a construction in Π. The measuring point M_s belonging to s can also be found in Π from F_s and the point O° obtained by rotating the eye O. With centre at F_s and radius $|F_sO^\circ|$ an arc of a circle is drawn, cutting h in M_s. The line joining M_s and A^c cuts l in A_l. $|SA_l|$ gives the true length of the segment $|SA|$ given by a central projection. If the image B^c of a further point B is prescribed on s^c, then with the help of M_s the true length of $|SB|$ can be constructed in a similar way. Furthermore, the segment $|A_lB_l|$ is equal to the true length of the segment $|AB|$ given in the image.

The method explained here for determining the true length of a horizontal segment from the central perspective image is known as the *measuring point method*, and the points M_s belonging to all lines parallel to s is known as the *measuring point* of s.

Furthermore, any given segment a can be repeatedly marked off on l and the points of division joined to M_s. The joining lines, through their intersections with s^c, generate a projective scale on these lines.

By these methods the central projection of a cube standing on Γ can be constructed, given the visible projection $|A^cB^c|$ of one edge a (Fig.). In the construction it is assumed that the central projection is given by the position line l, the horizon h, the principal point H and the eye distance d.

Perspective collineation. Suppose that a point A in Γ and its central projection A^c are given (Fig.). The planes Γ and Ω are rotated in the same sense about l and h, respectively, to Π. Then A goes into A° and O into O°. The chords AA° and OO° are parallel and span a plane, which cuts Π in $O^\circ A^\circ$. Since this plane contains the visual ray OA, A^c also lies on $O^\circ A^\circ$. Furthermore, the parallels

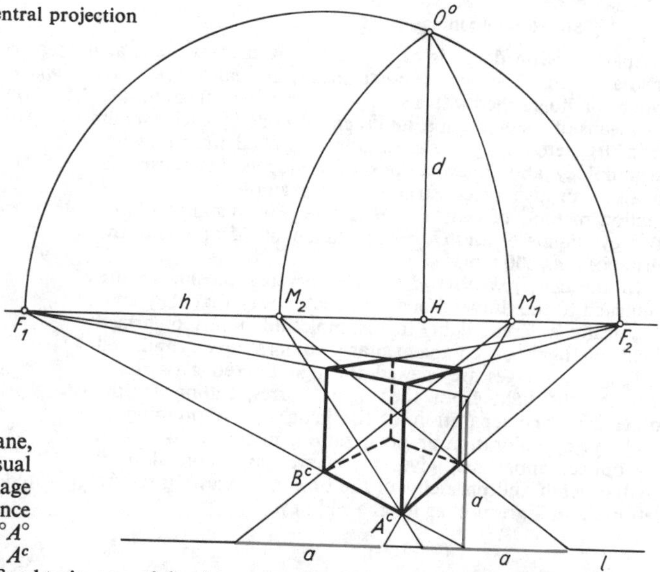

9.3-20 Construction of the central projection of a cube given an edge $A^c B^c$

OH and AA_l span a plane, which also contains the visual ray OA and cuts the image plane Π in the line HA_l. Hence the point of intersection of $O^\circ A^\circ$ and HA_l is the image point A^c of A. The construction, confined to its essentials, shows that there is a perspective collinear relation between A^c and A° with O° as centre of collineation, l as axis of collineation, and h as counteraxis.

The central perspective image of a figure in Γ and its rotation about l into the image plane Π are related by a perspective collineation (Fig.).

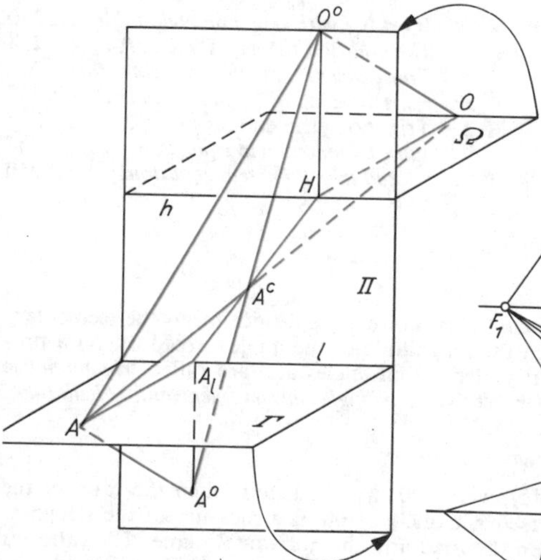

9.3-21 Perspective collinear point relation between the central projection of a point A in Γ and the result of rotating it

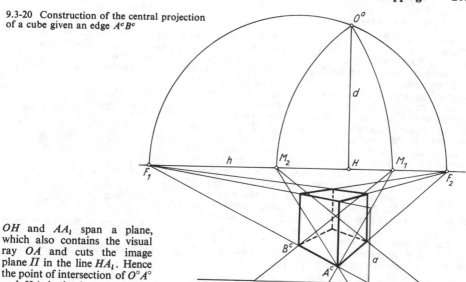

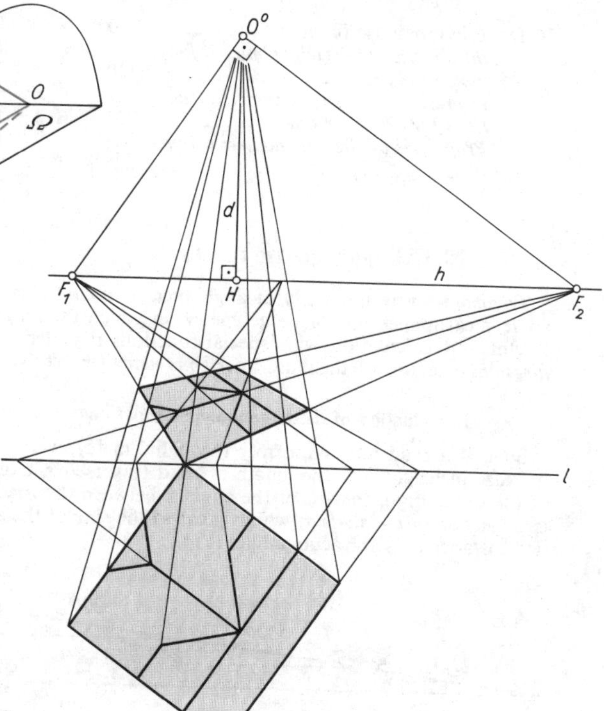

9.3-22 Perspective collineation between a plane figure in Γ and the result of rotating it into the image plane Π

Stereoscopic image pairs

Spatial vision depends on the fact that each eye sees a different central perspective image of a spatial object. The two different images are called *stereoscopic image pairs* or *stereo-images*. They can be photographed with a stereo-camera of about 65 mm ($\approx$ 2.56″) objective distance and separated by means of a stereoscope, and then seen one by each eye. Since a spatial object can be reconstructed from its stereo-images, *stereoscopy* is applied in surveying, criminology and investigations of accidents. Stereoscopic images can also be constructed, for example, by the intersection method of central perspective. The distance of the two eye-points O and $\bar{O}$ is then taken as 65 mm and the distance d as 200 mm.

In the *anaglyph method* the two stereoscopic images are returned to the drawing-board in such a way that they emit physically different light, for example, differently polarized light or light of complementary colours green and red (Fig.). The images are viewed through filtered glass. Each absorbs the light from the image corresponding to the other eye. The separation of the images corresponding to each eye by coloration or polarization, in conjunction with an optical apparatus, gives the viewer an impression of spatial depth and plasticity of the object. In viewing the figure a red filter should be used for the left eye and a green filter for the right eye.

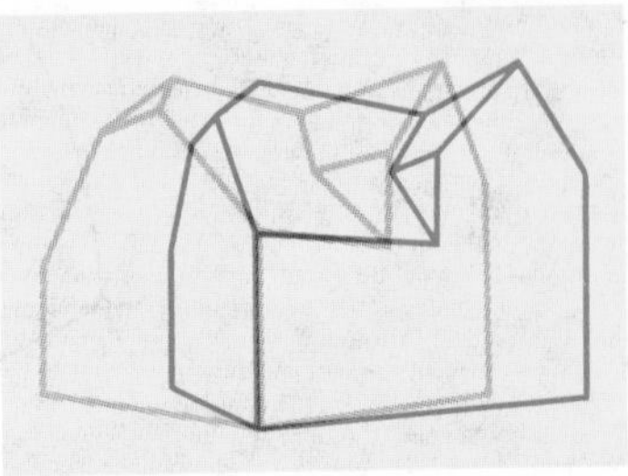

9.3-23 Two central perspective images of a model of a house

10. Trigonometry

10.1. Trigonometric functions

Trigonometry is the study of angle measurement. This, however, does not mean the elementary angle measurement of plane geometry, in which the magnitude of the angle is read off on a protractor, but calculation with special functions that depend on angles and are called *trigonometric functions* because of their use in *trigonometry* (the study of *triangle measurement and calculation*).

Introduction of the trigonometric functions

Sine. If a road rises uniformly through 3 m for every 100 m of its length, then the ratio of the increase in height h to the length of road traversed s, namely 3/100, is a measure of the steepness of the road (Fig.), that is, of the angle α between the road and the horizontal plane. The ratio h/s is a *function of the angle* α, which is called the sine of the angle α, and is accordingly defined in the first instance only for acute angles (Fig.).

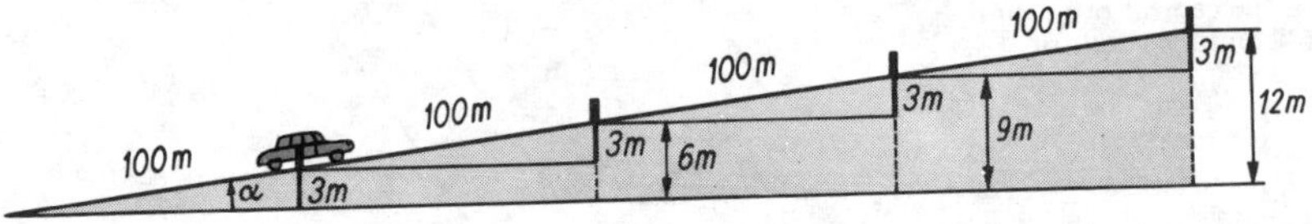

10.1-1 Ascent on an inclined plane (drawing exaggerated)

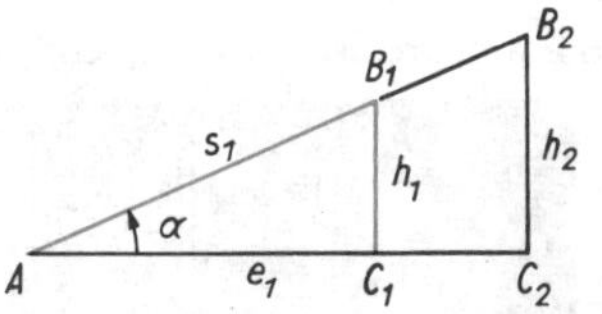

10.1-2 Sine and cosine of the angle α; $|AB_2| = s_2$, $|AC_2| = e_2$; $h_1/s_1 = h_2/s_2$; $e_1/s_1 = e_2/s_2$; $\triangle\ AB_1C_1$ is similar to $\triangle\ AB_2C_2$; $h/s = \sin\alpha$, $h = s\sin\alpha$, $e/s = \cos\alpha$, $e = s\cos\alpha$

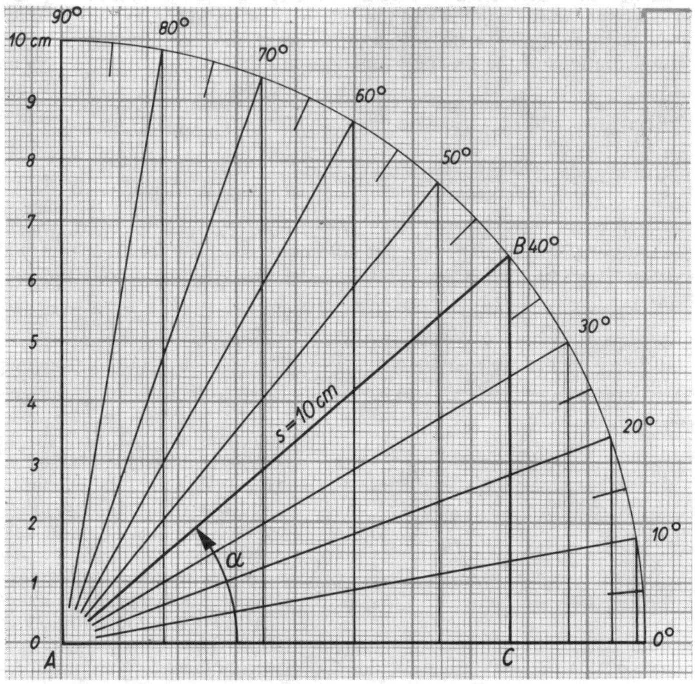

10.1-3 Graphical determination of the sine and cosine of an acute angle α from the ratios h/s and e/s; for example, $\sin 40° \approx 0.643$, $\cos 40° \approx 0.770$

From a sufficiently large drawing on squared paper the value of $h/s = \sin\alpha$ can be read off. The graphical determination of the sine is particularly simple if the divisor s is a power of 10, for example 10 cm (Fig.). For $\alpha = 40°$, for example, one obtains $h/s = 6.43\text{ cm}/10.00\text{ cm} = 0.643$. The accuracy of this method is not very high, but it can be increased by enlargement of the drawing. For every acute angle α the sine function has a fixed value, which is always less than 1 and for greater angles is greater than for smaller angles.

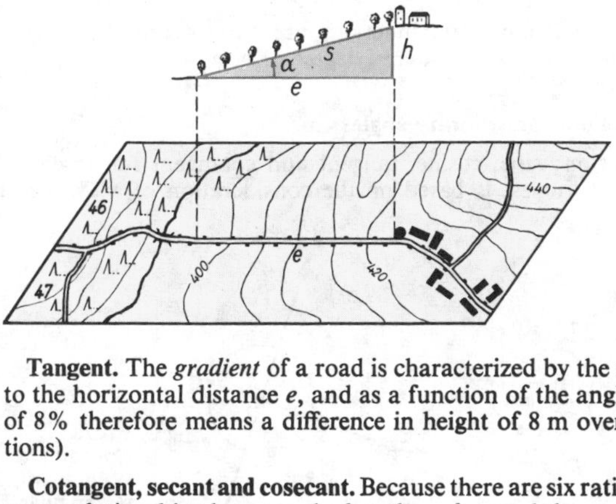

Cosine. On maps the *projections e on the horizontal plane* of the inclined segments s appear as map distances (Fig.), The ratio e/s is also a function of the angle α, called the cosine: $\cos\alpha = e/s$, $e = s\cos\alpha$. The values of the cosine function for an acute angle α decrease as the angle α increases. One obtains graphically $\cos 40° \approx 7.70\text{ cm}/10\text{ cm} = 0.770$ (Fig. 10.1-3).

10.1-4 Projection of an inclined line segment s onto the plane of the map (plan)

Tangent. The *gradient* of a road is characterized by the ratio $h/e = \tan\alpha$ of the increase in height h to the horizontal distance e, and as a function of the angle α it is called the tangent of α. A gradient of 8% therefore means a difference in height of 8 m over a map distance of 100 m (see linear functions).

Cotangent, secant and cosecant. Because there are six ratios, in general, between three distances, three more relationships between the lengths s, h, e and the angle α can be defined. Of these the cotangent is the reciprocal of the tangent and the remaining two, secant and cosecant, are used less frequently, for example, in astronomy or in navigation.

sine: $\sin\alpha = h/s$, cosine: $\cos\alpha = e/s$,
tangent: $\tan\alpha = h/e$, cotangent: $\cot\alpha = e/h$,
secant: $\sec\alpha = s/e$, cosecant: $\operatorname{cosec}\alpha = s/h$.

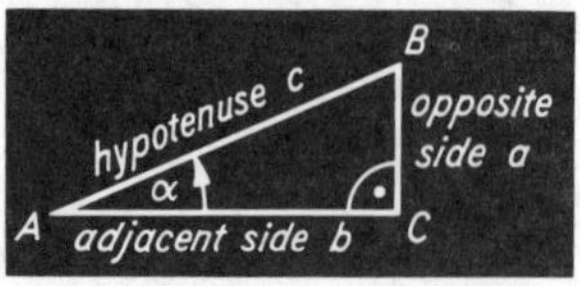

10.1-5 Angle α in the right-angled triangle

In the right-angled triangle *ABC* with hypotenuse *c* (Fig.), the side *a* opposite the angle α and the side *b* adjacent to the angle α are known as the *opposite side* and *adjacent side*, respectively. The definitions of the six angle functions are then:

$$\sin\alpha = \frac{a}{c} = \frac{\text{opposite side}}{\text{hypotenuse}}, \quad \tan\alpha = \frac{a}{b} = \frac{\text{opposite side}}{\text{adjacent side}}, \quad \sec\alpha = \frac{c}{b} = \frac{\text{hypotenuse}}{\text{adjacent side}},$$

$$\cos\alpha = \frac{b}{c} = \frac{\text{adjacent side}}{\text{hypotenuse}}, \quad \cot\alpha = \frac{b}{a} = \frac{\text{adjacent side}}{\text{opposite side}}, \quad \operatorname{cosec}\alpha = \frac{c}{a} = \frac{\text{hypotenuse}}{\text{opposite side}}.$$

Between these trigonometric functions several relationships hold, which can easily be verified in the right-angled triangle (Fig.), but hold quite generally for an arbitrary angle α:

$$\sin^2\alpha + \cos^2\alpha = 1 \qquad \tan\alpha\cot\alpha = 1 \qquad \sec\alpha = \frac{1}{\cos\alpha} \qquad \operatorname{cosec}\alpha = \frac{1}{\sin\alpha}$$

$$\tan\alpha = \frac{\sin\alpha}{\cos\alpha} = \frac{1}{\cot\alpha}, \quad \cot\alpha = \frac{\cos\alpha}{\sin\alpha} = \frac{1}{\tan\alpha}, \quad 1 + \tan^2\alpha = \frac{1}{\cos^2\alpha}, \quad 1 + \cot^2\alpha = \frac{1}{\sin^2\alpha}$$

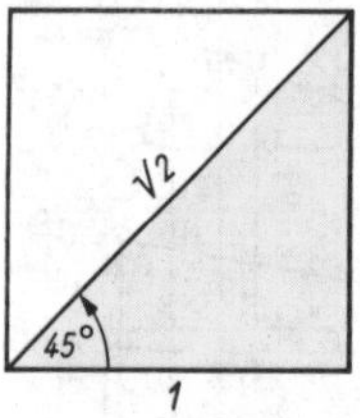

10.1-6 Square of side 1

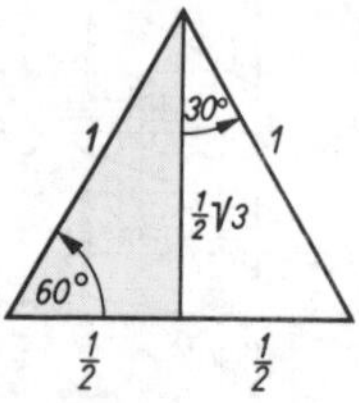

10.1-7 Equilateral triangle of side 1

The angle 45° occurs in a square of side 1 with the diagonal $d = \sqrt{2}$, and the angles 30° and 60° in an equilateral triangle of side 1 with the altitude $h = \frac{1}{2}\sqrt{3}$ (Fig.). For the four trigonometric functions commonly used one obtains the values given in the table. These values are also shown calculated to four decimal places.

A few of these values are *rational* and the remaining ones are *irrational*, but *algebraic*. Using properties derived from the addition theorems, one can obtain from these by algebraic operations the values of the trigonometric functions for $\varphi/2$, $\varphi/4$, ... and for 2φ,

Function	$\varphi = 30°$	$\varphi = 45°$	$\varphi = 60°$	Function	$\varphi = 30°$	$\varphi = 45°$	$\varphi = 60°$
$\sin\varphi$	$1/2$	$\frac{1}{2}\sqrt{2}$	$\frac{1}{2}\sqrt{3}$	$\sin\varphi$	0.5000	0.7071	0.8660
$\cos\varphi$	$\frac{1}{2}\sqrt{3}$	$\frac{1}{2}\sqrt{2}$	$1/2$	$\cos\varphi$	0.8660	0.7071	0.5000
$\tan\varphi$	$\frac{1}{3}\sqrt{3}$	1	$\sqrt{3}$	$\tan\varphi$	0.5774	1.0000	1.7321
$\cot\varphi$	$\sqrt{3}$	1	$\frac{1}{3}\sqrt{3}$	$\cot\varphi$	1.7321	1.0000	0.5774

3φ, 4φ, 5φ, ... In general, the values of the trigonometric functions are *transcendental numbers*, whose values can be calculated from infinite series to any desired degree of accuracy.

Definition of the trigonometric functions for arbitrary angles

The definition of the trigonometric functions sine, cosine, tangent and cotangent for angles of arbitrary magnitude, and not only for acute angles, is based on the consideration of a *Cartesian*

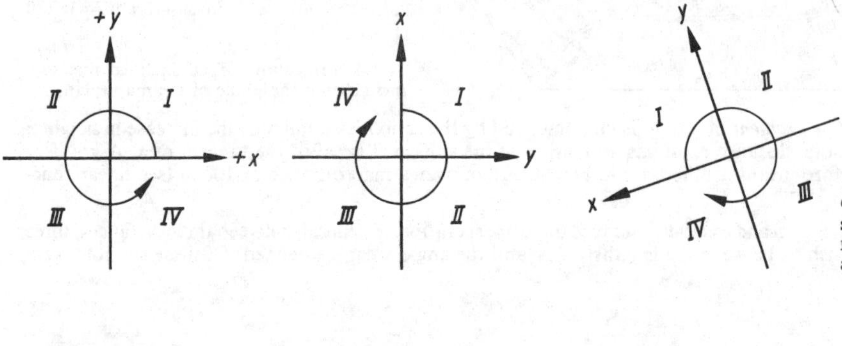

10.1-8 Coordinate systems with rectangular axes

coordinate system (Fig.), usually a left-handed system, in which an anticlockwise rotation is regarded as a *rotation in the positive sense.*

In *mining geometry* and in *geophysics right-handed systems* are often used, whose *x-axis points to the North* and *y-axis to the East* and in which the positive sense of rotation is clockwise. Some possible coordinate systems with rectangular axes are shown in the figure.

Definition on the unit circle. In a plane Cartesian coordinate system an angle φ can run through all four quadrants. Its magnitude can be measured in degrees, in new degrees or *grades*, or in radians (see Chapter 7.). Its moving arm cuts the circle with radius $r = 1$ and centre at the origin O, the so-called unit circle, in a point B_i (Fig.). For the intersection B_0 of the x-axis with the unit circle, φ has the value zero. During one complete revolution of the moving arm of φ about the origin, φ runs through all values from 0° to 360°, or 2π. The relations to be derived also hold for angles greater than 2π, since for them the point B_i assumes the same positions as for angles between 0 and 2π. The position of the points B_i, for example, B_1, B_2 or B_3, is determined by its coordinates. The abscissa is the *orthogonal projection of the particular radius* $r = 1$ *on the x-axis, and the ordinate is the orthogonal projection of this radius on the y-axis.* For the position B_3, for example, their signed numerical values are both negative, that is, $\overrightarrow{OC_3}$ is in the direction opposite to the x-axis and $\overrightarrow{C_3B_3}$ is opposite to the y-axis (Fig.). In the first quadrant the definitions already given of sine, cosine, tangent and cotangent, say from the triangle OC_1B_1, are valid. It is agreed that the same definitions shall remain valid for all quadrants, that is, for all positions of the point B_i:

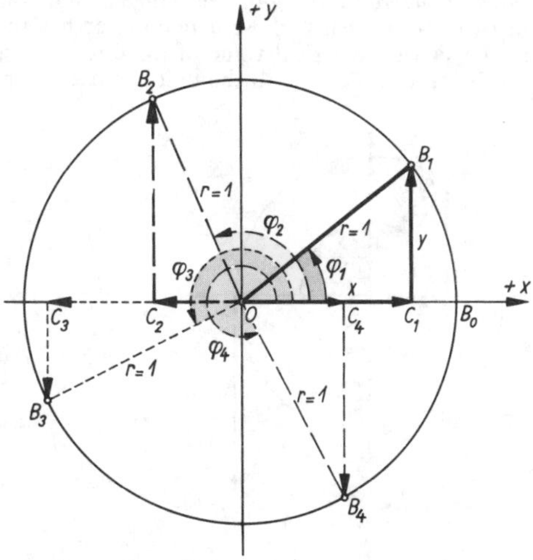

10.1-9 Definition of the trigonometric functions on a circle of radius $r = 1$

$$\sin\varphi = \frac{\text{ordinate}}{\text{radius}}, \quad \cos\varphi = \frac{\text{abscissa}}{\text{radius}},$$
$$\tan\varphi = \frac{\text{ordinate}}{\text{abscissa}}, \quad \cot\varphi = \frac{\text{abscissa}}{\text{ordinate}}.$$

Table of signs

Function	Quadrant I	II	III	IV
$\sin\varphi$	+	+	−	−
$\cos\varphi$	+	−	−	+
$\tan\varphi$	+	−	+	−
$\cot\varphi$	+	−	+	−

In these definitions the abscissae and ordinates have different signs in different quadrants, but the radius is always positive. Thus, in the figure $\sin\varphi_2$, $\tan\varphi_3$, $\cot\varphi_3$ and $\cos\varphi_4$ are positive, but $\cos\varphi_2$, $\tan\varphi_2$, $\cot\varphi_2$, $\sin\varphi_3$, $\cos\varphi_3$, $\sin\varphi_4$, $\tan\varphi_4$ and $\cot\varphi_4$ are negative. The table shows the signs of the four trigonometric functions in all the quadrants.

The procedure described for extending the domain of validity of a definition to a new region (the quadrants I, II, III, IV), in such a way that the relations holding in the original domain of definition (quadrant I) remain valid, is often used in mathematics (the *principle of permanence*). In particular, all the relations between the trigonometric functions that were found when they were first introduced now hold for all values of the angle φ.

The trigonometric functions can also be determined for the angles 0°, 90° ($\pi/2$), 180° (π), 270° ($3\pi/2$) and 360° (2π), since the abscissae and ordinates for these angles have one of the values 0, +1 or −1. *Discontinuities* occur for the tangent and cotangent functions if the denominator of the fraction tends to zero. For example, if the angle φ approaches the value 90° from below, then

$$\lim_{\varphi_1 \to 90^\circ} \tan\varphi_1 = \lim_{|OC_1| \to 0} \frac{|C_1B_1|}{|OC_1|} = +\infty;$$

on the other hand, if φ approaches 90° from above,

$$\lim_{\varphi_2 \to 90^\circ} \tan\varphi_2 = \lim_{|OC_2| \to 0} \frac{|C_2B_2|}{-|OC_2|} = -\infty.$$

Angle φ	0° 0	90° $\pi/2$	180° π	270° $3\pi/2$	360° 2π
$\sin\varphi$	0	+1	−1	−1	0
$\cos\varphi$	+1	0	0	0	+1
$\tan\varphi$	0	$\pm\infty$	0	$\pm\infty$	0
$\cot\varphi$	$\pm\infty$	0	$\pm\infty$	0	$\pm\infty$

As the increasing angle φ passes through the value $\varphi = 90°$, the value of the tangent function jumps from $+\infty$ to $-\infty$. For $\varphi = 90°$ itself the function is not defined. The notation for this situation is abbreviated to $\tan 90° = \pm\infty$. Similar *jump discontinuities* occur for the tangent function at $\varphi = 3\pi/2$ and for the cotangent function at $\varphi = 0$ and at $\varphi = \pi$. Since the radius of the unit circle has length $r = +1$, the sine and cosine are given by the *ordinate* and *abscissa*, respectively (*with the appropriate sign*). The tangent and cotangent functions can also be expressed as the ratio of two line segments whose denominator has the value 1. The value of the *tangent* can be read off as the signed numerical value of the directed segment intercepted between the arms of the angle φ on the tangent ($x = 1$) to the unit circle at the point B_0. For, by the intercept theorems (Fig.):

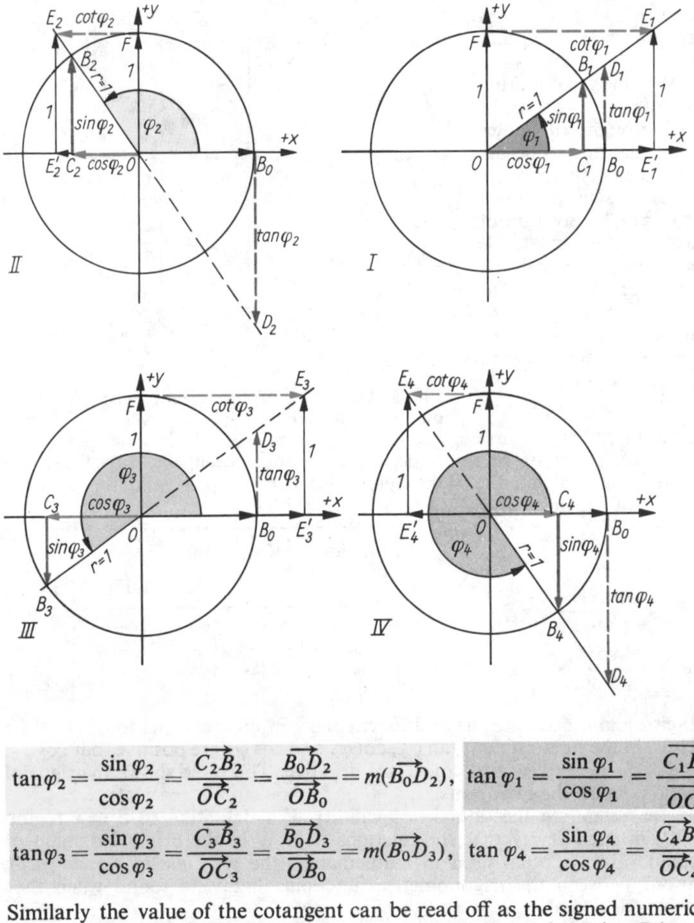

10.1-10 The trigonometric functions in the four quadrants

$$\tan\varphi_2 = \frac{\sin\varphi_2}{\cos\varphi_2} = \frac{\overrightarrow{C_2B_2}}{\overrightarrow{OC_2}} = \frac{\overrightarrow{B_0D_2}}{\overrightarrow{OB_0}} = m(\overrightarrow{B_0D_2}), \quad \tan\varphi_1 = \frac{\sin\varphi_1}{\cos\varphi_1} = \frac{\overrightarrow{C_1B_1}}{\overrightarrow{OC_1}} = \frac{\overrightarrow{B_0D_1}}{\overrightarrow{OB_0}} = m(\overrightarrow{B_0D_1})$$

$$\tan\varphi_3 = \frac{\sin\varphi_3}{\cos\varphi_3} = \frac{\overrightarrow{C_3B_3}}{\overrightarrow{OC_3}} = \frac{\overrightarrow{B_0D_3}}{\overrightarrow{OB_0}} = m(\overrightarrow{B_0D_3}), \quad \tan\varphi_4 = \frac{\sin\varphi_4}{\cos\varphi_4} = \frac{\overrightarrow{C_4B_4}}{\overrightarrow{OC_4}} = \frac{\overrightarrow{B_0D_4}}{\overrightarrow{OB_0}} = m(\overrightarrow{B_0D_4}).$$

Similarly the value of the cotangent can be read off as the signed numerical value m of the directed segment intercepted on the tangent to the unit circle at the point F ($y = 1$) by the positive y-axis and the moving arm of the angle φ. Again, by the intercept theorems:

$$\cot\varphi_2 = \frac{\cos\varphi_2}{\sin\varphi_2} = \frac{\overrightarrow{OC_2}}{\overrightarrow{C_2B_2}} = \frac{\overrightarrow{OE_2'}}{\overrightarrow{E_2'E_2}} = m(\overrightarrow{FE_2}), \quad \cot\varphi_1 = \frac{\cos\varphi_1}{\sin\varphi_1} = \frac{\overrightarrow{OC_1}}{\overrightarrow{C_1B_1}} = \frac{\overrightarrow{OE_1'}}{\overrightarrow{E_1'E_1}} = m(\overrightarrow{FE_1}),$$

$$\cot\varphi_3 = \frac{\cos\varphi_3}{\sin\varphi_3} = \frac{\overrightarrow{OC_3}}{\overrightarrow{C_3B_3}} = \frac{\overrightarrow{OE_3'}}{\overrightarrow{E_3'E_3}} = m(\overrightarrow{FE_3}), \quad \cot\varphi_4 = \frac{\cos\varphi_4}{\sin\varphi_4} = \frac{\overrightarrow{OC_4}}{\overrightarrow{C_4B_4}} = \frac{\overrightarrow{OE_4'}}{\overrightarrow{E_4'E_4}} = m(\overrightarrow{FE_4}).$$

The terms *tangent* and *cotangent* can accordingly be visualized as the *signed numerical values of segments on the tangent* (with point of contact at $x = 1$) and the *cotangent* (with point of contact at $y = 1$). The values of the functions for the angles 0, $\pi/2$, π, $3\pi/2$ and 2π can also be read off from the unit circle.

Graph of the angle functions in the four quadrants. A clear picture of the shape of the graphs of the trigonometric functions can be obtained by introducing a Cartesian coordinate system in which the argument φ in radians is taken as abscissa and the values of the respective trigonometric functions are taken as ordinates. The figure shows the pointwise construction of the graphs of the functions sine and tangent for angles at intervals of 15° ($\pi/12$ or $16^{2}/_{3}{}^{g}$) in the quadrants I and II. In the following figure the scale is reduced by half in order to show the graphs of the curves for all quadrants.

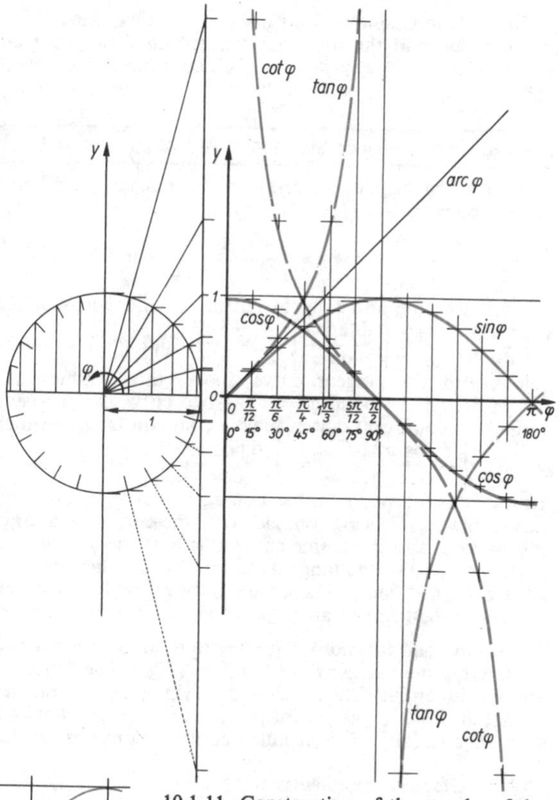

10.1-11 Construction of the graphs of the trigonometric functions

Properties of the trigonometric functions

From the two figures showing the graphical representation of the trigonometric functions a number of properties of these functions can be read off, whose validity can usually be proved from the unit circle. The angles can have arbitrary positive values and also, as will be seen, arbitrary negative values.

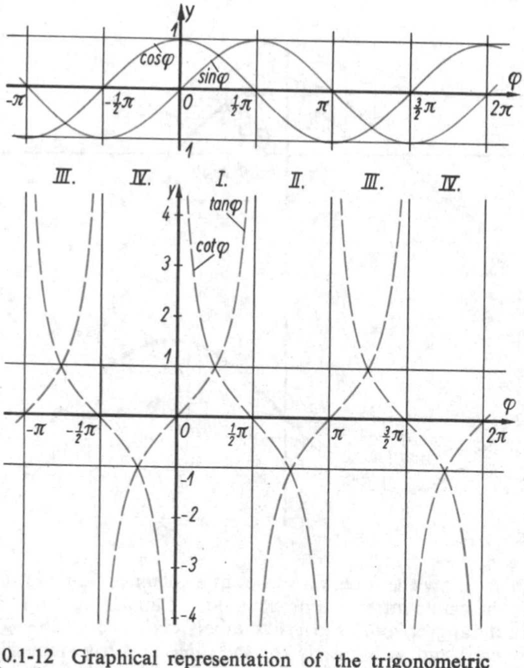

10.1-12 Graphical representation of the trigonometric functions in the four quadrants (arguments in radian measure)

Periodicity and range of values of the trigonometric functions. The trigonometric functions are periodic. The sine and the cosine functions have the period 2π (360°); the tangent and the cotangent functions have the period π (180°). In the unit circle the free arms of the angles $(\varphi \pm 2n\pi)$ all have the same position, and thus their trigonometric functions have the same values.

$$\sin(\varphi \pm 2n\pi) = \sin\varphi; \quad \cos(\varphi \pm 2n\pi) = \cos\varphi; \quad n = 1, 2, 3, \ldots$$

The unit circle representation shows further that the free arms of all the angles $(\varphi \pm n\pi)$ cut each of the tangents to the unit circle at the points where $x = 1$, $y = 1$, respectively, at a single point, and that the tangent and cotangent functions of these angles therefore all have the same value.

$$\tan(\varphi \pm n\pi) = \tan\varphi; \quad \cot(\varphi \pm n\pi) = \cot\varphi; \quad n = 1, 2, 3, \ldots$$

The functions sine and cosine take all their function values in a subinterval, for example, for $0 \leqslant \varphi \leqslant 2\pi$, and the functions tangent und cotangent take all their values in a smaller interval, for example, for $0 \leqslant \varphi \leqslant \pi$. In such an interval the functions $\sin\varphi$ and $\cos\varphi$ oscillate between the values -1 and $+1$; on the other hand, the functions $\tan\varphi$ and $\cot\varphi$ take all values between $-\infty$ and $+\infty$.

$$-1 \leqslant \sin\varphi \leqslant +1 \text{ or } |\sin\varphi| \leqslant 1, \; -1 \leqslant \cos\varphi \leqslant +1 \text{ or } |\cos\varphi| \leqslant 1$$

Slope of the tangent. According to the rules of the differential calculus, the derivative of a function at each point of its graph gives the slope of the tangent to the curve at this point.

$$\frac{\mathrm{d}\sin\varphi}{\mathrm{d}\varphi} = \cos\varphi, \quad \frac{\mathrm{d}\cos\varphi}{\mathrm{d}\varphi} = -\sin\varphi, \quad \frac{\mathrm{d}\tan\varphi}{\mathrm{d}\varphi} = \frac{1}{\cos^2\varphi}, \quad \frac{\mathrm{d}\cot\varphi}{\mathrm{d}\varphi} = -\frac{1}{\sin^2\varphi}.$$

Thus, the sine and tangent curves cut the φ-axis at the point $\varphi = 0$ at an angle of 45°, since $\left[\frac{\mathrm{d}\sin\varphi}{\mathrm{d}\varphi}\right]_{\varphi=0} = \left[\frac{\mathrm{d}\tan\varphi}{\mathrm{d}\varphi}\right]_{\varphi=0} = +1$. But as the angle increases, the sine curve moves away below, and the tangent curve above, this common tangent, because $\cos\varphi$ is decreasing for this angle. At the point $\varphi = \pi$ the two curves are perpendicular to one another. For $\varphi = \pi/2$ the cosine and cotangent curves have a common tangent which makes an angle $-45°$ with the positive φ-axis; $\left[\frac{\mathrm{d}\cos\varphi}{\mathrm{d}\varphi}\right]_{\varphi=\pi/2} = \left[\frac{\mathrm{d}\cot\varphi}{\mathrm{d}\varphi}\right]_{\varphi=\pi/2} = -1$. As the angle increases, the cosine curve moves away above, and the cotangent curve below, this common tangent. For $\varphi = 3\pi/2$ these curves intersect at a right angle. The sine curve has a tangent parallel to the φ-axis at $\varphi = \pi/2$ and at $\varphi = 3\pi/2$, and the cosine curve at $\varphi = 0$ and $\varphi = \pi$.

The progress of the tangents to the sine and tangent curves in quadrant I establishes the estimate $\sin\varphi < \operatorname{arc}\varphi < \tan\varphi$, where $\operatorname{arc}\varphi$ is the radian measure of the angle φ. In the graph $\operatorname{arc}\varphi$ is represented by a straight line at an angle of 45° to the positive φ-axis.

Even and odd functions. The function $\cos\varphi$ is even, because for positive and negative angles φ it has the same values; $f(-\varphi) = f(\varphi)$ (Fig.). The functions sine, tangent, and cotangent, however, are odd functions. Their curves are symmetrical about the origin (Fig.), because $f(-\varphi) = -f(\varphi)$, so that the function values for positive and negative angles have the same absolute value but opposite signs. The validity of these much used relationships can be seen from the unit circle.

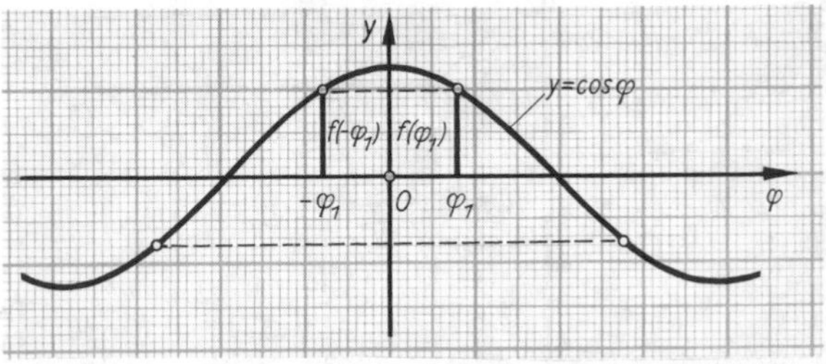

10.1-13 Graphical representation of the *even function* $y = \cos\varphi = \cos(-\varphi)$

even function
$\cos(-\varphi) = +\cos\varphi$
odd functions
$\sin(-\varphi) = -\sin\varphi$ $\tan(-\varphi) = -\tan\varphi$ $\cot(-\varphi) = -\cot\varphi$

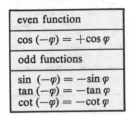

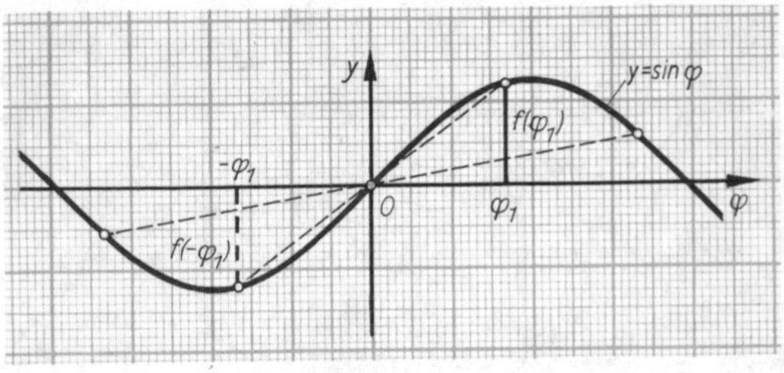

10.1-14 Graphical representation of the *odd function* $y = \sin\varphi = -\sin(-\varphi)$

Because of these properties it is sufficient to know the function values in a *subinterval of half the period length* in order to give the values for the whole interval. For example, for angles from 0 to π the cosine runs through the same values as for the angles from 2π to π; in symbols, $\cos\varphi = \cos(2\pi - \varphi)$, $\varphi \leq \pi$. Thus, it assumes all its values between 0 and π. Similarly for the three odd trigonometric functions the values in a subinterval are sufficient: from 0 to π for $\sin\varphi$, and from 0 to $\pi/2$ for $\tan\varphi$ and $\cot\varphi$.

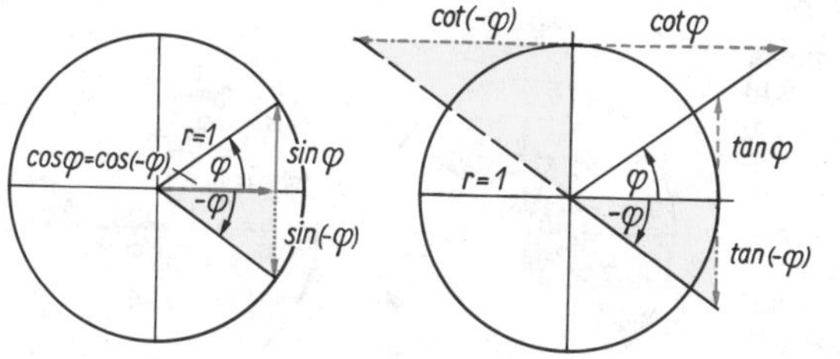

10.1-15 Sine, cosine, tangent and cotangent of a *negative angle*

From the relations between a function and its cofunction (Fig.) and those between the quadrants it suffices to know the values of $\sin\varphi$ for $0 \leqslant \varphi \leqslant \pi/2$ in order to calculate the values of all the other trigonometric functions. To simplify the calculations it is of practical convenience to give the values of the tangent together with those of the sine for $0 \leqslant \varphi \leqslant \pi/2$.

Relations between the trigonometric functions for the same angle. From the relations found in the introduction, each trigonometric function can be expressed in terms of every other one for the same argument. For example, if one wishes to express $\sin\varphi$ or $\cot\varphi$ in terms of $\cos\varphi$, one obtains:

1. $\sin\varphi = \pm\sqrt{(1-\cos^2\varphi)}$; 2. $\cot\varphi = \cos\varphi/\sin\varphi = \cos\varphi/[\pm\sqrt{(1-\cos^2\varphi)}]$.

The following table contains all the relations:

required \ given	$\sin\varphi$	$\cos\varphi$	$\tan\varphi$	$\cot\varphi$
$\sin\varphi =$	$\sin\varphi$	$\pm\sqrt{(1-\cos^2\varphi)}$	$\dfrac{\tan\varphi}{\pm\sqrt{(1+\tan^2\varphi)}}$	$\dfrac{1}{\pm\sqrt{(1+\cot^2\varphi)}}$
$\cos\varphi =$	$\pm\sqrt{(1-\sin^2\varphi)}$	$\cos\varphi$	$\dfrac{1}{\pm\sqrt{(1+\tan^2\varphi)}}$	$\dfrac{\cot\varphi}{\pm\sqrt{(1+\cot^2\varphi)}}$
$\tan\varphi =$	$\dfrac{\sin\varphi}{\pm\sqrt{(1-\sin^2\varphi)}}$	$\dfrac{\pm\sqrt{(1-\cos^2\varphi)}}{\cos\varphi}$	$\tan\varphi$	$\dfrac{1}{\cot\varphi}$
$\cot\varphi =$	$\dfrac{\pm\sqrt{(1-\sin^2\varphi)}}{\sin\varphi}$	$\dfrac{\cos\varphi}{\pm\sqrt{(1-\cos^2\varphi)}}$	$\dfrac{1}{\tan\varphi}$	$\cot\varphi$

For angles in the first quadrant the positive signs of the roots are valid. In the remaining quadrants the signs of the roots are determined from the table of signs or from the unit circle.

Example: In the third quadrant $\cos\varphi$ and $\sin\varphi$ are negative, but $\tan\varphi$ and $\cot\varphi$ are positive. Hence for $\pi < \varphi < 3\pi/2$ the second line of the table is

$$\cos\varphi = -\sqrt{(1-\sin^2\varphi)} = -\frac{1}{\sqrt{(1+\tan^2\varphi)}} = -\frac{\cot\varphi}{\sqrt{(1+\cot^2\varphi)}}.$$

Function and cofunction. The word cosine means *complementary sine*, that is, the sine of the complementary angle. Similarly cotangent and cosecant, respectively, mean *tangent* and *secant of the complementary angle*. The complement β of a given acute angle α is such that $\alpha + \beta$ is a right angel. The right angle can be measured in degrees, gons, or radians. Thus, if α and β are in radians, then $\alpha + \beta = \pi/2$. The expressions cosine, cotangent and cosecant therefore imply the mathematical statements

$$\cos\alpha = \sin(\pi/2 - \alpha) = \sin\beta,$$
$$\operatorname{cosec}\alpha = \sec(\pi/2 - \alpha) = \sec\beta.$$
$$\cot\alpha = \tan(\pi/2 - \alpha) = \tan\beta,$$

1 right angle = 90° = 100 gon = $\pi/2$ radians

In a right-angled triangle in which a and b are the sides opposite the angles α and β, respectively, it is immediately obvious that

$$\sin\alpha = a/c = \cos\beta, \quad \cos\alpha = b/c = \sin\beta, \quad \tan\alpha = a/b = \cot\beta, \quad \cot\alpha = b/a = \tan\beta.$$

In addition one sees that $\sin\alpha = \cos(\pi/2 - \alpha)$, $\tan\alpha = \cot(\pi/2 - \alpha)$, $\sec\alpha = \operatorname{cosec}(\pi/2 - \alpha)$, so that the *sine function is the cofunction of the cosine*, and the *tangent and secant*, respectively, are the *cofunctions of the cotangent and cosecant*.

Each trigonometric function assumes for the argument increasing from 0 *to* $\pi/2$ *the same values as its cofunction for the argument decreasing from* $\pi/2$ *to* 0.

Quadrant relations. Between trigonometric functions whose arguments differ by a right angle or a multiple of a right angle, certain relationships hold, the so-called quadrant relations.

The quarter turn theorem. Passage from one quadrant into the next follows from the rotation of the figure through a quarter of a complete rotation, or from the addition of a right angle to the argument φ (Fig.). In this rotation the abscissa and cosine value, respectively, become the ordinate and sine value of the same absolute value, and conversely $|\sin(\pi/2+\varphi)| = |\cos\varphi|$, $|\cos(\pi/2+\varphi)| = |\sin\varphi|$. A positive $\cos\varphi$ lies along the positive x-axis and thus, as $\sin(\pi/2+\varphi)$, lies along the positive y-axis after the rotation, hence is positive. Similarly a negative $\cos\varphi$ on the negative x-axis, after a rotation through a right angle, goes into a negative value $\sin(\pi/2+\varphi)$ on the negative y-axis. Hence $\cos\varphi = \sin(\pi/2+\varphi)$. On the other hand, a positive value $\sin\varphi$ goes from the positive y-axis to the negative x-axis under the rotation, and a negative $\sin\varphi$ from the negative y-axis to the positive x-axis. Consequently, $\sin\varphi = -\cos(\pi/2+\varphi)$.

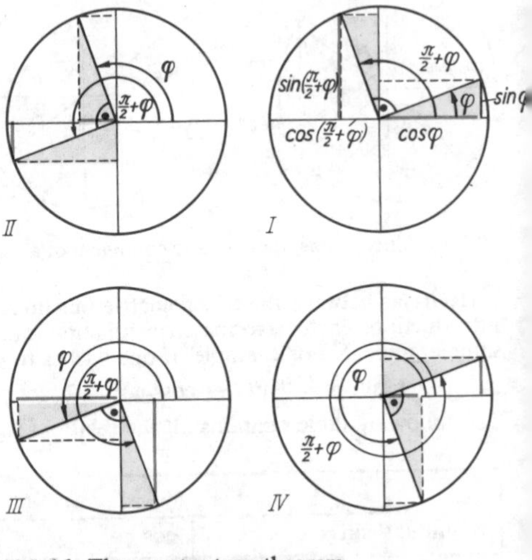

10.1-16 The quarter turn theorem
$\sin(\pi/2+\varphi) = \cos\varphi$, $\cos(\pi/2+\varphi) = -\sin\varphi$

In the coordinate system the radius of the unit circle has the same position for a positive angle φ as for a negative angle $-\psi$ if $\varphi + \psi = 4\pi/2 = 2\pi$. Hence the trigonometric functions also have the same value if φ is replaced by $-\psi$: $\sin(\pi/2-\psi) = \cos(-\psi) = \cos\psi$, $\cos(\pi/2-\psi) = -\sin(-\psi) = \sin\psi$. Because $\tan\varphi = \sin\varphi/\cos\varphi$ and $\cot\varphi = \cos\varphi/\sin\varphi$, it also follows that $\tan(\pi/2+\varphi) = -\cot\varphi$, $\cot(\pi/2+\varphi) = -\tan\varphi$, $\tan(\pi/2-\psi) = \cot\psi$, $\cot(\pi/2-\psi) = \tan\psi$.

In this way the relationships between function and cofunction have been generalized for an arbitrary angle ψ. Geometrically the equation $\sin(\pi/2+\beta) = \cos\beta$ means that the curve of the sine function may be regarded as that of the cosine curve translated by $90° = \pi/2$.

Further quadrant relations. Passage to the next quadrant but one is accomplished by the addition of $2\pi/2$, and to the next quadrant but two by the addition of $3\pi/2$. The relations that hold in these cases are obtained by substituting $\pi/2+\psi$ or $2\pi/2+\psi$, respectively, for φ in the quarter turn theorem.

Examples: 1. $\tan(2\pi/2+\psi) = \tan(\pi/2+\pi/2+\psi) = -\cot(\pi/2+\psi) = +\tan\psi$.
2. $\cos(3\pi/2+\psi) = \cos(\pi/2+2\pi/2+\psi) = -\sin(2\pi/2+\psi)$
$= -\sin(\pi/2+\pi/2+\psi) = -\cos(\pi/2+\psi) = +\sin\psi$.

One can also apply the formulae of the quarter turn theorem to a multiple of a right angle minus an arbitrary angle φ.

Example: $\sin(2\pi/2-\varphi) = \sin(\pi/2+\pi/2-\varphi) = \cos(\pi/2-\varphi) = \sin\varphi$.

Summary of the quadrant relations. By the quadrant relations a trigonometric function of an angle $(n\pi/2 \pm \delta)$ for $n = 1, 2, 3, 4$ can be expressed as a function of the angle δ, where δ is an arbitrary angle. Putting $(n\pi/2 \pm \delta) = \Phi$, the quadrant relations can be collected together in the following table:

$\Phi =$	$\pi/2-\delta$	$\pi/2+\delta$	$2\pi/2-\delta$	$2\pi/2+\delta$	$3\pi/2-\delta$	$3\pi/2+\delta$	$4\pi/2-\delta$	$4\pi/2+\delta$
$\sin\Phi$	$\cos\delta$	$\cos\delta$	$\sin\delta$	$-\sin\delta$	$-\cos\delta$	$-\cos\delta$	$-\sin\delta$	$\sin\delta$
$\cos\Phi$	$\sin\delta$	$-\sin\delta$	$-\cos\delta$	$-\cos\delta$	$-\sin\delta$	$+\sin\delta$	$+\cos\delta$	$\cos\delta$
$\tan\Phi$	$\cot\delta$	$-\cot\delta$	$-\tan\delta$	$+\tan\delta$	$+\cot\delta$	$-\cot\delta$	$-\tan\delta$	$\tan\delta$
$\cot\Phi$	$\tan\delta$	$-\tan\delta$	$-\cot\delta$	$+\cot\delta$	$+\tan\delta$	$-\tan\delta$	$-\cot\delta$	$\cot\delta$
quadrant	I	II	II	III	III	IV	IV	I

The table shows the following rules:

1. when the angle δ is added to, or subtracted from, an *odd multiple* of $\pi/2$, that is for $\Phi = \pi/2 \pm \delta$ or $\Phi = 3\pi/2 \pm \delta$, the cofunction of δ occurs in the expression for the required function of Φ;
2. when the angle δ is added to, or subtracted from, an *even multiple* of $\pi/2$, that is, for $\Phi = \pi \pm \delta$ or $\Phi = 2\pi \pm \delta$, the same function of δ occurs in the expression for the required function of Φ;

3. if δ is taken to be an acute angle, then the free arm of the angle Φ is in the quadrant given in the last line of the table. The sign of the function of Φ is determined for this quadrant from the unit circle.

By means of this table the trigonometric functions of an arbitrary angle Φ can be related to functions of an acute angle δ. Since in practical problems, especially if the angles are given in degrees, minutes and seconds, only positive increments are used, the columns $\Phi = \pi/2 + \delta$, $\Phi = 2\pi/2 + \delta$ and $\Phi = 3\pi/2 + \delta$ are specially marked. From the table, for example, $\sin(2\pi/2 + \delta) = -\sin\delta$, $\tan(\pi/2 + \delta) = -\cot\delta$, $\cot(3\pi/2 + \delta) = -\tan\delta$.

Inverse functions. If a directed segment $\overrightarrow{OF_1}$ (or $\overrightarrow{OF_3}$) of length less than unity is marked off from the origin on the y-axis (Fig.), then a line parallel to the x-axis through the point F_1 (or F_3) cuts the unit circle in two points B_1 and B_2 (or B_3 and B_4). The figure shows that OB_1 and OB_2, respectively, are the free arms of two angles φ_1 and φ_2 whose sine functions are given by the signed numerical values of the segment $\overrightarrow{OF_1}$, $\sin\varphi_1 = \sin\varphi_2 = m(\overrightarrow{OF_1})$.

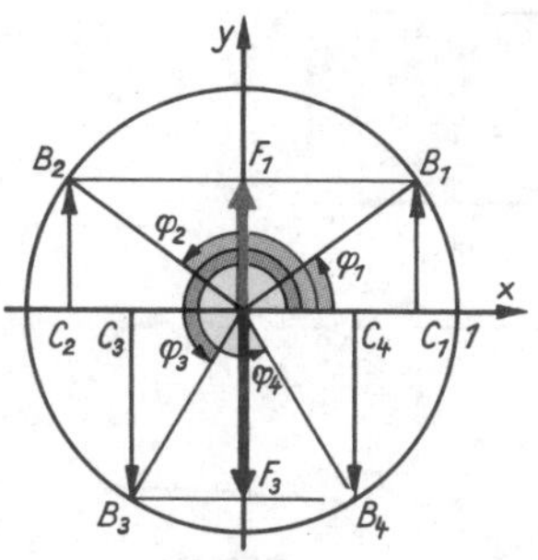

10.1-17 Construction of the angle for two given values of the sine

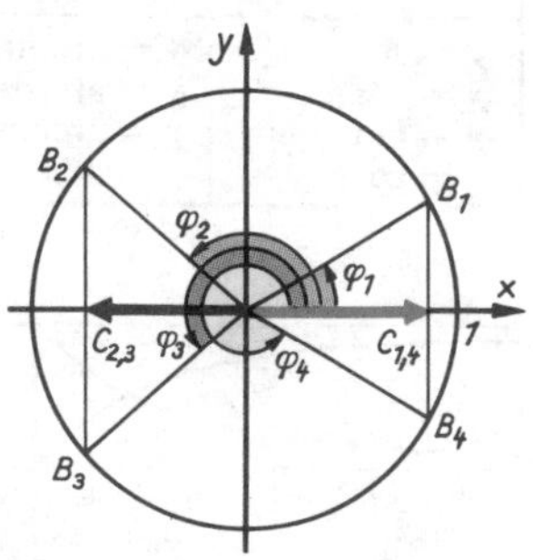

10.1-18 Construction of the angle for two given values of the cosine

Similarly OB_3 and OB_4, respectively, are the free arms of angles φ_3 and φ_4, where $\sin\varphi_3 = \sin\varphi_4 = m(\overrightarrow{OF_3})$. Thus, every number y with $|y| \leqslant 1$ can be expressed as the value of a sine function, and if $|y| < 1$, there are always two angles that satisfy the equation $y = \sin\psi$. Of course, it is clear from the figure that complete rotations of the free arm cannot be distinguished; thus, there are actually infinitely many solutions

$$\psi_1 = \varphi_1 \pm 2n\pi \quad \text{and} \quad \psi_2 = \varphi_2 \pm 2n\pi \quad \text{for} \quad y > 0,$$
$$\psi_1 = \varphi_3 \pm 2n\pi \quad \text{and} \quad \psi_2 = \varphi_4 \pm 2n\pi \quad \text{for} \quad y < 0, \qquad n = 0, 1, 2, \ldots$$

From the symmetry about the y-axis, these angles satisfy the equations

$$\varphi_1 + \varphi_2 = \pi, \qquad \varphi_3 + \varphi_4 = 3\pi.$$

Similarly, by constructing parallels to the y-axis through the end points $C_{1,4}$ and $C_{2,3}$ of directed segments on the x-axis of length x with $|x| < 1$, one obtains two angles φ_1 and φ_4, φ_2 and φ_3, respectively, as solutions of the equation $\cos\psi = x$ (Fig.).

$$\psi_1 = \varphi_1 \pm 2n\pi, \quad \psi_2 = \varphi_4 \pm 2n\pi, \quad n = 0, 1, 2, \ldots$$

or

$$\psi_1 = \varphi_2 \pm 2n\pi, \quad \psi_2 = \varphi_3 \pm 2n\pi, \quad n = 0, 1, 2, \ldots$$

These two solutions satisfy the condition $\varphi_1 + \varphi_4 = 2\pi$, or $\varphi_2 + \varphi_3 = 2\pi$.

From the value of the cosine or of the sine one obtains two values of the angle ψ in the interval $0 \leqslant \varphi < 2\pi$; from $y = \sin\psi$, for example, the values φ_1 and φ_2. From the unit circle one sees that the angle ψ is uniquely determined by one of these functions and the sign of the other. From the relationship $\tan\psi/2 = \sin\psi/(1 + \cos\psi)$, which will be derived in connection with the addition theorem, it follows that the value of the tangent of the half angle is sufficient to determine uniquely the angle ψ, $0 \leqslant \psi < 2\pi$.

The problem of finding angles for which the tangent function or cotangent function assume the given values y or x, respectively, can likewise be solved geometrically on the unit circle (Fig.). The directed segment $\overrightarrow{B_0D_{1,3}}$ corresponding to the number y is marked off on the tangent at B_0 ($x = 1$) to the unit circle, and the line joining the origin to its end-point $D_{1,3}$ meets the unit circle in the points B_1 and B_3: one sees that $\psi_1 = \varphi_1 \pm n\pi$, $n = 0, 1, 2, \ldots$, are solutions of the equation $\tan\psi = y$. Similarly the directed segment $\overrightarrow{FE_{2,4}}$ corresponding to the number x is marked off on the tangent at F ($y = 1$) to the unit circle, and the line joining the origin to its end-point $E_{2,4}$ meets the unit circle in the points B_2 and B_4; $\psi_2 = \varphi_2 \pm n\pi$,

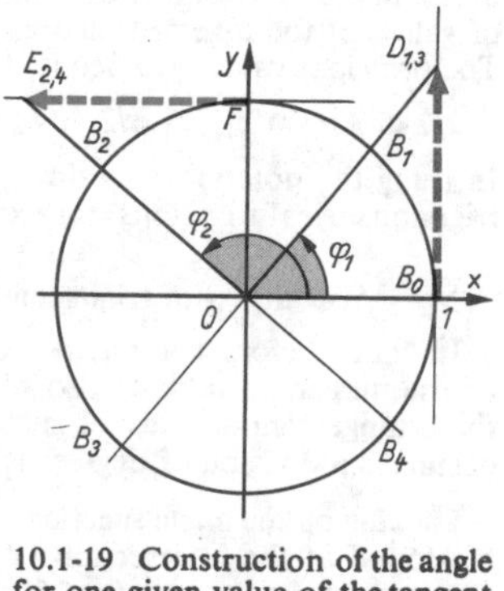

10.1-19 Construction of the angle for one given value of the tangent and of the cotangent

$n = 0, 1, 2, \ldots$, are solutions of the equation $\cot \psi = x$. A function that determines the angle, measured in radians, for which a trigonometric function assumes a given value is called a *circular* or *inverse trigonometric function* (see inverse functions in Chapter 5.).

After the usual interchange of x and y in the derivation of the inverse function, y represents the angle (in radians) whose sine has the value x. The Latin phrase *arcus cuius sinus x est* (the arc whose sine is x) has led to the symbol arcsin x. The notation for these functions is collected together

Trigonometric function	Inverse function
$y = \sin x$	$y = \arcsin x = \sin^{-1} x$
$y = \cos x$	$y = \arccos x = \cos^{-1} x$
$y = \tan x$	$y = \arctan x = \tan^{-1} x$
$y = \cot x$	$y = \text{arccot}\, x = \cot^{-1} x$

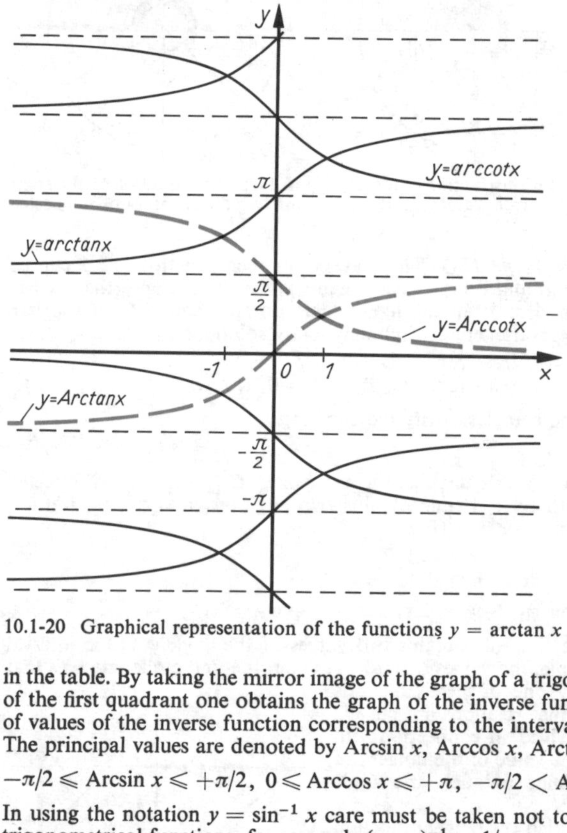

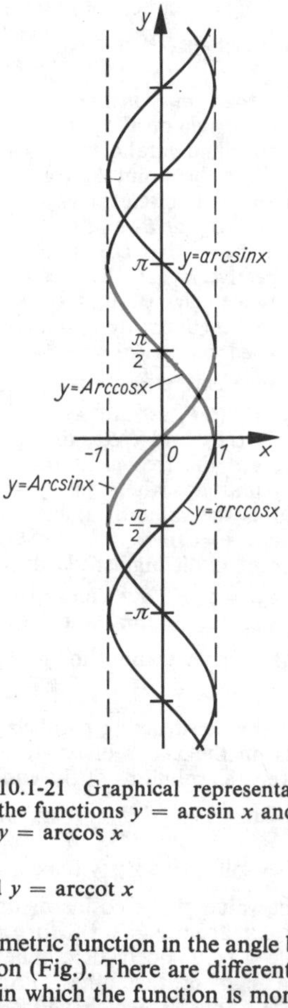

10.1-21 Graphical representation of the functions $y = \arcsin x$ and $y = \arccos x$

10.1-20 Graphical representation of the functions $y = \arctan x$ and $y = \text{arccot}\, x$

in the table. By taking the mirror image of the graph of a trigonometric function in the angle bisector of the first quadrant one obtains the graph of the inverse function (Fig.). There are different ranges of values of the inverse function corresponding to the intervals in which the function is monotonic. The principal values are denoted by Arcsin x, Arccos x, Arctan x and Arccot x, where

$$-\pi/2 \leqslant \text{Arcsin}\, x \leqslant +\pi/2,\ 0 \leqslant \text{Arccos}\, x \leqslant +\pi,\ -\pi/2 < \text{Arctan}\, x < +\pi/2,\ 0 < \text{Arccot}\, x < +\pi.$$

In using the notation $y = \sin^{-1} x$ care must be taken not to confuse it with the reciprocals of the trigonometrical functions, for example $(\cos x)^{-1} = 1/\cos x = \sec x$.

Working with trigonometric tables

In times before computers became widely used it was essential for trigonometric calculations to use numerical tables. Also today it may be useful to know how to use them. The principle of the arrangement and use of such tables is the same for all methods of subdividing the angle. Since decimal subdivision of degrees is now widely used, the explanation is based on it.

Looking up the angle functions. In the table reproduced here the values of the function represented are found at the intersection of a horizontal row with a vertical column (Fig.). In the figure, for example, where the row for '5' degrees meets the column for '0°.4' one finds that $\sin 5.4° = 0.0941$, as the heading indicates. Since the values of the sine are all less than 1 (except for the single value

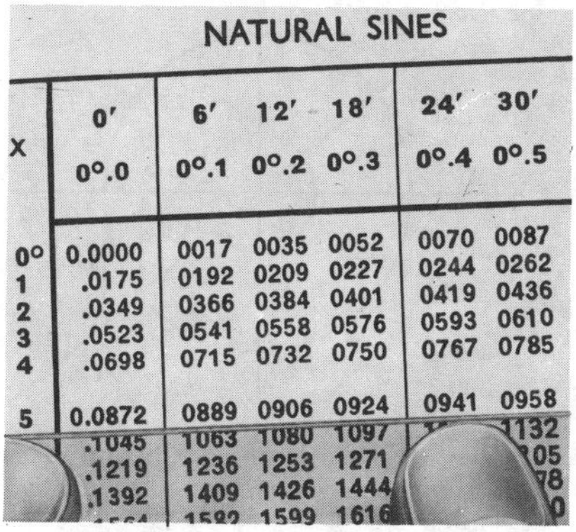

NATURAL SINES

x	0′ 0°.0	6′ 0°.1	12′ 0°.2	18′ 0°.3	24′ 0°.4	30′ 0°.5
0°	0.0000	0017	0035	0052	0070	0087
1	.0175	0192	0209	0227	0244	0262
2	.0349	0366	0384	0401	0419	0436
3	.0523	0541	0558	0576	0593	0610
4	.0698	0715	0732	0750	0767	0785
5	0.0872	0889	0906	0924	0941	0958
	.1045	1063	1080	1097	[illegible]	1132
	.1219	1236	1253	1271	[illegible]	[illegible]
	.1392	1409	1426	1444	[illegible]	[illegible]

10.1-22 Looking up sin α when α is given

$1 = \sin 90°$), frequently only the places after the decimal point are given. All tables of trigonometric functions have a *double entry*; they can be read from the left and above or from the right and below. This means that the rows can be counted from top to bottom and the columns from left to right, or conversely, the rows from bottom to top and the columns from right to left. In the figure the value $\sin 5.4° = 0.0941$ appears in row 5 and column 0°.4, counted from the left and above. From the right and below the same value appears in the row 84 and the column .6. But because $5.4° + 84.6° = 90°$, this gives the value of the cofunction, that is, $\cos 84.6° = 0.0941$. In this way the values of a function and its cofunction are contained in one and the same table, the sine and tangent values from the left and above, and the cosine and cotangent values from the right and below. By the quadrant relations only values in the first quadrant need be given; the sign for the required function is determined from the table of signs or from the unit circle.

Because the use of calculating machines is growing, the importance of tables of the *natural values* of the angle functions is increasing. Formerly *logarithmic trigonometric tables* were preferred for accurate calculations. In some of these the characteristic is increased by 10. For example, $\lg \sin 5.4° = 0.9736 - 2 = 8.9736 - 10$, and such a table gives 8.9736. Since there are no logarithms of negative numbers, these tables contain only the logarithms of the absolute values of the trigonometric functions. On the other hand, their signs are decisive for the magnitude of the angles to be calculated; the signs can then be indicated by a p (positive) or n (negative) placed after the logarithm (see examples of calculations with logarithms in Chapter 2.).

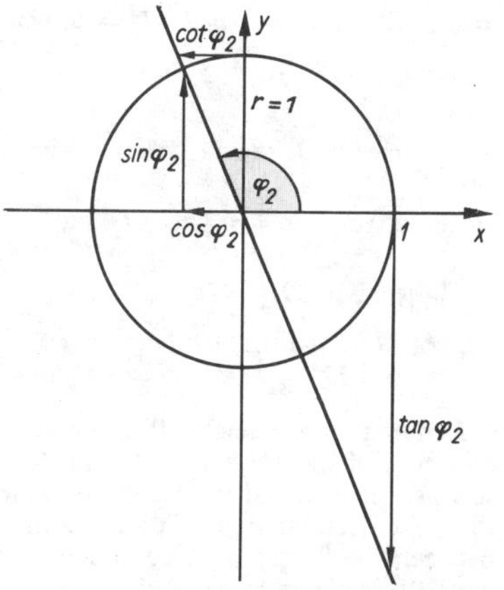

10.1-23 Values of the trigonometric functions for the angle $\varphi_2 = 113.4°$

Example 1: For the angle $\varphi_1 = 56.6°$ one obtains:

$\sin 56.6° = 0.8348$; $\cos 56.6° = 0.5505$;
$\tan 56.6° = 1.517$; $\cot 56.6° = 0.6594$;
$\lg \sin 56.6° = 9.9216$; $\lg \cos 56.6° = 9.7407$;
$\lg \tan 56.6° = 0.1809$; $\lg \cot 56.6° = 9.8191$.

Example 2: For the angle $\varphi_2 = 113.4°$ (Fig.) one obtains:

$$\sin 113.4° = \sin(90° + 23.4°) = +\cos 23.4° = +0.9178;$$
$$\cos 113.4° = \cos(90° + 23.4°) = -\sin 23.4° = -0.3971;$$
$$\tan 113.4° = -\cot 23.4° = -2.311;$$
$$\cot 113.4° = -\tan 23.4° = -0.4327;$$

$\lg \sin 113.4° = 9.9627p$; $\lg |\cos 113.4°| = 9.5990n$;
$\lg |\tan 113.4°| = 0.3638n$; $\lg |\cot 113.4°| = 9.6362n$.

Example 3: For the angle $\varphi_3 = 244.8°$ (Fig.) one obtains:

$$\sin 244.8° = \sin(180° + 64.8°) = -\sin 64.8° = -0.9048;$$
$$\cos 244.8° = \cos(180° + 64.8°) = -\cos 64.8° = -0.4258;$$
$$\tan 244.8° = \tan 64.8° = +2.125;$$
$$\cot 244.8° = \cot 64.8° = +0.4706;$$

$\lg |\sin 244.8°| = 9.9566n$; $\lg |\cos 244.8°| = 9.6292n$;
$\lg \tan 244.8° = 0.3274p$; $\lg \cot 244.8° = 9.6726p$.

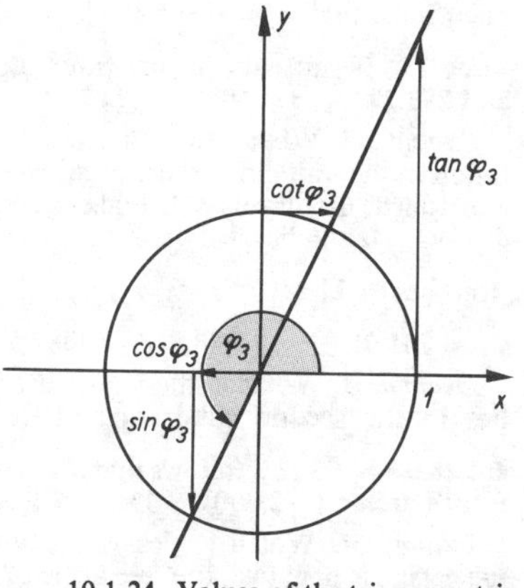

10.1-24 Values of the trigonometric functions for the angle $\varphi_3 = 244.8°$

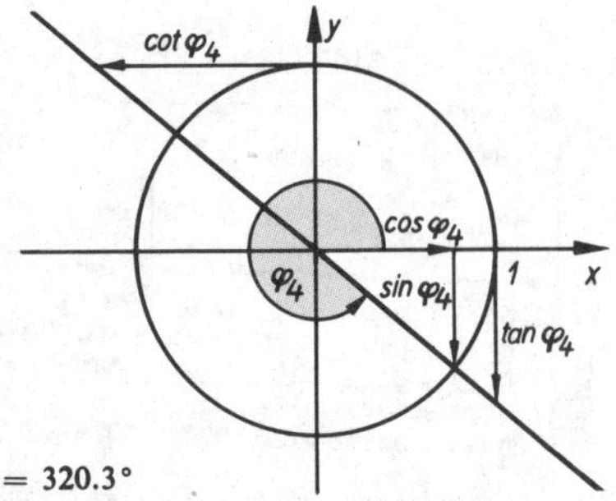

Example 4: For the angle $\varphi_4 = 320.3°$ (Fig.) one obtains:

$\sin 320.3° = \sin(270° + 50.3°) = -\cos 50.3° = -0.6388;$
$\cos 320.3° = \cos(270° + 50.3°) = +\sin 50.3° = +0.7694;$
$\tan 320.3° = -\cot 50.3° = -0.8302;$
$\cot 320.3° = -\tan 50.3° = -1.205;$
$\lg|\sin 320.3°| = 9.8053n;$ $\lg \cos 320.3° = 9.8862p;$
$\lg|\tan 320.3°| = 9.9192n;$ $\lg|\cot 320.3°| = 0.0808n.$

10.1-25 Values of the trigonometric functions for the angle $\varphi_4 = 320.3°$

If a further decimal digit z in the value of the angle is known, then the required function value lies between two values t_1 and t_2 given in the table, and by *linear interpolation* on the *table difference* $d = t_2 - t_1$ one obtains the *correction* $c = d \cdot z/10$ (see Chapter 2.).

Because the cofunctions cosine and cotangent decrease with increasing argument, the correction c for them must be subtracted from the table value.

The value of sin 5.47° lies between 0.0941 and 0.0958. The table difference d is $17 \cdot 10^{-4} = 0.0017$ and the next digit z is 7. For the correction one obtains $c = \frac{17 \cdot 7}{10} \cdot 10^{-4} = (11.9) \cdot 10^{-4} \approx 12 \cdot 10^{-4} = 0.0012$; that is, $\sin 5.47° = 0.0953$. The value of cos 56.64° lies between 0.5505 and 0.5490; here $d = -15 \cdot 10^{-4}$, $z = 4$, $c = \frac{-15 \cdot 4}{10} \cdot 10^{-4} = -(6.0) \cdot 10^{-4}$, that is, $\cos 56.64° = 0.5499$.

Examples: 1. $\tan 113.43° = -\cot 23.43° = -\left(2.311 - \frac{11 \cdot 3}{10} \cdot 10^{-3}\right) = -2.308.$

2. $\lg|\cos 244.86°| = \lg \cos 64.86°n = \left(\bar{1}.6292 - \frac{16 \cdot 6}{10} \cdot 10^{-4}\right) n = \bar{1}.6282n.$

3. $\lg|\sin 320.39°| = \lg \cos 50.39°n = \left(9.8053 - \frac{9 \cdot 9}{10} \cdot 10^{-4}\right) n = \bar{1}.8045n.$

4. $\cot 81.36° = 0.1519$, because $\cot 81.3° = 0.1530$ and $\cot 81.4° = 0.1512$.

5. $\tan 62° 37' = 1.931$, because $\tan 62° 30' = 1.921$ and $\tan 62° 40' = 1.935$.

Looking up the angle. If the given function value appears directly in the table, then the determination of the angle is just a matter of reading off the numbers of the row and column that intersect at this value. If the function value does not agree with any value in the table, then one obtains the next decimal digit z of the angle from the table difference d and the correction difference c between the function value and the nearest value in the table in the sense of increasing argument: $c/(d/10) = z$ or $z = c \cdot 10/d$.

Example 1: The cosine value 0.3950 lies between $\cos 66.7° = 0.3955$ and $\cos 66.8° = 0.3939$; one finds that $d = -16 \cdot 10^{-4}$, $c = -5 \cdot 10^{-4}$, so that $z = \frac{+5 \cdot 10}{16} \approx 3$, or $\arccos 0.3950 = 66.73°$. Because the inverse function is many-valued, $\varphi = -66.73° \mathrel{\hat{=}} 293.27°$; $66.73° \pm n \cdot 360°$ and $293.27° \pm n \cdot 360°$, $n = 1, 2, \dots$ are also valid solutions.

Example 2: What is the value of arcsin (−0.7777)? – From the table of signs or from the representation in the unit circle there exist, when the period is ignored, two solutions φ_1 and φ_2 in the third and fourth quadrants satisfying the equation $\varphi_1 + \varphi_2 = 3\pi = 540°$. From the quadrant relations $\delta = \varphi_1 - 2\pi/2$ lies between 51.0° and 51.1°, since $\sin 51.0° = 0.7771$ and $\sin 51.1° = 0.7782$; from $d = 11 \cdot 10^{-4}$, $c = 6 \cdot 10^{-4}$ it follows that $z = \frac{6 \cdot 10}{11} \approx 5$, that is, $\delta = 51.05°$, $\varphi_1 = 231.05° \pm n \cdot 360°$, $\varphi_2 = 308.95° \pm n \cdot 360°$, $n = 0, 1, 2, \dots$

Example 3: What is the value of arctan (−2.000)? – The angle φ for which $\tan \varphi = -2.000$ lies in the second quadrant and for $\delta = \varphi - \pi/2$, $\cot \delta = 2.000$. From $\cot 26.5° = 2.006$ and $\cot 26.6° = 1.997$ it follows that $d = -9 \cdot 10^{-3}$, $c = -6 \cdot 10^{-3}$, $z = \frac{6 \cdot 10}{9} \approx 7$, that is, $\delta = 26.57°$. Hence $\arctan(-2.000) = 116.57° \pm n \cdot 180°$, $n = 0, 1, 2, \dots$

Example 4: Which angles φ satisfy the equation $\lg|\cos \varphi| = \bar{1}.74435n$? – Because the value of the cosine is negative, the free arms of the angles φ_2 and φ_3 lie in the 2nd and 3rd quadrants, symmetrically with respect to the x-axis. By the quadrant relations $\delta = \varphi_3 - 2\pi/2$ is determined by lg cos $\delta = \bar{1}.74435p$. From a 5-figure table one finds that lg cos 56.28° = $\bar{1}.74440$ and lg cos 56.29° = $\bar{1}.74428$; from $d = -12 \cdot 10^{-5}$, $c = -5 \cdot 10^{-5}$, $z = \frac{5 \cdot 10}{12} \approx 4$ one obtains $\delta = 56.284°$. Hence $\varphi_2 = 123.716° \pm n \cdot 360°$, $\varphi_3 = 236.284° \pm n \cdot 360°$.

The addition theorems

The addition theorems show how the trigonometric functions of a sum or difference of two angles α and β can be expressed in terms of the trigonometric functions of the individual angles.

The addition theorems for sine and cosine. In the unit circle the values of the cosine and sine, respectively, of an angle φ are represented as the signed numerical values of the abscissa and ordinate of the radius in the direction of the free arm of the angle φ. The two segments are the orthogonal projections of this radius on the x- and y-axes. By a theorem of vector algebra, *the projection of this radius is equal to the sum of the projections of any two vectors of which it is the sum.* In the Fig. 10.1-26, for example, in each case $\overrightarrow{OQ} = \overrightarrow{OT} + \overrightarrow{TQ}$, where $\overrightarrow{OT}$ is the orthogonal projection of the free arm $\overrightarrow{OQ}$ of the angle β on the free arm $\overrightarrow{OP}$ of the angle α, and $\overrightarrow{TQ}$ is the orthogonal projection of the free arm $\overrightarrow{OQ}$ of the angle β on a direction S which is the $\bar{y}$-axis of a second Cartesian

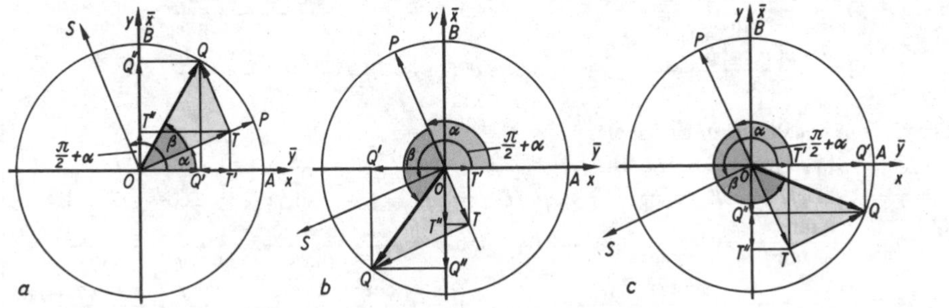

10.1-26 Examples of the addition theorem for the sine and the cosine function

coordinate system $(\bar{x}, \bar{y})$ determined by the $\bar{x}$-axis in direction of $\overrightarrow{OP}$ and $\sphericalangle (\bar{x}, \bar{y}) = \pi/2$. Between the systems (x, y) and $(\bar{x}, \bar{y})$ hold the angular relationships

$$\sphericalangle (x, \bar{x}) = \alpha, \quad \sphericalangle (x, \bar{y}) = \alpha + \pi/2, \quad \sphericalangle (y, \bar{x}) = -\pi/2 + \alpha, \quad \sphericalangle (y, \bar{y}) = -\pi/2 + \alpha + \pi/2 = \alpha.$$

If $m_{\bar{x}}$ denotes the signed numerical value of the orthogonal projection on the $\bar{x}$-axis and $m_{\bar{y}}$ that of the orthogonal projection on the $\bar{y}$-axis, than $\overrightarrow{OT} = m_{\bar{x}}\ (\overrightarrow{OQ}) = \cos\beta$ and $\overrightarrow{TQ} = m_{\bar{y}}(\overrightarrow{OQ}) = \sin\beta$. Correspondingly in the (x, y)-system are valid:

$$m_{\bar{x}}(\overrightarrow{OQ}) = m_x(\overrightarrow{OT}) + m_x(\overrightarrow{TQ}) = \cos(\alpha + \beta) \quad \text{and}$$
$$m_y(\overrightarrow{OQ}) = m_y(\overrightarrow{OT}) + m_y(\overrightarrow{TQ}) = \sin(\alpha + \beta).$$

Because of

$$m_x(\overrightarrow{OT}) = \cos\beta \cos(x, \bar{x}) = \cos\alpha\cos\beta,$$
$$m_x(\overrightarrow{TQ}) = \sin\beta \cos(x, \bar{y}) = -\sin\alpha\sin\beta,$$
$$m_y(\overrightarrow{OT}) = \cos\beta \cos(y, \bar{x}) = \sin\alpha\cos\beta,$$
$$m_y(\overrightarrow{TQ}) = \sin\beta \cos(y, \bar{y}) = \cos\alpha\sin\beta$$

it follows $\cos(\alpha + \beta) = \cos\alpha\cos\beta - \sin\alpha\sin\beta$ and $\sin(\alpha + \beta) = \sin\alpha\cos\beta + \cos\alpha\sin\beta$.

These arguments are valid for arbitrary angles α and β; the three figures are examples for three selected cases.

If at the same time one uses the fact, obvious from the periodicity, that every angle β_1 can be replaced by an angle $-\beta_2$, where $\beta_1 + \beta_2 = 2\pi$ (or 400^g or 360°), it follows that differences of angles can also occur in the addition theorem:

$$\sin(\alpha-\beta)=\sin\alpha\cos\beta-\cos\alpha\sin\beta,$$
$$\cos(\alpha-\beta)=\cos\alpha\cos\beta+\sin\alpha\sin\beta.$$

The addition theorems for tangent and cotangent. These are obtained at once in a universally valid form by division and suitable rearrangement:

$$\tan(\alpha+\beta)=\frac{\sin(\alpha+\beta)}{\cos(\alpha+\beta)}=\frac{\sin\alpha\cos\beta+\cos\alpha\sin\beta}{\cos\alpha\cos\beta-\sin\alpha\sin\beta}.$$

Both the numerator and the denominator are divided by $\cos\alpha\cos\beta$:

$$\tan(\alpha+\beta)=\frac{\tan\alpha+\tan\beta}{1-\tan\alpha\tan\beta};\quad \tan(\alpha-\beta)=\frac{\tan\alpha-\tan\beta}{1+\tan\alpha\tan\beta}.$$

Similarly one obtains

$$\cot(\alpha+\beta)=\frac{\cot\alpha\cot\beta-1}{\cot\alpha+\cot\beta};\quad \cot(\alpha-\beta)=\frac{\cot\alpha\cot\beta+1}{\cot\beta-\cot\alpha}.$$

$$\sin(\alpha+\beta)=\sin\alpha\cos\beta+\cos\alpha\sin\beta \qquad \sin(\alpha-\beta)=\sin\alpha\cos\beta-\cos\alpha\sin\beta$$
$$\cos(\alpha+\beta)=\cos\alpha\cos\beta-\sin\alpha\sin\beta \qquad \cos(\alpha-\beta)=\cos\alpha\cos\beta+\sin\alpha\sin\beta$$
$$\tan(\alpha+\beta)=\frac{\tan\alpha+\tan\beta}{1-\tan\alpha\tan\beta} \qquad \tan(\alpha-\beta)=\frac{\tan\alpha-\tan\beta}{1+\tan\alpha\tan\beta}$$
$$\cot(\alpha+\beta)=\frac{\cot\alpha\cot\beta-1}{\cot\beta+\cot\alpha} \qquad \cot(\alpha-\beta)=\frac{\cot\alpha\cot\beta+1}{\cot\beta-\cot\alpha}$$

Functions of double and of half angles

$$\sin 2\varphi=2\sin\varphi\cos\varphi \qquad \cos 2\varphi=\cos^2\varphi-\sin^2\varphi=1-2\sin^2\varphi=2\cos^2\varphi-1$$
$$\sin\varphi=2\sin\varphi/2\cos\varphi/2 \qquad \cos\varphi=\cos^2\varphi/2\,\sin^2\varphi/2=1-2\sin^2\varphi/2=2\cos^2\varphi/2-1$$
$$\tan 2\varphi=\frac{2\tan\varphi}{1-\tan^2\varphi}=\frac{2}{\cot\varphi-\tan\varphi} \qquad \tan\varphi=\frac{2\tan\varphi/2}{1-\tan^2\varphi/2}=\frac{2}{\cot\varphi/2-\tan\varphi/2}$$
$$\cot 2\varphi=\frac{\cot^2\varphi-1}{2\cot\varphi}=\frac{\cot\varphi-\tan\varphi}{2} \qquad \cot\varphi=\frac{\cot^2\varphi/2-1}{2\cot\varphi/2}=\frac{\cot\varphi/2-\tan\varphi/2}{2}$$
$$\sin\varphi=+\sqrt{\left(\frac{1-\cos 2\varphi}{2}\right)} \qquad \sin\varphi/2=\pm\sqrt{\left(\frac{1-\cos\varphi}{2}\right)}$$
$$\cos\varphi=\pm\sqrt{\left(\frac{1+\cos 2\varphi}{2}\right)} \qquad \cos\varphi/2=\pm\sqrt{\left(\frac{1+\cos\varphi}{2}\right)}$$
$$\tan\varphi=\pm\sqrt{\left(\frac{1-\cos 2\varphi}{1+\cos 2\varphi}\right)}=\frac{\sin 2\varphi}{1+\cos 2\varphi}=\frac{1-\cos 2\varphi}{\sin 2\varphi}$$
$$\tan\varphi/2=\pm\sqrt{\left(\frac{1-\cos\varphi}{1+\cos\varphi}\right)}=\frac{\sin\varphi}{1+\cos\varphi}=\frac{1-\cos\varphi}{\sin\varphi}$$
$$\cot\varphi=\pm\sqrt{\left(\frac{1+\cos 2\varphi}{1-\cos 2\varphi}\right)}=\frac{\sin 2\varphi}{1-\cos 2\varphi}=\frac{1+\cos 2\varphi}{\sin 2\varphi}$$
$$\cot\varphi/2=\pm\sqrt{\left(\frac{1+\cos\varphi}{1-\cos\varphi}\right)}=\frac{\sin\varphi}{1-\cos\varphi}=\frac{1+\cos\varphi}{\sin\varphi}$$
$$\sin 2\varphi=\frac{2\tan\varphi}{1+\tan^2\varphi},\ \cos 2\varphi=\frac{1-\tan^2\varphi}{1+\tan^2\varphi};\ \sin\varphi=\frac{2\tan\varphi/2}{1+\tan^2\varphi/2},\ \cos\varphi=\frac{1-\tan^2\varphi/2}{1+\tan^2\varphi/2}$$

Functions of multiple angles

$$\sin 3\varphi=3\sin\varphi-4\sin^3\varphi \qquad \cos 3\varphi=4\cos^3\varphi-3\cos\varphi$$
$$\sin 4\varphi=4\sin\varphi\cos\varphi-8\sin^3\varphi\cos\varphi \qquad \cos 4\varphi=8\cos^4\varphi-8\cos^2\varphi+1$$
$$\sin 5\varphi=5\sin\varphi-20\sin^3\varphi+16\sin^5\varphi \qquad \cos 5\varphi=16\cos^5\varphi-20\cos^3\varphi+5\cos\varphi$$
$$\tan 3\varphi=\frac{3\tan\varphi-\tan^3\varphi}{1-3\tan^2\varphi} \qquad \cot 3\varphi=\frac{\cot^3\varphi-3\cot\varphi}{3\cot^2\varphi-1}$$
$$\tan 4\varphi=\frac{4\tan\varphi-4\tan^3\varphi}{1-6\tan^2\varphi+\tan^4\varphi} \qquad \cot 4\varphi=\frac{\cot^4\varphi-6\cot^2\varphi+1}{4\cot^3\varphi-4\cot\varphi}$$

Consequences of the addition theorems

From the addition theorems many relationships between the trigonometric functions can be derived, and they are collected together in the following tables. A few examples illustrate the way in which they are derived.

$$\begin{aligned}&\sin(\alpha+\beta)\sin(\alpha-\beta)\\&=\sin^2\alpha\cos^2\beta-\cos^2\alpha\sin^2\beta\\&=\sin^2\alpha\cos^2\beta-\cos^2\alpha(1-\cos^2\beta)\\&=\cos^2\beta(\sin^2\alpha+\cos^2\alpha)-\cos^2\alpha\\&=\cos^2\beta-\cos^2\alpha.\end{aligned}$$

$$\begin{aligned}\sin 3\varphi&=\sin(2\varphi+\varphi)\\&=\sin 2\varphi\cos\varphi+\cos 2\varphi\sin\varphi\\&=2\sin\varphi\cos^2\varphi+(1-2\sin^2\varphi)\sin\varphi\\&=2\sin\varphi(1-\sin^2\varphi)+\sin\varphi-2\sin^3\varphi\\&=3\sin\varphi-4\sin^3\varphi.\end{aligned}$$

In the equation $\sin(\varphi+\psi)+\sin(\varphi-\psi)=2\sin\varphi\cos\psi$ one puts $\alpha=\varphi+\psi$, $\beta=\varphi-\psi$, so that $\varphi={}^1/_2(\alpha+\beta)$, $\psi={}^1/_2(\alpha-\beta)$, and one obtains $\sin\alpha+\sin\beta=2\sin{}^1/_2(\alpha+\beta)\cos{}^1/_2(\alpha-\beta)$.

$$\tan\alpha\pm\tan\beta=\frac{\sin\alpha}{\cos\alpha}\pm\frac{\sin\beta}{\cos\beta}=\frac{\sin\alpha\cos\beta\pm\cos\alpha\sin\beta}{\cos\alpha\cos\beta}=\frac{\sin(\alpha\pm\beta)}{\cos\alpha\cos\beta}.$$

Sums, differences and products of trigonometric functions

$$\sin\alpha+\sin\beta=2\sin\frac{\alpha+\beta}{2}\cos\frac{\alpha-\beta}{2}\qquad\cos\alpha+\cos\beta=2\cos\frac{\alpha+\beta}{2}\cos\frac{\alpha-\beta}{2}$$

$$\sin\alpha-\sin\beta=2\cos\frac{\alpha+\beta}{2}\sin\frac{\alpha-\beta}{2}\qquad\cos\alpha-\cos\beta=-2\sin\frac{\alpha+\beta}{2}\sin\frac{\alpha-\beta}{2}$$

$$\tan\alpha+\tan\beta=\frac{\sin(\alpha+\beta)}{\cos\alpha\cos\beta}\qquad\cot\alpha+\cot\beta=\frac{\sin(\alpha+\beta)}{\sin\alpha\sin\beta}$$

$$\tan\alpha-\tan\beta=\frac{\sin(\alpha-\beta)}{\cos\alpha\cos\beta}\qquad\cot\alpha-\cot\beta=\frac{-\sin(\alpha-\beta)}{\sin\alpha\sin\beta}$$

$$\cos\alpha+\sin\alpha=\sqrt{2}\sin(45^\circ+\alpha)=\sqrt{2}\cos(45^\circ-\alpha)$$

$$\cos\alpha-\sin\alpha=\sqrt{2}\cos(45^\circ+\alpha)=\sqrt{2}\sin(45^\circ-\alpha)$$

$$\sin(\alpha+\beta)\sin(\alpha-\beta)=\cos^2\beta-\cos^2\alpha\qquad\cos(\alpha+\beta)\cos(\alpha-\beta)=\cos^2\beta-\sin^2\alpha$$

$$\sin\alpha\sin\beta={}^1/_2[\cos(\alpha-\beta)-\cos(\alpha+\beta)]\qquad\cos\alpha\cos\beta={}^1/_2[\cos(\alpha-\beta)+\cos(\alpha+\beta)]$$

$$\sin\alpha\cos\beta={}^1/_2[\sin(\alpha-\beta)+\sin(\alpha+\beta)]\qquad\cos\alpha\sin\beta={}^1/_2[\sin(\alpha+\beta)-\sin(\alpha-\beta)]$$

$$\tan\alpha\tan\beta=\frac{\tan\alpha+\tan\beta}{\cot\alpha+\cot\beta}=-\frac{\tan\alpha-\tan\beta}{\cot\alpha-\cot\beta}\qquad\cot\alpha\cot\beta=\frac{\cot\alpha+\cot\beta}{\tan\alpha+\tan\beta}=-\frac{\cot\alpha-\cot\beta}{\tan\alpha-\tan\beta}$$

$$\tan\alpha\cot\beta=\frac{\tan\alpha+\cot\beta}{\cot\alpha+\tan\beta}=-\frac{\tan\alpha-\cot\beta}{\cot\alpha-\tan\beta}$$

$$\sin\alpha\sin\beta\sin\gamma={}^1/_4[\sin(\alpha+\beta-\gamma)+\sin(\beta+\gamma-\alpha)+\sin(\gamma+\alpha-\beta)-\sin(\alpha+\beta+\gamma)]$$

$$\cos\alpha\cos\beta\cos\gamma={}^1/_4[\cos(\alpha+\beta-\gamma)+\cos(\beta+\gamma-\alpha)+\cos(\gamma+\alpha-\beta)+\cos(\alpha+\beta+\gamma)]$$

$$\sin\alpha\sin\beta\cos\gamma={}^1/_4[-\cos(\alpha+\beta-\gamma)+\cos(\beta+\gamma-\alpha)+\cos(\gamma+\alpha-\beta)-\cos(\alpha+\beta+\gamma)]$$

$$\sin\alpha\cos\beta\cos\gamma={}^1/_4[\sin(\alpha+\beta-\gamma)-\sin(\beta+\gamma-\alpha)+\sin(\gamma+\alpha-\beta)+\sin(\alpha+\beta+\gamma)]$$

Powers of trigonometric functions

$$\sin^2\varphi={}^1/_2(1-\cos 2\varphi)\qquad\cos^2\varphi={}^1/_2(1+\cos 2\varphi)$$

$$\sin^3\varphi={}^1/_4(3\sin\varphi-\sin 3\varphi)\qquad\cos^3\varphi={}^1/_4(3\cos\varphi+\cos 3\varphi)$$

$$\sin^4\varphi={}^1/_8(\cos 4\varphi-4\cos 2\varphi+3)\qquad\cos^4\varphi={}^1/_8(\cos 4\varphi+4\cos 2\varphi+3)$$

$$\sin^5\varphi={}^1/_{16}(10\sin\varphi-5\sin 3\varphi+\sin 5\varphi)\qquad\cos^5\varphi={}^1/_{16}(10\cos\varphi+5\cos 3\varphi+\cos 5\varphi)$$

General formulae for the sine and cosine of a multiple angle. De Moivre's theorem in the theory of complex numbers states that $(\cos\varphi+i\sin\varphi)^n=\cos n\varphi+i\sin n\varphi$. Bearing in mind that $i^2=-1$ this can be proved for $n=1, 2, 3, \dots$ by the method of induction by means of the addition theorems. If the left-hand side is expanded by the binomial theorem, equating the real and imaginary parts gives

$$\cos n\varphi=\cos^n\varphi-\binom{n}{2}\cos^{n-2}\varphi\sin^2\varphi+\binom{n}{4}\cos^{n-4}\varphi\sin^4\varphi-\cdots$$

$$\sin n\varphi=\binom{n}{1}\cos^{n-1}\varphi\sin\varphi-\binom{n}{3}\cos^{n-3}\varphi\sin^3\varphi+\binom{n}{5}\cos^{n-5}\varphi\sin^5\varphi-\cdots$$

The general sine curve. In nature and technology the mathematical description of oscillations, for example, in high frequency technology, optics, acoustics or mechanics, is based on sine and cosine functions. In these oscillations the greatest displacement, the *amplitude a*, of a sine oscillation can be different from 1, its *wave length* λ different from 2π, and the ordinate at the zero point different from 0. The function $y = a \sin x$, for example, has the *amplitude a* (Fig.) and the function $y = \sin(2\pi x/\lambda)$ the *wave length* λ, because for $0 \leqslant x \leqslant \lambda$ the argument $2\pi x/\lambda$ runs through the values from 0 to 2π (Fig.). The function $y = \sin(nx)$, where n is an integer, has exactly n complete oscillations in the interval from 0 to 2π, since $\lambda = 2\pi/n$. Finally, the function $y = \sin(n\pi x/l)$ with $\lambda = 2l/n$ describes an oscillation of which n waves have length $2l$.

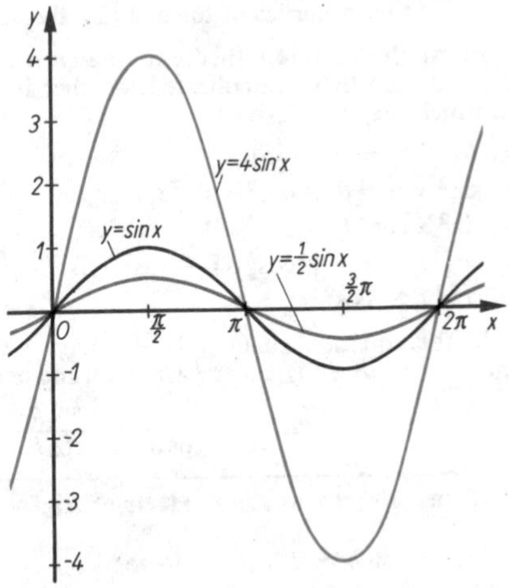

10.1-27 Graphs of the functions $y = \sin x$, $y = 4 \sin x$ and $y = {}^1/_2 \sin x$

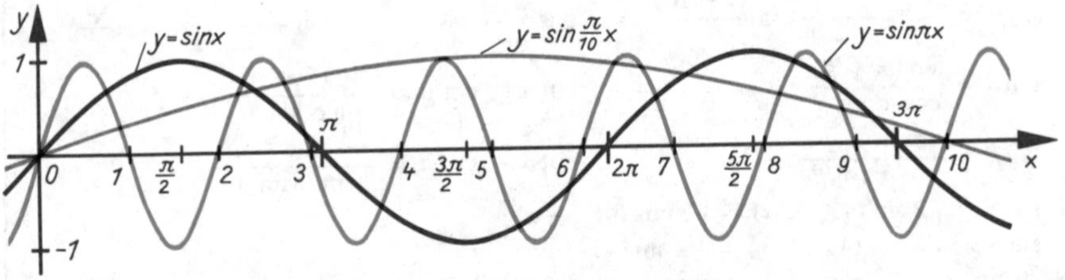

10.1-28 Graphs of the functions $y = \sin \pi x$ and $y = \sin(\pi x/10)$

Superposition. If several physical quantities that can be represented by oscillations act at a point, then the ordinates for this point are added. For example, $y_1 = 2 \sin x$ and $y_2 = -\cos 2x$ gives $y = y_1 + y_2 = 2 \sin x - \cos 2x$ (Fig.).

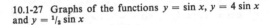

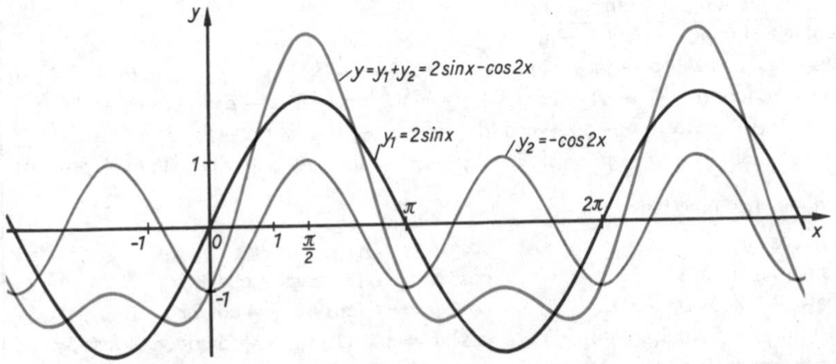

10.1-29 Graph of the function $y = y_1 + y_2 = 2 \sin x - \cos 2x$

Damped oscillations. If an oscillating system loses energy, then the amplitude decreases. For example, the function $a = 3e^{-2x/\pi}$ has the value $3/e$ for $x = \pi/2$, only $3/e^2$ for $x = 2\pi/2$, and so on. The figure shows the graph for $y = 3e^{-2x/\pi} \sin 4x$.

Angular frequency ω and phase difference φ. If the time t is regarded as the independent variable, the equation of the general sine curve has the form $y = a \sin(\omega t + \varphi)$. From the fact that for $\omega t = 2\pi$ a whole oscillation is completed it follows that the time for a *complete oscillation* (through wave peak and wave trough) is $t = 2\pi/\omega$. This time is called the *periodic time* of the oscillation and is denoted by T. If T is measured in seconds, then $1/T$ is the number of oscillations in one second, that is, the *frequency* f of the oscillation: $f = 1/T$. The angular frequency $\omega = 2\pi/T = 2\pi(1/T) = 2\pi f$ gives the number of oscillations in 2π seconds. Finally, the phase difference φ is the angle by which the given curve *leads* the sine curve (Fig.). For $t = 0$ the function y already has the value $y = \sin\varphi$. For a negative phase difference φ one speaks of *lagging*. For $a = 1$ and $\varphi = +\pi/2$ the function $y = a\sin(\omega t + \varphi)$ becomes the cosine function $\cos\omega t$, that is, the cosine curve leads the sine curve by $\pi/2$. If a general sinusoidal oscillation with angular frequency ω is given, then a and φ

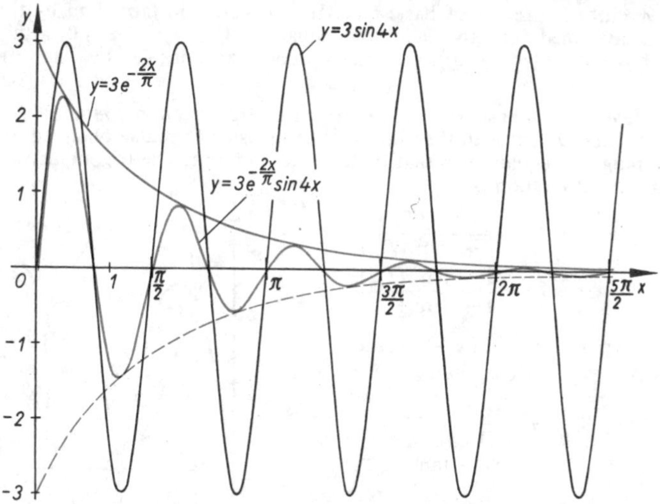

10.1-30 Graph of the function $y = 3e^{-2x/\pi}\sin 4x$

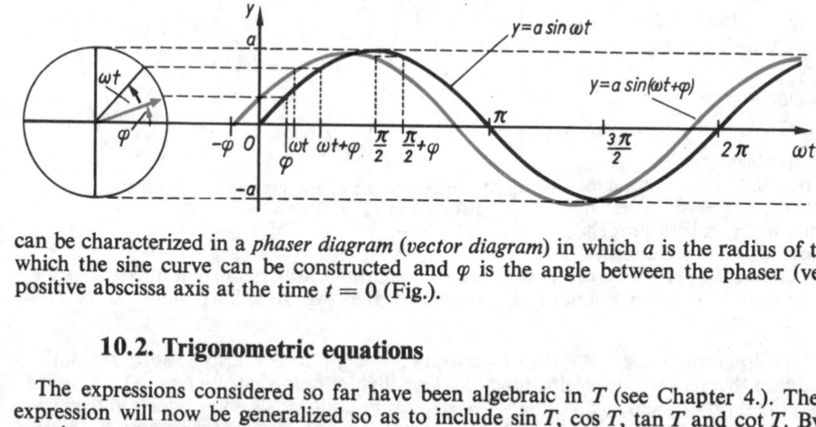

10.1-31 General sine curve $y = a\sin(\omega t + \varphi)$. Left, *phaser diagram* or *vector diagram*; right, curve representation or *line diagram*

can be characterized in a *phaser diagram* (*vector diagram*) in which a is the radius of the circle from which the sine curve can be constructed and φ is the angle between the phaser (vector) and the positive abscissa axis at the time $t = 0$ (Fig.).

10.2. Trigonometric equations

The expressions considered so far have been algebraic in T (see Chapter 4.). The notion of an expression will now be generalized so as to include $\sin T$, $\cos T$, $\tan T$ and $\cot T$. By equating expressions and at the same time taking into account the range of values of the variables, new equations are formed. In *trigonometric equations* with one variable, the variable x occurs in at least one such generalized expression. In *pure trigonometric equations* x occurs only in such expressions, for example, in $\sin(2x + \pi) - \sqrt{2}\cos x = 0$; in *mixed trigonometric equations* x also occurs in algebraic expressions, for example, in $\tan x - 3x = 0$.

Trigonometric equations are transcendental equations (see Chapter 4.). There is no general algorithm for their solution, but they can be solved graphically, or by numerical approximation methods, with arbitrary precision. For certain special types of pure trigonometric equations *solution algorithms* do exist. Because of the periodicity of the trigonometric functions the domain of the variable of a trigonometric equation is often confined to an interval whose length is a primitive period, say $0 \leqslant x < 2\pi$.

Pure trigonometric equations

Basic type. A pure trigonometric equation is said to be of basic type if the variable occurs only in expressions involving *one* trigonometric function, for example, in $\sin T$, and the equation is algebraic in this expression.

Example 1: The equation $\cos^3(2x) = b$, in which x is variable and b is a real parameter, is of basic type. Moreover, it is algebraic in $\cos 2x$, and the substitution $t = \cos 2x$ leads to $t^3 = b$ with the solution $t = \sqrt[3]{b}$. From $\cos 2x = \sqrt[3]{b}$ one can look up the solutions for x within the accuracy of the table.

Example 2: The equation $\tan^2 x + p \tan x + q = 0$ with the variable x and parameters p and q is likewise of basic type. It is algebraic in $\tan x$ and by the substitution $u = \tan x$ it is transformed into the quadratic equation $u^2 + pu + q = 0$ with the solutions $(\tan x)_{1,2} = {}^1/_2(-p \pm \sqrt{(p^2 - 4q)})$. With the help of a table one can then find the solutions for x.

Reduction to basic type. If the trigonometric equation contains several of the terms $\sin T$, $\cos T$, $\tan T$, $\cot T$, but with the same T, then by using formulae obtained in the previous section one can arrange the equation so that its terms contain only one trigonometric function. The most advantageous substitution is

$$\sin T = \frac{2\tan(T/2)}{1 + \tan^2(T/2)}, \qquad \cos T = \frac{1 - \tan^2(T/2)}{1 + \tan^2(T/2)}.$$

Example 3: $5 \sin x - 3 \cos x = 3$, $0 \leqslant x < 2\pi$.

$$5 \cdot \frac{2\tan(x/2)}{1 + \tan^2(x/2)} - 3 \cdot \frac{1 - \tan^2(x/2)}{1 + \tan^2(x/2)} = 3;$$

$$10\tan(x/2) - 3 + 3\tan^2(x/2) = 3 + 3\tan^2(x/2);$$

$$\tan(x/2) = 3/5, \quad x/2 = 30.96°; \quad x = 61.92°.$$

The transformation of the given equation is not valid for $x = \pi$, since $\tan(x/2)$ then does not exist. A test of the original equation by substituting the value $x = \pi$ shows that $x = \pi$ is a second solution in the given range of values of the variable. The solutions are obtained graphically as the abscissae of the points of intersection of the graphs of the two functions $y_1 = 5 \sin x$ and $y_2 = 3 \cos x + 3$ (Fig.).

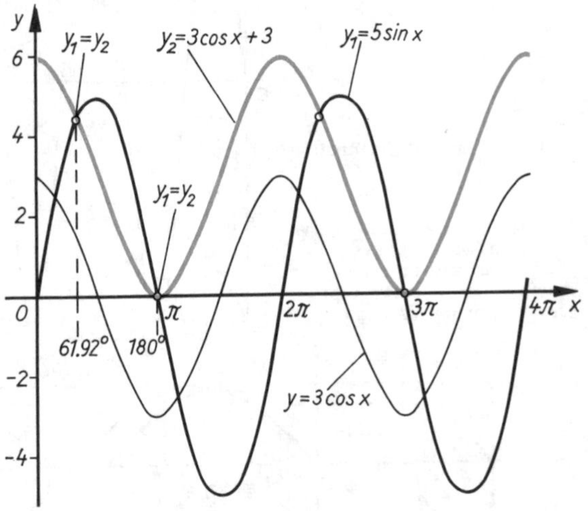

10.2-1 Intersections of the graphs of the functions $y_1 = 5 \sin x$ and $y_2 = 3 \cos x + 3$

Example 4: The equation $a \cos x + b \sin x = c$ with $c^2 \leqslant a^2 + b^2$ can also be solved with the help of the addition theorem for the cosine function. One divides both sides by $r = +\sqrt{(a^2 + b^2)}$ and puts $a/[+\sqrt{(a^2 + b^2)}] = \cos h$, $b/[+\sqrt{(a^2 + b^2)}] = \sin h$, $\tan h = b/a$. The equation then becomes $\cos h \cos x + \sin h \sin x = c/[+\sqrt{(a^2 + b^2)}]$ or $\cos(x - h) = c/[+\sqrt{(a^2 + b^2)}]$; $x + h = \arccos\{c/[+\sqrt{(a^2 + b^2)}]\}$. The *auxiliary angle h* is uniquely determined by $\tan h = b/a$. Hence x is also known (there are two solutions between 0 and 2π). For the numerical values $a = -3$, $b = 5$, $c = 3$ one obtains: $-3 \cos x + 5 \sin x = 3$, $\tan h = 5/(-3) = \sin h/\cos h$. Because $\sin h > 0$ and $\cos h < 0$, h lies in quadrant II; $h = 120.96°$. From $\cos(x - h) = 3/(+\sqrt{34}) = 0.5145$ it follows that $(x - h)_1 = 59.04°$ or $(x - h)_2 = -59.04°$. Thus $x_1 = 180°$, $x_2 = 61.92°$.

If the trigonometric equation consists of expressions in only one trigonometric function, say $\cot T_1$, $\cot T_2$, ..., with different $T_1, T_2, \ldots$, then in certain circumstances it can be reduced to basic type. For example, if all the T_i are integral multiples of a single term T, this can be done with the help of the addition theorems.

Example 5: $\dfrac{2\cot 2x}{1 - 3\cot x} = \dfrac{1}{2}$, or $4 \cot 2x = 1 - 3 \cot x$. Because $\cot 2x = \dfrac{\cot^2 x - 1}{2 \cot x}$, the equation is equivalent to $\dfrac{2(\cot^2 x - 1)}{\cot x} = 1 - 3\cot x$, or $5 \cot^2 x - \cot x - 2 = 0$ (Fig.).

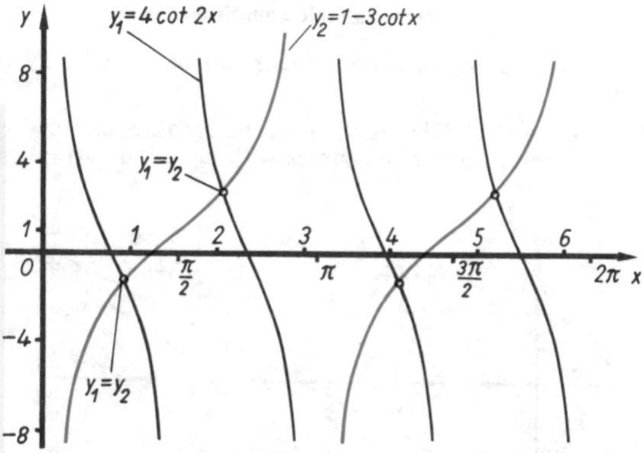

10.2-2 Intersections of the graphs of the functions $y_1 = 4 \cot 2x$ and $y_2 = 1 - 3 \cot x$

Putting $\cot x = u$ one obtains $u^2 - u/5 - 2/5 = 0$; $u = 1/10 + \sqrt{41}/10$, that is, $u_I = (\sqrt{41} + 1)/10$, $u_{II} = -(\sqrt{41} - 1)/10$.

Solutions for $0 \leqslant x < 2\pi$	
$(\cot x)_I = 0.7403$	$(\cot x)_{II} = -0.5403$
$x_1 = 0.9335$ (53,5°) $x_2 = 4.0751$ (233.5°)	$x_3 = 2.0662$ (118.4°) $x_4 = 5.2078$ (298.7°)

Test: all 4 values satisfy the equation.

The formula for $\cot 2x$ is not valid for the values 0 and π. However, one sees at once from the given equation that these values are not solutions.

Further examples show that a reduction to basic type is possible in other cases.

Example 6: $\sin(2x + \pi) - \sqrt{2}\cos x = 0$. Using the quadrant relations or an addition theorem one obtains $-\sin 2x - \sqrt{2}\cos x = 0$ or $2 \sin x \cos x + \sqrt{2} \cos x = 0$,

$(2\sin x + \sqrt{2})$ ·	$\cos x = 0$
$\sin x = -\sqrt{2}/2$ $x_1 = 5\pi/4$ $x_2 = 7\pi/4$	$\cos x = 0$, $x_3 = \pi/2$ $x_4 = 3\pi/2$

By testing one can verify that the solutions are correct (Fig.).

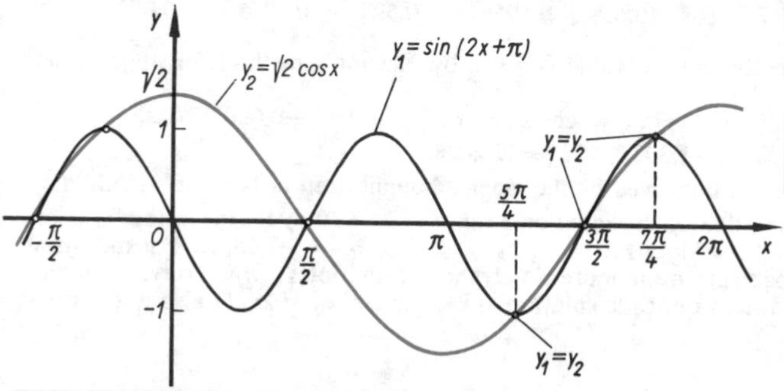

10.2-3 Intersections of the graphs of the functions $y_1 = \sin(2x + \pi)$ and $y_2 = \sqrt{2}\cos x$

Example 7: The equation $\cos(3x/7) + \sin x = 0$ can be simplified by writing $\sin x = \cos(\pi/2 - x)$ and using the formula $\cos\alpha + \cos\beta = 2\cos[(\alpha + \beta)/2]\cos[(\alpha - \beta)/2]$:

$$\cos(3x/7) + \cos(\pi/2 - x) = 0.$$

$2\cos(\pi/4 - 2x/7)$ ·	$\cos(5x/7 - \pi/4) = 0.$
$\cos(\pi/4 - 2x/7) = 0$, $\pi/4 - 2x/7 = \pi/2 + k\pi$, $x_1 = -7\pi/8 - 7k\pi/2$.	$\cos(5x/7 - \pi/4) = 0$, $5x/7 - \pi/4 = \pi/2 + k\pi$ or $5x/7 = 3\pi/4 + k\pi$, $x_2 = 21\pi/20 + 7k\pi/5$.

Since k can take the values $0, \pm 1, \pm 2, \ldots$ one can replace $-k$ by $+k$ in the formula for x_1:

$$x_1 = -7\pi/8 + 7k\pi/2; \quad x_2 = -21\pi/20 + 7k\pi/5.$$

Tests show that all the values satisfy the equation. It should be noted that the solutions for consecutive integers k differ not by 2π, but by $7\pi/2$ or $7\pi/5$, respectively (Fig.).

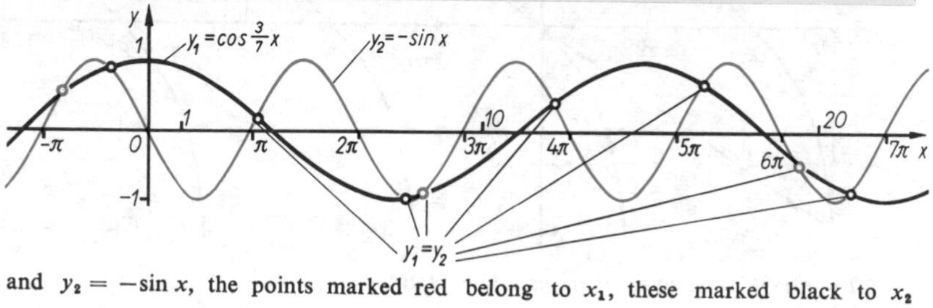

10.2-4 Intersections of the graphs of the functions $y_1 = \cos(3x/7)$ and $y_2 = -\sin x$, the points marked red belong to x_1, these marked black to x_2

Mixed trigonometric equations

Mixed trigonometric equations can be solved only by graphical or iterative methods (see Chapter 29.).

Example 1: The solutions of the equation $\cos x - x/2 + 1.7 = 0$ are the abscissae of the points of intersection of the curves with equations $y_1 = \cos x$, $y_2 = x/2 - 1.7$ (Fig.). They have only

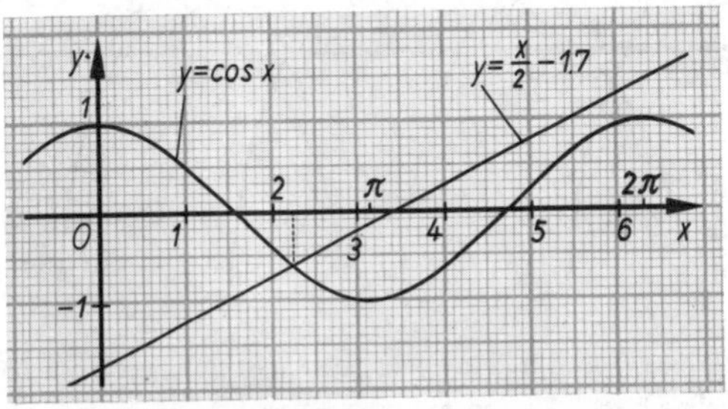

10.2-5 Graphical solution of the equation $\cos x = x/2 - 1.7$

one point of intersection with the abscissa $x_0 \approx 2.21$. If the graphs in the neighbourhood of this intersection are drawn on a larger scale, the accuracy of the reading can be improved. Here one obtains $x_0 \approx 2.209$.

Test: $\cos 2.209 - \dfrac{2.209}{2} + 1.7 = \cos 140.63^g + 0.5955 = -0.5958 + 0.5955 = -0.0003$.

A closer approximation x_1 to the correct value is given by Newton's method for approximate solutions;

$$x_1 = x_0 - f(x_0)/f'(x_0), \quad f(x_0) = \cos x_0 - x_0/2 + 1.7 = -0.0003,$$

$$f'(x_0) = -\sin x_0 - 1/2 = -1.3032, \quad x_1 = 2.2088.$$

The approximation can be further improved by the repeated application of Newton's method.

Example 2: The graphical solution of the equation $3 \tan x - 2x = 0$ by means of the functions $y_1 = \tan x$, $y_2 = 2x/3$, yields the solutions $x_1 = 0$, $x_2 = \pm 4.38$, $x_3 = \pm 7.65$, ... For increasing values of x the solutions approach more and more closely the odd multiples of $\pi/2$. To every solution x_0 there corresponds the equal and opposite solution $-x_0$; for $\tan x_0 = 2x_0/3$ also implies that $\tan(-x_0) = {}^2/_3(-x_0)$ (Fig.).

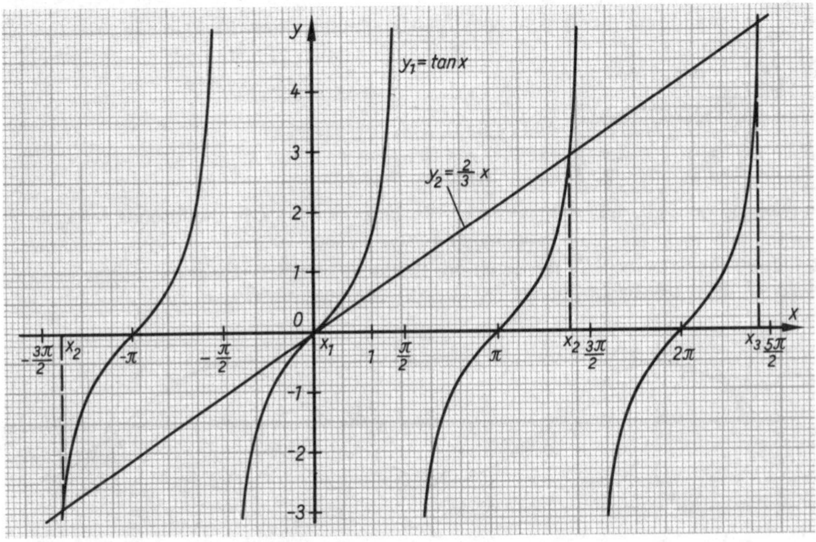

10.2-6 Graphical solution of the equation $3 \tan x - 2x = 0$

11. Plane trigonometry

The trigonometric functions already defined make it possible to use angles to calculate unknown quantities in plane rectilinear figures. Angles can often be measured with less effort and greater accuracy than lengths. As the name indicates, trigonometry is concerned with the measurement or calculation of triangles into which every figure bounded by straight lines can be subdivided by diagonals. In this one always has in mind the use of known angles.

11.1. Solution of right-angled triangles

General methods

The definition of the trigonometric functions was first given in the right-angled triangle and then extended to arbitrary angles with the help of the unit circle. These definitions contain all the relations between lengths and angles in the right-angled triangle and thus suffice to calculate all the rest when any two of the six quantities are given.

When the right angle is denoted by γ and the hypotenuse by c, two additional relationships in the right-angled triangle ABC (Fig.) are available from geometry:

I. the theorem of Pythagoras: $c^2 = a^2 + b^2$,
II. the fact that each of the angles with its vertex on the hypotenuse is the complement of the other: $\alpha + \beta = 90°$.

From these relationships or by re-lettering the triangle all possible cases in which two of the quantities a, b, c, α and β are given can be reduced to four cases, namely c, α; c, a; a, α and a, b, for which the solutions will now be stated.

I. Given the hypotenuse c and one adjacent angle, say α:
1. $\beta = 90° - \alpha$; 2. $\sin\alpha = a/c$, $a = c \sin\alpha$;
3. $\cos\alpha = b/c$, $b = c\cos\alpha$.

II. Given the hypotenuse c and one other side, say a:
1. $\sin\alpha = a/c$; 2. $\beta = 90° - \alpha$;
3a. $b = \sqrt{(c^2 - a^2)}$ or with the help of the calculated angle α:
3b. $\cot\alpha = b/a$, $b = a\cot\alpha$; or 3c. $\cos\alpha = b/c$, $b = c\cos\alpha$.

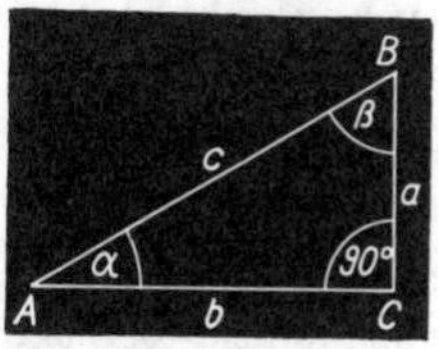

11.1-1
Right-angled triangle

III. Given an angle and the side opposite to it, say a and α:
1. $\beta = 90° - \alpha$; 2. $\cot\alpha = b/a$, $b = a\cot\alpha$;
3. $\sin\alpha = a/c$, $c = a/\sin\alpha$ or with the help of the calculated angle β:
2a. $\tan\beta = b/a$, $b = a\tan\beta$; 3a. $\cos\beta = a/c$, $c = a/\cos\beta$.

IV. Given the two sides a and b containing the right angle:
1. $\tan\alpha = a/b$, 2. $\beta = 90° - \alpha$; 3a. $c = \sqrt{(a^2 + b^2)}$ or with the help of the calculated angle α:
3b. $c = a/\sin\alpha$; or 3c. $c = b/\cos\alpha$.

Checks and accuracy. One usually tries to find the solution using only the given quantities. *Auxiliary solutions* with the help of quantities already calculated can be used as checks, because the same quantity calculated in different ways must theoretically have the same value. Another *check* is based on the theorem that the sum of the angles of a triangle is 180°. In surveying checks are provided for almost every trigonometric calculation. In this the permissible deviation of the value for the same quantity depends essentially on the tables used. In evaluating a possible deviation

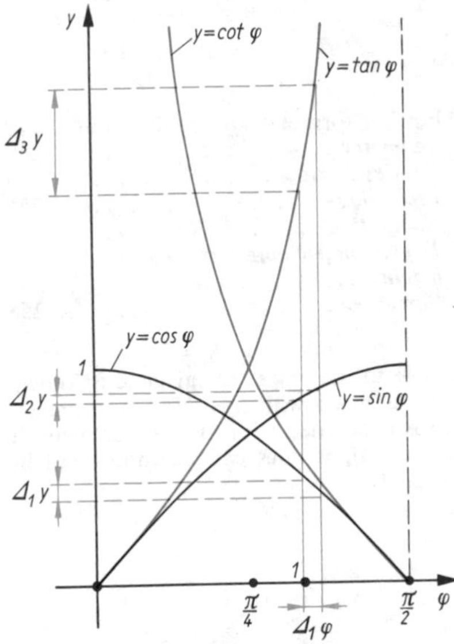

one must bear in mind that for a given small interval $\Delta_1\varphi$ of an angle φ the errors Δy in looking up values of different trigonometric functions are of *different magnitude*. In the figure, for example, $\Delta_3 y$ for $y = \tan\varphi$ is greater than $\Delta_1 y$ for $y = \cos\varphi$. Of course, conversely, for a given small interval $\Delta_2 y$ the *value of the angle* can be determined more accurately from the tangent function or from the cotangent function than from the other two functions. For the function $y = \sin\varphi$, in particular, the figure shows once more the dependence of the magnitude Δy of the interval of the function values upon the magnitude of the interval of the angle values. For small values of

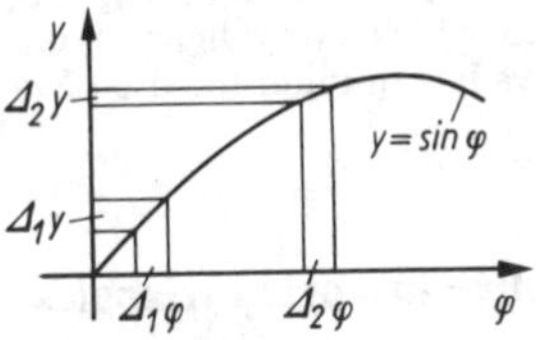

11.1-2 Accuracy in working with trigonometric functions

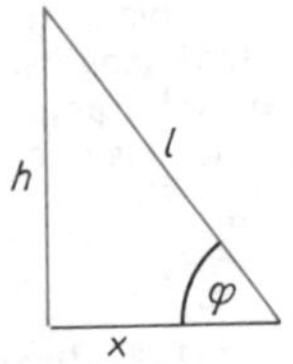

11.1-3 Inclination of a ladder leaning against a wall, $h = 1.2$, $l = 1.5$

the angle in the neighbourhood of $\varphi = 0°$, Δy is large; on the other hand, for large values in the neighbourhood of $\varphi = 90°$, Δy is small. The angle φ can be determined from the value found for the sine with greater precision in the first case than in the second. The accuracy of the *check* must, of course, be in agreement with the *measured value*. To calculate the angle made with the horizontal by a ladder of length $l = 1.50$ m leaning against a vertical wall at a height $h = 1.20$ m, one obtains $\sin\varphi = 1.2/1.5 = 0.8$ (Fig.). The distance x of the foot of the ladder from the wall is given by $x = \sqrt{(1.5)^2 - (1.2)^2} = 0.90$ m. As a check $x_s = 1.5\cos\varphi$ and $x_t = 1.2\cot\varphi$ are calculated. The round value $\varphi_1 = 53°$ taken from a 4-figure table without interpolation gives the values $x_{1s} = 0.903$ and $x_{1t} = 0.904$ which correspond to the accuracy of l and h. From a 7-figure table one obtains the less meaningful values $\varphi_2 = 53°7'48.4''$, $x_{2s} = 0.900\,000\,0$ and $x_{2t} = 0.899\,999\,6$. The distance of the ladder will hardly be measured to within 4 millimetres and certainly not to within 4 ten-thousandths of a millimetre. *The result cannot be more accurate than the given values.*

To increase the accuracy in surveying, additional measurements are made and the most probable value is calculated by the methods of errors and least squares.

Applications

Length of a chord of a circle. The angle subtended at the centre of a circle of radius r by the chord of length s is twice the angle subtended at the circumference by the same chord (Fig.). The perpendicular from the centre M of the circle to the chord s bisects both the angle at the centre and the chord and forms two congruent right-angled triangles. It then follows that: $\sin\gamma = s/2r$ or $s = 2r\sin\gamma$.

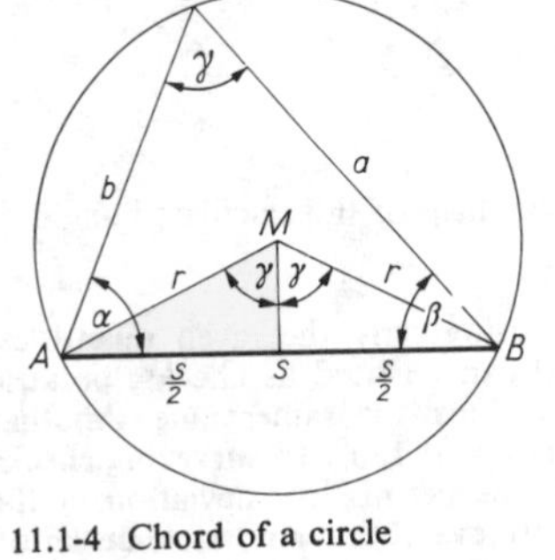

11.1-4 Chord of a circle

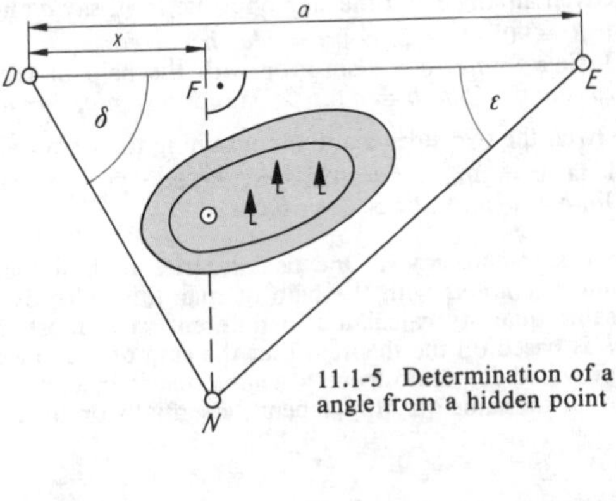

11.1-5 Determination of a right angle from a hidden point

Determination of a right angle from a hidden point. From a water pipe running in a straight line between the villages D and E (Fig.) a perpendicular branch pipe to a village N ist to be constructed and a water tower is to be built on the intervening ridge. N cannot be seen from the required point F at which the branch pipe leaves the main pipe, though it can be seen from D and E. The distance $a = |DE|$ and the angle δ are measured. The position of F on DE is determined by the distance $x = |DF|$. From the right-angled triangles DFN and EFN one obtains: $|FN| = x \tan \delta$, $|FN| = (a - x) \tan \varepsilon$, so that $x \tan \delta = (a - x) \tan \varepsilon$ and hence $x(\tan \delta + \tan \varepsilon) = a \tan \varepsilon$, $x = a \dfrac{\tan \varepsilon}{\tan \delta + \tan \varepsilon}$.
For the calculation of x using logarithms this expression is transformed using the addition theorem:

$$x = \frac{a \sin \varepsilon/\cos \varepsilon}{\sin \delta/\cos \delta + \sin \varepsilon/\cos \varepsilon} = \frac{a \sin \varepsilon \cos \delta \cos \varepsilon}{\cos \varepsilon(\sin \delta \cos \varepsilon + \cos \delta \sin \varepsilon)} = a \frac{\cos \delta \sin \varepsilon}{\sin(\delta + \varepsilon)}.$$

Determination of heights. The height of a tree can be determined (Fig.) by measuring the angle of elevation of the top of the tree from a point A, the distance s between the foot of the tree F and the base S of the point of observation, and the height h_2 of the measuring instrument (that is, the vertical distance $|AS|$). Then $h_1 = s \tan \psi$, and the actual height H of the tree is given by

$$H = h_1 + h_2 = s \tan \psi + h_2.$$

Approximate methods of determining heights. 1. Instead of measuring the angle of elevation ψ, the top of the tree can be sighted along the hypotenuse of an *isosceles right-angled triangle ABC* in which the side CB is held in a vertical line by a plumb line. The angle ψ is then 45° and $h_1 = s$, $H = s + h_2$.

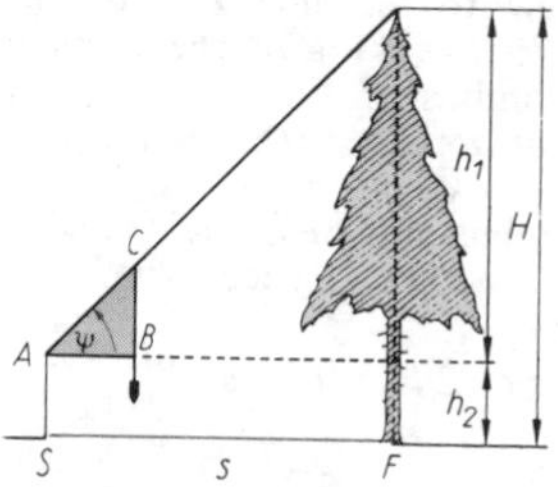

11.1-6 Determination of the height of a tree

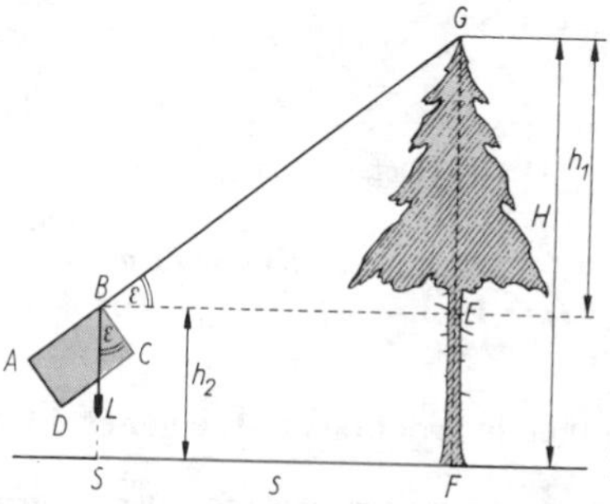

11.1-7 Method of measuring heights in forestry

This method can be used only when there is enough room to choose the point of observation suitably. Otherwise one can employ the following method, which is usual in forestry.

2. A *rectangle ABCD* (made of wood or cardboard) is held in such a position that the top G of the tree is sighted along the edge AB (Fig.). A plumb line suspended from the point B then cuts the side CD of the rectangle in the point L. The two angles marked ε are equal, since the arms of one are perpendicular to the corresponding arms of the other and the right-angled triangles BCL and BEG are similar. Then $|GE|/|BE| = \tan \varepsilon = |CL|/|BC|$. If one chooses $|BC| = 10''$ and subdivides the side $|CD|$ into inches, then $|CL|/|BC| = |CL|/10$ is always a decimal fraction whose value is $\tan \varepsilon$. The rectangle $ABCD$ 'calibrated' in this way is a disguised *table of tangents*, which is particularly simple to handle. From $h_1 = |GE| = s \tan \varepsilon$ it follows that the height of the tree $H = s \tan \varepsilon + h_2 = s(|CL|/10) + h_2$.

Determination of the altitude of the sun. From the length b of the *shadow* cast by a *vertical rod* of length s on a horizontal plane (Fig.) the angle φ between the rays of the sun and the horizontal can be determined. It is called the altitude of the sun. One obtains $\tan \varphi = s/b$ or $\cot \varphi = b/s$. If the rod is of length 1 yard, then the length of the shadow in yards gives the value of $\cot \varphi$ immediately.

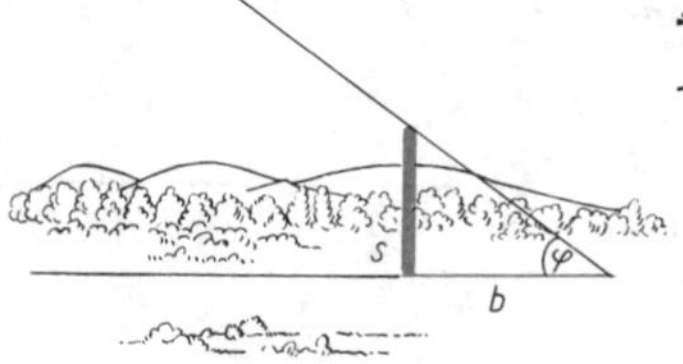

11.1-8 Altitude of the sun

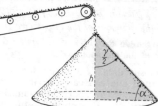

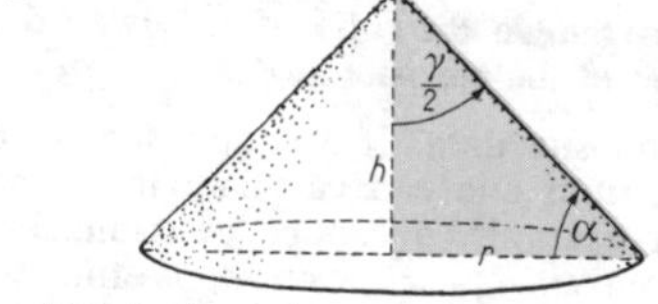

11.1-9 Sand tip

The angle of a tip. If sand is transported on a conveyor belt, then a conical heap or a *sand tip* is formed as it falls off (Fig.). Its content can be calculated from the diameter $d = 2r$ of its circular base and the *tip angle* α between a line in the curved surface of the cone and the horizontal. $V = \pi r^2 h/3$, where $h = r \tan \alpha$, so that $V = (\pi r^3/3) \tan \alpha$. If the vertical angle γ of the cone is used instead of the tip angle α, then $h = r \cot(\gamma/2)$ and $V = (\pi r^3/3) \cot(\gamma/2)$.

For sand the tip angle is approximately 33° and for vulcanite about 36°.

The angle between the plane faces of a regular tetrahedron and a regular octahedron. The regular *tetrahedron* is bounded by four congruent equilateral triangles and six edges of equal length k. The angle ν between two adjacent triangular faces can be seen in a plane section of the tetrahedron containing the edge BD, bisecting the edge AC skew to BD, and perpendicular to AC (Fig.). The section BDM is an isosceles triangle. Its equal sides are altitudes of faces of the tetrahedron and have length $h = {}^1/_2 k \sqrt{3}$. The height η of the tetrahedron is perpendicular to one of these equal sides and divides it in the ration $|MF| : |FB| = 1 : 2$, because the altitudes of the equilateral triangle ABC are also medians. In the right-angled triangle MFD, h is the hypotenuse and $|MF| = h/3$ the side adjacent to the angle ν. Hence $\cos \nu = {}^1/_3 h/h = {}^1/_3$, $\nu = 70°31'44''$.

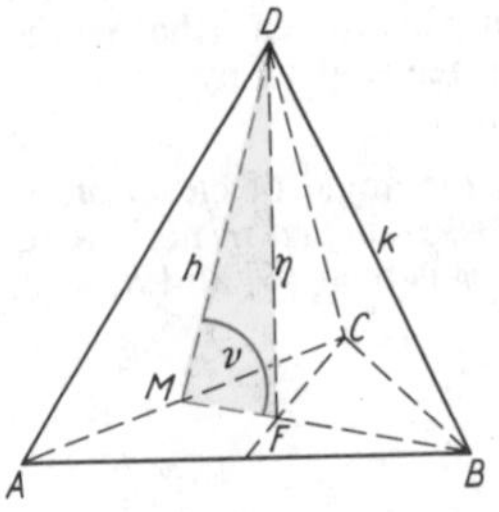

11.1-10 Tetrahedron

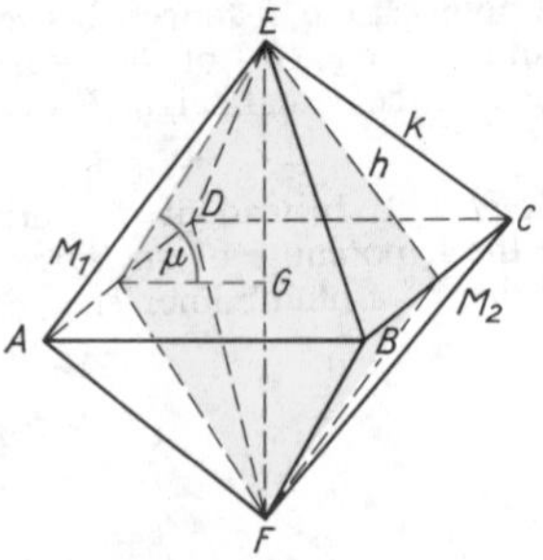

11.1-11 Octahedron

The regular *octahedron* is bounded by eight congruent equilateral triangles and twelve edges of equal length k. The angle 2μ between two adjacent triangular faces can be seen in a plane section through two opposite vertices E, F and through the midpoints M_1, M_2 of two parallel edges ($AD \parallel BC$) that are skew to the line EF joining these vertices (Fig.). The section is a rhombus of side $h = {}^1/_2 k \sqrt{3}$ whose diagonals, $|EF| = k\sqrt{2}$ and $|M_1 M_2| = k$, bisect the angles of the rhombus and are at right angles to one another. Hence from the right-angled triangle $M_1 GE$ it follows that the half-angle μ is given by:

$$\cos \mu = {}^1/_2 k / ({}^1/_2 k \sqrt{3}) = 1/\sqrt{3} = {}^1/_3 \sqrt{3}; \quad \mu = 54°44'07'' \quad \text{or} \quad 2\mu = 109°28'14''.$$

11.2. The trigonometric functions in the general triangle

In many cases the lengths and angles accessible for measurement do not lie in right-angled triangles. Relationships between the sides and angles of the general triangle were therefore derived. The most important are the sine rule and the ccsine rule. They are sufficient for every calculation. The cosine rule is less advantageous for calculations, especially when tables are used, because the formula contains a sum of squares and a product term. It can be replaced by the tangent or by the half-angle formula.

The formulae of plane trigonometry

The sine rule. Every triangle ABC (Fig.) has a *circumcircle* whose centre M is at the intersection of the perpendicular bisectors of the sides of the triangles. The sides of the triangle are *chords* of he circle and the opposite angles are angles at its circumference. If the radius of the circumcircle is denoted by R, then the sides can be calculated as chords of the circle: $a = 2R \sin \alpha$, $b = 2R \sin \beta$, $c = 2R \sin \gamma$. From these one obtains for the diameter $2R = a/\sin \alpha = b/\sin \beta = c/\sin \gamma$.

Sine rule
$a/\sin \alpha = b/\sin \beta = c/\sin \gamma$ or $a : b : c = \sin \alpha : \sin \beta : \sin \gamma$

In any triangle the ratio of each side to the sine of the opposite angle is a constant (*equal to the diameter of the circumcircle*).

The sine rule. In a plane triangle the ratio of any two sides is equal to the ratio of the sines of the opposite angles.

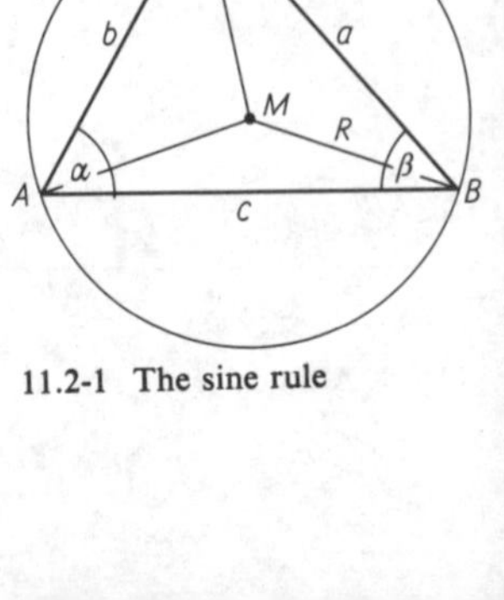

11.2-1 The sine rule

The sine rule connects opposite data. If two opposite data are given, then from any third datum one can calculate the opposite one. Given a, α and b, for example, β can be determined from $\sin \beta / \sin \alpha = b/a$, $\sin \beta = (b/a) \sin \alpha$; or given b, β and γ the side c can be determined from $c/b = \sin \gamma / \sin \beta$, $c = b \sin \gamma / \sin \beta$.

In calculating an angle by means of the sine rule one should, of course, observe that two angles φ_1 and φ_2 are given by $\sin\varphi$, as can be seen from the unit circle. One of these angles is acute and the other is the difference between the acute angle and 180°; $\varphi_1 + \varphi_2 = 180°$. One must distinguish in each particular case which of these angles corresponds to the given geometrical situation.

The cosine rule. In the triangle ABC let D be the foot of the altitude h_c and $|AD| = q$ the projection of the side b on the side c (Fig.). This projection $q = b\cos\alpha$ is positive for an acute angle α and negative for an obtuse angle. The segment DB ist thus of length $c - q = |DB|$ for arbitrary values of α. The altitude h_c always has the length $h_c = b\sin\alpha$. Applying the theorem of Pythagoras to the right-angled triangle DBC one obtains $a^2 = h_c^2 + (c-q)^2 = b^2\sin^2\alpha + c^2 + b^2\cos^2\alpha - 2cb\cos\alpha$, or $a^2 = b^2 + c^2 - 2bc\cos\alpha$. Corresponding relationships can be found using the altitudes h_a and h_b. These can be obtained formally by *a cyclic permutation* in which a is replaced by b, b by c and c by a; the same holds for the angles $\alpha \to \beta \to \gamma \to \alpha$.

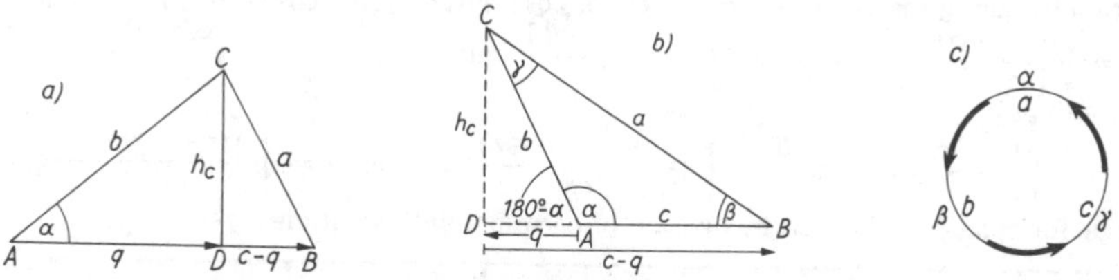

11.2-2 The cosine rule: a) for an *acute-angled* triangle, b) for an *obtuse-angled* triangle, c) cyclic permutation

Cosine rule	$a^2 = b^2 + c^2 - 2bc\cos\alpha,\ b^2 = c^2 + a^2 - 2ca\cos\beta,\ c^2 = a^2 + b^2 - 2ab\cos\gamma$

The cosine rule. In a plane triangle the square of one side is equal to the sum of the squares of the other two sides minus twice the product of these two sides and the cosine of the angle between them.

When two sides and the included angle are known, the third side can be calculated using the cosine rule, and when three sides are known any angle can be found:

$$\cos\alpha = \frac{b^2 + c^2 - a^2}{2bc}, \quad \cos\beta = \frac{c^2 + a^2 - b^2}{2ca}, \quad \cos\gamma = \frac{a^2 + b^2 - c^2}{2ab}.$$

The tangent formula. Using the rule for the ratios of corresponding sums and differences and applying the addition theorems one can deduce:

$$\frac{a}{b} = \frac{\sin\alpha}{\sin\beta}, \quad \frac{a-b}{a+b} = \frac{\sin\alpha - \sin\beta}{\sin\alpha + \sin\beta} = \frac{2\cos[(\alpha+\beta)/2]\sin[(\alpha-\beta)/2]}{2\sin[(\alpha+\beta)/2]\cos[(\alpha-\beta)/2]}.$$

Dividing both numerator and denominator by $\cos[(\alpha+\beta)/2]\cos[(\alpha-\beta)/2]$ one obtains the tangent formula for the sides a and b. The corresponding formulae for the remaining pairs of sides are obtained by a cyclic permutation:

$$\frac{a-b}{a+b} = \frac{\tan[(\alpha-\beta)/2]}{\tan[(\alpha+\beta)/2]}, \quad \frac{b-c}{b+c} = \frac{\tan[(\beta-\gamma)/2]}{\tan[(\beta+\gamma)/2]}, \quad \frac{c-a}{c+a} = \frac{\tan[(\gamma-\alpha)/2]}{\tan[(\gamma+\alpha)/2]}.$$

Tangent formulae	$\tan\dfrac{\alpha-\beta}{2} = \dfrac{a-b}{a+b}\tan\dfrac{\alpha+\beta}{2}, \quad \dfrac{\alpha+\beta}{2} = \dfrac{180° - \gamma}{2}$
	$\tan\dfrac{\beta-\gamma}{2} = \dfrac{b-c}{b+c}\tan\dfrac{\beta+\gamma}{2}, \quad \dfrac{\beta+\gamma}{2} = \dfrac{180° - \alpha}{2}$
	$\tan\dfrac{\gamma-\alpha}{2} = \dfrac{c-a}{c+a}\tan\dfrac{\gamma+\alpha}{2}, \quad \dfrac{\gamma+\alpha}{2} = \dfrac{180° - \beta}{2}$

From two sides (for example, a and b) and the included angle (γ) the other two angles (α and β) can be calculated by means of these formulae. Their half-sum $(\alpha+\beta)/2 = 90° - \gamma/2$ is given by the included angle and their half-difference $(\alpha-\beta)/2$ is given by the tangent formula: from $(\alpha+\beta)/2 = \xi$ and $(\alpha-\beta)/2 = \eta$ one obtains $\alpha = \xi + \eta$ and $\beta = \xi - \eta$.

The half-angle formulae. To obtain a formula that is suitable for logarithmic calculations in the case of three given sides, one substitutes the expression $\cos\alpha = \dfrac{b^2 + c^2 - a^2}{2bc}$ given by the cosine rule into the formula

$$\cos\frac{\alpha}{2} = \sqrt{\left[\frac{1+\cos\alpha}{2}\right]} \quad \text{(see Chapter 10):}$$

$$\cos\frac{\alpha}{2} = \sqrt{\left[\frac{2bc+b^2+c^2-a^2}{4bc}\right]} = \sqrt{\left[\frac{(b+c)^2-a^2}{4bc}\right]} = \sqrt{\left[\frac{b+c-a}{2}\cdot\frac{b+c+a}{2}\cdot\frac{1}{bc}\right]}.$$

Similar formulae hold for $\cos\frac{\beta}{2}$ and $\cos\frac{\gamma}{2}$. If one introduces the perimeter $2s$ of the triangle, so that $a+b+c=2s$ or $s=(a+b+c)/2$,
one obtains $s-a={}^1/_2(b+c-a)$, $\quad s-b={}^1/_2(c+a-b)$, $\quad s-c={}^1/_2(a+b-c)$
and hence $\cos\frac{\alpha}{2} = \sqrt{\left[\frac{(s-a)\,s}{bc}\right]}$, $\quad \cos\frac{\beta}{2} = \sqrt{\left[\frac{(s-b)\,s}{ca}\right]}$, $\quad \cos\frac{\gamma}{2} = \sqrt{\left[\frac{(s-c)\,s}{ab}\right]}$.

Similarly, by substituting the values for $\cos\alpha$, $\cos\beta$, $\cos\gamma$ given by the cosine rule into the formulae $\sin\frac{\alpha}{2} = \sqrt{\left[\frac{1-\cos\alpha}{2}\right]}$, $\sin\frac{\beta}{2} = \sqrt{\left[\frac{1-\cos\beta}{2}\right]}$ and $\sin\frac{\gamma}{2} = \sqrt{\left[\frac{1-\cos\gamma}{2}\right]}$ one obtains the relationships

$$\sin\frac{\alpha}{2} = \sqrt{\left[\frac{(s-b)(s-c)}{bc}\right]}, \quad \sin\frac{\beta}{2} = \sqrt{\left[\frac{(s-c)(s-a)}{ca}\right]}, \quad \sin\frac{\gamma}{2} = \sqrt{\left[\frac{(s-a)(s-b)}{ab}\right]}.$$

The half-angle formulae are obtained by division of corresponding equations.

Half-angle formulae	$2s = a+b+c$
$\tan\frac{\alpha}{2} = \sqrt{\left[\frac{(s-b)(s-c)}{s(s-a)}\right]}$,	$\tan\frac{\beta}{2} = \sqrt{\left[\frac{(s-c)(s-a)}{s(s-b)}\right]}$, $\quad \tan\frac{\gamma}{2} = \sqrt{\left[\frac{(s-a)(s-b)}{s(s-c)}\right]}$

For practical calculations it is advisable to calculate all three angles α, β, γ from the three sides a, b, c. The known angle sum of a triangle can then be used as a check.

The four main cases for the solution of a triangle

In a triangle the following data can be given: two angles and one side; two sides and one angle that is either opposite one of the two sides or included between them; three sides. The method of solution for these cases will be given.

I. Given two angles and a side. Since the angle sum of a triangle is 180°, the third angle is also known. By means of the sine rule the remaining sides can be calculated; from c, α, β, for example, it follows that $\gamma = 180° - (\alpha+\beta)$ and $a = c\sin\alpha/\sin\gamma$, $b = c\sin\beta/\sin\gamma$.

Example: A force $F = 130$ units is to be decomposed into two components F_1 and F_2 in such a way that F_1 makes an angle $\delta = 18°$ with F and the two components make an angle $\varepsilon = 65°$ with one another (Fig.). The diagonal $F = |AC|$ of the parallelogram $ABCD$ is given. The position of the point B is determined by the angles $\delta = 18°$ and $\omega = \varepsilon - \delta = 47°$. In the triangle ABC it follows that:

$$F_1 = F\frac{\sin\omega}{\sin(180° - \varepsilon)} = F\frac{\sin 47°}{\sin 65°}; \quad F_1 = 104.902 \text{ units},$$

$$F_2 = F\frac{\sin\delta}{\sin\varepsilon} = F\frac{\sin 18°}{\sin 65°}; \quad F_2 = 44.324 \text{ units}.$$

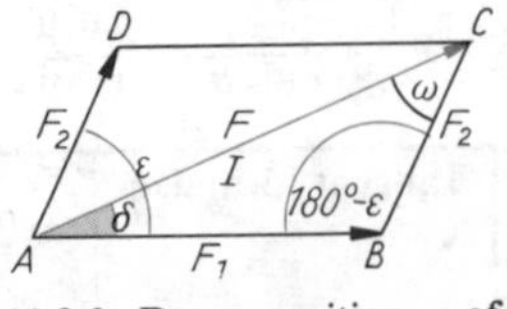

11.2-3 Decomposition of the force F into two components F_1 and F_2

II. Given two sides and the angle opposite to one of them. Let a, c and γ be given (Fig.); then one obtains
1. $\sin\alpha = (a/c)\sin\gamma$; $\quad$ 2. $\beta = 180° - (\alpha+\gamma)$; $\quad$ 3. $b = c\sin\beta/\sin\gamma$.

Of course, equation 1. can hold only if $(a/c)\sin\gamma \leqslant 1$. Because of this condition there are several possible cases.

II (1) $a < c$, with the given angle opposite the greater side. There always exists an angle α, which must be smaller than γ, since it is opposite the smaller side. Moreover, the solution is ***unique***; although the sine function has the same value for the angles α_1 and $\alpha_2 = 180° - \alpha_1$, only $\alpha_1 < \gamma_1$ is a solution of the problem.

Example: $a = 56.9$ m, $c = 68.0$ m, $\gamma = 63°57'$.
1. $\sin\alpha = a/c\sin\gamma = (56.9/68.0)\sin 63°57'$; $\alpha_1 = 48°45'$; $\alpha_2 = 180° - \alpha_1 = 131°15'$ is greater than γ.

2. $\beta = 180° - (\alpha_1 + \gamma)$; $\beta = 67°18'$.
3. $b = c\dfrac{\sin\beta}{\sin\gamma} = 68.0\text{ m}\dfrac{\sin 67°18'}{\sin 63°57'}$; $b = 69.8$ m.

II (2) $a = c$; the triangle is isosceles and hence $\alpha = \gamma$.

II (3) $a > c$, with the given angle opposite the smaller side. Then a can be so large that the condition $\sin\alpha \leqslant 1$ is not satisfied. II (3.1) : *no solution exists* and no triangle can be constructed from the given data; for example, if $c = 2''$, $a = 5''$, $\gamma = 75°$. II (3.2) : $\sin\alpha$ may be equal to 1 so that α is a right angle, because $\alpha_2 = 180° - \alpha_1 = \alpha_1$. The solution and the construction are *unique*, for example, if $a = 2''$, $c = 1''$, $\gamma = 30°$. II (3.3) : if $\sin\alpha < 1$, the angles α_1 and $\alpha_2 = 180° - \alpha_1$ can be calculated. Because $\sin\alpha > \sin\gamma$, it also follows that $\alpha_1 > \gamma$, so that $(180° - \alpha_1) + \gamma < 180°$ and the angle α_2 also satisfies the geometric conditions.

The problem has *two solutions*.

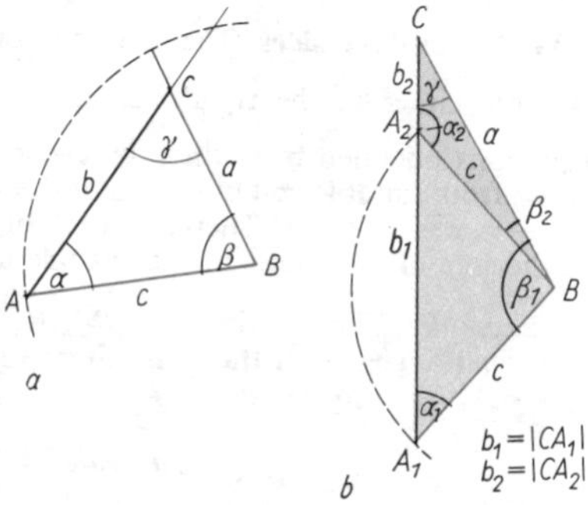

11.2-4 Solution of a triangle, given two sides and an angle opposite to one of them; a) *one solution*, b) *two solutions*

Example: $a = 87.23$ m, $c = 65.95$ m, $\gamma = 30.42°$.

1. $\sin\alpha = (87.23/65.95)\sin 30.42°$; $\alpha_1 = 42.04°$; $\alpha_2 = 180° - \alpha_1 = 137.96°$; $\alpha_1 > \gamma$, $\alpha_2 > \gamma$.
2. $\beta_1 = 180° - (\alpha_1 + \gamma)$; $\beta_1 = 107.54°$, $\beta_2 = 11.62°$.
3. $b_1 = 65.95\text{ m}\cdot(\sin 104.54°/\sin 30.42°) = 126.0$ m and
$b_2 = 65.95\text{ m}\cdot(\sin 11.62°/\sin 30.42°) = 26.23$ m.

III. Given two sides and the included angle. The solution comes from the cosine rule or the tangent formula. Given the values of b, c and α in the triangle ABC, then the *cosine* rule gives $a^2 = b^2 + c^2 - 2bc\cos\alpha$ and from this the unique value $a = \sqrt{(b^2 + c^2 - 2bc\cos\alpha)}$. The angle β can also be determined uniquely from the cosine rule, that is, from $\cos\beta = (c^2 + a^2 - b^2)/(2ca)$. However, it is usually preferable to use the sine rule and obtain $\sin\beta = (b/a)\sin\alpha$. Of the two angles β_1 and β_2 that satisfy this equation only one corresponds to the geometric conditions. From $(\gamma + \beta)/2 = 90° - \alpha/2$, and by the *tangent formula* one obtains: $\tan[(\gamma + \beta/2](c - b)/(c + b) = \tan[(\gamma - \beta)/2]$; from $(\gamma + \beta)/2$ and $(\gamma - \beta)/2$ the angles β and γ can be found. The third side can then be determined by the sine rule; $c = a\sin\gamma/\sin\alpha$.

Example: A cable is to be laid in a straight line through wooded country between two places R and S. They are not visible from one another, but a point A can be found from which the distances $d = |AR| = 2.473$ miles and $e = |AS| = 3.752$ miles and the angle $\tau = \sphericalangle RAS = 42°26'10''$ can be measured (Fig.). What must the length x of the cable be and at what angles ε, δ from R, S respectively must it be laid? – For comparison two methods of solution are given.

1. $x^2 = d^2 + e^2 - 2de\cos\tau$
$x^2 = 6.497313$
$x = 2.549$ miles
2. $\sin\varepsilon = (e/x)\sin\tau$
$\varepsilon_1 = 83°20'00''$
$\varepsilon_2 = 96°40'00''$
3. $\delta = 180° - (\varepsilon + \tau)$
$\delta_1 = 54°13'50''$
$\delta_2 = 40°53'50''$

Since $e > x > d$, it must also follow that $\varepsilon > \tau > \delta$; this condition is satisfied only by δ_2. Therefore the solution is x, ε_2, δ_2.

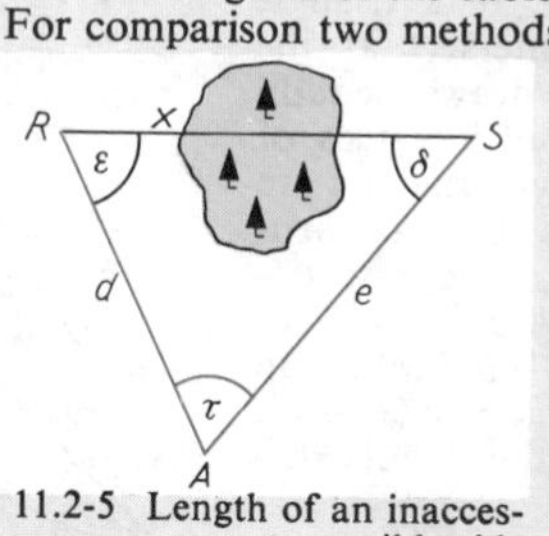

11.2-5 Length of an inaccessible side

$$\tan\frac{\varepsilon - \delta}{2} = \frac{e - d}{e + d}\tan\frac{\varepsilon + \delta}{2}$$

1. $\varepsilon + \delta = 180° - \tau = 137°33'50''$
$(\varepsilon + \delta)/2 = 68°46'55''$
$(\varepsilon - \delta)/2 = 27°53'18''$

$\varepsilon = 96°40'13''$
$\delta = 40°53'37''$
$\tau = 42°26'10''$

$\varepsilon + \delta + \tau = 180°00'00''$ (check)

2. $x = e\dfrac{\sin\tau}{\sin\varepsilon} = 2.549$ miles

The *agreement* between the two results is *unsatisfactory*. The reason for this (which was discussed fully in the introduction) is that the sine function was used to determine an angle in the neighbourhood of 90°: $\sin 96°40'00'' = 0.99324$, $\sin 96°40'10'' = 0.99323$, $\sin 96°40'20'' = 0.99323$; on the number of seconds nothing reliable can be said. A greater precision can be obtained in this case if the angle ε ist also calculated by the cosine rule, that is, from the equation $\cos\varepsilon = \dfrac{x^2 + d^2 - e^2}{2xd}$. One obtains the unique value $\cos\varepsilon = \dfrac{-1.464462}{2(2.549)(2.473)}$ or $\varepsilon_2' = 96°40'14''$, in sufficiently close agreement with the value found by the tangent formula. The solution found from the cosine rule is therefore: $x = 2.549$ miles, $\varepsilon_2' = 96°40'14''$, $\delta_2' = 40°53'36''$.

IV. Given three sides. The solution comes from the cosine rule or the half-angle formulae, that is, from either of the equations $\cos\alpha = \dfrac{b^2 + c^2 - a^2}{2bc}$, $\tan\dfrac{\alpha}{2} = \sqrt{\left[\dfrac{(s-b)(s-c)}{s(s-a)}\right]}$, and the equations obtained from these by cyclic permutation. Both solutions are unique and are obtained either from suitable combinations of the six numbers $a^2, b^2, c^2, 2ab, 2bc, 2ca$ or of the four numbers $s, s-a, s-b, s-c$. Therefore each of the three angles α, β, γ should be calculated and the value of the sum of the angles of the triangle used as a check.

Example: Three points R_1, R_2, R_3 on raised ground are to be connected by radar (Fig.). At what angles must the transmitter and receiver at each point R_1, R_2, R_3 be built? – $|R_1R_2| = c = 45.21$ miles; $|R_2R_3| = a = 52.46$ miles; $|R_3R_1| = b = 39.37$ miles.

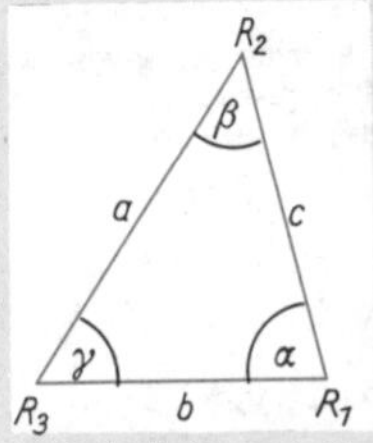

11.2-6 Solution of a triangle, given three sides

Cosine rule

$$a^2 = 2752.0516$$
$$b^2 = 1549.9969$$
$$c^2 = 2043.9441$$
$$b^2 + c^2 - a^2 = 841.8894$$
$$c^2 + a^2 - b^2 = 3245.9988$$
$$a^2 + b^2 - c^2 = 2258.1044$$
$$\alpha = 76°19'12''$$
$$\beta = 46°49'06''$$
$$\gamma = 56°51'42''$$
$$180°00'00''$$

Half-angle formula

		lg
$a = 52.46$	$s - a = 16.06$	1.20575
$b = 39.37$	$s - b = 29.15$	1.46464
$c = 45.21$	$s - c = 23.31$	1.36754
$2s = 137.04$ →	$s = 68.52$	1.83582

$$\alpha = 76°19'12''$$
$$\beta = 46°49'06''$$
$$\gamma = 56°51'42''$$
$$180°00'00''$$

11.3. Further formulae and applications

In many fields arguments are made precise with the aid of mathematical relations; for example, when directions and angles in plane rectilinear figures occur, then theorems of plane trigonometry are used. One of these fields, namely surveying, plays a special role. In this discipline the relationships in question rest more directly on these theorems than in other fields, and historically the requirements of surveying were responsible for the development of plane trigonometry. For this reason the possible applications in this field are dealt with in a special section.

Geometry

The radius r of the inscribed circle. In a triangle ABC the bisectors of the angles intersect at the centre M of the inscribed circle. If one draws the radii through the points of contact E, F, G of the sides of the triangle (Fig.), then six right-angled triangles are formed. They are congruent in pairs and, in particular, the pairs of sides marked x, y, z, respectively, are equal. Their lengths are $\cdot x = s - a$, $y = s - b$, $z = s - c$, where $s = (a + b + c)/2$. In the triangle AGM, $\tan(\alpha/2) = r/x = r/(s-a)$, but by the tangent formula for the whole triangle $\tan\dfrac{\alpha}{2} = \sqrt{\left[\dfrac{(s-b)(s-c)}{s(s-a)}\right]}$. Hence

$$\frac{r}{s-a} = \sqrt{\left[\frac{(s-b)(s-c)}{s(s-a)}\right]},$$

$$r = (s-a)\sqrt{\left[\frac{(s-b)(s-c)}{s(s-a)}\right]}.$$

The same result would have been obtained by considering $\tan(\beta/2)$ or $\tan(\gamma/2)$.

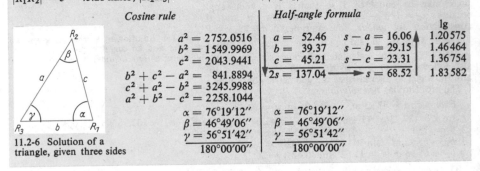

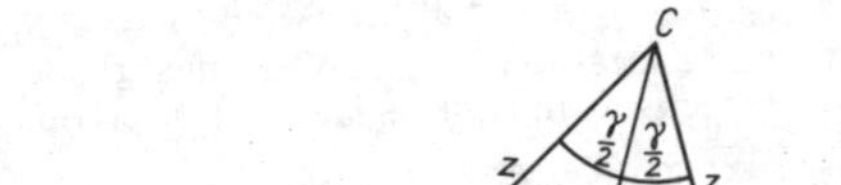

11.3-1 Inscribed circle of a triangle

Radius of the inscribed circle	$r = \sqrt{\left[\dfrac{(s-a)(s-b)(s-c)}{s}\right]}$

Marking out an arc of a circle whose centre is inaccessible. Between two points A and B, whose distance apart e is known, arbitrarily many points P_i are to be constructed, all lying on a circle through A and B with given radius r (Fig.). The centre of the circle is inaccessible. It is required to find the distance s from A of the points P_i and the angle φ between AP_i and AB. Let P be one of the required points. Then the triangle AMP is isosceles with base $s = 2r \sin(\sigma/2)$ subtending an angle σ at the centre of the circle. The angle $\sphericalangle PMB$ subtended at the centre by the chord PB is $\varepsilon - \sigma$ and the angle at the circumference is φ. Thus, $\varphi = (\varepsilon - \sigma)/2$ or $\sigma = \varepsilon - 2\varphi$. But the angle ε in the triangle ABM can be determined from $e = 2r \sin(\varepsilon/2)$, so that $\sin \varepsilon/2 = e/2r$. Hence the distance s in dependence on the angle φ is given by: $s = 2r \sin(\sigma/2) = 2r \sin(\varepsilon/2 - \varphi)$, where $\varepsilon/2 = \text{Arcsin}\,(e/2r)$.

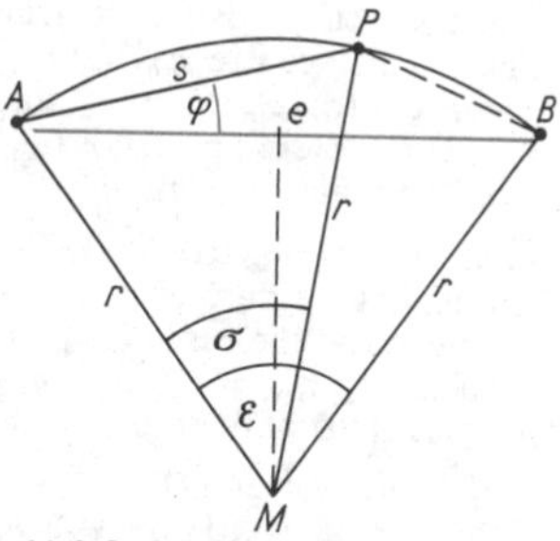

11.3-2 Marking out an arc of a circle

Area of a triangle. From the area formula $A = ch_c/2$ and $h_c = b \sin \alpha$ it follows that $A = (bc/2) \sin \alpha$. From the relationship $\sin \alpha = a/(2R)$, where R is the radius of the circumcircle (see Fig. 11.2-1), it follows that $A = abc/(4R)$, or from $b = 2R \sin \beta$ and $c = 2R \sin \gamma$, $A = 2R^2 \sin \alpha \sin \beta \sin \gamma$. Again, since $R = \dfrac{a}{2 \sin \alpha}$, $A = a^2 \dfrac{\sin \beta \sin \gamma}{2 \sin \alpha}$. In Fig. 11.3-1, by adding the areas of the partial triangles ABM, BCM and CAM with altitude equal to r, the radius of the inscribed circle, one obtains Heron's formula:

$$A = {}^1/_2(cr + br + ar) = rs = \sqrt{[s(s-a)(s-b)(s-c)]}, \quad \text{where} \quad 2s = a + b + c.$$

Area of a triangle	$A = (bc/2) \sin \alpha = (ca/2) \sin \beta = (ab/2) \sin \gamma = abc/(4R) = 2R^2 \sin \alpha \sin \beta \sin \gamma$ $A = a^2 \dfrac{\sin \beta \sin \gamma}{2 \sin \alpha} = b^2 \dfrac{\sin \gamma \sin \alpha}{2 \sin \beta} = c^2 \dfrac{\sin \alpha \sin \beta}{2 \sin \gamma}$
Heron's formula	$A = rs = \sqrt{[s(s-a)(s-b)(s-c)]}$

Example: It is required to calculate the area of a triangle with the sides $a = 345.8$, $b = 236.5$, $c = 497.3$. Using Heron's formula one finds that $s = 539.8$, $s - a = 194.0$, $s - b = 303.3$, $s - c = 42.5$; hence, with the help of 4-figure logarithm tables, $A = 36\,740$. A rough calculation is always to be recommended; in this case by using a slide rule one obtains:

$$\begin{aligned} A &= \sqrt{(539.8 \cdot 194 \cdot 303.3 \cdot 42.5)} \\ &\approx \sqrt{(5.40 \cdot 10^2 \cdot 1.94 \cdot 10^2 \cdot 3.03 \cdot 10^2 \cdot 4.25 \cdot 10)} \\ &= 10^3 \sqrt{(5.40 \cdot 1.94 \cdot 3.03 \cdot 42.5)} = 36\,700. \end{aligned}$$

The position of the decimal point was estimated as follows:

$$A \approx 10^3 \sqrt{(10 \cdot 3 \cdot 10 \cdot 4.2)} = 10^4 \sqrt{12.6} \approx 3.5 \cdot 10^4.$$

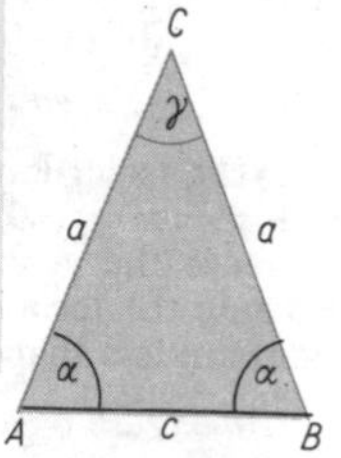

11.3-3 Area of an isosceles triangle

Isosceles triangle. If the equal sides are denoted by a, the base by c and the base angles by α (Fig.), then the area A ist given by:

1. $A = {}^1/_2 a^2 \sin \gamma$, where $\gamma = 180° - 2\alpha$;
2. $A = \dfrac{c^2 \sin^2 \alpha}{2 \sin \gamma} = \dfrac{c^2 \sin^2 \alpha}{2 \sin 2\alpha} = \dfrac{c^2 \sin^2 \alpha}{4 \sin \alpha \cos \alpha}$, $\quad A = \dfrac{c^2}{4} \tan \alpha$;
3. $s = a + c/2$, $s - a = c/2$, $s - c = a - c/2$, and hence $A = \sqrt{[(a + c/2)(c/2)(c/2)(a - c/2)]} = (c/2) \sqrt{(a^2 - c^2/4)} = (c/4) \sqrt{(4a^2 - c^2)}$.

Equilateral triangle. Each side is of length a.

1. $A = {}^1/_2 a^2 \sin 60°$, $A = (a^2/4) \sqrt{3}$.
2. By Heron's formula: $s = 3a/2$, $s - a = s - b = s - c = a/2$ and $A = \sqrt{[(3a/2) \cdot (a^3/8)]} = (a^2/4) \sqrt{3} = a^2 \sqrt{3}/4$.

Regular hexagon. This polygon is composed of six equilateral triangles of side R (the radius of the circumcircle) and it therefore follows that

$$A_6 = (6/4) R^2 \sqrt{3}, \quad A_6 = (3/2) R^2 \sqrt{3}.$$

Regular *n*-sided polygon. Its area is composed of n isosceles triangles in which the equal sides are radii of the circumcircle and the angle φ at the centre of the circle included between them is the nth part of the complete angle; $\varphi_n = 360°/n$ (Fig.).

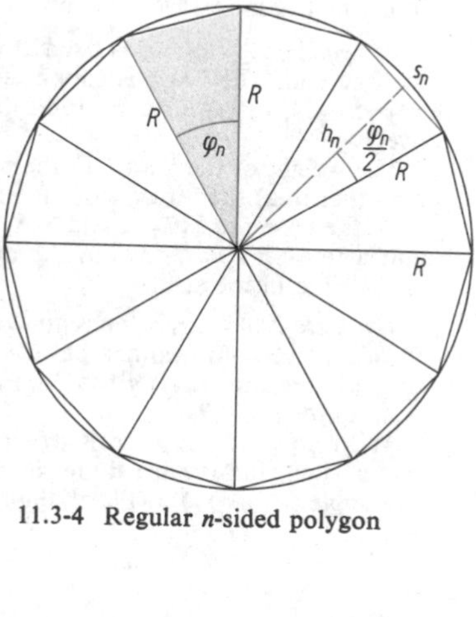

11.3-4 Regular *n*-sided polygon

In each single isosceles triangle the altitude h_n bisects the side s_n of the polygon and the angle at the centre φ_n. Thus, $s_n = 2R \sin(\varphi_n/2)$, $h_n = R\cos(\varphi_n/2)$, so that $A_n = (n/2)\, s_n h_n = (n/2)\, R^2 \cdot 2\sin(\varphi_n/2)\cos(\varphi_n/2) = (n/2)\, R^2 \sin\varphi_n = (n/2)\, R^2 \sin(360°/n)$.

Regular *n*-sided polygon	$A_n = \frac{n}{2} R^2 \sin\frac{360°}{n}$

The general quadrilateral. For the general quadrilateral *ABCD* a formula analogous to Heron's formula can be derived. Since a quadrilateral is determined by five data, one can take as given the four sides and the sum of a pair of opposite angles, for example, α and γ (Fig.). Denoting the semi-perimeter by s, $s = {}^1/_2(a+b+c+d)$, and the sum of the angles α and γ by 2ε, the areas of the triangles *ABD* and *BCD* and the area A_q of the quadrilateral are given by:

$$A_{\mathrm{I}} = {}^1/_2 ad \sin\alpha; \quad A_{\mathrm{II}} = {}^1/_2 bc \sin\gamma;$$

$$A_q = {}^1/_2(ad\sin\alpha + bc\sin\gamma).$$

Using the cosine rule in the two triangles one obtains

$$a^2 + d^2 - 2ad\cos\alpha = f^2 = b^2 + c^2 - 2bc\cos\gamma$$

or

$$a^2 + d^2 - b^2 - c^2 = 2(ad\cos\alpha - bc\cos\gamma).$$

Then

$$(4A_q)^2 + (a^2+d^2-b^2-c^2)^2 = 4(a^2d^2 + b^2c^2 - 2abcd\cos 2\varepsilon),$$ and finally,

$$16A_q^2 = (a+d+b-c)(a+d-b+c)(b+c+a-d)(b+c-a+d) - 16abcd\cos^2\varepsilon.$$

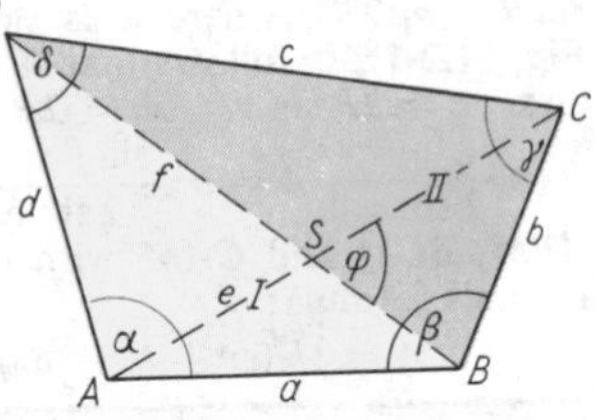

11.3-5 Area of a general quadrilateral

Area of a general quadrilateral
$A_q = \sqrt{[(s-a)(s-b)(s-c)(s-d) - abcd\cos^2\varepsilon]}$

If φ is the angle at which the diagonals of a quadrilateral intersect at the point *S*, then the area A_q af the quadrilateral can be expressed as the sum of the areas of the four triangles *ABS*, *BCS*, *CDS* and *DAS*, so that

$$A_q = {}^1/_2\,[|AS|\cdot|BS|\sin(180° - \varphi) + |BS|\cdot|CS|\sin\varphi + |CS|\cdot|DS|\sin(180° - \varphi) + |DS|\cdot|AS|\sin\varphi] = {}^1/_2\,[|AS|(|BS| + |DS|) + |CS|(|BS| + |DS|)]\sin\varphi$$
$$= {}^1/_2\,[|AS| + |CS|)(|BS| + |DS|)]\sin\varphi,$$

$$A_q = {}^1/_2\, ef\sin\varphi.$$

Thus, *the area is equal to half the product of the diagonals and the sine of the angle between them.*

Cyclic quadrilateral. In a cyclic quadrilateral the *sum of a pair of opposite angles* is 180^0, that is, $\alpha + \gamma = \pi = 180°$, $\varepsilon = 90°$, $\cos\varepsilon = 0$.

Hence the general formula for the area simplifies to $A_{cq} = \sqrt{[(s-a)(s-b)(s-c)(s-d)]}$. Because the term $abcd\cos^2\varepsilon$ is never negative, the areas of all other quadrilaterals with the same sides are less than that of the cyclic quadrilateral.

Among all quadrilaterals with sides a, b, c, d the cyclic quadrilateral has the greatest area.

Physics

All physical quantities that can be represented by vectors (for example, force or velocity) require the use of trigonometric functions for their calculation.

Example 1: A *beam B* is fixed at right angles to a wall (Fig.) and a load of *f* lb. wt. hangs from the free end. The beam is supported by either a) a *tie T* or b) a *strut S* making an angle α or β, respectively, with the beam. Find the forces (tension or thrust) occurring in *B* and *T* or *S*, respectively. –

The weight of the load *f* is the resultant of two forces, one in the direction of the beam *B* and the other in a) the direction of the tie *T* or b) the direction of the strut *S*. Since *f* is perpendicular to *B*, the triangles $T_1T_2T_3$ and $S_1S_2S_3$ are right-angled; a) there is a thrust $d = f\cot\alpha$ in the beam and a tension $t = f/\sin\alpha$ in the tie; b) there is a tension $z = f\cot\beta$ in the beam and a thrust $s = f/\sin\beta$ in the strut.

Example 2: An *aircraft* has an average speed $v_1 = 360$ m.p.h. and flies in the direction N 23.5° E from a place *A* to another place *B* distant 300 miles from *A*. The wind speed is $v_2 = 45$ m.p.h. towards the direction N 18° W. Find the course along which the aircraft must fly and the time it takes to reach *B*. –

With no wind the aircraft would reach *B* in 300/360 h $\doteq$ 5/6 h, that is, in 50 minutes. Because of the side-wind it must fly in the direction Nα_3E (Fig.). By using the parallelogram of velocities the angle $(\alpha_3 - \alpha_1)$ in the triangle *ACE* can be calculated from three given quantities, namely, the sides v_1 and v_2 and the angle $\sphericalangle AEC = \alpha = \alpha_1 + \alpha_2$. By the sine rule

$\sin(\alpha_3 - \alpha_1) = \sin(\alpha_1 + \alpha_2)\, v_2/v_1$,

and $\quad v = v_1 \dfrac{\sin(180° - \alpha_2 - \alpha_3)}{\sin(\alpha_1 + \alpha_2)}$,

since $\alpha_3 - \alpha_1 = 4.75°$,

$\alpha_3 = 28.25°$, $\quad v \approx 390.18$ m.p.h.

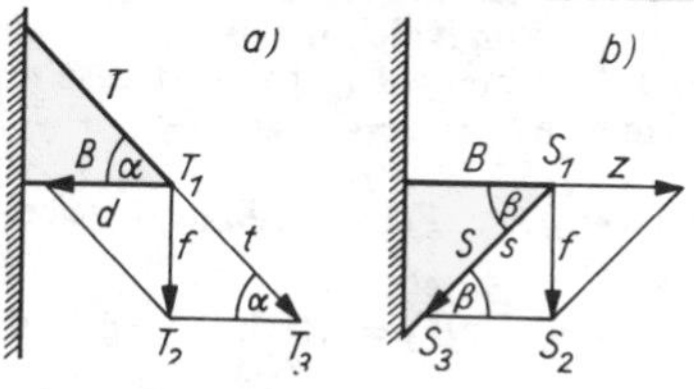

11.3-6 Beam: a) with a *tie*, b) with a *strut*

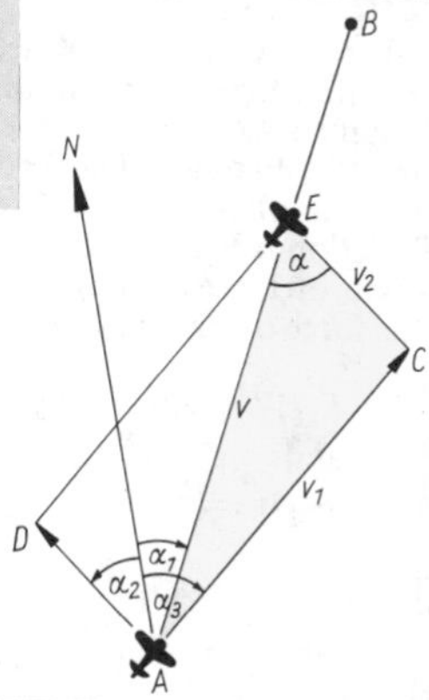

11.3-7 Path of an *aircraft* with a side-wind

11.3-8 Height and distance of a lightning flash

The aircraft flies in the direction N 28.25° E and at a speed of $v = 390.18$ miles per hour reaches B in approximately 46 minutes.

Example 3: A *lightning flash* is observed at an angle α to the horizontal and the *thunderclap* is heard t seconds later at the point of observation. It then follows that the flash occurs at a distance $e = 333t$ m and at a height $h = 333t \sin\alpha$ m (Fig.). Here 333 m/s is the speed of sound, and because the speed of light is $c = 300\,000$ km/s, the time taken for the light to travel to the point of observation can be neglected.

Technology

The laws of technology are applied laws of physics. The trigonometric functions and theorems occur in just the same way as soon as angles play a part.

Crankshafts. In a crankhaft the position of the *big end K* is a function of the angle of rotation φ of the crank (Fig.). If r is the radius of the crank and l the length of the connecting rod, then by the cosine rule:

$l^2 = x^2 + r^2 - 2xr\cos(180° - \varphi) \Rightarrow x^2 + 2rx\cos\varphi = l^2 - r^2$.

The solution of this quadratic equation gives

$$x = -r\cos\varphi + \sqrt{(r^2\cos^2\varphi + l^2 - r^2)}$$
$$= -r\cos\varphi + \sqrt{\{r^2(\cos^2\varphi - 1) + l^2\}},$$
$$x = -r\cos\varphi + \sqrt{(l^2 - r^2\sin^2\varphi)}.$$

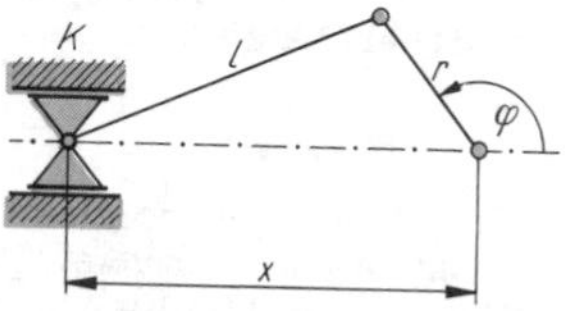

11.3-9 *Crankshaft*

Length of a driving belt. If two *pulleys* of radii R and r have axes at a distance a apart, one can calculate the length L of the driving belt that lies taut around them both (Fig.): $t^2 = a^2 - (R - r)^2$, $\cos\alpha = (R - r)/a$ or $\alpha = \text{Arccos}\,[(R - r)/a]$ in radians. From this it follows that

$$L = 2t + K + k = 2\sqrt{[a^2 - (R - r)^2]} + R(2\pi - 2\alpha) + r \cdot 2\alpha,$$
$$L = 2\{\sqrt{[a^2 - (R - r)^2]} - \alpha(R - r) + R\pi\}.$$

For the case $r = R/2$ and $a = 2R$ one obtains $L = 8.838R$.

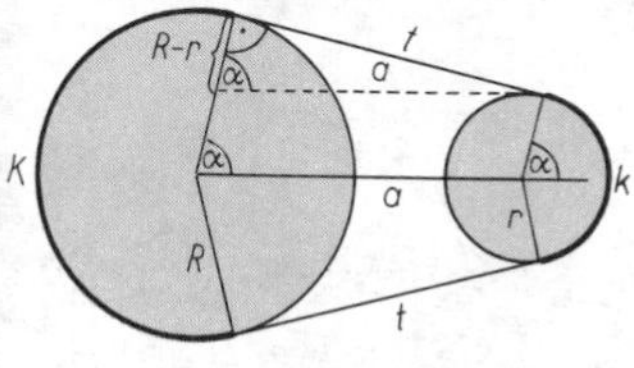

11.3-10 Length of a *driving belt*

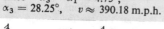

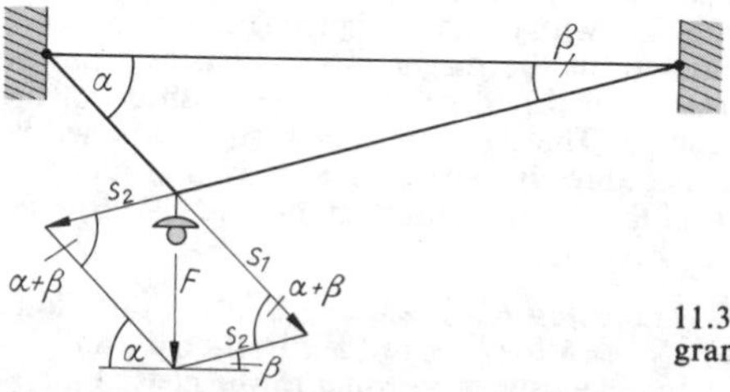

11.3-11 Parallelogram of forces

The parallelogram of forces. A *street lamp* is suspended from two ropes of unequal length that are inclined to the horizontal at angles α and β, respectively. If the sag of the ropes can be neglected, the tensions S_1 and S_2 in the ropes can be calculated using the sine rule (Fig.). Bearing in mind that $\sin(90° - x) = \cos x$, one obtains:

$$S_1 = F\cos\beta/\sin(\alpha+\beta), \quad S_2 = F\cos\alpha/\sin(\alpha+\beta).$$

Motion on an inclined plane. A body of weight W lies on a plane inclined at an angle α to the horizontal. It is required to find the force F_1 in the direction of the plane (Fig.) that will move the body up the plane with constant speed and the force F_2 in the direction of the plane that is necessary to prevent the body from sliding down. The force of friction R is proportional to the normal reaction N between the body and the plane; $R = \mu N$, where μ is called the *coefficient of friction*. Putting $\mu = \tan\varrho$, the angle of friction ϱ is the inclination of a plane on which the given body just fails to slide down. In the triangle of forces (Fig.) the force F ist increased in the first case and decreased in the second case by the force of friction $R = N\tan\varrho$ to give $F_1 = F + R$ and $F_2 = F - R$, respectively. By the sine rule:

$$F_1/W = \sin(\alpha+\varrho)/\sin(90° - \varrho), \quad F_1 = W\sin(\alpha+\varrho)/\cos\varrho$$

and

$$F_2/W = \sin(\alpha-\varrho)/\sin(90° + \varrho), \quad F_2 = W\sin(\alpha-\varrho)/\cos\varrho.$$

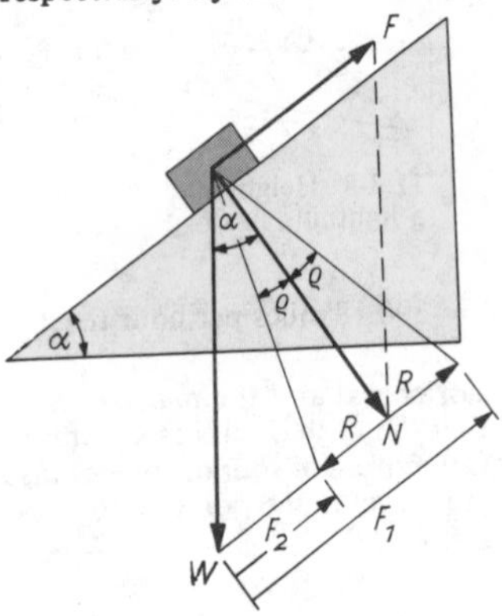

11.3-12 Motion on an *inclined plane* with the corresponding triangle of forces

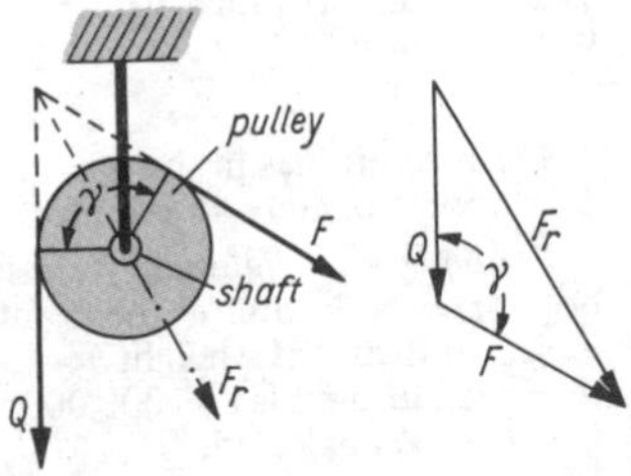

11.3-13 Forces on a fixed *pulley*

Forces on a fixed pulley. To calculate the friction between a fixed pulley and the shaft one must obtain the resultant force F_r of the load Q and the applied force F. If the rope wrapped round the pulley subtends an angle γ at the centre, then F_r is given by the cosine rule (Fig.):

$$F_r = \sqrt{(F^2 + Q^2 - 2FQ\cos\gamma)}.$$

For a smooth pulley the rope tensions F and Q are equal to one another, and since $2 - 2\cos\gamma = 2(1 - \cos\gamma) = 4\sin^2(\gamma/2)$, the equation can be simplified to give $F_r = 2Q\sin(\gamma/2)$. For $\gamma = 180°$, F and Q are parallel and $F_r = 2Q$.

Navigation

To determine the position of a ship and its path on the sea the earth is regarded as a sphere. The calculations must therefore be based on the laws of spherical trigonometry. Methods referring to stars or satellites for determining position are also based on them. Smaller regions, for example for journeys near a coast, may be taken as plane. The location and relative distances of characteristic points marked on a coastal map are assumed to be known.

The direction of motion of a ship, its *course*, is fixed by the angle between the direction of its keel and a fixed reference direction. The *true course* is measured from the geographic north towards the east up to 360°; the *magnetic course* is measured from the magnetic north in the same sense. The difference between the two, that is, the angle between the geographical and the magnetic meridian, is called the *declination*. On an iron ship with its own magnetic field the compass meridian deviates from the magnetic meridian by an angle called the *deviation*, which depends on the position of the ship and its course. The *compass course* so determined is, however, measured directly. The course is given as an angle between points of the compass, for example as N 35° E, read *north* 35 *degrees east*.

Example: From a *ship* F a *lighthouse* L in the direction S 55.3° E and a *church tower* K in the direction S 28.5° W are sighted simultaneously. According to the coastal map the distance $|KL| = s = 33.25$ km ≈ 17.95 nautical miles in the direction N 84.7° E (nautical mile = 1.852 km).

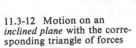

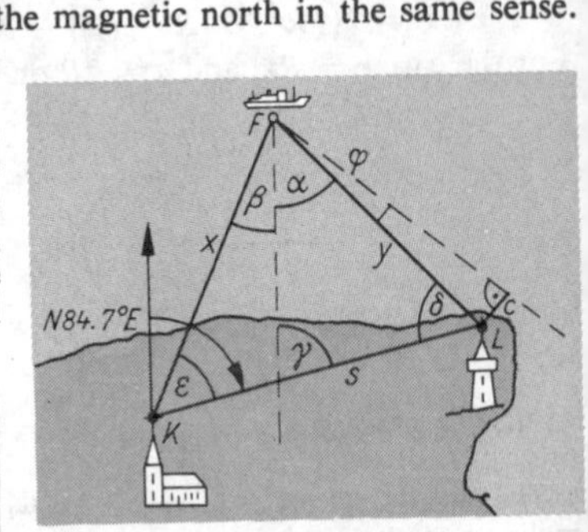

11.3-14 Ship's course

a) Find the *distances* $|FK| = x$ and $|FL| = y$ (Fig.);
b) what course must the ship maintain if it is to pass by the lighthouse at a distance $c = 4$ nautical miles $= 7.408$ km? (Calculation of the *circle of danger*). –

From the given values $\alpha = 55.3°$, $\beta = 28.5°$, $\gamma = 84.7°$, $s = 33.25$ km one obtains $\delta = 180° - (\alpha + \gamma) = 40°$, $\varepsilon = \gamma - \beta = 56.2°$, $x = s \dfrac{\sin \delta}{\sin(\alpha + \beta)} = 21.49 \text{ km} \approx 11.6$ nautical miles, $y = s \dfrac{\sin \varepsilon}{\sin(\alpha + \beta)} = 27.77 \text{ km} \approx 15$ nautical miles, $\sin \varphi = c/y$, $\varphi = 15.47°$. Ship's course: S $(\alpha + \varphi)°$ E, that is, S 70.77 ° E.

Trigonometric determination of heights

In practice angles in a horizontal plane can be measured with greater precision than those in a vertical plane, because light does not travel in an straight line through air of variable density. In addition to this *terrestrial refraction*, for distances over 200 m the curvature of the earth must also be taken into account.

Schematic construction of the theodolite. The theodolite is the instrument used for measuring angles in surveying. The many instruments used for special applications are all based on a simple principle. A vertical hollow spindle *hS* is attached to three *levelling screws Sc* which rest on a *base plate*, often fixed to a tripod (Fig.). The spindle *hS* carries a horizontal circular disc *D* with a ring calibrated in the clockwise sense, called the *limb L*. The *alidade A* is a circular disc, free to rotate in the spindle, having two diametrically opposite pointers and carrying a *spirit level sL* and two *supports Sp* for the telescope axis *aT*. Rigidly attached to this axis are the *telescope T* and the vertical circle *V* for measuring angles of height. The alidade ist made horizontal by means of the levelling screws and the spirit level, and in a good theodolite the axis *a* about which the alidade turns must then be *vertical*. The axis *aT* is then *horizontal* and the telescope *T* is at *right angles* to it. Methods exist to eliminate small deviations from these conditions by preliminary measurements (called *adjustments*) or to determine their magnitude in order to allow for them in the measurement proper. The quality of a theodolite depends essentially upon the accuracy of the two scales on the horizontal and vertical circles as well as on the *reading device R*, which could be a pointer, a vernier or a microscope. Accordingly the angles can be read, for example, to within minutes or seconds. To increase the precision of the angle measurement, a definite procedure is adopted and the readings repeated. The telescope is considered to point correctly towards the object in question if its image, or a characteristic portion of it, coincides with the *cross hairs* of the instrument (in the simplest case a horizontal and a vertical line). It follows from the construction of the theodolite that horizontal angles can also be measured even if the objects aimed at are at different heights. The reference position of the horizontal scale is immaterial, since the horizontal angle is always obtained as the difference of two readings. On the other hand, for the measurement of vertical angles the zero reference direction must be adjusted horizontally by using a spirit level.

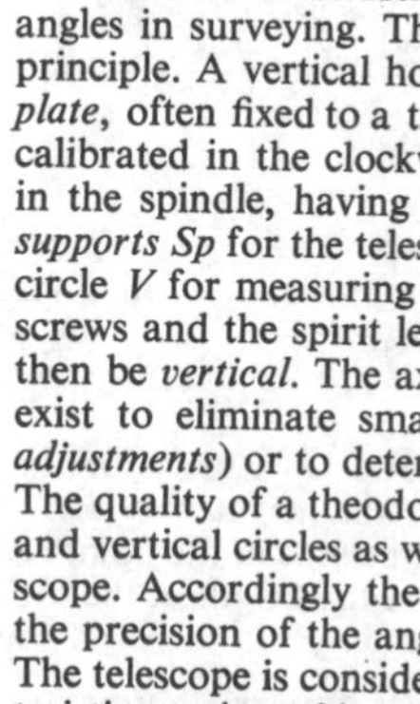

11.3-15 Theodolite (schematic)

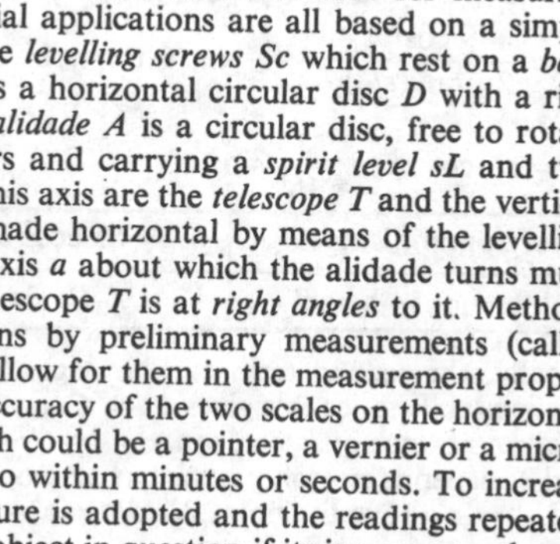

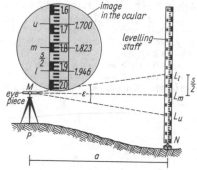

11.3-16 Tacheometrical levelling for a *horizontal sighting*

Tacheometrical levelling. Tacheometry means *fast measurement*. It ist used to determine from the position and height of a known point *P* the positions and heights of a whole series of new points merely from theodolite readings. It can, for example, be used to investigate the surface geometry of terrain as a basis for a building project. For this purpose there are two horizontal hairs, an upper one *u* and a lower one *l*, parallel to the horizontal cross hair *m* of the telescope and at equal distances $p/2$ above and below it, respectively. A rod (called a *levelling staff*) is set up vertically at the new point *N* and the images of the three hairs mark out three points L_m, L_l and L_u on the rod (Fig.).

The difference $L_l - L_u$ between the upper and lower readings is the rod section s. Because of the inversion of the image the greater numbers on the levelling staff lie at the top. The rod section s appears from the instrument to have an *angle of parallax* ε. Depending on the distance p between the horizontal hairs, the focal length f of the object lens and the path of the light in the telescope, the horizontal distance a of the staff from the theodolite for a horizontal sighting can be found in the form $a = Cs$, in which the instrument contant C usually has the round value 100. For the angle of parallax ε it then follows that $\tan(\varepsilon/2) = s/(2a) = 1/(2C)$. If the line of sight from the telescope F to the middle reading L_m is inclined at an angle α to the horizontal (Fig.), then the horizontal distance a' can be determined by the following trigonometric calculation:

$$s = |HL_l| - |HL_u| = a'[\tan(\alpha + \varepsilon/2) - \tan(\alpha - \varepsilon/2)]$$

$$= a' \frac{\sin(\alpha + \varepsilon/2)\cos(\alpha - \varepsilon/2) - \sin(\alpha - \varepsilon/2)\cos(\alpha + \varepsilon/2)}{\cos(\alpha + \varepsilon/2)\cos(\alpha - \varepsilon/2)}$$

$$= a' \frac{\sin\varepsilon}{\cos^2\alpha\cos^2(\varepsilon/2) - \sin^2\alpha\sin^2(\varepsilon/2)},$$

$$a' = s\frac{\cos^2\alpha\cos^2(\varepsilon/2) - \sin^2\alpha\sin^2(\varepsilon/2)}{2\sin(\varepsilon/2)\cos(\varepsilon/2)}$$

$$= \frac{s}{2}\frac{\cos^2\alpha}{\tan(\varepsilon/2)} - \frac{l}{2}\sin^2\alpha\tan(\varepsilon/2),$$

or

$$a' = sC\cos^2\alpha - \frac{s}{4C}\sin^2\alpha.$$

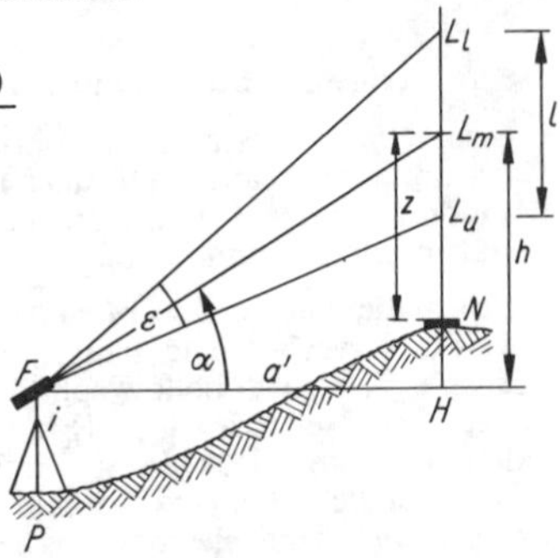

11.3-17 Tacheometrical levelling for an *inclined sighting*

Horizontal distance	$a' = Cs\cos^2\alpha$
Difference in height	$h = {}^1/_2 Cs\sin 2\alpha$

Finally, the second term can be neglected because s and $\sin^2\alpha$ are small and $C = 100$. For the horizontal distance this gives $a' = Cs\cos^2\alpha$ and from this approximate value the height $h = |HL_m|$ is given at once by $h = a'\tan\alpha = {}^1/_2 Cs\sin 2\alpha$. The difference Δh between the heights of the points P and N depends not only on h but also on the height i of the theodolite and on the length of the rod $|NL_m| = z$ above the point N: $\Delta h = i + h - z$.

Calculation of heights with the help of vertical angles. In the following examples terrestrial refraction will be neglected, either because the sighting distance does not exceed 200 m or because a lesser precision in the results (decimetres or even yards) is acceptable.

Horizontal base line and vertical angles. If a horizontal base line $|AB| = s$ in the direction to the point G in the terrain is measured, and also the vertical angles α and β of G from the end-points A and B (Fig.), then the height h of G can be calculated by the sine rule:

1. $\gamma = \beta - \alpha$; 2. $u = s\sin\alpha/\sin\gamma$ and 3. $h = u\sin\beta$, so that $h = s\sin\alpha\sin\beta/\sin(\beta - \alpha)$.

Inclined base line. The angle β of inclination of a base line $|AB| = s$ lies in a vertical plane passing through G, and α and γ are the vertical angles of G from A and B (Fig.). The difference in height between A and G can be calculated by the sine rule. If $|AG| = x$, then $h = x\sin\alpha$, where $x = s\sin\varepsilon/\sin\sigma$, $\varepsilon = \beta + (180° - \gamma)$ and $\sigma = \gamma - \alpha$. By substitution one obtains $h = s\sin\varepsilon\sin\alpha/\sin\sigma$ or $h = s\sin(\gamma - \beta)\sin\alpha/\sin(\gamma - \alpha)$.

Base line in an arbitrary direction relative to the line of sight. The vertical through G meets the horizontal plane through the horizontal base line $|AB| = s$ in the point F. From the end-points A and

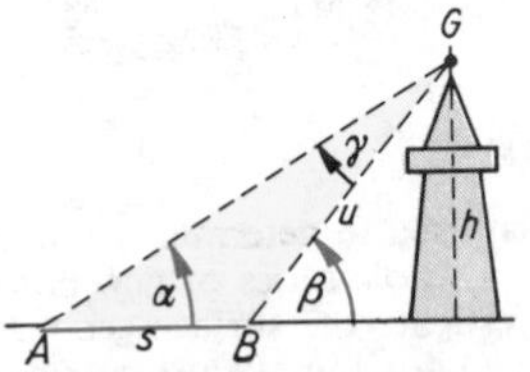

11.3-18 Trigonometrical determination of heights with a *horizontal base line* in a vertical plane through the summit

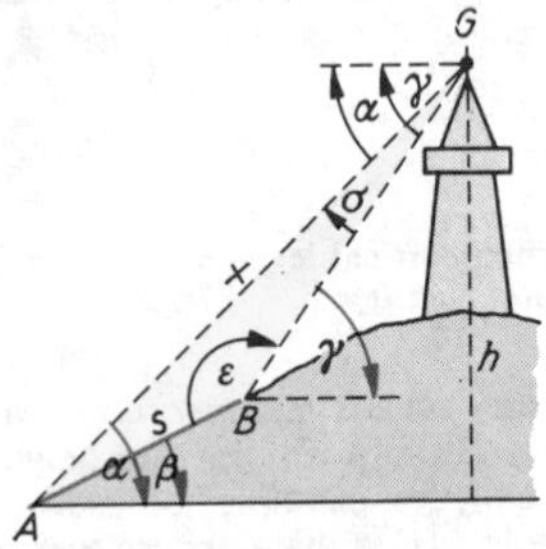

11.3-19 Trigonometrical determination of heights with an *inclined base line* in a vertical plane through the summit

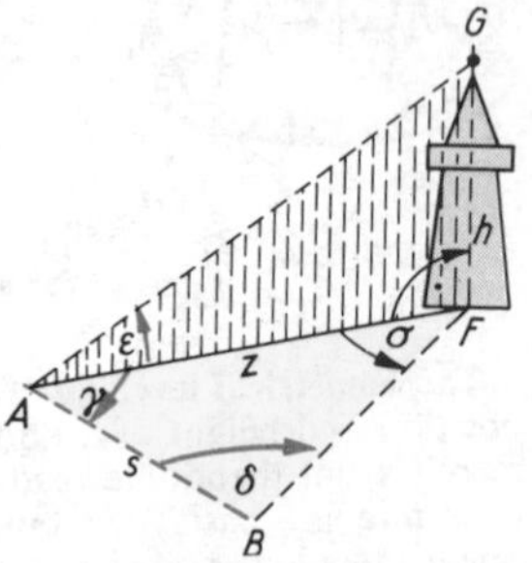

11.3-20 Trigonometrical determination of heights with a *horizontal base line*

B the angles $\sphericalangle FAB = \gamma$ and $\sphericalangle FBA = \delta$ are measured and also the angle of elevation ε of G from A (Fig.). The plane AFG is perpendicular to the horizontal plane FAB. If $|AF| = z$, then $h = z \tan \varepsilon$, where $z = s \sin \delta / \sin \sigma = s \sin \delta / \sin [180° - (\gamma + \delta)] = s \sin \delta / \sin (\gamma + \delta)$. By substitution one obtains: $h = s \sin \delta \tan \varepsilon / \sin (\gamma + \delta)$.

If the base line $|AB| = s$ is inclined to the horizontal at an angle ε_1 with B above A (Fig.), if α and β, respectively, are the horizontal angles between the base line and the point G measured from A and B, and if ε and ε_2 are the vertical angles of G measured from A and B, respectively, then the problem can be reduced to one of the type already solved. In the triangle AHB' in the horizontal plane through A the side $s' = s \cos \varepsilon_1$ and the angles α and β are known. By the sine rule it follows that $a' = s' \sin \alpha / \sin (\alpha + \beta)$; $b' = s' \sin \beta / \sin (\alpha + \beta)$. From the triangle AHG lying in a vertical plane one can then deducet hat: $h = b' \tan \varepsilon = s' \sin \beta \tan \varepsilon / \sin (\alpha + \beta) = s \cos \varepsilon_1 \sin \beta \tan \varepsilon / \sin (\alpha + \beta)$. As a check one has $h = h_1 + h_2$, where $h_1 = s \sin \varepsilon_1$ and $h_2 = a' \tan \varepsilon_2 = s' \sin \alpha \tan \varepsilon_2 / \sin (\alpha + \beta) = s \cos \varepsilon_1 \sin \alpha \tan \varepsilon_2 / \sin (\alpha + \beta)$.

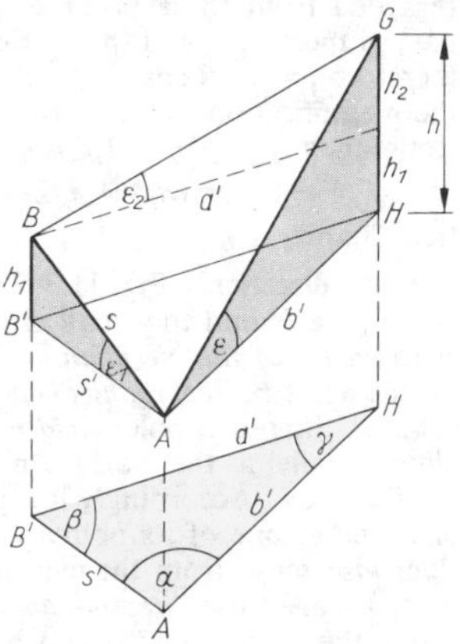

11.3-21 Trigonometrical determination of heights with an *inclined base line*

Surveying

The ultimate aim of surveying is to fix uniquely any desired point of the earth's surface. This is achieved by means of *coordinates* or in pictorial form in *maps*. As a first approximation the earth's surface is regarded as the surface of a sphere on which the position of a point is fixed by the *longitude* λ and the *latitude* φ, that is, by the intersection of a great circle through the north and south poles (a *meridian*) with a *latitute circle*. The meridian through Greenwich is taken as the *zero meridian* and all other meridians are measured by their angular distance λ from this one, where λ is between 0° and 180° towards the east or the west. The latitude circles are small circles parallel to the equator. Their distance φ from the equator is measured in degrees of angle on a meridian from 0° at the equator to 90° at the pole on each side, giving northern and southern latitudes. These coordinates are spherical coordinates and are determined by astronomical measurements, as will be shown in Chapter 12.

Gauss–Krüger or transverse Mercator projection. A cylindrical or a conical surface can be cut along a generator and rolled out, or *developed*, onto a plane. In this process all lengths, areas, and angles remain unchanged; they therefore appear in the map with their true size. The surface of a sphere is not developable. Hence, following C. F. Gauss and the one-time Director of the Potsdam Institute of Geodesy, J. H. L. Krüger (1857–1923), lunes (*meridian strips*) whose bounding meridians include an angle of 6° at the poles are mapped onto the surface of a cylinder that touches the lune along its mean meridian (making an angle of 3° with each bounding meridian). The figure shows approximately how narrow these strips which cover the whole earth are. It is therefore understandable that lengths and areas deviate only a little from their true values. A *plane map of the meridian strip* is then obtained by developing the cylinder. The *meridian of contact m* belongs both to the sphere and to the cylinder, and therefore appears in the plane *with its true length*. If the distance from the equator of a point on this meridian is given by the angle ξ measured in radians then the corresponding point in the plane has coordinate $x = \xi R$, where R is the radius of the earth. Great circles that cut the meridian of contact at right angles and therefore have as poles the points of intersection A_1, A_2 of the axis of the cylinder with the sphere (Fig.) give rise to lines on the map that cut the x-axis at right angles, that is, to generators of the cylinder. Distances of a point P of the sphere from the meridian of contact m are measured on these orthogonal great circles by the angle η; the image point P' has the corresponding distance y from the x-axis. The relationship between η and y can be derived from the requirement that the Gauss-Krüger projection is angle-preserving (or conformal,

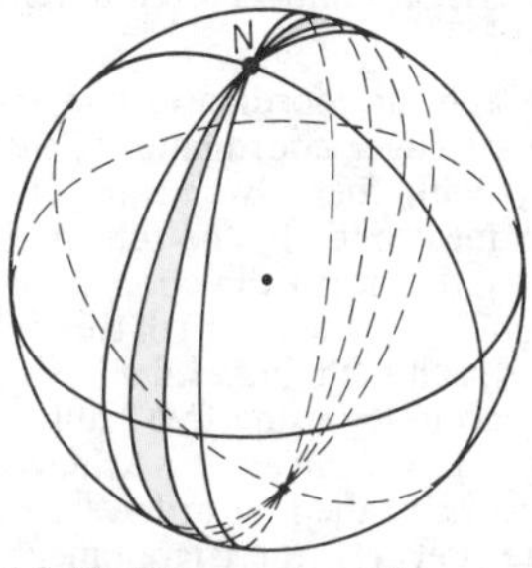

11.3-22 Three consecutive meridian strips

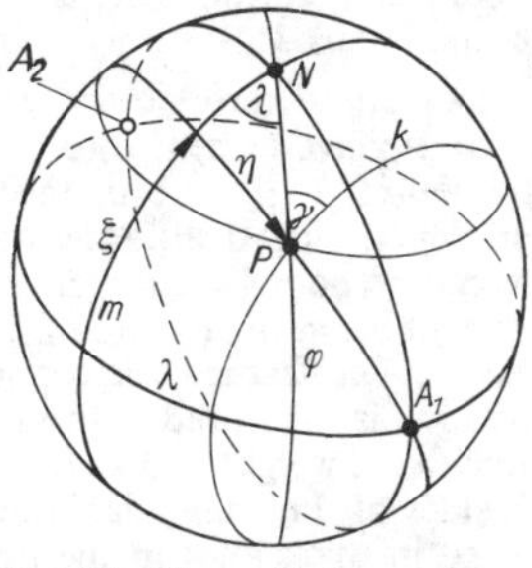

11.3-23 Gauss-Krüger coordinates

see Chapter 23.). The conformal property is ensured if triangles on the sphere are mapped onto similar triangles and the *magnification* for lines is the same in all directions. A small circle k through the point P that cuts all great circles through the poles A_1 and A_2 at right angles is regarded from these poles as a latitude circle of latitude η. The orthogonal great circles play the part of meridians and the meridian of contact that of the equator. The length of the small circle is therefore $l = 2\pi R \cos \eta$. On the other hand, on the cylinder and in the plane $L = 2\pi R$, so that the magnification is $L/l = 1/\cos \eta$. By the conformal property the magnification is equal to $dy/d\eta$. Consequently $dy/d\eta = 1/\cos \eta$, or $dy = d\eta/\cos \eta$. By integration it follows that:

$$y = \ln \tan(\pi/4 + \eta/2) = (1/M) \lg \tan(\pi/4 + \eta/2), \quad \text{where} \quad 1/M = 1/\lg e = 2.302\,585\,1$$

(see Chapter 2.).

North directions. Fig. 11.3–23 shows an angle γ at the point P between the meridian to the north pole N and the small circle k parallel to the meridian of contact m. This angle is called the *convergence of the meridian* and gives the deviation of the grid north from the geographical north for the point P. *Geographical north* (or true north) is the direction from the point P along the meridian towards the north pole. *Grid north* ist the direction from the image point P' in the Gauss-Krüger plane parallel to the x-axis. On the sphere it corresponds to the direction of the tangent at P to the small circle k. Accordingly it is possible to fix the direction of a line in different ways. The *azimuth* α of a line at one of its points P is the angle between the true north and the line, measured in the clockwise sense from the meridian. The angle between grid north and the line measured in the same sense is called the *direction angle* v and is usually denoted in terms of two points P_1 and P_2 of the line in the form $v = (P_1P_2)$, where $(P_1P_2) = (P_2P_1) \pm 180°$. For the sake of completeness it should be mentioned that sometimes a third north direction, magnetic north, is used; it differs from true north by the deviation of a compass needle. With respect to magnetic north one speaks of the *declination* of the line.

Latitudes and departures. The x-value in Gauss–Krüger coordinates is measured along the meridian of contact towards the north (or south); it is called the *latitude* and gives the *true distance from the equator*. The y-value is called the *departure*. Positive y-values denote points that lie to the east of the meridian of contact m. In order to avoid negative y-coordinates, the contact or mean meridian is not given the y-coordinate 0 m, but 500000 m; at the same time the meridian strip on the sphere from which the map in question has been derived is given by a characteristic number in front of this value. This characteristic number is 1 for the meridian of contact 3° (1500000 m), 2 for 9° (2500000 m), 3 for 15° (3500000 m) and so on: $(\lambda_m + 3)/6$. A point that lies 65370 m east of the meridian 3° thus has the y-coordinate 1500000 m + 65370 m = 1565370 m. This number is called the *right value*. For a point that lies 74250 m west of the meridian 9°, the right value is 2500000 m − 74250 m = 2425750 m. Conversely, one reaches a point with right value 4374981 m and latitude 5755899 m by going 5755899 m to the north from the equator on the 21°-meridian (because $4 \cdot 6 - 3 = 21$) and then in the perpendicular direction through 500000 m − 374981 m = 125019 m to the west. On *topographical maps* the right values and latitudes are given only in whole kilometres and the first two numbers are written as superscripts, for instance, as $^{43}74$ and $^{57}55$ for the last example. Since the distortions in lengths are greatest near the bounding meridians, the coordinates of important points are calculated in addition by using strips of width 0.5° on each side of each bounding meridian. Thus, for points of a strip 1° wide, about 70 km wide at latitude 52°, every pair of coordinates is available both for westerly and for easterly meridian strips.

On topographical maps and for geodesy meridian strips of width 3° instead of 6° are also used. The same considerations and notation hold for these; only the characteristic numbers are different. They are 1, 2, 3, ... $(\lambda_m/3)$ for the contact meridians 3°, 6°, 9°, ..., λ_m°.

Triangulation. Only for a few points does one determine the geographical coordinates and the azimuth of line segments between them and convert them into Gauss-Krüger coordinates. Other characteristic points, called *trigonometric points* (TP), are connected with them by means of a *triangular framework of the first order* in which almost all angles are measured. In the schematic figure the geographical coordinates of the four points characterized by the north direction N are determined and the azimuth a of a line segment between any two of them (Fig.). The length of this line is calculated by means of a *basis framework*. The framework connects each such line with a basis b from 4 to 10 km long on level ground, which is measured with great accuracy using free hanging invar wires of length 24 m at a tension equal to the weight of 5 kg (≈ 11 lb.). A mean error is achieved of 8 mm in 10 km, or 1 in 1250000. The sides of the triangular framework of the first order are on average 40 to 70 km long; they are marked in thick lines in the figure. Between the trigonometric points of the first order thus determined in Gauss-Krüger coordinates the points of the *framework of the second order* are superimposed, merely by angle measurements. Its sides are on average 20 km long, those of the *framework of the third order* 5 to 10 km long, and finally those of the *framework of the fourth order* 2 to 5 km long. These trigonometric points of the first to the fourth order are marked by a granite plate engraved with a cross and sunk in the ground, and a vertical rectangular

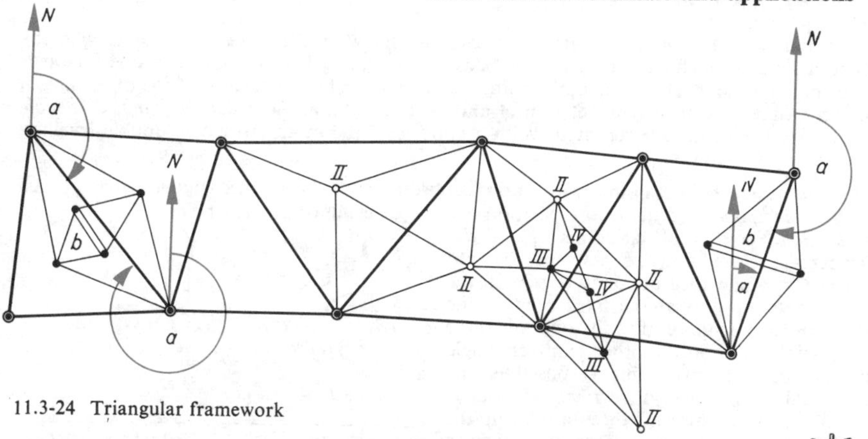

11.3-24 Triangular framework

granite pillar with a cross standing over it (Fig.). In order to be able to see the TP from a greater distance and perform observations upon it a *signal* is erected over it (Fig.) which at the same time forms a good protection from damage. Triangulation frameworks whose triangles are arranged in a strip are called triangular chains. A triangular chain along a meridian of the earth was once used to determine the shape of the earth. As a result of the development of radar and other methods, with the help of electromagnetic waves, of measuring distances with greater precision, *trilateration* has become possible as an alternative to triangulation. While astronomers up to now have tried to determine the distance to a planet, usually Venus, by measuring angles from two places on the earth separated by at most a distance equal to the diameter of the earth, they are now determining these distances from the time taken for a radar signal to travel. The aim of this investigation is to determine more accurately than before the astronomical unit, the distance between the earth and the sun.

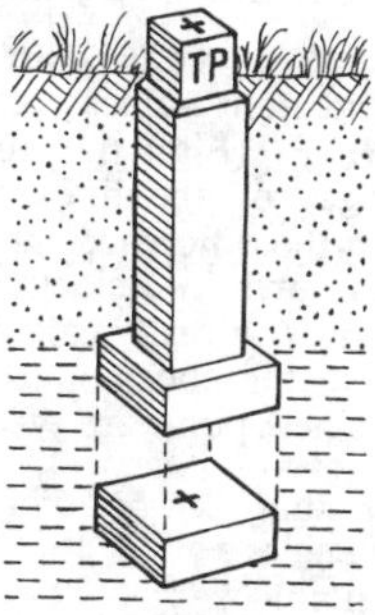

11.3-25 Trigonometric point (TP)

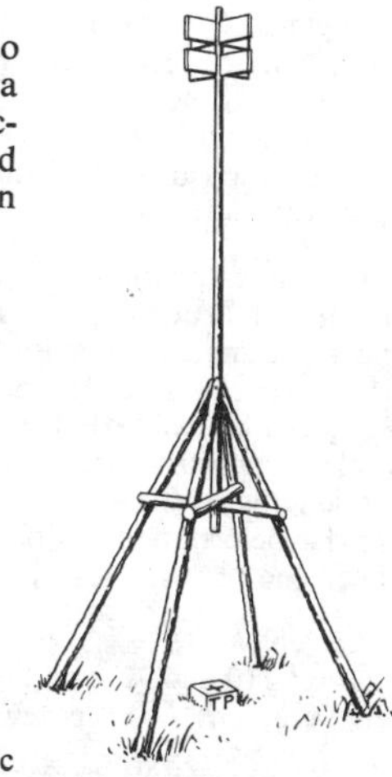

11.3-26 Signal

Bench mark systems. Two lines radiating from the centre of a series of concentric spheres intersect the surface of each sphere in two points. The greater the radius of the sphere, the longer the line joining the corresponding points of intersection. From this geometrical fact it follows that between two plump lines, each hanging in a deep shaft, one measures different distances according to the height at which the measurements are made. The distance between two earth radii is greater in the mountains than at sea level. Consequently in surveying all lengths must be calculated at the same height, at *sea level*. Hence the height of each calculated point above a zero level must be measured. For this purpose a network of points of known height, called *bench marks*, is determined. The datum of the Ordnance Survey is determined by 'the approximate mean water at Liverpool', and all levels on the maps of that survey are altitudes above this datum.

To measure differences of height instruments called *levels* are used. The telescope axis of the level must be exactly parallel to the axis of a *delicate spirit level*, and thus exactly horizontal. A backwards reading is then taken on a levelling staff L_r provided with a centimetre scale and held vertically at a point R, and then a forwards reading on a staff L_v at the point V (Fig.). The difference $r - v = d$ between these two readings then measures how much higher V lies than R. Centimetres are read off on the staff and millimetres are either estimated or measured as deviations on a parallel *sliding plate*.

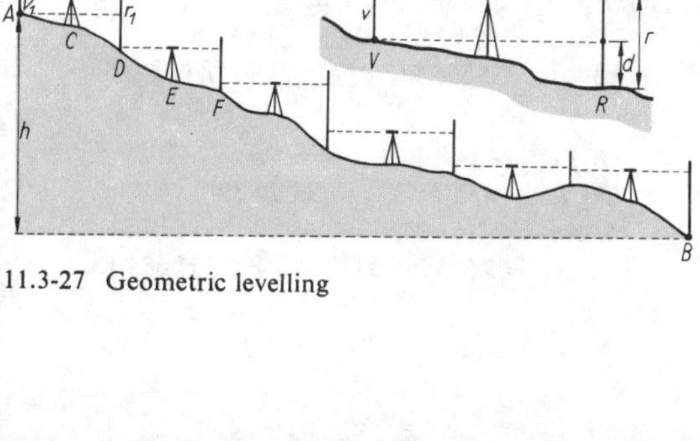

11.3-27 Geometric levelling

In the lower part of the figure a chain of measurements with the level is indicated, whereby at each intermediate point, for example D, one backwards reading is taken and then one forwards reading after moving the level, for example from C to E. By algebraic addition of the differences d one obtains the difference in height between A and B. Such a chain is called *levelling*, or *double levelling* if the measurements are repeated. With a good level the mean error of a double levelling of length 1 km is ± 0.4 mm.

Determination of new points. Points with known coordinates, for example, trigonometric points, are designated *fixed points*. Points whose position is to be determined are *new points*.

Forward section. From two fixed points F_1 and F_2, whose distance apart s is known, a new point N is to be determined by angle measurement (Fig.). If the theodolite can beset up only at the fixed points, one speaks of a *forward section*. If only one of these points is accessible, but on the other hand an angle measurement from the new point is possible, the procedure is called a *sideways section*. Of course, one usually tries to measure all three angles in the triangle F_1F_2N in order to use the angle sum as a check on the angle observations. Geometrically one side and three angles of the triangle F_1F_2N are always known, and the remaining sides s_1 and s_2 can be calculated using the sine rule:

$s_1 = s \sin\alpha/\sin\gamma, \quad s_2 = s \sin(360° - \beta')/\sin\gamma.$

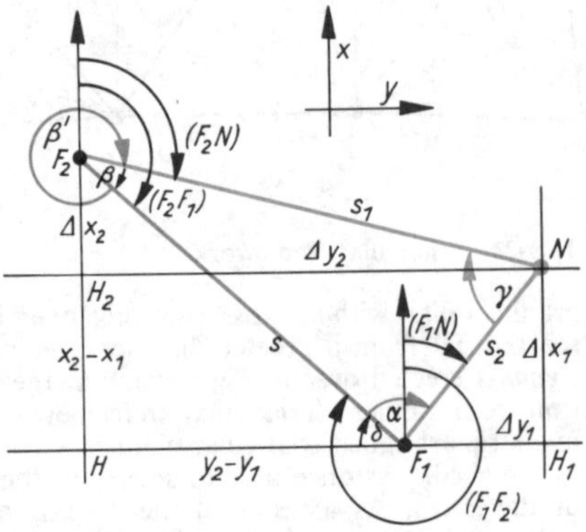

11.3-28 Forward section, the direction of x is grid north

Geodetically the fixed points F_1 and F_2 are given by their latitudes x_1, x_2 and right values y_1, y_2. In the right-angled triangle HF_1F_2 (Fig.) the direction angle (F_1F_2) and the length of the line segment F_1F_2 can be calculated from the differences $y_2 - y_1$ and $x_2 - x_1$ of the coordinates. The direction angle is measured from grid north in a clockwise sense. In the triangles F_2NH_2 and F_1H_1N, using the lengths of the sides s_1 and s_2 already calculated, the coordinate differences Δy_1, Δx_1, Δx_2, Δy_2 can be determined. Added to the coordinates of F_1 or F_2 these give the coordinates of N. As a check the coordinates of the new point are twice calculated.

Example: It is given that $x_1 = 2\,524\,950.98$, $y_1 = 5\,711\,619.35$ and $x_2 = 2\,525\,616.57$, $y_2 = 5\,710\,664.92$, and the angles $\alpha = 61°13'33''$ and $\beta' = 328°32'15''$ are measured from the given line segment s to the new point N.

1. *Angle* $\gamma = 180° - \alpha - (360° - \beta')$ $\qquad$ $\beta = 360° - \beta' = 31°27'45''$
$\gamma = 87°18'42''$

2. *Direction angle* (F_1F_2):

$$\tan(F_1F_2) = \frac{y_2 - y_1}{x_2 - x_1} \qquad y_2 - y_1 = -954.43$$
$$x_2 - x_1 = +665.59$$
$$F_1F_2 = \frac{x_2 - x_1}{\cos(F_1F_2)} = \frac{y_2 - y_1}{\sin(F_1F_2)} \qquad (F_1F_2) = 304°53'24'', \quad \delta = 34°53'24''$$
$$\tan(F_1F_2) = -\cot\delta$$
$$\cos(F_1F_2) = \sin\delta, \quad \sin(F_1F_2) = -\cos\delta$$
$$F_1F_2 = 1163.6$$

3. *Lengths of the sides* s_1 *and* s_2:
$s_1 = |F_1F_2| \sin\alpha/\sin\gamma = |F_2N|$ $\qquad$ $s_1 = 1021.0 = |F_2N|$
$s_2 = |F_1F_2| \sin\beta/\sin\gamma = |F_1N|$ $\qquad$ $s_2 = 608.9 = |F_1N|$

4. *From* F_1 *to the new point:*
$(F_1N) = (F_1F_2) + \alpha$ $\qquad$ $(F_1N) = 366°06'57'' = 6°06'57''$
$x_N - x_1 = |F_1N| \cos(F_1N) = \Delta x_1$ $\qquad$ $\Delta x_1 = +604.53, \quad \Delta y_1 = +64.78$
$y_N - y_1 = |F_1N| \sin(F_1N) = \Delta y_1$ $\qquad$ $x_N = {}^{25}25555.51, \quad y_N = {}^{57}11684.13$

5. *From* F_2 *to the new point:*
$(F_2N) = (F_2F_1) + \beta'$ $\qquad$ $(F_2F_1) = 124°53'24''$
$x_N - x_2 = |F_2N| \cos(F_2N) = \Delta x_2$ $\qquad$ $(F_2N) = 453°25'39'' = 93°25'39''$
$y_N - y_2 = |F_2N| \sin(F_2N) = \Delta y_2$ $\qquad$ $\Delta x_2 = -61.04, \quad \Delta y_2 = +1019.21$
$x_N = 2525555.53, \quad y_N = 5711684.13.$

Backward section. If three fixed points F_1, F_2, F_3 are given and observation is possible only from the new point N (Fig.), then one speaks of a backward section. The new point must be chosen in such a way that it *does not lie on the circumcircle of the triangle* $F_1F_2F_3$. The most accurate result is obtained if N lies in the interior of this triangle. Let the angles of the triangle to be calculated be φ_1 at F_1, φ_2 at F_2 and φ_3 at F_3, and let the angles measured from N be $\sphericalangle F_2NF_3 = \nu_1$, $\sphericalangle F_3NF_1 = \nu_2$ and $\sphericalangle F_1NF_2 = \nu_3$. A solution for machine calculation arises from the coordinates for the *centre of gravity S* of the given triangle: $s_x = (x_1 + x_2 + x_3)/3$, $s_y = (y_1 + y_2 + y_3)/3$. In these formulae the vertices have the same *weight with respect to the medians.* If one gives them different weights g_1, g_2, g_3, this gives rise to different transversals through the vertices, which can lie outside the triangle if some of the weights have negative values. The point of intersection N of the transversals then has the coordinates

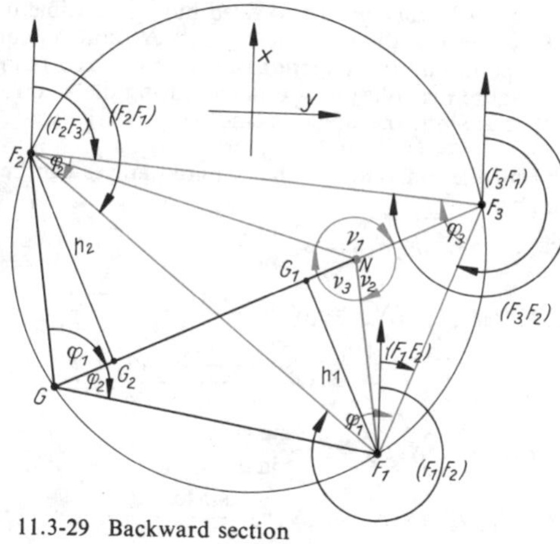

11.3-29 Backward section

$$x = (g_1x_1 + g_2x_2 + g_3x_3)/(g_1 + g_2 + g_3), \qquad y = (g_1y_1 + g_2y_2 + g_3y_3)/(g_1 + g_2 + g_3).$$

The weights can be obtained from the mechanical concept that with respect to a transversal through one vertex the *moments* of the other two vertices must be equal. Let G be the point of intersection of the transversal F_3N with the circumcircle, G_1 and G_2, respectively, the feet of the perpendiculars h_1 and h_2 from F_1 and F_2 to this transversal. The moments of the vertices F_1 and F_2 are equal: hence $g_1h_1 = g_2h_2$ or $g_1/g_2 = h_2/h_1 = \dfrac{h_2}{|GN|}\Big/\dfrac{h_1}{|GN|}$, where:

$$\frac{h_2}{|GN|} = \frac{h_2}{|GG_2| + |G_2N|} = \frac{1}{|GG_2|/h_2 + |G_2N|/h_2} = \frac{1}{\cot\varphi_1 - \cot\nu_1},$$

$$\frac{h_1}{|GN|} = \frac{h_1}{|GG_1| + |G_1N|} = \frac{1}{|GG_1|/h_1 + |G_1N|/h_1} = \frac{1}{\cot\varphi_2 - \cot\nu_2}.$$

By cyclic permutation and substitution one obtains:

$$g_1 : g_2 : g_3 = \frac{1}{\cot\varphi_1 - \cot\nu_1} = \frac{1}{\cot\varphi_2 - \cot\nu_2} = \frac{1}{\cot\varphi_3 - \cot\nu_3}.$$

Since an arbitrary proportionality factor makes no difference to the coordinates of the new point, it may be taken equal to 1. The weights are then:

$$g_1 = \frac{1}{\cot\varphi_1 - \cot\nu_1}, \quad g_2 = \frac{1}{\cot\varphi_2 - \cot\nu_2}, \quad g_3 = \frac{1}{\cot\varphi_3 - \cot\nu_3}.$$

From the coordinates of the fixed points F_1, F_2, F_3 (see forward section) one finds:

$x_1 = 2\,524\,950.98$, $\quad x_2 = 2\,525\,616.57$, $\quad x_3 = 2\,525\,555.51$,
$y_1 = 5\,711\,619.35$, $\quad y_2 = 5\,710\,664.92$, $\quad y_3 = 5\,711\,684.14$,
$(F_1F_2) = 304°53'24''$, $\quad (F_2F_3) + \varphi_2 = (F_2F_1)$, $\quad (F_2F_3) = 93°25'39''$,
$(F_3F_1) + \varphi_3 = (F_3F_2)$, $\quad (F_3F_1) = 186°06'57''$, $\quad (F_1F_2) + \varphi_1 = (F_1F_3)$.

Calculated	Cotangent	Measured	Cotangent
$\varphi_2 = 31°27'45''$	1.63425	$\nu_2 = 153°12'22''$	−1.98019
$\varphi_3 = 87°18'42''$	0.0469547	$\nu_3 = 97°20'08''$	−0.128734
$\varphi_1 = 61°13'33''$	0.549176	$\nu_1 = 109°27'30''$	−0.353300
180°00'00''		360°00'00''	

The numerical values of the weights are $g_1 = 1.10806$, $g_2 = 0.27667$, $g_3 = 5.69188$ and the coordinates of the new point are $x = 2525463.25$, $y = 5711634.13$.

Backward section has particular significance for ships and aircraft in determining their own positions.

Hansen's problem. If two fixed but inaccessible points F_1 and F_2 are given, for example, the tops of two towers, then two new points N_1 and N_2 can be determined if from each of them the other new point and the directions to the two fixed points can be observed (Fig.). With the notation of the figure the solution is obtained using the sine rule, provided that the angles φ and ψ can be calculated. Since the angles ϱ in the two triangles N_1N_2S and F_1F_2S are equal, it follows that: $(\varphi+\psi)/2 = (\alpha+\gamma)/2 = \varepsilon_1$.

Half the difference of the required angles can be found in the following way, using an auxiliary angle:

$$\triangle F_1F_2N_1 : |N_1F_1| = s\,\frac{\sin\psi}{\sin\beta};$$

$$\triangle F_1N_2N_1 : |N_1N_2| = |N_1F_1|\,\frac{\sin(\alpha+\beta+\gamma)}{\sin\gamma} = s\,\frac{\sin(\alpha+\beta+\gamma)\sin\psi}{\sin\beta\sin\gamma};$$

$$\triangle F_1F_2N_2 : |N_2F_2| = s\,\frac{\sin\varphi}{\sin\delta};$$

$$\triangle F_2N_2N_1 : |N_1N_2| = |N_2F_2|\,\frac{\sin(\alpha+\gamma+\delta)}{\sin\alpha} = s\,\frac{\sin(\alpha+\gamma+\delta)\sin\varphi}{\sin\alpha\sin\delta};$$

$$\frac{\sin\varphi}{\sin\psi} = \frac{\sin\alpha\sin\delta\sin(\alpha+\beta+\gamma)}{\sin\beta\sin\gamma\sin(\alpha+\gamma+\delta)} = \cot\eta.$$

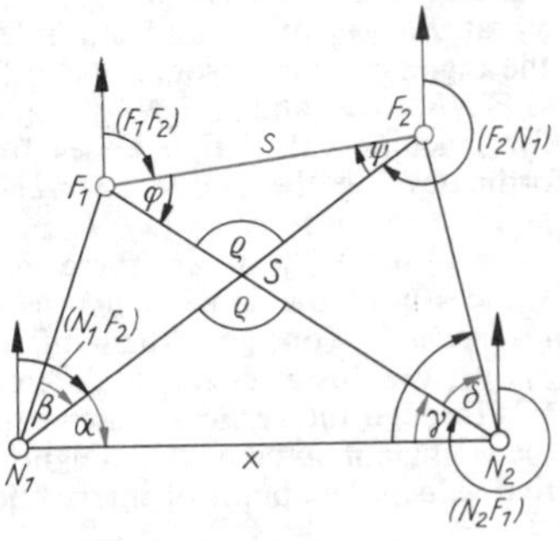

11.3-30 Hansen's problem

The auxiliary angle η is known up to a multiple of 180°. By addition and subtraction and using trigonometric relationships, remembering that $\cot 45° = 1$, one obtains:

$$\frac{\sin\varphi-\sin\psi}{\sin\varphi+\sin\psi} = \frac{\cot\eta-1}{\cot\eta+1},\quad \frac{2\cos[(\varphi+\psi)/2]\sin[(\varphi-\psi)/2]}{2\sin[(\varphi+\psi)/2]\cos[(\varphi-\psi)/2]} = \frac{\cot 45°\cot\eta-1}{\cot\eta+\cot 45°},$$

$$\tan[(\varphi-\psi)/2] = \tan[(\varphi+\psi)/2]\cot(45°+\eta),$$

and hence the value $(\varphi-\psi)/2 = \varepsilon_2$ is known. Then $\varphi = \varepsilon_1+\varepsilon_2$, $\psi = \varepsilon_1-\varepsilon_2$. Consequently the line segments $|F_1N_1|$, $|N_1N_2|$, and $|N_2F_2|$ can be calculated. For the direction angles one finds:

$$\begin{aligned}
(F_1N_2) &= (F_1F_2)+\varphi,\\
(N_2F_1) &= (F_1N_2)+180°,\\
(N_2F_2) &= (N_2F_1)+\delta,\\
(N_2N_1) &= (N_2F_1)-\gamma,\\
(F_2N_1) &= (F_2F_1)-\psi,\\
(N_1F_2) &= (F_2N_1)+180°,\\
(N_1F_1) &= (N_1F_2)-\beta,\\
(N_1N_2) &= (N_1F_2)+\alpha.
\end{aligned}$$

From the direction angles and the lengths the coordinate differences can be found (see forward section), and hence the series of coordinates for the points $F_1 \to N_1 \to N_2 \to F_2$, which must give the known values for the coordinates of F_2.

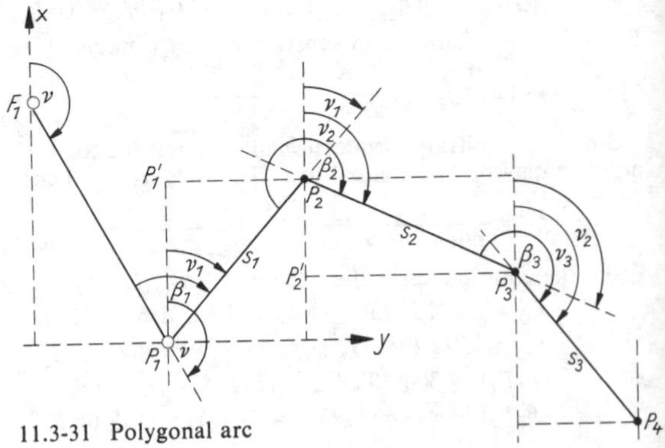

11.3-31 Polygonal arc

Polygonal arcs. In addition to points already determined trigonometrically, the coordinates of further points can be calculated by measuring lines and angles. If $P_1, P_2, P_3, \ldots, P_n$ are the vertices of a polygonal arc starting at the known point P_1, then the lines $s_1 = |P_1P_2|$, $s_2 = |P_2P_3|$, and so on, are measured with a measuring tape and at every point the angle of deviation $\beta_1, \beta_2, \ldots$ is measured. This angle is the difference between the direction of the preceding segment and that of the following one, measured in the clockwise sense (Fig.). For the first measurement at P_1 the direction towards another fixed point F_1 is taken as the direction of the preceding segment. By the measurement of the angle of deviation at P_1 the polygonal arc is connected to a known direction. The accuracy of the polygonal arc measurement can be appreciably increased if the coordinates of the last point P_n are known and a further fixed point F_2 can be sighted from P_n. The polygonal arc then connects the two given directions (F_1P_1) and (P_nF_2). The direction angles can then be calculated by adding the angles of deviation:

$(P_1F_1) = (F_1P_1) \pm 180°, \rightarrow (P_1P_2) = (P_1F_1) + \beta_1,$

$(P_2P_1) = (P_1P_2) \pm 180°, \rightarrow (P_2P_3) = (P_2P_1) + \beta_2,$

and so on. The coordinate differences Δx_i and Δy_i between the point P_i and the point P_{i+1} are given by transforming from polar coordinates (P_iP_{i+1}), s_i to Cartesian coordinates; for example,

$\Delta x_1 = x_2 - x_1 = |P_1P_1'| = s_1 \cos(P_1P_2)$ and $\Delta y_1 = y_2 - y_1 = |P_1'P_2| = s_1 \sin(P_1P_2)$.

The signs of the coordinate differences depend on the magnitudes of the direction angles; in the figure these are denoted by $\nu_i = (P_iP_{i+1})$; $\nu_1 = (P_1P_2)$ lies in the first quadrant, $\nu_2 = (P_2P_3)$ and $\nu_3 = (P_3P_4)$ in the second quadrant; thus, Δx_1 is positive, but Δx_2 and Δx_3 are negative.

12. Spherical trigonometry

As its name implies, spherical trigonometry is concerned with the solution of triangles on the surface of a sphere. It has developed from astronomy and navigation, with the task of determining the positions of points and the distances between them and also angles on the celestial sphere or on the surface of the earth, regarded as a sphere. The basis of the Gauss-Krüger coordinates, which are important in surveying, is also obtained from astronomical measurements.

12.1. Great circles, small circles and lunes

Every straight line through the centre M of a sphere cuts its surface in the extremities of a diameter whose length is twice the radius R of the sphere. Every plane perpendicular to a diameter and at a distance l (less than R) from the centre M cuts the sphere in a circle of radius $r = \sqrt{(R^2 - l^2)}$. If this plane contains M, then the intersection is a great circle with $r = R$. For $l = R$ one obtains a *tangent plane* that has only one point in common with the sphere since $r = 0$.

A *pencil of planes* may be made to pass through two points A and B on a sphere that do not lie on a diameter, and this cuts the sphere in a pencil of circles (Fig.). Among these circles there is a smallest one, for which the line AB is a diameter, and a largest one whose centre coincides with the centre of the sphere. This single circle having a radius equal to that of the sphere is called a *great circle*; all others are *small circles*. If all the planes of the pencil, together with their circles of intersection, are rotated about the line through A and B into the plane of the great circle, a family of coaxial circles through A and B is obtained. The smaller arc between A and B on each of these circles clearly is the smaller, the larger the radius r of the circles; thus it has its smallest value for the great circle with $r = R$. By means of differential geometry it can be shown that the arc $\widehat{AB}$ of the great circle is not only the shortest circular arc joining A and B, but also the shortest of all curves on the sphere connecting A and B. It is a portion of a *geodesic line*.

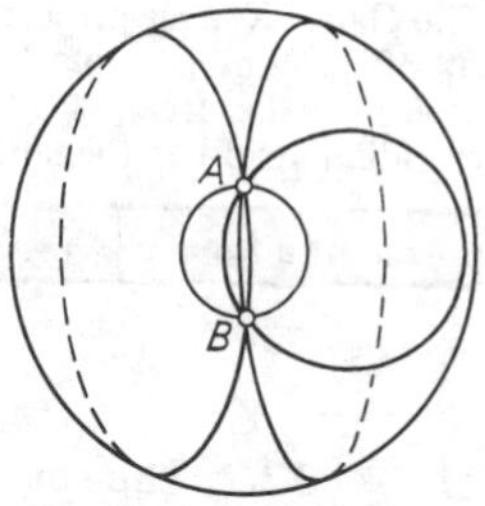

12.1-1 Circles through two points A and B on the sphere

Great circles. All distances between points on the sphere are measured along arcs of great circles. On spheres of sufficiently large radius these go over arbitrarily closely into the Euclidean distance along a straight line. From the theorems of plane geometry the length of the arc $\widehat{AB}$ of the great circle between the points A and B depends on the magnitude of the radius R and on the angle subtended by the arc at the centre, which can be given in radians or degrees and is usually denoted by, a small Latin letter, for example, by a or $a°$.

Arc of a great circle	$\widehat{AB} = R\hat{a} = \pi R a°/180°$

Two great circles intersect in two points A and B that are the ends of a diameter. Such points, in which a straight line through the centre of the sphere meets its surface, are called *diametrically opposite points*, or *poles*. The great circle whose plane is perpendicular to the straight line AB is called the *polar circle* of A (or B). If one describes the polar circle in a definite sense, one can distinguish between a left-hand and a right-hand pole.

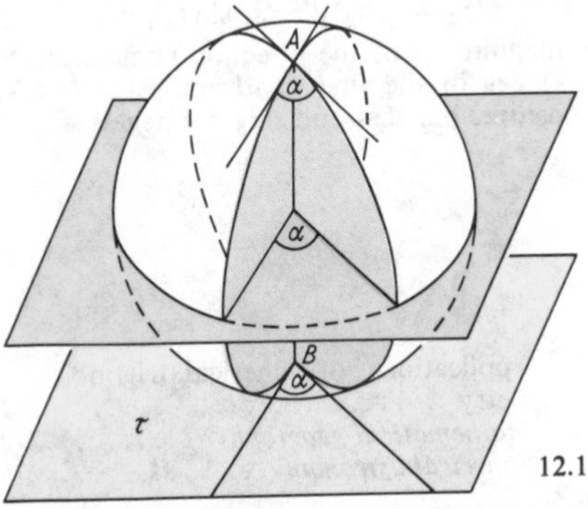

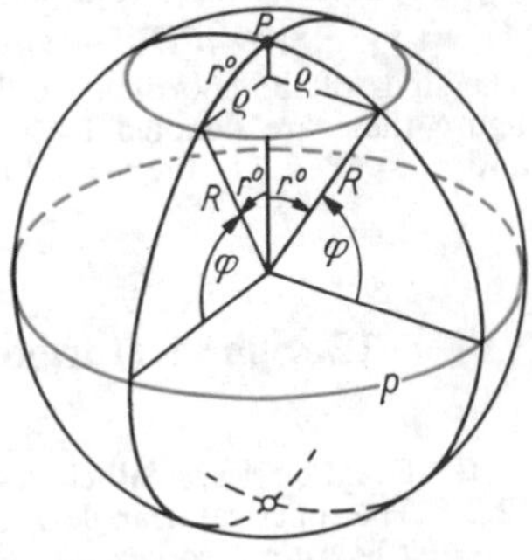

12.1-3 Spherical circle, latitude circle

12.1-2 Angle α between two great circles, τ tangent plane

Every plane perpendicular to the diameter AB (Fig.) cuts each of the planes of two great circles through A and B in a straight line, and the angle α between these two lines is the angle between the planes. The tangents to the two great circles at a pole are both perpendicular to the diameter AB, and the angle between them is also equal to α. This is the *angle α between the two great circles.*

Spherical circles. All points of a sphere that lie at the same distance from a point P, measured along a great circle through P, lie on a circle called a *spherical circle*. The constant spherical distance is called its *spherical radius* and the point P is its spherical centre or *pole*; for example, all latitude circles of latitude φ are spherical circles. They have spherical radius $(90° - \varphi)$ and the pole is their spherical centre. The greatest spherical circle is the polar circle p for which the pole is the spherical centre; its spherical radius is $\pi/2$ or 90°. The other spherical circles are small circles that are intersections with the surface of the sphere of planes parallel to the plane of the polar circle. If the spherical radius of a circle is $r°$ and the radius of the sphere R (Fig.), then the radius of the circle in the intersecting plane is $\varrho = R\cos(90° - r°)$ and the circumference is $2\pi\varrho = 2\pi R\cos(90° - r°)$. Thus, the circumference of a latitude circle is $2\pi\varrho = 2\pi R\cos\varphi$.

Lunes. Two great circles always have a pair of diametrically opposite points in common and divide the surface of the sphere into four lunes. Each of these has two equal sides of magnitude $s = 180°$ (or π). The magnitude of its area depends only upon the angle α between the great circles. The Gauss-Krüger projection uses lunes having an angle of 6°. For an angle of 90° (or $\pi/2$) the area A_0 of the lune is a quarter of the surface area of the sphere and is therefore πR^2. For an angle of $\alpha°$ (or $\hat{\alpha}$) the area is, by proportion, equal to $A = \pi R^2 \alpha°/90°$ (or $2R^2\hat{\alpha}$). Thus, a Gauss-Krüger meridian strip has the surface area

Area of a lune	$A = 2R^2\hat{\alpha} = \pi R^2\,\alpha°/90°$

$A = \pi R^2 \cdot 6°/90° = \pi R^2/15 = 8\,501\,665\ \text{km}^2$ (if R is taken to be 6371.221 km).

12.2. The spherical triangle

If three points A, B, C lying on a sphere are such that no two of them form a pair of diametrically opposite points and they do not all three lie on one great circle, then they determine three great circles, each of which joins two of the points, and which also intersect in pairs in the points $\bar{A}$, $\bar{B}$, $\bar{C}$ diametrically opposite to the given points. By these circles the surface of the sphere is divided into eight portions, each of which is bounded by arcs of the three great circles that are less than π (Fig.). These regions are called spherical triangles, in particular *Euler triangles* to distinguish them from triangles in which sides greater than π are possible, for example,

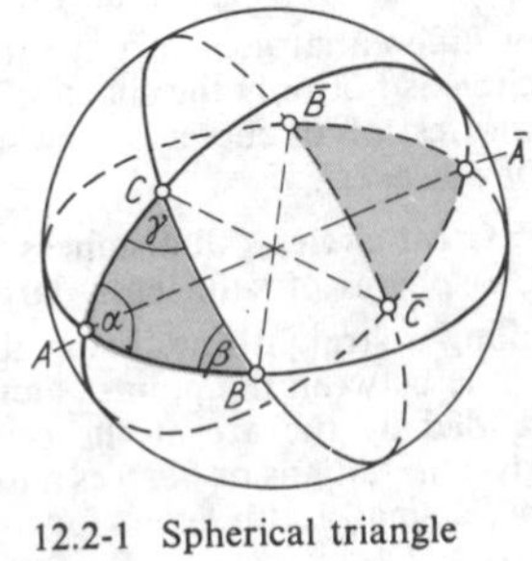

12.2-1 Spherical triangle

the triangle with sides AB, BC and $C\bar{A}\bar{C}A$ in the figure. This non-Euler triangle differs from the hemisphere bounded by the great circle $C\bar{A}\bar{C}A$ only by the Euler triangle ABC. For this reason only Euler triangles will be considered here. The angles α, β, γ of the triangle are the angles between the planes of the great circles that intersect in pairs at the vertices of these angles; they are also the angles between the tangents at the vertices to the great circles that intersect at these points. In Euler triangles no angle exceeds π.

Area of a spherical triangle. Each pair of the eight spherical triangles with vertices at diametrically opposite points is symmetrical about the centre of the sphere and therefore all their data and their areas are equal. For example, $\triangle ABC = \triangle \bar{A}\bar{B}\bar{C}$, or $\triangle AB\bar{C} = \triangle \bar{A}\bar{B}C$. Each triangle having a side in common with the triangle ABC forms with it a lune whose area can be stated. From

$$\triangle ABC + \triangle BC\bar{A} = 2R^2\hat{\alpha}, \qquad \triangle ABC + \triangle CA\bar{B} = 2R^2\hat{\beta},$$
$$\triangle ABC + \triangle AB\bar{C} = 2R^2\hat{\gamma}$$

it follows that

$$3\triangle ABC + [\triangle BC\bar{A} + \triangle CA\bar{B} + \triangle AB\bar{C}] = 2R^2(\hat{\alpha} + \hat{\beta} + \hat{\gamma}).$$

By the symmetry about the centre

$$\triangle ABC + [\triangle BC\bar{A} + \triangle CA\bar{B} + \triangle AB\bar{C}] = \triangle ABC + \triangle BC\bar{A} + \triangle CA\bar{B} + \triangle \bar{A}\bar{B}C = 2\pi R^2$$

(a hemisphere), and thus

$$2\triangle ABC + 2\pi R^2 = 2R^2(\hat{\alpha} + \hat{\beta} + \hat{\gamma}),$$

or

$$\triangle ABC = R^2(\hat{\alpha} + \hat{\beta} + \hat{\gamma} - \pi) = (\pi R^2/180°)(\alpha° + \beta° + \gamma° - 180°).$$

The excess of the sum of the angles of a spherical triangle over π (or 180°) is called the *spherical excess* ε. One obtains:

Area of a spherical triangle	$A = \hat{\varepsilon}R^2 = \pi R^2\varepsilon/180°$ $\hat{\varepsilon} = \hat{\alpha} + \hat{\beta} + \hat{\gamma} - \pi = \alpha° + \beta° + \gamma° - 180°$

It follows that in every spherical triangle with non-zero surface area the angle sum is greater than two right angles. For example, in an Euler triangle whose vertices are poles of the opposite sides the angle sum is 3 right angles $= 3\pi/2 = 270°$.

Polar triangle. Corresponding to each spherical triangle a three-sided solid angle can be determined by the vectors $\boldsymbol{A}$, $\boldsymbol{B}$, $\boldsymbol{C}$ (of length R) from the centre M of the sphere to its vertices A, B, C. Spheres with different radii and the same centre M cut the rays determined by the vectors $\boldsymbol{A}$, $\boldsymbol{B}$, $\boldsymbol{C}$ in similar spherical triangles that have the same sides and angles. It may therefore be assumed that the vectors are of length 1. Now P_a, P_b, P_c are the feet of the perpendiculars from a point P in the interior of the three-sided angle to its three bounding planes. These perpendiculars determine the *polar solid angle* of the given angle. The magnitudes of its sides are measured by the angles $\bar{c}$, $\bar{a}$ and $\bar{b}$; $\sphericalangle P_aPP_b = \bar{c}$, $\sphericalangle P_bPP_c = \bar{a}$, $\sphericalangle P_cPP_a = \bar{b}$ (Fig.). The plane face $PP_a\bar{B}P_c$, for example, is perpendicular to the faces MBC and MAB of the original solid angle, and is therefore also perpendicular to their line of intersection $\boldsymbol{B}$. The angle $P_a\bar{B}P_c$ is the angle β between the plane surfaces of the sides a and c. In the quadrilateral $PP_a\bar{B}P_c$ it therefore follows that $\bar{b} + \beta = 180°$, since the other two angles are right angles. Similarly one finds that $\bar{c} + \gamma = 180°$ and $\bar{a} + \alpha = 180°$. If one chooses a point within the polar solid angle, for simplicity the point M, then the vectors $\boldsymbol{A}$, $\boldsymbol{B}$, $\boldsymbol{C}$ are the perpendiculars from it to the sides $\bar{a}$, $\bar{b}$, $\bar{c}$ of the polar solid angle, and $\bar{A}$, $\bar{B}$, $\bar{C}$ are the feet of these perpendiculars. Thus, the original solid angle $MABC$ is the polar solid angle of its polar solid angle. Its sides a, b, c are perpendicular to the line segments PP_a, PP_b, PP_c, respectively. The angles $\bar{\alpha}$, $\bar{\beta}$, $\bar{\gamma}$ of the polar solid angle are contained in the rectangles $M\bar{B}P_a\bar{C}$, $M\bar{C}P_b\bar{A}$, $M\bar{A}P_c\bar{B}$, respectively ($\bar{\alpha} = \sphericalangle\bar{B}P_a\bar{C}$, $\bar{\beta} = \sphericalangle\bar{C}P_b\bar{A}$, $\bar{\gamma} = \sphericalangle\bar{A}P_c\bar{B}$), and thus $\bar{\alpha} + a = 180°$, $\bar{\beta} + b = 180°$, $\bar{\gamma} + c = 180°$, since in each case the other two angles are both right angles.

Relationships between sides and angles of a solid angle and its polar solid angle
$\bar{a} + \alpha = 180°$, $\bar{b} + \beta = 180°$, $\bar{c} + \gamma = 180°$ $\bar{\alpha} + a = 180°$, $\bar{\beta} + b = 180°$, $\bar{\gamma} + c = 180°$

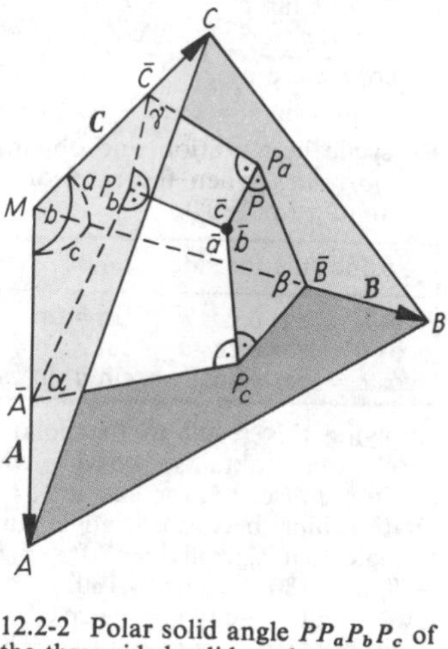

12.2-2 Polar solid angle $PP_aP_bP_c$ of the three-sided solid angle $MABC$

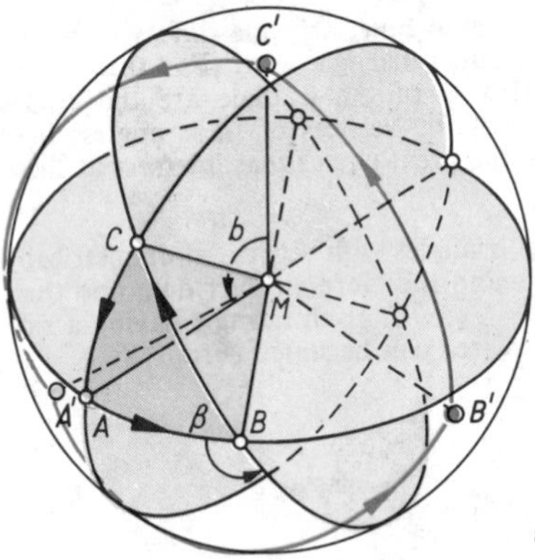

If the arbitrarily chosen point P approaches the point M, then PP_a, PP_b, PP_c become perpendiculars to the sides a, b, c, respectively, each cutting the sphere in two diametrically opposite points. The points A', B', C' are the left-hand poles of the sides of the given triangle described in the following sense: $A \to B$, $B \to C$, $C \to A$. Then the spherical triangle $A'B'C'$ is called the polar triangle of the given triangle; the sides and angles of the two triangles are connected by the relationship given above (Fig.).

12.2-3 Spherical triangle, three-sided solid angle, *polar triangle* and polar solid angle

The main theorems for the solution of the general spherical triangle

The cosine rules for the sides and for the angles. On a sphere with centre at M and radius 1, let the vertices of the triangle ABC be the end-points of the vectors $\boldsymbol{A}$, $\boldsymbol{B}$, $\boldsymbol{C}$ from M for which $|\boldsymbol{A}| = |\boldsymbol{B}| = |\boldsymbol{C}| = 1$, $\boldsymbol{AB} = \cos c$, $\boldsymbol{BC} = \cos a$, $\boldsymbol{CA} = \cos b$. Also, let $\boldsymbol{t}_{AC}$, $\boldsymbol{t}_{AB}$; $\boldsymbol{t}_{BA}$, $\boldsymbol{t}_{BC}$; $\boldsymbol{t}_{CB}$; $\boldsymbol{t}_{CA}$ be vectors of the length 1 on the tangents to the great circles at the vertices A, B, C. Each pair determines a tangent plane in which the corresponding angle of the triangle can be measured;

$$\sin\alpha = |\boldsymbol{t}_{AB} \times \boldsymbol{t}_{AC}|, \quad \sin\beta = |\boldsymbol{t}_{BC} \times \boldsymbol{t}_{BA}|, \quad \sin\gamma = |\boldsymbol{t}_{CA} \times \boldsymbol{t}_{CB}|.$$

In the figure the side $b = \widehat{AC}$ appears in its true magnitude and the tangent plane through $\boldsymbol{t}_{AC}$ and $\boldsymbol{t}_{AB}^{(0)}$ is perpendicular to the plane of the diagram. If this plane is rotated about $\boldsymbol{t}_{AC}$ into the tangent plane, then the magnitude of the angle $\alpha^{(0)}$ between $\boldsymbol{t}_{AC}$ and $\boldsymbol{t}_{AB}^{(0)}$ is equal to the true value of the angle α. In the plane through two vectors, for example through $\boldsymbol{A}$ and $\boldsymbol{C}$, the tangent $\boldsymbol{t}_{AC}$ at one point cuts the extension of the other vector $\boldsymbol{C}$ at H_1. In the figure the triangle AMH_1 lies in the plane of the diagram. Using the auxiliary point H_2 and the intercept theorems one finds that $|MH_1|/|MC| = |MA|/|MH_2|$, where $|MH_1| = 1/\cos b$ and $|AH_1| = \tan b$. By vector addition one obtains $\overrightarrow{MA} + \overrightarrow{AH_1} = \overrightarrow{MH_1}$ or $\boldsymbol{A} + \boldsymbol{t}_{AC} \tan b = \boldsymbol{C}/\cos b$, $\boldsymbol{t}_{AC} \tan b = \boldsymbol{C}/\cos b - \boldsymbol{A}$. Similarly in a plane through $\boldsymbol{A}$ and $\boldsymbol{B}$ one obtains $\boldsymbol{t}_{AB} \tan c = \boldsymbol{B}/\cos c - \boldsymbol{A}$. By equating the scalar product of the vectors on the left-hand side of the two equations with that of the vectors on the right-hand one derives

$$(\boldsymbol{t}_{AC} \cdot \boldsymbol{t}_{AB}) \tan b \cdot \tan c = \frac{\boldsymbol{C} \cdot \boldsymbol{B}}{\cos b \cos c} + \boldsymbol{A} \cdot \boldsymbol{A} - \frac{\boldsymbol{C} \cdot \boldsymbol{A}}{\cos b} - \frac{\boldsymbol{B} \cdot \boldsymbol{A}}{\cos c},$$

$$\cos\alpha \tan b \tan c = \frac{\cos a}{\cos b \cos c} + 1 - \frac{\cos b}{\cos b} - \frac{\cos c}{\cos c},$$

$$\cos\alpha \sin b \sin c = \cos a - \cos b \cos c.$$

By cyclic permutation one obtains the *cosine rule for sides,* when the sides and angles are less than π (or 180°).

Cosine rule for sides
$\cos a = \cos b \cos c + \sin b \sin c \cos\alpha$
$\cos b = \cos c \cos a + \sin c \sin a \cos\beta$
$\cos c = \cos a \cos b + \sin a \sin b \cos\gamma$

Applying this result to the polar triangle $\bar{A}\bar{B}\bar{C}$ one obtains $\cos\bar{a} = \cos\bar{b}\cos\bar{c} + \sin\bar{b}\sin\bar{c}\cos\bar{\alpha}$, for example. From the relationships between triangle and polar triangle, that is, from $\bar{a} = 180° - \alpha$, $\bar{b} = 180° - \beta$, $\bar{c} = 180° - \gamma$, $\bar{\alpha} = 180° - a$, it follows that $-\cos\alpha = (-\cos\beta)(-\cos\gamma) + \sin\beta\sin\gamma(-\cos a)$.

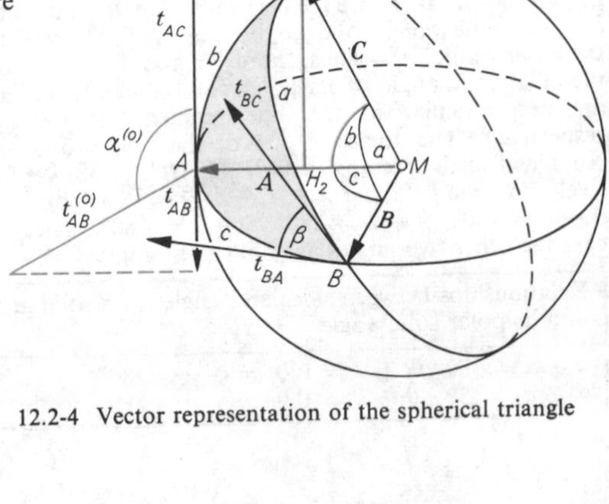

12.2-4 Vector representation of the spherical triangle

Hence the *cosine rule for angles* is obtained by cyclic permutation.

Cosine rule for angles	$\cos\alpha = -\cos\beta\cos\gamma + \sin\beta\sin\gamma\cos a$ $\cos\beta = -\cos\gamma\cos\alpha + \sin\gamma\sin\alpha\cos b$ $\cos\gamma = -\cos\alpha\cos\beta + \sin\alpha\sin\beta\cos c$

The sine rule. The relationship $\boldsymbol{t}_{AB}\tan c = \boldsymbol{B}/\cos c - \boldsymbol{A}$ and $\boldsymbol{t}_{AC}\tan b = \boldsymbol{C}/\cos b - \boldsymbol{A}$ are used to derive the cosine rule for sides. Multiplication by $\cos c$ and $\cos b$, respectively, leads to $\boldsymbol{t}_{AB}\sin c = \boldsymbol{B} - \boldsymbol{A}\cos c$ and $\boldsymbol{t}_{AC}\sin b = \boldsymbol{C} - \boldsymbol{A}\cos b$. Substituting these values in the vector product $\boldsymbol{t}_{AB}\times\boldsymbol{t}_{AC} = \sin\alpha\cdot\boldsymbol{A}$ one obtains

$$\sin b\sin c\cdot\boldsymbol{A}\sin\alpha = \boldsymbol{B}\times\boldsymbol{C} - \cos b(\boldsymbol{B}\times\boldsymbol{A}) - \cos c(\boldsymbol{A}\times\boldsymbol{C}) + \cos b\cos c(\boldsymbol{A}\times\boldsymbol{A}),$$

where $\boldsymbol{A}\times\boldsymbol{A} = 0$ and the vectors $\boldsymbol{B}\times\boldsymbol{A}$ and $\boldsymbol{A}\times\boldsymbol{C}$ are perpendicular to $\boldsymbol{A}$. Since $\boldsymbol{A}\cdot(\boldsymbol{B}\times\boldsymbol{A}) = 0$ and $\boldsymbol{A}\cdot(\boldsymbol{A}\times\boldsymbol{C}) = 0$, scalar multiplication by $\boldsymbol{A}$ and subsequent cyclic permutation gives the three relationships

$$\sin b\sin c\sin\alpha = \boldsymbol{A}\cdot(\boldsymbol{B}\times\boldsymbol{C}),\quad \sin c\sin a\sin\beta = \boldsymbol{B}\cdot(\boldsymbol{C}\times\boldsymbol{A}),\quad \sin a\sin b\sin\gamma = \boldsymbol{C}\cdot(\boldsymbol{A}\times\boldsymbol{B}).$$

Because the triple scalar product is unaltered by cyclic permutation of the vectors, the three right-hand sides have the same value. Equating the left-hand sides,

$$\sin b\sin c\sin\alpha = \sin c\sin a\sin\beta = \sin a\sin b\sin\gamma$$

leads to the sine rule.

Sine rule
$\sin a : \sin b : \sin c = \sin\alpha : \sin\beta : \sin\gamma$

Half-angle and half-side formulae. The formulae corresponding to the half-angle formula of plane trigonometry can be used in the same way, to calculate the angles from three given sides and, conversely, the sides from three given angles. From the cosine rule for sides and with the help of trigonometric relationships:

$$\cos^2\frac{\alpha}{2} = \frac{1}{2}(1 + \cos\alpha) = \frac{1}{2}\cdot\frac{\sin b\sin c + \cos a - \cos b\cos c}{\sin b\sin c}$$

$$= \frac{\cos a - \cos(b+c)}{2\sin b\sin c} = -\frac{\cos(b+c) - \cos a}{2\sin b\sin c}$$

$$= \frac{\sin[(b+c+a)/2]\ \sin[(b+c-a)/2]}{\sin b\sin c} = \frac{\sin s\sin(s-a)}{\sin b\sin c}$$

$$\sin^2\frac{\alpha}{2} = \frac{1}{2}(1 - \cos\alpha) = \frac{1}{2}\cdot\frac{\sin b\sin c + \cos b\cos c - \cos a}{\sin b\sin c}$$

$$= \frac{\cos(b-c) - \cos a}{2\sin b\sin c}$$

$$= \frac{\sin[(a+c-b)/2]\sin[(a+b-c)/2]}{\sin b\sin c} = \frac{\sin(s-b)\sin(s-c)}{\sin b\sin c}$$

These make use of the following facts:

$$\cos\alpha = \frac{\cos a - \cos b\cos c}{\sin b\sin c}$$

$$\cos\varphi - \cos\psi = -2\sin\frac{\varphi+\psi}{2}\sin\frac{\varphi-\psi}{2}$$

$$s = [(b+c+a)/2]$$
$$(s-a) = (b+c-a)/2$$
$$(s-b) = (c+a-b)/2$$
$$(s-c) = (a+b-c)/2$$

The half-angle formula is obtained by division, using the fact that $\tan(\alpha/2) = \sin(\alpha/2)/\cos(\alpha/2)$.

Half-angle formulae	$\tan\frac{\alpha}{2} = \sqrt{\left(\frac{\sin(s-b)\sin(s-c)}{\sin s\sin(s-a)}\right)}$, $\tan\frac{\beta}{2} = \sqrt{\left(\frac{\sin(s-c)\sin(s-a)}{\sin s\sin(s-b)}\right)}$, $\tan\frac{\gamma}{2} = \sqrt{\left(\frac{\sin(s-a)\sin(s-b)}{\sin s\sin(s-c)}\right)}$, where $2s = a+b+c$

The half-side formulae are polar to the half-angle formulae. In the polar triangle $\bar{A}\bar{B}\bar{C}$ the half-angle formula gives $\tan^2(\bar{\beta}/2) = \dfrac{\sin(\bar{s}-\bar{c})\sin(\bar{s}-\bar{a})}{\sin\bar{s}\sin(\bar{s}-\bar{b})}$. By substitution, using the relationships $\sigma = {}^1/_2(\alpha+\beta+\gamma)$, $\bar{\beta} = 180° - b$, $\bar{a} = 180° - \alpha$, $\bar{b} = 180° - \beta$, $\bar{c} = 180° - \gamma$, $\bar{s} = {}^1/_2(a+b+c) = 270° - \sigma$, $\bar{s} - \bar{a} = 90° - (\sigma - \alpha)$, $\bar{s} - \bar{b} = 90° - (\sigma - \beta)$, $\bar{s} - \bar{c} = 90° - (\sigma - \gamma)$, one obtains $\cot^2\dfrac{b}{2} = \dfrac{\cos(\sigma-\gamma)\cos(\sigma-\alpha)}{-\cos\sigma\cos(\sigma-\beta)}$.

Half-side formulae	$\tan\frac{b}{2} = \sqrt{\left(\frac{-\cos\sigma\cos(\sigma-\beta)}{\cos(\sigma-\gamma)\cos(\sigma-\alpha)}\right)}$, $\tan\frac{c}{2} = \sqrt{\left(\frac{-\cos\sigma\cos(\sigma-\gamma)}{\cos(\sigma-\alpha)\cos(\sigma-\beta)}\right)}$, $\tan\frac{a}{2} = \sqrt{\left(\frac{-\cos\sigma\cos(\sigma-\alpha)}{\cos(\sigma-\beta)\cos(\sigma-\gamma)}\right)}$, where $2\sigma = \alpha+\beta+\gamma$

Napier's analogies. For the complete solution by logarithms of a spherical triangle, given two sides and the included angle or two angles and the side between them, the so-called Napier's analogies are available. They can be derived from the half-angle or half-side formulae, using trigonometric relations, in particular those concerning sums and differences of trigonometric functions. It suffices here to give one of each of the sets of three formulae that follow from one another by cyclic permutation.

Napier's analogies	
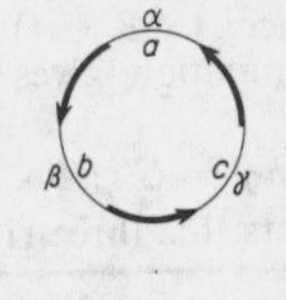	1a) $\tan\frac{a}{2}\cos\frac{\beta-\gamma}{2}=\tan\frac{b+c}{2}\cos\frac{\beta+\gamma}{2}$; 1b); 1c) 2a) $\tan\frac{a}{2}\sin\frac{\beta-\gamma}{2}=\tan\frac{b-c}{2}\sin\frac{\beta+\gamma}{2}$; 2b); 2c) 3a) $\cot\frac{\alpha}{2}\cos\frac{b-c}{2}=\tan\frac{\beta+\gamma}{2}\cos\frac{b+c}{2}$; 3b); 3c) 4a) $\cot\frac{\alpha}{2}\sin\frac{b-c}{2}=\tan\frac{\beta-\gamma}{2}\sin\frac{b+c}{2}$; 4b); 4c)

For frequent use of Napier's analogies the following mnemonic rules are suggested: all arguments are halved; if the tangents or cotangents have sides as arguments, then the sines and cosines have angles (and conversely); the function of the half-side is related to the half-sum or half-difference of the other two sides, and similarly for angles. It is easy to formulate other precise mnemonic rules.

The basic problems for the general spherical triangle

Unlike the plane triangle, the spherical triangle is also determined by three angles, so that there are six basic problems. For their solution one uses general relationships in the Euler triangle.

Limit passage to plane trigonometry. Three points A, B, C in space that do not lie in a straight line, determine a plane and in it one plane triangle. They can, however, form the vertices of a spherical triangle on infinitely many different spheres. If these are ordered according to increasing radius R, then as $R\to\infty$, the spherical triangle tends continuously to the plane triangle, each spherical angle tending to the corresponding plane angle, and the spherical excess becomes arbitrarily small. To the lengths of the sides $\bar{a}$, $\bar{b}$, $\bar{c}$, in the plane triangle there correspond the sides $\bar{a}/R$, $\bar{b}/R$, $\bar{c}/R$, measured in radians, of the spherical triangle. In the formula

$$\tan(\hat{\varepsilon}/4)=\sqrt{\{\tan(\hat{s}/2)\tan[(\hat{s}-\hat{a})/2]\tan[(\hat{s}-\hat{b})/2]\tan[(\hat{s}-\hat{c})/2]\}},$$

which was obtained by *L'Huilier*, the tangent of the angle may be replaced by its radian measure because the angle is small. This gives

$$\frac{\hat{\varepsilon}}{4}=\frac{1}{4}\sqrt{\left[\frac{\bar{s}}{R}\cdot\frac{(\bar{s}-\bar{a})}{R}\cdot\frac{(\bar{s}-\bar{b})}{R}\cdot\frac{(\bar{s}-\bar{c})}{R}\right]}$$

The area A of the spherical triangle then becomes the area given by Heron's formula for a plane triangle:

$$A=\hat{\varepsilon}R^2=(\hat{\varepsilon}/4)\cdot 4R^2=\sqrt{[\bar{s}(\bar{s}-\bar{a})(\bar{s}-\bar{b})(\bar{s}-\bar{c})]}.$$

For large, but still finite, values of R a theorem due to LEGENDRE holds:

Theorem of Legendre: A spherical triangle with small sides and therefore small spherical excess has approximately the same area as a plane triangle with sides of lengths of the same absolute value. Each angle of the plane triangle is smaller than the corresponding angle of the spherical triangle by approximately one third of the spherical excess.

Using L'Huilier's formula the spherical excess ε can be found for a triangle on the earth (of radius R) with sides $a = 31.075$ miles, $b = 37.290$ miles and $c = 43.505$ miles (for example, between Cambridge, Luton and Corby). The magnitudes of the sides are given in radian measure by $\hat{a}=\bar{a}/R$, $\hat{b}=\bar{b}/R$, $\hat{c}=\bar{c}/R$, or in degrees by $a^\circ=360^\circ\bar{a}/2\pi R)$, $b^\circ=360^\circ\bar{b}/(2\pi R)$, $c^\circ=360^\circ\bar{c}/(2\pi R)$. or alternatively in seconds by multiplying a by 206204.8'', since 1 radian corresponds to 206264.8''

Miles	Radians	Seconds	$s/2 = 24'16.85''$
31.075	0.0078479	1618.75''	$(s-a)/2 = 10'47.47''$
37.290	0.0094173	1942.46''	$(s-b)/2 = 8'5.62''$
43.505	0.0109869	2266.21''	$(s-c)/2 = 5'23.75''$

The results to the right show that the spherical excess is $\varepsilon = 7.6''$. By Legendre's theorem the triangle may be regarded as plane as long as the accuracy of the measured angles is not less than $\varepsilon/3\approx 2.5''$.

To derive the limiting form as $R\to\infty$ of the sine and cosine rules the trigonometric functions are expanded in convergent series. Writing $\bar{a}/R=q_a$, $\bar{b}/R=q_b$, $\bar{c}/R=q_c$, one obtains $\sin q_a$

$= q_a - q_a^3/3! + \cdots = q_a[1 - q_a^2/6 + \delta_1]$ and $\cos q_a = 1 - q_a^2/2! + \delta_2$, where δ_1 and δ_2 are of the order of $1/R^4$. For the sine rule of spherical trigonometry this gives:

$$\sin\alpha : \sin\beta : \sin\gamma = \bar{a}[1 - q_a^2/6 + \delta_1] : \bar{b}[1 - q_b^2/6 + \delta_3] : \bar{c}[1 - q_c^2/6 + \delta_5],$$

so that in the limiting case $\sin\alpha : \sin\beta : \sin\gamma = \bar{a} : \bar{b} : \bar{c}$. This is the sine rule of plane trigonometry. Similarly, for the spherical cosine rule one obtains the cosine rule of plane trigonometry: $\cos q_a = \cos q_b \cos q_c + \sin q_b \sin q_c \cos\alpha$ or

$$[1 - q_a^2/2 + \delta_2] = [1 - q_b^2/2 + \delta_4]\,[1 - q_c^2/2 + \delta_6] + q_b q_c \cos\alpha[1 - q_b^2/6 + \delta_3]\,[1 - q_c^2/6 + \delta_5],$$

$$-{}^1/_2 q_a^2 + \delta_2 = -{}^1/_2(q_b^2 + q_c^2) - \delta_{4,6} + q_b q_c \cos\alpha[1 - (q_b^2 + q_c^2)/6 + \delta_{3,5}],$$

$$\bar{a}^2 = \bar{b}^2 + \bar{c}^2 - 2\bar{b}\bar{c}\cos\alpha.$$

General relationships in Euler spherical triangles. Since no angle and no side can be greater than π (or 180°), the arguments are given uniquely by the tangent, cotangent and cosine functions; on the other hand, two values are given by the sine function. If two arguments are possible, the geometrically correct solutions are selected from the theoretically possible ones by means of inequalities.

1. In an Euler spherical triangle the sum of the angles lies between π and 3π and the sum of the sides between 0 and 2π:

$$\pi < \hat{\alpha} + \hat{\beta} + \hat{\gamma} < 3\pi \quad \text{and} \quad 0 < \hat{a} + \hat{b} + \hat{c} < 2\pi$$

or $$180° < \alpha + \beta + \gamma < 540° \quad \text{and} \quad 0 < a + b + c < 360°$$

2. The greater angle lies opposite the greater side.

If $a > b$, for example, in the Napier analogy 4c) $\cot(\gamma/2)\sin[(a-b)/2] = \tan[(\alpha-\beta)/2]\sin[(a+b)/2]$, that is, $\sin[(a-b)/2] > 0$, then because $\sin[(a+b)/2] > 0$ and $\cot(\gamma/2) > 0$ in Euler triangles, it follows that $\tan[(\alpha-\beta)/2] > 0$ also. But this means that $(\alpha - \beta) > 0$ or $\alpha > \beta$.

3. The sum of two sides is greater than the third. The difference between two sides is smaller than the third.

Corresponding to each spherical triangle there exists a solid angle. This degenerates into a plane circular sector when the sum of two sides is equal to the third side, and is impossible in space if the sum is smaller than the third side. If the difference between the two sides a and b is greater than or equal to the third side c, $a - b \geqslant c$, then it would follow that $a \geqslant b + c$, in contradiction to the first part of the theorem.

4. The sum of two angles is less than the third increased by π (or 180°).

As has just been shown, in the polar triangle $\bar{A}\bar{B}\bar{C}$:

$$\bar{a} + \bar{b} > \bar{c} \quad \text{and} \quad \bar{a} - \bar{b} < \bar{c}.$$

Because $\bar{a} = 180° - \alpha$, $\bar{b} = 180° - \beta$, $\bar{c} = 180° - \gamma$ this means that for the triangle ABC:

$$180° - \alpha + 180° - \beta > 180° - \gamma, \quad \text{and} \quad 180° - \alpha - 180° + \beta < 180° - \gamma,$$

$$180° + \gamma > \alpha + \beta, \quad \text{and} \quad \beta + \gamma < 180° + \alpha.$$

5. If the sum of two sides is greater (or less) than two right angles, then the sum of the two opposite angles is greater (or less) than two right angles.

In Napier's analogy 3c) $\cot(\gamma/2)\cos[(a-b)/2] = \tan[(\alpha+\beta)/2]\cos[(a+b)/2]$, let $a + b > \pi$, so that $\cos[(a+b)/2] < 0$. Because $\cot(\gamma/2)$ and $\cos[(a-b)/2]$ must be positive in the Euler triangle, it follows that $\tan[(\alpha+\beta)/2] < 0$. This means that $(\alpha+\beta)/2 > \pi/2$ or $\alpha + \beta > \pi$. Similarly it follows from $a + b < \pi$ that $\alpha + \beta < \pi$, and from $a + b = \pi$ that $\alpha + \beta = \pi$.

The basic problem 1a. In the spherical triangle ABC the three sides a, b, c are given and it is required to find the three angles α, β, γ. The sum of each pair of sides must be greater than the third and the sum of all three sides less than 360°. The solution is found by means of the cosine rule for sides (Fig.) $\cos\alpha = (\cos a - \cos b\cos c)/(\sin b \sin c)$ or by the half-angle-formulae

$$s = (a + b + c)/2, \quad \tan\frac{\alpha}{2} = \sqrt{\left(\frac{\sin(s-b)\sin(s-c)}{\sin s \sin(s-a)}\right)}.$$

The formulae for $\cos\beta$, $\cos\gamma$ or for $\tan(\beta/2)$ and $\tan(\gamma/2)$ are obtained by cyclic permutation.

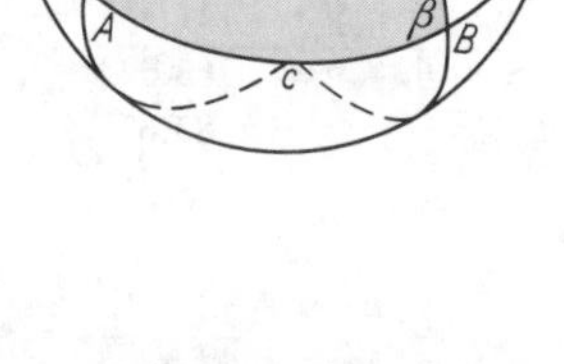

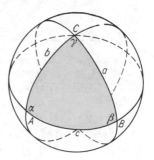

12.2-5 Solution of a spherical triangle, given three sides, or three angles

The basic problem 1b. In the spherical triangle ABC the three angles α, β, γ are given, such that the sum of two angles is less than the third increased by 180°, and the sum of all three angles lies between 180° and 540°. The sides can then be calculated either by means of the cosine rule for angles:

$$\cos a = (\cos\alpha + \cos\beta\cos\gamma)/(\sin\beta\sin\gamma), \ldots,$$

or by the half-side formulae:

$$2\sigma = \alpha + \beta + \gamma, \qquad \tan\frac{a}{2} = \sqrt{\left(\frac{-\cos\sigma\cos(\sigma-\alpha)}{\cos(\sigma-\beta)\cos(\sigma-\gamma)}\right)}, \ldots$$

The basic problem 2a. If in the spherical triangle ABC two sides and the included angle, say b, c and α, are given, the third side is found from the cosine rule for sides:

$$\cos a = \cos b\cos c + \sin b\sin c\cos\alpha.$$

With this side the remaining angles β and γ can then be found from the sine rule:

$$\sin\beta = \sin b\sin\alpha/\sin a \quad\text{and}\quad \sin\gamma = \sin c\sin\alpha/\sin a.$$

From each sine function one obtains two corresponding arguments, which are supplementary. However, using the theorem that the greater angle lies opposite the greater side, the angle β that corresponds to the given values of the problem can be uniquely determined as the angle that is greater or smaller than the angle α, according as the side b is greater or smaller than the side a. The angle γ is similarly chosen so that $\gamma \gtreqless \alpha$ according as $c \gtreqless a$, where either the two upper or the two lower inequalities hold simultaneously. For logarithmic calculation Napier's analogies are available. By 3a) and 4a) one obtains:

$$\tan\frac{\beta+\gamma}{2} = \cot\frac{\alpha}{2}\cos\frac{b-c}{2}\Big/\cos\frac{b+c}{2}$$

and

$$\tan\frac{\beta-\gamma}{2} = \cot\frac{\alpha}{2}\sin\frac{b-c}{2}\Big/\sin\frac{b+c}{2}.$$

From these the angles $(\beta+\gamma)/2$ and $(\beta-\gamma)/2$ and hence β and γ can be calculated. The remaining side a is given by the sine rule $\sin a = \sin\alpha\cdot\sin b/\sin\beta$; from the two values a of the Arcsin function, that one is chosen that is greater or less than c, according as α is greater or less than γ.

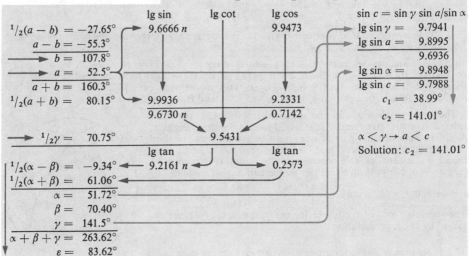

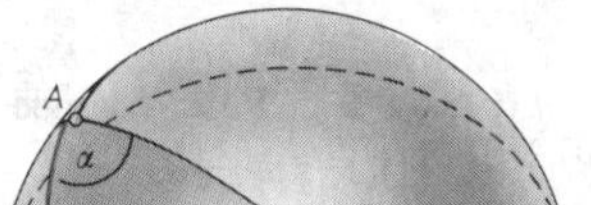

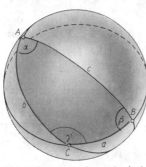

12.2-6 Solution of a spherical triangle with $a = 52.5°$, $b = 107.8°$ and $\gamma = 141.5°$

Example: It is given that $a = 52.5°$; $b = 107.8°$; $\gamma = 141.5°$ (Fig.).

By Napier's analogies 3c) and 4c):

$$\tan\frac{\alpha+\beta}{2} = \cot\frac{\gamma}{2}\cos\frac{a-b}{2}\Big/\cos\frac{a+b}{2},$$

$$\tan\frac{\alpha-\beta}{2} = \cot\frac{\gamma}{2}\sin\frac{a-b}{2}\Big/\sin\frac{a+b}{2}.$$

	lg sin	lg cot	lg cos	$\sin c = \sin\gamma\sin a/\sin\alpha$
$^1/_2(a-b) = -27.65°$	9.6666 n		9.9473	$\lg\sin\gamma =$ 9.7941
$a - b = -55.3°$				$\lg\sin a =$ 9.8995
$b =$ 107.8°				9.6936
$a =$ 52.5°				$\lg\sin\alpha =$ 9.8948
$a + b =$ 160.3°				$\lg\sin c =$ 9.7988
$^1/_2(a+b) =$ 80.15°	9.9936		9.2331	$c_1 =$ 38.99°
	9.6730 n		0.7142	$c_2 =$ 141.01°
$^1/_2\gamma =$ 70.75°		9.5431		$\alpha < \gamma \rightarrow a < c$
	lg tan		lg tan	Solution: $c_2 =$ 141.01°
$^1/_2(\alpha-\beta) =$ −9.34°	9.2161 n		0.2573	
$^1/_2(\alpha+\beta) =$ 61.06°				
$\alpha =$ 51.72°				
$\beta =$ 70.40°				
$\gamma =$ 141.5°				
$\alpha+\beta+\gamma =$ 263.62°				
$\varepsilon =$ 83.62°				

Basic problem 2b. Two angles of the spherical triangle and the side between them are now given, for example, β, γ and a. The problem and hence also the solution is polar to the basic problem 2a. It is therefore sufficient to compile the formulae.

I. The cosine rule for angles gives the angle α: $\cos\alpha = -\cos\beta\cos\gamma + \sin\beta\sin\gamma\cos a$. The sine rule gives b and c: $\sin b = \sin\beta\sin a/\sin\alpha$, $\sin c = \sin\gamma\sin a/\sin\alpha$, where $b \gtreqless a$ according as $\beta \gtreqless \alpha$, and $c \gtreqless a$ according as $\gamma \gtreqless \alpha$.

II. Napier's analogies 1a) and 2a) give the sides b and c:

$$\tan\frac{b+c}{2} = \tan\frac{a}{2}\cos\frac{\beta-\gamma}{2}\Big/\cos\frac{\beta+\gamma}{2} \quad\text{and}\quad \tan\frac{b-c}{2} = \tan\frac{a}{2}\sin\frac{\beta-\gamma}{2}\Big/\sin\frac{\beta+\gamma}{2}.$$

The angle α is given by the sine rule $\sin\alpha = \sin a\sin\beta/\sin b$, where $\alpha \gtreqless \beta$ according as $a \gtreqless b$.

Basic problem 3a. If two sides and one of the opposite angles of the spherical triangle ABC are given, for example, a, c and γ, then there is either no solution, or one or two solutions (Fig.). The three cases are illustrated in the figure by the three spherical circles k_1, k_2, and k_3 about the point B with different spherical radii c. The circle k_1 with radius $\widehat{BA_4}$ does not intersect the great circle through C and A_1; k_2 touches it at the point A_3, and k_3, on the other hand, cuts it at A_1 and A_2. With the length of side $\widehat{BA_3} = c$ there is one solution, the right-angled triangle A_3BC. With side $\widehat{BA_1} = \widehat{BA_2} = c$, however, one obtains the two triangles A_1BC and A_2BC. Because the triangle A_2BA_1 is isosceles, it follows that $\alpha_2 = \sphericalangle BA_1A_2$, or $\alpha_1 = 180° - \alpha_2$. By calculation one obtains these two angles α_1 and α_2 from the sine rule $\sin\alpha = \sin a\sin\gamma/\sin c$, since $\sin\alpha_1 = \sin\alpha_2$. For each value of the angle α the side b and the angle β are then given uniquely by Napier's analogies 2b) and 4b):

$$\tan\frac{b}{2} = \tan\frac{c-a}{2}\sin\frac{\gamma+\alpha}{2}\Big/\sin\frac{\gamma-\alpha}{2},$$

$$\cot\frac{\beta}{2} = \tan\frac{\gamma-\alpha}{2}\sin\frac{c+a}{2}\Big/\sin\frac{c-a}{2}.$$

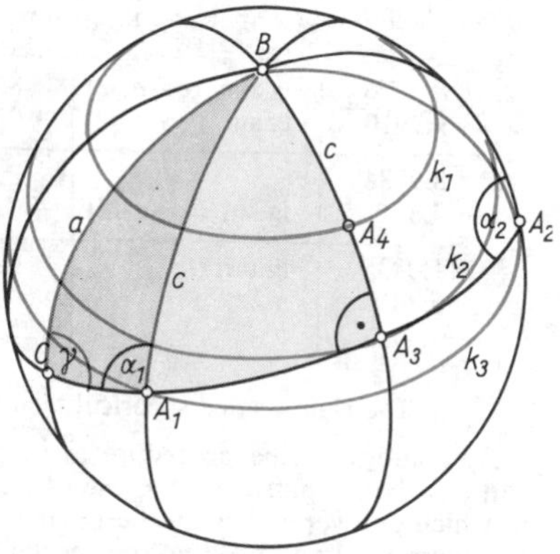

12.2-7 Solution of a spherical triangle, given two sides a and c, and one opposite angle γ

The analytical discussion of the possible cases is based on the relationship $\sin\alpha = \sin a\sin\gamma/\sin c$ and the procedure is analogous to that followed in plane trigonometry (Fig.).

I. $(\sin a\sin\gamma/\sin c) > 1$, so that $\sin\alpha > 1$; no real solution.

II. $(\sin a\sin\gamma/\sin c) = 1$, $\sin\alpha = 1$, $\alpha = \pi/2$; one solution, for example, the triangle A_3BC.

III. $\sin\alpha = (\sin a\sin\gamma/\sin c) < 1$.

III (1). $\sin a < \sin c \to \sin\alpha < \sin\gamma$; one solution, since for each given triple of values (a_i, c_i, γ_i), $i = 1, 2$, the value of α is uniquely determined, because $\alpha \gtreqless \gamma$ according as $a \gtreqless c$ (Fig.).

III (2). $\sin a = \sin c \to \sin\alpha = \sin\gamma$; one solution [see III (1).].

III (3). $\sin a > \sin c \to \sin\alpha > \sin\gamma$; two solutions. Either $a > c \to \alpha > \gamma$, that is, $c = c_1$ (acute) $\to \gamma = \gamma_1$ (acute) and α_1, $\alpha_2 = 180° - \alpha_1$ are solutions; or $a < c \to \alpha < \gamma$, that is, $c = c_2$ (obtuse) $\to \gamma = \gamma_2$ (obtuse) and $\alpha_1, \alpha_2 = 180° - \alpha_1$ are solutions (see Fig.).

Basic problem 3b. The polar problem, to solve a spherical triangle ABC from two angles and a side opposite one of them, for example, from α, γ and c, leads to the corresponding different cases. It is therefore sufficient to give the method of calculation without further discussion:

1. $\sin a = \sin\alpha\sin c/\sin\gamma$;
2. $\tan(b/2) = \tan[(c-a)/2]\sin[(\gamma+\alpha)/2]/\sin[(\gamma-\alpha)/2]$;
3. $\cot(\beta/2) = \tan[(\gamma-\alpha)/2]\sin[(c+a)/2]/\sin[(c-a)/2]$.

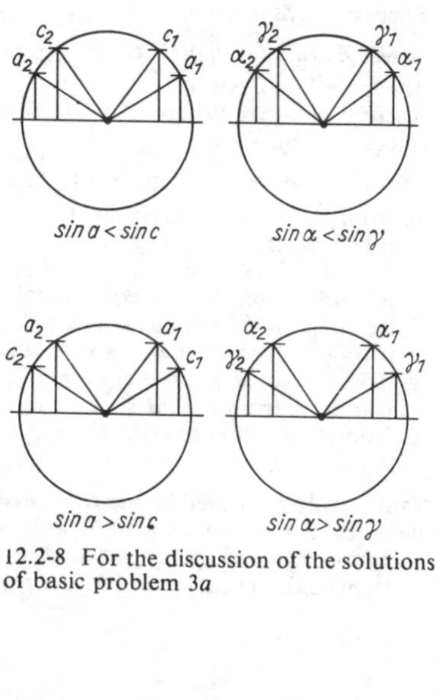

12.2-8 For the discussion of the solutions of basic problem 3a

Example: Given that $c = 96.5°$; $\alpha = 101.2°$; $\gamma = 102.1°$; to find a, b, β (Fig.).

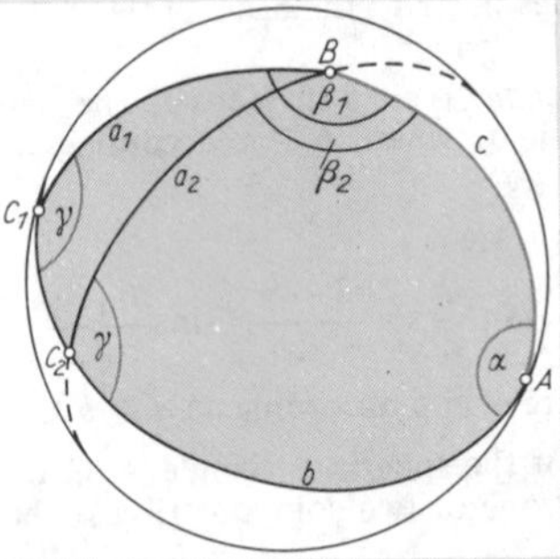

1.

$\sin a = \sin\alpha \sin c / \sin\gamma$

$$\begin{aligned} \lg \sin\alpha &= 9.9916 \\ +\lg \sin c &= 9.9972 \\ \hline & 9.9888 \\ -\lg \sin\gamma &= 9.9902 \\ \hline \lg \sin a &= 9.9986 \end{aligned}$$

$a_1 = 85.40°$, $a_2 = 94.60°$

Since $\sin\alpha > \sin\gamma$, case III (3) applies with $\alpha < \gamma$ and there are two solutions.

1a)

90.95°	$(c+a)/2$	95.55°
48.25° 42.70°	$c/2$ $a/2$	48.25° 47.30°
5.55° 101.65°	$(c-a)/2$ $(\gamma+\alpha)/2$	0.95° 101.65°
51.05° 50.60°	$\gamma/2$ $\gamma/2$	51.05° 50.60°
0.45°	$(\gamma-\alpha)/2$	0.45°

12.2-9 Solution of a spherical triangle with $c = 96.5°$, $\alpha = 101.2°$ and $\gamma = 102.1°$

2.

8.9876 +9.9910	$\lg\tan[(c-a)/2]$ $\lg\sin[(\gamma+\alpha)/2]$	8.2196 +9.9910
8.9786 −7.8951	$\lg\sin[(\gamma-\alpha)/2]$	8.2106 −7.8951
1.0835 85.28° 170.56°	$\lg\tan(b/2)$ $b/2$ b	0.3155 64.19° 128.38°

3.

7.8951 +9.9999	$\lg\tan[(\gamma-\alpha)/2]$ $\lg\sin[(c+a)/2]$	7.8951 +9.9980
7.8950 −8.9855	$\lg\sin[(c-a)/2]$	7.8931 −8.2196
8.9095 85.34° 170.68°	$\lg\cot(\beta/2)$ $\beta/2$ β	9.6735 64.75° 129.50°

The right-angled spherical triangle

By analogy to the procedure in plane trigonometry, the calculations in spherical trigonometry can also be simplified by the use of right-angled triangles. Polar to these are right-sided triangles, in which one vertex lies on the polar of a second vertex. However the right-sided spherical triangle is seldom used and need not be specially considered.

Napier's rules. In a spherical triangle ABC suppose that the angle γ is 90°; then c is the hypotenuse (Fig.). Since $\sin 90° = 1$ and $\cos 90° = 0$, the theorems for the general triangle simplify to the following forms:

Sine rule: $\sin a = \sin\alpha \sin c / \sin 90°$,

1. $\sin a = \sin\alpha \sin c$, **(1)** $\cos(90° - a) = \sin\alpha \sin c$,
2. $\sin b = \sin\beta \sin c$, **(2)** $\cos(90° - b) = \sin\beta \sin c$.

Cosine rule for sides: $\cos c = \cos a \cos b + \sin a \sin b \cos 90°$,

3. $\cos c = \cos a \cos b$, **(3)** $\cos c = \sin(90° - a)\sin(90° - b)$.

Cosine rule for angles: $\cos\alpha = -\cos\beta\cos 90° + \sin\beta \sin 90° \cos a$,

4. $\cos\alpha = \sin\beta\cos a$, **(4)** $\cos\alpha = \sin(90° - a)\sin\beta$,
5. $\cos\beta = -\cos 90° \cos\alpha + \sin 90° \sin\alpha\cos b$,
   $\cos\beta = \sin\alpha\cos b$, **(5)** $\cos\beta = \sin(90° - b)\sin\alpha$.

From these five relationships further ones can be found:

6. from 4.: $\cos a = \cot\alpha \cdot \sin\alpha/\sin\beta$ and 5.: $\cos b = \cot\beta \cdot \sin\beta/\sin\alpha$
   it follows from 3. that $\cos c = \cot\alpha\cot\beta$, **(6)** $\cos c = \cot\alpha\cot\beta$,
7. from 1.: $\sin\alpha = \sin a/\sin c$ and 3.: $\cos b = \cos c/\cos a$
   it follows from 5. that $\cos\beta = \tan a\cot c$, **(7)** $\cos\beta = \cot(90° - a)\cos c$,
8. from 2.: $\sin\beta = \sin b/\sin c$ and 3.: $\cos a = \cos c/\cos b$
   it follows from 4. that $\cos\alpha = \tan b\cot c$, **(8)** $\cos\alpha = \cot(90° - b)\cot c$,
9. from 5.: $\sin\alpha = \cos\beta/\cos b$ and 2.: $\sin c = \sin b/\sin\beta$
   it follows from 1. that $\sin a = \tan b\cot\beta$, **(9)** $\cos(90° - a) = \cot(90° - b)\cot\beta$,
10. from 4.: $\sin\beta = \cos\alpha/\cos a$ and 1.: $\sin c = \sin a/\sin\alpha$
    it follows from 2. that $\sin b = \tan a\cot\alpha$, **(10)** $\cos(90° - b) = \cot(90° - a)\cot\alpha$.

Napier collected together the relationships (1) to (10) in the rules that bear his name. To formulate the rules, one visualises a triangle in which γ is a right angle. Leaving the right angle out of consideration, the remaining two angles, the hypotenuse and the complements of the sides containing the right angle are called the *circular parts* of the triangle. These five parts, β, c, α, $(90° - b)$, $(90° - a)$,

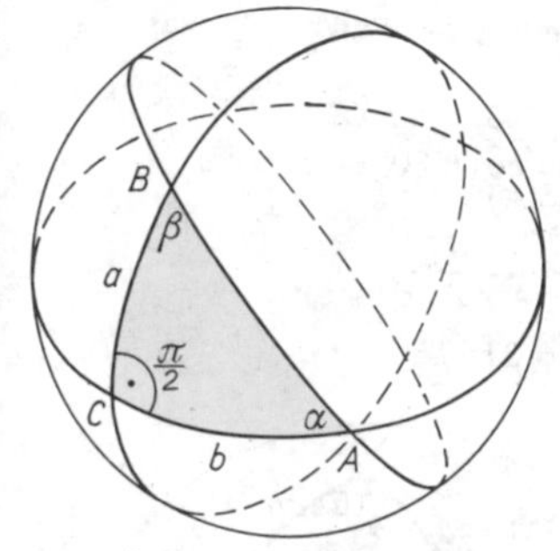

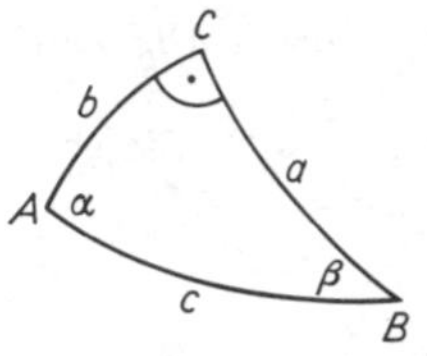

12.2-10 Right-angled spherical triangle *ABC*

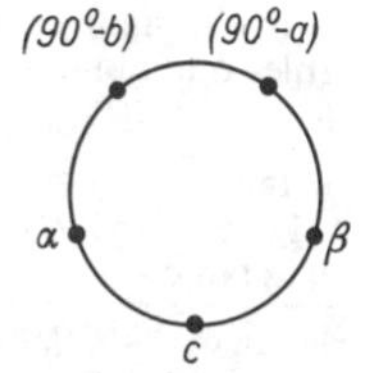

12.2-11 Position of the parts for *Napier's rules*

are arranged around a circle in the order in which they naturally occur in the triangle (Fig.). If any one part is selected, the two on either side of it are called the adjacent parts and the remaining two the opposite parts. Napier's rules then have the following form:

Napier's rules. In a right-angled spherical triangle the cosine of any part is equal to the product of the cotangents of the adjacent parts, and is also equal to the product of the sines of the opposite parts.

From the application of Napier's rules to Euler triangles the arguments of all trigonometric functions except the sine function are given uniquely and that of the sine function is two-valued. From the general relationships in the Euler triangle one can distinguish whether one or two solutions exist.

Example: If it is given that $a = 38.4°$, $\alpha = 42.9°$, then one finds two solutions. From $\sin b = \cot\alpha \tan a$, two values b_1 and $b_2 = 180° - b_1$ are obtained and from $\cos\alpha = \cos a \sin\beta$, two values for β, which can also be calculated from $\cos\beta = \sin\alpha\cos b$. Finally, the hypotenuse c can be determined from $\cos\alpha = \cot c \tan b$.

$\lg \cot\alpha =$	0.0319	$\lg\cos\alpha =$	9.8648
$\lg\tan a =$	9.8990	$\lg\cos a =$	9.8941
$\lg\sin b =$	9.9309	$\lg\sin\beta =$	9.9707
$b_1 =$	58.52°	$\beta_1 =$	69.2°
$b_2 =$	121.48°	$\beta_2 =$	110.8°

$\lg\cos\alpha =$	9.8648
$\lg\cot b_{1,2} =$	9.7870
$\lg\cot c =$	9.6518
$c_1 =$	65.85°
$c_2 =$	114.15°

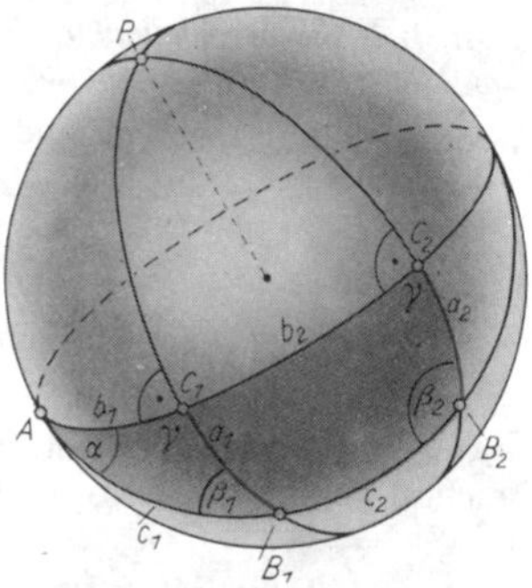

12.2-12 Solution of a right-angled spherical triangle, given a side a containing the right angle and the opposite angle α

Altitudes. With the help of Napier's rules altitudes of spherical triangles can be calculated. They are measured on a great circle through a vertex perpendicular to the opposite side. The altitude gives the spherical distance of the vertex from the side. An arbitrary spherical triangle is divided by an altitude into two right-angled triangles and can be solved using Napier's rules. In this way Napier's analogies can, in general, be avoided. Moreover, in applications the altitude often has a direct meaning. Suppose, for instance, that in the figure of the following example the great circle through A, C and F represents the earth's equator and B is the position of a ship. Then h is the geographical latitude of this position. Or suppose that B represents the North Pole of the earth and an aircraft or ship is moving along the great circle through A and C; then h is its shortest distance from the pole and the side $\widehat{CF}$ the path to the position at which this distance is reached.

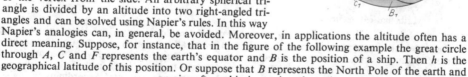

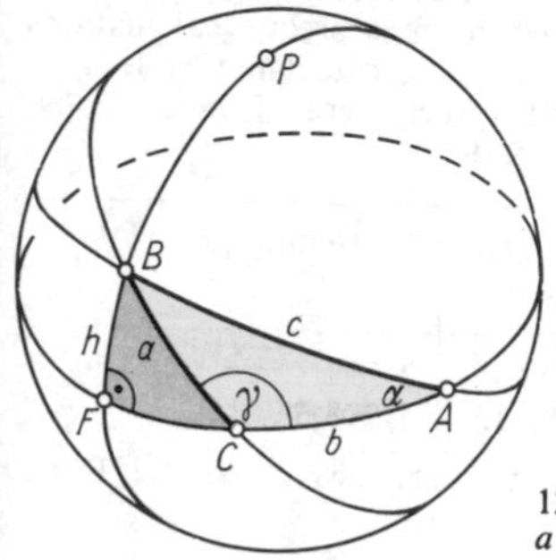

Example: In the triangle with sides $c = 84°$, $a = 42.7°$ and angle $\gamma = 135°$ (Fig.) the altitude $h = \widehat{BF}$ to the side b lies outside the triangle because the angle γ is obtuse. By the sine rule $\sin\alpha = \sin\gamma \sin a / \sin c$, two values α_1 and α_2 are obtained for α. Since the smaller angle lies opposite the smaller side, only α_1 can be a solution.

$\lg\sin\gamma =$	9.8495
$\lg\sin a =$	9.8313
	9.6808
$\lg\sin c =$	9.9976
$\lg\sin\alpha =$	9.6832
$\alpha_1 =$	28.83°
$\alpha_2 =$	151.17°

12.2-13 Solution of a spherical triangle ABC, given two sides $a = 42.7°$, $c = 84°$ and one opposite angle $\gamma = 135°$

In the right-angled triangles ABF and CBF the hypotenuse and an angle are given; by Napier's rules one obtains:

1. $\cos\alpha = \cot c \tan \widehat{AF}$ and

$$\tan\widehat{AF} = \cos\alpha \tan c$$
$$\lg\cos\alpha = 9.9425$$
$$\lg\tan c = 0.9784$$
$$\lg\tan\widehat{AF} = 0.9209$$
$$\widehat{AF} = 83.16°$$

2. $\cos(180° - \gamma) = \cot\alpha \tan\widehat{CF}$

$$\tan\widehat{CF} = \cos(180° - \gamma)\tan a$$
$$\lg\cos(180° - \gamma) = 9.8495$$
$$\lg\tan a = 9.9651$$
$$\lg\tan\widehat{CF} = 9.8146$$
$$\widehat{CF} = 33.12°$$

Thus, the side b has magnitude $\widehat{AF} - \widehat{CF} = 50.04°$

By the sine rule applied to the triangle ABC the angle β is given by

$$\sin\beta = \frac{\sin b \sin\gamma}{\sin c}.$$

Because β must be less than γ, $\beta_1 = 33.02°$ is the only solution.

$$\lg\sin b = 9.8845$$
$$\lg\sin\gamma = 9.8495$$
$$9.7340$$
$$\lg\sin c = 9.9976$$
$$\lg\sin\beta = 9.7364$$
$$\beta_1 = 33.02°$$
$$\beta_2 = 146.98°$$

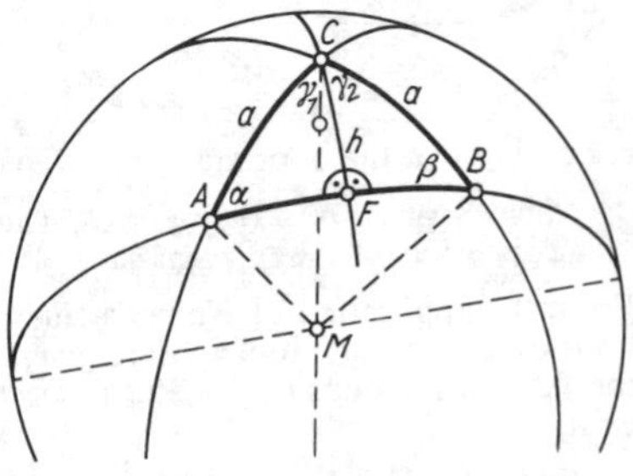

12.2-14 An isosceles spherical triangle

Isosceles spherical triangles. If two sides of the spherical triangle ABC are equal to one another, for example, if $a = b$, then the triangle is isosceles. Let F be the foot of the altitude h to the third side (Fig.). The calculation of the altitude h by applying Napier's rules to the right-angled triangle AFC with b and α must give the same value of h as from triangle BFC with a and β; since $a = b$ it follows that $\alpha = \beta$. With $a = b$ and $\alpha = \beta$, it then follows that $\widehat{AF}$ and $\widehat{BF}$ have the same value, that is, $\gamma_1 = \gamma_2$. Thus, the relationships known in plane geometry also hold here.

In an isosceles spherical triangle the altitude to the base bisects the base and the angle opposite to it. It is the perpendicular bisector of the base and a line of symmetry of the triangle. The base angles are equal to one another.

A corresponding theorem holds for a spherical triangle with two equal angles. Such a triangle is also isosceles.

12.3. Applications of spherical trigonometry

Among the applications of spherical trigonometry, two call for special attention because of their practical importance. They are the applications to mathematical geography and to astronomy.

Mathematical geography

The form of the earth is, in fact, irregular and is called a geoid. However the deviations from one of the bodies amenable to mathematical calculation are small in relation to their size. The analysis of the paths of the artificial earth satellites has shown that a suitable ellipsoid with three axes gives the best fit for the geoid. In fact, the difference between the two axes lying in the equatorial plane is so small that it has not so far been determined by earth measurements. Accordingly in higher geodesy the earth is regarded as a spheroid (ellipsoid of rotation). The first precise calculations were made by Friedrich Wilhelm Bessel (1784–1846). In 1924 the ellipsoid calculated by J. Hayford (1868–1925) was internationally recognized. The most recent values were given by F. N. Krasovskii (1878–1948); they are used for work in geodesy in the USSR.

Earth ellipsoid	Equatorial radius a		Polar radius b		Flattening $(a - b)/a$
	km	miles	km	miles	
Hayford	6378.388	3964.194	6356.912	3950.846	1/297
Krasovskii	6378.245	3964.105	6356.863	3950.816	1/298.3

In a first approximation the earth may be regarded as a sphere of mean radius $R = 6371.221$ km, [$\lg R = 3.804\,2227$] or $R = 3959.740$ miles.

Units of measurement on the earth sphere

1° on a great circle	111.20 km ≙ 69.111 miles
1° on the equator	111.32 km ≙ 69.186 miles
1 *geographical mile* = 1/15 equatorial degree	7.422 km ≙ 4.613 miles
arc length of a *meridian quadrant*	10 002.288 km ≙ 6 216.462 miles
mean arc length of a *meridian degree*	111.137 km ≙ 69.072 miles
1 *nautical mile* or 1 mean minute on a longitude circle	1.852 km
1 knot = 1 nautical mile/hour	1.852 km/h

A point on the earth's surface is determined by its longitude λ and its latitude φ, as has already been described in the derivation of the Gauss-Krüger coordinates in Chapter 11. The meridians are great circles, but the latitude circles are not; their radius ϱ is given by $\varrho = R\cos\varphi$. *Distances* on the earth's surface are measured along great circles, because they are geodesic lines and represent the shortest connections on the sphere. *Bearings* (or courses) are angles made with the meridian.

Determination of distance and course. If two places P_1 and P_2 on the earth are given by their longitude λ_1, λ_2 and latitude φ_1, φ_2 then the great circle distance between them and the angles between this circle and the meridians through P_1 and P_2 can be calculated. The formulae developed in the basic problems and Napier's rules are available for the solution.

Example: If an aircraft flies with air speed 800 km/h ≙ 497.2 m.p.h. from Leningrad ($\varphi_L = 59.9°$ N; $\lambda_L = 30.3°$ E $= -30.3°$) to San Francisco ($\varphi_F = 37.8°$ N; $\lambda_F = 122.4°$) by the shortest route, then its path is the arc $\widehat{LF}$ of the great circle through L and F (Fig.). On each of the meridians through the two places the arc from the equator to the place is given as the geographical latitude. The meridian arc from the place to the north pole N has magnitude $(90° - \varphi)$ and the two arcs $\widehat{LN} = 90° - \varphi_L$, $\widehat{FN} = 90° - \varphi_F$ form with the great circle arc $\widehat{LF}$ a spherical triangle in which the angle $\Delta\lambda$ between the two meridians is known; $\Delta\lambda = \lambda_F - \lambda_L = 122.4° + 30.3° = 152.7°$. In the spherical triangle two sides and the included angle are given. The great circle arc $g = \widehat{LF}$ is found by the cosine rule for sides:

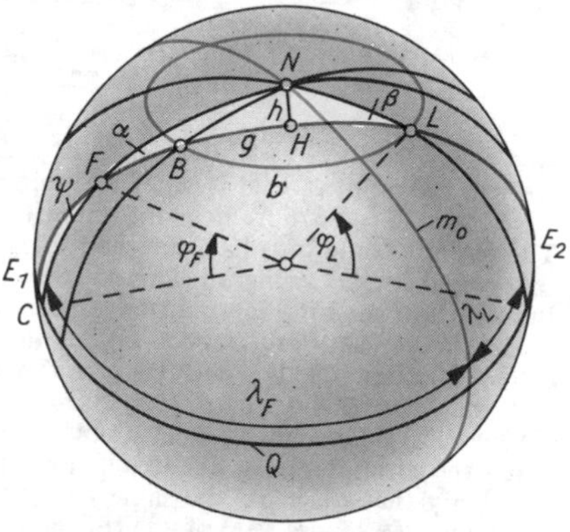

12.3-1 Flight path from Leningrad L to San Francisco F (schematic)

$$\cos g = \cos(90° - \varphi_L)\cos(90° - \varphi_F) + \sin(90° - \varphi_L)\sin(90° - \varphi_F)\cos\Delta\lambda,$$

$$\cos g = \sin\varphi_L \sin\varphi_F + \cos\varphi_L\cos\varphi_F\cos\Delta\lambda.$$

$\lg\sin\varphi_L = 9.9371$	$\lg\cos\varphi_L = 9.7003$	$\lg 2\pi = 0.7982$
$\lg\sin\varphi_F = 9.7874$	$\lg\cos\varphi_F = 9.8977$	$\lg R = 3.8042$
$\lg u = 9.7245$	$\lg\cos\Delta\lambda = 9.9487\,n$	$\lg g = 1.9017$
$u = 0.5304$	$\lg v = 9.5467\,n$	6.5041
$+v = -0.3522$	$v = -0.3522$	$\lg 360° = 2.5563$
$\cos g = u + v = +0.1782$		$\lg\bar{g} = 3.9478$
$g = 79.74°$		$\bar{g} = 8868$ km ≙ 5511.5 miles

This arc has the length $\bar{g} = 2\pi Rg/360° = 8868$ km, where the approximate value 6371 km (3960 miles) is taken for R, and could therefore be traversed at the given speed in about 11 hours (11.08 h).

The angles α and β in the spherical triangle are given by the sine rule. The aircraft leaves Leningrad on a course N 21.61° W and arrives at San Francisco on a course S 13.52° W. Each meridian cuts the flight path at a different angle. The course angle steadily increases by 144.87° from N 21.61° W to the final course S 13.52° W. At one point H of the flight path the aircraft is flying due west. It is then at its nearest point to the North Pole. The point H

$$\sin\alpha = \cos\varphi_L \sin\Delta\lambda/\sin g$$
$$\sin\beta = \cos\varphi_F \sin\Delta\lambda/\sin g$$

$\lg\sin\beta =$	$9.5662 \rightarrow \beta = 21.61°$
$\lg\cos\varphi_F =$	9.8977
$\lg\sin\Delta\lambda = 9.6615$, $\lg\sin g = 9.9930$	$\rightarrow 9.6685$
$\lg\cos\varphi_L =$	9.7003
$\lg\sin\alpha =$	$9.3688 \rightarrow \alpha = 13.52°$

ist the foot of the perpendicular h from the pole to the side $\widehat{LF}$. The altitude h divides the triangle LNF into two right-angled triangles. In the triangle LNH the distance h from the pole and the angle $\lambda_1 = \sphericalangle LNH$ can be determined. On the meridian $\lambda_H = 40.78°$ W the aircraft is flying due west; it is then at its closest position to the pole, 1183 km away from it. It crosses the latitude circle of Leningrad later at the point B, at the same angle as in Leningrad; its course at this point is therefore S 21.61° W. The meridian of the point B is given by $\lambda_B = \lambda_H + \lambda_1 = 40.78° + 71.08° = 111.86°$ W; thus, the geographical coordinates of the point B are $\varphi_B = 59.9°$ N and $\lambda_B = 111.86°$ W. The arc $\widehat{BH} = \widehat{LH}$ can be found from the right-angled triangle LNH using Napier's rules.

$\cos(90° - \varphi_L) = \cot\lambda_1 \cot\beta$, $\cot\lambda_1 = \tan\beta \sin\varphi_L$

$\cos(90° - h) = \sin(90° - \varphi_L)\sin\beta$, $\sin h = \cos\varphi_L \sin\beta$

$\lg\tan\beta = 9.5978$	$\lg\sin\beta = 9.5662$	$\cos\beta = \tan\widehat{LH}\tan\varphi_L$
$\lg\sin\varphi_L = 9.9371$	$\lg\cos\varphi_L = 9.7003$	$\tan\widehat{LH} = \cot\varphi_L\cos\beta$
$\lg\cot\lambda_1 = 9.5349$	$\lg\sin h = 9.2665$	$\lg\cot\varphi_L = 9.7632$
$\lambda_1 = 71.08°$	$h = 10.64°$	$\lg\cos\beta = 9.9684$
$-\lambda_L = -30.3°$	$\lg 2\pi R = 4.6024$	$\lg\tan\widehat{LH} = 9.7316$
$\lambda_H = 40.78°$ W	$-\lg 360° = 2.5563$	$\widehat{LH} = 28.33°$
	2.0461	$\lg(2\pi R/360°) = 2.0461$
	$\lg h = 1.0269$	$\lg\widehat{LH} = 1.4522$
	$\lg \bar{h} = 3.0730$	3.4983
	$\bar{h} = 1183$ km ≈ 735 miles	$\widehat{LH} = 3150$ km ≈ 1958 miles

Only at the point B, that is, after travelling a distance of $\widehat{LB} = 6300$ km or after a flight time of 7 h 52 min 30 s (7.875 h), does the aircraft turn for the first time towards more southerly latitudes. The aircraft could also have reached the point B by flying along the latitude circle $\varphi = 59.9°$ N that goes through Leningrad. This course would have cut all the meridians at right angles. However, the path b from L to B would have been longer, since it would not have been along a geodesic (great circle as shortest connection) but along a loxodrome (curve of constant bearing). The radius of the latitude circle is $\varrho = R\cos\varphi_L$. The arc b subtends an angle $\Delta\lambda$ at the centre of this circle, and thus $\bar{b} = 2\pi R\cos\varphi_L \Delta\lambda/360°$. One obtains $\bar{b} = 8516$ km (5293 miles) instead of 6300 km (3915 miles) along the geodesic, the difference being 2216 km (1337 miles). It would have taken the aircraft about 2 h 45 min longer to fly along the latitude circle.

A body that describes the same path, but with the speed $v = 8$ km/s (4.97 m.p.sec.) of an artificial earth satellite, needs 787.5 s = 13 min 7.5 s for the arc $\widehat{LB}$ and reaches San Francisco after 1108.5 s or 18 min 28.5 s, if friction is neglected. Its path cuts the equator Q in two points E_1 and E_2 which, as intersections of two great circles, lie on diameter of the sphere. Ift he intersection of the meridian of San Francisco with the equator Q is denoted by C, then the spherical triangle E_1CF is right-angled at C. In this triangle the angle $\psi = 13.52°$ and the side $\widehat{CF} = \varphi_F$ are known, and the side $\widehat{CE_1}$ can be found by Napier's rules. The point E_1 has coordinates $\varphi_{E_1} = 0$, $\lambda_{E_1} = \lambda_F + 8.39° = 130.79°$ W and consequently E_2 has coordinates $\varphi_{E_2} = 0$, $\lambda_{E_2} = (130.79° + 180°)$ W = 310.79° W or $\lambda_{E_2} = 49.21°$ E.

$\lg(2\pi R/360°) = 2.0461$
$\lg\cos\varphi_L = 9.7003$
$\lg\Delta\lambda = 2.1838$
$\lg\bar{b} = 3.9302$
$\bar{b} = 8516$ km ≈ 5293 miles

$\sin\varphi_F = \cot\alpha\tan\widehat{E_1C}$
$\tan\widehat{E_1C} = \tan\alpha\sin\varphi_F$
$\lg\tan\alpha = 9.3811$
$\lg\sin\varphi_F = 9.7874$
$\lg\tan\widehat{E_1C} = 9.1685$
$\widehat{E_1C} = 8.39°$

Loxodromes. The advantage for a ship or an aircraft of travelling along a geodesic to reach its destination in the shortest time contains the disadvantage that the course must be altered throughout the journey, strictly speaking at every instant. A curve that cuts all the meridians at the same bearing angle α is called a *loxodrome*. A latitude circle is a loxodrome for the bearing $\alpha = 90°$, a meridian one for $\alpha = 0°$. In the general case, for an arbitrary angle α, there is a curve for which a transcendental function gives the relationship between the latitude φ and the longitude λ at every point. If one considers two neighbouring points A and B (Fig.) with the coordinates (λ, φ) and $(\lambda + \Delta\lambda, \varphi + \Delta\varphi)$ on a loxodrome l, and the latitude circle with radius $\varrho = R\cos\varphi$ through the point A, then the arcs $\widehat{AC} = R\cos\varphi\,\Delta\lambda$, $\widehat{CB} = R\,\Delta\varphi$ and $\widehat{AB} = \Delta s$ form a right-angled triangle ABC. This, however, is not a spherical triangle (only $R\,\Delta\varphi$ lies along a great circle), but it may be regarded as plane if $\Delta\varphi$ and $\Delta\lambda$ are chosen sufficiently small. The following relationships can be read off in this triangle:

$$\tan\alpha = \Delta\lambda\cos\varphi/\Delta\varphi, \quad \Delta\lambda/\Delta\varphi = \tan\alpha/\cos\varphi \quad \text{and} \quad \Delta s\cos\alpha = R\,\Delta\varphi, \quad \Delta s/\Delta\varphi = R/\cos\alpha.$$

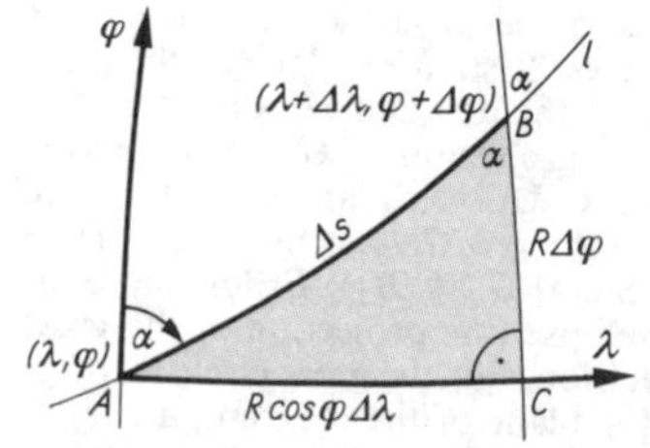

In the limit as $\Delta\varphi \to 0$ these equations tend to two differential equations $d\lambda/d\varphi = \tan\alpha/\cos\varphi$ and $ds/d\varphi = R/\cos\alpha$ in which the variables can easily be separated. By integration the first yields the equation of the loxodrome:

$$d\lambda = \tan\alpha(d\varphi/\cos\varphi), \quad \lambda = \tan\alpha[\ln\tan(\pi/4 + \varphi/2) + C],$$
$$\lambda_2 - \lambda_1 = \tan\alpha[\ln\tan(\pi/4 + \varphi_2/2) - \ln(\pi/4 + \varphi_1/2)].$$

12.3-2 Derivation of the loxodrome

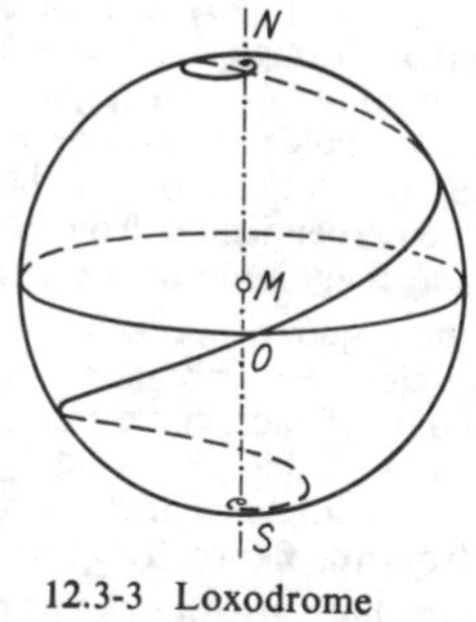

12.3-3 Loxodrome

The second gives the arc length s of the loxodrome:

$$ds = (R/\cos\alpha)\,d\varphi, \quad s = (R/\cos\alpha)(\varphi_2 - \varphi_1).$$

As a closer examination of the first equation shows, the loxodrome circulates about the pole of the great circle that cuts the initial meridian at right angles at the starting point, and spirals around it in ever smaller windings infinitely often, without reaching it (an *asymptotic point*). The change in the latitude during one rotation becomes steadily smaller (Fig.).

From the second equation, the flight path of an aircraft from a point on the equator where it cuts the meridian at the course angle α to the point at which it reaches latitude φ, has the length $s = R\varphi/\cos\alpha$. It is therefore the longer, the greater the angle α. For a flight to the north pole ($\varphi = \pi/2$) at a constant bearing $\alpha = 60°$, for example, it follows that $s = 2R\pi/2$ (since $\cos 60° = 1/2$), while the shortest route along a meridian is of length $s = R\pi/2$. Thus, the path along the loxodrome is twice as long.

Determination of position by a fix. By means of a fix, the position of a ship, aircraft, or other body is determined from the directions from which signals are received that are propagated in straight lines and are as a rule not optical. Geometrically this is based on the same scheme as forward and backward sections in plane trigonometry. There are two kinds of fix: in one the directions of signals emitted from the object to be located are determined by two fixed ground stations and from these the coordinates of the object are calculated, and in the other the signals are emitted by two known ground stations and their directions observed and the position calculated at the object itself.

In practice radio signals are almost always used. Whereas in surveying the precision of the angles can be increased by repeated measurements and the most probable value for the result calculated by the method of last squares, the fix rests upon a single measurement of the directions, which are of lesser precision. For this reason physical properties of the waves employed, for example, interference, oscillations, or other methods such as radar, are also made use of. Above all, one is almost always concerned with a moving object. The position must therefore be determined by means of tables, by graphical methods, or by electronic apparatus, so that the result is available while the object of the fix is still in the neighbourhood of the position required. Thus, in practice, making a fix has become a physical and technical problem. It is sufficient to describe here a simple graphical procedure.

A graphical method of making a fix. In order to obtain a clear picture, let the basis b be 60°. Let the measured angles β_1 and β_2 between the basis and the direction of the new point C at the points B_1 and B_2 be $\beta_1 = 60°$ and $\beta_2 = 110°$. The plane projection of the earth can be chosen in such a way that the image of the great circle through B_1 and B_2 is a circle (Fig.); $\sphericalangle B_2MB_1 = 60°$. The projection of the great circle through B_1 (and its diametrically opposite point B_1') and also through the required point C is an ellipse, whose major axis is $|B_1B_1'|$ and whose minor axis is therefore the per-

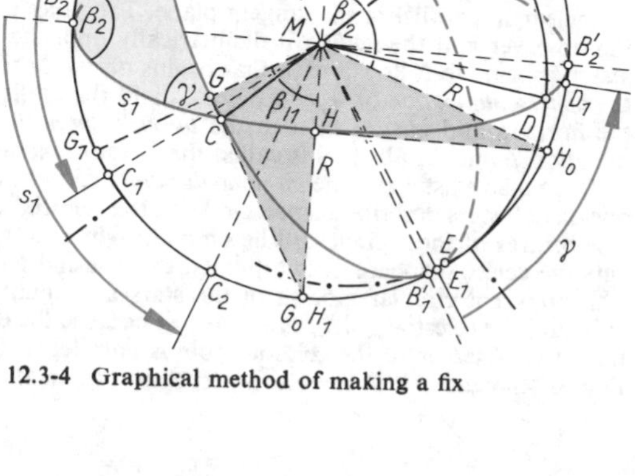

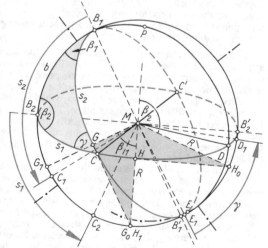

12.3-4 Graphical method of making a fix

pendicular to B_1B_1' through M. The length of the semi-minor axis is the projection of the radius $R = |MG'|$ of the sphere, and is given as the intersection of two planes: firstly, the plane of the great circle through B_1CB_1' that is inclined at an angle β_1 to the plane of the diagram Π, and secondly, the plane through M is perpendicular to Π and $B_1B'_1$. If G is the projection of G', then $\Delta MG'G$ is a right-angled triangle whose hypotenuse $|MG'| = R$ and angle $G'MG = \beta_1$ are known, where MG is perpendicular to B_1B_1'. In the figure this triangle is folded into $\triangle MGG_0$ in the plane of the diagram, and gives the length $|MG|$ of the semi-minor axis (in general $G_0 \neq H_1$). From the semi-major axis MB_1 and the semi-minor axis MG every point on the ellipse (the projection of the great circle through B_1, G and B_1') can be constructed with arbitrary precision. For the great circle through the points B_2 and B_2', whose plane is inclined at an angle β_2 to the plane of the diagram, a similar result holds. Thus, one makes the following sequence of constructions: the perpendicular at M to B_2B_2'; marking out of the angle β_2; the intersection H_0 of its free arm with the circle of radius R; the perpendicular from H_0 to the perpendicular at M to B_2B_2' gives H; then $|MH|$ gives the position and magnitude of the semi-minor axis of the required ellipse. The intersection C of two ellipses is the required point. To determine the true values of the sides $s_1 = \widehat{B_2C}$ and $s_2 = \widehat{B_1C}$ their great circles need only be rotated about a diameter into the plane of the diagram. The projection of each point of each circle moves on a perpendicular to the axis of rotation. Thus, C moves to C_1 or C_2 and $\sphericalangle B_1MC_1 = s_2$, $\sphericalangle B_2MC_2 = s_1$.

The angle of inclination γ of the planes of the two great circles indicated can also be deduced from the figure. The two planes intersect in the straight line CMC'. The polar circle with C as pole cuts both great circles at right angles, in the points D and E. The arc $\widehat{DE}$ corresponds to the angle γ. By rotating this polar circle about a diameter into the plane of the diagram the true magnitude of the angle γ can be read off: $\sphericalangle E_1MD_1 = \gamma$.

Spherical astronomy

Apart from the method of the fix, the positions of ships and aircraft are, even today, found by means of the stars. It was once the only method of navigation on the high seas. Explorers in unknown lands relied on them alone. The necessary measurements were made with the compass, the theodolite, a mirror sextant or similar angle-measuring instrument and an accurate clock. Later wireless telegraphy was used to transmit time signals to check the clocks. Knowledge of the most important constellations is enough for an approximate orientation. For precise determination of position one must know data concerning the position of easily located stars and the motion of the sun, the planets, the moon and Jupiter's moons, and the astronomical coordinate systems in which positions in the heavens are given. The data from spherical astronomy that are important for the purposes of navigation appear in the nautical and astronomical almanacs; the astronomical coordinate systems that are indispensable for navigation are the *horizontal* and the *equatorial* systems.

Like all astronomical coordinate systems, these are based on the fact that the starry sky appears to an observer as a portion of a gigantic sphere called the celestial sphere. The position of each point on it can be fixed by two numerical coordinates (corresponding to longitude and latitude on the earth's surface). Any great circle with its poles (pole and polar) is suitable as a reference system for these two coordinates. One angle is measured on this circle in a prescribed sense from a fixed point; the second is measured on a perpendicular great circle through the point whose position is to be fixed and the pole of the basic circle.

The horizontal system. To an observer O on the sea or in flat country the night sky appears as a hemisphere bounded by the horizon H (Fig.). Mathematically the (apparent) *horizon* is the circle in which a tangent plane to the earth at the point of observation cuts the celestial sphere. In relation to the distances of most of the stars the radius of the earth is negligibly small. The apparent horizon therefore coincides with the true horizon, which is the intersection of a plane through the centre of the earth, parallel to the tangent plane. The poles of the horizon are the *zenith Z* vertically above the observer and the *nadir Na* diametrically opposite to the zenith. To the observer the motion of a star appears to follow a path that begins on the horizon (it is said to rise at A), ascends to a peak, the *upper culmination* or *transit* point C, and then falls again to *descend* at D and finally sinks below the horizon and passes through the *lower culmination* point lC. The great circle through the culmination points of all stars is called the *celestial meridian m*.

There also exist stars, the *circumpolar stars cS*, whose path lies entirely above the horizon. During one day all stars describe a small circle on the celestial sphere. Their paths are parallel to one another. The centres of these circles all lie on a straight line that forms the axis of the celestial sphere. This cuts the celestial sphere in two points, the celestial North Pole P_N and the celestial South Pole P_S. This apparent circular motion of the stars is a consequence of the rotation of the earth about its axis, and the celestial poles remain at rest because the earth's axis points towards them. The direction from the observer to the celestial pole is parallel to the earth's axis; for an observer at the North Pole of the earth it is therefore perpendicular to the horizon and for an observer on the earth's equator it is horizontal. If one imagines that the tangent plane slides along an earth's meridian from the equator to the North Pole, then the altitude of the celestial pole increases steadily from 0° to 90° and is always equal to the geographical latitude. The *altitudes h* of a star *St* are measured on the great circles through zenith and nadir, the *verticals V*, which are perpendicular to the horizon, and vary from 0° on the horizon to +90° at the zenith and −90° at the nadir. Measurement of the altitude of the celestial pole gives the geographical latitude φ of the observer. The intersections of the vertical through the celestial pole with the horizon are called the *north point N* and the *south point S* diametrically opposite to it. If the observer looks towards the north point, then to his right, at right-angles to the line of vision, is the east point *E* and to his left, the west point *W*. These four points are called the *cardinal points* and their directions are the celestial directions north, south, east and west. They can be determined by dropping a perpendicular from the celestial North Pole or by determination of the vertical on which an arbitrary fixed star culminates; this bisects the angle between two verticals on which the fixed star has the same altitude.

Together with the altitude *h*, the *azimuth a* serves as second coordinate. This is measured at the position of the observer as the angle between the meridian plane and the vertical plane of the stars *St*, and varies from 0° at the south point to 360°, in the sense west, north, east of the (apparent) daily motion of the star. Consequently the azimuth appears also as an arc on the horizon and as an angle at the zenith point. In place of the altitude the complementary angle is often measured, the *zenith distance z* of the star from the zenith: $h + z = 90°$.

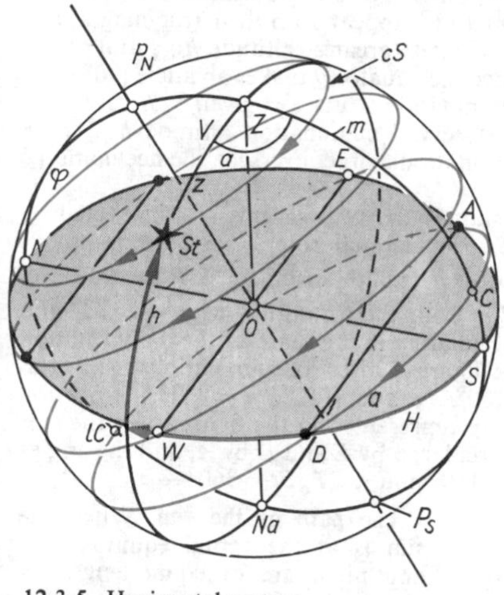

12.3-5 Horizontal system

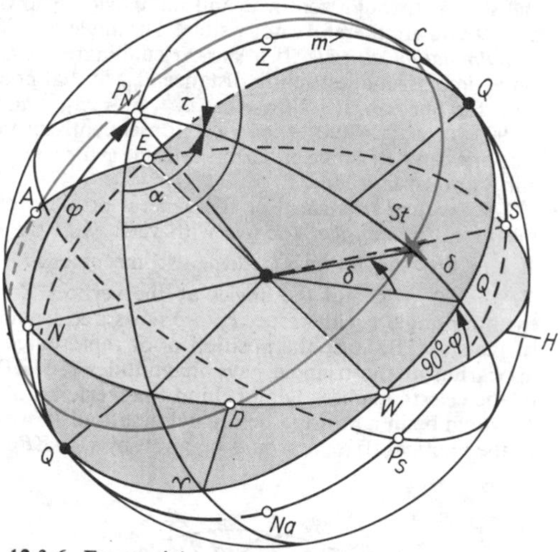

12.3-6 Equatorial system

Equatorial systems. Because all stars move on parallel circles about the celestial pole, their distance from each of these circles must remain constant. As circle of reference one chooses the great circle among them, which is polar to the celestial pole. It is called the *celestial equator Q*, because it is the line of intersection of the plane of the earth's equator with the celestial sphere. On the sphere it runs approximately through the constellation Pisces, the upper one of the three stars in Orion's belt and the star Altair in the constellation Aquila. The equator cuts the horizon at the west point and at the east point (Fig.) and is inclined at an angle $(90° - \varphi)$ to the horizon *H*. The altitude of a star *St* above the equator is called its declination δ; it is measured on a great circle called the *hour circle*, which passes through the *celestial poles* P_N and P_S and is therefore perpendicular to the equator. The *hour angle* τ is taken as the second coordinate. This is the angle between this circle and the *celestial meridian m* on which the star culminates. This meridian passes through the south and north points, the zenith *Z* and nadir *Na*, and through the celestial poles P_N and P_S. It represents the intersection of the plane of the observer's meridian with the celestial sphere. The hour angle is measured from the meridian in the sense of the apparent daily motion of the stars and takes values from 0° to 360°, or from 0^h to 24^h. Thus, the west point W has an hour angle of 90° or 6^h.

This *first equatorial* or *hour angle* system is independent of the geographical latitude of the observer's position, because the declination is referred to the equator. The zero direction from which the hour angle is measured is, however, determined by the meridian of the observer and thus depends on his geographical longitude. The hour angle of the same star at the same time is, for example, greater for Moscow than for London, since the star culminates about 2 h 29.2 min = 149.2 min earlier in Moscow than in London because of the rotation of the earth. Since 24 h corresponds to an angle of 360°, the difference between the hour angles is 149.2°/4 = 37.3°. Thus, Moscow lies further east than London by $\Delta\lambda = 37.3°$. To make the second coordinate in the equatorial system independent of the position of the observer, one selects a reference point on the celestial equator. This point, denoted by ♈, is called the *vernal equinox* (or *first point of Aries*). As a point of the equator it takes part in the apparent rotation of the celestial sphere. The angle measured from it along the equator in the opposite sense to that of the apparent rotation is therefore constant. It is called the *right ascension* α. Right ascension α and declination δ are the coordinates of the *second equatorial system* (or *right ascension system*). The approximate position of the vernal equinox is found by extending the hour circle from the pole star (P_N) through the right-hand end of the W-shaped constellation Cassiopeia to meet the celestial equator.

The relations between the horizontal and the hour angle systems. If one combines the two astronomical systems (Fig.), the horizon and the equator intersect at the east and west points. Through the star St pass the hour circle and the vertical. The path of the star runs parallel to the equator; it reaches its upper and lower culmination points at C and lC, respectively. A is the point at which it ascends, D is the point at which it descends. The altitude of the celestial pole P_N above the horizontal plane is the geographical latitude of the position O of the observer; $\widehat{NP_N} = \varphi$. The figure arises as the orthogonal projection of the figure *Equatorial systems* on the plane of the meridian m through N, P_N, Z, C, Q and S. The hour circle of the vernal equinox ♈ is not indicated, but the vertical of the star St through zenith Z and nadir Na is shown. The points E and A lie behind the points W and D and are therefore not visible. The angles φ, $(90° - \varphi)$ and δ appear with their true magnitude.

Culmination altitude. If a star St culminates at C, it reaches its greatest altitude h_{max} and at the same time its smallest zenith distance z_{min} at that point. Since the equator Q makes an angle $(90° - \varphi)$ with the horizon, it follows that $\varphi = \delta + z_{min}$ and the culmination altitude is given by $h_{max} + z_{min} = 90°$, or $h_{max} = 90° - \varphi + \delta$. Consequently from the observed culmination altitude h_{max} of a star one can determine either the latitude φ for a known declination δ or, conversely, the declination δ for a known latitude φ.

The nautical triangle. For the general position of the star St the two systems are connected by means of the *nautical triangle* with vertices at the star St, the celestial pole P_N and the zenith Z. It contains the following elements: the sides $\widehat{StZ} = 90° - h$ (zenith distance), $\widehat{StP_N} = 90° - \delta$, $\widehat{ZP_N} = 90° - \varphi$ and the angles at the vertices Z and P_N. Both the azimuth a with vertex Z and the hour angle τ with vertex P_N are measured from the meridian m in the sense of the daily rotation of the stars. Because the position of St represented in the figure is after its culmination, the angles appearing in the triangle have magnitudes $\sphericalangle StZP_N = 180°\ \ a$ and $\sphericalangle ZP_NSt = \tau$. If the portion of the celestial sphere lying behind the meridian plane were represented in the figure, then the star St would be in a position before culmination, W would be replaced by E and D by A, and the angles in the nautical triangle would be given by $\sphericalangle StZP_N = a\ \ 180°$ and $\sphericalangle ZP_NSt = 360° - \tau$.

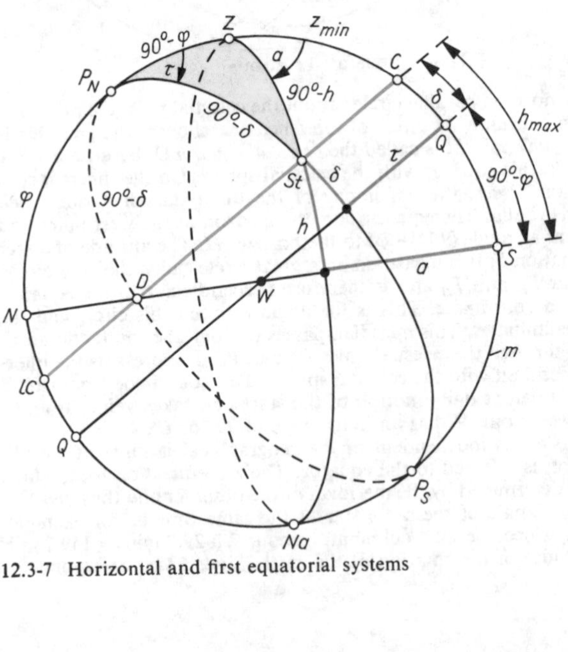

12.3-7 Horizontal and first equatorial systems

The path of the sun. When the sun is at the vernal equinox, day and night are of equal length; it rises at 6 o'clock in the morning at the east point, moves across the sky approximately along the celestial equator, and sets at 6 o'clock in the evening at the west point. Its right ascension $\alpha_\odot$ and declination $\delta_\odot$, however, unlike those for all fixed stars, are not constant. The right ascension increases steadily and the declination decreases steadily from December 22 to June 22. Because of the increasing right ascension (Fig.) the sun reaches the celestial meridian every day later than the vernal equinox. In the course of a year this time delay grows to one full day. Whereas the vernal equinox and all the fixed stars culminate 366 times, the sun culminates only 365 times. Because of the increasing declination of the sun, its ascending point A and descending point D shift to the north, to A_1 and D_1.

The days become longer until the summer solstice. The sun then has its greatest declination of $\delta = 23°26'$ (circle of rotation of Cancer). After this declination decreases, is zero at the autumnal equinox, at the winter solstice is $-23°26'$ (circle of rotation of Capricorn), and at the vernal equinox it is again zero. Altogether the apparent path of the sun in the sky is not a circle, as for the other fixed stars, but a spiral of 365 windings described twice, occupying a zone of width $2 \cdot 23°26'$. While each fixed star, almost without exception, has the same neighbouring stars throughout the course of a year (with noticeable deviations in only a few cases), the sun wanders through 13 constellations, which were reduced to 12 on account of the duodecimal system. These constellations lie on a great circle in the neighbourhood of the apparent yearly path of the sun, called the *ecliptic*. The constellations are Aries, the Ram ♈; Taurus, the Bull ♉; Gemini, the Twins ♊; Cancer, the Crab ♋; Leo, the Lion ♌; Virgo, the Virgin ♍; Libra, the Scales ♎; Scorpio, the Scorpion ♏; Sagittarius, the Archer ♐; Capricornus, the Goat ♑; Aquarius, the Watercarrier ♒, and Pisces, the Fishes ♓. The ecliptic cuts the equator at an angle of $\varepsilon = 23°26'$, at the vernal equinox and its diametrically opposite point, the autumnal equinox.

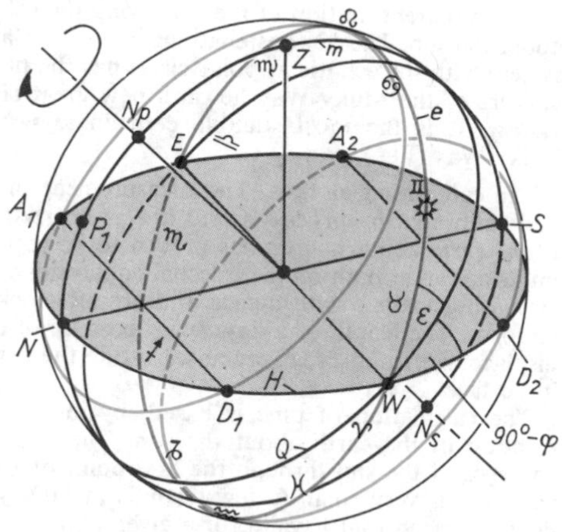

12.3-8 Apparent motion of the sun in the course of a year

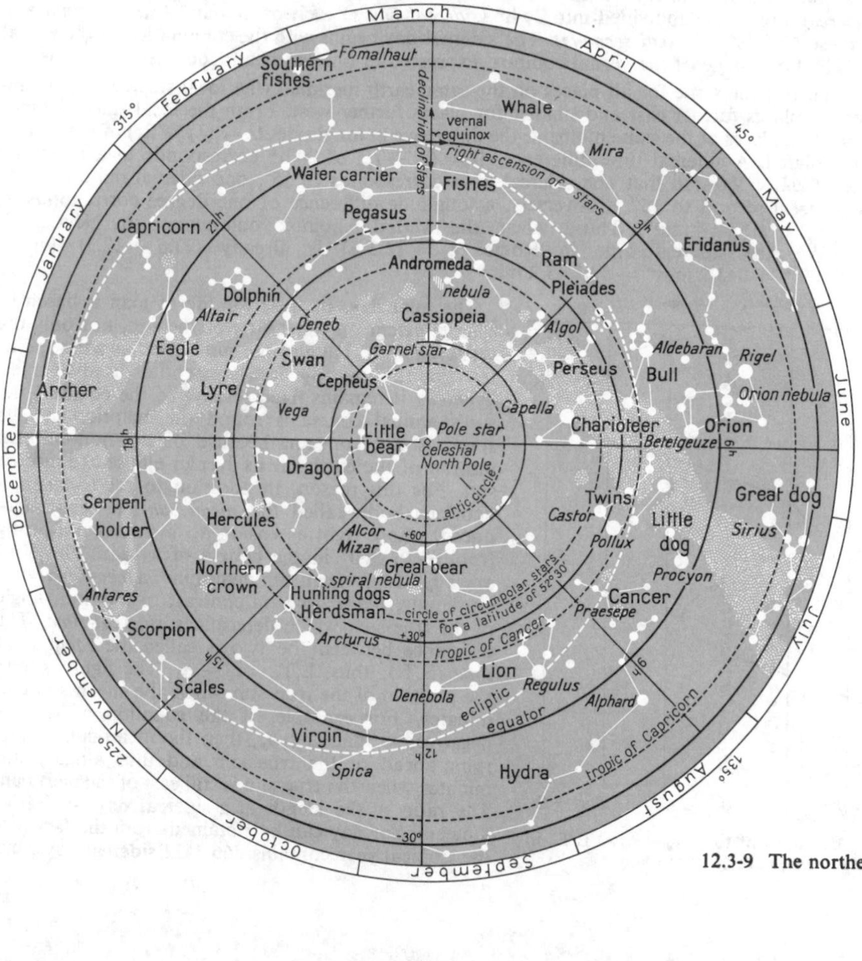

12.3-9 The northern sky

This apparent motion of the sun along the ecliptic is the consequence of the motion of the earth about the sun. The 12 constellations lie in the plane of the earth's orbit about the sun. A coordinate system with the ecliptic as polar circle has the plane of the earth's orbit as reference plane. The star clusters of the Milky Way lie on a new great circle that forms the reference plane of the *galactic system*. It is the most suitable coordinate system to describe the distribution of the stars of the Milky Way (Fig.).

The calculation of time. The measurement of time intervals requires clocks that are controlled and calibrated by processes that are as nearly as possible constant, mostly periodic. The *rotation of the earth* about its axis has proved to be very uniform; a fixed star or the first point of Aries ♈ on its apparent path on the celestial equator can serve as the pointer of a very accurate clock. However, observations with quartz and atomic clocks have shown that this rotation is not completely uniform. The length of a day varies because of tidal friction and changes irregularly through mass displacements and other processes inside the earth, as well as through meteorological processes on its surface.

The calculation of time is based, by international agreement, on the duration of the tropical rotation of the earth about the sun. The *tropical year* denotes the time between two successive passages of the sun through the first point of Aries. However, this period is also variable, but the variation is very small (a few seconds in 1000 years) and its magnitude is known. By choice of a definite period that is valid for a given point in time, a definite tropical year is chosen as a normal year. The time based on this is for purposes of calculation absolutely uniform and is called *Ephemeris* or *Newtonian time*, because it was used in astronomy for the calculation of the coordinates of the heavenly bodies called the Ephemerides. Accordingly the second, s, is fixed as the 31 556 925.9747th part of the tropical year for 1900, January 0, 12 o'clock Ephemeris time; according to the calendar 1900, January 0 is 31. 12. 1899.

Sidereal time. The time interval between two successive culminations of the first point of Aries ♈ is the *sidereal day*. It is subdivided into 24 h* (*sidereal hours*) each of 60 min* (*sidereal minutes*), each of these of 60 s* (*sidereal seconds*). The sidereal day begins with the culmination of the vernal equinox. The hour angle of the vernal equinox, expressed in time units and denoted by t_{Υ}, is the sidereal time. It is the same for all places on the same earth meridian (the local sidereal time), and is greater for places further east and smaller for places further west. From the local sidereal times t_1 and t_2 of two places at the same moment, the difference in longitude $\Delta\lambda = (\lambda_2 - \lambda_1)$ of two places can be calculated. A sidereal time difference $\Delta t = (t_2 - t_1)$ of 24 h* corresponds to a longitude difference of $\Delta\lambda = 360°$, so that one sidereal hour corresponds to 15°, one sidereal minute to 15′ and one sidereal second to 15″. Conversely, a longitude difference of one degree corresponds to 24 h*/360 = 1 h*/15 = 4 min*. Thus, when the vernal equinox culminates in New York ($\lambda_2 = 73.5$ W), the sidereal time in Rome ($\lambda_1 = 12.3$ E) is already (73.5 + 12.3) · 4 min* = 85.8 · 4 min* = 343.2 min* = 5 h* 43.2 min*.

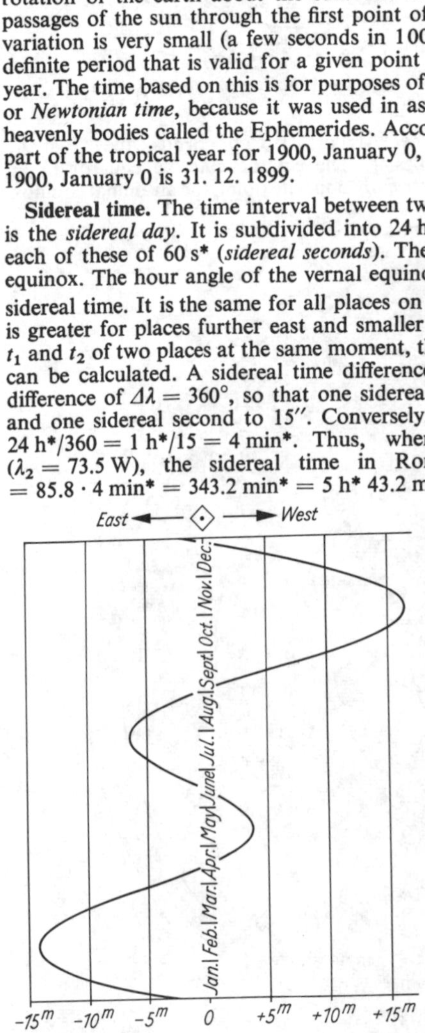

12.3-10 Equation of time, $t_s - t_m$ = E.T., in the course of a year; ⊙ mean sun

The solar day. Because the life of man is based to a large extent on the course of the sun, not only the vernal equinox, but also the sun is used as a time pointer. For this purpose it has essential disadvantages. Whereas it remains fixed relative to the equator, the sun's annual course is round the ecliptic, and its speed is not constant because of the non-uniform motion of the earth on its Kepler ellipse around the sun. For this reason, in addition to the true sun, a fictitious body called the *mean sun* has been introduced, whose right ascension α_m increases uniformly from 0° to 360° in the course of a year. The hour angle t_m of this mean sun determines a *mean solar time* or, simply, *mean time*, in contrast to the hour angle of the true sun, which determines *sidereal time*. The difference between the two is called the *equation of time* (E.T.): thus, E.T. = $t_s - t_m$. The figure shows the position of the true sun when the mean sun culminates. For example, if the equation of time is negative, so that $t_m > t_s$, then the mean sun is hurrying ahead of the true sun and thus already culminates when the true sun is still east of the meridian. The ratio of the length of a sidereal day to that of a mean solar day can be obtained from the fact that the tropical year contains 366.2422 sidereal days, but

365.2422 mean solar days. One obtains: 24 h mean solar time = 24 h* 3 min* 56.55536 s*; 24 h* = 23 h 56 min 4.09058 s mean solar time; 1 h mean solar time = 1.002737909 h*; 1 h* = 0.997269567 h mean solar time.

Time zones. Of course, both true and mean time are *local times*, and only places lying on the same earth meridian have the same local time. This fact of nature, so inconvenient for modern traffic, is made tolerable by dividing the earth's surface into zones bounded by meridians of longitude at intervals 15° apart, and using the local mean time appropriate to the central meridian at all places within a zone. The local mean time of the meridian of Greenwich, $\lambda = 0$, is called *Universal Time* or *Greenwich Mean Time* (G.M.T.); that of the meridian $\lambda = -15°$ or $\lambda = 15°$ E is Mid-European Time (M.E.T.).

Example 1: On the morning of November 18 a ship lies at latitude $\varphi = 54°57'$ N. The altitude of the sun is observed to be $h_s = 9°15'$. The ship's chronometer gives G.M.T. = $8^{h}58^{min}20^{s}$, the nautical almanac $\delta_s = -19°12'$ and the equation of time +14 min 50 s. On which meridian does the ship lie? –

In the nautical triangle zenith Z — pole P_N — sun S the three sides are known: $\widehat{ZS} = 90° - h = 80°45'$, $\widehat{ZP_N} = 90° - \varphi = 35°03'$, $\widehat{SP_N} = 90° - \delta = 109°12'$. For the difference t' between the hour angle t and 360° the half-angle formula gives

$$\sin\frac{1}{2}t' = \sqrt{\left[\frac{\sin(s - [90° - \varphi])\sin(s - [90° - \delta])}{\sin(90° - \varphi)\sin(90° - \delta)}\right]}.$$

		lg sin
$90° - h =$	$80°45'$	9.98950
$90° - \varphi =$	$35°03'$	+8.76015
$90° - \delta =$	$109°12'$	8.74965
$2s =$	$225°00'$	−9.75913
$s =$	$112°30'$	−9.97515
$s - [90° - \varphi] =$	$77°27'$	9.01537
$s - [90° - \delta] =$	$3°18'$	9.50768
$t'/2 =$	$18°46'34''$	
$t' =$	$37°33'8'' = 2$ h 30 min 13 s	

Thus, the observation fixes the time as 2 h 30 min 13 s before the culmination of the true sun, that is, at $12^{h} - 2^{h}30^{min}13^{s} = 9^{h}29^{min}47^{s}$. But the local mean time is 14 min 50 s less, or $9^{h}14^{min}57^{s}$. The difference between this and G.M.T. is $9^{h}14^{min}57^{s} - 8^{h}58^{min}20^{s} = 16$ min 37 s, or $(16°37')/4 = 4°9'15''$. The ship's position is $\lambda = 4°9'15''$ E and 54°57′ N.

Example 2: On a ship travelling on a calm sea north of the equator the sun's altitude of $h_1 = 21.7°$ is measured at $18^{h}50^{min}$ G.M.T. The declination of the sun is obtained from nautical almanac as $\delta_1 = -10.15°$ and the equation of time as E.T. = +15 min 3 s. After steaming for 15.2 nautical miles on the great circle determined by the course N 67.5° W, the ship observes that the sun culminates at an altitude of $h_2 = 35°$ with a declination $\delta_2 = -10.21°$ (Fig.). What are the coordinates of the two positions of observation? – The culmination altitude h_2 of the sun satisfies:

$$h_{max} = h_2 = 90° - \varphi_2 + \delta_2 \quad \text{or} \quad \varphi_2 = 90° + \delta_2 - h_2, \quad \text{that is,} \quad \varphi_2 = 44.79°.$$

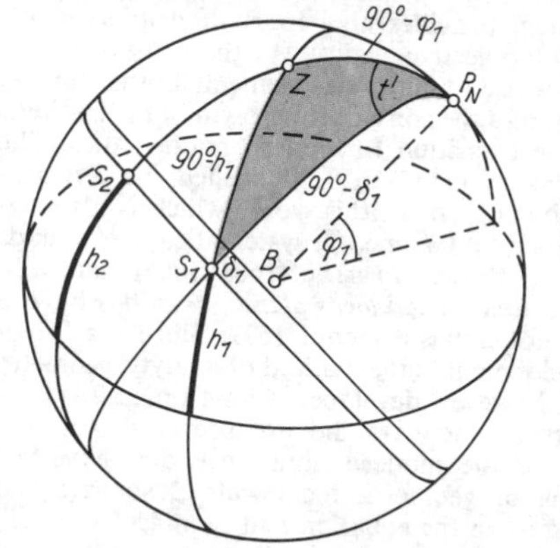

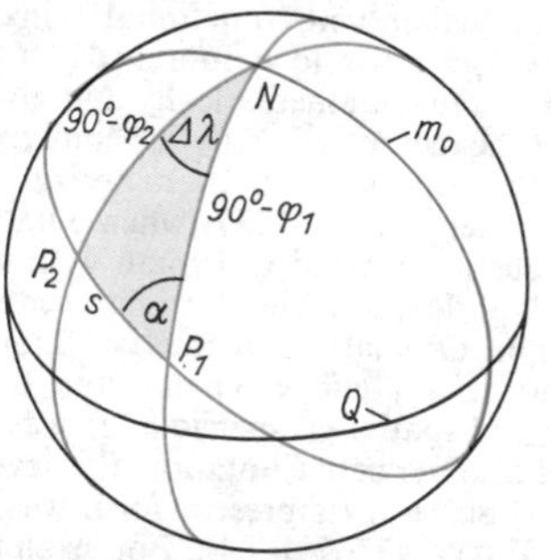

12.3-11 Schematic representation for Example 2 of the sky (left) and of the earth's surface (right), Q equator, m_0 meridian of Greenwich

The two observation points P_1 and P_2, together with the North Pole N determine a spherical triangle P_1NP_2 on the earth's surface. With a course angle $\alpha = 67.5°$ at the point P_1, the ship has travelled a distance $\widehat{P_1P_2} = 15.2$ nautical miles $= 15.2 \cdot 1.852$ km between the observation points, and this corresponds to an arc $s = \dfrac{360 \cdot 15.2 \cdot 1.852}{2\pi R} = 0.253°$. The side $\widehat{P_2N}$ opposite the course angle α is $90° - \varphi_2 = 45.21°$; the sine rule, $\sin\Delta\lambda = \sin s \sin\alpha / \sin(90° - \varphi_2)$, gives

$\Delta\lambda = 0.329°$. In the same triangle Napier's analogy 2a) gives:

$$\tan[(90° - \varphi_1)/2] = \tan[(90° - \varphi_2 - s)/2] \sin[(\alpha + \Delta\lambda)/2]/\sin[(\alpha - \Delta\lambda)/2],$$

and hence $90° - \varphi_1 = 45.3°$, $\varphi_1 = 44.7°$.

In the nautical triangle ZP_NS_1 of the first observation point the three sides $\widehat{ZS_1} = 90° - h_1$, $\widehat{ZP_N} = 90° - \varphi_1$ and $\widehat{P_NS_1} = 90° - \delta_1$ are known. By the cosine rule for sides the difference t' between 360° and the hour angle t can be calculated:

$$\cos t' = (\sin h_1 - \sin\varphi_1 \sin\delta_1)/(\cos\varphi_1 \cos\delta_1).$$

One obtains $t' = 45.13° = 3.01$ h = 3 h 0 min 36 s. At the first observation it was $12^h - 3^h0^{min}36^s = 8^h59^{min}24^s$ true local time, or $8^h44^{min}21^s$ local mean time, since $t_m = t_s -$ E.T. Relative to the local mean time of Greenwich the time difference is $18^h50^{min} - 8^h44^{min}21^s = 10$ h 05 min 39 s, or 10.094 h. Consequently the difference in longitude is $10.094 \cdot 15° = 151.41°$. Thus, Greenwich lies east of P_1 and the longitude of P_1 is $\lambda_1 = 151.41°$ W and that of P_2 is $\lambda_2 = \lambda_1 + \Delta\lambda = 151.74°$.

13. Analytic geometry of the plane

The main idea of analytic geometry is that geometric investigations can be carried out by means of algebraic calculations. This method has proved extraordinarily fruitful. The fusion of geometric and algebraic thinking, together with functional thinking, provides an important help to man's understanding of the exploration and comprehension of objective reality. At the same time the method is particularly attractive mathematically and gives rise to important elements in the training of the mind. The birth of the method of analytic geometry, and the consequent growth of the methods of the differential and integral calculus, characterize the transition to modern mathematics. The year of birth can be taken to be 1637, when DESCARTES (1596–1650) published his *Discours de la Méthode* anonymously, to avoid a dispute with the church. In this work, which is also significant for the history of philosophy, the third part, entitled La Géométrie, systematically expounds the fundamental principle of analytic geometry. Shortly before, FERMAT (1601–1665) had also worked out the method of analytic geometry, but his treatise *Ad locos planos et solidos isagoge* (Introduction to planar and spatial geometric loci) was not published until 1679. Since the 'Geometry' of Descartes had also the better notation, the development of the method of analytic geometry is usually attributed to Descartes. Its present form was, however, developed a long time after Descartes, particularly by EULER (1707–1783). For example, DESCARTES did not use two axes, and only since the time of Euler, to whom a large part of the modern notation is due, have far-reaching conclusions been drawn from the equations of geometric loci, while DESCARTES and FERMAT generally regarded their investigations as ending when the equation had been set up.

13.1. Plane coordinate systems

The fusion of geometric and algebraic thinking is attained by regarding geometric figures as sets of points and by assigning numerical quantities to each point, which distinguish it from other points. A curve or a line is then the carrier of a totality of points whose numerical quantities satisfy certain relations, which are called the equations of the figure, for example, the equation of an ellipse or a line. The graph of a linear equation in two variables is always a line, and that of a quadratic equation is a conic. The foundation of this construction of analytic geometry is the correspondence between points and numbers, which must be *one-to-one*. On a line, or more generally a curve, one number is sufficient to fix a point uniquely, on a plane or a surface a number pair, in space a number triple; conversely, a point on a curve uniquely determines one number, on a surface a number pair and in space a number triple. These numbers are called coordinates. They can be obtained in different ways; coordinate systems are the means of fixing them.

The number line. On a line the position of any point P is uniquely determined if a *zero point* O and a *unit segment* $u = O1$ are given on it. The integral multiples of the unit segment are obtained by repeatedly laying off u either from O beyond 1 in the *positive direction* or from 1 beyond O in the *negative direction* (Fig.). The end-points of the multiples correspond to the whole numbers, positive or negative. The point P is either an end-point, or it lies between two of the end-points, say n and $n+1$; there is always a real number x such that x times u is the distance $|OP|$ of the point P from the origin O. One has $n \leqslant x \leqslant n+1$ for positive x, and $-n' \geqslant x \geqslant -n'-1$ for negative x. The number x is the *coordinate* of the point P. Conversely, any real number x uniquely determines a point P of the number line by means of the equation $m(OP) = xu$, where $m(OP) = |OP|$ if $x > 0$ and $m(OP) = -|OP|$ if $x < 0$.

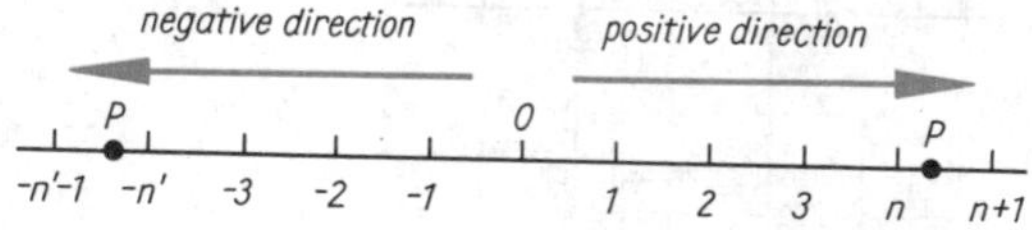

13.1-1 The number line

Parallel coordinate systems

Oblique parallel coordinates. To fix the position of a point in a plane two non-parallel number lines, with *origins* O and O' and unit segments $u = O1$ and $u' = O'1'$, are needed, because the plane has two dimensions. The lines are always arranged so that their zero points coincide, $O = O'$; they are called the *axes of the coordinate system*, and are usually called the x-axis, or axis of *abscissae*, and the y-axis, or *axis of ordinates* (latin *abscindere*, to cut off, *ordinare*, to order). If the axes enclose an angle $\alpha < 180°$, then in a *right-handed system* the notation for the axes is chosen so that a rotation of the $+x$-axis in the *mathematically positive sense* (anticlockwise) through the angle α leads to the $+y$-axis; in a *left-handed system* the opposite sense of rotation holds. The plane is divided by the coordinate axes into four regions. These *quadrants* are numbered I, II, III and IV, in the same sense of rotation as that of the coordinate system. If a point P lies in one of these quadrants, then a line can be drawn through it parallel to each coordinate axis, meeting the other axis in one point, the x-axis in P' and the y-axis in P'' (Fig.). The coordinates x and y of these points on the number lines are the *coordinates of the point P*. Different points lead to different number pairs (x, y). Conversely, for any number pair (a, b) there are two points P_a, P_b on the coordinate axes such that $m(OP_a) = au$ and $m(OP_b) = bu'$. The lines through these points P_a, P_b parallel to the other coordinate axes intersect in one point P, whose coordinates are a and b. Even if the point P lies on one of the coordinate axes, a number pair is necessary to determine its position as a point of the plane. It then coincides with P' or P'', while the point on the other axis coincides with the origin, and so its coordinate is zero. Points on the axis of abscissae have coordinates $(x, 0)$, and points on the axis of ordinates have coordinates $(0, y)$. In the number pair that characterizes the point P, the *x-coordinate* or *abscissa* always appears *in the first place*, and the *y-coordinate* or *ordinate in the second place*. To the origin there correspond the coordinates $(0, 0)$.

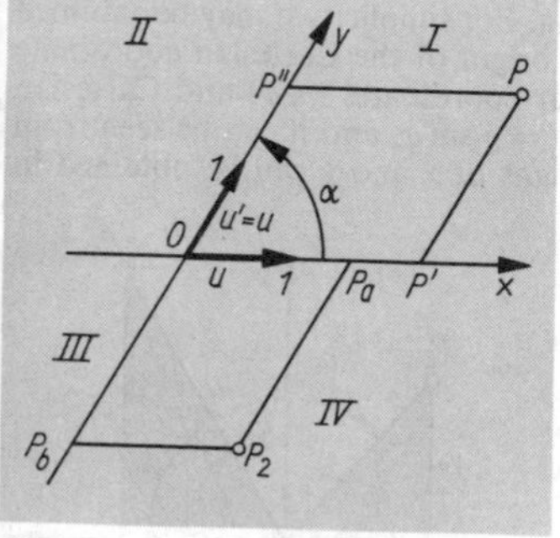

13.1-2 Oblique coordinate system

In each quadrant the coordinates have definite signs. For the given numbering of the quadrants, the corresponding signs are shown in the adjoining table.

Sign		Point lies
abscissa	ordinate	in quadrant
+	+	I
−	+	II
−	−	III
+	−	IV

Rectangular parallel coordinates, Cartesian coordinates. In a Cartesian coordinate system the coordinate axes are perpendicular to one another, and *the same unit of length* is chosen on the two axes. Also, the two parallels through a point P by means of which the corresponding coordinates are found are perpendicular to one another and to the coordinate axes. This rectangular coordinate system is used in the majority of cases, and as a rule a right-handed system, but in surveying a left-handed system is used (see Chapter 11.).

Example: In the figure the point P_1 has the coordinates $x_1 = +2$ and $y_1 = +3$. If a point P_2 is to be drawn with the coordinates $x_2 = -3/2$ and $y_2 = +5/4$, it can only have the given position. The origin has the coordinates (0, 0).

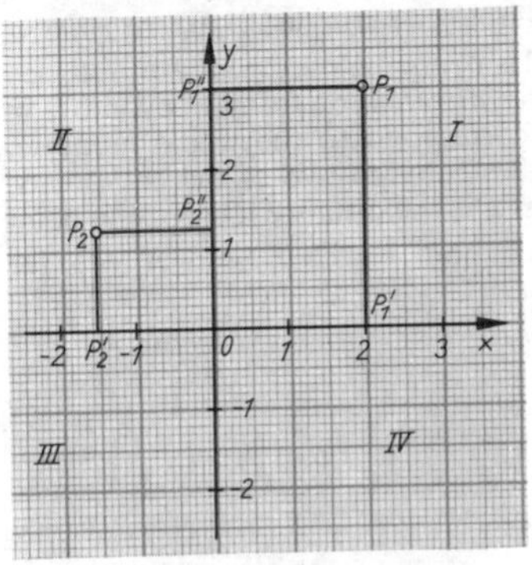

13.1-3 Rectangular parallel coordinates

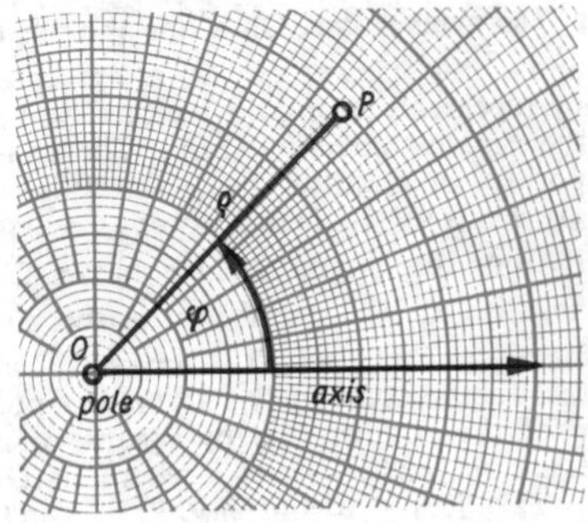

13.1-4 Polar coordinates of the point P: $\varphi = 45°$ and $\varrho = 4$

Polar coordinates

The polar coordinate system. A polar coordinate system is determined by a fixed point O, the *origin* or *pole*, and a *zero direction* or *axis* through it, on which positive lengths can be laid off and measured, as on a number line. An arbitrary point P of the plane can then be fixed firstly by the *angle* φ through which the axis must be rotated in the mathematically positive sense so as to pass through P, and secondly by the *positive distance* ϱ of the point P from the pole, measured along the number line (Fig.). The angle φ is called the ***argument, phase,*** or ***amplitude***; it can take values from 0° up to 360°; the length $|OP| = \varrho$ is called the ***radius***; it can only take non-negative values. For the point O itself, $\varrho = 0$ and φ is indeterminate.

Changing from one coordinate system to another

The same geometric figure, say a circle, can be described in two different coordinate systems C_1 and C_2, for example, in a Cartesian system and a polar coordinate system. For the same geometric properties two equations $f_1(x, y) = 0$ and $f_2(\xi, \eta) = 0$ are found. Instead of deriving each of the two functions from the geometric data, one can calculate one function from the other by means of the properties of the coordinate systems and their relative position. One then talks of a *transformation* from one system to the other. The equations of the transformation must obviously state how to calculate the coordinates (ξ, η) of a point in C_2 from the coordinates (x, y) of the same point in C_1, and conversely. If the ***equations of transformation*** are $x = t_1(\xi, \eta)$, $y = t_2(\xi, \eta)$, and their *inverses* $\xi = \tau_1(x, y)$, $\eta = \tau_2(x, y)$, then the equations $f_1(x, y) = 0$ and $f_2(\xi, \eta) = 0$ that describe the geometric figure are transformed into one another.

Transformation from polar coordinates to Cartesian and vice versa. For simplicity it may be assumed that the pole of the polar coordinate system coincides with the origin of the Cartesian coordinate system, and its axis with the x-axis. Then if a point P has polar coordinates (ϱ, φ) and Cartesian coordinates (x, y), the trigonometric relations give $x = \varrho \cos \varphi$, $y = \varrho \sin \varphi$, and it can be seen from the unit circle, in particular, that all possible combinations of signs of x and y can be obtained in the various quadrants by making φ take all values from zero to 2π (Fig.).

$x = \varrho \cos \varphi$ $y = \varrho \sin \varphi$	$x^2 + y^2 = \varrho^2$ $\varrho = \sqrt{(x^2 + y^2)}$	$\cos \varphi = x/\sqrt{(x^2 + y^2)}$ $\sin \varphi = y/\sqrt{(x^2 + y^2)}$

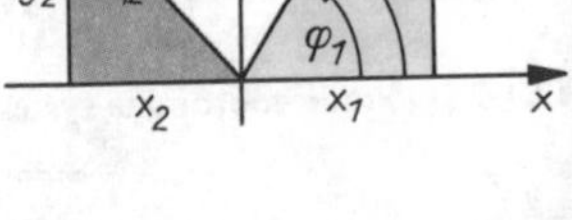

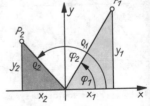

13.1-5 Relation between Cartesian and polar coordinates

Example 1: If P_1 has rectangular parallel coordinates (3, 4), then $\varrho_1 = \sqrt{(3^2 + 4^2)} = \sqrt{25} = 5$; $\cos \varphi_1 = 3/5 = 0.6$, $\sin \varphi_1 = 4/5 = 0.8$; from the trigonometric tables, $\varphi_1 = 53.13°$. Hence P_1 has polar coordinates $\varrho_1 = 5$ and $\varphi_1 = 53.13°$.

Example 2: P_2 has polar coordinates $\varrho_2 = 3$, $\varphi_2 = 120°$. Then the rectangular parallel coordinates of P_2 are $x_2 = 3 \cos 120°$, $y_2 = 3 \sin 120°$, and so from the values of the trigonometric tables, $x_2 = -3/2$, $y_2 = (3/2)\sqrt{3}$.

Example 3: In polar coordinates the equation of a circle with centre at the pole and radius r is given by $\varrho = r$, $0 \leqslant \varphi < 2\pi$. Without any further geometric considerations, the equation of the circle in Cartesian coordinates can be obtained by substitution from the equations of transformation: $\varrho = \sqrt{(x^2 + y^2)} = r$ or $x^2 + y^2 = r^2$.

Parallel displacement of a system of rectangular parallel coordinates. Two different Cartesian coordinate systems C_1 with coordinates x and y and C_2 with coordinates ξ and η are related in such a way that corresponding axes are parallel to one another and the origin O_2 of C_2 has coordinates (a, b) in C_1 (Fig.). The same point P then has coordinates (x, y) in C_1 and (ξ, η) in C_2, where $x = a + \xi$, $y = b + \eta$, or $\xi = x - a$, $\eta = y - b$.

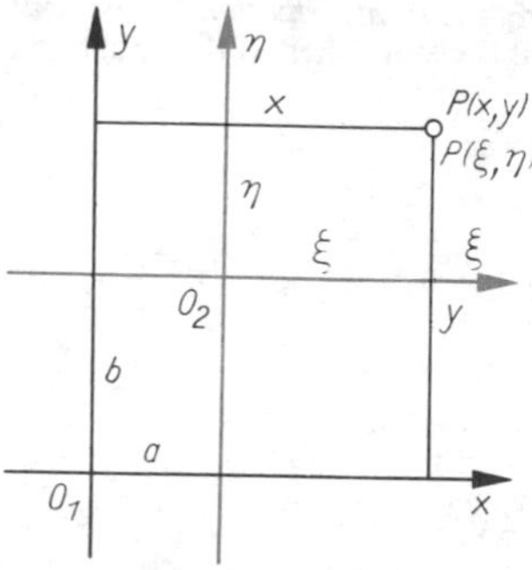

13.1-6 Two parallel rectangular coordinate systems displaced relative to one another

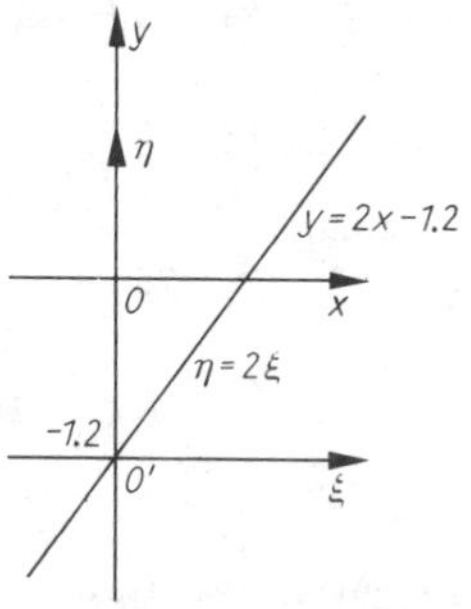

13.1-7 Transformation of the equation of a line

Coordinate transformation by parallel displacement	$x = a + \xi$ $y = b + \eta$
Inverse transformation	$\xi = x - a$ $\eta = y - b$

These transformation formulae always hold, irrespective of the quadrant in which the origin of the new system happens to lie; for example, if a and b are both positive, the displacement is upwards and to the right; if a and b are both negative, it is downwards and to the left.

Example 1: The (x, y)-system is to be transformed so that the origin of the (ξ, η)-system parallel to it is at the point $(4, -2.5)$, that is, $a = 4$, $b = -2.5$. The transformation equations are $x = 4 + \xi$, $y = -2.5 + \eta$.

Example 2: In the (x, y)-coordinate system there is a curve (a straight line) whose equation is $y = 2x - 1.2$ (Fig.). If one puts $x = 0$, one obtains its point of intersection with the y-axis, whose coordinates are $(0, -1.2)$. Let this point be the origin of a (ξ, η)-coordinate system, whose axes are parallel to those of the (x, y)-system. Since $a = 0$ and $b = -1.2$, the transformation equations are $x = \xi$, $y = \eta - 1.2$. They hold for every point of the plane. In the (ξ, η)-system the curve (line) therefore has the equation $\eta - 1.2 = 2\xi - 1.2$, or $\eta = 2\xi$. It can be seen that in this case the form of the equation is simplified by the transformation.

Rotation of a system of rectangular parallel coordinates. Suppose that the (x, y)-system of rectangular parallel coordinates is rotated (keeping the origin fixed) in the mathematically positive sense through an angle ψ into a (ξ, η)-system. Let a point P have the coordinates (x, y) in the old system, and (ξ, η) in the new system (Fig.).

For any angle ψ the projection of the ξ-coordinate on the x-axis has the value $\overrightarrow{OC} = \xi \cos \psi$. The η-axis is inclined to the x-axis at an angle $\psi + \pi/2$, and so the projection of the η-coordinate on the x-axis is $\overrightarrow{CA} = \eta \cos (\psi + \pi/2) = -\eta \sin \psi$, by a theorem of trigonometry. Hence, in the sense of *vector addition*: $x = \overrightarrow{OA} = \overrightarrow{OC} + \overrightarrow{CA} = \xi \cos \psi - \eta \sin \psi$.

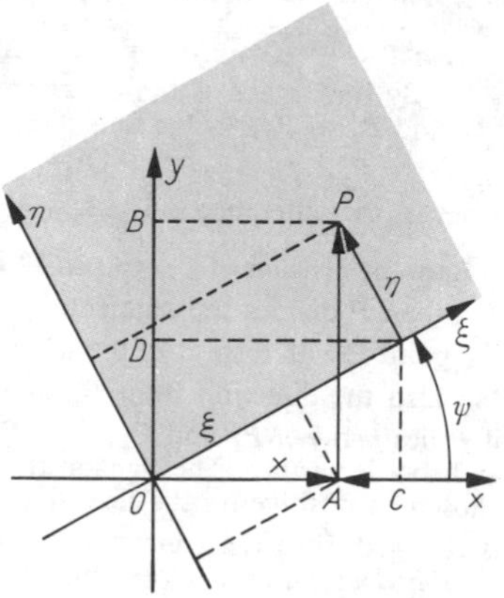

13.1-8 Rotation of the coordinate system

The inclination of the ξ-axis to the y-axis is $-(\pi/2-\psi)$ for any angle ψ; that of the η-axis to the y-axis is ψ. Hence, for the projection of the ξ-coordinate on the y-axis, $OD=\xi\cos(\psi-\pi/2)=\xi\sin\psi$ and for the projection of the η-coordinate on the y-axis, $DB=\eta\cos\psi$; hence the y-coordinate satisfies the transformation equation $y=OB=OD+DB=\xi\sin\psi+\eta\cos\psi$.

Coordinate transformation by rotation of the (x, y)-system through an angle ψ	$x=\xi\cos\psi-\eta\sin\psi$ $y=\xi\sin\psi+\eta\cos\psi$	$\xi=x\cos\psi+y\sin\psi$ $\eta=-x\sin\psi+y\cos\psi$

The formulae for ξ and η are obtained by rotating the (ξ,η)-system through an angle $-\psi$.

Example: What are the coordinates of the point $P(2, 4)$ in the coordinate system resulting from a rotation through 30°? – The old coordinates are $x=2$, $y=4$; since $\sin\psi=1/2$, $\cos\psi={}^1/_2\sqrt{3}$,

$$\xi=2\times{}^1/_2\sqrt{3}+4\times{}^1/_2=2+\sqrt{3},\quad \eta=-2\times{}^1/_2+4\times{}^1/_2\sqrt{3}=-1+2\sqrt{3}.$$

Hint. By parallel displacement of the coordinate system the absolute term of the equation of a curve can be eliminated, as in the example above. By means of a rotation it is always possible to *remove the mixed term xy* from an equation that is quadratic in the variables x and y (see Discussion of the general equation of the second degree). Here one is using the *transformation to principal axes*. Suppose, for example, that the equation $x^2+xy+y^2-3=0$ is given. Then, if the equations of the rotation through 45°,

$$x=\xi\cos 45°-\eta\sin 45°=(\xi-\eta)\cdot{}^1/_2\sqrt{2},$$
$$y=\xi\sin 45°+\eta\cos 45°=(\xi+\eta)\cdot{}^1/_2\sqrt{2}$$

are substituted into the equation, the new equation is

$${}^1/_2(\xi^2-2\xi\eta+\eta^2)+{}^1/_2(\xi^2-\eta^2)+{}^1/_2(\xi^2+2\xi\eta+\eta^2)-3=0,\quad\text{or}\quad 3\xi^2+\eta^2-6=0.$$

13.2. Point and line

Segment and ratio of division

Length of a segment. The length of a segment, that is, the distance between its two end-points, is measured in pure geometry by a ruler, but in analytic geometry it is calculated *from the coordinates of its end-points*. If the end-points P_1 and P_2 of the segment (Fig.) have rectangular parallel coordinates $P_1(x_1, y_1)$ and $P_2(x_2, y_2)$, then the length of the segment P_1P_2 is found by means of the theorem of Pythagoras.

Length of the segment P_1P_2, distance between the points P_1 and P_2
$\lvert P_1P_2\rvert=\sqrt{[(x_2-x_1)^2+(y_2-y_1)^2]}$

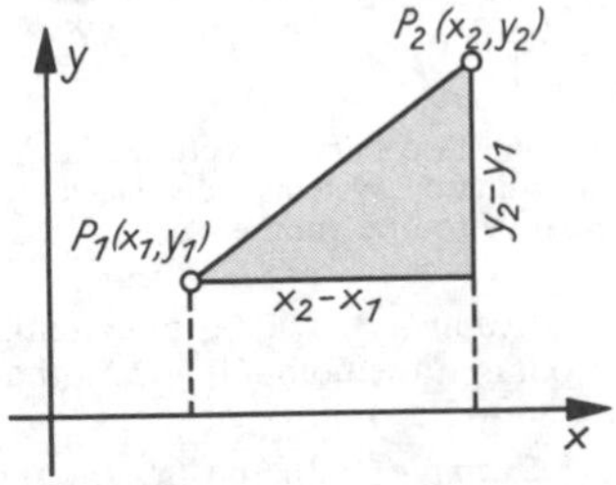

13.2-1 Distance between two points. Length of a segment

Examples: 1. Given $P_1(1, 8)$, $P_2(4, 2)$: $\lvert P_1P_2\rvert=\sqrt{[(4-1)^2+(2-8)^2]}=\sqrt{[3^2+(-6)^2]}=\sqrt{45}\approx 6.71$.

2. Given $P_3(-3, -2)$, $P_4(-6, -1)$: $\lvert P_3P_4\rvert=\sqrt{[(-6+3)^2+(-1+2)^2]}=\sqrt{10}\approx 3.16$.

3. By how much is the direct route from P_1 to P_4 shorter than the detour from P_1 through P_2 and P_3 to P_4? – One finds that $\lvert P_1P_4\rvert=\sqrt{130}\approx 11.40$ and

$$\lvert P_1P_2\rvert+\lvert P_2P_3\rvert+\lvert P_3P_4\rvert=\sqrt{45}+\sqrt{65}+\sqrt{10}\approx 6.71+8.06+3.16=17.93,$$

that is, the difference is $17.93-11.40=6.53$.

Ratio of division of a segment. If a point P lies on the segment P_1P_2 and does not coincide with P_2, then P divides the segment P_1P_2 in the ratio $P_1P: PP_2=\lambda$; λ is called the *ratio of division* of the point P with respect to the segment P_1P_2; P_1P denotes not only the length $\lvert P_1P\rvert$ of the segment, but also the direction from P_1 to P, that is, $P_1P=-PP_1$. The same holds for PP_2. Therefore, if P lies *between* P_1 and P_2, then P_1P and PP_2 have the same direction and hence the same sign, and so λ is *positive*. The signs of the two segments depend, as for any number line, on the direction chosen as positive on the line, that is, on the orientation of the line. If the orientation of the line is changed, then the segments P_1P and PP_2 have reversed, but still equal signs. *The orientation of the line does not change the ratio of division* λ and can therefore be ignored. If P lies outside the segment P_1P_2, then P_1P and PP_2 have opposite signs, and so λ is negative.

A more precise investigation shows that each position of the point P on the line can be characterized by one value of the ratio of division λ (Fig.). In fact, it can be seen that λ increases monotonically as P moves along the segment P_1P_2 from P_1, since in $\lambda = P_1P/PP_2$ the numerator always increases and the denominator decreases. For $P = P_1$, $\lambda = 0$; for the mid-point M of the segment P_1P_2 $\lambda_M = +1$; as P approaches the point P_2 arbitrarily closely, λ increases beyond any finite value.

If P is an *outer point of division* of the segment P_1P_2, then the difference of the lengths of P_1P and PP_2 is always $|P_1P_2|$, and it has less influence on the ratio $P_1P : PP_2 = \lambda$ the greater the segments P_1P and PP_2 are, that is, if P is sufficiently distant from the segment P_1P_2, then λ differs arbitrarily little from -1. It makes no difference whether P moves away in the direction P_1P_2 or in the direction P_2P_1. One says briefly: at the *improper* or *infinitely distant point* P of the line the ratio of division has the value $\lambda = -1$. If P approaches the point P_1 in the direction of P_1P_2, then $|PP_2| = |PP_1| + |P_1P_2| > |PP_1|$, and so the absolute value of the ratio of division is always less than 1, that is, λ increases from -1 to 0. If P approaches the point P_2 in the direction of P_2P_1, then $|P_1P| = |P_1P_2| + |P_2P| > |P_2P|$, that is $|\lambda| = |P_1P| : |PP_2| > 1$, and so $|PP_2| \to 0$ or $|\lambda| \to \infty$ as $P \to P_2$; λ therefore decreases monotonically from -1 to $-\infty$ as P moves, as an outer point of division, towards P_2 in the direction P_2P_1. For $P = P_2$, λ is not defined, but λ converges to $+\infty$ when P is an inner point and to $-\infty$ when P is an outer point.

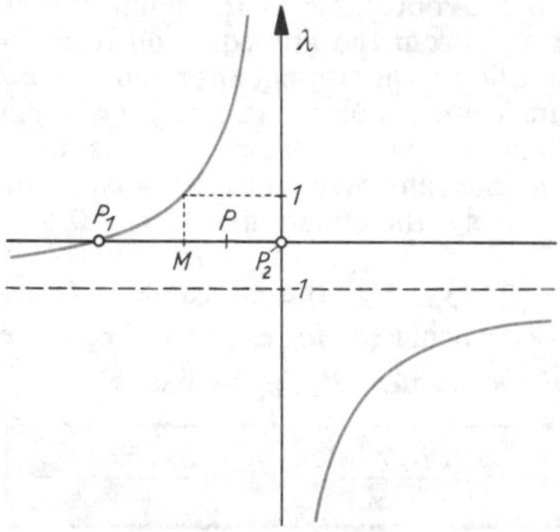

13.2-2 Value of the ratio of division λ as P describes a line

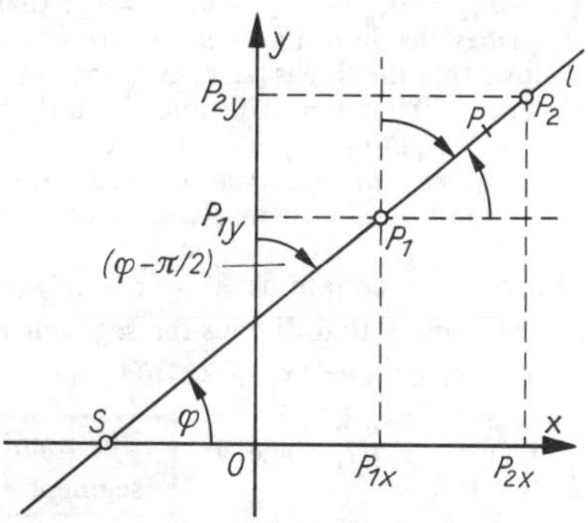

13.2-3 Ratio of division and the equation of a line

Equations of a line

Direction of a line. An oriented line l makes with the $+x$-axis an angle $\sphericalangle(x, l) = \varphi$, that is, the $+x$-axis moves into the direction of the line by means of a rotation about its point of intersection S with the line l in the mathematically positive sense (in a right-handed system, the opposite sense in a left-handed system). The line makes an angle $\sphericalangle(y, l) = \varphi - \pi/2$ with the $+y$-axis. Suppose that two points P_1 and P_2 are given on the line, so that P_1P_2 is positive. Lines are drawn through each of these points parallel to each coordinate axis (Fig.), cutting the axes in P_{1x}, P_{2x} and P_{1y}, P_{2y}. The *projections of the segment P_1P_2 on the axes* are then given for any angle φ by the formulae $P_{1x}P_{2x} = P_1P_2 \cos\varphi$ and $P_{1y}P_{2y} = P_1P_2 \cos(\varphi - \pi/2) = P_1P_2 \sin\varphi$. If (x_1, y_1) and (x_2, y_2) are the coordinates of P_1 and P_2, then

$$x_1 + P_{1x}P_{2x} = x_2, \quad x_2 - x_1 = P_{1x}P_{2x} \quad \text{and} \quad y_1 + P_{1y}P_{2y} = y_2, \quad y_2 - y_1 = P_{1y}P_{2y};$$

since $\quad |P_1P_2| = +\sqrt{[(x_2 - x_1)^2 + (y_2 - y_1)^2]}$,

$$\cos\varphi = \frac{x_2 - x_1}{\sqrt{[(x_2 - x_1)^2 + (y_2 - y_1)^2]}}, \qquad \sin\varphi = \frac{y_2 - y_1}{\sqrt{[(x_2 - x_1)^2 + (y_2 - y_1)^2]}}.$$

The *angle* φ is thus determined by the coordinates of the points P_1 and P_2; it can take values between 0 and 2π. However, in all cases where the orientation of the line l need not be taken into account, it is sufficient to determine the angle φ from the value of its tangent:

$$m = \tan\varphi = (y_2 - y_1)/(x_2 - x_1), \qquad \varphi = \tan^{-1}[(y_2 - y_1)/(x_2 - x_1)].$$

It is best to take the *principal value of the inverse tangent function*, that is, the value of φ in the interval $-\pi/2 < \varphi + \pi/2$. The value m is called the *gradient* or slope of the line l.

Equation of a line. If a point P divides the segment P_1P_2 in the ratio $\lambda = P_1P : PP_2$, then the equations found above for the segments P_1P and PP_2 hold, that is,

$$x - x_1 = P_1P\cos\varphi \quad \text{and} \quad y - y_1 = P_1P\sin\varphi, \quad x_2 - x = PP_2\cos\varphi \quad \text{and} \quad y_2 - y = PP_2\sin\varphi.$$

Hence, for the ratio of division,

$$\lambda = P_1P : PP_2 = (x - x_1)/\cos\varphi : (x_2 - x)/\cos\varphi = (x - x_1) : (x_2 - x)$$

or

$$\lambda = P_1P : PP_2 = (y - y_1)/\sin\varphi : (y_2 - y)/\sin\varphi = (y - y_1) : (y_2 - y).$$

If λ takes all values between $-\infty$ and $+\infty$, then the point P describes the line l; if the coordinates (x, y) of a point P satisfy the equation $(x - x_1) : (x_2 - x) = (y - y_1) : (y_2 - y)$, then the point P lies on the line. The coordinates (x, y) are often called the *current coordinates*. By corresponding addition, the equation $a : b = c : d$ can be put into the form $a : (a + b) = c : (c + d)$; hence the equation of the line is $(x - x_1) : (x_2 - x_1) = (y - y_1) : (y_2 - y_1)$. By interchanging the inner terms, one obtains the *two point form* of the equation.

Two point form	$(y - y_1)/(x - x_1) = (y_2 - y_1)/(x_2 - x_1)$

Point-direction form	$(y - y_1) = m(x - x_1)$

This emphasizes that two points P_1 and P_2 completely determine the line in the coordinate system. In the *point-direction form*, a point P_1 and the gradient $m = (y_2 - y_1) : (x_2 - x_1) = (y - y_1) : (x - x_1)$ determine the position of the line.

The right-hand side $(y_2 - y_1)/(x_2 - x_1)$ of the two point form corresponds to the meaning of $\tan\varphi$ and can take all values of this trigonometric function, according to the relative position of P_1 and P_2. Of particular interest are the *special cases* of the quotient for $(y_2 - y_1) = 0$ and $(x_2 - x_1) = 0$. In the first case, since $\tan\varphi = 0$, the line determined by the points P_1 and P_2 is parallel to the x-axis, and $\varphi = 0°$ or $\varphi = 180°$. From the line equation it follows that $(y - y_1)/(x - x_1) = 0$, $y - y_1 = 0$, $y = y_1$; that is, for any value of the x-coordinate of the point P that describes the line, its *y-coordinate has the constant value* $y = y_1$. From the line equation it again follows that the line is parallel to the x-axis. In the second case it follows from the equation $\tan\varphi = \infty$ that $\varphi = 90°$ or $\varphi = 270°$ and similarly from the identically transformed equation $(x - x_1)/(y - y_1) = (x_2 - x_1)/(y_2 - y_1) = 0$ or $x - x_1 = 0$; for any y-coordinate of the current point P its *x-coordinate must always have the constant value* $x = x_1$; that is, the line runs parallel to the y-axis. For the x- and y-axes themselves, since $y_1 = 0$ and $x_1 = 0$, respectively, the equations are $y = 0$ and $x = 0$.

From the equations $\lambda = (x - x_1)/(x_2 - x)$ and $\lambda = (y - y_1)/(y_2 - y)$ the coordinates (x, y) of the point P that divides the segment P_1P_2 in the ratio λ can be calculated; for example, $\lambda x_2 - \lambda x = x - x_1$ or $x = (x_1 + \lambda x_2)/(1 + \lambda)$. For the *mid-point M of the segment P_1P_2* one has $x = (x_1 + x_2)/2$, $y = (y_1 + y_2)/2$, since $\lambda = +1$.

The point P divides the segment P_1P_2 in the ratio λ	$x = \dfrac{x_1 + \lambda x_2}{1 + \lambda}$,	$y = \dfrac{y_1 + \lambda y_2}{1 + \lambda}$

Example 1: Direction angle φ of the segment P_1P_2.

a) Given $P_1(2, 3)$ and $P_2(7, 8)$:
$\cos\varphi = (7 - 2)/\sqrt{[(7 - 2)^2 + (8 - 3)^2]} = 5/\sqrt{[5^2 + 5^2]} = 1/\sqrt{2}$;
$\sin\varphi = (8 - 3)/(5\sqrt{2}) = 1/\sqrt{2}$. Direction angle $\varphi = 45°$.

b) Given $P_1(-1, -2)$ and $P_2(0, 8)$:
$\cos\varphi = (0 + 1)/\sqrt{[(0 + 1)^2 + (8 + 2)^2]} = 1/\sqrt{101}$; $\sin\varphi = (8 + 2)/\sqrt{101} = 10/\sqrt{101}$;
$\tan\varphi = 10$. Direction angle $\varphi = 84.3°$.

c) Given $P_1(2, -3)$ and $P_2(-3, +5)$:
$\cos\varphi = (-3 - 2)/\sqrt{[(-3 - 2)^2 + (5 + 3)^2]} = -5/\sqrt{89}$;
$\sin\varphi = 8/\sqrt{89}$; $\tan\varphi = -8/5 = -1.6$. Second quadrant: $\varphi = 180° - 58° = 122°$.

Example 2: Find the coordinates of the point T that divides the segment joining $P_1(3, -2)$ and $P_2(-5, 4)$ in such a way that $P_1T : TP_2 = 2 : 3$. – Since $\lambda = {}^2/_3$, it follows that

$$x = \frac{x_1 + \lambda x_2}{1 + \lambda} = \frac{3 + {}^2/_3(-5)}{1 + {}^2/_3} = \frac{9 - 10}{5} = -{}^1/_5;$$

$$y = \frac{y_1 + \lambda y_2}{1 + \lambda} = \frac{-2 + {}^2/_3 \cdot 4}{1 + {}^2/_3} = \frac{-6 + 8}{5} = {}^2/_5; \quad T(-{}^1/_5, {}^2/_5).$$

The segment is bisected ($\lambda = +1$) at the point $M(-1, 1)$.

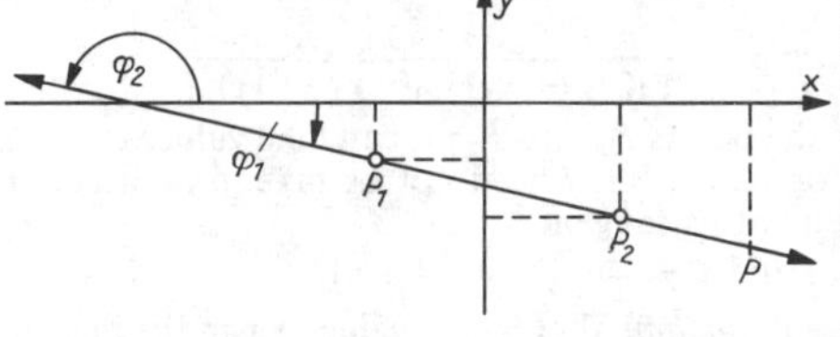

13.2-4 The line through $P_1(-4, -2)$ and $P_2(5, -4)$

Example 3: Find the equation of the line joining the points $P_1(-4, -2)$ and $P_2(5, -4)$ (Fig.). – By substituting the given values $x_1 = -4$, $y_1 = -2$, $x_2 = 5$, $y_2 = -4$, one obtains the equation

$(y + 2)/(x + 4) = (-4 + 2)/(5 + 4)$, which can be simplified to $y = -(2/9)\,x - 26/9$. The gradient m has the value $m = -2/9 = -0.2222\ldots = \tan\varphi$. The value of φ is $\varphi_1 = -12.53°$ or $\varphi_2 = 180° - 12.53° = 167.47°$; the principal value is $\varphi = -12.53°$. A *point* P_3 on the line at a distance $3\,|P_1P_2|$ from P_2 in the direction $\overrightarrow{P_1P_2}$ *divides the segment* $\overline{P_1P_2}$ *in the ratio* $\lambda = \overrightarrow{P_1P_3}/\overrightarrow{P_3P_2} = 4\,|P_1P_2|/(-3\,|P_1P_2|) = -4/3$. It therefore has the coordinates

$$x_3 = [x_1 - {}^4\!/_3 x_2]/(1 - {}^4\!/_3) = [-4 - {}^4\!/_3 \cdot 5]/(-{}^1\!/_3) = 12 + 20 = 32$$

and $$y_3 = [y_1 - {}^4\!/_3 y_2]/(1 - {}^4\!/_3) = [-2 - {}^4\!/_3(-4)]/(-{}^1\!/_3) = -16 + 6 = -10.$$

Example 4: To find the equation of the line that passes through the point $P_1(3, 4)$ and has the *direction angle* $\alpha = 60°$. Since $x_1 = 3$, $y_1 = 4$, $\tan\alpha = m = \tan 60° = \sqrt{3}$, the equation of the line is $y - 4 = \sqrt{3}(x - 3)$ or $y = x\sqrt{3} + 4 - 3\sqrt{3}$.

Cartesian normal form of the equation of a line. In the last example the equation of the line was simplified to the form $y = ax + b$. This can be done generally for the equation of any line, provided that it is not parallel to the y-axis. From the point-direction form, for example, one obtains

$$y - y_1 = m(x - x_1) = mx - mx_1,$$
$$y = mx + (y_1 - mx_1),$$
or $$y = mx + c, \quad \text{where} \quad c = y_1 - mx_1.$$

Cartesian normal form	$y = mx + c$

As was shown above, lines parallel to the y-axis have the equation $x = x_1$. Since $m = \tan\varphi$, mx_1 is obviously the difference of the ordinates of the given point P_1 and the point of intersection S of the line with the y-axis; $c = (y_1 - mx_1)$ is therefore the ordinate of S. This is confirmed by putting $x = 0$ into the normal form, which gives $y_S = c$.

If one takes the line $y = mx$, and gives x the value 1 (Fig.), then $y = m$; the line $y = mx + c$ is then obtained from this line by a *parallel displacement of c in the direction of the y-axis*. Hence, if one lays off a length m from the point (1, 0) along the line parallel to the y-axis at a distance $+1$, then the line from the origin to the end-point of this segment is the line $y = mx$. The line parallel to this through the end-point of the segment of length c from the origin along the y-axis is the required line l (Fig.).

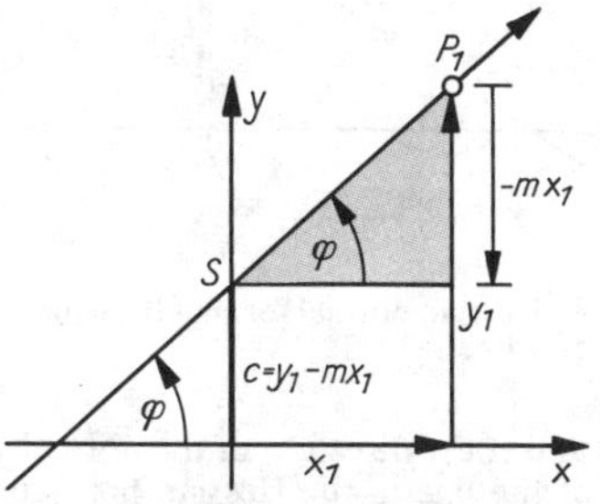

13.2-5 Derivation of the Cartesian normal form from the point-direction form

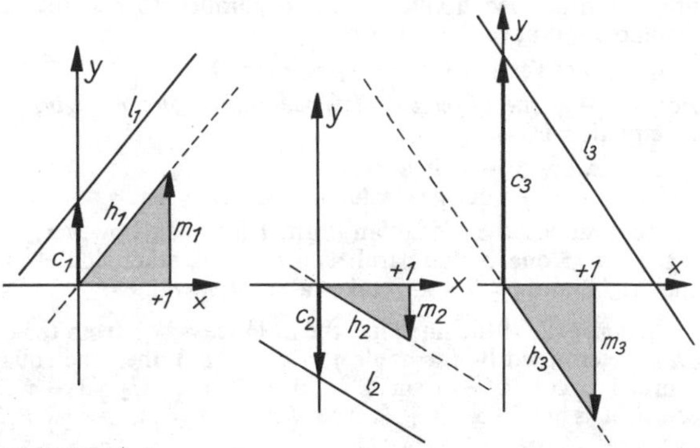

13.2-6 Three examples of the normal form $y = mx + c$

Intercept form of the equation of a line. A line that *does not pass through the origin* and is *not parallel to either coordinate axis* can be fixed in the coordinate system if the intercepts a and b that it cuts on the axes are given. If it cuts the axes in the points $P_1(a, 0)$ and $P_2(0, b)$, then the two-point form of the equation $(y - 0)/(x - a) = (b - 0)/(0 - a)$ can be expressed as $y/b = [x/(-a) + 1]$ or $x/a + y/b = 1$ (Fig.).

Intercept form of the equation of a line	$x/a + y/b = 1$

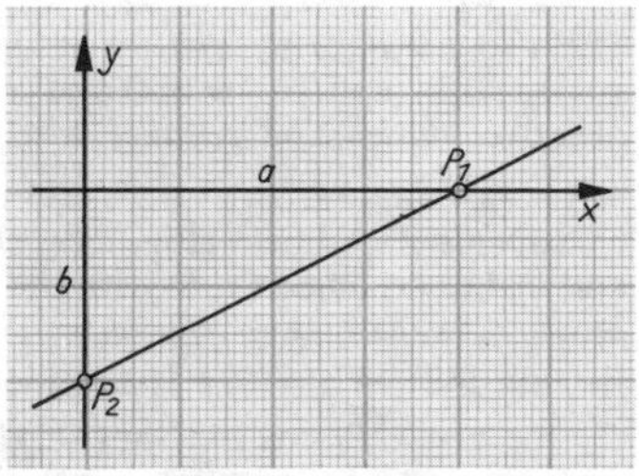

13.2-7 Derivation of the intercept form of the equation of a line

Examples: 1. The intercepts are $a = 4$, $b = -2$. The equation of the line in the intercept form is $x/4 + y/(-2) = 1$. This can be put into the normal form $y = x/2 - 2$.

2. The general intercept equation can also be put into the normal form. From the equation $x/a + y/b = 1$ one obtains $y = -(b/a)\,x + b$, that is, $m = -b/a$, $c = b$.

3. At what points does the line $y = -4x/3 + 8$ cut the axes? – From the formula of Example 2 it follows immediately that $b = 8$, and therefore $a = 6$.

The Hessian normal form of the equation of a line. This form of the equation is named after Otto HESSE (1811–1874). The (x, y)-plane is divided by an *oriented line* l into two half-planes, of which the one that lies to the left of l as it is described in the sense of the orientation is called *positive*. There is then a line n through the origin *normal* to l, oriented so that the angle (l, n), in the sense of rotation of the coordinate system, is $+90°$ (Fig.). By means of this normal n, the *distance p of the line l from O* can be determined. If the perpendicular from the origin O to the line l cuts it in L, then $\overrightarrow{OL} = p$. In the figure, this distance p is positive; if O were to lie in the positive half-plane, then the distance $\overrightarrow{OL}$ would be negative; if O and L coincide, then p has the value zero.

If the normal, and therefore the distance p, makes an angle φ with the x-axis, $\sphericalangle(x, n) = \varphi$, then the angle (x, l) that the line l makes with the x-axis is obtained by rotating the normal through $-\pi/2$, that is, $\sphericalangle(x, l) = \sphericalangle(x, n) - \pi/2 = \varphi - \pi/2$. The angle (y, n) that the normal makes with the y-axis is given by $\sphericalangle(y, n) = \varphi - \pi/2$. The angle $\varphi = \sphericalangle(x, n)$ can take all values between 0 and 2π. If a *point P* has the Cartesian coordinates $x = \overrightarrow{OR}$, $y = \overrightarrow{RP}$, and the *distance* $d = \overrightarrow{QP}$ *from the line l*, then there are two vector paths from the origin O to the point P, namely $\overrightarrow{OL} + \overrightarrow{LQ} + \overrightarrow{QP}$ and $\overrightarrow{OR} + \overrightarrow{RP}$. Their projections on the normal must be equal in magnitude and sense:

$$p + 0 + d = x\cos\varphi + y\cos(y, n) = x\cos\varphi + y\sin\varphi \quad \text{or} \quad d = x\cos\varphi + y\sin\varphi - p.$$

For points P in the positive half-plane the distance d from the line l is positive, and for points in the negative half-plane the distance is negative.

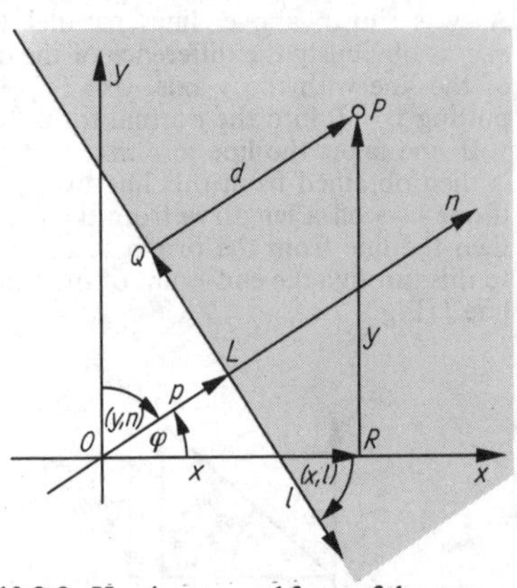

13.2-8 Hessian normal form of the equation of a line

Points P that lie on the line have the distance $d = 0$ from it; the equation of the line is therefore $x\cos\varphi + y\sin\varphi - p = 0$.

Hessian normal form	$x\cos\varphi + y\sin\varphi - p = 0$

The sign of p in the equation depends on the orientation, as described above. The two parallels to l at the distances $\pm\delta$ have the equations

$$x\cos\varphi + y\sin\varphi - (p \pm \delta) = 0.$$

For $\delta = -p$ *one of the parallels passes through the origin*, its equation is

$$x\cos\varphi + y\sin\varphi = 0$$

or $$y = -x\cot\varphi = x\tan(\varphi - \pi/2) = mx,$$

hence assumes the Cartesian normal form. If, however, $\delta > p$, then one of the parallels lies on the other side of the origin, and $p' = p - \delta$ takes a negative value.

Example 1: If the line l has the distance $p = 3$ from the origin and if the direction of the normal n is determined by the angle $\varphi = 30°$ (Fig.), then the equation of the line in the Hessian normal form is $x\cos 30° + y\sin 30° - 3 = 0$ or $x \cdot {}^1/_2\sqrt{3} + y \cdot {}^1/_2 - 3 = 0$; in the Cartesian normal form it is $y = -x\sqrt{3} + 6$. The *distances of the points* $P_1(5, 7)$ and $P_2(-1, -3)$ from the line l are:

$$d_1 = {}^5/_2\sqrt{3} + {}^7/_2 - 3 \approx 4.33 + 0.5 = 4.83;$$
$$d_2 = -{}^1/_2\sqrt{3} - {}^3/_2 - 3 \approx -(0.87 + 4.5) = -5.37.$$

The two *parallels* p_2 *and* p_1 at the *distances* $\delta = \pm 6$ from l have the following equations: $x \cdot {}^1/_2\sqrt{3} + y \cdot {}^1/_2 - 9 = 0$ and $x \cdot {}^1/_2\sqrt{3} + y \cdot {}^1/_2 + 3 = 0$; in the second equation, p has a negative value. The *parallel* with distance $p_1 > 0$ and normal $n' = -n$ has the equation: $-x \cdot {}^1/_2\sqrt{3} - y \cdot {}^1/_2 - 3 = 0$ or $x\cos 210° + y\sin 210° - 3 = 0$.

Example 2: A line h (Fig.) cuts the x- and y-axes in the points $P_1 = (-5, 0)$ and $P_2 = (0, +8)$ and makes an angle (x, h) with the x-axis for which $\tan(x, h) = {}^8/_5 = 1.6$, that is, $\sphericalangle(x, h) = 58°$. Since $\sphericalangle(x, h) = \varphi - \pi/2$, $\varphi = 58° + 90° = 148°$. In the *Hessian normal form* $x\cos 148° + y\sin 148° - p = 0$, p is obtained by projecting the segment $\overrightarrow{OP_1}$ or $\overrightarrow{OP_2}$ on the normal n:

$$p = \overrightarrow{OP_1}\cos 148° = (-5)(-\sin 58°) = 5 \cdot 0.8480 \approx 4.24$$

or $$p = \overrightarrow{OP_2}\cos(148° - 90°) = 8\cos 58° = 8 \cdot 0.5299 \approx 4.24.$$

The Hessian normal form is therefore $-x(0.85) + y(0.53) - 4.24 = 0$.
The *distance d* of the point $P_3(6, 5)$ from the line h is given by $d = -6 \times 0.85 + 5 \times 0.53 - 4.24 = -5.09 + 2.65 - 4.24 \approx -6.68$. This distance is larger than p by $p_1 = 6.68 - 4.24 = 2.44$. The parallel h_1 to h through P_3 therefore has the equation $-x(0.85) + y(0.53) + 2.44 = 0$, and the parallel h_1' with $n_1 = -n$ has the equation $x\cos(148° + 180°) + y\sin(148° + 180°) - 2.44 = 0$ or $x(0.85) - y(0.53) - 2.44 = 0$.

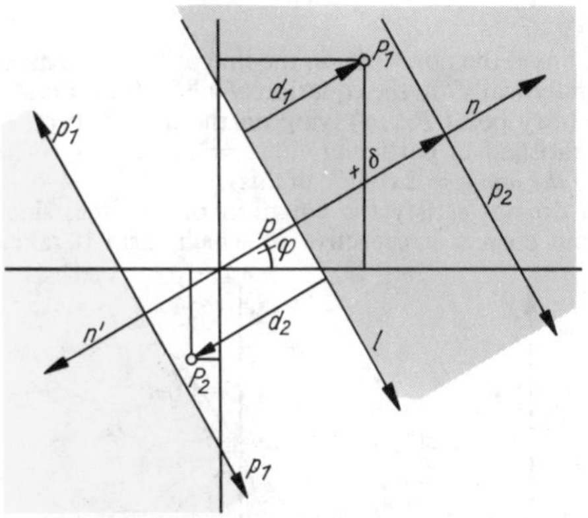

13.2-9 Example 1 of a line in Hessian normal form: $\varphi = 30°$, $p = 3$

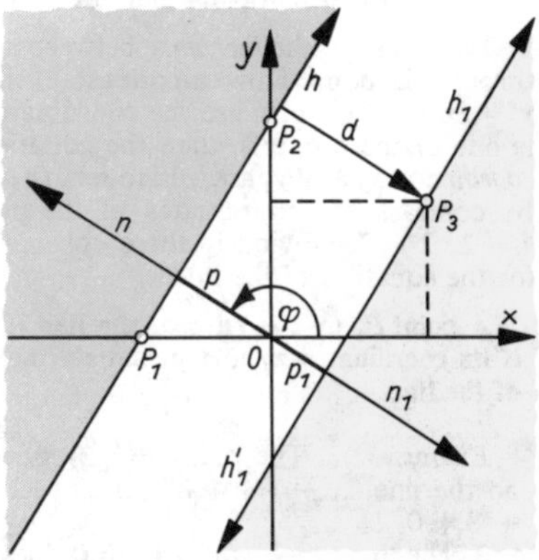

13.2-10 Example 2 of a line in Hessian normal form

The general form of the equation of a line. The general equation of a line is $Ax + By + C = 0$, where A, B, C are arbitrary real numbers, except that A and B are not both zero. Its graph is always a straight line. If either A or B is zero, say $A = 0, B \neq 0$, then the equation $By + C = 0$, $y = -C/B$, represents a line parallel to the x-axis at a distance $y = -C/B$; if $B = 0, A \neq 0$, the line is parallel to the y-axis at a distance $x = -C/A$. If A and B are both non-zero, then the equation $y = -(A/B)\,x - (C/B)$ is the Cartesian normal form with gradient $m = -A/B$ and intercept $c = -C/B$ on the y-axis. If $C = 0$ the line passes through the origin.

The *half-plane* that contains only those points $P(x, y)$ whose coordinates give positive values to the linear function $Ax + By + C = f(x, y)$ is called *positive*. The equation $y = (8/5)\,x + 8$ considered in Example 2 corresponds to the linear function $5y - 8x - 40$. For the point $P_3(6, 5)$ it has the value $25 - 48 - 40 = -63$, and so P_3 lies in the negative half-plane.

The line is oriented so that it is described in the positive sense when the positive half-plane lies to the left of it. If the equation $Ax + By + C = 0$ is multiplied by $\frac{\varepsilon}{\sqrt{(A^2 + B^2)}}$, where $\varepsilon = \pm 1$:

$$\frac{\varepsilon Ax}{\sqrt{(A^2 + B^2)}} + \frac{\varepsilon By}{\sqrt{(A^2 + B_2)}} + \frac{\varepsilon C}{\sqrt{(A^2 + B^2)}} = 0,$$

it is thereby *normalized*, that is, the sum of the squares of the coefficients of x and y is 1:

$$\left(\frac{\varepsilon A}{\sqrt{(A^2 + B^2)}}\right)^2 + \left(\frac{\varepsilon B}{\sqrt{(A^2 + B^2)}}\right)^2 = 1.$$

These coefficients can then be interpreted as the values of the cosine and sine of an angle φ (Fig.). If one puts

$$\frac{\varepsilon A}{\sqrt{(A^2 + B^2)}} = \cos\varphi, \qquad \frac{\varepsilon B}{\sqrt{(A^2 + B^2)}} = \sin\varphi,$$

$$\frac{\varepsilon C}{\sqrt{(A^2 + B^2)}} = -p,$$

then the equation is in the *Hessian normal form*; only for $\varepsilon = +1$ the distance d of a point P of the positive half-plane is positive.

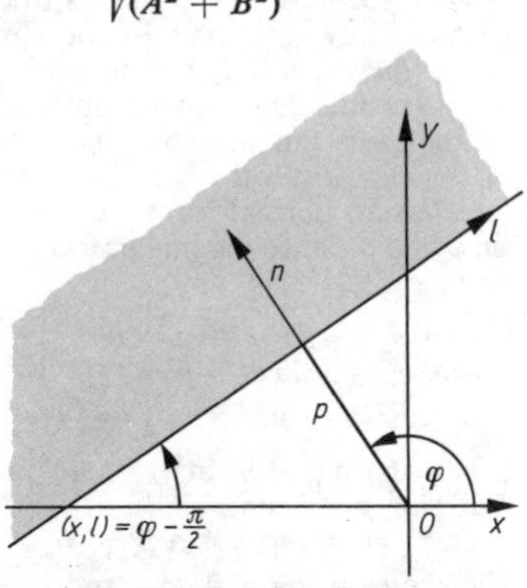

13.2-11 Normalizing the equation $3y - 2x - 4 = 0$

Example: To bring the equation $3y - 2x - 4 = 0$ to the Hessian normal form. – The origin belongs to the negative half-plane, since $3 \times 0 - 2 \times 0 - 4 = -4$. Since $A = -2$, $B = +3$, $C = -4$, each coefficient must be divided by $\sqrt{(4 + 9)} = \sqrt{13}$. This gives the equation of the line in the Hessian normal form

$$-(2/\sqrt{13})\,x + (3/\sqrt{13})\,y - 4/\sqrt{13} = 0,$$

that is, $\cos\varphi = -2/\sqrt{13}$, $\sin\varphi = +3/\sqrt{13}$, $\tan\varphi = -3/2 = -1.5$; $\varphi = 123.68°$; $p = +4/\sqrt{13}$. From the Cartesian normal form $y = {}^2/_3 x + {}^4/_3$ one has as a check $m = \tan(x, l) = {}^2/_3$, $(x, l) = 33.69°$, that is $\varphi - 90° = 33.69°$, $\varphi = 123.69°$.

Incidence of point and line

One speaks of the *incidence* between a point and a line if the point lies on the line or the line passes through the point. How can one establish incidence analytically? In the equation of a line, for example, $y = 2x - 7$, x and y are the coordinates of an arbitrary point $P(x, y)$ lying on the line. If $P_1(4, 1)$ is one of these points, then the equation must be satisfied in particular for $x = x_1$, $y = y_1$. The *coordinates* x_1 and y_1 are said to *satisfy the equation of the line* $y = 2x - 7$; in this case, $1 = 2 \cdot 4 - 7$. By contrast, the coordinates of the point $P_2(2, 4)$ do not satisfy the equation of the line, since $4 \neq 2 \cdot 2 - 7$. Obviously these considerations remain correct irrespective of which form is taken for the equation of the line.

A point P_1 (x_1, y_1) lies on the line if and only if its coordinates x_1 and y_1 satisfy the equation of the line.

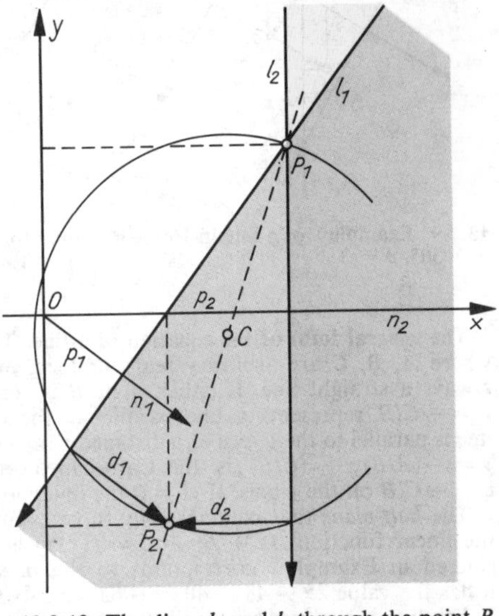

13.2-12 The lines l_1 and l_2 through the point P_1 have distances d_1 and d_2 from the point P_2

Examples: 1. The point $P(2, 3)$ does not lie on the line $2x - y/4 + 8 = 0$, since $2 \cdot 2 - 3/4 + 8 \neq 0$.

2. The line $x/2 + y/3 - 17 = 0$ does not pass through the origin, since $0/2 + 0/3 - 17 \neq 0$.

3. The point $P_1(57, 88)$ lies on the line $y - 8 = 2 \cdot (x - 17)$, since $88 - 8 = 2 \cdot (57 - 17)$.

4. The line through the points $P_1(0, 3/2)$ and $P_2(2, 5/2)$ has the equation $(y - 3/2)/(x - 0) = (5/2 - 3/2)/2$ or $y = x/2 + 3/2$. It cuts the x-axis in the point S whose ordinate is $y_0 = 0$. Its abscissa is then $x_0 = -3$. The point $S(-3, 0)$ is the intersection of the line $y = x/2 + 3/2$ with the x-axis; $x_0 = -3$ is the zero of the function.

5. If a point P_1 with the abscissa $x_1 = 5$ is to lie on the line $y = 2x/3 - 2$, then its ordinate y must have the value $y_1 = 2x_1/3 - 2 = 10/3 - 2 = 4/3$.

6. To find the line l_1 through the point $P_1(6, 4)$ that has the distance $d = 3$ from the point $P_2(3, -5)$. – The line can be constructed geometrically by means of the circle of Thales on the segment P_1P_2 as diameter (Fig.).

Two lines l_1 and l_2 are determined in this way; their distance, according to the orientation of the perpendicular drawn from the origin, is either positive ($d_1 = +3$) or negative ($d_2 = -3$). At the same time it should be noted that *the given distance d must be smaller than the length of the segment P_1P_2* if a solution is to be possible. It is advisable to take the equation of the line in the Hessian normal form $x \cos \varphi + y \sin \varphi - p = 0$, and to determine the three numbers $\cos \varphi$, $\sin \varphi$ and p. Since the line passes through the point $P_1(6, 4)$ and has the distance $d = \pm 3$ from the point $P_2(3, -5)$,

$$\left| \begin{array}{l} 6 \cos \varphi + 4 \sin \varphi - p = 0 \\ 3 \cos \varphi - 5 \sin \varphi - p = \pm 3 \end{array} \right| \begin{array}{l} + \\ - \end{array} \rightarrow 3 \cos \varphi + 9 \sin \varphi = \mp 3$$

$$\cos^2 \varphi + \sin^2 \varphi = 1 \leftarrow \cos \varphi = \mp 1 - 3 \sin \varphi$$

$$1 \pm 6 \sin \varphi + 9 \sin^2 \varphi + \sin^2 \varphi = 1$$

$$10 \sin^2 \varphi \pm 6 \sin \varphi = 0$$

$$\sin \varphi_1 = \mp 3/5, \quad \sin \varphi_2 = 0$$

$$\cos \varphi_1 = \pm 4/5, \quad \cos \varphi_2 = \mp 1$$

$$p_1 = \pm 2^2/_5, \quad p_2 = \mp 6$$

The choice of $p_1 = +2^2/_5$ and $p_2 = +6$ fixes the orientation of the two lines l_1 and l_2; putting $\cos \varphi_1 = +4/5$, $\sin \varphi_1 = -3/5$, $p_1 = +2^2/_5$ and $\cos \varphi_2 = +1$, $\sin \varphi_2 = 0$, $p_2 = +6$, one obtains the required equations $+4x/5 - 3y/5 - 2^2/_5 = 0$ and $x - 6 = 0$, or, in the Cartesian normal form, $y = 4x/3 - 4$ and $x = 6$. If the coordinates $x = 0$, $y = 0$ are substituted into the two functions $f_1(x, y) = -3y/5 + 4x/5 - 2^2/_5$ and $f_2(x, y) = x - 6$ one sees that the origin O lies in the negative half-plane for both lines.

13.3. Several lines

In plane geometry it is well known that the relative position of two lines in a plane can be described by means of the concepts parallel and distance, or point of intersection and angle. In analytic geometry the corresponding characterization of two lines can be read off from their equations.

In the Cartesian normal form of two lines $y = m_1x + c_1$ and $y = m_2x + c_2$, their directions are characterized by their gradients m_1 and m_2. The lines are parallel if and only if $m_1 = m_2$. If, on the other hand, they are given in the Hessian normal form, $x \cos \varphi_1 + y \sin \varphi_1 - p_1 = 0$ and $x \cos \varphi_2 + y \sin \varphi_2 - p_2 = 0$, then they are parallel if the coefficients of the linear terms are equal to within a common factor $\varkappa$, $\cos \varphi_1 = \varkappa \cos \varphi_2$ and $\sin \varphi_1 = \varkappa \sin \varphi_2$. Since these coefficients can be obtained by normalization from the general linear equations, the same condition holds for two parallel lines given in the form $A_1x+B_1y+C_1=0$ and $A_2x+B_2y+C_2=0$: $A_1B_2-A_2B_1=0$.

If in a linear equation, say $Ax + By + C = 0$, x and y are interpreted as coordinates of a given point, then the coefficients A, B, C are *parameters*, and the equation means that from the set of all points (x, y) a subset is picked out: it consists of those points whose coordinates satisfy the condition given by the ratios $A : B : C$ of the parameters. It has been shown that this subset has a line as carrier. Conversely, A, B, C, where A and B not both zero, can be regarded as homogeneous coordinates. Then x, y are parameters, which pick out from the set of all lines (A, B, C) or $(\cos \varphi, \sin \varphi, p)$ the subset of those that have the point (x, y) as carrier. They form a *pencil of lines*. If the relations are given that must hold between the point coordinates of a point P on a line dividing the segment between two other points P_1 and P_2 of the line in a given ratio, then one can immediately find the relations between the line coordinates that must hold if the two lines are parallel.

Point and angle of intersection

Determination of the point of intersection of two lines. Required are the coordinates x_0 and y_0 of the point of intersection $P_0(x_0, y_0)$ of two lines $A_1x + B_1y + C_1 = 0$ and $A_2x + B_2y + C_2 = 0$, given in the general form. P_0 as the point of intersection of the two lines must lie on both, hence its coordinates x_0 and y_0 must satisfy the equations of the two lines. The point of intersection is therefore obtained by solving the system of equations

$$\left| \begin{array}{l} A_1x_0 + B_1y_0 + C_1 = 0 \\ A_2x_0 + B_2y_0 + C_2 = 0 \end{array} \right|,$$

which can be done according to the usual rules (see Chapter 4.). If the system has a solution (x_0, y_0), this gives the coordinates of the point of intersection. If it has no solution because the equations are incompatible, then the lines are parallel. If the system has infinitely many solutions, because the equations are linearly dependent, then the two lines coincide.

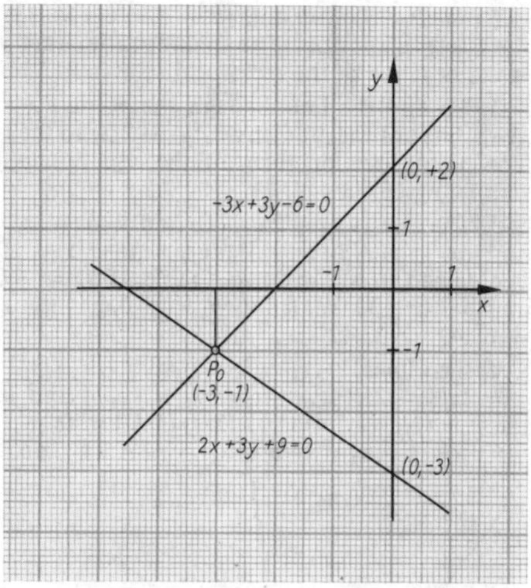

13.3-1 Determination of the point of intersection of two lines

Examples: 1. At which point do the two lines $-3x + 3y - 6 = 0$ and $2x + 3y + 9 = 0$ intersect? – The following scheme shows the solution of the system of equations.

$$\left| \begin{array}{l} -3x_0 + 3y_0 - 6 = 0 \\ 2x_0 + 3y_0 + 9 = 0 \end{array} \right| \begin{array}{l} \longrightarrow 5x_0 + 15 = 0 \\ + \nearrow \qquad x_0 = -3 \end{array}$$

$$-3(-3) + 3y_0 - 6 = 0$$

$$y_0 = -1$$

The lines intersect at the point $P_0(-3, -1)$ (Fig.).

2. To find the point of intersection P_0 of two lines given in the normal form, say $y=-3x+14$, $y = -x - 1$, one can most conveniently solve the system by equating the right-hand sides. In the given example, the lines intersect at $P_0(7.5, -8.5)$.

3. The lines $3x + y - 7 = 0$ and $2x - y - 3 = 0$ intersect at the point $P_0(2, 1)$.

4. The lines $2x - 3y + 5 = 0$ and $3y - 2x + 2 = 0$ are parallel. For each line the origin belongs to the positive half-plane. In the Hessian normal form the system of equations becomes

$$\left| \begin{array}{l} 2x_0 - 3y_0 + 5 = 0 \\ -2x_0 + 3y_0 + 2 = 0 \end{array} \right| \Longrightarrow \left| \begin{array}{l} +2x_0/\sqrt{13} - 3y_0/\sqrt{13} + 5/\sqrt{13} = 0 \\ -2x_0/\sqrt{13} + 3y_0/\sqrt{13} + 2/\sqrt{13} = 0 \end{array} \right|$$

The lines are at a distance $7/\sqrt{13}$ from one another.

5. The equations $0.8x + 0.4y - 1.2 = 0$ and $2x + y - 3 = 0$ represent the same line; the first equation is obtained from the second by dividing by 5/2. In the system of equations the two equations are dependent on one another. The lines coincide.

6. The lines $y = 2x - 8$ and $y = 2x + 12$ are parallel, since $m_1 = 2 = m_2$. Similarly the lines $x/4 + y/6 = 1$ and $x/2 + y/3 = 1$ are parallel, since their Cartesian normal forms are $y = -3x/2 + 6$ and $y = -3x/2 + 3$.

7. To find the equation of the line that passes through $P_1(2, -1)$ and is parallel to the line $y = 2x - 3$. – For the required line a point and the gradient $m = 2$ are given. By using the point-direction form, one obtains the equation of the line $y + 1 = 2(x - 2)$.

The angle of intersection of two lines. The angle $\psi = \sphericalangle(l_1, l_2)$ at which two lines l_1 and l_2 intersect is obtained most simply from the Hessian normal form; from $x \cos \varphi_1 + y \sin \varphi_1 - p_1 = 0$ and $x \cos \varphi_2 + y \sin \varphi_2 - p_2 = 0$ it follows immediately that $\psi = \sphericalangle(l_1, l_2) = \varphi_2 - \varphi_1$.

From the Cartesian normal form the angle ψ is obtained, by the restriction to the principal value of the inverse tangent function, only to within an additive constant $\pm\pi$. If $y = m_1 x + c_1$ and $y = m_2 x + c_2$ are the equations of the lines, then $m_1 = \tan \alpha_1$ and $m_2 = \tan \alpha_2$, where $\alpha_1 = \sphericalangle(x, l_1)$, $\alpha_2 = \sphericalangle(x, l_2)$, and so $\sphericalangle(x, l_1) + \sphericalangle(l_1, l_2) = \sphericalangle(x, l_2)$, or $\sphericalangle(l_1, l_2) = \psi = \alpha_2 - \alpha_1$. By the addition theorem for the tangent function it follows that

$$\tan \psi = \frac{\tan \alpha_2 - \tan \alpha_1}{1 + \tan \alpha_1 \tan \alpha_2} \quad \text{or} \quad \boxed{\tan \psi = \frac{m_2 - m_1}{1 + m_1 m_2}}$$

By interchanging the lines one obtains $\psi' = \sphericalangle(l_2, l_1) = \alpha_1 - \alpha_2 = -\psi$ or $\psi' = \pi - \psi$.

This condition includes the condition for *the lines to be parallel*: from $\psi = 0$ it follows that $m_1 = m_2$, which was found earlier. For $\psi = \pi/2$ one obtains the condition for the *two lines to be perpendicular*. Since $\tan \psi = \infty$, the denominator must be zero, that is, $1 + m_1 m_2 = 0$.

Condition for orthogonality	$m_2 = -1/m_1$

Examples: 1. The lines $y - 2 = 5(x - 13)$ and $y = -x/5 + 18$ are perpendicular, since in the Cartesian normal form their equations are $y = 5x - 63$ and $y = -1/5x + 18$, that is, $m_1 = 5$ is the negative of the reciprocal of $m_2 = -1/5$, or $m_2 = -1/m_1$.

2. To find the line through the point $P_1(1, 1)$ perpendicular to the line $y = -2x/3 + 3$. The given line has the gradient $m_1 = -2/3$, so the required line has the gradient $m_2 = -1/m_1 = +3/2$. The point-direction form of the equation is $(y - 1)/(x - 1) = +3/2$ or $2(y - 1) = 3(x - 1)$, that is, $y = 3x/2 - 1/2$.

3. The lines $y = -2x + 16$ and $y = -3x/5 + 3/5$ *cut at an angle* $\psi = 32.48°$, since $m_1 = -2$ and $m_2 = -3/5$ and so $\tan \psi = (-3/5 + 2)/(1 + 2 \times 3/5) = 7/11 = 0.6364$; the given value is obtained from the tangent table. If $\alpha_1 = -63.43°$ had been obtained from $m_1 = -2 = \tan \alpha_1$ and $\alpha_2 = -30.96°$ from $m_2 = -0.6 = \tan \alpha_2$, then ψ would have been $\alpha_2 - \alpha_1 = -30.96° + 63.43° = 32.47°$.

4. The lines $x/4 + y/5 = 1$ and $x/3 - y/2 = 1$ have the normal forms $y = -5x/4 + 5$ and $y = 2x/3 - 2$ (Fig.). Since $m_1 = -5/4$ and $m_2 = +2/3$, the *angle of intersection* ψ is given by $\tan \psi = (2/3 + 5/4)/[1 - (2/3)(5/4)] = (8 + 15)/(12 - 10) = 23/2 = 11.5$; $\psi = 85.03°$.

Check: $\tan \alpha_1 = -5/4$, $\alpha_1 = -51.34°$; $\tan \alpha_2 = +2/3$, $\alpha_2 = +33.69°$; $\psi = \alpha_2 - \alpha_1 = 85.03°$.

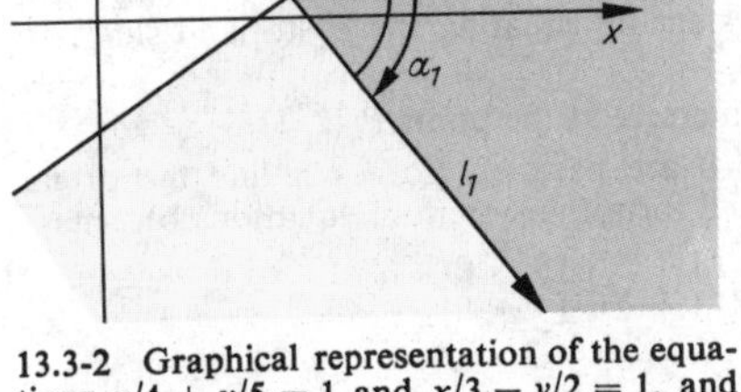

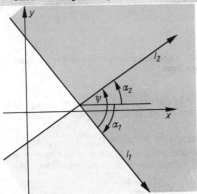

13.3-2 Graphical representation of the equations $x/4 + y/5 = 1$ and $x/3 - y/2 = 1$, and the angle of intersection of the lines

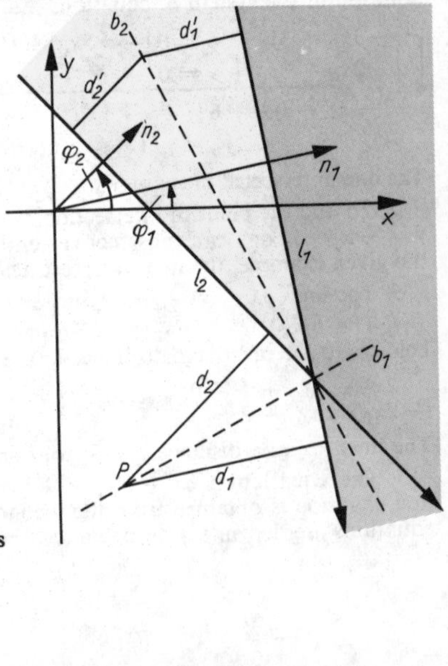

13.3-3 Determination of the angle-bisectors

If the equation of a line is to be found that passes through a given point and cuts a given line at a given angle ψ, then $\tan\psi$ and m_1 are given; by solving the formula one obtains the required value $m_2 = (m_1 + \tan\psi)/(1 - m_1 \tan\psi)$ and hence the required equation of the line by using the point-direction form.

The equations of the angle bisectors. Two intersecting lines l_1 and l_2 have two angle bisectors b_1 and b_2 (Fig.). They are defined as the locus of those points that have the same distance from the two lines. The Hessian normal form is recommended for this. The lines are given by

$$x \cos\varphi_1 + y \sin\varphi_1 - p_1 = 0 \text{ and } x \cos\varphi_2 + y \sin\varphi_2 - p_2 = 0.$$

They determine four sectors; *one of them belongs to the two positive half-planes*. It is bisected by b_1. In this sector each point of b_1 has positive distances d_1 from l_1 and d_2 from l_2. In the *vertically opposite sector* the distances of a point P of b_1 from l_1 and l_2 are both negative. In the equations $d_1 = \varepsilon_1(x \cos\varphi_1 + y \sin\varphi_1 - p_1)$ and $d_2 = \varepsilon_2(x \cos\varphi_2 + y \sin\varphi_2 - p_2)$, $\varepsilon_1 = \pm 1$ and $\varepsilon_2 = \pm 1$ have the same sign; it follows from $d_1 = d_2$ that the equation of the bisector b_1 is $x(\cos\varphi_1 - \cos\varphi_2) + y(\sin\varphi_1 - \sin\varphi_2) - (p_1 - p_2) = 0$. On the bisector b_2 of the remaining two sectors each point has positive distance from one line and negative distance from the other, that is, $\varepsilon_1 = -\varepsilon_2$ or $d_1' = -d_2'$; it follows that b_2 has the equation $x(\cos\varphi_1 + \cos\varphi_2) + y(\sin\varphi_1 + \sin\varphi_2) - (p_1 + p_2) = 0$.

Equations of the two angle bisectors of two intersecting lines	$x(\cos\varphi_1 \pm \cos\varphi_2) + y(\sin\varphi_1 \pm \sin\varphi_2) - (p_1 \pm p_2) = 0$. The choice of the negative sign gives the bisector that divides the sector belonging to the two positive half-planes.

Example: The equations of the lines $x + y - 2 = 0$ and $7x + y - 32 = 0$ are given in the Hessian normal form by $x/\sqrt{2} + y/\sqrt{2} - 2/\sqrt{2} = 0$ and $7x/(5\sqrt{2}) + y/(5\sqrt{2}) - 32/(5\sqrt{2}) = 0$. The equations of the two angle bisectors are

$$x(1/\sqrt{2} \pm 7/(5\sqrt{2})) + y(1/\sqrt{2} \pm 1/(5\sqrt{2})) - (2/\sqrt{2} \pm 32/(5\sqrt{2})) = 0,$$

or $x(5 \pm 7) + y(5 \pm 1) - (10 \pm 32) = 0$; in Cartesian normal form these are $y = x/2 - 11/2$ and $y = -2x + 7$.

Triangle and polygon

The area of a triangle. If $P_1(x_1, y_1)$, $P_2(x_2, y_2)$ and $P_3(x_3, y_3)$ are the vertices of the triangle (Fig.), then its area A is known to be $A = {}^1/_2\, |P_1P_2| \cdot h_3 = {}^1/_2\, |P_1P_2| \cdot |P_1P_3| \cdot \sin\alpha$, where h_3 is the distance of P_3 from P_1P_2 and α is the angle between the segments $\overrightarrow{P_1P_2}$ and $\overrightarrow{P_1P_3}$ *in the sense of rotation of the coordinate system*. If this sense is the mathematically positive, then P_3 lies to the left of the segment $\overrightarrow{P_1P_2}$; if the perimeter of the triangle is described in the sequence $P_1 \to P_2 \to P_3$, then the triangle lies to the left and the sign of the sine function is positive. For the *opposite sense of rotation* of the angle the triangle lies to the right; the oriented area of the triangle is counted as negative.

By the *parallel displacement* $x' = x - x_1$, $y' = y - y_1$ one can go over to a coordinate system in which P_1 is the origin; in this system, ϱ_2, φ_2 and ϱ_3, φ_3 are the *polar coordinates* of P_2 and P_3; then $2A = \varrho_2\varrho_3 \sin(\varphi_3 - \varphi_2) = \varrho_2 \cos\varphi_2 \cdot \varrho_3 \sin\varphi_3 - \varrho_2 \sin\varphi_2 \cdot \varrho_3 \cos\varrho_3 = x_2'y_3' - y_2'x_3'$

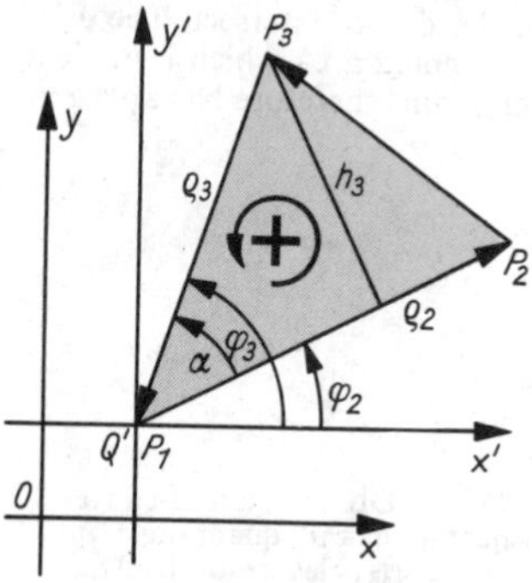

13.3-4 Area of a triangle

$$= \begin{vmatrix} x_2' & y_2' \\ x_3' & y_3' \end{vmatrix} = \begin{vmatrix} x_2 - x_1 & y_2 - y_1 \\ x_3 - x_1 & y_3 - y_1 \end{vmatrix}$$

$$= \begin{vmatrix} 1 & x_1 & y_1 \\ 0 & x_2 - x_1 & y_2 - y_1 \\ 0 & x_3 - x_1 & y_3 - y_1 \end{vmatrix} = \begin{vmatrix} 1 & x_1 & y_1 \\ 1 & x_2 & y_2 \\ 1 & x_3 & y_3 \end{vmatrix} = 2A.$$

Area A of a triangle with vertices $P_1(x_1, y_1)$, $P_2(x_2, y_2)$, $P_3(x_3, y_3)$
$A = {}^1/_2[x_1(y_2 - y_3) + x_2(y_3 - y_1) + x_3(y_1 - y_2)]$ $= {}^1/_2 \begin{vmatrix} 1 & x_1 & y_1 \\ 1 & x_2 & y_2 \\ 1 & x_3 & y_3 \end{vmatrix}$

If this determinant is expanded in terms of the second column or if the two-rowed determinant is multiplied out and the terms are reordered, an expression is obtained in which the indices in each of the three summands can be interchanged by a *cyclic permutation*. If P_3 lies on the line P_1P_2, *the area of the triangle is zero*; for $A = 0$ the equation gives

$$x_3(y_1 - y_2) - y_3(x_1 - x_2) + (x_1y_2 - x_2y_1) = 0$$

or $$y_3 = x_3(y_1 - y_2)/(x_1 - x_2) + (x_1y_2 - x_2y_1)/(x_1 - x_2),$$

that is, it becomes the *equation of the line* through P_1 and P_2 in the Cartesian normal form, which agrees with geometrical intuition. The condition that three points should lie on a line is $A = 0$.

Example: The triangle $P_1(2, 1)$, $P_2(6, 3)$, $P_3(4, 7)$ has the area

$$A = {}^1/_2 \begin{vmatrix} 1 & 2 & 1 \\ 1 & 6 & 3 \\ 1 & 4 & 7 \end{vmatrix} = {}^1/_2[2(3 - 7) + 6(7 - 1) + 4(1 - 3)] = {}^1/_2(-8 + 36 - 8) = 10.$$

The triangle $P_4(-4, -5)$, $P_5(5, -3)$, $P_6(6, 2)$ has the area

$$A = {}^1/_2[-4(-3 - 2) + 5(2 + 5) + 6(-5 + 3)] = {}^1/_2(20 + 35 - 12) = 21.5.$$

Area of a polygon. In a *convex polygon* the line segment P_xP_y joining two arbitrarily chosen interior points P_x and P_y contains only *interior* points. All the diagonals of an n-gon through one vertex lie wholly in the interior and *divide the polygon into* $(n - 2)$ *triangles*. Any two adjacent triangles have a diagonal as common side and together they cover the whole area of the polygon, irrespective of which vertex is chosen (Fig.). If each of the triangles is described in the sense fixed as positive, then the polygon is also described in this sense. Each diagonal is described once in one sense and then in the neighbouring triangle in the opposite sense. *The area of the polygon is the sum of the areas of the triangles.*

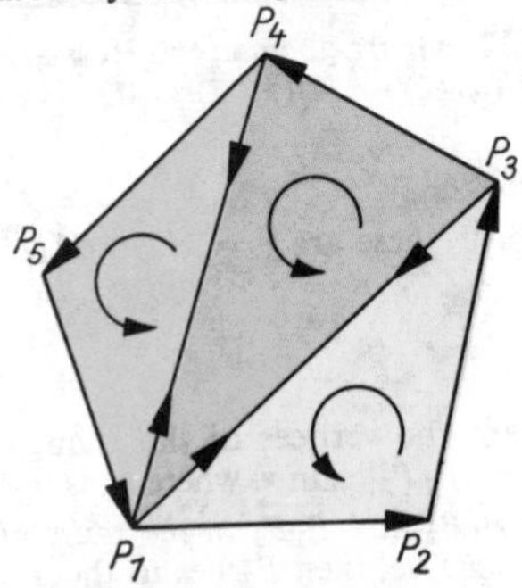

13.3-5 Dividing a convex n-gon into $(n - 2)$ triangles

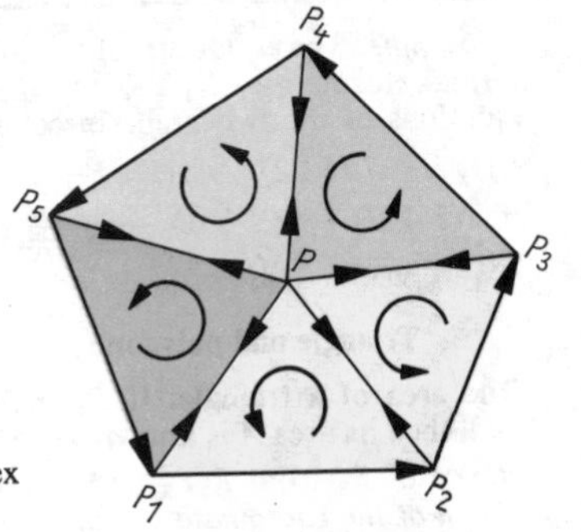

13.3-6 Dividing a convex n-gon into n triangles

If the area is divided *into n triangles* by lines joining an arbitrary point P inside the polygon to the vertices, then again the sense of description of the triangles corresponds to that of the polygon, and the interior sides are described twice in opposite senses (Fig.). A *non-convex polygon* can also be divided into triangles by either of the two methods. However, there arise diagonals and 'interior' sides which contain exterior points of the polygon (Fig.). If the area of a triangle described in the sense opposite to that of the polygon is counted as negative, for example, the triangle $P_1P_2P_3$ in the pentagon $P_1P_2P_3P_4P_5$, then again the *area of the polygon is the algebraic sum of those of the triangles*, as long as the polygon is not *folded*, that is, as long as its sides do not intersect. The quadrangle $P_1P_2P_3P_4$ in Fig. 13.3-7 is folded. One must ascribe to it an oriented area which is made up of one *positive* (red) area and one *negative* (blue); a folded 'parallelogram' therefore has area zero.

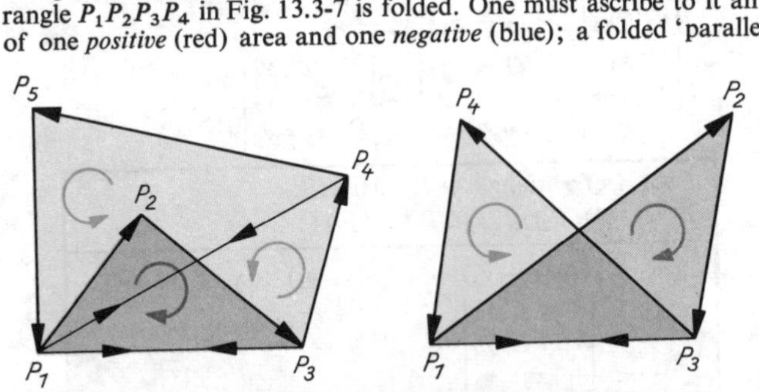

13.3-7 Dividing a non-convex pentagon and quadrangle into 3 and 2 triangles, respectively

Centre of gravity of a triangle. In a triangle with the vertices $P_1(x_1, y_1)$, $P_2(x_2, y_2)$, $P_3(x_3, y_3)$ the midpoints of the sides are $M_1[{}^1/_2(x_2 + x_3), {}^1/_2(y_2 + y_3)]$, $M_2[{}^1/_2(x_3 + x_1), {}^1/_2(y_3 + y_1)]$, $M_3[{}^1/_2(x_1 + x_2), {}^1/_2(y_1 + y_2)]$. On the medians $s_1 = |P_1M_1|$, $s_2 = |P_2M_2|$, $s_3 = |P_3M_3|$ points G_1, G_2, G_3 are fixed by means of the ratios $\lambda_1, \lambda_2, \lambda_3$ (Fig.); their coordinates are

$$\xi_1 = \frac{x_1 + \lambda_1(x_2 + x_3)/2}{1 + \lambda_1}, \quad \eta_1 = \frac{y_1 + \lambda_1(y_2 + y_3)/2}{1 + \lambda_1},$$
$$\xi_2 = \frac{x_2 + \lambda_2(x_3 + x_1)/2}{1 + \lambda_2}, \quad \eta_2 = \frac{y_2 + \lambda_2(y_3 + y_1)/2}{1 + \lambda_2},$$
$$\xi_3 = \frac{x_3 + \lambda_3(x_1 + x_2)/2}{1 + \lambda_3}, \quad \eta_3 = \frac{y_3 + \lambda_3(y_1 + y_2)/2}{1 + \lambda_3}.$$

It turns out that under the apparently arbitrary choice $\lambda_1 = \lambda_2 = \lambda_3 = 2$, the three pairs of coordinates become equal, that is, they represent the same point $G = G_1 = G_2 = G_3$:

The three medians of a triangle meet at a point G, which divides each of them in the ratio $|P_iG| : |GM_i| = 2:1$ and is called the centre of gravity of the triangle.

The coordinates of the centre of gravity are the arithmetic means of the coordinates of the vertices of the triangle.

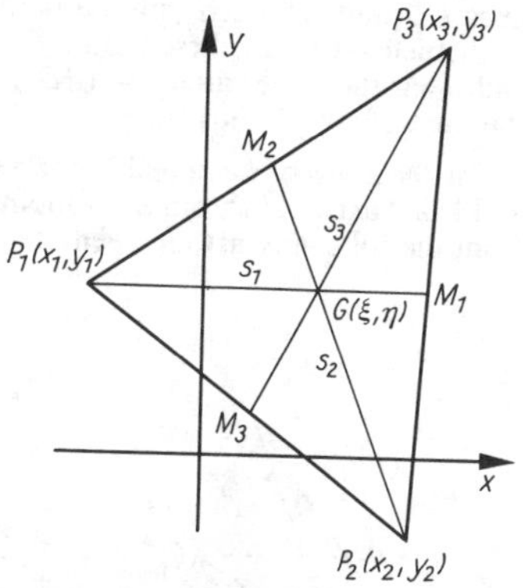

13.3-8 Centre of gravity of a triangle

Coordinates of the centre of gravity of a triangle	$\xi = \frac{x_1 + x_2 + x_3}{3}, \quad \eta = \frac{y_1 + y_2 + y_3}{3}$

Example: In the triangle $P_1(-5, 3)$, $P_2(-2, -1)$, $P_3(7, 8)$ the coordinates of the centre of gravity are $\xi = (-5 - 2 + 7)/3 = 0$; $\quad \eta = (3 - 1 + 8)/3 = 3^1/_3$.

Theorem of Apollonius. Before proving this theorem, a lemma on the angle bisectors of a triangle is derived.

In a triangle the bisector of an interior angle and the bisector of the corresponding exterior angle divide the opposite side in the ratio of the sides containing the angle.

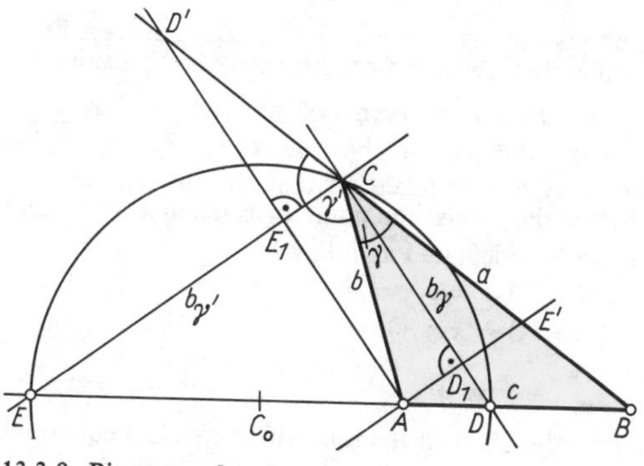

13.3-9 Bisectors of an interior angle γ and the corresponding exterior angle γ'

In the figure the *bisectors* of the angles γ and γ' of the triangle ABC are denoted by b_γ and $b_{\gamma'}$. They are *perpendicular*, since γ and γ' are supplementary angles. The lines AD' and AE' parallel to them through A cut the bisectors *at right angles* at D_1 and E_1. From the congruence of the two pairs of triangles $\triangle AD_1C \cong \triangle E'D_1C$ and $\triangle AE_1C \cong \triangle D'E_1C$ it follows that $|CE'| = |CA| = |CD'|$. The lines BD' and BE are cut by the parallels AD', DC and by the parallels CE, $E'A$. Hence by the intercept theorem:

1. $|AD| : |DB| = |D'C| : |CB| = |CA| : |CB|$ and
2. $|AE| : |EB| = |E'C| : |CB| = |CA| : |CB|$.

If one considers the *directions of segments* on the side c, then $\overrightarrow{AD}$, $\overrightarrow{DB}$ and $\overrightarrow{EB}$ have the same sign, and $\overrightarrow{AE}$ the opposite sign. The ratios λ in which the points D and E of the bisectors divide the segment $\overrightarrow{AB}$ have the same numerical value, but opposite sign: $\lambda_1 = (ABD) = \overrightarrow{AD}/\overrightarrow{DB} = +(b:a)$ and $\lambda_2 = (ABE) = \overrightarrow{AE}/\overrightarrow{EB} = -(b:a)$. Two points D and E that divide a segment $\overrightarrow{AB}$ internally and externally in the same ratio are called *harmonic points* and the ratio of the two ratios $\lambda_1 : \lambda_2$ is called the *cross-ratio* $(A, B;\ D, E)$. For harmonic points the cross-ratio therefore has the value $-1 = [+(b:a)] : [-(b:a)]$.

Theorem of Apollonius. The locus of the vertices C of all triangles ABC with a given side $|AB|$, whose other sides are in a constant ratio $|AC| : |BC| = \lambda$, is the circle on the segment $|DE|$ as diameter, whose end-points D and E divide the side $|AB|$ internally and externally in the ratio λ.

In the figure of the theorem proved above, if the segment AB and its points of division D and E are regarded as given, then, apart from the triangle ABC, there are other triangles ABC with the property that the bisectors of the angles γ and γ' pass through D and E. To be angle bisectors, the lines CD and CE need only be perpendicular, that is, C must lie on the circle on the segment DE as diameter. For all these points C the ratio $(b : a) = \lambda$ of its distances from the two fixed points A and B has the same value $\lambda = |AD| : |DB|$. The circle is therefore the locus of all the points C, as the theorem of Apollonius states.

The theorems of Ceva and Menelaus. These theorems are named after Giovanni CEVA (1648–1734) and MENELAUS of Alexandria (about 98 A. D.). Their dual character (see Chapter 25.) can be seen from the following arrangement of the questions (Fig.).

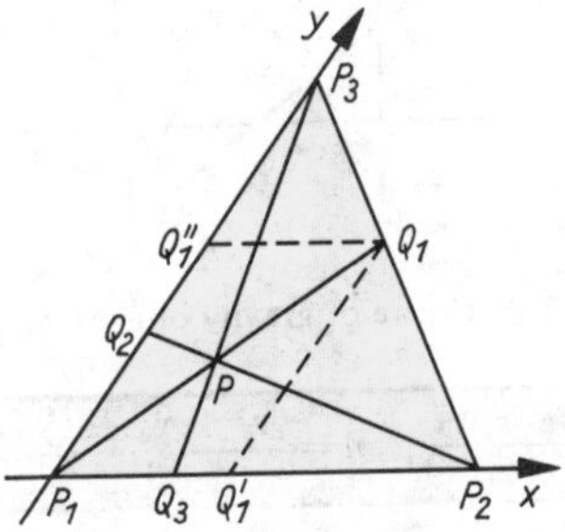

13.3-10 The theorem of Ceva

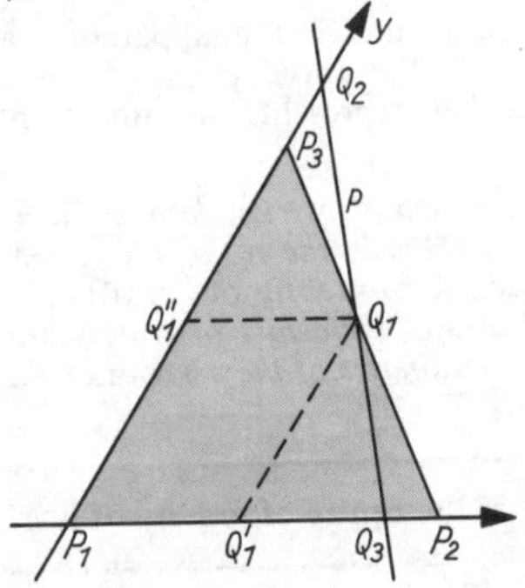

13.3-11 The theorem of Menelaus

Theorem of Ceva. Under what conditions do three lines, each of which passes through one vertex of the triangle but does not coincide with a side, meet in a point? –

Theorem of Menelaus. Under what conditions do three points, each of which lies on one side of the triangle but does not coincide with a vertex, lie on a line? –

Theorem of Ceva. Three lines, each of which passes through one vertex of a triangle, meet in a point if and only if the product of the ratios in which they divide the opposite sides has the value 1.

If the vertices of the triangle are denoted by P_1, P_2, P_3 and the points of intersection of the three lines with the opposite sides by Q_1, Q_2, Q_3, then they form the ratios: $\lambda_1 = \overrightarrow{P_2Q_1} : \overrightarrow{Q_1P_3}$; $\lambda_2 = \overrightarrow{P_3Q_2} : \overrightarrow{Q_2P_1}$; $\lambda_3 = \overrightarrow{P_1Q_3} : \overrightarrow{Q_3P_2}$. A coordinate system is taken so that the line through P_1 and P_2 is the x-axis and the line through P_1 and P_3 is the y-axis; the point P_2 is taken as $(1, 0)$ and P_3 as $(0, 1)$. Then the coordinates of Q_1, Q_2, Q_3 can be calculated as follows:

$$\overrightarrow{P_1Q_2} : \overrightarrow{Q_2P_3} = 1 : \lambda_2 \text{ or } \overrightarrow{P_1Q_2} : \overrightarrow{P_1P_3} = 1 : (1 + \lambda_2) = y_2;\ x_2 = 0;$$
$$\overrightarrow{P_1Q_3} : \overrightarrow{Q_3P_2} = \lambda_3 \text{ or } \overrightarrow{P_1Q_3} : \overrightarrow{P_1P_2} = \lambda_3 : (1 + \lambda_3) = x_3;\ y_3 = 0;$$
$$\overrightarrow{P_1Q_1'} : \overrightarrow{P_1P_2} = \overrightarrow{P_3Q_1} : \overrightarrow{P_3P_2} = 1 : (1 + \lambda_1) = x_1;$$
$$\overrightarrow{P_1Q_1''} : \overrightarrow{P_1P_3} = \overrightarrow{P_2Q_1} : \overrightarrow{P_2P_3} = \lambda_1 : (1 + \lambda_1) = y_1.$$

If x and y are the coordinates of the point of intersection P, then the following three line equations must hold simultaneously:

(1) line through Q_1, P and P_1: $(y_1 - 0)/(x_1 - 0) = y/x$ or $\lambda_1 = y/x$;
(2) line through Q_2, P and P_2: $(y_2 - 0)/(x_2 - 1) = y/(x - 1)$ or $[1/(1 + \lambda_2)] : (-1) = y/(x - 1)$;
(3) line through Q_3, P and P_3: $(y_3 - 1)/x_3 = (y - 1)/x$ or $(-1) : [\lambda_3/(1 + \lambda_3)] = (y - 1)/x$.

By eliminating y from (1), (2) and (3) one obtains

(2′) $-1/(1 + \lambda_2) = x\lambda_1/(x - 1)$; $1 - x = x(\lambda_1 + \lambda_1\lambda_2)$, $x = 1/(1 + \lambda_1 + \lambda_1\lambda_2)$;
(3′) $-(1 + \lambda_3)/\lambda_3 = (x\lambda_1 - 1)/x$, $x + x\lambda_3 = \lambda_3 - x\lambda_1\lambda_3$, $x = \lambda_3/(1 + \lambda_3 + \lambda_1\lambda_3)$.

By eliminating x from (2′) and (3′) one obtains

$$\lambda_3 + \lambda_1\lambda_3 + \lambda_1\lambda_2\lambda_3 = 1 + \lambda_3 + \lambda_1\lambda_3,$$
$$\lambda_1\lambda_2\lambda_3 = 1, \text{ which was to be proved.}$$

Theorem of Menelaus. A transversal cuts the sides of a triangle in such a way that the product of the ratios in which the points of intersection divide the three sides has the value -1.

Just as in Ceva's theorem, let

$$\lambda_1 = \overrightarrow{P_2Q_1} : \overrightarrow{Q_1P_3}, \quad \lambda_2 = \overrightarrow{P_3Q_2} : \overrightarrow{Q_2P_1}, \quad \lambda_3 = \overrightarrow{P_1Q_3} : \overrightarrow{Q_3P_2}$$

be the ratios in which the points of intersection divide the sides; the vertices of the triangle can again have the coordinates $P_1(0, 0)$, $P_2(1, 0)$, $P_3(0, 1)$. The coordinates of the points of intersection have the same values, except that one of the ratios (or all three), λ_2 in the figure, has a negative value. If the points Q_1, Q_2, Q_3 are to lie on a line, then

$$(y_2 - y_3):(x_2 - x_3) = (y_1 - y_3):(x_1 - x_3),$$
$$[1/(1+\lambda_2)]:[-\lambda_3/(1+\lambda_3)]$$
$$= [\lambda_1/(1+\lambda_1)]:[1/(1+\lambda_1) - \lambda_3/(1+\lambda_3)],$$
$$-(1+\lambda_3)/[\lambda_3(1+\lambda_2)]$$
$$= \lambda_1(1+\lambda_3)/[1+\lambda_3-\lambda_3(1+\lambda_1)],$$
$$-(1+\lambda_3)+\lambda_3(1+\lambda_1) = \lambda_1\lambda_3(1+\lambda_2),$$
$$-1-\lambda_3+\lambda_3+\lambda_1\lambda_3 = \lambda_1\lambda_3+\lambda_1\lambda_2\lambda_3,$$
$$-1 = \lambda_1\lambda_2\lambda_3,$$ which was to be proved.

	Q_1	Q_2	Q_3
x	$\frac{1}{1+\lambda_1}$	0	$\frac{\lambda_3}{1+\lambda_3}$
y	$\frac{\lambda_1}{1+\lambda_1}$	$\frac{1}{1+\lambda_2}$	0

13.4. The circle

Equations of a circle

Equations of a circle in rectangular coordinates. The circle is the locus of all points $P(x, y)$ of the plane that have a constant distance r from a fixed point $C(c, d)$; C is called the centre and r the radius of the circle. By the theorem of Pythagoras (Fig.) one obtains the required equation $(x-c)^2+(y-d)^2=r^2$. If the centre is at the origin, then $c=d=0$, and the equation of the circle is $x^2+y^2=r^2$.

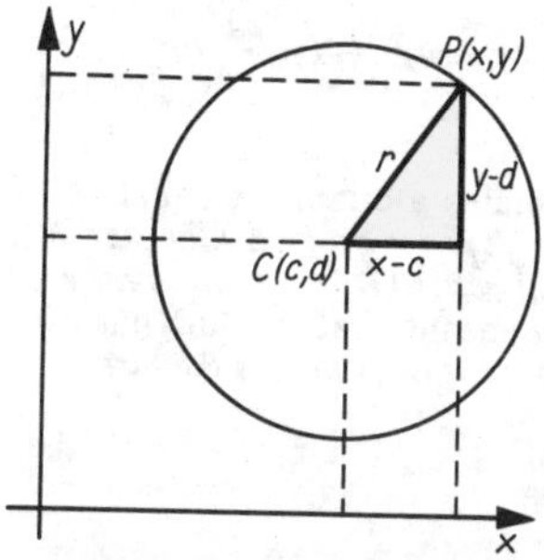

Equation of the circle in general position; centre $C(c, d)$, radius r	$(x-c)^2+(y-d)^2=r^2$
Equation of the circle with centre at the origin, radius r	$x^2+y^2=r^2$

13.4-1 Derivation of the equation of a circle

Example 1: The circle with centre at $C(4, 3)$ and radius 2 has the equation $(x-4)^2+(y-3)^2=2^2$.

Example 2: The point $P_0(1, 2)$ does not lie on the circle $(x-1/2)^2+(y-2)^2=5^2$, because its coordinates $x_0=1, y_0=2$ do not satisfy the equation of the circle. For $(1-1/2)^2+(2-2)^2 \neq 25$.

Example 3: What are the ordinates of the points P_1 and P_2 on the circle $(x-1)^2+(y-2)^2=61$ that have the abscissa 6? – Required are the ordinates y_1 and y_2 of the points $P_1(6, y_1)$ and $P_2(6, y_2)$. If one puts $x=6$ into the equation of the circle and solves for y, one obtains $y-2 = \pm\sqrt{[61-(6-1)^2]}$, that is, $y_1=8$, $y_2=-4$.

The equation of a circle in polar coordinates. The centre C of a circle of radius r has the polar coordinates $C(\varrho_0, \varphi_0)$. P is an arbitrary point on the circle with the coordinates (ϱ, φ). In the triangle OCP, $|CP|=r$ and $|OC|=\varrho_0$ have fixed values, and ϱ varies with the angle φ between the values $\varrho_{min}=|\varrho_0-r|$ and $\varrho_{max}=\varrho_0+r$ (Fig.). By the *cosine theorem*, $\varrho^2+\varrho_0^2-2\varrho_0\varrho\cos(\varphi-\varphi_0)=r^2$. If the centre of the circle lies on the polar axis and the circle passes through the origin one speaks of the *vertex position* – then since the angle in a semi-circle is a right angle, the simplified equation of the circle is $\varrho=2r\cos\varphi$.

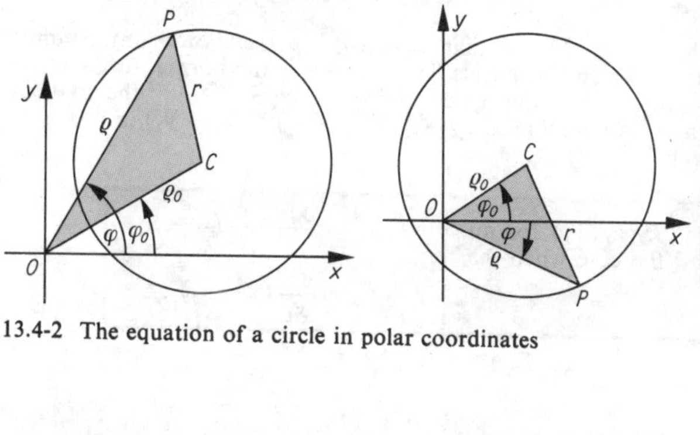

General equation of a circle with centre at $C(\varrho_0, \varphi_0)$ and radius r in polar coordinates
$\varrho^2+\varrho_0^2-2\varrho_0\varrho\cos(\varphi-\varphi_0)=r^2$
Special equation in the vertex position
$\varrho=2r\cos\varphi;\ \varrho \geqslant 0$

13.4-2 The equation of a circle in polar coordinates

Example: If C has the coordinates (4, 30°) and $r = 3$, the equation of the circle in polar coordinates is $\varrho^2 + 16 - 2\varrho \cdot 4 \cos(\varphi - 30°) = 9$, or $\varrho^2 - 8\varrho \cos(\varphi - 30°) + 7 = 0$.

Parametric representation of the circle. If the two coordinates x and y are regarded as functions $x = \varphi_1(t)$, $y = \varphi_2(t)$ of *one variable t*, then t is called a *parameter*, and one speaks of a parametric representation (Fig.). In physical applications the time is often taken as the parameter. For the circle the parametric representation is $x = c + r\cos t$, $y = d + r \sin t$ if the parameter t is taken to be the angle between the positive direction of the x-axis and the radius to the variable point $P(x, y)$.

Parametric representation of the circle with centre at $C(c, d)$ and radius r	$x = c + r\cos t$ $y = d + r\sin t$

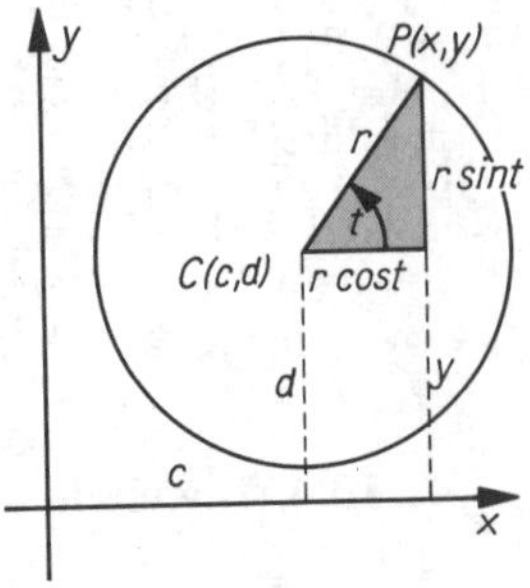

13.4-3 The parametric equations of a circle

Example: The circle with centre at $C(3, 4)$ and radius 2 has the parametric representation $x = 3 + 2\cos t$, $y = 4 + 2\sin t$.

Circle and line

Suppose that a circle $(x - c)^2 + (y - d)^2 = r^2$ and a line $y = mx + \bar{c}$ are given. The coordinates x_0, y_0 of a *point of intersection P_0* must satisfy the equation of the circle and the equation of the line. One obtains the system of equations

$$\left| \begin{aligned} (x_0 - a)^2 + (y_0 - b)^2 &= r^2 \\ mx_0 + \bar{c} &= y_0 \end{aligned} \right| .$$

By substituting for y_0, squaring and collecting terms together one obtains a quadratic equation of the form $x_0^2 + 2px_0 + q = 0$ with general solution $x_0 = -p \pm \sqrt{(p^2 - q)}$ (see Chapter 4.). According to the sign of its *discriminant* $D = p^2 - q$ it has *two real roots* ($D > 0$), *one real root* ($D = 0$), or *two conjugate complex roots* ($D < 0$). Geometrically this means that the line has two, one, or no points in common with the circle, that is, it is a *secant*, a *tangent* or it *misses the circle*.

Examples: 1. For the points of intersection of the circle $(x - 3)^2 + (y - 2)^2 = 40$ with the line $y = -x + 9$ one obtains

$$\left| \begin{aligned} (x_0 - 3)^2 + (y_0 - 2)^2 &= 40 \\ y_0 &= -x_0 + 9 \end{aligned} \right| \rightarrow (x_0 - 3)^2 + (-x_0 + 7)^2 = 40 \rightarrow 2x_0^2 - 20x_0 = -18$$

$$x_{01} = 9,\ x_{02} = 1 \rightarrow y_{01} = 0,\ y_{02} = 8$$

The two points of intersection are $P_1(9, 0)$ and $P_2(1, 8)$.

2. To find the points of intersection of the line $y = -x/2 + {}^5/_2 \sqrt{5}$ with the circle $x^2 + y^2 = 25$, one solves the quadratic equation obtained by substitution; often, for simplicity, the index that characterizes the points of intersection is emitted.

$$\begin{aligned} x^2 + x^2/4 + 125/4 - (5/2)\sqrt{5}x &= 25, \\ (5/4)x^2 - (5/2)\sqrt{5}x &= -25/4, \\ x^2 - 2x\sqrt{5} &= -5, \\ (x - \sqrt{5})^2 &= 0, \\ x_1 = x_2 &= \sqrt{5}. \end{aligned}$$

The discriminant $D = 5 - 5$ has the value zero, and the line touches the circle at the point $x_0 = \sqrt{5}$, $y_0 = 2\sqrt{5}$.

3. The line $x = 6$ has no point in common with the circle $x^2 + y^2 = 25$. By substitution one obtains a quadratic equation for y, $36 + y^2 = 25$, with discriminant $D = 25 - 36 = -11$, that is, there is no real solution.

Normals to the circle. Geometrically it is well known that a tangent is *perpendicular* to the radius through the point of contact. The line on which this radius lies is therefore the normal to the circle at the point of contact. For the point $P_1(x_1, y_1)$ on the circle $(x - c)^2 + (y - d)^2 = r^2$ the gradient of the normal is $(y_1 - d)/(x_1 - c)$ and its equation is $(y - y_1)/(x - x_1) = (d - y_1)/(c - x_1)$ or $y - y_1 = [(y_1 - d)/(x_1 - c)](x - x_1)$.

Equation of the normal through $P_1(x_1, y_1)$ for an arbitrary position of the circle and the case when the centre is at the origin	$y - y_1 = \dfrac{y_1 - d}{x_1 - c}(x - x_1)$ $y - y_1 = \dfrac{y_1}{x_1}(x - x_1)$

Example: The normal to the circle $(x-2)^2+(y-1)^2=25$ through $P_1(5,-3)$ has the equation $y+3=[(-3-1)/(5-2)](x-5)$ or in the Cartesian normal form $y=-(4/3)x+11/3$.

Tangents to the circle. If the *point of contact* $P_1(x_1, y_1)$ of a tangent to the circle $(x-c)^2+(y-d)^2=r^2$ is given, then $m_1=(y_1-d)/(x_1-c)$ is the gradient of the radius to the point of contact, and $m_2=-1/m_1=-(x_1-c)/(y_1-d)$ is the *gradient of the tangent* (Fig.).

In the *point-direction form* the equation of the tangent is $y-y_1=-(x_1-c)(x-x_1)/(y_1-d)$, or, by multiplying up and collecting terms together, $yy_1-y_1^2-yd+y_1d=-xx_1+x_1^2+cx-cx_1$, that is, $xx_1+yy_1-cx-dy=x_1^2+y_1^2-cx_1-dy_1$. If one adds to both sides the expression $(c^2+d^2-cx_1-dy_1)$ and bears in mind the fact that P_1 lies on the circle, so that $(x_1-c)^2+(y_1-d)^2=r^2$, one obtains the equation of the tangent in the form $(x-c)x_1-(x-c)c+(y-d)y_1-(y-d)d=r^2$ or $(x-c)(x_1-c)+(y-d)(y_1-d)=r^2$.

Equation of the tangent at $P_1(x_1, y_1)$ for an arbitrary position of the circle and the case when the centre is at the origin	
$(x-c)(x_1-c)+(y-d)(y_1-d)=r^2$	$xx_1+yy_1=r^2$

Example: The equation of the tangent to the circle $(x-2)^2+(y-1)^2=25$ at the point $P(5,-3)$ is $(x-2)(5-2)+(y-1)(-3-1)=25$ or $3(x-2)-4(y-1)=25$, or in the Cartesian normal form $y=(3/4)x-27/4$.

By means of the *differential calculus* one can find the gradient of the tangent by differentiating the equation of the circle. From $(x-c)^2+(y-d)^2=r^2$ it follows that $2(x-c)+2(y-d)y'=0$ or $y'=-(x-c)/(y-d)$. The gradient of the tangent at $P_1(x_1, y_1)$ is therefore $y_1'=-(x_1-c)/(y_1-d)$, which agrees with the value already found.

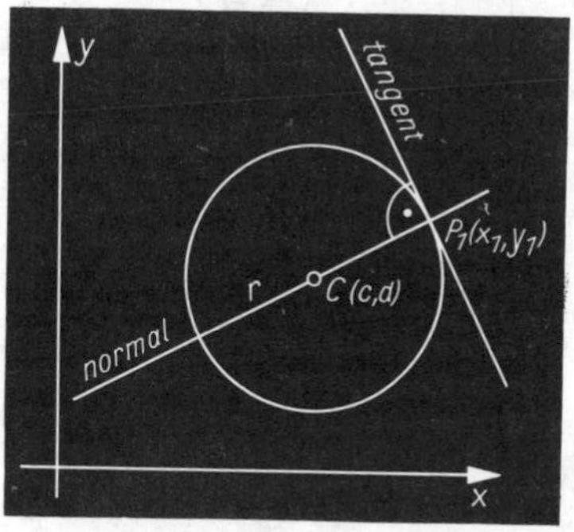

13.4-4 Tangent to a circle

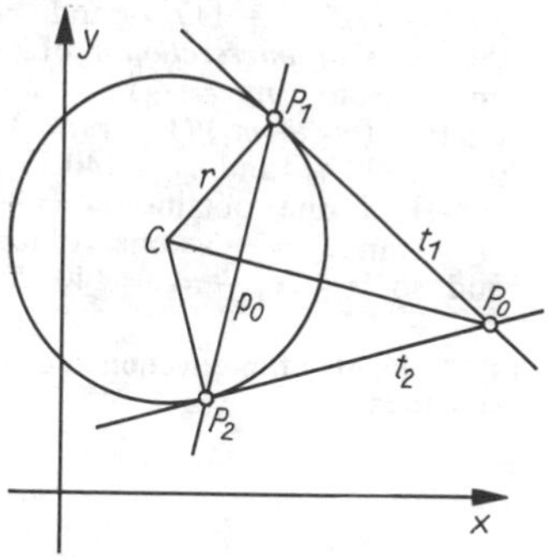

13.4-5 Tangents to a circle from a point outside the circle

The tangents from a point to a circle. If $C(c, d)$ is the centre of a circle of radius r and $P_0(x_0, y_0)$ is a point outside the circle, then there are two tangents from P_0 to the circle (Fig.). Their *points of contact* are $P_1(x_1, y_1)$ and $P_2(x_2, y_2)$. The equations

$$(x-c)(x_1-c)+(y-d)(y_1-d)=r^2 \quad \text{and} \quad (x-c)(x_2-c)+(y-d)(y_2-d)=r^2$$

are satisfied for the coordinates of $P_0(x_0, y_0)$:

$$(x_0-c)(x_1-c)+(y_0-d)(y_1-d)=r^2, \quad (x_0-c)(x_2-c)+(y_0-d)(y_2-d)=r^2.$$

The equation $(x_0-c)(x-c)+(y_0-d)(y-d)=r^2$

therefore holds for the coordinates of both points of contact $P_1(x_1, y_1)$ and $P_2(x_2, y_2)$. The equation therefore represents a line that passes through both points of contact. This line is called the *polar* p_0 of the *pole* P_0 and is determined by the coordinates (c, d) of the centre C and those of the pole $P_0(x_0, y_0)$, in addition to the radius r of the circle. *Its points of intersection with the circle are the points of contact of the two tangents from* P_0. From the coordinates of the pole (x_0, y_0) and one point of contact, one can always give the equation of the tangent, for example, in the two-point form.

Example: To find the tangents from $P_0(3, 5)$ to the circle with the equation $(x+2)^2+y^2=5$. The equation of the *polar* is $(3+2)(x+2)+(5-0)(y-0)=5$ or $y=-x-1$. For its *points of intersection* P_1 and P_2 with the circle one has $y=-x-1$ and $(x+2)^2+y^2=5$, and so $x^2+4x+4+x^2+2x+1=5$ or $x^2+3x=0$; $x_1=0$, $x_2=-3$. The *points of contact* are therefore $P_1(0,-1)$ and $P_2(-3, 2)$. The tangent that touches at P_1 has the equation $y=2x-1$, and the other is $y=x/2+7/2$.

Two circles

Points of intersection of two circles. Two circles, whose equations are $(x - c_1)^2 + (y - d_1)^2 = r_1^2$ and $(x - c_2)^2 + (y - d_2)^2 = r_2^2$, can lie in such a position that they *intersect* in two points, or they can *touch* at one point, or they can be *separate*, that is, have no common point.

To find a possible point of intersection $P_0(x_0, y_0)$ one has to solve the system of equations

$$\left| \begin{array}{l} (x_0 - c_1)^2 + (y_0 - d_1)^2 = r_1^2 \\ (x_0 - c_2)^2 + (y_0 - d_2)^2 = r_2^2 \end{array} \right| .$$

If it has *two real* distinct solutions x_{01}, y_{01} and x_{02}, y_{02}, then $P_1(x_{01}, y_{01})$ and $P_2(x_{02}, y_{02})$ are the two points of intersection; in the case of a *real double solution* the circles touch. If the system of equations has *no real solution*, the circles are separate.

Example: By subtracting the equations $(x+4)^2+(y+5)^2 = 194$ and $(x-3)^2+(y-2)^2 = 40$ of two circles one obtains the equation $x + y = 9$ of the line which is a common chord of the two circles. Its points of intersection with one of the circles are also the points of intersection of the two circles; one finds that these are $P_1(9, 0)$ and $P_2(1, 8)$.

The angle of intersection of two circles. The angle of intersection of two circles is defined as the angle between the *tangents* at each point of intersection; this has the same value for both points of intersection (Fig.).

Example: The circles $(x + 4)^2 + (y + 5)^2 = 194$ and $(x - 3)^2 + (y - 2)^2 = 40$ intersect at the point $P_1(9, 0)$. The tangents are

$$\begin{array}{ll} & (9 + 4)(x + 4) + 5(y + 5) = 194 \\ \text{and} & (9 - 3)(x - 3) + (-2)(y - 2) = 40 \end{array}$$

or, in the Cartesian normal form, $y = -(13/5)x + 117/5$ and $y = 3x - 27$. For the *angle of intersection* ψ of these two lines one finds from $m_1 = -13/5$ and $m_2 = 3$ that $\tan \psi = (m_2 - m_1)/(1 + m_1 m_2) = -14/17$ and so $\psi_2 = -39.47°$ and $\psi_1 = 140.53°$; from $m_1 = \tan \alpha_1 = -13/5$ one obtains $\alpha_1 = -68.96°$ and from $m_2 = \tan \alpha_2 = +3$ one obtains $\alpha_2 = +71.57°$ and so $\psi = \alpha_2 - \alpha_1 = 140.53°$.

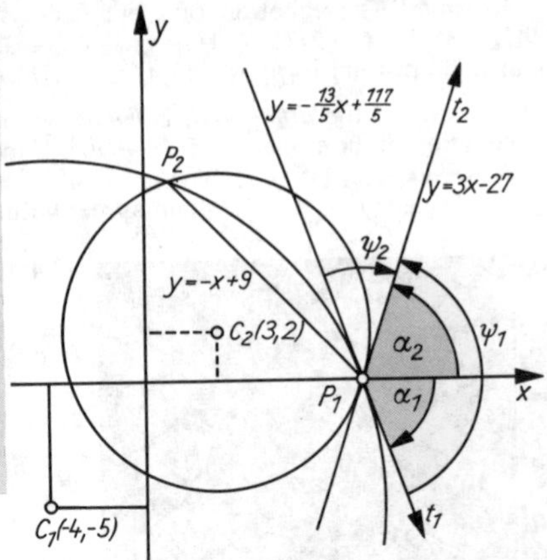

13.4-6 Points of intersection and angle of intersection of two circles

13.5. The conics

Conics as intersections of a circular cone with planes

In antiquity conics were defined as intersections of a plane E with a circular cone. The intersections are called circle, ellipse, hyperbola and parabola. If the intersecting plane E contains the vertex Z of the (double) cone, the intersection is either a *point*, the vertex Z, or a *generating line*, if E touches the cone, or *two generating lines* intersecting at Z, if the plane contains interior points of the cone. These intersections are known as *degenerate conics*.

If E does not contain Z, but is perpendicular to the axis of a right cone, then the intersection is a circle; if E is parallel to a tangent plane, the intersection is a parabola; if E is neither parallel to a tangent plane nor perpendicular to the axis, then the intersection is an ellipse if it intersects all the generators on the same side of Z, and a hyperbola otherwise.

All non-degenerate conics can be regarded as *perspective images* of one another; Z is the centre of perspective. *On each generator there lies one and only one point of each conic*, apart from three exceptions, which can be eliminated by means of the concept of the improper point or point at infinity of a line (just as in the ratio of division). For the *parabola*, the generator g_0 parallel to E has a point (at infinity) in common with the line through the vertex V of the parabola parallel to it, and this is a point of the parabola; the parallel line VF through the vertex of the parabola is called the *axis of the parabola*. In the case of the *hyperbola* there are two generators in which the plane E' parallel to E through Z cuts the cone. In projective geometry the two planes E and E'

have a line (at infinity) l in common, and each of the two generators cuts l in a point (at infinity), which is a point of each of the lines parallel to the generator and is a point of the hyperbola; in particular, the two lines parallel to these generators through the centre of the hyperbola are called its *asymptotes*.

The Dandelin spheres. Pierre DANDELIN (1794–1847) was the first to use the spheres that touch the cone and the intersecting plane E to derive properties of the conics.

The parabola. If E is parallel to a tangent plane of the cone that touches it along a generator g_0, then there is *only one Dandelin sphere* that touches the cone and E; its diameter is the distance of g_0 from E (Fig.). The sphere touches E at a point F and the cone along a circle c, which meets g_0 in a point D. The plane E_1 through c cuts E in a line l, which is called the directrix and is perpendicular to the plane Σ through g_0 and the axis of the cone. The plane Σ cuts E in the *axis of the parabola*; the (finite) point of the parabola on the axis is the vertex V of the parabola. By rotating Σ about g_0 one obtains lines of intersection with E, for example, BP, parallel to the axis of the parabola, that is, perpendicular to the directrix. The plane Σ_A obtained by rotation about g_0 cuts the cone in a second generator through the points Z, A and P; A lies on c and P lies on the line of intersection; P is therefore a point of the parabola. The segments PF and PA are of equal length as tangents from P to the sphere. The segments PA and PB are of equal length, since the lines ZP and DB intersect at A and are cut by the parallels BP and DZ, so $|BP|:|PA| = |DZ|:|ZA| = 1$, that is $|BP| = |PA|$.

The parabola is the locus of all points P of the plane that have the same distance from a fixed point F and a fixed line l. The ratio $|PF|:|PB| = \varepsilon$ has the value 1 and is called the numerical eccentricity.

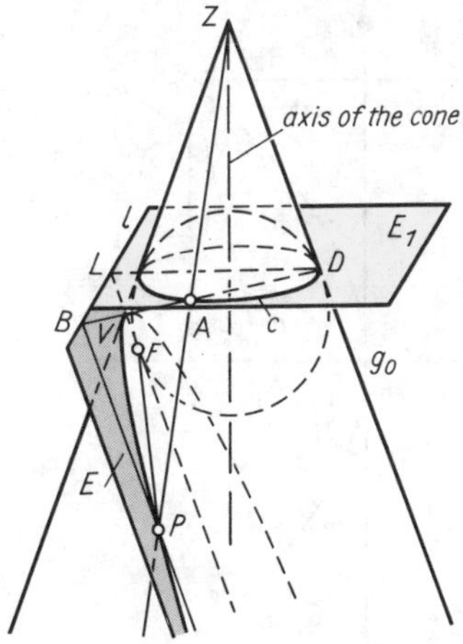

13.5-1 Parabola as section of a cone

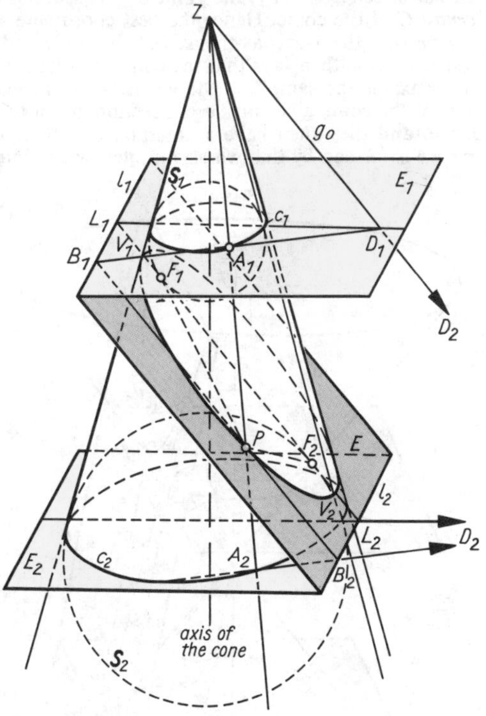

13.5-2 Ellipse as section of a cone

The ellipse and hyperbola. If E is not parallel to a tangent plane of the cone, there are two Dandelin spheres that touch E at points F_1 and F_2 and touch the cone along the circles c_1 and c_2, respectively. The planes E_1 and E_2 of these circles cut E in a pair of parallel directrices l_1 and l_2. The plane Σ perpendicular to the directrices through the axis of the cone cuts E in the *axis of the ellipse* or *hyperbola* (Fig.).

A line g_0 in this plane Σ parallel to the axis of the conic through the vertex Z of the cone cuts the planes E_1 in D_1 and E_2 in D_2. If Σ is rotated about g_0, then its line of intersection with E, say B_1B_2, remains parallel to the axis of the conic, that is, perpendicular to the directrices l_1 and l_2. The plane Σ_A contains a generator ZA_1A_2; the point of intersection P of B_1B_2 with A_1A_2 is a point of the conic. The segments PF_1 and PA_1 are of equal length as tangents from P to the sphere S_1; similarly, $|PF_2| = |PA_2|$. Because of the position of S_1 and S_2 on opposite sides or on the same side of E, one has, for the *ellipse* $|PF_1| + |PF_2| = |A_1A_2|$, and for the *hyperbola* $|PF_1| - |PF_2| = |A_1A_2|$.

The ellipse is the locus of all points P of the plane for which the sum of the distances from two fixed points F_1 and F_2 (the foci) is constant; by symmetry this constant ($2a$) is equal to the distance between the points V_1 and V_2 of the ellipse that lie in Σ, which are called vertices.

The hyperbola is the locus of all points P of the plane for which the difference of the distances from two fixed points F_1 and F_2 (the foci) is constant; for the vertices V_1 and V_2 in Σ one again has $|V_1V_2| = 2a = |A_1A_2|$.

In the plane Σ_A that arises by rotation about g_0 the lines ZP and B_1D_1 intersect at A_1, and B_1P is parallel to ZD_1. Hence $|PA_1| : |PB_1| = |ZA_1| : |ZD_1| = |PF_1| : |PB_1| = \varepsilon$.

The ratio of the distance $|PF_1|$ of a point P on a conic from a focus F_1 to its distance $|PB_1|$ from the corresponding directrix l_1 is a constant ε, the numerical eccentricity; for the ellipse $0 < \varepsilon < 1(|ZA_1| < |ZD_1|)$, and for the hyperbola $\varepsilon > 1(|ZA_1| > |ZD_1|)$.

The equations of the conics. To arrive at an analytic expression for the conics a suitable coordinate system must be chosen. From its definition, a conic is *symmetrical about its axis*. In addition, the ellipse and hyperbola, from the considerations above, must also be *symmetrical about the perpendicular bisector of* F_1F_2; the point of intersection of this perpendicular bisector with the axis is the *centre* C of the conic. Hence the best coordinate system for an ellipse or hyperbola is a *Cartesian system* with the x-axis as the axis of the conic and the y-axis as the perpendicular line through the centre C. One then says that the conic is in its *central position*. One speaks of the *vertex position* if the x-axis is the same, but the y-axis is the tangent at a vertex. Also *polar coordinates*, where the axis of the conic gives the zero direction and a focus is the pole, are suitable for all three types of conic, and they then have a common equation. For the hyperbola, a natural *oblique coordinate system* is formed by the two asymptotes, which intersect at the centre.

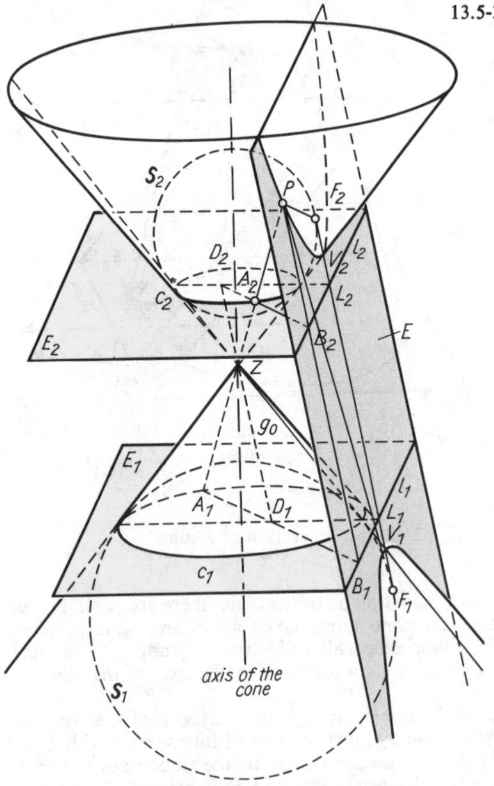

13.5-3 Hyperbola as section of a cone

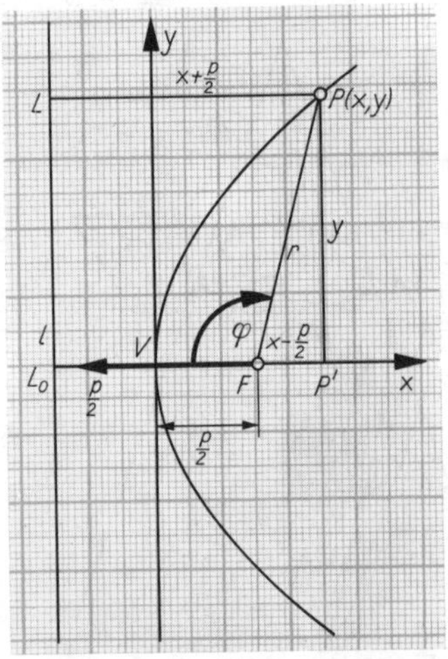

13.5-4 Derivation of the equation of a parabola

Equations of the parabola

The vertex equation. The Cartesian coordinate system is such that the x-axis is the *axis of the parabola* and the y-axis is the *tangent at the vertex* (Fig.). From the definition of the parabola, each of its points P has the same distance from the *focus* F and the directrix l. The *vertex* V must therefore bisect the perpendicular FL_0 from F to l. There are two points on the parabola whose ordinates are equal to the distance from the directrix. The absolute value p of this ordinate is called the *semi-parameter* of the parabola; $|L_0F| = p$. The focus F therefore has the coordinates $(p/2, 0)$. An arbitrary point $P(x, y)$ of the parabola has the distance $|FP| = \sqrt{[y^2 + (x - p/2)^2]}$ from the focus F and the distance $|PL| = p/2 + x$ from the directrix. By the definition of the parabola,

$(p/2 + x)^2 = y^2 + (x - p/2)^2$ or $y^2 = 2px$.

Vertex equation of the parabola	$y^2 = 2px$

The equation shows that the x-axis is an axis of symmetry; the vertex is at the origin; for each abscissa $x > 0$ there are two points of the parabola whose ordinates are equal and opposite. The semiparameter determines the *form of the parabola*. The smaller the value of p is, the nearer the focus and directrix come to the y-axis and the more slowly y increases. In the limit, as $p \to 0$, the parabola degenerates to the positive x-axis counted twice. On the other hand, if p takes a very large value, then the focus and directrix are at a large distance apart, and as $p \to \infty$, the parabola degenerates to the y-axis, since $x \to 0$.

The equations $x^2 = 2py$, $y^2 = -2px$ and $x^2 = -2py$ with $p > 0$ also represent parabolas, as can be seen from the diagram (Fig.), in which the parabola $\eta^2 = 2p\xi$ goes over into one of the given equations by a suitable rotation of the ξ, η-coordinate system through an angle ψ. The equations of transformation are $\xi = x \cos\psi - y \sin\psi$ and $\eta = x \sin\psi + y \cos\psi$.

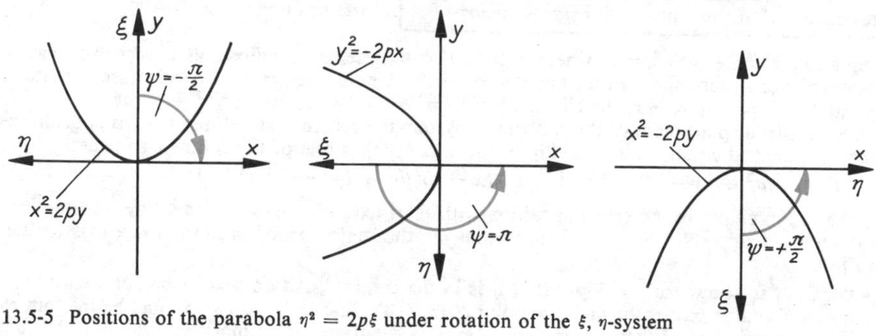

13.5-5 Positions of the parabola $\eta^2 = 2p\xi$ under rotation of the ξ, η-system

Parabola	ψ	Equations of transformation	Transformed equation	Fig.	Interval for x	Interval for y
$\eta^2 = 2p\xi$	$-\pi/2$	$\xi = y, \quad \eta = -x$	$x^2 = 2py$	a	$-\infty < x < +\infty$	$0 \leqslant y < +\infty$
$\eta^2 = 2p\xi$	$+\pi$	$\xi = -x, \eta = -y$	$y^2 = -2px$	b	$-\infty < x \leqslant 0$	$-\infty < y < +\infty$
$\eta^2 = 2p\xi$	$+\pi/2$	$\xi = -y, \eta = x$	$x^2 = -2py$	c	$-\infty < x < +\infty$	$-\infty < y \leqslant 0$

If, after a parallel displacement of the coordinate system, the vertex has the coordinates (c, d), then the equation of the parabola takes one of the following forms ($p > 0$):

$$(y - d)^2 = 2p(x - c); \qquad (x - c)^2 = 2p(y - d);$$
$$(y - d)^2 = -2p(x - c); \qquad (x - c)^2 = -2p(y - d).$$

Example 1: To find the equation of the parabola, in the vertex position with x-axis as the axis of the parabola, that passes through the point $P_0(2, 4)$. – The equation of the parabola must be satisfied by the coordinates of P_0: $4^2 = 2p \cdot 2$; hence $p = 4$. The equation of the parabola is therefore $y^2 = 8x$.

Example 2: The parabola that has its vertex at $V(2, 3)$, is concave downwards, and passes through the point $P_0(4, 1)$, must have an equation that is a transformation by parallel displacement of axes of the equation $x^2 = -2py$; its equation is $(x - 2)^2 = -2p(y - 3)$. Since it passes through the point $P_0(4, 1)$, one has $(4 - 2)^2 = -2p(1 - 3)$, or $p = +4/4 = +1$. The equation of the parabola is therefore $(x - 2)^2 = -2(y - 3)$. The focus is at a distance $^1/_2 p = 1/2$ from the vertex along the axis, that is, its coordinates are $F(2, 2.5)$.

Equations of the ellipse

The central equation of the ellipse. The x-axis coincides with the *axis of the ellipse*, and the y-axis with the perpendicular bisector of the segment V_1V_2 between the *vertices* (Fig.). The y-axis cuts the ellipse in two points N_1 and N_2, the *secondary vertices*. The length $|V_1V_2| = 2a$ is called the *major axis*, the length $|N_1N_2| = 2b$ the *minor axis* and $|F_1F_2|/2 = e$ the *linear eccentricity*. Since $|N_1F_1| + |N_1F_2| = 2a$, the segments a, b and e form a right-angled triangle, and so $e^2 + b^2 = a^2$.

The foci therefore have coordinates $F_1(+e, 0)$ and $F_2(-e, 0)$. An arbitrary point of the ellipse $P(x, y)$ has the distances $|PF_1| = r_1 = \sqrt{[y^2 + (e - x)^2]}$ and $|PF_2| = r_2 = \sqrt{[y^2 + (e + x)^2]}$ from the foci. From the definition of the ellipse, $r_1 + r_2 = 2a$ or $r_1 = 2a - r_2$. If one substitutes for r_1 and r_2 and squares both sides, one square root remains:

$$y^2 + (e - x)^2 = 4a^2 - 4a\sqrt{[y^2 + (e + x)^2]} + y^2 + (e + x)^2$$

or $\quad a\sqrt{[y^2 + (e + x)^2]} = a^2 + ex.$

By squaring both sides one obtains

$$a^2y^2 + a^2e^2 + 2a^2ex + a^2x^2 = a^4 + 2a^2ex + e^2x^2.$$

Since $e^2 = a^2 - b^2$ this equation can be simplified:

$$a^2y^2 + a^4 - a^2b^2 + a^2x^2 = a^4 + a^2x^2 - b^2x^2$$

or $\quad x^2/a^2 + y^2/b^2 = 1.$

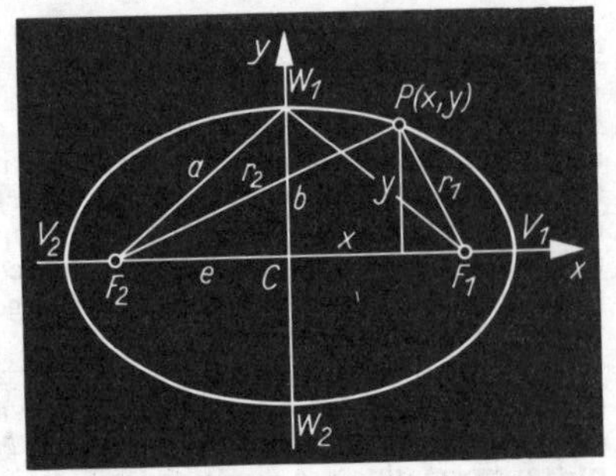

13.5-6 The equation of an ellipse in central position

Central equation of the ellipse	$x^2/a^2 + y^2/b^2 = 1$

The equation $x^2/b^2 + y^2/a^2 = 1$, where $a > b$, also represents an ellipse, as is seen by rotating the ξ, η-coordinate system through an angle $\psi = -\pi/2$. Under the transformation (see Equations of the parabola) $\xi = y, \eta = -x$, the ellipse $\xi^2/a^2 + \eta^2/b^2 = 1$ goes into $x^2/b^2 + y^2/a^2 = 1$.

If after a parallel displacement of the coordinate system the centre of the ellipse has the coordinates (c, d), then the central equation of the ellipse, for $a > b$, takes one of the following forms:

$$(x - c)^2/a^2 + (y - d)^2/b^2 = 1 \quad \text{or} \quad (x - c)^2/b^2 + (y - d)^2/a^2 = 1.$$

Example 1: An ellipse in the central position with semi-axes of lengths 3 and 4 has the equation $x^2/16 + y^2/9 = 1$ or $x^2/9 + y^2/16 = 1$ according as the major axis lies along the x-axis or the y-axis Fig.).

Example 2: It is known of an ellipse that it is in the central position, that it has one semi-axis of length 5, and that it passes through the point $P_0(3, -8)$. Since $8 > 5$, $b = 5$ must be the minor semi-axis. One can find a by substituting $x_0 = 3, y_0 = -8$ in the central equation: $3^2/5^2 + (-8)^2/a^2 = 1$ or $64/a^2 = (25 - 9)/25 = 16/25$, that is, $a = \sqrt{(25 \cdot 64/16)} = 10$. The equation of the ellipse is therefore $x^2/5^2 + y^2/10^2 = 1$.

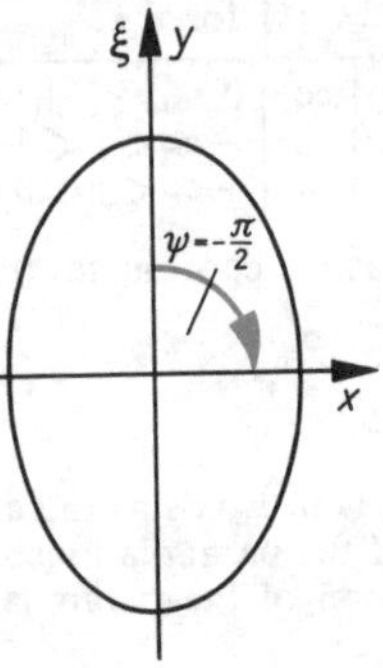

13.5-7 Rotation of the ξ, η-system and the ellipse $x^2/a^2 + y^2/b^2 = 1$

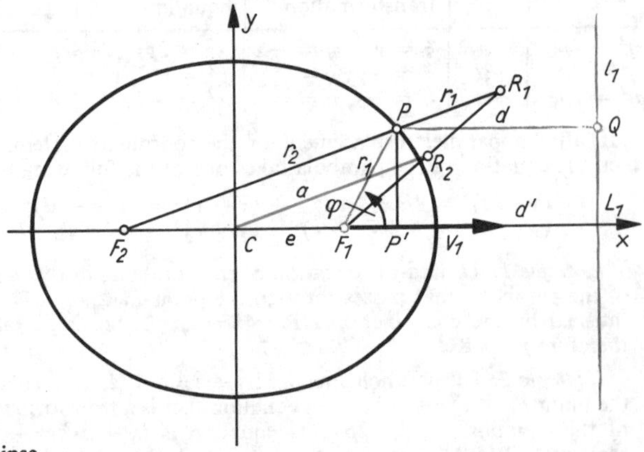

13.5-8 Numerical eccentricity of an ellipse

The numerical eccentricity. A line parallel to the y-axis of an ellipse at a distance $|CL_1| = a^2/e$ from it is called a directrix l_1. The vertex V_1 is at a distance d' from it, where $d' = |V_1L_1| = (a^2/e) - a = (a/e)(a - e)$, and an arbitrary point P of the ellipse is at a distance d from it, where $d = |PQ| = a^2/e - x$. For the distances r_1 and r_2 of the point P from the foci F_1 and F_2 one has, by the theorem of Pythagoras, $r_2^2 = r_1^2 + (2e)^2 - 2 \cdot 2e(e - x) = r_1^2 + 4e^2 - 4e^2 + 4ex$ or $r_2^2 - r_1^2 = 4ex$. Since $r_2 + r_1 = 2a$, it follows by division that $r_2 - r_1 = 2ex/a$, and so $r_2 = a + ex/a$ and $r_1 = a - ex/a$. If one then substitutes, from the last equation, $x = (a^2 - r_1a)/e$ in the expression for d, one obtains $d = r_1 \cdot (a/e)$, that is, the ratio $d : r_1 = a : e$ is independent of the chosen point P. Its reciprocal $\varepsilon = e/a$ is called the numerical eccentricity. In the figure it is shown how to construct ε

as the ratio of two segments: R_1 is determined on F_2P by $|PR_1| = r_1$, and R_2 on F_1R_1 by $CR_2 \parallel F_2R_1$; then $|CF_1| : |CR_2| = e : a = \varepsilon$.

The ellipse is the locus of all points P of the plane for which the ratio $r_1 : d$ of the distance r_1 of P from a focus F_1 to its distance d from the corresponding directrix l_1 has the constant value $\varepsilon = e : a$.

Parametric representation of the ellipse. The ellipse can be regarded as the affine image of the circle $\xi^2 + \eta^2 = a^2$ by reducing all the ordinates in the ratio $y : \eta = b : a$; in this case, the transformation $\xi = x$, $\eta = ya/b$ takes the circle into the ellipse $x^2/a^2 + y^2/b^2 = 1$. The construction is carried out by taking the points of intersection A and B of a ray through the origin with the circles c_a and c_b of radii a and b, drawing lines through A and B parallel to the axes and taking the point P of the ellipse as the point of intersection. One sees that the condition $y : \eta = b : a$ is satisfied. If t is the angle that an arbitrary ray makes with the x-axis, one obtains the parametric representation $x = a \cos t$, $y = b \sin t$, which satisfies the equation of the ellipse (Fig.).

Parametric representation of the ellipse	$x = a \cos t,\ y = b \sin t$

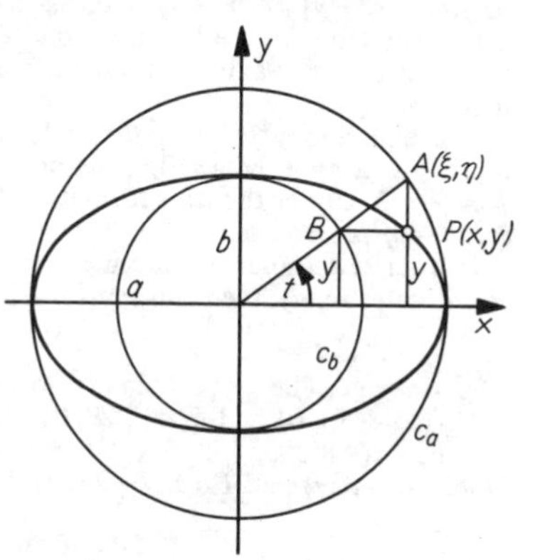

13.5-9 Parametric representation of an ellipse

Equations of the hyperbola

Like the ellipse, the hyperbola is *symmetrical* about the *axis* through the *vertices* V_1 and V_2 and about the line perpendicular to this axis through the *centre* C, $|CV_1| = |CV_2|$. The segment $|CV_1|$ is denoted by a, and the segments $|CF_1| = |CF_2|$ by e. The hyperbola does not have a minor semi-axis; since $e > a$, there is, however, a segment b given by $b^2 = e^2 - a^2$.

Central equation of the hyperbola. Because of the symmetry of the hyperbola, there is a particularly suitable Cartesian coordinate system in which the x-axis coincides with the axis of the hyperbola and the y-axis is the line perpendicular to it through C. The foci have the coordinates $F_1(+e, 0)$ and $F_2(-e, 0)$ (Fig.). The distances of an arbitrary point $P(x, y)$ of the hyperbola from the foci are

$$|PF_1| = r_1 = \sqrt{[y^2 + (x - e)^2]}$$

and $$|PF_2| = r_2 = \sqrt{[y^2 + (x + e)^2]}.$$

From the definition of the hyperbola, $r_2 - r_1 = 2a$ or $r_2 = 2a + r_1$. If one substitutes the expressions for r_1 and r_2 and squares both sides, one square root remains:

$$\cancel{y^2} + (x + e)^2 = 4a^2 + 4a\sqrt{[y^2 + (x - e)^2]} + \cancel{y^2} + (x - e)^2$$

or $$ex - a^2 = a\sqrt{[y^2 + (x - e)^2]}.$$

Squaring both sides again, one obtains

$$e^2x^2 + a^4 - \cancel{2a^2ex} = a^2y^2 + a^2x^2 - \cancel{2a^2ex} + a^2e^2.$$

Since $e^2 = a^2 + b^2$, this equation can be simplified:

$$\cancel{a^2x^2} + b^2x^2 + \cancel{a^4} = a^2y^2 + \cancel{a^2x^2} + \cancel{a^4} + a^2b^2$$

or $$x^2/a^2 - y^2/b^2 = 1.$$

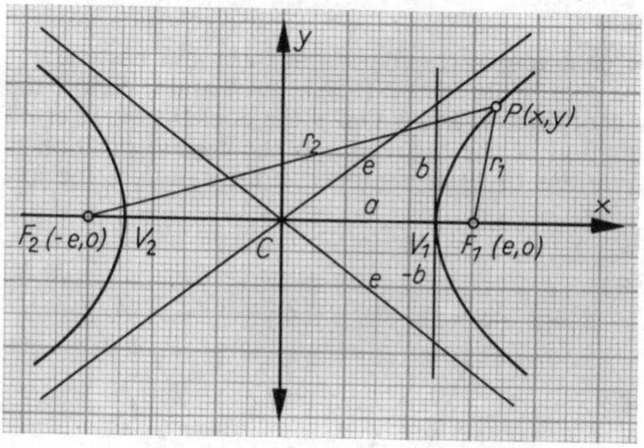

13.5-10 The equation of a hyperbola in central position

The *significance of the number b* can be realized from the following rearrangement of the equation:

$a^2y^2 = b^2(x^2 - a^2)$,
$y/x = \pm(b/a)\sqrt{[1 - a^2/x^2]}$.

Central equation of the hyperbola	$x^2/a^2 - y^2/b^2 = 1$

The limiting value of this expression, as $x \to \infty$, is $\lim_{x\to\infty} y/x = \pm b/a$. The lines $\eta = \pm(b/a)\,\xi$ having these limiting values as gradients are the asymptotes of the hyperbola.

Asymptotes of the hyperbola	$y = \pm(b/a)\,x$

By symmetry it is sufficient to consider the behaviour of the hyperbola $x^2/a^2 - y^2/b^2 = 1$ and the line $\eta = (b/a)\,\xi$ in the first quadrant. If the perpendicular is dropped from a point (ξ, η) with

$\xi > a$ onto the x-axis, it cuts the hyperbola in a point $P(x, y)$. Then $\xi = x$, and from $\eta = (b/a)\,x$ and $y = (b/a)\,x\sqrt{[1 - a^2/x^2]}$ it follows that $y < \eta$, because the factor $\sqrt{[1 - a^2/x^2]}$ is less than 1. The larger x, the smaller the difference $\eta - y$, since it follows from the equation of the hyperbola that $[x/a - y/b] = [x/a + y/b]^{-1}$ and since $[x/a + y/b] \to \infty$, as $x \to \infty$, $\lim\limits_{x\to\infty} [(x/a - (y/b)] = 0$ or $[(b/a)\,x - y] = (\eta - y) \to 0$, as $x \to \infty$. For large values of x the hyperbola therefore comes arbitrarily close to the line $\eta = (b/a)\,\xi$. The line is an asymptote of the hyperbola. Its angle of inclination to the x-axis is obtained from a right-angled triangle with the sides a and b and hypotenuse e.

The equation $y^2/a^2 - x^2/b^2 = 1$, where a is the principal semi-axis and lies along the y-axis, represents a hyperbola (Fig.), as one sees by rotating the ξ, η-coordinate system through an angle $\psi = -\pi/2$. Under the transformation $\xi = y, \eta = -x$ (see Equations of the Parabola) the hyperbola $\xi^2/a^2 - \eta^2/b^2 = 1$ goes into $y^2/a^2 - x^2/b^2 = 1$.

If after a parallel displacement of the coordinate system the centre of the hyperbola has the coordinates (c, d), then the central equation of the hyperbola takes one of the following forms:

$$(x - c)^2/a^2 - (y - d)^2/b^2 = 1 \quad \text{or} \quad (y - d)^2/a^2 - (x - c)^2/b^2 = 1.$$

Example: The hyperbola $x^2/25 - y^2/4 = 1$ has the *vertices* $V_1(5, 0)$ and $V_2(-5, 0)$, *foci* $F_1(\sqrt{(25 + 4)}, 0)$ and $F_2(-\sqrt{29}, 0)$ and *asymptotes* $y = \pm(2/5)\,x$.

The hyperbola $y^2/25 - x^2/4 = 1$, on the other hand, has the *vertices* $V_3(0, 5)$ and $V_4(0, -5)$, *foci* $F_3(0, \sqrt{29})$ and $F_4(0, -\sqrt{29})$ and *asymptotes* $y = \pm(5/2)\,x$.

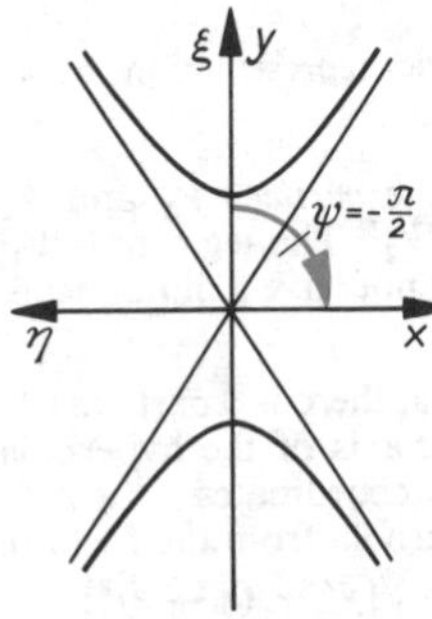

13.5-11 Rotation of the ξ, η-system and the hyperbola $x^2/a^2 - y^2/b^2 = 1$

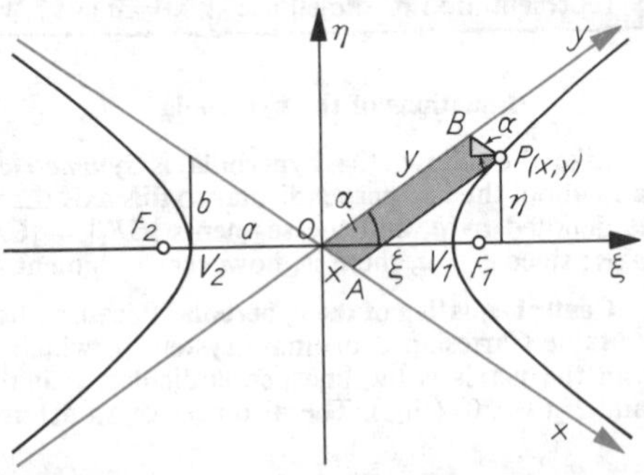

13.5-12 The asymptotic equation of a hyperbola

The asymptotic equation of the hyperbola. The equation of the hyperbola turns out particularly simple if its asymptotes ar taken as the axes of a coordinate system (Fig.). Let ξ and η denote the coordinates in the original system of rectangular coordinates, and x and y the coordinates referred to the asymptotes as oblique axes. Then the equation of the hyperbola in the central position is $\xi^2/a^2 - \eta^2/b^2 = 1$, and $\eta = \pm(b/a)\,\xi$ are the equations of its asymptotes. If $\tan\alpha = b/a$, the following relations hold between the coordinates:

$$\left|\begin{array}{l}\eta = y\sin\alpha - x\sin\alpha\\ \xi = x\cos\alpha + y\cos\alpha\end{array}\right| \quad \text{or} \quad \left|\begin{array}{l}\eta = (y - x)\sin\alpha\\ \xi = (y + x)\cos\alpha\end{array}\right|.$$

Substituting in the central equation, one obtains

$$[(y + x)^2\cos^2\alpha]/a^2 - [(y - x)^2\sin^2\alpha]/b^2 = 1$$

or $\;(y + x)^2\,b^2\cos^2\alpha - (y - x)^2\,a^2\sin^2\alpha = a^2b^2.$

Since $b^2\cos^2\alpha = a^2\sin^2\alpha$, this gives $2xy(b^2\cos^2\alpha + a^2\sin^2\alpha) = a^2b^2$ and so $4xy\cos^2\alpha = a^2$ and $4xy\sin^2\alpha = b^2$. By addition one obtains $4xy = a^2 + b^2$ and since $\sin 2\alpha = 2ab/(a^2 + b^2)$ one has finally $xy\sin 2\alpha = ab/2$. This equation states that the parallelogram $OAPB$ always has the same area.

The equation of a hyperbola referred to its asymptotes has the form $xy =$ const. *Conversely, any function of this type represents a hyperbola.*

Asymptotic equation of the hyperbola
$xy = (a^2 + b^2)/4 = e^2/4 = \text{const}$

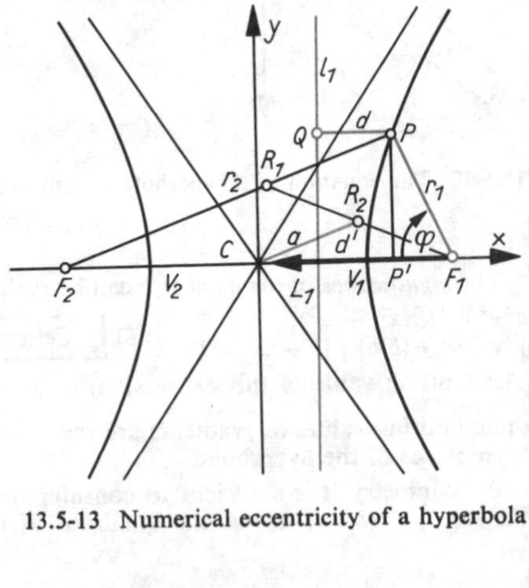

13.5-13 Numerical eccentricity of a hyperbola

The numerical eccentricity. A line parallel to the y-axis of a hyperbola at a distance $|CL_1| = a^2/e$ is called a directrix l_1 (Fig.). The vertex V_1 has the distance d' from it, where $d' = |V_1L_1| = a - a^2/e = a(e - a)/e$ and an arbitrary point P of the hyperbola has the distance d from it, where $d = x - a^2/e$. In the triangle F_1F_2P, by the theorem of Pythagoras, $r_2^2 = r_1^2 + (2e)^2 - 2 \cdot 2e(e - x)$ or $r_2^2 - r_1^2 = 4ex$. Since $r_2 - r_1 = 2a$, one has $r_2 + r_1 = 2ex/a$ and so $r_2 = a + ex/a$ and $r_1 = ex/a - a$. By substituting $x = (r_1a + a^2)/e$ in $d = x - a^2/e$ one obtains $d = r_1a/e$, that is, $r_1 : d = e : a = \varepsilon$. This constant ε is called the *numerical eccentricity*.

The hyperbola is the locus of all points P of the plane for which the ratio $r_1 : d$ of the distance r_1 of P from one focus F_1 to its distance d from the corresponding directrix l_1 has the constant value $\varepsilon = e : a$.

If R_1 is determined on F_2P by $|PR_1| = r_1$ and R_2 on R_1F_1 by $CR_2 \parallel F_2P$, then $|CF_1| : |CR_2| = e : a$

Conic and line

Points of intersection of conic and line. In the derivation of the conics by means of the Dandelin spheres as the plane section of a circular cone it is clear that *to each point P of the circle of contact c* of a Dandelin sphere there corresponds *a point P' of the conic*. To a line l in the plane E_1 or E_2 of the circle there corresponds a line l' in the plane of intersection E, which arises as the line of intersection of this plane E and a plane determined by the line l and the vertex Z of the cone. As in any projective mapping, points of intersection in E_1 or E_2 go into points of intersection in the image plane E. *According as the line l cuts, touches, or misses the circles c_1 or c_2, the image line l' is a secant of or a tangent of the conic or has no point in common with it*; there can also be points of the circle whose images are improper points of the conic, which must therefore be treated separately.

1. In the case of the *parabola*, this is the point D of c; it has the greatest distance from the directrix l (see Fig. 13.5-1). *Each secant through D of the circle c has a line parallel to the axis of the parabola as image* (for example, DA has the image BP), and therefore cuts the parabola in only one finite point P. The *tangent at D* to the circle c is parallel to the directrix, its image is *the line at infinity* of E.

2. In the case of a *hyperbola*, a plane through the vertex Z of the cone parallel to the plane E cuts the circle in two points P_2 and P_3 whose images are the points at infinity on the asymptotes. *The line through these two points has the line at infinity of E as image.* A *secant* of the circle c_1 or c_2 *through one of these points*, say P_2, has a *line parallel to the asymptote* as its image, which cuts the hyperbola in only one finite point. The *tangents* to the circle at P_2 and P_3 go into the *asymptotes*. Their point of intersection is therefore the inverse image of the centre C of the hyperbola.

In finding the points of intersection of a conic with a line it is easy to see from the Cartesian normal form of the line equation whether the special case of a secant with only one point of intersection occurs ($m_p = 0$ for a parabola, $m_h = \pm(b/a)$ for a hyperbola). In all other cases, by substituting for y from the line equation in the equation of the conic a quadratic equation is obtained, whose discriminant gives information about the number of points of intersection.

Example 1: For the coordinates of the points of intersection of the line $y = -x/2 + 2$ with the parabola $x^2 = 4y$ one has to solve the system of these two equations. By substitution one obtains

$$x^2 = -2x + 8 \quad \text{or} \quad x^2 + 2x + 1 = 9,$$
$$x_1 = 2,\ x_2 = -4 \quad \text{and} \quad y_1 = 1,\ y_2 = 4.$$

The points of intersection are therefore $P_1(2, 1)$ and $P_2(-4, 4)$.

Example 2: The conic $16x^2 + 25y^2 + 32x - 100y - 284 = 0$ lies parallel to the axes, since there is no mixed term in its equation. The equation can be written as

$$16x^2 + 32x + 16 + 25y^2 - 100y + 100 - 16 - 100 - 284 = 0$$
$$\text{or} \quad 16(x + 1)^2 + 25(y - 2)^2 = 400 \quad \text{or} \quad (x + 1)^2/25 + (y - 2)^2/16 = 1.$$

The centre C of the conic has the coordinates $(-1, 2)$. The line $5y = 28x - 62$ cuts the ellipse in the points $P_1(2, -6/5)$ and $P_2(3, 22/5)$, since by substituting the equation of the line into the equation of the ellipse one obtains the quadratic equation $x^2 - 5x + 6 = 0$ with the roots $x_1 = 2$, $x_2 = 3$.

Gradients of tangents. Equations of tangents to a conic. The equations of tangents to a parabola, ellipse and hyperbola can be obtained by the methods of analytic geometry, just as for a circle. However, it is much more advantageous to use the method of the *differential calculus*. The derivative of the equation of a conic at x_1 gives the *gradient* of the tangent to the conic at the point $P_1(x_1, y_1)$, where y_1 is the value of the function, that is, the ordinate of the conic at x_1. Taking into account the equation of the conic, the *point-direction form* of the line equation gives the equation of the tangent: for example, for the *parabola* $y^2 = 2px$ one obtains by differentiation $2yy' = 2p$ or $y' = p/y$

and so the gradient y_1' of the tangent (at the point P_1) is $y_1' = p/y_1$ and the equation of the tangent is

$$y_1' = (y - y_1)/(x - x_1) \quad \text{or} \quad (p/y_1)\,x - (p/y_1)\,x_1 = y - y_1,$$
$$px - px_1 = yy_1 - y_1^2 = yy_1 - 2px_1, \quad \text{that is,} \quad p(x + x_1) = yy_1.$$

For the *ellipse and hyperbola*, if $P_1(x_1, y_1)$ is the point of contact,

$$x_1^2/a^2 \pm y_1^2/b^2 = 1; \quad 2x_1/a^2 + 2y_1y_1'/b^2 = 0; \quad y_1' = \mp(b^2x_1)/(a^2y_1)$$

is the gradient, and the equation of the tangent is

$$y_1'(x - x_1) = y - y_1, \quad \mp(xx_1/a^2)(b^2/y_1) \pm (x^2/a^2)(b^2/y_1) = y - y_1,$$
$$(xx_1/a^2) \pm (yy_1/b^2) = x_1^2/a^2 \pm y_1^2/b^2 \quad \text{or} \quad xx_1/a^2 \pm yy_1/b^2 = 1.$$

Corresponding derivations can be made for the parabolas $x^2 = 2py$, $y^2 = -2px$, and $x^2 = -2py$, the ellipse $x^2/b^2 + y^2/a^2 = 1$ and the hyperbola $y^2/a^2 - x^2/b^2 = 1$. The following table contains the results for the most important cases.

Conic	Equation	Tangent at the point $P_1(x_1, y_1)$: gradient	equation
Parabola, vertex V			
$V(0, 0)$	$y^2 = 2px$	p/y_1	$yy_1 = p(x + x_1)$
$V(c, d)$	$(y - d)^2 = 2p(x - c)$	$p/(y_1 - d)$	$(y - d)(y_1 - d) = p(x - c + x_1 - c)$
Ellipse, centre C			
$C(0, 0)$	$\dfrac{x^2}{a^2} + \dfrac{y^2}{b^2} = 1$	$-\dfrac{b^2}{a^2}\dfrac{x_1}{y_1}$	$\dfrac{xx_1}{a^2} + \dfrac{yy_1}{b^2} = 1$
$C(c, d)$	$\dfrac{(x - c)^2}{a^2} + \dfrac{(y - d)^2}{b^2} = 1$	$-\dfrac{b^2}{a^2}\dfrac{x_1 - c}{y_1 - d}$	$\dfrac{(x - c)(x_1 - c)}{a^2} + \dfrac{(y - d)(y_1 - d)}{b^2} = 1$
Circle, centre C			
$C(0, 0)$	$x^2 + y^2 = r^2$	$-x_1/y_1$	$xx_1 + yy_1 = r^2$
$C(c, d)$	$(x - c)^2 + (y - d)^2 = r^2$	$-(x_1 - c)/(y_1 - d)$	$(x - c)(x_1 - c) + (y - d)(y_1 - d) = r^2$
Hyperbola, centre C			
$C(0, 0)$	$\dfrac{x^2}{a^2} - \dfrac{y^2}{b^2} = 1$	$\dfrac{b^2}{a^2}\dfrac{x_1}{y_1}$	$\dfrac{xx_1}{a^2} - \dfrac{yy_1}{b^2} = 1$
$C(c, d)$	$\dfrac{(x - c)^2}{a^2} - \dfrac{(y - d)^2}{b^2} = 1$	$\dfrac{b^2}{a^2}\dfrac{x_1 - c}{y_1 - d}$	$\dfrac{(x - c)(x_1 - c)}{a^2} - \dfrac{(y - d)(y_1 - d)}{b^2} = 1$

Example: The tangent to a conic at a point. The equation of the tangent to the ellipse

$$(x + 1)^2/25 + (y - 2)^2/16 = 1$$

at the point $P_1(2, -6/5)$ is

$$(x + 1)(x_1 + 1)/a^2 + (y - 2)(y_1 - 2)/b^2 = 1$$

or

$$16(x + 1)(2 + 1) + 25(y - 2)(-6/5 - 2) = 400,$$
$$48x + 48 - 80y + 160 = 400,$$
$$y = (3/5)\,x - 12/5.$$

The angle of intersection of a conic and a line. This angle of intersection is defined as the angle between the line and the tangent to the conic at the appropriate point of intersection; it can be calculated as the angle between two lines. The line $y = -x/2 + 2$ and the parabola $x^2 = 4y$ intersect at the point $P_1(2, 1)$. The tangent t_1 to the parabola

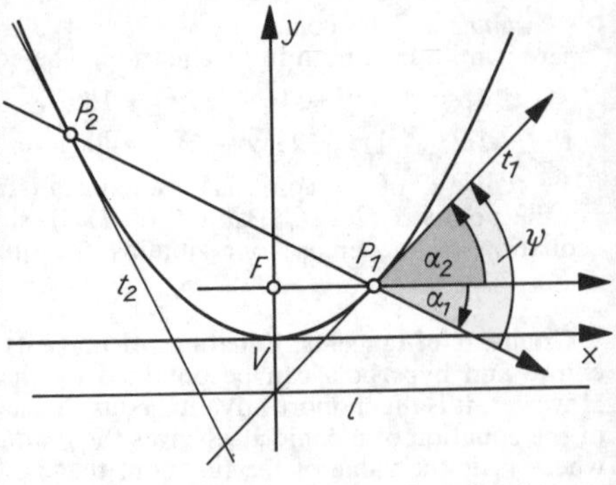

13.5-14 Point of intersection and angle of intersection of a line and a parabola

at P_1 has the equation $y = x - 1$; its angle of inclination α_2 to the x-axis has the value $\alpha_2 = 45°$, and that of the line is $\alpha_1 = -26.56°$; the angle ψ between them is therefore $\psi = \alpha_2 - \alpha_1 = 71.56°$ (Fig.).

Tangents to a conic having a prescribed gradient. The gradient with the given value m_1 must be equal to the gradient y_1' of the conic. In the case of the parabola the direction given by m_1 must not be that of the axis, since no tangent of the parabola has this direction. From $y_1' = m_1 = p/y_1$ one obtains one coordinate $y_1 = p/m_1$; the other can be calculated from the equation of the parabola, since the coordinates of the point of contact must satisfy it. In the case of an ellipse or hyperbola ($m_1 = \mp b^2x_1/a^2y_1$) one obtains the ratio of the coordinates x_1, y_1 of the point of contact and by substituting in the equation of the conic one obtains $x_1^2 = a^4m_1^2/(a^2m_1^2 \pm b^2)$, a pure quadratic equation which, for all ellipses and for those hyperbolas for which $|m_1| > b/a$, has two numerically equal roots. The result corresponds to the geometrical property that *for any direction there are two parallel tangents to an ellipse*, whose points of contact are symmetrical about the centre, but for a *hyperbola* this case only happens when the *line through the origin in the given direction lies outside the region between the asymptotes.* If the required tangent is *parallel* to a line $y = mx + c$, then m_1 is determined by $m_1 = m$; if the tangent is *perpendicular* to the line $y = mx + c$, then $m_1 = -1/m$; finally, if the two lines *enclose an angle* ψ, then from $\tan\psi = (m_1 - m)/(1 + m_1 m)$ one obtains the value $m_1 = (m + \tan\psi)/(1 - m\tan\psi)$.

Example: If the ellipse $36x^2 + 100y^2 = 9$ and the line $y = -(4/5)\,x$ are given, then $m = -4/5$. Since $\dfrac{x^2}{(9/36)} + \dfrac{y^2}{(9/100)} = 1$, the semi-axes of the ellipse have the lengths $a = 1/2$ and $b = 3/10$. For a tangent parallel to the line, $-4/5 = -(9/25)\,(x_1/y_1)$, that is, $y_1 = (9/20)\,x_1$. Further, the points of contact of the tangents lie on the ellipse, that is, $36x_1^2 + 100y_1^2 = 9$. From these two equations one obtains by substitution $36x_1^2 + (100 \cdot 81/400)\,x_1^2 = 9$ or $x_1^2 = 36/225$ and so $x_{1,2} = \pm 2/5, y_{1,2} = \pm 9/50$. The points of contact are therefore $B_1(2/5, 9/50)$ and $B_2(-2/5, -9/50)$; the equations of the tangents are $y = -(4/5)\,x + 1/2$ and $y = -(4/5)\,x - 1/2$.

Any tangent to a hyperbola forms with the asymptotes a triangle (P_2MP_3) of constant area $A = ab$.

The tangent $xx_1/a^2 - yy_1/b^2 = 1$ at the point $P_1(x_1, y_1)$ of the hyperbola $x^2/a^2 - y^2/b^2 = 1$ cuts the asymptotes $y = \pm(b/a)\,x$ at the points P_2 and P_3 (Fig.). For their coordinates one finds by substitution:

$x[x_1/a^2 \mp y_1/(ab)] = 1$ or $x_{2,3} = a^2b/(bx_1 \mp ay_1)$
and $y[\pm x_1/(ab) - y_1/b^2] = 1$ or
$y_{2,3} = ab^2/(\pm bx_1 - ay_1)$.

For the area A of the triangle MP_3P_2 one obtains:

$$A = {}^1/_2 \begin{vmatrix} 1 & 0 & 0 \\ 1 & x_3 & y_3 \\ 1 & x_2 & y_2 \end{vmatrix}$$
$$= {}^1/_2 \begin{vmatrix} 1 & 0 & 0 \\ 1 & a^2b/(bx_1 + ay_1) & ab^2/(-bx_1 - ay_1) \\ 1 & a^2b/(bx_1 - ay_1) & ab^2/(bx_1 - ay_1) \end{vmatrix}$$
$$= {}^1/_2\,[a^3b^3/(b^2x_1^2 - a^2y_1^2) + a^3b^3/(b^2x_1^2 - a^2y_1^2)]$$
$$= a^3b^3/(b^2x_1^2 - a^2y_1^2) = ab/[x_1^2/a^2 - y_1^2b^2] = ab.$$

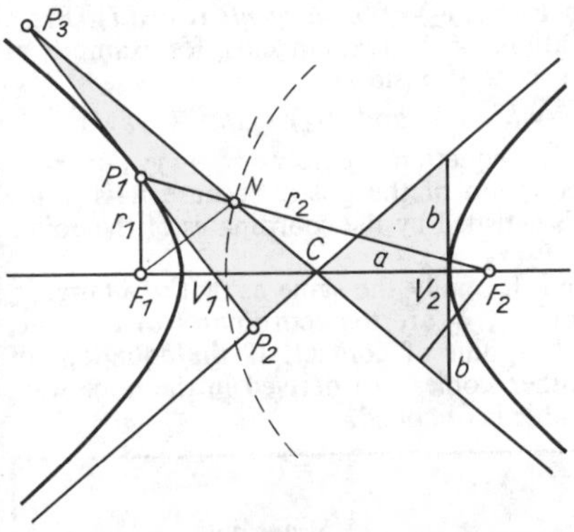

13.5-15 Triangles formed by tangents to a hyperbola and the asymptotes

Normal and Polar of a conic

Gradients of normals. Equations of normals. A normal is the line perpendicular to the tangent at its point of contact $P_1(x_1, y_1)$. One can therefore use the table of gradients of tangents to the conic to obtain the gradient of a normal and so, by means of the point-direction form, obtain the equation of a normal. In the most important cases one has:

Equation of conic	Gradient of the normal	Equation of the normal at the point $P_1(x_1, y_1)$
$y^2 = 2px$	$-y_1/p$	$y - y_1 = -(y_1/p)(x - x_1)$
$x^2/a^2 + y^2/b^2 = 1$	$[a^2y_1/(b^2x_1)]$	$y - y_1 = [a^2y_1/(b^2x_1)](x - x_1)$
$x^2/a^2 - y^2/b^2 = 1$	$-[a^2y_1/(b^2x_1)]$	$y - y_1 = -[a^2y_1/(b^2x_1)](x - x_1)$

Example: The normal to the hyperbola $x^2/16 - y^2/9 = 1$ at the point $P_1(5, -9/4)$ has the equation $y + \frac{9}{4} = -\frac{16(-9/4)}{9 \times 5}(x-5)$, that is, $y + \frac{9}{4} = \frac{4}{5}x - 4$ or $y = \frac{4}{5}x - \frac{25}{4}$ (Fig.).

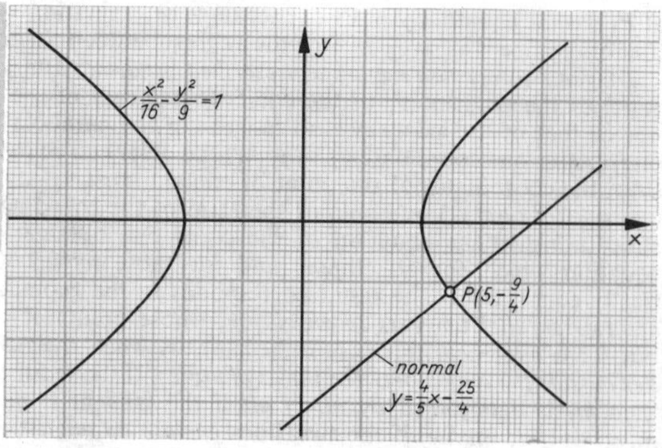

13.5-16 Normal to a hyperbola

Important theorems on normals to a conic.

The line joining a point P_1 of a parabola to the focus and the line through P_1 parallel to the axis make the same angle with the normal at P_1, since they make the same angle with the tangent at P_1.

The normal at a point P_1 of an ellipse bisects the angle between the lines joining P_1 to the foci, since these lines make the same angle with the tangent at P_1.

The tangent and normal to an ellipse at P_1 and the lines joining P_1 to the foci form a harmonic pencil, since in any triangle $F_1F_2P_1$ the internal and external bisectors of one angle (at P_1) are harmonically conjugate to the two sides that form the angle.

The tangent and normal to a hyperbola at P_1 and the lines joining P_1 to the foci form a harmonic pencil, since the tangent and normal at P_1 are the internal and external bisectors of the angle in P_1 of the triangle $F_1F_2P_1$. The normal bisects the angle supplementary to the angle between the lines joining P_1 to the foci.

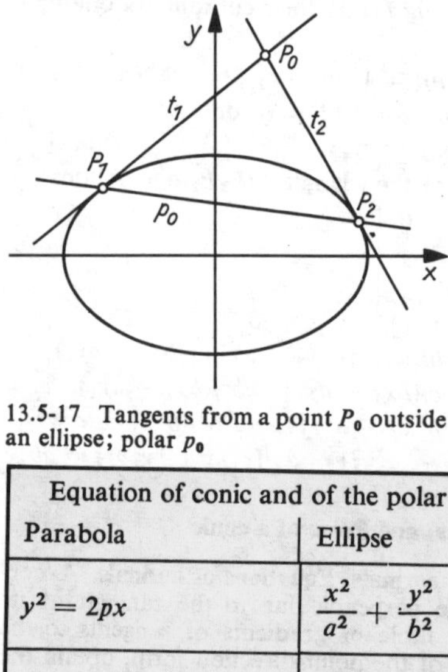

13.5-17 Tangents from a point P_0 outside an ellipse; polar p_0

The equation of the polar. Just as for a circle, one can find the tangents to a conic from a point $P_0(x_0, y_0)$ outside by means of the polar p_0 of the pole P_0. The polar p_0 is defined as *the line joining the points of contact $P_1(x_1, y_1)$ and $P_2(x_2, y_2)$ of the tangents t_1 and t_2* (Fig.). From the equations of the two tangents, for example, of an ellipse in the central position:

$(t_1)\; x_1x/a^2 + y_1y/b^2 = 1$ and $(t_2)\; x_2x/a^2 + y_2y/b^2 = 1$

one obtains the equation $(p_0)\; xx_0/a^2 + yy_0/b^2 = 1$, which is the equation of the polar, because it is a line equation that is satisfied by the coordinates of the points $P_1(x_1, y_1)$ and $P_2(x_2, y_2)$.

The equation is formally the same as that of a tangent, but the constants x_0, y_0 are the coordinates of the pole, not those of the point of contact. If the equations of the polars of other conics are derived in the same way, the following table is obtained.

Equation of conic and of the polar p_0 of the pole $P_0(x_0, y_0)$			
Parabola	Ellipse	Circle	Hyperbola
$y^2 = 2px$	$\frac{x^2}{a^2} + \frac{y^2}{b^2} = 1$	$x^2 + y^2 = r^2$	$\frac{x^2}{a^2} - \frac{y^2}{b^2} = 1$
$yy_0 = p(x + x_0)$	$\frac{xx_0}{a^2} + \frac{yy_0}{b^2} = 1$	$xx_0 + yy_0 = r^2$	$\frac{xx_0}{a^2} - \frac{yy_0}{b^2} = 1$

A line q that has no point in common with a conic, for example, a hyperbola, can also be interpreted as the polar of a point Q in the interior of the conic. If two distinct points $Q_1(\xi_1, \eta_1)$ and $Q_2(\xi_2, \eta_2)$ of q are taken as poles, then their polars q_1 and q_2 have the equations $\xi_1x/a^2 - \eta_1y/b^2 = 1$ and $\xi_2x/a^2 - \eta_2y/b^2 = 1$. From these equations one can calculate the point of intersection $Q(x_0, y_0)$ of the polars q_1 and q_2. This gives $xx_0/a^2 - yy_0/b^2 = 1$, which is the equation of a line that passes through Q_1 and Q_2 and is therefore the polar q of Q.

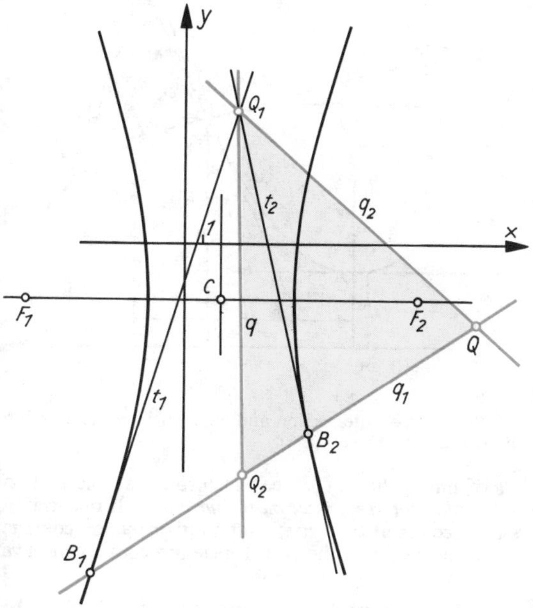

Example: The equations $(x-2)^2/16 - (y+3)^2/100 = 1$ and $y = -70x + 217$ determine a hyperbola and a line q, which do not intersect (Fig.). The points $Q_2(3^{31}/_{113}, -12^{23}/_{113})$ and $Q_1(3, 7)$ lie on q. The polar q_1 of Q_1 has the equation $(x-2)(\xi_1 - 2)/16 - (y+3)(\eta_1 + 3)/100 = 1$, and since $\xi_1 = 3$, $\eta_1 = 7$, this takes the form $y = {}^5/_8 x - {}^{114}/_8$.

Similarly the polar q_2 of Q_2 has the equation

$$100(x-2)(3^{31}/_{113} - 2) - 16(y+3)(-12^{23}/_{113} + 3) = 1600$$

or

$$y = -{}^{45}/_{52}x + 9^{31}/_{52}.$$

The coordinates of the point of intersection Q of q_1 and q_2 are given by these two equations; one finds that $x_0 = 16$, $y_0 = -4^1/_4$ are the coordinates of the pole Q of the line q. In the triangle QQ_1Q_2, each vertex is the pole of the opposite side.

13.5-18 The hyperbola given by $(x-2)^2/16 - (y+3)^2/100 = 1$

Tangents from a point to a conic. If the tangents from a point P outside a conic are to be drawn, it is convenient to take the points of intersection of the polar p of P with the conic. These points B_1 and B_2 are the *points of contact* of the tangents.

Example: The polar of the point $P(14, 1)$ with respect to the ellipse $x^2 + 4y^2 = 100$ (Fig.) has the equation $14x/100 + y/25 = 1$ or $14x + 4y = 100$, $y = -(7/2)x + 25$. It cuts the ellipse at $B_1(8, -3)$ and $B_2(6, 4)$, since $x^2 + 4(49x^2/4 - 175x + 625) = 100$, $x^2 - 14x = -48$, $x_1 = 6$, $x_2 = 8$; $y_1 = 4$, $y_2 = -3$.

Hence the equations of the tangents are $y = (2/3)x - 25/3$ and $y = -(3/8)x + 25/4$.

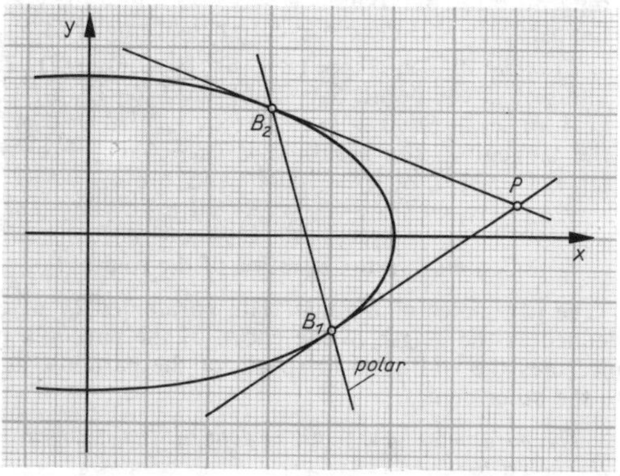

13.5-19 Tangents from a point P to an ellipse

Two conics

The points of intersection of two conics. To determine the points of intersection of two conics the corresponding system of equations must be solved. The real solutions give the coordinates of the points of intersection.

Example 1: The parabola $y^2 = 12x$ and the circle $(x+3)^2 + y^2 = 72$ intersect in the points $S_1(3, 6)$ and $S_2(3, -6)$, since their coordinates satisfy both equations. They are obtained from the system of equations

$$\left| \begin{aligned} (x_0+3)^2 + y_0^2 &= 72 \\ y_0^2 &= 12x_0 \end{aligned} \right|,$$

which has the solutions $x_1 = 3$, $y_1 = 6$ and $x_2 = 3$, $y_2 = -6$.

Example 2: To determine the points of intersection of the ellipse $(x+6)^2/80+(y-2)^2/20=1$ with the parabola $(x+6)^2=4(y-2)$ one has to solve the system of equations (Fig.)

$$\left|\begin{array}{l}(x_0+6)^2/80+(y_0-2)^2/20=1\\(x_0+6)^2=4(y_0-2)\end{array}\right|$$

Under the transformation $x_0+6=\xi$, $y_0-2=\eta$ the equations go into $20\xi^2+80\eta^2=1600$ and $\xi^2=4\eta$. By eliminating ξ one obtains $80\eta+80\eta^2=1600$, $\eta^2+\eta+1/4=81/4$, $\eta_{1,2}=-1/2\pm 9/2$, $\eta_1=4$, $\eta_2=-5$ and $\xi_{1,2}=\pm 4$, $\xi_{3,4}=\pm 2\sqrt{5}i$. Hence $x_1=\xi_1-6=-2$, $x_2=\xi_2-6=-10$, $y_1=\eta_1+2=6$, $y_2=\eta_1+2=6$. The conics therefore have two points of intersection $S_1(-2, 6)$ and $S_2(-10, 6)$.

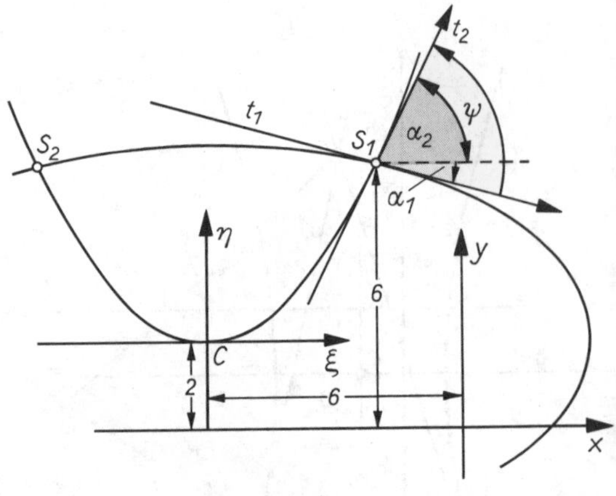

13.5-20 Point of intersection and angle of intersection of an ellipse and a parabola

Two conics need not have any (real) points of intersection. They can also touch at one or two points. It can be proved *that two non-degenerate conics intersect in at most four points.* If one section of a (double) cone consists of two generators intersecting at the vertex and another section consists of one generator, then, for a suitable position of the sections, the two degenerate conics can have infinitely many points in common.

The angle of intersection of two conics. The angle of intersection of two conics is defined as the angle between their tangents at the point of intersection. One must therefore find the equations of the tangents and calculate the angle of intersection.

Example 1: The parabola $y^2=12x$ and the circle $(x+3)^2+y^2=72$ intersect at the points $S_1(3, 6)$ and $S_2(3, -6)$. The tangent to the circle at S_1 has the equation $y=-x+9$ and the tangent to the parabola is $y=x+3$. Since the gradients are negative reciprocals of one another, the parabola and the circle cut at right angles at S_1, and similarly at S_2.

Example 2: The ellipse $(x+6)^2/80+(y-2)^2/20=1$ and the parabola $(x+6)^2=4(y-2)$ intersect at the points $S_1(-2, 6)$ and $S_2(-10, 6)$. The equation of the tangent to the ellipse at S_1 is $(x+6)(x_1+6)/80+(y-2)(y_1-2)/20=1$ or $y=-(1/4)x+11/2$, and the equation of the tangent to the parabola is $(x+6)(x_1+6)=2(y-2+y_1-2)$ or $y=2x+10$. Since $\tan\alpha_1=-1/4$, $\alpha_1=-14.04°$ and $\tan\alpha_2=+2$, $\alpha_2=+63.43°$, the angle of intersection is $\psi=\alpha_2-\alpha_1=77.47°$.

Common vertex equation of the conics

The parameter of a conic. The parameter $2p$ of a parabola $y^2=2px$ in the vertex position is defined as the length of the chord of the parabola perpendicular to the axis through the focus; it measures the width, so to speak, of the parabola at the focus. This definition can be carried over to the other conics.

The parameter of a conic is defined as the length of the chord perpendicular to the principal axis through a focus.

The parameter of a conic whose principal axis lies along the x-axis can be calculated by working out twice the positive ordinate y_F at a focus, that is, by substituting the abscissa x_F of the focus into the equation of the conic and solving this equation for y_F:

parabola: $y^2=2px$, $x_F=p/2$, so $y_F=p$,
ellipse: $x^2/a^2+y^2/b^2=1$, $x_F=e$,
$e^2/a^2+y_F^2/b^2=1$, so $y_F=\pm(b/a)\sqrt{[a^2-e^2]}$,
or, since $a^2-e^2=b^2$, $y_F=b^2/a$,
hyperbola: $x^2/a^2-y^2/b^2=1$,
$x_F=e$, $e^2/a^2-y_F^2/b^2=1$, so $y_F=\pm\sqrt{[e^2-a^2]}$,
or, since $e^2-a^2=b^2$, $y_F=b^2/a$.

Parameter	parabola	$2p$
	ellipse	$2p=2b^2/a$
	hyperbola	$2p=2b^2/a$

Vertex equations of the conics. The inner relationship between the conics is clear from their vertex equations. For the parabola this is $y^2=2px$, and for the ellipse and hyperbola it can be obtained from the central equation by a parallel displacement of the coordinate system.

Ellipse: From the central equation $\xi^2/a^2 + \eta^2/b^2 = 1$ in the ξ, η-system one obtains by transforming the origin to the vertex $V_2(-a, 0)$, that is, by the transformation $x = \xi + a$, $y = \eta$, the equation $(x - a)^2/a^2 + y^2/b^2 = 1$ in the new system; this equation can be rearranged into $y^2 = 2b^2x/a - b^2x^2/a^2$, or, by using the semiparameter $p = b^2/a$ of the ellipse, into $y^2 = 2px - (p/a)\,x^2$ (Fig.). The relation to the vertex equation of the parabola is obvious: *from the term $2px$ for the parabola the term $(p/a)\,x^2$ is subtracted* to obtain the ellipse. This explains the name ellipse: it refers to a deficiency (Greek: ***elleipsis***) compared with the parabola.

Vertex equation of the ellipse
$y^2 = 2px - (p/a)\,x^2,\ p = b^2/a$

Vertex equation of the hyperbola
$y^2 = 2px + (p/a)\,x^2,\ p = b^2/a$

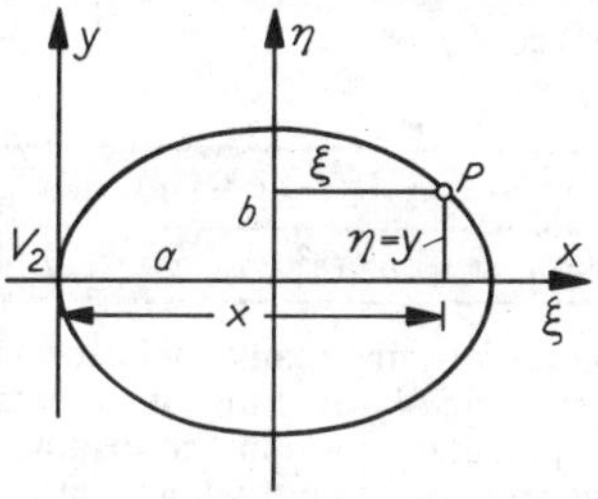

13.5-21 Transformation of an ellipse into the vertex position

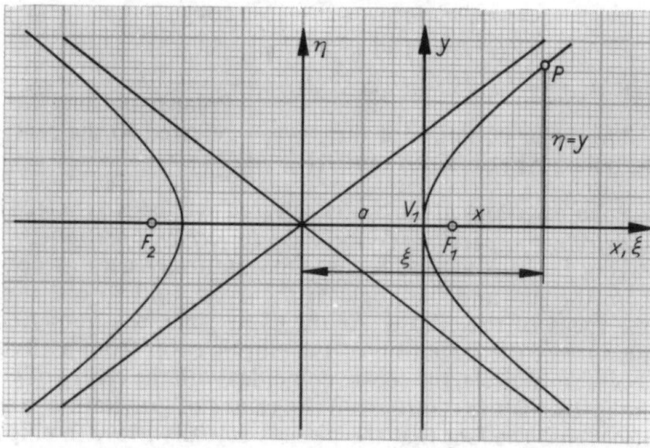

13.5-22 Transformation of a hyperbola into the vertex position

Hyperbola: From the central equation $\xi^2/a^2 - \eta^2/b^2 = 1$ in the ξ, η-system one obtains, by transforming the origin to the vertex $V_1(a, 0)$, that is, by the transformation $x = \xi - a$, $y = \eta$, the equation $(x + a)^2/a^2 - y^2/b^2 = 1$ in the new system; this equation can be rearranged into $y^2 = 2b^2/a + b^2x^2/a^2$, or, by using the semiparameter $p = b^2/a$ of the hyperbola, into $y^2 = 2px + (p/a)\,x^2$ (Fig.). Compared with the parabola $y^2 = 2px$, there is a term $(p/a)\,x^2$ in *excess* of the term $2px$. This explains the name hyperbola (Greek: ***hyperbole***, the excess).

Common vertex equation of the conics. By introducing the numerical eccentricity $\varepsilon = e/a$ for the ellipse $(0 < \varepsilon < 1)$ and the hyperbola $(\varepsilon > 1)$ and $\varepsilon = 1$ for the parabola, all three conics can be given a common vertex equation. For the ellipse, $p/a = b^2/a^2 = (a^2 - e^2)/a^2 = 1 - \varepsilon^2 > 0$ since $1 < \varepsilon < 1$; for the hyperbola, on the other hand, $p/a = b^2/a^2 = (e^2 - a^2)/a^2 = \varepsilon^2 - 1$, and so $0 - \varepsilon^2$ is always negative. For $\varepsilon = 1$ the term $(1 - \varepsilon^2)\,x^2$ obviously has the value zero; the equation $y^2 = 2px - (1 - \varepsilon^2)\,x^2$ therefore describes each of the three conics, depending on the value of ε.

Common vertex equation of the conics
$y^2 = 2px - (1 - \varepsilon^2)\,x^2$

The vertex equation of the circle is also included in this equation. If one puts $p = r$ and $\varepsilon = 0$, one obtains $y^2 = 2rx - x^2$ or $y^2 = x(2r - x)$; this relation is satisfied, by virtue of the altitude theorem for right-angled triangles.

In the common vertex equation a conic is determined by the parameter $2p$ and the numerical eccentricity ε. The quantities used up to now to characterize a conic, the semiaxes a and b and the linear eccentricity e, can be expressed in terms of p and ε if one considers that $y_0 = 0$ gives $x_0 = 2a$ and that $p = b^2/a$ for the ellipse and hyperbola and $p = r$ for the circle. One finds for the ellipse $a = p/(1 - \varepsilon^2)$, $b = p/\sqrt{(1 - \varepsilon^2)}$, $e = p\varepsilon/(1 - \varepsilon^2)$ and for the hyperbola $a = p/(\varepsilon^2 - 1)$, $b = p/\sqrt{(\varepsilon^2 - 1)}$, $e = p\varepsilon/(\varepsilon^2 - 1)$; if one chooses $p = 1$, then, for $\varepsilon = 0.8$, for example, the rounded-off values are $a = 2.78$, $b = 1.67$, $e = 2.22$, while for $\varepsilon = 1.5$, $a = 0.8$, $b = 0.89$, $e = 1.2$ (Fig.).

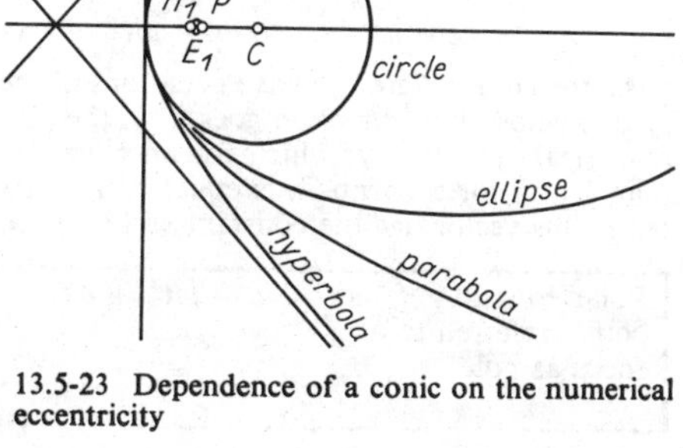

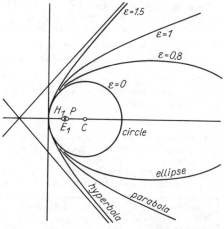

13.5-23 Dependence of a conic on the numerical eccentricity

Polar equations of the conics

To describe the conic in polar coordinates it is natural to take its axis as the zero direction; for the ellipse and hyperbola one could choose the centre as pole, but it is more usual to take a focus as pole.

Polar equations of conics referred to the centre as pole. The central equation of the ellipse $x^2/a^2 + y^2/b^2 = 1$ is transformed to polar coordinates by putting $x = r\cos\varphi$, $y = r\sin\varphi$, where the pole is the centre of the ellipse, so the polar equation is:

$$(r^2/a^2)\cos^2\varphi + (r^2/b^2)\sin^2\varphi = 1$$

or

$$\begin{aligned} 1 &= (r^2/b^2)(b^2\cos^2\varphi + a^2\sin^2\varphi)/a^2 = (r^2/b^2)(b^2\cos^2\varphi + a^2 - a^2\cos^2\varphi)/a^2 \\ &= (r^2/b^2)[a^2 - (a^2 - b^2)\cos^2\varphi]/a^2 = (r^2/b^2)[1 - (e^2/a^2)\cos^2\varphi] \\ &= (r^2/b^2)(1 - \varepsilon^2\cos^2\varphi), \end{aligned}$$

that is, $r^2 = b^2/(1 - \varepsilon^2\cos^2\varphi)$.

The polar equation of the hyperbola can be obtained similarly.

Polar equations, referred to the centre	ellipse $r^2 = b^2/(1 - \varepsilon^2\cos^2\varphi)$
	hyperbola $r^2 = b^2/(\varepsilon^2\cos^2\varphi - 1)$

Polar equations of the conics referred to a focus as pole. These conic equations find many applications in *astronomy*, particularly because of *Kepler's first law*, which states that the planets move in ellipses having the sun as one focus. One naturally uses as coordinates of planetary motion the distance from the sun and the angle in the orbit, and so one uses a polar coordinate system whose pole is one focus of the ellipse. At the same time, the numerical eccentricity ε is used in astronomy as a measure of the deviation of the elliptic orbit from a circular path. The word eccentricity is a happy choice: in a circle, the centre coincides with the centre of gravitation; the longer the ellipse is stretched, the further is the centre from the centre of gravitation, and so the more eccentric is the path. KEPLER discovered the fact that the planets actually move in ellipses, not circles, by considering Mars which, of all the planets then known, has the greatest eccentricity, $\varepsilon = 0.0933$. The eccentricity of the orbit of the Earth is only $\varepsilon = 0.0168$. Also meteors, comets and artificial satellites, if they have periodic motion inside the solar system, move in elliptical orbits. If they are not periodic, that is, their kinetic energy is sufficient to take them outside the solar system, then they move in parabolas or hyperbolas, provided that one neglects the disturbance caused by the force of attraction of the planets.

Polar equation of the ellipse. In Fig. 13.5–8 the focus F_1 is taken as the pole of a polar coordinate system, whose zero direction is that of the x-axis from F_1 to V_1. In the triangle F_1PF_2, since $r_2 = 2a - r_1$ and $|F_2F_1| = 2e$, the cosine law gives:

$$(2a - r_1)^2 = (2e)^2 + r_1^2 + 2\cdot 2er_1\cos\varphi \quad \text{or} \quad 4a^2 - 4ar_1 + r_1^2 = 4e^2 + r_1^2 + 4er_1\cos\varphi,$$
$$r_1 = (a^2 - e^2)/(a + e\cos\varphi) = a(1 - \varepsilon^2)/(1 + \varepsilon\cos\varphi) = b^2/[a(1 + \varepsilon\cos\varphi)] = p/(1 + \varepsilon\cos\varphi),$$

on putting $\varepsilon = e/a$ and $b^2/a = p$.

Polar equation of the hyperbola. In Fig. 13.5–13 the focus F_1 is taken as the pole of a polar coordinate system, whose zero direction is that of the $-x$-axis from F_1 to V_1. In the triangle F_1PF_2, since $r_2 = 2a + r_1$ and $|F_2F_1| = 2e$, the cosine law gives:

$$(2a + r_1)^2 = (2e)^2 + r_1^2 - 2\cdot 2er_1\cos\varphi \quad \text{or} \quad 4a^2 + 4ar_1 + r_1^2 = 4e^2 + r_1^2 - 4er_1\cos\varphi,$$
$$r_1 = (e^2 - a^2)/(a + e\cos\varphi) = b^2/[a(1 + \varepsilon\cos\varphi)] = p/(1 + \varepsilon\cos\varphi).$$

Polar equation of the parabola. In Fig. 13.5–4 the focus F is taken as the pole of a polar coordinate system whose zero direction is that of the $-x$-axis from F to V. Since $|L_0F| = p$, the definition of the parabola gives

$$p - r\cos\varphi = r \quad \text{or} \quad r = p/(1 + \cos\varphi).$$

All the conics therefore have equations of the same form $r = p/(1 + \varepsilon\cos\varphi)$ in a polar coordinate system whose zero direction goes from the pole to the nearest vertex; they differ in the values of the numerical eccentricity, which for an ellipse is positive but less than 1, for a hyperbola is greater than 1 and for a parabola is equal to 1. Also, the circle can be included by taking $\varepsilon = 0$, so that the radius vector has the constant value $r = p$.

Polar equation of the conics referred to a focus as pole	$r = p/(1 + \varepsilon\cos\varphi)$	$\varepsilon > 1$ hyperbola $\varepsilon = 1$ parabola $0 < \varepsilon < 1$ ellipse $\varepsilon = 0$ circle

For the parabola ($\varepsilon = 1$), r is not defined when $\varphi = \pi$. If $\varepsilon = 0$ or $0 < \varepsilon < 1$, that is, for the circle or the ellipse, to any value of the angle there corresponds a unique value of r. Finally, if $\varepsilon > 1$, r is not defined for any value φ_1 for which $1 + (e/a)\cos\varphi = 0$, that is, $\cos\varphi = -a/e$, when the free side of the angle φ_1 or $-\varphi_1$ is parallel to an asymptote.

Example: The *perihelion* is defined as the point of a planetary orbit nearest to the sun, and the *aphelion* the furthest point. What is the distance of the aphelion of Mars from the sun? – From astronomical observations it is known that the major semi-axis a of the orbit of Mars is, in round figures, 1.52 radii of the Earth's orbit (1 radius of the Earth's orbit is about 92.6 million miles) and its eccentricity $\varepsilon = 0.0933$. At the aphelion $\varphi = \pi$. Since $p = b^2/a = a(b^2/a^2) = a(a^2 - e^2)/a^2 = a(1 - \varepsilon^2)$, $r = a(1 - \varepsilon^2)/(1 - \varepsilon) = a(1 + \varepsilon) = 1.52 \times 1.0933 \approx 1.66$, measured in radii of the Earth's orbit. This means that the distance of Mars from the sun at aphelion is about $154 \cdot 10^6$ miles.

The eccentric anomaly. In astronomy and in the calculation of the elliptic paths or artificial satellites, the eccentric anomaly E, introduced by KEPLER, is used. This is the angle E measured from the zero direction to CP'', where C is the centre of the ellipse and P'' is the point of the auxiliary circle that corresponds to a point P of the ellipse (Fig.).
In plane geometry the construction of an ellipse is carried out from the auxiliary circle with radius a (half the major axis) and the concentric circle with radius $b = \sqrt{(a^2 - e^2)}$ (half the minor axis). As was shown in the parametric representation of the ellipse, all its chords perpendicular to the major axis V_2V_1 are in the ratio $b : a$ to the corresponding chords of the auxiliary circle. If P is a point of the ellipse, the segments $|P'P|$ and $|P'P''|$ are half these chords, so $|P'P| : |P'P''| = b : a$. In the right-angled triangle shown in the figure, $|P'P| = r\sin\varphi$, $|P'P''| = a\sin E$, and so $ba\sin E = ar\sin\varphi$ or $r\sin\varphi = b\sin E$. On the major axis, because $|CF_1| = e$ and $|CP'| = a\cos E$, one obtains $r\cos\varphi = a\cos E - e$. By using the equations $1 = \sin^2\varphi + \cos^2\varphi$ and $e^2 = a^2 - b^2$, r can be expressed as a function of E:

$$\begin{aligned} r^2 &= b^2\sin^2 E + (a^2\cos^2 E - 2ae\cos E + e^2) \\ &= b^2\sin^2 E + a^2\cos^2 E - 2ae\cos E + a^2 \\ &\quad - b^2\sin^2 E - b^2\cos^2 E \\ &= (a^2 - b^2)\cos^2 E - 2ae\cos E + a^2 = (a - e\cos E)^2, \end{aligned}$$

and since $a > e$ and $r > 0$, $r = a - e\cos E$.

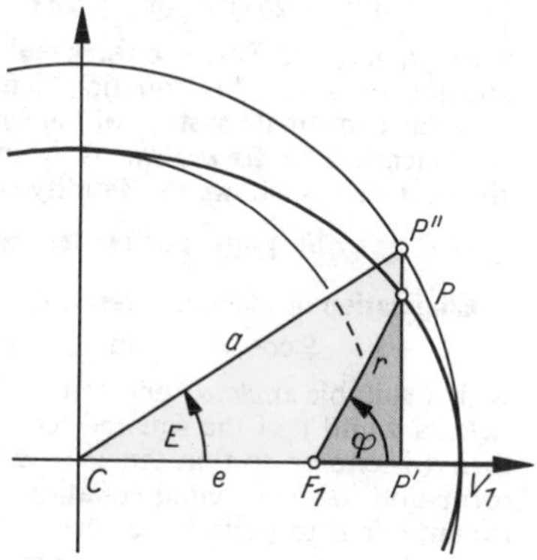

13.5-24 Eccentric anomaly

This equation contains *Kepler's first law*, according to which the planets move round the sun in elliptic orbits with the sun as one focus.

The relation between the anomaly φ and the eccentric anomaly E is given by the two equations

$$\cos\varphi = (1/r)(a\cos E - e) = (a\cos E - e)/(a - e\cos E),$$
$$\sin\varphi = (1/r)\, b\sin E = \sqrt{(a^2 - e^2)}\sin E/(a - e\cos E),$$

which can also be expressed in the form $\tan(\varphi/2) = \sqrt{[(a + e)/(a - e)]}\tan(E/2)$. To obtain the time t as a function of E, one of these equations, the second, for example, is differentiated with respect to t, where, as usual, differentiation with respect to t is denoted by a dot:

$$\begin{aligned} \cos\varphi \cdot \dot\varphi &= \frac{b\cos E \cdot \dot E(a - e\cos E) - e\sin E \cdot \dot E b\sin E}{(a - e\cos E)^2} \\ &= b \cdot \dot E \cdot \frac{a\cos E - e\cos^2 E - e\sin^2 E}{(a - e\cos E)^2} = b \cdot \dot E \cdot \frac{a\cos E - e}{(a - e\cos E)^2}. \end{aligned}$$

It follows that $\dot\varphi = \dfrac{d\varphi}{dt} = b\dot E \cdot \dfrac{a\cos E - e}{(a - e\cos E)^2} \cdot \dfrac{a - e\cos E}{a\cos E - e}$ or $\dot\varphi = \dfrac{d\varphi}{dt} = \dfrac{b\dot E}{a - e\cos E} = \dfrac{b\dot E}{r}$.

By *Kepler's second law*, the *area covered* by the radius vector in a given time is constant: $r^2\dot\varphi = C$. Introducing C into the last relation gives

$$\dot E = \frac{dE}{dt} = \frac{r^2\dot\varphi}{br} = \frac{C}{br} = \frac{C}{b(a - e\cos E)} \quad \text{or} \quad dt = \frac{b}{C}(a - e\cos E)\,dE,$$

and so the required function $t = t(E)$ is obtained by *integration*:

$$t = \frac{b}{C}(Ea - e\sin E) = \frac{\sqrt{(a^2 - e^2)}}{C}(Ea - e\sin E).$$

As E increases from 0 to 2π, the *orbital time* T is obtained:

$$T = \frac{b}{C} \cdot 2\pi a = \frac{2\pi a \sqrt{(a^2 - e^2)}}{C}.$$

By *Kepler's third law*, for each planet there exists a *constant* $\mu/(4\pi^2)$ for which $a^2/T^2 = \mu/(4\pi^2)$, or, by substituting the above value for T, $\mu = aC^2/(a^2 - e^2)$. Therefore three of the four constants are sufficient for all the relations; it is usual to choose $\varepsilon = e/a$, C and μ and to derive from $r = p/(1 + \varepsilon \cos \varphi) = b^2/[a(1 + \varepsilon \cos \varphi)]$:

$$r = C^2/[\mu(1 + \varepsilon \cos \varphi)] = C^2(1 - \varepsilon \cos E)/[\mu(1 - \varepsilon^2)],$$
$$\cos \varphi = (-\varepsilon + \cos E)/(1 - \varepsilon \cos E), \quad \sin \varphi = \sqrt{(1 - \varepsilon^2)} \sin E/(1 - \varepsilon \cos E),$$
$$t = C^3(E - \varepsilon \sin E)/[\mu^2(1 - \varepsilon^2)^{3/2}].$$

Discussion of the general equation of the second degree

The general equation of the second degree in two variables x and y has the form

$$ax^2 + 2bxy + cy^2 + 2dx + 2ey + f = 0,$$

where a, b, c, d, e, f are arbitrary real coefficients. It is actually of the second degree only when a, b, c are not all zero. This equation defines a curve in the x, y-coordinate system. Henceforth a rectangular coordinate system will always be assumed. The type of curve depends on the values of the coefficients. The *discussion*, by which one means the characterization of the curve depending on the coefficients, shows the validity of the following theorem.

The general equation of the second degree always represents a conic.

Elimination of the mixed term. By a rotation of the coordinate system, that is, by a transformation

$$x = \xi \cos \alpha - \eta \sin \alpha, \qquad y = \xi \sin \alpha + \eta \cos \alpha$$

with a suitable angle α, one can always arrange that the mixed term with $\xi\eta$ vanishes. If the coefficients a and c of the squared terms are equal ($a = c$), one chooses $\alpha = 45°$; if they are different, then α is chosen so that $\tan 2\alpha = 2b/(a - c)$, as one can see by substituting the equations of transformation in the original equation. In this way a transformed equation in ξ and η is obtained. It is convenient to write these variables as x and y again; the equation is then of the form

$$Ax^2 + Cy^2 + 2Dx + 2Ey + F = 0.$$

This means that *the axes of the conic are now parallel to the coordinate axes.*

Elimination of the linear terms. The central equations of the ellipse and hyperbola have no linear terms. One therefore looks for a parallel displacement $x = \xi + c$, $y = \eta + d$ of the coordinate system, where c and d are constants such that the linear terms vanish. By carrying out the transformation one obtains:

$$A\xi^2 + C\eta^2 + 2(Ac + D)\,\xi + 2(Cd + E)\,\eta + Ac^2 + Cd^2 + 2Dc + 2Ed + F = 0.$$

Discussion:

(1) If $A \neq 0$ and $C \neq 0$, both linear terms can be removed by choosing $c = -D/A$ and $d = -E/C$. The equation then takes the form $A\xi^2 + C\eta^2 = N$, where $N = D^2/A + E^2/C - F$. Three cases are possible for N: $N > 0$, $N = 0$, $N < 0$.

$N > 0$: Case 1: A and C both positive. The curve is an *ellipse* with the central equation $\frac{\xi^2}{(N/A)} + \frac{\eta^2}{(N/C)} = 1$, and with the semi-axes $\sqrt{(N/A)}$ and $\sqrt{(N/C)}$.
Case 2: A and C negative. There is *no real curve.*
Case 3: A and C of opposite signs. The curve is a *hyperbola.*

$N = 0$: Case 1: If A and C have the same sign, the equation is satisfied only for $\xi = \eta = 0$. The curve is a *single point.*
Case 2: A and C of opposite signs. The left-hand side of the equation then factorizes, and the curve is a *pair of intersecting lines.*

$N < 0$: The same conics are obtained as for $N > 0$ (except that Cases 1 and 2 are interchanged).

(2) If $AC = 0$, there are three possibilities.

$A = 0$, $C \neq 0$: Case 1: $D \neq 0$. Then c and d can be chosen so that $Cd + E = 0$, $Cd^2 + 2Dc + 2Ed + F = 0$. The equation becomes $\eta^2 = -2(D/C)\,\xi$, and the curve is a *parabola.*

Case 2: $D = 0$. The equation is a quadratic in η, and it therefore represents a *pair of parallel lines*. They coincide, and therefore represent a *double line*, if $E^2 - FC = 0$.

$A \neq 0, C = 0$: Case 1: $E \neq 0$. The curve is a *parabola*.
Case 2: $E = 0$. The curve is a *pair of parallel lines* or a *double line*.

$A = 0, C = 0$: Case 1: D and E not both zero. The curve is a *single line*.
Case 2: $D = E = 0$. Then F must also be zero.

Note: The case of a pair of parallel lines can be regarded as a special case of a section of a cone by a plane parallel to the axis, where the vertex of the cone is at infinity, and the cone is therefore a cylinder.

Discussion of the equation $Ax^2 + Cy^2 + 2Dx + 2Ey + F = 0$

$AC \neq 0$	$N = \dfrac{D^2}{A} + \dfrac{E^2}{C} - F$		
	$N > 0$	$A > 0, C > 0$	ellipse
		$A < 0, C < 0$	no real curve
		$AC < 0$	hyperbola
	$N = 0$	$AC > 0$	point
		$AC < 0$	pair of intersecting lines
	$N < 0$	$A > 0, C > 0$	no real curve
		$A < 0, C < 0$	ellipse
		$AC < 0$	hyperbola
$AC = 0$	$A = 0, C \neq 0$	$D \neq 0$	parabola
		$D = 0$	pair of parallel lines, coinciding if $E^2 - FC = 0$
	$A \neq 0, C = 0$	$E \neq 0$	parabola
		$E = 0$	pair of parallel lines, coinciding if $D^2 - FA = 0$
	$A = 0, C = 0$	D and E not both zero	line
		$D = E = 0$	(trivial)

Example 1: In the equation $3x^2 - 30x + 8y + 65 = 0$, the values of the coefficients are $A = 3 \neq 0$, $C = 0$, $D \neq 0$; the curve is therefore a *parabola*. To find the vertex, focus and parameter, one divides by 3 and completes the square; from the equation $x^2 - 10x + 25 = -(8/3)\,y - 65/3 + 75/3$ or $(x - 5)^2 = -(8/3)\,(y - 5/4)$ one sees that the parabola is concave downwards, that the vertex is at $V(5, 5/4)$ and that the parameter is $p = 4/3$.

Example 2: The equation $25x^2 + 49y^2 + 150x - 196y - 804 = 0$ describes an *ellipse*, whose principal axis is parallel to the x-axis, since $A = 25 \neq 0$, $C = 49 \neq 0$, $N > 0$. The equation can be brought to the central form by twice completing the square: $(x + 3)^2/49 + (y - 2)^2/25 = 1$. The centre of the ellipse is at $C(-3, 2)$. The major semi-axis is $a = 7$, and the minor semi-axis is $b = 5$.

Example 3: The equation $64x^2 - 25y^2 + 256x + 300y - 2244 = 0$ represents a hyperbola, as one sees immediately. From the equation it follows that $64(x^2 + 4x) - 25(y^2 - 12y) = 2244$. By completing the square twice, one finds that $64(x^2 + 4x + 4) - 25(y^2 - 12y + 36) = 2244 + 256 - 900$ or $64(x + 2)^2 - 25(y - 6)^2 = 1600$. The central equation is therefore $(x + 2)^2/25 - (y - 6)^2/64 = 1$. The major axis is parallel to the x-axis, the centre is at $C(-2, 6)$ and the semi-axes are 5 and 8.

Example 4: For the conic $9x^2 - 4y^2 = 0$, $AC \neq 0$ but $N = 0$. Since $AC < 0$, the conic is a *pair of intersecting lines*. In fact, $9x^2 - 4y^2 = (3x - 2y)\,(3x + 2y) = 0$. Each factor gives a line, and the equations of the lines are $y = (3/2)\,x$ and $y = -(3/2)\,x$. The two lines intersect at the origin.

II. Steps towards higher mathematics

14. Set theory

Set theory is the foundation stone of the edifice of modern mathematics. The precise definitions of all mathematical concepts are based on set theory. Furthermore, the methods of mathematical deduction are characterized by a combination of logical and set-theoretical arguments. To put it briefly, the language of set theory is the common idiom spoken and understood by mathematicians the world over. From all this it follows that if one is to make any progress in higher mathematics itself or in its practical applications, one has to become familiar with the basic concepts and results of set theory and with the language in which they are expressed.

The definition of a set quoted below gives the impression that the naive set concept is easy to grasp because of its apparent perspicuity. In actual fact it leads to great difficulties, which have only been overcome by the development of axiomatic systems for set theory.

When Georg CANTOR (1845–1918), who founded set theory, published his daring new concepts and arguments, their importance was recognized by only a few mathematicians. But in its further development the theory was to penetrate almost all branches of mathematics, having a profound influence on their development, and changing the appearance even of established theories. Indeed, the development of some disciplines, such as topology, was essentially dependent on the means of set theory. What is more, set theory proved a unifying force, giving all branches of mathematics a common basis, and their concepts a new clarity and precision.

The following sections emphasize those parts of set theory that have particularly important applications in the development of the various branches of mathematics.

14.1. The concept of a set

In colloquial usage the term 'set' is taken, as a rule, to mean a collection of things that in some sense or another belong together or are akin. This latter aspect is difficult to make precise and is therefore omitted from the mathematical concept.

> Cantor's definition of a set: A set is the result of collecting together certain well-determined objects of our perception or our thinking into a single whole; these objects are called the elements of the set.

In spite of the lack of precision of this definition — which actually leads to contradictions (see Example 5) — it is sufficient to introduce several important definitions and concepts.

> If an object a is an element of a set S, one writes $a \in S$ (read 'a belongs to S' or 'S contains a'); one writes $a \notin S$ if a is not an element of S. If S is the set of elements $a, b, c, \ldots$, one writes $S = \{a, b, c, \ldots\}$, for instance, $\{1, 2, \ldots\}$ is the set of the positive natural numbers. If S contains only one element a, then S is called a *singleton*, $S = \{a\}$. If S contains two distinct elements a and b, then S is called an unordered pair, $S = \{a, b\}$.

A *subset* T of a set S is any set whose elements all belong to S; this is denoted by $T \subseteq S$. The subsets T of S that are distinct from S itself are called *proper subsets* of S; in this case one writes $T \subset S$. The empty set is a set that has no elements at all. The introduction of this set has proved convenient to round off statements and arguments of set theory, just as the number 0 (historically a late invention) rounds off the statements and calculations of arithmetic. The usual symbol for the empty set is $\emptyset$.

Sets whose elements are themselves sets are called *families* or *systems*, for instance, a nation is a set of people and an element of the 'family' of nations. A very important system is the set of all subsets of a given set S; this is called the *power set* of S and is denoted by $P(S)$.

Example 1: The set S of all people in a certain building B at a certain time t. This set is well-defined even if nobody is in the building at the chosen time; in that case S is the empty set. The set W of all women in B at the time t is a subset of S, $W \subseteq S$; W is not necessarily a proper subset of S.

Example 2: The set of all prime numbers. This set is infinite, as was already proved by EUCLID, whereas the sets of Example 1 are always finite.

Example 3: The set of all regular polygons inscribed in the unit circle is used in the calculation of π.

Example 4: The set of all subsets of the natural numbers. This is also infinite, in fact, as will be shown later, 'more' infinite than the natural numbers themselves.

Example 5: The set of all sets that do not contain themselves as elements. This set, which is perfectly admissible under Cantor's definition, leads to the celebrated paradox of Bertrand RUSSELL (1872–1970). If the set is denoted by R and if one supposes that R is an element of itself ($R \in R$), then R is – like every other element of R – a set that does not contain itself as an element ($R \notin R$); that is, the assumption leads to a contradiction. If, on the other hand, R is not an element of itself ($R \notin R$), then, since R contains every set that does not contain itself, R cannot be one of those sets. Therefore R contains itself ($R \in R$), which is again a contradiction. Since one of the two assumptions must be true, the whole situation stands in contradiction to the laws of logic.

Example 5 shows that the construction of new sets must not be extended without bounds, if contradictions are to be avoided.

The examples make it clear how sets are to be constructed. A set is determined by the description of a property. To be a little more precise, the set consists of all objects ξ for which a statement $A(x)$ with the object variable x becomes true if x is replaced by ξ.

In Examples 1 to 4 these statements are (in the same order):

x is a human being and is in the building B at the time t,
x is a prime number,
x is a regular polygon inscribed in the unit circle,
x is a subset of the set of positive natural numbers.

If x is replaced by an arbitrary object, a statement results that is either true or false. In axiomatic systems of set theory it is of central importance to delineate precisely what logical form a statement must have if it is to be admissible for the definition of a set.

The set defined by the statement $H(x)$ is denoted by $\{x \mid H(x)\}$ (read: 'the set of all x such that $H(x)$').

Example 6: $\{x \mid x$ is a natural number and there is a natural number y with $x = y^2\}$ is the set of all squares. The notation can be abbreviated: $\{x \in \mathbf{N} \mid x = y^2 \text{ for some } y \in \mathbf{N}\}$.

The axiomatic systems of set theory developed in the first half of this century all have four basic principles in common: the principles of extensionality, of set construction, the existence of infinite sets, and the axiom of choice.

The principle of *extensionality* says that two sets S and T having the same elements (that is, being of the same extent) are identical ($S = T$). The word identical is taken here in Leibniz' sense, that is, in any statement S can be replaced by T and vice versa, without changing the truth or falsity of the statement.

The principle of *construction* asserts that certain restricted types of statements do define sets; a usual restriction is that the statement contains only object symbols, logical symbols, and the symbol $\in$.

The *existence of infinite sets* states just that. The meaning of infinite must, of course, be made precise. This principle is difficult to motivate by a direct reference to reality. But without it major parts of mathematics and theoretical science, such as the differential and integral calculus and classical mechanics, would become meaningless. One could not even give a set-theoretical foundation to the theory of natural numbers.

Finally, there is the *axiom of choice*, which is basic for many mathematical arguments. Nevertheless many authors regard this axiom with a doubt similar to that which Euclid's parallel postulate met in an earlier era.

Axiom of choice. If $\boldsymbol{S}$ is a system of non-empty sets, then there exists a set A having exactly one element in common with every set S of $\boldsymbol{S}$.

14.2. Operations on sets

Operations on sets are used to construct new sets from given ones. The most important are the intersection, union, and difference of sets S and T.

> Intersection: $S \cap T =_{\text{def}} \{x \mid x \in S \text{ and } x \in T\}$
> Union: $S \cup T =_{\text{def}} \{x \mid x \in S \text{ or } x \in T\}$
> Difference: $S \setminus T =_{\text{def}} \{x \mid x \in S \text{ and } x \notin T\}$

Example 1: $\{a, b, c\} \cap \{a, c, d\} = \{a, c\}$, $\{a, b, c\} \cup \{a, c, d\} = \{a, b, c, d\}$, $\{a, b, c\} \setminus \{a, c, d\} = \{b\}$.

Example 2: The intersection of the set of all rectangles and the set of all rhombi is the set of all squares.

Example 3: The union of the set of all rectangles and the set of all parallelograms is the set of all parallelograms, because every rectangle is a parallelogram, and so nothing is added to the set of all parallelograms.

It is important to distinguish between the union $S \cup T$ and the set of all elements belonging *either to S or to T* (but not to both). This latter set is called the symmetric difference of S and T and is only used occasionally for special purposes.

Sets S and T whose intersection is empty are called *disjoint*. If S is any subset of U, then $U \setminus S$ is called the *complement of S in U*.

The basic properties of the operations on sets in the following table can be illustrated by representing the sets by bounded areas of the plane.

Commutativity
$S \cap T = T \cap S$
$S \cup T = T \cup S$

Associativity
$S \cap (T \cap R) = (S \cap T) \cap R$
$S \cup (T \cup R) = (S \cup T) \cup R$

Distributivity
$S \cap (T \cup R) = (S \cap T) \cup (S \cap R)$
$S \cup (T \cap R) = (S \cup T) \cap (S \cup R)$

Idempotence
$S \cap S = S$
$S \cup S = S$

If S and T are subsets of U and their complements in U are written S' and T' for short, then *De Morgan's rules* hold: $(S \cap T)' = S' \cup T'$; $(S \cup T)' = S' \cap T'$.

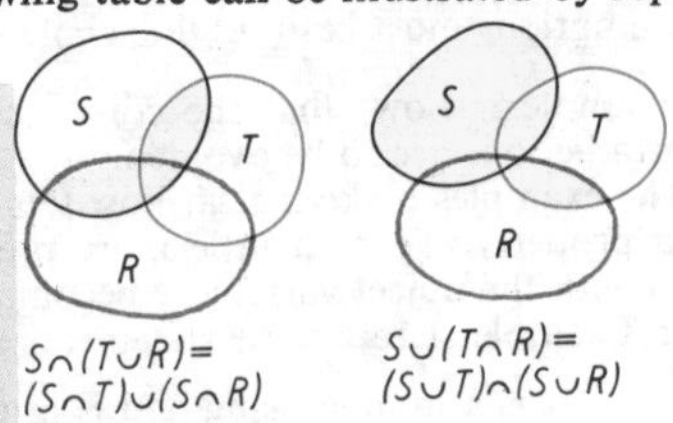

14.2-1 Distributivity of $\cap$ and $\cup$:

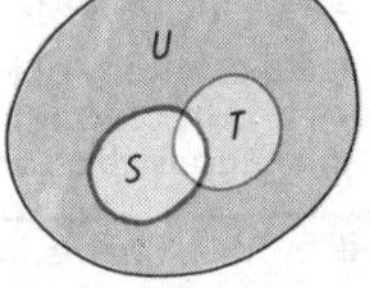

14.2-2 De Morgan's rule $(S \cap T)' = S' \cup T'$; S' is blue, T' is red, $S \cap T$ is left white

As an example, here is a proof of the first statement (Fig.). To show that $(S \cap T)' = S' \cup T'$ one proves the two statements (*i*) $(S \cap T)' \subseteq S' \cup T'$ and (*ii*) $(S \cap T)' \supseteq S' \cup T'$.

To prove (*i*), let $x \in (S \cap T)'$, that is, $x \in U$, but $x \notin S \cap T$. Now either $x \in S$ or $x \notin S$. If the latter, then $x \notin S'$ and therefore $x \in S' \cup T'$. If the former, then $x \notin T$, for otherwise x would be an element of $S \cap T$. Therefore $x \in T'$ and again $x \in S' \cup T'$. This completes the proof of (*i*). To prove (*ii*), let $x \in S' \cup T'$, that is, $x \in S'$ or $x \in T'$ (it is, of course, possible that both hold). In the first case $x \notin S$ and therefore $x \notin S \cap T$, and in the second case $x \notin T$ and again $x \notin S \cap T$.

Generalized operations on sets. The operations of intersection and union are initially defined as operations on two arguments. They can, however, be generalized not only to 3, 4, ... sets, but to arbitrary systems of sets. But first a few explanations.

Systems of sets are denoted in what follows by upper case bold face letters. The members $S, T, \ldots$ of a system $\mathbf{S}$ are sometimes labelled by subscripts or are made to depend on parameters. For instance, a finite system $\mathbf{S}$ may be written as $\{S_1, \ldots, S_k\}$ or $\mathbf{S} = \{S_i \mid i = 1, \ldots, k\}$. Frequently $\mathbf{S} = \{S_i\}_{i \leqslant k}$ is used. Thus, if $A_n = \{x \in \mathbf{N} \mid x \leqslant n\}$, then $\{A_n\}_{n \in \mathbf{N}}$ is the family of all initial segments of the sequence of natural numbers.

In general, $\{S_i\}_{i \in I}$ is called an *indexed family of sets* if a set I, the *index set*, is given and if to every $i \in I$ a set S_i of the family is assigned. Every set of the family must occur at least once, but it is not required that distinct indices give distinct sets. In the terminology of mappings (see 14.4.), an indexed family $\mathbf{S}'$ is a surjective mapping of I onto $\mathbf{S}$; here $\mathbf{S}$ itself is the range of the mapping. Every family $\mathbf{S}$ can be indexed by taking $\mathbf{S}$ itself as the index set.

> Definition of the intersection and union of an arbitrary system $\mathbf{S}$:
> $\bigcap \mathbf{S} =_{\text{def}} \{x \mid x \in S \text{ for all } S \in \mathbf{S}\}$; $\bigcup \mathbf{S} =_{\text{def}} \{x \mid x \in S \text{ for some } S \in \mathbf{S}\}$.
> If $\mathbf{S}$ is indexed, $\mathbf{S} = \{S_i\}_{i \in I}$, then one writes $\bigcap \mathbf{S} = \bigcap_{i \in I} S_i$; $\bigcup \mathbf{S} = \bigcup_{i \in I} S_i$.

These definitions also include the original case of two sets, when the system has only two members.

Generalizations of the distributive laws: $S \cap \bigcup_{i \in I} S_i = \bigcup_{i \in I} S \cap S_i$; $S \cup \bigcap_{i \in I} S_i = \bigcap_{i \in I} S \cup S_i$.

If all the sets are subsets of a set U, the following generalizations of De Morgan's rules hold:

$[\bigcap_{i \in I} S_i]' = \bigcup_{i \in I} S_i'$; $[\bigcup_{i \in I} S_i]' = \bigcap_{i \in I} S_i'$.

14.3. Relations

It is well known that if a and b are distinct real numbers, then $a < b$ or $b < a$. If $a < b$, one could also say that the relation 'less than' holds for the pair (a, b). This relation can be denoted by $R_<$ and is completely characterized by the set of all ordered pairs of real numbers for which it holds. An extension of this train of thought leads to the following basic definition.

A relation R on a set S is a set of ordered pairs of elements of S. If $(a, b) \in R$, one also says that R holds for the ordered pair (a, b) and sometimes one writes this in the form aRb.

In the definition above the term 'ordered pair' is used in a naive, intuitive sense as an aggregation of the objects a and b in such a way that a is distinguished as the first element of the ordered pair (a, b), and b as the second. In the further course of this section a rigorous set-theoretic definition of the notion of ordered pair will be given.

Example 1: In the set S of all humans living at the present moment a relation C can be defined by 'A is a parent of B' or 'B is a child of A'.

Example 2: The relation R_d on the set $S = \{1, 2, 3, 4, 6, 12\}$, defined by the statement 'x divides y' consists of the pairs $(1, 1), (1, 2), \ldots, (2, 2), (2, 4)$ and so on. In the Figure 14.3-1 the relation is represented by an arrow diagram in which the numbers, represented by dots, are connected by arrows if the relation holds between them. Since every number divides itself, every point is connected to itself by a circular arrow (loop).

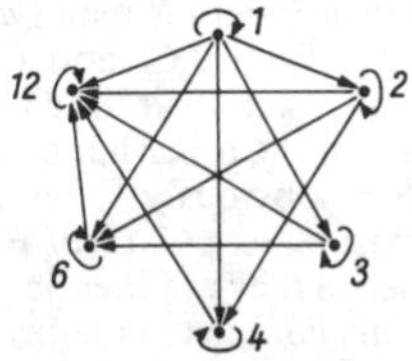

14.3-1 Arrow diagram for the divisibility relation on {1, 2, 3, 4, 6, 12}

The set $\{x \in S \mid (x, y) \in R$ for at least one y in $S\}$ is called the *support* of R. The set $\{x \in S \mid (y, x) \in R$ for at least one $y \in S\}$ is called the *range* of R. The support and range of R are denoted in what follows by *Supp R* and *Ran R*. The set *Dom R* = *Supp R* $\cup$ *Ran R* is called the *domain* of R. Obviously *Dom R* $\subseteq S$.

The support of C in Example 1, for instance, consists of all humans with at least one child, the range consists of all humans one of whose parents is still alive. In Example 2 all elements belong to both the support and the range.

In mathematics certain properties of relations play a particular role; some of the most important are given in the following table (where R denotes a relation on S).

R is *reflexive*	$=_{\text{def}}$ xRx holds for all $x \in S$.
R is *irreflexive*	$=_{\text{def}}$ there is no $x \in S$ for which xRx holds.
R is *symmetric*	$=_{\text{def}}$ for all $x, y \in S$, if xRy, then yRx.
R is *asymmetric*	$=_{\text{def}}$ there are no elements $x, y \in S$ with xRy and yRx.
R is *antisymmetric*	$=_{\text{def}}$ for all $x, y \in S$: if xRy and yRx, then $x = y$.
R is *transitive*	$=_{\text{def}}$ for all $x, y, z \in S$: if xRy and yRz, then xRz.
R is *connected*	$=_{\text{def}}$ for all $x, y \in S$: if $x \neq y$, then xRy or yRx.
R is *left unique*	$=_{\text{def}}$ for all $x, y, z \in S$: if xRz and yRz, then $x = y$.
R is *right unique*	$=_{\text{def}}$ for all $x, y, z \in S$: if xRy and xRz, then $y = z$.
R is *biunique*	$=_{\text{def}}$ R is left unique and right unique.

Restrictions of relations. If R is a relation on S and T a subset of S, then $\{(x, y) \in R \mid x, y \in T\}$ is a relation on T. It is called the *restriction* of R to T and is frequently denoted by $R_{|T}$. For instance, the relation 'less than' on the natural numbers $\mathbf{N}$ is the restriction to $\mathbf{N}$ of the relation 'less than' on the real numbers $\mathbf{R}$.

Equivalence relations. An equivalence relation on a set S is a relation that is reflexive, symmetric, transitive, and has support S. Equivalence relations are found not only in every corner of mathematics, but in almost all the sciences.

Example 3: A line l is *parallel* to a line l': $l \parallel l'$.
Example 4: A number a is *congruent* to a number b modulo m: $a \equiv b \pmod{m}$.
Example 5: A triangle ABC is *similar* to a triangle $A'B'C'$: $\triangle ABC \sim \triangle A'B'C'$ or a figure F is *homeomorphic* to a figure F' (see Chapter 34.).
Example 6: x is identical with y. The identity relation on S, id_S is the set $\{(x, x) \mid x \in S\}$.

An equivalence relation R on S induces a partition of S into classes, which consist of those elements between which the relation holds.

A partition of a set S is a family P of non-empty subsets of S, called the classes of the partition, with the following two properties: 1. Any two distinct classes are disjoint, 2. every element of S lies in one class (Fig.).

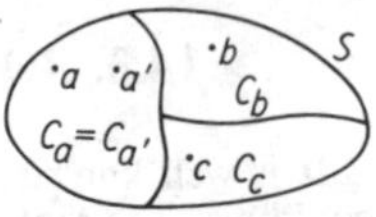

14.3-2 Partition of a set S into three classes

If P is a partition, then every element a of S lies in exactly one class $C \in P$, which is denoted by C_a. Obviously $C_a = C_b$ if and only if b lies in C_a.
The following simple theorem is of fundamental importance. It is the basis of the *principle of identification by abstraction.*

Main theorem on equivalence relations. If R is an equivalence relation on a set S, then there exists a partition P of S such that elements a, $b \in S$ lie in the same class of P if and only if aRb holds. Conversely, if P is a partition of S, then the relation $\{(a, b) \mid$ there is a class $C \in P$ with $a, b \in C\}$ is an equivalence relation.

Proof. Let R be given. Define $C_a =_{\text{def}} \{x \in S \mid aRx\}$, and call it the equivalence class of a. Let P be the family of equivalence classes of elements of S. Since aRa for all elements of S, $a \in C_a$. Thus, every element of S lies in a class of P. It remains to show that distinct classes of P are disjoint. Suppose that C_a and C_b are not disjoint, say $c \in C_a \cap C_b$, then aRc and bRc. Since R is symmetric, this implies cRb, and aRb by the transitivity of R. Now if $e \in C_b$, then bRe and again by transitivity aRe. So $e \in C_a$ and $C_b \subseteq C_a$. In the same way one shows that $C_a \subseteq C_b$, and therefore $C_a = C_b$. Thus, non-disjoint classes are identical, and P is the required partition.
On the other hand, let P be a partition of S and let R be defined as in the sta tement of the theorem. R is obviously reflexive and symmetric. Suppose that aRb and bRc, then by the definition there exist classes C, C' of P with $a, b \in C$ and $b, c \in C'$. These classes are not disjoint, because $b \in C \cap C'$, hence they are identical. But now $a, c \in C$ and so aRc by the definition of R. Therefore R is transitive. This completes the proof.
The equivalence classes of Example 3 are the *directions* in the plane or space. Those of Example 4 are the *residue classes* mod m and those of Example 6 are the *singletons* in S.

Order relations. A relation R on a set S is called a partial ordering on S if R is reflexive, transitive and antisymmetric. If R is also connected, it is called a total or linear ordering.

Example 7: The divisibility relation R_d is a partial ordering of the natural numbers.
Example 8: The relation 'S is a subset of T' is a partial ordering of the subsets of a set U.
Example 9: The relation $a \leqslant b$, 'a is less than or equal to b', is a partial ordering, in fact, a total ordering of the set of real numbers.

An ordered set is defined as a pair (S, R), where R is a partial ordering of the set S. It is usual to let the symbol S also stand for the ordered set (S, R). The restriction $R_{|T}$ of R to a subset T of S is again an ordering, in other words, subsets of ordered sets are also ordered. If S is an ordered set with a partial ordering $R_{\leqslant}$, then an element $u \in S$ is called an upper bound for a subset T of S, if $xR_{\leqslant}u$ holds for all $x \in T$. An element $m \in S$ is called maximal in S if there is no $x \neq m$ in S with $mR_{\leqslant}x$.

One of the most frequently used lemmas in the whole of mathematics is the following, which is equivalent to the axiom of choice.

Kuratowski-Zorn lemma. If every totally ordered subset of an ordered set (S, R) has an upper bound in S, then S has a maximal element.

An important example of the use of this lemma occurs in the section on cardinal numbers.

Set-theoretical definition of an ordered pair. Since a is distinguished as the left-hand member of the ordered pair (a, b), and b as the right-hand one, the pair cannot simply be defined as $\{a, b\}$. This difficulty is overcome by the following subtle definition.

Definition of an ordered pair	$(a, b) =_{\text{def}} \{\{a\}, \{a, b\}\}$

If $a \neq b$, one can distinguish the left-hand element of the ordered pair (a, b) as the element of the singleton of the set, while the right-hand element is that which is not in the singleton.

From this definition one can derive the following fundamental property of ordered pairs:

The statement $(a_1, a_2) = (b_1, b_2)$ holds if and only if $a_1 = b_1$ and $a_2 = b_2$.

The Cartesian product $S \times T$ of two sets S and T is the set of all ordered pairs (a, b) with $a \in S$ and $b \in T$. The product $S \times S$ is abbreviated as S^2, $S^2 \times S$ is abbreviated as S^3 and so on. The elements of S^n are called *n-tuples* of elements of S. The 3-tuple, or *triple* $((a, b), c)$ is abbreviated (a, b, c) and so forth.

Example 11: The set $\mathbf{C}$ of complex numbers can be regarded as the product $\mathbf{R} \times \mathbf{R} = \mathbf{R}^2$ of the set of all real numbers with itself.

A subset of S^n is called an *n-argument* (or *n-ary*) *relation* on S. If S^1 is defined as S itself, then the 1-argument relations on S are the subsets of S. *Binary* or *2-argument relations* are the ones so far considered in this section. Occasionally *n*-argument relations are called *predicates*.

Examples 12: The relation 'the point X lies between the points Y and Z' is a three-argument relation on the points of the plane.

Example 13: 'z is the sum of x and y' is a three-argument relation on the set of natural numbers, and also on other sets of numbers.

Example 14: 'The quadruple of points $[O, P, Q, R]$ forms a parallelogram in the plane' is a four-argument relation on the set of points in the plane.

14.4. Mappings

A *function* on a set S with values in a set T is a *right unique relation* with support in S and range in T. If the support is the whole of S, it is called a *mapping* from S *into* T. If the range of a mapping is the whole of T, it is called a mapping of S *onto* T, or *surjective*.

Remark. Obviously functions on S with values in T are particular subsets of $S \times T$. In some branches of mathematics (for example, complex analysis) functions are not necessarily assumed to be right unique. The definition given here corresponds to common usage in mathematics.

Functions and mappings pervade the whole of mathematics. Most frequently they are mappings from one set S into another T, which are also written $F\colon S \to T$. The set of all mappings of S into T is denoted T^S.

Since one frequently has to deal with several different types of mappings simultaneously, for instance, with mappings whose arguments are functions or mappings (see Example 3), there are a number of synonyms for mappings, or for particular types of mappings. The most frequent are *operation* (principally for mapping of S^2 into S), *operator*, *functional* (mainly for real-valued functions of functions), *functor*, and *morphism* (mainly for mappings that are in some sense structure preserving).

Image and inverse image. If F is a function on S with values in T, and if $(x, y) \in F$, then y is called the image of x under F, or the value of F at x. This can be denoted in several ways: $y = x^F$, $y = xF$, $y = F(x)$, or $y = F_x$. If $y = F(x)$, then x is called an inverse image of y under F. The set $F^{-1}(y) =_{\text{def}} \{x \in S \mid F(x) = y\}$ is the complete inverse image of y.

Particular functions. Functions on the set $\mathbf{R}$ of real numbers with values also in $\mathbf{R}$ are called *real functions*, or functions of a real variable. Functions of n real variables are functions on $\mathbf{R}^n$ with values in $\mathbf{R}$. Mappings of the set $\mathbf{N}$ of natural numbers into itself are called *arithmetical*, or *number-theoretical* functions.

Example 1: $y = x^2$ is a real function; the notation, though common, can easily give rise to misunderstandings, a better notation would be $F\colon x \to x^2$. For the moment the function will be called Sq. The support of Sq is obviously the whole of $\mathbf{R}$, its range is $\mathbf{R}^{\geqslant 0}$, the set of non-negative real numbers.

Mappings of the set $\{0, \ldots, n-1\}$ into a set S are called *n-term sequences* of elements of S. If $F(i) = a_i$ $(i = 0, \ldots, n-1)$, then F is written as $F = (a_0, \ldots, a_{n-1})$. Mappings of $\mathbf{N}$, the set of natural numbers, into S are simply called *sequences* of elements of S. The sequence F with $F(i) = a_i$ is written as $(a_1, a_2, \ldots)$ or $(a_i)_{i \in \mathbf{N}}$.

Restrictions of mappings. If F is a mapping of S into T, and if U is a subset of S, then $\{(x, y) \in F \mid x \in U\}$ is a mapping of U into T. It is called the restriction of F to U and denoted $F_{|U}$. For instance, the operation of addition of natural numbers is the restriction of the operation with the same name on real numbers. As the example shows, restrictions of mappings are frequently denoted by the same symbol as the unrestricted mapping.

Injective functions. A function on S with values in T is called *injective*, *invertible*, or *one-to-one*, if it is a left unique relation. In that case every element of the range of F has a unique inverse image, and the set $\{(y, x) \in T \times S \mid (x, y) \in F\}$ is a function on T with values in S. It is called the *inverse function* of F and is denoted by F^{-1}. If F is an injective mapping, then F^{-1} is a mapping if and only if F is surjective, and such mappings are called *bijective*. The inverse of a bijective mapping is again bijective.

Example 2: id_S is a bijective mapping of S onto itself; it is equal to its inverse.

Example 3: $S^{\{0,1\}}$ is the set of all mappings of $\{0, 1\}$ into S, that is, the set of all two-term sequences of elements of S: $\{(a_0, a_1) \mid a_0, a_1 \in S\}$. Let F: $M^{\{0,1\}} \to M^2$ be the mapping that associates with the sequence (a_0, a_1) the ordered pair (a_0, a_1). It is clear that F is bijective. Because of this there is no essential difference between n-term sequences of elements of S and n-tuples of elements of S.

Example 4: Let $\mathbf{S}$ be a family of sets and A a set having exactly one element in common with each member of the family. Associate with each member S of the family the unique element of $S \cap A$. The mapping ε from $\mathbf{S}$ onto A thus defined is called a *choice function* for $\mathbf{S}$. Choice functions are invertible only in special cases.

Combination of mappings. If F and G are functions, then the set

$$H =_{\text{def}} \{(x, z) \mid \text{there exists a } y \text{ with } (x, y) \in F \text{ and } (y, z) \in G\}$$

is again a function, whose support is contained in the support of F and whose range is contained in the range of G. This H is called the *combination*, *composition*, or *product* of F and G. If the images under the functions are written as $F(x)$ etc., then it is denoted by $G \cdot F$, because $z = H(x)$ means that $z = G(F(x))$. If the notation $y = x^F$ and $z = y^G$ is used, then the product $z = (x^F)^G = x^{FG}$ is written in the inverse order. Care must be taken to ensure that it is clear which notation is being used.

If F is a mapping from S to U, and G a mapping from U to T, then $H = G \cdot F$ (where the order is taken as in the preceding paragraph) is a mapping from S to T. Here $F \cdot G$ need not even contain a single ordered pair. The product of mappings is associative, that is, $F \cdot (G \cdot H) = (F \cdot G) \cdot H$ for any three mappings F, G and H.

Example 5: Parallel shifts are particular (bijective) mappings of the set of points of the plane onto itself. In this case the composition of two parallel shifts $\mathbf{p}$ and $\mathbf{q}$ is written as a sum $\mathbf{p} + \mathbf{q}$. The operation $+$ is commutative, but in general, the composition of functions is not commutative.

14.5. Infinite sets and cardinal numbers

Definitions of finiteness. In the naive sense a set S is finite if there is a natural number n such that the elements of S can be counted out by the numbers before n; more precisely, if there is a bijective mapping from the set of natural numbers less than n onto S.

This definition has the disadvantage that it takes the natural numbers as already given. On the other hand, it turns out that to define the natural numbers one needs the concept of a finite set. This difficulty was first clearly recognized by DEDEKIND. He overcame it by giving a definition of finiteness that avoids the use of the natural numbers and uses mappings instead.

Dedekind's definition of finiteness. A set S is finite if every injective mapping of S into itself is bijective.

It follows from this definition that S is infinite if and only if there exists an injective mapping of S into itself that is not surjective, in other words, if there exists a bijective mapping of S onto a proper subset of S.

A particularly convenient definition, perhaps the most frequently used, is due to RUSSELL.

Russell's definition of finiteness. A set S is finite if it belongs to every system $\mathbf{S}$ with the following properties: 1. $\emptyset \in \mathbf{S}$; 2. if $U \in \mathbf{S}$, then $U \cup \{a\} \in \mathbf{S}$ for all $a \in S$.

It is easy to show that a set that is finite in the sense of Russell's definition is also finite in Dedekind's sense. The converse can be shown using the axiom of choice.

Example 1: The set **N** of natural numbers is infinite, because there is an injective map of **N** onto a proper subset of **N**, for instance onto the set of even numbers (Fig.). An equally suitable mapping is $F: n \to n+1$.

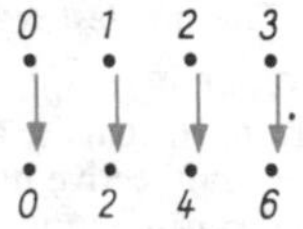

14.5-1 Bijective mapping of the set of natural numbers N onto a proper subset (after GALILEI)

Cardinal numbers

Two sets S and T are said to be *equipotent* or of the same *power* (written: $S \sim T$) if there is a bijective map from S to T.

It is easy to see that the relation defined above is an equivalence relation on any suitable family of sets. Hence it leads to a partition of the family into classes of equipotent sets.

A *cardinal number* is the class of sets equipotent to a given set. The cardinal numbers of *finite sets* are called the *natural numbers*. The cardinal numbers of *infinite sets* are called *transfinite*.

One cannot operate with the family of all sets, or even the family of all sets equipotent to a given set, because that would lead to Russell's paradox. To avoid this one usually restricts the definition above to a family F that is as large as possible and as necessary. In that case the cardinal numbers are themselves sets, namely families of sets. It may, however, become necessary to enlarge the family F later.

Comparison of cardinal numbers. The power or cardinal number of a set S is denoted by card S. Cardinal numbers are denoted by lower case bold type letters, $\mathbf{s}$, $\mathbf{t}$, etc.

$\mathbf{n} \leqslant \mathbf{s} =_{\text{def}} \mathbf{n}$ is the *cardinal number of a subset* of a set S with card $S = \mathbf{s}$.

This definition is independent of the choice of S.

Bernstein's theorem. If there are injective maps from S into T, and from T into S, then S and T are equipotent.

This theorem implies that the relation $\leqslant$ on cardinal numbers is anti-symmetric.

Theorem. The relation $\leqslant$ on cardinal numbers is an ordering.

At the end of this section it will be shown that any two cardinal numbers can be compared, that is, that $\leqslant$ is a total ordering. In 14.6. it will even be shown that any non-empty set of cardinal numbers contains a smallest element.

The existence of arbitrarily large cardinal numbers. The following theorem of CANTOR is basic for the theory of transfinite cardinals.

Cantor's theorem. To any set there exists a set of higher power, indeed, card $P(S) >$ card S.

The proof of this theorem is surprisingly short and elegant. On the one hand, it is clear that there is an injective mapping of S into $P(S)$, namely the one taking the elements $a \in S$ to the singletons $\{a\}$ in $P(S)$. It is now necessary to show that no injective map from S to $P(S)$ is surjective, in other words, that for every injective map φ from S into $P(S)$ there exist elements of $P(S)$ that have no inverse image. This is done by showing that the set $U =_{\text{def}} \{x \in S \mid x \notin \varphi(x)\}$ is never an image under φ. Suppose the contrary: that $U = \varphi(u)$, say, for some $u \in S$. Now either $u \in U$ or $u \notin U$. If $u \in U$, then $u \in \varphi(u) = U$, since $U = \varphi(u)$; but, by definition, U contains only those elements of S that are not elements of their images under φ. So this assumption leads to a contradiction. But the other assumption also leads to a contradiction, for $u \notin U$ means that $u \notin \varphi(u)$, and since U contains all elements of S that are not elements of their images, this implies that $u \in U$. Hence the original assumption is untenable (compare this proof with Russell's paradox. Here an assumption was made, which is shown to be untenable because it leads to a contradiction; in Russell's paradox the argument, applied to the set of all sets, is the same, but there is no previous assumption. Hence it leads to an undissolved paradox).

Countable sets

A set S is called countable (or *denumerable*) *if it is equipotent to the set* **N** *of natural numbers,* that is, if there is a bijective mapping $\varphi: n \to a_n$ from **N** onto S. The cardinal of the countable sets is denoted by $\aleph_0$.

The smallest transfinite cardinal is $\aleph_0$.

Proof. Example 1 shows that $\aleph_0$ is transfinite. It remains to show that $\aleph_0 \leqslant \boldsymbol{n}$ for all transfinite cardinals, that is, that every transfinite set contains a countable subset. So let S be infinite and let φ be an injective map of S into a proper subset T of S. Choose $a \in S \setminus T$ and put $a = a_0$, and define $a_{n+1} = \varphi(a_n)$ inductively. The set $\{a_i \mid i \in \mathbf{N}\}$ is a countable subset of S.

In the usual naive proof of this theorem one chooses an element a_0 of S, and then an element a_1 of $S_1 = S \setminus \{a_0\}$ and continues in this manner. This process does not break off because the fact that S is infinite implies that the sets S_i are not empty. This argument is a tacit application of the axiom of choice, but it can be made completely precise.

The union of countably many countable sets is countable.

Let $M_i = \{a_{i0}, a_{i1}, a_{i2}, \ldots\}$ and suppose that the elements of all the sets M_i are arrayed in an infinite matrix (Fig.). The counting can begin in the top left-hand corner and continue diagonally in the manner shown by the arrows. Elements that have been counted once are omitted on repetitions, for instance, if $a_{11} = a_{20}$, then a_{11} is omitted. It is clear that in this manner $\bigcup_{i=0}^{\infty} M_i$ is completely counted out.

This proof was used by Cantor to show that the set of rational numbers is countable. One can imagine them arrayed in the same manner as the elements a_{ij} in an infinite matrix.

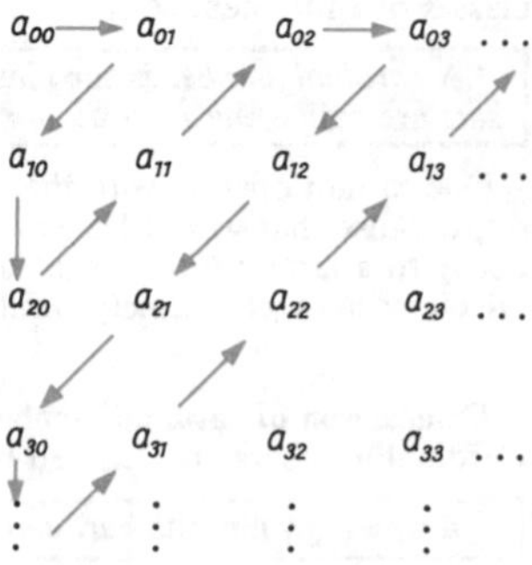

14.5-2 Matrix array to prove that the union of countably many countable sets $M_i = \{a_{i0}, a_{i1}, \ldots\}$ is countable

Sums and products of cardinal numbers. If S and T are disjoint representatives of the cardinals $\boldsymbol{s}$ and $\boldsymbol{t}$, then the sum of $\boldsymbol{s}$ and $\boldsymbol{t}$ is defined as $\boldsymbol{s} + \boldsymbol{t} =_{\text{def}} \text{card}\,(S \cup T)$, and the product as $\boldsymbol{s} \cdot \boldsymbol{t} =_{\text{def}} \text{card}\,(S \times T)$. These definitions can be extended to arbitrarily many cardinal numbers.

Let $\{\boldsymbol{m}_i\}_{i \in I}$ be a system of cardinal numbers and $\{M_i\}_{i \in I}$ be a system of mutually disjoint representatives, then one defines $\sum_{i \in I} \boldsymbol{m}_i =_{\text{def}} \text{card} \bigcup_{i \in I} M_i$.

To define the product, and finally powers, of cardinal numbers one needs a generalization of the Cartesian product to arbitrary systems of sets $\{S_i \mid i \in I\}$. This generalization is also useful in other parts of mathematics.

Generalization of the Cartesian product

$\bigtimes_{i \in I} S_i =_{\text{def}} \{f \mid f \text{ is a mapping from } I \text{ into } \bigcup_{i \in I} S_i \text{ with } f(i) \in S_i\}$ $\qquad \prod_{i \in I} \boldsymbol{m}_i =_{\text{def}} \text{card} \bigtimes_{i \in I} M_i$

If $S_i = S$ for all $i \in I$, one writes S^I for $\bigtimes_{i \in I} S_i$ and similarly $\boldsymbol{m}^{\boldsymbol{n}}$ if all the cardinals $\boldsymbol{m}_i$ are equal ($\boldsymbol{n} = \text{card}\, I$).

It can be shown that for transfinite cardinals $\boldsymbol{m} + \boldsymbol{n} = \boldsymbol{m} \cdot \boldsymbol{n} = \max(\boldsymbol{m}, \boldsymbol{n})$. In particular, $\boldsymbol{m} + \aleph_0 = \boldsymbol{m} \cdot \aleph_0 = \boldsymbol{m}$ for any transfinite cardinal $\boldsymbol{m}$. This means that the ordinary arithmetical operations become trivial when extended to transfinite cardinals. But this is not true for powers; for instance, the following theorem shows that $\boldsymbol{m} < 2^{\boldsymbol{m}}$.

If a set M has the cardinal $\boldsymbol{m}$, then $P(M)$ has the cardinal $2^{\boldsymbol{m}}$.

To prove this one associates with each subset T of M its *characteristic function* $\chi = \chi_T$, which is defined by

$$\chi(a) = \begin{cases} 1 & \text{if } a \in T \\ 0 & \text{if } a \in M \setminus T. \end{cases}$$

This gives a bijective mapping of $P(M)$ onto the set $\{0, 1\}^M$ of all mappings from M to $\{0, 1\}$. But the set of these mappings has the cardinal $2^{\boldsymbol{m}}$, by definition.

The continuum. The cardinal of the set of real numbers is called the cardinal of the continuum and denoted by $\aleph$ or $\boldsymbol{c}$. The cardinal of the set of real numbers in the open interval (0, 1) is also $\aleph$, because this interval is mapped bijectively to the set of all real numbers, for example by the function $y = (x - 1/2)/[x(1 - x)]$ (Fig.).

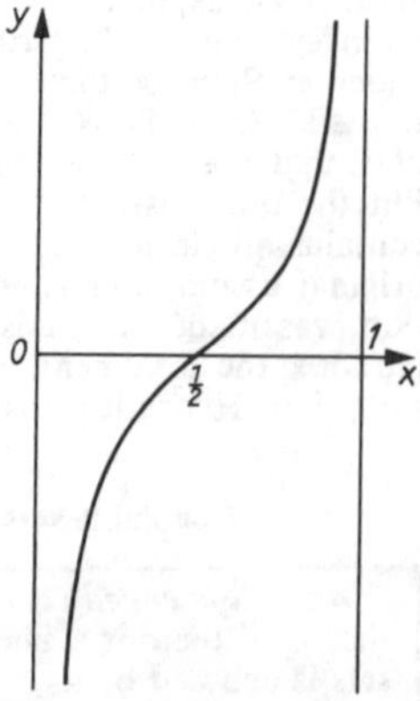

14.5-3 Mapping of the interval (0, 1) onto the whole real line by the function $y = (x - 1/2)/[x(1 - x)] = 1/(1 - x) - 1/[2x(1 - x)]$

The cardinals $\aleph_0$ and $\aleph$ are connected by the formula $\aleph = 2^{\aleph_0}$.

This is proved by defining two mappings. The first is an injective mapping from $P(\mathbf{N})$ into $\mathbf{R}$, the set of real numbers: if $M \in P(\mathbf{N})$, then M is mapped to the decimal $0.\, a_1 a_2 \ldots$, where $a_i = 1$ if $i \in M$ and $a_i = 0$ otherwise. This proves that $2^{\aleph_0} \leqslant \aleph$. The second maps the set of real numbers in (0, 1) injectively into $P(\mathbf{N})$: let $r = 0.\, a_1 a_2 \ldots$ $(0 \leqslant a_i \leqslant 9)$, and exclude periods of the digit 9. Then map r to the set $\{1a_1, 1a_1a_2, \ldots\}$ of natural numbers, for example $r = 0.1406\ldots$, is mapped to the set $\{11, 114, 1140, 11406, \ldots\}$. This proves that $\aleph \leqslant 2^{\aleph_0}$ and hence that $\aleph = 2^{\aleph_0}$.

The continuum hypothesis states that there is no cardinal number between $\aleph_0$ and $\aleph$, in other words, that an infinite set of real numbers is either countable or has the cardinal $\aleph$. In 1964 Cohen proved that it is impossible to prove the continuum hypothesis by means of the standard set-theoretical axioms; previously, in 1938, Gödel had shown that the continuum hypothesis does not contradict these axioms. Together the two results show that the continuum hypothesis is independent of the other set-theoretical axioms.

Comparison of cardinals.

For any two cardinal numbers $m \neq n$ one of the relations $m < n$ or $n > m$ holds.

It is sufficient to prove that for any pair of sets M and N there is an injective function φ on M with values in N such that the support of φ is M or the range of φ is N. For in the first case card $M \leqslant$ card N and in the second card $N \leqslant$ card M. The proof given here exhibits a use of the Zorn-Kuratowski lemma, which is typical for modern mathematics.

Let Φ be the set of injective functions on M with values in N. This set is not empty, because the empty function φ_0 with $Supp\, \varphi_0 = Ran\, \varphi_0 = \emptyset$ is contained in Φ. For elements φ, ψ of Φ one defines $\varphi \leqslant \psi$ if φ is a restriction of ψ, or equivalently, if $\varphi \subseteq \psi$, when these mappings are regarded as sets of ordered pairs. It is clear that $\leqslant$ is an ordering of Φ. Now if Ω is a *chain* (totally ordered subset) of Φ, then $\bigcup \Omega$ (with the functions again regarded as sets of pairs) is an injective function on M with values in N, and thus an upper bound of Ω in Φ. Now by Zorn's lemma Φ contains a maximal element φ^*. Suppose that $Supp\, \varphi^* \subset M$ and $Ran\, \varphi^* \subset N$, then let $a \in M \setminus Supp\, \varphi^*$ and $b \in N \setminus Ran\, \varphi^*$, and define $\varphi' = \varphi^* \cup \{(a, b)\}$. Since φ' is still injective, it is in Φ, but this contradicts the maximality of φ^*. Therefore $Supp\, \varphi^* = M$ or $Ran\, \varphi^* = N$, q.e.d.

14.6. Well-ordered sets and ordinal numbers

Order types.

Two ordered sets S and T are called *similar* if there is a bijective map φ from S to T such that $a < b$ if and only if $a^\varphi < b^\varphi$ for all $a, b \in S$.

Similarity is an equivalence relation on ordered sets. The equivalence classes are called *similarity types* of ordered sets.

The same difficulties apply to this relation as to the relation of equipotence for arbitrary sets. And they are avoided in the same way by restricting all arguments to a suitable large family of ordered sets.

Order types are the similarity classes of totally ordered sets.

Example 2: The *set of all real numbers* has the same order type as the set of real numbers in the open interval (0, 1), because the bijective mapping given in 14.5. preserves the ordering in both directions. This is called the order type of the linear continuum.

Example 3: The ordered *set of rational numbers* has the following properties: 1. it is *countable*, 2. it is *dense*, that is, between any two distinct elements there is a further element, and 3. it has *no initial or final element*. Cantor proved that all totally ordered sets with the properties 1., 2., and 3. have the same *order type* η. Thus, this is the order type of every open rational interval, but also of the set of all algebraic real numbers in their natural ordering, because this set is the union of countably many countable sets and thus countable.

Example 4: Two finite totally ordered sets have the same order type if and only if they have the same cardinal number. The similarity is constructed by first mapping the minimal element of one to that of the other, then the minimal element above that, and so on. The finite order types are thus in one-to-one correspondence with the finite cardinal numbers. The natural numbers can be regarded either as cardinals or as order types.

Example 5: The order type of the even numbers is the same as that of the natural numbers. Indeed, any countable set S can be given an ordering of the same type as the natural numbers by using a bijective mapping $\varphi: \mathbf{N} \to S$ to define the ordering on S: $m^\varphi \leqslant n^\varphi$ if and only if $m \leqslant n$.

Addition and multiplication of order types. Let A and B be disjoint representatives of the order types α and β, and therefore totally ordered sets. The *sum* $\alpha + \beta$ is defined as the order type of $A \cup B$, ordered by placing B behind A; that is, for all $a, b \in A \cup B$ one defines:

$$a < b \text{ in } A \cup B \text{ if and only if } \begin{cases} a \in A \text{ and } b \in B \text{ or} \\ a, b \in A \text{ (or } B) \text{ and } a < b \text{ in } A \text{ (or } B). \end{cases}$$

The *product* $\alpha \cdot \beta$ is defined as the order type of the product $A \times B$ with the following ordering:

$$(a, b) < (c, d) \text{ if and only if } \begin{cases} b < d \text{ or} \\ b = d \text{ and } a < c. \end{cases}$$

This is called the *anti-lexicographical ordering* of $A \times B$.

For natural numbers, regarded as order types, the sum and product are the same as those already defined. Addition and multiplication of order types are associative and distributive, but not, in general, commutative.

Well-ordered sets. The ordered set of the natural numbers has the following remarkable property: *every non-empty subset has a smallest element*. This property is used in counting the natural numbers by always taking the smallest 'unused' number, and is the basis of the principle of mathematical induction. CANTOR recognized the central importance of this property and used it to define well-ordered sets.

A totally ordered set $(S, <)$ is called well-ordered if every non-empty subset has a smallest (unique minimal) element. The order types of well-ordered sets are called ordinal numbers.

By definition every well-ordered set has a smallest element. The set of natural numbers is well-ordered in its natural ordering. Its (transfinite) ordinal number is denoted by ω (Fig.). A *segment* of a well-ordered set S is a proper subset T of S which contains every element of S that is smaller than some element of T. If A is a segment of S, then there is always an element $a \in S$, such that $A = \{x \in S \mid x < a\}$. With the help of this concept it is now possible to define a comparison of ordinal numbers.

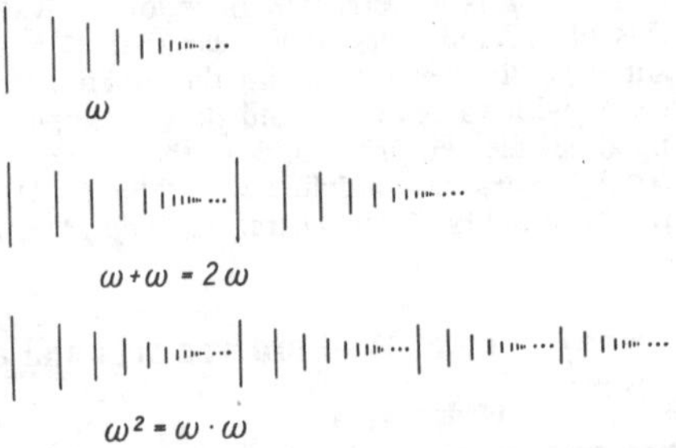

14.6-1 Schematic representation of some ordinal numbers

If α and β are ordinal numbers with representatives A and B, then $\alpha < \beta$ is defined to mean that A is similar to a segment of B, in other words, $\alpha < \beta$ if α is the ordinal number of a segment of B.

Every set of ordinal numbers is totally ordered by $<$.

This theorem cannot be reformulated as '*The set of all ordinal numbers is totally ordered*', because the concept 'set of all ordinal numbers' leads to a contradiction just as the 'set of all cardinals'.

It is immediate that $<$ is transitive. The statement that $<$ is irreflexive is equivalent to the statement that no well-ordered set is similar to one of its segments. The opposite assumption leads to a contradiction. For suppose that there is a similarity φ of S onto its segment A. Then there must be elements $x \in S$ such that $x^{\varphi} < x$. Let a be the smallest such element and $b = a^{\varphi}$. Since $b < a$, it then follows that $b^{\varphi} < a^{\varphi} = b$, thus $b^{\varphi} < b$, contradicting the minimality of a. The proof that any two ordinals can be compared is somewhat more complicated.

Every set of ordinal numbers is well-ordered, in other words, every set of ordinal numbers has a minimal element.

To prove this, let $W(\alpha)$ denote the set of all ordinal numbers less than a given ordinal number α. If A is a well-ordered set of type α, then A and $W(\alpha)$ are similar; for every ordinal number $\beta < \alpha$ corresponds to a segment S of A and this in its turn corresponds to an element b of A, with $S = \{x \in A \mid x < b\}$. Thus, $W(\alpha)$ is well-ordered. If now Z is any set of ordinal numbers, and α in Z is chosen arbitrarily, then $Z \cap W(\alpha)$ if not empty, has a smallest element, by the first part, and this must be the smallest element of Z.

The smallest transfinite ordinal number is ω.

Classes of ordinals. If m is a transfinite cardinal number, one can consider the class of all ordinal numbers whose representatives have the cardinal m. These sets are called the *transfinite classes of ordinals*. In every non-empty class there is a smallest ordinal number; this is called the *initial ordinal number* of the class. CANTOR combined the finite ordinal number into the *first class*, and the ordinal numbers of countable sets into the *second class* (see Well-ordering of the cardinals).

To every ordinal number α there is a larger one, for instance, its successor $\alpha + 1$; furthermore, to any set of ordinal numbers Z there is an ordinal number that is larger than any $\alpha \in Z$. For the set $\bigcup_{\alpha \in Z} W(\alpha + 1)$ is well-ordered, and its ordinal number β is larger than any $\alpha \in Z$. Indeed, β is the smallest such ordinal and is called the *supremum* of Z (sup Z).

Obviously $\sup W(\alpha) = \alpha$, in particular, $\sup \{0, 1, 2, \ldots\} = \omega$.

An ordinal number α is called a *limit number* if $W(\alpha)$ has no largest element. All other ordinal numbers have immediate predecessors and are called *isolated.*

Thus, ω is a limit ordinal, but all the finite ordinal numbers are isolated, and so is $\omega + 1$.

Recursion principle (transfinite induction). This is an important generalization of the principle of induction to arbitrary well-ordered sets.

Proof by recursion. Let S be well-ordered and suppose that a statement is true for the smallest element of S and, further, that it is true for any element of S if it is true for all smaller elements. Then the statement is true for every element of S.

This is very easy to prove. The assumption that the statement is not true for every element of S leads to a contradiction. For if a is the smallest element for which it is false (this must exist by the definition of well-ordering), then the statement is true for all elements less than a, and therefore, by hypothesis, for a, because a is not the smallest element of S. The following principle is more intricate.

Definition by recursion. If S is a well-ordered set and T is any set, then a unique mapping from S to T is defined if the image $f(a_0)$ of the smallest element a_0 of S is given and if $f(a)$ is determined by the values of f for all elements less than a.

This principle can be used to define powers of ordinal numbers α^β by a system of recursive equations.

(*i*) $\alpha^0 = 1$; (*ii*) $\alpha^{\beta+1} = \alpha^\beta \cdot \alpha$; (*iii*) $\alpha^\lambda = \sup \{\alpha^\xi \mid \xi < \lambda\}$ for limit ordinals λ.

Example 6: $\omega^1 = \omega^0 \cdot \omega = 1 \cdot \omega = \omega$, $\omega^2 = \omega^1 \cdot \omega = \omega \cdot \omega, \cdots, \omega^\omega = \sup \{\omega, \omega^2, \omega^3, \ldots\}$, and $\omega^{\omega^\omega} = \sup \{\omega^\omega, \omega^{\omega^2}, \omega^{\omega^3}, \ldots\}$.

All these numbers belong to the second class, as does every supremum of a countable set of numbers of the second class. The first number in the second class that cannot be expressed as a sum of powers of ω is

$$\varepsilon_0 = \sup \{\omega, \omega^\omega, \omega^{\omega^\omega}, \ldots\}.$$

This is called the *smallest ε-number*. It satisfies the equation $\omega^\varepsilon = \varepsilon$. An analogous continuation of this process leads to the numbers $\varepsilon_1, \varepsilon_2, \ldots, \varepsilon_\omega, \ldots, \varepsilon_\varepsilon, \ldots \varepsilon_{\varepsilon_\varepsilon}$ (ε with the subscript ε_ε) and so on. The naming of numbers of the second class can be continued ad infinitum, but it is impossible to define a universal notation, because it can be shown that there are uncountably many numbers in the class.

The well-ordering theorem. The preceding arguments do not exclude the possibility that certain transfinite classes of ordinal numbers might be empty; in other words, the question is still open, whether every set can be well-ordered in at least one way. This is indeed the case and forms the content of the well-ordering theorem. This theorem is equivalent to the axiom of choice. The first rigorous proof (using the axiom of choice) was given by Ernst ZERMELO (1871–1953) in 1904 in a letter to HILBERT. It started the controversy about the admissibility of the axiom of choice, which still has not been resolved.

Well-ordering theorem. On every set S there is a relation under which S is well-ordered.

CANTOR had considered this theorem as a principle of thought and had made it plausible in the following manner. Take any element a_0 of S, then a second, and so on. If S is infinite, one obtains a sequence $a_0, a_1, \ldots$ Now either S is exhausted or not; if not, repeat the process, as long as is necessary to exhaust the set. If S is ordered by the sequence in which the elements are chosen, then every subset has a smallest element, namely the one chosen first.

By present-day standards of mathematical rigour this argument can only be regarded as a heuristic first approach. Rigorous proofs are long and far too complicated to be included here.

Well-ordering of the cardinals. By the well-ordering theorem no transfinite cardinal $\boldsymbol{m}$ has an empty class of ordinals $Z_{\boldsymbol{m}}$. The well-ordering theorem can be used to show that any two cardinals are comparable, that is, that for any two sets S and T there is an injective mapping from S to T or from T to S. This is not surprising, since Zorn's lemma, the axiom of choice and the well-ordering theorem are all equivalent. By well-ordering S and T the statement is reduced to the comparability of ordinals.

In addition, it can now be shown that every non-empty set of cardinal numbers K has a smallest element; for the set of cardinal numbers is similar to the set of initial ordinal numbers of their classes (in fact, the cardinal numbers are often identified with these initial numbers). From this identification it also follows that to any set of cardinal numbers there exists a cardinal number larger than any element of the set. It is thus possible to index the cardinal numbers by the ordinal numbers in the following manner:

$\aleph_0$ = smallest infinite cardinal number,
$\aleph_{\alpha+1}$ = the smallest cardinal number greater than $\aleph_\alpha$,
$\aleph_\lambda$ = $\sup\{\aleph_\xi \mid \xi < \lambda\}$ for limit ordinal numbers λ.

This gives the famous Cantor sequence of cardinal numbers $\aleph_0, \aleph_1, \ldots, \aleph_\omega, \ldots$

Since $2^{\aleph_0} > \aleph_0$, the problem of the continuum hypothesis can be restated as the question where $\aleph$, the cardinal number of the continuum, occurs in this sequence. Cantor's continuum hypothesis is that $2^{\aleph_0} = \aleph_1$. The so-called generalized continuum hypothesis is that $2^{\aleph_\alpha} = \aleph_{\alpha+1}$.

15. The elements of mathematical logic

One of the main tasks of mathematical logic is the investigation of *formal thinking* and *inference* by means of mathematical methods taken, for example, from algebra or the theory of algorithms.

But this task, which has its origins in philosophy, is not its only one; nowadays mathematical logic comprises a multitude of questions and applications to the most diverse domains such as the natural sciences, switchwork algebra, the theory of data processing systems, linguistics, several branches of the social sciences like philosophy, laws, und ethics.

A decisive impetus towards the development of mathematical logic came from the situation of mathematics at the end of the 19th century. Up to that time mathematics had gathered together an abundance of individual results and had reached a high degree of abstraction, without having achieved a corresponding clarity about the contents of the fundamental concepts, which were used in an intuitive manner, for example, the concept of a set or of logical inference (see Chapter 42.). Apart from the need for an unquestionable foundation of the concept of a set, for the first time it became necessary to gain insight into the proper meaning of logic and logical deduction.

15.1. Propositional logic

Principles of classical propositional logic. *Propositions* is the name used for certain linguistic formations that serve to describe and to communicate facts. *Classical propositional logic* starts out from two assumptions. According to the *two-value principle* every proposition is either true or false. The concept of *truth* used here, which goes back to Aristotle, regards a proposition to be true if the statement asserted by it corresponds to a fact. The two-value principle really subsumes two principles:

1. the *principle of the excluded middle* according to which every proposition is true or false, and
2. the *principle of the excluded contradiction* according to which there is no proposition that ist both true and false. Therefore the class of all propositions splits into two disjoint subclasses, which are denoted by the symbols 1 (true) und 0 (false) and are called *truth values*.

By means of linguistic particles such as 'not', 'and', 'or' etc., given propositions can be combined into more complicated propositions. According to the second fundamental principle, the *principle of extensionality*, the truth value of a compound proposition is determined exclusively by the truth values of its components and does not depend on their meaning. Consequently such combinations can be regarded as functions that assign truth values to n-tuples of truth values.

The connective particles most frequently used in propositional logic correspond to *truth functions*: the function non corresponding to 'not', et to 'and', vel to 'or', seq to 'if ..., then ...', and aeq to 'if and only if ...'. These functions are determined as follows:

Truth function	Functorial notation	p 0	p 1	$p\,q$ 0 0	$p\,q$ 0 1	$p\,q$ 1 0	$p\,q$ 1 1
non p	$\sim p$	1	0				
et (p, q)	$p \wedge q$			0	0	0	1
vel (p, q)	$p \vee q$			0	1	1	1
seq (p, q)	$p \to q$			1	1	0	1
aeq (p, q)	$p \leftrightarrow q$			1	0	0	1

These definitions do not quite agree with the meaning of the particles in everyday language. For example, the following proposition is true:

'If $2 \cdot 2 = 5$, then the moon is inhabited by conscious living beings'; because in functorial notation $(0 \to 0) = 1$.

The task of propositional logic consists in the mathematical analysis of these concepts, which for the purpose are formalized within the framework of a calculus, the *propositional calculus*. To develop it one starts out from a collection of *fundamental symbols* of the following kinds:

(*i*) *variables* for propositions: $p_1, p_2, \ldots; p, q, r, s, \ldots$;
(*ii*) *functors* $\sim, \wedge, \vee, \to, \leftrightarrow$, denoting in this order the functions non, et, vel, seq, aeq;
(*iii*) *technical symbols*: (,).

Among the set of all sequences of symbols, the fundamental objects of the propositional calculus, the so-called *expressions*, are now selected by means of an inductive definition:

Definition of expressions

(*i*) The variables $p, q, \ldots$ are expressions.
(*ii*) If H and G are expressions, then so are $\sim H$, $(H \wedge G)$, $(H \vee G)$, $(H \to G)$, $(H \leftrightarrow G)$.
(*iii*) A sequence of symbols is an expression only if it is formed in accordance with (*i*) and (*ii*).

This definition makes it possible to decide in finitely many steps whether a given sequence of symbols is an expression or not.

Example 1: The following are expressions: $((p \to q) \wedge (r \vee s))$ and $((p \leftrightarrow q) \to (\sim q \to \sim p))$.

To simplify the presentation of expressions one uses *rules for saving brackets*:

(*i*) If the whole of an expression is included in brackets, these brackets may be omitted.
(*ii*) In the sequence $\sim, \wedge, \vee, \to, \leftrightarrow$ each functor separates stronger than the preceding one; for example, $p \wedge q \to r$ is to be read unambiguously as $(p \wedge q) \to r$.
(*iii*) A functor marked with a dot beneath separates stronger than one without a dot (see Examples 3., 4., 5., 6.; in 6. two dots separate stronger than one dot).

Semantics provides a link between the truth values and truth functions on the one hand, and the expressions on the other. This is done by means of the notion of a covering. A *covering of the propositional variables* is a function that assigns to each variable one of the two truth values 0 or 1. Such a covering f can be extended in a natural way to a function v_f that assigns a truth value to *every* expression. For a given f this function v_f is defined inductively:

(*i*) For variables p one has: $v_f(p) = f(p)$;
(*ii*) $v_f(\sim H) = \text{non}\,(v_f(H))$;
(*iii*) for expressions H and G one has:

$$\begin{aligned} v_f(H \wedge G) &= \text{et}\,(v_f(H), v_f(G)), \\ v_f(H \vee G) &= \text{vel}\,(v_f(H), v_f(G)), \\ v_f(H \to G) &= \text{seq}\,(v_f(H), v_f(G)), \\ v_f(H \leftrightarrow G) &= \text{aeq}\,(v_f(H), v_f(G)). \end{aligned}$$

The important concepts of semantic equivalence and universal validity can now be defined. Two expressions H and G are said to be *semantically equivalent*, in symbols $H \equiv G$, if $v_f(H) = v_f(G)$ for every covering f. An expression H is said to be *universally valid* or a *tautology* if $v_f(H) = 1$, in other words, if H is true for every covering f.

Example 2: $p \to (q \to p)$ is a tautology; $p \underset{\cdot}{\to} (p \to q) \to q$ is a tautology; $(p \to q) \wedge (p \to \sim q) \to \sim p$ is a tautology, by the principle of the excluded middle.

Logical inference serves to obtain new true propositions from given propositions that have already been established to be true. Therefore the *rules of inference* to be used must carry over the truth of an expression to the deduced expression. In the derivation of such rules of inference tautologies play a particular role: every tautology of the form $H \to G$ leads to a rule of inference. The conditions for the application of a rule, the *premisses*, are written down above a horizontal line, the result of the application, the *conclusion*, below. A system S of rules of inference determines a relationship 'A can be derived from S', in symbols $S \vdash A$.

Examples for rules of inference, in which H, G, F denote expressions and S a set of expressions.

3. $p \dot{\rightarrow} (p \rightarrow q) \rightarrow q$ leads to the rule:
$$\frac{S \vdash H \qquad S \vdash H \rightarrow G}{S \vdash G}$$

4. The *chain of inferences* $(p \rightarrow q) \dot{\rightarrow} (q \rightarrow r) \rightarrow p \rightarrow r$ leads to the rule:
$$\frac{S \vdash H \rightarrow G \qquad S \vdash G \rightarrow F}{S \vdash H \rightarrow F}$$

5. The *contraposition* $p \rightarrow \sim q \dot{\rightarrow} q \rightarrow \sim p$ leads to the rule:
$$\frac{S \vdash H \rightarrow \sim G}{S \vdash G \rightarrow \sim H}$$

6. The *principle of contradiction* $p \rightarrow q \dot{\Rightarrow} p \rightarrow \sim q \dot{\rightarrow} \sim p$ leads to the rule:
$$\frac{S \vdash H \rightarrow G \qquad S \vdash H \rightarrow \sim G}{S \vdash \sim H}$$

The rules of inference of the propositional calculus do not take account of the finer structure of propositions; further-reaching rules of inference are treated in the predicate calculus.

15.2. Predicate logic

The expressions of the propositional calculus are not sufficient to formulate the facts occurring in mathematics; a formalized version of the mathematical language must be considerably richer. A characteristic feature is the frequent use of variables and of special symbols for *functions* or for *relations*. *Variables* are preassigned symbols, which denote arbitrary objects of a previously delineated domain. Symbols whose meaning is fixed are called *constants*, such as 0 and + within the domain of natural numbers.

A further feature of the mathematical language is the possibility of *binding variables* by means of the *quantifiers* of predicate logic.

In the expression 'there exist prime numbers p and q such that $2n = p + q$' the symbols p and q are bound by the predicate-logical functor 'there exists ...', while the variable n is free. It has turned out that for the purposes of binding variables in mathematics the two operations of predicate logic $\exists$ '*there is* ...' and $\forall$ '*for all* ...' are sufficient. Therefore the languages of predicate logic are based on this kind of binding variables only.

In *predicate logic* the finer structure of mathematical statements is investigated; for example, the propositional calculus is incapable of grasping the statement about rational numbers

$$\forall x \, \forall y \, \exists z (x < y \rightarrow x < z < y).$$

Syntax of elementary languages. The propositions of a mathematical theory contain as fundamental concepts certain predicates and functions; for example, in the theory of sets the relationship $\in$ '... is element of ...', in geometry the relationship of incidence and of betweenness, in arithmetic addition, multiplication and the relationship of order. For these fundamental concepts symbols are introduced, which taken together form the *signature* of this theory. A signature therefore consists of symbols for relations, for functions, and for individuals. Each of these symbols has its appropriate *valency* or '*arity*'. In the signature $\Sigma = \{+, \cdot, <, 0, 1\}$ of elementary arithmetic + and $\cdot$ are binary operational symbols, $<$ is the symbol of a binary relation, and 0 and 1 are symbols for individuals.

Apart from the symbols in Σ a *mathematical theory* uses variables for individuals, such as the symbols $x, y, z, \ldots$, logical symbols such as $\sim, \wedge, \vee, \rightarrow, \leftrightarrow, =, \exists, \forall$ and auxiliary technical symbols.

Just as in the propositional calculus, an *elementary language* L_Σ (or language of the predicate calculus) can now be defined with reference to a given signature Σ by means of these fundamental symbols. Its elements are certain sequences of symbols called *expressions* or *propositional forms*. The construction of these expressions takes place after introducing the so-called terms.

Definition of terms

(*i*) Variables and constants for individuals are terms.
(*ii*) If F is an n-ary function symbol und $t_1, \ldots, t_n$ are terms, then so is $Ft_1 \ldots t_n$.
(*iii*) A sequence of symbols is a term only if it is formed in accordance with (*i*) and (*ii*).

Example 7: If sin, +, $\cdot$ are the usual function symbols interpreted in the domain of real numbers, then the following sequences of symbols are terms: $\sin x$, $x^2 \cdot y + y^3 + z^3$, $\sin (x + \sin (y^2 + x))$.

The expressions of the elementary language L_Σ are characterized inductively.

Definition of expressions and propositional forms

(*i*) If R is an n-ary relation symbol and $t_1, \ldots, t_n$ are terms, then $Rt_1 \ldots t_n$ is an expression, a so-called *atomic expression*.

(*ii*) If A and B are expressions, then so are $\sim A$, $(A \wedge B)$, $(A \vee B)$, $(A \rightarrow B)$, $(A \leftrightarrow B)$.
(*iii*) If $A(x)$ is an expression containing the variable x, but not the symbols $\exists x$ or $\forall x$, then so are $\exists x\, A(x)$ and $\forall x A(x)$.
(*iv*) A sequence of symbols is an expression only if it is formed in accordance with (*i*) – (*iii*).

Example 8: In the signature $\Sigma = \{P, Q, R, f, g, T\}$ the following sequences of symbols are expressions: $\forall x(Rxy \rightarrow Qxf(y))$, $\sim \exists x[Rxy \vee Qxg(y, x)]$, $\forall x[Px \wedge \exists y(Tyx \rightarrow Sxy)]$.

Just as in the propositional calculus, it can be decided in finitely many steps whether a given sequence of symbols is an expression or not.

A variable x occurs in an expression H *freely* if x occurs in H but not $\exists x$ or $\forall x$; x is *quantified* in H if $\exists x$ or $\forall x$ occurs in it. After every place of the form Θx, where Θ is $\exists$ or $\forall$, there begins a uniquely determined partial expression H' of H in which the variable x without this symbol Θx would be free. This partial expression H' of H is called the *domain of influence of the quantifier* Θ for the place in question. Within it the variable x is quantified. A variable x occurs freely at a certain place in an expression H if it occurs at that place and is neither quantified there nor within the domain of influence of a quantifier. If there is at least one place in H at which a variable x occurs freely, then x is said to occur freely in H.

Example 9: In the expression $\overset{1}{\exists} x[\overset{5}{P}x \wedge Q\overset{8}{\boxed{y}} \wedge \overset{10}{g}(\overset{12}{\boxed{y}}) = \overset{15}{z}] \rightarrow [\overset{20}{\exists} x \forall \overset{22}{\boxed{y}}\; Rx\overset{25}{\boxed{y}} \wedge f(x, z) \overset{30}{=} z]$ the places are indicated by numbers over the symbols. The variable y occurs freely at the places 8 and 12 and bound at 22 and 25; it is quantified at 22, and at 25 it is within the domain of influence of the place 21.

In general, expressions are not propositions. The expression $x < y$, in which $<$ denotes the order among natural numbers, only becomes a proposition if definite symbols are substituted for the variables x and y, for example, $0 < 1$, $3 < 2$, $5 < 7$, or when the free variables are bound by quantifiers, for example, $\forall x\, \exists y\, x < y$. *Propositions* can therefore be characterized as those expressions in which there are *no free variables.*

Examples of propositions:

10. The monotonic law for addition of natural numbers: $\forall x\, \forall y\, \forall z\; (x < y \rightarrow x + z < y + z)$.

11. Fermat's conjecture: $\sim \exists x\, \exists y\, \exists z\, \exists n\; (n > 2 \wedge x^n + y^n = z^n)$.

12. Goldbach's conjecture: $\forall x[2 \mid x \wedge x \neq 2 \wedge x \neq 0 \rightarrow \exists y\, \exists z\; (\text{prime } y \wedge \text{prime } z \wedge x = y + z)]$; here prime y is an abbreviation for $y \neq 1 \wedge \forall u\, \forall v\; (y = u \cdot v \rightarrow u = 1 \vee v = 1)$, and $2 \mid x$ is an abbreviation for $\exists y\; (y + y = x)$. In words the expression reads: 'For all natural numbers x, if x is even, non-zero, and not 2, then there exist prime numbers y and z such that x is the sum of y and z'.

A generalization of the elementary languages is obtained if one allows quantification of unary predicates, that is, if one treats them like individuals. In such languages, which are also called *monadic* of the second order, considerably more statements can be expressed than in the elementary languages.

Examples of propositions in monadic languages of the second order:

13. Peano's axiom for the natural numbers:

$$\forall P(P\,0 \wedge \forall x(Px \rightarrow Px') \rightarrow \forall x\; Px),$$

in words: 'If a unary predicate P holds for 0, and if P holds, together with an element x, also for its successor x', then P holds for all natural numbers'.

14. Axiom of the least upper bound for real numbers:

$$\forall P[\exists z\; Pz \wedge \exists u\, \forall v(Pv \rightarrow v \leq u) \rightarrow \exists y(\forall v(Pv \rightarrow v \leq y) \wedge \sim \exists y'(\forall v(Pv \rightarrow v \leq y') \wedge y' < y))];$$

in words: 'Every non-empty set of real numbers that is bounded above has a least upper bound'.

The *languages of predicate logic are descriptive*, that is, the expressions of such languages describe relationships prevailing in mathematical structures.

With the development of data processing by machines algorithmic languages gain in importance. *Algorithmic languages* have the purpose of making commands, initiating actions, and directing processes. Examples of algorithmic languages used in programming technology are ALGOL 60, PL 1, FORTRAN, COBOL and others.

Some algorithmic elements are contained even in elementary languages: a term can be regarded as a sequence of commands to be carried out; for example, $(x + 1) \cdot y$ denotes the sequence 'add 1 to x and then multiply the result by y'.

Semantics of elementary languages. Just as in the proportional calculus, semantics establishes a connection between the expressions of L_Σ and the realm of mathematical structures in which the expressions have a meaning.

Let Σ be a set of operational and functional symbols, and S a non-empty set. By an *interpretation* of Σ in S one understands a mapping δ that assigns to every n-ary relation symbol R in Σ an n-ary relation R^δ on S, that is, a subset of S^n, and to every n-ary operational symbol F an n-argument function F^δ on S, that is, a single-valued mapping of S^n into S. The equality symbol $=$ is always interpreted by the relationship of identity.

Let δ_Σ denote the sequence of the relations and operations assigned to the symbols in Σ.

A **Σ-structure, Σ-algebra,** or **Σ-model** is defined as an ordered pair $\boldsymbol{S} = (S, \delta_\Sigma)$. Let $\boldsymbol{K}_\Sigma$ denote the class of all Σ-structures. The symbols contained in Σ always refer to the class $\boldsymbol{K}_\Sigma$. A *truth concept* can now be set up for the elementary languages L_Σ, that is, a definition of the statement: 'the proposition H is true within the structure $\boldsymbol{S}$', symbolically $\boldsymbol{S} \models H$. This concept is fundamental for the whole of semantics. The truth concept in elementary languages can be made precise by introducing first a more general concept: 'the S-cover α satisfies the expression H in $\boldsymbol{S}$', symbolically $\boldsymbol{S} \models_\alpha H$.

By an S-cover α one means a function that assigns to every individual variable an element of S. Such a cover α can be extended in a natural way, just as in the propositional calculus, to a mapping α of all terms of L_Σ into S: $c^\alpha = c^\delta$, where c is a constant for an individual in Σ:

$$(F(t_1, \ldots, t_n))^\alpha = F^\delta(t_1^\alpha, \ldots, t_n^\alpha).$$

Example 15: Let $t = (x+1) \cdot y$, $\alpha(x) = 2$, $\alpha(y) = 3$. Then $t^\alpha = (x^\alpha + 1) \cdot y^\alpha = (2+1) \cdot 3 = 9$.

Definition of the relation 'α satisfies A in $\boldsymbol{S}$', in symbols, $\boldsymbol{S} \models_\alpha A$, in which **iff** stands for 'if and only if'.

(*i*) $\boldsymbol{S} \models_\alpha Rt_1 \ldots t_n$ **iff** $(t_1^\alpha, \ldots, t_n^\alpha) \in \boldsymbol{R}^\delta$, that is R holds for the n-tuple $(t_1^\alpha, \ldots, t_n^\alpha)$:

(*ii*) $\boldsymbol{S} \models_\alpha \sim A$ **iff** not $\boldsymbol{S} \models_\alpha A$;
$\boldsymbol{S} \models_\alpha A \wedge B$ **iff** $\boldsymbol{S} \models_\alpha A$ and $\boldsymbol{S} \models_\alpha B$;
$\boldsymbol{S} \models_\alpha A \vee B$ **iff** $\boldsymbol{S} \models_\alpha A$ or $\boldsymbol{S} \models_\alpha B$;
$\boldsymbol{S} \models_\alpha A \rightarrow B$ **iff** $\boldsymbol{S} \models_\alpha A$ implies $\boldsymbol{S} \models_\alpha B$;
$\boldsymbol{S} \models_\alpha A \leftrightarrow B$ **iff** $\boldsymbol{S} \models_\alpha A \rightarrow B$ and $\boldsymbol{S} \models_\alpha B \rightarrow A$;

(*iii*) $\boldsymbol{S} \models_\alpha \exists x\, A(x)$ **iff** the value of α for the variable x can be modified so that the modified cover α' satisfies the expression $A(x)$ in $\boldsymbol{S}$;
$\boldsymbol{S} \models_\alpha \forall x\, A(x)$ **iff** every cover α arising only by modifying the value for the variable x satisfies the expression $A(x)$ in $\boldsymbol{S}$.

Example 16: $\exists x(y = x \cdot x)$, where $\cdot$ denotes the multiplication of natural numbers. If α is a covering of all variables such that $y^\alpha = 4$, then α satisfies this expression; for the covering α that assigns to the variable x the value 2 and agrees with α for all the remaining variables satisfies the expression $y = x \cdot x$.

This example shows that whether $\boldsymbol{S} \models_\alpha A$ holds or not depends only on the variables occurring freely in A. If A is a proposition, that is, an expression without free variables, then $\boldsymbol{S} \models_\alpha A$ holds for all α or for no α.

Definition. (*i*) An expression A is valid in $\boldsymbol{S}$, in symbols $\boldsymbol{S} \models A$, if and only if every cover α satisfies A in $\boldsymbol{S}$, that is, if $\boldsymbol{S} \models_\alpha A$ holds for every S-cover α;

(*ii*) an expression $A \in L_\Sigma$ is called *universally valid* (or *valid in predicate logic*) if A is valid in every Σ-structure.

Examples: 17. The proposition $\forall x\, \exists y\, x < y$ is valid in the domain of natural numbers; but it is not universally valid, because it is false in a finite ordered set $(S, <)$.

18. The proposition $\forall x\, \forall y\, Rxy \vee \sim\forall x\, \forall y\, Rxy$ is universally valid; indeed, $\boldsymbol{S} \models H$ or $\boldsymbol{S} \models \sim H$ holds for every proposition H and every structure $\boldsymbol{S}$.

An expression H is said to be *indecomposable in propositional logic* if it begins with a quantifier, that is, if H is of the form $H = \Theta x H'$, where Θ stands for $\exists$ or $\forall$. Every expression is composed from indecomposable expressions by means of the functors $\sim$, $\wedge$, $\vee$, $\rightarrow$, and $\leftrightarrow$. When one substitutes propositional variables for the indecomposable components of an expression, one obtains an expression of the propositional calculus; for example, the expression $\forall x\, \forall y\, Rxy \vee \sim\forall x\, \forall y\, Rxy$ goes over into the tautology $p \vee \sim p$. An expression H is said to be *universally valid in propositional logic* if the corresponding expression of propositional logic is universally valid within the framework of the propositional calculus. If H is universally valid in propositional logic, then also in predicate logic. However, there are expressions that are universally valid in predicate logic, but not in propositional logic; an example is: $\forall x\, Px \rightarrow \exists x\, Px$. This is the reason for the occurrence in predicate logic of methods of inference that are not available in the propositional calculus.

Definition. Let S be a set of expressions in L_Σ. A Σ-structure $\boldsymbol{M}$ is a model of S if all the expressions $A \in S$ are valid in $\boldsymbol{M}$. Let Mod S be the class of all models of S.

Let $\Sigma = \{+, 0\}$, and let $S \subseteq L_\Sigma$ be the following set of expressions:

(i) $(x + y) + z = x + (y + z)$, *(iii)* $\forall x\, \exists y (x + y = 0)$,

(ii) $\forall x (x + 0 = x)$, *(iv)* $\forall x\, \forall y (x + y = y + x)$.

Then a Σ-structure $\mathbf{M} = (M, +, 0)$ is a model of S if and only if $\mathbf{M}$ is an additively written Abelian group, and Mod S *is the class of all Abelian groups.*

Two expressions H and G are called *semantically* or *logically equivalent* if the expression $H \leftrightarrow G$ is universally valid.

Examples of logical equivalences:

19. $\sim \exists x\, A(x) \equiv \forall x \sim A(x)$. *20.* $\sim \forall x\, A(x) \equiv \exists x \sim A(x)$.

21. $\Theta x A(x) \equiv \Theta y A(y)$ if y does not occur in $A(x)$ nor x in $A(y)$ und Θ is one of the quantifiers $\exists$ or $\forall$.

22. $\forall x [A(x) \wedge B(x)] \equiv \forall x\, A(x) \wedge \forall x\, B(x)$. *23.* $\exists x [A(x) \vee B(x)] \equiv \exists x\, A(x) \vee \exists x\, B(x)$.

Expressions of the form $\Theta_1 x_1 \ldots \Theta_n x_n A(x_1, \ldots, x_n)$, where each Θ_i is $\exists$ or $\forall$ and A is free from quantifiers, are said to be *in prenex form.*

Every expression is logically equivalent to an expression in prenex form of the same signature and with the same free variables.

Examples for the transformation of an expression into a prenex form logically equivalent to it.

24. $\forall x\, \forall y [x < y \to \exists z (x < z < y)] \equiv \forall x\, \forall y\, \exists z (x < y \to x < z < y)$.

25. $\forall x\, \exists y\, \forall z\, Qxyz \to \forall y\, \exists z \sim Ryz \equiv \forall u\, \exists v\, \exists x\, \forall y\, \exists z \sim (Qxyz \wedge Ruv)$.

Mathematical inference. Mathematical inference serves to obtain new true propositions from given true propositions. The backbone of mathematical inference is the *conclusion.* If S is a set of propositions that are true in a structure $\mathbf{S}$ and if a proposition A can be deduced from S, then A must preserve its truth in $\mathbf{S}$, that is, the proposition A must be true in the structure $\mathbf{S}$.

Definition of a conclusion. If S is a set of propositions of an elementary language L_Σ and if H is an expression in L_Σ, then H is said to follow from S, in symbols $S \Vdash H$ if every model of S is also a model of H, that is, if Mod $S \subseteq$ Mod H. The *set of consequences* of S is $S^{\Vdash} = \{H \in L_\Sigma, S \Vdash H\}$.

For example, if S is the following system of axioms:

$$\forall x\, \forall y\, \forall z (x \cdot y) \cdot z = x (y \cdot z), \quad \forall x\, \forall y\, x \cdot y = y \cdot x, \quad \forall x\, \forall y\, \forall z (y \cdot x = z \cdot x \to y = z),$$

then a proposition H follows from S if and only if H holds in every commutative cancellation semigroup. Therefore the set of consequences of S contains all the propositions of the elementary theory of the class of all commutative cancellation semigroups.

In mathematical inference one does not always go back to the definition of a conclusion, but one makes use of certain *rules of inference*, which are *hereditary under conclusion*, that is, are valid for the conclusion process.

Examples of conclusion-hereditary rules of inference.

26. Separation rule:
$$\frac{S \Vdash H \qquad S \Vdash H \to G}{S \Vdash G}$$

27. Derivation rule:
$$\frac{S \cup \{A\} \Vdash B}{S \Vdash A \to B}$$

28. Deduction theorem:
$$\frac{S \Vdash A \to B}{S \cup \{A\} \Vdash B}$$

29. Indirect conclusion:
$$\frac{S \cup \{A\} \Vdash B \qquad S \cup \{A\} \Vdash \sim B}{S \Vdash \sim A}$$

For every system R of rules of inference of this kind one can define a *relationship of derivability*: 'The expression A is derivable, provable, from the set S'. An expression A is derivable or provable from S by means of the R-rules if one can obtain A from certain initial expressions belonging to S by applying rules in R finitely many times. A proof or a derivation of A can be regarded as a finite sequence of expressions $(F_1, \ldots, F_n, A)$ which can be obtained successively from S by application of the rules in R. If the set R of rules and the initial set S are finite, then it can always be decided in finitely many steps whether or not a given finite sequence of expressions is a proof.

The conclusion relation, which is at the basis of every inference, can be characterized by a finite system of rules of inference.

In accordance with this fundamental fact, in what follows a system of rules of inference is indicated that is adapted as far as possible to the natural deduction process. For each of the functors $\sim$, $\wedge$, $\vee$, $\to$, $\leftrightarrow$, $\exists$, $\forall$ two rules of inference are given, one to *introduce* it and one to *remove* it. The *relationship of derivability* laid down by these rules of inference is denoted by the symbol $\vdash$.

Definition of a system of rules of inference.

(0a) $\dfrac{A \in S}{S \vdash A}$ (0b) $\dfrac{S \vdash A,\ S \subseteq S'}{S' \vdash A}$

(1a) $\dfrac{S, A \vdash B}{S \vdash A \rightarrow B}$ (1b) $\dfrac{S \vdash A,\ A \rightarrow B}{S \vdash B}$

The rule (1a) corresponds to the deduction theorem for conclusion (see Example 28).

(2a) $\dfrac{S, A \vdash B, \sim B}{S \vdash \sim A}$ (2b) $\dfrac{S, \sim A \vdash B, \sim B}{S \vdash A}$

(3a) $\dfrac{S \vdash A, B}{S \vdash A \wedge B}$ (3b) $\dfrac{S \vdash A \wedge B}{S \vdash A, B}$

(4a) $\dfrac{S \vdash A}{S \vdash A \vee B,\ B \vee A}$ (4b) $\dfrac{S \vdash A \vee B,\ A \rightarrow C,\ B \rightarrow C}{S \vdash C}$

(5a) $\dfrac{S \vdash A \rightarrow B,\ B \rightarrow A}{S \vdash A \leftrightarrow B}$ (5b) $\dfrac{S \vdash A \leftrightarrow B}{S \vdash A \rightarrow B,\ B \rightarrow A}$

(6a) $\dfrac{S \vdash A(t)}{S \vdash \exists x\, A(x)}$ (6b) $\dfrac{S \vdash \exists x\, A(x),\ A(y) \rightarrow B}{S \vdash B}$

In (6a) t is an arbitrary term, in (6b) y does not occur in B nor in S.

(7a) $\dfrac{S \vdash A(y)}{S \vdash \forall x\, A(x)}$ (7b) $\dfrac{S \vdash \forall x\, A(x)}{S \vdash A(t)}$

In (7a) y does not occur in S.

(8a) $S \vdash t = t$ (8b) $\dfrac{S \vdash A(t),\ t = t'}{S \vdash A(t')}$

In (8a, b) t and t' denote terms of the language L_Σ.

All these rules are conclusion-hereditary, that is, 'if $S \vdash A$, then $S \Vdash A$' holds.

Apart from these rules mathematical inference makes use of a number of other rules of inference that can be derived from the given ones. Examples are the following:

(1c) $\dfrac{S \vdash A \rightarrow B}{S, A \vdash B}$ (2c) $\dfrac{S \vdash \sim\sim A}{S \vdash A}$ (4c) $\dfrac{S \vdash A \vee B}{S, \sim A \vdash B}$

(6c) $\dfrac{S, A(x) \vdash B}{S, \exists x\, A(x) \vdash B}$ (x not free in B nor in S) (8c) $\dfrac{S \vdash A(t),\ t = t'}{S \vdash A(t\,/\!/\,t')}$

In (8c) $A(t\,/\!/\,t')$ means that the term must not necessarily be replaced by the term t' at all places where it occurs in $A(t)$.

The following important theorem holds for the relationship of derivability.

Theorem on the completeness of the relationship of derivability. For every set S of formulae in L_Σ and every formula $A \in L_\Sigma$ $S \Vdash A$ holds if and only if $S \vdash A$. In particular, $\emptyset \Vdash A$ if and only if $\emptyset \vdash A$, that is, A is universally valid if and only if A is derivable without axioms, in other words, from the empty set.

Example 30: Strictly formal derivation of a number-theoretical statement. By way of explanation, after the lines of the formal proof the meaning of the expressions and the natural conclusions corresponding to the formal derivations are added in square brackets.

$$\forall x\, (\sim \exists y\ x = 3 \cdot y \rightarrow \exists z\ x^2 - 1 = 3 \cdot z).$$

[For all integers x, if x is not divisible by 3, then $x^2 - 1$ is divisible by 3.]

Let $B(x, z)$ be an abbreviation for $x^2 - 1 = 3 \cdot z$. According to rule (7a) it is sufficient to show that $S \vdash \sim\exists y\ a = 3 \cdot y \rightarrow \exists z\, B(a, z)$.

[It is enough to prove the assertion for a fixed but arbitrary integer a.]

By (1a) it is sufficient to show that

$$S, \sim\exists y\ a = 3 \cdot y \vdash \exists z\, B(a, z).$$

[Assuming that a is not divisible by 3, it has to be shown that $a^2 - 1$ is divisible by 3.]

The fact that $S \vdash \exists x\ a = 3 \cdot x \vee \exists x\ a + 1 = 3 \cdot x \vee \exists x\ a - 1 = 3 \cdot x$ is taken to be known. By (4c) then

$$S, \sim\exists x\ a = 3 \cdot x \vdash \exists x\ a + 1 = 3 \cdot x \vee \exists x\ a - 1 = 3 \cdot x.$$

[Since a is not divisible by 3, either $a + 1$ or $a - 1$ is divisible by 3.]

By (4b) it is now enough to show that

(*i*) $S, \exists x\, a+1 = 3\cdot x \vdash \exists z\, B(a, z)$ and (*ii*) $S, \exists x\, a-1 = 3x \vdash \exists z\, B(a, z)$.

[Two cases are to be considered: (*i*) $a + 1$ is divisible by 3, (*ii*) $a - 1$ is divisible by 3.]

Only (*i*) will be proved, similar arguments hold for (*ii*). By (6c) it is enough to show that $S, a+1 = 3\cdot b \vdash \exists z\, B(a, z)$ and by (6a) $S, a+1 = 3\cdot b \vdash B(a, t)$ for a certain term t.

[Let b be one of the elements x for which $a + 1 = 3\cdot x$; it is sufficient to indicate a number t for which $a^2 - 1 = 3\cdot t$.]

Now $S \vdash (a+1)(a-1) = a^2 - 1$, hence by application of the rules (8a), (8b), and (8c)

$S, a+1 = 3\cdot b \vdash a^2 - 1 = 3\cdot b\cdot(a-1)$,

that is, the term $t = b\cdot(a-1)$ has the required properties.

15.3. Formalized theories

The formalization of a theory proceeds in several steps. First of all, the domain of objects and the appropriate relations have to be laid down. On this level, the first mathematical concepts are obtained, by *abstraction from real situations*; for example, the fundamental geometrical concepts such as point and line arose by abstraction from reality. In the second step the concept of a *proposition* is made *precise*, and an *interpretation of the propositions on the domain in question* is defined. Finally, an *axiom system* and a *relationship of derivability* are given. An axiom system should aim at *completeness*, that is, it should characterize the domain in question completely. This means: every proposition that is valid in this domain should be derivable from the system of axioms.

Most mathematical theories refer to certain classes of structures. The theory of a *class* K *of structures* can be identified with the set of propositions that are valid in every structure of this class.

Definition. Relative to a given class K of Σ-structures the elementary theory $T(K)$ of this class is defined by $T(K) = \{H \in L_\Sigma : K \models H\}$; here $K \models H$ means that $A \models H$ for every structure $A \in K$, that H is true in K.

$T(K)$ is the elementary theory of the class K of structures. A set X of propositions is an axiom system for a theory T if $X^\vdash = T$ and if X is decidable, that is, if for every expression $H \in L_\Sigma$ it can be decided in finitely many steps whether $H \in X$ or $H \notin X$.

Examples of formalized elementary theories.

31. *The theory of fields* with $\Sigma = \{+, \cdot, 0, 1\}$ is characterized by the following system of axioms:

$\forall x\,\forall y\,\forall z[x + (y + z) = (x + y) + z]$, $\forall x[x + 0 = 0 + x = x]$,
$\forall x\,\exists y(x + y = 0)$, $\forall x\,\forall y(x + y = y + x)$,
$\forall x\,\forall y\,\forall z[(x\cdot y)\cdot z = x\cdot(y\cdot z)]$, $\forall x\,\forall y(x\cdot y = y\cdot x)$,
$\forall x(x\cdot 1 = x)$, $\forall x[x \neq 0 \rightarrow \exists y(x\cdot y = 1)]$,
$\forall x\,\forall y\,\forall z[(x + y)\cdot z = x\cdot z + y\cdot z]$.

32. The *theory of linearly ordered sets* with $\Sigma = \{<\}$ is characterized by:

$\sim\exists x(x < x)$, $\forall x\,\forall y\,\forall z(x < y \wedge y < z \rightarrow x < z)$,
$\forall x\,\forall y(x = y \vee x < y \vee y < x)$.

33. The *theory of groups* with $\Sigma = \{\cdot, 1\}$ is characterized by:

$\forall x\,\forall y\,\forall z[(x\cdot y)\cdot z = x\cdot(y\cdot z)]$, $\forall x(x\cdot 1 = x)$,
$\forall x\,\exists y(x\cdot y = 1)$.

Definability in formalized theories. Frequently in a mathematical theory T, apart from the fundamental symbols given by the signature Σ, new concepts, predicates, and operations are defined. For example, in the arithmetic of natural numbers the relationship $x \mid y$ of divisibility 'x divides y' can be defined as follows: $x \mid y =_{\mathrm{def}} \exists z\ (y = x\cdot z)$ or the relation $a \leq b$ as follows: $a \leq b =_{\mathrm{def}} \exists x(a + x = b)$. Such explicit definitions themselves can be characterized as formal propositions of a particular kind. If in this example one extends the initial signature $\Sigma = \{+, \cdot, 0, 1\}$ of elementary arithmetic by adding the binary predicate symbol $|$, then to the arithmetical axioms the proposition $\forall x\,\forall y[x \mid y \leftrightarrow \exists z(y = x\cdot z)]$ can be added, which is called the definition of the predicate $x \mid y$.

Also functional and individual symbols can be introduced by definition. An *explicit definition of an n-argument function* F in a theory T has the following form:

$$\forall x_1 \dots \forall x_n\, \exists y[Fx_1 \dots x_n = y \leftrightarrow A(x_1, \dots, x_n, y)],$$

where it is assumed that in T the propositions

$\forall x_1 \dots \forall x_n\, \exists y\, A(x_1, \dots, x_n, y)$ and $\forall x_1 \dots \forall x_n\, \forall y\, \forall z[A(x_1, \dots, x_n, y) \wedge A(x_1, \dots, x_n, z) \rightarrow y = z]$

are derivable.

Definition. If in an elementary theory T with the signature Σ a relation R is an element of Σ and if $\Sigma' \subseteq \Sigma \setminus \{R\}$ is a subset of Σ, then the relation R is said to be explicitly definable in T if there is a definition for R derivable in T whose defining expression contains symbols of Σ' only.

If in a theory T a relation R is explicitly definable by means of the remaining relations, then every expression can be transformed equivalently in T into an expression that does not contain the symbol R. Consequently definable relations are dispensable. But from the methodological point of view the search for suitable definitions is just as important as the search for a suitable proof.

If a predicate R is definable within a theory T by means of predicates $Q_1, \ldots, Q_n$, then in every model of T the interpretation of R is uniquely determined by the interpretations of the Q_i. Hence the following principle is valid.

Padoa's principle. The fact that a predicate R cannot be defined within a theory T in terms of predicates $Q_1, \ldots, Q_n$ can be established by indicating two models $\mathbf{M}$ and $\mathbf{M}'$ that differ solely in the meaning of R.

Axiomatic definitions are of a different kind. They aim at grasping a concept or a relation of a domain of objects axiomatically, that is, to characterize them by means of a set of propositions. For a class $\boldsymbol{K}$ of structures this means that an axiom system is to be given for the elementary theory of $\boldsymbol{K}$.

15.4. Algorithms and recursive functions

Within the framework of mathematics and logic, algorithms made their appearance as general methods for the solution of all problems of a given class. Their purpose is to describe processes in such a way that afterwards they can be imitated or governed by a machine. Examples of algorithmic processes are logical inference and certain calculating processes occurring in mathematics, in particular, solution methods for various types of equations.

A characteristic feature of an *algorithm* is that it transforms given quantities (input data) into other quantities (output data) on the basis of a system of *transformation rules*. But it only makes sense to talk of an algorithm if certain additional conditions are satisfied:

(*i*) The *system of quantities* to be transformed into one another must be *given effectively*;

(*ii*) the algorithm must be describable by *finitely many rules* because no machine can store infinitely many rules;

(*iii*) the transformation of quantities, the working of the algorithm proceeds in the form of mechanical *working units*, each unit consisting in the application of one of the given rules.

During the period 1931–1947, within the framework of mathematical logic, a number of well delineated concepts of an 'algorithm' were developed, making the intuitive notion more precise. The most important are the *calculus of equations* (J. Herbrand, K. Gödel, S. C. Kleene, about 1931 to 1936), the Turing machine (A. M. Turing 1936), the λ-calculus (A. Church 1936) and the algorithmic concepts of E. L. Post (1936) and A. A. Markov (1947).

Of great significance is the fact that all these notions are equivalent in the sense that the same number-theoretical functions, the so-called *recursive functions*, can be calculated by each of them. Here a number-theoretical function is understood to be one that is defined in the domain of natural numbers. On the basis of this equivalence one can take the view that the intuitive concept of an algorithm is comprehended in the newly gained precision. This point of view was formulated in 1936 by Church and is known in the mathematical literature as *Church's hypothesis*.

A number-theoretical function f is said to be *calculable* if there is an algorithm by which the value $f(n)$ can be found for every value n of the argument.

Examples of calculable functions.

34. Let $f(x)$ be the xth *prime number*. Here the method of the sieve of Eratosthenes (see Chapter 1.) can be used to calculate the function.

35. Let $f(x, y)$ be the *greatest common divisor* of x and y. This function can be calculated by means of Euclid's algorithm (see Chapter 1.).

36. Let $f(x)$ be the xth digit in the decimal representation of $\pi = 3.14159\ldots$ Here a convergent series representation of π is suitable for the calculation of the function (see Chapter 21.).

The *class of recursive functions* arises by making the intuitive concept of a calculable function more precise. Certain *initial functions*, which can be regarded to be immediately calculable, are called recursive, and certain rules are specified by means of which new recursive functions can be generated from given ones. The rules are such that for every new recursive function one can indicate at once an algorithm to calculate the function values if such algorithms are available for the given recursive functions.

A. Initial functions

(*i*) The *identity functions* $I_n^m (1 \leqslant m \leqslant n)$ are defined by the equations $I_n^m(x_1, \ldots, x_n) = x_m$;
(*ii*) the *constant functions* F_c^m are defined by the equations $F_c^m(x_1, \ldots, x_m) = c$, in which c is a fixed natural number;
(*iii*) the *successor function* is defined by $f(x) = x + 1$.

B. Generating rules for functions

(*i*) *The substitution of functions.* If f is a k-argument function and $g_1, \ldots, g_k$ are n-argument functions, then the relation $g(x_1, \ldots, x_n) = f[g_1(x_1, \ldots, x_n), \ldots, g_k(x_1, \ldots, x_n)]$ determines an n-argument function.
(*ii*) *Primitive recursion.* If h is a $(k + 1)$-argument and g a $(k - 1)$-argument function, then the following system of equations determines a unique k-argument function:

$$f(x_1, \ldots, x_{k-1}, 0) = g(x_1, \ldots, x_{k-1}),$$
$$f(x_1, \ldots, x_{k-1}, y + 1) = h[x_1, \ldots, x_{k-1}, y, f(x_1, \ldots, x_{k-1}, y)].$$

Existence and uniqueness of this function is guaranteed by Dedekind's justification theorem (see Chapter 3.).
(*iii*) *The formation of a minimum.* If f is a $(k + 1)$-argument function such that for every k-tuple $(x_1, \ldots, x_k)$ of natural numbers there exists a number y with $f(x_1, \ldots, x_k, y) = 0$, then a new function g is determined by the stipulation that $g(x_1, \ldots, x_k)$ is the smallest y for which $f(x_1, \ldots, x_k, y) = 0$.

A number-theoretical function is said to be recursive if it is an initial function or if it can be generated from initial functions in finitely many steps by the application of the rules stated. If only the rules *B* (*i*) und *B* (*ii*) are admitted, then the resulting class of functions consists of the so-called *primitive recursive functions*.

Examples of primitive recursive functions.

37. The sequence of Fibonacci numbers: $f(0) = 1, \quad f(1) = 1, \quad f(x + 2) = f(x + 1) + f(x)$.

38. The function $f(x, y) = x + y$ is obtained by primitive recursion from the primitive recursive functions $h(x, y, z) = z + 1$ and $I_1^1(x) = x$:

$$x + 0 = I_1^1(x) = x, \quad x + (y + 1) = h(x, y, x + y).$$

39. The function $g(x, y) = x \cdot y$ results by primitive recursion from the primitive recursive functions $h'(x, y, z) = x + z$ and $C_0(x) = 0$:

$$g(x, 0) = x \cdot 0 = C_0(x) = 0, \quad g(x, y + 1) = h'(x, y, x \cdot y).$$

40. The function $e(x, y) = x^y$ results by primitive recursion from the primitive recursive functions $h''(x, y, z) = x \cdot z$, $C_1(x) = 1$:

$$e(x, 0) = C_1(x) = 1, \quad e(x, y + 1) = h''[x, y, e(x, y)].$$

Owing to the previously mentioned equivalence of the various concepts of algorithm the view can be taken that the class of calculable functions coincides with the class of recursive functions.

Church's hypothesis. A number-theoretical function is calculable if and only if it is recursive.

The decision problem. A precise concept of algorithm was a necessary prerequisite for the investigation of the question whether certain problems are algorithmically soluble. Such questions were discussed even in the Middle Ages. For example, around 1300 Raymundus LULLUS developed the idea of an *Ars magna*, by which he meant a general method of finding all possible truths. These ideas reached their first climax when LEIBNIZ (1646–1716) recognized that strictly speaking the concept of an Ars magna comprises two concepts, namely that of an *Ars iudicandi*, a *decision method*, and that of an *Ars inveniendi*, a *method of generating and axiomatizing*. After LEIBNIZ these ideas were not developed further. One of the reasons was that the formalizing and interpreting technique of mathematical logic, which is necessary for such investigations, did not exist yet.

But by means of the recursive functions a precise version of the decision and generating method can be indicated. These concepts are first defined for sets of natural numbers.

Definition of the decision and generating method.

(*i*) A set S of natural numbers is *recursively enumerable* if and only if there exists a recursive function f whose range of values coincides with S. This function f evidently provides a *generating method* for the set S.

(*ii*) A set S of natural numbers is *decidable* if and only if the characteristic function f_S of S is recursive; here the function f_S is defined as follows:

$$f_S(n) = \begin{cases} 1 \text{ if } n \in S, \\ 0 \text{ if } n \notin S. \end{cases}$$

If f_S is recursive, then it can be decided whether a given natural number n is an element of S or not.

The set of all even numbers is decidable.
The set of all Fibonacci numbers is decidable.
The set of all prime numbers is decidable.

The original unrestricted concept of an algorithm does not refer to natural numbers only, but also to more general objects, for example, the algorithm for the differentiation of polynomials.

Non-numerical algorithms can be reduced to recursive functions and recursive sets of natural numbers.

Let K be a class of non-numerical input and output data; suppose that a one-to-one mapping of this class into the set of natural numbers is fixed. This mapping, which is called a *codification*, is assumed to be chosen so that

(*i*) it is itself given by an algorithm;
(*ii*) there exists an algorithm to decide whether a number is an image of a non-numerical object in K, and if so, to construct this object;
(*iii*) such a codification is to be used only when there exists an algorithm to get hold of the non-numerical class K.

By identifying the objects of a non-numerical class with their code numbers the decision problem for subclasses of K can be reduced to the decision problem for certain sets of natural numbers.

Of particular importance are decision and axiomatization problems for mathematical theories, especially elementary theories. In the investigation of such questions one starts out from a codification Φ that assigns a natural number to every sequence of symbols taken from the set

$$A = \Sigma \cup \{\sim, \wedge, \vee, \rightarrow, \leftrightarrow, \exists, \forall, x_1, x_2, \ldots\}$$

of an elementary language. Such codifications are easy to find.

Definition of decidability and axiomatizability of an elementary theory. Let Φ be a codification of the sequences of symbols taken from an elementary language L_Σ. An elementary theory $T \subseteq L_\Sigma$ is *decidable* if and only if the set $\Phi(T^\vdash)$ is *recursive*. T is *axiomatizable* if and only if there exists a decidable set $S \subseteq L_\Sigma$ such that $S^\vdash = T^\vdash$.

By means of these definitions the medieval attempts to create an Ars magna obtain an exact meaning. The first important results in this direction are due to GÖDEL. He proved in 1930 that the universally valid expressions of an elementary language are axiomatizable, in other words, can be generated in the sense of the Ars inveniendi. Subsequently GÖDEL obtained an even more important result: he proved that the elementary theory of numbers is not axiomatizable, that is, there is no algorithm to produce precisely the propositions that are valid in the domain of natural numbers $\mathbf{N} = (\mathbf{N}, +, \cdot, 0, 1)$. Of course, such a proof can only be given when there is a general definition of the concept of algorithm. GÖDEL based his proof on the concept of a recursive function and so gave at the same time an example of an algorithmically unsolvable problem.

From then on a number of other elementary theories have been proved to be undecidable.

The elementary theory of groups is undecidable.
The elementary theory of fields is undecidable.
If the signature Σ contains an n-ary relation symbol with $n \geqslant 2$, then the set P_Σ of the logically valid expressions is undecidable.

In 1970 a famous problem was answered in the negative. This is Hilbert's tenth problem, which he proposed in 1900, at the First International Congress of Mathematicians in Paris: does there exist a universal algorithm to solve arbitrary Diophantine equations?

But there are also some decidable theories.

The elementary theory of the field of real numbers is decidable.
The elementary theory of Euclidean geometry is decidable.
The elementary theory of Abelian groups is decidable.

It has turned out that every sufficiently expressive theory is undecidable. This recognition of the limitations and of the scope of the axiomatic method must be regarded as one of the most important results of research on the foundations of mathematics.

16. Groups and fields

16.1. Groups and semigroups

Groups

Sets of elements or objects of which any two can be combined according to a specified rule and in a particular order to obtain a third element of the set occur frequently in all branches of mathematics.

> An *operation* on a set S is a mapping that associates with any ordered pair (a, b) of elements of S a third element c of the set. Operations are usually written as multiplication or addition. One writes $c = ab$ or $c = a + b$ and calls it the product or the sum of a and b, respectively.

Examples: Ordinary addition and multiplication are operations on the sets of integers, rational numbers, real numbers and complex numbers. Matrix multiplication is an operation on the set of all $(n \times n)$-matrices, on the set of $(n \times n)$-matrices with non-zero determinant, and on the set of $(n \times n)$-matrices whose determinant is 1.

An operation can be defined on the set of permutations of a certain fixed number of objects, by defining the product of two permutations to be the permutation obtained by performing one after the other (this is a special case of the composition of mappings). For the permutations $p_1 = \begin{pmatrix} 1 & 2 & 3 & 4 \\ 2 & 3 & 1 & 4 \end{pmatrix}$ and $p_2 = \begin{pmatrix} 1 & 2 & 3 & 4 \\ 4 & 1 & 3 & 2 \end{pmatrix}$ the product is $p_1 \cdot p_2 = \begin{pmatrix} 1 & 2 & 3 & 4 \\ 1 & 3 & 4 & 2 \end{pmatrix}$, as can be seen by the following scheme, in which the effect on the objects is made clear

$$\begin{pmatrix} 1 & 2 & 3 & 4 \\ 2 & 3 & 1 & 4 \end{pmatrix} \begin{pmatrix} 1 & 2 & 3 & 4 \\ 4 & 1 & 3 & 2 \end{pmatrix} = \begin{pmatrix} 1 & 2 & 3 & 4 \\ 1 & 3 & 4 & 2 \end{pmatrix}.$$

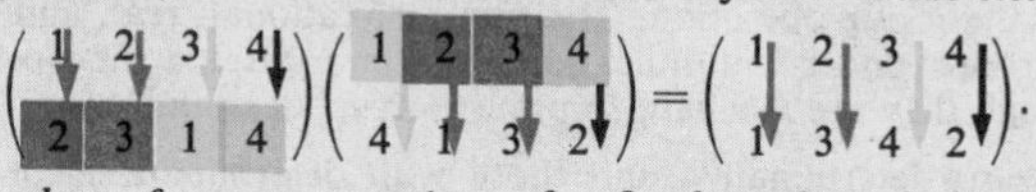

The product of two permutations of a fixed number of objects is another permutation of those objects.

A permutation $P = \begin{pmatrix} 1 & 2 & \dots & r & \dots & n \\ i_1 & i_2 & \dots & i_r & \dots & i_n \end{pmatrix}$ of n objects can also be written as a product of *cycles* by writing the image under P of each element after the element. The element i_r which follows r must itself occur in the top row and is then followed by its image i_r'. The next step gives a further element i_r'' and so on. This process breaks off after finitely many steps when the element r is reached again. This cycle can have at most n elements. If it does not contain all the elements, a new cycle is started; if $i_r = r$, the cycle can be written (r), but is usually omitted from the product.

For example, the permutations $A = \begin{pmatrix} 1 & 2 & 3 & 4 & 5 & 6 & 7 \\ 2 & 4 & 1 & 7 & 6 & 5 & 3 \end{pmatrix}$ and $B = \begin{pmatrix} 1 & 2 & 3 & 4 & 5 & 6 & 7 \\ 7 & 3 & 5 & 1 & 2 & 4 & 6 \end{pmatrix}$ can be written as $A = (1\ 2\ 4\ 7\ 3)(5\ 6)$ and $B = (1\ 7\ 6\ 4)(2\ 3\ 5)$. Their products are $AB = C = \begin{pmatrix} 1 & 2 & 3 & 4 & 5 & 6 & 7 \\ 3 & 1 & 7 & 6 & 4 & 2 & 5 \end{pmatrix}$ $= (1\ 3\ 7\ 5\ 4\ 6\ 2)$ and $BA = D = \begin{pmatrix} 1 & 2 & 3 & 4 & 5 & 6 & 7 \\ 3 & 1 & 6 & 2 & 4 & 7 & 5 \end{pmatrix} = (1\ 3\ 6\ 7\ 5\ 4\ 2)$.

The preceding examples are not all of the same kind. Some of the sets are infinite, and some finite, and closer examination of the operations reveals further differences. An operation on a set is called *associative* if for any three elements a, b, c, of S one has $(a\ b)\ c = a(b\ c)$ (if the operation is written as multiplication) and $(a + b) + c = a + (b + c)$ (if it is written as addition). The operation is called *commutative* if for any two elements one has $ab = ba$ or $a + b = b + a$, respectively. It is easy to check that the multiplication of matrices and permutations are examples of associative operations. The multiplication and addition of numbers are associative and commutative. However, multiplication of permutations and matrices is not commutative as is shown by the example above where $AB \neq BA$.

An element e of a set S is called a *neutral element* of an operation if for all elements a of S application of the operation to e and a in either order gives a. If the operation is written as multiplication, e is called a *unit element* and $ea = ae = a$. For example, the permutation $\begin{pmatrix}1&2&3&4\\1&2&3&4\end{pmatrix}$ is a unit element in the set of permutations of four objects, and the matrix $\begin{pmatrix}1&0\\0&1\end{pmatrix}$ is a unit element in the set of (2×2)-matrices. The same is true for the number 1 in the sets of integers, rational numbers, real numbers, and complex numbers. If S is a set with a multiplicatively written operation and with a unit element e, then an element $a' \in S$ is called an inverse of an element $a \in S$ if $a'a = aa' = e$. The element a' is denoted by a^{-1}; for instance, the inverse permutation to $p = \begin{pmatrix}1&2&3&4\\2&4&3&1\end{pmatrix}$ (arrows pointing down) is $p^{-1} = \begin{pmatrix}1&2&3&4\\2&4&3&1\end{pmatrix}$ (arrows pointing up) $= \begin{pmatrix}1&2&3&4\\4&1&3&2\end{pmatrix}$; in cycle notation, $p = (1\ 2\ 4)$ and $p^{-1} = (1\ 4\ 2)$.

By a process of abstraction from numerous examples one obtains the concept of a group.

A set G is called a group if the following four conditions are satisfied:

(I) An operation (multiplication) is defined on G.
(II) The operation is associative.
(III) The set G has a unit element e.
(IV) Every element a in G has an inverse a^{-1} in G.

It is very easy to see, by using (II), that the unit element e in (III) and the inverse element a^{-1} in (IV) are uniquely determined.

If the operation is also commutative, the group is called *commutative* or *Abelian*, in honour of N. H. Abel (1802–1829).

The use of the multiplicative notation for the group operation is purely a matter of convention and does not say anything about the nature of the operation. One could just as well use the additive notation. In that case the unit element is called the *zero* element and the inverse is called the *opposite element*. As a rule the additive notation is reserved for Abelian groups.

Following the concepts of set theory one distinguishes between *finite* and *infinite* groups. The number of elements in a group is called its *order*. For finite groups one can write out the multiplication table. This is a square array in which the element in the ith row and the jth column is the product of the ith group element and the jth group element. Examples are the tables given in the section on subgroups.

Examples of groups: The integers, the rational, the real, and the complex numbers form infinite Abelian groups, with addition as the group operation. The non-zero rational, real, and complex numbers form infinite Abelian groups under multiplication (for these groups the operation is written multiplicatively, even though they are Abelian, to prevent misunderstandings).

The $(n \times n)$-matrices with non-zero determinants, and those with determinant 1 form infinite non-Abelian groups under matrix multiplication. The first is called the *general linear* group $GL(n)$ and the second the *special linear group* $SL(n)$.

Permutation groups. The permutations of a certain fixed number n of objects form a finite group under the multiplication defined above, the *symmetric group* S_n. Its order is $n!$ For $n \geqslant 3$ it is non-Abelian.

If $p = \begin{pmatrix}1&2&\dots&n\\i_1&i_2&\dots&i_n\end{pmatrix}$ is a permutation, then the number of *inversions* counts how many times a larger number occurs before a smaller one in the sequence $i_1, i_2, \dots, i_n$. If the number of inversions is even, the permutation is called even, otherwise odd.

Example 1: The permutation $p = \begin{pmatrix}1&2&3&4&5\\4&3&1&5&2\end{pmatrix}$ has 6 inversions: 4 comes before 3, 2, and 1, 3 before 2, and 1, and 5 before 2.

S_n splits into $n!/2$ even permutations and $n!/2$ odd ones.

The product of two even permutations is even. The product of two odd permutations is also even. The product of an odd and an even permutation is odd.

On the basis of this fact a sign is introduced for permutations: $\operatorname{sgn} p = +1$ if p is even, and $\operatorname{sgn} p = -1$ if p is odd. It follows that the sign of the product of two permutations is the product of their signs. It is now easy to see that the inverse of an even permutation is even, because the identity permutation is even. Thus, the even permutations form a group of order $n!/2$, called the *alternating group* A_n.

Subgroups. A subset H of a group G is called a subgroup if H forms a group under the same operation as on G. By this definition the group itself and the unit subgroup, containing only the unit element, are subgroups of G. All groups having only one element are called *trivial*, and all subgroups of a group G other than G itself are called *proper*.

Some of the groups mentioned in the introduction are subgroups of others. Thus, the additive group of the integers is a subgroup of the additive group of the rational, and this, in its turn, is a subgroup of the additive group of the real numbers. The multiplicative group of the non-zero rational numbers is a subgroup of the multiplicative group of the non-zero real numbers. The alternating group $\boldsymbol{A}_n$ is a subgroup of the symmetric group $\boldsymbol{S}_n$. The special linear group SL(n) is a subgroup of the general linear group GL(n). The following proposition follows immediately from the definition of a subgroup:

The intersection of a family of subgroups is a subgroup.

If a is an element of a group G, there are subgroups containing that element, for instance G itself. The intersection of all these subgroups also contains a, by definition, and so is the smallest subgroup containing a. It is called the cyclic subgroup generated by a and written $\langle a\rangle$. Clearly, $\langle a\rangle$ consists of all the powers a^n, $n \in Z$ (negative powers are powers of the inverse; powers are, as usual, products of an element with itself).

If all the powers a^n are distinct, then $\langle a\rangle$ is called an *infinite cyclic group*. Otherwise there exists a smallest positive integer n such that $a^n = e$, and $\langle a\rangle$ is a *cyclic group of order* n: $\langle a\rangle$ consists of the elements $e, a, \ldots, a^{n-1} = a^{-1}$; $a^{n+1} = a$ etc. A group that coincides with one of its cyclic subgroups is called *cyclic*.

The set-theoretical *union* $U_1 \cup U_2$ *of two subgroups* of a group is not, in general, itself a subgroup, as can be seen by the example below. The union is merely a subset of G; but to every subset S of G one can define the *intersection* of all subgroups containing S (which itself contains S and is the smallest subgroup containing S) as the *subgroup* $\langle S\rangle$ *generated by* S. The subgroup $\langle U_1, U_2\rangle$ is then the smallest subgroup containing both U_1 and U_2. If $\langle S\rangle = G$ for a subset S of G, one says that G is *generated by* S.

Example 2: The elements of $\boldsymbol{S}_3$ in cycle notation are $p_1 = (1)$, $p_2 = (1\ 2\ 3)$, $p_3 = (1\ 3\ 2)$, $p_4 = (1\ 2)$, $p_5 = (1\ 3)$, and $p_6 = (2\ 3)$. The group table is:

	p_1	p_2	p_3	p_4	p_5	p_6
p_1	p_1	p_2	p_3	p_4	p_5	p_6
p_2	p_2	p_3	p_1	p_6	p_4	p_5
p_3	p_3	p_1	p_2	p_5	p_6	p_4
p_4	p_4	p_5	p_6	p_1	p_2	p_3
p_5	p_5	p_6	p_4	p_3	p_1	p_2
p_6	p_6	p_4	p_5	p_2	p_3	p_1

Using the group table it is easy to check that the sets $A = \{p_1, p_4\}$, $B = \{p_1, p_5\}$, $C = \{p_1, p_6\}$, and $D = \{p_1, p_2, p_3\}$ are subgroups of $\boldsymbol{S}_3$. Further $A = \langle p_4\rangle$, $B = \langle p_5\rangle$, $C = \langle p_6\rangle$ are cyclic of order 2 and $D = \langle p_2\rangle = \langle p_3\rangle$ is cyclic of order 3. The union of A and D is not a subgroup, for then the product $p_5 = p_3 p_4$ would lie in $\{p_1, p_4\} \cup \{p_1, p_2, p_3\}$, which is obviously not the case. The group generated by the union is the whole of $\boldsymbol{S}_3$.

Homomorphisms

Homomorphism. The concept of a *homomorphism* occupies a central position in the whole of group theory. It is characterized by two statements. One relates to the sets of elements of the groups and the other to the group operations.

A mapping f of a group G into a group G' is called a homomorphism if for any two elements $a, b \in G$ the relation (H) $\quad f(a \cdot b) = f(a) \cdot f(b)$ *holds.*

Here the product on the left is taken in G and on the right in G'. The image of G under f is always a subgroup of G'. If there is a surjective homomorphism of G onto G', that is, if every element of G' occurs as an image under f, then G' is called a *homomorphic image* of G. It is perfectly possible that under a homomorphism distinct elements of G are mapped to the *same element* of G'. It is not a requirement that homomorphisms should be injective.

Example 1: Let G be the group of all real (2×2)-matrices with non-zero determinant, and G' the multiplicative group of non-zero real numbers. The mapping taking each matrix to its determinant satisfies the requirement (H) and is a homomorphism. What is more, it is surjective, because every real number r is the determinant of the matrix $\begin{pmatrix} 1 & 0 \\ 0 & r \end{pmatrix}$.

$$\begin{pmatrix} a & b \\ c & d \end{pmatrix} \rightarrow \begin{vmatrix} a & b \\ c & d \end{vmatrix} = (ad - bc)$$

Example 2: The mapping taking every permutation $p \in \boldsymbol{S}_n$ to its sign sgn p is homomorphism of $\boldsymbol{S}_n$ into the multiplicative group of real numbers, by the multiplication rules for signs: $\operatorname{sgn}(p_1 \cdot p_2) = \operatorname{sgn} p_1 \cdot \operatorname{sgn} p_2$. Its image is the subgroup consisting of the numbers $+1$ and -1.

The condition (H) means that in a certain sense a homomorphism must preserve the structure of the original group. The image is in general 'smaller' than the original group. For instance, all matrices of the form $\begin{pmatrix}\lambda a & \lambda b\\ \mu c & \mu d\end{pmatrix}$, where λ and μ are numbers with $\lambda\mu = 1$, have the same determinant as the matrix $\begin{pmatrix}a & b\\ c & d\end{pmatrix}$. The set of those elements that are mapped to the unit element of the image is a measure of the shrinkage of G. These elements form a special kind of subgroup of the original group and are called the *kernel* of the homomorphism. The kernel of the homomorphism in Example 1 is the special linear group SL(2) of (2×2)-matrices, and the kernel of the homomorphism in Example 2 is the alternating group $\boldsymbol{A}_n$ of all the even permutations in $\boldsymbol{S}_n$.

Isomorphisms. *A bijective homomorphism is called an isomorphism.*

If f is an isomorphism of G onto G', then its image is G', and its kernel is the unit subgroup of G. It can be shown that these two conditions are sufficient for f to be an isomorphism. The inverse map from G' to G also satisfies (H) and is thus also an isomorphism. If there is an isomorphism from G to G', then the groups are called *isomorphic*, in symbols: $G \cong G'$.

Example 3: Let V_4 be the *Klein four group* consisting of the permutations $e = (1), a = (1\ 2)(3\ 4)$, $b = (1\ 3)(2\ 4)$ and $c = (1\ 4)(2\ 3)$. Let G be the group consisting of the matrices $e' = \begin{pmatrix}1 & 0\\ 0 & 1\end{pmatrix}$, $a' = \begin{pmatrix}-1 & 0\\ 0 & -1\end{pmatrix}$, $b' = \begin{pmatrix}0 & 1\\ 1 & 0\end{pmatrix}$, and $c' = \begin{pmatrix}0 & -1\\ -1 & 0\end{pmatrix}$. Define the bijective mapping f: $e \longrightarrow e'$, $a \longrightarrow a'$, $b \longrightarrow b'$, $c \longrightarrow c'$.

V_4	e	a	b	c
e	e	a	b	c
a	a	e	c	b
b	b	c	e	a
c	c	b	a	e

G	e'	a'	b'	c'
e'	e'	a'	b'	c'
a'	a'	e'	c'	b'
b'	b'	c'	e'	a'
c'	c'	b'	a'	e'

From the group tables of the two groups V_4 and G one can read off that the relation (H) is satisfied for arbitrary elements of V_4. Thus, V_4 is isomorphic to the group G of matrices.

As can be seen from the example, isomorphic groups have identical structures, even though their elements may be of completely different kinds, here permutations and matrices. Homomorphisms and isomorphisms are not restricted to finite groups. Isomorphic groups always have the same cardinal number.

Example 4: Let $R^{\times}$ be the multiplicative group of positive real numbers, and R^{+} the additive group of all real numbers. The mapping f: $a \to \ln a$ taking every positive real number to its natural logarithm is an isomorphism of the two groups; as is well known, $\ln(a \cdot b) = \ln a + \ln b$, in other words, $f(a \cdot b) = f(a) + f(b)$, so that $R^{\times} \cong R^{+}$.

Isomorphism is an *equivalence relation* between groups, so that the class of all groups is partitioned into isomorphism classes. Emphasizing that one is only interested in the isomorphism class of a group and not in its particular representation, one speaks of an *abstract* group.

Isomorphic groups have the same structure, and calculations in them follow the same laws and rules, even though they may have different kinds of elements, and the operations may be defined in different ways.

Normal subgroups. It has already been mentioned that kernels of homomorphisms form a special type of subgroup. A subgroup N that occurs as the kernel of a homomorphism of a group G into some other group is called *normal*. Thus, in Example 1 (of the section on homorphisms) the special linear group is a normal subgroup of the general linear group. In Example 2 the alternating group is a normal subgroup of the symmetric group. A group whose only normal subgroups are the whole group and the trivial one (these are always normal) is called *simple*. A homomorphism from a simple group G onto a group H is either an isomorphism, or H is trivial. The following result on permutation groups is mentioned here because of its applications in *Galois theory*.

For $n > 4$ the alternating group $\boldsymbol{A}_n$ is the only non-trivial proper subgroup of $\boldsymbol{S}_n$. For $n > 4$ the alternating group $\boldsymbol{A}_n$ is simple.

A proper normal subgroup M of G is called *maximal* if for any normal subgroup N of G with $M \subseteq N \subseteq G$, either $M = N$ or $N = G$ holds. The Klein four group V_4 is a maximal normal subgroup of $\boldsymbol{A}_4$.

Factor groups. If S is a subset of a group G and a is an element of G, the set aS is defined to be $\{as \mid s \in S\}$. A similar definition is made for multiplication on the right. If H is a subgroup of G, the sets aH for $a \in G$ are called the *left cosets* of H in G. It is easy to show that they form a partition of G. Naturally the same definitions can be made for right cosets, and they also form a partition of G. However, these two partitions are not, in general, the same. Now if f is a homomorphism of G and its kernel is N, then N has the remarkable property that for any a in G the cosets aN and Na are identical, because they both consist exactly of those elements of G that are mapped by the homomorphism f to the same element as a. This property is so important that for the moment it will be given a special name.

A subgroup $\bar{N}$ of G for which the equation $a\bar{N} = \bar{N}a$ holds for any element a of G is called invariant. The condition is equivalent to $a\bar{N}a^{-1} = \bar{N}$ for any a in G.

Since the left and right cosets of invariant subgroups are equal, the words left and right can be omitted. As they form a partition, two cosets are either equal or they have no elements in common. Generalizing now the product notation still further and defining for subsets S and T of G the product $ST = \{st \mid s \in S \text{ and } t \in T\}$, then for subgroups H of G the relation $HH = H$ holds. For invariant subgroups $\bar{N}$ one can go even further. If $a\bar{N}$ and $b\bar{N}$ are two cosets then $(a\bar{N})(b\bar{N}) = (a\bar{N})(\bar{N}b) = a((\bar{N}\bar{N})\,b) = a(\bar{N}b) = a(b\bar{N}) = (ab)\,\bar{N}$. Thus the product of two cosets of an invariant subgroup is a coset and contains the products of elements of the two cosets. One can now verify that the cosets of the invariant group $\bar{N}$ actually form a group under this multiplication, whose identity element is $e\bar{N} = \bar{N}$, and in which the inverse of $a\bar{N}$ is $a^{-1}\bar{N}$. What is more, from the equations above the mapping π: $a \to a\bar{N}$, that takes every element of G to its coset is a homomorphism, which is called the *canonical homomorphism*. The kernel of π is, by its very definition, the invariant subgroup $\bar{N}$. The concepts *normal* and *invariant* subgroup *are identical*. The group of the cosets of N is called the factor group or quotient group of G by N, and is denoted by G/N.

Example 5: The factor group of $\boldsymbol{S}_n$ by $\boldsymbol{A}_n$ consists of the elements $\boldsymbol{A}_n$ and $p_0\boldsymbol{A}_n$, where p_0 is an *odd* permutation. The mapping $\pi : p \to \begin{cases} p\boldsymbol{A}_n = \boldsymbol{A}_n \text{ if } p \text{ is even,} \\ p\boldsymbol{A}_n = p_0\boldsymbol{A}_n \text{ if } p \text{ is odd,} \end{cases}$ is easily seen to be the canonical homomorphism π from $\boldsymbol{S}_n$ to $\boldsymbol{S}_n/\boldsymbol{A}_n$.

The homomorphism theorem. If f is a homomorphism from G onto a group G' (that is, f is surjective), then there is a natural mapping from the cosets of N to G', because all the elements of a coset have the same image under f. The map φ taking the coset aN to the common image $f(a)$ of its elements is an isomorphism of G/N and the homomorphic image G'. This is the content of the homomorphism theorem (Fig.).

Homomorphism theorem. Every surjective homomorphism of a group G onto a group G' can be split into the product of the canonical homomorphism π from G onto $G/\text{Ker} f$ and an isomorphism φ of $G/\text{Ker} f$ onto G'.

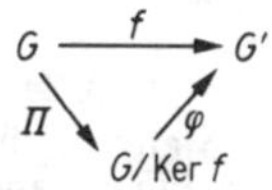

16.1-1 The homomorphism theorem

The homomorphism theorem implies that the investigation of arbitrary homomorphisms can be reduced to the investigation of factor groups, isomorphisms, and subgroups.

Example 6: It has already been shown that the mapping f: $p \to \text{sgn}\, p$ is a homomorphism of the symmetric group $\boldsymbol{S}_n$ onto the subgroup $\{+1, -1\}$ of the real numbers with kernel $\boldsymbol{A}_n$. In Example 5 the canonical homomorphism $\pi : p \to \begin{cases} p\boldsymbol{A}_n = \boldsymbol{A}_n \text{ if } p \text{ is even,} \\ p\boldsymbol{A}_n = p_0\boldsymbol{A}_n \text{ if } p \text{ is odd,} \end{cases}$ was introduced. The splitting asserted in the homomorphism theorem is now $f = \pi \cdot \varphi$, where φ is the mapping $\varphi : \begin{cases} \boldsymbol{A}_n \to +1 \\ p_0\boldsymbol{A}_n \to -1 \end{cases}$. It is obvious that φ is indeed an isomorphism of the factor group $\boldsymbol{S}_n/\boldsymbol{A}_n$ onto the group $\{+1, -1\}$.

The relation between maximal normal subgroups and simple groups is given by the following theorem.

A normal subgroup N of a group G is maximal if and only if the factor group G/N is simple.

Automorphisms. An *automorphism* is an isomorphism of a group G onto itself. The product of two automorphisms of a group G is again an automorphism of G. The identity mapping of G onto itself, which leaves every element fixed, is an automorphism, and if f is an automorphism of G onto itself, so is the inverse mapping f^{-1}. The automorphisms of G form a group under the product of mappings as operation. It is called the *automorphism group* of G.

Finite groups

The theory of finite groups was originally developed as a tool for dealing with the problem of solving algebraic equations, and dealt at first only with *permutation groups*, that is, subgroups of symmetric groups. The importance of these groups is apparent from the following two theorems.

Cayley's theorem. Every group of order n is isomorphic to a subgroup of the symmetric group S_n.
Lagrange's theorem. For any subgroup H of a finite group G, the index $[G:H]$ of H in G is defined as the number of left cosets of H in G. If E is the unit subgroup, then $[G:E]$ is the order of G. Now for any subgroup H of G the order of H divides the order of G. More precisely: $[G:E] = [G:H]\,[H:E]$.

The *order of an element a* of a group G is the *order of the cyclic subgroup generated* by a. Obviously, if G is finite, then every element has finite order, indeed by Lagrange's theorem its order divides the order of the group. There are, however, infinite groups, whose elements all have finite order, for example, the multiplicative group of all complex roots of unity, that is, the group of all solutions of the equations $x^n - 1 = 0$, where $n = 1, 2, \ldots$ W. BURNSIDE raised the problem whether a group in which all elements have finite order and which is generated by finitely many elements is necessarily finite. The question was only settled (in the negative) in 1967 by NOVIKOV and ADYAN.

In the theory of solutions of algebraic equations the concept of a composition series of a group plays an important role. A *composition series* of a finite group G is a sequence of subgroups $G = G_0 \supset G_1 \supset G_2 \cdots \supset G_l = E$, each containing the next, such that each group is a maximal normal subgroup of its immediate predecessor. The simple groups $G_0/G_1, G_1/G_2, \ldots, G_{l-2}/G_{l-1}, G_{l-1}/E = G_{l-1}$ are called the *composition factors*. Groups with composition factors of prime order are called soluble, because of their relation to the solubility of algebraic equations. The composition factors of a finite group are uniquely determined up to isomorphism and the order in which they occur (Theorem of *Jordan-Hölder*).

Applications. Apart from the applications of group theory to geometry, one can say roughly that groups play a role anywhere where mappings, transformations, and symmetries (in some sense or another) occur, and concepts are investigated that are invariant under the mappings, transformations, or symmetries. In particular, the theory of finite groups is applied in the theory of solutions of algebraic equations (see Galois theory). In physics group theory is important in relativity theory, where the group of *Lorentz transformations* is used, and in quantum mechanics. The division of physics into relativistic and non-relativistic physics, is a division by group-theoretical criteria. In crystallography group theory makes it possible to determine all possible crystal forms by investigating their symmetry groups. It is perhaps also of interest that all two- and three-dimensional ornaments can be classified by means of group theory.

Topological groups

Many of the groups that are important in applications, particularly in geometry and physics, are infinite and carry, apart from their algebraic structure, also a structure as a topological space.

A *topological group* is a set of elements that is, on the one hand, a group and, on the other, a topological space, such that these two structures are compatible, in the sense that *multiplication and inversion are continuous mappings*. Examples are various matrix groups, the transformation groups of the several branches of geometry, and the *Lorentz group*.

For instance, it makes sense to ask whether two (2×2)-matrices with real entries are only slightly different, whether their entries are close together. This leads to a topology for (2×2)-matrices derived from the topology of the real line. It can be shown that in this topology multiplication of matrices is continuous. Thus, the general linear group GL(2) of (2×2)-matrices with non-zero determinant is a topological group. While the fact that topological groups are always infinite makes it harder to analyze them, their topological structure makes it possible to use new, not necessarily algebraic, methods, which have led to excellent results in the theory of Abelian topological groups and in the theory of compact topological groups. The interplay between topological and algebraic ideas gives this branch of group theory its particular attraction.

$$\begin{pmatrix} \cos\varphi & -\sin\varphi \\ \sin\varphi & \cos\varphi \end{pmatrix}$$

Lie groups. The rotations of the plane about a fixed point form a group. Each rotation is determined by its angle φ of rotation and can be described by a matrix of the adjacent form. It can be shown that these matrices form a group. Since φ varies continuously between 0 and 2π, a topology can be defined on the group. But this group differs from other topological groups in that its elements are dependent on a parameter, and that this dependence is described by differentiable functions. This makes it possible to define not only continuous but also differentiable functions on the group, by calling a function differentiable if it is *differentiable as a function of the real parameter* φ. A topological space on which differentiable functions can be defined is called a *differentiable manifold*, and the group of rotations about a fixed point is not just a topological space, but actually a differentiable manifold. Such groups are called *Lie groups*, after Sophus LIE; they form a special part of the class of topological groups and are in many respects easier to deal with than arbitrary topological groups, because the tools of analysis can also be utilized in their investigation.

Applications. Lie groups and their representations (see Chapter 33.) are particularly important in the theory of special functions (*spherical functions, Bessel functions* etc.) and in the theory of *almost periodic functions*. LIE used his theory to classify and solve differential equations. In quantum theory the LIE group of rotations of a sphere and the Lorentz group, which is also a LIE group, play an important part.

Semigroups

A *semigroup* is a non-empty set with an *associative operation*. If the operation is also *commutative*, the semigroup is called *commutative*. If H is a multiplicative semigroup and contains an element e such that $ea = ae = a$ for all elements a of H, then H is called a *semigroup with unit element*. Examples of commutative semigroups with unit element that are not groups are the integers under multiplication and the non-negative integers under addition (in which 0 is the unit element). Naturally every group is a semigroup. Theorems that are true for all semigroups are also true for all groups. By analogy to groups one distinguishes *finite* and *infinite* semigroups and calls the number of elements of a semigroup its *order*. If the equations $ax = b$ and $ya = b$ each have at most one solution for arbitrary pairs of elements a and b of a semigroup H, then H is called *regular*. The non-zero integers form a regular semigroup under multiplication.
For finite semigroups the following theorem holds: *A regular semigroup of finite order is a group.*

Example 7: If a function $f(t)$ describes a time-dependent process, and $f(t+\alpha)$ the process delayed by the time interval α, then the set of mappings $T_\alpha: f(t) \to f(t+\alpha)$ forms a semigroup.

Example 8: The power set of an arbitrary set S (that is, the set of all subsets of S) forms a commutative semigroup with unit element under intersection as operation, and also under union. The whole set S is the unit element for intersections, and the empty set for unions.

16.2. Fields and algebraic equations

Up to the beginning of the nineteenth century algebra could be described as the theory of solutions of algebraic equations. Its purpose was to find methods, as general as possible, of computing such solutions. The actual solutions were of less interest than the methods used in finding them. The ideas of nineteenth century mathematicians in connection with these problems led to the definition of groups and fields as essential tools in the theory of solutions of algebraic equations. Later these concepts attained an interest and importance of their own, principally because applications were found in quite different areas. Group theory and field theory are now extensive branches of algebra. Further objects with similar structures were also found in many parts of mathematics and this led to the definitions of such algebraic structures as rings, algebras, lattices, and integral domains (see Chapter 33.).

Fields and integral domains

Fields. A *field* is a set K of elements satisfying the following *field axioms*:

Field axioms. Axiom 1. Two *operations*, called addition and multiplication, are defined on K.
Axiom 2. Under addition the elements of K form an *Abelian group*, with zero element 0.
Axiom 3. Under multiplication the elements of K different from 0 form an *Abelian group*.
Axiom 4. Addition and multiplication are linked by the distributive law, that is, for any three elements, a, b and c of K one has: $a(b+c) = ab + ca$.

The rational numbers, the real numbers, and the complex numbers are the most important examples of fields. Intuitively one can say that a field is a set of elements in which one can do arithmetic in the usual way. Just as for groups, one distinguishes finite and infinite fields. A subset P of a field Ω is called a *subfield* if it satisfies the field axioms for the operations defined in Ω, in particular, sum and product of elements in P must lie in P, and so must negatives and inverses of elements of P. The field Ω is also called an *extension field* of P. If K is an extension field of P and also a subfield of Ω, then K is called an *intermediate field* between P and Ω: $P \subseteq K \subseteq \Omega$.

Every field can be regarded as a *vector space* (see Chapter 17.) over any subfield P as set of scalars, by taking field addition as the vector space addition and field multiplication with elements of the subfield as multiplication by scalars. The dimension of Ω over P regarded as a vector space is called the *degree of the extension field* Ω over P. If Ω is a finite-dimensional vector space over P, the extension is called *finite*. Its degree is denoted by $n = [\Omega : P]$. In this case one can find elements $\beta_1, \beta_2, \ldots, \beta_n$ in Ω such that every element $\beta \in \Omega$ can be written uniquely in the form $\beta = c_1\beta_1 + c_2\beta_2 + \cdots + c_n\beta_n$ with elements $c_1, c_2, \ldots, c_n$ of P. Such elements $\beta_1, \beta_2, \ldots, \beta_n$ are said to form a *basis* of Ω over P.

If Ω is an extension field of P and $\alpha_1, \alpha_2, \ldots, \alpha_m$ are arbitrary elements of Ω, then $P(\alpha_1, \alpha_2, \ldots, \alpha_m)$ denotes the smallest subfield of Ω containing P and $\alpha_1, \alpha_2, \ldots, \alpha_m$. It consists of all the elements of Ω that can be obtained from P and $\alpha_1, \alpha_2, \ldots, \alpha_m$ by means of the elementary arithmetical operations. One says that the field $P(\alpha_1, \alpha_2, \ldots, \alpha_m)$ is obtained from P by *adjoining* $\alpha_1, \alpha_2, \ldots, \alpha_m$ to P. It can be obtained by first adjoining α_1 to P to obtain a field $K_1 = P(\alpha_1)$, then adjoining α_2 to K_1 to obtain $K_2 = K_1(\alpha_2) = P(\alpha_1, \alpha_2)$, and so on. After m steps one has $K_m = K_{m-1}(\alpha_m) = P(\alpha_1, \alpha_2, \ldots, \alpha_m)$. An extension field $P(\alpha)$ that is obtained by adjoining a single element is called a *simple extension* of P.

If K_1 and K_2 are fields, then a bijective mapping f from K_1 onto K_2 such that $f(a+b) = f(a) + f(b)$ and $f(ab) = f(a)f(b)$, for arbitrary elements a and b of K_1, is called an *isomorphism* of K_1 onto K_2; K_1 and K_2 are called *isomorphic*, and one writes $K_1 \cong K_2$. An isomorphism of a field K onto itself is called an automorphism of K. If K_1 and K_2 are fields and if P is a subfield of K_1 and K_2, then an isomorphism of K_1 onto K_2 leaving P elementwise fixed is called a *relative isomorphism*. For $K_1 = K_2$ one speaks of a relative automorphism. Numerous examples of these concepts can be found in the following sections.

Integral domains. The field $\mathbb{Q}$ of rational numbers has a subset satisfying the field axioms 1, 2, and 4, but not axiom 3: the integers. In the set of integers division is not universally possible, but at least they satisfy the weaker *cancellation law*: If $ab = ac$ and $a \neq 0$, then $b = c$ (see Semigroups).

An integral domain is a set I in which axioms 1, 2, and 4 for fields are satisfied, and axiom 3 is replaced by axiom 3′: The non-zero elements of I form a regular commutative semigroup under multiplication.

All fields are integral domains. The integers are an example of an integral domain that is not a field. A further important example of an integral domain is the set of polynomials with coefficients in a field (see Polynomials). If the set I is finite, then I is called a finite integral domain. From the theorem on finite regular semigroups mentioned above one obtains at once:

Every finite integral domain is a field.

Two integral domains are called *isomorphic* ($I_1 \cong I_2$) if there is a bijective mapping f from I_1 to I_2 that is compatible with the operations in the same sense as a field isomorphism. The mapping is called an *isomorphism* (see Field of fractions).

Examples: Apart from the fields and integral domains already mentioned, the *Gaussian numbers* form a further important example. They consist of all complex numbers of the form $a + bi$, where a and b are rational numbers and $i^2 = -1$. The subset of Gaussian numbers for which a and b are integers is called the set of *Gaussian integers* and is an integral domain. The Gaussian numbers are obtained from the rationals by adjoining i.

The smallest finite field consists of two elements 0 and e with adjacent addition and multiplication tables.

+	0	e
0	0	e
e	e	0

·	0	e
0	0	0
e	0	e

The fields of *residues modulo a prime number* are also important examples (see Chapter 31.).

Polynomials. An expression of the form $f(x) = a_n x^n + a_{n-1}x^{n-1} + \cdots + a_1 x + a_0$, in which n is a natural number and $a_0, \ldots, a_n$ are elements of a field K, is called a *polynomial in the indeterminate* x over K. The quantities $a_0, \ldots, a_n$ are called its *coefficients*. A polynomial whose coefficients are all zero is called the *zero polynomial*. If the coefficient a_n is not zero, then the *degree* of the polynomial is n. If the indeterminate x is replaced by a field element, one obtains a function defined on K with values in K. The functions are called *polynomial functions* or *integral rational functions* on the field K; conversely, if the field is infinite, each integral rational function determines a unique polynomial. In dealing with polynomials it is frequently convenient to use the simplified notation $f(x) = \sum_{i=0}^{\infty} a_i x^i$, where it is understood that only finitely many a_i are different from zero. One can then write the sum and product of polynomials $f(x) = \sum_{i=0}^{\infty} a_i x^i$ and $g(x) = \sum_{j=0}^{\infty} b_j x^j$ as $f(x) + g(x) = \sum_{k=0}^{\infty} c_k x^k$ and $f(x) \cdot g(x) = \sum_{k=0}^{\infty} d_k x^k$, where the coefficients c_k and d_k satisfy the equations $c_k = a_k + b_k$ and $d_k = \sum_{i+j=k} a_i b_j$. Polynomials with coefficients in a certain fixed field K form an integral domain, which is denoted by $K[x]$. In this integral domain $K[x]$ one can perform *division with remainder* just as for the integers. This means that if $f(x)$ and $g(x)$ with $g(x) \neq 0$ are two given polynomials, there exist unique polynomials $h(x)$ and $r(x)$ such that $f(x) = h(x) \cdot g(x) + r(x)$, and $r(x) = 0$ or the degree of $r(x)$ is less than that of $g(x)$. By analogy to number theory (see Chapter 1.) one calls two polynomials $f_1(x)$ and $f_2(x)$ *congruent* modulo $g(x)$ and writes $f_1(x) \equiv f_2(x) \bmod g(x)$ if they leave the same remainder $r(x)$ on division by $g(x)$. Congruence modulo $g(x)$ is an equivalence relation and leads to a partition of the integral domain $K[x]$ into classes. These classes are added and multiplied by choosing a polynomial from each class, adding or multiplying the polynomials, and defining as sum or product of the classes the class of the sum or product of the polynomials. It can be shown that the same class is always obtained even if the polynomials chosen from the original two classes are changed. If $g(x)$ is an irreducible polynomial, these classes, which are called the *residue classes* modulo $g(x)$, form a field, which is denoted by $K[x]/(g(x))$ and is called the *residue class field* of $K[x]$ modulo $g(x)$.

A non-constant polynomial is called *irreducible* over K if it cannot be written as a product of two polynomials over K of smaller but positive degree. A polynomial of degree n is called *monic* if its highest coefficient a_n is 1. The following theorem holds for polynomials over a field K:

Every polynomial $f(x)$ over K has a representation $f(x) = cp_1(x) \cdots p_s(x)$ as a product of a field element c and irreducible monic polynomials $p_1(x), \ldots, p_s(x)$ in $K(x)$. This representation is unique up to the order of the irreducible polynomials.

Field of fractions. The concept of a field of fractions arises from the question, 'What is the smallest field containing a given integral domain?' The question can be reformulated: 'Given an integral domain I, is there a field K containing I, whose elements can all be written as fractions of elements of I?' – To obtain an idea of such a field one first assumes that it exists. One can operate on the fraction a/b in the usual manner. The elements a and $b \neq 0$ are in I and so: (1) $a/b = a'/b'$ if $ab' = a'b$, (2) $a/b + c/d = (ad + bc)/bd$, (3) $(a/b) \cdot (c/d) = ac/bd$, where, of course, all the denominators must be different from zero. If the field exists, the rules (1), (2), and (3) must hold. One uses these rules to *construct* a field K, by first replacing the required fractions by *ordered pairs* (a, b) of elements a and $b \neq 0$ of I. An equivalence relation is defined on these pairs by $(a, b) = (a', b')$ if $ab' = a'b$ in I. The classes are denoted by square brackets [ ]. Addition and multiplication of the classes are defined using the rules (2) and (3): by (2) $[a, b] + [c, d] = [ad + bc, bd]$ and by (3) $[a, b] \cdot [c, d] = [ac, bd]$. One checks that these operations are independent of the particular representative of the class $[a, b]$ etc. chosen, and that the set of classes K' forms a field under this addition and multiplication, in which every element $[a, b]$ can be represented as a fraction $[a, e]/[b, e]$, where e is the unit element of I. It can now be shown that the elements $[a, e]$ of K' form an integral domain I' that is isomorphic to I by the mapping $[a, e] \to a$. Thus, a field of the required type has been constructed; however, it does not contain I itself, but an integral domain isomorphic to I. Now if the elements of I' are replaced by their isomorphic images in I, and the elementary operations are redefined suitably, one obtains a field K containing I, in which every element can be represented as a fraction a/b of elements of I. This field is called the *field of fractions of the integral domain I*. It is the *smallest field containing I*. If I is the set of integers, then this process coincides with the one used to construct the rational numbers (see Chapter 3.).

The field of fractions of the integers is the set of rational numbers.
If I is the integral domain of polynomials $K[x]$ over a field K, its field of fractions is called the field of rational functions over K and is denoted by $K(x)$.

Algebraic equations and field extensions. An *algebraic equation of degree n* is an equation $f(x) = 0$ whose left-hand side is a polynomial of degree n. In particular, the equation is called irreducible over K if the polynomial $f(x)$ is irreducible over K. The solutions of the equation $f(x) = 0$ are called the *roots of the polynomial* $f(x)$ or of the equation $f(x) = 0$. In general, algebraic equations can only be solved in an extension field; for example, the need to define a root for every quadratic polynomial has led to a considerable enlargement of the number system: the complex numbers. Frequently a smaller extension is sufficient. For instance, the coefficients of the equations $x^2 - 2 = 0$ and $x^2 + 4 = 0$ lie in the field of rational numbers and their solutions lie in the field $\mathbb{Q}(\sqrt{2})$ and in the Gaussian number field, respectively. If $f(x) = 0$ is an irreducible equation over K of degree greater than 1, then it has no solution in K and an extension is necessary to find a solution.

If α is a solution of an irreducible equation $f(x) = 0$ over K, then α is called *algebraic* over K. If Ω is an extension field of K and every element of Ω is algebraic over K, then Ω is called an *algebraic extension* of K, and an algebraic extension of $\mathbb{Q}$ is called a *number field*. Extensions that are not algebraic are called *transcendental*. If $f(x) = 0$ is an irreducible equation with coefficients in a number field K, then by the *fundamental theorem of algebra* (see Chapter 4.) it has a solution α in the field of complex numbers. The field $K(\alpha)$ obtained by adjoining α to K is the smallest subfield of the complex numbers containing K and the root α of the equation. It is not possible to construct a suitable extension of this type for the general equation of degree n, whose coefficients are unknowns in the meaning of algebra, because there is no field available that must contain at least one solution of the equation. Since even in the case of a particular equation the fundamental theorem of algebra only asserts the existence of a solution in the field of the complex numbers and gives no method of constructing it, it seems plausible in all cases to try and find a general method of constructing an extension field containing a root of a given equation, a so-called *root field*.

Construction of a root field. Let $f(x)$ be a monic irreducible polynomial of degree n with coefficients in a field K. The residue class field $K[x]/(f(x))$ contains a subfield $\bar{K}$, consisting of the congruence classes $\bar{a}$ of the elements a of K. The class $\bar{a}$ of a consists of all polynomials leaving the remainder a on division by $f(x)$. The mapping $a \to \bar{a}$ is an isomorphism of K onto $\bar{K}$: $K \cong \bar{K}$. If $\bar{K}$ is replaced by K in $K[x]/(f(x))$ leaving *all the operations and relations the same*, one obtains a field K' containing K and isomorphic to $K[x]/(f(x))$. Now let $f(x) = x^n + a_{n-1}x^{n-1} + \cdots + a_0$ and let α be the class of all polynomials leaving the remainder x on division by $f(x)$. Then $\alpha^n + a_{n-1}\alpha^{n-1} + \cdots + a_0$ is, by the rules of addition and multiplication of residue classes, the class of $f(x)$, but this is the class of all polynomials leaving the remainder 0 on division by $f(x)$, and in K' this has been replaced by 0 itself. Thus, $\alpha^n + a_{n-1}\alpha^{n-1} + \cdots + a_1\alpha + a_0 = 0$, and α is a solution of the equation $f(x) = 0$. The field K' contains a solution of the equation $f(x) = 0$ and is called a *root field* of it.

The root field $K' = K(\alpha)$ is a *simple algebraic extension* of K, and every element of K' can be written in the form $b_0 + b_1\alpha + \cdots + b_{n-1}\alpha^{n-1}$, where $b_0, b_1, \ldots, b_{n-1}$ are elements of K. The degree of the extension $K(\alpha)$ over K is equal to the degree of the irreducible polynomial $f(x)$: $[K(\alpha):K] = \text{degree}\,(f(x)) = n$.

Example 1: A root field obtained for the equation $x^2 + 1 = 0$ over the rational numbers $\mathbb{Q}$ is the simple extension $\mathbb{Q}(\mathrm{i})$ in which every element can be written $a + b\mathrm{i}$ where a and b are rational numbers and $\mathrm{i}^2 + 1 = 0$. This field is isomorphic to the field of Gaussian numbers.

Splitting fields. Let $f(x)$ be a monic polynomial of degree n with coefficients in a field K. The *splitting field* of $f(x)$ over K is the *smallest* extension field L of K over which $f(x)$ splits into linear factors: $f(x) = (x - \alpha_1)(x - \alpha_2) \cdots (x - \alpha_n)$, where $\alpha_1, \alpha_2, \ldots, \alpha_n$ are the roots of $f(x)$ in L. The splitting field of a polynomial or algebraic equation over K is the smallest extension of K containing *all* the solutions of the algebraic equation. It is unique up to isomorphism. One can construct the splitting field by repeated applications of the construction of a root field. First $f(x)$ is factorized into irreducible polynomials over K. If all the irreducible factors have degree 1, then K itself is the splitting field. If not, one constructs a root field K' for one of the irreducible factors of degree >1 and factorizes $f(x)$ into irreducible factors over K'. If all the factors now are of degree 1, then K' is the required splitting field. Otherwise one chooses a factor of degree >1, and constructs a root field K'' over K' by the same process. Again $f(x)$ is factorized into irreducible polynomials over K'', and so on. As the degree of at least one irreducible factor is reduced at each stage, the process must end with a splitting field of $f(x)$.

Example 2: Consider the polynomial $f(x) = x^3 - 2$ over $\mathbb{Q}$. Its roots are $\alpha_1 = \sqrt[3]{2}$, $\alpha_2 = \omega\sqrt[3]{2}$, $\alpha_3 = \bar{\omega}\sqrt[3]{2}$, where $\omega = {}^1/_2(-1 + \mathrm{i}\sqrt{3})$ and $\bar{\omega} = {}^1/_2(-1 - \mathrm{i}\sqrt{3})$. The splitting field is $L = \mathbb{Q}(\sqrt[3]{2}, \omega\sqrt[3]{2}, \bar{\omega}\sqrt[3]{2}) = \mathbb{Q}(\sqrt[3]{2}, \omega)$, since $\bar{\omega} = -1 - \omega$.

Galois theory

The Galois group and the fundamental theorem of Galois theory. The connection between solutions of algebraic equations and group theory discovered by E. GALOIS leads to particularly beautiful results in the theory of finite field extensions. For this reason the *central part of field theory* dealing with solutions of algebraic equations is called *Galois theory*. If N is the splitting field of a polynomial over P *without repeated roots*, then N has the important property that every relative automorphism of any extension L_1 of P containing N maps N onto itself, that is, induces an automorphism of N. A field extension with this property is called *normal*. If K is an extension of P contained in L_1 but not normal over P, then it is mapped by the relative automorphisms of L_1 over P onto isomorphic copies $K', K'', \ldots$, which are called the *conjugates* of K in L_1. The relative automorphisms of L_1 determine relative isomorphisms of K over P. A normal field extension K is distinguished by the property that K is *equal to all its conjugates*.

Let $f(x)$ be an irreducible polynomial of degree n over a field P. Suppose that an extension L of P contains all the solutions $\alpha_1, \ldots, \alpha_n$ of the equation $f(x) = 0$, and that they are all distinct. If α is one of these solutions, then the conjugates of the simple extension $P(\alpha)$ are $P(\alpha_1), \ldots, P(\alpha_n)$, of which one is identical with $P(\alpha)$. The n relative isomorphisms of $P(\alpha)$ map α to $\alpha_1, \ldots, \alpha_n$. Since every relative isomorphism must take a root of $f(x)$ to another root of $f(x)$, there can be no further relative isomorphisms of $P(\alpha)$.

The number of relative isomorphisms of a simple extension $P(\alpha)$ of P is equal to the degree $[P(\alpha):P]$.

This statement can be generalized. A *finite* extension $K = P(\beta_1, \ldots, \beta_m)$ can always be obtained by adjunction of a single element: $K = P(\vartheta)$, provided that the irreducible equations whose roots $\beta_1, \ldots, \beta_m$ are adjoined have no repeated roots. If $N = P(\vartheta)$ is a normal extension of P, then the number of relative automorphisms of N is equal to the degree of the extension $[N:P]$. The relative automorphisms of a normal extension N form a group of order $[N:P]$ under multiplication of mappings. This group is called the *Galois group* G of the normal extension N over P: $[G:E] = [N:P]$.

If K is an intermediate field between P and N, then N is also normal over K, and those relative automorphisms of the Galois group G of N over P that leave all the elements of K fixed, in other words, the relative automorphisms of N over K, form a subgroup H of G, which is just the Galois group G of N over K.

In the above manner a subgroup H of the Galois group G is associated with every intermediate field K. This correspondence can be inverted. If H is a subgroup of G, then the elements of N that are left fixed by all the relative automorphisms in H form an intermediate field K. Thus, the investigation of the intermediate fields between N and P is reduced to that of the subgroups of the Galois group. The methods of group theory can now be applied to field theory. If the subgroups of the Galois group G of N over P are known, one is in a position to survey completely all the extensions between P and N, and the intermediate fields, and their relationships to one another. If P is a field containing the rational numbers, then the fundamental theorem of Galois theory holds.

Fundamental theorem of Galois theory. Let N be a finite normal extension of P and G its Galois group. (I) There is a one-to-one correspondence between the subgroups H of G and the intermediate fields K of the extension.

(II) If K and H correspond to each other, then H consists of all the relative automorphisms of N that leave the elements of K fixed; and K consists of all the elements of N that are fixed under the relative automorphisms in H.

(III) An intermediate field K is normal over P if the associated subgroup H is normal in G. In that case the Galois group of K over P is isomorphic to the factor group G/H.

(IV) The following relations hold:

$$[H:E] = [N:K] \qquad [G:H] = [K:P]$$

$$\begin{array}{ccccc} & N & \longleftrightarrow & E & \\ [N:K] \rightarrow & | & & | & \rightarrow [H:E] \\ & K & \longleftrightarrow & H & \\ [K:P] \rightarrow & | & & | & \rightarrow [G:H] \\ & P & \longleftrightarrow & G & \end{array}$$

Example 3: Consider the finite normal extension $L = \mathbb{Q}(\sqrt[3]{2}, \omega\sqrt[3]{2}, \bar{\omega}\sqrt[3]{2}) = \mathbb{Q}(\sqrt[3]{2}, \omega)$, the splitting field of the polynomial $x^3 - 2$ over $\mathbb{Q}$. To find the Galois group of L over $\mathbb{Q}$, one first finds all relative isomorphisms of $\mathbb{Q}(\sqrt[3]{2})$ in L. Here: $\sqrt[3]{2} \to \sqrt[3]{2}$, $\sqrt[3]{2} \to \omega\sqrt[3]{2}$, or $\sqrt[3]{2} \to \bar{\omega}\sqrt[3]{2}$. Therefore the relative automorphisms of $\mathbb{Q}(\sqrt[3]{2}, \omega)$ act on $\sqrt[3]{2}$ and ω in the following six possible ways:

$\sqrt[3]{2} \to \sqrt[3]{2}$, $\quad \omega \to \omega \sim E$
$\sqrt[3]{2} \to \omega\sqrt[3]{2}$, $\quad \omega \to \omega \sim A$
$\sqrt[3]{2} \to \bar{\omega}\sqrt[3]{2}$, $\quad \omega \to \omega \sim A^2$
$\sqrt[3]{2} \to \sqrt[3]{2}$, $\quad \omega \to \bar{\omega} \sim B$
$\sqrt[3]{2} \to \omega\sqrt[3]{2}$, $\quad \omega \to \bar{\omega} \sim A^2B$
$\sqrt[3]{2} \to \bar{\omega}\sqrt[3]{2}$, $\quad \omega \to \bar{\omega} \sim AB$.

These six relative automorphisms form the Galois group G of L over $\mathbb{Q}$, with multiplication defined as performing the two mappings in succession. If one notes that $\omega^2 = \bar{\omega}$ and $\omega\bar{\omega} = 1$, it is easy to verify the relations above. Furthermore one has: $A^3 = E$, $B^2 = E$, and $BA = A^2B$. The subgroups $\{E, A, A^2\}$, $\{E, B\}$, $\{E, AB\}$, and $\{E, A^2B\}$ correspond to the intermediate fields $\mathbb{Q}(\omega)$, $\mathbb{Q}(\sqrt[3]{2})$, $\mathbb{Q}(\omega\sqrt[3]{2})$, and $\mathbb{Q}(\bar{\omega}\sqrt[3]{2})$.

$$\begin{array}{cccccc} \mathbb{Q}(\sqrt[3]{2}, \omega), & \mathbb{Q}(\omega), & \mathbb{Q}(\sqrt[3]{2}), & \mathbb{Q}(\omega\sqrt[3]{2}), & \mathbb{Q}(\bar{\omega}\sqrt[3]{2}), & \mathbb{Q} \\ \updownarrow & \updownarrow & \updownarrow & \updownarrow & \updownarrow & \updownarrow \\ \langle E\rangle & \langle A\rangle & \langle B\rangle & \langle AB\rangle & \langle A^2B\rangle & G \end{array}$$

The Galois group of an equation. If $f(x) = x^n + a_{n-1}x^{n-1} + \cdots + a_1x + a_0 = 0$ is an equation with coefficients in $\mathbb{Q}$, then there are certain rational relations $H(\alpha_1, \ldots, \alpha_n) = 0$ between the solutions $\alpha_1, \ldots, \alpha_n$ of the equation; for example the adjacent equations (see Chapter 4.).

$$\begin{aligned} \alpha_1 + \alpha_2 + \cdots + \alpha_n &= -a_{n-1} \\ \alpha_1\alpha_2 + \alpha_1\alpha_3 + \cdots + \alpha_{n-1}\alpha_n &= a_{n-2} \\ &\vdots \\ \alpha_1\alpha_2 \ldots \alpha_n &= (-1)^n a_0 \end{aligned}$$

These equations are independent of the order in which the solutions are numbered. In other words: these relations between the solutions *remain invariant under all permutations of the solutions.* It can happen that there are further relations that depend on the particular equation in question, and that these may or may not be invariant under certain permutations. If one considers the set of those permutations that do not destroy any of the relations among the solutions, then the following result holds:

The set of those permutations that leave all relations $H(\alpha_1, \ldots, \alpha_n) = 0$ with coefficients in $\mathbb{Q}$ invariant is a subgroup of S_n. It is called the Galois group of the equation.

Example 4: To find the Galois group of the equation $f(x) = (x^2-2)(x^2-3) = 0$. The roots of this equation are $\alpha_1 = \sqrt{2}, \alpha_2 = -\sqrt{2}, \alpha_3 = \sqrt{3}, \alpha_4 = -\sqrt{3}$. It is required to find all the permutations of four elements that leave all relations between these roots invariant. It is sufficient to consider the relations $H_1(\alpha_1, \ldots, \alpha_4) = \alpha_1\alpha_2 = 2$, and $H_2(\alpha_1, \ldots, \alpha_4) = \alpha_3\alpha_4 = 3$, from which it is easy to see that the permutations

$$e = \begin{pmatrix}1 & 2 & 3 & 4\\ 1 & 2 & 3 & 4\end{pmatrix}, \quad p_1 = \begin{pmatrix}1 & 2 & 3 & 4\\ 2 & 1 & 3 & 4\end{pmatrix}, \quad p_2 = \begin{pmatrix}1 & 2 & 3 & 4\\ 1 & 2 & 4 & 3\end{pmatrix} \quad \text{and} \quad p_3 = \begin{pmatrix}1 & 2 & 3 & 4\\ 2 & 1 & 4 & 3\end{pmatrix}$$

are exactly the elements of the Galois group G, because every permutation that takes α_1 or α_2 to α_3 or α_4, or vice versa, would destroy one of the relations H_1 or H_2.

If $f(x) = 0$ is the *general equation of degree n*, that is, $f(x) = 0$ is an equation with *indeterminate coefficients*, which can be arbitrarily replaced by elements of any field, then there are no further relations apart from the ones given above, the so-called elementary symmetric relations and their consequences.

The Galois group of the general equation of degree n is the symmetric group S_n.

The Galois groups of an equation and of its splitting field. To find a connection between the Galois group of an equation $f(x) = 0$ and that of a field extension, one considers the splitting field L of the polynomial $f(x)$ over $\mathbb{Q}$. Let $\alpha_1, \ldots, \alpha_n$ be the roots of the polynomial in L, and assume that they are all distinct. If A is a relative automorphism of L over $\mathbb{Q}$, then every rational relation $H(\alpha_1, \ldots, \alpha_n) = 0$ with coefficients in $\mathbb{Q}$ is taken by A to $H(A\alpha_1, \ldots, A\alpha_n) = 0$. Further, since every relative automorphism of L takes roots of $f(x)$ to roots of $f(x)$, one has $A\alpha_1 = \alpha_{i_1}, A\alpha_2 = \alpha_{i_2}, \ldots, A\alpha_n = \alpha_{i_n}$, say. Thus, every relative automorphism A, in other words, every element of the Galois group of L over $\mathbb{Q}$, defines a permutation $p = \begin{pmatrix}1 & 2 & \ldots & n\\ i_1 & i_2 & \ldots & i_n\end{pmatrix}$ of the roots of the equation $f(x) = 0$, under which every valid relation $H(\alpha_1, \ldots, \alpha_n) = 0$ goes over into a valid relation $H(\alpha_{i_1}, \ldots, \alpha_{i_n}) = 0$, and the permutation belongs to the Galois group of the equation. It can be shown that the mapping defined by $A \to p$ is, in fact, an isomorphism of the Galois group of the extension L over $\mathbb{Q}$ onto the Galois group of the equation $f(x) = 0$.

Example 5: The Galois group of the equation $f(x) = (x^2-2)(x^2-3)$ consists of the elements (see Example 4):

$$e = \begin{pmatrix}1 & 2 & 3 & 4\\ 1 & 2 & 3 & 4\end{pmatrix}, \quad p_1 = \begin{pmatrix}1 & 2 & 3 & 4\\ 2 & 1 & 3 & 4\end{pmatrix}, \quad p_2 = \begin{pmatrix}1 & 2 & 3 & 4\\ 1 & 2 & 4 & 3\end{pmatrix} \quad \text{and} \quad p_3 = \begin{pmatrix}1 & 2 & 3 & 4\\ 2 & 1 & 4 & 3\end{pmatrix}.$$

The Galois group of the corresponding field extension $\mathbb{Q}(\sqrt{2}, \sqrt{3})$ consists of the elements e', p_1', p_2', and p_3', which have the following effect on $\sqrt{2}$ and $\sqrt{3}$:

$$\begin{array}{lll}
e' \sim \sqrt{2} \to \sqrt{2}, & \sqrt{3} \to \sqrt{3} & e \to e'\\
p_1' \sim \sqrt{2} \to -\sqrt{2}, & \sqrt{3} \to \sqrt{3} & p_1 \to p_1'\\
p_2' \sim \sqrt{2} \to \sqrt{2}, & \sqrt{3} \to -\sqrt{3} & p_2 \to p_2'\\
p_3' \sim \sqrt{2} \to -\sqrt{2}, & \sqrt{3} \to -\sqrt{3} & p_3 \to p_3'.
\end{array}$$

It is very easy to check that this mapping is an isomorphism between the two Galois groups.

Solution of equations by radicals. Apart from the problem of the *existence* of solutions of an equation $f(x) = 0$, which has been solved by the construction of the splitting field, there is also the problem of *determining* them, that is, of finding a method of giving their precise values. For quadratic equations the formula has been known for a very long time. The *formula of Cardano* can be used to solve cubic equations, and FERRARI, a pupil of CARDANO, was able to give a corresponding formula for quartic equations, that is, equations of the fourth degree. These formulae all use only the four basic arithmetical operations and the extraction of roots. After FERRARI all attempts to find a general formula using only these operations for equations of the *fifth or higher degrees* remained unsuccessful. The 'power of radicals' was greatly overestimated, as will be shown by the following argument. As a justification of these efforts it should be remarked that the expectation of finding corresponding solution formulae for the general equation of degree higher than 4 was fostered by the fact that one can very well find special equations whose solutions can be expressed by radicals. Such equations are called soluble, or *soluble by radicals*. A *radical* is a solution of a pure equation of the nth degree $x^n - a = 0$ and is denoted by $\sqrt[n]{a}$.

A *radical expression* over a field P is defined in the following manner: there is an element $g_1 \in P$, and finitely many polynomials $g_2(x_1), g_3(x_1, x_2), \ldots, g_m(x_1, \ldots, x_{m-1})$, and $g(x_1, \ldots, x_m)$ and positive integers $n_1, \ldots, n_m$ such that $\beta = g(\beta_1, \ldots, \beta_m)$ with $\beta_1 = \sqrt[n_1]{g_1}, \beta_2 = \sqrt[n_2]{g_2(\beta_1)}, \beta_3 = \sqrt[n_3]{(g_3(\beta_1, \beta_2))}, \ldots, \beta_m = \sqrt[n_m]{(g_m(\beta_1, \ldots, \beta_{m-1}))}$.

Example 6: With $g_1 = 2$, $g_2 = 6x_1^3 + 5x_1 + 3$, and $g(x_1, x_2) = (1 + 3x_1)^3 x_2 + 7x_1 + 2$, and $n_1 = 2$ and $n_2 = 4$, one obtains the radical expression β:

$$\beta = g(\beta_1, \beta_2) = (1 + 3\sqrt{2})^3 \sqrt[4]{[6(\sqrt{2})^3 + 5\sqrt{2} + 3]} + 7\sqrt{2} + 2$$

over the field of rational numbers. Here one has

$$\beta_1 = \sqrt{2} \quad \text{and} \quad \beta_2 = \sqrt[4]{[3 + 5\sqrt{2} + 6(\sqrt{2})^3]}.$$

An equation $f(x) = 0$ is called *soluble by radicals* over P if its solutions are *radical expressions* over P. In the language of field theory, a radical expression corresponds to a *tower of fields*:

$$P = K_0 \subseteq K_0(\beta_1) = K_1 \subseteq K_1(\beta_2) = K_2 \subseteq K_2(\beta_3) = K_3 \cdots = K_{m-1}(\beta_m) = K_m = K,$$

where β is an element of K and every extension K_{i+1} over K_i is obtained by solving a pure equation $x^{n_i} - g_i(\beta_1, \ldots, \beta_{i-1}) = 0$. In this case the field K is called soluble. Thus, in the language of field theory the solubility of an equation can be expressed in the following manner:

The equation $f(x) = 0$ is soluble by radicals if and only if there exists a soluble field K containing the splitting field L of the polynomial $f(x)$.

It can be proved that under these circumstances the splitting field L of $f(x)$ must itself be soluble:

The equation $f(x) = 0$ is soluble by radicals if and only if the splitting field L of the polynomial $f(x)$ can be reached by a tower of fields $P = K_0 \subseteq K_0(\beta_1) = K_1 \subseteq \cdots \subseteq K_{m-1}(\beta_m) = K_m = L$, in which each field $K_{i+1} = K_i(\beta_{i+1})$ is obtained from K_i by adjunction of a solution of a pure equation $x^{n_i} - b_i = 0$.

It can be shown that the existence of such a tower implies the existence of a tower in which each field is the splitting field of a pure equation over its predecessor. By the fundamental theorem of Galois theory this implies the existence of series of subgroups in the Galois group G:

$$G = H_m \supseteq H_{m-1} \supseteq \cdots \supseteq H_1 \supseteq H_0 = E,$$

in which each subgroup is normal in its predecessor, and the factor groups are isomorphic to the Galois groups of the extensions in the tower. These are always Abelian for pure equations, if the base field contains all the nth roots of unity. From this it can be shown that the Galois group has a composition series with factors of prime order. The converse is also true, and so one obtains the definitive criterion:

The equation $f(x) = 0$ is soluble by radicals if and only if its Galois group is soluble.

Since the symmetric groups of degree 2, 3, and 4 are soluble, this implies the solubility of the general equations of degree 2, 3, and 4 by radicals. On the other hand, the symmetric groups of degree 5 and higher are not soluble, and therefore there can be no formulae for the solutions of the general equations of degree 5 and higher. This result was discovered independently by GALOIS and by ABEL; the latter did not, however, succeed in giving Galois' general criterion for the solubility of particular equations by radicals.

Cubic equations. The reduced form of the cubic equation (see Chapter 4.) is: $x^3 + px + q = 0$, where p and q are elements of a field P. The Galois group of the equation is S_3. The composition series $S_3 \supseteq A_3 \supseteq E$ of the Galois group corresponds to a tower of fields $P \subseteq K \subseteq N$. Furthermore one has the relations $[S_3 : A_3] = [K : P] = 2$ and $[A_3 : E] = [N : K] = 3$. To make matters simpler, assume that P already contains the cube roots of unity. To get from P to K one need only adjoin a square root $\sqrt{D}$ to P, where $\sqrt{D}$ must remain fixed under all permutations of A_3. From this condition one obtains $\sqrt{D} = \sqrt{(-4p^3 - 27q^2)}$. If one denotes the solutions of the original equation by $\alpha_1, \alpha_2, \alpha_3$ and forms *Lagrange's resolvent*, $r = \alpha_1 + \omega\alpha_2 + \bar{\omega}\alpha_3$, where ω, $\bar{\omega}$ are the cube roots of unity, then it can be shown that $r^3 = 27q/2 + (3/2)\sqrt{(-3D)} = s$ lies in the field $K = P(\sqrt{D})$. Now one obtains the extension N by adjoining $r = \sqrt[3]{s}$ to K. The roots $\alpha_1, \alpha_2, \alpha_3$ can now be computed from the equations $\alpha_1 + \alpha_2 + \alpha_3 = 0$, $\alpha_1 + \omega\alpha_2 + \bar{\omega}\alpha_3 = r$, and $\alpha_1 + \bar{\omega}\alpha_2 + \omega\alpha_3 = -3p/r$.

Constructions by ruler and compass. The problem of constructions by ruler and compass alone can be formulated in the following way; from finitely many given points in the plane, to construct in *finitely many steps* a required point, where each step is of one of the following types:

(1) The ruler may be used only to draw the line joining two given or previously constructed points.

(2) The compass may only be used to draw a circle whose centre is a given or previously constructed point and whose radius is the distance between two given or previously constructed points.

(3) New points can be constructed by intersecting two straight lines, a line and a circle, or two circles, that have been constructed by the rules (1) and (2).

To obtain a survey of the points that can be constructed one translates the geometrical problem into algebraic language. This is done by introducing a rectangular Cartesian coordinate system to describe the points $P_1, P_2, \ldots, P_n$, in which is P_1, say, the origin and the point (0, 1) is P_2. If K is the smallest field containing the coordinates of all the given points, then it can be shown by means of analytic geometry that all points constructible in a single step of type (1), (2), or (3) must have coordinates in K or in a field obtained from K by adjoining square roots. On the other hand, it can also be shown that all the rational operations and the extracting of square roots can be performed by sequences of steps of the types (1), (2), and (3). Generally, one obtains the following important criterion:

A point can be constructed by ruler and compass alone if and only if its coordinates lie in a finite normal extension field of K whose degree over K is a power of 2.

Since in many cases the problem is to construct a quantity x that is given by an equation $f(x) = 0$, the criterion can be reworded: A quantity x can be constructed by ruler and compass if and only if the equation $f(x) = 0$ can be solved by quadratic radicals over K.

The famous problems of classical Greek mathematics, to construct by ruler and compass alone a square of area equal to that of a given circle (*the squaring of the circle*), to divide an angle into three equal parts (*trisection of the angle*), and to find a cube of double the volume of a given cube (*the doubling of the cube*) all turn out to be unsoluble. The algebraic formulation of the doubling of the cube is $x^3 - 2 = 0$, where x is the edge of the required cube. This equation is irreducible over the field of rational numbers. Each of its roots generates a field extension of degree 3. Such a field can never be contained in a field extension whose degree is a power of 2.

The trisection of the angle α is equivalent to constructing a segment of length $\cos(\alpha/3)$, where $\cos\alpha$ is given. The resulting equation is $4[\cos(\alpha/3)]^3 - 3\cos(\alpha/3) - \cos\alpha = 0$. The question is whether the roots of the equation $4x^3 - 3x - \cos\alpha = 0$ lie in an extension field of degree 2^m (m a natural number) of $\mathbb{Q}(\cos\alpha)$, where $\mathbb{Q}$ is the field of rational numbers. It can be shown that the equation is, in general, irreducible, and then just as with the doubling of the cube each root generates a field extension of degree 3, and there cannot be a general ruler and compass construction method for the trisection of an angle.

The squaring of the circle requires the construction of a straight line segment of length $\sqrt{\pi}$. Since π is transcendental over the rational numbers, that is, does not satisfy any algebraic equation, this problem also is insoluble.

The construction of a regular n-gon. The nth roots of unity divide the unit circle into n equal parts. A regular n-gon inscribed in the unit circle can be constructed by ruler and compass alone if and only if n is of the form $2^l p_1 p_2 \ldots p_k$, where l is a non-negative integer and $p_1, p_2, \ldots, p_k$ are distinct *Fermat primes*, that is, primes of the form $2^m + 1$. Thus, regular n-gons can be constructed for $n = 3, 4, 5, 6, 8, 10, 12, 15, 16, 17, 20, 24, \ldots, 257, \ldots$, using only ruler and compass.

Applications

Field theory has found manifold applications in other parts of mathematics. Its methods are used in Galois theory and *algebraic number theory*. Many classes of functions of interest in complex analysis (see Chapter 23.) form fields, for example, the rational functions and the *elliptic* functions. On the other hand, the methods of complex analysis are sometimes used in field theory for instance, in the proof of the so-called fundamental theorem of algebra, and other investigations on the field of complex numbers.

Algebraic varieties are discussed in Chapter 32. and in Chapter 33.

17. Linear algebra

17.1. Systems of linear equations

In Chapter 4. it was examined under what circumstances, and by what means, equations of the form $ax = b$ or $ax + by = c$ can be solved, when a, b, and c are rational, real, or complex numbers. The first equation contains the variable x, and the second the two variables x and y. In both cases, the variables occur only to the first power; such equations are called *linear* in one or two variables, respectively. In general, a *linear equation in n variables* (or *unknowns*) $x_1, x_2, \ldots, x_n$ is an equation of the form $a_1x_1 + a_2x_2 + \cdots + a_nx_n = b$.

The numbers $a_1, a_2, \ldots, a_n$ are called the *coefficients* and b is called the *constant* or *absolute term* of the equation. For $n = 1$ and $n = 2$ one obtains the cases mentioned above.

Many problems in mathematics lead not to a single linear equation, but to a whole system of such equations.

Systems of linear equations. A simple example of a system (or a set) of simultaneous linear equations is given by the two adjacent equations. Here a_1, a_2, b_1, b_2, c_1, and c_2 are given numbers. A solution of such a system of equations is a pair of numbers $(\bar{x}, \bar{y})$ such that when x is replaced by $\bar{x}$ and y by $\bar{y}$, both equations are satisfied simultaneously, that is, both become true propositions on equality. Generalizing this situation one understands by a *system of m linear equations in n variables (unknowns)* $x_1, x_2, \ldots, x_n$ a system of the adjacent form. In this system the a_{ij} and b_i are given numbers for $i = 1, \ldots, m$, and $j = 1, \ldots, n$. The index i indicates in which equation of the system the number occurs, and the index j of the coefficient a_{ij} indicates the unknown with which it is associated; for instance, a_{23} (read 'a two three', 'a sub two three', not 'a twenty three') is the coefficient of the unknown x_3 in the second equation. A single linear equation forms a special case of a system, characterized by $m = 1$.

$$a_1x + b_1y = c_1,$$
$$a_2x + b_2y = c_2.$$

$$\begin{aligned} a_{11}x_1 + a_{12}x_2 + \cdots + a_{1n}x_n &= b_1, \\ a_{21}x_1 + a_{22}x_2 + \cdots + a_{2n}x_n &= b_2, \\ &\vdots \\ a_{m1}x_1 + a_{m2}x_2 + \cdots + a_{mn}x_n &= b_m. \end{aligned}$$

A system of linear equations is called *homogeneous* if the constant terms $b_1, b_2, \ldots, b_m$ are all zero; otherwise, that is, if even one b_i is not zero, the system is called *inhomogeneous*. If in an inhomogeneous system all the constant terms are replaced by zeros, then the resulting system is called the *associated homogeneous system*.

A *solution* of a system of m linear equations in n unknowns is a sequence of numbers $\bar{x}_1, \bar{x}_2, \ldots, \bar{x}_n$ such that when all the unknowns are replaced by the corresponding numbers, then all the equations are satisfied simultaneously. A sequence of numbers $c_1, c_2, \ldots, c_n$ is called an *n-tuple* and is written $(c_1, c_2, \ldots, c_n)$. Two n-tuples $(c_1, c_2, \ldots, c_n)$ and $(d_1, d_2, \ldots, d_n)$ are equal if and only if $c_1 = d_1$, $c_2 = d_2$, ..., and $c_n = d_n$. If the n-tuple $(\bar{x}_1, \bar{x}_2, \ldots, \bar{x}_n)$ is a solution of the given system of linear equations, it may be necessary to check whether the values $\bar{x}_i$ lie in the permitted fundamental domain of variability. For simplicity it will be assumed that this domain is the set of real (complex) numbers, provided that the coefficients and constants of the system are real (or complex) numbers, respectively.

The investigation of the solutions of a system of linear equations leads to three problems, which have to be treated each in its own manner. The first is the question of the *existence* of solutions. This asks under what conditions a system has solutions; for even the single equation $0 \cdot x = 1$ has no solution. The second problem is to find a *method* that gives a solution of a given system of linear equations. Finally, the third problem is to describe the *totality* of *all* solutions of the given system of equations.

The existence of solutions. The following manipulations of a system of linear equations do not alter the solubility or non-solubility nor the solutions of the system:

1. Addition of equations of the system to other equations of the system.
2. Multiplication of equations of the system by non-zero factors.
3. Changing the sequence of the equations.

Example 1:
$$\left.\begin{aligned} 2x_1 + x_2 &= 1 \\ x_1 - 3x_2 &= 4 \end{aligned}\right| \begin{matrix} +1 \\ -2 \end{matrix} \rightarrow \left.\begin{aligned} 2x_1 + x_2 &= 1 \\ -2x_1 + 6x_2 &= -8 \end{aligned}\right| + \rightarrow \left.\begin{aligned} 2x_1 + x_2 &= 1 \\ 7x_2 &= -7 \end{aligned}\right|$$

All the systems have the unique solution $\bar{x}_1 = 1$, $\bar{x}_2 = -1$.

The following criterion gives a theoretical insight into the solubility of a system. It can also be used in practice to prove that a given system has no solutions.

A system of linear equations has a solution if and only if the following condition holds: whenever repeated application of the operations 1. and 2. leads to an equation in which all the coefficients are zero, then the constant term of that equation is also zero.

Example 2: The adjacent system of equations has no solution. As described by the numbers in red, the equations are multiplied by -1, 1, and -1, respectively, and then added; the resulting equation is

$$0 \cdot x_1 + 0 \cdot x_2 + 0 \cdot x_3 = 1.$$

$$\left.\begin{aligned} 2x_1 + x_2 + x_3 &= 1 \\ x_1 + 2x_2 + x_3 &= 1 \\ -x_1 + x_2 \phantom{{}+x_3} &= -1 \end{aligned}\right| \begin{matrix} -1 \\ +1 \\ -1 \end{matrix}$$

Homogeneous equations and the complete set of solutions. The problem of finding all the solutions of a given inhomogeneous system can be reduced in part to the simpler problem of finding all the solutions of the associated homogeneous system. This is a consequence of the following easily verified properties of solutions of homogeneous systems (note the analogy to the operations 1. and 2.).

1. If $(\bar{x}_1, \bar{x}_2, \dots, \bar{x}_n)$ and $(\bar{y}_1, \bar{y}_2, \dots, \bar{y}_n)$ are both solutions of a homogeneous system of linear equations, then so is their sum $(\bar{x}_1 + \bar{y}_1, \bar{x}_2 + \bar{y}_2, \dots, \bar{x}_n + \bar{y}_n)$, which is defined componentwise.
2. If $(\bar{x}_1, \bar{x}_2, \dots, \bar{x}_n)$ is a solution, then so is its multiple by a factor c, $(c\bar{x}_1, c\bar{x}_2, \dots, c\bar{x}_n)$.
3. The n-tuple $(0, 0, \dots, 0)$ is always a solution of any homogeneous system and is called the *trivial* solution.

From the properties 1. and 2. it follows that if $(\bar{x}_1^{(1)}, \bar{x}_2^{(1)}, \dots, \bar{x}_n^{(1)})$; $(\bar{x}_1^{(2)}, \bar{x}_2^{(2)}, \dots, \bar{x}_n^{(2)})$, ..., $(\bar{x}_1^{(m)}, \bar{x}_2^{(m)}, \dots, \bar{x}_n^{(m)})$ are m solutions of a homogeneous system of linear equations, then so is

$$\begin{aligned}&\lambda_1(\bar{x}_1^{(1)}, \bar{x}_2^{(1)}, \dots, \bar{x}_n^{(1)}) + \lambda_2(\bar{x}_1^{(2)}, \bar{x}_2^{(2)}, \dots, \bar{x}_n^{(2)}) + \dots + \lambda_m(\bar{x}_1^{(m)}, \bar{x}_2^{(m)}, \dots, \bar{x}_n^{(m)})\\ &= (\lambda_1\bar{x}_1^{(1)} + \lambda_2\bar{x}_1^{(2)} + \dots + \lambda_m\bar{x}_1^{(m)}, \dots, \lambda_1\bar{x}_n^{(1)} + \lambda_2\bar{x}_n^{(2)} + \dots + \lambda_m\bar{x}_n^{(m)})\end{aligned}$$

for any real numbers λ_i $(i = 1, 2, \dots, m)$. Such a sum of multiples is called a *linear combination*. Thus, the statements under 1. and 2. together say that any linear combination of solutions of a system of homogeneous linear equations is again a solution.

These properties do not hold for inhomogeneous systems. However, the following theorem shows the connection between the solutions of an inhomogeneous system and those of the associated homogeneous system.

If a solution of the associated homogeneous system is added to an arbitrary solution of the inhomogeneous system, then the result is again a solution of the inhomogeneous system. If an arbitrary but fixed solution of the inhomogeneous system is chosen, then every solution of the inhomogeneous system can be obtained by adding a solution of the homogeneous system to the chosen solution.

Solution by elimination and Gauss's algorithm. This method can be used to find just one solution, or the whole manifold of solutions, for a given system of linear equations.

The algorithm is particularly suitable for use on computers, and for this reason it has gained in significance in recent years. The basic idea of the method is the following: using the operations 1. to 3. the given set of m equations in n unknowns is transformed into a new set of m equations in n unknowns, in which one of the unknowns, say x_1, only occurs in a single equation. One says that x_1 has been *eliminated* from the other $m - 1$ equations. By the same method these $m - 1$ equations are transformed so that another unknown, say x_2, only occurs in a single one of them. Repeating this process a system is finally obtained in which x_1 occurs only in the first equation, x_2 occurs only there and in the second, and so on. This kind of system is easy to solve.

The following very simple example should clarify the working of Gauss's algorithm.

Example 3:

$$\begin{aligned}3x_1 - 3x_2 + x_3 &= 0 \quad &(1)\\ 4x_2 - x_3 &= 5 \quad &(2)\\ 2x_1 - 2x_2 + x_3 &= 1 \quad &(3)\end{aligned}$$

$$\begin{aligned}3x_1 - 3x_2 + x_3 &= 0 \quad &(1')\\ 4x_2 - x_3 &= 5 \quad &(2')\\ {}^1/_3 x_3 &= 1 \quad &(3')\end{aligned}$$

Equation (2) does not contain x_1; this occurs only in equations (1) and (3). Multiplication of (1) by $-2/3$ and subsequent addition to (3) yields a new set of equations (1′), (2′), and (3′).

Since x_2 is already absent from (3′), it is unnecessary to eliminate it. From (3′) it can be seen that $\bar{x}_3 = 3$, then by substituting this value in (2′) the value $\bar{x}_2 = 2$ is obtained and now (1′) yields $\bar{x}_1 = 1$.

A more difficult example is the case of three equations in four unknowns with general coefficients. By applying operation 3., if necessary, it can be assumed that a_{11} is not zero in the adjacent system S.

$$S:\quad\begin{aligned}a_{11}x_1 + a_{12}x_2 + a_{13}x_3 + a_{14}x_4 &= b_1 \quad &(1)\\ a_{21}x_1 + a_{22}x_2 + a_{23}x_3 + a_{24}x_4 &= b_2 \quad &(2)\\ a_{31}x_1 + a_{32}x_2 + a_{33}x_3 + a_{34}x_4 &= b_3 \quad &(3)\end{aligned}$$

Then the system S_1 can be obtained from S by subtracting a_{21}/a_{11} and a_{31}/a_{11} times equation (1) from equations (2) and (3), respectively.

$$S_1:\quad\begin{aligned}a_{11}x_1 + a_{12}x_2 + a_{13}x_3 + a_{14}x_4 &= b_1 \quad &(1')\\ a'_{22}x_2 + a'_{23}x_3 + a'_{24}x_4 &= b'_2 \quad &(2')\\ a'_{32}x_2 + a'_{33}x_3 + a'_{34}x_4 &= b'_4 \quad &(3')\end{aligned}$$

Here $a'_{ij} = a_{ij} - a_{i1}a_{1j}/a_{11}$, and $b'_i = b_i - a_{i1}b_1/a_{11}$ $(i = 2, 3; j = 2, 3, 4)$. If all the coefficients of (2′) and (3′) are zero and b'_2 or b'_3 is not, then by the criterion of the preceding paragraph the system has no solution. If, on the other hand, $b'_2 = b'_3 = 0$, then x_2, x_3, and x_4 can be chosen arbitrarily and x_1 determined by equation (1). If neither of these cases occurs, then by interchanging (2′) and (3′), if necessary, and possibly renaming and interchanging the unknowns, it can be arranged that a'_{22} is not zero. To avoid overloading the notation it is assumed that this is already the case in the system S_1 as it stands (the reader is referred to the subsequent examples).

The system S_2 is obtained from S_1 by subtracting a'_{32}/a'_{22} times equation (2′) from equation (3′). Here $a''_{3j} = a'_{3j} - a'_{32}a'_{2j}/a'_{22}$, and $b''_3 = b'_3 - a'_{32}b'_2/a'_{22}$ ($j = 3, 4$). As at the previous stage if $a''_{33} = a''_{34} = 0$, there are two cases to distinguish. If $b''_3 = 0$, then x_3 and x_4 can be chosen arbitrarily and x_1 and x_2 determined from (1″) and (2″).

$$S_2 \quad \begin{array}{rl} a_{11}x_1 + a_{12}x_2 + a_{13}x_3 + a_{14}x_4 = b_1 & (1'') \\ a'_{22}x_2 + a'_{23}x_3 + a'_{24}x_4 = b'_2 & (2'') \\ a''_{33}x_3 + a''_{34}x_4 = b''_3 & (3'') \end{array}$$

Now let $a''_{33} \neq 0$. If x_4 is taken as an arbitrary number d, then x_3 is determined from (3″), x_2 from (2″) and finally x_1 from (1″). There is a solution for each d and all solutions can be obtained in this way. If $a''_{33} = 0$ but $a''_{34} \neq 0$, then the solutions are obtained in the same manner by interchanging x_3 and x_4.

Example 4: From the given system S_1 the new system S_2 is obtained by interchanging the first two rows. Now in S_2 $0/1 = 0$ times equation (1) is subtracted from (2) and $3/1 = 3$ times (1) from (3). This yields the system S_3. S_4 is obtained from S_3 by interchanging x_2 and x_4 and renaming them x'_4 and x'_2, respectively. Now in S_4 $-4/4$ times (2′) is subtracted from (3′), that is, the two equations are added, yielding S_5. If in S_5 one puts $\bar{x}'_4 = \bar{x}_2 = d$, then $\bar{x}_3 = 2$, $\bar{x}'_2 = \bar{x}_4 = 1$, and $\bar{x}_1 = -7 + 2d$. Thus, the set **S** of solutions of S_1 is $\mathbf{S} = \{(-7 + 2d, d, 2, 1);\ d \text{ real}\}$.

$$S_1 \quad \begin{array}{r} -\ x_3 + 4x_4 = 2 \\ x_1 - 2x_2 + 4x_3 + 3x_4 = 4 \\ 3x_1 - 6x_2 + 8x_3 + 5x_4 = 0 \end{array}$$

$$S_2 \quad \begin{array}{rl} x_1 - 2x_2 + 4x_3 + 3x_4 = 4 & (1) \\ -\ x_3 + 4x_4 = 2 & (2) \\ 3x_1 - 6x_2 + 8x_3 + 5x_4 = 0 & (3) \end{array}$$

$$S_3 \quad \begin{array}{r} x_1 - 2x_2 + 4x_3 + 3x_4 = 4 \\ -\ x_3 + 4x_4 = 2 \\ -\ 4x_3 - 4x_4 = -12 \end{array}$$

$$S_4 \quad \begin{array}{rl} x_1 + 3x'_2 + 4x_3 - 2x'_4 = 4 & (1') \\ 4x'_2 - x_3 = 2 & (2') \\ -4x'_2 - 4x_3 = -12 & (3') \end{array} \longrightarrow S_5 \quad \begin{array}{r} x_1 + 3x'_2 + 4x_3 - 2x'_4 = 4 \\ 4x'_2 - x_3 = 2 \\ -\ 5x_3 = -10 \end{array}$$

Example 5:

Two systems of four equations in three variables with the same coefficients but different right-hand sides.

$$\begin{array}{r} x_1 + x_2 + x_3 = 6 \\ 2x_1 + x_2 - x_3 = 1 \\ 4x_1 - x_2 + 2x_3 = 8 \\ -x_1 + x_2 + 2x_3 = 7 \end{array} \qquad \begin{array}{r} x_1 + x_2 + x_3 = 6 \\ 2x_1 + x_2 - x_3 = 0 \\ 4x_1 - x_2 + 2x_3 = 8 \\ -x_1 + x_2 + 2x_3 = 7 \end{array}$$

Application of the algorithm leads to the following two systems:

$$\begin{array}{r} x_1 + x_2 + x_3 = 6 \\ -x_2 - 3x_3 = -11 \\ 13x_3 = 39 \\ 0 = 0 \end{array} \qquad \begin{array}{r} x_1 + x_2 + x_3 = 6 \\ -x_2 - 3x_3 = -12 \\ 13x_3 = 44 \\ 0 = -11/13 \end{array}$$

While the left-hand system has the unique solution $\bar{x}_3 = 3$, $\bar{x}_2 = 2$, and $\bar{x}_1 = 1$, the right-hand system has no solutions at all because of the contradiction in the last equation.

Geometrical interpretation. It is well known from analytic geometry that a linear equation in two unknowns defines a *straight line* in the plane. For if one associates with each solution $(\bar{x}, \bar{y})$ the point with the coordinates $(\bar{x}, \bar{y})$, then the image of the set of solutions is a line. Hence two equations in two unknowns determine a pair of lines in the plane, and solutions of the system, if they exist, must be the *points of intersection of* the two lines. The following cases can occur (see Fig. 4.2-2, 4.2-3):

(*i*) There are infinitely many solutions and infinitely many points of intersection of the two lines. In this case the two lines coincide; one of the unknowns can be given an arbitrary value, and the other is determined.

(*ii*) The system of equations has a unique solution. In this case the lines intersect in a single point.

(*iii*) The system has no solution. The lines are distinct and parallel.

The situation is similar with equations in three unknowns x, y, z. Each single equation defines a *plane* in three-dimensional space. Again the solutions of the system are the points contained in all the planes.

17.2. Determinants

In methods of solving systems of n equations in n unknowns, other than Gauss's, certain functions of the coefficients, the determinants, play a decisive role. These functions are also important in other branches of mathematics, such as the differential and integral calculus of several variables.

A *determinant* is a function of n^2 variables, usually written as a square scheme of the adjacent form. The numbers a_{ij} are called the *elements* or *entries* of the determinant. The ith *row* of the determinant is the n-tuple $(a_{i1}, \ldots, a_{in})$ of entries with i as their first subscript. The jth *column* is the n-tuple $(a_{1j}, \ldots, a_{nj})$ of entries with j as their second subscript. The *value* of the determinant is $\Sigma\,(-1)^k\, a_{1s_1}\, a_{2s_2} \cdots a_{ns_n}$, where the indices $s_1, \ldots, s_n$ form a permutation of the numbers $1, \ldots, n$ and are therefore distinct. The sum is taken over all possible permutations of $1, \ldots, n$, that is, the summands are all possible products containing exactly one entry from each row and from each column. Since there are $n!$ permutations, the sum has $n!$ terms. The sign $(-1)^k$ is determined by the number k of inversions in the permutation; for example, in the product $a_{13}a_{21}a_{34}a_{42}$ the permutation $\begin{pmatrix}1&2&3&4\\3&1&4&2\end{pmatrix}$ has $k=3$ inversions (namely 3 before 1, 3 before 2 and 4 before 2), so that the sign is -1.

$$\begin{vmatrix} a_{11} & a_{12} & \cdots & a_{1n} \\ a_{21} & a_{22} & \cdots & a_{2n} \\ \vdots & \vdots & & \vdots \\ a_{n1} & a_{n2} & & a_{nn} \end{vmatrix}$$

Example 1: Computation of 2×2 determinants. The permutations $\begin{pmatrix}1&2\\1&2\end{pmatrix}$ and $\begin{pmatrix}1&2\\2&1\end{pmatrix}$ have 0 and 1 inversions, respectively.

$$\begin{vmatrix} a_{11} & a_{12} \\ a_{21} & a_{22} \end{vmatrix} = a_{11}a_{22} - a_{12}a_{21}.$$

Example 2: Computation of 3×3 determinants. The permutations π of three elements contain k inversions

$$D = \begin{vmatrix} a_{11} & a_{12} & a_{13} \\ a_{21} & a_{22} & a_{23} \\ a_{31} & a_{32} & a_{33} \end{vmatrix}$$

π	$\begin{pmatrix}1&2&3\\1&2&3\end{pmatrix}$	$\begin{pmatrix}1&2&3\\1&3&2\end{pmatrix}$	$\begin{pmatrix}1&2&3\\2&1&3\end{pmatrix}$	$\begin{pmatrix}1&2&3\\2&3&1\end{pmatrix}$	$\begin{pmatrix}1&2&3\\3&1&2\end{pmatrix}$	$\begin{pmatrix}1&2&3\\3&2&1\end{pmatrix}$
k	0	1	1	2	2	3

$$D = a_{11}a_{22}a_{33} - a_{11}a_{23}a_{32} - a_{12}a_{21}a_{33} + a_{12}a_{23}a_{31} + a_{13}a_{21}a_{32} - a_{13}a_{22}a_{31}.$$

These rules for calculating 2×2 and 3×3 determinants can be easily remembered in the following way. Write out the determinant and supplement the 3×3 determinant with the first two columns (Fig.).

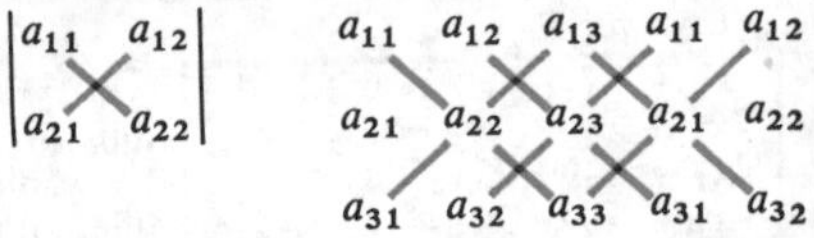

The numbers connected by red lines are multiplied and the products are added; from this the sum of the products of numbers connected by blue lines is subtracted. The result is the value of the determinant. For 3×3 determinants this is called *Sarrus' rule*.

Properties of determinants. The following statements can be derived from the definition of the determinant:

1. **The determinant is a linear function of the entries of each row.**
2. **If two rows are interchanged, the determinant changes sign.**
3. **The value of the determinant is zero if one of the rows is a linear combination of the others, in particular, if the entries in one row are all zero or if two rows are identical.**
4. **The value of the determinant does not change if to a given row a linear combination of the other rows is added.**
5. **The value of the determinant does not change if rows are made into columns and vice versa.**

The first statement means: (*i*) A factor common to all the entries of a row can be taken before the whole determinant. (*ii*) If the entries of a row can all be written as a sum of m elements, then the whole determinant can be written as a sum of m determinants in which the other rows remain unchanged. For example:

$$\begin{vmatrix} a_{11} & a_{12} & a_{13} \\ (a_{21}+b_{21}) & (a_{22}+b_{22}) & (a_{23}+b_{23}) \\ a_{31} & a_{32} & a_{33} \end{vmatrix} = \begin{vmatrix} a_{11} & a_{12} & a_{13} \\ a_{21} & a_{22} & a_{23} \\ a_{31} & a_{32} & a_{33} \end{vmatrix} + \begin{vmatrix} a_{11} & a_{12} & a_{13} \\ b_{21} & b_{22} & b_{23} \\ a_{31} & a_{32} & a_{33} \end{vmatrix}$$

The statements 3. and 4. are immediate consequences of 1. and 2., and 5. means that all theorems on rows of determinants are also true for their columns.

The rules 3. and 4. are frequently used in the practical computation of determinants.

Example 3: $\begin{vmatrix} 5 & 3 & -1 \\ 0 & 0 & 0 \\ 7 & 9 & 8 \end{vmatrix} = 0;$ $\begin{vmatrix} 1 & 1 & 1 \\ 2 & 4 & 5 \\ 1 & 1 & 1 \end{vmatrix} = 0;$ $\begin{vmatrix} 1 & 1 & 1 \\ 2 & 4 & 5 \\ 4 & 6 & 7 \end{vmatrix} = 0;$

$$\begin{vmatrix} 1 & 2 & 3 & 5 \\ 2 & -2 & 8 & 4 \\ 1 & 1 & -1 & 3 \\ 7 & 0 & 2 & 1 \end{vmatrix} = 2 \begin{vmatrix} 0+1 & 1+1 & 4-1 & 2+3 \\ 1 & -1 & 4 & 2 \\ 1 & 1 & -1 & 3 \\ 7 & 0 & 2 & 1 \end{vmatrix} = 2 \begin{vmatrix} 0 & 1 & 4 & 2 \\ 1 & -1 & 4 & 2 \\ 1 & 1 & -1 & 3 \\ 7 & 0 & 2 & 1 \end{vmatrix} +$$

$$+2 \begin{vmatrix} 1 & 1 & -1 & 3 \\ 1 & -1 & 4 & 2 \\ 1 & 1 & -1 & 3 \\ 7 & 0 & 2 & 1 \end{vmatrix} = 2 \begin{vmatrix} 0 & 1 & 4 & 2 \\ (1-0) & (-1-1) & (4-4) & (2-2) \\ 1 & 1 & -1 & 3 \\ 7 & 0 & 2 & 1 \end{vmatrix} = 2 \begin{vmatrix} 0 & 1 & 4 & 2 \\ 1 & -2 & 0 & 0 \\ 1 & 1 & -1 & 3 \\ 7 & 0 & 2 & 1 \end{vmatrix} \quad \text{(see Example 5)}$$

Minors. If any m rows and any m columns of an $n \times n$ determinant are deleted, the resulting $(n-m) \times (n-m)$ determinant is called a *minor* of the original one.

Example 4: Deleting the 2nd and 5th rows and 2nd and 4th columns of the larger determinant yields the 3×3 minor.

$$\begin{vmatrix} a_{11} & a_{12} & a_{13} & a_{14} & a_{15} \\ a_{21} & a_{22} & a_{23} & a_{24} & a_{25} \\ a_{31} & a_{32} & a_{33} & a_{34} & a_{35} \\ a_{41} & a_{42} & a_{43} & a_{44} & a_{45} \\ a_{51} & a_{52} & a_{53} & a_{54} & a_{55} \end{vmatrix} \longrightarrow \begin{vmatrix} a_{11} & a_{13} & a_{15} \\ a_{31} & a_{33} & a_{35} \\ a_{41} & a_{43} & a_{45} \end{vmatrix}$$

If only the ith row and the jth column are deleted, one obtains an $(n-1) \times (n-1)$ minor. If this determinant is multiplied by the sign $(-1)^{i+j}$, then its value is called the *cofactor* or *algebraic complement* of the element a_{ij} and is written A_{ij}.

Computation of determinants. The cofactors play an important part in the computation of determinants as the following theorem shows:

Theorem (development of a determinant by rows): Every determinant D can be computed from the elements of any fixed row and their cofactors: $D = a_{i1}A_{i1} + a_{i2}A_{i2} + \cdots + a_{in}A_{in}$

By 5. the analogous statement for columns is also true.

Example 5: Development by the second row.

$$\begin{vmatrix} 0 & 1 & 4 & 2 \\ 1 & -2 & 0 & 0 \\ 1 & 1 & -1 & 3 \\ 7 & 0 & 2 & 1 \end{vmatrix} = 1 \cdot (-1)^{2+1} \begin{vmatrix} 1 & 4 & 2 \\ 1 & -1 & 3 \\ 0 & 2 & 1 \end{vmatrix} + (-2)\,(-1)^{2+2} \begin{vmatrix} 0 & 4 & 2 \\ 1 & -1 & 3 \\ 7 & 2 & 1 \end{vmatrix}$$

$$+\,0 \cdot (-1)^{2+3} \begin{vmatrix} 0 & 1 & 2 \\ 1 & 1 & 3 \\ 7 & 0 & 1 \end{vmatrix} + 0 \cdot (-1)^{2+4} \begin{vmatrix} 0 & 1 & 4 \\ 1 & 1 & -1 \\ 7 & 0 & 2 \end{vmatrix} = -189.$$

Thus, the development theorem reduces the calculation of $n \times n$ determinants to the calculation of $(n-1) \times (n-1)$ determinants. Determinants with 2 or 3 rows can be computed directly by this method.

The example shows that it is advantageous to develop by a row with many zeros. The rules 1. and 5. can frequently be used to obtain such a row.

Solution of systems of linear equations by determinants. If the coefficients a_{ij} of a system of n linear equations in n unknowns are written (in the order in which they appear in the system) as the entries of a determinant D, then the determinant D_j is defined as the one obtained from D by deleting the jth column and replacing it by the column of constants on the right-hand side of the system. The determinants D and D_j can be used to solve the equations provided that the value of D is not zero.

Cramer's rule. If the determinant D of the coefficients of a system of n linear equations in n unknowns is not zero, then $\bar{x}_j = D_j/D$ ($j = 1, 2, \ldots, n$) is the only solution of that system.

Example 6: (Example 3 of solution by elimination):

$$\left.\begin{aligned} 3x_1 - 3x_2 + x_3 &= 0 \\ 4x_2 - x_3 &= 5 \\ 2x_1 - 2x_2 + x_3 &= 1 \end{aligned}\right|, \quad D = \begin{vmatrix} 3 & -3 & 1 \\ 0 & 4 & -1 \\ 2 & -2 & 1 \end{vmatrix} = 4, \quad D_1 = \begin{vmatrix} 0 & -3 & 1 \\ 5 & 4 & -1 \\ 1 & -2 & 1 \end{vmatrix} = 4,$$

$$D_2 = \begin{vmatrix} 3 & 0 & 1 \\ 0 & 5 & -1 \\ 2 & 1 & 1 \end{vmatrix} = 8, \quad D_3 = \begin{vmatrix} 3 & -3 & 0 \\ 0 & 4 & 5 \\ 2 & -2 & 1 \end{vmatrix} = 12; \quad \begin{aligned} \bar{x}_1 &= D_1/D = 1, \\ \bar{x}_2 &= D_2/D = 2, \\ \bar{x}_3 &= D_3/D = 3. \end{aligned}$$

In a homogenous system all the determinants D_j are necessarily zero. Therefore such a system has non-trivial solutions only if D is 0. The converse is also true.

A homogeneous system of n linear equations in n unknowns has non-trivial solutions if and only if the determinant of its coefficients is zero.

17.3. Vector spaces

Introduction. The exposition in the section on 'Homogeneous equations' shows that the solutions of a homogeneous system form an example of a set of objects that can be added together or multiplied by numbers without leaving the set. A generalization and abstract definition of these properties leads to the concept of a *vector space*, which is central for the whole of linear algebra.

A vector space is a set of objects or elements that can be added together and multiplied by numbers (the result being an element of the set), in such a way that the usual rules of calculation hold.

Linear algebra can be regarded as the theory of vector spaces. The elements of a vector space are called *vectors*, the numbers by which they can be multiplied are called *scalars*. The set of scalars can be the rational, real or complex numbers. Other more general structures can also be used (fields; see Chapter 16.). In what follows the set of scalars will always be taken to be the real numbers. The characteristic rules in a vector space V are the following ($\mathbf{x}, \mathbf{y}, \mathbf{z}$ are elements of the vector space V, that is, vectors, and a and b are scalars).

1. *Associative law of addition:* $(\mathbf{x} + \mathbf{y}) + \mathbf{z} = \mathbf{x} + (\mathbf{y} + \mathbf{z})$.
2. *Commutative law of addition:* $\mathbf{x} + \mathbf{y} = \mathbf{y} + \mathbf{x}$.
3. *Existence of zero:* There exists an element $\mathbf{o}$ in V such that $\mathbf{x} + \mathbf{o} = \mathbf{x}$ for all $\mathbf{x}$ in V.
4. *Existence of inverses:* To every $\mathbf{x}$ in V there exists an element $-\mathbf{x}$ in V such that $\mathbf{x} + (-\mathbf{x}) = \mathbf{o}$.
5. *Associative law of multiplication:* $a(b\mathbf{x}) = (ab)\,\mathbf{x}$.
6. *Unital law:* $1\mathbf{x} = \mathbf{x}$.
7. *First distributive law:* $a(\mathbf{x} + \mathbf{y}) = a\mathbf{x} + a\mathbf{y}$.
8. *Second distributive law:* $(a + b)\,\mathbf{x} = a\mathbf{x} + b\mathbf{x}$.

Every set in which an addition of the elements of the set and a multiplication by scalars are defined so that the results always lie in the set and the laws 1. to 8. hold is a vector space.

Examples of vector spaces. 1. The set of all *polynomial functions* (*integral rational functions*) forms a vector space. If $f(x) = a_nx^n + \cdots + a_1x + a_0$ and $g(x) = b_mx^m + \cdots + b_1x + b_0$ are polynomial functions and $n \geq m$, say, then their sum is $f(x) + g(x) = a_nx^n + \cdots + a_{m+1}x^{m+1} + (a_m + b_m)\,x^m + \cdots + (a_1 + b_1)\,x + (a_0 + b_0)$. The product $a \cdot f(x)$ of a real number and a polynomial function is $a \cdot f(x) = (aa_n)\,x^n + \cdots + (aa_1)\,x + (aa_0)$. It is now easy to verify the rules 1. to 8. The zero element $\mathbf{o}$ is the polynomial function $f(x) = 0$.

2. The set of all *differentiable functions* and also the set of *integrable functions* form vector spaces. The zero element is again the function $f(x) = 0$. The functions are added by adding their values and multiplied by a number by multiplying their values by that number.

3. The sets of *real numbers* and *complex numbers* form vector spaces with the usual multiplication and addition.

4. The set of n-tuples $(a_1, a_2, \ldots, a_n)$ with real entries a_i form a vector space $\mathbf{R}^n$ for every natural number n. For $n = 2$ they are also called *ordered pairs*, for $n = 3$ *triples* and for $n = 4$ *quadruples*. Addition is defined by $(a_1, a_2, \ldots, a_n) + (b_1, b_2, \ldots, b_n) = (a_1 + b_1, a_2 + b_2, \ldots, a_n + b_n)$ and multiplication by $a(a_1, a_2, \ldots, a_n) = (aa_1, aa_2, \ldots, aa_n)$.

Vector algebra

The vector space V_3. In this paragraph the properties of a particulary important vector space are investigated. It plays a central part in physics and technology; and it clarifies the importance of vector spaces and thus of linear algebra in practical applications. The name vector was first used for elements of this particular space and later generalized to the present terminology.

The vectors will first be described geometrically, starting with the three-dimensional space, in which length, breadth, and height are defined. A *shift* of the space consists in associating with each point P a point Q such that the (oriented) line segments connecting points with their images are parallel and all have the same length. Such a shift is called a *translation or vector*. From the definition it follows that a vector is completely determined if its effect on a single point P is known, that is, if the point Q associated with P is known. Therefore the vector can be characterized by drawing the line segment $\overrightarrow{PQ}$ and putting an arrow-head at Q to indicate that P goes to Q. Such an oriented segment is called a *representative* of the vector. Here P is called the *initial point* or *point of application* of the vector and Q is called its *end-point* (Fig.).

A vector is a shift of three-dimensional space.

For every point P there is a representative of any given vector with P as its initial point, and every point Q also occurs as end-point of a suitable representative. Different representatives of the same vector are parallel and have the same length. Hence it makes sense to define the *length* or *modulus* or *norm* of a vector $\boldsymbol{a}$ as the distance between the points P and Q of any representative of $\boldsymbol{a}$. The length of $\boldsymbol{a}$ is denoted by $|\boldsymbol{a}|$ or $|\overrightarrow{PQ}|$. It is always non-negative. Vectors of length 1 are called *unit vectors*.

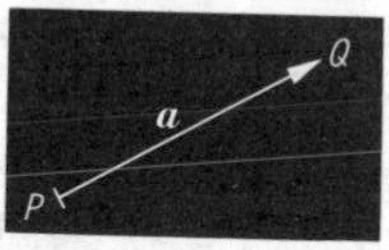

17.3-1 Representative of a vector

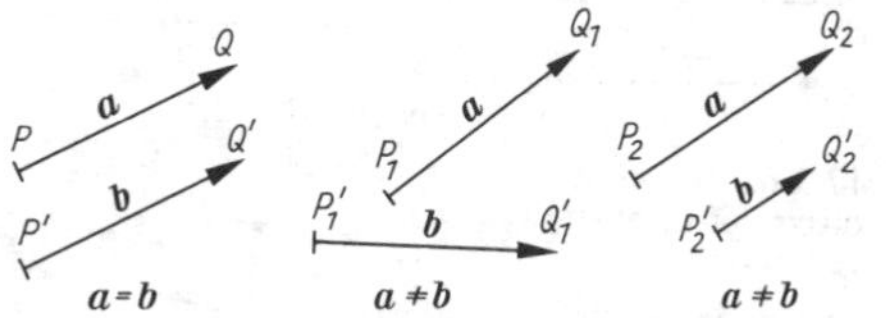

17.3-2 Equality of vectors

In what follows most of the other concepts of vector algebra will be introduced by representatives. But it is incorrect to identify a vector with a single representative. If this is done, one usually attempts to avoid the resulting difficulties by some sentence such as 'vectors can be shifted in any direction'. But this sentence really means that by a parallel displacement of a representative of a vector one obtains another representative of the same vector. This leads to the following proposition.

Two vectors are equal if and only if any two of their representatives are equal in length and direction. All parallel segments of equal length and orientation are representatives of one and the same vector (Fig.).

To obtain a vector space, addition of vectors and multiplication by scalars must still be defined.

Vector addition. The sum of two vectors $\boldsymbol{a}$ and $\boldsymbol{b}$ is the shift $\boldsymbol{a} + \boldsymbol{b}$ obtained by performing the shifts $\boldsymbol{a}$ and $\boldsymbol{b}$ in succession. Using representatives the sum can be defined in the following way: If $\overrightarrow{PQ}$ is a representative of $\boldsymbol{a}$ in P and $\overrightarrow{QR}$ is a representative of $\boldsymbol{b}$ in Q, then $\overrightarrow{PR}$ is a representative of $\boldsymbol{a} + \boldsymbol{b}$ (Fig.). It is easy to see that this definition does not depend on the choice of representatives.

If one considers the representatives $\overrightarrow{PQ}$, $\overrightarrow{PQ'}$, $\overrightarrow{Q'R}$ and $\overrightarrow{QR}$ of $\boldsymbol{a}$ and $\boldsymbol{b}$ in P, $\boldsymbol{a}$ in Q', and $\boldsymbol{b}$ in Q, one obtains a parallelogram with its diagonal $\overrightarrow{PR}$ representing both $\boldsymbol{a} + \boldsymbol{b}$ and $\boldsymbol{b} + \boldsymbol{a}$ (Fig.). Consequently $\boldsymbol{a} + \boldsymbol{b} = \boldsymbol{b} + \boldsymbol{a}$ and the *commutative law of addition* holds. It is just as easy to verify the *associative law of addition* $(\boldsymbol{a} + \boldsymbol{b}) + \boldsymbol{c} = \boldsymbol{a} + (\boldsymbol{b} + \boldsymbol{c})$ (Fig.).

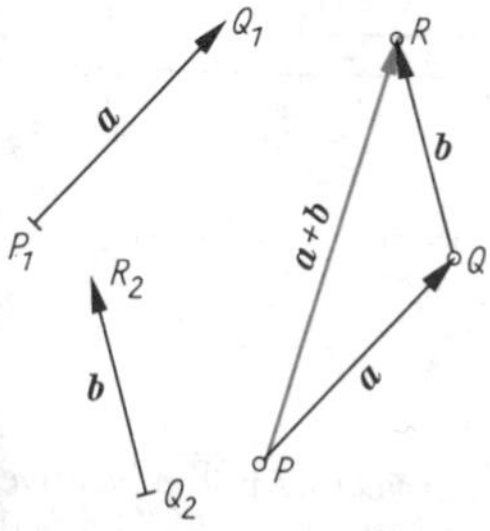

17.3-3 Addition of vectors

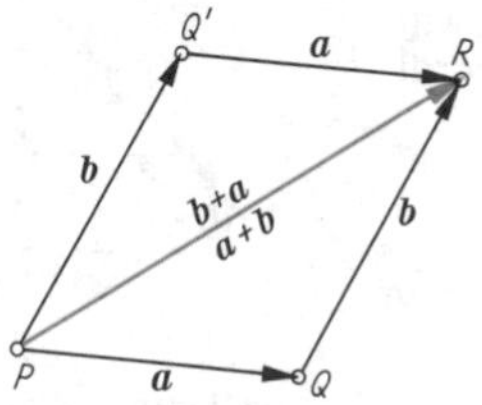

17.3-4 Commutativity of vector addition

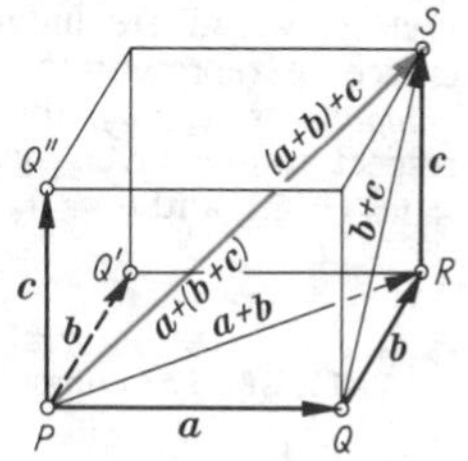

17.3-5 Associativity of vector addition
$\boldsymbol{a} + (\boldsymbol{b} + \boldsymbol{c}) = (\boldsymbol{a} + \boldsymbol{b}) + \boldsymbol{c}$

The validity of these laws implies the following rules for adding more than two vectors:

Several vectors can be added by choosing any sequence of representatives, one of each vector, such that the end-point of each representative is the initial point of the next. The sum or resultant is the vector represented by the segment going from the initial point of the first representative to the end-point of the last.

The null (or zero) vector. The translation that shifts a point P to P itself, and thus leaves all points of space fixed is the *null vector*, written $\boldsymbol{o}$. It cannot be given a particular direction, its length is 0. It has the characteristic property that $\boldsymbol{a} + \boldsymbol{o} = \boldsymbol{a}$ for all vectors $\boldsymbol{a}$.

Subtraction. To define subtraction of vectors one uses the existence of a unique inverse to every vector. If $\boldsymbol{a} + \boldsymbol{b} = \boldsymbol{o}$, then $\boldsymbol{b} = -\boldsymbol{a}$ and a representative of $\boldsymbol{b}$ is obtained by interchanging the initial and end-points of a representative of $\boldsymbol{a}$ (Fig.). Thus, $-\boldsymbol{a}$ has the same length as $\boldsymbol{a}$, but the opposite direction. In particular $\boldsymbol{o} = -\boldsymbol{o}$.

It now makes sense to say that the difference $\boldsymbol{a} - \boldsymbol{b}$ of two vectors is the sum of $\boldsymbol{a}$ and $-\boldsymbol{b}$, $\boldsymbol{a} - \boldsymbol{b} = \boldsymbol{a} + (-\boldsymbol{b})$.

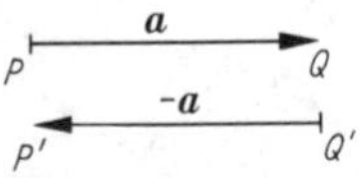

17.3-6 Representatives of opposite vectors

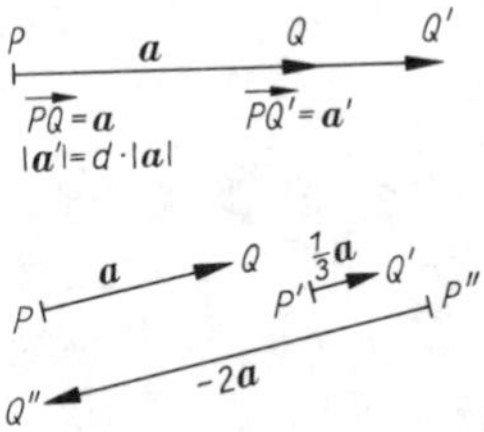

17.3-7 Multiplication of vectors by scalars

Multiplication of vectors by scalars. If the points P, Q and Q' ($P \neq Q$, $P \neq Q'$) lie on a line, then $\overrightarrow{PQ}$ and $\overrightarrow{PQ'}$ are representatives of vectors $\boldsymbol{a}$ and $\boldsymbol{a}'$ in the same direction or in opposite directions. However, in general, the length of $\boldsymbol{a}$ and $\boldsymbol{a}'$ will be different. But there is a real number $d > 0$ such that $|\boldsymbol{a}'| = d \cdot |\boldsymbol{a}|$, namely $d = |\boldsymbol{a}'|/|\boldsymbol{a}|$ (Fig.).

If $\boldsymbol{a}$ and $\boldsymbol{a}'$ have the same direction, one defines $\boldsymbol{a}' = d\boldsymbol{a}$, with $d = |\boldsymbol{a}'|/|\boldsymbol{a}|$, if they have opposite directions, one defines $\boldsymbol{a}' = -d\boldsymbol{a}$. This leads to the following definition, which also covers the cases $\boldsymbol{a} = \boldsymbol{o}$ or $d = 0$. The *product* $d \cdot \boldsymbol{a}$ of a vector $\boldsymbol{a}$ by a real number d is the vector of length $d \cdot |\boldsymbol{a}|$, in the same direction as $\boldsymbol{a}$ if $d > 0$, and in the opposite direction if $d < 0$. For $d = 0$ one defines $0 \cdot \boldsymbol{a} = \boldsymbol{o}$.

In particular, it follows that $1 \cdot \boldsymbol{a} = \boldsymbol{a}$, $(-1) \cdot \boldsymbol{a} = -\boldsymbol{a}$, $d \cdot \boldsymbol{o} = \boldsymbol{o}$ for all d and $n \cdot \boldsymbol{a} = \underbrace{\boldsymbol{a} + \boldsymbol{a} + \cdots + \boldsymbol{a}}_{n \text{ times}}$ for every natural number n. If $\boldsymbol{a} \neq \boldsymbol{o}$, then vector $\boldsymbol{a}/|\boldsymbol{a}|$ has the length 1, hence is a unit vector, which in what follows will be denoted by $\boldsymbol{a}^0$. Thus, $\boldsymbol{a} = |\boldsymbol{a}| \cdot \boldsymbol{a}^0$. The *associative* law of multiplication and the two *distributive laws* are easily verified. In this way a *vector space* has been corstructed *consisting of the translations of three-dimensional space. It will be denoted by* V_3.

Components and Coordinates in V_3. To make geometry amenable to computational methods one introduces a system of coordinates, for example, an orthogonal (Cartesian) system with x-, y-, and z-axes. The perpendicular projections onto the axes of a representative $\overrightarrow{PQ}$ of a vector $\boldsymbol{a}$ are again representatives of certain vectors. These vectors, which are independent of the choice of representative, are called the *components* $\boldsymbol{a}_x$, $\boldsymbol{a}_y$ and $\boldsymbol{a}_z$ of $\boldsymbol{a}$ with respect to the given system of coordinates (Fig.), and $\boldsymbol{a} = \boldsymbol{a}_x + \boldsymbol{a}_y + \boldsymbol{a}_z$.

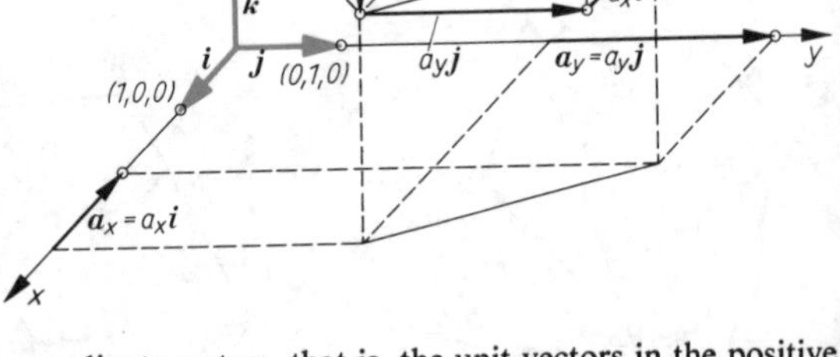

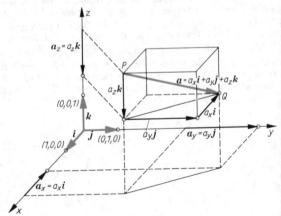

17.3-8 The components of a vector

If $\boldsymbol{i}$, $\boldsymbol{j}$, and $\boldsymbol{k}$ are the *basis vectors* of the coordinate system, that is, the unit vectors in the positive directions of the x-, y-, and z-axes, respectively, then $\boldsymbol{a}_x = a_x \cdot \boldsymbol{i}$, $\boldsymbol{a}_y = a_y \cdot \boldsymbol{j}$, $\boldsymbol{a}_z = a_z \cdot \boldsymbol{k}$. The real numbers a_x, a_y, and a_z are called the *coordinates* of $\boldsymbol{a}$ with respect to the given coordinate system.

Components of $\boldsymbol{a}$: $\boldsymbol{a}_x, \boldsymbol{a}_y, \boldsymbol{a}_z$ (vectors)	$\boldsymbol{a} = \boldsymbol{a}_x + \boldsymbol{a}_y + \boldsymbol{a}_z$
Coordinates of $\boldsymbol{a}$: a_x, a_y, a_z (numbers)	$\boldsymbol{a} = a_x\boldsymbol{i} + a_y\boldsymbol{j} + a_z\boldsymbol{k}$
Norm of $\boldsymbol{a}$ with coordinates a_x, a_y, a_z	$\lvert\boldsymbol{a}\rvert = \sqrt{(a_x^2 + a_y^2 + a_z^2)}$
Components and norm of the vector $\boldsymbol{a}$ represented by $\overrightarrow{PQ}$, $P = (x_0, y_0, z_0)$, $Q = (x_1, y_1, z_1)$	$\boldsymbol{a} = (x_1 - x_0)\,\boldsymbol{i} + (y_1 - y_0)\,\boldsymbol{j} + (z_1 - z_0)\,\boldsymbol{k}$ $\lvert\boldsymbol{a}\rvert = \sqrt{[(x_1 - x_0)^2 + (y_1 - y_0)^2 + (z_1 - z_0)^2]}$

If $\overrightarrow{PQ}$ represents $\boldsymbol{a}$ and P and Q have the coordinates (x_0, y_0, z_0) and (x_1, y_1, z_1), respectively, then the components of $\boldsymbol{a}$ are $(x_1 - x_0)\,\boldsymbol{i}$, $(y_1 - y_0)\,\boldsymbol{j}$, and $(z_1 - z_0)\,\boldsymbol{k}$, respectively. Thus, the coordinates of $\boldsymbol{a}$ are the differences between the coordinates of the end-point and initial point of an arbitrary representative of $\boldsymbol{a}$.

Since every vector determines a triple of coordinates and, on the other hand, every triple (a_1, a_2, a_3) determines a unique vector $\boldsymbol{a}$ with $a_x = a_1$, $a_y = a_2$, and $a_z = a_3$, the vector space V_3 can be identified with the vector space of triples of real numbers $\mathbf{R}^3$. But for this to make sense it must be shown that addition and scalar multiplication in V^3 corresponds to addition and scalar multiplication in $\mathbf{R}^3$, in other words, that the coordinates of $\boldsymbol{a} + \boldsymbol{b}$ and $d \cdot \boldsymbol{a}$ are $(a_x + b_x, a_y + b_y, a_z + b_z)$ and (da_x, da_y, da_z), respectively. And indeed, from $\boldsymbol{a} = a_x\boldsymbol{i} + a_y\boldsymbol{j} + a_z\boldsymbol{k}$ and $\boldsymbol{b} = b_x\boldsymbol{i} + b_y\boldsymbol{j} + b_z\boldsymbol{k}$ it follows that $\boldsymbol{a} + \boldsymbol{b} = (a_x + b_x)\,\boldsymbol{i} + (a_y + b_y)\,\boldsymbol{j} + (a_z + b_z)\,\boldsymbol{k}$ and $d \cdot \boldsymbol{a} = (d \cdot a_x)\,\boldsymbol{i} + (d \cdot a_y)\boldsymbol{j} + (d \cdot a_z)\,\boldsymbol{k}$.

In the first case the commutative and associative laws of addition and the second distributive law are required; in the second the first distributive law and the associative law of multiplication. One says that addition and multiplication are performed *componentwise* or *coordinatewise*. Since $-\boldsymbol{a} = -1\boldsymbol{a}$ and thus has the coordinates $-a_x$, $-a_y$, and $-a_z$, subtraction is also performed componentwise.

Vectors are added (or subtracted) by adding (or subtracting) their coordinates; a vector is multiplied by a scalar by multiplying its coordinates by that scalar.

Example 1: For $\boldsymbol{a} = 2\boldsymbol{i} + (1/2)\,\boldsymbol{j} - \boldsymbol{k}$; $\boldsymbol{b} = -3\boldsymbol{i} + 2\boldsymbol{j} + 5\boldsymbol{k}$ and $d = 2$ one obtains $\boldsymbol{a} + \boldsymbol{b} = -\boldsymbol{i} + (5/2)\,\boldsymbol{j} + 4\boldsymbol{k}$; $-\boldsymbol{b} = 3\boldsymbol{i} - 2\boldsymbol{j} - 5\boldsymbol{k}$; $\boldsymbol{a} - \boldsymbol{b} = 5\boldsymbol{i} - (3/2)\,\boldsymbol{j} - 6\boldsymbol{k}$; and $d\boldsymbol{a} = 4\boldsymbol{i} + \boldsymbol{j} - 2\boldsymbol{k}$.

Thus, there is a bijective (one-to-one and onto) map of V_3 onto $\mathbf{R}^3$, which preserves addition and scalar multiplication. The vectors $\boldsymbol{i}, \boldsymbol{j}, \boldsymbol{k}$ are taken to the triples (1, 0, 0). (0, 1, 0) and (0, 0, 1) and the arbitrary vector $\boldsymbol{a} = a_x\boldsymbol{i} + a_y\boldsymbol{j} + a_y\boldsymbol{k}$ goes to (a_x, a_y, a_z). Although the space V_3 is more intuitively defined, calculations are more convenient in $\mathbf{R}^3$, because they give a much clearer picture of the operations in V_3. V_3 and $\mathbf{R}^3$ have the same structure as vector spaces, but different sets of objects. In what follows they will not be regarded as the same space (see Linear maps).

The inner and vector product in V_3. In V_3 there are two further natural operations. One of them associates a scalar with any pair of vectors and is called the *inner* or *dot product*, because the product of $\boldsymbol{a}$ and $\boldsymbol{b}$ is written as $\boldsymbol{a} \cdot \boldsymbol{b}$. The other produces a vector and is called the *vector* or *cross product*, because this product of $\boldsymbol{a}$ and $\boldsymbol{b}$ is written as $\boldsymbol{a} \times \boldsymbol{b}$. The inner product can be generalized to other vector spaces, but this is not immediately possible for the vector product.

Both products have physical applications. For instance, the work done by a force $\boldsymbol{F}$ moving along a straight path $\boldsymbol{s}$ is calculated by the dot product $\boldsymbol{F} \cdot \boldsymbol{s}$, and the velocity $\boldsymbol{v}$ of a point P on a body rotating about an axis is calculated as the vector product of the radius from the axis to P and the angular velocity.

The inner product. The representatives with the same initial point P of two non-zero vectors $\boldsymbol{a}$ and $\boldsymbol{b}$ enclose an angle α between 0° and 180° and an angle β between 180° and 360° such that $\alpha + \beta = 360°$. The angle $\measuredangle(\boldsymbol{a}, \boldsymbol{b})$ between the vectors $\boldsymbol{a}$ and $\boldsymbol{b}$, is defined as the smaller angle α.

The inner or dot product $\boldsymbol{a} \cdot \boldsymbol{b}$ (read '$\boldsymbol{a}$ dot $\boldsymbol{b}$') of two non-zero vectors $\boldsymbol{a}$ and $\boldsymbol{b}$ is defined as the real number $\lvert\boldsymbol{a}\rvert\,\lvert\boldsymbol{b}\rvert \cos \measuredangle(\boldsymbol{a}, \boldsymbol{b})$.

The *commutative law* $\boldsymbol{a} \cdot \boldsymbol{b} = \boldsymbol{b} \cdot \boldsymbol{a}$ holds; one says the inner product is *symmetric*. The *associative law* does not hold, in general, for the product of three vectors, because $(\boldsymbol{a} \cdot \boldsymbol{b}) \cdot \boldsymbol{c}$ is a multiple of $\boldsymbol{c}$, whereas $\boldsymbol{a} \cdot (\boldsymbol{b} \cdot \boldsymbol{c})$ is, if anything, a multiple of $\boldsymbol{a}$. On the other hand, there is a *composition law* for the multiplication by scalars and the dot product. Using the commutative law as well one obtains the following equation:

$$(a\boldsymbol{b} \cdot \boldsymbol{c}) = a(\boldsymbol{b} \cdot \boldsymbol{c}) = a(\boldsymbol{c} \cdot \boldsymbol{b}) = (a\boldsymbol{c} \cdot \boldsymbol{b}) = (\boldsymbol{b} \cdot a\boldsymbol{c}).$$

An operation on vectors satisfying this law and the distributive laws is called *bilinear*. The *distributive law* always holds $\boldsymbol{a} \cdot (\boldsymbol{b} + \boldsymbol{c}) = \boldsymbol{a} \cdot \boldsymbol{b} + \boldsymbol{a} \cdot \boldsymbol{c}$.

Two vectors $\boldsymbol{a}$ and $\boldsymbol{b}$ are called *orthogonal* if $\boldsymbol{a} \cdot \boldsymbol{b} = 0$. If both $\boldsymbol{a}$ and $\boldsymbol{b}$ are non-zero, this means that $\cos \measuredangle(\boldsymbol{a}, \boldsymbol{b}) = 0$ or $\alpha = 90°$, in other words, the representatives of $\boldsymbol{a}$ and $\boldsymbol{b}$ are perpendicular to one another.

The dot product cannot be inverted, for one cannot define a unique vector $\boldsymbol{a}$ in a meaningful way such that $\boldsymbol{a}\cdot\boldsymbol{b}=c$ (or such that '$\boldsymbol{a}$ is the quotient of the real number c and the vector $\boldsymbol{b}$'). For given $\boldsymbol{b}$ and c there are always infinitely many vectors $\boldsymbol{a}$ satisfying the equation $\boldsymbol{a}\cdot\boldsymbol{b}=c$; for example, if $c=0$, then any multiple of a vector $\boldsymbol{a}\neq\boldsymbol{o}$ orthogonal to $\boldsymbol{b}$ will do. *Division by vectors is not permissible.*

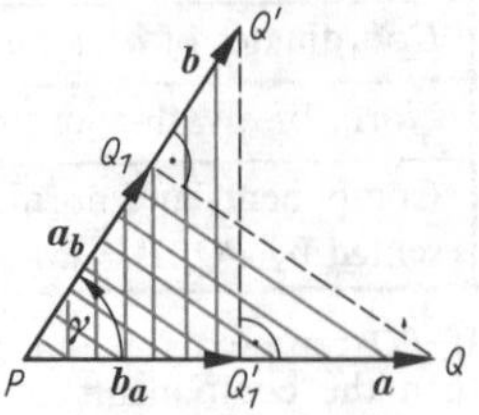

17.3-9 Projection of a vector onto another

In the dot product $\boldsymbol{a}\cdot\boldsymbol{b}$, the vector $\boldsymbol{a}$ can be replaced by the vector $\boldsymbol{a}_b$ of length $|\boldsymbol{a}|\cdot\cos\sphericalangle(\boldsymbol{a},\boldsymbol{b})$ and in the same or the opposite direction to $\boldsymbol{b}$ according as $\sphericalangle(\boldsymbol{a},\boldsymbol{b})$ is less than or greater than 90°. A representative of $\boldsymbol{a}_b$ can be obtained by taking the perpendicular projection of a representative of $\boldsymbol{a}$ onto the line through a representative of $\boldsymbol{b}$, with the same initial point (Fig.). The same can be done for $\boldsymbol{b}$. Thus, $\boldsymbol{a}\cdot\boldsymbol{b}=\boldsymbol{a}_b\cdot\boldsymbol{b}=\boldsymbol{b}_a\cdot\boldsymbol{a}$. The product $\boldsymbol{a}_b\cdot\boldsymbol{b}$ has the desirable property that $\boldsymbol{a}_b\cdot\boldsymbol{b}=|\boldsymbol{a}_b|\cdot|\boldsymbol{b}|$, whereas $\boldsymbol{a}\cdot\boldsymbol{b}=|\boldsymbol{a}|\cdot|\boldsymbol{b}|\cdot\cos\sphericalangle(\boldsymbol{a},\boldsymbol{b})$.

The dot product in coordinates. To calculate the dot product in coordinates all one needs are the values of the dot products $\boldsymbol{i}\cdot\boldsymbol{i}$, $\boldsymbol{i}\cdot\boldsymbol{j}$, ..., $\boldsymbol{k}\cdot\boldsymbol{k}$ (and the distributive and associative laws). From the definition these are easily seen to be 1 if the two factors are equal, and 0 otherwise. One can thus compute the dot product of $\boldsymbol{a}$ and $\boldsymbol{b}$ from their coordinates without knowing the angle between them, and indeed one can use this to obtain the angle between two vectors from their dot product.

Inner products of the basis vectors	$\boldsymbol{i}\cdot\boldsymbol{i}=\boldsymbol{j}\cdot\boldsymbol{j}=\boldsymbol{k}\cdot\boldsymbol{k}=1$	$\boldsymbol{i}\cdot\boldsymbol{j}=\boldsymbol{i}\cdot\boldsymbol{k}=\boldsymbol{j}\cdot\boldsymbol{k}=0$
Inner product $\boldsymbol{a}\cdot\boldsymbol{b}$ in coordinates	$\boldsymbol{a}=a_x\boldsymbol{i}+a_y\boldsymbol{j}+a_z\boldsymbol{k}$, $\boldsymbol{b}=b_x\boldsymbol{i}+b_y\boldsymbol{j}+b_z\boldsymbol{k}$	$\boldsymbol{a}\cdot\boldsymbol{b}=a_xb_x+a_yb_y+a_zb_z$
Angles, if $\lvert\boldsymbol{a}\rvert$ and $\lvert\boldsymbol{b}\rvert$ are $\neq 0$	$\cos\sphericalangle(\boldsymbol{a},\boldsymbol{b})=\dfrac{\boldsymbol{a}\cdot\boldsymbol{b}}{\lvert\boldsymbol{a}\rvert\,\lvert\boldsymbol{b}\rvert}=$	$\dfrac{a_xb_x+a_yb_y+a_zb_z}{\sqrt{[(a_x^2+a_y^2+a_z^2)(b_x^2+b_y^2+b_z^2)]}}$

Example 2: From $\boldsymbol{a}=3\boldsymbol{i}-4\boldsymbol{j}$ and $\boldsymbol{b}=\boldsymbol{i}+2\boldsymbol{j}-2\boldsymbol{k}$ one obtains $\cos\sphericalangle(\boldsymbol{a},\boldsymbol{b})=-1/3$ and hence $\sphericalangle(\boldsymbol{a},\boldsymbol{b})\approx 109°28'$.

The vector product. The vector product $\boldsymbol{a}\times\boldsymbol{b}$ (read '$\boldsymbol{a}$ cross $\boldsymbol{b}$') of two non-zero vectors is defined to be the vector $\boldsymbol{c}$ that has the following properties 1. $\boldsymbol{a}\cdot\boldsymbol{c}=\boldsymbol{b}\cdot\boldsymbol{c}=0$, that is, $\boldsymbol{c}$ is orthogonal to both $\boldsymbol{a}$ and $\boldsymbol{b}$. 2. $|\boldsymbol{c}|=\boldsymbol{a}\cdot\boldsymbol{b}\sin\sphericalangle(\boldsymbol{a},\boldsymbol{b})$ and 3. the determinant $\begin{vmatrix}a_x & a_y & a_z\\ b_x & b_y & b_z\\ c_x & c_y & c_z\end{vmatrix}$, formed from the coordinates of $\boldsymbol{a}$, $\boldsymbol{b}$ and $\boldsymbol{c}$ in the given way, is non-negative. If $\boldsymbol{a}$ or $\boldsymbol{b}$ is the null vector, then $\boldsymbol{a}\times\boldsymbol{b}$ is also the null vector.

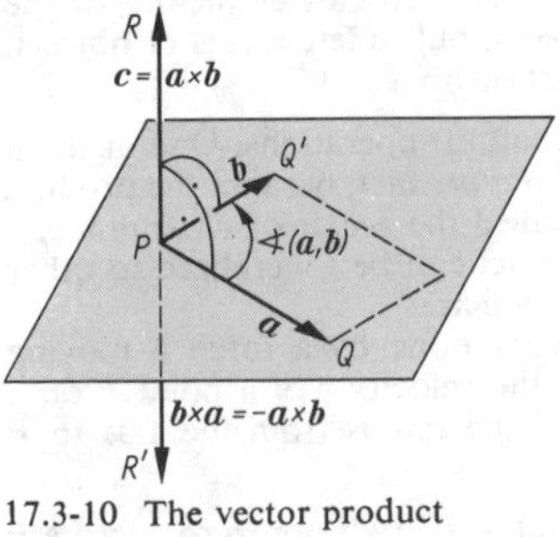

17.3-10 The vector product

To find a geometrical interpretation of the vector product of two non-zero vectors that are not scalar multiples of one another, consider the plane spanned by two of their respective representatives, $\overrightarrow{PQ}$ and $\overrightarrow{PQ'}$ (Fig.). Then the properties given above have the following meaning for the representative $\overrightarrow{PR}$ of $\boldsymbol{c}$:

1. $\overrightarrow{PR}$ is perpendicular to the plane spanned by $\overrightarrow{PQ}$ and $\overrightarrow{PQ'}$.
2. The length of $\overrightarrow{PR}$ is $|\boldsymbol{a}|\,|\boldsymbol{b}|\sin\sphericalangle(\boldsymbol{a},\boldsymbol{b})$, which is the area of the parallelogram with the sides $\overrightarrow{PQ}$ and $\overrightarrow{PQ'}$.
3. $\overrightarrow{PQ}$, $\overrightarrow{PQ'}$, $\overrightarrow{PR}$ form a *right-handed* system. This means: viewed from R, the shorter of the two possible rotations taking $\overrightarrow{PQ}$ to $\overrightarrow{PQ'}$ is anti-clockwise (if the thumb of the right hand points in the direction of $\overrightarrow{PQ}$ and the first finger in the direction of $\overrightarrow{PQ'}$ then the palm faces the direction of $\overrightarrow{PR}$).

If $\overrightarrow{PQ}$, $\overrightarrow{PQ'}$, $\overrightarrow{PR}$, $\overrightarrow{PR'}$, are representatives of the vectors $\boldsymbol{a}$, $\boldsymbol{b}$, $\boldsymbol{a}\times\boldsymbol{b}$, and $\boldsymbol{b}\times\boldsymbol{a}$, then $\overrightarrow{PR}$ and $\overrightarrow{PR'}$ are both perpendicular to the plane spanned by $\overrightarrow{PQ}$ and $\overrightarrow{PQ'}$, and they both have the same length. But since both $\overrightarrow{PQ}$, $\overrightarrow{PQ'}$, $\overrightarrow{PR}$ and $\overrightarrow{PQ'}$, $\overrightarrow{PQ}$, $\overrightarrow{PR'}$ form right-handed systems in the given orders, $\overrightarrow{PR}$ and $\overrightarrow{PR'}$ must point in opposite directions. Thus, $\boldsymbol{a}\times\boldsymbol{b}=-\boldsymbol{b}\times\boldsymbol{a}$. This law is known as *anti-commutativity*. The vector product is not commutative or associative, but it is *bilinear*, that is, if a is a scalar and $\boldsymbol{x}$, $\boldsymbol{y}$, and $\boldsymbol{z}$ are vectors, then $a(\boldsymbol{x}\times\boldsymbol{y})=a\boldsymbol{x}\times\boldsymbol{y}=\boldsymbol{x}\times a\boldsymbol{y}$ and $\boldsymbol{x}\times(\boldsymbol{y}+\boldsymbol{z})=(\boldsymbol{x}\times\boldsymbol{y})+(\boldsymbol{x}\times\boldsymbol{z})$ and $(\boldsymbol{x}+\boldsymbol{y})\times\boldsymbol{z}=(\boldsymbol{x}\times\boldsymbol{z})+(\boldsymbol{y}\times\boldsymbol{z})$.

The vector product in coordinates. The definition immediately gives the values for the vector products of the basis vectors $\boldsymbol{i}$, $\boldsymbol{j}$, and $\boldsymbol{k}$. The bilinearity of the product then makes it possible to calculate the coordinates of $\boldsymbol{a} \times \boldsymbol{b}$. They are the determinants

$$\begin{vmatrix} a_y & a_z \\ b_y & b_z \end{vmatrix}, \quad -\begin{vmatrix} a_x & a_z \\ b_x & b_z \end{vmatrix}, \quad \text{and} \quad \begin{vmatrix} a_x & a_y \\ b_x & b_y \end{vmatrix}.$$

This can be easily remembered by using the following mnemonic device. Write down a 3×3 determinant in which the first row is $\boldsymbol{i}$, $\boldsymbol{j}$, $\boldsymbol{k}$, and the second and third rows are coefficients of $\boldsymbol{a}$ and $\boldsymbol{b}$. If the value is calculated and the terms containing $\boldsymbol{i}$, $\boldsymbol{j}$, and $\boldsymbol{k}$ are sorted, then these are the components of $\boldsymbol{a} \times \boldsymbol{b}$.

Vector products of the basis vectors	$\boldsymbol{i} \times \boldsymbol{i} = \boldsymbol{j} \times \boldsymbol{j} = \boldsymbol{k} \times \boldsymbol{k} = \boldsymbol{o}$ $\boldsymbol{i} \times \boldsymbol{j} = \boldsymbol{k}, \quad \boldsymbol{j} \times \boldsymbol{k} = \boldsymbol{i}, \quad \boldsymbol{k} \times \boldsymbol{i} = \boldsymbol{j}$
Components of $\boldsymbol{a} \times \boldsymbol{b}$	$\boldsymbol{a} \times \boldsymbol{b} = (a_y b_z - b_y a_z)\,\boldsymbol{i} + (a_z b_x - a_x b_z)\,\boldsymbol{j} + (a_x b_y - a_y b_x)\,\boldsymbol{k}$ $= \begin{vmatrix} \boldsymbol{i} & \boldsymbol{j} & \boldsymbol{k} \\ a_x & a_y & a_z \\ b_x & b_y & b_z \end{vmatrix}$

Examples 3: The vector product of $\boldsymbol{a} = 5\boldsymbol{i} - 3\boldsymbol{j} + \boldsymbol{k}$ and $\boldsymbol{b} = -\boldsymbol{i} - \boldsymbol{j} - 2\boldsymbol{k}$ is

$$\boldsymbol{a} \times \boldsymbol{b} = \begin{vmatrix} \boldsymbol{i} & \boldsymbol{j} & \boldsymbol{k} \\ 5 & -3 & 1 \\ -1 & -1 & 2 \end{vmatrix} = -5\boldsymbol{i} - 11\boldsymbol{j} - 8\boldsymbol{k}.$$

Basis and dimension. Several concepts that played a part in the discussion of V_3 can also be useful in the analysis of other vector spaces. Examples are the introduction of coordinates and representation of operations in coordinates. However, some ideas, such as the vector product, cannot be generalized.

Linearly dependent and independent vectors. If $\boldsymbol{x}_1, \boldsymbol{x}_2, \ldots, \boldsymbol{x}_n$ are vectors of a vector space V and $\boldsymbol{x}$ is a vector in V, $\boldsymbol{x}$ is said to be *linearly dependent* on $\boldsymbol{x}_1, \boldsymbol{x}_2, \ldots, \boldsymbol{x}_n$ if there are numbers $a_1, a_2, \ldots, a_n$, such that $\boldsymbol{x} = a_1\boldsymbol{x}_1 + a_2\boldsymbol{x}_2 + \cdots + a_n\boldsymbol{x}_n$. One also says that $\boldsymbol{x}$ depends on $\boldsymbol{x}_1, \boldsymbol{x}_2, \ldots, \boldsymbol{x}_n$ or that $\boldsymbol{x}$ is a *linear combination* of $\boldsymbol{x}_1, \boldsymbol{x}_2, \ldots, \boldsymbol{x}_n$. For instance, every vector $\boldsymbol{a}$ of V_3 depends on the system $\boldsymbol{x}_1 = \boldsymbol{i}$, $\boldsymbol{x}_2 = \boldsymbol{j}$, and $\boldsymbol{x}_3 = \boldsymbol{k}$: $\boldsymbol{a} = a_x\boldsymbol{i} + a_y\boldsymbol{j} + a_z\boldsymbol{k}$.

Obviously the zero vector $\boldsymbol{o}$ depends on any system of vectors $\boldsymbol{x}_1, \boldsymbol{x}_2, \ldots, \boldsymbol{x}_n$. One need only choose $a_1 = a_2 = \cdots = a_n = 0$.

If $\boldsymbol{x}$ is linearly dependent on $\boldsymbol{x}_1, \boldsymbol{x}_2, \ldots, \boldsymbol{x}_n$, then there are numbers $a_1, a_2, \ldots, a_n$, and $a = -1$ such that $\boldsymbol{o} = a_1\boldsymbol{x}_1 + a_2\boldsymbol{x}_2 + \cdots + a_n\boldsymbol{x}_n + a\boldsymbol{x}$. On the other hand, this equation does not state that $\boldsymbol{x}$ is dependent on $\boldsymbol{x}_1, \boldsymbol{x}_2, \ldots, \boldsymbol{x}_n$. What it means is that if at least one of the coefficients $a_1, a_2, \ldots, a_n$, or a is not 0, then the corresponding vector is dependent on the rest. This leads to the following definition.

A system $\boldsymbol{x}_1, \boldsymbol{x}_2, \ldots, \boldsymbol{x}_n$ of vectors is called *linearly dependent* if there are numbers $a_1, a_2, \ldots, a_n$, *not all zero*, such that $\boldsymbol{o} = a_1\boldsymbol{x}_1 + a_2\boldsymbol{x}_2 + \cdots + a_n\boldsymbol{x}_n$.
The system is called *linearly independent* if this equation can only be satisfied by $a_1 = a_2 = \cdots a_n = 0$.

The concept of linear independence is particularly important, because it is a necessary and sufficient condition for the solution of the equation

$$\boldsymbol{x} = a_1\boldsymbol{x}_1 + a_2\boldsymbol{x}_2 + \cdots + a_n\boldsymbol{x}_n$$

to be unique for all $\boldsymbol{x}$ that depend on $\boldsymbol{x}_1, \boldsymbol{x}_2, \ldots, \boldsymbol{x}_n$. In other words: $\boldsymbol{x}_1, \boldsymbol{x}_2, \ldots, \boldsymbol{x}_n$ are linearly independent if and only if every vector $\boldsymbol{x}$ can be written in one and only one way as a linear combination of $\boldsymbol{x}_1, \boldsymbol{x}_2, \ldots, \boldsymbol{x}_n$, or not at all.

Example 4: The vectors $\boldsymbol{i}$, $\boldsymbol{j}$ of V_3 form a linearly independent system. For if $\boldsymbol{o} = a_1\boldsymbol{i} + a_2\boldsymbol{j}$, then $0 = \boldsymbol{o} \cdot \boldsymbol{i} = (a_1\boldsymbol{i} + a_2\boldsymbol{j}) \cdot \boldsymbol{i} = a_1$ and similarly $0 = a_2$.

The three vectors $\boldsymbol{i}$, $\boldsymbol{j}$, $\boldsymbol{k}$ are also linearly independent, as can be proved in the same way. They have the further property that every vector is dependent on them: $\boldsymbol{a} = a_x\boldsymbol{i} + a_y\boldsymbol{j} + a_z\boldsymbol{k}$. This leads to the following definition.

A basis of a vector space V is a system $\boldsymbol{B}$ of vectors of V such that every vector in V can be represented in exactly one way as a linear combination of vectors of $\boldsymbol{B}$.

From the above it follows that this is equivalent to the following definition:

A basis of a vector space V is a linearly independent system $\boldsymbol{B}$ of vectors of V such that every vector in V is linearly dependent on $\boldsymbol{B}$.

Thus, the system $\boldsymbol{i}, \boldsymbol{j}, \boldsymbol{k}$ forms a basis of V_3.

If there is a finite basis in a vector space V, it is called *finite-dimensional*, otherwise *infinite-dimensional*. For finite-dimensional vector spaces the following theorem holds:

If V is finite-dimensional, then any two bases have the same number of elements. This number is called the dimension of V.

The dimension of V_3 is 3, since $\boldsymbol{i}, \boldsymbol{j}, \boldsymbol{k}$ is a basis, thus, every basis of V_3 contains exactly 3 elements:

Subspaces. A non-empty subset S of a vector space V is called a *subspace* if with the same addition and multiplication by scalars it satisfies the vector space axioms. This means, in particular, that the sum of two elements of S lies in S and any scalar multiple of an element of S lies in S. In fact, these two properties are the only ones that need to be verified, as the others all hold automatically.

A subset V' of V containing at least one element is a subspace if and only if the sum of $\boldsymbol{x} + \boldsymbol{y}$ of any two elements $\boldsymbol{x}$ and $\boldsymbol{y}$ of V' lies in V' and the scalar multiples $a\boldsymbol{x}$ of an element $\boldsymbol{x}$ of V' all lie in V'.

Example 5: The set consisting only of $\boldsymbol{o}$ is a subspace of every vector space. Every vector space is a subspace of itself. These are the *trivial subspaces*. The first one has dimension 0.

Example 6: If $\boldsymbol{x} \neq \boldsymbol{o}$ is a vector of V, the set V' of all scalar multiples of $\boldsymbol{x}$ is a subspace of V, the subspace *spanned* (or *generated*) by $\boldsymbol{x}$. It has $\boldsymbol{x}$ as a basis and its dimension is 1.

Coordinates. If $\boldsymbol{x}_1, \boldsymbol{x}_2, \ldots, \boldsymbol{x}_n$ is a basis of V, then by definition every vector $\boldsymbol{x}$ can be written in just one way in the form $\boldsymbol{x} = a_1\boldsymbol{x}_1 + a_2\boldsymbol{x}_2 + \cdots + a_n\boldsymbol{x}_n$. The real numbers $a_1, a_2, \ldots, a_n$ are called the *coordinates* of $\boldsymbol{x}$ with respect to the given basis. The coordinates of $\boldsymbol{x}$ change if the basis is changed, but their number is always equal to the dimension of V.

If two vectors $\boldsymbol{x}$ and $\boldsymbol{y}$ are given by their coordinates with respect to the same basis, then it follows from the vector space axioms that they can be added *coordinatewise*. Similarly $\boldsymbol{x}$ can be multiplied coordinatewise by a scalar c.

$\boldsymbol{x} = a_1\boldsymbol{x}_1 + a_2\boldsymbol{x}_2 + \cdots + a_n\boldsymbol{x}_n$ $\boldsymbol{y} = b_1\boldsymbol{x}_1 + b_2\boldsymbol{x}_2 + \cdots + b_n\boldsymbol{x}_n$	$\boldsymbol{x} + \boldsymbol{y} = (a_1 + b_1)\,\boldsymbol{x}_1 + (a_2 + b_2)\,\boldsymbol{x}_2 + \cdots + (a_n + b_n)\,\boldsymbol{x}_n$ $c\boldsymbol{x} = (ca_1)\,\boldsymbol{x}_1 + (ca_2)\,\boldsymbol{x}_2 + \cdots + (ca_n)\,\boldsymbol{x}_n.$

Thus, given a basis in a n-dimensional vector space V, one can associate uniquely with every vector $\boldsymbol{x}$ an n-tuple $(a_1, a_2, \ldots, a_n)$ in $\mathbf{R}^n$ and vice versa; what is more, this association preserves addition and multiplication by scalars. Such a mapping (association) is an *isomorphism* (see Linear maps) of V onto $\mathbf{R}^n$. Nevertheless, in the investigation of arbitrary vector spaces it is not advisable to exploit this isomorphism with $\mathbf{R}^n$, for it depends on the choice of a basis in V; this introduces an element of arbitrariness, and many investigations can be considerably complicated if the chosen basis is inconvenient.

$\mathbf{R}^n$ has a *standard basis*, namely $\boldsymbol{e}_1 = (1, 0, \ldots, 0)$, $\boldsymbol{e}_2 = (0, 1, 0, \ldots, 0)$, ..., $\boldsymbol{e}_n = (0, \ldots, 0, 1)$. For the subsequent discussion this is always taken to be chosen once and for all. In this basis the vector $\boldsymbol{x} = (a_1, a_2, \ldots, a_n)$ has the representation $\boldsymbol{x} = a_1\boldsymbol{e}_1 + a_2\boldsymbol{e}_2 + \cdots + a_n\boldsymbol{e}_n$.

The inner product. Generalizing the inner product of V_3 one defines the inner or dot product in $\mathbf{R}^n$ to be

$$\boldsymbol{x} \cdot \boldsymbol{y} = (a_1, a_2, \ldots, a_n) \cdot (b_1, b_2, \ldots, b_n) = a_1b_1 + a_2b_2 + \cdots a_nb_n.$$

The dot product of $\boldsymbol{x}$ with itself is written $\boldsymbol{x}^2$ for convenience. Just as in V_3, the dot product is *symmetric* and *bilinear*.

The *length, norm,* or *modulus* $|\boldsymbol{x}|$ of a vector $\boldsymbol{x} = (a_1, a_2, \ldots, a_n)$ is also defined by generalizing from V_3: $|\boldsymbol{x}| = \sqrt{\boldsymbol{x}^2} = \sqrt{(a_1^2 + a_2^2 + \cdots + a_n^2)}$.

In V_3, if $\overrightarrow{PQ}$ represents $\boldsymbol{a}$ and $\overrightarrow{QR}$ represents $\boldsymbol{b}$, then the third side $\overrightarrow{PR}$ of the triangle PQR represents $\boldsymbol{a} + \boldsymbol{b}$. It is an axiom that in a triangle no side is longer than the sum of the other two, so that $|\overrightarrow{PR}| \leqslant |\overrightarrow{PQ}| + |\overrightarrow{QR}|$ or $|\boldsymbol{a} + \boldsymbol{b}| \leqslant |\boldsymbol{a}| + |\boldsymbol{b}|$.

This is therefore called the *triangle inequality*; it can be proved (by induction) for $\mathbf{R}^n$ as well. It can also be extended (by a very easy induction) to the form

$$|\boldsymbol{x}_1 + \boldsymbol{x}_2 + \cdots + \boldsymbol{x}_m| \leqslant |\boldsymbol{x}_1| + |\boldsymbol{x}_2| + \cdots + |\boldsymbol{x}_m|.$$

The triangle inequality implies also that $||\boldsymbol{x}| - |\boldsymbol{y}|| \leqslant |\boldsymbol{x} - \boldsymbol{y}|$ and, for the dot product, that $|\boldsymbol{x} \cdot \boldsymbol{y}| \leqslant |\boldsymbol{x}|\,|\boldsymbol{y}|$. If this is written in coordinates, it becomes the *Cauchy-Schwarz inequality*, which is occasionally called the *Bunyakovskii* inequality.

In the proof of the Cauchy-Schwarz inequality one uses the bilinearity and symmetry of the dot product to prove $(x \pm y)^2 = x^2 \pm 2x \cdot y + y^2$.
Similarly one proves $(x + y) \cdot (x - y) = x^2 - y^2$.

Angles. In the section on V_3 a formula was found for the cosine of the angle between two vectors in terms of their inner product. In $\mathbf{R}^n$ this same formula is used to give an analytical definition of angles. The angle $\sphericalangle(x, y)$ between two non-zero vectors x and y is that angle between 0 and π whose cosine satisfies the adjacent formula

$$\cos \sphericalangle(x, y) = \frac{x \cdot y}{|x| \cdot |y|}.$$

This condition determines $\sphericalangle(x, y)$ uniquely. This definition of an angle makes it plausible to call x and y *orthogonal* if $x \cdot y = 0$.
In the basis $e_1, \ldots, e_n$ of $\mathbf{R}^n$ all the vectors have length 1 and any two distinct vectors are orthogonal. This is a fundamental property of this basis.

Inner product $x \cdot y = (a_1, \ldots, a_n) \cdot (b_1, \ldots, b_n)$ **Rules**	$x \cdot y = a_1 b_1 + \cdots + a_n b_n = y \cdot x$ $x \cdot (y + z) = x \cdot y + x \cdot z$ $d(x \cdot y) = (dx) \cdot y = x \cdot (dy)$.
Norm of x	$\lvert x \rvert = \sqrt{(a_1^2 + \cdots + a_n^2)}$
Generalized triangle inequality	$\lvert x_1 + \cdots + x_m \rvert \leqslant \lvert x_1 \rvert + \cdots + \lvert x_m \rvert$
Cauchy-Schwarz inequality	$\lvert x \cdot y \rvert \leqslant \lvert x \rvert \lvert y \rvert \rightarrow \sum_{i=1}^{n} a_i b_i \leqslant \sum_{i=1}^{n} a_i^2 \cdot \sum_{i=1}^{n} b_i^2$
Angle between x **and** y	$\cos \sphericalangle (x, y) = \frac{x \cdot y}{\lvert x \rvert \lvert y \rvert}, \quad 0 \leqslant \sphericalangle (x, y) \leqslant \pi$

Euclidean vector spaces. The introduction of an inner product gives $\mathbf{R}^n$ a structure additional to that of an ordinary vector space. The inner product associates with each pair of vectors x and y a real number $x \cdot y$ and can therefore be regarded as a function of the variables x and y satisfying certain properties.
In $\mathbf{R}^n$ coordinates were used in the definition. But the concept of an inner product can be generalized to arbitrary vector spaces as follows:

> *If V is a vector space and q a function that associates with every pair of vectors x and y in V a real number, then q is called an inner product on V if the following rules hold:*
>
> 1. $q(x, y) = q(y, x)$,
> 2. $q(x + x', y) = q(x, y) + q(x', y)$,
> 3. $q(ax, y) = aq(x, y)$,
> 4. $q(x, x) \geqslant 0$, *and* $q(x, x) = 0$ *if and only if* $x = o$.
>
> *A vector space equipped with such an inner product is called a Euclidean vector space. If no confusion is to be feared, the function $q(x, y)$ is written as (x, y).*

The inner product on $\mathbf{R}^n$ has these properties, hence $\mathbf{R}^n$ is a Euclidean vector space. A function satisfying 1. is said to be *symmetric*; if it satisfies 2. and 3. and the corresponding laws in the second term y (which follow automatically for symmetric functions), it is called *bilinear*. The last property 4. is *called non-singularity*. It indicates that $q(x, y) = x \cdot y$ is positive definite, that is, $x \cdot y = 0$ for all y is only possible when $x = o$. This is equivalent to the statement that the matrix of the $e_i \cdot e_j$ is non-zero, where the e_i form an arbitrary basis.
If V is a Euclidean vector space with a given inner product q, one can define length and angle by generalizing the definitions in $\mathbf{R}^n$. The *length, norm, or modulus* of a vector x is $|x| = \sqrt{q(x, x)}$. Property 4. of q states that every non-zero vector has positive length. If x and y are non-zero vectors, then the angle φ between 0 and π satisfying $\cos \varphi = \frac{q(x, y)}{|x|\,|y|}$ is called the *angle between* x *and* y.
Vectors of length 1 are called *unit vectors*. If $q(x, y) = 0$, then x and y are called *orthogonal* (with respect to q).
The properties of the basis $e_1, \ldots, e_n$ suggest the following definition:

> A basis of a Euclidean vector space is called *orthonormal* if all its vectors are unit vectors and if any two distinct vectors of the basis are orthogonal.

In the investigation of Euclidean vector spaces one usually tries to find an orthonormal basis, because its properties simplify many computations (in particular, the inner product can be computed by the ordinary formula for the dot product using the coordinates with respect to an orthonormal basis).

17.4. Linear maps

Properties of linear maps. A map A from a vector space V to a vector space V' is called linear if for any vectors $\boldsymbol{x}$ and $\boldsymbol{y}$ of V and any real number a the equations $A(\boldsymbol{x}+\boldsymbol{y}) = A(\boldsymbol{x}) + A(\boldsymbol{y})$ and $A(a\boldsymbol{x}) = aA(\boldsymbol{x})$ hold. This means that it does not matter whether one performs calculations on vectors in V and then applies the map to the result or first applies the map to the vectors and then does the corresponding calculations with their images in V'. The final result will be the same vector in both cases. In other words, the equations express the compatibility of the map with the fundamental vector space operations in V and V'. The map A is sometimes expressed by an arrow $A\colon V \to V'$ or $\boldsymbol{x} \to A(\boldsymbol{x})$. Here the uniquely determined vector $A(\boldsymbol{x})$ is called the *image* of $\boldsymbol{x}$, and $\boldsymbol{x}$ is called a *pre-image* or *inverse image* of $A(\boldsymbol{x})$. A vector in V' may have no pre-images, one pre-image, or more than one (see Chapter 5.).

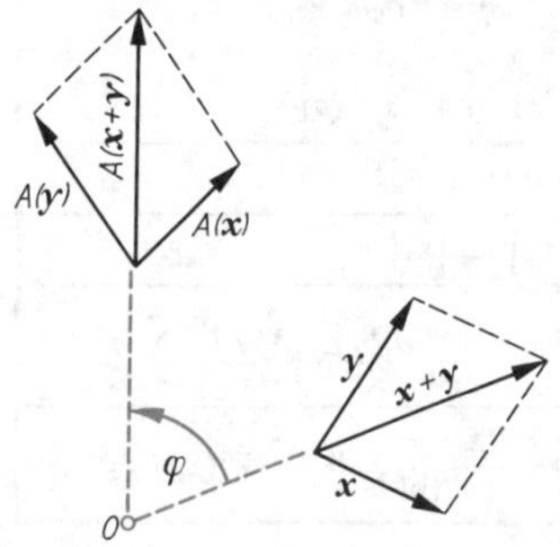

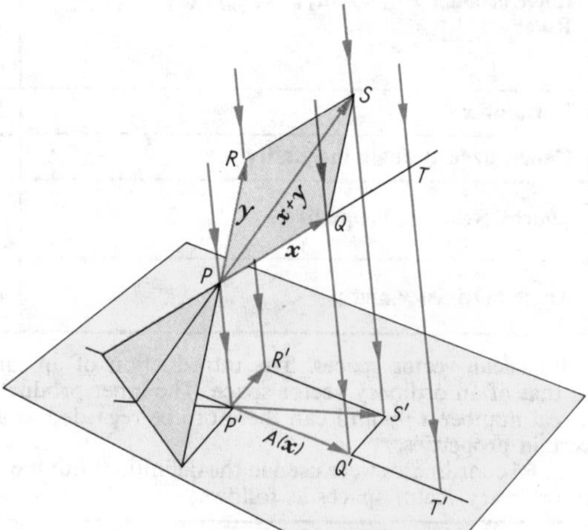

17.4-1 Rotation of a plane about a fixed point O through the angle φ

17.4-2 Parallel projection of the three-dimensional space onto a plane, and linearity of the parallel projection,
$\overrightarrow{PQ} = \boldsymbol{x}$, $\overrightarrow{PR} = \boldsymbol{y}$, $\overrightarrow{PS} = \boldsymbol{x}+\boldsymbol{y}$, $\overrightarrow{PT} = a\boldsymbol{x}$,
$\overrightarrow{P'Q'} = A(\boldsymbol{x})$, $\overrightarrow{P'R'} = A(\boldsymbol{y})$,
$\overrightarrow{P'S'} = A(\boldsymbol{x}) + A(\boldsymbol{y})$, $\overrightarrow{P'T'} = aA(\boldsymbol{x})$

Example 1: If a plane is rotated through an angle φ about a point O (Fig.), then every class of directly parallel oriented segments of the same length is taken to another such class. The rotation induces a map A of vectors by defining the image of $\boldsymbol{x}$ to be the vector associated with the class obtained after rotating the class of representatives of $\boldsymbol{x}$. The adjacent drawing in which the rotation is shown for vectors $\boldsymbol{x}$, $\boldsymbol{y}$ and $\boldsymbol{x}+\boldsymbol{y}$ shows that A is linear, because the parallelogram given by $\boldsymbol{x}$ and $\boldsymbol{y}$ is rotated as a whole. The equation $A(\boldsymbol{x}+\boldsymbol{y}) = A(\boldsymbol{x}) + A(\boldsymbol{y})$ holds. The relation $A(a\boldsymbol{x}) = aA(\boldsymbol{x})$ follows immediately from the fact that rotations preserve length.

Example 2: Similarly any parallel projection of three-dimensional space onto a plane also gives a linear map of the corresponding vector spaces (Fig.). The equation $A(\boldsymbol{x}+\boldsymbol{y}) = A(\boldsymbol{x}) + A(\boldsymbol{y})$ holds, because the parallelogram defining the sum $\boldsymbol{x}+\boldsymbol{y}$ is mapped to the parallelogram defining $A(\boldsymbol{x})+A(\boldsymbol{y})$. The equation $A(a\boldsymbol{x}) = aA(\boldsymbol{x})$ follows from the proportion $|PQ| : |PT| = |P'Q'| : |P'T'|$.

Kernel and image of a linear map. With each linear map A from V to V', there are associated two distinguished subspaces of V and V', respectively, its kernel and its image. The *kernel* of A is the subspace of V consisting of all those vectors that are mapped by A to the null vector of V'. The *image* of A is the subspace of V' consisting of all those vectors of V' that are images of vectors in V. In Example 2 the image of A is the plane, onto which the space is projected, and the kernel is the set of those vectors in three-dimensional space that are parallel to the direction of projection (and of course the null vector). If V is finite-dimensional and A is a linear map from V to V', then both the kernel and the image of A are also finite-dimensional. The dimension of the kernel is called the *nullity* of A, and the dimension of the image is called the *rank* of A. An important theorem on linear maps then states that

nullity of A + rank of A = dimension of V.

From this theorem it follows that the dimension of the image of A is at most equal to the dimension of V. The nullity measures the degree by which A differs from a one-to-one map. If the nullity is 0, then A is one-to-one.

Example 3: The left-hand side of a system of linear equations defines a linear map A from the vector space $\mathbf{R}^n$ to the vector space $\mathbf{R}^m$, by associating with each n-tuple $x = (x_1, \ldots, x_n)$ as its image the m-tuple

$$\begin{array}{c} a_{11}x_1 + \cdots + a_{1n}x_n = b_1 \\ \vdots \\ a_{m1}x_1 + \cdots + a_{mn}x_n = b_m \end{array}$$

$A(x) = (a_{11}x_1 + \cdots + a_{1n}x_n, \ldots, a_{m1}x_1 + \cdots + a_{mn}x_n)$. It is easy to check that this map is indeed linear. The problem of solving the system of equations can now be interpreted in the following manner: given a vector (or m-tuple) $(b_1, \ldots, b_m)$ in $\mathbf{R}^m$, to find all the vectors (n-tuples) in $\mathbf{R}^n$ that are mapped onto $(b_1, \ldots, b_m)$ by A. The associated homogeneous system is obtained by asking for the vectors mapped onto $(0, \ldots, 0)$ by A. *The vector space of solutions of the homogeneous system is the kernel of A.* If the nullity of A is 0, then the homogeneous system has only the trivial solution, and the inhomogeneous system has at most one solution, because A is one-to-one. The image of A is the set of vectors $(b_1, \ldots, b_m)$ for which solutions of the system exist. The rank of A can be computed from the coefficients of the equations (see Rank of a matrix).

One-to-one linear maps from a vector space V onto a vector space V' play an important role in linear algebra. Such maps are called *isomorphisms*. If A is an isomorphism from V to V', then the inverse map is an isomorphism from V' to V. The spaces V and V' are called isomorphic, in symbols $V \cong V'$. Isomorphic vector spaces have identical algebraic properties. Isomorphism is an equivalence relation on vector spaces, that is, it is reflexive, symmetric, and transitive.

Examples of linear maps: 4. On any vector space the *identity* map I, taking every vector to itself, is linear and indeed an isomorphism.

$$I(x + y) = x + y = I(x) + I(y) \quad \text{and} \quad I(ax) = a \cdot x = a \cdot I(x).$$

5: If V is an n-dimensional space with a basis $e_1, \ldots, e_n$, then the *coordinate* map Φ from V to $\mathbf{R}^n$ takes the vector $x = a_1e_1 + \cdots + a_ne_n$ to the n-tuple $(a_1, \ldots, a_n)$ of its coordinates with respect to the given basis. The map Φ is linear, one-to-one, and every n-tuple of real numbers is the image of a vector in V. Hence Φ is an isomorphism, and hence: *Every n-dimensional vector space is isomorphic to $\mathbf{R}^n$.*

The significance of linear maps. Since linear maps are compatible with the operations in a vector space, they make it possible to transfer an algebraic situation or problem from one space to another. Of particular importance are the isomorphisms, because the algebraic properties of subsets of vector spaces, such as linear dependence and independence or dimension, are invariant. Under these maps any theorem involving only these concepts is true for all spaces isomorphic to V, once it has been proved for V. In particular, one can exploit the isomorphism of n-dimensional spaces with $\mathbf{R}^n$.

The coordinate map translates relationships between vectors into equations involving real numbers, namely the coefficients of these vectors. However, the coordinate map depends essentially on the choice of a basis, hence different sets of equations may reflect the same set of relationships. From this point of view linear maps are important, because they describe the relationship between one set of coordinates and another.

Operations on linear maps. It is remarkable that the set of linear maps from a vector space V to another V' can itself be made into a vector space in a natural way. If A and B are linear maps from V to V', their sum $A + B$ is defined by $(A + B)(x) = A(x) + B(x)$ for all vectors x in V. Similarly, the scalar multiple $a \cdot A$ is defined by $(a \cdot A)(x) = a \cdot A(x)$. It is easy to check that $A + B$ and $a \cdot A$ are again linear and that the characteristic properties of vector spaces hold. If the dimensions of V and V' are m and n, respectively, the dimension of the vector space of linear maps from V to V' is $m \cdot n$.

The *product of two maps* is defined as the result of carrying out the maps in succession. For this to have a meaning, the first map must be defined on the image of the second; in other words, if B is a linear map from V to V' and A one from V' to a vector space V'', then one obtains the map $A \cdot B$ by applying B to the vectors of V and then A to the result. Thus, $A \cdot B$ is a linear map from V to V'' (Fig.).

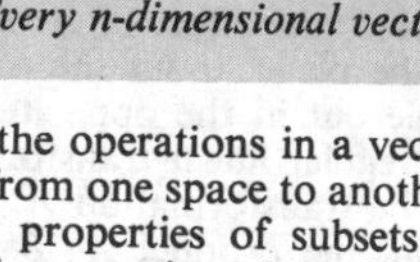

17.4-3 The product $A \cdot B$ of two linear maps

The fact that $A \cdot B$ is defined does not mean, in general, that $B \cdot A$ is also defined. And even if they are both defined, they need not necessarily be equal (Fig. 17.4-4). Thus, multiplication of maps is not commutative. It is, however, associative and the distributive laws also hold:

$$A \cdot (B + C) = A \cdot B + A \cdot C \quad \text{and} \quad (A + B) \cdot C = A \cdot C + B \cdot C.$$

Special linear maps. Of particular interest in the investigation of the structure of a vector space V are linear maps of V into itself. These are called *linear operators* or *linear transformations* on V. *The linear transformations of an n-dimensional vector space V form a vector space of dimension n^2.*

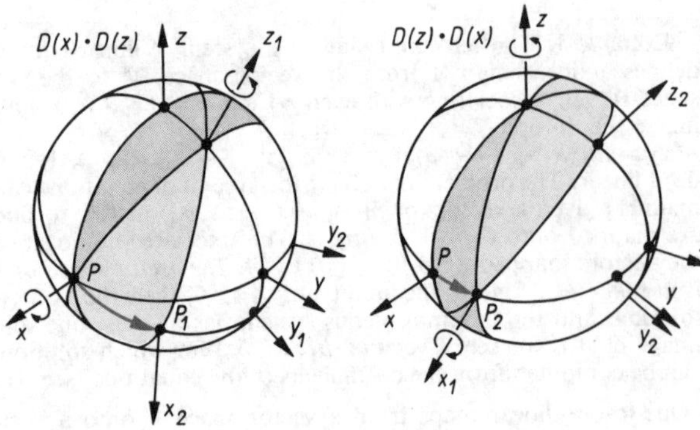

17.4-4 Rotations of the sphere about the x-axis ($D(x)$) and about the z-axis ($D(z)$); $D(x) \cdot D(z)$ takes the point P to the point P_1, on the other hand, $D(z) \cdot D(x)$ takes P to P_2, which is different from P_1

Furthermore, for two linear transformations A and B of the same space V the condition for the products $A \cdot B$ and $B \cdot A$ to exist is always satisfied. Therefore a (non-commutative) multiplication is also defined on the set of linear transformations of V. An example of a transformation is the identity map I on a space V, which is called the *identity transformation*. This transformation is an isomorphism of V onto itself. Linear transformations that are isomorphisms of V onto itself are called *regular* (or *non-singular*); if a linear transformation is not an isomorphism, it is called *singular*. Regular transformations have the property that the image of a basis is again a basis. They can also be characterized in the following way: *A linear transformation is regular if and only if there exists a linear transformation such that* $A \cdot B = B \cdot A = I$. In this case B is uniquely determined; it is called the *inverse transformation* to A and is denoted by A^{-1}. Thus, A^{-1} is the transformation that reverses the effect of A. For instance, $I^{-1} = I$. Examples of regular transformations are rotations of the plane about the origin. The inverse transformation is then the rotation through the same angle but in the opposite direction. In the same way rotations about an axis through the origin are regular linear transformations of the three-dimensional space V_3.

The transformation $A + B$ need not, in general, be regular even if A and B are. On the other hand, the product $A \cdot B$ or $B \cdot A$ of two regular transformations is always regular. The inverse transformation to $A \cdot B$ is $B^{-1} \cdot A^{-1}$. The set of regular transformations therefore has under multiplication properties similar to those of the non-zero real numbers (except that the commutative law $A \cdot B = B \cdot A$ is *not* satisfied). The identity transformation plays the same role as the number 1, since $A \cdot I = I \cdot A = A$ for any transformation A. In abstract algebra a set with these properties is called a group (see Chapter 16.). The set of regular transformations is called the *general linear group* on V and is denoted by $\mathrm{GL}(V)$.

If the vector space is Euclidean, one can associate with every linear transformation A a second transformation A^*, which is uniquely determined by the condition that $A(\boldsymbol{x}) \cdot \boldsymbol{y} = \boldsymbol{x} \cdot A^*(\boldsymbol{y})$ for all vectors $\boldsymbol{x}$ and $\boldsymbol{y}$ in V. This A^* is called the *adjoint transformation* to A. The following rules hold for adjoint transformations

$$(A + B)^* = A^* + B^*, \qquad (A \cdot B)^* = B^* \cdot A^*, \qquad (a \cdot A)^* = a \cdot A^*, \qquad (A^*)^* = A.$$

Of particular importance are *self-adjoint* or *symmetric transformations*. They are characterized by the property that $A^* = A$. These transformations occur frequently in physical problems and have in certain respects a very simple structure. The analogous transformations for infinite-dimensional complex vector spaces, the so-called *Hermitian transformations*, play an important part in quantum mechanics. Trivial examples of symmetric transformations are the multiples $a \cdot I$ of the identity transformation.

In a Euclidean vector space V the inner product can be used to define the lengths of vectors and the angles between them. Thus, in the investigation of such a space those linear transformations are particularly useful that are compatible with these additional properties of V. Such transformations are called *orthogonal transformations*. A linear transformation A on V is orthogonal if it leaves the inner product invariant, that is, if $\boldsymbol{x} \cdot \boldsymbol{y} = A(\boldsymbol{x}) \cdot A(\boldsymbol{y})$ for all vectors $\boldsymbol{x}$ and $\boldsymbol{y}$ in V. An orthogonal transformation preserves the lengths of vectors and the angles between them. The rotations of the plane and of three-dimensional space are again examples of orthogonal transformations; they obviously preserve lengths and angles. Orthogonal transformations can be characterized in the following way: *A linear transformation is orthogonal if and only if the image of an orthonormal basis is again an orthonormal basis*. Shorter and algebraically more succinct is the following description: A linear transformation A is orthogonal if and only if $A^* = A^{-1}$. Since according to this definition every orthogonal transformation has an inverse, orthogonal transformations are regular. The inverse of an orthogonal transformation is orthogonal, and the product of two orthogonal transformations is also orthogonal: the set of orthogonal transformations forms a group. This is called the *orthogonal group* on V. All the orthogonal transformations of V_3 can be obtained as rotations or as products of rotations and reflections in planes. A rotation is completely determined by the rotation induced in a plane perpendicular to the axis. This fact can be used to find special matrices representing orthogonal transformations on V_3.

17.5. Matrices

$$a_{11}x_1 + \cdots + a_{1n}x_n = b_1$$
$$\vdots$$
$$a_{m1}x_1 + \cdots + a_{mn}x_n = b_m$$

$$A = \begin{pmatrix} a_{11} & a_{12} & \dots & a_{1n} \\ \vdots & \vdots & & \vdots \\ a_{m1} & a_{m2} & \dots & a_{mn} \end{pmatrix}$$

The properties of the solution set of a system of linear equations depend essentially on the coefficients a_{ij} of the system. The rectangular array of these coefficients in their m rows and n columns is called an *$m \times n$-matrix* A (read 'm-by-n matrix'). The numbers in such an array are called the *entries* or *elements* of the matrix A. The n-tuple of entries with the same first subscript i is called the *ith row* and the m-tuple of entries with the same second subscript j is called the *jth column* of the matrix.

An $m \times n$-matrix has m rows and n columns. If $m = n$, the matrix is called *square*. The notation for the matrix A is often abbreviated as $A = (a_{ij})$ and sometimes as $A = (a_{ij})_{m.n}$ to indicate the number of rows and columns.

Operations on matrices. Matrices of the same shape (that is, the same number of rows and the same number of columns) can be added. The sum of two such matrices is defined as the matrix whose entries are just the sums of the corresponding entries of the original matrices. A matrix can be multiplied by a real number by multiplying every entry by that number.

Addition of matrices of same shape

$$\begin{pmatrix} a_{11} & a_{12} & \dots & a_{1n} \\ \vdots & \vdots & & \vdots \\ a_{m1} & a_{m2} & \dots & a_{mn} \end{pmatrix} + \begin{pmatrix} b_{11} & b_{12} & \dots & b_{1n} \\ \vdots & \vdots & & \vdots \\ b_{m1} & b_{m2} & \dots & b_{mn} \end{pmatrix} = \begin{pmatrix} a_{11}+b_{11} & a_{12}+b_{12} & \dots & a_{1n}+b_{1n} \\ \vdots & \vdots & & \vdots \\ a_{m1}+b_{m1} & a_{m2}+b_{m2} & \dots & a_{mn}+b_{mn} \end{pmatrix}$$

Multiplication of a matrix by a real number

$$c\begin{pmatrix} a_{11} & a_{12} & \dots & a_{1n} \\ \vdots & \vdots & & \vdots \\ a_{m1} & a_{m2} & \dots & a_{mn} \end{pmatrix} = \begin{pmatrix} ca_{11} & ca_{12} & \dots & ca_{1n} \\ \vdots & \vdots & & \vdots \\ ca_{m1} & ca_{m2} & \dots & ca_{mn} \end{pmatrix}$$

Example 1:

$$\begin{pmatrix} 1 & -2 & 0 \\ -1 & 1 & 2 \end{pmatrix} + (-2)\begin{pmatrix} 2 & 1 & -2 \\ 3 & -2 & -2 \end{pmatrix} = \begin{pmatrix} 1 & -2 & 0 \\ -1 & 1 & 2 \end{pmatrix} + \begin{pmatrix} -4 & -2 & 4 \\ -6 & 4 & 4 \end{pmatrix} = \begin{pmatrix} -3 & -4 & 4 \\ -7 & 5 & 6 \end{pmatrix}$$

The usual rules hold for addition of $m \times n$ matrices and multiplication by scalars.

The set of $m \times n$-matrices forms a vector space of dimension mn.

The null vector of this space is the *null matrix* or *zero matrix*, whose entries are all zero.

Matrices cannot always be multiplied. The product AB of an $m \times n$-matrix A by an $r \times s$-matrix B is defined only if $n = r$. In that case the product is an $m \times s$-matrix $C = (c_{ij})$, with entries defined in the following manner:

$$c_{ij} = a_{i1}b_{1j} + a_{i2}b_{2j} + \cdots + a_{in}b_{nj} = \sum_{k=1}^{n} a_{ik}b_{kj}.$$

The entry c_{ij} can be interpreted as the inner product of the ith row of A with the jth column of B.

Example 2:

$$\begin{pmatrix} 1 & -1 \\ 2 & 0 \end{pmatrix}\begin{pmatrix} 2 & 1 & 0 \\ 2 & 0 & -1 \end{pmatrix} = \begin{pmatrix} 1\cdot 2+(-1)\cdot 2 & 1\cdot 1+(-1)\cdot 0 & 1\cdot 0+(-1)\cdot(-1) \\ 2\cdot 2+\quad 0\cdot 2 & 2\cdot 1+\quad 0\cdot 0 & 2\cdot 0+\quad 0\cdot(-1) \end{pmatrix} = \begin{pmatrix} 0 & 1 & 1 \\ 4 & 2 & 0 \end{pmatrix}$$

In general, the existence of the product $A \cdot B$ does not imply that $B \cdot A$ is also defined, and even if both are defined, they need not be equal, as is shown by the following calculation:

$$\begin{pmatrix} 1 & 2 \\ 0 & 1 \end{pmatrix}\cdot\begin{pmatrix} 2 & 2 \\ 1 & 0 \end{pmatrix} = \begin{pmatrix} 4 & 2 \\ 1 & 0 \end{pmatrix}, \quad \text{but} \quad \begin{pmatrix} 2 & 2 \\ 1 & 0 \end{pmatrix}\cdot\begin{pmatrix} 1 & 2 \\ 0 & 1 \end{pmatrix} = \begin{pmatrix} 2 & 6 \\ 1 & 2 \end{pmatrix}.$$

Like the multiplication of linear maps, so the multiplication of matrices is not commutative. Apart from this exception the usual rules (such as the associative and distributive laws) hold for the multiplication of matrices.

Multiplication of matrices	Condition for feasibility: $n = r$	Rules
$A = (a_{ij})_{m,n}$ $B = (b_{ij})_{r,s}$	$(a_{ij})_{m,n} \cdot (b_{ij})_{n,s} = (c_{ij})_{m,s}$ with $c_{ij} = a_{i1}b_{1j} + \cdots + a_{in}b_{nj} = \sum_{i=1}^{n} a_{ik} b_{kj}$	$(AB)\,C = A(BC)$ $A(B + C) = AB + AC;$ $(A + B)\,C = AC + BC$

Square matrices of the same size can always be multiplied. There is a special $n \times n$-matrix, the *unit matrix* or *identity matrix* I, which leaves any $n \times n$-matrix fixed under multiplication on the left or on the right: $I \cdot A = A \cdot I = A$.

$$I = \begin{pmatrix} 1 & 0 & \dots & 0 \\ 0 & 1 & \dots & 0 \\ \vdots & \vdots & & \vdots \\ 0 & 0 & \dots & 1 \end{pmatrix}.$$

The set of $n \times n$-matrices thus has properties similar to the set of transformations of a vector space. By analogy to the definition of a regular transformation, a square matrix A is called *regular* (or *non-singular*) if there is a (necessarily square) matrix B such that $AB = BA = I$; otherwise A is *singular*. The matrix B is then uniquely determined, it is called the *inverse* matrix of A and is denoted by A^{-1}. Rules for calculating the inverse are discussed below.

Example 3: The inverse of

$$A = \begin{pmatrix} 2 & 1 \\ -1 & 1 \end{pmatrix} \quad \text{is} \quad A^{-1} = \begin{pmatrix} 1/3 & -1/3 \\ 1/3 & 2/3 \end{pmatrix}, \quad \text{for} \quad \begin{pmatrix} 2 & 1 \\ -1 & 1 \end{pmatrix} \cdot \begin{pmatrix} 1/3 & -1/3 \\ 1/3 & 2/3 \end{pmatrix} = \begin{pmatrix} 1 & 0 \\ 0 & 1 \end{pmatrix}.$$

Just as with linear transformations, inverses and products of regular matrices are regular, and the equations $(AB)^{-1} = B^{-1}A^{-1}$ and $(A^{-1})^{-1} = A$ both hold.

The set of all regular $n \times n$-matrices is a group under matrix multiplication. It is called the general linear group of degree n and is denoted by GL(n). It is isomorphic too, but conceptually different from, the group GL (V).

With every $m \times n$-matrix A one can associate an $n \times m$-matrix A^T, the transposed of A. It is obtained from A by interchanging the rows and the columns:

$$A = \begin{pmatrix} a_{11} & \dots & a_{1n} \\ \vdots & & \vdots \\ a_{m1} & \dots & a_{mn} \end{pmatrix} \rightarrow A^T = \begin{pmatrix} a_{11} & \dots & a_{m1} \\ \vdots & & \vdots \\ a_{1n} & \dots & a_{mn} \end{pmatrix}.$$

For a square matrix this process can be easily visualized as reflection in the *main diagonal*, from the top left-hand to the bottom right-hand corner.

Examples: 4.

$$A_1 = \begin{pmatrix} 1 & 0 & 1 \\ 2 & 0 & -1 \\ 2 & 1 & 2 \end{pmatrix}; \quad A_1^T = \begin{pmatrix} 1 & 2 & 2 \\ 0 & 0 & 1 \\ 1 & -1 & 2 \end{pmatrix}.$$

5.

$$A_2 = \begin{pmatrix} 1 & 2 \\ 0 & 1 \\ 1 & 2 \end{pmatrix}; \quad A_2^T = \begin{pmatrix} 1 & 0 & 1 \\ 2 & 1 & 2 \end{pmatrix}.$$

The rules governing the transposition of a matrix are similar to those for taking the adjoint of a linear transformation. $(A + B)^T = A^T + B^T$; $(aA)^T = a \cdot A^T$; $(A \cdot B)^T = B^T \cdot A^T$; $(A^T)^T = A$.

If A is a regular matrix, then so is A^T, and the inverse of A^T is the transposed of A^{-1}: $(A^{-1})^T = (A^T)^{-1}$. The matrix $(A^T)^{-1}$ is called the *contragredient* matrix to A.

The determinant of a matrix. Computation of the inverse matrix. With every square matrix A there is associated a real number, the *determinant* of A (see Determinants):

$$A = \begin{pmatrix} a_{11} & \dots & a_{1n} \\ \vdots & & \vdots \\ a_{n1} & \dots & a_{nn} \end{pmatrix}, \quad \det A = \begin{vmatrix} a_{11} & \dots & a_{1n} \\ \vdots & & \vdots \\ a_{n1} & \dots & a_{nn} \end{vmatrix}$$

There is a remarkable connection between matrix multiplication and determinants, the *product theorem*.

Product theorem	$\det(A \cdot B) = (\det A) \cdot (\det B)$

Since the rules for calculating determinants immediately give $\det I = 1$ when I is the identity matrix, it follows that if A is regular, then $(\det A)(\det A)^{-1} = 1$. Therefore the determinant of a regular matrix is non-zero. The converse is also true. *If the determinant of a matrix is not zero, then A is regular*. This is made explicit by a formula for computing A^{-1}.

The inverse A^{-1} of a matrix A	$A = \begin{pmatrix} a_{11} & \dots & a_{1n} \\ \vdots & & \vdots \\ a_{n1} & \dots & a_{nn} \end{pmatrix}$; $\quad A^{-1} = \frac{1}{\det A} \begin{pmatrix} A_{11} & \dots & A_{n1} \\ \vdots & & \vdots \\ A_{1n} & \dots & A_{nn} \end{pmatrix}$	A_{ij} is the cofactor of a_{ij} in A.

Example 6:

$$A = \begin{pmatrix} 1 & 0 & 2 \\ 2 & -1 & 1 \\ -2 & 0 & 1 \end{pmatrix}; \quad A^{-1} = \frac{-1}{5} \begin{pmatrix} -1 & 0 & 2 \\ -4 & 5 & 3 \\ -2 & 0 & -1 \end{pmatrix}; \quad A_{12} = - \begin{vmatrix} 2 & 1 \\ -2 & 1 \end{vmatrix} = -4.$$

The number $-4 = A_{12}$, say, in the right hand matrix is the cofactor of $a_{12} = 0$ in A.

Example 7: Computation of the inverse of a regular 2 × 2 matrix

$$A = \begin{pmatrix} a_{11} & a_{12} \\ a_{21} & a_{22} \end{pmatrix}; \quad A^{-1} = \frac{1}{a_{11}a_{22} - a_{12}a_{21}} \begin{pmatrix} a_{22} & -a_{12} \\ -a_{21} & a_{11} \end{pmatrix}.$$

Apart from this method of finding the inverse matrix by the cofactors, it can also be done by solving a suitable set of equations. The equations are obtained by considering the entries of A^{-1} as unknowns in the matrix equation ($AA^{-1} = I$)

$$\begin{pmatrix} a_{11} & \cdots & a_{1n} \\ \vdots & & \vdots \\ a_{n1} & \cdots & a_{nn} \end{pmatrix} \begin{pmatrix} x_{11} & \cdots & x_{1n} \\ \vdots & & \vdots \\ x_{n1} & \cdots & x_{nn} \end{pmatrix} = \begin{pmatrix} 1 & \cdots & 0 \\ \vdots & & \vdots \\ 0 & & 1 \end{pmatrix}.$$

After multiplication of the two matrices on the left one obtains a system of n^2 linear equations to determine the n^2 unknowns x_{ij}. The solution of these equations by Cramer's rule gives the cofactor formula above for A^{-1}.

It is also possible to compute the inverse A^{-1} by considering the system of n equations in the $2n$ variables $x_1, \ldots, x_n, y_1, \ldots, y_n$:

$$\begin{matrix} a_{11}x_1 + \cdots + a_{1n}x_n = y_1 \\ \vdots \\ a_{n1}x + \cdots + a_{nn}x = y_n. \end{matrix}$$

This system can be solved by Cramer's rule or by Gauss's algorithm:

$$\begin{matrix} x_1 = b_{11}y_1 + \cdots + b_{1n}y_n \\ \vdots \\ x_n = b_{n1}y + \cdots + b_{nn}y. \end{matrix}$$

The matrix $B = (b_{ij})$ of the coefficients on the right-hand side is then the inverse of A. This method requires fewer equations.

Representation of linear maps by matrices

The operations on matrices show remarkable similarities to those on linear maps. This is true not only of the relatively simple operations of addition and multiplication by scalars, but also of the multiplication of the linear maps and matrices themselves, in particular, the conditions for the existence of the product, inverse etc. These similarities, already emphasized by a similar terminology, are not accidental. Indeed, the importance of matrices lies to a large extent in the fact that they can be used to describe linear maps numerically. This aspect subsumes the use of matrices in describing systems of linear equations. A linear map A from an n-dimensional vector space V to an m-dimensional vector space V' can be represented by an $m \times n$-matrix in the following way: If $x_1, \ldots, x_n$ and $y_1, \ldots, y_m$ are bases of V and V', respectively, then the images of $x_1, \ldots, x_n$ can be expressed in terms of the basis $y_1, \ldots, y_n$:

$$\begin{matrix} A(x_1) = a_{11}y_1 + \cdots + a_{m1}y_n \\ \vdots \\ A(x_n) = a_{1n}y_1 + \cdots + a_{mn}y_n \end{matrix} \quad \text{or} \quad A(x_j) = \sum_{i=1}^{m} a_{ij}y_i \quad \text{for} \quad j = 1, \ldots, n.$$

Now the linear map A is completely determined by the images $A(x_j)$ of the basis vectors $x_1, \ldots, x_n$; for an arbitrary vector $x = a_1x_1 + \cdots + a_nx_n$ of V then has the image $A(x) = A(a_1x + \cdots + a_nx_n) = a_1A(x_1) + \cdots + a_nA(x_n)$. Thus, the linear map is completely characterized by the $m \cdot n$ numbers a_{ij}.

It turns out to be more convenient to use as the matrix representing A the transposed of the coefficient matrix above

$$A \to A = \begin{pmatrix} a_{11} & \cdots & a_{1n} \\ \vdots & & \vdots \\ a_{m1} & \cdots & a_{mn} \end{pmatrix}.$$

The jth column of A is simply the set of coordinates of $A(x_j)$ with respect to the basis $y_1, \ldots, y_m$. It is important to remember that the choice of the matrix representing A depends on the choice of bases in V and V'.

If the bases in V and V' are fixed, then the correspondence between linear maps and matrices has the adjacent properties:

If $A \to A$ and $B \to B$, then $A + B \to A + B$ and $a \cdot A \to a \cdot A$.

Under the same condition there is a unique $m \times n$-matrix associated with every linear map, and vice versa. These statements are summarized in the following theorem:

The vector space of linear maps from V to V' is isomorphic to the vector space of $m \times n$-matrices.

A similar fact emerges for multiplication. If A is a linear map from V' to V'' and B a linear map from V to V', then by choosing bases in V, V', and V'' matrices are associated with A and B, and it can be shown that:

If $A \to A$ and $B \to B$, then $A \cdot B \to A \cdot B$.

The representation of linear maps by matrices is completely analogous to the representation of vectors by n-tuples with respect to a basis. The coordinates of a linear map are only arranged in a special way to make up a matrix. The analogies in the operations now appear as consequences of the fact that the operations for matrices are defined to correspond to the operations on the linear maps they represent.

If a linear map from V to V' is represented by the $m \times n$-matrix $\mathbf{A} = (a_{ij})$ with respect to fixed bases in V and V', then the vector equation $A(\boldsymbol{x}) = \boldsymbol{x}_0$ can be solved. Here it is required to find all vectors $\boldsymbol{x}$ in V that are mapped by A to a given vector $\boldsymbol{x}_0$ of V'.

If $b_1, \ldots, b_m$ are the coefficients of $\boldsymbol{x}_0$ in V', then the problem of finding the coordinates, $x_1, \ldots, x_n$, of such a vector $\boldsymbol{x}$ is simply that of solving the system of equations

$$\begin{matrix} a_{11}x_1 + \cdots + a_{1n}x_n = b_1 \\ \vdots \qquad \vdots \qquad \vdots \\ a_{m1}x_1 + \cdots + a_{mn}x_n = b_m. \end{matrix} \Longrightarrow \begin{pmatrix} a_{11} & \cdots & a_{1n} \\ \vdots & & \vdots \\ a_{m1} & \cdots & a_{mn} \end{pmatrix} \begin{pmatrix} x_1 \\ \vdots \\ x_n \end{pmatrix} = \begin{pmatrix} b_1 \\ \vdots \\ b_m \end{pmatrix}.$$

This shows the connection between the equations given by linear maps and systems of linear equations. If the system is written in matrix form, it becomes evident that it is merely the vector equation $A(\boldsymbol{x}) = \boldsymbol{x}_0$ in coordinates.

Here the coordinates of $\boldsymbol{x}$ and $\boldsymbol{x}_0$ are written as matrices with a single column.

Representations of linear transformations. To associate a matrix with a linear transformation of an n-dimensional vector space V, it is sufficient to choose a basis $\boldsymbol{x}_1, \ldots, \boldsymbol{x}_n$. From the equations

$$\begin{matrix} A(\boldsymbol{x}_1) = a_{11}\boldsymbol{x}_1 + \cdots + a_{n1}\boldsymbol{x}_n \\ \vdots \qquad \vdots \qquad \vdots \\ A(\boldsymbol{x}_n) = a_{1n}\boldsymbol{x}_1 + \cdots + a_{nn}\boldsymbol{x}_n \end{matrix} \quad \text{or} \quad A(\boldsymbol{x}_j) = \sum_{i=1}^{n} a_{ij}\boldsymbol{x}_i \quad \text{for } j = 1, \ldots, n$$

one obtains the matrix representing the transformation $A \to \mathbf{A} = \begin{pmatrix} a_{11} & \cdots & a_{1n} \\ \vdots & & \vdots \\ a_{n1} & \cdots & a_{nn} \end{pmatrix}$.

Linear transformations are always represented by square matrices.

If a linear transformation is regular, then so is the matrix representing it, and vice versa. The inverse transformation is represented by the inverse matrix.

If $A \to \mathbf{A}$, then $A^{-1} \to \mathbf{A}^{-1}$.

If $A \to \mathbf{A}$ and $B \to \mathbf{B}$, then $A + B \to \mathbf{A} + \mathbf{B}$; $a \cdot A \to a \cdot \mathbf{A}$; and $A \cdot B \to \mathbf{A} \cdot \mathbf{B}$.

Example 8: If A is the transformation already mentioned repeatedly that is obtained by rotating the plane about the origin O through the angle φ, and if $\boldsymbol{x}_1$ and $\boldsymbol{x}_2$ are two orthogonal basis vectors of length 1, then their representatives in the point O are mapped to representatives of their images $A(\boldsymbol{x}_1)$ and $A(\boldsymbol{x}_2)$ (Fig.). Obviously the adjacent equations hold for $A(\boldsymbol{x}_1)$ and $A(\boldsymbol{x}_2)$. The operator A is thus represented by the matrix $\mathbf{A}$. The operator A^{-1} is simply the rotation through φ in the opposite direction, that is, a rotation through $-\varphi$.

$$A(\boldsymbol{x}_1) = \cos\varphi\,\boldsymbol{x}_1 + \sin\varphi\,\boldsymbol{x}_2$$
$$A(\boldsymbol{x}_2) = -\sin\varphi\,\boldsymbol{x}_1 + \cos\varphi\,\boldsymbol{x}_2$$
$$\mathbf{A} = \begin{pmatrix} \cos\varphi & -\sin\varphi \\ \sin\varphi & \cos\varphi \end{pmatrix}$$

$$A^{-1} \to \mathbf{A}^{-1} = \begin{pmatrix} \cos(-\varphi) & -\sin(-\varphi) \\ \sin(-\varphi) & \cos(-\varphi) \end{pmatrix} = \begin{pmatrix} \cos\varphi & \sin\varphi \\ -\sin\varphi & \cos\varphi \end{pmatrix}.$$

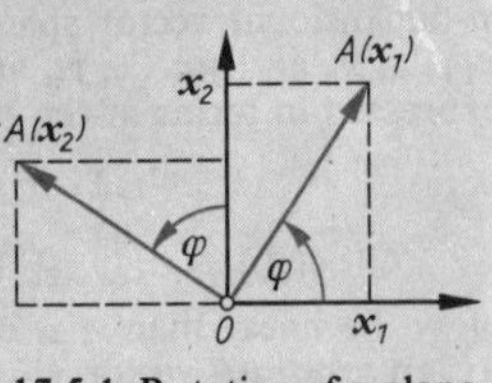

17.5-1 Rotation of a plane about the fixed point O through the angle φ

Example 9: If I is the identity transformation on V and $\boldsymbol{x}_1, \ldots, \boldsymbol{x}_n$ is any basis of V, then

$$\begin{aligned} I(\boldsymbol{x}_1) &= \boldsymbol{x}_1 = 1 \cdot \boldsymbol{x}_1 + 0 \cdot \boldsymbol{x}_2 + \cdots + 0 \cdot \boldsymbol{x}_n \\ I(\boldsymbol{x}_2) &= \boldsymbol{x}_2 = 0 \cdot \boldsymbol{x}_1 + 1 \cdot \boldsymbol{x}_2 + \cdots + 0 \cdot \boldsymbol{x}_n \\ &\ \vdots \\ I(\boldsymbol{x}_n) &= \boldsymbol{x}_n = 0 \cdot \boldsymbol{x}_1 + 0 \cdot \boldsymbol{x}_2 + \cdots + 1 \cdot \boldsymbol{x}_n \end{aligned} \qquad I \to \mathbf{I} = \begin{pmatrix} 1 & 0 & \cdots & 0 \\ 0 & 1 & \cdots & 0 \\ \vdots & \vdots & & \vdots \\ 0 & 0 & \cdots & 1 \end{pmatrix}$$

For any basis the identity transformation is represented by the identity matrix.

In general, the matrix $\mathbf{A}$ representing a linear transformation A depends on the choice of basis. If $\boldsymbol{x}_1, \ldots, \boldsymbol{x}_n$ and $\boldsymbol{x}'_1, \ldots, \boldsymbol{x}'_n$ are two bases of V, then, say,

$A \to \mathbf{A}$ with respect to the basis $\boldsymbol{x}_1, \ldots, \boldsymbol{x}_n$, and

$A \to \mathbf{A}'$ with respect to the basis $\boldsymbol{x}'_1, \ldots, \boldsymbol{x}'_n$.

A new linear transformation C can now be defined by means of the two bases: $C(\boldsymbol{x}_1) = \boldsymbol{x}'_1, \ldots, C(\boldsymbol{x}_n) = \boldsymbol{x}'_n$, that is, C is the transformation taking one basis to the other. If the transformation C is represented by the matrix $\mathbf{C}$ with respect to the basis $\boldsymbol{x}_1, \ldots, \boldsymbol{x}_n$, then the relation $\mathbf{A}' = \mathbf{C}^{-1}\mathbf{A}\mathbf{C}$ holds. This is the *rule of transformation* for matrices representing the same operator with respect to different bases. Matrices for which the above relation holds are called *similar*. A natural question in this context is that of the existence of a basis for which the matrix representing a given transformation is as simple as possible. This is the problem of finding *normal forms* for transformations and is closely connected with the theory of *eigenvalues* (see Transformation to principal axes in Eigenvalues).

Change of coordinates. If in a vector space V two bases $x_1, \ldots, x_n$ and $y_1, \ldots, y_n$ are given, then a vector x has coordinates with respect to each of these bases:

$$x = x_1 x_1 + \cdots + x_n x_n = y_1 y_1 + \cdots + y_n y_n.$$

The change from one coordinate system to the other is described by the equations:

$$\begin{aligned} y_1 &= a_{11} x_1 + \cdots + a_{n1} x_n \\ &\vdots \\ y_n &= a_{1n} x_1 + \cdots + a_{nn} x_n \end{aligned} \quad \text{or} \quad y_j = \sum_{i=1}^{n} a_{ij} x_i \quad \text{for} \quad j = 1, \ldots, n.$$

The coordinates $x_1, \ldots, x_n$ and $y_1, \ldots, y_n$ now satisfy the relations

$$x_j = \sum_{i=1}^{n} a_{ji} y_i \quad \text{for} \quad j = 1, \ldots, n.$$

The inverse formulae are obtained by going from the matrix $A = (a_{ij})$ to its inverse $A^{-1} = (a'_{ij})$

Change from basis $x_1, \ldots, x_n$ to basis $y_1, \ldots, y_n$			
Transformation of basis vectors		**Transformation of coordinates**	
$y_j = \sum_{i=1}^{n} a_{ij} x_i$	$(a_{ij}) = A$	$y_i = \sum_{i=1}^{n} a'_{ji} x_i$	$(a'_{ji}) = (A^{-1})^T$
$x_j = \sum_{i=1}^{n} a'_{ij} y_i$	$(a'_{ij}) = A^{-1}$	$x_j = \sum_{i=1}^{n} a_{ji} y_i$	$(a_{ji}) = A^T$

The difference in the way they are transformed is expressed by calling the transformation of the coordinates *contragredient* to that of the bases. For if the basis $y_1, \ldots, y_n$ is represented by the matrix A with respect to the basis $x_1, \ldots, x_n$, then the coordinates of x with respect to the basis $y_1, \ldots, y_n$ are obtained from those with respect to $x_1, \ldots, x_n$ by the contragredient matrix $(A^{-1})^T$, as is shown by the equations above.

For orthonormal bases in a Euclidean vector space the transformation matrix is orthogonal and hence equal to its contragredient. In this particular case coordinates are transformed in the same way as bases.

The rank of a matrix. For any $m \times n$-matrix A one can determine the maximal number of linearly independent columns or rows by considering the rows and columns as elements of $\mathbf{R}^n$ and $\mathbf{R}^m$, respectively. These two numbers are always equal and are called the *rank of the matrix*. If the matrix A represents the linear map A, then the rank of A is the same as the rank of A. The rank can be computed by using the following facts:

The rank of a matrix remains unchanged if 1. a multiple of one row (column) is added to another row (column), or 2. rows (or columns) are interchanged.

By using these rules a matrix can be brought into a form in which only entries with the same row and column index can be different from zero. The rank of A is the number of such non-zero entries. This method is very similar to Gauss's algorithm of solving systems of linear equations. For a quadratic matrix it is sufficient to transform it to *triangular form* in which all the entries below (or above) the main diagonal are zero. If this is done so that as many diagonal elements as possible are non-zero, then again the number of such elements is the rank.

Example 10: The matrix A is transformed into A_2 by adding the second column to the first and third. By subtracting three times the first from the third one obtains A_3. Interchanging the first two columns gives A_4. The rank of A is 2.

$$A = \begin{pmatrix} -1 & 1 & -1 \\ 1 & 0 & 3 \end{pmatrix} \longrightarrow A_2 = \begin{pmatrix} 0 & 1 & 0 \\ 1 & 0 & 3 \end{pmatrix} \longrightarrow A_3 = \begin{pmatrix} 0 & 1 & 0 \\ 1 & 0 & 0 \end{pmatrix} \longrightarrow A_4 = \begin{pmatrix} 1 & 0 & 0 \\ 0 & 1 & 0 \end{pmatrix}$$

Example 11: In the initial matrix A, three times the first row is added to the second, and twice the first row to the third. In the transformed matrix the second and third column are interchanged. The rank of A is 3.

$$A = \begin{pmatrix} 1 & 1 & 0 \\ -3 & -3 & 1 \\ -2 & 1 & 0 \end{pmatrix} \to \begin{pmatrix} 1 & 1 & 0 \\ 0 & 0 & 1 \\ 0 & 3 & 0 \end{pmatrix} \to \begin{pmatrix} 1 & 0 & 1 \\ 0 & 1 & 0 \\ 0 & 0 & 3 \end{pmatrix}.$$

Special types of matrices. Corresponding to the special types of linear transformations there are special types of matrices. If V is a Euclidean vector space and if a transformation A is represented by the matrix A with respect to an orthonormal basis, then the adjoint transformation A^* is represented by the transposed matrix A^T. Hence symmetric transformations, for which $A = A^*$, are represented by *symmetric matrices*, for which $A = A^T$.

Of particular importance are *orthogonal matrices,* because they transform orthonormal bases into one another. Expressed in terms of coordinates this says that: *The coordinates with respect to one rectangular coordinate system are transformed into those with respect to another by means of an orthogonal matrix.* A matrix is *orthogonal* if $A^T = A^{-1}$. This equation can also be written in the form $A \cdot A^T = I$ and interpreted thus: *In an orthogonal matrix the inner product of different rows is zero, the inner product of a row with itself is one.* The same statements are true for the columns of A, and either set is a sufficient condition for the matrix to be orthogonal. For instance, every 2×2 orthogonal matrix can be written in the form:

$$\begin{pmatrix} \cos\varphi & -\sin\varphi \\ \sin\varphi & \cos\varphi \end{pmatrix} \quad \text{or} \quad \begin{pmatrix} \cos\varphi & \sin\varphi \\ \sin\varphi & -\cos\varphi \end{pmatrix}.$$

In the first case the matrix represents a rotation of the plane through the angle φ, in the second case there is an additional reflection in a line. Matrices of the second type can be distinguished from those of the first by the fact that their determinant is -1, whereas the determinant of a rotation is always $+1$. In general, the determinant of a orthogonal matrix is always $+1$ or -1. If an orthogonal matrix has determinant $+1$, it is sometimes called *proper,* in general they correspond to orientation-preserving orthogonal transformations of a Euclidean vector space. The following matrices are of this type:

$$A_{12}(\varphi) = \begin{pmatrix} \cos\varphi & -\sin\varphi & 0 \\ \sin\varphi & \cos\varphi & 0 \\ 0 & 0 & 1 \end{pmatrix}, \quad A_{13}(\psi) = \begin{pmatrix} \cos\psi & 0 & -\sin\psi \\ 0 & 1 & 0 \\ \sin\psi & 0 & \cos\psi \end{pmatrix}, \quad A_{23}(\vartheta) = \begin{pmatrix} 1 & 0 & 0 \\ 0 & \cos\vartheta & -\sin\vartheta \\ 0 & \sin\vartheta & \cos\vartheta \end{pmatrix}.$$

Here φ, ψ, and ϑ are arbitrary angles. If a fixed order of the basis vectors e_1, e_2, e_3 is chosen, then $A_{12}(\varphi)$ represents a rotation of space about the e_3-axis. The e_1, e_2-plane is rotated through φ while e_3 is left unchanged. This fact gives rise to the special form of the matrix. Every proper orthogonal 3×3 matrix A can be written as a product $A = A_{23}(\vartheta) \cdot A_{13}(\psi) \cdot A_{12}(\varphi)$ for suitable choices of φ, ψ, and ϑ.

Just as for the orthogonal transformations, so the set of orthogonal $n \times n$ matrices forms a group. The proper orthogonal matrices form a subgroup of this group.

17.6. Eigenvalues

Eigenvalues and eigenvectors. A number λ is called an *eigenvalue* (*or characteristic value*) *of a linear transformation* A if there exists a vector $x \neq o$ such that $A(x) = \lambda \cdot x$. The vector x is then called an *eigenvector of the transformation* A belonging to λ. The eigenvectors belonging to λ together with the null vector form a subspace, called an *eigenspace* of A.

If the equation $A(x) = \lambda \cdot x$ is rewritten in the form $(A - \lambda I)\, x = o$, then it can be stated that:

A number λ is an eigenvalue of the operator A if and only if the operator $A - \lambda I$ is singular.

In this formulation it is possible to define an eigenvalue in terms of a matrix A representing the transformation A: *A number λ is an eigenvalue of the matrix A if $A - \lambda I$ is singular.*

Example 1: Let A be a singular transformation; then there exists a non-zero vector x such that $A(x) = o = 0 \cdot x$. Hence $\lambda = 0$ is an eigenvalue of A, and the non-zero vectors of the kernel are the eigenvectors belonging to 0.

Example 2: Suppose that the matrix A representing the operator A with respect to a basis $x_1, \dots, x_n$ is diagonal:

$$\begin{matrix} A(x_1) = \lambda_1 x_1 \\ \vdots \qquad \vdots \\ A(x_n) = \lambda_n x_n \end{matrix} \qquad A \to A = \begin{pmatrix} \lambda_1 & 0 & \dots & 0 \\ 0 & \lambda_2 & \dots & 0 \\ \vdots & & & \vdots \\ 0 & 0 & \dots & \lambda_n \end{pmatrix}.$$

Then the basis vectors are all eigenvectors of A. Such transformations are particularly easy to describe, because they change the basis vectors only by multiplying them by scalars. They are called *diagonal* (or *diagonalizable*) transformations. Every transformation of an n-dimensional space with n distinct eigenvalues is diagonalizable.

The significance of eigenvalues in physics. Eigenvalue problems are important in many branches of physics. They make it possible to find coordinate systems in which the transformations in question take on their simplest forms. In mechanics for instance, the principal moments of a rigid body are found with the help of the eigenvalues of the symmetric matrix representing its inertia tensor. The situation is similar in the mechanics of continua, where the rotations and deformations of a body in the principal directions are found with the help of the eigenvalues of a symmetric matrix. Eigenvalues are of central importance in quantum mechanics, in which the measured values of physical 'observables' appear as the eigenvalues of certain operators. The term 'transformation' is used predominantly in pure mathematical (geometrical) context, whereas 'operator' is more customary in applications (physics, technology).

Computation of eigenvalues and eigenvectors. If a basis in a vector space V is chosen, then the equation $(A - \lambda I)(x) = o$ is represented by the following system of equations for the coordinates $x_1, \ldots, x_n$ of x:

$$\begin{aligned} (a_{11} - \lambda)\, x_1 + a_{12} \quad & x_2 + \cdots + a_{1n} \quad x_n = 0 \\ a_{21}x_1 \quad + (a_{22} - \lambda)\, & x_2 + \cdots + a_{2n} \quad x_n = 0 \\ \vdots \qquad\qquad & \vdots \qquad\qquad \vdots \\ a_{n1}x_1 \quad + a_{n2} \quad & x_2 + \cdots + (a_{nn} - \lambda)\, x_n = 0. \end{aligned}$$

The coefficient matrix is the matrix $\boldsymbol{A} - \lambda \boldsymbol{I}$ representing the transformation $A - \lambda I$. Since only non-zero vectors can be eigenvectors, the problem is to find non-zero solutions of this homogeneous system. A necessary and sufficient condition for the existence of such solutions is that the determinant of the matrix of coefficients should vanish: $\det(\boldsymbol{A} - \lambda \boldsymbol{I}) = 0$. This is the case if and only if $A - \lambda I$ is singular, that is, if λ is an eigenvalue of A. The determinant can be seen to be a polynomial of degree n in λ:

$$\det(\boldsymbol{A} - \lambda \boldsymbol{I}) = a_0 + a_1\lambda + \cdots + a_n\lambda^n.$$

This is called the *characteristic polynomial* of the matrix $\boldsymbol{A}$. If $\boldsymbol{A}'$ is another matrix representing A, then $\boldsymbol{A}' = \boldsymbol{C}^{-1}\boldsymbol{A}\boldsymbol{C}$ for some matrix $\boldsymbol{C}$, and its associated polynomial is the same:

$$\det(\boldsymbol{A}' - \lambda \boldsymbol{I}) = \det(\boldsymbol{C}^{-1}\boldsymbol{A}\boldsymbol{C} - \lambda \boldsymbol{I}) = \det(\boldsymbol{C}^{-1}(\boldsymbol{A} - \lambda \boldsymbol{I})\,\boldsymbol{C}) = \det(\boldsymbol{A} - \lambda \boldsymbol{I}).$$

To find an eigenvector x one must therefore first find a root of the characteristic polynomial of $\boldsymbol{A}$. The coordinates $x_1, x_2, \ldots, x_n$ of x can then be found as a non-trivial solution of the homogeneous system given above.

Example 3: For $n = 2$ and $\boldsymbol{A} = \begin{pmatrix} 2 & 3 \\ -1 & -2 \end{pmatrix}$ the eigenvalues are roots of the equation:

$$\det(\boldsymbol{A} - \lambda \boldsymbol{I}) = \begin{vmatrix} 2-\lambda & 3 \\ -1 & -2-\lambda \end{vmatrix} = \lambda^2 - 1 = 0.$$

Thus, they are $+1$ and -1. The coordinates x_1, x_2 of the eigenvectors belonging to the eigenvalue $+1$ are the solutions of the system:

$$\begin{aligned} 1x_1 + 3x_2 &= 0 \\ -1x_1 - 3x_2 &= 0 \end{aligned} \;\longrightarrow\; (x_1, x_2) = \tau \cdot (-3, 1),$$

where τ is an arbitrary non-zero number. In general, eigenvectors are determined only up to scalar multiples.

The transformation to principal axes

For symmetric transformations the theory leads to a particularly simple result. All the eigenvalues of a symmetric transformation are real and there exists an orthonormal basis of eigenvectors. If A is represented by the matrix $\boldsymbol{A}$, this means that there exists an orthogonal matrix $\boldsymbol{C}$ such that $\boldsymbol{A}' = \boldsymbol{C}^{-1}\boldsymbol{A}\boldsymbol{C}$ is diagonal, with the eigenvalues on the main diagonal. $\boldsymbol{A}'$ is called the *normal form* of $\boldsymbol{A}$ and the change of basis represented by $\boldsymbol{C}$ is called the *transformation to principal axes*. The matrix $\boldsymbol{C}$ is the matrix of the coordinates of an orthonormal basis of eigenvectors with respect to the basis under which A is represented by the matrix $\boldsymbol{A}$.

Example 4: For $A \to \boldsymbol{A} = \begin{pmatrix} 3 & -1 \\ -1 & 3 \end{pmatrix}$ the eigenvalues are $+2$ and $+4$. The eigenvectors belonging to $+2$ are $(x_1, x_2) = \tau_1 \cdot (1, 1)$ and the eigenvectors belonging to $+4$ are $(x_1, x_2) = \tau_2(-1, 1)$. The numbers τ_1 and τ_2 can be chosen to give the vectors the length 1. The eigenvectors $(1/\sqrt{2}, 1/\sqrt{2})$ and $(-1/\sqrt{2}, 1/\sqrt{2})$ form an orthonormal basis, and $\boldsymbol{C}$ is the matrix.

$$\boldsymbol{C} = \begin{pmatrix} 1/\sqrt{2} & -1/\sqrt{2} \\ 1/\sqrt{2} & 1/\sqrt{2} \end{pmatrix}; \quad \boldsymbol{C}^{-1} = \boldsymbol{C}^T = \begin{pmatrix} 1/\sqrt{2} & 1/\sqrt{2} \\ -1/\sqrt{2} & 1/\sqrt{2} \end{pmatrix}; \quad \boldsymbol{C}^{-1}\boldsymbol{A}\boldsymbol{C} = \begin{pmatrix} 2 & 0 \\ 0 & 4 \end{pmatrix}.$$

By means of the transformation to principal axes the equation of centred conics or quadrics can be considerably simplified, by changing the Cartesian coordinate system to one consisting of symmetry axes of the curve, or surface. These are the principal axes of the figure, which explains the name transformation to principal axes.

Example 5: If $ax^2 + 2bxy + cy^2 = d$ is the equation of a conic section, then the coordinates on the left-hand side are arranged in a symmetric matrix $\boldsymbol{A}$. Under a coordinate transformation to new rectangular coordinates (x', y'), by

$$\boldsymbol{A} = \begin{pmatrix} a & b \\ b & c \end{pmatrix}$$

an orthogonal matrix $C = (c_{ij})$ the matrix A is transformed to $A' = C^T A C = C^{-1} A C$. Thus, by choosing a suitable matrix C, A' can be made diagonal

$$\begin{aligned} x' &= c_{11}x + c_{21}y \\ y' &= c_{12}x + c_{22}y \end{aligned}; \quad C = \begin{pmatrix} c_{11} & c_{12} \\ c_{21} & c_{22} \end{pmatrix}; \quad A' = C^{-1}AC = \begin{pmatrix} \lambda_1 & 0 \\ 0 & \lambda_2 \end{pmatrix}.$$

This means that in the new coordinate system the curve is described by the equation $\lambda_1 x'^2 + \lambda_2 y'^2 = d$.

For example, let $3x^2 - 2xy + 3y^2 = 2$ be the equation of a curve. The transformation matrix C of the corresponding symmetric matrix A was found in Example 4:

$$A = \begin{pmatrix} 3 & -1 \\ -1 & 3 \end{pmatrix}; \quad C = \begin{pmatrix} \frac{1}{\sqrt{2}} & -\frac{1}{\sqrt{2}} \\ \frac{1}{\sqrt{2}} & \frac{1}{\sqrt{2}} \end{pmatrix},$$

$$\left.\begin{aligned} x' &= \frac{1}{\sqrt{2}}x + \frac{1}{\sqrt{2}}y \\ y' &= \frac{-1}{\sqrt{2}}x + \frac{1}{\sqrt{2}}y \end{aligned} \;\right|\; \left.\begin{aligned} x &= \frac{1}{\sqrt{2}}x' - \frac{1}{\sqrt{2}}y' \\ y &= \frac{1}{\sqrt{2}}x' + \frac{1}{\sqrt{2}}y' \end{aligned} \right| .$$

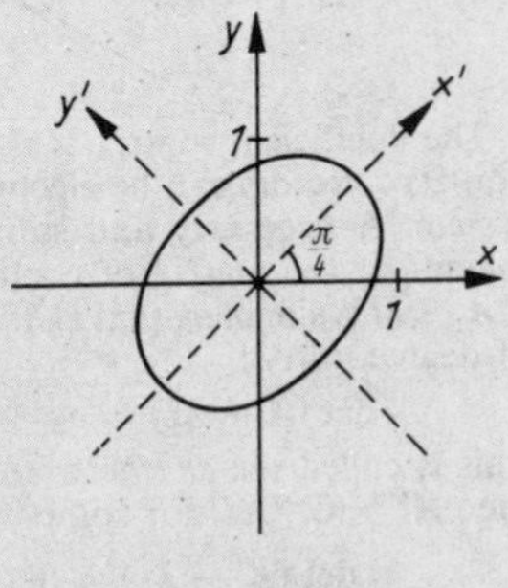

17.6-1 Transformation to principal axes for $3x^2 - 2xy + 3y = 2x'^2 + 4y'^2 = 2$

The coefficients of the last two equations are the entries of $C^{-1} = C^T$. If the expressions for x and y are substituted in the equation, the resulting equation for the curve in the new coordinate system is $2x'^2 + 4y'^2 = 2$.

The matrix C describes a rotation of the plane about the origin through an angle of $\pi/4$, which takes the old coordinate axes into the new ones (Fig.).

17.7. Multilinear algebra

The principal object of multilinear algebra is the investigation of *multilinear forms*, which are generalizations of linear forms. A multilinear form on a vector space V is a function that associates with any r vectors a number and is linear with respect to each variable. This means that if any $r - 1$ vectors are fixed, the mapping so defined is linear in the last vector.

Bilinear forms. If $r = 2$ the form is called *bilinear*. An example of bilinear form is the inner product of vectors. If a basis of the space V is chosen, the bilinear form can be expressed in coordinates; for example, in the case of a two-dimensional space the general expression for a bilinear form is

$$B(x, y) = a_{11}x_1y_1 + a_{12}x_1y_2 + a_{21}x_2y_1 + a_{22}x_2y_2 .$$

If one puts $x = y$, one obtains a *quadratic form* $a_{11}x_1^2 + a_{12}x_1x_2 + a_{21}x_2x_1 + a_{22}x_2^2$. The most important problem in the theory of quadratic forms is to express the given form in the simplest possible way, for instance, in a form without mixed terms. This can always be done by means of the transformation to principal axes.

Tensors. The coefficients of a bilinear form exhibit a regular behaviour under transformations of that form, which is characteristic of tensor coordinates. By generalizing the concept of a vector space in linear algebra one defines *tensor spaces*, whose elements are then called *tensors*.

Applications. *Tensor algebra*, the investigation of tensor spaces, has an important application in differential geometry. There the curvature of a surface or of a space is described by a tensor, the *curvature tensor*. In the theory of relativity the impossibility of separating the energy and impulse of a particle is reflected by the existence of a tensor whose components are the energy and the components of the impulse, the so-called *energy-impulse tensor*. Tensors are also useful in other areas of physics, for instance, in crystal optics and elasticity theory. Thus, the deformation or tension of an elastic medium is described by the *deformation or tension tensor*.

The theory of bilinear and quadratic forms is used in analytic geometry to arrive at the standard classification of conics and quadrics. It is also used in physics, particularly in the description of physical systems subject to small vibrations.

18. Sequences, series, limits

18.1. Sequences

From every non-empty set S of real numbers sequences can be selected by choosing from S in succession a first number a_1, a second number a_2, a third number a_3, and so on, and by considering a_1 to be the first term of the sequence, a_2 the second, a_3 the third, and so on. For example, if from the set of positive integers one selects in their natural order the numbers that are divisible by 2, one obtains the sequence of even numbers, whose first five terms are 2, 4, 6, 8, 10. In the formation of sequences an element of S may be chosen more than once, as for example the number 2 in the sequence 2, 4, 2, 6, 2, 8, 2, 10. If the same number a is always chosen, one obtains a *constant sequence* $a, a, \ldots, a, \ldots$

A *finite sequence* consists of finitely many, say N, terms; a_N is then its last term. The sequence 2, 4, 2, 6, 2, 8, 2, 10 defined above is a finite sequence of eight terms; $a_8 = 10$ is its last term. On the other hand, the sequence of even numbers has no last term, because every term is followed by another one. Such sequences are called *infinite*.

An infinite sequence is given when to every natural number $n \geqslant 1$ there corresponds exactly one real number a_n; a_n is called the nth term of the sequence. If this correspondence exists only for each natural number n between 1 and N ($1 \leqslant n \leqslant N$), then one obtains a finite sequence.

A tabulated representation of this correspondence, for example,

Term number n	1	2	3	4	5	...
Term a_n of the sequence	↓ 2	↓ 4	↓ 6	↓ 8	↓ 10	...

for the sequence of even numbers, shows that one can regard every sequence as a set of ordered pairs of numbers (n, a_n) whose first component n is a natural number, and whose second component, the therm a_n, is a real number. Since the correspondence is single-valued, sequences can also be defined as functions.

Sequences are functions whose domain of definition is a set of natural numbers and whose range consists of real numbers.

Of course, the plausible graphical representation of a sequence, for example, by the sequence of discrete points with the coordinates (n, a_n) in a Cartesian coordinate system, or a tabulated representation, is as unsuitable for the complete description of an infinite sequence as is the enumeration of some of the initial terms of the sequence. For instance, the terms $a_1 = 2$, $a_2 = 3$, $a_3 = 5$ can be continued in a sequence in many, even in infinitely many ways. Examples of such continuations are the sequence of prime numbers, the finite sequence of all the factors of 210, or the sequence 2, 3, 5, 8, 13, 21, ..., in which the kth term for $k > 2$ is the sum of the two preceding terms.

For the complete description of an infinite sequence one tries, therefore, to represent the unique correspondence between the term number n and the corresponding term a_n of the sequence by a *defining law*. In most cases it is possible to state the defining law by means of an *analytical expression* $a_n = f(n)$, $n = 1, 2, 3, \ldots$ One can then denote the sequence $a_1, a_2, a_3, \ldots$ by $\{a_n\} = \{f(n)\}$.

Examples of sequences whose defining law can be stated by means of an analytical expression.

1. The sequence 2, 4, 6, ... of even numbers has the defining law $a_n = 2n$.

2. The sequence 1, 4, 9, ... of perfect squares: $\{a_n\} = \{n^2\}$.

3. The seventh term of the sequence $\{a_n\} = \{n/(n+1)\}$ is obtained by substituting $n = 7$ in the analytical expression to give $a_7 = 7/(7+1) = 7/8$.

4. The sequence $\{a_n\} = \{2^n\}$ for $1 \leqslant n \leqslant 10$ is a finite sequence; its last term is $a_{10} = 2^{10} = 1024$.

5. The defining law $a_n = (-1)^{n+1}\, n$ leads to the sequence $1, -2, 3, -4, 5, -6, \ldots$ This sequence is *alternating*, that is, neighbouring terms have opposite signs. This example also shows that an infinite sequence does not necessarily have a largest or a smallest term.

Sometimes the defining law of a sequence can be given by means of a *recurrence relation*, from which a term a_n can be calculated only when the preceding terms a_i with $i < n$ are already known. For example, the sequence 0, 1, 1, 2, 3, 5, 8, 13, 21, ... of the *Fibonacci numbers* is defined by $a_1 = 0$, $a_2 = 1$, and for $n \geqslant 3$ by the recurrence relation $a_n = a_{n-1} + a_{n-2}$.

However, there are sequences for which neither an analytical expression nor a recursive law can be given, for example the sequence of prime numbers, or the sequence 3, 1, 4, 1, 5, 9, 2, 6, 5, ... whose nth term is the nth digit of the decimal expansion of the number π. From the terms of a sequence one can obtain further sequences, for example, from 1, 1/2, 1/3, ..., $1/n$, ... the sequence $s_1 = 1$, $s_2 = 1 + 1/2$, $s_3 = 1 + 1/2 + 1/3$, ..., $s_n = 1 + 1/2 + 1/3 + \cdots + 1/n$, ..., whose nth term is the sum of the first n terms of the given sequence. In this case the sequence $s_1, s_2, \ldots, s_n, \ldots$ is given by an indirect rule.

The main interest lies in infinite sequences. Of particular interest are properties that follow from the relationship between successive terms.

Monotonic sequences. These are sequences whose terms steadily increase (or steadily decrease) with increasing term number (see Chapter 5.).

A sequence $\{a_n\}$ is called monotonic increasing if each of its terms is greater than its predecessor, that is, $a_{n+1} > a_n$ for all n. It is called monotonic decreasing if $a_{n+1} < a_n$ for all n.

Sometimes in this definition equality is also allowed, and a sequence is called monotonic increasing if $a_{n+1} \geqslant a_n$ and monotonic decreasing if $a_{n+1} \leqslant a_n$. To distinguish between them sequences with the property $a_{n+1} > a_n (a_{n+1} < a_n)$ are then called *strictly monotonic increasing* (*strictly monotonic decreasing*).

The sequence 1, 1/2, 1/3, ..., $1/n$, ... of fractions, for example, is (strictly) monotonic decreasing, and the sequence $-12, -9, -6, -3, 0, \ldots, [-12 + 3(n - 1)], \ldots$ is (strictly) monotonic increasing. Most sequences are neither monotonic increasing nor monotonic decreasing, for example, the sequence 1, 1/2, 2, 1/3, 3, 1/4, 4, ...

Bounded sequences. The sequence $-1/2, 0, 1/6, 2/8, \ldots$ with the defining law $a_n = (n-2)/(2n)$ has the property that none of its terms is greater than 1, and also that none is less than $-1/2$, so that the inequality $-1/2 = a_n < 1$ holds for all n. Such sequences are called bounded.

A sequence $\{a_n\}$ is said to be bounded if there exist two numbers k and K such that the inequality $k \leqslant a_n \leqslant K$ holds for every term a_n of the sequence.

Such a k is called a *lower bound* and K an *upper bound* for the sequence. If $k \leqslant a_n \leqslant K$ for every term a_n of a sequence, then $|a_n| \leqslant M = \text{Max}\,(|k|, |K|)$. Conversely, if a bound M exists for the absolute values $|a_n|$ of the terms, $|a_n| \leqslant M$, then $-M \leqslant a_n \leqslant M$, that is, the sequence is bounded. The definition can therefore also be stated as follows:

A sequence $\{a_n\}$ is bounded if there exists a positive number M that is not exceeded by the absolute value of any element of the sequence: $|a_n| \leqslant M$ for all n.

The numbers k, K, M are not uniquely determined. Clearly, if k is a lower bound of the sequence, so is every smaller number $k' < k$, and if K is an upper bound, so is every greater number $K' > K$. A finite sequence is always bounded; the smallest term of the sequence can be chosen as a lower bound k, and the greatest term as an upper bound K. Infinite sequences can be *unbounded*, for example, the sequence of squares 1, 4, 9, ..., n^2, ... The least upper bound is called the *supremum* G; every smaller number $G - \varepsilon$, where ε is arbitrarily small and positive, is exceeded by at least one term a_m of the sequence $\{a_n\}$, that is, $a_m > G - \varepsilon$. Similarly the greatest lower bound is called the *infimum* g; every greater number $g + \varepsilon (\varepsilon > 0$, arbitrary) exceeds at least one term a_k of the sequence, that is, $a_k < g + \varepsilon$. It can be shown that every bounded sequence has a uniquely determined supremum and a uniquely determined infimum.

These considerations can also be applied generally to number sets, if one replaces 'sequence' by 'set' and 'term' by 'element'.

Arithmetic sequences. In an arithmetic sequence the *difference* d between two consecutive terms is *constant* and non-zero: $a_n - a_{n-1} = d$. For example, the sequence of even numbers 2, 4, 6, 8, ... has the common difference $d = 2$. If one chooses $d = -3$ and the first term $a_1 = 25$, one obtains the sequence 25, 22, 19, 16, 13, ... If d is positive, the arithmetic sequence increases monotonically, and if d is negative, it decreases monotonically. Every infinite arithmetic sequence is unbounded.

Arithmetic sequence	$a_1, a_2 = a_1 + d, a_3 = a_1 + 2d, \ldots, a_n = a_1 + (n - 1)\,d, \ldots$

The name is derived from the fact that every term a_k $(k \geqslant 2)$ is the *arithmetic mean* of its two neighbouring terms: clearly the mean of $a_{k-1} = a_k - d$ and $a_{k+1} = a_k + d$ is $(a_{k-1} + a_{k+1})/2 = (2a_k)/2 = a_k$.

Example 1: The arithmetic sequence with the first term $a_1 = 33$ and the common difference $d = 8$ has the 100th term $a_{100} = a_1 + (n - 1)\,d = 33 + 99 \cdot 8 = 825$.

Example 2: If $a_{10} = 15$ is the 10th term of an arithmetic sequence whose common difference is 2, then the first term is $a_1 = a_n - (n - 1)\,d = 15 - 9 \cdot 2 = -3$.

The process of *linear interpolation* consists in inserting m further terms between two terms a_k and a_{k+1} of an arithmetic sequence with the common difference d, so that they again form an arithmetic sequence. Let d' be the common difference of the required sequence; then

$$a_{k+1} = a_k + (m+1)\,d' = a_k + d, \quad \text{so that} \quad d' = d/(m+1).$$

Example: To interpolate 6 terms between each pair of terms of the arithmetic sequence $\boxed{3}$, $\boxed{17}$, $\boxed{31}$, 45, 59, $\cdots$. Since $d = 14$, the common difference d' of the new sequence is given by $d' = 14/7 = 2$, and this gives the sequence $\boxed{3}$, 5, 7, 9, 11, 13, 15, $\boxed{17}$, 19, 21, 23, 25, 27, 29, $\boxed{31}$, ...

From a given sequence the *difference sequence* is formed by taking the difference between consecutive terms. Thus, an arithmetic sequence can also be described as one whose first difference sequence is constant. In practical mathematics and in the calculus of errors and approximations, *arithmetic sequence of higher order*, for example, of the nth order, are used. In these the nth difference sequence Δ^n is the first one that is constant.

Example: The sequence 1, 8, 27, 64, 125, 216, ... is arithmetic of the third order, since its third difference sequence Δ^3 is constant.

Sequence	1		8		27		64		125		216		...
Δ^1		7		19		37		61		91		...	
Δ^2			12		18		24		30		...		
Δ^3				6		6		6		...			

Geometric sequences. In a geometric sequence the *ratio* $q \neq 1$ of two neighbouring terms is constant: $a_n = a_{n-1}q$. For example, the sequence 9, 3, 1, 1/3, 1/9, ... has the first term $a_1 = 9$ and the common ratio $q = 1/3$. For $a_1 = -1/2$, $q = -2$, one obtains the sequence $-1/2, 1, -2, 4, \ldots$ and for $a_1 = -24$, $q = 1/2$, the sequence $-24, -12, -6, -3, \ldots$ If q is positive, all the terms have the same sign as a_1; if q is negative, the sequence is *alternating*. Geometric sequences are bounded if $|q| \leqslant 1$, and are otherwise unbounded. They increase monotonically for $a_1 > 0$, $q > 1$, and also for $a_1 < 0$, $0 < q < 1$. They decrease monotonically for $a_1 > 0$, $0 < q < 1$, and also for $a_1 < 0$, $q > 1$.

Geometric sequence	$a_1, a_2 = a_1q, a_3 = a_1q^2, \ldots, a_n = a_1q^{n-1}, \ldots$

The name of the sequence is derived from the fact that every term a_k ($k \geqslant 2$) is numerically equal to the *geometric mean* of its two neighbouring terms: for $a_{k-1} = a_k/q$ and $a_{k+1} = a_kq$ have the geometric mean $\sqrt{[(a_k/q)(a_kq)]} = \sqrt{(a_k^2)} = |a_k|$.

Example 1: The geometric sequence with the first term $a_1 = 2$ and the common ratio $q = 1/2$ has the 10th term $a_{10} = a_1q^9 = 2 \cdot (1/2)^9 = 1/256$.

Example 2: If the first term of a geometric sequence is $a_1 = 2/3$ and its 10th term is $a_{10} = a_1q^9 = 13122$, then the common ratio is given by $q = \sqrt[9]{(a_{10}/a_1)} = \sqrt[9]{(3 \cdot 13122/2)} = 3$.

Example 3: If a sufficiently large piece of paper of thickness $a_1 = 0.1$ mm (0.003937″) is folded 40 times, a layer of paper is obtained of thickness $d = a_{41} = 0.1 \text{ mm} \times 2^{40} = 109951162777.6$ mm $\approx$ 109951 km $\approx$ 68335 miles.

Example 4: In passing through a glass plate a light ray loses 1/12 of its intensity L by reflection at the boundary surfaces and by inhomogeneity of the material. After passing through the first plate it has the intensity $a_1 = L - 1/12L = 11/12L$; after passing through the second one it has the intensity $a_2 = 11/12L - 1/12(11/12)L = (11/12)^2 L$; and after passing through the nth plate it has the intensity $a_n = (11/12)^n L$. If it is established by measurement that the intensity a_n is only half the original value, from $a_n = (11/12)^n L = 1/2L$ the number n of plates can be calculated. It is found that $n = \lg 2/(\lg 12 - \lg 11) \approx 8$. Thus, the light ray has penetrated eight plates.

Between any two terms a_k and $a_{k+1} = a_kq$ of a geometric sequence, m numbers can be *interpolated* in such a way that the resulting sequence is again geometric. If q' is the common ratio of this sequence to be determined, then $a_{k+1} = a_k(q')^{m+1} = a_kq$. From this it follows that $q' = \sqrt[m+1]{q}$.

Example 1· Interpolate four terms between each pair of terms of the sequence $\boxed{32}$, $\boxed{1}$, $\boxed{1/32}$, 1/1024 ... For the given sequence $q = 1/32$, and the common ratio of the interpolated sequence is $q' = \sqrt[5]{(1/32)} = 1/2$. Thus, one obtains the sequence $\boxed{32}$, 16, 8, 4, 2, $\boxed{1}$, 1/2, 1/4, 1/8, 1/16, $\boxed{1/32}$, ...

Example 2: In tuning by equal temperament, 11 intermediate notes are arranged at equal distances between the notes of an octave. In C major, for example, the notes are C#, D, D#, E, F, F#, G, G#, A, A#, B. The frequencies of the tones form a geometric sequence between the tones of an octave with frequency ratio $q = 2$. The ratio q' of the required tones is obtained from $q' = \sqrt[12]{2} = 1.059463$, giving the sequence of frequency ratios: C = 1, C# = 1.05946, D = 1.122 44, D# = 1.189 21, E = 1.259 92, F = 1.337 92, F# = 1.414 21, G = 1.498 31, G# = 1.587 40, A = 1.681 79, A# = 1.781 80, B = 1.887 75, C′ = 2.

Basic series of norm numbers

	R 5	*R* 10	*R* 20	*R* 40	*R* 80
q	$\sqrt[5]{10} \approx 1.6$	$\sqrt[10]{10} \approx 1.25$	$\sqrt[20]{10} \approx 1.12$	$\sqrt[40]{10} \approx 1.06$	$\sqrt[80]{10} \approx 1.03$

In normalizing one tries to find gradations of magnitudes that satisfy practical requirements with a minimum number of steps. One uses so-called *decimal geometric sequences*. These are geometric sequences with the *step* or *common ratio* $q = \sqrt[n]{10}$, which are called in technology the basic series of norm numbers.

Accordingly every decimal region is subdivided into n steps. From the basic series one can select further series by using only every second, every third, or every mth step of the series.

Series *R* 10	1	1.25	1.6	2	2.5	3.15	4	5	6.3	8	10	12.5
Selected series *R* 10/2	1		1.6		2.5		4		6.3		10	
		1.25		2		3.15		5		8		12.5

Technological products, machines and machine parts, tools etc., are manufactured according to these basic series. Pressures in presses, lifting forces and heights of cranes and winches, numbers of revolutions, cutting speeds and power of turbines are likewise graded. Internationally agreed paper formats are also geometrically graded, and coins and banknotes are often based on geometric sequences with $q = \sqrt[3]{10} \approx 2.2$, giving the very approximate sequence 1, 2, 5, 10, 20, 50, ... The decimal coinage in Great Britain and the United States coinage conform to this scheme.

Convergence and divergence of sequences. The terms of the sequence 1, 3/4, 4/6, 5/8, ... with the defining law $a_n = (n + 1)/(2n)$ differ from 1/2 by less and less as the term number n becomes greater. The difference $|a_n - 1/2|$ between the terms of the sequence and 1/2 can be made *arbitrarily* small; that is, a suitable value of the index n can be chosen, so that *from this value of n onwards all* the differences $|a_n - 1/2|$ are smaller than an arbitrarily small given positive number ε. If it is required, for example, that the deviation from 1/2 shall be at most $\varepsilon = 0.001$, then from $|a_n - 1/2| = |(n+1)/(2n) - 1/2| = |1/2 + 1/(2n) - 1/2| = 1/(2n) < 0.001$, it follows that all terms a_n with $n > 500$ have the required property. At most 500 terms have a greater deviation from 1/2. If the required precision is increased to $\varepsilon = 0.000001$, then only 500000 terms have a greater deviation from 1/2, and $|a_n - 1/2| < 0.000001$ for all terms a_n with $n > 500000$. In general, $|a_n - 1/2| < \varepsilon$ for *all* $n > 1/(2\varepsilon)$. Thus, no matter how small ε is chosen, it is always possible to choose an index such that from this term onwards all the terms of the sequence differ from 1/2 by less than ε. The sequence $\{a_n\}$ is then said to *converge* to the *limit* 1/2.

A sequence $\{a_n\}$ is said to be convergent to the limit a if to every arbitrarily small positive number ε there corresponds a number $N(\varepsilon)$ such that the inequality $|a_n - a| < \varepsilon$ is satisfied for all terms a_n of the sequence with $n > N(\varepsilon)$.

The number N beyond which $|a_n - a| < \varepsilon$ depends, in general, on ε; the smaller ε is chosen, the greater is N. For this reason it is denoted more precisely by $N(\varepsilon)$. For a sequence $\{a_n\}$ converging to the limit a, to every $\varepsilon > 0$ there corresponds, of course, a number $N_2(\varepsilon)$ beyond which $|a_n - a| < \varepsilon/2$, a number $N_k(\varepsilon)$ beyond which $|a_n - a| < \varepsilon/k$, a number $N'(\varepsilon)$ beyond which $|a_n - a| < \varepsilon^\alpha$, and so on. Abundant use will be made of this. If the sequence $\{a_n\}$ converges to the limit a, one writes $\{a_n\} \to a$ as $n \to \infty$, or $\lim\limits_{n \to \infty} a_n = a$ (read *a_n converges to a as n tends to infinity*, or *the limit of a_n, as n tends to infinity, is a*). Pictured geometrically, this means that only finitely many terms of the sequence lie outside the ε-neighbourhood $a - \varepsilon \cdots a + \varepsilon$ of the limit a, whilst all other terms lie within this ε-neighbourhood. Thus, one says that *almost all* the terms of the sequence lie in the ε-neighbourhood of a, no matter how small ε may be.

Example 1: The sequence 0.3, 0.33, 0.333, ... with the defining law $a_n = 3/10 + 3/10^2 + \cdots + 3/10^n$ converges to 1/3. For the magnitudes of the deviations from the limit one finds

$$\begin{aligned} |a_n - 1/3| &= |3(10^{n-1} + \cdots + 10^1 + 1)/10^n - 1/3| \\ &= |[9(10^n - 1)/(10 - 1) - 10^n]/(3 \cdot 10^n)| = 1/(3 \cdot 10^n) < \varepsilon. \end{aligned}$$

For an arbitrary given positive ε, this inequality is satisfied for all $n > N(\varepsilon) = \lg[1/(3\varepsilon)]$. For $\varepsilon = 10^{-12}$, for example, $N(\varepsilon) = 12 - \lg 3$; thus, in this case only 12 terms differ from 1/3 by more than $\varepsilon = 10^{-12}$. In general, every infinite decimal fraction $0.z_1z_2z_3 \ldots$ with digits z_i can be regarded as a convergent sequence $\{a_n\}$ with $a_n = z_1/10 + z_2/10^2 \cdots + z_n/10^n$. The limit of the sequence is the real number represented by the decimal fraction.

Example 2: The sequence 1, 1/4, 1/9, 1/16, ... of reciprocals of squares has the limit zero, because for arbitrary $\varepsilon > 0$, $|a_n - 0| = |1/n^2 - 0| = 1/n^2 < \varepsilon$ for all $n > N(\varepsilon) = 1/\sqrt{\varepsilon}$.

Sequences with the limit zero are called *null sequences*. From every null sequence $\{b_n\}$ a sequence $\{b_n + b\}$ with the limit b can be constructed. Conversely, if the sequence $\{a_n\}$ converges to the limit a, then $\{a_n - a\}$ is a null sequence.

Convergence behaviour of arithmetic and geometric sequences. Sequences that do not converge are called *divergent*. For example, every arithmetic sequence is divergent. Since the difference between two consecutive terms is always d, it is never possible for almost all its terms to lie in a neighbourhood of a fixed value. For positive values of d the terms a_n of the sequence are all eventually greater than every arbitrary large number. One therefore writes symbolically $\lim_{n\to\infty} a_n = \infty$ and calls such a sequence *definitely divergent*. For negative values of d the terms are eventually less than every negative number of arbitrarily large absolute value. This sequence is also definitely divergent; one writes $\lim_{n\to\infty} a_n = -\infty$.

The infinite geometric sequence with the defining law $a_n = a_1 q^{n-1}$ converges to zero if the absolute value $|q|$ of the common ratio is less than 1. If $|q|$ is greater than 1, then the sequence $\{a_n\}$ is divergent, in particular, definitely divergent for $q > 1$.

Subsequences. If $p_1, p_2, p_3, \ldots, p_n, \ldots$ is any strictly monotonic increasing infinite sequence of natural numbers, then $\{p_n\}$ is called a *subsequence* of the sequence of natural numbers; for example, the sequence 1, 3, 7, 9, 13, 14, 27, ... If such a sequence $\{p_n\}$ of indices is chosen, this determines from any sequence $\{a_n\}$ one of its subsequences $\{a_{p_n}\}$. For example, 1, 1/8, 1/64, ... is a subsequence of the sequence 1, 1/2, 1/4, 1/8, 1/16, ... If the terms a_n for all $n > N(\varepsilon)$ lie in the ε-neighbourhood of the limit a, so that $|a_n - a| < \varepsilon$, then the terms a_{p_n} of the subsequence with $p_n > N(\varepsilon)$ also lie in this neighbourhood. Hence the following theorem holds.

Every subsequence $\{a_{p_n}\}$ of a convergent sequence $\{a_n\} \to a$ converges to the same limit a.

Theorems about convergent sequences. The convergence of the sequence $\{a_n\}$ is decided by the *existence* of a subscript $N(\varepsilon)$ beyond which $|a_n - a| < \varepsilon$; the size of $N(\varepsilon)$ is completely immaterial. For this reason finitely many terms can be removed or added without altering the convergence or the limit of the sequence, since this affects at most the size of $N(\varepsilon)$. Such properties, which depend only on the behaviour of all terms 'beyond the place $N(\varepsilon)$', are called *infinitary properties* of a sequence. The convergence of a sequence is an infinitary property.

Convergent sequences are bounded.

If a sequence $\{a_n\}$ has the limit a, then almost all its terms lie in the interval from $a - \varepsilon$ to $a + \varepsilon$; but the set of terms that lie outside this interval is finite and therefore is also bounded.

If the convergent sequence $\{a_n\}$ has the upper bound K, then its limit a is also not greater than K. Otherwise infinitely many terms of the sequence would have to fall in a neighbourhood of a lying entirely to the right of K, in contradiction to the bounding property of K. Similarly, the limit a of the sequence cannot be less than any of its lower bounds.

A convergent sequence has exactly one limit.

If $\{a_n\}$ has two different limits a and a', then ε can be chosen so small that the ε-neighbourhoods of a and a' have no point in common. From some place $N(\varepsilon)$ onwards, infinitely many terms of the sequence lie outside the ε-neighbourhood of a, and infinitely many lie outside the ε-neighbourhood of a', which contradicts the limit property of a and of a'.

If the sequences $\{a_n\}$ and $\{b_n\}$ have the limits a and b, then the sequences $\{a_n + b_n\}$, $\{a_n - b_n\}$, $\{a_n b_n\}$ converge to the limits $a + b$, $a - b$, ab, respectively, and if b_n and b are different from zero, then $\{a_n/b_n\}$ converges to the limit a/b.

Suppose, for example, that it is required to show for an arbitrarily prescribed ε that $|(a_n + b_n) - (a + b)| < \varepsilon$. From the convergence of the sequence $\{a_n\}$, an index N_1 can be determined so that $|a_n - a| < \varepsilon/2$ if $n > N_1$, and similarly, for the sequence $\{b_n\}$, an index N_2 so that $|b_n - b| < \varepsilon/2$ if $n > N_2$. For all $n > \max(N_1, N_2)$, it follows from the triangle inequality that $|(a_n + b_n) - (a + b)| = |(a_n - a) + (b_n - b)| \leqslant |a_n - a| + |b_n - b| < \varepsilon$, as required.

To show that $|a_n/b_n - a/b|$ can be made smaller than any given positive number ε, one first notes that $|a_n/b_n - a/b| = |[b(a_n - a) - a(b_n - b)]/(b \cdot b_n)| \leqslant [|b| \cdot |a_n - a| + |a| \cdot |b_n - b|]/(|b| \cdot |b_n|)$. N_3 can be determined so that $|b_n| \geqslant g > 0$ for all $n > N_3$; this is always possible since $b \neq 0$. Finally N_1 can be determined so that $|a_n - a| < g\varepsilon/2$ for all $n > N_1$, and N_2 so that $|b_n - b|$

$< g|b|\,\varepsilon/(2|a|)$ for all $n > N_2$. It then follows that $|a_n/b_n - a/b| < \varepsilon$ when $n > \max(N_1, N_2, N_3)$.

The following statements are important special cases of the last theorem.

1. If c, c_1, and c_2 are constants, and $\{a_n\} \to a$, $\{b_n\} \to b$, then $\{ca_n\} \to ca$ and $\{c_1 a_n + c_2 b_n\} \to c_1 a + c_2 b$.

2. Since the sequence of the products of the terms of two convergent sequences converges to the product of their limits, it follows that $\{a_n^k\} \to a^k$ for every positive integer k whenever $\{a_n\} \to a$. If $a_n \neq 0$ and $a \neq 0$, this also holds for every negative integer k. One can even deduce that $\{a_n^\alpha\} \to a^\alpha$ for every real number α if $a_n \neq 0$, $a \neq 0$.

3. If $\{a_n\}$ and $\{b_n\}$ are null sequences, so are the sequences $\{a_n + b_n\}$, $\{a_n - b_n\}$ and $\{a_n b_n\}$.

The sequence $\{a_n/b_n\}$ formed from the null sequences $\{a_n\}$ and $\{b_n\}$ is not, in general, a null sequence. For example, $\{a_n\} = \{1/2^n\}$ and $\{b_n\} = \{1/4^n\}$ are null sequences, but $\{a_n/b_n\} = \{2^n\}$ is definitely divergent.

If the sequences $\{a_n'\}$ and $\{a_n''\}$ converge to the same limit a, and the relation $a_n' \leqslant a_n \leqslant a_n''$ holds for almost all terms of the sequence $\{a_n\}$, then $\{a_n\}$ also converges to the limit a.

Corresponding to an arbitrary $\varepsilon > 0$ there exists an $N(\varepsilon)$ beyond which all terms of the sequence $\{a_n'\}$ and all terms of the sequence $\{a_n''\}$ lie in the ε-neighbourhood of a. Since $a_n' \leqslant a_n \leqslant a_n''$, almost all terms of the sequence $\{a_n\}$ also lie in this neighbourhood, so that $\lim_{n\to\infty} a_n = a$.

Limits of some important convergent sequences

$\lim_{n\to\infty} \sqrt[n]{q} = 1$, for arbitrary $q > 0$	$\lim_{n\to\infty} \sqrt[n]{n} = 1$	$\lim_{n\to\infty} (\log_b n)/n = 0$, $b > 0$, $b \neq 1$

1. For arbitrary positive values of q, $\{x_n\} = \{\sqrt[n]{q} - 1\}$ is a null sequence. For $q = 1$ every term has the value zero. For $q > 1$, $\sqrt[n]{q} > 1$, so that the numbers x_n are positive. Hence $q = (1 + x_n)^n > 1 + nx_n > nx_n > 0$, or $0 < x_n < q/n$. But $\{q/n\}$ is a null sequence, so that $\{x_n\}$ is a null sequence. For $q < 1$, $1/q > 1$ and hence $\{\sqrt[n]{(1/q)} - 1\}$ has the limit zero. If this null sequence is multiplied term-by-term by the sequence $\{\sqrt[n]{q}\}$, which is bounded since $\sqrt[n]{q} < 1$, then the product sequence $\{1 - \sqrt[n]{q}\}$, and hence the sequence $\{\sqrt[n]{q} - 1\}$, is a null sequence.

2. As was shown above, the terms of the sequence $\{q^{1/n}\}$ have the limit 1 as n tends to infinity, where q is an anbitrary positive number. Hence a number N can be found, such that for all $m > N$ both values $q^{\pm 1/m}$ lie between $1 - \varepsilon$ and $1 + \varepsilon$. For a null sequence $\{a_n\}$, an index N_1 can always be found so that a_n lies between $-1/m$ and $+1/m$ for all $n > N_1$, and thus for all $n > N_1$ the powers q^{a_n} lie between $1 - \varepsilon$ and $1 + \varepsilon$. Hence $q^{a_n} - 1$ lies between $-\varepsilon$ and $+\varepsilon$, that is, $\{q^{a_n} - 1\}$ is a null sequence if $\{a_n\}$ is a null sequence. From this it follows that $\{q^{a_n}\}$ converges to the limit q^a if $\{a_n\} \to a$. For $q^{a_n} - q^a = q^a(q^{a_n - a} - 1)$, where $\{a_n - a\}$ is a null sequence and hence $\{q^{a_n - a} - 1\}$ is also one.

$\{q^{a_n}\} \to 1$ if $\{a_n\} \to 0$ and q is positive $\{q^{a_n}\} \to q^a$ if q is positive and $\{a_n\} \to a$ $\{(a_n)^\alpha\} \to a^\alpha$ if $\{a_n\} \to a$, $a_n > 0$, $a > 0$, α real	

It will be shown below under 4. that for an arbitrary basis $b > 1$ of a system of logarithms, $\{\log_b a_n\} \to \log_b a$ if $\{a_n\} \to a$. If α is an arbitrary real constant, then $\{\alpha \log_b a_n\} \to \alpha \log_b a$ and also, as has just been shown, $\{b^{\alpha \log_b a_n}\} \to b^{\alpha \log_b a}$, hence $\{a_n^\alpha\} \to a^\alpha$.

3. The sequence $\sqrt[n]{n}$ converges to 1, that is, $\{x_n\} = \{\sqrt[n]{n} - 1\}$ is a null sequence. For $n \geqslant 2$ its terms x_n are positive. From $(1 + x_n)^n = n$ one obtains from the binomial theorem $n(n-1)\,x_n^2/2 \leqslant n$, or $|x_n| \leqslant \sqrt{[2/(n-1)]}$. If corresponding to the prescribed number $\varepsilon > 0$, the number $N(\varepsilon) = 2/\varepsilon^2 + 1$ is chosen, then $|x_n| < \varepsilon$ for all $n > N(\varepsilon)$.

4. For an arbitrary basis $b > 1$ of logarithms, $\{(\log n)/n\}$ is a null sequence; in other words, for an arbitrary ε there must exist a number $N(\varepsilon)$ such that $(\log n)/n < \varepsilon$ for all $n > N(\varepsilon)$. But

$$(\log n)/n < \varepsilon \Longleftrightarrow \log n < n\varepsilon \Longleftrightarrow n < b^{\varepsilon n} \Longleftrightarrow \sqrt[n]{n} < b^{\varepsilon}.$$

Since b^ε is greater than 1, $\sqrt[n]{n}$ converges to 1, and the above argument is reversible, the result follows. Moreover, since $\log_{1/b} n = -\log_b n$, the sequence $\{(\log n)/n\}$ also converges to zero for $0 < b < 1$.

Convergence criteria for sequences. From the definition of convergence one can test whether a number a is, in fact, the limit of the sequence $\{a_n\}$. On the other hand, if no such number a is known, one uses *convergence criteria* (or tests for convergence) that allow one to determine the convergence or divergence of a sequence from properties that are, in general, easily verified. However, there is no general method for the determination of the limit; this can be found only by a variety of methods, specially constructed for particular sequences.

The first test for convergence. The terms of a monotonic increasing, unbounded sequence assume arbitrarily large values; the sequence is *definitely divergent*. But if a monotonic sequence is bounded, it can be shown to have a limit.

The first test for convergence: A monotonic and bounded sequence is always convergent.

The number e *as a limit.* The sequence $\{a_n\}$ with $a_n = (1 + 1/n)^n$ increases monotonically, since for $n \geqslant 2$

$$a_{n-1} = [1 + 1/(n-1)]^{n-1} = [n/(n-1)]^{n-1} = [n/(n-1)]^n \cdot (1 - 1/n) < [n/(n-1)]^n \cdot (1 - 1/n^2)^n = [(n+1)/n]^n = (1 + 1/n)^n = a_n .$$

The inequality results for $a = -1/n^2$ from the use of *Bernoulli's inequality* $1 + na < (1 + a)^n$, which holds for $a > -1$, $a \neq 0$, $n \geqslant 2$. The sequence $\{(1 + 1/n)^n\}$ is bounded. Since all the terms are positive, zero is a lower bound. The binomial theorem gives

$$a_n = (1 + 1/n)^n = 1 + \binom{n}{1}\Big/ n + \cdots + \binom{n}{k}\Big/ n^k + \cdots \binom{n}{n}\Big/ n^n .$$

One can obtain an estimate for each term in this sum by

$$\binom{n}{k}\Big/ n^k = (1/k!)\,(1 - 1/n)\,(1 - 2/n) \cdots [1 - (k-1)/n] \leqslant 1/(2 \cdot 3 \cdots k) \leqslant 1/2^{k-1},$$

so that

$$a_n = (1 + 1/n)^n < 1 + 1 + 1/2 + 1/2^2 + \cdots + 1/2^{n-1} < 1 + 1/(1 - 1/2) = 3 .$$

Thus, the sequence is also bounded above. It is therefore convergent, and following Leonhard EULER its limit is denoted by e. The number e is sandwiched between the terms of the sequence considered and those of the monotonic decreasing sequence $\{[1 + 1/(n-1)]^n\}$, which likewise converges to e. However, e is usually calculated by means of the series

$$\mathrm{e} = 1 + 1/1! + 1/2! + 1/3! + \cdots .$$

$\lim_{n\to\infty} (1 + 1/n)^n = \mathrm{e}$	$\mathrm{e} = 2.718281828459045 23536...$

The second (or Cauchy) test for convergence. While the first test for convergence applies only to monotonic sequences, the Cauchy test holds for arbitrary sequences. If the differences of all possible pairs of terms from some place $N(\varepsilon)$ onwards are less than a given positive number ε, then almost all terms of the sequence lie in an ε-neighbourhood. At most the finitely many terms a_i with $i \leqslant N(\varepsilon)$ can lie outside it. The expression 'if and only if' in the following indicates that the test is both necessary and sufficient.

The second test for convergence: A sequence $\{a_n\}$ is convergent if and only if corresponding to every arbitrary positive number ε, a number $N(\varepsilon)$ can always be chosen so that $|a_n - a_m| < \varepsilon$ for all indices n and m greater than $N(\varepsilon)$.

Example 1: The sequence 1/2, 5/4, 5/6, 9/8, 9/10, 13/12, 13/14, ... with the general term $a_n = 1 + (-1)^n/(2n)$ is bounded, but not monotonic. The first convergence test is not applicable. The Cauchy test establishes the convergence of the sequence, for

$$\begin{aligned} |a_{n+1} - a_n| &= |1 + (-1)^{n+1}/(2n+2) - 1 - (-1)^n/(2n)| \\ &= |[(-1)^{n+1} \cdot 2n - (-1)^n\,(2n+2)]/[2n(2n+2)]| \leqslant |[2n + 2n + 2]/[2n(2n+2)]| \\ &= |[4n+2]/[4n^2 + 4n]| \leqslant [4n+4]/[4n^2 + 4n] = 1/n < \varepsilon \text{ for all } n > 1/\varepsilon . \end{aligned}$$

As is obvious from the defining law, all the elements following a_{n+1} lie between a_n and a_{n+1}; thus, for arbitrary $n, m > 1/\varepsilon$, $|a_n - a_m| \leqslant |a_{n+1} - a_n| < \varepsilon$.

Example 2: The sequence 1, $1 + 1/2$, $1 + 1/2 + 1/3$, $1 + 1/2 + 1/3 + 1/4$, ... with the general term $a_n = 1 + 1/2 + 1/3 + \cdots + 1/n$ does not satisfy the Cauchy convergence test, since if one chooses an $\varepsilon < 1/2$, there always exist two numbers $n, m > N(\varepsilon)$, for which $|a_n - a_m| > \varepsilon$, no matter how large $N(\varepsilon)$ is. Suppose that $m > N$ and $n = 2m > N$. Then one obtains

$$\begin{aligned} |a_n - a_m| &= 1/(m+1) + 1/(m+2) + 1/(m+3) + \cdots + 1/(2m) \\ &> 1/(2m) + 1/(2m) + \cdots + 1/(2m) = m \cdot 1/(2m) = 1/2 > \varepsilon , \end{aligned}$$

where each of the fractions $1/(m+i)$ $(i = 1, 2, \ldots, n)$ has been replaced by the smaller, or at most equally large, fraction $1/(2m)$.

Accumulation point of a sequence. The sequence $1 + 1/2, 2 + 1/2, 3 + 1/2, 1 + 1/3, 2 + 1/3, 3 + 1/3, \ldots, 1 + 1/n, 2 + 1/n, 3 + 1/n, \ldots$ has the property that infinitely many terms of the sequence lie in every neighbourhood of each of the numbers 1, 2, and 3. The terms of the sequence accumulate in the neighbourhood of the points 1, 2, and 3, which are therefore called *accumulation points* of the sequence.

A number A is called an accumulation point of the sequence $\{a_n\}$ if for every arbitrary positive number ε the inequality $|a_n - A| < \varepsilon$ is satisfied for infinitely many distinct terms a_n.

From this it follows that the limit of a sequence is always one of its accumulation points. On the other hand, an accumulation point is not necessarily a limit, because for an accumulation point A the inequality $|a_n - A| < \varepsilon$ has only to be satisfied for *infinitely many n*, but for a limit A it must be satisfied for *all n* from a particular place $N(\varepsilon)$ onwards. Consequently, a convergent sequence can have only one accumulation point, since only finitely many terms of the sequence lie outside each ε-neighbourhood of the limit L, and in particular, it is impossible for infinitely many terms to lie in every ε-neighbourhood of $L' \neq L$. The following theorem shows that the converse of this statement is also true.

A bounded sequence with exactly one finite accumulation point is convergent. But if a sequence has no finite accumulation point, or more than one, then it is divergent.

The Bolzano-Weierstrass theorem: Every bounded infinite sequence has at least one accumulation point.

If k is a lower bound and K an upper bound of the sequence, then all its terms lie in the interval $J_0 = [k, K]$. This interval is bisected and the half in which infinitely many terms of the sequence lie (or the left-hand half if both contain infinitely many) is denoted by J_1. The same procedure of bisecting and choosing one half interval applied to the interval J_1 gives J_2, and so on. The *nest of intervals* so constructed contains exactly one real number A, which is an accumulation point of the sequence. For since the lengths of the intervals of the nested set converge to zero, corresponding to every ε-neighbourhood of A there is an interval of the set that lies entirely in this neighbourhood and, by construction, contains in addition infinitely many terms of the sequence.

The concept of the accumulation point can be extended to *arbitrary sets of numbers*; the Bolzano-Weierstrass theorem ensures the existence of at least one accumulation point for bounded infinite sets of numbers. This need not itself be an element of the set; for example, the accumulation points 1, 2, 3 in the example considered above do not belong to the set.

18.2. Series

Series are of special significance, as much for the inner structure of mathematics as for practical applications. Many numerical methods are based on the theory of series; for example, the construction of tables of logarithms and of trigonometric functions and the calculation of important constants such as e and π are best accomplished with the help of series (see Chapter 21.).

The concept of a series. The Greek sophist ZENON (5th cent. B. C.) posed the question whether Achilles, who runs twelve times as fast as a tortoise, can overtake it if he gives it a start of 1 stadion (an ancient measure of length, approximately 200 yards). While the tortoise crawls a distance of 1/11 stadion, Achilles with twelve times the speed covers $12 \cdot 1/11 = 1 + 1/11$ stadion, that is, the head start and the path of the tortoise; thus, he has overtaken it. On the other hand Zenon argued: when Achilles has covered 1 stadion, the tortoise has crawled 1/12 stadion; when he has run this twelfth, the tortoise still has a lead of $1/12^2$ stadion; when he has covered this, the tortoise is still $1/12^3$ stadion ahead, and so on. The distance covered by Achilles to the point of overtaking can therefore be expressed in the form $1 + 1/12 + 1/12^2 + 1/12^3 + \cdots$, where the dots denote that every term $a_k = 1/12^{k-1}$ is followed by another one $a_{k+1} = 1/12^k$, so that the expression does not terminate. Such an expression is called an *infinite series*. ZENON believed that in this example he had found a contradiction in the formal thinking, since it seemed certain to him that the value of the infinite series is greater than every quantity so that Achilles could never catch up with the tortoise. However, the series correctly set up by him is geometric and has the sum 12/11, as follows from the rules derived for these series.

By an infinite series (or just a series) one understands an expression of the form $a_1 + a_2 + a_3 + \cdots$, abbreviated to $\sum_{i=1}^{\infty} a_i$, where the a_i are terms of an infinite number sequence $\{a_n\}$.

Use of the summation sign. To write a sum in an abbreviated form one uses the Greek letter Σ and writes, for example, $b_1 + b_2 + b_3 + \cdots + b_n = \sum_{i=1}^{n} b_i$ (read sum of b_i for i equals 1 to n). The addition to the sign Σ of the condition 'i equals 1 to n' implies that the terms of the sum are given by letting the *summation index* i assume in succession the values of all the natural numbers from 1 to n. For example, $\sum_{i=1}^{5} 1/i^2 = 1/1^2 + 1/2^2 + 1/3^2 + 1/4^2 + 1/5^2$. The sign is also used to write infinite series in abbreviated form. For example, the series obtained by ZENON is written $1 + 1/12 + 1/12^2 + \cdots = \sum_{i=0}^{\infty} 1/12^i$. The symbol ∞ means that the series does not terminate. The summation index no longer occurs when the series is written as the sum of its terms, and it is therefore immaterial whether it is denoted by i, k, or any other letter. It is often convenient to give the first term of a series the subscript 0, so that the series has the form $\sum_{n=0}^{\infty} a_n$.

Convergence and divergence. Sum of a series. As has already been mentioned, one can associate the value 12/11 with the series $\sum_{i=0}^{\infty} 1/12^i$. In order to be able to decide, in general, whether a value can be associated with a series $\sum_{i=1}^{\infty} a_i$, one forms from the terms a_i of the series the sequence $\{s_n\}$ of its *partial sums*. If and only if this sequence of partial sums converges, to the limit S say, will a value be ascribed to the series, namely the value S. One says: the series *converges* and has the *sum* S.

An infinite series $\sum_{i=1}^{\infty} a_i$ is said to be convergent if and only if the sequence of its partial sums converges. The limit S of the sequence of partial sums is called the sum of the series

$$S = a_1 + a_2 + a_3 + \cdots \quad \text{or} \quad S = \sum_{i=1}^{\infty} a_i .$$

On the other hand, if the sequence of partial sums of the given series diverges, the series is said to be divergent; it has no sum.

	Sequence of partial sums
a_1	$s_1 = a_1 = \sum_{i=1}^{1} a_i$
a_2	$s_2 = a_1 + a_2 = \sum_{i=1}^{2} a_i$
a_3	$s_3 = a_1 + a_2 + a_3 = \sum_{i=1}^{3} a_i$
$\vdots$	$\vdots$
a_n	$s_n = a_1 + a_2 + a_3 + \cdots + a_n = \sum_{i=1}^{n} a_i$

The word *sum* of a series is chosen only on the grounds of the formal analogy with sums of finitely many terms and is simply a synonym for the concept of *limit of the sequence of partial sums.* In ZENON's series the general term s_n of its sequence of partial sums has the value $s_n = (12/11)(1 - 1/12^n)$ (see Sum of a finite geometric series), and this converges to the limit $S = \lim_{n\to\infty} s_n = \lim_{n\to\infty} (12/11)(1 - 1/12^n) = 12/11$.

Example: If a square of unit area is repeatedly halved, as indicated in the accompanying figure, then the area of the resulting rectangles can be considered as terms of the infinite series $1/2 + 1/4 + 1/8 + \cdots + 1/2^n + \cdots$.

The geometrical aspect leads one to suppose that the series has the sum 1. Since the sequence of partial sums of the given series is 1/2, 3/4, 7/8, ..., $(2^n - 1)/2^n$, ..., it does indeed converge to the limit $s = \lim_{n\to\infty} s_n = \lim_{n\to\infty} (2^n - 1)/2^n = \lim_{n\to\infty} (1 - 1/2^n) = 1$.

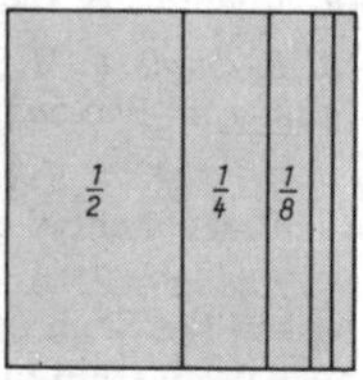

18.2-1 The convergence of the series $1/2 + 1/4 + 1/8 + \cdots$

Even in the 18th century these concepts had not been clarified. For example, the infinite series $1 - 1 + 1 - 1 + 1 \ldots$ was written either $(1 - 1) + (1 - 1) + (1 - 1) + \cdots$ or $1 - (1 - 1) - (1 - 1) - \cdots$ with the corresponding sums 0 or 1. But the sequence of its partial sums $s_1 = 1$, $s_2 = 0$, $s_3 = 1$, $s_4 = 0$, ... diverges, and the series has no sum.

Arithmetic series. In an *arithmetic series* $\sum_{i=1}^{\infty} a_i$ the a_i are the terms of an arithmetic sequence $\{a_n\}$. Clearly every infinite arithmetic series is divergent. Only its nth partial sum $s_n = \sum_{i=1}^{n} a_i$, often called a *finite arithmetic series*, is of any interest. When Carl Friedrich GAUSS was nine years old, his school teacher gave to the class the task of adding together all the whole numbers from 1 to 100. He had hardly sat down, however, when little GAUSS put his slate down on the desk with the words: 'There it is'. The teacher was all the more astonished when all the slates were finally given in to find on the first one only one number, just the right answer 5050, which GAUSS had calculated in his head by means of the scheme indicated below.

The teacher recognized that little Carl Friedrich could not learn much in his arithmetic class, and procured a special arithmetic book for him from Hamburg: Remer's Arithmetica. The idea of the nine-year-old Gauss can be applied to every finite arithmetic series.

$$\begin{array}{l} s_n = a_1 + (a_1 + d) + (a_1 + 2d) + \cdots + (a_1 + [n-1]\,d) \\ s_n = a_n + (a_n - d) + (a_n - 2d) + \cdots + (a_n - [n-1]\,d) \\ \hline 2s_n = n(a_1 + a_n) \quad \text{or} \quad s_n = n(a_1 + a_n)/2. \end{array}$$

$$\begin{array}{c} 1 + 2 + 3 + \cdots + 50 + \\ 100 + 99 + 98 + \cdots + 51 \\ \hline 50 \times 101 = 5050 \end{array}$$

Finite arithmetic series	first term a_1, last term $a_n = a_1 + (n-1)\,d$, common difference d, sum $s_n = n(a_1 + a_n)/2 = na_1 + dn(n-1)/2$

From the formulae for the sum it can be seen that three of the numbers a_1, a_n, d, n and s_n must be given; the remaining ones can then be calculated from linear or quadratic equations.

Example 1: If for a finite arithmetic series $a_1 = 3$, $a_n = 43$ and $d = 5$ are known, then the number of terms n and the sum s_n can be calculated.

$$a_n = a_1 + (n-1)\,d \rightarrow 43 = 3 + (n-1)\cdot 5 \rightarrow n = 9;$$
$$s_n = n(a_1 + a_n)/2 \qquad s_9 = 9(3 + 43)/2 = 207.$$

Example 2: From $d = 12$, $s_n = 180$, and $a_n = 60$ the first term a_1 and the number of terms n can be found

$$a_1 = a_n - (n-1)\,d; \quad s_n = (n/2)\,(a_1 + a_n) \rightarrow s_n = (n/2)\,[2a_n - (n-1)\,d] \rightarrow$$
$$180 = (n/2)\,[120 - (n-1)\cdot 12] \rightarrow n^2 - 11n + 30 = 0.$$

This quadratic equation has the solutions $n_1 = 6$, $n_2 = 5$. From $n_1 = 6$ it follows that $a_1 = 0$ and from $n_2 = 5$ that $a_1 = 12$. The finite series $12 + 24 + 36 + 48 + 60$ and $0 + 12 + 24 + 36 + 48 + 60$ both correspond to the given values.

Geometric series. In a *geometric series* $\sum_{i=1}^{\infty} a_i$, the a_i are terms of a geometric sequence $\{a_n\}$. The nth term s_n of the sequence of partial sums, often denoted also as the *sum of the finite geometric series* $\sum_{i=1}^{n} a_i$, is obtained by means of the adjacent scheme:

$$\begin{array}{rl} s_n &= a_1 + a_1q + a_1q^2 + \cdots + a_1q^{n-1} \\ qs_n &= \quad a_1q + a_1q^2 + \cdots + a_1q^{n-1} + a_1q^n \\ \hline s_n - qs_n &= a_1 - a_1q^n, \quad \text{or} \quad s_n = a_1(1-q^n)/(1-q). \end{array}$$

Finite geometric series	first term a_1, last term $a_n = a_1q^{n-1}$, common ratio q, sum $s_n = a_1(1-q^n)/(1-q) = a_1(q^n-1)/(q-1)$ for $q \neq 1$

From the formula for the sum it can be seen that three of the numbers a_1, a_n, q, n and s_n must be given and the remaining ones can then be calculated. However, in the process exponential equations or equations of the nth degree can occur.

Example 1: If in a geometric series the first term $a_1 = 2$, the common ratio $q = 5$, and the sum $s_n = 976\,562$ are given, then the number of terms n can be calculated

$$s_n = a_1(q^n - 1)/(q-1) \rightarrow 976562 = 2(5^n - 1)/(5-1) \rightarrow 5^n = 1953125.$$

This exponential equation has the solution $n = 9$. The series has nine terms.

Example 2: According to the Arabian historian Ja'qubi, the inventor of the game of chess asked the Shah of Persia as a reward for the number of grains of wheat that would result from placing 1 grain on the first of the 64 squares of the chess board, 2 on the second, 4 on the third, and so on, placing on each square twice as many as on the previous one. The total number of grains is given by the formula

$$s_n = a_1(q^n - 1)/(q-1), \quad \text{so that} \quad s_{64} = 1(2^{64} - 1)/(2-1) = 2^{64} - 1 \approx 1.84 \times 10^{19}.$$

Assuming that the surface of the earth (roughly 13×10^{10} acres) forms a single wheatfield with a yield of 1.6 tons per acre, and assuming 20 million grains to the ton, then four harvests would still not be enough to yield the required quantity.

Infinite geometric series. The sum $s_n = a_1(1-q^n)/(1-q)$ for $q \neq 1$ is the nth term of the sequence of partial sums. The numbers a_1 and q are constants and the convergence of the sequence depends only on the magnitude of $(1 - q^n)$. For $q > 1$ and for $q < -1$ the sequence $\{q^n\}$ is divergent, so that the geometric series also has *no sum*. For $|q| < 1$, $\{q^n\}$ is a null sequence, so that $\{1 - q^n\}$ has the limit 1. In this case the geometric series converges and has the sum $s = a_1/(1-q)$.

Infinite geometric series	$s = a_1/(1-q)$ for $\lvert q\rvert < 1$

Example 1: Every periodic decimal fraction can clearly be represented by a convergent geometric series. The formula for the sum makes it possible to transform the decimal fraction into a vulgar fraction. For example, the decimal fraction 0.2525... corresponds to the series $25/100 + 25/10000 + \cdots$ with the first term $a_1 = 25/100$ and the common ratio $q = 1/100$; it has the sum $s = 25/99$.

Example 2: Six lines can be drawn through a point O in such a way that each pair of neighbouring lines includes an angle $\alpha = 30° \triangleq \pi/6$. From the point P_1 on one of the lines at a distance a from the point O, the perpendicular to a neighbouring line is drawn. From its foot P_2 the perpendicular to the following line is drawn, and so on (Fig.). The consecutive perpendiculars form a polygonal arc $|P_1P_2| + |P_2P_3| + \cdots$ that spirals around and shrinks towards the point O. The perpendicular $|P_iP_{i+1}|$ has the length l_i, where $l_1 = a \sin(\pi/6)$, $l_2 = a \sin(\pi/6) \cos(\pi/6)$, $l_3 = a \sin(\pi/6)[\cos(\pi/6)]^2, \ldots$ The series $l_1 + l_2 + l_3 + \cdots$ is geometric with $a_1 = a \sin(\pi/6)$ and $q = \cos(\pi/6)$; it converges since $\cos(\pi/6) < 1$ and has the sum

$$s = a \sin(\pi/6)/[1 - \cos(\pi/6)]$$
$$= (a/2)/(1 - {}^1/_2\sqrt{3}) = a(2 + \sqrt{3}).$$

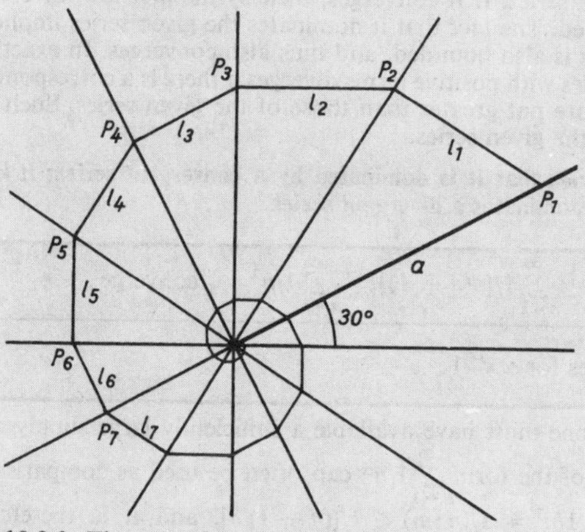

18.2-2 The sum of a geometric series

Convergence tests for series with positive terms. Here again, as for sequences, the questions whether a given series converges, and if so what is its sum, play a particularly important role. Theorems by means of which the convergence behaviour of a series can be decided are called *convergence criteria*, or *convergence tests*. One distinguishes between necessary conditions, sufficient conditions, and those that are both necessary and sufficient. A *necessary condition* leads to several possibilities. Series that do not satisfy it are certainly divergent. On the other hand, if the condition *is* satisfied, then the series *can*, but *need not* converge. A series certainly converges if it satisfies a *sufficient condition*; but if it fails to satisfy this, it may *nevertheless converge*. Definite conclusions can be drawn only if a sufficient condition is satisfied, or a necessary condition is not satisfied. Consequently the most useful criteria are those that are *necessary and sufficient*, because they allow one to distinguish at once between convergence and divergence.

A series converges if the sequence of its partial sums converges, in other words, if all its partial sums from some index n_0 onwards lie in an ε-neighbourhood of the limit s. But the partial sum s_{n+1} arises from s_n by the addition of a_{n+1}. A necessary condition for the convergence of the sequence $\{s_n\}$ of partial sums is therefore that the sequence $\{a_n\}$ of the terms of the series is a null sequence.

For the series $\sum_{i=1}^{\infty} a_i$ to converge it is necessary, but not in general sufficient, that its terms form a null sequence, or that $\lim_{n\to\infty} a_n = 0$.

The first main test for convergence: For the series $\sum_{i=1}^{\infty} a_i$ of positive terms to converge it is necessary and sufficient that the sequence of its partial sums is bounded.

Since the series has only positive terms, the sequence of partial sums increases monotonically. If it is also bounded, then it must be convergent, by the first convergence test for sequences. Since this criterion does not require *strict monotonic* behaviour, the statement also holds for series with non-negative terms.

Example 1: The terms of the *harmonic series* $\sum_{i=1}^{\infty} 1/i = 1 + 1/2 + 1/3 + 1/4 + \cdots$ form a null sequence; but the series itself *diverges*, since the sequence $\{s_n\}$ of its partial sums is not bounded. For s_n exceeds every number C when $n > 2^m$ and $m > 2C$

$$s_n > (1 + 1/2) + (1/3 + 1/4) + (1/5 + \cdots + 1/8) + \cdots + (\cdots + 1/2^m) > 1/2 + 2 \cdot (1/4)$$
$$+ 4 \cdot (1/8) + \cdots + 2^{m-1}(1/2^m) = m/2; \quad s_n > C.$$

Example 2: The series $\sum_{n=2}^{\infty} 1/[(n-1)\,n] = 1/(1 \cdot 2) + 1/(2 \cdot 3) + 1/(3 \cdot 4) + \cdots$ has the nth partial sum

$$s_n = 1/(1 \cdot 2) + 1/(2 \cdot 3) + \cdots + 1/[n(n+1)]$$
$$= (1 - 1/2) + (1/2 - 1/3) + (1/3 - 1/4) + \cdots + [1/n - 1/(n+1)] = 1 - 1/(n+1).$$

Since $\{1/(n+1)\}$ is a null sequence, the sequence $\{s_n\} = \{1 - 1/(n+1)\}$ is bounded. The given series converges and has the sum 1.

Comparison test. A series whose terms are not smaller than those of a given series with positive terms is said to *dominate* or *majorize* the given series. If it converges, then by the first test for convergence its sequence of partial sums is bounded. The fact that it dominates the given series implies that the sequence of partial sums of that series is also bounded, and thus also converges. In exactly the same way one can conclude that a given series with positive terms diverges if there is a corresponding divergent comparison series whose terms are not greater than those of the given series. Such a comparison series is said to be *subordinate to* the given series.

It is sufficient for the convergence of a series that it is dominated by a convergent series; it is sufficient for the divergence of a series that it dominates a divergent series.

$\sum_{n=1}^{\infty} 1/n,\quad \sum_{n=1}^{\infty} 1/\sqrt{n}$ diverge	$\sum_{n=1}^{\infty} 1/[n(n+1)],\quad \sum_{n=1}^{\infty} 1/n^2$ converge
$\sum_{n=1}^{\infty} 1/n^\alpha$ converges for $\alpha > 1$, diverges for $\alpha \leqslant 1$	

In order to be able to use comparison tests, one must have available a sufficiently large supply of known convergent or divergent series. Series of the form $\sum_{n=1}^{\infty} 1/n^\alpha$ can often be used as comparison series. The series $\sum_{n=1}^{\infty} 1/n^2$ converges, since $1/n^2 = 1/(n \cdot n) < 1/[(n-1)\,n]$, and it is therefore dominated by the convergent series $\sum_{n=2}^{\infty} 1/[(n-1)\,n]$ (see Example 2). For $\alpha \geqslant 2$, $\sum_{n=1}^{\infty} 1/n^\alpha$ converges, since $1/n^\alpha \leqslant 1/n^2$. Because $1/n \leqslant 1/\sqrt{n}$, the series $\sum_{n=1}^{\infty} 1/\sqrt{n} = 1 + 1/\sqrt{2} + 1/\sqrt{3} + \cdots$ dominates the divergent harmonic series $\sum_{n=1}^{\infty} 1/n$, and therefore diverges. Since $1/n \leqslant 1/n^\alpha$ for all $\alpha \leqslant 1$, the series $\sum_{n=1}^{\infty} 1/n^\alpha$ diverges for every $\alpha \leqslant 1$. For $\alpha \geqslant 2$ the convergence of the series has already been established. For $1 < \alpha < 2$ a dominating geometric series can be found. If k is an integer such that $2^k > n$, then

$$\begin{aligned} s_n \leq s_{2^k-1} &= 1 + (1/2^\alpha + 1/3^\alpha) + (1/4^\alpha + 1/5^\alpha + 1/6^\alpha + 1/7^\alpha) + \cdots \\ &\quad + (1/(2^{k-1})^\alpha + \cdots + 1/(2^k - 1)^\alpha) \\ &\leqslant 1 + 2/2^\alpha + 4/4^\alpha + \cdots + 2^{k-1}/(2^{k-1})^\alpha \\ &= 1 + 1/2^{\alpha-1} + 1/(2^{\alpha-1})^2 + \cdots + 1/(2^{\alpha-1})^{k-1}. \end{aligned}$$

This is a geometric series with common ratio $q = 1/2^{\alpha-1}$, which is less than 1 for $1 < \alpha < 2$. Thus, the geometric series converges, and therefore the given series also converges.

The ratio test: If $\sum_{i=1}^{\infty} a_i$ is a series of positive terms and if there exists a positive number q less than 1 such that $a_{n+1}/a_n \leqslant q$ from an index n_0 onwards, then the series converges. On the other hand, if from an n_0 onwards $a_{n+1}/a_n \geqslant 1$, then the series diverges.

By hypothesis:

$$\begin{aligned} a_{n_0+1}/a_{n_0} &\leqslant q \longrightarrow a_{n_0+1} \leqslant q a_{n_0}, \\ a_{n_0+2}/a_{n_0+1} &\leqslant q \longrightarrow a_{n_0+2} \leqslant q\; a_{n_0+1} \leqslant q^2 a_{n_0} \quad \text{etc.} \end{aligned}$$

It follows that the series formed by the terms of the remainder of the given series from the n_0th term onwards is dominated by the geometric series $\sum_{i=1}^{\infty} a_{n_0} q^i$. But the convergence of this remainder series decides the convergence of the given series, which therefore converges for $q < 1$. On the other hand, if $a_{n+1}/a_n \geqslant 1$ it follows that $a_{n+1} \geqslant a_n > 0$; the terms of the series do not form a null sequence, and the series must diverge. The condition is *sufficient*. If it is not satisfied, nothing can be concluded about the convergence or divergence of the series. This is the case, for example, if the quotient is less than 1, but not less than a fixed number q less than 1. The ratio of two successive terms of the harmonic series, known to be divergent, is $a_{n+1}/a_n = n/(n+1) < 1$, but it is not true that $n/(n+1) \leqslant q < 1$. For the convergent series $\sum_{n=1}^{\infty} 1/n^2$, however, $a_{n+1}/a_n = (n/(n+1))^2 < 1$, but again it is not true that $(n/(n+1))^2 \leqslant q < 1$. In both cases $\lim_{n\to\infty} a_{n+1}/a_n = 1$.

If $\lim_{n\to\infty} a_{n+1}/a_n = k$, then the series $\sum_{n=1}^{\infty} a_n$ $\begin{cases} \text{converges for } k < 1 \\ \text{diverges for } k > 1 \end{cases}$	The ratio test fails if the sequence $\{a_{n+1}/a_n\}$ tends to 1 'from the left'.

Example: The series $\sum_{n=1}^{\infty} n!/n^n = 1!/1 + 2!/2^2 + 3!/3^3 + 4!/4^4 + \cdots$ converges, since $a_n = n!/n^n$, $a_{n+1} = (n+1)!/(n+1)^{n+1}$ gives $a_{n+1}/a_n = [(n+1)!\, n^n]/[(n+1)^{n+1}\, n!] = [(n+1)\, n^n]/[(n+1)^{n+1}] = [n/(n+1)]^n = 1/(1+1/n)^n \leqslant 1/2 < 1$ for all n.

$\sum_{n=1}^{\infty} n!/n^n$ converges

The root test: If $\sum_{n=1}^{\infty} a_n$ is a series of positive terms, and if there exists a positive number $q < 1$ such that $\sqrt[n]{a_n} \leqslant q$ from some index n_0 onwards, then the series converges; on the other hand, if $\sqrt[n]{a_n} \geqslant 1$ from some index n_0 onwards, then the series diverges.

If $\sqrt[n]{a_n} \leqslant q < 1$ for all $n \geqslant n_0$, then the convergent geometric series $\sum_{n=n_0}^{\infty} q^n$ dominates the remainder $\sum_{n=n_0}^{\infty} a_n$ of the given series. If however $\sqrt[n]{a_n} \geqslant 1$, the terms of the series do not form a null sequence. The condition is *sufficient*. It fails, for example, if $\sqrt[n]{a_n}$ tends to 1 'from the left'.

If $\lim_{n\to\infty} \sqrt[n]{a_n} = k$, the series $\sum_{n=1}^{\infty} a_n$ $\begin{cases}\text{converges for } k < 1\\ \text{diverges for } k > 1\end{cases}$

Example 1: The series $\sum_{n=1}^{\infty} \alpha^n/n^n$ converges for every fixed $\alpha > 0$. $\sqrt[n]{(\alpha^n/n^n)} = \alpha/n$, which is less than $1/2$ for all $n > 2\alpha$. By the root test $\sum_{n=1}^{\infty} (\alpha^n/n^n)$ is convergent.

Example 2: For the series $\sum_{n=1}^{\infty} (1 - 1/n)^n$ the root test fails, since $\lim_{n\to\infty} (1 - 1/n) = 1$. However, one can show that $\lim_{n\to\infty} a_n = \lim_{n\to\infty} (1 - 1/n)^n = 1/e$. Thus the terms a_n of the series do not form a null sequence and the series diverges.

Convergence tests for series with arbitrary terms. Applying the second, or Cauchy, convergence test for sequences to the sequence $\{s_n\}$ of partial sums of the series $\sum a_n$, one obtains the *second test* for convergence of series.

The second main test for convergence: The series $\sum_{n=1}^{\infty} a_n$ converges if and only if to every arbitrary positive number ε there corresponds an integer $N(\varepsilon)$ such that $|s_{n+p} - s_n| = |a_{n+1} + a_{n+2} + \cdots + a_{n+p}| < \varepsilon$ for all $n > N(\varepsilon)$ and all $p \geqslant 1$.

Of course, this test is not easy to handle. It is simpler to prove the convergence of $\sum a_n$ by establishing the convergence of $\sum |a_n|$, which dominates it. On the other hand, if the dominating series diverges, one can draw no conclusion about the given series. In spite of this the following statements about divergence are correct, because the terms of series that satisfy one of these conditions do not form a null sequence, and therefore violate a necessary condition for convergence.

A series $\sum_{n=1}^{\infty} a_n$ with arbitrary terms converges if it satisfies one of the following conditions: $|a_{n+1}/a_n| \leqslant q < 1$ for all $n \geqslant n_0$; $\lim_{n\to\infty} |a_{n+1}/a_n| < 1$; $\sqrt[n]{|a_n|} \leqslant q < 1$ for all $n \geqslant n_0$; $\lim_{n\to\infty} \sqrt[n]{|a_n|} < 1$.
On the other hand, the series diverges if it satisfies one of the following conditions:

$$|a_{n+1}/a_n| \geqslant 1 \text{ for all } n \geqslant n_0; \quad \sqrt[n]{|a_n|} \geqslant 1$$

for all $n \geqslant n_0$; $\lim_{n\to\infty} |a_{n+1}/a_n| > 1$; $\lim_{n\to\infty} \sqrt[n]{|a_n|} > 1$.

Leibniz' test for convergence: An alternating series converges if the absolute values of its terms form a monotonic null sequence.

If the alternating series is written in the form $a_1 - a_2 + a_3 - a_4 \ldots$, the a_i denote the absolute values of the terms. The subsequence $s_2, s_4, s_6, \ldots, s_{2n}, \ldots$ of its partial sums is monotonic increasing, since $s_{2k+2} = s_{2k} + (a_{2k+1} - a_{2k+2}) \geqslant s_{2k}$.
The expression in the bracket is non-negative, since the sequence a_i of absolute values of the terms decreases monotonically. For the same reason it follows from

$$s_{2n} = a_1 - (a_2 - a_3) - (a_4 - a_5) - \cdots - (a_{2n-2} - a_{2n-1}) - a_{2n}$$

that $s_{2n} < a_1$. As a monotonic increasing and bounded sequence $\{s_{2n}\}$ has a limit s. It then follows that the subsequence $s_1, s_3, s_5, \ldots, s_{2n+1}, \ldots$ has the same limit. For $\lim_{n\to\infty} s_{2n+1} = \lim_{n\to\infty} (s_{2n} + a_{2n+1}) = s$, because $\lim_{n\to\infty} a_{2n+1} = 0$. Thus, the sequence $\{s_n\}$ of partial sums converges. The arguments also show that the sequence $\{|a_n|\}$ need not be *strictly* decreasing.

Example 1: The series $\sum_{n=1}^{\infty} (-1)^{n-1}/n = 1 - 1/2 + 1/3 - 1/4 + \cdots$ is convergent (its sum is ln 2).

Example 2: The alternating series $1 - 1/5 + 1/2 - 1/5^2 + 1/3 - 1/5^3 + 1/4 - \cdots$ diverges. The absolute values of its terms form a null sequence, but this sequence is not monotonic. The $(2n)$th partial sum can be arranged as follows:

$$s_{2n} = (1 + 1/2 + 1/3 + \cdots + 1/n) - (1/5 + 1/5^2 + \cdots + 1/5^n).$$

As n increases, the first part exceeds every finite number, while the second part, as a geometric series, tends to a finite limit.

Calculations with convergent series. The rules that are valid for calculations with finite sums can be applied only partially to convergent infinite series (see the following Theorems 1, 2, 3); some can be applied only to series that satisfy stronger convergence conditions.

1. The convergence and the limit of a convergent series are not altered if the terms are bracketed together arbitrarily, but without altering their order.

If $S = \sum_{i=1}^{\infty} a_i = a_1 + a_2 + a_3 + \cdots$, then it is also true that $S = \sum_{i=1}^{\infty} A_i$, where $A_1 = (a_1 + a_2 + \cdots + a_{r_1})$, $A_2 = (a_{r_1+1} + a_{r_1+2} + \cdots + a_{r_2})$, $A_3 = (a_{r_2+1} + \cdots + a_{r_3})$, ... This follows because the sequence $\{s_n'\}$ of partial sums of the series $\sum A_i$ is a subsequence of the sequence $\{s_n\}$ of partial sums of the series $\sum a_i$, and this is known to have the same limit as $\{s_n\}$.

2. If every term a_i of a series converging to the sum s is multiplied by a constant c, the resulting series converges to the sum cs.

$$\sum_{i=1}^{\infty} ca_i = cs \text{ if } c \text{ is constant and } \sum_{i=1}^{\infty} a_i = s$$

By a theorem on sequences, $\{s_n\} \to s$ implies that $c\{s_n\} \to cs$.

3. Term-by-term addition of the convergent series $\sum_{i=1}^{\infty} a_i = A$ and $\sum_{i=1}^{\infty} b_i = B$ gives the series $\sum_{i=1}^{\infty} (a_i + b_i)$, which also converges and has the sum $A + B$.

If $\{A_n\}$ and $\{B_n\}$ are the sequences of partial sums of the series $\sum a_i$ and $\sum b_i$, then by a theorem on sequences, $\{A_n\} \to A$ and $\{B_n\} \to B$ implies that $\{A_n + B_n\} \to A + B$.

Absolute convergence. A series $\sum_{i=1}^{\infty} a_i$ with arbitrary terms is said to be absolutely convergent if the series $\sum_{i=1}^{\infty} |a_i|$ of their absolute values converges.

The series $1 - 1/2 + 1/3 - 1/4 + \cdots$ is not absolutely convergent, since the series of absolute values is the divergent harmonic series. If the series $\sum a_i$ and $\sum |a_i|$ have the sums s and S, then since $|s_n| = |a_1 + a_2 + \cdots + a_n| \leqslant |a_1| + |a_2| + \cdots + |a_n| < S$, it clearly follows that $|s| \leqslant S$.

Rearrangement of series. The question now arises to what extent is the commutative law for finite sums also valid for infinite series. Let $\sum_{n=1}^{\infty} a_n$ be a series and $(k_1, k_2, \ldots, k_n, \ldots)$ a sequence of natural numbers with the property that it contains every natural number exactly once. The series $\sum_{n=1}^{\infty} a_{k_n}$ is then called a rearrangement of the series $\sum_{n=1}^{\infty} a_n$. For example, the series

$$1 - 1/2 + 1/3 - 1/4 + 1/5 - \cdots \quad \text{and} \quad 1 + 1/3 - 1/2 + 1/5 + 1/7 - 1/4 + \cdots$$

are obtained from one another by rearrangement. Both converge, but to different sums s_1 and s_2. This shows that one has to use care in rearranging series.

$$\begin{aligned} s_1 &= 1 - 1/2 + 1/3 - 1/4 + 1/5 - 1/6 + 1/7 - \cdots \\ &= 1 - 1/2 + 1/3 - (1/4 - 1/5) - (1/6 - 1/7) - \cdots \\ &= 10/12 - (1/4 - 1/5) - (1/6 - 1/7) - \cdots < 10/12 \end{aligned}$$

and

$$\begin{aligned} s_2 &= 1 + 1/3 - 1/2 + 1/5 + 1/7 - 1/4 + \cdots \\ &= (1 + 1/3 - 1/2) + (1/5 + 1/7 - 1/4) + (1/9 + 1/11 - 1/6) + \cdots \\ &= 5/6 + 13/140 + (1/9 + 1/11 - 1/6) + \cdots > 11/12, \end{aligned}$$

since the expression $[1/(2n-3) + 1/(2n-1) - 1/n] = (4n-3)/[(2n-3)(2n-1)n]$ is always positive.

Such convergent series whose sum depends on the ordering of the terms are called *conditionally convergent*. Series that remain convergent and have the same sum, no matter how they are rearranged, are called *unconditionally convergent*. The important question for calculation with series, how one can distinguish unconditionally convergent series from conditionally convergent ones, or in other words, how one recognizes whether one may disregard the ordering of the terms of a series or not, is answered in a surprisingly simple way by the following theorem:

Every absolutely convergent series is also unconditionally convergent; every series that is convergent but not absolutely convergent is only conditionally convergent.

The first part of the theorem is easily established. Firstly, if $\sum a_n$ is an absolutely convergent series with non-negative terms and $\sum a_{k_n}$ is a series obtained from it by rearrangement, then the partial sums s_n of the first series and s_n' of the rearranged series satisfy the inequality

$$s_n' = a_{k_1} + a_{k_2} + \cdots + a_{k_n} \leqslant a_1 + a_2 + \cdots + a_N = s_N < s$$

if N is chosen so large that all the k_i $(i = 1, 2, \ldots, n)$ occur among the numbers $1, 2, \ldots, N$. Thus, the sequence $\{s_n'\}$ of the partial sums of the rearranged series $\sum a_{k_n}$ is bounded, and its convergence follows from the first main test for convergence. If now $\sum a_n$ is an absolutely convergent series with arbitrary terms and $\sum a_{k_n}$ is a rearrangement of it, then $\sum |a_n|$ is an absolutely convergent series with non-negative terms; the convergence of $\sum |a_{k_n}|$ follows, as has already been shown, and this implies the convergence of $\sum a_{k_n}$.

The rearrangement also does not influence the sum. Because $\sum a_n$ is assumed to be absolutely convergent, corresponding to an arbitrary $\varepsilon > 0$ a number m can be chosen so that for all $k \geqslant 1$

$$|a_{m+1}| + |a_{m+2}| + \cdots + |a_{m+k}| < \varepsilon.$$

If N is now chosen so large that all the indices $1, 2, \ldots, m$ occur among the numbers $k_1, k_2, \ldots, k_N$, then the difference $|s_n' - s_n|$ for $n > N$ contains only terms a_i with $i > m$, and from this it follows that $|s_n' - s_n| < \varepsilon$ for all $n > N$. Consequently

$$s' = \lim_{n\to\infty} s_n' = \lim_{n\to\infty} [s_n + (s_n' - s_n)] = \lim_{n\to\infty} s_n + \lim_{n\to\infty} (s_n' - s_n) = s + 0 = s.$$

To prove the second part of the theorem one can show that from the convergent, but not absolutely convergent series $\sum a_n$, divergent series or convergent series with arbitrarily prescribed sums can be formed by suitable rearrangement.

$$\begin{matrix} a_{11} + a_{12} + \cdots + a_{1i} + \cdots = z_1 \\ a_{21} + a_{22} + \cdots + a_{2i} + \cdots = z_2 \\ \vdots \qquad \vdots \qquad \cdots \qquad \vdots \qquad \cdots \qquad \vdots \\ a_{k1} + a_{k2} + \cdots + a_{ki} + \cdots = z_k \\ \vdots \qquad \vdots \qquad \cdots \qquad \vdots \qquad \cdots \qquad \vdots \end{matrix}$$

On the other hand, absolutely convergent series can be rearranged in an essentially more general sense, without affecting their convergence or their sums. Let $\sum a_n$ be an absolutely convergent series and the adjacent scheme denote an infinite sequence of partial series of the given series $\sum a_n$, with the property that every term of the series $\sum a_n$ occurs in exactly one of the partial series (a_{ki}, for example, is the ith term in the kth partial series). Then the series $\sum_{k=1}^{\infty} z_k$ is obtained from the given absolutely convergent series $\sum_{n=1}^{\infty} a_n$ by a 'rearrangement in an extended sense'. $\sum_{k=1}^{\infty} z_k$ is likewise absolutely convergent and has the same sum s as the series $\sum_{n=1}^{\infty} a_n$.

Very remarkable is the fact that, under certain conditions, the converse of this theorem holds. That is, the terms of the absolutely convergent partial series can be put together in an arbitrary way to form a series, and all possible series resulting in this way are convergent and have the same sum. The information about this is given by the major rearrangement theorem going back to CAUCHY.

$$\begin{matrix} a_{11} + a_{12} + \cdots + a_{1i} + \cdots = z_1 \\ a_{21} + a_{22} + \cdots + a_{2i} + \cdots = z_2 \\ \vdots \qquad \vdots \qquad \cdots \qquad \vdots \qquad \cdots \qquad \vdots \\ a_{k1} + a_{k2} + \cdots + a_{ki} + \cdots = z_k \\ \vdots \qquad \vdots \qquad \cdots \qquad \vdots \qquad \cdots \qquad \vdots \end{matrix}$$

Major rearrangement theorem. Suppose that the adjacent array is a sequence of absolutely convergent series, that is, for $k = 1, 2, \ldots$, each of the series $\sum_{i=1}^{\infty} |a_{ki}|$ converges and has a sum denoted by ζ_k. If, in addition, $\sum_{k=1}^{\infty} \zeta_k$ converges, then the terms occurring one below another in the same column of the given array likewise form absolutely convergent series. If one writes $\sum_{k=1}^{\infty} a_{ki} = s_i$, then the series $\sum_{i=1}^{\infty} s_i$ also converges absolutely, and $\sum_{i=1}^{\infty} s_i = \sum_{k=1}^{\infty} z_k$. Thus, the series of the row sums and the series of the column sums are both absolutely convergent and have the same sum.

To prove this one puts together all the terms a_{ki} occurring in the given array in any manner to form a sequence which one denotes by $a_1, a_2, ..., a_n, ...$ Then the series $\sum a_n$ converges absolutely, because if N is chosen so large that the terms $a_1, a_2, ..., a_n$ all occur in the first N rows of the array,

$$|a_1| + |a_2| + \cdots + |a_n| \leqslant \zeta_1 + \zeta_2 + \cdots + \zeta_N.$$

Because $\sum \zeta_k$ is assumed to be convergent, the right-hand side is bounded, and it follows that the nth partial sum of the series $\sum |a_n|$ on the left-hand side is also bounded, and consequently $\sum a_n$ is absolutely convergent. The 'column series' $\sum\limits_{i=1}^{\infty} a_{ki} = s_i$, as partial series of $\sum a_n$, are also absolutely convergent, and $|s_i| = |\sum\limits_{i=1}^{\infty} a_{ki}| \leqslant \sum\limits_{i=1}^{\infty} |a_{ki}|$. From this it follows that the nth partial sum of $\sum\limits_{i=1}^{\infty} s_i$ certainly does not exceed the sum of the series $\sum |a_n|$; this means, however, that $\sum s_i$ is absolutely convergent. Finally, the series $\sum s_i$ and $\sum z_k$ have the same sum, since each sum is equal to the sum of $\sum a_n$. Because of the absolute convergence of $\sum a_n$, according to the second main test one can choose the index m so that for all $k \geqslant 1$, $|a_{m+1}| + |a_{m+2}| + \cdots + |a_{m+k}| < \varepsilon$. One now determines N so that the terms $a_1, a_2, ..., a_m$ all occur in the first N rows of the above array. Denoting the nth partial sum of the series $\sum a_n$ by σ_n, the difference $|\sum\limits_{k=1}^{n} z_k - \sigma_n|$ is less than ε for all $n \geqslant N$, since only terms $\pm a_r$ with $r > m$ occur in this expression. Thus, $\lim\limits_{n\to\infty} \sum\limits_{k=1}^{n} z_k = \lim\limits_{n\to\infty} \sigma_n = s$. An analogous calculation for the column series yields $\lim\limits_{n\to\infty} \sum\limits_{i=1}^{n} s_i = \lim\limits_{n\to\infty} \sigma_n = s$.

Multiplication of series. If one multiplies every term of the series $\sum\limits_{i=1}^{\infty} a_i$ by every term of the series $\sum\limits_{i=1}^{\infty} b_i$, one obtains the partial products indicated in the array below. Each row of the array contains infinitely many terms all having the same a_i as a factor, and each column contains infinitely many terms all having the same b_j as a factor. The product of the two series is now defined to be the series $\sum\limits_{k=1}^{\infty} c_k$, where c_k is the sum of the partial products in the kth diagonal of the array as indicated. For example, $c_1 = a_1 b_1, c_2 = a_1 b_2 + a_2 b_1, c_3 = a_1 b_3 + a_2 b_2 + a_3 b_1, ..., c_k = \sum\limits_{i+j=k+1} a_i b_j, ...$ These partial products can be found by a translation method, in which one series is written in reverse order and the other is written on a strip of paper which is moved along above the first one. The diagram shows the position for the third term of two product series $c_3 = a_1 b_3 + a_2 b_2 + a_3 b_1$.

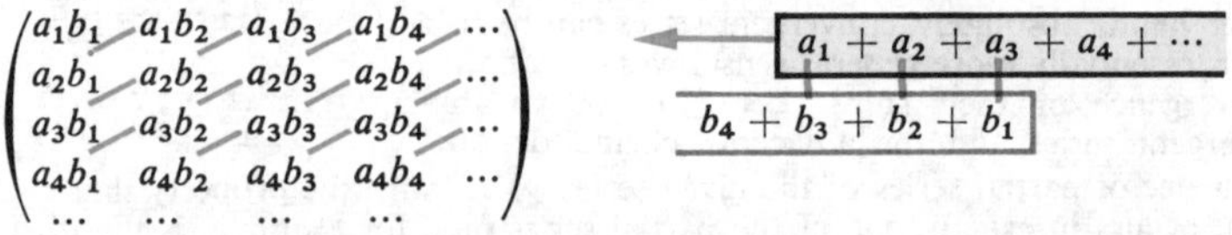

$$\begin{pmatrix} a_1 b_1 & a_1 b_2 & a_1 b_3 & a_1 b_4 & \cdots \\ a_2 b_1 & a_2 b_2 & a_2 b_3 & a_2 b_4 & \cdots \\ a_3 b_1 & a_3 b_2 & a_3 b_3 & a_3 b_4 & \cdots \\ a_4 b_1 & a_4 b_2 & a_4 b_3 & a_4 b_4 & \cdots \\ \cdots & \cdots & \cdots & \cdots & \end{pmatrix}$$

$$a_1 + a_2 + a_3 + a_4 + \cdots$$
$$b_4 + b_3 + b_2 + b_1$$

If the series $\sum\limits_{i=1}^{\infty} a_i = A$ and $\sum\limits_{j=1}^{\infty} b_j = B$ are both absolutely convergent, then the product series $\sum\limits_{k=1}^{\infty} c_k = C$, with $c_k = \sum\limits_{i+j=k+1} a_i b_j$, is also absolutely convergent and has the sum $C = AB$.

The following example shows that the convergence of the two 'factor series' is not enough to ensure the convergence of the product series.

Example: The square of the convergent, but not absolutely convergent, series $1 - 1/\sqrt{2} + 1/\sqrt{3} - 1/\sqrt{4} + \cdots$ is divergent, because its terms do not form a null sequence. The general term c_n of the product of the series with itself satisfies

$$|c_n| = 1 \cdot [1/\sqrt{n}] + [1/\sqrt{2}] \cdot [1/\sqrt{(n-1)}] + [1/\sqrt{3}] \cdot [1/\sqrt{(n-2)}] + \cdots + [1/\sqrt{n}] \cdot 1$$
$$\geqslant [1/\sqrt{n}] \cdot [1/\sqrt{n}] + [1/\sqrt{n}] \cdot [1/\sqrt{n}] + \cdots + [1/\sqrt{n}] \cdot [1/\sqrt{n}] = 1.$$

18.3. Limit of a function – Continuity

Limit and continuity of a function are concepts without which a rigorous construction of higher analysis is impossible. If a function describes a physical situation, then the concepts of limit and continuity often have a physical meaning also.

Limit of a function

Limit at a point. The concept of the limit of a function $y = f(x)$ can be related to the concept of the limit of a sequence. This is done by allowing the independent variable x to run through a convergent sequence of numbers $\{x_n\}$ tending to the limit a (the *abscissa sequence*), and considering the *ordinate sequence* $\{f(x_n)\}$ of the values $f(x_i)$ of the function corresponding to the x_i. If the convergence behaviour of the ordinate sequence $\{f(x_n)\}$ depends upon the choice of abscissa sequence, that is, if two different abscissa sequences both converging to a have corresponding ordinate sequences converging to different limits, or if an ordinate sequence diverges, then the function $f(x)$ does not tend to a limit as x tends to a. On the other hand, if the ordinate sequence $\{f(x_n)\}$ tends to L for *every* abscissa sequence $\{x_n\}$ tending to a, one says that the function $f(x)$ has the *limit* L as $x \to a$. This means that the values $f(x)$ of the function come the nearer to the number L the nearer the argument x comes to the value a. The difference $|f(x) - L|$ between the value of the function and the limit is less than every *arbitrarily chosen* positive number ε, provided that the value of x differs from a by less than a *suitably chosen* number $\delta = \delta(\varepsilon)$ depending on ε, that is, provided that $0 < |x - a| < \delta(\varepsilon)$ (Fig.). The number $\delta(\varepsilon)$ is by no means uniquely determined, for if one $\delta(\varepsilon)$ with the required properties has been found, then clearly every smaller number $\delta' < \delta(\varepsilon)$ will also serve.

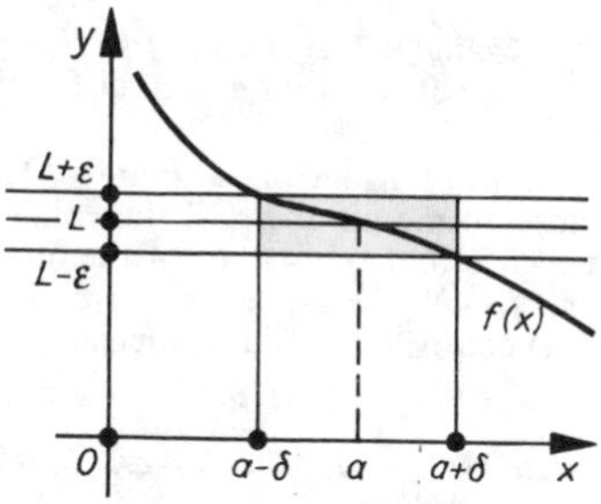

18.3-1 Geometrical illustration of the limit concept

The function $f(x)$ has the limit L as $x \to a$, $\lim\limits_{x \to a} f(x) = L$, if to every $\varepsilon > 0$, however small, there corresponds a number $\delta(\varepsilon) > 0$, such that the inequality $|f(x) - L| < \varepsilon$ holds for every x satisfying the condition $0 < |x - a| < \delta(\varepsilon)$.

Example 1: The function x^2 has the limit zero as the argument x tends to zero, $\lim\limits_{x \to 0} x^2 = 0$, since $|x^2 - 0| < \varepsilon$ for all x such that $|x - 0| < \delta(\varepsilon) \leqslant \sqrt{\varepsilon}$.

Example 2: The function $1/x$ tends to the limit $1/a$ as x tends to $a \neq 0$, $\lim\limits_{x \to a} 1/x = 1/a$. This follows from a theorem on number sequences, since for every abscissa sequence $\{x_n\} \to a \neq 0$, the corresponding ordinate sequence $\{1/x_n\} \to 1/a$.

Example 3: From the existence of the value $f(a)$ of the function one can certainly not conclude that the limit $\lim\limits_{x \to a} f(x)$ must also exist and be equal to $f(a)$, though this is very often the case. The function $f(x) = \{+1 \text{ for } x \neq 0, 0 \text{ for } x = 0\}$, for example, has the limit 1 as $x \to 0$, but the value of the function is $f(0) = 0$ (Fig.).

Example 4: The function $f(x) = (x^2 - 4)/(x - 2)$ is not defined for $x = 2$, since the numerator and denominator vanish simultaneously. But for $x \neq 2$, $f(x) = x + 2$, and so, $x \to 2$, the function tends to the limit 4, $\lim\limits_{x \to 2} [(x^2 - 4)/(x - 2)] = 4$; for $|(x^2 - 4)/(x - 2) - 4| = |x + 2 - 4| = |x - 2| < \varepsilon$ for all x for which $|x - 2| < \delta(\varepsilon) = \varepsilon$.

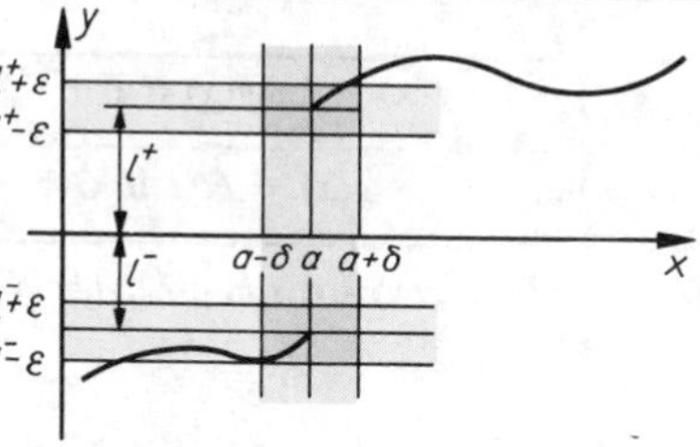

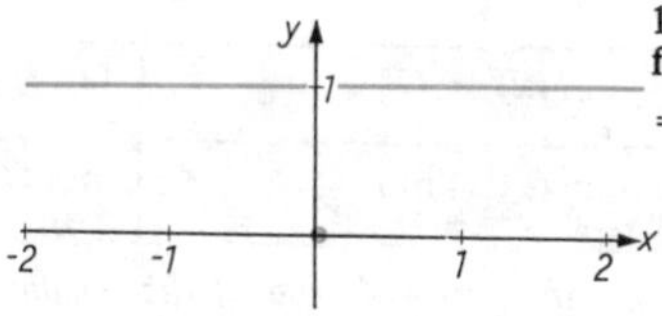

18.3-2 Graph of the function $f(x) = \begin{cases} +1 \text{ for } x \neq 0 \\ 0 \text{ for } x = 0 \end{cases}$

18.3-3 Left-hand limit l^- and right-hand limit l^+ as $x \to a$ are different

One-sided limits. It may be important in the passage to the limit whether the independent variable approaches the value a in the sense of increasing values of x, that is, from the left, or in the sense of decreasing values of x, that is, from the right. One speaks of a left-hand limit l^- if $|f(x) - l^-| < \varepsilon$ for all x with $a - \delta(\varepsilon) < x < a$, and of a right-hand limit l^+ *if* $|f(x) - l^+| < \varepsilon$ for all x with $a < x < a + \delta(\varepsilon)$. One writes $\lim\limits_{x \to a-0} f(x) = l^-$, $\lim\limits_{x \to a+0} f(x) = l^+$, respectively, indicating symbolically by $a - 0$, $a + 0$ from which side x converges to a. These two limits l^- and l^+ may differ from one another, for example, at a jump discontinuity of a function $f(x)$ (Fig.). The function has no two-sided limit there. On the other hand, the function has a limit as $x \to a$ if and only if the left-hand

and right-hand limits as $x \to a$ are equal.

$$\lim_{x\to a-0} f(x) = l^- = \lim_{x\to a+0} f(x) = l^+ = L \longleftrightarrow \lim_{x\to a} f(x) = L.$$

Infinite limits. As x tends to zero in any manner whatsoever, the values of the function $f(x) = 1/x^2$ ultimately exceed every number, however large. One writes $\lim_{x\to 0} 1/x^2 = +\infty$ and says that the function tends to the *limit plus infinity* as $x \to 0$. Similarly, $\lim_{x\to 0} (-1/x^2) = -\infty$; the function $f(x) = -1/x^2$ tends to the *limit minus infinity* as $x \to 0$, since it is ultimately less than any number $-N$ $(N > 0)$ no matter how large N is.

$\lim_{x\to a} f(x) = +\infty$ *(or* $\lim_{x\to a} f(x) = -\infty$*) if to every positive number N, however large, there corresponds a number* $\delta(N)$*, such that* $f(x) > N$ *(or* $f(x) < -N$*) for all x with* $0 < |x - a| < \delta(N)$.

Example: The *tangent function* $y = \tan x$ is not defined for $x = \pi/2$, but has both a right-hand and a left-hand infinite limit for this value; $\lim_{x\to \pi/2-0} \tan x = +\infty$, $\lim_{x\to \pi/2+0} \tan x = -\infty$.

Limit of a function at infinity. The values of the function $f(x) = 1/x + b$ clearly come arbitrarily close to the number b if the value of x is chosen sufficiently large. For example, the difference between b and the values of the function is less than 0.000 001 for all x larger than 10^6. In general, $|f(x) - b| < \varepsilon$ for all $x > 1/\varepsilon$. This example shows that the concept of the limit L of a function $f(x)$ can be extended to the case of unbounded increasing (or decreasing) abscissae.

$\lim_{x\to\infty} f(x) = L$ *if to every arbitrary* $\varepsilon > 0$ *there corresponds a sufficiently large* $\omega(\varepsilon) > 0$ *such that* $|f(x) - L| < \varepsilon$ *for all* $x > \omega(\varepsilon)$. *Similarly,* $\lim_{x\to-\infty} f(x) = L$ *if to every arbitrary* $\varepsilon > 0$ *there corresponds a sufficiently large* $\omega(\varepsilon) > 0$ *such that* $|f(x) - L| < \varepsilon$ *for all* $x < -\omega(\varepsilon)$.

The limits $\lim_{x\to\infty} f(x)$ and $\lim_{x\to-\infty} f(x)$ of the function $f(x)$, if they exist, describe the *behaviour of the function* at infinity, that is, for very large positive and very large negative values of x.

Example 1: $\lim_{x\to\infty} 1/x = 0$, since $|1/x - 0| = |1/x| < \varepsilon$ for all x satisfying the condition $x > \omega(\varepsilon) = 1/\varepsilon$.

Example 2: The limit $\lim_{x\to\infty} \sin x$ does not exist. No matter how large a value of x, say x_0, is chosen, because of the periodicity of the sine function there are always infinitely many abscissae greater than x_0 for which the function takes any prescribed value between -1 and $+1$.

The behaviour of rational functions at infinity is dealt with in Chapter 5.

Calculations with limits. The rules drawn up for calculation with limits of sequences can be carried over word-for-word to calculations with limits of functions. These rules, which have already been applied in the examples of the previous section, state that the *operation of forming the limit* can be interchanged with addition, subtraction, multiplication and division (if $L \neq 0$), provided that all the limits occurring exist and are finite. The first two rules hold also for sums and products of several functions, but not necessarily for infinite sums. A function $h(x)$ whose values in a neighbourhood of the point a lie between those of two functions $f(x)$ and $g(x)$ that both have the limit L as $x \to a$ also has the limit L.

$\lim_{x\to a} f(x) = F$ and $\lim_{x\to a} g(x) = G$	$\lim_{x\to a} [f(x) \pm g(x)] = \lim_{x\to a} f(x) \pm \lim_{x\to a} g(x) = F \pm G,$
$\lim_{x\to a} [f(x)\cdot g(x)] = \lim_{x\to a} f(x) \cdot \lim_{x\to a} g(x) = F\cdot G,$	$\lim_{x\to a} f(x)/g(x) = \lim_{x\to a} f(x)/\lim_{x\to a} g(x) = F/G$ if $G \neq 0$

If $\lim_{x\to a} f(x) = L$ *and* $\lim_{x\to a} g(x) = L$*, and if the inequality* $f(x) \leqslant h(x) \leqslant g(x)$ *holds in a neighbourhood of a, then* $\lim_{x\to a} h(x) = L$ *also.*

Example 1: $\lim_{x\to\infty} (\sin x)/x = 0$. Because $\sin x$ lies between -1 and $+1$, for $x > 0$ holds the inequality $-1/x \leqslant (\sin x)/x \leqslant 1/x$. The result follows, since $\lim_{x\to\infty} 1/x = \lim_{x\to\infty} (-1/x) = 0$ (Fig.).

Example 2: $\lim_{x\to 0} x \sin 1/x = 0$, since $-|x| \leqslant x \sin 1/x \leqslant |x|$ and $\lim_{x\to 0} |x| = \lim_{x\to 0} (-|x|) = 0$.

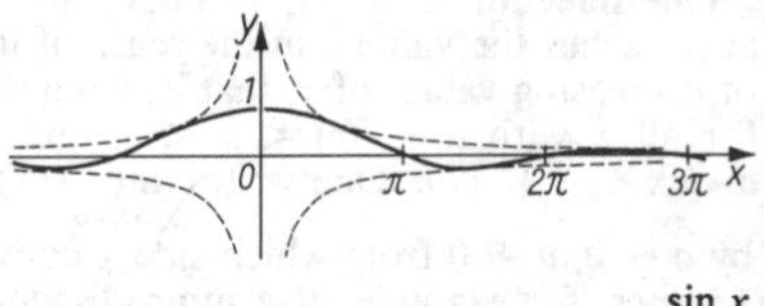

18.3-4 Graph of the function $y = \dfrac{\sin x}{x}$

Some important limits

For the determination of the limit of a function there are hardly any generally applicable methods. Some frequently used limits will be derived here, with the help of knowledge of convergent number sequences.

1. $a^x \to 1$ as $x \to 0$ if $a > 0$. $\lim_{x\to 0} a^x = 1$ for $a > 0$

It has already been shown that the sequence $\{\sqrt[n]{a}\}$ for $a > 0$ converges to 1. The sequence $\{1/\sqrt[n]{a}\}$ then has the reciprocal limit, likewise equal to 1. It follows that to every arbitrary $\varepsilon > 0$ a positive integer N can be determined, so that for all $n \geqslant N$, the numbers $a^{1/n}$ and $a^{-1/n}$ lie in the interval from $1 - \varepsilon$ to $1 + \varepsilon$. Since the exponential function is monotonic, all a^x with $-1/N < x < 1/N$ also lie in this interval. Thus, $1 - \varepsilon < a^x < 1 + \varepsilon$, or $|a^x - 1| < \varepsilon$ if $|x| < \delta(\varepsilon) = 1/N$.

2. $(1 + 1/x)^x \to \text{e}$ as $x \to \infty$. $\lim_{x\to\infty} (1 + 1/x)^x = \text{e}$ $\lim_{y\to 0} (1 + y)^{1/y} = \text{e}$

It is sufficient to show, that for an arbitrary abscissa sequence $\{x_n\} \to \infty$, the corresponding ordinate sequence $\{(1 + 1/x_n)^{x_n}\}$ has the limit e. To this end one chooses natural numbers p_n such that $p_n \leqslant x_n \leqslant p_n + 1$ for every n. It follows that

$$[1 + 1/(p_n + 1)]^{p_n} \leqslant [1 + 1/x_n]^{x_n} \leqslant [1 + 1/p_n]^{p_n+1}.$$

Now $\lim_{p_n\to\infty} [1 + 1/(p_n + 1)]^{p_n} = \text{e}$ and also $\lim_{p_n\to\infty} [1 + 1/p_n]^{p_n+1} = \text{e}$. The ordinate sequence under investigation is enclosed between two sequences, both converging to e, and hence has the same limit.

In the above relation $1/x$ may be replaced by y.

3. $(1 + a/x)^x \to \text{e}^a$ as $x \to \infty$. $\lim_{x\to\infty} (1 + a/x)^x = \text{e}^a$

For $a = 0$ the statement is trivial. For $a \neq 0$,

$$\lim_{x\to\infty} (1 + a/x)^x = \lim_{x\to\infty} [(1 + 1/(x/a))^{(x/a)}]^a = \lim_{z\to\infty} [(1 + 1/z)^z]^a = [\lim_{z\to\infty} (1 + 1/z)^z]^a = \text{e}^a.$$

In the last step the continuity of the function $x^\alpha (x > 0)$ is used.

4. $\log_b x \to \log_b a$ as $x \to a$ if $a > 0$, $b > 1$. $\lim_{x\to a} \log_b x = \log_b a$; $a > 0$, $b > 1$

It must be shown that to every $\varepsilon > 0$, there corresponds a $\delta(\varepsilon) > 0$, such that the inequality $|\log_b x - \log_b a| < \varepsilon$ is satisfied for all x with $|x - a| < \delta(\varepsilon)$. Firstly, since $b > 1$, for positive ε the numbers $\varepsilon_1 = b^\varepsilon - 1$ and $\varepsilon_2 = 1 - b^{-\varepsilon}$ are also positive. With $b^\varepsilon > 1$ it follows further that $\varepsilon_2 = 1 - b^{-\varepsilon} < b^\varepsilon(1 - b^{-\varepsilon}) = \varepsilon_1$. Now let $\varepsilon > 0$ be arbitrary. Then one can choose $\delta(\varepsilon) = a\varepsilon_2$ and obtains:

$$|x - a| < a\varepsilon_2 \Rightarrow \left|\frac{x-a}{a}\right| < \varepsilon_2 \Rightarrow -\varepsilon_2 < \frac{x-a}{a} < \varepsilon_2 < \varepsilon_1 \Rightarrow$$

$$-1 + b^{-\varepsilon} < \frac{x-a}{a} < b^\varepsilon - 1 \Rightarrow b^{-\varepsilon} < 1 + \frac{x-a}{a} < b^\varepsilon \Rightarrow b^{-\varepsilon} < \frac{x}{a} < b^\varepsilon.$$

From the definition of the logarithm, and because of its monotonicity, this gives

$$-\varepsilon < \log_b (x/a) < \varepsilon \Rightarrow |\log_b x - \log_b a| < \varepsilon.$$

Thus, the limit of a logarithm can be determined as the logarithm of the limit. Accordingly, as $x \to 0$, $[\log_b (1 + x)]/x = \log_b [(1 + x)^{1/x}] \to \log_b \text{e}$, and this limit takes the value $\ln \text{e} = 1$ for $b = \text{e}$.

$\lim_{x\to 0} [\log_b (1 + x)]/x = \log_b \text{e}$; $b > 1$ | $\lim_{x\to 0} [\ln (1 + x)]/x = 1$

5. $(a^x - 1)/x \to \ln a$ as $x \to 0$ if $a > 0$.

$\lim_{x\to 0} (a^x - 1)/x = \ln a$; $a > 0$ | $\lim_{x\to 0} (\text{e}^x - 1)/x = 1$

If $a = 1$, then the numerator is zero, and the statement is true since $\ln 1 = 0$. If $a \neq 1$, put $a^x = 1 + y$. Since $a^x \to 1$ as $x \to 0$, one has $y \to 0$. But $x \ln a = \ln (1 + y)$, and so $(a^x - 1)/x = y \ln a/\ln (1 + y) = \ln a/\ln [(1 + y)^{1/y}]$, whose denominator tends to 1. The most important special case is $a = \text{e}$ with $\ln \text{e} = 1$.

6. $\cos x \to 1 \quad \text{as} \quad x \to 0.$

$\lim_{x \to 0} \cos x = 1$

Since $|\cos x - 1| = 2|\sin^2(x/2)| = 2|\sin(x/2)| \cdot |\sin(x/2)| \leqslant 2|x/2| \cdot |x/2| = x^2/2 < \varepsilon$

for $|x| < \sqrt{(2\varepsilon)}$,

$|\cos x - 1|$ converges to zero as $x \to 0$.

7. $(\sin x)/x \to 1 \quad \text{as} \quad x \to 0.$

$\lim_{x \to 0} (\sin x)/x = 1$

The function $h(x) = (\sin x)/x$ can be included between the two functions $f(x) = 1$ and $g(x) = \cos x$, which both tend to the limit 1 as $x \to 0$. From the figure one can see that the area A'_{OEB} of the sector OEB of the circle lies between the areas of the triangles OEB and OED:

$$A_{OEB} < A'_{OEB} < A_{OED} \longrightarrow 1 \cdot \sin x < 1 \cdot x < \tan x \cdot 1$$
$$\longrightarrow 1 < x/\sin x < 1/\cos x \longrightarrow 1 > (\sin x)/x > \cos x.$$

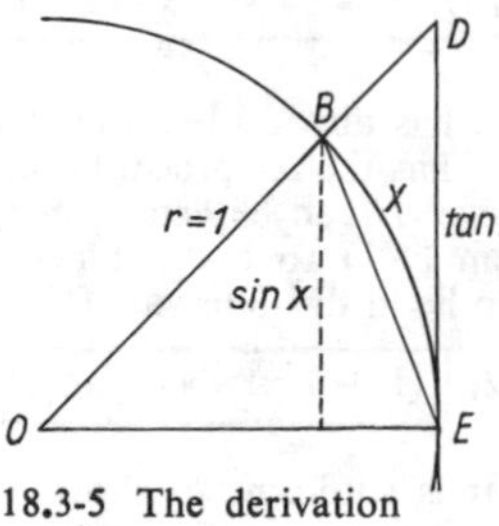

18.3-5 The derivation of $\dfrac{\sin x}{x} \to 1$ as $x \to 0$

The inequalities hold only for positive x; but as $x \to 0$, the right-hand and left-hand limits exist, both with the value 1, since $[\sin(-x)]/(-x) = (\sin x)/x$.

8. $(\tan x)/x \to 1 \quad \text{as} \quad x \to 0.$

$\lim_{x \to 0} (\tan x)/x = 1$

For $(\tan x)/x = [(\sin x)/x] \cdot [1/\cos x]$, and each factor tends to the limit 1 as $x \to 0$.

The rule of Bernoulli and L'Hospital

It is known that one can interchange the operations of addition, subtraction, multiplication and division with the passage to the limit only if all the limits occurring exist, are finite, and in the case of division are different from zero. On the other hand, if by an uncritical acceptance of the interchangeability there arise meaningless expressions of the form $0/0$, ∞/∞, $0 \cdot \infty$, 0^0, ∞^0 or 1^∞, then it is necessary to determine the given limit directly. One speaks of an *indeterminate form* if one of these expressions arises formally for $x \to a$. For the limit $\lim_{x \to 0} (\sin x)/x$ the indeterminate form $0/0$ results if one replaces the limit of the quotient by the quotient of the limits. However, since the limit of the denominator is zero, this procedure is not permissible. By other means it has already been shown that $\lim_{x \to 0} (\sin x)/x = 1$.

The indeterminate form 0/0. For the determination of the limit $\lim_{x \to a} f(x)/g(x)$ in the case $\lim_{x \to a} f(x) = \lim_{x \to a} g(x) = 0$, Johann Bernoulli (1667–1748) developed a rule which the Marquis de L'Hospital (1661–1704) published.

The rule of Bernoulli and L'Hospital: If both the numerator $f(x)$ and the denominator $g(x)$ of a quotient tend to the limit zero as $x \to a$, and if the derivatives $f'(x)$ and $g'(x) \neq 0$ of the functions $f(x)$ and $g(x)$ exist in a neighbourhood of $x = a$ and the limit $\lim_{x \to a} f'(x)/g'(x)$ of the quotient of the derivatives also exists, then this is equal to the limit $\lim_{x \to a} f(x)/g(x)$ of the quotient of the functions.

The rule uses the concept of the derivative, which is explained in Chapter 19. It can be deduced from the extended mean-value theorem mentioned there. In the expression

$$f(x)/g(x) = [f(x) - f(a)]/[g(x) - g(a)] = f'(\xi)/g'(\xi),$$

ξ lies between a and x, and thus also converges to a as $x \to a$. If $\lim_{x \to a} f'(\xi)/g'(\xi) = L$, the theorem holds. It can also be used in the case $x \to \infty$.

If an indeterminate expression of the form $0/0$ again arises, the rule can be applied to the quotient $f'(x)/g'(x)$ and the limit $\lim_{x \to a} f''(x)/g''(x)$ investigated. However, it can happen that one always obtains in this way an indeterminate form, or that the limit of the derivatives does not exist, although the given quotient does have a limit. The rule is then not applicable to the given function, and the limit must be found by other methods.

Example 1: $\lim\limits_{x\to 1} \dfrac{\ln x}{x-1} = \lim\limits_{x\to 1} \dfrac{1/x}{1} = 1.$

Example 2: $\lim\limits_{x\to 0} \dfrac{x^3}{2e^x - x^2 - 2x - 2} = \lim\limits_{x\to 0} \dfrac{3x^2}{2e^x - 2x - 2} = \lim\limits_{x\to 0} \dfrac{6x}{2e^x - 2} = \lim\limits_{x\to 0} \dfrac{6}{2e^x} = 3.$

Example 3: $\lim\limits_{x\to 0} \dfrac{\cos x - 1}{x^2} = \lim\limits_{x\to 0} \dfrac{-\sin x}{2x} = \lim\limits_{x\to 0} \dfrac{-\cos x}{2} = -\dfrac{1}{2}.$

Example 4: $\lim\limits_{x\to \pi/2} \dfrac{\sin 2x}{\cos^2 x} = \lim\limits_{x\to \pi/2} \dfrac{2\cos 2x}{-\sin 2x} = \pm\infty$, according as x tends to the value $\pi/2$ from the left or from the right.

Example 5: $\lim\limits_{x\to \infty} \dfrac{\ln [x/(x-1)]}{5/x} = \lim\limits_{x\to \infty} \dfrac{[(x-1)/x]\cdot[-1/(x-1)^2]}{-5/x^2} = \lim\limits_{x\to\infty} \dfrac{x}{5(x-1)}$

$$= \lim_{x\to\infty} \frac{1}{5(1-1/x)} = \frac{1}{5}.$$

Example 6: For the determination of the limit $\lim\limits_{x\to +0} [\sqrt{(x^2 + \sin^2 x)}]/x$ L'Hospital's rule fails. By other methods it is easy to show that the limit is $\sqrt{2}$.

The indeterminate form ∞/∞. If the numerator $f(x)$ and the denominator $g(x)$ of a quotient both tend to infinity as $x \to a$, then the functions $1/f(x)$ and $1/g(x)$ both tend to zero. If $f(x)$ and $g(x)$ are differentiable in a neighbourhood of $x = a$, and if $g'(x)$ is different from zero and the quotient $f'(x)/g'(x)$ tends to a limit, then L'HOSPITAL's rule can be applied: $\lim\limits_{x\to a} f(x)/g(x) = \lim\limits_{x\to a} f'(x)/g'(x)$. This also holds for $x \to \infty$.

Example 1: $\lim\limits_{x\to 1+0} \dfrac{-\ln (x-1)}{1/(x-1)} = \lim\limits_{x\to 1+0} \dfrac{-1/(x-1)}{-1/(x-1)^2} = \lim\limits_{x\to 1+0} (x-1) = 0.$

Example 2: $\lim\limits_{x\to \pi/2} \dfrac{\tan 3x}{\tan x} = \lim\limits_{x\to \pi/2} \dfrac{3/\cos^2 3x}{1/\cos^2 x} = \lim\limits_{x\to \pi/2} \dfrac{3\cos^2 x}{\cos^2 3x} = \lim\limits_{x\to \pi/2} \dfrac{-6\cos x \sin x}{-6 \cos 3x \sin 3x}$

$$= \lim_{x\to \pi/2} \frac{\sin 2x}{\sin 6x} = \lim_{x\to \pi/2} \frac{2\cos 2x}{6\cos 6x} = \frac{1}{3}.$$

Example 3: $\lim\limits_{x\to\infty} \dfrac{x + \sin x}{x}$ cannot be treated by L'HOSPITAL's rule, since $\lim\limits_{x\to\infty} \cos x$ does not exist. However,

$$\lim_{x\to\infty} (x + \sin x)/x = \lim_{x\to\infty} [1 + (\sin x)/x] = 1.$$

Example 4: $\lim\limits_{x\to\infty} \dfrac{e^x}{x^4} = \lim\limits_{x\to\infty} \dfrac{e^x}{4x^3} = \lim\limits_{x\to\infty} \dfrac{e^x}{12x^2} = \lim\limits_{x\to\infty} \dfrac{e^x}{24x} = \lim\limits_{x\to\infty} \dfrac{e^x}{24} = \infty.$

Example 5: $\lim\limits_{x\to\infty} \dfrac{x^n}{a^x} = \lim\limits_{x\to\infty} \dfrac{nx^{n-1}}{a^x \ln a} = \cdots = \lim\limits_{x\to\infty} \dfrac{n!}{a^x (\ln a)^n} = 0$ for n a positive integer and $a > 1$.

Example 6: $\lim\limits_{x\to\infty} \dfrac{\ln x}{x^n} = \lim\limits_{x\to\infty} \dfrac{1/x}{nx^{n-1}} = \lim\limits_{x\to\infty} \dfrac{1}{nx^n} = 0$ for n a positive integer.

The last two examples show that the exponential function a^x tends to infinity faster than every power x^n, but every power tends to infinity faster then the logarithm.

The remaining indeterminate forms. With the help of L'HOSPITAL's rule the remaining indeterminate forms can be treated, by expressing the functions in a form that leads to one of the indeterminate forms $0/0$ or ∞/∞ for the critical point. To calculate $\lim\limits_{x\to a} [f(x)\cdot g(x)]$ for the case $\lim\limits_{x\to a} f(x) = 0$, $\lim\limits_{x\to a} g(x) = \infty$, one writes $f(x)\cdot g(x) = \dfrac{f(x)}{1/g(x)}$ or $f(x)\cdot g(x) = \dfrac{g(x)}{1/f(x)}$, and one then has the case $0/0$ or ∞/∞.

Example: $\lim\limits_{x\to\infty} x \operatorname{arccot} x = \lim\limits_{x\to\infty} \dfrac{\operatorname{arccot} x}{1/x} = \lim\limits_{x\to\infty} \dfrac{-1/(1+x^2)}{-1/x^2} = \lim\limits_{x\to\infty} \dfrac{x^2}{x^2+1} = 1.$

To calculate $\lim\limits_{x\to a} [f(x) - g(x)]$ when $\lim\limits_{x\to a} f(x) = \lim\limits_{x\to a} g(x) = \infty$, one writes

$$f(x) - g(x) = \frac{1}{1/f(x)} - \frac{1}{1/g(x)} = \frac{1/g(x) - 1/f(x)}{1/[f(x)\cdot g(x)]} = \frac{\varphi(x)}{\psi(x)}$$

where $\lim\limits_{x\to a} \varphi(x) = \lim\limits_{x\to a} \psi(x) = 0.$

Example: $\lim_{x\to 0}[1/\sin x - 1/(x+x^2)] = \lim_{x\to 0}\frac{x+x^2-\sin x}{(x+x^2)\sin x}$

$= \lim_{x\to 0}\frac{1+2x-\cos x}{(1+2x)\sin x + (x+x^2)\cos x}$

$= \lim_{x\to 0}\frac{2+\sin x}{2\sin x - (x+x^2)\sin x + (2+4x)\cos x} = 1.$

To calculate $\lim_{x\to a} f(x)^{g(x)}$ in one of the cases $\lim_{x\to a} f(x) = \lim_{x\to a} g(x) = 0$; $\lim_{x\to a} f(x) = \infty$, $\lim_{x\to a} g(x) = 0$; or $\lim_{x\to a} f(x) = 1$, $\lim_{x\to a} g(x) = \infty$, one notes that in each of these cases $\ln f(x)^{g(x)} = g(x)\ln f(x)$ is a product of which one factor tends to zero and the other to infinity. Hence $\lim_{x\to a}[g(x)\ln f(x)]$ can be determined by the method already known.

Example 1: To calculate $\lim_{x\to +0} x^x$ one notes that $\ln x^x = x\ln x$ and $\lim_{x\to +0}\frac{\ln x}{1/x} = \lim_{x\to +0}\frac{1/x}{-1/x^2}$ $= \lim_{x\to +0}(-x) = 0$. It follows that $\lim_{x\to +0} x^x = \lim_{x\to +0} e^{x\ln x} = 1$, since $\lim_{\xi\to 0} a^\xi = 1$.

Example 2: To calculate $\lim_{x\to\infty}\sqrt[x]{x}$, note that $\ln\sqrt[x]{x} = \frac{1}{x}\ln x$ and $\lim_{x\to\infty}\frac{\ln x}{x} = \lim_{x\to\infty}\frac{1/x}{1} = 0$. It follows that $\lim_{x\to\infty}\sqrt[x]{x} = 1$.

Continuity of a function

Intuitively one regards the picture of a function that is continuous on an interval I as a smooth curve that is nowhere broken, that one can draw 'without taking the pencil off the paper'. This means that the function is defined at every point $x = \xi$ of the interval and that it changes very little for small changes of the argument (Fig.). This idea can be made precise.

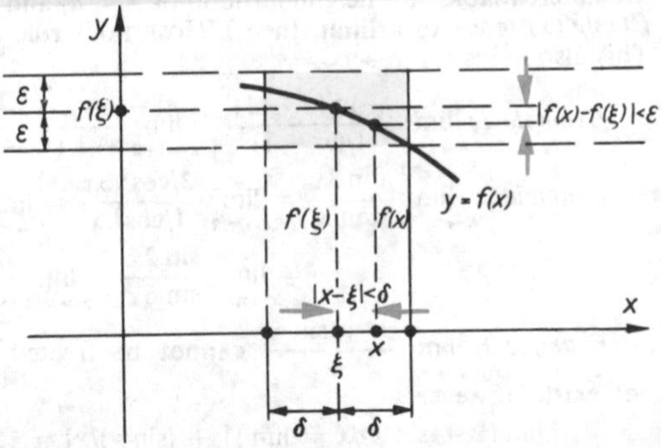

18.3-6 Geometrical illustration of continuity

A function $f(x)$, defined in a neighbourhood of $x = \xi$ and at ξ itself, is said to be continuous at ξ if to every small $\varepsilon > 0$ a suitable number $\delta(\varepsilon) > 0$ can always be chosen so that $|f(x) - f(\xi)| < \varepsilon$ for all x such that $|x - \xi| < \delta(\varepsilon)$.

This statement is equivalent to saying that the limit $\lim_{x\to\xi} f(x)$ exists and is equal to the value $f(\xi)$ of the function: $\lim_{x\to\xi} f(x) = f(\xi)$.

Example: $f(x) = 3x^2 - 1$ is continuous at every point $x = \xi$. First suppose that $|x - \xi| < 1$; then $|x + \xi| = |2\xi + (x - \xi)| \leqslant 2|\xi| + 1$. From this one obtains
$|f(x) - f(\xi)| = |3x^2 - 1 - (3\xi^2 - 1)| = 3|x^2 - \xi^2| = 3|x + \xi|\,|x - \xi| \leqslant 3(2|\xi| + 1)|x - \xi| < \varepsilon$
for all x with $|x - \xi| < \frac{\varepsilon}{3(2|\xi| + 1)}$. If for a given $\varepsilon > 0$ one chooses $\delta(\varepsilon) = \mathrm{Min}\,[1, \varepsilon/[3(2|\xi| + 1)]]$, then one has established the continuity of the given function.

One-sided continuity. If as $x \to \xi$ only the right-hand (or only the left-hand) limit exists and is equal to the value $f(\xi)$ of the function, one speaks of *right-hand* (or *left-hand*) *continuity*. For example, $f(x) = \sqrt{x}$ is continuous from the right at $x = 0$, since $\lim_{x\to +0}\sqrt{x} = 0 = f(0)$. If a function is continuous at $x = \xi$, then it is continuous both from the right and from the left. The converse is also true.

Continuity in an interval. A function $f(x)$ is continuous in an interval if it is continuous at every interior point of the interval and, if the interval is closed on the left (right), the function is continuous from the right (left) at the end-point.

The function $f(x) = 3x^2 - 1$, for example, is continuous in every interval. On the other hand, the function $f(x) = 1/(2 - x)$ is not continuous in the whole interval $1 \leqslant x \leqslant 5$. For $x = 2$ the function is *discontinuous*; its value does not exist there.

Uniform continuity. The example of the function $f(x) = 3x^2 - 1$ shows that the $\delta(\varepsilon)$ corresponding to a given $\varepsilon > 0$ depends in general on the value of ξ. If a function $f(x)$ is such that for a given $\varepsilon > 0$ a *single* value $\delta(\varepsilon)$ can be chosen for *all* ξ in an interval in order to guarantee that $|f(x) - f(\xi)| < \varepsilon$, then the function is said to be *uniformly continuous* in this interval. Such a value of $\delta(\varepsilon)$, valid for a whole interval, exists precisely when the set of all the values of δ, corresponding to all the ξ in the interval, has a positive lower limit. For example, the function $f(x) = 3x^2 - 1$ is uniformly continuous in the interval $2 \leqslant x \leqslant 5$, since $\delta(\varepsilon, \xi) = \text{Min}\left[1, \dfrac{\varepsilon}{3(2|\xi| + 1)}\right] \geqslant \dfrac{\varepsilon}{33}$, so that $\delta(\varepsilon) = \dfrac{\varepsilon}{33}$ serves for every ξ in this interval. On the other hand, the function $f(x) = \tan x$ is continuous, but not uniformly continuous, in the interval $0 \leqslant x < \pi/2$. For a given $\varepsilon > 0$, the nearer ξ approaches to the value $\pi/2$, the smaller must the corresponding value $\delta(\varepsilon)$ be. As $\xi \to \pi/2$, the values of δ tend to zero. The following theorem holds quite generally.

A function $f(x)$ that is continuous in a closed interval $[a, b]$ is also uniformly continuous in the interval.

The example $f(x) = \tan x$ shows that the condition on the interval to be closed is essential.

Points of discontinuity. A point $x = \xi$ at which the function $f(x)$ is not continuous is called a *point of discontinuity*. At such a point, either the function value or the limit fails to exist, or both exist but are not equal to one another. *Poles* of rational functions are examples of points of discontinuity; they are investigated in Chapter 5.

At an *indeterminate point* a function assumes formally an indeterminate form. For example, the function $f(x) = (\sin x)/x$ has an indeterminate point at $x = 0$. However, since as $x \to 0$ the left-hand and right-hand limits both exist and are equal to 1, one can consider a replacement function $f^*(x)$ that takes the value $(\sin x)/x$ for $x \neq 0$, and the value of the limit 1 for $x = 0$; $f^*(x)$ is then continuous for $x = 0$, and the discontinuity has been 'removed'. One therefore speaks of a *removable discontinuity*. The discontinuity of the function $f(x)$ at an indeterminate point $x = \xi$ is removable if the one-sided limits $\lim\limits_{x \to \xi+0} f(x) = \lim\limits_{x \to \xi-0} f(x) = L$ exist, and are finite and equal. One can then replace $f(x)$ by the function $f^*(x) = f(x)$ for $x \neq \xi$; $f(x) = L$ for $x = \xi$, which is continuous at $x = \xi$.

If the numerator $p(x)$ and the denominator $q(x)$ of a *rational* function $p(x)/q(x)$ have the common linear factor $x - x_0$, then $x = x_0$ is an indeterminate point of the function. If $p(x) = (x - x_0)^i\, p_1(x)$, $q(x) = (x - x_0)^k\, q_1(x)$, where $p_1(x_0) \neq 0$, $q_1(x_0) \neq 0$ and $i > k$, then x_0 is a *removable discontinuity*. The replacement function $f^*(x)$, continuous for $x = x_0$, is equal to $(x - x_0)^{i-k} \cdot p_1(x)/q_1(x)$ for $x \neq x_0$, and is zero for $x = x_0$. For $i = k$ the discontinuity is also removable, and the replacement function is $p_1(x)/q_1(x)$. For $i < k$, however, the replacement function $(x - x_0)^{i-k} \cdot p_1(x)/q_1(x)$ has a pole of order $(k - i)$ at the point $x = x_0$ and is therefore not continuous there.

Jump discontinuity. At a jump discontinuity the left-hand and right-hand limits are different from one another and the function cannot be continuous there.

Heat supplied to a solid body raises its temperature t. Its heat content H at the melting point $t = t_m$ is not a continuous function of the temperature, since at this temperature the heat content of the molten substance is greater than that of the solid (Fig.).

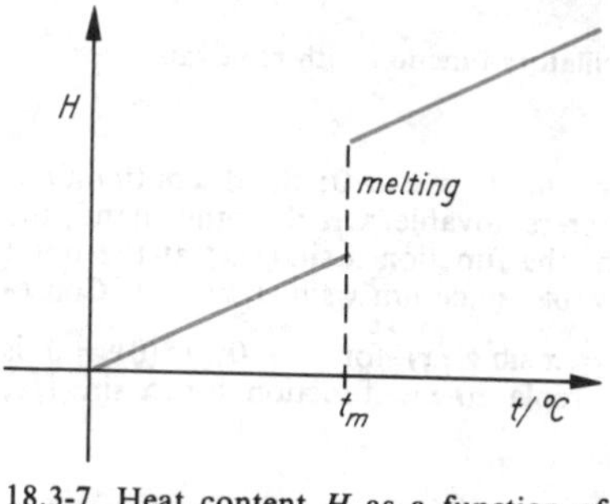

18.3-7 Heat content H as a function of the temperature t; t_m is the melting point

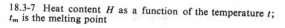

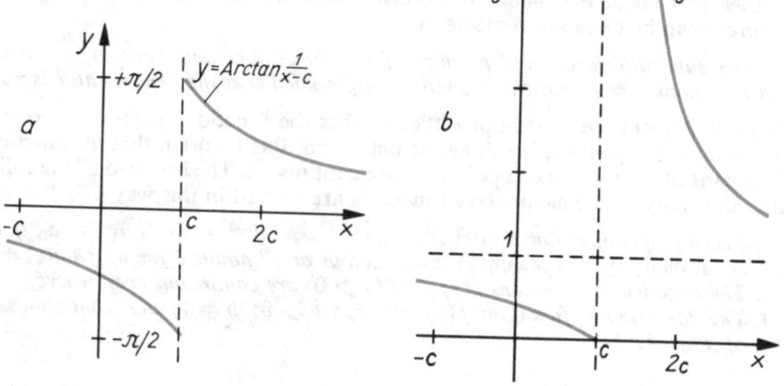

18.3-8 Graph of a function a) with finite, b) with infinite jump discontinuity

Example 1: The function $f(x) = \text{Arctan}\,[1/(x-c)]$ has a jump discontinuity of magnitude π at the point $x = c$ (Fig. 18.3-8 a), since

$$\lim_{x\to c-0} \text{Arctan}\,[1/(x-c)] = \lim_{z\to-\infty} \text{Arctan}\, z = -\pi/2 \text{ and } \lim_{x\to c+0} \text{Arctan}\,[1/(x-c)] = \lim_{z\to+\infty} \text{Arctan}\, z = +\pi/2.$$

Example 2: The function $f(x) = e^{1/(x-c)}$ has an infinite discontinuity at the point $x = c$ (see Fig. 18.3-8 b), since $\lim_{x\to c-0} e^{1/(x-c)} = \lim_{z\to-\infty} e^z = 0$ and $\lim_{x\to c+0} e^{1/(x-c)} = \lim_{z\to+\infty} e^z = +\infty$.

Example 3: The function $f(x) = 1/\cos x$ has infinite discontinuities at the points $\pi/2 + k\pi$ $(k = 0, \pm1, \pm2, \ldots)$, from $+\infty$ to $-\infty$ for even k and from $-\infty$ to $+\infty$ for odd k (Fig.).

Oscillatory functions with discontinuities. The function $f(x) = \sin(1/x)$ is not defined at the point $x = 0$ and is therefore not continuous there. If a positive number δ, however small, is chosen, there always exist in the interval $-\delta < x < +\delta$ infinitely many points $x = 2/(\pi n)$, that is $1/x = \pi n/2$, $[n > 2/(\pi\delta)]$ with the following property: for $n_1 = 2k$, $n_2 = 4k + 1$, $n_3 = 4k + 3$ (k an integer) the function $\sin(1/x) = \sin \pi n/2$ takes the values $0, +1, -1$ (Fig.). The function oscillates between $+1$ and -1 more and more rapidly, the larger n is or the nearer x approaches to zero. The function

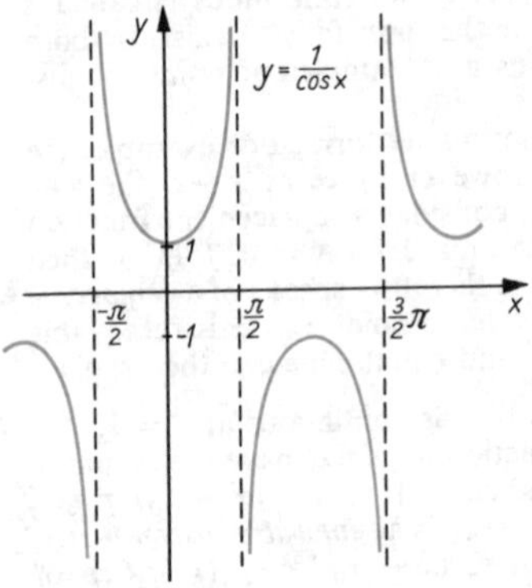

18.3-9 Jump from $-\infty$ to $+\infty$

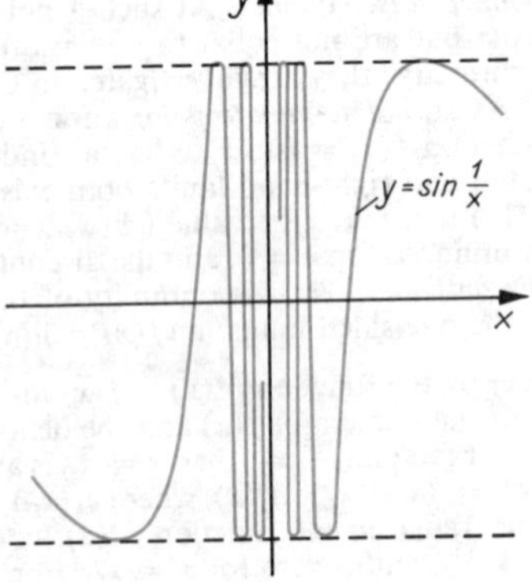

18.3-10 A discontinuous oscillatory function

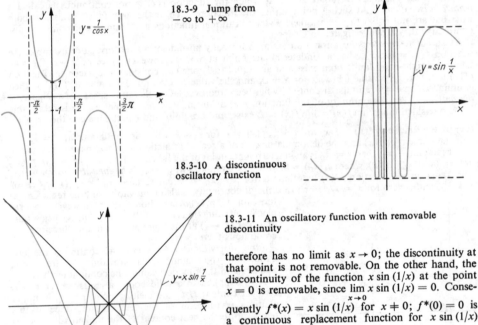

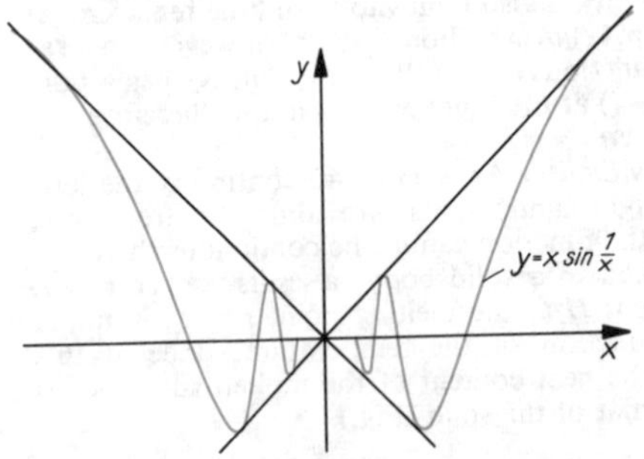

18.3-11 An oscillatory function with removable discontinuity

therefore has no limit as $x \to 0$; the discontinuity at that point is not removable. On the other hand, the discontinuity of the function $x \sin(1/x)$ at the point $x = 0$ is removable, since $\lim_{x\to0} x \sin(1/x) = 0$. Consequently $f^*(x) = x \sin(1/x)$ for $x \neq 0$; $f^*(0) = 0$ is a continuous replacement function for $x \sin(1/x)$ (Fig.).

Theorems about continuous functions. From the rules for calculating with limits the following theorem can be deduced immediately.

The sum, difference, and product of two functions continuous at $x = \xi$ are likewise continuous at this point. Their quotient is continuous provided that the denominator is not zero for $x = \xi$.

Since it is recognized without difficulty that the functions $g(x) = c$ = constant and $h(x) = x$ are continuous everywhere, it follows at once from this theorem that all functions obtained from them by means of the four basic operations are continuous. The first two of the following statements about the continuity of the elementary functions are proved in this way.

1. *Every polynomial function $f(x) = a_n x^n + a_{n-1} x^{n-1} + \cdots + a_1 x + a_0$ is continuous everywhere.*
2. *A rational function $p(x)/q(x)$ is continuous at all points ξ for which $q(\xi) \neq 0$.*
3. *The exponential functions $f(x) = a^x (a > 0)$ are continuous everywhere.*
4. *The logarithmic functions $f(x) = \log_b x (b > 0;\ b \neq 1)$ are continuous for all positive values of x.*

5. The trigonometric functions $\sin x$ *and* $\cos x$ *are continuous everywhere; the function* $\tan x = \sin x/\cos x$ *is continuous for all* $\xi \neq (2k+1)\pi/2$ *(k an integer) and the function* $\cot x = \cos x/\sin x$ *for all* $\xi \neq k\pi$ *(k an integer).*

With the help of the limit $\lim_{x\to 0} a^x = 1$ already obtained, one deduces that $\lim_{x\to\xi} a^x = \lim_{x\to\xi}(a^\xi \cdot a^{x-\xi}) = a^\xi \cdot \lim_{x\to\xi} a^{x-\xi} = a^\xi \cdot \lim_{h\to 0} a^h = a^\xi \cdot 1 = a^\xi$. Thus, the exponential function is continuous. The continuity of the logarithmic function follows similarly from $\lim_{x\to\xi} \log_b x = \log_b \xi (\xi > 0, b > 1)$. Since $\log_b x = -\log_{1/b} x$, the result holds also for $0 < b < 1$, and thus for all admissible bases.

Finally $\lim_{x\to 0} \sin x = 0$ since $-|x| \leqslant \sin x \leqslant |x|$, and $\lim_{x\to 0} \cos x = 1$ as has already been shown. From this one obtains

$$\sin x = \sin(\xi + x - \xi) = \sin\xi\cos(x-\xi) + \cos\xi\sin(x-\xi) \to \sin\xi \quad \text{as} \quad x\to\xi,$$
$$\cos x = \cos(\xi + x - \xi) = \cos\xi\cos(x-\xi) - \sin\xi\sin(x-\xi) \to \cos\xi \quad \text{as} \quad x\to\xi.$$

For the continuity of the functions $\tan x = \sin x/\cos x$ and $\cot x = \cos x/\sin x$, only the zeros of the denominators must be excluded.

Continuity of the inverse functions. The *circular functions* Arcsin x, Arccos x, Arctan x, Arccot x, as inverse functions of the continuous trigonometric functions, are likewise continuous, since the following theorem is true:

If a function $f(x)$ *has an inverse function* $\varphi(x)$ *in an interval I, then the continuity of* $f(x)$ *at the point* $x = \xi$ *implies the continuity of* $\varphi(x)$ *at the point* $x = f(\xi)$.

Accordingly the root functions $\sqrt[n]{x}$ are continuous for all positive x, since they are the inverse functions of the functions x^n for $x > 0$.

Continuity of composite functions. Let $y = f[\varphi(x)]$ be a composite function, whose *inner function* $t = \varphi(x)$ is continuous at the point $x = \xi$, and whose *outer function* $y = f(t)$ is continuous at the point $t = \tau = \varphi(\xi)$. Then the composite function $y = f[\varphi(x)]$ is continuous at $x = \xi$.

Every continuous function of a continuous function is again continuous.

Since $\lim_{t\to\tau} f(t) = f(\tau)$, to an arbitrary $\varepsilon > 0$ there always corresponds a suitable number $\delta_1(\varepsilon) > 0$, such that $|f(t) - f(\tau)| < \varepsilon$ for all $|t - \tau| < \delta_1(\varepsilon)$. Further, since $\lim_{x\to\xi} \varphi(x) = \varphi(\xi) = \tau$, to every arbitrary positive number, say to $\delta_1(\varepsilon)$, there corresponds a suitable number $\delta(\delta_1(\varepsilon)) = \delta_2(\varepsilon)$ such that $|\varphi(x) - \varphi(\xi)| < \delta_1(\varepsilon)$ for all x with $|x - \xi| < \delta_2(\varepsilon)$. Consequently, to every arbitrary $\varepsilon > 0$ one can choose a number $\delta_2(\varepsilon)$, such that $|f[\varphi(x)] - f[\varphi(\xi)]| < \varepsilon$ for all x satisfying $|x - \xi| < \delta_2(\varepsilon)$.

With the help of this theorem one can establish the continuity of many functions. For example, all functions $\sqrt[n]{p(x)}$, in which $p(x)$ denotes a polynomial, are continuous for all values of x for which $p(x) \geqslant 0$, since the polynomial $p(x)$ is continuous for all x, and the function $\sqrt[n]{t}$ is continuous for all $t \geqslant 0$. The function $f(x) = e^{\sin x}$ is continuous everywhere, since $t = \sin x$ and $y = e^t$ are everywhere continuous functions. Similarly the functions Arctan (x^2), $\cos(5x^2 - e^{4x+1})$, $\sin(1/x)$ $(x \neq 0)$ are continuous everywhere.

Properties of continuous functions. Functions that are continuous in an interval form a class of functions with noteworthy properties, such as the following, which GAUSS and other leading mathematicians of his time regarded as obvious, and for which Bernard BOLZANO (1781–1848) published the first proof.

Bolzano's theorem: If a function $f(x)$, continuous in a closed interval, assumes values with opposite signs at two points a and b in this interval, then there exists at least one point ξ between a and b at which the function vanishes.

The proof proceeds by enclosing such a point ξ for which $f(\xi) = 0$ under the given assumptions, in a nest of intervals.

Bolzano's theorem is the basis of many approximation methods for the solution of equations. For example, it follows from this theorem that a polynomial $p(x)$ of odd degree has at least one real zero, since $p(\omega)$ and $p(-\omega)$ certainly have opposite signs for sufficiently large values of ω.

The following are consequences of Bolzano's theorem:

1. A continuous function that does not vanish in an interval I must have the same sign everywhere in that interval.

2. If a function $f(x)$*, continuous in a closed interval, takes the values* $f(a) = A$ *and* $f(b) = B$ $(A \neq B)$ *at two points in the interval, then it takes every value between A and B at least once.*

Further fundamental properties of continuous functions are:

If a function $f(x)$ is continuous at x_0, where $f(x_0) \neq 0$, then $f(x)$ has the same sign in a certain neighbourhood of x_0 as it has at x_0 itself.

Theorem of Weierstrass. A function that is continuous in a closed interval is bounded there. A function that is continuous in a closed interval takes both a greatest and a least value in the interval.

In these theorems the assumption that the interval is closed cannot be omitted. For example, the function $1/x$ is continuous in the interval $0 < x \leqslant 1$, which is open on the left, but it takes larger and larger values as x approaches 0; it is unbounded. However, it is bounded in every closed interval $a \leqslant x \leqslant 1$ $(a > 0)$; $1 \leqslant f(x) \leqslant 1/a$. The theorem of Weierstrass also need not hold in an interval that is not closed. For example, the function $f(x) = x$ is continuous in the interval $0 \leqslant x < 1$ which is open on the right, but since $\lim_{x \to 1} f(x) = 1$, there is no point of the interval at which the value of the function is greater than at all other points.

19. Differential calculus

The differential and integral calculus, jointly known as the infinitesimal calculus, are basic disciplines of higher analysis. The objects of the differential calculus are functions, and its methods are the investigation and calculation of limiting values. Its central concept, the derivative of a function $f(x)$, is a measure of the sensitivity with which $f(x)$ reacts to a change in its argument. Many geometrical problems also, such as the calculation of the gradient of the tangent to a curve or the determination of the curvature of a curve, can be solved with the help of the differential calculus.

Because the relationships between quantities in the physical world can frequently be expressed by continuous and differentiable functions, only the differential calculus makes it possible in the natural sciences and technological disciplines to express mathematically not only states but also processes. For example, if $s = f(t)$ describes the dependence of the distance s described by a moving point mass on the time t, then the derivative of this function represents the instantaneous speed of this point mass. As an extension of this idea the concept of speed can be carried over to other circumstances in which time plays the part of the independent variable. The concepts of heating or cooling of a body, of reaction speed of a chemical process, rate of decay of a radioactive process and rate of growth of a biological organism can be defined and calculated. For mathematics itself the methods and results of the differential calculus have become the basis of higher analysis. The development of many disciplines is unthinkable without it, for example, the investigation of the relationships between functions and their derivatives, the expansion of functions in infinite series, the treatment of differential equations or differential geometry.

The beginnings of the infinitesimal calculus go back to the end of the 16th century; the theory was developed in the second half of the 17th century simultaneously, but independently, by Gottfried Wilhelm LEIBNIZ (1646–1716) and Isaac NEWTON (1643–1727) as a calculus, that is, an easily manageable method. Whereas LEIBNIZ started with the tangent problem, NEWTON arrived at the differential calculus by investigating physical problems. NEWTON also recognized as early as 1665 that differentiation and integration (see Chapter 20.) are inverse problems to one another.

19.1. The derivative of a function

To analyse the journey of a train from a town A to a town B, at a distance $s_A - s_B$ apart, a distance-time diagram (Fig.) can serve as a graph of the function $s = f(t)$, in which each point of time t_i corresponds to the distance s_i travelled by the train. The train departs from A at time t_0, brakes at t_3, because the signal S is at red, but does not come to rest because the signal just changes to green. By increasing its speed, the train reaches B on time.

The ratio of the distance travelled $(s_n - s_m)$ to the time $(t_n - t_m)$ taken to travel this distance, where $n > m$, is a measure of the speed. In a graphical railway timetable the points A and B are joined by a straight line. Consequently a uniform *average speed* $\bar{v} = (s_B - s_A)/(t_8 - t_0)$ is assumed. If tangents to the curve parallel to the line AB are drawn, then by inspection the speed of the train is $v = \bar{v}$ at times t_1, t_3, t_5 and t_7. In the time intervals (t_0, t_1) and (t_4, t_6) the speed v *increases*, and in (t_3, t_4) and (t_7, t_8) it *decreases*. The larger v is, the *steeper* the curve. At the points (t_1, s_1) and (t_5, s_5) it is *curved* to the left, and at (t_3, s_3) and (t_7, s_7) to the right.

Of course, each of these statements must be made more precise geometrically, but above all this process of making precise leads to purely analytical statements about a function $s = f(t)$, which are valid without any reference to a geometrical meaning.

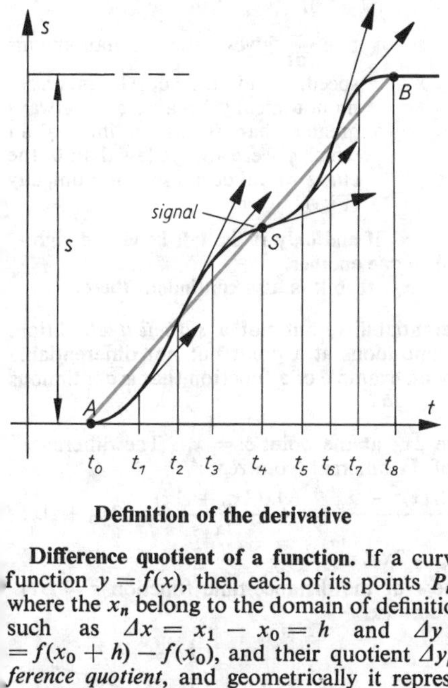

19.1-1 Schematic distance-time diagram for the journey of a train from town A to town B, s distance, t time

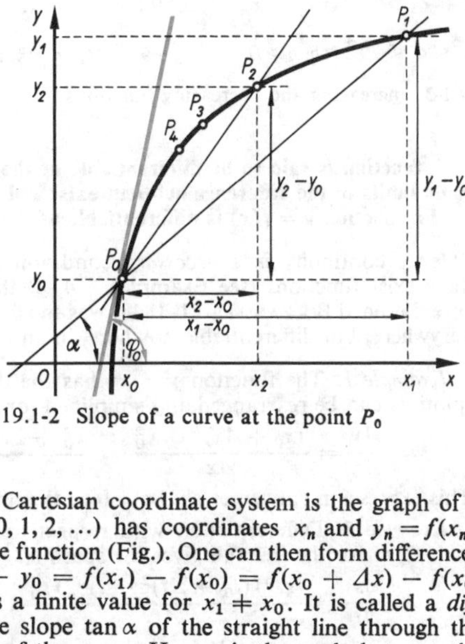

19.1-2 Slope of a curve at the point P_0

Definition of the derivative

Difference quotient of a function. If a curve in a Cartesian coordinate system is the graph of a function $y = f(x)$, then each of its points P_n $(n = 0, 1, 2, \ldots)$ has coordinates x_n and $y_n = f(x_n)$, where the x_n belong to the domain of definition of the function (Fig.). One can then form differences such as $\Delta x = x_1 - x_0 = h$ and $\Delta y = y_1 - y_0 = f(x_1) - f(x_0) = f(x_0 + \Delta x) - f(x_0) = f(x_0 + h) - f(x_0)$, and their quotient $\Delta y/\Delta x$ has a finite value for $x_1 \neq x_0$. It is called a *difference quotient*, and geometrically it represents the slope $\tan\alpha$ of the straight line through the points $P_0(x_0, y_0)$ and $P_1(x_1, y_1)$, which is a secant of the curve. Here α is the angle between the positive x-axis and the secant, measured in the positive sense

$$\frac{\Delta y}{\Delta x} = \frac{y_1 - y_0}{x_1 - x_0} = \frac{f(x_1) - f(x_0)}{x_1 - x_0} = \frac{f(x_0 + \Delta x) - f(x_0)}{\Delta x} = \frac{f(x_0 + h) - f(x_0)}{h} = \tan\alpha .$$

Difference quotients are frequently used, for example, in numerical mathematics as divided differences, in physics for average speeds or average temperature gradients.

Derivative. If the point $P_0(x_0, y_0)$ is kept fixed, but the point $P_1(x_1, y_1)$ moves along the curve towards the point P_0, then the secants change their position, and the difference quotient, hence the angle α, change their values. If the difference quotient $\Delta y/\Delta x$ tends to a limit as $x_n \to x_0$, this limit is called the *derivative* $\left(\frac{dy}{dx}\right)_{x=x_0}$ of the function $y = f(x)$ at the point x_0. The curve of this function then has a tangent at the point $P_0(x_0, y_0)$, whose position is determined by the limit φ of the angle α, given by $\lim\limits_{\Delta x \to 0} \frac{\Delta y}{\Delta x} = \left(\frac{dy}{dx}\right)_{x=x_0} = \tan\varphi$. The derivative can also be denoted by $f'(x_0)$ or $y'_{x=x_0}$ (read *f dashed of* x_0, *y′ at the point* x_0 or dy *by* dx *for* $x = x_0$).

From the consideration of limits (see Chapter 18.) it follows that the left- and right-hand limits should be equal to one another.

Derivative at the point x_0

$$f'(x_0) = y'_{x=x_0} = \left(\frac{dy}{dx}\right)_{x=x_0} = \lim_{\Delta x \to 0} \frac{\Delta y}{\Delta x} = \lim_{x_n \to x_0} \frac{y_n - y_0}{x_n - x_0} = \lim_{x_n \to x_0} \frac{f(x_n) - f(x_0)}{x_n - x_0}$$
$$= \lim_{\Delta x \to 0} \frac{f(x_0 + \Delta x) - f(x_0)}{\Delta x} = \lim_{h \to 0} \frac{f(x_0 + h) - f(x_0)}{h}$$

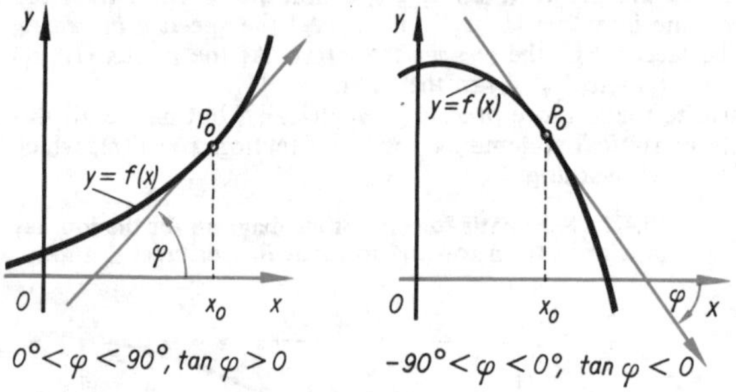

19.1-3 Increasing and decreasing functions

In connection with the analysis of the journey of a train from a town A to a town B it follows that the curve of the function $s = s(t)$ has a tangent for precisely those points of time t for which the limit $\lim_{\Delta t \to 0} \frac{\Delta s}{\Delta t} = \frac{ds}{dt} = \tan \varphi$ exists. The derivative $\frac{ds}{dt}$ gives the instantaneous speed, and the curve increases monotonically because φ is always greater than 0. In an interval in which φ were always less than 0, the curve would decrease monotonically (Fig.).

A function is said to be differentiable at the point $x = x_0$ if and only if the left-hand and right-hand limits of the difference quotient exist and are equal to one another.
If a function $y = f(x)$ is differentiable at the point $x = x_0$, then it is also continuous there.

Hence continuity is a *necessary* condition for differentiability, but not a *sufficient* condition. There exist functions (see Examples 3, 4, 5) that are continuous at a point but not differentiable there. Bernard BOLZANO (1781–1848) was the first to give an example of a function that is continuous everywhere, but differentiable nowhere, in an interval.

Example 1: The function $y = x^2$ has the derivative $2x_0$ at the point $x = x_0$. The difference quotient can be rearranged and simplified for values of Δx different from zero:

$$\frac{\Delta y}{\Delta x} = \frac{(x_0 + \Delta x)^2 - x_0^2}{\Delta x} = \frac{x_0^2 + 2x_0\,\Delta x + (\Delta x)^2 - x_0^2}{\Delta x} = \frac{\Delta x(2x_0 + \Delta x)}{\Delta x} = 2x_0 + \Delta x.$$

This expression converges as $\Delta x \to 0$ to the limit $y'_{x=x_0} = 2x_0$.

Example 2: Differentiation with respect to the time t of the distance-time function $s = f(t) = (g/2)\,t^2$ for the free fall gives the speed $v_{t=t_0} = gt_0$, because

$$\frac{\Delta s}{\Delta t} = \frac{(g/2)\,(t_0 + \Delta t)^2 - (g/2)\,t_0^2}{\Delta t}$$
$$= \frac{g}{2} \cdot \frac{\Delta t(2t_0 + \Delta t)}{\Delta t} = gt_0 + (g/2)\,\Delta t,$$
$$\lim_{\Delta t \to 0} \frac{\Delta s}{\Delta t} = \lim_{\Delta t \to 0} (gt_0 + (g/2)\,\Delta t) = gt_0.$$

Example 3: The function $y = x^{1/3}$ is not differentiable at the point $x = 0$ (Fig.). The difference quotient

$$\frac{\Delta y}{\Delta x} = \frac{(0 + \Delta x)^{1/3} - 0^{1/3}}{\Delta x} = \frac{(\Delta x)^{1/3}}{\Delta x} = \frac{1}{(\Delta x)^{2/3}}$$

does not tend to finite limit as $\Delta x \to 0$, but increases beyond every bound or tends to infinity. The tangent to the curve of the function $y = x^{1/3}$ is perpendicular to the x-axis at the point $x = 0$.

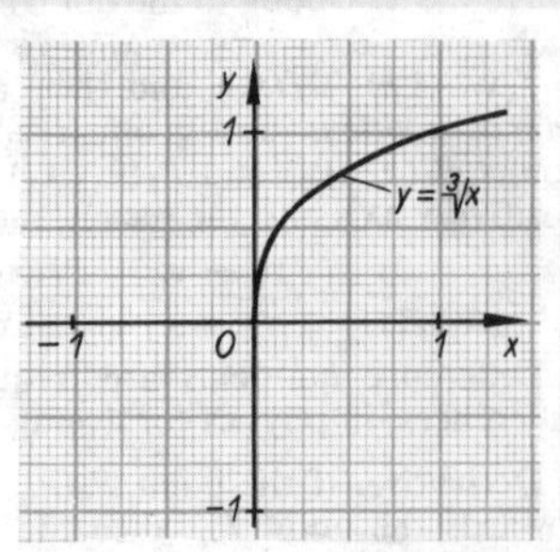

19.1-4 Graph of the function $y = \sqrt[3]{x}$

Example 4: The function $y = e^{|x-2|}$ is continuous at the point $x = 2$, but not differentiable there (Fig.). Its difference quotient is

$$\frac{\Delta y}{\Delta x} = \frac{e^{|2+\Delta x-2|} - e^{|2-2|}}{\Delta x} = \frac{e^{\Delta x} - 1}{\Delta x}$$

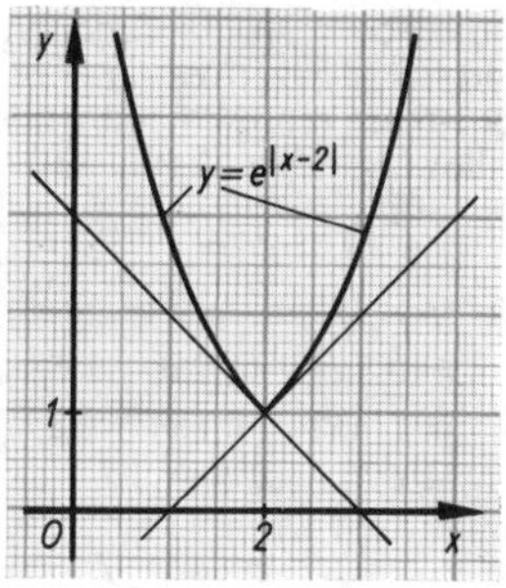

and this tends to the value +1 or −1 according to the sign of Δx (see Chapter 18.). At this point (2, 1) the curve has two tangents.

Example 5: At the point $x = 0$ the function $y = +x^{3/2}/2$ is only right-hand differentiable, because negative abscissae do not belong to its domain of definition. The value of its right-hand derivative at $x = 0$ is zero, because

$$\frac{\Delta y}{\Delta x} = \frac{(1/2)(0 + \Delta x)^{3/2} - (1/2)\cdot 0^{3/2}}{\Delta x} = \frac{1}{2}\frac{(\Delta x)^{3/2}}{\Delta x} = \frac{1}{2}(\Delta x)^{1/2}$$

and $\lim\limits_{\Delta x \to 0} (1/2)(\Delta x)^{1/2} = 0$.

19.1-5 Graph of the function $y = e^{|x-2|}$

The derivative as a function

If the derivative of a function $y = f(x)$ exists for all points in an interval $x_0 \leqslant x \leqslant x_1$, then the function is *differentiable in the whole interval.* To each value x in the interval there corresponds the derivative $f'(x)$ of the function at the point x; thus, $f'(x)$ is a function of x, the *derived function* or the *derivative.*

Example: For all values of x the function $y = x^2$ has the derivative $y' = 2x$. At the points $x_1 = 3$ and $x_2 = -2$ it has the values $y_1' = 6$ and $y_2' = -4$, respectively.

Higher derivatives. The derivative $y' = f'(x)$ of a function $y = f(x)$ is a function of x. Assuming that this is again differentiable, as is almost always the case for elementary functions, then the derivative of the first derivative is called the *second derivative* or the *derivative of the second order,* and is denoted by $y'' = f''(x) = \frac{d^2y}{dx^2}$ (read *y double dashed, f double dashed of x, or* d *two y by* d*x squared*). Similarly there can be a *third, fourth, nth derivative,* or a *derivative of the nth order.* Expressions such as 'existence of the derivative of the *n*th order' or 'differentiable arbitrarily often' are to be understood in this sense.

The following examples can be calculated by the rules derived in 19.2.

Example 1: $y = f(x) = x^5 + x^4/2 - 5x^3/6 + x^2 + 5x + 2;$

$$y' = f'(x) = 5x^4 + 2x^3 - 5x^2/2 + 2x + 5;$$

$$y'' = f''(x) = 20x^3 + 6x^2 - 5x + 2; \quad y''' = f'''(x) = 60x^2 + 12x - 5;$$

$$y^{IV} = y^{(4)} = f^{IV}(x) = f^{(4)}(x) = 120x + 12;$$

$$y^{(5)} = f^{(5)}(x) = 120; \quad y^{(6)} = f^{(6)}(x) = 0.$$

Example 2: $y = f(x) = \frac{x^2}{(x-1)^2}; \quad y' = f'(x) = -\frac{2x}{(x-1)^3};$

$$y'' = f''(x) = \frac{2(2x+1)}{(x-1)^4}; \quad y''' = f'''(x) = -\frac{12(x+1)}{(x-1)^5}; \ \dots$$

Example 3: $y = f(x) = \sin x; \quad \frac{dy}{dx} = \frac{d}{dx}\sin x = \cos x; \quad \frac{d^2y}{dx^2} = \frac{d^2}{dx^2}\sin x = -\sin x;$

$$\frac{d^3y}{dx^3} = \frac{d^3}{dx^3}\sin x = -\cos x; \quad \frac{d^4y}{dx^4} = \frac{d^4}{dx^4}\sin x = \sin x; \ \dots$$

All these are examples of functions that are differentiable arbitrarily often.

Example 4: From the distance-time law of the free fall $s = (g/2)\,t^2$ one obtains by differentiation $s' = gt$ and $s'' = g$, where g is a constant, the acceleration due to gravity. Consequently the free fall is a motion with constant acceleration.

Physically $s'' = \frac{d^2s}{dt^2}$ denotes the derivative of the speed $s' = \frac{ds}{dt}$, that is, the *acceleration.* In the example of the journey of a train from a town A to a town B described in the introduction, the time intervals during which the train is accelerating can be deduced. In these the angle φ between the x-axis and the tangent is increasing ($s'' > 0$); they are the intervals t_0 to t_2 and t_4 to t_6. In the intervals t_2 to t_4 and t_6 to t_8, however, this angle φ is getting smaller and the speed of the train is decreasing ($s'' < 0$).

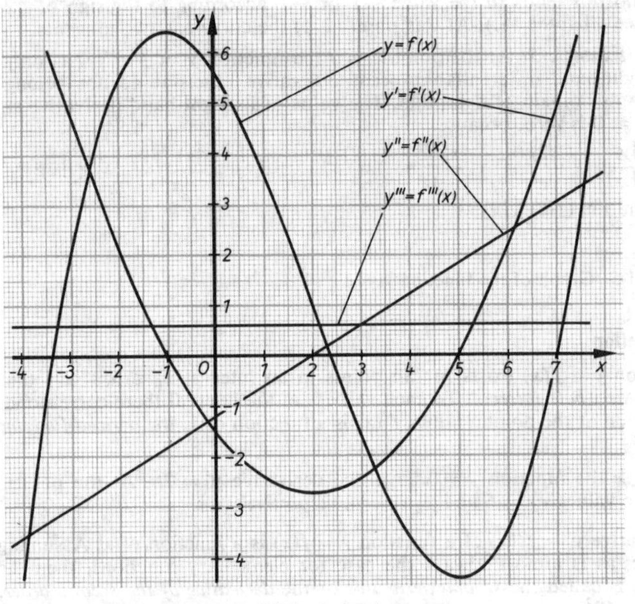

19.1-6 Curve of a function with its derived curves

The curves of a function and its derivatives in Fig. 19.1-6 are represented by the following table of values:

$$y = f(x) = 0.1x^3 - 0.6x^2 - 1.5x + 5.6;$$
$$y' = f'(x) = 0.3x^2 - 1.2x - 1.5;$$
$$y'' = f''(x) = 0.6x - 1.2;$$
$$y''' = f'''(x) = 0.6.$$

x	y	y'
−4	−4.4	
−3	2	4.8
−2	5.4	2.1
−1	6.4	0
0	5.6	−1.5
1	3.6	−2.4
2	1	−2.7
3	−1.6	−2.4
4	−3.6	−1.5
5	−4.4	0
6	−3.4	2.1
7	0	4.8
8	6.4	

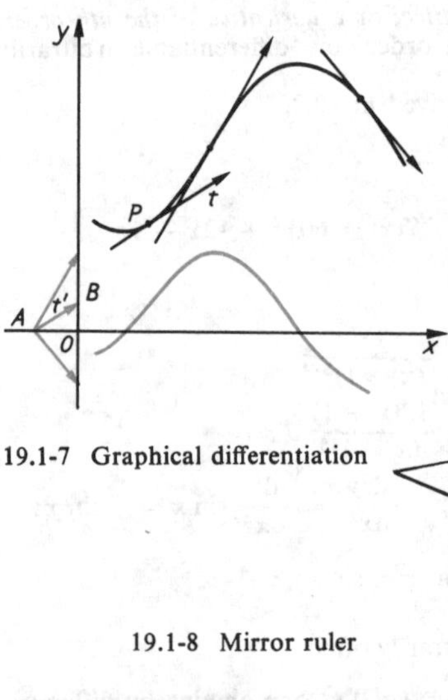

19.1-7 Graphical differentiation

Graphical differentiation. The *derivative* at a point P of the curve given by the function $y = f(x)$ is the value of the *tangent* function of the angle φ that the tangent t at the point P makes with the $+x$-axis. If a parallel t' to this tangent through the point $A(-1, 0)$ cuts the y-axis at the point B, then $\tan\varphi = |OB|/|AO| = y'$ (Fig.). The direction of the tangent at the point P can be determined with a *mirror ruler* (Fig.). The plane mirror of the ruler stands at right angles to the plane of the graph. The visible part of the curve goes over into its mirror image without a kink only if the ruler cuts the curve at right angles at the point P. The line through P perpendicular to this *normal to the curve* is the tangent t.

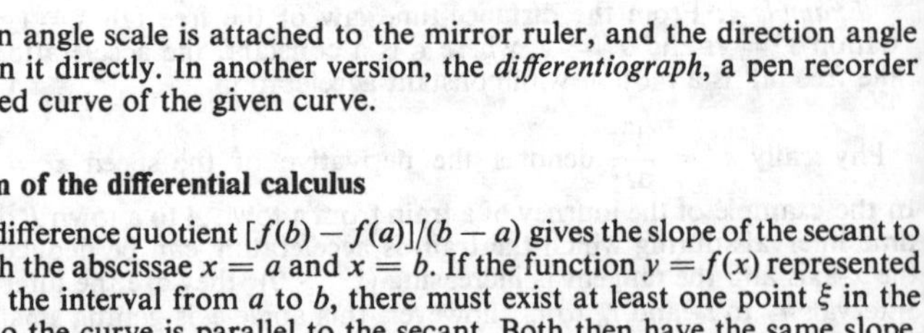

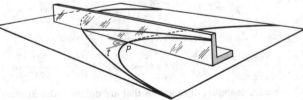

19.1-8 Mirror ruler

In a more complicated tool an angle scale is attached to the mirror ruler, and the direction angle of the tangent can be read off on it directly. In another version, the *differentiograph*, a pen recorder is attached that draws the derived curve of the given curve.

The mean value theorem of the differential calculus

The mean value theorem. The difference quotient $[f(b) - f(a)]/(b - a)$ gives the slope of the secant to the curve through the points with the abscissae $x = a$ and $x = b$. If the function $y = f(x)$ represented by the curve is differentiable in the interval from a to b, there must exist at least one point ξ in the interval for which the tangent to the curve is parallel to the secant. Both then have the same slope, that is, $f'(\xi) = [f(b) - f(a)]/(b - a)$. If one denotes the values a and b by $a = x$ and $b = x + h$, then ξ can be expressed in the form $\xi = x + \vartheta h$, where ϑ is a positive number less than 1, $0 < \vartheta < 1$ (Fig.). The mean value theorem then has the form

$$\frac{f(x+h) - f(x)}{h} = f'(x + \vartheta h), \quad \text{where} \quad 0 < \vartheta < 1.$$

Mean value theorem: If a function $y = f(x)$ is continuous in the closed interval $a \leqslant x \leqslant b$ and differentiable in the open interval $a < x < b$, then there exists in the interior of the interval at least one intermediate value ξ (mean value) for which $[f(b) - f(a)]/(b - a) = f'(\xi)$, where $a < \xi < b$.

With the help of the mean value theorem numerical calculations can be performed, for example, the estimation of a value of a function from a known neighbouring value.

Example: From $f(x) = \ln 690 = 6.53669$, $f(x + h) = \ln 691$ can be determined to five places of decimals. From $f(x + h) = f(x) + hf'(x + \vartheta h)$, $hf'(x + \vartheta h)$ is the increment to be added to $f(x) = \ln 690$. Because $x = 690$ and $x + h = 691$, it follows that $h = 1$, and because $f'(x) = \frac{\mathrm{d}}{\mathrm{d}x} \ln x = 1/x$ the increment is $1 \cdot f'(x + \vartheta) = 1/(690 + \vartheta)$, which lies between $1/690 = 0.0014492\ldots$ and $1/691 = 0.0014471\ldots$ To five decimal places both these numbers have the value 0.00145. One therefore obtains $\ln 691 = 6.53669 + 0.00145 = 6.53814$.

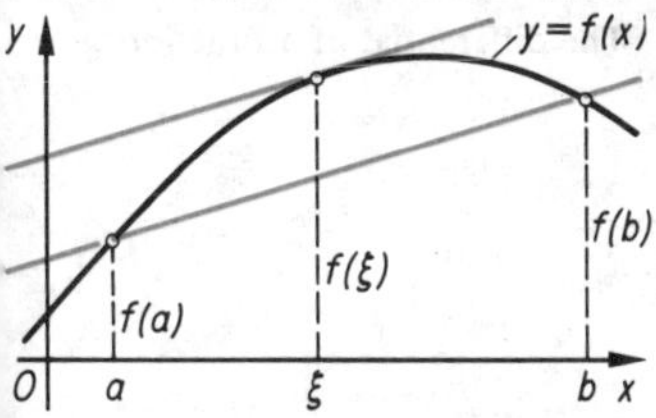

19.1-9 The mean value theorem of the differential calculus

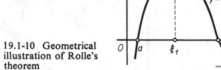

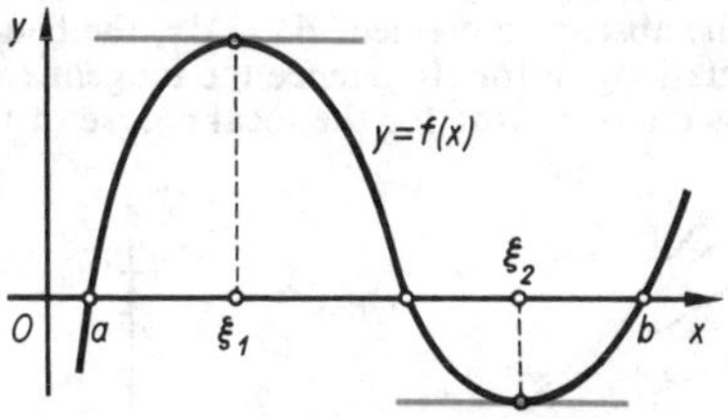

19.1-10 Geometrical illustration of Rolle's theorem

Rolle's theorem. If in the mean value theorem the values $f(a)$ and $f(b)$ of the function are equal then there exists a value ξ with $a < \xi < b$ for which $f'(\xi) = 0$, that is, there exists in this interval a tangent parallel to the x-axis. In the theorem named after Michel ROLLE (1652–1719) (Fig.) the additional condition $f(a) = f(b)$ is imposed.

Rolle's theorem: If a function $y = f(x)$ is continuous in the closed interval $a \leqslant x \leqslant b$ and differentiable in the open interval $a < x < b$ and if $f(a) = f(b)$, then there exists in the interior of the interval at least one intermediate value ξ such that $f'(\xi) = 0$ with $a < \xi < b$.

The extended mean value theorem. For the sake of completeness an extension of the mean value theorem is stated here, which is useful for many purposes:

If two functions $f(x)$ and $g(x)$ are continuous in the closed interval $a \leqslant x \leqslant b$, differentiable in the open interval $a < x < b$, and if $g'(x) \neq 0$ in the interval, then there exists in the interior of the interval at least one intermediate value ξ such that $[f(b) - f(a)]/[g(b) - g(a)] = f'(\xi)/g'(\xi)$, where $a < \xi < b$.

Consequences of the mean value theorem. If the derivative of a function is zero for all points of an interval and if $x_1 < x_2$ are two of these points, then $f'(\xi) = [f(x_2) - f(x_1)]/(x_2 - x_1) = 0$, or $f(x_2) = f(x_1)$. The function is a constant.

A function that is differentiable in an interval and whose derivative $f'(x)$ vanishes everywhere in the interval, is constant in that interval.

If the derivatives of the functions $\varphi(x)$ and $\psi(x)$ have the same values in an interval, then the derivative of $f(x) = \varphi(x) - \psi(x)$ has the constant value zero, so that $f(x)$ is a constant.

Two functions, that are differentiable and whose derivatives are equal in an interval, differ in that interval only by an additive constant.

The theorem already obtained intuitively, that a function $f(x)$ increases in the interval $a < x < b$ if its derivative is positive there, and decreases if its derivative is negative, can be proved rigorously with the help of the mean value theorem.

Differential

Differential of a function. For a function $f(x)$, differentiable in an interval, the difference between the difference quotient and the derivative at the point x_0 is a function $\varphi(\Delta x)$ of Δx. From $[f(x_0 + \Delta x) - f(x_0)]/\Delta x - f'(x_0) = \varphi(\Delta x)$ the increment in the function is given by Δy

$= f(x_0 + \Delta x) - f(x_0) = f'(x_0) \cdot \Delta x + \varphi(\Delta x) \cdot \Delta x$. It consists of the part $f'(x_0) \cdot \Delta x$, which is linear in Δx and proportional to it and tends to zero as $\Delta x \to 0$, and of the part $\varphi(\Delta x) \cdot \Delta x$, which tends to zero 'of a higher order' than Δx as $\Delta x \to 0$. The linear part of the increment Δy is called the *differential of the function* at the point x_0 and is denoted by $dy = df(x_0) = f'(x_0) \cdot dx$. The quantity $dx = \Delta x$ is called the differential of the independent variable.

Differential of the function $y = f(x)$ at the point x_0	$dy = f'(x_0) \cdot dx$

Example: The differential of the function $y = f(x) = x^2$ at the point x_0 is $dy = 2x_0 \cdot dx$.

After the introduction of the concept of the differential, the derivative 'dy by dx' can be represented as the quotient 'dy over dx' of two finite quantities.

In a geometrical illustration (Fig.) the points $P_0[x_0, f(x_0)]$ and $P_1[x_0 + \Delta x, f(x_0 + \Delta x)]$ lie on the curve of the function $y = f(x)$. If the increment $\Delta y = |RP_1|$ corresponding to the increment Δx is cut by the tangent to the curve at the point P_0 in the point T, then $|RT| = dy$ is the differential. Clearly the smaller the abscissa increment $\Delta x = dx$, the better the approximation dy for Δy. Hence the tangent can be regarded as characteristic for the local course of the curve.

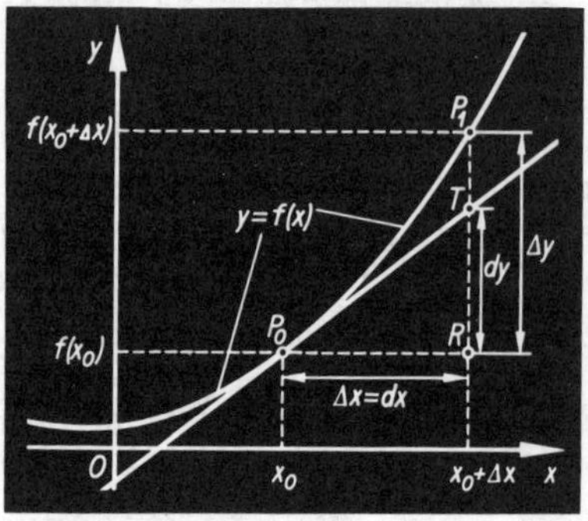

19.1-11 The differential of a function

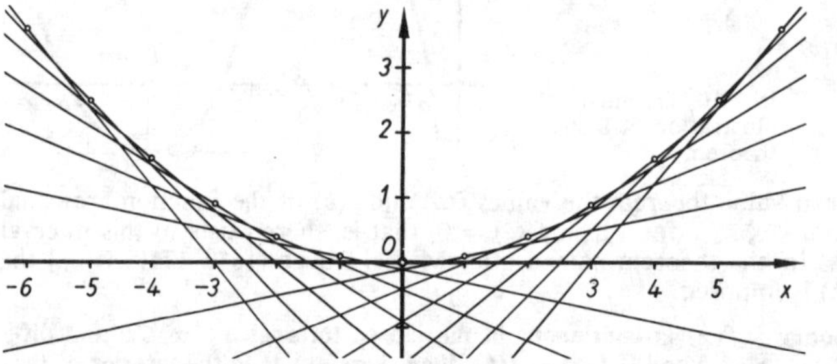

19.1-12 Tangents to the curve with the equation $y = x^2/10$

The graph of the function $y = x^2/10$ is given as the envelope of the tangents drawn at individual points (Fig.).

The approximation $\Delta y \approx dy$. In the calculation of approximations one makes use of the fact that for small $|\Delta x| = |dx|$ the increment Δy of the function in the neighbourhood of the point x_0 can be replaced by the differential dy of the function at this point with good precision: the approximation $\Delta y \approx dy$ is valid. For the function $y = \sin x$ in the neighbourhood of $x_0 = 0$, $\Delta y = \sin \Delta x - \sin 0 = \sin \Delta x = \sin dx$, $dy = \cos 0 \cdot dx = dx$, and consequently the approximation formula $\sin dx \approx dx$ holds, or $\sin h \approx h$ for small h if one writes h for dx.

Occasionally differentials are said to be *infinitely small*. This is an unprecise and misleading statement, because one considers throughout finite non-zero quantities that are chosen only sufficiently small for the problem under consideration, that is, small enough to correspond to the required degree of accuracy.

Differentials of higher order. Let the function $y = f(x)$ be n times differentiable ($n > 1$). Then its *differential of the first order* $dy = f'(x)\,dx$ is a differentiable function of x with the derivative $(dy)' = f''(x)\,(dx)^2$, because dx is independent of x and in the differentiation must be treated as a constant factor. The differential of dy is called the *differential of the second order* and is written d^2y (read d two y). Similarly one can form a differential of the third order, $d^3y = d(d^2y) = f'''(x)\,(dx)^3$, and so on.

Differential of the second, ..., nth order	$d^2y = d(dy) = f''(x)\,(dx)^2$ $d^ny = d(d^{n-1}y) = f^{(n)}(x)\,(dx)^n$

From this definition the nth derivative $f^{(n)}(x)$ of the function $y = f(x)$ can also be written as the quotient of two differentials: $f^{(n)}(x) = d^ny/dx^n$ (read dn y by dx to the nth).

19.2. The technique of differentiation

On the basis of the definition of the derivative, the following steps must be taken in differentiating a function: form the difference quotient, rearrange it suitably, and then find its limit. But one often obtains the result more quickly by using *general formulae* for the derivatives of typical composite functions and special functions. To prove these, however, the steps described above must, in general, be followed.

Derivatives of typical composite functions

Derivative of a product with a constant factor. If $y = cf(x)$, where the factor c is a constant, then the factor c can be taken out of the difference quotient; $\Delta y/\Delta x = c(\Delta f(x)/\Delta x)$. But if a quantity a tends to the limit a_0, then ca_0 is the limit of ca.

Factor rule	$\frac{d}{dx}[cf(x)] = c\left[\frac{d}{dx}f(x)\right]$

Examples: $y = 6x^2$, $y' = 6 \cdot 2x = 12x$; $y = \pi \sin x$, $y' = \pi \cos x$;

$y = (2/3)\,x^3$, $y' = (2/3) \cdot 3x^2 = 2x^2$; $y = 2\sqrt{x}$, $y' = 2 \cdot \frac{1}{2\sqrt{x}} = \frac{1}{\sqrt{x}}$.

Derivative of a sum. The difference quotient of a sum $f(x) = u(x) + v(x)$ can be rearranged as follows:

$$\frac{\Delta y}{\Delta x} = \frac{[u(x+\Delta x) + v(x+\Delta x)] - [u(x) + v(x)]}{\Delta x} = \frac{u(x+\Delta x) - u(x)}{\Delta x} + \frac{v(x+\Delta x) - v(x)}{\Delta x}.$$

Because the limit of a sum is equal to the sum of the limits, it follows that

$$\lim_{\Delta x \to 0} \frac{\Delta y}{\Delta x} = u'(x) + v'(x).$$

Sum rule	$\frac{d}{dx}(u + v) = \frac{du}{dx} + \frac{dv}{dx}$

In this derivation each of the elements in the sum can again be a sum; thus, the theorem holds also for the sum of finitely many terms.

The derivative of a sum of finitely many functions is equal to the sum of the derivatives of the individual functions.

The rule holds also for differences, since subtraction of a function can be regarded as addition of the same function multiplied by the constant factor (−1).

Derivative of a product. The difference quotient of a product $y = f(x) = u(x) \cdot v(x)$ can be rearranged as follows:

$$\frac{\Delta y}{\Delta x} = \frac{u(x+\Delta x) \cdot v(x+\Delta x) - u(x)\,v(x+\Delta x) + u(x)\,v(x+\Delta x) - u(x)\,v(x)}{\Delta x}$$

$$= \frac{u(x+\Delta x) - u(x)}{\Delta x} \cdot v(x+\Delta x) + u(x) \cdot \frac{v(x+\Delta x) - v(x)}{\Delta x}.$$

The operation of forming a limit can be interchanged with the operations of addition and multiplication; thus, the derivative is given by $f'(x) = u'(x) \cdot v(x) + u(x) \cdot v'(x)$.
For three factors $(v_1v_2v_3)' = (v_1v_2)'\,v_3 + v_1v_2v_3' = v_1'v_2v_3 + v_1v_2'v_3 + v_1v_2v_3'$.

Product rule	
$\frac{d}{dx}(uv) = \frac{du}{dx} \cdot v + \frac{dv}{dx} \cdot u$	$(uv)' = u'v + v'u$

Power rule
$\frac{dx^n}{dx} = nx^{n-1}$; n a positive integer

This rule can be generalized to n factors by induction. In the special case when each factor is equal to x, it follows that $(x^n)' = 1 \cdot x^{n-1} + 1 \cdot x^{n-1} + \cdots = nx^{n-1}$. With the help of the sum, the product and the power rules, every polynomial function can be differentiated.

The derivative of a polynomial function of degree n is a polynomial function of degree $n - 1$.

Example 1: $y = (3x^2 - 5x + 6)(4x^2 + 3x - 7) = u \cdot v$,
$y' = (6x - 5)(4x^2 + 3x - 7) + (8x + 3)(3x^2 - 5x + 6) = u' \cdot v + v' \cdot u$,
$y' = 48x^3 - 33x^2 - 24x + 53$.

One arrives at the same result if one performs the multiplication first, giving $y = 12x^4 - 11x^3 - 12x^2 + 53x - 42$, and then differentiates using the sum rule.

If non-rational, for example, transcendental functions occur as terms in a sum or as factors, then derivatives that will be derived later must be used.

Example 2: $y = x^2 \cdot \sin x$, $\quad y' = 2x \sin x + x^2 \cos x$.

Example 3: $y = x^2 \cdot \ln x$; $\quad y' = 2x \ln x + x^2 \cdot (1/x) = x(2 \ln x + 1)$;

$$y'' = 1 \cdot (2 \ln x + 1) + x \cdot (2/x) = 2 \ln x + 3.$$

Example 4: $y = x \sin x \cos x = uvw$;

$$y' = u'vw + uv'w + uvw' = 1 \cdot \sin x \cos x + x \cos x \cos x + x \sin x(-\sin x);$$

$$y' = \sin x \cos x + x \cos 2x.$$

Derivative of a quotient. Under the assumption that a quotient $y = f(x) = u(x)/v(x)$, $v(x) \neq 0$, is differentiable, its derivative can be deduced from the product rule. From $y = u/v$ it follows that $yv = u$ or $u' = y'v + yv'$. Thus,

$$y' = (1/v) \cdot (u' - yv') = (1/v)\,[u' - (u/v) \cdot v'] = [u'v - uv']/v^2.$$

Quotient rule	$\frac{d}{dx}\left(\frac{u}{v}\right) = \frac{1}{v^2}\left(v \cdot \frac{du}{dx} - u \cdot \frac{dv}{dx}\right)$	$\left(\frac{u}{v}\right)' = \frac{u' \cdot v - u \cdot v'}{v^2}$

It is not hard to prove, starting out from $\frac{\Delta y}{\Delta x}$ as in the case of a product, that a quotient of two differentiable functions is, in fact, differentiable.

From the quotient rule one can establish the validity of the formula found for the derivative of $y = x^n$ for negative integer exponents also, $n = -m$, m positive. Since $y = x^{-m} = 1/x^m$, one puts $u = 1$, $v = x^m$ and obtains:

$$y' = \frac{0 - mx^{m-1}}{x^{2m}} = -mx^{-m-1} = nx^{n-1}.$$

$$\frac{dx^n}{dx} = nx^{n-1}, \ n \text{ a negative integer}$$

Example 1: In the function $y = \frac{3x^2 - 5}{x^4 + 2}$, take $u = 3x^2 - 5$, $v = x^4 + 2$. Because $u' = 6x$ and $v' = 4x^3$ one obtains

$$y' = \frac{6x(x^4 + 2) - 4x^3(3x^2 - 5)}{(x^4 + 2)^2} = \frac{2x(6 + 10x^2 - 3x^4)}{(x^4 + 2)^2}.$$

Example 2: In the function $y = \frac{x^3}{x^2 - 1}$, take $u = x^3$ and $v = x^2 - 1$. Because $u' = 3x^2$ and $v' = 2x$ one obtains

$$y' = \frac{3x^2(x^2 - 1) - 2x \cdot x^3}{(x^2 - 1)^2} = \frac{x^2(x^2 - 3)}{(x^2 - 1)^2}.$$

For the second derivative one must take $u = x^4 - 3x^2$ and $v = (x^2 - 1)^2$. Because $u' = 4x^3 - 6x$ and $v' = 4x(x^2 - 1)$ one obtains

$$y'' = \frac{(4x^3 - 6x)(x^2 - 1)^2 - 4x(x^2 - 1)(x^4 - 3x^2)}{(x^2 - 1)^4} = \frac{2x(x^2 + 3)}{(x^2 - 1)^3}.$$

Example 3: The derivative of the function $y = \frac{1 + \tan x}{1 - \tan x} = \frac{u}{v}$, with $u' = 1/\cos^2 x$, $v' = -1/(\cos^2 x)$ is found to be

$$y' = \frac{(1 - \tan x)/\cos^2 x + (1 + \tan x)/\cos^2 x}{(1 - \tan x)^2} = \frac{2}{\cos^2 x(1 - \tan x)^2} = \frac{2}{1 - \sin 2x}.$$

Chain rule. As already described (see Chapter 5.), $y = f[\varphi(x)]$ is a composite function if the domain of definition of the function consists of values of x for which the values t of the function $t = \varphi(x)$ belong to the domain of definition of the function $y = f(t)$. The difference quotient may be replaced by $\frac{\Delta y}{\Delta x} = \frac{\Delta y}{\Delta t} \cdot \frac{\Delta t}{\Delta x}$. If the function $t = \varphi(x)$ is differentiable at the point ξ, so that $\frac{dt}{dx} = \varphi'(\xi)$ exists, and if further the function $y = f(t)$ has a derivative $\frac{df}{dt} = f'(\tau)$ at the point $\tau = \varphi(\xi)$, then the composite function $y = f[\varphi(x)]$ is also differentiable at the point ξ. One obtains $\frac{dy}{dx} = \frac{dy}{dt} \cdot \frac{dt}{dx}$, $\frac{df}{dx} = \frac{df}{d\varphi} \cdot \frac{d\varphi}{dx}$, or $f'[\varphi(x)] = f'(t)\,\varphi'(x)$, where $t = \varphi(x)$.

This proof is valid only when $\Delta t \neq 0$, but the result always holds.

Chain rule	$\dfrac{dy}{dx} = \dfrac{dy}{dt} \cdot \dfrac{dt}{dx}$ or $f'(x) = f'(t)\varphi'(x)$

Example 1: The function $y = (3x^2 + 5)^4$ is of the form $y = f(t) = t^4$, where $t = \varphi(x) = 3x^2 + 5$. It follows from the chain rule that

$$y' = \frac{dy}{dt} = \frac{df}{dt} \cdot \frac{d\varphi}{dx} = 4t^3 \cdot 6x = 4(3x^2 + 5)^3 \cdot 6x = 24x(3x^2 + 5)^3.$$

Example 2: In the function $y = \sqrt{(5x^3 - 7x + 8)}$, $y = f(t) = t^{1/2}$ and $t = \varphi(x) = 5x^3 - 7x + 8$. The chain rule gives

$$y' = \frac{df}{dt} \cdot \frac{d\varphi}{dx} = \frac{1}{2} t^{-1/2}(15x^2 - 7) = \frac{15x^2 - 7}{2\sqrt{(5x^3 - 7x + 8)}}.$$

Example 3: In the function $y = \sin 2x$, $y = f(t) = \sin t$ and $t = \varphi(x) = 2x$. One finds that $y' = \dfrac{df}{dt} \cdot \dfrac{d\varphi}{dx} = \cos 2x \cdot 2 = 2 \cos 2x$.

Example 4: To differentiate the function $y = \ln \sin \sqrt{(a + bx)}$, the chain rule must be used several times. Putting $y = f(t) = \ln t$, $t = \varphi(u) = \sin u$, $u = \psi(v) = \sqrt{v}$ and $v = a + bx$, one obtains in succession

$$y' = \frac{df}{dt} \cdot \frac{d\varphi}{du} \cdot \frac{d\psi}{dv} \cdot \frac{dv}{dx} = \frac{1}{t} \cdot \cos u \cdot \frac{1}{2\sqrt{v}} \cdot b$$

$$= \frac{1}{\sin \sqrt{(a + bx)}} \cdot \cos \sqrt{(a + bx)} \cdot \frac{b}{2\sqrt{(a + bx)}} = \frac{b \cot \sqrt{(a + bx)}}{2\sqrt{(a + bx)}}.$$

Logarithmic differentiation. In certain cases it is more advantageous to differentiate not the given function $y = f(x)$, but its natural logarithm $\ln f(x)$. The chain rule gives

$$\frac{d}{dx} \ln f(x) = \frac{1}{f(x)} \cdot \frac{d}{dx} f(x) \quad \text{or} \quad f'(x) = f(x) \cdot \frac{d}{dx} \ln f(x).$$

$\dfrac{d}{dx} f(x) = f(x) \cdot \dfrac{d}{dx} \ln f(x)$

Example 1: The natural logarithm of the function $y = x^x$ for positive values of x is $\ln x^x = x \ln x$. The product rule gives $\dfrac{d}{dx} \ln x^x = 1 \cdot \ln x + \dfrac{x}{x}$. Thus, the derivative of the given function is $y' = x^x(\ln x + 1)$.

Example 2: For positive values of x the function $y = x^{1/x}$ has the derivative

$$y' = x^{1/x} \cdot \frac{d}{dx}(\ln x^{1/x}) = x^{1/x} \frac{d}{dx}\left(\frac{1}{x} \cdot \ln x\right) = x^{1/x} \cdot \frac{-\ln x + 1}{x^2} = x^{(1-2x)/x}(1 - \ln x).$$

Example 3: The function $y = x^n[\varphi(x)^{1/m}] \sin^2 x$ has three factors. Its natural logarithm is $\ln y = n \ln x + 1/m \cdot \ln \varphi(x) + 2 \ln \sin x$, whose derivative is $\dfrac{d}{dx} \ln y = \dfrac{n}{x} + \dfrac{\varphi'(x)}{m\varphi(x)} + \dfrac{2 \cos x}{\sin x}$.

From this the derivative of the given function is obtained as

$$y' = x^n[\varphi(x)]^{1/m} \sin^2 x \left[\frac{n}{x} + \frac{\varphi'(x)}{m\varphi(x)} + 2 \cot x\right].$$

Example 4: For $y = (\sin x)^x$ one obtains

$$y' = (\sin x)^x (\ln \sin x + x \cos x/\sin x) = (\sin x)^x (\ln \sin x + x \cot x).$$

Derivatives of mutually inverse functions. If a function $y = f(x)$ is monotonic and continuous in an interval $a < x < b$ and has a finite and non-zero derivative $f'(x)$ for every x in the interval, then the function $x = \varphi(y)$ inverse to $y = f(x)$ is also differentiable in the corresponding y-interval, and $f'(x) \cdot \varphi'(y) = 1$.

Inverse function rule	$\varphi'(y) = \dfrac{1}{f'(x)}, \quad \dfrac{dx}{dy} = 1 \Big/ \dfrac{dy}{dx}$

By the assumptions made, Δx and Δy in each of the two difference quotients $\Delta x/\Delta y$ and $\Delta y/\Delta x$ have the same values, so that $\dfrac{\Delta y}{\Delta x} \cdot \dfrac{\Delta x}{\Delta y} = 1$. By the assumption that $\lim\limits_{\Delta x \to 0} \dfrac{\Delta y}{\Delta x} = f'(x)$ exists and is different from zero, the limit $\lim\limits_{\Delta y \to 0} \dfrac{\Delta x}{\Delta y}$ also exists and has the value $\dfrac{1}{f'(x)}$. For the geometrical interpretation one interchanges the variables in $x = \varphi(y)$. The curve $y = \varphi(x)$ is then obtained by taking the mirror image of the curve of $y = f(x)$ in the line of symmetry $x = y$ of the coordinate system. If a tangent to the curve of $y = f(x)$ makes an angle α with the $+x$-axis, then the corresponding tangent to the curve of $y = \varphi(x)$ makes the same angle α with the $+y$-axis, that is, the angle $\beta = \pi/2 - \alpha$ with the $+x$-axis. But for these complementary angles $\tan\alpha \cdot \tan\beta = 1$, or $f'(x) \cdot \varphi'(x) = 1$ (Fig.).

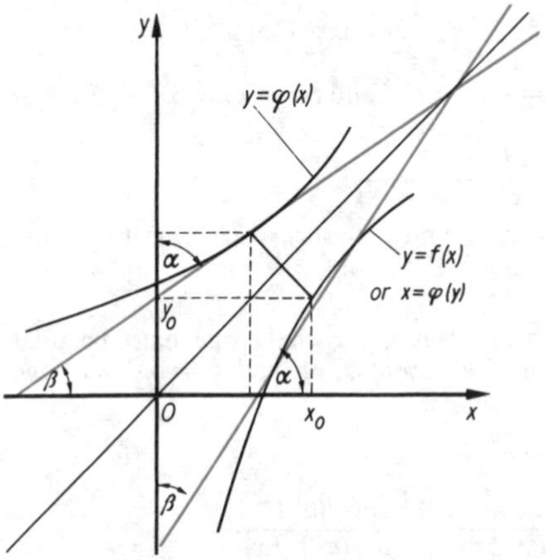

19.2-1 Slope of the curves of mutually inverse functions

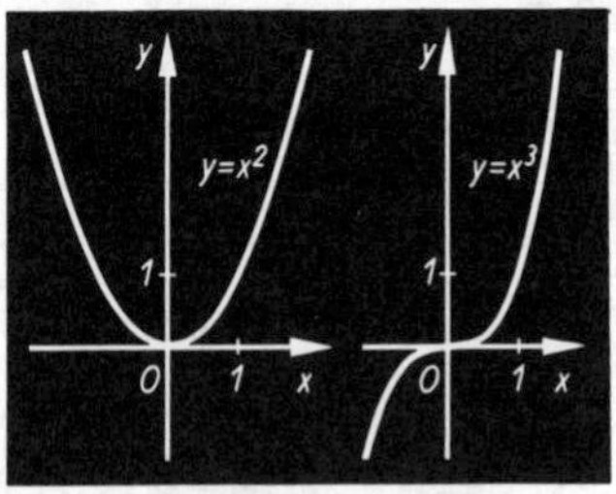

19.2-2 Graphs of the functions $y = x^2$ and $y = x^3$, whose inverse functions are not differentiable for $x = 0$

If $f'(x) \equiv 0$ in an interval, then the function $f(x)$ certainly does not have a unique inverse there, because in that case a single value of y corresponds to all the values of x in the interval. But if $f'(x_i) = 0$ only for individual points x_i in the interval in which the inverse function $x = \varphi(y)$ corresponding to the function $f(x)$ is determined, then because $f(x)$ is monotonic, $f'(x)$ cannot change its sign in passing through these points x_i. If, on the other hand, $f'(x_i) = 0$ and $f'(x)$ always has the same sign in a neighbourhood of the point x_i, then $f(x)$ certainly has an inverse in this neighbourhood, but the inverse is not differentiable at the point x_i: for example, the function $y = x^3$ at the point $x_i = 0$ (Fig.).

The inverse function rule for differentiation is used to find the derivative of a function when the derivative of its inverse function is already known; for example, those of the logarithm function, of the circular functions, and of the inverse hyperbolic functions.

Differentiation of functions in parametric form. A parametric representation of a function $y = f(x)$ is given by $x = \varphi(t)$ and $y = \psi(t)$. One can then also express y as a composite function $y = f[\varphi(t)]$ of the parameter t, and the chain rule for differentiation yields $\dfrac{dy}{dt} = \dfrac{dy}{dx} \cdot \dfrac{dx}{dt}$. In this calculation it is assumed that the functions $\varphi(t)$ and $\psi(t)$ are differentiable with respect to the parameter t and that $\varphi'(t) \neq 0$.

Derivative of a function in parametric form	$\dfrac{dy}{dx} = \dfrac{dy}{dt} \Big/ \dfrac{dx}{dt}$ or $f'(x) = \psi'(t)/\varphi'(t)$

Example 1: The ellipse with the equation $x^2/a^2 + y^2/b^2 = 1$ has the parametric representation $x = a\cos t$ and $y = b \sin t$. From the derivatives with respect to the parameter, $\dfrac{dx}{dt} = -a \sin t$ and $\dfrac{dy}{dt} = b\cos t$, the derivative $\dfrac{dy}{dx} = -\dfrac{b\cos t}{a \sin t} = (-b/a)\cot t$ is found. Because $\cos t = x/a$ and $\sin t = y/b$, one obtains $\dfrac{dy}{dx} = -b^2x/(a^2y)$ as the slope of the tangent at the point $P(x, y)$ to the given ellipse.

Example 2: The common cycloid has the parametric representation $x = a(t - \sin t)$ and $y = a(1 - \cos t)$. Because $\dfrac{dx}{dt} = a(1 - \cos t) = 2a\sin^2(t/2)$ and $\dfrac{dy}{dt} = a \sin t = 2a \sin(t/2)\cos(t/2)$, its derivative $\dfrac{dy}{dx}$ is given by $\dfrac{dy}{dx} = \cot(t/2)$. It follows from this result that at each of the points with $t = 2k\pi$, $(k = 0, \pm 1, \pm 2, \ldots)$ at which it meets the x-axis the common cycloid has a cusp with a tangent perpendicular to the x-axis.

Differentiation of functions in polar coordinates. If $r = r(\vartheta)$ is the representation of a function in polar coordinates, then by means of the relations $x = r\cos\vartheta$ and $y = r\sin\vartheta$ between the polar coordinates and the Cartesian coordinates one can go over to a parametric representation of the function with the parameter ϑ: $x = r(\vartheta)\cos\vartheta$ and $y = r(\vartheta)\sin\vartheta$. Thus, its derivative is given by $\frac{dy}{dx} = \frac{dy}{d\vartheta}\Big/\frac{dx}{d\vartheta}$. If a dot denotes differentiation with respect to the parameter, $\frac{dr}{d\vartheta} = \dot{r}$, then:

$$\frac{dy}{d\vartheta} = \dot{r}\sin\vartheta + r\cos\vartheta$$

and $$\frac{dx}{d\vartheta} = \dot{r}\cos\vartheta - r\sin\vartheta.$$

Derivative of a function $r = r(\vartheta)$ in polar coordinates	$\frac{dy}{dx} = \frac{\dot{r}\sin\vartheta + r\cos\vartheta}{\dot{r}\cos\vartheta - r\sin\vartheta}$

Example: The logarithmic spiral has the equation $r = ae^{k\vartheta}$. By the above rule its derivative is:

$$\frac{dy}{dx} = \frac{ak\,e^{k\vartheta}\sin\vartheta + a\,e^{k\vartheta}\cos\vartheta}{ak\,e^{k\vartheta}\cos\vartheta - a\,e^{k\vartheta}\sin\vartheta} = \frac{k\sin\vartheta + \cos\vartheta}{k\cos\vartheta - \sin\vartheta}.$$

This result shows that the direction of the tangent depends only on ϑ, so that an arbitrary position vector making an angle ϑ_0 with the positive x-axis cuts the spiral at a constant angle φ_0.

In order to calculate the angle φ between the tangent and the position vector $\overrightarrow{OP}$ in the general case, one takes the relation $\varphi = \alpha - \vartheta$ (Fig.) and deduces that

$$\tan\varphi = \tan(\alpha - \vartheta) = \frac{\tan\alpha - \tan\vartheta}{1 + \tan\alpha\tan\vartheta} = \frac{\frac{dy}{dx} - \tan\vartheta}{1 + \frac{dy}{dx}\tan\vartheta}$$

$$= \frac{y'\cos\vartheta - \sin\vartheta}{y'\sin\vartheta + \cos\vartheta} = \frac{r}{\dot{r}},$$

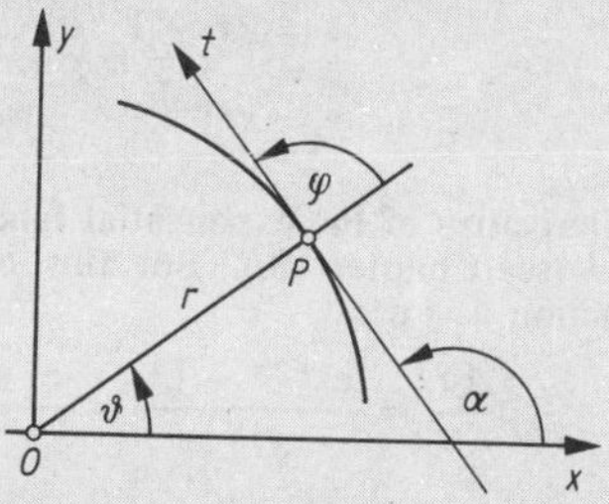

19.2-3 Angle φ between the tangent to a curve and the position vector $\overrightarrow{OP}$

where the last expression is obtained by applying the rule for the derivative of a function in polar coordinates and then solving for $r/\dot{r}$.

If one applies this result to the logarithmic spiral (see 19.5. – Special curves), one obtains $\tan\varphi = (a\,e^{k\vartheta})/(ak\,e^{k\vartheta}) = 1/k$.

This means that the logarithmic spiral cuts all radii vectors at the same angle $\varphi = \arctan(1/k)$. For this reason the cutting edges of the knife discs of certain cutting machines have the form of a logarithmic (or equiangular) spiral, to ensure a constant cutting angle.

Differentiation of implicit functions. It is often necessary to differentiate a function defined implicitly by $F(x, y) = 0$. For this purpose the expression $F(x, y)$, as a function of two variables, must be differentiable with respect to y for fixed x, and with respect to x for fixed y. Furthermore, if there exists a continuous explicit form $y = f(x)$ for the given function, then for $\frac{\partial F(x, y)}{\partial y} \neq 0$, $y = f(x)$ is also differentiable and its derivative $y' = f'(x)$ can be obtained by means of the following formula (see Derivatives of functions of several variables) without first finding the explicit form $y = f(x)$. The round letters ∂ are used to indicate that the partial derivatives of the function $F(x, y)$ are intended.

Implicit differentiation (see p. 422)	$\frac{dy}{dx} = -\frac{\partial F(x, y)}{\partial x}\Big/\frac{\partial F(x, y)}{\partial y}$ or $y' = -F_x/F_y$

Example: The slope of the hyperbola given by $F(x, y) = 2x^2 - y^2 + 12x - 2y + 3 = 0$ at the point $P_0(2, 5)$ is given by the derivative of the function $y = f(x)$ at the point $x_0 = 2$. From $\frac{\partial F}{\partial x} = 4x + 12$, $\frac{\partial F}{\partial y} = -2y - 2$, it follows that $f'(x) = -(4x + 12)/(-2y - 2) = (2x + 6)/(y + 1)$ and $f'(2) = 5/3$.

Derivatives of special functions

Derivative of the constants and of the power functions. Because every difference quotient of a constant function vanishes, its derivative is also equal to zero. For the derivative of the power function $y = x^n$, the product rule gives $y' = nx^{n-1}$ if n is a *positive integer*. Combining this with the quotient rule, this result can be extended to *negative integer* exponents. The exponent n of the power function may also be a rational number p/q, or in general a *real number* (see Chapter 2.). One then defines $y = x^{p/q} = (x^p)^{1/q}$ or $y = x^\alpha = e^{\alpha \ln x}$, where the variable x is restricted to positive values. Using the chain rule, the derivative of an exponential function for arbitrary α is found to be

$$\frac{dy}{dx} = e^{\alpha \ln x} \cdot \alpha \cdot (1/x) = \alpha \cdot (x^\alpha/x) = \alpha x^{\alpha-1}.$$

Independently of this result, the derivative of a power with rational exponent p/q, in which the integers p and q have no common factor, can be obtained directly in the following steps, using the inverse function rule:

$$y_1 = x^{1/q} \rightarrow y_1^q = x \rightarrow \frac{dy_1}{dx} = 1 \bigg/ \frac{dx}{dy_1} = 1/(q y_1^{q-1});$$

$$\frac{dx^n}{dx} = nx^{n-1}, \ n \text{ real}$$

$$y = x^{p/q} = y_1^p \rightarrow \frac{dy}{dx} = p y_1^{p-1} \cdot \frac{dy_1}{dx} = (p/q) \cdot (y_1^{p-q}) = (p/q)\, x^{(p-q)/q} = (p/q)\, x^{(p/q)-1}.$$

Examples: function	derivative	function	derivative
$y = x$	$y' = 1$	$y = x^{-4}$	$y' = -4x^{-5}$
$y = 2x - 1$	$y' = 2$	$y = x^{1/3}$	$y' = 1/(3x^{2/3})$
$y = -x/2 + 2$	$y' = -1/2$	$y = x^{2\sqrt{2}}$	$y' = 2\sqrt{2}x^{2\sqrt{2}-1}$
$y = x^{15}$	$y' = 15x^{14}$	$y = x^\pi$	$y' = \pi x^{\pi-1}$

Derivative of the exponential function. The limit, as x tends to zero, of $(e^x - 1)/x$ was found to be 1 (see Chapter 18.). But this expression occurs in the difference quotient of the exponential function $y = e^x$:

$$\frac{\Delta y}{\Delta x} = \frac{e^{x+\Delta x} - e^x}{\Delta x} = \frac{e^x e^{\Delta x} - e^x}{\Delta x} = e^x \left(\frac{e^{\Delta x} - 1}{\Delta x}\right).$$

$$\frac{d\,e^x}{dx} = e^x$$

Here Δx tends to zero, but e^x is a constant for each arbitrary, fixed value of x. The difference quotient therefore tends to the limit e^x. It is the only derivative that is *equal to the function.* For this reason the exponential function $y = e^x$ is appropriate for the description of natural events, for which the variation y' of the given variable y is equal or proportional to y, for example, in the decay of radioactive material. From the chain rule it follows that $\frac{d}{dx} e^{kx} = k\, e^{kx}$ (k is the factor of proportionality).

The derivative of the general exponential function $y = a^x = e^{x \ln a}$ is given by the chain rule:

$$\frac{dy}{dx} = \ln a \cdot e^{x \ln a}.$$

$$\frac{d}{dx} a^x = a^x \ln a$$

Derivative of the logarithmic function. The inverse function of the logarithmic function $y = \log_a x$ is the function $x = a^y$ whose derivative is $\frac{dx}{dy} = a^y \ln a$. The function $x = a^y$ is monotonic and its derivative is never zero for finite values of the variable. It follows that the reciprocal of its derivative is the derivative of the logarithm (see Chapter 2.):

$$\frac{dy}{dx} = 1 \bigg/ \frac{dx}{dy} = 1/(a^y \ln a) = 1/(x \ln a) = (1/x) \log_a e.$$

$$\frac{d}{dx} \log_a x = \frac{1}{x \ln a} = \frac{1}{x} \cdot \log_a e$$

$$\frac{d}{dx} \ln x = \frac{1}{x}$$

Derivatives of the trigonometric functions. The difference quotient is formed and rearranged in preparation for the passage to the limit:

$$\frac{\Delta y}{\Delta x} = \frac{\sin(x + \Delta x) - \sin x}{\Delta x} = \frac{2\cos(x + \Delta x/2)\sin(\Delta x/2)}{\Delta x} = \cos(x + \Delta x/2) \cdot \frac{\sin(\Delta x/2)}{\Delta x/2}.$$

Here the formula of trigonometry

$$\sin\alpha - \sin\beta = 2\cos[(\alpha + \beta)/2]\sin[(\alpha - \beta)/2]$$

is used, with $x + \Delta x = \alpha$ and $x = \beta$.

As $\Delta x \rightarrow 0$, the quotient $\sin(\Delta x/2)/(\Delta x/2)$ tends to the limit 1. Thus, $\Delta y/\Delta x$ tends to the limit $\cos x$, as $\Delta x \rightarrow 0$, since the cosine function is continuous.

$\dfrac{d \sin x}{dx} = \cos x$	$\dfrac{d \tan x}{dx} = \dfrac{1}{\cos^2 x} = 1 + \tan^2 x = \sec^2 x$
$\dfrac{d \cos x}{dx} = -\sin x$	$\dfrac{d \cot x}{dx} = -\dfrac{1}{\sin^2 x} = -(1 + \cot^2 x) = -\operatorname{cosec}^2 x$

The derivative of the function $y = \cos x$ is obtained by a corresponding rearrangement. Because $y = \tan x = \sin x/\cos x$ and $y = \cot x = \cos x/\sin x$, the derivatives of these functions can be found using the quotient rule. The derivatives are valid, however, only for values of x for which $\cos x$ or $\sin x$ are different from zero, and hence not for the values $x = (2k + 1)\,\pi/2$, or $x = 2k \cdot \pi/2 = k\pi$, where k can be any integer.

Derivatives of the circular functions. The function $y = \operatorname{Arcsin} x$ with $-1 \leqslant x \leqslant +1$ and $-\pi/2 \leqslant y \leqslant +\pi/2$ is the inverse of the function $x = \sin y$, which is continuous and monotonic in the given interval. Hence its derivative is given by $\dfrac{dy}{dx} = 1 \Big/ \dfrac{dx}{dy} = 1/\cos y = 1/\sqrt{(1-x^2)}$. Because of the condition $\dfrac{dx}{dy} \neq 0$ it is necessary to restrict the result to the open interval $-1 < x < +1$. If the function $x = \sin y$ is inverted in another of its intervals of monotonicity, for example, in $-\pi/2 + k\pi \leqslant y \leqslant +\pi/2 + k\pi$ with k an integer, then $y = (-1)^k \operatorname{Arcsin} x + k\pi$ is the inverse

$\dfrac{d \operatorname{Arcsin} x}{dx} = \dfrac{1}{\sqrt{(1 - x^2)}}$; $\lvert x\rvert < 1$	$\dfrac{d \operatorname{Arctan} x}{dx} = \dfrac{1}{1 + x^2}$
$\dfrac{d \operatorname{Arccos} x}{dx} = -\dfrac{1}{\sqrt{(1 - x^2)}}$; $\lvert x\rvert < 1$	$\dfrac{d \operatorname{Arccot} x}{dx} = -\dfrac{1}{1 + x^2}$

function and its derivative is $\dfrac{dy}{dx} = \dfrac{(-1)^k}{\sqrt{(1 - x^2)}}$. Similarly the function $y = \operatorname{Arccos} x$ with $-1 \leqslant x \leqslant +1$ and $0 \leqslant y \leqslant \pi$ is the inverse of the function $x = \cos y$, and its derivative in the interval $-1 < x < +1$ is $\dfrac{dy}{dx} = \dfrac{1}{-\sin y} = -\dfrac{1}{\sqrt{(1 - x^2)}}$. The inverse of the function $x = \cos y$ in the interval $k\pi \leqslant y \leqslant (k + 1)\,\pi$, where k is an integer, in which it is monotonic, is $y = (-1)^k \operatorname{Arccos} x + k\pi$. This has the derivative $\dfrac{dy}{dx} = \dfrac{(-1)^{k+1}}{\sqrt{(1 - x^2)}}$. In a similar manner the derivatives of $y = \operatorname{Arctan} x$ for $-\pi/2 < y < +\pi/2$, $y = \operatorname{Arccot} x$ for $0 < y < \pi$ can be obtained, and the method is also valid for the intervals $-\pi/2 + k\pi < y < +\pi/2 + k\pi$, $k\pi < y < (k + 1)\,\pi$, respectively.

Derivatives of the hyperbolic functions. These functions are defined as rational functions of the exponential function, and can therefore be differentiated using the sum and quotient rules. For example,

$$\frac{d \sinh x}{dx} = \frac{d}{dx}(e^x - e^{-x})/2 = (e^x + e^{-x})/2 = \cosh x.$$

$\dfrac{d \sinh x}{dx} = \cosh x$	$\dfrac{d \cosh x}{dx} = \sinh x$	$\dfrac{d \tanh x}{dx} = \dfrac{1}{\cosh^2 x}$	$\dfrac{d \coth x}{dx} = -\dfrac{1}{\sinh^2 x}$

From $y = \tanh x = \sinh x/\cosh x$, $\dfrac{d}{dx}\dfrac{\sinh x}{\cosh x} = \dfrac{\cosh^2 x - \sinh^2 x}{\cosh x^2}$, where $\cosh^2 x - \sinh^2 x = 1$.

Derivatives of the inverse hyperbolic functions. By the inverse function rule, the derivatives of the inverse hyperbolic functions can be found from the relation $\dfrac{dy}{dx} = 1 \Big/ \dfrac{dx}{dy}$; for example, $\dfrac{d \sinh^{-1} x}{dx} = 1 \Big/ \dfrac{d \sinh y}{dy} = 1/\cosh y = 1/\sqrt{(1 + x^2)}$. A similar procedure is used for $y = \tanh^{-1} x$ in the domain of definition $\lvert x\rvert < 1$, and for $y = \coth^{-1} x$ in $\lvert x\rvert > 1$. One obtains the derivatives displayed here, which represent different functions in spite of their formal equality, because they

$\dfrac{d \sinh^{-1} x}{dx} = \dfrac{1}{\sqrt{(x^2 + 1)}}$	$\dfrac{d \tanh^{-1} x}{dx} = \dfrac{1}{1 - x^2}$; $\lvert x\rvert < 1$
$\dfrac{d \cosh^{-1} x}{dx} = \dfrac{1}{\sqrt{(x^2 - 1)}}$; $x > 1$	$\dfrac{d \coth^{-1} x}{dx} = \dfrac{1}{1 - x^2}$; $\lvert x\rvert > 1$

have different domains of definition. The function $y = \cosh^{-1} x$ is the inverse function of $x = \cosh y$ in the interval of monotonicity $0 \leqslant y < +\infty$; hence $\frac{\mathrm{d} \cosh^{-1} x}{\mathrm{d}x} = 1 \Big/ \frac{\mathrm{d} \cosh y}{\mathrm{d}y} = 1/\sinh y$ $= 1/\sqrt{(x^2 - 1)}$. The function is differentiable for all x in the domain of definition $x \geqslant 1$ with the exception of $x = 1$. In the interval of monotonicity $-\infty < y \leqslant 0$, $x = \cosh y$ has the inverse function $y = -\cosh^{-1} x$, and consequently its derivative is $\frac{\mathrm{d}y}{\mathrm{d}x} = -\frac{1}{\sqrt{(x^2 - 1)}}$.

Derivative of an integral with respect to a limit. In the integral $\int_a^x f(\xi)\,\mathrm{d}\xi$, the number a is fixed and the upper limit x is variable; consequently the integral is a function $\Phi(x)$ of its upper limit. The derivative of the integral with respect to this variable upper limit is equal to the function value $f(x)$ of the integrand at this upper limit (see Chapter 20.).

19.3. Derivatives of functions of several variables

Partial derivatives of a function

In the function $z = f(x_1, x_2, \ldots, x_n)$, $x_1, x_2, \ldots, x_n$ denote variables that are independent of one another; for example, in $z = f(x, y)$ the variables are $x_1 = x$ and $x_2 = y$, and in $z = f(u, v, w)$, $x_1 = u$, $x_2 = v$, $x_3 = w$. If one regards all the variables except one, x_i say, as constants, $x_{1,0}$, $x_{2,0}, \ldots, x_{i-1,0}, x_{i+1,0}, \ldots, x_{n,0}$, then the function becomes a function of one variable. If this function is differentiable, one can form a *partial derivative*, denoted by

$$\frac{\partial z}{\partial x_i} = \frac{\partial}{\partial x_i} f(x_{1,0}, \ldots, x_i, \ldots, x_{n,0}) = f_{x_i};$$

the round letters ∂ indicate the partial derivative (read *partial* ∂f *by* ∂x_i). For $z = f(x, y)$, for example,

$$\frac{\partial}{\partial x} f(x, y_0) = \lim_{\Delta x \to 0} \frac{f(x + \Delta x, y_0) - f(x, y_0)}{\Delta x}, \qquad \frac{\partial}{\partial y} f(x_0, y) = \lim_{\Delta y \to 0} \frac{f(x_0, y + \Delta y) - f(x_0, y)}{\Delta y},$$

whenever the two limits exist. The rules for differentiation of a function of one variable hold with respect to the single non-fixed variable x_i.

Example 1: $z = f(x, y) = x^3 + 7x^2y + 3xy^5 - 5y^6$;

$$\frac{\partial z}{\partial x} = f_x(x, y) = 3x^2 + 14xy + 3y^5, \qquad \frac{\partial z}{\partial y} = f_y(x, y) = 7x^2 + 15xy^4 - 30y^5.$$

Example 2: $z = f(x, y) = \arctan (x/y)$; $\frac{\partial z}{\partial x} = z_x = \frac{1}{1 + (x/y)^2} \cdot \frac{1}{y} = \frac{y}{x^2 + y^2}$;

$$\frac{\partial z}{\partial y} = z_y = \frac{1}{1 + (x/y)^2} \cdot (-xy^{-2}) = -\frac{x}{x^2 + y^2}.$$

Example 3: $w = f(x, y, z) = \sqrt{(x^2 + y^2 + z^2)}$;

$$\frac{\partial w}{\partial x} = w_x = \frac{x}{\sqrt{(x^2 + y^2 + z^2)}},$$

$$\frac{\partial w}{\partial y} = w_y = \frac{y}{\sqrt{(x^2 + y^2 + z^2)}},$$

$$\frac{\partial w}{\partial z} = w_z = \frac{z}{\sqrt{(x^2 + y^2 + z^2)}}.$$

Geometrical significance of the partial derivatives of a function of two variables. A function $z = f(x, y)$ of two variables can, in general, be represented by a surface in space. The assumption $y = y_0 = \text{const}$ selects the points of this surface that lie at the same time in the plane $y = y_0$ parallel to the x, z-plane. They form a plane curve, and $\frac{\partial}{\partial x} f(x, y_0)$ is the slope of the tangent t_1 to

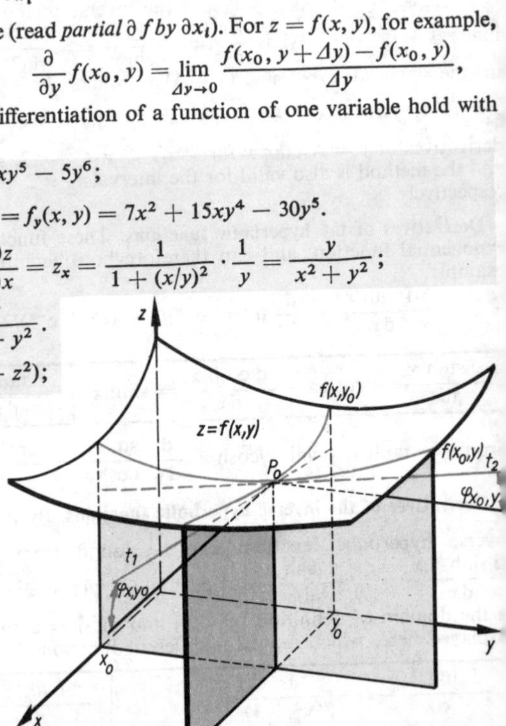

19.3-1 Geometrical significance of the partial derivative of a function of two variables

this curve at the point (x, y_0): $z_x = \tan \varphi_{x, y_0}$. The angle φ_{x, y_0} describes the slope of t_1 to the $+x$-axis (Fig.). Similarly $z_y = \frac{\partial}{\partial y} f(x_0, y) = \tan \varphi_{x_0, y}$ is the slope with respect to the $+y$-axis of the tangent t_2 to the curve in which the surface $z = f(x, y)$ is cut by the plane $x = x_0$ parallel to the y, z-plane. At each point P of the surface both a tangent t_1 determined by z_x and a tangent t_2 determined by z_y are defined. Under certain assumptions, which are usually satisfied in practice, the two tangents span a plane to the surface at the point P.

Partial derivatives of higher order. Each partial derivative is again a function of the same variables, and can itself have partial derivatives if the limits of the corresponding difference quotients exist. These are called *partial derivatives of higher order*, for example, of the second, third, ... nth order. Partial derivatives with respect to different variables are called *mixed derivatives*. From $z = f(x, y)$, for example, by differentiating $\frac{\partial z}{\partial x} = f_x$ and $\frac{\partial z}{\partial y} = f_y$ one obtains four derivatives of the second order:

$$f_{xx} = \frac{\partial f_x}{\partial x} = \frac{\partial^2 z}{\partial x^2}, \quad f_{xy} = \frac{\partial f_x}{\partial y} = \frac{\partial^2 z}{\partial x\, \partial y}, \quad f_{yx} = \frac{\partial f_y}{\partial x} = \frac{\partial^2 z}{\partial y\, \partial x} \quad \text{and} \quad f_{yy} = \frac{\partial f_y}{\partial y} = \frac{\partial^2 z}{\partial y^2}.$$

From the way in which they are formed, the functions f_{xy} and f_{yx} are different from one another. But Leonhard EULER (1707–1783) already knew conditions under which they are equal; Hermann Amandus SCHWARZ (1843–1921) proved the theorem named after him.

Theorem of Schwarz: If the mixed partial derivatives of the second order f_{xy} and f_{yx} of a function $f(x, y)$ are continuous functions of x and y in a domain D, then they are equal to one another in the interior of this domain.

$$f_{xy} = f_{yx}$$

The *continuity* of a function $u = f(x, y)$ at the point (x_0, y_0) means that corresponding to an arbitrary prescribed $\varepsilon > 0$ there always exists a positive number $\delta = \delta(\varepsilon)$ such that $|f(x, y) - f(x_0, y_0)| \leqslant \varepsilon$ for all pairs of numbers (x, y) satisfying $(x - x_0)^2 + (y - y_0)^2 \leqslant \delta^2$. Geometrically this means that the values of the function $f(x, y)$ differ by an arbitrarily small amount from $f(x_0, y_0)$ provided that the argument (x, y) is chosen within a sufficiently small circle with centre at (x_0, y_0). If $f(x, y)$ is continuous at every point of a domain D, then the *function is continuous* in D. The theorem of Schwarz holds also for partial derivatives of higher order and also for functions of more than two variables; for example, for $z = f(x, y)$, $f_{xxy} = f_{xyx} = f_{yxx}$ and $f_{xyy} = f_{yxy} = f_{yyx}$.

Example 1:

$$z = f(x, y) = x^3 + 7x^2y + 3xy^5 - 5y^6, \quad f_x = 3x^2 + 14xy + 3y^5;$$
$$f_y = 7x^2 + 15xy^4 - 30y^5; \quad f_{xx} = 6x + 14y; \quad f_{xy} = 14x + 15y^4 = f_{yx};$$
$$f_{yy} = 60xy^3 - 150y^4; \quad f_{xxx} = 6; \quad f_{xxy} = f_{xyx} = f_{yxx} = 14;$$
$$f_{xyy} = f_{yxy} = f_{yyx} = 60y^3; \quad f_{yyy} = 180xy^2 - 600y^3.$$

Example 2:

$$z = f(x, y) = \arctan x/y; \quad f_x = \frac{y}{x^2 + y^2}; \quad f_y = -\frac{x}{x^2 + y^2};$$
$$f_{xx} = -\frac{2xy}{(x^2 + y^2)^2}; \quad f_{xy} = \frac{x^2 - y^2}{(x^2 + y^2)^2} = f_{yx}; \quad f_{yy} = \frac{2xy}{(x^2 + y^2)^2}.$$

Total differential

Total differential of the first order. If $z = f(x, y)$ is a function of the two variables x and y and if its partial derivatives of the first order f_x and f_y exist, then these are the limits of the difference quotients $\frac{\Delta_x z}{\Delta x}$ and $\frac{\Delta_y z}{\Delta y}$. Thus, $\frac{\Delta_x z}{\Delta x} = \frac{\partial z}{\partial x} + \varphi_1(\Delta x)$ and $\frac{\Delta_y z}{\Delta y} = \frac{\partial z}{\partial y} + \varphi_2(\Delta y)$, where $\varphi_1(\Delta x)$ and $\varphi_2(\Delta y)$ denote the differences between the corresponding difference quotients and the partial derivatives, as in the case of the total differential of a function $f(x)$ of one variable. If these equations are solved in the same way for the increments of the function $z = f(x, y)$, the equations $\Delta_x z = \frac{\partial z}{\partial x} \cdot \Delta x + \varphi_1(\Delta x) \cdot \Delta x$ and $\Delta_y z = \frac{\partial z}{\partial y} \cdot \Delta y + \varphi_2(\Delta y) \cdot \Delta y$ are obtained. In these expressions the terms $\varphi_1(\Delta x) \cdot \Delta x$ and $\varphi_2(\Delta y) \cdot \Delta y$ are small 'to a higher order' than the partial differentials $\frac{\partial z}{\partial x}\,\mathrm{d}x$ and $\frac{\partial z}{\partial y}\,\mathrm{d}y$. Consequently, from the total increment Δz of the function $z = f(x, y)$, if $\frac{\partial z}{\partial x}$ and $\frac{\partial z}{\partial y}$ are continuous and terms of higher order of smallness are neglected, one obtains the total differential dz of the first order

$$\Delta z = f(x+\Delta x, y+\Delta y) - f(x, y+\Delta y) + f(x, y+\Delta y) - f(x, y),$$

$$\Delta z = \left[\frac{f(x+\Delta x, y+\Delta y) - f(x, y+\Delta y)}{\Delta x}\right]\Delta x + \left[\frac{f(x, y+\Delta y) - f(x, y)}{\Delta y}\right]\Delta y$$

$$= \frac{\partial f}{\partial x}\Delta x + \varphi_1(\Delta x)\cdot\Delta x + \frac{\partial f}{\partial y}\Delta y + \varphi_2(\Delta y)\cdot\Delta y$$

and $dz = \frac{\partial f}{\partial x}dx + \frac{\partial f}{\partial y}dy$.

Total differential of the first order	$dz = \frac{\partial z}{\partial x_1}dx_1 + \frac{\partial z}{\partial x_2}dx_2$

The larger the chosen values of Δx and Δy, the larger is the difference between the total differential dz and the total increment $\Delta z = f(x+\Delta x, y+\Delta y) - f(x, y)$. For the function $z = x^2 - y^2$ one has $dz = 2(x\,dx - y\,dy)$. The table shows the difference $\Delta z - dz$ in the neighbourhood of the point $x = 2$, $y = 1$ for each of the pairs of values 2, 1 and 0.2, 0.1 for Δx and Δy.

x	2	2	y	1	1	$(\Delta z - dz)/\Delta z$	33%	5%
$dx, \Delta x$	2	0.2	$dy, \Delta y$	1	0.1	$\Delta z - dz$	3	0.03
$x\,dx$ →	4	0.4	$y\,dy$ →	1	0.1	→ dz	6	0.60
$x + \Delta x$	4	2.2	$y + \Delta y$	2	1.1	Δz	9	0.63
x^2 →	4	4	y^2 →	1	1	→ $f(x, y)$	3	3
$(x + \Delta x)^2$ →	16	4.84	$(y + \Delta y)^2$ →	4	1.21	→ $f(x, y) + \Delta z$	12	3.63

Geometrical significance of the differential of a function of two variables. The partial differential $d_x z$ represents the increment in the ordinate of the tangent to the curve $z = f(x, y_0)$ and the partial differential $d_y z$ the increment in the ordinate of the tangent to the curve $z = f(x_0, y)$. The total differential $dz = \frac{\partial z}{\partial x}dx + \frac{\partial z}{\partial y}dy$ is a function of four variables (x, y, dx, dy) and represents geometrically the increase that results in the ordinate of the point of contact of the tangent plane with the surface $z = f(x, y)$ if x is increased by $\Delta x = h = dx$ and y by $\Delta y = k = dy$ (Fig.).

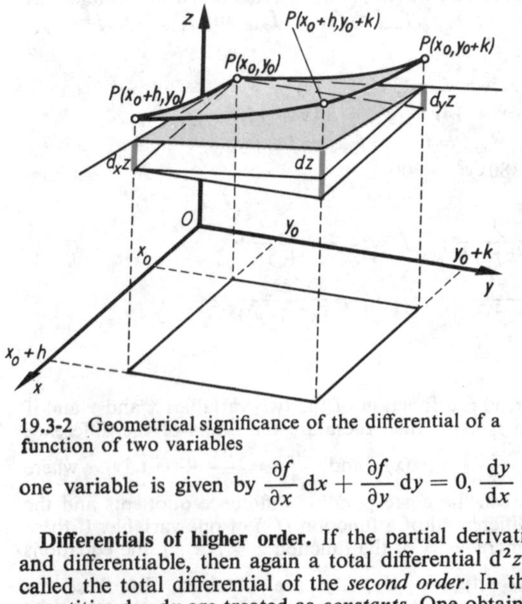

19.3-2 Geometrical significance of the differential of a function of two variables

Example: For the function $z = f(x, y) = x^3 + 7x^2y + 3xy^5 - 5y^6$, $z_x = 3x^2 + 14xy + 3y^5$ and $z_y = 7x^2 + 15xy^4 - 30y^5$. It has the total differential

$$dz = (3x^2 + 14xy + 3y^5)\,dx + (7x^2 + 15xy^4 - 30y^5)\,dy.$$

If $z = f(x, y) \equiv 0$, then the function can be regarded as an implicit form of a function of one variable. Because $dz = 0$, the derivative $\frac{dy}{dx}$ of the *implicit function* of one variable is given by $\frac{\partial f}{\partial x}dx + \frac{\partial f}{\partial y}dy = 0$, $\frac{dy}{dx} = -\frac{\partial f}{\partial x}\Big/\frac{\partial f}{\partial y}$ (see p. 417).

Differentials of higher order. If the partial derivatives of a function are themselves continuous and differentiable, then again a total differential d^2z of the total differential can be formed. It is called the total differential of the *second order*. In the differentiation the finite, arbitrarily chosen quantities dx, dy are treated as *constants*. One obtains

$$d^2z = \frac{\partial}{\partial x}\left(\frac{\partial z}{\partial x}dx + \frac{\partial z}{\partial y}dy\right)dx + \frac{\partial}{\partial y}\left(\frac{\partial z}{\partial x}dx + \frac{\partial z}{\partial y}dy\right)dy$$

$$= \frac{\partial^2 z}{\partial x^2}dx^2 + \frac{\partial^2 z}{\partial y\,\partial x}dx\,dy + \frac{\partial^2 z}{\partial x\,\partial y}dx\,dy + \frac{\partial^2 z}{\partial y^2}dy^2.$$

Since $z_{xy} = z_{yx}$ by the theorem of Schwarz, the differential assumes a form in which the coefficients and the products of dx and dy are formally given by the *binomial theorem.*

Total differential of the second order	$d^2z = \frac{\partial^2 z}{\partial x^2} \cdot dx^2 + 2\frac{\partial^2 z}{\partial x\, \partial y} \cdot dx\, dy + \frac{\partial^2 z}{\partial y^2} \cdot dy^2$

For the total differential of the second order, for example, one obtains $d^2z = \left(\frac{\partial}{\partial x}\, dx + \frac{\partial}{\partial y}\, dy\right)^{(2)} z.$ The expression inside the brackets is multiplied by z in the sense of an *operator*. It can be shown that the total differentials of higher order, for example, of the nth order, are given by this formal relation.

Total differential of the nth order	$d^nz = \left(\frac{\partial}{\partial x}\, dx + \frac{\partial}{\partial y}\, dy\right)^{(n)} z$
of the 3rd order	$d^3z = \frac{\partial^3 z}{\partial x^3}\, dx^3 + 3\frac{\partial^3 z}{\partial x^2\, \partial y}\, dx^2\, dy + 3\frac{\partial^3 z}{\partial x\, \partial y^2}\, dx\, dy^2 + \frac{\partial^3 z}{\partial y^3}\, dy^3$

Example: The function $z = f(x, y) = x^3 + 7x^2y + 3xy^5 - 5y^6$ has the partial derivatives $z_x = 3x^2 + 14xy + 3y^5$; $z_y = 7x^2 + 15xy^4 - 30y^5$; $z_{xx} = 6x + 14y$; $z_{xy} = 14x + 15y^4$; $z_{yy} = 60xy^3 - 150y^4$. Consequently its total differential of the second order is

$$d^2z = (6x + 14y)\, dx^2 + 2(14x + 15y^4)\, dx\, dy + (60xy^3 - 150y^4)\, dy^2.$$

Solution of equations in several variables. The equation $3x - 4y + 5 = 0$ can be regarded as an implicit form of the equation of a function whose explicit form $y = 3x/4 + 5/4$ is easily obtained. In general, from a given equation $F(x, y) = 0$ it is required to find a function $y = y(x)$ of one variable for which the equation $F[x, y(x)] = 0$ is satisfied identically. This *solution* may be possible by means of elementary functions or by the application of limiting processes, such as infinite series. For $x^2 + y^2 + 1 = 0$, for example, it is not possible. It may also happen that in the neighbourhood of different points (x_0, y_0) with $F(x_0, y_0) = 0$ different solutions for y exist. For example, the equation $F(x, y) = 5x^2 + y^2 - 9 = 0$ has the solution $y = \sqrt{(9 - 5x^2)}$ in the neighbourhood of the point (1,2), and the solution $y = -\sqrt{(9 - 5x^2)}$ in the neighbourhood of (0, −3).

The equation $F(x, y) = 0$ determines in a neighbourhood $U(x_0, y_0)$ of the point (x_0, y_0) with $F(x_0, y_0) = 0$ exactly one continuous function $y = y(x)$ with the properties $y_0 = y(x_0)$ and $F[x, y(x)] = 0$ for all $x \in U$ if the following conditions are satisfied: 1. the function $F(x, y)$ is continuous in $U(x_0, y_0)$; 2. the partial derivatives F_x and F_y exist and are continuous; 3. $F_y(x_0, y_0) \neq 0$. The function $y = y(x)$ is then also differentiable and $y' = y'(x) = -F_x/F_y$. If the given function $F(x, y)$ has continuous partial derivatives up to the kth order, then $y = y(x)$ is also k times continuously differentiable.

These results can be extended at once to functions of more than two variables. If $F(x_1, x_2, \dots, x_k)$ is a continuous function with continuous partial derivatives F_{x_i} in a neighbourhood of $(x_1^0, x_2^0, \dots, x_k^0)$ and $F(x_1^0, x_2^0, \dots, x_k^0) = 0$ with $F_{x_j}(x_1^0, x_2^0, \dots, x_k^0) \neq 0$ for a fixed j, then there exists in a neighbourhood of $(x_1^0, x_2^0, \dots, x_k^0)$ a continuous function $x_j = f(x_1, \dots, x_{j-1}, x_{j+1}, \dots, x_k)$ with $x_j^0 = f(x_1^0, \dots, x_{j-1}^0, x_{j+1}^0, \dots, x_k^0)$ and $F(x_1, \dots, x_{j-1}, f, x_{j+1}, \dots, x_k) = 0$.

Example 1: The equation $F(x, y) = e^y - e^{-y} - 2x = 0$ can be solved for y in a neighbourhood of (0, 0), because $F(x, y)$ is a continuous function with the continuous derivatives $F_x = -2$, $F_y = e^y + e^{-y}$ and $F(0, 0) = 0$, $F_y(0, 0) = 2 \neq 0$. The solution is $y = \ln(x + \sqrt{x^2 + 1}) = \sinh^{-1} x$.

Example 2: The equation $F(x, y) = x^3 + y^3 - 3axy = 0$ for the *folium of Descartes* cannot be solved for y in the neighbourhood of (0, 0) because $F_y = 3y^2 - 3ax$, and hence $F_y(0, 0) = 0$. One gathers this intuitively from the graph of the function (see Fig. 19.5-6).

The following theorem gives conditions under which a *system of m equations* $F_i(x_1, \dots, x_n; y_1, \dots, y_m) = 0$, $i = 1, 2, \dots, m$, can be solved for the m functions $y_1, y_2, \dots, y_m$ in a neighbourhood of the point $(x_1^0, \dots, x_n^0; y_1^0, \dots, y_m^0)$.

If the functions $F_i(x_1, \dots, x_n; y_1, \dots, y_m)$ for $i = 1, 2, \dots, m$ are continuous in a neighbourhood U of the point $(x_1^0, \dots, x_n^0; y_1^0, \dots, y_m^0)$ and have continuous partial derivatives $\frac{\partial F_i}{\partial x_j}, \frac{\partial F_i}{\partial y_k}$ at that point, and if the functional determinant $\text{Det}\left[\frac{\partial F_i}{\partial y_k}(x_1^0, \dots, x_n^0; y_1^0, \dots, y_m^0)\right]$ formed from the partial derivatives $\frac{\partial F_i}{\partial y}$ at the point $(x_1^0, \dots, x_n^0; y_1^0, \dots, y_m^0)$ is different from zero, then there exists in U *exactly one system of m differentiable functions* $y_i = y_i(x_1, \ldots, x_n)$ *with the properties that* $y_i^0 = y_i(x_1^0, \ldots, x_n^0)$

and $F_i[x_1, \ldots, x_n, y_1(x_1, \ldots, x_n), \ldots, y_m(x_1, \ldots, x_n)] = 0$.

Functional determinant or Jacobian

$$\begin{vmatrix} \frac{\partial F_1}{\partial y_1} & \frac{\partial F_1}{\partial y_2} & \cdots & \frac{\partial F_1}{\partial y_m} \\ \frac{\partial F_2}{\partial y_1} & \frac{\partial F_2}{\partial y_2} & \cdots & \frac{\partial F_2}{\partial y_m} \\ \vdots & \vdots & & \vdots \\ \frac{\partial F_m}{\partial y_1} & \frac{\partial F_m}{\partial y_2} & \cdots & \frac{\partial F_m}{\partial y_m} \end{vmatrix}$$

Example 1: The system of three equations given here represents the connection between the *Cartesian* coordinates (x, y, z) of a point and its spherical polar coordinates (r, ϑ, φ). The Jacobian is given by

$$x = r \cos \varphi \sin \vartheta$$
$$y = r \sin \varphi \sin \vartheta$$
$$z = r \cos \vartheta$$

$$D = \begin{vmatrix} \sin\vartheta \cos\varphi & \sin\vartheta \sin\varphi & \cos\vartheta \\ r\cos\vartheta\cos\varphi & r\cos\vartheta\sin\varphi & -r\sin\vartheta \\ -r\sin\vartheta\sin\varphi & r\sin\vartheta\cos\varphi & 0 \end{vmatrix} = r^2 \sin\vartheta$$

and $D \neq 0$ for all points that do not lie on the z-axis. For these points the system can be solved for r, ϑ, φ, giving

$$r = \sqrt{(x^2 + y^2 + z^2)}; \quad \varphi = \arccos \frac{x}{\sqrt{(x^2 + y^2)}}; \quad \vartheta = \arccos \frac{z}{\sqrt{(x^2 + y^2 + z^2)}}.$$

Example 2: Generalization of *1.* If the functions $y_k = f_k(x_1, \ldots, x_n)$ for $k = 1, 2, \ldots, n$ and their partial derivatives of the first order are continuous in a neighbourhood of the point $(x_1^0, \ldots, x_n^0)$ and the Jacobian $\operatorname{Det}\left[\frac{\partial f_i}{\partial x_k}(x_1^0, \ldots, x_n^0)\right] \neq 0$, then there exist continuous functions $x_k = x_k(y_1, y_2, \ldots, y_n)$ with $x_k(y_1^0, \ldots, y_n^0) = x_k^0$ and $f_k[x_1(y_1, \ldots, y_n), x_2(y_1, \ldots, y_n), \ldots, x_n(y_1, \ldots, y_n)] = y_k$ for $k = 1, 2, \ldots, n$. Thus, under the conditions stated, a system of equations has (speaking intuitively) an 'inverse system'.

19.4. Extreme values of functions

Extreme values of functions of one variable

From the graph (Fig.) of the function

$$y = f(x) = {}^1/_6(x^3 - 3x^2 - 9x + 17)$$

one recognizes that in the range between $-\infty$ and -1 the ordinate value increases as the abscissa increases. On the other hand, from $x = -1$ to $x = +3$ the ordinates decrease steadily, and for $x > 3$ they again increase steadily. In a suitable chosen neighbourhood of the point $x_{max} = -1$, the value $f(x)$ of the function is less than $f(x_{max})$ for all values of the abscissa x different from x_{max}. The function is said to have a *local maximum* at the point x_{max}. At the point $x_{min} = +3$, the function is said to have a *local minimum*, because the value $f(x)$ of the function is greater than $f(x_{min})$ for all values of x different from x_{min} in a suitably chosen neighbourhood of this point. Both values, maximum and minimum, are called *local extreme values* (or *extrema*); local, because there are places at which the function assumes greater values than $f(-1) = +11/3$ and smaller values than $f(+3) = -5/3$. In the closed interval $-2 \leqslant x \leqslant 4$, however, they are the *absolute* or *global* extreme values

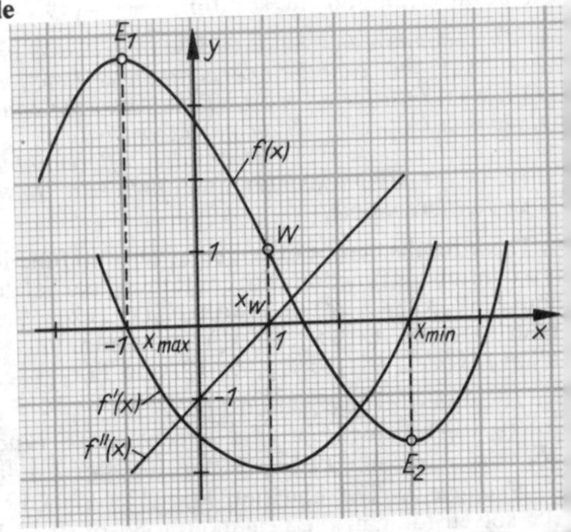

19.4-1 Extrema and points of inflection of the grap of the function $y = f(x) = (1/6)(x^3 - 3x^2 - 9x + 1$

Local maximum $f(x_{max}) > f(x)$ for $x \neq x_{max}$ Local minimum $f(x_{min}) < f(x)$ for $x \neq x_{min}$	in a sufficiently small neighbourhood

By the *theorem of Weierstrass*, a function that is continuous in a closed interval assumes i supremum and its infimum in that interval. These absolute extrema, however, can occur at th boundary points of the interval, for example, for the function considered above in the interv $-5 \leqslant x \leqslant 10$.

Conditions for the occurrence of local extrema. As the argument x of a differentiable function $f(x)$ passes through the point x_m of a local extremum, the sign of its derivative $f'(x)$ changes. If $x_m = x_{max}$ is the position of a local maximum, then $f'(x) > 0$ for $x < x_{max}$, because for these arguments the function $f(x)$ is increasing, and $f'(x) < 0$ for $x_{max} < x$, because $f(x)$ is decreasing. If the derivative is continuous, then $f'(x_{max}) = 0$ must hold. Similarly for the position x_{min} of a local minimum, $f'(x) < 0$ for $x < x_{min}$ and $f'(x) > 0$ for $x_{min} < x$, and consequently for a continuous derivative $f'(x_{min}) = 0$.

It follows that a *necessary condition* for a local extremum is that the derivative $f'(x)$ vanishes. Only the sign change of the first derivative $f'(x)$ discussed above, expressed analytically by $f''(x_{max}) < 0$ or $f''(x_{min}) > 0$, guarantees the existence of a local maximum or minimum. Thus, it is *sufficient* for their existence that $f'(x_{max}) = 0$ and $f''(x_{max}) < 0$ or $f'(x_{min}) = 0$ and $f''(x_{min}) > 0$. In the case when $f''(x_m) = 0$, the higher derivatives of the function $f(x)$ can be taken into consideration.

$f'(x_0) = 0$ and $f''(x_0) < 0 \rightarrow x_0$ is a local maximum
$f'(x_0) = 0$ and $f''(x_0) > 0 \rightarrow x_0$ is a local minimum

Extrema at zeros of the second and higher derivatives. If $f'''(x)$ is the first non-vanishing derivative for $x = x_m$, then it follows from the discussion above that the function $f'(x)$ has a minimum at this point if $f'''(x_m) > 0$, or a maximum if $f'''(x_m) < 0$. In both cases, since $f'(x_m) = 0$, the curve of $f'(x)$ touches the x-axis at the point $x = x_m$. For the graph of the function $f(x)$ this means: when the argument x passes through the point x_m in the sense of x increasing, then if $f'''(x_m) > 0$, the slope $\tan\varphi$ of the tangent decreases from positive values to $\varphi = 0$ for $x = x_m$, and then increases again, and if $f'''(x_m) < 0$, $\tan\varphi$ increases from negative values to $\varphi = 0$ for $x = x_m$, and then decreases again (Fig.). Such points are called *horizontal points of inflection.*

A function $f(x)$ that is at least n times differentiable ($n \geqslant 2$) at the point ξ has a local extremum at that point if n is even and $f'(\xi) = f''(\xi) = \cdots = f^{(n-1)}(\xi) = 0$, but $f^{(n)}(\xi) \neq 0$; if $f^{(n)}(\xi) < 0$, a local maximum occurs, and if $f^{(n)}(\xi) > 0$, a local minimum.

If one expands $f(x)$ about the point ξ using *Taylor's theorem*, then since $f'(\xi) = f''(\xi) = \cdots = f^{(n-2)}(\xi) = 0$, $f(x) - f(\xi) = (h^{n-1}/(n-1)!) \cdot f^{(n-1)}(\xi + \vartheta h)$ with $0 < \vartheta < 1$.

Because $f^{(n)}(x)$ is the derivative of $f^{(n-1)}(x)$, for $f^{(n)}(\xi) < 0$, $f^{(n-1)}(x)$ decreases monotonically through the value $f^{(n-1)}(\xi)$, but for $f^{(n)}(\xi) > 0$ it increases monotonically. The sign of h, on the other hand, is always negative to the left of ξ and positive to the right; the same holds for h^{n-1}, since $n-1$ is odd. Consequently, as the following table shows, the remainder term, and hence the difference $f(x) - f(\xi)$ is negative on both sides of the point $x = \xi$ for $f^{(n)}(\xi) < 0$, so that $f(\xi) > f(x)$, and the function has a *local maximum*, whilst $f(\xi) < f(x)$ on both sides of ξ if $f^{(n)}(\xi) > 0$, and the function has a *local minimum.*

		$x < \xi$	ξ	$\xi < x$
	h^{n-1}	−	0	+
$f^{(n)}(\xi) < 0$	$f^{(n-1)}(x)$	+	0	−
$f^{(n)}(\xi) > 0$	$f^{(n-1)}(x)$	−	0	+

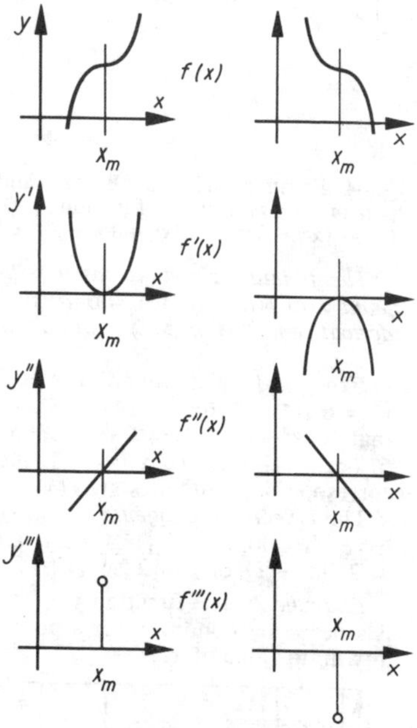

19.4-2 Schematic representation for $f'(x) = 0$; $f''(x) = 0$; $f'''(x) \neq 0$

Points of inflection. The tangent to the curve of the function $y = (1/6)(x^3 - 3x^2 - 9x + 17)$ (see Fig. 19.4-1) at the point $(-3, -5/3)$ has the slope $f'(-3) = +6$. The slope decreases to the value $f'(-1) = 0$ at the maximum $(-1, +11/3)$, and on passing through this point decreases still further as far as the point $x_w = +1$, where it has the value $f'(1) = -2$. From there on the slope increases monotonically. The function $f'(x)$ has a local minimum at $x_w = 1$. In the interval $-\infty < x \leqslant +1$ the direction of a tangent moves into a direction of a following tangent, in the sense of increasing values of x, by a *rotation to the right*, in the mathematically negative sense. In the interval $+1 \leqslant x < +\infty$ there is a corresponding *rotation to the left*, in the mathematically positive sense. At the *point of inflection* the sense of rotation changes from right to left.

If one imagines driving a vehicle along the curve, then the road lies to the right of the tangent as far as the point of inflection, but after one has passed through this point it lies to the left of the tangent. The *curvature* of the curve before the point of inflection is opposite in sign to that after it. If a portion of a curve through three points situated sufficiently close to one another is replaced by a circular arc, then the centre of this *circle of curvature* lies to the right of the curve before the point of inflection and to the left of it afterwards. The tangent at the point of inflection or *inflectional tangent* thus separates portions of the curve whose curvatures are in opposite senses.

From these considerations it follows that the function has a point of inflection where its first derivative assumes an extreme value. If in addition $f'(x_w) = 0$, then the inflectional tangent is horizontal; one speaks of a *horizontal point of inflection* (Fig.).

If $f''(x_w) = 0$ and $f'''(x_w) \neq 0$ then x_w is a point of inflection. Equation of the inflectional tangent: $(y - y_w) = f'(x_w)(x - x_w)$.

The criteria for extrema can be applied to the first derivative $f'(x)$, regarded as a function $\varphi(x)$. Consequently a sufficient condition for a local maximum or minimum of $f'(x)$ at the point x_w is $\varphi'(x_w) = f''(x_w) = 0$ and $\varphi''(x_w) = f'''(x_w) < 0$ or $f'''(x_w) > 0$. If $f'''(x_w) = 0$, then the last theorem above holds, since a derivative of odd order of $f(x)$ is one of even order of $\varphi(x)$.

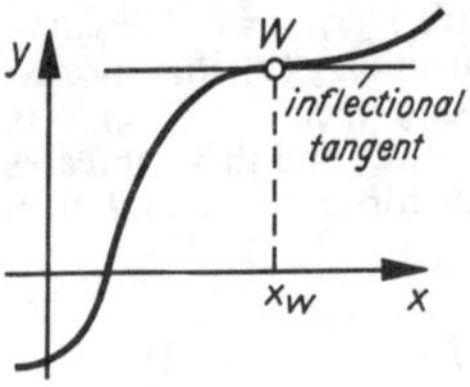

19.4-3 Graph of a function with a horizontal point of inflection

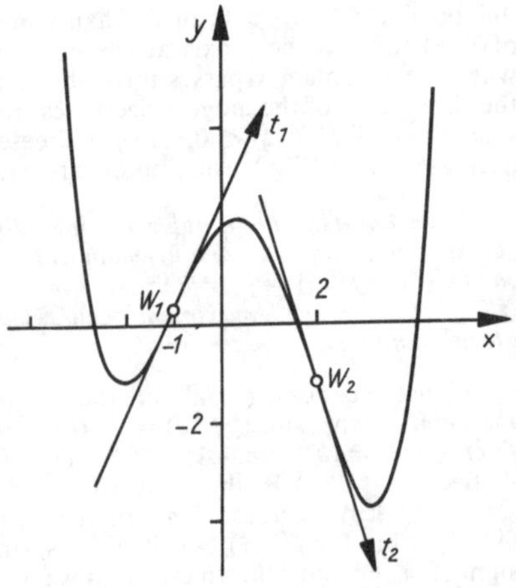

19.4-4 Points of inflection W_1, W_2 and inflectional tangents t_1, t_2 of the graph of the function $f(x) = (x^4 - 2x^3 - 12x^2 + 8x + 20)/10$

The function $y = f(x)$ has a point of inflection at a point ξ at which $f''(\xi) = 0$ if the first non-vanishing derivative $f^{(n)}(\xi)$ $(n > 2)$ is of odd order.

Example 1: The function $y = f(x) = 0.1(x^4 - 2x^3 - 12x^2 + 8x + 20)$ has the derivatives $y' = 0.1(4x^3 - 6x^2 - 24x + 8)$, $y'' = 1.2x^2 - 1.2x - 2.4$ and $y''' = 2.4x - 1.2$. From $y'' = 0$, that is, $x^2 - x - 2 = 0$ one obtains $x_1 = -1$ and $x_2 = +2$. Because $f'''(x_1) = -3.6 \neq 0$ and $f'''(x_2) = +3.6 \neq 0$, $x_1 = -1$ and $x_2 = +2$ are abscissae of *points of inflection* (Fig.). The corresponding ordinates are $f(x_1) = +0.3$ and $f(x_2) = -1.2$.

The *inflectional tangents* t_1 and t_2 at the points of inflection $W_1(-1, +0.3)$ and $W_2(+2, -1.2)$ have the slopes $f'(x_1) = +2.2$ and $f'(x_2) = -3.2$ and hence have the equations $(y - 0.3) = 2.2(x + 1)$, or $y = 2.2x + 2.5$ for t_1 and $(y + 1.2) = -3.2(x - 2)$ or $y = -3.2x + 5.2$ for t_2.

Example 2: The function $y = f(x) = (x^2 - 4)/x$ has no point of inflection, because $f''(x) = 0$ is a necessary condition for a point of inflection, but $y'' = -8/x^3$ cannot have the value zero for any finite value of x.

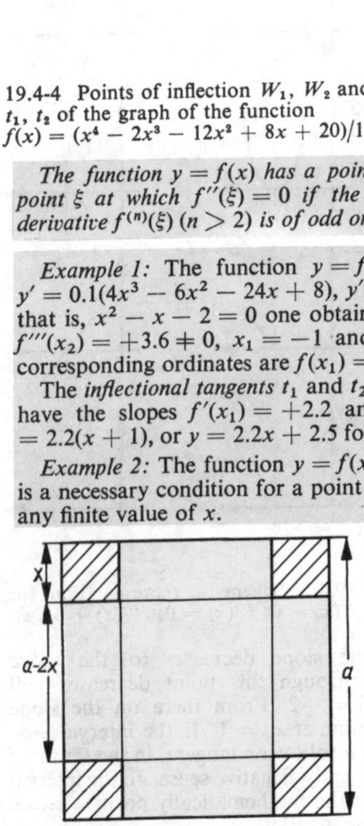

19.4-5 Box of greatest volume made out of a square

Applications. If one succeeds in expressing a variable f as a continuous and differentiable function of a variable x, then one can calculate for which value of x the variable f has an extreme value. From the given conditions it can be determined whether this value is a local maximum or minimum.

In applications, however, it is usually required to find absolute extrema. If the function $f(x)$ is continuous in the closed interval $a \leqslant x \leqslant b$ and differentiable in the open interval $a < x < b$, then its absolute minimum (or maximum) is either the smallest local minimum, the greatest local maximum or one of the boundary values $f(a)$ or $f(b)$.

Example 1: A box is constructed from a square of cardboard of side a by cutting away four squares from its corners and folding the resulting rectangles (Fig.). The four shaded equal squares can be used to stick the carton together. How big must these squares be for the volume V of the carton to be as large as possible? –

By the formula for the volume of a rectangular box V is given by

$$V = y = f(x) = x(a-2x)^2 = 4x^3 - 4ax^2 + a^2x.$$

This function can have extreme values only for $y' = 12x^2 - 8ax + a^2 = 0$, that is, $x^2 - 2ax/3 + a^2/12 = 0$; thus, $x_1 = a/6$, because for $x_2 = a/2$ the cardboard would fall apart. From $y'' = 24x - 8a$ it follows that $f''(x_1) = -4a < 0$, and thus $x_1 = a/6$ is the value of the abscissa of a local maximum. This is also the absolute maximum; the cut must be 1/6 of the length of side a.

Example 2: What must be the dimensions of a cylindrical preserve tin so that for a given content V the smallest possible amount of sheet metal is required for its construction? – A right circular cylinder is determined by the radius r of its circular base and its height h. Its surface area S must be expressed as a function of one variable. The second variable in $S = 2\pi r^2 + 2\pi rh$ can be eliminated using the given additional condition $V = \pi r^2 h$. With $h = V/(\pi r^2)$ one obtains:

$$S = y = f(r) = 2\pi r^2 + 2V\cdot(1/r), \quad y' = 4\pi r - 2V/r^2, \quad y'' = 4\pi + 4V/r^3.$$

A local extremum can occur only for $4\pi r_1 = 2V/r_1^2$ or $r_1 = (V/2\pi)^{1/3}$.

Because $f''(r_1) = 12\pi > 0$, the surface area has a local minimum at the point r_1; this is, moreover, the absolute minimum. The height h_1 of the cylinder is given by $h_1 = V/(\pi r_1^2) = 2r_1$.

If one substitutes for V a given value, say $V = 50$ cubic inches, one obtains $r_1 \approx 2''$ and $h_1 \approx 4''$. Of all cylindrical tins with the same volume, the one whose surface area is the smallest is that whose diameter $2r_1$ is equal to its height h_1.

Example 3: A beam of rectangular cross section is to be cut from a log, whose cross section may be assumed to be circular with diameter d (Fig.). For what measurements does its load carrying capacity T reach a maximum, if T is proportional to the breadth b and the square of the height h; $T = cbh^2$ (c = const)? – The theorem of Pythagoras yields the additional condition $h^2 = d^2 - b^2$, and hence one obtains T as a function of one variable $T = f(b) = cd^2b - cb^3$. Thus, $f'(b) = cd^2 - 3cb^2$, $f''(b) = -6cb$. A local extremum can occur only for $f'(b_1) = 0 = cd^2 - 3cb_1^2$, that is, $b_1 = (d/3)\sqrt{3}$. Because $f''(b_1) = -2cd\sqrt{3} < 0$, this extremum is a maximum of the load carrying capacity. Finally from $h^2 = d^2 - b^2$, $h = (d/3)\sqrt{6}$. The ratio $h/b = \sqrt{2}/1$ is independent of the diameter of the log.

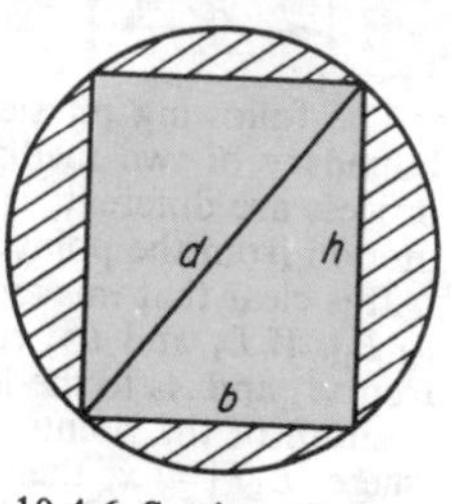

19.4-6 Section of a beam cut from a log

Denoting the sides of a rectangle by a and b, its perimeter by P and its area by A, then the following table shows the validity of two theorems:

Of all rectangles with given perimeter the square has the greatest area.
Of all rectangles with given area the square has the smallest perimeter.

Given	$P = 2(a+b)$	$b = P/2 - a$	$A = ab$	$b = A/a$
Required	$A = ab = f(a)$ $= Pa/2 - a^2$		$P = 2(a+b) = f(a)$ $= 2(a + A/a)$	
1st derivative	$f'(a_1) = P/2 - 2a_1 = 0$	$a_1 = P/4$	$f'(a_1) = 2 - 2A/a_1^2 = 0$	$a_1 = \sqrt{A}$
2nd derivative	$f''(a_1) = -2 < 0$	maximum	$f''(a_1) = +4/\sqrt{A} > 0$	minimum
Solution	$b_1 = P/4 = a_1$	square	$b_1 = \sqrt{A} = a_1$	square

Example 4: A sector is to be cut out of a circular piece of sheet metal of radius R, and the remainder bent together to form a conical funnel (Fig.). For what angle ε at the centre does the funnel have the greatest capacity? –

The formula $V = (\pi/3)\,r^2h$ for the volume of a cone, together with the additional condition $r^2 = R^2 - h^2$, gives the equation $V = f(h) = (\pi/3)\,(R^2h - h^3)$. For extrema, $f'(h_1) = \pi(R^2 - 3h_1^2)/3 = 0$; $h_1 = (R/3)\sqrt{3}$, $f''(h) = -2\pi h$, $f''(h_1) = -(2/3)\,\pi R\sqrt{3} < 0$. Thus, h_1 gives a maximum. From the additional condition, $r_1 = (R/3)\sqrt{6}$. In bending the sheet the circular arc of length $b = \hat{\varepsilon}R$

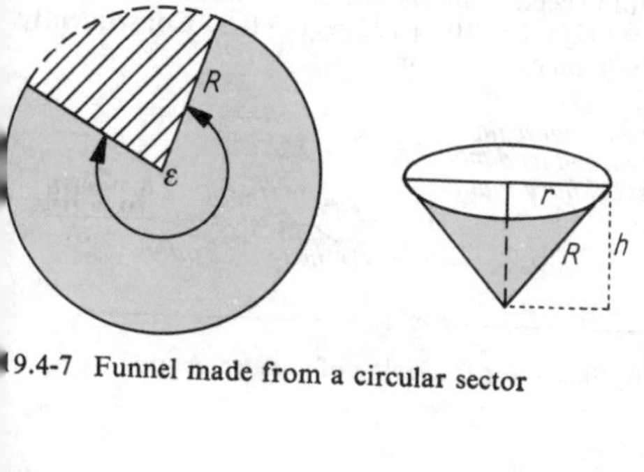

19.4-7 Funnel made from a circular sector

becomes the circumference $2\pi r$ of the circular base. From $R\hat{\varepsilon} = 2\pi r_1$ it follows that $\hat{\varepsilon} = (2/3)\,\pi\,\sqrt{6}$, or $\varepsilon \approx 294°$.

Example 5: Of all cylinders that can be inscribed in a right circular cone of radius R and height H, the one with the greatest volume is required (Fig.).
The volume V of the cylinder is given by $V = \pi r^2 h$. The additional condition comes in this case from the intercept theorem: $h/(R - r) = H/R$, $h = (H/R)\,(R - r)$. Hence $V = f(r) = \pi(H/R)\,(Rr^2 - r^3)$; $f'(r_1) = \pi(H/R)\,(2Rr_1 - 3r_1^2) = 0$ gives $r_1 = 2R/3$, because the solution $r_2 = 0$ corresponds to the volume $V = 0$. Because $f''(r_1) = \pi(H/R)\,(2R - 6r_1) = -2\pi H < 0$, the volume is a maximum for $r = r_1$.

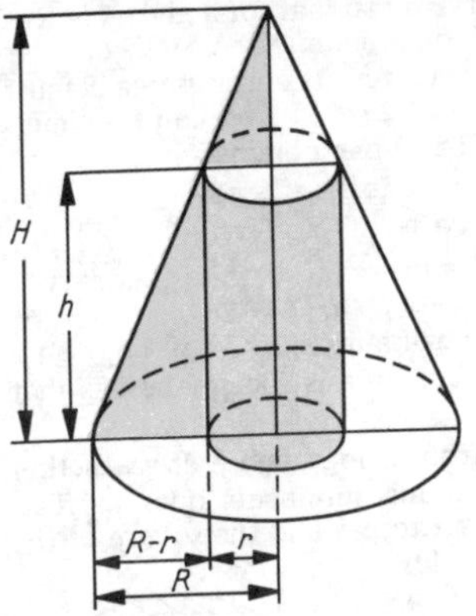

19.4-8 Cylinder inscribed in a right circular cone

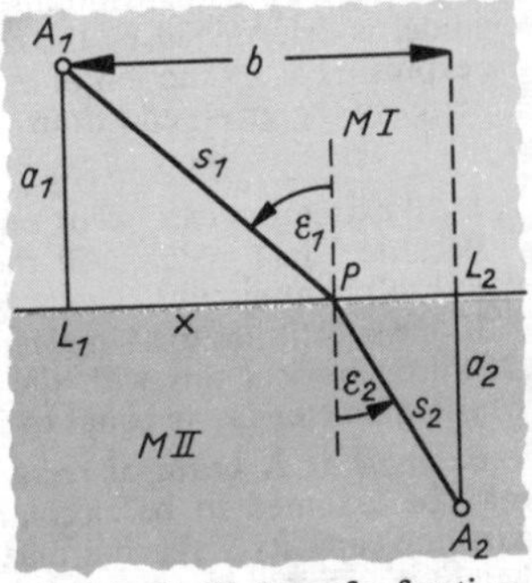

19.4-9 Snell's law of refraction

The following physical problem leads to *Snell's law of refraction*. A plane E_1 is the common boundary of two media M I and M II, in which the velocities of propagation of a body or of a process are different, v_1 in M I and v_2 in M II. Under what conditions is the time required for the motion from the point A_1 in M I to A_2 in M II the smallest possible (Fig.)? –
It is clear that this motion takes place in a plane E_2 passing through A_1 and A_2 and perpendicular to E_1. If L_1 and L_2, respectively, are the feet of the perpendiculars $|A_1L_1| = a_1$ and $|A_2L_2| = a_2$ from A_1 and A_2 to the line of intersection of the two planes E_1 and E_2, and if $|L_1L_2| = b$, then the position of the points A_1 and A_2 is fixed. If the path of the motion cuts the boundary line at P, where $|L_1P| = x$, then the length of the path s_1 from A_1 to P is given by $s_1 = \sqrt{(a_1^2 + x^2)}$, and $|PA_2| = s_2 = \sqrt{\{a_2^2 + (b - x)^2\}}$. The time t required to describe the whole path A_1PA_2 is the sum of the individual times $t_1 = s_1/v_1$ and $t_2 = s_2/v_2$. This gives the function $t(x)$, and the condition for the extremum is obtained from this:

$$t = t(x) = t_1 + t_2 = (1/v_1)\cdot\sqrt{(a_1^2 + x^2)} + (1/v_2)\cdot\sqrt{\{a_2^2 + (b - x)^2\}},$$

$$t'(x) = (1/v_1)\,x/\sqrt{(a_1^2 + x^2)} - (1/v_2)\,(b - x)/\sqrt{\{a_2^2 + (b - x)^2\}} = 0 = x/(v_1 s_1) - (b - x)/(v_2 s_2).$$

This means geometrically that $\sin\varepsilon_1/\sin\varepsilon_2 = v_1/v_2$, because $x/s_1 = \sin\varepsilon_1$ and $(b - x)/s_2 = \sin\varepsilon_2$.

Example 6: What maximum speed may an express train have if braking produces a uniform retardation of $b = 2.2$ ft/s^2, and the braking distance may not exceed $s_1 = 3000$ ft? –
Substituting the braking distance s_1 into the distance-time equation $s = vt - (b/2)\,t^2$ of uniformly retarded motion gives the braking time t_1 as a function of the maximum speed v. The speed $s' = v - bt$ must be zero after the braking time t_1. In both formulae v denotes the speed when the brakes are first applied, and hence the maximum speed. Elimination of t_1 from the equations $3000 = vt_1^2 - 1.1t_1^2$ and $0 = v - 2.2t_1$ gives $v = \sqrt{(2bs_1)} = 10\cdot\sqrt{132} \approx 115$ ft/s. Consequently the speed of the express train may not exceed $v \approx 78$ miles per hour.

Example 7: A water main is to be laid from a water tower W to the main buildings H (Fig.). In addition, some buildings S at some distance from the main are to be supplied with water by means of a subsidiary main. S is at a distance of 1 mile from the principal main and the foot of the perpendicular from S to the main is at a distance of

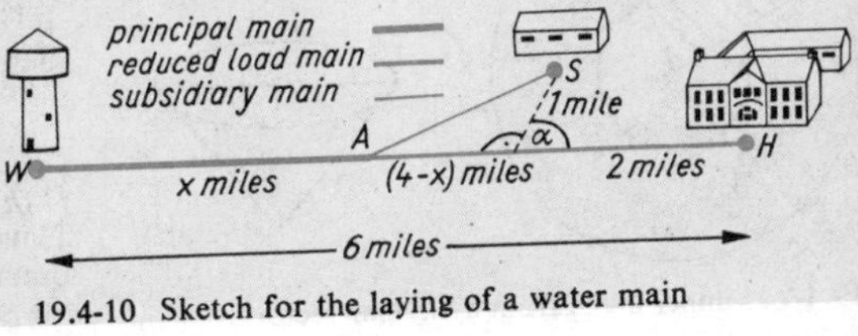

19.4-10 Sketch for the laying of a water main

2 miles from the buildings H. The distance from H to the water tower is 6 miles. The cost of one mile of water main is estimated as follows: principal main 30 units, reduced load main 22 units, subsidiary main 12 units. All mains are laid in straight lines. At what distance from the water tower must the subsidiary main branch off from the principal main, in order that the cost of laying them shall be as small as possible? –

Introducing the variable x for the distance between the water tower W and the branch point A, the length $|AS|$ of the subsidiary main is given by $|AS| = \sqrt{1 + (4 - x)^2}$. The total cost C is then made up as follows: $C = 30x + 22(6 - x) + 12\sqrt{1 + (4 - x)^2}$. An extreme value for the function $C = f(x) = 132 + 8x + 12\sqrt{17 - 8x + x^2}$ must now be calculated. $f'(x) = 8 + (12x - 48)/\sqrt{17 - 8x + x^2}$; hence the necessary condition $f'(x) = 0$ leads to the quadratic equation $x^2 - 8x + 76/5 = 0$ with the solutions $x_{1,2} = 4 \pm (2/5)\sqrt{5}$. The value $x_1 = 4 + (2/5)\sqrt{5}$ does not satisfy the equation in the form containing the square root (check). The second derivative $f''(x) = 12/(17 - 8x + x^2)^{3/2}$ is greater than zero for x_2 and hence indicates a local minimum that is at the same time an absolute minimum. The subsidiary main must therefore branch the principal main at a distance of 3.11 miles from the water tower.

Extreme values of functions of several variables

The k variables $\xi_1, \xi_2, \dots, \xi_i, \dots, \xi_k$ of the function $y = f(\xi_1, \dots, \xi_k)$ can be regarded as an ordered k-tuple $\boldsymbol{x} = (\xi_1, \xi_2, \dots, \xi_k)$ of real numbers in a k-dimensional Euclidian space (see Chapter 40.). The element $\boldsymbol{x}$ lies in a neighbourhood of the element $\boldsymbol{x}_m = (\xi_1^{(m)}, \dots, \xi_k^{(m)})$ if positive numbers h_i can be found such that each of the variables ξ_i $(i = 1, 2, \dots, k)$ lies in an interval $\xi_i^{(m)} - h_i < \xi_i < \xi_i^{(m)} + h_i$. Each k-tuple $\boldsymbol{x}$ corresponds uniquely to a function value $f(\boldsymbol{x}) = y$, and a local maximum of the function $f(\boldsymbol{x}) = y$ at the point $\boldsymbol{x}_m$ can occur only if the ordinate values $f(\boldsymbol{x})$ for all elements different from $\boldsymbol{x}_m$ in a neighbourhood of $\boldsymbol{x}_m$ are less than $f(\boldsymbol{x}_m)$. Similarly $f(\boldsymbol{x}) > f(\boldsymbol{x}_m)$ must hold in a neighbourhood of a local minimum.

Necessary condition for the occurrence of local extrema. For functions of two variables one puts $\xi_1 = x$, $\xi_2 = y$ and $f(x, y) = z$. The graph of the function is a surface in three-dimensional space, and the conditions for a local extremum have an intuitive meaning: $f(\boldsymbol{x}) < f(\boldsymbol{x}_m)$ means that in a neighbourhood of the maximum $P_{max} = (x_{max}, y_{max}, z_{max})$ all other points of the surface lie below a horizontal plane through this point. Similarly $f(\boldsymbol{x}) > f(\boldsymbol{x}_m)$ means that at the point $P_{min} = (x_{min}, y_{min}, z_{min})$ all points of the surface in a neighbourhood lie above a horizontal plane. These planes are tangent planes and are spanned by the two tangents determined by $z_x = \dfrac{\partial f(x, y)}{\partial x}$ and $z_y = \dfrac{\partial f(x, y)}{\partial y}$ (see 19.3. – Partial derivatives of a function), which are parallel to the x, y-plane only if $z_x = 0$ and $z_y = 0$.

$$\text{Local extremum for } (x_m, y_m) \text{ only if } \frac{\partial f(x_m, y_m)}{\partial x} = 0 \text{ and } \frac{\partial f(x_m, y_m)}{\partial y} = 0$$

This condition is necessary; a *saddle point* (Fig.) shows, however, that it is not sufficient. Although both tangents are horizontal there, no matter how small a neighbourhood of the saddle point is chosen, two points of the surface can always be found within it and on opposite sides of the tangent plane at the point.

The necessary condition already found can be generalized to differentiable functions of k variables.

A function $y = f(\xi_1, \xi_2, \dots, \xi_k) = f(\boldsymbol{x})$ can have a local extremum at the point $\boldsymbol{x} = \boldsymbol{x}_m$ only if each partial derivative of the first order vanishes at $\boldsymbol{x}_m$.

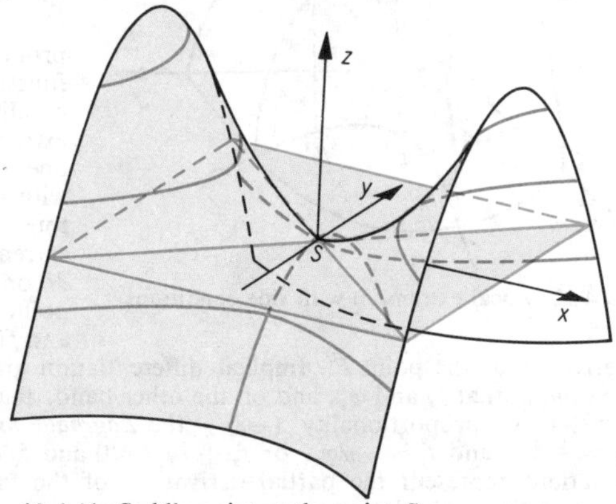

19.4-11 Saddle point at the point S

Sufficient condition for the occurrence of a local extremum. If one expands the function $z = f(x, y)$ in the neighbourhood $x_m - h_1 < x < x_m + h_1$, $y_m - h_2 < y < y_m + h_2$ of a local extreme value at the point (x_m, y_m) by Taylor's theorem and breaks off the expansion after the term $n = 1$, then because $f_x = 0$ and $f_y = 0$, one obtains: $2!\Delta = 2![f(x_m + h_1, y_m + h_2) - f(x_m, y_m)] = h_1^2 f_{xx}(x_m + \vartheta_1 h_1, y_m + \vartheta_2 h_2) + 2h_1h_2 f_{xy}(x_m + \vartheta_1 h_1, y_m + \vartheta_2 h_2) + h_2^2 f_{yy}(x_m + \vartheta_1 h_1, y_m + \vartheta_2 h_2)$, $0 < \vartheta_1, \vartheta_2 < 1$. If the second derivatives f_{xx}, f_{xy} and f_{yy} are continuous functions, then they have the same sign at the point $(x_m + h_1, y_m + h_2)$ as at (x_m, y_m), provided that h_1 and h_2 are chosen sufficiently small. In particular, if $f_{xx} \neq 0$, then the difference Δ can be expressed in the form:

$$2!\Delta = h_1^2 f_{xx} + 2h_1h_2 f_{xy} + h_2^2 f_{yy} = (1/f_{xx})\,[(h_1 f_{xx} + h_2 f_{xy})^2 + h_2^2(f_{xx}f_{yy} - f_{xy}^2)].$$

In the square bracket, besides the square there occurs only the expression $(f_{xx}f_{yy} - f_{xy}^2)$. If it is positive, then the square bracket is positive; Δ is different from zero and has the same sign as f_{xx}. Thus, for $f_{xx} < 0$, the ordinate difference Δ is always negative in a neighbourhood of the point (x_m, y_m) and the function has a *local maximum*; for $f_{xx} > 0$ however, it has a *local minimum*. From $(f_{xx}f_{yy} - f_{xy}^2) > 0$ it also follows that f_{yy} has the same sign as f_{xx} in the neighbourhood considered.

$f(x, y)$ has a local extremum at (x_m, y_m) if $f_x = 0$, $f_y = 0$ and $(f_{xx}f_{yy} - f_{xy}^2) > 0$ at this point; for $f_{xx} < 0$ it is a maximum, and for $f_{xx} > 0$ a minimum.

It can be shown that no extremum can occur for $(f_{xx}f_{yy} - f_{xy}^2) < 0$; for example, f_{xx} and f_{yy} have different signs at a saddle point. However, when $(f_{xx}f_{yy} - f_{xy}^2) = 0$, no conclusion can be reached as to whether an extremum occurs or not.

When the first and second partial derivatives of a function $y = f(x)$ are denoted by $p_i = \dfrac{\partial y}{\partial \xi_i}$ $(i = 1, 2, \ldots, k)$ and $p_{ij} = \dfrac{\partial^2 y}{\partial \xi_i\, \partial \xi_j}$ $(i, j = 1, 2, \ldots, k)$, then the function $y = f(x)$ has a local extremum at the point x_m if all minors of even order of the determinant displayed here are positive and the signs of the minors of odd order agree with the sign of p_{11}; for $p_{11} < 0$ the local extremum is a maximum, and for $p_{11} > 0$ a minimum.

$$\begin{vmatrix} p_{11} & p_{12} \cdots p_{1k} \\ p_{21} & p_{22} \cdots p_{2k} \\ \vdots & \vdots \quad\; \vdots \\ p_{k1} & p_{k2} \cdots p_{kk} \end{vmatrix}$$

Maxima and minima with side conditions. In many problems of applications extrema of functions of several variables are to be determined in which the variables are not independent of one another, but are connected by side conditions. Such problems can often be solved by reducing the number of variables by elimination with the help of the side conditions. Some extreme value problems were solved above by this procedure. This means, however, that preference, not always justifiable, is given to one variable over the others. A treatment giving equal weight to all variables, the *Lagrange method of undetermined multipliers*, will be derived intuitively in the following for functions of two variables.

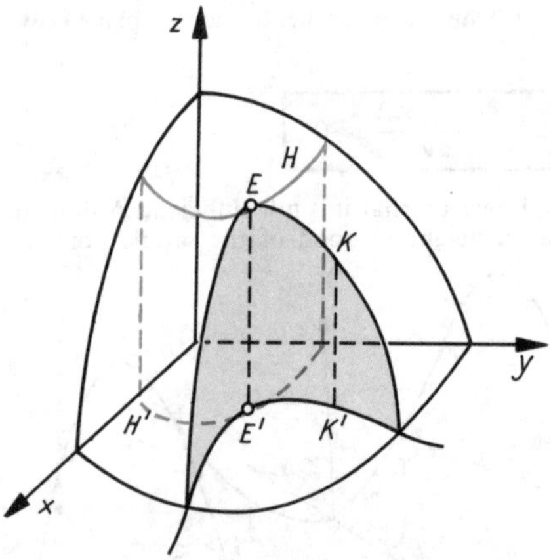

19.4-12 Local extremum with side conditions

The two variables x and y of a function $z = f(x, y)$ are connected by means of a side condition $\varphi(x, y) = 0$. The function $z = f(x, y)$ represents a surface in space, and the equation giving the side condition $\varphi(x, y) = 0$ defines a curve K' in the x, y-plane (Fig.). In calculating extrema the only values x, y of the function $z = f(x, y)$ that are of interest are those that satisfy the side condition, that is, the points of that curve K on the surface $z = f(x, y)$ whose projection on the x, y-plane is precisely the curve K'. Thus, the problem of determining the local extrema of the function $z = f(x, y)$, taking account of the side condition $\varphi(x, y) = 0$, means finding the local extrema of the space curve K. For this purpose one considers the family of curves $c = f(x, y)$ with $c = \text{const}$. One of these contour lines H will touch the space curve K at a point E; E is an extremum. The projection H' of this contour line H on the x, y-plane touches the curve K' at the point E'. The functions defined by $\varphi(x, y) = 0$ and $f(x, y) - c = 0$ must therefore have the same derivative at the point E'. Implicit differentiation gives $f_x/f_y = \varphi_x/\varphi_y$. From this it follows on the one hand, that f_x and φ_x, and on the other hand, that f_y and φ_y are proportional. Introducing the constant of proportionality $(-\lambda)$ – the *Lagrange multiplier* – one obtains the two equations $f_x = -\lambda\varphi_x$ and $f_y = -\lambda\varphi_y$, or $f_x + \lambda\varphi_x = 0$ and $f_y + \lambda\varphi_y = 0$. But the left-hand sides of these equations represent the partial derivatives of the function $F(x, y) = f(x, y) + \lambda\varphi(x, y)$. All this leads to the method of *Lagrangian multipliers*:

To determine the local extrema of a function $z = f(x, y)$, subject to the side condition $\varphi(x, y) = 0$, an auxiliary function $F(x, y) = f(x, y) + \lambda\varphi(x, y)$ with the undetermined multiplier λ is formed, and the first partial derivatives of this function are found. From the system of equations $F_x = f_x + \lambda\varphi_x = 0$, $F_y = f_y + \lambda\varphi_y = 0$, $\varphi(x, y) = 0$ the coordinates of possible extrema and the multiplier λ are calculated.

This rule simply gives an elegant way of calculating those points at which local extrema can occur. The investigation to determine whether an extremum really occurs, and of what nature, is in general complicated. In a concrete example it is often clear from the formulation of the problem whether a maximum or a minimum is to be expected with certainty.

Example: Among all right-angled triangles with given hypotenuse c it is required to find the one with the greatest area. Denoting the sides containing the right angle by x and y, $A = f(x, y) = xy/2$ is to be a maximum. Since the triangle is right-angled, the side condition $x^2 + y^2 = c^2$ or $\varphi(x, y) = x^2 + y^2 - c^2 = 0$ holds. Thus, one forms the auxiliary function $F(x, y) = xy/2 + \lambda(x^2 + y^2 - c^2)$. From the system of equations

$$\frac{\partial F}{\partial x} = y/2 + 2\lambda x = 0 \longrightarrow \lambda = -y/(4x), \quad \frac{\partial F}{\partial y} = x/2 + 2\lambda y = 0 \longrightarrow x^2 = y^2,$$

$$\varphi(x, y) = x^2 + y^2 - c^2 = 0 \longrightarrow x^2 = c^2/2$$

one obtains $x = y = c/\sqrt{2}$. Hence the isosceles right-angled triangle is a possible solution, and it can be proved that for this the area is an absolute maximum.

Similarly the method of undetermined multipliers can be extended to the determination of extrema of a function of n variables with side conditions.

19.5. Applications to plane curves

By means of the differential calculus the important points on the curve given by a function can be determined; in particular, it can be decided when singular points occur. The properties of the evolute and the involute arise out of the clarification of the concept of curvature.

Discussion of the curve defined by an explicit function

It is required to investigate the function $y = f(x)$ with the *domain of definition* $D(f)$. If $D(f)$ is not given explicitly, then it is taken to consist of all real numbers x for which the analytical expression $f(x)$ is defined. If the domain of definition extends on one or on both sides to infinity, then the behaviour of the function for large values of $|x|$ can be investigated by means of limiting processes, that is, for $x \to +\infty$ or for $x \to -\infty$. One therefore speaks also of the *behaviour of the function at infinity*. For rational functions this was done in Chapter 5. The coordinates of the *points of intersection with the coordinate axes* are obtained from the equation $y = f(x)$ by putting $x = 0$ or $y = 0$, respectively. For the determination from the equation $f(x) = 0$ of the *zeros* of the function, that is, the values of the abscissae of the intersections of its curve with the x-axis, approximation methods are applied if necessary. For rational functions Sturm's theorem (see Chapter 5.2. – Zeros) can be used to find intervals in which a zero lies.

The behaviour of the function in the neighbourhood of its *points of discontinuity* is examined with the help of limiting arguments and the type of each discontinuity, for example, pole, indeterminate point, jump discontinuity or oscillation, is determined.

Finally, the function is tested by the methods of the differential calculus for *extrema* and *points of inflection*. If the zeros of the first and second derivatives of the function are determined for this purpose, then for twice continuously differentiable functions one also knows the intervals in which the first derivative has a constant sign and the intervals in which the second derivative does not change sign. Hence one has found the intervals in which the function is *monotonic decreasing*, *monotonic increasing*, *convex or concave*.

Examples of curves for discussion. In what follows the results are given of the discussion of the curves of typical examples. In particular: y_{0i} are the ordinates of the points of intersection with the y-axis; x_{0i} the zeros; x_{mi} the abscissae of extrema; x_{wi} the abscissae of points of inflection; M_i a maximum; m_i a minimum; W_i a point of inflection.

Example 1: The function $y = f(x) = (x/300)(x^2 - 45)(x^2 - 10)$ is defined for all x (Fig.).

Its derivatives are

$$y' = f'(x) = (1/60)(x^2 - 30)(x^2 - 3); \quad y'' = f''(x) = (x/30)(2x^2 - 33);$$
$$y''' = f'''(x) = x^2/5 - 11/10.$$

Behaviour at infinity: $f(x) \to \pm\infty$, as $x \to \pm\infty$. Intersections with the axes:

$$y_0 = 0; \quad x_{01} = 0; \quad x_{02} = -3\sqrt{5} \approx -6.71; \quad x_{03} = +3\sqrt{5} \approx +6.71;$$
$$x_{04} = +\sqrt{10} \approx 3.16; \quad x_{05} = -\sqrt{10} \approx -3.16.$$

Local extrema:

$$f'(x_m) = 0 \to x_{m1} = -\sqrt{30}; \quad x_{m2} = -\sqrt{3}; \quad x_{m3} = +\sqrt{3}; \quad x_{m4} = +\sqrt{30};$$
$$M_1 \equiv (-5.48, 5.48); \quad M_2 \equiv (1.73, 1.7); \quad m_1 \equiv (-1.73, -1.7);$$
$$m_2 \equiv (5.48, -5.48).$$

Points of inflection:

$$f''(x_w) = 0 \to x_{w1} = -4.06; \quad x_{w2} = 0; \quad x_{w3} = +4.06;$$
$$W_1 \equiv (-4.06, 2.58); \quad W_2 \equiv (0, 0); \quad W_3 \equiv (4.06, -2.58).$$

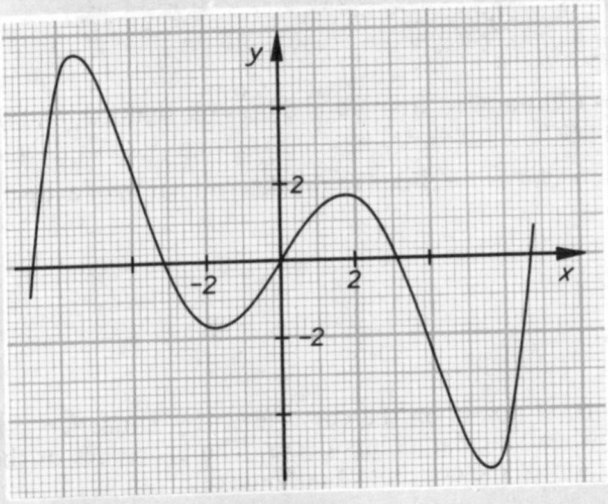

19.5-1 Graph of the function
$y = (x/300)(x^2 - 45)(x^2 - 10)$

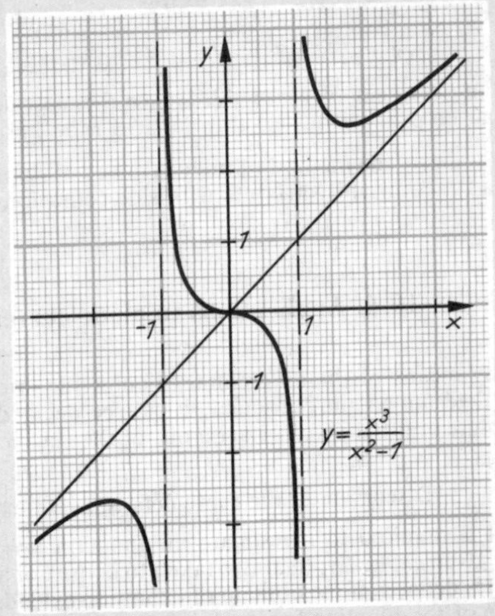

19.5-2 Graph of the function $y = x^3/(x^2 - 1)$

Example 2: The function $y = f(x) = x^3/(x^2 - 1)$ (Fig.) is defined for all x with the exception of the values $x = \pm 1$, for which the denominator is zero. Its derivatives are

$$y' = f'(x) = \frac{x^2(x^2 - 3)}{(x^2 - 1)^2}; \quad y'' = f''(x) = \frac{2x(x^2 + 3)}{(x^2 - 1)^3}; \quad y''' = f'''(x) = \frac{-6(x^4 + 6x^2 + 1)}{(x^2 - 1)^4}.$$

Behaviour at infinity: $\dfrac{x^3}{x^2 - 1} = x + \dfrac{x}{x^2 - 1} \to \pm\infty$, as $x \to \pm\infty$.

Asymptote $y = x$ (see Chapter 5.).
Intersections with the axes: $y_0 = 0$; $x_0 = 0$.
Discontinuities: Poles for $x_1 = 1$ and $x_2 = -1$ with vertical asymptotes.
Local extrema: $f'(x_m) = 0 \to x_{m1} = -\sqrt{3}$; $x_{m2} = +\sqrt{3}$; $x_{m3} = 0 = x_w$; $M \equiv (-1.73, -2.6)$; $m \equiv (1.73, 2.6)$.
Points of inflection: $f''(x_w) = 0 \to x = 0$, $W \equiv (0, 0)$. Since $f'(x_w) = 0$, this is a horizontal point of inflection.

Example 3: The function $y = f(x) = x\sqrt{(9 - x^2)}$ is defined only in the interval $-3 \leqslant x \leqslant +3$ (Fig.). Its derivatives are

$$y' = f'(x) = \frac{9 - 2x^2}{\sqrt{(9 - x^2)}}; \quad y'' = f''(x) = \frac{x(2x^2 - 27)}{\sqrt{\{(9 - x^2)^3\}}}; \quad y''' = f'''(x) = \frac{-243}{\sqrt{\{(9 - x^2)^5\}}}.$$

Intersections with the axes: $y_0 = 0$; $x_{01} = 0$; $x_{02} = 3$; $x_{03} = -3$.
Local extrema: $f'(x_m) = 0 \to x_{m1} = -(3/2)\sqrt{2}$; $x_{m2} = +(3/2)\sqrt{2}$; $m \equiv (-2.12, -4.5)$; $M \equiv (2.12, 4.5)$.

Points of inflection: $f''(x_w) = 0 \to x_{w1} = 0$; $x_{w2} = +(3/2)\sqrt{6}$; $x_{w3} = -(3/2)\sqrt{6}$; x_{w2} and x_{w3} are outside the domain of definition. $W_1 = (0, 0)$.

The mirror image of this curve in the x-axis is the curve defined by the function $y = -x\sqrt{(9-x^2)}$. Both functions are defined by the algebraic equation $x^4 - 9x^2 + y^2 = 0$. $P(0, 0)$, as a double point, is a singular point.

Example 4: The function $y = f(x) = \sin 2x + 2\cos x$ is defined for all x; because of its periodicity the investigation will be restricted to the interval $0 \leqslant x \leqslant 2\pi$ (Fig.). Its derivatives are

$$y' = f'(x) = 2\cos 2x - 2\sin x;$$
$$y'' = f''(x) = -4\sin 2x - 2\cos x;$$
$$y''' = f'''(x) = -8\cos 2x + 2\sin x.$$

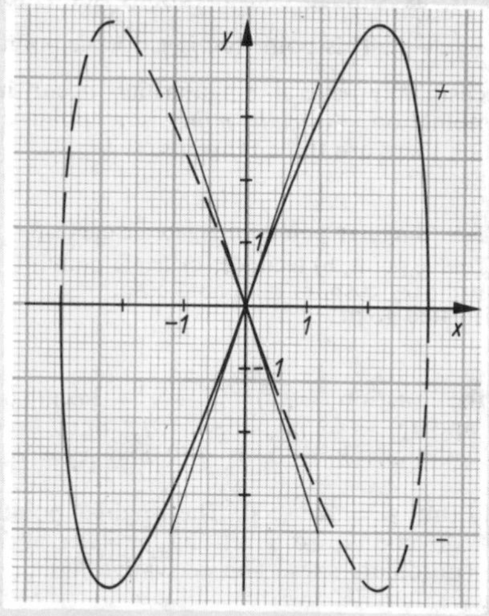

19.5-3 Discussion of the function defined by the equation $x^4 - 9x^2 + y^2 = 0$

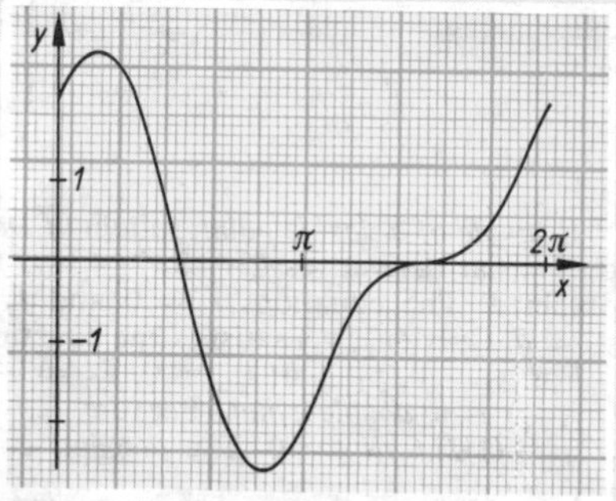

19.5-4 Graph of the function $y = \sin 2x + 2\cos x$

Intersections with the axes: $y_0 = 2$; $x_{01} = \pi/2 \approx 1.57$; $x_{02} = 3\pi/2 \approx 4.71$.

Local extrema: $f'(x_m) = 0 \to x_{m1} = \pi/6$; $x_{m2} = 5\pi/6$; $x_{m3} = 3\pi/2 = x_{w3}$. $M \equiv (0.52, 2.6)$; $m \equiv (2.62, -2.6)$.

Points of inflection: $f''(x_w) = 0 \to x_{w1} = \pi/2$; $x_{w2} \approx 3.39$; $x_{w3} = 3\pi/2$; $x_{w4} \approx 6.03$; $W_1 \equiv (1.57, 0)$; $W_2 \equiv (3.39, -1.45)$; $W_3 \equiv (4.71, 0)$, a horizontal point of inflection since $f'(x_{w3}) = 0$; $W_4 \equiv (6.03, 1.45)$.

Singular points

A survey of singular points can be made by investigating the possible tangent directions at these points. If the equation of the curve is given in implicit form by $f(x, y) = 0$, and if φ is the angle between a tangent and the $+x$-axis, then

$$y' = \tan\varphi = \frac{dy}{dx} = -\frac{\partial f}{\partial x} \Big/ \frac{\partial f}{\partial y} = -f_x/f_y.$$

A tangent parallel to the x-axis is then characterized by $f_x = 0$, and one perpendicular to the x-axis ($\varphi = \pi/2$) by $f_y = 0$. If both partial derivatives have the value zero, however, then a singular point is to be expected.

Singular points of an algebraic curve. In a neighbourhood $x_s - h_1 < x < x_s + h_1$, $y_s - h_2 < y < y_s + h_2$ of a singular point (x_s, y_s), which is supposed to be a zero, because $f(x_s, y_s) = 0$, $f_x(x_s, y_s) = 0$ and $f_y(x_s, y_s) = 0$, the Taylor expansion with the remainder term R_3 has the form:

$$f(x_s + h_1, y_s + h_2) = {}^1/_2(h_1^2 f_{xx} + 2h_1h_2 f_{xy} + h_2^2 f_{yy}) + R_3.$$

If $h_1 = \Delta x$ and $h_2 = \Delta y$ tend to zero, then (R_3/h_1^2) is likewise a null sequence, because R_3 contains only the third and higher powers of h_1 and h_2.

The value of $y' = \tan\varphi$ can then be determined from the quadratic equation $f_{xx} + 2y'f_{xy} + y'^2 f_{yy} = 0$ if the three partial derivatives do not all vanish. If $f_{yy} \neq 0$, then the number of solutions of this quadratic equation depends on the value of $\Delta = f_{xy}^2 - f_{xx}f_{yy}$. For $\Delta > 0$ two distinct tangents exist; the curve has a *double point* in which two branches of the curve intersect. For $\Delta = 0$ two coincident tangents exist; two branches of the curve have a common tangent and touch one another either at a *tacnode* or at a *cusp*. At an *ordinary cusp* the two branches of the curve lie on opposite

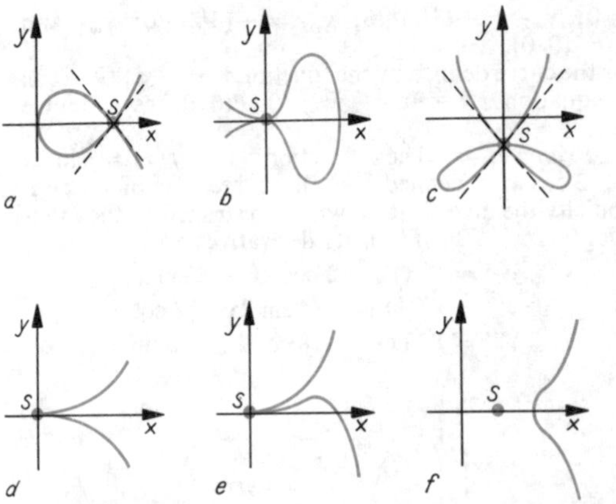

sides of the tangent, and at a *ramphoid cusp* they lie on the same side. For $\Delta < 0$ there are no real tangents and the curve has an *isolated point* there (Fig.). If $f_{yy} = 0$, then one tangent has the slope $y' = -(1/2) f_{xx}/f_{yy}$, whilst a second tangent is parallel to the y-axis, as similar considerations show for $\frac{dx}{dy} = \frac{1}{y'}$. This point too is a double point. Occasionally several singularities occur in combination. A *triple point* is a point at which there are three tangents.

19.5-5 Singular points of algebraic curves; a) double point, b) tacnode, c) triple point, d) ordinary cusp, e) ramphoid cusp, f) isolated point

Example of a double point, folium of Descartes: From the equation $f(x, y) = x^3 + y^3 - 3axy = 0$ the partial derivatives $f_x = 3x^2 - 3ay$, $f_y = 3y^2 - 3ax$ and $f_{xx} = 6x$, $f_{xy} = -3a$, $f_{yy} = 6y$ are obtained. The equations $f_x = 0$ and $f_y = 0$ have two solutions $x_1 = 0$, $y_1 = 0$ and $x_2 = a$, $y_2 = a$. Only (x_1, y_1) satisfies the equation $f = 0$.

For $x_1 = 0$, $y_1 = 0$: $f_{xx} = 0$, $f_{xy} = -3a$, $f_{yy} = 0$, and thus $\Delta = 9a^2 > 0$. The singular point (x_1, y_1) is a *double point*. Because $f_{yy} = 0$ the tangent directions are given by $y' = \infty$ and $y' = -(1/2) f_{xx}/f_{yy} = 0$. The *tangents* coincide with the coordinate axes. The derivative has the value $\frac{dy}{dx} = -f_x/f_y = -(x^2 - ay)/(y^2 - ax)$.

For $f_x = 0$ but $f_y \neq 0$ there is a tangent t_x parallel to the x-axis; for $f_y = 0$ but $f_x \neq 0$ there is a tangent t_y parallel to the y-axis. The calculation gives: $f_y = 0$, $f_x \neq 0$: $y^2 = ax \rightarrow y^6/a^3 + y^3 - 3y^3 = 0$ or $y^6 = 2y^3a^3$; with $y \neq 0$ $\rightarrow y_3 = a\sqrt[3]{2} \approx 1.26a$ and $x_3 = a\sqrt[3]{4} \approx 1.59a$. Similarly for $f_x = 0$, $f_y \neq 0$, $x_4 = a\sqrt[3]{2} \approx 1.26a$ and $y_4 = a\sqrt[3]{4} \approx 1.59a$.

Substituting $y = mx$ in the equation $f(x, y) = 0$ one obtains $x^3(1 + m^3) - 3amx^2 = 0$ or $1 + m^3 - 3am/x = 0$. As $x \rightarrow \pm\infty$ this gives $1 + m^3 = 0$ or $m = -1$. The folium of Descartes (Fig.) therefore has an *asymptote* with slope $m = -1$. If its equation is $y = -x + c$, the value of c is obtained by substituting for y in the equation $f(x, y) = 0$: $x^3 + (c - x)^3 - 3ax(c - x) = 0$ or $3x^2(a + c) - 3x(c^2 + ac)^3 + c^3 = 0$. Hence, letting $x \rightarrow \pm\infty$, $a + c = 0$ or $c = -a$. The equation of the asymptote is $y = -x - a$.

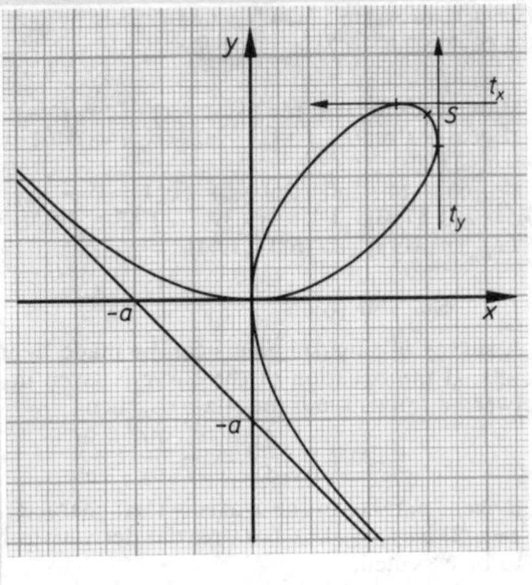

19.5-6 Folium of Descartes

Curvature, evolute and involute

Curvature. In the discussion of the point of inflection the concept of curvature was used to characterize the different course of the curve of the function $y = f(x)$ before and after a point of inflection. It was established there that the same property of the curve is also characterized by the variation of the direction angle τ of two successive tangents to the curve. Because it is clear that the larger the increment of arc length Δs required for a given variation $\Delta\tau$ of the direction angle, the smaller the curvature $\varkappa$, the latter is defined to be the rate of change of this angle as a function of the arc length s:

$$\varkappa = \lim_{\Delta s \to 0} \frac{\Delta\tau}{\Delta s} = \frac{d\tau}{ds}$$

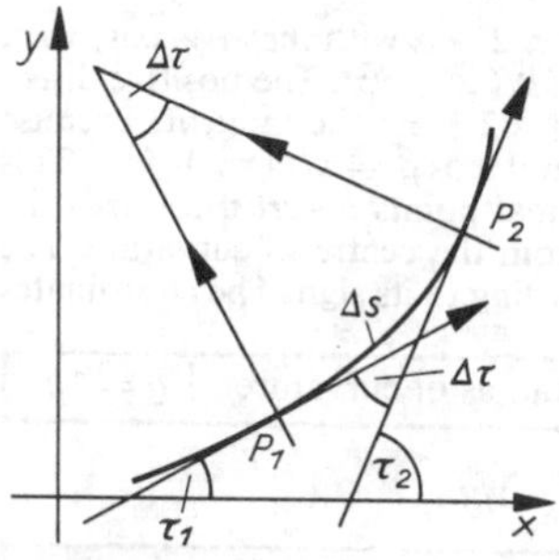

19.5-7 Curvature of a plane curve with small left-curvature

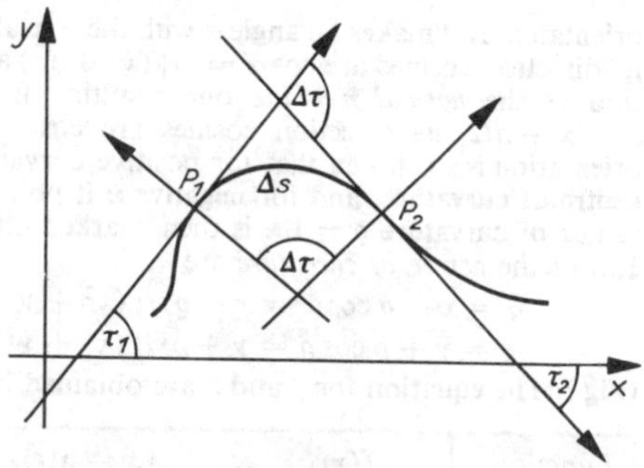

19.5-8 Curvature of a plane curve with large right-curvature

(Fig.). But the angle τ is defined by $\tau = \arctan y'$; by the chain rule one obtains

$$\frac{d\tau}{ds} = \frac{d\tau}{dx}\bigg/\frac{ds}{dx} = y''/(1+y'^2)\cdot[1/\sqrt{(1+y'^2)}] = y''/(1+y'^2)^{3/2} = \varkappa.$$

If x and y are regarded as functions of a parameter and differentiation with respect to this parameter is denoted by a dot, then $\tau = \arctan(\dot{y}/\dot{x})$, $\varkappa = (\dot{x}\ddot{y} - \dot{y}\ddot{x})/(\dot{x}^2 + \dot{y}^2)^{3/2}$. The curvature can also be expressed in polar coordinates $x = r\cos\vartheta$, $y = r\sin\vartheta$. One obtains

$$\dot{x} = \cos\vartheta\,\dot{r} - r\sin\vartheta\,\dot{\vartheta}; \qquad \dot{y} = \sin\vartheta\,\dot{r} + r\cos\vartheta\,\dot{\vartheta}$$

and

$$\ddot{x} = \ddot{r}\cos\vartheta - 2\dot{r}\sin\vartheta\,\dot{\vartheta} - r\cos\vartheta\,\dot{\vartheta}^2 - r\sin\vartheta\,\ddot{\vartheta},$$
$$\ddot{y} = \ddot{r}\sin\vartheta + 2\dot{r}\cos\vartheta\,\dot{\vartheta} - r\sin\vartheta\,\dot{\vartheta}^2 + r\cos\vartheta\,\ddot{\vartheta}$$

and hence

$$\varkappa = (2\dot{r}^2\dot{\vartheta} + r\dot{r}\ddot{\vartheta} - r\ddot{r}\dot{\vartheta} + r^2\dot{\vartheta}^3)/(\dot{r}^2 + r^2\dot{\vartheta}^2)^{3/2} = (2r'^2 - rr'' + r^2)/(r'^2 + r^2)^{3/2},$$

where

$$r' = \frac{dr}{d\vartheta} = \frac{\dot{r}}{\dot{\vartheta}} \quad\text{and}\quad r'' = \frac{dr'}{dt}\cdot\frac{dt}{d\vartheta} = \frac{\ddot{r}\dot{\vartheta} - \dot{r}\ddot{\vartheta}}{\dot{\vartheta}^3}.$$

From the first of these formulae it follows that the curvature has the same sign as y''. Thus, if the derivative increases as the curve is described in the sense of increasing abscissa, then the second derivative y'' is positive and so $\varkappa$ is also positive; regarded from a point P with a very large ordinate, the curve appears *concave*. A road that is curved in this way has a *left-curve*. If the curve appears *convex* from P, or if a similarly curved road has a *right-curve*, then the derivative constantly decreases, and y'' and $\varkappa$ are *negative*.

Circle of curvature. The circle is a curve with constant curvature; from the parametric representation $x = \varrho\cos t$, $y = \varrho\sin t$ one obtains $\varkappa = 1/\varrho$. This result corresponds to the intuitive property that the smaller the *radius of curvature* ϱ, the greater the curvature of a circle (Fig.). To every point (x, y) of the curve of the function $y = f(x)$ a circle of curvature $y = g(x)$ is assigned in accordance with the following rules: both curves pass through the *same point*, $f(x) = g(x)$, have the *same tangent* there, $f'(x) = g'(x)$, and the *same curvature*, so that $f''(x) = g''(x)$. The tangent t at the point (x, y) is

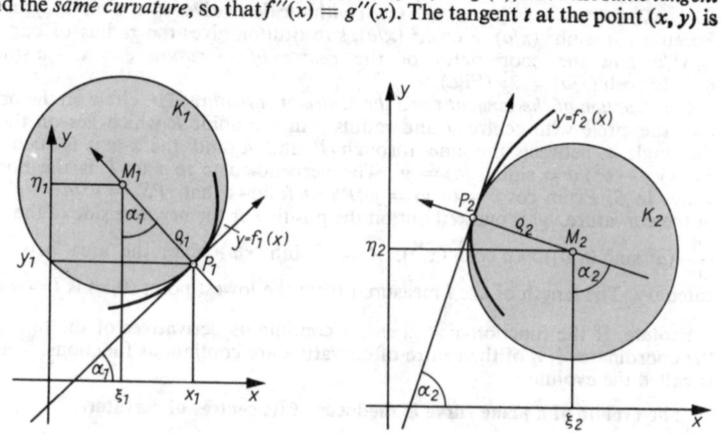

19.5-9 Circle of curvature K, radius of curvature ϱ and centre of curvature M; $\varrho_1 > 0$, $\varrho_2 < 0$

orientated. If it makes an angle α with the $+x$-axis and an angle $\beta = \pi/2 - \alpha$ with the $+y$-axis, then its direction cosines are $\cos\alpha = \dot{x}/\sqrt{(\dot{x}^2 + \dot{y}^2)}$ and $\cos\beta = \sin\alpha = \dot{y}/\sqrt{(\dot{x}^2 + \dot{y}^2)}$. The positive direction of the *normal* n is the one resulting from a rotation of $+\pi/2$ from the tangent. Because $\bar{\alpha} = \alpha + \pi/2$, its direction cosines are $\cos\bar{\alpha} = -\dot{y}/\sqrt{(\dot{x}^2 + \dot{y}^2)}$ and $\cos\bar{\beta} = \dot{x}/\sqrt{(\dot{x}^2 + \dot{y}^2)}$. This orientation is so chosen that for positive curvature the positive normal points *inwards*, towards the centre of curvature, and for negative $\varkappa$ it points *outwards*, away from the centre of curvature. The radius of curvature $\varrho = 1/\varkappa$ is then marked off on the normal according to its sign. The coordinates ξ, η of the *centre of curvature* are

Radius of curvature	$\varrho = 1/\varkappa$

$$\xi = x + \varrho\cos\bar{\alpha} = x - \varrho\dot{y}/\sqrt{(\dot{x}^2 + \dot{y}^2)},$$
$$\eta = y + \varrho\cos\bar{\beta} = y + \varrho\dot{x}/\sqrt{(\dot{x}^2 + \dot{y}^2)}$$

(Fig.). The equation for ξ and η are obtained by substituting for $\varrho = 1/\varkappa$.

Function	$y = f(x)$	$x = x(t),\ y = y(t)$	$r = f(\vartheta)$
Curvature	$\varkappa = \dfrac{y''}{(1 + y'^2)^{3/2}}$	$\varkappa = \dfrac{\dot{x}\ddot{y} - \dot{y}\ddot{x}}{(\dot{x}^2 + \dot{y}^2)^{3/2}}$	$\varkappa = \dfrac{r^2 + 2r'^2 - rr''}{(r^2 + r'^2)^{3/2}}$
Centre of curvature	$\xi = x - \dfrac{1 + y'^2}{y''}\cdot y'$, $\eta = y + \dfrac{1 + y'^2}{y''}$	$\xi = x - \dfrac{\dot{x}^2 + \dot{y}^2}{\dot{x}\ddot{y} - \dot{y}\ddot{x}}\cdot\dot{y}$, $\eta = y + \dfrac{\dot{x}^2 + \dot{y}^2}{\dot{x}\ddot{y} - \dot{y}\ddot{x}}\cdot\dot{x}$	$\xi = r\cos\vartheta - \dfrac{(r^2 + r'^2)(r\cos\vartheta + r'\sin\vartheta)}{r^2 + 2r'^2 - rr''}$, $\eta = r\sin\vartheta - \dfrac{(r^2 + r'^2)(r\sin\vartheta - r'\cos\vartheta)}{r^2 + 2r'^2 - rr''}$

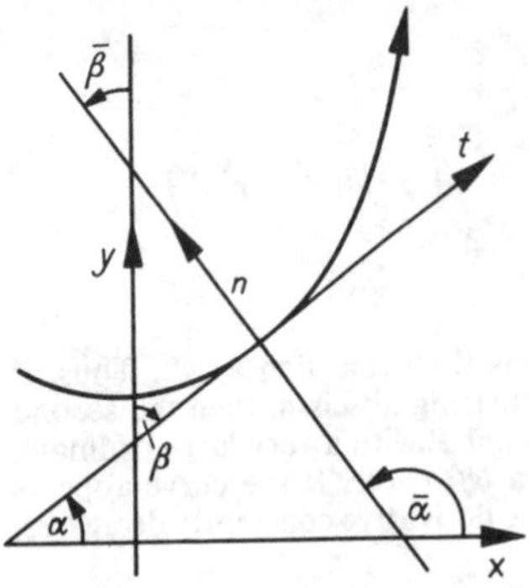

19.5-10 Direction cosines of the tangent *t* and the normal *n*

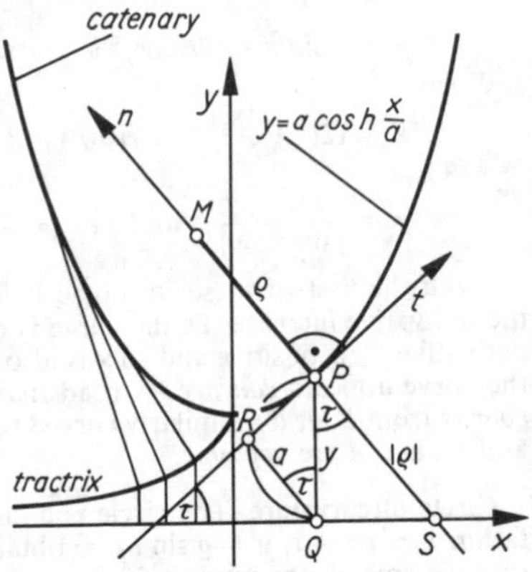

19.5-11 Centre of curvature of the catenary

Catenary. A completely flexible heavy thread, suspended from two points, assumes in equilibrium the form of the catenary. It is the evolute of the *tractrix* and is the graph of the function $y = (a/2)\,(e^{x/a} + e^{-x/a}) = a\cosh(x/a)$ with the derivatives $y' = \sinh(x/a)$ and $y'' = (1/a)\cosh(x/a)$. Because $1 + \sinh^2(x/a) = \cosh^2(x/a)$, substitution gives the radius of curvature $\varrho = a\cosh^2(x/a) = y^2/a$ and the coordinates of the *centre of curvature* $\xi = x - a\sinh(x/a) = x - ay'$ and $\eta = 2a\cosh(x/a) = 2y$ (Fig.).

Construction of the tangent t and the centre of curvature. The circle on the ordinate $|PQ|$ as diameter cuts the circle with centre Q and radius a in the point R which lies on the tangent at P, because the angle τ between the line through P and R and the x-axis is such that $\tan\tau = |RP|/|RQ| = \sqrt{(y^2 - a^2)}/a = \sinh(x/a) = y'$. The perpendicular to t at P is the normal n, which cuts the x-axis in S. From $\cos\tau = (a/y) = y/|PS|$ it follows that $|PS| = y^2/a = |\varrho|$. According to the sign of the curvature, $|\varrho|$ is marked out on the positive or the negative side of the normal from P. Because $\dfrac{d}{dx}(a^2\sinh(x/a)) = a\cosh(x/a)$, $A = a^2\sinh(x/a)$ gives the area between the x-axis and the catenary. The length of arc l measured from the lowest point $(0, a)$ is $l = a\sinh(x/a)$.

Evolute. If the function $y = f(x)$ has continuous derivatives of the first and second order, then the coordinates ξ, η of the centre of curvature are continuous functions. The curve defined by them is called the evolute.

The evolute of a plane curve is the locus of its centres of curvature.

The evolute can also be constructed as the envelope of the normals; thus, the normals to the original curve are tangents to its evolute (Fig.). Because the centres of curvature of the original curve lie on the evolute, the formulae for the coordinates of the centre of curvature give at the same time a parametric representation for the evolute. One need only regard ξ and η as running coordinates.

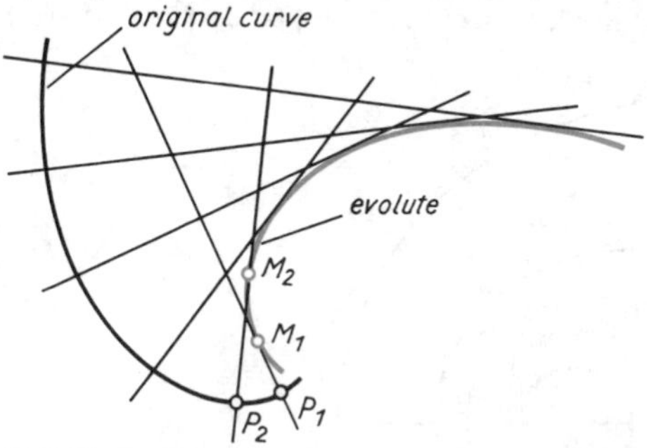

19.5-12 Evolute

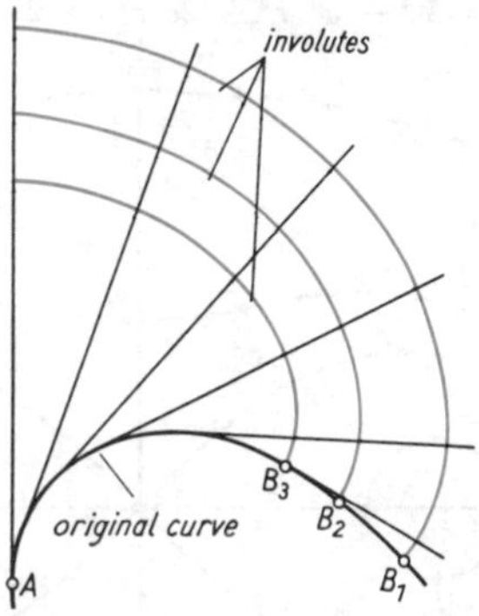

19.5-13 Family of involutes of a plane curve

Involute. The *involute* is a developed curve. One imagines a curve with an inextensible thread laid along it (Fig.). The thread is attached to a point A on the curve. If one then considers a point B_1 of the thread and unwinds the tautly held thread from the curve, then the point B_1 describes a new curve, an involute of the original curve. Since each point B describes such an involute, a whole family of involutes belongs to one given curve. Because the thread is always held taut during the unwinding, the unwound portion of it is always a tangent to the original curve. The point B describes about the instantaneous point of the tangent an infinitesimal circular arc as element of arc of the involute; but this means that the unwound portion of the thread is always normal to the involute. Thus, the tangents of the original curve cut the involute at a right angle. From this the following theorems arise:

The involutes of a plane curve are the orthogonal trajectories (curves cutting at right angles) of the tangents to the original curve.
Every curve is the evolute of each of its involutes.
Every curve is an involute of its evolute.

Example: In machine design the involute of the circle finds application as the profile curve of the teeth of involute gears (Fig.). From the diagram the coordinates of a point P of the involute can be read off as $\xi = x + s \sin t$ and $\eta = y - s \cos t$. From the parametric representation of the circle $x = r \cos t$; $y = r \sin t$ together with the formula $s = rt$ for the length of the unwound circular arc, the parametric representation $\xi = r(\cos t + t \sin t)$; $\eta = r(\sin t - t \cos t)$ for the involute of the circle is obtained.

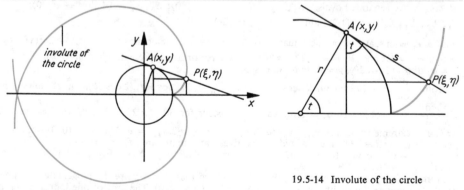

19.5-14 Involute of the circle

Special curves

In the discussion of double points and the centre of curvature in the previous section, the most important properties of the *folium of Descartes* and of the *catenary* were stated incidentally. Properties of certain other curves are given in the following.

Cassinian ovals. Cassinian ovals are defined as the loci of all points P for which the product of the distances $r_1 = |F_1P|$ and $r_2 = |F_2P|$ from two fixed points F_1 and F_2 has a constant value a^2. If the two points F_1 and F_2 lie on the x-axis of a Cartesian coordinate system at distances $+e$ and $-e$ from the origin, then $r_1^2 = (x-e)^2 + y^2$; $r_2^2 = (x+e)^2 + y^2$; $r_1^2 r_2^2 = a^4$ or $(x^2+y^2)^2 - 2e^2(x^2 - y^2) = a^4 - e^4$, $r^4 - 2e^2r^2 \cos 2\vartheta = a^4 - e^4$, $r^2 = e^2 \cos 2\vartheta \pm \sqrt{(e^4 \cos^2 2\vartheta + a^4 - e^4)}$.

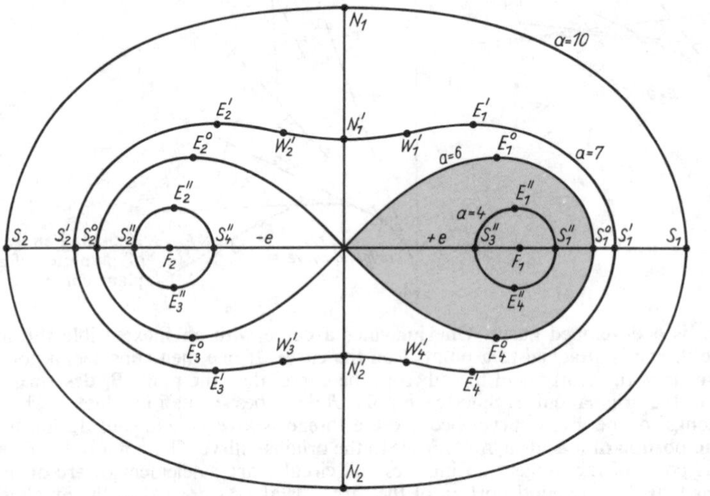

19.5-15 Cassinian ovals for $e = 6$, $a = 10, 7, 6$ and 4; $e = a = 6$ is the lemniscate

Cassinian ovals of different form are obtained according to the ratio of the two constants a and e (Fig.). In the following survey they are characterized by the intersections S_1, S_2, S_3, S_4 with the x-axis, N_1, N_2 with the y-axis, by the extrema E_1, E_2, E_3, E_4 and by the points of inflection W_1, W_2, W_3, W_4.

1. $a > e\sqrt{2}$, the curve resembles an ellipse. $S_1, S_2 \equiv (\pm\sqrt{(a^2+e^2)}, 0)$; $N_1, N_2 \equiv (0, \pm\sqrt{(a^2 - e^2)})$. For $a = e\sqrt{2}$ also the curve has a form like that of the ellipse, $S_1, S_2 \equiv (\pm e\sqrt{3}, 0)$; $N_1, N_2 \equiv (0, \pm e)$, but at N_1 and N_2 the curvature is zero.

2. $e < a < e\sqrt{2}$, indented oval. $S_1', S_2' \equiv (\pm\sqrt{(a^2+e^2)}, 0)$; $N_1', N_2' \equiv (0, \pm\sqrt{(a^2-e^2)})$; $E_1', E_2', E_3', E_4' \equiv (\pm(1/2e)\sqrt{(4e^4 - a^4)}, \pm(a^2/2e))$; $W_1', W_2', W_3', W_4' \equiv (\pm\sqrt{\{(v-u)/2\}}, \pm\sqrt{\{(u+v)/2\}})$, where $u = (a^4 - e^4)/(3e^2)$ and $v = \sqrt{\{(a^4 - e^4)/3\}}$.

3. $a < e$, two separated ovals. $S_1'', S_2'' \equiv (\pm\sqrt{(a^2+e^2)}, 0)$; $S_3'', S_4'' \equiv (\pm\sqrt{(e^2 - a^2)}, 0)$; $E_1'', E_2'', E_3'', E_4'' \equiv (\pm(1/2e)\sqrt{(4e^4 - a^4)}, \pm(a^2/2e))$.

4. $a = e$, lemniscate. For the equation of the lemniscate one obtains $(x^2+y^2)^2 - 2a^2(x^2 - y^2) = 0$ or $r^2 = 2a^2 \cos 2\vartheta$, $r = a\sqrt{(2\cos 2\vartheta)}$. Thus, a parametric representation is given by $x = a \cos\vartheta \sqrt{(2\cos 2\vartheta)}$, $y = a \sin\vartheta \sqrt{(2 \cos 2\vartheta)}$.

From $\dfrac{dx}{d\vartheta} = -2a \cdot \sin 3\vartheta/\sqrt{(2\cos 2\vartheta)}$ and $\dfrac{dy}{d\vartheta} = 2a \cdot \cos 3\vartheta/\sqrt{(2 \cos 2\vartheta)}$, it follows that $y' = \dfrac{dy}{dx} = -\cot 3\vartheta$. Thus, extreme values can occur for $3\vartheta = \pi/2, 3\pi/2, 5\pi/2$ or $\vartheta = \pi/6, \pi/2, 5\pi/6$. The extreme values are $x_{1,2} = \pm(a/2)\sqrt{3}$, $y_{1,2} = \pm a/2$, $r_{1,2} = a$. The point (0, 0) is a double point. The values of the partial derivatives at (0, 0) are $f_x = 4(x^3 + xy^2 - a^2x) = 0$; $f_y = 4(x^2y + y^3 + a^2y) = 0$; $f_{xx} = 4(3x^2 + y^2 - a^2) = -4a^2$; $f_{xy} = 8xy = 0$; $f_{yy} = 4(x^2 + 3y^2 + a^2) = 4a^2$; $\Delta = +16a^4$.

From $-a^2 + y'^2a^2 = 0$ it follows that $y' = \pm 1$, so that $y = \pm x$ are the tangents at the point (0, 0). The intersections with the x-axis are $S_1^0, S_2^0 = (\pm a\sqrt{2}, 0)$. The area of one loop is given by

$$A = {}^1/_2 \int r^2(\vartheta)\, d\vartheta = a^2 \int_{-\pi/4}^{+\pi/4} \cos 2\vartheta\, d\vartheta = \frac{a^2}{2} \sin 2\vartheta \Big|_{-\pi/4}^{+\pi/4} = a^2.$$

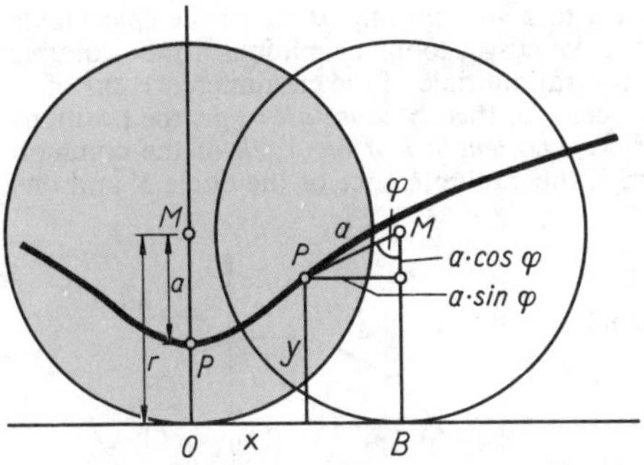

19.5-16 Derivation of the equation of a cycloid

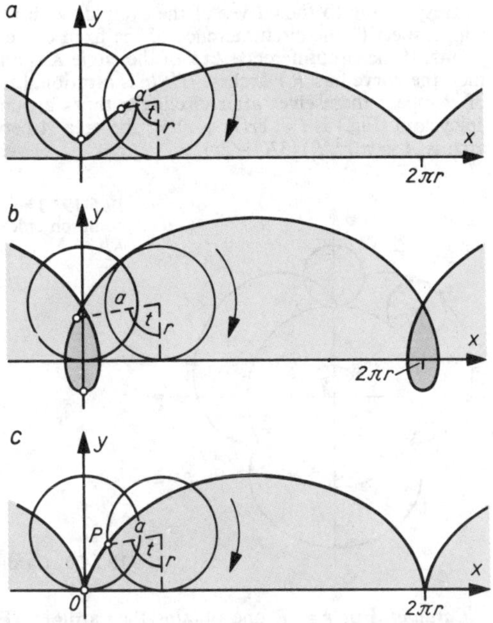

19.5-17 Contracted, extended and common cycloids

Cycloids. A cycloid arises mechanically as the locus of a point P rigidly connected to a circle of radius r at a distance a from its centre M, if this circle rolls on a straight line without slipping. If one lets the circle roll on the x-axis of a Cartesian coordinate system, in which the abscissae are measured from the position in which P is at its lowest point, and denotes the *angle of rotation* by φ, then the unrolled arc $|OB| = r\varphi$ is longer than the x-coordinate of P by an amount $a \sin \varphi$ and r is longer than its ordinate by $a \cos \varphi$ (Fig.): $x = r\varphi - a \sin \varphi$; $y = r - a \cos \varphi$. According to the ratio a/r one distinguishes the *contracted* $(a < r)$, the *extended* $(a > r)$ and the *common cycloid* $(a = r)$ (Fig.).

The *contracted cycloid* has minima for $\varphi = 0, 2\pi, 4\pi, \ldots$ with $y_m = r - a$; the *extended cycloid* has for the corresponding values of x two points with the same abscissae. One is the minimum with $y_m = r - a$. The φ-value of the other is given by the trigonometric equation $r\varphi = a \sin \varphi$. The *common cycloid* has cusps at these places. Its element of arc is $ds = 2r \sin \varphi/2 \, d\varphi$, and thus the length s of the complete *cycloidal arch* is $s = \int_{\varphi=0}^{2\pi} ds = 8r$. The area under this complete arch is $A = \int_{\varphi=0}^{2\pi} y \, dx = r^2 \int_0^{2\pi} (1 - 2 \cos \varphi + \cos^2 \varphi) \, d\varphi = 2\pi r^2 + 0 + \pi r^2 = 3\pi r^2$, and is therefore equal to three times the area of the rolling circle.

Epicycloids. An epicycloid arises mechanically as the locus of a point P, rigidly fixed to a circle k of radius r, if this circle k rolls on the outside of a fixed circle K of radius R (Fig.). According to the distance a of the point P from the centre M of the circle k, one distinguishes the *contracted* $(a < r)$, the *extended* $(a > r)$, and the *common epicycloid* $(a = r)$. If the radius $|OM| = R + r$ turns through an angle φ, then the circle k turns through an angle ψ, where $\varphi R = \psi r$. The perpendicular MB from M to the x-axis cuts off from ψ the angle $(\pi/2 - \varphi)$, and the remaining angle ϑ is $\vartheta = \psi + \varphi - \pi/2 = [(R + r)/r] \cdot \varphi - \pi/2$. The coordinates of the point P are given by

$$x = (R + r) \cos \varphi - a \cos [\varphi \cdot (R + r)/r];$$
$$y = (R + r) \sin \varphi - a \sin [\varphi(R + r)/r].$$

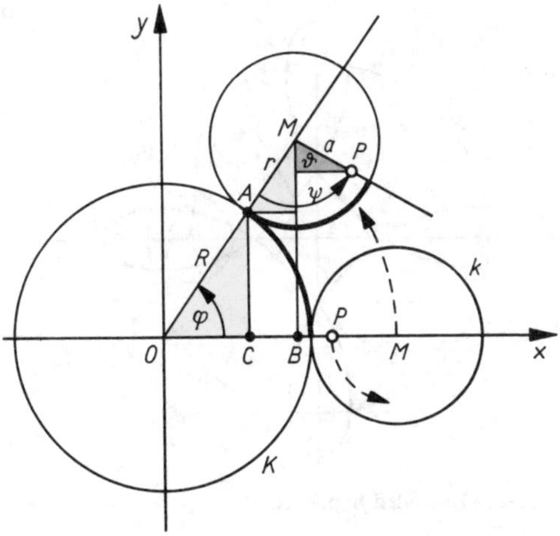

19.5-18 Derivation of the equation of an epicycloid

Corresponding to the curves of the cycloids with respect to a straight line, those of the epicycloids with respect to the circumference of the fixed circle K have cusps, loops or minima without double points. If the circumference $2\pi R$ of the circle K is an integral multiple of the circumference $2\pi r$ of k, then the curve has R/r arches. If R/r is a rational number p/q, then because $qR = pr$, the positions of P repeat themselves after circling q times around K. The *length l of one arch* of the common epicycloid (Fig.) is $l = 8r(R + r)/R$; the area A between the circumference of the circle K and one arch is $A = \pi(r^2/R)(3R + 2r)$.

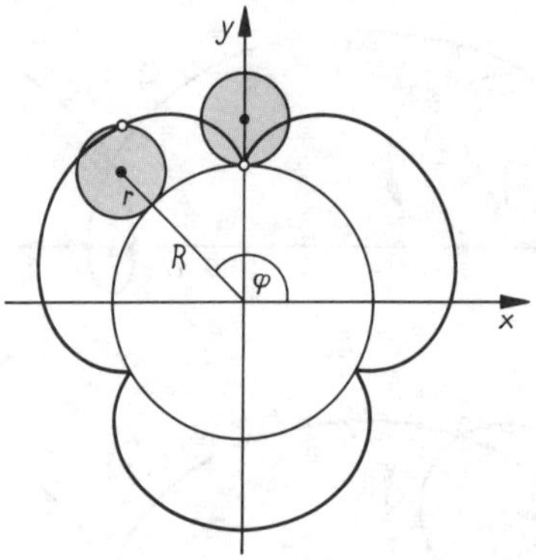

19.5-19 The common epicycloid; $R/r = 3$

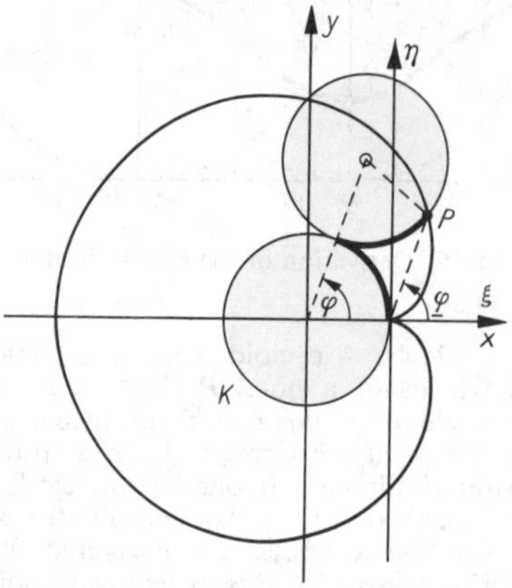

19.5-20 Cardioid

Cardioid. For $r = R$ one obtains the cardioid (Fig.) with the parametric representation $x = R(2\cos\varphi - \cos 2\varphi)$; $y = R(2\sin\varphi - \sin 2\varphi)$. By eliminating φ one obtains the algebraic equation $(x^2 + y^2 - R^2)^2 = 4R^2[(x - R)^2 + y^2]$. The length l of the curve is $l = 8R$, and its area is $A = 6\pi R^2$, that is, six times the area of the fixed circle K.

In the ξ, η-coordinate system the equation of the cardioid becomes particularly simple: $\xi = x - R$, $\eta = y$, so that $\xi = 2R\cos\varphi(1 - \cos\varphi)$, $\eta = 2R\sin\varphi(1 - \cos\varphi)$. Taking polar coordinates r, ϑ, where $r = 2R(1 - \cos\varphi)$ and $\cos\vartheta = \xi/r = \cos\varphi$ and $\sin\vartheta = \eta/r = \sin\varphi$, the polar equation of the cardioid is $r = 2R(1 - \cos\vartheta)$.

Hypocycloids. In contrast to the epicycloid, the hypocycloid arises mechanically when a circle k rolls without slipping on the inside of a fixed circle K. One can think of the moving circle k as being rotated about the tangent. The segments r and a, together with the angle of rotation ψ change their signs. The parametric representation then takes the form:

$$x = (R - r)\cos\varphi + a\cos[\varphi(R - r)/r]; \quad y = (R - r)\sin\varphi - a\sin[\varphi(R - r)/r].$$

In a *contracted hypocycloid* $a < r$, in an *extended* one (Fig.) $a > r$, and in the *common hypocycloid* $a = r$. The corresponding curves have rounded cusps (minima related to the fixed circle), loops or cusps.

The form of the hypocycloid depends on the ratio R/r. If it is an integer, then the curve closes up after a single rotation of the moving circle about the fixed one. If it is not integral, but rational,

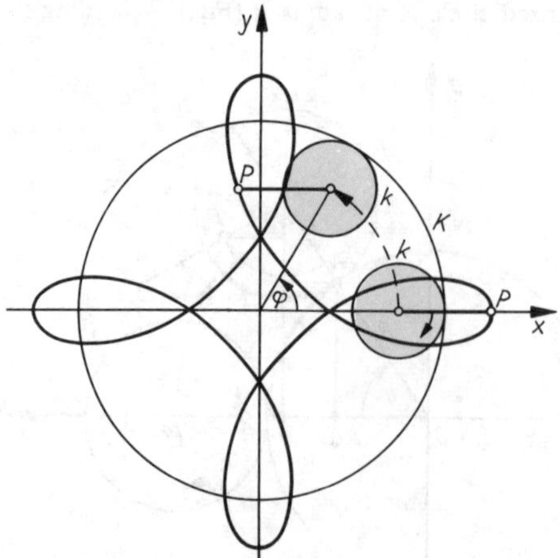

19.5-21 Extended hypocycloid

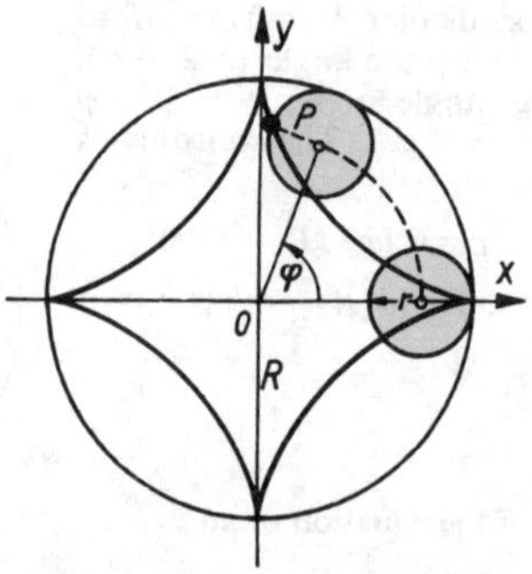

19.5-22 Astroid

$R/r = m/n$, (m and n having no common factor), then the curve closes up only after n circuits. For irrational values of the ratio R/r the curve is not closed. If R/r is an integer, then the common hypocycloid has the length $l = 8(R - r)$, and the area A between one complete arch and the fixed circle K is $A = \pi(r^2/R) \cdot (3R - 2r)$.

The astroid or star curve is a common hypocycloid with $4r = R$. Its parametric representation is therefore $x = 4r \cos^3 \varphi = R \cos^3 \varphi$, $y = 4r \sin^3 \varphi = R \sin^3 \varphi$, because $\sin 3\varphi = 3 \sin \varphi - 4 \sin^3 \varphi$ and $\cos 3\varphi = 4 \cos^3 \varphi - 3 \cos \varphi$. In Cartesian coordinates one obtains the equation $x^{2/3} + y^{2/3} = R^{2/3}$ (Fig.).

Tractrix. A heavy point P at the end of an inextensible thread of length a describes a tractrix if the end-point K of the thread moves along the x-axis (Fig.). Thus, the thread is stretched in the direction of a tangent to the curve, so that $\dfrac{dy}{dx} = \mp y/\sqrt{(a^2 - y^2)}$. Integration gives the equation $x = a \ln |[a \pm \sqrt{(a^2 - y^2)}]/y| \mp \sqrt{(a^2 - y^2)} = \cosh^{-1}(a/y) \mp \sqrt{(a^2 - y^2)}$. The point A is a cusp. The length of arc l measured from A is $l = a \ln (a/y)$.

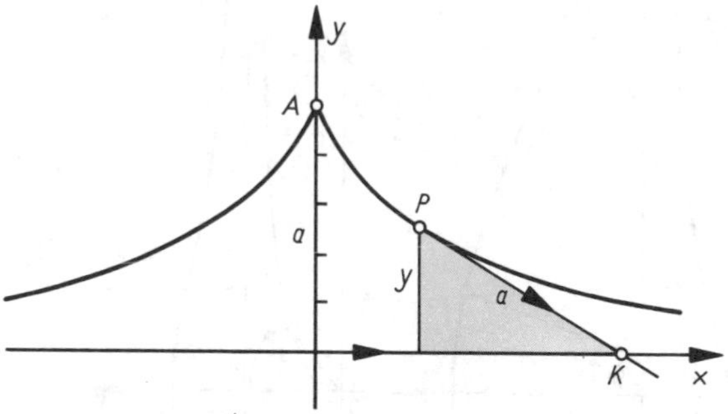

19.5-23 Tractrix, $a = 5$

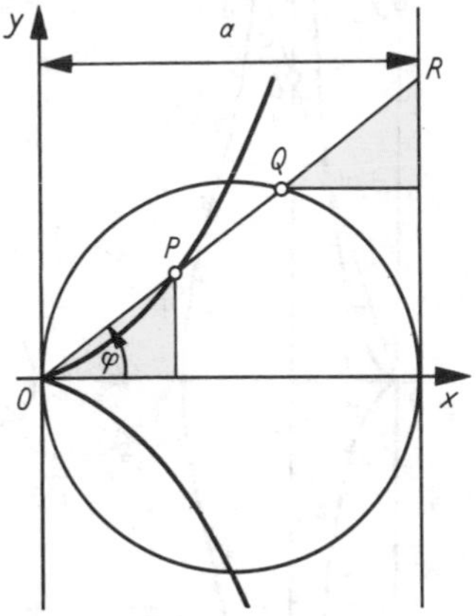

19.5-24 Cissoid

Cissoid. Let a circle touch two parallel lines at a distance a apart. On a secant through one point of contact O the circle and the other parallel line cut off the segment $|QR|$. The point P on the secant is then constructed so that $|OP| = |QR|$, and the cissoid is the geometric locus of all points P constructed on each secant through O in this way (Fig.). The secant makes the angle φ with the x-axis in the Cartesian coordinate system introduced into the figure; $m = \tan \varphi$ serves as a parameter. The point R has the coordinates $x_1 = a$ and $y_1 = am$; from $(x - r)^2 + m^2x^2 = r^2$ the coordinates of Q are obtained as $x_2 = a/(1 + m^2)$, $y_2 = am/(1 + m^2)$. Thus, those of P are given by $x = x_1 - x_2 = am^2/(1 + m^2)$, $y = y_1 - y_2 = am^3/(1 + m^2)$. With the help of $m^2 = x/(a - x)$ and $1 + m^2 = a/(a - x)$ the parameter can be eliminated, giving $y^2 = x^3/(a - x)$. In polar coordinates this gives $r^2 = x^2 + y^2 = a^2m^4/(1 + m^2)$ or $r = a \sin^2 \varphi/\cos \varphi$. The point P is a cusp and the parallel $x = a$ is the asymptote of the cissoid. The area A between the curve and the asymptote is $A = 3\pi a^2/4$, and hence is three times the area of the given circle with the radius $a/2$.

Strophoid. In a Cartesian coordinate system the point $A(-a, 0)$ is the vertex of a pencil of lines. If a line of the pencil cuts the y-axis at the point $B(0, b)$, then the points P and P' on it for which $|BP| = |BP'| = |OB|$ are points of the strophoid (Fig.). From the equation of the line $y = (b/a)\,x + b$ and

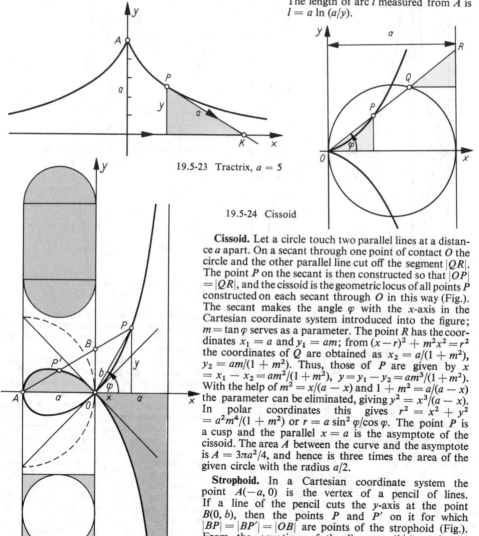

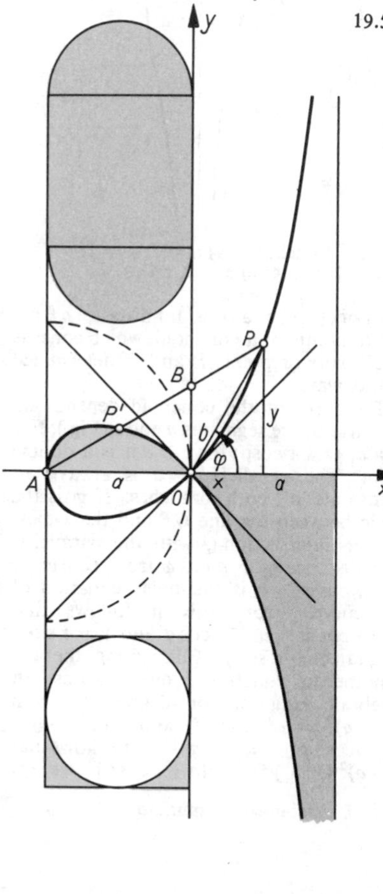

19.5-25 Strophoid: portions of the same colour are of equal area

the distance condition $(y-b)^2+x^2=b^2$, b is eliminated; from $b=ya/(a+x)$ one obtains $y^2/x^2=(a+x)/(a-x)$. It follows that $|AP'|=(a-x)\sqrt{(1+b^2/a^2)}$, $|AP|=(a+x)\sqrt{(1+b^2/a^2)}$, and because $OB \perp AO$, the secant-tangent theorem gives $|AP|\cdot|AP'|=a^2$. The points of the strophoid go into one another under the transformation by reciprocal radii. If φ is the angle between the $+x$-axis and the line OP, $m=\tan\varphi=y/x$ serves as a parameter. The equation of the strophoid gives $m^2=(a+x)/(a-x)$ or $x=a\cdot(m^2-1)/(m^2+1)$, $y=am\cdot(m^2-1)/(m^2+1)$ and the polar equation follows: $r^2=x^2+y^2=a^2\cdot(m^2-1)^2/(m^2+1)$, or $r=-a\cdot\cos 2\varphi/\cos\varphi$. The point (0, 0) is a double point with the tangents $y=+x$. The line $x=a$ is an asymptote. Half the area of the loop is $A'=a^2-\pi a^2/4$, and half the area between the curve and the asymptote is $A=a^2+\pi a^2/4$.

Conchoids of the line. A line passes through the origin O of a Cartesian coordinate system and cuts the line $x=a$ parallel to the y-axis in the point Q. The two points P and P' on OP, on opposite

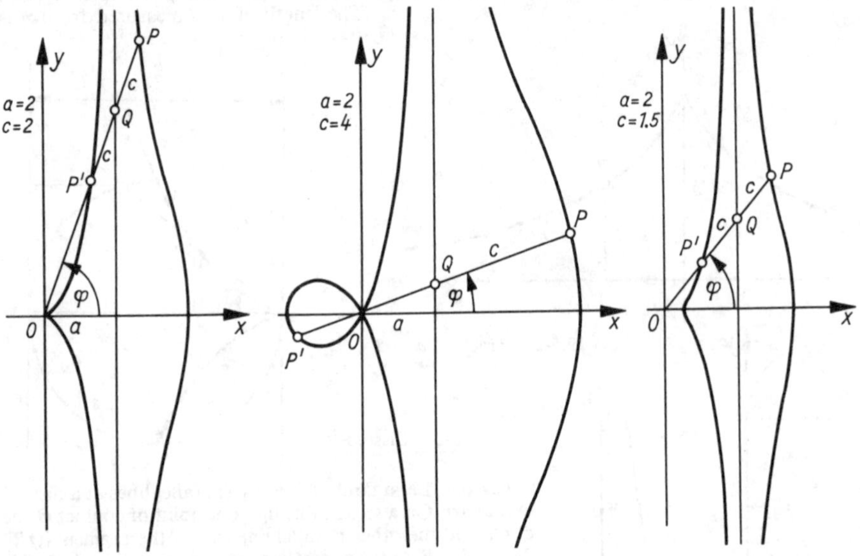

19.5-26 Conchoids: a) $a=2$, $c=2$; b) $a=2$, $c=4$; c) $a=2$, $c=1.5$

sides of Q and at a constant distance c from it, are points of a conchoid, which consists of all pairs of points P and P' determined in this way.

The form of the conchoid depends on the ratio of the segments a and c (Fig.). For $c=a$, O is a cusp; for $c>a$ it is a double point. The parallel $x=a$ is always an asymptote of both branches. If φ is the angle between the line OP and the x-axis, then its intersection Q with the parallel is at a distance $r_Q=a/\cos\varphi$ from O. Hence $r=a/\cos\varphi\pm c$ is the polar equation of the curve. From this it follows that $x=r\cos\varphi=a\pm c\cos\varphi$ and $y=r\sin\varphi=a\tan\varphi\pm c\sin\varphi$. Eliminating the trigonometric functions, one obtains an algebraic equation of degree 4: from $(x-a)^2=c^2\cos^2\varphi$ and $(x^2+y^2)=a^2/\cos^2\varphi\pm 2ac/\cos\varphi+c^2$ it follows that $(x-a)^2(x^2+y^2)=c^2(a\pm c\cos\varphi)^2=c^2x^2$.

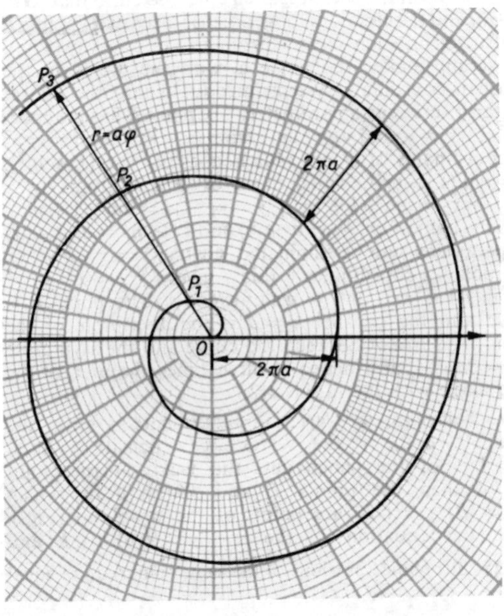

19.5-27 Archimedean spiral, $a=2$

Spirals. The term spirals denotes curves whose radius vector r is a single-valued function of the vectorial angle φ, $r = f(\varphi)$, where φ goes from 0 or from $-\infty$ to $+\infty$ and $r(\varphi)$ can be different from $r(\varphi + 2\pi)$. Examples are $r = a\varphi$ or $r = a\,e^{k\varphi}$.

Archimedean spiral. In polar coordinates this spiral (Fig.) has the equation $r = a\varphi$. Points $P_1, P_2, \ldots$ on the same radius vector are at a constant distance $2\pi a$ apart, because $r_2 = a(\varphi + 2\pi) = a\varphi + 2\pi a = r_1 + 2\pi a$. The element of arc ds has the value $ds = a\sqrt{(1 + \varphi^2)}\,d\varphi$; consequently the arc length is given by

$$s = a \int_0^1 \sqrt{(1 + \varphi^2)}\,d\varphi = (a/2)\left(\varphi_1 \sqrt{(1 + \varphi_1^2)} + \sinh^{-1} \varphi_1\right).$$

For large values of φ_1 the approximation $s \approx (a/2)\,\varphi_1^2$ holds. The area of a sector between two radius vectors $r_1 = a\varphi_1$ and $r_2 = a\varphi_2$ is $A = (a^2/6)(\varphi_2^3 - \varphi_1^3)$.

Logarithmic spiral. In polar coordinates this spiral has the equation $r = a\,e^{k\varphi}$, $k > 0$. For negative values of φ the curve winds with decreasing radius vector about the pole O, ever closer to it: this is an *asymptotic point*. It was shown in the treatment of differentiation of functions in polar coordinates that every line through the pole O cuts the logarithmic spiral at the same angle $\tau_0 = \operatorname{arccot} k$ (Fig.), and that the tangents at these points of intersection are parallel to one another. In addition it was shown that

$$\frac{dr}{d\varphi} = r' = r/\tan\tau = rk \quad \text{or} \quad d\varphi = \frac{dr}{rk}.$$

Using a relation derived in Chapter 20., the arc length s can be calculated:

$$\begin{aligned} ds &= \sqrt{\left(r^2 + \left(\frac{dr}{d\varphi}\right)^2\right)}\,d\varphi \\ &= \sqrt{(r^2 + r^2k^2)}\,d\varphi \\ &= r\sqrt{(1 + k^2)}\,d\varphi \\ &= (1/k)\sqrt{(1 + k^2)}\,dr, \end{aligned}$$

or $s = (1/k)\sqrt{(1 + k^2)}\,(r_2 - r_1)$.

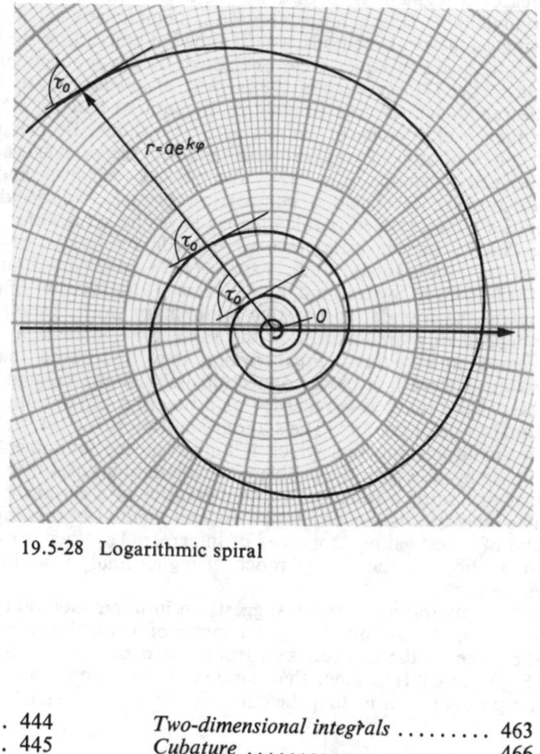

19.5-28 Logarithmic spiral

20. Integral calculus

It is the task of the differential calculus to study the properties of the derivative of a function, which gives a limit representation, for example, for the slope of the tangent at a point on the graph of a function. This value of the derivative at a point depends only on the values of the function in an arbitrarily small neighbourhood, that is, on *local properties* of the function. The problem of defining rigorously and of calculating the area of a region bounded by a closed curve, or the volume of a region bounded by a closed surface, led to the discussion of a limiting process of an entirely different kind, the study of which is the object of the so-called *integral calculus* (*integer*, lat. whole). The limiting process arises when the given plane or solid region is approximated more and more closely by regions whose area or volume can be calculated by elementary methods. While the *tangent problem* refers to a local property, the *area* or *volume problem* requires a knowledge of the bounding curve or surface as a whole.

20.1. The definite integral

Already ARCHIMEDES (287–212 B. C.) succeeded in proving that the area of a segment ASB of a parabola is (4/3) times the area of the triangle ABC (Fig.). Since Greek mathematicians calculated areas by using geometrical methods, which aimed at converting the given region into a square of equal area, one spoke of the *problem of quadrature*, and in the 17th century of the *method of exhaustion*, to indicate that the region whose area or volume is required can be exhausted by a sequence of regions of known area or volume. By means of suitable decompositions of regions into smaller pieces KEPLER (1571–1630) obtained formulae for the volume of barrels, and CAVALIERI (1598–1647) developed a comparison principle to decide when two bodies lying between parallel planes have the same volume (see Chapter 8.). Further investigations were made by DESCARTES (1596–1656), FERMAT (1601–1665), and PASCAL (1623–1662) in France, GULDIN (1577–1643) in Switzerland, and WALLIS (1616–1703) in England. Building on the foundations of this preliminary work, LEIBNIZ (1646–1716) and NEWTON (1643–1727), independently and almost at the same time, created a satisfactory calculus for the computation of areas and volumes. Furthermore, they discovered that despite the different limiting processes involved there is a close connection between the tangent problem and the quadrature problem. If the definite integral is regarded as a function of the upper limit of integration, then its derivative is equal to the integrand. The standard notation for integrals, which is due to LEIBNIZ, corresponds to this important fact, which is known as the fundamental theorem of the calculus.

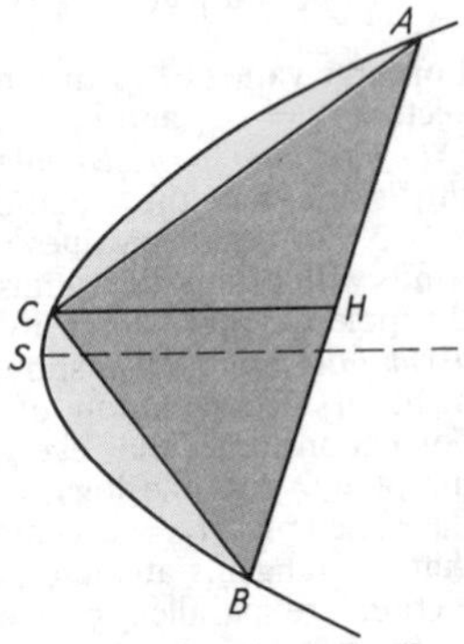

20.1-1 Quadrature of a segment of a parabola

Side by side with the problem of calculating areas then stands the second main problem, namely that of finding a function whose derivative in a certain interval is equal to another given function. The integral calculus develops a general method of treating these apparently very different problems and of investigating properties of integrals. Together the differential and integral calculus form the foundation for the entire branch of higher analysis and are indispensable for modern science and technology.

The quadrature problem suggests an intuitive method of defining the definite integral as the area F_a^b of the region bounded by the curve of a function $y = f(x)$ between the ordinates $f(a)$ and $f(b)$, the piece of the abscissa axis from $x = a$ to $x = b$, and the two straight lines $x = a$ and $x = b$ (Fig.). Here it is assumed that the curve is the graph of a continuous function and that all its ordinates are positive (that is, that the curve lies above the x-axis).

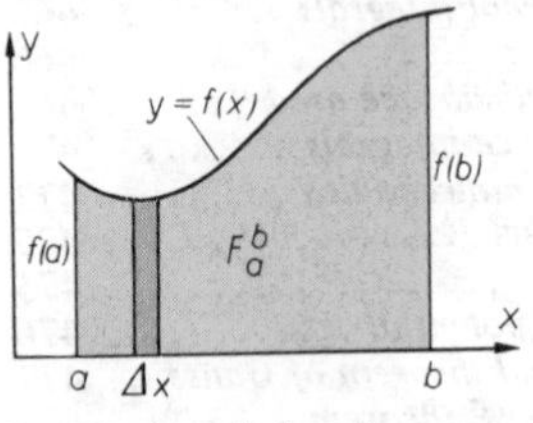

20.1-2 Area below the curve $y = f(x)$ between $x = a$ and $x = b$

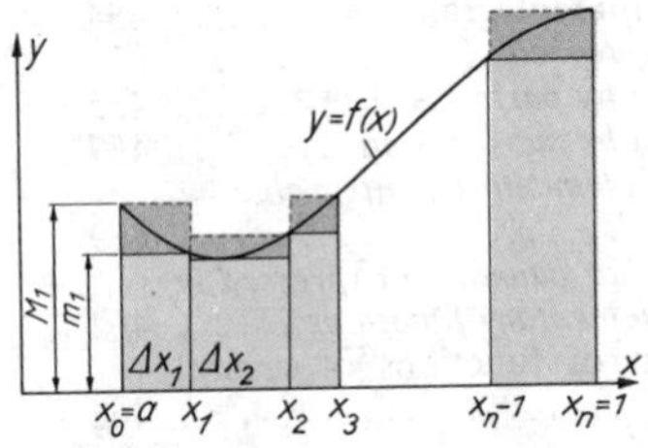

20.1-3 Upper sum and lower sum

The area F_a^b can be approximated from below and above by sequences of suitable figures, for example, by *step polygons*, which are included in, or include, the given region (Fig.). These areas can be calculated elementarily as sums of areas of rectangles. It is intuitively clear that the approximation to F_a^b will improve as the size of the steps gets smaller, so that the sequences formed by the areas of the inscribed and circumscribed step polygons should approach a common limit. The area F_a^b is defined to be this limit, and the limit (assuming that it exists) is denoted by $\int_a^b f(x)\,dx$ (read: integral from a to b of $f(x)$) and called the definite integral.

The definite integral as a limit

Limit of upper and lower sums. One divides the interval from a to b into n subintervals Δx_i, $i = 1, \ldots, n$, for example, into n equal intervals Δx, so that $n\,\Delta x = b - a$; for simplicity both the interval and its length are denoted by Δx. The ordinates at any two adjacent points of division bound a strip of width Δx. If one chooses as heights m_i, the smallest, and M_i, the largest ordinate in the ith interval and compares the area of the strip with the areas $m_i\Delta x$, $M_i\Delta x$ of the corresponding rectangles, then the sum $\sum m_i\Delta x$ is smaller, and the sum $\sum M_i\Delta x$ is larger, than the area required. These are called upper sums and lower sums (Fig.). If one refines the subdivision by splitting up the subintervals, then the new smallest ordinate will be at least as large as the original m_i, and the new largest ordinates not larger than the original M_i. Thus, the lower sum increases and remains less than F_a^b, the upper sum decreases and remains greater than F_a^b. The lower sums form a monotonically increasing bounded sequence, the upper sums a monotonically decreasing bounded sequence, and therefore both have a limit. If these limits are identical, that is, if the sequence of their differences $\sum(M_i - m_i)\,\Delta x$ tends to zero as $\Delta x \to 0$ and $n \to \infty$, then (and only then) does the integral $\int_a^b f(x)\,dx$ exist. This always happens for continuous functions.

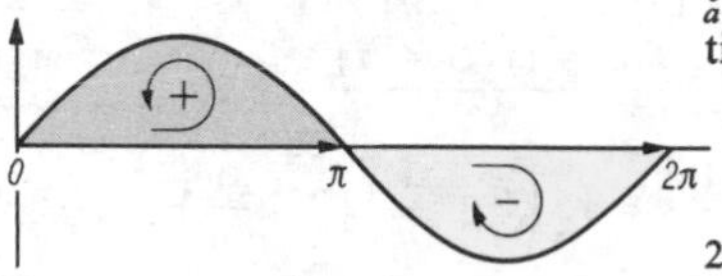

20.1-4 The integral $I = \int_0^{2\pi} \sin\varphi\,d\varphi = 0$ as an area

The original assumption that the curve lies above the x-axis, or that the function has only positive values, is not necessary. For parts of the curve below the x-axis the oriented area takes negative values because in the corresponding parts of the sums the ordinates are negative, while the intervals Δx_i are still positive. The diagram shows by the example of the integral of the sine function from 0 to 2π that the area can have the value zero. Running along the x-axis with increasing x, positive areas lie to the left and negative areas to the right — in accordance with the definition of oriented area in plane geometry.

Analytical definition of the definite integral. The geometrical significance of the definite integral as an area can be dispensed with. Let $f(x)$ be a bounded function in the interval $a \leqslant x \leqslant b$; again one divides this interval by division points $a = x_0 < x_1 < x_2 < \cdots < x_{n-1} < x_n = b$ into n subintervals of lengths $\Delta x_i = x_i - x_{i-1}$ $(i = 1, 2, \ldots, n)$. Since a bounded function need not have a greatest or least value in an interval, one forms upper sums by using G_i, the least upper bound (supremum) of the function in the subinterval, and lower sums by using g_i, the greatest lower bound (infimum). If the difference $\sum(G_i - g_i)\,\Delta x_i$ tends to zero as the length of the largest interval Δx_i tends to zero when $n \to \infty$, then the upper and lower sums both tend to the same limit. Also every sequence $F_n = \sum_{i=1}^{n} f(\xi_i)\,\Delta x_i$, where ξ_i is an arbitrary point in the subinterval Δx_i, then tends to this limit.

If the limit $\lim_{n\to\infty;\,\Delta x_i \to 0} F_n = I$ of the sums $F_n = \sum_{i=1}^{n} f(\xi_i)\,\Delta x_i$ exists and is independent of the choice of division points x_i and intermediate points ξ_i, then the function $f(x)$ is said to be integrable over the interval $[a, b]$ and the limit I is called the definite integral $\int_a^b f(x)\,dx$ of $f(x)$ from $x = a$ to $x = b$.

This concept of integral goes back to Bernhard RIEMANN (1826–1866); the *Lebesgue integral* is a modern extension of this concept. The function $f(x)$ is called the *integrand*, and x is called the *variable of integration*; a and b are the lower and upper *limits of integration*. The value I of the integral of a given integrand f does not depend on the name given to the variable of integration.

Definite integral of $f(x)$ between the limits $x = a$ and $x = b$	$\lim_{n\to\infty} \sum_{i=1}^{n} f(\xi_i)\,\Delta x_i = \int_a^b f(x)\,dx$

If the integrand is continuous in the closed interval $[a, b]$, then the integral certainly exists.

Every continuous function is integrable.

All these arguments remain correct if the function has finitely many discontinuities in the interval $[a, b]$, but is bounded (Fig.).

Every function that is bounded and has only finitely many discontinuities in $[a, b]$ *is integrable.*

More complicated functions, such as the function which for $0 \leqslant x \leqslant 1$ has the value 1 for rational x and 0 for irrational x, are treated in measure theory (see Chapter 35.).

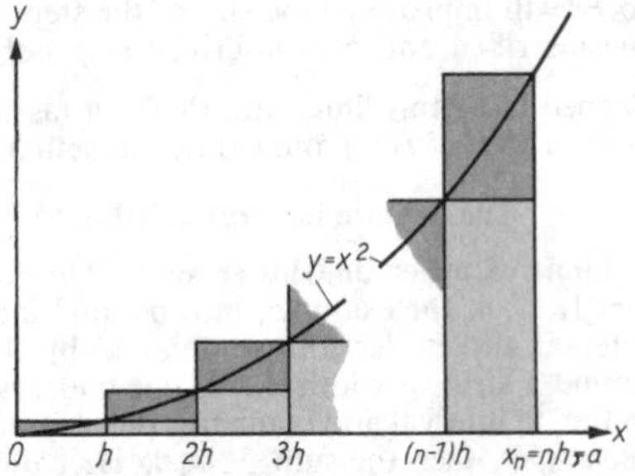

20.1-5 Area under the parabola $y = x^2$

Example: To calculate the area under the parabola $y = x^2$ between $x = 0$ and $x = a$, one divides the interval $[0, a]$ into n equal parts of length $h = a/n$ (Fig.). The n left end-points of sub-intervals are $x_0 = 0, x_1 = h, \dots, x_{n-1} = (n-1)h$, the n right end-points are $x_1 = h, x_2 = 2h, \dots, x_n = nh = a$. Since the function $y = x^2$ is monotonically increasing in $[0, a]$, the lower sum $\underline{F_n}$ and upper sum $\overline{F_n}$ are:

$$\underline{F_n} = 0 + 1^2h^2 \cdot h + 2^2h^2 \cdot h + \cdots + (n-1)^2 h^2 h \qquad \overline{F_n} = 1^2h^2 \cdot h + 2^2h^2 \cdot h + \cdots + n^2h^2 \cdot h$$

$$= h^3 \cdot [1^2 + 2^2 + \cdots + (n-1)^2] \qquad = h^3[1^2 + 2^2 + \cdots + n^2]$$

$$= h^3 \cdot \frac{(n-1)\, n(2n-1)}{6} \qquad = h^3 \frac{n(n+1)(2n+1)}{6}$$

$$\underline{F_n} = \frac{1}{6} a^3 \left(1 - \frac{1}{n}\right) \cdot 1 \cdot \left(2 - \frac{1}{n}\right) \qquad \overline{F_n} = \frac{1}{6} a^3 \cdot 1 \cdot \left(1 + \frac{1}{n}\right)\left(2 + \frac{1}{n}\right)$$

Both sequences have the same limit $F = {}^1/_3 a^3$ as $n \to \infty$,

so that the area under the parabola between $x = 0$ and $x = a$ is ${}^1/_3 a^3$. The adjacent table, for $a = 6$ and $F = 72$, shows how the upper and lower sums vary with n.

n	h	$\underline{F_n}$	$\overline{F_n}$
6	1	55	91
12	1/2	$63^1/_4$	$81^1/_4$
24	1/4	$67^9/_{16}$	$76^9/_{16}$
48	1/8	$69^{49}/_{64}$	$74^{17}/_{64}$
96	1/16	$70^{225}/_{256}$	$73^{33}/_{256}$

Properties of the definite integral

If one interchanges the limits of integration, one alters the direction of integration; the factors Δx_i change sign as oriented segments, and so do all the sums $F_n = \sum f(\xi_i)\,\Delta x_i$ and the integral, because the ordinates $f(\xi_i)$ do not change their sign. To preserve the formula below, for $a = b$ one defines $\int_a^a f(x)\,dx = 0$.

$$\int_a^b f(x)\,dx = -\int_b^a f(x)\,dx$$

Interchange of limits of integration changes the sign of the integral.

Sums of integrals. If $f(x)$ is integrable over the interval from $x = a$ to $x = b$, and if c lies in this interval, then c can be made a division point of each subdivision, and so any sum $\sum f(\xi_i)\,\Delta x_i$ can be split into two sums, one for the interval from a to c and the other from c to b. Since the limit of a sum is equal to the sum of the limits, the definite integral can be written as the sum of two others. By the same theorem about limits:

$$\int_a^c f(x)\,dx + \int_c^b f(x)\,dx = \int_a^b f(x)\,dx \qquad \int_a^b [f(x) + g(x)]\,dx = \int_a^b f(x)\,dx + \int_a^b g(x)\,dx$$

The integral of the sum of two integrable functions $f(x)$ and $g(x)$ is equal to the sum of the integrals of the separate functions.

The splitting of the interval of integration can be used to separate out positive and negative contributions to signed areas by using the zeros of the integrand as splitting points. For instance,

$$I = \int_0^{2\pi} \sin\varphi\,d\varphi = \int_0^{\pi} \sin\varphi\,d\varphi + \int_{\pi}^{2\pi} \sin\varphi\,d\varphi = 2 - 2 = 0.$$

For integrands with finite jump discontinuities it is also necessary to split the range of integration at the point of discontinuity (Fig.). But one must beware of integrating across points where the integrand becomes infinite; for instance, the integral $\int_{-1}^{1} \frac{dx}{x}$ does not make sense because the interval of integration contains $x = 0$, where the integrand is infinite. Such cases will be treated in greater detail in the section on improper integrals.

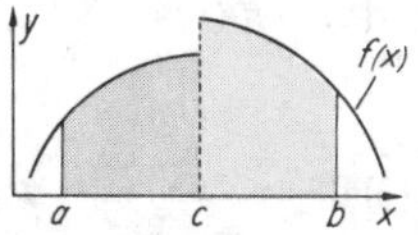

20.1-6 Integral of a discontinuous function

$$\int_a^c f(x)\,dx + \int_c^b f(x)\,dx = \int_a^b f(x)\,dx$$

Integration of a function with a constant factor. If the integrand is a product $cf(x)$, where c is constant, then $\sum cf(\xi_i)\Delta x_i = c\sum f(\xi_i)\Delta x_i$, and so c appears as a factor outside the integral.

Example: $\int_0^3 4x^2\,dx = 4\int_0^3 x^2\,dx = 4 \cdot 27/3 = 36.$

$$\int_a^b cf(x)\,dx = c\int_a^b f(x)\,dx$$

Mean-value theorem of the integral calculus. Every continuous function $f(x)$ attains, in a closed interval $[a, b]$, its maximum M and minimum m. It follows from the definition of the definite integral that

$$m(b-a) \leqslant \int_a^b f(x)\,dx \leqslant M(b-a);$$

the area under the curve lies between the areas of rectangles with base $b - a$ and heights m or M. Therefore there must be a μ between m and M such that the rectangle of base $b - a$ and height μ, hence of area $\mu(b-a)$, has equal area with the region under the curve (Fig.). By the continuity of the function $f(x)$ one can always find a place ξ in $[a, b]$ with $\mu = f(\xi)$; hence it follows that $\int_a^b f(x)\,dx = (b-a)f(\xi)$. As in the mean value theorem of the differential calculus, ξ can also be written as $a + \vartheta(b-a)$ for a suitable ϑ between 0 and 1.

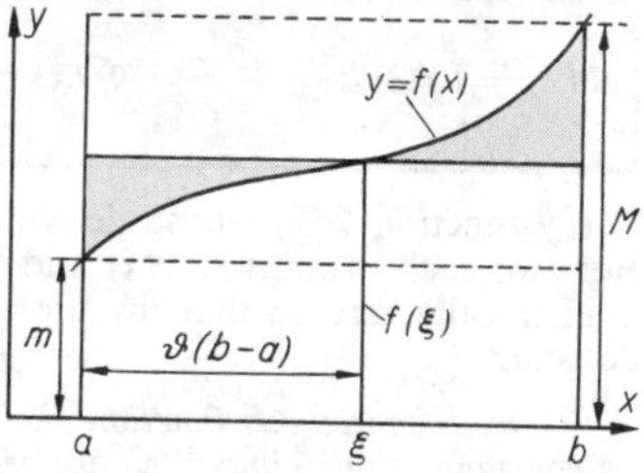

20.1-7 Mean value theorem of the integral calculus

$\int_a^b f(x)\,dx = (b-a)f(\xi)$ with ξ in $[a, b]$	$\int_a^b f(x)\,dx = (b-a)f[a + \vartheta(b-a)],\ 0 \leqslant \vartheta \leqslant 1$

If the function $f(x)$ is continuous in the closed interval $[a, b]$, then the definite integral of this function from $x = a$ to $x = b$ can be expressed as the product of the length of the interval and the value of the function at some intermediate point of the interval.

The mean value theorem is used, for example, in estimating definite integrals of functions that cannot be integrated in elementary terms or whose integral is difficult to find. If $f(x)$, $g(x)$, $h(x)$ are integrable functions on the interval $a \leqslant x \leqslant b$ which satisfy the inequalities $f(x) \leqslant g(x) \leqslant h(x)$, then

$$\int_a^b f(x)\,dx \leqslant \int_a^b g(x)\,dx \leqslant \int_a^b h(x)\,dx.$$

Example: In the interval $0 \leqslant x \leqslant 1/2$ the function e^{-x^2}, which cannot be integrated in elementary terms, satisfies the inequalities $1 - x^2 \leqslant e^{-x^2} \leqslant 1/(1 + x^2)$. It follows that

$$0.458 = [x - x^3/3]_0^{1/2} = \int_0^{1/2} (1 - x^2)\,dx \leqslant \int_0^{1/2} e^{-x^2}\,dx \leqslant \int_0^{1/2} dx/(1 + x^2) = [\arctan \alpha]_0^{1/2} = 0.464.$$

For completeness' sake, the following *generalized mean value theorem* is quoted without proof:

If $f(x)$ and $g(x)$ are continuous in the closed interval $[a, b]$ and if $g(x)$ does not change sign there, then for a suitable ξ in $[a, b]$,

$$\int_a^b f(x)\,g(x)\,dx = f(\xi)\int_a^b g(x)\,dx.$$

Integration as inverse to differentiation. If one of the limits of integration of a definite integral, say the upper limit, is regarded as a variable x, then to each value of x there corresponds the value

$\Phi(x)$ of the integral. The integral is a function of this limit, $\Phi(x) = \int_a^x f(\xi)\,d\xi$. Now on the one hand, $\Phi(x + \Delta x) - \Phi(x) = \int_x^{x+\Delta x} f(\xi)\,d\xi$; on the other hand, by the mean value theorem of the integral calculus,

$$\int_x^{x+\Delta x} f(\xi)\,d\xi = \Delta x \cdot f(x + \vartheta\,\Delta x); \text{ hence } \frac{\Phi(x+\Delta x) - \Phi(x)}{\Delta x} = \frac{1}{\Delta x}\int_x^{x+\Delta x} f(\xi)\,d\xi = f(x + \vartheta\,\Delta x).$$

As $\Delta x \to 0$, the expression $f(x + \vartheta\,\Delta x)$ tends to $f(x)$ provided that $f(x)$ is continuous. It then follows that: $\longrightarrow \lim_{\Delta x \to 0} \dfrac{\Phi(x+\Delta x) - \Phi(x)}{\Delta x} = f(x)$.

This relation represents a fundamental connection between the differential and integral calculus.

If the function $f(x)$ is continuous, then the function $\Phi(x) = \int_a^x f(\xi)\,d\xi$ is differentiable, and its derivative is equal to the value of the integrand at the upper limit of integration

$$\Phi'(x) = \frac{d}{dx}\int_a^x f(\xi)\,d\xi = f(x).$$

$$\Phi(x) = \int_x^a f(\xi)\,d\xi \longrightarrow \Phi'(x) = \frac{d}{dx}\int_a^x f(\xi)\,d\xi = f(x)$$

Every function $\Phi(x)$ whose derivative is equal to the integrand $f(x)$ is called a *primitive* of $f(x)$. For two such primitives $\Phi(x)$ and $\Psi(x)$ of the same integrand $f(x)$ the derivative of $\Psi(x) - \Phi(x)$ is identically zero so that, by the mean value theorem of the differential calculus, $\Psi(x) - \Phi(x)$ is constant.

To each continuous function $f(x)$ there belongs a family of primitives, any two of which differ by a constant. If $\Phi(x)$ and $\Psi(x)$ are primitives of the same function $f(x)$, then $\Phi(x) = \Psi(x) +$ const.

y
x
$\Phi(x)=\frac{1}{12}x^3+C$
with C=3
$\Phi(x)=\frac{1}{12}x^3+C$
with C=0
$f(x)=\frac{1}{4}x^2$
8
7
6
5
4
2
1
-1
-2
-3
-4
-5
-4
-3
-2
-1
1
2
3
4

20.1-8 Integral curves of $\int (x^2/4)\,dx = x^3/12 + C$

The graphs of the functions $\Phi(x)$ therefore form a family of parallel curves (Fig.). The class of all primitives is called the *indefinite integral* of $f(x)$ and is denoted by $\int f(x)\,dx$.

Examples: 1. The indefinite integral of the function $f(x) = 3x^2$ is $\Phi(x) = x^3 +$ const, because x^3 has the derivative $3x^2$.

2. The derivative of $\sin x$ is $\cos x$, so that the integral of $\cos x$ is $\sin x$. Written as an indefinite integral this is $\int \cos x\,dx = \sin x + C$.

3. The function $f(x) = 2x^2$ is the derivative of its indefinite integral $\Phi(x)$. If it is also required that $\Phi(1) = 1$, then the constant C in the indefinite integral $\int 2x^2\,dx = (2/3)\,x^3 + C$ can be determined from $1 = \Phi(1) = 2/3 + C$. Thus, $C = 1/3$ and $\Phi(x) = (2/3)\,x^3 + 1/3$.

4. For a function $\Phi(x)$, its derivative $\Phi'(x) = 3x^4$ and the value $\Phi(5) = 0$ are known. From the indefinite integral $\int 3x^4\,dx = (3/5)\,x^5 + C$ and the condition $\Phi(5) = 0$ one derives $(3/5)\cdot(5^5) + C = 0$, so that $C = -1875$ and therefore $\Phi(x) = (3/5)\,x^5 - 1875$.

The definite integral as ordinate difference. Knowledge of the indefinite integral makes it easier to calculate the definite integral. Let $\Phi(x)$ be any primitive of $f(x)$, and note that $\Psi(x) = \int_a^x f(\xi)\,d\xi$ is also a primitive, so that $\Psi(x) = \Phi(x) + C$. But $\Psi(a) = \int_a^a f(\xi)\,d\xi = 0$, so that $C = -\Phi(a)$, and therefore

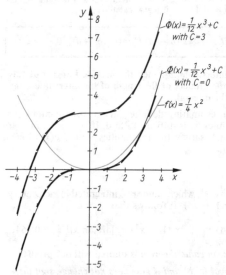

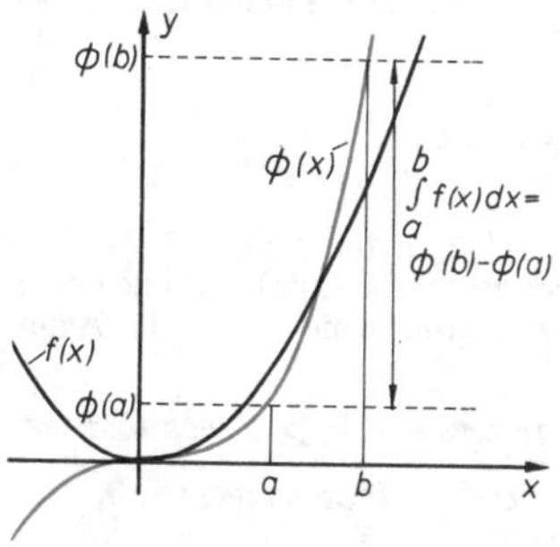

$$\int_a^b f(\xi)\,d\xi = \Psi(b) = \Phi(b) + C = \Phi(b) - \Phi(a) \quad \text{(Fig.)}$$

Examples: 1. $\int_1^2 x^2\,dx = \left[x^3/3\right]_{x=1}^{x=2} = 8/3 - 1/3 = 7/3.$

2. $\int_{-\pi/2}^{+\pi/2} \cos x\,dx = \sin x \Big|_{-\pi/2}^{\pi/2}$
$= \sin(\pi/2) - \sin(-\pi/2) = 1 - (-1) = 2.$

20.1-9 The definite integral as ordinate difference

Improper integral. The integral defined as the limit of sums $\sum f(\xi_i)\,\Delta x_i$ is called a *proper integral,* in contrast to an *improper integral,* in which either the integrand becomes infinite at some point p of the interval of integration or at least one of the limits of integration is infinite. The improper integral is defined as a limit of proper integrals. The interval of integration is first restricted; in the first case one integrates only to $p + \varepsilon$ or $p - \varepsilon$, in the second case only to ω, and then considers the behaviour of the corresponding integrals $I(\varepsilon)$ or $I(\omega)$, as $\varepsilon \to 0$ or $\omega \to \pm\infty$. If $I(\varepsilon)$ or $I(\omega)$ tends to a finite limit, one speaks of a *convergent improper integral,* and if not, the improper integral is said to *diverge.*

Integrand with an infinity. In the following integral the integrand has a pole at $x = 0$ (see Chapter 5.2. – Power functions with negative exponents). Excluding this place, one obtains in the interval from $\varepsilon > 0$ to 1 a proper integral (see Standard integrals) $I(\varepsilon)$ and investigates the limit of $I(\varepsilon)$ as $\varepsilon \to 0$:

$$\int_0^1 \frac{1}{x^\alpha}\,dx = \lim_{\varepsilon\to 0} I(\varepsilon) = \lim_{\varepsilon\to 0} \int_\varepsilon^1 \frac{1}{x^\alpha}\,dx = \lim_{\varepsilon\to 0}\left[\frac{1}{1-\alpha}(1-\varepsilon^{1-\alpha})\right].$$

$$\int_0^1 \frac{1}{x^\alpha}\,dx = \frac{1}{1-\alpha}; \quad 0 < \alpha < 1$$

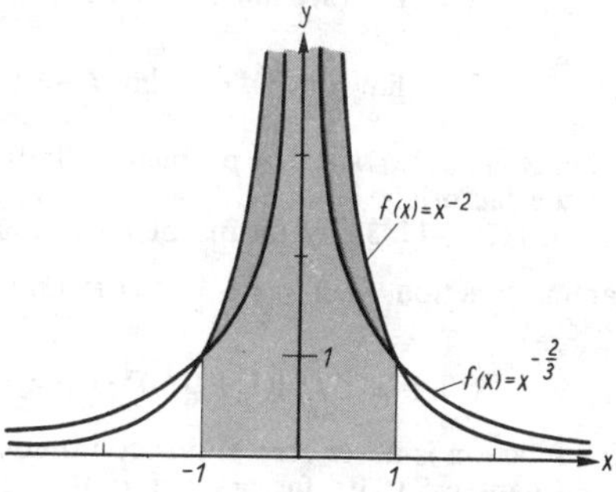

20.1-10 Behaviour of the functions $f(x) = x^{-2/3}$ and $f(x) = x^{-2}$ near $x = 0$

The existence of the limit depends on the positive exponent $\alpha \neq 1$. If $\alpha > 1$, then $\varepsilon^{1-\alpha}$ has a negative exponent and certainly diverges as $\varepsilon \to 0$. On the other hand, if $\alpha < 1$, then $\varepsilon^{1-\alpha} \to 0$ as $\varepsilon \to 0$; the improper integral then converges to $1/(1-\alpha)$. The figure displays the two cases $\alpha = 2/3 < 1$ and $\alpha = 2 > 1$. More generally, the following comparison principle (majorant criterion) holds: Suppose that the function $f(x)$ is integrable in the interval $a \leqslant x \leqslant p - \varepsilon$ for every $\varepsilon > 0$, but is unbounded in the interval $p - \varepsilon \leqslant x < p$. If there exists a number $\alpha < 1$ such that the function $(p - x)^\alpha f(x)$ is bounded for $a \leqslant x < p$, then the improper integral $\int_a^p |f(x)|\,dx$ converges and therefore so does $\int_a^p f(x)\,dx$. If there exist an $\alpha < 1$ and a bound K such that $|f(x)| \leqslant \dfrac{K}{|p-x|^\alpha}$ in the interval $[a, p)$, then one has the estimate

$$\int_a^p |f(x)|\,dx = \lim_{\varepsilon\to 0} \int_a^{p-\varepsilon} |f(x)|\,dx = \lim_{\varepsilon\to 0} \int_a^{p-\varepsilon} \frac{K}{|p-x|^\alpha}\,dx = \frac{K}{1-\alpha}(p-a)^{1-\alpha}.$$

Example: By Standard integrals one has

$$\int_0^1 \frac{dx}{\sqrt{(1-x^2)}} = \lim_{\varepsilon\to 0} \int_0^{1-\varepsilon} \frac{dx}{\sqrt{(1-x^2)}} = \lim_{\varepsilon\to 0} \arcsin(1-\varepsilon) = \frac{\pi}{2}.$$

Infinite interval of integration. The contrast to an improper integral in which the integrand becomes infinite can again be clearly illustrated by the example $f(x) = 1/x^\alpha$.

$$\int_1^\infty \frac{1}{x^\alpha}\,dx = \lim_{\omega\to\infty} I(\omega) = \lim_{\omega\to\infty} \int_1^\omega \frac{dx}{x^\alpha} = \lim_{\omega\to\infty} \left[\frac{1}{1-\alpha}\,(\omega^{1-\alpha} - 1)\right].$$

If $\alpha \neq 1$ is positive, then the behaviour of $I(\omega)$ depends on that of $\omega^{1-\alpha}$; if the exponent $1 - \alpha$ is positive, that is, $\alpha < 1$, then $I(\omega) \to \infty$ as $\omega \to \infty$ and the integral diverges; if the exponent is negative, that is, $\alpha > 1$, then $\omega^{1-\alpha} \to 0$ as $\omega \to \infty$ and the integral converges to the value $1/(\alpha - 1)$. Again there is a comparison principle (majorant criterion):

If $f(x)$ is integrable on any finite sub-interval of $x \geq a$, and if there exists an $\alpha > 1$ such that the function $x^\alpha f(x)$ in bounded for all $x > a$, then the improper integral $\int_a^\infty f(x)\,dx$ is convergent.

If K is a bound for the function $x^\alpha|f(x)|$, and $\alpha > 1$, then one has the estimate $\int_a^\infty |f(x)|\,dx = \lim_{\omega\to\infty} \int_a^\omega |f(x)|\,dx \leq \lim_{\omega\to\infty} \int_a^\omega \frac{K}{x^\alpha}\,dx = \frac{Ka^{1-\alpha}}{\alpha - 1}.$

Examples: 1. In the following improper integral the integrand vanishes to the second order as $x \to \infty$ (Fig.)

$$\int_0^\infty \frac{dx}{1+x^2} = \lim_{\omega\to\infty} \int_0^\omega \frac{dx}{1+x^2} = \lim_{\omega\to\infty} (\arctan \omega - \arctan 0) = \pi/2.$$

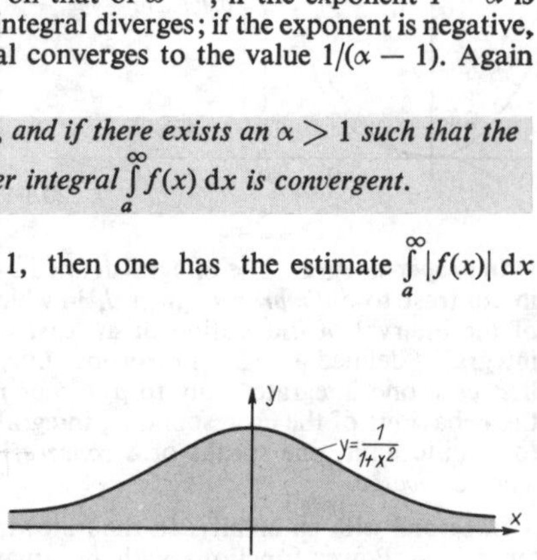

20.1-11 Area under the graph of the function $f(x) = 1/(1 + x^2)$

2. $\int_0^\infty e^{-x}\,dx = \lim_{\omega\to\infty} \int_0^\omega e^{-x}\,dx = \lim_{\omega\to\infty} (-e^{-\omega} + 1) = 1.$

The gamma function. The problem of finding a function whose values for positive integral arguments are the factorials $1! = 1$, $2! = 1 \cdot 2 = 2$, $3! = 1 \cdot 2 \cdot 3 = 6$, ..., $n! = 1 \cdot 2 \cdot \cdots \cdot n$ was solved by EULER (1707–1783) by means of an improper integral. LEGENDRE (1752–1833) called this Euler's gamma function $\Gamma(x) = \int_0^\infty e^{-t} t^{x-1}\,dt$; GAUSS (1777–1855) gave a definition of $\Gamma(x)$ as an infinite product

$$\Gamma(x) = x^{-1} \prod_{n=1}^\infty [(1 + 1/n)^x (1 + x/n)^{-1}].$$

The function is never zero and is continuous except at $x = 0, -1, -2, \ldots$ where it has simple poles (see Chapter 5.). Its factors are obtained by substituting $n = 1, 2, 3, \ldots$ in the square brackets. From the functional equation $\Gamma(x + 1) = x\Gamma(x)$ and the value $\Gamma(1) = 1$ it follows that $\Gamma(n + 1) = n\Gamma(n) = n!$ for integral arguments $n = 1, 2, 3, \ldots$

Euler's gamma function	$\Gamma(x) = \int_0^\infty e^{-t} t^{x-1}\,dt, \quad x > 0$
Gauss's definition of $\Gamma(x)$	$\Gamma(x) = \lim_{n\to\infty} \frac{n!\,n^{x-1}}{x(x+1)(x+2)\cdots(x+n-1)}$, $x \neq 0, -1, -2, \ldots$ $\Gamma(x) = x^{-1} \prod_{n=1}^\infty [(1 + 1/n)^x \cdot (1 + x/n)^{-1}]$
Functional equations for $\Gamma(x)$	$\Gamma(x + 1) = x\Gamma(x)$; $\Gamma(x)\,\Gamma(1 - x) = \pi/\sin(\pi x)$; $\Gamma(1/2 + x)\,\Gamma(1/2 - x) = \pi/\cos(\pi x)$; $\Gamma(x)\,\Gamma(-x) = -\pi/[x \sin(\pi x)]$

Quadrature

By quadrature one understands the calculation of areas of plane regions with curved boundaries.

Area under a curve. The area F above the x-axis and below the curve with the equation $y = f(x)$, between $x = a$ and $x = b$, is found from the integral $F = \int_a^b f(x)\,dx$ if all the values $f(x)$ are positive in the interval $a \leqslant x \leqslant b$. If $f(x)$ changes sign in $[a, b]$, one may split this integral into pieces in which $f(x)$ takes positive and negative values, and the integral into a sum of positive and negative contributions corresponding to oriented areas. If one ignores orientation, the total area is the sum of the absolute values of these contributions.

Example 1: Quadrature of Neil's parabola $y = ax^{3/2}$ (Fig.). $F = a \int_0^g x^{3/2}\,dx = 2ag^{5/2}/5$. If $h = a \cdot g^{3/2}$, then $F = 2gh/5$, so that the area F is by $gh/10$ less than the area of the right-angled triangle with base g and height h.

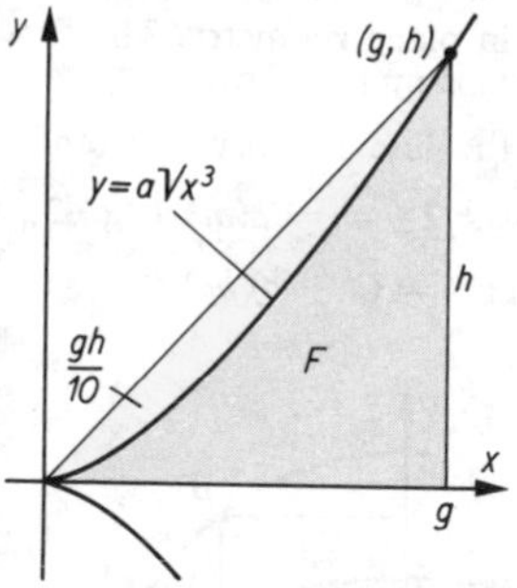

20.1-12 Area under the positive branch of Neil's parabola

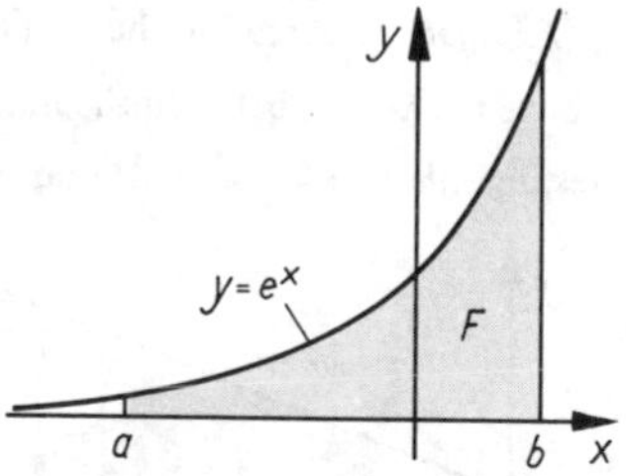

20.1-13 Area under the exponential curve $y = e^x$

Example 2: Quadrature of the *exponential curve* $y = e^x$ from $x = a$ to $x = b$. If a is finite, $\int_a^b e^x\,dx = e^b - e^a$. As $a \to -\infty$, F tends to the finite limit $F = e^b$; the improper integral $\int_{-\infty}^b e^x\,dx$ converges and (Fig.) the region extending to infinity has finite area.

Example 3: Quadrature of the *rectangular hyperbola* $y = k^2/x$ from $x = a$ to $x = b$; $k^2 \int_a^b \frac{dx}{x} = k^2(\ln b - \ln a)$; $F = k^2 \ln(b/a)$ (Fig.). Here one has a 'rational' curve for which the calculation of area involves a transcendental function.

Example 4: Quadrature of the *sine curve* $y = \sin x$ from $x = 0$ to $x = \pi$; $\int_0^\pi \sin x\,dx = \cos 0 - \cos \pi$; $F = 2$. Here one has a 'transcendental' curve with rational area.

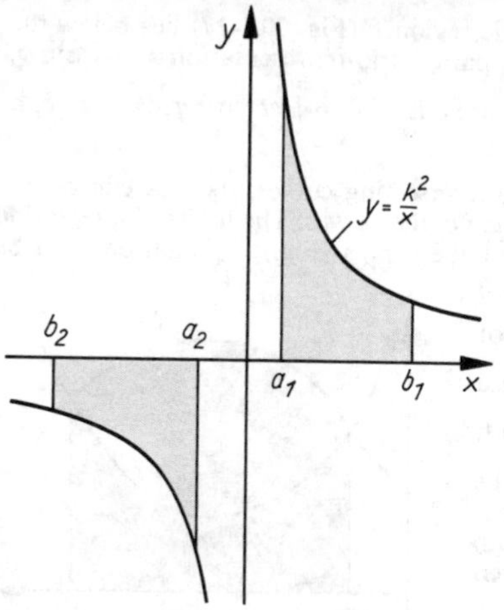

20.1-14 Area under the hyperbola $y = k^2/x$

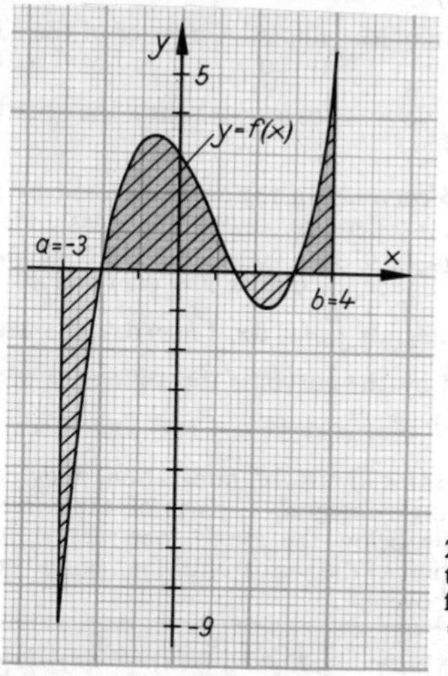

20.1-15 Area under the curve of the function $y = f(x) = x^3/3 - 5x^2/6 - 3x/2 + 3$

Example 5: To calculate the area of the region bounded by the curve $y = f(x) = x^3/3 - 5x^2/6 - 3x/2 + 3$, the x-axis, and the lines $x = -3$ and $x = 4$ (Fig.). In the interval $-3 \leqslant x \leqslant 4$ the function has zeros at $x_1 = -2$, $x_2 = 3/2$, $x_3 = 3$. If one ignores orientation, the area is the

sum of the absolute values of the integrals over the subintervals between zeros. Since
$\int (x^3/3 - 5x^2/6 - 3x/2 + 3)\,dx = F(x) = x^4/12 - 5x^3/18 - 3x^2/4 + 3x,$
this leads to the following evaluation of the area:

$$\begin{aligned} F &= \left|\int_{-3}^{-2} f(x)\,dx\right| + \left|\int_{-2}^{3/2} f(x)\,dx\right| + \left|\int_{3/2}^{3} f(x)\,dx\right| + \left|\int_{3}^{4} f(x)\,dx\right| \\ &= |[F(x)]_{-3}^{-2}| + |[F(x)]_{-2}^{3/2}| + |[F(x)]_{3/2}^{3}| + |[F(x)]_{3}^{4}| \\ &= |-5.444 + 1.5| + |2.297 + 5.444| + |1.5 - 2.297| + |3.556 - 1.5| \\ &= |-3.944| + |7.741| + |-0.797| + |2.056| = 14.358. \end{aligned}$$

The integral calculus may be used to derive the area formulae known in plane geometry. The formulae for the trapezium, the circle, and the ellipse will be developed as examples below.

Example 1: Area of the trapezium: To calculate the area below the line $y = mx + a$, and above the x-axis, between the limits $x = 0$ and $x = h$: $\int_0^h (mx + a)\,dx = mh^2/2 + ah = h(mh + 2a)/2$; setting $mh + a = b$ (Fig.), one obtains the familiar trapezium formulae $A = (a + b)\,h/2$.

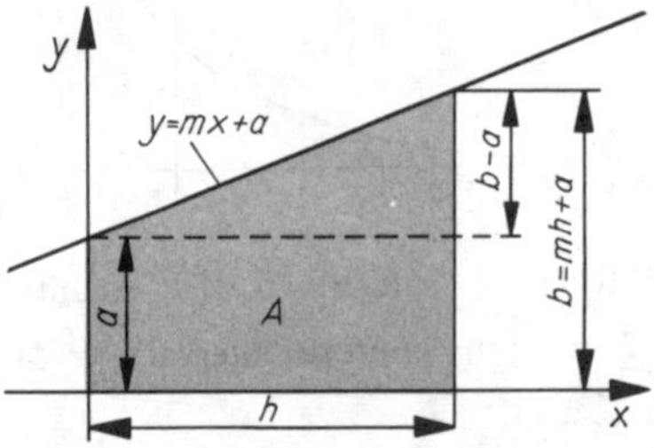

20.1-16 The area formula for a trapezium

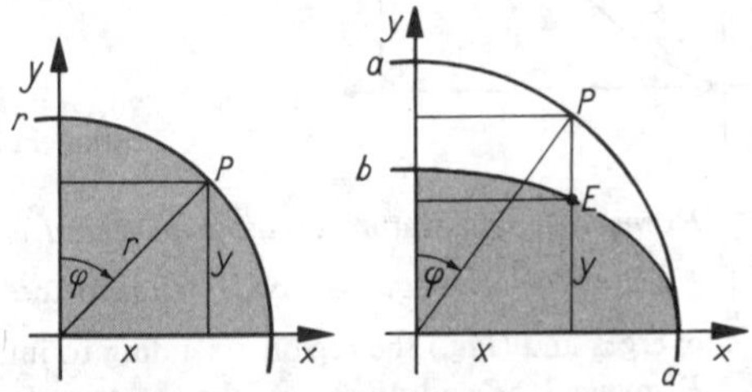

20.1-17 The area formulae for circle and ellipse

Example 2: Area of the circle. A quarter of the circular region A (Fig.) lies below the arc $y = \sqrt{(r^2 - x^2)}$ between the limits $x = 0$ and $x = r$. Substituting $x = r \sin\varphi$, $dx = r\cos\varphi\,d\varphi$ one obtains $A/4 = \int_0^r \sqrt{(r^2 - x^2)}\,dx = \int_0^{\pi/2} r^2 \cos^2\varphi\,d\varphi = (r^2/2)\,[\sin\varphi\cos\varphi + \varphi]_0^{\pi/2} = (r^2/2)\cdot(\pi/2) = \pi r^2/4$, or $A = \pi r^2$.

Example 3: Area of the ellipse. A quarter of the elliptic region F (Fig. 20.1-17) lies below the arc $y = (b/a)\sqrt{(a^2 - x^2)}$ between $x = 0$ and $x = a$. The parametric representation $x = a\sin\varphi$, $y = b\cos\varphi$, gives $dx = a\cos\varphi\,d\varphi$; hence the quarter area is $F/4 = \int_0^{\pi/2} ab\cos^2\varphi\,d\varphi = \pi ab/4$, that is, $F = \pi ab$.

Area between two curves. If a region is enclosed by two intersecting curves, its area can be calculated as the absolute difference between the areas under each of the curves. The limits of integration are the abscissae x_1, x_2 of two consecutive intersections of the curves (Fig.); orientation is to be ignored. The areas under the curves $y = g(x)$ and $y = h(x)$ are given by the integrals $\int_{x_1}^{x_2} g(x)\,dx$, $\int_{x_1}^{x_2} h(x)\,dx$; the area of the region enclosed by the curves is the difference of these integrals, that is, $F = \left|\int_{x_1}^{x_2} g(x)\,dx - \int_{x_1}^{x_2} h(x)\,dx\right|$. Since both integrals have the same limits of integration, they can be combined into a single integral. If parts of the curves between x_1 and x_2 lie below the x-axis, one may shift both curves in the direction of the y-axis until the region between them lies entirely above

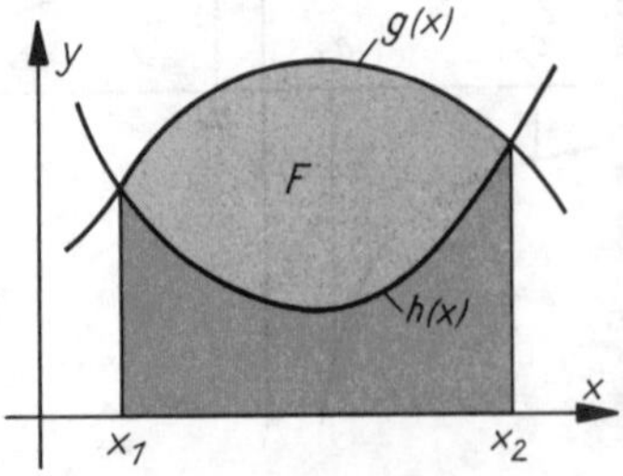

20.1-18 Area between two curves

Area between two curves	$F = \left\vert\int_{x_1}^{x_2} [g(x) - h(x)]\,dx\right\vert$

the x-axis. This changes both functions by the same additive constant, which cancels out after subtraction.

Example 1: The curves determined by $g(x) = 3\sqrt{x}$ and $f(x) = x^2 - 4x + 6$ (Fig.) intersect at the points (1, 3) and (4, 6). Now $g(x) - h(x) = (3\sqrt{x} - x^2 + 4x - 6)$, so that $F = \int_1^4 (3\sqrt{x} - x^2 + 4x - 6)\,dx = [2x\sqrt{x} - x^3/3 + 2x^2 - 6x]_1^4 = 5$.

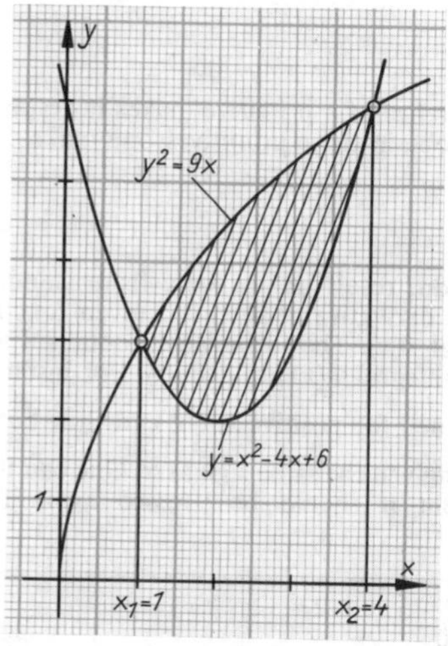

20.1-19 Area between the curves of the functions $y^2 = 9x$ and $y = x^2 - 4x + 6$

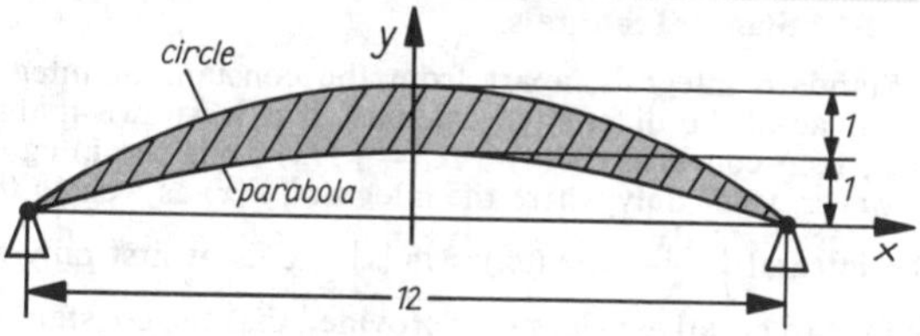

20.1-20 Cross cut trough a steel girder

Example 2: Fig. 20.1-20 represents a cross-cut through a steel girder. The upper boundary is an arc of a circle, the lower that of a parabola. This makes it possible to calculate the mass of the girder when the thickness d of the steel sheet and the density ϱ of the material are known. If A is the area of the cross-cut, then its mass is $M = A \cdot d \cdot \varrho$. The cross-cut area A can be calculated by means of integrals. For a suitable choice of the coordinate system one has the general equations $x^2 + (y - b)^2 = r^2$ for the circle, and $y = ax^2 + c$ for the parabola. The given data lead to the special equations $x^2 + (y + 8)^2 = 100$ for the circle and $y = -x^2/36 + 1$ for the parabola.

Substituting $g(x) - h(x) = \sqrt{(100 - x^2)} - 8 + x^2/36 - 1$ one obtains for the required area

$$A = 2\int_0^6 (\sqrt{(100 - x^2)} + x^2/36 - 9)\,dx$$
$$= 2[{}^1/_2(100 \arcsin(x/10) + x\sqrt{(100 - x^2)} - x^3/108 - 9x]_0^6$$
$$= 2[{}^1/_2(100 \cdot 0.6435 + 48) - 52] = 8.36.$$

Graphical integration. Just as one can determine graphically the derivative of a given curve, so one can construct, conversely, from the given graph of a derivative an appropriate integral curve. One begins by selecting by means of an initial condition from the family of integral curves one, say the curve passing through a point P_0 with the coordinates $x_0 = 1$, $y_0 = 0$. Every ordinate of the derivative $f(x)$ represents the slope of the integral curve to be constructed at the point in question. Therefore, if one drops a perpendicular from $f(x_0) = f(1)$ to the y-axis and joins the foot B_0 of the perpendicular to the point $A = (-1, 0)$, then the line through A and B_0, on account of $\tan\alpha_0 = |B_0O|/1 = f(1)$, gives the direction of the integral curve at the point P_0.

A parallel to this direction through P_0 is tangent to the integral curve and represents it approximately in a small neighbourhood of P_0. Since no further points of the integral curve are known, one chooses a point x_1, halves the interval from x_0 to x_1 by a parallel to the y-axis, and shifts the tangential direction for the point P_1 to be constructed so that it intersects the tangent at P_0 on the mid-line of the interval. Continuing in this manner one obtains a polygonal arc, which represents the integral curve approximately (Fig.). The drawing of an integral curve for a given derivative can also be done mechanically by means of special tools.

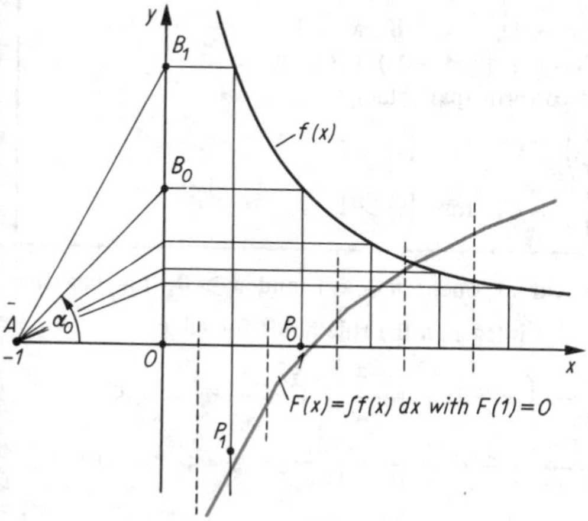

20.1-21 Graphical integration

20.2. The indefinite integral

The expression of the definite integral as an ordinate difference of a primitive makes it desirable to obtain *primitives* for as many functions as possible. To obtain these primitives one makes use of the fact that integration is the inverse of differentiation.

Standard integrals

Standard integrals, apart from the constant of integration, result immediately from standard formulae of the differential calculus. If it is known that the derivative of $\Phi(x)$ is a given function $f(x)$, then conversely $\Phi(x) + C = \int f(x)\,dx$ is the integral of $f(x)$. The formulae so obtained are, however, valid only where the integrand $f(x)$ as well as the integral $\Phi(x)$ are defined. For instance, the integral $\int \frac{dx}{x} = \ln(cx) = \ln|x| + C$ is at first only defined for positive x, but can then be extended to negative values of x provided that the constant c is given the same sign as x. One can then write $\ln(cx) = \ln(|c|\,|x|) = \ln|x| + \ln|c| = \ln|x| + C$. But it should be stressed again that $x = 0$ must never belong to the interval of integration.

Table of standard integrals

$$\int dx = x + C \qquad \int x^n\,dx = \frac{x^{n+1}}{n+1} + C; \quad n \neq -1$$

$$\int e^x\,dx = e^x + C \qquad \int \frac{dx}{x} = \ln(cx) = \ln|x| + C; \quad x \neq 0$$

$$\int a^x\,dx = \frac{a^x}{\ln a} + C = a^x \log_a e + C; \quad 0 < a \neq 1$$

$$\int \cos x\,dx = \sin x + C \qquad \int \sec^2 x\,dx = \int \frac{dx}{\cos^2 x} = \tan x + C; \quad x \neq \frac{(2k+1)\pi}{2}, \; k \in \mathbf{Z}$$

$$\int \sin x\,dx = -\cos x + C \qquad \int \operatorname{cosec}^2 x\,dx = \int \frac{dx}{\sin^2 x} = -\cot x + C; \quad x \neq k\pi, \; k \in \mathbf{Z}$$

$$\int \cosh x\,dx = \sinh x + C \qquad \int \frac{dx}{\cosh^2 x} = \tanh x + C$$

$$\int \sinh x\,dx = \cosh x + C \qquad \int \frac{dx}{\sinh^2 x} = -\coth x + C; \quad x \neq 0$$

$$\int \frac{dx}{\sqrt{(1-x^2)}} = \operatorname{Arcsin} x + C = -\operatorname{Arccos} x + C'; \quad |x| < 1$$

$$\int \frac{dx}{1+x^2} = \arctan x + C = -\operatorname{arccot} x + C'$$

$$\int \frac{dx}{\sqrt{(1+x^2)}} = \sinh^{-1} x + C = \ln|x + \sqrt{(1+x^2)}| + C'$$

$$\int \frac{dx}{\sqrt{(x^2-1)}} = \begin{cases} \cosh^{-1} x + C = \ln(x + \sqrt{(x^2-1)}) + C' & \text{if } x > 1 \\ -\cosh^{-1}(-x) + C = -\ln(-x + \sqrt{(x^2-1)}) + C' & \text{if } x < -1 \end{cases}$$

with $0 \leq \cosh^{-1} x < \infty$ (principal value)

$$\int \frac{dx}{1-x^2} = \begin{cases} \tanh^{-1} x + C = \ln \sqrt{\left(\frac{1+x}{1-x}\right)} + C' & \text{for } |x| < 1 \\ \coth^{-1} x + C_1 = \ln \sqrt{\left(\frac{x+1}{x-1}\right)} + C_1' & \text{for } |x| > 1 \end{cases}$$

The rule $\int x^n\,dx = \frac{x^{n+1}}{n+1} + C$ holds for any real exponent $n \neq -1$ and $x > 0$; for negative integers it is enough to have $x \neq 0$, and for positive integers n the rule holds for all x.

Examples: *1.* $\int x^3\,dx = \frac{x^4}{4} + C$. 2. $\int \frac{dx}{x^3} = \int x^{-3}\,dx = \frac{x^{-2}}{-2} + C = -\frac{1}{2x^2} + C$.

3. $\int \frac{dx}{x^r} = \int x^{-r} dx = \frac{x^{-r+1}}{-r+1} + C = \frac{x^{-(r-1)}}{-(r-1)} + C = \frac{-1}{(r-1)\,x^{(r-1)}} + C; \; (r \neq 1)$.

4. $\int \sqrt[3]{x}\, dx = \int x^{1/3}\, dx = \frac{x^{4/3}}{4/3} + C = 3/4 \sqrt[3]{x^4} + C = 3/4x\sqrt[3]{x} + C.$

5. $\int \sqrt[m]{x}\, dx = \int x^{1/m}\, dx = \frac{x^{1/m+1}}{1/m+1} + C = \frac{x^{(1+m)/m}}{(1+m)/m} = \frac{m\sqrt[m]{x^{m+1}}}{m+1} + C = \frac{m\,x\sqrt[m]{x}}{m+1} + C.$

6. $\int \frac{dx}{\sqrt[3]{x^2}} = \int x^{-2/3}\, dx = \frac{x^{1/3}}{1/3} + C = 3\sqrt[3]{x} + C.$

7. $\int (5x^3 + 4x^2 - 3x + 2)\, dx = 5\int x^3\, dx + 4\int x^2\, dx - 3\int x\, dx + 2\int dx$
$= 5x^4/4 + 4x^3/3 - 3x^2/2 + 2x + C.$

8. $\int [ax^2 + 1/x - b/x^2 + 1/(1+x^2)]\, dx = ax^3/3 + \ln|x| + b/x + \arctan x + C.$

9. $\int \frac{x^3 + 2x^2 - x + 3}{x}\, dx = \int (x^2 + 2x - 1 + 3/x)\, dx = x^3/3 + x^2 - x + 3\ln|x| + C.$

10. $\int_{-\pi/4}^{+\pi/4} (\cos x - \sin x + 1/\cos^2 x)\, dx = [\sin x + \cos x + \tan x]_{-\pi/4}^{+\pi/4}$
$= (^1/_2\sqrt{2} + {^1/_2}\sqrt{2} + 1) - (-^1/_2\sqrt{2} + {^1/_2}\sqrt{2} - 1) = 2 + \sqrt{2}.$

Integration by parts

Sometimes a difficult integral can be found more easily by 'partial integration'. It is assumed that the integrand can be written as the product of two functions of which one is easy to integrate. The rule for integration by parts is a consequence of the rule for differentiating a product:

$$\frac{d(uv)}{dx} = v\frac{du}{dx} + u\frac{dv}{dx}.$$

Integration on both sides leads to $\int \frac{d(uv)}{dx}\, dx = uv = \int v\frac{du}{dx}\, dx + \int u\frac{dv}{dx}\, dx$, and so to the rule for integration by parts. 'By parts' indicates that one writes the integrand as a product uv' of two parts u and v', where the integral of the part v' is known and the new integral with the integrand $u'v$ is easier to find.

Integration by parts	$\int u\, dv = uv - \int v\, du$	$\int uv'\, dx = uv - \int u'v\, dx$

Example: To calculate the integral $\int x\, e^x\, dx$, one puts $u = x$ and $v' = e^x$, so that $u' = 1$ and $v = e^x$. The rule for the integration by parts yields

$$\int x\, e^x\, dx = x\, e^x - \int e^x\, dx = x\, e^x - e^x + C = (x-1)\, e^x + C.$$

Recurrence formulae. The integral $\int x^6 e^x\, dx$ cannot be found from a single integration by parts, but in this case and similar cases one can often make several successive integrations by parts which lead step-by-step to simplifications of the integral, until one reaches one of the standard integrals. Then one has found a recurrence formula.

Since e^x is equal to its derivative, the factorization $u = x^n$, $v' = e^x$ for the integrand $x^n e^x$ seems promising; this gives $v = e^x$, $uv = x^n e^x$, $u'v = nx^{n-1} e^x$ and hence the adjacent recursion formula.

1. $\int x^n e^x\, dx = x^n e^x - n\int x^{n-1} e^x\, dx$	n integer

Example: $\int x^2 e^x\, dx = x^2 e^x - 2\int x\, e^x\, dx = x^2 e^x - 2[x\, e^x - \int e^x\, dx]$
$= x^2 e^x - 2[x\, e^x - e^x + C] = x^2 e^x - 2x\, e^x + 2\, e^x - 2C$
$= (x^2 - 2x + 2)\, e^x + C_1$, where $C_1 = -2C$.

2. $\int x^n \sin x\, dx = -x^n \cos x + n\int x^{n-1} \cos x\, dx$ $\int x^n \cos x\, dx = +x^n \sin x - n\int x^{n-1} \sin x\, dx$	n integer

Here the appropriate factorizations of the integrand are easily guessed, because the integrals of $\sin x$ and $\cos x$ are known.

Examples: 1. $\int x^2 \sin x\, dx = -x^2 \cos x - \int (-2x \cos x)\, dx$
$= -x^2 \cos x + 2[x \sin x - \int \sin x\, dx]$
$= -x^2 \cos x + 2[x \sin x + \cos x + C] = -x^2 \cos x + 2x \sin x + 2\cos x + 2C.$

2. $\int x^3 \cos x \, dx = x^3 \sin x - 3 \int x^2 \sin x \, dx$
$= x^3 \sin x - 3[-x^2 \cos x + 2x \sin x + 2 \cos x + 2C]$
$= x^3 \sin x + 3x^2 \cos x - 6x \sin x - 6 \cos x - 6C.$

3. $\int (\ln x)^n \, dx = x (\ln x)^n - n \int (\ln x)^{n-1} \, dx$	$n \neq -1$, integer

One factorizes the integrand as $1 \cdot (\ln x)^n$, that is, one puts $v' = 1$, $u = (\ln x)^n$, thus, $v = x$, $u' = n(\ln x)^{n-1} \cdot (1/x)$ and $u'v = n(\ln x)^{n-1}$. If the exponent n is a natural number, then $n - 1$ partial integrations lead to the integral $\int \ln x \, dx$, whose integrand can again be written as a product $1 \cdot \ln x$, which leads to $\int 1 \cdot \ln x \, dx = x \ln x - \int x \cdot (1/x) \, dx = x \ln x - x + C$. For $n = -1$ one obtains the logarithmic integral (see Chapter 21.).

4. $\displaystyle\int x^n \ln x \, dx = \ln x \cdot \frac{x^{n+1}}{n+1} - \int \frac{x^n}{n+1} \, dx = \frac{x^{n+1}}{n+1}\left(\ln x - \frac{1}{n+1}\right)$	$n \neq -1$

Here one puts $v' = x^n$ and $u = \ln x$, so that the integrand $v \cdot u' = \dfrac{x^{n+1}}{n+1} \cdot \dfrac{1}{x} = \dfrac{x^n}{n+1}$ no longer contains $\ln x$ and can be integrated at once; a recurrence formula is not needed here.

5. $\displaystyle\int \sin^n x \, dx = -\frac{\cos x \sin^{n-1} x}{n} + \frac{n-1}{n} \int \sin^{n-2} x \, dx$	n integer
$\displaystyle\int \cos^n x \, dx = \frac{\sin x \cos^{n-1} x}{n} + \frac{n-1}{n} \int \cos^{n-2} x \, dx$	$n \neq 0$

The integrand, for example $\sin^n x$, is written as a product $\sin x \cdot \sin^{n-1} x$; thus, one may put $u = \sin^{n-1} x$, $v' = \sin x$ to get $v = -\cos x$, $u' = (n-1) \sin^{n-2} x \cos x$, hence $u'v = -(n-1) \sin^{n-2} x \cos^2 x = -(n-1) \sin^{n-2} x (1 - \sin^2 x) = -(n-1) \sin^{n-2} x + (n-1) \sin^n x$. Therefore $\int \sin^n x \, dx = -\sin^{n-1} x \cos x + (n-1) \int \sin^{n-2} x \, dx - (n-1) \int \sin^n x \, dx$, $(1 + n - 1) \int \sin^n x \, dx = -\sin^{n-1} x \cos x + (n-1) \int \sin^{n-2} x \, dx$. Division of each side by n leads to the first recursion formula; the second is obtained similarly. If n is a negative integer, one uses the recurrence formula to express the integral on the right in terms of the integral on the left.

Examples: 1. $\int \sin^2 x \, dx = -(\cos x \sin x)/2 + (1/2) \int dx = -(\cos x \sin x)/2 + x/2 + C.$

2. $\displaystyle\int_0^{2\pi} \cos^2 x \, dx = (1/2) [\sin x \cos x + x]_0^{2\pi} = \pi.$

3. $\int \cos^3 x \, dx = (1/3) \sin x \cos^2 x + (2/3) \int \cos x \, dx = (1/3) \sin x \cos^2 x + (2/3) \sin x + C.$

4. $\displaystyle\int_0^{2\pi} \sin^3 x \, dx = -(1/3) [\cos x(\sin^2 x + 2)]_0^{2\pi} = 0.$

Wallis' product formula. The previous recurrence formulae lead to a representation of $\pi/2$ which was found by John WALLIS (1616–1703). Since $0 \leqslant \sin x < 1$ in the interval $0 \leqslant x < \pi/2$, the inequality $\sin^{2k+1} x \leqslant \sin^{2k} x \leqslant \sin^{2k-1} x$ holds for natural numbers $k \geqslant 1$. On the other hand, the recurrence formulae lead to:

$$\int_0^{\pi/2} \sin^{2k} x \, dx = \frac{2k-1}{2k} \int_0^{\pi/2} \sin^{2k-2} x \, dx = \frac{(2k-1)(2k-3) \cdots 1}{2k(2k-2) \cdots 2} \cdot \frac{\pi}{2},$$

$$\int_0^{\pi/2} \sin^{2k+1} x \, dx = \frac{2k}{2k+1} \int_0^{\pi/2} \sin^{2k-1} x \, dx = \frac{2k(2k-2) \cdots 2}{(2k+1)(2k-1) \cdots 3}$$

and therefore to the inequalities:

$$\frac{2 \cdot 4 \cdots 2k}{3 \cdot 5 \cdots (2k+1)} \leqslant \frac{1 \cdot 3 \cdots (2k-1)}{2 \cdot 4 \cdots 2k} \cdot \frac{\pi}{2} \leqslant \frac{2 \cdot 4 \cdots (2k-2)}{3 \cdot 5 \cdots (2k-1)},$$

or

$$1 \leqslant (2k+1) \left[\frac{1 \cdot 3 \cdots (2k-1)}{2 \cdot 4 \cdots 2k}\right]^2 \cdot \frac{\pi}{2} \leqslant \frac{2k+1}{2k} = 1 + \frac{1}{2k}.$$

Since $\lim\limits_{k \to \infty} \left(1 + \dfrac{1}{2k}\right) = 1$, it follows that $\lim\limits_{k \to \infty} (2k+1) \left[\dfrac{1 \cdot 3 \cdots (2k-1)}{2 \cdot 4 \cdots 2k}\right]^2 \cdot \dfrac{\pi}{2} = 1.$

Wallis' product formula	$\frac{\pi}{2} = \lim_{k\to\infty} \frac{2^2 \cdot 4^2 \dots (2k)^2}{1^2 \cdot 3^2 \cdot 5^2 \dots (2k-1)^2 \cdot (2k+1)}$

For $k = 10$, this gives the (rather poor) approximation $\pi/2 \approx 1.5339$ or $\pi \approx 3.0678$.

Integration by substitution

An integral may become easier or simpler to find if the variable x is replaced by a new variable z by the substitution $x = \varphi(z)$, or if some part $\varphi(x)$ of the integrand is introduced as a new variable z. In all cases the connection between the differentials dx and dz of the given variable x and the new variable z must be taken into account.

Integrand as function $f[\varphi(x)]$ of a linear function $\varphi(x) = mx + c$. One substitutes $\varphi(x) = mx + c = z$ and notes that $m\,dx = dz$ or $dx = dz/m$. The substitution succeeds if $f(z)$ can be integrated.

Examples: 1. If the integrand is $(ax + b)^5$, put $ax + b = z$, $dx = dz/a$, to obtain
$\int (ax + b)^5\,dx = (1/a)\int z^5\,dz = (1/6a)\cdot z^6 + C = (1/6a)\cdot(ax + b)^6 + C.$

2. The integrand $\sqrt{(3x - 4)}$ becomes $\sqrt{z}$ if one puts $3x - 4 = z$; then $dx = dz/3$ and for the integral one obtains

$$\int \sqrt{(3x-4)}\,dx = {}^1/_3 \int \sqrt{z}\,dz = {}^1/_3 \cdot {}^2/_3 \cdot z^{3/2} + C = {}^2/_9 z\sqrt{z} + C$$
$$= {}^2/_9 (3x - 4)\sqrt{(3x-4)} + C.$$

3. The substitution $\omega t + \pi/2 = z$, $dt = dz/\omega$, leads to
$\int \sin(\omega t + \pi/2)\,dt = (1/\omega)\int \sin z\,dz = -(1/\omega)\cos z + C = -(1/\omega)\cos(\omega t + \pi/2) + C.$

4. In the integrand e^{-3x} one substitutes $-3x = z$, $dx = -dz/3$, to obtain
$\int e^{-3x}\,dx = -{}^1/_3 \int e^z\,dz = -{}^1/_3 e^z + C = -{}^1/_3 e^{-3x} + C.$

Integrand of the form $\varphi'(x)/\varphi(x)$. If the integrand is a quotient in which the numerator is the derivative of the denominator, one substitutes $\varphi(x) = z$. Then $\varphi'(x)\,dx = dz$ or $dx = \frac{dz}{\varphi'(x)}$, so that the integral becomes $\int \frac{dz}{z} = \ln z + C$.

$$\int \frac{\varphi'(x)}{\varphi(x)}\,dx = \ln|\varphi(x)| + C \qquad \int \frac{x^{n-1}}{x^n + a}\,dx = \frac{1}{n}\int \frac{nx^{n-1}}{x^n + a}\,dx = \frac{1}{n}\ln|x^n + a| + C$$

For instance, the denominator $\varphi(x) = x^n + a$ has the derivative $\varphi'(x) = nx^{n-1}$, so that if the numerator is $x^{n-1} = \frac{nx^{n-1}}{n}$, one may take the constant factor $\frac{1}{n}$ outside the integral, to give the integrand the form $\varphi'(x)/\varphi(x)$.

Examples: 1. $\int \frac{3x^2 - 4}{x^3 - 4x + 7}\,dx = \ln[c(x^3 - 4x + 7)] = \ln|x^3 - 4x + 7| + C.$

2. $\int \frac{x^3}{x^4 - 5}\,dx = {}^1/_4 \int \frac{4x^3}{x^4 - 5}\,dx = {}^1/_4 \ln[c(x^4 - 5)] = {}^1/_4 \ln|x^4 - 5| + C.$

3. $\int \frac{3 - 5x}{1 + x^2}\,dx = 3\int \frac{dx}{1 + x^2} - {}^5/_2 \int \frac{2x}{1 + x^2}\,dx = 3\arctan x - {}^5/_2 \ln(1 + x^2) + C.$

The integrand $\frac{1}{\sqrt{(x^2 + a^2)}} = \frac{1}{D}$ can be rearranged into the form $\varphi'(x)/\varphi(x)$ as follows:

$$\frac{1}{D} = \frac{(x + D)}{D(x + D)} = \frac{(x + D)/D}{(x + D)} = \frac{1 + x/D}{x + D} = \frac{1 + x/\sqrt{(x^2 + a^2)}}{x + \sqrt{(x^2 + a^2)}}.$$

Now the numerator is the derivative of the denominator.

$$\int \frac{dx}{\sqrt{(x^2 + a^2)}} = \int \frac{1 + x/\sqrt{(x^2 + a^2)}}{x + \sqrt{(x^2 + a^2)}}\,dx = \ln|x + \sqrt{(x^2 + a^2)}| + C$$

Integrals of the function $\tan x$, $\cot x$, $\tanh x$ and $\coth x$. Each of these integrands can be written as a quotient in which the numerator is the derivative of the denominator, and can therefore be integrated by the method described. For example, $\int \tan x\,dx = -\int \frac{-\sin x}{\cos x}\,dx = -\ln|\cos x| + C.$

Integrals of the functions $\arctan x$, $\operatorname{arccot} x$, $\tanh^{-1} x$ *and* $\coth^{-1} x$. For these integrands, partial integration leads first to an integral of the form $\int \frac{\varphi'(x)}{\varphi(x)}\,dx$; for example, if $\arctan x$ is written

$$\int \tan x \, dx = -\ln|\cos x| + C \qquad \int \cot x \, dx = \ln|\sin x| + C$$
$$\int \tanh x \, dx = \ln|\cosh x| + C \qquad \int \coth x \, dx = \ln|\sinh x| + C$$

$$\int \arctan x \, dx = x \arctan x - \tfrac{1}{2} \ln(1 + x^2) + C$$
$$\int \operatorname{arccot} x \, dx = x \operatorname{arccot} x + \tfrac{1}{2} \ln(1 + x^2) + C$$
$$\int \tanh^{-1} x \, dx = x \tanh^{-1} x + \tfrac{1}{2} \ln(1 - x^2) + C, \quad |x| < 1$$
$$\int \coth^{-1} x \, dx = x \coth^{-1} x + \tfrac{1}{2} \ln(x^2 - 1) + C, \quad |x| > 1$$

as a product $1 \cdot \arctan x$, then $u = \arctan x$, $v' = 1$, so that $v = x$, $u' = \dfrac{1}{1 + x^2}$, $vu' = \tfrac{1}{2} \cdot \dfrac{2x}{1 + x^2}$ and $\int \arctan x \, dx = x \arctan x - \tfrac{1}{2} \int \dfrac{2x}{1 + x^2} dx = x \arctan x - \tfrac{1}{2} \ln(1 + x^2) + C$. The other three functions can be integrated similarly.

Integrand of the form $f[\varphi(x)] \cdot \varphi'(x)$, with $\varphi'(x) \neq 0$. If the integrand can be written as the product of a function $\varphi(x)$ and its derivative $\varphi'(x)$, then the substitution $\varphi(x) = z$, $\varphi'(x) \, dx = dz$ also leads to an easily integrated function of z and to the integral $\int z \, dz = z^2/2 + C$. More generally, if the

$$\int \varphi(x) \varphi'(x) \, dx = \tfrac{1}{2} [\varphi(x)]^2 + C \qquad \int \varphi^n(x) \cdot \varphi'(x) \, dx = \int z^n \, dz = \frac{1}{n + 1} \varphi^{n+1}(x) + C$$

integrand has the form $f[\varphi(x)] \cdot \varphi'(x)$, the substitution $\varphi(x) = z$, $dx = \dfrac{dz}{\varphi'(x)}$, leads to an integrand $f(z)$, so that if $f(z)$ has a known integral (for example, $f(z) = z^n$), the integral of $f[\varphi(x)] \cdot \varphi'(x)$ can be found.

Examples: 1. $\int \sin x \cos x \, dx = \int \sin x \, d(\sin x) = \int z \, dz = z^2/2 + C = \tfrac{1}{2} \sin^2 x + C.$

2. $\displaystyle\int \frac{\ln x}{x} dx = \int \ln x \, d(\ln x) = \int z \, dz = [\ln x]^2/2 + C.$

3. $\displaystyle\int \frac{\arctan^5 x}{1 + x^2} dx = \int \arctan^5 x \, d(\arctan x) = \int z^5 \, dz = z^6/6 + C = \tfrac{1}{6} \arctan^6 x + C.$

4. $\int (1 - x^4)^7 x^3 \, dx = -\tfrac{1}{4} \int (1 - x^4)^7 (-4x^3 \, dx) = -\tfrac{1}{4} \int (1 - x^4)^7 \, d(1 - x^4)$
$= -\tfrac{1}{4} \int z^7 \, dz = -z^8/32 + C = -(1 - x^4)^8/32 + C.$

5. $\displaystyle\int \frac{x \, dx}{(x^2 + 1)\sqrt{(x^2 + 1)}} = \int \frac{x \, dx}{\sqrt{(x^2 + 1)^3}} = \tfrac{1}{2} \int \frac{2x \, dx}{(x^2 + 1)^{3/2}} = \tfrac{1}{2} \int \frac{d(x^2 + 1)}{(x^2 + 1)^{3/2}}$
$\displaystyle= \tfrac{1}{2} \int z^{-3/2} \, dz = -\tfrac{1}{2} \cdot 2z^{-1/2} + C = -1/\sqrt{(x^2 + 1)} + C.$

Integrals of the functions $\arcsin x$, $\arccos x$, $\sinh^{-1} x$ *and* $\cosh^{-1} x$. For these integrals, integration by parts leads to integrals of the form $\int \varphi(x) \varphi'(x) \, dx$, which have the value $\varphi^2/2$, apart from the constant of integration. For instance, writing $\arcsin x$ as a product $1 \cdot \arcsin x$, that is, taking $u = \arcsin x$, $v' = 1$, $v = x$,

$u' = \dfrac{1}{\sqrt{(1 - x^2)}}$, $vu' = \dfrac{x}{\sqrt{(1 - x^2)}} = -\tfrac{1}{2} \left(\dfrac{-2x}{\sqrt{(1 - x^2)}} \right)$, integration by parts gives

$$\int \arcsin x \, dx = x \arcsin x + \sqrt{(1 - x^2)} + C$$
$$\int \arccos x \, dx = x \arccos x - \sqrt{(1 - x^2)} + C$$
$$\int \sinh^{-1} x \, dx = x \sinh^{-1} x - \sqrt{(1 + x^2)} + C$$
$$\int \cosh^{-1} x \, dx = x \cosh^{-1} x - \sqrt{(x^2 - 1)} + C$$

$\displaystyle\int \arcsin x \, dx = x \arcsin x + \int \frac{-x \, dx}{\sqrt{(1 - x^2)}}$
$\displaystyle= x \arcsin x + \tfrac{1}{2} \int \frac{dz}{z} = x \arcsin x + \sqrt{z} + C$
$= x \arcsin x + \sqrt{(1 - x^2)} + C.$ The other three functions can be treated similarly.

The substitution of a new variable by $x = \varphi(z)$. This substitution converts the integrand $f(x)$ into the composite function $f[\varphi(z)]$; the differentials are related by $dx = \varphi'(z) \, dz$, and the integral takes the form $\int f(x) \, dx = \int f[\varphi(z)] \varphi'(z) \, dz$. For a definite integral one must, however, also convert the limits of the x-interval of integration into the corresponding limits of z-integration by means of the inverse function $z = \psi(x)$. The function $\varphi(z)$ must have an inverse function and must be differentiable, with $\varphi'(z) \neq 0$. The substitution $x = |a| z$, $dx = |a| \, dx$ or $x = (|a|/b) z$, $dx = (|a|/b) \, dz$ lead to the following integrals:

$$\int \frac{dx}{\sqrt{(a^2 - x^2)}} = \operatorname{Arcsin} \frac{x}{|a|} + C \quad \text{for} \quad |x| < |a|; \qquad \int \frac{dx}{a^2 + x^2} = \frac{1}{a} \arctan \frac{x}{a} + C;$$
$$\int \frac{dx}{\sqrt{(a^2 - b^2x^2)}} = \frac{1}{b} \operatorname{Arcsin} \frac{bx}{|a|} + C \quad \text{for} \quad |x| < \left|\frac{a}{b}\right|; \qquad \int \frac{dx}{a^2 + b^2x^2} = \frac{1}{ab} \arctan \frac{b}{a} x + C$$

Examples: 1. To remove the square root $\sqrt{x}$ in the following integral one puts $x = \varphi(z) = z^2$, $dx = 2z \, dz$. Integration by parts leads to evaluation of the resulting integral in z; using the inverse function $z = \psi(x) = \sqrt{x}$ to $x = \varphi(z)$, one obtains

$$\int e^{\sqrt{x}} \, dx = \int e^z \cdot 2z \, dz = 2 \int z e^z \, dz = 2 e^z(z - 1) + C = 2 e^{\sqrt{x}}(\sqrt{x} - 1) + C.$$

2. $\int_1^4 e^{\sqrt{x}}\,dx = 2[e^z(z-1)]_1^2 = 2\,e^2$; the new limits of integration are obtained from $z = +\sqrt{x}$ for $x_1 = 1$ and $x_2 = 4$.

3. $\int_{-\pi/2}^{\pi} \sin 2x\,dx = (1/2)\int_{-\pi}^{2\pi} \sin z\,dz = -(1/2)\,[\cos z]_{-\pi}^{2\pi}$
$= -(1/2)\,(1+1) = -1.$

Substitution: $2x = z$, $dx = dz/2$; $z_1 = -2\pi/2 = -\pi$; $z_2 = 2\pi$.

4. $\int_0^{\sqrt{3}} \frac{5x}{4-x^2}\,dx = -(5/2)\int_4^1 \frac{dz}{z} = -(5/2)\ln z\,|_4^1$
$= -(5/2)(\ln 1 - \ln 4) = (5/2)\ln 4 = 5\ln 2.$

Substitution: $4 - x^2 = z$, $-2x\,dx = dz$; $x_1 = 0 \to z_1 = 4$, $x_2 = \sqrt{3} \to z_2 = 4 - 3 = 1$.

5. $\int_{\sqrt{e}}^{e} \frac{dx}{x\sqrt{[\ln x(1-\ln x)]}} = \int_{1/2}^{1} \frac{dz}{\sqrt{[z(1-z)]}}$
$= \int_{\pi/4}^{\pi/2} \frac{2\sin u\cos u}{\sin u\cos u}\,du = 2[u]_{\pi/4}^{\pi/2} = 2(\pi/2 - \pi/4) = \pi/2.$

Substitution: $z = \ln x$, $dx = x\,dz$; $x_1 = \sqrt{e} \to z_1 = 1/2$, $x_2 = e \to z_2 = 1$. $z = \sin^2 u$, $dz = 2\sin u\cos u\,du$; $z_1 = 1/2 \to u_1 = \pi/4$, $z_2 = 1 \to u_2 = \pi/2$.

Classes of elementarily integrable functions

Integrable functions $f(x)$, such as x^n, $\sin x$, e^x, whose indefinite integrals can be expressed in closed form in terms of elementary functions are called *elementarily integrable* (or integrable in elementary terms). The following section describes the most important types of elementarily integrable functions together with methods of finding their indefinite integrals.

Rational functions $R(x)$, partial fractions. Every rational function is elementarily integrable: Since every power of x has an elementary integral, so does any polynomial. Rational functions that are not polynomials can be written as a sum of partial fractions (see Chapter 5.) and can then be integrated because for all natural numbers $k > 1$ each of the following fractions can be integrated:

$$\frac{A}{x-x_1},\quad \frac{A}{(x-x_1)^k},\quad \frac{Ax+B}{x^2+px+q},\quad \frac{Ax+B}{(x^2+px+q)^k},\ \text{with } p^2 < 4q \text{ and } A \neq 0.$$

The integrals of the first two expressions are standard integrals; the numerators of the last two can always be written as a sum $Ax + B = (2x + p)\,A/2 + (B - Ap/2)$. The first term leads to an

$$\int \frac{A}{x-x_1}\,dx = A\ln|x-x_1| + C;\qquad \int \frac{A\,dx}{(x-x_1)^k} = -\frac{A}{(k-1)(x-x_1)^{k-1}} + C$$

integral of the form $\int \frac{\varphi'(x)}{[\varphi(x)]^k}\,dx$, which is elementary; the second term is a constant which can be taken outside the integral. It remains only to show that the integral $\int \frac{dx}{(x^2+px+q)^k}$ can be found elementarily for $k = 1, 2, \ldots$

For $k = 1$, the denominator $x^2 + px + q$ can be written as $(x + p/2)^2 + (q - p^2/4)$ by completing the square. The substitution $x + p/2 = \sqrt{(q - p^2/4)}\,u$, $dx = \sqrt{(q - p^2/4)}\,du$ gives

$$\frac{1}{\sqrt{(q-p^2/4)}}\int \frac{du}{u^2+1} = \frac{1}{\sqrt{(q-p^2/4)}}\arctan u.$$

$$\int \frac{dx}{x^2+qx+q} = \frac{2}{\sqrt{(4q-p^2)}}\arctan\frac{2x+p}{\sqrt{(4q-p^2)}} + C$$

$$\int \frac{Ax+B}{x^2+px+q}\,dx = \frac{A}{2}\ln|x^2+px+q| + \frac{2B-Ap}{\sqrt{(4q-p^2)}}\cdot\arctan\frac{2x+p}{\sqrt{(4q-p^2)}} + C$$

For $k > 1$ one can get a recurrence formula of the type

$$\int \frac{dx}{(x^2+px+p)^k} = \frac{c_1x+c_2}{(x^2+px+q)^{k-1}} + c_3\int \frac{dx}{(x^2+px+q)^{k-1}}$$

by finding the undetermined coefficients c_1, c_2, c_3 as follows: one differentiates both sides, then clears fractions by multiplying through by $(x^2 + px + q)^k$ and so obtains an identity

$$1 = -(k-1)(c_1 x + c_2)(2x + p) + (c_1 + c_3)(x^2 + px + q);$$

now one equates coefficients:

Coefficients of x^2: $-2c_1(k-1) + c_1 + c_3 = 0$.
Coefficients of x: $-2c_2(k-1) - c_1 p(k-1) + (c_1 + c_3)p = 0$.
Constant terms: $-c_2 p(k-1) + (c_1 + c_3)q = 1$.

This gives $c_1 = \dfrac{2}{(k-1)(4q-p^2)}$, $c_2 = \dfrac{p}{(k-1)(4q-p^2)}$, $c_3 = \dfrac{2(2k-3)}{(k-1)(4q-p^2)}$.

$$\int \frac{dx}{(x^2+px+q)^k} = \frac{2x+p}{(k-1)(4q-p^2)(x^2+px+q)^{k-1}} + \frac{2(2k-3)}{(k-1)(4q-p^2)} \int \frac{dx}{(x^2+px+q)^{k-1}}$$

These results are often also displayed for the denominator $(ax^2 + bx + c)^k$.

$$\int \frac{Ax+B}{(ax^2+bx+c)^k}\,dx = \frac{-A}{2a(k-1)(ax^2+bx+c)^{k-1}} + \left(B - \frac{Ab}{2a}\right) \int \frac{dx}{(ax^2+bx+c)^k}$$

$$\int \frac{dx}{(ax^2+bx+c)^k} = \frac{2ax+b}{(k-1)(4ac-b^2)(ax^2+bx+c)^{k-1}} + \frac{2(2k-3)a}{(k-1)(4ac-b^2)} \int \frac{dx}{(ax^2+bx+c)^{k-1}}$$

$$\int \frac{dx}{ax^2+bx+c} \begin{cases} = \dfrac{2}{\sqrt{(4ac-b^2)}} \cdot \arctan \dfrac{2ax+b}{\sqrt{(4ac-b^2)}} + C \text{ if } b^2 - 4ac < 0 \\ = \dfrac{1}{\sqrt{(b^2-4ac)}} \cdot \ln \left| \dfrac{2ax+b-\sqrt{(b^2-4ac)}}{2ax+b+\sqrt{(b^2-4ac)}} \right| + C \text{ if } b_2 - 4ac > 0 \\ = \dfrac{-2}{2ax+b} + C \text{ if } b^2 - 4ac = 0 \end{cases}$$

Examples: (the decomposition into partial fractions is assumed to have been carried out)

1. $\displaystyle\int \frac{4x^2 - 7x + 25}{x^3 - 6x^2 + 3x + 10}\,dx = 2\int \frac{dx}{x+1} - 3\int \frac{dx}{x-2} + 5\int \frac{dx}{x-5}$
$\displaystyle = 2\ln|x+1| - 3\ln|x-2| + 5\ln|x-5| + C = \ln \left| \frac{(x+1)^2(x-5)^5}{(x-2)^3} \right| + C.$

2. $\displaystyle\int \frac{3x^2 - 20x + 20}{(x-2)^3(x-4)}\,dx = \frac{3}{2}\int \frac{dx}{x-2} + 6\int \frac{dx}{(x-2)^2} + 4\int \frac{dx}{(x-2)^3} - \frac{3}{2}\int \frac{dx}{x-4}$
$\displaystyle = \frac{3}{2}\ln|x-2| - \frac{6}{x-2} - \frac{2}{(x-2)^2} - \frac{3}{2}\ln|x-4| + C = \frac{2(5-3x)}{(x-2)^2} + 3\ln \sqrt{\left(\frac{x-2}{x-4}\right)} + C.$

3. $\displaystyle\int \frac{3x^2 - 3x - 10}{x^3 - 5x^2 + 11x - 15}\,dx = \int \frac{dx}{x-3} + \int \frac{2x+5}{x^2 - 2x + 5}\,dx$
$\displaystyle = \ln|x-3| + \ln|x^2 - 2x + 5| + \frac{7}{2}\arctan \frac{x-1}{2} + C$
$\displaystyle = \ln|(x-3)(x^2 - 2x + 5)| + \frac{7}{2}\arctan \frac{x-1}{2} + C.$

4. $\displaystyle\int \frac{-3x^3 + x - 4}{(x+1)(x^2+x+1)^2}\,dx = -2\int \frac{dx}{x+1} + \int \frac{2x-3}{x^2+x+1}\,dx + \int \frac{8x+1}{(x^2+x+1)^2}\,dx$
$\displaystyle = -2\ln|x+1| + \ln|x^2+x+1| - \frac{8}{3}\sqrt{3}\arctan \frac{2x+1}{\sqrt{3}} - \frac{4}{x^2+x+1} - \frac{2x+1}{x^2+x+1}$
$\displaystyle - \frac{4}{3}\sqrt{3}\arctan \frac{2x+1}{\sqrt{3}} + C = \ln \left| \frac{x^2+x+1}{(x+1)^2} \right| - 4\sqrt{3}\arctan \frac{2x+1}{\sqrt{3}} - \frac{2x+5}{x^2+x+1} + C.$

Integrals of functions $R[x, \sqrt[n]{(ax+b)}]$ or $R[x, \sqrt[n]{[(ax+b)/(cx+d)]}]$. The integrands are rational functions of x and of $\sqrt[n]{(ax+b)}$ or $\sqrt[n]{[(ax+b)/(cx+d)]}$, that is, they can be obtained from x

and $\sqrt[n]{(ax+b)}$ or $\sqrt[n]{[(ax+b)/(cx+d)]}$ by finitely many additions, subtractions, multiplications and divisions. The substitutions $z=\sqrt[n]{(ax+b)}$ for which $x=(z^n-b)/a$, or $z=\sqrt[n]{[(ax+b)/(cx+d)]}$ for which $x=(b-dz^n)/(cz^n-a)$, lead to rational functions of z which, as has just been shown, can be integrated elementarily. For special forms R the integration can be simplified by using particular devices.

Integrals of functions $R[x, \sqrt{(ax^2+bx+c)}]$. Various substitutions, depending on the nature of the coefficients a, b, and c, lead to a rational function in a new variable z.

1. If a and ax^2+bx+c are positive, and $b^2 \neq 4ac$, one makes the substitution $\sqrt{(ax^2+bx+c)} = x\sqrt{a}+z$.
2. If c is non-negative, one puts $\sqrt{(ax^2+bx+c)} = xz+\sqrt{c}$.
3. If a is negative, and if the equation $ax^2+bx+c=0$ has two distinct real roots x_1 and x_2, one makes the substitution $\sqrt{(ax^2+bx+c)} = z(x-x_1)$.

There exist tables of integrals for special functions R.

$$I=\int\frac{dx}{\sqrt{(ax^2+bx+c)}}=\frac{1}{\sqrt{a}}\ln\left|\frac{2ax+b}{2\sqrt{a}}+\sqrt{(ax^2+bx+c)}\right|+C \text{ for } a>0,\quad b^2-4ac\neq 0$$

$I=-\dfrac{1}{\sqrt{(-a)}}\arcsin\dfrac{2ax+b}{\sqrt{(b^2-4ac)}}+C$ for $a<0$, $b^2-4ac>0$	$I=\dfrac{1}{\sqrt{a}}\arcsin\dfrac{2ax+b}{\sqrt{(4ac-b^2)}}+C$ for $a>0$, $b^2-4ac<0$	$I=\dfrac{1}{\sqrt{a}}\ln(2ax+b)+C$ for $a>0$, $b^2-4ac=0$

Another method of integration consists in completing the square in the integrand $ax^2+bx+c=a[[x+b/(2a)]^2+(4ac-b^2)/(4a^2)]$, so as to bring it into one of the forms displayed in the adjacent table, and then using the displayed substitutions to obtain rational expressions involving trigonometric or hyperbolic functions.

Integrand	Substitution
$R[x, \sqrt{(\alpha^2-x^2)}]$	$x=\alpha\sin z$
$R[x, \sqrt{(\alpha^2+x^2)}]$	$x=\alpha\sinh z$
$R[x, \sqrt{(x^2-\alpha^2)}]$	$x=\alpha\cosh z$

Examples: 1. The substitution $x=\sin z$, $dx=\cos z\,dz$, $z=\arcsin x$ yields

$$\int\sqrt{(1-x^2)}\,dx=\int\sqrt{(1-\sin^2 z)}\cos z\,dz=\int\cos^2 z\,dz=(1/2)(z+\sin z\cos z)+C$$
$$=(1/2)(\arcsin x+x\sqrt{(1-x^2)})+C.$$

2. From $x=r\sin z$, $dx=r\cos z\,dz$ one obtains similarly

$$\int\sqrt{(r^2-x^2)}\,dx=(1/2)(r^2\arcsin x/r+x\sqrt{(r^2-x^2)})+C.$$

Integral	Result	Substitution
$\int\sqrt{(1+x^2)}\,dx$	$^1/_2[\sinh^{-1}x+x\sqrt{(1+x^2)}]+C$	$x=\sinh z$
$\int\sqrt{(a^2+x^2)}\,dx$	$\pm(a^2/2)\sinh^{-1}x/a+(x/2)\sqrt{(a^2+x^2)}+C$ for $a\lesseqgtr 0$	$x=a\sinh z$
$\displaystyle\int\frac{dx}{a^2-b^2x^2}$	$(1/ab)\tanh^{-1}(bx/a)+C$ for $\lvert x\rvert<\lvert a/b\rvert$	$x=(a/b)\tanh z$

Integrals of functions $R(\sin x, \cos x, \tan x, \cot x)$. Transformation into a rational function of z can be achieved by the substitution $z=\tan(x/2)$:

$$\sin x=2\sin(x/2)\cos(x/2)=\frac{2\tan(x/2)\cos^2(x/2)}{\sin^2(x/2)+\cos^2(x/2)}=\frac{2z}{1+z^2},$$

$$\cos x=\frac{\cos^2(x/2)-\sin^2(x/2)}{\cos^2(x/2)+\sin^2(x/2)}=\frac{1-z^2}{1+z^2},\qquad \tan x=\frac{2\tan(x/2)}{1-\tan^2(x/2)}=\frac{2z}{1-z^2},$$

$$\cot x=\frac{1-z^2}{2z},\qquad \frac{dz}{dx}=\frac{1}{2\cos^2(x/2)}=\frac{1+z^2}{2},\qquad \frac{dx}{dz}=\frac{2}{1+z^2}.$$

$$\int R(\sin x,\cos x,\tan x,\cot x)\,dx=\int R\left(\frac{2z}{1+z^2},\frac{1-z^2}{1+z^2},\frac{2z}{1-z^2},\frac{1-z^2}{2z}\right)\frac{2\,dz}{1+z^2}$$

Examples: 1. $\int \frac{dx}{\sin x} = 2\int \frac{1+z^2}{2z(1+z^2)}\,dz = \int \frac{dz}{z} = \ln(cz) = \ln(c \cdot \tan x/2)$.

2. $\int \frac{1-\sin x}{\sin x(1-\cos x)}\,dx = \int \frac{[1-2z/(1+z^2)]\cdot[2/(1+z^2)]}{[2z/(1+z^2)]\,[1-(1-z^2)/(1+z^2)]}\,dz$

$= (1/2)\int \frac{z^2-2z+1}{z^3}\,dz = (1/2)\int [1/z - 2/z^2 + 1/z^3]\,dz = (1/2)\,[\ln|z| + 2/z - 1/(2z^2)] + C$

$= (1/2)\,[\ln|\tan(x/2)| + 2\cot(x/2) - (1/2)\cot^2(x/2)] + C$.

Integrals of functions $R(\sinh x, \cosh x, \tanh x, \coth x)$. According to the definitions of the hyperbolic functions, these integrals may be converted into integrals of rational functions by means of the substitution $e^x = t$, for example, $\sinh x = [t - 1/t]/2$. By analogy with the trigonometric case the substitution $z = \tanh(x/2)$ is also successful:

$$\sinh x = 2\sinh(x/2)\cosh(x/2) = \frac{2\tanh(x/2)\cosh^2(x/2)}{\cosh^2(x/2) - \sinh^2(x/2)} = \frac{2z}{1-z^2},$$

$$\cosh x = \frac{\sinh^2(x/2) + \cosh^2(x/2)}{\cosh^2(x/2) - \sinh^2(x/2)} = \frac{1+z^2}{1-z^2},$$

$$\tanh x = \frac{2\tanh(x/2)}{1+\tanh^2(x/2)} = \frac{2z}{1+z^2}, \qquad \coth x = \frac{1+z^2}{2z},$$

$$\frac{dz}{dx} = \frac{1}{2\cosh^2(x/2)} = \frac{\cosh^2(x/2) - \sinh^2(x/2)}{2\cosh^2(x/2)} = \frac{1-z^2}{2}, \qquad \frac{dx}{dz} = \frac{2}{1-z^2}.$$

Binomial integrals $\int x^m(a+bx^n)^p\,dx$. Here the coefficients a and b are real numbers and the exponents m, n and p are rational numbers. A theorem of P. L. Chebyshev (1821–1894) states that these integrals can be expressed as elementary functions when at least one of the numbers p, $(m+1)/n$, or $(m+1)/n + p$ is an integer. If p is an integer, the integrand is a sum of powers with rational exponents which can be integrated. If $(m+1)/n$ is an integer and $p = s/r$, one puts $z = \sqrt[r]{(a+bx^n)}$; if $(m+1)/n + p$ is an integer, one puts $z = \sqrt[r]{[(a+bx^n)/x^n]}$.

Integrals that cannot be expressed in terms of elementary functions

The calculation of the length of an elliptical arc, of the period of oscillation of a circular pendulum, and of other problems lead to *elliptic integrals*. These are integrals whose integrand contains the square root of a cubic or quartic polynomial with no repeated root.

Elliptic integrals of the first and second kind	$\int \frac{dx}{\sqrt{[(1-x^2)(1-k^2x^2)]}}, \quad \int \frac{x^2\,dx}{\sqrt{[(1-x^2)(1-k^2x^2)]}}, \quad 0 < k^2 < 1$

Joseph Liouville (1809–1882) proved that they belong to the class of those integrals that cannot be expressed in closed form in terms of elementary functions. There are other integrals of this type with comparatively simple integrands, such as $\frac{1}{\sqrt{(\cos\alpha - \cos x)}}$, $\frac{\sin x}{x}$, or $\frac{1}{\sqrt{(1+x^4)}}$. But this does not mean that these integrals do not exist: as indefinite integrals of continuous functions they are, as has been shown, differentiable functions of the upper limit of integration. On the contrary, integrals that cannot be expressed by elementary functions are accepted into mathematics as *new, higher, non-elementary functions*. They are often treated by first expanding the integrand as an infinite series, which is then integrated term by term (see Chapter 21.).

If the infinite series $\sum_{n=0}^{\infty} f_n(x)$ converges uniformly on the interval $a \leqslant x \leqslant b$ and if each term $f_n(x)$ is integrable, then the series obtained by integrating termwise over $[a, b]$ also converges.

20.3. Integration of functions of several variables

Since the definite integral is particularly useful for calculating areas of plane regions, it is natural to look for a generalization to facilitate the calculations of volumes of spatial regions. If a bounded continuous function $z = f(x_1, x_2, \ldots, x_n)$ is defined on a measurable bounded region G in n-dimensional space, one divides G up into a finite number of measurable subsets and forms, just as in the definition of a simple definite integral, upper and lower sums involving the volumes of these subsets and the maxima and minima of $f(x_1, x_2, \ldots, x_n)$ in each of the subsets. If these sums approach the same limit as the subdivision is refined, this limit is called the *n-fold volume integral* of f over G. The two-fold volume integral, called the *double integral*, will be discussed in greater detail; it can be used to calculate the volume of solid bodies that are bounded by curved surfaces. However, the range of integration of an n-fold integral can be restricted to a manifold of lower dimension. For example, one speaks of a *line integral* for $n = 3$ when this manifold is a 1-dimensional curve, or a *surface integral* when it is a 2-dimensional surface.

Two-dimensional integrals

Double integral. The definite integral was defined as a limit of sums in which each term is the product of two factors, the lengths Δx_i of subintervals and the ordinates $f(\xi_i)$ at a point ξ_i of the subinterval, the number n of subintervals tending to infinity and the length of the longest subinterval tending to zero. The interval $[a, b]$ of integration on the x-axis is now replaced by a plane region G on which a function $z = f(x, y)$ is defined, and G is divided into n subregions ΔG_i, $i = 1, 2, \ldots, n$. To simplify the notation, one writes ΔG_i for the subregion and also for its area.

Suppose that the function is continuous and bounded in the region G. Then one can form lower sums with the infimum m_i in ΔG_i, and upper sums with the supremum M_i in ΔG_i. If the subdivision of G is refined, then as $n \to \infty$ and $\Delta G_i \to 0$, the sequence of lower sums tends to the same limit as the sequence of upper sums, and any sequence of intermediate sums $\sum_{i=1}^{n} f(\xi_i, \eta_i)\,\Delta G_i$ tends to the same limit whatever intermediate point (ξ_i, η_i) is chosen in ΔG_i. The integral of the function $z = f(x, y)$ over the region G is defined to be this common limit and is called a double integral, because there are two variables of integration.

Double integral	$\iint_G f(x, y)\,dG = \lim_{\substack{n\to\infty \\ \Delta G_i \to 0}} \sum_{i=1}^{n} m_i \Delta G_i = \lim_{\substack{n\to\infty \\ \Delta G_i \to 0}} \sum_{i=1}^{n} M_i \Delta G_i$

The existence of such a double integral can also be guaranteed if the function $z = f(x, y)$ is bounded and piecewise continuous in G. If the double integral exists, the function is said to be integrable over G.

Geometrical interpretation of the double integral. The simple definite integral may be regarded as the area of a plane region below a curve. Similarly, the double integral of a continuous function of two variables may be interpreted as the volume below a surface $z = f(x, y)$ (Fig.) provided that $z = f(x, y)$ takes only positive values in G. The ΔG_i are elements of area in the x, y-plane which, after the limiting process, are denoted by $dx\,dy$ in Cartesian coordinates and by $r\,dr\,d\varphi$ in polar coordinates. Every product $m_i \Delta G_i$ is the volume of a cylinder with base area ΔG_i and height m_i, and likewise for $M_i \Delta G_i$ and the similar cylinder of height M_i. For the volume V below the surface $z = f(x, y)$ one therefore has $\sum_{i=1}^{n} m_i \Delta G_i \leqslant V \leqslant \sum_{i=1}^{n} M_i \Delta G_i$. Refinement of the subdivision leads to a monotonically increasing sequence of lower sums and monotonically decreasing sequence of upper sums; both sequences have the same limit because the sequence of differences of corresponding upper and lower sums tends to zero.

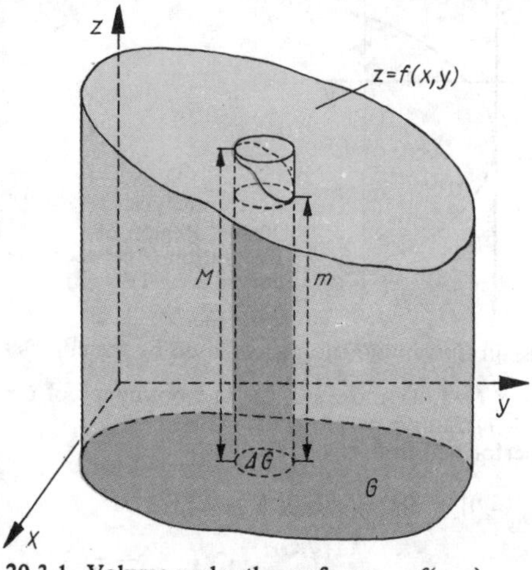

20.3-1 Volume under the surface $z = f(x, y)$ above G

Calculations of the double integral. A double integral may be calculated by making two successive integrations over each variable in turn. Suppose that the region G of integration has a simple boundary which meets the rectangle $a_1 \leqslant x \leqslant a_2$, $b_1 \leqslant y \leqslant b_2$ at the points A_1, A_2, B_1, B_2 (Fig.). The points A_1, A_2 separate the boundary curve of G into two arcs: $A_1B_1A_2$, which is the graph of a function $y = y_1(x)$, and $A_1B_2A_2$, the graph of $y = y_2(x)$. Similarly B_1, B_2 separate the boundary curve into $B_1A_1B_2$ and $B_1A_2B_2$, given by $x = x_1(y)$ and $x = x_2(y)$. For fixed $x = \xi_i$, $y_1(\xi_i)$ and $y_2(\xi_i)$ are end-points of an interval $y_1(\xi_i) \leqslant y \leqslant y_2(\xi_i)$ over which the function $f(\xi_i, y)$ of the single variable y must be integrated; $\varphi(\xi_i) = \int_{y_1(\xi_i)}^{y_2(\xi_i)} f(\xi_i, y)\,dy$ is, for fixed $x = \xi_i$, a constant which depends on x, that is, a function $\varphi(x)$ of x on the interval $a_1 \leqslant x \leqslant a_2$. With suitable assumptions about the boundary curve of G, $\varphi(x)$ is a continuous function of x and is therefore integrable over the interval $[a_1, a_2]$. Similarly $\psi(\eta_i) = \int_{x_1(\eta_i)}^{x_2(\eta_i)} f(x, \eta_i)\,dx$ is, for fixed $y = \eta_i$, a constant which depends on y and as a continuous function $\psi(y)$ on the interval $b_1 \leqslant y \leqslant b_2$, is integrable over this interval. It can be shown that both repeated integrations lead to the same value, which is also equal to the value of the double integral; this agrees with the geometrical idea that the $\varphi(x)$ are areas of plane sections parallel to the y, z-plane, and the $\psi(y)$ areas of the plane sections parallel to the x, z-plane, of the same solid body lying below the surface $z = f(x, y)$ and above the plane region G.

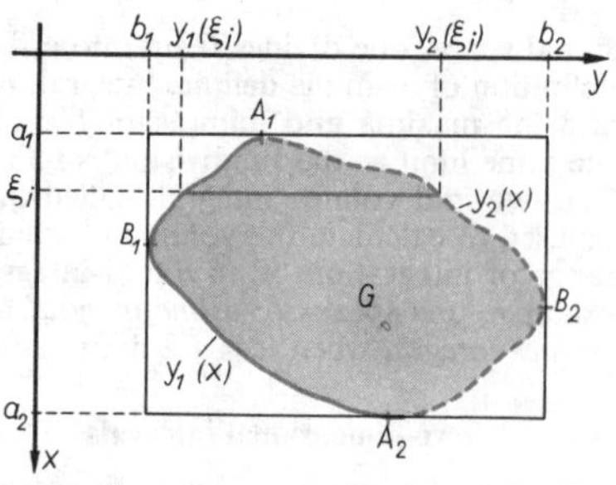

20.3-2 Decomposition of the boundary of the region G of integration

$$\iint_G f(x, y)\,dG = \int_{x=a_1}^{x=a_2}\left(\int_{y_1(x)}^{y_2(x)} f(x, y)\,dy\right)dx = \int_{y=b_1}^{y=b_2}\left(\int_{x_1(y)}^{x_2(y)} f(x, y)\,dx\right)dy$$

For a function $\Phi(r, \varphi)$ given in polar coordinates $x = r\cos\varphi$, $y = r\sin\varphi$, the element of area dG takes the form $dG = r\,dr\,d\varphi$, as follows from calculating the Jacobian (see Transformation of multiple integrals).

$$\begin{vmatrix} \dfrac{\partial x}{\partial r} & \dfrac{\partial y}{\partial r} \\[2ex] \dfrac{\partial x}{\partial \varphi} & \dfrac{\partial y}{\partial \varphi} \end{vmatrix} = \begin{vmatrix} \cos\varphi & \sin\varphi \\ -r\sin\varphi & +\, r\cos\varphi \end{vmatrix} = r.$$

$$\iint_G \Phi(r, \varphi)\,dG = \int_{\varphi_1}^{\varphi_2}\int_{r_1(\varphi)}^{r_2(\varphi)} \Phi(r, \varphi)\,r\,dr\,d\varphi$$

Example 1: In the double integral $\iint_G (x + y)\,dG$, let G be the region between the lines $x = 0$, $y = 1$ and $x + y = 3$ (Fig.). For fixed x, the y-integration goes from the constant limit $y_1 = 1$ up to the variable limit $y_2 = 3 - x$

$$\varphi(x) = \int_1^{3-x} (x + y)\,dy = [xy + y^2/2]_1^{3-x} = x(3 - x) + (3 - x)^2/2 - (x + 1/2) = 4 - x - x^2/2.$$

This function $\varphi(x)$ must now be integrated in the x-direction; the limits of integration are $x = a_1 = 0$ and $x = a_2 = 2$, and the value of the integral is

$$\int_0^2 (4 - x - x^2/2)\,dx = [4x - x^2/2 - x^3/6]_0^2 = 8 - 2 - 4/3 = 14/3; \qquad \int_0^2\int_1^{3-x} (x + y)\,dy\,dx = 14/3.$$

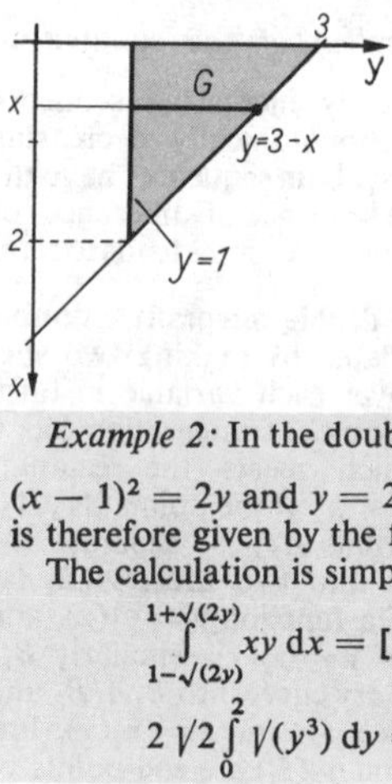

20.3-3 Region G between the lines $x = 0$, $y = 1$ and $x + y = 3$

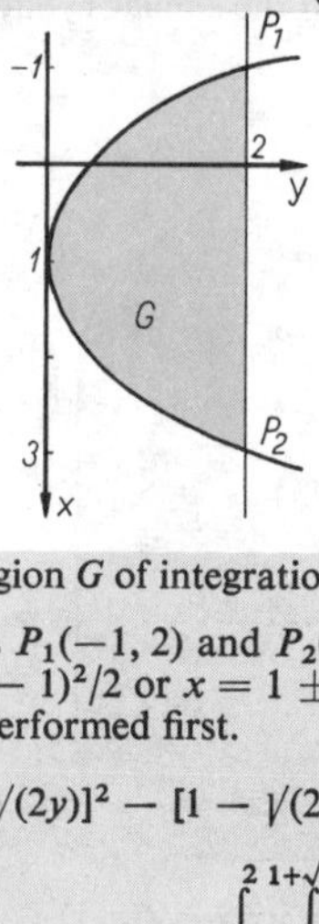

20.3-4 Region of integration G between the curves $(x - 1)^2 = 2y$ and $y = 2$

Example 2: In the double integral $\iint_G xy\,dG$, the region G of integration is enclosed by the curves $(x - 1)^2 = 2y$ and $y = 2$, which meet at the points $P_1(-1, 2)$ and $P_2(3, 2)$. The boundary of G is therefore given by the functions $y_1 = 2$, $y_2 = (x - 1)^2/2$ or $x = 1 \pm \sqrt{(2y)}$ (Fig.).

The calculation is simpler if the x-integration is performed first.

$$\int_{1-\sqrt{(2y)}}^{1+\sqrt{(2y)}} xy\,dx = [yx^2/2]_{1-\sqrt{(2y)}}^{1+\sqrt{(2y)}} = {}^1\!/_2 y\{[1 + \sqrt{(2y)}]^2 - [1 - \sqrt{(2y)}]^2\} = 2\sqrt{(2y^3)};$$

$$2\sqrt{2}\int_0^2 \sqrt{(y^3)}\,dy = 2\sqrt{2}[{}^2\!/_5 y^2\sqrt{y}]_0^2 = 32/5; \qquad \int_0^2\int_{1-\sqrt{(2y)}}^{1+\sqrt{(2y)}} xy\,dx\,dy = 32/5.$$

The calculation for the opposite order of integrations is

$$\int_{(x-1)^2/2}^{2} xy\,dy = [xy^2/2]_{(x-1)^2/2}^{2} = 2x - (x-1)^4 \cdot x/8$$

$$= -x^5/8 + x^4/2 - 3x^3/4 + x^2/2 + 15x/8 = \varphi(x),$$

$$\int_{-1}^{3} \varphi(x)\,dx$$

$$= [-x^6/48 + x^5/10 - 3x^4/16 + x^3/6 + 15x^2/16]_{-1}^{3}$$

$$= 32/5.$$

Example 3: A vertical cylinder is erected above the ellipse $x^2/a^2 + y^2/b^2 = 1$ in the x, y-plane and is cut off obliquely by the plane $z = f(x, y) = mx + ny + c$, where c is so large that the plane $z = f(x, y)$ cuts the x, y-plane in a line outside the ellipse (Fig.). The volume of this truncated cylinder is given by the double integral $\iint_G (mx + ny + c)\,dG$, where the region G is bounded by the ellipse $y = \pm(b/a)\sqrt{(a^2 - x^2)}$. Then

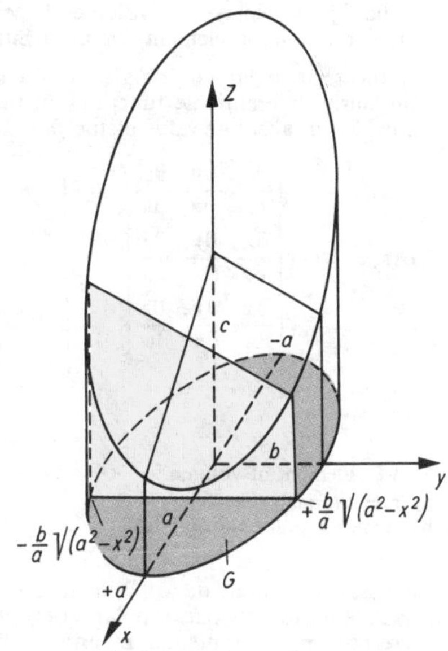

20.3-5 Obliquely truncated elliptic cylinder

$$V = \int_{-a}^{+a} \left[\int_{-(b/a)\sqrt{(a^2-x^2)}}^{+(b/a)\sqrt{(a^2-x^2)}} (mx + ny + c)\,dy \right] dx$$

$$= \int_{-a}^{a} [mxy + ny^2/2 + cy]_{-(b/a)\sqrt{(a^2-x^2)}}^{+(b/a)\sqrt{(a^2-x^2)}}\,dx$$

$$= \int_{-a}^{a} 2(b/a)(mx + c)\sqrt{(a^2 - x^2)}\,dx$$

$$= 2(b/a)\,[m \int_{-a}^{a} x\sqrt{(a^2 - x^2)}\,dx + c\int_{-a}^{a} \sqrt{(a^2 - x^2)}\,dx].$$

The first integral with the factor m outside is zero (one may, for instance, use the substitution $a^2 - x^2 = z$, $-2x\,dx = dz$), and the substitution $x = a \sin z$, $dx = a \cos z\,dz$ leads to the value $a^2 c\pi/2$ for the second integral. Hence the volume V has the value $V = abc\pi$.

Multiple integrals. Just as integration of functions of two variables leads to double integrals, so the integration of functions of three or more variables leads to triple or multiple integrals. If one considers a function of three variables, defined in a bounded three-dimensional region R, and subdivides R into parts ΔR_i, one can again form lower sums $\sum_{i=1}^{n} m_i \Delta R_i$, upper sums $\sum_{i=1}^{n} M_i \Delta R_i$, and intermediate sums $\sum_{i=1}^{n} f(\xi_i, \eta_i, \zeta_i)\,\Delta R_i$. Here m_i is the infimum and M_i the supremum of the function f in the subregion ΔR_i, and $f(\xi_i, \eta_i, \zeta_i)$ is a function value at some point in ΔR_i. If the sequences of lower and upper sums tend to a common limiting value as $n \to \infty$ and $\Delta R_i \to 0$, then so do the intermediate sums, and the common limit is defined to be the triple integral of the function $f(x, y, z)$ over the region R.

Triple integral	$\iiint_R f(x, y, z)\,dR = \lim_{\substack{n\to\infty \\ \Delta R_i \to 0}} \sum_{i=1}^{n} m_i \Delta R_i = \lim_{\substack{n\to\infty \\ \Delta R_i \to 0}} \sum_{i=1}^{n} M_i \Delta R_i$

Any function f that is bounded and continuous in R is integrable in this sense. Whereas the region of integration can still be thought of geometrically as a region in space, a geometrical interpretation of the integral is no longer possible; in the contexts of mechanics it could be interpreted as the total mass of the region if $f(x, y, z)$ were the density at (x, y, z) in the region R. For integrals of functions of more than three variables, defined analogously, even the region of integration no longer has a direct geometrical meaning. Just like double integrals, so multiple integrals may be calculated, under suitable assumptions about the region of integration, by the appropriate number of successive integrations over each of the variables; the limits of integration depend on the nature of the boundary of R.

$$\iiint_R f(x, y, z)\,dR = \int_{x=a_1}^{x=a_2} \left\{ \int_{y_1(x)}^{y_2(x)} \left[\int_{z_1(y,z)}^{z_2(y,z)} f(x, y, z)\,dz \right] dy \right\} dx$$

Transformation of multiple integrals. In many cases it is convenient not to use rectangular (Cartesian) coordinates to describe the region R, but other coordinate systems. The most usual ones, depending on the particular problem being studied, are cylindrical and spherical polar coordinates.

The figure displays the volume element ΔR for cylindrical and spherical polar coordinates. To derive the volume element for an arbitrary coordinate system, the following theorem is quoted:

If the rectangular coordinates $x = x(u, v, w)$, $y = y(u, v, w)$, $z = z(u, v, w)$ are one-to-one continuously differentiable functions of the coordinates u, v, w, then the volume element dR is multiplied by the absolute value of the *functional determinant* (Jacobian) $D(u, v, w)$ displayed below.

$$D(u, v, w) = \begin{vmatrix} \dfrac{\partial x}{\partial u} & \dfrac{\partial y}{\partial u} & \dfrac{\partial z}{\partial u} \\ \dfrac{\partial x}{\partial v} & \dfrac{\partial y}{\partial v} & \dfrac{\partial z}{\partial v} \\ \dfrac{\partial x}{\partial w} & \dfrac{\partial y}{\partial w} & \dfrac{\partial z}{\partial w} \end{vmatrix}$$

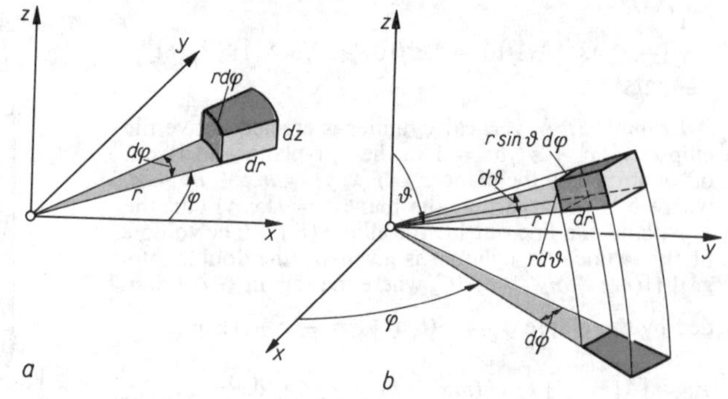

20.3-6 Element of volume
a) in cylindrical,
b) in special polar coordinates

That is, $dR = dx\,dy\,dz = |D|\,du\,dv\,dw$. For *cylindrical* polar coordinates $x = r\cos\varphi$, $y = r\sin\varphi$, $z = z$, and for *spherical* polar coordinates $x = r\sin\vartheta\cos\varphi$, $y = r\sin\vartheta\sin\varphi$, $z = r\cos\vartheta$, the corresponding determinants D_c and D_s turn out to be

$$D_c = \begin{vmatrix} \cos\varphi & \sin\varphi & 0 \\ -r\sin\varphi & +r\cos\varphi & 0 \\ 0 & 0 & 1 \end{vmatrix}, \quad D_s = \begin{vmatrix} \sin\vartheta\cos\varphi & \sin\vartheta\sin\varphi & \cos\vartheta \\ r\cos\vartheta\cos\varphi & r\cos\vartheta\sin\varphi & -r\sin\vartheta \\ -r\sin\vartheta\sin\varphi & r\sin\vartheta\cos\varphi & 0 \end{vmatrix},$$

that is, $D_c = r$ and $D_s = r^2\sin\vartheta$, so that dR becomes $r\,dr\,d\varphi\,dz$ and $r^2\sin\vartheta\,dr\,d\vartheta\,d\varphi$, respectively.

$$\iiint_R f(x, y, z)\,dR = \int_{z_1}^{z_2}\int_{\varphi_1(z)}^{\varphi_2(z)}\int_{r_1(\varphi, z)}^{r_2(\varphi, z)} F(r, \varphi, z)\,r\,dr\,d\varphi\,dz = \int_{\varphi_1}^{\varphi_2}\int_{\vartheta_1(\varphi)}^{\vartheta_2(\varphi)}\int_{r_1(\vartheta, \varphi)}^{r_2(\vartheta, \varphi)} \Phi(r, \vartheta, \varphi)\,r^2\sin^2\vartheta\,dr\,d\vartheta\,d\varphi$$

Cubature

Multiple integrals have important applications in the calculation of the volume V of a solid body B. It has already been observed that the double integral $\iint_G f(x, y)\,dx\,dy$ represents the volume of a cylinder with base G and upper boundary $z = f(x, y)$. It follows that a cylindrical body with cross-section G and upper and lower boundaries $z = f_1(x, y)$, $z = f_2(x, y)$ has the volume $V = \iint_G [f_1(x, y) - f_2(x, y)]\,dx\,dy$. In this way one can find the volume of any body that can be pieced together from a finite number of such cylindrical bodies; most bodies B occurring in practical applications are of this form. The volume V of B can also be expressed as the triple integral $\iiint_B dV$, where the shape of the boundary surface of B determines the limits of integration; in particular, if one inserts the limits for z and performs the z-integration, one is left with a double integral of the type discussed earlier. A further method of calculating the volume of B, if B has a piecewise smooth boundary surface, is furnished by the integral theorem of Gauss. This formula states that

$$V = \iiint_B dV = {}^1/_3 \iint_{\partial B} \boldsymbol{rn}\,dS,$$

where ∂B is the boundary of B, dS is an element of surface, $\boldsymbol{n}$ is the outward normal, and $\boldsymbol{r} = x\boldsymbol{i} + y\boldsymbol{j} + z\boldsymbol{k}$ is the vector field; this reduces the calculation of the volume to the evaluation of a surface integral.

Calculation of volume from areas of cross-sections. Suppose that the solid is referred to Cartesian coordinates x, y, z and lies between the two planes $x = a$, $x = b$ perpendicular to the x-axis. Suppose also that the areas of cross-sections of the body by planes perpendicular to the x-axis are known, and are given by a continuous function $q(x)$. One may then think of the body as being made of slices of thickness Δx_i (Fig.). In each slice there is a smallest cross-sectional area, q_i, and a largest, Q_i; the volume V_i of the ith slice lies between that of a cylinder of height Δx_i and base area q_i, and of a similar cylinder of base area Q_i. Just as for areas one obtains, as approximations to the volume V, lower sums $v(n)$ and upper sums $V(n)$

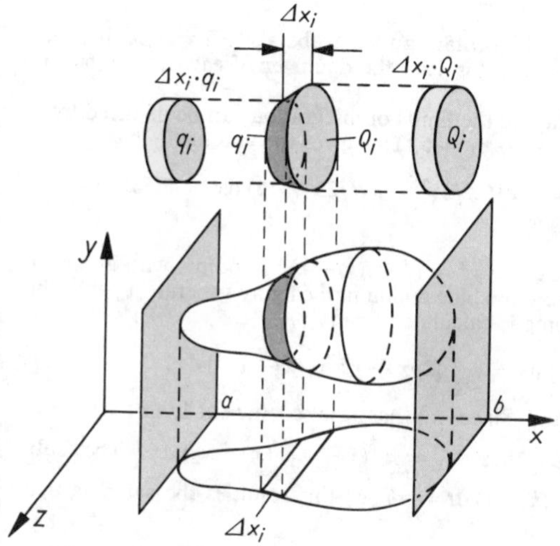

20.3-7 Cubature of a solid

$$v(n) = \sum_{i=1}^{n} q_i \Delta x_i \leqslant V \leqslant \sum_{i=1}^{n} Q_i \Delta x_i = V(n),$$

which have the same limiting value as $n \to \infty$ and $\Delta x_i \to 0$. Hence the volume V may be represented as a definite integral.

$$V = \int_a^b q(x)\,dx$$

Cavalieri's principle. If another body has cross-sectional areas $\bar{q}(x)$ which in $[a, b]$ for each x are the same as $q(x)$, so that $\bar{q}(x) \equiv q(x)$, then the volumes $\bar{V}$ and V of the bodies are equal. This principle was formulated by CAVALIERI before the methods of the integral calculus had been developed.

Two bodies lying between two parallel planes have the same volume if their cross-sections by any plane parallel to these planes have equal areas.

Volume of a solid of revolution. Solids with certain symmetry properties can often be regarded as having boundaries that are generated by the rotation of a curve; for instance, the surface of a sphere is obtained by rotating a semicircle about its diameter. Such a body is called a *solid of revolution*. If its surface is obtained by rotating the continuous curve $y = f(x)$ about the x-axis, or $x = \varphi(y)$ about the y-axis, then the cross-sections by planes perpendicular to the axis are circular regions with areas $q(x) = \pi[f(x)]^2$ or $q(y) = \pi[\varphi(y)]^2$. If the solid of revolution is bounded by planes $x = x_1$ and $x = x_2$, or $y = y_1$ and $y = y_2$, the formulae displayed below give their volumes.

If the solid is obtained by rotation of a continuous curve that consists of several arcs joined together, it is best to add up the volumes of the individual pieces. It may also be appropriate, just as in the calculation of the area between two curves, to integrate the difference of the squares of two suitable functions

$$\pi \int_{x_1}^{x_2} \{[g(x)]^2 - [h(x)]^2\}\,dx.$$

Axis of rotation	Area of cross-section	Volume
x-axis	$q(x) = \pi[f(x)]^2$	$V_x = \pi \int_{x_1}^{x_2} [f(x)]^2\,dx$
y-axis	$q(y) = \pi[\varphi(y)]^2$	$V_y = \pi \int_{y_2}^{y_1} [\varphi(y)]^2\,dy$

Example 1: The curve $y = f(x) = x^2/36$ is rotated between the limits $x_1 = 0$ and $x_2 = 12$ a) about the x-axis, b) about the y-axis (Fig.). The volumes of the corresponding solids are to be calculated:

a) $V_x = \pi \int_{x_1}^{x_2} [f(x)]^2\,dx = \pi \int_0^{12} (x^4/1296)\,dx = 192\pi/5 \approx 120.6.$

b) Using $x = \varphi(y) = 6\sqrt{y}$ and $y_1 = f(0) = 0$, $y_2 = f(12) = 4$, one obtains

$$V_y = \pi \int_{y_1}^{y_2} [\varphi(y)]^2\,dy = \pi \int_0^4 36y\,dy = 288\pi \approx 908.4.$$

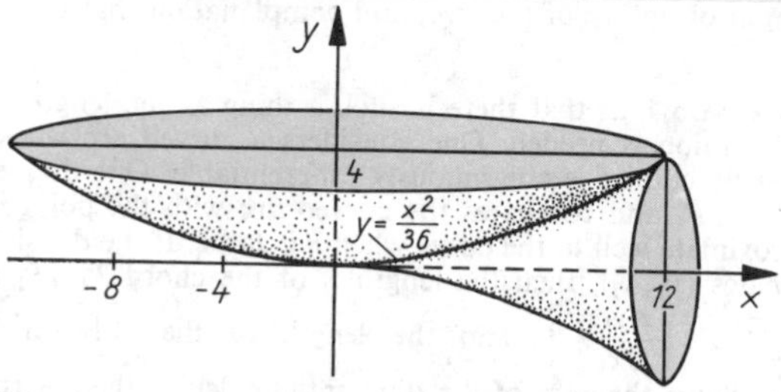

20.3-8 Rotation about x-axis and about y-axis

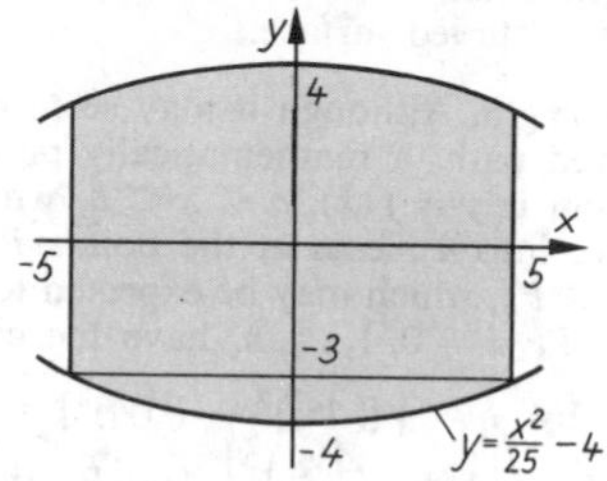

20.3-9 Barrel as paraboloid of rotation

Example 2: The surface of a barrel is described by rotating the parabola $y = ax^2 + c$, between two limits, about the x-axis. The length of the barrel is 1 m, the diameter of each end is 60 cm, the largest diameter is 80 cm (Fig.).

The constants in the equation of the parabola and the limits of integration can be deduced from the given measurements: $y = x^2/25 - 4$, $x_1 = -5$, $x_2 = 5$. This gives the volume

$$V_x = \pi \int_{-5}^{5} (x^4/25^2 - 8x^2/25 + 16)\,\mathrm{d}x = 2\pi \int_{0}^{5} (x^4/25^2 - 8x^2/25 + 16)\,\mathrm{d}x \approx 425.2.$$

The barrel holds 425.2 l.

Example 3: The parabola $y^2 = 2px$ cuts the circle $y^2 = r^2 - (x - c)^2$ in points with abscissae x_1 and x_2 (Fig.). Rotation about the x-axis makes the blue region in the figure generate a parabolic spherical ring of height $h = x_2 - x_1$. Its volume is calculated from

$$V_x = \pi \int_{x_1}^{x_2} 2px\,\mathrm{d}x - \pi \int_{x_1}^{x_2} [r^2 - (r - c)^2]\,\mathrm{d}x = \pi \int_{x_1}^{x_2} [2px - r^2 + (x - c)^2]\,\mathrm{d}x.$$

Since the integrand vanishes at x_1 and x_2, and since x^2 has the coefficient 1, one can put $2px - r^2 + (x - c)^2 = (x - x_1)(x - x_2)$, so that $V_x = \pi \int_{x_1}^{x_2} (x - x_1)(x - x_2)\,\mathrm{d}x$. The substitution $x - x_1 = t$ then leads to $V_x = \pi \int_0^h t(h - t)\,\mathrm{d}t = \pi h^3/6$; this result is the same as that for the cylindrical spherical ring.

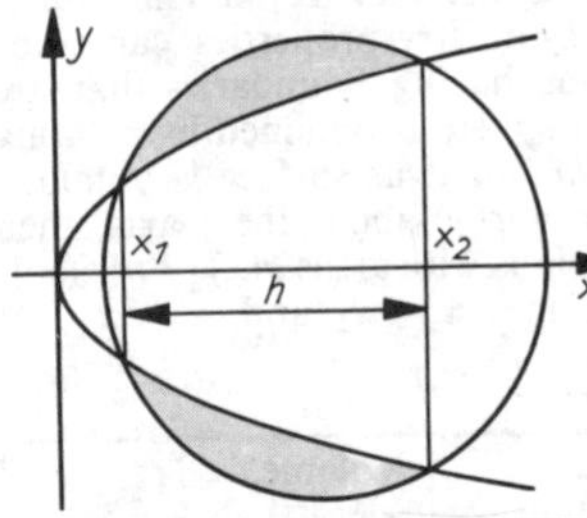

20.3-10 Parabolic spherical ring

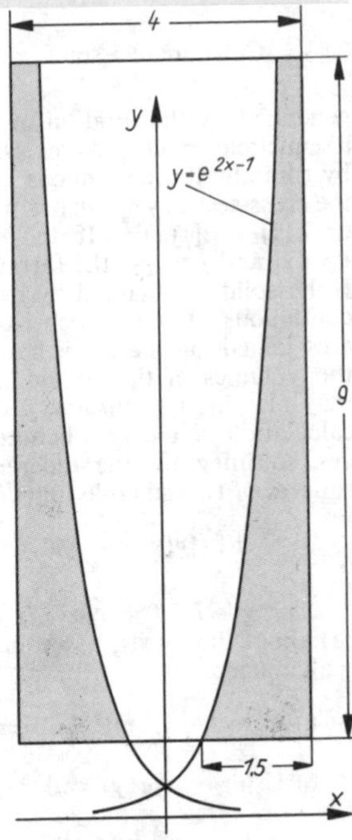

20.3-11 Rotation about the y-axis

Example 4: The diagram (Fig.) indicates a steel cylinder into which a hole has been bored by rotating the curve $y = e^{2x-1}$ about the y-axis. The volume of the hole is $V_y = \pi \int_{y_1}^{y_2} x^2\,\mathrm{d}y$, the limits of integration are obtained from the measurements given in the diagram as $y_1 = 1$, $y_2 = 10$. By solving the equation for x in terms of y, and squaring, one gets

$$x^2 = (1/4)(\ln^2 y + 2\ln y + 1).$$

Hence the required volume is

$$V_y = (\pi/4) \int_1^{10} (\ln^2 y + 2\ln y + 1)\,\mathrm{d}y = (\pi/4)\,[y \ln^2 y + y]_1^{10}$$
$$\approx 48.73.$$

Arc length and surface area

Rectification is the calculation of the length of an arc of a curve, and complanation that of the area of a curved surface.

Arc length. Although it may seem intuitively obvious that there is such a thing as the length of a curved path, a mathematically precise definition is needed. One considers a curved arc whose equation is $y = f(x)$, $a \leqslant x \leqslant b$, where the function f is continuously differentiable. One divides this arc into n pieces at the points $P_0, P_1, \ldots, P_n$ and compares the curved arc with the polygon $P_0P_1 \ldots P_n$, which may be expected to approximate well to the curve when n is large. If the division points P_i, $i = 0, 1, \ldots, n$, have the coordinates (x_i, y_i), then the length l_i of the chord $P_{i-1}P_i$ is given by $l_i = \sqrt{[(\Delta x_i)^2 + (\Delta y_i)^2]} = \Delta x_i \sqrt{\left[1 + \left(\frac{\Delta y_i}{\Delta x_i}\right)^2\right]}$, and the length of the polygon is $s_n = \sum_{i=1}^{n} \sqrt{\left[1 + \left(\frac{\Delta y_i}{\Delta x_i}\right)^2\right]}\,\Delta x_i$. By the mean value theorem of the differential calculus there exists a place ξ_i in the interval $x_{i-1} \leqslant x \leqslant x_i$ such that the derivative $f'(\xi_i)$ is equal to the difference quotient $\Delta y_i/\Delta x_i$, so that $l_i = \Delta x_i \sqrt{(1 + [f'(\xi_i)]^2)}$ (Fig.). Each subdivision of the interval from $x_0 = a$ to $x_n = b$ produces a polygon of length s_n. If one refines the subdivision so that the number n of subintervals tends to infinity and the length Δx_i of the longest subinterval tends to zero, then the sequence s_n of lengths of inscribed polygons has a limiting value $s = \int_a^b \sqrt{(1 + [f'(x)]^2)}\, dx$ because of the continuity of $f'(x)$ and the definition of the definite integral as a limit of sums.

A continuous curve is called *rectifiable* if the lengths s_n of inscribed polygons remain bounded above for all possible subdivisions of $[a, b]$, and then the supremum s of such lengths is called the *arc length* of the curve; it can also be shown that s is the limit of the s_n as the subdivision is refined, that is,

$$s = \lim_{\substack{n \to \infty \\ \Delta x_i \to 0}} s_n .$$

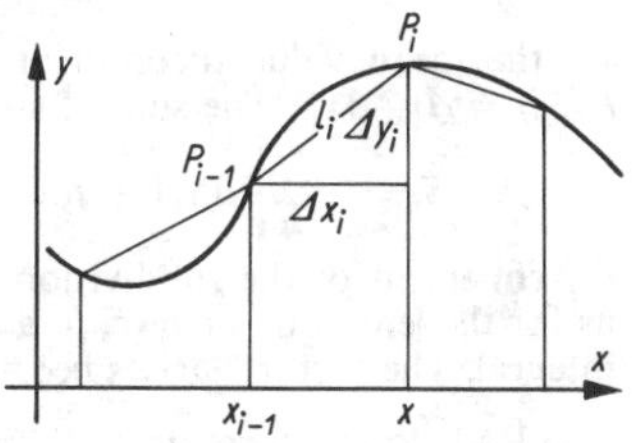

20.3-12 Length of a plane curve

The above derivation has proved that a curve is rectifiable if $y = f(x)$ is continuously differentiable in $[a, b]$. If the continuous curve is rectifiable, but $y = f(x)$ is not continuously differentiable in $[a, b]$, then a direct calculation of the arc length is generally impossible.

Arc length	in rectangular coordinates	in parametric form	in polar coordinates
	$s = \int_a^b \sqrt{(1 + y'^2)}\, dx$	$s = \int_{t_1}^{t_2} \sqrt{(\dot{x}^2 + \dot{y}^2)}\, dt$	$s = \int_{\varphi_1}^{\varphi_2} \sqrt{\left[\left(\frac{dr}{d\varphi}\right)^2 + r^2\right]}\, d\varphi$

Element of length. For a fixed lower limit and a variable upper limit, the length of a curve is a function of the upper limit, $s(x) = \int_a^x \sqrt{(1 + y'^2)}\, dx$. The differential ds of this function is often called *element of length* of the curve, so that the length is the integral of the element of length (Fig.).

Element of length
$ds = \sqrt{(1 + y'^2)}\, dx = \sqrt{[(dx)^2 + (dy)^2]} = \sqrt{(\dot{x}^2 + \dot{y}^2)}\, dt$

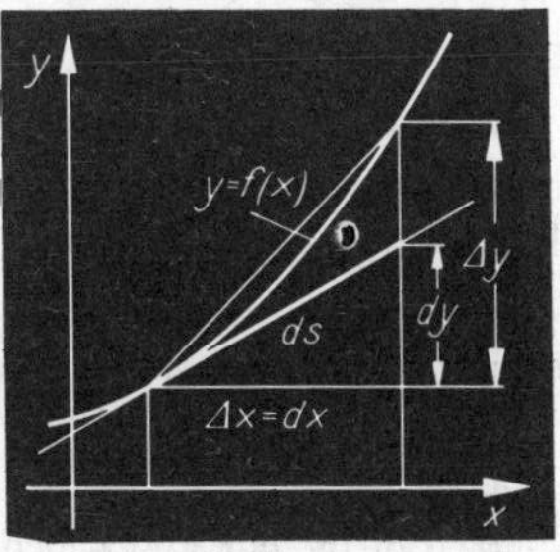

20.3-13 Element of length

Example 1: For the circumference of a circle of radius r, one obtains $y = \sqrt{(r^2 - x^2)}$; $y' = -x/\sqrt{(r^2 - x^2)}$; $1 + y'^2 = r^2/(r^2 - x^2)$. Therefore the circumference C is given by

$$C = 4 \int_0^r \frac{r}{\sqrt{(r^2 - x^2)}}\, dx = 4r \int_0^1 \frac{dt}{\sqrt{(1 - t^2)}} = 4r[\text{Arcsin}\, t]_0^1 = 2\pi r.$$

Example 2: For the cycloid in parametric form $x = a(t - \sin t)$, $y = a(1 - \cos t)$ one has $\dot{x} = a(1 - \cos t)$, $\dot{y} = a \sin t$ so that the element of length is

$$\begin{aligned} ds &= \sqrt{(\dot{x}^2 + \dot{y}^2)}\, dt = \sqrt{[a^2(1 - \cos t)^2 + a^2 \sin^2 t]}\, dt \\ &= a \sqrt{(1 - 2 \cos t + \cos^2 t + \sin^2 t)}\, dt = a\sqrt{2} \cdot \sqrt{(1 - \cos t)}\, dt \\ &= a\sqrt{2}\sqrt{(2 \sin^2 t/2)}\, dt = 2a \sin t/2\, dt. \end{aligned}$$

This gives arc length

$$s = \int_0^{2\pi} \sqrt{(\dot{x}^2 + \dot{y}^2)}\, dt = 2a \int_0^{2\pi} \sin t/2\, dt = -4a[\cos t/2]_0^{2\pi} = 8a.$$

The length of a full arc of the pointed cycloid is therefore four times the diameter of the rolling circle generating the cycloid.

Surface area. The area of a curved surface is given by a surface integral. In what follows, a formula for the area of a surface of revolution will be derived. Let P_1 and P_n be points on a continuously differentiable curve $y = f(x)$, corresponding to $x_1 = a$ and $x_n = b$; then rotation about the x-axis makes the arc of the curve from P_1 to P_n describe a surface of revolution. If, as in the discussion of length, the arc is replaced by a polygon with $n-1$ sides, then the corresponding surface of revolution is a sum of lateral surfaces of frustums of cones. The surface area of a typical such lateral surface is

$$\sigma_\nu = \pi[f(x_\nu) + f(x_{\nu+1})] \sqrt{[(\Delta x_\nu)^2 + (\Delta y_\nu)]^2} = \pi[f(x_\nu) + f(x_{\nu+1})] \Delta x_\nu \sqrt{\left[1 + \frac{\Delta y_\nu}{\Delta x_\nu}\right]^2}.$$

By the mean value theorem of the differential calculus there exists a ξ_ν in $(x_\nu, x_{\nu+1})$ for which $f'(\xi_\nu) = \Delta y_\nu / \Delta x_\nu$. The sum of surface areas is therefore

$$S_n = \pi \sum_{\nu=1}^{n-1} [f(x_\nu) + f(x_{\nu+1})] \sqrt{[1 + (f'(\xi_\nu))^2]} \, \Delta x_\nu .$$

A refinement of the subdivision of the interval $a \leqslant x \leqslant b$ leads to improved approximations and, as for the length of a curve, to a limiting value for the sum S_n, which may be expressed as a definite integral. The factor 2 arises because both $f(x_\nu)$ and $f(x_{\nu+1})$ occur in S_n. Use of the element of length also leads to the formula $S = 2\pi \int_{s_1}^{s_2} y \, ds$.

Area of a surface of revolution	$S = 2\pi \int_a^b y \sqrt{(1 + y'^2)} \, dx$

Example: The formulae for the surface areas of a sphere, spherical cap, and belt of a sphere – because of $y = \sqrt{(r^2 - x^2)}$; $1 + y'^2 = r^2/(r^2 - x^2)$ are:

Sphere: $S = 2\pi \int_{-r}^{r} \sqrt{(r^2 - x^2)} \, r/\sqrt{(r^2 - x^2)} \, dx = 4\pi r \int_0^r dx = 4\pi r^2$.

Spherical cap: $S_c = 2\pi r \int_\xi^r dx = 2\pi r(r - \xi) = 2\pi rh$ with $h = r - \xi$.

Spherical belt: $S_b = 2\pi r \int_{\xi_1}^{\xi_2} dx = 2\pi r(\xi_2 - \xi_1) = 2\pi rh$ with $h = \xi_2 - \xi_1$.

Line and surface integrals

Line integrals. In order to put such physical notions as work, potential, etc. into a mathematical form, it is appropriate to generalize the original concept of integral by considering limits of sums whose summands depend in a certain way on a curve, the path of integration. This leads to the concept of a *line integral.*

Suppose that a smooth curve **C** in three-dimensional space is given in parametric form by the functions $x = x(s)$, $y = y(s)$ and $z = z(s)$ with continuous first derivatives. The parameter may, for example, be the arc length s. Further, let $f(x, y, z)$ be a continuous function whose domain of definition includes the arc AB of the curve corresponding to the parameter interval $\sigma_1 \leqslant s \leqslant \sigma_2$. Then to each s in the interval $[\sigma_1, \sigma_2]$ corresponds a point $P[x(s), y(s), z(s)]$ of the curve at which the function takes the value $f[x(s), y(s), z(s)]$, so that one now has a function of the parameter s. If one subdivides the curve into n arcs or, what amounts to the same thing, divides $[\sigma_1, \sigma_2]$ into n subintervals Δs_i, and forms the sum $\sum_{i=1}^{n} f[x(s_i), y(s_i), z(s_i)] \Delta s_i$, where s_i is an arbitrary parameter value from the subinterval Δs_i, one obtains a sequence of sums. If this sequence tends to a limit when the length of the largest subinterval approaches zero and the number n of subintervals increases to infinity, and if this limit does not depend on the choice of subdivisions and intermediate points s_i, then this limit is called the line integral of the first kind of the function $f(x, y, z)$ along the curve **C** from A to B.

Line integral	$\int_{\mathbf{C}} f(x, y, z) \, ds = \lim_{\substack{\Delta s_i \to 0 \\ n \to \infty}} \sum_{i=1}^{n} f[x(s_i), y(s_i), z(s_i)] \Delta s_i$

The calculation of the integral can be reduced to that of a definite integral. If $x = x(t)$, $y = y(t)$, $z = z(t)$ is any parametric representation of the curve **C** (the arc AB corresponding to the parameter interval $t_1 \leqslant t \leqslant t_2$) then because $\frac{ds}{dt} = \sqrt{[\dot{x}(t)^2 + \dot{y}(t)^2 + \dot{z}(t)^2]}$ one has:

$$\int_{\mathbf{C}} f(x, y, z) \, ds = \int_{t_1}^{t_2} f[x(t), y(t), z(t)] \frac{ds}{dt} \, dt = \int_{t_1}^{t_2} f[x(t), y(t), z(t)] \sqrt{[\dot{x}(t)^2 + \dot{y}(t)^2 + \dot{z}(t)^2]} \, dt$$

If $P(x, y, z)$, $Q(x, y, z)$ and $R(x, y, z)$ are continuous functions, one can define similarly other types of line integral $\int_C P(x, y, z)\,dx$, $\int_C Q(x, y, z)\,dy$, $\int_C R(x, y, z)\,dz$; for example, the first of these is the limit of sums $\sum_{i=1}^{n} P[x_i(s), y_i(s), z_i(s)]\,\Delta x_i$, where Δx_i is the projection on the x-axis of the ith arc of subdivision of the curve. If these three integrals are added together, one obtains the line *integral of the second kind*

$$\int_C [P(x, y, z)\,dx + Q(x, y, z)\,dy + R(x, y, z)\,dz].$$

The calculation of such an integral in two dimensions can sometimes be simplified by applying the following theorem.

If $P(x, y)\,dx + Q(x, y)\,dy$ is the total differential $dF(x, y)$ of a function $F(x, y)$, and if $P(x, y)$ and $Q(x, y)$ are continuous in a connected region G, then the value of the line integral $\int_C [P(x, y)\,dx + Q(x, y)\,dy]$ depends only on the end-points A and B of the path of integration in G and not on the particular path joining A to B.

This follows from the fact that

$$\int_C [P(x, y)\,dx + Q(x, y)\,dy] = \int_C dF(x, y) = \int_{t_1}^{t_2} dF(x, y) = F[x(t_1), y(t_1)] - F[x(t_2), y(t_2)]$$

depends only on the limits of integration. An equivalent statement is that the line integral $\int_C [P\,dx + Q\,dy]$ is zero for every closed curve $\mathbf{C}$ in the region G.

The following theorem, which is easy to deduce from Gauss's theorem (see Divergence and theorem of Gauss), provides a criterion for $P\,dx + Q\,dy$ to be a total differential.

If the region G in question is simply-connected and if the functions $P(x, y)$ and $Q(x, y)$ are continuously differentiable in G, then the integrability condition $\dfrac{\partial P}{\partial y} = \dfrac{\partial Q}{\partial x}$ is necessary and sufficient for $P(x, y)\,dx + Q(x, y)\,dy$ to be a total differential.

In Chapter 22. it will be shown that if $P\,dx + Q\,dy$ is not a total differential, then it is always possible to find an integrating factor $\mu(x, y)$ such that the product $\mu(x, y)(P\,dx + Q\,dy)$ is a total differential.

Integrability condition	$\dfrac{\partial P}{\partial y} = \dfrac{\partial Q}{\partial x}$

Example: To calculate $\int_C (x\,dx + y\,dy)$ along the parabola $y = x^2$ from $A(0, 0)$ to $B(2, 4)$, take x as a parameter. Then $dy = 2x\,dx$, and

$$\int_C (x\,dx + y\,dy) = \int_0^2 (x\,dx + x^2 \cdot 2x\,dx) = \int_0^2 (x + 2x^3)\,dx = [x^2/2 + x^4/2]_0^2 = 10.$$

Since the integrability condition $\dfrac{\partial P}{\partial y} = \dfrac{\partial Q}{\partial x}$ is satisfied, the integral is independent of the path of integration; integration, for example, along the curve $y = 4\sin(\pi x/4)$ from $A(0, 0)$ to $B(2, 4)$ yields the same result as before.

Surface integrals. Just as the line integral generalizes the simple definite integral, so surface integrals are the analogous generalization of the double integral over plane regions. Suppose that $\mathbf{S}$ is a smooth surface, bounded by a piecewise smooth curve, in three-dimensional space with Cartesian coordinates (x, y, z); here 'smooth' means that the tangent plane at interior points of $\mathbf{S}$ depends continuously on its point of contact. Let $\mathbf{S}$ have the parametric representation (see Chapter 26.) $\boldsymbol{x} = \boldsymbol{x}(u, v)$, where the parameters u and v range over the region $U = \{u_1 \leqslant u \leqslant u_2, v_1 \leqslant v \leqslant v_2\}$, and let $f(x, y, z)$ be a continuous function defined on $\mathbf{S}$. One divides $\mathbf{S}$ into small portions S_i formed from a network of smooth curves on $\mathbf{S}$, chooses an arbitrary point $P_i(x_i, y_i, z_i)$ in S_i, and forms the sum $\sum_{i=1}^{n} f(x_i, y_i, z_i)\,\Delta S_i$, where ΔS_i is the surface area of S_i. If this sum approaches a limiting value as $n \to \infty$ and $\Delta S_i \to 0$, independent of the choice of the points P_i, this limit is called the *surface integral* of the function $f(x, y, z)$ over the *surface* $\mathbf{S}$ and is denoted by $\int_S f(x, y, z)\,dS$.

Surface integral	$\int_S f(x, y, z)\,dS = \lim\limits_{\substack{\Delta S_i \to 0 \\ n \to \infty}} \sum_{i=1}^{n} f(x_i, y_i, z_i)\,\Delta S_i$

Calculation of a surface integral can be reduced to that of a double integral as follows: one inserts the parametric representation of the coordinates x, y, z into $f(x, y, z)$; the surface element dS (see

Chapter 26.) has the form

$$dS = \sqrt{(EG - F^2)}, \quad \text{where} \quad E = x_u \cdot x_u, \quad F = x_u \cdot x_v, \quad G = x_v \cdot x_v.$$

Therefore

$$\int_S f(x, y, z)\, dS = \int_U f[x(u, v), y(u, v), z(u, v)] \sqrt{(EG - F^2)}\, du\, dv.$$

For $f = 1$, the integral yields the surface area of **S**.

Surface integrals of the second kind are defined analogously to line integrals of the second kind.

Applications in mechanics

Work. The concept 'work done by a force' is defined in terms of a line integral of the second kind. The force $\boldsymbol{F}$ is a special vector field (see Vector analysis); let F_x, F_y, F_z be its components referred to a Cartesian (x, y, z)-coordinate system. If $P(x, y, z)$, the point at which the force $\boldsymbol{F}$ is applied, moves along a smooth path **C**, then $W = \int_C (F_x\, dx + F_y\, dy + F_z\, dz)$ is called the *work done* by the force $\boldsymbol{F}$ along the path **C** (Fig.). If the vector with components dx, dy, dz is denoted by $d\boldsymbol{r}$, then the integral for work done can be written vectorially, using the scalar product $\boldsymbol{F} \cdot d\boldsymbol{r}$.

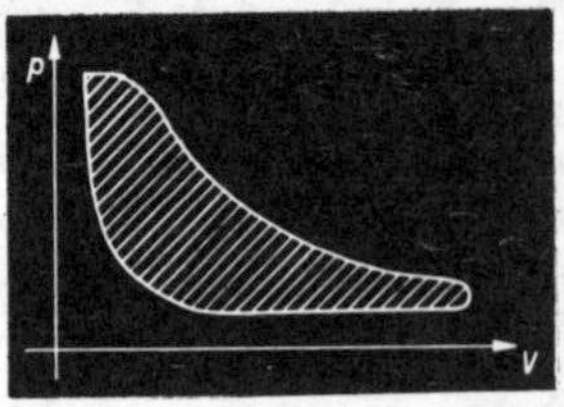

20.3-14 Indicator diagram of a steam engine

Work done is the line integral of the force.

Work as integral	$W = \int_C \boldsymbol{F} \cdot d\boldsymbol{r}$

For instance, if the force $\boldsymbol{F}$ is constant and the path **C** is a segment starting at the origin and represented by a vector $\boldsymbol{r}$ of length $|\boldsymbol{r}|$, then the angle φ between the directions of $\boldsymbol{F}$ and $\boldsymbol{r}$ is also constant and the work integral leads to the formula $W = |\boldsymbol{F}| \cdot |\boldsymbol{r}| \cdot \cos\varphi$; in this case the work done is the product of the component of force $|\boldsymbol{F}| \cos\varphi$ along the direction of the path and the path length $|\boldsymbol{r}|$. In physical problems the components of force are usually the partial derivatives of a function V called the *potential*. Work done is then the line integral of a total differential and depends only on the end-points of the path. The field of force is then called *conservative*.

Example 1: What is the work done in extending a spiral spring by l units of length if the force acts in the direction of the spring? – Denoting the spring constant by D, the force is $F = Dx$ and the work done $W = \int F\, dx = D \int_0^l x\, dx = Dl^2/2$.

Example 2: To accelerate a body of mass m from the velocity v_1 to the velocity v_2 requires work W; since $F = m \dfrac{dv}{dt}$, one obtains

$$W = \int_{s_1}^{s_2} F\, ds = \int_{s_1}^{s_2} m \frac{dv}{dt}\, ds = m \int_{v_1}^{v_2} v\, dv = (v_2^2 - v_1^2) \cdot m/2.$$

The work done is equal to the increase in kinetic energy.

Static moment. The *static moment* M of a point mass about an axis is defined to be the product of the distance l of the point mass from the axis and of the mass m.

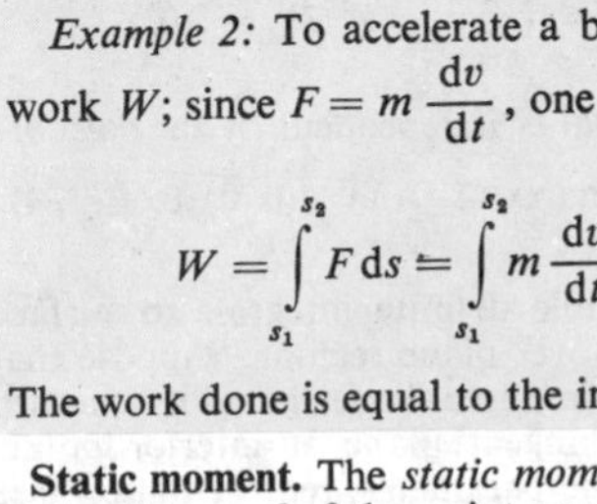

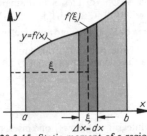

20.3-15 Static moment of a region

Static moment of a continuous mass distribution
$M = \int l\, dm = \varrho \int_V l\, dV$

For the static moment dM of an element of mass dm of a continuous mass distribution, the differential expression $dM = l \cdot dm$ holds, and the static moment of the entire mass is obtained by integration. If ϱ is the constant density and dV the volume element, then $dm = \varrho\, dV$. To define the static moments about each of the coordinate axes for the region below the curve $y = f(x)$, one simply calculates the moments for a continuous mass distribution of density $\varrho = 1$ over a body of thickness $d = 1$ lying above the region (Fig.). To calculate the moment about the y-axis of the region under the curve $y = f(x)$, one decomposes the region into strips of width $\Delta x = \mathrm{d}x$. The area of the strip, by the mean value theorem of the integral calculus, is $f(\xi)\,\mathrm{d}x$ for some intermediate value ξ in Δx, and the moment of the strip is therefore $\mathrm{d}M = \xi f(\xi)\,\mathrm{d}x$. Integration then gives the moment of the region. To obtain the moment about the x-axis, one decomposes each of the above strips into elements of breadth $\Delta y = \mathrm{d}y$. This element has the moment $\mathrm{d}M = \eta\,\mathrm{d}y\,\mathrm{d}x$, where η is an intermediate value in Δy, and integration in the y-direction yields $\mathrm{d}M = \int_0^y \eta\,\mathrm{d}y\,\mathrm{d}x$ for the moment of the strip about the x-axis. Finally, integration over x yields the total moment of the region.

Static moment about the y-axis of the region below the curve $y = f(x)$ between a and b	$M_y = \int_a^b xy\,\mathrm{d}x$
Static moment about the x-axis of the region beneath the curve $y = f(x)$ between a and b	$M_x = \int_a^b \int_0^y \eta\,\mathrm{d}y\,\mathrm{d}x = \frac{1}{2}\int_a^b y^2\,\mathrm{d}x$

Similarly the static moment of a curve can be obtained by considering a uniform mass distribution of density $\varrho = 1$ along the curve; the moments about the x-axis and the y-axis are

$$M_x = \int_a^b y\sqrt{(1 + y'^2)}\,\mathrm{d}x \quad \text{and} \quad M_y = \int_a^b x\sqrt{(1 + y'^2)}\,\mathrm{d}x,$$

respectively. For a body of revolution about the x-axis, one can calculate the static moment with respect to the plane through the origin perpendicular to the x-axis as

$$M = \pi \int_a^b xy^2\,\mathrm{d}x.$$

$$\boxed{M = \pi \int_a^b xy^2\,\mathrm{d}x}$$

Centre of mass. Every solid body can be regarded as a system of point masses, and there is always one point, the *centre of mass*, at which one can imagine the entire mass of the body to be concentrated. The static moment of a continuous mass distribution is equal to the static moment of the centre of mass with respect to the same axis: $M = \int_m l\,\mathrm{d}m = l_c\,m$. By using the appropriate static moments, one can obtain the coordinates of the centre of mass for uniform distributions along a plane curve, over a plane region, and a solid of revolution:

$$x_c = M_y/s \quad \text{and} \quad y_c = M_x/s; \qquad x_c = M_y/A \quad \text{and} \quad y_c = M_x/A; \qquad x_c = M/V.$$

Coordinates (x_c, y_c) of the centre of mass		
a) uniform distribution on a plane curve	b) uniform distribution below the curve $y = f(x)$	c) uniform solid of revolution about the x-axis
$x_c = \dfrac{M_y}{s} = \dfrac{\int_a^b x\sqrt{(1 + y'^2)}\,\mathrm{d}x}{\int_a^b \sqrt{(1 + y'^2)}\,\mathrm{d}x}$;	$x_c = \dfrac{M_y}{A} = \dfrac{\int_a^b xy\,\mathrm{d}x}{\int_a^b y\,\mathrm{d}x}$;	$x_c = \dfrac{M}{V} = \dfrac{\int_a^b xy^2\,\mathrm{d}x}{\int_a^b y^2\,\mathrm{d}x}$;
$y_c = \dfrac{M_x}{s} = \dfrac{\int_a^b y\sqrt{(1 + y'^2)}\,\mathrm{d}x}{\int_a^b \sqrt{(1 + y'^2)}\,\mathrm{d}x}$	$y_c = \dfrac{M_x}{A} = \dfrac{\int_a^b y^2\,\mathrm{d}x}{2\int_a^b y\,\mathrm{d}x}$	$y_c = z_c = 0$

For the solid of revolution about the x-axis, the centre of mass lies on the x-axis, that is, $y_c = z_c = 0$.

Example: To calculate the coordinate of the centre of mass of the region below the curve $y = f(x) = \cos x$ between 0 and $\pi/2$. – The area is $A = \int_0^{\pi/2} \cos x\,\mathrm{d}x = 1$. The required integrals can be found after an integration by parts in each case: $\int_0^{\pi/2} x\cos x\,\mathrm{d}x = \pi/2 - 1$, $\int_0^{\pi/2} \cos^2 x\,\mathrm{d}x = \pi/4$. The formulae for the coordinates of the centre of mass then lead to $x_c = \pi/2 - 1$, $y_c = \pi/8$.

The formulae lead to *Pappus' rules*. The generating region for the solid of revolution about the x-axis has the static moment $M_x = (1/2)\int_a^b y^2\,\mathrm{d}x$ and the ordinate of the centre of mass is $y_c = M_x/A$.

This gives, for the volume of the solid of revolution, the relation

$$V_x = \pi \int_a^b y^2\,dx = 2\pi \cdot {}^1/_2 \int_a^b y^2\,dx = 2\pi M_x = 2\pi y_c A.$$

Pappus' rule for the volume of a solid of revolution: the volume is the product of the area revolved and the length of the path described by its centre of mass.

The surface of the solid of revolution is generated by revolving the curve $y = f(x)$ about the x-axis; this curve has the static moment $M_x = \int_a^b y\sqrt{(1+y'^2)}\,dx$ and its centre of mass has the coordinate $y_c = M_x/s$. It follows that the surface area S_x is given by $S_x = 2\pi \int_a^b y\sqrt{(1+y'^2)}\,dx = 2\pi M_x = 2\pi y_c s$.

Pappus' rule for the area of a surface of revolution: the surface area is the product of the length of the generating curve and the length of the path described by its centre of mass.

Moment of inertia. The kinetic energy W of a body of mass M and velocity v is $W = v^2M/2$. If a rigid body rotates about a fixed axis A, its various portions have different velocities. If ω denotes the constant angular velocity and x the distance of the element of mass dm from the axis of rotation, then this element has the velocity $v = x\omega$ and the kinetic energy $dW = (1/2)\,x^2\omega^2\,dm$. The kinetic energy of the whole body is obtained by integration as $W = (1/2)\,\omega^2 \int_m x^2\,dm$, where the integration is taken over all elements of mass.

axial Moment of inertia polar	$I_A = \int_m x^2\,dm$ $I_P = \int_m r^2\,dm$	dm mass element; x or r distance from axis A of rotation or point P of reference

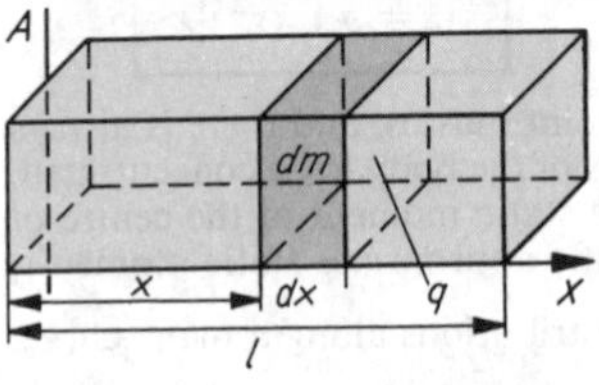

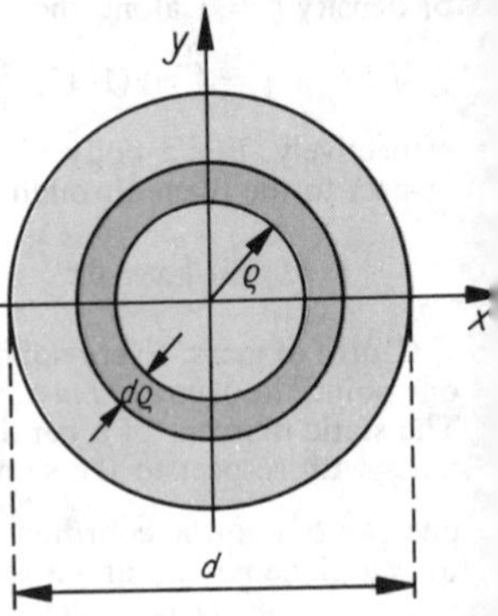

20.3-16 Moment of inertia

If one compares the two expressions for kinetic energy, one notices that the mass M has been replaced by the integral $\int_m x^2\,dm$; this is called the *axial moment of inertia* I_A with respect to the axis of rotation A (Fig.). If the moment of inertia is defined not with respect to a reference axis A but with respect to a reference point P, one obtains the *polar moment of inertia* I_P.

An important relation between the polar moment of inertia Ip with respect to the origin O of a rectangular Cartesian coordinate system and the axial moments of inertia I_x and I_y with respect to the two coordinate axes can be obtained by using the relation $r^2 = x^2 + y^2$, where r, x, y are the distances of a mass element from the origin and from the axes. Thus,

$$I_P = \int_m r^2 dm = \int_m (x^2 + y^2)\,dm = \int_m x^2\,dm + \int_m y^2\,dm = I_x + I_y.$$

Connection between polar and axial moments of inertia	$I_P = I_x + I_y$

Example 1: To find the moment of inertia of a thin straight rod of length l, cross-sectional area q and uniform density ϱ, with respect to an axis passing through one end of the rod and at right angles to it (Fig. 20.3-16). If dx is an element of length of the rod, the corresponding element of mass is $dm = \varrho q\,dx$. Hence $I_A = \int_m x^2\,dm = \int_0^l x^2 \varrho q\,dm = \varrho q \int_0^l x^2\,dm = q\varrho l^3/3 = Ml^2/3$, since the total mass of the rod is $M = \varrho q l$.

Example 2: To find the moments of inertia of a thin circular plate of diameter d with respect to the centre and with respect to one of the lines through the centre (Fig. 20.3-16). To simplify matters, it is assumed the mass per unit area is 1. First one calculates the polar moment of inertia I_P. The mass dm of the dark circular ring in the diagram is $dm = 2\pi\varrho\,d\varrho$, so that $I_P = \int_0^r \varrho^2\,dm = 2\pi \int_0^r \varrho^3\,d\varrho = \pi r^4/2 = \pi d^4/32$. For reasons of symmetry, $I_x = I_y$ and therefore $I_P = I_x + I_y = 2I_x$, $I_x = I_P/2$, so that the axial moments are $I_x = I_y = \pi d^4/64$.

Steiner's theorem. Let I_C be the moment of inertia of a body about an axis C passing through the centre of mass. The moment of inertia I_A about an axis A parallel to C and at a distance a from C is evidently

$$I_A = \int_m (x+a)^2 \, \mathrm{d}m = \int_m (x^2 + 2xa + a^2) \, \mathrm{d}m = I_C + 2a \int_m x \, \mathrm{d}m + a^2 m .$$

But x denotes distance from the axis C, so that the integral $\int_m x \, \mathrm{d}m$ is the static moment about this axis and is zero because the axis C passes through the centre of mass.

Steiner's theorem. The moment of inertia I_A of a body about an arbitrary axis is equal to its moment of inertia about the axis C through the centre of mass and parallel to A plus the product of its mass and the square of the distance from A to C.

Steiner's theorem	$I_A = I_C + a^2 m$

20.4. Vector analysis

In vector analysis one considers vector-valued functions of one or several variables and applies the concepts and methods of the differential and integral calculus. Its applications lie mainly in the fields of mathematical physics and of differential geometry.

Fields

Scalar fields. A scalar function φ in space is called a scalar field if, in a given region, a scalar $\varphi(x, y, z) = \varphi(\boldsymbol{r})$ is assigned to each point $P(x, y, z)$ with position vector $\boldsymbol{r}$; for instance, temperature or density in a body are scalar fields. They can be visualized through the *level surfaces* $\varphi(x, y, z) = \text{const}$ in space or the *level curves* $\varphi(x, y) = \text{const}$ in the plane; for example, there are maps giving lines of constant height above sea level (contours) and lines of constant temperature (isotherms). The function φ changes the more rapidly the closer the level surfaces or level curves are to each other.

Example: The level surfaces of the scalar field $\varphi = x^2 + y^2 + z^2 = r^2$ are spheres with centre at the origin.

Vector fields. If a function $\boldsymbol{v} = \boldsymbol{v}(x, y, z) = \boldsymbol{v}(\boldsymbol{r})$ assigns a vector $\boldsymbol{v}$ to each point in a region, then $\boldsymbol{v} = \boldsymbol{v}(\boldsymbol{r})$ describes a vector field; for instance, fields of force $\boldsymbol{F}(\boldsymbol{r})$ or electric fields $\boldsymbol{E}(\boldsymbol{r})$ are vector fields. These can be visualized by attaching arrows to different points $\boldsymbol{r}$ whose lengths and directions represent the corresponding vector $\boldsymbol{v}(\boldsymbol{r})$ (Fig.).

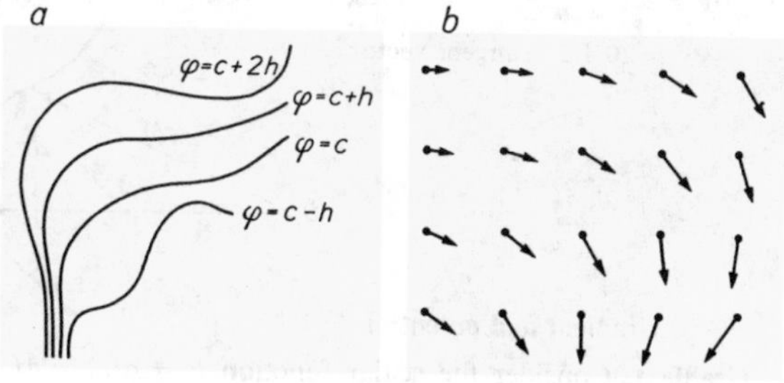

20.4-1 a) Scalar field b) vector field

One usually writes the vector functions of a vector field in the form

$$\boldsymbol{a} = \boldsymbol{a}(x, y, z, t) = u(x, y, z, t)\, \boldsymbol{i} + v(x, y, z, t)\, \boldsymbol{j} + w(x, y, z, t)\, \boldsymbol{k},$$

that is, the field vector $\boldsymbol{a}$ depends on the time t and on the space coordinates x, y, z, which may themselves be functions of t. Differentiation of this vector function is reduced to differentiation of the scalar functions u, v, w by the definition

$$\mathrm{d}\boldsymbol{a} = \mathrm{d}u\, \boldsymbol{i} + \mathrm{d}v\, \boldsymbol{j} + \mathrm{d}w\, \boldsymbol{k},$$

where $\mathrm{d}u = \dfrac{\partial u}{\partial x}\mathrm{d}x + \dfrac{\partial u}{\partial y}\mathrm{d}y + \dfrac{\partial u}{\partial z}\mathrm{d}z + \dfrac{\partial u}{\partial t}\mathrm{d}t$ with analogous expressions for $\mathrm{d}v$ and $\mathrm{d}w$. An equivalent definition is the following:

$$\frac{\partial \boldsymbol{a}}{\partial x} = \lim_{\Delta x \to 0} \frac{\boldsymbol{a}(x + \Delta x, y, z, t) - \boldsymbol{a}(x, y, z, t)}{\Delta x} = \frac{\partial u}{\partial x}\boldsymbol{i} + \frac{\partial v}{\partial x}\boldsymbol{j} + \frac{\partial w}{\partial x}\boldsymbol{k},$$

and analogously

$$\frac{\partial \boldsymbol{a}}{\partial y} = \frac{\partial u}{\partial y}\boldsymbol{i} + \frac{\partial v}{\partial y}\boldsymbol{j} + \frac{\partial w}{\partial y}\boldsymbol{k}, \quad \frac{\partial \boldsymbol{a}}{\partial z} = \frac{\partial u}{\partial z}\boldsymbol{i} + \frac{\partial v}{\partial z}\boldsymbol{j} + \frac{\partial w}{\partial z}\boldsymbol{k}, \quad \frac{\partial \boldsymbol{a}}{\partial t} = \frac{\partial u}{\partial t}\boldsymbol{i} + \frac{\partial y}{\partial t}\boldsymbol{j} + \frac{\partial w}{\partial t}\boldsymbol{k}.$$

That is to say, *differentiation* is carried out *componentwise* according to the usual rules. For example, let $\boldsymbol{a}_1$ and $\boldsymbol{a}_2$ be vector functions and φ a scalar function; then the following rules for differentiating products hold:

1) $\frac{\partial}{\partial x}(\varphi \boldsymbol{a}) = \varphi \frac{\partial \boldsymbol{a}}{\partial x} + \frac{\partial \varphi}{\partial x}\boldsymbol{a}$ and similarly for $\frac{\partial}{\partial y}(\varphi \boldsymbol{a})$, $\frac{\partial}{\partial z}(\varphi \boldsymbol{a})$, $\frac{\partial}{\partial t}(\varphi \boldsymbol{a})$;

2) $\frac{\partial}{\partial x}(\boldsymbol{a}_1 \cdot \boldsymbol{a}_2) = \boldsymbol{a}_1 \cdot \frac{\partial \boldsymbol{a}_2}{\partial x} + \frac{\partial \boldsymbol{a}_1}{\partial x} \cdot \boldsymbol{a}_2$ and similarly for $\frac{\partial}{\partial y}(\boldsymbol{a}_1 \cdot \boldsymbol{a}_2)$, $\frac{\partial}{\partial z}(\boldsymbol{a}_1 \cdot \boldsymbol{a}_2)$, $\frac{\partial}{\partial t}(\boldsymbol{a}_1 \cdot \boldsymbol{a}_2)$;

3) $\frac{\partial}{\partial x}(\boldsymbol{a}_1 \times \boldsymbol{a}_2) = \boldsymbol{a}_1 \times \frac{\partial \boldsymbol{a}_2}{\partial x} + \frac{\partial \boldsymbol{a}_1}{\partial x} \times \boldsymbol{a}_2$ and similarly for

$\frac{\partial}{\partial y}(\boldsymbol{a}_1 \times \boldsymbol{a}_2)$, $\frac{\partial}{\partial z}(\boldsymbol{a}_1 \times \boldsymbol{a}_2)$, $\frac{\partial}{\partial t}(\boldsymbol{a}_1 \times \boldsymbol{a}_2)$.

In what follows, it will be assumed, to simplify matters, that all functions and the partial derivatives that occur are continuous, so that the order of partial differentiations may be interchanged; for instance, $\frac{\partial^2}{\partial x\, \partial y} = \frac{\partial^2}{\partial y\, \partial x}$ (see Chapter 19.).

The most important special cases of vector functions are:

1. The field vector $\boldsymbol{a}$ does not depend explicitly on the time variable t and so has the form

$$\boldsymbol{a}(x, y, z) = u(x, y, z)\,\boldsymbol{i} + v(x, y, z)\,\boldsymbol{j} + w(x, y, z)\,\boldsymbol{k};$$

the field is then said to be *time-independent*.

2. The field depends on a scalar parameter t, not necessarily identical with the time variable; thus $u = x(t)$, $v = y(t)$, $w = z(t)$, or

$$\boldsymbol{a} = \boldsymbol{r}(t) = x(t)\,\boldsymbol{i} + y(t)\,\boldsymbol{j} + z(t)\,\boldsymbol{k}.$$

The position vector $\boldsymbol{r} = \boldsymbol{r}(t)$, as t varies, describes a path in space; if now t is the time variable and $\boldsymbol{r}(t)$ the position of a particle, then the derivative $\frac{d\boldsymbol{r}}{dt} = \frac{dx}{dt}\boldsymbol{i} + \frac{dy}{dt}\boldsymbol{j} + \frac{dz}{dt}\boldsymbol{k} = \dot{x}\boldsymbol{i} + \dot{y}\boldsymbol{j} + \dot{z}\boldsymbol{k}$ represents the *velocity* of the particle and $\frac{d^2\boldsymbol{r}}{dt^2}$ its *acceleration*. The vector $\frac{d\boldsymbol{r}}{dt}$ is tangent to the path $\boldsymbol{r}(t)$ (Fig.). If $\boldsymbol{r}(t)$ has constant length, $|\boldsymbol{r}| = \text{const}$, then also $\boldsymbol{r}^2 = \text{const}$ and hence $\boldsymbol{r} \cdot \frac{d\boldsymbol{r}}{dt} + \frac{d\boldsymbol{r}}{dt} \cdot \boldsymbol{r} = 0$, so that $\boldsymbol{r}$ and $\frac{d\boldsymbol{r}}{dt}$ are perpendicular.

20.4-2 Tangent vector

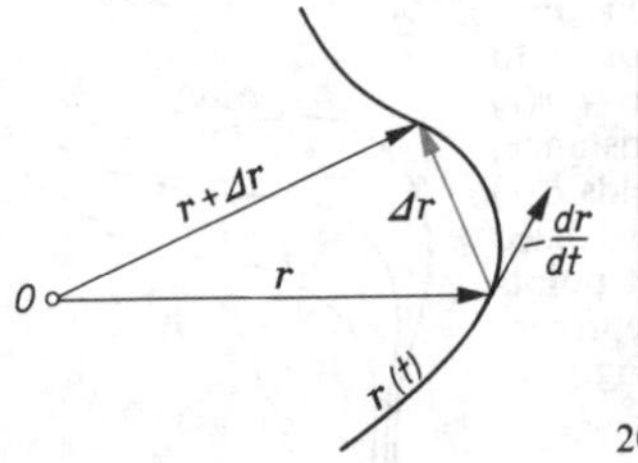

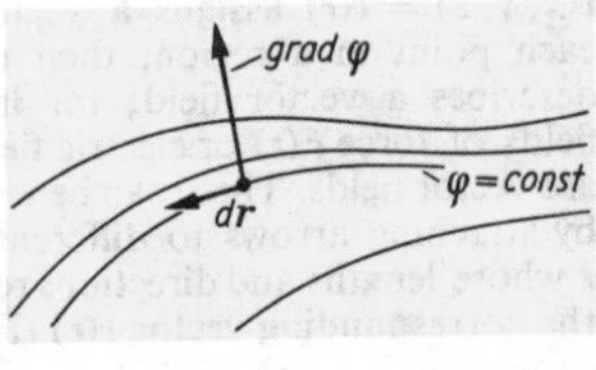

20.4-3 Level surfaces and gradient

Gradient and potential

Gradient. Consider the scalar function $\varphi = \varphi(\boldsymbol{r}) = \varphi(x, y, z)$. If $\boldsymbol{r}$ is increased by $d\boldsymbol{r} = dx\,\boldsymbol{i} + dy\,\boldsymbol{j} + dz\,\boldsymbol{k}$, then the change $d\varphi$ in φ is given by $d\varphi = \frac{\partial \varphi}{\partial x}dx + \frac{\partial \varphi}{\partial y}dy + \frac{\partial \varphi}{\partial z}dz$. This expression can be regarded as the scalar product of the vector $d\boldsymbol{r}$, describing the small movement in space, and a vector $\text{grad}\,\varphi = \frac{\partial \varphi}{\partial x}\boldsymbol{i} + \frac{\partial \varphi}{\partial y}\boldsymbol{j} + \frac{\partial \varphi}{\partial z}\boldsymbol{k}$, the *gradient* of φ. The change in the direction $d\boldsymbol{r}$ is then $d\varphi = \text{grad}\,\varphi \cdot d\boldsymbol{r}$. If, in particular, one chooses $d\boldsymbol{r}$ so that it lies in a level surface $\varphi = \text{const}$, then φ does not change, so that $d\varphi = 0$; hence the gradient $\text{grad}\,\varphi$ is *perpendicular to the level surface* $\varphi = \text{const}$ (Fig.). Now suppose that $d\boldsymbol{r}$ is at an angle ϑ to $\text{grad}\,\varphi$; then $d\varphi = \text{grad}\,\varphi \cdot d\boldsymbol{r} = |\text{grad}\,\varphi| \cdot |d\boldsymbol{r}| \cos\vartheta$, and hence the function *increases most rapidly* when $\vartheta = 0$, that is, *in the direction of the gradient*. If one also sets $|d\boldsymbol{r}| = ds$, then $\frac{d\varphi}{ds} = |\text{grad}\,\varphi| \cos\vartheta$, which means:

Gradient	$\text{grad}\,\varphi = \frac{\partial \varphi}{\partial x}\boldsymbol{i} + \frac{\partial \varphi}{\partial y}\boldsymbol{j} + \frac{\partial \varphi}{\partial z}\boldsymbol{k}$

The derivative of φ in any direction is equal to the projection of the gradient onto this direction.

Potential. The operation of forming the gradient led from a scalar field $\varphi = \varphi(\boldsymbol{r})$ to a vector field $\text{grad}\,\varphi$. In general, one cannot go in the opposite direction, that is, not every vector field is the gradient of a scalar field; vector fields $\boldsymbol{a} = \frac{\partial \varphi}{\partial x}\boldsymbol{i} + \frac{\partial \varphi}{\partial y}\boldsymbol{j} + \frac{\partial \varphi}{\partial z}\boldsymbol{k}$ which are gradients are called *conservative*, and φ is called the potential of $\boldsymbol{a}$ (see Chapter 37.).

Consider the line integral $\int_{P_0}^{P} \boldsymbol{a}\,d\boldsymbol{r} = \int_{P_0}^{P}(u\,dx + v\,dy + w\,dz)$, and let the path of integration be given in the parametric form $\boldsymbol{r} = \boldsymbol{r}(t) = x(t)\,\boldsymbol{i} + y(t)\,\boldsymbol{j} + z(t)\,\boldsymbol{k}$, so that the line integral can be written as an ordinary integral

$$\int_{t_0}^{t}\left(\boldsymbol{a}\,\frac{d\boldsymbol{r}}{dt}\right)dt = \int_{t_0}^{t}\left[u(t)\,\frac{dx}{dt} + v(t)\,\frac{dy}{dt} + w(t)\,\frac{dz}{dt}\right]dt.$$

In general, this integral depends not only on the end-points P_0 and P of the path of integration, but also on the whole path. However, if the vector field $\boldsymbol{a} = u(x, y, z)\,\boldsymbol{i} + v(x, y, z)\,\boldsymbol{j} + w(x, y, z)\,\boldsymbol{k}$ and the path of integration $\boldsymbol{r} = x(t)\,\boldsymbol{i} + y(t)\,\boldsymbol{j} + z(t)\,\boldsymbol{k}$ are defined in a simply-connected region G and have continuous derivatives there, then $\boldsymbol{a} = \operatorname{grad}\varphi$ is a necessary and sufficient condition for the line integral to be independent of the path. Then the following theorem holds:

The line integral of a conservative vector does not depend on the path of integration, and is equal to the potential difference between the initial and final points of the path. Conversely, if the line integral $\int \boldsymbol{a}\,d\boldsymbol{r}$ depends only on the end-points of the path, then $\boldsymbol{a}$ is the gradient of a potential φ.

An equivalent statement is:

In a simply-connected region G the line integral $\oint \boldsymbol{a}\,d\boldsymbol{r}$ vanishes for every closed path in G if and only if $\boldsymbol{a} = \operatorname{grad}\varphi$ for some scalar function φ.

A necessary and sufficient condition for $\boldsymbol{a} = \operatorname{grad}\varphi$ in a simply-connected region G is that

$$\frac{\partial u}{\partial y} = \frac{\partial v}{\partial x}, \qquad \frac{\partial v}{\partial z} = \frac{\partial w}{\partial y}, \qquad \frac{\partial w}{\partial x} = \frac{\partial u}{\partial z},$$

that is, $\operatorname{curl}\boldsymbol{a} \equiv 0$ (see Curl and Stokes' theorem).

Example: To calculate $\oint(-y\,dx + x\,dy)$ round the unit circle with centre at the origin. – The unit circle has the parametric representation $x = \cos t$, $y = \sin t$. Hence $\oint(-y\,dx + x\,dy) = \int_0^{2\pi}[-\sin t(-\sin t) + \cos t(\cos t)]\,dt = \int_0^{2\pi}[\sin^2 t + \cos^2 t]\,dt = 2\pi$. The integral does not vanish, that is, the vector field $\boldsymbol{a} = -x\boldsymbol{i} + y\boldsymbol{j}$ is not conservative; this also follows from $\dfrac{\partial u}{\partial y} = -1$, whereas $\dfrac{\partial v}{\partial x} = +1$.

Divergence and theorem of Gauss

Divergence. The divergence is a scalar field that can be derived from a vector field

$$\bar{\boldsymbol{a}} = \bar{u}(x, y, z)\,\boldsymbol{i} + \bar{v}(x, y, z)\,\boldsymbol{j} + \bar{w}(x, y, z)\,\boldsymbol{k}.$$

As an aid to interpretation one thinks of the field $\boldsymbol{a}$ as the *velocity field* of a fluid flow of density $\varrho = \varrho(x, y, z)$. One assumes that the flow of the fluid is *steady*, that is, $\boldsymbol{a}$ and ϱ do not depend explicitly on time. The x-component $\bar{u}\boldsymbol{i}$ represents the distance per unit time in the x-direction covered by a fluid particle. If one considers a small cuboid (Fig.), with edges parallel to the axes, then the volume of fluid entering across the surface $dA_1 = dy\,dz$ perpendicular to the x-axis in unit time is $dA_1\bar{u} = \bar{u}\,dy\,dz$, and therefore the mass entering per unit time is $\varrho\bar{u}\,dy\,dz$. The mass leaving the volume element $dA_2 = dy\,dz$ is

$$\varrho(x + dx, y, z)\,\bar{u}(x + dx, dy, dz)\,dy\,dz = \left[\varrho\bar{u} + \frac{\partial(\varrho\bar{u})}{\partial x}\,dx\right]dy\,dz;$$

the difference $\dfrac{\partial(\varrho\bar{u})}{\partial x}\,dx\,dy\,dz$ gives the rate at which mass is lost from the volume element across the faces dA_1 and dA_2. The total loss per unit time is obtained by adding contributions from the other two pairs of opposite faces and is equal to

$$\left[\frac{\partial(\varrho\bar{u})}{\partial x} + \frac{\partial(\varrho\bar{v})}{\partial y} + \frac{\partial(\varrho\bar{w})}{\partial z}\right]dx\,dy\,dz.$$

The *loss of mass per unit of time and volume*, the *divergence* of the vector field, is therefore given by $\operatorname{div}\boldsymbol{a}$ after putting $\boldsymbol{a} = \varrho\bar{\boldsymbol{a}}$, $u = \varrho\bar{u}$, $v = \varrho\bar{v}$, $w = \varrho\bar{w}$.

$$\operatorname{div}\boldsymbol{a} = \frac{\partial u}{\partial x} + \frac{\partial v}{\partial y} + \frac{\partial w}{\partial z}$$

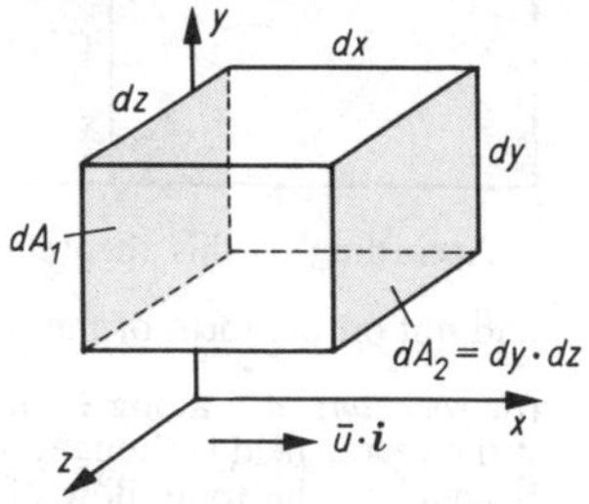

20.4-4 Interpretation of divergence

If there is no loss or gain of mass, then $\operatorname{div}\boldsymbol{a} = 0$ and the field is called *source-free*. Points where $\operatorname{div}\boldsymbol{a} > 0$ are called *sources* (more mass flows out than in); points with $\operatorname{div}\boldsymbol{a} < 0$ are called *sinks* (more flows in than out).

Theorem of Gauss. The *total loss of mass* from a finite region G can be calculated from the volume integral $\iiint_G \operatorname{div} \boldsymbol{a}\, d\tau$. This mass must have flowed out from G across its boundary surface **S**. If $d\boldsymbol{s} = \boldsymbol{n}\, d\sigma$ is a directed surface element, that is, a vector in the direction of the outward normal $\boldsymbol{n}$ ($|\boldsymbol{n}| = 1$) of magnitude equal to the area $d\sigma$ of the surface element, then the previous arguments show that the outward mass flow per unit time across $d\sigma$ is $\boldsymbol{a} \cdot \boldsymbol{n}\, d\sigma$. The total outward flow across **S** is therefore equal to the surface integral $\iint_S (\boldsymbol{a} \cdot \boldsymbol{n})\, d\sigma = \iint_S a_n\, d\sigma$. By equating the two expressions for mass flows, one obtains the theorem of Gauss.

Theorem of Gauss	$\iiint_G \operatorname{div} \boldsymbol{a}\, d\tau = \iint_S \boldsymbol{a} \cdot \boldsymbol{n}\, d\sigma = \iint_S a_n\, d\sigma$
in components	$\iiint_G \left(\frac{\partial u}{\partial x} + \frac{\partial v}{\partial y} + \frac{\partial w}{\partial z}\right) d\tau = \iint_S [u \cos(x, \boldsymbol{n}) + v \cos(y, \boldsymbol{n}) + w \cos(z, \boldsymbol{n})]\, d\sigma$

This theorem, so intuitively clear in the hydrodynamical example, is true quite generally for continuously differentiable vector fields when G is a bounded closed region whose boundary **S** can be divided into pieces with continuously varying normals, apart from finitely many curves or points. The theorem of Gauss makes it possible to convert volume integrals into surface integrals; it also leads to the intuitively obvious result:

If $\boldsymbol{a}$ is source-free in a region, that is, $\operatorname{div} \boldsymbol{a} = 0$, then the total flow across the boundary of the region vanishes.

Curl and Stokes' theorem

Curl. The curl is an operation of differentation, which converts the vector field $\boldsymbol{a} = u\boldsymbol{i} + v\boldsymbol{j} + w\boldsymbol{k}$ into another vector field.

$$\operatorname{curl} \boldsymbol{a} = \left(\frac{\partial w}{\partial y} - \frac{\partial v}{\partial z}\right)\boldsymbol{i} + \left(\frac{\partial u}{\partial z} - \frac{\partial w}{\partial x}\right)\boldsymbol{j} + \left(\frac{\partial v}{\partial x} - \frac{\partial u}{\partial y}\right)\boldsymbol{k}$$

Stokes' theorem. Suppose that **C** is a closed, not necessarily plane, curve bounding a piece of surface **S**. Suppose that **S** has a continuously varying unit normal $\boldsymbol{n}$, except at finitely many curves or points, and that **C** has a continuously varying tangent, except at finitely many points. The sense of direction for **C** shall be that for which **S** lies on the left if one looks at **S** from the side indicated by $\boldsymbol{n}$ (Fig.). Then Stokes' theorem holds for any continuously differentiable vector field.

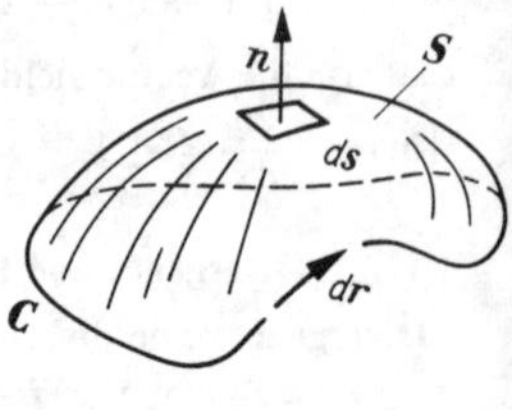

20.4-5 Stokes' theorem

Theorem of Stokes	$\oint_C \boldsymbol{a}\, d\boldsymbol{r} = \iint_S (\boldsymbol{n} \cdot \operatorname{curl} \boldsymbol{a})\, d\sigma = \iint_S \operatorname{curl}_n \boldsymbol{a}\, d\sigma$
in components	$\oint_C (u\, dx + v\, dy + w\, dz) = \iint_S \left\{\left(\frac{\partial w}{\partial y} - \frac{\partial v}{\partial z}\right) \cos(x, \boldsymbol{n}) + \left(\frac{\partial u}{\partial z} - \frac{\partial w}{\partial x}\right) \cos(y, \boldsymbol{n}) + \left(\frac{\partial v}{\partial x} - \frac{\partial u}{\partial y}\right) \cos(z, \boldsymbol{n})\right\} d\sigma$

According to this theorem the integral $\iint_S (\boldsymbol{n} \cdot \operatorname{curl} \boldsymbol{a})\, d\sigma$ depends only on the boundary curve **C** and not on the form of the surface **S** bounded by **C**. The line integral $\int_C \boldsymbol{a}\, d\boldsymbol{r}$ is called the *circulation* (or *rotation*) of $\boldsymbol{a}$ along **C**; it measures the strength of the rotational movement of a fluid element if the vector field is thought of as describing a fluid flow. Stokes' theorem states that this circulation is equal to the total flow of the normal component of the vector field $\operatorname{curl} \boldsymbol{a}$ across a surface **S** spanned by **C**.

If one interprets $\boldsymbol{a}$ as a field of force, then $-\boldsymbol{a}\, d\boldsymbol{r}$ is the work done against the force $\boldsymbol{a}$ along $d\boldsymbol{r}$, and $\oint \boldsymbol{a}\, d\boldsymbol{r}$ is the total work done after a complete circuit round **C**. It is zero, and therefore independent of the path **C**, only if $\operatorname{curl} \boldsymbol{a} \equiv 0$. Fields for which $\operatorname{curl} \boldsymbol{a} \equiv 0$ are called *irrotational*. An irrotational field $\boldsymbol{a}$ can always be represented as a gradient $\boldsymbol{a} = \operatorname{grad} \varphi$. Hence $\operatorname{curl}(\operatorname{grad} \varphi) = 0$, a relationship that can also be verified directly.

The operator nabla, rules of calculation

The three differential operators grad, div, curl can be written in terms of a single operator, the *nabla operator*, introduced by HAMILTON and denoted by ∇. The name nabla is that of a Hebrew stringed instrument whose shape resembles that of the symbol.

The operator ∇ is defined as

$$\nabla = \boldsymbol{i}\frac{\partial}{\partial x} + \boldsymbol{j}\frac{\partial}{\partial y} + \boldsymbol{k}\frac{\partial}{\partial z}.$$

If one interprets the 'product' $\frac{\partial}{\partial x}\cdot\varphi$ as $\frac{\partial\varphi}{\partial x}$, one can write $\nabla\varphi = \operatorname{grad}\varphi$. The scalar product gives $\nabla\cdot\boldsymbol{a} = \operatorname{div}\boldsymbol{a}$, and the vector product gives $\nabla\times\boldsymbol{a} = \operatorname{curl}\boldsymbol{a}$.

Another differential operator, introduced by LAPLACE denoted by $\triangle$ and called delta, is defined for scalar fields φ by

$$\triangle\varphi = \operatorname{div}\operatorname{grad}\varphi = \nabla\cdot(\nabla\varphi) = \frac{\partial^2\varphi}{\partial x^2} + \frac{\partial^2\varphi}{\partial y^2} + \frac{\partial^2\varphi}{\partial z^2},$$

but for vector fields $\boldsymbol{a}(x, y, z)$ by

$$\triangle\boldsymbol{a} = \operatorname{grad}\operatorname{div}\boldsymbol{a} - \operatorname{curl}\operatorname{curl}\boldsymbol{a} = \nabla(\nabla\cdot\boldsymbol{a}) - \nabla\times(\nabla\times\boldsymbol{a}) = \frac{\partial^2\boldsymbol{a}}{\partial x^2} + \frac{\partial^2\boldsymbol{a}}{\partial y^2} + \frac{\partial^2\boldsymbol{a}}{\partial z^2}.$$

The following identities hold:

$$\operatorname{grad}(\varphi_1\varphi_2) = \varphi_1\operatorname{grad}\varphi_2 + \varphi_2\operatorname{grad}\varphi_1, \qquad \operatorname{div}(\varphi\boldsymbol{a}) = \varphi\operatorname{div}\boldsymbol{a} + \boldsymbol{a}\cdot\operatorname{grad}\varphi,$$

$$\operatorname{curl}(\varphi\boldsymbol{a}) = \varphi\operatorname{curl}\boldsymbol{a} - \boldsymbol{a}\times\operatorname{grad}\varphi, \qquad \operatorname{curl}\operatorname{grad}\varphi = 0,$$

$$\operatorname{div}(\boldsymbol{a}_1\times\boldsymbol{a}_2) = \boldsymbol{a}_2\cdot\operatorname{curl}\boldsymbol{a}_1 - \boldsymbol{a}_1\cdot\operatorname{curl}\boldsymbol{a}_2, \qquad \operatorname{div}\operatorname{curl}\boldsymbol{a} = 0.$$

Finally, it should at least be mentioned that gradient, divergence, and curl of a field are objects that are independent of the coordinate system used. One says that these quantities are *invariant* under coordinate transformations.

21. Series of functions

Modern mathematics cannot do without the theory and applications of series of functions, which are very important in the development of analysis and the theory of functions. It was already pointed out in Chapter 20. that some functions can be integrated only by using their expansion in a series. Such expansions are also often useful in practical applications. They can be used to investigate properties of a function when only a few of its values are known, to calculate approximate values of functions, and to give rapid and reliable estimates for the accuracy of methods of calculation.

In Chapter 18. the properties of infinite series with constant terms are treated. Now the terms of a series are functions of some variable; of special importance are power series, whose nth term is a function of the form a_nx^n, and Fourier series, whose general term has the form $a_n\cos nx + b_n\sin nx$.

21.1. Series of functions

Properties of a series of constant terms were derived by considering sequences of numbers. These arguments will now be extended by considering expressions $F(x) = \sum_{n=0}^{\infty} f_n(x)$, which have the following interpretation:

1. For each natural number $n = 0, 1, 2, \ldots$ a function of the sequence $f_0(x), f_1(x), \ldots, f_n(x), \ldots$ is given, each of the functions is defined for x ranging over an interval I, that is, for each x in this domain of definition it assumes a unique value of its range.

2. The sequence $F_n(x)$ of *partial sums* $F_n(x) = f_0(x) + f_1(x) + \cdots + f_n(x)$, $n = 0, 1, 2, \ldots$, consists of functions that are defined, for each n, in the interval I.

3. For each x in I the sequence $F_n(x)$ tends to a limit, which is denoted by $F(x) = \lim_{n\to\infty} F_n(x)$. This *limit function* exists in the interval I, which is called the *convergence interval*. The difference $R_n(x) = F(x) - F_n(x)$ between the limit function $F(x)$ and the approximation $F_n(x)$ is called the remainder and must tend to zero, as $n \to \infty$, if convergence is to take place.

Uniform convergence. In the series of functions $F(x) = x^2 + x^2(1 - x^2) + x^2(1 - x^2)^2 + \cdots = \sum_{n=0}^{\infty} x^2(1 - x^2)^n$, the terms $f_n(x) = x^2(1 - x^2)^n$ are continuous. The sequence of partial sums $F_0(x) = x^2$, $F_1(x) = x^2 + x^2(1 - x^2)$, ... converges for all $x \neq 0$ in the interval $-1 \leqslant x \leqslant 1$ because the series $F(x) = x^2[1 + (1 - x^2) + (1 - x^2)^2 + \cdots]$ is a geometric series with the ratio $1 - x^2 = q < 1$ and its sum is $F(x) = x^2\{1/[1 - (1 - x^2)]\} = 1$. On the other hand, $F(0) = 0$. Thus, the limit function $F(x)$, unlike the functions $f_n(x)$, is not continuous at $x = 0$.

This raises the question under what conditions properties, such as continuity or differentiability, of the terms $f_n(x)$ of the series can be transferred to the sum function $F(x)$. The diagram of the partial sum curves for the series $F(x) = \sum x^2(1 - x^2)^n$ gives a hint. For $n > 10$, the curves can hardly be separated from each other when $|x| > 0.6$ but for $x = 0.2$, for instance, they differ quite appreciably (Fig.). That is to say: the index N beyond which the size of the remainder $R_n(x)$ lies below a given positive number ε depends, in general, on the choice of x within the interval I. These indices $N(\varepsilon, x)$, in the present example, increase indefinitely as $x \to 0$ for any given $\varepsilon > 0$. If, however, one can find an N that does not depend on x, one speaks of *uniform convergence* of the function series $F(x)$. This is always the case if the set of numbers $N(\varepsilon, x)$ is bounded above whenever ε is fixed and x varies over I. It will be shown in what follows that in this case $F(x)$ is continuous, and also that the series $F(x)$ can be integrated 'term-by-term' and can be differentiated in this way if the functions $f_n(x)$ are all differentiable over I and the series resulting from term-by-term differentiation is uniformly convergent. On the other hand, the series $\sum x^2(1 - x^2)^n$ considered earlier is convergent, but not uniformly convergent, in the interval $-1 \leqslant x \leqslant 1$.

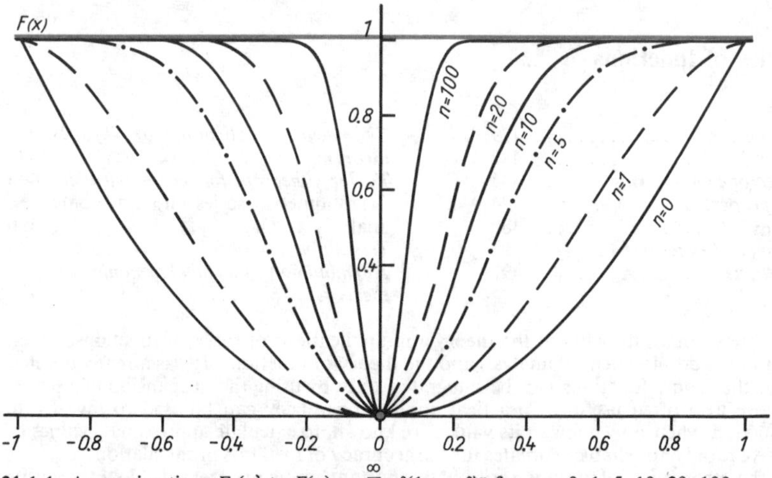

21.1-1 Approximations $F_n(x)$ to $F(x) = \sum_{n=0}^{\infty} x^2(1 - x^2)^n$ for $n = 0, 1, 5, 10, 20, 100$

A function series $F(x) = \sum_{n=0}^{\infty} f_n(x)$ is called uniformly convergent in an interval I if, for each given $\varepsilon > 0$, there exists an $N = N(\varepsilon)$, depending only on ε but not on x, such that $|R_n(x)| = |f_{n+1}(x) + f_{n+2}(x) + \cdots| < \varepsilon$ for every x in I provided that $n \geqslant N$.

The concept of uniform convergence was introduced by Karl WEIERSTRASS (1815–1897) and others; it can be extended, as can the following criterion, to the complex case.

The Weierstrass majorant criterion ('M-test'). If each function $f_n(x)$ in the series $F(x) = \sum_{n=0}^{\infty} f_n(x)$ is bounded, with $|f_n(x)| \leqslant M_n$ for all x in I, and if the series $\sum_{n=0}^{\infty} M_n$ converges, then the series

$F(x) = \sum_{n=0}^{\infty} f_n(x)$ is uniformly convergent in the interval I. The series $\sum_{n=0}^{\infty} M_n$ is then called a convergent majorant for $\sum_{n=0}^{\infty} f_n(x)$.

Example 1: The series $\sum_{n=1}^{\infty} \sin(nx)/n^2$ is uniformly convergent for all x, because $|\sin(nx)/n^2| \leqslant 1/n^2$ for all x and the majorant series $\sum_{n=1}^{\infty} 1/n^2$ converges.

Example 2: The geometric series $x + x^2 + x^3 + \cdots$ converges in the interval $-1 < x < 1$. For a fixed x_0 with $0 < x_0 < 1$ and given $\varepsilon > 0$ one can always find an $N(\varepsilon)$ such that for $n > N(\varepsilon)$ $|R_n(x_0)| = |x_0^{n+1} + x_0^{n+2} + \cdots| = x_0^{n+1}/(1 - x_0) < \varepsilon$. But if x_0 increases towards $+1$, then $|R_n|$ increases indefinitely for any fixed $n > N(\varepsilon)$, that is, $\lim_{x_0 \to 1} x_0^{n+1}/(1 - x_0) = \infty$. This shows that the geometric series converges uniformly in any closed portion $|x| \leqslant x_0 < 1$ of the convergence interval $-1 < x < 1$, but does not converge uniformly in the whole open convergence interval.

It can be shown that instead of the remainder $R_n(x)$ one may consider an arbitrary section $F_{n+k}(x) - F_n(x)$ of the series; the condition for uniform convergence then reads

$$|F_{n+k}(x) - F_n(x)| = |f_{n+1}(x) + f_{n+2}(x) + \cdots + f_{n+k}(x)| < \varepsilon$$

for all x in I, all $n \geqslant N(\varepsilon)$ and all $k \geqslant 1$.

Limits of a series of functions. Suppose that the series $F(x) = \sum_{n=0}^{\infty} f_n(x)$ converges uniformly in the interval $a < x < x_0$. Then, for given $\varepsilon > 0$ one can find an index n_1 such that for all x in the interval, all $n > n_1$ and all $k \geqslant 1$,

$$|f_{n+1}(x) + f_{n+2}(x) + \cdots + f_{n+k}(x)| < \varepsilon.$$

Suppose, in addition, that each function $f_n(x)$ has a left-hand limit a_n as $x \to x_0 - 0$; then it may be substituted in the inequality $|a_{n+1} + a_{n+2} + \cdots + a_{n+k}| < \varepsilon$, but this means that the series $\sum a_n$ formed from the limits converges. If its sum is s and its partial sums are s_n, an index n_2 may be chosen so that $|s_m - s| < \varepsilon/3$ for all $m > n_2$ and also $|R_m(x)| < \varepsilon/3$. This can be used to show that the limit function $F(x)$ has the limit s as $x \to x_0 - 0$. Indeed, for the chosen but fixed m and all x in $a < x < x_0$

$$|F(x) - s| = |(F_m(x) - s_m) + (s_m - s) + R_m(x)| < |F_m(x) - s_m| + \varepsilon/3 + \varepsilon/3.$$

But here the function $F_m(x)$, being the sum of a fixed number of functions $f_n(x)$, has the limit s_m as $x \to x_0 - 0$; this means that for a positive $\delta < (x_0 - a)$ a subinterval $x_0 - \delta < x < x_0$ can be found such that $|F_m(x) - s_m| < \varepsilon/3$ and hence $|F(x) - s| < \varepsilon$ for all such x. The sum s is therefore the limit of $F(x)$ as $x \to x_0 - 0$. The result can be summarized in the following form:

$$\lim_{x \to x_0 - 0} \left[\sum_{n=0}^{\infty} f_n(x)\right] = \sum_{n=0}^{\infty} \left[\lim_{x \to x_0 - 0} f_n(x)\right],$$

which means that in a uniformly convergent series of functions passages to a limit can be carried out term-by-term.

Analogous arguments hold for right-hand limits. If the functions $f_n(x)$ are continuous at x_0, then their left-hand and right-hand limits are equal to their values $f_n(x_0)$, that is, the sum function $F(x)$ is continuous at x_0.

If the series $F(x) = \sum_{n=0}^{\infty} f_n(x)$ is uniformly convergent in an interval I and if the terms $f_n(x)$ are continuous at $x = x_0$, then $F(x)$ is also continuous at $x = x_0$.

Differentiation and integration term-by-term. If the functions $f_n(x)$ are differentiable in I, if the series $f(x) = \sum_{n=0}^{\infty} f_n'(x)$ formed with the derivatives is uniformly convergent in I, and if $\sum_{n=0}^{\infty} f_n(x)$ converges for at least one $x = x_0$ in I, then this series converges uniformly for all x in I, and the derivative of its sum $F(x) = \sum_{n=0}^{\infty} f_n(x)$ is the sum of the differentiated series, that is, $F'(x) = \sum_{n=0}^{\infty} f_n'(x)$. The detailed proof of these statements is omitted; they are often stated in the less precise form:

A series of functions may be differentiated term-by-term if the resulting series is uniformly convergent.

$$\frac{d}{dx}\left[\sum_{n=0}^{\infty} f_n(x)\right] = \sum_{n=0}^{\infty} f_n'(x)$$

Although he did not have the concept of uniform convergence, ABEL (1802–1829) gave the following examples to show that termwise differentiation need not always lead to correct answers:

1. The series $\sum_{n=1}^{\infty} \sin(nx)/n$ converges for all real x, but the series obtained by termwise differentiation is $\sum_{n=1}^{\infty} \cos nx$ which diverges for all x. 2. The series $\sum_{n=1}^{\infty} f_n(x) = \sum_{n=1}^{\infty} \sin(nx)/n^2$ converges uniformly for all x, but the differentiated series $\sum_{n=1}^{\infty} f_n'(x) = \sum_{n=1}^{\infty} \cos(nx)/n$ diverges for $x = 0$.

If the terms $f_n(x)$ of a series are integrable on an interval I and if the series is uniformly convergent in I, then the series obtained by termwise integration also converges in I and represents the integral $\int F(x)\,dx$ of the limit function $F(x) = \sum_{n=0}^{\infty} f_n(x)$, that is, $\int F(x)\,dx = \sum_{n=0}^{\infty} \int f_n(x)\,dx$.

A uniformly convergent series of functions may be integrated term-by-term.

Arc length and circumference of an ellipse can be obtained by termwise integration of the series for the element of arc. For the arc s (Fig.) of the ellipse, let $x = a \sin\varphi$, $y = b\cos\varphi$, so that $dx = a\cos\varphi\,d\varphi$, $dy = -b\sin\varphi\,d\varphi$, and let $\varepsilon^2 = 1 - b^2/a^2$ (ε is called the eccentricity). Then

$$s = \int_0^{\varphi} \sqrt{(dx^2 + dy^2)} = a\int_0^{\varphi} \sqrt{(1 - \varepsilon^2 \sin^2\varphi)}\,d\varphi,$$

where use has been made of the fact that

$$\begin{aligned} a^2\cos^2\varphi + b^2\sin^2\varphi &= a^2 - (a^2 - b^2)\sin^2\varphi \\ &= a^2(1 - \varepsilon^2\sin^2\varphi). \end{aligned}$$

Since $|\varepsilon| < 1$, the square root may be expanded in a uniformly convergent binomial series

$$\sqrt{(1 - \varepsilon^2\sin^2\varphi)}$$
$$= 1 - \frac{\varepsilon^2}{2}\sin^2\varphi - \frac{\varepsilon^4}{2\cdot 4}\sin^4\varphi - \frac{1\cdot 3\cdot \varepsilon^6}{2\cdot 4\cdot 6}\sin^6\varphi - \cdots.$$

Term-by-term integration leads to

$$s = a\left(\varphi - \frac{\varepsilon^2}{2}\int_0^{\varphi}\sin^2\varphi\,d\varphi - \frac{\varepsilon^4}{2\cdot 4}\int_0^{\varphi}\sin^4\varphi\,d\varphi - \cdots\right),$$

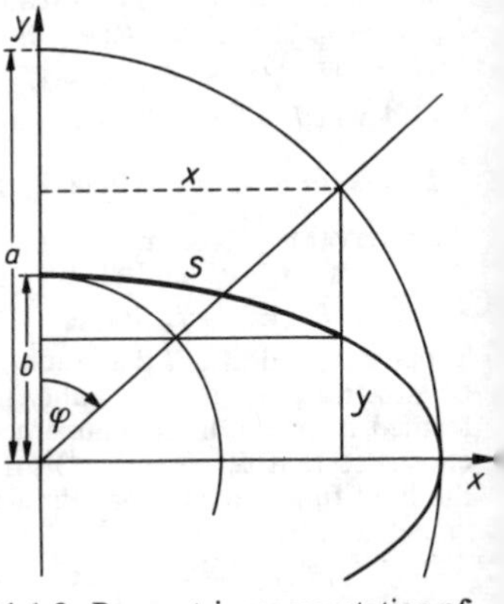

21.1-2 Parametric representation of an ellipse

which makes it possible to calculate the arc length s for any angle φ. To find the circumference, put $\varphi = \pi/2$ and calculate the length of a quarter of the circumference, using the recurrence formulae in Chapter 20. for the derivation of Wallis' product to obtain:

Circumference of ellipse	$U = 2\pi a\left[1 - \left(\frac{1}{2}\right)^2 \varepsilon^2 - \left(\frac{1\cdot 3}{2\cdot 4}\right)^2 \frac{\varepsilon^4}{3} - \left(\frac{1\cdot 3\cdot 5}{2\cdot 4\cdot 6}\right)^2 \frac{\varepsilon^6}{5} - \cdots\right]$	
Approximation formula	$U \approx \pi\left[\frac{3}{2}(a + b) - \sqrt{(ab)}\right]$	$\varepsilon^2 = 1 - b^2/a^2$

To estimate the error Δ caused by using the approximation formula, one first uses the relations $(a + b)/2 = (a/2)\,[1 + \sqrt{(1 - \varepsilon^2)}]$, $\sqrt{(ab)} = a\sqrt[4]{(1 - \varepsilon^2)}$, to expand these expressions in binomial series with remainders R_3' and R_3'' for which upper estimates r' and r'' can be found. From these series one can then find a series for $\pi[3(a + b)/2 - \sqrt{(ab)}]$, which differs from the exact series only by terms in ε^8 and beyond. If r is an upper estimate for the accuracy of the exact series, then one gets the error estimate $\Delta < r + 3r' + r''$, and if one carries through the calculations, it turns out that $\Delta < 0.4\varepsilon^8/(1 - \varepsilon^2)$.

21.2. Power series

Power series are special series of functions, in which the functions are powers of the variable multiplied by a coefficient, $f_n(x) = a_n x^n$. The partial sums $F_n(x) = a_0 + a_1 x + \cdots + a_n x^n$ are polynomials, which are defined for all x. The range of convergence of the power series $F(x) = \lim_{n\to\infty} F_n(x) = \sum_{n=0}^{\infty} a_n x^n = a_0 + a_1 x + a_2 x^2 + \cdots$ must be examined in each particular case. It can happen that the series is *always convergent* that is, convergent for all x, or is *never convergent* except when $x = 0$.

Examples: 1. The series $F(x) = \sum_{n=1}^{\infty} n^n x^n = x + 4x^2 + 27x^3 + 256x^4 + \cdots$ is never convergent.

2. The series $F(x) = \sum_{n=1}^{\infty} \frac{x^n}{n!} = x + \frac{x^2}{2} + \frac{x^3}{6} + \frac{x^4}{24} + \cdots$ is always convergent.

Convergence of power series

For each fixed value $x = x_0$ the terms of the power series can be regarded as constants so that the results in Chapter 18. can be used. In particular, the concept of absolute convergence, that is, convergence of the series of absolute values, is applicable to power series. It can be proved, though the proof will not be given here, that the power series $\sum a_n x^n$ converges absolutely whenever $|x| < |x_1|$ if the series $\sum a_n x_1^n$ converges.

The radius of convergence of power series. A positive number r is called the radius of convergence of a power series if the series converges for every x with $|x| < r$, but diverges for $|x| > r$; the interval from $-r$ to $+r$ is the range or interval of convergence (Fig.). One may put $r = \infty$ for an always convergent series and $r = 0$ for a never convergent series.

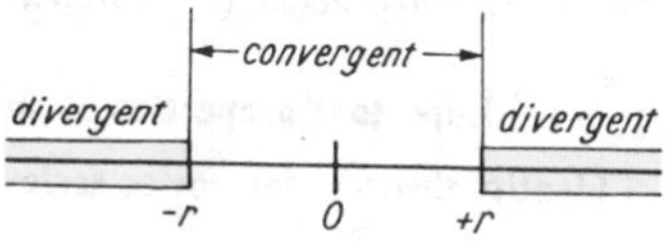

21.2-1 Convergence interval of a power series

Theorem of Abel. For every power series that is neither always convergent nor never convergent, there exists an $r > 0$ such that the series converges for $|x| < r$ and diverges for $|x| > r$.

The Cauchy-Hadamard formula. A formula for finding the radius of convergence was stated in 1821 by CAUCHY (1789–1857) but attracted no attention. It was only rediscovered 70 years later by HADAMARD (1865–1963). One considers the upper limit $\mu = \overline{\lim} \sqrt[n]{|a_n|}$ of the sequence

$$|a_1|, \sqrt[2]{|a_2|}, \sqrt[3]{|a_3|}, \ldots, \sqrt[n]{|a_n|},$$

that is, the number μ with the property that for every $\varepsilon > 0$ infinitely many terms of the sequence are greater than $\mu - \varepsilon$, but only finitely many are greater than $\mu + \varepsilon$.

If μ is finite and positive, $0 < \mu < +\infty$, then $1/\mu$ is also finite and positive, and one can find x_1 and ϱ such that $|x_1| < \varrho < 1/\mu$, so that $1/\varrho > \mu$. This means that $\sqrt[n]{|a_n|} < 1/\varrho$ or $\sqrt[n]{|a_n x_1^n|} < |x_1|/\varrho < 1$ for all $n > N_1$. The power series therefore converges absolutely at x_1. On the other hand, if $|x_2| > 1/\mu$, then $\sqrt[n]{|a_n|} > 1/|x_2|$ or $|a_n x_2^n| > 1$ for infinitely many n, so that the series diverges at x_2. Thus, the number $r = 1/\mu$ is the radius of convergence.

Radius of convergence	$r = \frac{1}{\mu} = \frac{1}{\overline{\lim} \sqrt[n]{\vert a_n\vert}}$

The following result, stated without proof, sometimes makes it possible to use the sequenec $(|a_{n+1}/a_n|)$ rather than $(\sqrt[n]{|a_n|})$.

If the sequence $\{|a_{n+1}/a_n|\}$ converges to a limit, then the sequence $\{\sqrt[n]{|a_n|}\}$ also converges and to the same limit.

Example: The power series $\sum_{n=1}^{\infty} x^n, \sum_{n=1}^{\infty} x^n/n, \sum_{n=1}^{\infty} x^n/n^2, \ldots, \sum_{n=1}^{\infty} x^n/n^p$ $(p \geqslant 0$ fixed$)$ all have the same radius of convergence $r = 1$. It suffices to calculate, for each $p = 0, 1, 2, \ldots$, the following limit: $\lim_{n\to\infty} |a_{n+1}/a_n| = \lim_{n\to\infty} |n^p/(n+1)^p| = \lim_{n\to\infty} [1 - 1/(n+1)]^p = 1$.

It is not possible to make any generally valid assertions about the behaviour of a power series at the ends $x = +r$ and $x = -r$ of the interval of convergence; a separate investigation must be made in each particular case. For instance, in the first three series of the preceding example one finds that:

1. the series $\sum_{n=1}^{\infty} x^n$ diverges for $x = -1$ and $x = +1$;
2. the series $\sum_{n=1}^{\infty} x^n/n$ converges for $x = -1$ and diverges for $x = +1$;
3. the series $\sum_{n=1}^{\infty} x^n/n^2$ converges for $x = -1$ and $x = +1$.

If a power series $\sum_{n=0}^{\infty} a_n x^n$ has the radius of convergence r, then it converges absolutely for any x with $|x| < r$.

Uniform convergence of power series. A theorem due to ABEL states:

A power series converges uniformly in every closed interval that lies entirely inside the interval of convergence.

According to this theorem all results on series of functions obtained on the assumption of uniform convergence are valid for power series. Hence in every closed interval inside the interval of convergence $f(x) = \sum_{n=0}^{\infty} a_n x^n$ is a continuous function whose integral can be obtained by term-by-term integration. Its derivative can be obtained by term-by-term differentiation, as will be shown later.

Power series with a complex variable. For power series with complex coefficients and a complex variable, the interval of convergence is replaced by a circular disc, whose radius is again called the radius of convergence (see Chapter 23.).

Important properties of power series

Identity theorem for power series. The power series $f(x) = \sum_{n=0}^{\infty} a_n x^n$ is a continuous function inside its convergence interval $|x| < r$, and in particular at $x = 0$. If the power series $g(x) = \sum_{n=0}^{\infty} b_n x^n$ is defined in the same interval, and hence continuous there, and if there is a sequence x_k with infinitely many non-zero-terms and $x = 0$ as an accumulation point, then it follows from $f(x_k) = g(x_k)$ for all x_k and from $\lim_{k\to\infty} f(x_k) = a_0$, $\lim_{k\to\infty} g(x_k) = b_0$, that $a_0 = b_0$. Since $x_k \neq 0$, one can now consider two new functions

$$f_1(x_k) = (f(x_k) - a_0)/x_k = a_1 + a_2 x_k + a_3 x_k^2 + \cdots,$$
$$g_1(x_k) = (g(x_k) - b_0)/x_k = b_1 + b_2 x_k + b_3 x_k^2 + \cdots,$$

for which again $f_1(x_k) = g_1(x_k)$, so that one obtains $a_1 = b_1$ letting $k \to \infty$. The procedure can be repeated to obtain $a_2 = b_2$, and by induction it follows that $a_n = b_n$ for all n, so that the two power series are identical.

If the power series $\sum_{n=0}^{\infty} a_n x^n$ and $\sum_{n=0}^{\infty} b_n x^n$ converge for $|x| < r$ and if their sums coincide on a sequence of points x_k with $x_k \neq 0$ and $x_k \to 0$, then the series are identical, that is, $a_n = b_n$ for all n.

The identity theorem also holds for power series of the form $\sum_{n=0}^{\infty} a_n (x - x_0)^n$. If a function $f(x)$ can be represented, in a neighbourhood of x_0, by such a power series, then this representation is unique: if two methods of calculation lead to two power series representing a given function, then the coefficients of corresponding powers must be equal. The method of equating coefficients, which was derived in Chapter 5. is therefore applicable to power series.

Example: For arbitrary real numbers a and b, $(1 + x)^a (1 + x)^b = (1 + x)^{a+b}$. In the domain of convergence $|x| < 1$ each factor can be represented by the binomial series

$$(1 + x)^a = \sum_{n=0}^{\infty} \binom{a}{n} x^n; \quad (1 + x)^b = \sum_{n=0}^{\infty} \binom{b}{n} x^n; \quad (1 + x)^{a+b} = \sum_{n=0}^{\infty} \binom{a+b}{n} x^n.$$

If one now uses the theorem about multiplication of power series, one obtains

$$\sum_{n=0}^{\infty} \binom{a+b}{n} x^n = \sum_{n=0}^{\infty} \left[\binom{a}{0}\binom{b}{n} + \binom{a}{1}\binom{b}{n-1} + \cdots + \binom{a}{n}\binom{b}{0}\right] x^n.$$

Comparison of coefficients leads to a surprisingly simple proof of the addition theorem for binomial coefficients.

Addition theorem for binomial coefficients	
$\binom{a}{0}\binom{b}{n} + \binom{a}{1}\binom{b}{n-1} + \cdots + \binom{a}{n-1}\binom{b}{1} + \binom{a}{n}\binom{b}{0} = \binom{a+b}{n}$	a, b real $n = 0, 1, 2, \ldots$

Transformation to a new centre. All the results found for power series remain valid if one uses $(x - x_0)$ instead of x as a variable. The function $f(x) = \sum_{n=0}^{\infty} a_n (x - x_0)^n$ is continuous inside the interval $|x - x_0| < r$. Now if x_1 lies in this interval, it can be taken as the centre for a new expansion

$f(x) = \sum_{k=0}^{\infty} b_k(x - x_1)^k$ of the same function $f(x)$. Its radius of convergence r_1 is at least $r_1 = r - |x_1 - x_0|$. For each point x in the interval $|x - x_1| < r_1$ the relation $|x_1 - x_0| + |x - x_1| < r$ holds (Fig.). Therefore if one substitutes $x - x_0 = (x_1 - x_0) + (x - x_1)$ in the series in powers of $x - x_0$, then not only is this series $f(x) = \sum_{n=0}^{\infty} a_n[(x_1 - x_0) + (x - x_1)]^n$ absolutely convergent, but also the series $\sum_{n=0}^{\infty} |a_n| (|x_1 - x_0| + |x - x_1|)^n$ converges. Under these assumptions the major rearrangements theorem holds (see Chapter 18.): If one expands each term $[(x_1 - x_0) + (x - x_1)]^n$ by the binomial theorem for $n = 0, 1, 2, \ldots$ and collects together powers of $(x - x_1)$ as follows:

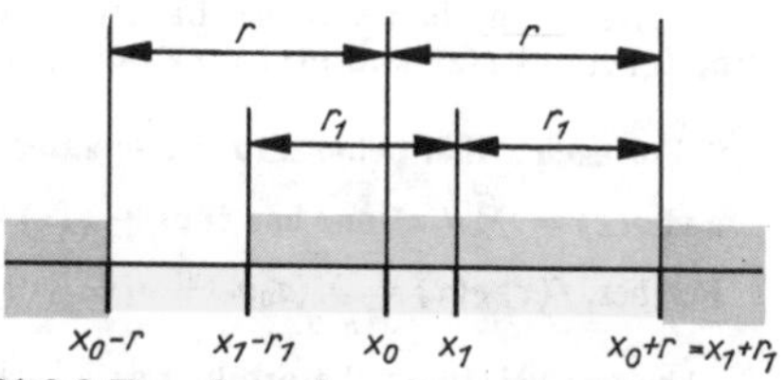

21.2-2 Transformation of a power series to a new centre

$$
\begin{aligned}
f(x) = {} & a_0(x_1 - x_0)^0 \\
& + a_1(x_1 - x_0)^1 + a_1\binom{1}{1}(x_1 - x_0)^{1-1}(x - x_1) \\
& + a_2(x_1 - x_0)^2 + a_2\binom{2}{1}(x_1 - x_0)^{2-1}(x - x_1) + a_2\binom{2}{2}(x_1 - x_0)^{2-2}(x - x_1)^2 + \cdots \\
& + a_3(x_1 - x_0)^3 + a_3\binom{3}{1}(x_1 - x_0)^{3-1}(x - x_1) + a_3\binom{3}{2}(x_1 - x_0)^{3-2}(x - x_1)^2 + \cdots \\
& \cdots\cdots\cdots\cdots\cdots\cdots\cdots\cdots\cdots\cdots\cdots\cdots \\
& + a_n(x_1 - x_0)^n + a_n\binom{n}{1}(x_1 - x_0)^{n-1}(x - x_1) + a_n\binom{n}{2}(x_1 - x_0)^{n-2}(x - x_1)^2 + \cdots \\
\hline
f(x) = {} & b_0 \qquad + \qquad b_1 \cdot (x - x_1) \qquad + \qquad b_2 \cdot (x - x_1)^2 + \cdots
\end{aligned}
$$

where $b_0 = \sum_{n=0}^{\infty} \binom{n}{0} a_n(x_1 - x_0)^n$, $b_1 = \sum_{n=0}^{\infty} \binom{n+1}{1} a_{n+1}(x_1 - x_0)^n$, $b_2 = \sum_{n=0}^{\infty} \binom{n+2}{2} a_{n+2}(x_1 - x_0)^n, \ldots,$

and each column contains an absolutely convergent series, the column sums form an absolutely convergent series, and for each x in the interval $|x - x_1| < r_1$ the function value $f(x)$ is given by $\sum b_k(x - x_1)^k$. Hence $f(x) = \sum_{k=0}^{\infty} b_k(x - x_1)^k$, with $b_k = \sum_{n=0}^{\infty} \binom{n+k}{k} a_{n+k}(x_1 - x_0)^n$.

Term-by-term differentiation of power series. The transformation just discussed serves to obtain power series expansions centred on an arbitrary point x_1 in the interval of convergence $|x - x_0| < r$ of the original power series (centred on x_0). The representation $f(x) = \sum_{k=0}^{\infty} b_k(x - x_1)^k$ shows that the function $f(x)$ is differentiable at $x = x_1$ because $f(x_1) = b_0$, and so

$$[f(x) - f(x_1)]/(x - x_1) = b_1 + b_2(x - x_1) + b_3(x - x_1)^2 + \cdots,$$

and $$\lim_{x \to x_1} [f(x) - f(x_1)]/(x - x_1) = f'(x_1) = b_1 = \sum_{n=0}^{\infty} \binom{n+1}{1} a_{n+1}(x_1 - x_0)^0.$$

This holds for every x_1 inside the convergence interval, and therefore

$$f'(x) = \sum_{n=0}^{\infty} (n + 1)\, a_{n+1}(x - x_0)^n.$$

This is precisely the series obtained by term-by-term differentiation of the series $f(x) = \sum_{n=0}^{\infty} a_n(x - x_0)^n$. This procedure can be repeated, and the next step leads to

$$[f'(x) - f'(x_1)]/(x - x_1) = 2!\, b_2 + 3!\, b_3(x - x_1) + \cdots$$

and $$\frac{1}{2!} f''(x_1) = b_2, \quad \text{where} \quad b_2 = \sum_{n=0}^{\infty} \binom{n+2}{2} a_{n+2}(x_1 - x_0)^n.$$

This series again being absolutely convergent. Proof by induction leads to the theorem:

A function represented by a power series is differentiable arbitrarily often at every interior point of the interval of convergence. The derivatives may be calculated by term-by-term differentiation of the power series.

Sum, difference and product of two power series. For any point that lies inside the convergence interval of each of two power series $f(x) = \sum_{n=0}^{\infty} a_n x^n$ and $g(x) = \sum_{n=0}^{\infty} b_n x^n$, each series converges absolutely. From theorems in Chapter 18. one can deduce that the sum, difference, and product of the functions $f(x)$ and $g(x)$ may be represented by power series with appropriate coefficients.

For each x that belongs to the domains of convergence of each of the power series $f(x) = \sum_{n=0}^{\infty} a_n x^n$ and $g(x) = \sum_{n=0}^{\infty} b_n x^n$ one has $f(x) \pm g(x) = \sum_{n=0}^{\infty} (a_n \pm b_n)\, x^n$.
Further, $f(x)\, g(x) = \sum_{n=0}^{\infty} (a_0 b_n + a_1 b_{n-1} + \cdots + a_n b_0)\, x^n$, where this series converges absolutely.

The coefficients in the product series can be found as diagonal sums in the array of products or by the method of sliding strips (see Chapter 18.).

Example: The geometric series $1/(1-x) = 1 + x + x^2 + x^3 + x^4 + \cdots$ converges for $|x| < 1$. As one can see from the position of the sliding strip for the third or fourth term of the product, multiplication of the series by itself leads successively to

$$1/(1-x)^2 = 1 + 2x + 3x^2 + 4x^2 + \cdots \quad \text{and} \quad 1/(1-x)^3 = 1 + 3x + 6x^2 + 10x^3 + \cdots$$

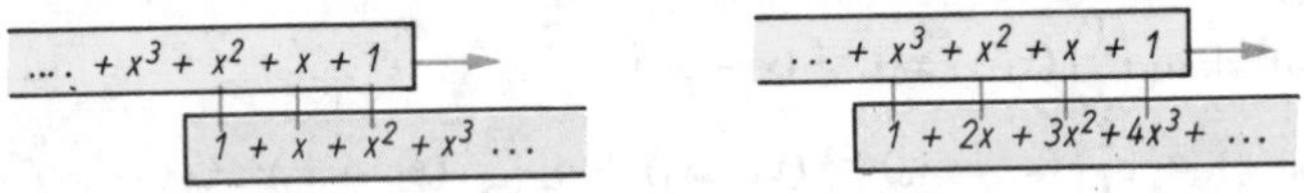

Further examples concerning powers of the sine and cosine series will be given after these series have been derived.

Substitution of one power series into another. If in a composite function (see Chapter 19.), for instance, in $y = f[\varphi(x)]$, the inner function $z = \varphi(x) = \sum_{n=0}^{\infty} a_n x^n$ is represented by a power series and if, for x inside the domain of convergence, this series assumes values z that lie in the domain of convergence of a power series $f(z) = \sum_{n=0}^{\infty} b_n z^n$ which represents the function $y = f(z)$, then this defines a composite function $y = F(x) = f[\varphi(x)]$. One may then ask whether this function $F(x)$ can be represented by a power series $F(x) = \sum_{n=0}^{\infty} c_n x^n$, and how the coefficients c_n can be found in terms of the coefficients a_n and b_n. Now

$$F(x) = b_0 + b_1(a_0 + a_1 x + \cdots) + b_2(a_0 + a_1 x + \cdots)^2 + \cdots,$$

and the powers z^k of $z = \sum a_n x^n$ can, by multiplication of power series, be written as power series $z^k = a_{k0} + a_{k1} x + \cdots + a_{kn} x^n + \cdots$.

If one now substitutes these expressions in the series $f(z) = \sum b_k z^k$ and collects terms involving the same power of x to obtain a power series $\sum c_n x^n$, it can be proved that this series converges absolutely and represents the function $F(x) = f[\varphi(x)]$.

Division by a power series. The task of dividing a power series $\sum b_n x^n$ by a power series $\sum a_n x^n$ can be reduced to finding a product, provided that $1/\sum a_n x^n$ can be represented by a power series. If one assumes that $\sum a_n x^n = a_0 + (a_1 x + a_2 x^2 + \cdots) = a_0 + z$ has the radius of convergence $r > 0$ and that a_0 is not zero, one can find part of the range of convergence in which $|z| = |a_1 x + a_2 x^2 + \cdots| < |a_0|$, and then the reciprocal of $\sum a_n x^n$ can be expanded in a geometric series

$$\frac{1}{\sum a_n x^n} = \frac{1}{a_0} \cdot \frac{1}{1 + z/a_0} = \frac{1}{a_0} - \frac{z}{a_0^2} + \frac{z^2}{a_0^3} - \frac{z^3}{a_0^4} + \cdots,$$

which converges in an x-interval within which $z = a_1 x + a_2 x^2 + \cdots$ also converges. By substituting the power series for z in the geometric series and collecting terms, as was done earlier, one obtains a power series $\sum c_n x^n$, which converges absolutely and represents the function $1/\sum a_n x^n$. Once one knows conditions under which the power series $\sum c_n x^n$ exists, it is easier to find the coefficients c_n from the identity $\sum a_n x^n \sum c_n x^n = 1$ and comparison of coefficients. This leads to a system of equations from which the unknowns $c_0, c_1, c_2, \ldots$ can be calculated step-by-step:

$$a_0 c_0 = 1, \quad a_0 c_1 + a_1 c_0 = 0, \ldots, \quad a_0 c_n + a_0 c_{n-1} + \cdots + a_n c_0 = 0, \ldots$$

More generally, if $\sum b_n x^n$ is to be divided by $\sum a_n x^n$, the identity $\sum b_n x^n = \sum a_n x^n \cdot \sum c_n x^n$ and comparison of coefficients leads to equations to determine the c_n.

Example: The series expansions

$$\sin x = x - \frac{x^3}{3!} + \frac{x^5}{5!} - \cdots, \qquad \cos x = 1 - \frac{x^2}{2!} + \frac{x^4}{4!} - \cdots,$$

to be derived later, can be used to find a power series for $\tan x = \sin x/\cos x$ by division. The conditions for the existence of the power series $\sum c_n x^n$ imply that division is possible only in that part of the domain $|x| < r = \infty$ of the cosine series in which $\cos x \neq 0$, that is, in the interval $|x| < \pi/2$. The identity for finding coefficients reads $\sin x = \cos x \cdot \sum c_n x^n$.

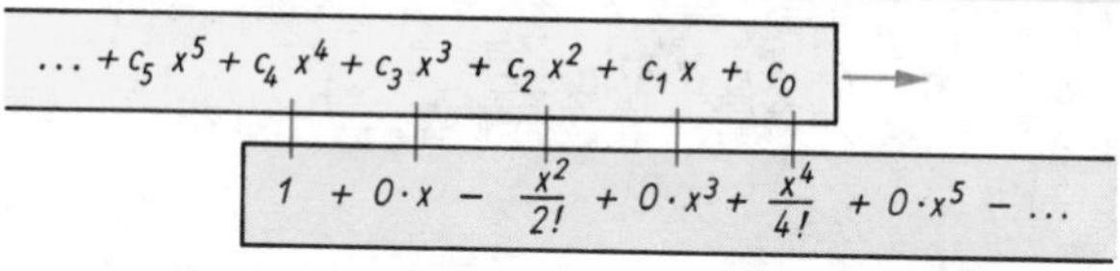

The position of the sliding strips for c_4 is shown above. One obtains

$$0 = c_0, \quad 1 = c_1, \quad 0 = c_2 - \frac{c_0}{2!}, \quad -\frac{1}{3!} = c_3 - \frac{c_1}{2!}, \quad 0 = c_4 - \frac{c_2}{2!} + \frac{c_0}{4!}, \ldots$$

and, step-by-step,

$$c_0 = 0, \ c_1 = 1, \ c_2 = 0, \ c_3 = \frac{1}{2!} - \frac{1}{3!} = \frac{1}{3}, \ c_4 = 0, \ c_5 = \frac{1}{5!} - \frac{1}{4!} + \frac{1}{3 \cdot 2!} = \frac{2}{15},$$

$$c_6 = 0, \ c_7 = \frac{1}{7!} + \frac{1}{6!} - \frac{1}{3 \cdot 4!} + \frac{2}{15 \cdot 2!} = \frac{17}{315}, \ldots$$

that is, the expansion, valid for $|x| < \pi/2$: $\tan x = x + \frac{x^3}{3} + \frac{2x^5}{15} + \frac{17x^7}{315} + \cdots$.

Bernoulli numbers. If one uses the expansion, to be derived later, $e^x = 1 + \frac{x}{1!} + \frac{x^2}{2!} + \cdots$ for the exponential function, then the function

$$f(x) = \frac{x}{e^x - 1} = \frac{1}{1 + x/2! + x^2/3! + x^3/4! + \cdots} = B_0 + B_1 x/1! + B_2 x^2/2! + \cdots$$

satisfies the conditions that permit the division, hence the expansion in a power series. As indicated in the above equation, the coefficients are put into the form $B_n/n!$. The numbers B_n are called Bernoulli numbers and can be calculated from the identity

$$1 = (1 + x/2! + x^2/3! + \cdots)\,[B_0 + (B_1/1!)\,x + (B_2/2!)\,x^2 + \cdots].$$

Comparison of coefficients (Fig.) leads to the relations

$B_0 = 1, \quad B_0/2! + B_1 = 0,$
$B_0/3! + B_1/(1!\,2!) + B_2/2! = 0,$
$B_0/4! + B_1/(1!\,3!) + B_2/(2!\,2!) + B_3/3! = 0, \ldots,$
or $B_0 = 1, \quad 2B_1 + B_0 = 0,$
$3B_2 + 3B_1 + B_0 = 0,$
$4B_3 + 6B_2 + 4B_1 + B_0 = 0,$
$5B_4 + 10B_3 + 10B_2 + 5B_1 + B_0 = 0, \ldots$

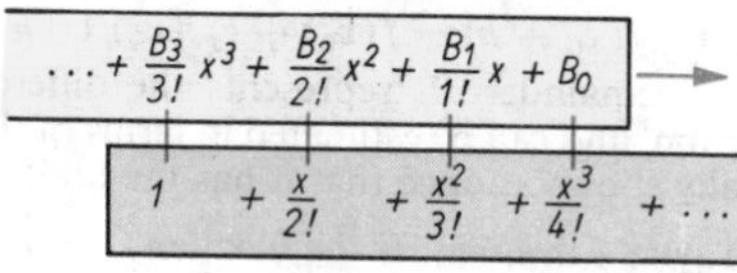

This leads, step-by-step, to

$$B_0 = 1, \quad B_1 = -1/2, \quad B_2 = 1/6, \quad B_3 = 0, \quad B_4 = -1/30, \ldots$$

From $n = 3$ onwards all B_n with odd index n are zero.

Inversion theorem for power series. Under suitable conditions of monotonicity, a function $y = f(x)$ has an inverse function $x = \varphi(y)$ (see Chapter 5.). For power series an analogous theorem, which will not be proved here, can be derived.

For a power series $y = f(x) = a_1 x + a_2 x^2 + a_3 x^3 + \cdots$ with radius of convergence r and $a_1 \neq 0$ there exists exactly one power series $x = \varphi(y) = b_1 y + b_2 y^2 + b_3 y^3 + \cdots$ that is convergent in a neighbourhood of $y = 0$ and such that $y \equiv f[\varphi(y)]$.

Once it has been shown that the power series $x = b_1 y + b_2 y^2 + \cdots$ has a positive radius of convergence r_1, its coefficients can be found by substituting this series in each term of the given power series $y = a_1 x + a_2 x^2 + \cdots$ and equating coefficients of powers of y. This determines the coefficients uniquely, so that there is only one power series expansion for $x = \varphi(y)$.

Example: From the power series $y = \sin x = x - \frac{x^3}{3!} + \frac{x^5}{5!} - \cdots$ the coefficients b_n in $x = \text{Arcsin}\, y = b_1 y + b_2 y^2 + b_3 y^3 + \cdots$ can be obtained by substitution as follows:

$$\begin{aligned} y = {} & b_1 y + b_2 y^2 + b_3 y^3 + b_4 y^4 + b_5 y^5 \\ & - (1/6)\,[b_1^3 y^3 + 3b_1^2 b_2 y^4 + 3b_1 b_2^2 y^5 + 3b_1^2 b_3 y^5 + \cdots] \\ & + (1/120)\,[b_1^5 y^5 + \cdots], \end{aligned}$$

so that hence

$$1 = b_1, \quad 0 = b_2, \quad 0 = b_3 - b_1^3/6, \quad 0 = b_4 - b_2 b_1^2/2,$$
$$0 = b_5 - b_1 b_2^2/2 + b_1^2 b_3/2 + b_1^5/120, \ldots$$

This leads step-by-step to

$$b_1 = 1, \quad b_2 = 0, \quad b_3 = 1/6, \quad b_4 = 0, \quad b_5 = 3/40, \ldots$$

or

$$x = \text{Arcsin}\, y = y + y^3/6 + 3y^5/40 + \cdots.$$

Taylor series

A power series $\sum a_n x^n$ with positive radius of convergence r defines a function $f(x) = \sum a_n x^n$, which is a continuous function of x for $|x| < r$. Repeated differentiation term-by-term of the series gives the derivatives of arbitrary order of the function $f(x)$. On the other hand, if a function $f(x)$, such as $\sin x$, $\sqrt{(1 + x^2)}$ or $\arctan x$, is given, it still remains to be shown that this function can be expanded in a convergent power series and how the coefficients in this expansion can be determined. This problem was solved by Brook TAYLOR (1685–1731) and Colin MACLAURIN (1698–1741).

If the given function $f(x)$ can be expanded at all as a power series, $f(x) = a_0 + a_1 x + a_2 x^2 + \cdots + a_n x^n + \cdots$, then $f(x)$ must be differentiable arbitrarily often. Differentiation term-by-term, and then setting $x = 0$ leads successively to

$$f'(x) = a_1 + 2a_2 x + \cdots + n a_n x^{n-1} + \cdots \longrightarrow f'(0) = a_1,$$
$$f''(x) = 2a_2 + 3 \cdot 2 x a_3 + \cdots + n(n-1)\, a_n x^{n-2} + \cdots \longrightarrow f''(0) = 2a_2,$$
$$f'''(x) = 3 \cdot 2 \cdot 1 a_3 + \cdots + n(n-1)(n-2)\, a_n x^{n-3} + \cdots \longrightarrow f'''(0) = 3!\, a_3,$$
$$\cdots$$
$$f^{(n)}(x) = n!\, a_n + (n+1) \cdots 2 a_{n+1} x + \cdots \longrightarrow f^{(n)}(0) = n!\, a_n.$$

The power series, provided that it converges, then takes the form of a so-called MACLAURIN series $f(x) = f(0) + (f'(0)/1!)\, x + (f''(0)/2!)\, x^2 + \cdots$. Similar considerations hold for a power series $\sum a_n (x - x_0)^n$ with centre x_0 and lead to the so-called TAYLOR series

$$f(x) = f(x_0) + (f'(x_0)/1!)(x - x_0) + (f''(x_0)/2!)(x - x_0)^2 + \cdots.$$

If one replaces x by $x_0 + h$, one obtains

$$f(x_0 + h) = f(x_0) + (f'(x_0)/1!)\, h + (f''(x_0)/2!)\, h^2 + \cdots.$$

Taylor's theorem. To investigate the convergence of these series one introduces the partial sums and the remainder R_n (see Taylor's form in Chapter 18.):

$$f(x_0 + h) = f(x_0) + (f'(x_0)/1!)\, h + \cdots + (f^{(n)}(x_0)/n!)\, h^n + R_n.$$

The remainder R_n represents the difference between the given function and an approximating function, and can be estimated in terms of the $(n + 1)$th derivative of $f(x)$. This form of the remainder usually shows at once that it has the limit zero as $n \to \infty$ so that the series converges to $f(x)$.

Taylor's theorem. If the function $f(x)$ has a continuous nth derivative $f^{(n)}(x)$ in the closed interval from x_0 to $x_0 + h$, and if its $(n + 1)$th derivative exists at least inside this interval, then the remainder R_n in $f(x_0 + h) = f(x_0) + (f'(x_0)/1!)\, h + (f''(x_0)/2!)\, h^2 + \cdots + (f^{(n)}(x_0)/n!)\, h^n + R_n$ can be written in

a) Lagrange's form: there exists a number ϑ with $0 < \vartheta < 1$ such that

$$R_n = \frac{h^{n+1}}{(n+1)!} f^{(n+1)}(x_0 + \vartheta h), \text{ or}$$

b) Cauchy's form: there exists a number ϑ' with $0 < \vartheta' < 1$ such that

$$R_n = \frac{h^{n+1}}{n!} (1 - \vartheta')^n f^{(n+1)}(x_0 + \vartheta' h).$$

To establish these forms of the remainder one uses an extension of the mean value theorem of the differential calculus, which states that for two continuous functions $F(x)$ and $\varphi(x)$ that are continuously differentiable inside the interval $[x_0, x_0 + h]$ with $\varphi'(x) \neq 0$ there exists a number ϑ with $0 < \vartheta < 1$ such that

$$\frac{F(x_0 + h) - F(x_0)}{\varphi(x_0 + h) - \varphi(x_0)} = \frac{F'(x_0 + \vartheta h)}{\varphi'(x_0 + \vartheta h)}.$$

Choosing the function $\varphi(x)$ as

$$\varphi(x) = (x_0 + h - x)^{n+1}, \quad \text{one has} \quad \varphi(x_0) = h^{n+1}, \quad \varphi(x_0 + h) = 0,$$

and

$$\varphi'(x) = -(n+1)(x_0 + h - x)^n.$$

The function $F(x)$ is obtained from the remainder

$$R_n = f(x_0 + h) - f(x_0) - hf'(x_0)/1! - \cdots - h^n f^{(n)}(x_0)/n!$$

by putting x_1 for $(x_0 + h)$ and $(x_1 - x_0)$ for h and then making x_0 a variable x. This function $F(x) = f(x_1) - f(x) - (x_1 - x)\cdot f'(x)/1! - (x_1 - x)^2 \cdot f''(x)/2! - \cdots - (x_1 - x)^n \cdot f^{(n)}(x)/n!$ is continuous in the interval $[x_0, x_0 + h]$, differentiable in its interior, and takes the values

$$F(x_0) = R_n, \quad F(x_0 + h) = 0, \quad F'(x) = -(x_1 - x)^n f^{(n+1)}(x)/n!.$$

Consequently the generalized mean value theorem yields

$$\frac{-R_n}{-h^{n+1}} = \frac{-[h^n(1-\vartheta)^n/n!]\, f^{(n+1)}(x_0 + \vartheta h)}{-(n+1)\, h^n (1-\vartheta)^n},$$

or

$$R_n = [h^{n+1}/(n+1)!]\, f^{(n+1)}(x_0 + \vartheta h),$$

that is, Lagrange's form of the remainder.

Different choices of the auxiliary function $\varphi(x)$ lead to other forms of the remainder. Cauchy's form is obtained by choosing $\varphi(x) = x_0 + h - x$.

Remainder in the Maclaurin series. Taylor's theorem also holds for this series. The remainder takes the forms $R_n = [x^{n+1}/(n+1)!] f^{(n+1)}(\vartheta x)$ (Lagrange), $R_n = [x^{n+1}/n!]\,(1-\vartheta')^n f^{(n+1)}(\vartheta' x)$ (Cauchy).

Trigonometric functions. The functions sin x and cos c have the following derivatives at $x = 0$

sin x			
$f(0)$	$= f^{(4k)}(0)$	$= \sin 0 =$	0
$f'(0)$	$= f^{(4k+1)}(0) =$	$\cos 0 =$	1
$f''(0)$	$= f^{(4k+2)}(0) =$	$-\sin 0 =$	0
$f'''(0)$	$= f^{(4k+3)}(0) =$	$-\cos 0 =$	-1

cos x			
$f(0)$	$= f^{(4k)}(0)$	$= \cos 0 =$	1
$f'(0)$	$= f^{(4k+1)}(0) =$	$-\sin 0 =$	0
$f''(0)$	$= f^{(4k+2)}(0) =$	$-\cos 0 =$	-1
$f'''(0)$	$= f^{(4k+3)}(0) =$	$\sin 0 =$	$0.$

Putting $n = 2m$ one obtains for arbitrary x and with $0 < \vartheta < 1$ or $0 < \vartheta' < 1$

$$\sin x = x - \frac{x^3}{3!} + \frac{x^5}{5!} - \frac{x^7}{7!} + \cdots + (-1)^{m-1}\frac{x^{2m-1}}{(2m-1)!} + (-1)^m \frac{x^{2m+1}}{(2m+1)!}\cos(\vartheta x),$$

$$\cos x = 1 - \frac{x^2}{2!} + \frac{x^4}{4!} - \frac{x^6}{6!} + \cdots + (-1)^{m-1}\frac{x^{2m-2}}{(2m-2)!} + (-1)^m \frac{x^{2m}}{(2m)!}\cos(\vartheta' x).$$

Both remainders tend to zero as $m \to \infty$ for any x, so that both series always converge.

Multiplication of these series by themselves leads to series for powers of the sine and cosine functions. The addition theorems can also be used to obtain these, for example,

$$\sin^2 x = \frac{1}{2}(1 - \cos 2x) = \frac{1}{2}\left[\frac{(2x)^2}{2!} - \frac{(2x)^4}{4!} + \frac{(2x)^6}{6!} - \cdots\right].$$

From sin x/cos x a series for tan x has already been obtained by division. Series for $1/\cos x$, $x/\sin x$ and $x \cot x$ can be obtained similarly by division.

$$\sin x = \frac{x}{1!} - \frac{x^3}{3!} + \frac{x^5}{5!} - + \cdots + (-1)^n \frac{x^{2n+1}}{(2n+1)!} + \cdots, \quad r = \infty$$

$$\cos x = 1 - \frac{x^2}{2!} + \frac{x^4}{4!} - + \cdots + (-1)^n \frac{x^{2n}}{(2n)!} + \cdots, \quad r = \infty$$

$\sin^2 x = x^2 - x^4/3 + 2x^6/45 - \cdots$ $\cos^2 x = 1 - x^2 + x^4/3 - 2x^6/45 + \cdots$

$\sin^3 x = x^3 - x^5/2 + 13x^7/120 - \cdots$ $\cos^3 x = 1 - 3x^2/2 + 7x^4/8 - 61x^6/240 + \cdots$

$\tan x = x + x^3/3 + 2x^5/15 + 17x^7/315 + \cdots, \quad r = \pi/2$

$x \cot x = 1 - x^2/3 - x^4/45 - 2x^6/945 - x^8/4725 - \cdots, \quad r = \pi$

$x \operatorname{cosec} x = x/\sin x = 1 + x^2/6 + 7x^4/360 + 31x^6/15120 + \cdots, \quad r = \pi$

$\sec x = 1/\cos x = 1 + x^2/2 + 5x^4/24 + 61x^6/720 + \cdots, \quad r = \pi/2$

Exponential and hyperbolic functions. Since all derivatives of the function e^x are equal to e^x and therefore are 1 at $x = 0$, one obtains $e^x = 1 + \dfrac{x}{1!} + \cdots + \dfrac{x^n}{n!} + \dfrac{x^{n+1}}{(n+1)!} e^{\vartheta x}$ with $0 < \vartheta < 1$.

For the general exponential $a^x = e^{x \ln a}$ an analogous series, convergent for all x, can be obtained.

The Taylor series can also be used to obtain the addition theorem for the exponential function, because

$$e^{x_0+h} = e^{x_0} + e^{x_0}h/1! + e^{x_0}h^2/2! + \cdots$$

gives

$$e^{x_0+h} = e^{x_0}(1 + h/1! + h^2/2! + \cdots) = e^{x_0}e^h.$$

The definitions $\sinh x = (e^x - e^{-x})/2$ and $\cosh x = (e^x + e^{-x})/2$, together with the power series for the exponential function, lead to power series for these hyperbolic functions. The series for

$\tanh x = \sinh x/\cosh x$ and $\coth x = \cosh x/\sinh x$ can then be obtained by division of the power series occurring in the numerator and denominator.

$$e^x = 1 + \frac{x}{1!} + \frac{x^2}{2!} + \frac{x^3}{3!} + \cdots + \frac{x^n}{n!} + \cdots, \quad r = \infty$$
$$a^x = e^{x \ln a} = 1 + \frac{x \ln a}{1!} + \frac{(x \ln a)^2}{2!} + \cdots, \quad r = \infty, \quad a > 0$$
$$\sinh x = x + x^3/3! + x^5/5! + \cdots + x^{2n+1}/(2n+1)! + \cdots, \quad r = \infty$$
$$\cosh x = 1 + x^2/2! + x^4/4! + \cdots + x^{2n}/(2n)! + \cdots, \quad r = \infty$$
$$\tanh x = x - x^3/3 + 2x^5/15 - 17x^7/315 + - \cdots, \quad r = \pi/2$$
$$x \coth x = 1 + x^2/3 - x^4/45 + 2x^6/945 - x^8/4725 + - \cdots, \quad r = \pi$$

Logarithm. For the logarithmic function $\ln(1+x)$, $f(1) = 0$ and $\frac{d^n \ln x}{dx^n} = (-1)^{n-1}(n-1)!/x^n$, so that $f^{(n)}(1)/n! = (-1)^{n-1}/n$. Taylor's theorem therefore yields $\ln(1+x) = x - x^2/2 + x^3/3 - + \cdots + (-1)^{n-1} x^n/n + (-1)^n x^{n+1}/[(n+1)(1+\vartheta x)^{n+1}]$, where $0 < \vartheta < 1$. The remainder term tends to zero as $n \to \infty$ when $0 \leqslant x \leqslant 1$. The same series can be derived, for the interval $|x| < 1$, by termwise integration of the geometric series $1/(1+x) = 1 - x + x^2 - x^3 + \cdots$.

Logarithmic function	$\ln(1+x) = x - \frac{x^2}{2} + \frac{x^3}{3} - \frac{x^4}{4} + \cdots, \quad \lvert x\rvert < 1$

This series is usually unsuitable for calculations because it converges too slowly unless x is very small.

From $\ln(1-x) = -x - x^2/2 - x^3/3 - \cdots$ and $\ln[(1+x)/(1-x)] = \ln(1+x) - \ln(1-x)$ one obtains a series which converges when $|x| < 1$; for $\xi > 1$, $1/\xi = x < 1$ one obtains the second formula.

$$\ln \frac{1+x}{1-x} = 2\left(x + \frac{x^3}{3} + \frac{x^5}{5} + \cdots\right)$$
$$\ln \sqrt{\frac{\xi+1}{\xi-1}} = \frac{1}{\xi} + \frac{1}{3\xi^3} + \frac{1}{5\xi^5} + \cdots$$

To calculate logarithms by using series, it is desirable to combine rapidly convergent series. Thus, $\ln 2 = 7\ln(10/9) - 2\ln(25/24) + 3\ln(81/80)$, because $\frac{81^3 \cdot 24^2 \cdot 10^7}{80^3 \cdot 25^2 \cdot 9^7} = 2$. The most slowly convergent series involved here is

$$\ln\left(\frac{10}{9}\right) = -\ln\left(1 - \frac{1}{10}\right) = \frac{1}{10} + \frac{1}{2 \cdot 100} + \frac{1}{3 \cdot 1000} + \cdots.$$

Similarly

$$\ln 3 = 11\ln(10/9) - 3\ln(25/24) + 5\ln(81/80),$$
$$\ln 5 = 16\ln(10/9) - 4\ln(25/24) + 7\ln(81/80).$$

It may have been the diversity of mathematical methods available that led GAUSS to remark: 'There is a kind of poetry in the calculation of logarithmic tables'.

Binomial series. For positive integral m,

$$f(x) = (1+x)^m = 1 + \binom{m}{1}x + \binom{m}{2}x^2 + \cdots + \binom{m}{m}x^m.$$

If m is not a positive integer, the function $f(x) = (1+x)^m$ can be expanded in a Maclaurin series, convergent for $|x| < 1$. Here $f(0) = 1$, $f'(0) = m$, $f''(0) = m(m-1)$, ..., $f^{(n)}(0) = \binom{m}{n} \cdot n!$

Binomial series	$(1+x)^m = 1 + \binom{m}{1}x + \binom{m}{2}x^2 + \binom{m}{3}x^3 + \cdots; \quad r = 1$

This series was discovered by NEWTON in 1676, but derived correctly by EULER only about 100 years later. It is useful for calculating approximate value of roots and powers with arbitrary exponents. For $m = 1/2, 1/3, -1/2$ and $-1/3$, and $|x| < 1$ one obtains:

$$\sqrt{(1+x)} = 1 + \frac{1}{2}x - \frac{1}{2 \cdot 4}x^2 + \frac{1 \cdot 3}{2 \cdot 4 \cdot 6}x^3 - + \cdots + (-1)^{n-1}\frac{1 \cdot 3 \cdot 5 \cdots (2n-3)}{2 \cdot 4 \cdot 6 \cdots 2n}x^n + \cdots$$
$$\sqrt[3]{(1+x)} = 1 + \frac{1}{3}x - \frac{1 \cdot 2}{3 \cdot 6}x^2 + \frac{1 \cdot 2 \cdot 5}{3 \cdot 6 \cdot 9}x^3 - + \cdots + (-1)^{n-1}\frac{1 \cdot 2 \cdot 5 \cdot 8 \cdots (3n-4)}{3 \cdot 6 \cdot 9 \cdot 12 \cdots 3n}x^n + \cdots$$

$$\frac{1}{\sqrt{(1+x)}} = 1 - \frac{1}{2}x + \frac{1\cdot 3}{2\cdot 4}x^2 - \frac{1\cdot 3\cdot 5}{2\cdot 4\cdot 6}x^3 + - \cdots + (-1)^n \frac{1\cdot 3\cdot 5\cdots(2n-1)}{2\cdot 4\cdot 6\cdots 2n}x^n + \cdots$$

$$\frac{1}{\sqrt[3]{(1-x)}} = 1 - \frac{1}{3}x + \frac{1\cdot 4}{3\cdot 6}x^2 - \frac{1\cdot 4\cdot 7}{3\cdot 6\cdot 9}x^3 + - \cdots + (-1)^n \frac{1\cdot 4\cdot 7\cdots(3n-2)}{3\cdot 6\cdot 9\cdots 3n}x^n + \cdots$$

Powers of the geometric series, that is, negative integral powers of $1/(1-x)$, can be obtained very simply by differentiation for $|x| < 1$:

$$1/(1-x) = 1 + x + x^2 + x^3 + x^4 + \cdots + x^n + \cdots$$
$$1/(1-x)^2 = 1 + 2x + 3x^2 + 4x^3 + \cdots + (n+1)\,x^n + \cdots$$
$$1/(1-x)^3 = 1 + 3x + 6x^2 + 10x^3 + \cdots + (1/2)\,(n+1)\,(n+2)\,x^n + \cdots$$

Inverse trigonometric and hyperbolic functions. Since the first derivatives of these functions are simple algebraic functions, which can be expanded as binomial series, expansions for the functions themselves can be obtained by termwise integration. For example, $\frac{d(\arctan x)}{dx} = 1/(1+x^2) = 1 - x^2 + x^4 - x^6 + \cdots$, $|x| < 1$, leads to $\int dx/(1+x^2) = x - x^3/3 + x^5/5 - x^7/7 + \cdots + c$, and the constant of integration is $c = \arctan 0 = 0$. Since $\operatorname{arccot} x = \pi/2 - \arctan x$, this also leads to an expansion for $\operatorname{arccot} x$. Integration of the binomial expansion for $1/\sqrt{(1-x^2)}$ leads to a power series for $\arcsin x$, and also for $\arccos x = \pi/2 - \arcsin x$. Analogous results hold for inverse hyperbolic functions, but not all the resulting series are power series.

$$\arcsin x = x + \frac{1}{2}\cdot\frac{x^3}{3} + \frac{1\cdot 3\cdot x^5}{2\cdot 4\cdot 5} + \cdots + \frac{1\cdot 3\cdots(2n-3)\,x^{2n-1}}{2\cdot 4\cdots(2n-2)\,(2n-1)} + \cdots, \quad r = 1$$

$$\arccos x = \pi/2 - x - \frac{1}{2}\cdot\frac{x^3}{3} - \frac{1\cdot 3\cdot x^5}{2\cdot 4\cdot 5} - \cdots, \quad r = 1$$

$$\arctan x = x - x^3/3 + x^5/5 - x^7/7 + - \cdots + (-1)^n x^{2n+1}/(2n+1) + \cdots, \quad r = 1, \quad x = +1$$

$$\operatorname{arccot} x = \pi/2 - x + x^3/3 - x^5/5 + x^7/7 - + \cdots, \quad r = 1, \quad x = 1$$

$$\sinh^{-1} x = x - \frac{1\cdot x^3}{2\cdot 3} + \frac{1\cdot 3\cdot x^5}{2\cdot 4\cdot 6} - + \cdots + (-1)^n \frac{1\cdot 3\cdots(2n-3)\,x^{2n-1}}{2\cdot 4\cdots(2n-2)\,(2n-1)}, \quad r = 1$$

$$\cosh^{-1} x = \pm\left[\ln(2x) - \frac{1}{2\cdot 2x^2} - \frac{1\cdot 3}{2\cdot 4\cdot 4x^4} - \cdots\right], \quad x > 1$$

$$\tanh^{-1} x = x + x^3/3 + x^5/5 + x^7/7 \cdots + x^{2n+1}/(2n+1) + \cdots, \quad r = 1$$

$$\coth^{-1} x = 1/x + 1/(3x^3) + 1/(5x^5) + 1/(7x^7) + \cdots, \quad |x| > 1$$

Series for special functions

Examples for integrals that can be evaluated explicitly only by expansions in series are Gauss's error integral, the cosine and sine integrals, and the logarithmic integral.

Example: For the sine integral, one obtains a convergent series by termwise integration of a uniformly convergent series for the integrand:

$$\int_0^x \frac{\sin t}{t}\,dt = \int_0^x \left(1 - \frac{t^2}{3!} + \frac{t^4}{5!} - \cdots\right) dt = x - \frac{x^3}{3\cdot 3!} + \frac{x^5}{5\cdot 5!} - \frac{x^7}{7\cdot 7!} + \cdots$$

Gauss's error integral, $r = \infty$, $\lim_{x\to\infty} \varphi(x) = 1$	$\varphi(x) = \frac{2}{\sqrt{\pi}} \int_{-\infty}^{x} e^{-t^2}\,dt = \frac{2}{\sqrt{\pi}}\left(\frac{x}{1} - \frac{x^3}{1!\,3} + \frac{x^5}{2!\,5} - \frac{x^7}{3!\,7} + \cdots\right)$
Sine integral, $r = \infty$	$\mathrm{Si}(x) = \int_0^x \frac{\sin t}{t}\,dt = x - \frac{x^3}{3!\,3} + \frac{x^5}{5!\,5} - \frac{x^7}{7!\,7} + \cdots$
Cosine integral, $r = \infty$	$\mathrm{Ci}(x) = \int_0^x \frac{\cos t}{t}\,dt = \ln\gamma + \ln x - \frac{x^2}{2!\,2} + \frac{x^4}{4!\,4} - \cdots$

Euler's (or Mascheroni's) constant	$\ln\gamma = \lim_{n\to\infty}\left(1 + \frac{1}{2} + \frac{1}{3} + \cdots + \frac{1}{n} - \ln n\right)$ $\ln\gamma = 0.57722\ldots$
Logarithmic integral,	$\mathrm{Li}(x) = \int_0^x \frac{\mathrm{d}t}{\ln t}, \quad t = \mathrm{e}^{-u}$
$r = \infty$	$\mathrm{Li}(\mathrm{e}^{-u}) = \ln\gamma + \ln\lvert u\rvert - u + \frac{u^2}{2!2} - \frac{u^3}{3!3} + \frac{u^4}{4!4} - + \cdots$

Approximations

Examples of applications of Taylor's theorem. Taylor's theorem is often used for calculating approximate values of a function $f(x)$. The remainder may be used to decide how many terms of the series are needed to achieve a prescribed accuracy and to give a bound for the error when a fixed number of terms is used and the variable is confined to a specific range.

Calculation of the number e. Taylor's theorem applied to e^x at $x = 1$ gives $\mathrm{e} = 1 + 1 + 1/2! + \cdots + 1/n! + \mathrm{e}^{\vartheta}/(n+1)!$, $0 < \vartheta < 1$. If e is to be calculated correctly to 7 decimal places it can be decided quickly how n should be chosen to attain the required accuracy. The remainder satisfies the inequality $1/(n+1)! < R_n < 3/(n+1)!$ because $\mathrm{e}^0 = 1 < \mathrm{e}^{\vartheta} < \mathrm{e}^1 < 3$. To avoid rounding errors it is best to require that $R_n < 10^{-8}$. Then n must satisfy $3/(n+1)! < 10^{-8}$ or $(n+1)! > 3.10^8$. Since $12! \approx 4.8 \cdot 10^8$ it suffices to take $n = 11$. Then $1 + 1 + 1/2! + \cdots + 1/11! = 2.718281826\ldots$ The remainder term can be estimated from $1/12! \approx 2.10^{-9}$ and $3/12! \approx 6.10^{-9}$. Thus, one obtains the inequality $2.718281828\cdots < \mathrm{e} < 2.718281832$ and e, correct to 7 decimal places, is given by $\mathrm{e} = 2.7182818\ldots$

21.2-3 The number e, base of the natural logarithms.

If one carries out this calculation, one realizes that not very much effort would be needed to increase the accuracy. However, one encounters a problem that often arises in practical calculations, the problem of rounding errors. The individual terms of a series used for numerical calculation are almost always periodic decimals, which have to be truncated and rounded off. This rounding off process may make it impossible to predict with absolute certainty the accuracy reached. In practice it has been found reliable to carry one or two decimal places more than required in the final result, and also to note for each term whether the rounding is downwards or upwards, so that the maximum possible rounding error can be stated before the final result is rounded off. To illustrate this point, here is the calculation of $\mathrm{e}^{-0.1}$ using Taylor's theorem: $\mathrm{e}^{-0.1} = 1 - 0.1 + 0.005 - \cdots + R_n$, where $R_n = (0.1)^{n+1} \cdot \mathrm{e}^{-\vartheta \cdot 0.1}\,(n+1)!$ with $0 < \vartheta < 1$. Note that $0.905 < \mathrm{e}^{-\vartheta \cdot 0.1} < 1$. The first four terms of the series lead to the adjacent calculation. For R_3 the inequality $0.000004167 > R_3 > 0.000003770$ holds, so that $\mathrm{e}^{-0.1}$ satisfies $0.904837103 < \mathrm{e}^{-0.1} < 0.904837500$. Therefore $\mathrm{e}^{-0.1} = 0.904837\ldots$ has been determined correct to six places. The most that could have been expected from using only four terms, together with the remainder, is accuracy to four or perhaps five places; the accuracy of the calculation, with an error of at most four units in the seventh place, is surprising. The use of Taylor's theorem always leads to favourable results when $f^{(n+1)}(x_0)$ and $f^{(n+1)}(x_0 + h)$ do not differ much and the function $f^{(n+1)}(x)$ is monotone in the interval from x_0 to $x_0 + h$.

1.000	000	000
−0.100	000	000
+0.005	000	000
−0.000	166	667
0.904	833	333

Calculation of the number π. The arctan series can be used for the calculation of π. It reads $\arctan x = x - x^3/3 + x^5/5 - x^7/7 + \cdots$. In this series, the remainder R_n cannot be used immediately in the form given in Taylor's theorem. Although the function $f(x) = \arctan x$ has derivatives of all orders, the way in which successive derivatives are formed is rather complicated and can hardly be used to give an explicit formula. However, one can obtain the following formula for the remainder by direct use of the mean value theorem:

$$\arctan x = x - x^3/3 + x^5/5 - + \cdots + (-1)^{k-1}\,x^{2k-1}/(2k-1)$$
$$+ (-1)^k\,x^{2k+1}/[(2k+1)(1 + \vartheta x^2)], \quad 0 < \vartheta < 1.$$

For $x = 1$, this yields

$$\pi/4 = 1 - 1/3 + 1/5 - + \cdots + (-1)^{2k-1}/(2k-1) + (-1)^k/[(2k+1)(1+\vartheta)], \quad 0 < \vartheta < 1.$$

This equation was found by James Gregory (1638–1675) and by Leibniz and is therefore called the *Gregory-Leibniz equation.* The remainder lies between $1/[2(2k+1)]$ and $1/(2k+1)$, so that one can see that the formula is not very suitable for practical calculation of π. One would need 100000 terms for calculating π to 5 decimal places. One therefore tries to find more suitable values of the independent variable to substitute in the arctan series. For instance, $\arctan(1/\sqrt{3}) = \pi/6$ leads to

$$\frac{\pi}{6} = \frac{1}{\sqrt{3}}\left[1 - \frac{1}{3 \cdot 3} + \frac{1}{5 \cdot 3^2} + \cdots + \frac{(-1)^k}{(2k+1)\,3^k} \cdot \frac{1}{1+\vartheta}\right], \quad 0 < \vartheta < 1.$$

Combination of several arctan series with different arguments ultimately leads to particularly convenient formulae, of which some are quoted, which have been used to calculate π by means of electronic computers:

GAUSS $\pi = 48 \arctan(1/18) + 32 \arctan(1/57) - 20 \arctan(1/239)$;
STÖRMER (1896) $\pi = 24 \arctan(1/8) + 8 \arctan(1/57) + 4 \arctan(1/239)$.

The last two formulae were used in 1961 for the calculation of π to 100265 decimal places. Two machines calculated in the binary system, as a check each by a different formula. Gauss's formula required 4 hours and 22 minutes, Störmer's 8 hours and 43 minutes. The transfer to the decimal system took 42 minutes; the print-out occupies 20 pages.

In practice such 'accuracy' is without significance. A knowledge of π to 14 decimal places suffices to calculate the circumference of a circle of radius 6400 km (the radius of the earth!) with an error of less than 0.001 mm. For such a bound of the error to be meaningful the radius would have to be determined with a similar accuracy; but in the present state of measuring technique this cannot be achieved by any means.

The use of the binomial series. The binomial series is often used for approximate calculation of roots. From Taylor's theorem,

$$(1+x)^a = 1 + \binom{a}{1}x + \binom{a}{2}x^2 + \cdots + \binom{a}{n}x^n + \binom{a}{n+1}x^{n+1}(1+\vartheta x)^{a-n-1}$$

with $0 < \vartheta < 1$ and $|x| < 1$.

For instance, to calculate $\sqrt[3]{999}$ to 12 places, one has to put in the formula above $x = -1/1000$ and $a = 1/3$, because $\sqrt[3]{999} = 10(1 - 1/1000)^{1/3}$. To reach the required accuracy it is enough to take $n = 2$ because the values of the $(n+1)$th derivative at $x_0 = 0$ and $x_0 + h = -0.001$ differ by very little. One has $1/(1.62 \times 10^{10}) < R_2 < 1.003/(1.62 \times 10^{10})$ and so

$$9.996665556173 < \sqrt[3]{999} < 9.996665556175, \text{ or } \sqrt[3]{999} = 9.99666555617\ldots$$

This degree of accuracy is hard to attain with many-figure tables, which shows the power of the method; greater accuracy would involve comparatively little extra effort.

If the binomial series is to be used for the approximate calculation of roots, it is necessary to bring the argument into the form $1 \pm \delta$, where in general δ should not exceed 0.1. To do this, it is often necessary to use a number of devices which are best illustrated by examples. If $\sqrt[n]{a}$ is to be found and if there is an integer b such that $b^n \approx a$, one needs to do no more than to write $\sqrt[n]{a} = \sqrt[n]{[b^n \cdot (a/b^n)]} = b\sqrt[n]{[1 + (a - b^n)/b^n]}$; for instance, $\sqrt[5]{33} = 2\sqrt[5]{(1 + 1/32)}$. If this method is inappropriate, it may be possible to find a rational number whose nth power is near to a. A classical example is $\sqrt{2}$. Here $\sqrt{2} \approx 1.4 = 7/5$, so that one can write

$$\sqrt{2} = \sqrt{\left(\frac{7^2 \cdot 5^2 \cdot 2}{5^2 \cdot 7^2}\right)} = \frac{7}{5}\sqrt{\left(\frac{50}{49}\right)} = \frac{7}{5}\sqrt{\left(1 + \frac{1}{49}\right)}.$$

Similarly $\sqrt[3]{92} \approx 4.5 = \dfrac{9}{2}$, so that one can write

$$\sqrt[3]{92} = \sqrt[3]{\left(\frac{9^3 \cdot 2^3 \cdot 92}{2^3 \cdot 9^3}\right)} = \frac{9}{2}\sqrt[3]{\left(\frac{736}{729}\right)} = \frac{9}{2}\sqrt[3]{\left(1 + \frac{7}{729}\right)}.$$

These examples indicate how the binomial series may be used to calculate roots very accurately. The series can also be used for any fractional exponent, and is particularly useful when 4-, 5- or even 7-place tables do not give enough accuracy.

Approximations. In rough calculations it has proved advantageous to use the first few terms of a power series expansion. There are familiar applications of this in science and technology. To neglect the square and higher powers of small quantities is quite common, for example in thermodynamics, when the cubic coefficient of expansion is set to be three times the linear coefficient. Frequently $\sin x$ for small angles x is replaced by x. The table that follows gives a survey of the more frequently used formulae and their range of validity. The cited values of $|x|$ should not be exceeded if the error in using the approximation is not to exceed 0.001 or 0.01. It will be seen that for practical purposes it is frequently unnecessary to use tables of function values, since it is then often sufficient to keep the error below 0.1% or 1%. In that case it is, for example, permissible to replace $\arcsin x$

by x in the range up to $10°$; this leads to a remarkable saving of effort. It is also possible to dispense with searching for tables of rare special functions if suitable approximate formulae can be derived from expansions in series; approximate calculation of integrals can also often be made from suitable approximation formulae quickly and with good accuracy.

Examples:

1. $\sqrt[4]{258.3} = 4\sqrt[4]{\left(\frac{258.3}{256}\right)} \approx 4\left(1 + \frac{2.3}{4 \times 256}\right) = 4.009$. The exact value to five places is 4.00895.

2. $e^{-0.1} \approx 1 - 0.1 + \frac{(0.1)^2}{2} = 0.905$ (exact value 0.90484 ...).

3. $e^{-0.023} \approx 1 - 0.023 = 0.977$ (exact value 0.977262 ...).

Frequently used approximations. The figure at the bottom gives the relation between angles x in degrees and x in rad.

Function	1st approximation	Error $\leqslant 10^{-3}$ for $\lvert x\rvert \leqslant$	Error $\leqslant 10^{-2}$ for $\lvert x\rvert \leqslant$	2nd approximation	Error $\leqslant 10^{-3}$ for $\lvert x\rvert \leqslant$	Error $\leqslant 10^{-2}$ for $\lvert x\rvert \leqslant$
$1/(1+x)$	$1-x$	0.031	0.099	$1-x+x^2$	0.096	0.20
$1/(1+x)^2$	$1-2x$	0.018	0.055	$1-2x+3x^2$	0.063	0.12
$1/(1+x)^3$	$1-3x$	0.012	0.039	$1-3x+6x^2$	0.046	0.095
$\sqrt{(1+x)}$	$1+x/2$	0.087	0.25	$1+x/2-x^2/8$	0.25	0.48
$\sqrt[3]{(1+x)}$	$1+x/3$	0.095	0.27	$1+x/3-x^2/9$	0.25	0.47
$\sqrt[4]{(1+x)}$	$1+x/4$	0.10	0.29	$1+x/4-3x^2/32$	0.24	0.49
$1/\sqrt{(1+x)}$	$1-x/2$	0.050	0.15	$1-x/2+3x^2/8$	0.14	0.28
$1/\sqrt[3]{(1+x)}$	$1-x/3$	0.065	0.19	$1-x/3+2x^2/9$	0.17	0.34
$(1+x)/(1-x)$	$1+2x$	0.022	0.068	$1+2x+2x^2$	0.077	0.16
$[(1+x)/(1-x)]^2$	$1+4x$	0.011	0.034	$1+4x+8x^2$	0.043	0.090
$\sqrt{[(1+x)/(1-x)]}$	$1+x$	0.043	0.13	$1+x+x^2/2$	0.12	0.25
$\sin x$	x	0.18	0.39	$x-x^3/6$	0.63	1.04
$\sin^2 x$	0	0.031	0.10	x^2	0.23	0.41
$\cos x$	1	0.044	0.14	$1-x^2/2$	0.39	0.70
$\cos^2 x$	1	0.031	0.10	$1-x^2$	0.23	0.42
$\tan x$	x	0.14	0.30	$x+x^3/3$	0.38	0.58
$\arcsin x$	x	0.18	0.38	$x+x^3/6$	0.42	0.63
$\arccos x$	$\pi/2-x$	0.18	0.38	$\pi/2-x-x^3/6$	0.42	0.63
$\arctan x$	x	0.14	0.31	$x-x^3/3$	0.35	0.57
$\operatorname{arccot} x$	$\pi/2-x$	0.14	0.31	$\pi/2-x+x^3/3$	0.35	0.57
e^x	$1+x$	0.044	0.13	$1+x+x^2/2$	0.17	0.38
$\ln(1+x)$	x	0.044	0.14	$x-x^2/2$	0.14	0.33
$\lg(1+x)$	$0.4343x$	0.069	0.23	$0.4343x+0.2171x^2$	0.20	0.45
$\sinh x$	x	0.18	0.39	$x+x^3/6$	0.65	1.03
$\cosh x$	1	0.044	0.14	$1+x^2/2$	0.39	0.70
$\tanh x$	x	0.14	0.31	$x-x^3/3$	0.38	0.61
$\sinh^{-1} x$	x	0.18	0.40	$x-x^3/6$	0.43	0.70
$\tanh^{-1} x$	x	0.14	0.30	$x+x^3/3$	0.37	0.52

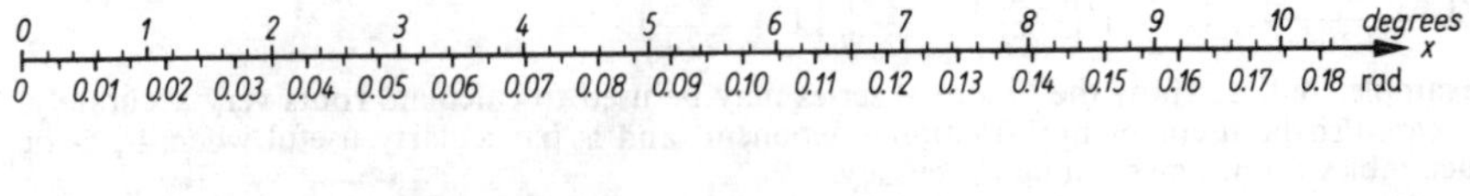

Geometrical applications of Taylor's theorem

Osculating parabola. Suppose that a curve has the equation $y = a_0 + a_1x + a_2x^2 + \cdots$ in a given coordinate system. If one introduces a new coordinate system in which the x-axis is tangent to the curve at a point P and the y-axis is along the normal at the same point, then for the curve referred to the new coordinate system the following values holds at the point P: $x = 0, f(x) = 0, f'(x) = 0$. The curvature $\varkappa$ is given, in general, by $\varkappa = f''/(1 + f'^2)^{3/2}$, and therefore $f''(0) = \varkappa$. It follows that $f(0) = a_0 = 0, f'(0) = a_1 = 0, f''(0) = 2a_2 = \varkappa$, and that the equation of the curve in the new coordinates is $f(x) = (1/2)\,\varkappa x^2 + \cdots$. The curve near P is therefore approximated well by the parabola $g(x) = (1/2)\varkappa x^2$, the osculating parabola. This is a second order approximation (Fig.).

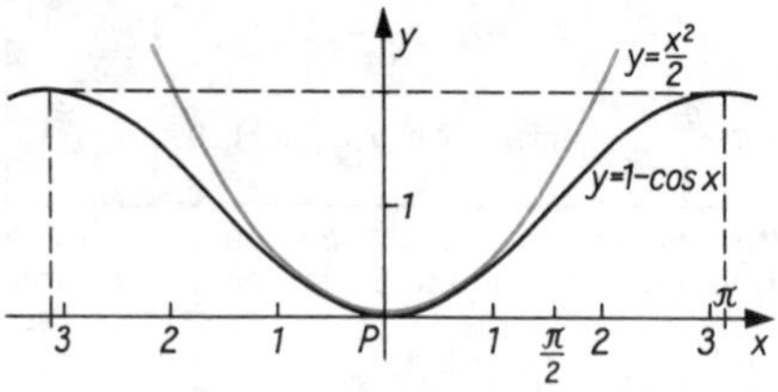

21.2-4 Osculating parabola for the function $y = 1 - \cos x$

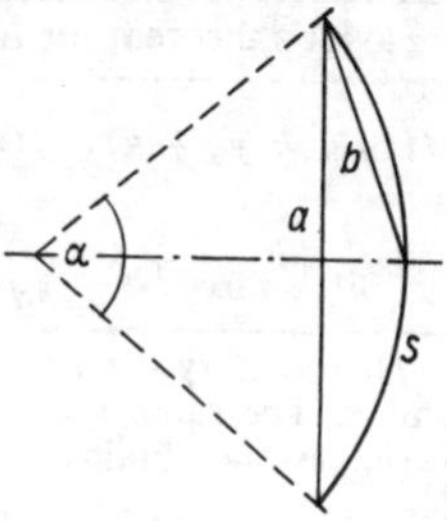

21.2-5 Determination of the length of a circular arc

Determination of the length of a circular arc. The arc length s of a circular arc with central angle α can be found from the approximation $s = (8b - a)/3$, where a is the chord of this arc and b the chord for half the arc (Fig.). The exact value is $s = r\alpha$. Since $a = 2r \sin(\alpha/2)$, $b = 2r \sin(\alpha/4)$, the series expansion for $\sin x$ leads to

$$\frac{8b-a}{3} = \frac{2r}{3}\left\{8\left[\frac{\alpha}{4} - \frac{(\alpha/4)^3}{3!} + \frac{(\alpha/4)^5}{5!} - \frac{(\alpha/4)^7}{7!}\cos\frac{\alpha\vartheta}{4}\right]\right.$$
$$\left. - \left[\frac{\alpha}{2} - \frac{(\alpha/2)^3}{3!} + \frac{(\alpha/2)^5}{5!} - \frac{(\alpha/2)^7}{7!}\cos\frac{\alpha\vartheta'}{2}\right]\right\}$$
$$= \frac{2r}{3}\left[\frac{3\alpha}{2} - \frac{3\alpha^5}{2^7 \cdot 5!} + \frac{\alpha^7}{2^7 \cdot 7!}\left(\cos\frac{\alpha\vartheta'}{2} - \frac{\cos(\alpha\vartheta/4)}{2^4}\right)\right]$$

or

$$\frac{8b-a}{3} = s - \frac{r\alpha^5}{5! \cdot 64}\left[1 - \frac{\alpha^2}{126}\left(\cos\frac{\alpha\vartheta'}{2} - \frac{1}{16}\cos\frac{\alpha\vartheta}{4}\right)\right].$$

If the term in square brackets in the last expression is replaced by 1, the error is increased, so that the error in the approximation is at most $r\alpha^5/7680$. For $r = 1$ unit and $\alpha = 30°$ the error is less than $5 \cdot 10^{-6}$ units.

Bending of a beam. The bending moment $M(x)$ of a beam is $M(x) = EJ\varkappa$, where E is the module of elasticity, J the moment of inertia of the cross section, and $\varkappa$ the curvature of the centre line of the beam at the position x (Fig.). In practice the angles α is very small, and one may expand the denominator in the equation $\varkappa = y''/(1 + y'^2)^{3/2}$ in powers of $y'^2 = \tan^2 \alpha$. This gives $\varkappa = y''(x)\,[1 - (3/2)\,y'(x)^2 + (15/8)\,y'(x)^4 + \cdots]$. For a slightly bent beam, a first approximation is $\varkappa = y''(x)$ and one obtains the *differential equation* $\dfrac{d^2y}{dx^2} = \dfrac{M(x)}{EJ}$ *for the bending of a beam.*

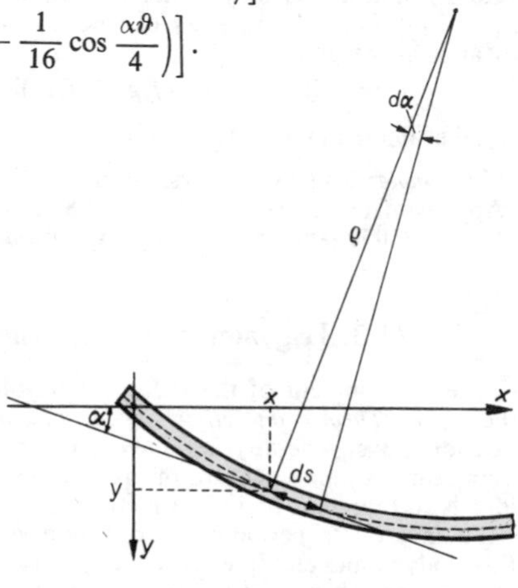

21.2-6 Bending of a beam

Taylor's theorem for several variables

Taylor's theorem can also be stated for functions of several variables. For a function $f(x, y)$ of two variables the case $n = 0$ reads:

$$f(x_0 + h, y_0 + k) = f(x_0, y_0) + hf_x(x_0 + \vartheta h, y_0 + \vartheta k) + kf_y(x_0 + \vartheta h, y_0 + \vartheta k), \qquad 0 < \vartheta < 1.$$

This is simply the mean value theorem for functions of two variables. The case $n = 1$ reads, with $0 < \vartheta < 1$,

$$f(x_0 + h, y_0 + k) = f(x_0, y_0) + hf_x(x_0, y_0) + kf_y(x_0, y_0)$$
$$+ (1/2!)\,[h^2 f_{xx}(x_0 + \vartheta h, y_0 + \vartheta k) + 2hkf_{xy}(x_0 + \vartheta h, y_0 + \vartheta k) + k^2 f_{yy}(x_0 + \vartheta h, y_0 + \vartheta k)].$$

The space required for higher values of n increases rapidly with n, for instance, $n = 5$ requires 28 different terms, and each of the highest derivatives has 6 indices. It is therefore helpful to use a shorthand symbolic notation by first writing

$$hf_x(x_0, y_0) + kf_y(x_0, y_0) = \left(h\frac{\partial}{\partial x} + k\frac{\partial}{\partial y}\right) f(x_0, y_0)$$

and then using powers of the symbolic operator $\left(h\dfrac{\partial}{\partial x} + k\dfrac{\partial}{\partial y}\right)$ to write down the terms of higher order.

Taylor's theorem for functions of two variables
$f(x_0+h, y_0+k) = f(x_0, y_0) + \left(h\frac{\partial}{\partial x} + k\frac{\partial}{\partial y}\right) f(x_0, y_0) + \frac{1}{2!}\left(h\frac{\partial}{\partial x} + k\frac{\partial}{\partial y}\right)^2 f(x_0, y_0) + \cdots + \frac{1}{n!}\left(h\frac{\partial}{\partial x} + k\frac{\partial}{\partial y}\right)^n f(x_0, y_0) + \frac{1}{(n+1)!}\left(h\frac{\partial}{\partial x} + k\frac{\partial}{\partial y}\right)^{n+1} f(x_0+\vartheta h, y_0+\vartheta k),\ 0<\vartheta<1$

The quantity ϑ takes the same value in all terms of the remainder, and depends on n, x_0, y_0, h and k. The same symbolism can be utilized for functions of three or more variables. For three variables one obtains

$$f(x_0+h, y_0+k, z_0+l) = f(x_0, y_0, z_0) + \sum_{\nu=1}^{n} \frac{1}{\nu!}\left(h\frac{\partial}{\partial x} + k\frac{\partial}{\partial y} + l\frac{\partial}{\partial z}\right)^{\nu} f(x_0, y_0, z_0)$$
$$+ \frac{1}{(n+1)!}\left(h\frac{\partial}{\partial x} + k\frac{\partial}{\partial y} + l\frac{\partial}{\partial z}\right)^{n+1} f(x_0+\vartheta h, y_0+\vartheta k, z_0+\vartheta l), \quad 0<\vartheta<1.$$

An extension of Newton's method. If one wishes to solve the pair of equations $f(x, y) = 0$, $g(x, y) = 0$ and if approximate values x_0, y_0 are known, one puts $f(x_0+h, y_0+k) = 0$, $g(x_0+h, y_0+k) = 0$, and expands by Taylor's theorem. For $n = 0$ this gives

$$f(x_0, y_0) + hf_x(x_0+\vartheta h, y_0+\vartheta k) + kf_y(x_0+\vartheta h, y_0+\vartheta k) = 0,$$
$$g(x_0, y_0) + hg_x(x_0+\vartheta' h, y_0+\vartheta' k) + kg_y(x_0+\vartheta' h, y_0+\vartheta' k) = 0.$$

If one puts $\vartheta = 0$, $\vartheta' = 0$ as an approximation, one commits an error, and instead of the exact values h and k one only obtains approximate values h_1 and k_1 from which new approximations $x_1 = x_0 + h_1$, $y_1 = y_0 + k_1$ can be calculated. The procedure can be repeated if necessary. For h_1 and k_1 one obtains

$$h_1 = -[(fg_y - gf_y)/(f_x g_y - f_y g_x)]_{\substack{x=x_0\\y=y_0}}, \qquad k_1 = [(fg_x - gf_x)/(f_x g_y - f_y g_x)]_{\substack{x=x_0\\y=y_0}}$$

to be evaluated at $x = x_0$, $y = y_0$.

Example: To solve the system of equations $f(x, y) = x^2 + y - 2 = 0$, $g(x, y) = xy - 2 = 0$. – Approximate values are $x_0 = -1.8$ and $y_0 = -1.1$. For these values one gets $h_1 = 0.031$, $k_1 = -0.030$, and the new approximation is $x_1 = -1.769$, $y_1 = -1.130$.

21.3. Trigonometric series and harmonic analysis

The development of the theory of trigonometric series began with the publication, in 1822, of the book '*Théorie analytique de la chaleur*' by Joseph DE FOURIER (1768–1830). His researches, extending over several years, have led to the development of an extensive theory for the series that now bear his name and are of great importance in mathematics, science and technology. Its basic idea is to represent periodic functions by series of particular (trigonometric) periodic functions.

To investigate periodic motions Fourier series are used in acoustics, electrodynamics, optics, thermodynamics etc. In electrical engineering problems such as the frequency behaviour of switching elements or the transfer of impulses can be solved by means of Fourier series. Prediction of the tides is important for navigation; since they are periodic phenomena, one utilizes Fourier series and constructs mechanical instruments, the tide predictors and waterlevel predictors, for all important harbours. Today there is hardly a branch of physics, mathematics, or technology in which Fourier series are not used.

Trigonometric series

Series of functions $\sum_{n=0}^{\infty} f_n(x)$ in which the general term is $f_n(x) = a_n \cos nx + b_n \sin nx$, with constant coefficients a_n and b_n, are called *trigonometric series*. If this series converges in an interval of length, 2π, then, since the trigonometric functions are periodic, it converges for all x and represents a periodic function $f(x)$. But this function is not necessarily continuous, indeed, it often has discontinuities between which it is given by different formulae (Fig.). On the other hand, if the series converges uniformly, then its sum $f(x)$ is continuous. In this case a connection can be established between the coefficients a_n, b_n and the sum function

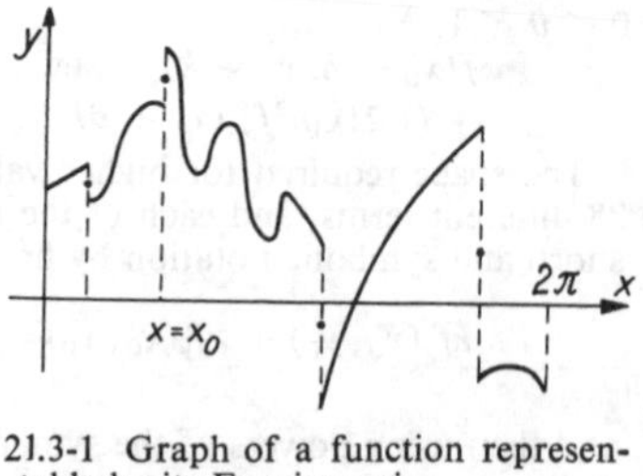

21.3-1 Graph of a function representable by its Fourier series

$f(x)$. Multiplication of the series

$$f(x) = \sum_{n=0}^{\infty} f_n(x) = \sum_{n=0}^{\infty}(a_n \cos nx + b_n \sin nx)$$

by the bounded factors $\cos px$ or $\sin px$, where p is a non-negative integer, does not disturb uniform convergence, so that one may calculate

$$\int_0^{2\pi} f(x) \cos px \, dx \quad \text{and} \quad \int_0^{2\pi} f(x) \sin px \, dx$$

by termwise integration of the series $\sum f_n(x) \cos px$ or $\sum f_n(x) \sin px$. These integrations involve the integrals over the interval $(0, 2\pi)$ of the functions $\cos nx \cos px$, $\sin nx \cos px$, $\cos nx \sin px$, $\sin nx \sin px$. One finds by partial integration that these integrals have the value 0 when $n \neq p$; for $p = n$ they are

$$\int_0^{2\pi} \cos^2 nx \, dx = \int_0^{2\pi} \sin^2 nx \, dx = \pi \quad \text{for} \quad n > 0,$$

and

$$\int_0^{2\pi} \cos^2 nx \, dx = 2\pi, \quad \int_0^{2\pi} \sin^2 nx \, dx = 0 \quad \text{for} \quad n = 0.$$

Because of the exceptional behaviour of $n = 0$, it has now become conventional to write the trigonometric series as

$$f(x) = {}^1/_2 a_0 + \sum_{n=1}^{\infty}(a_n \cos nx + b_n \sin nx)$$

so that the coefficients can be written for all $n \geqslant 0$ thus:

Euler-Fourier formulae	$a_n = \frac{1}{\pi} \int_0^{2\pi} f(x) \cos nx \, dx, \quad b_n = \frac{1}{\pi} \int_0^{2\pi} f(x) \sin nx \, dx$

Fourier series. One may well ask what functions $f(x)$ can be represented by trigonometric series. If $f(x)$ is integrable, one can at least use the Euler-Fourier formulae to calculate the numbers a_n and b_n and then write down the formal series ${}^1/_2 a_0 + \sum_{n=1}^{\infty} (a_n \cos nx + b_n \sin nx)$.

One calls this the Fourier series of $f(x)$, and a_n, b_n the *Fourier coefficients* of the function $f(x)$. However, it may happen either that the Fourier series of $f(x)$ does not converge at all, or that it converges, but that its sum is not equal to $f(x)$; this can occur even if $f(x)$ is continuous. Also it is conceivable that $f(x)$ has other representations by a trigonometric series.

However, if the Fourier series of a continuous function $f(x)$ turns out to be uniformly convergent, then its sum must be $f(x)$, and $f(x)$ has no other representation by a uniformly convergent trigonometric series. This is only a sufficient condition: the problem of finding necessary and sufficient conditions for the convergence of the Fourier series of $f(x)$ is still not completely settled.

Since the terms $f_n(x)$ of the Fourier series are periodic functions of period 2π, the sum function also has the period 2π, so that it makes sense to consider Fourier series for periodic functions of period 2π. From a function of period $2l$, one can obtain a function of period 2π by replacing the variable x by $\pi x/l$. If one needs the Fourier expansion for a function $f(x)$ defined in some interval I of length 2π, it is appropriate to extend the function outside this interval by requiring that $f(x + 2k\pi) = f(x)$ ($x \in I$; k an integer) so that the extended function has the period 2π.

Dirichlet's condition. A further sufficient condition for the convergence to $f(x)$ of the Fourier series of $f(x)$ is due to DIRICHLET (1805–1859). It suffices for practical purposes and covers a wide class of functions including functions of the type described in the picture below. The functions $f(x)$ may, without loss of generality, be assumed to be periodic of period 2π.

Suppose that $f(x)$ is a periodic function of period 2π and is defined and bounded for $0 \leqslant x < 2\pi$, and suppose that the interval $(0, 2\pi)$ can be split into finitely many subintervals in each of which the function is continuous and monotonic. Then the Fourier series of $f(x)$ converges at each point of continuity x_0 to $f(x_0)$, and at a point a jump discontinuity x to the mean value $(1/2)[\lim_{x \to x^*-0} f(x) + \lim_{x \to x^*+0} f(x)]$ of its left and right limiting values.

Hence, if one prescribes at the points of 'jump' x^* with

$$\lim_{x \to x^*-0} f(x) \neq \lim_{x \to x^*+0} f(x) \quad \text{that} \quad f(x^*) = (1/2)[\lim_{x \to x^*-0} f(x) + \lim_{x \to x^*+0} f(x)],$$

then the Fourier series of $f(x)$ converges to $f(x)$ at all points in the domain of definition. The requirement that the interval $(0, 2\pi)$ can be split into finitely many subintervals in each of which $f(x)$ is continuous and monotonic means that the function has only finitely many discontinuities and only finitely many extrema.

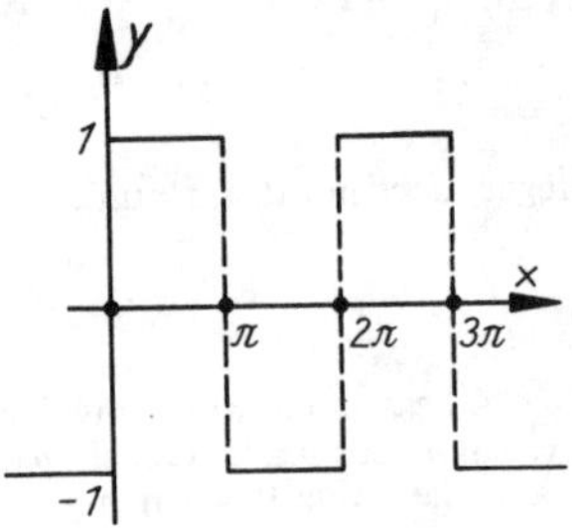

Example: Let $f(x)$ be given by $f(x) = 1$ for $0 < x < \pi, f(x) = -1$ for $\pi < x < 2\pi, f(x + 2k\pi) = f(x), k = \pm 1, \pm 2, \pm 3, \ldots$ At the jumps, let $f(0) = f(k\pi) = 0$. Dirichlet's condition obviously holds (Fig.).

The integrations in the Euler-Fourier formulae lead to $a_n = 0$ for all n, $b_{2n} = 0$ $(n = 1, 2, \ldots)$, $b_{2n+1} = 4/[\pi(2n+1)]$ if $n = 0, 1, 2, \ldots$

The Fourier series is $f(x) = \dfrac{4}{\pi}\left[\sin x + \dfrac{\sin 3x}{3} + \dfrac{\sin 5x}{5} + \cdots\right]$.

21.3-2 Rectangular curve

1 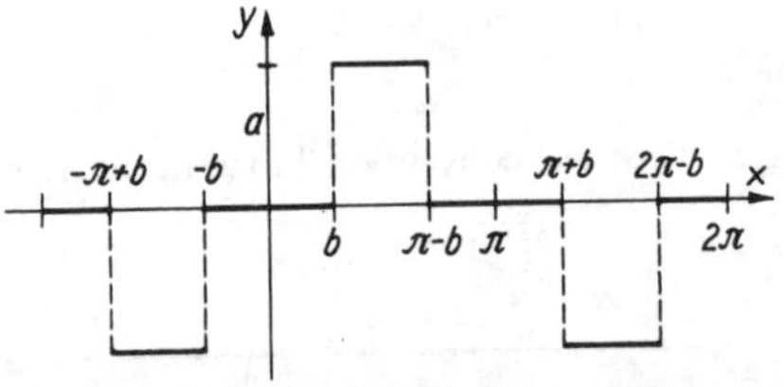

2 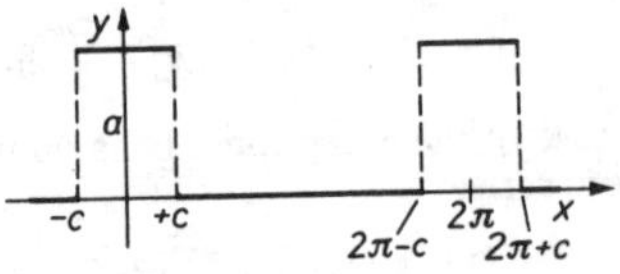

1. *Rectangular impulse* of the first kind:

$$f(x) = \frac{4a}{\pi}\left[\frac{\cos b}{1} \cdot \sin x + \frac{\cos 3b}{3} \cdot \sin 3x + \frac{\cos 5b}{5} \cdot \sin 5x + \cdots\right].$$

2. *Rectangular impulse* of the second kind:

$$f(x) = \frac{2a}{\pi}\left[\frac{c}{2} + \frac{\sin c}{1} \cdot \cos x + \frac{\sin 2c}{2} \cdot \cos 2x + \frac{\sin 3c}{3} \cdot \cos 3x + \cdots\right].$$

3 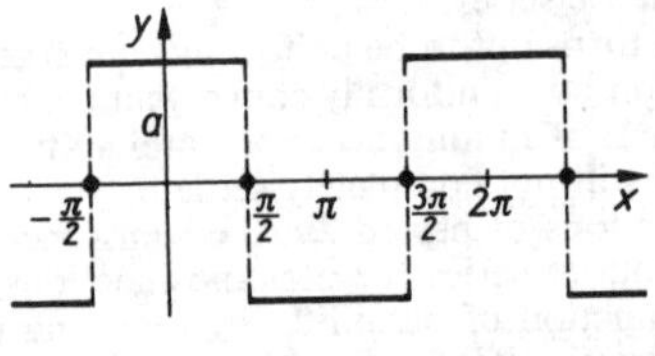

4 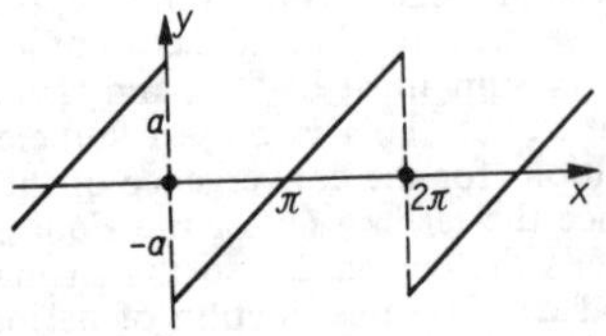

3. *Rectangular curve:* $f(\pi/2) = f(3\pi/2) = \cdots = 0, f(x) = \dfrac{4a}{\pi}\left[\cos x - \dfrac{\cos 3x}{3} + \dfrac{\cos 5x}{5} - + \cdots\right]$.

4. *Sawtooth curve:* $f(0) = f(2\pi) = \cdots = 0$, $f(x) = -\dfrac{2a}{\pi}\left[\sin x + \dfrac{\sin 2x}{2} + \dfrac{\sin 3x}{3} + \cdots\right]$.

5 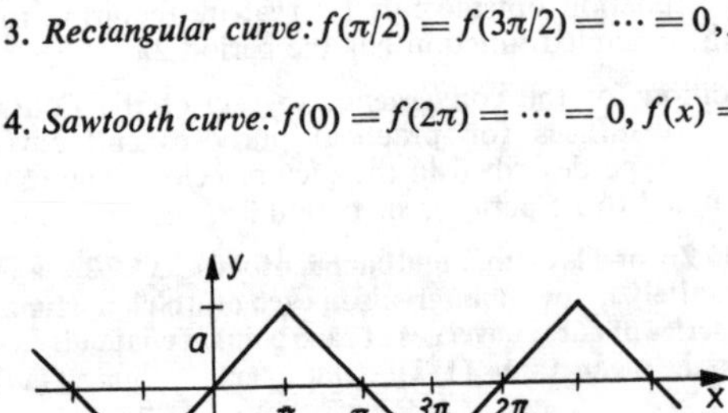

6 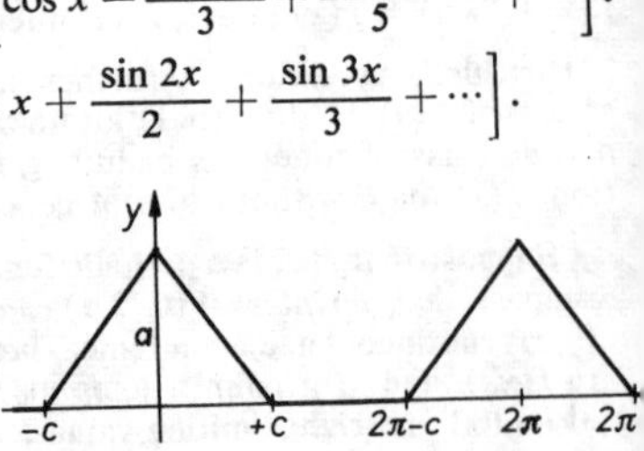

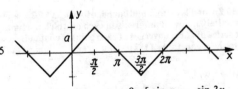

5. *Triangular curve:* $f(x) = \dfrac{8a}{\pi^2}\left[\dfrac{\sin x}{1^2} - \dfrac{\sin 3x}{3^2} + \dfrac{\sin 5x}{5^2} - + \cdots\right]$.

6. *Triangular impulse:*

$$f(x) = \frac{ac}{2\pi} + \frac{2a}{\pi^2}\left[\frac{1 - \cos c}{1^2}\cos x + \frac{1 - \cos 2c}{2^2}\cos 2x + \frac{1 - \cos 3c}{3^2}\cos 3x + \cdots\right].$$

7 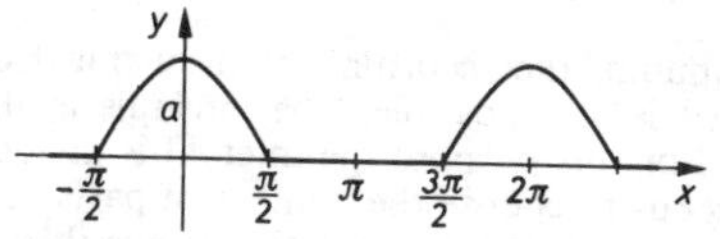

8 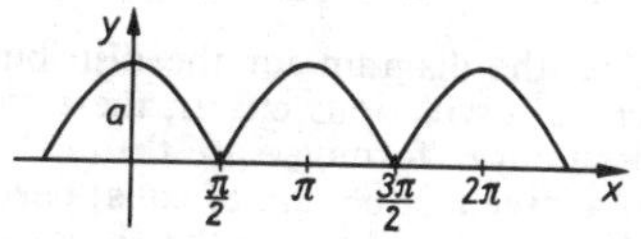

7. *Alternating current rectified in one direction, half waves of a cosine curve:*

$$f(x) = \frac{a}{\pi}\left[1 + \frac{\pi}{2}\cdot\cos x + \frac{2}{1\cdot 3}\cdot\cos 2x - \frac{2}{3\cdot 5}\cdot\cos 4x + \frac{2}{5\cdot 7}\cdot\cos 6x - + \cdots\right].$$

8. *Alternating current rectified in two directions:*

$$f(x) = |\cos x|,\ f(x) = \frac{2a}{\pi}\left[1 + \frac{2}{1\cdot 3}\cdot\cos 2x - \frac{2}{3\cdot 5}\cdot\cos 4x + \frac{2}{5\cdot 7}\cdot\cos 6x - + \cdots\right].$$

Harmonic analysis and harmonic synthesis

Harmonic analysis. This is the determination of the Fourier coefficients $a_0, a_1, a_2, \ldots, b_1, b_2, \ldots$ In technology it is frequently used to analyse periodic phenomena. An oscillation is split up by harmonic analysis into a sum of pure sine oscillations (harmonic oscillations) and a constant part. Apart from the *fundamental oscillation* there occur the so-called 'harmonics' whose frequency is twice, three times etc. the fundamental frequency. As a rule, the phase of an individual harmonic is shifted by comparison with the fundamental oscillation. One can always set $a_n \cos nx + b_n \sin nx = c_n \cos(nx - x_n)$; this leads to $a_n = c_n \cos x_n$ and $b_n = c_n \sin x_n$ and hence to $c_n = \sqrt{(a_n^2 + b_n^2)}$, $x_n = \arctan(b_n/a_n)$.

The process of setting up the Fourier coefficients for the rectangular curve is an example of harmonic analysis.

Much labour can be saved in harmonic analysis if one observes certain symmetry properties of the function $f(x)$ to be analysed:

In the Fourier expansion of an even function $f(x) = f(-x)$ all the sine terms are absent, that is, all the $b_n = 0$. For an odd function $f(x) = -f(-x)$ all the cosine terms are absent, that is, all the $a_n = 0$ (including a_0). For a function with the property $f(x+\pi) = -f(x)$ the absolute term is $a_0 = 0$, and only coefficients with an odd index occur ($a_2 = a_4 = \cdots = b_2 = b_4 = \cdots = 0$).

If one looks for a best possible approximation to a periodic function $f(x)$ by a finite sum $\Phi_n(x)$ of sine and cosine functions, $\Phi_n(x) = \sum_{j=0}^{n}(a_j \cos jx + b_j \sin jx)$, one chooses by analogy to the method of least squares the integral $\frac{1}{2\pi}\int_0^{2\pi} [f(x) - \Phi_n(x)]^2\, dx$ as a measure for the difference $f(x) - \Phi_n(x)$. This assumes its minimum when the a_j and b_j are the Fourier coefficients of the function $f(x)$. This is another important property of the Fourier coefficients.

Harmonic synthesis. This is the inverse process to harmonic analysis. The individual pure oscillations are added and yield a resultant. The figure shows the first three terms of the Fourier expansion of the rectangular curve and the sum curve y allows a comparison with the original curve y_R.

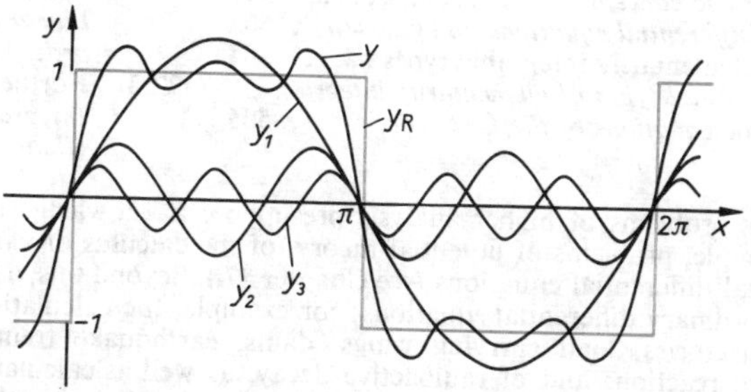

21.3-4 Graphical representation of the harmonic synthesis of a rectangular curve

Approximate calculation of the Fourier coefficients. In practice the functions to be expanded in a Fourier series are frequently not given by an analytic expression. As a rule they are curves drawn by a measuring instrument equipped with a pen, such as the tangential force diagram of a piston engine, the diagram for the distribution of pressure in a pump, the recording of mechanical or electrical oscillations etc. In these cases the Fourier analysis is also possible. The integrals in the Euler-Fourier formulae are then calculated approximately. For this purpose the interval is divided into a large number $2m$ of equal parts (Fig.). It is advantageous to choose the number of parts as a multiple of 4 and to use the values 12, 24, 36, 72, ..., because such a division makes it possible to utilize the symmetry properties of the sine and cosine functions, and this saves calculating labour. After fixing a coordinate system the function values at the places $x_0, x_1, x_2, \ldots, x_{2m-1}$ are measured; they are denoted by $y_0, y_1, y_2, \ldots, y_{2m-1}$. Then

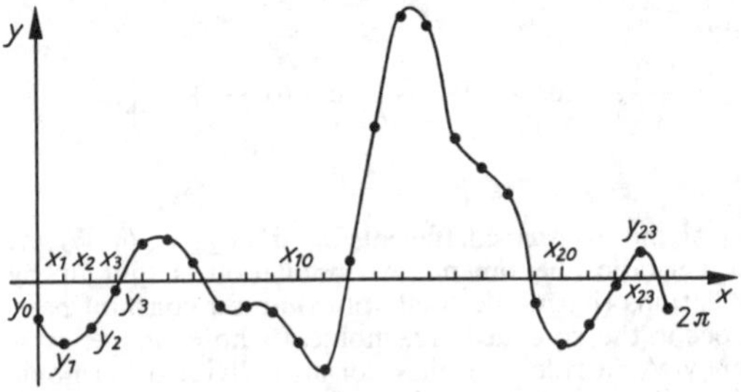

$$a_0 = \frac{1}{2m}\sum_{i=0}^{2m-1} y_i, \quad a_m = \frac{1}{2m}\sum_{i=0}^{2m-1} y_i \cos(i\pi),$$

$$a_n = \frac{1}{m}\sum_{i=0}^{2m-1} y_i \cos\frac{ni\pi}{m},$$

$$b_n = \frac{1}{m}\sum_{i=0}^{2m-1} y_i \sin\frac{ni\pi}{m}$$

for $n = 1, 2, \ldots, (m-1)$.

21.3-5 Fourier analysis of a curve given empirically

If one chooses $2m = 24$, one obtains the 24 coefficients $a_0, a_1, a_2, \ldots, a_{12}, b_1, b_2, \ldots, b_{11}$. The resulting function $a_0 + \sum_{n=1}^{11} (a_n \cos nx + b_n \sin nx) + a_{12} \cos 12x = f(x)$ has the values $f(x_i) = y_i$ at the places x_i $(i = 0, 1, \ldots, 23)$.

The amount of work to be done in harmonic analysis is considerable. With the help of an electric calculating machine and special techniques a trained operator can carry out a harmonic analysis with 12 points in about half an hour, with 24 points in about 2 hours, with 36 points in about 6 hours and with 72 points in about 16 hours. Without resorting to the special techniques one has to form for 72 points about 5000 products, which have to be combined in 72 sums. An electric computer of medium speed performs the calculations for 36 points in about 2 minutes. The time required to print out the result is usually larger than the calculating time.

Harmonic analysers. The large amount of time required for the Fourier analysis of curves has led to the development of mechanical tools and devices. One operates with them as with a planimeter. The given curve is traced with a moving pen, and the value of a Fourier coefficient or a value proportional to it can be read off the calculating works. Instruments of this kind are called *harmonic analysers*.

22. Ordinary differential equations

Many problems of higher analysis presuppose a knowledge of ordinary differential equations; for example, problems of potential theory, of the calculus of variations, of theoretical physics and of partial differential equations (see Chapter 37.). Beyond this, a wide field of applications is opened up by ordinary differential equations; for example, the calculation of pendulum oscillations, satellite trajectories, load carrying wings, dams, earthquake tremors, heat propagation, speeds of chemical reactions and of radioactive decay, as well as calculations in electrotechnology and ship building. Only differential equations for real variables and real-valued functions will be examined here and, renouncing full mathematical rigour, methods of solution will be given that occur frequently in practice. A first glance will also be given at typical problems in this field, at its vast and often difficult theory.

22.1. Preliminary survey

Basic concepts

Differential equation. If a relation exists between a function of one or more variables and some of its derivatives, in the form of an equation in which the independent variables can also occur, then one speaks of a *differential equation*. Every solution of the differential equation is called a *solution* or an *integral*; for example, the differential equation $\left(\frac{dy}{dx}\right)^2 + y^2 = 1$ has the solution $y = \sin x$, since substitution gives the identity $\cos^2 x + \sin^2 x = 1$, which holds for all x. Conversely, for a function $z = f(x, y)$ of the two independent variables x and y one can set up a differential equation that has, for example, the solution $z = xy$. Because $\frac{\partial z}{\partial x} = y, \frac{\partial z}{\partial y} = x$ in this case, $z = xy$ satisfies the differential equation $\frac{\partial z}{\partial x} y + \frac{\partial z}{\partial y} x = x^2 + y^2$.

If the functions occurring in the differential equation depend on only one independent variable, and thus also derivatives with respect to only one variable occur, then one speaks of an *ordinary differential equation*. To these belong, for example,

$$\frac{dy}{dx} = \cos x, \quad \frac{d^2y}{dx^2} + y^2 = 3xy, \quad y'^3 - y'xy = 0.$$

On the other hand, if the required functions depend on several independent variables and accordingly partial derivatives occur, one speaks of *partial differential equations*. Examples are

$$\frac{\partial^2 z}{\partial x\, \partial y} + z\frac{\partial z}{\partial x} = 0 \quad \text{and} \quad \frac{\partial^2 z}{\partial x^2} + \frac{\partial^2 z}{\partial y^2} = 4xy\frac{\partial z}{\partial x};$$

it is required to find functions $z = f(x, y)$ of x and y. Only ordinary differential equations will be dealt with in the following.

Order and degree of a differential equation. The *order* of a differential equation is defined as the highest order of the derivatives contained in it. A differential equation of the nth order can be expressed in the form $F(x, y, y', y'', \ldots, y^{(n)}) = 0$, where F denotes a function of the arguments in the bracket. In particular, $y' = f(x, y)$ is the general *explicit* and $F(x, y, y') = 0$ the general *implicit differential equation* of the first order. If F is a polynomial function of the arguments $y, y', \ldots, y^{(n)}$, then its *degree* is equal to that of the differential equation; the dependence upon x plays no part in this. However, in the case of the differential equation $y' = x + \sin y'$ one cannot speak of a degree.

Differential equation	Order	Degree
$y' = x + \sin y'$	1	—
$y'^2 = x \sin x$	1	2
$y'' = 3x^2 y$	2	1
$y'' + 3y' + y \cos x = \sin x$	2	1
$y'''y'' = y$	3	2

Differential equations of the first degree, or linear differential equations, are particularly important for applications. In these the unknown function and its derivatives occur only to the first power and also not multiplied together. Consequently, the general linear differential equation of the nth order has the form $f + f_0 y + f_1 y' + f_2 y'' + \cdots + f_n y^{(n)} = 0$, where $f, f_0, f_1, \ldots, f_n$ denote given functions of x.

The integral of a differential equation. If the equation $F(x, y, y', \ldots, y^{(n)}) = 0$, after the substitution of a function $y = \varphi(x)$ and its derivatives $y', y'', \ldots, y^{(n)}$, becomes an identity in x valid for all x in an interval, then $y = \varphi(x)$ is called a *solution* or *integral*; the process of obtaining it is called *integration*, and the graph of $y = \varphi(x)$ in the x, y-plane is an *integral curve*. The solutions are often not elementary functions or even closed forms of these functions. On the contrary, certain non-elementary functions that are important for applications are defined precisely as solutions of special types of differential equations. For example, in 1785 in investigating the force of attraction of an ellipsoid at a point outside it, LEGENDRE hit upon a differential equation still called after him, whose solution is represented by the *Legendre polynomials*. It is often sufficient, without insisting on the complete solution, to determine the analytic properties of a solution in a neighbourhood of a point x_0 and to investigate the shape of the integral curves, the uniqueness of the solution, or other questions. Finally, *existence theorems* are concerned with the properties of a differential equation from which it can be deduced with certainty that solutions exist at all.

Preliminary survey of the nature of the integrals of differential equations. One distinguishes between the *general integral* and *particular* and *singular integrals*. The nature of the solution can be summarized crudely and somewhat imprecisely as follows:

The general integral of a differential equation of the nth order contains exactly n arbitrary constants $C_1, C_2, \ldots, C_n$; it is determined only to within these constants.

Correspondingly, in the integral calculus one obtains as solution of the differential equation $y' = f(x)$ the integral $y = \int f(x)\,dx + C$. If one assigns to the $C_1, C_2, \ldots, C_n$ arbitrary fixed numerical values, then one obtains a *particular integral*. Consequently, all the particular integrals are, so to speak, contained in the general integral.

Example: The differential equation $y'^2 + y^2 = 1$ has $y = \sin(x + C)$ as its general integral. For $C = \pi/2$ one obtains the particular integral $y = \sin(x + \pi/2) = \cos x$. By substituting in the differential equation one can easily see that it is, in fact, a solution.

Besides the general integral and particular integrals, a differential equation may also have *singular integrals*, which usually correspond to certain discontinuities of the given equation. Singular integrals cannot be obtained from the general integral by a choice of the constants. For example, the differential equation $y'^2 + y^2 = 1$ already mentioned has the singular integral $y = \pm 1$, as one can see by differentiation and substitution.

Example: The second order differential equation $y'' + y = 0$ has the general integral $y = C_1 \sin x + C_2 \cos x$. By suitable choice of the constants C_1 and C_2 one obtains the particular integrals $y = 0$, $y = \cos x$, $y = 2\cos x$, $y = \sin x$, $y = \pi \sin x$. Singular integrals do not exist.

Differential equations and geometry

The direction field of a differential equation of the first order. In the implicit form $F(x, y, y') = 0$, and more particularly in the explicit form $y' = f(x, y)$, a differential equation assigns to the points of the x, y-plane for which $f(x, y)$ is defined, a value $p = y' = f(x, y)$ of the derivative of the required function $y(x)$, which gives the direction of the tangent to the curve representing the function $y(x)$. The *direction field* of the differential equation of the first order arises in this way. The number triple x, y, p is called a *line element*; the point (x, y) is its carrier. At least an approximate idea of the course of the integral curves of a first order differential equation can be obtained with the help of the direction field, in which the direction of the tangent at the point (x, y) is marked by a short line (Fig.). Geometrically expressed, the problem of the integration of the differential equation of the first order consists of finding all curves that *fit the direction field*, that is, have a tangent at every point and contain only those line elements that agree with the values given by $y' = f(x, y)$.

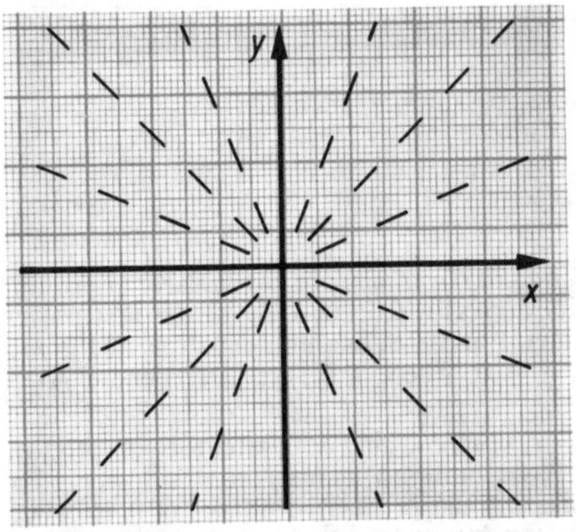

22.1-1 Direction field of the differential equation $y' = y/x$

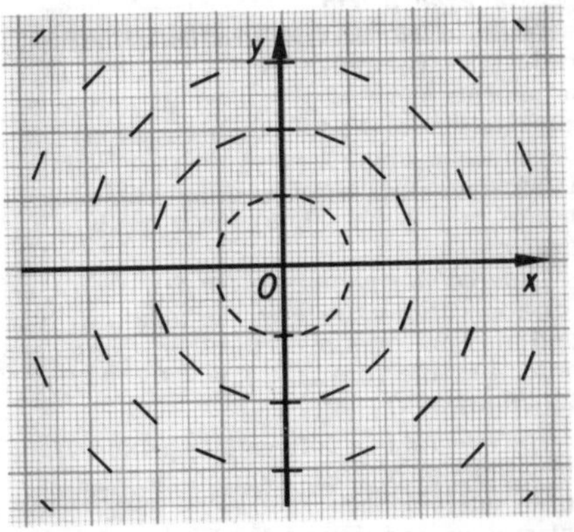

22.1-2 Direction field of the differential equation $y' = -x/y$

Differential equation and family of curves. The result stated above, that the solution of a differential equation of the first order contains one arbitrary constant, can be interpreted geometrically; the solution consists of a *one-parameter family of curves*. The converse also holds: A one-parameter family of curves $y = \varphi(x, C)$ is represented analytically by a differential equation of the first order. This is obtained by eliminating C from the system of equations $y = \varphi(x, C)$; $\dfrac{dy}{dx} = \varphi'(x, C)$.

Example: The family of all straight lines through the origin has the equation $y = Cx$. Then $y' = C$. From this the differential equation $y = y'x$, or $y' = y/x$ is obtained (see Fig. 22.1-1).

An n-parameter family of curves can be represented analytically by a differential equation of the nth order. Conversely, the general solution of a differential equation of the nth order represents an n-parameter family of curves.

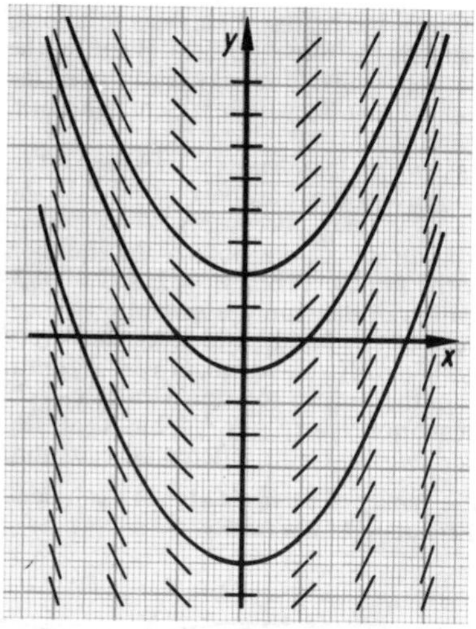

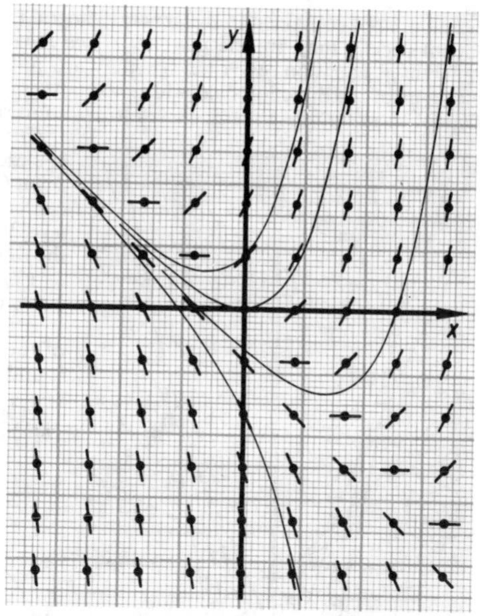

22.1-3 Direction field of the differential equation $y' = x$

22.1-4 Direction field of the differential equation $y' = x + y$

The second part of this theorem clearly follows immediately from the nature of the general integral of a differential equation of the nth order. On the other hand, from the equation of a family of curves that contains n parameters, the corresponding differential equation can be found: one differentiates the equation of the family sufficiently often until one succeeds in eliminating the parameters from the original equation and the equations obtained from it by differentiation, and in obtaining a differential equation free of parameters.

Example 1: $y = C_1 x + C_2$ is the two-parameter equation of the family of all straight lines in the plane not parallel to the y-axis. By differentiating twice, $y'' = 0$; an elimination is not necessary. The differential equation states, in fact, that it is concerned with all curves whose curvature is everywhere zero; these are precisely the straight lines.

Example 2: The family of all circles of fixed radius a has the equation $(x - C_1)^2 + (y - C_2)^2 = a^2$. By differentiation one obtains $x = C_1 + (y - C_2)\,y' = 0$, and a second differentiation gives $1 + y'^2 + (y - C_2)\,y'' = 0$. From these one obtains by elimination $C_2 = (1/y'')\,(1 + y'^2 + yy'')$ and $C_1 = x - (1 + y'^2)\,(y'/y'')$, and then by substitution the differential equation $y''^2 a^2 = (1 + y'^2)^3$.

All curves of the family are contained among the solutions of the corresponding differential equation. However, it may very well happen that the solutions of the differential equation contain additional curves that do not belong to the original family; for example, the family of curves $y = C_1 x + C_2$, $C_1 > 0$, consisting of all straight lines with positive slope, leads to the differential equation $y'' = 0$; but among the solutions of this equation are not only all straight lines with positive slope, but also all those with negative slope.

Singular solutions, envelopes of families of curves. The *family of all circles* $y^2 + (x - C)^2 = 1$ of radius 1 whose centres lie on the x-axis satisfies the differential equation $y^2 y'^2 + y^2 - 1 = 0$, because $yy' + x - C = 0$, $C = x + yy'$ (Fig.). It is also satisfied by the functions $y = 1$ and $y = -1$,

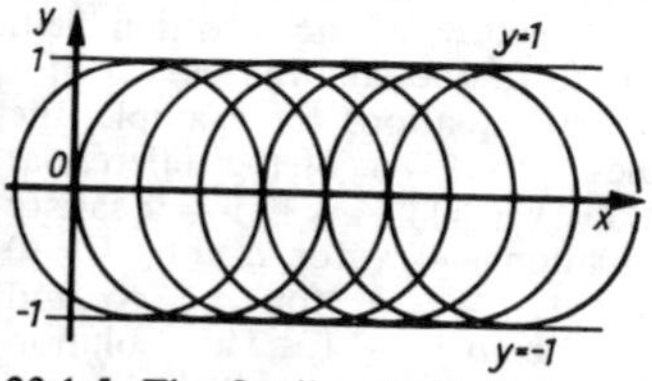

22.1-5 The family of all circles of radius 1 whose centres lie on the x-axis

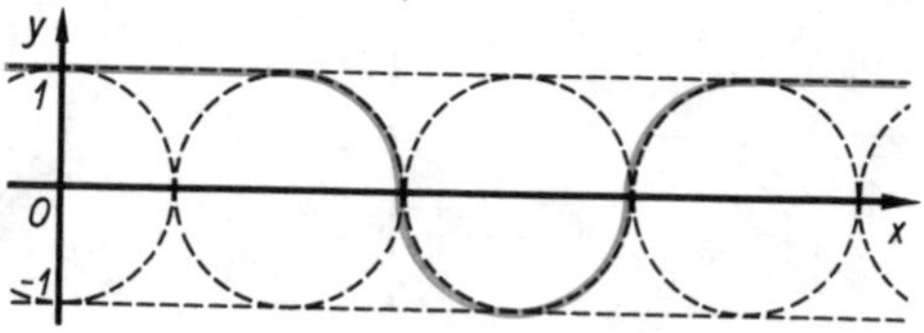

22.1-6 Composite solution curve which fits the direction field of the differential equation $y^2 y'^2 + y^2 - 1 = 0$

which are not contained in the general integral $(x - C)^2 + y^2 = 1$, but represent singular solutions. Geometrically they are the *tangents to the family of circles* and fit the direction field given by the differential equation, even though they are not contained in the family of circles. From the line elements of the direction field additional curves can be constructed that likewise represent solutions. One of these infinitely many curves is drawn in red in the figure.

Family of tangents to the parabola $y = x^2$(Fig.). The equation of the tangent to the parabola $y = x^2$ at the point (x_1, y_0) is $y + y_0 = 2xx_0$. Because $y_0 = x_0^2$ and x_0 can be regarded as parameter C, one obtains the equation $y = 2Cx - C^2$ for the family. From $y' = 2C$, $C = y'/2$, the differential equation of the family is given by $y = xy' - y'^2/4$. The *envelope* of this family, which touches every curve of the family, is clearly the parabola $y = x^2$ itself. It is not contained in the general solution $y = 2Cx - C^2$ of the differential equation $y = xy' - y'^2/4$, but it satisfies it and is the *singular solution* of the differential equation.

The envelope of a family of curves is always a solution of the differential equation of the family.

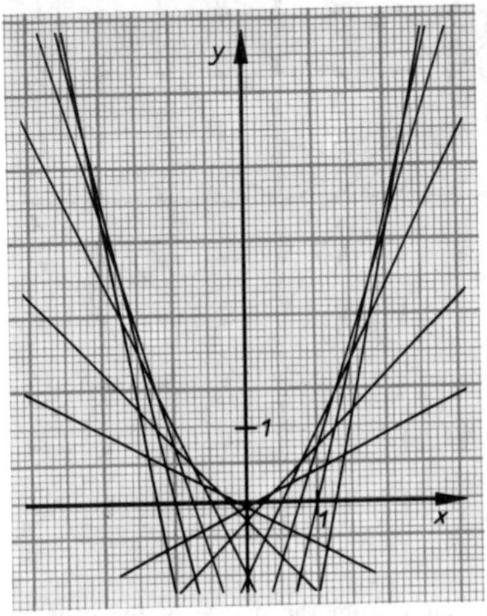

22.1-7 Family of tangents to the parabola $y = x^2$

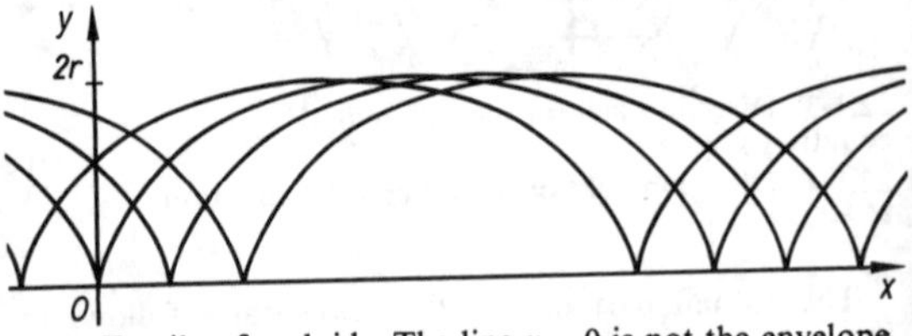

22.1-8 Family of cycloids. The line $y = 0$ is not the envelope

From this fact follows also a method, stated here without proof, of finding the envelope of a one-parameter family of curves when it exists. If one knows the general solution $\Phi(x, y, C) = 0$ of the differential equation corresponding to the family, then one eliminates the parameter C from the equation $\Phi(x, y, C) = 0$ and the equation $\dfrac{\partial\Phi(x, y, C)}{\partial C} = 0$ obtained by differentiating it partially with respect to C. Occasionally this procedure also yields other curves, besides the envelope, that have geometrical significance for the family of curves; for example, the cusp locus for the family of cycloids (Fig.) or the node locus in the case in which the individual curves of the family intersect themselves.

Isoclines, orthogonal trajectories. Points of a direction field of a differential equation of the first order having the same field direction lie on a curve called an *isocline*. The equation of an isocline is obtained by substituting $y' = \text{constant} = a$ in the equation $y' = f(x, y)$. From the isoclines one can obtain a picture of the direction field and hence of the solution curves of a differential equation; for example, the isoclines with $y' = a$ of the differential equation $(x + y)y' + x - y = 0$ satisfy the equation $y = x$ for $a = 0$, $x = 0$ for $a = 1$, $y = -x$ for $a = \infty$ and $y = 0$ for $a = -1$. The solution curves form a so-called *vortex* (Fig.).

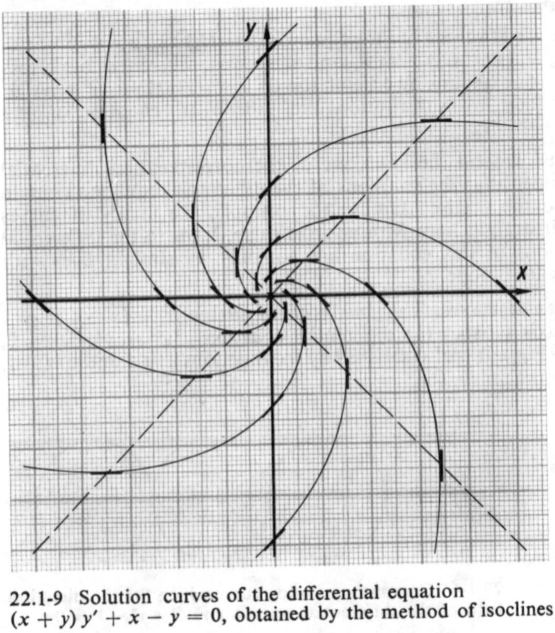

22.1-9 Solution curves of the differential equation $(x + y)y' + x - y = 0$, obtained by the method of isoclines

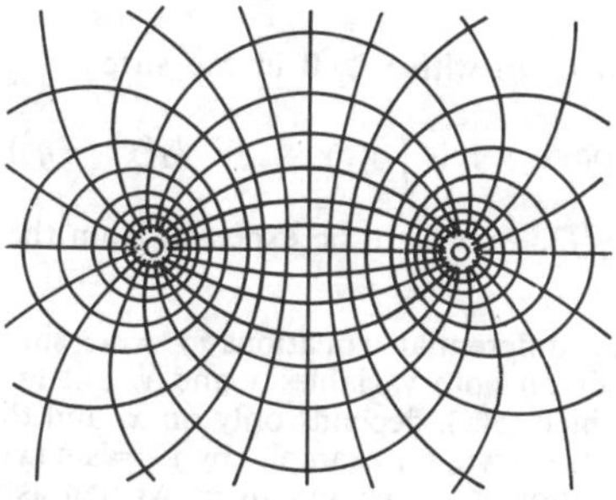

In geometry and above all in physics the problem often arises of finding the *family of orthogonal trajectories* of a family of curves. These are the curves that cut every curve of the first family at right angles. The lines of force of a magnetic or electric dipole form the family of field lines that are cut orthogonally by the equipotential lines (Fig.). Analytically one obtains the differential equation of the orthogonal trajectories by replacing y' by $-1/y'$ in the differential equation $y' = f(x, y)$ belonging to the family of curves $\varphi(x, y, C) = 0$. This method is based on the fact that product of the slopes of two orthogonal curves is -1.

22.1-10 Lines of force of a dipole cutting the equipotential lines at right angles. One family of curves consists of the orthogonal trajectories of the other

Example: The *orthogonal trajectories of the family of parabolas* $y^2 = -2(x + C)$, whose differential equation is $yy' = -1$, satisfy the differential equation $y' = y$ with the general solution $y = C\,e^x$. The family of exponential curves forms the family of orthogonal trajectories of the family of parabolas, and vice versa.

22.2. Elementarily integrable types

A differential equation is called *elementarily integrable* if its general solution can be obtained by ordinary integrations (quadratures) as a combination of finitely many elementary functions. This is possible only for certain types of differential equations that occur frequently in applications. By the solution procedures dealt with in the following, the question of the existence of solutions is decided positively by actually giving them.

Special types of elementarily integrable differential equations of the first order

The general *implicit* differential equation of the first order $F(x, y, y') = 0$ can be solved for y' in the neighbourhood of a point (x_0, y_0, y_0') by the implicit function theorem, provided that $\frac{\partial F}{\partial y'} \neq 0$ at that point; one obtains the *explicit form* $y' = f(x, y)$.

Differential equation of the type $y' = g(x)$. In this type of differential equation the right-hand side depends only on x. If $g(x)$ is integrable in the open interval (a, b), for example, if it is continuous, then for an arbitrary but fixed ξ in the interval (a, b), the functions

$$y = \int_{\xi}^{x} g(t)\,dt + C, \quad a < x < b,$$

satisfy the differential equation $y' = g(x)$ for arbitrary values of the constant C. The integral calculus shows that these are all the functions that satisfy it; thus, y represents the general integral.

Differential equation of the type $y' = h(y)$. If the function $h(y)$ depending on y only is continuous for $c < y < d$ and is *nowhere equal to zero* in this open interval, then this differential equation can be reduced to the type just considered. If $y = y(x)$ is a solution of the differential equation $y' = h(y)$, then the *inverse function* $x = \psi(y)$ satisfies the differential equation $\psi' = \frac{dx}{dy} = \frac{1}{h(y)}$. Then $x = \int \frac{1}{h(y)}\,dy$, in an interval for the inverse function corresponding to the interval (c, d), yields the function $x = \psi(y)$ that is inverse to the solution $y = y(x)$.

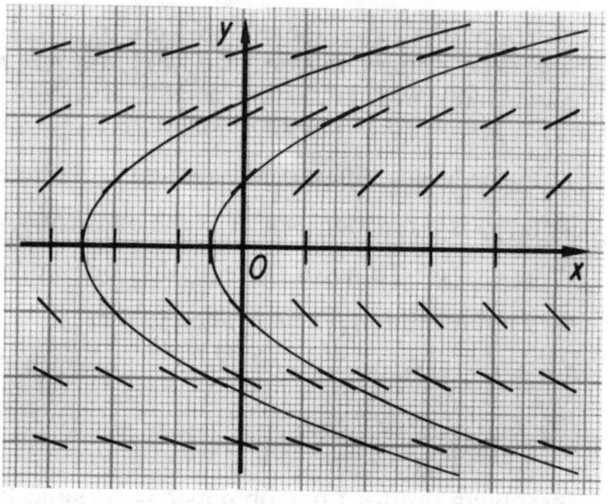

22.2-1 Direction field of the differential equation $y' = 1/y$

Example: The differential equation $y' = 1/y$ has a solution in the interval $0 < c < y < d$, because all the conditions are satisfied with $h(y) = 1/y$. Its *isoclines* are lines parallel to the x-axis.

The integral curve of $\frac{dx}{dy} = y$ passing through the point (ξ, η) with $\eta > 0$ in the strip $\{-\infty < x < +\infty; c < y < d\}$ is obtained by solving for y from $x = \xi + \int_\eta^y y\,dy = \xi + {}^1/_2(y^2 - \eta^2)$ for $y > 0$. This gives $y = \sqrt{\{\eta^2 + 2(x - \xi)\}}$ for $x > \xi - \eta^2/2$. As was to be expected from the direction field, the integral curves are parabolas (Fig.).

Differential equations with variables separable. In the differential equations $y' = e^x \sin y$, $y' = y/x^2$, $y' = (y+1)/(x-1)$, the right-hand side depends on both variables x and y, but in a particular way; it is the product of two functions, one of which, $g(x)$, depends only on x, and the other, $h(y)$, depends only on y. This is not always the case, as is shown, for example, by $y' = \sin(xy)$ or $y' = x + y$. If the right-hand side of the differential equation $y' = f(x, y)$ can be written as a product $g(x) \cdot h(y)$, the *variables* are said to be *separable*. In this case the differential equation $y' = g(x)\,h(y)$ can be solved easily, if $g(x)$ and $h(y)$ are continuous functions and $h(y)$ is different from zero in a whole interval (c, d). From $\frac{dy}{dx} = g(x)\,h(y)$ one obtains $dy/h(y) = g(x)\,dx$, and after integrating both sides

$$\int dy/h(y) = \int g(x)\,dx + C;$$

the general solution in $c < y < d$ is obtained by solving for y.

Example 1: $y' = -y/x$ for $x > 0$, $y > 0$.
Here $g(x) = -1/x$, $h(y) = y$. One obtains

$$\begin{aligned} \int dy/y &= -\int dx/x + C, \\ \ln y + \ln x &= C, \\ \ln xy &= C, \\ xy &= e^C = c. \end{aligned}$$

The integral curves are hyperbolas.

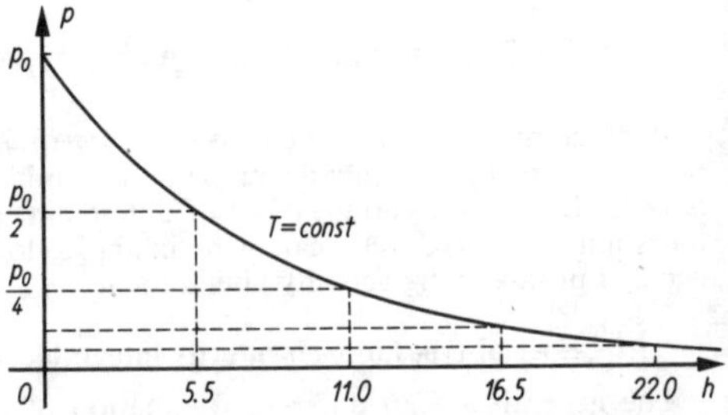

22.2-2 Decrease of the pressure p in the atmosphere at constant temperature as a function of the distance h in km above the ground

Example 2: The atmospheric pressure p varies with the height h above the ground (Fig.). For an increase in height dh, p increases by $dp = -\varrho g\,dh$, where ϱ is the density of the atmosphere and g the acceleration due to gravity. By Boyle's law the ratio $\varrho/p = \varrho_0/p_0 = a$ is constant, and hence

$$dp = -pag\,dh, \quad \int dp/p = -\int ag\,dh + C,$$
$$\ln p = -agh + C.$$

Barometric height formula
$p = p_0 e^{-\varrho_0 gh/p_0}$

For $h = 0$ the atmospheric pressure is p_0, the pressure at ground level, so that $C = \ln p_0$. One therefore obtains $\ln(p/p_0) = -agh$, or $p = p_0 e^{-agh} = p_0 e^{-\varrho_0 gh/p_0}$. The pressure decreases exponentially with increasing height; assuming a uniform atmospheric temperature, it is reduced by half every 5.54 km approximately.

Homogeneous differential equation. A differential equation $y' = f(x, y)$ is called homogeneous if $f(x, y)$ is a function $\varphi(y/x)$ of the quotient y/x; for example, $y' = \sin(y/x)$, $y' = \frac{(y/x - 1)\,x}{y}$ and $y' = -x^2/y^2$. To solve the equation one introduces a *new variable* into the equation $y' = \varphi(y/x)$ by the substitution $y/x = t$. Then $y = tx$, $y' = \frac{dy}{dx} = t'x + t$. This leads to the differential equation $t'x + t = \varphi(t)$, or $\frac{dt}{dx} = \frac{\varphi(t) - t}{x}$, in which the variables are separable. The general integral is

$$\int \frac{dt}{\varphi(t) - t} = \ln x + C$$

By solving this for t one obtains $t = t(x)$ and from this the required function $y = y(x)$. The method fails if the *denominator* $(\varphi(t) - t)$ of the integrand *vanishes*, if $\varphi(t) = t$, that is, if the given equation is $y' = y/x$. In this case, however, it could have been treated in the first place as a differential equation with variables separable.

Example: In order to find all curves $y(x)$ that cut every radius vector at the same angle α, one selects such a vector making an angle φ with the x-axis. At its intersection with the required curve

$y(x)$ the slope of the tangent to the curve is

$$y' = \tan(\varphi + \alpha) = \frac{\tan\varphi + \tan\alpha}{1 - \tan\varphi\tan\alpha} = \frac{(y/x) + \tan\alpha}{1 - (y/x)\tan\alpha}.$$

Writing $\tan\alpha = a$, one obtains the differential equation $y' = \dfrac{a + (y/x)}{1 - a\cdot(y/x)}$, which is homogeneous and has the solution $(2/a)\arctan y/x + C = \ln(x^2 + y^2)$.

In polar coordinates $r = \sqrt{(x^2 + y^2)}$ and $\varphi = \arctan y/x$ the solution has the equation $\varphi = a\ln r - (a/2)\cdot C$ or $r = e^{\varphi/a + C/2}$.

The required curves are *equiangular spirals* (Fig.).

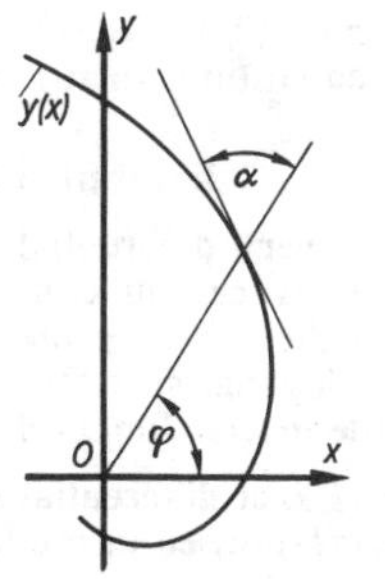

22.2-3 The derivation of the differential equation of an equiangular spiral

Linear differential equation $y' + p(x)\,y + q(x) = 0$. In this equation $p(x)$ and $q(x)$ are given functions of x that are assumed to be continuous. It is called a *linear homogeneous differential equation* if $q(x) \equiv 0$, that is, if there is no term not involving y or y'. It can be solved by treating it as a differential equation with variables separable. From $\dfrac{dy}{dx} + p(x)\,y = 0$ one obtains $\ln y = -\int p(x)\,dx + c_1$ or $y = Ce^{-\int p(x)\,dx}$. To obtain from this a solution of the original *inhomogeneous* differential equation $y' + p(x)\,y + q(x) = 0$ one uses the method of *variation of the parameter*, due to LAGRANGE. One regards C not as a constant, but as a function of x, $C = C(x)$. From $y = C(x)\,e^{-\int p(x)\,dx} = C(x)\,\psi(x)$ one obtains $y' = C'(x)\,\psi(x) + C(x)\,\psi'(x)$ and by substituting in the inhomogeneous equation

$$C'\psi + q + C[\psi' + p\psi] = 0.$$

The expression inside the square bracket is zero, because ψ satisfies the homogeneous equation. Thus, one obtains the differential equation $C'(x)\,\psi(x) + q(x) = 0$ for the determination of $C(x)$. From this one obtains $C(x) = C_1 - \int q(x)\,e^{\int p(x)\,dx}\,dx$.

General solution of the linear differential equation $y' + p(x)\,y + q(x) = 0$	$y = e^{-\int p(x)\,dx}\left[C_1 - \int q(x)\,e^{\int p(x)\,dx}\,dx\right]$

One should not remember the final formula, but the method: setting up the homogeneous equation, separation of the variables, variation of the parameter.

Example: $xy' - y = x^2\cos x$, that is, $y' - y/x - x\cos x = 0$; $x \neq 0$. The homogeneous equation $y' - (1/x)\,y = 0$ has the solution $y = Cx$. Variation of the parameter gives $y' = C'(x)\,x + C(x)$ and substitution in the given equation gives

$$C'(x)\,x + C(x) - C(x) - x\cos x = 0,$$
$$C'(x) - \cos x = 0,$$
$$C(x) = \sin x + C_1.$$

Hence the general solution is $y = x\sin x + C_1x$, as one can verify by substitution.

The Bernoulli differential equation $y' + p(x)\,y + q(x)\,y^n = 0$. This equation is called after the brothers Jakob and Johann BERNOULLI, who occupied themselves with it in 1695 and 1697 in competition with LEIBNIZ. For $n = 0$ this differential equation is linear; for $n = 1$ the variables are separable. One can therefore assume that $n \neq 0$, $n \neq 1$, and also that $y \neq 0$, for example, $y > 0$. Finally, the functions $p(x)$ and $q(x)$ are assumed to be continuous in an interval $a < x < b$. The following differential equations are of this kind:

$y' - (x^2 + 1)\,y - y^2 = 0$ with $n = 2$, $p(x) = -x^2 - 1$, $q(x) = -1$

or $xy' - y^3\ln x + y = 0$ with $n = 3$, $p(x) = 1/x$, $q(x) = -(\ln x)/x$.

To solve the equation one introduces a new function $z = z(x)$ by means of the substitution $y = z^{1/(1-n)}$. One obtains

$$y' = 1/(1-n)\cdot z^{n/(1-n)}z'(x),$$

and by substitution in the given equation, a linear differential equation for $z(x)$

$$z' + (1-n)\,p(x)\,z + (1-n)\,q(x) = 0.$$

From the general integral $z(x)$ of this equation the solution $y = y(x)$ of the given equation can be obtained.

Example: In the equation $y' - 4y/x - x\sqrt{y} = 0$ one has $n = 1/2$, $x \neq 0$, $y > 0$. Substituting $y = z^{1/(1-1/2)} = z^2$, from which $y' = 2zz'$, one obtains the linear differential equation

$z' - (2z)/x - x/2 = 0$ with the general solution $z = x^2[{}^1/_2 \ln x + C]$. The integral of the given equation is then $y = z^2 = x^4[{}^1/_2 \ln x + C]^2$.

Integration of an arbitrary differential equation of the first order

Every differential equation of the first order can be integrated if the functions contained in it satisfy certain conditions, for example, concerning continuity, which are stated more precisely in the existence theorems. It will be assumed that all the operations to be performed in the following, such as solution of given implicit functions, differentiation, integration, formation of the inverse function, etc. are possible.

Exact differential equation. The explicit differential equation of the first order $y' = f(x, y)$ can be expressed in the form $y' = -h(x, y)/g(x, y)$ or, in order to avoid fractions, in the form

$$y'g(x, y) + h(x, y) = 0.$$

If the left-hand side is the perfect derivative of a function $F(x, y)$, that is, if $y'g(x, y) + h(x, y) = \frac{d}{dx} F(x, y)$, the differential equation is said to be *exact*. The equation can then be integrated easily. From $\frac{d}{dx} F(x, y) = 0$ it follows that $F(x, y) = C$, and the general integral $y = y(x, C)$ is obtained by solving for y. If $h(x, y) + g(x, y)\, y' = \frac{d}{dx} F(x, y) = \frac{\partial F(x, y)}{\partial x} + \frac{\partial F(x, y)}{\partial y} y'$ is a perfect derivative, then one has necessarily

$$\frac{\partial F}{\partial x} = h(x, y) \quad \text{and} \quad \frac{\partial F}{\partial y} = g(x, y).$$

Condition of integrability	$\frac{\partial h}{\partial y} = \frac{\partial g}{\partial x}$

From $\frac{\partial^2 F}{\partial x\, \partial y} = \frac{\partial^2 F}{\partial y\, \partial x}$ one obtains the *condition of integrability*, a *necessary and sufficient* condition for the differential equation $y'g(x, y) + h(x, y) = 0$ to be exact.

Example: The equation $y'(6xy + x^2 + 3) + 3y^2 + 2xy + 2x = 0$ is given. Here $g(x, y) = 6xy + x^2 + 3$, $\frac{\partial g}{\partial x} = 6y + 2x$; $h(x, y) = 3y^2 + 2xy + 2x$, $\frac{\partial h}{\partial y} = 6y + 2x$. Because $\frac{\partial g}{\partial x} = \frac{\partial h}{\partial y}$ the differential equation is exact.

Method of solution. If a differential equation $y'g(x, y) + h(x, y) = 0$ is given, one first tests with the help of the *condition of integrability* whether it is exact. If this is the case, then there exists a corresponding function $F(x, y)$. By solving $F(x, y) = C$ for y one obtains the general integral of the differential equation. In the following it will be shown how to find the function $F(x, y)$.

General case	Example:
$y'g(x, y) + h(x, y) = 0$	$y'(6xy + x^2 + 3) + 3y^2 + 2xy + 2x = 0$
$\frac{\partial F}{\partial x} = h(x, y)$	$\frac{\partial F}{\partial x} = 3y^2 + 2xy + 2x$
$F = \int h(x, y)\, dx + \varphi(y).$	$F = 3y^2x + x^2y + x^2 + \varphi(y).$

The result of the integration with respect to x is determined only to within a function φ that is unknown for the time being and depends on y alone.

General case	Example
$\frac{\partial F}{\partial y} = \frac{\partial}{\partial y} \int h(x, y)\, dx + \varphi'(y)$	$\frac{\partial F}{\partial y} = = 6xy + x^2 + \varphi'(y);$
$\frac{\partial F}{\partial y} = g(x, y)$	but $\frac{\partial F}{\partial y} = 6xy + x^2 + 3$ also,
$\varphi'(y) = g(x, y) - \frac{\partial}{\partial y} \int h(x, y)\, dy$	that is, $6xy + x^2 + 3 = 6xy + x^2 + \varphi'(y).$

This is a differential equation for $\varphi(y)$.

General case	Example
Because the condition of integrability is satisfied, one can prove that the right-hand side does not depend on x; thus,	Clearly $\varphi'(y) = 3$ does not depend on x.
$\varphi(y) = \int \left[g - \frac{\partial}{\partial y} \int h\, dx\right] dy,$	$\varphi(y) = 3y + \text{constant},$
$F(x, y) = \int h(x, y)\, dx + \varphi(y).$	$F(x, y) = 3y^2x + x^2y + x^2 + 3y.$

Hence the general integral is

General case	Example
$\int h(x, y)\, dx + \varphi(y) = C$	$3y^2x + x^2y + x^2 + 3y = C.$

Integrating factor method. If a differential equation of the form $y'g(x, y) + h(x, y) = 0$ is not exact, one follows a method proposed by EULER and multiplies it by a function $\mu(x, y)$, which is chosen so that the equation becomes exact, in other words, that the lefthand side of

$$y'g(x, y)\,\mu(x, y) + h(x, y)\,\mu(x, y) = 0$$

is a perfect derivative. Such a function $\mu(x, y)$ is called an *Euler multiplier* or an *integrating factor*.

Example: The differential equation $y'(xy - x^2) + y^2 - 3xy - 2x^2 = 0$ is not exact, because $\frac{\partial g}{\partial x} = y - 2x$ and $\frac{\partial h}{\partial y} = 2y - 3x$. However, the simple function $\mu(x, y) = 2x$ is an integrating factor, because after multiplication by $2x$ one obtains

$$y'(xy - x^2)\,2x + (y^2 - 3xy - 2x^2)\,2x = 0,$$

and now

$$\frac{\partial}{\partial x}(xy - x^2)\,2x = 4xy - 6x^2 \quad \text{and} \quad \frac{\partial}{\partial y}(y^2 - 3xy - 2x^2)\,2x = 4xy - 6x^2.$$

By the above method, integration of this exact differential equation gives the general integral $y^2x^2 - 2x^3y - x^4 = C$.

The condition for $y'g\mu + h\mu$ to be a perfect derivative is clearly $\frac{\partial(g\mu)}{\partial x} = \frac{\partial(\mu h)}{\partial y}$, or $h\frac{\partial\mu}{\partial y} - g\frac{\partial\mu}{\partial x} = \mu\left(\frac{\partial g}{\partial x} - \frac{\partial h}{\partial y}\right)$. This is a partial differential equation for the determination of $\mu(x, y)$. It appears that the problem of integration has only been made harder. However, since one needs only a single particular integral of this partial differential equation, a real advantage has nevertheless been achieved. It can even be shown that it always has an integral, that is, that always at least one integrating factor exists for the equation $y'g + h = 0$.

Linear differential equations of higher order

Linear differential equations of higher order occur frequently in applications. In the general linear differential equation of the nth order $b_0(x)\,y + b_1(x)\,y' + b_2(x)\,y'' + \cdots + b_n(x)\,y^{(n)} = g(x)$, the *coefficients* $b_i(x)$ and the *perturbation function* $g(x)$ are taken to be real continuous and bounded functions of x, and $b_n(x)$ is assumed not to vanish in the interval considered. Dividing by $b_n(x)$ one then obtains the form $a_0(x)\,y + a_1(x)\,y' + a_2(x)\,y'' + \cdots + y^{(n)} = f(x)$, in which the $a_i(x)$ and $f(x)$ are likewise continuous and bounded. If $f(x)$ is identically equal to zero, the equation is said to be *homogeneous*; otherwise it is *inhomogeneous*. One begins by solving the homogeneous differential equation. This is achieved most easily when the coefficients $a_i(x)$ are constant numbers. The linear differential equation of the second order will serve as a pattern for linear differential equations of arbitrary order.

Linear homogeneous differential equation of the second order $a_0(x)\,y + a_1(x)\,y' + y'' = 0$. One disregards the *trivial solution* $y \equiv 0$.

Because this differential equation is linear and homogeneous in y and its derivatives, if $y_1(x)$ and $y_2(x)$ are any two particular integrals, then $C_1y_1(x)$ and $C_2y_2(x)$ and every linear combination $C_1y_1(x) + C_2y_2(x)$ are also solutions, where C_1 and C_2 denote arbitrary constants.

For a linear combination $C_1y_1 + C_2y_2$ of two particular solutions of the differential equation to represent the general integral, y_1 and y_2 must be *linearly independent*. If they were linearly dependent, then two constants α_1 and α_2 could be found, not both zero, for which $\alpha_1y_1 + \alpha_2y_2 = 0$. If $\alpha_1 \neq 0$, then $y_1 = -(\alpha_2/\alpha_1)\,y_2 = \bar{\alpha}_1y_2$ and if $\alpha_2 \neq 0$, then $y_2 = -(\alpha_1/\alpha_2)\,y_1 = \bar{\alpha}_2y_1$. Thus, the two functions y_1 and y_2 would represent only the same particular integral, since one is just a multiple of the other.

Example: $y_1 = \cos^2 x - \cos 2x$ and $y_2 = {}^1/_2 \sin^2 x$ are *linearly dependent*, because $y_1 - 2y_2 = 0$ for all x. But $y_1 = x$ and $y_2 = x^2$, or $y_1 = \sin x$ and $y_2 = \cos x$ are *linearly independent*.

If two particular solutions $y_1(x)$ and $y_2(x)$ are linearly independent, they form a *fundamental system* for the differential equation. In this case the quotient y_1/y_2 is not constant, and consequently its derivative $\frac{\mathrm{d}}{\mathrm{d}x}\left(\frac{y_1}{y_2}\right) = \frac{y_1'y_2 - y_2'y_1}{y_2^2}$ is not identically zero. The determinant

$$\Delta = \begin{vmatrix} y_1' & y_1 \\ y_2' & y_2 \end{vmatrix} = y_1'y_2 - y_2'y_1$$

is called the *Wronskian determinant*. The following theorem holds:

Two particular solutions y_1 and y_2 form a fundamental system and their linear combination $y = C_1y_1 + C_2y_2$ represents the general integral of the differential equation $a_0(x)\, y + a_1(x)\, y' + y'' = 0$ if and only if the Wronskian determinant formed from them is different from zero.

Example: The linear differential equation $xy'' + 2y' + axy = 0$, in which a denotes an arbitrary number, is transformed by the substitution $u = xy$ into a differential equation with constant coefficients. By differentiation, from $u = xy$ one obtains in succession $u' = y + xy'$, or $y' = u'/x - u/x^2$, and hence $y'' = (u''x - u')/x^2 - (u'x - 2u)/x^3$. Substituting the expressions for y, y' and y'' into the given equation, one obtains $u'' + au = 0$. As will be shown later, for $a = -1$ one obtains $u_1 = e^x$ and $u_2 = e^{-x}$ as solutions, and it follows that $y_1 = e^x/x$ and $y_2 = e^{-x}/x$ represent a fundamental system for the differential equation.

For arbitrary coefficients $a_0(x)$ and $a_1(x)$ there is no general procedure of finding a fundamental system for the differential equation

$$a_0(x)\, y + a_1(x)\, y' + y'' = 0.$$

But there are reference works in which one can look up solutions or suitable methods of solution. However, if the coefficients are constant numbers, there is a method that is always successful in setting up a fundamental system.

Linear homogeneous differential equations of the second order with constant coefficients. The differential equation has the form $y'' + c_1y' + c_2y = 0$. By the substitution $y(x) = e^{rx}$, so that $y' = r\,e^{rx}, y'' = r^2\,e^{rx}$, the equation becomes $(r^2 + c_1r + c_2)\,e^{rx} = 0$. Since the exponential function vanishes nowhere, the value of r can be determined from the quadratic equation.

Characteristic equation	$r^2 + c_1r + c_2 = 0$

If r_1 and r_2 are its roots, then $y_1 = e^{r_1x}$ and $y_2 = e^{r_2x}$ are *particular solutions* of the differential equation. The *general integral* is obtained as in one of the following cases.

1. The roots r_1 and r_2 are *real* and *distinct*. Then $y_1/y_2 = e^{(r_1-r_2)x}$ is not constant, y_1 and y_2 are linearly independent, and $y = C_1\,e^{r_1x} + C_2\,e^{r_2x}$, where C_1 and C_2 are arbitrary constants, represents the general integral.

2. The characteristic equation has the repeated root $r_1 = r_2 = -c_1/2$; y_1 and y_2 are then linearly dependent. By substitution one finds that $y_2 = x\,e^{r_1x}$ also satisfies the differential equation $y'' + c_1y' + c_2y = 0$. Because the quotient $y_2/y_1 = x$ is not constant, the particular solutions y_1 and y_2 form a fundamental system, and $y = C_1\,e^{r_1x} + C_2\,xe^{r_1x}$, where C_1 and C_2 are arbitrary constants, is the general integral.

3. The roots r_1 and r_2 are *complex*. Because c_1 and c_2 are real by hypothesis, r_1 and r_2 are *conjugate complex* numbers: $r_1 = \alpha + i\beta$, $r_2 = \alpha - i\beta$. The two particular solutions $y_1 = e^{(\alpha+i\beta)x} = e^{\alpha x}(\cos\beta x + i\sin\beta x)$ and $y_2 = e^{(\alpha-i\beta)x} = e^{\alpha x}(\cos\beta x - i\sin\beta x)$ form a fundamental system, and the general integral is $y^* = y_1C_1 + y_2C_2$ or, if one substitutes $C_1^* = C_1 + C_2$ and $C_2^* = i(C_1 - C_2)$, $y^* = e^{\alpha x}(C_1^*\cos\beta x + C_2^*\sin\beta x)$. Differential equations of this kind occur in oscillation problems.

Example: The differential equation $y'' - 3y' + 2y = 0$ has the characteristic equation $r^2 - 3r + 2 = 0$ with the roots $r_1 = 1$ and $r_2 = 2$. It follows that its general integral is $y = C_1\,e^x + C_2\,e^{2x}$.

Example: Mathematical pendulum. A pendulum, whose total mass m is assumed to be concentrated at the point A (Fig.), is suspended from a point O by a thread of length $l = |OA|$. It executes oscillations under the influence of gravity, in which friction and further influences are neglected. If the angle between the pendulum and the vertical at the time t is φ, then the force mg acts vertically downwards on the mass m, so that the force $mg\sin\varphi$ acts in the direction of the tangent; g denotes the acceleration due to gravity. By Newton's second law, this force is equal to the product of the mass m and the acceleration $l\cdot\dfrac{d^2\varphi}{dt^2}$; thus, one obtains for the angle of inclination $\varphi(t)$ the differential equation

$$ml\cdot\frac{d^2\varphi}{dt^2} = -mg\sin\varphi, \quad \text{or} \quad \frac{d^2\varphi}{dt^2} + \frac{g}{l}\sin\varphi = 0.$$

This is not linear, and by separation of the variables leads to an *elliptic integral*, which can be evaluated by series expansion or from tables. By

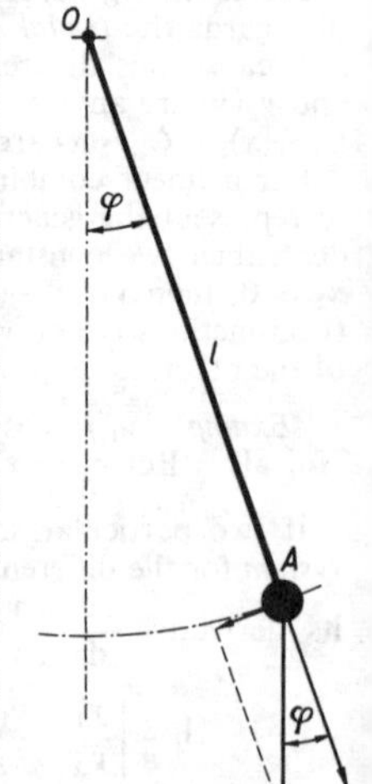

22.2-4 The mathematical pendulum

linearization, which is frequently applied, above all in physics, one obtains a linear differential equation that is far easier to solve. One limits oneself to small deviations φ from the vertical, so that one can take $\sin\varphi \approx \varphi$ and obtains $\frac{d^2\varphi}{dt^2} + \frac{g}{l} \cdot \varphi = 0$, with the solution

$$\varphi = \alpha \cos(\omega t + \delta),$$

where $\omega = \sqrt{(g/l)}$ is the angular frequency, hence $\tau = 2\pi/\omega$ is the *periodic time*, and α and δ denote the constants of integration. This formula for the solution expresses the fact, which was already noticed by GALILEI, that the periodic time is independent of the magnitude of the oscillation. Of course, this holds only approximately; for oscillations of greater magnitude the periodic time is given by

$$\tau = 2\pi \sqrt{(l/g)}\,[1 + (1/2)^2 \sin^2(\varphi_0/2) + [(1 \cdot 3)/(2 \cdot 4)]^2 \sin^4(\varphi_0/2) + \cdots],$$

where φ_0 denotes the greatest inclination, the *amplitude of the oscillation*. Compared with this precise formula, which is obtained by solving the non-linearized differential equation, the error is only 0.002% for $\varphi_0 = 1°$ and only 0.05% for $\varphi_0 = 5°$.

Instead of the circle, a curve chosen in such a way that a body oscillating on it always has a periodic time independent of the magnitude of the oscillations is called a *tautochrone*. In 1673 HUYGENS found that the *cycloid* has this property, and in the light of this he was able to construct a pendulum clock. The thread of the *cycloidal pendulum* constructed by him wrapped itself in oscillating around two cycloidal shaped surfaces. The pendulum bob then described a cycloid, because the evolute of a cycloid is also a cycloid. The pendulum oscillates tautochronously.

Linear inhomogeneous differential equation of the second order $a_0(x)\,y + a_1(x)\,y' + y'' = f(x)$.

The general solution of the inhomogeneous linear differential equation of the second order is equal to the sum of the general solution of the corresponding homogeneous differential equation and any particular solution of the inhomogeneous equation.

Thus, if $C_1y_1(x) + C_2y_2(x)$ is the general solution of the homogeneous differential equation and $p(x)$ a particular integral of the inhomogeneous equation, then $y = C_1y_1(x) + C_2y_2(x) + p(x)$ represents the general integral of the inhomogeneous differential equation. A *particular integral* $p(x)$ of the inhomogeneous differential equation can be obtained from the general integral of the homogeneous equation by the method of *variation of parameters* due to LAGRANGE. In the expression

$$p(x) = C_1(x)\,y_1(x) + C_2(x)\,y_2(x)$$

the coefficients C_1 and C_2 are regarded as functions of x. Because two functions $C_1(x)$ and $C_2(x)$ then have to be determined, they can be made to satisfy an *additional condition*; one chooses them so that $C_1'y_1 + C_2'y_2 = 0$. Substituting for $p(x)$, $p'(x)$ and $p''(x)$ in the inhomogeneous differential equation and taking account of the assumption that y_1 and y_2 are solutions of the homogeneous equation, one obtains the equation

$$C_1'y_1' + C_2'y_2' = f(x).$$

$$C_1'y_1 + C_2'y_2 = 0,$$
$$C_1'y_1' + C_2'y_2' = f(x)$$

This, together with the additional condition, gives a system of equations for the determination of C_1' and C_2'; this system always has a solution, since the determinant of its coefficients, the *Wronskian determinant*, never vanishes because y_1 and y_2 are linearly independent. $C_1(x)$ and $C_2(x)$ are obtained by integration, and hence the general integral of the inhomogeneous differential equation. However, in practice it is usually quite awkward to carry out the procedure of variation of the parameters, above all because it leads, in general, to integrals that cannot be evaluated in closed form.

Example: For the differential equation $xy'' + 2y' - xy = e^x$ the functions $y_1 = e^x/x$ and $y_2 = e^{-x}/x$ form a *fundamental system* for the homogeneous equation. Variation of the parameters requires a certain amount of calculation and yields $p(x) = {}^1/_2 e^x$ as a *particular integral* of the inhomogeneous equation. Hence the *general solution* is

$$y(x) = C_1\,e^x/x + C_2\,e^{-x}/x + {}^1/_2 e^x.$$

Particular solutions of the inhomogeneous linear differential equation of the second order with constant coefficients for special perturbation functions. If the coefficients c_1 and c_2 are constants, a particular integral of the differential equation $y'' + c_1y' + c_2y = f(x)$ can be found for certain types of perturbation function $f(x)$, without using the method of variation of the parameters.

Type 1: If the *perturbation function is a polynomial*, $f(x) = a_0 + a_1x + a_2x^2 + \cdots + a_nx^n$, $a_n \neq 0$, then one sets $p(x) = b_0 + b_1x + \cdots + b_{n-1}x^{n-1} + b_nx^n$ if $c_2 \neq 0$; if $c_2 = 0$, however, one also introduces into the expression the term $b_{n+1}x^{n+1}$. The required coefficients $b_0, b_1, \ldots$ are obtained by equating coefficients.

Example: $y'' + y = x^2$, in which $a_0 = 0$, $a_1 = 0$, $a_2 = 1$, $c_2 = 1$. One sets $p(x) = b_0 + b_1x + b_2x^2$, so that $p' = b_1 + 2b_2x$, $p'' = 2b_2$. Substitution in the equation gives $(b_0 + 2b_2) + b_1x + b_2x^2 = x^2$, and equating coefficients, $b_0 = -2$, $b_1 = 0$, $b_2 = 1$. Hence $p(x) = -2 + x^2$ is a particular integral and $y = C_1 \cos x + C_2 \sin x + x^2 - 2$ is the general solution of the differential equation.

Type 2: If the *perturbation function is an exponential function*, $f(x) = a\,e^{kx}$, one makes the substitution $p(x) = b\,e^{kx}$, which contains only those exponential functions that occur in the perturbation function. The value of b is to be determined.

Example: $y'' + y = 2e^{3x}$, in which $a = 2$, $k = 3$. Set $p(x) = b\,e^{3x}$. By substitution one obtains the equation $9b + b = 2$ for the determination of b, and hence the particular integral $p(x) = {}^1/_5 e^{3x}$ and the general integral $y = {}^1/_5 e^{3x} + C_1 \cos x + C_2 \sin x$.

Type 3: If the *perturbation function is a trigonometric function* $f(x) = a \cos mx + b \sin mx$, one sets $p(x) = a^* \cos mx + b^* \sin mx$, into which both the cosine and the sine function enter, even when only one of these functions occurs in the perturbation function, that is, if a or b is zero. Here a^* and b^* are determined by equating coefficients.

Example: $y'' + y = 2 \sin 3x$, in which $a = 0$, $b = 2$, $m = 3$. Set $p(x) = a^* \cos 3x + b^* \sin 3x$. Substituting and equating coefficients one obtains $a^* = 0$, $b^* = -{}^1/_4$. It follows that the general integral is $y = -{}^1/_4 \sin 3x + C_1 \cos x + C_2 \sin x$.

Type 4: If the *perturbation function is a linear combination* of functions that occur in types 1, 2 and 3, then the particular solution is a linear combination of the corresponding individual *particular solutions*. In other words, to the general solution of the homogeneous equation must be added in succession the particular solutions obtained in turn by ignoring all but one of the terms of the perturbation function.

Example: $y'' + y = x^2 + 2e^{3x} + 2 \sin 3x$. It follows from the previous examples that the general integral is $y = C_1 \cos x + C_2 \sin x + x^2 - 2 + {}^1/_5 e^{3x} - {}^1/_4 \sin 3x$.

All these procedures for determining particular solutions of the inhomogeneous differential equation fail in the case of *resonance*, that is, if the perturbation function or one of its terms is at the same time an integral of the homogeneous differential equation. Resonance occurs, for example, in the differential equations $y'' + y = \cos x$ and $y'' - y' = e^x$.

22.3. Further considerations

Integration procedures in practice

As already emphasized, it is exceptional when a differential equation is elementarily integrable. But there are also methods for finding integrals in the most difficult cases, at least approximately, for example, by approximating to the differential equation by means of a difference equation and applying the methods of the calculus of difference equations. In the sequel certain other important procedures will be described.

Integration by means of power series. One expresses the required solution $y = y(x)$ of the differential equation $y' = f(x, y)$ in the form of a power series $y = a_0 + a_1x + a_2x^2 + \cdots$ with coefficients a_i that are initially undetermined, and substitutes for y and its derivatives in the differential equation. Under certain conditions the coefficients a_i of the series can then be calculated by *equating coefficients*.

Example: $y' = x^2 + y$. Set $y = a_0 + a_1x + a_2x^2 + a_3x^3 + a_4x^4 + \cdots$, $y' = a_1 + 2a_2x + 3a_3x^2 + 4a_4x^3 + \cdots$; it follows that $a_1 + 2a_2x + 3a_3x^2 + 4a_4x^3 + \cdots = a_0 + a_1x + (a_2 + 1)x^2 + a_3x^3 + a_4x^4 + \cdots$. Equating coefficients yields $a_1 = a_0$, $a_2 = a_0/2$, $a_3 = (a_0 + 2)/6$, $a_4 = (a_0 + 2)/24, \ldots$ Hence one obtains as the integral $y = a_0 + a_0x + a_0x^2/2 + (a_0 + 2)\,x^3/6 + (a_0 + 2)\,x^4/24 + \cdots$ and from this by rearrangement $y = (a_0 + 2)/1 + (a_0 + 2)\,x/1! + (a_0 + 2)\,x^2/2! + (a_0 + 2)\,x^3/3! + \cdots - 2 - 2x - x^2$, or $y = (a_0 + 2)\,e^x - 2 - 2x - x^2$.

In this particular example the final rearrangement has led to a closed form for the integral. By differentiation one can show that this function does indeed satisfy the differental equation. If no closed form can be found, then the series is terminated at a suitable place according to the accuracy required. The convergence of the series obtained is a consequence of the following theorem, stated without proof. The theorem is applicable because the right-hand side of the differential equation is a polynomial in x and y.

The solution of the differential equation $y' = f(x, y)$ can always be represented by a convergent power series if the right-hand side itself can be expanded as a power series $f(x, y) = c_{00} + c_{10}x + c_{01}y + c_{20}x^2 + c_{11}xy + c_{02}y^2 + \cdots = \sum c_{\lambda\mu}x^\lambda y^\mu$ that is absolutely convergent in a certain domain of the x, y-plane.

The method of solution by power series can also be applied to differential equations of higher order, as shown by the following two important differential equations.

Gaussian differential equation, hypergeometric series. In 1812 GAUSS studied thoroughly a particular differential equation, which he called *hypergeometric*. It contains *several parameters* to which arbitrary values can be assigned, and it can therefore readily be made to fit special conditions in applications. It is

$$x(x-1)\,y'' + [(\alpha+\beta+1)\,x - \gamma]\,y' + \alpha\beta y = 0,$$

where α, β, γ are the parameters. For $\gamma \neq 0, -1, -2, \ldots$ it has a power series solution that can be obtained by taking $y = \sum_{k=1}^{\infty} a_k x^k$. Let it be stated here without proof that one obtains the *recursion formula* $(k+1)(k+\gamma)\,a_{k+1} = (k+\alpha)(k+\beta)\,a_k$ and hence finally the series

$$y = a_0\left\{1 + \frac{\alpha}{1!}\cdot\frac{\beta}{\gamma}x + \frac{\alpha(\alpha+1)}{2!}\cdot\frac{\beta(\beta+1)}{\gamma(\gamma+1)}x^2 + \cdots\right\} = a_0F(\alpha, \beta, \gamma, x)$$

with *radius of convergence* 1. The series $F(\alpha, \beta, \gamma, x)$ in the brackets, which depends on the 3 parameters and the variable x, is called a *hypergeometric series*. It includes many functions well known in analysis, which arise by special choices of the parameters; for example,

$F(1, \beta, \beta, x) = 1/(1-x)$, the geometric series;
$F(-n, \beta, \beta, -x) = (1+x)^n$; $\quad xF(1, 1, 2, -x) = \ln(1+x)$;
$\lim_{\beta\to\infty} F(1, \beta, 1, x) = e^x$; $\quad \lim_{\alpha,\beta\to\infty} xF[\alpha, \beta, 3/2, -x^2/(4\alpha\beta)] = \sin x$.

In the general case, however, $F(\alpha, \beta, \gamma, x)$ cannot be expressed in terms of finitely many elementary functions.

Bessel's differential equation, cylinder functions. Following the studies of D. BERNOULLI and EULER, BESSEL investigated a special differential equation of the second order, which occurs in many problems of physics and technology, particularly those concerned with oscillations. It is the differential equation

$$xy'' + (1+n)\,y' - y = 0, \quad n \text{ constant}.$$

The power series expansion $y = \sum_{k=0}^{\infty} a_k x^k$ yields here $a_{k+1} = \dfrac{a_k}{(k+1)(k+1+n)}$.

and hence $y = a_0\left(1 + \dfrac{x}{1!\,(1+n)} + \dfrac{x^2}{2!\,(1+n)(2+n)} + \cdots\right) = a_0 j_n(x)$.

The series $j_n(x)$ depending on n (appearing in the brackets) are always convergent for $n \neq -1$, $n \neq -2, \ldots$ They are called *Bessel functions (cylinder functions) of the first kind.*

Graphical methods of integration. For the graphical solution of differential equations numerous integration procedures have been developed that are suitable for special types of equation and meet the required degree of accuracy. There is space here for only basic considerations and even then only for differential equations of the first order.

Method of polygonal arcs. If one wants to draw an integral curve of the differential equation $y' = f(x, y)$ that passes through a fixed point $P_0(x_0, y_0)$, and consequently satisfies an initial condition, then $y_0' = f(x_0, y_0)$ gives the *direction of the tangent* to the required curve at this point. On the tangent a point $P_1(x_1, y_1)$ is taken at a certain distance from P_0 and the procedure is repeated; the smaller the distance chosen, the more accurate the solution curve will be. So one obtains

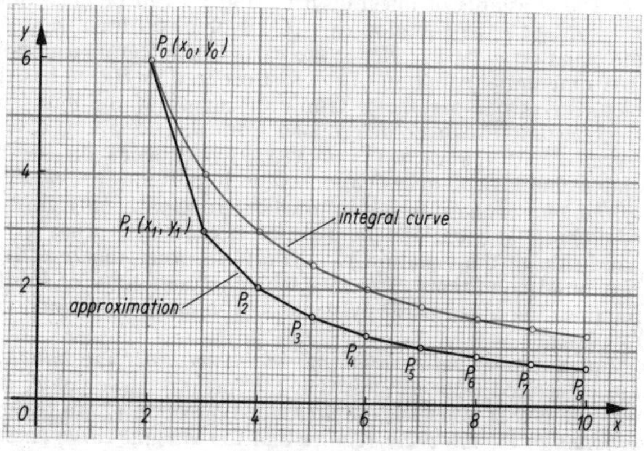

22.3-1 Graphical integration of the differential equation $y' = -y/x$ by the method of polygonal arcs

an approximation for the integral curve. The figure shows the procedure for the differential equation $y' = -y/x$ with the initial conditions $x_0 = 2$, $y_0 = 6$. It also shows points marked by red circles that lie on the true integral curve. The deviation is considerable. Essentially more accurate is the *method of interpolated half steps* (Fig.). In this method the direction $y'_0 = f(x_0, y_0)$ calculated for the point $P_0(x_0, y_0)$ is used only to calculate the direction $\eta'_1 = f(\xi_1, \eta_1)$ for an *intermediate point* $\Pi_1(\xi_1, \eta_1)$ (*half step*). The first *whole step* from P_0 to the first point $P_1(x_1, y_1)$ of the polygon is taken in this direction η'_1. The next half step leads from P_1 with $y'_1 = f(x_1, y_1)$ to $\Pi_2(\xi_2, \eta_2)$ and the next whole step from P_1 to $P_2(x_2, y_2)$ with $\eta'_2 = f(\xi_2, \eta_2)$. As can be seen by comparing the two polygonal arcs, the approximation to the true integral curve marked by red circles is significantly better in the second method. It could be further improved by reducing the size of the steps, above all in those domains in which the curve changes direction rapidly.

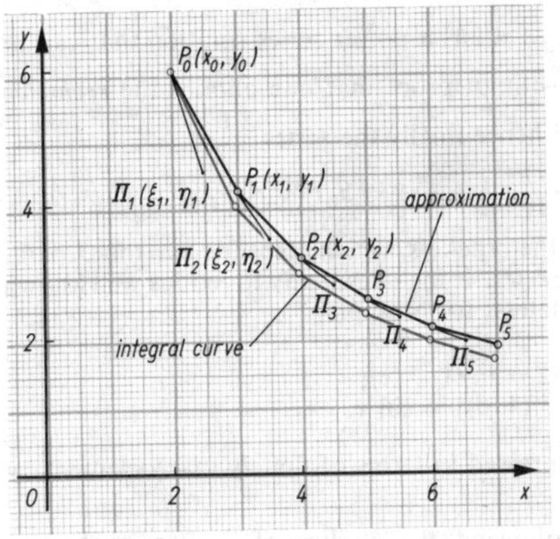

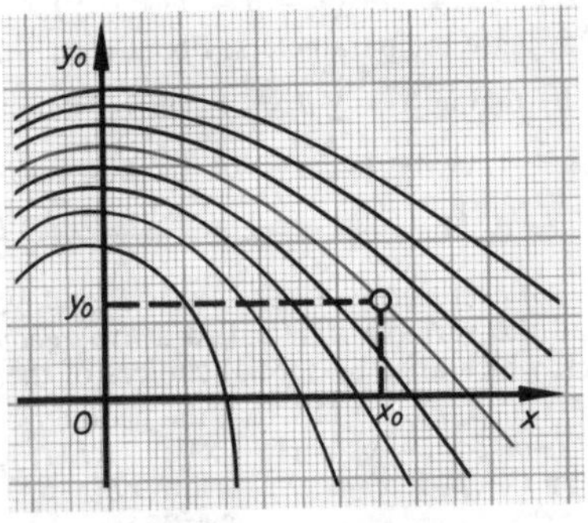

22.3-3 The special integral curve selected to take account of the initial condition $y(x_0) = y_0$

22.3-2 Graphical solution of the differential equation $y' = -\dfrac{y}{x}$ by the method of interpolated half-steps with the initial condition $x_0 = 2$, $y_0 = 6$. The integral curve is shown in red

Glances at the theory

Initial value problem, boundary value problem. The picture of the general integral of a differential equation of the second order is a two-parameter family of curves. In the case of an application, the *integral curve of a special solution* (that is, of a particular integral) must be extracted from these taking care of the special circumstances. For example, if g denotes the acceleration due to gravity, then $\dfrac{d^2y}{dt^2} = g$ is the differential equation of free fall. By integration one obtains $\dfrac{dy}{dt} = gt + C_1$, $y = {}^1/_2 gt^2 + C_1t + C_2$. But one wants to know where the falling body will be after a time t if it was *at a height* y_0 at the beginning of the fall, that is, *at time* $t_0 = 0$, and if its initial speed was v_0. The constants of integration C_1 and C_2 can be determined from these *initial conditions.* From $y' = gt + C_1$ the equation $v_0 = C_1$ is obtained for $t = t_0 = 0$, and from $y = {}^1/_2 gt^2 + C_1t + C_2$ the equation $y_0 = C_2$. Hence the particular integral required is $y = {}^1/_2 gt^2 + v_0t + y_0$.

A differential equation of the first order $y' = f(x, y)$ has the general integral $y = \varphi(x, C)$, whose picture is a one-parameter family of curves. If the initial condition $y_0 = \varphi(x_0, C)$ is prescribed, the constant C can be determined from this equation, $C = \psi(x_0, y_0)$. The required particular integral $y = \varphi[x; \psi(x_0, y_0)] = \Phi(x; x_0, y_0)$ depends on the initial conditions (Fig.). For physical and technological investigations it is important that $\Phi(x; x_0, y_0)$ shall be a *continuous function of the initial values,* which in practice are known only approximately. Consequently, in the theory of differential equations the question is examined: what requirements must be placed on the right-hand side of the equation $y' = f(x, y)$ in order that $y = \Phi(x; x_0, y_0)$ is a continuous function of the initial conditions. Together with the continuity of the function $f(x, y)$, a sufficient condition for this (and also for the existence and uniqueness theorem) is that a *Lipschitz condition* is satisfied (see later).

For a *differential equation of the second order* the *initial conditions* $y(x_0) = y_0$ and $y'(x_0) = y'_0$ are prescribed; this requires that the integral curve passes through the point (x_0, y_0) and that the tangent at that point has the prescribed direction y'_0. In general, the *initial conditions of a differential equation of the nth order* require that the integral curve passes through a point (x_0, y_0) and that at this point the first $(n - 1)$ derivatives assume prescribed values. For differential equations of higher order also the continuous dependence of the solutions on the initial conditions must be investigated.

For differential equations of order higher than the first, not only initial conditions, but also *boundary values* can be prescribed. Quite new types of problems then occur. Suppose, for example, that the solution $y(x)$ of the differential equation $y'' + \lambda y = 0$ in the interval $0 \leqslant x \leqslant \pi$ is required. It deals with an oscillation problem; one can think of a string stretched between the points 0 and π. At these points the string cannot be displaced from the position of rest; one must impose the *boundary conditions (of the first kind)* $y(0) = 0$, $y(\pi) = 0$. From the solution $y(x) = C_1 \sin(x\sqrt{\lambda} + C_2)$ one obtains the system of equations

$$y(0) = C_1 \sin C_2 = 0, \quad y(\pi) = C_1 \sin(\pi\sqrt{\lambda} + C_2) = 0.$$

If $\sqrt{\lambda}$ is not an integer, it follows that $C_1 = C_2 = 0$, and then the solution is $y \equiv 0$. The string would *not oscillate*, but would remain at rest. If, however, $\sqrt{\lambda}$ is an integer, it follows that λ is a perfect square, $\lambda = 1, 4, 9, \ldots$, the system of equations is satisfied, and one obtains infinitely many solutions $y(x) = C_1 \sin(x\sqrt{\lambda})$. The string moves between the boundaries in the sinusoidal oscillations $y = C_1 \sin x$, $y = C_1 \sin 2x$, $y = C_1 \sin 3x$, ...; these are the *fundamental oscillation* and the possible higher harmonics for the string, in which it can oscillate without external influences, once it is disturbed. These oscillations, which can be superimposed and can again be separated by harmonic analysis, are therefore called *characteristic (or eigen-) oscillations* and the numbers $\lambda_1 = 1$, $\lambda_2 = 4$, $\lambda_3 = 9, \ldots, \lambda_n = n^2, \ldots$ are called the *eigenvalues* of this problem.

In *boundary value problems of the second kind* the values at the boundary points x_1 and x_2 of the derivatives $y'(x_1) = a$, $y'(x_2) = b$ are prescribed. Theory and practice require the determination of *eigenfunctions* and *eigenvalues*. It is characteristic of these problems that non-trivial solutions $y \not\equiv 0$ exist only for certain discrete values of a parameter, namely, for the eigenvalues.

Existence theorem, uniqueness theorem. The geometrical interpretation of a differential equation of the first kind $y' = f(x, y)$ by means of the direction field suggests the validity of the *existence* theorem:

If $f(x, y)$ is a continuous function of both variables, then an integral curve passes through every point (x_0, y_0) of its domain of continuity D.

This theorem was, in fact, proved by PEANO. The question of the existence of an integral, in general, was first raised in the years 1820 to 1830 by CAUCHY and was also proved by him under the assumptions that $f(x, y)$ is *continuous* and has a *continuous partial derivative* $f_y(x, y)$. One recognizes that the existence theorem of Peano, in comparison with that of Cauchy, represents an essential sharpening, since the existence of the integral can be deduced from weaker assumptions.

It is for theory and practice an equally important question, under what assumptions about the right-hand side of the differential equation $y' = f(x, y)$ can the *uniqueness* of the solution also be deduced. Geometrically expressed, the uniqueness would be violated, for example, if the integral curve were to branch at a point (x_0, y_0). Physically the circumstance that, in spite of the same initial conditions, the process continued in different ways, would correspond to a violation of the principle of causality.

Both the existence and the uniqueness of the solution are guaranteed when the continuous function $f(x, y)$ satisfies a so-called *Lipschitz condition* in a domain D of the x, y-plane. A function $f(t)$ of a single variable t is said to satisfy a Lipschitz condition with a constant L in an interval $[a, b]$ when its difference quotients are bounded by L: $|f(t_1) - f(t_2)| \leqslant L|t_1 - t_2|$ for all t_1, t_2 in $[a, b]$. In the present situation $f(x, y)$ is a function of two variables x and y, but x is to be treated as a constant parameter.

Lipschitz condition	$|f(x, y_1) - f(x, y_2)| \leqslant L|y_1 - y_2|$ for all x, y_1 and y_2 in D

Geometrically this states that on every ordinate $f(x, y)$ has a bounded difference quotient with respect to y. Consequently the Lipschitz condition is certainly satisfied if the partial derivative $f_y(x, y)$ is bounded. With this concept the *existence and uniqueness theorem* has the form:

Suppose that the function $f(x, y)$ is continuous and satisfies a Lipschitz condition in a domain D. Then for every point (x_0, y_0) in D there is exactly one integral curve $y = \varphi(x)$ passing through (x_0, y_0).

The proof is based on the method of iteration, by which a sequence of successive approximations $\varphi_n(x)$ arises if one begins with the function $\varphi_0(x) = y_0$ and defines the approximations by

$$\varphi_{n+1}(x) = y_0 + \int_{x_0}^{x} f(t, \varphi_n(t))\,dt, \quad n = 0, 1, 2, \ldots$$

It can be shown, by making essential use of the Lipschitz condition, that the approximations $\varphi_n(x)$ converge uniformly to the solution $y = \varphi(x)$: $\lim_{n\to\infty} \varphi_n(x) = \varphi(x)$.

This *proof* is even *constructive*; it allows one in practice to obtain a solution, for example, that of the differential equation $y' = yx$ with the initial conditions $x_0 = 0$, $y_0 = 1$, although this is easier to solve by another method. One obtains in succession:

$$\varphi_0 = 1,$$

$$\varphi_1(x) = 1 + \int_0^x x\,dx = 1 + \frac{x^2}{2}, \quad \varphi_2(x) = 1 + \int_0^x \left(x + \frac{x^3}{2}\right) dx = 1 + \frac{x^2}{2} + \frac{x^4}{2\cdot 4}, \dots$$

$$\varphi_n(x) = 1 + \frac{x^2}{2} + \frac{x^4}{2\cdot 4} + \dots + \frac{x^{2n}}{2\cdot 4\dots(2n)} = 1 + (x^2/2) + (x^2/2)^2/2! + \dots + (x^2/2)^n/n!,$$

and finally,

$$\varphi(x) = \lim_{n\to\infty} \varphi_n(x) = \sum_{k=0}^{\infty} (x^2/2)^k/k! = e^{x^2/2}.$$

The question of conditions that ensure the existence and uniqueness of the integral of a differential equation is more difficult for the *implicit differential equation of the first order* $F(x, y, y') = 0$. It is still more difficult to answer generally for differential equations of higher order.

If the conditions of the existence and uniqueness theorems are not satisfied at a point (x_0, y_0), then there may be several or even infinitely many integral curves passing through (x_0, y_0), or none at all. Such a point is called a *singular point* of the differential equation (Fig.). It is of theoretical interest that the integral curves in its neighbourhood can behave in quite an unusual way. But a singular point is also important for the physicist or technologist. It characterizes mathematically a transition point of the physical event and the critical technical data: the fracture of a beam under bending, the breaking of a rope under tension, a change of state etc.

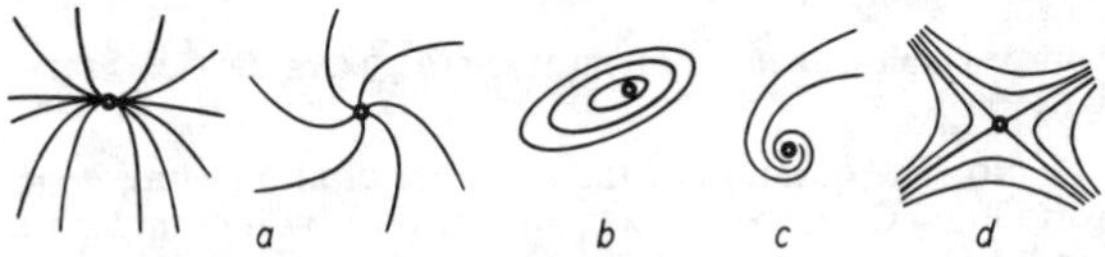

22.3-4 The behaviour of the integral curves in the neighbourhood of a singular point; a) nodes, b) whirl, c) vortex, d) saddle point

Differential equations of higher order and systems of differential equations. The differential equation of the nth order

$$F(x, y, y', \dots, y^{(n-1)}, y^{(n)}) = 0$$

can always be written as a system of n differential equations of the first order, by introducing new functions $y_1 = y', y_2 = y'', \dots, y_{n-1} = y^{(n-1)}$. One obtains the system $y_1 = y', y_2 = y_1', y_3 = y_2', \dots, F(x, y, y_1, \dots, y_{n-1}, y_{n-1}') = 0$.

Similarly a system of differential equations of higher order can also always be written as a system of differential equations of the first order. Consequently the proof of the existence and uniqueness of the integral of differential equations of higher order can also be reduced to that of the integral of a system of differential equations of the first order and made considerably easier in this way.

Applications in mechanics. These ideas are of great significance in mechanics. The fundamental concept of acceleration is expressed mathematically by the second derivative with respect to the time t of the space coordinates $x(t)$, $y(t)$, $z(t)$ of the point mass. It follows that in order to determine the motion of a point of mass m one must integrate the system of equations

$$m\frac{d^2x}{dt^2} = P(x, y, z), \quad m\frac{d^2y}{dt^2} = Q(x, y, z), \quad m\frac{d^2z}{dt^2} = R(x, y, z),$$

where P, Q and R, depending on the space coordinates (x, y, z) of the point, denote the components of the force acting on the point mass. If one reduces this system to a system of differential equations of the first order, one obtains a system of 6 differential equations

$$u = \dot{x}, \quad v = \dot{y}, \quad w = \dot{z}, \quad m\dot{u} = P(x, y, z), \quad m\dot{v} = Q(x, y, z), \quad m\dot{w} = R(x, y, z),$$

in which the new functions u, v, w denote the velocities. Systems of differential equations occur most frequently in which the number of functions required agrees with the number of differential equations. In general, a *system of n differential equations for n functions* of the variable t, in which the equations are solved for the derivatives, has the form

$$(dx_i)/(dt) = f_i(t, x_1, x_2, \dots, x_n); \quad i = 1, 2, \dots, n.$$

A *solution* or *integral* of this system is a system $x_1(t), x_2(t), \dots, x_n(t)$ which when substituted in the given system makes each individual equation an identity in t. The real difficulty of integration lies as one can see, in the fact that one cannot integrate one differential equation after another, because one required function x_i is also contained in the right-hand side of another differential equation

Physically speaking, the individual motions expressed by the $x_i(t)$ influence one another; the physicist speaks of *coupling*. One can think, for example, of two pendulums oscillating, not independently of one another, but coupled by means of a spring attached to each of the pendulum rods (Fig.).

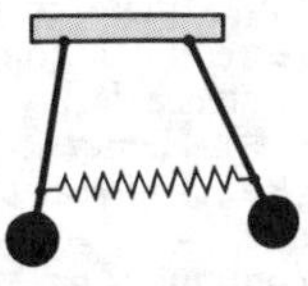

22.3-5 Coupling of two oscillating pendulums leads to a system of differential equations

To simplify the statement of the problem and to help the intuition one prefers to use the concepts of multidimensional geometry in the theory of differential equations. If the integral of a differential equation of the first order is represented geometrically by an integral curve in the x, y-plane, then the integrals of the system $(dx_i)/(dt) = f_i$, $i = 1, 2, ..., n$, can be regarded as integral curves in n-dimensional space, in which the $x_i(t)$, $i = 1, 2, ..., n$, describe the coordinates of the motion of a point.

The theory of first integrals. With the help of this nomenclature one can sketch the *theory of first integrals*, which is also of special importance for physics. An equation $F(x_1, ..., x_n) = C$, where C is constant, defines in the n-dimensional coordinate space of the x_i an $(n-1)$-dimensional hypersurface, and for variable C a one-parameter family of hypersurfaces. If $n-1$ families of hypersurfaces

$$F_i(x_1, ..., x_n) = C_i; \quad i = 1, 2, ..., n-1$$

are given, and from each family one hypersurface is selected, then, in general, these will intersect in a curve of the n-dimensional space. Altogether in this way a family of curves is obtained that depends on the $n-1$ parameters C_i, $i = 1, 2, ..., n-1$. When does this family represent the integral curves of the system $(dx_i)/(dt) = f_i$, $i = 1, 2, ..., n$? – For this every integral curve of the family must lie completely in a particular hypersurface of each family of hypersurfaces; thus, each $F_i(x_1, ..., x_n)$ must be constant. Every function $F(x_1, ..., x_n)$ that is constant along all the integral curves of the system is called a *first integral* of this system. For a function $F(x_1, ..., x_n)$ having a total differential to be a first integral, it is necessary and sufficient that

$$\frac{\partial F}{\partial x_1} f_1 + \frac{\partial F}{\partial x_2} f_2 + \cdots + \frac{\partial F}{\partial x_n} f_n = 0.$$

All the functions F_i, $i = 1, 2, ..., n-1$, must satisfy this condition; hence one has to solve the system of equations

$$\sum_{k=1}^{n} \frac{\partial F_i}{\partial x_k} f_k = 0, \quad i = 1, 2, ..., n-1.$$

The $f_k = \frac{dx_k}{dt}$ are known. Consequently it is possible, in general, to integrate the system with a knowledge of $n-1$ first integrals. If not all $n-1$ first integrals are known, but only $m < n-1$ of them, then an $(n-m)$-dimensional manifold is determined by the fixed

$$F_i(x_1, x_2, ..., x_n) = C_i; \quad i = 1, 2, ..., m.$$

The integral curves lying in this manifold are to be determined. One then has a system of $n-1-m$ differential equations of the first order; a knowledge of m first integrals reduces the number of equations of the system by m.

This possibility is of great significance in mechanics, for example in celestial mechanics. The celebrated *three body problem* investigates the motion of three mutually attracting masses, for example, the sun and two planets. It leads to 18 differential equations with 18 unknown functions, namely 9 coordinate functions and 9 velocity components. With the help of 12 known first integrals this problem is reduced to the integration of 6 differential equations of the first order.

23. Complex analysis

23.1. Differentiation and integration of complex-valued functions

Complex-valued functions. Two *real-valued functions* u and v defined in a set M of the x, y-plane assign to every point $(x, y) \in M$ a point (u, v) of the u, v-plane. If each of these points (x, y) and (u, v) is regarded as a complex number $z = x + iy$ and $w = u + iv$, then to every complex number

$z \in M$ there corresponds a complex number $w = f(z) = u(x, y) + iv(x, y)$. This correspondence represents a complex-valued function f (Fig.). It is called continuous at a point $z_0 \in M$ if for every sequence $\{z_n\}$ with $z_n \in M$ that converges to z_0 for $n = 1, 2, \ldots$ the sequence $f(z_n)$ converges to $f(z_0)$. Here a sequence $\{z_n\}$, $n = 1, 2, \ldots$, of complex numbers is said to converge if the sequences $\{\mathrm{Re}\, z_n\}$ of real parts and $\{\mathrm{Im}\, z_n\}$ of imaginary parts converge. But this means that f is continuous at $z_0 = x_0 + iy_0$ if and only if u and v are continuous at (x_0, y_0). A function defined in M is called continuous *on* M if it is continuous at every point of M.

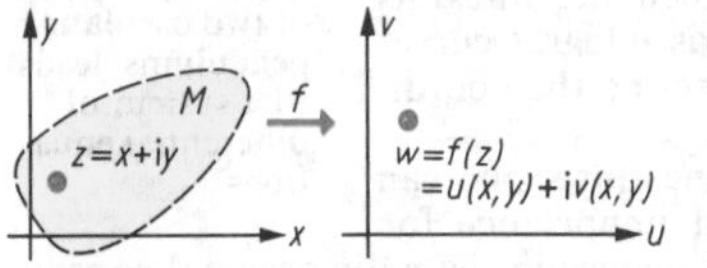

23.1-1 Correspondence between points $z = x + iy$ and points $w = u + iv$ under a complex-valued function $w = f(z)$

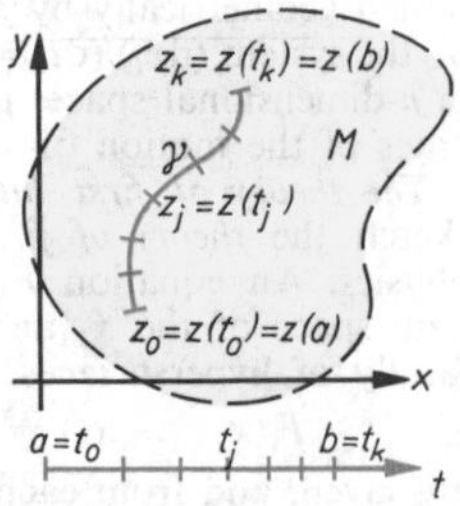

23.1-2 Subdivision of a curve γ represented by $z(t) = x(t) + iy(t)$

Complex curvilinear integrals. A continuous curve in the z-plane is a point set γ that can be represented in the form $z = z(t) = x(t) + i(yt)$, where $a \leqslant t \leqslant b$ and $x(t)$, $y(t)$ are continuous real-valued functions of t. A curve is called continuously differentiable if the functions $x(t)$ and $y(t)$ have continuous first derivatives; the curve then has a finite length $l(\gamma)$ (see Chapter 20.3. – Arc length and surface area). Suppose now that the interval $[a, b]$ is divided into k subintervals $[a, t_1]$, $[t_1, t_2]$, ..., $[t_{k-1}, b]$ by points t_j, $j = 0, 1, \ldots, k$, $t_0 = a$, $t_k = b$, to which on the curve there correspond the points $z_j = z(t_j)$ (Fig.). If the set is contained in the domain of definition M of a complex-valued function f that is continuous on M, then for distinguished subdivisions the sums $\sum_{j=1}^{k} f(z_j)(z_j - z_{j-1})$ converge to a complex number, which is called the complex curvilinear integral $\int_\gamma f(z)\,dz$ of f along γ. Here a sequence of subdivisions is called *distinguished* if the lengths of the longest subintervals form a null sequence for $k \to \infty$. The limit is independent of the choice of the distinguished sequence of subdivisions. By $f = u + iv$ and $z_j = x(t_j) + iy(t_j)$ one obtains for the products in the sum

$$f(z_j)(z_j - z_{j-1}) = u(x_j, y_j)(x_j - x_{j-1}) - v(x_j, y_j)(y_j - y_{j-1}) + i[v(x_j, y_j)(x_j - x_{j-1}) + u(x_j, y_j)(y_j - y_{j-1})]$$

and according to the definition of a real curvilinear integral of the second kind (see Chapter 20.3. – Line and surface integrals)

$$\int_\gamma f(z)\,dz = \int_\gamma (u\,dx - v\,dy) + i\int_\gamma (u\,dy + v\,dx)$$

$$= \int_{t=a}^{b} \left(u\frac{dx}{dt} - v\frac{dy}{dt}\right) dt + i\int_{t=a}^{b} \left(v\frac{dx}{dt} + u\frac{dy}{dt}\right) dt,$$

because, for example, $\int_\gamma u(x, y)\,dx = \int_{t=a}^{b} u(x(t), y(t))\frac{dx(t)}{dt}\,dt$. By the definition $\frac{dx(t)}{dt} + i\frac{dy(t)}{dt} = \frac{dz(t)}{dt}$ one finally obtains $\int_\gamma f(z)\,dz = \int_{t=a}^{b} f(z(t))\frac{dz(t)}{dt}\,dt$. Let $\bar{z} = \alpha - i\beta$ be the complex number conjugate to $z = \alpha + i\beta$; this means that $\alpha = (z + \bar{z})/2$ and $\beta = (z - \bar{z})/(2i)$. Correspondingly one obtains from sums $\sum_{j=1}^{k} f(z_j)(\bar{z}_j - \bar{z}_{j-1})$ the curvilinear integral $\int_\gamma f(z)\,d\bar{z} = \int_{t=a}^{b} f(z(t))\overline{\left(\frac{dz(t)}{dt}\right)}\,dt$.

Example: The circle γ with centre at $z = 0$ and radius $r = 1$ (Fig.) traversed in the positive sense can be represented by $z(t) = \cos t + i \sin t$ with $0 \leqslant t \leqslant 2\pi$. For this representation one has $\frac{dz(t)}{dt} = -\sin t + i\cos t$ and one obtains

$$\int_\gamma \frac{1}{z}\,dz = \int_{t=0}^{2\pi} \frac{1}{z(t)} \cdot \frac{dz(t)}{dt}\,dt = \int_{t=0}^{2\pi} \frac{-\sin t + i\cos t}{\cos t + i\sin t}\,dt = i\int_{t=0}^{2\pi} dt = 2\pi i,$$

because $i(\cos t + i\sin t) = -\sin t + i\cos t$.

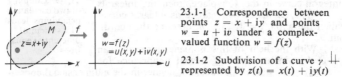

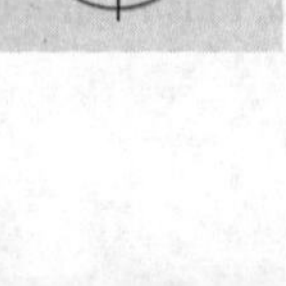

23.1-3 The integral $\int_\gamma \frac{dz}{z} = 2\pi i$, where the unit circle γ is traversed in the positive sense

Complex partial differentiations. Two real-valued functions u and v defined in an open set M of the z-plane and having continuous partial derivatives of the first order can be linearized at $(x_0, y_0) \in M$, that is, approximated by polynomials of the first degree:

$$\tilde{u}(x, y) = u(x_0, y_0) + \frac{\partial u(x_0, y_0)}{\partial x}(x - x_0) + \frac{\partial u(x_0, y_0)}{\partial y}(y - y_0)$$

and
$$\tilde{v}(x, y) = v(x_0, y_0) + \frac{\partial v(x_0, y_0)}{\partial x}(x - x_0) + \frac{\partial v(x_0, y_0)}{\partial y}(y - y_0).$$

Consequently $f = u + iv$ can be linearized at $z_0 = x_0 + iy_0$ by

$$\tilde{f}(z) = f(z_0) + \left[\frac{\partial u(x_0, y_0)}{\partial x} + i\frac{\partial v(x_0, y_0)}{\partial x}\right](x - x_0) + \left[\frac{\partial u(x_0, y_0)}{\partial y} + i\frac{\partial v(x_0, y_0)}{\partial y}\right](y - y_0).$$

After substituting $(x - x_0) = [(z - z_0) + (\overline{z - z_0})]/2$ and $y - y_0 = [(z - z_0) - (\overline{z - z_0})]/(2i)$ and observing the relations $\frac{\partial u}{\partial x} + i\frac{\partial v}{\partial x} = \frac{\partial f}{\partial x}$ and $\frac{\partial u}{\partial y} + i\frac{\partial v}{\partial y} = \frac{\partial f}{\partial y}$ this linearization becomes

$$\tilde{f}(z) = f(z_0) + \frac{1}{2}\left[\frac{\partial f(z_0)}{\partial x} - i\frac{\partial f(z_0)}{\partial y}\right](z - z_0) + \frac{1}{2}\left[\frac{\partial f(z_0)}{\partial x} + i\frac{\partial f(z_0)}{\partial y}\right](\overline{z - z_0}).$$

Starting out from this one defines the complex partial derivatives of the first order of f relative to z and $\bar{z}$ at the point z_0:

$$\frac{\partial f(z_0)}{\partial z} = \frac{1}{2}\left[\frac{\partial f(z_0)}{\partial x} - i\frac{\partial f(z_0)}{\partial y}\right] \quad \text{and} \quad \frac{\partial f(z_0)}{\partial \bar{z}} = \frac{1}{2}\left[\frac{\partial f(z_0)}{\partial x} + i\frac{\partial f(z_0)}{\partial y}\right].$$

For these differentiations the standard rules for real-valued functions are valid, for example,

$$\frac{\partial[f(z) + g(z)]}{\partial z} = \frac{\partial f(z)}{\partial z} + \frac{\partial g(z)}{\partial z} \quad \text{or} \quad \frac{\partial[f(z)\cdot g(z)]}{\partial \bar{z}} = f(z)\cdot\frac{\partial g(z)}{\partial \bar{z}} + g(z)\cdot\frac{\partial f(z)}{\partial \bar{z}}.$$

Examples: 1. For $f(z) = \text{const}$ it follows that $\frac{\partial f(z)}{\partial x} = 0$ and $\frac{\partial f(z)}{\partial y} = 0$, hence $\frac{\partial f(z)}{\partial z} = 0$ and $\frac{\partial f(z)}{\partial \bar{z}} = 0$.

2. For $f(z) = z = x + iy$ it follows that $\frac{\partial f(z)}{\partial x} = 1$ and $\frac{\partial f(z)}{\partial y} = i$, hence $\frac{\partial f(z)}{\partial z} = 1$ and $\frac{\partial f(z)}{\partial \bar{z}} = 0$.

3. For $f(z) = \bar{z} = x - iy$ it follows that $\frac{\partial f(z)}{\partial x} = 1$ and $\frac{\partial f(z)}{\partial y} = -i$, hence $\frac{\partial f(z)}{\partial z} = 0$ and $\frac{\partial f(z)}{\partial \bar{z}} = 1$.

4. The rule for the differentiation of a product applied to $z^2, z^3, \ldots, z^n$ yields by induction $\frac{\partial z^2}{\partial z} = 2z$, $\frac{\partial z^3}{\partial z} = 3z^2$, ..., $\frac{\partial z^n}{\partial z} = nz^{n-1}$, as well as $\frac{\partial z^2}{\partial \bar{z}} = 0$, $\frac{\partial z^3}{\partial \bar{z}} = 0$, ..., $\frac{\partial z^n}{\partial \bar{z}} = 0$.

5. For a polynomial $f(z) = a_0 + a_1 z + a_2 z^2 + \cdots + a_n z^n$ with constant coefficients a_j it follows that $\frac{\partial f(z)}{\partial z} = 0 + a_1 + 2a_2 z + \cdots + na_n z^{n-1}$ and $\frac{\partial f(z)}{\partial \bar{z}} = 0$.

The sequence $\{s_n(z)\}$ of partial sums $s_n(z) = \sum_{j=0}^{n} a_j(z - z_0)^j$ of the power series $\sum_{j=0}^{\infty} a_j(z - z_0)^j$ converges, as $n \to \infty$, either at $z = z_0$ only, or in a circular disc $|z - z_0| < R$, or in the whole plane. The limit function f satisfies $\frac{\partial f(z)}{\partial z} = \sum_{j=1}^{\infty} ja_j(z - z_0)^{j-1}$ and $\frac{\partial f(z)}{\partial \bar{z}} = 0$ (see Chapter 21.). A special case is the definition of the complex exponential function $\exp z = \sum_{j=0}^{n} \frac{z^j}{j!}$; its circle of convergence is the whole z-plane, and its partial derivative is $\frac{\partial \exp z}{\partial z} = \exp z$.

The exponential function satisfies Euler's formula $\exp(i\varphi) = \cos\varphi + i\sin\varphi$. The functional equation $\exp(z_1 + z_2) = \exp z_1 \cdot \exp z_2$ or $\exp z_0 = \exp(z_0 - z)\cdot\exp z$ results from

$$\frac{\partial}{\partial z}[\exp(z_0 - z)\cdot\exp z] = \exp(z_0 - z)\exp z + \exp(z_0 - z)\exp z = 0,$$

because this shows that $\exp(z_0 - z)\exp z = \text{const} = \exp z_0$.

Holomorphic functions. A function f defined in an open set M is called holomorphic if $\frac{\partial f(z)}{\partial \bar{z}} = 0$ for every point $z \in M$. For holomorphic functions one writes $\frac{df}{dz}$ or f' instead of $\frac{\partial f}{\partial z}$. As already

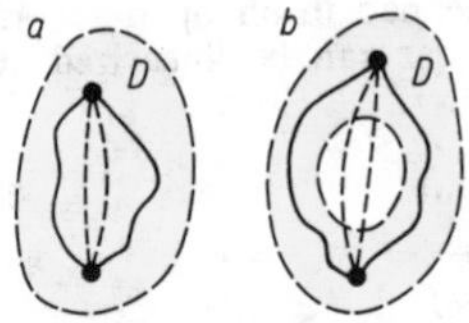

mentioned, the limit function of a power series is holomorphic. A domain D is an open point set in which any two points can be connected by a curve lying entirely in D. In a *simply-connected domain* D two such curves with the same initial point and the same end-point can always be deformed into each other continuously so that the domain D is never left (Fig.).

23.1-4 a) Simply-connected, b) multiply-connected domain D

For every holomorphic function f defined in a simply-connected domain D there exists one and, apart from an additive constant, only one holomorphic *primitive function* F for which $f(z) = \dfrac{dF(z)}{dz}$. Along any curve γ leading from z_1 to z_2 in D one has $\int_\gamma f(z)\,dz = F(z_2) - F(z_1)$.

Example: For $f(z) = z^2$ a primitive F is $F(z) = (1/3)\,z^3$. If γ joins the point $z_1 = 1$ to $z_2 = 2 + i$, then

$$\int_\gamma z^2\,dz = (1/3)\,(2 + i)^3 - (1/3)\cdot 1^3 = {}^1\!/_3 + {}^{11}\!/_3 i.$$

Let γ be the boundary of a part B of D. Then the *theorem of Ostrogradskii-Gauss* holds:

$$\int_\gamma (u\,dx + v\,dy) = \iint_B \left(\frac{\partial v}{\partial x} - \frac{\partial u}{\partial y}\right) dx\,dy.$$

From this theorem it follows that $\int_\gamma f(z)\,dz = \int_\gamma (u\,dx - v\,dy) + i\int_\gamma (u\,dy + v\,dx)$

$$= -\iint_B \left(\frac{\partial v}{\partial x} + \frac{\partial u}{\partial y}\right) dx\,dy + i\iint_B \left(\frac{\partial u}{\partial x} - \frac{\partial v}{\partial y}\right) dx\,dy$$

$$= \iint_B \left[i\left(\frac{\partial u}{\partial x} + i\frac{\partial v}{\partial x}\right) + i^2\left(\frac{\partial u}{\partial y} + i\frac{\partial v}{\partial y}\right)\right] dx\,dy = i\iint_B \left[\frac{\partial f}{\partial x} + i\frac{\partial f}{\partial y}\right] dx\,dy = 2i\iint_B \frac{\partial f(z)}{\partial \bar{z}}\,dx\,dy.$$

Since for holomorphic functions $\dfrac{\partial f(z)}{\partial \bar{z}} = 0$, it follows that $\int_\gamma f(z)\,dz = 0$. This is a way of proving the Cauchy integral theorem.

Cauchy's integral theorem: If γ_0 is a closed curve in a simply-connected domain D, then $\int_{\gamma_0} f(z)\,dz = 0$ for every function f holomorphic in D.

Cauchy's integral theorem also follows from the theorem of the existence of a primitive function $F(z)$ for f in a simply-connected domain. For if γ is closed, then the initial point z_1 is the same as the end-point z_2, and therefore $\int_\gamma f(z)\,dz = F(z_2) - F(z_1) = 0$.

The values of a holomorphic function f in the interior of a circular disc $|z - z_0| \leqslant R$ entirely contained in M are determined, according to Cauchy's integral formula, by the values $f(\zeta)$ which f assumes on the boundary γ_0 of this disc, traversed in the positive direction (Fig.). For by Cauchy's integral theorem

$$\int_{\gamma_{01}} \frac{f(\zeta)}{\zeta - z}\,d\zeta = 0, \qquad \int_{\gamma_{02}} \frac{f(\zeta)}{\zeta - z}\,d\zeta = 0$$

along the closed curves γ_{01} and γ_{02}. If one adds the integrals and lets the radius ϱ tend to zero, one obtains Cauchy's integral formula, using the example on complex curvilinear integrals (on page 518).

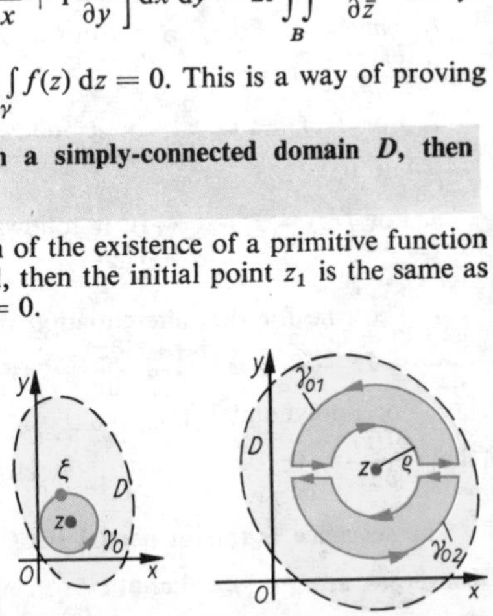

23.1-5 Cauchy's integral formula

Cauchy's integral formula	$f(z) = \dfrac{1}{2\pi i}\displaystyle\int_{\gamma_0} \frac{f(\zeta)}{\zeta - z}\,d\zeta$

A function f holomorphic in the disc $|z - z_0| < R$ can be represented as limit function by a uniquely determined power series $f(z) = \sum_{j=0}^{\infty} a_j (z - z_0)^j$ with the circle of convergence R and has the derivative $\dfrac{df(z)}{dz} = \sum_{j=1}^{\infty} j a_j (z - z_0)^{j-1}$, which is also holomorphic. By repeated differentiation one obtains holomorphic derivatives of every order k, which can also be obtained from Cauchy's integral formula by differentiation. Setting $z = z_0$ one can determine by comparison of the two results the coefficients a_j of the series for $f(z)$, beginning with $a_0 = f(z_0)$.

A function f holomorphic in $|z - z_0| < R$ and uniquely represented by the power series $f(z) = \sum_{j=0}^{\infty} a_j(z - z_0)^j$ has holomorphic derivatives $\frac{\mathrm{d}^k f(z)}{\mathrm{d}z^k} = \frac{k!}{2\pi \mathrm{i}} \int_{\gamma_0} \frac{f(\zeta)\,\mathrm{d}\zeta}{(\zeta - z)^{k+1}}$ of every order k, and the coefficients of the power series are determined by $a_j = \frac{1}{2\pi \mathrm{i}} \int_{\gamma_0} \frac{f(\zeta)}{(\zeta - z_0)^{j+1}}\,\mathrm{d}\zeta$, where γ_0 is a curve in the interior of the disc of convergence that goes around z or z_0 once in the positive direction.

Isolated singularities of holomorphic functions. Let M be an open set containing the punctured disc $0 < |z - z_0| < R$ (Fig.), but not necessarily the centre z_0 of the disc. Let f be a function holomorphic in M. Then f can be represented in the disc by a uniquely determined *Laurent expansion*, whose *principal part* consists of the terms with negative powers of the coefficients. The coefficient of its first term $a_{-1} = \mathrm{Res}\, f$ is called the *residue* of f at z_0. The point z_0 is called an *isolated singularity* of f.

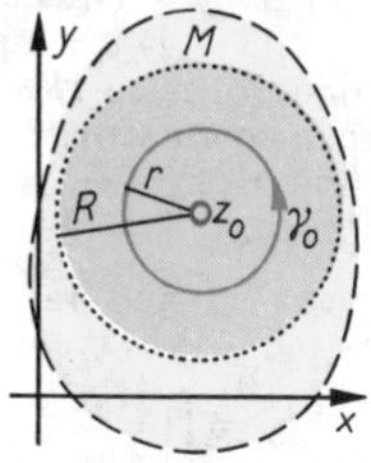

23.1-6 Laurent expansion in a punctured open disc

Laurent expansion: $f(z) = \sum_{j=-\infty}^{+\infty} a_j(z - z_0)^j = \dots + \frac{a_{-2}}{(z - z_0)^2} + \frac{a_{-1}}{z - z_0} + a_0 + a_1(z - z_0) + a_2(z - z_0)^2 + \dots$ with

$$a_j = \frac{1}{2\pi \mathrm{i}} \int_{\gamma_0} \frac{f(\zeta)\,\mathrm{d}\zeta}{(\zeta - z_0)^{j+1}}, \qquad a_{-1} = \frac{1}{2\pi \mathrm{i}} \int_{\gamma_0} f(\zeta)\,\mathrm{d}\zeta;$$

where γ_0 is the circle $|\zeta - z_0| = r$, $0 < r < R$, traversed in the positive direction.

If $a_j = 0$ for all negative j, that is, if the principal part is zero, then the Laurent series reduces to a power series, and f is holomorphic in the entire disc $|z - z_0| < R$ if one sets $f(z_0) = a_0$. In this case z_0 is called a *removable singularity* of f. It is called a *pole* of order n of f if $a_{-n} \neq 0$, but $a_{-n-1} = 0$, $a_{-n-2} = 0$, ..., that is, if only finitely many a_j with negative j differ from zero. Then $|f(z)|$ becomes arbitrarily large for places z sufficiently near to z_0. An *essential singularity* z_0 of f occurs if $a_j \neq 0$ for infinitely many negative j. By a theorem of Casorati-Weierstrass f then comes arbitrarily near to every complex value in every neighbourhood of z_0.

A function f is called *meromorphic* in an open set M if it is holomorphic in M apart from removable singularities or poles.

Example: The function f represented by $f(z) = \frac{1}{z+1} + \frac{z}{1+z^2} = \frac{1}{z+1} + \frac{z}{2\mathrm{i}(z - \mathrm{i})} - \frac{z}{2\mathrm{i}(z + \mathrm{i})}$ is meromorphic in the whole plane; it has a pole of the first order at each of the places $z_1 = -1$, $z_2 = \mathrm{i}$, and $z_3 = -\mathrm{i}$.

If one multiplies a meromorphic function f having a pole of order at most n at z_0 by $(z - z_0)^n$, then a power series arises in which a_{-1} is the coefficient of the term $a_{-1}(z - z_0)^{n-1}$. By repeated differentiation one obtains from this the residue a_{-1} of the function f

$$\operatorname*{Res}_{z_0} f = a_{-1} = \frac{1}{(n-1)!} \lim_{z \to z_0} \frac{\mathrm{d}^{n-1}[(z - z_0)^n f(z)]}{\mathrm{d}z^{n-1}}.$$

If γ_0 is a closed curve within the domain of definition M of a meromorphic function f and if z_0 is an isolated singularity of f that does not lie on γ_0, then for every point ζ on γ_0 one can calculate by $\zeta - z_0 = |\zeta - z_0| \cdot (\cos\varphi + \mathrm{i}\sin\varphi)$ (Fig.) the angle φ which the direction from z_0 to ζ makes with the positive real axis. This angle is determined only up to integral multiples $2k\pi$ of 2π, but k can be chosen so that φ changes continuously when ζ runs continuously through γ_0. When ζ after

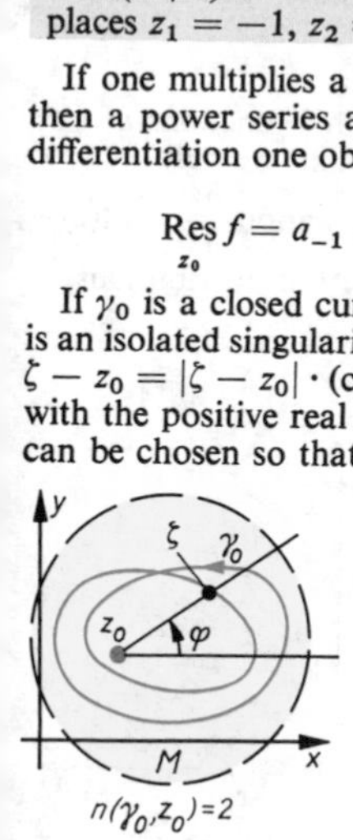

23.1-7 Definition of the winding number $n(\gamma_0, z_0)$

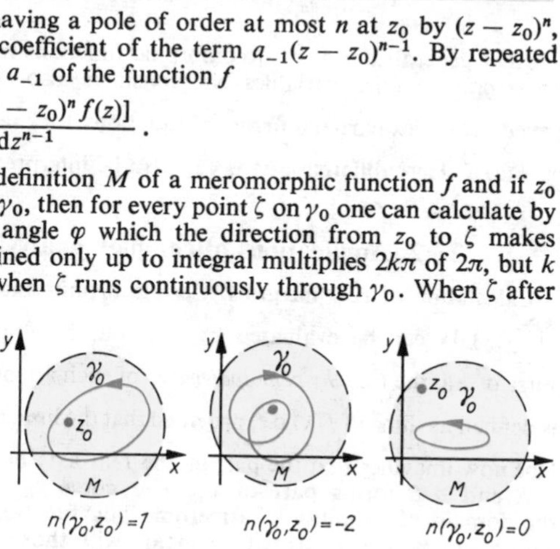

23.1-8 Examples of winding numbers

traversing γ_0 returns to the initial point, φ has changed by $2n\pi$. The *winding number* $n = n(\gamma_0, z_0)$ is an integer and depends on the curve γ_0, the sense of direction, and the position of z_0 relative to it (Fig.). If γ_0 surrounds z_0, has no double point, and is traversed in the positive direction, then by the Laurent expansion $\int_{\gamma_0} f(\zeta)\,d\zeta = 2\pi i a_{-1}$. More generally, the following theorem can be derived.

Residue theorem. $\int_{\gamma_0} f(\zeta)\,d\zeta = 2\pi i \sum_{z_0} n(\gamma_0, z_0) \cdot \operatorname{Res}_{z_0} f$, *provided that* γ_0 *is a closed curve in the simply-connected domain M and f is holomorphic in M apart from isolated singularities* z_0. *The sum is taken over the* z_0.

Example: The meromorphic function f represented by $f(z) = 1/(z+1) - 1/(z-1)$ has a pole each at $z_1 = -1$ and $z_2 = +1$. In a neighbourhood of z_1, $-1/(z-1)$ is a holomorphic function and can be expanded in a power series P_1, similarly $1/(z+1)$ in a neighbourhood of z_2 in a power series P_2. From $f(z) = 1/(z+1) + P_1 = -1/(z-1) + P_2$ one obtains $\operatorname{Res}_{z_0=-1} f = +1$ and $\operatorname{Res}_{z_0=+1} f = -1$. If γ_0 is the circle with centre at $z = 2$ and radius 2 traversed in the negative direction (Fig.), then $n(\gamma_0, -1) = 0$ and $n(\gamma_0, +1) = -1$; hence the residue theorem yields

$$\int_{\gamma_0} \left(\frac{1}{\zeta+1} - \frac{1}{\zeta-1}\right) d\zeta = 2\pi i[1 \cdot 0 + (-1)(-1)] = 2\pi i.$$

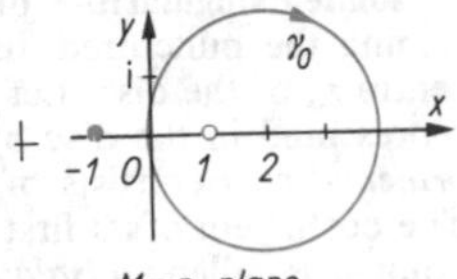

23.1-9 Application of the residue theorem to $f(z) = \dfrac{1}{z+1} - \dfrac{1}{z-1}$

Holomorphic functions of several complex variables. A function f defined in an open set D in the set $\mathbf{C}^n$ of all ordered n-tuples $(z_1, \ldots, z_n)$ of complex numbers is called holomorphic if every function f is holomorphic in which only one of the z_j is variable and the other $n-1$ are fixed. Hence they satisfy the differential equations $\dfrac{df}{d\bar{z}_1} = 0, \ldots, \dfrac{df}{d\bar{z}_n} = 0$. The point set $\{(z_1, \ldots, z_n) : |z_j - z_j^0| \leq R_j, j = 1, \ldots, n\}$ is called a *closed polycylinder*, where $(z_1^0, \ldots, z_n^0)$ is a fixed chosen point in $\mathbf{C}^n$. If it lies entirely in D, then at all points in its interior $\{(z_1, \ldots, z_n) : |z_j - z_j^0| < R_j, j = 1, \ldots, n\}$ the function f can be represented by the *generalized Cauchy integral formula* in which the determining surface $\mathbf{S}$: $\{(z_1, \ldots, z_n) : |z_j - z_j^0| = R_j, j = 1, \ldots, n\}$ is a subset of the boundary of the polycylinder.

Generalized Cauchy integral formula

$$f(z_1, \ldots, z_n) = \frac{1}{(2\pi i)^n} \int_{\mathbf{S}} \frac{f(\zeta_1, \ldots, \zeta_n)}{(\zeta_1 - z_1) \cdots (\zeta_n - z_n)} d\zeta_1, \ldots, d\zeta_n.$$

If D is a domain, that is, a connected open point set, then two functions holomorphic in D are equal at every point of D if their values coincide on the determining surface of a polycylinder situated in D.

Locally the holomorphic functions can be represented as the limit functions of a power series

$$\sum_{\nu_1, \ldots, \nu_n} c_{\nu_1, \ldots, \nu_n} (z_1 - z_1^0)^{\nu_1} \ldots (z_n - z_n^0)^{\nu_n}.$$

Further generalizations of holomorphic functions arise if instead of holomorphic functions of one or several complex variables one considers also complex valued functions that are solutions of general complex partial differential equations, for example, Vekua's differential equation $\dfrac{\partial w}{\partial \bar{z}} = A(z)\,w + B(z)\,\bar{w}$. Here differentiations can also be interpreted in the sense of the theory of distributions.

23.2. Applications of complex analysis

Calculation of real integrals. Certain definite integrals $\int_{-\infty}^{+\infty} f(x)\,dx$ can be evaluated by 'contour integration'. First of all, the *Cauchy principal value* of such an integral is defined as $\lim_{R\to\infty} \int_{-R}^{+R} f(x)\,dx$, provided that the limit exists. One now imagines that the part of the real axis between $-R$ and $+R$ forms part of a closed curve γ_R within the domain of definition of a meromorphic function $f(z)$ whose values, when z is real, coincide with those of the given function $f(x)$, and that the curvilinear integral

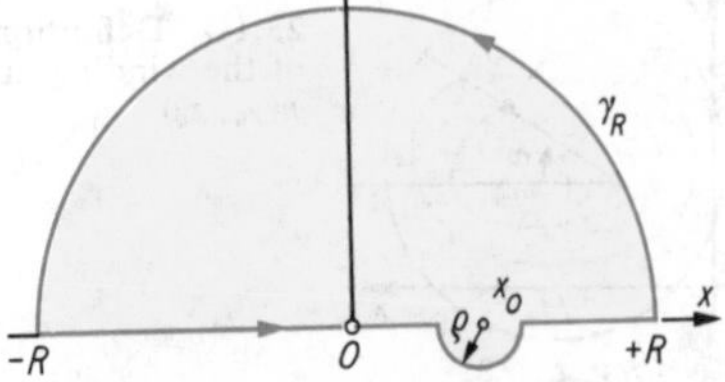

23.2-1 Path of contour integration

along the remaining part of γ_R has the limit zero as $R \to \infty$. If f has poles on the real axis, one of which at x_0, say, then the point x_0 can be excluded from, or included in, the interior of γ_R by a semicircle of radius ϱ (small) and the integral along the contour γ_R must be evaluated for $\varrho \to 0$ (Fig.). For such poles with $\operatorname{Im} z_0 = 0$ the factor $\mp\pi i$ therefore occurs in place of $2\pi i$, which holds for poles z_0 with positive imaginary part $\operatorname{Im} z_0 > 0$.

$$\int_{-\infty}^{+\infty} [p_1(x)/p_2(x)]\,dx = 2\pi i \sum_{\operatorname{Im} z_0 > 0\; z_0} \operatorname{Res}\,[p_1(z)/p_2(z)] + \pi i \sum_{\operatorname{Im} z_0 = 0\; z_0} \operatorname{Res}\,[p_1(z)/p_2(z)],$$

$$\int_{-\infty}^{+\infty} [p_3(x)/p_4(x)] \cos x\,dx + i \int_{-\infty}^{+\infty} [p_3(x)/p_4(x)] \sin x\,dx$$

$$= 2\pi i \sum_{\operatorname{Im} z_0 > 0\; z_0} \operatorname{Res}\,\{[p_3(z)/p_4(z)] \exp iz\} + \pi i \sum_{\operatorname{Im} z_0 = 0\; z_0} \operatorname{Res}\,\{[p_3(z)/p_4(z)] \exp iz\},$$

provided that the functions defined in the z-plane by $p_1(z)/p_2(z)$ and $p_3(z)/p_4(z)$ have only poles of order 1 on the real axis, that the degree of p_2 exceeds that of p_1 by at least 2, and the degree of p_4 exceeds that of p_3 by at least 1.

Example 1: $\displaystyle\int_0^{+\infty} \frac{1}{1+x^2}\,dx = \frac{1}{2}\int_{-\infty}^{+\infty} \frac{1}{1+x^2}\,dx = \frac{\pi}{2}$. If one sets in the formula above $p_1(z) = 1$, $p_2(z) = 1 + z^2 = (z+i)(z-i)$, one finds $\displaystyle\operatorname*{Res}_{z_0 = i} \frac{1}{1+z^2} = \lim_{z \to i} \frac{z-i}{1+z^2} = \lim_{z \to i} \frac{1}{z+i} = \frac{1}{2i}$ $= -\dfrac{i}{2}$, and so the result.

Example 2: $\displaystyle\int_{-\infty}^{+\infty} \frac{\sin x}{x}\,dx = \pi$. The residue theorem is applied to the meromorphic function represented by $(1/z) \exp iz$, which only at $z_0 = 0$ has a pole of the first order. Its residue at $z_0 = 0$ is $\displaystyle\operatorname*{Res}_{z_0}\left[\frac{\exp iz}{z}\right] = \lim_{z\to 0}\left[z\,\frac{\exp iz}{z}\right] = \lim_{z \to 0} \exp iz = 1$. From $\displaystyle\int_{-\infty}^{+\infty} \frac{\cos x}{x}\,dx + i\int_{-\infty}^{+\infty} \frac{\sin x}{x}\,dx = \pi i \cdot 1$ one derives $\displaystyle\int_{-\infty}^{+\infty} \frac{\cos x}{x}\,dx = 0$ and the assertion made above.

Connections between complex analysis and partial differential equations. For a holomorphic function $f = u + iv$ one has, by definition, $\dfrac{\partial f(z)}{\partial \bar z} = \dfrac{1}{2}\left[\dfrac{\partial f(z)}{\partial x} + i\,\dfrac{\partial f(z)}{\partial y}\right] = 0$, $\dfrac{\partial f(z)}{\partial x} = -i\,\dfrac{\partial f(z)}{\partial y}$, respectively, $\dfrac{\partial u}{\partial x} + i\,\dfrac{\partial v}{\partial x} = -i\,\dfrac{\partial u}{\partial y} + \dfrac{\partial v}{\partial y}$. This leads fo the Cauchy-Riemann differential equations.

Cauchy-Riemann differential equations	$\dfrac{\partial u}{\partial x} = \dfrac{\partial v}{\partial y}$	$-\dfrac{\partial v}{\partial x} = \dfrac{\partial u}{\partial y}$

Example: The function defined by $f(z) = z^2 = (x^2 - y^2) + 2ixy$ is holomorphic, therefore $u(x, y) = x^2 - y^2$ and $v(x, y) = 2xy$ is a solution of the Cauchy-Riemann differential equations.

Converse theorem. If the partial derivatives of the first order of the real-valued functions u and v exist and satisfy the Cauchy-Riemann differential equations, then $f = u + iv$ is holomorphic.

If one differentiates the first Cauchy-Riemann differential equation with respect to x and the second with respect to y, one obtains the *Laplace differential equation* $\dfrac{\partial^2 u}{\partial x^2} + \dfrac{\partial^2 u}{\partial y^2} = 0$. Since a holomorphic function f has derivatives of every order, the partial derivatives of arbitrarily high order of u and of v also exist.

The real part u and the imaginary part v of a holomorphic function $f = u + iv$ satisfy the Laplace differential equation $\Delta u = \dfrac{\partial^2 u}{\partial x^2} + \dfrac{\partial^2 u}{\partial y^2} = 0$, respectively, $\Delta v = \dfrac{\partial^2 v}{\partial x^2} + \dfrac{\partial^2 v}{\partial y^2} = 0$. Conversely, in a simply-connected domain D for a function u that is a solution in D of the Laplace differential equation there is a function v, uniquely determined apart from an additive constant, which together with u determines in D the holomorphic function $f = u + iv$.

Example: The real part $u(x, y) = x^2 - y^2$ of the holomorphic function f defined by $f(z) = z^2$ is a solution of Laplace's differential equation.

Conformal mappings. A holomorphic function f defined in a domain D by $w = f(z)$ assigns to every point z of D a point w of the w-plane. If γ is represented by $z = z(t)$, $a \leqslant t \leqslant b$, and if $\frac{dz(t)}{dt} = \varrho(t) \exp(i\beta(t))$, then the tangent to the curve at the point $z(a)$ makes the angle $\beta(a)$ with the positive real axis. If $\left.\frac{df(z)}{dz}\right|_{z=z_0} = \tilde{\varrho} \exp(i\alpha)$, then since $\left.\frac{df(z(t))}{dt}\right|_{t=a} = \left.\frac{df(z)}{dz}\right|_{z=z_0} \cdot \left.\frac{dz(t)}{dt}\right|_{t=a}$ $= \tilde{\varrho}\varrho(a) \exp[i(\beta(a) + \alpha)]$, the tangent to the image curve at the point $f(z(a))$ makes the angle $\beta(a) + \alpha$ with the positive real axis, that is, all angles are rotated by α. Consequently the angle φ between two curves remains unchanged (Fig.). Hence the mapping induced by f is called *angle-preserving* or *conformal*, more accurately, *directly conformal*, because the sense of rotation is also preserved. If $r(t)$ is the distance between the points $z(t)$ and z_0, and $\tilde{r}(t)$ that between $f(z(t))$ and $f(z_0)$, then $\lim_{t \to 0} \frac{\tilde{r}(t)}{r(t)} = |f'(z_0)|$ if $f'(z_0) \neq 0$. This means that in the limit $t \to 0$ distances are multiplied by the factor $|f'(z_0)|$ (Fig.).

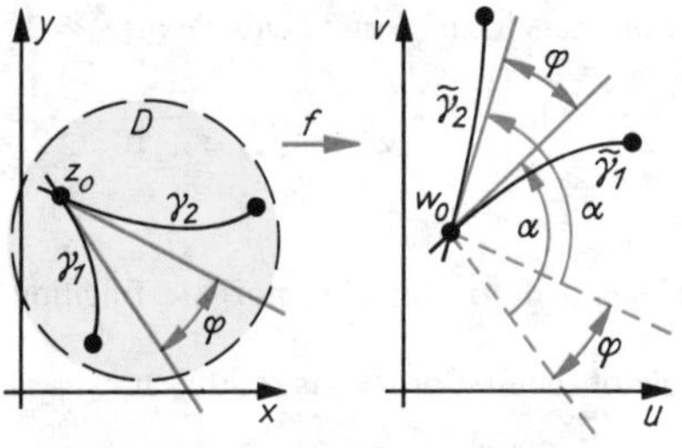

23.2-2 The mapping by a holomorphic function $w = f(z)$ is conformal

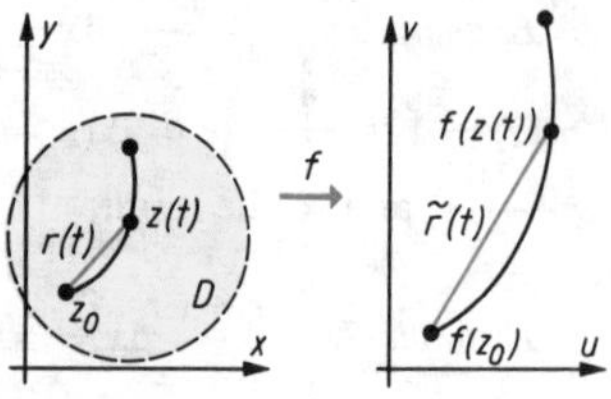

23.2-3 Distance ratios for conformal mappings

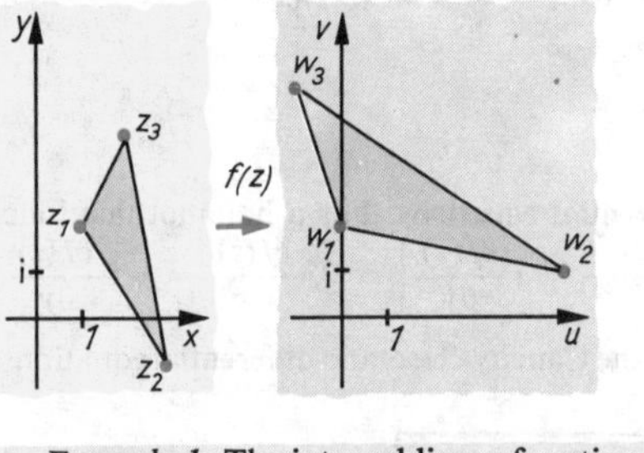

23.2-4 Similarity transformation of a triangle by means of $w = (1 + i) z + (1 - i)$

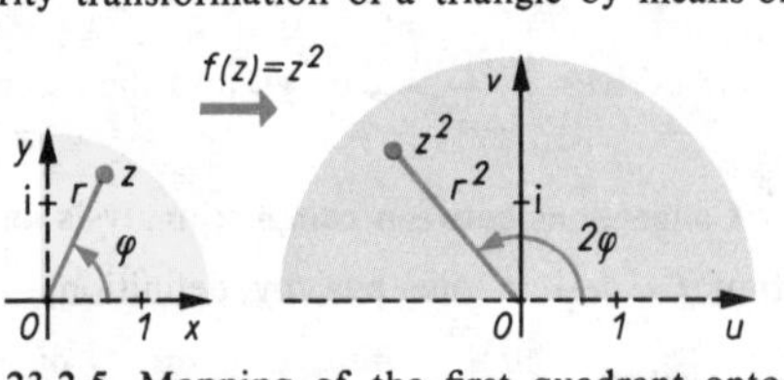

23.2-5 Mapping of the first quadrant onto the upper halfplane by $w = z^2$

Example 1: The integral linear function f defined by $w = f(z) = az + b$ with complex constants $a \neq 0$ and b maps every figure of the z-plane, for example, a triangle, to a similar figure in the w-plane (Fig.).

Example 2: Since $z = r \exp(i\varphi)$, one has $w = z^2 = r^2 \exp(2i\varphi)$; consequently the function f defined by $w = z^2$ maps the first quadrant ($\operatorname{Re} z > 0$, $\operatorname{Im} z > 0$) of the z-plane to the upper half-plane ($\operatorname{Im} w > 0$) of the w-plane (Fig.).

Example 3: If $w = f(z) = \exp(ic) \frac{z - z_0}{1 - \bar{z}_0 z}$, where c is a real constant and z_0 is fixed with $|z_0| < 1$, then

$$|w|^2 = |\exp(ic)|^2 \frac{(z - z_0)(\bar{z} - \bar{z}_0)}{(1 - \bar{z}_0 z)(1 - z_0 \bar{z})} = 1 \cdot \frac{z\bar{z} + z_0\bar{z}_0 - z_0\bar{z} - z\bar{z}_0}{1 + z_0\bar{z}_0 z\bar{z} - \bar{z}_0 z - z_0\bar{z}},$$

because the absolute value $|\zeta|^2$ of a complex number ζ satisfies $|\zeta|^2 = \zeta\bar{\zeta}$. Hence $|w| = 1$ if and only if $z\bar{z} + z_0\bar{z}_0 - z_0\bar{z} - z\bar{z}_0 = 1 + z_0\bar{z}_0 z\bar{z} - \bar{z}_0 z - z_0\bar{z}$, that is, $|z|^2 (1 - |z_0|^2) = 1 - |z_0|^2$, or $|z| = 1$. Since $|f(z_0)| = 0$, continuity of arguments shows that $|f(z)| < 1$ for all z with $|z| < 1$. Conversely, since $z = \exp(-ic) \frac{w + \exp(ic) z_0}{1 + \exp(-ic) \bar{z}_0 w}$ every w with $|w| < 1$ is the image of precisely one z with $|z| < 1$. Altogether this means that f maps the open unit disc $\equiv |z| < 1$ one-to-one onto itself. On account of $\frac{df(z)}{dz} \neq 0$ the mapping is conformal. If $z_0 = 0$, then $f(z) = \exp(ic)\, z$ is a rotation around $z = 0$ by the angle c measured in radian (Fig.).

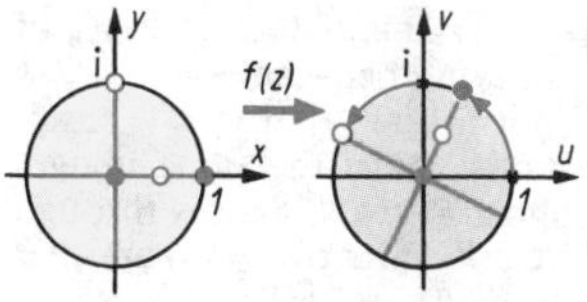

23.2-6 Rotation about $z = 0$

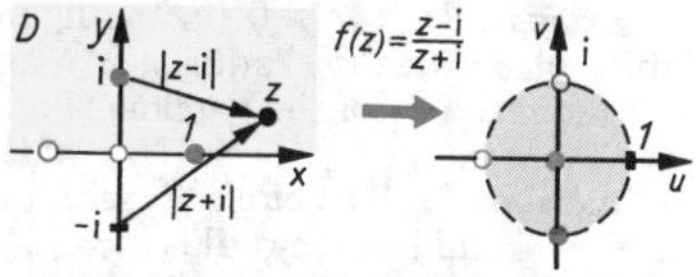

23.2-7 Mapping of the upper half-plane onto the unit circle by $w = (z - i)/(z + i)$

Example 4: The holomorphic function defined by $w = f(z) = \dfrac{z - i}{z + i}$ in the upper half-plane $\operatorname{Im} z > 0$ maps the latter conformally to the open disc $|w| < 1$. Since $|z - i| < |z + i|$ for all z with $\operatorname{Im} z > 0$, it is true that $|w| < 1$. The images of real z are points on the unit circle because $|z - i| = |z + i|$. The images of the points $0, 1, \infty, -1$ of the z-plane are the points $-1, -i, 1, i$ of the w-plane (Fig.).

Riemann mapping theorem. Let D be a simply-connected proper part of the complex z-plane. Given any point z_0 of D and any direction at z_0, there is one and only one conformal mapping of D onto the unit disc $|w| < 1$ by a holomorphic function $w = f(z)$ with non-vanishing derivative such that z_0 goes over into the centre $w = 0$ and the given direction at z_0 into that of the positive real axis (Fig.).

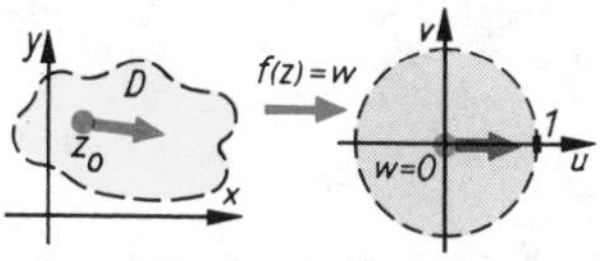

23.2-8 Riemann's mapping theorem

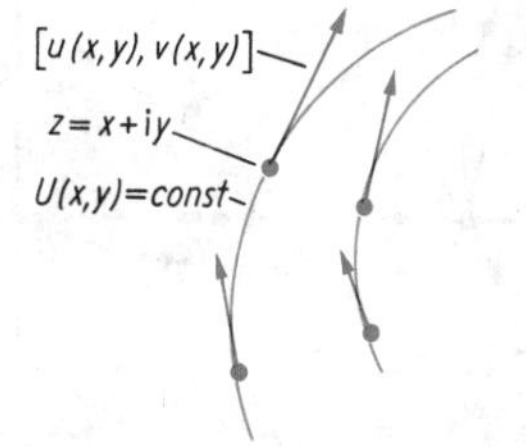

23.2-9 Stream lines $U(x, y) = \text{const}$

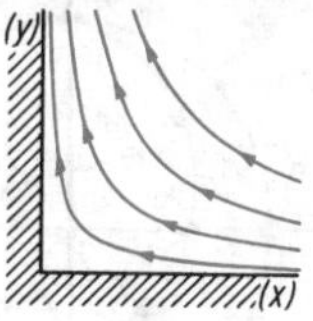

23.2-10 Flow in a rectangular bend

Problems of stream flow. A stationary flow, that is, one independent of the time, in a domain of the x, y-plane can be characterized by the velocity vector $[u(x, y), v(x, y)]$ of a particle following the flow. In a flow free of sources and vortices one has $\dfrac{\partial v}{\partial y} = \dfrac{\partial u}{\partial x}$ and $-\dfrac{\partial v}{\partial x} = \dfrac{\partial u}{\partial y}$, so that the components u and v form a holomorphic function $f = u + iv$. In a simply-connected domain there always exists a primitive $F = U + iV$, which is then holomorphic. With $U(x, y) = \text{const}$ it describes the streamlines along which the particles move (Fig.). The velocity vectors are tangents to the streamlines. Since conformal mappings carry holomorphic functions into holomorphic functions, they are a suitable means of describing the course of streamlines.

Example 1: In the upper w-half-plane with $\operatorname{Im} w > 0$ the streamlines are $\operatorname{Im} w = \text{const}$. If $w = z^2 = x^2 - y^2 + 2ixy$, this half-plane is the biunique image of the first quadrant in the z-plane (see Example 2 on conformal mappings). This means that the hyperbolae $\operatorname{Im} w = 2xy = \text{const}$ are also streamlines, namely those of the flow in a rectangular bend (Fig.).

Example 2: The function $w = (1/2)(z + 1/z)$ maps the circle $|z| = 1$ to the segment from $+1$ to -1 in the w-plane, traversed twice, so that the z-points $+1, i, -1, -i$ have the images $+1, 0, -1, 0$. With the exception of this segment the w-plane is the image of the exterior of the unit circle $|z| > 1$. In it the parallels $\operatorname{Im} w = \text{const}$ are streamlines. Their originals give the stream around the unit circle. Since $z = r(\cos\varphi + i \sin\varphi)$, $1/z = (1/r)(\cos\varphi - i \sin\varphi)$, and $w = (1/2)(r + 1/r)\cos\varphi + (i/2)(r - 1/r)\sin\varphi$, it can be seen that $(r - 1/r)\sin\varphi = c = \text{const}$ are the equations of these streamlines (Fig.).

The flow around other contours is obtained by mapping the exterior of the contour conformally to the exterior of a circle.

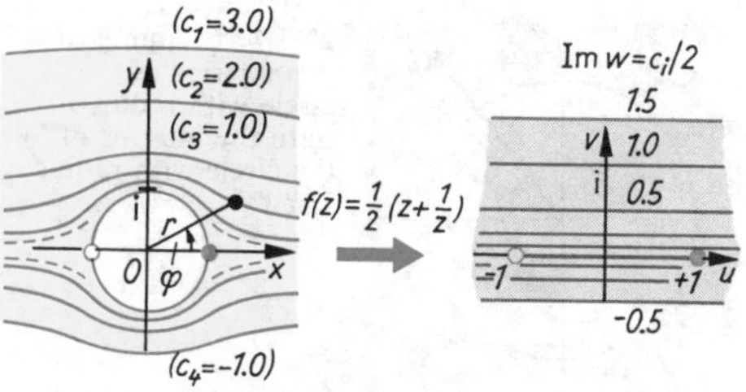

23.2-11 Conformal mapping of the streamlines $(r - 1/r)\sin\varphi = \text{const}$ of the z-plane onto the streamlines of the parallel flow $\operatorname{Im} w = \text{const}$ of the w-plane

Example 3: If $h > 0$ (in the figure $h = 2.75$), then $w = (1/2)(z + h^2/z)$ maps the circle K_h of the z-plane with the radius $r = h$ and the centre $z = 0$ to the segment lying between $-h$ and $+h$ on the real axis of the w-plane (Fig.); for from $z = h(\cos\varphi + i\sin\varphi)$ it follows that $w = h\cos\varphi$. If $\zeta = h^2/z$, then $w(\zeta) = (1/2)(h^2/z + h^2 \cdot z/h^2) = w(z)$, that is, z and ζ have the same images in the w-plane. The circle K_ϱ with the centre $M(-1, 1)$ passing through $z = h$ also passes through $z = -hi$ and has the radius $\varrho = \sqrt{[(h+1)^2 + 1]}$. By $\zeta = h^2/z$ the circle K_ϱ goes over into another circle K_η with the centre $H(h/(h+2), h/(h+2))$ and the radius $\eta = h\varrho/(h+2)$. The circles K_ϱ and K_η are mapped to the so-called *Zhukovskii profile*. Here the point set lying between K_ϱ and K_η goes over into the point set bounded by the Zhukovskii profile; in general, pairs of points lying between K_ϱ and K_η go over into one and the same point. Incidentally, the sickle-shaped point set lying in the first quadrant between K_h and K_η, and also that lying in the fourth quadrant between K_h and K_ϱ, goes over into the point set between the real axis of the w-plane and the part $G' - I' - A'$ of the Zhukovskii profile. The images of circles around $(-1, +1)$ with increasing radii $\varrho_i = 5, 6, \ldots$ become more and more circular for sufficiently large $|z|$, because $|h^2/z| < \varepsilon$.

In general, such mappings are defined by $a_0/z + a_1 z + a_2 z^2 + \cdots$.

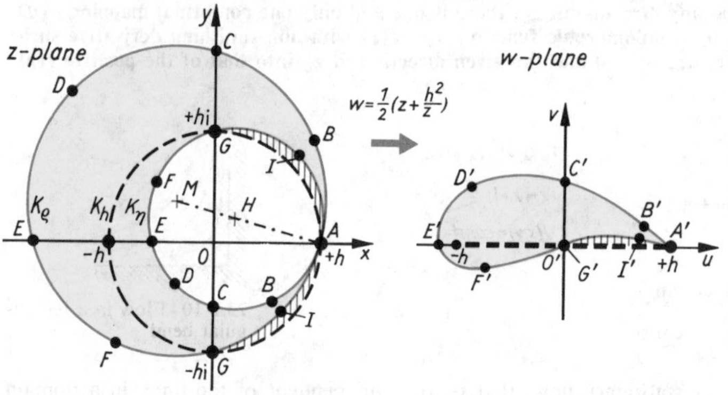

23.2-12 Conformal mapping of the circles K_ϱ and K_η to the Zhukovskii profile by $w = (1/2)(z + h^2/z)$

23.3. The total course of complex-valued functions

The Riemann number sphere. The holomorphic function defined for $z \neq 0$ by $\zeta = 1/z$ maps the exterior of the circle $|z| > R$ conformally to the punctured disc $0 < |\zeta| < 1/R$, which does not contain the point $\zeta = 0$ (Fig.). But the images $\zeta = 1/r \exp(-i\varphi)$ of the points $z = r\exp(i\varphi)$ approach it arbitrarily closely as $r \to \infty$. One therefore interprets $\zeta = 0$ as the image of the point $z = \infty$ of the z-plane. An intuitive idea of the point at infinity of the z-plane can be given by means of *Riemann's number sphere*. The sphere touches at its point S the z-plane at $z = 0$. The line joining the point N diametrically opposite to S to a point z pierces the surface of the sphere at the image point P of z. The mapping $z \to P$ is one-to-one, and the point N is regarded as the image of the point at infinity in the z-plane (Fig.).

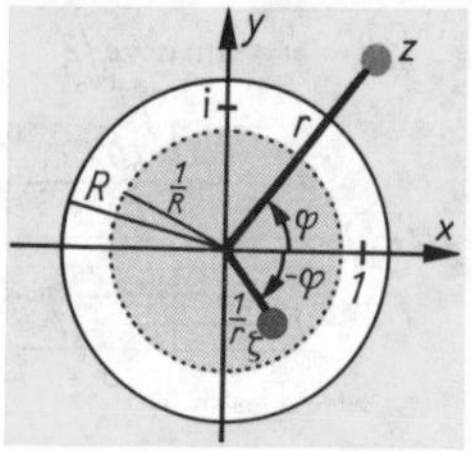

23.3-1 Mapping of the exterior of the circle with radius R onto the interior of the circle with radius $1/R$ by $\zeta = 1/z$

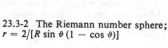

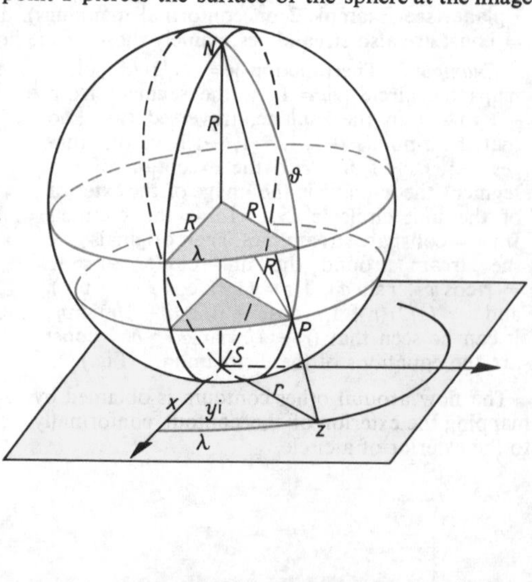

23.3-2 The Riemann number sphere; $r = 2/[R\sin\vartheta(1 - \cos\vartheta)]$

The rays on which a point $z = r \exp(i\varphi)$ and its image $\zeta = 1/r \exp(-i\varphi)$ lie go over one into one another by the reflection $Z = \bar{z}$. This is an *indirect conformal mapping* under which the sense of rotation of the argument φ is reversed. The *transformation by reciprocal radii* $\zeta = 1/Z = 1/\bar{z} = 1/r \exp(i\varphi)$ is therefore *indirectly conformal*. Original and image point lie on the same ray. The quantity $1/r$ can easily be constructed elementarily (Fig.).

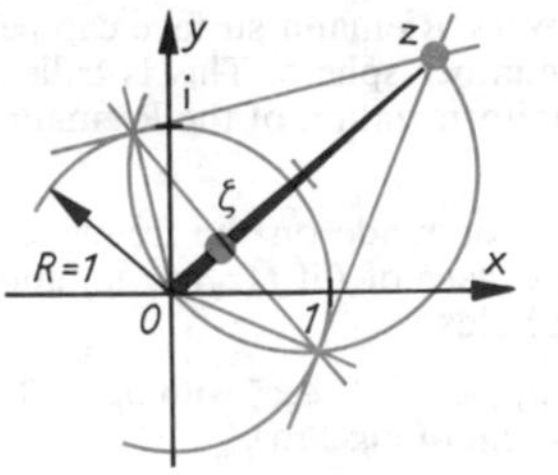

23.3-3 Mapping $\zeta = 1/z$ by reciprocal radii; $|z| = r$ implies that $|\zeta| = 1/r$ and $|\bar{z}| \cdot |\zeta| = 1$

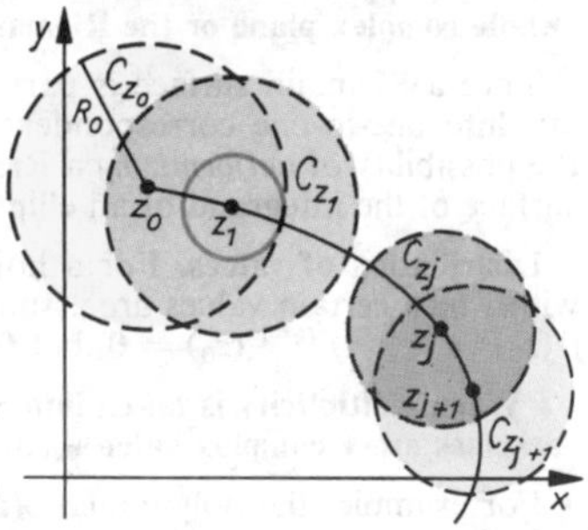

23.3-4 Analytic continuation

Riemann surfaces. If the power series $P_{z_0} = \sum_{j=0}^{\infty} a_j(z - z_0)^j$ converges in the disc K_{z_0}: $|z - z_0| < R_0$, then it defines in it a holomorphic function f. If z_1 lies in K_{z_0}, then in accordance with $(z - z_0) = [(z - z_1) + (z_1 - z_0)]$ the power series P_{z_0} can be rearranged into a power series P_{z_1}, which also converges and represents the same function in the intersection $K_{z_0} \cap K_{z_1}$. If P_{z_1} converges not only in K_{z_0}, but also at points z in a disc K_{z_1} lying partly outside K_{z_0}, then f has been continued analytically by P_{z_1} (Fig.). By analytic continuation in every possible way one obtains the *complete analytic function* generated by the power series P_{z_0}. It can happen that this assigns to every point of the z-plane exactly one function value $f(z)$; but there can also be points z for which one obtains different function elements according to the path on which they are approached. To avoid this ambiguity one imagines that every function element is defined in an individual copy of the plane, a sheet, so that over such places z the complete analytic function is defined uniquely in a corresponding number of sheets. This covering surface or sphere is called a *Riemann surface* $\boldsymbol{R}$.

For example, the function defined by $w = \sqrt{z} = r \exp(i\varphi/2)$ can be continued analytically from the positive real axis in the sense of increasing φ or the negative sense of decreasing φ. At the negative real axis one obtains, according to the sense of rotation, the values $w^+(\pi) = r \exp(i\pi/2)$ and $w^-(-\pi) = r \exp(-i\pi/2)$, respectively (Fig.). After a further complete circuit these values interchange, because $w^+(3\pi) = \exp(3i\pi/2) = r \exp(-i\pi/2)$ and $w^-(-3\pi) = r \exp(-3i\pi/2) = r \exp(i\pi/2)$, respectively. The *Riemann surface* of this function therefore has two sheets (Fig.) which are cut each along the negative real axis and are then pasted together crosswise so that the upper boundary of the cut in each sheet is connected with the lower boundary of the other sheet. Then the function values on the Riemann surface go continuously into one another. In the branch points the two sheets hang together; from the z-sphere one sees that for $w = \sqrt{z}$ both $z = 0$ and $z = \infty$ are branch points.

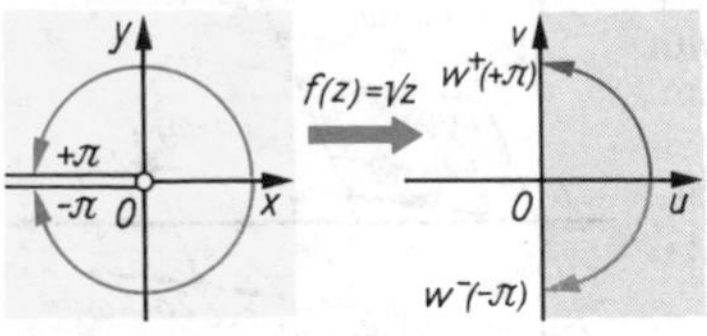

23.3-5 The values $w^+(\pi)$ and $w^-(\pi)$ of the function $w = \sqrt{z}$ at the negative real axis

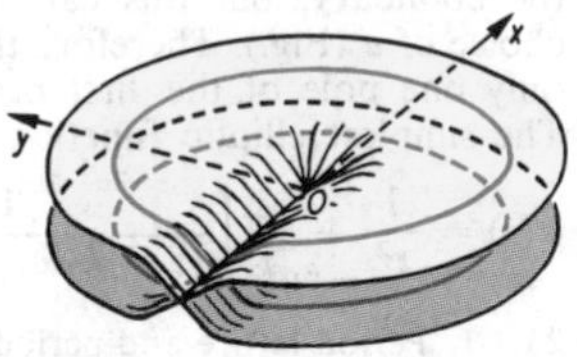

23.3-6 Riemann surface of $w = \sqrt{z}$

Uniformization. A Riemann surface $\boldsymbol{R}$ is a covering surface of the z-plane or the z-sphere. For $\boldsymbol{R}$ a further covering can be constructed, the *universal covering surface*. If one starts out from a definite point P_0 on $\boldsymbol{R}$, all the points P of the universal covering surface are to be the end-points of all possible curves beginning at P_0; but two curves γ_1 and γ_2 are to lead to the same point P only when they can be carried continuously into one another within $\boldsymbol{R}$; in other words, if there is a curve going from P_0 to P that cannot be carried continuously into γ_1 or γ_2, then it defines another point of the universal covering surface. The example of the annulus shows that such curves $\bar{\gamma}$ can occur (Fig.).

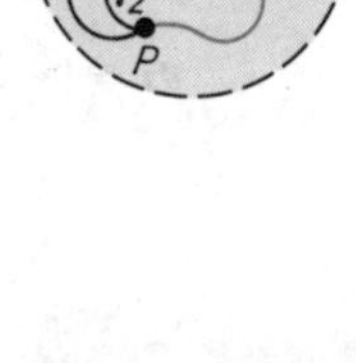

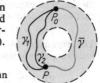

23.3-7 Two curves γ_1 and γ_2 define the same point of the universal covering surface if and only if they can be carried continuously into one another on the Riemann surface $\boldsymbol{R}$

It can then be shown that the universal covering surface is simply-connected and that the following generalization of Riemann's mapping theorem holds:

Generalized Riemann mapping theorem. The universal covering surface of every Riemann surface can be mapped one-to-one and conformally onto the interior of the unit disc, respectively, onto the whole complex plane or the Riemann sphere.

Since a Riemann surface is part of its universal covering surface, every Riemann surface can be put into one-to-one correspondence with a subset of the Riemann number sphere. This is called the possibility of *uniformizing* a Riemann surface; an example is the uniformization of the Riemann surface of the integrand of an elliptic integral (see Elliptic integrals).

Distribution of values. For a holomorphic function f statements can be made on the frequency with which certain values are assumed. A point z_0 is called a k-fold w_0-place of f if $f(z_0) = w_0$ and $f'(z_0) = 0, \dots, f^{(k-1)}(z_0) = 0$, but $f^{(k)}(z_0) \neq 0$. The following theorem holds:

When multiplicity is taken into account, a polynomial $p(z) = a_0 + a_1 z + \cdots + a_n z^n$ with $a_n \neq 0$ assumes every complex value w_0 at exactly n places (Fundamental theorem of algebra).

For example, the polynomial $p(z) = z^2$ assumes the value $w_0 = +1$ at $z = 1$ and $z = -1$; the value $w_0 = 0$ is assumed at $z = 0$ only, but with multiplicity 2, because $p'(z) = 2z$ and $p''(z) = 2$, so that $p'(0) = 0$ and $p''(0) \neq 0$. A stronger statement than that of the theorem of Casorati-Weierstrass is the theorem of Picard.

Theorem of Picard: A complex-valued function having an essential singularity at z_0 assumes in every neighbourhood of z_0 every complex value with at most one exception.

23.4. Elliptic integrals

The Weierstrass $\wp$-function. A function f defined for all z is said to be *periodic* of period ω if $f(z + \omega) = f(z)$ for all z. A function f is said to be *doubly-periodic* if there are two complex numbers ω_1, ω_2, for which ω_2/ω_1 is not real, such that all numbers $\omega = k_1\omega_1 + k_2\omega_2$ with arbitrary integers k_1, k_2 determine the set of all periods, that is, the *lattice of periods* with $f(z + \omega) = f(z)$ for all z.

Let a be any complex number. Then the numbers $a, a + \omega_1, a + \omega_1 + \omega_2, a + \omega_2$ are the vertices of a so-called *period parallelogramm*. A doubly-periodic function f assumes all its values within any period parallelogram. An *elliptic function* is a doubly-periodic meromorphic function. An elliptic function that is not a constant must have poles. But the sum of the residues inside any periodic parallelogram is always zero. For if the integral is taken around the boundary of such a parallelogram, then

$$\oint f(z)\,dz = \int_a^{a+\omega_1} f(z)\,dz + \int_{a+\omega_1}^{a+\omega_1+\omega_2} f(z)\,dz + \int_{a+\omega_1+\omega_2}^{a+\omega_2} f(z)\,dz + \int_{a+\omega_2}^{a} f(z)\,dz = 0;$$

for the substitution of $z + \omega$ for z in the third integral shows that its value is the negative of that of the first, and similarly for the other pair. Here it is assumed that there are no poles on the boundary, but this can always be achieved by a suitable choice of a (Fig.). Therefore there are no elliptic integrals with only one pole of the first order in the period parallelogram. The simplest elliptic function is the Weierstrass $\wp$-function

$$\wp(z) = \frac{1}{z^2} + \sum_{k_1, k_2}{}' \left[\frac{1}{(z - k_1\omega_1 - k_2\omega_2)^2} - \frac{1}{(k_1\omega_1 + k_2\omega_2)^2}\right];$$

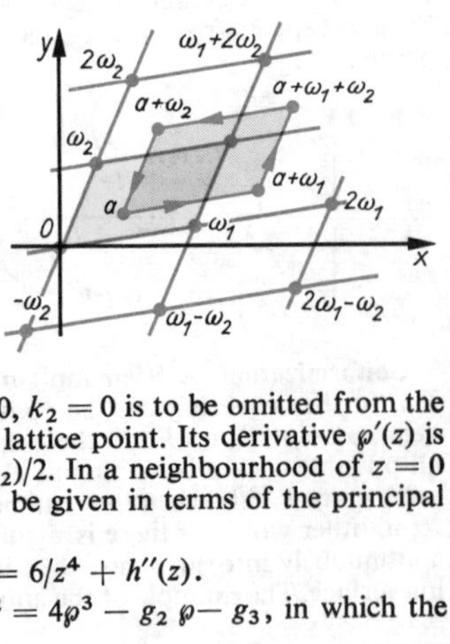

23.4-1 Period lattice and period parallelogram with the vertex a

here the dash at the symbol $\sum$ indicates that the term with $k_1 = 0$, $k_2 = 0$ is to be omitted from the summation. It has a pole of order 2 with the residue 0 at every lattice point. Its derivative $\wp'(z)$ is also elliptic and has zeros of order 1 at $\omega_1/2$, $\omega_2/2$, and $(\omega_1 + \omega_2)/2$. In a neighbourhood of $z = 0$ the Laurent expansions of the $\wp$-function and its derivatives can be given in terms of the principal part and a holomorphic function h:

$$\wp(z) = 1/z^2 + h(z), \quad \wp'(z) = -2/z^3 + h'(z), \quad \wp''(z) = 6/z^4 + h''(z).$$

The expansion in a series leads to the differential equation $\wp'^2 = 4\wp^3 - g_2\wp - g_3$, in which the following abbreviations are used:

$$g_2 = 60 \sum_{k_1, k_2}{}' \frac{1}{(k_1\omega_1 + k_2\omega_2)^4} \quad \text{and} \quad g_3 = 140 \sum_{k_1, k_2}{}' \frac{1}{(k_1\omega_1 + k_2\omega_2)^6}.$$

The *inversion problem* for the $\wp$-function requires, for given values g_2 and g_3 with $g_2^3 - 27g_3^2 \neq 0$, to find a periodic lattice such that for the associated $\wp$-function the quantities g_2 and g_3 assume the prescribed values.

If at N places z_j, $j = 1, \ldots, N$, inside a period parallelogram the principal parts of the Laurent expansions are prescribed so that the residues satisfy $\sum_{j=1}^{N} a_{-1}^{(j)} = 0$, then apart from an additive constant there exists one and only one elliptic function f having these principal parts. Apart from this constant, it is a linear combination of the $\wp$-function, its derivatives, and the primitive ζ of $-\wp$. But this function ζ is not elliptic, because it has poles of order 1 at the lattice points and is otherwise holomorphic.

The Weierstrass normal form of an elliptic integral. In an elliptic integral the integrand rat (z, w) is a rational function of z and w; here w is the square root of a polynomial $p_4(z)$ of degree 4 or $p_3(z)$ of degree 3 in z, and the 4 or 3 zeros of these polynomials are simple, that is, distinct. In the z-plane the integrand is determined to within a factor ± 1, and only on the two-sheeted Riemann surface of $w = \sqrt{p(z)}$ it is uniquely determined. Its four branch points are the four zeros e_1, e_2, e_3, e_4 of $p_4(z)$, or the three zeros of $p_3(z)$ together with the point $z = \infty$. The path of integration γ is situated on this Riemann surface. By the substitution $z' = 1/(z - e_4)$ one reduces $p_4(z)$ to $p_3(z)$. By a translation one can achieve that the centre of gravity of the triangle formed from the remaining 3 zeros e_1, e_2, e_3 lies at $z = 0$ (Fig.). Then $e_1 + e_2 + e_3 = 0$, and by Vieta's root theorem, apart from a constant factor, $p(z) = 4z^3 + c_1 z + c_2$. This leads to the *Weierstrass normal form* of the elliptic integral. By solving the inversion problem of the $\wp$-function for these values $-c_1$ and $-c_2$, two periods ω_1 and ω_2 can be found in a $\tilde{z}$-plane such that the relevant period lattice determines a $\wp(\tilde{z})$-function for which $g_2 = -c_1$ and $g_3 = -c_2$. By $z = \wp(\tilde{z})$ every period parallelogram of the $\tilde{z}$-plane is mapped one to one onto a sheet, and the whole $\tilde{z}$-plane onto the universal covering surface of $w = \sqrt{p(z)}$. According to the differential equation for the $\wp$-function and since $z = \wp(\tilde{z})$ it follows that $\wp'^2 = 4\wp^3 - g_2\wp - g_3 = w^2$, that is, $\wp'(\tilde{z}) = w$. If $\tilde{\gamma}$ is the inverse image in the $\tilde{z}$-plane of the curve γ in the z-plane, then the elliptic integral in the $\tilde{z}$-plane is:

$$\int_\gamma \text{rat}\,(z, w)\,\mathrm{d}z = \int_{\tilde{\gamma}} \text{rat}\,(\wp, \wp')\,\wp'(\tilde{z})\,\mathrm{d}\tilde{z}.$$

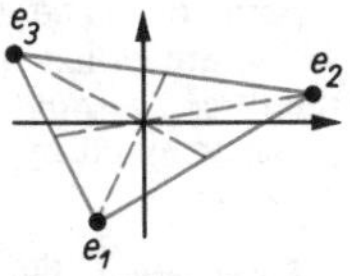

23.4-2 Zeros e_1, e_2, e_3 with $e_1 + e_2 + e_3 = 0$ of $\wp(z) = 4z^3 + c_1 z + c_2$

The integrand is said to be of the *a*) first, *b*) second, or *c*) third kind according as the integrand in the $\tilde{z}$-plane is an elliptic function *a*) without poles, *b*) having poles but such that their residues are all zero, *c*) otherwise. In case *a*) the integrand is therefore a constant in the $\tilde{z}$-plane, so that rat $(\wp, \wp') = \text{const}/\wp'$ and rat $(z, w) = \text{const}/w$. Since $w^2 = 4z^3 - g_2 z - g_3$, one obtains for the elliptic integrals:

Elliptic integrals: $\text{const} \displaystyle\int_\gamma \frac{\mathrm{d}z}{w} = \text{const} \int_\gamma \frac{\mathrm{d}z}{\sqrt{(4z^3 - g_2 z - g_3)}}$ first kind,

$\displaystyle\int_\gamma \frac{z\,\mathrm{d}z}{\sqrt{(4z^3 - g_2 z - g_3)}}$ second kind, $\displaystyle\int_\gamma \frac{\mathrm{d}z}{(z - z_0)\sqrt{(4z^3 - g_2 z - g_3)}}$ third kind.

The Legendre form of an elliptic integral arises when $w^2 = (1 - z^2)(1 - k^2 z^2)$ instead of $w^2 = 4z^3 - g_2 z - g_3$; k is called the *modulus*.

24. Analytic geometry of space

The essence of analytic geometry of space consists in setting up a correspondence between the points of the space and real numbers. Curves (1-dimensional manifolds) and surfaces (2-dimensional manifolds) then correspond to solution sets of equations, and geometrical constructions can be replaced by algebraic and analytic methods. Since these methods form the basis of analytic geometry, the subject did not arise until progress was made in algebra and analysis.

24.1. Coordinate systems

Rectangular coordinates

Setting up a system. Coordinate systems are the 'middlemen' between points and numbers. To set up a rectangular or Cartesian coordinate system in space, the first thing to do is to choose a point of space as the *origin*. Through this point three mutually perpendicular lines are drawn. These are called the *coordinate axes*, usually the x-, y-, and z-axis. The three coordinate axes span the three *coordinate planes* in space, the x,y-, x,z-, and y,z-plane. Any two axes divide the coordinate plane spanned by them into four quadrants, and the three coordinate planes divide space into eight *octants*.

Orientation. On each coordinate axis a unit vector is fixed: on the x-axis the vector $\boldsymbol{i}$, on the y-axis the vector $\boldsymbol{j}$, on the z-axis the vector $\boldsymbol{k}$. The coordinate axes are directed by these vectors, and the coordinate system is *oriented* (Fig.).

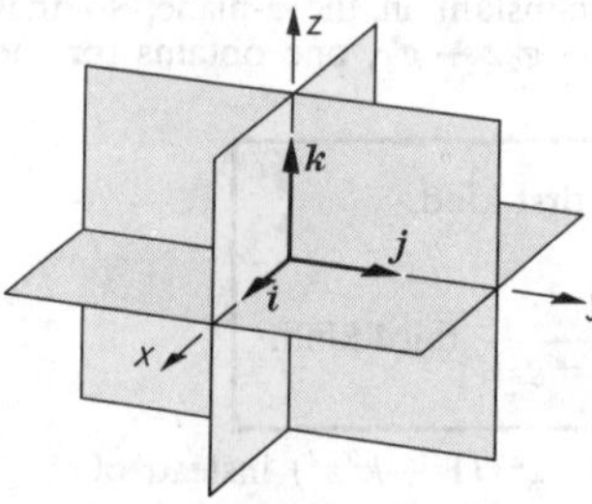

24.1-1 Rectangular coordinate system, right-oriented

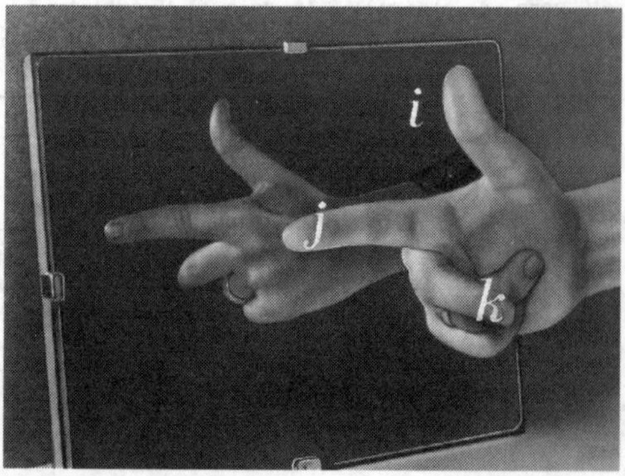

24.1-2 Orientation is reversed in a mirror

That part of each coordinate axis which contains the end-point of the unit vector from the origin on that axis is called the positive axis, the other one the negative. Any two positive axes bound a principal quadrant, and the three principal quadrants bound the principal octant.

One can always point the thumb, index finger, and middle finger of one hand in the directions of $\boldsymbol{i}$ (thumb), $\boldsymbol{j}$ (index finger), and $\boldsymbol{k}$ (middle finger). If this is done on the right hand, the system is called right-oriented or a *right-system*, otherwise a *left-system*. By reversal of one axis or by reflection in a plane a right-system goes into a left-system and vice versa (Fig.).

Points in space. If a rectangular coordinate system is chosen, then to every point of space there corresponds uniquely a triple of real numbers, and conversely, to every triple of real numbers there corresponds a unique point of space. The three numbers corresponding to a point of space are called the *rectangular* or *Cartesian coordinates* of the point. To determine the rectangular coordinates of a given point P one drops perpendiculars from P to each of the coordinate axes and measures the oriented lengths of the projections in units corresponding to the lengths of the fixed unit vectors. The values so obtained are the coordinates of P. To determine a point P from given coordinates x, y, z, one uses the vector notation. Starting from the origin, the vector $\boldsymbol{x} = x\boldsymbol{i} + y\boldsymbol{j} + z\boldsymbol{k}$ leads directly to the required point P (Fig.). Its distance from the origin can be calculated by means of the theorem of Pythagoras. It is $|\boldsymbol{x}| = \sqrt{(x^2 + y^2 + z^2)}$.

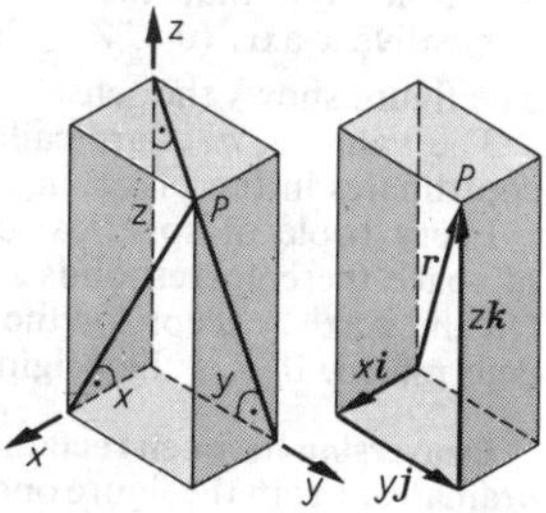

24.1-3 a) The rectangular coordinates of a point P in space, b) the vector $\boldsymbol{x} = x\boldsymbol{i} + y\boldsymbol{j} + z\boldsymbol{k}$ ends at P

Example: The rectangular coordinates $x = 3$, $y = 4$, $z = 12$ of a point P are given. This is written briefly as $P(3, 4, 12)$. The distance of this point from the origin is

$|\boldsymbol{x}| = \sqrt{(3^2 + 4^2 + 12^2)} = \sqrt{(9 + 16 + 144)} = \sqrt{169} = 13.$

The distance of P from the origin is 13 units.

Oblique coordinates

An oblique coordinate system is a generalization of a rectangular one. To set it up one takes any three lines through the origin that do not lie in one plane and prescribes a vector on each of these lines. Then all the results for a rectangular coordinate system hold word for word, with the following exceptions:

1. The numbers corresponding to a point of space are no longer called rectangular coordinates, but more generally *parallel coordinates*.
2. To determine the parallel coordinates of a given point P, one draws a parallelepiped whose edges are parallel to the coordinate axes and which has the origin and P as vertices. The oriented measures of the lengths of the edges lying along the coordinate axes are the parallel coordinates of P.
3. The vector $\boldsymbol{x} = x\boldsymbol{i} + y\boldsymbol{j} + z\boldsymbol{k}$, starting from the origin leads to the point P, as in a rectangular coordinate system, but as a rule $|\boldsymbol{x}| \neq \sqrt{(x^2 + y^2 + z^2)}$, since the theorem of Pythagoras does not hold in a general triangle.

Homogeneous coordinates

In *projective geometry* it is required that two lines in a plane always have a point of intersection, and also in space, if the lines are not skew. Hence an *improper point*, or *point at infinity*, is introduced as the 'point of intersection' of parallel lines. Analytic geometry, in the form established up to now, cannot cope with points at infinity. One possibility of satisfying the requirements of projective geometry and also of analytic geometry consists in the introduction of *homogeneous coordinates.*

If x', y', z' are the parallel coordinates of a point P in space, then numbers x, y, z, t determined by the equations $x' = x/t$, $y' = y/t$, $z' = z/t$ are called homogeneous coordinates of P. This quadruple is not uniquely determined. If x, y, z, t are homogeneous coordinates of a point of space, then for any real number $\varrho \neq 0$, $\varrho x, \varrho y, \varrho z, \varrho t$ are homogeneous coordinates of the same point. Conversely, to any homogeneous coordinates x, y, z, t with $t \neq 0$ there corresponds a unique triple of parallel coordinates. The reverse transformation results by simply going over to $x/t, y/t, z/t$.

Example. The point $P(2, 3, -1)$ has homogeneous coordinates $x = 2s$, $y = 3s$, $z = -s$, $t = s$ with an arbitrary $s \neq 0$.

Some problems are soluble in homogeneous coordinates, but not in parallel coordinates. For example, if $y' = ax' + b_1$ and $y' = ax' + b_2$ with $b_1 \neq b_2$ are two parallel lines in the x', y'-plane, then no point of intersection exists in parallel coordinates. In homogeneous coordinates the same lines have the equations $y = ax + b_1t$ and $y = ax + b_2t$.

This system of equations is soluble: in fact, it has infinitely many solutions $x = \varrho$, $y = a\varrho$, $t = 0$. For any $\varrho \neq 0$ the triple $\varrho, a\varrho, 0$ represents the homogeneous coordinates of a point of the x', y'-plane, a 'point at infinity', which is common to the two lines and characterizes their common direction.

Spherical coordinates

Setting up spherical coordinates. It is convenient for certain problems, for example, those concerned with the surface of a sphere, to introduce non-parallel coordinates. Instead of determining an arbitrary point P of space by rectangular coordinates x, y, z, it can also be determined by

1. the distance $r \geqslant 0$ of P from the origin O,
2. the angle φ that the segment OP makes with the x, y-plane $(-\pi/2 \leqslant \varphi \leqslant +\pi/2)$,

3. the angle λ that the projection OP' of the segment OP onto the x, y-plane makes with the positive x-axis ($0 \leqslant \lambda < 2\pi$).

The figure shows the sense of rotation of angle measurement.

The values r, φ, λ are called the *spherical coordinates* of the point P. They correspond to polar coordinates in the plane and are therefore also called *spatial polar coordinates.*

Every triple of spherical coordinates corresponds to exactly one point of space. To a point P of space there corresponds a unique triple of spherical coordinates if P does not lie on the z-axis. On the z-axis, except for the origin, only r and φ ($= \pm\pi/2$) are uniquely determined, and λ is undetermined. If P is the origin, only $r = 0$ is uniquely determined, and φ and λ are undetermined.

Conversion between rectangular and spherical coordinates. From the figure one obtains the relations

$x = |OP'| \cos\lambda, \quad y = |OP'| \sin\lambda, \quad |OP'| = r\cos\varphi.$

The rectangular coordinates of a point of space can therefore be calculated from the spherical coordinates by the adjacent formulae.

$$\begin{aligned} x &= r\cos\varphi\cos\lambda \\ y &= r\cos\varphi\sin\lambda \\ z &= r\sin\varphi \end{aligned}$$

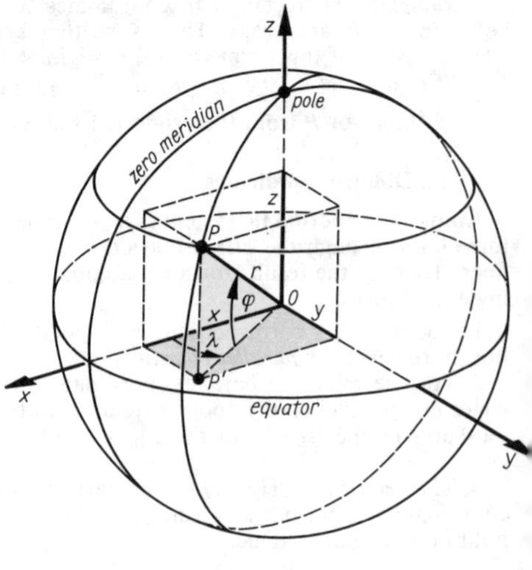

24.1-4 Spherical coordinates r, φ, λ of a point P in space

It follows that

$x^2 + y^2 + z^2 = r^2,$

$x/\sqrt{(x^2+y^2)} = \cos\lambda, \quad y/\sqrt{(x^2+y^2)} = \sin\lambda,$

$z/\sqrt{(x^2+y^2)} = \sin\varphi/\cos\varphi = \tan\varphi,$

$y/x = \sin\lambda/\cos\lambda = \tan\lambda.$

Hence the spherical coordinates of a point of space can be obtained from the rectangular coordinates by the formulae

$r = \sqrt{(x^2+y^2+z^2)},$

$\varphi = \operatorname{Arctan} z/\sqrt{(x^2+y^2)} \qquad (\text{for } x^2+y^2 \neq 0),$

$\lambda = \operatorname{Arctan}(y/x) \qquad (\text{for } x > 0, y > 0),$

$\lambda = \pi + \operatorname{Arctan}(y/x) \qquad (\text{for } x < 0),$

$\lambda = 2\pi + \operatorname{Arctan}(y/x) \qquad (\text{for } x > 0, y < 0).$

Furthermore,

$\varphi = \pi/2$ for $x = y = 0, z > 0$; $\qquad \lambda = \pi/2$ for $x = 0, y > 0$;

$\varphi = -\pi/2$ for $x = y = 0, z < 0$; $\qquad \lambda = 3\pi/2$ for $x = 0, y < 0$;

φ is undetermined for $x = y = 0, z = 0$; $\qquad \lambda$ is undetermined for $x = y = 0$.

Arctan, as always, is the principal value.

Example. What are the spherical coordinates of the point $P(3, -4, -12)$? –

$r = \sqrt{(3^2 + 4^2 + 12^2)} = 13;$

$\varphi = \operatorname{Arctan} -12/\sqrt{(3^2+4^2)} = \operatorname{Arctan}(-12/5) \approx -67.38°;$

$\lambda = 360° + \operatorname{Arctan}(-4/3) \approx 360° - 53.13° = 306.87°.$

The spherical coordinates of P are therefore $r = 13$, $\varphi \approx -67.38°$, and $\lambda \approx 306.87°$.

Cylindrical coordinates

For problems on the surface of a cylinder it is convenient to introduce *cylindrical coordinates* (Fig.). Starting from a rectangular coordinate system an arbitrary point P of space can be determined by

1. the distance $r \geqslant 0$ of P' from the origin O, where OP' is the projection of the segment OP onto the x, y-plane,
2. the angle φ that the segment OP' makes with the positive x-axis ($0 \leqslant \varphi < 2\pi$),
3. the oriented distance z of the point P from the x, y-plane ($-\infty < z < +\infty$).

To every triple of cylindrical coordinates there corresponds exactly one point of space. Again, to a point P of space there corresponds a unique triple of cylindrical coordinates if P does not lie on the z-axis. For points on the z-axis, $r = 0$ and z is determined, but φ is undetermined.

Cylindrical coordinates are often used in physics when cylindrically formed bodies are to be investigated, for example, in the calculation of the moment of inertia of a cylinder or in problems of heat conduction in cylindrical bodies.

The cylindrical coordinates r, φ, z of P coincide with the polar coordinates of the point P' in the x, y-plane and the rectangular z-coordinate of P. This gives the conversion formulae. Those for φ hold only if $x^2 + y^2 \neq 0$; φ is undetermined if $x = y = 0$.

$x = r \cos \varphi$ $y = r \sin \varphi$ $z = z$	$r = \sqrt{(x^2 + y^2)}$ $\cos \varphi = x/\sqrt{(x^2 + y^2)}$, $\sin \varphi = y/\sqrt{(x^2 + y^2)}$ $z = z$

Example. Given the cylindrical coordinates $r = 3$, $\varphi = -30°$, $z = 1$ of a point P, to find the rectangular coordinates.

$$x = 3 \cos (-30°) = 3 \cos 30° = (3/2) \sqrt{3} \approx 2.598,$$
$$y = 3 \sin (-30°) = -3 \sin 30° = -3/2 = -1.5,$$
$$z = 1.$$

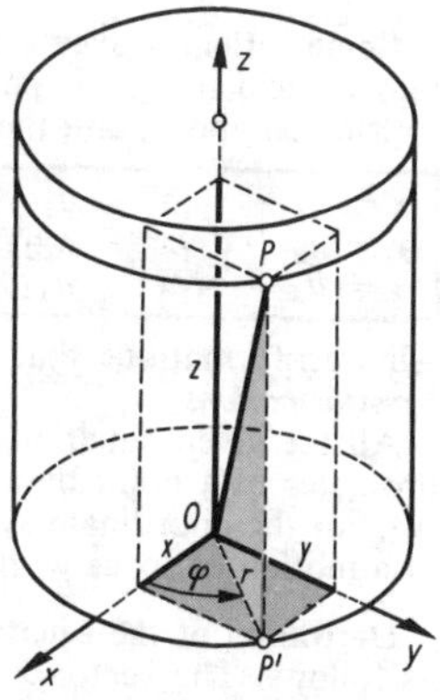

24.1-5 Cylindrical coordinates r, φ, z of a point P in space

Transformations of coordinates

If two coordinate systems are given in space (henceforth they will both be rectangular right-systems with the same unit of length) and they do not coincide, the problem often arises of calculating the coordinates x^*, y^*, z^* of a point P in one system from the coordinates x, y, z of P in the other system. Such a conversion of coordinates is called a *transformation of coordinates*. Three cases can be distinguished: translation, rotation and a combination of the two.

Translation. The two coordinate systems are so situated in space that one can be brought into coincidence with the other by means of a *parallel shift* (Fig.).

If the origin O^* of the second system has coordinates a_1, a_2, a_3 with respect to the first system with origin O, then the given relations hold between the coordinates x, y, z of a point P of space with respect to the first system and the coordinates x^*, y^*, z^* of P with respect to the second system.

$x = x^* + a_1$ $y = y^* + a_2$ $z = z^* + a_3$	$x^* = x - a_1$ $y^* = y - a_2$ $z^* = z - a_3$

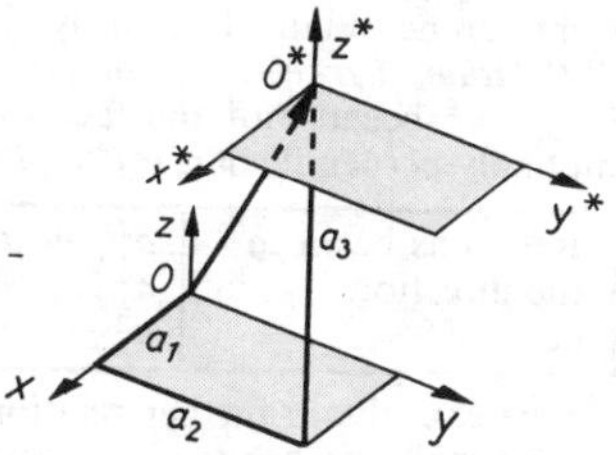

24.1-6 Translation of a coordinate system

Example. All points whose rectangular coordinates satisfy the equation $3x + 2y - z = 5$ lie in a plane in space. What is the equation of this plane with respect to a coordinate system whose origin has the coordinates $a_1 = -5$, $a_2 = 2$, $a_3 = 7$ with respect to the first system? – $x = x^* - 5$, $y = y^* + 2$, $z = z^* + 7$, so that $3x + 2y - z = 5$ goes into $3x^* + 2y^* - z^* = 5 + 15 - 4 + 7$. With respect to the new system the equation of the plane is $3x^* + 2y^* - z^* = 23$.

Rotation. The two coordinate systems have the same point of space as origin ($O^* = O$), but their axes have different directions.

In this case each axis of one system makes an angle with each axis of the other system. The cosines of these angles are denoted by a_{ik}, where i and k run through the values 1, 2 and 3. The first index always refers to the x, y, z-system and the second index to the x^*, y^*, z^*-system. The index 1 corresponds to the x- or x^*-axis, 2 to the y- or y^*-axis and 3 to the z- or z^*-axis; that is,

$$a_{11} = \cos (x, x^*) \quad a_{12} = \cos (x, y^*) \quad a_{13} = \cos (x, z^*)$$
$$a_{21} = \cos (y, x^*) \quad a_{22} = \cos (y, y^*) \quad a_{23} = \cos (y, z^*)$$
$$a_{31} = \cos (z, x^*) \quad a_{32} = \cos (z, y^*) \quad a_{33} = \cos (z, z^*).$$

The coordinates of an arbitrary point then transform according to the following equations. The a_{ik} are called *direction cosines*. The given equations of transformation are derived and further discussed below. It should be noted that the system of equations on the right contains the same coefficients as that on the left. Their positions are interchanged by a reflection in the main diagonal (top left to bottom right) of the coefficient matrix.

$x = a_{11}x^* + a_{12}y^* + a_{13}z^*$ $y = a_{21}x^* + a_{22}y^* + a_{23}z^*$ $z = a_{31}x^* + a_{32}y^* + a_{33}z^*$	$x^* = a_{11}x + a_{21}y + a_{31}z$ $y^* = a_{12}x + a_{22}y + a_{32}z$ $z^* = a_{13}x + a_{23}y + a_{33}z$

Combination. The two coordinate systems do not have the same origin and cannot be brought into coincidence by a parallel displacement alone. This case is a combination of the two cases considered above, and therefore leads to the following equations of transformation.

$x = a_1 + a_{11}x^* + a_{12}y^* + a_{13}z^*$	$x^* = a_{11}(x - a_1) + a_{21}(y - a_2) + a_{31}(z - a_3)$
$y = a_2 + a_{21}x^* + a_{22}y^* + a_{23}z^*$	$y^* = a_{12}(x - a_1) + a_{22}(y - a_2) + a_{32}(z - a_3)$
$z = a_3 + a_{31}x^* + a_{32}y^* + a_{33}z^*$	$z^* = a_{13}(x - a_1) + a_{23}(y - a_2) + a_{33}(z - a_3)$

All transformations that lead to a uniquely soluble system of linear equations are called *affine transformations.*

All the given equations of transformation can be interpreted as formulae for changing the coordinates of a point by a motion in the fixed space (translation, rotation or a combination of the two) of the coordinate system. However, they can also be regarded as the analytic representation of a motion of space with the coordinate system fixed.

Derivation of the equations of a rotation. The system of equations for a rotation can be derived as follows. The vector $\boldsymbol{x}$ of an arbitrary point P is given in the first system by $\boldsymbol{x} = x\boldsymbol{i} + y\boldsymbol{j} + z\boldsymbol{k}$ and in the second by $\boldsymbol{x} = x^*\boldsymbol{i}^* + y^*\boldsymbol{j}^* + z^*\boldsymbol{k}^*$. If one writes $\boldsymbol{x}$ in the first system in the form $\boldsymbol{x} = |\boldsymbol{x}|\ [(x/|\boldsymbol{x}|)\ \boldsymbol{i} + (y/|\boldsymbol{x}|)\ \boldsymbol{j} + (z/|\boldsymbol{x}|)\ \boldsymbol{k}]$ and first treats the special case where this vector is equal to $\boldsymbol{i}^*$ in the second system, that is, $x^* = 1$, $y^* = 0$, $z^* = 0$, then, by definition, $x/|\boldsymbol{x}| = a_{11}$, $y/|\boldsymbol{x}| = a_{21}$, $z/|\boldsymbol{x}| = a_{31}$.

Similar results are obtained from the special cases $\boldsymbol{x} = \boldsymbol{j}^*$ and $\boldsymbol{x} = \boldsymbol{k}^*$. This gives

$$\boldsymbol{i}^* = a_{11}\boldsymbol{i} + a_{21}\boldsymbol{j} + a_{31}\boldsymbol{k},\quad \boldsymbol{j}^* = a_{12}\boldsymbol{i} + a_{22}\boldsymbol{j} + a_{32}\boldsymbol{k},\quad \boldsymbol{k}^* = a_{13}\boldsymbol{i} + a_{23}\boldsymbol{j} + a_{33}\boldsymbol{k}.$$

If these expressions are substituted in $\boldsymbol{x} = x^*\boldsymbol{i}^* + y^*\boldsymbol{j}^* + z^*\boldsymbol{k}^*$ and the result is equated to $\boldsymbol{x} = x\boldsymbol{i} + y\boldsymbol{j} + z\boldsymbol{k}$, the system of equations on the left for a rotation is obtained. The system on the right can be obtained similarly.

Relations between the direction cosines. These relations are obtained from the expressions for $\boldsymbol{i}^*$, $\boldsymbol{j}^*$, $\boldsymbol{k}^*$ because of the fact that these vectors are unit vectors, $|\boldsymbol{i}^*| = |\boldsymbol{j}^*| = |\boldsymbol{k}^*| = 1$, and are mutually perpendicular, $\boldsymbol{i}^*\boldsymbol{j}^* = \boldsymbol{i}^*\boldsymbol{k}^* = \boldsymbol{j}^*\boldsymbol{k}^* = 0$.

Relations between the direction cosines	$a_{11}^2 + a_{21}^2 + a_{31}^2 = 1$	$a_{11}a_{12} + a_{21}a_{22} + a_{31}a_{32} = 0$
	$a_{12}^2 + a_{22}^2 + a_{32}^2 = 1$	$a_{11}a_{13} + a_{21}a_{23} + a_{31}a_{33} = 0$
	$a_{13}^2 + a_{23}^2 + a_{33}^2 = 1$	$a_{12}a_{13} + a_{22}a_{23} + a_{32}a_{33} = 0$

One can obtain further relations by taking into account the fact that $\boldsymbol{i}, \boldsymbol{j}, \boldsymbol{k}$ are unit vectors and are mutually perpendicular; however, it can be shown that there are only *six independent relations* between these direction cosines.

Since there are six independent equations connecting the nine direction cosines that characterize a rotation, a rotation can be completely described by means of *three quantities*. This was shown quite generally by CAYLEY. This is the basis for two particularly intuitive ways of characterizing a general rotation, namely by three angles (the Euler angles) or by an axis and one angle (Euler's theorem).

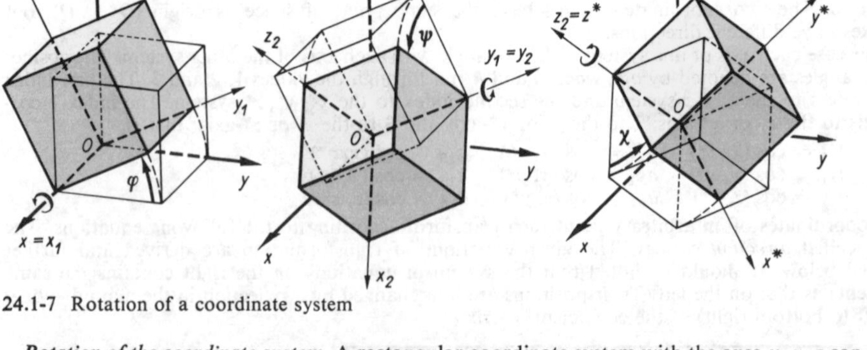

24.1-7 Rotation of a coordinate system

Rotation of the coordinate system. A rectangular coordinate system with the axes x, y, z can always be brought into coincidence with a second rectangular coordinate system with the same origin and axes x^*, y^*, z^* by first rotating about the x-axis through an angle φ, then about the y-axis through an angle ψ and finally about the z-axis through an angle χ (Fig.).

Relations between the direction cosines a_{ik} and the angles φ, ψ and χ			
a_{ik}	$k = 1$	$k = 2$	$k = 3$
$i = 1$	$\cos\psi\cos\chi$	$-\cos\psi\sin\chi$	$\sin\psi$
$i = 2$	$\cos\varphi\sin\chi + \sin\varphi\sin\psi\cos\chi$	$\cos\varphi\cos\chi - \sin\varphi\sin\psi\sin\chi$	$-\sin\varphi\cos\psi$
$i = 3$	$\sin\varphi\sin\chi - \cos\varphi\sin\psi\cos\chi$	$\sin\varphi\cos\chi + \cos\varphi\sin\psi\sin\chi$	$\cos\varphi\cos\psi$

Example of a rotation. On the surface of a sphere, whose centre is chosen as the origin of a rectangular coordinate system, there lies a point $P(-4, 8, -16)$. If the coordinate system is rotated anticlockwise about the x-axis through an angle $\varphi = 30°$, about the y-axis through an angle $\psi = 45°$ and about the z-axis through an angle $\chi = 60°$, one obtains the direction cosines by the given formulae and therefore the new coordinates of P.

$$a_{11} = \sqrt{2}/4, \quad a_{12} = -\sqrt{6}/4, \quad a_{13} = \sqrt{2}/2,$$
$$a_{21} = 3/4 + \sqrt{2}/8, \quad a_{22} = \sqrt{3}/4 - \sqrt{6}/8, \quad a_{23} = -\sqrt{2}/4,$$
$$a_{31} = \sqrt{3}/4 - \sqrt{6}/8, \quad a_{32} = 1/4 + 3\sqrt{2}/8, \quad a_{33} = \sqrt{6}/4,$$
$$x^* = 6 - 4\sqrt{3} + 2\sqrt{6} \approx 3.97, \quad y^* = -4 - 6\sqrt{2} + 2\sqrt{3} \approx -9.02,$$
$$z^* = -4\sqrt{2} - 4\sqrt{6} \approx -15.5.$$

The following theorem, which is given without proof, is of great importance in mechanics.

Euler's theorem. If two rectangular coordinate systems with the same origin and arbitrary directions of axes are given in space, one can always specify a line through the origin such that one coordinate system goes into the other by a rotation about this line.

Applied to a rigid body, Euler's theorem can be stated as follows:

For a rigid body, of which one point O is fixed relative to a system of reference, if a possible initial position is given and any other as final position, one can always specify an axis through O such that the body can be taken from the initial position to the final position by a rotation about this axis.

It is impossible to move a sphere whose centre is fixed in space so that at the end of the motion the position of all points on the surface of the sphere is different from their original position. In fact, either two or all the points must lie in their original positions.

Any (x, y, z)-coordinate system can be brought into coincidence with a second (x^*, y^*, z^*)-coordinate system with the same origin by rotation through angles ψ, φ, ϑ (Fig.), where k is the line of intersection of the x, y-plane and the x^*, y^*-plane. These angles are called the *Euler angles*.

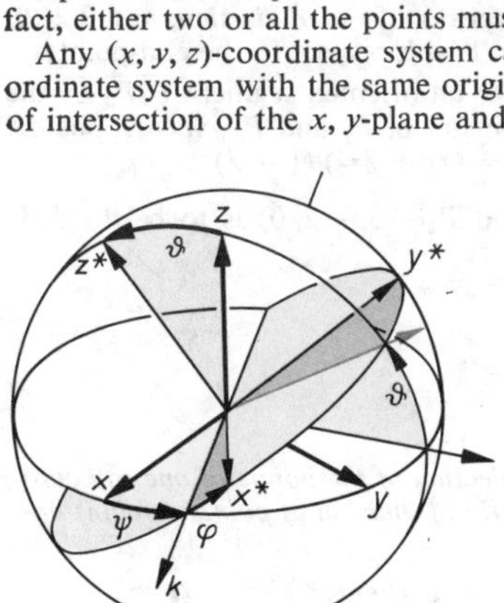

24.1-8 The Euler angles ψ, φ, ϑ

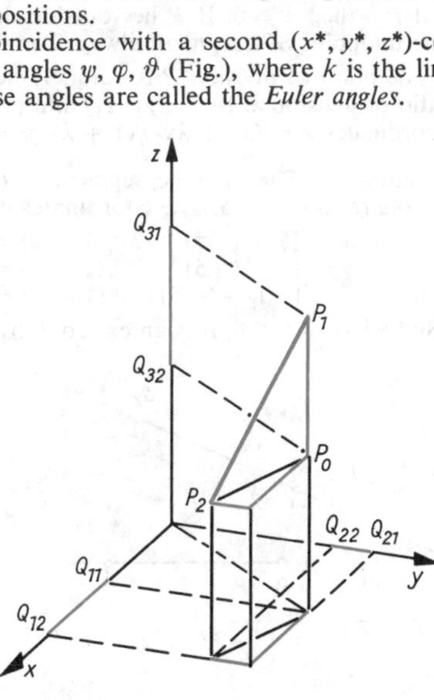

24.2-1 Components of a segment in space

24.2. Linear spaces

Segment

Length and orientation. The segment P_1P_2 on the line through two points P_1 and P_2 is the set of all points lying between P_1 and P_2 inclusive (see Chapter 7.). If the points lie on an oriented line, such as Q_{11} und Q_{12} on the x-axis of a spatial *Cartesian* coordinate system (Fig.), one regards the segment $Q_{11}Q_{12}$ as a directed quantity or vector, which is taken to be positive or negative according as its direction agrees or does not agree with the direction of the x-axis, in other words, one sets $|\overrightarrow{Q_{11}Q_{12}}| = x_2 - x_1$ and $|\overrightarrow{Q_{12}Q_{11}}| = x_1 - x_2$ so that $|\overrightarrow{Q_{11}Q_{12}}| = -|\overrightarrow{Q_{12}Q_{11}}|$. The length of the segment is denoted by $|Q_{11}Q_{12}|$.

If two points P_1 and P_2 in space are given by their Cartesian coordinates (x_1, y_1, z_1) and (x_2, y_2, z_2), then the three planes parallel to the coordinate planes through P_1 and P_2, respectively, cut off on the coordinate axes the segments $Q_{11}Q_{12}$, $Q_{21}Q_{22}$ and $Q_{31}Q_{32}$, which are called the *components* of the segment P_1P_2. If the segment is oriented as $\overrightarrow{P_1P_2}$, say, then its components are $\overrightarrow{Q_{11}Q_{12}}$, $\overrightarrow{Q_{21}Q_{22}}$ and $\overrightarrow{Q_{31}Q_{32}}$. In the right-angled triangle $P_0P_1P_2$ the length $|P_1P_2|$ of the segment can be calculated by Pythagoras' theorem.

Length of segment P_1P_2	$\lvert P_1P_2\rvert = \sqrt{[(x_2 - x_1)^2 + (y_2 - y_1)^2 + (z_2 - z_1)^2]}$

Example: The length $|P_1P_2|$ of the segment P_1P_2 between the points $P_1(5, 2, -1)$ and $P_2(-3, -2, 0)$ is calculated to be

$$|P_1P_2| = \sqrt{[(-3-5)^2 + (-2-2)^2 + (0+1)^2]} = \sqrt{(8^2 + 4^2 + 1^2)} = \sqrt{81} = 9.$$

The segment P_1P_2 is 9 units long.

The distances between three arbitrary points P_1, P_2 and P_3 in space satisfy the triangle in equality.

Triangle inequality	$\lvert P_1P_3\rvert \leqslant \lvert P_1P_2\rvert + \lvert P_2P_3\rvert$

The sign of a directed segment on an oriented line is obtained analytically as the scalar product of its vector with the unit vector on the oriented line: for example, $\overrightarrow{Q_{11}Q_{12}} \cdot \boldsymbol{i} = x_2 - x_1 > 0$ or $\overrightarrow{Q_{21}Q_{22}} \cdot \boldsymbol{j} = y_2 - y_1 < 0$. If the line determined by two points P_1 and P_2 has the orientation $\overrightarrow{P_1P_2}$, then its unit vector is $\boldsymbol{e} = \overrightarrow{P_1P_2}/|P_1P_2|$. If Q_1 and Q_2 are two arbitrary points of a line oriented by $\overrightarrow{P_1P_2}$, then an *oriented distance* $|\overrightarrow{Q_1Q_2}|$ is fixed by the following agreement: $|\overrightarrow{Q_1Q_2}| = +|Q_1Q_2|$ if the orientation $\overrightarrow{Q_1Q_2}$ corresponds to that of $\overrightarrow{P_1P_2}$, but $|\overrightarrow{Q_1Q_2}| = -|Q_1Q_2|$ if the orientation $\overrightarrow{Q_1Q_2}$ corresponds to that of $\overrightarrow{P_2P_1}$.

Division. Suppose that a given line is oriented by $\overrightarrow{P_1P_2}$ and that P is an arbitrary point on it other than P_2. One says that P divides the oriented segment $\overrightarrow{P_1P_2}$ in the ratio $\lambda = |\overrightarrow{P_1P}| : |\overrightarrow{PP_2}|$ and calls λ the *ratio of division.* In particular, one speaks of *inner division* when P lies between P_1 and P_2: then $\lambda > 0$. If P lies outside the segment $\overrightarrow{P_1P_2}$, one speaks of *outer division,* and $\lambda < 0$. The midpoint of a segment always has the ratio of division $\lambda = 1$ with respect to the end-points.

The point of division P is uniquely determined with respect to an oriented segment $\overrightarrow{P_1P_2}$ by the ratio of division λ. If (x_1, y_1, z_1) and (x_2, y_2, z_2) are the coordinates of P_1 and P_2, then P has the coordinates $x = (x_1 + \lambda x_2)/(1 + \lambda)$, $y = (y_1 + \lambda y_2)/(1 + \lambda)$, $z = (z_1 + \lambda z_2)/(1 + \lambda)$.

Example: The oriented segment $\overrightarrow{P_1P_2}$ with $P_1(5, 2, -1)$ and $P_2(-3, -2, 0)$ is to be divided in the ratio $\lambda = -5$. The coordinates of the point of division are

$$x = [5 + (-5)(-3)]/(1 - 5) = 20/(-4) = -5,$$
$$y = [2 + (-5)(-2)]/(1 - 5) = 12/(-4) = -3,$$
$$z = [-1 + (-5)(0)]/(1 - 5) = -1/(-4) = 1/4.$$

Since $\lambda = -5 < 0$, it is an external division.

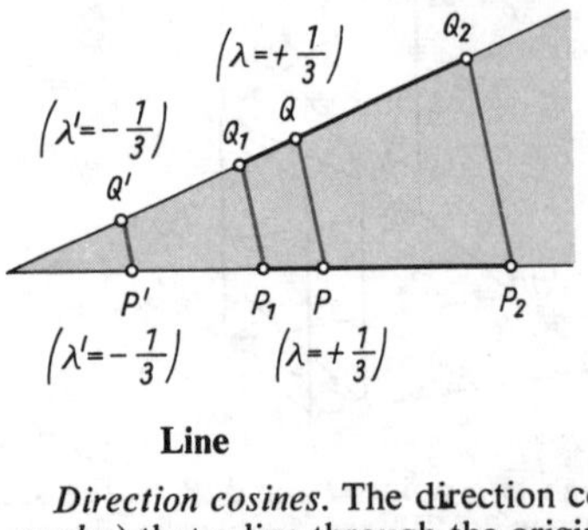

Under parallel projection of the points of one ray onto any other ray the ratio of division of points remains unchanged (Fig.).

24.2-2 Ratio of division remains unchanged by a parallel projection

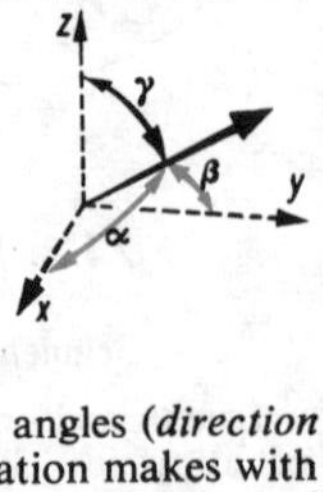

24.2-3 The direction angles of a line

Line

Direction cosines. The direction cosines of an oriented line are the cosines of the angles (*direction angles*) that a line through the origin parallel to the given line with the same orientation makes with the positive coordinate axes (Fig.). These three angles are two-valued, according to the sense of rotation, but their cosines are uniquely determined, since $\cos\alpha = \cos(2\pi - \alpha)$.

Relations. If an oriented line with the direction angles α, β and γ (with respect to the x-, y- and z-axes) passes through a point P_1 with the coordinates (x_1, y_1, z_1), and if $P(x, y, z)$ is an arbitrary point of the line, then $x = x_1 + |\overrightarrow{P_1P}| \cos\alpha$, $y = y_1 + |\overrightarrow{P_1P}| \cos\beta$, $z = z_1 + |\overrightarrow{P_1P}| \cos\gamma$. The proof is simple. One first takes a rectangular coordinate system with P_1 as origin and then carries out a translation. It follows that $(x - x_1)^2 + (y - y_1)^2 + (z - z_1)^2 = |P_1P_2|^2 (\cos^2\alpha + \cos^2\beta + \cos^2\gamma)$. By taking into account the standard formula for $|P_1P_2|$ one obtains the fundamental relation between the direction cosines of an oriented line.

$$\cos^2\alpha + \cos^2\beta + \cos^2\gamma = 1$$

Conversely, any three numbers a, b, c for which $a^2 + b^2 + c^2 = 1$ can be regarded as the cosines of the direction angles of an oriented line in space.

Calculation of the direction cosines from two given points. If P_1 and P_2 are two points of space with the coordinates (x_1, y_1, z_1) and (x_2, y_2, z_2), then the cosines of the direction angles α, β and γ of the oriented line that goes from P_1 to P_2 are given by the following formulae.

$$\cos\alpha = (x_2 - x_1)/|P_1P_2|, \quad \cos\beta = (y_2 - y_1)/|P_1P_2|, \quad \cos\gamma = (z_2 - z_1)/|P_1P_2|$$
$$\text{with } |P_1P_2| = \sqrt{[(x_2 - x_1)^2 + (y_2 - y_1)^2 + (z_2 - z_1)^2]}$$

Example: To determine the direction cosines of the oriented line that goes from $P_1(5, 2, -1)$ to $P_2(-3, -2, 0)$. One has $\sqrt{[(x_2 - x_1)^2 + (y_2 - y_1)^2 + (z_2 - z_1)^2]} = \sqrt{(8^2 + 4^2 + 1^2)} = 9$; hence $\cos\alpha = (-3 - 5)/9 = -8/9$, $\cos\beta = (-2 - 2)/9 = -4/9$, $\cos\gamma = [0 - (-1)]/9 = 1/9$. As a check one can verify that the sum of the squares of the cosines is equal to 1.

Equations of a line. By introducing the vectors $\boldsymbol{x} = x\boldsymbol{i} + y\boldsymbol{j} + z\boldsymbol{k}$, $\boldsymbol{x}_1 = x_1\boldsymbol{i} + y_1\boldsymbol{j} + z_1\boldsymbol{k}$, $\boldsymbol{e} = \cos\alpha\, \boldsymbol{i} + \cos\beta\, \boldsymbol{j} + \cos\gamma\, \boldsymbol{k}$ the equations $x = x_1 + |\overrightarrow{P_1P}| \cos\alpha$, $y = y_1 + |\overrightarrow{P_1P}| \cos\beta$, $z = z_1 + |\overrightarrow{P_1P}| \cos\gamma$ can be put into the simple form $\boldsymbol{x} = \boldsymbol{x}_1 + |\overrightarrow{P_1P}|\, \boldsymbol{e}$. The vector $\boldsymbol{e}$ is called a *direction vector* and is a unit vector. Occasionally a multiple of $\boldsymbol{e}$ is taken instead of $\boldsymbol{e}$. Then, instead of $\boldsymbol{e}$, the letter $\boldsymbol{a}$ will be used.

If one writes generally $\boldsymbol{x} = \boldsymbol{x}_1 + t\boldsymbol{a}$, then for given $\boldsymbol{x}_1$ and $\boldsymbol{a}$ to any real number t there corresponds a vector $\boldsymbol{x}$, which goes from the origin to a point of the line. As t runs through all numbers from $-\infty$ to $+\infty$, one obtains all the points of the line. Conversely, to any point of the line there corresponds a number t such that the vector $\boldsymbol{x} = \boldsymbol{x}_1 + t\boldsymbol{a}$ ends at the point. Therefore $\boldsymbol{x} = \boldsymbol{x}_1 + t\boldsymbol{a}$ is called the *point-direction equation* of the line or, since t is called a parameter, a parametric representation of the line (Fig.).

Parametric representation of the line	$\boldsymbol{x} = \boldsymbol{x}_1 + t\boldsymbol{a}$

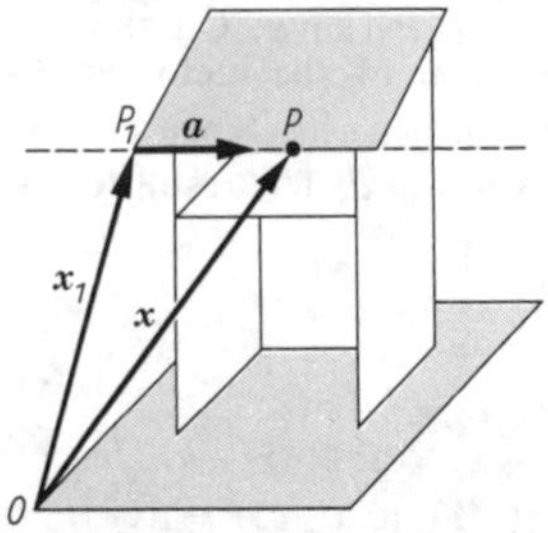

For a parametric representation of a line it is important only that to each value of the parameter in $(-\infty, +\infty)$ there corresponds a unique point and conversely, and not in which sense the line is described as t goes from $-\infty$ to $+\infty$. Hence the sense of direction of $\boldsymbol{a}$ plays no part, and $\boldsymbol{a}$ need not be a unit vector.

24.2-4 The point-direction equation of a line

Example: The point-direction equation of the line in the previous example is to be found. Hence P_1 and the direction cosines are to be regarded as given

$$\boldsymbol{x}_1 = 5\boldsymbol{i} + 2\boldsymbol{j} - \boldsymbol{k} \quad \text{and} \quad \boldsymbol{e} = -(1/9)(8\boldsymbol{i} + 4\boldsymbol{j} - \boldsymbol{k}).$$

The point-direction equation is therefore

$$\boldsymbol{x} = (5\boldsymbol{i} + 2\boldsymbol{j} - \boldsymbol{k}) - (t/9)(8\boldsymbol{i} + 4\boldsymbol{j} - \boldsymbol{k}).$$

If a new parameter $u = -t/9$ is taken, then the equation is

$$\boldsymbol{x} = (5\boldsymbol{i} + 2\boldsymbol{j} - \boldsymbol{k}) + u(8\boldsymbol{i} + 4\boldsymbol{j} - \boldsymbol{k}).$$

It is still a parametric representation of the given line, but its direction vector is no longer a unit vector and it no longer has the orientation of $\boldsymbol{e}$.

Two points are given. If P_1 and P_2 with the coordinates (x_1, y_1, z_1) and (x_2, y_2, z_2) are two given points of a line, let $\boldsymbol{x}_1 = x_1\boldsymbol{i} + y_1\boldsymbol{k} + z_1\boldsymbol{k}$ and $\boldsymbol{x}_2 = x_2\boldsymbol{i} + y_2\boldsymbol{j} + z_2\boldsymbol{k}$. As direction vector one can take $\boldsymbol{a} = \boldsymbol{x}_2 - \boldsymbol{x}_1$. If this expression is substituted for $\boldsymbol{a}$ in the point-direction equation, one obtains a *two-point equation* of the line.

Two-point equation of the line	$\boldsymbol{x} = \boldsymbol{x}_1 + t(\boldsymbol{x}_2 - \boldsymbol{x}_1)$

Example: To find a two-point equation of the line through $P_1(5, 2, -1)$ and $P_2(-3, -2, 0)$. From $\boldsymbol{x}_1 = 5\boldsymbol{i} + 2\boldsymbol{j} - \boldsymbol{k}$ and $\boldsymbol{x}_2 = -3\boldsymbol{i} - 2\boldsymbol{j}$ it follows that $\boldsymbol{x}_2 - \boldsymbol{x}_1 = -8\boldsymbol{i} - 4\boldsymbol{j} + \boldsymbol{k}$. A two-point equation is $\boldsymbol{x} = (5\boldsymbol{i} + 2\boldsymbol{j} - \boldsymbol{k}) + t(-8\boldsymbol{i} - 4\boldsymbol{j} + \boldsymbol{k})$ or, by choosing a new parameter, $u = -t$, $\boldsymbol{x} = (5\boldsymbol{i} + 2\boldsymbol{j} - \boldsymbol{k}) + u(8\boldsymbol{i} + 4\boldsymbol{j} - \boldsymbol{k})$.

Basic geometric problems. Some formulae are now derived that help to solve the most important geometric problems.

Angle between two lines. One says that two lines oriented by their direction vectors $\boldsymbol{a}$ and $\boldsymbol{a}^*$ enclose an angle φ if the lines through the origin parallel to them and with the same orientation enclose this angle. From the definition of the inner product (scalar product), $\boldsymbol{a} \cdot \boldsymbol{a}^* = |\boldsymbol{a}|\,|\boldsymbol{a}^*| \cos\varphi$. Two lines are perpendicular to one another if $\boldsymbol{a} \cdot \boldsymbol{a}^* = 0$. Since $\boldsymbol{e} = \boldsymbol{a}/|\boldsymbol{a}|$ and $\boldsymbol{e}^* = \boldsymbol{a}^*/|\boldsymbol{a}^*|$, one obtains $\cos\varphi = \boldsymbol{e} \cdot \boldsymbol{e}^*$.

$\cos\varphi = \boldsymbol{e} \cdot \boldsymbol{e}^*$

Example: Two oriented lines are given by $\boldsymbol{x} = (2\boldsymbol{i} - 3\boldsymbol{j} + 4\boldsymbol{k}) + t(3\boldsymbol{i} - 4\boldsymbol{j} + 12\boldsymbol{k})$ and $\boldsymbol{x}^* = (\boldsymbol{i} + 5\boldsymbol{j} - 3\boldsymbol{k}) + t^*(4\boldsymbol{i} + 3\boldsymbol{k})$. What angle $\varphi < \pi$ do they enclose if their orientations correspond to the given direction vectors? – The given direction vectors must first be normalized. This gives $\boldsymbol{e} = (3\boldsymbol{i} - 4\boldsymbol{j} + 12\boldsymbol{k})/\sqrt{(9 + 16 + 144)} = (1/13)(3\boldsymbol{i} - 4\boldsymbol{j} + 12\boldsymbol{k})$
and $\quad \boldsymbol{e}^* = (4\boldsymbol{i} + 3\boldsymbol{k})/\sqrt{(16 + 9)} = (1/5)(4\boldsymbol{i} + 3\boldsymbol{k})$ and so

$$\cos\varphi = [1/(5 \times 13)](3\boldsymbol{i} - 4\boldsymbol{j} + 12\boldsymbol{k}) \cdot (4\boldsymbol{i} + 3\boldsymbol{k}) = [1/(5 \times 13)](3 \times 4 - 4 \times 0 + 12 \times 3)$$
$$= 48/65 \approx 0.738\ldots$$

The angle enclosed by the given lines is therefore $\varphi \approx 42.4°$. This is not to say that the lines intersect!

Distance of a point from a line. If $\boldsymbol{x} = \boldsymbol{x}_1 + t\boldsymbol{e}$ is the equation of a given line and (x_2, y_2, z_2) the coordinates of a given point P_2, then $\overrightarrow{OP_2} = \boldsymbol{x}_2$ is determined, and $d = |(\boldsymbol{x}_2 - \boldsymbol{x}_1) \times \boldsymbol{e}|$ is the distance of the point P_2 from the given line, that is, the length of the perpendicular from P_2 to the given line (Fig.). In dealing with the line, the *distance* is regarded as a non-negative number, in accordance with the elementary notion; the concept of *oriented distance* will be taken up in dealing with the plane.

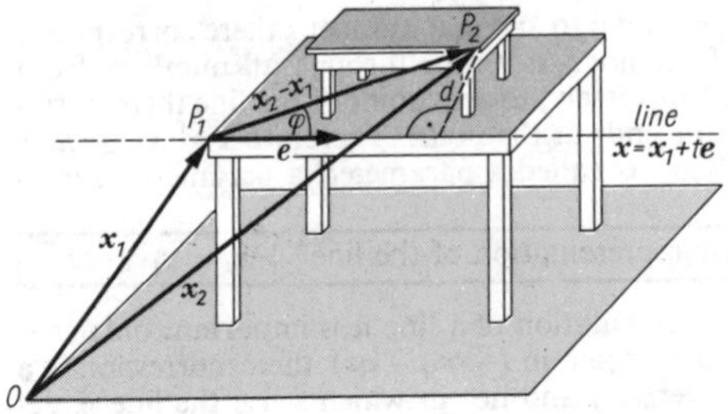

24.2-5 Distance of a point from a line

Point-line distance	$d = \lvert(\boldsymbol{x}_2 - \boldsymbol{x}_1) \times \boldsymbol{e}\rvert$

Proof. Let $\varphi \leqslant \pi$ be the angle enclosed by $\boldsymbol{e}$ and $\overrightarrow{P_1P_2}$. Then $d = |P_1P_2| \sin\varphi$. On the other hand, from the definition of the vector product $|\overrightarrow{P_1P_2} \times \boldsymbol{e}| = |P_1P_2| \cdot |\boldsymbol{e}| \sin\varphi$. Since $|\boldsymbol{e}| = 1$, $|\overrightarrow{P_1P_2} \times \boldsymbol{e}| = |P_1P_2| \sin\varphi = d$. The result follows, because $\overrightarrow{P_1P_2} = \boldsymbol{x}_2 - \boldsymbol{x}_1$.

Example: To find the distance of the point $P_2(3, 1, 5)$ from the line

$$\boldsymbol{x} = (2\boldsymbol{i} - 3\boldsymbol{j} + 4\boldsymbol{k}) + (t/13)(3\boldsymbol{i} - 4\boldsymbol{j} + 12\boldsymbol{k}).$$

Since $\boldsymbol{x}_1 = 2\boldsymbol{i} - 3\boldsymbol{j} + 4\boldsymbol{k}$ and $\boldsymbol{x}_2 = 3\boldsymbol{i} + \boldsymbol{j} + 5\boldsymbol{k}$, it follows that $\boldsymbol{x}_2 - \boldsymbol{x}_1 = \boldsymbol{i} + 4\boldsymbol{j} + \boldsymbol{k}$. Hence

$$(\boldsymbol{x}_2 - \boldsymbol{x}_1) \times \boldsymbol{e} = (\boldsymbol{i} + 4\boldsymbol{j} + \boldsymbol{k}) \times (1/13)(3\boldsymbol{i} - 4\boldsymbol{j} + 12\boldsymbol{k}) = (1/13)(52\boldsymbol{i} - 9\boldsymbol{j} - 16\boldsymbol{k}).$$

The magnitude of this vector is the required distance

$$d = (1/13)\sqrt{(52^2 + 9^2 + 16^2)} = (1/13)\sqrt{(2704 + 81 + 256)} = (1/13)\sqrt{3041} \approx 4.24.$$

The point P_2 is at a distance of approximately 4.24 units from the given line.

Distance between two skew lines. Two lines that have no point in common and are not parallel are called *skew*. If l and l^* are two skew lines, there is always exactly one point Q on l and exactly one point Q^* on l^* such that the vector $\overrightarrow{QQ^*}$ is perpendicular to both lines (see Chapter 9.). The length of this vector is the shortest distance that any two points of l and l^* can have from one another. It is called the distance between the two lines.

The distance d can be calculated from the equation $\boldsymbol{x} = \boldsymbol{x}_1 + t\boldsymbol{a}$ of l and the equation $\boldsymbol{x}^* = \boldsymbol{x}_1^* + \tau\boldsymbol{a}^*$ of l^*. There is a parameter t_1 such that $\boldsymbol{x}_1 + t_1\boldsymbol{a} = \overrightarrow{OQ}$ and a parameter τ_1 such that $\boldsymbol{x}_1^* + \tau_1\boldsymbol{a}^* = \overrightarrow{OQ^*}$. Since $\overrightarrow{QQ^*}$ is perpendicular to $\boldsymbol{a}$ and $\boldsymbol{a}^*$, $\overrightarrow{QQ^*} = d \cdot (\boldsymbol{a} \times \boldsymbol{a}^*)/|\boldsymbol{a} \times \boldsymbol{a}^*|$. If this expression is substituted in $\overrightarrow{OQ^*} = \overrightarrow{OQ} + \overrightarrow{QQ^*}$ and the scalar product of both sides with $(\boldsymbol{a} \times \boldsymbol{a}^*)$ is taken, then the solution for d is the distance d between the skew lines l and l^*.

Distance between two skew lines	$d = \lvert(\boldsymbol{x}_1 - \boldsymbol{x}_1^*) \cdot (\boldsymbol{a} \times \boldsymbol{a}^*)\rvert / \lvert(\boldsymbol{a} \times \boldsymbol{a}^*)\rvert$

Example: For the lines $\boldsymbol{x} = (\boldsymbol{i}+\boldsymbol{j}+\boldsymbol{k}) + t(\boldsymbol{i}-\boldsymbol{j}+\boldsymbol{k})$ and $\boldsymbol{x}^* = (\boldsymbol{i}-\boldsymbol{j}+\boldsymbol{k}) + \tau(-\boldsymbol{i}+\boldsymbol{j}+\boldsymbol{k})$, $\boldsymbol{x}_1 - \boldsymbol{x}_1^* = 2\boldsymbol{j}$ and $\boldsymbol{a} \times \boldsymbol{a}^* = -2\boldsymbol{i} - 2\boldsymbol{j}$ and so the distance d is given by

$$d = |2\boldsymbol{j} \cdot (-2\boldsymbol{i} - 2\boldsymbol{j})|/\sqrt{(2^2+2^2)} = 4/\sqrt{8} = \sqrt{2} \approx 1.414.$$

The distance between the lines is approximately 1.414 units.

Intersection of two lines. In space two lines generally have no point in common. For if l and l^* are two lines with the equations $\boldsymbol{x} = \boldsymbol{x}(t) = \boldsymbol{x}_1 + t\boldsymbol{a}$ and $\boldsymbol{x}^* = \boldsymbol{x}^*(\tau) = \boldsymbol{x}_1^* + \tau\boldsymbol{a}^*$ and they have (at least) one common point, then there must be (at least) one pair of values t, τ such that $\boldsymbol{x}(t) = \boldsymbol{x}^*(\tau)$. To this vector equation there corresponds a system of three linear equations for the two unknowns, which is not generally soluble.

The adjacent conditions are necessary and sufficient for the existence of a unique solution, that is, for two lines in space to have exactly one point in common.

$$\boldsymbol{a} \times \boldsymbol{a}^* \neq \boldsymbol{o} \quad \text{and} \quad (\boldsymbol{x}_1 - \boldsymbol{x}_1^*) \cdot (\boldsymbol{a} \times \boldsymbol{a}^*) = 0$$

The first condition says that the lines are not parallel and therefore cannot coincide, and the second follows immediately from the formula for the distance between two skew lines, since two intersecting lines must have zero distance. If there exist two parameters t and τ as the unique solution of the system of equations, then these substituted into the equations of the lines give the *point of intersection* of l and l^*. Otherwise there is either no point of intersection (no solution) or l and l^* coincide (infinitely many solutions).

Example: For the lines $\boldsymbol{x} = (2\boldsymbol{i} - 3\boldsymbol{j} + 4\boldsymbol{k}) + t(3\boldsymbol{i} - 4\boldsymbol{j} + 12\boldsymbol{k})$ and $\boldsymbol{x}^* = (\boldsymbol{i} + 5\boldsymbol{j} - 3\boldsymbol{k}) + \tau(36\boldsymbol{i} + 212\boldsymbol{j} + 27\boldsymbol{k})$ the distance is found to be $d = 0$. Since $\boldsymbol{a} \times \boldsymbol{a}^* \neq \boldsymbol{o}$, the lines have exactly one point of intersection. There must then be one t and one τ such that

$$(2\boldsymbol{i} - 3\boldsymbol{j} + 4\boldsymbol{k}) + t(3\boldsymbol{i} - 4\boldsymbol{j} + 12\boldsymbol{k}) = (\boldsymbol{i} + 5\boldsymbol{j} - 3\boldsymbol{k}) + \tau(36\boldsymbol{i} + 212\boldsymbol{j} + 27\boldsymbol{k}),$$

hence $(1 + 3t - 36\tau)\,\boldsymbol{i} - (8 + 4t + 212\tau)\,\boldsymbol{j} + (7 + 12t - 27\tau)\,\boldsymbol{k} = \boldsymbol{o}$. Since a vector is the null vector only if all its components vanish, the adjacent system of equations is obtained, which has the unique solution $t = -25/39$, $\tau = -1/39$. If these parameter values are substituted in the equations of the lines, then $\boldsymbol{x} = \boldsymbol{x}^*$ and this vector ends at the point of intersection of the lines. This gives $\boldsymbol{x} = \boldsymbol{x}^* = (1/39)(3\boldsymbol{i} - 17\boldsymbol{j} - 144\boldsymbol{k})$. The coordinates of the point of intersection are $x \approx 0.077$, $y \approx -0.436$, $z \approx -3.692$.

$$\begin{aligned} 3t - 36\tau &= -1, \\ 4t + 212\tau &= -8, \\ 12t - 27\tau &= -7. \end{aligned}$$

A system of lines that pass through a fixed point is called a *bundle of lines*. If, in addition, the lines all lie in one plane, one speaks of a *pencil of lines* or, for oriented half-lines, a *pencil of rays*.

Plane

Equations of a plane. A plane can be fixed in space by three points that do not lie on a line or by two points and a direction vector not parallel to the line joining the points or by a point and two non-parallel direction vectors.

Parametric representation. Suppose that a point P_1 with coordinates (x_1, y_1, z_1) and two nonparallel direction vectors $\boldsymbol{a}$ and $\boldsymbol{a}^*$ are given. If the vector $\overrightarrow{OP_1}$ is denoted by $\boldsymbol{x}_1$, then $\boldsymbol{x}^* = \boldsymbol{x}_1 + t\boldsymbol{a}$ ends at a point P^* on the line determined by $\boldsymbol{x}_1$ and $\boldsymbol{a}$. The vector $\boldsymbol{x} = \boldsymbol{x}^* + \tau\boldsymbol{a}^*$ ends at a point P in the plane determined by $\boldsymbol{x}_1$, $\boldsymbol{a}$ and $\boldsymbol{a}^*$. Hence for any pair of parameters t, τ a point of the plane is determined by $\boldsymbol{x} = \boldsymbol{x}_1 + t\boldsymbol{a} + \tau\boldsymbol{a}^*$. Conversely, for any point P of the plane there are two numbers t and τ for which such a representation holds. This is a *parametric representation* of the plane (Fig.).

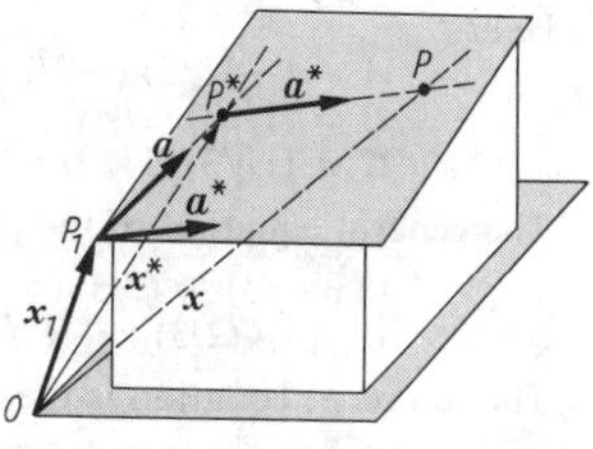

24.2-6 Parametric representation of a plane

Parametric representation of the plane	$\boldsymbol{x} = \boldsymbol{x}_1 + t\boldsymbol{a} + \tau\boldsymbol{a}^*$

If two points P_1 and P_2 and a direction vector $\boldsymbol{a}$ are given, then $\boldsymbol{a}^*$ can be determined as the direction vector of the line through P_1 and P_2. If three points are given, then two lines can be drawn through them and the direction vectors $\boldsymbol{a}$ and $\boldsymbol{a}^*$ can be calculated. In each case a parametric representation of the given form can be obtained.

Example: The points $P_1(0, 1, 1)$, $P_2(1, 0, 1)$ and $P_3(1, 1, 0)$ determine a plane in space. What is its parametric representation? – One first calculates two direction vectors, which need not be normalized, for example,

$$\overrightarrow{P_1P_2} = \overrightarrow{OP_2} - \overrightarrow{OP_1} = \boldsymbol{i} - \boldsymbol{j} \quad \text{and} \quad \overrightarrow{P_1P_3} = \overrightarrow{OP_3} - \overrightarrow{OP_1} = \boldsymbol{i} - \boldsymbol{k}.$$

If $x_1 = \overrightarrow{OP_1}$, for example ($O$ being the origin), then the parametric representation of the plane is

$$x = (i + k) + u(i - j) + v(i - k).$$

The parameters are denoted by u and v. To obtain a representation with normalized direction vectors one introduces the parameters $t = u|i - j| = u\sqrt{2}$ and $\tau = v|i - k| = v\sqrt{2}$. In terms of t and τ,

$$x = (i + k) + (t/\sqrt{2})(i - j) + (\tau/\sqrt{2})(i - k).$$

General equation. The given vector equation, as parametric representation of the plane, represents the system of three linear equations written out in detail below, where $a = \lambda e$, $a^* = \lambda^* e^*$. If one multiplies the first by $A = \cos\beta\cos\gamma^* - \cos\beta^*\cos\gamma$, the second by $B = \cos\gamma\cos\alpha^* - \cos\gamma^*\cos\alpha$, the third by $C = \cos\alpha\cos\beta^* - \cos\alpha^*\cos\beta$ and adds all three equations, one obtains

$$x = x_1 + \lambda\cos\alpha\cdot t + \lambda^*\cos\alpha^*\cdot\tau$$
$$y = y_1 + \lambda\cos\beta\cdot t + \lambda^*\cos\beta^*\cdot\tau$$
$$z = z_1 + \lambda\cos\gamma\cdot t + \lambda^*\cos\gamma^*\cdot\tau.$$

$Ax + By + Cz = Ax_1 + By_1 + Cz_1$ or $A(x - x_1) + B(y - y_1) + C(z - z_1) = 0$ as the equation of a plane through P_1. If in the first form the terms on the right-hand side are written as a constant $-D$, then the *general equation of a plane* is obtained.

General equation of a plane	$Ax + By + Cz + D = 0$

All points of space whose coordinates (x, y, z) satisfy an equation of this form, where A, B, C are not all zero, lie in a plane, and to each plane there is an equation of this kind which is satisfied by all points of the plane. More precisely, to any plane there are infinitely many such equations, since such an equation can be multiplied by any non-zero number without it representing a different plane. Hence the actual values of A, B, C and D have no geometrical significance, but only their ratios.

The *intercept form* of the equation of a plane is obtained from the general equation by bringing D to the right-hand side, dividing both sides by $-D$ and putting $a = -D/A$, $b = -D/B$, $c = -D/C$. This assumes that the numbers A, B, C, D are all non-zero; if not, an intercept equation is found by carrying out the corresponding operations as far as one can (see the following example). From the intercept equation it can be seen that the plane cuts off a segment a from the x-axis, b from the y-axis and c from the z-axis (Fig.). A plane passing through the origin has no intercept equation.

Intercept equation of a plane	$\dfrac{x}{a} + \dfrac{y}{b} + \dfrac{z}{c} = 1$

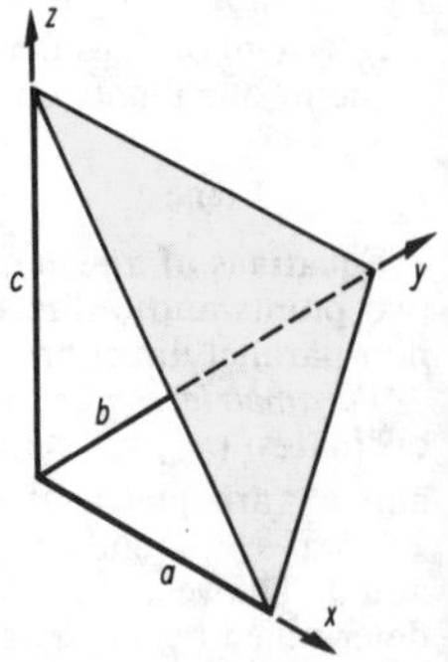

24.2-7 Intercept equation of a plane

Example: If $x = (3i + j - 2k) + (t/\sqrt{3})(i - j + k) + (\tau/\sqrt{3})(i + j - k)$ is the parametric representation of a plane, then the direction vectors a and a^* have the direction cosines

$$\cos\alpha = 1/\sqrt{3}, \qquad \cos\alpha^* = 1/\sqrt{3}$$
$$\cos\beta = -1/\sqrt{3}, \qquad \cos\beta^* = 1/\sqrt{3}$$
$$\cos\gamma = 1/\sqrt{3}, \qquad \cos\gamma^* = -1/\sqrt{3}.$$

Hence

$$A = (-1/\sqrt{3})(-1/\sqrt{3}) - (1/\sqrt{3})(1/\sqrt{3}) = 0$$
$$B = (1/\sqrt{3})(1/\sqrt{3}) - (-1/\sqrt{3})(1/\sqrt{3}) = 2/3$$
$$C = (1/\sqrt{3})(1/\sqrt{3}) - (1/\sqrt{3})(-1/\sqrt{3}) = 2/3.$$

The general equation of the plane is therefore

$$0(x - 3) + (2/3)(y - 1) + (2/3)(z + 2) = 0$$

or

$$0\cdot x + (2/3)y + (2/3)z = -2/3.$$

The intercept equation is therefore

$$0x - y - z = 1 \quad\text{or}\quad y/(-1) + z/(-1) = 1.$$

The plane cuts the x-axis 'at infinity', that is, it is parallel to the x-axis. The plane cuts the y-axis at $y = -1$ and the z-axis at $z = -1$.

Hessian normal form. If one divides the general equation by $\sqrt{(A^2 + B^2 + C^2)}$ and puts $A/\sqrt{(A^2 + B^2 + C^2)} = n_1$, $B/\sqrt{(A^2 + B^2 + C^2)} = n_2$, $C/\sqrt{(A^2 + B^2 + C^2)} = n_3$ and $D/\sqrt{(A^2 + B^2 + C^2)} = p$, one obtains the *Hessian normal form* of the equation of a plane.

Hessian normal form of the equation of a plane
$n_1x + n_2y + n_2z + p = 0$

Hessian normal form in vectors	$n\cdot x = -p$

By introducing the vectors $x = xi + yj + zk$ and $n = n_1i + n_2j + n_3k$ the Hessian normal form can be represented in a very simple way.

The vector $\boldsymbol{n}$ is perpendicular to the plane and is called the *normal vector* of the plane. It is a unit vector. Starting from the general equation of the plane, the orientation of $\boldsymbol{n}$ is determined by the sign of $\sqrt{(A^2 + B^2 + C^2)}$. It is usual to take the positive square root. Then the side of the plane that lies in the direction of $\boldsymbol{n}$ is defined as the positive side and the other as the negative side, and one speaks of the positive and negative half-spaces. The plane is oriented so that on the positive side the anticlockwise sense is taken as positive. Just like the oriented distance between two points on an oriented line, the *oriented distance* of a point from an oriented plane is introduced and applied in what follows. The distance of the origin from the plane given by $\boldsymbol{n} \cdot \boldsymbol{x} = -p$ is p. If $p > 0$, then the origin lies in the positive half-space, and if $p < 0$ it lies in the negative half-space. The figure gives an illustration of the Hessian normal form. The yellow surface represents an arbitrary plane E in space, and the red surface represents the plane through the origin O parallel to it. Let P be an arbitrary point of E, p the distance of E from O and $\boldsymbol{n}$ the normal vector of E. If φ denotes the angle enclosed by $\overrightarrow{OP} = \boldsymbol{x}$ and $\boldsymbol{n}$, then from the definition of the inner product $\boldsymbol{n} \cdot \boldsymbol{x} = |\boldsymbol{n}|\,|\boldsymbol{x}| \cos\varphi$ and $|\boldsymbol{n}| = 1$ it follows that $\boldsymbol{n} \cdot \boldsymbol{x} = |\boldsymbol{x}| \cos\varphi = -|\boldsymbol{x}| \cos(180° - \varphi)$.

From the right-angled triangle OPP' it follows that $|\boldsymbol{x}| \cos(180° - \varphi) = p$, that is $\boldsymbol{n} \cdot \boldsymbol{x} = -p$.

Example: To find the Hessian normal form of the plane given in the previous example.
From $A = 0$, $B = 2/3$, $C = 2/3$ it follows that $\sqrt{(A^2 + B^2 + C^2)} = \sqrt{(8/9)} = (2/3)\sqrt{2}$. If the general equation of the plane is divided by $(2/3)\sqrt{2}$, the Hessian normal form is obtained. It is $0 \cdot x + (\sqrt{2}/2)\,y + (\sqrt{2}/2)\,z + \sqrt{2}/2 = 0$ or, in vector form, $(\sqrt{2}/2)(\boldsymbol{j} + \boldsymbol{k}) \cdot \boldsymbol{x} = -\sqrt{2}/2$.

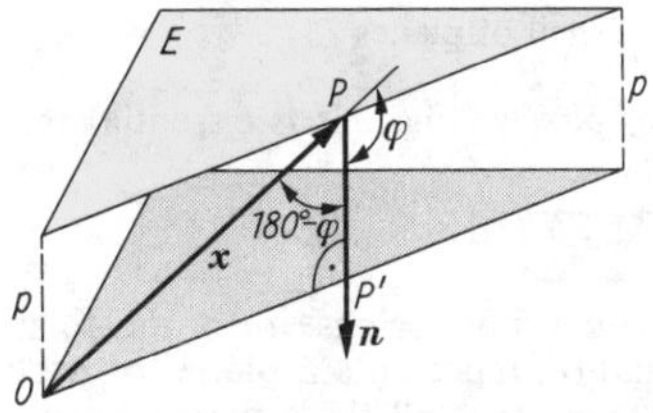

24.2-8 Illustration of the Hessian normal form of a plane in space

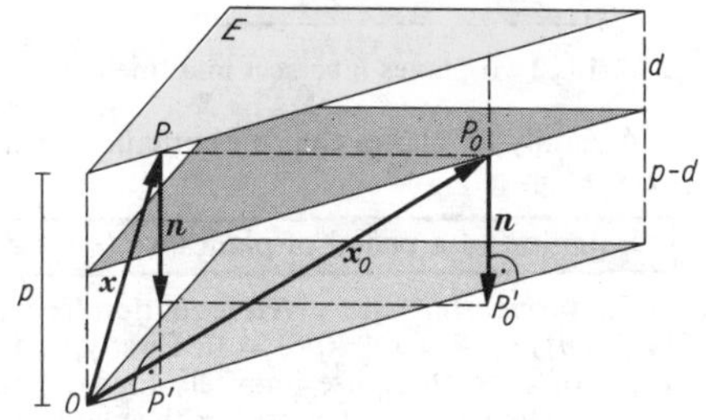

24.2-9 Distance of a point from a plane

Basic geometrical problems. Some basic problems can be solved particularly elegantly by using the vector notation.

Distance of a point from a plane. If $\boldsymbol{n} \cdot \boldsymbol{x} = -p$ is the Hessian normal form of a plane and $P_0(x_0, y_0, z_0)$ is an arbitrary point of space, then $d = \boldsymbol{n} \cdot \boldsymbol{x}_0 + p$ is the distance of P_0 from the plane. This is easily seen from the figure. The yellow surface again represents the given plane E and the red surface represents the plane through O parallel to E. The grey plane is the plane through P_0 parallel to E. Just as $\boldsymbol{n} \cdot \boldsymbol{x} = -p$ is obtained for an arbitrary point P of the plane by means of the triangle OPP', so if P_0 is an arbitrary point of space, by means of the triangle OP_0P_0' one obtains $\boldsymbol{n} \cdot \boldsymbol{x}_0 = -(p - d)$ and hence
$d = \boldsymbol{n} \cdot \boldsymbol{x}_0 + p$.

Distance of a point from a plane	$d = \boldsymbol{n} \cdot \boldsymbol{x}_0 + p$

Example: To find the distance of the point $P_0(3, -1, 2)$ from the plane given in the previous example. The distance formula gives

$$d = (\sqrt{2}/2)(\boldsymbol{j} + \boldsymbol{k}) \cdot (3\boldsymbol{i} - \boldsymbol{j} + 2\boldsymbol{k}) + (\sqrt{2}/2) = (\sqrt{2}/2)(0 \times 3 - 1 \times 1 + 1 \times 2) + (\sqrt{2}/2)$$
$$= \sqrt{2} \approx 1.414.$$

The required distance is approximately 1.414 units.

The angle between two planes. If $\boldsymbol{n} \cdot \boldsymbol{x} = -p$ and $\boldsymbol{n}^* \cdot \boldsymbol{x} = -p^*$ are the Hessian normal forms of two planes, then the angle φ between them is equal to the angle between their normal vectors $\boldsymbol{n}$ and $\boldsymbol{n}^*$. Hence $\cos\varphi = \boldsymbol{n} \cdot \boldsymbol{n}^*$. In particular, the two planes are perpendicular to one another if and only if $\boldsymbol{n} \cdot \boldsymbol{n}^* = 0$.

$\cos\varphi = \boldsymbol{n} \cdot \boldsymbol{n}^*$

Example: Are the planes $5x + 3y - z = 10$ and $2x - y + 7z = 5$ perpendicular to one another? – The normal vectors of the planes are $\boldsymbol{n} = (1/\sqrt{35})(5\boldsymbol{i} + 3\boldsymbol{j} - \boldsymbol{k})$ and $\boldsymbol{n}^* = (1/\sqrt{54})(2\boldsymbol{i} - \boldsymbol{j} + 7\boldsymbol{k})$. Hence $\cos\varphi = (1/\sqrt{35})(1/\sqrt{54})(5 \times 2 - 3 \times 1 - 1 \times 7) = 0$, that is $\varphi = 90°$. The two given planes are perpendicular to one another.

Intersection of two planes. Two planes always intersect in a line as long as they are not parallel. Hence $|\boldsymbol{n} \cdot \boldsymbol{n}^*| < 1$ or $\boldsymbol{n} \times \boldsymbol{n}^* \neq \boldsymbol{o}$ is a necessary and sufficient condition for two planes to intersect.

Two non-parallel planes always have a line in common. This is called the *line of intersection.* Since it is perpendicular to $\boldsymbol{n}$ and $\boldsymbol{n}^*$, its direction vector can be expressed as $\boldsymbol{a} = \boldsymbol{n} \times \boldsymbol{n}^*$. If any point that satisfies both the equations $\boldsymbol{n} \cdot \boldsymbol{x} = -p$ and $\boldsymbol{n}^* \cdot \boldsymbol{x} = -p^*$ is determined, then the parametric representation of the line is obtained from this and $\boldsymbol{a}$. In detail, a point is to be found whose coordinates satisfy the given system of equations, where n_1, n_2, n_3 are the components of $\boldsymbol{n}$ and n_1^*, n_2^*, n_3^* are the components of $\boldsymbol{n}^*$. This point together with the direction vector $\boldsymbol{a}$ gives the point-direction equation of the line of intersection. For example, if $n_1 n_2^* - n_1^* n_2 \neq 0$, then

$$n_1 x + n_2 y + n_3 z = -p,$$
$$n_1^* x + n_2^* y + n_3^* z = -p^*.$$

$$x = (p^* n_2 - p n_2^*)/(n_1 n_2^* - n_1^* n_2),\; y = (p n_1^* - p^* n_1)/(n_1 n_2^* - n_1^* n_2),\; z = 0$$

is a solution of the above system of equations (Fig.).

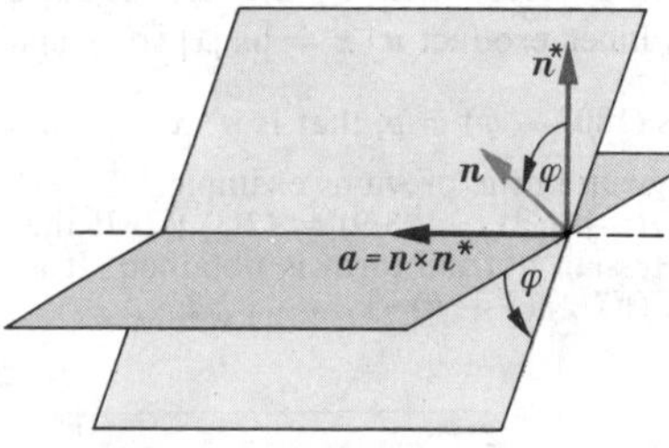

24.2-10 Two planes intersect in a line

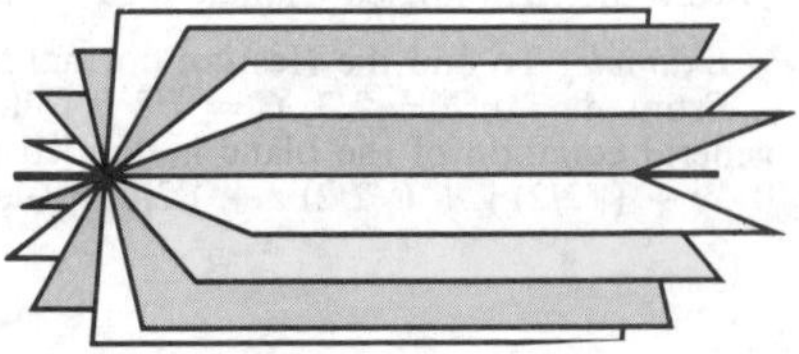

24.2-11 Pencil of planes

A family of planes that all contain one line is called a *pencil of planes* (Fig.). It is e ssential that $\boldsymbol{n} \times \boldsymbol{n}^* \neq \boldsymbol{o}$.

Equation of a pencil of planes	$(\boldsymbol{n} \cdot \boldsymbol{x} + p) + \lambda(\boldsymbol{n}^* \cdot \boldsymbol{x} + p^*) = 0$

To prove that the given equation represents a pencil of planes, one brings it to the form $(\boldsymbol{n} + \lambda \boldsymbol{n}^*) \cdot \boldsymbol{x} + (p + \lambda p^*) = 0$. Firstly, it is obvious that the equation represents a plane for each value of λ, so there are infinitely many planes. It must now be shown that all these planes have a common line. One first considers the planes given by values λ_1 and λ_2 of $\lambda(\lambda_1 \neq \lambda_2)$. Since $\boldsymbol{n} \times \boldsymbol{n}^* \neq \boldsymbol{o}$, these planes are not parallel, for $(\boldsymbol{n} + \lambda_1 \boldsymbol{n}^*) \times (\boldsymbol{n} + \lambda_2 \boldsymbol{n}^*) = (\boldsymbol{n} \times \boldsymbol{n}^*)(\lambda_2 - \lambda_1) \neq \boldsymbol{o}$. Hence they have a line of intersection with the direction vector $\boldsymbol{a} = \boldsymbol{n} \times \boldsymbol{n}^*$. To set up a point-direction equation of the line of intersection a point must be determined that satisfies both the equations $(\boldsymbol{n} + \lambda_1 \boldsymbol{n}^*) \cdot \boldsymbol{x} + (p + \lambda_1 p^*) = 0$ and $(\boldsymbol{n} + \lambda_2 \boldsymbol{n}^*) \cdot \boldsymbol{x} + (p + \lambda_2 p^*) = 0$. Since $\lambda_1 \neq \lambda_2$, this is possible if and only if the given system of equations is satisfied. It is clear that $\boldsymbol{a}$ and $\boldsymbol{x}$ can be determined independently of λ_1 and λ_2. This means that the line of intersection of the planes given by λ_1 and λ_2 is common to all the planes, which are therefore expressible by the equation given at the beginning.

$$\boldsymbol{n} \cdot \boldsymbol{x} + p = 0$$
$$\boldsymbol{n}^* \cdot \boldsymbol{x} + p^* = 0$$

It should be noted that the equation of a pencil of planes, expressed in the given form, does not contain one of the planes through the line, namely the plane $\boldsymbol{n}^* \cdot \boldsymbol{x} + p^* = 0$. This can be avoided by introducing a homogeneous parameter. If $\lambda = \varkappa^*/\varkappa$, then the equation can be written $\varkappa(\boldsymbol{n} \cdot \boldsymbol{x} + p) + \varkappa^*(\boldsymbol{n}^* \cdot \boldsymbol{x} + p^*) = 0$, and the plane $\boldsymbol{n}^* \cdot \boldsymbol{x} + p^* = 0$ is given by $\varkappa = 0, \varkappa^* = 1$.

Example: To find a point-direction equation for the line of intersection of the pencil of planes

$$[(1/\sqrt{14})(3\boldsymbol{i} - 2\boldsymbol{j} + \boldsymbol{k}) \cdot \boldsymbol{x} + 1] + \lambda[(1/\sqrt{14})(2\boldsymbol{i} + \boldsymbol{j} - 3\boldsymbol{k}) \cdot \boldsymbol{x} - 1] = 0.$$

A direction vector of the line of intersection is given by

$$\boldsymbol{a} = (3\boldsymbol{i} - 2\boldsymbol{j} + \boldsymbol{k}) \times (2\boldsymbol{i} + \boldsymbol{j} - 3\boldsymbol{k}) = 5\boldsymbol{i} + 11\boldsymbol{j} + 7\boldsymbol{k}.$$

$$3x - 2y + z = -1$$
$$2x + y - 3z = 1$$

A point of the line is obtained as a solution of the system of equations above.

One solution is $x = 1/7$, $y = 5/7$, $z = 0$, as one can check by substitution. A point-direction equation of the line of intersection is therefore

$$\boldsymbol{x} = (1/7)(\boldsymbol{i} + 5\boldsymbol{j}) + t(5\boldsymbol{i} + 11\boldsymbol{j} + 7\boldsymbol{k}).$$

Intersection of three planes. If three planes are given, then their equations $\boldsymbol{n} \cdot \boldsymbol{x} + p = 0$, $\boldsymbol{n}^* \cdot \boldsymbol{x} + p^* = 0$ and $\boldsymbol{n}^{**} \cdot \boldsymbol{x} + p^{**} = 0$ form a system of three linear equations for the three components x, y, z of $\boldsymbol{x}$. If this system is uniquely soluble, then the three planes have exactly one point in common. The adjacent condition is a necessary and sufficient condition for this.

$$\begin{vmatrix} n_1 & n_2 & n_3 \\ n_1^* & n_2^* & n_3^* \\ n_1^{**} & n_2^{**} & n_3^{**} \end{vmatrix} \neq 0$$

Otherwise the planes either have no common point or they have a common line or they coincide. The first is the case when either two of the planes are parallel or the three planes taken in pairs have distinct parallel lines of intersection. The second is the case when the three planes belong to a pencil. All the planes that have one point in common form a *bundle of planes*.

Equations of a bundle of planes	$(\boldsymbol{n}\cdot\boldsymbol{x}+p)+\lambda_1(\boldsymbol{n}^*\cdot\boldsymbol{x}+p^*)+\lambda_2(\boldsymbol{n}^{**}\cdot\boldsymbol{x}+p^{**})=0$

By introducing homogeneous parameters ($\lambda_1=\varkappa^*/\varkappa$, $\lambda_2=\varkappa^{**}/\varkappa$ one obtains

$$\varkappa(\boldsymbol{n}\cdot\boldsymbol{x}+p)+\varkappa^*(\boldsymbol{n}^*\cdot\boldsymbol{x}+p^*)+\varkappa^{**}(\boldsymbol{n}^{**}\cdot\boldsymbol{x}+p^{**})=0.$$

24.3. Quadrics

Principal axes

The set of all points whose rectangular coordinates satisfy an equation of the form $F(x, y, z) = 0$ is, under certain conditions, called a *surface*. The condition might be, for example, that the function $F(x, y, z)$ should be continuous in all the variables. According to the conditions that are imposed, different concepts of a surface are obtained.

If $F(x, y, z)$ is a linear function of the three variables x, y, z, that is, of the form $Ax+By+Cz+D$, where the coefficients A, B, C are not all zero, then the equation $F(x, y, z) = 0$ represents a plane.

Henceforth, let $F(x, y, z)$ be a quadratic function. Then $F(x, y, z) = 0$ is an algebraic equation of the second degree, that is, an equation of the form

$$a_{11}x^2+2a_{12}xy+2a_{13}xz+a_{22}y^2+2a_{23}yz+a_{33}z^2+2a_{14}x+2a_{24}y+2a_{34}z+a_{44}=0.$$

A surface representable by an equation of this form (where the first six coefficients must not all be zero) is called a *quadric* or a *surface of the second order*.

Under a linear transformation of coordinates (translation, rotation or a combination of the two) an algebraic equation in the rectangular coordinates x, y, z with coefficients a_{11} up to a_{44} goes into an algebraic equation of the second degree in the rectangular coordinates x^*, y^*, z^* with coefficients a^*_{11} up to a^*_{44}. Of fundamental importance is the fact that *in every case a rotation can be found* so that $a^*_{12}=a^*_{13}=a^*_{23}=0$. This transformation is called a *transformation to principal axes*.

Example: Under the rotation

$$x=(\sqrt{2}/2)\,x^*+(\sqrt{2}/2)\,y^*,\quad y=-(\sqrt{2}/2)\,x^*+(\sqrt{2}/2)\,y^*,\quad z=z^*$$

the equation $x^2+y^2+z^2+xy-1=0$ goes into

$$x^{*2}/a^2+y^{*2}/b^2+z^{*2}/c^2-1=0 \quad \text{with} \quad a=\sqrt{2},\quad b=\sqrt{(2/3)},\quad c=1.$$

Thanks to the transformation to principal axes, the discussion of geometric figures representable by algebraic equations of the second degree can be reduced to the discussion of equations of the following form.

$$a_{11}x^2+a_{22}y^2+a_{33}z^2+2a_{14}x+2a_{24}y+2a_{34}z+a_{44}=0$$

However, such an equation in which the first three coefficients are not all zero, can in general be simplified still further by a transformation of coordinates (in fact, a translation). The kind of translation needed in a particular case, and the form of simplified equation it leads to, depend on the nature of the coefficients. By considering all possible cases one arrives at the result that an arbitrary equation of the second degree can be reduced to one of 17 different special equations, each of which consists of at most 4 terms.

Three of these equations have the form $x^2/a^2+y^2/b^2+z^2/c^2+1=0$, $x^2/a^2+y^2/b^2+1=0$, $x^2/a^2+1=0$, where a, b and c are non-zero. These have no real solution and therefore do not represent a geometrical figure. The other 14 equations represent 14 different geometrical figures. The following nine are *degenerate* or *improper quadrics*:

1. $x^2/a^2+y^2/b^2+z^2/c^2=0$, the point (0, 0, 0).
2. $x^2/a^2+y^2/b^2=0$, a line, the z-axis.
3. $x^2/a^2=0$, a plane, the y, z-plane.
4. $x^2/a^2=1$, the two planes parallel to the y, z-plane at distances $x=\pm a$.
5. $x^2/a^2-y^2/b^2=0$, the two planes that cut the x, y-plane at right angles in the lines $y=\pm(b/a)\,x$.
6. $x^2/a^2-y^2/b^2=1$, the surface of a cylinder that is cut by planes perpendicular to the z-axis in hyperbolas; these are parallel and congruent to the hyperbola $x^2/a^2-y^2/b^2=1$ in the x, y-plane.
7. $x^2/a^2+y^2/b^2=1$, the surface of a cylinder that is cut by planes perpendicular to the z-axis in ellipses; these are parallel and congruent to the ellipse $x^2/a^2+y^2/b^2=1$ in the x, y-plane; if $a=b$ the ellipses are circles.
8. $x^2-2py=0$, the surface of a cylinder that is cut by planes perpendicular to the z-axis in parabolas; these are parallel and congruent to the parabola $x^2-2py=0$ in the x, y-plane.

9. $x^2/a^2 + y^2/b^2 - z^2/c^2 = 0$, the surface of a (double) cone that is cut by planes perpendicular to the z-axis in ellipses (or circles, if $b^2 = a^2$).

These figures are either not surfaces in the usual sense (1. and 2.), or they reduce to one plane (or two planes) and are really surfaces of the first order (3. to 5.), or they can be developed into a plane (6. to 9.).

There remain finally just five geometric figures that are called *proper quadrics*.

Proper quadrics

Classification. After carrying out a transformation to principal axes the axes of the coordinate system are in the directions of the principal axes of the surface. The characterization of the surface depends on one distinguished principal axis (either such an axis exists, or it is immaterial which principal axis is taken) and a section of the surface by a plane perpendicular to it.

According to whether the first section is an ellipse, parabola or hyperbola, the surface is called an *ellipsoid*, *paraboloid* or *hyperboloid*.

The form of the second section, if it is necessary to make a distinction, determines the adjective *elliptic* or *hyperbolic*. The adjective parabolic is not used, since there is no proper quadric that has a parabolic transverse section.

No distinction is made between elliptic and circular sections. An ellipsoid can therefore be a sphere, and an elliptic paraboloid can have circular transverse sections. For hyperboloids another distinction is essential. One distinguishes between hyperboloids of *one sheet* and *two sheets*.

Altogether there are five proper quadrics, which are called ellipsoid, elliptic paraboloid, hyperbolic paraboloid, hyperboloid of one sheet and hyperboloid of two sheets.

Ellipsoid. In rectangular coordinates the simplest form of the equation of an ellipsoid is as follows.

Ellipsoid	$\frac{x^2}{a^2} + \frac{y^2}{b^2} + \frac{z^2}{c^2} = 1$

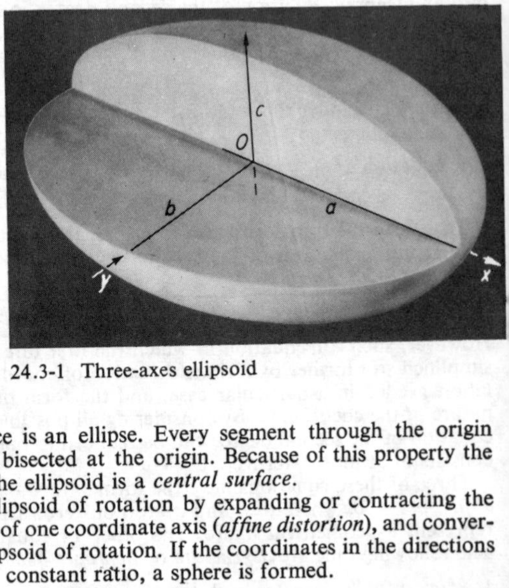

24.3-1 Three-axes ellipsoid

Here a, b, c are the lengths of half the principal axes of the ellipsoid (Fig.). If $a = b = c$, the ellipsoid is a sphere. If two of the numbers a, b, c are equal, it is an *ellipsoid of revolution* or *two-axes ellipsoid* and it is *stretched* (prolate) if the two equal axes are shorter than the third and *flattened* (oblate) if they are longer. If the numbers a, b, c are all different, one speaks of a *three-axes ellipsoid*.

The geometric figure represented by the above equation is a connected finite surface. It lies symmetrically about the three coordinate planes. Every plane section of the surface is an ellipse. Every segment through the origin joining two points of the surface (*diameter*) is bisected at the origin. Because of this property the origin is called the *centre* of the ellipsoid and the ellipsoid is a *central surface*.

Any ellipsoid can be transformed into an ellipsoid of rotation by expanding or contracting the coordinates in a constant ratio in the direction of one coordinate axis (*affine distortion*), and conversely, any ellipsoid can be formed from an ellipsoid of rotation. If the coordinates in the directions of two coordinate axes are suitably altered in a constant ratio, a sphere is formed.

Elliptic paraboloid. One of the three principal axes of the elliptic paraboloid is distinguished. In a rectangular coordinate system whose z-axis is in the direction of the distinguished principal axis, the simplest form of the equation is as follows.

Elliptic paraboloid	$\frac{x^2}{a^2} + \frac{y^2}{b^2} - 2z = 0$

Here a and b are half the lengths of the principal axes of the ellipse cut by the plane parallel to the x, y-plane at a distance $z = 1/2$ (Fig.). Another way of putting it is to say that a^2 is the semi-parameter of the parabola cut on the elliptic paraboloid by the x, z-plane and similarly b^2 for the y, z-plane. If $a = b$, the elliptic paraboloid is a *paraboloid of revolution*.

The geometric figure represented by the above equation is a connected infinite surface in the positive half-space determined by the x, y-plane. It is symmetrical about the x, z- and y, z-planes. The section of the surface by every plane parallel to the z-axis is a parabola, and the section of the surface by every plane perpendicular to the z-axis (and meeting it on the positive side) is an ellipse. The z-axis is called the *axis* for short, and the origin is called the *vertex*. The surface has no centre.

By means of an affine distortion in the x- or y-direction the elliptic paraboloid can be transformed into a paraboloid of revolution, and conversely, any elliptic paraboloid can be formed from a paraboloid of rotation.

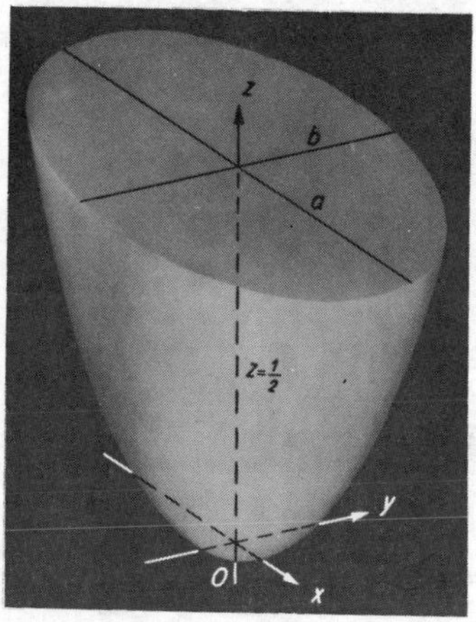

24.3-2 Elliptic paraboloid

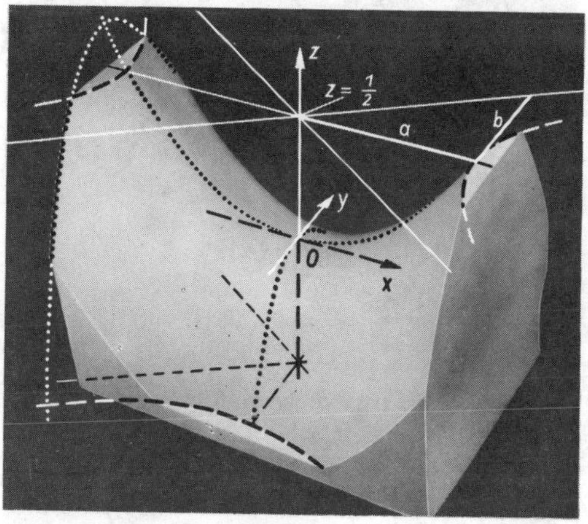

24.3-3 Hyperbolic paraboloid

Hyperbolic paraboloid. One of the principal axes of the hyperbolic paraboloid is distinguished. In a rectangular coordinate system whose z-axis is in the direction of the distinguished principal axis, the simplest form of the equation is as follows.

Hyperbolic paraboloid	$\frac{x^2}{a^2} - \frac{y^2}{b^2} - 2z = 0$

Here a and b are half the lengths of the principal axes of the hyperbola cut by the plane parallel to the x, y-plane at a distance $z = {}^1/_2$ (Fig.). Another way of putting it is to say that a^2 is the semi-parameter of the parabola cut by the x, z-plane and similarly $-b^2$ for the y, z-plane.

The geometric figure represented by the above equation is a connected infinite surface that lies in each octant. It is symmetrical about the x, z- and y, z-planes. The section of the surface by every plane parallel to the z-axis is a parabola, and the section of the surface by every plane perpendicular to the z-axis that does not pass through the origin is a hyperbola; the vertices of each hyperbola lie on a line parallel to the x-axis if the plane of the hyperbola meets the z-axis on the positive side and on a line parallel to the y-axis if the plane of the hyperbola meets the z-axis on the negative side. The x, y-plane cuts the surface in a pair of lines $x/a + y/b = 0$ and $x/a - y/b = 0$. The planes through each of these lines and the z-axis cut on planes perpendicular to the z-axis the asymptotes of the corresponding hyperbola. The z-axis is called the *axis* for short, and the origin is called the *vertex* of the hyperbolic paraboloid. It is a *saddle point*. The surface has no centre.

The hyperbolic paraboloid is the only proper quadric that can never be a surface of rotation and therefore cannot be transformed into such a surface by an affine distortion. This is due essentially to the fact that no plane cuts it in an ellipse.

There are other interesting possible ways of generating a hyperbolic paraboloid; one of these consists of translating a parabola that is convex downwards along a parabola that is convex upwards. For this reason, the hyperbolic paraboloid counts as a *translation surface*.

Finally, a hyperbolic paraboloid can be generated by families of lines. If one writes the above equation in the form $(x/a + y/b)(x/a - y/b) = 2z$ and puts $z/(x/a - y/b) = u$ and $2/(x/a - y/b) = v$, then the following two pairs of equations can be derived from the equation of the hyperbolic paraboloid:

1. $(x/a + y/b) = 2u$, $\quad (x/a + y/b) = vz$, $\qquad$ 2. $(x/a - y/b) = z/u$, $\quad (x/a - y/b) = 2/v$.

Each of these equations represents a family of planes, and each pair of equations determines a family of lines. These families of lines lie on the hyperbolic paraboloid. The lines are the *generators* of the hyperbolic paraboloid (Fig. 24.3-4).

Every surface generated by a family of lines is called a *ruled surface*. Among surfaces of the second order the elliptic, parabolic and hyperbolic cylinders, the (double) cone, the hyperbolic paraboloid and the hyperboloid of one sheet are ruled surfaces.

Since the cylinders and the cone can be developed into a plane, they are called *developable surfaces*. The example of the hyperbolic paraboloid shows that not every ruled surface is developable. The hyperboloid of one sheet is not developable either.

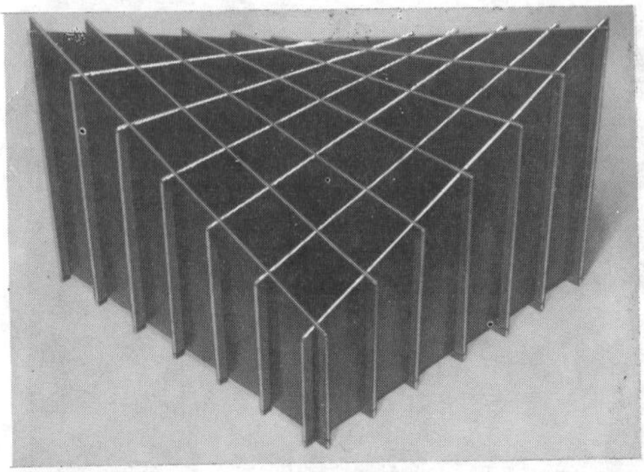

24.3-4 Hyperbolic paraboloid with its two families of generators

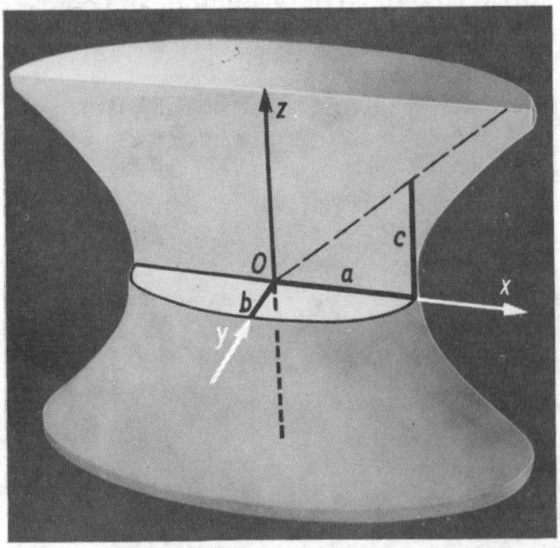

24.3-5 Hyperboloid of one sheet

The hyperboloid of one sheet. One of the three principal axes of a hyperboloid of one sheet is distinguished. In a rectangular coordinate system whose z-axis is in the direction of the distinguished principal axis, the simplest form of the equation is as follows:

Hyperboloid of one sheet	$\frac{x^2}{a^2}+\frac{y^2}{b^2}-\frac{z^2}{c^2}=1$

Here a and b are half the lengths of the principal axes of the ellipse cut by the x, y-plane. Similarly b and c are half the lengths of the principal axes of the hyperbola cut by the y, z-plane (Fig.). If $a = b$, the hyperboloid of one sheet is a *one-sheet hyperboloid of rotation.*

The geometric figure represented by the above equation is a connected infinite surface that lies in each octant. It lies symmetrically about the three coordinate planes. The section of the surface by every plane parallel to the z-axis is a *hyperbola*, and the section by any plane perpendicular to the z-axis is an *ellipse*. The z-axis is called the *axis* for short. The origin is the centre, hence the hyperboloid of one sheet is a central surface.

By an affine distortion in the x- or y-direction the hyperboloid of one sheet can be transformed into a hyperboloid of revolution, and conversely, it can be generated from such a surface.

The hyperboloid of one sheet can also be generated by families of lines. If one writes the above equation in the form $(x/a + z/c)(x/a - z/c) = (1 + y/b)(1 - y/b)$ and puts $(1 - y/b)/(x/a - z/c) = u$ and $(1 + y/b)/(x/a - z/c) = v$, then two pairs of equations can be derived from the equation of the hyperboloid of one sheet:

1. $x/a + z/c = u(1 + y/b)$, $\quad x/a + z/c = v(1 - y/b)$,
2. $x/a - z/c = (1 - y/b)/u$, $\quad x/a - z/c = (1 + y/b)/v$.

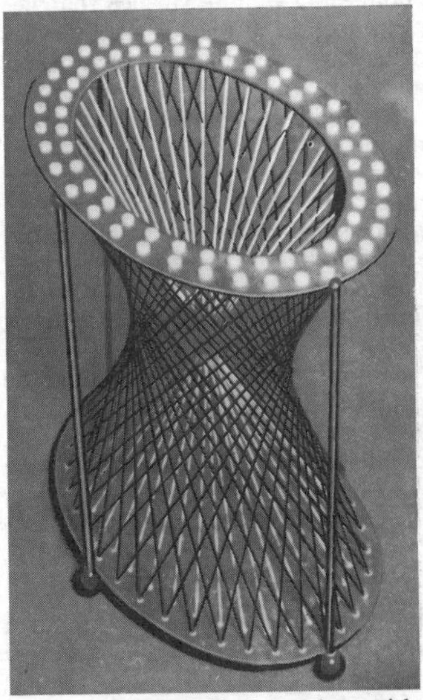

24.3-6 Hyperboloid of one sheet with its two families of generators and asymptotic cone

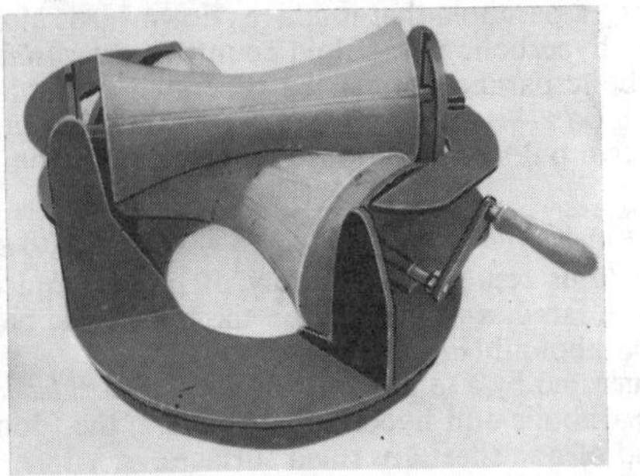

24.3-7 Model of the transmission of rotations by means of two hyperboloids of one sheet

Each of these equations represents a family of planes, and each pair of equations represents a family of lines. These are the *generators* of the hyperboloid of one sheet. The hyperboloid of one sheet is therefore a *ruled surface* (Fig.). Owing to this property, two hyperboloids of one sheet can be used in technology, like two cones, to transmit a rotation about one axis into a rotation about another arbitrarily directed axis (*hyperbolic cog-wheels*, Fig.).

If the generators are translated so as to pass through the origin, they form the *asymptotic cone* of the hyperboloid of one sheet. Its equation is $x^2/a^2 + y^2/b^2 - z^2/c^2 = 0$.

The hyperboloid of two sheets. One of the three principal axes of the hyperboloid of two sheets is distinguished. In a rectangular coordinate system whose z-axis is in the direction of the distinguished principal axis, the simplest form of the equation is as follows:

Hyperboloid of two sheets	$-\dfrac{x^2}{a^2} + \dfrac{y^2}{b^2} + \dfrac{z^2}{c^2} = 1$

Here a and b are half the lengths of the principal axes of the ellipse cut on the hyperboloid of two sheets by the planes parallel to the x, y-plane at a distance $z = \pm c\sqrt{2}$. Also, c and a are half the lengths of the principal axes of the hyperbola cut by the x, z-plane, and similarly c and b for the y, z-plane (Fig.). If $a = b$ the hyperboloid of two sheets is a two-sheeted *hyperboloid of revolution*. The geometric figure represented by the above equation is an infinite disconnected surface, consisting of two parts, with points in each octant. The two parts lie symmetrically about the coordinate planes. The section by any plane parallel to the z-axis is a *hyperbola*, and for $|z| > c$ the section by any plane perpendicular to the z-axis is an *ellipse*. The z-axis is called the *axis* for short, and the origin is the centre of the hyperboloid of two sheets. It is therefore a central surface. By an affine distortion in the x- or y-direction the hyperboloid of two sheets can be transformed into a hyperboloid of revolution, and conversely, it can be generated from such a surface. As for the hyperboloid of one sheet, there exists an asymptotic cone.

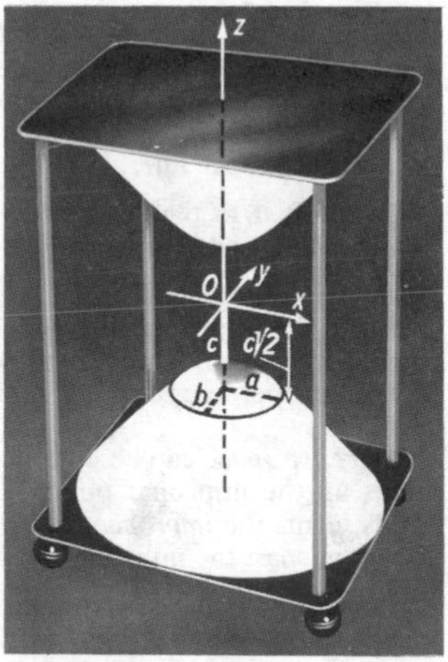

24.3-8 Hyperboloid of two sheets

25. Projective geometry

Projective geometry investigates those properties of geometrical figures that are unaltered by projection. The impetus for these investigations was provided by the study of perspective in painting and architecture. Following the development of descriptive geometry, principally by Gaspard Monge (1746–1818), Victor Poncelet (1788–1867) gave a first outline of projective geometry in his '*Traité des propriétés projectives des figures*'. *Analytical methods* in projective geometry were introduced mainly by August Ferdinand Möbius (1790–1868) and Julius Plücker (1801–1868), while Jacob Steiner (1796–1863) and Christian von Staudt (1798–1867) perfected a development of projective geometry without these methods. The first beginnings of this *synthetic approach* are to be found in the work of Pappus (250–300? B. C.), who introduced the cross-ratio, referring to a lost work of Apollonius of Perga (265–180 B. C.?). The connection between projective and Euclidean geometry was clarified by Felix Klein (1849–1925). He also introduced the idea of a geometry as the invariant theory of a certain group of mappings.

25.1. The basic elements of projective geometry

Improper elements. In general, a *parallel projection* between two lines of a plane maps all the points of one line one to one onto all the points of the other. In a *central projection*, or *perspectivity*, the correspondence between the points of two lines l_1 and l_2 is determined by means of lines p

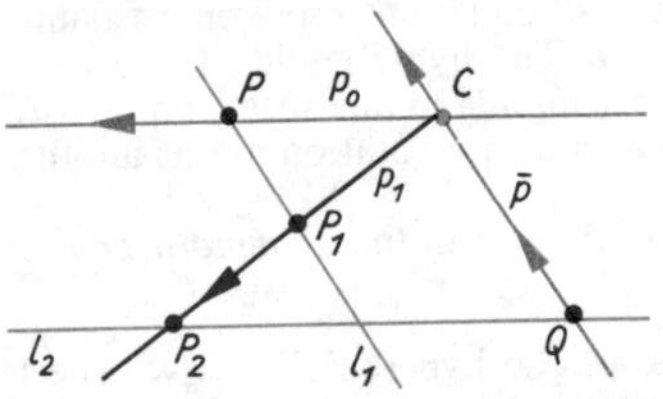

25.1-1 Introduction of improper points

through one point, the *centre C*; for example, if P_1 is a point of l_1, then the line $p_1 = (CP_1)$ cuts the line l_2 in P_2, the image of P_1 (Fig.). This mapping does not cover all the points of l_1 and l_2: to the point P for which $p_0 = (CP) \parallel l_2$ there corresponds no point on l_2, and to the point Q for which $\bar{p} = (CQ) \parallel l_1$ there corresponds no point on l_1. In order to preserve the one-to-one mapping, one adjoins to the *proper* points of the plane all the directions of lines of the plane as *improper points*; the image of the point P is then the direction of l_2, and the inverse image of Q is the direction of l_1. One proper and one improper point determine exactly one proper line through the proper point in the direction of the improper point, for example, the line p_0 through P and its image. By contrast, two improper points determine the *improper line*, which consists of all the improper points. Two parallel lines cut in an improper point, which corresponds to the common direction of the two lines, while a proper line and the improper line cut in the improper point that corresponds to the direction of the proper line.

In contrast to the geometry of the Euclidean plane, two lines in the projective plane always cut in one point, and through two points there passes exactly one line.

The *projective plane* consists of all proper and improper points. Its linear subspaces are the proper lines and the improper line. Since proper and improper points can be mapped into each other by a central projection, it makes no sense to distinguish between them. In the same way, the distinction between proper lines and the improper line loses its meaning, and it makes no sense to talk of parallel lines.

Projective space can be obtained from Euclidean space in the same way, by adjoining to the proper points all the improper points, that is, directions of the lines in space. The set of these improper points forms the *improper plane*. The improper lines are cut on this plane by the proper planes, and therefore span the improper plane.

In projective space two lines are either skew or cut in exactly one point. Two planes cut in exactly one line. A line and a point not lying on it span exactly one plane.

If one takes the lines as the basic elements of the projective plane, instead of the points, then the points, as vertices of pencils of lines, are the linear subspaces.

25.2. Projective coordinates

If S is a point outside the projective plane Π, then there is a one-to-one correspondence in which each line through S corresponds to its point of intersection P with the plane Π (Fig.). If the line is represented by a non-zero vector $\boldsymbol{x}$, then $\varrho\boldsymbol{x}(\varrho \neq 0)$ gives the same point P of Π, which is thus characterized by a 1-dimensional vector space. Referred to a basis $\boldsymbol{e}_0$, $\boldsymbol{e}_1$, $\boldsymbol{e}_2$ of the vector space, $\boldsymbol{x}$ has the representation $\boldsymbol{x} = \boldsymbol{e}_0 x_0 + \boldsymbol{e}_1 x_1 + \boldsymbol{e}_2 x_2 = \sum_{i=0}^{2} \boldsymbol{e}_i x_i$ or $\varrho\boldsymbol{x} = \sum_{i=0}^{2} \boldsymbol{e}_i \varrho x_i$. The three numbers (x_0, x_1, x_2)

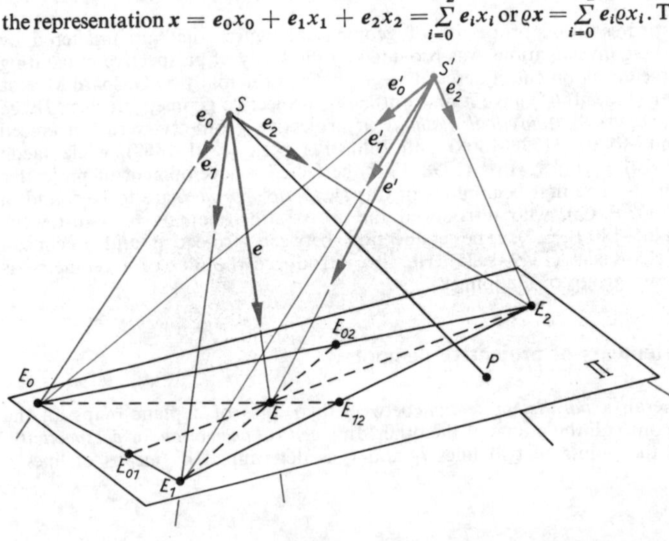

25.2-1 Introduction of projective coordinates

so determined by the point P are its *homogeneous projective coordinates*, homogeneous because $(\varrho x_0, \varrho x_1, \varrho x_2)$ $(\varrho \neq 0)$ also represent the same point. The *base-points* E_i with $\boldsymbol{x} = \boldsymbol{e}_i$ $(i = 0, 1, 2)$ have the homogeneous coordinates (1, 0, 0), (0, 1, 0) and (0, 0, 1), and the *unit point* E with $\boldsymbol{x} = \boldsymbol{e} = \boldsymbol{e}_0 + \boldsymbol{e}_1 + \boldsymbol{e}_2$ has the homogeneous coordinates (1, 1, 1). The base-points and unit-point form a *basis* for the projective coordinate system. If instead of the vectors $\boldsymbol{e}_i$ and $\boldsymbol{e}$, one chooses the vectors $\boldsymbol{e}'_i = \varrho_i \boldsymbol{e}_i$ and $\boldsymbol{e}' = \varrho \boldsymbol{e}$, then $\boldsymbol{e}' = \varrho \boldsymbol{e} = \varrho \sum_{i=0}^{2} \boldsymbol{e}_i$ and also $\boldsymbol{e}' = \sum_{i=0}^{2} \boldsymbol{e}'_i = \sum_{i=0}^{2} \boldsymbol{e}_i \varrho_i$. It follows that $\varrho_i = \varrho$ and $x_i = \varrho x'_i$, that is, the homogeneous coordinates remain unaltered if the ratios $x_0 : x_1 : x_2$ remain constant.

If S' is a point outside Π, distinct from S, with the vector $\sum_{i=0}^{2} s_i \boldsymbol{e}_i$ relative to S, then, relative to S', the points E_i and E have vectors of the form $\boldsymbol{e}'_i = \sum_{j=0}^{2} a_{ij} \boldsymbol{e}_j$ and $\boldsymbol{e}' = \sum_{i=0}^{2} \boldsymbol{e}'_i$. The affine mapping which makes each point X, with the vector $\sum_{j=0}^{2} x_j \boldsymbol{e}_j$ relative to S, correspond uniquely to the point X', with the vector $\sum_{j=0}^{2} (s_j + \sum_{i=0}^{2} x_i a_{ij}) \boldsymbol{e}_j$ relative to S, and each vector $\boldsymbol{x} = \sum_{i=0}^{2} x_i \boldsymbol{e}_i$ to the vector $\boldsymbol{x}' = \sum_{j=0}^{2} (\sum_{i=0}^{2} x_i a_{ij}) \boldsymbol{e}_j$, is uniquely determined by the conditions that S goes into S', and the vectors $\boldsymbol{e}_i$ into $\boldsymbol{e}'_i$ and $\boldsymbol{e}$ into $\boldsymbol{e}'$. If then the vector $\boldsymbol{p} = \sum_{i=0}^{2} p_i \boldsymbol{e}_i$ represents the point P, its image P' is represented by $\boldsymbol{p}' = \sum_{j=0}^{2} (\sum_{i=0}^{2} p_i a_{ij}) \boldsymbol{e}_j = \sum_{i=0}^{2} p_i \boldsymbol{e}'_i$ and it therefore has, relative to the basis $\boldsymbol{e}'_i$, the same coordinates as P relative to the basis $\boldsymbol{e}_i$. The projective coordinates therefore depend only on the basis, and not on the points S, S' of the surrounding space, which are only used to derive them.

Conversely, four points of a plane, E_0, E_1, E_2, E, *no three of which are collinear*, determine a projective coordinate system, for which these points are the basis. One need only join one point S outside the plane by lines to the given points and so choose vectors $\boldsymbol{e}_i$ and $\boldsymbol{e}$ on SE_i and SE so that $\sum_{i=0}^{2} \boldsymbol{e}_i = \boldsymbol{e}$. The vectors $\boldsymbol{e}_i$ are linearly independent, since the points E_i do not lie on one line, and sherefore they form a basis of the vector space.

In space a projective coordinate system is correspondingly determined by a basis of five points E_0, E_1, E_2, E_3, E, no four of which lie in one plane. From a point S outside the space, the base-points are represented by vectors $\boldsymbol{e}_i$ $(i = 0, 1, 2, 3)$, which form a basis of the 4-dimensional vector space, and the unit point by the vector $\boldsymbol{e} = \sum_{i=0}^{3} \boldsymbol{e}_i$. The coordinate ratios $x_0 : x_1 : x_2 : x_3$ do not depend on the choice of the point S nor the choice of the vectors $\boldsymbol{e}_i$, $\boldsymbol{e}$ on the corresponding lines through S.

On a projective line, three distinct points suffice as a basis of a coordinate system, two as base-points and one as unit-point.

Given any coordinate system in the plane, determined by four points E_0, E_1, E_2, E, one can *restrict* it to the three coordinate axes E_0E_1, E_0E_2 and E_1E_2 by taking as unit-points E_{01} on E_0E_1 with coordinates (1, 1, 0) corresponding to the vector $\boldsymbol{e}_0 + \boldsymbol{e}_1$, E_{02} on E_0E_2 with coordinates (1, 0, 1) corresponding to $\boldsymbol{e}_0 + \boldsymbol{e}_2$, and E_{12} on E_1E_2 with coordinates (0, 1, 1) corresponding to $\boldsymbol{e}_1 + \boldsymbol{e}_2$ (Fig.).

Conversely, any projective coordinate system on a line can be extended to a coordinate system of a plane containing the line by taking the base-points E_0 and E_1 and unit-point E_{01} on the line, choosing a point E_2 not on the line as third base-point, and choosing as a new unit point E a point of the line $E_{01}E_2$ distinct from E_{01} and E_2. The original coordinate system on the line is then a restriction of the coordinate system in the plane to the coordinate axis E_0E_1.

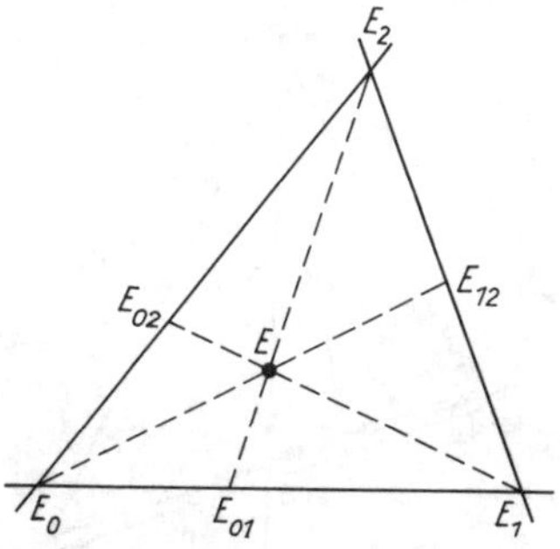

25.2-2 Restriction of the coordinate system of the plane onto any one of the three coordinate axes

25.3. Cross-ratios

Let A, B, C, D be four points of a projective line l (Fig.) such that no three of them coincide and that A and B are distinct. The line l and a point S not on l determine a plane, and with respect to a basis $\boldsymbol{e}_0$, $\boldsymbol{e}_1$ with origin S the four points correspond to vectors $\boldsymbol{a} = a_0\boldsymbol{e}_0 + a_1\boldsymbol{e}_1$, $\boldsymbol{b} = b_0\boldsymbol{e}_0 + b_1\boldsymbol{e}_1$, $\boldsymbol{c} = c_0\boldsymbol{e}_0 + c_1\boldsymbol{e}_1$, $\boldsymbol{d} = d_0\boldsymbol{e}_0 + d_1\boldsymbol{e}_1$. Since $\boldsymbol{a}$ and $\boldsymbol{b}$ are linearly independent, $\boldsymbol{c}$ and $\boldsymbol{d}$ can be expressed in terms of $\boldsymbol{a}$ and $\boldsymbol{b}$; one then has $\boldsymbol{c} = \lambda_0\boldsymbol{a} + \mu_0\boldsymbol{b}$ $(c_i = \lambda_0 a_i + \mu_0 b_i)$ and $\boldsymbol{d} = \lambda_1\boldsymbol{a} + \mu_1\boldsymbol{b}$ $(d_i = \lambda_1 a_i + \mu_1 b_i)$.

One can now define the cross-ratio of the four points:

Cross-ratio	$(A, B; C, D) = (\mu_0/\lambda_0):(\mu_1/\lambda_1)$

The cross-ratio does not depend on the choice of the vectors $\boldsymbol{a}, \boldsymbol{b}, \boldsymbol{c}, \boldsymbol{d}$ nor on the choice of S. The first part of this statement follows from a calculation of the cross-ratio for the new vectors $\boldsymbol{a}' = \alpha\boldsymbol{a}$, $\boldsymbol{b}' = \beta\boldsymbol{b}$, $\boldsymbol{c}' = \gamma\boldsymbol{c} = \gamma(\lambda_0\boldsymbol{a} + \mu_0\boldsymbol{b}) = \gamma[(\lambda_0/\alpha)\boldsymbol{a}' + (\mu_0/\beta)\boldsymbol{b}']$ and $\boldsymbol{d}' = \delta\boldsymbol{d} = \delta(\lambda_1\boldsymbol{a} + \mu_1\boldsymbol{b}) = \delta[(\lambda_1/\alpha)\boldsymbol{a}' + (\mu_1/\beta)\boldsymbol{b}']$. This gives $\lambda_0' = (\gamma/\alpha)\lambda_0$, $\mu_0' = (\gamma/\beta)\mu_0$, $\lambda_1' = (\delta/\alpha)\lambda_1$, $\mu_1' = (\delta/\beta)\mu_1$ or $(A, B; C, D) = (\mu_0'/\lambda_0'):(\mu_1'/\lambda_1') = (\mu_0/\lambda_0):(\mu_1/\lambda_1)$. If, on the other hand, S' is a point not on l and distinct from S, and if the vectors $\boldsymbol{a}', \boldsymbol{b}', \boldsymbol{c}', \boldsymbol{d}'$, with respect to S', correspond to the points A, B, C, D, then, as was shown for the projective plane, there is an affine mapping connecting the plane determined by l and S with the plane determined by l and S', which takes S into S' and the vectors $\boldsymbol{a}, \boldsymbol{b}$ into $\boldsymbol{a}', \boldsymbol{b}'$. Because of the linearity of the mapping $\boldsymbol{c}' = \lambda_0\boldsymbol{a}' + \mu_0\boldsymbol{b}'$ and $\boldsymbol{d}' = \lambda_1\boldsymbol{a}' + \mu_1\boldsymbol{b}'$, and so with respect to S' one has $(A, B; C, D) = (\mu_0/\lambda_0):(\mu_1/\lambda_1)$.

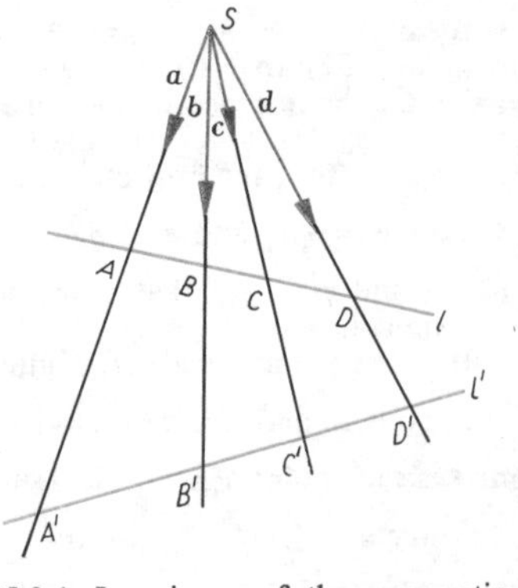

25.3-1 Invariance of the cross-ratio under projection

The cross-ratio remains invariant under central projection.

The invariance of the cross-ratio follows from the fact that the vectors $\boldsymbol{a}, \boldsymbol{b}, \boldsymbol{c}, \boldsymbol{d}$, relative to S, represent not only the points A, B, C, D as points of intersection of the four lines with the line l, but also the points of intersection A', B', C', D' with the line l'.

Representation of the cross-ratio in projective coordinates. In an arbitrary coordinate system on the line l, let a_i, b_i, c_i, d_i $(i = 0, 1)$ be the coordinates of the points A, B, C, D, where $c_i = \lambda_0 a_i + \mu_0 b_i$ and $d_i = \lambda_1 a_i + \mu_1 b_i$. Since $A \neq B$, the determinant $\Delta = |a_i b_i| = \begin{vmatrix} a_0 & b_0 \\ a_1 & b_1 \end{vmatrix} = a_0 b_1 - a_1 b_0$ is nonzero, so that from the system of equations for c_i and d_i the real numbers λ_i and μ_i can be calculated:

$$\lambda_0 = -\frac{1}{\Delta}\begin{vmatrix} b_0 & c_0 \\ b_1 & c_1 \end{vmatrix} = -|b_i c_i|/|a_i b_i|, \quad \mu_0 = \frac{1}{\Delta}\begin{vmatrix} a_0 & c_0 \\ a_1 & c_1 \end{vmatrix} = |a_i c_i|/|a_i b_i|,$$

$$\lambda_1 = -\frac{1}{\Delta}\begin{vmatrix} b_0 & d_0 \\ b_1 & d_1 \end{vmatrix} = -|b_i d_i|/|a_i b_i|, \quad \mu_1 = \frac{1}{\Delta}\begin{vmatrix} a_0 & d_0 \\ a_1 & d_1 \end{vmatrix} = |a_i d_i|/|a_i b_i|.$$

Hence

$$(A, B; C, D) = \frac{\mu_0}{\lambda_0} : \frac{\mu_1}{\lambda_1} = \frac{|a_i c_i|}{|b_i c_i|} : \frac{|a_i d_i|}{|b_i d_i|} = \frac{F(\boldsymbol{a}, \boldsymbol{c})}{F(\boldsymbol{b}, \boldsymbol{c})} : \frac{F(\boldsymbol{a}, \boldsymbol{d})}{F(\boldsymbol{b}, \boldsymbol{d})}.$$

Here $F(\boldsymbol{x}, \boldsymbol{y})$ denotes the area of the parallelogram determined by the vectors $\boldsymbol{x}$ and $\boldsymbol{y}$. This relation clarifies the connection with the usual definition of the cross-ratio of four proper points on an affine line l. By introducing a suitable factor ϱ one can arrange that A, B, C, D are the end-points of vectors $\boldsymbol{a}, \boldsymbol{b}, \boldsymbol{c}, \boldsymbol{d}$, drawn from S. The ratios of the areas of the parallelograms are those of triangles with the same height h (Fig.), that is, the ratios of the lengths along the line l. Treating these as directed segments, one has

$$(A, B; C, D) = (\overrightarrow{AC}/\overrightarrow{BC}):(\overrightarrow{AD}/\overrightarrow{BD}) = (\overrightarrow{AC}/\overrightarrow{CB}):(\overrightarrow{AD}/\overrightarrow{DB}),$$

that is, the quotient of the two ratios

$$(\overrightarrow{AC}/\overrightarrow{CB}) \text{ and } (\overrightarrow{AD}/\overrightarrow{DB}).$$

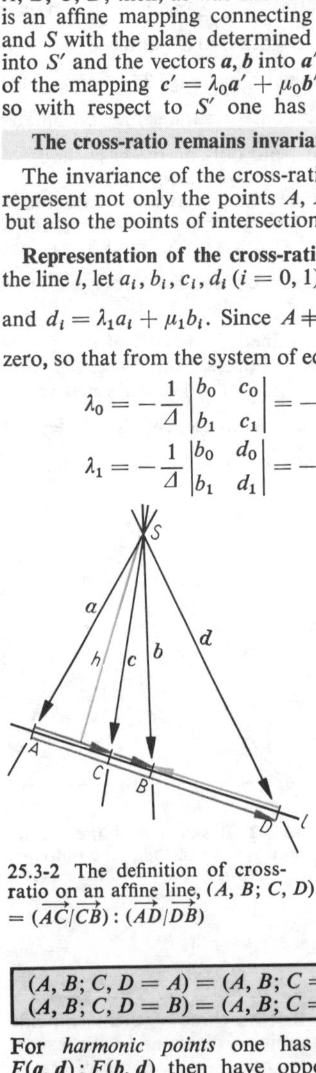

25.3-2 The definition of cross-ratio on an affine line, $(A, B; C, D) = (\overrightarrow{AC}/\overrightarrow{CB}):(\overrightarrow{AD}/\overrightarrow{DB})$

Special positions of the four points. If two of the points coincide, then the determinants, and therefore the cross-ratio, take special values; for example, $|a_i d_i| = 0$ if $A = D$, $|b_i d_i| = 0$ if $B = D$, $|b_i c_i| = 0$ if $C = B$ and $|a_i c_i| = 0$ if $A = C$.

$(A, B; C, D = A) = (A, B; C = B, D) = \infty$	$(A, B; C, D = C) = 1$
$(A, B; C, D = B) = (A, B; C = A, D) = 0$	

For *harmonic points* one has $(A, B; C, D) = -1$. The ratios of areas $F(\boldsymbol{a}, \boldsymbol{c}):F(\boldsymbol{b}, \boldsymbol{c})$ and $F(\boldsymbol{a}, \boldsymbol{d}):F(\boldsymbol{b}, \boldsymbol{d})$ then have opposite signs. From the definition of the vector product it follows that for either sense of rotation in the plane determined by l and S only one of the vectors $\boldsymbol{c}$ or $\boldsymbol{d}$ can lie between $\boldsymbol{a}$ and $\boldsymbol{b}$. One says that the pairs of points A, B and C, D separate each other.

There are 24 permutations of four points. If the two pairs are interchanged, or if the two points in each pair are interchanged, then the cross-ratio is unaltered:

$$(A, B; C, D) = (C, D; A, B) = (B, A; D, C) = (D, C; B, A).$$

$$\begin{array}{lll} (A, B; C, D) = k, & (A, B; D, C) = 1/k, & (A, C; B, D) = 1 - k \\ (A, D; B, C) = (k-1)/k, & (A, D; C, B) = k/(k-1), & (A, C; D, B) = 1/(1-k) \end{array}$$

Hence the 24 possible permutations give six values of the cross-ratio. In calculating these cross-ratios one must go back to the definition of a determinant; for example,

$$|a_i b_i| \cdot |c_i d_i| - |c_i b_i| \cdot |a_i d_i| = (a_0 b_1 - a_1 b_0)(c_0 d_1 - c_1 d_0) - (c_0 b_1 - c_1 b_0)(a_0 d_1 - a_1 d_0)$$
$$= (a_0 c_1 - a_1 c_0)(b_0 d_1 - b_1 d_0) = |a_i c_i| \cdot |b_i d_i|.$$

It follows immediately from the definition that if the points of the second pair are interchanged, one obtains the reciprocal of the cross-ratio. If two points of different pairs are interchanged, one obtains, for example,

$$(A, C; B, D) = \frac{|a_i b_i| \cdot |c_i d_i|}{|c_i b_i| \cdot |a_i d_i|} = \frac{|c_i b_i| \cdot |a_i d_i| + |a_i b_i| \cdot |c_i d_i| - |c_i b_i| \cdot |a_i d_i|}{|c_i b_i| \cdot |a_i d_i|}$$
$$= 1 + \frac{|a_i c_i| \cdot |b_i d_i|}{|c_i b_i| \cdot |a_i d_i|} = 1 - k, \quad \text{since} \quad |c_i b_i| = -|b_i c_i|.$$

Similarly, $(A, D; B, C) = 1 - 1/k = (k-1)/k$.

The *cross-ratio of four lines* of a pencil in a projective plane is defined as the cross-ratio of the four points of intersection of these lines with an arbitrary line l. The fact that this is independent of the choice of l follows from the invariance of the cross-ratio under central projection (Fig.).

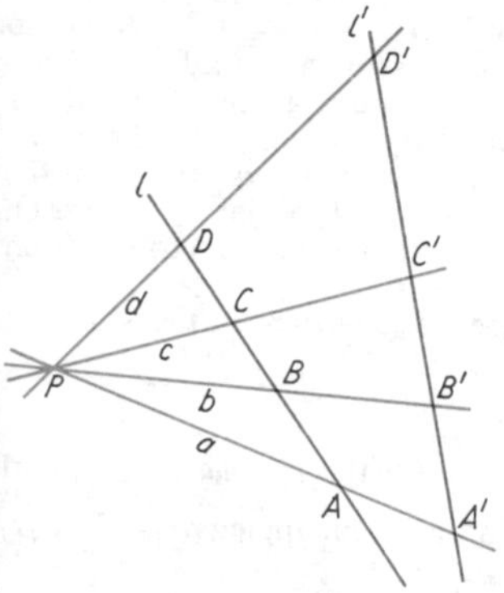

25.3-3 The cross-ratio of four lines is independent of the choice of the line l

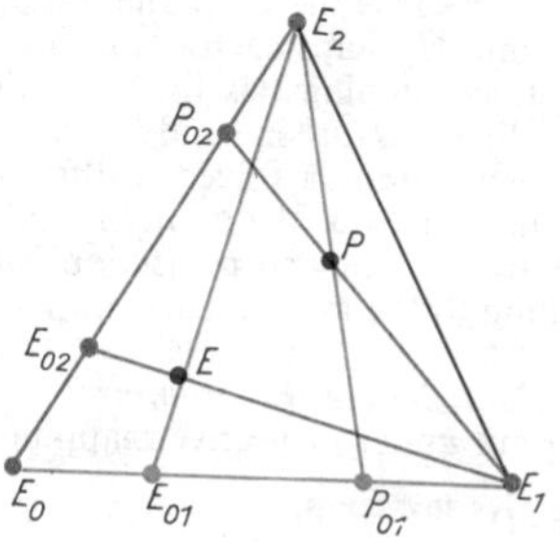

25.3-4 Introduction of projective coordinates by means of cross-ratio

Introduction of projective coordinates by means of cross-ratios. If with respect to a basis E_i $(i = 0, 1, 2)$ and E a point P has the projective coordinates (x_0, x_1, x_2), and if one projects the points E and P from E_2 onto the coordinate axis E_0E_1 (Fig.), then E_{01} and P_{01} have the coordinates $(1, 1)$ and (x_0, x_1) in the coordinate system restricted to the line E_0E_1. One then has $(E_0, E_1; E_{01}, P_{01}) = x_1 : x_0$. Similarly, $(E_0, E_2; E_{02}, P_{02}) = x_2 : x_0$ for the projections of E and P onto the line E_0E_2. One can therefore define the coordinate ratios $x_0 : x_1 : x_2$ by means of cross-ratios, that is, by projective quantities only. In the special case when E_1 and E_2 are improper points and the lines E_1P and E_2P are parallel to the coordinate axes E_0E_1 and E_0E_2, the coordinate ratios $x_1 : x_0$ and $x_2 : x_0$ represent *affine parallel coordinates.*

Given three points of a line, one can always choose a fourth point of the line so that the cross-ratio of these four points takes a given real value λ; for one can treat the three given points as the basis E_0, E_1, E_{01} of a coordinate system on the line, and then the fourth point is determined by

$$(E_0, E_1; E_{01}, P_{01}) = x_1 : x_0 = \lambda.$$

25.4. Projective mappings

Under a central projection A the projective coordinates of the image point P are the same as those of the original point, referred to the images $\overline{E}_i$, $\overline{E}$ of the basis E_i, E. With respect to an arbitrary basis E'_i, E', with corresponding basis vectors e'_i, e' connected to the basis vectors $\overline{e}_i$ by equations

$\bar{e}_i = \sum_{j=0}^{2} a_{ij} e'_j$ with $\det(a_{ij}) \neq 0$, the point P has the coordinates $x'_i = \sum_{j=0}^{2} a_{ij} x_j$. Hence a central projection is described with respect to an arbitrary basis by means of a linear coordinate transformation A with non-singular matrix (a_{ij}): $\varrho x'_i = \sum_{j=0}^{2} a_{ij} x_j$. If A is followed by another central projection B with the coordinate transformation B: $\varrho x''_k = \sum_{j=0}^{2} b_{kj} x'_j$ with $\det(b_{kj}) \neq 0$, then the resultant $\Gamma = BA$ of the two central projections is described by the coordinate transformation C: $\varrho x''_k = \sum_{i=0}^{2} b_{ki} \sum_{j=0}^{2} a_{ij} x_j = \sum_{j=0}^{2} c_{kj} x_j$ with $\det(c_{kj}) \neq 0$. As a generalization of the central projection one defines a *projective mapping* as a one-to-one mapping of a projective plane onto itself or another plane, described by a *regular linear* coordinate transformation $\varrho x'_i = \sum_{j=0}^{2} a_{ij} x_j$ $(i = 0, 1, 2)$ with $\det(a_{ij}) \neq 0$. It can be shown that any projective mapping arises from finitely many successive central projections.

Main theorem of projective geometry. There is exactly one projective mapping of a plane Π onto a plane Π' that takes four given points of Π no three of which are collinear into four points of Π' no three of which are collinear.

To prove this one takes as bases in Π and Π' the four given points, so that in the corresponding coordinate systems the points of each of the two sets have the projective coordinates $(1, 0, 0)$, $(0, 1, 0)$, $(0, 0, 1)$ and $(1, 1, 1)$. If one substitutes these coordinates for the four points and their images in $\varrho x'_i = \sum_{j=0}^{2} a_{ij} x_j$, it follows that $a_{ij} = 0$ for $i \neq j$ and $a_{ij} = 1$ for $i = j$, that is, $\varrho x'_i = x_i$. Conversely, the mapping described by the coordinate transformation $\varrho x'_i = x_i$ takes the four points of the basis in Π into those in Π'. It is therefore the unique projective mapping of this kind.

A central projection of Π onto Π' maps each point on the line of intersection $s = (\Pi \cap \Pi')$ onto itself. If an arbitrary projective mapping has this property, then it must be a central projection. For if l' in Π' is the image of a line l in Π, then l and l' meet in the same point of s, since by hypothesis the point of intersection of l with s remains fixed, and the lines l and l' therefore determine a plane. Then if P' and Q' on l' are the images of the points P and Q on l, the point of intersection C of the lines PP' and QQ' is the centre of a central projection, which takes P into P', Q into Q' and each point of s into itself. According to the main theorem, this central projection is identical to the given projective mapping.

According to Felix KLEIN's *Erlangen Programme* (1872), projective geometry is the study of properties that remain invariant under projective mappings.

The cross-ratio is a projective invariant.

If the cross-ratio $(A, B; C, D)$ of four points A, B, C, D on a line l with $c_i = \lambda_0 a_i + \mu_0 b_i$ and $d_i = \lambda_1 a_i + \mu_1 b_i$ is determined by $(\mu_0/\lambda_0) : (\mu_1/\lambda_1)$, then under the projective mapping $\varrho x'_i = \sum_{j=0}^{2} a_{ij} x_j$ $(i = 0, 1, 2)$ with $\det(a_{ij}) \neq 0$, the images A', B', C', D' satisfy the relations $\varrho a'_i = \sum_{j=0}^{2} a_{ij} a_j$, $\varrho b'_i = \sum_{j=0}^{2} a_{ij} b_j$ and $\varrho c'_i = \sum_{j=0}^{2} a_{ij} c_j = \varrho(\lambda_0 a'_i + \mu_0 b'_i)$, $\varrho d'_i = \sum_{j=0}^{2} a_{ij} d_j = \varrho(\lambda_1 a'_i + \mu_1 b'_i)$, from which one obtains $(A', B'; C', D') = (\mu'_0/\lambda'_0) : (\mu'_1/\lambda'_1) = (\mu_0/\lambda_0) : (\mu_1/\lambda_1) = (A, B; C, D)$.

All the projective mappings form a *group*, which is characterized by the invariance of the cross-ratio. Those projective mappings that leave a line fixed, but not necessarily pointwise, form a subgroup. The group of *affine mappings* is a subgroup for which this fixed line is the improper line of the plane. This, in turn, has a subgroup of *similarity mappings*, which take orthogonal lines into orthogonal lines. The subgroup of *congruence mappings* also leaves distances between two points invariant.

On a projective line three basis points determine the projective coordinates, and a projective mapping is described by a linear transformation $\varrho x'_0 = a x_0 + b x_1$, $\varrho x'_1 = c x_0 + d x_1$ with $ad - bc \neq 0$.

Main theorem on projective mappings of a line. There is exactly one projective mapping between two lines that takes three distinct points of the original line into three distinct points of the image line.

Central projections are then projective mappings of lines whose point of intersection is a *fixed point*.

The equation of a line. A point P of a projective line is characterized relative to two points A and B of the line with corresponding vectors $\boldsymbol{a}$ and $\boldsymbol{b}$ by the ratio $t_1 : t_2$ of the parameters in the expression $\varrho \boldsymbol{x} = t_1 \boldsymbol{a} + t_2 \boldsymbol{b}$. In homogeneous coordinates (x_0, x_1, x_2) one therefore has the following system of homogeneous equations, from which the values of t_1, t_2, ϱ can be found, to within a non-zero factor λ. If one assumes, without loss of generality, that $x_2 \neq 0$, one obtains $t_1 = \lambda(x_0 b_1 - x_1 b_0)$, $t_2 = \lambda(x_1 a_0 - x_0 a_1)$ and $\varrho = (\lambda/x_2)\{(x_0 b_1 - x_1 b_0)\, a_2 + (x_1 a_0 - x_0 a_1)\, b_2\}$. If one substitutes these values in the first equation, one obtains $x_0 u_0 + x_1 u_1 + x_2 u_2 = 0$, where $u_0 = a_2 b_1 - a_1 b_2$, $u_1 = a_0 b_2 - a_2 b_0$ and $u_2 = a_1 b_0 - a_0 b_1$.

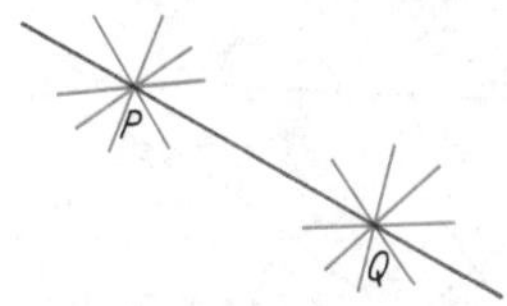

25.4-1 The connection between two pencils

$$\begin{aligned} t_1 a_0 + t_2 b_0 - \varrho x_0 &= 0 \\ t_1 a_1 + t_2 b_1 - \varrho x_1 &= 0 \\ t_1 a_2 + t_2 b_2 - \varrho x_2 &= 0 \end{aligned}$$

Equation of a line	$x_0 u_0 + x_1 u_1 + x_2 u_2 = 0$

The *Plücker line coordinates* (u_0, u_1, u_2), like the point coordinates (x_0, x_1, x_2), are determined only to within a non-zero scalar λ. The formal equivalence of the two triples in the equation of a line makes it clear that for a given triple (u_0, u_1, u_2) the equation determines a line l as the locus of all points whose coordinate triple (x_0, x_1, x_2) satisfies the equation, while for a given triple (x_0, x_1, x_2) the equation determines a point P as the vertex of a pencil of lines whose triple (u_0, u_1, u_2) satisfies the equation. The system of homogeneous linear equations $x_0 u_0 + x_1 u_1 + x_2 u_2 = 0$, $x_0 v_0 + x_1 v_1 + x_2 v_2 = 0$ determines the point (x_0, x_1, x_2) common to the two lines (u_0, u_1, u_2) and (v_0, v_1, v_2), their point of intersection. Similarly, the equations $x_0 u_0 + x_1 u_1 + x_2 u_2 = 0$ and $y_0 u_0 + y_1 u_1 + y_2 u_2 = 0$ determine the line (u_0, u_1, u_2) common to the two pencils with vertex at (x_0, x_1, x_2) and at (y_0, y_1, y_2), that is, the line joining their vertices (Fig.).

Principle of duality. The pairs of concepts 'point of a line' and 'line of a pencil', 'join' and 'intersect', 'lie on' and 'pass through' can be interchanged, since they are represented by equivalent algebraic operations or equations. There is, in this sense, a *principle of duality*: true statements of projective geometry are transformed by the interchange into true statements; for example, the theorem 'two distinct points lie on exactly one line' goes over into the true theorem 'two distinct lines pass through exactly one point'. To every theorem of projective geometry there is, therefore, a dual form, whose proof follows from that of the original theorem.

Theorem of Desargues. If the lines joining corresponding vertices of two triangles pass through a point S, then the points of intersection of corresponding sides lie on a line s.

Dual form: If the points of intersection of corresponding sides of two triangles lie on a line s, then the lines joining corresponding vertices pass through a point S.

The dual form of the theorem of Desargues is also its converse. For a proof of the theorem itself one assumes that the triangles are $A_1B_1C_1$ and $A_2B_2C_2$, where the lines A_1A_2, B_1B_2 and C_1C_2 meet in the point S (Fig.). It is required to prove that the points $A = (B_1C_1 \cap B_2C_2)$, $B = (C_1A_1 \cap C_2A_2)$ and $C = (A_1B_1 \cap A_2B_2)$ lie on a line s. One can apply a projective mapping to the two triangles in such a way that A and B go over into two improper points. Then in the image the sides $B_1'C_1'$ and $B_2'C_2'$ are parallel, and so are $C_1'A_1'$ and $C_2'A_2'$. By the intercept theorem it follows that $A_1'B_1'$ and $A_2'B_2'$ are parallel, that is, C' also lies on the improper line, and so in the original figure C lies on the line $s = AB$.

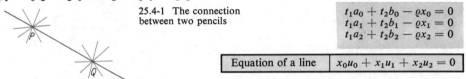

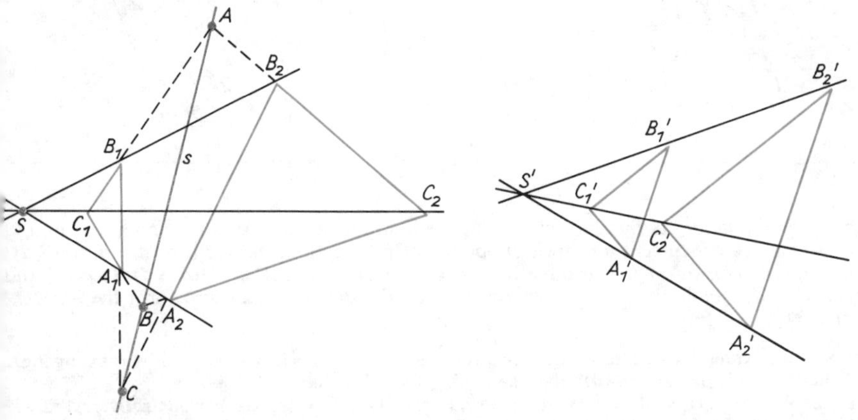

25.4-2 The theorem of Desargues

Theorem of Pappus. Given two lines l_1 and l_2, let A_1, B_1, C_1 be three points on l_1 and A_2, B_2, C_2 three points on l_2, distinct from the point of intersection O of the two lines. Then the points of intersection $A = (B_1C_2 \cap B_2C_1)$, $B = (C_1A_2 \cap C_2A_1)$, $C \cap (A_1B_2 \cap A_2B_1)$ of their cross-joins lie on a line l (Fig.).

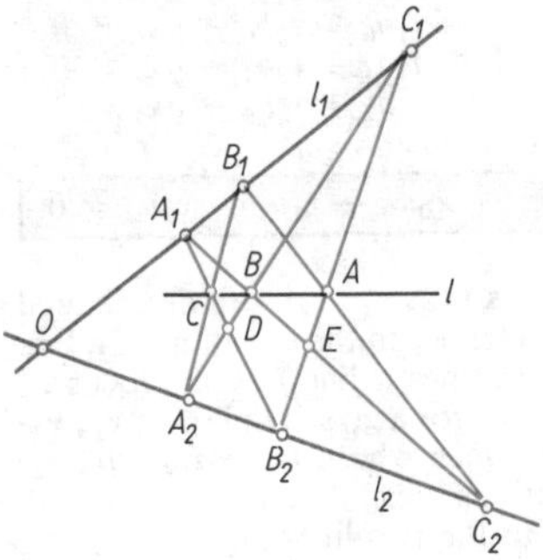

25.4-3 The theorem of Pappus

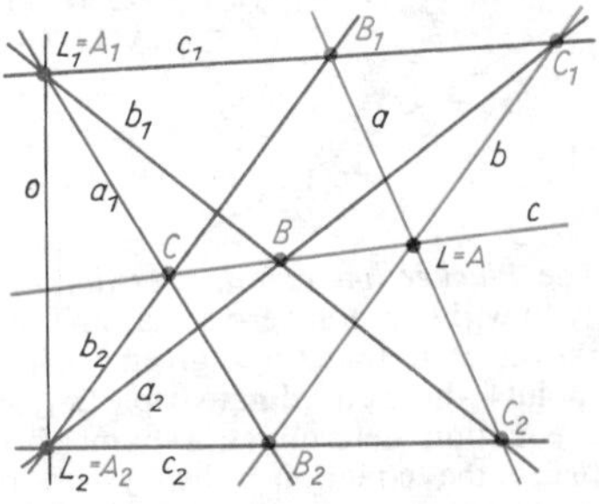

25.4-4 Dualizing Pappus' theorem

Dual form: Given two points L_1 and L_2, let a_1, b_1, c_1 be three lines through L_1 and a_2, b_2, c_2 three lines through L_2, distinct from the line o joining the two points. Then the lines $a = [(b_1 \cap c_2), (b_2 \cap c_1)]$, $b = [(c_1 \cap a_2), (c_2 \cap a_1)]$, $c = [(a_1 \cap b_2), (a_2 \cap b_1)]$ meet in a point L.

For a proof of the theorem of PAPPUS one introduces the points of intersection $D = A_1B_2 \cap A_2C_1$ and $E = B_2C_1 \cap A_1C_2$ (Fig). If one projects the line A_1B_2 from A_2 onto the line l_1, the ordered triple (C, D, B_2) goes into the triple (B_1, C_1, O), while the point A_1 remains fixed. A similar central projection of l_1 onto the line B_2C_1 from C_2 takes the ordered triple (A_1, B_1, O) into the triple (E, A, B_2) and leaves the point C_1 fixed. The two central projections together take the ordered quadruple (A_1, C, D, B_2) into the quadruple (E, A, C_1, B_2), and, since the point B_2 is fixed, this represents a central projection, whose centre B is the point of intersection of the lines C_1A_2 and C_2A_1. But then CA must also pass through B, that is, A, B and C lie on a line.

Complete quadrangle and complete quadrilateral. A *complete quadrangle* consists of four vertices A, B, C, D no three of which are collinear and the six lines joining them in pairs, the *six sides* AB, CD; AD, BC; AC, BD. The three points of intersection $P = (AB \cap CD)$, $Q = (AD \cap BC)$ and $R = (AC \cap BD)$ of these six lines are called the *diagonal points* and are the vertices of the *diagonal triangle* PQR (Fig.).

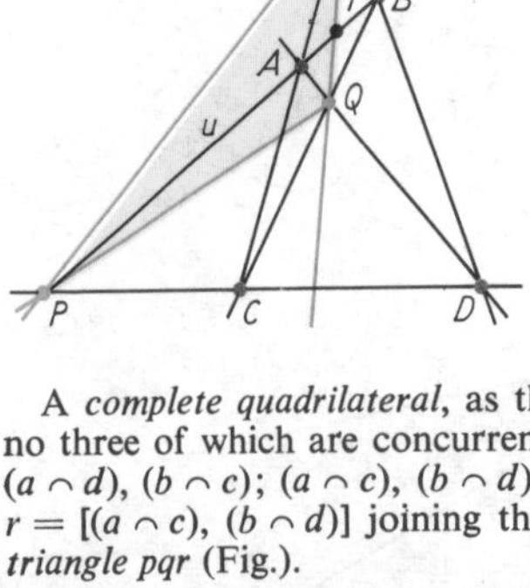

25.4-5 The complete quadrangle

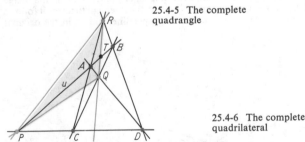

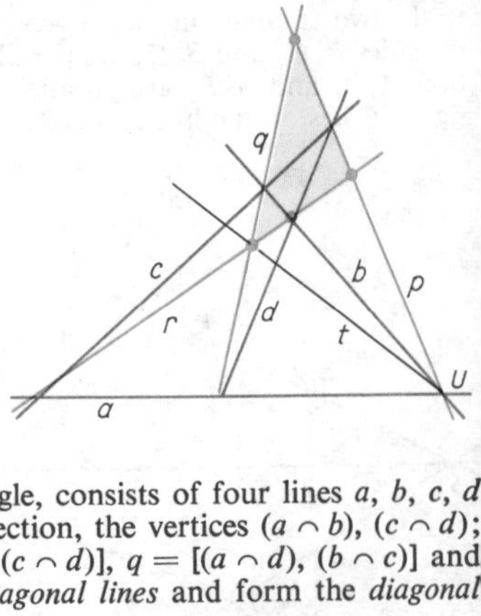

25.4-6 The complete quadrilateral

A *complete quadrilateral*, as the dual of a complete quadrangle, consists of four lines a, b, c, d no three of which are concurrent and their six points of intersection, the vertices $(a \cap b)$, $(c \cap d)$; $(a \cap d)$, $(b \cap c)$; $(a \cap c)$, $(b \cap d)$. The three lines $p = [(a \cap b), (c \cap d)]$, $q = [(a \cap d), (b \cap c)]$ and $r = [(a \cap c), (b \cap d)]$ joining these six points are called the *diagonal lines* and form the *diagonal triangle pqr* (Fig.).

On any side of a complete quadrangle, the two vertices are harmonically separated by the diagonal point and the point of intersection with the line joining the other diagonal points.

Dual form: At any vertex of a complete quadrilateral, the two sides are harmonically separated by the diagonal and the line joining the vertex to the point of intersection of the other diagonals.

If, for example, one picks out the side $AB = u$ of the complete quadrangle (↑ Fig. 25.4-5), the vertices A and B are harmonically separated by the diagonal point P and the point of intersection T of this side with the line QR joining the other diagonal points. Correspondingly, in the dual figure, the sides a and b that meet at the vertex U are harmonically separated by the diagonal p and the line t joining U to the point of intersection $(q \frown r)$ of the other two diagonals.

The proof of the theorem on the complete quadrangle depends on the fact that any point X of a projective plane can be characterized by the vector $\boldsymbol{x}$ drawn to it from a point S outside the plane. The vectors $\boldsymbol{a}, \boldsymbol{b}, \boldsymbol{c}, \boldsymbol{d}$ corresponding to the vertices A, B, C, D are linearly dependent, but any three of them are linearly independent, since no three of the points are collinear. Therefore there exist four real non-zero numbers $\alpha, \beta, \gamma, \delta$ such that $\alpha\boldsymbol{a} + \beta\boldsymbol{b} + \gamma\boldsymbol{c} + \delta\boldsymbol{d} = \boldsymbol{o}$. The diagonal point P on the sides AB and CD can be characterized by a vector $\boldsymbol{p} = \alpha\boldsymbol{a} + \beta\boldsymbol{b} = -(\gamma\boldsymbol{c} + \delta\boldsymbol{d})$, which is a linear combination of the vectors $\boldsymbol{a}$ and $\boldsymbol{b}$, and also of the vectors $\boldsymbol{c}$ and $\boldsymbol{d}$. Similarly, the other diagonal points Q and R are characterized by the vectors $\boldsymbol{q} = \alpha\boldsymbol{a} + \delta\boldsymbol{d} = -(\beta\boldsymbol{b} + \gamma\boldsymbol{c})$ and $\boldsymbol{r} = \alpha\boldsymbol{a} + \gamma\boldsymbol{c} = -(\beta\boldsymbol{b} + \delta\boldsymbol{d})$. If T is the point of intersection of QR with AB, then the corresponding vector $\boldsymbol{t}$ is given by $\boldsymbol{t} = \boldsymbol{q} + \boldsymbol{r} = \alpha\boldsymbol{a} - \beta\boldsymbol{b}$. The cross-ratio of the four points A, B, P, T is then given by $(A, B; P, T) = [\alpha/\beta] : [\alpha/(-\beta)] = -1$.

By means of the complete quadrangle one can construct, using a ruler only, the fourth harmonic point T of a point pair A, B and a third point P on a line u (Fig.). If one chooses on a line through P the remaining vertices C and D of a complete quadrangle, distinct from each other and from P, then the diagonal points Q and R are determined by $Q = (AD \frown BC)$ and $R = (AC \frown BD)$. The diagonal QR then cuts the side AB in the required fourth harmonic point T.

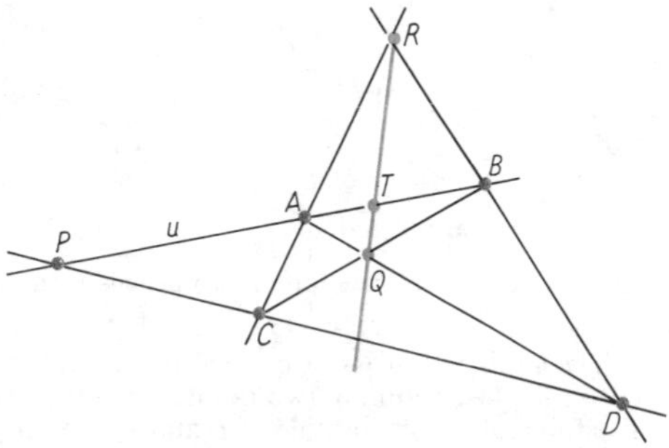

25.4-7 Construction of the fourth harmonic point

Duality in space. To the equation of a line $\sum_{i=0}^{2} x_i u_i = 0$ with Plücker coordinates u_i in the plane there corresponds in space the equation $\sum_{i=0}^{3} x_i u_i = 0$ of a plane. The pairs of dual concepts in space are therefore 'point' and 'plane', and a line through two points in space corresponds to the line common to two planes, so that the concept 'line' is self-dual. The theorem 'A point and a line not through it determine exactly one plane' has the dual form 'A plane and a line not in it determine exactly one point'.

In n-dimensional space the dual of a point is a *hyperplane*, and generally the dual of an m-dimensional subspace is a $(n - m - 1)$-dimensional subspace.

Collineations. Any projective mapping, as a one-to-one correspondence between the points of two projective planes, can be regarded as a one-to-one correspondence between the lines of the planes, if the lines, as duals of the points, are regarded as the basic elements of the plane. In this interpretation, a projective mapping is described relative to the coordinates u_i and u_i' by a linear transformation of the form $\varrho u_i' = \sum_{j=0}^{2} \alpha_{ij} u_j$ $(i = 0, 1, 2)$ with $\det(\alpha_{ij}) \neq 0$.

Dual form of the main theorem on projective mappings. There is exactly one projective mapping of the lines of a plane Π onto the lines of a plane Π' that takes four given lines of Π no three of which are concurrent into four given lines of Π' no three of which are concurrent.

The cross-ratio of four lines of a pencil is an invariant of a projective mapping.

Since projective mappings take collinear points into collinear points and concurrent lines, that is, lines of a pencil, into concurrent lines, they are called *collineations*.

Dual form of the main theorem on projective mappings of a line. There is exactly one collineation of a pencil with vertex P into a pencil with vertex P' that maps three distinct lines of the first pencil into three distinct lines of the second.

Under a central projection the correspondence between the points of two lines l and l' is by means of lines through the centre C, and the point of intersection $F = (l \frown l')$ is a *fixed point* of the mapping. Conversely, any collineation of two lines having a fixed point is a central projection or *central collineation*. Dually, the correspondence between the lines of two pencils L and L' under a central projection is by means of points on a fixed line c; for example, if p is a line of the pencil L, its image p' is the line joining the point $P = (p \frown c)$ to the vertex of the pencil L'. The line $f = (L, L')$ is then a *fixed line* of the mapping (Fig.). Conversely, any collineation between two pencils having a fixed line is a central collineation.

Since the cross-ratio of four lines p_i $(i = 1, 2, 3, 4)$ of a pencil L is equal to that of their points of intersection $P_i = (p_i \cap l)$ with an arbitrary line l that does not pass through L, and since collineations are characterized by the invariance of the cross-ratio, for any collineation λ between two pencils L and L' there is also a collineation λ^* between two lines l and l' that do not belong to the pencils L' and L, respectively. Any point P of l is the point of intersection of l with some line p of the pencil L, and then λ^* maps P into the point $P' = (p' \cap l')$ of the line l'. In the same way, for any collineation λ^* between two lines l and l' one can construct a collineation λ between two pencils L and L' in which any line $p = PL$ corresponds to the line $p' = P'L'$.

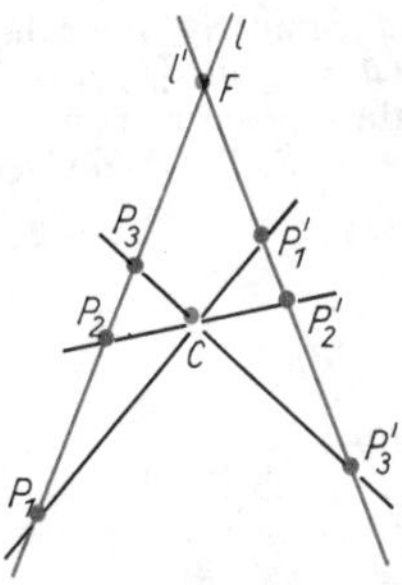

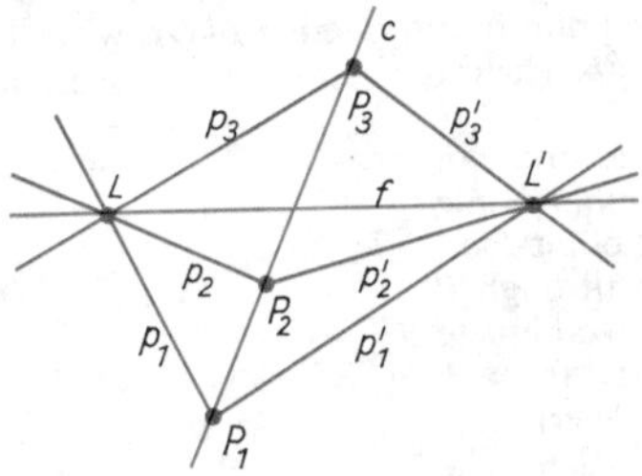

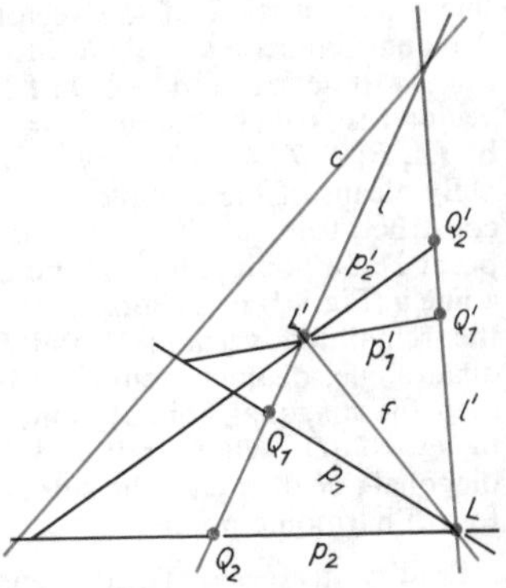

25.4-8 Central projection

25.4-9 Central collineation of two pencils with vertices L and L'

The collineation λ so obtained is a central collineation if the vertices L and L' of the two pencils lie on the line f joining two points that correspond in the collineation λ^*, because this line is then a fixed line of λ, for example, if L and L' are corresponding points of l and l' with respect to λ (Fig.).

Similarly, the collineation λ^* between two lines l and l' that arises from a collineation λ between two pencils L and L' is central if l and l' meet in the point of intersection $F = (p \cap p')$ of two lines p and p' that correspond under λ, because this point F is then a fixed point of λ^* (Fig.).

If p_1, p_2 and p_1', p_2' are two further pairs of lines that correspond under λ, and if they cut the lines l and l', respectively, in the points P_1, P_2 and P_1', P_2', then the vertex $C = (s_1 \cap s_2)$ of the central collineation is determined as the point of intersection of the two lines $s_1 = P_1P_1'$ and $s_2 = P_2P_2'$ (Fig.).

A *correlation* maps the points of a projective plane Π one to one onto the lines of a plane Π' and the lines of Π onto the points of Π'. In point and line coordinates, correlations are given by linear transformations

$$\varrho u_i' = \sum_{j=0}^{2} a_{ij} x_j \quad (i = 0, 1, 2) \quad \text{with} \quad \det(a_{ij}) \neq 0,$$

$$\varrho x_i' = \sum_{j=0}^{2} b_{ij} u_j \quad (i = 0, 1, 2) \quad \text{with} \quad \det(b_{ij}) \neq 0.$$

A correlation takes the point of intersection of two lines l_1, l_2 into the line joining the points L_1, L_2, the images of the lines l_1, l_2, and the line joining two points P_1, P_2 into the point of intersection of the lines p_1, p_2, the images of the points P_1, P_2.

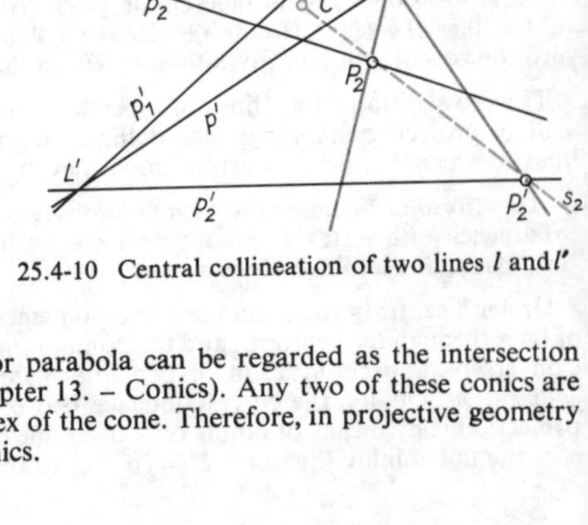

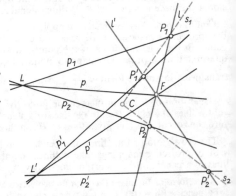

25.4-10 Central collineation of two lines l and l'

25.5. Conics

The equation of a conic. An ellipse, hyperbola, or parabola can be regarded as the intersection of a cone with a suitable projective plane (see Chapter 13. – Conics). Any two of these conics are mapped into each other by projection from the vertex of the cone. Therefore, in projective geometry there is no longer any distinction between these conics.

In homogeneous projective coordinates the equation of a conic is

$$a_{00}x_0^2 + a_{11}x_1^2 + a_{22}x_2^2 + 2a_{01}x_0x_1 + 2a_{02}x_0x_2 + 2a_{12}x_1x_2 = 0$$

or

$$\sum_{i,j=0}^{2} a_{ij}x_ix_j = 0 \quad \text{with} \quad a_{ij} = a_{ji} \quad \text{and} \quad \det(a_{ij}) \neq 0.$$

By a suitable linear transformation a non-degenerate conic can be reduced to the form $x_0^2 - x_1^2 - x_2^2 = 0$. A quadratic form with non-vanishing determinant does not necessarily represent a conic. The equation $x_0^2 + x_1^2 + x_2^2 = 0$ of the *empty conic* cannot represent a real conic under any transformation. A quadratic form $\sum_{i,j=0}^{2} a_{ij}x_ix_j$ with $\det(a_{ij}) = 0$ represents a *degenerate* or *singular conic*, which can be a pair of lines, a point, or a line counted twice.

Polarity. For any regular conic $\sum_{i,j=0}^{2} a_{ij}x_ix_j = 0$ with $\det(a_{ij}) \neq 0$ there is a special correlation $u_i = \sum_{j=0}^{2} a_{ij}x_j$ $(i = 0, 1, 2)$ of the plane onto itself, a so-called *polarity*. With any point P as *pole* it associates a line p as *polar*. The coordinates y_i of a point Q of the polar of P then satisfy the equation $\sum_{i=0}^{2} y_iu_i = \sum_{i,j=0}^{2} a_{ij}y_ix_j = 0$, where x_j are the coordinates of the pole P. All points Q of the polar of P are *conjugate* to P. The equation of the polar of a point P of the conic is, however, that of the tangent at P.

The polar p of a point P of a conic is the tangent t to the conic at P.

Any point P of the conic is therefore conjugate to every point of the tangent t at P, and therefore conjugate to itself. The points of the conic are the only points of the plane that are conjugate to themselves in this way.

If the tangents t_1, t_2 at two points B_1, B_2 of the conic meet at a point P, this point is conjugate to B_1 and to B_2, and therefore, since all the points conjugate to P lie on a line, it is conjugate to all points of the line B_1B_2, which is thus the polar p of P (Fig.).

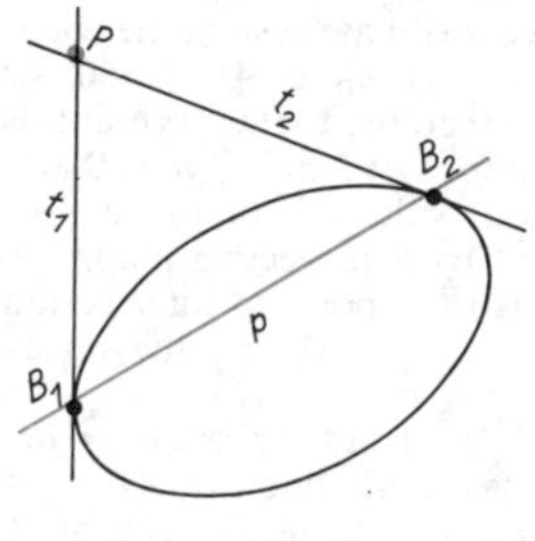

25.5-1 Pole P and polar p when there are real tangents from P to the conic

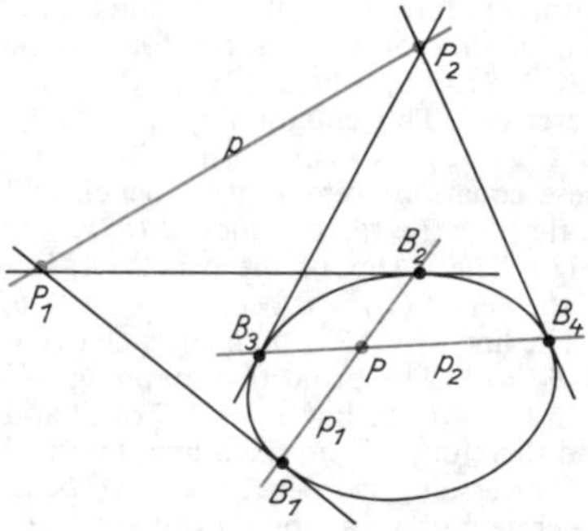

25.5-2 Pole P and polar p when there are no real tangents from P to the conic

This conclusion makes it possible to construct the polar p of any point P, even if no tangents to the conic pass through P. Let p_1, p_2 be two lines through P that cut the conic in B_1, B_2 and B_3, B_4, respectively. Then the point of intersection P_1 of the tangents at B_1 and B_2 and the point of intersection P_2 of the tangents at B_3 and B_4 are conjugate to any points of the lines p_1 and p_2, respectively, and therefore to their point of intersection P. Hence the line $p = P_1P_2$ is the polar of the point P (Fig.).

The line l through the two points P and Q with coordinates x_i and y_i cuts the conic $\sum_{i,j=0}^{2} a_{ij}x_ix_j = 0$ at the points R, R' with the coordinates $r_i = x_it_1 + y_it_2$, which are determined by the equation $\sum_{i,j=0}^{2} a_{ij}r_ir_j = \sum_{i,j=0}^{2} a_{ij}(x_it_1 + y_it_2)(x_jt_1 + y_jt_2) = t_1^2\alpha + t_2^2\beta + 2t_1t_2\gamma = 0$, where $\alpha = \sum_{i,j}^{2} a_{ij}x_ix_j$, $\beta = \sum_{i,j=0}^{2} a_{ij}y_iy_j$ and $\gamma = \sum_{i,j}^{2} a_{ij}x_iy_j$. The two solutions $t_1 : t_2$ and $t_1' : t_2'$ of this quadratic equation differ only in sign if $\gamma = 0$, that is, if P and Q are conjugate. The four points P, Q, R, R' are then harmonic, since $(P, Q; R, R') = (t_2/t_1) : (t_2'/t_1') = -1$.

The polar p of a point P is the locus of points Q such that the points of intersection R and R' of PQ with the conic separate P, Q harmonically (Fig.).

Since conics, as sets of points, are *curves of the second order*, their duals are *curves of the second class*, regarded as envelopes of their tangents. If P is a point of the conic $\sum_{i,j=0}^{2} a_{ij}x_i x_j = 0$, then the tangent at P (as the polar) has the coordinates $u_i = \sum_{j=0}^{2} a_{ij}x_j$. Conversely, from the coordinates u_j of the polar p one obtains the coordinates of P as $x_i = \sum_{j=0}^{2} b_{ij}u_j$ $(i = 0, 1, 2)$; here the matrix (b_{ij}) is the inverse of the matrix (a_{ij}), because it describes the inverse mapping. If one puts these coordinates into the equation of a curve of the second order, one obtains the equation $\sum_{i,j=0}^{2} b_{ij}u_i u_j = 0$ of a curve of the second class.

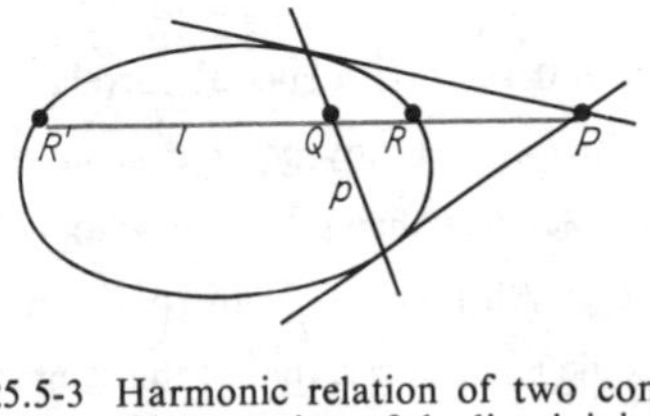

25.5-3 Harmonic relation of two conjugate points and the points of intersection of the line joining them with the conic

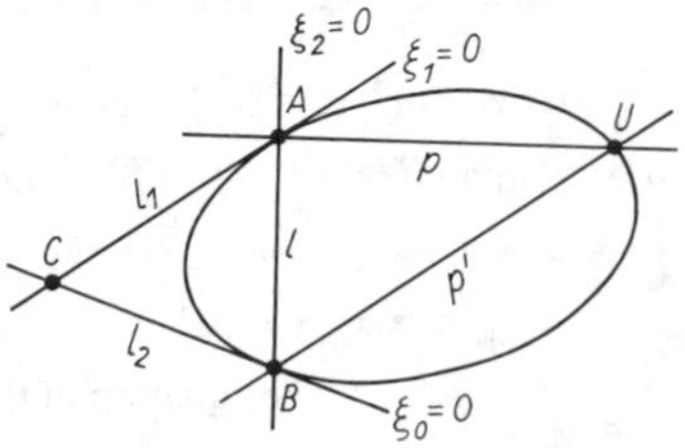

25.5-4 Projective generation of the conic

Projective generation of a conic. The equation $\sum_{i,j=0}^{2} a_{ij}x_i x_j = 0$ of a conic takes a simple form if two points A and B of the conic and the point of intersection C of the tangents at A and B are taken as base-points of a system of coordinates, and therefore have the coordinates $A(1, 0, 0)$, $B(0, 1, 0)$ and $C(0, 0, 1)$. If one puts these coordinates a_i of A and b_i of B into the equation of the conic, one obtains $a_{00} = 0$ and $a_{11} = 0$. Since C is conjugate to A and B, one has $\sum_{i,j=0}^{2} a_{ij}a_i c_j = 0$ and $\sum_{i,j=0}^{2} a_{ij}b_i c_j = 0$. Putting in the values of a_i, b_i and c_i, one obtains $a_{02} = a_{12} = 0$. In this coordinate system the equation of the conic takes the form $2a_{01}x_0x_1 + a_{22}x_2^2 = 0$, that ist, $(2x_0/a_{22}) \cdot a_{01}x_1 + x_2^2 = 0$, or $\xi_0\xi_1 + \xi_2^2 = 0$, where $(2x_0/a_{22}) = \xi_0$, $a_{01}x_1 = \xi_1$ and $x_2 = \xi_2$, and thereby the unit point is prescribed. This equation $\xi_2/\xi_0 + \xi_1/\xi_2 = 0$ can be split into two equations (A) $\xi_1 v + \xi_2 u = 0$ and (B) $\xi_0 u - \xi_2 v = 0$ by putting $\xi_2/\xi_0 = u/v$ and $\xi_1/\xi_2 = -u/v$. For various values of (u, v), each of these equations represents a pencil of lines, which are therefore related by a projective mapping, so that for the same values of (u, v) a point of the conic is determined as their point of intersection (Fig.). The vertex of the pencil (A) is A, since $\xi_1 = 0$, $\xi_2 = 0$ for $a_1 = 0$, $a_2 = 0$, and the vertex of the pencil (B) is B because $b_0 = 0$, $b_2 = 0$.

The line $l = AB$ is then characterized by $\xi_2 = 0$, the line $l_1 = AC$ by $\xi_1 = 0$ and the line $l_2 = BC$ by $\xi_0 = 0$. The projective mapping of the pencil A onto the pencil B takes the line $l = AB$ of A $(v = 0)$ into the line $l_2 = BC$ of B and the line $l_1 = AC$ of A $(u = 0)$ into the line $l = AB$ of B, and therefore has no fixed line, that is, it is not perspective.

Conversely, two pencils with vertices A and B that are projectively, but not perspectively, related generate a conic as the set of points of intersection of corresponding lines. If the line $l = AB$ of the pencil A goes into the line l_2 of the pencil B, and if l as a line of the pencil B is the image of the line l_1 of the pencil A, then one can take the points A, B and $C = (l_1 \cap l_2)$ as base-points of a coordinate system, whose unit point U is the point of intersection of any two corresponding lines p and p'. In this coordinate system the line l is characterized by the equation $\xi_2 = 0$, the line l_1 by $\xi_1 = 0$, the line l_2 by $\xi_0 = 0$ and the lines p and p' by $\xi_1 + \xi_2 = 0$ and $\xi_0 - \xi_2 = 0$. However, the collineation of the pencil A into the pencil B that is determined by the correspondence of lines $\xi_1 v + \xi_2 u = 0$ and $\xi_0 u - \xi_2 v = 0$ with the same ratio $u : v$, has the property that the three lines l, l_1 and p of the first pencil go over into the three lines l_2, l and p' of the second pencil. According to the main theorem, this coincides with the given collineation. By eliminating u and v from the equations of the lines, one sees that the point of intersection of corresponding lines describes the conic $\xi_0\xi_1 + \xi_2^2 = 0$.

In perspectively related pencils, corresponding lines meet on the line from which the projection takes place, that is, a degenerate conic is formed.

To construct a conic that passes through three given non-collinear points A, B, P and has two lines l_1 and l_2 as tangents at A and B, it is sufficient to form a projective mapping of the pencil A onto the pencil B. According to the main theorem, such a collineation is uniquely determined by the condition that the lines l_1, $l = AB$ and $p = AP$ of the pencil A correspond to the lines l, l_2 and $p' = BP$ of the pencil B. The points of intersection of corresponding lines in this collineation uniquely generate a conic, which passes through the points A, B, P and has l_1 and l_2 as tangents at A and B (Fig.).

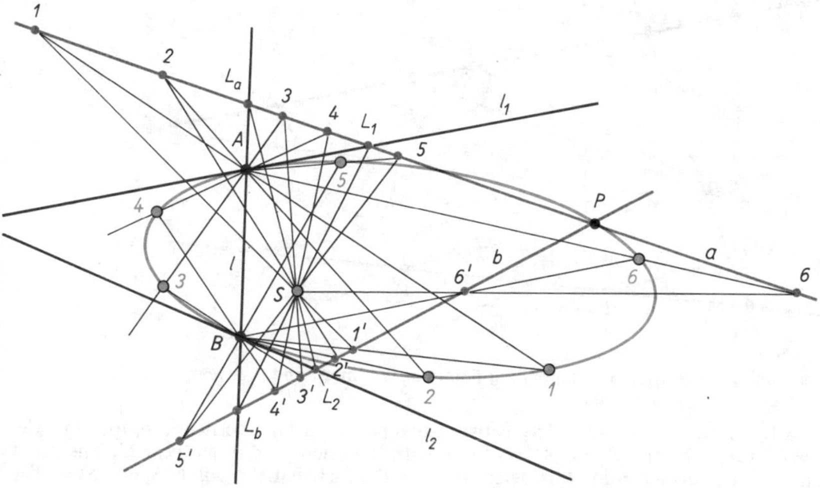

25.5-5 Construction of a conic from three points and two tangents

25.5-6 Construction of a conic from five points no three of which are collinear

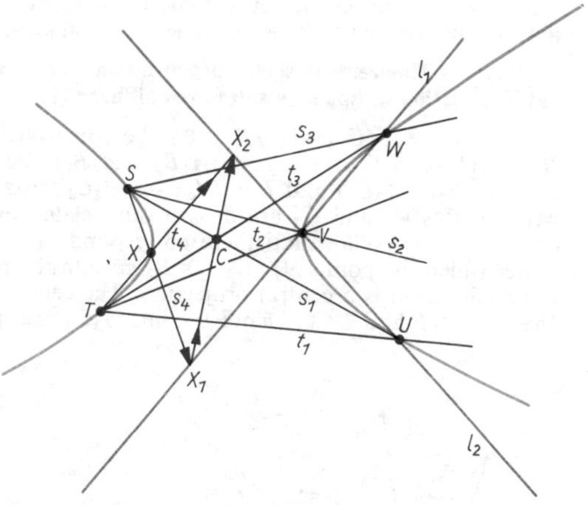

This collineation can be constructed by means of a central collineation between two lines a and b intersecting at P, which cut the lines l_1 and l in points L_1 and L_a and the lines l and l_2 in points L_b and L_2, with centre $S = (L_1L_b \cap L_aL_2)$.

If a conic is to be constructed through five given points, S, T, U, V, W, no three of which are collinear, one can pick out two of them as vertices of pencils that are projectively, but not perspectively related (Fig.), for example, S with $s_1 = SU$, $s_2 = SV$, $s_3 = SW$ and T with $t_1 = TU$, $t_2 = TV$, $t_3 = TW$. The collineation of the pencil S onto the pencil T that is uniquely determined by the condition that the lines s_i go into the lines t_i, then defines a conic, which consists of the points of intersection of corresponding lines. This collineation can be represented as a central collineation between two lines l_1 and l_2 intersecting at the point V. For example, if $l_1 = VW$ and $l_2 = UV$ are chosen, then the centre $C = (s_1 \cap t_3)$ of this perspectivity is the point of intersection of the lines s_1 and t_3. To a line s_4 cutting l_1 in a point X_1 there corresponds the line $t_4 = TX_2$, where $X_2 = (l_2 \cap X_1C)$ and the two lines cut in a point X of the conic.

Since the collineation is uniquely determined by the five given points, and the conic is in turn uniquely determined by this collineation, this construction gives a unique conic through these five points.

There is exactly one conic that passes through five given points no three of which are collinear.

As the dual of the generation by projective pencils, a conic can be generated by means of projectively related lines. The lines joining pairs of corresponding points of two lines l and l' envelop a non-degenerate conic if the two lines are not perspectively related (Fig.). If the two lines are projectively related, the *degenerate conic envelope* consists of only two points, the centre of projection and the point of intersection of the two lines.

The projective correspondence of the points of the lines l and l' is determined by three pairs of points. If A, B, C on l and their images A', B', C' on l' are given, then this correspondence can be

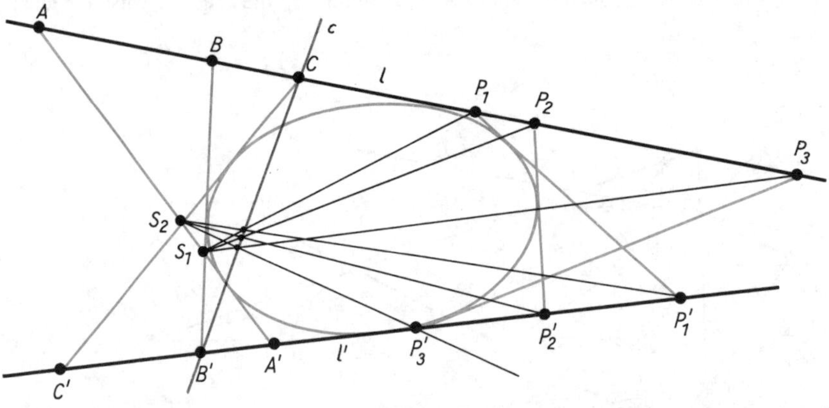

25.5-7 The lines joining projectively corresponding points of the lines l and l' envelop a conic

similarly realized by means of a perspectivity between two pencils, whose vertices are, for example, the points $S_1 = (AA' \frown BB')$ and $S_2 = (AA' \frown CC')$, with the line $c = CB'$ as 'centre'. The point P' corresponding to any further point P is determined by the fact that the point $(S_1P \frown S_2P')$ lies on c. The lines AA', BB', CC', $P_1P'_1$, ... are tangents of the conic.

Just as one can construct a conic through five points, so one can construct exactly one conic that touches five given lines no three of which are concurrent.

Pascal's Theorem. If six points of a conic are regarded as vertices of a hexagon, then the points of intersection of opposite sides are collinear (Fig.).

Let A_1, B_1, C_1, A_2, B_2, C_2 be the vertices of the hexagon, $A_3 = (B_1C_2 \frown B_2C_1)$, $B_3 = (C_1A_2 \frown C_2A_1)$, $C_3 = (A_1B_2 \frown A_2B_1)$ the points of intersection of opposite sides and $D = (A_1B_2 \frown A_2C_1)$ and $E = (A_1C_2 \frown B_2C_1)$ two further points of intersection. Then the pencils with vertices A_2 and C_2 are projectively related by the fact that corresponding lines meet on the conic. To this collineation there corresponds a collineation of the line A_1B_2 onto the line C_1B_2 under which the points A_1, C_3, D, B_2 go into the points E, A_3, C_1, B_2. Since the point B_2 is fixed, the collineation is a central projection. The centre of this projection is the point of intersection of the lines A_1E and DC_1. Hence the line A_3C_3 also passes through the point B_3.

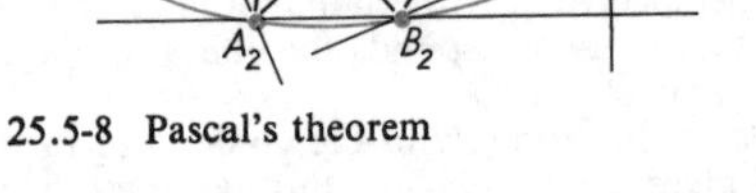

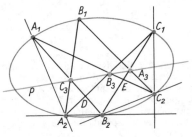

25.5-8 Pascal's theorem

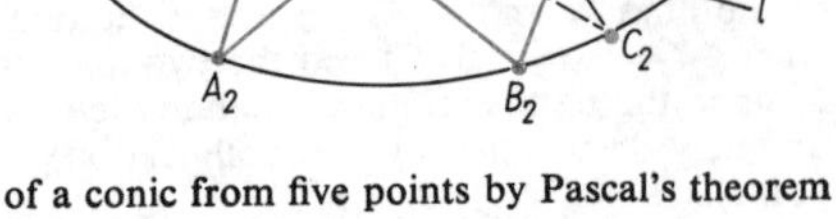

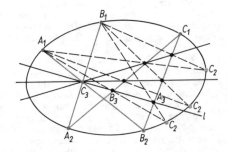

25.5-9 Construction of a conic from five points by Pascal's theorem

Pascal's theorem gives another possible construction for a conic through five points. Let A_1, B_1, C_1, A_2, B_2 be these five points no three of which are collinear. Then the lines A_1B_2 and A_2B_1 meet in the point C_3. An arbitrary line l through C_3 meets the lines C_1A_2 and C_1B_2 in the points B_3 and A_3. Then the lines A_1B_3 and B_1A_3 meet in a point C_2, which describes a conic through A_1, B_1, C_1, A_2, B_2 when l runs through all the lines through C_3 (Fig.).

Dual form of Pascal's theorem: Brianchon's theorem. If six tangents of a conic are regarded as the sides of a hexagon, then the lines joining opposite vertices are concurrent (Fig.).

Dual to the points of intersection A_3, B_3, C_3 of opposite sides of the Pascal hexagon, one has now the lines a_3, b_3, c_3 joining opposite vertices of the Brianchon hexagon. Dual to the Pascal line p through A_3, B_3 and C_3 one has the Brianchon point P as the point of intersection of a_3, b_3, c_3.

Brianchon's theorem gives a method, suitable for a draughtsman, of constructing a conic with five given lines as tangents. No three of the five given lines a_1, a_2, b_1, b_2, c_1 should be concurrent (Fig.). Of the three lines joining the points of intersection of opposite tangents, only one, $c_3 = [(a_1 \frown b_2), (b_1 \frown a_2)]$ is fixed. The other two lines a_3 and b_3 are determined by any position of the Brianchon point P on c_3, and by means of their points of intersection with the tangents b_1 and a_1 the sixth tangent c_2 is fixed, as the line through these points. This construction gives the conic as the envelope of all lines which, together with the five given lines, make up a Brianchon hexagon.

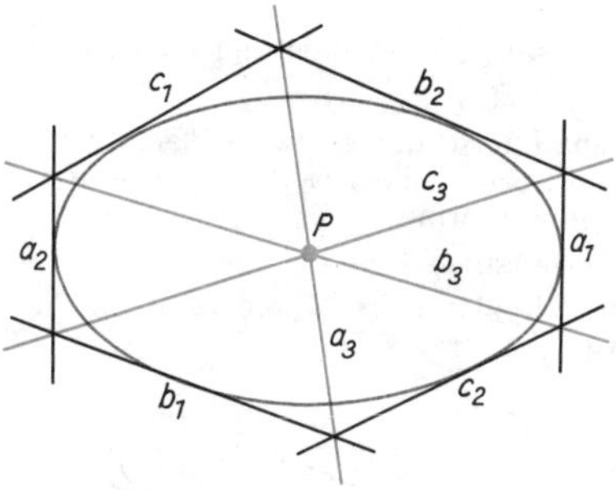

25.5-10 Brianchon's theorem

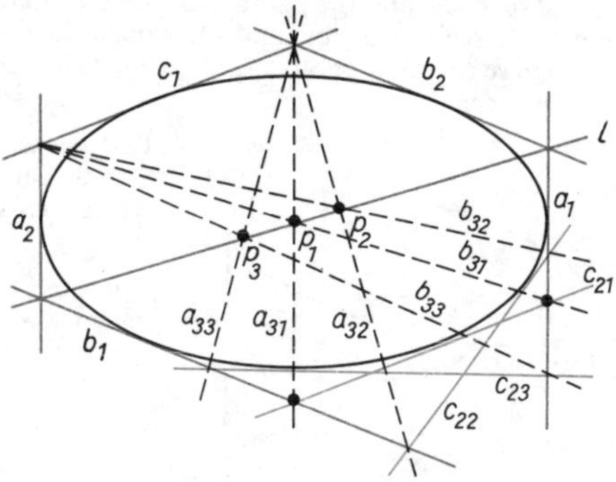

25.5-11 Construction of a conic from five tangents by Brianchon's theorem

26. Differential geometry, convex bodies, integral geometry

26.1. Differential geometry

In differential geometry the concepts and methods of analysis, particularly of differential calculus and the theory of differential equations, are applied to the study of geometric figures. The underlying *geometrical spaces* or *manifolds* must, as in analytical geometry, be referred to *coordinates*. Other geometrical figures are embedded in these spaces, for example, general curves or curved surfaces, which are characterized by *sufficiently differentiable equations* or *functions*. To understand the more advanced parts of differential geometry, one must be fully conversant with the tensor calculus; furthermore, a knowledge of topology and some other branches of mathematics is essential.

Theory of curves in Euclidean space

Definition of a curve. Let e_i $(i = 1, 2, 3)$ be three pairwise orthogonal unit vectors, which form an orthonormal trihedron of 3-dimensional Euclidean space E_3, and let x_i $(i = 1, 2, 3)$ be *Cartesian coordinates* with respect to this trihedron. By a *parametric representation of a portion of a curve* one understands the specification of the coordinates x_i of a point of the curve as functions of a real parameter t on an interval $a \leqslant t \leqslant b: x_i = f_i(t)$ $(i = 1, 2, 3)$. These three equations are usually summed up in one vector equation

$$\boldsymbol{x} = \boldsymbol{x}(t) = \sum_{i=1}^{3} f_i(t)\, \boldsymbol{e}_i.$$

The functions $f_i(t)$ must be continuously differentiable sufficiently often; it is usually sufficient to postulate the existence and continuity of the first three derivatives. By a *curve* one understands a connected set of points C such that for any point P of C there is a neighbourhood U such that the points of C lying in U can be represented as a portion of a curve. The parameter of a portion of a curve can be chosen almost arbitrarily; if t is a parameter, one obtains any other parameter t' by means of a *parameter transformation* $t' = \varphi(t)$, where the function $\varphi(t)$ is continuously differentiable sufficiently often and its derivative $\frac{d\varphi}{dt}$ is never zero. Of importance are only the geometrical properties of the curve that are independent of the special choice of the parameter, and not the more incidental analytical form of the representation. Often the coordinate system in Euclidean 3-dimensional space E_3 and the parameter t on the curve C can be chosen so that the functions representing C are as simple as possible and the calculations become easier.

A curve can also be given by an *implicit representation*, by means of two independent equations of the form

$$g(x_1, x_2, x_3) = 0, \quad h(x_1, x_2, x_3) = 0,$$

that is, geometrically as the intersection of two surfaces $g = 0$, $h = 0$. One of the simplest space curves is the *circular helix*, which can be represented in the form

$$\boldsymbol{x}(t) = a(\boldsymbol{e}_1 \cos t + \boldsymbol{e}_2 \sin t) + bt\boldsymbol{e}_3 .$$

The thread of a screw is a circular helix, where $2a$ is the diameter and $2\pi b$ the pitch of the screw (Fig.).

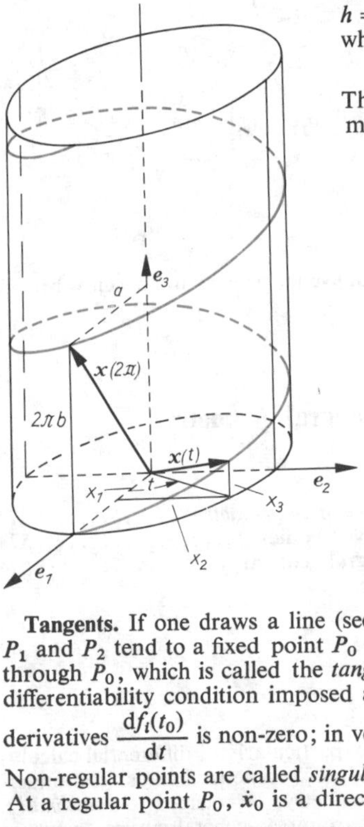

26.1-1 Circular helix

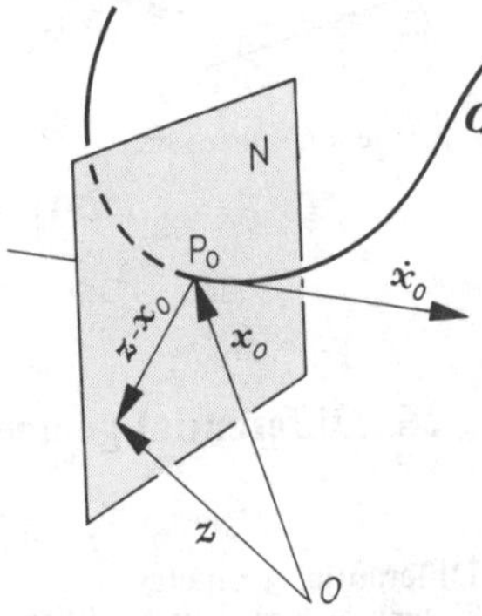

26.1-2 Normal plane N at a point P_0

Tangents. If one draws a line (secant) through two points P_1, P_2 of the curve C and then lets P_1 and P_2 tend to a fixed point P_0 of C with position vector $\boldsymbol{x}_0 = \boldsymbol{x}(t_0)$, the secant tends to a line through P_0, which is called the *tangent to* C at P_0. The existence of the tangent is ensured by the differentiability condition imposed above if the point P_0 is *regular*, that is, if at least one of the derivatives $\frac{df_i(t_0)}{dt}$ is non-zero; in vector form, $\dot{\boldsymbol{x}}_0 = \frac{d\boldsymbol{x}(t_0)}{dt} = \sum_{i=1}^{3} \boldsymbol{e}_i \frac{df_i(t_0)}{dt} \neq \boldsymbol{o}$.
Non-regular points are called *singular*, and their properties must always be investigated separately. At a regular point P_0, $\dot{\boldsymbol{x}}_0$ is a direction vector of the tangent at P_0. For the position vector $\boldsymbol{y}$ of a point on the tangent one obtains the given equation by introducing a parameter $\tau(-\infty < \tau < \infty)$.

Equation of the tangent	$\boldsymbol{y} = \boldsymbol{x}_0 + \dot{\boldsymbol{x}}_0\tau$

The plane through P_0 perpendicular to the tangent is called the *normal plane* N of C at P_0 (Fig.). If $\boldsymbol{z}$ is the position vector of a point of it, and $\boldsymbol{a} \cdot \boldsymbol{b}$ denotes the scalar product of the vectors $\boldsymbol{a}$ and $\boldsymbol{b}$, the equation of the normal plane N is:

Equation of the normal plane	$\dot{\boldsymbol{x}}_0 \cdot (\boldsymbol{z} - \boldsymbol{x}_0) = 0$

Osculating plane. Suppose that the curve C is not a straight line. Then, in general, three arbitrary points P_1, P_2, P_3 of it do not lie on one line. Any three such points therefore determine a plane. If P_1, P_2, P_3 tend to the same point P_0 of C, then in the limit their plane converges to a plane through P_0, which is called the *osculating plane* T *of* C *at* P_0 (Fig.). Its existence is ensured if the first two derivatives of the position vector $\boldsymbol{x}(t)$ are linearly independent for $t = t_0$, that is, if $\dot{\boldsymbol{x}}_0 \times \ddot{\boldsymbol{x}}_0 \neq \boldsymbol{o}$. Here $\ddot{\boldsymbol{x}}_0 = \frac{d^2\boldsymbol{x}(t_0)}{dt^2} = \sum_{i=1}^{3} \boldsymbol{e}_i \frac{d^2 f_i(t_0)}{dt^2}$; $\boldsymbol{a} \times \boldsymbol{b}$ denotes the *vector product* of the vectors $\boldsymbol{a}$ and $\boldsymbol{b}$.

If z is the position vector of a point of the osculating plane T and (a, b, c) denotes the scalar triple product of the vectors a, b and c, that is, $(a, b, c) = (a \times b) \cdot c$, then the equation of the osculating plane is

Equation of the osculating plane	$(\dot{x}_0 \times \ddot{x}_0) \cdot (z - x_0) = 0$ or $(\dot{x}_0, \ddot{x}_0, z - x_0) = 0$

The curve has *contact of the first order* with its tangents, that is, for a suitable choice of the parameter the first derivatives of the curve and the tangent coincide at the point of contact. The osculating plane can be defined as the plane that has contact of the second order with the curve at the point P_0, that is, the first two derivatives $\dot{x}_0$, $\ddot{x}_0$ must lie in it. If $\dot{x}_0 \times \ddot{x}_0 \neq o$, the osculating plane is uniquely determined. It is the plane that is spanned by the vectors $\dot{x}(t_0)$ and $\ddot{x}(t_0)$ at P_0. The plane perpendicular to the osculating plane and normal plane is called the *rectifying plane* R at P_0. If z is the position vector of a point of it, then one obtains:

Equation of the rectifying plane	$(\dot{x}_0, \dot{x}_0 \times \ddot{x}_0, z - x_0) = 0$

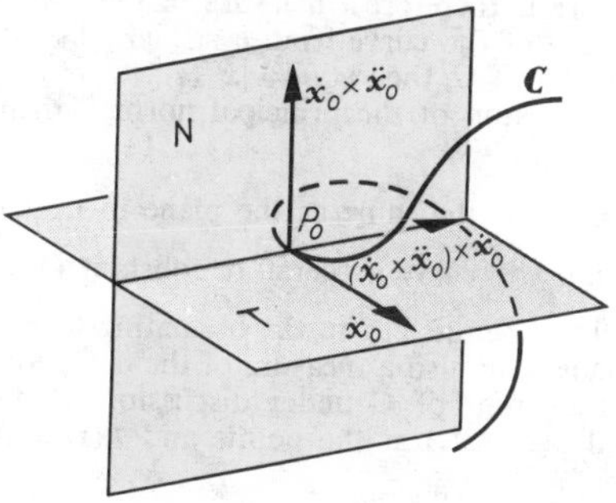

26.1-3 Osculating plane T at a point P_0

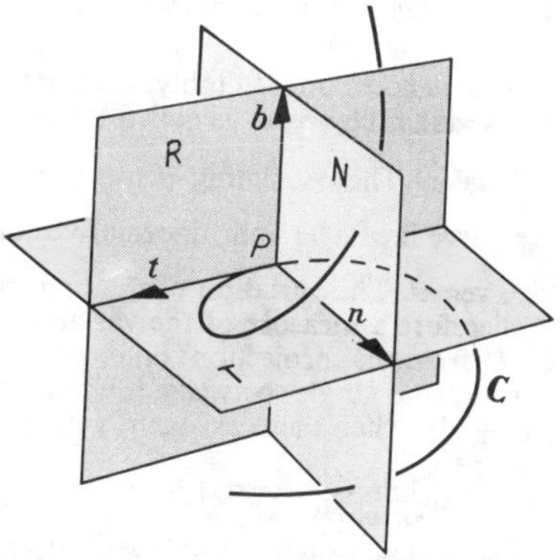

26.1-4 Moving trihedron of a curve

Normals. Any line that lies in the normal plane and passes through P_0 is called a *normal of C at P_0*. The normal lying in the osculating plane is called the *principal normal to C at P_0*, and the normal lying in the rectifying plane is called the *binormal*. A direction vector of the principal normal is $(\dot{x}_0 \times \ddot{x}_0) \times \dot{x}_0$, and a direction vector of the binormal is $\dot{x}_0 \times \ddot{x}_0$ (if $\dot{x}_0 \times \ddot{x}_0 \neq o$). If at each point P of the curve one draws three vectors t, n, b of length 1 in the directions of the tangent, principal normal and binormal to C, then one has an orthonormal trihedron, which is called the (uniquely determined) *moving trihedron of the curve*. The idea of a moving trihedron (or n-hedron) has proved very fruitful not only in the theory of curves, but in differential geometry generally (Fig.).

Arc length. The length of a polygon in E_3 can be defined as the sum of the lengths of its segments Δx. The curves considered in differential geometry can be approximated arbitrarily closely by polygons. Then the length $\Sigma \Delta x$ of the approximating polygon tends to a limiting value l, which is called the *length of the curve*. For a portion of a curve C with the parametric representation $x = x(t)$, $0 \leqslant t \leqslant a$, one has $\Delta x = x(t + \Delta t) - x(t)$, and it can be proved, under the usual differentiability conditions, that the length of a portion of a curve, as described above, is equal to the integral

$$l = \int_0^a |\dot{x}| \, dt = \int_0^a [\sum_{i=1}^{3} (\dot{f}_i(t))^2]^{1/2} \, dt;$$

here $|\dot{x}| = \sqrt{(\dot{x} \cdot \dot{x})}$ denotes the length of the vector $\dot{x}$. If one considers only the arc C_t from the point with the parameter 0 to the point with the parameter t, then the length s of this arc is a function of t:

$$s = s(t) = \int_0^t |\dot{x}(\tau)| \, d\tau.$$

If C is regular, it follows that $ds/dt = |\dot{x}(t)| > 0$, and s can be introduced as a new parameter, which is characterized geometrically. This parameter s is called the *arc length* of C or the *natural parameter* (Fig.). The derivative of the position vector with respect to arc length is the unit tangent vector

$$t = x' = dx/ds, \quad |x'| = 1.$$

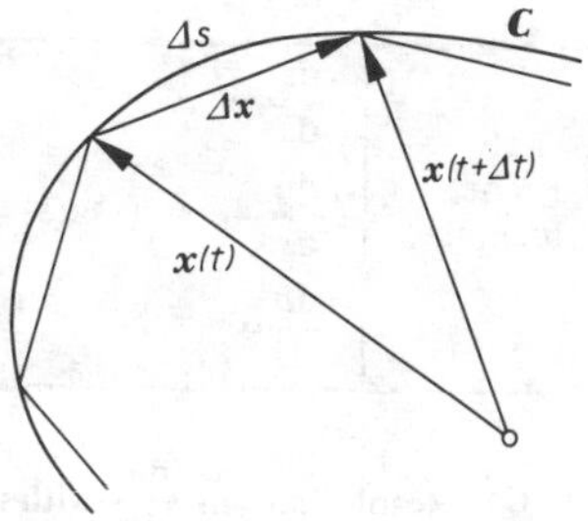

26.1-5 Definition of arc length

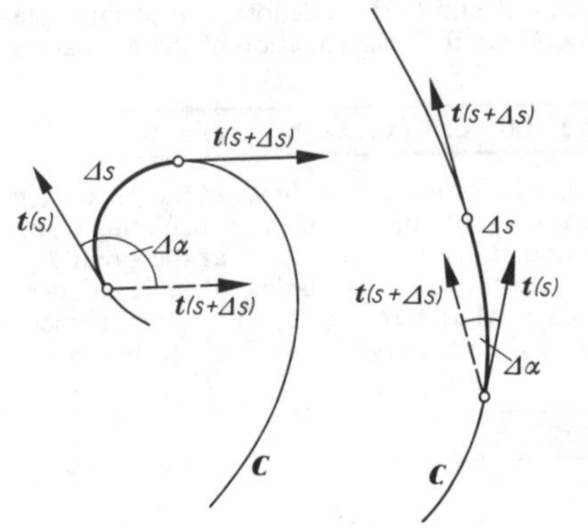

26.1-6 Definition of curvature

It follows that $x' \cdot x'' = 0$, that is, $x'' = \dfrac{d^2x}{ds^2}$ is perpendicular to x' and is therefore a direction vector of the principal normal, provided that $x'' \neq o$.

Curvature. Let $x = x(s)$, $0 \leqslant s \leqslant l$, be a portion of a curve C in terms of the arc length. The tangent vectors $t(s)$ and $t(s + \Delta s)$ at the points $P(s)$ and $P(s + \Delta s)$ with parameters s and $s + \Delta s$ form an angle $\Delta\alpha$ with one another. If now Δs tends to zero, then there exists the limiting value

$$\lim_{\Delta s \to 0} \left| \frac{\Delta\alpha}{\Delta s} \right| = \varkappa(s),$$

which is called the *curvature of C at the point P(s)* (Fig.). For a straight line one always has $\Delta\alpha = 0$, that is, the curvature $\varkappa(s)$ is identically zero.

The curvature is therefore a measure of the deviation of the form of the curve from a straight line. If s is the arc length of C, then $\varkappa(s) = |x''(s)|$.

If n denotes the suitably oriented unit vector in the direction of the principal normal, then $x'' = \varkappa(s)\, n$. Hence x'' is called the *curvature vector.*

Torsion. The osculating plane of a plane curve at any point is the same as the plane in which the curve lies. The unit binormal vector $b = \dfrac{x' \times x''}{|x' \times x''|}$ of a plane curve is therefore constant (and vice versa). The variation of the binormal vector b, which is perpendicular to the osculating plane, is therefore a measure of the variation of the osculating plane, and also a measure of the deviation of C from its projection onto the osculating plane at the point of C under discussion. If $\Delta\beta$ denotes the angle between the binormal vectors $b(s)$ and $b(s + \Delta s)$ at the points and $P(s)$ and $P(s + \Delta s)$, then there exists, in general, the limiting value

$$\lim_{\Delta s \to 0} \frac{\Delta\beta}{\Delta s} = \tau(s),$$

which is called the *torsion of C at the point P(s).*

Natural equations. Curvature, torsion, and arc length are *invariants* for a curve under Euclidean motions, that is, if a curve (made of wire, for example) is moved as a rigid body in space, then the curvature, torsion, and arc length do not change. In addition, these quantities do not depend on the arbitrary choice of parametric representation $x = x(t)$, hence are also invariant under parameter transformations. These two invariance properties follow immediately from the definitions given above. The three quantities s, $\varkappa$, τ are connected by the two equations

$$\varkappa = \varkappa(s) \geqslant 0, \quad \tau = \tau(s),$$

which are called the *natural equations of the curve.* The following theorem can be regarded as the main result of curve theory; it states that $\varkappa(s)$ and $\tau(s)$ form a *complete system of invariants* for C:

For any given continuous functions $\varkappa = \varkappa(s) > 0$ and $\tau = \tau(s)$, there is one, and, to within a Euclidean motion, only one curve C such that $\varkappa(s)$ is the curvature and $\tau(s)$ the torsion of C.

The Frenet formulae. The proof of this theorem is conducted by the method of the moving trihedron (Fig.). For the variation of the vectors $t(s)$, $n(s)$, $b(s)$ of the moving trihedron the following *Frenet formulae* hold:

Frenet formulae	$\dfrac{dt}{ds} =$		$\varkappa(s)\, n(s)$	
	$\dfrac{dn}{ds} =$	$-\varkappa(s)\, t(s)$		$+\tau(s)\, b(s)$
	$\dfrac{db}{ds} =$		$-\tau(s)\, n(s)$	

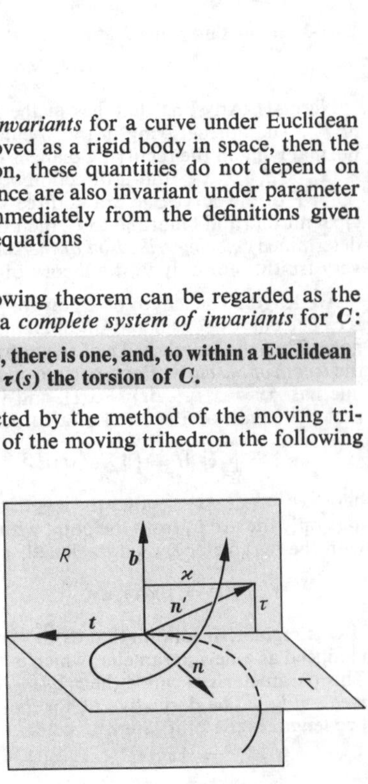

26.1-7 Resolution of $\dfrac{dn}{ds}$ with respect to the moving trihedron t, n, b

The first of these equations has already been proved, because $\frac{dt}{ds} = x''$ is the curvature vector. Since n, t, b are pairwise perpendicular unit vectors, one has $n \cdot n = b \cdot b = t \cdot t = 1$ and $n \cdot b = n \cdot t = b \cdot t = 0$; by differentiation $b' \cdot b = 0$, $n \cdot n' = 0$, $t' \cdot b = -t \cdot b'$, $n' \cdot t = -t' \cdot n$, $n' \cdot b = -b' \cdot n$. By scalar multiplication of $b' = \alpha_3 t + \beta_3 n + \gamma_3 b$ and $n' = \alpha_2 t + \beta_2 n + \gamma_2 b$ by suitable vectors n, t or b, one can determine the components α_i, β_i, γ_i ($i = 3, 2$). One obtains: $b' \cdot b = \gamma_3 = 0$; $b' \cdot t = \alpha_3 = -t' \cdot b = -\varkappa n \cdot b = 0$; $b' = \beta_3 n$. By the definition of torsion it follows that $|b'| = |\beta_3| = |\tau|$. For consistency with $\tau(s) = \lim\limits_{\Delta s \to 0} \frac{\Delta \beta}{\Delta s}$, one must put $b' = -\tau(s)\, n$. For the second equation: $\beta_2 = n \cdot n' = 0$; $\alpha_2 = n' \cdot t = -t' \cdot n = -\varkappa(s)$ and $\gamma_2 = n' \cdot b = -b' \cdot n = \tau(s)$, that is, $n' = -\varkappa t + \tau b$.

For given continuous functions $\varkappa(s) > 0$ and $\tau(s)$ the Frenet formulae are a system of linear differential equations to determine t, n, b. Once $t(s)$ has been obtained, the curve can be found by integration of $\frac{dx}{ds} = t(s)$; for example, the circular helices are characterized as the curves whose curvature and torsion are constant.

Theory of surfaces in Euclidean space

Definition of a surface. If the three coordinates x_i ($i = 1, 2, 3$) of a point of E_3 are given as functions of two parameters u and v, $x_i = f_i(u, v)$ ($i = 1, 2, 3$), or in vector form, $x = x(u, v) = \sum_{i=1}^{3} f_i(u, v)\, e_i$, where u and v vary in a certain domain D of a plane, then this is called a *parametric representation of a portion of a surface.*

A connected point set S of E_3 is called a *surface* if for each point P of S there is a neighbourhood U such that the points of S that lie in U have a parametric representation as a portion of a surface. If the values of the parameters u and v (also known as *coordinates on the surface*) are given, then the position of the point on the portion of the surface is uniquely determined; for example, a point on the earth's surface is fixed by its *latitude* and *longitude*. Within wide limits, the parameters of a portion of a surface can be chosen arbitrarily; instead of u and v one can take as parameters u' and v' in a domain D' of the plane if there is a one-to-one transformation of the form

$$\begin{aligned} u' &= u'(u, v) \\ v' &= v'(u, v) \end{aligned} \quad \text{with the determinant} \begin{vmatrix} \frac{\partial u'}{\partial u} & \frac{\partial u'}{\partial v} \\ \frac{\partial v'}{\partial u} & \frac{\partial v'}{\partial v} \end{vmatrix} \neq 0;$$

this is called a *parameter transformation.* The geometrical concepts of the theory of surfaces must be *invariant* under Euclidean motions and under parameter transformations. A surface can also be given *implicitly* by an equation $g(x_1, x_2, x_3) = 0$.

Tangent plane. In order to investigate properties of a surface in the neighbourhood of a point P_0 with the parameters u_0 and v_0, one considers curves that lie on the surface and pass through P_0. An arbitrary curve of this kind can be given in a parametric representation of the form

$$x(t) = x(u_0 + u(t), v_0 + v(t)),$$

where $u(0) = v(0) = 0$. In the special case $u = u_0 + t$, $v = v_0$, that is, $v(t) = 0$ for all t, one obtains the *parameter curve* through P_0, along which $v = v_0$ is constant. Similarly, if $u = u_0$, $v = v_0 + t$, one has the other parameter curve, along which $u = u_0$ is constant. The point P_0 is the point of intersection of these parametric curves. In geographical coordinates on the earth's surface, the meridians and circles of latitude are the corresponding parameter curves.

The tangent vector of a curve through P_0 at the point $t = 0$, that is, at P_0, is obtained by differentiation of the parametric representation

$$\left.\frac{dx}{dt}\right|_{t=0} = \frac{\partial x_0}{\partial u} \cdot \frac{du(0)}{dt} + \frac{\partial x_0}{\partial v} \cdot \frac{dv(0)}{dt},$$

where $\frac{\partial x_0}{\partial u} = \frac{\partial x(u_0, v_0)}{\partial u}$, $\frac{\partial x_0}{\partial v} = \frac{\partial x(u_0, v_0)}{\partial v}$. From these formulae one sees that: the vectors $\frac{\partial x_0}{\partial u}$ and $\frac{\partial x_0}{\partial v}$ are tangent vectors to the parameter curves. If these two vectors are linearly independent, then all the tangent vectors to curves that lie on the surface and pass through P_0 lie in the plane through P_0 spanned by $\frac{\partial x_0}{\partial u}$ and $\frac{\partial x_0}{\partial v}$. This plane is called the *tangent plane* to the surface at P_0. If a and b are parameters of points of the tangent plane and z is its position vector, then one obtains a *parametric representation of the tangent plane*:

$$z = x_0 + a\,\frac{\partial x_0}{\partial u} + b\,\frac{\partial x_0}{\partial v}\,.$$

Equation of the tangent plane	$\left(\frac{\partial x_0}{\partial u} \times \frac{\partial x_0}{\partial v}\right) \cdot (z - x_0) = 0$

These formulae make sense only if $\frac{\partial x_0}{\partial u}$ and $\frac{\partial x_0}{\partial v}$ are linearly independent; points at which this is so are called *regular*; if it is not so, the point is called *singular*. Obviously, a point P_0 of S is regular if and only if $\frac{\partial x_0}{\partial u} \times \frac{\partial x_0}{\partial v} \neq o$.

In geographical coordinates on the earth's surface, the poles are singular. The vertex of a circular cone is singular for any parametric representation, because no tangent plane to the cone exists at this point. In differential geometry one deals almost exclusively with regular surfaces; singular points must always be treated separately.

The line through P_0 perpendicular to the tangent plane is called the *normal* (Fig.).

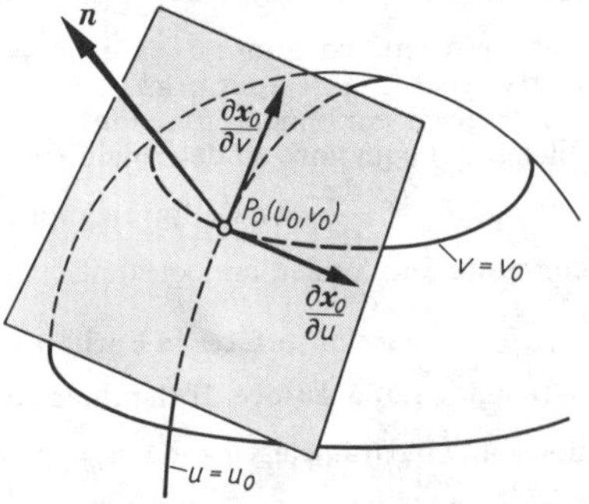

26.1-8 Tangent plane and surface normal

Normal unit vector	$n = \left(\frac{\partial x_0}{\partial u} \times \frac{\partial x_0}{\partial v}\right) \Big/ \left\lvert \frac{\partial x_0}{\partial u} \times \frac{\partial x_0}{\partial v} \right\rvert$

Intrinsic geometry. In the early years of differential geometry, surfaces were regarded as the outer boundaries of solid bodies, or as 'infinitely thin' solid bodies, embedded in 3-dimensional Euclidean space. The geometer Gaspard MONGE (1746–1818) can be regarded as a founder of this way of thinking; he wrote the first textbook on differential geometry (*Application de l'Analyse à la Géométrie*, Paris 1809). In connection with practical questions of geodesy, GAUSS posed the question: how can one draw conclusions about the spatial form of a surface from measurements on the surface itself? – This problem is of great importance for the determination of the form of the earth, which was regarded originally as a sphere, then as a flattened ellipsoid of rotation and today as a surface that cannot be represented in an elementary way, a so-called *geoid*. The investigation of these questions led GAUSS to the *intrinsic* geometry of the surface; he described this in his treatise *Disquisitiones generales circa superficies curvas* (1827). In this part of surface theory, surfaces are regarded not as solid bodies, but as skins, which can be *bent*, but not *stretched*. By a *bending* of a surface one understands a continuous deformation of the surface under which the lengths of all curves on the surface remain fixed. More generally, two surfaces S and S' are called *isometric* if there is a one-to-one correspondence $P' = \varphi(P)$ between the points P of S and the points P' of S' such that curves that are transformed into one another by the correspondence have the same length. The correspondence φ is then called an *isometric* mapping or *isometry*; for example, a cone can be isometrically mapped onto a region of the plane if it is cut along one of its generators. A bending of a surface S into a surface S' is also an isometric mapping of S onto S'; however, two isometric surfaces need not have a continuous bending from one to the other. Properties of surfaces that do not change under isometric mappings can be established by measurements on the surface; they form the content of the *intrinsic geometry of the surface*. In this sense, *plane geometry* is the intrinsic geometry of the plane and *spherical trigonometry* is the intrinsic geometry of the sphere.

Element of arc of a surface. The intrinsic geometry is completely governed by the element of arc of the surface. Let $x = x(t) = x(u(t), v(t))$, $t_0 \leqslant t \leqslant t_1$, be a curve C lying on the surface S. By differentiating with respect to t one obtains the tangent vector

$$\dot{x} = \frac{dx}{dt} = \frac{\partial x}{\partial u}\frac{du}{dt} + \frac{\partial x}{\partial v}\frac{dv}{dt}\,.$$

From the definition of arc length $s(t)$ of C one has $\left(\frac{ds}{dt}\right)^2 = |\dot{x}|^2 = \dot{x} \cdot \dot{x}$. Substituting the above expression for x one obtains

$$\left(\frac{ds}{dt}\right)^2 = \frac{\partial x}{\partial u} \cdot \frac{\partial x}{\partial u}\left(\frac{du}{dt}\right)^2 + 2\,\frac{\partial x}{\partial u} \cdot \frac{\partial x}{\partial v}\,\frac{du}{dt}\,\frac{dv}{dt} + \frac{\partial x}{\partial v} \cdot \frac{\partial x}{\partial v}\left(\frac{dv}{dt}\right)^2.$$

If, following GAUSS, one introduces the notation

$$E(u, v) = \frac{\partial x}{\partial u} \cdot \frac{\partial x}{\partial u}\,, \qquad F(u, v) = \frac{\partial x}{\partial u} \cdot \frac{\partial x}{\partial v}\,, \qquad G(u, v) = \frac{\partial x}{\partial v} \cdot \frac{\partial x}{\partial v}$$

and instead of writing the derivatives with respect to t one writes only the differentials, then one obtains the *element of arc* or the *first fundamental form* of the surface in the form

First fundamental form of the surface	$ds^2 = E(u, v)\,du^2 + 2F(u, v)\,du\,dv + G(u, v)\,dv^2$

The length l of the curve **C** is expressed in terms of the element of arc as follows:

$$l = \int_{t_0}^{t_1} \sqrt{(\dot{x}\cdot\dot{x})}\,dt = \int_{t_0}^{t_1} \sqrt{\left[E\left(\frac{du}{dt}\right)^2 + 2F\frac{du}{dt}\frac{dv}{dt} + G\left(\frac{dv}{dt}\right)^2\right]}\,dt;$$

in the integration naturally one has to substitute for the arguments u and v of E, F, G the equations $u = u(t)$, $v = v(t)$ of the curve.

By means of the first fundamental form one can not only calculate arc lengths, but also define and determine all the quantities that can be found by measurements on the surface; for example, the angle between two curves that lie on the surface **S** and meet at a point P_0 of **S**, and also the area of a point set lying on **S** can be defined in this way.

The area $A(U)$ of a domain U of **S** is

$$A(U) = \iint (EG - F^2)^{1/2}\,du\,dv.$$

The integrand $dA = (EG - F^2)^{1/2}\,du\,dv$ is called the *element of area* of **S**, and can be thought of intuitively as the area of an infinitely small mesh of parameter curves (Fig.); $(EG - F^2)^{1/2}\,\Delta u\,\Delta v$ is the area of the parallelogram spanned by the vectors $\frac{\partial \boldsymbol{x}}{\partial u}\Delta u$ and $\frac{\partial \boldsymbol{x}}{\partial v}\Delta v$. In the calculation of $A(U)$, the integration is to be taken over all parameters u and v for which $\boldsymbol{x}(u, v)$ lies in U.

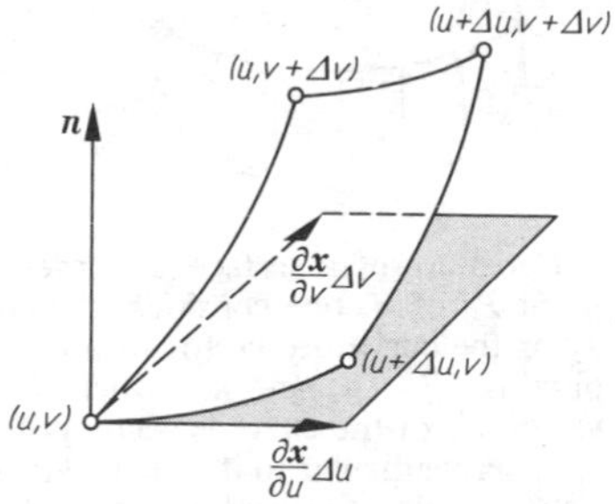

26.1-9 Definition of an element of area

The element of arc determines the intrinsic geometry of the surface completely: two surfaces S and S′ are isometric if and only if one can find parametric representations of them for which the elements of arc coincide.

Geodesics. If among all curves that lie on the surface and pass through two points P_1 and P_2 there is one of least length, it is called a *shortest curve*. The determination of shortest curves of a surface is one of the oldest problems of differential geometry and the calculus of variations. Given two points in a plane, there is only one shortest curve through them, namely the line segment joining them. There may be pairs of points on a surface that cannot be joined by a shortest curve. On the other hand, it can happen that through two points there is more than one shortest curve, even infinitely many; for example, for two diametrically opposite points of a sphere, any great semicircle through the points is a shortest curve. However:

If U is a sufficiently small neighbourhood of a point P_1 of the surface, and if P_2 is another point of U, then there is one shortest curve connecting P_1 and P_2.

A curve C lying on a surface **S** is called a *geodesic* if it is a shortest curve between any two of its points that are sufficiently close. On a sphere the great circles are geodesics but obviously need not be shortest curves; a great circle is divided into two arcs by two of its points, which in general have different lengths, so that only the smaller one is a shortest curve. An arc of a great circle whose length is greater than πR, where R is the radius of the sphere, is therefore not a shortest curve, but it is a geodesic. The differential equation of geodesics of an arbitrary surface is one of the second order that depends only on the first fundamental form.

Through any point of a regular surface there is exactly one geodesic in any given direction. Two points of a complete (intuitively, 'rimless') surface can be joined by a shortest curve, and so also by a geodesic.

Parallel displacement. The idea of parallel displacement can also be carried over to an arbitrary curved surface. At a point P_0 of a geodesic g, if a tangent vector $\boldsymbol{a}_0 = \boldsymbol{a}(P_0)$ to the surface **S** forms with the tangent vector $\boldsymbol{t}_0 = \boldsymbol{t}(P_0)$ to the geodesic an angle α, then at a point P of the geodesic one obtains the vector *$\boldsymbol{a}(P)$ parallel to $\boldsymbol{a}(P_0)$ along g* by constructing from P in the tangent plane the vector $\boldsymbol{a}$ of length $|\boldsymbol{a}_0|$ which forms the same angle α with the tangent vector $\boldsymbol{t} = \boldsymbol{t}(P)$ of g. It follows from this definition that tangent vectors of constant length of a geodesic (as for a straight line) are given a parallel displacement along it; here $\alpha = 0$. If this definition is also applied to curved polygons, whose components consist of geodesics, and if one thinks of an arbitrary curve on **S** as approximated by such *geodesic polygons*, then one obtains an intuitive idea of parallel displacement of a tangent vector along an arbitrary curve of the surface. The most important difference between parallel displacement on a curved surface and that in affine (or Euclidean) space is that *in the first case the parallel displacement depends on the curve* along which it takes place. If one displaces a vector around a closed path on a surface, then, in general, one does not get back to the original position (Fig.). In the figure, there is an angle of $\pi/2$ between the original vector $\boldsymbol{a}_0$ and the vector $\boldsymbol{a}_3$ displaced around a spherical triangle with three right angles.

26.1-10 Parallel displacement along a spherical triangle

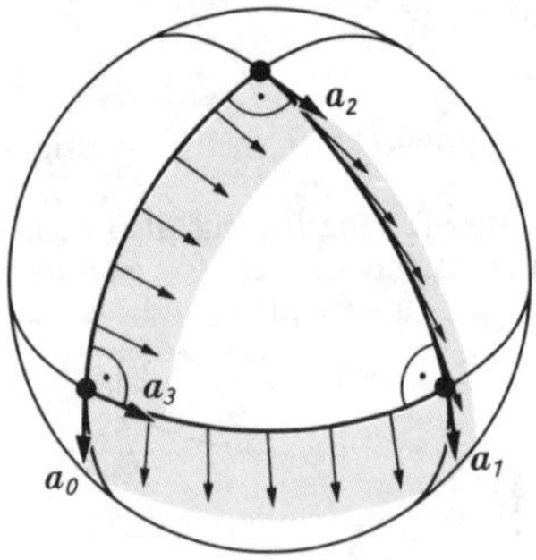

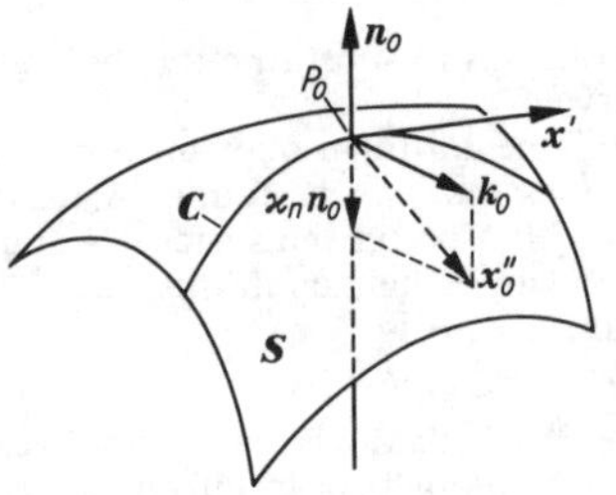

26.1-11 Normal curvature and geodesic curvature

Curvature of a surface. In order to investigate curvature properties in the neighbourhood of a point P_0 of $\boldsymbol{S}$, one considers the curvature of curves that lie on $\boldsymbol{S}$ and pass through P_0 (Fig.). If $\boldsymbol{x}_0''$ is the curvature vector of a curve $\boldsymbol{C}$ at P_0, one projects it onto the normal to the surface and obtains $\boldsymbol{x}_0'' = \varkappa_n \boldsymbol{n}_0 + \boldsymbol{k}_0$, where $\boldsymbol{k}_0 \cdot \boldsymbol{n}_0 = 0$, so that $\boldsymbol{k}_0$ is also a tangent vector; the curvature vector $\boldsymbol{x}_0''$ of the curve is decomposed into its tangential component $\boldsymbol{k}_0$ and the normal component $\varkappa_n \boldsymbol{n}_0$ perpendicular to it. The length $\varkappa_n$ of the projection onto the normal, taken with the appropriate sign, is called the *normal curvature* of $\boldsymbol{C}$ at P_0. The length $\varkappa_g = |\boldsymbol{k}_0|$ of $\boldsymbol{k}_0$ is called the *geodesic curvature*. It follows immediately from the resolution of the curvature vectoı $\boldsymbol{x}_0''$ into the normal component $\varkappa_n \boldsymbol{n}_0$ and the tangential component $\boldsymbol{k}_0$ that the (total) curvature $\varkappa(s) = |\boldsymbol{x}''(s)|$, the normal curvature $\varkappa_n = \boldsymbol{x}''(s) \cdot \boldsymbol{k}_0$, and the geodesic curvature $\varkappa_g = |\boldsymbol{k}_0|$ are linked by the relation $\varkappa^2 = \varkappa_n^2 + \varkappa_g^2$. The geodesic curvature is *invariant* under bending, and is therefore a concept of the intrinsic geometry, while the normal curvature depends on the embedding of the surface in space; for example, in a plane the normal curvature of every curve is obviously equal to zero. If one now bends a strip of the plane into part of a circular cylinder of radius r, then every generating circle of the cylinder has the normal curvature $1/r$.

The geodesic curvature of a curve on a surface can be defined by means of parallel displacement, just like the curvature of a space curve. *Geodesics* can be characterized as curves on a surface whose geodesic curvature vanishes. On a circular cylinder, the circular helices, the generating lines, and the circles perpendicular to the generators are the geodesics; when the cylinder is developed into a plane, by the isometric mapping obtained by cutting along a generator, the geodesics go into segments or straight lines.

The normal curvature is given by the *second fundamental form* of surface theory; if $u = u(s)$, $v = v(s)$ are the equations of the curve $\boldsymbol{C}$, where s is its arc length, then the second fundamental form of the surface is:

Second fundamental form of the surface	$\varkappa_n = L(u, v)\left(\frac{du}{ds}\right)^2 + 2M(u, v)\frac{du}{ds}\cdot\frac{dv}{ds} + N(u, v)\left(\frac{dv}{ds}\right)^2$

where $L = \boldsymbol{n} \cdot \dfrac{\partial^2 \boldsymbol{x}}{\partial u^2}$, $M = \boldsymbol{n} \cdot \dfrac{\partial^2 \boldsymbol{x}}{\partial u\, \partial v}$, $N = \boldsymbol{n} \cdot \dfrac{\partial^2 \boldsymbol{x}}{\partial v^2}$.

It follows that the normal curvature depends only on the direction of the curve at the point P_0.

All curves on $\boldsymbol{S}$ having the same tangent at P_0 also have the same normal curvature there.

A more precise investigation of the normal curvature leads to a classification of points of the surface. Firstly, if all the quantities L, M, N vanish at P_0, as is the case for every point of a plane, then P_0 is called a *flat point*. If this is not the case, one distinguishes three types of point.

Gaussian curvature	$K(P) = (LN - M^2)/(EG - F^2)$

If the *Gaussian curvature* of the surface at a point P with the coordinates (u, v) has the value $K(P) > 0$, then P is called *elliptic*, if $K(P) < 0$, then P is called *hyperbolic*, and if $K(P) = 0$, P is called *parabolic*

(Fig.). This purely formal division has a close connection with the shape of the surface. On a bicycle tube (torus), for example, the points towards the inside are hyperbolic, and the points toward the outside are elliptic; these two point sets are separated from one another by two circles, which consist of parabolic points. An ellipsoid has only elliptic points, a hyperbolic paraboloid (saddle surface) has only hyperbolic points, and a circular cylinder has only parabolic points (Fig.).

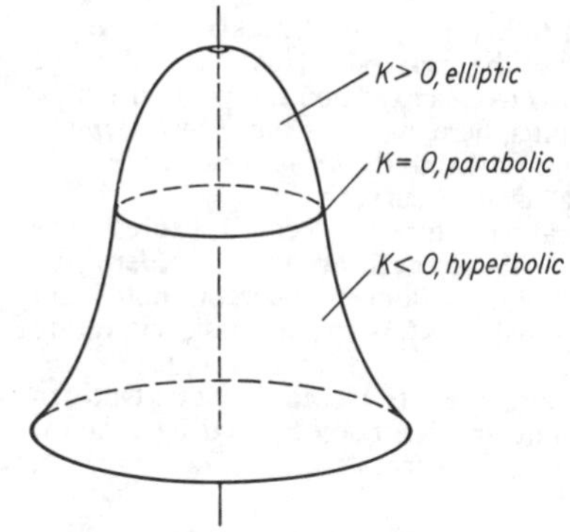

26.1-12 Classification of the points of a surface of revolution (bell)

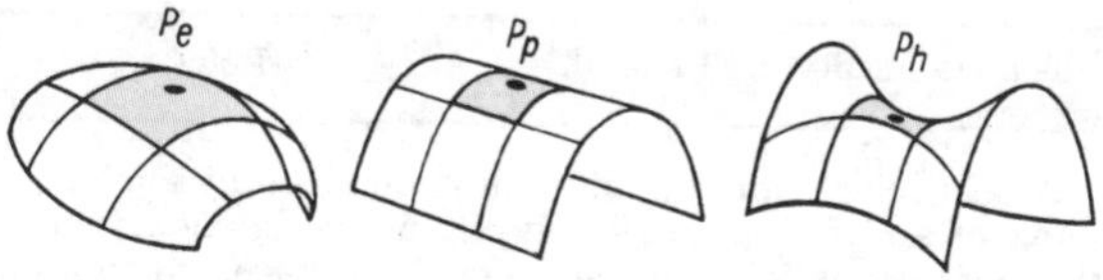

26.1-13 Elliptic (P_e), parabolic (P_p) and hyperbolic (P_h) point

Theorema egregium. The first and second fundamental forms assigned to a surface are *invariant* under motions, that is, if the surface is moved in space as a rigid body (without altering its shape), then the fundamental forms do not change. If the surface is bent, that is, deformed *isometrically*, then the first fundamental form remains unchanged, while the second fundamental form, which determines the normal curvature, changes. The first fundamental form is a *bending invariant*.

GAUSS showed that the Gaussian curvature ist not only invariant under motions and parameter transformations, but also under bending. He called this striking and unexpected result the Theorema egregium (remarkable theorem).

Theorema egregium. The Gaussian curvature K remains invariant under isometric mappings.

For a proof of this theorem one derives a formula for K in which only the coefficients of the first fundamental form and their derivatives appear. Since these are bending invariants, K must also be a bending invariant. By a suitable choice of parameters u and v it can be arranged that the parameter curves of the surface cut at right angles, that is, $\frac{\partial \mathbf{x}}{\partial u} \cdot \frac{\partial \mathbf{x}}{\partial v} = F = 0$. If this is assumed, then the Theorema egregium is expressed by the formula

$$K = \frac{LN - M^2}{EG} = -\frac{1}{\sqrt{(EG)}} \left[\frac{\partial}{\partial u} \left\{ \frac{1}{\sqrt{E}} \frac{\partial \sqrt{G}}{\partial u} \right\} + \frac{\partial}{\partial v} \left\{ \frac{1}{\sqrt{G}} \frac{\partial \sqrt{E}}{\partial v} \right\} \right].$$

It follows, for example, that a sphere of radius r, for which the Gaussian curvature at every point is equal to $1/r^2$, cannot be mapped isometrically onto a plane, for which $K \equiv 0$.

It is therefore impossible to draw a faithful map of part of the earth's surface; only by restriction to a sufficiently small domain can one obtain an approximately accurate representation.

Determination of a surface from the fundamental forms. The Theorema egregium is directly connected with the following question: let

$$\varphi_1 = E\xi^2 + 2F\xi\eta + G\eta^2, \quad \varphi_2 = L\xi^2 + 2M\xi\eta + N\eta^2,$$

be two quadratic forms whose coefficients are functions of the two variables u and v; furthermore, suppose that φ_1 is positive definite. Is there always a surface $\mathbf{S}$ for which φ_1 is the first and φ_2 the second fundamental form? – This problem is analogous to the determination of a curve with given curvature and torsion. But in contrast to this simpler problem, where curvature and torsion can be given independently of one another, in the case of a surface the two fundamental forms cannot be chosen independently of one another; their coefficients are connected by three conditions, the so-called *integrability conditions*, which hold on every surface. One of these conditions is given in the previous section, the equation that expresses the Theorema egregium (under the assumption that $F = 0$); the other two integrability conditions are called the *Codazzi-Mainardi formulae*. If these three conditions are satisfied for φ_1 and φ_2, then, at least for a sufficiently small region U of the variables u and v, there is always a portion of a surface that has φ_1 and φ_2 as fundamental forms; two distinct such portions of surfaces defined on U are congruent.

The Gauss-Bonnet theorem. By integrating the Gaussian curvature K, multiplied by the element of area dA, over a domain U of the surface $\boldsymbol{S}$, one obtains the so-called *integral curvature* $K(U)$ of the region

$$K(U) = \iint_U K \, dA = \iint K(EG - F^2)^{1/2} \, du \, dv,$$

which is naturally also a bending invariant.

An intuitive interpretation of the integral curvature, and therefore of the Gaussian curvature, is obtained by investigating the *spherical image* of a domain U of the surface $\boldsymbol{S}$. This spherical image is obtained by drawing the normal unit vector $\boldsymbol{n}$ of a point P of U from a fixed point, say the origin O. The ends of these vectors then describe a domain V of the unit sphere, which is the *spherical image* of U. The area of the spherical image is then (apart from the sign) equal to the *integral curvature* of U. It is intuitively obvious that this area is the larger the more sharply $\boldsymbol{S}$ curves.

If U is bounded by a simple closed curve $\boldsymbol{C}$, then the integral curvature $K(U)$ can be expressed as an integral around $\boldsymbol{C}$. Here the Gauss-Bonnet theorem holds, in which $\varkappa_g$ is the geodesic curvature and s is arc length along $\boldsymbol{C}$.

The Gauss-Bonnet theorem	$\iint_U K \, dA + \oint_C \varkappa_g \, ds = 2\pi$

A particularly interesting result is obtained by applying this theorem to closed surfaces. One can think of a *closed surface* intuitively as the boundary of a finite smooth body pierced by g holes; the number g is called the *genus* of the surface; examples of closed surfaces are the sphere ($g = 0$), the torus ($g = 1$), and the 'pretzel' ($g = 2$) (Fig.).

The integral curvature of a closed surface S of genus g does not depend on the shape of the surface and is equal to

$$K(S) = \iint_S K \, dA = 4\pi(1 - g).$$

This result is of great importance, since it makes it possible to express *topological* properties of the surface, in this case the genus g, which remain invariant even under arbitrary continuous deformations, in terms of quantities of differential geometry (here the integral curvature). Its generalizations and similar questions have led in recent decades to the development of one of the most interesting and difficult branches of modern geometry in which connections between properties of geometrical forms in differential geometry and in topology are also investigated in higher dimensions.

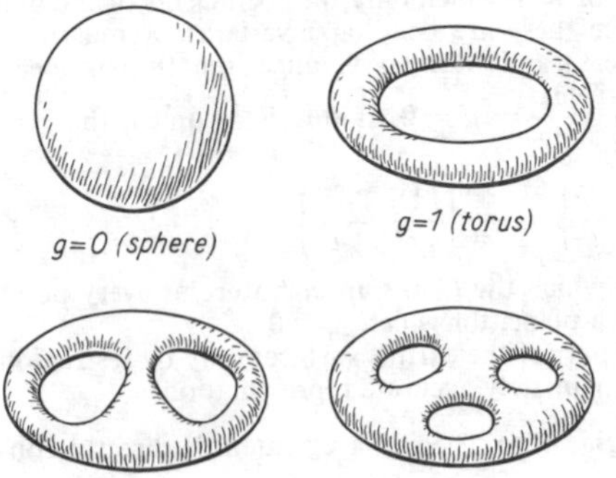

26.1-14 Surfaces of different genus g

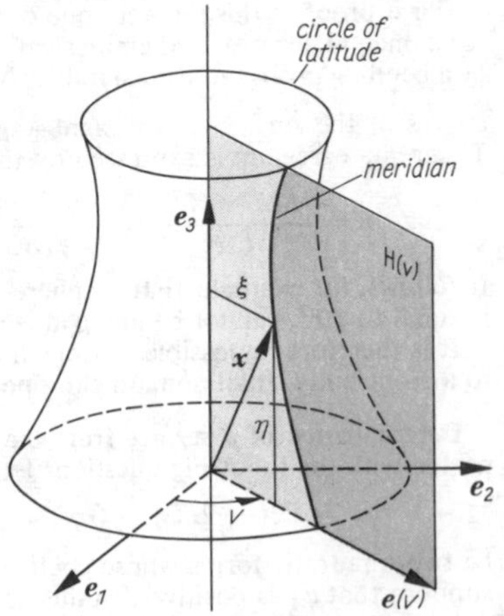

26.1-15 Definition of a surface of revolution

Surfaces of revolution. By a *surface of revolution* one understands a surface that arises by rotation of a plane curve about an axis lying in the plane of the curve. Such surfaces with rotational symmetry occur frequently in practice. To obtain a parametric representation of a surface of revolution one takes the axis of rotation as the $\boldsymbol{e}_3$-axis of a rectangular coordinate system in space (Fig.). In the $\boldsymbol{e}_1, \boldsymbol{e}_2$-plane perpendicular to it, a unit vector $\boldsymbol{e}(v)$ is defined by

$$\boldsymbol{e}(v) = \boldsymbol{e}_1 \cos v + \boldsymbol{e}_2 \sin v, \quad \text{whose derivative is} \quad \boldsymbol{e}^*(v) = \frac{d\boldsymbol{e}}{dv} = -\boldsymbol{e}_1 \sin v + \boldsymbol{e}_2 \cos v = \boldsymbol{e}(v + \pi/2).$$

It follows immediately that for all values of v the vectors $\boldsymbol{e}(v)$, $\boldsymbol{e}^*(v)$, $\boldsymbol{e}_3$ form a right-handed orthonormal trihedron. A plane $H(v)$ passing through the origin and spanned by $\boldsymbol{e}(v)$ and $\boldsymbol{e}_3$ rotates about the $\boldsymbol{e}_3$-axis as v varies. Any fixed curve $\boldsymbol{x}(u) = \xi(u)\,\boldsymbol{e} + \eta(u)\,\boldsymbol{e}_3$ in $H(v)$, for which u is the arc length, so that $\frac{\mathrm{d}\boldsymbol{x}}{\mathrm{d}u}\cdot\frac{\mathrm{d}\boldsymbol{x}}{\mathrm{d}u} = 1$, generates a surface of revolution as the plane $H(v)$ rotates about the $\boldsymbol{e}_3$-axis. Hence one obtains a parametric representation of the surface of revolution generated by $\boldsymbol{x}(u)$ in the form

$$\boldsymbol{x}(u, v) = \xi(u)\,\boldsymbol{e}(v) + \eta(u)\,\boldsymbol{e}_3\,.$$

The parameters of the surface are u and v. The parameter curves are the *meridians*, which are the generating curves $\boldsymbol{x}(u, v_0)$, $v_0 = \text{const}$, and the *circles of latitude* $\boldsymbol{x}(u_0, v)$, $u_0 = \text{const}$, which are the circles of intersection of the surface with planes perpendicular to the axis of rotation. To find the singular points of the surface – the generating curve is assumed to be regular – one calculates

$$\frac{\partial \boldsymbol{x}}{\partial u}\times\frac{\partial \boldsymbol{x}}{\partial v} = (\xi'\boldsymbol{e} + \eta'\boldsymbol{e}_3)\times\xi\boldsymbol{e}^* = \xi(-\eta'\boldsymbol{e} + \xi'\boldsymbol{e}_3),$$

where the dash denotes differentiation with respect to the arc length u of the generating curve. The vector $\boldsymbol{n} = -\eta'\boldsymbol{e} + \xi'\boldsymbol{e}_3$ is the normal unit vector of this curve and therefore a normal vector of the surface. A point of the surface is singular if and only if $\xi = 0$, that is, if the point lies on the axis of rotation. For the coefficients of the first fundamental form one obtains immediately $E = \frac{\partial \boldsymbol{x}}{\partial u}\cdot\frac{\partial \boldsymbol{x}}{\partial u} = |\boldsymbol{x}'|^2 = 1$, because u is the arc length on the meridian, $F = \frac{\partial \boldsymbol{x}}{\partial u}\cdot\frac{\partial \boldsymbol{x}}{\partial v} = 0$ (the meridians and circles of latitude cut orthogonally), and finally $G = \frac{\partial \boldsymbol{x}}{\partial v}\cdot\frac{\partial \boldsymbol{x}}{\partial v} = \xi^2(u)$; the first fundamental form is therefore

$$\mathrm{d}s^2 = \mathrm{d}u^2 + \xi^2(u)\,\mathrm{d}v^2.$$

To calculate the second fundamental form one needs the second derivatives of the parametric representation

$$\frac{\partial^2 \boldsymbol{x}}{\partial u^2} = \boldsymbol{x}'' = \varkappa_r\boldsymbol{n}, \qquad \frac{\partial^2 \boldsymbol{x}}{\partial u\,\partial v} = \xi'\boldsymbol{e}^*, \qquad \frac{\partial^2 \boldsymbol{x}}{\partial v^2} = -\xi\boldsymbol{e}.$$

The quantity $\varkappa_r$ is called the *relative* curvature of the meridian. Obviously, $|\varkappa_r| = (\boldsymbol{x}''\cdot\boldsymbol{x}'')^{1/2}$, because $\boldsymbol{n}\cdot\boldsymbol{n} = 1$, that is, the value of $\varkappa_r$ is equal to the curvature of the generating curve; $\varkappa_r$ is positive when the curve curves in the direction of the normal vector $\boldsymbol{n}$, and negative when it curves in the opposite direction; in the figure, $\varkappa_r < 0$. By scalar multiplication with the normal vector it follows that

$$L = \varkappa_r(u), \qquad M = 0, \qquad N = \xi\eta',$$

so that the second fundamental form is

$$\varkappa_n = \varkappa_r(u)\left(\frac{\mathrm{d}u}{\mathrm{d}s}\right)^2 + \xi(u)\,\eta'(u)\left(\frac{\mathrm{d}v}{\mathrm{d}s}\right)^2.$$

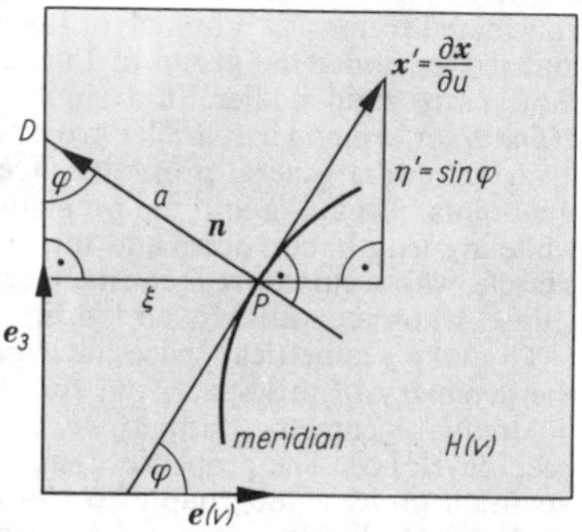

26.1-16 Curvature of a surface of revolution

From this one can easily calculate the normal curvature of the meridians and circles of latitude. For a meridian $v = v_0$, $\mathrm{d}v = 0$ and $\mathrm{d}u = \mathrm{d}s$ from the first fundamental form. It follows that $\varkappa_{n(\mathrm{Mer})} = \varkappa_r$; the normal curvature of a meridian is equal to its relative curvature. For a circle of latitude $u = u_0$, $\mathrm{d}u = 0$, that is, $\mathrm{d}s^2 = \xi^2\,\mathrm{d}v^2$ from the first fundamental form. It follows that $\varkappa_{n(\mathrm{Lat})} = \eta'/\xi$. These formulae can be interpreted geometrically. Since $\boldsymbol{x}' = \frac{\partial \boldsymbol{x}}{\partial u}$ is a unit vector, $\boldsymbol{x}' = \xi'\boldsymbol{e} + \eta'\boldsymbol{e}_3 = \boldsymbol{e}\cos\varphi + \boldsymbol{e}_3\sin\varphi$, where φ is the angle that $\boldsymbol{x}'$ makes with $\boldsymbol{e}$. From the figure one sees immediately that $\xi/a = \sin\varphi = \eta'$, so that $\eta'/\xi = 1/a$ is the reciprocal of the length a of the segment PD of the normal from the point P of the surface to the axis of rotation. It can be shown that the calculated values are just the maximum and minimum of the normal curvature for arbitrary curves of the surface through P. The extreme values of the normal curvature are called the *principal curvatures* of the surface at P. Curves which at every point have one of the principal curvatures as their normal curvature are called *lines of curvature*. In general, through each regular point of a surface there pass two lines of curvature, which cut at right angles; in the case of a surface of revolution, these are just the meridians and the circles of latitude. For the Gaussian curvature one obtains immediately $K = (LN - M^2)/(EG - F^2) = \varkappa_r\xi\eta'/\xi^2 = \varkappa_r/a$.

The Gaussian curvature is the product of the principal curvatures.

It follows from this that for a surface of revolution $K < 0$ when the curve arches towards the axis ($\varkappa_r < 0$), and $K > 0$ when the curve arches away from the axis ($\varkappa_r > 0$).

Mean curvature and minimal surfaces. If by a suitable choice of an orthonormal basis in the tangent plane at P the second fundamental form of the surface is brought to the normal form $\varkappa_n = \lambda_1 \left(\frac{du}{ds}\right)^2 + \lambda_2 \left(\frac{dv}{ds}\right)^2$, then one obtains the *principal curvatures* $\lambda_1(u, v)$ and $\lambda_2(u, v)$ as important invariants of the surface. A point at which the two normal curvatures coincide and are non-zero is called an *umbilic* of the surface. On a sphere every point is an umbilic; conversely, a surface on which every point is an umbilic is part of a sphere. If both principal curvatures are equal to zero, one speaks of a *flat point*; a surface on which every point is a flat point is part of a plane. At an umbilic or a flat point the normal curvature does not depend on the direction of the curve. The elementary symmetric functions of the principal curvatures are the *Gaussian curvature* $K = \lambda_1 \lambda_2$ and the *mean curvature* $H = (\lambda_1 + \lambda_2)/2$ of the surface. The problem of drawing a surface through a simple closed curve in space, which can be regarded as a continuously deformed circle, such that the surface has the least possible area, has as a necessary condition the equation $H = 0$, which was discovered as early as 1760 by LAGRANGE (1736–1813) and is a special case of the Euler-Ostrogradskii differential equation (see Chapter 38.). The non-planar solutions of this equation are called *minimal surfaces*. Since it follows from $H = 0$ and $K \neq 0$ that $K < 0$, minimal surfaces have negative Gaussian curvature. Further global results can only be mentioned.

The only closed surfaces with constant Gaussian curvature are spheres.

If $K > 0$ always holds for a closed regular surface, then it is an ovoid, that is, it bounds a finite convex body.

The only regular closed surfaces of genus zero with constant mean curvature are spheres.

Klein's Erlangen programme

According to Felix KLEIN (1849–1925), the various *geometries are to be regarded as invariant theories of the corresponding groups of transformations.* Thus, the Euclidean differential geometry considered so far – as a branch of Euclidean geometry – is the theory of invariants of curved surfaces and curves under the group of Euclidean motions (or *transformations*), which can be imagined as motions of rigid bodies. In a similar way, affine geometry is the theory of the invariants under *affine transformations* (parallel projections), and projective geometry studies properties that remain invariant under general projections (central projections). For example, the correspondence between the points of a curve and the osculating planes is not only a Euclidean, but a projective invariant, while arc-length, curvature and torsion are only Euclidean invariants, and not even affine. Indeed, a circle, whose curvature is constant, can be transformed by an affine transformation into an arbitrary ellipse, whose curvature is no longer constant.

To every geometrical space having a *Lie group of transformations* there belongs, as a branch of the geometry of this space, the corresponding differential geometry. Today, apart from *Euclidean* differential geometry, also *affine*, *projective*, *elliptic*, *hyperbolic* differential geometry, etc. have been developed. The properties that are invariant under a group G of transformations are *a fortiori* invariant under a subgroup of G; for example, it turns out that the classification of the points of a surface into elliptic, hyperbolic and parabolic is not only a Euclidean, but also an affine and even a projective invariant.

In addition, in differential geometry the interesting properties must be invariant under differentiable parameter transformations. Quite generally, one can ask what properties of geometrical forms remain invariant under sufficiently differentiable mappings of the space onto itself. The property of being a straight line or a plane is obviously invariant under projective mappings, but a plane can be transformed into a fairly arbitrarily curved surface by a suitable differentiable mapping. *Orders of contact* of curves or surfaces are invariant under sufficiently differentiable mappings. Also, the *geometry of webs* is invariant under these mappings. Questions about properties invariant under differentiable mappings can be very fruitful, even though the set of these mappings no longer forms a group, in general.

These considerations lead to a classification of the properties of differential geometry according to the principles of Klein's Erlangen programme. One can consider, for example, all twice continuously differentiable one-to-one mappings of surfaces of E_3 onto surfaces of E_3. The intrinsic geometry was defined above as the theory of properties of surfaces that remain invariant under isometric mappings. Here the set of isometric mappings is a proper subset of the set of differentiable mappings of surfaces onto one another.

A mapping is called *conformal* if angles between curves remain invariant; for example, *stereographic projection* of a sphere onto a plane is a conformal mapping. Every isometric mapping is conformal, but not conversely. The properties that remain invariant under conformal mappings form the subject matter of the *conformal geometry* of the surface. Similarly one can consider *area-preserving* and other classes of mappings and their corresponding geometries.

Riemannian geometry

Manifolds. All the geometrical figures studied in differential geometry are regarded as point sets, referred to *parameters* or *coordinates*. The dimension of a figure is defined as the number of coordinates that are necessary to fix a point of it. Thus, a curve is one-dimensional; for its points can be characterized by the values of one parameter t. Correspondingly, a surface is two-dimensional, and the space surrounding us is three-dimensional. Spaces of higher dimension occur in physical and technical applications. For example, if one wishes to describe the path of an aeroplane, that is, not only its route, but also its progress in time, one must know at any time t its longitude u, its latitude v and its height h. One thus obtains a curve in the four-dimensional space of the variables t, u, v, h. If one wishes to follow the aeroplane more precisely, one must add the components of instantaneous velocity $\dot{u} = \frac{du}{dt}$, $\dot{v} = \frac{dv}{dt}$, $\dot{h} = \frac{dh}{dt}$, and thus obtains a curve in the seven-dimensional space of the variables $t, u, v, h, \dot{u}, \dot{v}, \dot{h}$. If one now considers N aeroplanes simultaneously, one must specify, apart from the time t, for each aeroplane the 6 position and velocity coordinates $u_i, v_i, h_i, \dot{u}_i, \dot{v}_i, \dot{h}_i$ $(i = 1, \ldots, N)$, in order to describe the 'system', here the aggregate of N aeroplanes; one obtains therefore a $(6N + 1)$-dimensional space. In statistical mechanics an ideal gas is represented as a system of N molecules, which move in space independently of one another (like our aeroplanes); to describe the gas, again $6N + 1$ coordinates are necessary. Here N is very large, of the order of magnitude of the Loschmidt number: $N = 6.02 \cdot 10^{23}$. All these examples have one thing in common: each of the spaces is a point set whose points are in one-to-one correspondence with n-tuples $(x_1, \ldots, x_n)$ of real numbers, the *coordinates* of the point. Such a point set is called an *n-dimensional differentiable manifold.* These are manifolds in which the coordinates can – like the parameters of curves and surfaces – again be subjected to sufficiently differentiable coordinate transformations. The only properties that have a geometrical meaning are those that do not depend on the choice of coordinate system. Since the real numbers are a mathematical model of our intuitive geometrical idea of the *continuum* (number line), it is not surprising that important geometrical considerations can be carried out in a differentiable manifold; for example, the idea of *tangent vector* and *tangent space* can be introduced, and a *theory* of *contact* of submanifolds (curves, surfaces, m-dimensional submanifolds of the given n-dimensional manifold) can be developed. On these foundations one can then build a *geometrical theory of partial differential equations.* Also, a geometrical theory of the calculus of variations, the so-called *Finsler geometry*, can be created.

Riemannian geometry. Although a geometry can be developed for differentiable manifolds, it is in comparison with Euclidean geometry, very meagre, because concepts such as length, angle, area parallel displacement and curvature are completely missing. In 1854, Bernhard RIEMANN (1826–1866 in his inaugural lecture 'Über die Hypothesen, welche der Geometrie zugrunde liegen' (On the hypotheses which underlie geometry) developed the basic ideas of a geometry, which much later found an important physical application as the mathematical foundation of *Einstein's general theory of relativity*. An n-dimensional manifold is called a *Riemannian space* if a quadratic differential form is given in it as *element of arc*:

$$ds^2 = \sum_{i,k=1}^{n} g_{ik}(x_1, \ldots, x_n)\, dx_i\, dx_k.$$

The simplest non-trivial special case of a Riemannian geometry is the intrinsic geometry of a surface, which is determined by its element of arc (first fundamental form) alone and does not depend on the actual embedding in Euclidean space. The first fundamental form goes over into the form given above for the element of arc of a 2-dimensional Riemannian space if one introduces the new notation $u = x_1$, $v = x_2$, $E = g_{11}$, $F = g_{12} = g_{21}$, $G = g_{22}$. Here x_1 and x_2 must not be confused with the space coordinates x_1, x_2, x_3 in E_3. It does not matter which of the mutually isometric surfaces in E_3 one considers.

Riemannian geometry is precisely a generalization of the intrinsic geometry of a surface into n dimensions. All the concepts mentioned above, which are missing from the theory of differentiable manifolds, can be defined by means of the element of arc as a meaningful analogy to intrinsic geometry. For example, if $x_i = x_i(t)$, $0 < t < 1$, is a representation of a curve in Riemannian space, then along it $dx_i = \dot{x}_i\, dt$, and one obtains again as invariant parameter $s = s(t)$ the arc length

$$s(t) = \int_0^t \sqrt{\left[\sum_{i,k=1}^{n} g_{ik}(x_1(t), \ldots, x_n(t))\, \dot{x}_i(t)\, \dot{x}_k(t)\right]}\, dt.$$

While in intrinsic geometry the form $\sum_{i,k=1}^{n} g_{ik}\dot{x}_i\dot{x}_k$ is always *positive definite*, in Riemannian geometry also *indefinite* forms are admitted, so that the arc length can occasionally be zero or imaginary. It is just such Riemannian spaces that are applied in the theory of relativity.

Also, Euclidean space is a special case of a Riemannian space; its element of arc in orthonormal Cartesian coordinates was $g_{ii} = 1$ and $g_{ik} = 0$ for $i \neq k$. One can say that a portion of a Riemannian space arises by distorting a portion of a Euclidean space of the same dimension, in the same way as, for example, a piece of car bodywork is formed from a flat piece of metal by cutting out and pressing. In a similar way, *manifolds with affine connection* arise by 'distorting' affine spaces: in these manifolds, length, angle and area are no longer defined, but only a parallel displacement, depending on the path, determines the geometry of the manifold. The curvature of a Riemannian space (and also of a manifold with affine connection) indicates the deviation of the geometry of the space from that of the Euclidean (or affine) space of the same dimension: it is measured by means of the *Riemann-Christoffel curvature tensor*.

26.2. Convex bodies

A body $\boldsymbol{B}$ in Euclidean space is called *convex*, or an *ovoid*, if the line-segment joining any two points of $\boldsymbol{B}$ lies in $\boldsymbol{B}$. Convex bodies have been investigated in geometry for a long time. A proper theory of convex bodies arose towards the end of the 19th century in the works of BRUNN and MINKOWSKI (1864–1909); it has been generalized to n-dimensional Euclidean space and to non-Euclidean spaces. Examples for convex bodies are the sphere, ellipsoid, cylinder, cone, cube, tetrahedron and rectangular parallelepiped. The last three bodies are *convex polyhedra*, that is, bodies whose boundary consists of finitely many convex polygons. In the *theory of convex polyhedra* the following questions, for example, have been considered: By how many conditions (on the edges, vertices, faces, area, and so on) is a convex polyhedron uniquely determined (to within motions)? – When does there exist a convex polyhedron with certain prescribed conditions?

In the theory of convex bodies extremal problems are treated frequently. The oldest of these is the *isoperimetric problem* (see Chapter 38.).

By a *convex surface* one understands the boundary of a convex body. A convex surface can have faces and vertices, as in the example of a convex polyhedron. Nevertheless, the most important results of the differential geometry of regular surfaces of Euclidean space, especially their intrinsic geometry, can be generalized to arbitrary convex surfaces. In this way more far-reaching results than in classical differential geometry have been obtained. An essential feature of the theory of convex bodies is that one operates directly with geometrical objects, points, lines, and so on, and avoids to a large extent the dependence on analytical methods such as coordinates or parametric representations. Recently a modern branch of geometry, the very general *geometry of sets*, has been built up on the foundation of these methods.

The theory of convex bodies finds applications in many other branches of mathematics. For example, the *geometry of numbers*, whose foundations were also laid by MINKOWSKI, was developed in conjunction with this theory; certain results of the theory of convex bodies were applied to problems of number theory. Connected with this is the very attractive and intuitive *theory of packing*. A typical problem in this theory is the following: How should one place pennies on a very large table so as to accomodate as many pennies as possible? – The pennies must not overlap. The solution is that each penny must touch six others. The analogous problem about the densest packing of spheres in space is still unsolved. The theory of convex bodies also has many connections with *integral geometry*.

26.3. Integral geometry

Integral geometry developed from problems in geometrical probability. The first such problem was posed in 1777 by Count George DE BUFFON (1707–1788) (Buffon's needle problem): Parallel lines are drawn on a plane at equal distances a apart. A needle of length $l < a$ is thrown at random onto the plane. What is the probability p that it should meet one of the lines? – The answer is that $p = 2l/(\pi a)$. Since l and a are known, and p can be estimated by statistical methods, this gives a possibility of (approximately) determining π *experimentally*.

Subsequently many similar examples were considered, and important partial results obtained. Wilhelm BLASCHKE (1885–1962) and his school were the first to found integral geometry as a proper geometrical discipline. The foundations of integral geometry of a geometrical space are certain *measures*, which are assigned invariantly to sets of geometrical objects; in the Euclidean plane, for example, with every elementary plane figure one associates its area. This measure is invariant, that is, congruent figures have the same area. Every figure can be regarded as a *set of geometrical objects*, namely as the set of points belonging to it. The *dual* geometrical objects to points in a plane are the lines. There now arises the obvious question whether one can define in an invariant way a measure for the set of lines. In this case it is possible. For example, one considers the set L of all lines l that meet a circle C of radius r. The position of such a line is fixed if its direction angle φ (the angle it makes with a fixed line, say the x-axis) and its distance p from the centre of the circle are given. Here φ must run through all directions in the interval $0 \leqslant \varphi < 2\pi$ and p must run through all distances $0 \leqslant p \leqslant r$. As a measure for the set of lines L one therefore chooses the product of the lengths of the two intervals. It can be shown that this measure is invariant and can be defined in a similar way for much more general sets of lines. In the given example, the measure for the set L is $2\pi r$, which is the circumference of the circle. More generally, the measure of all lines that meet a convex plane figure is equal to the circumference of the figure.

If one considers two convex figures that do not overlap, one can ask for the measure of all the lines that simultaneously meet both figures. M. W. CROFTON deduced that this measure is equal to the length of the crossed belt that encircles the two figures, minus the length of the uncrossed belt around the two figures (Fig.).

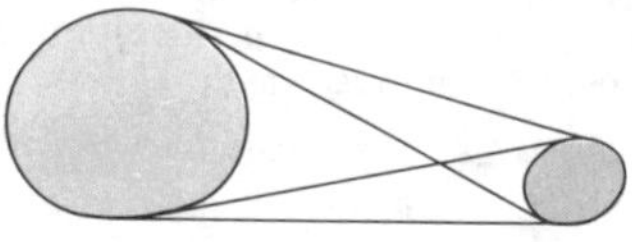

26.3-1 Crofton's belt theorem

Apart from the measures of sets of lines one can define a *kinematic* measure for sets of mutually congruent figures; for example, one can calculate the measure of all equilateral triangles of side 1 that meet the given figure. The kinematic measure was introduced by Henri POINCARÉ (1854–1912).

Just as in the plane, so one can develop a more or less meaningful integral geometry in other geometries determined by *Lie groups of transformation* (in the sense of *Klein's* Erlangen programme). In particular, the integral geometry of *n-dimensional Euclidean space* has been studied extensively. Also, for *curved surfaces* and for the *Finsler geometry* of the calculus of variations, attempts have been made to set up an integral geometry.

Integral geometry finds applications in other branches of geometry, particularly in the theory of convex bodies. Also, in practical problems its methods have been applied in connection with mathematical statistics; for example, a method has been worked out for the statistical determination of the surface of a lung.

27. Probability theory and statistics

27.1. Combinatorial analysis

Combinatorial analysis investigates the different possibilities for the arrangement of objects, for example, the questions: 'How many ways are there of arranging four letters in a row?', or 'In how many different ways can five different numbers be selected from 90 numbers?' – The objects of the investigation can be, for instance, numbers, letters, persons, tests. They are called *elements* and are denoted by numerals or letters. If two arrangements do not contain the same elements, orr example, *ab*, *cd*, or if they do còntain the same elements but not the same number of times for each, for example, *aab*, *abb*, then they are regarded as different. Arrangements such as *aabb*, *abab* fae regarded as different only if the ordering is taken into account.

Permutations

Every arrangement of finitely many elements in any order in which all the elements are used is called a *permutation* of the given elements. For example, the arrangements *acdbe*, *dbcae* are permutations of the elements a, b, c, d, e.

Number of permutations. The number of permutations of distinct elements can be obtained by inductive reasoning as follows: From two elements a and b the two permutations ab, ba can be formed. Of the three elements a, b, c, each can stand in the first place, and at the same time the other two can be ordered in two different ways:

$$abc \quad acb \quad bac \quad bca \quad cab \quad cba.$$

It follows that there are $3 \cdot 2 = 6$ permutations. For four elements the corresponding argument leads to $4 \cdot 3 \cdot 2 = 24$ permutations. In general, n elements can be arranged in $1 \cdot 2 \cdot 3 \cdot \cdots \cdot (n-1)\, n = n!$ (read *n factorial*) ways.

There are n! permutations of n distinct elements

Number of permutations of n distinct elements
$n! = 1 \cdot 2 \cdot 3 \cdot \cdots \cdot n$

Example: From the nine digits 1, 2, 3, ..., 9 one can form $9! = 362\,880$ nine-figure numbers such that in every one each of the nine digits occurs only once.

Table of factorials from 1! *to* 20!

n	$n!$	n	$n!$	n	$n!$	n	$n!$
1	1	6	720	11	39 916 800	16	20 922 789 888 000
2	2	7	5 040	12	479 001 600	17	355 687 428 096 000
3	6	8	40 320	13	6 227 020 800	18	6 402 373 705 728 000
4	24	9	362 880	14	87 178 291 200	19	121 645 100 408 832 000
5	120	10	3 628 800	15	1 307 674 368 000	20	2 432 902 008 176 640 000

If groups of identical elements occur in a number of elements, then the number of permutations is smaller than if all the elements are distinct. For example, in permuting the five elements $e_1 = a$, $e_2 = a$, $e_3 = b$, $e_4 = b$, $e_5 = b$, all orderings of the elements e_1 and e_2, and similarly all those of the elements e_3, e_4 and e_5 must be regarded as identical. Since the numbers of permutations of these elements are 2!, 3! repectively, the total number of distinct permutations is only $\dfrac{5!}{2!\,3!} = 10$. In general, if n elements consist of m groups containing $p_1, p_2, \ldots, p_m$ identical elements, respectively, and if the $p_i!$ permutations of the elements p_i $(i = 1, 2, \ldots, m)$ are regarded as being the same, then the total number of permutations of the n elements is given by:

$$\frac{n!}{p_1!\,p_2!\,\ldots\,p_m!}; \qquad p_1 + p_2 + \cdots + p_m = n$$

Example: Could a passionate bridge player play all possible hands in the course of his life? – The number of possible hands is $\dfrac{52!}{13!\,13!\,13!\,13!}$, because permutations of the 13 cards of each of the players do not lead to a different hand. This number exceeds $5.36 \cdot 10^{28}$. If he played 200 hands every day for 100 years, that is, 7 300 000 hands, the player could play only a tiny fraction of the total.

Lexicographical ordering. The search for all $n!$ permutations of n elements is greatly simplified by *lexicographical ordering*. In this one first fixes some natural order, for example, for numbers their order of magnitude, or for letters their alphabetical order. Permutations are then said to be in lexicographical order if of two different permutations the one whose first element is in its natural order comes first; if they have the same first element, then that with the second element in its natural order comes first, and if the first two elements are the same for both, then one distinguishes them according to the third element, and so on. The first two of the following pairs of permutations are in lexicographical order, as are also the six permutations of the three elements given above, but the third pair is not.

1.	2.	3.
$a\ b\ c\ f\ g$	$a\ b\ c\ h\ i$	$a\ b\ d\ f\ e$
$a\ b\ c\ g\ f$ ↓	$a\ c\ b\ i\ h$ ↓	$a\ b\ d\ e\ f$ ↓

Inversion. Two elements in a permutation are said to form an *inversion* if their ordering is opposite to their natural order. In the permutation *cdbea* formed from the elements a, b, c, d, e, the element c precedes b, c precedes a, d precedes b, d precedes a, b precedes a and e precedes a, and they each form an inversion. The given permutation thus contains six inversions. If the number of inversions of a permutation is even, the permutation is said to be *even*; otherwise it is called *odd*.

Combinations

Every collection of k elements taken from n elements is called a *combination of the kth class* or kth *order*.

Example: $ab, ac, ad, bc, bd, cd, bb, dd, \ldots$ are combinations of the second class from the four elements a, b, c, d.

If only distinct elements are chosen for each collection, one speaks of *combinations without repetition*; otherwise they are called *combinations with repetition.* If two combinations containing the same elements, but in a different order, are regarded as being distinct, one speaks of *permutations* in an extended sense.

The number of permutations. The number of permutations of k elements selected from n elements will be denoted by ${}^nP_k$. The first element can be chosen in n different ways, the second can then be chosen in $(n-1)$ ways, the third in $(n-2)$ ways, and the kth in $(n-k+1)$ ways. Hence

$${}^nP_k = n(n-1)\ldots(n-k+1) = \frac{n!}{(n-k)!}.$$

Number of permutations of k elements selected from n elements without repetition	${}^nP_k = \dfrac{n!}{(n-k)!}$

Thus, for $n = 4$, the number of permutations of the 4 elements a, b, c, d taken 3 at a time is ${}^4P_3 = 4 \cdot 3 \cdot 2 = 24$. These are

abc → *abd* → *acb* → *acd* → *adb* → *adc* → *bac* → *bad* → *bca* → *bcd* →
bda → *bdc* → *cab* → *cad* → *cba* → *cbd* → *cda* → *cdb* → *dab* → *dac* →
dba → *dbc* → *dca* → *dcb*.

If repetitions are allowed, then the second element can also be chosen in n ways, and so can the third, and so on. Thus, for $n = 3$ elements, the number of permutations taken 2 at a time with repetition is ${}^3P_2^{(r)} = 3^2 = 9$. These are

aa → *ab* → *ac* → *ba* → *bb* → *bc* → *ca* → *cb* → *cc*.

More generally, ${}^nP_k^{(r)} = n^k$.

Number of permutations of k elements selected from n elements with repetition	${}^nP_k^{(r)} = n^k$

Example 1: From the digits 1, 2,..., 9, 0 one can form ${}^{10}P_3^{(r)} = 10^3 = 1000$ permutations of the third class with repetition. These are precisely the numbers 000, 001, 002, up to 999.

Example 2: By means of *Braille* the blind are able to feel letters, numbers and punctuation, which are printed on paper as arrangements of six points that appear either raised or non-raised. *Point* and *no point* are the two variable elements, and the number of possible signs represented by them is the number of permutations with repetition of these two elements taken 6 at a time. This gives $2^6 = 64$ signs. These possibilities are enough to represent the blind alphabet, together with numbers and punctuation (Fig.).

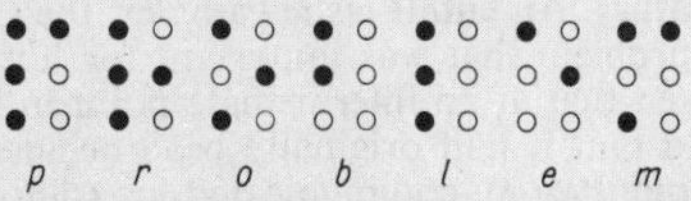

27.1-1 Braille for 'problem'

The number of combinations. In combinations the order in which the elements occur is not taken into account. The number of combinations of k elements taken from n elements is denoted by ${}^nC_k$. From the four elements a, b, c, d the following combinations of the second class can be formed: *ab* → *ac* → *ad* → *bc* → *bd* *cd*. If one permutes the elements of each combination, one obtains the number of permutations ${}^4P_2$ of two elements taken from four, which is 2! times as large. Similarly, the number of permutations ${}^nP_k$ of k elements taken from n elements is $k!$ times the number of combinations ${}^nC_k$. Thus,

$$k! \cdot {}^nC_k = {}^nP_k = \frac{n!}{(n-k)!}, \text{ or } {}^nC_k = \binom{n}{k} = \frac{n!}{(n-k)!\,k!}.$$ The $\binom{n}{k}$ are the *binomial coefficients.*

Number of combinations ${}^nC_k$ of k elements taken from n elements without repetition	${}^nC_k = \dbinom{n}{k} = \dfrac{n!}{(n-k)!\,k!}$

Example: In the game of *Bingo* $k = 5$ different numbers can be chosen from $n = 90$ numbers in $\binom{90}{5} = 43\,949\,268$ ways. Only if one has cards with all these possibilities is one certain to have a line of five. The number of fours and threes can be calculated in a similar way. From the five (correct) numbers one is always missing; thus, there are $\binom{5}{4} = 5$ combinations of four. For three correct numbers, with two always missing, there are $\binom{5}{3} = 10$ combinations. Each of the five combinations of four occurs $\binom{90-5}{1} = 85$ times among the possible incorrect lines of five, since this is the number of possible ways of adding a fifth number from the remaining 85 incorrect ones. Thus, the number of lines with four correct numbers is $\binom{5}{4}\binom{90-5}{1} = \binom{5}{4}\binom{85}{1}$ $= 5 \cdot 85 = 425$. Each combination of three numbers can be combined with two of the remaining numbers in $\binom{85}{2} = 85 \cdot 42$ ways so that neither a line of four correct nor one of five correct results. Thus, the number of lines with three correct numbers is $\binom{5}{2} \cdot \binom{85}{2} = 35\,700$.

If repetitions are allowed, the number of combinations of n elements of the kth class is denoted by ${}^nC_k^{(r)}$. For the three elements a, b, c and for the n elements $a_1, a_2, \ldots, a_n$ one obtains the following combinations of the second class:

$$\begin{array}{ccc} aa & ab & ac \\ & bb & bc \\ & & cc \end{array} \qquad \begin{array}{ccccc} a_1a_1 & a_1a_2 & a_1a_3 & \ldots & a_1a_n \\ & a_2a_2 & a_2a_3 & \ldots & a_2a_n \\ & & a_3a_3 & \ldots & a_3a_n \\ & & & & \vdots \\ & & & & a_na_n \end{array}$$

Thus, ${}^3C_2^{(r)} = 3 + 2 + 1 = 6$ and ${}^nC_2^{(r)} = n + (n-1) + (n-2) + \cdots + 1 = \binom{n+1}{2}$. In general, ${}^nC_k^{(r)} = \binom{n+k-1}{k}$; this statement is true for $k = 2$ as has been shown above, and can be established for every natural number $k > 2$ by the method of mathematical induction.

Number of combinations of n elements taken k at a time with repetition	${}^nC_k^{(r)} = \binom{n+k-1}{k}$

27.2. Probability theory

Historical background. The beginning of probability theory goes back to the middle of the 17th century. An enthusiastic gambler, the Chevalier DE MÉRÉ, asked Blaise PASCAL (1623–1662) to solve a problem that was important for him: to describe the distribution of the wins of two gamblers, given that at an intermediate point in the game one had won $n < m$ rounds and the other $p < m$, and that it had originally been decided that the first one to win m rounds should win the whole game. PASCAL communicated his solution to Pierre FERMAT (1601–1665), who also found a method of solution. A third one came from Christian HUYGENS (1629–1695). These learned men recognized the significance of the question for the investigation of the laws governing random events. The concepts and the first methods of the new science developed from problems of games of chance. Much later, in the 19th century, the rapidly increasing interest in natural sciences made it necessary to extend the theory of probability beyond the framework of games of chance. This development is closely linked with the names of Jacob BERNOULLI (1654–1705), Abraham de MOIVRE (1667–1754), Pierre-Simon de LAPLACE (1749–1827), Carl Friedrich GAUSS (1777–1855), Simon Denis POISSON (1781–1840), Pafnuti Lvovich CHEBYSHEV (1821–1894), Andrei Andreevich MARKOV (1856–1922), and most recently with those of Alexander Yakovlevich KHINCHINE (1894–1959) and Andrei Nikolaevich KOLMOGOROV (b. 1903). Connected with the investigation of the laws governing random events is that of mass events. For example, the production of an article in everyday use is a mass event and the occurrence of a faulty article among them is a random event. Probability theory today is connected with many other branches of mathematics and with many fields of natural science, technology, and economics.

Probability of random events

Event. An event E, in the sense of a random event, is the result of a trial that can, but need not occur. A *trial* can be an observation or an experiment and is characterized by a set of conditions to be satisfied and by its repeatability. The limiting cases are also regarded as events: the *certain event* S, which definitely occurs, and the *impossible event* $\emptyset$, which never occurs. In the trial 'throw of a die', for example, E_3 is the event '3 turns up', S is '1 or 2 or 3 or 4 or 5 or 6 turns up', and '7 turns up' is $\emptyset$.

Events are *mutually exclusive* or *incompatible* if as the result of a trial only one of them can occur; for example, in the trial 'throw of a die' the events E_i (i turns up), $i = 1, 2, 3, 4, 5, 6$, are mutually exclusive, since only one of them can occur. In drawing a ball out of an urn that contains red and black balls, E_1 'drawing a red ball' and E_2 'drawing a black ball' are incompatible, since they cannot occur simultaneously. All events that are mutually exclusive in pairs form a *complete* system of events if, as the result of a trial, one of them must necessarily occur, for example, the events E_i ($i = 1, 2, 3, 4, 5, 6$) for the trial 'throw of a die'.

If two events E_1 and E_2 form a *complete system* of events, each of them is *complementary* to the other. In tossing a coin, for example, 'heads' and 'tails' are complementary.

One speaks of the *sum C of the two events A and B*, denoted by $C = A \cup B$ or $C = A + B$, if in a trial at least one of the events A or B occurs. This idea can be extended to more than two events. For example, for the trial 'throw of a die' the event 'an even number turns up' is equivalent to the sum of the events $E_2 + E_4 + E_6$.

One speaks of the *product* C of the events A and B, denoted by $C = A \cap B$ or $C = A \cdot B$ (or simply $C = AB$) if in a trial the events A and B both occur. For the trial 'throw of two dice', for example, the event C, 'throw of 12', occurs when a 6 is thrown with each die. The idea of a product can also be extended to more than two events.

The classical definition of probability. Although there exists an axiomatic theory of probability, the important laws can be derived from the classical definition.

Classical definition of probability. If a trial can result in n equally likely events and if m of these are favourable to the occurrence of an event E, then the probability $P(E)$ of the event E occurring is

$$P(E) = \frac{m}{n} = \frac{\text{number of favourable events}}{\text{number of possible events}}.$$

Thus, $P(E)$ is always a number between 0 and 1; $0 \leqslant P(E) \leqslant 1$. For the sure event S, $P(S) = 1$. The probability of an event is often expressed as a percentage.

For the trial 'throw of a die' the events E_i (i turns up) with $i = 1, 2, 3, 4, 5, 6$ are all possible. If the throw of a 3 is regarded as favourable, then the probability of its occurrence is $P(E) = 1/6$. This assumes an *ideal die* that is geometrically and mechanically homogeneous, so that no side is favoured more than another on account of the form and mass distribution. The events E_i are then equally probable. They form a complete system, and it follows that their sum is the certain event of throwing one of the numbers 1 to 6. But the probability of this is 1, and therefore the probability of each E_i is 1/6. For the event $E = E_2 + E_4 + E_6$ (an even number turns up) one obtains $P(E) = 3/6 = 1/2$.

The addition law in probability theory. If as the result of a trial n events are possible, and if m_i of these are favourable to the occurrence of the event E_i for $i = 1, 2, \ldots, k$, then $m = \sum_{i=1}^{k} m_i$ are favourable for the occurrence of the event $E = \sum_{i=1}^{k} E_i$, provided that the events E_i ($i = 1, 2, \ldots, k$) are mutually exclusive. It then follows that $P(E_i) = m_i/n$ for $i = 1, 2, \ldots, k$, and $P(E) = m/n = (m_1 + m_2 + \cdots + m_k)/n = \sum_{i=1}^{k} m_i/n = \sum_{i=1}^{k} P(E_i)$.

The probability of the sum of a number of mutually exclusive events is equal to the sum of the probabilities of these events $P(E_1 + E_2 + \cdots + E_k) = P(E_1) + P(E_2) + \cdots + P(E_k)$.

Example: As a result of throwing an ideal die, let E_4 and E_5 be the events of 4 and 5 turning up. Then for $E = E_4 + E_5$ (a 4 *or* a 5 turns up) it follows that $P(E) = P(E_4 + E_5) = P(E_4) + P(E_5) = 2/6 = 1/3$ (Fig.).

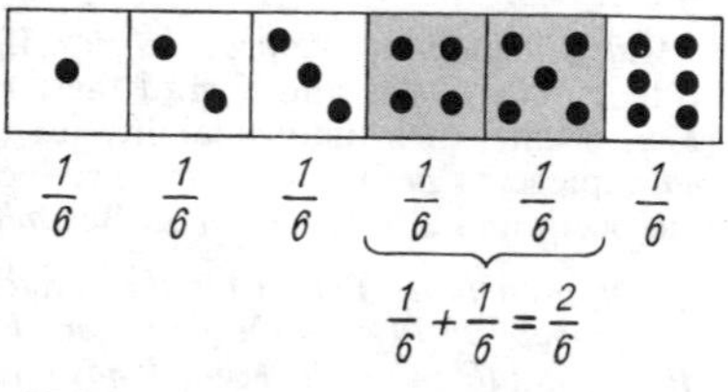

27.2-1 The addition law for an ideal die

If in a trial only the k events E_i with $i = 1, 2, \ldots, k$ are possible, then they form a *complete system* of events; since $n = m = m_1 + m_2 + \cdots + m_k$ in this case, it follows that $P(E) = 1$. For two mutually exclusive events E_1 and E_2, $P(E_1 + E_2) = P(E_1) + P(E_2) = 1$, or $P(E_2) = 1 - P(E_1)$; for example, if E_1 is the event 'birth of a boy' and E_2 the event 'birth of a girl'.

If in a trial with n possible outcomes the event E_1 occurs m_1 times and the event E_2 occurs m_2 times, but these two events are *not mutually exclusive*, the addition theorem can still be used if one considers the l cases when the event E_1E_2 occurs. There are three groups of mutually exclusive

events: $\bar{m}_1 = (m_1 - l)$ cases in which only the event E_1 occurs, $\bar{m}_2 = (m_2 - l)$ cases with E_2 alone, and l cases in which both occur. By the addition law it therefore follows that
$P(E_1 + E_2) = (m_1 - l)/n + (m_2 - l)/n + l/n = m_1/n + m_2/n - l/n = P(E_1) + P(E_2) - P(E_1E_2)$.
If $P(E_1E_2)$ is not known, the estimate $P(E_1 + E_2) \leqslant P(E_1) + P(E_2)$ holds. This addition theorem can be extended to more than two events.

Conditional probabilities. An *unconditional probability* depends only upon the set of conditions initially fixed for the trial; for example, that every die used is ideal, so that each number is equally likely at each throw. A *conditional* probability depends, in addition, on at least one further condition. The probability of the occurrence of the event E on the assumption that the event F has already occurred is then denoted by $P(E/F)$.

Example 1: If an urn contains n balls, of which m are black and $(n - m)$ white, then for the trial 'drawing with replacement' two events are possible, F_1 'drawing a black ball' and F_2 'drawing a white ball'. For these two one obtains the *unconditional probabilities* $P(F_1)$ and $P(F_2)$. For the trial 'drawing without replacement' the number of balls available for the second withdrawal depends upon the outcome of the first one. If the event F_1 occurred, then there are $(m - 1)$ black and $(n - m)$ white balls in the urn; if F_2 occurred, there are m black and $(n - m - 1)$ white balls. For the events E_1 'a black ball at the second withdrawal' and E_2 'a white ball at the second withdrawal' there are therefore four conditional probabilities $P(E_1/F_1)$, $P(E_2/F_1)$, $P(E_1/F_2)$ and $P(E_2/F_2)$.

Event	F_1	F_1	F_2	F_2	first withdrawal
Probability	$\frac{m}{n}$	$\frac{m}{n}$	$\frac{n-m}{n}$	$\frac{n-m}{n}$	
Event	(E_1/F)	(E_2/F_1)	(E_1/F_2)	(E_2/F_2)	second withdrawal
Probability	$\frac{m-1}{n-1}$	$\frac{n-m}{n-1}$	$\frac{m}{n-1}$	$\frac{n-m-1}{n-1}$	

Example 2: In throwing two ideal dice 36 events $E_{a,b} = (a, b)$ with $a = 1, 2, \ldots, 6$ and $b = 1, 2, \ldots, 6$ occur, and they are mutually exclusive in pairs. In this notation, as with ordered number pairs in general, (a, b) is to be regarded as different from (b, a). One therefore obtains the unconditional probability $P(E_{a,b}) = 1/36$. For the event 'total score 8 at one throw' the addition theorem gives $P(a + b = 8) = 5/36$, because there are five favourable events corresponding to $a + b = 2 + 6 = 3 + 5 = 4 + 4 = 5 + 3 = 6 + 2$. Under the additional condition S_e 'total score is even' one obtains on the other hand $P(E_{a,b}/S_e) = 1/18$ and $P(a + b = 8/S_e) = 5/18$. The condition 'score of 4 with the second die' gives another conditional probability. For $b = 4$, (a, b) can be one of the six number pairs (1, 4), (2, 4), (3, 4), (4, 4), (5, 4), (6, 4), so that $P(E_{a,b}/(b = 4)) = 6/36 = 1/6$. The same numerical value 1/6 is obtained for the probability that out of the six number pairs with $a = 4$, the one with $b = 4$ occurs (Fig.).

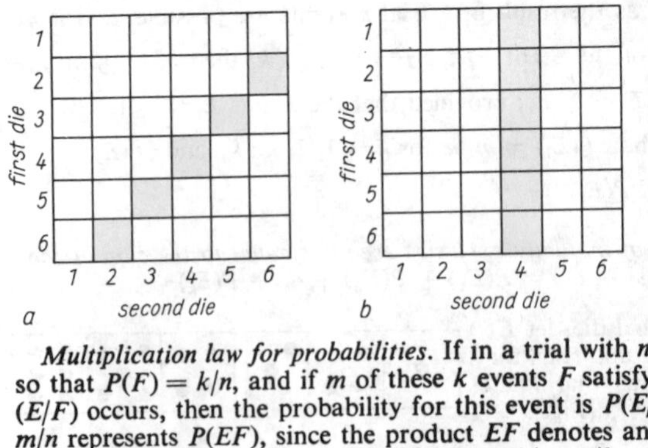

27.2-2 Probabilities in throwing two dice
a) 5/36 for total score 8
b) 6/36 = 1/6 for score of 4 with second die

Multiplication law for probabilities. If in a trial with n possible results the event F occurs k times, so that $P(F) = k/n$, and if m of these k events F satisfy a further condition under which the event (E/F) occurs, then the probability for this event is $P(E/F) = m/k$. On the other hand, the fraction m/n represents $P(EF)$, since the product EF denotes an event that satisfies both the conditions for F to occur and also those for E. But $m/n = k/n \cdot m/k$, and it follows that $P(EF) = P(F) \cdot P(E/F)$.

The probability $P(EF)$ for the simultaneous occurrence of two events E and F is the product of the probability $P(F)$ of the first event F and the conditional probability $P(E/F)$ of the event E under the assumption that the event F has already occurred.

Example: When two dice are thrown, the 36 events $E_{a,b}$ are possible. Let F be the event 'score (a, b) such that $a + b$ is divisible by 2' and (E/F) the events among these for which $a + b$ is also divisible by 3. Consequently EF are the events for which $a + b$ is divisible by 6. One obtains $P(F) = 18/36$, $P(E/F) = 6/18$ and $P(EF) = 6/36 = 1/6$, because EF consists precisely of the

6 events (1, 5), (2, 4), (3, 3), (4, 2), (5, 1), and (6, 6). Alternatively, $P(EF) = 6/36 = (18/36) \cdot (6/18) = P(F) \cdot P(E/F)$.

The law of total probability. If the events F_i for $i = 1, 2, \ldots, n$ form a *complete system* and if E is a further event, then the events $E \cdot F_i$ for $i = 1, \ldots, n$ are mutually exclusive in pairs. The sum of these events is equivalent to the event E. By the addition law for probabilities the probability $P(E)$ is given by the sum of the $P(EF_i)$ for $i = 1, 2, \ldots, n$. But for each term in the sum one obtains $P(EF_i) = P(F_i)\,P(E/F_i)$, by the multiplication law for probabilities. Thus, combining these results, $P(E) = \sum_{i=1}^{n} P(F_i)\,P(E/F_i)$. This unconditional probability $P(E)$ is called the *total probability*.

Example: Two urns contain black and white balls in proportions that can be different for each urn. Then the event 'drawing a white ball from one of the two urns' can be represented as the sum of two mutually exclusive events $EF_1 + EF_2$, where F_1 denotes 'drawing a ball from urn 1' and F_2 'drawing a ball from urn 2'. By the law of total probability the probability for the event E is given by $P(E) = P(EF_1) + P(EF_2) = P(F_1)\,P(E/F_1) + P(F_2)\,P(E/F_2)$. Thus, $P(E)$ can be calculated from $P(F_1)$, $P(F_2)$, $P(E/F_1)$ and $P(E/F_2)$.

Independent events. Two events E and F are independent of one another if the occurrence or non-occurrence of one has no influence on the occurrence or non-occurrence of the other; for example, in throwing two ideal dice the score thrown with one does not depend on that thrown with the other. If E denotes 'a score of 4' and the dice are distinguished by the indices 1 and 2, then $P(E_1) = P(E_2) = 1/6$, but $P(E_2/E_1) = 1/6$ also. For the event $E_1 \cap E_2 = E_1 \cdot E_2$, that is, 'a score of 4 with each die', it then follows that $P(E_1 \cap E_2) = 1/36 = 1/6 \cdot 1/6$. This result can be generalized.

Multiplication law for the probability of independent events	$P(E_1 \cap E_2 \cap E_3) = P(E_1) \cdot P(E_2) \cdot P(E_3)$

The axiomatic definition of probability. The development of the natural sciences and technology led to problems to which the classical definition of probability could no longer be applied uncritically. One cannot always assume that the number of possible cases is finite and that the individual cases are equally likely. For example, it is difficult, purely from arguments of symmetry, to determine the probability that during a specified time interval m conversations out of a total of n will be taking place over a telephone cable. The *statistical definition* of probability is superior to the classical one for these problems, but its character is more descriptive than formal-mathematical. It became necessary to investigate systematically the basic concepts of the calculus of probability and to establish the conditions for its applicability in an axiomatic structure. Of the different approaches that were proposed, the one generally followed today is that which KOLMOGOROV developed at the beginning of the 1930's for the solution of the new problems. He linked the concept of probability with modern set theory, measure theory, and functional analysis. His method proceeds from the main properties of probability that are valid whether they are based on the classical or the statistical definition. KOLMOGOROV created an axiomatic foundation for the concept of probability, which includes both the classical and the statistical definition and, in addition, satisfies the more stringent requirements of modern natural science and technology.

This axiomatic development is based on a set S of *elementary events* and a system $\boldsymbol{B}$ of subsets of S. The elements of the system $\boldsymbol{B}$, that is, the subsets of S, are called *random events*. If the system $\boldsymbol{B}$ of random events satisfies the following conditions, then it is called a *Borel field of events*.

Borel field:

1. **The set S is an element of $\boldsymbol{B}$.**
2. **If two sets E_1 and E_2 are elements of $\boldsymbol{B}$, then their union $E_1 \cup E_2$, their intersection $E_1 \cap E_2$, and their complements $\overline{E_1}$ and $\overline{E_2}$ are also elements of $\boldsymbol{B}$.**
3. **If sets $E_1, E_2, \ldots, E_n, \ldots$ are elements of $\boldsymbol{B}$, then their union $E_1 \cup E_2 \cup \cdots \cup E_n \cup \cdots$ and their intersection $E_1 \cap E_2 \cap \cdots \cap E_n \cap \cdots$ are also elements of $\boldsymbol{B}$.**

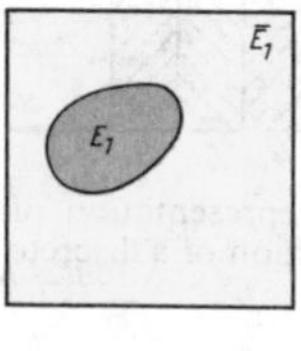

27.2-3 Event E_1 and event $\overline{E_1}$

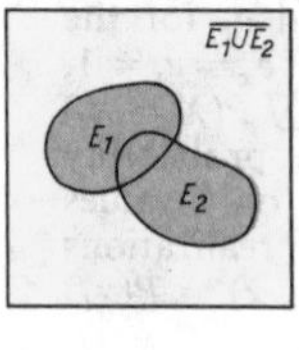

27.2-4 Event $E_1 \cup E_2$ and event $\overline{E_1 \cup E_2}$

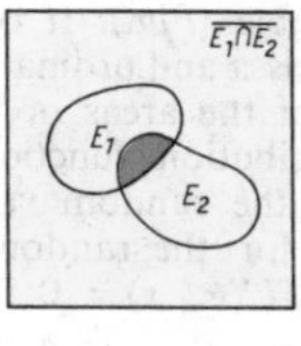

27.2-5 Event $E_1 \cap E_2$ and event $\overline{E_1 \cap E_2}$

If only conditions 1. and 2. are satisfied, one speaks of a *field of events.*

By the second condition $\bar{S}$, that is, the empty set Ø, must be an element of **B**. It is called the *impossible event*. The random events E_1, $\bar{E}_1$, $E_1 \cup E_2$, $\overline{E_1 \cup E_2}$, $E_1 \cap E_2$ and $\overline{E_1 \cap E_2}$ are illustrated in the figures, in which random events are represented by points of a square. Each point set then represents a random event.

The explanation will be helped by an example. If a die is thrown, then the set of elementary events consists of the six elements e_i ($i = 1, 2, \ldots, 6$), where e_i denotes that a throw results in a score i. The system of random events **B** then consists of $2^6 = 64$ elements: $(e_1), (e_2), \ldots, (e_6), (e_1, e_2), \ldots, (e_5, e_6), (e_1, e_2, e_3), \ldots, (e_4, e_5, e_6), \ldots, (e_1, e_2, e_3, e_4, e_5, e_6)$ and the empty set Ø. Enclosed in each pair of brackets stand the elements of S of which the corresponding subset of S is composed.

On the basis of the system **B** of random events, in which S denotes the certain event, $\bar{S}$ the impossible event and E and $\bar{E}$ opposite events, the probability of the occurrence of an event is defined by Kolmogorov's system of axioms.

Kolmogorov's system of axioms:

1. **Axiom: To every random event E in the field of events there is assigned a non-negative real number $P(E)$, called the probability of E.**
2. **Axiom: The probability of the certain event S is 1, $P(S) = 1$.**
3. **Axiom: If the events $E_1, E_2, \ldots, E_n$ are mutually exclusive in pairs, then $P(E_1 \cup E_2 \cup \cdots \cup E_n) = P(E_1) + P(E_2) + \cdots + P(E_n)$.**

The system is supplemented by the following extended addition axiom, which makes it possible to take account of those events (of frequent occurrence in probability theory) that are composed of infinitely many partial events.

Extended addition axiom: If the occurrence of an event E is equivalent to the occurrence of an arbitrary one of the events $E_1, E_2, \ldots, E_n, \ldots$, mutually exclusive in pairs, then $P(E) = P(E_1) + P(E_2) + \cdots + P(E_n) + \cdots$

From these axioms it follows as a first consequence that $P(E) \leqslant 1$ for every event E in **B**. The axiom system is free from contradictions but not complete. The structure of probability theory is based on it. The measure-theoretical concept of probability, in conjunction with a sufficiently wide interpretation of frequency, is the basis of mathematical statistics.

Random variables and distribution

A *random variable* X is a variable which, in different experiments carried out under the same conditions, assumes different values x, each of which then represents a *random event*. In what follows only *discrete* random variables X will be considered, which assume a finite number or at most a countably infinite number of values x_i, or *continuous* random variables X, which can take all values in a finite or infinite interval. The numerical value of the score obtained by throwing a die is a discrete random variable; the random events, or *realizations*, are $x_i = i$ for $i = 1, 2, \ldots, 6$. On the other hand, the instantaneous speed X of a molecule in a gas is a continuous random variable, which can take every value in an interval. Random variables are completely characterized by their probability, density, and distribution functions.

Probability and distribution function of a discrete random variable. The random events x_i are regarded as *discontinuities* and their probabilities $P(x_i)$ as *magnitudes of the discontinuities.* The probability function then relates the magnitudes to the discontinuities; for example, for a loaded die the probabilities $P(X = x_i)$ correspond to the events x_i 'score of i'. Since one of these events must occur, $\sum_{i=1}^{6} P(X = x_i) = 1$.

x_i	1	2	3	4	5	6
$P(X = x_i)$	1/6	1/6	1/8	1/6	1/8	1/4

Graphically one represents the probability function by vertical strips, rectangles of width e_x and of height $e_y \cdot P(X = x_i)$. For a loaded die the values $P(X = x_i)$ are given in the table (Fig.). If one chooses e_x and e_y for the units in the abscissa and ordinate directions, so that $e_x = e_y = 1$, then the sum of the areas of the rectangles is $\sum P(X = x_i) = 1$. The distribution function $F(x)$ gives the probability $P(X < x)$ that the random variable X assumes only values $x_i < x$. Thus, for the random variable X with realizations $i = 1, 2, \ldots, n$, $P(X < x) = 0$ for $x < x_1$, $P(X < x) = P(x_1)$

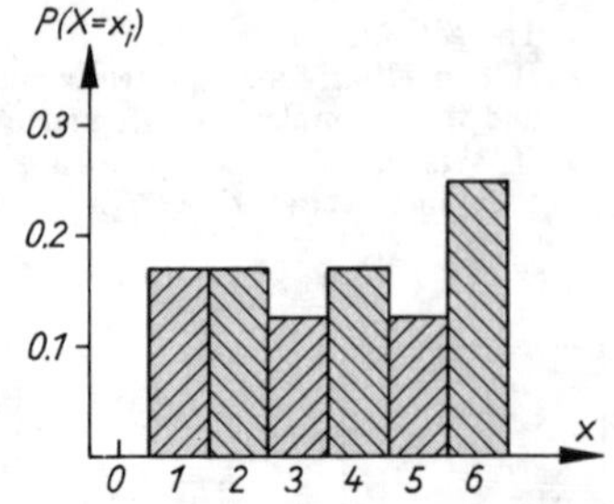

27.2-6 Graphical representation of the probability function of a discrete random variable

for $x_1 \leqslant x < x_2$, $P(X < x) = P(x_1) + P(x_2)$ for $x_2 \leqslant x < x_3$, and $P(X < x) = 1$ for $x > x_n$. Hence the distribution function $F(x)$ increases monotonically from $F(-\infty) = 0$ to $F(+\infty) = 1$. For the given probability function of a loaded die, for example, because $x_i = i$ one obtains:

$F(1) = P(X < 1) = 0; \quad F(2) = P(X < 2) = P(X = 1) = 1/6;$

$F(3) = P(X < 3) = P(X = 1) + P(X = 2) = 1/3;$

$F(4) = P(X < 4) = \sum_{i=1}^{3} P(X = i) = 11/24;$

$F(5) = P(X < 5) = \sum_{i=1}^{4} P(X = i) = 5/8;$

$F(6) = P(X < 6) = \sum_{i=1}^{5} P(X = i) = 3/4,$

$F(x > 6) = P(X < x) = \sum_{i=1}^{6} P(X = i) = 1.$

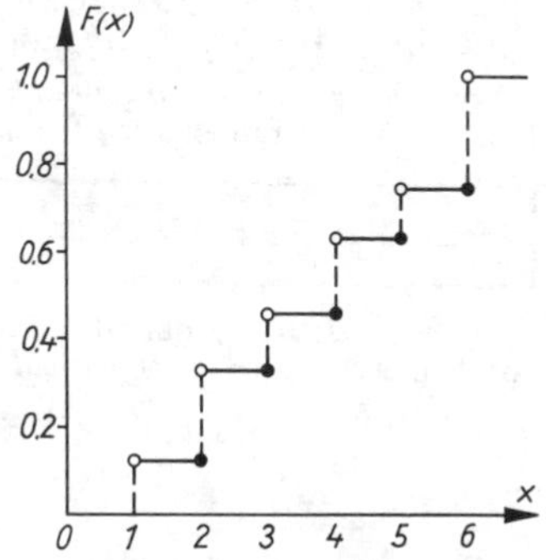

27.2-7 Graphical representation of the distribution function $F(x)$ of a discrete random variable

The graph of the distribution function $F(x)$ is a step function (Fig.), where the x_i are taken as the abscissae and the corresponding $F(x)$ as the ordinates in a Cartesian coordinate system.

Density and distribution function of a continuous random variable X. For a continuous random variable X every value x in an interval is a random event for which the probability of its occurrence is zero. But to each value x there corresponds a value $f(x)$ of the *probability density* or *density function.* For this function $f(x) = 0$ and $\int_{-\infty}^{\infty} f(t)\,dt = 1$ (Fig.); that is, the area between the x-axis and the graph of $f(x)$ is of magnitude 1.

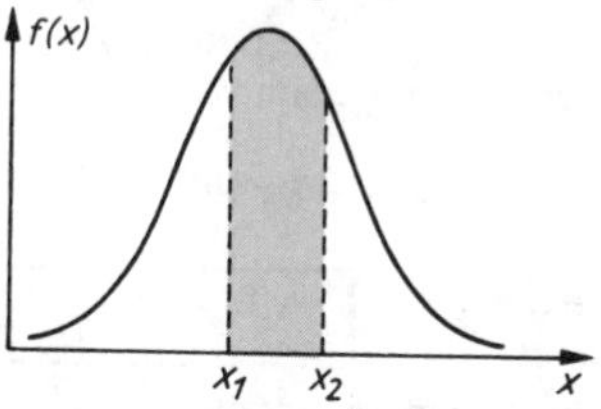

27.2-8 Representation of a density function

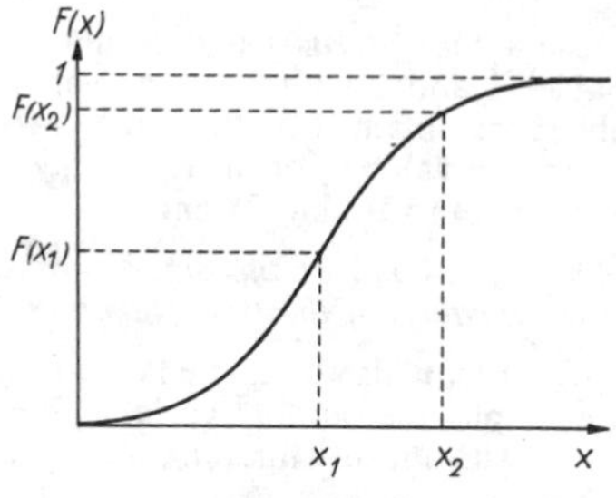

27.2-9 Distribution function of a continuous random variable

The *distribution function* $F(x)$ again represents the probability $P(X < x)$ that the continuous random variable assumes values less than x. Because $f(t) \geqslant 0$ it therefore follows from $F(x) = P(X < x) = \int_{-\infty}^{x} f(t)\,dt$ that the function $F(x)$ increases monotonically from $F(-\infty) = 0$ to $F(+\infty) = 1$ (Fig.).

The probability that the random variable X assumes a value in the interval $x_1 \leqslant x < x_2$, given by $P(x_1 \leqslant X < x) = P(X < x_2) - P(X < x_1) = F(x_2) - F(x_1) = \int_{x_1}^{x_2} f(t)\,dt$, is represented by the blue area in the graph of the density function. A well-known continuous random variable X is the *normal distribution,* which will be investigated later.

Mean value or expectation and variance

A random variable is completely described by the probability function if it is discrete, and by the density function if it is continuous. From these functions parameters can be calculated that characterize the random variable. The most important are the mean value or expectation and the variance.

Mean value or expectation. The mean value μ of a *discrete random variable* X is obtained by multiplying each of its possible values by the corresponding probability and forming the sum of all these products. This mean value need not occur among the values of the discrete random variable X.

If X is described by the probability function $P(X = x_i) = p_i$ $(i = 1, 2, \ldots, n)$, then the mean value μ is determined by $\mu = x_1 p_1 + x_2 p_2 + \cdots + x_n p_n$.

Mean value of a discrete random variable	$\mu = \sum_{i=1}^{n} x_i p_i$

Example 1: For an *ideal* die, $p_i = 1/6$, $i = 1, 2, \ldots, 6$; thus, the mean value is $\mu = 1 \cdot 1/6 + 2 \cdot 1/6 + 3 \cdot 1/6 + 4 \cdot 1/6 + 5 \cdot 1/6 + 6 \cdot 1/6 = 3.5$.

For a *continuous random variable* X the mean value μ is calculated by multiplying x by the density function $f(x)$ and integrating the product from $-\infty$ to $+\infty$.

Mean value of a continuous random variable	$\mu = \int_{-\infty}^{+\infty} x f(x)\, dx$

Example 2: It is required to find the mean value of the continuous random variable X with the probability density of the *normal* or *Gaussian* distribution (Fig.)

$$p(x) = 1/[a\sqrt{(2\pi)}]\, e^{-(x-b)^2/(2a^2)}.$$

Then $\mu = \int_{-\infty}^{+\infty} x/[a\sqrt{(2\pi)}]\, e^{-(x-b)^2/(2a^2)}\, dx$.

By means of the substitution $(x - b)/(a\sqrt{2}) = z$, $dx/(a\sqrt{2}) = dz$ and with the help of the integral formulae $\int_{-\infty}^{+\infty} e^{-x^2}\, dx = \sqrt{\pi}$ and $\int_{-\infty}^{+\infty} x\, e^{-x^2}\, dx = 0$ one obtains

$$\mu = a\sqrt{2}/\sqrt{\pi} \int_{-\infty}^{+\infty} [z + b/(a\sqrt{2})]\, e^{-z^2}\, dz$$

$$= b/\sqrt{\pi} \int_{-\infty}^{+\infty} e^{-z^2}\, dz = b.$$

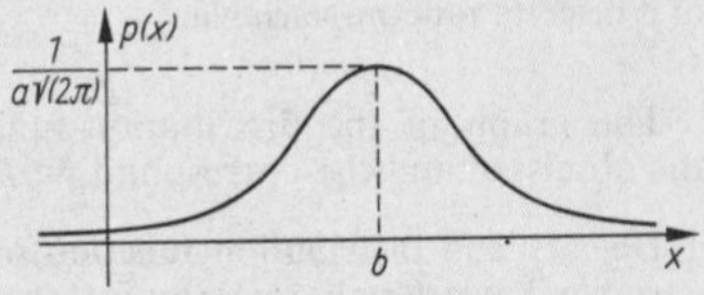

27.2-10 Mean value of a normal distribution

Mean values of sums and products of random variables. The sum $Z = X + Y$ of the random variables X and Y with the mean values μ_x and μ_y is also a random variable; for example, the value of the score obtained by throwing two dice.

The mean value μ_z of the random variable Z is determined from the mean values μ_x and μ_y of the single random variables X and Y.

The mean value of the sum of two random variables is equal to the sum of the mean values of the two random variables.

$$\mu_z = \mu_x + \mu_y$$

This rule can also be carried over to the mean value of the sum of three or more random variables. For example, the random variable $Z = U + X + Y$ formed from the sum of the random variables U, X, Y with mean values μ_u, μ_x, μ_y has the mean value $\mu_z = \mu_u + \mu_x + \mu_y$.

Example 3: If two ideal dice are thrown, then the random variables X and Y are the values of the score obtained with the first die and second die, respectively. The mean value of each is known: $\mu_x = 3.5$; $\mu_y = 3.5$. The mean value of the values of the total score thrown with both dice is then $\mu_z = 3.5 + 3.5 = 7$.

Example 4: The output of a factory in a small unit of time, say in one day, can be regarded as a random variable, because it is subject to small variations due to disturbances that cannot always be predicted and also cannot be eliminated by technical means. In two factories A and B the number of articles (random variables X in A and Y in B) has the mean value $\mu_x = 260$ in A and $\mu_y = 90$ in B. Then the production in both factories (random variable Z) has the mean value

$$\mu_z = 260 + 90 = 350.$$

Likewise for the product of two random variables X and Y a simple rule holds if X and Y are independent, that is, if the equation

$$P(X < x;\ Y < y) = P(X < x) \cdot P(Y < y)$$

holds for arbitrary x and y. Here $P(X < x;\ Y < y)$ is the probability that X is less than x and Y less than y. If μ_x and μ_y are the mean values of the two independent random variables X and Y, then the mean value μ_z of the random variable $Z = X \cdot Y$ is given by $\mu_z = \mu_x \cdot \mu_y$.

The mean value of the product of two independent random variables is equal to the product of the mean values of the two variables.

$$\mu_z = \mu_x \cdot \mu_y$$

Just as for the sum of more than two random variables, so this rule can be carried over to the product of more than two independent random variables.

Example 5: In the making of rectangular plates both the length (in inches) and the width (in inches) are random variables (X and Y, respectively). It follows that the area is also a random variable ($Z = X \cdot Y$). Let the mean values of X and Y be $\mu_x = 120$ in. and $\mu_y = 80$ in. Then the mean value μ_z of the area is $\mu_z = 120 \cdot 80 = 9600\ \text{in}^2$.

Variance. In many cases the mean value does not suffice to characterize a random variable X. In the production of bolts, for example, the diameter is an important measurement and is a random variable X in the sense of probability theory. With the best adjustment of the machine its mean value μ is equal to the desired value. During production, however, it is found that many diameters are greater and many are smaller than the desired value. For the same mean value the deviation can be large with one machine and small with another, but they must lie within the tolerance limits.

The variance σ^2 of the random variable X is used to describe them, and its square root σ is called the *standard deviation* or *root mean square deviation.* It is a measure of the magnitudes of the deviations from the mean value.

The variance σ^2 of a discrete random variable X is obtained by multiplying the square of each deviation $(x_i - \mu)$ from the mean value by the corresponding probability and adding all these products.

The same probability function applies to the magnitudes $(x_i - \mu)^2$ as to X, namely $\dfrac{x_1 x_2 \cdots x_n}{p_1 p_2 \cdots p_n}$ with $\sum_{i=1}^{n} p_i = 1$.

Variance of a discrete random variable	$\sigma^2 = \sum_i (x_i - \mu)^2 p_i$

Example 6: When an ideal die is thrown, the score obtained is a random variable X with mean value $\mu = 3.5$ and probability function

x_i	1	2	3	4	5	6
p_i	1/6	1/6	1/6	1/6	1/6	1/6

Hence the variance is given by:

$$\sigma^2 = (1 - 3.5)^2 \cdot 1/6 + (2 - 3.5)^2 \cdot 1/6 + (3 - 3.5)^2 \cdot 1/6 + (4 - 3.5)^2 \cdot 1/6 + (5 - 3.5)^2 \cdot 1/6 + (6 - 3.5)^2 \cdot 1/6 = 2.92 \quad \text{and} \quad \sigma = \sqrt{2.92} = 1.71.$$

The variance σ^2 of a continuous random variable X is obtained by multiplying the square of the deviation $(x - \mu)$ from the mean value by the density function $f(x)$ and integrating the product from $-\infty$ to $+\infty$.

Variance of a continuous random variable	$\sigma^2 = \int_{-\infty}^{+\infty} (x - \mu)^2 f(x)\,dx$

Example 7: It is required to find the variance of the continuous random variable X with mean value b and probability density

$$p(x) = 1/[a\sqrt{(2\pi)}]\, e^{-(x-b)^2/(2a^2)} \qquad \textit{(Gaussian or normal distribution).}$$

The definition of the variance gives: $\sigma^2 = \int_{-\infty}^{+\infty} (x - b)^2/[a\sqrt{(2\pi)}]\, e^{-(x-b)^2/(2a^2)}\,dx$.

Using the substitution given above one obtains the result $\sigma^2 = a^2$.

Variance of the sum of two independent random variables. If X and Y are two independent random variables with variance σ_x^2 and σ_y^2, respectively, then $Z = X + Y$ is also a random variable (for example, the score obtained by throwing two dice) and the variance σ_z^2 of the random variable Z is determined by the variances of the two variables X and Y.

The variance of the sum of two independent random variables is equal to the sum of their variances.

$$\sigma_z^2 = \sigma_x^2 + \sigma_y^2$$

Example 8: The score obtained by throwing two ideal dice is a random variable Z. It is the sum of the random variables X and Y, these being the separate scores obtained with each of the two dice at each throw. The variance of each of these is $\sigma_x^2 = \sigma_y^2 = 2.92$. Thus, the variance σ_z^2 of the random variable Z is given by $\sigma_z^2 = 2.92 + 2.92 = 5.84$.

Chebyshev's inequality

In the preceding section it was shown that a general idea of the behaviour of a random variable can be obtained from the mean value and the variance. However, when these are given, it is still not possible to answer the question, what is the probability for a particular deviation from the mean value μ. *Chebyshev's inequality* gives a simple estimate for this.

One begins with a discrete or continuous random variable X with values x, mean value μ, and variance σ^2. Chebyshev's inequality, which will not be derived here, is then as follows:

Chebyshev's inequality
$P\{\lvert x-\mu\rvert \geqslant \varepsilon\} \leqslant \sigma^2/\varepsilon^2$

The probability that the absolute value of the difference $(x-\mu)$ is greater than or equal to an arbitrary number $\varepsilon > 0$ is less than or equal to the variance divided by ε^2.

With the help of this inequality the probabilities for the different deviations from the mean value can be estimated. If, for example, in measuring a length, a mean length of 300 yd. and a variance of 36 is established, then the probability that a deviation of more than 30 yd. occurs is estimated as follows: $P(\lvert x-300\rvert \geqslant 30) \leqslant 36/900 = 0.04$; that is, the probability is at most 0.04.

The law of large numbers

In everyday life and in theoretical investigations events whose probabilities lie close to one or to zero play an important role. One is interested, for example, that the probability for the safe transport of passengers shall be hardly distinguishable from one, or that the probability for the collapse of a bridge shall be practically zero. Thus, it is an essential task of probability theory to find instances whose probabilities lie near to one. Among these laws, the *law of large numbers* is of particular significance and will be explained in two forms.

The law according to CHEBYSHEV. Given are n pairwise independent random variables $X_1, X_2, \ldots, X_n$ with mean values $\mu_1, \mu_2, \ldots, \mu_n$ and variances that are all less than b^2. Let $A = (\mu_1 + \mu_2 + \cdots + \mu_n)/n$ denote the arithmetic mean of the mean values. From Chebyshev's inequality it follows that $P(\lvert(1/n)\sum_{i=1}^{n} X_i - A\rvert < \varepsilon) \geqslant 1 - b^2/(n\varepsilon^2)$, where ε is an arbitrary positive number.

The law of large numbers according to Chebyshev. For sufficiently large values of n the arithmetic mean A of the mean values of n pairwise independent random variables differs from the arithmetic mean of these variables by less than ε with a probability that is arbitrarily close to 1.

Chebyshev's law of large numbers justifies the rule that the mean value of n measurements is more reliable than any single measurement.

The law according to Jakob BERNOULLI: The probability for the occurrence of an event E is p, and in n independent trials the event E occurs n_1 times. Then for an arbitrary positive number ε, $P(\lvert n_1/n - p\rvert < \varepsilon) \geqslant 1 - 1/(4\varepsilon^2 n)$.

The law of large numbers according to Bernoulli. For sufficiently large values of n, the relative frequency of the occurrence of the event E in n observations differs from the probability p for the occurrence of the event by less than ε, with a probability that is arbitrarily close to 1.

Bernoulli's law of large numbers is a special case of Chebyshev's law. It establishes that the relative frequency can be used for the estimation of unknown probabilities.

	n	n_1 heads	n_1/n	$1-1/(4\varepsilon^2 n)$
Buffon	4 040	2 048	0.507	0.993 8
K. Pearson	12 000	6 019	0.501 6	0.997 9
K. Pearson	24 000	12 012	0.500 5	0.999 0

Example: In tossing a coin one distinguishes between two events *heads* and *tails*. The result of several long series of tosses are to be given, using the law of large numbers with $\varepsilon = 0.1$. One sees that as the number of trials increases, the probability that the relative frequency n_1/n deviates from the probability $p = 0.5$ by less than 0.1, tends to 1.

Some important distributions

A random variable X is completely characterized by its probability function or density function, or by its distribution function; for short one says: by its distribution. Some types of distribution have acquired great significance in practice.

The binomial distribution. The binomial distribution (sometimes called the *Bernoulli distribution*) can be used for problems based on the following scheme of trials: There are black and white balls in an urn, and the probability for the event E 'drawing a black ball' is p. The probability for the event $\bar{E}$, 'drawing a white ball', is then $(1-p)$. The trial, that in a series of n draws with replacement the event E occurs k times and fails to occur $(n-k)$ times, determines the random variable X. Its distribution is the *binomial distribution*. Its *probability function* $P_n(k)$ can be determined as follows: Since each ball is replaced after being drawn, every draw is an independent event, and by the multiplication law the probability of drawing k black and $(n-k)$ white balls is $p^k(1-p)^{n-k}$.

There are $\binom{n}{k} = \frac{n!}{k!\,(n-k)!}$ different permutations of the sequence in which k black and $(n-k)$ white balls can be drawn, each of which leads to the same result of the trial. Thus, $P_n(k) = \binom{n}{k} p^k (1-p)^{n-k}$. The probabilities that belong to the individual values of k give the probability function. By summing over all $k < x$ one obtains the distribution function $F_n(x) = \sum_k P_n(k)$.

Example 1: A ball is drawn out of an urn and then replaced into the urn. The probability of drawing a black ball is $p = 1/4$. Each 10 such draws form a group. If the trials are continued, the number of black balls in the individual groups will vary: it is a random variable. It is required to find the probability function and the distribution function of this random variable. The 'number of black balls among 10 balls' is required. With the help of the formula $P_{10}(k) = \binom{10}{k}\left(\frac{1}{4}\right)^k\left(\frac{3}{4}\right)^{10-k}$, where $k = 0, 1, \ldots, 10$, one obtains the probability function (Fig. 27.2-11):

k	0	1	2	3	4	5	6	7	8	9	10
$P_{10}(k)$	0.056	0.188	0.282	0.250	0.146	0.058	0.016	0.003	0.001	0.0	0.0

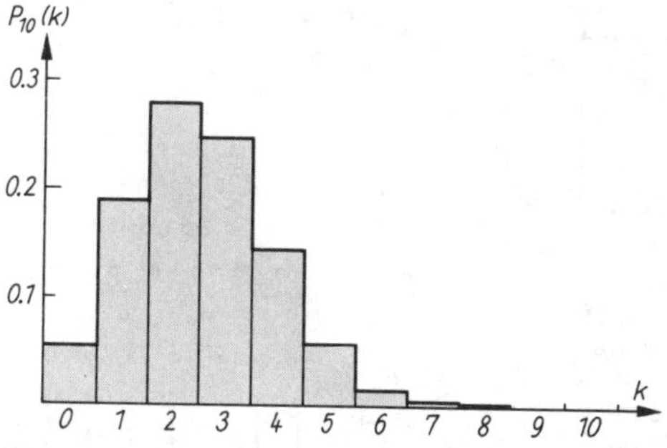

27.2-11 Probability function $P_{10}(k)$ of a binomial distribution

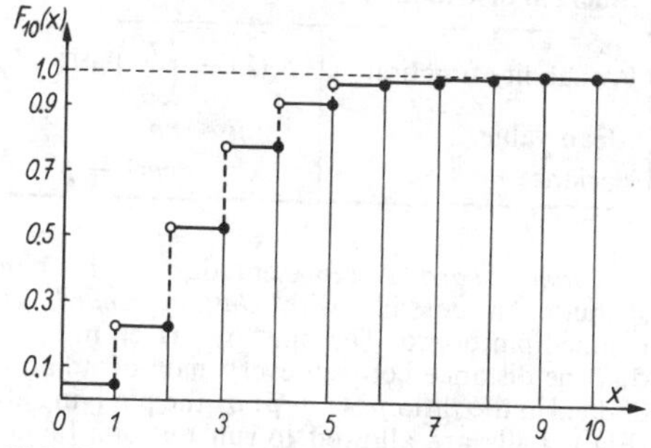

27.2-12 Representation of the distribution function $F_{10}(x)$

and from $F_{10}(x) = \sum_k P_{10}(k)$, where the summation extends over all $k < x$, the distribution function (Fig. 27.2-12):

k	0	1	2	3	4	5	6	7	8	9	10	>10
$F_{10}(k)$	0.0	0.056	0.244	0.526	0.776	0.922	0.980	0.996	0.999	1.0	1.0	1.0

One obtains an impression of the probability function and the distribution function from graphical representations. In these k and x are taken as abscissae, $P_{10}(k)$ and $F_{10}(x)$, respectively, as ordinates.

Example 2: From the probability $p = 0.515$ of a male birth, one can calculate with the help of a binomial distribution the probability that a family of, say, 6 children contains 0, 1, 2, 3, 4, 5 or 6 boys:

Boys	6	5	4	3	2	1	0
Probability	0.019	0.105	0.248	0.312	0.220	0.083	0.013

In order to simplify the tedious calculation of the probabilities $P_n(k)$, one derives the recursion formula:

Recursion formula	$P_n(k+1) = \frac{n-k}{k+1} \cdot \frac{p}{1-p} \cdot P_n(k)$

Mean value and variance of the binomial distribution. Substituting the corresponding magnitudes in the formula for the mean value of a discrete random variable one obtains

$$\mu = \sum_{m=0}^{n} mP_n(m) = \sum_{m=0}^{n} m\binom{n}{m} p^m(1-p)^{n-m} = np\sum_{m=0}^{n-1}\binom{n-1}{m} p^m(1-p)^{n-1-m}.$$

When the binomial theorem is observed, it follows that $\mu - np[p + (1-p)]^{n-1} = np$.

With the mean value np, the variance is then given by

$$\sigma^2 = \sum_{m=0}^{n}(m - np)^2 P_n(m)$$

$$= \sum_{m=0}^{n} m^2 \binom{n}{m} p^m(1-p)^{n-m} - 2np \sum_{m=0}^{n} m \binom{n}{m} p^m(1-p)^{n-m} + n^2p^2 \sum_{m=0}^{n} \binom{n}{m} p^m(1-p)^{n-m}$$

$$= \sum_{m=0}^{n} m^2 \binom{n}{m} p^m(1-p)^{n-m} - n^2p^2.$$

If the first sum is calculated in the same way as for the mean value, it follows finally that

$$\sigma^2 = pn[(1-p) + pn] - n^2p^2 = np(1-p).$$

For the Example 2 above the mean value and variance are given by

$$\mu = 10 \cdot 1/4 = 2.5; \quad \sigma^2 = 10 \cdot (1/4) \cdot (3/4) = 1.875.$$

The binomial distribution is easy to use for small values of n and k. For large values, however, the calculations become very troublesome and either the Poisson distribution or the normal distribution is used, according to the nature of the problem.

Binomial distribution	
Probability function	$P_n(k) = \binom{n}{k} p^k(1-p)^{n-k}$
Mean value	$\mu = np$
Variance	$\sigma^2 = np(1-p)$

Galton's board. A representation of the binomial distribution is possible with *Galton's board.* This is an inclined pin board. The pins are driven in in such a way that the distance between every pair of adjacent pins is divided in the ratio $p:(1-p)$ by the pin lying above them (Fig.). Balls are allowed to run out of a funnel through the rows of pins. Each ball hits a nail from row to row and can be deflected to the right or to the left. After the balls have run through n rows, they are caught in $(n+1)$ compartments. The contents of the compartments show the distribution of the balls in the form of a histogram. If N balls are allowed to roll through a Galton board with n rows of pins, then $N \cdot P_n(m)$ balls are to be expected in the mth compartment. Various binomial distributions can be demonstrated by various arrangements of the rows of pins.

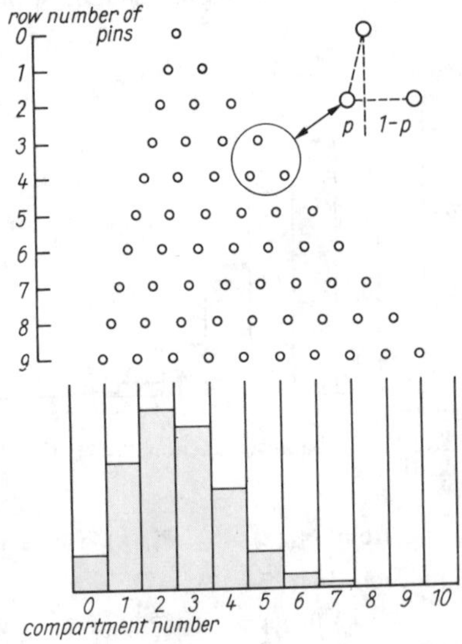

27.2-13 Schematic representation of a Galton board

Poisson distribution. This distribution is based on essentially the same problem as the binomial distribution. It differs only in that the number n of balls drawn from the urn is very large and the probability p for the drawing of a black ball is very small. In other words: the Poisson distribution is the limiting case of the binomial distribution as $n \to \infty$ and $p \to 0$, with the additional assumption that the product $np = a$ is constant. The distribution is applied, therefore, if an event occurs very rarely. With these assumptions, taking the limit $\lim_{n\to\infty} P_n(k)$, the probability of drawing k black balls in n draws is given by:

$$\psi_n(k) = a^k e^{-a}/k!,$$

where $np = a$. The Poisson distribution is determined by the quantity a alone. The probabilities for the individual values of k give the probability function, and summing the individual probabilities over all $k < x$ one obtains the distribution function

$$F_n(x) = \sum_k \psi_n(k).$$

Example: One ball is drawn from an urn repeatedly and replaced into the urn. The probability of drawing a black ball is $p = 0.01$. Each 60 such draws form a group. If the trials are continued, the number of black balls in the individual groups will vary; it is a random variable. The probability function and the distribution function of this random variable 'number of black balls among 60 balls' are required. From $\psi_{60}(k) = (0.6)^k e^{-0.6}/k!$, where $a = 60 \cdot 0.01 = 0.6$ and k can take the values 0, 1, 2, 3, ..., 60, one obtains the probability function

k	0	1	2	3	4	5	...	60
$\psi_{60}(k)$	0.549	0.329	0.099	0.020	0.003	0.000	...	0.000

and from $F_{60}(x) = \sum_k \psi_{60}(k)$, where the sum extends over all $k < x$, the distribution function:

x	0	1	2	3	4	5	6	...	60	60
$F_{60}(x)$	0.0	0.549	0.878	0.977	0.997	1.000	1.000	...	1.000	1.000

One obtains graphical representations in the same way as for the binomial distribution.

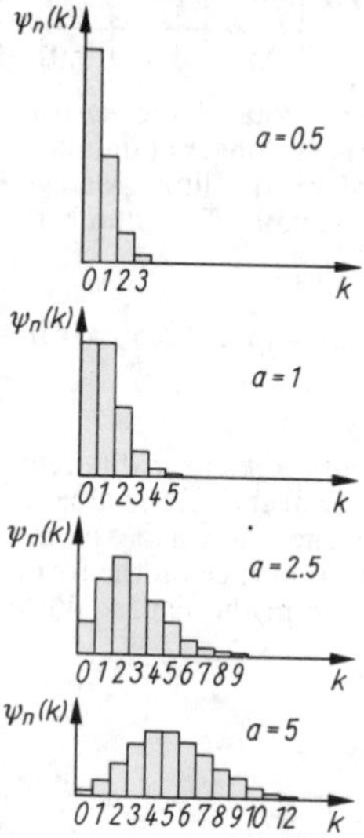

27.2-14 Poisson distribution for different values of a

In Fig. 27.2-14 Poisson distributions with different values of a are drawn. It shows that the peak of the distribution for increasing values of a moves further to the right and reduces the asymmetry of the curve.

For the calculation of individual probabilities one uses the appropriate recursion formula obtained from the ratio

$\psi_n(k+1)/\psi_n(k) = [k!a^{k+1}e^{-a}]/[(k+1)!\,a^k e^{-a}] = a/(k+1)$.

Recursion formula	$\psi_n(k+1) = \dfrac{a}{k+1}\psi_n(k)$

Mean value and variance of the Poisson distribution. These are calculated from the corresponding quantities for the binomial distribution, in which np is put equal to a and p tends to 0. One obtains $\mu = np = a$; $\sigma^2 = np = a$. Thus, the mean value and the variance are equal.

Poisson distribution	
Probability function	$\psi_n(k) = \dfrac{a^k e^{-a}}{k!}$
Mean value	$\mu = a$
Variance	$\sigma^2 = a$

The field of application of the Poisson distribution was at one time limited to very rare events, for example, to child suicides or to deaths from horse kicks in an army. In recent decades, however, it has acquired considerable importance. Today it plays an important role, for example, in telecommunications, in statistical quality control, for the description of the decay of radioactive substances, in the textile industry, in biology, and in meteorology. In addition, the Poisson distribution is used in many cases as an approximation for the binomial distribution, since in practice the agreement is adequate for sufficiently large values of n and small values of p.

The normal or Gaussian distribution. The normal distribution is one of the most important distributions of probability theory and was discovered by GAUSS in connection with the application of the method of least squares to surveying.

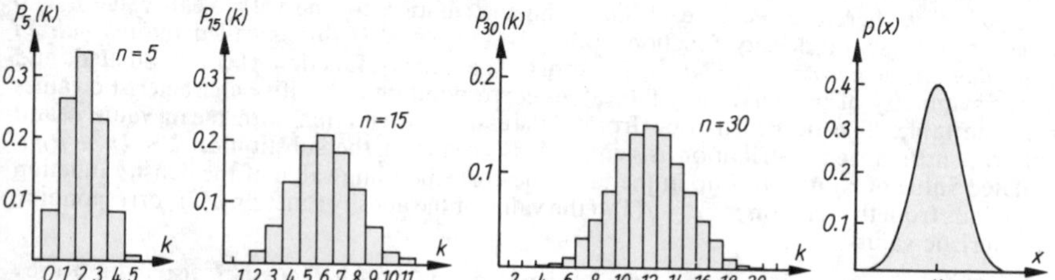

27.2-15 Graphical representation of the probability function for an increasing number of draws n

For the binomial distribution the probability function $P_n(k) = \binom{n}{k} p^k(1-p)^{n-k}$ was given, where the probability p for the event 'drawing a black ball' has a fixed value between 0 and 1. For an increasing number n of draws (see the binomial distribution) the probability function loses its asymmetry (Fig.). For $p = 0.2$ and $n = 5$, 15 and 30 one obtains

k	0	1	2	3	4	5
$P_5(k)$	0.08	0.26	0.34	0.23	0.08	0.01

k	2	3	4	5	6	7	8	9	10	11
$P_{15}(k)$	0.02	0.06	0.13	0.19	0.21	0.18	0.12	0.06	0.02	0.01

k	6	7	8	9	10	11	12	13	14	15	16	17	18
$P_{30}(k)$	0.01	0.03	0.05	0.08	0.12	0.14	0.15	0.14	0.11	0.08	0.05	0.03	0.01

As $n \to \infty$, when the number of draws n increases beyond every limit, one obtains the *normal distribution*. The number of characteristic values x of the random variable X is no longer countable; the random variable X is continuous. Its *density function* $p(x)$ is obtained by the limit passage $\lim_{n\to\infty} P_n(k) = p(x) = 1/[a\sqrt{(2\pi)}]\, e^{-(x-b)^2/(2a^2)}$, as can be shown in greater detail. The numbers a and b are constants. For the mean value μ and the variance σ^2 one obtains $\mu = \int_{-\infty}^{+\infty} x/[a\sqrt{(2\pi)}]\, e^{-(x-b)^2/(2a^2)}\,dx = b$, $\sigma^2 = \int_{-\infty}^{+\infty}(x-b)^2/[a\sqrt{(2\pi)}] \cdot e^{-(x-b)^2/(2a^2)}\,dx = a^2$. (see Mean value or expectation and variance, Examples 2 and 7).

The mean value and the variance describe this distribution completely.

Fig. 27.2-16 shows the graphical representation of the frequency distribution $p(x)$ for different values of $a^2 = \sigma^2$. The curves are bell-shaped. The peak of each distribution lies at the mean value μ. The curve falls from this value symmetrically on both sides of it and approaches the x-axis asymptotically. The curve has inflections at distances $\pm\sigma$ from the mean value. The influence on the form of the bell-shaped curve of the magnitude of the variance is easy to recognize in the figure. With increasing σ the curves become flatter and broader.

Gaussian or normal distribution
Density function $p(x) = \dfrac{1}{a\sqrt{(2\pi)}}\, e^{-(x-b)^2/(2a^2)}$
Mean value $\mu = b$
Variance $\sigma^2 = a^2$

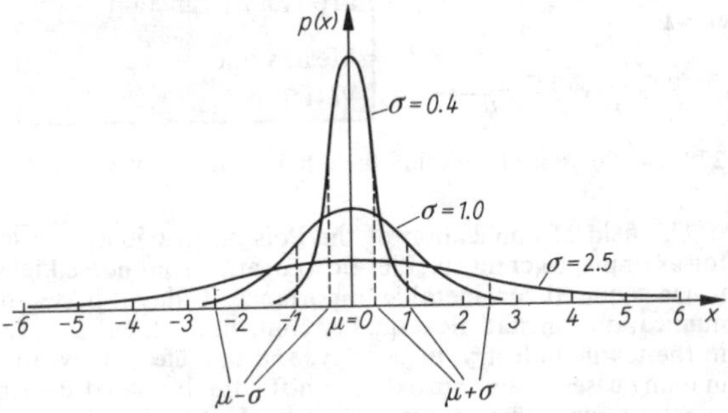

27.2-16 Frequency distribution of the Gaussian distribution for different σ

Normalized Gaussian distribution. It is tedious to calculate individual values of the density function $p(x)$ of a random variable X that can be described by a Gaussian distribution with mean value μ and variance σ^2. One therefore relates each Gaussian distribution to one with mean value $\mu = 0$ and variance $\sigma^2 = 1$, whose density function $\varphi(\lambda) = 1/\sqrt{(2\pi)} \cdot e^{-\lambda^2/2}$; this is called the *normalized Gaussian distribution*, or *normal distribution*, for short. The density function $\varphi(\lambda)$ is tabulated, and because of the symmetry of the curve only the values corresponding to positive characteristic values are given in the table. The transformation from a Gaussian distribution with mean value μ and variance σ^2 to a normalized distribution is achieved by means of the substitution $\lambda = (x - \mu)/\sigma$. For a calculated value of λ one looks up in the table the corresponding value of the density function $\varphi(\lambda)$ and then finds from the relation $p(x) = \varphi(\lambda)/\sigma$ the value of the density function $p(x)$ corresponding to the characteristic value x.

Example: The mean value $\mu = 20$ and the vaiiance $\sigma^2 = 25$ of a Gaussian distribution are known. One first calculates $\lambda = (x - 20)/5$, then looks up $\varphi(\lambda)$ in the table and finally calculates $p(x) = \varphi(\lambda)/5$.

x	20	21	22	...	25	...	30
λ	0	0.2	0.4	...	1	...	2
$\varphi(\lambda)$	0.3989	0.3910	0.3683	...	0.2420	...	0.054
$p(x)$	0.0798	0.0782	0.0737	...	0.0584	...	0.0108

Ordinates of the normal distribution $\varphi(\lambda) = 1/\sqrt{(2\pi)} \cdot e^{-\lambda^2/2}$

λ	0	1	2	3	4	5	6	7	8	9
0.0	0.3989	3989	3989	3988	3986	3984	3982	3980	3977	3973
0.1	0.3970	3965	3961	3956	3951	3945	3939	3932	3925	3918
0.2	0.3910	3902	3894	3885	3876	3867	3857	3847	3836	3825
0.3	0.3814	3802	3790	3778	3765	3752	3739	3726	3712	3697
0.4	0.3683	3668	3653	3637	3621	3605	3589	3572	3555	3538
0.5	0.3521	3503	3485	3467	3448	3429	3411	3391	3372	3352
0.6	0.3332	3312	3292	3271	3251	3230	3209	3187	3166	3144
0·7	0.3123	3101	3079	3056	3034	3011	2989	2966	2943	2920
0.8	0.2897	2874	2850	2827	2803	2780	2756	2732	2709	2685
0.9	0.2661	2637	2613	2589	2565	2541	2516	2492	2468	2444
1.0	0.2420	2396	2371	2347	2323	2299	2275	2251	2227	2203
1.1	0.2179	2155	2131	2107	2083	2059	2036	2012	1989	1965
1.2	0.1942	1919	1895	1872	1849	1827	1804	1781	1759	1736
1.3	0.1714	1692	1669	1647	1626	1604	1582	1561	1540	1518
1.4	0.1497	1476	1456	1435	1415	1394	1374	1354	1334	1315
1.5	0.1295	1276	1257	1238	1219	1200	1182	1163	1145	1127
1.6	0.1109	1092	1074	1057	1040	1023	1006	0989	0973	0957
1.7	0.0941	925	909	893	878	863	848	833	818	804
1.8	0.0790	775	761	748	734	721	707	694	681	669
1.9	0.0656	644	632	620	608	596	584	573	562	551
2.0	0.0540	529	519	508	498	488	478	468	459	449
2.1	0.0440	431	422	413	404	396	387	379	371	363
2.2	0.0355	347	339	332	325	317	310	303	297	290
2.3	0.0283	277	271	264	258	252	246	241	235	229
2.4	0.0224	219	213	208	203	198	194	189	184	180
2.5	0.0175	171	167	163	159	155	151	147	143	139
2.6	0.0136	132	129	126	122	119	116	113	110	107
2.7	0.0104	101	99	96	94	91	89	86	84	81
2.8	0.0079	77	75	73	71	69	67	65	63	61
2.9	0.0060	58	56	55	53	51	50	49	47	46
3.0	0.0044	43	42	41	39	38	37	36	35	34
3.1	0.0033	32	31	30	29	28	27	26	25	25
3.2	0.0024	23	22	22	21	20	20	19	18	18
3.3	0.0017	17	16	16	15	15	14	14	13	13
3.4	0.0012	12	12	11	11	10	10	10	9	9
3.5	0.0009									
4.0	0.0001									

Distribution function of the Gaussian distribution. The distribution function of the Gaussian distribution

$$F(x) = \int_{-\infty}^{x} p(t)\, dt = 1/[a\sqrt{(2\pi)}] \int_{-\infty}^{x} e^{-(t-b)^2/(2a^2)}\, dt$$

is called the *Gaussian* or *error integral.*

Values of
$\Phi(\lambda) = \int_0^\lambda \varphi(t)\,dt$

λ	$\Phi(\lambda)$
0.0	0.0000
0.1	0.0398
0.2	0.0793
0.3	0.1179
0.4	0.1554
0.5	0.1915
0.6	0.2257
0.7	0.2580
0.8	0.2881
0.9	0.3159
1.0	0.3413
1.1	0.3643
1.2	0.3849
1.3	0.4032
1.4	0.4192
1.5	0.4332
1.6	0.4452
1.7	0.4554
1.8	0.4641
1.9	0.4713
2.0	0.4772
2.1	0.4821
2.2	0.4861
2.3	0.4893
2.4	0.4918
2.5	0.4938
2.6	0.4953
2.7	0.4965
2.8	0.4974
2.9	0.4981
3.0	0.4987
3.1	0.4990
3.2	0.4993
3.3	0.4995
3.4	0.4997

It represents the area under the curve $p(x)$ between $-\infty$ and x (Fig.). The function $F(x)$ has the x-axis and the line $F(x) = 1$ as asymptotes and has a point of inflection at $x = \mu$. It gives the probability that a characteristic value is less than x. Bearing in mind the symmetry of the bell-shaped curve, the distribution function $F(x)$ for $\mu = 0$ and $\sigma^2 = 1$ is tabulated in the following form:

$$\Phi(\lambda) = \int_0^\lambda \varphi(t)\,dt = 1/\sqrt{(2\pi)} \int_0^\lambda e^{-t^2/2}\,dt.$$

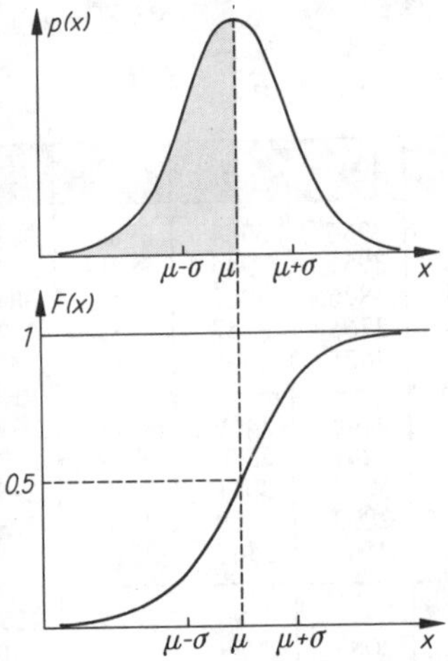

27.2-17 Representation of the density function and the distribution function of the Gaussian distribution

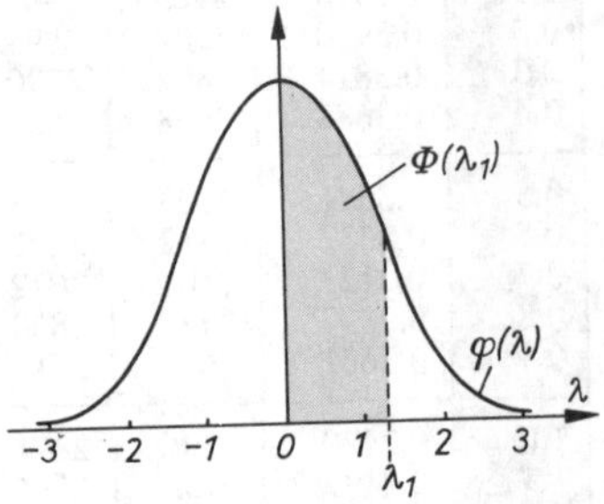

27.2-18 Geometric meaning of the function $\Phi(\lambda)$

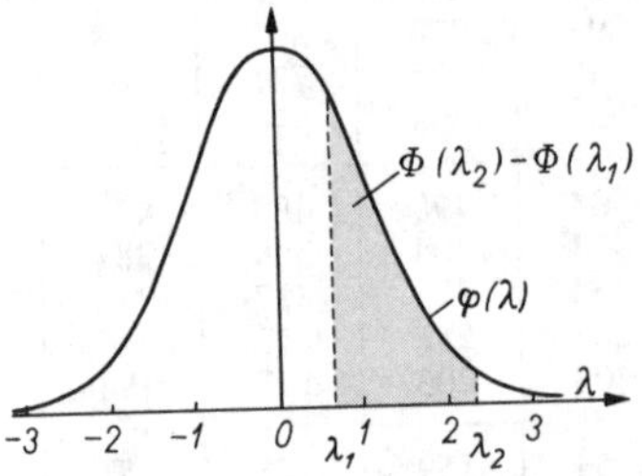

27.2-19 Area given by $\Phi(\lambda_2) - \Phi(\lambda_1)$

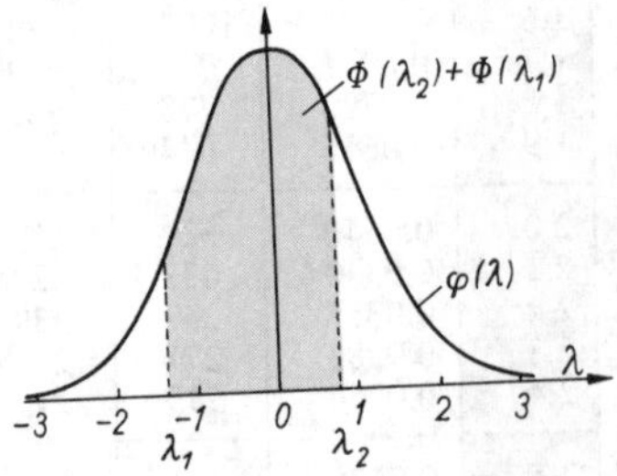

27.2-20 Area given by $\Phi(\lambda_2) + \Phi(\lambda_1)$

Fig. 27.2-18 shows the area represented by this function. Each Gaussian distribution with mean value μ and variance σ^2 is connected with $\Phi(\lambda)$ by $\lambda = (x - \mu)/\sigma$. With the help of this relation the areas corresponding to the characteristic values x_1 and x_2 can be calculated. There are different cases to be considered:

(*i*) If λ_1 and λ_2 both lie to the right of zero and $\lambda_2 > \lambda_1$, then the area under the curve between these values is $\Phi(\lambda_2) - \Phi(\lambda_1)$ (Fig.). A similar result holds if both values lie to the left of zero.

(*ii*) If λ_1 and λ_2 lie on opposite sides of zero, then the area is given by $\Phi(\lambda_1) + \Phi(\lambda_2)$ (Fig.).

In both cases the area calculated represents the probability with which an observed value is expected to lie in the interval bounded by x_1 and x_2.

Example: The mean value $\mu = 20$ and the variance $\sigma^2 = 25$ of a Gaussian distribution are known. It is required to find the probability that an observed value lies between
(*i*) $x_1 = 25$ and $x_2 = 35$; (*ii*) $x_1 = 5$ and $x_2 = 35$.
One calculates
(*i*) $\lambda_1 = (25 - 20)/5 = 1$, $\lambda_2 = (35 - 20)/5 = 3$, (*ii*) $\lambda_1 = (5 - 20)/5 = -3$, $\lambda_2 = (35 - 20)/5 = 3$
and obtains from the table
(*i*) $\Phi(\lambda_1) = 0.3413$, $\Phi(\lambda_2) = 0.4987$, (*ii*) $\Phi(\lambda_1) = 0.4987$, $\Phi(\lambda_2) = 0.4987$.
Subtraction in (*i*) and addition in (*ii*) gives the required probabilities (*i*) 0.1574; (*ii*) 0.9974.

With the help of the error integral, the proportion of the whole area under the bell-shaped curve between the limits x_1 and x_2 can be determined. By subtraction from 1 the corresponding proportion of the remaining area is obtained. Important partial areas that have great significance in mathematical statistics are collected together in the accompanying table.

Whole area and remainder from the error integral

Limits on the x-axis		Proportion of the total area	Remaining area
$\mu - \lambda\sigma$	$\mu + \lambda\sigma$	(%)	(%)
$\mu - 1\sigma$	$\mu + 1\sigma$	68.26	31.74
$\mu - 1.96\sigma$	$\mu + 1.96\sigma$	95	5
$\mu - 2\sigma$	$\mu + 2\sigma$	95.44	4.56
$\mu - 2.58\sigma$	$\mu + 2.58\sigma$	99	1
$\mu - 3\sigma$	$\mu + 3\sigma$	99.73	0.27
$\mu - 3.29\sigma$	$\mu + 3.29\sigma$	99.9	0.1

Limit theorems for sums of independent random variables

Many processes in the natural sciences, technology and economics are described on the assumption that they are influenced by a large number of random factors independent of one another, of which each alters the course of the processes only a little. In general, only the sum of their effects is observed in investigating the processes; for example, the error in a measurement forms such a random variable which is the sum of many independent random variables. Probability theory has established limiting value theorems concerning the rules governing the behaviour of these sums.

The de Moivre-Laplace integral limit theorem. If for each of n trials, p is the probability that the event E occurs and $q = 1 - p$ the probability that E does not occur, then a random variable X_k can be determined so that $X_k = 1$ if E occurs in the kth trial and $X_k = 0$ if E does not occur. The random variable $\sum_{k=1}^{n} X_k$ then determines how often E occurred in n consecutive trials. Because of the distribution of the terms in the sum, the *probability function* of $\sum_{k=1}^{n} X_k$ is a *binomial distribution* with expected value $\mu = np$ and variance $\sigma^2 = np(1 - p) = npq$. By the *de Moivre-Laplace theorem* the *distribution function* of this sum $\sum_k X_k$ does not tend to a *limiting distribution function* as $n \to \infty$, but that for the random variable $\sum_{k=1}^{n} \frac{X_k - np}{\sqrt{(npq)}}$ does, and its limiting distribution function is the *normalized Gaussian distribution*. This means that for arbitrary numbers $a < b$ the following relation holds as $n \to \infty$.

De Moivre-Laplace theorem	$P\left\{a \leqslant \sum_{k=1}^{n} \frac{m_b - \mu}{\sqrt{(npq)}} < b\right\} \to 1/\sqrt{(2\pi)} \int_a^b e^{-x^2/2}\, dx$

The de Moivre-Laplace theorem raises the question whether the relation obtained is dependent upon the choice of method of summation, and whether this relation is still valid if fewer conditions are placed on the distribution function of the terms of the sum. The *central limit theorem* gives a partial answer in its simplest form, which can be generalized considerably.

The central limit theorem. If the pairwise independent random variables $X_1, X_2, \ldots, X_n$ have the same distribution, and if $\mu = E(X_n)$ and $\sigma^2 = D^2(X_n) > 0$ exist, then the random variable

$$\frac{\frac{1}{n}\sum_{k=1}^{n} X_k - E\left(\frac{1}{n}\sum_{k=1}^{n} X_k\right)}{D\left(\frac{1}{n^2}\sum_{k=1}^{n} X_k\right)}$$

has the normalized Gaussian distribution as limiting distribution function.

For a long time the main task of the classical side of this theory was to find the most general conditions under which the distribution function of sums of independent random variables would tend towards the normal distribution with increasing number of terms. Parallel with the conclusion of this classical side a further direction was developed in the theory of limit theorems for sums of independent random variables, that is closely bound up with the stochastic processes sketched later. The question in this direction is as follows: What distributions, other than the normal, can be limiting distributions of sums of independent random variables? – In these investigations it turned out that not only the normal distribution occurs as a limiting distribution. One considered the problem of finding conditions on the summands such that the distribution of the sum shall approach one or other limiting distribution for a sufficiently large number of terms. This modern aspect of the limit theorems for sums of independent random variables has been developed strongly in the last thirty years and is closely connected with the names of KOLMOGOROV, KHINCHINE, GNEDENKO and others. The limit theorems have practical significance, for example, in the development of mathematical statistics and in the theory of errors of observation.

Stochastic processes

Random or *stochastic processes* are described by random variables that depend on at least one parameter. Such a parameter can either assume only a discrete set of values or can vary continuously. The degree of wear of a car tyre, for example, depends on the number t of miles it has been driven, but according to the initial conditions it is a random function of t. Also in the development of the number of inhabitants of a town over a lengthy period of time, besides the time t as parameter, systematic and random influences must be taken into account.

For the case of a parameter t the stochastic process is denoted by $X(t, \omega)$. In this the parameter ω expresses the random nature, $\omega \in \Omega$, where Ω is the set of all possible events that can occur; the parameter t expresses a systematic dependence of the random variable, usually on the time, but it could also be dependence on a sequence in the case of a numbering, or on a distance. For a fixed t, for example, for a 'snapshot' taken at time $t = t_0$, $X(t, \omega)$ is a random variable. For a fixed ω, $X(t, \omega)$ is a function of t that is called a *realization of the process*. For example, $X(t, \omega)$ can give the number of individuals or parts, the temperatures or the velocity vectors, depending on time.

Important types of stochastic processes are Markov and stationary processes.

Markov processes and Markov chains. A *Markov process* or *process without after-effect* is a stochastic process in which the knowledge of the future development is completely determined by the present state. That is, if the distribution functions of the random variables $X(t_0, \omega)$, $X(t_1, \omega)$, ..., $X(t_m, \omega)$ at different points of time $t_0 < t_1 < \cdots < t_m$ are known, then the distribution function of the random variable $X(t, \omega)$ at a point of time $t > t_m$ can be calculated from that at the time t_m alone. For example, suppose that a dam contains the quantity of water $X(t_m, \omega)$ at the beginning t_m of the time interval $(t_m, t_m + \Delta t)$ and that the amount $Z(t_m, \omega)$ of water flows into the dam in this interval and the constant amount M flows out. Then the amount of water in the dam at time $t = t_m + \Delta t$ is $X(t_m + \Delta t, \omega) = X(t_m, \omega) + Z(t_m, \omega) - M$ and the probability can be given that this amount of water does not exceed a definite capacity y of the dam. The amounts of water $Z(t, \omega)$ flowing in are random, but are independent of one another for different points of time t.

In a *Markov chain* the parameter t runs through only a discrete set of values t_i with $i = \ldots, -1, 0, +1, \ldots$

Ergodic theorems are concerned with the properties of limits of Markov processes in which the parameter increases beyond all bounds.

Poisson processes, a special class of Markov processes, play a role in the description of radioactive decay or in servicing processes, for example, in the work of a telephone exchange to calculate the waiting time for a call or for machine repair.

Stationary processes. Stochastic processes in which the causes of the variations are *independent of the time* are called stationary. The local atmospheric temperature at a point in a room, for example, varies irregularly about a fixed mean value (Fig.) if one chooses the time interval of observation so small that the variations depending on the time of day can be neglected. The variation of the diameter of a thread drawn from a spinning jet also has a time-independent mean value. If one describes these processes by means of the function $X(t, \omega)$, in which the variable t is interpreted as the time, then for each t there exist a mean value m and a variance σ^2.

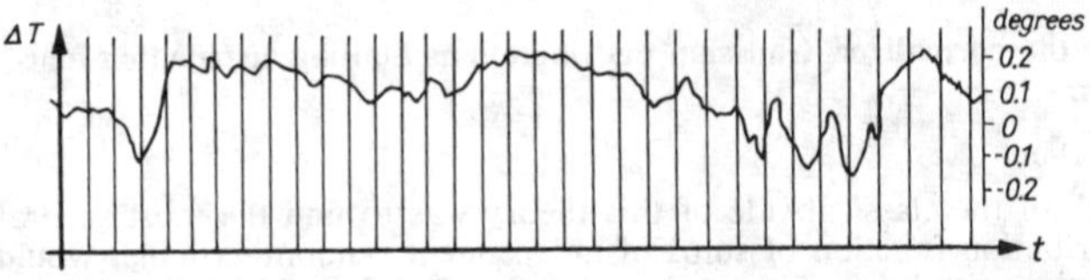

27.2-21 Curve showing local atmospheric temperature variations

Important in practice are those stochastic processes that are stationary in the sense of KHINCHINE. One postulates that their *mean value* $m = E(X(t, \omega))$ and *variance* $\sigma^2 = D^2(X(t, \omega))$ assume constant finite values and that the *correlation function* $R(t-s) = E([X(t, \omega) - m]\,[X(s, \omega) - m])$ depends only on the difference $t - s$, where t and s are two arbitrary points of time and $t > s$. These stationary processes are used, for example, in electrotechnology, in information technology, in the investigation of turbulent currents in the atmosphere, in the treatment of economic problems, and in medicine.

27.3. Statistics

The early beginnings of statistics are to be found in census counts before and around the beginning of the first century A. D. However, only in the 18th century did it begin to develop as an independent scientific discipline, by serving to describe the features that characterize the condition of a state. The concept of *statistics* is derived from the Latin word *status*, meaning condition. For a long time it was limited to work in this field, and only in recent decades did it depart from this exclusive character and, with the help of probability theory, begin to work out methods of analysing statistical data and proving statistical hypotheses. The methods of this *mathematical statistics* – or simply statistics, for short, in what follows – became an effective tool in natural science and technology by revealing new laws.

Population and sample. The *population* of a statistical investigation has as its *elements* observations or experiments under the same conditions. Each element can be examined with respect to different characteristics, which can be regarded as random variables $X, Y, \ldots$ If the characteristic X under consideration has the distribution function $F(x)$ in the population, then one says that the population has the distribution $F(x)$ with respect to the characteristic X. In statistical investigations one always considers a finite subset of elements from the population. It is called a *sample*, and the number of elements n contained in it is called the *sample size*.

Example: If the weight of 10-year old boys is the random variable X, then all boys of this age form the population. Measurements of the weights of boys in a number of places form a sample and each boy is an element of the population. The weight is a characteristic of the elements. Other characteristics, for example, are height and chest measurement.

Design of experiments

For working out a problem by statistical methods a plan of experiment must be set up that includes the method of collecting the data, the size of the sample and the method of solution of the problem. The more thorough the planning of the experiment, the better will be the results obtained by the methods of statistics. In particular, it must ensure that no measurements that are important for the conclusions are omitted or are incomplete. But it can also avoid achieving with a very expensive test series only as much as could have been achieved with an insignificant proportion of the cost. The following points are important in this connection.

(*i*) The *material* investigated should be *homogeneous*; that is, during the investigation the method of testing must remain the same. No changes must be made in the apparatus or conditions of production, and measuring instruments of different precision must not be used.

(*ii*) *Systematic errors* or influences must be *excluded* as far as possible. If one wishes, for example, to compare two materials, one must manufacture both on the same machine, otherwise differences in the machines enter into the results of the investigation. In agriculture, in testing different fertilizers the land must be divided into parallel strips in order to equalize the influence of the type of soil and its position.

(*iii*) A *control* must be provided. Either standard values exist for the characteristic under consideration, which can be compared with the results of the test, or control tests must be carried out. In experiments with fertilizers, for example, one has to assess the influence of a fertilizer from the difference between plants that grew with and without it, but otherwise under the same environmental conditions.

(*iv*) The choice of the *sample* must be *random* or *representative*. A *random choice* is one in which every element has the same probability of being or not being a member of the sample. In a consignment of screws, for example, the sample to be tested must not be chosen all from one place, but must be distributed over the whole consignment. In measuring the thickness of wires the measured points must be randomly distributed over the whole length of the wire. The random choice of elements can be made with the help of *tables of random numbers*. A *representative choice* of sample can be made when the material under investigation can be uniquely subdivided into parts. It is possible to subdivide a consignment of screws, for example, in such a way that each part contains the product of only one machine. Then from every part a number of pieces proportional to the size of the part can be chosen at random, and together these form the sample. In this way one obtains a picture of the consignment on a reduced scale.

(*v*) With regard to the size of the test sample, one has to consider that the bigger it is, the better the deductions about the population that can be made from it; but on the other hand, for reasons of time and effort, the size must usually be kept small, so that one has to reckon with a random deviation of the results. When deductions are made about a population by statistical methods, the size of the test sample is taken into account.

The collection and evaluation of material

The set of raw values that result from the experiments will be called the *original population.* They can be collected in lists, on charts, on bordered punched cards or on data processing cards, according to the size of the sample and the number of characteristics for each element. In the case of a single characteristic or a small sample one is content with a list; for several characteristics and larger sample size a chart or *bordered punched* card is prepared for each element in order to reduce the work of sorting in the evaluation process. Marks are punched at places on the border according to a predetermined key. In the case of many characteristics and a large number of elements, data processing cards are preferred for recording the values obtained, since the subsequent evaluation process is lightened by these preparations and this type of storing.

The preparation of the material. In order to obtain a preliminary survey of a given material one orders the values of the characteristics in the original population according to their magnitude and determines how frequently each value occurs. A *frequency distribution* arises in this way. Both the continuous and the discrete random variables that were explained in the section on probability theory appear as discrete variables, because the values are rounded off by the given or required accuracy.

Division into classes. In the case of a large sample the range of the characteristic values is subdivided into *classes* of equal size; in this way several values are grouped together to form a class. The choice of the size of the individual classes depends on the size of the sample and on the *scatter R*, that is, the difference between the greatest and the smallest values in the sample. The number of classes must not be too small in order that the character of the distribution shall not be blurred. On the other hand, if the number of classes is too large, then abnormal values are exaggerated and the given distribution is difficult to recognize. A class is characterized either by its limits or by its mean value. The width d of the class is the difference between its upper and lower limits; the class mean x_{M_i}, in the case of characteristics that are described by *discrete random variables*, is the arithmetic mean of the characteristic values in the particular class, and in the case of characteristics that are described by *continuous random variables,* is the arithmetic mean of the upper and lower limits of the class.

Example 1: Frequency distribution of a sample of size $n = 80$ for a characteristic that is described by a continuous random variable; x_i is the characteristic value, h_i the frequency.
(*i*) without subdivision into classes

x_i	h_i		x_i	h_i		x_i	h_i	
31.1	I	1	40.9	III	3	43.8	II	2
35.2	I	1	41.1	II	2	43.9	III	3
36.6	I	1	41.3	II	2	44.2	II	2
37.2	I	1	41.4	I	1	44.3	II	2
37.6	II	2	41.7	III	3	44.7	I	1
37.9	I	1	41.9	III	3	44.9	I	1
38.2	II	2	42.1	IIII	4	45.2	II	2
38.8	II	2	42.2	II	2	45.3	I	1
39.0	I	1	42.5	II	2	45.5	II	2
39.2	I	1	42.6	II	2	45.6	II	2
39.3	II	2	42.8	II	2	45.7	III	3
39.4	I	1	42.9	II	2	45.8	II	2
39.7	I	1	43.0	I	1	45.9	I	1
40.1	II	2	43.2	I	1	47.4	I	1
40.3	I	1	43.5	II	2	47.8	I	1
40.7	I	1	43.6	I	1			

(*ii*) with subdivision into classes; x_{M_i} is the class mean

Class	x_{M_i}	h_i	
From 33 up to and excluding 35	34	I	1
From 35 up to and excluding 37	36	II	2
From 37 up to and excluding 39	38	卌 III	8
From 39 up to and excluding 41	40	卌 卌 III	13
From 41 up to and excluding 43	42	卌 卌 卌 卌 卌	25
From 43 up to and excluding 45	44	卌 卌 卌 I	16
From 45 up to and excluding 47	46	卌 卌 III	13
From 47 up to and excluding 49	48	II	2

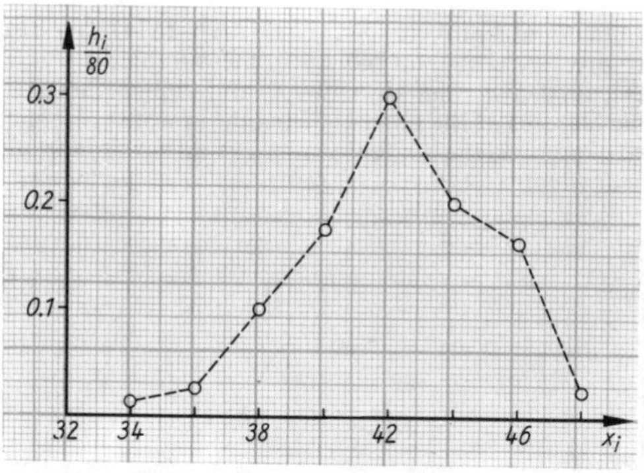

27.3-1 Representation of a distribution by a line diagram

Graphical representation of a frequency distribution. After preparing the material it is advisable to give a graphical representation of the empirical frequency distribution (Fig.). This can be done in different ways according to the purpose of the investigation and the characteristic considered, as can be seen from the examples.

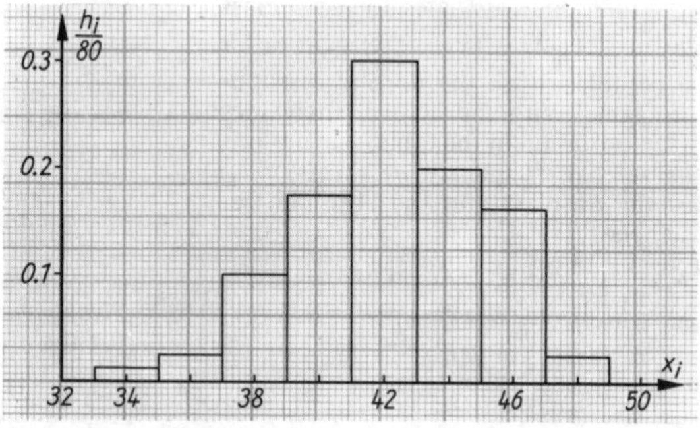

27.3-2 Representation of a distribution by a block diagram

Mean value and variance of the sample. A sample of size n can be characterized by the values of the mean $\bar{x}$ and the variance s^2, which are considered as estimates of the corresponding values μ and σ^2 for the population.

Mean value. The mean value, the arithmetic mean $\bar{x}$, is given by $\bar{x} = \frac{1}{n}\sum_{i=1}^{n} x_i$, where x_i $(i = 1, 2, \ldots, n)$ are the individual measured characteristic values. In frequency distributions the mean value is calculated by $\bar{x} = \frac{1}{n}\sum_{i=1}^{k} h_i x_i$, where h_i are the frequencies, x_i the characteristic values (or x_{M_i} the class means) and k the number of characteristic values (or the number of classes). Besides the arithmetic mean $\bar{x}$, the median value $\tilde{x}$ is also used in practice as a mean value. For odd values of n it is the characteristic value that stands in the $(n + 1)/2$th place in the series of values arranged in order of magnitude. For even values of n the median $\tilde{x}$ is the arithmetic mean of the characteristic values in the $n/2$th and the $(n/2 + 1)$th places.

Variance. For the n individual values x_i $(i = 1, 2, \ldots, n)$ of a sample the variance s^2 is given by

$$s^2 = \frac{1}{n-1}\sum_{i=1}^{n}(x_i - \bar{x})^2 = \frac{1}{n-1}\left[\sum_{i=1}^{n} x_i^2 - \frac{1}{n}\left(\sum_{i=1}^{n} x_i\right)^2\right];$$

s is called the *mean-square deviation* or the *standard deviation*. For a given frequency distribution with k characteristic values x_i (or k classes with class means x_{M_i}) and frequencies h_i the variance s^2 is calculated as follows:

$$s^2 = \frac{1}{n-1}\sum_{i=1}^{k} h_i \cdot (x_i - \bar{x})^2 = \frac{1}{n-1}\left[\sum_{i=1}^{k} h_i x_i^2 - \frac{1}{n}\left(\sum_{i=1}^{k} h_i x_i\right)^2\right].$$

Sample of size n	Mean value	Variance	Range of variation
	$\bar{x} = \frac{1}{n}\sum_{i=1}^{n} x_i$	$s^2 = \frac{1}{n-1}\sum_{i=1}^{n}(x_i - \bar{x})^2$	$R = x_{max} - x_{min}$

Besides the variance s^2 another quantity is used to characterize the range over which the characteristic values extend. This is the *scatter* or *range of variation* R, which is the difference between the greatest value x_{max} and the smallest value x_{min} of the characteristic values: $R = x_{max} - x_{min}$.

Example 2: For the frequency distribution of size $n = 80$ given above:

Subdivision into classes	Mean value	Variance	Median	Range of variation
without	$\bar{x} = 42.14$	$s^2 = 6.84$	$\tilde{x} = 42.2$	$R = 16.7$
with	$\bar{x} = 42.23$	$s^2 = 8.30$		

The discrepancies in the mean values and variance are due to the subdivision into classes for a comparatively small sample. For increasing n they become more and more nearly equal.

Normal distribution. Anthropological measurements of Lambert QUETELET (1796–1874) gave rise to the assumption that as far as their frequencies are concerned, all biological measurements follow a *Gaussian distribution*. For this reason it was called a *normal distribution* and the methods of statistics were built on this assumption. The basic features of the normal distribution were presented in the section on probability theory. For the sake of completeness a few examples that are important in statistical practice will be introduced in the following.

Because the normal distribution is determined by its mean value and variance alone, it can be calculated from the mean value $\bar{x}$ and variance s^2 of a sample. In this way it is possible to decide whether a particular characteristic is based on such a distribution.

(*i*) If the sample consists of n characteristic values that are subdivided into k classes of width d with class mean x_{M_i}, then one calculates for each class the number $\lambda_i = (x_{M_i} - \bar{x})/s$, in order to be able to use the normal distribution table. With the help of the values $\varphi(\lambda_i)$ looked up in the table, one obtains the relative frequency $q_i = (d/s)\,\varphi(\lambda_i)$ for the ith class and the absolute frequencies

$$k_i = nq_i \qquad (i = 1, 2, \ldots, k).$$

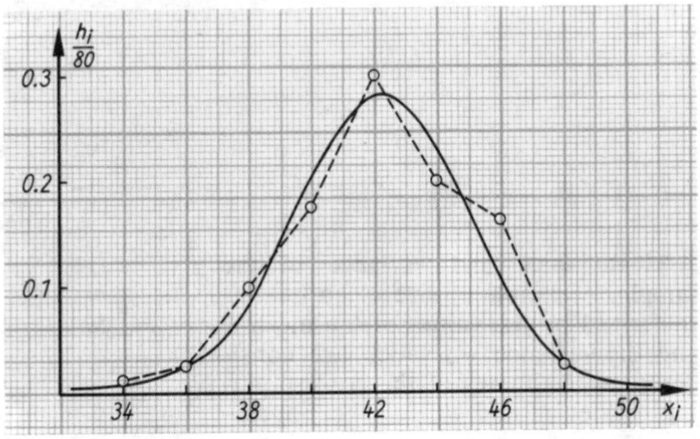

27.3-3 Example of a frequency distribution with the normal distribution drawn in

Example 3: Calculation of the normal distribution for the above sample (see Example *1*):

x_{M_i}	h_i	λ_i	$\varphi(\lambda_i)$	q_i	k_i
34	1	−2.86	0.0067	0.0046	0.4
36	2	−2.16	0.0387	0.0267	2.1
38	8	−1.47	0.1354	0.0934	7.5
40	13	−0.77	0.2966	0.2047	16.4
42	25	−0.08	0.3977	0.2744	22.0
44	16	0.61	0.3312	0.2285	18.3
46	13	1.31	0.1692	0.1167	9.3
48	2	2.00	0.0540	0.0373	3.0
	80			0.9863	79.0

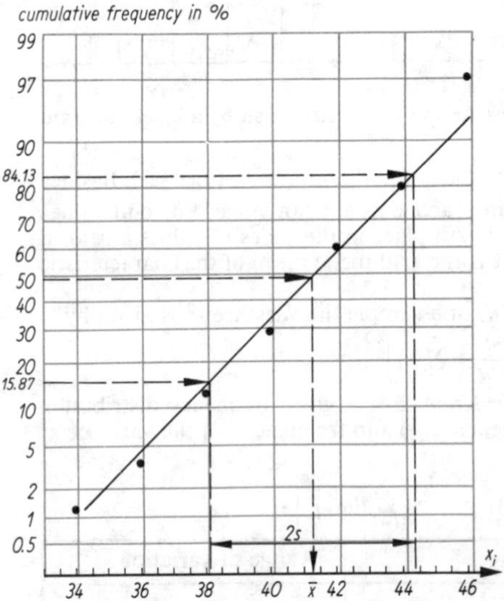

27.3-4 Cumulative percentage curve of a frequency distribution

(*ii*) If one is content with the graphical representation of the normal distribution, one uses the following rule of thumb: with the help of the formula $q = (d/s)\,\varphi(\lambda)$, as given in (*i*), the peak y_{max} of the normal distribution for $y = 0$ is calculated (Fig.). Additional ordinates can be found by the following rule:

x	$\bar{x}$	$\bar{x} \pm 0.5s$	$\bar{x} \pm s$	$\bar{x} \pm 2s$	$\bar{x} \pm 3s$
y	y_{max}	$7y_{max}/8$	$5y_{max}/8$	$y_{max}/8$	$y_{max}/80$

If one wants to have the representation in absolute frequencies, every value is multiplied by n. In the figure the frequency distribution of the above example is shown with that of the normal distribution calculated according to this rule.

(*iii*) It can also be tested with the help of *probability paper* whether the distribution of a characteristic under investigation fits a normal distribution and in addition what the mean value $\bar{x}$ and standard deviation s are. The probability net is coordinate paper whose ordinate scale is constructed so that the cumulative percentage curve of the normal distribution is a straight line (Fig.).

Regression and correlation

A large and important field in statistics is the analysis of regression and correlation. They are concerned with exposing and describing the dependence of two or more characteristics (random variables). Whilst the analysis of regression is concerned with the nature of the correspondence between the characteristics, it is the task of the analysis of correlation to determine the degree of this dependence. Only the basic concepts for the case of *linear dependence* of two characteristics (random variables X and Y) can be described here.

Regression. In an investigation in a school the height (random variable X) and weight (random variable Y) of the children is measured. It is required to determine whether on average a greater weight corresponds to a greater height, whether the correspondence is linear and what average weight corresponds to a given height. In this case the measurements show that the answer to the first question is in the affirmative. However, it is not possible to answer the other two questions without further investigation. The following regression calculation serves to provide an answer.

The individual pairs of values (x, y), where (x) is the height and (y) the weight of a child, are plotted as points with coordinates (x, y) relative to a pair of rectangular axes. The points plotted in this way form a *point set* that either has no particular form or more or less fits a curve. If the point set approximates to a line – only this case will be considered –, then the relation between the random variables X and Y is described by two lines, the regression lines. For the dependence of the weight (y) on the height (x) there exists a line of regression $Y = a_x + b_x x$, where the unknown coefficients a_x and b_x are calculated by means of Gauss's *method of least squares*. For a sample of n pairs of values (x_i, y_i) $(i = 1, 2, \ldots, n)$ one requires that

$$\sum_{i=1}^{n}(y_i - Y_i)^2 = \sum_{i=1}^{n}[y_i - (a_x + b_x x_i)]^2$$

shall be a minimum. This leads to

$$b_x = \frac{\sum_{i=1}^{n}(x_i - \bar{x})(y_i - \bar{y})}{\sum_{i=1}^{n}(x_i - \bar{x})^2} = \frac{\sum_{i=1}^{n} x_i y_i - \frac{1}{n}\left[\sum_{i=1}^{n} x_i \sum_{i=1}^{n} y_i\right]}{\sum_{i=1}^{n} x_i^2 - \frac{1}{n}\left(\sum_{i=1}^{n} x_i\right)^2},$$

$$a_x = \bar{y} - b_x\bar{x}.$$

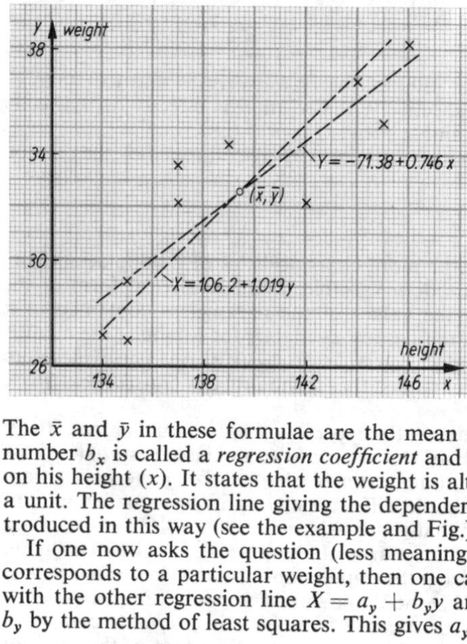

27.3-5 Point set and regression lines for the example (here the intersection of the axes is not the zero point of the coordinate system)

The $\bar{x}$ and $\bar{y}$ in these formulae are the mean values of the numbers x_i and y_i, respectively. The number b_x is called a *regression coefficient* and refers to the dependence of the weight (y) of a child on his height (x). It states that the weight is altered on average by b_x when the height increases by a unit. The regression line giving the dependence of the weight of a child on his height can be introduced in this way (see the example and Fig.).

If one now asks the question (less meaningful in this particular example) what average height corresponds to a particular weight, then one cannot use the equation given above, but must begin with the other regression line $X = a_y + b_y y$ and again calculate the unknown coefficients a_y and b_y by the method of least squares. This gives $a_y = \bar{x} - b_y\bar{y}$, where

$$b_y = \frac{\sum_{i=1}^{n}(x_i - \bar{x})(y_i - \bar{y})}{\sum_{i=1}^{n}(y_i - \bar{y})^2} = \frac{\sum_{i=1}^{n} x_i y_i - \frac{1}{n}\left[\sum_{i=1}^{n} x_i \sum_{i=1}^{n} y_i\right]}{\sum_{i=1}^{n} y_i^2 - \frac{1}{n}\left(\sum_{i=1}^{n} y_i\right)^2}$$

b_y is also called a regression coefficient and refers to the dependence of the height (x) of a child on his weight (y). This coefficient states that the height is altered on average by b_y when the weight increases by a unit. If this regression line is likewise introduced, then one observes that the two lines intersect in the centre of gravity $(\bar{x}, \bar{y})$ of the point set and make a scissor form. The more nearly closed these scissors are, the closer is the stochastic dependence between the random variables X and Y. They close up completely if a strictly linear, hence a *functional relation* exists.

Example: The height x (in cm) and the weight y (in kg. wt.) of 10 school children measured. Fig. 27.3-5 shows the point set with the two regression lines; all the necessary calculations are contained in the table following:

x	y	$(x-\bar{x})$	$(y-\bar{y})$	$(x-\bar{x})^2$	$(y-\bar{y})^2$	$(x-\bar{x})(y-\bar{y})$
135	29.30	−4.4	−3.31	19.36	10.9561	14.5640
145	35.20	5.6	2.59	31.36	6.7081	14.5040
139	34.50	−0.4	1.89	0.16	3.5721	−0.7560
142	32.10	2.6	−0.51	6.76	0.2601	−1.3260
137	33.60	−2.4	0.99	5.76	0.9801	−2.3760
137	32.30	−2.4	−0.31	5.76	0.0961	0.7440
134	27.20	−5.4	−5.41	29.16	29.2681	29.2140
144	36.70	4.6	4.09	21.16	16.7281	18.8140
135	26.90	−4.4	−5.71	19.36	32.6041	25.1240
146	38.20	6.6	5.69	43.56	32.3761	37.5540
1394	326.1			182.40	133.5490	136.0600

$\bar{x} = 139.4$,
$\bar{y} = 32.61$,
$b_x = 0.746$,
$b_y = 1.019$,
$a_x = -71.38$,
$a_y = 106.2$,
$Y = -71.38 + 0.746x$,
$X = 106.2 + 1.019y$.

Correlation. The degree of the dependence, an impression of which is given by the regression lines, is measured quantitatively by the *correlation coefficient* r_{xy}.

Correlation coefficient

$$r_{xy} = \frac{\sum_{i=1}^{n} (x_i - \bar{x})(y_i - \bar{y})}{\sqrt{\left[\sum_{i=1}^{n} (x_1 - \bar{x})^2 \cdot \sum_{i=1}^{n} (y_1 - \bar{y})^2\right]}} = \frac{\sum_{i=1}^{n} x_i y_i - \frac{1}{n}\left[\sum_{i=1}^{n} x_i \sum_{i=1}^{n} y_i\right]}{\sqrt{\left[\sum_{i=1}^{n} x_i^2 - \frac{1}{n}\left(\sum_{i=1}^{n} x_i\right)^2\right]} \cdot \sqrt{\left[\sum_{i=1}^{n} y_i^2 - \frac{1}{n}\left(\sum_{i=1}^{n} y_i\right)^2\right]}}$$

This correlation coefficient is independent of the units of the characteristics and can take all values between -1 and $+1$. If $r_{xy} = +1$ or -1, respectively, the relation is directly or indirectly linear. If $r_{xy} = 0$, then there is no relation. In the example above, $r_{xy} = +0.87$.

The correlation coefficient r_{xy} and the regression coefficients b_x and b_y are connected by the relation $r_{xy}^2 = b_x \cdot b_y$.

Methods of statistical estimation

It is often possible to draw conclusions about one or more characteristics of a population from the values of a sample. This is characterized by a random variable. If the analytical form of the corresponding distribution function is known, then the values of the parameters contained in it must be estimated. For such an estimation many possibilities are available, for example, the median or the arithmetic mean for the expectation of a random variable. In 1930 R. A. FISHER therefore drew up criteria for the goodness of an estimate; he demanded that it should be *unbiassed, consistent,* and *effective.*

For an *unbiassed estimate* $\hat{\Theta}$ of an unknown parameter Θ, the expectation of $\hat{\Theta}$ should agree with Θ; for example, the arithmetic mean $\bar{x}$ or the variance s^2 of the sample is an unbiassed estimate of the expectation μ or variance σ^2, respectively, of the random variable characterizing the population. A *consistent* estimate $\hat{\Theta}$ of an unknown parameter Θ should differ from Θ by an ever smaller amount with increasing probability for increasing sample size; that is, for arbitrary $\varepsilon > 0$, $P(|\hat{\Theta} - \Theta| < \varepsilon) \to 1$ for sufficiently large samples. For example, the arithmetic mean $\bar{x}$ of a sample is a consistent estimate of the expectation μ of the random variable characterizing the population. For an *effective estimate* $\hat{\Theta}$ of the parameter Θ the variance of the random variable $\hat{\Theta}$ should be small compared with the variance of other possible estimates; for example, the arithmetic mean $\bar{x}$ is an effective estimate compared with the median $\tilde{x}$, since the variance of the random variable x is smaller than that of the random variable $\tilde{x}$.

The estimate of a parameter can be either a point or an interval estimate. In a *point estimation* the true value of the parameter of the random variable is taken to be equal to the estimate of the value obtained from a sample. It will, however, agree with the true value with only a small probability; thus, little is known about the accuracy of the estimate. One therefore seeks by *interval estimation* to determine an interval $(\hat{\Theta} - \delta, \hat{\Theta} + \delta)$ containing the estimate $\hat{\Theta}$, in such a way that this includes the unknown parameter with probability $(1 - \alpha)$. The number $(1 - \alpha)$ is called the *confidence coefficient* and α is a number such that $0 < \alpha < 1$, from which the interval width 2δ can be calculated.

The most useful method of point estimation for a parameter is the *method of maximum likelihood.* For the normal distribution this was already developed by GAUSS. The name, the justification and the further development of the method, however, go back to FISHER. Its principle consists in choosing the estimate $\hat{\Theta}$ of a parameter Θ in such a way that the *likelihood function* for the given sample has a maximum. This likelihood function shall be given for a continuous random variable X

with known probability density $f(x;\Theta)$, where the parameter Θ is to be estimated from a test sample of n independent values $x_1, x_2, \ldots, x_n$.

One considers the likelihood function $L(x_1, x_2, \ldots, x_n; \Theta) = f(x_1;\Theta) f(x_2;\Theta) \cdots f(x_n;\Theta)$ as a function of the unknown parameter Θ and chooses as an estimate $\hat{\Theta}$ for it that value for which the function L assumes a maximum. Thus, one determines $\hat{\Theta}$ as a solution of the equation $\frac{dL}{d\Theta} = 0$. In practical calculations the equation is replaced by $\frac{d(\ln L)}{d\Theta} = \frac{1}{L} \cdot \frac{dL}{d\Theta} = 0$. If the density $f(x;\Theta_1,\Theta_2)$ of the continuous random variable X depends on two parameters Θ_1 and Θ_2, then the estimates $\hat{\Theta}_1$ and $\hat{\Theta}_2$ for them are given as solutions of the system of equations $\frac{\partial(\ln L)}{\partial\Theta_i} = \frac{1}{L} \cdot \frac{\partial L}{\partial \Theta_i} = 0$ for $i = 1, 2$.

Example: The parameters $\Theta_1 = \mu$ and $\Theta_2 = \sigma^2$ of a normally distributed random variable X can be estimated from a sample with the values $x_1, x_2, \ldots, x_n$. From the likelihood function

$$L(x_1, x_2, \ldots, x_n; \mu, \sigma^2) = [1/\sqrt{(2\pi\sigma^2)}]^n \cdot \exp\left[-1/(2\sigma^2) \sum_{k=1}^{n} (x_k - \mu)^2\right]$$

one obtains $\ln L = -(n/2) \ln 2\pi - (n/2) \ln \sigma^2 - 1/(2\sigma^2) \sum_{k=1}^{n} (x_k - \mu)^2$ and hence $\frac{\partial \ln L}{\partial \mu} = 1/\sigma^2 \sum_{k=1}^{n} (x_k - \mu) = 0$ and $\frac{\partial \ln L}{\partial \sigma^2} = -n/(2\sigma^2) + 1/(2\sigma^4) \sum_{k=1}^{n} (x_k - \mu)^2 = 0$. From these one obtains the estimates $\hat{\mu} = \frac{1}{n} \sum_{k=1}^{n} x_k = \bar{x}$ and $\hat{\sigma}^2 = \frac{1}{n} \sum_{k=1}^{n} (x_k - \bar{x})^2$.

Finally, the procedure for *interval estimation* will be given for a simple case in which X is a normally distributed random variable of whose parameters μ and σ^2 only σ^2 is known. For μ a confidence estimate is to be given from a sample with values $x_1, x_2, \ldots, x_n$. The arithmetic mean $\bar{x} = \frac{1}{n} \sum_{i=1}^{n} x_i$ is chosen as the estimate. This is known to be normally distributed with parameters μ and σ^2/n. For each α with $0 < \alpha < 1$, a λ_α can be determined from the table for the normal distribution in such a way that $P(|\bar{x} - \mu| < \lambda_\alpha \sigma/\sqrt{n}) = 1 - \alpha$ or $P(\bar{x} - \lambda_\alpha \sigma/\sqrt{n} < \mu < \bar{x} + \lambda_\alpha \sigma/\sqrt{n}) = 1 - \alpha$. The confidence interval $(\bar{x} - \lambda_\alpha \sigma/\sqrt{n}, \bar{x} + \lambda_\alpha \sigma/\sqrt{n})$ is a confidence estimate of μ with confidence coefficient $(1 - \alpha)$.

Example: If X is a normally distributed random variable, then one looks up the value $\lambda_\alpha = 1.96$ for $\alpha = 0.05$ and confidence coefficient $1 - \alpha = 0.95$ in the table for the normal distribution. For a sample of size $n = 16$ and with standard deviation $\sigma = 1.5$ one obtains for the parameter μ the confidence interval $P(\bar{x} - 1.96 \cdot 1.5/4 < \mu < \bar{x} + 1.96 \cdot 1.5/4) = 0.95$; $\bar{x}$ is the estimate obtained from the sample. The parameter μ lies in the interval $(\bar{x} - 0.74, \bar{x} + 0.74)$ with a probability of 0.95.

Statistical testing procedures

In many statistical problem s it is not enough to describe the available material by means of a frequency distribution or by numerical values, for example, if questions of the following kind are to be answered:

(*i*) In one neighbourhood 10-year old boys were found to have a greater mean weight than is usual. Is this deviation only a random one, or can the difference be due to other causes? –

(*ii*) In feeding experiments a number of rats were given a standard feed and others were fed with a food to be tested. If at the end of the series of tests a difference of mean weight of the two groups is found to exist, then it is required to establish whether the food under test really causes greater increase in weight.

In these questions one wants to know whether the apparent deviations are of a random or a significant nature. This is decided by test procedures, which all depend on a comparison; for example, either the corresponding measured values for two samples are compared with one another, or the values for one sample are compared with the corresponding known quantities for the population. In the test procedures one starts with the assumption or *hypothesis* in the first case that both samples examined belong to the same population, and in the second case that the sample belongs to the particular population considered, that is, that in both cases the differences are only random. This hypothesis is called the *null hypothesis* (H_0). Correspondingly the other possibility is called the *alternative hypothesis* (H_1).

With the help of the test procedure one decides whether to accept or reject the null hypothesis. In doing so one must bear in mind that the acceptance of the null hypothesis is nothing more than preferring this to the alternative hypothesis. One does not claim that this decision is correct in every case, because it rests on a sample of size n and an *error* is possible. Accordingly one makes the decision with an *error probability* α, which is generally chosen to be 0.05, 0.01 or 0.001. The error that occurs when the null hypothesis is rejected, although it is, in fact, correct, is called an *error of the first kind.* The other possible false decision, that of accepting the null hypothesis although it is false, is called an *error of the second kind.* For example, if as a result of a comparison one has concluded that a new medicine is better than an old one, although in reality they are both of equal value, then one has made an error of the first kind. If, however, one has concluded that both are equally good, although the new one is actually better, then one has made an error of the second kind.

Test distributions. To test the null hypothesis one uses test variables that are random variables and consequently can be described by means of a distribution. For the tests given in what follows the test variables are either normally distributed or are based on other distributions, the t-, the F- and the χ^2-distributions. Tables of these distributions are given in the worked examples.

Normal distribution. If the test variables are based on a normal distribution, then the error probability α has an intuitive meaning, if one considers the tables of the cumulative and remainder areas given by the error integral. In these the remainder area α, which corresponds to the error probability, is given as a percentage. To an error probability there corresponds a $\lambda = |(x - \mu)/\sigma|$; in general, μ and σ may be assumed to be known and only to be estimated by $\bar{x}$ and s for large samples. An error probability of $\alpha = 0.05$ or 5%, for example, corresponds to a $\lambda_\alpha = 1.96$; thus, the null hypothesis will be rejected if the value of λ calculated from the sample satisfies $\lambda > \lambda_\alpha = 1.96$, and accepted if $\lambda < \lambda_\alpha = 1.96$.

t-distribution. The procedure described for the normal distribution is no longer possible if μ and σ are unknown and have to be estimated from $\bar{x}$ and s for a sample of small size n. In this case s can deviate considerably from σ and does not then serve as a good estimate. It can be used only if the error probability belonging to λ is correspondingly increased. This occurs in STUDENT's t-distribution, which takes into account the size of the sample as well as the error probability α. For increasing values of n it becomes ever closer to the normal distribution and goes over to it in the limit as $n \to \infty$. The t-distribution is tabulated for different values of the error probability and of the degrees of freedom f, in place of the sample size. The *degree of freedom* is defined as the difference between the sample size n and the number m of characteristic measurements used in the calculation; $f = n - m$. In each of the following tests the degree of freedom is given. In Fig. 27.3-6 the normal distribution and the t-distribution for $f = 5$ degrees of freedom and an error probability $\alpha = 0.05$ are shown.

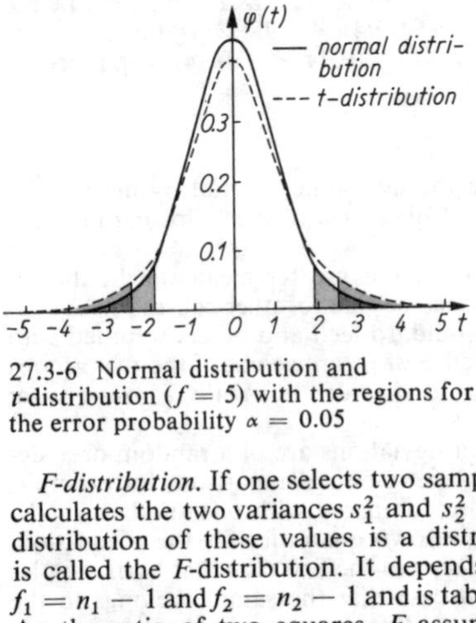

27.3-6 Normal distribution and t-distribution ($f = 5$) with the regions for the error probability $\alpha = 0.05$

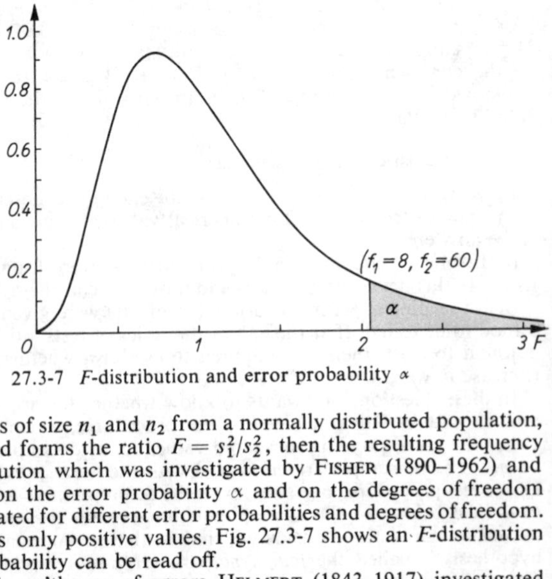

27.3-7 F-distribution and error probability α

F-distribution. If one selects two samples of size n_1 and n_2 from a normally distributed population, calculates the two variances s_1^2 and s_2^2 and forms the ratio $F = s_1^2/s_2^2$, then the resulting frequency distribution of these values is a distribution which was investigated by FISHER (1890–1962) and is called the F-distribution. It depends on the error probability α and on the degrees of freedom $f_1 = n_1 - 1$ and $f_2 = n_2 - 1$ and is tabulated for different error probabilities and degrees of freedom. As the ratio of two squares, F assumes only positive values. Fig. 27.3-7 shows an F-distribution from which the meaning of the error probability can be read off.

χ^2-distribution. In connection with GAUSS'theory of errors HELMERT (1843–1917) investigated the sum of squares of variables that are normally distributed. The distribution obtained in this way was later called by Karl PEARSON (1857–1936) the χ^2-distribution. This is based on the following assumptions: $X_1, \ldots, X_n$ are n random variables that are mutually independent and based on the same normal distribution with parameters μ and σ^2. The distribution of the sum of squares

$\chi^2 = \frac{1}{\sigma^2}\sum_{k=1}^{n}(x_k - \mu)^2$, where $x_1, \ldots, x_n$ are values of the random variables $X_1, \ldots, X_n$, is called the χ^2-distribution. It depends on the error probability α and the degree of freedom f and is tabulated for these values. Fig. 27.3-8 shows a χ^2-distribution and the meaning of the error probability.

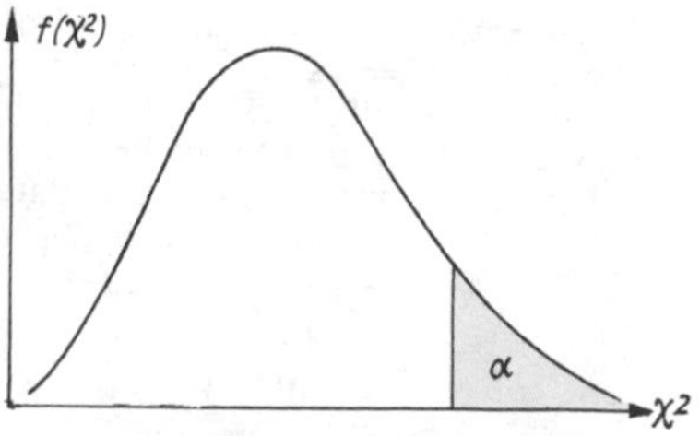

27.3-8 χ^2-distribution and error probability α for $f = 3$

Procedures for the testing of hypotheses. In the following some test procedures for frequently recurring problems are collected together. The choice of error probability depends on the nature of the problem and is determined accordingly. In industry and in agriculture, in general, an error probability of 0.05 is customary, and in medicine one of 0.01 or 0.001.

1. *Comparison of the mean values $\bar{x}$ and μ.* The mean value $\bar{x}$ of a sample of size n and variance s^2 taken from a normally distributed population is to be compared with the mean value of a normally distributed population. The null hypothesis, that the difference between the two is only random, is to be accepted with an error probability α if the calculated value t_c of the test variable $t = (1/s)/|\bar{x} - \mu| \sqrt{n}$ is less than the tabulated value t_T of the t-distribution for α and $f = n - 1$ degrees of freedom.

Example: From the examination of $n = 49$ samples of a material the proportion of a chemical element is determined, with mean $\bar{x} = 2.4\%$ and variance $s^2 = 0.4$. The value of μ is assumed to be 3%. Are the deviations that occur random with an error probability of $\alpha = 0.05$? –
The value of the test variable is $t_c = \frac{|2.4 - 3| \cdot \sqrt{(49)}}{\sqrt{(0.4)}} = 6.7$. The tabulated value of the t-distribution for an error probability of $\alpha = 0.05$ and $f = n - 1 = 48$ degrees of freedom is $t_T = 2.01$. Because $t_c > t_T$, the null hypothesis must be rejected. There are significant (essential) differences between the actual mean value and the assumed value.

Table of the t-distribution

Degrees of freedom f	Error probability $\alpha = 0.05$	Error probability $\alpha = 0.01$	Degrees of freedom f	Error probability $\alpha = 0.05$	Error probability $\alpha = 0.01$	Degrees of freedom f	Error probability $\alpha = 0.05$	Error probability $\alpha = 0.01$
1	12.71	63.66	17	2.11	2.90	45	2.01	2.69
2	4.30	9.92	18	2.10	2.88	50	2.01	2.68
3	3.18	5.84	19	2.09	2.86	60	2.00	2.66
4	2.78	4.60	20	2.09	2.85	70	1.99	2.65
5	2.57	4.03	21	2.08	2.83	80	1.99	2.64
6	2.45	3.71	22	2.07	2.82	90	1.99	2.63
7	2.37	3.50	23	2.07	2.81	100	1.98	2.63
8	2.31	3.36	24	2.06	2.80	120	1.98	2.62
9	2.26	3.25	25	2.06	2.79	140	1.98	2.61
10	2.23	3.17	26	2.06	2.78	160	1.98	2.61
11	2.20	3.11	27	2.05	2.77	180	1.97	2.60
12	2.18	3.06	28	2.05	2.76	200	1.97	2.60
13	2.16	3.01	29	2.05	2.76	300	1.97	2.59
14	2.15	2.98	30	2.04	2.75	400	1.97	2.59
15	2.13	2.95	35	2.03	2.72	500	1.97	2.59
16	2.12	2.92	40	2.02	2.70	1000	1.96	2.58
						∞	1.96	2.58

2. *Comparison of two mean values $\bar{x}_1$ and $\bar{x}_2$.* Two samples of size n_1 and n_2, respectively, are independent of one another and are assumed to come from normally distributed populations; in addition the deviations of their variances s_1^2 and s_2^2 are assumed to be random. The null hypothesis, that their mean values $\bar{x}_1$ and $\bar{x}_2$ differ only randomly with an error probability α, is accepted if the calculated value t_c for the test variable

$$t = \frac{|\bar{x}_1 - \bar{x}_2|}{s_d} \sqrt{\left(\frac{n_1 n_2}{n_1 + n_2}\right)} \quad \text{with } s_d^2 = \frac{s_1^2(n_1 - 1) + s_2^2(n_2 - 1)}{n_1 + n_2 - 2}$$

is less than the tabulated value t_T for α and $f = n_1 + n_2 - 2$ degrees of freedom.

Example: For two materials, from $n_1 = 20$ and $n_2 = 32$ tests, respectively, a mean tensile strength of $\bar{x}_1 = 18 \cdot 10^7$ N/m², $\bar{x}_2 = 24 \cdot 10^7$ N/m² with variance $s_1^2 = 4 \cdot 10^{14}$ and $s_2^2 = 6 \cdot 10^{14}$ is determined. With respect to an error probability of $\alpha = 0.05$, are the two materials essentially different with respect to the breaking strength? –

For the test variables one calculates:

$$t_b = \frac{|18 \cdot 10^7 - 24 \cdot 10^7|}{\sqrt{(5.24)} \cdot 10^7} \cdot \sqrt{\left(\frac{20 \cdot 32}{20 + 32}\right)} = 9.20,$$

where

$$s_d^2 = \frac{4 \cdot 10^{14} \cdot 19 + 6 \cdot 10^{14} \cdot 31}{20 + 32 - 2} = 5.24 \cdot 10^{14}.$$

The tabulated value of the t-distribution for an error probability of $\alpha = 0.05$ and $f = n_1 + n_2 - 2 = 50$ degrees of freedom is $t_T = 2.01$. Because $t_b > t_T$, the null hypothesis must be rejected; this means that there are significant differences between the two means.

3. *Comparison of two variances s_1^2 and s_2^2.* Two samples of size n_1 and n_2 are assumed to be independent of one another and to be taken from a normally distributed population. The null hypothesis, that their variances s_1^2 and s_2^2 differ from one another only randomly, is assumed with an error probability α if the calculated value F_c for the test variable $F = s_1^2/s_2^2, s_1^2 > s_2^2$, is less than the tabulated value F_T for α and $f_1 = n_1 - 1$ and $f_2 = n_2 - 1$ degrees of freedom.

Example: Two machines are to be compared with respect to the tolerances maintained by them in a given manufacturing process, to determine whether, with respect to an error probability of $\alpha = 0.05$, they differ essentially. For this purpose $n_1 = 25$ and $n_2 = 31$ tests are carried out on the first and second machine, respectively, and the variances $s_1^2 = 17.9$ and $s_2^2 = 17.5$ for the results obtained are calculated. For the test variable one obtains $F_c = (17.9)/(17.5) = 1.023$. The tabulated value of the F-distribution for an error probability of $\alpha = 0.05$ and $f_1 = 24, f_2 = 30$ degrees of freedom is $F_T = 1.89$. Because $F_c < F_T$, the null hypothesis must be accepted, that is, the differences in the tolerances of the two machines are only random.

F-distribution for $\alpha = 0.05$ (f_1 = degrees of freedom of the greater scatter)

f_2	$f_1=1$	$f_1=2$	$f_1=3$	$f_1=4$	$f_1=5$	$f_1=6$	$f_1=8$	$f_1=12$	$f_1=24$	$f_1=\infty$	f_2
1	161.4	199.5	215.7	224.6	230.2	234.0	238.9	243.9	249.0	254.3	1
2	18.51	19.00	19.16	19.25	19.30	19.33	19.37	19.41	19.45	19.50	2
3	10.13	9.55	9.28	9.12	9.01	8.94	8.84	8.74	8.64	8.53	3
4	7.71	6.94	6.59	6.39	6.26	6.16	6.04	5.91	5.77	5.63	4
5	6.61	5.79	5.41	5.19	5.05	4.95	4.82	4.68	4.53	4.36	5
10	4.96	4.10	3.71	3.48	3.33	3.22	3.07	2.91	2.74	2.54	10
20	4.35	3.49	3.10	2.87	2.71	2.60	2.45	2.28	2.08	1.84	20
30	4.17	3.32	2.92	2.69	2.53	2.42	2.27	2.09	1.89	1.62	30
40	4.08	3.23	2.84	2.61	2.45	2.34	2.18	2.00	1.79	1.51	40
60	4.00	3.15	2.76	2.52	2.37	2.25	2.10	1.92	1.70	1.39	60
120	3.92	3.07	2.68	2.45	2.29	2.17	2.02	1.83	1.61	1.25	120
∞	3.84	2.99	2.60	2.37	2.21	2.09	1.94	1.75	1.52	1.00	∞

4. *Comparison of frequencies.* If an event occurs z times in a sample of size n whose elements are independent of one another, but occurs in the population with probability p, then the deviation of the relative frequency z/n from p is only random, with an error probability α, provided that the calculated value t_c of the test variable

$$t = \frac{|z - np|}{\sqrt{\{np(1-p)\}}} \quad \text{or} \quad t = \frac{|z/n - p|}{\sqrt{\{p(1-p)\}}} \cdot \sqrt{n}$$

is less than the tabulated value t_T for α and $f = n - 1$ degrees of freedom.

Example: On the basis of observations over a long period the mortality rate for a certain animal disease is $p = 0.4$. A new drug is tested on $n = 71$ animals that contract this disease, and $z = 20$ of them die. Is the drug a suitable treatment, with respect to an error probability of $\alpha = 0.01$? –

According to the null hypothesis, that the deviation of the relative frequency $z/n = 20/71$ from $p = 0.4$ is only random, one calculates $t_c = \frac{|20 - 71 \cdot 0.4|}{\sqrt{(71 \cdot 0.4 \cdot 0.6)}} = 2.035$. The tabulated value of the t-distribution for an error probability of $\alpha = 0.01$ and $f = n - 1 = 70$ degrees of freedom is $t_T = 2.65$. Because $t_c < t_T$, the null hypothesis can be accepted.

5. *Testing of distributions.* An empirical distribution deviates only randomly from a theoretical distribution with an error probability α if the calculated value χ_c^2 of the test variable $\chi^2 = \sum_{i=1}^{k} [(h_i - k_i)^2/k_i]$ is less than the tabulated value χ_T^2 for α and $f = k - m - 1$ degrees of freedom, where m is the number of unknown parameters estimated from the sample. Here the material under investigation is divided into k classes, and h_i is the observed value, k_i the theoretical absolute frequency in the ith class ($i = 1, 2, \ldots, k$). The theoretical absolute frequency in each class is required to be at least 5. This can be achieved by combining several classes if necessary.

Example: A certain characteristic is measured on 80 articles made on a machine. The resulting measurements are divided into classes and the frequencies h_i given in the table are obtained. It is required to test whether, with respect to an error probability of $\alpha = 0.05$, the measured values correspond to a normal distribution. For this purpose the theoretical frequency k_i belonging to each class is calculated with the help of the Gaussian distribution and the null hypothesis is tested, that the deviation of the empirical distribution from the theoretical one is only random. The test variable is calculated with the help of the following table, which contains the individual steps in the calculation:

h_i	k_i	$h_i - k_i$	$(h_i - k_i)^2$	$(h_i - k_i)^2/k_i$
1, 2, 8 } 11	0.4, 2.1, 7.5 } 10.0	1.0	1.0	0.10
13	16.4	−3.4	11.56	0.70
25	22.0	−3.0	9.00	0.41
16	18.3	−2.3	5.29	0.29
13, 2 } 15	9.3, 3.0 } 12.3	2.7	7.29	0.59
80	79.0			2.09

Corresponding to the calculated value $\chi_c^2 = 2.09$ the tabulated value of χ^2 for an error probability $\alpha = 0.05$ and 2 degrees of freedom (since 2 parameters were estimated) is looked up, giving $\chi_T^2 = 5.99$. Because $\chi_c^2 < \chi_T^2$, the null hypothesis is accepted.

Table of the χ^2-distribution

Degrees of freedom f	Error probability $\alpha = 0.05$	$\alpha = 0.01$
1	3.84	6.64
2	5.99	9.21
3	7.82	11.35
4	9.49	13.28
5	11.07	15.09
6	12.59	16.81
7	14.07	18.48
8	15.51	20.09
9	16.92	21.67
10	18.31	23.21
20	31.41	37.57
30	43.77	50.89
40	55.76	63.69
60	79.08	88.38
120	146.57	158.95

Fields of application of statistics

From the numerous fields of application *technological statistics* and *biometry* will be selected here.

Technological statistics. The first beginnings of a statistical treatment of problems of technology go back to the beginning of the nineteen-twenties. At that time Karl DAEVES (b. 1893) recognized that the mass production of modern industry brings with it aspects of measurement that follow certain regularities. He collected together his examination procedure under the concept of *large number research.* But only in the last 25 to 30 years has statistics been applied to any considerable extent to industrial questions, for example, to the evaluation of test samples, to estimation from a series of measurements, or to current control of production. In the course of this development the term *technological statistics* has evolved. By this one understands a collection of all statistical methods that can be applied in technology and that are specially tailored for it.

These methods can be divided into two groups:

1. Methods for the statistical examination and evaluation of self-contained observation material. This is concerned essentially with statistical estimation and test procedures and with regression and correlation analyses to reveal and describe relationships. For this group the following problems are typical: the life of electric bulbs; the influence of measuring inaccuracy in the case of mechanical instruments of high precision; bending strength of natural wool fibre; tearing strength of certain fabrics; determination of the dependence of the tensile strength of a steel on various factors; comparison of the properties of two working materials.

2. Methods for the initial and final control and for the running control of a production process, briefly referred to as *statistical quality control.* These methods are based on the idea of controlling

the production process by statistical methods in such a way that the rejects are discovered as soon as they occur and their causes can be eliminated.

For this control there are two possibilities:

(*i*) the manufacture is regulated by means of control charts in such a way that the number of rejects and the finishing work are greatly reduced;

(*ii*) the finished and half-finished products are examined by means of test sampling procedures to see whether they satisfy certain quality requirements. In this case the finishing cannot be influenced; one can only draw inferences for future production.

Control charts. A finishing process is controlled by means of a control chart in such a way that characteristics of the product considered can be judged and it can be decided whether deviations from the required value are random or systematic. According to the kind of judgement one distinguishes:

(*i*) control charts for measurable characteristics,

(*ii*) control charts for non-measurable characteristics.

The design and use of a control chart is illustrated by a *single-value chart* (Fig.); a corresponding procedure holds for other control charts. For the control of a characteristic in a particular production process, objects are chosen at random, that were finished at the end of certain time intervals, and the characteristic value in question is measured. Instead of being entered in a book, this value is plotted above a time mark in the control chart. If one imagines the set of data entered as a frequency distribution, then in many cases the resulting distribution is normal, at least approximately. If one draws its mean value $\bar{x}$ as mean line (ML) and the value $\bar{x} + 3s$, $\bar{x} - 3s$ as upper control limit (UCL) and lower control limit (LCL), respectively, then 99.73% of all measured values must lie in the region bounded by the control lines. If the three lines have been determined by preliminary tests, then the production process can be controlled with the chart prepared in this way. If a measured value lies within the region, then the deviations from the mean may be regarded as random. The deviation is systematic if an entry lies outside the control lines (indicated in the figure by an arrow). In this case one has to look for the reason for the disturbance before continuing production. The graphical picture of the control chart always provides a better survey than a list. It shows the development of the characteristic considered during the production process and indicates when faults should be eliminated and new adjustments made to the machine. Because of these properties the control chart is very effective in operation and, in connection with the search for the origin of faults, it indicates frequently recurring faults and hence leads to changes in the technology.

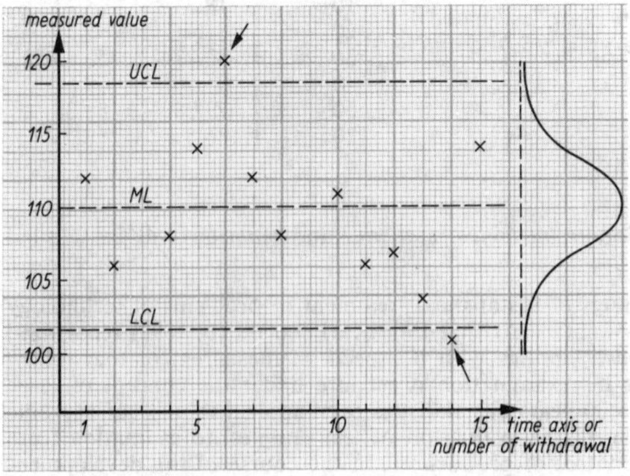

27.3-9 Single-value chart

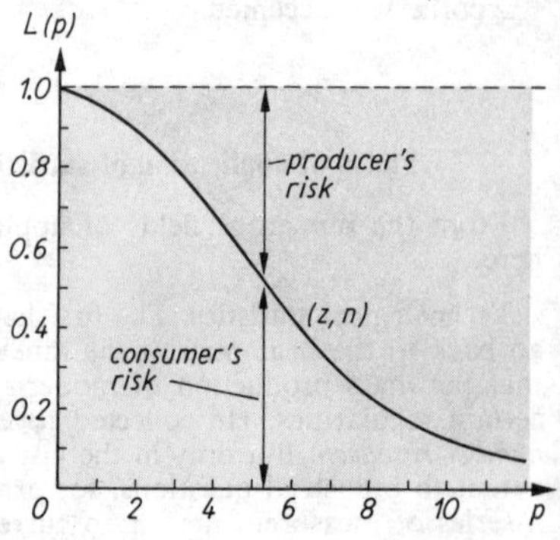

27.3-10 Operation characteristic

Sampling plans. Control charts are used when a running statistical control is to take place during a finishing process. But these methods fail, in general, if material of unknown quality is supplied, or if one forgoes a control during the finishing process and requires a quality examination to take place afterwards. In both these cases, in initial and final control, one could exercise 100% control. However, this is very costly. Besides, even with 100% control, there is no guarantee that all defective parts will be found, as experience confirms. One is therefore content with sample testing, whereby the acceptance or rejection of a consignment is decided by the quality of a sample drawn from the consignment. The tests, here for example good-bad-tests, are carried out according to a *sampling plan.* From a consignment of N items a sample of size n is withdrawn and tested. If it contains more than z bad parts, then the consignment will be rejected; it will be accepted with at most z bad parts.

Thus, the *sampling plan* is characterized by the number pair (z, n). It is based on the assumption that the percentage of faults in the sample agrees with that in the consignment. This occurs only with a certain probability, so that the producer and the consumer both take a risk in accepting the sampling plan, which can be seen from the operation characteristic (Fig.). It represents the acceptance probability $L(p)$ for the consignment in dependence upon its percentage rejection rate p, and its form depends upon the sampling plan. It is calculated, in general, by means of the binomial or Poisson distribution. From the figure the risk of the producer or of the consumer can be seen, that is, one can read off the probability that a consignment with a percentage rejection rate p shall be rejected or accepted. It is a matter for agreement by contract to choose a sampling plan and with the help of the operation characteristic to determine the permissible percentage rejection rate p. In practice it has become usual to choose a plan in such a way that a consignment with p_1% rejection rate will be accepted with a probability of 95% and one with p_2% rejection rate ($p_1 < p_2$) will be rejected with a probability of 90%.

Biometry. K. PEARSON defined biometry as the *study of the application of mathematical (statistical) methods to the examination of the multiplicity of life*. In the investigation of the laws and phenomena of living things one meets an incomparably more difficult situation than, for example, in physics. There it is possible to plan experiments that are reproducible. The conditions of the test can be kept constant and only the variable under investigation varies in order to find the required law. In *biology*, however, very many factors operate that cannot be influenced by the investigator (for example, the effect of weather on the cultivation of plants). In *medicine* the methods are still more problematical, since experiment is usually excluded (on people, for example, on ethical grounds) and often only pure observation remains. To this must be added the multiplicity of biological measures and values, which is called *biological variability*. For these reasons a well thought-out test plan for the arrangement, execution and evaluation of the tests is absolutely essential. In the course of the development of biometry special methods for the planning of tests or observations have evolved. For the evaluation of tests or observations special methods were created which particularly take into account:

(*i*) the usually small size of samples governed by the difficulty of satisfying the homogeneity requirement;

(*ii*) the frequency distributions that cannot be reduced to a normal distribution.

The difficulties and at the same time the attraction of biometry lie in finding statistical methods that are best suited for the given problems of reality.

28. Calculus of errors, adjustment of data, approximation theory

28.1. Calculus of errors

The calculus of errors is concerned with the precision of numerical data and the results of calculations. Errors that rest upon false mathematical reasoning, failure to pay attention to the laws of calculation, haste and lack of care in calculating, are not the subject of the calculus of errors. It does not excuse the calculator in any way from exercising extreme care in carrying out the operations of calculation. From the sides $a = 7.49$, $b = 5.32$ of a triangle and the included angle $\gamma = 30°$ a schoolboy calculates the length of the third

a^2	=	56.1001
b^2	=	28.3024
$a^2 + b^2$	=	84.4025
$\cos\gamma$	=	0.87
$2ab\cos\gamma$	=	69.3334
c^2	=	15.0691
c	=	3.88

side c from the formula $c = \sqrt{(a^2 + b^2 - 2ab \cos \gamma)}$ (see the accompanying calculation). The teacher finds the result too inaccurate; he had expected the solution $c = 3.92$. In looking up $\cos \gamma$ the boy has clearly taken too few decimal places into consideration. The questions arise, what error occurs as the result of the neglected decimal places, and how many places of $\cos \gamma$ would have had to be retained in order to give the required accuracy (see Accuracy of the result obtained by calculating with approximate values, Example 4).

Approximate values. In practical applications the numerical values of measured quantities are known only approximately. A man wants to drive his car to London. A road sign gives the distance as 75 miles. His car covers on average 30 miles per gallon of petrol. He therefore calculates that his fuel requirement for his journey will be (75/30 = 2.5) gallons. However, the road sign does not give the 'true value' of the length of his journey, but only an *approximate value*. For such a calculation a more accurate statement of distance would have no particular value. The average petrol consumption of his car is also an approximate value (Fig.).

28.1-1 This distance sign is rounded to whole miles

Even for pure numbers often only approximate values can be used in calculations, because most numbers can be represented in the decimal system only as infinite decimals, for example, $\sqrt{2}$, π, lg 3.

If one wishes to indicate that a is an approximate value for a quantity x, one usually writes $x \approx a$; x is the *exact value*, a the *approximate value*. For example, $\sqrt{2} \approx 1.41$; $\pi \approx 3.14$; $\lg 3 \approx 0.4771$.

Absolute and relative error

Error and absolute error. Each approximate value a is judged by its deviation from the exact value x, the difference $\varepsilon = a - x$ is called *error* $\varepsilon = a - x$, its absolute value $|\varepsilon| = |a - x|$ *absolute error*. The smaller $|\varepsilon|$ is, the more accurate is an approximate value a. For example, the approximate value $a_1 = 0.66667$ for $x = 2/3$ is a hundred times more accurate than the approximate value $a_2 = 0.667$.

approximate value	exact value	error	absolute error	relative error
a	x	$\varepsilon = a - x$	$\|\varepsilon\| = \|a - x\|$	$\|\varepsilon/x\|$

Correction. If one wishes to obtain the exact value x of a quantity from the approximate value a, one must add to a the correction c: $c = x - a = -\varepsilon$.

Relative error. Instead of the absolute error $|\varepsilon|$ of an approximate value a, its *relative* error $|\varepsilon/x|$ is often given. It is usually expressed as a percentage. The accuracy of approximate values for different quantities can then be compared with one another in this way.

Example: For the exact values $x = 2/3$, $y = 1/15$ respectively, the approximate values $a_1 = 0.67$, $a_2 = 0.07$ are used. Then the errors are $\varepsilon_1 = a_1 - x = 0.67 - 2/3 = 1/300$ and $\varepsilon_2 = a_2 - y = 0.07 - 1/15 = 1/300$; for the relative errors one obtains $|\varepsilon_1/x| = \dfrac{1/300}{2/3} = 1/200 = 0.005 = 0.5\%$ and $|\varepsilon_2/y| = \dfrac{1/300}{1/15} = 1/20 = 0.05 = 5\%$. Although the absolute errors are equal, a_1 is a ten times more accurate approximation for x than a_2 for y.

Bounds for the absolute error. Every statement concerning the magnitude of the absolute or relative error of an approximate value represents a statement about the *accuracy* of this approximation. However, the exact value is usually unknown, for example, for approximate values obtained from measurements. Then neither the absolute nor the relative error of the approximation can be calculated. In such a case bounds for the error or the relative error of the approximation ought to be given. By a bound for the (absolute) error of an approximate value a one understands a positive number Δa that is never exceeded in value by the absolute error. The inequality

$$-\Delta a \leqslant \varepsilon \leqslant \Delta a \quad \text{or} \quad a - \Delta a \leqslant x \leqslant a + \Delta a$$

always holds. If a bound Δa is stated, then at the same time both a lower and an upper *bound* for x are given. This is expressed for short in the form $x \approx a(\pm \Delta a)$, or $x = a \pm \Delta a$; Δa gives information about the accuracy of a. The smaller Δa is, the more accurate is the approximation a.

a approximate value for x, Δa bound for the absolute error of a	$x \approx a(\pm \Delta a)$ or $x = a \pm \Delta a$

Alternatively, if two bounds x_1 and x_2 for a quantity x are known, such that $x_1 \leqslant x \leqslant x_2$, then $a = (x_1 + x_2)/2$ is an approximation for x with $\Delta a = (x_2 + x_1)/2$.

Bounds for the relative error. In technical data the accuracy is often given in the form $x \approx a(\pm \delta \cdot 100\%)$ or also $x = a \pm a \cdot \delta \cdot 100\%$. The quantity $\delta = |\Delta a/a|$ is a bound for the relative error of a.

$x \approx a(\pm\delta \cdot 100\%)$ or $x = a \pm a \cdot \delta \cdot 100\%$	a approximate value for x; $\delta = \|\Delta a/a\|$ bound for the relative error of a; Δa bound for the absolute error

Example: The capacitance of a capacitor is given as 250 pF $\pm$ 10%. The relative error of the approximate value $a = 250$ pF is $\delta = 0.1$. From this it follows that $\Delta a = a \cdot \delta = 25$ pF is a bound for the absolute error. Consequently the exact value of the capacitance lies between 225 pF and 275 pF.

If approximate values for numbers such as π, $\sqrt{2}$, lg 3 are given, then as a rule one dispenses with a special statement of the accuracy. The representation of such approximate values is subject to certain rules that permit to infer at once the accuracy of the given tables.

Truncation. The number π can be represented only by an infinite decimal expansion. In a table, for example, one finds the number given as $\pi = 3.141\,592\,653\,589\ldots$ In this way the table gives as approximation for π a decimal truncated after 12 places. In *truncating* an infinite decimal, its sequence of digits is cut off completely at a particular place. The three dots at the end indicate that further digits would follow. In this all the digits given are *valid digits*, that is, the sequence of digits in the truncated decimal agrees completely with the sequence of digits in the non-truncated decimal, up to the place after which they were discarded.

The number π, truncated after 4 decimal places, is therefore $\pi = 3.141\,5\ldots$ Any of the digits from 0 to 9 can follow as the next one after the last digit of a truncated decimal. Hence if one uses a number truncated after k digits as an approximate value, then the absolute error of this approximation is negative and its absolute value is less than a unit of the order of the last digit included. A number truncated after k decimal places therefore has an error less than 10^{-k}; for example, for $\pi \approx 3.141\,5$ the absolute error is less than $10^{-4} = 1/10\,000$.

Rounding off. A usual method of curtailing decimal places is by *rounding off*. In this the last digit retained is unchanged, as in the method of truncating, if it is followed by a 0, 1, 2, 3 or 4 (*rounding down*). The last digit retained is increased by 1, however, if it is immediately followed by a 5, 6, 7, 8 or 9 (*rounding up*). Consequently an approximation for π rounded to four decimal places is $\pi \approx 3.1416$; its absolute error is less than $10^{-4}/2$. If one follows this rule, then one has the guarantee that the absolute error of a rounded number has absolute value less than half of a unit of the order of the last digit given. It can, however, be positive or negative. Only in the case for which the first digit neglected is a 5 followed by zeros is the rounding error exactly equal to half of a unit of the order of the last digit retained. It is customary in this case to round in such a way that the last digit retained is even. For example, $1/8 = 0.125\,00$ would be rounded to 0.12, but $7/40 = 0.175\,00$ to 0.18.

Reliable digits. The digits of a rounded number need not all be valid digits, since they may be the result of rounding up. But a correctly rounded number has only *reliable digits*. All digits of an approximate value are called reliable if the absolute error of this approximation is at most a half unit of the order of the last digit retained.

Example: In a five-figure table of square roots the value 6.245 00 is given for $\sqrt{39}$. This value has only reliable digits, because its absolute error is less than $5 \cdot 10^{-6}$. Its last three digits, however, are not valid digits, because $\sqrt{39} = 6.244\,997\,9\ldots$ Thus, if an approximate value contains only reliable digits, a statement of accuracy need not follow at the same time. On the other hand, if the accuracy of an approximation is not given, one must assume that all its digits are reliable. This occurs, in particular, for all numbers given in mathematical tables.

If there is no risk of misunderstanding, one often writes $x = a$ instead of $x \approx a$, if a is an approximate value for x resulting from rounding off.

Significant and non-significant digits. A difficulty arises in rounding large numbers. For example, if the number 1778 is rounded to the nearest hundred, one obtains 1800. This number is correctly rounded, but does not contain only reliable digits, because the absolute error is greater than 0.5. In place of the neglected digits 7 and 8, zeros are introduced. They serve only to fix the order of magnitude of the rounded number. They are called *non-significant digits*. The introduction of non-essential digits (zeros) can give rise to misunderstanding in the consideration of accuracy. One therefore uses another notation with the help of powers of ten. In the case in question one writes $18 \cdot 10^2$ for the rounded number. But if the number to be rounded is 1 799.7, then both the zeros in 1800 are *significant digits*. They are included, for example, in the form $1.800 \cdot 10^3$.

The rounding of rounded numbers. A further difficulty is encountered if numbers that are already rounded have to be rounded yet again. For example, if the number 0.4747 is rounded to two decimal places, this gives 0.47; if however the number is first rounded to three places, and this number is again rounded to two places, one obtains first 0.475 and finally 0.48. The last digit is now no longer

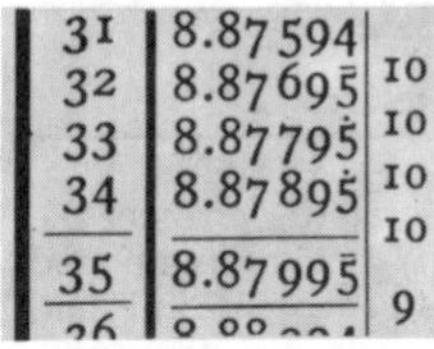

31	8.87594	
		10
32	$8.8769\bar{5}$	
		10
33	$8.8779\dot{5}$	
		10
34	$8.8789\dot{5}$	
		10
35	$8.8799\bar{5}$	
		9

28.1-2 Characterization of a 5 in the last place in a table of logarithms

reliable. The uncertainty resulting from repeated rounding always occurs when the last digit has to be rounded to a 5. It is therefore useful for possible further rounding to notice whether a 5 in the last place of a rounded number is genuine, or the result of rounding up or rounding down. A 5 is sometimes characterized by a bar or a point, respectively, above it, according as it is the result of rounding up or the digits following it are discarded (Fig.); $2.6146 \approx 2.61\bar{5} \approx 2.61$; $2.6153 \approx 2.61\dot{5} \approx 2.62$.

Accuracy of the result obtained by calculating with approximate values

Initial error and error of calculation. If a calculation is carried out with approximate values, then, in general, the result will likewise be only approximately correct. In the first place the inaccuracy of the result depends on the errors of the approximations entering into the calculation. The error in the result caused in this way is called the *initial error*. Further, a certain error occurs in the course of the calculation itself, for example, through rounding up or down. This is called the *error of calculation*. The error of calculation must always be smaller than the initial error, or else the accuracy of the initial data is not fully utilized.

Method of bounding values. The *method of bounding values* yields the most exact determination of the accuracy of the result of a calculation. In this one finds a lower and an upper bound for the result from the lower and upper bounds of the initial values. For the basic methods of calculation simple rules can be given. Let $L(x)$ and $U(x)$ be the lower and upper bounds, respectively, for the value of x, $L(y)$ and $U(y)$ lower and upper bounds for the value of y. Then

$$L(-x) = -U(x), \quad L(x+y) = L(x) + L(y), \quad L(x-y) = L(x) - U(y),$$
$$U(-x) = -L(x), \quad U(x+y) = U(x) + U(y), \quad U(x-y) = U(x) - L(y).$$

These relations can be derived from the inequalities $L(x) \leqslant x \leqslant U(x), L(y) \leqslant y \leqslant U(y)$; for example, from the inequality $L(x) \leqslant x \leqslant U(x)$ it follows on multiplying by -1 that $-U(x) \leqslant -x \leqslant -L(x)$, so that $-U(x)$ is a lower bound for $-x$ and $-L(x)$ an upper bound for $-x$.

From	$x \leqslant U(x)$	$L(x) \leqslant x$	$L(x) \leqslant x$	$x \leqslant U(x)$
follows	$x+y \leqslant U(x)+y;$	$L(x)+y \leqslant x+y;$	$L(x)-y \leqslant x-y;$	$x-y \leqslant U(x)-y;$
and since	$y \leqslant U(y)$	$L(y) \leqslant y$	$y \leqslant U(y)$	$L(y) \leqslant y$
it follows	$x+y \leqslant U(x)+U(y)$	$L(x)+L(y) \leqslant x+y$	$L(x)-U(y) \leqslant x-y$	$x-y \leqslant U(x)-L(y)$

In determining the bounds of xy and x/y, the signs of the bounds play a part. If x and y have only positive bounds, then

$$L(xy) = L(x)\,L(y), \quad U(xy) = U(x)\,U(y),$$
$$L(x/y) = L(x)/U(y), \quad U(x/y) = U(x)/L(y).$$

When rounding off in the course of calculation care must be taken that lower bounds may only be reduced as a result of the rounding, and upper bounds only increased.

Example 1: The height h of a frustrum of a right circular cone is to be calculated from its upper radius $r_2 \approx 61(\pm 0.5)$ in., its lower radius $r_1 \approx 74(\pm 0.5)$ in. and its slant height $s \approx 82(\pm 0.5)$ in. The formula is $h = \sqrt{[s^2-(r_1-r_2)^2]}$. The adjacent calculation gives 80.28 in. as a lower bound for the result and 81.63 in. as an upper bound. The result can be expressed more concisely in the form $h \approx 80.955(\pm 0.675)$ in. or as a somewhat coarser approximation $h \approx 81.0(\pm 0.8)$ in.

	Lower bound	Upper bound
s	81.5	82.5
r_1	73.5	74.5
r_2	60.5	61.5
$r_1 - r_2$	12.0	14.0
$(r_1 - r_2)^2$	144.0	196.0
s^2	6642.25	6806.25
h^2	6446.25	6662.25
h	80.28	81.63

This method is rather pessimistic, because in connection with arithmetic operations the operands are considered as independent and the worst combinations get the same weight as the most probable ones. Therefore probabilistic and socalled fuzzy methods were proposed to get more realistic estimations for error bounds. But these more sophisticated methods are in the concrete applications advisable only if the interval bounds lead to undiscussable consequences.

Method of limiting error. The method of bounding values takes account both of the initial error and of the error of calculation. Its application is, however, very time consuming, since each calculation must be carried out twice.

If one is interested only in the initial error, then the *method of limiting error* leads more quickly to the goal. It rests on the following principle.

A function $f(x_1, x_2, \ldots, x_k)$ of k variables is to be calculated. For the values $x_1, x_2, \ldots, x_k$ needed for the calculation, only approximate values $a_1, a_2, \ldots, a_k$ are available. It is required to estimate the error associated with the result if the calculation is based on the approximate values $a_1, a_2, \ldots, a_k$. Suppose that the approximations have the absolute errors $\varepsilon_1 = a_1 - x_1$, $\varepsilon_2 = a_2 - x_2$, ..., $\varepsilon_k = a_k - x_k$, which are very small compared with the values a_i. The exact result would be

$$f(x_1, x_2, \ldots, x_k) = f(a_1 - \varepsilon_1, a_2 - \varepsilon_2, \ldots, a_k - \varepsilon_k).$$

If one expands the right-hand side of this equation in a series, by the method known from the differential calculus, neglecting terms of higher order in the absolute errors ε_i, one obtains

$$f(x_1, \ldots, x_k) = f(a_1, \ldots, a_k) - \varepsilon_1 \frac{\partial f}{\partial x_1} - \varepsilon_2 \frac{\partial f}{\partial x_2} - \cdots - \varepsilon_k \frac{\partial f}{\partial x_k}.$$

The values of the partial derivatives of $f(x_1, \ldots, x_k)$ at the point $x_1 = a_1$, $x_2 = a_2, \ldots, x_k = a_k$ are to be taken here. Up to terms of higher order in the ε_i, this equation gives for the error of the result the expression

$$\begin{aligned}\varepsilon_f &= f(a_1, \ldots, a_k) - f(x_1, \ldots, x_k) \\ &= \varepsilon_1 f_{x_1}(a_1, \ldots, a_k) + \varepsilon_2 f_{x_2}(a_1, \ldots, a_k) + \cdots + \varepsilon_k f_{x_k}(a_1, \ldots, a_k).\end{aligned}$$

Thus, the absolute error can be estimated as follows:

$$|\varepsilon_f| \leqslant |\varepsilon_1|\,|f_{x_1}(a_1, \ldots, a_k)| + |\varepsilon_2|\,|f_{x_2}(a_1, \ldots, a_k)| + \cdots + |\varepsilon_k|\,|f_{x_k}(a_1, \ldots, a_k)|.$$

If bounds $\Delta a_1, \Delta a_2, \ldots, \Delta a_k$ for the errors $\varepsilon_1, \varepsilon_2, \ldots, \varepsilon_k$ are known, this inequality can be sharpened to

$$|\varepsilon_f| \leq \Delta a_1 |f_{x_1}(a_1, \ldots, a_k)| + \Delta a_2 |f_{x_2}(a_1, \ldots, a_k)| + \cdots + \Delta a_k |f_{x_k}(a_1, \ldots, a_k)| = \Delta f.$$

The value of Δf gives a bound, to a good approximation, for the absolute error of the result.

Basic equation for the estimation of the accuracy of the results of calculation
$\Delta f = \Delta a_1 |f_{x_1}(a_1, \ldots, a_k)| + \Delta a_2 |f_{x_2}(a_1, \ldots, a_k)| + \cdots + \Delta a_k |f_{x_k}(a_1, \ldots, a_k)|$

From this equation it is possible to calculate a limiting error for the result from limiting errors of the approximate values entering into the calculation. Neglecting in its derivation the terms of higher order in the errors ε_i $(i = 1, \ldots, k)$ makes hardly any difference in practice.

Application of the method of limiting error to the elementary rules of calculation. If a and b are approximations with limiting errors Δa and Δb for the quantities x and y, respectively, then the basic equation assumes the following forms:

Addition: $f(x, y) = x + y$; $|f_x| = 1$; $|f_y| = 1$; $\Delta f = \Delta a + \Delta b$.
Subtraction: $f(x, y) = x - y$; $|f_x| = 1$; $|f_y| = 1$; $\Delta f = \Delta a + \Delta b$.

The sum of the limiting errors of two approximate values represents a bound for the absolute error of the sum and of the difference of the two approximate values.

Multiplication: $f(x, y) = xy$; $|f_x| = |y|$; $|f_y| = |x|$; $\Delta f = |a|\,\Delta b + |b|\,\Delta a$; division by $|f(a, b)| = |ab|$ gives $\Delta f/|f| = \Delta a/|a| + \Delta b/|b|$.

Division: $f(x, y) = x/y$; $|f_x| = 1/|y|$; $|f_y| = |x|/y^2$; $\Delta f = \Delta a(1/|b|) + \Delta b|a|/b^2$; division by $|f(a, b)| = |a/b|$ gives $\Delta f/|f| = \Delta a/|a| + \Delta b/|b|$.

The sum of the bounds of the relative errors of two approximate values represents an approximate bound for the relative error of the product and of the quotient of the approximate values.

Raising to a power: $f(x) = x^n$; $|f_x| = |nx^{n-1}|$; $\Delta f = \Delta a|na^{n-1}|$; division by $|f(a)| = |a^n|$ gives $\Delta f/|f| = |n| \cdot \Delta a/|a|$.

n times the bound for the relative error of an approximate value is an approximate bound for the relative error of the nth power of this approximate value.

The method of limiting errors can also be applied to more complicated calculations that are composed of sums, differences, products and quotients.

Formulae for limiting error
x and y exact values, a and b their approximations, Δa and Δb their limiting errors

Type of calculation	$f(x, y)$	Bounds for the absolute error Δf	Bounds for the relative error $\Delta f/\lvert f\rvert$
Addition	$x+y$	$\Delta a + \Delta b$	$(\Delta a + \Delta b)/\lvert a+b\rvert$
Subtraction	$x-y$	$\Delta a + \Delta b$	$(\Delta a + \Delta b)/\lvert a-b\rvert$
Multiplication	xy	$\Delta a\lvert b\rvert + \Delta b\lvert a\rvert$	$\Delta a/\lvert a\rvert + \Delta b/\lvert b\rvert$
Division	x/y	$(\Delta a\lvert b\rvert + \Delta b\lvert a\rvert)/b^2$	$\Delta a/\lvert a\rvert + \Delta b/\lvert b\rvert$
Raising to a power	x^n	$\Delta a\lvert na^{n-1}\rvert$	$\lvert n\rvert\, \Delta a/\lvert a\rvert$
General	$f(x, y)$	$\Delta a\lvert f_x(a, b)\rvert + \Delta b\lvert f_y(a, b)\rvert$	$\dfrac{\Delta a\lvert f_x(a, b)\rvert + \Delta b\lvert f_y(a, b)\rvert}{\lvert f(a, b)\rvert}$

Example 2: The calculation of the expression $f = ab/c$ for $a = 2 \pm 0.1$, $b = 4 \pm 0.2$, $c = 2.5 \pm 0.1$ gives $f = (2 \cdot 4)/2.5 = 3.2$ with a relative error $\Delta f/|f| = \Delta a/|a| + \Delta b/|b| + \Delta c/|c| = 0.1/2 + 0.2/4 + 0.1/2.5 = 0.14 \mathrel{\hat{=}} 14\%$ and an absolute error $\Delta f = 3.2 \cdot 0.14 = 0.448$. The result is $f = 3.2 \pm 0.4$.

Example 3: The area of a triangle with two sides and the included angle given by $a \approx 5.2(\pm 0.05)$ $b \approx 3.4(\pm 0.05)$ and $\gamma = 35°(\pm 10')$ (Fig.) is

$$A = {}^1/_2 ab \sin\gamma \approx 5.070.$$

The estimate of the error is

$$\Delta A = {}^1/_2 \Delta a|b|\,|\sin\gamma| + {}^1/_2\,|a|\,\Delta b\,|\sin\gamma| + {}^1/_2\,|a|\,|b|\,|\cos\gamma|\,\Delta\gamma,$$

$$\Delta A/|A| = \Delta a/|a| + \Delta b/|b| + |\cos\gamma/\sin\gamma|\,\Delta\gamma = 0.05/5.2 + 0.05/3.4 + 1.428 \cdot 0.0029 = 0.0096 + 0.0147 \cdot 0.0042 = 0.0285 \mathrel{\hat{=}} 2.85\%.$$

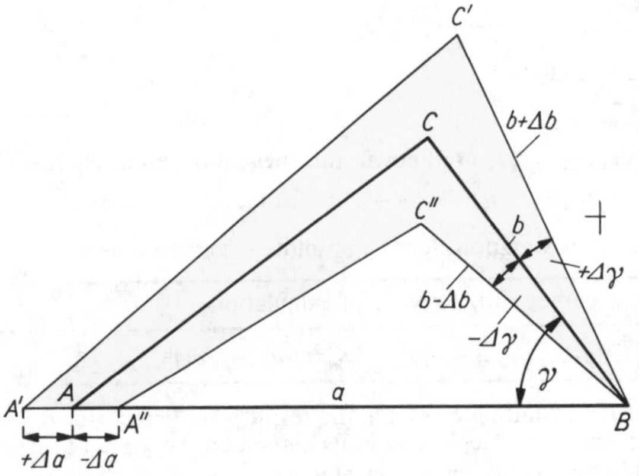

28.1-3 The exact triangle to be calculated lies between the triangles $A''BC''$ and $A'BC'$

Consequently $\Delta A = 5.070 \cdot 0.0285 = 0.144$ or $A \approx 5.070(\pm 0.144)$, that is, $A \approx 5.070(\pm 2.85\%)$.

The basic equation of the method of limiting error connects the bounds for the errors of the initial values with the bound for the error of the result of a calculation. If only a single approximate value enters into the calculation, then from this basic equation one can calculate the accuracy that must be chosen for this approximation in order to ensure a desired accuracy for the result. In this case the basic equation has the form $\Delta f = |f'(a)|\,\Delta a$, where a and Δa denote the approximation and its limiting error. If the result is to have an error not exceeding Δ_0, then one must have $\Delta f < \Delta_0$ or $\Delta a < \Delta_0/|f'(a)|$.

Example 4: How many decimal places of $\cos\gamma$ must be taken into account so that the absolute error in the length c of the third side of a triangle, calculated from the two sides $a = 7.49$ and $b = 5.32$ and the included angle $\gamma = 30°$, is less than 0.005? –

$$c = \sqrt{(a^2 + b^2 - 2ab\cos\gamma)};\ \text{hence}\ \frac{\partial c}{\partial(\cos\gamma)} = -\frac{ab}{\sqrt{(a^2 + b^2 - 2ab\cos\gamma)}} = -\frac{ab}{c}$$

and $\Delta c = \Delta(\cos\gamma)\,(ab/c) = \Delta(\cos\gamma) \cdot 10.2$.

Because $\Delta c < 0.005$ must hold, $\Delta(\cos\gamma) < 0.005/10.2 = 4.9 \cdot 10^{-4}$. Hence the value of $\cos\gamma$ must be accurate to at least *three places of decimals*.

If the result of a calculation depends on several initial values, then the problem of determining from the basic equation for the limiting error the accuracy that must be required of the initial data in order to ensure a given accuracy for the result is, of course, indeterminate. There is only one linear equation available for the calculation of several variables. However, by means of the basic equation one can estimate the magnitudes of the influence of individual errors on the result to recognize which initial values must be chosen with particular accuracy.

Example 5: The volume of a right circular cone is to be determined. The diameter of the circular base $d \approx 16$ and the height $h \approx 32$ are measured. How accurate must the measurements be and how many decimal places of π must be taken into account in the calculation so that the relative error of the result does not exceed 1%? – The volume of the cone is given by $V = (\pi/12)\,hd^2$. If $\Delta\pi$, Δh, Δd are the bounds for the absolute errors of π, h, d respectively, then a bound for the relative error of the volume is given by $\Delta V/V = \Delta\pi/\pi + \Delta h/h + 2\Delta d/d$. From the condition $\Delta V/V < 0.01$ one obtains the inequality

$$0.318\,31\Delta\pi + 0.031\,25\Delta h + 0.125\Delta d < 0.01\,.$$

It is not possible to give a unique estimation of $\Delta\pi$, Δh and Δd using this single relation. One can see, however, that an error in determining the diameter has an effect on the error of the result about four times as great as an error in measuring the height. The diameter must therefore be measured with particular care. An accuracy of $\Delta d = 0.1$ in the measurement of the diameter is not enough. The relative error of the result because of this error alone could be 1.25%. If the diameter is measured with an accuracy of 0.05, then $\Delta d = 0.05$ and bounds for the other two errors must satisfy the condition

$$0.318\,31\Delta\pi + 0.031\,25\Delta h < 0.003\,75\,.$$

The error in the measurement of the height could then be at most 0.12, and π would have to be free of error. If the accuracy of the height measurement is increased to $\Delta h = 0.1$, then the bound for the error of π must satisfy $0.318\,31\Delta\pi < 0.000\,625$ or $\Delta\pi < 0.002$. One can satisfy this condition by taking the value 3.14 for π, rounded to two places of decimals. Hence one can calculate the volume of the cone with an accuracy of at least 1% if one measures the diameter with an accuracy of $\Delta d = 0.05$ and the height with an accuracy $\Delta h = 0.1$ and takes $\pi = 3.14$ (Fig.).

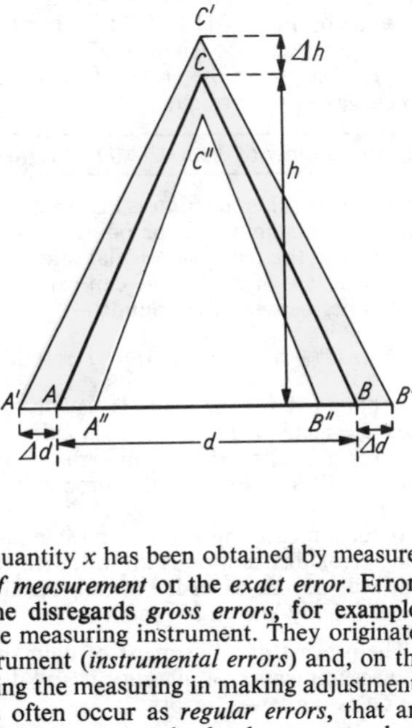

28.1-4 The exact section of the right circular cone lies between the figures $A'B'C'$ and $A''B''C''$

Errors of measurement and observation

Errors of measurement. If the approximate value a for a quantity x has been obtained by measurement, in this case the error $\varepsilon = a - x$ is called the *error of measurement* or the *exact error*. Errors of measurement are those that are unavoidable, when one disregards *gross errors*, for example, those that can result from inattention or mishandling of the measuring instrument. They originate, on the one hand, from the precision of the measuring instrument (*instrumental errors*) and, on the other hand, from involuntary errors made by the person doing the measuring in making adjustments and taking readings (*personal errors*). Instrumental errors often occur as *regular errors*, that are either *constant* or *systematic*. For example, if one reads the time on an absolutely accurate clock that is wrongly set, then this time measurement has a constant error, namely the precise amount by which the clock is wrongly set. But if one knows that a clock gains five minutes in the course of a day, then the time read on this clock contains a systematic error. Its magnitude depends on how much time has elapsed since the clock was last put right. Many constant and systematic errors must be taken into account in measurements. Because of their regularity, however, they can always be determined and eliminated.

Errors of observation. The situation is different for *irregular* or *random errors of measurement*. They are likewise unavoidable, but it is not always possible to eliminate them. For the most part the personal errors of the observer must be regarded as random errors. They are then called *errors of observation*. But random errors can also be produced by uncontrollable, random influences during the measuring process.

A single measurement suffices in itself to provide an approximate value a for a quantity x, but from this single measurement nothing can be said about the random error of measurement $\varepsilon = a - x$. It can be at one time larger, at another time smaller, positive or negative. Naturally, from a knowledge of the measuring instrument and the care and experience of the observer, a bound Δa for the measurement error can be given that will certainly not be exceeded, but will often be somewhat coarse. For this reason the measurement is carried out not only once, but a number of times, and if possible the individual measurements are undertaken by different observers. If n measurements of a quantity x are made, the n results $a_1, a_2, \ldots, a_n$ of the measurement do not, as a rule, agree completely, and especially not if exacting demands are placed upon the accuracy of the readings and values between the calibration marks of the measuring scale must be estimated. Such estimations always contain an element personal to the observer. Moreover, the accuracy of the adjustments always varies from measurement to measurement. A purely physiological source of error also arises from the fact that the determination of coincidences with the naked eye is not possible without ambiguity. The n measurements a_i give rise to n equations

$$\varepsilon_1 = a_1 - x, \quad \varepsilon_2 = a_2 - x, \ldots, \quad \varepsilon_n = a_n - x$$

for the n exact errors of measurement ε_i and the unknown exact value x, that is, for $(n+1)$ unknowns. In the *adjustment of data* methods are developed for finding the best possible approximation a for x and calculating its accuracy. The possibility of a solution of this problem rests on the fact that although the errors of observation ε_i in individual cases can be uncontrollably larger or smaller, positive or negative, seen as a whole they follow a strict law.

The error distribution law of Gauss. Gauss was one of the first to draw attention to the law governing errors of observation. The *density function of the normal distribution* of a continuous random variable is given by $p(x) = \{1/[a\sqrt{(2\pi)}]\} \cdot \exp[-(x-b)^2/(2a^2)]$, where $a^2 = \sigma^2$ is its *variance* and $\mu = b$ its expectation (see Chapter 27.). The Gaussian law of errors is obtained from this as a *density distribution* $\varphi(\varepsilon)$ if the *error of observation* $\varepsilon = x - b$ is chosen as the abscissa and the *relative frequency* as the ordinate.

Gaussian law of errors	$\varphi(\varepsilon) = \{1/[\sigma\sqrt{(2\pi)}]\} \cdot \exp[-\varepsilon^2/(2\sigma^2)]$

The graph of this function is bell-shaped, extending over the whole abscissa axis $(-\infty < \varepsilon < +\infty)$, and has a maximum at the point $\varepsilon = 0$ and points of inflection for $\varepsilon = -\sigma$ and $\varepsilon = +\sigma$. For large values of σ the curve $\varphi(\varepsilon)$ is flat and wide, and for small σ it is steep and narrow. By means of the Gaussian law of errors one can calculate the probability that the magnitude of an error of observation lies between the bounds $-\Delta$ and $+\Delta$. This probability is

$$P(-\Delta \leqslant \varepsilon \leqslant +\Delta) = \int_{-\Delta}^{+\Delta} \varphi(\varepsilon)\,d\varepsilon.$$

Error bound $\Delta = \lambda\sigma$	Probability P
0.67σ	0.500
1.00σ	0.683
1.96σ	0.950
2.00σ	0.954
2.58σ	0.990
3.00σ	0.997

The bound Δ for the error is usually expressed in units of σ. One puts $\Delta = \lambda\sigma (\lambda > 0, \sigma > 0)$. Evaluation of the *error integral* shows that the absolute value of an error of observation ε does not exceed the bound $\Delta = \lambda\sigma$ with the probability P shown in the adjacent table.

If for a measuring process the standard deviation σ of the underlying error distribution is known, then bounds $\Delta = \lambda\sigma$ can be given that will not be exceeded with a certain probability by the error of observation. Unfortunately, σ is not usually known in practice.

The process of adjustment of data shows how one can estimate the value of σ from a number of measurements of a quantity x, and draw conclusions about the error of observation by means of the Gaussian law of errors.

28.2. Adjustment of data

The process of adjustment was developed essentially by Gauss and applied to the calculation of the orbits of comets and to triangulations, for which he himself had carried out the measurements. Even today these methods are indispensible in the treatment of astronomical and geodetical measurements and, moreover, are applied with advantage in all fields in which exact calculations are to be made with the results of observation and measurement. With the help of adjustment it is possible from measurements containing errors to determine estimations (approximations) for the quantities to be measured, and to state their accuracy.

The method of least squares

Likelihood function. If n independent measurements $a_1, a_2, \ldots, a_n$ are made to determine the n quantities $y_1, y_2, \ldots, y_n$ then the Gaussian law of errors applies to each error of observation $\varepsilon_1 = a_1 - y_1$, $\varepsilon_2 = a_2 - y_2, \ldots, \varepsilon_n = a_n - y_n$. With $d\varepsilon_i = da_i$,

$$\varphi(a_i - y_i)\, da_i = \{1/[\sigma_i \sqrt{(2\pi)}]\} \exp[-(a_i - y_i)^2/(2\sigma_i^2)]\, da_i \quad (i = 1, \ldots, n)$$

gives the probability that the observed value lies in the differential interval $(a_i, a_i + da_i)$ or, for short, that *the measured value of y_i is a_i*. Each standard deviation σ_i $(i = 1, \ldots, n)$ depends on the precision of the corresponding measurement. By the multiplication law of the probability calculus, one calculates that the probability for the measured value of y_1 to be a_1, and at the same time the measured value of y_2 to be $a_2, \ldots$, and the measured value of y_n to be a_n, is given by

$$P = \frac{1}{\sqrt{(2\pi)^n}} \cdot \frac{1}{\sigma_1} \cdot \frac{1}{\sigma_2} \cdots \frac{1}{\sigma_n} \cdot e^{-(1/2)\left[\left(\frac{a_1 - y_1}{\sigma_1}\right)^2 + \left(\frac{a_2 - y_2}{\sigma_2}\right)^2 + \cdots + \left(\frac{a_n - y_n}{\sigma_n}\right)^2\right]} da_1\, da_2 \cdots da_n.$$

This equation may be written in the simpler form $P = L\, da_1\, da_2 \cdots da_n$ if L denotes the *likelihood function*; the expression S in it is usually called the *sum of the squares of the errors*, or simply the *sum of the squares*:

$$L = [1/\sqrt{(2\pi)}]^n (1/\sigma_1) \cdot (1/\sigma_2) \cdots (1/\sigma_n) \exp[-S/2]$$

with $$S = \sum_{i=1}^{n} [(a_i - y_i)/\sigma_i]^2 = \sum_{i=1}^{n} [\varepsilon_i/\sigma_i]^2.$$

$$S = \sum_{i=1}^{n} [(a_i - y_i)/\sigma_i]^2 \to \text{minimum}$$

Gauss's principle of least squares, the maximum likelihood principle. If the quantities $y_1, \ldots, y_n$ are measured, then the measured values $a_1, \ldots, a_n$ are known. The exact values of the quantities $y_1, \ldots, y_n$ remain unknown. According to GAUSS, estimations for the values of $y_1, \ldots, y_n$ appear *plausible* if the measurements $a_1, \ldots, a_n$ for them arise with the greatest probability. Hence $y_1, \ldots, y_n$ are determined in such a way that the probability P is a maximum if the measured values obtained are substituted for $a_1, \ldots, a_n$. The values for $y_1, \ldots, y_n$ obtained in this way are therefore also called the *most probable estimations* for the quantities to be measured. If the probability P assumes a maximum, then the likelihood function L must also be a maximum. This principle of estimation is therefore called the *maximum likelihood principle* and the estimated values for $y_1, \ldots, y_n$ given by it are called *maximum likelihood estimations*. The likelihood function L assumes a maximum precisely if the sum S assumes a minimum. Thus, by the maximum likelihood principle the estimated values for the quantities $y_1, \ldots, y_n$ to be measured are determined in such a way that the sum S of the squares of the errors is made a minimum. This is the *method of least squares*, which was developed by GAUSS for the estimation of exact values from a set of observations containing errors. More precisely, one should say the method of the least sum of the squares of the errors. This method forms the basis of the entire calculus of data adjustment (or smoothing). By its application the errors of observation are more or less smoothed out.

The practice of the method of least squares. If the quantities $y_1, \ldots, y_n$ are all different and if no relations exist between them, then the *method of least squares* leads to the solution $y_i = a_i$ $(i = 1, \ldots, n)$, that is, each quantity is estimated by its single observed value and the sum S is then exactly equal to zero. There can be no adjustment of the observations. This case hardly ever occurs in practice. As a rule, either the quantities $y_1, \ldots, y_n$ have the same value that is measured repeatedly, or there exist relations between them. In the latter case (which includes the former) there are fewer unknowns $t_1, t_2, \ldots, t_k$ $(k < n)$ in terms of which the quantities $y_1, \ldots, y_n$ can be expressed, $y_i = f_i(t_1, \ldots, t_k)$, for example, $y_i = c_{i1}t_1 + \cdots + c_{ik}t_k$ $(i = 1, \ldots, n)$. A representation is usually possible in the form of linear equations in which the coefficients $c_{i\varrho}$ $(\varrho = 1, \ldots, k)$ are known; for example, if a quantity is measured n times, then one has only a single unknown t, and the equations become $y_1 = t$, $y_2 = t, \ldots, y_n = t$.

Normal equations. If the number of unknowns is smaller than the number of measurements n, then there are surplus measurements. In the sum S of the squares of the errors the quantities y_i $(i = 1, \ldots, n)$ are also expressed in terms of the unknowns $t_1, \ldots, t_k$ and the partial derivatives of S with respect to these unknowns are formed. To determine the minimum of S these derivatives are put equal to zero. This produces a system of equations for $t_1, \ldots, t_k$, the so-called *normal equations*, and they are solved for the unknowns $t_1, \ldots, t_k$. With these solutions the estimations for the measured quantities $y_1, \ldots, y_n$ are calculated. One usually writes the maximum likelihood estimations for the measured quantities in the form $\hat{y}_1, \hat{y}_2, \ldots, \hat{y}_n$, to distinguish them from the unknown values $y_1, y_2, \ldots, y_n$. Likewise the value for $t_1, \ldots, t_k$ given by the normal equations are denoted by $\hat{t}_1, \hat{t}_2, \ldots, \hat{t}_k$. For the unknowns $t_1, \ldots, t_k$ other more suitable letters are often chosen for the particular problem. If the functions $f_i(t_1, \ldots, t_k)$ are linear, then the normal equations also form a system of linear equations for $t_1, \ldots, t_k$. If, however, the functions $f_i(t_1, \ldots, t_k)$ are not linear, then the solution of the normal equations can present considerable difficulties. It is then useful to *linearize* the problem. One first assumes rough approximate values $N_{t_1}, N_{t_2}, \ldots, N_{t_k}$ for $t_1, t_2, \ldots, t_k$. Let $t_1 = N_{t_1} + \delta t_1$, $t_2 = N_{t_2} + \delta t_2$, ..., $t_k = N_{t_k} + \delta t_k$. One expands the functions $f_i(t_1, \ldots, t_k)$ in Taylor series and breaks them off after the linear terms:

$$f_i(t_1, \ldots, t_k) = f_i(N_{t_1}, \ldots, N_{t_k}) + \frac{\partial f_i}{\partial t_1} \cdot \delta t_1 + \frac{\partial f_i}{\partial t_2} \cdot \delta t_2 + \cdots + \frac{\partial f_i}{\partial t_k} \cdot \delta t_k \quad \text{for } i = 1, \ldots, n.$$

One now has only to determine the unknown corrections $\delta t_1, \ldots, \delta t_k$, by means of the method of least squares.

Mean error and law of propagation of errors

Individual measurement. The precision of a measurement is given by the standard deviation σ appearing in the error distribution law. Instead of σ the quantity $h = 1/(\sigma\sqrt{2})$ is often introduced as a *measure of precision*. It was found from the Gaussian law of errors that the magnitude of the exact error of observation lies within an error bound of 0.674σ with a probability of 50%, and is less than $1 \cdot \sigma$ with a probability of 68.3%. In the adjustment of data σ is called the *mean error* and 0.67σ, or more precisely 0.674σ, the *probable error*.

Mean error	σ
Probable error	$0.674\,\sigma$

A number of measurements. If $a_1, a_2, \ldots, a_n$ are the measured values of the quantities $y_1, y_2, \ldots, y_n$ and $h_i = 1/(\sigma_i\sqrt{2})$ is the measure of the precision of each measurement, then for the sum of the squares of the errors one has

$$S = \sum_{i=1}^{n} [(a_i - y_i)/\sigma_i]^2 = 2\sum_{i=1}^{n} h_i^2 (a_i - y_i)^2 = 2\sum_{i=1}^{n} h_i^2 \varepsilon_i^2.$$

Weights of measurements. The individual errors ε_i do not contribute equally to the formation of the sum. In the formation of S a greater *weight* is attached to the square of the error if the measurement was dealt with greater precision h_i than to the square of the error of a less accurate measurement with smaller precision. One can attach directly to each individual measurement a weight p_i, which states to what extent, in relation to the other measurements, its error of observation ε_i enters into the calculation of the sum of squares of the errors. These weights must be in the ratio of the squares of their corresponding measures of precision, that is, $p_1 : p_2 : \cdots : p_n = h_1^2 : h_2^2 : \cdots : h_n^2$. As ratios, they are pure numbers and can be determined from the precision measures h_i only up to an arbitrary constant factor. If one chooses this factor in such a way that the weight $p = 1$ is attached to the precision h, then $p_i : 1 = h_i^2 : h^2$, or $h_i^2 = p_i h^2$ for $i = 1, \ldots, n$. For measurements of the same accuracy it is convenient to attach to each measurement the weight $p_i = 1$ $(i = 1, \ldots, n)$. Because the mean error σ_i of a measurement can be calculated from the precision h_i by the formula $\sigma_i = 1/(h_i\sqrt{2})$, $\sigma = 1/(h\sqrt{2})$ gives the mean error of an individual measurement with weight $p = 1$. Writing $h_i = h\sqrt{p_i}$ one obtains $\sigma_i = \sigma/\sqrt{p_i}$, $i = 1, \ldots, n$, and consequently from the mean error σ of an individual measurement of weight 1 one can calculate the mean error of an individual measurement with arbitrary weight p_i. Using the weight p_i and the mean error σ of an individual measurement of weight $p = 1$, the sum S of the squares of the errors takes the above form.

Mean error	$\sigma_i = \dfrac{\sigma}{\sqrt{p_i}}$

Sum of the squares of the errors	$S = (1/\sigma^2) \sum_{i=1}^{n} p_i (a_i - y_i)^2$

The mean square error method with weighting coefficients h_i is often preferrable, if in different parts of a data series to be fitted different error conditions exist. Sometimes we cannot assume an additive superposition of the errors to the true values. If we are confronted rather with a multiplicative influence of perturbances it is useful to choose for the weights the expressions $h_i = 1/y_i$.

Standard deviation of a linear combination of absolute errors. For the Gaussian law of errors

$$\varphi(\varepsilon) = [1/\sigma\sqrt{(2\pi)}] \exp[-(1/2)(\varepsilon/\sigma)^2]$$

the following three integrals hold:

$$(1) \int_{-\infty}^{+\infty} \varphi(\varepsilon)\, d\varepsilon = 1; \quad (2) \int_{-\infty}^{+\infty} \varepsilon\varphi(\varepsilon)\, d\varepsilon = 0; \quad (3) \int_{-\infty}^{+\infty} \varepsilon^2\varphi(\varepsilon)\, d\varepsilon = \sigma^2.$$

(1) and (3) are dealt with in the theory of probability, and (2) follows because the integrand is an odd function. If an error of observation ε is a linear combination of two independent individual errors

ε_1 and ε_2 of the form $\varepsilon = c_1\varepsilon_1 + c_2\varepsilon_2$, where c_1 and c_2 are constants, then using these three integrals one can show that the adjacent relation holds between the standard deviation σ of ε and the standard deviations σ_1 and σ_2 of ε_1 and ε_2, respectively. This result can be generalized:

$$\sigma^2 = c_1^2\sigma_1^2 + c_2^2\sigma_2^2$$

If an error of observation ε can be expressed as a linear combination of n independent individual errors $\varepsilon_1, \varepsilon_2, \ldots, \varepsilon_n$ with standard deviations $\sigma_1, \sigma_2, \ldots, \sigma_n$ in the form $\varepsilon = c_1\varepsilon_1 + c_2\varepsilon_2 + \cdots + c_n\varepsilon_n$ (where the c_i are constants), then the standard deviation σ of ε is given by

$$\sigma^2 = c_1^2\sigma_1^2 + c_2^2\sigma_2^2 + \cdots + c_n^2\sigma_n^2 .$$

Law of propagation of errors. If a function $y = f(x_1, \ldots, x_n)$ of the quantities $x_1, \ldots, x_n$ is to be calculated, and if only measured values $a_1, \ldots, a_n$ with exact errors $\varepsilon_1, \ldots, \varepsilon_n$ are available for $x_1, \ldots, x_n$, then the exact error ε of y is given, up to quantities of higher order in the ε_i, by the expression

$$\varepsilon = \frac{\partial f}{\partial x_1}\varepsilon_1 + \frac{\partial f}{\partial x_2}\varepsilon_2 + \cdots + \frac{\partial f}{\partial x_n}\varepsilon_n$$

(see Calculus of errors). The exact error ε of the result can be expressed as a linear form in the ε_i. It follows from this, by the above considerations, that the standard deviation σ of ε can be calculated from the standard deviations $\sigma_1, \sigma_2, \ldots, \sigma_n$ of the exact errors $\varepsilon_1, \varepsilon_2, \ldots, \varepsilon_n$ by Gauss's law of propagation of errors. Because the standard deviation σ of ε corresponds to the mean error of the result and the standard deviations $\sigma_1, \sigma_2, \ldots, \sigma_n$ are the mean errors of the measured values $a_1, a_2, \ldots, a_n$, *Gauss's law of propagation of errors* gives the mean error of the result in terms of the mean errors of the initial data.

Gauss's law of propagation of errors	$\sigma = \sqrt{\left[\left(\frac{\partial f}{\partial x_1}\right)^2 \sigma_1^2 + \left(\frac{\partial f}{\partial x_2}\right)^2 \sigma_2^2 + \cdots + \left(\frac{\partial f}{\partial x_n}\right)^2 \sigma_n^2\right]}$

This propagation law of the mean statistical error is not necessarily bound to the assumption, that the input variables are normally distributed. It also holds under the condition, that the input variables possess arbitrary random distributions, but they must be statistically independent from each other.

Mean error of mean value. The law of propagation of errors assumes a particularly simple form in the case when the function is the mean value of the quantities $x_1, \ldots, x_n$, $y = \bar{x} = (x_1 + x_2 + \cdots + x_n)/n$.

Because $\dfrac{\partial f}{\partial x_i} = \dfrac{1}{n}$, one obtains $\sigma_{\bar{x}}^2 = \dfrac{1}{n^2}(\sigma_1^2 + \sigma_2^2 + \cdots + \sigma_n^2)$.

If the measured values $a_1, a_2, \ldots, a_n$ of the quantities $x_1, x_2, \ldots, x_n$ have the same precision, then $\sigma_1^2 = \sigma_2^2 = \cdots = \sigma_n^2 = \sigma_x^2$ and one obtains

$$\sigma_{\bar{x}}^2 = \sigma_x^2/n, \quad \sigma_{\bar{x}} = \sigma_x/\sqrt{n}.$$

The mean error of the mean of n measurements made with equal precision is equal to the mean error of an individual measurement divided by $\sqrt{n}$.

Estimation of the mean error from the observations. In general, the mean errors σ_i of the individual measurements in a measuring process are not known, but only the weight p_i to be ascribed to the measurements $a_1, \ldots, a_n$ of the quantities $y_1, \ldots, y_n$. One is then faced with the problem of estimating the mean errors of the individual measurements from the available observed values $a_1, \ldots, a_n$. Because one can determine the mean error σ_i of an individual measurement of weight p_1 from the mean error σ of a measurement of weight $p = 1$ by means of the formula $\sigma_i = \sigma/\sqrt{p_i}$, it is sufficient to estimate the mean error σ. This estimation is denoted by m.

If the quantities $y_1, \ldots, y_n$ to be measured can be expressed in terms of exactly $k < n$ unknowns $t_1, t_2, \ldots, t_k$, then an estimate m for σ can be derived by the methods of mathematical statistics. The quantities $\hat{y}_i$ ($i = 1, \ldots, n$) are the maximum likelihood estimations of the quantities $y_1, \ldots, y_n$ to be measured.

Estimated mean error of an individual measurement of weight $p = 1$	
$m = \sqrt{\left\{\frac{1}{n-k}\left[\sum_i^n p_i(a_i - \hat{y}_i)^2\right]\right\}}$	p_i weights of measurements, a_i measured values, $\hat{y}_i$ adjusted measured values, $n - k$ number of surplus measurements

In the adjustment it is customary to denote the *mean error of the individual measurement of weight* $p = 1$ by m itself, and the probable error by $0.674\,m$, although m is only an estimation for σ, that can itself be subject to random variations. These variations can be considerable, especially for a small number n of observations. The statements about the bounds of the exact error of observation resulting from the Gaussian law of errors are therefore only approximately correct if one uses the estimate m for the mean error in place of σ. Mathematical statistics show how one can arrive at exact bounds for the exact error of observations with the help of m. To characterize the accuracy of measurements one must always state the (estimated) mean error m of the individual observation of weight $p = 1$. From m one can obtain the mean error of an individual measurement of weight p_i, using the adjacent formula, which corresponds to the formula $\sigma_i = \sigma/\sqrt{p_i}$. With the

Mean error of an individual measurement of weight p_i	$m_i = m/\sqrt{p_i}$

help of Gauss's law of propagation of errors one is in a position to calculate also the mean error of each quantity formed from the observed values $a_1, \ldots, a_n$. This arises, in particular, for the maximum likelihood estimation of the quantities $y_1, \ldots, y_n$ to be measured. If required, bounds for the exact error of observation, that will not be exceeded with prescribed probabilities, can be obtained from these mean errors.

Adjustment of direct measurements

A quantity y is measured directly n times with the same precision. Let the measured values be $a_1, a_2, \ldots, a_n$. The single unknown in this measuring process is y ($k = 1$). Hence the equations $y_1 = y, y_2 = y, \ldots, y_n = y$ hold. For individual measurements of equal precision these measurements have the same weight $p_1 = p_2 = \cdots = p_n = 1$. The sum of squares of the errors and its derivative with respect to the unknown y are given by

$$S = (1/\sigma^2)\sum_{i=1}^{n}(a_i - y)^2; \qquad \frac{dS}{dy} = -(2/\sigma^2)\sum_{i=1}^{n}(a_i - y).$$

Equating this derivative to zero one obtains the normal equation $\sum_{i=1}^{n}(a_i - y) = 0$, whose solution for y gives the estimation $\hat{y}$.

The mean value of the individual measurements serves as an estimation for the quantity to be measured.

Estimation by direct measurements of equal precision	$\hat{y} = (a_1 + a_2 + \cdots + a_n)/n = \bar{a}$

The approximation for the quantity y to be measured is given in the form $y \approx \hat{y}(\pm m_{\hat{y}})$ or $y = \hat{y} + m_{\hat{y}}$; the mean error $m_{\hat{y}}$ of the estimation is calculated from the mean error m of the individual measurement by the relation $m_{\hat{y}} = m/\sqrt{n}$.

Mean error of the individual measurement for measurements of equal precision	$m = \sqrt{\left\{\left[\sum_{i=1}^{n}(a_i - \bar{a})^2\right] \Big/ (n-1)\right\}}$
Mean error of the estimation	$m_{\hat{y}} = \sqrt{\left\{\left[\sum_{i=1}^{n}(a_i - \bar{a})^2\right] \Big/ [n(n-1)]\right\}}$

Example: Each of five school children measures the length y of the edge of a model cube. The estimation obtained from the results of the measurements (see table) is

$$\hat{y} = [(12.2 + 12.1 + 12.5 + 12.3 + 12.4)/5] \text{ in.} = 12.30 \text{ in.}$$

Results of measurement
$a_1 = 12.2$ in.
$a_2 = 12.1$ in.
$a_3 = 12.5$ in.
$a_4 = 12.3$ in.
$a_5 = 12.4$ in.

The mean error of the individual measurement is

$$m = \sqrt{\{[(12.2-12.3)^2+(12.1-12.3)^2+(12.5-12.3)^2+(12.3-12.3)^2+(12.4-12.3)^2]/4\}} \text{ in.},$$
$m = 0.158$ in.

This gives for $\hat{y}$ a mean error $m_{\hat{y}} = 0.158/\sqrt{5} = 0.071$ in. and consequently the approximation for the length of the edge of the cube is $y \approx 12.3$ in. (± 0.071 in.), or $y \approx (12.30 \pm 0.07)$ in.

Adjustment of direct measurements of unequal precision. A quantity y is measured directly n times Let the measured value be $a_1, a_2, \ldots, a_n$. Because of the unequal precision of the measurements they must be taken with their weights $p_1, p_2, \ldots, p_n$. The single unknown in this measuring process is y ($k = 1$) and the relations $y_1 = y, y_2 = y, \ldots, y_n = y$ hold. If one forms the sum S of the squares of the errors, differentiates it with respect to y, and equates the derivative to zero, one obtains the normal equations

$$S = (1/\sigma^2) \sum_{i=1}^{n} p_i(a_i - y)^2 \longrightarrow \frac{dS}{dy} = -(2/\sigma^2) \sum_{i=1}^{n} p_i(a_i - y) = 0 \longrightarrow \sum_{i=1}^{n} p_i(a_i - y) = 0,$$

from which one finds the estimation $\hat{y}$.

The weighted mean of the individual measurements with weights p_i serves as an estimation for the quantity to be measured.

Mean error of the individual measurement of weight $p = 1$	$m = \sqrt{\left[\left(\sum_{i=1}^{n} p_i(a_i - \hat{y})^2\right) \Big/ (n-1)\right]}$
Estimation by direct measurements of unequal precision	$\hat{y} = \left(\sum_{i=1}^{n} p_i a_i\right) \Big/ \left(\sum_{i=1}^{n} p_i\right)$

The mean error m of an individual measurement of weight $p = 1$ is needed to calculate easily the mean error of other relevant quantities. From it, for example, the mean error $m_i = m/\sqrt{p_i}$ can be found of the individual measurement a_i of weight p_i. Further, it can be used to calculate the mean error $m_{\hat{y}}$ for the estimation $\hat{y}$ according to the law of propagation of errors. Because $\dfrac{\partial \hat{y}}{\partial a_i} = \dfrac{p_i}{p_1 + p_2 + \cdots + p_n}$, it follows that

$$m_{\hat{y}} = \sqrt{\left\{\sum_{i=1}^{n}\left[m_i^2 p_i^2 \Big/ \left(\sum_{i=1}^{n} p_i\right)^2\right]\right\}} = \sqrt{\left\{\sum_{i=1}^{n}\left[m^2 p_i \Big/ \left(\sum_{i=1}^{n} p_i\right)^2\right]\right\}}.$$

The approximate value of the quantity to be measured can be given in the form $y \approx \hat{y}(\pm m_{\hat{y}})$ or $y = \hat{y} \pm m_{\hat{y}}$.

Mean error of the estimation by measurements of unequal precision
$m_{\hat{y}} = m \Big/ \sqrt{\left[\sum_{i=1}^{n} p_i\right]}$

Example: A length l is measured five times, and afterwards three more times with greater precision. Because of the more accurate measuring procedure the measurements a_6, a_7, a_8 (see table) of the second group must be given five times the weight of the first group. Hence $p_1 = p_2 = p_3 = p_4 = p_5 = 1$ and $p_6 = p_7 = p_8 = 5$. The estimation $\hat{l}$ for l is given by $\hat{l} = 12.34$ in. $= {}^1/_{20}\,[1 \cdot (12.35 + 12.40 + 12.25 + 12.30 + 12.35) + 5 \cdot (12.37 + 12.32 + 12.34)]$ in. Since $[0.01^2 + 0.06^2 + 0.09^2 + 0.04^2 + 0.01^2 + 5 \cdot (0.03^2 + 0.02^2 + 0^2)] = 0.0200$ and $\sqrt{(0.0200/7)} = 0.0535$ the mean error m of the individual measurement has the calculated value $m = 0.0535$ in. Because $p_1 = p_2 = p_3 = p_4 = p_5 = 1$, this is at the same time the mean error of the measurement of the first group, $m_1 = 0.0535$ in. The mean error of the measurement of the second group is given by $m_2 = m/\sqrt{5}$ in. $= 0.0239$ in.

$a_1 = 12.35$ in.
$a_2 = 12.40$ in.
$a_3 = 12.25$ in.
$a_4 = 12.30$ in.
$a_5 = 12.35$ in.
$a_6 = 12.37$ in.
$a_7 = 12.32$ in.
$a_8 = 12.34$ in.

The mean error of the estimated value is $m_{\hat{l}} = m/\sqrt{20}$ in. $= 0.0120$ in. Hence from the measurements the approximate value of the required length l is calculated to be $l \approx 12.34$ in. (± 0.01 in.).

If the mean errors σ_i for measurements $a_1, a_2, \ldots, a_n$ are themselves known instead of the weights p_i, then the weights p_i can be determined up to a constant factor by the ratios

$$p_1 : p_2 : \ldots : p_n = (1/\sigma_1^2) : (1/\sigma_2^2) : \ldots : (1/\sigma_n^2).$$

This factor is chosen so that either the $p_1, \ldots, p_n$ are easily manageable numbers, or that $p_1 + p_2 + \cdots + p_n = 1$. Denoting the arbitrary factor by λ^2, so that $p_i = \lambda^2/\sigma_i^2$, the mean error of an observation of weight $p = 1$ is $\sigma = \sigma_i \sqrt{p_i} = \lambda$. Hence if m is calculated from the observed values with the prescribed weights, the result must be approximately equal to λ, since m is an estimation for σ. If there is a very large difference between m and λ, then it can be concluded that systematic errors occur in some measurements.

Adjustment of observations

Conditional observations. The angles α, β, γ of a triangle are to be determined by measurement. Each angle is measured repeatedly. The measurements of the angle α give the values $a_1, a_2, \ldots, a_{n_1}$ (n_1 measurements). For the angle β the measured values are $b_1, b_2, \ldots, b_{n_2}$ (n_2 measurements) and for the angle γ they are $c_1, c_2, \ldots, c_{n_3}$ (n_3 measurements). Altogether $n = n_1 + n_2 + n_3$ measurements are carried out. The individual measurements are made with equal precision. Again denoting the exact values of the quantities to be determined by measurement by $y_1, y_2, \ldots, y_{n_1}$ (measurements of α), $y_{n_1+1}, y_{n_1+2}, \ldots, y_{n_1+n_2}$ (measurements of β), $y_{n_1+n_2+1}, y_{n_1+n_2+2}, \ldots, y_{n_1+n_2+n_3}$ (measurements of γ), then these quantities can be expressed in terms of the unknowns α, β, γ by

$$\begin{aligned} y_1 &= y_2 = \cdots = y_{n_1} = \alpha, \\ y_{n_1+1} &= y_{n_1+2} = \cdots = y_{n_1+n_2} = \beta, \\ y_{n_1+n_2+1} &= y_{n_1+n_2+2} = \cdots = y_{n_1+n_2+n_3} = \gamma. \end{aligned}$$

The sum of the squares of the errors is

$$S = \frac{1}{\sigma^2}\left[\sum_{i=1}^{n}(a_i-\alpha)^2 + \sum_{j=1}^{n}(b_j-\beta)^2 + \sum_{k=1}^{n}(c_k-\gamma)^2\right].$$

However, the method of least squares cannot be applied immediately, because the unknowns α, β, γ are subject to a *condition*. The sum of the angles of a triangle is 180°. Hence α, β, γ must satisfy the *equation of condition* $\varphi(\alpha, \beta, \gamma) = \alpha + \beta + \gamma - 180° = 0$. The adjustment of the observations must be achieved, taking this condition into account. In such a case one speaks of the *adjustment of conditional observations*. Strictly speaking, not three, but only two unknowns occur here. For if α and β have been determined, the value of γ follows from the conditional equation. One can deal with such a case in two ways. One can either use the conditional equation $\varphi(\alpha, \beta, \gamma) = 0$ to express one angle, say γ, in terms of the other two, substitute this expression for γ in the sum S and then apply the *method of least squares*, or one can determine directly the minimum of S subject to the side condition $\varphi(\alpha, \beta, \gamma) = 0$. To do this one makes use of the *method of Lagrangian multipliers* (see Chapter 19.4. – Extreme values of functions of several variables); that is, one determines α, β, γ and λ so that the expression

$$T = S + \lambda\varphi(\alpha, \beta, \gamma) = \frac{1}{\sigma^2}\left[\sum_{i=1}^{n}(a_i-\alpha)^2 + \sum_{j=1}^{n}(b_j-\beta)^2 + \sum_{k=1}^{n}(c_k-\gamma)^2\right] + \lambda(\alpha+\beta+\gamma-180°)$$

assumes a minimum. The quantity λ is the Lagrangian multiplier belonging to the side condition. The normal equation are then

$$\frac{\partial T}{\partial \alpha} = -\frac{2}{\sigma^2}\sum_{i=1}^{n_1}(a_i-\alpha) + \lambda = 0; \qquad \frac{\partial T}{\partial \gamma} = -\frac{2}{\sigma^2}\sum_{k=1}^{n_3}(c_k-\gamma) + \lambda = 0;$$

$$\frac{\partial T}{\partial \beta} = -\frac{2}{\sigma^2}\sum_{j=1}^{n_2}(b_j-\beta) + \lambda = 0; \qquad \frac{\partial T}{\partial \lambda} = \alpha + \beta + \gamma - 180° = 0.$$

The solutions of these normal equations are

$$\hat{\alpha} = \bar{a} - K/n_1; \quad \hat{\beta} = \bar{b} - K/n_2; \quad \hat{\gamma} = \bar{c} - K/n_3; \quad \hat{\lambda} = 2K/\sigma^2$$

with the correction $K = (\bar{a} + \bar{b} + \bar{c} - 180°)/(1/n_1 + 1/n_2 + 1/n_3)$. The mean error of the individual measurements is calculated, as usual, from

$$m^2 = \left[\sum_{i=1}^{n_1}(a_i-\hat{\alpha})^2 + \sum_{j=1}^{n_2}(b_j-\hat{\beta})^2 + \sum_{k=1}^{n_3}(c_k-\hat{\gamma})^2\right] \Big/ (n-2).$$

The mean errors of the estimations $\hat{\alpha}, \hat{\beta}, \hat{\gamma}$ are given by the law of propagation of errors by

$$m_{\hat{\alpha}} = (m/n_1)\sqrt{[n_1 - 1/(1/n_1 + 1/n_2 + 1/n_3)]};$$
$$m_{\hat{\beta}} = (m/n_2)\sqrt{[n_2 - 1/(1/n_1 + 1/n_2 + 1/n_3)]};$$
$$m_{\hat{\gamma}} = (m/n_3)\sqrt{[n_3 - 1/(1/n_1 + 1/n_2 + 1/n_3)]}.$$

Example: The angles α, β, γ are measured four, three, four times, respectively, $n_1 = 4$, $n_2 = 3$, $n_3 = 4$ (see table), and the mean values $\bar{a}$, $\bar{b}$, and $\bar{c}$ are calculated. Because $1/(1/n_1 + 1/n_2 + 1/n_3) = 1.2$ and $\bar{a} + \bar{b} + \bar{c} - 180° = 3''$, the correction K is given by $K = 3'' \cdot 1.2 = 3.6''$, and hence the estimations $\hat{\alpha} = \bar{a} - K/4$, $\hat{\beta} = \bar{b} - K/3$ and $\hat{\gamma} = \bar{c} - K/4$ can be calculated.

Angle α	$a_i - \hat{\alpha}$	Angle β	$b_j - \hat{\beta}$	Angle γ	$c_k - \hat{\gamma}$
$a_1 = 62°17'14''$	$+0.9''$	$b_1 = 73°20'25''$	$+1.2''$	$c_1 = 44°22'25''$	$+1.9''$
$a_2 = 62°17'11''$	$-2.1''$	$b_2 = 73°20'27''$	$+3.2''$	$c_2 = 44°22'26''$	$+2.9''$
$a_3 = 62°17'16''$	$+2.9''$	$b_3 = 73°20'23''$	$-0.8''$	$c_3 = 44°22'22''$	$-1.1''$
$a_4 = 62°17'15''$	$+1.9''$			$c_4 = 44°22'23''$	$-0.1''$
$\bar{a} = 62°17'14''$		$\bar{b} = 73°20'25''$		$\bar{c} = 44°22'24''$	
$\hat{\alpha} = 62°17'13.1''$		$\hat{\beta} = 73°20'23.8''$		$\hat{\gamma} = 44°22'23.1''$	

The mean error of an individual measurement is

$$m = \sqrt{\{[(0.9^2+2.1^2+2.9^2+1.9^2)+(1.2^2+3.2^2+0.8^2)+(1.9^2+2.9^2+1.1^2+0.1^2)]/9\}''} = 2.181''.$$

From this the mean errors

$$m_{\hat{\alpha}} = m \cdot 0.418 = 0.912''; \quad m_{\hat{\beta}} = m \cdot 0.447 = 0.975''; \quad m_{\hat{\gamma}} = m \cdot 0.418 = 0.912''$$

are obtained. The approximate values for the three angles calculated from the measurements are

$$\alpha \approx 62°17'13.1''(\pm 0.91''); \quad \beta \approx 73°20'23.8''(\pm 0.97''); \quad \gamma \approx 44°22'23.1''(\pm 0.91'').$$

The method described in the example of the measurement of the angles of a triangle can be applied quite generally for the adjustment of conditional observations. If the quantities $y_1, y_2, \ldots, y_n$ to be measured can be expressed in terms of the k unknowns $t_1, \ldots, t_k$, and if these unknowns satisfy r

independent conditional equations

$$\varphi_1(t_1, \ldots, t_k) = 0; \quad \varphi_2(t_1, \ldots, t_k) = 0; \quad \ldots; \quad \varphi_r(t_1, \ldots, t_k) = 0,$$

then the unknowns $t_1, \ldots, t_k$ and the Lagrangian multipliers $\lambda_1, \ldots, \lambda_r$ are determined so that the expression

$$T = S + \lambda_1 \varphi_1(t_1, \ldots, t_k) + \cdots + \lambda_r \varphi_r(t_1, \ldots, t_k)$$

has a minimum. The remaining steps in the calculation again correspond completely to the method of least squares. Because of the r conditional equations there are actually not k unknowns, but only $k - r$ unknowns to determine from the observations. This effective number of unknowns is to be taken with weight $p = 1$ in the calculation of the mean error m of the individual measurements.

Adjustment of indirect observations. Often the quantities to be determined are not accessible for direct measurement; for example, to determine the density of a body its weight G and its volume V are measured; G and V are *indirect observations* for the determination of the density. In the sense of the adjustment of data, the interest in indirect observations lies not so much in finding the exact values of the measured quantities $y_1, y_2, \ldots, y_n$, but in the determination of the unknowns $t_1, t_2, \ldots, t_k$, in terms of which $y_1, \ldots, y_n$ can be represented. The method of least squares again yields the estimations $\hat{t}_1, \hat{t}_2, \ldots, \hat{t}_k$ for the unknowns. From the mean error m of the individual observation, the mean error of the estimations $\hat{t}_1, \hat{t}_2, \ldots, \hat{t}_k$ can be calculated, using the law of propagation of errors.

Example: A ring consists of a silver-gold-alloy. To determine the weight of gold and silver the ring is weighed a number of times with a Jolly spring balance, in air G and in water W (see table). Letting g_1 denote the weight of gold, g_2 that of silver and ϱ_1 and ϱ_2 the densities of gold and silver, respectively:
$G = g_1 + g_2$; $W = (1 - 1/\varrho_1) g_1 + (1 - 1/\varrho_2) g_2$.

Weight	
in air G	in water W
$a_1 = 4.01$ g	$b_1 = 3.72$ g
$a_2 = 3.98$ g	$b_2 = 3.72$ g
$a_3 = 4.03$ g	$b_3 = 3.69$ g
$a_4 = 4.02$ g	

The sum of the squares of the errors is

$$S = (1/\sigma^2)\left[\sum_{i=1}^{4}(a_i - g_1 - g_2)^2 + \sum_{j=1}^{3}[b_j - (\varrho_1 - 1) g_1/\varrho_1 - (\varrho_2 - 1) g_2/\varrho_2]^2\right].$$

This leads to the normal equations:

$$\frac{\partial S}{\partial g_1} = -(2/\sigma^2)\left[\sum_{i=1}^{4}(a_i - g_1 - g_2) + \sum_{j=1}^{3}[b_j - (\varrho_1 - 1) g_1/\varrho_1 - (\varrho_2 - 1) g_2/\varrho_2](\varrho_1 - 1)/\varrho_1\right] = 0,$$

$$\frac{\partial S}{\partial g_2} = -(2/\sigma^2)\left[\sum_{i=1}^{4}(a_i - g_1 - g_2) + \sum_{j=1}^{3}[b_j - (\varrho_1 - 1) g_1/\varrho_1 - (\varrho_2 - 1) g_2/\varrho_2](\varrho_2 - 1)/\varrho_2\right] = 0.$$

The estimations given by these are

$$\hat{g}_1 = -\bar{a}\varrho_1 \cdot (\varrho_2 - 1)/(\varrho_1 - \varrho_2) + \bar{b}\varrho_1\varrho_2/(\varrho_1 - \varrho_2); \quad \bar{a} = {}^1/_4 \sum_{i=1}^{4} a_i; \quad \bar{b} = {}^1/_3 \sum_{j=1}^{3} b_j;$$

$$\hat{g}_2 = +\bar{a}\varrho_2 \cdot (\varrho_1 - 1)/(\varrho_1 - \varrho_2) - \bar{b}\varrho_1\varrho_2/(\varrho_1 - \varrho_2); \quad \hat{G} = \hat{g}_1 + \hat{g}_2;$$

$$\hat{W} = (1 - 1/\varrho_1)\hat{g}_1 + (1 - 1/\varrho_2)\hat{g}_2.$$

The mean error of the individual measurement is

$$m = \sqrt{\left\{\left[\sum_{i=1}^{4}(a_i - \bar{a})^2 + \sum_{j=1}^{3}(b_j - \bar{b})^2\right] \Big/ (7 - 2)\right\}}.$$

Because $\dfrac{\partial \hat{g}_1}{\partial a_i} = -{}^1/_4 \cdot \varrho_1 \cdot (\varrho_2 - 1)/(\varrho_1 - \varrho_2)$ and $\dfrac{\partial \hat{g}_1}{\partial b_j} = {}^1/_3 \cdot \varrho_1\varrho_2/(\varrho_1 - \varrho_2)$, the law of propagation of errors gives

$$m_{\hat{g}_1} = m\sqrt{[(1/4) \cdot \varrho_1^2((\varrho_2 - 1)^2/(\varrho_1 - \varrho_2)^2) + (1/3) \cdot \varrho_1^2\varrho_2^2/(\varrho_1 - \varrho_2)^2]};$$

$$m_{\hat{g}_2} = m\sqrt{[(1/4) \cdot \varrho_2^2(\varrho_1 - 1)^2/(\varrho_1 - \varrho_2)^2 + (1/3) \cdot \varrho_1^2\varrho_2^2/(\varrho_1 - \varrho_2)^2]}.$$

The numerical calculation with $\varrho_1 = 19.3$ g/cm³ and $\varrho_2 = 10.5$ g/cm³ gives:

$$\bar{a} = 4.01\text{ g}; \quad \bar{b} = 3.71\text{ g}; \quad \hat{g}_1 = 1.89\text{ g}; \quad \hat{g}_2 = 2.12\text{ g};$$

$$m = 0.02\text{ g}; \quad m_{\hat{g}_1} = 16.89m = 0.34\text{ g}; \quad m_{\hat{g}_2} = 17.20m = 0.34\text{ g}.$$

The weights given by the measurements are

$$g_1 \approx 1.89\text{ g}(\pm 0.34\text{ g}) \quad \text{and} \quad g_2 \approx 2.12\text{ g}(\pm 0.34\text{ g}).$$

Adjustment of relations

The relation between a quantity y and an independent variable x on which it depends is often given in the form of a *linear equation* $y = \alpha + \beta x$. The quantity y is observed for different values

$x_1, x_2, \ldots, x_n$ of the variable x. Hence the exact values of the quantities to be determined by the individual measurements are

$$y_i = \alpha + \beta x_i \qquad (i = 1, \ldots, n).$$

Let the measured values again be $a_1, a_2, \ldots, a_n$, and let $p_1, p_2, \ldots, p_n$ be the weights associated with them. Because the values $x_1, \ldots, x_n$ of the variable x are given, α and β are the only unknowns in this measuring process. The estimation for α and β can be made by adjustment of the data. As in the case of indirect observations, the interest lies not so much in finding the exact values y_i, but in determining the unknowns α and β. Because the unknowns occur as constants in a linear equation, one speaks of an *estimation of constants.* The quantity β is called a *coefficient of regression.*

In applying the method of least squares one again begins with the sum of squares of the errors,

$$S = (1/\sigma^2) \sum_{i=1}^{n} p_i(a_i - y_i) = (1/\sigma^2) \sum_{i=1}^{n} p_i(a_i - \alpha - \beta x_i)^2.$$

Equating to zero the partial derivatives of S with respect to the unknowns α and β leads to the normal equations

$$\frac{\partial S}{\partial \alpha} = -(2/\sigma^2) \sum_{i=1}^{n} p_i(a_i - \alpha - \beta x_i) = 0; \qquad \frac{\partial S}{\partial \beta} = -(2/\sigma^2) \sum_{i=1}^{n} p_i(a_i - \alpha - \beta x_i) \cdot x_i = 0.$$

In a notation due to GAUSS the normal equations assume the form:

Notation due to GAUSS	$\sum_{i=1}^{n} z_i = [z]$

$$\alpha[p] + \beta[px] = [pa], \quad \alpha[px] + \beta[px^2] = [pax].$$

Estimations of α and β in the regression equation	$\hat{\alpha} = \frac{[pa]}{[p]} - \beta \frac{[px]}{[p]}; \ \hat{\beta} = \frac{[pax]\,[p] - [pa]\,[px]}{[px^2]\,[p] - [px]^2}$

From these equations the estimations $\hat{\alpha}$ and $\hat{\beta}$ for the unknowns α and β can be found. From these one can also calculate the estimations $\hat{y}_i = \hat{\alpha} + \hat{\beta} x_i$ $(i = 1, \ldots, n)$ for the exact values of the quantities $y_1, y_2, \ldots, y_n$. The mean error for the individual measurements of weight $p = 1$ is given by

$$m = \sqrt{\left\{\left[\sum_{i=1}^{n} p_i(a_i - \hat{y}_i)^2\right] \Big/ (n-2)\right\}},$$

because two unknowns are to be determined from the observations. Then $m_i = m/\sqrt{p_i}$ is the mean error of a measurement of weight p_i. With the help of the law of propagation of errors one calculates

$$m_{\hat{\alpha}} = m \frac{1}{[p]} \sqrt{\left[[p] + \frac{[p]\,[px]^2}{[p]\,[px^2] - [px]^2}\right]} = m \sqrt{\left[\frac{[px^2]}{[p]\,[px^2] - [px]^2}\right]},$$

$$m_{\hat{\beta}} = m \sqrt{\left[\frac{[p]}{[p]\,[px^2] - [px]^2}\right]}.$$

Example: For ten plantations of pine trees of different ages the mean diameter x of the tree trunks and the mean height y are determined. The measured values, arranged in order of magnitude with respect to x (see table) have the same precision ($p_i = 1$; $i = 1, 2, \ldots, 10$). A graphical plotting

Measured values, say x in centimetres and y in metres

i	x_i	y_i
1	10.6	8.6
2	14.0	11.5
3	18.1	12.4
4	23.2	15.6
5	25.0	15.1
6	26.4	17.7
7	30.5	18.9
8	32.5	18.6
9	36.6	21.3
10	40.1	24.3

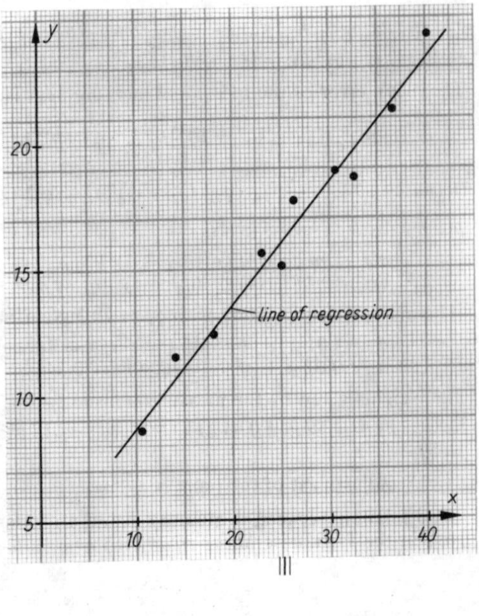

28.2-1 The mean height plotted against the mean diameter for ten pine plantations. Line of regression $\hat{y} = 3.837 + 0.4888x$

of the connection between diameter and height gives the representation of the dependence of y upon x (Fig.). This relation is to be described by means of a linear equation $y = \alpha + \beta x$.

From the measured values one calculates

$$[p] = 10; \quad [px] = 257.0; \quad [pa] = 164.0;$$
$$[px^2] = 7430.24; \quad [pxa] = 4618.26.$$

It follows that

$$\hat{\beta} = (4618.26 \cdot 10 - 164.0 \cdot 257.0)/(7430.24 \cdot 10 - 257.0^2) = 0.4888;$$
$$\hat{\alpha} = 164.0/10 - \hat{\beta} \cdot 257.0/10 = 3.837.$$

The line of regression $\hat{y} = 3.837 + 0.4888x$ is drawn in the figure. Further calculations give $\sum_{i=1}^{10} p_i(a_i - \hat{y}_i)^2 = 5.1522$, and the mean error of an individual measurement is $m = 0.8025$. From this the mean errors for $\hat{\alpha}$ and $\hat{\beta}$ are found to be $m_{\hat{\alpha}} = 0.7614$ and $m_{\hat{\beta}} = 0.0279$.

Adjustment of non-linear relations. A quantity y may depend linearly on several variables $z_0, z_1, z_2, \dots, z_k$. The representation for y then has the form

$$y = \beta_0 z_0 + \beta_1 z_1 + \beta_2 z_2 + \cdots + \beta_k z_k.$$

To determine the regression coefficients $\beta_0, \beta_1, \beta_2, \dots, \beta_k$, the quality y is measured for different values $z_{0i}, z_{1i}, z_{2i}, \dots, z_{ki}$ $(i = 1, \dots, n)$ of the variables on which it depends. The exact values of the quantities to be determined by the individual measurements are then

$$y_i = \beta_0 z_{0i} + \beta_1 z_{1i} + \beta_2 z_{2i} + \cdots + \beta_k z_{ki} \quad (i = 1, \dots, n).$$

Let the measured values be $a_1, a_2, \dots, a_n$, and let $p_1, p_2, \dots, p_n$ be the weights associated with them. If the number of measurements n is greater than the number of unknowns $k + 1$, then there are again surplus measurements available. The unknowns $\beta_0, \beta_1, \beta_2, \dots, \beta_k$ can be determined by the method of adjustment of data. Starting from the sum of the squares of the errors

$$S = (1/\sigma^2) \sum_{i=1}^{n} p_i(a_i - \beta_0 z_{0i} - \beta_1 z_{1i} - \cdots - \beta_k z_{ki})^2$$

one obtains the normal equations

$$\begin{aligned}
\beta_0[pz_0^2] + \beta_1[pz_0z_1] + \beta_2[pz_0z_2] + \cdots + \beta_k[pz_0z_k] &= [paz_0]\\
\beta_0[pz_1z_0] + \beta_1[pz_1^2] + \beta_2[pz_1z_2] + \cdots + \beta_k[pz_1z_k] &= [paz_1]\\
\vdots \qquad\qquad \vdots \qquad\qquad \vdots \qquad\qquad\qquad \vdots \qquad & \quad \vdots\\
\beta_0[pz_kz_0] + \beta_1[pz_kz_1] + \beta_2[pz_kz_2] + \cdots + \beta_k[pz_k^2] &= [paz_k].
\end{aligned}$$

One solves these equations for the unknowns and finds the estimations $\hat{\beta}_0, \hat{\beta}_1, \dots, \hat{\beta}_k$. With these one can calculate the estimations $\hat{y}_i = \hat{\beta}_0 z_{0i} + \cdots + \hat{\beta}_k z_{ki}$ $(i = 1, \dots, n)$ for the exact values of the measured quantities $y_1, y_2, \dots, y_n$. The mean error of the individual measurement of weight $p = 1$ is given by

$$m = \sqrt{\left\{\left[\sum_{i=1}^{n} p_i(a_i - \hat{y}_i)^2\right] \Big/ (n - k - 1)\right\}}.$$

With the help of m one can find in the usual way the mean error of an individual measurement of weight p_i and, by the law of propagation of errors, the mean error of the estimations $\hat{\beta}_0, \hat{\beta}_1, \dots, \hat{\beta}_k$.

If the quantity y can be expressed as a polynomial in x,

$$y = \beta_0 + \beta_1 x + \beta_2 x^2 + \cdots + \beta_k x^k,$$

then a non-linear relation exists between y and the variable x on which it depends. This is a special case of the linear dependence on several variables dealt with above. One need only put

$$\begin{aligned}
z_0 &= 1, & z_1 &= x, & z_2 &= x^2, \dots, z_k = x^k,\\
z_{0i} &= 1, & z_{1i} &= x_i, & z_{2i} &= x_i^2, \dots, z_{ki} = x_i^k \quad (i = 1, \dots, n)
\end{aligned}$$

and the determination of $\hat{\beta}_0, \dots, \hat{\beta}_k$ and of the mean error can proceed according to the formulae already given.

Example: In pine plantations with different mean trunk diameters d at a height of 1.30 m above the ground the average amount of wood V in each trunk is observed (see table). The determination of the amount of wood V_i is carried out with equal precision. The relation between V and d is to be adjusted by means of a cubical parabola. To simplify the calculation the variable $x = (d - 30)/5$ is introduced as a new independent variable. The quantity V is to be represented by the cubic equation $V = \beta_0 + \beta_1 x + \beta_2 x^2 + \beta_3 x^3$, whose coefficients $\beta_0, \beta_1, \beta_2, \beta_3$ are to be determined by adjustment of the observed data. The sum of the squares of the errors is

$$S = (1/\sigma^2) \sum_{i=1}^{9} (V_i - \beta_0 - \beta_1 x_i - \beta_2 x_i^2 - \beta_3 x_i^3)^2.$$

The normal equations have the form

$$9\cdot\beta_0+[x]\,\beta_1+[x^2]\,\beta_2+[x^3]\,\beta_3=[V],$$
$$[x]\,\beta_0+[x^2]\,\beta_1+[x^3]\,\beta_2+[x^4]\,\beta_3=[xV],$$
$$[x^2]\,\beta_0+[x^3]\,\beta_1+[x^4]\,\beta_2+[x^5]\,\beta_3=[x^2V],$$
$$[x^3]\,\beta_0+[x^4]\,\beta_1+[x^5]\,\beta_2+[x^6]\,\beta_3=[x^3V].$$

From the observed data one calculates

$$[x]=0,\quad [x^2]=60,\quad [x^3]=0,\quad [x^4]=708,$$
$$[x^5]=0,\quad [x^6]=9\,780,$$
$$[V]=8.382,\qquad [xV]=19.058,$$
$$[x^2V]=69.090,\quad [x^3V]=227.456.$$

The solutions of the normal equations are given by

$$\hat\beta_0=0.645\,40,\qquad \hat\beta_1=0.296\,348,$$
$$\hat\beta_2=0.042\,890,\qquad \hat\beta_3=0.001\,8039.$$

Table of the

i	observation d_i (cm)	observation V_i (m^3)	calculation x_i	calculation $\hat V_i$ (m^3)
1	10	0.030	−4	0.031
2	15	0.094	−3	0.093
3	20	0.213	−2	0.210
4	25	0.388	−1	0.390
5	30	0.643	0	0.645
6	35	0.987	+1	0.986
7	40	1.426	+2	1.424
8	45	1.969	+3	1.969
9	50	2.632	+4	2.632

Consequently the parabolic regression curve has the equation

$$V=0.645\,40+0.296\,348[(d-30)/5]+0.042\,890[(d-30)/5]^2+0.001\,8039[(d-30)/5]^3.$$

The adjusted values $\hat V_i$ are shown in the table.

Representation of a function with the help of simpler functions

The method of least squares is applied in mathematics not only for the adjustment of observations. If one wishes, for example, to find an approximation to a complicated function $y=f(x)$ in terms of simpler functions $\varphi_0(x), \varphi_1(x), \dots, \varphi_k(x)$, one can determine the coefficients $\beta_0, \beta_1, \dots, \beta_k$ in the linear expression $y=f(x)=\beta_0\varphi_0(x)+\beta_1\varphi_1(x)+\dots+\beta_k\varphi_k(x)$ by the method of least squares.

If the function $y=f(x)$ is known only at the discrete points x_i ($i=1, \dots, n$; $n>k$), $y_i=f(x_i)$, then one begins with the sum of the squares of the deviations

$$S=\sum_{i=1}^{n}[y_i-\beta_0\varphi_0(x_i)-\beta_1\varphi_1(x_i)-\dots-\beta_k\varphi_k(x_i)]^2.$$

If, on the other hand, the whole course of the function $y=f(x)$ is known in an interval $a\leqslant x\leqslant b$, then one begins with the integral of the square of the deviation

$$S=\int_a^b[f(x)-\beta_0\varphi_0(x)-\beta_1\varphi_1(x)-\dots-\beta_k\varphi_k(x)]^2\,dx.$$

In both cases the coefficients $\beta_0, \dots, \beta_k$ are determined in such a way that S assumes a minimum. Denoting the sum $\sum_{i=1}^{n}\varphi_j(x_i)\,\varphi_k(x_i)$, or the integral $\int_a^b\varphi_j(x)\,\varphi_k(x)\,dx$, as the case may be, by $[\varphi_j\varphi_k]$ and the sum $\sum_{i=1}^{n}y(x_i)\,\varphi_k(x_i)$ or the integral $\int_a^b y(x)\,\varphi_k(x)\,dx$ by $[y\varphi_k]$, one obtains the normal equations in the form

$$\begin{aligned}
\beta_0[\varphi_0^2]\ &+\beta_1[\varphi_0\varphi_1]+\beta_2[\varphi_0\varphi_2]+\dots+\beta_k[\varphi_0\varphi_k]=[y\varphi_0],\\
\beta_0[\varphi_1\varphi_0]&+\beta_1[\varphi_1^2]\ \ +\beta_2[\varphi_1\varphi_2]+\dots+\beta_k[\varphi_1\varphi_k]=[y\varphi_1],\\
&\ \vdots\\
\beta_0[\varphi_k\varphi_0]&+\beta_1[\varphi_k\varphi_1]+\beta_2[\varphi_k\varphi_2]+\dots+\beta_k[\varphi_k^2]\ \ =[y\varphi_k].
\end{aligned}$$

From this system of linear equations $\beta_0, \dots, \beta_k$ are to be calculated. The solution of the normal equations takes a particularly simple form if $[\varphi_i\varphi_k]$ has the value 0 for $i\neq k$, and 1 for $i=k$. The solutions are then $\beta_i=[y\varphi_i]$. This case occurs if the functions $\varphi_k(x)$ form an *orthonormal* (*normed orthogonal*) *system of functions*. The best known systems of orthogonal functions are the trigonometric functions $\sin n\varphi$, $\cos n\varphi$ ($n=0, 1, 2, 3, \dots$) and the Legendre polynomials.

28.3. Approximation theory

Every calculation with approximate values can be said to belong to approximation theory. In the stricter sense, however, one understands by approximation theory certain mathematical procedures that make it possible to replace complicated calculations by simpler ones. One must accept the fact that in this way one obtains not exact solutions, but only approximate ones. On the one hand, these *approximation methods* serve to save work in calculation. On the other hand, they enable one to obtain numerical solutions for very many mathematical problems by means of approximate methods alone; to obtain, for example, the numerical value of a given integral that cannot be expressed in a closed form, one must resort to approximate methods of integration. Approximation procedures have been worked out for widely different types of mathematical problems. Each approximation method must allow an estimation of the error perpetrated by its application.

Approximation methods for calculating the values of a function

The whole of numerical calculation takes place exclusively in the field of the four basic operations addition, subtraction, multiplication and division. If a more complicated mathematical function $f(x)$ is to be calculated, one must express it in such a way that only these four basic operations have to be applied. This is usually achieved by means of power series expansions (see Chapter 21.).

Asymptotic representations for large values of the argument. If one has to calculate values of a function $F(x)$ for very large values of the argument, one possibility is to obtain an approximation formula by substituting $z = 1/x$ in the function, and expanding the function $f(z) = F(1/z)$ in a Taylor series in $z = 1/x$. Because z is very small, the expansion can, in general, be restricted to a few terms. Another possibility is to determine for $F(x)$ an *asymptotic representation* or *asymptotic expansion*. A function $\varphi(x)$ is said to be an asymptotic representation of $F(x)$ if $\lim_{x\to\infty} [F(x) - \varphi(x)] = 0$. One then writes $F(x) \sim \varphi(x)$. If one sets $F(x) = \varphi(x) + R(x)$, the remainder term $R(x)$ must satisfy $\lim_{x\to\infty} R(x) = 0$. A function $\varphi(x)$ is often also called an asymptotic representation of $F(x)$, $F(x) \sim \varphi(x)$, if $F(x) = \varphi(x) \cdot [1 + r(x)]$, where $\lim_{x\to\infty} r(x) = 0$, that is, if the quotient $F(x)/\varphi(x)$ tends to 1 for large values of x.

Asymptotic representation of the Gaussian error integral. By two partial integrations one obtains for the Gaussian error integral the representation

$$\begin{aligned}\Phi(x) &= [1/\sqrt{(2\pi)}] \int_{-\infty}^{x} \exp[-t^2/2]\, dt = 1 - [1/\sqrt{(2\pi)}] \int_{x}^{\infty} \exp[-t^2/2]\, dt \\ &= 1 - [1/\sqrt{(2\pi)}] \{(1/x) \exp[-x^2/2] - \int_{x}^{\infty} (1/t^2) \exp[-t^2/2]\, dt\} \\ &= 1 - [1/\sqrt{(2\pi)}] \{(1/x) \exp[-x^2/2] - (1/x^3) \exp[-x^2/2] + \int_{x}^{\infty} (3/t^4) \exp[-t^2/2]\, dt\}.\end{aligned}$$

With the first three terms one has already found a good asymptotic representation $\Phi(x) \approx 1 - [1/\sqrt{(2\pi)}] \{(1/x) \exp[-x^2/2] - (1/x^3) \exp[-x^2/2]\}$. The remainder term $R_3(x) = -[1/\sqrt{(2\pi)}] \int_{x}^{\infty} (3/t^4) \exp[-t^2/2]\, dt$ can be estimated from

$$|R_3(x)| \leqslant [1/\sqrt{(2\pi)}] \cdot (3/x^5) \int_{x}^{\infty} t \exp[-t^2/2]\, dt = [1/\sqrt{(2\pi)}] \cdot (3/x^5) \exp[-x^2/2].$$

It tends to zero as $x \to \infty$. Even for $x = 2$ the error of the asymptotic formula is $< 5 \cdot 10^{-3}$.

The Euler sum formula. If the function $F(x)$ for which an asymptotic representation is required can be represented as a sum $F(x) = f(1) + f(2) + \cdots + f(x-1) + f(x)$, where $f(z)$ is a given function and x a positive integer, then an asymptotic representation can be obtained from the Euler sum formula.

Euler sum formula	$F(x) = \int_{1}^{x} f(t)\, dt + \frac{1}{2} [f(x) + f(1)] + \sum_{k=1}^{n} \frac{B_{2k}}{(2k)!} [f^{(2k-1)}(x) - f^{(2k-1)}(1)] + R_n(x)$

In this formula the B_{2k} are the *Bernoulli numbers* of which the first few are $B_2 = 1/6$, $B_4 = -1/30$, $B_6 = 1/42$, $B_8 = -1/30$, $B_{10} = 5/66$, $B_{12} = -691/2730$ (see Chapter 21.). The remainder $R_n(x)$ can be estimated by $|R_n(x)| \leqslant [4/(2\pi)^{2n}] \int_{1}^{x} |f^{(2n)}(t)|\, dt$. If $R_n(x)$ tends to zero as $x \to \infty$, then by neglecting the remainder in the Euler sum formula one has already found an asymptotic representation for $F(x)$. If, however, $R_n(x)$ tends to a limit C_n as $x \to \infty$, then an asymptotic representation is obtained by replacing $R_n(x)$ by the limit C_n in the Euler sum formula.

Asymptotic representation for the factorial function $x!$ Taking the natural logarithm of $x!$ one has $F(x) = \ln x! = \ln 1 + \ln 2 + \cdots + \ln x$. Retaining only the terms up to the first derivative ($n = 1$) in the Euler sum formula one obtains

$$\begin{aligned}F(x) &= \int_{1}^{x} \ln t\, dt + (1/2)(\ln x + \ln 1) + (1/6) \cdot (1/2)(1/x - 1) + R_1(x) \\ &= x \ln x - x + \ln x/2 + 1/(12x) + 1 - 1/12 + R_1(x).\end{aligned}$$

The remainder $R_1(x)$ can be estimated by $|R_1(x)| \leqslant (1/\pi^2)(1 - 1/x^2)$. A more precise investigation gives $\lim_{x\to\infty} R_1(x) = C_1 = 1/12 - 1 + \ln \sqrt{(2\pi)}$. Replacing $R_1(x)$ by this limit in the Euler sum formula, one obtains the asymptotic representation

$$F(x) \approx (x + 1/2) \ln x - x + 1/(12x) + \ln \sqrt{(2\pi)}.$$

From this it follows that

$$x! = e^{F(x)} \approx \sqrt{(2\pi x)}\, x^x \exp\,[-x + 1/(12x)].$$

For very large values of x the term $1/(12x)$ in the exponential can be neglected, and then Stirling's formula is obtained.

Stirling's formula	$x! \approx \sqrt{(2\pi x)}\, x^x e^{-x}$

Approximation of functions by means of polynomials

If for a function $y = f(x)$ the function values at the arguments $x = x_0$, $x = x_1, \ldots, x = x_n$ are known, these points are called *basic points* and the corresponding function values $y_0 = f(x_0)$, $y_1 = f(x_1), \ldots, y_n = f(x_n)$ are called *basic values*. The problem consists in calculating the function value $y = f(x)$ for an arbitrary value of x lying between two adjacent basic points. If the exact determination of $y = f(x)$ involves very extensive calculations, then one tries to calculate the required function value y approximately from the known function values $y_0, y_1, \ldots, y_n$, as one says, to determine it by interpolation.

Linear interpolation. One becomes acquainted with the simplest interpolation procedure in working with angle functions or logarithms, when values are to be determined that lie between those given in the table. The values given in the table are then the basic values and the required value is usually found by *linear interpolation*. For this one needs only two basic points x_0 and x_1 with the basic values $y_0 = f(x_0)$, $y_1 = f(x_1)$. The value $y = f(x)$ is required, where $x_0 < x < x_1$. By linear interpolation one finds from the ratio equation $(y - y_0)/(x - x_0) = (y_1 - y_0)/(x_1 - x_0)$ the value $y = y_0 + (x - x_0)(y_1 - y_0)/(x_1 - x_0)$. Thus, one replaces the arc of the function $y = f(x)$ in the interval (x_0, x_1) by a straight line approximation that passes through the points (x_0, y_0) and (x_1, y_1) (Fig.).

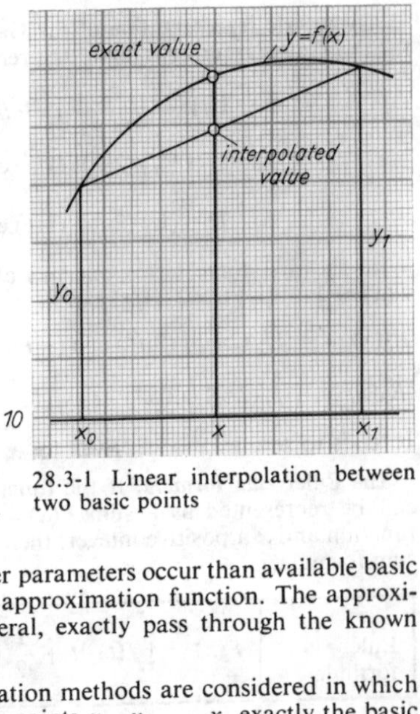

28.3-1 Linear interpolation between two basic points

Interpolation in the wider sense. In general, all interpolation methods consist in replacing the function $y = f(x)$ in a neighbourhood of the basic points $x_0, x_1, \ldots, x_n$ by simpler functions that are best possible approximations to the function $y = f(x)$ in this neighbourhood. One method of determining such approximation functions is the *method of least squares*. Fourier analysis and smoothing of data, among other procedures, rest upon this. They can be applied if fewer parameters occur than available basic points and have to be determined for the specification of the approximation function. The approximation functions determined in this way do not, in general, exactly pass through the known basic values $(x_0, y_0), (x_1, y_1), \ldots, (x_n, y_n)$.

Interpolation in the stricter sense. In what follows interpolation methods are considered in which the approximation functions for $y = f(x)$ assume at the basic points $x_0, x_1, \ldots, x_n$ exactly the basic values $y_0, y_1, \ldots, y_n$. Because polynomials are the simplest functions available, one tries to find a polynomial approximation to the function $y = f(x)$ in a neighbourhood of $x_0, x_1, \ldots, x_n$. From algebra it is well known that exactly one polynomial $P_n(x)$ of degree n passes through $n + 1$ given points $(x_0, y_0), (x_1, y_1), \ldots, (x_n, y_n)$. This polynomial is chosen as an approximation function for $y = f(x)$. There are various methods for determining it. They all lead to the same polynomial $P_n(x)$; its external form varies, but this is the only difference between the different interpolation methods.

Polynomial form. If the polynomial is written in the form $P_n(x) = a_0 + a_1 x + a_2 x^2 + \cdots + a_n x^n$ with undetermined coefficients $a_0, a_1, \ldots, a_n$ and with the condition that it passes through the points $(x_0, y_0), (x_1, y_1), \ldots, (x_n, y_n)$, then the adjacent equations must be satisfied.

$$\begin{aligned} y_0 &= a_0 + a_1 x_0 + a_2 x_0^2 + \cdots + a_n x_0^n \\ y_1 &= a_0 + a_1 x_1 + a_2 x_1^2 + \cdots + a_n x_1^n \\ &\vdots \\ y_n &= a_0 + a_1 x_n + a_2 x_n^2 + \cdots + a_n x_n^n \end{aligned}$$

These are $n + 1$ equations for the determination of $a_0, a_1, \ldots, a_n$. They have a unique solution if the basic points $x_0, x_1, \ldots, x_n$ are all distinct.

Example: The function $y = \sqrt{x}$ is to be approximated by a polynomial of degree 2 that passes through the points $(x_0 = 1,\ y_0 = 1)$, $(x_1 = 1.21,\ y_1 = 1.1)$ and $(x_2 = 1.44,\ y_2 = 1.2)$. Let $P_2(x) = a_0 + a_1 x + a_2 x^2$. From the adjacent equations one finds that $a_0 = 0.4099$, $a_1 = 0.6842$, $a_2 = -0.0941$. Substitution of the value $x = 1.3$ in the interpolation polynomial $y = \sqrt{x} \approx 0.4099 + 0.6842x - 0.0941x^2$ gives $\sqrt{1.3} \approx 1.1403$. The exact value for $\sqrt{1.3}$ is 1.140 175.

$$\begin{aligned} 1.0 &= a_0 + a_1 + a_2 \\ 1.1 &= a_0 + 1.21a_1 + 1.4641a_2 \\ 1.2 &= a_0 + 1.44a_1 + 2.0736a_2 \end{aligned}$$

This is a simple example of an *inverse interpolation* in a table of squares. Although the form assumed in this method is very simple, the final determination of the interpolation polynomial requires a considerable amount of calculation, especially if a large number of basic points is to be taken into account. For this reason LAGRANGE and NEWTON chose the form of the polynomial $P_n(x)$ rather differently and hence arrived at formulae that are simpler for calculation.

Lagrange's interpolation polynomial. LAGRANGE begins by assuming the form

$$P_n(x) = L_0(x)\, y_0 + L_1(x)\, y_1 + \cdots + L_n(x)\, y_n$$

for the approximating polynomial, in which the coefficients $L_i(x)$ of the basic values y_i are polynomials in x of degree n. They are calculated from the basic points x_j $(j = 0, 1, \ldots, n)$ alone and take the values $L_i(x_j)$ at these points. The polynomial approximation $P_n(x)$ certainly passes exactly through the points (x_0, y_0), (x_1, y_1), ..., (x_n, y_n) if the polynomials $L_i(x)$ can be determined in such a way that $L_i(x_j)$ has the value 1 for $i = j$, and 0 for $i \neq j$. The polynomials defined by Lagrange satisfy this condition.

Lagrange's interpolation polynomial and Lagrange polynomials

$$y = f(x) \approx P_n(x) = L_0(x)\, y_0 + L_1(x) y_1 + \cdots + L_n(x)\, y_n$$

$$L_i(x) = \frac{(x - x_0)(x - x_1) \cdots (x - x_{i-1})(x - x_{i+1}) \cdots (x - x_n)}{(x_i - x_0)(x_i - x_1) \cdots (x_i - x_{i-1})(x_i - x_{i+1}) \cdots (x_i - x_n)}; \quad i = 0, 1, \ldots, n$$

If one of the values $x_0, x_1, \ldots, x_{i-1}, x_{i+1}, \ldots, x_n$ is substituted for x, then there is always just one factor of the numerator that vanishes; but for $x = x_i$ the numerator is equal to the denominator. Introducing these polynomials into the form assumed for $P_n(x)$ one obtains Lagrange's interpolation polynomial.

Example: The parabolic approximation for the function $y = \sqrt{x}$, passing through the points $(x_0 = 1,\ y_0 = 1)$, $(x_1 = 1.21,\ y_1 = 1.1)$, $(x_2 = 1.44,\ y_2 = 1.2)$ is to be determined with the help of Lagrange's interpolation method. From the adjacent Lagrange polynomials the polynomial approximation is

$$L_0(x) = \frac{(x - x_1)(x - x_2)}{(x_0 - x_1)(x_0 - x_2)} = \frac{(x - 1.21)(x - 1.44)}{0.0924}$$

$$L_1(x) = \frac{(x - x_0)(x - x_2)}{(x_1 - x_0)(x_1 - x_2)} = \frac{(x - 1.0)(x - 1.44)}{-0.0483}$$

$$L_2(x) = \frac{(x - x_0)(x - x_1)}{(x_2 - x_0)(x_2 - x_1)} = \frac{(x - 1.0)(x - 1.21)}{0.1012}$$

$$\begin{aligned} P_2(x) = {} & (x - 1.21)(x - 1.44)(1.0/0.924) - (x - 1.0)(x - 1.44)(1.1/0.483) \\ & + (x - 1.0)(x - 1.21)(1.2/0.1012); \end{aligned}$$

$$\begin{aligned} P_2(x) = {} & (x - 1.21)(x - 1.44) \cdot 10.8225 - (x - 1.0)(x - 1.44) \cdot 22.7743 \\ & + (x - 1.0)(x - 1.21) \cdot 11.8577. \end{aligned}$$

In this form the polynomial can already be used for numerical calculations. Multiplying out the brackets and collecting together terms in like powers of x one finds again the polynomial determined earlier $P_2(x) = 0.4099 + 0.6842x - 0.0941x^2$.

Newton's interpolation polynomial. If a polynomial of degree n passing through the basic points (x_0, y_0), (x_1, y_1), ..., (x_n, y_n) has already been determined with the help of Lagrange's formula, and a further basic point (x_{n+1}, y_{n+1}) is added, then if one wishes to determine a polynomial approximation of degree $n + 1$ passing through all $n + 2$ basic points, in applying the Lagrange formula the whole calculation begins anew. All the Lagrange polynomials $L_0(x), L_1(x), \ldots, L_n(x)$ must be calculated again. For *Newton's interpolation method*, on the other hand, in this case only one extra term has to be added. The method starts by assuming the form

$$P_n(x) = b_0 + b_1(x - x_0) + b_2(x - x_0)(x - x_1) + \cdots + b_n(x - x_0)(x - x_1) \cdots (x - x_{n-1})$$

for the polynomial approximation. The coefficients $b_0, b_1, \dots, b_n$ are again determined in such a way that the polynomial passes through the points $(x_0, y_0), (x_1, y_1), \dots, (x_n, y_n)$. Substituting the values $x_0, x_1, \dots, x_n$ for x in Newton's formula one obtains the *stepped system of equations*

$$\begin{aligned} y_0 &= b_0 \\ y_1 &= b_0 + b_1(x_1 - x_0) \\ y_2 &= b_0 + b_1(x_2 - x_0) + b_2(x_2 - x_0)(x_2 - x_1) \\ &\vdots \\ y_n &= b_0 + b_1(x_n - x_0) + b_2(x_n - x_0)(x_n - x_1) + \cdots + b_n(x_n - x_1) \cdots (x_n - x_{n-1}). \end{aligned}$$

This system can be solved step-by-step for $b_0, b_1, \dots, b_n$. Using the so-called divided differences (see Chapter 29.2.) one can, beginning with $b_0 = y_0$, give a formula for each coefficient b_i.

From the second, third, and following equations one finds step-by-step:

$$b_1 = (y_1 - y_0)/(x_1 - x_0) = [x_1 x_0];$$
$$y_2 = y_0 + [x_1 x_0](x_2 - x_0) + b_2(x_2 - x_0)(x_2 - x_1)$$

or

$$[x_2 x_0] = [x_1 x_0] + b_2(x_2 - x_1)$$

and finally

$$b_2 = [x_2 x_1 x_0], \dots, \qquad b_k = [x_k x_{k-1} \cdots x_1 x_0].$$

Substituting these coefficients in the assumed formula one obtains Newton's interpolation polynomial.

Newton's interpolation polynomial	$y = f(x) \approx y_0 + [x_1 x_0](x - x_0) + [x_2 x_1 x_0](x - x_0)(x - x_1) + \cdots + [x_n x_{n-1} \cdots x_2 x_1 x_0](x - x_0)(x - x_1) \cdots (x - x_{n-1})$

If a further basic point (x_{n+1}, y_{n+1}) is to be taken into account, then one simply introduces into the polynomial already calculated the term $[x_{n+1} x_n x_{n-1} \cdots x_2 x_1 x_0](x - x_0)(x - x_1) \cdots (x - x_n)$ and thus obtains a polynomial of degree $n + 1$ that passes through all the points $(x_0, y_0), (x_1, y_1), \dots, (x_n, y_n), (x_{n+1}, y_{n+1})$.

The *decreasing divided differences* (singly underlined in 29.2.) are used in Newton's formula. In deriving it, however, it is not essential that the basic points are taken in the order $x_0, x_1, x_2, \dots, x_n$. If they are arranged in an arbitrary order $x_{i_0}, x_{i_1}, \dots, x_{i_n}$ and the procedure described is applied, then Newton's interpolation polynomial in the general form is obtained.

$$y = f(x) \approx P_n(x) = y_{i_0} + (x - x_{i_0})[x_{i_1} x_{i_0}] + (x - x_{i_0})(x - x_{i_1})[x_{i_2} x_{i_1} x_{i_0}] + \cdots + (x - x_{i_0})(x - x_{i_1}) \cdots (x - x_{i_{n-1}})[x_{i_n} x_{i_{n-1}} \cdots x_{i_1} x_{i_0}]$$

Arranging the basic points in the order $x_n, x_{n-1}, \dots, x_1, x_0$, one obtains

$$y = f(x) \approx P_n(x) = y_n + (x - x_n)[x_{n-1} x_n] + (x - x_n)(x - x_{n-1})[x_{n-2} x_{n-1} x_n] + \cdots + (x - x_n)(x - x_{n-1}) \cdots (x - x_1)[x_0 x_1 \cdots x_n].$$

This formula uses the *increasing divided differences* (doubly underlined in the scheme). The formula can also be rearranged in such a way that the *central divided differences* (in the middle of the scheme) are used to form the polynomial approximation. Whatever form one chooses for the representation, one always obtains the uniquely determined polynomial of degree n that passes through the points $(x_0, y_0), (x_1, y_1), \dots, (x_n, y_n)$.

Example: To determine the *parabolic approximation for the function* $y = \sqrt{x}$ passing through the points $(x_0 = 1, y_0 = 1), (x_1 = 1.21, y_1 = 1.1), (x_2 = 1.44, y_2 = 1.2)$ by means of Newton's interpolation method, one first calculates the divided differences.

$x_{i+2} - x_i$	$x_{i+1} - x_i$	x_i	y_i	Δ	$[x_{i+1} x_i]$	Δ	$[x_2 x_1 x_0]$
		1	1				
	0.21			0.1	0.476190		
0.44		1.21	1.1			−0.041408	−0.0941
	0.23			0.1	0.434782		
		1.44	1.2				

When *decreasing* divided differences are used, Newton's interpolation polynomial is

$$P_2(x) = 1 + (x - 1) \cdot 0.476190 - (x - 1)(x - 1.21) \cdot 0.0941;$$

when *increasing* divided differences are used, it is

$$P_2(x) = 1.2 + (x - 1.44) \cdot 0.434782 - (x - 1.44)(x - 1.21) \cdot 0.0941$$

or with *central* divided differences

$$P_2(x) = 1.1 + (x - 1.21) \cdot 0.434782 - (x - 1.21)(x - 1.44) \cdot 0.0941 .$$

Collecting together terms containing like powers of x, all three cases lead to the previously determined polynomial

$$P_2(x) = 0.4099 + 0.6842x - 0.0941x^2 .$$

Equidistant basic points. If the basic points $x_0, x_1, \ldots, x_n$ are equidistant (with spacing h), then in Newton's interpolation polynomial

$$P_n(x) = y_0 + (x - x_0)[x_1 x_0] + \cdots + (x - x_0)(x - x_1) \cdots (x - x_{n-1})[x_n x_{n-1} \cdots x_1 x_0],$$

which passes through the points $(x_0, y_0), \ldots, (x_n, y_n)$, the divided differences can be expressed in terms of the simple differences if one puts $x = x_0 + t\,h$ (see Chapter 29.2.). In the scheme introduced there, the differences used here lie on a diagonal line falling from left to right.

Newton's interpolation formula with decreasing differences
$P_n(x_0 + t\,h) = y_0 + t\,\Delta^1 y_{1/2} + \dfrac{t(t-1)}{2!}\Delta^2 y_1 + \cdots + \dfrac{t(t-1)(t-2)\cdots(t-n+1)}{n!}\Delta^n y_{n/2}$

If one also takes account of basic points $x_{-1} = x_0 - h,\ x_{-2} = x_0 - 2h, \ldots, x_{-n} = x_0 - nh$ and determines the Newton interpolation polynomial through the points $(x_0, y_0), (x_{-1}, y_{-1}), \ldots, (x_{-n}, y_{-n})$, one obtains

$$P_n(x) = y_0 + (x - x_0)[x_0 x_{-1}] + (x - x_0)(x - x_{-1})[x_0 x_{-1} x_{-2}] + \cdots + (x - x_0)(x - x_{-1}) \ldots (x - x_{-n+1}) \cdot [x_0 x_{-1} \ldots x_{-n}].$$

In this polynomial too the divided differences can be replaced by simple differences.

Newton's interpolation formula with increasing differences
$P_n(x_0 + t\,h) = y_0 + t\,\Delta^1 y_{-1/2} + \dfrac{t(t+1)}{2!}\Delta^2 y_{-1} + \cdots + \dfrac{t(t+1)(t+2)\cdots(t+n-1)}{n!}\Delta^n y_{-n/2}$

Finally, the basic points can be arranged in the alternating sequence $x_0, x_1, x_{-1}, x_2, x_{-2}, x_3, \ldots$ The corresponding Newton interpolation polynomial is

$$P_n(x) = y_0 + (x - x_0)[x_1 x_0] + (x - x_0)(x - x_1)[x_1 x_0 x_{-1}] + (x - x_0)(x - x_1)(x - x_{-1})[x_2 x_1 x_0 x_{-1}] + \cdots.$$

According as the number of basic points available is even ($n = 2k$) or odd ($n = 2k + 1$), the polynomial ends with the term

$$(x - x_0)(x - x_1)(x - x_{-1}) \ldots (x - x_{k-1})[x_k x_{k-1} \ldots x_0 \ldots x_{-k+1}]$$

or

$$(x - x_0)(x - x_1)(x - x_{-1}) \ldots (x - x_k)[x_k x_{k-1} \ldots x_0 \ldots x_{-k+1} x_{-k}].$$

If the divided differences are replaced by the ordinary differences, one obtains Gauss's interpolation formula.

Gauss's interpolation formula
$P_n(x_0 + t\,h) = y_0 + t\,\Delta^1 y_{1/2} + \dfrac{t(t-1)}{2!} \cdot \Delta^2 y_0 + \dfrac{t(t-1)(t+1)}{3!} \cdot \Delta^3 y_{1/2} + \dfrac{t(t-1)(t+1)(t-2)}{4!} \cdot \Delta^4 y_0 \ldots$

This uses the differences that stand near the middle of the difference scheme. There is a second Gaussian formula in which the basic points are arranged in the order $x_0, x_{-1}, x_1, x_{-2}, x_2, \ldots$ Stirling, Laplace, Bessel, Everett and others have given further interpolation formulae, obtained by various choices of the basic points and by suitable combinations of the formulae of Newton and Gauss.

Interpolation in tables

If the basic points $x_0, x_1, \dots, x_n$ are arranged in order of magnitude $x_0 < x_1 < \dots < x_n$, and if the function $y = f(x)$ is replaced in the interval $x_0 \dots x_n$ by an interpolation polynomial $P_n(x)$ of degree n that passes through the points $(x_0, y_0), (x_1, y_1), \dots, (x_n, y_n)$, then the approximate error is determined by the remainder term R_{n+1}. The quantity ξ is, in general, an unknown value

$$R_{n+1}(x) = f(x) - P_n(x) = (x - x_0)(x - x_1) \dots (x - x_n)\,[y^{n+1}(\xi)/(n+1)!]$$

in the interval (x_0, x_n). For example, if one determines by linear interpolation the function value $y = f(x)$ for an argument x lying between two tabulated values x_0 and $x_0 + h$ (h the spacing of the table), then the interpolation error is given by $R_2(x) = (x - x_0)(x - x_0 - h)\, y''(\xi)/2$. The product $(x - x_0)(x - x_0 - h)$ has its greatest absolute value for $x = x_0 + h/2$. The interpolation error can therefore be estimated by $|R_2(x)| \leqslant (h^2/8)\,|y''(x)|$, in which the maximum value of $|y''(x)|$ in the interval (x_0, x_1) is to be substituted. By interpolation in a k-figure table the interpolation error should not exceed a half unit of the kth place. Hence:

In a k-figure table of spacing h a linear interpolation is permissible if $(h^2/8)\,|y''(x)| < 0.5 \cdot 10^{-k}$.

Example: A five-figure table for $\lg \sin x$ $(0° \leqslant x \leqslant 45°)$ has spacing $h = 0.01°$. Because $y''(x) = -M/\sin^2 x$, where $M = \lg e$, x must satisfy the inequality $(h^2/8)(M/\sin^2 x) < 0.5 \cdot 10^{-5}$, where h is in radians. From this it follows that $\sin x > 0.01819$. This condition is satisfied for $x > 1.04°$. Consequently one can interpolate linearly in this table for $x > 1°$.

If linear interpolation in a table is not permissible, then interpolation formulae of higher order must be applied. In these one can keep the interpolation error determined by means of the remainder term small if one chooses the basic points in such a way that the interpolation takes place somewhere near the middle of the region encompassed by the basic points. This is achieved with interpolation formulae that work with central differences, such as Gauss's interpolation formulae. Only if one has to interpolate at the beginning or at the end of a table, where the central differences are not available, then one falls back on Newton's interpolation formula with decreasing or increasing differences.

29. Numerical analysis

Mathematics in the strict sense uses the continuum of the real and complex numbers for its quantitative statements and also relations between numbers or objects such as vectors or matrices that are based on this number system. *In numerical mathematics* all statements must be arrived at with the aid of rational numbers and as a rule, for example, when a digital computer is used, with only finitely many. In numerical analysis the formulation of a procedure for the solution of a mathematical problem therefore requires the setting up of a model, and this gives rise to rounding errors and errors of procedure.

29.1. Introduction

Rounding errors arise in the mapping of the real numbers that occur in a given procedure into the domain of permissible rational numbers. In doing this, not only the initial data of the problem, but also the intermediate results after every step in the calculation are falsified. *Errors of procedure* occur because every transcendental operation must be replaced by a finite chain of realizable operations, such as addition, subtraction, multiplication and division. The model procedure is *numerically stable* if its quantitative result differs from that of the exact procedure by only a specified small amount.

Estimation of accuracy of numerical procedures. The estimation of the accuracy of numerical procedure is a very important practical problem, but is not easy to solve. Its solution depends very much on the type of problem in question. In applications, above all two classes of problems can be distinguished: approximation of a mathematical object that can be described by means of a formula, and approximation of a mathematical object that is known to exist, but can be determined only approximately by means of measurements.

The problem of accuracy for mathematical objects given by formulae. Numbers, vectors, functions, functionals and operators are mathematical objects given by formulae. Numbers, vectors or functions are regarded as points of a space and approximated by sequences or, what amounts to the same thing, by series $x = \sum_{i=0}^{\infty} c_i e_i$ with coordinate elements e_i and components c_i.

In numerical analysis every *expansion* of this type must be broken off after finitely many approximating steps, that is, instead of x one is satisfied with the approximation $x^* = \sum_{i=1}^{n} c_i e_i$. To this there belong two *quality criteria* Q_1 and Q_2: $Q_1 = n$ is a measure of the extent of the expansion, and Q_2 a measure of the closeness of the approximation, for example, $|x - x^*|$ for numbers or $\sup |x(t) - x^*(t)|$ for functions.

One is interested in keeping Q_1 as well as Q_2 small. The smallness of Q_1, however, contradicts that of Q_2, and vice versa. If the coefficients c_i are calculated by means of a definite rule, for example, by the Taylor series expansion, then for a given $Q_1 = n$ there is no room to play with for Q_2, but Q_2 is fixed by the element x to be approximated. Q_2, however, is not known, and one must therefore be satisfied in practice with a more or less accurate estimate for Q_2, for example, with estimates for the remainder term in series expansions.

More favourable is the situation in which c_i is not obtained by a definite rule, but is determined for a given $Q_1 = n$ in such a way that Q_2 becomes minimal. A rule for this determination represents an *optimal expansion process.* In this case one usually finds the rule that $Q_{2\,\min}(n)$ is a monotonic decreasing function.

If one considers, for example, orthogonal series expansions, whose coordinate elements e_i with a suitable moment operation M satisfy the orthogonality requirement $M(e_i e_j) = 0$ for $i \neq j$, and if one chooses $Q_2 = M[(x - x^*)(x - x^*)]$ as a measure of the approximation, then it is required to find values of c_i that guarantee a minimum of Q_2 for a given n. For this one obtains

$c_i = M(xe_i)/M(e_i e_i)$ and the minimal value is given by $Q_{2\,\min} = M(xx) - \sum_{i=1}^{n} M(xe_i)^2/M(e_i e_i)$.

Obviously $Q_{2\,\min}$ is a monotonic decreasing function of n. For numerical series one takes the moment operation $M(xy) = x \cdot y$, for vector series the scalar product $M(xy) = (x, y)$, for random variables the expectation operation, and for functions the scalar product $M(xy) = \int_a^b x(t)\, y(t)\, p(t)\, dt$ in the function space. In the case of functionals or operations given by formulae, the estimation of accuracy is made difficult by the fact that an approximation to the given operation is required that produces almost the same effect as that operation for a comprehensive set of original elements y.

If for the given operation (functional or operator) the relation $x = Fy$ holds, then one obtains for the approximating operation F^* the corresponding relation $x^* = F^*y$. The usual procedure is the approximation of F by a linear combination $F^* = \sum_{i=0}^{n} c_i f_i$ of certain linear basic operations f_i. One can proceed here in a similar way as for the expansion of elements x; the difficulty lies solely in the fact that one has to eliminate the dependence on y in a suitable manner. This can be achieved by an additional averaging over y or by the method of moments, where now $\sup_{y \in Y} M((x - x^*)^2)$ is made a minimum. However, one comes up against the same difficulties here as in the Chebyshev approximation, that is, the approximation of functions by uniformly convergent series.

The accuracy problem for given mathematical objects that have to be approximated by measurements. Measurements necessarily always provide incomplete information about the objects to be determined. One must attempt to produce from the measurements the best possible approximation of the mathematical object. This is often the case for problems involving a search for extreme values, or for the problem of approximating an operation F over the measurements x_i for a certain set of original elements y_i.

In these problems two phases must be distinguished, the *learning phase* and the *execution phase*. In the learning phase measurements, which are subject to a criterion of effort, are carried out to a certain extent. From the results of these measurements one has to decide, in the best possible manner, an approximation to the mathematical object, usually by means of an *estimation of parameters*. The expectation operation M then converts the measured values obtained into estimates c_i^* of the parameters c_i. In addition to the measure of effort Q_1 and the measure of the approximation Q_2, formulated with the operation M, there now enters a measure Q_3, which estimates the randomness of the c_i^* on the basis of the choice of tests that have led to the measured values.

In this connection one speaks of the *design of experiments* if one endeavours to carry out the trial tests in such a way that the random errors introduced by the estimation of c_i^* are as small as possible. The execution phase has the character of an extrapolation. Here the model of the mathematical object with the estimated values c_i^* is applied under arbitrary admissible conditions. If strong deviations occur from the $Q_{2\,\min}$ that was obtained in the learning phase, then one can try by subsequent learning phases to improve step-by-step the model so far achieved. Such a procedure is of special significance for statistical methods of numerical analysis, for example, for *regression analysis*.

Representation of numbers. In a *positional system* with base $q > 0$ a real number z is represented in the form

$$z = \pm(a_k q^k + a_{k-1} q^{k-1} + \cdots + a_0 q^0 + a_{-1} q^{-1} + a_{-2} q^{-2} + \cdots),$$

where each of the numbers a_i is one of the non-negative integers $0, 1, \ldots, q - 1$, and in which the *integerpart* with exponents $l \geqslant 0$ of q is distinguished from the *fractional part with* $l < 0$. In digital computers only numbers of *word length* L can be represented, with negative powers up to the maximum order L. For *floating point representation* the normal form is used,

$$z = \pm q^e(b_{-1} q^{-1} + b_{-2} q^{-2} + \cdots + b_{-L} q^{-L}) \quad \text{with} \quad b_{-1} \neq 0.$$

With $q = 10$, for example, the number -36.12 is represented as

$$-10^2(3 \cdot 10^{-1} + 6 \cdot 10^{-2} + 1 \cdot 10^{-3} + 2 \cdot 10^{-4}) = -10^2 \cdot 0.3612.$$

In this the exponent e varies, as a rule, between $-L$ and $+L$ and is also represented in the positional system with the basis q. With automatic calculating machines negative exponents can be avoided by using, instead of the external number in normed form, the internal number $z' = \pm q^{e+L}(b_{-1} q^{-1} + \cdots + b_{-L} q^{-L})$, whose exponent is too high by an amount L. For a given e, an equidistant grid of $2 \cdot (q^L - 1)$ numbers can thus be realized; its grid distance is q^{e-L}. The totality of all realizable numbers is given by $-L \leqslant e \leqslant +L$. The basic arithmetical operations may lead to a departure from the region of admissible numbers. Once the basis q has been agreed, it is sufficient to state the sequence of numbers $z = \pm a_k a_{k-1} \cdots a_0 \cdot a_{-1} a_{-2} \cdots a_{-L}$ or the sequence of numbers consisting of the exponent e and those of the normed fractional part.

Example: The decimal number 132[10] is converted by division by 2[10]: $132/2 = 66 + 0/2$ with $b_0 = 0$; $66/2 = 33 + 0/2$ with $b_1 = 0$; $33/2 = 16 + 1/2$ with $b_2 = 1$; $16/2 = 8 + 0/2$ with $b_3 = 0$; $8/2 = 4 + 0/2$ with $b_4 = 0$; $4/2 = 2 + 0/2$ with $b_5 = 0$; $2/2 = 1 + 0/2$ with $b_6 = 0$ and $1/2 = 0 + 1/2$ with $b_7 = 1$, so that for 132 one obtains the *binary number* 10000100. To obtain the decimal number again from this by conversion one divides by 10[2] = 1010. This gives

$$\begin{array}{lll}
10000100/1010 = 1101 + 10/1010 \quad \text{with} \quad b_0 = 10[2] = 2[10]; & & \\
\underline{-1010} & & \\
\quad 1101 & 1101/1010 = 1 + 11/1010 \quad \text{with} \quad b_1 = 11[2] = 3[10]; & \\
\underline{\;-1010} & \underline{-1010} & \\
\quad 1100 & \quad\;\; 11 \quad 1/1010 = 0 + 1/1010 \quad \text{with} \quad b_2 = 1[2] = 1[10], & \\
\underline{\;-1010} & & \\
\qquad 10 & \text{that is, } b_2 \cdot 10^2 + b_1 \cdot 10 + b_0 \cdot 10^0 = 132. &
\end{array}$$

For a real number z simple recursive procedures give the binary representation of its integer part int (z) and of its fractional part frac (z). Assume that

$$\text{int}(z) = r_k \cdot 2^k + r_{k-1} \cdot 2^{k-1} + \cdots + r_1 \cdot 2 + r_0$$

with $r_i = 0$ or $r_i = 1$. Write w for int (z). Then one obtains $r_0 = w - 2 \cdot \text{int}(w/2)$. *Now write w for int (w/2). With this "new" value w one obtains* $r_1 = w - 2 \cdot \mathit{int}(w/2)$. Repeating this renaming and calculation procedure recursively gives th sequences $r_0, r_1, \ldots, r_k$.

The fractional part can be written as

$$\text{frac}(z) = s_1/2 + s_2/2^2 + \cdots + s_k/2^k + \text{rem}_{k+1}/2^{k+1}$$

with $s_i = 0$ or $s_i = 1$ and a $(k + 1)$st remainder rem_{k+1}.

Again write u for frac (z). Then one obtains $s_1 = \text{int}(2 \cdot u)$. *Now write u for $2 \cdot u - \mathit{int}(2 \cdot u)$. With this "new" value u one obtains* $s_2 = \mathit{int}(2 \cdot u)$. Repeating this procedure recursively gives the sequence $s_1, s_2, \ldots, s_k$.

Besides the *decimal system* with $q = 10$, other positional systems are used. The *binary system* with $q = 2$ has the advantage for computers that only two physical states, which are denoted by 0 and 1, are required for the representation. By *conversion* it is possible to change a number $z[q] = a_k q^k + a_{k-1} q^{k-1} + \cdots + a_0 q^0$ from the q-representation into the p-representation $z[p] = b_l p^l + b_{l-1} p^{l-1} + \cdots + b_0 p^0$, on dividing $z[q]$ by $p[q]$. One can see from $z[p]$ that this gives an integer g_1 and a fractional part $r_1/p[q]$, to which the value $b_0 p^0$ corresponds in the representation $z[p]$, so that $r_1 = b_0$. Similarly from $g_1/p[q]$ one obtains the coefficients b_1, and then $b_2, b_3, \ldots, b_l$.

Interval calculus. In order to determine the inaccuracies in the results of the basic arithmetical operations when the initial data are rounded off, MOORE developed the interval calculus; every number z is replaced by the smallest closed rational interval $[a, b]$ in which it must lie. As a generalization of the operations on numbers, which in every possible case can be visualized on the number line, one obtains the *arithmetical operations on intervals.* If $z_1 = [a, b]$ and $z_2 = [c, d]$, then $z_1 + z_2 = [a + c, b + d]$ and, since $-[c, d] = [-d, -c]$, $z_1 - z_2 = [a - d, b - c]$ (Fig.). For operations of the second kind one obtains

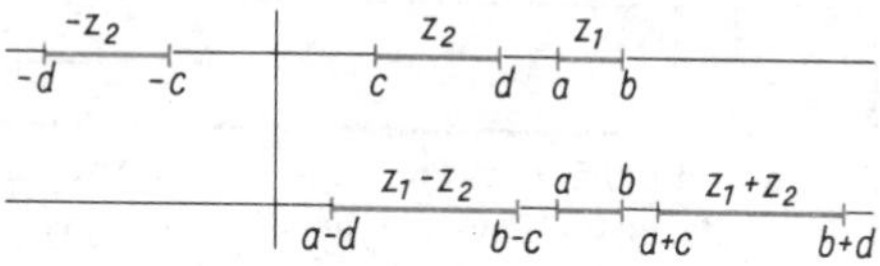

29.1-1 Intervals for $z_1 + z_2$, for $-z_2$ and for $z_1 - z_2$

$z_1 \cdot z_2 = [a, b] \cdot [c, d] = [\min(ac, ad, bc, bd), \max(ac, ad, bc, bd)]$

and $z_1/z_2 = [a, b]/[c, d] = [\min(a/c, a/d, b/c, b/d), \max(a/c, a/d, b/c, b/d)]$. In this way *interval functions* can be defined, in particular, rational interval functions, which must replace other functions in the numerical model.

Example: The power function $f(x) = x^k$ with positive integral exponent k is defined as the k-fold product of x with itself; thus, one obtains for $x = [x_1, x_2]$ as interval for the power function: $x^k = [x_1, x_2]^k = [\min(x_1^k, x_1^{k-1}x_2, x_1^{k-2}x_2^2, \ldots, x_2^k), \max(x_1^k, x_1^{k-1}x_2, \cdots, x_2^k)]$.

29.2. Interpolation and calculus of differences

The basic idea of interpolation is to replace a function $f(x)$, of which the values $y_i = f(x_i)$ are given at finitely many points $x_1, x_2, \ldots, x_n$ or possibly the derivatives $y_i^{(j)} = f^{(j)}(x_i)$ up to the order $m_i \geqslant j$, by an approximation consisting of a superposition $\sum_j A_j \varphi_j(x) = f^*(x) \approx f(x)$ of standard functions $\varphi_j(x)$. The form of the functions $\varphi_j(x)$ for a given class of functions and the *superposition coefficients* A_j must be uniquely determined from the given values in such a way that the replacement function $f^*(x)$ assumes the values y_i or $y_i^{(j)}$, $i = 1, \ldots, n$, at the given points $\{x_i\}$. For points other than those of the set $\{x_i\}$ the *goodness of the interpolation formula* is given by an estimation of the remainder $R = f(x) - f^*(x)$. If x lies in the interior of the smallest interval that contains the set $\{x_i\}$, $i = 1, \ldots, n$, then one speaks of a proper *interpolation*, and if x lies outside this interval, of an *extrapolation*.

Many numerical procedures can be carried out more readily by the superposition of standard functions, for example, the determination of roots, integration, differentiation, or integration of differential equations. As standard functions *polynomials* are often used. The Taylor and Lagrange interpolations are two limiting cases of practical importance.

Taylor interpolation. The value of the function $f(x_1)$ and the values of the derivatives $f^{(j)}(x_1)$ with $m_1 \geqslant j$ are given at only one point x_1. The *standard polynomials* are $\varphi_j(x) = (x - x_1)^j/j!$ for $j = 0, 1, \ldots, m_1$ and the *superposition coefficients* are $A_j = f^{(j)}(x_1)$. The remainder is $R = f^{(m_1+1)}(\xi) \cdot (x - x_1)^{m_1+1}/(m_1 + 1)!$, where ξ is a point between x_1 and x (see Chapter 21.).

Example: The Taylor interpolation of the function $\sin x$ at the point $x = 0$ consists of finitely many terms of its Taylor series expansion $\sin x = x - x^3/3! + x^5/5! - \ldots$

Lagrangian interpolation. At the points $x_1, x_2, \ldots, x_n$ only the values of the function $y_i = f(x_i)$ are given. The *standard polynomials* are $\varphi_j(x) = \dfrac{\Phi_n(x)}{(x - x_j) \cdot \Phi_n'(x_j)}$ with $\Phi_n(x) = \prod_{i=1}^{n} (x - x_i)$, and the *superposition coefficients* are $A_j = y_j = f(x_j)$. The remainder is $R = \dfrac{f^{(n)}(\xi)}{n!} \Phi_n(x)$, where ξ is a point in the smallest interval that contains the set $\{x_i\}$, $i = 1, \ldots, n$.

Divided differences. If for a function $y = f(x)$ the values at the $n + 1$ *basic points* $x_0, x_1, \ldots, x_n$, $y_0 = f(x_0), \ldots, y_n = f(x_n)$ are given, then divided differences – also called *gradients* – of orders 0 to n can be determined:

0. $[x_0] = y_0,\ [x_1] = y_1, \ldots,\ [x_n] = y_n$;
1. $[x_i x_j] = (y_i - y_j)/(x_i - x_j)$, for example, $[x_1 x_0] = (y_1 - y_0)/(x_1 - x_0)$,
   $[x_n x_{n-1}] = (y_n - y_{n-1})/(x_n - x_{n-1})$;
2. $[x_i x_j x_k] = ([x_i x_j] - [x_j x_k])/(x_i - x_k)$, for example,
   $[x_2 x_1 x_0] = ([x_2 x_1] - [x_1 x_0])/(x_2 - x_0)$;

$(r+1)$. $[x_i x_j x_{k_1} x_{k_2} \cdots x_{k_r}] = ([x_i x_j x_{k_1} \cdots x_{k_{r-1}}] - [x_j x_{k_1} \cdots x_{k_r}])/(x_i - x_{k_r})$.

$$n.\quad [x_n x_{n-1} \cdots x_0] = \frac{[x_n x_{n-1} \cdots x_1] - [x_{n-1} x_{n-2} \cdots x_0]}{x_n - x_0}$$

All divided differences are symmetric in their arguments; for example,

$$[x_i x_j] = (y_i - y_j)/(x_i - x_j) = (y_j - y_i)/(x_j - x_i) = [x_j x_i],$$
$$[x_i x_j x_k] = ([x_k x_j] - [x_j x_i])/(x_k - x_i) = [x_k x_j x_i],$$

and similarly it can be shown that

$$[x_i x_j x_k] = [x_i x_k x_j] = [x_j x_i x_k] = [x_j x_k x_i] = [x_k x_i x_j].$$

One can therefore rearrange the arguments in divided differences in an arbitrary way. For calculating divided differences the following *gradient scheme* is used.

Scheme for the calculation of divided differences

			x_0	y_0					
		$x_1 - x_0$			$y_1 - y_0$	$\underline{[x_1 x_0]}$			
	$x_2 - x_0$		x_1	y_1			$[x_2 x_1] - [x_1 x_0]$	$\underline{[x_2 x_1 x_0]}$	
…		$x_2 - x_1$			$y_2 - y_1$	$[x_2 x_1]$			…
	$x_3 - x_1$		x_2	y_2			$[x_3 x_2] - [x_2 x_1]$	$[x_3 x_2 x_1]$	
		$x_3 - x_2$			$y_3 - y_2$	$[x_3 x_2]$			…
⋮	⋮		x_3	y_3			⋮	⋮	⋮
…		⋮			⋮	⋮			…
			⋮	⋮					
	$x_n - x_{n-2}$						$[x_n x_{n-1}] - [x_{n-1} x_{n-2}]$	$\underline{\underline{[x_n x_{n-1} x_{n-2}]}}$	
		$x_n - x_{n-1}$			$y_n - y_{n-1}$	$\underline{\underline{[x_n x_{n-1}]}}$			
			x_n	$\underline{\underline{y_n}}$					

Single underlining indicates the value of *decreasing divided differences*, double underlining those of *increasing divided differences*, and in the middle of the scheme lie the *central divided differences*.

Example: $y = x^3$; basic points: $x_0 = 1$, $x_1 = 3$, $x_2 = 4$. Gradient scheme:

$x_{i+2} - x_i$	$x_{i+1} - x_i$	x_i	y_i	Δ	$[x_{i+1} x_i]$	Δ	$[x_2 x_1 x_0]$
		1	$\underline{1}$				
	2			26	$\underline{13}$		
3		3	27			24	$\underline{8}$
	1			37	37		
		4	64				

Result: $[x_0] = 1$, $[x_1 x_0] = 13$, $[x_2 x_1 x_0] = 8$.

Properties of the divided differences. If $f(x) = f_1(x) + f_2(x)$ and one denotes the divided differences of the functions $f_1(x)$ and $f_2(x)$ by attaching the suffixes 1 and 2, then

$$[x_n x_{n-1} \ldots x_0] = [x_n x_{n-1} \ldots x_0]_1 + [x_n x_{n-1} \ldots x_0]_2 .$$

For a function $f(x) = c f_1(x)$, where c is a constant,

$$[x_n x_{n-1} \ldots x_0] = c[x_n x_{n-1} \ldots x_0]_1 .$$

For divided differences an independent expression can be given that does not presuppose the formation of divided differences of lower order:

$$[x_n x_{n-1} \ldots x_0] = \sum_{i=0}^{n} \frac{f(x_i)}{(x_i - x_0) \cdots (x_i - x_{i-1})(x_i - x_{i+1}) \cdots (x_i - x_n)} .$$

If the function $f(x)$ is n times continuously differentiable in an interval containing the basic points $x_0, x_1, \ldots, x_n$, then its divided differences can be expressed in terms of the derivatives of the function: $[x_n x_{n-1} \cdots x_0] = f^{(n)}(\xi)/n!$. Here ξ denotes a suitable point in the interval containing the basic points. It follows from this that all divided differences of the nth order of a polynomial function of the nth degree are equal.

Difference table for equidistant basic points. The scheme for the calculation of divided differences becomes particularly simple if the basic points $x_0, x_1, \ldots, x_n$, arranged in order of magnitude, are chosen to be *equidistant*. Then for a given *spacing* of width h, $x_1 = x_0 + h$, $x_2 = x_0 + 2h, \ldots,$ $x_n = x_0 + nh$. For the differences of the arguments in the left-hand part of the scheme one obtains $x_{i+k} - x_i = kh$. Further, if one introduces

the first difference $y_{i+1} - y_i = \Delta^1 y_{i+1/2}$,
the second difference $\Delta^1 y_{i+1} - \Delta^1 y_i = \Delta^2 y_{i+1/2}$,
$\vdots$
the nth difference $\Delta^{n-1} y_{i+1} - \Delta^{n-1} y_i = \Delta^n y_{i+1/2}$,

then a simple relation exists between these ordinary differences and the divided differences. The adjacent formula holds.

$$[x_{i+k}x_{i+k-1} \ldots x_{i+1}x_i] = \frac{1}{h^k} \cdot \frac{1}{k!} \cdot \Delta^k y_{i+k/2}$$

It follows from this that not only all the nth divided differences, but also all the nth ordinary differences of a polynomial function of degree n are equal among themselves.

If an auxiliary variable t is introduced by the equation $x = x_0 + th$ and if the basic points $x_{-1} = x_0 - h$, $x_{-2} = x_0 - 2h, \ldots$ preceding the basic point x_0 are also taken into account, then one obtains the following *difference table*.

Difference table

t	$x = x_0 + th$	y	Δ^1	Δ^2	Δ^3	Δ^4	Δ^5	Δ^6
$\vdots$	$\vdots$	$\vdots$	$\vdots$	$\vdots$	$\vdots$	$\vdots$	$\vdots$	$\vdots$
-2	x_{-2}	y_{-2}		$\Delta^2 y_{-2}$				
			$\Delta^1 y_{-3/2}$		$\Delta^3 y_{-3/2}$			
-1	x_{-1}	y_{-1}		$\Delta^2 y_{-1}$		$\Delta^4 y_{-1}$		
			$\Delta^1 y_{-1/2}$		$\Delta^3 y_{-1/2}$		$\Delta^5 y_{-1/2}$	
0	x_0	y_0		$\Delta^2 y_0$		$\Delta^4 y_0$		$\Delta^6 y_0$
			$\Delta^1 y_{1/2}$		$\Delta^3 y_{1/2}$		$\Delta^5 y_{1/2}$	
1	x_1	y_1		$\Delta^2 y_1$		$\Delta^4 y_1$		
			$\Delta^1 y_{3/2}$		$\Delta^3 y_{3/2}$			
2	x_2	y_2		$\Delta^2 y_2$				
$\vdots$	$\vdots$	$\vdots$	$\vdots$	$\vdots$	$\vdots$	$\vdots$	$\vdots$	$\vdots$

In calculating the difference table one begins with a table of the basic values, forms the difference series for the function values y_i, and thus obtains the first differences. For these one again forms the difference series to obtain the second differences, and so on.

Aitken's interpolation. The *recursive interpolation process* due to AITKEN can be applied with advantage if the point x at which the function $f(x)$ is to be interpolated is given. By linear interpolation (Fig.) one first determines $h_{i,i+1}$ in such a way that $h_{i,i+1}(x_{i+1} - x_i) = y_i(x_{i+1} - x) + y_{i+1}(x - x_i)$, that is, by

$$h_{i,\,i+1} = \frac{1}{(x_{i+1} - x_i)} \det \begin{vmatrix} y_i & (x_i - x) \\ y_{i+1} & (x_{i+1} - x) \end{vmatrix}.$$

By adding a further point x_{i+2} one arrives at a polynomial $h_{i,\,i+1,\,i+2}$ of the second degree for the interpolation and hence to a higher accuracy

$$h_{i,\,i+1,\,i+2} = \frac{1}{(x_{i+2} - x_i)} \det \begin{vmatrix} h_{i,\,i+1} & (x_i - x) \\ h_{i+1,\,i+2} & (x_{i+2} - x) \end{vmatrix}.$$

One proceeds in this way as required until the values of successive approximations differ only by an amount lying within the limits of accuracy that are in any case to be expected.

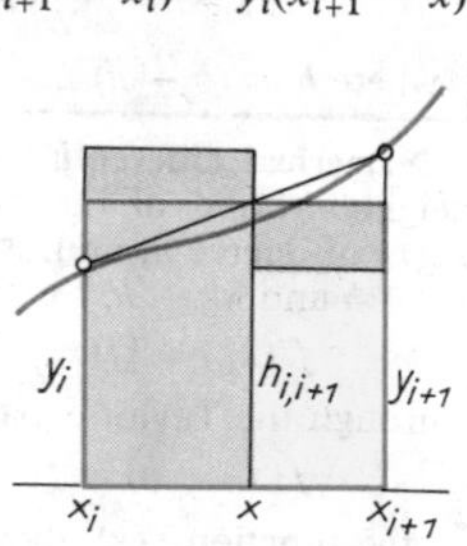

29.2-1 Aitken's interpolation; $h_{i,i+1}(x_{i+1} - x_i) = y_i(x_{i+1} - x) + y_{i+1}(x - x_i)$

29.3. Numerical models for integration and differentiation

Numerical integration. By a quadrature formula one understands a model with values of the given function $f(x)$ or its

Quadrature formula	$\int_a^b f(x)\,dx \approx \sum_{i,j} A_{ij} f^{(j)}(x_{ij})$

derivatives at basic points x_{ij} in which the superposition coefficients A_{ij} or the basic points x_{ij} are to be determined by the demands on the goodness of the model. In a quadrature formula of *amplitude type*, the basic points are given and the A_{ij} are to be determined; in one of *argument type*, on the other hand, the A_{ij} are given and suitable basic points are to be determined. The basic points need not belong to the interval of integration.

Example: A quadrature formula of the third order that provides exact values for polynomials up to the third degree is $\int_a^b f(x)\,dx \approx [(b-a)/2]\{f(a)+f(b)-[(b-a)/6]\,[f'(b)-f'(a)]\}$.

Interpolation quadrature formulae are obtained by replacing the function to be integrated in the interval of integration $[a, b]$ by a Lagrangian interpolation polynomial of the nth degree with the basic points $x_i = a + ih$, $i = 0, 1, 2, \ldots, n$. By the goodness requirement that the quadrature formula shall be exact for every polynomial of the nth degree, one obtains the model

$$\int_a^b f(x)\,dx \approx \sum_{i=1}^{n} A_i f(a+ih) \quad \text{with} \quad A_i = \frac{b-a}{n}\cdot\frac{(-1)^{n-i}}{i!\,(n-i)!}\int_0^n \frac{t^{[n+1]}}{t-i}\,dt,$$

where $t^{[n+1]} = t(t-1)(t-2)\ldots(t-n)$. If $n = 2m-1-d$, with $d = 0$ for odd values of n and $d = 1$ for even values of n, then the error of the model amounts to

$$R_n = -M_n(b-a)^{2m+1} f^{(2m)}(\xi), \quad \text{where} \quad a < \xi < b$$

and

$$M_1 \approx 8.333\cdot 10^{-2}, \quad M_2 \approx 3.472\cdot 10^{-4}, \quad M_3 \approx 1.543\cdot 10^{-4},$$
$$M_4 \approx 5.167\cdot 10^{-7}, \quad M_5 \approx 2.910\cdot 10^{-7}, \quad M_6 \approx 6.379\cdot 10^{-10},$$
$$M_7 \approx 3.912\cdot 10^{-10}, \quad M_8 \approx 5.133\cdot 10^{-13}.$$

The trapezoidal rule and Simpson's rule are often used.

Trapezoidal rule.

$$\int_a^b f(x)\,dx = [(b-a)/n]\,[f(0)/2 + f(h) + f(2h) + \cdots + f((n-1)\,h) + f(nh)/2] - [(b-a)^3/(12n^2)]\cdot f''(\xi),$$

or for $n = 1$ in the case of linear interpolation between the ends of the interval:

$$\int_a^b f(x)\,dx \approx [(b-a)/2]\,[f(a)+f(b)] - [(b-a)^3/12]\,f''(\xi).$$

Simpson's rule for even n.

$$\int_a^b f(x)\,dx = [(b-a)/(3n)]\,\{f(0) + f(nh) + 2[f(2h) + f(4h) + \cdots + f((n-2)\,h)] + 4[f(h) + f(3h) + \cdots + f((n-1)\,h)]\} - [(b-a)^5/(180n^4)]\,f^{(4)}(\xi),$$

or for $n = 2$ and interpolation by means of a quadratic polynomial between the ends a and b of the interval:

$$\int_a^b f(x)\,dx \approx (h/3)\,[f(a) + 4f(a+h) + f(b)] - h^5\cdot f^{(4)}(\xi)/90,$$

where $h = (b-a)/2$.

Numerical differentiation. As for numerical integration, one interpolates the function $f(x)$ in a neighbourhood of the point x_0 at which $f'(x_0)$ is to be formed by an interpolation polynomial $P_n(x)$ of degree n, and uses the model $f'(x_0) \approx P_n'(x_0)$. For linear interpolation between the points $x_0 - h$ and $x_0 + h$, for example, one obtains

$$f'(x_0) \approx [f(x_0+h) - f(x_0-h)]/(2h).$$

Through the Taylor expansion

$$f(x_0+h) = f(x_0) + f'(x_0)\,h + f''(x_0)\,(h^2/2!) + \cdots + f^{(n)}(x_0)\,(h^n/n!) + \cdots = e^{(hd/dx)}f(x_0)$$

of the function $f(x)$ about the point x_0 one obtains the following universally applicable model of the *differential operator*

$$\frac{d}{dx} \approx (1/h)\ln(1 + h\Delta) = (1/h)\,[(h\Delta) - (h\Delta)^2/2 + (h\Delta)^3/3 - (h\Delta)^4/4 + \cdots];$$

where Δ is the difference operator with $\Delta f(x) = [f(x+h) - f(x)]/h$. Truncating this series after the nth power of the operator Δh results in a model which for polynomials up to and including degree n gives exact values for the derivative; for it turns out to be identical with the model obtained by the substitution $f'(x_0) \approx P_n'(x_0)$.

Example: For $P(x) = a_0 + a_1 x + a_2 x^2$, $(h\Delta)\,P(x) = P(x+h) - P(x) = a_1 h + a_2(2xh + h^2)$ and $(h\Delta)^2 = a_1 h + a_2[2(x+h)\,h + h^2] - a_1 h - a_2(2xh + h^2) = 2a_2 h^2$. Hence the differentiation operator becomes

$$(1/h)\,[(h\Delta) - (h\Delta)^2/2]\,P(x) = a_1 + 2a_2 x + a_2 h - a_2 h = a_1 + 2a_2 x = P'(x).$$

Numerical solution of ordinary differential equations

Many practical problems require the integration of a differential equation $y' = f(x, y)$ with a continuous function $f(x, y)$, starting from the initial value $y(x_0) = y_0$ and proceeding in the direction of x increasing (or the integration of a system of such equations). With $x_i = x_0 + ih$ one determines approximately the values $y_i = y(x_i)$. In many numerical methods for the solution the integral equation $y(x) = y_0 + \int_{x_0}^{x} f(t, y(t))\,dt$ equivalent to the differential equation is used.

Adams' method. For the sequence of solutions $\{y_i\}$ the integral equation gives the increment $y_{i+1} - y_i = \int_{x_i}^{x_{i+1}} f(t, y(t))\,dt$; in this the function $f(t, y(t))$ is replaced by a *Newton's interpolation polynomial* of degree n and the required integration is performed exactly.

The *Adams' interpolation formula* uses the basic points $x_i, x_{i-1}, \ldots, x_{i-n}$

The *Adams' extrapolation formula* uses the basic points $x_{i+1}, x_i, \ldots, x_{i-n+1}$

and with $f_i = f(x_i, y_i)$ and $\Delta f_i = f_{i+1} - f_i$ one obtains

(interpolation formula)

$$y_{i+1} - y_i = h \sum_{r=0}^{n} B_r \Delta^r f_{i-r}, \quad \text{where } B_0 = 1,$$

$$B_r = \frac{1}{r!} \int_0^1 u(u+1)(u+2)\ldots(u+r-1)\,du;$$

for example, $B_1 = 0.5$; $B_2 = 0.41$; $B_3 = 0.375$; $B_4 = 0.3486$; $B_5 = 0.32986$.

(extrapolation formula)

$$y_{i+1} - y_i = h \sum_{r=0}^{n} B'_r \Delta^r f_{i+1-r},$$

where $B'_r = B_r - B_{r-1}$.

To start this process one has to work first of all with interpolation polynomials of lower order, or

(interpolation formula) the starting values $y_0, y_1, \ldots, y_n$ must already be known with sufficient accuracy.

(extrapolation formula) the starting values $y_0, y_1, \ldots, y_{n-1}$ must be known. The subsequent value is given by a transcendental equation which can be solved by *iteration.*

Runge-Kutta method. In the integration of the differential equation $y' = f(x, y)$ the interpolation theory or the difference calculus require a finite set $\{y_i\}$ ofstored *backward values.* The Runge-Kutta method, on the other hand, uses only the properties of the continuous function $f(x, y)$; hence it requires a smaller store and moreover, apart from providing an independent solution, can also be used to calculate starting values for other methods. Furthermore, the *spacing* h can be changed during the course of the calculation, and *numerical stability* is more readily secured than in the case of methods based on the calculus of differences. In the Runge-Kutta method an *approximate value* $\tilde{y}_{i+1}$ is calculated for

$$y_{i+1} = y_i + \int_{x_i}^{x_i+h} f(t, y(t))\,dt, \quad \text{where} \quad \tilde{y}'_{i+1} = y_i + k$$

and the increment k is obtained via the intermediate steps:

$$k_0 = 0, \qquad k_j = hf(x_i + a_j h, y_i + \sum_{r=0}^{j-1} b_{jr} k_r) \quad \text{for} \quad j = 1, 2, \ldots, r$$

and $k = \sum_{j=1}^{r} g_j k_j$. The parameters a_j, b_{jr} and g_j contained in these steps are determined by the goodness required by the method. It is customary to demand that $\varphi(h) = \int_{x_i}^{x_i+h} f(t, y(t))\,dt - \sum_{j=1}^{r} g_j k_j$ possesses for $h = 0$ vanishing derivatives $\varphi(0) = \varphi'(0) = \cdots = \varphi^{(l)}(0)$ of as high an order l as possible, and l is called the *order of the Runge-Kutta method.*

The Runge-Kutta method is an improvement of the *Euler method* $\tilde{y}'_{i+1} = y_1 + hf(x_i, y_i)$, which has an unfavourable error propagation.

Examples of proven Runge-Kutta methods of order 4:

1. $\tilde{y}'_{i+1} = y_i + (1/6)(k_1 + 2k_2 + 2k_3 + k_4)$, where
$k_1 = hf(x_i, y_i)$, $k_2 = hf(x_i + h/2, y_i + k_1/2)$,
$k_3 = hf(x_i + h/2, y_i + k_2/2)$, $k_4 = hf(x_i + h, y_i + k_3)$.

2. $\tilde{y}'_{i+1} = y_i + (1/3)(k_1/2 + 3k_2/2 + k_3/2 + k_4/2)$, where
$k_1 = hf(x_i, y_i)$, $k_2 = hf(x_i + h/2, y_i + k_1/2)$,
$k_3 = hf(x_i + h/2, y_i - k_1/2 + k_2)$, $k_4 = hf(x_i + h, y_i + k_2/2 + k_3/2)$.

Determination of roots

From the range of values of a function $y = f(x)$ one obtains indications of those arguments x for which $f(x) = 0$. Accordingly if one determines an interpolation polynomial, then its roots are approximate values for the required roots. One tries to use these values as a step in the estimation by an iterative process.

Method of false position. Two different points $y_1 = f(x_1)$ and $y_2 = f(x_2)$ are connected by a straight line, which serves as an interpolation polynomial of the first degree (Fig.). From $(x_2 - x)/(x_2 - x_1) = (y_2 - y)/(y_2 - y_1)$ one obtains
$P_1(x) = y = y_1(x - x_2)/(x_1 - x_2) + y_2(x - x_1)/(x_2 - x_1)$ with the root x' for $P_1(x) = 0$.

Method of false position	$x' = \dfrac{y_1x_2 - y_2x_1}{y_1 - y_2}$

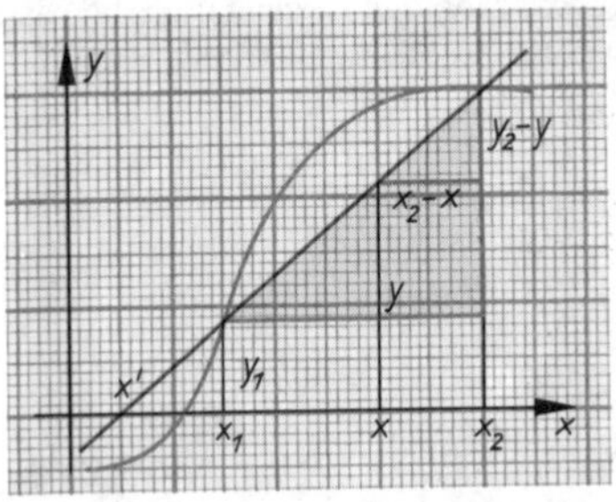

29.3-1 Method of false position

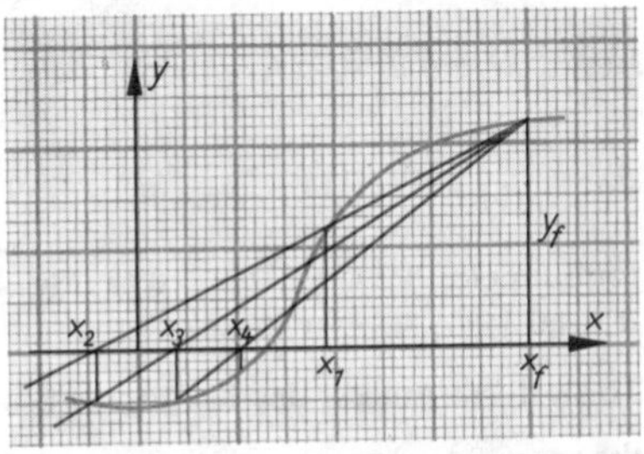

29.3-2 Method of false position: fixed point method

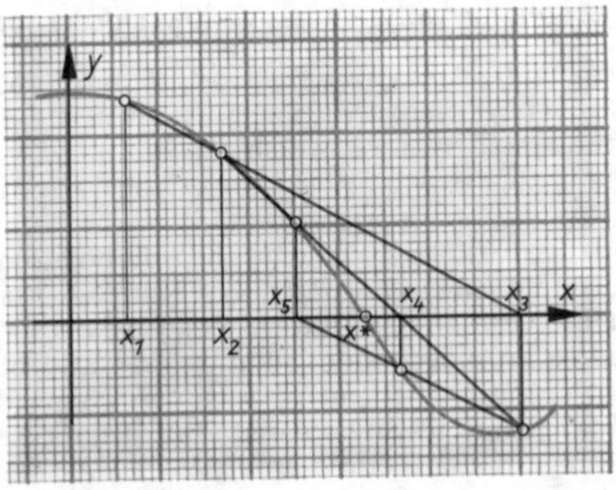

29.3-3 Method of false position: secant method

If one determines $y' = f(x')$ by substitution, then iterative processes can be found. According to the *fixed point method* (Fig.) one fixes a pair of values $x_2 = x_f$ and $y_2 = y_f$ as the fixed point. Writing $x_1 = x_i$, $y_1 = y_i$ and $x' = x_{i+1}$, $y' = y_{i+1}$, then $x_{i+1} = (x_f y_i - x_i y_f)/(y_i - y_f)$. According to the *secant method* one writes $x_2 \to x_{i-1}$, $y_2 \to y_{i-1}$, $x_1 \to x_i$, $y_1 \to y_i$ and $x' \to x_{i+1}$ and obtains $x_{i+1} = (y_i x_{i-1} - y_{i-1}x_i)/(y_i - y_{i-1}) = [x_{i-1}f(x_i) - x_i f(x_{i-1})]/[f(x_i) - f(x_{i-1})]$. This process (Fig.) converges very rapidly to x^*, where $f(x^*) = 0$, provided that the following conditions are satisfied for $f(x)$: from the lower bound m_1 of $f'(x)$ and the upper bounds M_1 of $|f'(x)|$ and M_2 of $|f''(x)|$ one calculates $K = M_2M_1^2/(2m_1^3)$, and the inequalities $K|x^* - x_0| < 1$ and $K|x^* - x_1| < 1$ must then hold.

Example: For the root $x^* = 2.094\,551\,481\,542\,3\ldots$ of the function $f(x) = x^3 - 2x - 5$ one obtains by the fixed point method with $x_f = 2$ and $x_1 = 3$ the following approximations:

$x_2 = 2.058\,823\,529\,4$; $x_3 = 2.096\,558\,636\,2$; $x_4 = 2.094\,440\,519\,3$;
$x_5 = 2.094\,557\,621\,8$; $x_6 = 2.094\,551\,139\,9$; $x_7 = 2.094\,551\,500\,6$.

Newton's method. As a Newton's interpolation polynomial of the first degree a straight line is fixed by a point $y_0 = f(x_0)$ on the graph of the function $f(x)$ and by the gradient y'_0 of the tangent at that point. From the equation of the tangent $y = P_1(x) = y_0 + y'_0(x - x_0)$ an estimation x' of the zero can be made.

Newton's method	$x' = x_0 - y_0/y'_0$

Iterative processes can be derived from this. In the case of *Newton's method with fixed gradient*,

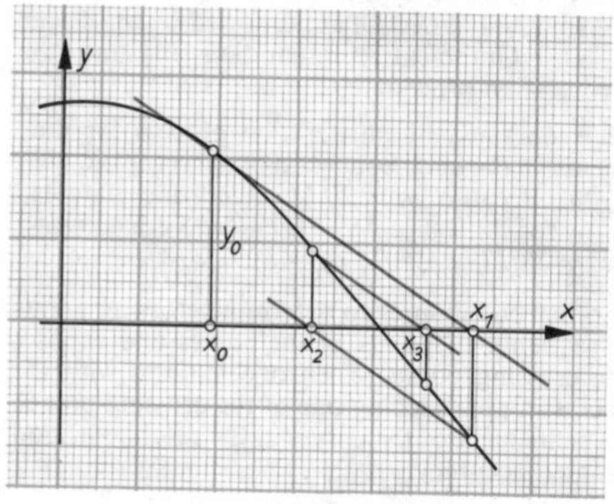

29.3-4 Newton's method with fixed gradient

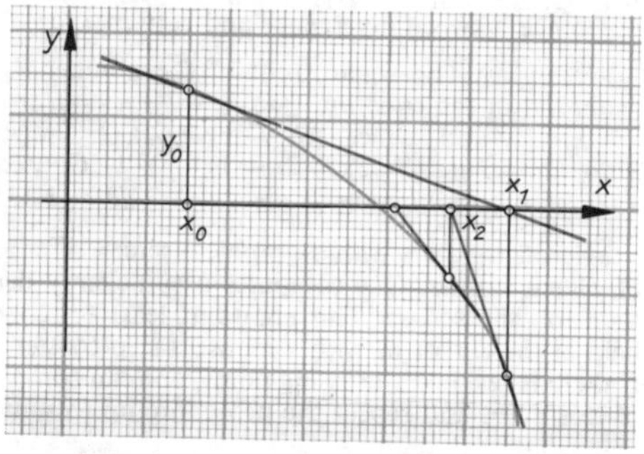

29.3-5 Newton's method with variable gradient

$y'_0 = f'(x_f)$ is left unchanged and the point $x' \to x_{i+1}$ with $y_{i+1} = f(x_{i+1})$ is obtained from the point $x_0 \to x_i$ with $y_0 \to f(x_i)$; this step $x_{i+1} = x_i - f(x_i)/f'(x_f)$ is iteratively repeated (Fig.). In the case of *Newton's method with variable gradient*, the derivative $f'(x_i)$ is calculated afresh at each point $(x_i, f(x_i))$, so that $x_{i+1} = x_i - f(x_i)/f'(x_i)$ (Fig.).

Example: The kth root of the number a is obtained as a root of the function $f(x) = x^k - a$ by means of the iteration equation

$$x_{i+1} = x_i - (x_i^k - a)/(kx_i^{k-1}) = x_i(1 - 1/k) + a/(kx_i^{k-1}).$$

For the *square root* it becomes $x_{i+1} = (x_i + a/x_i)/2$.

Newton's method converges rapidly if $K|x^* - x_1| < 1$ holds for the root x^* and K has the value described in the method of false position.

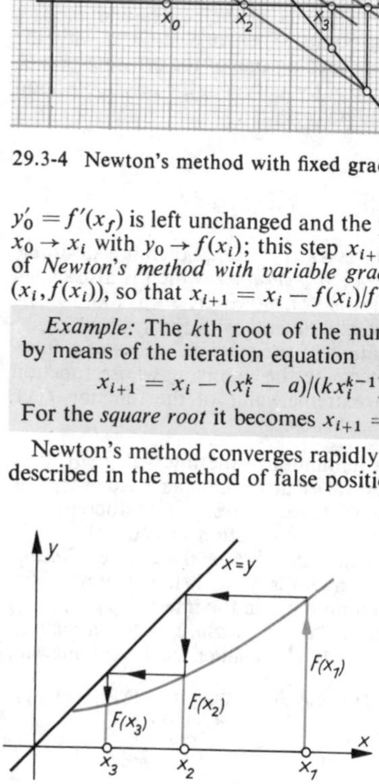

29.3-6 Start of the iteration process $x_{i+1} = F(x_i)$

Method of iteration. In general, for the determination of a root the equation $f(x) = 0$ is expressed in the form $x_{i+1} = F(x_i)$, which is capable of iteration (Fig.). If one chooses the function $F(x) = x - cf(x)$, where $c > 0$, then an optimal value for c can be chosen with reference to the lower bound m_1 and the upper bound M_1 for the derivative $f'(x)$, that is, $m_1 < f'(x) < M_1$. For the derivative $F'(x) = 1 - cf'(x)$ it then follows that $1 - cm_1 > 1 - cf'(x) > 1 - cM_1$, that is, $F'(x)$ lies within bounds that are narrower, the smaller the value of $\max(|1 - cm_1|, |1 - cM_1|)$. For $c = 2/(M_1 + m_1)$ one obtains $|1 - cm_1| = |1 - cM_1| = (M_1 - m_1)/(M_1 + m_1) = a < 1$. By the mean value theorem of the differential calculus it then follows that $|F(x_{i+1}) - F(x_i)| \leqslant |F'(\xi)|\,|x_{i+1} - x_i| < a \cdot |x_{i+1} - x_i|$.

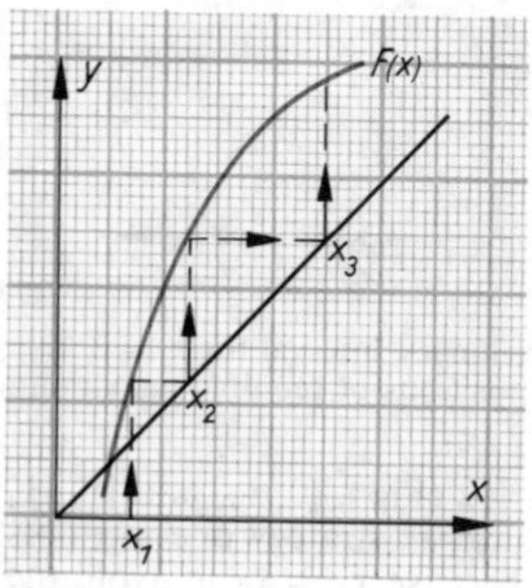

29.3-7 Graphical illustration of a divergent iteration

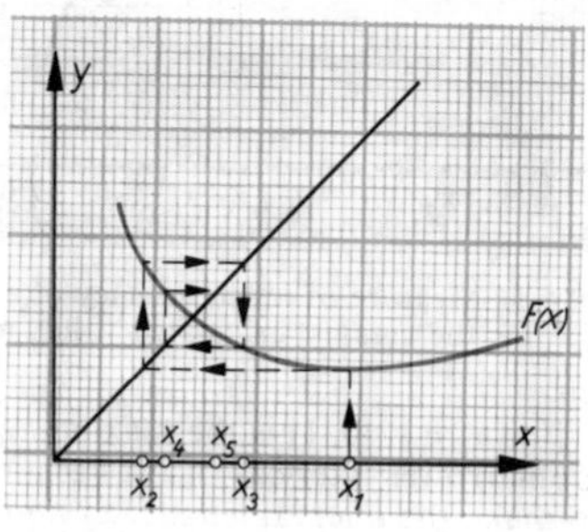

29.3-8 Graphical illustration of a convergent iteration

The smaller the value of a the better the convergence of the iteration process. For the fixed point *method of false position* it is clear that $F(x) = [x_f f(x) - xf(x_f)]/[f(x) - f(x_f)]$ and for *Newton's method* $F(x) = x - f(x)/f'(x)$. The convergence can be improved by the *δ^2-process of Aitken*. In this case two normal iteration steps $x_{3i+1} = F(x_{3i})$ and $x_{3i+2} = F(x_{3i+1})$ are followed by an *Aitken step* $x_{3i+3} = x_{3i} - (x_{3i+1} - x_{3i})^2/(x_{3i+2} - 2x_{3i+1} + x_{3i})$. Graphical representations illustrate the difference between divergence and convergence of the iteration procedure (Fig.).

Example: The square root of the number $a > 1$ is obtained by an iteration process for the solution of the equation $f(x) = x^2 - a = 0$. In general, in the neighbourhood of the zero, $1 < \sqrt{a} = x < a$, and hence for $f'(x) = 2x$, $2 < f'(x) < 2a$. Because $m = 2$, $M = 2a$, one obtains $c = (2a - 2)/(2a + 2) = (a - 1)/(a + 1)$. Consequently for $\sqrt{2}$, $x_{i+1} = x_i - (x_i^2 - 2)/3$ is a convergent iteration function. It gives $x_1 = 4/3 = 1.333$; $x_2 = 38/27 \approx 1.407$; $x_3 = 3092/2187 \approx 1.412$.

Microcomputers with a graphical screen offer also other possibilities to determine crossing points of planar curves. One computes a sufficiently dense sequence of points of the curves and exposes them on the screen. This shows at least roughly the crossing points. Then one initiates with the help of the cursor or with a mouse a step-by-step process in approaching one of the crossing-points. If there is reached a crossing point, one can compute a refinement of the curves in the neighbourhood of this point and repeat the procedure up to some wanted accuracy.

29.4. Search for extreme values

One-dimensional processes

If the mode of operation of a system depends on the parameters $x_1, x_2, \ldots, x_n$, then it is desirable to have a criterion $F(x_1, x_2, \ldots, x_n)$ for the *goodness of the mode of operation*, which is also called the *objective* or *cost function*. This function of several variables is not, however, always known by **means of a formula. Its values then have to be determined by experiments with the system.**

A simple case of such a mode of operation is the determination of the extreme values of a function $f(x)$ of one variable. The parameters $x_1, \ldots, x_n$ now determine the points x_i whose function values $f(x_i)$ provide information about the proximity of an extreme value of the function $f(x)$. In the search for these points several strategies have proved practicable.

Overall strategies. If the function $f(x)$ in question has one minimum in the interval $[a, b]$, then the function $-f(x)$ has precisely one maximum; if it has several relative maxima, then every one of the strategies described leads to an approximation for one of these values. It may therefore be assumed that $f(x)$ has exactly one maximum, and that after the transformation $u = (x - a)/(b - a)$, the interval is $[0, 1]$. An overall strategy $Z_n = (x_1, x_2, \ldots, x_n)$ then consists in the choice of n different points x_i of the interval $[0, 1]$, with $x_i < x_j$ for $i < j$. If $f(x_k)$ for $x_i = x_k$ is the greatest of the calculated values of the function, then the argument of the maximum lies in the interval $[x_{k-1}, x_{k+1}]$ (Fig.). *The indeterminacy of the interval* $L_n = x_{k+1} - x_{k-1}$ leads to the *measure of indeterminacy of the strategy* $L_n = \max_{1 \leqslant i \leqslant n} (x_{i+1} - x_{i-1})$, where $x_0 = 0$ and $x_{n+1} = 1$. The smaller the largest interval, the better the strategy. The optimum measure of indeterminacy $L_{n\,\text{opt}} = \min_{Z_n} \{\max_{1 \leqslant i \leqslant n} (x_{i+1} - x_{i-1})\}$ is therefore the characteristic of the *minimax strategy*.

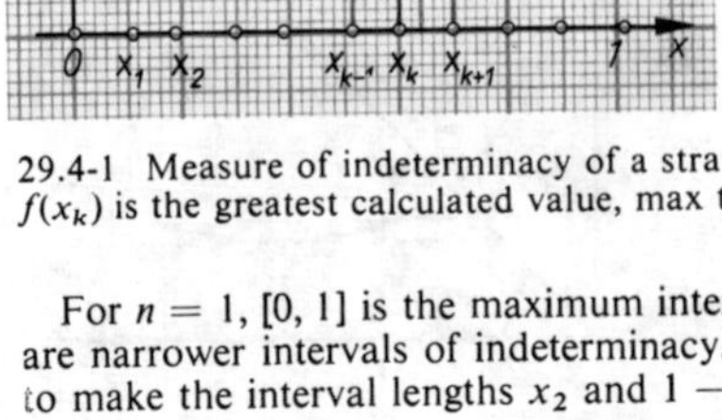

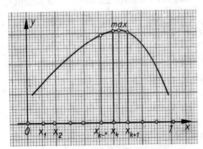

29.4-1 Measure of indeterminacy of a strategy; $f(x_k)$ is the greatest calculated value, max the maximum

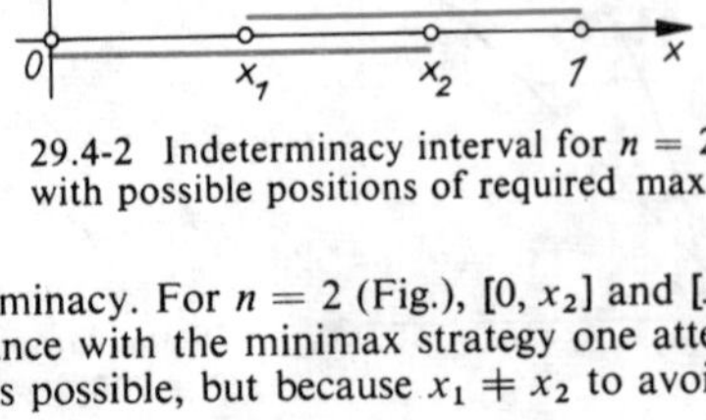

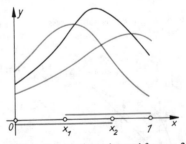

29.4-2 Indeterminacy interval for $n = 2$ with possible positions of required maximum

For $n = 1$, $[0, 1]$ is the maximum interval of indeterminacy. For $n = 2$ (Fig.), $[0, x_2]$ and $[x_1, 1]$ are narrower intervals of indeterminacy. If in accordance with the minimax strategy one attempts to make the interval lengths x_2 and $1 - x_1$ as small as possible, but because $x_1 \neq x_2$ to avoid the unrealizable value $x_1 = x_2 = 0.5$, then the ε-optimal overall strategy for $n = 2$ gives the values $x_1 = 0.5 - \varepsilon/2$ and $x_2 = 0.5 + \varepsilon/2$ with a sufficiently small $\varepsilon > 0$. Here the choice of ε also depends on the error variation of the function values $f(x)$, since it is not possible to determine of two values $f(x)$ and $f(x + \varepsilon)$ differing by less than the width of the variation which one is the greater.

For $n = 3$ the new third point can at most increase the sharpness of the separation. A narrowing of the indeterminacy interval can only be achieved by a pair of new points. Optimal is an arrangement of equidistant pairs which, for even n, is given by the points $x_k = (1 + \varepsilon) \cdot [(k + 1)/2]/\{(n/2) + 1\} - \{[(k + 1)/2] - [k/2]\}\,\varepsilon$, where $[x]$ denotes the largest integer less than or equal to x. The length of the optimal uncertainty interval is then $L_{n\,\text{opt}} = (1 + \varepsilon)/\{(n/2) + 1\}$.

Example: For $n = 4$ one obtains the partition points $x_1 = 1/3 - 2\varepsilon/3$; $x_2 = 1/3 + \varepsilon/3$; $x_3 = 2/3 - \varepsilon/3$; $x_4 = 2/3 + 2\varepsilon/3$ and the optimal uncertainty interval $L_{4\text{opt}} = 1/3 + \varepsilon/3$.

Sequential strategies. As the name implies, every new step in this strategy starts from the preceding one, so that the uncertainty interval obtained from that step is made the new interval to be examined. In this way one avoids too many partition points.

In the case of the *dichotomic sequential search*, the ε-optimal overall strategy for $n = 2$ is applied repeatedly. One generalizes $L_{2\,\text{opt}} = (1 + \varepsilon)/2$ to the recurrence relation

$$L_{2k\,\text{opt}} = (L_{2(k-1)\,\text{opt}} + \varepsilon)/2$$

and obtains for $n = 2k$ points, $L_{n\,\text{opt}} = 2^{-n/2} + \varepsilon(1 - 2^{-n/2})$. One can see that for the same n the length of the interval is less than for the optimal minimax overall strategy.

Example: The calculation of the first 12 partition points in the search for the minimum of the function $f(x) = |x^2 - 2|$ with $\varepsilon = 10^{-4}$ shows the effort required: $x_1 = 1 - \varepsilon/2$; $x_2 = 1 + \varepsilon/2$; $x_3 = 1.5 - 3\varepsilon/4$; $x_4 = 1.5 + \varepsilon/4$; $x_5 = 1.25 - 5\varepsilon/8$; $x_6 = 1.25 + 3\varepsilon/8$; $x_7 = 1.375 - 11\varepsilon/16$; $x_8 = 1.375 + 5\varepsilon/16$; $x_9 = 1.4375 - 23\varepsilon/32$; $x_{10} = 1.4375 + 9\varepsilon/32$; $x_{11} = 1.406\,25 - 45\varepsilon/64$; $x_{12} = 1.406\,25 + 19\varepsilon/64$.

The Fibonacci search procedure. The number of tests n that are to be carried out during the search is fixed. Starting from the initial search interval $[a_1, b_1]$ the subsequent search intervals are fixed by means of a sequence of numbers d_i, which are determined from the Fibonacci numbers. The *Fibonacci numbers* $F_0 = 1$, $F_1 = 1$, $F_2 = 2$, $F_3 = 3$, $F_4 = 5$, $F_5 = 8$, $F_6 = 13$, $F_7 = 21$, $F_8 = 34$, $F_9 = 55$, $F_{10} = 89$, $F_{11} = 144$, $F_{12} = 233$, $F_{13} = 377$, $F_{14} = 610$ satisfy the recurrence relation $F_i = F_{i-1} + F_{i-2}$. This gives $1 = F_{i-1}/F_i + F_{i-2}/F_i$, and because $F_{i-1} > F_{i-2}$ it follows that $F_{i-2}/F_i < 1/2$. One puts $L_1 = b_1 - a_1$, $d_1 = L_2 = L_1 F_{n-1}/F_n$, $d_2 = L_3 = L_1 F_{n-2}/F_n$, $d_3 = L_2 F_{n-3}/F_{n-1} = L_1 F_{n-3}/F_n$, $d_4 = L_3 F_{n-4}/F_{n-2} = L_1 F_{n-4}/F_n, \ldots,$ $d_{n-1} = L_n = L_{n-2} F_1/F_3 = \cdots = L_1/F_n$. Since $F_n > 2^{n/2}$ for $n \geqslant 3$, the length of the interval L_n is less than that for the dichotomic search for the same n.

With the help of these values the x_i are fixed. From $x_1 = a_1 + d_2$, $x_2 = b_1 - d_2$ it follows that $x_2 - x_1 = L_1 - 2d_2 > 0$ or $x_2 > x_1$, since $d_2 < {}^1/_2 L_1$. For the search intervals one has $[a_1, b_1 - d_2]$ or $[a_1 + d_2, b_1]$ of the same length $L_1 - d_2 = L_1[(F_n - F_{n-2})/F_n] = L_1 F_{n-1}/F_n = L_2 = d_1$. The point x_3 and the new search interval depend on the function values $f(x_1)$ and $f(x_2)$:

for $f(x_1) \geqslant f(x_2)$ one puts $a_2 = a_1$, $b_2 = x_2$ and $x_3 = a_2 + d_3$, where $a_2 = a_1 < x_3 < x_1 < x_2 = b_2$. Comparison of the function values at the points x_3 and x_1 gives two possible new uncertainty intervals $[a_1, x_1]$, $[x_3, x_2]$ of length $L_3 = d_2$.

for $f(x_1) < f(x_2)$ one puts $a_2 = x_1$, $b_2 = b_1$ and $x_3 = b_2 - d_3$, where $a_2 \leqslant x_1 < x_2 < x_3 < b_1 = b_2$. Comparison of the function values at the points x_2 and x_3 gives two possible new uncertainty intervals $[x_1, x_3]$, $[x_2, b_1]$ of length $L_3 = d_2$.

For the determination of x_4 one applies to the interval $[a_2, b_2]$ the same arguments that have led to the point x_3. The subsequent points up to x_n are found similarly.

Example: If the minimum of the function $|x^2 - 2| = f(x)$ is determined by the Fibonacci search procedure, one obtains for the lengths of the uncertainty intervals (to three decimal places) $L_1 = 2.000$, $L_2 = 1.236$, $L_3 = 0.764$, $L_4 = 0.472$, $L_5 = 0.292$, $L_6 = 0.180$, $L_7 = 0.113$, $L_8 = 0.068$, $L_9 = 0.045$, $L_{10} = 0.023$ and the points of subdivision $x_1 = 0.764$, $x_2 = 1.236$, $x_3 = 1.528$, $x_4 = 1.708$, $x_5 = 1.416$, $x_6 = 1.348$, $x_7 = 1.461$, $x_8 = 1.393$, $x_9 = 1.438$, $x_{10} = 1.415$ (Fig.).

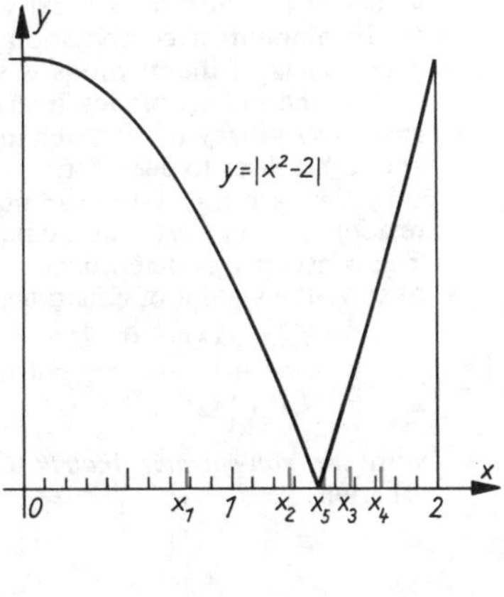

29.4-3 Intervals of a Fibonacci search procedure for $y = |x^2 - 2|$

The golden section search procedure. This procedure is only a little less effective than the Fibonacci procedure, but it does not require the number of search steps to be fixed in advance.

In the search interval $[a, b]$ two points x and x' are fixed by a parameter τ still to be determined. From $\tau = (b - a)/(b - x)$ one obtains $x = a/\tau + b(1 - 1/\tau)$ and hence from $\tau = (b - x)/(b - x')$, the value $x' = a/\tau^2 + b(1 - 1/\tau^2)$; for $a = 0$ and $b = 1$, for example, one obtains $x = 2/3$, $x' = 8/9$. A point configuration equivalent to (a, x, x', b) in a reduced search interval depends on the values of the function at the points x and x'; for $f(x') \geqslant f(x)$ one chooses $a := x$, $x := x'$ and $x' := b(1 - 1/\tau^2) + a/\tau^2$, and for $f(x') < f(x)$, on the other hand, $b := x'$, $x' := x$, $x := b(1 - 1/\tau) + a/\tau$. The length L of the uncertainty interval then changes for $f(x') \geqslant f(x)$ by $L := L(1 - 1/\tau^2)$ and for $f(x') < f(x)$ by $L := L/\tau$. Every point in an interval is regarded as being equally likely to be an extremum. The probability of an interval to contain an extremum is thus proportional to the length of the interval; for the two intervals it is in the ratio $(1 - 1/\tau^2) : 1/\tau$. But the most favourable chain of decision is the one in which one has to distinguish between two equally probable cases. For this one has $1 - 1/\tau^2 = 1/\tau$, or the optimal τ-value $\tau = 1/2 + (1/2)\sqrt{5} = 1.618\,033\,989\ldots$ (see Chapter 7.). For the uncertainty intervals the recurrence relation $L := L/\tau$ holds, so that after n search steps $L_n = L_1/\tau^n$. The ratio of the effectiveness of this procedure to that of the dichotomic search is as τ is to $\sqrt{2} \approx 1.142\ldots$, so that it is approximately 14% higher.

The connection with the Fibonacci procedure follows from the relation $F_i = (1/\sqrt{5})[\tau^{i+1} - 1/(-\tau)^{i+1}]$. For $i = 1$ and $i = 2$ one obtains by substitution $F_1 = 1$ and $F_2 = 2$. In general, however, the recursion formula $F_{i+1} = F_i + F_{i-1}$ holds also for the right-hand side of the relation; for if one multiplies $\tau^{i+1} - 1/(-\tau)^{i+1} = \tau^i - 1/(-\tau)^i + \tau^{i-1} - 1/(-\tau)^{i-1}$ by $(-\tau)^{i+1}$ on boths sides, then, whether i is even or odd, one obtains the relation $\tau^{2i}(\tau^2 - \tau - 1) = \tau^2 - \tau - 1$, which is always correct because τ was determined from $\tau^2 - \tau - 1 = 0$.

The same result is obtained by the *z-transformation method* for linear recurrence relations. One introduces the auxiliary function $F(z) = \sum_{i=0}^{\infty} F_i z^i$ and takes into account the fact that from the recurrence relation $F_i = F_{i-1} + F_{i-2}$, $\sum_{i=2}^{\infty} F_i z^i = z \sum_{i=2}^{\infty} F_{i-1} z^{i-1} + z^2 \sum_{i=2}^{\infty} F_{i-2} z^{i-2}$. One then obtains $F(z) = F_0 + F_1 z + \sum_{i=2}^{\infty} F_i z^i = F_0 + F_1 z - zF_0 + zF_0 + z \sum_{i=0}^{\infty} F_i z^i + z^2 \sum_{i=0}^{\infty} F_i z^i$ or $-F_0 + z(F_0 - F_1) = z^2 F(z) + zF(z) - F(z)$, giving

$F(z) = [(F_0 - F_1)\, z - F_0]/(z^2 + z - 1) = \dfrac{-1}{z^2 + z - 1}$. For $z_{1,2} = -(1/2) \pm (1/2)\sqrt{5}$ the partial fraction expansion is $F(z) = \dfrac{1}{\sqrt{5}}\left[\dfrac{1}{z_1 - z} - \dfrac{1}{z_2 - z}\right]$; each term of the sum can be expanded as a geometric series in powers of z, and comparison of coefficients with the series $F(z) = \sum_{i=0}^{\infty} F_i z^i$ gives the relation stated.

Example: The determination of the minimum of the function $f(x) = |x^2 - 2|$ by the golden section search procedure gives for the initial interval $a_1 = 0$, $b_1 = 2$ the search points: $x_1 = 0.764$, $x_2 = 1.236$, $x_3 = 1.528$, $x_4 = 1.708$, $x_5 = 1.415$, $x_6 = 1.348$, $x_7 = 1.459$, $x_8 = 1.391$, $x_9 = 1.373$, $x_{10} = 1.399$, $x_{11} = 1.405$ (Fig.).

Compared with the iteration method $x_{i+1} = x_i - (x_i^2 - 2)/3$ the golden section procedure tends to the required solution noticeably more slowly. However, it is a procedure for more general functions, whilst the iteration process is suited for special functions only.

Multi-dimensional search processes and systems of non-linear equations

By a generalization of the one-dimensional procedure one attempts to determine the extreme values of functions of several variables or the solution of systems of non-linear equations. Owing to the amount of computation modern digital computers are necessary for this purpose and the *effectiveness* of the methods is smaller. If in the case of $n = 1$ variable 90% has been eliminated, so that the indeterminacy interval is only 10% of the original one, then because $0.9 \cdot 0.9 = 0.81$ an indeterminacy of 19% remains for $n = 2$, for $n = 3$ it amounts to 27%, because $(0.9)^3 \approx 0.73$, and it increases to 34% for $n = 4$, to 41% for $n = 5$, 47% for $n = 6$ and 52% for $n = 7$.

In the case $n = 1$ for the non-linear equation $f(x) = 0$, one arrives by way of the equivalent relation $x = x - cf(x)$ at the iterative process $x^k = x^{k-1} - cf(x^{k-1})$, where k and $k - 1$ are indices. The constant c is determined optimally by a goodness criterion. Generalizing to $n = 2$, one seeks to solve the system of equations

$$\begin{aligned} f_1(x_1, x_2) &= 0 \\ f_2(x_1, x_2) &= 0 \end{aligned} \quad \text{by putting} \quad \begin{aligned} x_1 &= x_1 - c_{11} f_1(x_1, x_2) - c_{12} f_2(x_1, x_2) \\ x_2 &= x_2 - c_{21} f_1(x_1, x_2) - c_{22} f_2(x_1, x_2) \end{aligned}$$

with the *non-singular matrix* $\mathbf{C} = (c_{ij}) = \begin{pmatrix} c_{11} & c_{12} \\ c_{21} & c_{22} \end{pmatrix}$, which is chosen according to a goodness criterion.

In an *iterative process* not only do the x_1^k and x_2^k depend on the values of x_1^{k-1} and x_2^{k-1} of the preceding iterative step, but also the constants $c_{ij}(x_1^k, x_2^k)$ on $c_{ij}(x_1^{k-1}, x_2^{k-1})$. Their values determine the convergence behaviour, which can be improved by *multi-step iteration algorithms*, in which the approximation in the kth iteration step depends on the finitely many preceding approximations.

For the multi-dimensional search for extreme values, for example, for the maximization of the function $f(x_1, x_2)$, there is the necessary condition $f_{x_1}(x_1, x_2) = \frac{\partial}{\partial x_1} f(x_1, x_2) = 0$ and $f_{x_2}(x_1, x_2) = \frac{\partial}{\partial x_2} f(x_1, x_2) = 0$ in the interior of the domain of definition. By solving this non-linear system of equations one obtains not only maxima, but also minima or saddle points.

In setting up the iteration sequence for the determination of roots, the choice of the matrix $\boldsymbol{C}$ should secure not only a rapid convergence, but also the approach to the position of the required extremum.

No generally satisfactory recommendations can yet be given for the choice of the search matrix $\boldsymbol{C}$. In *stochastic search procedures* the $\boldsymbol{C}$-matrix is dependent on chance; random improvement steps are carried out, but the sequence of these steps should converge with probability 1, that is with certainty, to the required point.

Newton's method. Generalizing the one-dimensional Newton's method, in the case $n = 2$ one replaces the functions $f_1(x_1, x_2)$ and $f_2(x_1, x_2)$ by their tangent planes in the neighbourhood of the point determined, and takes as the next approximation the point at which the line of intersection of the two planes cuts the plane $z = 0$.

With k used again as an index, the tangent planes are clearly

$$z = f_1(x_1^{k-1}, x_2^{k-1}) + \frac{\partial f_1(x_1^{k-1}, x_2^{k-1})}{\partial x_1}(x_1 - x_1^{k-1}) + \frac{\partial f_1(x_1^{k-1}, x_2^{k-1})}{\partial x_2}(x_2 - x_2^{k-1}),$$

$$z = f_2(x_1^{k-1}, x_2^{k-1}) + \frac{\partial f_2(x_1^{k-1}, x_2^{k-1})}{\partial x_1}(x_1 - x_1^{k-1}) + \frac{\partial f_2(x_1^{k-1}, x_2^{k-1})}{\partial x_2}(x_2 - x_2^{k-1}).$$

Putting $z = 0$ and solving the resulting linear system of equations for the factors $(x_1 - x_1^{k-1})$ and $(x_2 - x_2^{k-1})$ one obtains the *C-matrix* as the inverse of the *Jacobian matrix* of the system of equations.

Example: If a solution of the system of equations

$$f_1(x_1, x_2) = x_1 + 3 \lg x_1 - x_2^2 = 0,$$
$$f_2(x_1, x_2) = 2x_1^2 - x_1 x_2 - 5x_1 + 1 = 0$$

is required, then one determines the intersection of the curves $f_1 = \text{const}$ and $f_2 = \text{const}$, and obtains approximately the points (1.4, −1.5) and (3.4, 2.2). As initial approximation one chooses the point (3.4, 2.2). From the Jacobian matrix

$$\begin{pmatrix} \frac{\partial f_1}{\partial x_1} & \frac{\partial f_1}{\partial x_2} \\ \frac{\partial f_2}{\partial x_1} & \frac{\partial f_2}{\partial x_2} \end{pmatrix} = \begin{pmatrix} 1 + \frac{3 \cdot 0.434\,29}{x_1} & -2x_2 \\ 4x_1 - x_2 - 5 & -x_1 \end{pmatrix}$$

one obtains the $\boldsymbol{C}$-matrix as its inverse, and hence the recursion scheme. From this one obtains in succession the adjacent approximations:

	x_1	x_2
$k = 1$	3.4899	2.2633
$k = 2$	3.4891	2.2621
$k = 3$	3.4875	2.2626

For these values $f_1(x_1^3, x_2^3) = 0.0002$ and $f_2(x_1^3, x_2^3) = 0.0000$.

The gradient method or method of steepest descent. The direction of the greatest increase of a function $f(x_1, x_2)$ is given by the direction of the *gradient* $(f_{x_1}(x_1, x_2), f_{x_2}(x_1, x_2))$.

If it is required to find the minimum of the function $f(x_1, x_2)$, then an improvement must result if one moves from the point (x_1^{k-1}, x_2^{k-1}) already obtained in the direction opposite to that of the gradient. This means making the following attempt for the iteration process:

$$x_1^k = x_1^{k-1} - l f_{x_1}(x_1^{k-1}, x_2^{k-1}) \quad \text{and} \quad x_2^k = x_2^{k-1} - l f_{x_2}(x_1^{k-1}, x_2^{k-1}) \quad \text{with} \quad l > 0.$$

Thus, in this case the $\boldsymbol{C}$-matrix is taken to be a diagonal matrix with diagonal elements l.

If for a fixed l a trial step is carried out, after which one examines whether an improvement has occurred and if, depending on the success, a new spacing l is chosen, then one speaks of the *gradient method*.

One can, however, attempt to choose the factor l optimally at every step. A natural requirement for l would be to make the vâlue of the function

$$F(l) = f[x_1^{k-1} - l f_{x_1}(x_1^{k-1}, x_2^{k-1}), x_2^{k-1} - l f_{x_2}(x_1^{k-1}, x_2^{k-1})]$$

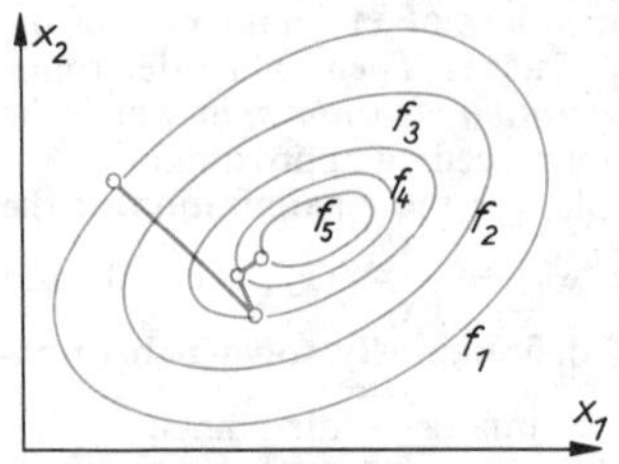

as small as possible. The required value for l can, for example, be determined by a one-dimensional search process (Fig.).

If this optimal value for l is chosen at every step, then one speaks of the *method of steepest descent*.

29.4-4 Gradient method; contour lines $f_1 > f_2 > f_3 > f_4 > f_5$ of $f(x_1, x_2)$

Example: The method will be illustrated by determining the minimum of the function

$$f(x_1, x_2) = 2x_1^2 - 2x_1 + x_2^2 - x_2 = 2(x_1 - 1/2)^2 + (x_2 - 1/2)^2 - 3/4;$$

in this simple case the minimum $x_1 = x_2 = 0.5$ is known. By the *gradient method* one obtains the recursion formulae

$$x_1^k = x_1^{k-1} - 2l(2x_1^{k-1} - 1) \quad \text{and} \quad x_2^k = x_2^{k-1} - l(2x_2^{k-1} - 1).$$

According to the *method of steepest descent* the optimal l is obtained from the equation

$$l = [4(2x_1^{k-1} - 1)^2 + (2x_2^{k-1} - 1)^2]/[16(2x_1^{k-1} - 1)^2 + 2(2x_2^{k-1} - 1)^2].$$

If one begins with the initial approximation $x_1^0 = x_2^0 = 0$, then one obtains in succession the adjacent l-values and improved approximations:

$l^0 = 0.278$	$x_1^1 = 0.556$	$x_2^1 = 0.278$
$l^1 = 0.414$	$x_1^2 = 0.457$	$x_2^2 = 0.461$
$l^2 = 0.283$	$x_1^3 = 0.504$	$x_2^3 = 0.482$
$l^3 = 0.429$	$x_1^4 = 0.497$	$x_2^4 = 0.497$

29.5. Numerical methods for the solution of linear equations and inequalities

Linear equations

A solution of a system of m linear equations $y_i = \sum_{j=1}^{n} a_{ij}x_j$ with $i = 1, 2, \ldots, m$ consists of n numbers x_j with $j = 1, 2, \ldots, n$. In n-dimensional space each one of these equations with a fixed y_i can be interpreted as a *hyperplane* with the normal vector $\boldsymbol{a}_i = (a_{i1}, a_{i2}, \ldots, a_{in})$. If for $m = n$ these normal vectors are linearly independent, that is, if the equation $\sum_{i=1}^{m} l_i \boldsymbol{a}_i = 0$ is satisfied only when all the l_i are zero, then the $n = m$ hyperplanes have a common point of intersection with the uniquely determined coordinates $x_j, j = 1, 2, \ldots, n$. If $m < n$ and the m normal vectors are linearly independent, then the corresponding result holds for the m-dimensional subspace determined by them. For $m > n$, the m vectors $\boldsymbol{a}_i$ are certainly linearly dependent. If the vector $\boldsymbol{a}_{i_0}$ is a linear combination of some of the other vectors $\boldsymbol{a}_j$, and if in addition y_{i_0} is the same linear combination of the y_j, and the hyperplane belonging to $\boldsymbol{a}_{i_0}$ contains the intersection of the hyperplanes belonging to these $\boldsymbol{a}_j$, then the hyperplane belonging to $\boldsymbol{a}_{i_0}$ does not represent additional conditions, and its equation need not be taken into account. A contradiction occurs if y_{i_0} is not the same linear combination of the y_j as $\boldsymbol{a}_{i_0}$ is of the $\boldsymbol{a}_j$. In this case the given linear system of equations is insoluble.

For $n = 2$ the hyperplanes are straight lines, which intersect, are parallel to one another, or coincide (Fig.).

Jordan's elimination. The system of equations is arranged in the form of a table, so that the rth row contains the coefficients a_{rj} of the x_j with $j = 1, 2, \ldots, n$ and the sth column the coefficients a_{is} of x_s with $i = 1, 2, \ldots, m$.

	x_1	x_2	$\cdots$	x_s	$\cdots$	x_n	
y_1	a_{11}	a_{12}	$\cdots$	a_{1s}	$\cdots$	a_{1n}	1
y_2	a_{21}	a_{22}	$\cdots$	a_{2s}	$\cdots$	a_{2n}	2
$\vdots$							
y_r	a_{r1}	a_{r2}	$\cdots$	a_{rs}	$\cdots$	a_{rn}	r
$\vdots$							
y_m	a_{m1}	a_{m2}	$\cdots$	a_{ms}	$\cdots$	a_{mn}	m
	1	2		s		n	

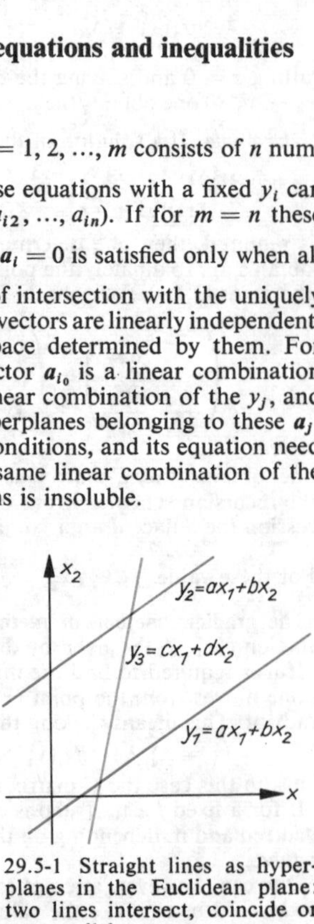

29.5-1 Straight lines as hyperplanes in the Euclidean plane: two lines intersect, coincide or are parallel

If then one of the coefficients is different from zero, for example, $a_{rs} \neq 0$, then x_s can be eliminated using y_r. From $y_r = a_{r1}x_1 + a_{r2}x_2 + \cdots + a_{rs}x_s + \cdots + a_{rn}x_n$ it follows that $x_s = (1/a_{rs})[-a_{r1}x_1 - a_{r2}x_2 - \cdots + y_r - \cdots - a_{rn}x_n]$. Substitution of this value for x_s changes all coefficients of the table. Those of the sth column, that is, those of y_r, then become $a_{1s}/a_{rs}, a_{2s}/a_{rs}, \ldots, a_{ms}/a_{rs}$. For the remaining b_{ij} with $i \neq r$ and $j \neq s$, $b_{ij} = a_{ij} - (a_{is} \cdot a_{rj})/a_{rs}$. The table then assumes the form:

	x_1	x_2	$\cdots$	y_r	$\cdots$	x_n	
y_1	b_{11}	b_{12}	$\cdots$	$+a_{1s}/a_{rs}$	$\cdots$	b_{1n}	1
y_2	b_{21}	b_{22}	$\cdots$	$+a_{2s}/a_{rs}$	$\cdots$	b_{2n}	2
$\vdots$							
x_s	$-a_{r1}/a_{rs}$	$-a_{r2}/a_{rs}$	$\cdots$	$+1/a_{rs}$	$\cdots$	$-a_{rn}/a_{rs}$	r
$\vdots$							
y_m	b_{m1}	b_{m2}	$\cdots$	$+a_{ms}/a_{rs}$	$\cdots$	b_{mn}	m
	1	2		s		n	

In this way one attempts as far as possible to exchange every x_s for a y_r. This process comes to an end when every coefficient at the intersection of the row of a not yet exchanged y_j with the column of a remaining x_i is zero. The exchanged x_s are then linear combinations of the exchanged y_r and the not exchanged variables x_i. These x_i do not obey any further conditions and can, as free parameters, be chosen arbitrarily. The not exchanged y_j are then linear combinations of the exchanged variables y_r alone. If the prescribed values for y_j do not satisfy these conditions, then the system of equations has no solution.

If the process does not come to an end, so that all x_s can be exchanged for variables y_r, then by the final table the x_s are unique functions of the exchanged y_r. Any not exchanged y_j that are still present are then unique linear combinations of the y_r. If the prescribed values for y_j do not satisfy these conditions, then contradictions occur and the system of equations has no solution.

If for $m = n$ all the x_s can be exchanged for the y_r, then the final table represents the inverse matrix A^{-1} of the matrix A of the original table.

Example 1: The Jordan elimination applied to the system of equations results in tables, with the coefficient $a_{rs} \neq 0$ in brackets, from whose rows the next elimination step follows. The elimination equation is placed in red underneath the table.

$$x_1 + x_2 - x_3 = y_1 = 2$$
$$x_1 - x_2 + x_3 = y_2 = 4$$
$$x_1 - x_2 - x_3 = y_3 = 8$$

	x_1	x_2	x_3
y_1	(1)	1	−1
y_2	1	−1	1
y_3	1	−1	−1

$x_1 = y_1 - x_2 + x_3$

	y_1	x_2	x_3
x_1	1	−1	1
y_2	1	−2	(2)
y_3	1	−2	0

$x_3 = -y_1/2 + x_2 + y_2/2$

	y_1	x_2	y_2
x_1	1/2	0	1/2
x_3	−1/2	1	1/2
y_3	1	(−2)	0

$x_2 = y_1/2 - y_3/2$

	y_1	y_3	y_2
x_1	1/2	0	1/2
x_3	0	−1/2	1/2
x_2	1/2	−1/2	0

This system of equations has a unique solution. Substitution gives $x_1 = 3$, $x_2 = -3$, $x_3 = -2$.

Example 2: For the system of equations

$$x_1 + x_2 - x_3 = y_1 = 2$$
$$x_1 - x_2 + x_3 = y_2 = 4$$
$$3x_1 - x_2 + x_3 = y_3 = 8$$

one obtains similarly

	x_1	x_2	x_3
y_1	1	1	−1
y_2	1	−1	1
y_3	3	−1	(1)

$x_3 = -3x_1 + x_2 + y_3$

	x_1	x_2	y_3
y_1	(4)	0	−1
y_2	−2	0	1
x_3	−3	1	1

$x_1 = y_1/4 + y_3/4$

	y_1	x_2	y_3
x_1	1/4	0	1/4
y_2	−1/2	(0)	1/2
x_3	−3/4	1	1/4

The missing exchange of the variable x_2 ist not possible, since the coefficient in brackets is zero. Its row gives the condition $y_2 = (y_3 - y_1)/2$ for the existence of a solution. It is not satisfied for the given numerical values and the system of equations has no solution. If the given values were $y_1 = 2$, $y_2 = 3$, $y_3 = 8$, then $y_2 = (y_3 - y_1)/2$ would be satisfied and would lead to the solution $x_1 = 2.5$, $x_3 = 0.5 + x_2$, in which the value of x_2 can be chosen arbitrarily.

The modified system of linear equations $y_i = 0 = \sum_{j=1}^{n} a_{ij}x_j - b_i$ leads to another form of the Jordan exchange problem. The initial table has an additional column for the b_i.

	x_1	x_2	$\dots$	x_n	$-b_i$	
y_1	a_{11}	a_{12}	$\dots$	a_{1n}	$-b_1$	1
y_2	a_{21}	a_{22}	$\dots$	a_{2n}	$-b_2$	2
$\vdots$						r
y_m	a_{m1}	a_{m2}	$\dots$	a_{mn}	$-b_m$	m
	1	2	s	n	$n+1$	

Because after every elimination of x_s by y_r the coefficients of the sth column have $y_r = 0$ as a factor, this column can be discarded.

Example: For the system

$$x_1 + x_2 - x_3 - 2 = y_1 = 0$$
$$x_1 - x_2 + x_3 - 4 = y_2 = 0$$
$$x_1 - x_2 - x_3 - 8 = y_3 = 0$$

one obtains

	x_1	x_2	x_3	1
y_1	(1)	1	-1	-2
y_2	1	-1	1	-4
y_3	1	-1	-1	-8

$x_1 = y_1 - x_2 + x_3 + 2$

	x_2	x_3	1
x_1	-1	1	2
y_2	-2	(2)	-2
y_3	-2	0	-6

$x_3 = x_2 + y_2/2 + 1$

	x_2	1
x_1	0	3
x_3	1	1
y_3	(-2)	-6

$x_2 = -y_3/2 - 3$

	1
x_1	3
x_3	-2
x_2	-3

Hence the solution is $x_1 = 3$, $x_2 = -3$, $x_3 = -2$.

Gauss's method. *This method of solution* is obtained from the Jordan elimination procedure. The rows of any of the exchanged variables x_i are not again entered in the table, but are separately noted. In this way the size of the table is steadily reduced and in the end one obtains a very easily soluble linear system of equations with a triangular matrix.

Example: For the above example one obtains in succession the following tables and equations, from which the solution is obtained recursively:

	x_1	x_2	x_3	1
y_1	(1)	1	-1	-2
y_2	1	-1	1	-4
y_3	1	-1	-1	-8

$x_1 = -x_2 + x_3 + 2$

	x_2	x_3	1
y_2	-2	(2)	-2
y_3	-2	0	-6

$x_3 = x_2 + 1$

	x_2	1
y_3–	-2	-6

$x_2 = -3$

or $x_2 = -3$, $x_3 = -2$, $x_1 = 3$.

Iteration process for the solution of systems of linear equations

By the substitution $a_{ij} = h_{ij} + c_{ij}$ the given system of equations $\sum_{j=1}^{n} a_{ij}x_j = b_i$ is decomposed into $\sum_{j=1}^{n} h_{ij}x_j + \sum_{j=1}^{n} c_{ij}x_j = b_i$. The h_{ij} are chosen in such a way that their matrix $\boldsymbol{H} = (h_{ij})$ has an inverse matrix $\boldsymbol{H}^{-1} = (h_{ri}^{-1})$ that is easily formed, for example, as a diagonal matrix. It then follows that $x_r = \sum_i h_{ri}^{-1}b_i - \sum_{i,j} h_{ri}^{-1}c_{ij}x_j$. This system of equations can be written in iterative form. If one takes $x_j^0 = \sum_i h_{ri}^{-1}b_i$ as initial value, then one obtains $x_r^k = x_j^0 - \sum_{i,j} h_{ri}^{-1}c_{ij}x_j^{k-1}$. This process converges to the solution of the linear system of equations if and only if the absolute value of all the eigenvalues of the matrix $\sum_i (h_{ri}^{-1}c_{ij}) = (k_{ij})$ is less than 1. An *eigenvalue* l of the matrix (k_{ij}) is defined to be a number l for which the system of equations $\sum_j k_{ij}x_j = lx_i$ has a non-trivial solution, that is, one for which not all the x_j vanish. This means that either $\sum_j |\sum_i h_{ri}^{-1}c_{ij}| < 1$ or $\sum_r |\sum_i h_{ri}^{-1}c_{ij}| < 1$. These conditions are satisfied, for example, if the condition $|k_{ii}| \geqslant \sum_{i \neq j} |k_{ij}|$ holds for the matrix $\boldsymbol{K} = (k_{ij})$. The increment of successive approximations is then given by

$$(x_k^r - x_r^{k-1}) = \sum_{i,j} h_{ri}^{-1}c_{ij}(x_j^{k-1} - x_j^{k-2}).$$

Example: For the solution by the iteration method the equations of the given system are expressed in a form suitable for iteration:

$$\begin{array}{l} 10x_1 - x_2 + x_3 = 10 \\ x_1 + 5x_2 + x_3 = 5 \\ x_1 - x_2 + 10x_3 = 10 \end{array} \longrightarrow \begin{array}{l} x_1^k = 0.1x_2^{k-1} - 0.1x_3^{k-1} + 1 \\ x_2^k = -0.2x_1^{k-1} - 0.2x_3^{k-1} + 1 \\ x_3^k = -0.1x_1^{k-1} + 0.1x_2^{k-1} + 1 \end{array}$$

For the initial approximation $x_1^0 = 1$, $x_2^0 = 1$, $x_3^0 = 1$ one then obtains the increments

$$\begin{pmatrix} 0 \\ -0.4 \\ 0 \end{pmatrix} \begin{pmatrix} -0.04 \\ 0 \\ -0.04 \end{pmatrix} \begin{pmatrix} 0.004 \\ 0.016 \\ 0.004 \end{pmatrix} \begin{pmatrix} 0.0012 \\ -0.0016 \\ 0.0012 \end{pmatrix} \begin{matrix} x_1 \\ x_2 \\ x_3 \end{matrix}$$

and adding these to the initial approximation gives the approximate solution after four iteration steps $x_1^4 = 0.9652$, $x_2^4 = 0.6144$, $x_3^4 = 0.9652$.

Eigenvalue problems. If one regards the linear system of equations $\sum_{j=1}^{n} a_{ij}y_j = x_i$ for $i = 1, 2, \ldots, n$ as a description of a linear system with the y_j as input and the x_i as output variables, whose cause-effect relationship is given by the matrix $A = (a_{ij})$, then the existence of an eigenvalue l according to the equation $\sum_{j=1}^{n} a_{ij}x_j = lx_i$ means that an *eigenvector* $(x_1, x_2, \ldots, x_n)$ used as an input variable remains unchanged except for the factor of proportionality l. The *eigenvalue equation* $\sum_{j=1}^{n} a_{ij}x_j = lx_i$ is a homogeneous linear system, which has solutions different from zero if the determinant of the matrix $A - lI$, in which I is the unit matrix, vanishes, that is,

$$\det|A - lI| = \begin{vmatrix} a_{11} - l & a_{12} & \cdots & a_{1n} \\ a_{21} & a_{22} - l & \cdots & a_{2n} \\ \vdots & & & \vdots \\ a_{n1} & a_{n2} & \cdots & a_{nn} - l \end{vmatrix} = 0.$$

This *characteristic equation* is a polynomial of degree n for the determination of the eigenvalues. For each eigenvalue l one obtains from the eigenvalue equation the corresponding eigenvector. With eigenvectors a decoupling of the system is possible, that is, it can be achieved that every output variable is dependent only on one input variable. In this way *normal oscillations* can be introduced in oscillatory mechanical systems. In the *theory of the gyroscope* one uses three axes placed in the directions of three independent eigenvectors of the matrix of the moment of inertia, in order to obtain a simple form of the dynamic equations. The theory of electric *n-ports* or *2-ports*, using wave *parameters*, rests on a representation of the linear transformation that corresponds to $A = (a_{ij})$ solely by means of the eigenvalues and eigenvectors.

The *eigenvalue of greatest absolute value* and an associated eigenvector can be obtained by a rule that has proved practical in electrical transmission theory. According to this an approximation to the eigenvector and an estimation of the eigenvalue of greatest absolute value are obtained from the output vector of a chain that begins with an arbitrary input and uses the output vector of one step in the chain as the input vector of the next step sufficiently often, in accordance with the equations $\sum_{j=1}^{n} a_{ij}x_j = lx_i$ for $i = 1, 2, \ldots, n$ (Fig.). The last output vector represents an eigenvector approximately, and the quotient of corresponding components of successive approximations represents the eigenvalue. With the essentially arbitrary initial vector $x_i^0 = b_i$, the iterative process $x_i^k = \sum_{j=1}^{n} a_{ij}x_j^{k-1}$ is valid. The quotients x_i^k/x_i^{k-1}, which approach equality with one another, serve as an estimation of the eigenvalue of greatest absolute value.

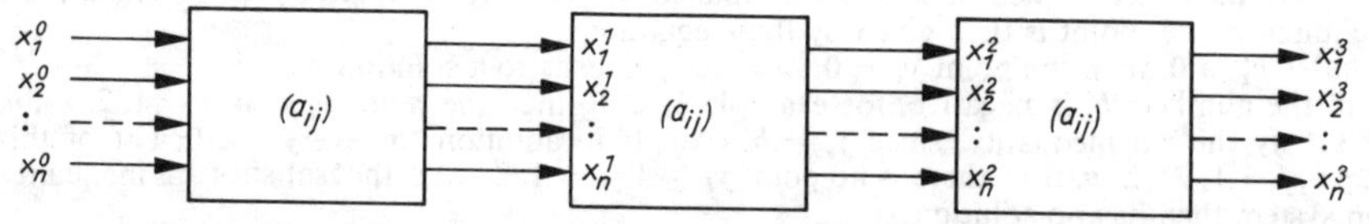

29.5-2 Scheme of a chain for the determination of the eigenvalue of greatest absolute value

Example: For the system of equations

$$\begin{array}{r} x_1 - x_2 = lx_1 \\ 2x_2 + x_3 = lx_2 \\ x_2 + x_3 = lx_3 \end{array}$$

by iteration with the initial values $x_1^0 = 1$, $x_2^0 = 1$, $x_3^0 = 1$ one obtains successively

x_1	0	-3	-11	-32	-87	-231	-608
x_2	3	8	21	55	144	377	987
x_3	2	5	13	34	89	233	610
$k =$	1	2	3	4	5	6	7

For x_i^7/x_i^6 one obtains 2.633, 2.62, 2.62 as an approximation for the eigenvalue of greatest absolute value. The true eigenvalues are obtained from the matrix A and its characteristic equation:

$$A = \begin{pmatrix} 1 & -1 & 0 \\ 0 & 2 & 1 \\ 0 & 1 & 1 \end{pmatrix} \rightarrow \begin{vmatrix} 1-l & -1 & 0 \\ 0 & 2-l & 1 \\ 0 & 1 & 1-l \end{vmatrix} = (1-l)^2(2-l) - (1-l) = 0$$

with $(1-l) = 0$ or $l_1 = 1$ and $l^2 - 3l + 2 = 1$ or $l_2 = (3 + \sqrt{5})/2 \approx 2.6173$ and $l_3 = (3 - \sqrt{5})/2 \approx 0.3825$.

Linear inequalities

In the application of mathematical methods to economics and planning, n-tuples $(x_1, x_2, \ldots, x_n)$ have to be determined from systems of inequalities $y_i = -\sum_{j=1}^{n} a_{ij}x_j + b_i \geqslant 0$ for $i = 1, 2, \ldots, m$. If the n-tuple is regarded as a point of an n-dimensional space, then every one of the m inequalities fixes a half-space bordered by a hyperplane $-\sum_{j=1}^{n} a_{ij}x_j + b_i = 0$. For $m \leqslant n$ the hyperplanes intersect in subspaces of dimension at least $n - m$. For the case $m > n$, which is important in practice, the configuration of the intersection contains points that are *corner points* of the intersection of the half-spaces. This intersection is the required solution region of the given inequalities. It is an n-dimensional convex polyhedron; together with any two points within it or on its boundary it contains all the points of the line joining these two points. A finite polyhedron, which does not contain infinitely distant points, is completely determined by its corner points, or vertices.

Under the assumptions that $m > n$ and that the matrix of the coefficients (a_{ij}) has *rank n* it can be decided by means of an algorithm whether there exist n-tuples that are solutions of the inequalities, and if so, how the corner points of the solution region are obtained.

Initial information

	$-x_1$	$-x_2$	$\ldots$	$-x_n$	1
y_1	a_{11}	a_{12}		a_{1n}	b_1
y_2	a_{21}	a_{22}		a_{2n}	b_2
$\vdots$					
y_m	a_{m1}	a_{m2}		a_{mn}	b_m

$$x_1 = -b_{11}y_1 - b_{12}y_2 - \cdots - b_{1n}y_n + b_1',$$
$$x_2 = -b_{21}y_1 - b_{22}y_2 - \cdots - b_{2n}y_n + b_2',$$
$$\cdots$$
$$x_n = -b_{n1}y_1 - b_{n2}y_2 - \cdots - b_{nn}y_n + b_n'$$

	$-y_1$	$-y_2$	$\ldots$	$-y_n$	1
y_{n+1}	$b_{n+1,1}$	$b_{n+1,2}$	$\ldots$	$b_{n+1,n}$	b_{n+1}'
y_{n+2}	$b_{n+2,1}$	$b_{n+2,2}$	$\ldots$	$b_{n+2,n}$	b_{n+2}'
$\cdot$	$\cdot$	$\cdot$	$\ldots$	$\cdot$	$\cdot$
$\cdot$	$\cdot$	$\cdot$	$\ldots$	$\cdot$	$\cdot$
y_r	b_{r1}	b_{r2}	$\ldots$	b_{rn}	b_r'
$\cdot$	$\cdot$	$\cdot$	$\ldots$	$\cdot$	$\cdot$
$\cdot$	$\cdot$	$\cdot$	$\ldots$	$\cdot$	$\cdot$
y_m	b_{m1}	b_{m2}	$\ldots$	b_{mn}	b_m'

Of the *initial information* presented in the table, it can be assumed as a result of the assumptions, that the first n rows of the coefficients a_{ij} are linearly independent. By Jordan elimination every variable x_i can then be exchanged against a variable y_j. One obtains a *standard form* of the system of inequalities consisting of n equations and a table, from which the variables y_i, $i = 1, 2, \ldots, n$ with $y_i \geqslant 0$ are to be determined in such a way that for $i = n + 1, \ldots, m$ also, $y_i \geqslant 0$. The n-tuple for the required corner point is then given by the n equations.

If in this table $b_i' \geqslant 0$, then the point $y_i = 0$, $i = 1, \ldots, n$, leads to a solution.

If one of the numbers b_i' is negative, for example $b_r' < 0$, then the point $y_i = 0$, $i = 1, 2, \ldots, n$, does not satisfy the rth inequality, since $y_r = b_r' < 0$. If in addition for every coefficient of this row $b_{rj} \geq 0$, $j = 1, 2, \ldots, n$, then there is no point $y_i \geq 0$, $i = 1, 2, \ldots, n$, that satisfies this inequality. The given system then has no solution.

If, however, for $b_r' < 0$ there exists a coefficient $b_{rs} < 0$ in the rth row, then one forms the quotients b_i'/b_{is}, $i = n + 1, \ldots, m$, with the coefficients of the sth column. If besides b_r'/b_{rs} there are other non-negative quotients, then one chooses the smallest. If this occurs in the i_0th row, then one chooses b_{i_0s} as exchange element for a Jordan step, which exchanges the variables y_{i_0} with y_s. From the equation $y_{i_0} = -\sum_{j=s} b_{i_0j}y_j - b_{i_0s}y_s + b_{i_0}'$ this gives the value $y_s = (-\sum_{j=s} b_{i_0j}y_j - y_{i_0})/b_{i_0s} + b_{i_0}'/b_{i_0s}$.

The term b'_{i_0}/b_{i_0s} is non-negative. If $i_0 = r$, then by the Jordan step one has achieved that all the elements of the new column 1 are non-negative and a corner point has been found.

If, however, $i_0 \neq r$, then the terms of the other rows $i \neq i_0$ have to be estimated. After the Jordan step they satisfy $-b_{is} \cdot b'_{i_0}/b_{i_0s} + b'_i = b_{is}[(b'_i/b_{is}) - (b'_{i_0}/b_{i_0s})]$. For rows i with $b'_i/b_{is} > b'_{i_0}/b_{i_0s} \geqslant 0$ an improvement is achieved in the case $b_{is} < 0$, since this negative term has a smaller absolute value after the Jordan step. A positive term $b_{is} > 0$ remains positive. For rows i with $b'_i/b_{is} < 0$, in the case $b_{is} < 0$ the term b'_i is positive and remains positive; in the case $b_{is} > 0$, b'_i was negative, remains negative, and its absolute value even increases.

It can be shown that for the cases $b'_i/b_{is} \geqslant 0$ but $b_{is} < 0$, for $b'_i/b_{is} < 0$ but $b_{is} > 0$, and for other special cases such as $b'_{i_0}/b_{i_0s} = 0$, a table with only positive elements results after finitely many exchange steps, leading to a corner point $(x_1, x_2, \ldots, x_n)$ of the solution region.

Example: To determine one corner point of the solution region of the inequalities

$$-x_1 + 2x_2 - 3x_3 - 2 \geqslant 0, \quad 4x_1 - x_2 + 4x_3 - 5 \geqslant 0,$$
$$-3x_1 + x_2 - 4x_3 + 3 \geqslant 0, \quad x_1 \geqslant 0, \quad x_2 \geqslant 0, \quad x_3 \geqslant 0.$$

This system of inequalities is already in standard form. According to the above algorithm one obtains successively the tables

	$-x_1$	$-x_2$	$-x_3$	1
y_1	1	−2	3	−2
y_2	−4	1	−4	−5
y_3	3	−1	4	3

	$-x_1$	$-x_2$	$-y_3$	1
y_1	$-5/4$	$-5/4$	$-3/4$	$-17/4$
y_2	−1	0	1	−2
x_3	$3/4$	$-1/4$	$1/4$	$3/4$

	$-x_3$	$-x_2$	$-y_3$	1
y_1	$20/12$	$-20/12$	$-4/12$	−3
y_2	$4/3$	$-1/3$	$4/3$	−1
x_1	$4/3$	$-1/3$	$1/3$	1

	$-x_3$	$-y_1$	$-y_3$	1
x_2	−1	$-12/20$	$4/20$	$36/20$
y_2	1	$-4/20$	$84/60$	$-8/20$
x_1	1	$-4/20$	$24/60$	$32/20$

	$-x_3$	$-y_2$	$-y_3$	1
x_2	−4	−3	−4	3
y_1	−5	−5	−7	2
x_1	0	−1	−1	2

29.6. Nomographical procedures

Nomograms represent the functional dependence of several variables graphically in such a way that the value of one of them can be obtained from the given values of the others by a simple geometrical construction.

Nomograms for two variables

For the functional relationship $y = f(x)$ its graphical representation in a Cartesian coordinate system already forms a nomogram, which consists of the two coordinate axes and, in general, a curve. Its graph can be changed if along the x- and y-axes one plots not multiples of a unit distance, but lengths $\xi = \varphi(x)$ and $\eta = \psi(y)$, respectively, which are given by invertible monotonic functions φ and ψ. For a suitable choice of these functions one can obtain graph paper in which the *scale carrier* $\eta = g(\xi)$ represents the given relation $y = \psi^{-1}g(\varphi(x)) = f(x)$, so that $f = \psi^{-1}g\varphi$, where ψ^{-1} is the inverse function of ψ. The *scale carrier is a straight line* if $\eta = \alpha + \beta\xi$, or $\psi(y) = \alpha + \beta\varphi(x)$.

Examples: 1. For *semi-logarithmic paper*, $\xi = x$ and $\eta = \log_a x$. Functions $y = Ka^{Lx}$ have as scale carrier a straight line with $\alpha = \log_a K$ and $\beta = L$.

2. For doubly logarithmic paper, $\xi = \log_a x$ and $\eta = m \log_a y$. Functions $y = Kx^L$ have as scale carrier a straight line with $\alpha = m \log_a K$ and $\beta = mL$.

3. For probability paper, $\xi = x$ and $\eta = F^{-1}(y)$, where F^{-1} is the inverse function of the Gaussian error function $F(w) = \dfrac{1}{\sqrt{(2\pi)}} \displaystyle\int_{-\infty}^{w} \exp\,[-x^2/2]\,dx$. The scale carrier is then a straight line for the functions $y = F(K + Lx)$, that is, the distribution functions of all normal distributions.

Double scales are scale carriers on which, corresponding to the x-values of a sufficient number of points, the associated y-values are arranged immediately opposite. One can imagine these value scales to be transferred by parallel projection from the x-axis and the y-axis, so that these axes need no longer be given (Fig.). One obtains a *functional scale* or a *curved scale* of a variable u, if every point of a curve marked with a parameter value u is determined in a fixed x, y-system by

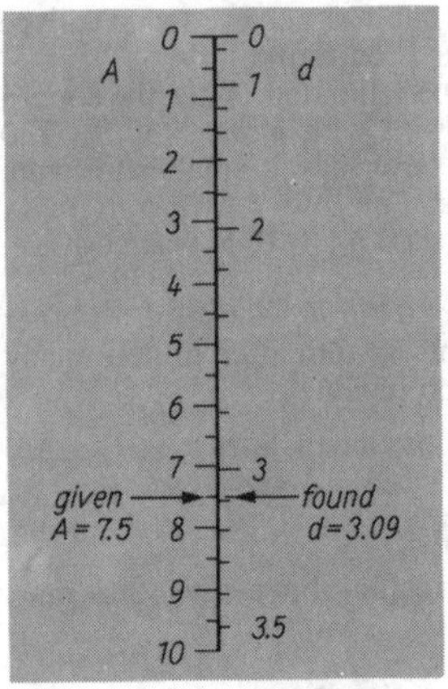

29.6-1 Double scale for the relation between the area of a circle $A = \pi d^2/4$ and its diameter d

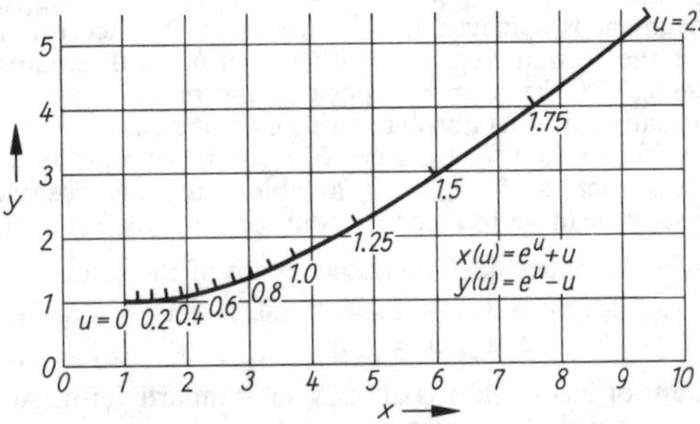

29.6-2 Scale holder with the equation $(x - y)/2 = \ln [(x + y)/2]$

$x = \varphi(u)$ and $y = \psi(u)$. The curve of this function scale then represents the relation between the functions $\varphi(u)$ and $\psi(u)$.

Example: For the functions $x = \varphi(u) = e^u + u$ and $y = \psi(u) = e^u - u$, because $u = (1/2) \ln (x + y)$, one obtains the equation $(x - y)/2 = \ln (x + y)/2$ of the scale carrier (Fig.).

Nomograms for three variables

For a functional relation $F(u, v, w) = 0$, to be able to read easily the value of one variable from those of the two remaining ones, one usually uses collineation nomograms or alignment charts.

Collineation nomograms. If one regards each of the three variables as a parameter, then by means of six functions φ_i, ψ_i with $i = 1, 2, 3$ one can find three functional scales in a unique x, y-coordinate system given by the equations $x_1 = \varphi_1(u), y_1 = \psi_1(u); x_2 = \varphi_2(v), y_2 = \psi_2(v); x_3 = \varphi_3(w), y_3 = \psi_3(w)$. To facilitate the reading it will additionally be required of the functions φ_i, ψ_i that value triples (u_0, v_0, w_0) belonging together in accordance with the equation $F(u_0, v_0, w_0) = 0$ lie on a straight line (Fig.), that is, that the triangle with vertices (x_1, y_1), (x_2, y_2), (x_3, y_3) has zero area. The Soreau equation is a necessary and sufficient condition for this.

Soreau equation	$\begin{vmatrix} 1 & \varphi_1(u) & \psi_1(u) \\ 1 & \varphi_2(v) & \psi_2(v) \\ 1 & \varphi_3(w) & \psi_3(w) \end{vmatrix} = 0$

If one has functions φ_i, ψ_i that satisfy this equation, then the function scales are determined by $x_i = \varphi_i$ and $y_i = \psi_i$; of course not uniquely, since every transformation of the common coordinate plane that transforms straight lines again into straight lines gives rise to a new collineation chart. Such a transformation can be used to improve the size of the variation intervals of the variables and thus the accuracy of the readings.

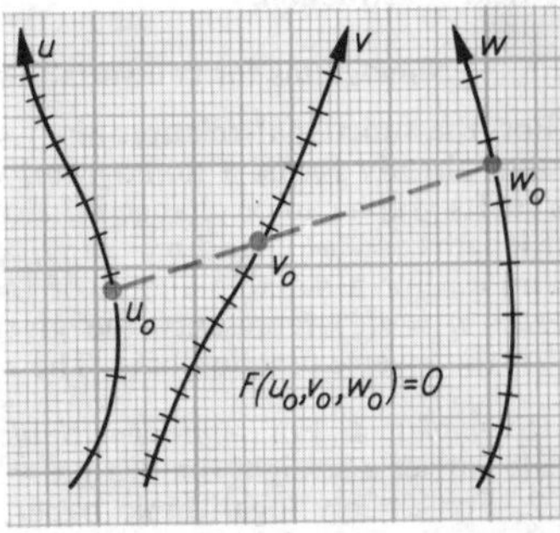

29.6-3 The associated value triple (u_0, v_0, w_0) of the collineation nomogram lies on a straight line

Basic forms and scale equations for collineation nomograms.

(1) The three points (x_1, y_1), (x_2, y_2), (x_3, y_3) lie on a straight line if $(y_1 - y_2)/(x_1 - x_2) = (y_1 - y_3)/(x_1 - x_3)$. If one puts $x_1 = g_1(u)$, $y_1 = f_1(u)$; $x_2 = -g_2(v)$, $y_2 = -f_2(v)$ and $x_3 = -g_3(w)$, $y_3 = -f_3(w)$, then no linear relation exists between the f_i, g_i; three curvilinear scales result and the basic form

$$[f_1(u) + f_2(v)]/[g_1(u) + g_2(v)] = [f_1(u) + f_3(w)]/[g_1(u) + g_3(w)].$$

(2) For $x_1 = 0$ the conditional equation (1) becomes

$$(y_1 - y_2)/(-x_2) = (y_1 - y_3)/(-x_3) \quad \text{or} \quad y_1 = (y_3 x_2 - y_2 x_3)/(x_2 - x_3).$$

If then the scales S_1, S_2, S_3 are determined by $x_1 = 0$, $y_1 = f_1(u)$, $x_2 = -1/g_2(v)$, $y_2 = f_2(v)/g_2(v)$ and $x_3 = 1/g_3(w)$, $y_3 = f_3(w)/g_3(w)$, then S_1 is rectilinear, whereas no linear relations exist between f_2, g_2 and f_3, g_3. The basic form is $f_1(u) = [f_2(v) + f_3(w)]/[g_2(v) + g_3(w)]$.

(3) If one substitutes $y_2 = px_2 + q$ into the conditional equation (2), so that $x_1 = 0$, $y_1 = f_1(u)$; $x_2 = -1/g_2(v)$, $y_2 = px_2 + q$; $x_3 = 1/g_3(w)$ and $y_3 = f_3(w)/g_3(w)$, then S_1 and S_2 are rectilinear, p can be chosen arbitrarily, and only between f_3 and g_3 there is no linear relation; the basic form is $f_1(u) = [-p + qy_2(v) + f_3(w)]/[g_2(v) + g_3(w)]$.

(4) If through $y_3 = mx_3 + c$ the scale S_3 also becomes linear, where m can be freely chosen, then under otherwise identical conditions as in (3) the basic form becomes

$$f_1(u) = [-p + qg_2(v) + m + cg_3(w)]/[g_2(v) + g_3(w)].$$

(5) If $x_2 = 1$ is substituted into the conditional equation (2), it becomes $y_1 = (y_3 - y_2x_3)/(1 - x_3)$ or $y_1(1 - x_3) + y_2x_3 - y_3 = 0$. If one now introduces: $x_1 = 0$, $y_1 = f_1(u)$; $x_2 = 1$, $y_2 = f_2(v)$ and $x_3 = g_3(w)/[f_3(w) + g_3(w)]$, $y_3 = -h_3(w)/[f_3(w) + g_3(w)]$, then the scales S_1 and S_2 are linear and these lines are parallel, whereas the scale S_3 is rectilinear only if a linear relation exists between f_3 and g_3. By substitution one obtains the basic form $f_1(u)f_3(w) + f_2(v)g_3(w) + h_3(u) = 0$.

(6) One substitutes $x_1 = a$, $x_2 = b$, $x_3 = c$ into the conditional equation (2) and obtains $y_1 = (by_3 - cy_2)/(b - c)$ or $y_1(b - c) = y_3(b - a) + y_2(a - c)$, that is $y_3(a - b) = y_1(c - b) + y_2(a - c)$ using condition (1). With $y_1 = f_1(u)/(c - b)$, $y_2 = f_2(v)/(a - c)$ and $y_3 = f_3(w)/(a - b)$ the basic form $f_3(w) = f_1(u) + f_2(v)$ results, where the scales S_1, S_2, S_3 are parallel straight lines.

(7) General investigations have shown, that in addition to the ones derived, the following three basic forms are also possible:

$f_3(w) = f_1(u)f_2(v)$,

$f_1(u)f_2(v)f_3(w) = f_1(u) + f_2(v) + f_3(w)$

and

$f_1(u)f_2(v)f_3(w) + f_1(u) + f_2(v)g_3(w) + h_3(w) = 0$.

Example: The real solutions of the reduced cubic equation $w^3 - 3pw - 2q = 0$ can be obtained from a collineation nomogram with two rectilinear scales: $x_1 = 0$, $y_1 = -3p$; $x_2 = 1$, $y_2 = -2q$ and $x_3 = 1/(1 + w)$, $y_3 = -w^3/(1 + w)$, as can be seen from the basic form $f_1(p)f_3(w) + f_2(q)g_3(w) + h_3(w) = 0$. From this one can read off:

$f_1(p) = -3p$, $f_2(q) = -2q$, $f_3(w) = w$,

$g_3(w) = 1$, $h_3(w) = w^3$ (Fig.).

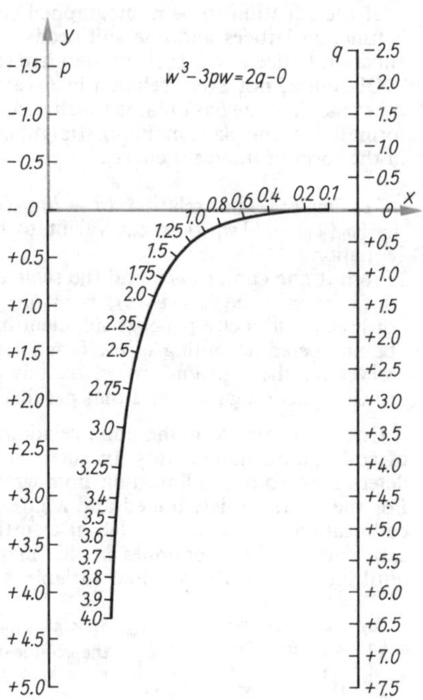

29.6-4 Nomogram for the real solutions of the equation $w^3 - 3pw - 2q = 0$

Nomograms for more than three variables

Instead of associating three points, one of each function scale, with one another by means of a straight line, one can look for other constructions for the association of a greater number of points, for example, one can associate with three points of the unique x, y-plane the centre of the circle determined by them, or of a triangle.

For the functional dependence $F(u_1, u_2, v_1, v_2, w_1, w_2) = 0$, of at most six variables, *intercept charts* have been constructed, which connect three *function lattices* by straight lines. By a function lattice one understands two families of coordinate curves, for example, the curves $r = \text{const} = \sqrt{(x^2 + y^2)}$ and the curves $\varphi = \text{const} = \arctan y/x$ of a polar coordinate system. Two values of the variables that are associated by this lattice determine a point with the coordinates $x_1 = \varphi_1(u_1, u_2)$, $y_1 = \psi_1(u_1, u_2)$. For the other two function lattices one then obtains in addition $x_2 = \varphi_2(v_1, v_2)$, $y_2 = \psi_2(v_1, v_2)$ and $x_3 = \varphi_3(w_1, w_2)$, $y_3 = \psi_3(w_1, w_2)$.

It is essential that for a function lattice there is a unique invertible correspondence between the points (u_1, u_2) and (x_1, y_1). As in the case of collineation nomograms one requires that three points (u_1^0, u_2^0), (v_1^0, v_2^0) and (w_1^0, w_2^0) of the three function lattices (u_1, u_2), (v_1, v_2) and (w_1, w_2), that satisfy

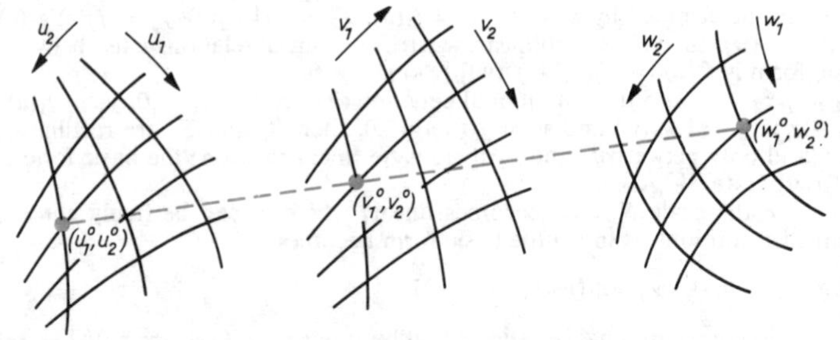

29.6-5 Corresponding points of three function lattices lie on a straight line

the given condition $F(u_1^0, u_2^0, v_1^0, v_2^0, w_1^0, w_2^0) = 0$, shall lie on a straight line (Fig.). In the Cartesian coordinate system common to all the function lattices the adjacent condition must then hold.

$$\begin{vmatrix} 1 & \varphi_1(u_1, u_2) & \psi_1(u_1, u_2) \\ 1 & \varphi_2(v_1, v_2) & \psi_2(v_1, v_2) \\ 1 & \varphi_3(w_1, w_2) & \psi_3(w_1, w_2) \end{vmatrix} = 0$$

If the equation to be nomographed contains 5 variables, then 4 variables can be represented by 2 function lattices and one still needs one scale carrier for the fifth variable; for 4 variables, one function lattice and 2 scale carriers are sufficient.

Of course, not every relation in 6 variables can be nomographed by an intercept chart. On the other hand, if one has obtained such a chart for a relation, then by application of an arbitrary transformation of the plane mapping straight lines again into straight lines one can obtain further solutions in the form of intercept charts.

Example: The relation $(a - b)\,\psi_3(w_2) = [a - \varphi_3(w_1)]\,\psi_2(v) + [\varphi_3(w_1) - b]\,\psi_1(u)$ is equivalent to the adjacent determinant equation.

$$\begin{vmatrix} 1 & a & \psi_1(u) \\ 1 & b & \psi_2(v) \\ 1 & \varphi_3(w_1) & \psi_3(w_2) \end{vmatrix} = 0.$$

From it one can at once find the scale equations, and the equations of the required function lattices are $x_1 = a$, $y_1 = \psi_1(u)$; $x_2 = b$, $y_2 = \psi_2(v)$; $x_3 = \varphi_3(w_1)$, $y_3 = \psi_3(w_2)$. For the realization ordinary millimetre paper is sufficient in which two straight lines at a distance $b - a$ apart must be numbered according to the functional scales $\psi_1(u)$ and $\psi_2(v)$. The same millimetre paper grid serves for the function lattice. Its axis perpendicular to the scale must be numbered according to the function $\varphi_3(w_1)$ and that parallel to the scale according to $\psi_3(w_2)$.

One often tries to nomograph relations with more than three variables by working with a chain of collineation nomograms. In such nomograms, for the given values of two variables one first determines from a collineation nomogram a value of an auxiliary variable, for example ξ (Fig.). For the value so determined and a given value of a third variable one determines from a second collineation nomogram a value of a further auxiliary variable, or of the solution variables, as the case may be. One continues in this manner, always by calculating values of auxiliary variables, until the value of the required variable is obtained from a final collineation nomogram.

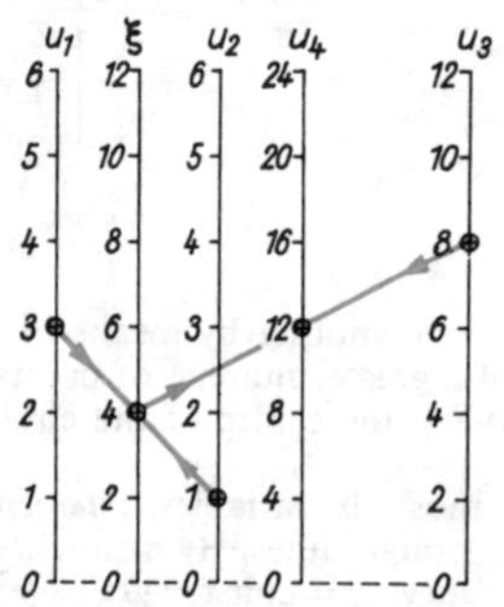

29.6-6 Auxiliary variable ξ in the collineation nomogram for $u_4 = u_1 + u_2 + u_3$

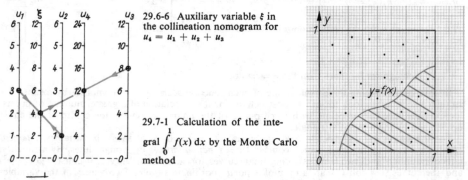

29.7-1 Calculation of the integral $\int_0^1 f(x)\,dx$ by the Monte Carlo method

29.7. Monte-Carlo methods

The name *Monte-Carlo methods* is given to all procedures that make use of the concept of randomness for the solution of deterministic problems, for example, the evaluation of integrals, the determination of extreme values, the solution of systems of equations and the solution of ordinary and partial differential equations. The example of the integral $\int_0^1 f(x)\,dx$ used to illustrate the method can be generalized directly to n-fold integrals; indeed, the value of Monte-Carlo methods only becomes evident for multi-dimensional problems. A finite interval of integration $[a, b]$ can be reduced by the linear transformation $u = (x - a)/(b - a)$ to the form $[0, 1]$.

Probability procedure. The function $f(x)$ is assumed to be bounded above and below so that it can be transformed to satisfy the condition $0 \leqslant f(x) \leqslant 1$.

The required integral is the value of the area of the region bounded by the curve $f(x)$, the abscissa and possibly also by straight line segments parallel to the ordinate axis (Fig.). If the area and that of the unit square were uniformly covered with mass, for example, cut from cardboard of uniform thickness, then the ratio of the masses of the two could be regarded as an estimate of the interval. If one imagines the two areas uniformly covered by n points of equal mass, and if m_f of them lie within the required area, then by counting one obtains m_f/n as an estimate of the required area. Here the number of points n is at least 10^4. Their *uniform distribution* must be truly random in both the x- and the y-directions, so that one deals with n mutually independent trials. Uniformly distributed random numbers make it possible to break off the procedure at a value n for which the successive estimates differ by less than a prescribed limit of accuracy. If $(k - 1)$ is the number of counting steps executed and I_{k-1} their result, then the recursive counting scheme

$$I_k = I_{k-1} + (\xi_k - I_{k-1})/k = [(k - 1)\, I_{k-1} + \xi_k]/k$$

has proved useful, where $\xi_k = 1$ if the kth point falls in the region of $f(x)$, and $\xi_k = 0$ otherwise.

Mean value procedure. If only uniformly distributed random numbers x_i in the interval $[0, 1]$ are chosen for the argument, and $f(x_i)$ is calculated for each, then the statistical mean $M[f(x)]$, multiplied by the width of the interval 1, is an estimate for the required integral. Because the arithmetic mean is an effective estimate for $M[f(x)]$, one obtains $(1/n) \sum f(x_i) = \int_0^1 f(x)\,dx$. Here the recursive formula $I_k = I_{k-1} + [f(x_k) - I_{k-1}]/k = [I_{k-1}(k - 1) + f(x_k)]/k$ has proved suitable.

Generation of uniformly distributed random numbers. When digital computers are used, it is not customary to build a random generator into the computer for the generation of random numbers. Such a truly random generator has relatively little flexibility if great complications are to be avoided. Moreover, the stationariness of the generated sequence of random numbers is not guaranteed over long periods of time, that is, their statistical properties change in the course of the test.

For this reason the random numbers are generated from deterministic recursive formulae. So that these *pseudo-random numbers* differ as little as possible from a sequence of truly random numbers one requires a *quasi-independence* of successive pseudo-random numbers and the non-occurrence of periodic number sequences. Recursive formulae based on elementary number theory have proved to be particularly favourable.

1. Reduced Fibonacci numbers $[F_k = F_{k-1} + F_{k-2}] \bmod m$;
2. $y_{k+1} = [(2^r + 1)\, y_k + c] \bmod 2^m$ with $r > 2$ and c even;
3. $y_k = s \cdot y_{k-1} \bmod m$, for example, $s = 23$ and $m = 10^8 + 1$;
4. $y_k = \sum_{j=1}^{r} c_j y_{k-j} \bmod m$, where the $y_0, y_1, \ldots, y_{r-1}$ are suitable numbers between 0 and $m - 1$ and the c_j for $j = 1, 2, \ldots, r$ are suitable constants.

Since all these pseudo-random numbers are determined modulo m or modulo 2^m, they are reduced by $x_k = y_k/m$ or $x_k = y_k/2^m$ to numbers in the interval $[0, 1]$.

30. Mathematical optimization

Optimization problems were already formulated by Euclid, but only with the development of the differential calculus and the calculus of variations in the 17th and 18th centuries was a mathematical tool forged for the solution of such problems. *Optimization problems in economics* are extreme value problems with auxiliary conditions, which are often characterized by the fact that the number of variables is very large and that non-negative solutions are sought.

In general, an economic occurrence to be studied is regarded as a process composed of different activities, and it is the aim to obtain by abstraction a corresponding mathematical model. For every activity several variants exist and their realization is dependent on constraints $g_j(x_i) = 0$ imposed by the available capacity, so that not for every activity can the most favourable variant be chosen. Instead a combination of possible variants is sought for which a given *objective function* $f(x_i)$ for the overall process assumes a maximum or minimum value; $g_j(x_i) = 0$ and $x_i \geqslant 0$ are called *constraints*.

Optimization problem	$\max\{f(x_i) \mid g_j(x_i) = 0,\ x_i \geqslant 0\}$	$i = 1, \ldots, n;\ j = 1, \ldots, m;\ n > m$

For *general non-linear optimization* there are no restrictions on the given functions f and g_j; for *quadratic optimization*, $f(x_i)$ is quadratic in the x_i and $g_j(x_i)$ is linear; for *linear optimization*, f and g_j are linear functions.

30.1. Linear optimization

In optimization it is usual to define the relation $<$ 'smaller than' also for matrices $\boldsymbol{A} = (a_{ij})$ and $\boldsymbol{B} = (b_{ij})$ of the same order (see Chapter 17.) by setting $\boldsymbol{A} < \boldsymbol{B}$ or $\boldsymbol{A} \leqslant \boldsymbol{B}$ if and only if $a_{ij} < b_{ij}$ or $a_{ij} \leqslant b_{ij}$, respectively, for each i and j. Correspondingly $\boldsymbol{A} > \boldsymbol{B}$ or $\boldsymbol{A} \geqslant \boldsymbol{B}$ are defined by $a_{ij} > b_{ij}$ or $a_{ij} \geqslant b_{ij}$, respectively. It is worth noticing that there may be two matrices of the same order for which none of the three relations $<, >, =$ holds, whereas for two rational or real numbers in any case exactly one of these relations holds.

If $\boldsymbol{c}$ and $\boldsymbol{x}$ are matrices of order $n \times 1$ with n rows and 1 column, $\boldsymbol{A} = (a_{ji})$ a matrix of order $m \times n$, $\boldsymbol{b}$ one of order $m \times 1$, $\boldsymbol{O}$ the zero matrix of order $m \times n$, $\boldsymbol{o}$ the zero matrix of order $n \times 1$, and $\boldsymbol{c}^T$ the transpose of the matrix $\boldsymbol{c}$, obtained from $\boldsymbol{c}$ by interchanging its rows and columns, then for linear objective functions and constraints, $f(x_i)$ can be represented by $\boldsymbol{c}^T\boldsymbol{x}$ and $g_j(x_i) = 0$ by $\boldsymbol{A}\boldsymbol{x} = \boldsymbol{b}$. When one puts $\boldsymbol{c} = -\boldsymbol{d}$, $\boldsymbol{A} = -\boldsymbol{B}$ and $\boldsymbol{b} = -\boldsymbol{h}$, the problem of maximization changes into one of minimization. For a geometrical interpretation, $\boldsymbol{x}$ may be regarded as a vector in n-dimensional Euclidean space R_n.

Linear optimization	$\max\{\boldsymbol{c}^T\boldsymbol{x} \mid \boldsymbol{A}\boldsymbol{x} \leqslant \boldsymbol{b},\ \boldsymbol{x} > \boldsymbol{o}\}$ or $\min\{\boldsymbol{d}^T\boldsymbol{x} \mid \boldsymbol{B}\boldsymbol{x} \geqslant \boldsymbol{h},\ \boldsymbol{x} \geqslant \boldsymbol{o}\}$

According to the elements of the matrices $\boldsymbol{c}, \boldsymbol{A}, \boldsymbol{b}$ or $\boldsymbol{d}, \boldsymbol{B}, \boldsymbol{h}$, various problems can be distinguished: *deterministic* problems if these coefficients are known constants, *parametric* problems if the coefficients (or some of them) can vary over known intervals, and *stochastic* problems if the coefficients (or some of them) are random variables.

Examples: 1. Maximum gain. If the components x_i of $\boldsymbol{x}$ are the piece numbers of a product or commodity in a manufacturing process and c_i the yield corresponding to one piece of the product i, then $\boldsymbol{x}$ represents a manufacturing programme and $\boldsymbol{c}^T\boldsymbol{x}$ its *total yield*, such as, for example, gain or proceeds in foreign exchange. Further, if k is one of the m activities, for example, a group of machines, b_k the available capacity (store capital), and if the coefficients a_{ki} of the matrix $\boldsymbol{A}$ represent the level of activity per piece of commodity, then the problem of maximizing the manufacturing programme is to calculate the maximum proceeds, taking into account the availability of the given capacities. The assumptions on which the model is based imply that both the proceeds and activities are proportional to the amount of production and that the vector $\boldsymbol{x}$ of the piece numbers can have only integral components $x_i \geqslant 0$. It is further assumed that the demand for the commodity is unlimited. If this is not the case, limits d_i in the sales can be introduced by the additional constraints $x_i \leqslant d_i$.

2. Diet problem. Let i be a nutriment of which the amount x_i is contained in a food combination to be determined, and let d_i be the cost per unit amount. Let k be a vitamin or nutrient substance of which the minimum amount h_k must occur and of which the nutriment i contains an amount b_{ki}. The solution leads to a minimization problem, which in this simple form, however, is applicable only to the cheapest combinations of animal feeding stuff. A different model for minimizing the cost of a food sequence of a hotel is obtained by additional refined and detailed assumptions, such as the *daily distribution of meals*, that is, breakfast, luncheon and dinner, the *structure of a meal*, for example, hors d'oevre, main dish and dessert, the *offer of a choice*, for example, of three dishes, and a *smallest period of sequence of dishes*, for example, of two weeks.

The maximization problem in linear optimization was formulated in 1939 by Kantorovich, who solved it by the method of *solution factors*. The diet problem was solved approximately in 1941 by Cornfield and in 1945 by Stigler. The problem of *linear optimization*, which was formulated quite generally by Wood and Dantzig, was solved by Dantzig by the *simplex method*, which has been further developed in many respects.

Simplex method. For the linear optimization the constraints consist of $Ax \leqslant b$ and $x \geqslant o$, or $a_{j1}x_1 + \cdots + a_{ji}x_i + \cdots + a_{jn}x_n \leqslant b_j$ and $x_i \geqslant 0$ for $j = 1, 2, \ldots, m$ and $i = 1, 2, \ldots, n$ (see Chapter 29.). Just as the condition $2x_1 + 3x_2 \leqslant 4$ defines a closed half-plane, so exactly $(n + m)$ closed half-spaces are determined by the above constraints if x is interpreted as a *point* or *vector* in an n-dimensional space R_n. If the $(n + m)$ constraints are consistent, then the intersection R of the $(n + m)$ half-spaces contains at least one point. Every point of R is a *feasible solution* or a *feasible vector*.

The *feasible region* R of the problem, as an intersection of $(n + m)$ half-spaces, is a *convex polyhedron* (Fig.). It is assumed that R is not empty and is bounded.

The objective function $f(x) = c^T x$ can be interpreted geometrically by considering the surface $f(x) = \text{const}$, which represents a family of parallel hyperplanes $c^T x = k$ in R_n. It is required to find the hyperplane with the greatest k having a non-empty intersection with the convex polyhedron R. Clearly this forms a *plane of support* of R for this family, that is, a hyperplane that has a point in common with R. Accordingly the maximum of f in R can occur only at boundary points.

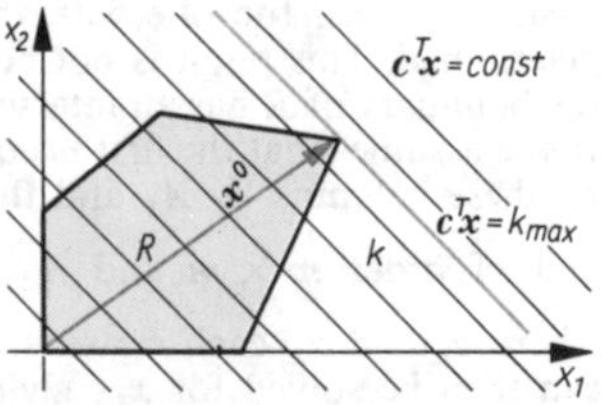

30.1-1 Geometrical representation of a maximum problem in a two-dimensional space R_2; R is the feasible region, $c^T x = k_{\max}$ a plane of support, x^o a feasible basic solution

With the above assumptions, R is the convex hull of its vertices. Let x^l $(l = 1, \ldots, s)$ be the vertices of R, of which there are at most $\binom{n+m}{n}$. Then every $x \in R$ can be represented by $x = \sum_{l=1}^{s} \lambda_l x^l$ with $\lambda_l \geqslant 0$ and $\sum_{l=1}^{s} \lambda_l = 1$. But it then follows that $f(x) = c^T x = c^T(\sum_{l=1}^{s} \lambda_l x^l) = \sum_{l=1}^{s} \lambda_l c^T x^l = \sum_{l=1}^{s} \lambda_l f(x^l)$. Among the s values $f(x^l)$ there is a greatest, say $f(x^{l_0})$. It is then certainly true that $f(x) = \sum_{l=1}^{s} \lambda_l f(x^l) \leqslant \sum_{l=1}^{s} \lambda_l f(x^{l_0}) = f(x^{l_0})$. When R is bounded and non-empty, the optimization problem thus reduces to the determination of the *vertices* x^l of R. In any case the solution is to be found among them.

The m inequalities $\sum_{i=1}^{n} a_{ji}x_i \leqslant b_j$ can be written in the form of equations $\sum_{i=1}^{n} a_{ji}x_i + x_{n+j} = b_j$, by introducing the m (so-called) slack variables $\bar{x} = \begin{pmatrix} x_{n+1} \\ \vdots \\ x_{n+m} \end{pmatrix}$. If I is the unit matrix, one obtains a further form of the LO-problem.

Linear optimization with slack variables	$\max \left\{ (c^T, o) \begin{pmatrix} x \\ \bar{x} \end{pmatrix} \middle\vert\, Ax + I\bar{x} = b,\ x \geq o,\ \bar{x} \geq o \right\}$

For the sake of simplicity one writes again $\max \{c^T x \mid Ax = b, x \geqslant o\}$ (with suitably enlarged matrices), where A is of order $m \times (m + n)$ and x is of order $(n + m) \times 1$. It can be assumed that A is of rank m, since otherwise either the equations $Ax = b$ would be inconsistent and no feasible vector would exist, or some of the equations would be superfluous, being linear combinations of the remaining ones.

A vector x having exactly m positive components that belong to m linearly independent columns of the matrix A is called a *feasible basic solution*.

The feasible basic solutions are precisely the vertices of the feasible region R.

For the analytical proof of this theorem one uses the *convex linear combination* $x = \lambda x^1 + (1 - \lambda) x^2 = \lambda(x^1 - x^2) + x^2$ with $0 < \lambda < 1$, which determines intermediate points x on the straight line segment connecting the points x^1 and x^2. The vertices of R alone cannot be represented by a convex linear combination of two different points of R. If A has m linearly independent columns $a_1, \ldots, a_m$ and if a feasible basic solution is x^1 with $x_1^1 > 0, \ldots, x_m^1 > 0$, $x_{m+1}^1 = x_{m+2}^1 = \cdots = x_{m+n}^1 = 0$, then a convex linear combination $x^1 = \lambda x^2 + (1 - \lambda)\, x^3$ with two different feasible points x^2 and x^3 is impossible. Because $x_{m+r}^t = 0$ for $t = 1, 2, 3$ and $r = 1, \ldots, n$, it would follow from $Ax^2 = b$ and $Ax^3 = b$ that $A(x^2 - x^3) = o$ with the trivial solution $x^2 - x^3 = o$; but this means that x^1 must be a vertex.

On the other hand, if it is assumed that x^1 is a vertex with positive components $x_1^1, \ldots, x_k^1$, then the corresponding columns $a_1, \ldots, a_k$ of A must be linearly independent. Since A has m rows, $k \leqslant m$ must hold, and it follows that x^1 is a feasible basic solution. For if the columns $a_1, \ldots, a_k$ were linearly dependent, then numbers $y_1, \ldots, y_k$, not all zero, could be found, such that $\sum_{j=1}^{k} y_j a_j = o$ and for $y > 0$ also $y \sum_{j=1}^{k} y_j a_j = o$. Consequently, since of $\sum_{j=1}^{k} x_j^1 a_j \pm y \sum_{j=1}^{k} y_j a_j = b$, by choosing a sufficiently small number y, two vectors $x^2 = (x_1^1 + yy_1, \ldots, x_k^1 + yy_k, 0, \ldots, 0)$ and $x^3 = (x_1^1 - yy_1, \ldots, x_k^1 - yy_k, 0, \ldots, 0)$ could be constructed whose first k components are positive. But because of the representation $x^1 = x^2/2 + x^3/2$ with $\lambda = 1/2$, then x^1 could not be a vertex, contrary to the first assumption.

The *degenerate case* $k < m$ is possible, but will be excluded here in these considerations. The degenerate case can be dealt with by means of the simplex method without special difficulties.

Thus, R has at most $\binom{m+n}{n} = \binom{m+n}{m}$ vertices. Among these finitely many vertices or feasible basic solutions, the one with the greatest function value k_{max} of the objective function must be determined. This point is not necessarily uniquely determined. This is the case, for example, when the boundary of R has an intersection of dimension $d \geqslant 1$ with the required hyperplane $c^T x = k_{max}$. If one assumes that the first m columns of A are linearly independent and denotes the matrix formed by these columns by A_1 and the remainder by A_2, then $A = (A_1, A_2)$, where A_1 is non-singular and of order $m \times m$ and A_2 is of order $m \times n$. Similarly one partitions $c = \begin{pmatrix} c_1 \\ c_2 \end{pmatrix}$, $x = \begin{pmatrix} x_1 \\ x_2 \end{pmatrix}$, where c_1 and x_1 each consists of the first m components. The equation $Ax = A_1 x_1 + A_2 x_2 = b$ can then be solved for x_1, giving $x_1 = A_1^{-1} b + A_1^{-1} A_2(-x_2)$. Assuming that $A_1^{-1} b > o$, together with $x_2 = o$, this gives a feasible basic solution x^1. Substitution into the objective function gives

$$f(x) = c_1^T A_1^{-1} b + [c_1^T A_1^{-1} A_2 - c_2^T](-x_2).$$

For $x_2 = o$ the value of the objective function becomes $f(x^1) = c_1^T A_1^{-1} b$. These relations are presented in the so-called *simplex tableau.*

Simplex tableau	$A_1^{-1} b$	$A_1^{-1} A_2$
	$c_1^T A_1^{-1} b$	$c_1^T A_1^{-1} A_2 - c_2^T$

The feasible basic solutions stand in the first column, and the last row gives the values of the corresponding objective function.

Three mutually exclusive cases may be distinguished in the simplex tableau:

1. The n elements of $c_1^T A_1^{-1} A_2 - c_2^T$ are non-negative. In this case an *optimal* solution exists, because if any element of x_2 is made positive, then the value of the objective function becomes at most smaller.

2. $c_1^T A_1^{-1} A_2 - c_2^T$ contains a negative element, say the kth; suppose that all the elements of the kth column of $A_1^{-1} A_2$ are non-positive. The kth component of x_2 can then be arbitrarily increased. If because of $x_1 = A_1^{-1} b + A_1^{-1} A_2(-x_2)$ the components of x_1 are changed at the same time, one always obtains feasible solutions for which the objective function increases beyond all bounds; $f(x)$ is not bounded in the feasible region and the problem has no solution.

3. The kth element of $c_1^T A_1^{-1} A_2 - c_2^T$ is again negative, but for each k the kth column of $A_1^{-1} A_2$ contains at least one positive element. In this case, too, one can increase the objective function by enlarging the kth component of x_2. One may do this, however, only until the first of the decreasing components of $x_1 = A_1^{-1} b + A_1^{-1} A_2(-x_2)$ assumes the value zero. The remaining (changed) components of x_1 and the kth component of x_2 determined in this way form a new feasible basic solution with a greater value for the objective function. The linear independence of the corresponding columns of A can be deduced. Since only finitely many feasible solutions exist and since because of the increase of the objective function at every simplex step one obtains new basic solutions, one arrives after finitely many steps at case 1 (optimal solution) or case 2.

Derivation of a first feasible basic solution. If $b > o$, then introducing slack variables and starting from the requirement $\max (c^T, o) \begin{pmatrix} x \\ \bar{x} \end{pmatrix}$ with $\bar{x} = b$ one obtains a feasible basic solution. If no slack variables can be introduced, then one can definitely achieve $b \geqslant o$ (and in every practical problem even $b > o$). By means of so-called *artificial variables* $y = (y_1, \ldots, y_m)$ one then first of all solves the problem

$$\min \{\sum_{j=1}^{m} y_j \mid Ax + Iy = b, x \geqslant o, y \geqslant o\}$$

for which $y = b$ provides a first feasible basic solution. If the minimum of $\sum_{j=1}^{m} y_j$ is positive, then the initial problem has *no feasible solution.* If the minimum is zero, then the optimal solution $(x^o, y^o) = (x^o, o)$ of the last problem is a feasible basic solution of the initial problem. For a computation programme one chooses the last described procedure, which is *problem independent.*

Duality. The problem $\min \{b^T y \mid A^T y \geqslant c, y \geqslant o\}$ is called the *dual problem* to the problem $\max \{c^T x \mid Ax \leqslant b, x \geqslant o\}$. The latter is also known as the *primal problem.* Here y is a matrix of order $m \times 1$ or a vector in R_m. A vector y with $A^T y \geqslant c$ and $y \geqslant o$ is called *feasible.*

Primal problem	$\max\{c^Tx \mid Ax \leqslant b, x \geqslant o\}$	Dual problem	$\min\{b^Ty \mid A^Ty \geqslant c, y \geqslant o\}$

If x and y (*primal* and *dual*, respectively) are feasible, then $c^Tx \leqslant b^Ty$.

Proof: x feasible means that $Ax \leqslant b$ and $x \geqslant o$; y feasible means that $A^Ty \geqslant c$ and $y \geqslant o$. Thus, $c^Tx \leqslant (A^Ty)^T x = (y^TA)\,x = y^T(Ax) \leqslant y^Tb = b^Ty$.
From this it follows without difficulty that:

If x^o and y^o are feasible and if $c^Tx^o = b^Ty^o$, then x^o and y^o are optimal for the primal and dual problem, respectively.

Proof: By the above theorem, for every feasible x, $c^Tx \leqslant b^Ty^o = c^Tx^o$, and on the other hand, for every feasible y, $b^Ty \geqslant c^Tx^o = b^Ty^o$. The first inequality shows that x^o is the solution of the primal problem and the second that y^o is the solution of the dual problem.
GALE, KUHN and TUCKER have proved the following duality theorem.

Duality theorem: x^o is a solution of the primal problem if and only if there exists a feasible y^o such that $c^Tx^o = b^Ty^o$; y^o is a solution of the dual problem if and only if there exists a feasible x^o such that $b^Ty^o = c^Tx^o$. The primal and the dual problem are soluble if and only if both have simultaneously feasible vectors.

These statements are of use especially if a problem can be solved only approximately and one would like to have an estimate of how far one is away from the optimum. This can also be important when a computer is used and to keep computation costs within economic limits one has to break off the computations.

Example 3: To maximize $x_1 + 4x_2$ under the constraints shown. – Using the slack variables x_3 and x_4 these and the objective function $f(x)$ become the equations shown below.
From these one obtains:

$$2x_1 + 3x_2 \leqslant 4,\quad 3x_1 + x_2 \leqslant 3,\quad x_1 \geqslant 0,\quad x_2 \geqslant 0$$

$$\begin{aligned} 1\cdot x_3 + 0\cdot x_4 + 2x_1 + 3x_2 &= 4,\\ 0\cdot x_3 + 1\cdot x_4 + 3x_1 + 1x_2 &= 3,\\ 0\cdot x_3 + 0\cdot x_4 + x_1 + 4x_2 &= f(x)\end{aligned}$$

$$A = \begin{pmatrix}1 & 0 & 2 & 3\\ 0 & 1 & 3 & 1\end{pmatrix},\quad A_1 = A_1^{-1} = \begin{pmatrix}1 & 0\\ 0 & 1\end{pmatrix},\quad A_2 = \begin{pmatrix}2 & 3\\ 3 & 1\end{pmatrix},$$

$$c_1^T = (0, 0),\ c_2^T = (1, 4),\ x_1^T = (x_3, x_4),\ x_2^T = (x_1, x_2)$$

and hence the new equations and the simplex tableau S_1. Since the inclusion of x_2 into the basis resulted in the greatest increase in the value of the objective function, x_2 will be used. Through the equation $x_3 = 4 - 2x_1 - 3x_2$, $x_2 = 4/3$ is determined, since this implies $x_3 = 0$, which thus leaves the basis. From the modified equations for x_2, x_4 and the value of $A_1^{-1}A_2$ as well as $c_1^T = (4, 0)$ and $c_2^T = (1, 0)$, a new form of the objective function $f(x) = 4x_2 + 0\cdot x_4 + x_1 + 0\cdot x_3$ can be determined:

$$\begin{aligned} x_3 &= 4 - 3x_1 - 3x_2,\\ x_4 &= 3 - 3x_1 - x_2,\\ f(x_1^T) &= 0 + (0 - 1)(-x_1) + (0 - 4)(-x_2).\end{aligned}$$

S_1			
	4	2	3
	3	3	1
	0	−1	−4

$$(x_1^T) = 16/3 + [8/3 - 1](-x_1) + [4/3 - 0](-x_3).$$

$$\begin{aligned} x_2 &= 4/3 - 2x_1/3 - x_3/3,\\ x_4 &= 5/3 - 7x_1/3 + x_3/3,\end{aligned}\qquad A_1^{-1}A_2 = \begin{pmatrix}2/3 & 1/3\\ 7/3 & -1/3\end{pmatrix}$$

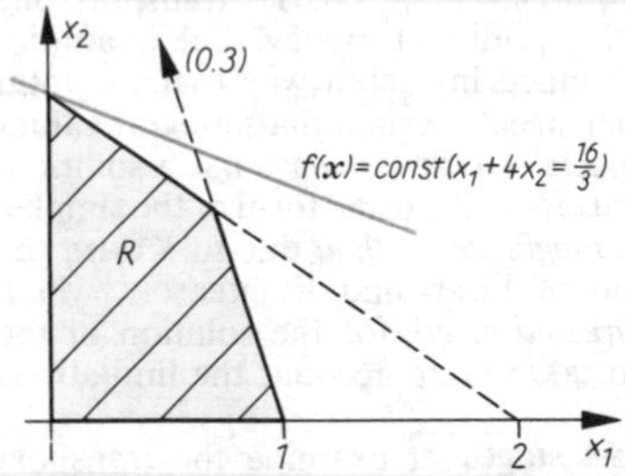

30.1-2 Solution of a primal problem with the objective function $f(x) = x_1 + 4x_2$

S_2			
	4/3	2/3	1/3
	5/3	7/3	−1/3
	16/3	5/3	4/3

One obtains the simplex tableau S_2. The optimal solution of the problem occurs for $x_2 = 4/3$, $x_4 = 5/3$, $x_1 = x_3 = 0$. In Fig. 30.1-2 the feasible region R of the original problem without slack variables is shown, as well as the straight line $x_1 + 4x_2 = 16/3$.

Shadow prices. In the form of the objective function that corresponds to the optimal solution, the coefficients $c_1^TA_1^{-1}A_2 - c_2^T$ belonging to the slack variables form the solution vector y^o of the dual problem. In the example $y^{oT} = (4/3, 0)$, and hence $b^Ty^o = 16/3 = f(x^o)$. The components of this dual solution vector are also called *shadow prices*. They indicate the extent to which the objective function is increased when the corresponding component of $\boldsymbol{b}$ is increased by unity. In the example an increase of $b_2 = 3$ would not achieve anything, since in the optimal solution $x_4 > 0$, and accordingly $3x_1 + x_2 < 3$ is not affected. On the other hand, an increase of 4/3 would result if $b_1 = 4$ were replaced by $b_1 = 5$, as can easily be verified.

Application of the simplex method. The simplex method has been improved in many respects with a view to reducing the rounding error, the required computer store and computer time.

LEMKE in 1954 developed the dual simplex method, in which he solved the primal problem via the solution of the dual problem. To save computer time one has also combined the primal and dual simplex. The *revised simplex method* is frequently used in conjunction with a representation in product form of the inverse matrix. For extensive problems one reverts by re-inversion to the data of the original matrix after a certain number of simplex steps, to reduce the rounding error.

For primal problems with an upper bound for the variables ($\boldsymbol{x} \leqslant \boldsymbol{d}$) DANTZIG developed a special algorithm whereby the amount of computation is comparable to that of the problem without such bounds. Finally, for problems with a special matrix structure, where only the cross-hatched area is occupied with non-zero elements (Fig.), DANTZIG and WOLFE developed in 1960 a *decomposition method*, in which the total problem is decomposed into a main and several auxiliary problems. In this way problems with 32000 constraints and 2 million variables have already been solved in justifiable computer times before 1963.

The method of the *fictitious game* suggested by BROWN and ROBINSON, which requires less store space than the simplex method, converges too slowly to become practicable.

The *transport problem*, which is an important special case of the primal problem, was formulated independently by HITCHCOCK in 1941 and KANTOROVICH in 1942.

Transport problem	$\min \{\sum_{i=1}^{m} \sum_{j=1}^{n} c_{ij}x_{ij} \mid \sum_{i=1}^{m} x_{ij} = b_j, \sum_{j=1}^{n} x_{ij} = a_i; x_{ij} \geqslant 0\}$

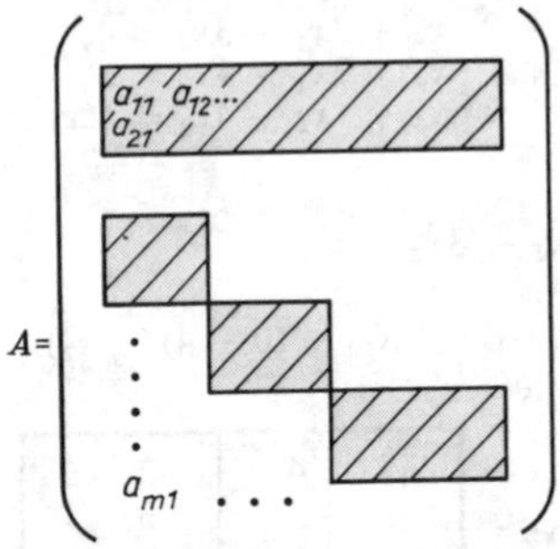

30.1-3 Decomposition method; scheme of a matrix in which only the elements indicated are different from zero

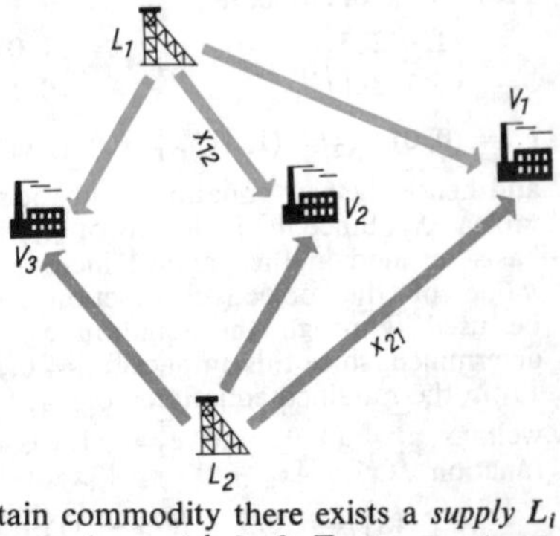

30.1-4 Transport optimization

The following meaning can be ascribed to it: For a certain commodity there exists a *supply* L_i with a *supply capacity* $a_i \geqslant 0$ and a *consumer* V_j with the *requirement* $b_j \geqslant 0$. To transport one unit of the commodity from the supplier to the consumer costs c_{ij} units of money. The quantities x_{ij}, which are transported from L_i to V_j (Fig.), are to be determined in such a way that the total transport is as cheap as possible. The matrix $\boldsymbol{A}$ now has a special structure such that integral values of b_j and a_i result in integral solutions for x_{ij}. Moreover, this special problem always has a solution. It has been used in practice very effectively and in many ways. Apart from the usual form of the simplex algorithm, one should mention the solution procedure by the *Hungarian method* due to KUHN, the *stepping-stone method* of CHARNES and COOPER and a method of FORD and FULKERSON, which makes use of the determination of the maximum flow in a *directed graph* for the solution of the problem. In this way one can solve the transport problem, also taking into account the limitations of the capacity of the transport routes.

Further generalizations distinguish between several transport stages or examine the transport of several commodities.

Example 4: Let $b_1 = 4$, $b_2 = 8$, $b_3 = 2$, $b_4 = 8$ be the requirements of 4 consumers and $a_1 = 10$, $a_2 = 7$, $a_3 = 5$ the capacity of 3 suppliers, so that $\sum a_i = \sum b_j = 22$. Should one of the cases $\sum a_i = a \gtrless b = \sum b_j$ occur, one assumes a fictitious supplier for the amount $(b - a)$ or a fictitious consumer for $(a - b)$, as the case may be. The coefficients c_{ij} of the matrix are the costs per piece for the transport from i to j.

$c_{11} = 20$	$c_{12} = 6$	$c_{13} = 4$	$c_{14} = 3$
$c_{21} = 10$	$c_{22} = 1$	$c_{23} = 5$	$c_{24} = 8$
$c_{31} = 9$	$c_{32} = 3$	$c_{33} = 8$	$c_{34} = 2$

By the *North-West corner rule* one attempts, starting from the North-West corner on the top left-hand side, to satisfy the given requirements according to the sums of b_j and a_i, respectively, from field to field in a maximal manner. The sequence of the fields with decision is indicated by red numbers, and those fields where at the same time the piece number has also been decided are marked by red arrows.

North-West corner rule

					a_i
	4 1	6 2	0	0	10
	0	2 3	2 4	3 5	7
	0	0	0	5	5
b_j	4	8	2	8	

Minimum matrix procedure

					a_i
	4	1	2	3 3	10
	0	7 1	0	0	7
	0	0	0	5 2	5
b_j	4	8	2	8	

Using the *minimum matrix procedure* one also satisfies maximally according to the sums a_i and b_j, but the fields with decision are fixed in each case according to the smallest value of c_{ij}, in order to save transportation costs. As is to be expected, the value of the objective function is now smaller; the North-West corner rule gives $f_1 = 162$, and the matrix minimum $f_2 = 120$. For the optimal solution $x_{13} = 2$, $x_{14} = 8$, $x_{22} = 7$, $x_{31} = 4$ and $x_{32} = 1$ one obtains $f_{\text{opt}} = 78$. The solution is degenerate, since only $5 = n + m - 2$ positive components occur, whilst because of the additional condition $a = b$, a total of $n + m - 1 = 6$ positive components is to be expected.

Integral optimization. In connection with the determination of a production programme which, in the presence of capacity restrictions, secures maximum proceeds, the problem of an integral solution of $\max \{c^T x \mid Ax \leqslant b, x \geqslant o\}$ occurs. One searches for the maximum of the objective function no longer among all the points of the feasible region R, but only among the *lattice points* contained in R.

For each concrete problem one must, of course, examine whether a treatment as an integral problem is really necessary. For example, if the coefficients of the problem are only estimates, then the additional labour does not pay. The inaccuracies introduced into the problem by the initial data are not appreciably increased by the normal procedure and the rounding off of the non-integral solution to a neighbouring lattice point.

For integral optimization the *method of the cutting plane* was developed by GOMORY in 1958. The problem is first of all treated in a normal manner until an optimal solution is obtained. If by chance this solution is not integral, then an additional constraint is introduced in such a way that the resulting solution is no longer feasible in the primal sense, but the remainder of the feasible region $R_1 \subseteq R$ contains all lattice points of R. The hyperplane, which cuts something off from R, has given the method its name. By its introduction the solution is not feasible as a primal solution, but remains feasible as a dual solution. In this way one can, by the dual simplex method, arrive relatively quickly at an optimal solution relative to R_1. If this is not integral, a second cutting plane is introduced, giving $R_2 \subseteq R_1$. This procedure leads to the desired integral solution after a finite number of steps.

In practical applications some problems arise because of rounding errors resulting from the finite calculations. Since the cutting planes that pass through the lattice points can be determined only approximately, certain limits of variation must be admitted when testing for the integral character of the solution. In spite of this it can happen that lattice points are cut off from R. This has led to many further studies of the problem, and the application of the method remains problematic as far as the technique of computation is concerned.

Parametric optimization. All coefficients in linear optimization can be dependent upon parameters. The simplest cases are

Parametric optimization	$\max \{(c + \lambda d)^T x \mid Ax \leqslant b, x \geqslant o\}$	$\max \{c^T x \mid Ax \leqslant b + \lambda f, x \geqslant o\}$

This problem arises when a production programme has to be determined which maximizes the gain for a given capacity, and when only bounds can be given for the gain per item of production, or when the question has to be answered as to how the optimal solution changes as a result of changes in production gains, for example, in sales prices. This is the first formulation. The second formulation answers the question as to changes in the production programme as a result of change in capacity, either through extensions or limitations. Here one can obtain, as was pointed out above, certain information from the components of the dual problem, the shadow prices.

The two formulations are mutually *dual*: the problem dual to the first formulation has the structure of the second. Hence the first formulation only will be considered in some detail. It is assumed that the feasible region R is bounded and not empty and that $\boldsymbol{c}$ and $\boldsymbol{d}$ are linearly independent. (For linearly dependent $\boldsymbol{c}$ and $\boldsymbol{d}$ a general primal problem results.) The hyperplanes $(\boldsymbol{c} + \lambda \boldsymbol{d})^T \boldsymbol{x} = k$ form for every *fixed* λ a family of parallel planes with parameter k. For a fixed k and variable λ one obtains a pencil of hyperplanes. Fig. 30.1-5 illustrates for two dimensions the situation, which can be established quite generally: If one draws the vector normal to the hyperplane $(\boldsymbol{c} + \lambda \boldsymbol{d})^T \boldsymbol{x} = k$ on the side of increasing k, then for $-\infty < \lambda < +\infty$ this vector sweeps out an angle of magnitude π. For the example of the figure, the λ-interval $(-\infty, +\infty)$ can be split up into four parts $-\infty < \lambda \leqslant \lambda_1 < 0$, $\lambda_1 \leqslant \lambda \leqslant \lambda_2 < 0$, $\lambda_2 \leqslant \lambda \leqslant \lambda_3 (\lambda_3 > 0)$, $\lambda_3 \leqslant \lambda < +\infty$, for which E_4, E_3, E_2, E_1, respectively, is the optimal solution. One can find quite generally a decomposition of $(-\infty, +\infty)$ into finitely many partial intervals such that for each λ-interval a feasible basic solution is optimal. For unbounded R no solutions need exist in the intervals $-\infty < \lambda \leqslant \lambda_1$ or $\lambda_r \leqslant \lambda < +\infty$.

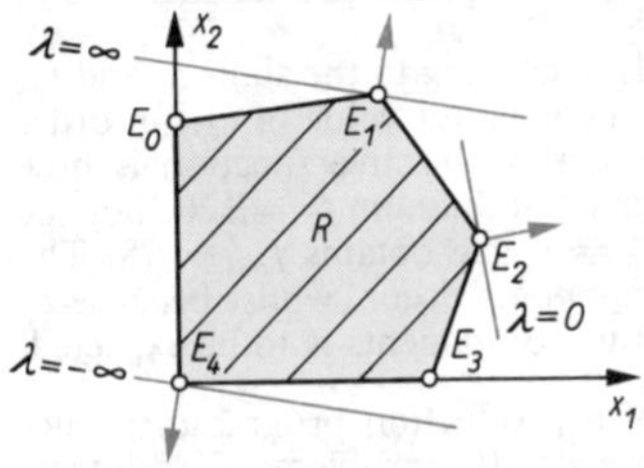

30.1-5 Parametric optimization in the plane

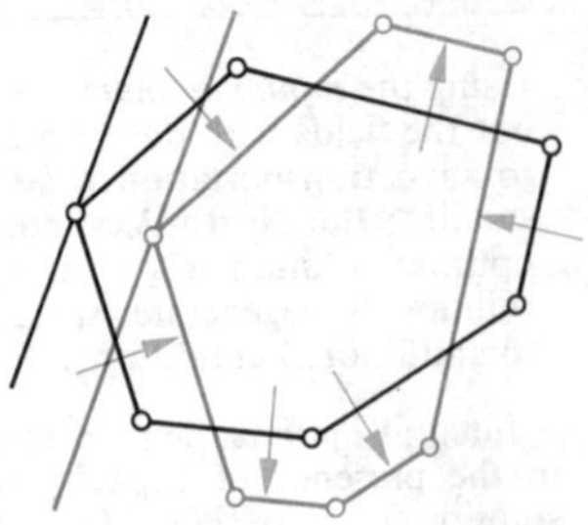

30.1-6 Geometric interpretation of the dual parametric problem

In practice one solves the problem for a certain λ by the simplex method and then determines the λ-interval for which the optimal solution obtained remains optimal. At the boundaries of this interval one can determine those components of $\boldsymbol{x}$ that for increasing or decreasing λ move into the basis, and likewise those that leave the basis. For this new basis one again obtains a λ-interval, and so on. Several procedures exist that are particularly suitable for this, for example, one proposed by SPURKLAND (1964).

For the dual problem a variation of λ means a parallel displacement of the hyperplanes enclosing the feasible region R (Fig.). It can be shown here that for certain λ-intervals the same components of $\boldsymbol{x}$ are in the basis, that is, are positive. The solution itself, however, changes with λ. The problem of the maximization of the gain of a production programme for given capacities suggests a further parametric problem. In Example 1 the element a_{ki} of the matrix $\boldsymbol{A}$ has represented the activity per unit of commodity i in the activity group k. These coefficients could also change, for example, by an increase in productivity or through the introduction of new technologies. Thus, the study of problems with more than one parameter is of interest, and a number of recent investigations show certain generalizations of the results that are known from one-parameter optimization.

Of course, with every practical problem the question arises of how a change in the coefficients influences the solution. In the case of bounded capacities the *shadow prices* give information about this, but the question can be answered in full generality only by *parametric optimization.*

Further applications of linear optimization. Linear optimization can be applied in many areas of natural sciences, technology, and economics. It is one of the effective mathematical methods of operational research. A few special cases of application will now be indicated.

Coordination problem. n agents are to be associated with exactly n tasks in such a way that each agent is associated with exactly one task and the total effort of the total cost is to be minimized. For example, a mechanical production process consists of n tasks and there are n workmen each of whom can, in principle, do each of the tasks, but in varying times. The times required by the workmen form a square matrix with elements c_{ij}. Each workman is to complete exactly one task. The mathematical formulation then assumes the form

$$\min \{\sum_{i=1}^{n} \sum_{j=1}^{n} c_{ij} x_{ij} \mid \sum_{i=1}^{n} x_{ij} = \sum_{j=1}^{n} x_{ij} = 1, x_{ij} \geqslant 0\}.$$

This is a special case of the transport problem. Exactly n of the x_{ij} are equal to 1 and the remainder 0. The basic solution contains, instead of $2n - 1$, only n positive components, that is, the problem is strongly degenerate. Kuhn's method is therefore more suitable for the solution than the simplex method.

Mixing problem. A typical mixing problem has already been mentioned in connection with the diet problem. The loading of a blast furnace for the production of cast iron is a second example.

One looks for the cheapest mixture of ore for the production of cast iron with definite properties. The production of a gas with prescribed calorific value by the mixing of gases of different manufacturing costs and known calorific values also leads to a problem of linear optimization.

Cutting problem. If sheets of given size are to be cut into different kinds of smaller sheets, then the question of reducing the waste to a minimum leads to a minimization problem. Such problems arise in the manufacture of metal, wood, textile and leather goods.

Stochastic linear optimization. A linear optimization problem whose coefficients are random quantities, max $\{\boldsymbol{c}^T\boldsymbol{x} \mid \boldsymbol{A}\boldsymbol{x} \leqslant \boldsymbol{b}, \boldsymbol{x} \geqslant \boldsymbol{o}\}$, is meaningful only in the sense of the *maximum expectation* of the objective function. If the random character of a problem is fully taken into account, it becomes in most cases either *non-linear*, or it must be represented or approximated by problems with piecewise-linear objective functions and linear constraints. There is only one very special case in which the resulting problem remains linear, namely when the components of $\boldsymbol{c}$, that is, the gains, or for a minimum problem the costs, are random quantities whose distribution functions are independent of the values of x_i. It is then permissible to replace the random components c_i of $\boldsymbol{c}$ by their *expectation values* $\bar{c}_i = E(c_i)$ and to treat the primal problem in a deterministic manner with the vector $\bar{\boldsymbol{c}}$ formed from the $\bar{c}_i$. This is possible, for example, in the treatment of the problem of *minimal cost of fodder* (diet problem) if one has to assume for the year under consideration such a randomness with known distribution for the prices of the individual types of food. The problem becomes more complicated if the components of $\boldsymbol{b}$ are random quantities, for example, if in the case of the optimum costs of a store of spare parts the demand is random, or if the elements of the matrix $\boldsymbol{A}$ are random.

30.2. Non-linear optimization

Among the *non-linear problems, convex optimization* can claim a certain completeness, at least as far as the theory is concerned.

Convex optimization. A region B of the n-dimensional Euclidean space R_n is called convex if for every two points of B all points that are *convex linear combinations* of them also belong to B (Fig.). A function $f(x)$ defined on a convex region B is called convex if for $\boldsymbol{x}^1$ and $\boldsymbol{x}^2$ in B and $0 < \lambda < 1$

$$f(\lambda \boldsymbol{x}^1 + (1-\lambda)\,\boldsymbol{x}^2) \leqslant \lambda f(\boldsymbol{x}^1) + (1-\lambda)\,f(\boldsymbol{x}^2)$$

always holds (Fig.). If for $\boldsymbol{x}^1 \neq \boldsymbol{x}^2$ the sign of equality is excluded, then $f(x)$ is called *strictly* or *properly convex*. A cube, parallelopiped, tetrahedron, sphere and ellipsoid are examples of convex regions in R_3.

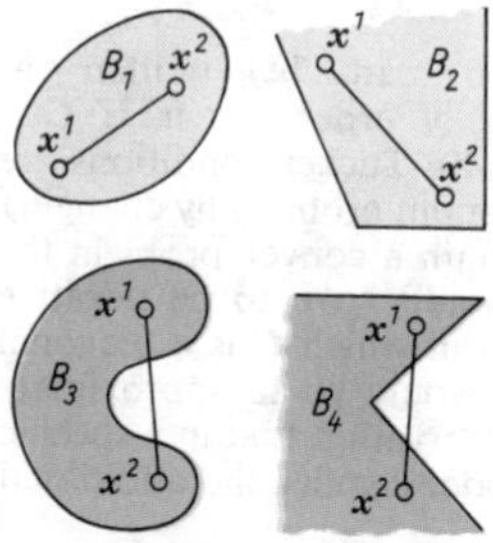

30.2-1 Convex sets M_1, M_2 and non-convex sets M_3, M_4

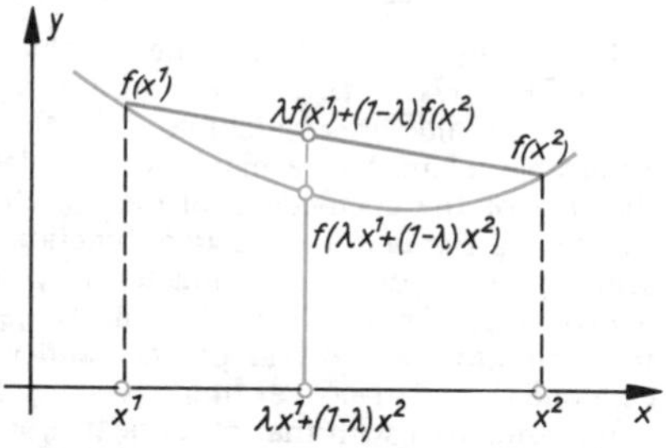

30.2-2 Graph of a convex function of one variable

The intersection of convex regions is a convex region, a property that has been used already in considering the feasible region R of the maximization problem. If the functions f and g_j in min $\{f(\boldsymbol{x}) \mid g_j(\boldsymbol{x}) \leqslant 0, x_i \geqslant 0\}$ $(i = 1 \ldots n, j = 1 \ldots m < n)$ are convex functions, then one speaks of a problem of *convex optimization.* The *feasible region* R of the problem is a *convex region* in R_n. For the existence of a solution Kuhn, Tucker and Slater have proved a fundamental theorem, the *saddle point theorem.* It refers to the saddle point $(\boldsymbol{x}^o, \boldsymbol{u}^o)$ of a function $F(\boldsymbol{x}, \boldsymbol{u})$ of two variables which for $\boldsymbol{u}$ assumes a maximum at this point, so that for values in the neighbourhood of $\boldsymbol{u}^o$ it decreases, but for $\boldsymbol{x}$ has a minimum, and thus increases for values in the neighbourhood of $\boldsymbol{x}^o$. This function $F(\boldsymbol{x}, \boldsymbol{u})$ is obtained by the *method of Lagrange multipliers* for extremal problems with side conditions. Using the Lagrange multipliers u_j with $j = 1, \ldots, m$ the *generalized Lagrange function* $F(\boldsymbol{x}, \boldsymbol{u}) = f(\boldsymbol{x}) + \sum_{j=1}^{m} u_j \cdot g_j(\boldsymbol{x})$ is defined in which $\boldsymbol{u}$ is the $m \times 1$ matrix formed from the u_j.

Saddle point theorem. If an $x^1 \geqslant o$ exists with $g_j(x^1) < 0$ for $j = 1, \ldots, m$, then $x^o \geqslant o$ is a solution of $\min \{f(x) \mid g_j(x) \leqslant 0, x_i \geqslant 0\}$ if and only if for all $x \geqslant o$, $u \geqslant o$ a $u^o \geqslant o$ exists with $F(x^o, u) \leqslant F(x^o, u^o) \leqslant F(x, u^o)$.

Accordingly the function $F(x, u)$ has at (x^o, u^o) a non-negative saddle point. It will now be shown that this is sufficient for $x^o \geqslant o$ to be a solution of the convex problem. One obtains

$$f(x^o) + \sum_{j=1}^{m} u_j g_j(x^o) \leqslant f(x^o) + \sum_{j=1}^{m} u_j^o g_j(x^o) \leqslant f(x) + \sum_{j=1}^{m} u_j^o g_j(x) \quad \text{for all } x \geqslant o \text{ and all } u \geqslant o.$$

From the left-hand inequality it follows that $g_j(x^o) \leqslant 0$ for $j = 1, \ldots, m$; for with a positive $g_{j_0}(x^o)$, taking $u_{j_0} > 0$, $u_j = 0$ for $j \neq j_0$, the left-hand side could be made to increase without limit. Hence $x^o \geqslant o$ is feasible. Furthermore, $\sum_{j=1}^{m} u_j^o g_j(x^o) = 0$, since otherwise the left-hand inequality would not hold for $u = o$, because all $g_j(x^o) < 0$ and $u^o \geqslant o$. It thus follows that $f(x^o) \leqslant f(x) + \sum_{j=1}^{m} u_j^o g_j(x)$ for all $x \geqslant o$, and consequently $f(x^o) \leqslant f(x)$ for all $x \geqslant o$ with $g_j(x) \leqslant 0$ $(j = 1, \ldots, m)$, since $u^o \geqslant o$. But this means that $x^o \geqslant o$ is a solution.

A complete proof of the saddle point theorem as stated here is due to SLATER. KUHN and TUCKER have proved it for differentiable functions, and for such functions the *local Kuhn-Tucker conditions*, which are equivalent to the conditions of the theorem, are stated below. These necessary and sufficient conditions for a solution of the convex optimization problem are utilized in many applications – especially in the method of quadratic optimization.

Local Kuhn-Tucker conditions. For convex differentiable functions $f(x)$, $g_j(x)$ the existence of an $x^o \geqslant o$ and a $u^o \geqslant o$ with

$$\frac{\partial F(x^o, u^o)}{\partial x} \geqslant o, \; x^{oT} \frac{\partial F(x^o, u^o)}{\partial x} = 0, \quad \frac{\partial F(x^o, u^o)}{\partial u} \leqslant o, \; u^{oT} \frac{\partial F(x^o, u^o)}{\partial u} = 0$$

is necessary and sufficient for $x^o \geqslant o$ to be a solution of the convex problem.

Quadratic optimization. The example given above of the determination of a manufacturing programme with maximum gain for a given capacity was a problem of linear optimization. The components c_i of c were the gain per unit produce i. The gain is the difference between the achieved price p_i and the costs k_i. The composition of k_i, as well as that of influencing factors that could alter p_i, will not be considered here in detail. The assumption that both p_i and k_i are independent of the piece number x_i of the product i is a great simplification. If it is assumed that a discount is offered for selling large numbers and that the cost per item decreases with increasing number, then the situation can be expressed approximately by $p_i = \bar{p}_i - r_i x_i$, $k_i = \bar{k}_i - s_i x_i$. One thus obtains the quadratic objective function $c^T x = \sum_{i=1}^{n} (p_i - k_i) x_i + \sum_{i=1}^{n} (s_i - r_i) x_i^2$.

In its complete form the problem of quadratic optimization can be written as $\min \{c^T x + x^T C x \mid Ax \leqslant b, x \geqslant o\}$, where C is a symmetric square matrix of order $n \times n$. If C is positive definite or semidefinite, then the problem is convex and the Kuhn-Tucker conditions are applicable. Here, too, a maximum problem can be changed into a minimum problem by changing the sign of the coefficients of the objective function. In this case, to obtain a convex problem the square matrix in the objective function for the maximum problem would have to be negative semi-definite. For the manufacturing programme the special case arises in which C is a diagonal matrix, which for $s_i - r_i \leqslant 0$ would be negative semi-definite. This, too, would be an approximate representation of the real process, and it may be useful to repeat the observation that in a specific case one should always examine whether the advantage of a 'better' model justifies the additional work compared with that of a linear model.

The Lagrange function for quadratic optimization assumes the form

$$F(x, u) = c^T x + x^T C x + u^T (Ax - b) \quad \text{so that} \quad \frac{\partial F}{\partial x} = c + 2Cx + A^T u, \quad \frac{\partial F}{\partial u} = Ax - b.$$

With $\frac{\partial F}{\partial x} = v$ and $-\frac{\partial F}{\partial u} = y$ one obtains the conditions:

(1) $Ax + y = b$; (2) $2Cx - v + A^T u = -c$;

(3) $x \geqslant o$, $v \geqslant o$, $y \geqslant o$, $u \geqslant o$; (4) $x^T v = 0$, $y^T u = 0$.

A vector $x \in R_n$ is a solution if and only if, together with a $v \in R_n$, a $u \in R_m$ and a $y \in R_m$, it satisfies the conditions (1) to (4). (1) to (3) form a linear system. The condition (4) can also be written as (4a) $x^T v + y^T u = 0$, since by (3) the vanishing of each individual term in the scalar product is required either by (4) or by (4a). Hence this condition states that a feasible solution is required for the linear system (1) to (3) for which at most one of the corresponding components of x and v, and similarly at most one of the corresponding components of y and u, may be positive. Altogether, at most $n + m$ components of the four vectors may be positive, that is, exactly the same number as there are equations in (1) and (2). The solutions of the system (1) to (4) are thus contained among the *feasible basic solutions* of the first three conditions. These feasible basic solutions can be determined by means of the *simplex method.*

For taking the last condition into account there are two possibilities:

The method of Wolfe. One introduces additional variables into (1) to (4) in such a way that a feasible basic solution for (1) to (3) satisfying (4) can be found without difficulty. One then proceeds with the simplex method in such a way that this condition remains satisfied and the additional variables are removed from the basis.

In the methods of Barankin-Dorfman and of Frank-Wolfe one begins with a feasible basic solution that does not satisfy the last condition and proceeds with the simplex method with the aim of minimizing the expression $x^Tv + y^Tu$.

Method of Frank-Wolfe. With $z^T = (x^T, y^T, v^T, u^T)$ and $\bar{z}^T = (v^T, u^T, x^T, y^T)$ one can write the Kuhn-Tucker conditions in the form $\min \{\bar{z}^T z \mid \bar{A}z = \bar{b},\ z \geqslant o\}$. Here $\bar{A} = \begin{pmatrix} A & I_m & O & O \\ 2C & O & -I_n & A^T \end{pmatrix}$ and $\bar{b} = \begin{pmatrix} b \\ -c \end{pmatrix}$, where I_m and I_n are unit matrices of order m and n, respectively. Because $\bar{z}^T z = 2(x^T v + y^T u)$, a solution of the original problem can be obtained as the x-portion of a solution z_0 of the transformed problem with $\bar{z}^T z_0 = 0$.

A feasible basic solution z_1 can be determined by methods known from linear optimization. Frank and Wolfe then consider $\bar{z}_1^T z$ with this fixed z_1 as objective function of the transformed problem. In this way the problem is linearized and the simplex method can be applied. If one obtains an optimal solution z_2 of this problem with $\bar{z}_1^T z_2 = 0$, then one has finished. Otherwise the following procedure is suggested. One continues with the method until one obtains with the basis z_k either the relation $\bar{z}_1^T z_k = 0$, in which case one has finished, or else $\bar{z}_1^T z_k \leqslant {}^1/_2 \bar{z}_1^T z_1$. In the second case Frank and Wolfe give a method for the construction of a *new* z_1. They show that one of the cases always occurs and that by the use of their method with a positive semi-definite C, the first case always occurs after a finite number of steps so that one obtains a solution.

Method of gradients. From the definitions $v = \dfrac{\partial F}{\partial x}$, $y = \dfrac{\partial F}{\partial u}$ it follows that $G(z) = \bar{z}^T z$ is a convex function. The linearization at the point z_1 is performed in such a way that $G(z)$ and the linear replacement function $H(z) = \bar{z}_1^T z$ have the *same gradient* at the point z_1. In this way one arrives at a new group of procedures, the *method of gradients*, which can be applied in quadratic as well as non-linear optimization. For a differentiable function $f(x)$ with $x \in R_n$, the gradient vector $\operatorname{grad} f = \dfrac{\partial f}{\partial x}$ is perpendicular to the surface $f(x) = \text{const}$ and points in the direction of maximum increase of the function $f(x)$. In order to minimize a given function without side conditions one starts from a given point x_0 and proceeds in the direction of $\operatorname{grad} f(x_0)$. If the function has a unique minimum, as for example for strictly convex functions with a finite minimum, then an iterative application of this method will certainly lead to the desired result. For a problem of convex optimization care must be taken, of course, that one remains in the feasible region.

The method of gradients can be described quite generally as follows: Starting from a feasible point x_0 one determines a direction such that, at least initially, one remains in the feasible region and that the value of the objective function decreases as rapidly as possible. One proceeds in this direction until either the objective function ceases to decrease or the boundary of the feasible region is reached. The point x_1 reached is used as the starting point for the next step. The various methods differ only in the way the direction of the steps is chosen.

Method of feasible directions of Zoutendijk. Let the problem be $\min \{f(x) \mid Ax \leqslant b\}$. Suppose that the given constraints also contain sign limitations for x. If a_j^T are the rows of the matrix A and b_j the components of b, then the constraints can be expressed by $a_j^T x \leqslant b_j$ for $j = 1, \ldots, m$. Let the objective function $f(x)$ be convex with continuous partial derivatives in the feasible region R. The method described above of proceeding from a point x^k in the feasible region to the next point x^{k+1} is carried out in the following form: One travels in the direction determined by a vector r^k along the ray $x^k + \lambda r^k$, where the direction of r^k is determined in such a way that for $\lambda > 0$ the ray remains at first in the feasible region. If x^k is an interior point of R, then this is no limitation. If x^k lies on the boundary of R, and if J is that index set for which $a_j^T x^k = b_j$ holds, then it is necessary and sufficient for the choice of r^k in the above sense that $a_j^T r^k \leqslant 0$ for $j \in J$. Such a direction is called *feasible*. In addition one aims to reduce the function $f(x)$ along the ray as much as possible; $f(x)$ will be reduced for all r^k with $\operatorname{grad}^T f(x^k) \cdot r^k < 0$. One writes $\operatorname{grad} f(x^k) = c$ and determines

r^k as the solution of the linear optimization problem min $\{c^T r \mid a_j^T r < 0$ for $j \in J\}$. Since, in general, $c^T r$ is not thereby bounded below, a feasible condition is still added. If one chooses $-1 \leqslant r_i \leqslant 1$ $(i = 1, \ldots, n)$ for the components r_i of r, then by the simplex method an r^k favourable for x^k can be determined. One goes along this ray $x^k + \lambda r^k$ until either $f(x)$ becomes minimal, that is, as far as λ_1 with grad $^T f(x^k + \lambda_1 r^k) \cdot r^k = 0$, or until a point is reached at which the ray leaves the feasible region. The corresponding value of λ_2 is determined from $\lambda_2 = \max \{\lambda \mid a_j^T(x^k + \lambda r^k) = b_j\}$ for $j = 1, \ldots, m$. By repeating the procedure described, the next point of approximation $x^{k+1} = x^k + \lambda_k r^k$ with $\lambda_k = \min(\lambda_1, \lambda_2)$ is obtained. If λ_k is not finite, then $f(x)$ has no finite minimum. ZOUTENDIJK has established the convergence of the method. In the special case of a quadratic function $f(x)$ it can be achieved by an additional rule that the procedure comes to a conclusion after finitely many steps.

Besides quadratic optimization there is a special form of non-linear optimization for which in recent years satisfactory methods of solution have been given and for which also a duality principle holds. These are problems in which the objective function is a *quotient of two linear functions* and the constraints are linear inequalities.

30.3. Dynamic optimization

The basic idea of dynamic optimization will first be illustrated by a simple example. A particular vehicle (lorry, goods wagon) is to be loaded with different objects of different kind; n is the number of kinds s_i $(i = 1, \ldots, n)$ of objects, $v_i > 0$ the price of the objects, $w_i > 0$ the weight of the objects, $u_i > 0$ the number of loaded objects of kind i, and z the total capacity of the vehicle in question, where $z \geqslant \min\limits_{i=1,\ldots,n} \{w_i\}$ must hold. The problem consists of determining the numbers u_i in such a way that a load with maximum price is achieved.

The problem thus leads to the following exercise in optimization: to determine the maximum of the objective function

$$f(u_1, \ldots, u_n) = \sum_{i=1}^{n} v_i u_i \quad \text{with the constraints}$$

$$u_i \geqslant 0 \quad (i = 1, \ldots, n) \quad \text{are all integers and} \quad \sum_{i=1}^{n} w_i u_i \leqslant z.$$

This problem may be regarded as an n-step process, in which at every step a u_i is determined, such that the required maximum is reached in the last step. The whole optimization problem is thus transformed into an event in time, that is, into a *process*.

Discrete deterministic processes. Let S be a system, for example, of an economic, mechanical or chemical nature, the state of which in the time interval $[t', t'']$ can be described by the n functions $x_i = x_i(t)$ $(i = 1, \ldots, n)$, where the set of the possible states $x^T = (x_1, \ldots, x_n)$ (in the time interval in question) lies in a given point set X of the n-dimensional Euclidean space. The components of x^T are called *state variables*.

For the definition of a discrete *deterministic process* the following assumptions are needed:

Let $t' = t_1 < \cdots < t_i < t_{i+1} < \cdots < t_{n+1} = t''$ be a given partition of the interval $[t', t'']$; then the state $x^{i+1} = (x_1(t_i), \ldots, x_n(t_i))$ of the system remains unchanged in the right-open time interval $[t_i, t_{i+1})$ $(i = 1, \ldots, n)$. The state x^{i+1} depends only on x^i and a certain decision e^i, that is, $x^{i+1} = T^i(x^i, e^i)$ $(i = 1, \ldots, n)$, where T^i (the transformation of the state in the ith step) is independent of earlier states.

The decision e^i is uniquely characterized by a certain n-dimensional vector $u^i = (u_1, \ldots, u_n)$, where the points u^i must lie in a prescribed region U^i. Each point $u^i \in U^i$ is a *permissible decision vector of the* i*th step* of the decision process. Every sequence $P = (u^1, \ldots, u^n)$ with $u^i \in U^i$ $(i = 1, \ldots, n)$ and $x^{i+1} = T^i(x^i, u^i) \in X$ for $i = 1, \ldots, n$ is called a *permissible strategy* (or *steering*) of the n-step decision process in question. Because the changes of state of the system S occur only at discrete points of time, and because the quantities involved are not random, all processes of this kind are called *discrete deterministic processes*.

Optimal strategy. For the discrete deterministic process of n steps a certain function $f(x^1, \ldots, x^n, x^{n+1}, u^1, \ldots, u^n)$ is given, which is defined over the domain $x^i \in X$ $(i = 1, \ldots, n)$, $u^i \in U^i$ $(i = 1, \ldots, n)$ and is called the *objective function*. If the initial state x^1 is given, then the objective function f can be expressed as a function of $x^1, u^1, \ldots, u^n$. This follows from the assumptions on the process, so that

$$f(x^1, T^1(x^1, u^1), T^2(T^1(x^1, u^1), u^2), \ldots, u^1, \ldots, u^n) = \tilde{f}(x^1, u^1, \ldots, u^n).$$

The optimization problem then consists of finding a permissible strategy $P_0 = (u_0^1, \ldots, u_0^n)$ for a given initial state x^1 with the property $\tilde{f}(x^1, u_0^1, \ldots, u_0^n) = \max\limits_{\{u^1, \ldots, u^n\}} \tilde{f}(x^1, u^1, \ldots, u^n)$, where $\{u^1, \ldots, u^n\}$

runs through the set of all permissible strategies. If such a strategy P_0 exists, it is called an *optimal strategy*. The method of dynamic optimization presupposes a certain property of the objective function, the so-called Markov property.

Markov property	**The function f is defined for each n, that is, a sequence of functions $f(x^1), f(x^1, x^2, u^1), f(x^1, x^2, x^3, u^1, u^2), \ldots$ is given that can be calculated recursively. The function $f(x^1, \ldots, x^n, x^{n+1}, u^1, \ldots, u^n)$ can be defined with the aid of the function $f(x^1, \ldots, x^n, u^1, \ldots, u^{n-1})$ and of x^{n+1} and u^n.**

It happens that for most decision processes in practical applications the class of *separable functions*, also called the class of *functions of additive character*, is adequate. They are functions that can be written in the form

$$f(x^1, \ldots, x^{n+1}, u^1, \ldots, u^n) = \sum_{i=1}^{n} g_i(x^i, u^i).$$

These functions then have the Markov property.

Under the assumption that for the (deterministic, discrete) n-step process described above with the separable objective function f the optimal strategy exists and hence also the optimal solution of the corresponding optimization problem, one introduces the notation $f_n(x^1) = \max_{\{u^1, \ldots, u^n\}} \sum_{i=1}^{n} g_i(x^i, u^i)$.

Bellman's principle of optimality. The method of dynamic optimization is based on the fact that instead of the original problem with fixed initial state x^1 and fixed number of steps n, a number of problems is considered. The value $f_n(x^1)$ is thus regarded as a function of x^1 and n. If one imagines that the value $f_n(x^1)$ has been calculated by any method whatsoever, then it is easy on the basis of the definition of $f_n(x^1)$ and the separable character of the objective function to derive recursively the formula

$$f_n(x^1) = \max_{u^1 \in U^1} \max_{u^2 \in U^2} \ldots \max_{u^n \in U^n} \{\sum_{i=1}^{n} g_i(x^i, u^i)\} = \max_{u^1 \in U^1} \{g_1(x^1, u^1) + f_{n-1}(x^2)\}.$$

But since $x^2 = T^1(x^1, u^1)$, the recursive formula goes over into the relation

$$f_n(x^1) = \max_{u^1 \in U^1} \{g_1(x^1, u^1) + f_{n-1}(T^1(x^1, u^1))\}.$$

This relation can also be derived on the basis of *Bellman's principle of optimality*, which contains the basic idea of dynamic optimization.

Bellman's principle of optimality	**If $u_0^1, \ldots, u_0^n$ represents the optimal strategy of a given n-step process with initial state x^1, then the decision sequence $u_0^2, \ldots, u_0^n$ represents the optimal strategy of the $(n-1)$-step process with initial state x^2. Here x^2 is that state into which the system S has changed from the initial state x^1 by the decision u^1.**

In the literature on dynamic optimization a reverse numbering has emerged, with x^{n+1} denoting the initial state and u^n the first decision in an n-step process. With this notation the transformation of states is described by $x^i = T^i(x^{i+1}, u^i)$, $i = 1, \ldots, n$, with $x^n = T^n(x, u^n)$, $x = x^{n+1}$, and the recursive formula, which corresponds to Bellman's optimality principle, goes over into the relation

$$f_n(x) = \max_{u^n \in U^n} \{g_n(x, u^n) + f_{n-1}(x^n)\} = \max_{u^n \in U^n} \{g_n(x, u^n) + f_{n-1}(T^n(x, u^n))\}.$$

Example 5: The problem mentioned at the beginning of this section can be regarded as a discrete, deterministic process of n steps, where with the natural numbering $u^i = (u_i)$ denotes the decision in the ith step, $(u^1, \ldots, u^n) = (u_1, \ldots, u_n)$ a permissible strategy with the given properties, and $x^{i+1} = x^i - u_i w_i = z - \sum_{j=1}^{i} w_j u_j$ $(i = 1, \ldots, n)$, where $x^1 = z$, the state in the ith step (free loading space). Then by the principle of optimality

$$f_n(z) = f_n(x^1) = \max_{\substack{u_1 \in \{0,1,\ldots\} \\ u_1 \cdot w_1 \leqslant z}} \{u_1 v_1 + f_{n-1}(z - u_1 w_1)\}, \quad \ldots \quad f_1(z) = \max_{\substack{u_1 \in \{0,1,\ldots\} \\ u_1 \cdot w_1 \leqslant z}} u_1 v_1$$

or, changing to the reverse numbering, as is customary in dynamic optimization,

$$f_n(z) = \max_{\substack{u_n \in \{0,1,\ldots\} \\ u_n w_n \leqslant z}} \{u_n v_n + f_{n-1}(z - u_n w_n)\}, \quad \ldots \quad f_1(z) = \max_{\substack{u_n \in \{0,1,\ldots\} \\ u_n w_n \leqslant z}} u_n v_n.$$

Clearly $g_i(u_i) = u_i v_i$.

Let the adjacent numerical values be given:

$n = 3$, $z = 100$,
$w_1 = 40$, $w_2 = 45$, $w_3 = 60$,
$v_1 = 20$, $v_2 = 75$, $v_3 = 102$.

In this case

$$g_1(u_1) = 20u_1, \quad g_2(u_2) = 75u_2, \quad g_3(u_3) = 102u_3, \quad f = 20u_1 + 75u_2 + 102u_3,$$

and the constraints are represented by $40u_1 + 45u_2 + 60u_3 \leqslant 100$, where u_1, u_2, u_3 are non-negative integers.

One first tabulates the functions $g_i(u_i)$ $(i = 1, 2, 3)$, where the condition $0 \leqslant u_i \leqslant z/w_i$ $(i = 1, 2, 3)$ must be observed with each u_i an integer.

z	u_1	$g_1(u_1)$	z	u_2	$g_2(u_2)$	z	u_3	$g_3(u_3)$
0–39	0	0	0–44	0	0	0–59	0	0
40–79	0	0	45–89	0	0	60–100	0	0
	1	20		1	75		1	102
80–100	0	0	90–100	0	0			
	1	20		1	75			
	2	40		2	150			

(1) (2) (3)

Using the formula $f_1(z) = \max\limits_{u_1} 20u_1$, with the aid of step (1), one calculates the adjacent table, in which $\bar{u}_1(z)$ denotes that value of u_1 for which $f_1(z)$ is obtained.

u_1	z	$f_1(z)$	$\bar{u}_1(z)$
0	0–39	0	0
1	40–79	20	1
2	80–100	40	2

(4)

For $n = 2$ one obtains $f_2(z) = \max\limits_{u_2} \{g_2(u_2) + f_1(z - u_2w_2)\}$ and by means of Tables (2) and (4) one arrives at the Table (5) where $\bar{u}_2(z)$ is that value of u_2 for which $f_2(z)$ is obtained. For $n = 3$ one has $f_3 = \max\limits_{u_3} \{g_3(u_3) + f_2(z - w_3u_3)\}$ and on the basis of this relation one arrives at the final Table (6).

$\bar{u}_1$	$\bar{u}_2$	z	$f_2(z)$	$\bar{u}_2(z)$
0	0	0–39	0	0
1	0	40–44	20	0
0	1	45–79	75	1
1	1	80–89	95	1
0	2	90–100	150	2

(5)

$\bar{u}_1$	$\bar{u}_2$	$\bar{u}_3$	z	$f_3(z)$	$\bar{u}_3(z)$
0	0	0	0–39	0	0
1	0	0	40–44	20	0
0	1	0	45–59	75	0
0	0	1	60–89	102	1
0	2	0	90–100	150	0

(6)

For $z = 100$ Table (6) shows that $\bar{u}_3 = 0$. Then $\bar{u}_2 = \bar{u}_2(z - \bar{u}_3w_3) = \bar{u}_2(100) = 2$, by Table (5). From Table (4) one obtains $\bar{u}_1 = \bar{u}_1(z - \bar{u}_2w_2 - \bar{u}_3w_3) = \bar{u}_1(100 - 2 \cdot 45) = \bar{u}_1(10) = 0$. Thus the optimal solution $(\bar{u}_1, \bar{u}_2, \bar{u}_3) = (0, 2, 0)$ is here unique. The optimal value of the load is then $2v_2 = 150$.

Method of functional equations. For this method, too, a description of a suitable example will first be presented.

An amount x of money is available and there are two possibilities for investing this money. The amount $u_1 (0 \leqslant u_1 \leqslant x)$ is invested in the first way and the amount $x - u_1$ in the second. In a given time interval, for example, in one year, the gain $g_1(u_1)$ is expected from the investment u_1, and the gain $g_2(x - u_1)$ from the investment $x - u_1$. At the end of the time interval the means that had to be employed to achieve the gains g_1 and g_2 will have lost some of their effectiveness, by amortization and the need for maintenance, so that after one year the state will be

$$x_1 = au_1 + b(x - u_1), \quad 0 < a < 1, \quad 0 < b < 1.$$

Starting with the amount x_1 at the beginning of the second year, the sum u_2 with $0 \leqslant u_2 \leqslant x_1$ will again be invested in the first way and the sum $x_1 - u_2$ in the second, so that the gain in two years is equal to

$$g_1(u_1) + g_2(x - u_1) + g_1(u_2) + g_2(x_1 - u_2).$$

In the same way, money will also be invested at the beginning of the third year, when the money disposed of will be $x_2 = au_2 + b(x_1 - u_2)$. After n years the gain achieved is

$$\sum_{i=1}^{n} \{g_1(u_i) + g_2(x_{i-1} - u_i)\} = \sum_{i=1}^{n} g(x_{i-1}, u_i)$$

with $x_0 = x$, where at the end of the nth year the sum available for disposal is $x_n = au_n + b(x_{n-1} - u_n)$. In this way a definite n-step process is described with the objective function

$$\sum_{i=1} g_i(x_{i-1}, u_i) = \sum_{i=1} g(x_{i-1}, u_i) = \sum_{i=1} \{g_1(u_i) + g_2(x_{i-1}, u_i)\}$$

with $x_0 = x$ and with the constraints $0 \leqslant u_i \leqslant x_{i-1}$ $(i = 1, \ldots, n)$, where the state transformation

$$x_i = T^i(x_{i-1}, u_i) = T(x_{i-1}, u_i) = au_i + b(x_{i-1} - u_i), \quad i = 1, \ldots, n,$$

does not depend on the step. If one chooses the *inverse* numbering of the state- and decision quantities, that is, if $x = x_{n+1}$ is regarded as the initial state, then one arrives at the following optimization problem: to determine the maximum of the objective function

$$\sum_{i=1}^{n} g(x_{i+1}, u_i) = \sum_{i=1}^{n} \{g_1(u_i) + g_2(x_{i+1} - u_i)\} \text{ with the constraints } 0 \leqslant u_i \leqslant x_{i+1} \quad (i = 1, \ldots, n),$$

$$x_i = T^i(x_{i+1}, u_i) = T(x_{i+1}, u_i) = au_i + b(x_{i+1} - u_i), \ i = 1, \ldots, n.$$

According to the optimality principle one obtains

$$f_n(x) = f_n(x_{n+1}) = \max_{0 \leqslant u_n \leqslant x_{n+1}} \{g_1(u_n) + g_2(x_{n+1} - u_n) + f_{n-1}(x_n)\}$$

$$= \max_{0 \leqslant u_n \leqslant x} \{g_1(u_n) + g_2(x - u_n) + f_{n-1}(T(x_{n+1}, u_n))\}$$

$$= \max_{0 \leqslant u \leqslant x} \{g_1(u) + g_2(x - u) + f_{n-1}(au + b(x - u))\}, \ n \geqslant 1, \text{ where } f_0 \text{ is defined by } f_0 = 0.$$

For $n = 1, 2, \ldots$ this system represents a system of functional equations for the unknown functions $f_1(x), \ldots, f_n(x)$. To solve the problem in a specific case, that is, when g_1, g_2 and the numbers a, b are prescribed, one determines recursively the solution of the last system of equations, that is, for each $n = 1, 2, \ldots$, the value $\bar{u}_n = \bar{u}_n(x)$ for which

$$g_1(\bar{u}_n) + g_2(x - \bar{u}_n) + f_{n-1}(a\bar{u}_n + b(x - \bar{u}_n)) = f_n(x)$$

with respect to $u \in [0, x]$ and $\bar{x}_i = \bar{x}_i(x) = a\bar{u}_i + b(\bar{x}_{i+1} - u_i)$, $i = 1, \ldots, n$. This is known as the *method of functional equations*.

Example 6: Let the gain functions of the introductory example be $g_1(u) = \alpha \sqrt{u}$, $g_2(x - u) = \beta \sqrt{(x - u)}$, where $\alpha > 0$, $\beta > 0$ are arbitrary numbers and $0 < a = b < 1$. Then in this case $x_i = au_i + b(x_{i+1} - u_i) = ax_{i+1}$, $i = 1, \ldots, n$, and since $x_{n+1} = x$, it follows that $x_i = a^{n+1-i}x$ $(i = 1, \ldots, n)$. The equations for f_n lead in this case to the relations

$$f_n(x) = \max_{0 \leqslant u \leqslant x} \{\alpha \sqrt{u} + \beta \sqrt{(x - u)} + f_{n-1}(ax), \quad n \geqslant 1, \quad f_0 = 0.$$

If for a fixed choice of $x > 0$ one defines the function $\varphi_{n,x}(u) = \alpha \sqrt{u} + \beta \sqrt{(x - u)} + f_{n-1}(ax)$, then $\frac{\mathrm{d}}{\mathrm{d}u}\varphi_{n,x} = \alpha/(2\sqrt{u}) - \beta/[2\sqrt{(x - u)}]$ for $u \in (0, x)$ and $\frac{\mathrm{d}}{\mathrm{d}u}\varphi_{n,x} = 0$ for the single point $0 < \bar{u}_n(x) = [\alpha^2/(\alpha^2 + \beta^2)] \cdot x < x$, at which the maximum of $\varphi_{n,x}(u)$ with respect to $u \in [0, x]$ is reached. Therefore $f_n(x) = \alpha \sqrt{\bar{u}} + \beta \sqrt{(x - \bar{u})} + f_{n-1}(ax) = \sqrt{(\alpha^2 + \beta^2)}\sqrt{x} + f_{n-1}(ax)$ for $n > 1$ with $f_0(x) = 0$.

From this it follows easily that:

$$f_1(x) = \sqrt{[(\alpha^2 + \beta^2)\, x]},$$

$$f_2(x) = \sqrt{[(\alpha^2 + \beta^2)\, x]} + \sqrt{(\alpha^2 + \beta^2)\, ax]} = \sqrt{[(\alpha^2 + \beta^2)\, x]}\,(1 + a^{1/2}),$$

$$\vdots$$

$$f_n(x) = \sqrt{[(\alpha^2 + \beta^2)\, x]}\,(1 + a^{1/2} + \cdots + a^{(n-1)/2}),$$

where $f_n(x)$ denotes the gain in the nth step.

Because $0 \leqslant u_i \leqslant x_{i+1}$ $(i = 1, \ldots, n)$,

$$\bar{u}_i(x_{i+1}) = [\alpha^2/(\alpha^2 + \beta^2)]\, x_{i+1} = [\alpha^2/(\alpha^2 + \beta^2)]\, a^{n-i}x \qquad (i = 1, \ldots, n).$$

Since the reverse numbering has been used, the sequence $[\alpha^2/(\alpha^2 + \beta^2)]\, x$, $[\alpha^2/(\alpha^2 + \beta^2)]\, ax, \ldots,$ $[\alpha^2/(\alpha^2 + \beta^2)]\, a^{n-i}x$ is the optimal strategy of the n-step process considered with the gain $f_n(x) = \sqrt{[(\alpha^2 + \beta^2)\, x]} \sum_{s=0}^{n-1} a^{s/2}$, where the amount $x_n = a^n x$ is available after the nth step.

Future developments. In contrast to the loading problem, the maximization of gain represents an n-step discrete process for which the permissible region of the decision quantities forms a connected compact set – in this case a closed interval. This is therefore a special case of those discrete n-step processes for which the region of the decision quantities is represented generally by a closed and bounded region of space of corresponding dimension. In such a simple case the method of functional equations can often be applied with success. In more complicated cases, especially when the decision at every step is characterized by a decision vector $\boldsymbol{u} = (u_1, \ldots, u_m)$, $m \geqslant 2$, further methods, such as the *method of multipliers* or the *method of successive approximations*, are used with the aim of reducing the original problem to finitely many simpler problems for which the store of a computer is adequate. The discrete deterministic problems form only a part of the decision problems that occur in dynamic optimization. It is sometimes advantageous to consider processes with infinitely many steps, although such a process never occurs in practice. For if a discrete process with very many steps is given and the passage to the limit $n \to \infty$ in the f_n-equations is possible, then instead of these relations one obtains a *single* functional equation

$$f(x) = \lim_{n\to\infty} f_n(x) = \max_{0 \leqslant u \leqslant x} \{g(x, u) + f_{n-1}(T(x, u))\}.$$

In general, this functional equation is easier to solve than the original problem and for large n it gives a good approximation to the required solution. The *theory of stationary processes* examines under what conditions this method is applicable.

Another class of optimization problems in dynamic optimization deals with decision processes in which, at every instant of time in the given time interval, a decision is possible and even required. One then speaks of *continuous decision processes*. The corresponding theory is closely linked with variational calculus and with the theory of optimal processes according to PONTRYAGIN. The mathematical requirements are quite demanding.

In contrast to the discrete deterministic processes are the *discrete stochastic processes*, for which the state at the end of a step is known only in the form of a probability distribution. These processes are often closer to the multifarious problems in economics than are deterministic models, and the methods of dynamic optimization have also been extended in this direction.

III. Brief reports on selected topics

31. Number theory

The original task of number theory was the investigation of the properties of the integers. Its systematic development as a branch of mathematics came rather late. Individual results were known in antiquity, for example to EUCLID (about 300 B.C.) and DIOPHANTOS (about 250 A.D.). In the 17th century remarkable discoveries of scientific significance occurred, above all, in the investigations of Pierre FERMAT (1601–1666). Great steps forward were taken in the many works of Leonhard EULER (1707–1783), which are full of fruitful far-reaching ideas. At last Carl Friedrich GAUSS (1777–1855) set up a uniform theory. In 1801 he published his *Disquisitiones arithmeticae*, a monumental work, which was the foundation of *higher arithmetic* in the strict sense.

Nowadays number theory is a widely ramified theory, making extensive use of both abstract algebra (mainly in *algebraic number theory*) and of profound methods of analysis (*in analytic number theory*). This leads to problems and to new branches that only have indirect connections with the integers.

In contrast to other parts of mathematics, many questions and results of number theory are comprehensible to the mathematical layman without specialized knowledge. But it turns out that proofs of theorems frequently require an extensive mathematical apparatus.

GAUSS called mathematics the *queen of the sciences*, and in 1808 (in a letter to his friend BÓLYAI) he said: 'It is remarkable that all those who study this science seriously develop a kind of passion for it'.

Rings and fields. The basic facts of *elementary number theory* have been treated in Chapter 1. From the theory of divisibility it is known that the quotient of two integers may, but need not be, an integer; for example, $15/3 = 5$, but $15/7$ is not an integer. One says: division, the inverse operation to multiplication, cannot always be performed within the domain of integers. Number systems in which the operations of addition, subtraction and multiplication can be performed without restriction are called *rings*. If division (except by 0) can also always be carried out in a number system, one speaks of a *field*; for example, the rational numbers form a field. In what follows, the ring of integers $0, \pm 1, \pm 2, \ldots$ is denoted by $\mathbf{Z}$ and the field of rational numbers by $\mathbf{Q}$. If a and b are two numbers in $\mathbf{Z}$ and $b \neq 0$, the quotient a/b lies in $\mathbf{Q}$, but not necessarily in $\mathbf{Z}$. If the latter happens, then a is said to be *divisible* by b or a *multiple* of b.

Ideals. Apart from $\mathbf{Z}$ one needs rings R whose elements may be real or complex numbers. Of great importance are subsets I of a ring R that have the following properties:

(*i*) if a and b are numbers in I, then so is $a - b$;

(*ii*) for every number r in R and every number a in I the product ra also lies in I.

These subsets I of R are called *ideals* in R.

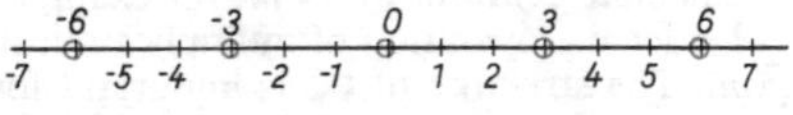

31-1 The ideal (3) on the number line

For example, if m is a natural number, the totality of numbers $0, \pm m, \pm 2m, \pm 3m, \ldots$ is an ideal in $\mathbf{Z}$ (Fig.). Clearly, the difference of two integral multiples of m is again a multiple of m (Property (*i*)) and every multiple of a multiple of m is itself a multiple of m (Property (*ii*)). In such a case one denotes the ideal by $M = (m)$ to indicate that it consists of all the multiples of the number m. Ideals consisting of all the multiples of a single element of a ring R, in the present case m, are called *principal* ideals. It is easy to verify that in $\mathbf{Z}$ every ideal is principal, so that all ideals (m) in $\mathbf{Z}$ are obtained by setting, in turn, $m = 0, 1, 2, \ldots$

A concept of divisibility can also be defined for ideals. One says that an ideal A is divisible by an ideal B if every element of A is also an element of B, in other words, if $A \subseteq B$ in the set-theoretical sense. Apparently the naive sense of the word 'divisible' is turned into its opposite, but the connection with the theory of divisibility immediately clarifies the reason for the nomenclature. An application of the definition to two ideals $\boldsymbol{A} = (a)$ and $\boldsymbol{B} = (b)$ in $\mathbf{Z}$ shows that $\boldsymbol{A}$ is divisible by $\boldsymbol{B}$ if and only if a is divisible by b. For example, the ideal (2) consists of the numbers $0, \pm 2, \pm 4, \pm 6, \pm 8, \ldots$ and the ideal (4) of the numbers $0, \pm 4, \pm 8, \pm 12, \ldots$ Hence $(4) \subseteq (2)$ set-theoretically; but this means that the ideal (4) is divisible by the ideal (2), because 4 is divisible by 2. Next, one defines a product $\boldsymbol{AB}$ of two ideals $\boldsymbol{A}$ and $\boldsymbol{B}$, namely as the ideal consisting of all sums of finitely many products ab, where now a and b are *arbitrary* elements of $\boldsymbol{A}$ and $\boldsymbol{B}$, respectively. In $\mathbf{Z}$ it follows that for $\boldsymbol{A} = (a)$ and $\boldsymbol{B} = (b)$ the product $\boldsymbol{AB} = (ab)$, for example $(2) \cdot (4) = (8)$.

The concept of an ideal was created in the development of algebraic number theory. *Ideal theory* studies the structure of rings and their ideals. In order to simplify the statement of results it is convenient to use the following mode of speaking: let a and b be numbers of a ring R and $\boldsymbol{I}$ an ideal in R; one says that $a \equiv b \pmod{\boldsymbol{I}}$ (read: a congruent to b modulo $\boldsymbol{I}$) if $a - b$ is a number in $\boldsymbol{I}$. This relation of *congruence* is an equivalence relation, being transitive, symmetric and reflexive. On the basis of these properties all the numbers of R can be divided into disjoint residue classes mod $\boldsymbol{I}$, in such a way that all the numbers congruent mod $\boldsymbol{I}$ belong to one and the same class. The significance of the formal way of writing lies in the fact that most calculating rules for equations also hold for congruences with respect to a fixed modulus.

Historically, the concept of congruence was first created by GAUSS for the ring $\mathbf{Z}$. Here $a \equiv b \pmod{m}$ means that the difference of the two integers a and b lies in the ideal $\boldsymbol{M} = (m)$, so that $a - b$ is divisible by m, or that a and b have the same remainder on division by m.

Examples: $88 \equiv -10 \pmod{14}$ because $88 - (-10) = 98$ is divisible by 14; $3^7 \equiv 1 \pmod{1093}$; $2^{32} \equiv -1 \pmod{641}$.

The integers

Certain properties of the residue classes in the ring of integers $\mathbf{Z}$ can be studied in relation to divisibility of numbers. Let (a, b) denote the greatest common divisor (gcd) of two numbers a and b and let p denote a prime number.

The residue class ring mod *m*. The residue classes mod m can be made into a ring by defining an addition and a multiplication of residue classes. This will be illustrated by an example. Let r_1 be the residue class $\bar{2}$ mod 6 consisting of the numbers $\ldots -10, -4, 2, 8, 14, \ldots$, let r_2 be the residue class $\bar{5}$ mod 6 consisting of the numbers $\ldots -7, -1, 5, 11, \ldots$ Then $r_1 + r_2$ is defined as the residue class which contains $2 + 5 = 7$, hence also, in fact, all the numbers that on division by 6 leave the remainder 1. One writes $\bar{2} + \bar{5} = \bar{1}$; other examples are $\bar{5} + \bar{0} = \bar{5}$, $\bar{3} + \bar{3} = \bar{0}$. The product is defined by $\bar{2} \cdot \bar{5} = \overline{10} = \bar{4}$; other examples mod 6 are $\bar{5} \cdot \bar{0} = \bar{0}$, $\bar{3} \cdot \bar{3} = \bar{3}$, $\bar{2} \cdot \bar{3} = \bar{0}$.

Under the addition and multiplication just defined the residue classes mod m form the so-called *residue class ring* mod m. The structure of this ring can be described in precise statements. It is a field if and only if m is a prime number.

The group of prime residue classes. By selecting from the m distinct residue classes $\bar{0}, \bar{1}, \bar{2}, \ldots, \overline{m-1}$ mod m all those whose numbers are prime to m one obtains the prime residue classes mod m. For $m = 6$ there are two prime residue classes $\bar{1}$ and $\bar{5}$; for $m = p$ there are always $p - 1$, namely $\bar{1}, \bar{2}, \ldots, \overline{p-1}$.

The number of prime residue classes mod m is denoted by $\varphi(m)$ (*Euler's function*). For example, $\varphi(6) = 2$, $\varphi(p) = p - 1$; $\varphi(m)$ is a number-theoretical function, that is, a function defined for integer arguments. It is a multiplicative function, that is, $\varphi(a, b) = \varphi(a)\,\varphi(b)$, provided that $(a, b) = 1$. It is easy to count out that $\varphi(p^k) = p^k - p^{k-1} = p^{k-1}(p - 1)$. The rule just stated makes it possible to calculate $\varphi(m)$ for every m; for example, $\varphi(3240) = \varphi(2^3)\,\varphi(3^4)\,\varphi(5) = 4 \cdot 54 \cdot 4 = 864$.

Under the operation of multiplication the prime residue classes mod m form a group G_m of order $\varphi(m)$. The structure of G_m is important for all m, but here only the case $m = p$ will be treated. G_p is cyclic, which means that every prime residue class mod p can be written as a power of a fixed residue class $\bar{g}$; such a g is called a *primitive root* mod p. For example, for $p = 11$ the residue class group G_{11} can be generated by $\bar{g} = \bar{2}$ (or by $\bar{6}, \bar{7}, \bar{8}$) for the powers $2^0 \equiv 1, 2^1 \equiv 2, 2^2 \equiv 4, 2^3 \equiv 8, 2^4 \equiv 5, 2^5 \equiv 10, 2^6 \equiv 9, 2^7 \equiv 7, 2^8 \equiv 3, 2^9 \equiv 6 \pmod{11}$ gives all the $p - 1 = 10$ prime residue classes $\bar{1}, \bar{2}, \ldots, \overline{10}$. Since every element a of a finite group G of order n satisfies the equality $a^n = e$ (e the unit element), one has for $G = G_m$ *Euler's theorem*:

Euler's theorem. *$a^{\varphi(m)} \equiv 1 \pmod{m}$ when $(a, m) = 1$. In the special case $m = p$ this becomes* **Fermat's theorem:** *$a^{p-1} \equiv 1 \pmod{p}$ when a is not divisible by p.*

Congruences with unknowns. In the residue class ring and residue class group certain algebraic problems can be solved. For example, one can ask what residue classes $\bar{x}$ mod m satisfy given equations, say $\overline{ax} = \bar{b}$. Such questions lead to congruences mod m with unknowns. The *linear congruence* $ax \equiv b \pmod{m}$ cannot always be solved, for example $3x \equiv 2 \pmod{12}$ is insoluble because no integer multiple of 3 leaves the remainder 2 on division by 12. In fact, $ax \equiv b \pmod{m}$

is soluble if and only if b is divisible by the greatest common divisor (a, m). A *congruence of degree n*, $x^n + a_1x^{n-1} + \cdots + a_n \equiv 0 \pmod m$ can have more incongruent solutions than its degree n indicates, for example $x^2 \equiv 1 \pmod 8$ has the solutions $x \equiv 1, 3, 5, 7$. If the modulus m is a prime number p, it need not have solutions at all and cannot have more than n residue classes as solutions.

Power residues. In the *binomial congruence* $x^n \equiv a \pmod p$ is a not divisible by p. Those residue classes $a \pmod p$ for which this congruence is soluble are called nth power residues mod p. Two fundamental questions arise:

1. What numbers are nth power residues for a given prime?
2. For what primes p is a given number a a nth power residue?

The first question is answered by *Euler's criterion* $a^{(p-1)/d} \equiv 1 \pmod p$ where $d = (p-1, n)$. Those and only those residue classes a that satisfy this condition are nth power residues. The answer to the second question leads to *reciprocity laws*, which are among the most beautiful and profound results of number theory. For $n = 2$ the residue classes $a \pmod p$ with $(a, p) = 1$ for which the congruence $x^2 \equiv a \pmod p$ is soluble are called *(quadratic) residues* and those for which the congruence is insoluble *(quadratic) non-residues*. For odd p there are equally many residues and non-residues mod p, namely $(p-1)/2$ each. For example, mod 17 one has $1^2 \equiv 1$, $2^2 \equiv 4$, $3^2 \equiv 9$, $4^2 \equiv 16$, $5^2 \equiv 8$, $6^2 \equiv 2$, $7^2 \equiv 15$, $8^2 \equiv 13$. Hence 1, 2, 4, 8, 9, 13, 15, 16 are the eight residues, and 3, 5, 6, 7, 10, 11, 12, 14 the eight non-residues mod 17.

The law of reciprocity. The investigation of the question for what moduli p a given number a is quadratic residue has led to the discovery of the celebrated quadratic reciprocity law, which was set up by EULER on the basis of extensive numerical material. Let p and q be odd prime numbers. If at least one of them is of the form $4k + 1$, then the congruences $x^2 \equiv p \pmod q$ and $y^2 \equiv q \pmod p$ are both soluble or both unsoluble (in other words, p is residue mod q if and only if q is residue mod p). But if both p and q are of the form $4k + 3$, then only one of the two congruences is soluble, the other insoluble. The first complete proof of the law was found by GAUSS when he was 18 years old. Later he gave six further proofs, of what he called *Theorema fundamentale*, and altogether more than fifty different proofs have been found since. The diverse principles employed and the endeavour to find reciprocity laws for higher power residues have given a powerful impetus to number theory.

About ten years befor Gauss's proof, a proof of the quadratic reciprocity law was published by LEGENDRE (1752–1833), but it contained a gap. Legendre introduced a useful symbol $\left(\frac{a}{p}\right)$ *(the Legendre symbol)*, which has the value $+1, -1$, or 0 according as a is residue (mod p), non-residue (mod p), or divisible by p. This makes it possible to express the law in a formula:

$$\left(\frac{p}{q}\right) \cdot \left(\frac{q}{p}\right) = (-1)^{\frac{p-1}{2} \cdot \frac{q-1}{2}}.$$

Supplements to the law of reciprocity are the propositions

$$\left(\frac{-1}{p}\right) = (-1)^{\frac{p-1}{2}} \quad \text{and} \quad \left(\frac{2}{p}\right) = (-1)^{\frac{p^2-1}{8}}.$$

The first of these states that the congruences $x^2 \equiv -1 \pmod p$ is soluble when p is of the form $4k + 1$, hence $(p-1)/2$ is even, and it is unsoluble when p has the form $4k + 3$, hence $(p-1)/2$ is odd.

Diophantine equations. Let $f(x_1, x_2, \ldots, x_n)$ be a polynomial in $x_1, x_2, \ldots, x_n$ with integer coefficients. An equation $f(x_1, x_2, \ldots, x_n) = b$ is called *Diophantine* if a solution is required to consist of integers. This kind of equation is named after the Greek mathematician DIOPHANTOS of Alexandria. A system of Diophantine equations $f_1 = b_1$, $f_2 = b_2, \ldots, f_k = b_k$ is equivalent to the single equation $(f_1 - b_1)^2 + (f_2 - b_2)^2 + \cdots + (f_k - b_k)^2 = 0$, that is, the solution sets in $\mathbf{Z}$ of the system and of the single equation are the same. A linear *Diophantine* equation
$a_1x_1 + a_2x_2 + \cdots + a_nx_n = b$ $(a_1, a_2, \ldots, a_n, b$ integers)
is soluble if and only if b is divisible by the gcd $(a_1, a_2, \ldots, a_n)$.

If it is soluble at all, then there are infinitely many solving n-tuples. The case $n = 2$ is easy to survey. If a_1 and a_2 are coprime and x_1', x_2' is any solution of $a_1x_1 + a_2x_2 = b$, then the totality of solutions can be represented in the form $x_1 = x_1' + a_2t$, $x_2 = x_2' - a_1t$, where t is a arbitrary integer. A special solving pair can be obtained from the last but one approximating fraction for the continued fraction expression of a_1/a_2.

Example: $43x_1 + 19x_2 = b$. The approximating fractions (see Chapter 3.6. – Continued fractions) of 43/19 are 7/3, 9/4, 43/19. The fraction 9/4 yields $x_1' = 4b$, $x_2' = -9b$, so that the general solution can be written in the form $x_1 = 4b + 19t$, $x_2 = -9b - 43t$.

For $n \geqslant 3$ more general continued fraction methods have been developed. If a linear system of m independent equations is given, $\sum_{j=1}^{n} a_{ij}x_j = b_i$, $i = 1, 2, \dots, m$, and if $m > n$, the system is overdetermined and does not have solutions, in general. But if $m < n$, the system is underdetermined and has, in general, infinitely many solutions. More accurately, in the last case the system is soluble in integers $x_1, x_2, \dots, x_n$ if and only if the greatest common divisor of the m-rowed determinants of the coefficient matrix (a_{ij}) is equal to the gcd of the m-rowed determinants of the augmented coefficient matrix, which arises from (a_{ij}) by adding the column of the b_i.

The problem of solving the most general Diophantine equation of the second degree in two unknowns x_1, x_2,

$$c_{11}x_1^2 + c_{12}x_1x_2 + c_{22}x_2^2 + c_{13}x_1 + c_{23}x_2 + c_{33} = 0,$$

with integers c_{ij} can be reduced by a transformation of variables to the same problem for an equation of the special form $y_1^2 - Dy_2^2 = b$ with integers D and b. Two cases are to be distinguished. If $D < 0$, then there are no solutions or finitely many solutions y_1, y_2. If $D > 0$ (D not a square) and $b = 1$, then the so-called *Pell's equation* $y_1^2 - Dy_2^2 = 1$ has, apart from the trivial solutions $y_1 = \pm 1$, $y_2 = 0$, infinitely many solutions, which can be obtained from a minimal solution. It is easy to see that the more general equation $y_1^2 - Dy_2^2 = b$ cannot have solutions at all if b is quadratic non-residue (mod D).

Examples: 1. $x_1^2 + x_1x_2 + x_2^2 = 19$ has exactly 12 solution pairs 2, 3; −2, −3; 3, 2; −3, −2; 2, −5; −2, 5; 5, −2; −5, 2; 3, −5; −3, 5; 5, −3; −5, 3.

2. $y_1^2 - 5y_2^2 = 1$. From the minimal solution $y_1' = 9$, $y_2' = 4$ one obtains for $n = 0, 1, 2, \dots$ the totality of solutions $y_1 = \pm[(9+4\sqrt{5})^n + (9-4\sqrt{5})^n]/2$, $y_2 = \pm[(9+4\sqrt{5})^n - (9-4\sqrt{5})^n]/(2\sqrt{5})$.

The problem of finding all right-angled triangles for which the adjacent sides x, y of the right angle and the hypotenuse z are integral multiples of the unit of length leads to the Diophantine equation $x^2 + y^2 = z^2$. Its solutions, the *Pythagorean numbers*, can be represented by the formulae $x = u^2 - v^2$, $y = 2uv$, $z = u^2 + v^2$, where u and v are arbitrary integers. The smallest solutions are $3^2 + 4^2 = 5^2$ and $5^2 + 12^2 = 13^2$.

There is a surprising difference between the cases $n = 2$ and $n \geqslant 3$. A *theorem of Thue* states that an equation $a_1x^n + a_2x^{n-1}y + \dots + a_ny^n = b$ ($n \geqslant 3$, $a_1 \neq 0$) with integers $a_1, a_2, \dots, a_n$ has only finitely many solutions unless the left-hand side can be split into homogeneous factors of lower degree with integer coefficients. Diophantine equations of higher degree lead to deep problems whose solution requires a knowledge of the theory of algebraic number fields.

Among them *Fermat's conjecture* (also called Fermat's last theorem) has met with particular interest: for any integer exponent $n > 2$ there are no non-zero integers x, y, z satisfying the equation $x^n + y^n = z^n$. FERMAT had written (about 1637) in the margin of his own copy of the works of DIOPHANTOS: 'I have discovered a truly wonderful proof, but this margin is too small to contain it'. His proof has never been found, and in spite of the efforts of famous mathematicians no one yet has succeeded in proving or disproving the conjecture. Interesting partial results have been obtained; for example, it is known that the conjecture is correct for all exponents n up to 125,000.

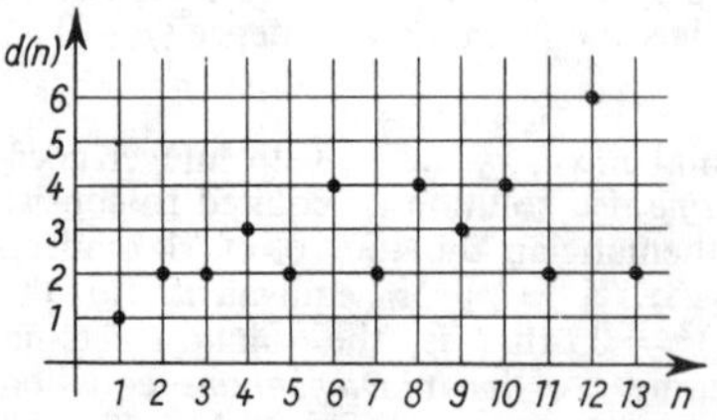

31-2 The number-theoretical function $d(n)$: total number of positive divisors of n

Analytic number theory. Apart from Euler's function $\varphi(n)$, many other number-theoretical functions $f(n)$, with the set of natural numbers as domain of definition, have been investigated.

Examples are: $\pi(x)$, the number of primes $\leqslant x$, $d(n)$, the number of divisors (Fig.), $\sigma(n)$, the sum of the positive divisors, of n; $r(n)$, the number of integer solution pairs x, y of the equation $x^2 + y^2 = n$. Some of these functions show a very erratic behaviour, for example, $d(1) = 1$, $d(2) = 2$, $d(3) = 2$, $d(4) = 3$, $d(5) = 2$, $d(6) = 4$, $d(7) = 2$, $d(8) = 4$ etc. Nevertheless, frequently a regularity can be found for the *average* of the first n function values of f. The function

$$[f(1) + f(2) + \dots + f(n)]/n$$

in many cases behaves asymptotically for increasing n like an analytic function of n. For example, $(1/n) \sum_{k \leqslant n} d(k)$ grows like the natural logarithm of n. Setting $[d(1) + d(2) + \dots + d(n)]/n = \ln n + R(n)$ one finds that $R(n)$ is of smaller order of magnitude in n than $\ln n$ is. The investigation of averages of number-theoretical functions and, above all, of their remainders $R(n)$, requires the finest tools of analysis. This branch of mathematics, *analytic number theory*, has occupied the minds of famous mathematicians, beginning with EULER in the 18th century and going right to the present day. It is still in a state of flux. Particular attention has been paid to the function $\pi(x)$, the number of primes up to and including x.

GAUSS conjectured the asymptotic law $\pi(x) \sim x/\ln x$, whose validity was not proved until about 1900 by HADAMARD and de la VALLÉE-POUSSIN. Their investigations showed that the best approximation to $\pi(x)$ is the *integral logarithm* $\text{li}(x) = \int_0^x \frac{dt}{\ln t}$ and that the estimate of the remainder $R(x) = \pi(x) - \text{li}(x)$ is closely connected with the position of the complex zeros of the *Riemann zeta-funktion* $\zeta(s) = \Sigma(1/n^s)$ with $s = \sigma + it$ and $\sigma > 1$.

Among the numerous other distribution problems there is the celebrated theorem, first proved by Lejeune DIRICHLET (1805–1859): Every arithmetic progression in which the initial term and the common difference are coprime contains infinitely many primes.

Additive number theory. The questions of additive number theory will be illustrated by a few special theorems and problems.

Fermat's theorem: Every prime number p with $p \equiv 1 \pmod 4$ is the sum of the squares of two natural numbers. The representation is unique apart from the order of the summands.

Example: $233 = 8^2 + 13^2$.

Lagrange's theorem: Every natural number can be represented as the sum of the squares of at most four natural numbers.

Example: 11 can be written as the sum of three squares: $11 = 3^2 + 1^2 + 1^2$, 7 requires four squares: $7 = 2^2 + 1^2 + 1^2 + 1^2$.

Waring's problem (raised in 1770 by WARING, first solved in 1909 by HILBERT). There is a number-theoretical function $g(k)$ such that every natural number can be represented as a sum of at most $g(k)$ kth powers of natural numbers. For example, $g(2) = 4$, by the theorem of Lagrange, and $g(3) = 9$. It is interesting that for $k \geqslant 3$ every sufficiently large n requires fewer than $g(k)$ summands. It is known that $239 = 4^3 + 4^3 + 3^3 + 3^3 + 3^3 + 3^3 + 1^3 + 1^3 + 1^3$ is the largest number that requires 9 cubes; all larger numbers require at most 8 cubes, and it has been proved that from a certain natural number onwards all subsequent numbers admit a decomposition into at most 7 cubes.

Goldbach's conjecture. In a letter to EULER in 1742, GOLDBACH conjectured that every even number $n > 6$ is the sum of two odd primes. This conjecture has so far neither been proved nor refuted. The best result in this direction, due to VINOGRADOV (born 1891) is the three primes theorem: Every sufficiently large odd integer is a sum of three odd primes. The proof uses very subtle analytic methods.

Partitions. By a partition of a natural number n one understands a representation of n as a sum of natural numbers. The total number of partitions of n is denoted by $p(n)$, where the number of terms is not restricted, equal summands are admitted, and the order of the summands is ignored. Thus $5 = 4 + 1 = 3 + 2 = 3 + 1 + 1 = 2 + 2 + 1 = 2 + 1 + 1 + 1 = 1 + 1 + 1 + 1 + 1$ are the seven partitions of 5, so that $p(5) = 7$. The function $p(n)$ has many interesting properties, for example, $p(5n + 4) \equiv 0 \pmod 5$ and $p(n) \sim [1/(4n\sqrt{3})] \exp[\pi\sqrt{(2n/3)}]$ for large n. More generally, one of the fundamental questions of additive number theory can be stated as follows. Let A be a set of natural numbers and $s > 2$ a given natural number. Can every natural number be represented as a sum of at most s elements of A? – By new methods and concepts (*density* and *order* of A) such problems have been tackled successfully from 1930 onwards.

Algebraic numbers

Algebraic number fields. An *algebraic number* α is a complex number that is a root of an algebraic equation $f(x) = 0$. Here $f(x) = a_0x^m + \cdots + a_m$ is a polynomial over the field $\mathbf{Q}$ of rational numbers, with $a_0 \neq 0$, $m \geqslant 1$, in other words, the coefficients $a_0, a_1, \ldots, a_m$ are rational. The number α is a root of infinitely many equations of various degrees; for example, $\alpha = \sqrt{3}$ satisfies the equations $x^2 - 3 = 0$, $x^3 - x^2 - 3x + 3 = 0$, $x^4 - 9 = 0$ etc. But the polynomials for the last two equations can be factored into polynomial of lower degree:

$$x^3 - x^2 - 3x + 3 = (x^2 - 3)(x - 1) \quad \text{and} \quad x^4 - 9 = (x^2 - 3)(x^2 + 3).$$

Such polynomials are called *reducible*. If it is impossible to factor a polynomial f over $\mathbf{Q}$ into non-constant factors of lower degree, again with coefficients in $\mathbf{Q}$, then f is called *irreducible* over $\mathbf{Q}$; for example $f(x) = x^2 - 3$ is irreducible.

Now for an algebraic number α there is exactly one rationally irreducible polynomial $\varphi(x)$ with leading coefficient 1 such that $\varphi(\alpha) = 0$. The degree of the algebraic number α is defined as the degree of $\varphi(x)$. For example, every rational number r is algebraic of the first degree, being the root of $x - r = 0$; $(1 + i\sqrt{3})/2$ is of degree 2, being a root of $x^2 - x + 1 = 0$; and $\sqrt[n]{2}$ is of degree n, as a root of $x^n - 2 = 0$.

The roots $\alpha^{(1)}, \alpha^{(2)}, \ldots, \alpha^{(n)}$ of $\varphi(x) = 0$ (one of which is α) are called the *conjugates* of α and are all distinct. If in $\varphi(x) = 0$ all the coefficients are integers, then α is called an *algebraic integer*.

Let ϑ be an algebraic number. The smallest field $k = \mathbf{Q}(\vartheta)$ that contains $\mathbf{Q}$ and also ϑ is called an *algebraic number field*. For example, $\mathbf{Q}(\sqrt{3})$ consists of all numbers of the form $a + b\sqrt{3}$, where a and b are rational numbers. It is easy to check that these numbers do, in fact, form a field; for example, $1/(a + b\sqrt{3}) = (a - b\sqrt{3})/(a^2 - 3b^2) = A + B\sqrt{3}$. The *degree* of k is defined as the (uniquely determined) degree n of any number ϑ for which $k = \mathbf{Q}(\vartheta)$.

Quadratic number fields. Among the fields of special type the *quadratic number fields* have been studied most thoroughly. Taking $\vartheta = \sqrt{d}$ one may assume that d is an integer not divisible by a square. Under this condition one distinguishes between quadratic number fields $\mathbf{Q}(\sqrt{d})$ with $d \equiv 1 \pmod 4$ and with $d \equiv 2$ or $3 \pmod 4$. In the former case an *integral basis* of the field is given by $\omega_1 = 1$, $\omega_2 = (-1 + \sqrt{d})/2$, in the latter by $\omega_1 = 1$, $\omega_2 = \sqrt{d}$. By this one means that every algebraic integer of $\mathbf{Q}(\sqrt{d})$ can be represented uniquely in the form $c_1\omega_1 + c_2\omega_2$ with rational integers c_1 and c_2.

Cyclotomic fields. For every natural number n the n roots of the equation $z^n - 1 = 0$, a so-called *cyclotomic* equation, are called the nth *roots of unity*, because when represented in the complex plane they divide the unit circle around the origin into n equal parts. The $n - 1$ roots other than $z = 1$ satisfy for $n \geqslant 2$ the equation $f(z) = 0$, where

$$f(z) = (z^n - 1)/(z - 1) = z^{n-1} + z^{n-2} + \cdots + z + 1.$$

When n is a prime number p, then the polynomial $f(z)$ is irreducible over $\mathbf{Q}$. It is easy to see that in this case the numbers $\omega, \omega^2, \ldots, \omega^{p-1}$ are all the roots of $f(z) = 0$, where ω is any one root. The field $\mathbf{Q}(\omega)$ is called a *cyclomatic field*. Its conjugates $\mathbf{Q}(\omega), \mathbf{Q}(\omega^2), \ldots, \mathbf{Q}(\omega^{p-1})$, all of degree $p - 1$, coincide among each other. For a general n the cyclotomic fields were studied intensively by KUMMER in connection with his investigation on Fermat's last theorem. His results were pioneering in the development of algebraic number theory. Even before him GAUSS had thought out a method of solving cyclotomic equations. His theory also made it possible to indicate all the regular n-gons that can be constructed by ruler and compass. Thus, in the theory of cyclotomy (Greek: division of the circle) three domains of mathematics, namely geometry, algebra, and number theory, are in close interaction in a wonderful manner.

Units. Algebraic integers of a special kind in an algebraic number field are the units ε, which divide the number 1. For them the reciprocal ε^{-1} is also an algebraic integer. In $\mathbf{Q}$ the only units are ± 1. But as a rule, an algebraic number field contains infinitely many units. They can all be derived from finitely many among them (fundamental units) by multiplication and exponentiation (*Dirichlet's unit theorem*). Besides $\mathbf{Q}$ the imaginary quadratic number fields (for which the generating number ϑ is complex) have finitely many units. The study of algebraic number fields has led to results of great interest.

Ideal theory. One could be tempted to assume on the basis of the laws of arithmetic in $\mathbf{Q}$ and $\mathbf{Z}$ that every algebraic integer can be split into a product of prime factors, which themselves are algebraic integers, uniquely apart from the order of the factors and from unit factors. But it was recognized in the 19th century that this assumption is false. For example, in the quadratic number field $\mathbf{Q}(\sqrt{-5})$ the number 6 has two distinct decompositions $6 = 2 \cdot 3 = (1 + \sqrt{-5})(1 - \sqrt{-5})$. It must and can be shown that, apart from unit factors, the numbers $2, 3, 1 + \sqrt{5}, 1 - \sqrt{-5}$ cannot be factored further in $\mathbf{Q}(\sqrt{-5})$.

This appeared to indicate that no simple arithmetic theory of the algebraic integers could be possible. But then KUMMER found a way, which was later developed independently by KRONECKER and DEDEKIND. DEDEKIND created the theory of ideals. He replaced the algebraic integers by the ideals of the ring R of algebraic integers in an algebraic number field k and proved the main theorem:

Every ideal in R, other than R itself and (0), *can be represented as a product of prime ideals, uniquely apart from the order of the factors.*

For the ring $\mathbf{Z}$ of rational integers this means that every principal ideal (m) is uniquely, apart from the order of the factors, a product of prime ideals $(p_1)(p_2) \cdots (p_m)$. But this is only another way of stating the fundamental theorem of elementary number theory $m = \pm p_1 p_2 \cdots p_n$.

Ideal classes. Two ideals A and B of the ring R in k are said to be equivalent if there are two principal ideals (α) and (β) such that $(\alpha)A = (\beta)B$. On the basis of this concept of equivalence one can split all the ideals in R into disjoint classes, and it turns out that the number h of resulting classes is *finite*. The principal ideals form a class by themselves. In $\mathbf{Z}$ this is the only class, so that here $h = 1$. The determination of h is a difficult task, but can be achieved with the help of analysis. There is a transcendental formula for the class number, due to DIRICHLET. For special types of field an arithmetic representation of h is known.

Transcendental numbers

A number γ that is not algebraic is called *transcendental*. It satisfies no algebraic equation with integer coefficients. It is not immediately clear that transcendental numbers exist. LIOUVILLE was the first to construct some explicitly, for example, $\sum_{n=1}^{\infty}(1/2^{n!})$. His method was based on theorems which state that algebraic numbers cannot be approximated 'arbitrarily well' by rationals. He proved, for example, that if α is an algebraic number of degree n, then a positive real constant c can be found such that $|\alpha - r/s| > c \cdot s^{-n}$ for all rational numbers r, s ($s > 0$). The following theorem of Thue-Siegel-Roth is very deep: the inequality $|\alpha - r/s| < s^{-\mu}$, where $\mu > 2$, α is any algebraic number, and r, s are rational integers ($s > 0$) has only finitely many solutions r, s.

It is of great interest to know whether a particular number or a value of a given analytic function is transcendental or not. HERMITE was the first to prove (in 1873) that the basis e of the natural logarithms is transcendental. Shortly afterwards LINDEMANN succeeded in proving (in 1882) that the area π of the unit circle is also a transcendental number. This showed that the quadrature of the circle is impossible, namely to draw with ruler and compass a square whose area is equal to that of a circle of given radius.

More generally, if $\alpha \neq 0$, then α and e^{α} cannot both be algebraic. Consequently, the functions e^x (for $x \neq 0$) and $\ln x$ ($x \neq 0, 1$) have transcendental values for algebraic arguments x. This result is proved with the aid of complex analysis, which is also used in showing, for example, that e^{π} is transcendental. It is still not known whether e^e or $e + \pi$ or Euler's constant γ are transcendental.

Around the turn of the century David HILBERT (1862–1943) made a list of 23 important mathematical problems. Among many others, his seventh problem has been solved: α^{β} is transcendental whenever $\alpha \neq 0, 1$ is algebraic and β is algebraic and irrational. This shows that the quotient of the logarithm of two algebraic numbers is either rational or transcendental.

Many transcendency statements refer to elliptic integrals and elliptic functions. For example, the perimeter of an ellipse for which the length of the axes are algebraic is transcendental. Questions of algebraic independence of transcendental numbers play an important role in the theory of transcendental numbers. Among them is the theorem of LINDEMANN.

If $\alpha_1, \alpha_2, \ldots, \alpha_n$ are algebraic numbers that are linearly independent over the field $\mathbf{Q}$ *of rational numbers, then there exists no relation $\beta_1 e^{\alpha_1} + \beta_2 e^{\alpha_2} + \cdots + \beta_n e^{\alpha_n} = 0$ with algebraic coefficients $\beta_1, \beta_2, \ldots, \beta_n$, unless they all vanish.*

32. Algebraic geometry

Algebraic geometry developed from the theory of *algebraic curves and surfaces* and the n-dimensional geometry of the Italian school. The first contributions to the theory of plane algebraic curves were made by Isaac NEWTON (1643–1727), Colin MACLAURIN (1698–1746), Leonhard EULER (1707–1783) and Gabriel CRAMER (1704–1752). The founder of algebraic geometry in the strict sense was Max NOETHER (1844–1921). The Italian geometers, principally Corrado SEGRE (1863–1924), Francesco SEVERI (1879–1961) and Federigo ENRIQUES (1871–1946) brought this discipline to complete development. In this century an investigation of the foundations of the subject from the algebraic point of view was undertaken by the German school, particularly by Emmy NOETHER (1882–1935), the daughter of Max Noether, Bartel L. VAN DER WAERDEN (b. 1903) and Wolfgang GRÖBNER (b. 1899).

Algebraic curves and surfaces. The central concept of algebraic geometry is the algebraic variety (AV) in n-dimensional *projective space* S_n, in which each point is given by the ratios $x_0 : x_1 : \cdots : x_n$ of $n + 1$ coordinates.

To explain this concept one first treats the case $n = 2$; then S_2 is a *projective plane*. A (*plane*) *algebraic curve* of S_2 is defined by a homogeneous *algebraic equation* $F(x_0, x_1, x_2) = 0$, that is, the curve is the totality of points (ξ_0, ξ_1, ξ_2) that satisfy the equation. For example, the equation of the parabola $y^2 = 2p(x - a)$ can be made homogeneous: $x_2^2/x_0^2 = 2p(x_1/x_0 - a)$ or $2apx_0^2 - 2px_0x_1 + x_2^2 = 0$; on the left-hand side there is a *homogeneous polynomial* (*form*) $F(x_0, x_1, x_2)$. From two homogeneous algebraic equations $F_1(x_0, x_1, x_2) = 0$ and $F_2(x_0, x_1, x_2) = 0$ the common points of the curves $F_1 = 0$ and $F_2 = 0$ are obtained, namely all points (ξ_0, ξ_1, ξ_2) that satisfy both equations. If $F_1(x_0, x_1, x_2)$ and $F_2(x_0, x_1, x_2)$ have no common factor, then there are only *finitely many* such points. Both cases (an algebraic curve; finitely many points) are examples of algebraic varieties in the projective plane.

In the case $n = 3$, that is, in *projective space* S_3, a form $F(x_0, x_1, x_2, x_3)$ gives an *algebraic surface*; two forms $F_1(x_0, x_1, x_2, x_3)$, $F_2(x_0, x_1, x_2, x_3)$ with no common factor give a planar or spatial algebraic *curve*. Three or more forms $F_1, F_2, F_3, \dots$ with no common factor can still give a curve. The points $(\xi_0, \xi_1, \xi_2, \xi_3)$ of a curve of S_3 satisfy other equations, apart from $F_i = 0$ ($i = 1, 2$ or $i = 1, 2, 3, \dots$); for example, $GF_i = 0$, where G can be a constant or, more generally, an arbitrary form, and (if F_i and F_k have the same degree) $F_i \pm F_k = 0$.

Polynomial ideals, sets of zeros, and algebraic varieties. In order to survey all these equations it is convenient to use the concept of an *ideal* in a commutative ring R.

Definition. A set of elements of a ring R is called an *ideal* $\boldsymbol{a}$ if, given two elements a and b of $\boldsymbol{a}$, $a - b$ is in $\boldsymbol{a}$, and furthermore if a is any element of $\boldsymbol{a}$, then every product ra is in $\boldsymbol{a}$, where r is any element of the ring. A simple example is the set of all elements ra and ar, where a is a fixed element of R. An ideal of this kind is called a *principal ideal* and denoted by (a). The concept of ideal was first used historically in number theory, in connection with Fermat's problem. In particular, if R is a ring of polynomials $f, g, \dots$, then its ideals are called *polynomial ideals.* If one is considering only forms in R, and therefore in an ideal of R, then one has a *homogeneous ideal* (*H-ideal*). A point $(\xi) = (\xi_0, \xi_1, \dots, \xi_n)$ of S_n is called a *zero* of the H-ideal $\boldsymbol{a}$ if every form $F(x) = F(x_0, x_1, \dots, x_n)$ of $\boldsymbol{a}$ vanishes for $x_0 = \xi_0$, $x_1 = \xi_1, \dots, x_n = \xi_n$. The totality of all zeros of $\boldsymbol{a}$ is called the *set of zeros* (SZ) of the H-ideal (see the examples above).

In what follows, further theorems on ideals in commutative rings R, which are important for algebraic geometry, are quotet.

The *intersection* $\boldsymbol{a} \cap \boldsymbol{b}$ of two ideals $\boldsymbol{a}$ and $\boldsymbol{b}$ of R is the totality of those elements a of R that belong to both $\boldsymbol{a}$ and $\boldsymbol{b}$. The intersection of two ideals is again an ideal. The *sum* $(\boldsymbol{a}, \boldsymbol{b})$ of $\boldsymbol{a}$ and $\boldsymbol{b}$ is the totality of all elements of the form $a + b$, where $a \in \boldsymbol{a}$, $b \in \boldsymbol{b}$. The sum of two ideals is again an ideal. The *intersection* $\mathrm{SZ}(\boldsymbol{a}) \cap \mathrm{SZ}(\boldsymbol{b})$ *of two sets of zeros* is the set of those points (ξ) that belong to both $\mathrm{SZ}(\boldsymbol{a})$ and $\mathrm{SZ}(\boldsymbol{b})$. The *join* or *sum*

$$\mathrm{SZ}(\boldsymbol{a}) \cup \mathrm{SZ}(\boldsymbol{b}) = \mathrm{SZ}(\boldsymbol{a}) + \mathrm{SZ}(\boldsymbol{b})$$

is the set of all points that belong to at least one of the SZ. The following two theorems hold: (1) $\mathrm{SZ}(\boldsymbol{a} \cap \boldsymbol{b}) = \mathrm{SZ}(\boldsymbol{a}) + \mathrm{SZ}(\boldsymbol{b})$, (2) $\mathrm{SZ}(\boldsymbol{a}, \boldsymbol{b}) = \mathrm{SZ}(\boldsymbol{a}) \cap \mathrm{SZ}(\boldsymbol{b})$. These definitions and theorems can be extended to the case of finitely many ideals.

An ideal $\boldsymbol{a}$ is called *reducible* if it is the intersection of two ideals distinct from $\boldsymbol{a}$; otherwise it is called *irreducible.* Correspondingly, $\mathrm{SZ}(\boldsymbol{a})$ is called *reducible* or *irreducible* according as $\boldsymbol{a}$ is reducible or irreducible.

Example: $\boldsymbol{a} = (x_1^2 - x_2^2) = (x_1 + x_2) \cap (x_1 - x_2)$; the SZ of $\boldsymbol{a}$ in S_2 is the pair of lines $x_1 + x_2 = 0$, $x_1 - x_2 = 0$.

However, a polynomial ideal is not uniquely determined by its SZ. For example, the H-ideals $\boldsymbol{a} = (x_1 + x_2)$ and $\boldsymbol{b} = ((x_1 + x_2)^2)$ have the same SZ in S_2, namely the points of the line $x_1 + x_2 = 0$. Hence it is convenient to agree that an *algebraic variety* $\mathrm{AV}(\boldsymbol{a})$ is determined not by $\mathrm{SZ}(\boldsymbol{a})$ alone but by further data. One could characterize $\mathrm{AV}(\boldsymbol{a})$ by the H-ideal $\boldsymbol{a}$ itself. In this sense, in the example above $\mathrm{AV}(\boldsymbol{a}) \neq \mathrm{AV}(\boldsymbol{b})$. Now every *polynomial ideal* can be expressed as the intersection of finitely many irreducible ideals (the Lasker-Noether theorem): (3) $\boldsymbol{a} = \boldsymbol{q}_1 \cap \boldsymbol{q}_2 \cap \dots \cap \boldsymbol{q}_s$. By definition, this gives the partition (4) $\mathrm{AV}(\boldsymbol{a}) = \mathrm{AV}(\boldsymbol{q}_1) + \mathrm{AV}(\boldsymbol{q}_2) + \dots + \mathrm{AV}(\boldsymbol{q}_s)$. It should be specially noted that the representation (3), and hence the partition (4), are not unique for all ideals $\boldsymbol{a}$.

For further investigation of irreducible ideals the following definitions are necessary. An ideal $\boldsymbol{p}$ of the ring R is called a *prime ideal* if $ab \in \boldsymbol{p}$ and $a \notin \boldsymbol{p}$ imply that $b \in \boldsymbol{p}$. An ideal $\boldsymbol{q}$ of R is called a *primary ideal* if $ab \in \boldsymbol{q}$ and $a \notin \boldsymbol{q}$, $b \notin \boldsymbol{q}$ imply that $a^\varrho \in \boldsymbol{q}$ and $b^\sigma \in \boldsymbol{q}$, where ϱ and σ are suitable natural numbers. For polynomial ideals the following theorems hold:

(5) Every prime ideal is irreducible.
(6) Every irreducible ideal is primary, but not every primary ideal is irreducible.
(7) For every primary ideal $\boldsymbol{q}$ there is exactly one prime ideal $\boldsymbol{p}$ such that $\mathrm{SZ}(\boldsymbol{q}) = \mathrm{SZ}(\boldsymbol{p})$.
(8) The prime ideals $\boldsymbol{p}_1, \boldsymbol{p}_2, \dots, \boldsymbol{p}_s$ belonging to the irreducible components $\boldsymbol{q}_1, \boldsymbol{q}_2, \dots, \boldsymbol{q}_s$ in (3) are uniquely determined by $\boldsymbol{a}$; they are called the prime ideals belonging to $\boldsymbol{a}$.

It follows (see (6), (7)) that to each SZ of an irreducible ideal $\boldsymbol{q}$ there is exactly one prime ideal $\boldsymbol{p}$ such that $\mathrm{SZ}(\boldsymbol{q}) = \mathrm{SZ}(\boldsymbol{p})$. Accordingly a prime ideal is uniquely determined by its SZ. Hence, by definition, $\mathrm{AV}(\boldsymbol{p}) = \mathrm{SZ}(\boldsymbol{p})$.

Generic zeros. According to VAN DER WAERDEN every $\mathrm{SZ}(\boldsymbol{p})$ can be characterized in the following way. Apart from the so-called *special* points (ξ), whose coordinates are elements of an algebraic extension of the coefficient field K, one can also consider points (9) $(\xi(t)) = (\xi_0(t_0, \dots, t_k), \dots,$

$\xi_n(t_0, \ldots, t_k))$, whose coordinates are elements of an algebraic extension of $K(t_0, \ldots, t_k)$, the field of rational functions in the indeterminates $t_0, \ldots, t_k$ with coefficients in K.

Definition: $(\xi(t))$ is called a *generic zero* of the ideal $\boldsymbol{a}$ or a *generic point* of SZ($\boldsymbol{a}$) if $F(x) \in \boldsymbol{a}$ is necessary and sufficient for $F(\xi(t)) = 0$. For example, the circle (10) $x_1^2 + x_2^2 = x_0^2$ (in inhomogeneous coordinates (11) $x^2 + y^2 = 1$) has the generic point (12) $x_0 = t_0(1 + t_1^2)$, $x_1 = t_0(1 - t_1^2)$, $x_2 = 2t_0t_1$.

An ideal has exactly one generic zero if it is a prime ideal $\boldsymbol{p}$.

From (12) one can obtain the generic point for (11) by putting (13) $x = x_1/x_0 = (1 - t_1^2)/(1 + t_1^2)$, $y = x_2/x_0 = 2t_1/(1 + t_1^2)$. Similarly, from (9) one can obtain the generic point in inhomogeneous coordinates by dividing all the coordinates by the first non-zero coordinate. The maximal number of algebraically independent *non-homogeneous coordinates* obtained in this way is called the *dimension* of $\boldsymbol{p}$ or of SZ($\boldsymbol{p}$); s numbers $u_1, \ldots, u_s$ are called *algebraically independent* over the field K if $f(u_1, \ldots, u_s) = 0$, where f is a polynomial with coefficients in K, implies that all the coefficients vanish. From (9) one obtains all points of SZ($\boldsymbol{p}$) (with the possible exception of components of lower dimension) by substituting all possible special values for $t_0, \ldots, t_k$ (*specialization of parameters*) in $(\xi(t))$. For example, the generic zero (12) of the circle (10) gives all its points except the point S for which $x_0 : x_1 : x_2 = 1 : -1 : 0$. Similarly, (13) gives all points of (11) except $S(x, y) = (-1, 0)$.

If in the Fig. 32-1 one takes all the lines (14) $y = m(x + 1)$, then each of them cuts the circle (11) in S and one further point P, whose coordinates (x, y) are uniquely determined by the gradient m. By substituting for y from (14) into (11), one obtains $(1 + m^2)x^2 + 2m^2x - (1 - m^2) = 0$. This quadratic equation has roots -1 and $(1 - m^2)/(1 + m^2)$; in conjunction with (14) this gives the two points $S(-1, 0)$ and $P((1 - m^2)/(1 + m^2), 2m/(1 + m^2))$. All the points P obtained in this way are distinct from S, since the equation $(1 - m^2)/(1 + m^2) = -1$ is not satisfied by any m. This corresponds to the fact that the line $x = -1$ cannot be expressed in the form (14). Conversely, any point $P(x, y)$ of the circle distinct from S corresponds to the value $m = y/(x + 1)$. If the gradient m is chosen as the parameter t_1, then the coordinates of P take the form (13). In contrast to $x = \cos t$, $y = \sin t$, this parametric representation of the circle is rational.

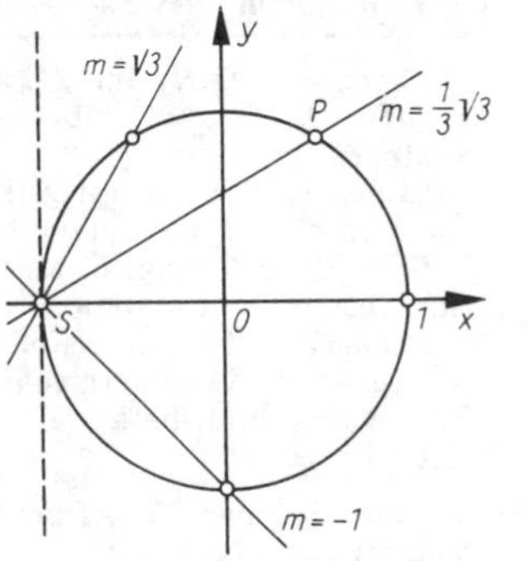

32-1 Rational parametric representation of the unit circle

Multiplicity. Instead of characterizing AV($\boldsymbol{a}$) by means of the ideal $\boldsymbol{a}$, as above, an AV can be uniquely defined if the prime ideals $\boldsymbol{p}_\sigma$ belonging to $\boldsymbol{a}$ are given (see (8)) and a non-negative number μ_σ is associated with each of them as its *multiplicity*. This can be written symbolically as

$$\mathrm{AV}(\boldsymbol{a}) = \mu_1 \mathrm{SZ}(\boldsymbol{p}_1) + \cdots + \mu_s \mathrm{SZ}(\boldsymbol{p}_s).$$

For example, the ideal $\boldsymbol{b} = ((x_1 + x_2)^2)$ considered above determines an AV consisting of the line $x_1 + x_2 = 0$ with multiplicity 2. The concept of multiplicity is particular important in the investigation of the *intersection* of algebraic varieties. Two definitions of multiplicity have been given, which coincide for S_2 and S_3 but can lead to different results in spaces of higher dimension. The first definition is based on the *principle of conservation of number*, postulated by the earlier geometers, by which the number of points of intersection in a *special case* is equal to the number in the *general case*. An exact formulation of this principle, and the limitations on its applicability, were given by VAN DER WAERDEN in 1927 by means of the concept of *relation-true specialization*. On the other hand, the second *ideal-theoretic*, definition of multiplicity is given as the *length of an ideal*. If this definition is used, then in contrast to the fierst, the generalized Bezout theorem is no longer unrestricted; in other problems, however, the use of the length of an ideal is appropriate.

Recent methods. In order to fix the concept of variety, Oscar ZARISKI (b. 1899) first used in 1938 the method of *valuations*, in addition to the SZ and multiplicity; this in turn has connections with Krull's theory of *local rings* (Wolfgang KRULL, 1899–1971), the *theory of functions* and *set-theoretical topology*.

Later, in 1946, André WEIL (b. 1906) gave a new foundation of algebraic geometry by using *topological* methods (*sheaf theory*, *cohomology theory*), which are often referred to in recent works.

33. Further algebraic structures

The characteristic features of an algebraic structure have been discussed in detail in Chapter 16. and in Chapter 17., with reference to a vector space. Since this concept is of central importance for contemporary algebra, there now follows a brief account of further developments leading from groups and fields to lattices, to rings and algebras, and to representation theory.

Lattices

The concept of a lattice was formed with a view to generalize and unify certain relationships that occur between subsets of a set, but also between substructures of certain structures, such as groups, fields, topological spaces, and so on. The development of the theory of lattices started about 1930 and was influenced by the work of Garrett BIRKHOFF.

A set L with two operations, called intersection ($\frown$) and union ($\smile$), is called a lattice if for arbitrary elements $a, b, \ldots$ of L the following axioms hold
(1) Commutativity $a \frown b = b \frown a$, $a \smile b = b \smile a$,
(2) Associativity $(a \frown b) \frown c = a \frown (b \frown c)$; $(a \smile b) \smile c = a \smile (b \smile c)$, and
(3) Absorption rules $a \smile (a \frown b) = a$ and $a \frown (a \smile b) = a$.

Example 1: If H_1 and H_2 are arbitrary subgroups of a group G, then $H_1 \frown H_2$ and $H_1 \smile H_2 = \langle H_1 \smile H_2 \rangle$ are both subgroups of G. With these operations the set of all subgroups of G becomes a lattice.
Example 2: The non-negative integers form a lattice with intersection defined as the greatest common divisor, and union defined as the least common multiple.
Example 3: Certain classes of logical propositions form a lattice under the operations *and* (intersection) and *or* (union).
Example 4: The subsets of a set form a lattice under the ordinary intersection and union.
Example 5: The intermediate fields between two given fields form a lattice, with the ordinary intersection, and the union of two fields defined as the smallest field containing both of them (see Chapter 16.).

A bijective mapping φ from one lattice L_1 onto a lattice L_2 is called an *isomorphism* if for arbitrary elements a and bo f L_1:

$$\varphi(a \frown b) = \varphi(a) \frown \varphi(b) \quad \text{and} \quad \varphi(a \smile b) = \varphi(a) \smile \varphi(b).$$

If the operations $\frown$ and $\smile$ are interchanged in a lattice L_2, one obtains a new lattice $D(L_2)$, the *dual lattice* of L_2. A bijective mapping φ from a lattice L_1 to a lattice L_2 is called a *dual isomorphism* if it is an isomorphism of L_1 onto $D(L_2)$.

The concepts of lattice theory allow the following reformulation of the fundamental theorem of Galois theory (see Chapter 16.).

Fundamental theorem of Galois theory: The mapping associating with each intermediate field the corresponding Galois group is a dual isomorphism from the lattice of intermediate fields to the lattice of subgroups of the Galois group.

Partially ordered sets. A set S with elements $a, b, c, \ldots$ on which a relation $a \subseteq b$ is defined is called a *partially ordered set* (with respect to that relation) if the relation is *reflexive*, *transitive*, and *anti-symmetric* (see Chapter 14.); anti-symmetric means that $a \subseteq b$, and $b \subseteq a$ together imply that $a = b$.

It should be emphasized that the relation $a \subseteq b$ or $b \subseteq a$ need not hold for all pairs of elements of S.

Example 6: The set of natural numbers 1, 2, 3, ... is partially ordered by the relation 'a divides b'.
Example 7: The set of all subsets of a given set is partially ordered by the set-theoretic relation 'A is a subset of B'.
Example 8: The set of all continuous real-valued functions on the interval [0, 1] is partially ordered by the relation $f \subseteq g$, meaning that $f(x) \leqslant g(x)$ for all x in [0, 1].

It is usual to interpret the relation $\subseteq$ as 'being contained in' and to represent partially ordered sets by diagrams. Such a diagram is obtained by associating with each element a a small circle K_a in the plane, and connecting the circles K_a and K_b by a line if a and b are comparable. If $a \subset b$, then the circle K_a lies below the circle K_b.

Example 9: S consists of the subsets of the set $M = \{a, b, c\}$: $M_0 = M$, $M_1 = \{a, b\}$, $M_2 = \{a, c\}$, $M_3 = \{b, c\}$, $M_4 = \{a\}$, $M_5 = \{b\}$, $M_6 = \{c\}$, $M_7 = \varnothing$. It is represented by the adjacent diagram (Fig.).

Example 10: The possible partially ordered sets of one, two, or three elements are given in the adjacent diagrams (Fig.).

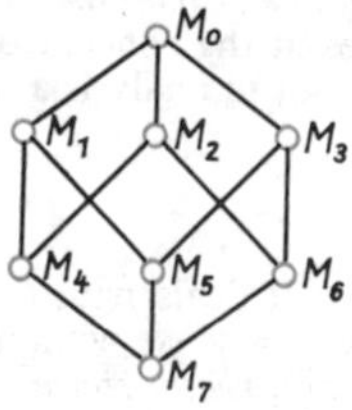

33-1 Diagram of the partially ordered set S (see Example 9)

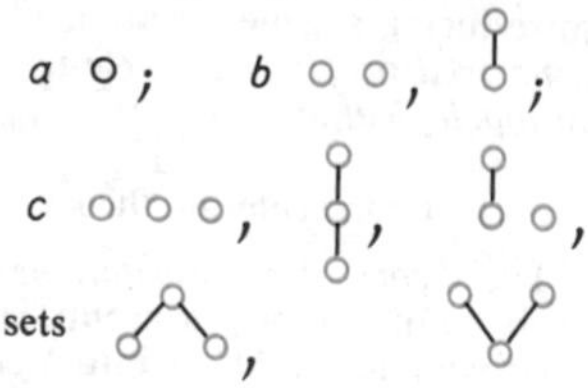

33-2 All possible diagrams of partially ordered sets of a) one, b) two, and c) three elements

Any lattice L *defines* a partially ordered set $S(L)$ with the *same* elements as L. One defines $a \subseteq b$ if $a \frown b = a$. For certain partially ordered sets this statement can be inverted.

Applications. The range of applications of lattice theory is extraordinarily wide on account of the generality of the theory. The most important examples are mathematical logic, foundations of mathematics, algebra, topology, and the theory of integration. Only by means of lattice-theoretical concepts was it possible to develop a sufficiently general concept of the integral for the needs of modern mathematics.

Rings and algebras

Rings. A set R is called a *ring* if two operations, addition and multiplication, are defined on it. Under addition R must be an Abelian group – the *additive group* of R – and addition and multiplication must be linked by the two distributive laws: $a(b + c) = ab + ac$, and $(a + b)\,c = ac + bc$.

If multiplication is associative, the ring is called an *associative ring* and one speaks of its *multiplicative semigroup*. If the multiplication is also commutative, the ring is called a *commutative ring*. Integral domains and fields are special kinds of commutative rings.

Further *examples* of rings are 1. the non-associative ring of vectors in three-dimensional Euclidean space, with vector addition and vector multiplication as the operations; 2. the associative, but non-commutative ring of $(n \times n)$-matrices under matrix addition and multiplication.

The investigation of rings, a particularly important part of current research in algebra, was decisive for the development of abstract algebra in our century. The analysis of algebraic structure, which today is standard practice in algebraic research, was suggested by Emmy Noether (1882–1935) and first put into practice by her and her pupils in important examples. Their investigations gave algebra completely new impulse and led to new areas of application.

Algebras. The concept of an algebra was developed in connection with rings that are also *vector spaces.*

An *algebra* is an associative ring A whose additive group A^+ is a vector space over a field K and in which multiplication by scalars of the field commutes with ring multiplication: $(\alpha u)\,v = u(\alpha v)$, $\alpha \in K$ and $u, v \in A$.

The dimension of an algebra A as a vector space is called its *rank*.

Examples: 1. The *continuous* and the *differentiable* real or complex valued functions on an interval form an algebra.

2. The real or complex $(n \times n)$-matrices form an algebra of rank n^2.

Structure constants. Since every element of an algebra of finite rank n can be represented as a linear combination of basis elements $u_1, \ldots, u_n$, one can write the product of two elements $u = \sum_{i=1}^{n} \alpha_i u_i$ and $v = \sum_{j=1}^{n} \beta_j u_j$ as $uv = \sum_{i,j}^{n} \alpha_i \beta_j (u_i u_j)$. From this it follows that all products uv can be computed if the products $u_i u_j$ of the basis elements are known. These products must themselves be linear combinations of the basis elements $u_i u_j = \sum_{k=1}^{n} \gamma_{ij}^k u_k$.

The n^3 constants γ_{ij}^k are called the structure constants of the algebra. They determine the multiplication in the algebra completely by means of the equations for uv and $u_i u_j$.

Example: The algebra H of rank 4 with the basis elements 1, i, j, k, and the products $1^2 = 1$, $1i = i$, $1j = j$, $1k = k$, $i^2 = j^2 = k^2 = -1$; $ij = k$, $jk = i$, $ki = j$ and $ji = -k$, $kj = -i$, $ik = -j$ is called *Hamilton's quaternion algebra*. It contains the complex numbers $a + bi$ as a subfield.

Applications. Apart from the applications mentioned under field theory, in which ring theory plays a decisive part, there have been important applications recently in *functional analysis*. By introducing a generalization of the absolute value (see Chapter 40.) one arrives at the concept of a *normed* or *Hilbert algebra*. The theory of normed algebras is an important tool in analysis and in *topological algebra*.

Representation theory

The *theory of representations* is closely connected with the theory of algebras. It deals with the problem of mapping a group, ring, or algebra homomorphically, that is, in a certain sense without destroying its structure, into a *group or ring of matrices or linear transformations of a vector space*.

This vector space is called the *representation space*. The determination of the representations of a group or algebra is of importance not only in analyzing its structure, but also for many applications in physics and chemistry, for instance, in *quantum mechanics*. Furthermore, representation theory can be regarded as an ordering principle in geometry and generalizes the theory of invariants, which flourished at the beginning of our century.

Representation of a group. Consider as representation space a complex vector space V. A representation of a group G over V is a homomorphism of G into the *general linear group* $GL(V)$ of *non-singular linear transformations* of V. If the dimension of V is n, one speaks of an *n-dimensional* representation. In that case every linear transformation can be represented after choosing a basis in V by an $(n \times n)$-matrix, and one obtains a homomorphism of G into $GL(n)$. Such a homomorphism is called a *matrix representation* of the group G.

The concrete description of representations can be particularly difficult. For certain important groups methods of finding all possible representations have therefore been specially developed; for example, for the group of all permutations of a fixed number of elements. For infinite groups, such as *topological* groups, the problems become particularly difficult. However, many questions have been solved for particular *Lie groups*, such as the rotation group and the Lorentz group.

Applications. Apart from the applications already mentioned, representation theory is frequently used in analysis. Thus, the representations of the rotation group, that is, the group of rotations of the sphere in 3-dimensional space, lead to a deeper theory of *spherical functions*, while representations of other groups are used to find important properties of the Bessel functions and others.

Naturally, the representations of the Lorentz group are important in physics.

Conclusion

Summing up one can say that today algebra is the theory of *algebraic structures*. An algebraic structure is a set on which certain operations (addition, multiplication, intersection etc.) are defined, where the exact nature of the objects of the set is irrelevant to the investigation. This concept of structure, which was developed at the beginning of this century, has been of the greatest importance for algebra, and has moulded algebraic thinking for the past 50 years. It has been modified and extended to other areas of mathematics, such as topological, or differentiable structures.

In recent years a series of new concepts have been developed in connection with other mathematical disciplines; however, their investigation is still in a state of flux, and their importance in many cases is not yet sufficiently clarified.

34. Topology

The topology of point sets

In several theorems of analysis and geometry the *connectivity* of a figure may play an essential part; for instance, the simple theorem that a function is one-to-one if its derivative is everywhere non-zero is only valid if the domain of definition of the function is connected. It is easy to find a function that is defined on the open intervals (0, 1) and (2, 3), is not one-to-one, but has the derivative 1 at every point of the intervals (Fig.). The situation is more complicated when connectivity in the plane is discussed. The theorem that a vector field with zero curl has a potential holds, in general, only if the domain where the field is defined contains no holes; such a domain is called simply-connected. For example, one can choose the lengths of the vectors of a vector field in such a way that the curl becomes zero (Fig.), although it has no potential; this can be seen by integrating along the curve marked in the illustration. Investigations of the connectivity properties of figures suggested by these and similar examples form a small but characteristic part of topology.

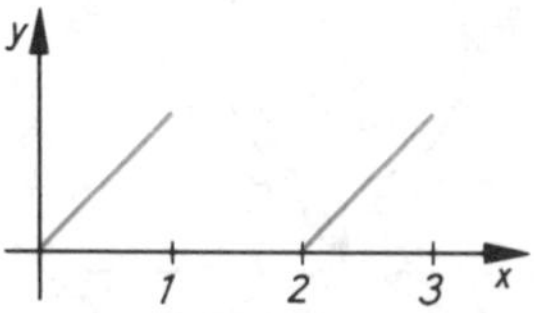

34-1 The represented function is not one-to-one

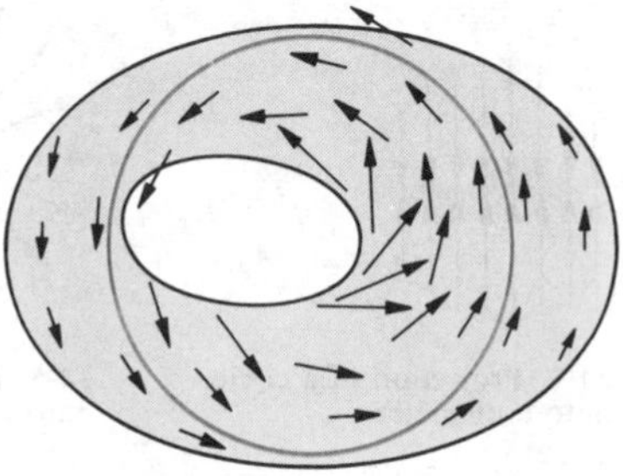
34-2 Vector field with zero curl and no potential

The concept of a figure. To begin with, the figures to be investigated lie on a straight line, in a plane, or in three-dimensional Euclidean space E^3. It is to be expected that the situation becomes more complicated as the dimension increases. On the line all that matters is how many parts the figure has, but in the plane one must also consider how many holes each of the parts encloses; and in three dimensions there are, in fact, two different kinds of holes, *cavities* like the holes in a Swiss cheese and *channels* like the holes in a sieve.

A *figure* is generally defined to be a *set of points* in the space under consideration. Therefore figures are also called point sets. The definition leads to very complicated examples, difficult to visualize, such as the set of all points in the plane for which one coordinate in a Cartesian coordinate system is rational, the other irrational. Although the following remarks are also valid for such complicated point sets, it is sufficient to focus attention on 'sensible' figures, for example, intervals on the line or surfaces in the plane or surfaces and solids in three dimensions.

Homeomorphic point sets. Before statements can be made about the connectivity of figures, there must be a precise definition of what it means for two figures to have the same connectivity; such figures are called *homeomorphic*. Intuitively two figures X and Y are homeomorphic if X can be bent and stretched and deformed into Y, without any tearing or pasting together any parts of X (Fig.). If X is transformed into Y in this manner, then with each point p of X there is associated a unique point $f(p)$ in Y and vice versa, that is, the map f that associates with each point p of X its transform $f(p)$ is a bijection of X onto Y. The condition that there should be no tears implies for f that if two points p and q of X are sufficiently close together, then their images $f(p)$ and $f(q)$ are also close together. This condition can be made precise by using the distance $d(p, q)$ between the points and then becomes analogous to the condition for continuity of functions of a real variable.

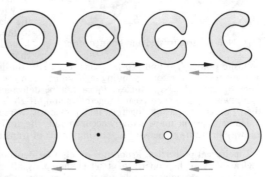

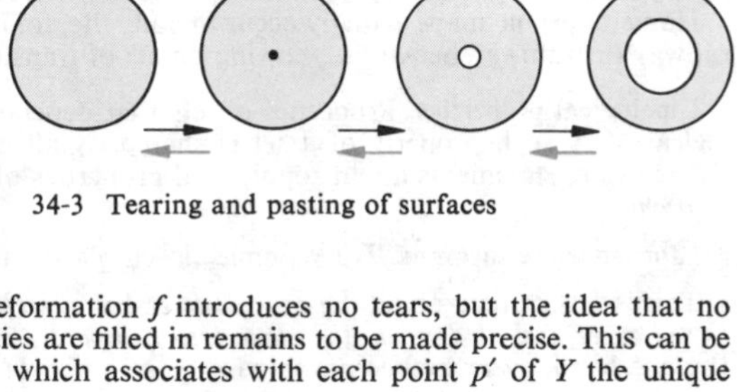
34-3 Tearing and pasting of surfaces

> A map f from a figure X to a figure Y is called continuous if for every point p in X and every positive number ε there exists a positive number δ such that for any point q in X it follows from $d(p, q) < \delta$ that $d(f(p), f(q)) < \varepsilon$.

Continuity expresses the fact that the deformation f introduces no tears, but the idea that no points are pasted together and that no cavities are filled in remains to be made precise. This can be done by considering the inverse map f^{-1}, which associates with each point p' of Y the unique point p of X (which exists because f is a bijection) such that $f(p) = p'$. To say that f does not past anything together is the same as saying that f^{-1} introduces no tears, or that f^{-1} is continuous. Thus, it is now possible to give a precise definition of homeomorphism.

Two figures X and Y are called homeomorphic if there exists a continuous bijective map f of X onto Y such that the inverse map f^{-1} is also continuous. The map f is then called a homeomorphism or topological map.

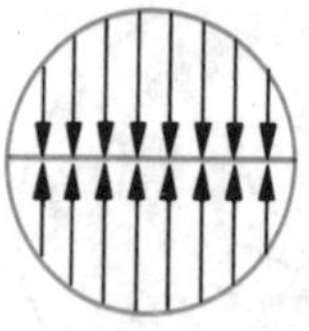

34-4 Projection of a circle onto a diameter

34-5 Projection of a square onto a circle

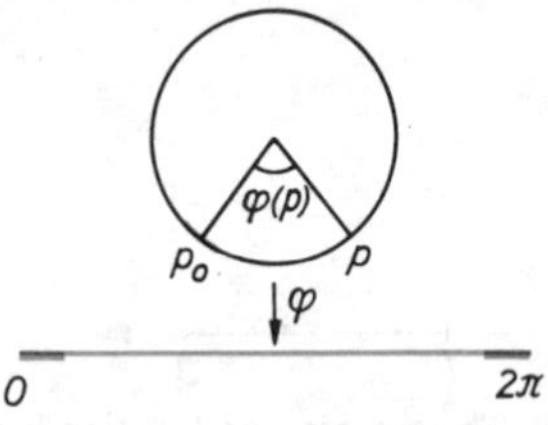

34-6 Mapping of a circle onto a half-open interval

Examples: 1. The perpendicular projection of a circle onto a diameter is a continuous map, but not a homeomorphism, since it is not one-to-one (Fig.).

2. The central projection of the boundary of a square onto a circle is a homeomorphism (Fig.).

3. The map from a circle C, to the half-open interval $[0, 2\pi)$ that takes each point p of C to its associated angle $\varphi(p)$, is not continuous at the point p_0 with angle 0; for the images of points near p_0 lie far apart, at opposite ends of the interval (Fig.).

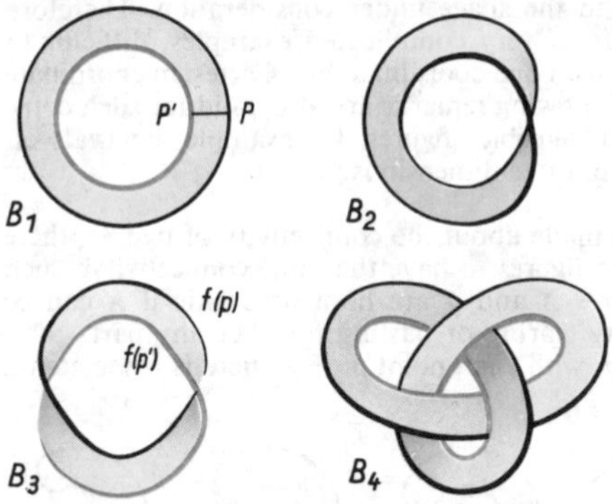

34-7 Non-twisted, twisted and knotted strips

The definition of a homeomorphism does not exactly correspond to the intuitive idea of deforming one figure into another without tearing or bending. It must also be permitted to cut the figure, on condition that after the deformation has taken place the cut is pasted together again point for point, exactly as it was initially. Of the four strips B_1, B_2, B_3, B_4 in the figure, B_3 and B_4 are homeomorphic to B_1, but the *Möbius strip* B_2 is not. B_2 is obtained from B_1 by cutting, twisting once, and pasting together again, B_3 is twisted twice, and in B_4 the strip is stretched and then knotted. B_1 and B_2 are not homeomorphic, since B_1 has two boundary curves, whereas B_2 has only one. B_1 and B_3 are homeomorphic, since the black boundary curve of B_1 can be mapped to the black boundary curve of B_3 and the red boundary of B_1 to the red one of B_3 in such a way that opposite points p and p' are mapped to opposite points $f(p)$ and $f(p')$. If then each straight-line segment between opposite points p and p' is mapped to the corresponding segment between $f(p)$ and $f(p')$, the result is a homeomorphism from B_1 onto B_3. Nevertheless B_1 cannot be deformed into B_3, it must be cut along a segment p, p', twisted, and then pasted together again. If B_1 is transformed into B_2 in a similar manner, then points are pasted together that previously were not close to each other.

Homeomorphic maps actually occur in daily life, for example, in schematic maps for underground railway or tramway networks, showing points of transfer.

Topological properties. Properties of sets that depend only on their connectivity are called topological. Any such property of a set is shared by all its homeomorphic images. Thus, *topological theorems* are statements about topological properties of point sets. An example is the *Jordan curve theorem.*

Jordan curve theorem. Every simple closed plane curve divides the plane into two parts.

This is a statement about the properties of being a *closed curve without self-intersections* or *simple closed curve* and of *dividing the plane into two parts.* Both these properties are topological; for on the one hand, every homeomorphic image of a simple closed curve must again be a simple closed curve – indeed one can define a simple closed curve to be a homeomorphic image of the circle; on the other hand, division into two parts means that the complement to the curve consists of two disconnected parts, which is also a topological property. The theorem itself, which seems almost obvious is not at all easy to prove. One begins to appreciate this if one realizes that a curve C in a plane can be extremely complicated, so that at first sight it is very difficult to decide whether a given point P is inside it or not (Fig.).

34-8 Simple closed curve $\boldsymbol{C}$ with exterior point P

Apart from theorems that are statements about topological properties, other theorems are also regarded as belonging to topology if they are concerned mainly with topological concepts, such as continuous maps in *Brouwer's fixed point theorem.*

The **Brouwer fixed point theorem.** If C is a closed circular disc including the perimeter, then every continuous map of K into itself has a fixed point, that is, a point that is mapped to itself.

Thus, if the disc is distorted in such a way that the resulting figure lies entirely inside it, then at least one point occupies its original position.

One of the main tasks of topology is to decide whether two given figures X and Y are homeomorphic. If this is true, it can frequently be proved by exhibiting a specific homeomorphism from X to Y by trial and error. If they are not homeomorphic, the proof is usually more complicated. The basic method for such proofs is to find a topological property that is satisfied by one of the figures, but not by the other. If this is the case, then X and Y cannot be homeomorphic, for if a point set has a topological property, then so does every point set homeomorphic to it. This method requires familiarity with a large number of topological properties; here is a list of some of the simplest and most important.

A point set Z is called *connected* (more accurately: *path-connected*) if any two points p and q in Z can be joined by a path in Z, that is, if there is a continuous map of an interval into Z that maps the end-points of the interval to p and q, respectively. Thus, figures are connected if they do not consist of several disjoint parts. Connected figures in the plane and in space can still have *holes*. Certain types of these are excluded by requiring the set to be simply-connected. A set Z is *simply-connected* if any closed curve in Z can be contracted inside Z to a point. It is intuitively clear that simply-connected plane figures can have no holes, since a curve going around such a hole cannot be contracted to a point inside Z. On the other hand, for three-dimensional figures the condition excludes channels, but not cavities; for instance, a hollow ball is simply-connected, even though it is hollow and thus has a cavity. A sieve, however, is not simply-connected.

This type of topological property, relating to holes in figures and their connectivity, was the starting point of *algebraic topology*, in which the connectivity of figures of arbitrarily high dimension is described by certain algebraic structures associated with the figures, such as the *homology* and *homotopy groups*.

With any point set X one can associate a natural number dim X, the so-called *dimension of the point set*. For familiar figures, such as a curve $\boldsymbol{C}$, a surface $\boldsymbol{S}$, or a body $\boldsymbol{B}$, for which there is an intuitive notion of what the dimension should be, dim X has the expected value, namely dim $\boldsymbol{C} = 1$, dim $\boldsymbol{S} = 2$, and dim $\boldsymbol{B} = 3$. In *dimension theory* a precise definition of dimension makes it possible to prove that homeomorphic point sets X and Y have the same dimension, dim $X =$ dim Y. Thus, the dimension of a point set is a topological property, and, for example, a curve can never be homeomorphic to a surface.

Neighbourhood of a point. Apart from connectivity and dimension, topology includes the study of further properties of point sets, some of which are also important in other branches of mathematics such as the differential calculus. In explaining the concept of the derivative at a point x of a function defined on a closed interval I it matters very much whether x is an end-point or not. In the former case one can only talk of a left or a right derivative. The situation is similar for a function of two variables $f(x, y)$ defined on a plane domain D. In defining the partial derivatives or the total differential of f at a point here also it is necessary to distinguish whether the point lies in the interior or on the boundary of D. Indeed, one frequently has to distinguish between interior points and boundary points of a figure. The first step in making these concepts precise is to introduce the idea of a neighbourhood of a point. One defines the ε-neighbourhood $U_\varepsilon(p)$ of a point p as the set of all points whose distance from p is less than ε, where ε is an arbitrary positive number. In this context it is important to distinguish whether the point p is regarded as an element of a line, a plane, or three-dimensional space. In the first case the ε-neighbourhood is an open interval (that is, without the end-points) of length 2ε, in the second it is an open disc of radius ε, and in the third a solid ball of the same radius. Note that the circumference of the disc and the surface of the ball are not counted as belonging to them.

In any figure F one distinguishes between *interior points p*, which have an ε-neighbourhood entirely contained in F, and *boundary points q*, for which every ε-neighbourhood contains some points not in F (Fig.). One sees that the dimension of the space in which F lies plays a decisive role in these definitions. For instance, in the plane the centre M of a disc K is an interior point, but in three dimensions the disc K consists only of boundary points, since no ball of positive radius lies inside K. *Open sets* are sets without boundary points, they consist only of interior points, for example, a ball in space without its boundary sphere. In the theory of functions of a complex variable open and connected sets in the plane play a special part; they are called *domains*.

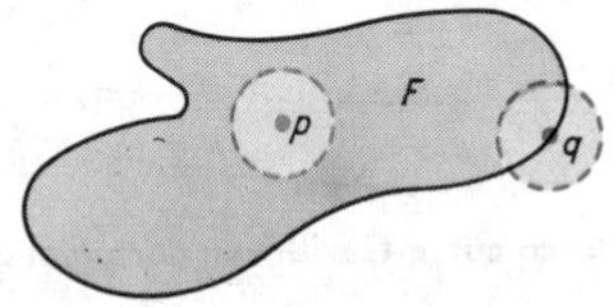

34-9 Boundary point q and interior point p

A point p is said to be *adherent* to a set X if every ε-neighbourhood of p contains at least one point of X. Intuitively, the set of points adherent to X consists of those points which, if not actually contained in X, are at least 'infinitely close' to X; the end-points of an open interval, for example, do not belong to the interval, but they are 'infinitely close' to it. The set of points adherent to an open disc K of radius r about a point z are, apart from the points p of the disc themselves, the points q on the perimeter of the disc, for which $d(q, z) = r$. If the only points adherent to a set F are the points of F itself, then F is called *closed*. The open disc K above can be made into a closed set by adjoining all the boundary points q with $d(q, z) = r$. The essential relationship between open and closed sets is the fact that a set G on a line, in a plane, or in three-dimensional space is open if and only if its complement, that is, the set of all points of the space in question that are not in G, ist closed.

For the purpose of later generalizations one defines relative ε-neighbourhoods $U_\varepsilon(p)$ of a point p and relative open sets with respect to a subset X. The relative *ε-neighbourhood* of p in X consists of the points q in X with $d(q, p) < \varepsilon$, that is, it is $X \cap U_\varepsilon(p)$. Similarly, a subset G of X is called *open in X* if every point p in G has a relative ε-neighbourhood in X entirely contained in G. The empty set, which is a subset of every set and hence also of X, is also regarded as open in X.

The following statements hold for the system of subsets of X that are open in X: 1. X itself and the empty set are both members of the system, that is, are open in X. 2. The union of arbitrarily many sets that are open in X is also open in X. 3. The intersection of finitely many sets that are open in X is open in X.

With these concepts of relative ε-neighbourhoods and open sets a new definition of continuity for maps f from one set X to another Y can be given.

A map f from a point set X to another point set Y is continuous if and only if for every point p in X and every (relative) ε-neighbourhood V of f(p) in Y one can find a δ-neighbourhood U of p in X, that is mapped by f to a subset of V (Fig.).

A map f from X to Y is continuous if and only if the inverse image of every open set in Y is open in X.

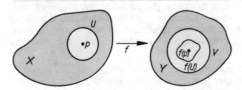

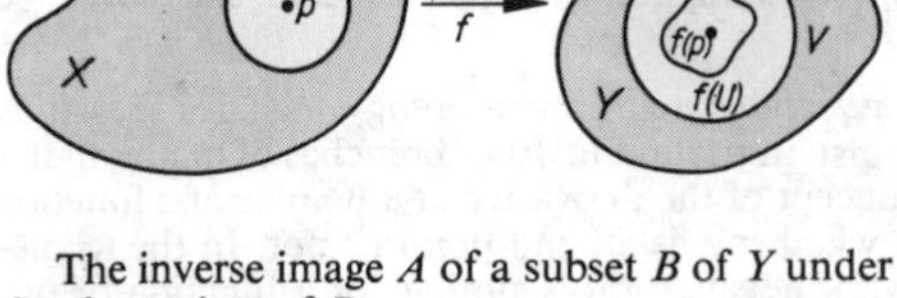

34-10 Continuous map

The inverse image A of a subset B of Y under a map f is the set of all points in X that are mapped by f to points of B.

n-dimensional spaces

The exposition so far has been restricted to figures on a line, in a plane or in three dimensions, that is, to the Euclidean spaces E^1, E^2 and E^3. The starting point for their generalization to n-dimensional Euclidean space E^n is the observation that every point p in E^3 can be described by a triple of coordinates (x_1, x_2, x_3). The so-defined map from E^3 to the set of all triples of real numbers is one-to-one and onto, and as long as a fixed Cartesian coordinate system is retained, the points of E^3 can be identified with their coordinate triples. The space E^n is defined similarly.

E^n is the set of all n-tuples of real numbers $(x_1, x_2, \ldots, x_n)$; an n-tuple regarded as an element of this set is usually called a point.

The distance function $d(p, q) = [(y_1 - x_1)^2 + (y_2 - x_2)^2 + (y_3 - x_3)^2]^{1/2}$ for $p = (x_1, x_2, x_3)$ and $q = (y_1, y_2, y_3)$ in E^3 can be generalized to E^n; this function retains the three essential properties of the distance in E^3, where the *triangle inequality* states that in a triangle with the vertices p, q and r each side is at most as long as the other two together.

Distance between points p, q in E^n	$d(p, q) = \left[\sum_{i=1}^{n} (y_i - x_i)^2\right]^{1/2}$	1. $d(p, q) = d(q, p)$, 2. $d(p, q) \geqslant 0$, $d(p, q) = 0$ if and only if $p = q$ 3. Triangle inequality: $d(p, r) \leqslant d(p, q) + d(q, r)$

Using this distance $d(p, q)$ it is now possible to define topological properties for subsets X and Y of E^n similarly to the way this has been done for E^n with $n = 1, 2, 3$. A *mapping* f from a set X to a set Y is *continuous* if for every point p in X and every positive number ε there exists a positive number δ such that $d(p, q) < \delta$ implies that $d(f(p), f(q)) < \varepsilon$. Such a mapping is a *homeomorphism* if it is bijective (that is, one-to-one and onto) and both f and f^{-1} are continuous. Two sets X and Y are called homeomorphic if there exists a homeomorphism from X to Y. Finally, the ε-neighbourhood of a point in X can be defined in exactly the same way as above, so that all the basic concepts have been extended to cover the more general sets now under consideration.

Higher-dimensional spaces are of use in the examination of objects or conditions that cannot be described by at most three coordinates, but can be described by finitely many. For instance, in physics an event is fixed not just by its three coordinates in space, but also by a time coordinate. Thus, every such event corresponds to a point in E^4. If it is desired to describe not just one event but several, such as in a continuing process, then one obtains a subset of E^4. A similar situation is encountered in physical systems with several degrees of freedom. But of course, this way of looking at things is only worthwhile if one succeeds in stating and proving interesting and practically useful geometrical or topological theorems in higher-dimensional spaces. Two examples of such theorems follow.

The Jordan-Brouwer theorem. In generalizing to E^3 the Jordan curve theorem, which was stated for E^2, the closed curve has to be replaced by a *2-dimensional topological sphere*. This is any subset of E^3 that is homeomorphic to the surface of a sphere, or intuitively speaking, a deformed sphere. Every two-dimensional topological sphere divides E^3 into two parts.

In E^n an *n-dimensional ball* is defined by analogy to that in three dimensions as the set of points p whose distance from a given fixed point z is at most a fixed number r, $d(p, z) < r$. The surface of the ball ist the set of those points p for which $d(p, z) = r$. An *$(n - 1)$-dimensional sphere* is a subset of E^n that ist homeomorphic to this surface; again this can be described intuitively by saying that it is the deformed surface of an n-dimensional ball. The Jordan-Brouwer theorem and the Brouwer fixed point theorem can now be stated in full generality.

The Jordan-Brouwer theorem. Every $(n - 1)$-dimensional topological sphere divides E^n into two parts.

The Brouwer fixed point theorem. If f is a continuous map of an n-dimensional ball into itself, then f has a fixed point.

Topological structures

Incomparably more far-reaching generalizations of the intuitive topological concepts can be made by introducing into topology the idea of a *structure* on a set, which was first developed in algebra. There structures such as rings and fields are obtained from number systems by a process of abstraction in which properties of numbers that are inessential for algebra are ignored and only those are retained that are needed as foundation for algebraic investigations. If a similar course is to be followed in topology, then it is first necessary to determine what properties of point sets form the basis of topological arguments, and then to define a *general structure* for which the only requirement is that these prerequisites for topological investigations are realized. One of the essential properties of point sets leading to topological concepts is the possibility of defining *continuity* for maps between the point sets, because many other topological concepts, such as homeomorphisms and connectivity, are defined in terms of this fundamental idea of continuity.

A *topological structure* will therefore be defined as a set T with certain properties that make it possible to declare for a map f from T to another such structure T' whether f is continuous or not. It is plausible to regard the *metric spaces* of *functional analysis* as a suitable structure, because continuity can be defined as soon as one has a concept of distance (see n-dimensional spaces), and metric spaces are by definition sets in which a distance $d(p, q)$ is defined for all pairs of elements p, q. Although this is a possible course, for some topological investigations metric spaces have been found to be too restricted. The final definition is arrived at by observing that continuity of a map f from a point set X to a point set Y can be defined as soon as the open sets of X and Y (see Topological properties) are known. For f is continuous if and only if the inverse image of every open set in Y is open in X.

Therefore the following definition of a *topological space* is a suitable structure for the study of topology.

A topological space T is a set together with a distinguished system O of subsets of T satisfying the following requirements: 1. T and the empty set belong to O; 2. the union of arbitrarily many sets belonging to O also belongs to O; 3. the intersection of finitely many sets belonging to O also belongs to O.

The sets in O are called the open sets of T.

These three axioms correspond exactly to the properties of the system of open subsets of a point set listed above, so that any point set together with its (relatively) open subsets forms a topological space T, the definition of continuity just repeated at the end of the previous paragraph makes sense for maps from a topological space T to a topological space T'. A map f from T to T' is called a homeomorphism if it is bijective and f and f^{-1} are both continuous. Two topological spaces T and T' are called homeomorphic if there is a homeomorphism mapping T to T'. A topological space T is (path-)connected if to any pair (p, q) of elements of T there exists a continuous map of an interval into T such that the end-points of the interval are mapped to p and q, respectively.

These examples show how concepts defined for point sets can be extended to arbitrary topological spaces. However, general statements about topological spaces tend to be far less intuitive in geometric terms than those about point sets. Further, there are concepts for point sets that elude generalization to arbitrary topological spaces; for instance, it is not possible to give a satisfactory definition of dimension for all topological spaces. *General or set-theoretical topology*, which is the study of arbitrary topological spaces, can in many of its parts hardly be regarded as belonging to geometry. Rather, it assumes the character of a *structure theory* comparable to group theory in algebra. Just as in group theory special classes of groups such as Abelian groups, are examined, so in general topology topological spaces are investigated that satisfy apart from 1., 2. and 3. above, further axioms for example, the *Hausdorff separation axiom*: To any two elements p and q of T there exist disjoint open subsets X and Y of T such that $p \in X$ and $q \in Y$ (Fig.).

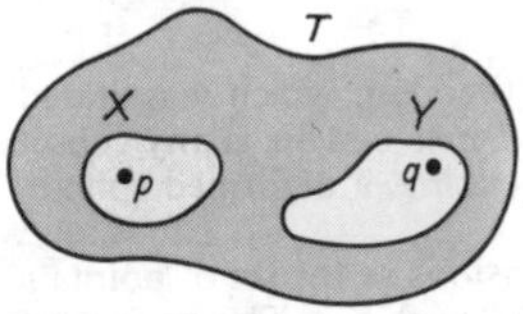

34-11 The Hausdorff separation axiom

Topological spaces are more general than *metric spaces*, so that, in particular, every metric space is a topological space. That is to say, one can distinguish in any metric space the system of open sets. These open sets are defined just as in Euclidean spaces: Let M be a metric space, p an element of M, and ε a positive number; the ε-neighbourhood of p in M is the set of all elements of M whose distance from p is less than ε. A subset X of M is open if for every element p of X there is an ε-neighbourhood of p entirely contained in X. It is not difficult to prove that the system of open sets so defined has the properties 1., 2. and 3., and that in this way M indeed becomes a topological space. This remark has the important consequence that theorems and concepts of general topology are applicable to metric spaces and, in particular, to the investigations of functional analysis.

As an example of an important problem of general topology the *metrization problem* may be quoted: under what conditions on the system of open sets O of a topological space T is T metrizable, that is, when can a distance function d be defined on T that makes T into a metric space whose open sets are precisely the sets in O? – It is easy to see that a necessary condition is the Hausdorff separation axiom, but this is not sufficient. The *metrization theorem* of NAGATA and SMIRNOV gives necessary and sufficient conditions for the existence of such a metric on a topological space. The precise formulation of these conditions, however, exceeds the scope of this chapter.

35. Measure theory

Measure theory deals with the determination of the content of geometrical configurations, or more generally, of point sets. It is directly connected with the integral calculus and set theory and finds important applications in many branches of analysis and in the foundation of probability theory. In contrast to the calculation of the areas of triangles, rectangles and other figures bounded by straight lines, figures bounded by curved lines or even more complicated ones present difficulties. Even to explain what one understands by the *content of a point set* is a problem. Its first solution, the *concept of Riemann content*, leaning heavily on the concept of the Riemann integral, was given in 1890 by Giuseppe PEANO (1858–1932) and Marie Ennemond Camille JORDAN (1838–1922).

In order to arrive at the content of a point set (for example, in a plane), a square grid is laid over the plane and the given figure is *approximated from within* by a region consisting only of squares of the grid (yellow in Fig. 35-1). An *outer approximation region* contains the figure in its interior (yellow and blue). If by halving one proceeds to a finer grid, then the new inner approximation contains the old one and is usually larger by a certain number of the new squares, whilst the new outer region results by deleting new squares from the old region. Thus, the difference between the areas of these approximate regions can only become smaller. If now with continuing refinement the *inner* and *outer contents* approach one another arbitrarily closely, then their common limiting value is called the *Peano-Jordan content*, or just the content, of the given figure.

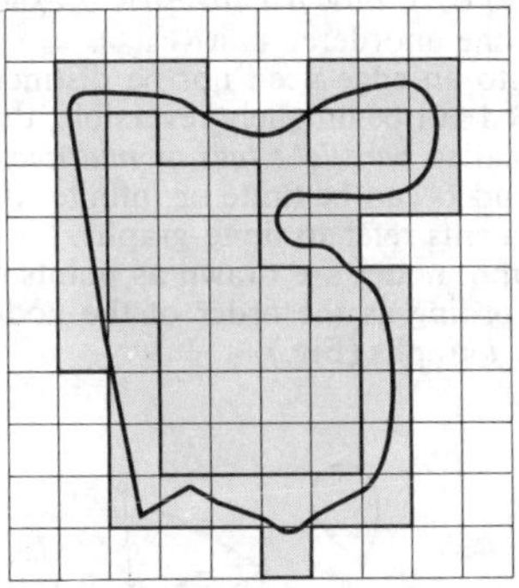

35-1 Approximations with respect to Peano-Jordan content

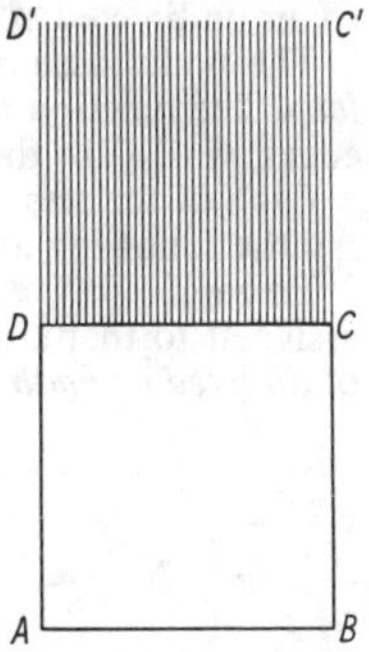

35-2 Configuration without Peano-Jordan content

This concept of content yields the well-known area formulae for figures bounded by straight lines, as well as for the circle and ellipse among others. However, there exist *point sets to which one can ascribe no Peano-Jordan content*. Fig. 35-2 represents a square *ABCD* on whose upper side *CD* a perpendicular of length equal to the side of the square is erected at every one of the infinitely many points whose distance from the vertex *C* is a *rational number*. In this case all the outer contents are at least twice as large as the inner ones and the two do not tend to a common limit as the grid is refined, because the whole square *CC'D'D* always belongs to the outer approximation and only to that one. Every grid square lying in *CC'D'D*, no matter how small, contains both points that do and others that do not belong to the figure.

Lebesgue measure. In modern mathematics precisely such point sets, which at first sight seemed exceptional, gained considerable significance. In very many cases a more comprehensive concept of content was successful, the *Lebesgue measure*, which was developed in 1902 by Henri Léon LEBESGUE (1875–1941). In contrast to the Peano-Jordan content, the approximation figures may also consist of infinitely many elementary areas of different magnitudes.

Point sets that have a content are also measurable, and their measure is numerically equal to their content. On the other hand, there exist point sets to which no content can be ascribed, but which do have a measure; for example, the measure of the configuration in Fig. 35-2 is equal to the content of the square *ABCD*; the subset consisting of the perpendiculars is of measure zero. *Sets of measure zero* play a particularly important part both in pure mathematics and also generally in the mathematical description of natural processes; they characterize, so to speak, the inessential. The considerations in the case of spaces of other dimensions are completely analogous; for example, for dimension three one obtains the ordinary volume, or *space measure*.

In the *integral calculus* the use of measure in place of content leads to the *Lebesgue integral*. It represents an extension of the concept of the Riemann integral, just as the Lebesgue measure is an extension of the concept of the Peano-Jordan content.

In the process of further abstraction, in general measure theory one understands by a measure on a set Ω a real-valued function $m(A)$ whose argument runs through certain subsets A of Ω and which has properties corresponding to the simplest geometrical interpretations. In the first place, $m(A) \geqslant 0$ and $m(A \cup B) = m(A) + m(B)$ for disjoint point sets A, B. The sets A belonging to the domain of definition of m are said to be measurable with respect to m. This approach makes it possible, for example, to apply measure-theoretical theorems directly to probability theory; in this a *random event* is regarded as a subset A of the 'point' set Ω of all elementary events, and the measure $m(A)$ as the *probability of the event A*.

36. Graph theory

Foundations

Directed and undirected graphs. A *graph* $G = [X, U, f]$ is a combination of two sets of elementary figures, the set X of *nodes* x and the set U of *edges* u, and a function f defined on U, the *incidence function*. This assigns to each *directed edge* $u \in U$ exactly one *ordered pair* of nodes x_i, $x_k \in X$ and to an undirected edge one unordered pair (Fig.).

The nodes assigned to an edge need not be distinct. If $x_i = x_k$, then $f(u) = (x_k, x_k)$ is called a *loop*. The function f need not be uniquely reversible, that is, a pair (x_i, x_k) can be assigned to several edges, which are then called *parallel edges* or *multiple edges*.

Each of the sets X and U can be finite or infinite. If X and U are finite, the graph is called *finite*. All the following arguments refer to finite graphs.

To represent the graph, nodes are drawn as points and edges as arcs of curves joining the nodes assigned to them. According as the order of the nodes in a pair matters or does not, one speaks of *directed* or *undirected* graphs (Fig.).

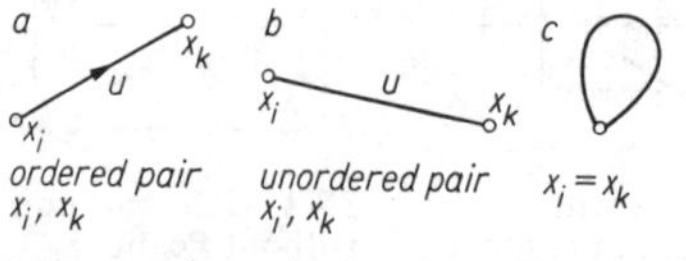

36-1 a) Directed edge, b) undirected edge c) loop

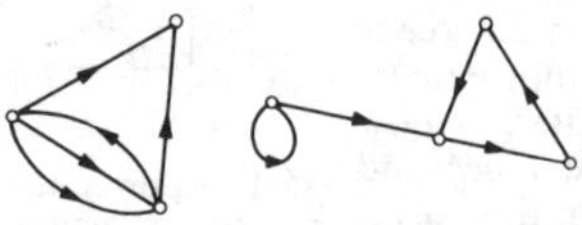

36-2 Directed graphs

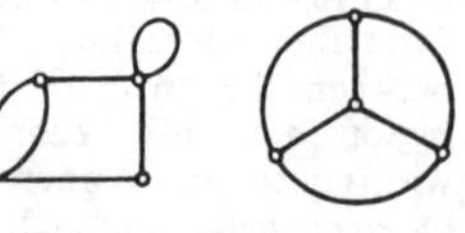

36-3 Undirected connected graphs; the right-hand one is complete

Example: The street plan of a town is usually represented on a map by an undirected graph. This is completely sufficient for pedestrians. However, if there are many one-way streets, a car driver needs a representation of the street plan as a directed graph.

Applications. The five *Platonic solids* (tetrahedron, cube, octahedron, dodecahedron, icosahedron), like all other polyhedra, represent graphs with their vertices and edges as the nodes and edges of the graph. The boundaries between countries on a geographical map form a graph, and so does the railway system, the shipping routes and the airlines.

All communication networks, such as telephone or teleprinter networks, can be represented as graphs, and electrical, water and central heating networks can also be described by means of graphs. In systems theory and cybernetics, complex systems are considered whose structure can be represented with the help of graphs, for example, block switch diagrams, signal flow diagrams and the production structure of a business. A practical branch of graph theory consists of *network techniques*. The starting times and connections of the parts of a complete process are transferred to a network and can then be calculated and directed (see Network techniques).

Generally it can be said that graphs are a tool in solving combinatorial problems.

Example: A ferryman F wishes to take a wolf W, a goat G and a cabbage C from the left bank of a river to the right by means of a boat which can only hold two out of F, W, G, C. The wolf and the goat must never be left together unguarded, nor the goat and the cabbage. How can this be achieved! –

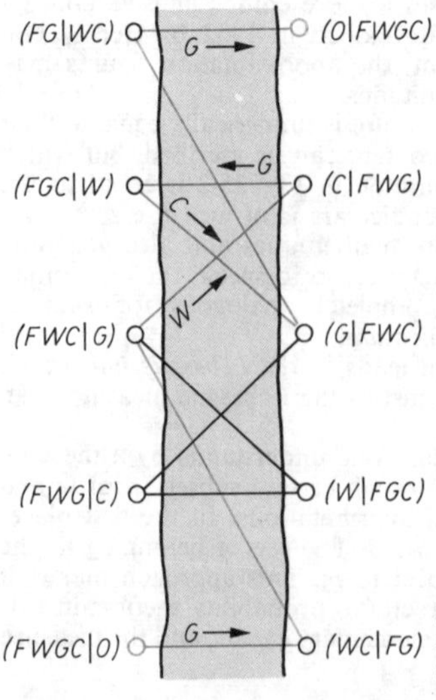

36-4 Combinations on both banks for the transport by a ferryman F of a wolf W, a goat G and a cabbage C; one of the possible sequences of connected edges is coloured

One first classifies the admissible combinations on the two banks. For example, one is (FWG/C), which means that F, W and G are on the left bank and C on the right. A zero means that none of the four things is on the corresponding bank. A combination in which the boat is on the right bank is now connected to a combination in which the boat is on the left if the ferryman can get from one to the other in one journey. The combinations and the connections are the nodes and edges of a graph (Fig.). The problem can now be solved by looking for a connected sequence of edges that begins at $(FWGC/0)$ and ends at $(0/FWGC)$. There are several of these.

Special graphs. A graph in which any two distinct nodes are joined by exactly one edge is called *complete*. If the graph consists only of isolated nodes, that is, the set of edges is empty, it is called a *null graph*. If one can go from any node to any other node along the edges, the graph is called *connected*. A complete graph is connected.

By repeatedly going from one edge to another, through a node common to the two edges, one obtains a *sequence of edges*. If each edge in a sequence of edges occurs only once, one speaks of a *path*. This is called *closed* if the starting point and end-point coincide. A closed path in which no node except the starting point occurs twice is called a (topological) *circle*. A connected non-empty graph is called a *tree* if it contains no closed path (Fig.). Trees are used, for example, in structure formulae of chemistry for chains of carbohydrates. Finally, a graph is called *planar* if it can be drawn in a plane or, what is topologically the same thing, embedded in the surface of a sphere, without the edges intersecting. For example, trees are planar graphs.

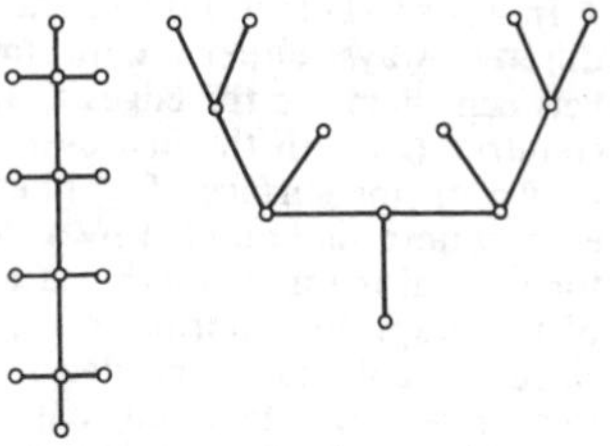

36-5 Trees

Combinatorial structures. Graph theory is concerned with combinatorial matters. The *object of graph theory* is not only the determination of numbers, as in elementary combinatorial theory, but the *combinatorial structure itself*. Investigations of combinatorial structure, that is, graph-theoretical considerations, were first used by Leonhard EULER in 1736. He started with the *problem of the seven bridges of Königsberg*, that is, the problem whether one can take a walk in which each of the seven bridges is crossed exactly once (Fig.). It can be seen immediately that the walk neither begins nor ends in at least two of the four regions I to IV. These regions are entered and left again. However, since an odd number of bridges leads to *each* region, such a walk is not possible. EULER investigated more generally under what conditions a given connected graph can be described in a closed path in such a way that each edge is covered exactly once. An *Euler path* of this kind exists if and only if an even number of edges meet at each node.

The graphs one comes across in practice often have a very general structure, and the main question is that of an algorithm for the effective solution of an optimization problem connected with the graph. This is exemplified by the *problem of the cheapest telephone network*.

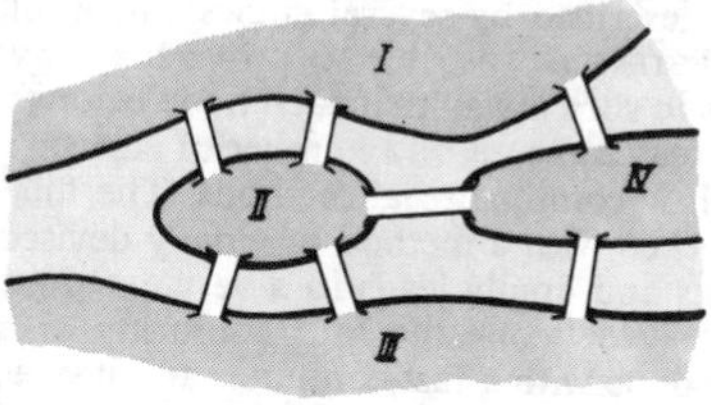

36-6 The seven bridges of Königsberg

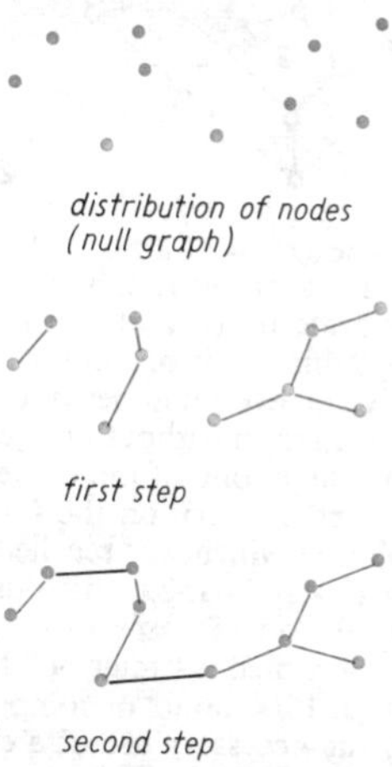

36-7 Minimal tree

Example: To connect n places by a telephone network at minimum cost, where the branch points occur only at the places themselves. The costs of the direct connections between any two places are known.

The required network is obviously a *tree*. A simple algorithm serves to construct it. *First step:* one joins each node to the one for which the cost is least, thus obtaining a *system of trees*. *Second step:* one contracts each tree to a point and repeats the process. If one continues in this way, one breaks off the process when only one tree remains. This is the required *minimal tree*. In Fig. 36-7 the process is carried out for 10 places in such a way that the distance between any two points are proportional to the costs.

If one wishes to solve the problem by trying all possibilities, one would have to test n^{n-2} possibilities for n places, that is, 10^8 possibilities for $n = 10$ places.

The four colour problem

Mapmakers know that any political map can be drawn with four colours in such a way that any two countries with a common frontier (not just a point) are coloured differently. It is fairly easy to prove that five colours are always sufficient to colour any map. The problem of whether four colours are always sufficient proved to be very hard, was unsolved for about 100 years and therefore of stimulating influence for the development of graph theory.

In a map (1) (Fig.) the countries and their frontiers can always be represented by a graph in two different ways: either the meeting-points of three or more frontiers are the nodes and the frontiers between them are the edges (2 a) or the countries are the nodes and the edges denote neighbouring countries (2 b). In the first case the areas are to be coloured, in the second case the nodes. A map drawn on the surface of a sphere is called *normal* if exactly three frontiers meet at each node and each country is bounded by a (topological) circle. The four-colour problem can be reduced to that for normal maps. If there is a topological circle or *Hamilton circuit* (Fig.) containing all the nodes of the graph of a normal map, then the countries of the map can be coloured with four colours, since two colours are needed for the inside and two for the outside of the Hamilton circuit. For a long time it was believed that a Hamilton circuit always exists. Counterexamples were given only in 1965, so this method does not solve the four colour problem.

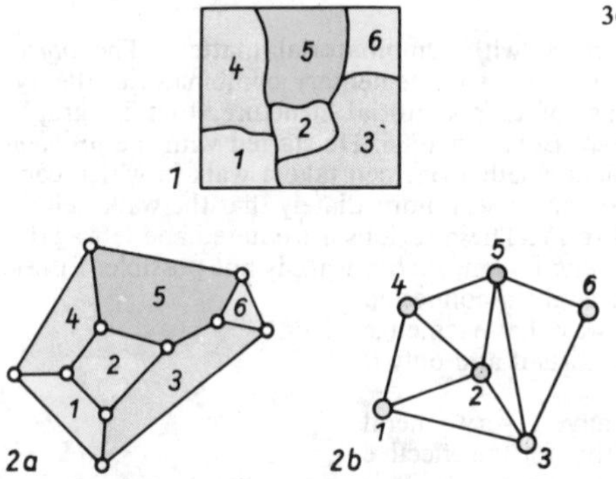

36-8 Map (1) with its graph (2a, 2b)

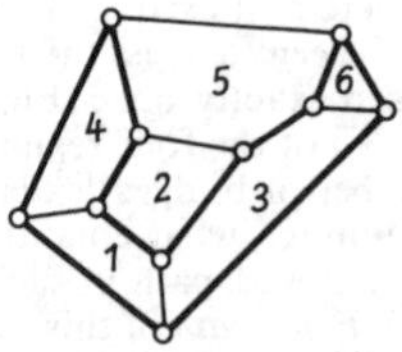

36-9 Graph of a normal map with Hamilton circuit (see Fig. 36-8, 2a)

The graph of a normal map is a *cubic graph*, that is, exactly three undirected edges meet at each node. Moreover, this graph contains no edges whose deletion would divide the graph into two separate parts, that is, the graph has no *bridges*. For a proof of the four colour problem it would be sufficient if one could show that any cubic graph can be described by several circles which all have an even number of edges. Without assuming planarity PETERSEN was able to prove that any cubic graph without bridges can be described by several circles in such a way that each node belongs to exactly one of the circles. Of course, some of these circles may have an odd number of edges.

Further work on the four colour problem used topological or combinatorial methods. The final solution which was reached in 1976 by APPELL and HAKEN worked with a method originally devised by KEMPE in 1879 which uses a reduction procedure for graphs and finally leads to a very extended distinction of cases. About 1800 cases had to be considered – each one the four colouring of a graph which is "interior" to a circuit of length $\leqslant 14$. This was done with a fast computer and needed about 1200 hours of computing time, thus providing a new type of mathematical proof characterized by the necessary use of a computer.

Network techniques

Network techniques are applied to represent, analyze and optimize the progress of complicated processes, for example, the erection of large buildings, which are composed of several partial processes. The aims of network techniques are: the planning of the finishing time and intermediate times and the search for spare time for the partial processes; the determination of the most advantageous sequence for the partial processes to shorten the total time, lower the cost and improve the utilization of capacity; the development of a control system and limitation of responsibility.

Activities and events. A *network* is a directed graph whose elements are expressed as activities and events. *Activities* are partial processes or parts of the work; to them there correspond *durations of time*; *events* are the attainments of individual steps of the process or occurrences of individual stages of completion; to them there correspond *points of time. Fictitious activities* are those of zero duration, which only express the dependence of actual activities. If the activities and events are represented by the edges and nodes of the network, then it is an *event-oriented network*. Conversely, if the activities are represented by the nodes and the interdependence of the activities by the edges of the network, then it is an *activity-oriented network*. Here the edges essentially correspond to the events, while the dependence of the activities usually consists of the fact that one activity must be finished before the next can begin. The following arguments refer to an event-oriented network.

Example: For the building of a machine shop with access road and grounds, the following work has to be done if u signifies a unit of time:

1.1. Laying the foundations	5 u	2.2. Laying out the grounds	10 u
1.2. Brickwork	11 u	3.1. Delivery time for the machines	24 u
1.3. Roof construction	4 u	3.2. Mounting the machines	3 u
1.4. Interior work	10 u	3.3. Other equipment	8 u
2.1. Building the access road	9 u		

The items in the first group (1.1. to 1.4.) must be carried out in succession. The building of the access road (2.1.) can start after laying the foundations (1.1.) and must be finished before mounting the machines (3.2.) (fictitious activity ②–④), while the interior work (1.4.) follows the mounting of the machines (fictitious activity ⑤–⑥) (Fig.). The fictitious activity ③–④ is necessary because the mounting of the machines (3.2.) cannot begin before the brickwork (1.2.).

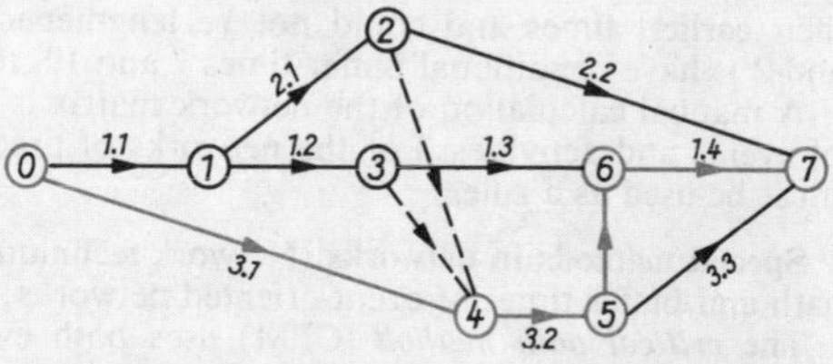

36-10 Network of the example; the critical path is marked in red

Critical path. In a network one is interested in the duration of the whole process, that is, the time between the first event and the last. This can be determined, since there is at least one path along the edges from the first event to the last for which the sum of the durations of the activities is a maximum. A path of this kind is called a *critical path*, and the activities on it are called *critical activities*. In Fig. 36-10 the critical path has 37 units of time. Any lengthening of a critical activity leads to a lengthening of the total duration, while this is not the case if a non-critical activity is lengthened within certain limits.

Activity times. For each activity in the network there is an earliest and latest starting time and an earliest and latest finishing time. The difference between the starting and finishing times is the *duration of the activity*. For critical activities the earliest and latest times coincide.

Network matrix. To determine the critical path of a network and to obtain the activity times one uses a network matrix (Fig.). Its rows and columns correspond to the events of the network, and its elements give the number of units of time for the duration of the activities joining the events (see the table in the example). To arrive at an algorithm for calculation, the successive events and the successive numbers must be obtained. If t_n is the earliest and T_n the latest time for the event n, then:

$t_0 = T_0$, where 0 denotes the first event;
$t_e = T_e$, where e denotes the last event;
$$\left.\begin{aligned} t_n &= \max(t_v + a_{vn}),\ v < n \\ T_n &= \min(T_v - a_{nv}),\ v > n \end{aligned}\right\}$$
for all other events, where a_{vn} is the duration of the activity that joins the events v and n.

t_n	0	1	2	3	4	5	6	7	events
0		5			24				0
5			9	11					1
14					0			10	2
16					0		4		3
24						3			4
27							0	8	5
27								10	6
37									7
$T_n =$	0	12	24	23	24	27	27	37	
$T_n - t_n =$	0	7	10	7	0	0	0	0	

36-11 Network matrix

To calculate the earliest times t_n one has to work with the columns of the network matrix, starting with the first event. To calculate t_4, for example, one forms the sums $t_0 + a_{04} = 0 + 24$, $t_2 + a_{24} = 14 + 0$, $t_3 + a_{34} = 16 + 0$, and takes the largest, $t_4 = 24$. To calculate the latest times T_n one begins with the last event and works with the rows of the matrix. For example, $T_5 = 27$ is the minimum of the differences $T_7 - a_{57} = 37 - 8$ and

$T_6 - a_{56} = 27 - 0$. To obtain the critical path, one marks on the network those events for which $t_n = T_n$ in the network matrix, that is, $n = 0, 4, 5, 6$ and 7.

Buffer times. In a network it is possible, within certain limits, to move or extend non-critical activities, without increasing the total time of the process, because of *buffer times* (Fig.). The duration of the activity Ⓝ→ Ⓜ is denoted by a_{nm}. By using the *total buffer time* $S_G = T_m - t_n - a_{nm}$ the buffer times of the preceding and following activities are reduced. If the activity is critical, then $t_n = T_n$, $t_m = T_m$ and $T_m - t_n = a_{nm}$, so $S_G = 0$. Using the *independent buffer time* $S_U = \max\{0, t_m - T_n - a_{nm}\}$ has no effect on the buffer times of the preceding and following activities. Using the *free buffer time* $S_F = t_m - t_n - a_{nm}$ only has an effect on the buffer times of the preceding activities, while using the *conditional buffer time* $S_B = T_m - t_m$ only has an effect on the buffer times of the following activities. Since $S_G \geqslant S_F \geqslant S_U \geqslant 0$ and $S_G = S_F + S_B$, all the buffer times are zero for the critical activities.

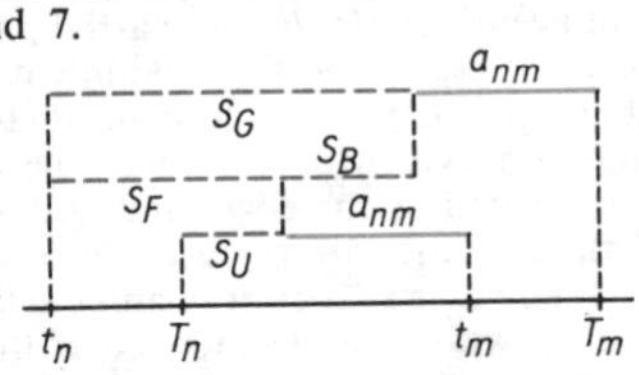

36-12 Buffer times

In the example, the activity 2.2. that leads from ② to ⑦ has buffer times $S_G = 13$, $S_U = 3$, $S_F = 13$, $S_B = 0$. Since $S_U = 3$, 2.2. can be lengthened by three units of time without setting back the completion of building. However, a lengthening of 13 units of time would mean that the path ⓪ → ① → ② → ⑦ would be critical, and so the activities 1.1. and 2.1. would have to begin at their earliest times and could not be lengthened, which would otherwise be possible, since 1.1. and 2.1. have conditional buffer times 7 and 10, respectively.

A manual calculation of the network matrix is possible only for a network with a small number of events and activities. For the networks of processes that arise in practice, electronic calculators must be used as a rule.

Special methods in networks. Network techniques are mainly concerned with finding the critical path and buffer times of event-oriented networks; the following methods have been developed.

The *critical path method* (CPM) uses both event-oriented and activity-oriented networks and evaluates only the activities by means of a (deterministically obtained) duration. Its aim is to obtain the critical path and to calculate the buffer times by means of a programme of calculation derived from the network matrix.

In the *metra-potential method* (MPM) both the activities (nodes) and the dependencies (edges) are evaluated in an activity-oriented network. For the activities a duration is evaluated, while the evaluation of the dependence expresses a *coupling distance*, for example, the time interval between the starting times of the activities joined by the edge. Coupling distances can take negative values. According to the relation between the coupling distance and the duration of the activity, one distinguishes between delayed completion (coupling distance greater than the duration of the activity), normal completion (coupling distance equal to the duration of the activity) and overlapping completion (coupling distance less than the duration of the activity). If normal completion holds in the complete network, one has a CPM-network.

The *programme evaluation and research task* (PERT) mostly uses an event-oriented network, but fixes the durations of activities not deterministically, but by means of stochastic statements. For the duration d of an activity there is an optimistic estimate d_o, a pessimistic estimate d_p and a most probable estimate d_m. The duration of the activity is given by $d = (d_o + d_p + 4d_m)/6$. Apart from the critical path, the expected values and the variance of the times are calculated.

The methods given up to now assume that the connectivity and dependence in the network, which is abstracted from the actual process, are essentially given, that is, these methods work with a prescribed topological structure of the network. In the *combination network* (CNW) the topological structure is no longer given, and the determination of an optimal structure is the aim of the method. It is now assumed that in the set of activities there is a relation such that $A \to B$ means that B must be carried out after A; $C \leftrightarrow D$ means that C and D must not proceed simultaneously. These conditions must be formulated for all the activities that occur in the network, and a structure of the network is to be found for which the critical path is minimal.

The above methods are connected with *calculations of resources.* This implies an understanding and consideration of the resources with which the activities are carried out. As resources one counts machines, labour and materials, and also the prices and costs of the activities. One distinguishes two groups of problems: 1. In optimal distribution of resources, given the time for the whole process one is trying to achieve the most equable loading of resources by using the *buffer times* and splitting the activities into sections. 2. In time-optimal distribution of limited resources upper limits are given for the availability of resources, and one is trying to make the total duration of the process as small as possible allowing for the limitations on the resources. Complete solutions of the resource problem are still not known, but quite effective approximation methods have been applied.

Finally, attempts are being made to produce network optimization algorithms concerned with complicated criteria for effectiveness and optimal working.

37. Potential theory and partial differential equations

Partial differential equations

Order, linearity, homogeneity. Ordinary differential equations contain only functions of one independent variable. By contrast, one speaks of a *partial* differential equation if the unknown function $u = u(x_1, x_2, \ldots, x_n)$ depends on several independent variables $x_1, x_2, \ldots, x_n$ and the equation contains partial derivatives $\dfrac{\partial u}{\partial x_i}$, $\dfrac{\partial^2 u}{\partial x_i \, \partial x_j}$ etc. for $i, j = 1, 2, \ldots, n$. The order of the highest derivative that appears in the equation determines the *order* of the equation. The differential equation is called *linear* if the unknown function and its derivatives occur linearly and are not multiplied together. A linear partial differential equation is called *homogeneous* if it contains no term free from the unknown function and its derivatives, otherwise *inhomogeneous*. For linear partial differential equations, as for ordinary ones, the *principle of superposition* holds: if u_1 and u_2 are solutions, then every linear combination $u = C_1 u_1 + C_2 u_2$, where C_1 and C_2 are constants, is also a solution.

Partial differential equations of the first order. The integration of a partial differential equation of the first order can always be reduced to the integration of a system of ordinary differential equations, the *characteristic system*. For the differential equation $F(x_0, \ldots, x_n, u, p_0, \ldots, p_n) = 0$, where $p_i = \dfrac{\partial u}{\partial x_i}$, this system has the form

$$x_i' = \frac{\partial F}{\partial p_i}, \quad p_i' = -\frac{\partial F}{\partial x_i} - p_i \frac{\partial F}{\partial u}, \quad u' = \sum_{j=0}^{n} \frac{\partial F}{\partial p_j} p_j$$

where the x_i and p_i are regarded as functions of a new parameter t and the dash denotes the derivative with respect to t.

If the differential equation does not depend explicitly on u, it can be brought to the form $\dfrac{\partial u}{\partial t} + H(t, x_1, \ldots, x_n, p_1, \ldots, p_n) = 0$, where $x_0 = t$, $p_0 = \dfrac{\partial u}{\partial t}$ and the variables are possibly renumbered. An equation of this form is called a *Hamilton-Jacobi differential equation*; the function H in it is called the *Hamiltonian*. The characteristic system then has the *canonical form* $\dfrac{dx_i}{dt} = \dfrac{\partial H}{\partial p_i}$, $\dfrac{dp_i}{dt} = -\dfrac{\partial H}{\partial x_i}$. The motions of point masses of certain mechanical systems are described by such equations. In this case, the x_i and p_i are generalized coordinates of position and impulse, and the Hamiltonian H is equal to the total energy (see Chapter 38).

Partial differential equations of higher order. There is no corresponding closed theory of integration for partial differential equations of higher order. Even though the general integral cannot be derived, one can often find particular solutions by means of a suitable trial solution in the form of a product or a sum of functions, each of which depends only on part of the set of variables: this is called *separation of the variables*. The given differential equation then splits into a number of simpler differential equations for these functions.

The properties of a special *linear partial differential equation of the second order* are investigated in potential theory.

Potential theory

Originally arising from problems in mechanics, potential theory has developed into an independent, extensive branch of mathematics. Its results are applied in numerous physical disciplines, particularly in the treatment of problems in mechanics, electrostatics, magnetostatics, electrodynamics, hydrodynamics and thermodynamics. Potential theory has also proved to be fruitful in the development of the theory of ordinary and partial differential equations, complex analysis, the theory of conformal mappings and differential geometry.

The Newtonian potential. The simplest concept of potential is that discovered by Newton to explain the mutual attraction of material bodies.

Newton's law of gravitation	$F = k \cdot (m\mu)/r^2$

Potential of a point. Newton's law of gravitation states that two bodies in three-dimensional space exert an attraction on each other which is directly proportional to their masses and inversely proportional to the square of the distance between them. If the bodies are regarded as idealized, so that their whole mass is concentrated at a point, the mass-point, say the mass m of one body at a point P and the mass μ of the other at a point Q, then the *above* formula is obtained. It becomes Coulomb's law if the masses are replaced by electric charges. Here r is the distance between the points and k is a factor of proportionality, for example, the constant of gravitation. To simplify the calculations it will be assumed that in the following $km = 1$. If one supposes that the mass at P is attracted by the mass at Q, then the force F is directed from P to Q. If this direction of force makes angles α, β, γ with the axes of a Cartesian coordinate system, in which P and Q have the coordinates (x, y, z) and (ξ, η, ζ), then, by the theorems of analytic geometry (Fig.), $r = \sqrt{[(x-\xi)^2 + (y-\eta)^2 + (z-\zeta)^2]}$, $\cos\alpha = (\xi - x)/r$, $\cos\beta = (\eta - y)/r$, $\cos\gamma = (\zeta - z)/r$, and so the components X, Y, Z of the force F are given by

$$X = F\cos\alpha = \mu\cdot(\xi - x)/r^3,\quad Y = F\cos\beta = \mu\cdot(\eta - y)/r^3,\quad Z = F\cos\gamma = \mu\cdot(\zeta - z)/r^3.$$

However, as LAGRANGE discovered in 1773, these three components are the partial derivatives of a function $U(x, y, z)$, which GAUSS in 1840 called the *potential* of the mass μ at Q for the point $P(x, y, z)$.

For if $U = \mu/r = \mu[(x-\xi)^2 + (y-\eta)^2 + (z-\zeta)^2]^{-1/2}$, and if Q is regarded as fixed and P as variable, then the partial derivative of U with respect to x is

$$\frac{\partial U}{\partial x} = -(\mu/2)\,[(x-\xi)^2 + (y-\eta)^2 + (z-\zeta)^2]^{-3/2}\cdot 2(x-\xi) = \mu\cdot(\xi - x)/r^3 = X.$$

Similarly the partial derivatives of U with respect to y and z are $\dfrac{\partial U}{\partial y} = Y$, $\dfrac{\partial U}{\partial y} = Z$.

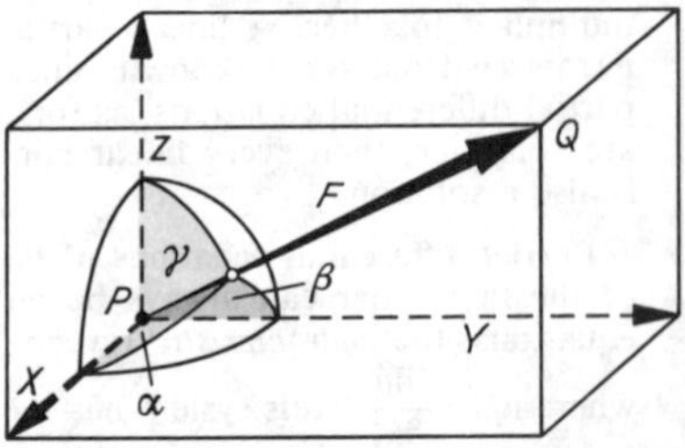

37-1 Splitting a force into components

Potential
$U = U(x, y, z) = \mu/\sqrt{[(x-\xi)^2 + (y-\eta)^2 + (z-\zeta)^2]}$

The potential U is defined in the whole three-dimensional space with the exception of Q; if $P = Q$, the denominator vanishes, and the expression μ/r is not defined for $r = 0$.

The value $-U(x, y, z)$ is equal to the potential energy of the system consisting of the two mass points.

Potential of finitely many points. If the mass at P is attracted by finitely many mass-points Q_s $(s = 1, 2, \dots, n)$ with masses μ_s, then the components X, Y, Z of the total force acting at P are the sums of the components X_s, Y_s, Z_s of the individual forces:

$$X_s = \mu_s\cdot(\xi - x)/r^3,\quad Y_s = \mu_s\cdot(\eta - y)/r^3,\quad Z_s = \mu_s\cdot(\zeta - z)/r^3;$$

$$X = \sum_{s=1}^{n} X_s,\quad Y = \sum_{s=1}^{n} Y_s,\quad Z = \sum_{s=1}^{n} Z_s.$$

Potential	$U = \sum_{s=1}^{n} \mu_s/r_s$

Similarly the potential U is the sum of the individual potentials, as long as P does not coincide with any of the points Q_s.

Potential of a continuously distributed mass. To generalize further, it is natural to abandon the abstraction of a mass point and to investigate the attraction that a *continuously distributed* mass exerts at a point P lying outside this mass. One thinks of the mass, which fills a region T, as divided into infinitesimal elements of volume $d\tau = d\xi\, d\eta\, d\zeta$ with mass $d\mu$ and density $\varrho = \dfrac{d\mu}{d\tau}$. The element of volume at $Q(\xi, \eta, \zeta)$ exerts at $P(x, y, z)$ a force of attraction with components dX, dY, dZ. The components of the force of attraction of the whole mass in T is obtained by summing over all the infinitely many elements of volume, that is, by integrating over T (Fig.):

$$dX = [(\xi - x)/r^3]\, d\mu,\qquad dY = [(\eta - y)/r^3]\, d\mu,$$
$$dZ = [(\zeta - z)/r^3]\, d\mu;$$

$$X = \iiint_T [(\xi - x)/r^3]\, d\mu,\qquad Y = \iiint_T [(\eta - y)/r^3]\, d\mu,$$
$$Z = \iiint_T [(\zeta - z)/r^3]\, d\mu.$$

These components are again partial derivatives of a potential U, as one can show by partially differentiating under the integral sign.

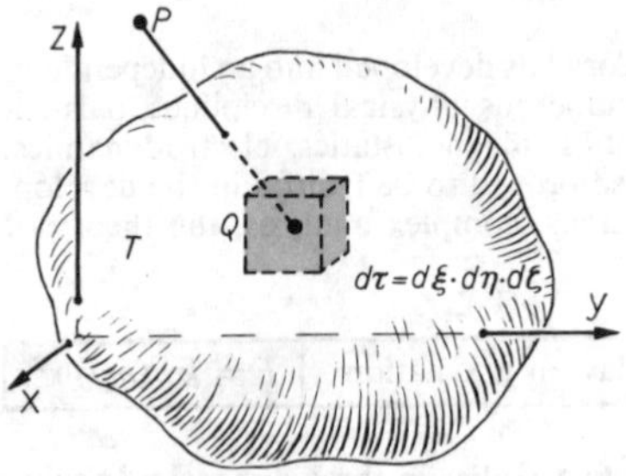

37-2 Derivation of the Newtonian potential

Newtonian potential	$U = \iiint_T \dfrac{d\mu}{r} = \iiint_T \dfrac{\varrho}{r}\, d\tau$

Equipotential surfaces. One way of giving a geometrical interpretation of potential is by means of *equipotential surfaces*. The potentials are defined for each point P of three-dimensional space, as long as P does not coincide with the attracting mass point or lie on or inside the attracting continuous mass. If all the points P at which the potential has the same value a are joined, the equipotential surface $U(x, y, z) = a$ is obtained. By varying a the given equation represents a 1-parameter family of surfaces.

In the case of the point potential $U = \mu/r$ the family is given by $\mu/r = a$. These are obviously concentric spheres with centre at Q, whose radii decrease as a increases. Figures 37-3 and 37-4 represent the family $\mu/r = a$ and show how the potential U depends on the distance r.

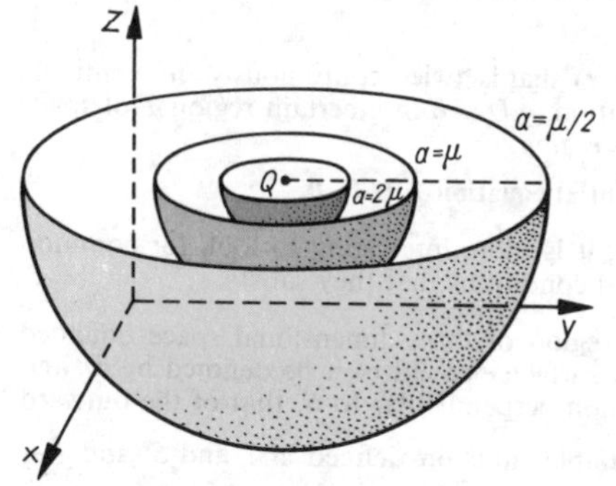

37-3 Equipotential surfaces defined by $a = \mu/r$

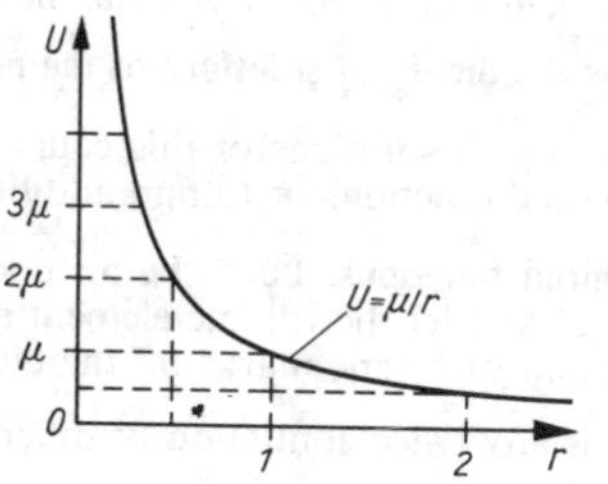

37-4 Dependence of the point potential U of the distance r

Differential equation for potentials. If one partially differentiates

$$1/r = [(x-\xi)^2 + (y-\eta)^2 + (z-\zeta)^2]^{-1/2}$$

twice with respect to x, one obtains

$$\frac{\partial}{\partial x}(1/r) = -[(x-\xi)^2 + (y-\eta)^2 + (z-\zeta)^2]^{-3/2}(x-\xi)$$

and

$$\frac{\partial^2}{\partial x^2}(1/r) = 3[(x-\xi)^2 + (y-\eta)^2 + (z-\zeta)^2]^{-5/2}(x-\xi)^2 - [(x-\xi)^2 + (y-\eta)^2 + (z-\zeta)^2]^{-3/2} = 3(x-\xi)^2/r^5 - 1/r^3.$$

Similarly for the partial derivatives with respect to y and z:

$$\frac{\partial^2}{\partial y^2}(1/r) = 3(y-\eta)^2/r^5 - 1/r^3, \quad \frac{\partial^2}{\partial z^2}(1/r) = 3(z-\zeta)^2/r^5 - 1/r^3.$$

If these three partial derivatives are added, the right-hand sides cancel:

$$\frac{\partial^2}{\partial x^2}(1/r) + \frac{\partial^2}{\partial y^2}(1/r) + \frac{\partial^2}{\partial z^2}(1/r) = 0.$$

The function $u = 1/r$ satisfies the differential equation $\frac{\partial^2 u}{\partial x^2} + \frac{\partial^2 u}{\partial y^2} + \frac{\partial^2 u}{\partial z^2} = 0$, which was first given by Laplace in 1782 and named after him.

This linear homogeneous partial differential equation of the second order is written for short as $\triangle u = 0$.

Since the differential equation is homogeneous and linear, it remains true if it is multiplied by a constant μ. Since $\triangle(1/r) = 0$, it follows that $\triangle(\mu/r) = 0$. The point potential $U = \mu/r$ is therefore a solution of Laplace's equation. Since the equation $\triangle(\mu_s/r_s) = 0$ is satisfied for each term in the sum $\sum_{s=1}^{n} \mu_s/r_s$, the differential equation is also satisfied for the sum. Finally, by differentiation under the integral sign one obtains

Laplace operator	$\frac{\partial^2}{\partial x^2} + \frac{\partial^2}{\partial y^2} + \frac{\partial^2}{\partial z^2} = \triangle$

$$\triangle U = \iiint \triangle(1/r)\,\varrho\,d\tau = 0.$$

All three potentials considered up to now are therefore solutions of Laplace's equation. This gives a new and interesting approach to potential theory: one takes Laplace's equation as starting point and calls the solutions potentials.

The case excluded so far, when P lies inside the attracting mass, leads to a non-homogeneous differential equation of the form $\triangle u = -4\pi\varrho$, where ϱ is the density of the mass; this was discovered by POISSON in 1813. Moreover, the *Laplace operator* occurs in further important partial differential equations of theoretical physics. Examples are:

1. the *Helmholtz oscillation equation* $\triangle u + k^2 u = 0$;
2. the *heat conduction equation* $\triangle u = \frac{1}{c^2} \cdot \frac{\partial u}{\partial t}$, which is also applied to diffusion problems;
3. the *wave equation* $\triangle u = \frac{1}{c^2} \cdot \frac{\partial^2 u}{\partial t^2}$ for electromagnetic and water waves, sound transmission and oscillations of strings;
4. the *telegraph equation* $\triangle u = a \frac{\partial^2 u}{\partial t^2} + b \frac{\partial u}{\partial t} + cu$ for the transmission of electromagnetic waves in cables.

The general potential function. Any function $U(x, y, z)$ that is twice continuously differentiable with respect to all three variables and satisfies the equation $\triangle U = 0$ in a certain region T of space is called a *potential function* or *harmonic function in this region.*

Potential theory is the theory of solutions of the potential equation $\triangle U = 0$.

Rather than collecting all solutions of this equation, it is more interesting to look for common properties of all potential functions or to find additional conditions that they satisfy.

Properties of potential functions. Let T be an open region of three-dimensional space bounded by a smooth surface S, and let the volume element and surface element of it be denoted by $d\tau$ and $d\sigma$ (Fig.). At every point of S, one marks off the direction perpendicular to S, that of the outward normal $\boldsymbol{n}$. Now if V is any twice continuously differentiable function defined in T and S, and $\frac{\partial V}{\partial n}$ is its partial derivative in the normal direction for all points of S, then, by Gauss's integral theorem:

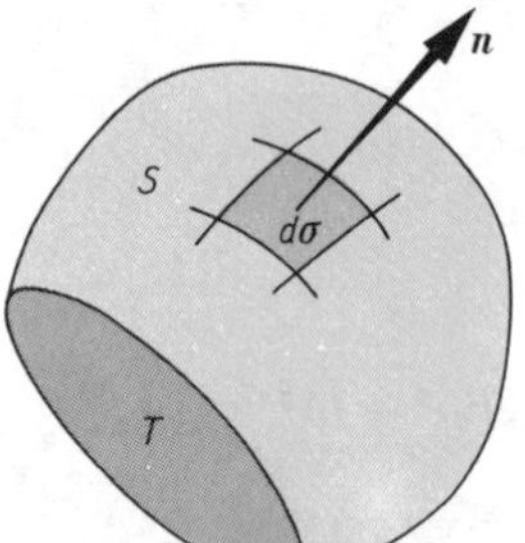

$$\iint_S \frac{\partial V}{\partial n}\, d\sigma = \iiint_T \triangle V\, d\tau.$$

If V is a potential function U, then $\triangle U = 0$ everywhere in T, and so:

For every potential function U, $\displaystyle\iint_S \frac{\partial U}{\partial n}\, d\sigma = 0$

37-5 Surface element and normal

This statement characterizes potential functions; if it holds on the surface S' of any region T' lying in T, then U is a potential function. According to another integral theorem, Green's theorem, for any two twice continuously differentiable functions V and W, defined in T and S,

$$\iint_S \left(W \frac{\partial V}{\partial n} - V \frac{\partial W}{\partial n}\right) d\sigma = \iiint_T (W \triangle V - V \triangle W)\, d\tau.$$

If one chooses $W = 1/r$, where r is the distance of P from a fixed point P_0, then $\triangle W = \triangle(1/r) = 0$ except at $P = P_0$. This point is excluded in the first instance from the region of integration T, since W has a singularity there. If one wishes to admit P_0 in T, one has to carry out a limiting process, which leads to $\iiint_T (1/r) \triangle V\, d\tau = -4\pi V(P_0)$ for $P_0 \in T$. If a potential function U is chosen for V, one obtains:

Green's representation formula for a general potential function	$U(P) = \frac{1}{4\pi} \iint_S \left(\frac{1}{r} \frac{\partial U}{\partial n} - U \frac{\partial(1/r)}{\partial n}\right) d\sigma$

U is therefore completely determined at every point P_0 of T as long as the values of the function U and its normal derivative $\frac{\partial U}{\partial n}$ are known on the boundary S.

If one takes for S a sphere C with centre at P_0 and radius R, then $\frac{\partial(1/r)}{\partial n} = \frac{\partial(1/r)}{\partial r} = -1/r^2$. Now $r = R = \text{const}$ on C, and by taking account of $\iint_C \frac{\partial U}{\partial n}\, d\sigma = 0$ one obtains:

Gauss's mean value property for potential functions	$U(P_0) = 1/(4\pi R^2) \iint_C U \, d\sigma$

The value of the function at the centre of the sphere is always equal to the average of the values of the function at points of the surface of the sphere. A potential function cannot therefore have a relative maximum or minimum at an interior point of T.

Boundary value problems. Green's formula leads to the following question: under what assumptions can one determine the potential function U inside a region T from given values of U and $\frac{\partial U}{\partial n}$ on the boundary S of T? – Problems of this kind are called boundary value problems. They occur in many branches of physics, for example in electrostatics, hydrodynamics and the theory of heat conduction.

Now U and $\frac{\partial U}{\partial n}$ cannot both be chosen arbitrarily on S. If the boundary values of U are given, then the function is uniquely determined; the difference $(U - \bar{U})$ of two functions with the same boundary values is zero over the whole boundary and therefore in the interior, by Gauss's mean value property.

The problem of determining the potential function in the interior from given boundary values of U is called the *first boundary value problem of potential theory* or the *Dirichlet problem*, after the mathematician who first worked on it.

The *second boundary value problem*, or *Neumann problem*, consists of finding a potential function that has a given normal derivative $\frac{\partial U}{\partial n}$ at all boundary points. Naturally the boundary values must be prescribed in such a way that the condition $\iint_S \frac{\partial U}{\partial n} \, \partial\sigma = 0$ is satisfied.

The *third boundary value problem* consists of finding solutions of the potential equation for which a linear combination $\frac{\partial U}{\partial n} + hU$, where h is a positive constant, takes prescribed values at the boundary points.

Simple solutions of the potential equation. Potential functions U in three-dimensional space are functions of three independent variables and have the form $U = U(x, y, z)$ in Cartesian coordinates, $U = U(\varrho, \varphi, z)$ in cylindrical coordinates and $U = U(r, \vartheta, \varphi)$ in spherical coordinates. One is often interested in solutions for which U can be written as a product of three functions of one variable: $U(x, y, z) = X(x)\, Y(y)\, Z(z)$ or $U(\varrho, \varphi, z) = P(\varrho)\, \Phi(\varphi)\, Z(z)$ or $U(r, \vartheta, \varphi) = R(r)\, \Theta(\vartheta)\, \Phi(\varphi)$. In this case one obtains from the partial differential equation $\triangle U = 0$ three ordinary differential equations, which can usually be solved directly. This procedure is known as *separation of the variables*. For example, if $U(x, y, z) = X(x)\, Y(y)\, Z(z)$, then the equation $\triangle U = \frac{\partial^2 U}{\partial x^2} + \frac{\partial^2 U}{\partial y^2} + \frac{\partial^2 U}{\partial z^2} = 0$ gives the three differential equations $\frac{1}{X} \cdot \frac{d^2 X}{dx^2} = k^2$, $\frac{1}{Y} \cdot \frac{d^2 Y}{dy^2} = l^2$, $\frac{1}{Z} \cdot \frac{d^2 Z}{dz^2} = -(k^2 + l^2)$ and so the solution is $U_{klm}(x, y, z) = e^{kx} e^{ly} e^{mz}$; $m^2 = -(k^2 + l^2)$; k, l, m complex.

Transformations. Certain mappings of three-dimensional space leave the property of being a potential function unaltered. These are inversions with respect to spheres and are called *Thomson transformations.* For example, after inversion with respect to the unit sphere $|r| = 1$, $U(r, \vartheta, \varphi)$ becomes $(1/r)\, U[(1/r), \vartheta, \varphi]$, which is also a potential function. The Newtonian potential $U = 1/r$ gives the constant potential $U = 1$, and conversely.

Potentials in the plane. Two-dimensional potentials are solutions of the differential equation $\triangle U = \frac{\partial^2 U}{\partial x^2} + \frac{\partial^2 U}{\partial y^2} = 0$. They often arise in physical problems when the solution is independent of the third coordinate; for example, the force of attraction of a very long uniform rod in the direction of the z-axis is the same at two points $P_1(x, y, z_1)$ and $P_2(x, y, z_2)$, as long as the coordinates z_1, z_2 are small in comparison with the length $2L$. For an approximate solution one may assume that U does not depend on z, and start with $U = U(x, y)$ (Fig.).

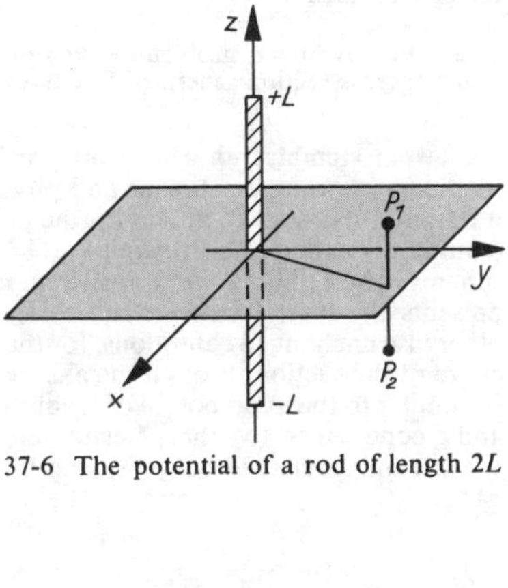

37-6 The potential of a rod of length $2L$

The solutions of the two-dimensional potential equation are closely connected with *complex analysis*: a function $w = u + iv$ of the complex variable $z = x + iy$ is analytic if and only if its real part $u(x, y)$ and its imaginary part $v(x, y)$ satisfy the *Cauchy-Riemann differential equations* $\frac{\partial u}{\partial x} = \frac{\partial v}{\partial y}, \frac{\partial u}{\partial y} = -\frac{\partial v}{\partial x}$. Then $\frac{\partial^2 u}{\partial x^2} = \frac{\partial^2 v}{\partial y\, \partial x} = \frac{\partial^2 v}{\partial x\, \partial y} = -\frac{\partial^2 u}{\partial y^2}$ and $\frac{\partial^2 u}{\partial x^2} + \frac{\partial^2 u}{\partial y^2} = 0$. The same holds for v.

The real and imaginary parts of any analytic function are potential functions. They are said to be conjugate.

Example: From $w = \ln z = \ln (re^{i\varphi}) = \ln r + i\varphi$ one obtains two conjugate potentials in the plane. They are $u(r, \varphi) = \ln r$ and $v(r, \varphi) = \varphi$, or in Cartesian coordinates $U(x, y) = \ln \sqrt{(x^2 + y^2)}$, $V(x, y) = \arctan (y/x)$.

The potential $U = \ln r$, the *logarithmic potential*, plays the same role in the plane as the Newtonian potential $U = 1/r$ does in space. It is the potential of the field of force of an attracting point, only the force of attraction is proportional to $1/r$ instead of $1/r^2$.

38. Calculus of variations

The methods of the calculus of variations are applied in the solution of many problems of geometry, theoretical physics and technology. Questions that led to problems in the calculus of variations had already emerged in antiquity, for example, the problem of finding among all plane areas with equal perimeters that with the greatest area. ZENODOROS (about 180 B. C.) recognized the isoperimetric problem. The same problem confronted the peasant Pakhom in Tolstoy's story 'How much earth does a man need?', when the Bashkiri Elder called to him: 'As much land as you can walk around in a day is yours.'

Isoperimetric problem: Of all plane figures with equal perimeters (isoperimetric figures) the circle has the greatest area (Fig.), and in space, of all bodies with equal surface areas the sphere has the maximum volume.

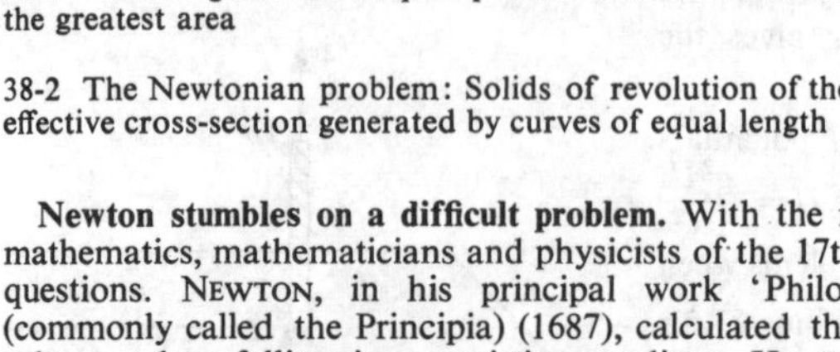

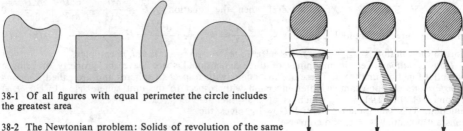

38-1 Of all figures with equal perimeter the circle includes the greatest area

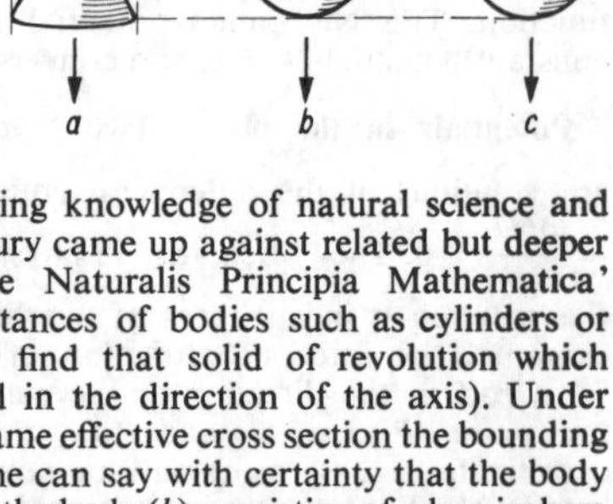

38-2 The Newtonian problem: Solids of revolution of the same effective cross-section generated by curves of equal length

Newton stumbles on a difficult problem. With the increasing knowledge of natural science and mathematics, mathematicians and physicists of the 17th century came up against related but deeper questions. NEWTON, in his principal work 'Philosophiae Naturalis Principia Mathematica' (commonly called the Principia) (1687), calculated the resistances of bodies such as cylinders or spheres when falling in a resisting medium. He tried to find that solid of revolution which presents the least resistance in falling (with the same speed in the direction of the axis). Under otherwise unchanged conditions, for the same length and the same effective cross section the bounding curve of the section through the axis is required. Although one can say with certainty that the body (*a*) similar to the hyperboloid of revolution is less suitable than the body (*b*) consisting of a hemisphere and a cone joined together, Newton and his contemporaries were not yet able to give as the solution the streamline-shaped body (*c*) (Fig.).

The brachystochrone problem. Still more fruitful and celebrated was the problem of the *brachystochrone*, which was raised publicly by Johann BERNOULLI in 1696: If two points P_1 and P_2 are given, at different heights but not lying one above the other, it is required to find among all possible curves connecting them, that one along which a material point slides from P_1 to P_2 under the influence of gravity (neglecting friction) in the shortest possible time. This problem occupied at the time the leading mathematicians in the whole of Europe: NEWTON, LEIBNIZ, Jakob BERNOULLI, L'HOSPITAL, HUDDE, FATIO and others. From then on, the calculus of variations developed as a special mathematical discipline.

In a suitably chosen coordinate system some such curves $y = f(x)$ joining the points P_1 and P_2 are drawn as possible curves for the fall (Fig.). Because the distance s, the time t and the instantaneous speed v at every point of this curve are connected by the relation $v = \dfrac{ds}{dt}$, because the speed v under the acceleration due to gravity g has the value $v = \sqrt{(2gy)}$, and finally because the element of arc ds is the well-known function $ds = \sqrt{(1 + y'^2)}\, dx$ of y' and x, one obtains the time T required for the motion as a definite integral between the limits x_1 and x_2; from

$$dt = [1/\sqrt{(2gy)}]\, ds = [1/\sqrt{(2gy)}]\, \sqrt{(1 + y'^2)}\, dx$$

it follows that

$$T = [1/\sqrt{(2g)}] \int_{x_1}^{x_2} \sqrt{[(1 + y'^2)/y]}\, dx.$$

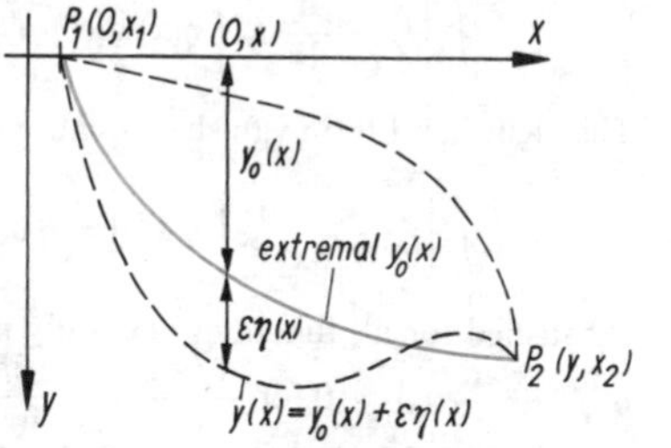

38-3 The brachystochrone problem

This integral is to have the smallest value for the *required function* $y_0 = y_0(x)$, that is, its value is larger for all functions $y(x)$ different from y_0. By the method sketched in what follows (solution of the Euler differential equation) one obtains as solution, with α as a parameter and two constants C_1 and C_2, the *cycloid*

$$x_0 = (C_1/2)(\alpha - \sin\alpha) + C_2, \qquad y_0 = (C_1/2)(1 - \cos\alpha).$$

Variation problem without side conditions, Euler's differential equation

The investigation of the brachystochrone has led to the problem of finding a function $y(x)$ for which the integral of a second function $f(x, y, y')$ has a smallest or a largest value; the function $f(x, y, y')$ is determined by the geometrical, technological or physical situation and is called the *basic function*. In the brachystochrone problem $f(x, y, y') = \sqrt{[(1 + y'^2)/y]}$ is the basic function. Characterizing the requirement of an extreme value by an exclamation mark, the condition

$$J = \int_{x_1}^{x_2} f(x, y, y')\, dx = \text{extreme value!}$$

is to hold. The basic function depends on the independent variable x and on the required function $y(x)$ and its derivative $y'(x)$. The required function $y_0(x)$ is called an *extremal.*

The basic problem of the calculus of variations is a maximum or minimum problem, but of a more difficult kind than in the differential calculus. It is to find a function for which a certain integral assumes a greatest or a smallest value.

Whereas the BERNOULLI brothers, NEWTON and others solved the brachystochrone problem by means of special tricks, EULER, LAGRANGE, WEIERSTRASS, OSTROGRADSKII, CARATHÉODORY and others were able to develop in the 19th and 20th centuries a method that always leads to a solution.

EULER succeeded in reducing the variation problem to differential equations. He began by assuming that all the admissible functions $y(x)$ must not differ from the required extremal $y_0(x)$ at the points P_1 and P_2. He thought of $y(x)$ as being always a combination of the extremal $y_0(x)$ and a variation function $\varepsilon\eta(x)$. Then $y(x) = y_0(x) + \varepsilon\eta(x)$ becomes the extremal for the value $\varepsilon = 0$ of the free parameter. At the same time $\eta(x)$ must have the value zero at the points P_1 and P_2, that is, for the values x_1 and x_2, so that the *boundary conditions* $\eta(x_1) = \eta(x_2) = 0$ must hold. For $y = y(x)$ the integral J is a function of ε, where

$$J(\varepsilon) = \int_{x_1}^{x_2} f(x, y_0 + \varepsilon\eta, y_0' + \varepsilon\eta')\, dx.$$

This function $J(\varepsilon)$ has an extremum for $\varepsilon = 0$ and consequently, by the rules of the differential calculus, its derivative must vanish for $\varepsilon = 0$. Since the limits of integration are fixed, one may differentiate under the integral sign:

$$J'(\varepsilon) = \int_{x_1}^{x_2} [f_y(x, y_0 + \varepsilon\eta, y_0' + \varepsilon\eta')\,\eta + f_{y'}(x, y_0 + \varepsilon\eta, y_0' + \varepsilon\eta')\,\eta']\,dx.$$

For every function $\eta(x)$ that is continuously differentiable between x_1 and x_2 and vanishes at x_1 and x_2,

$$J'(0) = \int_{x_1}^{x_2} [f_y(x, y_0, y_0')\,\eta + f_{y'}(x, y_0, y_0')\,\eta']\,dx = 0.$$

Integrating the second term by parts one obtains

$$\int_{x_1}^{x_2} \left[f_y - \frac{d}{dx} f_{y'}\right] \eta(x)\,dx + [f_{y'}\eta(x)]_{x_2}^{x_1} = 0.$$

The right-hand term vanishes because of the boundary conditions. But the relation

$$\int_{x_1}^{x_2} \left[f_y - \frac{d}{dx} f_{y'}\right] \eta(x)\,dx = 0$$

is satisfied for all functions $\eta(x)$ only if the expression in square brackets vanishes identically, that is, $f_y - \frac{d}{dx} f_{y'} = 0$, or $\frac{\partial f}{\partial y} - \frac{d}{dx}\frac{\partial f}{\partial y'} = 0$. If one carries out the total differentiation with respect to x one obtains the celebrated Euler differential equation of the calculus of variations.

Euler's differential equation	$f_{y'y'}y'' + f_{yy'}y' + f_{xy'} - f_y = 0$

This represents an ordinary differential equation of the second order.

The requirement that a given integral shall assume an extremum for a function is replaced by a differential equation for this function.

The first variation. The variation function is called the *variation δy of the function $y(x)$*, according to a notation introduced by LAGRANGE in 1755:

$$\delta y = \varepsilon\eta(x); \quad y = y_0 + \delta y.$$

Because the integral J has an extremum for the extremal, the difference $\Delta J = J(y) - J(y_0)$ can never be positive for a maximum, and never negative for a minimum of the integral. For arbitrarily small values of ε the variation of the value of the integral can be expressed as the differential of the function $J(\varepsilon)$ for $\varepsilon = 0$. The product $J'(0)\,\varepsilon$ is called the first variation of J and is denoted by δJ:

$$\delta J = J'(0)\,\varepsilon = [f_{y'} \cdot \delta y]_{x_1}^{x_2} + \int_{x_1}^{x_2} \left(f_y - \frac{d}{dx} f_{y'}\right) \delta y\,dx.$$

The necessary condition for the existence of an extremum of the integral that was found in the derivation of the Euler differential equation can thus be expressed by saying that the first variation must vanish: $\delta J = 0$.

Generalizations. The basic function $f(x, y, y')$, and consequently the integral J, can depend on several (finitely many) functions $y_1(x), y_2(x), \ldots, y_n(x)$ and their derivatives, instead of on one function $y(x)$. In place of one Euler differential equation there is then a system of n differential equations. On the other hand, with one function $y(x)$, *higher derivatives* of $y(x)$ can occur in the basic function, besides $y'(x)$ also $y''(x), \ldots, y^{(n)}(x)$. The Euler differential equation is then of order $2n$. Finally, one can enquire into the extremal properties of surfaces in space. A *soap film* stretched across a closed, non-plane loop of wire, for example, always assumes the smallest possible surface area because of the surface tension (see Plate 55). Such surfaces with minimum area are called *minimal surfaces.* The integral J is then a double integral, x and y are independent variables in the basic function, and the function $z = z(x, y)$ determining the area is required:

$$\iint_B f(x, y, z, z_x, z_y)\,dx\,dy = \text{extreme value!}$$

As a condition for the occurrence of an extremum one obtains the Ostrogradskii differential equation, a partial differential equation of the second order:

Ostrogradskii differential equation	$f_z - \frac{\partial}{\partial x} f_{z_x} - \frac{\partial}{\partial y} f_{z_y} = 0$

Necessary and sufficient conditions for the occurrence of an extremum

Under the assumption that an extremum exists, the first derivative $J'(0)$ must vanish and the Euler differential equation must hold. Such a condition, which must be satisfied for every solution, is called a *necessary* condition. Whether a solution exists cannot be decided by a necessary condition alone. WEIERSTRASS was able to find *sufficient* conditions for the existence of an extremum by means of arguments that were new in principle. It is not possible here, however, to go into this, nor into the proof of an existence theorem, which shows that a solution exists at all. In many cases one is content to investigate subsequently whether the solution of the Euler differential equation really has extremal properties under the given geometrical, technological or physical conditions; for example, it is intuitively obvious that between two points of the plane there is no shortest, but non-rectilinear, connecting curve, because for every connecting line one can imagine a shorter one, provided that the straight line is excluded by hypothesis (Fig.).

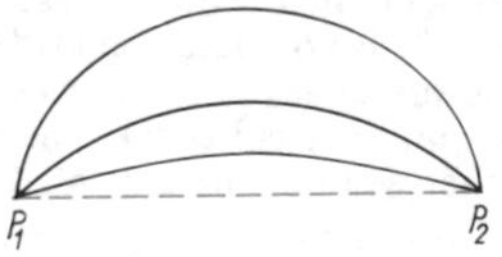

38-4 Between two points P_1 and P_2 there is no shortest non-rectilinear connecting curve

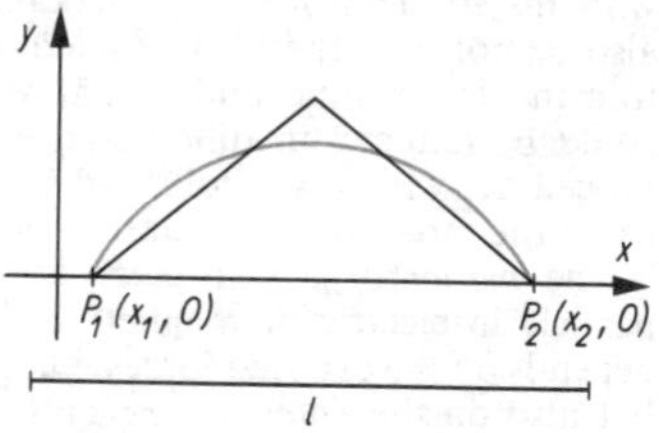

38-5 The area bounded by the line segment P_1P_2 and a curve of prescribed length l between these points is greatest when the curve is a circular arc

Variation problems with side conditions

Frequently variation problems occur in which not only the integral $\int_{x_1}^{x_2} f(x, y, y')\,dx$ must be made an extremum, but additional conditions must also be satisfied. Such conditions are called *side conditions.*

Isoperimetric problem. If one wishes the area $\int_{x_1}^{x_2} y\,dx$, $x_2 > x_1$, under the required curve $y(x)$ to be an extremum, then, for example, the arc length $\int_{x_1}^{x_2} \sqrt{(1 + y'^2)}\,dx = l$ may be prescribed, where l cannot be less than $x_2 - x_1$, $l \geqslant x_2 - x_1$. The solution of this problem of enclosing the greatest possible area by a given perimeter, which consists here of the line segment P_1P_2 and of l (Fig.), is a segment of a circle.

In general, in the isoperimetric problem, together with the requirement $\int_{x_1}^{x_2} f(x, y, y')\,dx =$ extreme value! there occurs a *side condition in integral form,* requiring that the integral of a function $g(x, y, y')$ depending on y and y' shall have a fixed, prescribed value a: $\int_{x_1}^{x_2} g(x, y, y')\,dx = a$.

This *general isoperimetric problem* can be reduced to a variation problem without side conditions by the *method of undetermined multipliers* developed by LAGRANGE. From the given functions $f(x, y, y')$ and $g(x, y, y')$ one forms with a constant multiplier λ the extended basic function $h(x, y, y') = f(x, y, y') + \lambda g(x, y, y')$ and solves the Euler differential equation for this function:

$$h_y - \frac{d}{dx} h_{y'} = 0.$$

In the example of the prescribed arc length l one puts $h(x, y, y') = y + \lambda \sqrt{(1 + y'^2)}$ and obtains by integrating the Euler differential equation $(x - \xi)^2 + (y - \eta)^2 = \lambda^2$; thus, the extremals are indeed circular arcs of radius $|\lambda|$.

Side conditions in equation form. In the isoperimetric problem, in spite of the formally complicated exterior form, there is basically only one prescribed number, for example a length. A side condition in the form of an equation, however, which may well appear simpler, leaves open many more possibilities for the required extremals, for example, the arbitrary path of the given curve on a surface. How many possible ways are there of arriving at one point from another on the surface of a sphere alone!

The *geodesic lines* on a surface are the shortest curves connecting two points of the surface. If one thinks of the coordinates of the surface point $x(t)$, $y(t)$, $z(t)$ as depending on a parameter t $\left(\frac{dx}{dt} = \dot{x}, \frac{dy}{dt} = \dot{y}, \frac{dz}{dt} = \dot{z}\right)$, then the arc length, that is, the integral $\int_{t_1}^{t_2} \sqrt{(\dot{x}^2 + \dot{y}^2 + \dot{z}^2)}\,dt$, has to be made a minimum. At the same time, however, it must be guaranteed that the curves really lie on the surface; the coordinates x, y, z must satisfy the equation $g(x, y, z) = g[x(t), y(t), z(t)] = 0$ of the surface as a side condition.

Minimal principles of theoretical physics

It was noticed relatively early by mathematicians and physicists that a *light ray* travels from one point to another in space along a path that requires a shorter time to traverse than any neighbouring path (see Chapter 19.4. – Extreme values of functions). If one makes this *principle of Fermat* the basis of geometrical optics, the laws of refraction and reflection, for example, can be derived deductively from it.

Such minimal principles were interpreted teleologically in the 18th century; indeed, MAUPERTUIS even tried to construct a proof of the existence of God from his *principle of least action*! VOLTAIRE with his ironic story of Dr. Akakia (1752) exposed Maupertuis to the ridicule of Europe and so also demolished the idea of teleology. It turned out that the light path could also on occasion lead to a maximum time and – what was more important – that the variational principles of mechanics could be reduced to differential equations, which no longer sounded or looked teleological. This procedure, which was followed by LAGRANGE, GAUSS, HAMILTON and JACOBI will now be indicated using the modern nomenclature of mathematical physics.

The *motion of a point-mass*, for example, in the gravitational field of the earth, or of a charged particle in electric or magnetic fields, is determined not only by its instantaneous velocity, which depends on the external forces, but also on potentials. It thus depends not only on the *kinetic energy T*, but also on the *potential energy U*. According to LAGRANGE one considers the *Lagrangian function* $L = T - U$ as a function of the time t, and of the space coordinates x, y, z and their derivatives $\dot{x}, \dot{y}, \dot{z}$. If one has not a single point-mass, but a *system of N point-masses*, then L is a function of the time t and of $3N$ coordinates and $3N$ velocity components. For various physical problems *generalized coordinates* q_k, $k = 1, 2, \ldots, 3N$, are introduced so that $L = L(t, q_k, \dot{q}_k)$, $k = 1, 2, \ldots, 3N$, is a function of these coordinates. The motion of the point-masses follows from the *Lagrangian equations of motion of the second kind*

$$\frac{\partial L}{\partial q_k} - \frac{\mathrm{d}}{\mathrm{d}t}\frac{\partial L}{\partial \dot{q}_k} = 0, \qquad k = 1, 2, \ldots, 3N.$$

These equations can be derived from *Hamilton's principle*. Among all imaginable conditions under which the system could change during the time interval $t_2 - t_1$ from a state 1 which it occupied at time t_1 to a given state 2, the motion that actually occurs is that for which the integral $J = \int_{t_1}^{t_2} L(t, q_k, \dot{q}_k)\,\mathrm{d}t$ is the smallest. This most important integral principle of classical mechanics leads in this way to a variation problem, and the Euler differential equations belonging to it are the Lagrangian equations of motion of the second kind.

Direct methods

Elegant as the methods of the calculus of variations sketched above appear to be, considerable difficulties can stand in the way of their practical application. In particular, for many problems the exact solution of the Euler differential equation is difficult or not possible at all. For this reason *approximation methods* have been developed which, because they circumvent the Euler differential equation, are called direct methods of the calculus of variations.

The method of Ritz (1909). For $J = \int_{x_{21}}^{x_2} f(x, y, y')\,\mathrm{d}x$ = extreme value! one assumes for the required function $y(x)$ the approximation

$$y = c_1\varphi_1(x) + \cdots + c_n\varphi_n(x),$$

where the $\varphi_i(x)$ must satisfy the boundary conditions. The problem consists in determining the constant coefficients c_i. One substitutes for y in J and obtains $J(c_1, \ldots, c_n)$ = extreme value! The c_i are given by the necessary conditions $\dfrac{\partial J}{\partial c_i} = 0$, $i = 1, \ldots, n$, for the occurrence of an extremum.

Example: If $\int_0^1 (y'^2 - y^2 - 2xy)\,\mathrm{d}x$ = extreme value! is to hold, where the solution satisfies the boundary conditions $y(0) = y(1) = 0$, one makes the assumption, for example, that $\varphi_1(x) = x(1 - x)$ and $\varphi_2(x) = x^2(1 - x)$. One then obtains the approximate solution $y = -7x^3/41 - 8x^2/369 + 71x/369$. As a check one finds in this example, from the Euler differential equation, the solution $y = \sin x/\sin 1 - x$. The differences between the exact and the approximate solution are only of the order of magnitude 10^{-4}.

39. Integral equations

An equation serving to determine a function is called an integral equation if the required function occurs in the integrand of an integral. A very simple example is the equation

(1) $$\int_a^s y(t)\,\mathrm{d}t = f(s) - f(a), \quad a \leqslant s \leqslant b.$$

The function $f(s)$ is given and it is required to find the function $y(t)$. Clearly the solution is $y(t) = \dfrac{\mathrm{d}f(t)}{\mathrm{d}t}$.

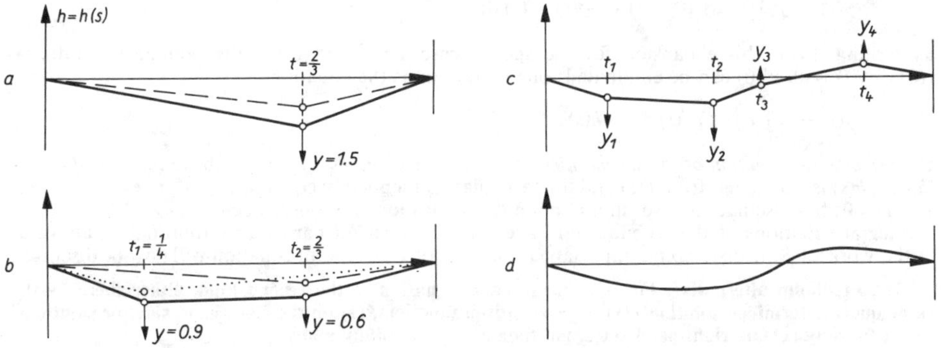

39-1 Plucked string

Integral equations frequently arise in the mathematical treatment of physical or technological problems; to this category there belong problems concerning elastic bending and oscillations (building of bridges) and concerning heat propagation processes. As an example consider a stretched string of length l, described by the interval $0 \leqslant s \leqslant l$. At the point with the coordinate t it is loaded with a force 1. Let the deflected string be represented by the function $E(s, t)$; in Fig. 39-1a (broken line) the deflected curve, for example, $h(s) = E(s, 2/3)$ for $t = 2/3$ is given. If the force is of magnitude y, the deflection is $h(s) = E(s, t)\,y$ (continuous line). If two forces y_1 and y_2 act at the points t_1 and t_2, then a total deflection $h(s) = E(s, t_1)\,y_1 + E(s, t_2)\,y_2$ arises which is composed of the individual deflections $E(s, 1/4) \cdot 0.9$ (dotted line) and $E(s, 2/3) \cdot 0.6$ (broken line) shown in Fig. 39-1b. Similarly, a loading with n forces $y_1, y_2, \ldots, y_n$ at the points $t_1, t_2, \ldots, t_n$ leads to the deflection

$$h(s) = E(s, t_1)\,y_1 + E(s, t_2)\,y_2 + \cdots + E(s, t_n)\,y_n$$

(see Fig. 39-1c). In particular, the deflection at the kth point of application $s = t_k$ has the value

(2) $$h_k = h(t_k) = E(t_k, t_1)\,y_1 + \cdots + E(t_k, t_n)\,y_n$$

or $$h_k = \sum_{i=1}^{n} E(t_k, t_i)\,y_i \qquad (k = 1, 2, \ldots, n).$$

For a continuously distributed force acting along the whole string one obtains similarly

(3) $$h(s) = \int_0^l E(s, t)\,y(t)\,\mathrm{d}t.$$

The function $y(t)$ is the *force density* or *force per unit length*, so that $y\,\mathrm{d}t$ is the differential force acting on the element of length $\mathrm{d}t$; correspondingly the summation sign in (2) goes over into the integral of (3). The deflected string forms a 'smooth' curve $h = h(s)$ (see Fig. 39-1d).

Conversely, if the form of the deflected string, that is, the function $h(s)$, is known and the loading $y(t)$ of the string is to be found, then the relation (3) becomes a *linear integral equation of the first kind* for the required function $y(t)$. The function $E(s, t)$ of two variables is called the *kernel* of the integral equation. To *solve the integral equation* one can proceed the opposite way round: the equation (3) is approximated by the equation (2), that is, the integral is approximated by a finite sum with the desired accuracy, in which one imagines the force density $y(t)$ replaced by sufficiently many individual forces y_i. The calculation of the y_i in (2) is, however, nothing but the *solution of a system of linear equations*, with n equations in n unknowns y_i.

In fact, the theory of linear integral equations has much in common with that of systems of linear equations; one could regard integral equations as *linear equations with infinitely many unknowns.*

FREDHOLM used this connection when he developed (about 1900) the first general theory. However, he considered a somewhat different type, the so-called *linear integral equations of the second kind* or *Fredholm integral equations*. These occur frequently, although mostly indirectly as the result of the rearrangement of differential equations; for example, the forced harmonic oscillation of a string is described by a boundary value problem for a differential equation of the second order:

(4) $$y'' + \lambda y = f(s), \quad y(0) = y(l) = 0.$$

It is required to find the deflection $y = y(s)$ of the string at the point s in a particular *phase of the oscillation*; the constant λ is given by the *frequency of the oscillation*, and the function $f(s)$ by the *external force* acting with the same frequency and in the same phase. For the rearrangement one substitutes $f(t) - \lambda y(t)$ for y'' from (4) in the Taylor expansion

$$y(s) = y(0) + sy'(0) + \int_0^s (s - t)\, y''(t)\, dt.$$

By writing down this expansion for the special case $s = l$ and using the boundary conditions $y(0) = y(l) = 0$, $y'(0)$ can be eliminated; one then obtains the equation

(5) $$y(s) - \lambda \int_0^l K(s, t)\, y(t)\, dt = h(s),$$

a *linear integral equation of the second kind for* $y(s)$. The *kernel* K is like the above *influence function* $E(s, t)$; $h(s)$ is calculated from $f(s)$ and, in particular, is identically equal to zero if no external forces act ($f = 0$: free oscillation). Additional boundary conditions no longer occur.

Integral equations of the second kind have been investigated particularly thoroughly, above all in the work of SCHMIDT. Some important properties of this type of equation will now be discussed.

The Fredholm alternative. For a linear integral equation of the second kind. either there exists a uniquely determined solution $y(s)$ for every given function $h(s)$ on the right-hand side, or solutions exist for only certain right-hand sides, but then always infinitely many.

The two cases are distinguished according to whether the *homogeneous* equation

(6) $$\bar{y}(s) - \lambda \int_0^l K(s, t)\, \bar{y}(t)\, dt = 0$$

belonging to (5) has only the *trivial solution* $\bar{y} = 0$, or has non-trivial, so-called null solutions $\bar{y}(s)$. The physical significance of the first case $\bar{y} \equiv 0$ in the example considered is that no *free oscillations*, called *characteristic oscillations*, are possible for the frequency determined by λ. Then there always exists a well-defined form of oscillation, no matter how the external forces are distributed. In the second case, with null solutions, on the other hand, characteristic oscillations exist. For a given force distribution $h(s)$ an oscillation $y(s)$ of the string can then exist, but an arbitrary characteristic oscillation $\bar{y}(s)$ can be superimposed on this: $y(s) + \bar{y}(s)$ is also a solution of the problem. However, there need not be any solution at all; this case occurs physically if the applied force is chosen in such a way that it 'builds up' a *characteristic oscillation* (*resonance*) and theoretically leads to infinitely large deflections of the string.

Eigenvalues. In general, those values of the parameter λ for which (6) has non-trivial solutions are relatively rare exceptions. They are called *eigenvalues* and the corresponding null solutions are called *eigenfunctions*; for example, for the oscillation of the string the eigenvalues are the numbers $\lambda_n = (\pi^2/l^2) \cdot n^2$ $(n = 1, 2, \ldots)$ and the corresponding eigenfunctions are $\bar{y}_n = \sin(s \cdot \sqrt{\lambda_n})$. Eigenvalues and eigenfunctions play a significant role in the theory and practice of integral equations; for example, series expansions of given functions in terms of eigenfunctions are important aids in the solution of differential and integral equations: the well-known Fourier series belong to this category.

The resolvent. If (5) has a unique solution, then the solution $y(s)$ can be represented with the help of the *solving kernel* or the *resolvent* $\Gamma(s, t)$:

(7) $$h(s) + \lambda \int_0^l \Gamma(s, t)\, h(t)\, dt = y(s).$$

For sufficiently small λ the resolvent can be calculated by an *iteration procedure*. For this purpose one substitutes for $y(t)$ under the integral sign in (5) the value

$$h(t) + \lambda \int_0^l K(t, r)\, y(r)\, dr$$

given by (5) itself; this leads to an equation of the form

$$y(s) - \lambda^2 \int_0^l K_2(s, r)\, y(r)\, dr = h(s) + \lambda \int_0^l K(s, t)\, h(t)\, dt.$$

In this the y under the integral sign is again replaced by the expression given above, and so on. Finally comparison with (7) yields the expansion

(8) $$\Gamma(s,t) = K(s,t) + \lambda K_2(s,t) + \lambda^2 K_3(s,t) + \cdots,$$

the so-called *Neumann series*. The *iteration kernels* K_2, K_3, ... are calculated from K by repeated integration.

Other types of equation. Besides the equations (3) and (5) there are also *linear integral equations of the third kind*:

(9) $$g(s)\,y(s) - \lambda \int_0^l K(s,t)\,y(t)\,\mathrm{d}t = h(s),$$

in which $g(s)$ and $h(s)$ are given functions. Integral equations of the first and second kinds are special cases of this type; they arise from (9) if $g(s)$ is a constant.

Further there is the extensive field of *non-linear* integral equations. Here there is no general theory yet. In these equations the required function y does not occur under the integral sign as a simple factor, but in a much more general, and usually more complicated, way. For example,

(10) $$y(s) - \int_0^l g(s,t)\,[y(t)]^2\,\mathrm{d}t = h(s)$$

is a non-linear integral equation.

40. Functional analysis

Functional analysis has been developed essentially in the last 40 years. Its beginnings lie in the recognition that widely different kinds of *mathematical operations*, from the basic operations of arithmetic to differentiation and integration, have strikingly many features in common and that the *mathematical objects* subjected to these operations exhibit the same or similar properties in relation to the operations, although they come from quite different fields of mathematics. The same rules of addition hold for addition of angles, of numbers, of vectors and so on. In this sense functional analysis formed originally a cross-section of certain branches of analysis, for example, of the theory of integral equations, of the calculus of variations, and of linear algebra.

The search for and recognition of such *deep-seated common properties*, the striving for the most general possible statements, which are independent of special mathematical objects and are determined only by *abstract relations*, have led to numerous new concepts, which have become the basis of functional analysis and are frequently used in modern mathematics.

Abstract spaces

Deviating from the usage of everyday language, the concept of *space* in functional analysis bears no direct relation to geometry or even to the space of our experience. Because of certain similarities to geometry, especially to analytical geometry and linear algebra, the word space was carried over to objects of functional analysis. Similarly other concepts also, such as *distance* or *length*, taken from the vocabulary of analytical geometry, have lost their original geometrical meaning.

The concept of an abstract space. In functional analysis a set of elements is called an *abstract space* if a limiting process is defined within the set, if the statement '*a sequence* $x_1, x_2, x_3, \ldots$ *of elements of the space tends to a limit* $x = \lim_{n\to\infty} x_n$' has a well-defined meaning.

Example 1: The elements of the *k-dimensional Euclidean space* $\mathbf{R}^k$ are the ordered k-tuples of real numbers $x = (\xi_1, \xi_2, \ldots, \xi_k)$. A sequence of elements $x_n = (\xi_1^{(n)}, \xi_2^{(n)}, \ldots, \xi_k^{(n)})$, $n = 1, 2, \ldots$, tends to the element $x = (\xi_1, \xi_2, \ldots, \xi_k)$ if for each $i = 1, 2, \ldots, k$ the number sequence $\{\xi_i^{(n)}\}$ tends to the corresponding ξ_i as $n \to \infty$. For $k = 3$, when $\mathbf{R}^3$ consists of the totality of triples $x = (\xi_1, \xi_2, \xi_3)$ of real numbers, the ξ_1, ξ_2, ξ_3 can be regarded as coordinates and x as a point in the sense of solid analytical geometry.

Example 2: The space of *polynomials of degree at most m* in a variable t has the elements $x = x(t) = \alpha_0 + \alpha_1 t + \alpha_2 t^2 + \cdots + \alpha_m t^m$, where the coefficients $\alpha_0, \alpha_1, \ldots, \alpha_m$ denote complex numbers. The set of these polynomials represents a space, once a concept of limit is defined in it.

A polynomial of degree m is determined uniquely by the value of the function and the values of its first m derivatives at a point $t = t_0$ according to Taylor's formula:

$$x(t) = x(t_0) + x'(t_0)(t - t_0) + \frac{x''(t_0)}{2!}(t - t_0)^2 + \cdots + \frac{x^m(t_0)}{m!}(t - t_0)^m.$$

One therefore defines: A sequence of polynomials $x_n = x_n(t)$ is *convergent to the polynomial* $x = x(t)$ if the values of the function x_n and its derivatives converge individually to the values of the function $x(t_0)$ and of its derivatives $x'(t_0), \ldots, x^{(m)}(t_0)$ at the point $t = t_0$, that is, if $\lim_{n\to\infty} x_n(t_0) = x(t_0)$, $\lim_{n\to\infty} x_n'(t_0) = x'(t_0)$, ..., $\lim_{n\to\infty} x_n^{(m)}(t_0) = x^{(m)}(t_0)$.

Linear spaces. In linear spaces a *multiplication* of the elements $x, y, z, \ldots$ by real or complex numbers $\lambda, \mu, \ldots$ and an addition of any pair of elements of the space must be defined. To each number λ and each element x of the space there corresponds a unique element denoted by λx, and likewise to each pair of elements (x, y) a unique element $x + y$ of the space. It is, however, left completely undetermined what these multiplications and additions look like in special cases; they must simply satisfy the following conditions:

Conditions for multiplication and addition in linear spaces:

(1) $\lambda(\mu x) = (\lambda\mu)x$, (2) $1 \cdot x = x$, (3) $(\lambda + \mu)x = \lambda x + \mu x$, (4) $x + y = y + x$,
(5) $(x + y) + z = x + (y + z)$, (6) $\lambda(x + y) = \lambda x + \lambda y$.
(7) There exists a zero element O for which $0 \cdot x = 0 \cdot y = O$ (with the real number 0) always holds.

Further, addition and multiplication must be *continuous operations*, that is, they must satisfy the following additional conditions:
(8) From $\lim_{n\to\infty} x_n = x$ and $\lim_{n\to\infty} y_n = y$ it follows that $\lim_{n\to\infty}(x_n + y_n) = x + y$.
(9) From $\lim_{n\to\infty} \lambda_n = \lambda$ and $\lim_{n\to\infty} x_n = x$ it follows that $\lim_{n\to\infty} \lambda_n x_n = \lambda x$.

Examples in which all the axioms are satisfied:

1. The k-dimensional Euclidean space $\mathbf{R}^k$ becomes a linear space if multiplication is defined by

$$\lambda x = \lambda(\xi_1, \xi_2, \ldots, \xi_k) = (\lambda\xi_1, \lambda\xi_2, \ldots, \lambda\xi_k)$$

and addition of two elements $x = (\xi_1, \xi_2, \ldots, \xi_k)$ and $y = (\eta_1, \eta_2, \ldots, \eta_k)$ by $x + y = (\xi_1 + \eta_1, \xi_2 + \eta_2, \ldots, \xi_k + \eta_k)$. The zero element is $O = (0, 0, \ldots, 0)$.

2. The space of polynomials of degree at most m, becomes a linear space by the definitions $\lambda x = \lambda\alpha_0 + \lambda\alpha_1 t + \cdots + \lambda\alpha_m t^m$ and $x + y = x(t) + y(t) = (\alpha_0 + \beta_0) + (\alpha_1 + \beta_1)t + \cdots + (\alpha_m + \beta_m)t^m$. The zero element is the polynomial $O = O(t) = 0$.

The concept of a metric and of a metric space. Two points P and Q of the three-dimensional geometrical space have a certain distance from one another, which is different from zero if they do not coincide; this distance is measured by the length $|PQ|$ of the line segment joining P and Q, where $|PQ| > 0$ for $P \neq Q$. The distance between P and Q is equal to the distance between Q and P, that is $|PQ| = |QP|$. If one takes a third point R not on the line passing through P and Q, then one obtains a triangle PQR. By a theorem of elementary geometry, any side of a triangle, say PQ, is less than the sum of the other two sides PR and QR, that is, $|PQ| < |PR| + |QR|$. This relation is called the *triangle inequality*. In the form $|PQ| \leqslant |PR| + |QR|$ this relation holds without limitation, even when P, Q and R no longer form a proper triangle, but lie on a straight line (Fig.).

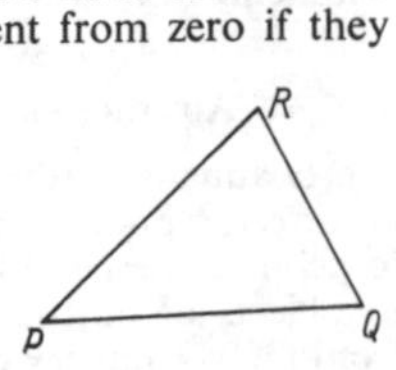

40-1 The triangle inequality

In analysis too one often has to measure *distances*, figuratively speaking, between the elements $x, y, z, \ldots$ considered, to decide whether two elements x, y are at a 'great' or 'small' distance from one another. In order to measure distances, a *distance function* or *metric* must be defined, that is, a real-valued function $d(x, y) \geqslant 0$ defined for all pairs of elements x, y.

Axioms for a metric: (1) $d(y, x) = d(x, y)$, (2) $d(x, y) = 0$ if and only if $x = y$,
(3) $d(x, y) \leqslant d(x, z) + d(z, y)$ (*triangle inequality*)

In this the analytical form of the distance function remains completely open. A space for whose pairs of elements a distance function is defined is called a *metric space*. The limiting process $\lim_{n\to\infty} x_n = x$, which is assumed in an abstract space, is defined by $\lim_{n\to\infty} d(x, x_n) = 0$.

Examples of metrics in the k-dimensional Euclidean space:

1. $d(x, y) = \sqrt{\left[\sum_{i=1}^{k} (\xi_i - \eta_i)^2\right]}$ is a generalization of the formula for the length in analytical geometry;

2. $d(x, y) = \max_{i=1,\dots,k} |\xi_i - \eta_i|$. 3. $d(x, y) = \sum_{i=1}^{k} |\xi_i - \eta_i|$.

For each of these metrics it has to be shown that it satisfies the three given axioms. This is quite easy; only for the first metric it is a little difficult to establish the validity of the triangle inequality.

Normed spaces. It is well known that to every complex number $\zeta = \xi + i\eta$ there corresponds the non-negative real number $|\zeta| = \sqrt{(\xi^2 + \eta^2)}$ as *absolute value* or *modulus* of ζ. In working with functions, vectors, matrices and so on, the form of the problem often suggests a way of ascribing a non-negative real number to each of the objects considered as a measure of its 'magnitude'. Such a numerical measure associated with the elements $x, y, \dots$ of a space is called a norm and is denoted by $\|x\|, \|y\|, \dots$, provided that it has the following properties.

Properties of the norms $\|x\|$. (1) $\|x\| > 0$ for $x \neq O$, $\|O\| = 0$; (2) $\|\lambda x\| = |\lambda| \cdot \|x\|$, where λ is an arbitrary real or complex number; (3) $\|x + y\| \leqslant \|x\| + \|y\|$ (triangle inequality).

These properties hold, for example, for the absolute value or modulus of a complex number.
A space to whose elements a norm is ascribed is called a *normed space*.

From a norm a metric can be derived, by defining a distance function $d(x, y)$ between two elements x, y to be the norm of their difference: $d(x, y) = \|x - y\|$.

In the k-dimensional Euclidean space $\mathbf{R}^k$ the following norms have all the required properties:

1. $\|x\| = \sqrt{\left(\sum_{i=1}^{k} \xi_i^2\right)}$; 2. $\|x\| = \max_{i=1,\dots,k} |\xi_i|$; 3. $\|x\| = \sum_{i=1}^{k} |\xi_i|$.

From these the metrics on $\mathbf{R}^k$ in the examples above are obtained. It often depends on the purpose of the investigation, which definition of the norm is appropriate to use for a particular space.

Complete metric spaces. If a sequence $x_1, x_2, \dots$ of elements of a metric space X converges to an element x, then the distances $d(x_1, x), d(x_2, x), \dots$ form a null sequence. By the triangle inequality the distances $d(x_i, x_k)$ between two arbitrary elements x_i, x_k of the sequence also tend to zero as the indices i and k increase; they form a Cauchy sequence.

A sequence $\{x_n\}$ is called a Cauchy sequence if for every positive number ε an index $n(\varepsilon)$ can be found such that $d(x_i, x_k) \leqslant \varepsilon$ for all indices $i, k \geqslant n(\varepsilon)$.

Every convergent sequence is a Cauchy sequence, but not every Cauchy sequence is a convergent sequence. Spaces in which a limit element x can be found for every Cauchy sequence $\{x_n\}$ are said to be *complete*, and complete normed linear spaces are called *Banach spaces*, after Stefan BANACH (1892–1945), who was one of the founders of functional analysis. All finite-dimensional spaces, for example, the space of polynomials of degree at most m, are complete. The space $L_2(a, b)$ (see Hilbert spaces) is likewise complete. In general, one requires a scalar product space to be complete before designating it as a Hilbert space.

Hilbert spaces. The spaces, called after David HILBERT (1862–1943), are important special cases of normed linear spaces. In these, for every pair of elements x, y a complex-valued function (x, y), called a *scalar product*, is defined, having the following properties (the bars above denote the conjugate complex number):

(1) $(x, y) = \overline{(y, x)}$;
(2) $(\lambda x, y) = \bar{\lambda}(x, y)$, where λ is an arbitrary complex number;
(3) $(x, x) \geqslant 0$, with equality if and only if $x = O$;
(4) $(x + y, z) = (x, z) + (y, z)$.

The space is normed; a norm is introduced with the help of the scalar product by the equation $\|x\| = \sqrt{(x, x)}$.

Examples of Hilbert spaces: 1. The space $\mathbf{C}^k$ of complex k-tuples with the scalar product $(x, y) = \sum_{i=1}^{k} \xi_i \bar{\eta}_i$ and the corresponding norm $\|x\| = \sqrt{\left(\sum_{i=1}^{k} |\xi_i|^2\right)}$, ($\xi_i, \eta_i$ complex).

2. The space $L_2(a, b)$. Its elements consist of complex-valued functions $x = x(t)$, defined in $a \leqslant t \leqslant b$, for which the integral $\int_a^b |x(t)|^2 \, dt$ exists. The scalar product defined by $(x, y) = \int_a^b x(t)\overline{y(t)} \, dt$ has all the requisite properties.

Example of functional-analytical arguments. The insight that can be gained through functional-analytical concepts is shown by some of the consequences of the *Schwarz inequality*.

The Schwarz inequality $\lvert(x, y)\rvert \leqslant \lVert x\rVert \cdot \lVert y\rVert$ holds in a Hilbert space.

The proof makes use first of all of the third property of the scalar product, according to which $(x + \lambda y, x + \lambda y) \geqslant 0$ for every arbitrary complex number λ. Using the other properties one obtains

$$(x + \lambda y, x + \lambda y) = (x, x + \lambda y) + \lambda(y, x + \lambda y) = \overline{(x + \lambda y, x)} + \lambda\overline{(x + \lambda y, y)}$$
$$= (x, x) + \overline{\lambda(y, x)} + \overline{\lambda(x, y)} + \lambda\overline{\lambda(y, y)} = (x, x) + \bar{\lambda}(x, y) + \lambda(y, x) + \lambda\bar{\lambda}(y, y) \geq 0.$$

This holds, in particular, for $\lambda = -(x, y)/(y, y)$; consequently

$$(x, x) - \overline{(x, y)}\,(x, y)/(y, y) - (x, y)\,(y, x)/(y, y) + (x, y)\,\overline{(x, y)}/(y, y) \geq 0.$$

From this it follows that

$$(x, y)\,(y, x) = (x, y)\,\overline{(x, y)} = \lvert(x, y)\rvert^2 \leq \lVert x\rVert^2 \cdot \lVert y\rVert^2, \text{ as required.}$$

The usefulness of this general result lies in the following. Once one recognizes that a numerical operation on arbitrary pairs of elements x and y of a space satisfies the conditions required of a scalar product for Hilbert spaces, then from the validity of the Schwarz inequality for this special space one immediately obtains important relations; for example, the inequalities

$\left\lvert\sum_{i=1}^{k} \xi_i\eta_i\right\rvert \leqslant \sqrt{\left(\sum_{i=1}^{k} \xi_i^2\right)} \cdot \sqrt{\left(\sum_{i=1}^{k} \eta_i^2\right)}$	for the space $\mathbf{R}^k$ and
$\left\lvert\int_a^b x(t) \cdot \overline{y(t)}\,\mathrm{d}t\right\rvert \leqslant \sqrt{\left(\int_a^b \lvert x(t)\rvert^2\,\mathrm{d}t\right)} \cdot \sqrt{\left(\int_a^b \lvert y(t)\rvert^2\,\mathrm{d}t\right)}$	for the space $L_2(a, b)$

These are relations which, besides a number of similar formulae, had been found earlier and independently for the individual spaces by CAUCHY, BUNYAKOVSKII, SCHWARZ, and others. By introducing functional-analytical concepts in this way essential properties common to different branches of analysis are discovered and worked out.

Similarly many relations already known are obtained quite simply as interpretations of a theorem of functional analysis; for example, the triangle inequality for norms follows from the *Schwarz inequality*. Because $\lVert x\rVert = \sqrt{(x, x)}$, $\lVert x + y\rVert^2 = (x + y, x + y) = (x, x) + (x, y) + (y, x) + (y, y) \leqslant \lVert x\rVert^2 + \lVert y\rVert^2 + \lvert(x, y) + (y, x)\rvert \leqslant \lVert x\rVert^2 + \lVert y\rVert^2 + 2\lVert x\rVert \cdot \lVert y\rVert = (\lVert x\rVert + \lVert y\rVert)^2$, and the result is proved.

For the spaces $\mathbf{R}^k$ and $L_2(a, b)$ this yields the Cauchy and the Minkowski inequalities, which were known earlier.

Cauchy's inequality	$\sqrt{\left(\sum_{i=1}^{k} (\xi_i + \eta_i)^2\right)} \leqslant \sqrt{\left(\sum_{i=1}^{k} \xi_i^2\right)} + \sqrt{\left(\sum_{i=1}^{k} \eta_i^2\right)}$
Minkowski's inequality	$\sqrt{\left(\int_a^b \lvert x(t) + y(t)\rvert^2\,\mathrm{d}t\right)} \leqslant \sqrt{\left(\int_a^b \lvert x(t)\rvert^2\,\mathrm{d}t\right)} + \sqrt{\left(\int_a^b \lvert y(t)\rvert^2\,\mathrm{d}t\right)}$

Operators

Whilst by means of the space concept the objects of a mathematical investigation are essentially only typified, an operator characterizes a definite mathematical operation that can be performed on the elements of the space. Almost every mathematical operation can be regarded as a *correspondence determined by a definite rule of calculation, mapping every element x of an abstract space X uniquely to an element y of a space Y*, which may, but need not, be different from X. The correspondence is also called a *mapping of X into Y*, and the law of correspondence is called an operator $A, B, \ldots$ or F; the correspondence is written in the form $y = Ax$ or $y = A(x)$ (Fig.).

The real functions F of a real variable x are special operators; they map the space of the real numbers $\mathbf{R}^1$ or a subspace X of it into $Y = \mathbf{R}^1$.

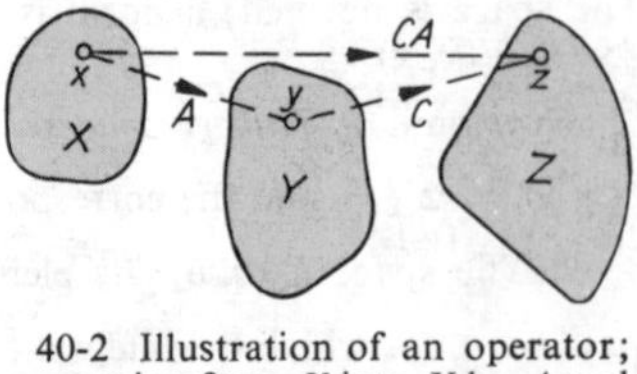

40-2 Illustration of an operator; mapping from X into Y by A and from Y into Z by C

If one assigns to each polynomial $x(t)$ of the space X of polynomials of degree at most m the polynomial

$$y = Ax = x''(t) - 3x'(t) - \alpha x^2(t),$$

then $y = Ax$ is a mapping from X into the space Y of polynomials of degree at most $2m$.

Linear operators. As far as applications are concerned, these operators form the most important class. They are defined by the properties (1) $A(\lambda x) = \lambda Ax$ for every arbitrary number λ, and (2) $A(x + y) = Ax + Ay$. As an example, the operator given above is a linear operator if $\alpha = 0$, and is otherwise non-linear.

Composition of operators. If A and B are two operators, each being a mapping from X into Y, and λ is an arbitrary number, then by the product λA one understands the operator that maps x into $\lambda(Ax)$, so that $(\lambda A)\,x = \lambda(Ax)$; the sum $A + B$, on the other hand, maps x into $Ax + Bx$, so that $(A + B)\,x = Ax + Bx$.

Finally, if a third operator C is given mapping the space Y into a space Z, then the *operator* CA maps each element x into that element $C(Ax)$ in Z into which x goes over by successive applications of the operators A and C; this is expressed by the formula $(CA)\,x = C(Ax)$.

Bounded linear operators. A linear operator A mapping a normed linear space X into a normed linear space Y is said to be *bounded* if an inequality of the form

$$\|y\| = \|Ax\| \leqslant K\|x\| \quad \text{with} \quad K > 0$$

holds for all x in X; the smallest number K having this property is called the *norm* of the operator A and is denoted by $\|A\|$. It goes without saying that the norm $\|y\|$ in the space Y may be different from the norm $\|x\|$ in X.

Example: Let X be $\mathbf{R}^3$ with the norm $\|x\| = \max|\xi_i|$ and let Y be $\mathbf{R}^2$ with the norm $\|y\| = \max|\eta_i|$ for $i = 1, 2$. By the equations $\eta_1 = a_{11}\xi_1 + a_{12}\xi_2 + a_{13}\xi_3$, $\eta_2 = a_{21}\xi_1 + a_{22}\xi_2 + a_{23}\xi_3$, a certain $y = (\eta_1, \eta_2)$ is made to correspond to each $x = (\xi_1, \xi_2, \xi_3)$. The operator defined by this is bounded, because

$$|\eta_i| \leqslant |a_{i1}| \cdot |\xi_1| + |a_{i2}| \cdot |\xi_2| + |a_{i3}| \cdot |\xi_3| \leqslant \sum_{k=1}^{3} |a_{ik}|\,\|x\|,$$

that is,

$$\|y\| = \max_{i=1,2} |\eta_i| \leqslant (\max_{i=1,2} \sum_{k=1}^{3} |a_{ik}|)\,\|x\|.$$

A closer investigation shows that $\max\limits_{i=1,2} \sum\limits_{k=1}^{3} |a_{ik}|$ is precisely the smallest number K, and therefore represents the norm of the operator A.

With the definitions given above of the addition of operators, the multiplication of an operator by a number, and the norm of an operator, the *totality of bounded linear operators A, B, ...,* that map a space X into a space Y, themselves form a normed linear space. This is of unusual significance for functional analysis and the application of functional-analytical methods. The circle of consideration again closes as it were: classes of operators, which as intermediaries between two spaces appear to stand outside the theory of spaces, themselves fall into the category of abstract spaces.

Functionals. Among the mappings of a space the *numerical functions* occupy a special place; these are mappings into the set of the real or of the complex numbers. They are called *functionals* and have given their name to functional analysis.

In normed linear spaces the norm, for example, is already a functional. For the sake of simplicity only normed linear spaces will be considered in the following.

An exceptional place is again occupied by the *linear functionals f*, which assign to every element $x, y, \ldots$ of the space X a real or complex number $f(x), f(y), \ldots$, so that the *linearity conditions* $f(x + y) = f(x) + f(y)$, $f(\alpha x) = \alpha f(x)$ hold for all elements x, y of the space X and all admissible real or complex numbers α. A linear functional is said to be *bounded*, or also *continuous*, if the *norm* $\|f\|$ of f satisfies the condition $\|f\| = \sup\limits_{x \in X} (|f(x)|/\|x\|) < \infty$ (where $x \neq 0$).

The totality of all continuous linear functionals defined on X forms the *dual space* X^*. If X is a normed linear space, this is likewise a normed linear space.

An important problem of functional analysis consists of determining the properties of continuous linear functionals, or of representing them and their values $f(x)$, $x \in X$, as a sum or an integral, and of characterizing sets and mappings of the original space X by elements and mappings of elements of the dual space X^*. From the point of view of this problem functional analysis is a further development of a geometrical discipline, *linear geometry*.

The theory of continuous linear functionals plays a significant role, for example, in the theory of *linear operator equations* or integral equations, in the theory of *approximate integration*, in the theory of *distributions* or *generalized functions*, and in the theory of the *Lagrange method of undetermined multipliers*. Here are some examples of results for specific spaces.

1. In $\mathbf{R}^k$, the space of k-tuples $x = (x_1, \ldots, x_k)$ of real numbers x_i, $i = 1, \ldots, k$, corresponding to each linear functional f there exist k real numbers $f_1, \ldots, f_k$ such that the value $f(x)$ of the functional can be represented in the form $f(x) = f_1 \cdot x_1 + f_2 \cdot x_2 + \cdots + f_k \cdot x_k$.

One speaks of a *representation* of f by means of this relation. Conversely, corresponding to each arbitrary k-tuple of real numbers $f_1, \ldots, f_k$, a continuous linear functional can be defined in this way. Depending on the norm (see Normed spaces) by which the elements $x \in \mathbf{R}^k$ are normed, one finds:

1. $\|f\| = \sqrt{\left(\sum_{i=1}^{k} f_i^2\right)}$, 2. $\|f\| = \sum_{i=1}^{k} |f_i|$ or 3. $\|f\| = \max_{i=1,\ldots,k} |f_i|$.

2. In the space $L_2(a, b)$ of all functions of Lebesgue integrable square over the interval $[a, b]$, the *Riesz representation theorem* states: *Corresponding to each continuous linear functional f there exists a uniquely determined function g in $L_2(a, b)$ such that the value $f(x)$ of the functional for $x \in L_2(a, b)$ can be represented in the form* $f(x) = \int_a^b x(t)\,\bar{g}(t)\,dt = (x, g)$.

More generally, in every (complete) Hilbert space X the values of the functional $f(x)$ can be represented as a scalar product (x, g). Conversely, by means of an arbitrary element $g \in X$ one can define a continuous linear functional $f(x) = (x, g)$. It can be shown, moreover, that the norm $\|f\|$ of the functional f is equal to the norm $\|g\|$ of the generating element.

3. The space X of all polynomials in one variable of degree at most m can be regarded as a normed linear space with the norm $\|x\| = \sum_{i=0}^{m} |x^{(i)}(t_0)|$.

A linear functional f on X must assign to the polynomials $1, t, t^2, \ldots, t^m$ certain complex numerical values $f_0, f_1, f_2, \ldots, f_m$ and to the polynomial x with the function values

$$x(t) = x(t_0) + x'(t_0)\,t + \cdots + \frac{x^{(m)}(t_0)}{m!}\,t^m$$

the numerical values

$$f(x) = x(t_0) f_0 + x'(t_0) f_1 + \cdots + \frac{x^{(m)}(t_0)}{m!} f_m.$$

Conversely, by means of this relation with arbitrary numbers $f_0, \ldots, f_m$ one can define a linear functional. This is also continuous, because $\|f\| = \max_{i=0,\ldots,m} (|f_i|/i!)$.

4. *Hyperplanes.* A linear equation in the variables x_1, x_2, x_3 determines a subset, namely a plane, in the three-dimensional space $\mathbf{R}^3$. As an extension of this situation, the totality of elements x of a linear space X that satisfy an equation $f(x) = \alpha$ is called a *hyperplane* H; $f(x)$ denotes a continuous linear functional and α a number. The *distance* $d(y, H)$ of an element $y \in X$ from the hyperplane H is defined to be the greatest lower bound of all distances $\|y - x\|$, $x \in H$, that is, $d(y, H) = \inf_{x \in H} \|y - x\|$.
In three-dimensional geometry, in which the absolute value of the distance measurement is used, an expression for the distance is given by the Hesse normal form. Similarly, in a general normed linear space X one has $d(y, H) = |f(y) - \alpha|/\|f\|$. If X is a complete space, then, as in the three-dimensional space, there exists an element $x_0 \in H$ whose distance $\|y - x_0\|$ is equal to the distance of the element y from the hyperplane H.

Optimization problems of control theory frequently involve the determination of the distance of a given element y from a hyperplane.

5. *Corresponding to an element u of a normed linear space there exists a continuous linear functional f of norm 1 such that* $\|u\| = f(u)$.

The norm of an element u, which is sometimes difficult to handle, can therefore be represented as the value of a linear functional. For example, if $x(t) = [\xi_1(t), \ldots, \xi_n(t)]$ is a k-dimensional vector-valued function with real differentiable components $\xi_i(t)$, where t is a real variable, and if s is a further value of the independent variable, then an upper estimate can be given for the norm $\|x(t) - x(s)\| := \sum_{i=1}^{k} |\xi_i(t) - \xi_i(s)|$ in terms of the argument difference $(s - t)$. By the theorem mentioned, one imagines a functional f on the space $\mathbf{R}^k$ chosen in such a way that $f(x(t) - x(s)) = \|x(t) - x(s)\|$ and $\|f\| = 1$; that is, one imagines k real numbers $f_1, \ldots, f_k$, satisfying the conditions $\|x(t) - x(s)\| = \sum_{i=1}^{k} f_i[\xi_i(t) - \xi_i(s)]$ and $\max_{i=1,\ldots,k} |f_i| = 1$. Taking $\varphi(t) = \sum_{i=1}^{k} f_i \cdot \xi_i(t)$ and using the first mean value theorem of the differential calculus one finds that $\|x(t) - x(s)\| = \varphi(t) - \varphi(s) = \varphi'(\tau) \cdot (t - s) \leqslant \sum_{i=1}^{k} |\xi_i'(\tau)| \cdot |t - s|$ for some point τ between s and t.

6. *Extension of functionals.* Occasionally a continuous linear functional is defined only on a linear subspace, and the problem then arises of defining the functional on the rest of the space in such a way that it remains linear and continuous and, if possible, also preserves the norm. Theorems concerning such an extension have been proved, for example, by HAHN, BANACH, KREIN and RUTMAN.

The use of functional-analytical methods in approximation theory

Approximation theory is concerned with the problem of giving methods for the approximate solution of equations of widely differing kinds, for example, of differential or integral equations. In an abstract scheme for such equations an operator A is given transforming the element x of a complete normed space X into the element y of a normed space Y, and an element x^* in X is sought that is mapped onto the zero element O of Y, and therefore satisfies $A(x^*) = O$. In many cases a method for determining x^* approximately proceeds by an *iteration process*: one rearranges the equation $A(x) = O$ in the equivalent form $x = B(x)$, chooses a first approximation x_0 more or less arbitrarily and then forms the further approximations $x_1 = B(x_0)$, $x_2 = B(x_1)$, $x_3 = B(x_2)$, ..., $x_n = B(x_{n-1})$, ... By the *Banach fixed point theorem* it can be decided, as a rule, whether the sequence $\{x_n\}$ converges to x^*.

Banach fixed point theorem: *If B is a mapping of a subset M of a Banach space into itself and if B satisfies, for all elements $x, y \in M$, a Lipschitz condition $\|B(x) - B(y)\| \leqslant L\|x - y\|$ with a Lipschitz constant $L < 1$, then for every arbitrary initial approximation x_0 in M, the sequence of approximations $x_1 = B(x_0)$, $x_2 = B(x_1)$, ..., $x_n = B(x_{n-1})$, ... converges to the unique solution x^* in M of the equation $x = B(x)$, and an estimate of the error is given by*

$$\|x^* - x_n\| \leqslant [L/(1 - L)]\, \|x_n - x_{n-1}\| \leqslant [L^n/(1 - L)]\, \|x_1 - x_0\|.$$

Example 1: Let $X = Y$ be the set of the real numbers, $A(x) \equiv x - \sin x - 1$, $B(x) \equiv \sin x + 1$, the subset M the interval $\pi/2 \leqslant x \leqslant 2$. M is mapped by B onto the interval $2 \geqslant x \geqslant 1 + \sin 2 (> \pi/2)$. For all $x, y \in M$ the Lipschitz condition $|B(x) - B(y)| = |\sin x - \sin y| \leqslant L \cdot |x - y|$ holds with $L = |\cos 2|$. Starting from $x_0 = \pi/2 \approx 1.571$ one obtains $x_1 = 2$, $x_2 = \sin 2 + 1 = 1.909$ ... The estimate of the error gives $|x^* - x_2| \leqslant 0.066$. The exact solution, correct to three decimal places, is $x^* = 1.935$ ...

Example 2: Let $X = Y = L_2(0, 1)$, $[A(x)](s) = x(s) - (1/2) \int_0^1 \frac{x(t)}{1 + s + t}\, dt - 2$ $(= 0$ for $0 \leqslant s \leqslant 1)$, $[B(x)](s) = (1/2) \int_0^1 \frac{x(t)}{1 + s + t}\, dt + 2$, $M = L_2(0, 1)$. A square integrable function x is mapped by B again into a square integrable function. For all $x, t \in L_2(0, 1)$ holds

$$\|B(x) - B(y)\|^2 = \int_0^1 \left| \frac{1}{2} \int_0^1 \frac{x(t) - y(t)}{1 + s + t}\, dt \right|^2 ds \leq \frac{1}{4} \int_0^1 |x(t) - y(t)|^2\, dt = \frac{1}{4} \|x - y\|^2.$$

The conditions of the Banach fixed point theorem are therefore satisfied with $L = 1/2$ and an arbitrary square integrable initial function, say $x_0(s) = 0$. One then obtains the sequence of approximations $x_1(s) = 2$, $x_2(s) = \ln [(2 + s)/(1 + s)] + 2$, ... The error estimation gives for the function x_2 a mean square error $\|x^* - x_2\| = (\int_0^1 |\ln [(2 + s)/(1 + s)]|^2\, ds)^{1/2} = 0.48$. The exact solution $x^*(s)$ is not known.

As can be seen, functional-analytical methods are of great value even in numerical problems, such as arise every day in engineering practice.

41. Foundations of geometry – Euclidean and non-Euclidean geometry

Euclidean geometry is the oldest and historically most important example of a deductive scientific discipline. Down to modern times it has been a model of an exact science and it became the starting point for a systematic development of the foundations of geometry. This development began at the turn of the 19th century with the discovery of non-Euclidean geometry, reached its zenith in the investigations of HILBERT, and now covers a wide field of inquiry.

Foundations of geometry

Euclid's Elements. In his *Elements* (the *Stoicheia*), EUCLID of Alexandria (c. 365–300 B.C.) gave a synopsis of the mathematical knowledge of his time. They contain propositions from *number theory*, for example, the Euclidean algorithm and a proof of the existence of infinitely many prime numbers, from *solid geometry* the theory of regular polyhedra, also the theory of proportion and of *similarity* together with a discussion of incommensurable quantities, and problems of *plane geometry*. The importance of the work lies in the fact that in it the theorems of geometry – with certain restrictions, according to present-day knowledge – are proved without recourse to the real world, but purely by *logical deductions* from a set of axioms.

ARISTOTLE [c. 384–322 B.C.] regarded *axioms* as statements that are self-evident, stemming directly from experience, and containing only concepts about whose meaning there can be no doubt. For this reason EUCLID gave *definitions* of his basic concepts, for example, *a point is what has no parts*. But in the subsequent deductions no use is made of these definitions.

It was only realized in the 19th century that EUCLID tacitly uses properties of order without having stated them as axioms. These inadequacies of Euclid's system of axioms were removed by HILBERT in 1899 in his book *Grundlagen der Geometrie* (Foundations of geometry), which at the same time answered scientific questions of a new and fundamental character. According to *Hilbert's axiomatics*, questions concerning the nature of the basic concepts or their relation to real objects do not belong to the mathematical theory concerned, but to its *metatheory*. The axioms only lay down certain relationships between the fundamental concepts. Many of Euclid's theorems are in principle just as easily verifiable as the axioms themselves; thus, to the modern way of thinking it is merely a matter of convenience in the systematic development of the theory that a particular axiomatic development is used.

The parallel axiom. Euclid's own formulation of this axiom or, as he calls it, postulate already makes it seem less self-evident than the others.

Euclid's version of the parallel axiom: If a straight line intersects two others in such a manner that the interior angles on one side of this line are together less than two right angles, then the two lines sufficiently extended intersect on that side (Fig.).

A whole series of apparently self-evident statements therefore led to fallacious 'proofs' of this axiom. The majority of these statements turned out to be equivalent to the axiom itself.

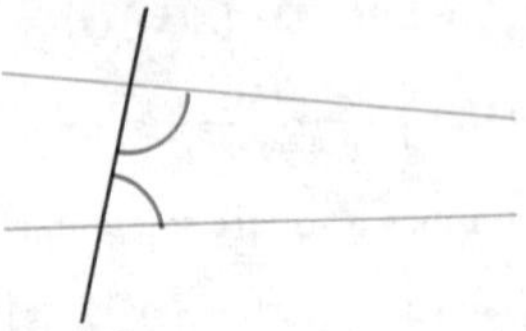

41-1 The parallel postulate of EUCLID

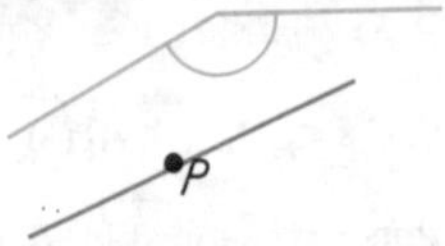

41-2 Statement of LEGENDRE

Statements equivalent to the parallel postulate:

1. Poseidonius (c. 135–51 B.C.): Two parallel lines are equidistant. – 2. Proclus (c. 500 A.D.): If a line intersects one of two parallel lines, then it also intersects the other. – 3. Saccheri (c. 1700): The sum of the interior angles of a triangle is two right angles. – 4. Legendre (1752–1833): A line through a point in the interior of an angle other than a straight angle intersects at least one of the arms of the angle (Fig.). – 5. Farkas Bólyai (1775–1856): There is a circle through every set of three non-collinear points.

The importance of this axiom was made clear, and the foundation was laid for both the modern interpretation of geometric axioms and of axiomatic systems in general, when GAUSS (1777–1855), LOBACHEVSKII (1792–1856) and Janos BÓLYAI (1802–1850) constructed, independently of one another, a geometry in which the parallel postulate does not hold. This non-Euclidean geometry showed a similar high degree of internal harmony and consistency as Euclidean geometry. Ten years after the death of LOBACHEVSKII, BELTRAMI succeeded in finding the first realization of the essential parts of this geometry on a curved surface, and later KLEIN fitted both Euclidean and non-Euclidean geometry into the larger framework of projective geometry.

Axiomatic characterization of Euclidean geometry. Euclidean geometry is a *categorical theory*, in which every statement is either true or disprovable in the sense that the assumption of its truth leads to a contradiction. On the other hand, it is intuitively related to direct experience, and many theorems are the result of experiments with ruler and compass and similar instruments. This highly empirical aspect declined with the growing importance of analytical geometry, in which the Euclidean plane is identified with the set of pairs of real numbers. HILBERT clarified this connection completely and proved that his system of axioms is categorical, by showing that any two of its models are isomorphic; the isomorphism type is that of the Euclidean plane over the field of real numbers. By extending his axiomatic interpretation to other number systems, HILBERT was able to use an axiomatic characterization of the real numbers, essentially based on discoveries of DEDEKIND, to prove that his system of axioms characterizing Euclidean geometry is *complete*.

The basic concepts of *Hilbert's system of axioms* for the Euclidean plane are *point, line, incidence* as a relation between points and lines, *betweenness* as a relation of triples of points, and *congruence* of *line-segments* and *angles*. The axioms are grouped in four sections: A *axioms of incidence*, B *axioms of order*, C *axioms of congruence*, D *axioms of continuity*, and in each section the concepts of the previous sections are used. For solid geometry a further concept, that of a *plane* must be introduced and the axioms must be modified. The notion of a categorical system of axioms was considerably deepened by TARSKI's result on the *completeness* and *decidability* of Euclidean geometry, provided that one regards it as an *elementary theory* in which set variables do not occur, or are at least avoidable in principle. Apart from certain considerations connected with continuity this can be done. The full Dedekind axiom of continuity has to be replaced by a continuity scheme that requires the existence of intersections only for those *Dedekind cuts* of lines that can be defined by expressions in the basic geometric concepts (see Chapter 15). TARSKI's result is that every statement of elementary geometry can be proved or disproved from these axioms by formal logic, and that there exists an algorithm to decide whether a given statement can be proved from the axioms and thus whether it is true or false. Such an algorithm, which could be realized on a machine, could even have practical uses, since there are many non-trivial problems of elementary geometry, such as tessellating problems or partition of polygons into simpler ones, that could then be solved mechanically. Just as decidability of elementary geometry reduces to that of the arithmetic of the real numbers by way of analytic geometry, so decidability has also been proved for other geometric theories, for example, for non-Euclidean (hyperbolic) geometry.

Since the selection of the basic concepts and axioms of Euclidean geometry is to a large extent arbitrary and a matter of convenience or even personal taste, the question arose whether HILBERT's choice could be simplified.

If one expresses the concepts of *line* and *incidence* in terms of a three-variable relation, *collinearity* of points, col (A, B, C), whose intuitive meaning is that A, B and C lie on a single line, then the parallel axiom can be rephrased in the following manner.

Parallel axiom in terms of collinearity: If A, B and P are non-collinear distinct points, then there exists a point Q such that col (A, B, R) *and* col (P, Q, R) *cannot both hold for any point R; further, if Q′ is any other such point, then* col (P, Q, Q') *holds.*

Collinearity itself can easily be reduced to the relation of *betweenness*, because three points are obviously collinear if and only if one of them lies between the other two. It is somewhat harder to show that betweenness can be expressed in terms of collinearity. The *metrical relations* can also be reduced; indeed, all of HILBERT's basic concepts can be formulated in terms of a single three-argument relation, for example, cir $(A, B; C)$, which means that A and B are equidistant from C. On the other hand, it can be shown that it is impossible to base Euclidean geometry on a single two-argument point relation.

System of axioms for plane geometry based on motions. The group-theoretical foundation of Euclidean geometry, which is based on the concept of motion, differs from other axiomatic systems of plane geometry more or less closely related to Hilbert's system in that the axioms of congruence are replaced by statements about motions. In proving the congruence propositions a number of metatheoretical statements are used. The basic concepts are *point, line* (as a distinguished set of points), *betweenness*, and motion. The *axioms* of *incidence* and of *betweenness* are retained initially, but the axioms of congruence are replaced by statements about *properties of the group of motions*. The following system of axioms leads to *absolute geometry* if the parallel axiom is omitted, and to non-Euclidean geometry if it is replaced by its negation. The symbol $P \mid l$ means that the point P and the line l are incident, in other words: 'l goes through P', 'P lies on l', or 'l contains P'.

Axioms of incidence: I_1. To any two points there exists exactly one line passing through both of them; every line contains at least two points. – I_2. Not all points lie on a single line. – I_3. Parallel postulate: To any line l and any point P not on l there exists exactly one line through P that has no point in common with l.

Axioms of betweenness: B_1. If R lies between P and Q, then R also lies between Q and P, and P, Q, R are distinct points on a single line (Fig.). – B_2. Of three distinct points on a line exactly one lies between the other two. – B_3. If R lies between P and Q, and Q lies between P and S, then R lies between P and S (Fig.). – B_4. If P, Q and R are not collinear and if the line l intersects PQ in a point Z_1 between P and Q, then l contains R or a point S between R and P or between R and Q (Fig.).

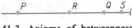

41-3 Axioms of betweenness B_2 and B_3

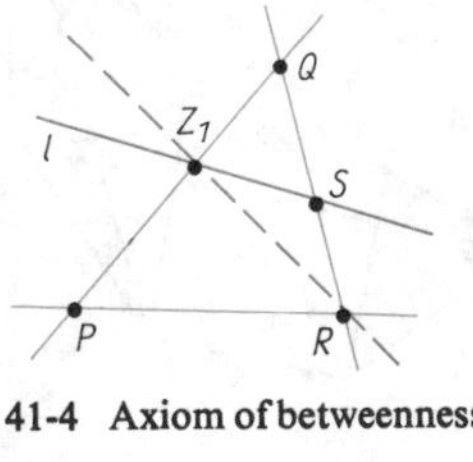

41-4 Axiom of betweenness B_4

Definitions: D_1. P and Q are said to *lie on the same side of a point O* relative to a line l if $P, Q, O \mid l$ and O is not between P and Q. The two equivalence classes of this relation for given l and $O \mid l$ are called the two *half-lines* of l with vertex O. – D_2. $P, Q \nmid l$ *lie on the same side of the line l* if no point of l lies between P and Q. The two equivalence classes of this relation for a given line l are called the half-planes bounded by l (Fig.). – D_3. A triple (P, h, H) consisting of a half-plane H, a half-line h on its boundary, and the vertex P of h is called a *flag* (Fig.). – D_4. An *automorphism* of a plane is a bi-unique map α of the plane onto itself such that a point R lies between points P and Q if and only if R^α lies between P^α and Q^α.

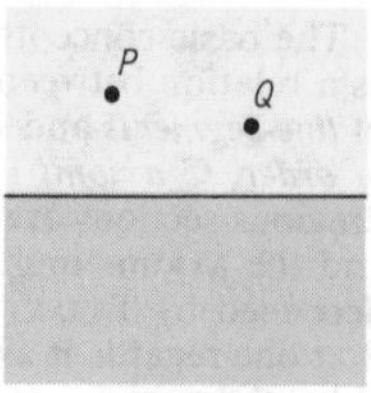

41-5 The half-planes bounded by l

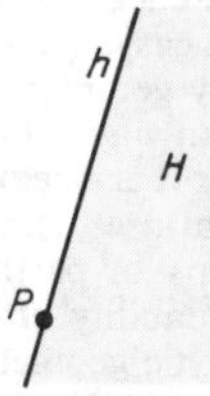

41-6 The flag defined by the triple (P, h, H)

It is easy to show that under order automorphisms lines, half-lines and half-planes are mapped to lines, half-lines, and half-planes, respectively. In particular, flags are mapped to flags. Certain automorphisms are called *motions*, and these are required to satisfy the following axioms.

Axioms of motion: M_1. If α and β are motions, then the combination $\alpha \cdot \beta$ (first α, then β) is also a motion. – M_2. The identity map 1 is a motion. – M_3. If F and F' are two flags, then there exists exactly one motion taking F to F'. – M_4. For any two points P and Q, there exists a motion interchanging them; for any two half-lines with a common vertex there exists a motion interchanging them.

A consequence of these axioms is that the motions form a *group*, and in particular, that if α is a motion, then so is α^{-1}; for if α maps an arbitrary flag F to F', then by M_2 there exists a motion β mapping F' to F. The map $\gamma = \alpha \cdot \beta$ maps F to itself, but then γ must be the identity, because otherwise $\gamma^2 \neq \gamma$ would be a second motion mapping F onto itself. Another consequence is that a motion is uniquely determined by three non-collinear points and their images.

Definitions of congruence and reflection: D_5. A pair of points P, Q is called *congruent* to a second pair P', Q' if there exists a motion taking P to P' and Q to Q'. – D_6. A motion that leaves exactly one line fixed is called a *reflection* (*in the line*). Reflections ϱ are involutions, that is, $\varrho \neq 1, \varrho^2 = 1$

The *main theorem on reflections* states that the product of three reflections in lines having a point or a perpendicular in common, is again a reflection (Fig.). These and all other statements can be proved without the use of the parallel postulate, and are therefore also valid in non-Euclidean geometry. For proofs avoiding the use of the parallel postulate it is particularly convenient to introduce a *calculus of reflections*. Instead of the partly group-theoretical system of concepts introduced above, one now starts out exclusively from the system Γ of reflections, which is regarded as a generating set consisting of *involutions* for a group (the group of motions). Lines are identified with elements of Γ, so that the elements of Γ are also called lines. Points are those products lm of two lines l and m that are again involutions, that is, for which $(l \cdot m)(l \cdot m) = 1$. This is equivalent to $l \cdot m = m \cdot l$. Thus, intuitively, points are identified with the reflections in them. A point P is incident with a line l if $P \cdot l = l \cdot P$. When all the other basic concepts are defined in the same manner, then geometric theorems

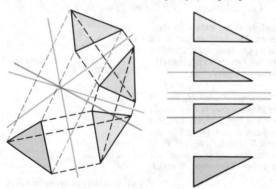

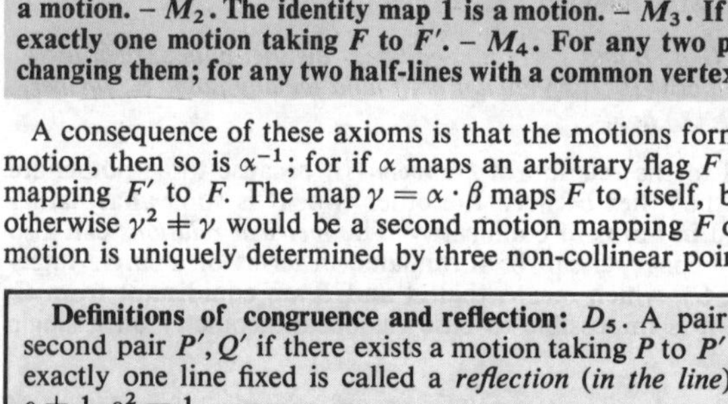
41-7 Three successive reflections whose axes pass through a common point or have a common perpendicular give together a reflection

simply become calculating rules for the group generated by Γ. This calculus of reflections represents a new process of transforming geometry into algebra apart from classical analytical geometry.

The introduction of coordinates. Apart from the complete axiomatic characterization of a single specific model of a geometry, the relations between the classes of models for geometric statements and their algebraic characterizations are also important. This kind of problem leads to questions about minimal axioms necessary to ensure the existence of certain standard procedures from model theory or mapping theory, for example, for the introduction of coordinates. The theories that have been created to tackle such problems have a geometrical terminology, but in their methodological development they resemble those of group theory or lattice theory; examples are the theory of vector spaces, or those of affine or projective planes.

The models of the axioms I_1, I_2 and I_3 form the class of *affine planes*, which have not yet been completely classified. Among them the *translation planes* are distinguished by the fact that to any two points there exists at least one translation, or parallel shift, carrying one point to the other. They also are required to satisfy the *minor theorem* d *of Desargues.*

Minor theorem d *of Desargues: If the corresponding pairs of vertices of two triangles lie on parallel lines and two pairs of corresponding sides are parallel, then so is the third pair* (*Fig.*).

It can be shown that the translations of a translation plane T form a vector space over a skew field K (the scalar field of T), whose dimension over K is even or infinite. The most elegant way of defining the scalars is as special linear mapping of the group of translations.

The dimension of the vector space of translations is precisely 2 if the plane T is a *Desarguian plane*, that is, if the *major theorem* D *of Desargues holds.*

Theorem D *of Desargues: If the lines through the corresponding vertices of a pair of triangles intersect in a single point, and two corresponding pairs of sides are parallel, then so is the third pair* (*Fig.*).

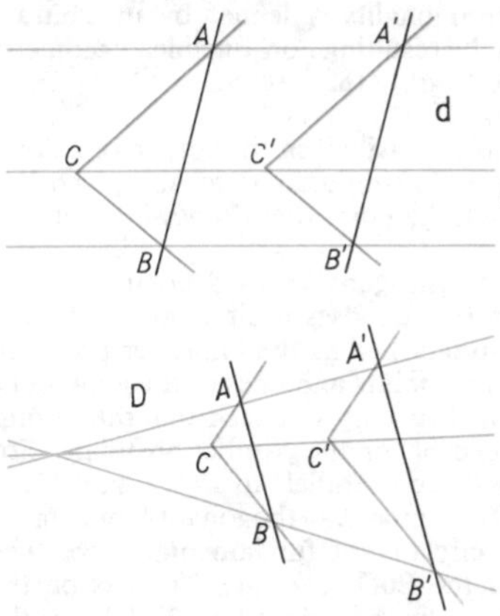

41-8 Desargues' theorem D and the minor version d

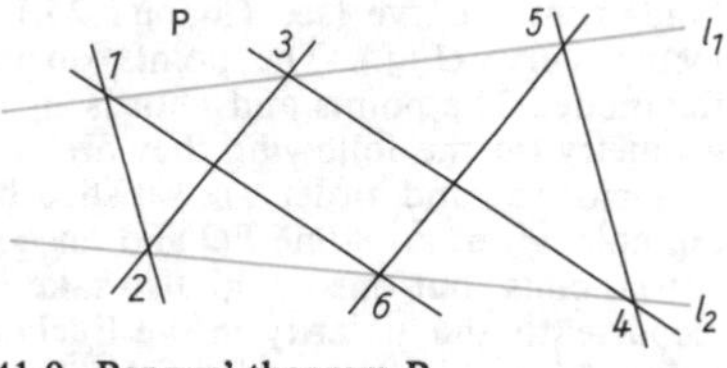

41-9 Pappus' theorem P

This statement D is required only for the proof that for all $\boldsymbol{a}, \boldsymbol{b} \in T$ with $\boldsymbol{a} \neq \boldsymbol{o}$ and $\boldsymbol{a} \parallel \boldsymbol{b}$ there is at least one linear mapping α of the vector group that takes $\boldsymbol{a}$ to $\boldsymbol{b}$. Thus, if an origin O and two coordinate axes l and m through O are chosen, any point P can be represented uniquely by its position vector, and this vector can be decomposed into its components relative to l and m. It is clear that one obtains coordinates by associating elements of the skew field (scalars) with the vectors on a line, and thus with the points on it. If one chooses an arbitrary vector $\boldsymbol{e} \neq \boldsymbol{o}$ on the line, then to every parallel vector $\boldsymbol{a}$ is associated that scalar which takes $\boldsymbol{e}$ into $\boldsymbol{a}$.

The propositions d and D are examples of so-called *closure theorems*. Another pair of closure theorems are the *theorems* P *and* p *of Pappus*. The validity of P is a necessary and sufficient condition for the field of scalars K to be commutative.

Major theorem P *of Pappus: If alternate vertices of a closed hexagon lie on two lines l_1 and l_2, and if none of them lies on both lines, and if the lines of two pairs of opposite sides are parallel, then the lines of the third pair are also parallel* (*Fig.*). *The minor theorem* p *of Pappus makes the same statement under the added condition that l_1 and l_2 are parallel.*

If the axioms of incidence hold, then these closure theorems are logically dependent on one another in the following order: $\mathrm{P} \rightarrow \mathrm{D} \rightarrow \mathrm{d} \rightarrow \mathrm{p}$. It is an open problem whether the last arrow can be reversed; the first three are not reversible. However, for *finite affine planes* one has $\mathrm{P} \Leftrightarrow \mathrm{D}$, because by Wedderburn's theorem every finite field is commutative. A geometrical proof of this fact has not yet been found.

Non-Euclidean geometry

Decisive connections between Euclidean, non-Euclidean and projective geometry were discovered around 1860 by Arthur CAYLEY (1821–1895) and were further developed a decade later by Felix KLEIN (1849–1925). They justify Cayley's statement: 'Projective geometry is all geometry'.

Projective geometry can be described by a system of axioms of incidence, order and continuity, that differs from the axioms of Euclidean geometry principally in the following points: any two lines intersect; the axioms of betweenness are replaced by axioms of a four-argument separation relation between pairs of points, because a projective line must always be regarded as cyclically closed.

The transition from projective to Euclidean or non-Euclidean geometry is effected by the introduction of the concepts of parallelism and orthogonality.

Two lines or planes are called parallel if they intersect in the improper or ideal plane of the space. Thus, in Euclidean geometry there is only one parallel to a given line through a point not on the line, because there is only one line through the point and the improper point on the given line. If an *absolute polarity*, that is, a polarity without a fundamental curve, is introduced in the improper plane, orthogonality can be defined for Euclidean space.

Now instead of a plane, an arbitrary set can be declared to be improper, for example, a quadric in projective space, which then splits into the inside and outside of this surface, for example, relative to a sphere or ellipsoid. A polarity of the whole space with the improper surface as its distinguished quadric defines 'orthogonality' for lines and planes in the interior of the surface, by defining a line to be orthogonal to a plane if it goes through the pole of the plane. One sees further that in a plane, which now consists only of the part on the interior of the fundamental surface, there are several parallels to a given line through a point not on the line. In this way one obtains the *hyperbolic geometry* discovered by GAUSS, BÓLYAI and LOBACHEVSKII.

If no surface at all is distinguished in projective space and orthogonality is defined by an arbitrary polarity on the whole space without a fundamental surface, then the resulting non-Euclidean geometry is *elliptic geometry*, which was first investigated by Bernhard RIEMANN (1826–1866).

Hyperbolic geometry. Plane hyperbolic geometry is obtained most easily if in the system of axioms for Euclidean geometry the parallel axiom is replaced by the following axiom: to any given line and any point not on that line there are at least two lines through the point that do not intersect the given line.

To obtain a model for this geometry one attempts to define orthogonality by a polarity with a fundamental curve (see Chapter 25.). In the case of a plane one chooses a circle as the fundamental curve (Fig.). The points on the circumference are distinguished as the improper points of the model. The points and chords in the interior are the proper points and lines of the hyperbolic geometry (in the following they are called h-points and h-lines). It is easy to check that the axioms of incidence and order are satisfied by h-points and h-lines, except for the parallel postulate. For example, given an h-line PQ and an h-point R both RU and RV are 'parallel' to PQ, U and V are not h-points, but one could also take the h-lines a, b and c. To define h-orthogonality one takes recourse to the polarity in the Euclidean plane for which the circle is the fundamental curve. The h-line ST is h-orthogonal to the h-line $P'Q'$ if the pole A of the Euclidean line $P'Q'$ lies on the Euclidean line ST, and the pole B of the Euclidean line ST lies on the Euclidean line $P'Q'$. Since the polar of any point on the Euclidean line $P'Q'$ passes through A, one need only draw a line through S and A or through T and A in order to drop an h-perpendicular to $P'Q'$ or to erect an h-perpendicular in S. The h-congruence is defined as follows: two h-segments PQ and $P'Q'$ are called h-congruent if the absolute values of the logarithms of the cross-ratios formed by the intersections of their lines with the fundamental circle are equal: PQ is h-congruent to $P'Q'$ if $|\ln D(P, Q; U, V)| = |\ln D(P', Q'; U', V')|$. This definition also determines, in the presence of continuity, the way of applying h-congruent segments. Having developed on this basis a definition for the h-congruence of h-angles one can show that the sum of the angles in an h-triangle is less than two right angles and that two h-triangles are h-congruent if they agree in the three h-angles.

The preceding arguments give essential information on the group of transformations which underlies hyperbolic geometry. According to KLEIN the question is: under what transformation groups are the concepts just defined invariant? – To answer this, the model is again interpreted as an object in the Euclidean plane. First of all it is clear that the relevant mappings must be collineations that carry both the circle and its interior onto itself. Examples of such mappings are rotations of the circle about its centre and reflections in a diameter; but translations or dilations and contractions are to be excluded. Now the collineations just characterized form a group. It is enough to note, in addition, that the mappings of this group preserve congruence of segments and angles. The invariance of congruence of segments is just a matter of definition by means of the cross ratio D. A pair of h-orthogonal lines goes into another such pair, because polarity is preserved. In these *automorphic collineations* of the circle one has the congruence group of the model in question.

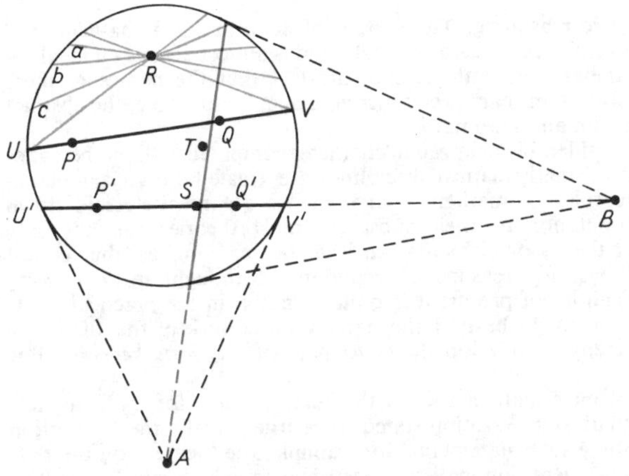

41-10 Model of a hyperbolic geometry

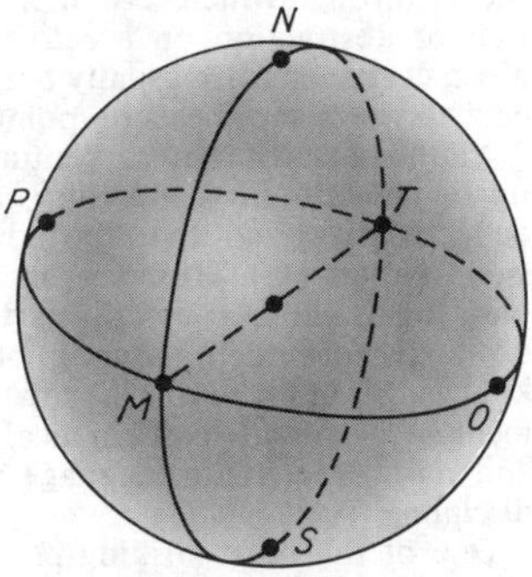

41-11 Model of an elliptic geometry

Elliptic geometry. This geometry cannot be obtained from Euclidean geometry in the same simple way as hyperbolic geometry by omitting or modifying a single axiom. For it follows from the axioms of incidence, order and congruence of plane Euclidean geometry that to any line there exists another one that does not intersect it. Thus, these sets of axioms must be changed; in particular, any two lines should always intersect. Elliptic geometry is essentially identical with *spherical geometry*. One regards the surface of a sphere as the elliptic plane, the *el*-lines are the great circles of the sphere and an *el*-point is a pair of diametrically opposite points of the sphere (Fig.).

Two *el*-lines always have an *el*-point in common, because any two great circles on the sphere intersect in two diametrically opposite points. The *el*-points (*N*, *S*) and (*M*, *T*) determine a unique *el*-line, namely (*NMST*), but one can drop infinitely many *el*-perpendiculars from the *el*-point (*N*, *S*) to the *el*-line (*MOTP*). If one chooses as elliptical distance between two points the length of the shorter arc on the great circle connecting them, then $\pi/2$ is the greatest distance in elliptic geometry. The sum of the angles in an *el*-triangle is always greater than two right angles.

In Chapter 17. it was shown that the rotations of a sphere about its centre form a group. In the present case this group of transformations of the sphere is the group of congruence transformations of elliptic plane geometry.

42. Foundations of mathematics

The modern development of the foundations of mathematics began, together with that of mathematical logic, towards the end of the last century. Today the two fields of study are still closely connected.

The problems of the foundations of mathematics or 'meta-mathematics' include a broad spectrum of questions, from the scientific investigation of special mathematical disciplines to philosophical questions on the nature of mathematical statements and mathematical knowledge. As far as the kind of question permits, the analysis and clarification of the relevant problems proceed with the same precision and rigour as is necessary for the successful treatment of mathematical problems. Of course, mathematical tools cannot always be used here without restriction, since the matter under discussion is often precisely the clarification of the foundations of the very tools. The following presentation can give only a first introduction to this type of problem, and can only take account of certain points of view. The reader especially interested in questions on the foundations of mathematics is referred to the comprehensive report by Andrzej Mostowski (1913–1975) 'Thirty years of foundational studies' in Acta Philosophica Fennica (Fasc. XVII, 1965).

It may be asserted that meta-mathematics presents the most advanced scientific theory of a specific scientific discipline. This is due to the fact that mathematics, to a greater extent than other sciences, was faced at an early stage with the problem of a critical analysis of its foundations, above all owing to the high degree of abstraction from concrete objectivity and the resulting demand for precision in mathematical definitions and deductive reasoning. The system of concepts of a mathematical theory through which certain properties of objective reality are reflected is almost always an idealization or abstraction on a set-theoretical basis, notwithstanding the fact that the names of these abstract concepts are usually borrowed from their narrower concrete usage, for example, the abstract concepts of a set, measure, point, algorithm and automaton.

Mathematics of the 20th century has considerably changed in character compared with mathematics up to the turn of the century. Almost all mathematical disciplines are treated in an axiomatic-deductive way, so that the admissible rules of mathematical inference are precisely laid down (see Chapter 15.). However, in the presentation of their substance most theories use not only pure logic, but also certain parts of set theory or elementary arithmetic, as long as this is not expressly forbidden. Among other things, for reasons of economy of thought in discovering knowledge and carrying out the technicalities of proofs, it is quite common in the practical treatment of a special mathematical discipline to go beyond the narrow framework of this discipline and its characteristic language and thereby to develop de facto part of the *metatheory* of this discipline.

One of the most striking problems of meta-mathematics is the *truth problem* for mathematical statements. Is it a fact that each mathematical proposition stated to be true is really the description of an objective feature, or are at least some such statements, for example, the theorem on the possibility of well-ordering the real numbers, mere nominalistic constructions, which can be justified only by a number of useful applications? – This problem will later be treated in more detail.

The traditional trends of thought in the philosophy of mathematics

To a certain extent, mathematical investigations have always been connected with a critical analysis of their foundations, corresponding to the state of knowledge at the time; this applies not only to the Greek period, but also to the mathematics of the Middle Ages and the early bourgeois society. Since it is customary nowadays to refer mainly to the *results of the 19th century* in the foundations of mathematics, these will now be sketched briefly.

After a period of rapid development, about the middle of the last century a critical scrutiny of the foundations of analysis took place. Apart from A. CAUCHY (1789–1857), C. F. GAUSS (1777 to 1855), K. WEIERSTRASS (1815–1897) and B. BOLZANO (1781–1848), the names of R. DEDEKIND (1831–1916) and G. CANTOR (1845–1918) must be mentioned. One main problem was an exact definition of the concept of a real number. The clarification of the number concept given by DEDEKIND, WEIERSTRASS, and CANTOR is part of the stock in trade of mathematicians nowadays.

1. Logicism. Independently of each other, G. FREGE (1848–1925) and DEDEKIND founded their theories of the natural numbers 'logically', or as we should say today, on a set-theoretical basis. Above all, FREGE aimed at founding the theory of natural numbers, and then successively the whole of mathematics, on the laws of pure thought and therefore on those of logic. In the terminology of KANT this would mean proving the analytical character of mathematical propositions. FREGE established the connection between logic and mathematics. He laid the foundation of a programme which is usually known as *logicism*. B. RUSSELL (1872–1970) observed that Frege's structure was inconsistent and developed an improved system, which is set forth in the well-known work *Principia Mathematica* (jointly with A. N. WHITEHEAD). Its essential content is ultimately the proof that the whole of mathematics can be developed on the basis of the set-theoretical (ramified) theory of types.

A variant of logicism, a radical interpretation which aims at regarding mathematics, as it were, as a result of rational thought, is *mathematical Platonism*; it appears, for example, in the ideas of CANTOR on the foundation of set theory. In this interpretation sets are ideal objects, which exist independently of intellectual activity. The task of the mathematical investigator is to track down the laws that prevail in this very general world of objects, the *Cantor universe.*

In a certain sense logicism is absorbed in the set-theoretical foundation of mathematics, which will later be described in more detail.

2. Formalism. The formalistic interpretation arose as the answer to the difficulties in the theory of knowledge expressed in logicism. A decisive step in this direction was the book, published in 1899, *Grundlagen der Geometrie* (*Foundations of geometry*) by D. HILBERT (1862–1943). Here it was shown for the first time by the example of geometry what is to be understood by formal axiomatics and its metalogical analysis. The programme of a formalistic foundation of mathematics was finally formulated by HILBERT in 1920 and taken in hand by him and his school. According to this programme, even such mathematical domains as number theory, analysis, and set theory which, by their nature, are at first sight specified by their contents, are to be understood as formal theories. The first problem for a student of the foundations is to establish formally that the system is consistent that is, to show conclusively that a statement and its negation are not both derivable from the axioms by means of the laws of logic. This task is to be accomplished by methods whose reliability is beyond all doubt. To these methods, which HILBERT called *finite*, there belong elementary combinatorial methods, in particular, the principle of proof by mathematical induction, but not those of transfinite set theory, the so-called *infinitistic methods*. The results of these endeavours (up to 1938) were recorded in the two-volume work *Grundlagen der Mathematik* (*Foundations of Mathematics*) by HILBERT and P. BERNAYS, which next to Principia Mathematica ranks among the most significant works on the foundations of mathematics of this century.

HILBERT's original attempt had to be revised on account of the results of K. GÖDEL (1906–1978). One of these results states that any proof that a formal system is free from contradictions necessarily requires methods beyond those provided by the system itself; accordingly, one cannot prove that number theory is free from contradictions by means of finite methods in the strict sense.

Today there is still no clear agreement on the type and the range of admissible extensions of finite methods; one of these possibilities is the incorporation of recursive functionals. It is not possible to go further into these questions here, except to say that finite investigations are not limited to the question of freedom from contradiction, but refer, for example, to decision problems and generally to the analysis of the finite core in fundamental mathematical and meta-mathematical results of infinitistic nature.

3. Intuitionism. This point of view, which is totally opposite to the logicistic and formalistic interpretations, was founded by L. E. J. BROUWER (1881–1966). Similar ideas had been put forward earlier by L. KRONECKER (1823–1891) and H. POINCARÉ (1854–1912). The following points of view are characteristic of intuitionism: 1. Rejection of the *actual infinite*. 2. The postulate of *effective constructability* as the only means of defining mathematical objects. 3. The original material of the construction consists of the *natural* numbers, and these are to be regarded only as a potentially (uncompleted) given aggregate. 4. Limitation of the classical logical principles in their application to infinite aggregates.

To explain at least one of these points of view, define

$$a_n = \begin{cases} 0 \text{ if } 2(n+1) \text{ is the sum of two prime numbers,} \\ 1 \text{ otherwise.} \end{cases}$$

The real number $g = 0.\,a_1a_2 \ldots$ could be called the *Goldbach number*; it can be calculated to any degree of accuracy, yet it is not known whether $g = 0$ or not. Since one cannot be certain whether Goldbach's problem (see Chapter 31.) will ever be solved, there is some justification for the argument that it makes no sense to say that either $g = 0$ or $g \neq 0$. However, this obviously implies a certain limitation on the *tertium non datur*, the law of the excluded middle.

4. The present situation. None of the schools of thought mentioned above has been able to achieve its original aim. In spite of this, the treatment of meta-mathematical questions from different points of view has brought to light valuable insights and results, which were not originally intended. The questions of decidability posed by meta-mathematics has led at an early stage to a precise formulation of computability and the *concept of an algorithm*. The formal mathematical languages, which were made precise by HILBERT and his school, are fundamental for the construction of algorithmic languages (for example, ALGOL or FORTRAN) and these examples could be increased indefinitely. Today an individual scholar can rarely be classified as belonging to a definite direction, rather he follows a dialectic course, by studying questions and results from different, partially contradictory, standpoints. The observance of certain differentiated constructivity postulates in the course of an investigation is not so much the expression of a certain philosophical position as of the methodological principle of not unnecessarily overstepping the bounds of secure knowledge.

Some main results in the foundation of mathematics

1. The set-theoretical foundation of mathematics. What remains of the original programme of logicism is the knowledge that the whole of mathematics can be built up on the basis of *axiomatic set theory*. This means that today every existing mathematical theory, irrespective of whether it has an axiomatic character or concerns a definite domain of objects, can be regarded as a partial domain of axiomatic set theory, suitably limited by the aims of the relevant theory.

The essential contributions to the axiomatic foundation of set theory are due to B. RUSSELL (1872 to 1970), E. ZERMELO (1871–1953), J. VON NEUMANN (1903–1957), A. FRAENKEL (1891–1965), P. BERNAYS (1888–1977) and K. GÖDEL (1906–1978). The *Bourbaki school* has undertaken a well thought-out methodical arrangement of mathematics from the set-theoretical point of view and has thereby essentially contributed to its popularization.

The language of axiomatic set theory in, say, the Zermelo-Fraenkel system is a very simple predicate language with the single predicate sign $\in$. The axioms correspond to the few so-called principles of the contents of set theory listed in the chapter on Set Theory. The rules of deduction are the formal rules of natural inference given in the chapter on Elements of Mathematical Logic – or rules reducible to them. The *definition of concepts* proceeds only by the rules of explicit definition; other, recursive or implicit, definitions are reducible to explicit definitions within the framework of axiomatic set theory.

The mathematician is not obliged on principle to go beyond the frame of formal set theory, but this naturally applies only to mathematical investigations as such and not to their application to physical or other extra-mathematical processes. The question of a suitable mathematical model for processes of this kind is not, strictly speaking, a mathematical question.

It should be observed that many mathematical theories have a complicated formal underlying apparatus; it is not always clear how they can be reduced to set theory and how their language can be adequately 'coded' in the simple language of set theory. This applies as a rule even to the presentation of the contents of set theory itself. Compare, for instance, the examples of forming sets in the chapter on Set Theory, most of which need to be made precise within the frame of axiomatic set theory. To form the set of all subsets of the set of real numbers, one has to have a definition of the concept of real number inside axiomatic set theory. This inturn assumes a definition of the concept of the set of natural numbers and leads first and foremost to an axiomatic analysis of the concept of finiteness within the frame of general set theory.

In addition, all the concepts used in the semantics of formal languages can be defined set-theoretically; this holds, in particular, for the linguistic objects themselves. These have been explained as finite sequences of certain 'symbols'. Just as in set-theoretical topology the elements of a set structured in a certain way are called *points*, so the elements of an arbitrary (usually denumerable) set can be designated as *symbols*.

2. Criticism of the set-theoretical foundation based on results in the foundations of mathematics. The question arises whether the possibility of a set-theoretical basis for mathematics as a whole can be regarded as throwing sufficient light on the problem of a meta-mathematical foundation of mathematics. Although by the reduction of the whole of mathematics to set theory the meta-mathematical problems appear to be reduced to a large extent to those of axiomatic set theory with its simple language and easily comprehensible axioms, this would in fact be a fallacy, for various reasons some of which will now be briefly discussed here.

(*i*) One immediate objection is the question of *consistency of the formal system of set theory*. In respect of actual experience the set-theoretical axiom system is on too high a level of abstraction to be able to speak of a direct verification. A kind of empirical control exists at best for certain consequences of these axioms, for example, for existence statements on solutions of differential equations with certain boundary conditions.

Hence it is not very surprising that set theory, just at the beginning of its development, had to eliminate a number of serious paradoxes in its system of concepts, which despite their removal here caused a permanent mistrust on the part of many mathematicians in too free a use of infinitistic methods.

(*ii*) A further objection concerns the *incompleteness in principle of the set-theoretical axiom system*, to be discussed in the next section, in the sense that in any far-reaching (recursive) axiom system there are statements that are independent of this axiom system. There is therefore no hope of completely grasping the intuitive universe of sets even approximately, through a chosen fixed axiom system.

(*iii*) Despite the facts mentioned above it could still be assumed that a certain object domain U (the intuitive set universe) corresponds to an accepted set-theoretical axiom system A, and that any set-theoretical statement is either true (that is, valid in U) or not. It is then possible to speak of the structure $\langle A, U\rangle$ within a suitable meta-language L^*. In L^* a well-known result of T. SKOLEM (1887–1963) – known as the *Skolem paradox* – can now be formulated; it states that there are several non-isomorphic models not only of the axiom system A, but even of the syntactically complete system of all statements valid in U (see Chapter 15.). But then the idea of a standard model, that is, a model of the axiomatic set theory that is distinguished in a certain way, becomes completely doubtful.

These objections make it clear that the aim of a logical-empirical foundation of mathematics, particularly in the classical form of Cantor's Platonism, is unattainably far removed. One is therefore justified in asking whether a universal set-theoretical foundation of mathematics is a factual requirement, or whether principles of a more constructive character are perhaps sufficient for this purpose; as regards the actual application of mathematical methods in extra-mathematical fields, on closer inspection essentially only *constructive methods* prove practicable. In the present state of things it must certainly be said that the *infinitistic methods* of set theory cannot be abandoned. Metaphorically speaking, the guns of infinitistic set theory have so far an unsurpassed range. This applies particularly to the capture of the constructive methods of the applications of mathematics. Furthermore, it can be said that – despite the fact that the Cantor universe proves to be a fiction on closer analysis – the mathematician, particularly one who studies the foundations, obtains his results, as a rule, only on the basis of a certain intuitive idea of an abstract mathematical reality.

3. Incompleteness of axiomatic theories and the indefinability of the concept of truth. In judging a mathematical theory T created with the object of providing a model for a certain domain of objects, for example, physical space or certain physical or economic processes, the only significant thing is the success. Since T represents only one consciously chosen idealization of a real process, the question of the truth of statements in T is of secondary importance. However, the truth problem becomes relevant for the whole of mathematics, being a closed science. The same is true for any branch, for example, the theory of natural numbers or set theory, that by its origins is not an axiomatic theory, but the description of a certain possibly abstract domain of objects.

To be sure, a great many mathematical statements, in spite of their abstract character, have an immediate relationship with reality. Consider, for example, the following theorem, whose validity is evident: '*if there exists a division of a finite set S into n disjoint classes* $C_1, \ldots, C_n$ *in which each class contains exactly m elements, then there also exists a division of S into m disjoint classes with n elements each.*'

The situation is quite different with respect to the statement, widely, accepted nowadays, that '*there exists a well-ordering relation on the set of real numbers*', and generally for existence statements in which nothing is said about the method of construction of the object in question.

If U is a certain domain of objects (universe of discourse), and L a formalized language over U, then it is known to be possible (see Chapter 15.) within the frame of a metatheory over L and U to make precise the concept of *validity* or *truth* of a statement of L in U. The first question is whether there is a codifiable axiom system A such that the set of statements derivable from A by the rules of formalized reasoning coincides with the set of true statements over U. In some cases this is in fact possible, for example, when U is a finite universe of discourse or when the language L is so lacking in expressive power that it does not even permit the formulation of complicated properties of U.

The following refers to a domain U that contains the natural numbers and to a language L in which the arithmetic of the natural numbers is expressible. Under these assumptions the *first incompleteness theorem of Gödel* holds, according to which any axiom system formulated in L that consists of finitely many or, more generally, of a recursive set of axioms is incomplete in the sense that not all true statements in U can be derived from A. A further fundamental result is the theorem of A. TARSKI (b. 1901) that under the given assumptions no predicate $W(x)$ is definable in L such that for an object a of U the statement $W(a)$ is true in U if and only if a is the code number of a true statement in U.

For the proofs of both theorems a *codification*, also called an *arithmetization* or *Gödelization*, of the language L by natural numbers is carried out. This is done in such a way that firstly natural numbers are assigned to the fundamental symbols so that sequences of symbols correspond to certain finite sequences of natural numbers. In the second stage the finite sequences of natural numbers are put in one-to-one correspondence with the natural numbers. The natural number that corresponds in this way to an expression H is called the *code* or *Gödel number* of H and denoted by H^*.

Let L, U and their semantic correlation be included in a new universe of discourse $\hat{U}$, and let $\hat{L}$ be an adequate language for $\hat{U}$. Thus L is called the *object language* of U, and $\hat{L}$ the *meta-language* of the system $\langle L, U\rangle$ (Fig.).

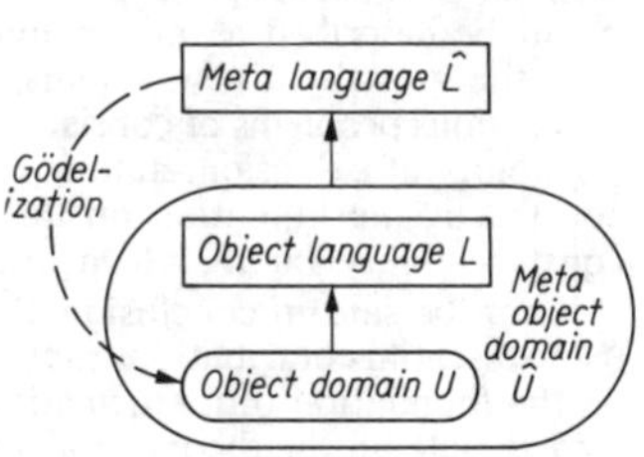

The coding of L makes it possible to project certain predicates of U which in the first instance are only metalinguistically expressible, into the object language L.

An example of a predicate in the metaobject domain is the one-place predicate 'the statement H is provable (from A)'. To this there corresponds a certain arithmetical predicate $B(n)$, which is true for a natural number n if and only if n is the code number of a provable statement. Under the assumptions made on U an expression $Nb(v)$ can now be constructed so that on substitution of a natural number n for the variable v the statement $Nb(n)$ says: 'the statement with the code number n is unprovable (from A)'. By means of a further device, a so-called *diagonal argument*, one can now find a natural number m so that $m = Nb(m)^*$. The statement $Nb(m)$ can then be regarded as a *self-referring* statement with the meaning 'I am unprovable'.

$Nb(m)$ is valid in U, otherwise its negation 'I am provable' would be valid in U; since each statement provable from A is naturally also valid in U, the statement $Nb(n)$ with the code number m in U would be both valid and invalid, which would be a contradiction. However, in accordance with the meaning of $Nb(m)$, the validity of $Nb(m)$ also indicates its unprovability. Hence the axiom system A turns out to be incomplete.

The result of TARSKI is obtained in a similar way. One assumes that an expression $W(v)$ exists with the meaning 'v is true'; the negation $Nw(v)$ of this expression then represents the predicate 'v is untrue'. A self-referring statement $Nw(m)$ ($Nw(m)^* = m$), as constructed above, would then mean 'I am an invalid statement'. This statement would be true if it is false and false if it is true. One can escape from this contradiction only by dropping the assumption that the predicate 'v is true in U' is expressible in L.

This kind of argument concerns deep analysis of a paradox already known in antiquity, which can be put in the following form: 'The theorem printed in red type on page n of this book is false.' This statement too is false if it is true and true if it is false, and therefore infringes the principle of excluded contradiction.

4. Relative consistency and the independence of the continuum hypothesis. A theory T is called relatively consistent with respect to a theory T' if the fundamental concepts of T can be defined in the language of T' in such a way that the axioms of T correspond to certain statements valid in T'; the theory T is then said to be *interpreted* in the theory T'.

The definitions just given are of a metatheoretical nature. Frequently the proof of this interpretability can be conducted completely inside the language of T', although model-theoretical arguments about T and T' succeed more quickly.

Of particular interest is the special case when T' is an extension of T within the same language L. In particular, a statement A of L is called consistent with respect to T if the theory $T \cup \{A\}$ is relatively consistent with respect to T; for example, it can be shown that Euclidean and non-Euclidean geometry are relatively consistent with respect to absolute geometry, that is, that the parallel axiom and also its negation are consistent with respect to the other geometrical axioms, or briefly that the parallel axiom is independent of the other axioms. The fact that the proof of this can be carried out entirely within absolute geometry is by no means trivial; however, by modern standards, the proof of independence is almost a banality if it is carried out by model-theoretic means, that is, in this case by means of analytic geometry.

GÖDEL showed in 1938 that the continuum hypothesis and the axiom of choice are relatively consistent with respect to the other axioms of set theory. Twenty-five years later P. COHEN showed that the negation of the continuum hypothesis is also consistent with respect to the other axioms.

Although these results have a formal analogy with geometry, the situation is quite different, since it is possible to set up the different kinds of geometry from a unified standpoint, namely that of general set theory. However, there is no unified principle for founding the different, mutually exclusive, systems of set theory. According to the present state of affairs, such principles of a mathematical nature do not even seem to exist, because a higher mathematical abstraction than that of set theory is absolutely inconceivable.

GÖDEL himself has expressed the view that the development of set theory will lead to new axioms, which will allow the continuum hypothesis to be disproved. The axioms so far taken into discussion for the extension of the usual bounds of set theory, for example, the axiom of TARSKI on the existence of inaccessible cardinal numbers, are not likely to suffice for this.

Tarski's axiom is an example of an axiom that ensures the existence of further sets, beyond the domain produced by the principles of set formation and choice. The acceptance of such axioms could be described as an unlimited extension of mathematics. It must be remembered, however, that the growth of new axioms of unlimited character is not a cogent demand and would cause new serious problems of consistency. There are certain limited ways of attaining the above mentioned possibility of extension, among them all constructivistically oriented statements. One kind of semi-constructivistic limitation on the boundless formation of sets would be the acceptance of Gödel's constructability axiom, which would imply the validity of the continuum hypothesis.

It can be said in conclusion that the result of research on the foundations of mathematics has made essential contributions to the clarification of the range and the bounds of the classical statements on the foundation of mathematics, and has, in addition, provided numerous practical applications, for example, in the theory of algorithms and the theory of formal systems.

43. Game theory

As early as the 17th century attempts were made to analyze games of chance and parlour games. A multitude of these games continue to be with us today, and in some (such as roulette) the outcome is purely accidental, in others (such as bridge) it depends on chance and the players' behaviour, while a third group (such as chess) is completely controlled by skill.

In 1943 J. von Neumann and O. Morgenstern were the first to provide a general description of the links between economic problems (competitive situations) and games, thus establishing the theory of games as we know it today. Nowadays it is seen as a discipline in the wider field of mathematical operations research.

Conflict situations

Nature and society are replete with cases in which the parties involved have conflicting interests and pursue them in different ways. Such a competitive situation is easily recognizable in a parlour game, military confrontation or economic competition. However, in some problems occurring in game theory it is necessary to construe a conflict situation.

Such a situation is mathematically modelled as a game where the *players* may be the natural persons attending a party or, in a more general sense, companies, armies, ships, nature etc. These players can choose to follow a *specific course of action* as embodied, for example, in the rules of a parlour game. They will then do their best to use this leeway skillfully to achieve their goals. One of these goals may be to win a game of chess.

The theory of games is concerned with describing and modelling these relations in mathematical terms and finding the best possible strategy for a player.

Matrix games

The most basic type of game is one involving only two people who act on the understanding that the gains made by one player are equal to the other's losses. Gains and losses may be measured in sums of money, and the total amount paid out in this case will be zero. Hence the name – *two-person zero-sum games*, also known as *matrix games* since they can be fully described by a matrix $\boldsymbol{A}$.

A matrix game means that the m courses of action H_i open to player P_1 are assigned to the m rows of a matrix $\boldsymbol{A}$. Similarly, the n columns of the same matrix $\boldsymbol{A}$ are assigned to the n courses of action h_k open to player P_2. This will make him the *column player*, while P_1 is known as the *row player*.

For actual play P_1 selects a row i and P_2 a column k of the matrix $\boldsymbol{A} = (a_{ik})$, both players proceeding independently of each other and without informing the other. A pay-off is then made by P_2 to P_1 at the rate of a_{ik}. P_1 gains something if $a_{ik} > 0$, there is no pay-off if $a_{ik} = 0$, and he loses if $a_{ik} < 0$. The matrix $\boldsymbol{A}$ is known as the pay-off matrix for P_1, and in many examples and applications the pay-off is merely symbolic.

> The pay-off matrix $\boldsymbol{A}$ for the row player P_1 tells us that this is a matrix game.

As play proceeds, the following conflict situation arises: P_1 aims to maximize his gains by selecting suitable rows, whereas P_2 seeks to minimize his losses by choosing the right columns. In mathematical terms, these goals may be described by introducing the concept of strategy. A strategy indicates the probability x_1 with which the course of action H_i (the row i) is selected in a particular game.

> Any vector $\boldsymbol{x}$ defined as $\boldsymbol{x}^T = (x_1 x_2 \dots x_m)$, $0 \leqslant x_i = 1$, $x_1 + x_2 + \cdots + x_m = 1$ is a strategy for P_1, $\boldsymbol{x}$ being the column vector and $\boldsymbol{x}^T$ the row vector.

If a component x_i equals 1 (which means all other components are zero), then only the row i will be selected in any game. This strategy $\boldsymbol{x}$ is known as a pure *strategy*; all other strategies are *mixed strategies*. By analogy, any vector $\boldsymbol{y}$ where $\boldsymbol{y}^T = (y_1 y_2 \dots y_n)$, $0 \leqslant y_k \leqslant 1$, $y_1 + y_2 + \cdots + y_n = 1$ is a strategy for P_2. Pure and mixed strategies are defined on the analogy of P_1.

In terms of probability theory the average pay-off P_2 makes to P_1 is $\boldsymbol{x}^T\boldsymbol{A}\boldsymbol{y}$. When a fixed strategy $\boldsymbol{x}$ is chosen, then P_1 is certain to gain $\min_{\boldsymbol{y}} \boldsymbol{x}^T\boldsymbol{A}\boldsymbol{y}$. This means that P_1 must select a strategy $\boldsymbol{x} = \boldsymbol{x}_0$ to maximize his gain.

A strategy x_0 is *optimal* for P_1 if $\min_y x_0^T Ay = \max_x \min_y x^T Ay$. By analogy, y_0 is an optimal strategy for P_2 if $\max_x x^T Ay_0 = \min_y \max_x x^T Ay$.

Evidently, $\max_x \min_y x^T Ay = \min_y \max_x x^T Ay$. J. VON NEUMANN has proved the following essential result.

Main theorem for a matrix game: $\max_x \min_y x^T Ay = \min_y \max_x x^T Ay = v$.

The number v is known as the *value* of the game and, together with the optimal strategies x_0 and y_0, it represents the *solution to the matrix game.*

If P_1 abandons an optimal strategy x_0 his gains may be less than the value of the game v; if P_2 strays from an optimal strategy y_0 his losses may exceed v. The optimal strategies x_0 and y_0 therefore describe the best possible approach the two players can take. The average pay-off made by P_2 to P_1 is then v, and a game where $v = 0$ is a *fair game.*

Pure optimal strategies. A special case among the matrix games is the *saddle-point game* with *pure optimal strategies* x_0 and y_0, as illustrated in the following example.

Each of two players P_1 and P_2 has the four aces from a pack of cards. The idea of the game is for both to put an ace face upwards on the table at the same time, followed by a pay-off under a previously agreed rule. That is the end of the game, and the amounts P_2 pays to P_1 can be seen from the following *pay-off matrix A*.

P_1 \ P_2	Clubs	Spades	Hearts	Diamonds	Row minimum
Clubs	3	0	2	−1	−1
Spades	−1	−1	0	1	−1
Hearts	2	0	1	0	0
Diamonds	2	−1	−2	3	−2
Column maximum	3	0	2	3	

The figures in the pay-off matrix A represent the following: If P_1 lays down the ace of clubs and P_2 the ace of hearts, then P_2 has to pay two units to P_1. If both players present the ace of spades, P_1 will pay one unit to P_2. Here the course of action open to a player is to choose from his aces. If this game were to go on for some time, P_1 will appear to be favoured because there are more positive figures in the pay-off matrix than negative ones, but this is a fallacy.

Let us now determine the optimal strategies for both players, assuming that neither wants to take a risk. If P_1 selects the first row (ace of clubs), he will be worst off if P_2 chooses the fourth column (ace of diamonds) as he loses one unit. P_1 will also lose if he opts for the second or fourth row but is certain to lose nothing by consistently playing the third row (ace of hearts). In that case he may even win if P_2 makes a mistake. This means that P_2 can at least be sure of gaining the amount that is the *greatest row minimum*, i. e. $w_1 = \max(-1, -1, 0, -2) = 0$. This applies by analogy to P_2. If he plays the first column (ace of clubs), the worst response he can get is the first row (ace of clubs), which will lose him three units. He also loses if he opts for the third or fourth column, but can be certain to lose nothing by playing the second column (ace of spades) consistently. He may in turn win if his opponent plays badly, and he can limit his losses to the *least column maximum*, i. e. $w_2 = \min(3, 0, 2, 3) = 0$.

As $w_1 = w_2 = 0$, neither side has an advantage, which makes this a fair game. The optimal strategy is for P_1 to always show the ace of hearts and for P_2 to stick to the ace of spades. In the pay-off matrix this takes the form of the saddle point shown in red. The solution to the game is $x_0^T = (0, 0, 1, 0)$, $y_0^T = (0, 1, 0, 0)$, $v = 0$.

There is no saddle point if $w_1 \neq w_2$, and this type of game becomes uninteresting once the optimal strategies are known, as in our case.

Examples of saddle-point games include chess, go, checkers and mill. In each, there would only be three possibilities if the players followed optimal strategies: a) white always wins, b) black always wins, c) there is always a draw. In the case of chess, go and checkers it is still unknown which of the three applies. The reason is that we cannot define the optimal behaviour of a player because of the many courses of action open to him under the rules of the game (but this is not the case with mill). This also explains why a chess computer will lose on occasion. By contrast, the options in an end game are a simple kind of chess problem and identical with the moves to be made.

Mixed optimal strategies. These occur in matrix games without a saddle point, and the optimal strategy of a player can be determined in a diagram if only two courses of action are open to him.

A case in point is the gentleman (P_1) who wishes to encounter a lady (P_2) as often as possible over a weekend, whereas she wants to see as little of him as possible. The options for him are to go out either on Saturday or Sunday, while she may do so in the morning or afternoon. He then draws up a table illustrating his chances of meeting her.

P_1 \ P_2	Morning	Afternoon	Row minimum
Saturday	2	−0.5	−0.5
Sunday	0	1	0
Column maximum	2	1	

She is of the same opinion, as regards the chances of an encounter. The 2 in the table means he is very hopeful of meeting her while doing his shopping on Saturday morning, while the −0.5 tells us she is rather certain of not running into him on the afternoon of the same day because of his football craze.

This is not a saddle-point game because the greatest row minimum $w_1 = \max(-0.5, 0) = 0$ differs from the least column maximum $w_2 = \min(2, 1) = 1$, w_1 is less than w_2. The game con-

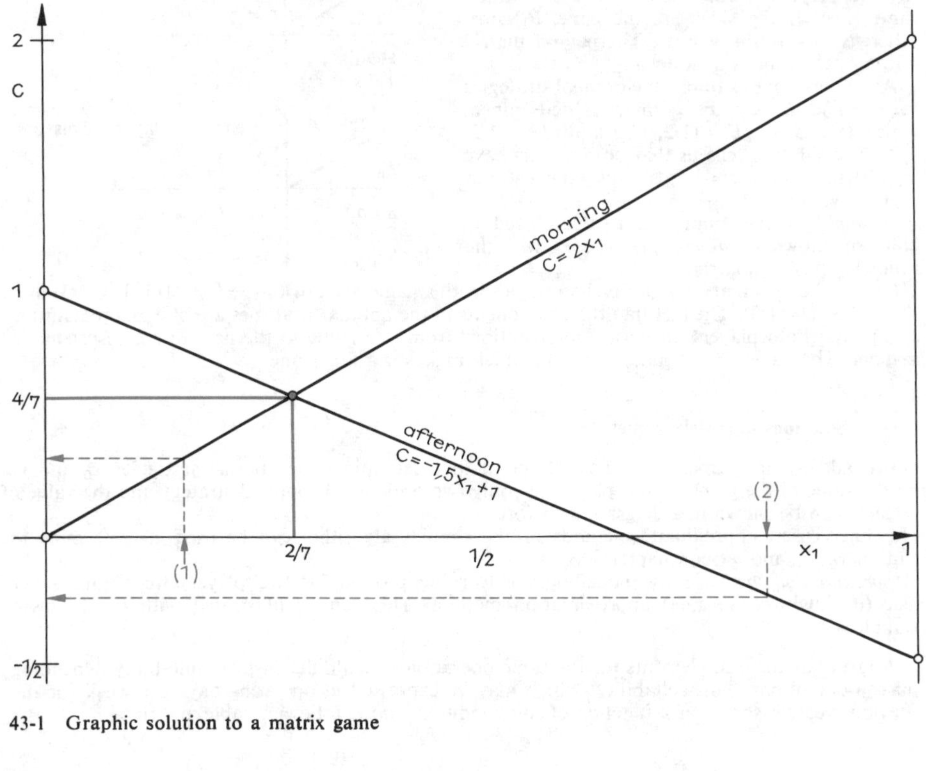

43-1 Graphic solution to a matrix game

tinues through several weekends a total of M times, with P_1 choosing Saturday m_1 times and Sunday m_2 times so that $m_1 + m_2 = M$ or $x_1 + x_2 = 1$ if $x_1 = m_1/M$ and $x_2 = m_2/M$. P_2 accordingly chooses the morning n_1 times, the afternoon n_2 times, and $y_1 + y_2 = 1$ if $y_1 = n_1/M$ and $y_2 = n_2/M$. The vectors $x^T = (x_1 x_2$ and $y^T = (y_1 y_2)$ are strategies for P_1 and P_2.

The optimal strategies x_0 and y_0 can now be graphically determined. The average chances for P_1 per weekend, if P_2 always chooses the morning, are: $C = 2x_1 + 0x_2 = 2x_1$. If P_2 invariably opts for the afternoon P_1's chances are $C = 0.5x_1 + 1x_2 = -0.5x_1 + 1$ because $x_1 + x_2 = 1$. The two equations correspona to straight lines in a rectangular Cartesian coordinate system, where the chance C is the ordinate and x_1 the abscissa. It must be borne in mind in this connection that x_1 is a relative frequency with a value that can only range from 0 to 1 (Fig. 43-1). The coordinates for the point of intersection of the two straight lines are $x_1 = 2/7$, $C = 4/7$. On the C axis the chances of P_1 can be read off as follows:

(1) If P_1 opts for Saturday with a relative frequency of $0 \leqslant x_1 < 2/7$, his chances are less than 4/7 in the worst possible case (P_2 goes out in the afternoon).

(2) If, on the other hand, P_1 chooses Saturday with a relative frequency of $2/7 < x_1 \leqslant 1$, then the chances are again less than 4/7 in the worst possible case (P_2 goes out in the afternoon).

The optimal strategy for P_1 is therefore $x_0^T = (2/7, 5/7)$, and the value of the game $C = 4/7$. This means that our gentleman has to adhere to a 2:5 ratio of Saturday to Sunday without any regularity and the game is *unfair* as far as P_1 is concerned.

Now P_2 tries to find a strategy limiting her mean loss to 4/7. Her average chances per weekend, if P_1 consistently opts for Saturday, are $C = 2y_1 - 0.5y_2 = 2y_1 - 0.5(1 - y_1) = 2.5y_1 - 0.5 \leqslant 4/7$. If P_1 always goes out on a Sunday her chances are $C = 0 \cdot y_1 + y_2 = 1 - y_1 \leqslant 4/7$. From the first inequality it follows that $y_1 \leqslant 3/7$ and from the second, $y_1 \geqslant 3/7$. The optimal strategy for P_2 is then $y_0^T = (3/7, 4/7)$ so that P_2 selects a morning to afternoon ratio of 3:4 with no regularity. So much for a complete discussion of this game. If one of the players abandons his optimal strategy, the other is favoured.

Two other well-known *examples* may be given for matrix games without a saddle point.

Example 1 is the *penny game* where each of the two players reveals the side of a coin at the same time. If the sides are the same, P_1 wins, otherwise P_2 is the winner. The pay-off matrix A for P_1 then looks as follows:

P_1 \ P_2	Heads	Tails
Heads	1	−1
Tails	−1	1

As in the above example, the optimal strategies for P_1 and P_2 can be graphically determined, which results in $x_0^T = (1/2, 1/2)$ and $y_0^T = (1/2, 1/2)$. The solution tells us that both players have to alternate between heads and tails without regularity.

Example 2 is the counting-out game played by children known as *stone-paper-scissors*, with the following pay-off matrix.

P_1 \ P_2	Stone	Paper	Scissors
Stone	0	−1	1
Paper	1	0	−1
Scissors	−1	1	0

Due to the symmetry the optimal strategies in this game are obvious: $x_0^T = (1/3, 1/3, 1/3)$ and $y_0^T = (1/3, 1/3, 1/3)$. The fact that the components in the optimal strategies are of equal magnitude means that both players must vary their options from one game to the next in an unsystematic manner. The value of the game is $v = 0$, which makes it a fair game.

Solutions to matrix games

In a saddle-point game it is the saddle point that determines the optimal strategies x_0 and y_0, and the value of the game, v. If a player has only two options, his optimal strategy and the value of the game can be shown in a diagram as before.

In other cases, approximate methods or the simplex algorithm can be used to determine the solution x_0, y_0 and v (see Chapter 30.1.).

Approximation methods. By modelling the learning process for two players we obtain a very basic (if slowly converging) approximation method. This can be illustrated with the following example.

A farm has three implements for the same operation which differ in their suitability depending on soil conditions. This suitability (which may be expressed in operating days per week for the previous year) is shown as a function of soil conditions in the following table, matrix A.

P_1 \ P_2	dry	normal	wet
Implement 1	1	0	2
Implement 2	3	0	0
Implement 3	0	2	1

Player P_1 is the farm, player P_2 the weather, and the game itself belongs in the category of games against nature. The solution tells us how often we can expect to use each of the three implements and what "action" nature may take.

The approximation procedure consists of the following steps:

(1) P_1 arbitrarily selects a row from matrix $\boldsymbol{A}$, for example the first row, and writes it under the matrix.

(2) P_2 then chooses that column of matrix $\boldsymbol{A}$ which contains the lowest figure in that row (column 2) and writes it next to the matrix.

(3) P_1 selects the row that has the highest figure in it (the third row) and writes the sum of the last and the newly choosen row under the matrix.

(4) P_2 again chooses the column which contains the lowest figure in that row and writes the sum of the last and the new row next to the matrix.

	$\boldsymbol{A}$			(2)	(4)	(6)	(8)	(10)	
	1	0	2	0	1	1	3	4	$x_1 \approx 0/5$
	3	0	0	0	**3**	3	3	**6**	$x_2 \approx 2/5$
	0	2	1	**2**	2	**4**	**5**	5	$x_3 \approx 3/5$
(1)	1	**0**	2						
(3)	**1**	2	3						
(5)	4	**2**	3						
(7)	4	4	**4**						
(9)	**4**	6	5						
	$y_1 \approx 2/5$	$y_2 \approx 2/5$	$y_3 \approx 1/5$						

The procedure continues, and if the smallest figure in a column or the highest figure in a row occur more than once, an arbitrary choice is made. In addition, the figures selected each time are marked in red when searching for the next row or column.

Approximations for the optimal strategies are obtained when this algorithm is followed: count the red figures in each of the three rows (next to $\boldsymbol{A}$) formed by the columns and in each of the columns (under $\boldsymbol{A}$) formed by the rows and divide by the number of steps. For example, three figures are marked in the third row (2, 4, 5) and the number of steps is five, so that $x_3 \approx 3/5$.

The value of the game, v, is contained between the last marked figure in the columns (4 in our example) and the last marked figure in the rows (6 in the example), divided by the number of steps in each case. The total result obtained after five steps for the optimal strategies $\boldsymbol{x}_0$ and $\boldsymbol{y}_0$, and for the value of the game, v, is: $\boldsymbol{x}_0^T \approx (0, 2/5, 3/5)$, $\boldsymbol{y}_0^T \approx (2/5, 2/5, 1/5)$, $4/5 \leqslant v \leqslant 6/5$.

This approximate solution is still far too vague, but it becomes more practical after ten steps: $\boldsymbol{x}_0^T \approx (1/10, 3/10, 6/10)$, $\boldsymbol{y}_0^T \approx (3/10, 4/10, 3/10)$, $9/10 \leqslant v \leqslant 11/10$.

The last approximate solution tells us that the implements will be used at the ratio of about 1:3:6 and that the "options" of nature (dry – normal – wet) are about 3:4:3.

The more steps that are carried out in the procedure, the closer we will get to the solution of the matrix game. To compare, let us look at the accurate solution, obtained with the aid of the simplex algorithm (see below): $\boldsymbol{x}_0^T = (1/4, 1/4, 1/2)$, $\boldsymbol{y}_0^T = (1/3, 1/3, 1/3)$, $v = 1$.

The conclusion is that implement no. 3 should be ready for use at all times if possible. The value of the game is only symbolic.

Simplex algorithm

Each matrix game can be transformed into a problem of linear programming and solved with the aid of the simplex algorithm (see Chapter 30.1.).

For this purpose, the matrix game with the pay-off matrix $\boldsymbol{A}$ is assigned the two dual linear optimization problems (see Chapter 30.) max $\{\boldsymbol{e}^T\boldsymbol{v} \mid \boldsymbol{A}\boldsymbol{v} \leqslant \boldsymbol{e},\ \boldsymbol{v} \geqslant 0\}$ and min $\{\boldsymbol{e}^T\boldsymbol{w} \mid \boldsymbol{A}^T\boldsymbol{w} \geqslant \boldsymbol{e},\ \boldsymbol{w} \geqslant 0\}$. In the vectors $\boldsymbol{e}$ all components are equal to 1. The vectors $\boldsymbol{v}$ and $\boldsymbol{w}$ contain the variables of the optimization problems.

If we let the optimal solutions $\boldsymbol{y}_0$ and $\boldsymbol{w}_0$ and the common optimal value of the objective function c of the two optimization functions be determined with the simplex algorithm, then the optimal strategies $\boldsymbol{x}_0$ and $\boldsymbol{y}_0$ and the value v of the game are obtained as follows:

$$v = \frac{1}{c},\ \boldsymbol{x}_0 = \frac{1}{c}\boldsymbol{w}_0,\ \boldsymbol{y}_0 = \frac{1}{c}\boldsymbol{c}_0.$$

The following example illustrates how a game can be solved using the simplex algorithm. The model is one of a military conflict in the Middle Ages in which General BLOTTO is defending a town with two gates. His forces consist of three troops, while the enemy is attacking with two; all troops are of equal strength. The town is considered to have fallen to the attackers (value of the game = 1) as soon as the troops attacking at either gate outnumber the defending troops. A ceasefire results if equal numbers of troops attack and defend (value of the game = 0), and the town has been held if the defenders outnumber the attackers (value of the game = −1). The courses of action open to the two opponents are to assign an optimal number of troops to the gates. For the attacker this results in the pay-off matrix $\boldsymbol{A}$ as follows.

Attackers Gate 1	Attackers Gate 2	Blotto Gate 1: 3 / Gate 2: 0	2 / 1	1 / 2	0 / 3
2	0	−1	0	1	1
1	1	1	0	0	1
0	2	1	1	0	−1

The attacks continue unremittingly. What are the optimal strategies for both sides? Who has any advantage?

According to Chapter 30.1. the following simplex tables can be drawn up.

(1)	v_1	v_2	v_3	v_4	
w_1	−1	0	1	1	1
w_2	1	0	0	1	1
w_3	1	1	0	−1	1
	−1	−1	−1	−1	0

(2)	v_1	w_3	v_3	v_4	
w_1	−1	0	1	1	1
w_2	1	0	0	1	1
v_2	1	1	0	−1	1
	0	1	−1	−2	1

(3)	v_1	w_3	v_3	w_2	
w_1	−2	0	1	−1	0
v_4	1	0	0	1	1
v_2	2	1	0	1	2
	2	1	−1	2	2

(4)	v_1	w_3	w_1	w_2	
v_3	−2	0	1	−1	0
v_4	1	0	0	1	1
v_2	2	1	0	1	2
	0	1	1	1	3

This gives the optimal solutions $\boldsymbol{v}_0^T = (0, 2, 1, 0)$, $\boldsymbol{w}_0^T = (1, 1, 1)$ and $c = 3$, and the optimal strategies $\boldsymbol{x}_0^T = (1/3, 1/3, 1/3)$ and $\boldsymbol{y}_0^T = (0, 2/3, 1/3, 0)$. Hence the attacker must make equal use of the three options open to him, whereas General BLOTTO has to deploy his second and third defense options at the rate of 2:1. The value of the game, $v = 1/3$, tells us that the attacker has any advantage.

Linear programming with several objective functions

Many economic applications involve solving a linear optimization problem with a number of competing objective functions in the presence of identical constraints. In cases where no assessment of the various objective functions is possible, a *compromise solution* can be put together from the optimal solutions to these functions, using an arrangement derived from game theory. It follows a line of thought whereby s objective functions need to be maximized as $z_1 = \boldsymbol{c}_1^T\boldsymbol{x}$, $z_2 = \boldsymbol{c}_2^T\boldsymbol{x}$, ..., $z_s = \boldsymbol{c}_s^T\boldsymbol{x}$ in the presence of identical constraints $\boldsymbol{Bx} \leqslant \boldsymbol{b}$, $\boldsymbol{x} \geqslant 0$.

Let each of the s linear programming problems be solvable and let the optimal solutions be determined, for example, with the simplex algorithm (see Chapter 30.1.) and specified as $\boldsymbol{x}_1, \boldsymbol{x}_2, \ldots, \boldsymbol{x}_s$. Let all the objective function maxima $\boldsymbol{c}_1^T\boldsymbol{x}_1, \boldsymbol{c}_2^T\boldsymbol{x}_2, \ldots, \boldsymbol{c}_s^T\boldsymbol{x}_s$ be positive. An assessment of the optimal solution $\boldsymbol{x}_i$ is possible with regard to all s objective functions, where $a_{ik} = \boldsymbol{c}_k^T\boldsymbol{x}_i/\boldsymbol{c}_k^T\boldsymbol{x}_k$. The following then applies: $0 < a_{ik} \leqslant 1$. The percentage fulfillment of the k-th objective function by the optimal solution $\boldsymbol{x}_i$ is embodied in $a_{ik} \cdot 100\%$. This percentage fulfillment may vary considerably, and the minimum can be low. The projected compromise solution will not have this disadvantage. Now the game is considered according to the pay-off matrix $A = (a_{ik})$, $i, k = 1, 2, \ldots, s$, and solved. The optimal strategy of the player P_1 is $\boldsymbol{x}_0$, and the value of the game is v.

If $\boldsymbol{x}_0^T = (x_{01}, x_{02}, \ldots, x_{0s})$, the following applies.

The compromise solution $\boldsymbol{k} = x_{01}\boldsymbol{x}_1 + x_{02}\boldsymbol{x}_2 + \cdots + x_{0s}\boldsymbol{x}_s$ ensures that the minimum fulfillment of all objective functions is at a maximum with $v \cdot 100\%$.

For *example*, if $s = 3$ objective functions $z_1 = 10x_1 + x_2$, $z_2 = 5x_1 + 5x_2$, $z_3 = -5x_1 + x_2 + 30$ and these are to be maximized with the constraints $5x_1 + x_2 \leqslant 40$, $x_1 + x_2 \geqslant 16$, $x_2 \leqslant 15$ and $x_1 \geqslant 0$, $x_2 \geqslant 0$, then the optimal solutions (see Chapter 30.1.) after the first/second/third objective function are $\boldsymbol{x}_1^T = (6, 10)$, $\boldsymbol{x}_2^T = (5, 15)$ and $\boldsymbol{x}_3^T = (1, 15)$.

In the following table the left side gives the values of $\boldsymbol{c}_k^T\boldsymbol{x}_i$, the centre is the game matrix $A = (a_{ik})$, and the right side indicates the values of $a_{ik} \cdot 100\%$.

	z_1	z_2	z_3	z_i/z_1	z_i/z_2	z_i/z_3	1st objective function	2nd objective function	3rd objective function
$\boldsymbol{x}_1$	70	80	10	1	80/100	10/40	100%	80%	25%
$\boldsymbol{x}_2$	65	100	15	65/70	1	15/40	92.86%	100%	37.5%
$\boldsymbol{x}_3$	25	80	40	25/70	80/100	1	35.71%	80%	100%

The 10 marked in red (left side) is obtained when the solution $\boldsymbol{x}_1$ is substituted into the objective function z_3: $z_3(\boldsymbol{x}_1) = -5 \cdot 6 + 10 + 30 = 10$. The maximum value of the objective function z_3 is achieved with the optimal solution $\boldsymbol{x}_3$: $z_3(\boldsymbol{x}_3) = -5 \cdot 1 + 15 + 30 = 40$. The figure $a_{13} = \boldsymbol{c}_3^T\boldsymbol{x}_1/\boldsymbol{c}_3^T\boldsymbol{x}_3 = z_3(\boldsymbol{x}_1)/z_3(\boldsymbol{x}_3) = 10/40$ provides an assessment of the solution $\boldsymbol{x}_1$ with regard to the optimal solution $\boldsymbol{x}_3$ for the third objective function (centre of the table). The figure $a_{13} \cdot 100\% = 25\%$ gives a percentage assessment and is at the same time the lowest figure in the right-hand table. This means that the third objective function is least fulfilled by the solution $\boldsymbol{x}_1$, the fulfillment being only 25%. Now that we have defined the game with the pay-off matrix A we can solve it using the simplex algorithm. The following is obtained for the value of the game, v, and the optimal strategy $\boldsymbol{x}_0$ of the row player ($\boldsymbol{y}_0$ is irrelevant):

$$v = \frac{623}{938} \approx 0.6642; \qquad \boldsymbol{x}_0^T = \frac{623}{938}\left(0 \ \frac{504}{623} \ \frac{434}{623}\right) = \left(0 \ \frac{252}{469} \ \frac{217}{469}\right).$$

This leads to the compromise solution

$$\boldsymbol{k} = 0\begin{pmatrix}6\\10\end{pmatrix} + \frac{252}{469}\begin{pmatrix}5\\15\end{pmatrix} + \frac{217}{469}\begin{pmatrix}1\\15\end{pmatrix} = \begin{pmatrix}1466/469\\15\end{pmatrix} \approx \begin{pmatrix}3.1493\\15\end{pmatrix}$$

in which the minimum fulfillment for all objective functions is at least $v \cdot 100\% = 66.42\%$; this minimum fulfillment is a maximum. When $\boldsymbol{k}$ is substitued into the objective function it can be seen that the fulfillment of z_1 is exactly 66.42%, while it is 90.75% for z_2 and 73.13% for z_3.

Other games

It is characteristic of matrix games that they involve only *two* players, each of whom has a *finite number* of options, and that the *gains* of one side are *equalled by the losses* of the other. Any of these three conditions may not apply, leading to *n-person games, infinite games* and *non-zero-sum games*.

Non-zero-sum games. Here the sum of all pay-offs made during a game is not zero, and the players may or may not enter into *coalitions*. Accordingly, a game may be called *cooperative* or *non-cooperative*. Parlour games are normally of the latter type, while economic systems are mostly cooperative (because arrangements will benefit all sides).

For example, a two-person non-zero-sum game can be described by a matrix in which the first figures are the gains of P_1 and the second figures the gains of P_2. Both players have two options each.

P_1 \ P_2	Option 1	Option 2
Option 1	2; 1	−1; −1
Option 2	−1; −1	1; 2

In the *non-cooperative version* P_1 makes no allowance for the interests of P_2 (the figures in second place in the matrix) and sees this as a zero-sum game even though it is not. A graphic solution, for example, will show the value of the game for P_1 (i.e. his average gains per game) to be 1/5, and his optimal strategy $y_0^T = (3/5, 2/5)$.

In a *cooperative game*, on the other hand, the players may agree only to take actions bearing equal numbers. Then the average gains for both will be 3/2 and their optimal strategies $x_0^T = (1/2, 1/2)$ and $y_0^T = (1/2, 1/2)$. This indicates that the cooperative version is the one from which both can expect greater benefits.

***n*-person games.** Among n players, each has a finite number of options but coalitions are prohibited. For each player a separate pay-off matrix A_i has been defined which depends on n arguments (courses of action). One game consists of each player selecting a course of action, followed by payment of the amount A_i to P_i.

These games are normally represented by game trees, and the *value* of the matrix game gives way to the more general concept of the *equilibrium point*. The strategies associated with an equilibrium point are optimal, and if one player abandons the optimal strategy while the others keep to it he may not normally expect larger gains but rather a reduction. The equilibrium point and related strategies therefore illustrate rational behaviour on the part of the players. The following applies: there is at least one equilibrium point in each finite non-cooperative n-person game.

This existence theorem, which is analogous to the minimax theorem, does not lend itself general calculation. Such calculation has so far only been possible in the case of specific three-person and four-person games.

Our *example* is a specific *take-away game* with the following rules: from a total of five matches, P_1 must first draw two or three. P_1, P_2 and P_3 alway take turns, one after another. The game is over when no more matches can be drawn, and the last person to take a match will pay the amount 1 to the previous player.

This game can be solved immediately by drawing up a game tree (Fig. 43-2). The figures given in the tree are the numbers of matches left at any one time. The line in red is the optimal strategy, and a player may expect to gain less if he departs from it. For example, the far left branch indicates that P_2 is the last to draw, and loses. The game is unfair to P_3, who will always lose and have to pay to P_2 if his opponents chooses an optimal strategy.

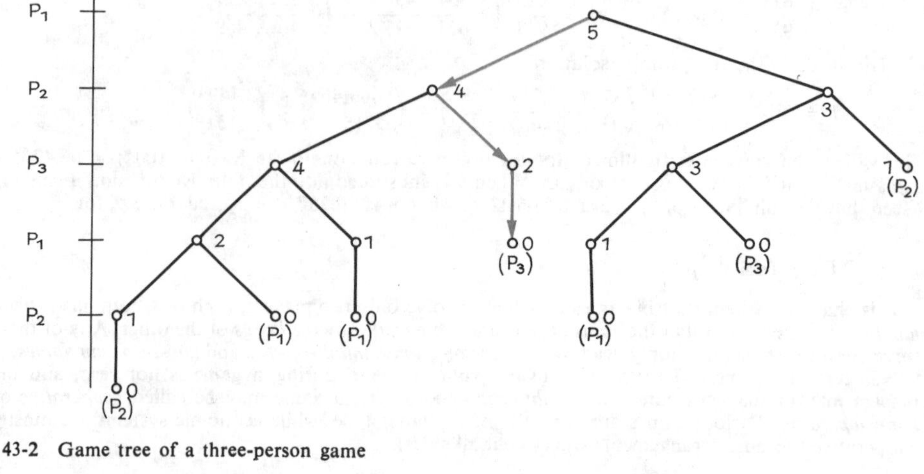

43-2 Game tree of a three-person game

44. Perturbation theory

The perturbation calculation is an approximation method which is widely used in science and technology.

In most cases the solution of an equation cannot be indicated explicitly. When the equation is slightly modified so that the solution can be indicated, this solution generally differs from the solution of the equation originally presented.

The entirety of methods, the calculation of corrections and the substantiation of these calculations is called the perturbation theory.

Example: Consider the equation $x = a + \varepsilon x^2$ for a given real a and ε, $|\varepsilon|$ small. For $\varepsilon = 0$, $x = a$ is the solution of the equation. Try to find the solution x in the form of a power series $x = a + \varepsilon x_1 + \varepsilon^2 x_2 + \dots$ with still indeterminate coefficients $x_1, x_2, \dots$

This power series is substituted into the equation:

$$a + \varepsilon x_1 + \varepsilon^2 x_2 + \dots = a + \varepsilon(a + \varepsilon x_1 + \varepsilon^2 x_2 + \dots)^2.$$

When equating the coefficients at the left side to those coefficients at the right side which are at the same powers of ε, $x_1 = a^2$, $x_2 = 2ax_1 = 2a^3$, ... follows.

Application to integral equations. Consider for a small $|\varepsilon|$ the integral equation

$$y(s) = y_0(s) + \varepsilon \int_0^1 K(s, t)\, y(t)\, dt$$

in the interval (0, 1) with given $y_0(s)$ and given function $K(s, t)$ of two variables s and t, then the solution is $y(s) = y_0(s)$ for $\varepsilon = 0$. The following power series can be considered

$$y(s) = y_0(s) + \varepsilon y_1(s) + \varepsilon^2 y_2(s) + \dots$$

with the functions $y_1, y_2, \dots$ to be determined.

Substitution into the integral equation and comparison of coefficients result in

$$y_1(s) = \int_0^1 K(s, t)\, y_0(t)\, dt, \qquad y_2(s) = \int_0^1 K(s, t)\, y_1(t)\, dt, \dots$$

Example: $y_0(s) = 1$, $K(s, t) = st$, $y_1(s) = s \int_0^1 t\, dt = \frac{1}{2}s$, $y_2(s) = s \int_0^1 t\left(\frac{1}{2}t\right) dt = \frac{s}{2}\int_0^1 t^2\, dt = \frac{s}{6}$.

The perturbation theory is applied to eigenvalue problems.

Eigenvalue problems for matrices. Let A_0 and A_1 be two real and symmetrical matrices with n lines and n columns. For a real ε, $|\varepsilon|$ small, consider $A(\varepsilon) = A_0 + \varepsilon A_1$.

When λ_0 is an eigenvalue of $A(0) = A_0$, the following is true: If the system of equations $(A_0 - \lambda_0 I)\, x = 0$ has exactly m linearly independent solutions, then there are exactly m eigenvectors $x^{(j)}(\varepsilon)$ of $A(\varepsilon)$ and m eigenvalues $\lambda_j(\varepsilon)$, $j = 1, 2, 3 \dots$ with $\lambda_j(0) = \lambda_0$ and

$$x^{(j)}(\varepsilon) \cdot x^{(k)}(\varepsilon) = \begin{cases} 1 & \text{for } j = k \\ 0 & \text{for } j \neq k \end{cases}$$

Furthermore it is true that $x^{(j)}(\varepsilon)$ and $\lambda_j(\varepsilon)$ are convergent power series in ε for $|\varepsilon| < \varepsilon_0$, ε_0 sufficiently small.

Example:

$$A_0 = \begin{pmatrix} 3 & -1 \\ -1 & 3 \end{pmatrix}, \quad A_1 = \begin{pmatrix} 2 & 3 \\ 3 & 2 \end{pmatrix}, \quad \lambda_0 = 2, \quad m = 1.$$

You can start with the power series $\lambda(\varepsilon) = \lambda_0 + \lambda_1 \varepsilon + \lambda_2 \varepsilon^2 + \dots$ and $x(\varepsilon) = x^{(0)} + \varepsilon x^{(1)} + \dots$ with $x^{(0)} = \frac{1}{\sqrt{2}}(1, 1)^T$.

Substitution into $A(\varepsilon)\, x(\varepsilon) = \lambda(\varepsilon)\, x(\varepsilon)$ and comparison of coefficients yield

$$(A_0 - \lambda_0 I)\, x^{(0)} = 0, \quad (A_0 - \lambda_0 I)\, x^{(1)} = -A_1 x^{(0)} + \lambda_1 x^{(0)}, \dots$$

λ_1, i.e. the first correction for λ can be calculated from the second equation. Since λ_0 is eigenvalue the following must be true: $(-A_1x^{(0)} + \lambda_1x^{(0)}) \cdot x^{(0)} = 0$.

From this results $\lambda_1 = A_1x^{(0)} \cdot x^{(0)} = \frac{1}{2}\begin{pmatrix}2 & 3\\3 & 2\end{pmatrix}\begin{pmatrix}1\\1\end{pmatrix} \cdot \begin{pmatrix}1\\1\end{pmatrix} = 5$.

Then $x^{(1)}, \lambda_2, x^{(2)}, \lambda_3, \ldots$ can be calculated successively.

Eigenvalue problems for ordinary differential equations. Oscillations result in eigenvalue problems in ordinary and partial differential equations. Frequently, the eigenvalues cannot be indicated explicitly. The disturbance theory often provides useful approximations.

Consider, for example, the eigenvalue problem $U''(x) + \varepsilon q(x)\,U(x) + \lambda U(x) = 0$ in the interval $(0, \pi)$ with the boundary conditions $u(0) = u(\pi) = 0$ and with the given function $q(x)$, the eigenvalues for $\varepsilon = 0$ are given by $\lambda_n = n^2$, $n = 1, 2, \ldots$ The (normed) eigenfunctions belonging to it are $U_n^{(0)}(x) = \sqrt{(2/\pi)}\sin(\lambda_n x)$.

For real ε, $|\varepsilon|$ is small – the set-ups $\lambda_n(\varepsilon) = \lambda_n + \lambda_n^{(1)}\varepsilon + \lambda_n^{(2)}\varepsilon^2 + \ldots$ and $U_n^{(x)} = U_n^{(0)}(x) + U_n^{(1)}(x)\,\varepsilon + U_n^{(2)}(x)\,\varepsilon^2 + \ldots$ with $\int_0^\pi U_n^2(x)\,dx = 1$ can be made.

For the first correction of the eigenvalue

$$\lambda_n^{(1)} = -\int_0^\pi q(x)[U_n^{(0)}(x)]^2\,dx = -\frac{2}{\pi}\int_0^\pi q(x)\sin^2 nx\,dx$$

is true.

Example: Let $q(x) = x$, then $\lambda_n^{(1)} = -\frac{2}{\pi}\int_0^\pi x\sin^2 nx\,dx = 1 - \pi$.

45. The pocket calculator

History. The forerunners of the pocket calculator were on the one hand the mechanical desk calculators, which were later enhanced with electromechanical functions, and, on the other, the first electronic computers. The use of transistors made it possible in 1962 to manufacture electronic calculators that were no bigger than conventional electromechanical types but which could perform the same operations. In some cases, the number of digits available on the new machines was smaller, but the effect of this was negligible in practice. Soon it was becoming clear that the electronic desk calculators were superior: they are quiet, much faster and soon capable of functions that were outside the range of electromechanical devices. The first such calculators were, however, quite expensive and prone to defects because they consisted of a multitude of components and soldering points.

Around 1970 it became possible to accommodate the inner workings of a calculator on a small and reliable silicon chip which could be mass-produced at a low cost. The result was a *pocket-sized calculator* whose price was steadily reduced in the years that followed, making it an article for mass consumption. Programmable calculators appeared around 1975, and some of the more recent models have functions similar to those of minicomputers. The pocket calculator has now completely superseded mechanical and electromechanical calculators, and the slide rule. Production costs are low, it is extremely handy, fast and quiet, and much more accurate than the slide rule, thus eliminating the need for rough calculation and logarithmic scales.

Types of calculators. There are calculators with *simple* and *extended* functions, others for *economic*, *economic/scientific* and *scientific/technical* calculations, *programmable calculators*, and special-purpose calculators. Simple versions can only perform the four fundamental operations, and this is enough for many users. These functions can be extended by a memory, a very appropriate addition, and basic operations such as percentage calculation ([%] key), squares and roots ([x^2] and [$\sqrt{x}$] keys), and inversion ([1/x] key). Economic calculators have additional facilities for computing interest; scientific calculators can do statistical evaluations (mean values [$\bar{x}$], standard deviation [σ], trend and regression to some extent); and scientific/technical versions incorporate functions for exponential and trigonometric calculations and the inversions thereof: [e^x], [ln x], [lg x], [10^x], [sin x], [cos x], [tan x], [arc sin x], [arc cos x], [arc tan x], [x^y].

From the viewpoint of computing logic there are calculators with *algebraic* and *arithmetic logic*, with and without *brackets*, with and without *hierarchy* and with *reverse Polish notation*. Characteristic of calculators with arithmetic logic is the key [±] (or [+=]), while the key [=] denotes models with algebraic logic. Instead of the = key, versions with inverse Polish notation have one with the symbol [↑] or the word [ENTER] (see below). Bracket keys are marked [[(] and [)]]. Whether a calculator uses a hierarchy or not is not obvious from the keyboard but can be determined with a test calculation, based e.g. upon the fact that $14 = 2 + (3 \cdot 4)$ but $20 = (2 + 3) \cdot 4$.

Example: Testing a calculator for hierarchy behaviour

Input	Output for calculators without hierarchy	Output for calculators with hierarchy
[2] [+] [3] [×] [4] [=]	20	14

Characteristic keys on programmable calculators include [GOTO], [LOAD], [LD], [LEARN], and [PROGR].

Outer appearance. The display has a maximum of 6–12 digits and a number of special characters, particularly signs (minus only) and a decimal point. On the keyboard, the only features that have been standardized are the relative positions of the figure keys [0] to [9] and the key for the decimal point. The keys for entry, operations, functions and cancel, where demarcation is not always consistent, are normally given different colours. Some keys have double/multiple lettering/symbols for more than one function, and the user has to press a *prefix key* (*function key*, often F) to select the particular function which, in the case of programmable calculators, also depends on whether the calculator is in the *loading* or *execution mode*. The only parts accessible to the user *inside* are the batteries or small accumulators.

Entering figures. These and the decimal point are entered from left to right, the negative sign being added afterwards by pressing the [+/−] key. For example, −3.1 is entered as follows: [3] [.] [1] [+/−].

Example: Entering the number −3.1

Input	Display
	0 .
[3]	3 .
[.]	3 .
[1]	3 . 1
[+/−]	−3 . 1

Further input (of figures) is blocked when all the free digits in the display have been taken up. Before entry starts the display may show, instead of 0., the result of an earlier calculation, which is normally cleared as soon as the first figure is entered. However, there may be figures left in invisible *registers* (memory cells), and these could lead to unpleasant surprises in the further course of the calculation. In some cases, all the registers and the display will be in an undefined state when the calculator is switched on. It is therefore advisable to activate the cancel key [C] once or the key [CE–C] twice to clear all registers. Pressing [CE] or [CE–C] once will clear the display and sometimes only the last digit entered. To cancel a wrong entry, press [CE]; to cancel a wrong negative sign activate [+/−] again.

More sophisticated calculators feature *floating-point* representation. Here, $58 \cdot 10^{27}$ will be entered as [5] [8] [EEX] [2] [7], the display showing 5 8 . 2 7 (or, better, 5 8 . $^{2\ 7}$), but the same figure will be displayed in the standard form 5 . 8 2 8 after activation of an operation key (or of [=]).

Example: Entering the number $58 \cdot 10^{27}$

Input	Display
	0 .
[5]	5 .
[8]	5 8 .
[EEX]	5 8 . 0 0
[2]	5 8 . 0 2
[7]	5 8 . 2 7
[=]	5 . 8 2 8

For the number -0.0027 the following or similar input sequences are possible: [0] [.] [0] [0] [2] [7] [+/−] or [.] [0] [0] [2] [7] [+/−] or [2] [7] [+/−] [EEX] [4] [+/−] or [2] [.] [7] [+/−] [EEX] [+/−] [0] [3] .
The number entered will then be stored in the "X register", and display continues to be linked to this throughout the rest of the calculation. However, the figure 0.6666667×10^{-4} can be entered as [.] [6] [6] [6] [6] [6] [6] [7] [EEX] [+/−] [4] or, at least, as [.] [6] [6] [6] [6] [6] [6] [7] [×] [1] [EEX] [+/−] [4] [=] . The readout will then only be 0 . 6 6 6 6 − 0 4 or (standardized) 6 . 6 6 6 6 − 0 5. In the làtter case, the X register contains 0.6666667-4 or 6.666667-05, which can only be displayed in part. Registers often contain other digits which cannot be directly displayed; see also the section on "hidden digits".

Simple operations. To calculate 19 + 85 one has to enter, on most calculators, [1] [9] [+] [8] [5] [=] , for which the readout on the display is 1 0 4. To arrive at this result, the calculator goes through the following steps:

The number 19 is first entered into the X register and then copied into a second register, the Y register, when the operation key + is pressed. This is where the calculator also remembers the addition operation. The X register is now free for the second operand 85 to be entered, and the addition proper is performed when the result key [=] is pressed. The sum, 104, is then contained in the X register and appears on the display. We can now explain what happens if we enter [1] [9] [+] [=]. The result key causes the values in the X and Y registers to be added together and the readout is 3 8. (Some calculators, however, show the result 1 9 for [1] [9] [+] [=] , indicating that here the first operand is not copied into the Y register until the other has been entered.) No uniform results can be expected for the entry [1] [9] [+] [8] [5] [=] [2] [0] [=] , where some calculators only show the 20 entered last because the Y register has been cleared after the addition. Other calculators give the result 39 (= 19 + 20), and in this case the content 19 + of the Y register has been retained after the addition and the second = begins the addition all over again (calculators with automatic constant for the first operand). Still other types (with more functions) keep +85 in the Y register after the first addition, so that [1] [9] [+] [8] [5] [=] [2] [0] [=] gives 1 0 5. This automatic constant for the second operand has an advantage when it comes to division. Computing sequences such as 3:57, 7:57, 11:57 ... which occur frequently in practice can then be entered in the form [3] [÷] [5] [7] [=] [7] [=] [1] [1] [=] ... (with the [÷] key for division and the [×] key for multiplication). In some calculators, however, the automatic constant for the various fundamental operations is not consistent and may sometimes be switched on/off. It is convenient for raising whole numbers to a power: 3^4 is then entered as [3] [×] [=] [=] [=] , the result is 8 1.

Example: Different kinds of automatic constant

Input	X register I	X register II	X register III	Y register I	Y register II	Y register III	Display I	Display II	Display III
[1]	1	1	1				1 .	1 .	1 .
[9]	19	19	19				1 9 .	1 9 .	1 9 .
[+]	19	19	19	19+	19+	19+	1 9 .	1 9 .	1 9 .
[8]	8	8	8	19+	19+	19+	8 .	8 .	8 .
[5]	85	85	85	19+	19+	19+	8 5 .	8 5 .	8 5 .
[=]	104	104	104	0	19+	+85	1 0 4 .	1 0 4 .	1 0 4 .
[2]	2	2	2	0	19+	+85	2 .	2 .	2 .
[0]	20	20	20	0	19+	+85	2 0 .	2 0 .	2 0 .
[=]	20	39	105	0	19+	+85	2 0 .	3 9 .	1 0 5 .

I: calculators without automatic constant,
II: calculators with automatic constant for the 1st operand,
III: calculators with automatic constant for the 2nd operand, with algebraic logic in each case

Calculators with arithmetic logic perform multiplication and division in the same way as models with algebraic logic, but without automatic constant; whereas addition + is coupled with the result key [=] as [±] or [+=]. To add 19 + 85 one has to enter [1] [9] [±] [8] [5] [±]. Each time the [±] key is activated the contents of the X and Y registers will be added into the X register, and result then copied into the Y register. For the difference, the key [−] or [≡] or [−=] has the same effect as the combination [+/−] [±], so that entering [1] [9] [±] [8] [5] [≡] gives the result −66. When [1] [9] [−] [8] [5] [±] is entered the readout on the display will be 66 (−19 + 85 = 66). This makes it clear that we are basically dealing with operations that follow the operands.

For a calculator with arithmetic logic, before the input process starts the Y register must be cleared (press [C] key). For multiplication and division operations must be stored in the Y register, but not for addition and subtraction. [+=] and [−=] trigger several operations in quick succession.

Example: Doing 19 + 85 on a calculator with arithmetic logic

Input	X register	Y register	Display
[1]	1	0	1 .
[9]	19	0	1 9 .
[+=]	19	0	1 9 .
	19	19	1 9 .
[8]	8	19	8 .
[5]	85	19	8 5 .
[−=]	−85	19	8 5 .
	−66	19	− 6 6 .
	−66	−66	− 6 6 .

Calculators with reverse Polish notation consistently use the principle of subsequent operation, and pressing the keys [1] [9] [↑] [8] [5] [+] gives the result 104. Entering the separation sign [↑] (or [ENTER]) copies the entry 19 into the Y register, no operation need be stored, and [+] starts the operation.

Example: Doing 19 + 85 on an calculator with reverse Polish notation

Input	X register	Y register	Display
[1]	1		1 .
[9]	19		1 9 .
[↑]	19	19	1 9 .
[8]	8	19	8 .
[5]	85	19	8 5 .
[+]	104		1 0 4 .

Hidden digits, overflow. Press [9] [÷] [7] [=]. The accurate result is 1.285714. An eight-digit calculator will show this as 1 . 2 8 5 7 1 4 2 or 1 . 2 8 5 7 1 4 3. In the first case the other digits were cut out and in the second rounding up took place. Now enter [−] [1] [.] [2] [8] [5] [7] [1] [4] [2] [=]. Either the result will be zero, if the calculator has no hidden digits, or else one or two more digits will be shown, e.g. 9 . − 0 8, in which case the calculator uses 9 digits internally instead of the 8 shown, the 9th being correctly rounded up. In other cases where the display might be 8 . − 0 8, for example, the 10th digit has been cut. But even if the 9th digit is correctly rounded up in the division 9:7, the calculation $1 + 9 \cdot 10^{-9} - 1$ (entry [1] [+] [9] [EEX] [9] [+/−] [−] [1] [=]) sometimes does not give the rounded up result of zero. In other models, a hidden 10th digit can be coaxed out with display 9 . − 0 9, while the intermediate pressing of the result key (entry [1] [+] [9] [EEX] [9] [+/−] [=] [−] [1] [=]) gives the rounded up readout 1 . − 0 8.

When the largest possible number is entered, e.g. 9999999 or $99999 \cdot 10^{99}$ and multiplied by 2, the calculator will signal an error (E, ERROR, several decimal points, flashing display etc.) or show the first digits of the result (1 9 9 9 9 9 9 9) with or withour signalling an error (e.g. missing decimal point). For a downward overflow, e.g. [1] [÷] (largest number) [=] [÷] (largest number) [=] the calculator will also show an error or, correctly rounded up, zero. The second alternative tells us that, just as in the case of an upward overflow, one cannot have blind faith in pocket calculators, and a check is in order from time to time.

Percent key. The principal function here is that pressing the [%] key causes the amount in the X register to be divided by 100, and one may proceed as follows:
[6][5][0][×][3][%][=] , where the readout is 1 9 . 5 which is 3% of 650, or
[6][5][0][−][3][%][=] , where the readout is 2 1 6 6 6 . 6 6 7 which means that 100% is 3%, or
[3][÷][6][5][0][%][=] , where the readout is 4 . 6 1 5 3 − 0 1 which explains how much 3 is in relation to 650.

Example: How much is 3 of 650 in percent?

Input	X register	Y register	Display
[3]	3		3 .
[+]	3	3+	3 .
[6]	6	3+	6 .
[5]	65	3+	6 5 .
[0]	650	3+	6 5 0 .
[%]	6 . 5	3+	6 . 5
[=]	0,461538462	3+	4 . 6 1 5 3 − 0 1

Some calculators also proceed as follows: [6][5][0] [+] [3] [%] [=] with the readout 6 6 9 . 5, which is 103% of 650, or
[6][5][0] [−] [3] [%] [=] with the result 6 3 0 . 5, which is 97% of 650.

Example: A way to calculate 97% of 650

Input	X register	Y register	Display
[6]	6		6 .
[5]	65		6 5 .
[0]	650		6 5 0 .
[−]	650	650−	6 5 0 .
[3]	3	650−	3 .
[%]	19.5	650−	1 9 5 .
[=]	630.5		6 3 0 . 5

Multiple operations. Let us enter 2 · 3 + 4. It will strike the non-expert as quite unusual to do this in reverse Polish notation, but it is indeed very logical to press [2][↑][3] [×] [4] [+] . The [4] need not be preceded by a [↑]. By entering [4] after the operation [×] the intermediate result 6 is to the Y register at the same time transfered. After [2] the [↑] is needed as a separation sign to avoid entering 23.

Example: Calculating 2 · 3 + 4 on a calculator with reverse Polish notation

Input	X register	Y register	Display
[2]	2		2 .
[↑]	2	2	2 .
[3]	3	2	3 .
[×]	6		6 .
[4]	4	6	4 .
[+]	10		1 0 .

Almost all calculators with algebraic logic give a correct result when [2] [×] [3] [=] [+] [4] [=] is entered, and the second [=] can be left out in most cases. Entering [+] leads both to calculation of the intermediate result $2 \cdot 3 = 6$, and storage of 6+ in the Y register. This is known as the *short-cut technique*.

Example: Calculating $2 \cdot 3 + 4$ on a calculator with algebraic logic

Input	X register	Y register	Display
[2]	2		2 .
[×]	2	2×	2 .
[3]	3	2×	3 .
[+]	6 6	 6+	6 . 6 .
[4]	4	6+	4 .
[=]	10		1 0 .

The picture is less clear when we enter [2] [+] [3] [×] [4]. In this case, most algebraic calculators with short-cut facilities will show the intermediate result 5 after [×] is pressed, and the final result 20. This means that the calculator is performing the operation $(2 + 3 \cdot 4)$. This can be avoided by entering [3] [×] [4] [+] [2] [=] ; the correct result is then 14. Some calculators use brackets, and one can enter [2] [+] [[(] [3] [×] [4] [)]] [=] . By pressing the bracket key [[(], the contents of the Y register are transferred to a third, "Z" register. The X and Y registers are then free for transfer operations (display shows 0. or the X register is cleared by entering 3). After entering [3] [×] [4] there is 3 in the Y register and 4 in the X register. When the bracket key is pressed the intermediate result $3 \cdot 4$ is calculated and displayed in the X register, then the 2+ stored in the Z register is transferred back to the Y register. Entering [=] begins the addition $2 + 12$. The number of bracket nestings is not standardized, and there may often be 2, 7 or 15.

Example: Calculation of $2 + 3 \cdot 4$ using a calculator with algebraic logic and brackets

Input	X register	Y register	Z register	Display
[2]	2			2 .
[+]	2	2+		2 .
[[(]	2 oder 0	2+	2+	2 . oder 0 .
[3]	3		2+	3 .
[×]	3	3	2+	3 .
[4]	4	3	2+	4 .
[)]]	12	2+		1 2 .
[=]	14			1 4 .

Similar things happen in hierarchic calculators when [2] [+] [3] [×] [4] [=] is entered. As soon as [×] is pressed the calculator gives it a higher rank than the [+] that is stored in the Y register. This causes 2+ to be transferred to the Z register while 3× goes into the Y register. The result key first connects the Y and X registers, followed by the transfer of the Z register contents to the Y register and then by connection of the X and Y registers again. In some cases the Z register is only used for addition and subtraction and the Y register just for multiplication and division.

Example: Calculating $2 + 3 \cdot 4$ using a hierarchic calculator

Input	X register	Y register	Z register	Display
[2]	2			2 .
[+]	2	2+		2 .
[3]	3	2+		3 .
[×]	3	3×	2+	3 .
[4]	4	3×	2+	4 .
[=]	12 14	2+		1 4 .

For inverse Polish notation the entry is [2][↑][3][↑][4][×] [+] . Each [↑] not only copies the X register into the Y register but also causes the contents that have accumulated in the Y register to be transferred to the Z register, and the contents of any following registers to be moved the next one, so that whatever was in the previous register is now lost. When [×] is entered, the product is written into the X register, and at the same time the contents of the Z register are transferred back into the Y register. The same happens with any following registers, the contents of the last one being retained. The procedure is similar for [+] .

Example: Calculating $2 + 3 \cdot 4$ using a *calculator with reverse Polish notation.* Where a fourth register is not available, $y' = z' = 2$ applies; for a calculator with exactly four registers, $y' = y = z'$ applies, and for one with at least five registers, $y' = y$ and $z' = z$.

Input	X register	Y register	Z register	Display
[2]	2	y	z	2 .
[↑]	2	2	y	2 .
[3]	3	2	y	3 .
[↑]	3	3	2	3 .
[4]	4	3	2	4 .
[×]	12	2	y'	1 2 .
[+]	14	y'	z'	1 4 .

An algebraic calculator of the hierarchic type or with brackets will easily solve problems of the kind $2 \cdot 3 + 4 \cdot 5$, the entry in the former case being [2] [×] [3] [+] [4] [×] [5] [=] and, in the latter, [2] [×] [3] [+] [[(] [4] [×] [5] [)]] = ; for a calculator with reverse Polish notation the entry would be [2] [↑] [3] [×] [4] [↑] [5] [×] [+] while other calculators would need a memory.

Problems of the kind $(2 + 3) \cdot (4 + 5)$ are insoluble even for hierarchic calculators unless they have a memory or additional brackets, but present no difficulties for calculators with brackets or reverse Polish notation.

It may therefore be said that the least amount of rethinking occurs in hierarchic calculators with all other functions equal (i.e., entry is as required by the problem, the minus for negative numbers follows after); whereas other types have greater capacities, and models with reverse Polish notation involve the smallest number of keys.

Memory. [X → M] or simply [M] will write the contents of the X register into the memory, while [RX] (read X) writes the contents of the memory back into the X register without clearing the memory. The contents of the memory often displayed if they do not equal zero. Some calculators have a key [X ↔ M] which causes the contents of the X register and the memory to be exchanged. Pressing [M+] or [M−] either adds the contents of the X register to the memory or substracts them from it. [CM] clears the memory. Where this key is absent an alternative is [O] [M] , but this will also clear the X register. The problem $2 \cdot 3 + 4 \cdot 5 + 6 \cdot 7$ can be solved by pressing

[CM] [2] [×] [3] [=] [M+] [4] [×] [5] [=] [M+] [6] [×] [7] [=] [M+] [RM] .

In the absence of M+ one can proceed as follows:

[2] [×] [3] [=] [M] [4] [×] [5] [+] [RM] [=] [M] [6] [×] [7] [+] [RM] [=] .

Example: Calculating $2 \cdot 3 + 4 \cdot 5 + 6 \cdot 7$ using M+ − key

Input	X register	Y register	Memory	Display
CM			0	
2	2		0	2 .
×	2	2×	0	2 .
3	3	2×	0	3 .
=	6		0	6 .
M+	6		6	6 . M
4	4		6	4 . M
×	4	4×	6	4 . M
5	5	4×	6	5 . M
=	20		6	2 0 . M
M+	20		26	2 0 . M
6	6		26	6 . M
×	6	6×	26	6 . M
7	7	6×	26	7 . M
=	42		26	4 2 . M
M+	42		68	4 2 . M
RM	68		68	6 8 . M

Example: Calculating $2 \cdot 3 + 4 \cdot 5 + 6 \cdot 7$ without M+ key

Input	X register	Y register	Memory	Display
2	2			2 .
×	2	2×		2 .
3	3	2×		3 .
=	6			6 .
M	6		6	6 . ?
4	4		6	4 . M
×	4	4×	6	4 . M
5	5	4×	6	5 . M
+	20	20+	6	2 0 . M
RM	6	20+	6	6 . M
=	26		6	2 6 . M
M	26		26	2 6 . M
6	6		26	6 . M
×	6	6×	26	6 . M
7	7	6×	26	7 . M
+	42	42+	26	4 2 . M
RM	26	42+	26	2 6 . M
=	68		26	6 8 . M

Simple functions. When function keys such as [+/−], [1/x], [x^2] are pressed (possibly preceded by actuating the prefix key F), then normally only the contents of the X register will change; the operands in the Y register are retained, thus eliminating the need for a memory. This takes precedence over the small saving that would occur in the number of key operations. Pressing [1/x] repeatedly is a way of checking the repetitive accuracy of the calculator, for the numbers must not drift away. The key [1/x] is useful for calculations of the type $\frac{1}{1/2 + 1/3 + 1/4}$, which can be performed by pressing

[2][1/x] [+] [3][1/x] [+] [4][1/x] [=] [1/x].

Please note that the arguments must appear in the display, either as a result or as an entry, before the function keys can be actuated. If the entry is [2][+] [1/x][3][=], the result will be 5 instead of the 2.3333333 you may have expected.

Example: Calculator with algebraic logic, wrong entry

Entry	X register	Y register	Display
[2]	2		2.
[+]	2	2+	2.
[1/x]	0.5	2+	0.5
[3]	3	2+	3.
[=]	5		5.

This tells us that all calculators use reverse Polish notation as far as functions are concerned. A key which triggers off a nullary function, in the mathematical sense, is [π], which brings π into the X register. If there is a key [arc tan x], than π can also be calculated by pressing [4][×][1] [arc tan x][=]. The subtraction of the "two" π tells us something about the accuracy of the calculator.

Other functions. Scientific calculators are also equipped with the root function, trigonometric functions, exponential functions, the inverse functions thereof and, possibly, hyperbolic functions, conversion functions for polar and spherical coordinates, and statistical functions. When these function keys are pressed, fixed stored programmes are operated for the approximate calculation of the function values, and one has to wait one or two seconds for the result. For example, if the entry is [π][√x][x^2] [−] [π] [=], there will be a remainder which to some extent indicates the accuracy of the root programme stored in the calculator.

Example: Test for the accuracy of calculators root program

Input	X register	Y register	Display
[π]	3.14159265		3.1415927
[√x]	1.77245384		1.7724538
[x^2]	3.14159261		3.1415926
[−]	3.14159261	3.14159261 −	3.1415926
[π]	3.14159265	3.14159261 −	3.1415927
[=]	0.00000004		− 4. − 0 8

This example also shows that the last digits indicated are not always reliable. Nevertheless, a test of two calculators solving the problem $e^{2\pi}$ ([2][×][π][=][e^x]) gave the results of 535.49164 and 535.49165 respectively, which is surprisingly accurate when one considers that the value found in a table is 535.491656. The error is greater, however, when a comparison is made between sin 0.0005 and cos ($\pi/2 - 0.0005$), which is the same from the mathematical point of view but which gives the results $4.9999999 \cdot 10^{-4}$ and $5.003681 \cdot 10^{-4}$, and $4.9999998 \cdot 10^{-4}$ and $5.000019 \cdot 10^{-4}$ respectively. But this would be a rather extreme case of course.

When normal levels of accuracy are required, the calculator completely surpasses the conventional tables, and interpolation is no longer necessary.

A particular feature is the [y^x] key which is used to calculate $e^x \cdot \ln y$ so that an error is indicated when y is negative. Some calculators have an [x^y] key instead (which could simply be the same, or else which requires either an exchange in the argument entry *or* actuation of the register exchange key [$x \leftrightarrow y$]). In some (but not all) hierarchic calculators, raising to a power takes precedence over multiplication and division, and these models have (at least) four registers.

Example: Calculation of $5 + 4 \cdot 3^2$ using a calculator with double hierarchy

Entry	X register	Y register	Z register	T register	Display
[5]	5				5.
[+]	5	5+			5.
[4]	4	5+			4.
[×]	4	4×	5+		4.
[3]	3	4×	5+		3.
[y^x]	3	3˙	4×	5+	3.
[2]	2,	3˙	4×	5+	2.
[=]	9	4×	5+		
	36	5+			
	41				41.

The unit of measurement is important for trigonometric functions and their inversions, and most calculators have a slide control for DEG (degree)/RAD (radian measure) and, possibly, GRD (grade). To convert these into one another, one can use a simple trick. If, for example, 40° is to be converted into radian measure, one can enter, with the slide control on DEG, [4][0][sin x], then set the slide to RAD and enter arc sin x (also known as [$\sin^{-1} x$]), and the result will be 0.698. An alternative is, of course, [4][0][÷][1][8][0][×][π][=].

All functions make it necessary to enter the argument first, possibly followed by pressing a prefix key and finally the function key. As a rule, the [=] key need not be actuated in these calculations.

Statistical calculations are possible with some calculators, using the memory and the Z and T registers. When an sequence of numbers is entered, separated by [x], one of the three memories stores the number of entries (which can be recalled with [n]), the second stores the sum (this can be recalled with [Σx]), and the third contains the quadratic sum ([Σx^2]). One can then recall the average $\Sigma \frac{x}{n}$ with $\bar{x}$, and the standard deviations $\sqrt{\Sigma \frac{(x-\bar{x})^2}{n-1}}$ and $\sqrt{\Sigma \frac{(x-\bar{x})^2}{n}}$ with [σ_{n-1}] and [σ_n], and sometimes with [σ^2] the square of one of those numbers too.

Several addressable memories. Where several memories are present the contents of the X register, for example, will be stored in memory 4 by pressing [STO][4]; with [RCL] (recall) [4] the number can be returned to the X register. Several stores are useful in such application to statistical evaluation When a sequence of numbers such as $a_1, a_2, \ldots, a_n, \ldots$ occurs one can store the partial sums $\sum_{i=1}^{n} a_i$ in memory 1, the related square sums $\sum_{i=1}^{n} a_i^2$ in memory 2, and possibly count the summands in memory 3 (n); all three memories are required to calculate the scatter for the sequence of numbers.

Sometimes a computing plan is useful, as is shown here for solving the quadratic equation $x^2 + px + q = 0$.

In this way one can establish small programme libraries which will be particularly useful for programmable calculators.

Example: Plan for solving a quadratic equation $x^2 + px + q = 0$

Step	Operation	Remarks
1	Enter p	
2	[STO] [1]	
3	Enter q	
4	[STO] [2]	
5	[RCL] [1] [÷] [2] [=] [+/−]	$-p/2$
6	[STO] [3]	
7	[x^2] [−] [RCL] [2] [=] [$\sqrt{x}$]	Calculators with several memories have root functions
8	[STO] [4]	
9	[+] [RCL] [3] [=]	First solution
10	[RCL] [4] [+/−] [+] [RCL] [3] [=]	Second solution

Programmable calculators. These have an additional programme memory with a capacity ranging from about 30 to more than 1,000 steps. For example, the abovementioned programme for calculating the roots of quadratic equations can be as shown in the following. It is assumed that the coefficients p and q are in the memories 1 and 2.

Example: A program for solving a quadradic equation

	[LOAD]	Transition to programming mode
	[START]	Set instruction counter to zero
00	[RCL] [1]	The individual instructions are accompanied in the display by a check computation or a code number for the key or combination of keys that has just been depressed, and by a serial number shown left here)
01	[+]	
02	[2]	
03	[=]	
04	[+/−]	
05	[STO] [3]	
06	[x^2]	
07	[−]	
08	[RCL] [2]	
09	[=]	
10	[×]	
11	[STO] [4]	
12	[+]	
13	[RCL] [3]	
14	[=]	

15	[HALT]	To display the value for x_1
16	[RCL] [4]	
17	[+/−]	
18	[+]	
19	[RCL] [3]	
20	[=]	
21	[STOP]	The value for x_2 is indicated

If you want to run the programme, make sure that the values for p and q are in memories 1 and 2. (Alternatively, one can use the [HALT] and [STOP] instructions to schedule a programme so that the calculator stops for the entry of certain values and then proceeds to store these automatically after a certain period or when the [RUN] key is actuated). If this is the case, press [START] (to set the instruction counter from 22 to 00) and then [RUN]. The calculator will then run as if the user had pressed the keys numbered 00 to 14. It will then stop so that x_1 can be read off. When [RUN] is pressed again, the programme continues through steps 16–20. At that stage the calculator comes to a final stop, and the value for x_2 can be read off. In our example, the instruction counter is reset to 00 after [STOP] but not after [HALT], so that pressing [RUN] again would repeat the calculation. Prior to that, one can change p and q in memories 1 and 2.

It will now be clear that a small programme such as this consisting of 22 steps utilizes a considerable part of the programme store in a small programmable calculator. It is therefore desirable for the programme store to have a minimum of 100 locations. Our programme will be slightly shorter if we use reverse Polish notation. Not all calculators will indicate situations where the capacity of the programme store has been exceeded before overwriting it from the beginning.

It is important to check the accuracy of the programme entered and to test it thoroughly before use. For this purpose one can press the [SST] key (single step) repeatedly to run the programme sections individually, which will be particularly necessary in cases where a faulty programme would run endlessly. One can then interrupt the programme with [STOP] and find the defect with [SST].

In the course of time, the user will establish a small programme library, and the fact that switching the calculator off erases all programme and data stores is then all the more regrettable. It means that programme and data in question must once again be entered and tested the next time. In the case of longer programme, this is laborious, and errors cannot always be avoided. Some calculators therefore have hold circuits, which preserve the contents of programme and data stores for a certain length of time (possibly until the batteries need changing). Other models can print out the programme and data on small permanent magnetic cards, from which they can be read in again at a later date without any errors or loss of time. In some cases cassette recorders can be connected as permanent memories.

In the above programme, that which is stored is merely a sequence of key operations, the only true programme instructions being [HALT] and [STOP]. In most programmable calculators one needs a greater diversity of contents and a larger capacity. For example, such commands as "if $x \geqslant 0$, go to nn, otherwise carry out the next instruction".

Let us consider an example where a certain amount of money (debts) yields 4.5% interest and is repaid at a rate of 1.5% per year plus the saving in interest payments. The question could then be, how many years are needed to repay the debt. The result (32 years) appears in the X register at the end of the operation and is shown in the display. The amount of money initially owed is irrelevant and is therefore assumed to be 1.

Often jumps to subprogrammes are possible. These are jumps to particular programme sections together with jumps back to the next instruction after "jump to subprogramme". The calculator must therefore remember the number of this follow-up instruction: the return address.

From the viewpoint of computer science, the programming of programmable calculators is carried out in a code that is very close to machine language. There are also calculators with facilities for programming in a high-level language (BASIC), but these are closer to microcomputers under functional aspects and are therefore not dealt with here separately.

Example: A program to calculate the time of repayment for a debt

	[LOAD]	
	[START]	
00	[1]	Initial sum
01	[STO][1]	The money owed at a particular time is in memory 1
02	[x]	
03	[.]	6% of the initial debt = annual repayment (annuity)
04	[0]	
05	[6]	
06	[=]	
07	[STO][2]	Annuity in memory 2
08	[0]	
09	[STO][3]	Annual counter
10	[RCL][1]	Debts to X register
11	[$x \leq 0$]	Test to see if contents of the X register (remaining debt) are still positive
12	[3][0]	If so, i.e. $x \leqslant 0$, continue with step 30 (result output), otherwise proceed to the next step (13). The destination address 30 cannot be inserted until the rest of the program is complete, or a large enough number is selected to be on the safe side (this is suitable also for modifications).
13	[x]	
14	[1]	
15	[.]	
16	[.]	
17	[4]	
18	[5]	Debt multiplied by 1.045 = debt plus interest
19	[−]	
20	[RCL][2]	
21	[=]	Debt plus interest minus annuity = new debt in memory 1
22	[STO][1]	
23	[RCL][3]	
24	[+]	
25	[1]	
26	[=]	
27	[STO][3]	Annual counter set one year forward

28	[GOTO]	Unconditional (return) jump
29	[1][1]	to step 11 to test and select program cycle repeat or result output
30	[RCL][3]	Result output
31	[STOP]	
	[START] [RUN]	Program running, final display: 32

In some calculators the steps 23 to 27 can be summarized.

46. Microcomputers

Historical background. The microcomputer, on the one hand, is a logical step in the development of microelectronics which has already given us pocket calculators, particularly of the programmable type and, on the other hand, embodies a variety of concepts and system architectures as realized especially in third generation computers.

The first integrated circuit (one transistor, three resistors, one capacitor on one germanium chip) was put together in a laboratory in 1958 and commercial sales started in 1962 (with eight transistors on one silicon chip). In 1971, one chip carrying 2,300 transistors represented the first central processing unit of a functional if not very versatile computer. The 4-bit microprocessor was born in which a group of 4 bits (one tetrad) is the smallest information unit that is processed at a time. Hence the frequently used classification as n-bit computers (with $n = 8, 16, 32$). As the integration of circuits increased (reaching 8,000 to 10,000 transistors per chip after 1974), both the capacity and functions of computers grew at a rapid pace.

All the concepts developed up to the third generation were again applied to these much smaller systems, and new technical features were incorporated.

In 1975 the first set of components became available for a small home-built computer, followed in 1977 by the first complete microcomputer. The decade thereafter saw an unprecedented massive spread of these systems amounting to a genuine scientific and technical revolution. No longer need one carry a problem (after preparation) to a computer (that is not always accessible in a computer centre where it requires special attendance), but the computer (with all its accessories) can be "close to the problem" because

- systems are small (500 g to 5 kg for desk top models),
- have a low power consumption ($\leqslant$ 25 W),
- work at ever-increasing speeds,
- continue to go down in price, and
- are extremely versatile (in conjunction with a variety of electronic assemblies).

Computer systems have penetrated all spheres of daily life, on and off the job. The microcomputers of our time defy an exact classification as to design or performance, leaving only the broad criteria of use as home computers and personal computers.

Construction – This is a useful way of differentiating between systems.

The **pocket computer** is a direct extension of the programmable pocket calculator but no longer key-programmed. It uses a higher-level programming language instead (which is often BASIC), is portable and receives its power from different types of batteries.

The **video computer** works in conjunction with other home appliances (such as TV sets and cassette recorders) and (in the simplest configuration) consists of the computer proper and a keyboard (similar to that of a typewriter). It is mains-connected and programmable in at least one programming language.

The **personal computer** (workplace/office computer) is an autonomous system and has its own screen, auxiliary stores (often with floppy-disk connection), printer etc. for a growing range of professional uses.

Video and personal computers are both expandable and configurable for use with many other devices such as memory extension assemblies, joysticks, light pens, bar code readers, interfaces for process signals, voice input and output devices, telephone transmission facilities etc.

Setup and functions. Let us consider a minimum configuration consisting of a computer, keyboard, cassette recorder and TV set with the following functions:

- through the keyboard the user can communicate with the computer and load a program/data (for the first time) or enter control statements;
- programs (stored/to be processed) can be shown on the TV screen, input data can be checked and results (output data) displayed; in dialogue-oriented programs (mostly games) the screen is a work area; most systems also have an acoustic output (a beep in the simple versions but also sound patterns from built-in synthesizers heard over the TV loudspeaker);
- the cassette recorder permits to input programs which are stored on (commercial) cassettes and to output programs and data from the memory on cassettes; these may be vital for the operation of the system (software), as in an operating system. They may also serve as an interpreter for a higher-level programming language or as application programs assisting the user in learning, playing, etc;
- the computer as such controls the system and all information processing.

These components are physically there and represent the hardware of the system.

Once a decision has been made for a model to be purchased one must (despite all standardization) make sure that the parts of the system are compatible especially if existing devices (TV set, recorder) are to be included in the configuration. This saves time and money. It is equally important to be aware of the demands to be made on the system that is being set up (purchased), and whether it is sufficiently flexible (and reasonably priced) for extension. Basic components are often quite cheap but adding on may then become much more expensive than a higher initial investment. It is recommendable to get as much advice as possible before making a start. When all components for the system have been acquired, read the instructions for mounting and installation carefully. Many disappointments can be avoided with an accurate and planned approach (minimum distance between the TV set and cassette recorder, switching-on sequence, etc.). Plug-and-socket connections can be marked, free-hanging cables should be avoïded (at home). Give the equipment the attendance and maintenance it needs (clean contacts, keyboards etc.).

Whether you use "only" prepared software (programs) which is fully sufficient for many purposes, or whether you write your own programs (which, of course, can be quite fascinating), observe the logic of interaction between your key components.

Even in the minimum configuration, the hardware components can come into four functional categories:

a) the central processor with ROM and RAM,
b) input devices,
c) output devices,
d) auxiliary memories.

The interaction of your hardware, all types of input and output and the processing of data (characters) in the computer are controlled by software using a variety of programs for different jobs.

The operating system is a key element of your software and absolutely necessary for operating the system regardless of the user's intentions. At the heart of it is a ROM (*R*ead *O*nly *M*emory) which controls the system after start-up.

Using a RESET function (a key in most cases), the user can always return to this point and start anew in a defined manner. At the worst, if all keys and functions fail and the computer has "crashed", switch it off and on again. These crashes may be caused by faulty machine programs.

It is absolutely necessary to know the functions of the operating system. In many cases (in the simpler configurations) the whole system is accomodated in the ROM. More comfortable setups (especially where floppy-disk systems are connected) have facilities for loading other parts of the operating system into the RAM (*R*andom-*A*ccess *M*emory).

All the information in the ROM is preserved when the computer is switched off and remains available. It can not, however, be modified by the user. The RAM, on the other hand, is a read-write memory to be filled with programs and data (sometimes automatically) when the system has been started. With instructions at system level the user can expand the capacity of his system. In the simplest case, when using application software, the respective programs can be loaded into the main data memory (LOAD), started (RUN) and filed in external memories after completion (or possible modifications) (SAVE). The system core often has programs to check the correct function of the system (AUTOTEST). The more efficient and comfortable the selected computer system, the more comprehensive and sophisticated its operating system. The recommendation can only be repeated here that a close study should be made of the available functions. One can then use prepared software such as programs for games, writing systems for daily correspondence etc.

The efficiency of a system increases with a printer (available in many types) and the connection of floppy-disk systems. Floppy disks are flexible plastic disks with a magnetic coating permanently encased in a plastic envelope which give random access to large amounts of information.

By connecting printers and floppy-disk systems the transition is made to office and personal computers. Even though the user need not concern himself with the internal functions of a mini-computer system and the way the hardware operates, it is worthwhile picturing the arrangement of the RAM.

It consists of a systematic sequence of memory cells numbered from 0 to N-1, mostly with a width of 8 bits and then called *bytes.*

0. byte								1. byte									(N-2). byte			(N-1). byte		
7	6	5	4	3	2	1	0	7	6	5	4	3	2	1	0	....	7	....	0	7	....	0

Often the bits within a byte are numbered from 0 to 7 from right to left. Since each bit can only by 0 or 1, exactly $2^8 = 256$ different binary combinations are possible for the contents of a byte.

If the 2^i is assigned to the digit i of a byte, then its contents can be directly interpreted as a binary number.

Quite often the content of a byte is interpreted as two consecutive hexadecimal digits by combining 4 bits into a tetrad and assigning the value 16^0 to the right tetrad and the value 16^1 to the left tetrad.

Example: Representation as decimal number of the byte contents read as binary number:

Binary number	Decimal number
0000 0000	0
0000 0001	1
0000 0010	2
0000 0011	3
0000 0100	4
.	
.	
.	
0000 1111	15
.	
.	
.	
1111 0000	240
1111 0001	241
.	
.	
.	
1111 1111	255

Binary number	Hexadecimal representation		Conversion	Decimal number
0000 0000	0	0	$0.16^1 + 0.16^0$	0
0101 0001	5	1	$5.16^1 + 1.16^0$	81
1001 1001	9	9	$9.16^1 + 9.16^0$	153

Since the numbers 10, 11 ... 15 can also be represented by 4 bits, six additional digits are required for the hexadecimal representation which are assigned in the following way:

Tetrad	Hexadecimal digit	Decimal number
1010	A	10
1011	B	11
1100	C	12
1101	D	13
1110	E	14
1111	F	15

Example: The hexadecimal number

$$EB = 14 \cdot 16^1 + 11 \cdot 16^0 = 1 \cdot 2^7 + 1 \cdot 2^6 + 1 \cdot 2^5 + 0 \cdot 2^4 + 1 \cdot 2^3 + 0 \cdot 2^2 + 1 \cdot 2^1 + 1 \cdot 2^0 = 235$$

corresponds to the binary number 1110 1011.

From these considerations it can be seen that it does not matter at all whether the binary number is directly converted into the appertaining decimal equivalent as sum of binary powers or split up into two tetrads and the conversion performed hexadecimally. Quite often the hexadecimal representation is only a short and easy description of the binary contents of a byte without special numerical references.

Example: The representations 38, AB, 7C describe the byte contents

0011	1000	1010	1011	0111	1100
3	8	A	B	7	C

This representation is frequently used.

The interpretation of the contents of a byte as character is of fundamental importance. This assignment is called code and in most cases ASCII (*A*merican *S*tandard *C*ode for *I*nformation *I*nterchange) is used in microcomputers, see Table.

Coding to ASCII

Hexadecimal code	Character	Hexadecimal code	Character	Hexadecimal code	Character
20	Space	5A	Z	4B	K
21	!	5B	[	4C	L
22	"	5C	\	4D	M
23	#	5D	]	4E	N
24	$	5E	^	4F	O
25	%	5F	_	50	P
26	&	60	‘	51	Q
27	’	61	a	52	R
28	(	62	b	53	S
29	)	63	c	69	i
2A	*	64	d	6A	j
2B	+	65	e	6B	k
2C	,	66	f	6C	l
2D	—	67	g	6D	m
2E	.	68	h	6E	n
2F	/	3A	:	6F	o
30	0	3B	;	70	p
31	1	3C	<	71	q
32	2	3D	=	72	r
33	3	3E	>	73	s
34	4	3F	?	74	t
35	5	40	@	75	u
36	6	41	A	76	v
37	7	42	B	77	w
38	8	43	C	78	x
39	9	44	D	79	y
54	T	45	E	7A	z
55	U	46	F	7B	{
56	V	47	G	7C	|
57	W	48	H	7D	}
58	X	49	I	7E	~
59	Y	4A	J		

Unprintable control characters (such as cursor motions on the screen) which are not uniform on microcomputers and which often vary from the ASCII Standard correspond to the hexadecimal values 00 to 1F. The values 7F to FF are frequently used for coding graphic symbols and other control characters (e.g. for the colours).

With a code of this kind, the basis for text processing has been created which is amazing for the first moment. The input, handling, storage, and output of texts is a handling of bytes and byte contents inside the computer. At the same time, it can be seen that the representation and processing of numbers is a special part of this possibility of processing characters.

These considerations also show how efficient the assemblies must be which are arranged between keyboard and memory or between memory and printer. Operation of a certain key must cause the transfer of the corresponding byte contents, the availability of byte contents, the drive of a certain print character etc.

Memory sizes are indicated in KBytes (i.e. Kilo-Bytes):

1 KByte = 4 pages = 1024 bytes,
1 page = 256 bytes,
1 byte = 8 bits,
1 word = 2 bytes = 16 bits.

When 2 bytes (16 bits) are used for numbering (addressing) a memory, 64 KBytes can be directly addressed. For example, the capacity of a floppy-disk is about 640 KBytes.

Programming of microcomputers. The final form of representing a program to be read by the computer is the *machine language*. Instruction formats, which are exactly determined and typical of the respective computer, show which instruction has to be executed, viz., which operations are applied to which operands. The set of all available instructions is the instruction list.

Each instruction contains

Operation code	Operand part

Both the operation code and the operand part are sequences of bits and occupy together a certain number of bytes.

The first possibility of programming is to immediately enter the bit combinations required into the corresponding memory cells. This type of programming is very expensive, error-prone and hardly used at present (not at all by beginners).

The socalled *assembly languages* are of greater use. The operations are coded by mnemonic operation codes, addresses can be represented by symbols; moreover, auxiliary means for program organization (use of constants, fixing of start addresses, definition of memory areas) are specified.

Example: A part of the program in a possible assembly language:

```
LDA OP1
MOV B, A
LDA OP2
ADD B
JZ ZERO
.
.
.
ZERO: ...
```

An operand OP1 is loaded into an arithmetic register (accumulator A), and then moved into a register B (MOVE !). Another operand OP2 is put into the accumulator and a summation of register B and accumulator is made (ADD !). When the result is zero, the instruction is continued with the address ZERO (JUMP ZERO) etc.

This language level also requires extensive knowledge of the function and design of the hardware. This type of programming will (still) be applied when time-critical sequences must be performed having optimal design. Higher-level programming languages frequently offer the possibility to include parts which have been set up in assembler language.

The fact that conversion into the machine language of a program written in the assembly language is automatically executed by the system is importent for the user. For this purpose, a translation program (compiler, in this case assembler) which transforms the symbolic representation into the machine language should be available in the operating system or at least loadable. This is done before the actual running of the program. If a program is required more frequently, it could be advisable to store both the assembler text (for possible corrections) and the program in machine language and to keep it.

The use of higher-level programming languages is the "most agreeable" form of programming for the user.

Here the range is extremely wide and is continuously extended. At present, mainly BASIC and PASCAL are of importance for microcomputers. Other interesting languages, mainly those with methodical and didactic intentions, must be reserved for more special representations.

The difference to the level of the assembly language is obvious when expressing in PASCAL and BASIC the instruction sequence indicated in the assembly language in the last example.

Example: Addition of two numbers and checking the result

```
A: = OP1 + OP2;                  10 LET A = OP1 + OP2
if A = 0 then goto 10;           20 if A = 0 then goto 100
.                                .
.                                .
.                                .
10: ...                          100 ...
PASCAL                           BASIC
```

The names occurring here (A, OP1, OP2) are substitutional for the contents of certain memory locations, the operational symbol + designates the addition; the symbol: = or LET ... = designates the destination of the result. The meaning of the second instruction directly results from the translation of the English keywords

if A = 0, **then goto** ...

A program text in such a form can be handled much more easily by the user than machine and assembly languages and should always be the starting point for one's own software efforts. The transformation of higher-level programming languages to the machine level is possible by means of *compilers* and *interpreters.*

Compiling operating mode. An *interpreter* processes the original text instruction by instruction. It specifies an equivalent sequence of machine instructions for each instruction, executes them and processes the next instruction. A *compiler* processes the complete program text and establishes a complete machine program which is usable completely independent of the source text.

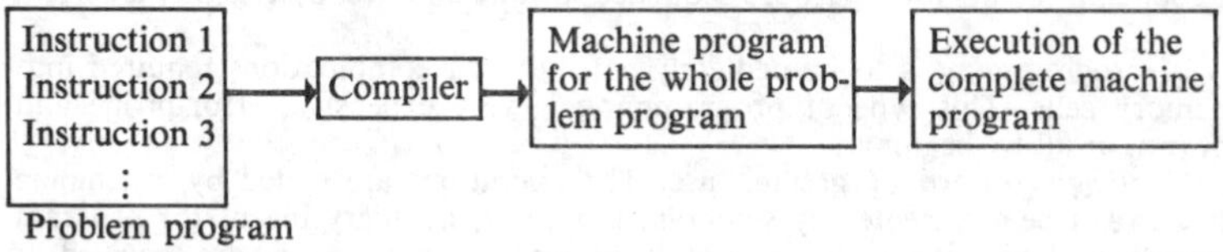

Interpreting operating mode

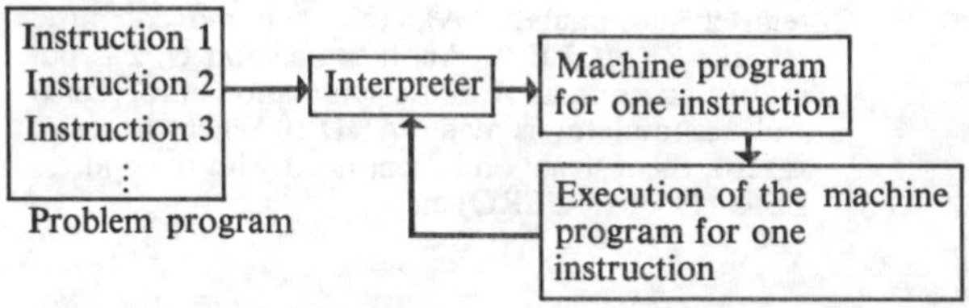

Generally, compilers are used for PASCAL, both compilers and interpreters are available for BASIC, however, preference is given to interpreters for microcomputers.

Thus another important system component is given which can be considered a part of the operating system for example, the BASIC interpreter can be available as software (loadable from cassette or floppy-disk) or on an (additional) slip-on ROM. When working with a certain program, you proceed from the system level to the language level.

Independent of the decision taken for a programming language, it can be seen that after some more or less successful attempts it is absolutely necessary to proceed systematically, carefully, disciplined and deliberately in the program design to succeed as quickly and safely as possible. For this purpose the whole process of problem solution, the design of the program text, is subdivided into single operational steps which should be executed completely and correctly. Any incorrectness, omission and negligence within a step are frequently of serious negative consequences for the subsequent stages.

The following should be performed in the correct sequence:

1. Problem. The task must be clearly defined without any misinterpretation; both the aims to be achieved and all conditions and prerequisites should be established.

2. Modelling. The problem is studied so thoroughly that there are starting points for the investigation of solving the problem and for solution. Very often this will be formulae for solutions (for mathematical problems), but also search methods, specifications for text and image configuration etc. are suitable. Moreover, data types and structures which are adapted to the problem must be selected. They play a major role in the effective solution of a problem. Of course, the availability of the respective structures in the case in question must already be taken into account (knowledge of the corresponding language version !).

3. Description of the input and output data. The following attributes for all input and output data must be defined:

Identifier:	Name for the data object used in the program
Meaning:	Description of the object which is represented by the data object
Unit of measurement:	Unit of measurement of the object
Type:	Type of the data object
Structure:	Structure of the data object

Example: Data object for performance

Identifier	Meaning	Unit of measurement	Type	Structure
P	performance	kWh	real number	single variable

4. Organization of input and output. Detailed statements on the sequence of input and the design of the output must be made:

a) Fixing of the objects for input,	or	a) Fixing of the objects for output,
b) Representation of the values,		b) Configuration of the (printed or screen) image,
c) Sequence of the data objects,		c) Sequence of data objects.

5. Program and data design. After elaborating steps 1 to 4 the actual program design can be carried out which uses a top-down strategy. The progress from a verbal description of the block solution through various refinement steps up to the ready program text is understood by this. The basic conception is to decompose complex steps (operations, tasks) into a number of "smaller" partial steps (operations, tasks) whose execution as a whole is functionally equivalent to the subdivided problem.

It is important to select the subdivisions (refinements) in such a way that the functional equivalence is not lost, but on the other hand to make a successive approximation to the terms of the target language. An idea of the available elementary operations and functions of the target language is necessary; the more exact and complete the knowledge is, the more effectively programming can be performed. If not all possibilities of the language are known you will certainly give away something, and if you know the language only insufficiently this can result in the tasks not being solved. Resemblance to the degree of mastering a foreign language is not accidental and illustrates these problems very impressively.

Elements of BASIC programming. In this section, important elements of the programming language BASIC (*B*eginners' *A*ll – Purpose *S*ymbolic *I*nstruction *C*ode) are shown which, at present, has become a dominating language in all microcomputers and, since the sixties, one of the most popular languages. Its main advantage is that the most important terms can be learned easily which mainly enables beginners to establish their own small programs "quickly". Furthermore, it can be extended easily; even operations which are oriented to the colours of the screen, the generation of sounds, the hardware and especially to backing memories can be easily included in the language.

At the same time this advantage is a great disadvantage since it resulted in a great number of "dialects" which are incompatible. Like in other programming languages, increased standardization efforts are being made for the BASIC language.

A main problem when using BASIC for greater problems is that the structured programming is not assisted or enforced by the language. A careful study of the methodology of programming and its consistent optional application is required.

Data and data type. The basis for understanding programming languages and their possibilities is the use of names (designations) which can be chosen by the user and allow access to memory locations. A memory location is reserved for each name chosen without the necessity to care for the specific computer-internal realization. The content of the memory location is accessible exactly via this name.

All names occuring in a program can be considered as being compiled to form a storage block.

A distinction is made between *constants* which remain permanently unchanged during the execution of this program and *variables* which can (perpetually) be changed.

Example: A storage block of names

Five variables are used, Pi and N remaining constant, while **sum, I** and **St** are the variables.

Sum	Pi	N	I	St
	3.14	100		

A certain *data type* is connected with each of these variables and even characteristics admissible for the respective variable. Important numerical data in BASIC are *integer* or *real data.* Integer data are integral numbers which must lie within a certain range of values and can be used in the normal way. Real data are used for real numbers with a finite number of places (i.e. exactly for rational numbers) and are also subject to certain specifications (they must be represented in a certain way, be within an interval, ...).

A certain number of admissible operations (operators) belongs to each data type. They are partly self-evident and must partly be tried by the user when studying the respective BASIC manual. Important operations for numerical data are addition, subtraction, division, multiplication, raising to a power and operations for the comparison of quantities. Operators always influence data and link them to a certain result. To achieve the definite use of the operators and to exclude brackets in the formulation of operation sequences, a certain order for the use of the operators is specified. The following specification is applicable for arithmetic operators:

1. Execution of raising to a power,
2. Execution of multiplication or division,
3. Execution of addition or subtraction.

Operations of the same kind and of the same stage are processed from left to right in their order of occurrence.

Operands and operators are compiled to *correct* terms and can be used accordingly.

Storage locations correspond to the variables occurring there which, at the time of execution of the program line, must in any case (!) contain a value. Many program errors can be attributed to the missing initialization of variables.

Example: Assume that the following BASIC line is given:

X1 = (A * A – 2 * A * B + B* B)/(A * A + B * B)

First, the term right of the equality symbol is considered, and evaluated in the following steps:

1. A * A
2. 2 * A * B
3. B * B
4. A * A – 2 * A * B
5. (A * A – 2 * A * B + B * B)
6. A * A
7. B * B
8. (A * A + B * B)
9. (A * A – 2 * A * B + B * B)/(A * A + B * B)

The function of the equality symbol is noteworthy for the variables to the left of it. An *assignment statement* is characterized by the equality symbol: the value of the term to the right of it becomes the value of the left variables. In the last example, the value which has been calculated by means of steps 1. to 9. is transported to place X1 and filed there (under this name).

SUM = 0
PI = 3.14
N = 100

are also simple assignment statements in this sense. The evaluation of the constants directly results in the indicated value and this value is stored. In this way, the initialization of variables can be carried out.

The compilation of data to vectors, matrices, ..., which are jointly designated *field* is very easy-to-use and handy. A special instruction DIM is used for the assignment of storage locations.

Example: The definitions of fields

DIM A(3, 3)
DIM B(10)
DIM C(2, 5)

give rise to the assignment of a matrix A with 3 rows and 3 columns, of a vector B with 10 components and of a matrix C with 2 rows and 5 columns.

A

A(1, 1)	A(1, 2)	A(1, 3)
A(2, 1)	A(2, 2)	A(2, 3)
A(3, 1)	A(3, 2)	A(3, 3)

B

B(1)	B(2)	...	B(9)	B(10)

C

C(1, 1)				
				C(2, 5)

The use of strings as data types is very typical of BASIC. Sequences of characters enclosed in quotation marks are understood by this. Variables for strings are specially marked by placing the character $ behind.

Example:

A $ = "BASIC"
B $ = "DESCRIPTION"

gives rise to the assignment of the indicated sequence of characters to the corresponding variables. For example, a typical operation for strings is their chaining.

C $ = A $ + B $

results in the value "BASICDESCRIPTION" for C $.

The operations available should be carefully studied and used in their entirety.

The use of predefined functions and the definition of own functions is very comfortable and expressive. While the latter requires some exercise and knowledge (and can be implemented by the DEF FN instruction) predefined functions can clearly be used with their name and the corresponding arguments.

Example:

X1 = −P/2 + SQR (P * P/4 − Q)
X2 = −P/2 − SQR (P * P/4 − Q)

These two instructions calculate the roots of a quadratic equation $x^2 + px - q = 0$ in case $p^2/4 - q \geqslant 0$. SQR designates the root function, the argument follows in brackets and is an arithmetic term. The programmer himself must take care that $p^2/4 - q$ is actually not negative.

The BASIC system can also be used as an (expensive, comfortable) pocket computer. Strictly speaking, this means that in the command mode (direct mode) each instruction is immediately executed after termination (by a special key which is frequently designated ENTER).

```
PRINT SQR (5)       ENTER
PRINT SIN (PI/2)    ENTER
PRINT 3 * 5 * 7 * 9 ENTER
```

results in the corresponding value being calculated and shown on the screen (PRINT) after each ENTER. This possibility will, of course, be used only seldom. The common application is the program mode.

These are the specifications to be observed:

1. The text of a BASIC program consists of a sequence of lines.
2. Each line starts with a line number; then the actual instruction follows, no number can be used twice.
3. The order of the lines is performed in accordance with the numbering chosen and in an ascending sequence.

A numbering of the lines in tens steps is useful for programming. If lines are added later on there is a reserve of 9 lines each which are frequently sufficient for this purpose. After this the required order (in tens steps) can automatically be restored by a special instruction (RENUMBER).

For the program text, imagine a storage block divided into squares in which each square contains a BASIC instruction. The interpreting mode already described opens a square (in accordance with the numbering) and realizes the specification contained in it (by access to the memory of variables, to the peripheral devices, ...).

The user must consider the basic principles shown and study the BASIC description thoroughly to realize the mode of action of all operations available. Then you will be able to understand written programs, to work them through manually (to interpret them) and to move over to writing one's own programs. The following example demonstrates the steps required.

Example: The computer shall calculate the mean value and the spread of a series of measured values.

$$m = \frac{1}{n} \sum_{i=1}^{n} a_i, \qquad st = \sqrt{\frac{1}{n-1} \sum_{i=1}^{n} (m - a_i)^2}$$

$$n \leqslant 100$$

Rough design of the algorithm	Data
(I) 1. Input request and input of n	n integer
2. Field input	field A for 100 real numbers
3. Computation m	m real number
4. Computation st	st real number
5. Output m, st	

Refined design of the algorithm	Data
(II) 1.1. Print: 'Input: ...'	
1.2. Input: n	
1.3. Initial values S1, S2	S1, S2 real numbers
2. Repeat for i from 1 to n	i integer control variable
2.1. Input: ai	
2.2. Compute: $S1 = S1 + ai$	
3. Compute: $m = S1/n$	
4.1. Repeat for i from 1 to n	
$S2 = S2 + (m - ai)^2$	
4.2. Compute: $st = \sqrt{1/(n-1) \cdot S2}$	
5. Print:	
5.1. 'The mean value is: ', m	
5.2. 'The spread is: ', st	

BASIC text

```
10 WINDOW: CLS
20 PRINT "Program mean value and spread"
30 PRINT "**************************
   ***"
40 DIM A(100): ! Specify a field of 100 elements
50 INPUT "How many values will you input?"; n
60 PRINT "Input the measured values!"
70 Su = 0
80 FOR i = 1 TO n
90 INPUT A(i)
100 Su = Su + A(i)
110 NEXT i
120 m = Su/n
130 Su = 0
140 FOR i = 1 TO n
150 Su = Su + (m – A(i))^2
160 NEXT i
170 st = SQR(Su/(n – 1))
180 PRINT "The mean value is"; m;" . "
190 PRINT "The spread was"; st; " . "
200 END
```

Generally, the study of the available language elements can be carried out according to the following instruction groups:

1. Mathematical operations
2. Assignment operations
3. Logical operations
4. Comparison operations
5. Program loops
6. Mathematical functions
7. Character string functions
8. Input of data
9. Output of data
10. Data file operation (operation with external data carriers)
11. Graphics and screen control
12. Access to hardware
13. Subroutine technology
14. Sundries

The available language elements of group 1. to 9. are absolutely necessary. The following table must be considered the core of a BASIC Standard. It was defined in 1978 by the American National Standards Institute.

BASIC Standard

Instruction or character	Meaning
"	Identification of the beginning or end of a character string
$	Identification of a variable as character string variable
.	Decimal point in REAL numbers
E	Identification of the exponent in REAL numbers (e.g. 3.55 E-6 for the number $3.55 \cdot 10^{-6} = 0.00000355$)
:	Separation of instructions in a line
END	Identification of the physical end of the program
REM	Start of comment line; the text up to the end of the line has only an explanatory meaning and is omitted in the program execution
DATA	Definition of a sequence of constants which can be assigned to certain variables by the READ instruction
INPUT	Identification of a data input (from the console)
READ	Reading of data from a DATA instruction, assignment of the values to certain variables
RESTORE	Setting of a pointer to the first element of the first DATA list; READ instructions always read the value to which the pointer points; after each reading operation the pointer is advanced
?	Abbreviation of PRINT

PRINT	Display of data on the screen; in this connection format controls are possible (PRINT A; B; C or PRINT A, B, C – PRINT AT – PRINT USING – PRINT # – PRINT # USING)
,	Separator in data lists, in PRINT instructions the leave blank of a tabulator step is effected
;	The next output is immediately connected to the previous output if there is a semicolon between the variables to be output
TAB	The output is performed at the point indicated (TAB (1Ø) effects the output as Item 1Ø)
*	Character for multiplication
+	Character for addition
−	Character for substraction
/	Character for division
^	Character for raising to a power
<	Character for "less than" (A < B) checks whether the relation "less than" is applicable (the occurrence of A < B is answered with "Yes" or "No" at a certain point)
< =	Character for "less than or equal to"
< >	Character for "not equal to"
>	Character for "greater than"
> =	Character for "greater than or equal to"
=	Character for assignment statement
LET	Identification of an assignment (can be omitted in most cases – LET A = B corresponds to A = B)
IF – LET	conditional assignment, the assignment behind LET is realized if the condition behind IF is met
ABS (*x*)	$\lvert x \rvert$
ATN (*x*)	arctan (*x*)
COS (*x*)	cos *x*
EXP (*x*)	e^x
INT	[*x*], the greatest integer $\leqslant x$
LOG (*x*)	log *x*
RANDOMIZE	Fixing of a start value for the generation of a random number sequence with the RND instruction
RND	Generation of a (pseudo) random number in the range between 0 and 1
SIN (*x*)	sin *x*
SQR	$\sqrt{x}$
TAN (*x*)	tan *x*
DEF FN ...	Definition of a user-owned function which can be called later on with FN ...
FEND or FNEND	Identification of the end of the functional definition
LEN	Determination of the number of characters in a character string
GOTO ...	Jump to the indicated line number
IF ... THEN ... ELSE ...	Branching in the program; if the condition after IF has been met, the instruction behind THEN is executed, otherwise the instruction behind ELSE. ELSE and the subsequent instruction can be missing; then a jump to the next line is made
IF ... GOTO ...	conditional jump
BASE	defines whether counting starts at 0 or 1 in indices of fields
DIM	Definition of fields
OPTION BASE	see BASE
CALL ...	Branching to a machine program starting at a certain address
GOSUB	Branching to a subroutine; execution of this program until a RETURN instruction occurs; this subroutine is then terminated and a jump to the line following GOSUB is made
RETURN	Leaving a subroutine, return to the (calling) main routine
FOR ... TO ... STEP ... ⋮ NEXT ...	Running instruction; the assignment statement which is written behind FOR assigns an initial value to a running variable; the final value is written behind TO; all instructions up to NEXT are repeated for each value of the running variables between the initial and the final value; the running variable is always increased by 1 if STEP has been omitted; if another step size is required, it can be indicated behind STEP.

Other instructions require more profound knowledge and should be carefully studied. In particular, hardware and graphics-oriented instructions can only be used efficiently if you know their purpose, and then they are a great attraction. Finally, it should be noted that every good BASIC system does not only include a compiler or interpreter, but represents a complete programming system. While the instructions described are means of expression which effect certain actions within a program, the level can be raised by means of other commands and effectively support the process of programming and the handling of the program in interactive communication with the system.

The following functions which can be controlled by commands are available:

- Input of programs (CLOAD, ...),
- Correction of programs (BREAK, EDIT, ...),
- Display of programs (LIST, ...),
- Execution of programs (RUN, ...),
- Operation with secondary data memories (CLOAD, CSAVE, ...),
- Support of program testing (TRACE, ...),
- Indication and alteration of memory mapping (TOP, ...),
- Erasure of the main memory (ERASE, NEW, ...).

Finally, there is an instruction (BYE) by which you can leave the BASIC system and return to the operating system.

Subject index

E

F

G

P

Q

T

U

V

1 Archimedes Poster of the town of Syracuse (Italy) on the occasion of the commemorating conference in honour of the ancient Greek mathematician Archimedes, which was held from 11 to 16 April 1961 and was attended by mathematicians, physicists, and engineers from all over the world

2 Mathematics in school I Introduction of the number seven

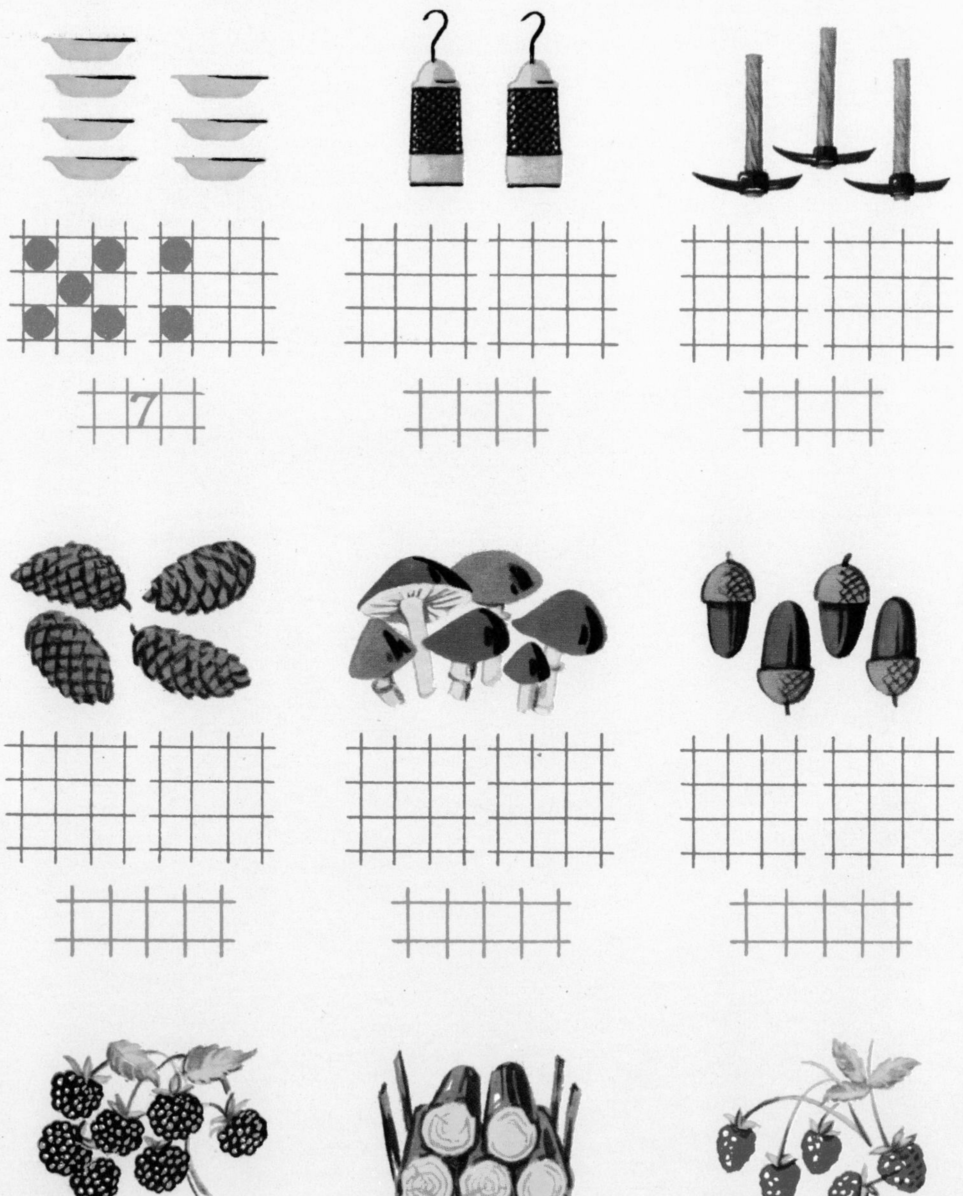

3 Mathematics in school II Revision exercises

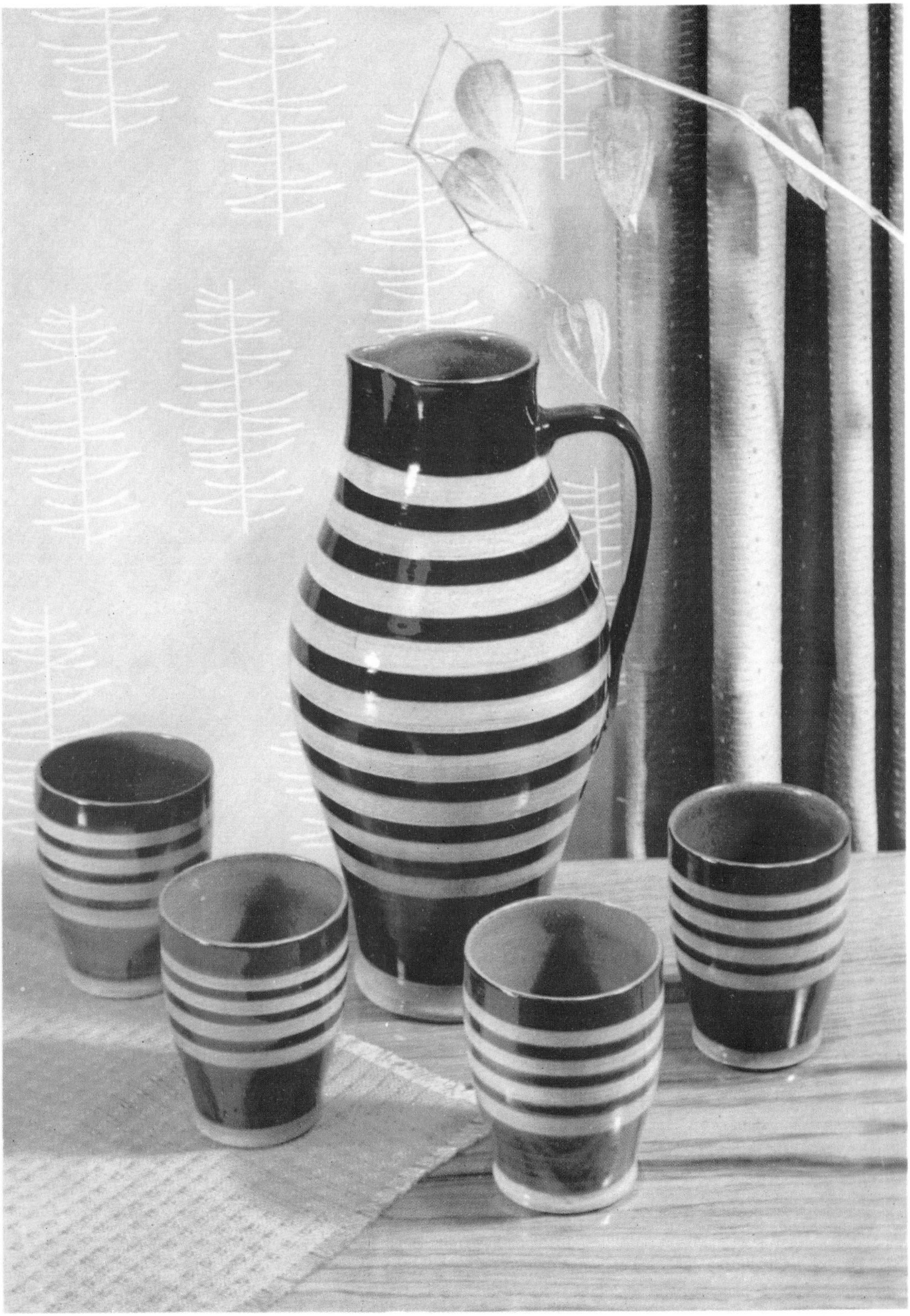

4 Mathematics in industrial arts Surfaces of revolution in the design of a pottery set

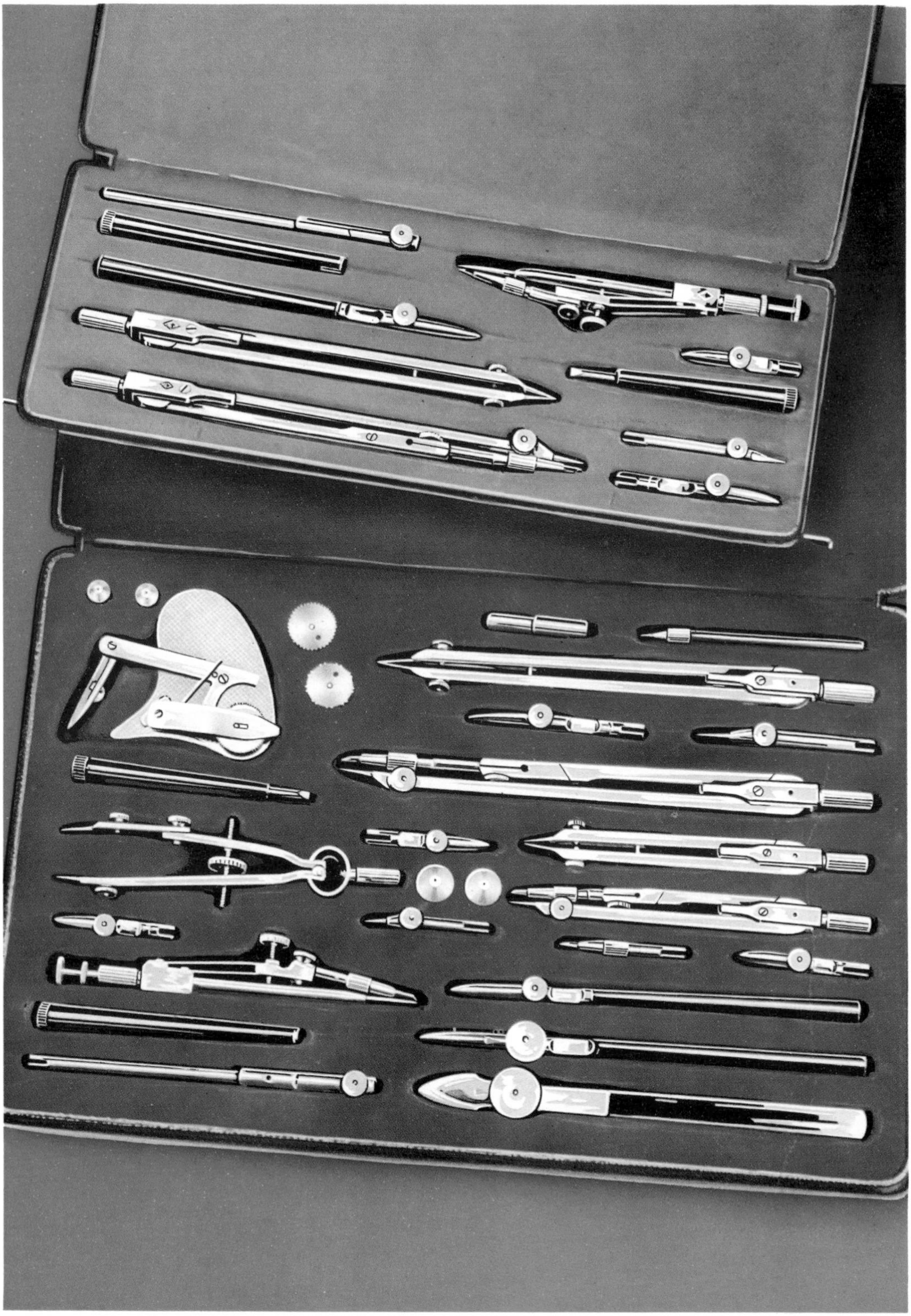

5 Drawing instruments I Geometry sets

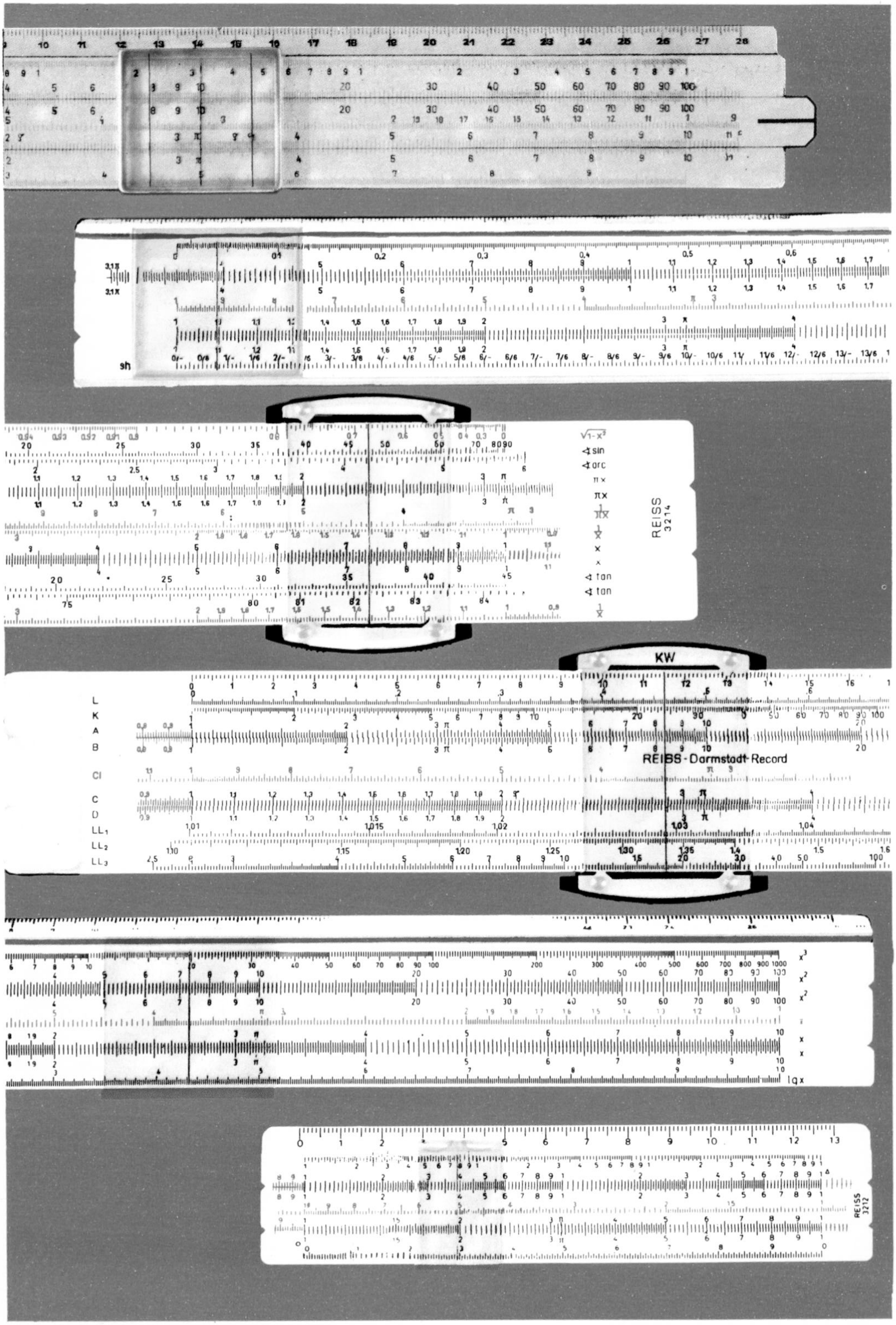

6 Drawing instruments II Slide rules

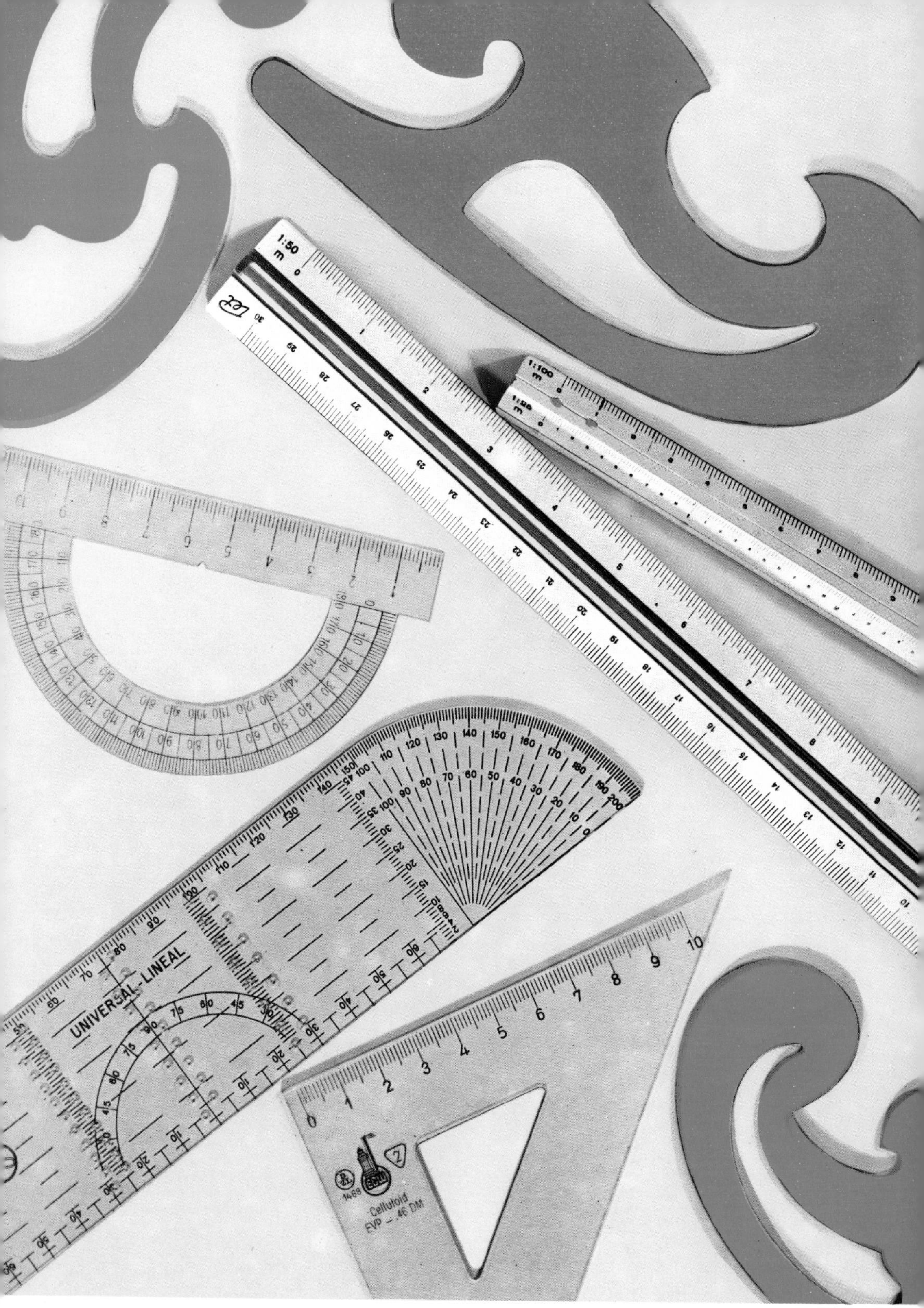

7 Drawing instruments III Rulers, protractors, and french curves

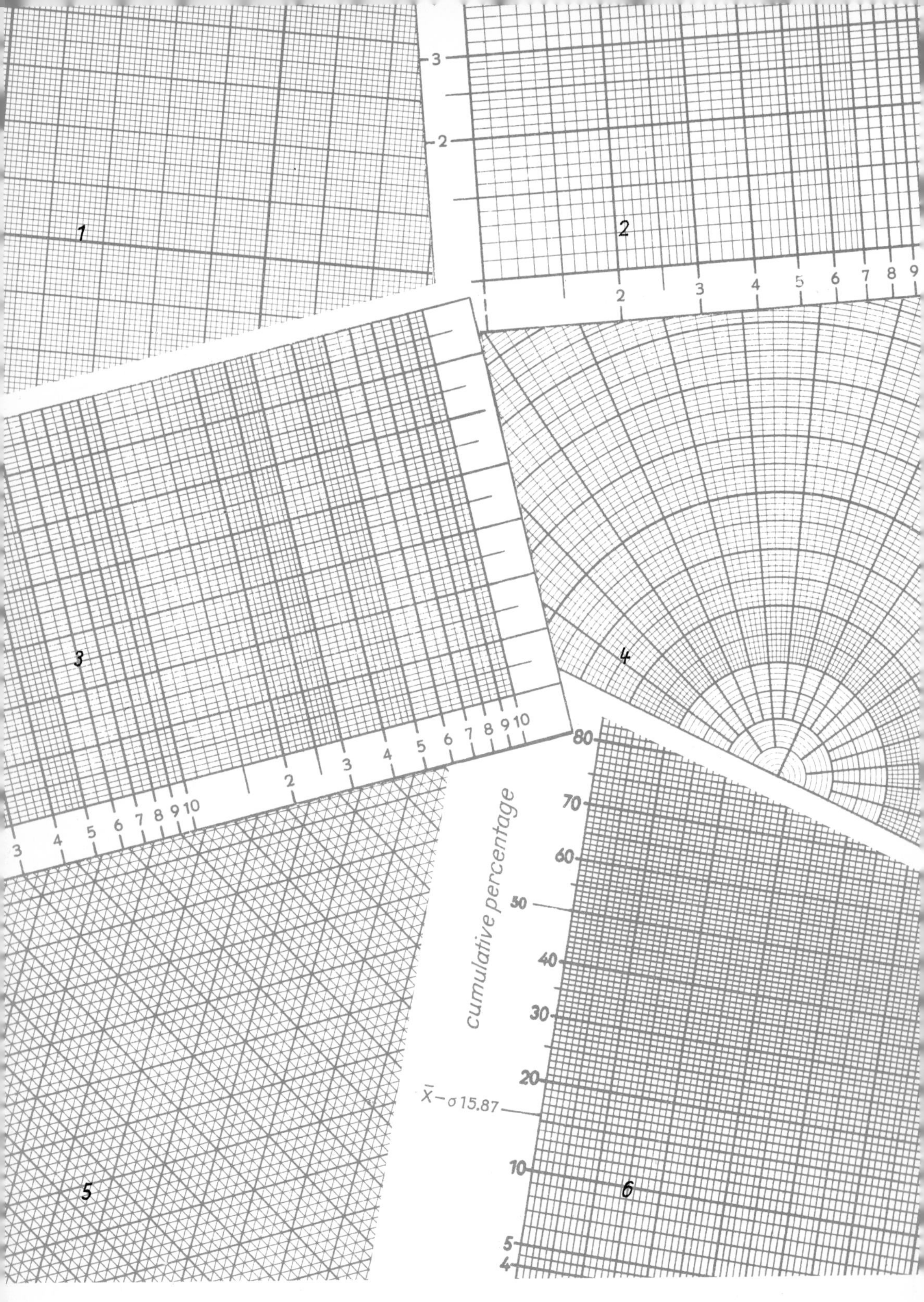

8 Graph papers

1 Millimetre paper
2 Doubly logarithmic paper
3 Simply logarithmic paper
4 Polar coordinate paper
5 Triangular net paper
6 Probability paper

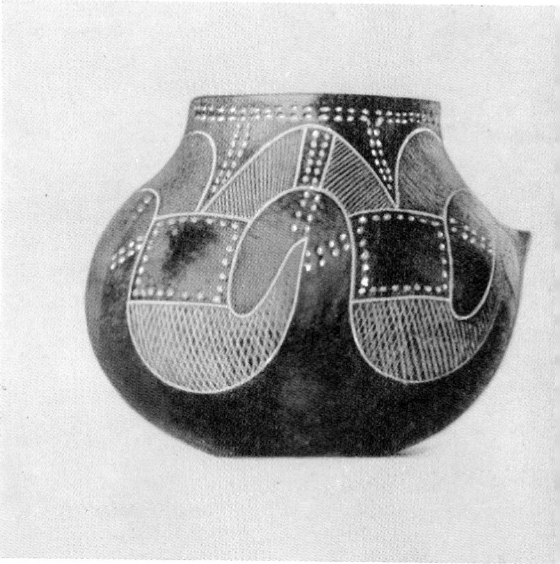

above and right
Clay vessels of the new stone age with mathematical ornamentations

9 From the earliest period of mathematics

below
Early Egyptian surveying (about 1415 B. C.; painting on plaster)

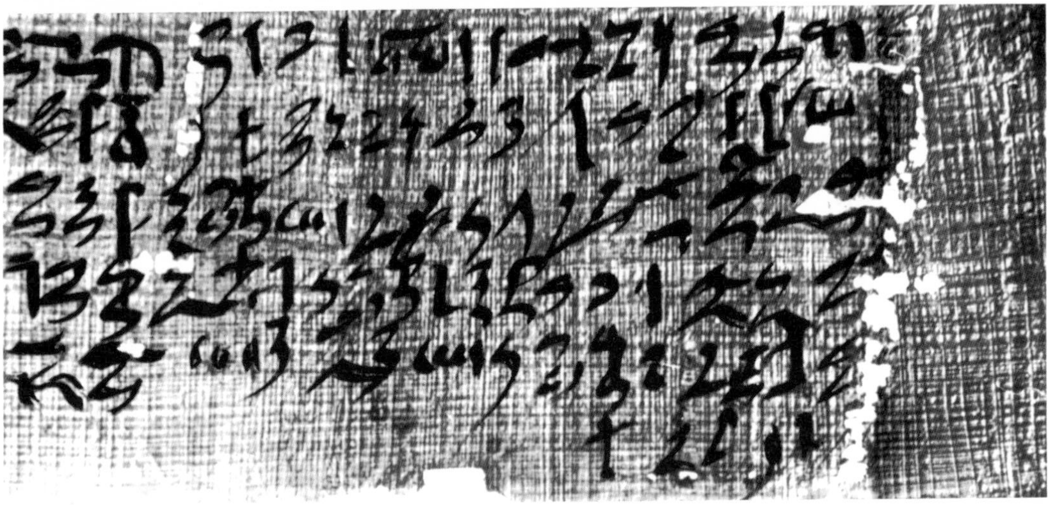

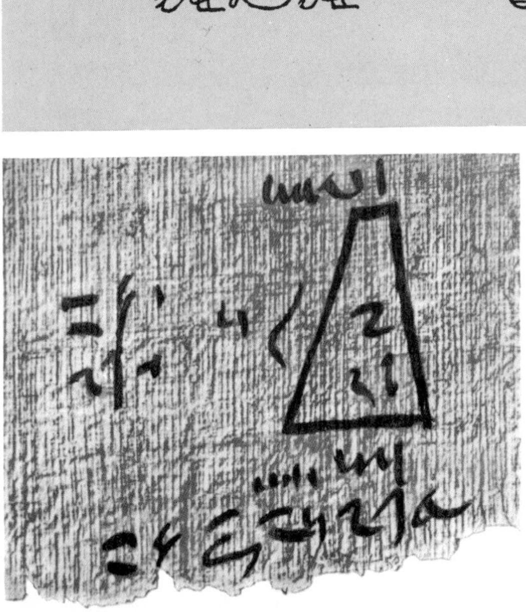

above
Original text of the Hau problem in Demotic writing
centre
Transcription of the same text into hieroglyphics
Calculation of a frustum of a pyramid

10 Ancient Egyptian mathematics
(The Moscow papyrus)

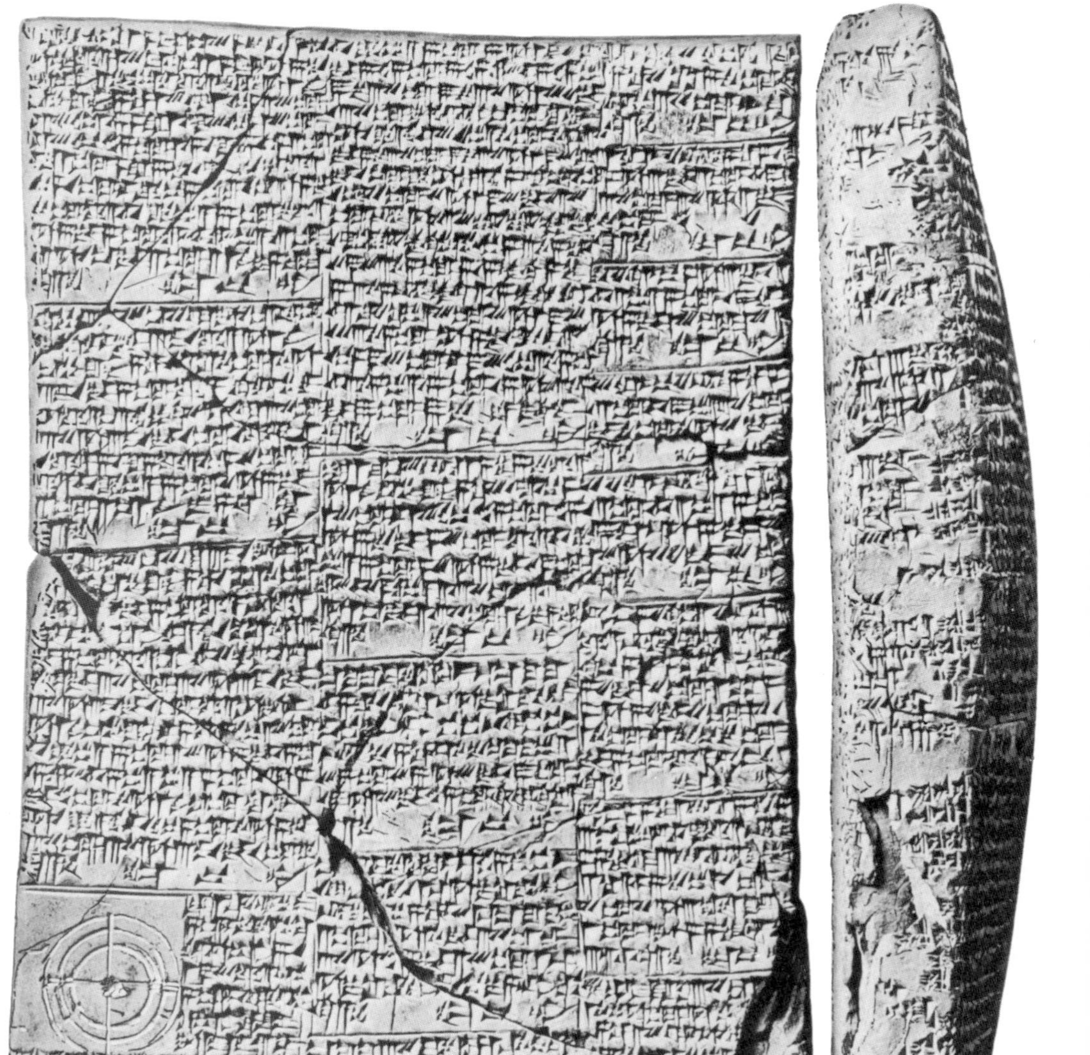

Cuneiform tablet with calculations of areas

11 Babylonian mathematics

Section of the tablet above

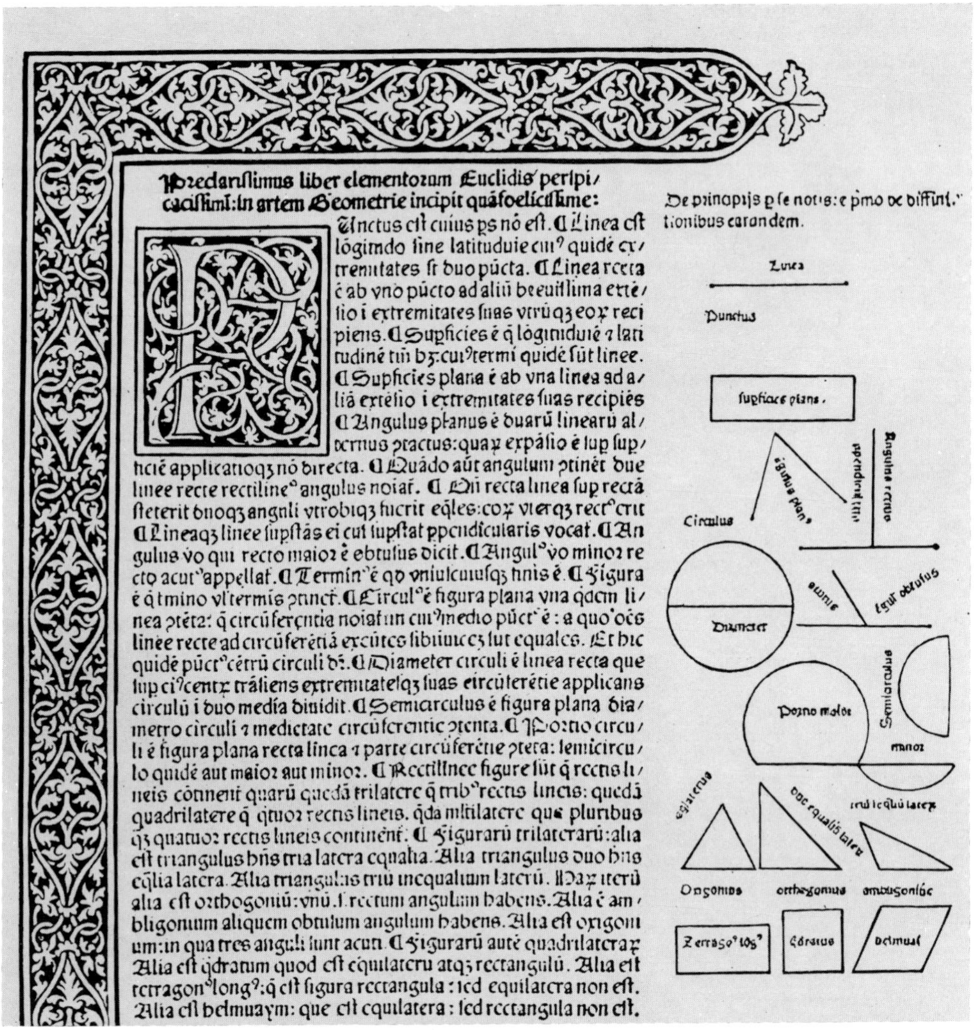

Preclarissimus liber elementorum Euclidis perspi/cacissimi: in artem Geometrie incipit quãfoelicissime:

Punctus est cuius ps nõ est. ¶ Linea est lõgitudo sine latitudine cui⁹ quidẽ ex/tremitates sũt duo pũcta. ¶ Linea recta ẽ ab vno pũcto ad aliũ breuissima extẽ/sio i extremitates suas vtrũq₃ eo℞ reci piens. ¶ Supficies ẽ q̃ lõgitudinẽ ⁊ lati tudinẽ tm̃ h₃: cui⁹ termi quidẽ sũt linee. ¶ Supficies plana ẽ ab vna linea ad a/liã extẽsio i extremitates suas recipiẽs ¶ Angulus planus ẽ duarũ linearũ al/ternus ptactus: qua℞ expãsio ẽ sup sup/ficiẽ applicatioq₃ nõ directa. ¶ Quãdo aũt angulum ptinẽt due linee recte rectiline⁹ angulus noiat. ¶ Qñ recta linea sup rectã steterit duoq₃ anguli vtrobiq₃ fuerit eq̃les: eo℞ vterq₃ rect⁹ erit ¶ Lineaq₃ linee supstãs ei cui supstat ppendicularis vocat. ¶ An gulus vo qui recto maior ẽ obtusus dicit. ¶ Angul⁹ vo minor re cto acut⁹ appellat. ¶ Termin⁹ ẽ qd vniuscuiusq₃ finis ẽ. ¶ Figura ẽ q̃ tmino vl' termis ptinet. ¶ Circul⁹ ẽ figura plana vna q̃dem li/nea ptẽta: q̃ circũferentia noiat in cui⁹ medio pũct⁹ ẽ: a quo oẽs linee recte ad circũferẽtiã exeũtes sibiinuice₃ sũt equales. Et hic quidẽ pũct⁹ cẽtrũ circuli dr̃. ¶ Diameter circuli ẽ linea recta que sup ci⁹ centrũ trãsiens extremitatesq₃ suas circũferẽtie applicans circulũ i duo media diuidit. ¶ Semicirculus ẽ figura plana dia/metro circuli ⁊ medietate circũferentie ptenta. ¶ Portio circu/li ẽ figura plana recta linea ⁊ parte circũferẽtie ptẽta: semicircu/lo quidẽ aut maior aut minor. ¶ Rectilinee figure sũt q̃ rectis li/neis cõtinent quarũ quedã trilatere q̃ trib⁹ rectis lineis: quedã quadrilatere q̃ qtuor rectis lineis. q̃da mltilatere que pluribus q₃ quatuor rectis lineis continẽt. ¶ Figurarũ trilaterarũ: alia est triangulus hñs tria latera equalia. Alia triangulus duo hñs eq̃lia latera. Alia triangulus triũ inequalium laterũ. Ha℞ iterũ alia est orthogoniũ: vnũ .s. rectum angulum habens. Alia ẽ am/bligonium aliquem obtusum angulum habens. Alia est oxigoni um: in qua tres anguli sunt acuti. ¶ Figurarũ autẽ quadrilatera℞ Alia est q̃dratum quod est equilaterũ atq₃ rectangulũ. Alia est tetragon⁹ long⁹: q̃ est figura rectangula: sed equilatera non est. Alia est helmuaym: que est equilatera: sed rectangula non est.

De principiis p se notis: ẽ pmo de diffini/tionibus earundem.

The Elements of Euclid, first printed edition 1482

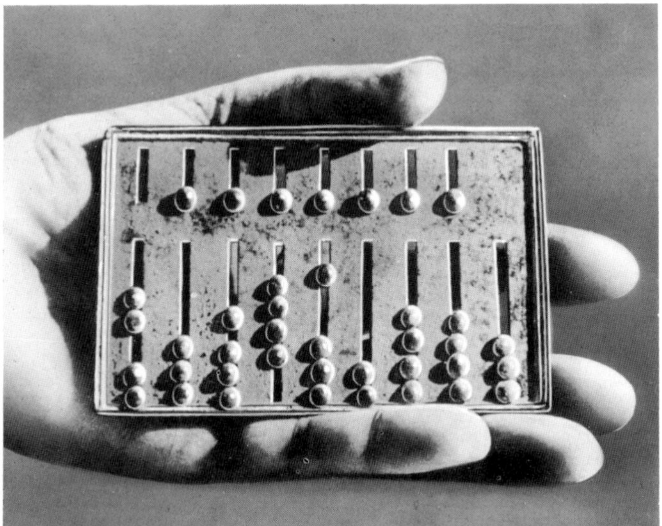

12 Graeco-Roman mathematics

Roman hand abacus

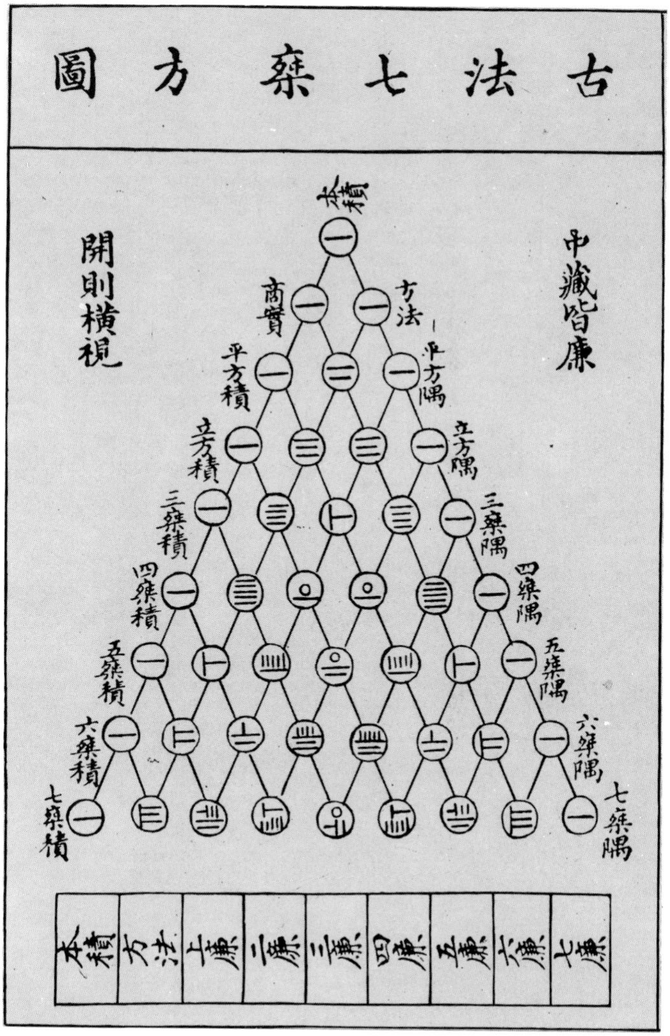

From a manuscript dated 1303, the triangle of numbers later named after Pascal

Bamboo digits

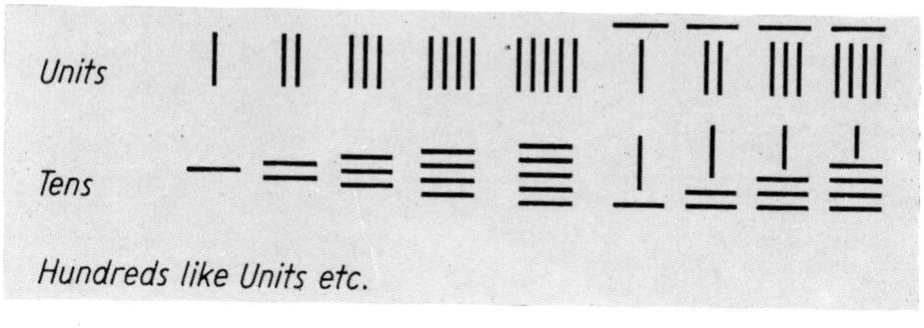

Chinese slide rule (about 1600)

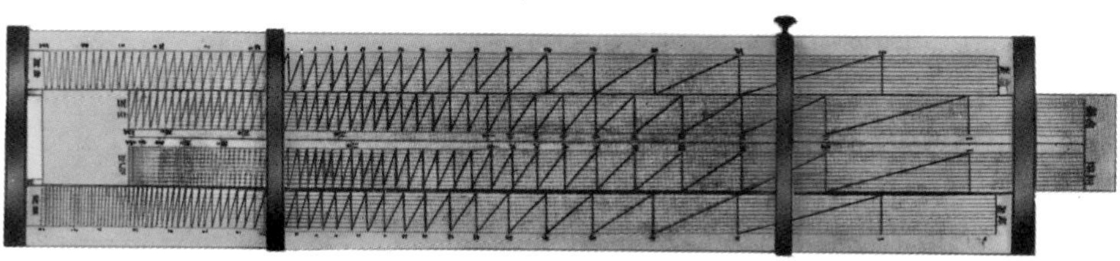

Mathematical-astronomical buildings of the 17th century for the determination of time, declination, and hour angle (near Delhi)

14 Ancient Hindu mathematics

Mathematical manuscript of the 16th century (copy of a manuscript of Bhaskara)

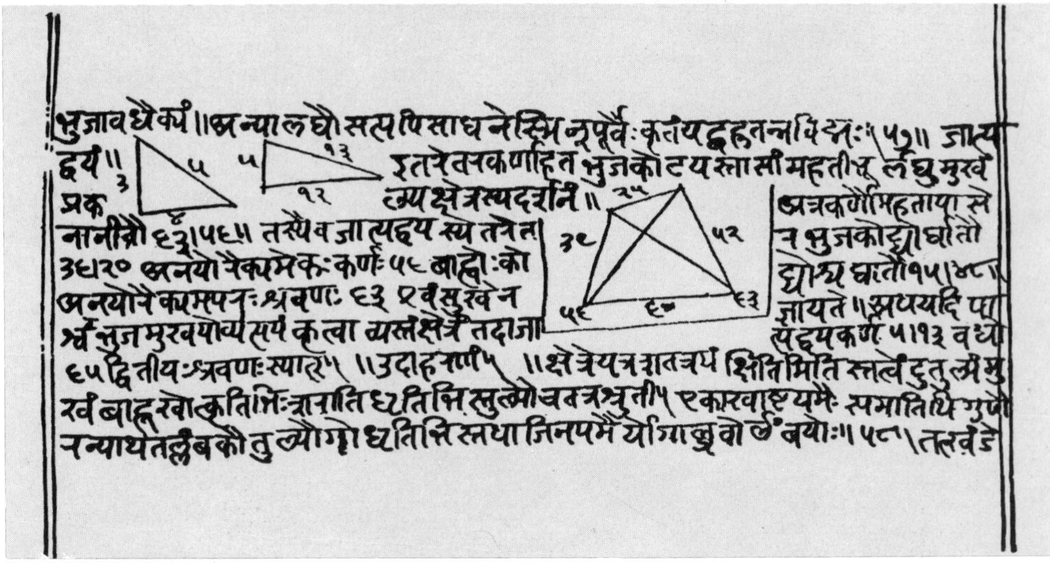

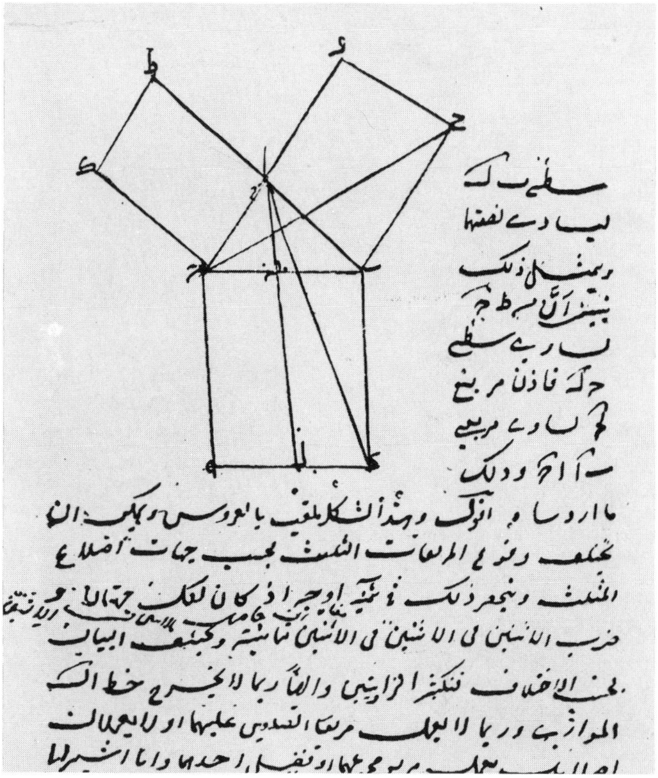

Theorem of Pythagoras in an Arabic mathematical manuscript of the 14th century

Arabic astrolabe for the measurement of heights and for astronomical calculations, beginning of the 15th century

16 Mathematics in Europe, 15th to 17th century

Triumph of the modern algorithm (digital calculation) over the ancient counter reckoning (abacus). Contemporary illustration dated 1504. Pythagoras, reputed to be the inventor of the abacus, sits sulkily on the right over his calculations, while on the left Boethius, regarded as the inventor of written calculations with Arabic numerals, is already finished. The goddess Arithmetica in the background supervises the contest

The use of Jacob's staff. Contemporary illustration of the 16th century

Ancient Egyptian mural: catching fish and hunting birds in a papyrus thicket

Projections in plan of characteristic cross-sections, see the position of head and shoulders of the principal persons

17 Mathematics and the visual arts I

Examples of the representation of spatial objects in a plane

Painting by Melozzo da Forli (1438–1494): Pope Sixtus IV appoints Platina as Prefect of the Vatican (Rome, Museum of the Vatican). The discovery of the vanishing point at the beginning of the 15th century marks a turning point in the history of painting

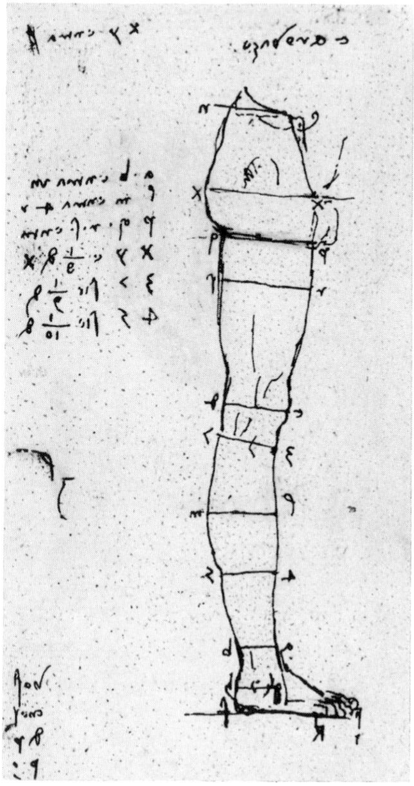

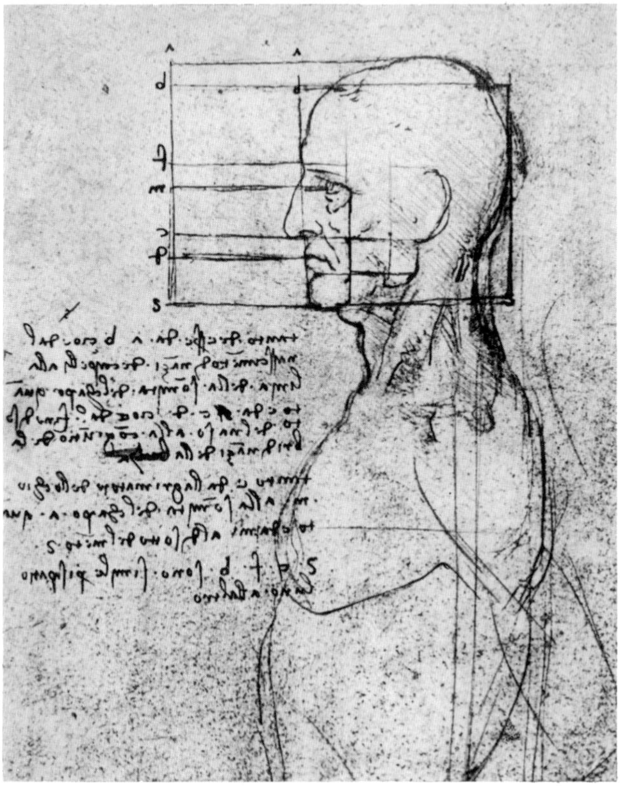

Drawing by Leonardo da Vinci

18 Mathematics and the visual arts II
Proportions of the human body

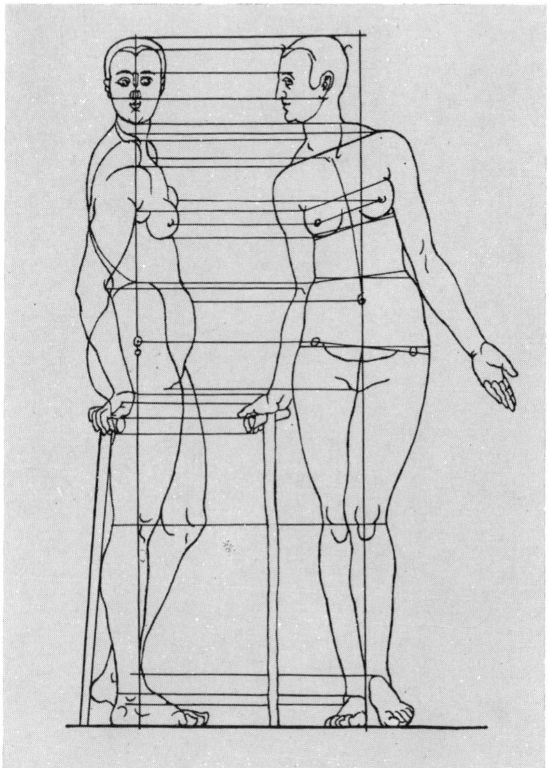

Scheme of proportions. Sketch by Albrecht Dürer

19 Mathematics and the visual arts III

Melancholia, copper engraving by Albrecht Dürer, with magic square; made in 1514 (see last row of square)

Egyptian pyramids near Giza (rigth square pyramid)

20 Geometric forms in architecture and technology I

Tower of city walls (rigth circular cylinder and cone)

The old town hall of Leipzig, 16th century. The tower divides the façade in the ratio of the golden section

21 Geometric forms in architecture and technology II

Modern water tower; its conical shape presents an unusual sight

Cooling towers of a generating plant (hyperboloids of revolution)

Obelisk in the great temple of Amun at Karnak (Ancient Egypt)

Wedge as a cleaving tool

22 Geometric forms in architecture and technology III

Hyperbolic paraboloid shells as roofs of an exhibition hall

1

2

3

4

5

6

7

23 Famous mathematicians of the 15th/16th century

1 Johannes Regiomontanus (1436 to 1476)
2 Simon Stevin (1548–1620)
3 Albrecht Dürer (1471–1528) (detail of his self-portrait)
4 Niccolo Tartaglia (about 1500–1557)
5 Geronimo Cardano (1501–1576)
6 Jost Bürgi (1552–1632)
7 Luca Pacioli (1445–1514) (after a painting by Jacopo de'Barbari)

Rechenung nach der
lenge/ auff den Linihen
vnd Feder.
Dar zu forteil vnd behendigkeit durch die Proportio=
nes/Practica genant/Mit grüntlichem
vnterricht des visierens.
Durch Adam Riesen.
im 1550. Jar.

Cum gratia & priuilegio
Cæsareo.

Adam Risen.
Vihekauff.

Item/einer hat 100. fl. dafür wil er 100.
haupt Vihes kauffen / nemlich / Ochsen/
Schwein/Kälber/ vnd Geissen/ kost ein Ochs
4 fl. ein Schwein anderthalben fl. ein Kalb
einen halben fl. vnd ein Geiß ein ort von einem
fl. wie viel sol er jeglicher haben für die 100. fl?
Machs nach den vorigen/mach eines jeglichen
kosten zu örtern/deßgleichen die 100. fl. vnd setz
als dann also:

	16	15	
	6	5	
100			400
	2	1	
	1		

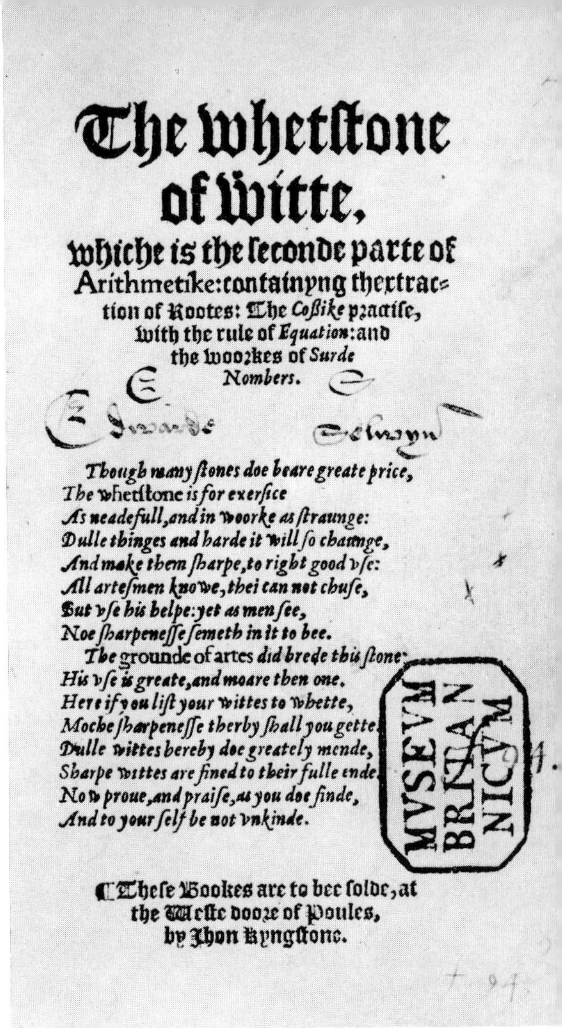

The whetstone
of witte,
whiche is the seconde parte of
Arithmetike: containyng thextrac-
tion of Rootes: The Coßike practise,
with the rule of Equation: and
the woorkes of Surde
Nombers.

Though many stones doe beare greate price,
The whetstone is for exersice
As neadefull, and in woorke as straunge:
Dulle thinges and harde it will so chaunge,
And make them sharpe, to right good vse:
All artesmen knowe, thei can not chuse,
But vse his helpe: yet as men see,
Noe sharpenesse semeth in it to bee.
The grounde of artes did brede this stone:
His vse is greate, and moare then one.
Here if you list your wittes to whette,
Moche sharpenesse therby shall you gette.
Dulle wittes hereby doe greately mende,
Sharpe wittes are fined to their fulle ende.
Now proue, and praise, as you doe finde,
And to your self be not vnkinde.

These Bookes are to bee solde, at
the Weste doore of Poules,
by Jhon Kyngstone.

right
Title page of Robert Recorde's 'Algebra' of 1557 (by courtesy of the Trustees of the British Museum)

top left
Title page of Adam Ries's 'Rechnung auff der Linihen und Feder …' of 1550. The words refer to calculations by means of the abacus and on paper

24 Famous mathematicians of the 16th century

below
A problem out of this book concerning the purchase of livestock

Conclusion of a business deal at a calculating desk on which lines and a distribution of coins are marked (old woodcut)

25 From old arithmetic books

Calculation of the capacity of a cask; title page of a booklet by Johann Frey printed in Nuremberg in 1531

The mathematics room of the National and University library in Prague

26 Two libraries

Entrance to the Science library of Erfurt (Boyneburg portal)

27 Old mathematical aids I

Pedometer, 1741

Slit bamboo as counting stick (from Sumatra)

below
Tally stick. On conclusion of a business deal both parts of the tally stick were marked at the appropriate notches. Each partner received one half as legal evidence

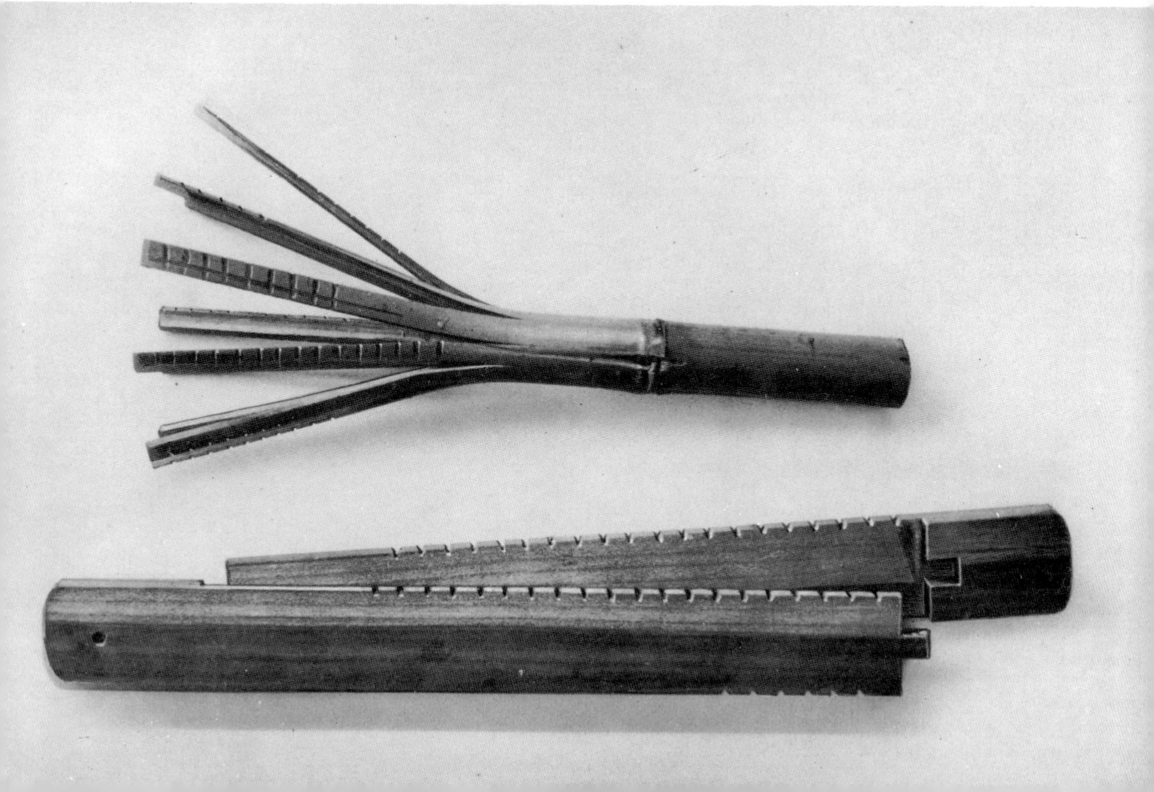

28 Old mathematical aids II
Counters or markers for arithmetic (not money) and an elaborate box, 16th century

29 Old mathematical aids III
Surveyor's compass, for the measurement and graphical determination of distances and angles in the field, about 1600

Illustration of a rod, by juxtaposition of 16 feet. The 19th century English rod (also perch or pole) is $16^1/_2$ feet long. From Jacob Köbel's 'Geometrie', Frankfurt 1616

30 Old measures I

16th century measuring rods with various graduations

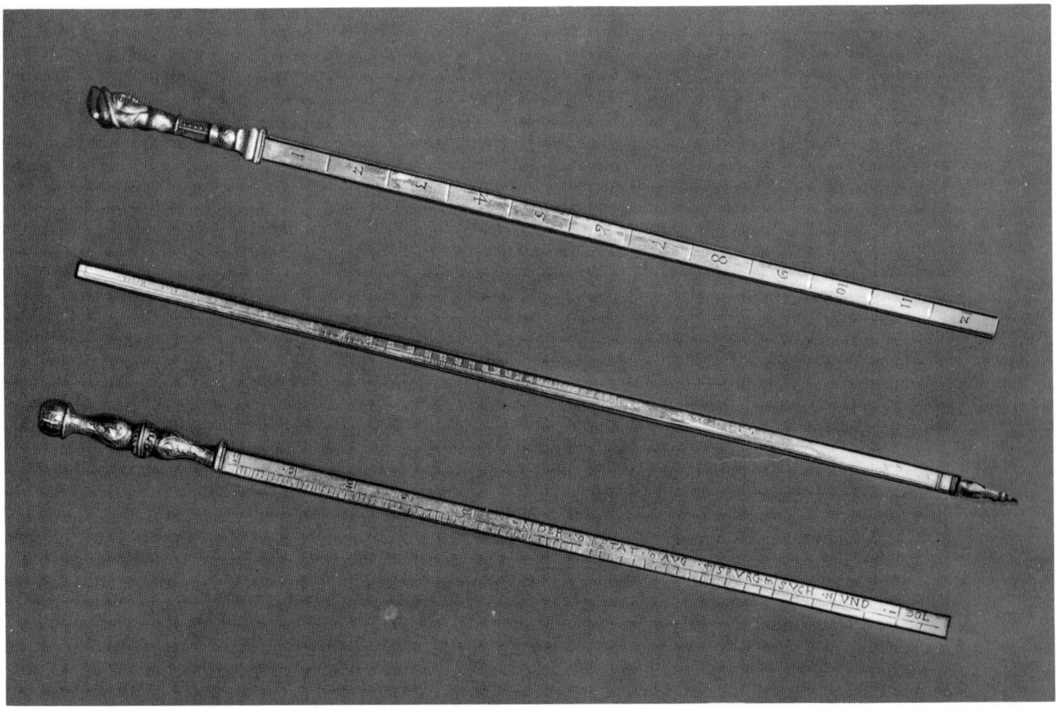

Hinged sun dial, ivory

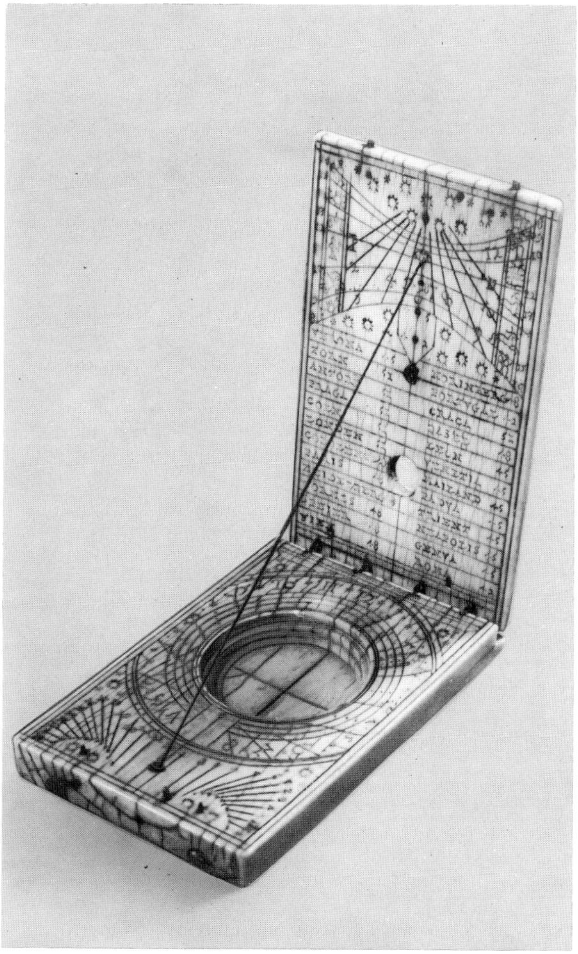

31 Old measures II

Set of weigths, Nuremberg 1588

32 Famous mathematicians of the 17th century I

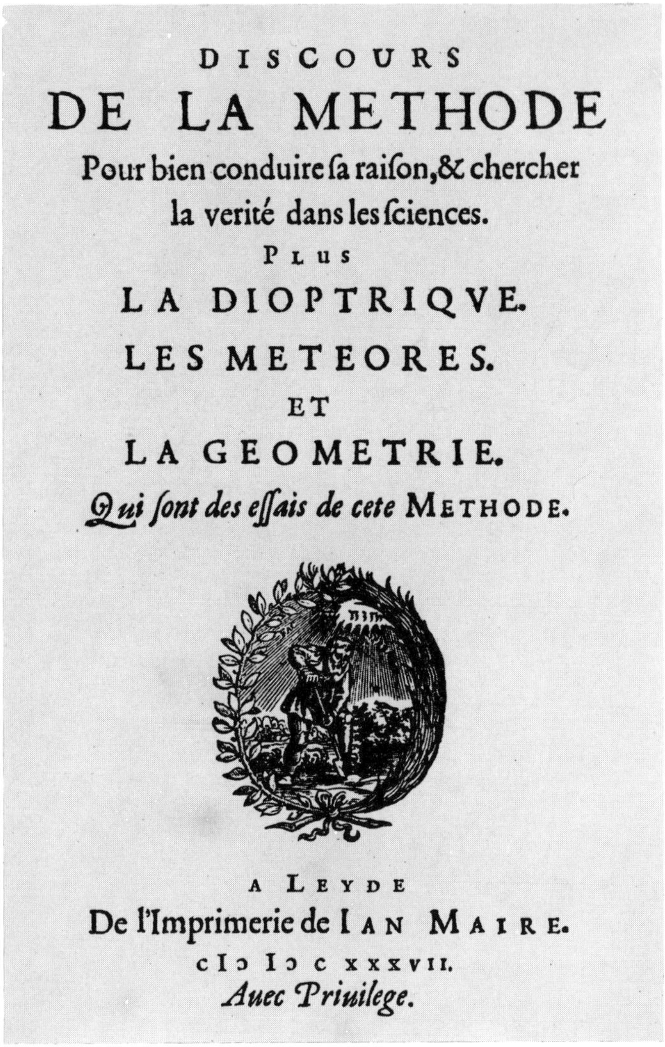
DISCOURS
DE LA METHODE
Pour bien conduire ſa raiſon,& chercher la verité dans les ſciences.
PLUS
LA DIOPTRIQVE.
LES METEORES.
ET
LA GEOMETRIE.
Qui ſont des eſſais de cete METHODE.

A LEYDE
De l'Imprimerie de IAN MAIRE.
cIↄ Iↄ c xxxvii.
Auec Priuilege.

Title page of Descartes' 'Discours de la méthode' whose third part 'La géométrie' contains the foundations of analytic geometry

René Descartes (1596–1650)

1

2

3

4

5

6

7

33 Famous mathematicians of the 17th century II

1 François Vieta (Viète; 1540–1603)
2 John Napier (Neper; 1550–1617)
3 Galileo Galilei (1564–1642)
4 Johannes Kepler (1571–1630)
5 Bonaventura Cavalieri (1598–1647)
6 Pierre de Fermat (1601–1665)
7 James Gregory (1638–1675)

top left
Gottfried Wilhelm Leibniz (1646–1716)

above
Blaise Pascal (1623–1662)

34 Famous mathematicians of the 17th/18th century I

The infinitesimal calculus was invented independently by Leibniz and Newton

Isaac Newton (1643–1727)

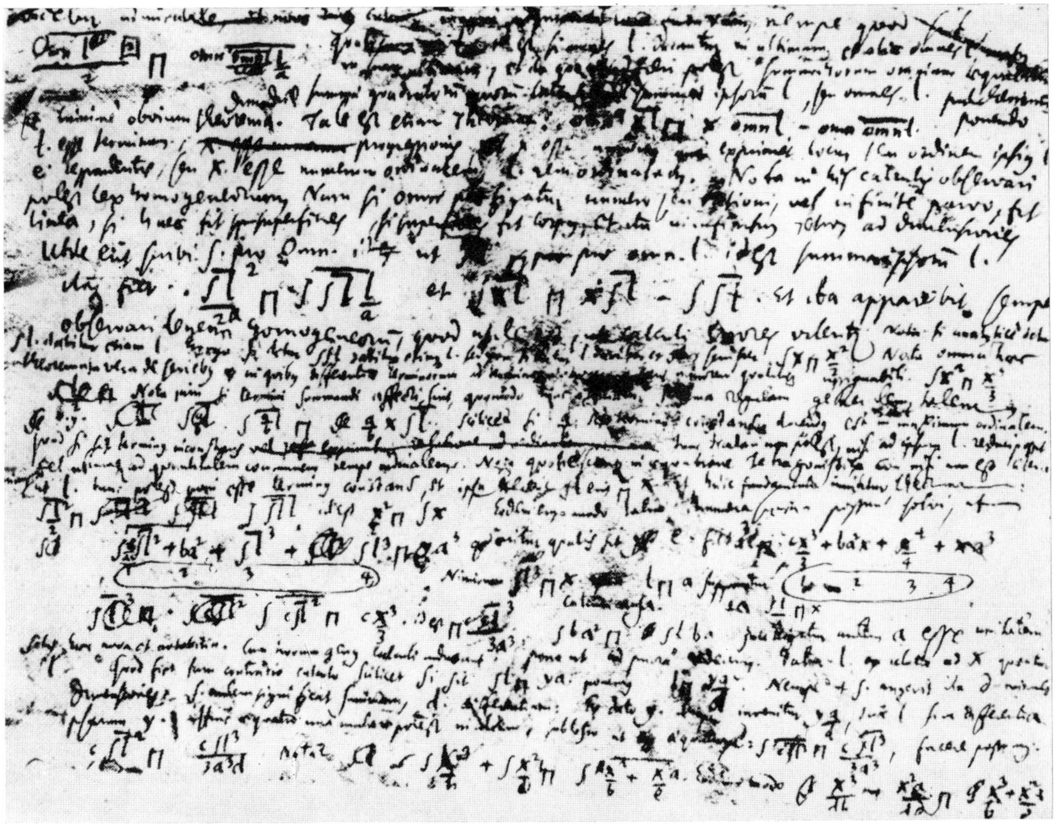

Extract from a manuscript of Leibniz dated 29 October 1673, in which the integral sign appears for the first time

35 Famous mathematicians of the 17th/18th century II

The mechanical calculator constructed by Pascal in 1642

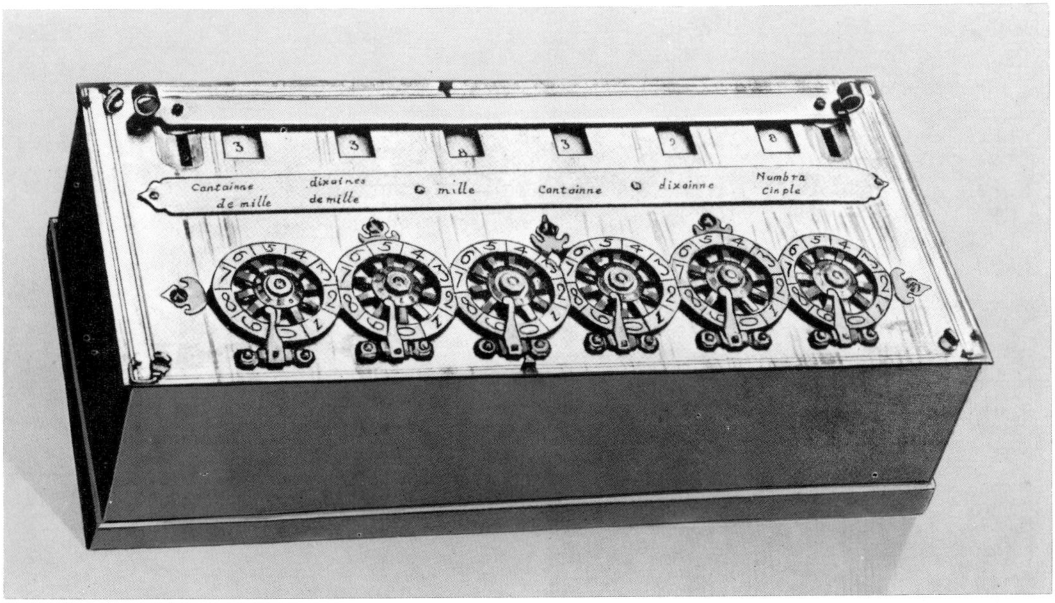

top left
Jakob Bernoulli (1654–1705)

above
Johann Bernoulli (1667–1748)

36 Famous mathematicians of the 17th/18th century III

The Swiss family of scholars produced within three generations eight eminent mathematicians. The three most important are Jakob, his brother Johann, and Daniel, the son of the latter

Daniel Bernoulli (1700–1782)

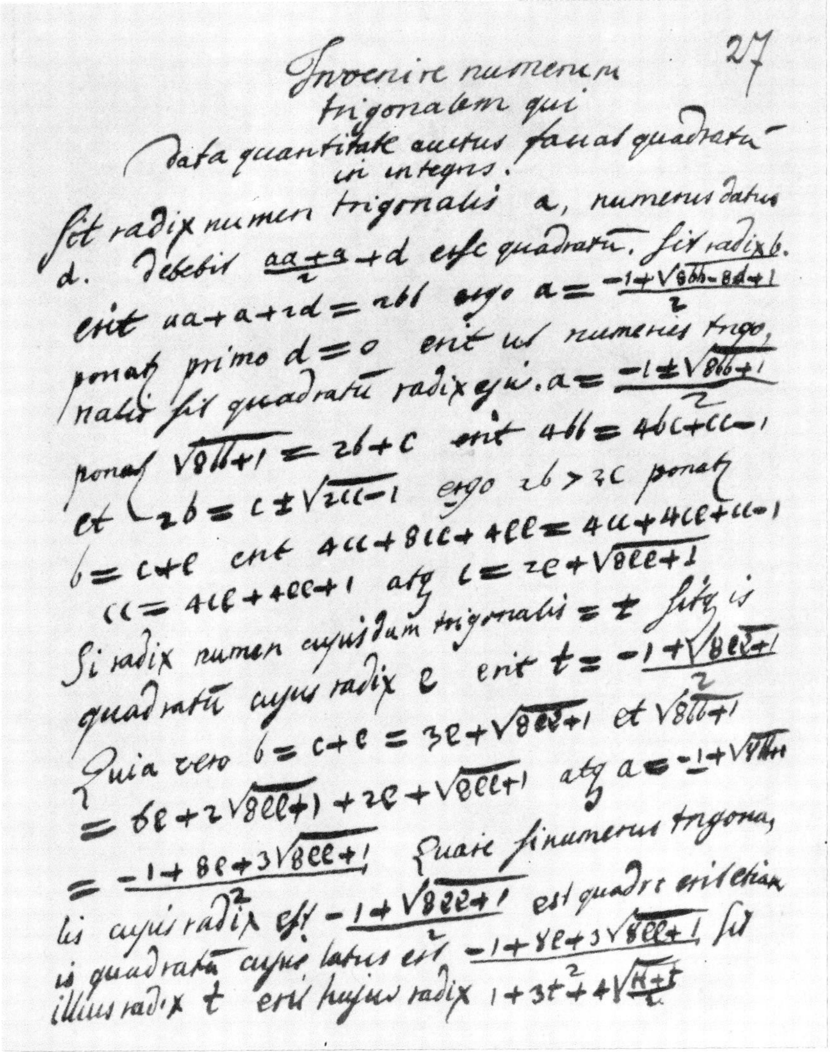

27

Invenire numerum trigonalem qui data quantitate auctus faciat quadratum in integris.

Sit radix numeri trigonalis a, numerus datus d. Debebit $\frac{aa+a}{2}+d$ esse quadratum. Sit radix b. erit $aa+a+2d=2bb$ ergo $a=\frac{-1+\sqrt{8bb-8d+1}}{2}$

ponatur primo $d=0$ erit ut numerus trigonalis sit quadratum radix ejus. $a=\frac{-1\pm\sqrt{8bb+1}}{2}$

ponatur $\sqrt{8bb+1}=2b+c$ erit $4bb=4bc+cc-1$ et $2b=c\pm\sqrt{2cc-1}$ ergo $2b>2c$ ponatur $b=c+e$ erit $4cc+8ce+4ee=4cc+4ce+cc-1$

$cc=4ce+4ee+1$ atque $c=2e+\sqrt{8ee+1}$

Si radix numeri cujusdam trigonalis $=t$ sitque is quadratum cujus radix e erit $t=\frac{-1+\sqrt{8ee+1}}{2}$

Quia vero $b=c+e=3e+\sqrt{8ee+1}$ et $\sqrt{8bb+1}$ $=6e+2\sqrt{8ee+1}+2e+\sqrt{8ee+1}$ atque $a=\frac{-1+\sqrt{8bb+1}}{2}$ $=\frac{-1+8e+3\sqrt{8ee+1}}{2}$ Quare si numerus trigonalis cujus radix est $\frac{-1+\sqrt{8ee+1}}{2}$ est quadratum erit etiam is quadratum cujus latus erit $\frac{-1+8e+3\sqrt{8ee+1}}{2}$ sit illius radix t erit hujus radix $1+3t+4\sqrt{\frac{tt+t}{2}}$

Page from a manuscript by Euler

37 Famous mathematicians of the 18th century I

Leonhard Euler (1707–1783)

1

2

3

4

5

6

7

38 Famous mathematicians of the 18th century II

1 Brook Taylor (1685–1731)
2 Moreau Maupertuis (1698–1759)
3 Johann Heinrich Lambert (1728–1777)
4 Joseph Louis Lagrange (1736–1813)
5 Gaspard Monge (1746–1818)
6 Adrien Marie Legendre (1752–1833)
7 Jean Baptiste Joseph de Fourier (1768–1830)

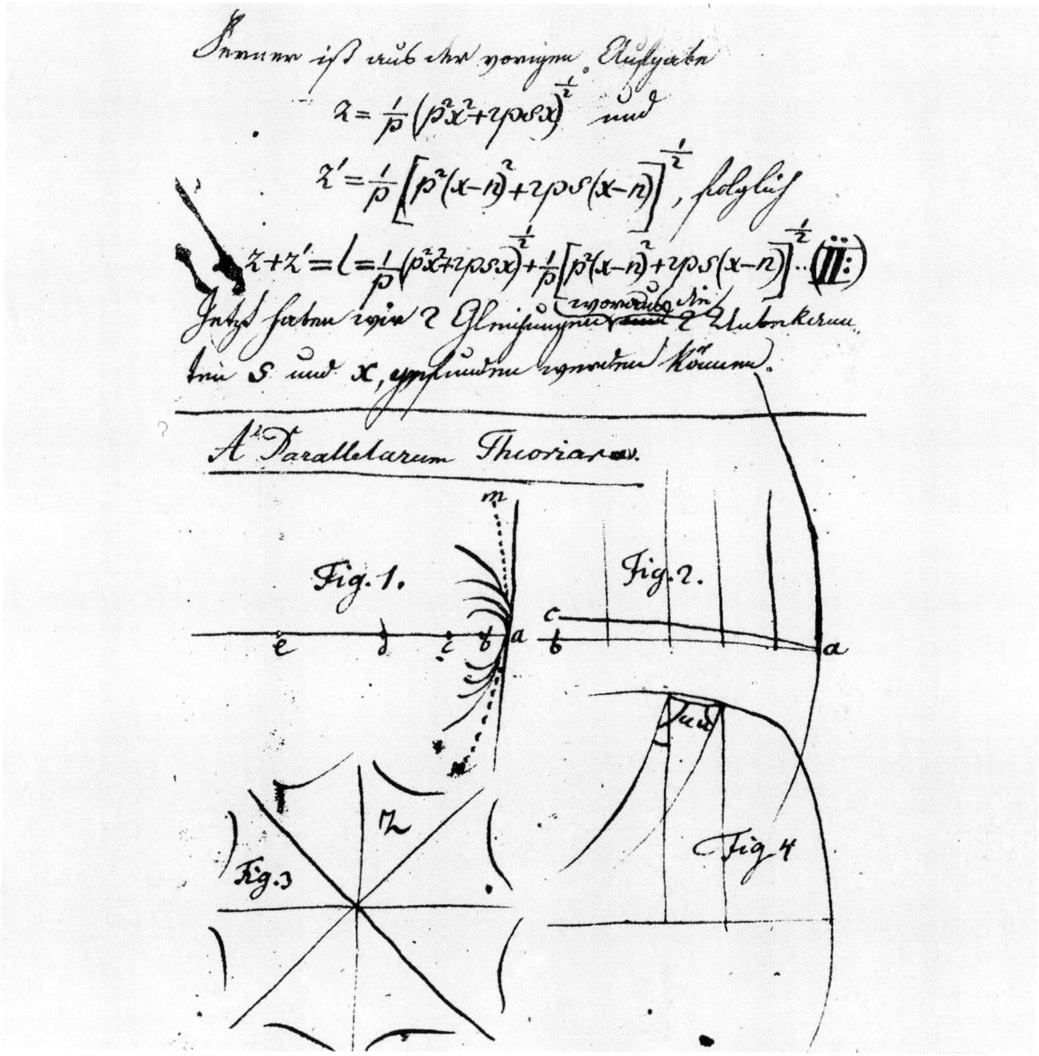

Drawing by
János Bólyai
on non-Euclidean
geometry (1820)

39 Famous mathematicians of the 19th century I

Nikolai Ivanovich Lobachevskii
(1792–1856)
He developed independently and almost simultanuously a non-Euclidean (hyperbolic) geometry

left Portrait of the young Gauss
right Gauss in his old age

Gauss's signature

40 Famous mathematicians of the 19th century II
One of the greatest mathematicians of all times was Carl Friedrich Gauss, born 30 April 1777 in Braunschweig, died 23 February 1855 in Göttingen

The University in Göttingen, where Gauss worked for nearly fifty years

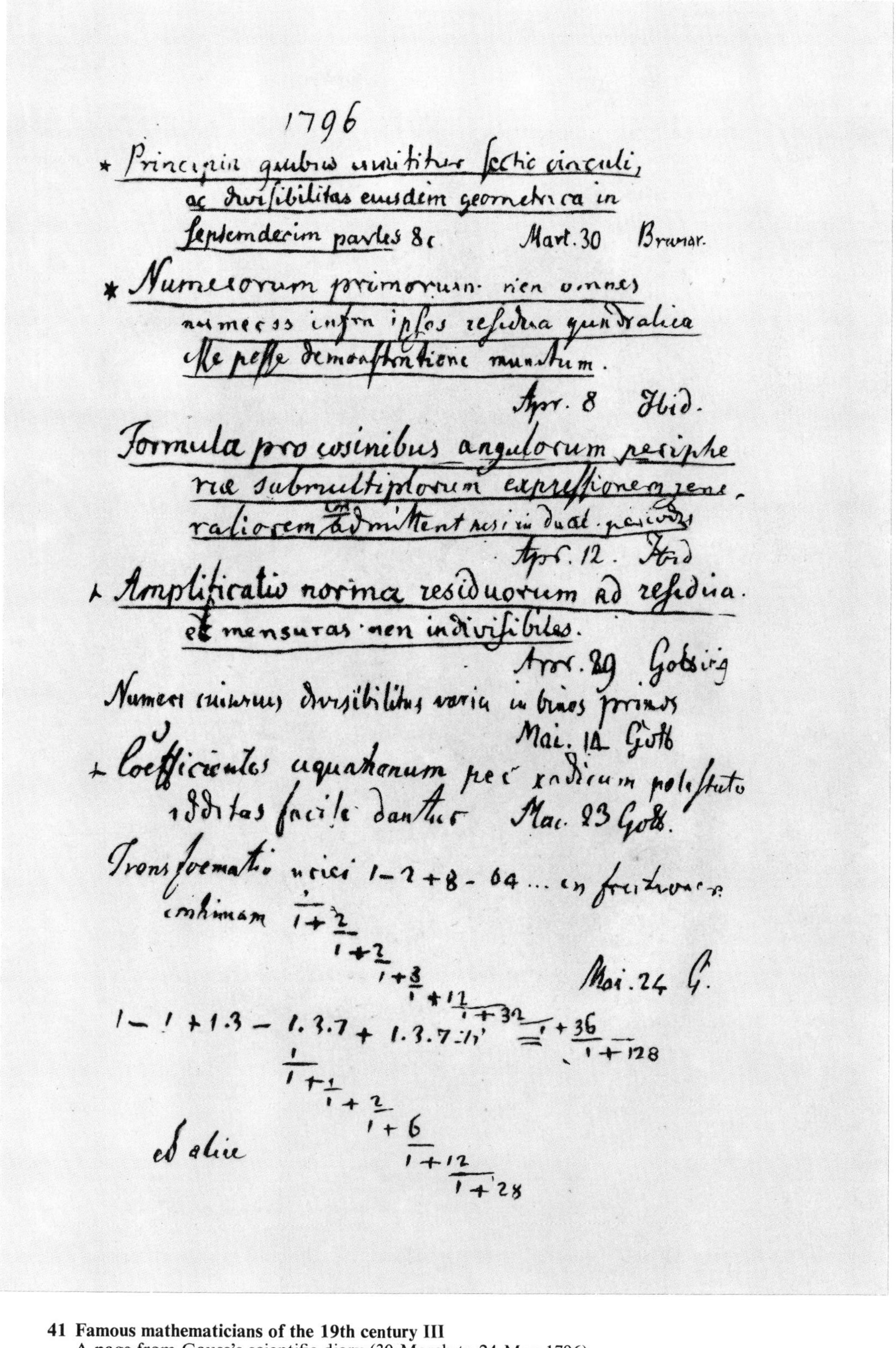

1796

* Principia quibus innititur sectio circuli, ac divisibilitas eiusdem geometrica in septemdecim partes &c. Mart. 30 Brunsv.

* Numerorum primorum non omnes numeros infra ipsos residua quadratica esse posse demonstratione munitum. Apr. 8 Ibid.

Formula pro cosinibus angulorum periphe rie submultiplorum expressionem gene raliorem non admittent nisi in duas periodos Apr. 12. Ibid

+ Amplificatio norma residuorum ad residua et mensuras non indivisibiles. Apr. 29 Gotting

Numeri cuiusvis divisibilitas varia in binos primos Mai. 14 Gott

+ Coefficientes aequationum per radicum potestatum summas facile dantur Mai. 23 Gott.

Transformatio seriei 1 − 2 + 8 − 64 ... in fractionem continuam

$$\cfrac{1}{1+\cfrac{2}{1+\cfrac{2}{1+\cfrac{8}{1+\cfrac{12}{1+32}}}}}$$

Mai. 24 G.

$$1 - 1 + 1.3 - 1.3.7 + 1.3.7.15 = \cfrac{1}{1+\cfrac{1}{1+\cfrac{2}{1+\cfrac{6}{1+\cfrac{12}{1+28}}}}} \qquad + \cfrac{36}{1+128}$$

et alia

41 Famous mathematicians of the 19th century III
A page from Gauss's scientific diary (30 March to 24 May 1796)

1

2

3

4

5

6

7

42 Famous mathematicians of the 19th century IV

1 Friedrich Wilhelm Bessel (1784–1846)
2 Augustin Louis Cauchy (1789–1857)
3 Jakob Steiner (1796–1863)
4 Niels Henrik Abel (1802–1829)
5 Peter Gustav Lejeune Dirichlet (1805–1859)
6 Évariste Galois (1811–1832)
7 Pafnuti Lvovich Chebyshev (1821–1894)

1

2

3

4

5

6

7

43 Famous mathematicians of the 19th century V

1 Carl Gustav Jacob Jacobi (1804–1851)
2 Bernhard Riemann (1826–1866)
3 Leopold Kronecker (1823–1891)
4 Karl Weierstraß (1815–1897)
5 Arthur Cayley (1821–1895)
6 Sophus Lie (1842–1899)
7 Sonya Kovalevski (1850–1891)

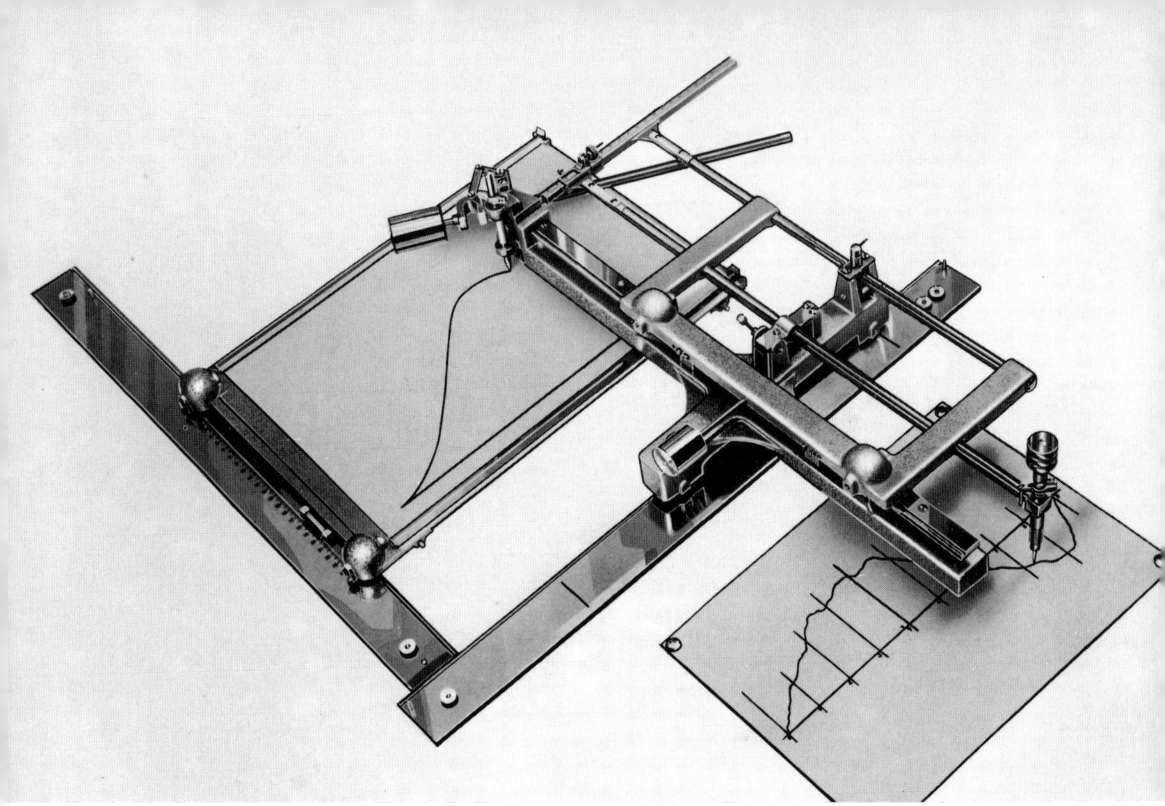

Instrument for drawing an integral curve of a given function or differential equation

44 Mathematical instruments I

Instrument to evaluate the integral of a function whose graph is given

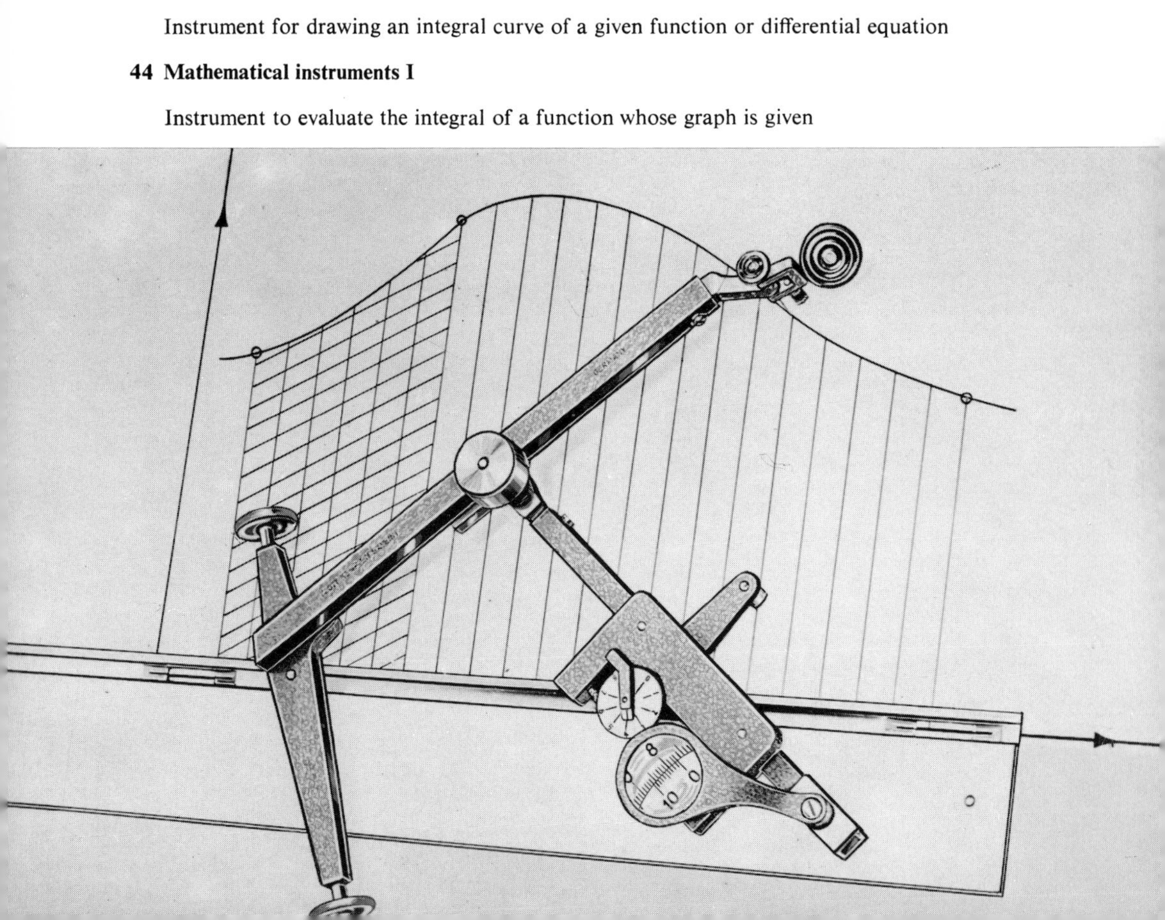

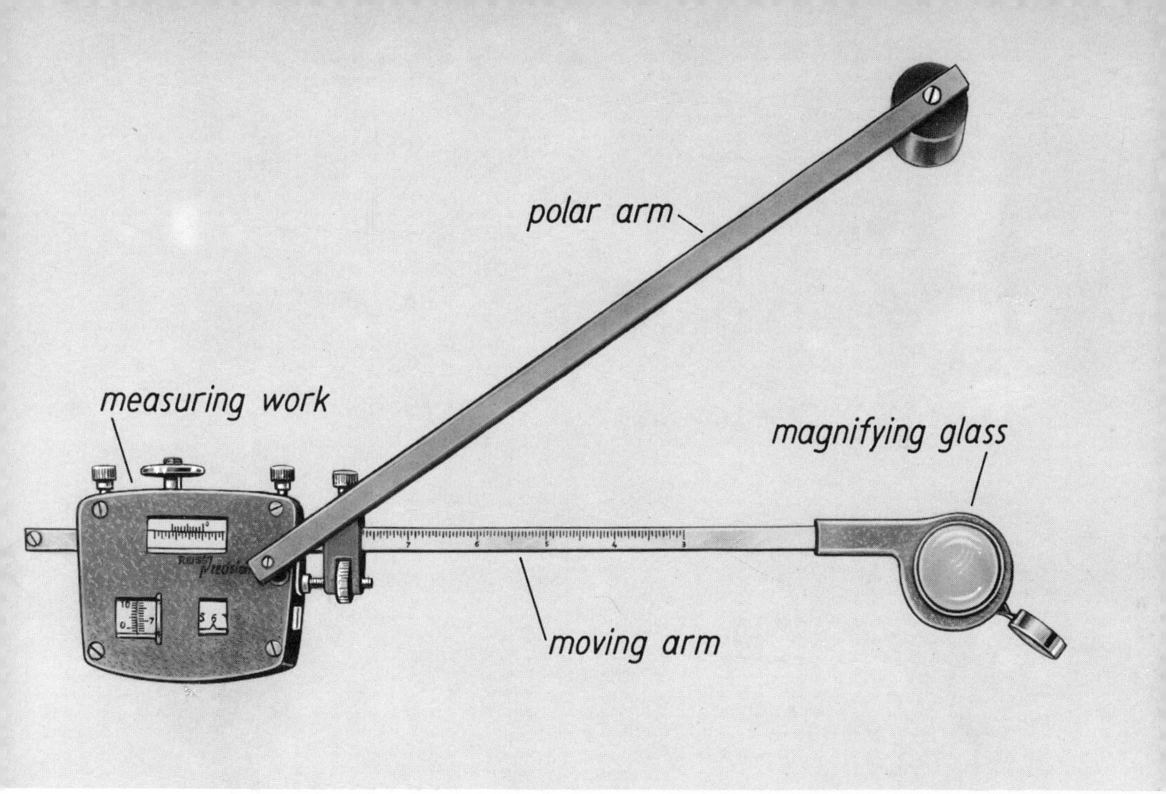

Compensating polar planimeter with polar am, for the measurement of areas in maps and plans; the moving point is equipped with a magnifying glass

45 Mathematical instruments II

Compensating polar planimeter with polar carriage for the evaluation of strip diagrams

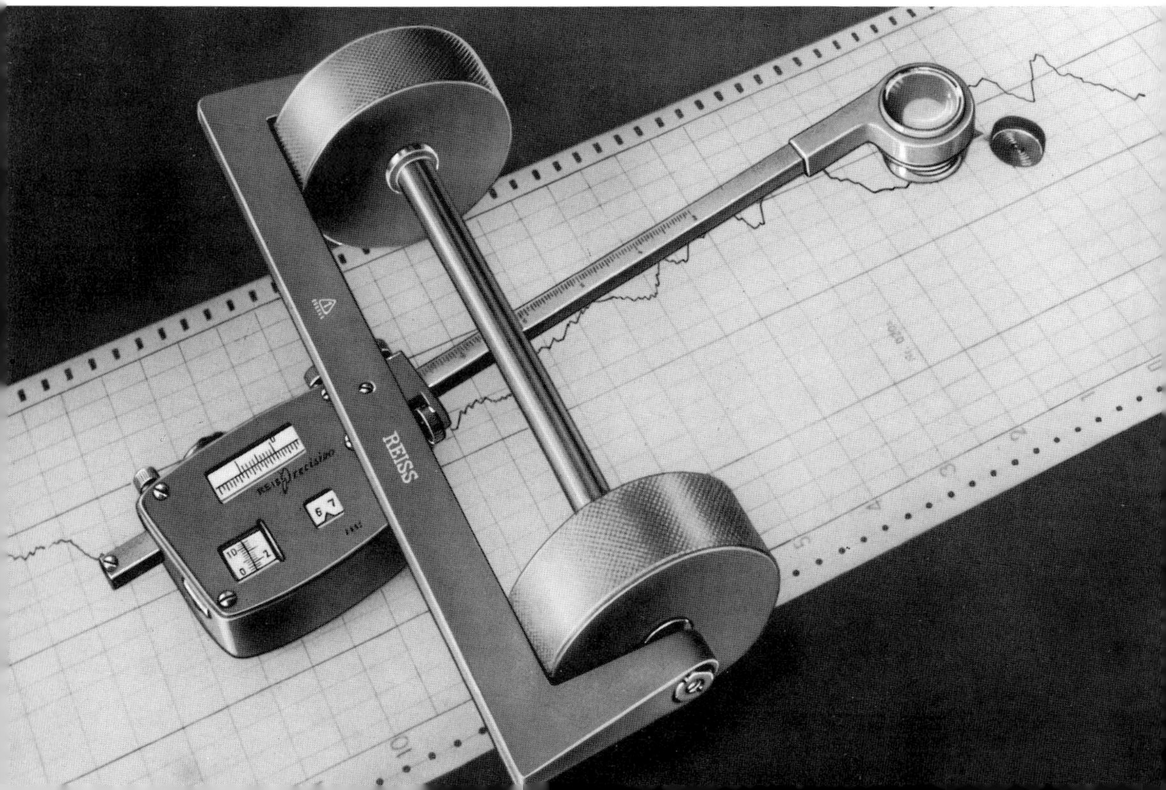

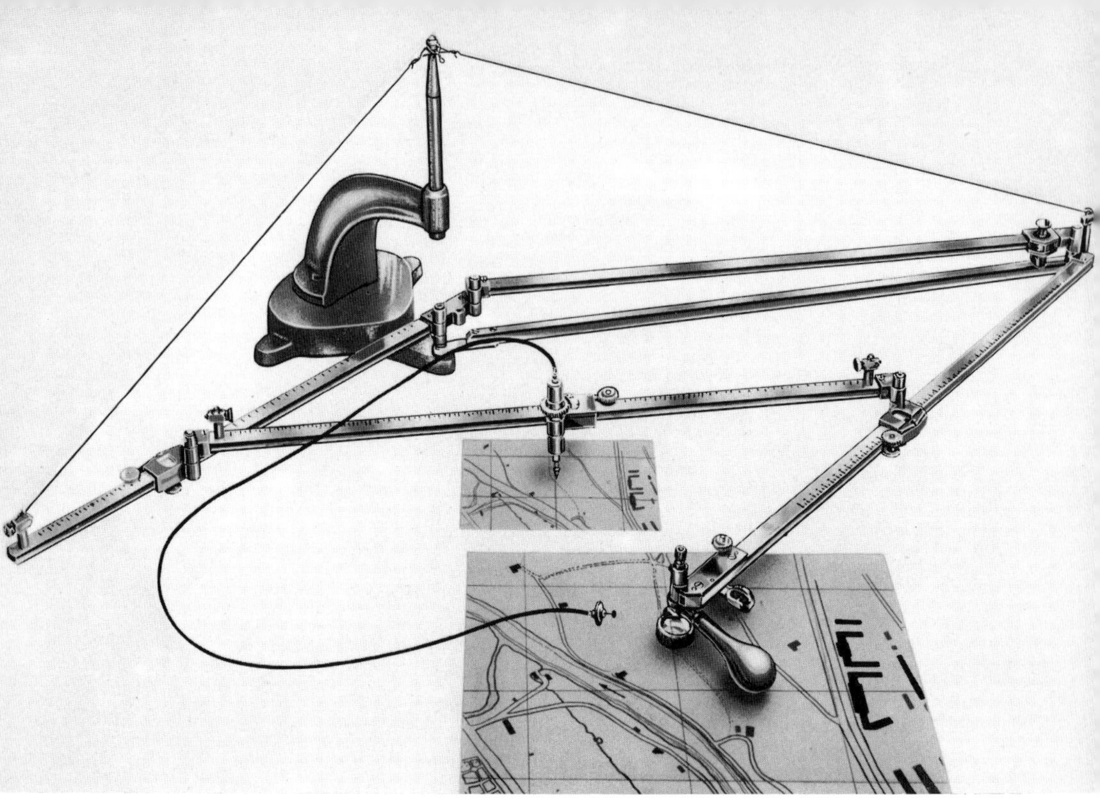

Precision pantograph, to reduce, magnify, or copy maps, plans etc.

46 Mathematical instruments III

Instrument for measurement of rectangular coordinates or the drawing of points with given coordinates

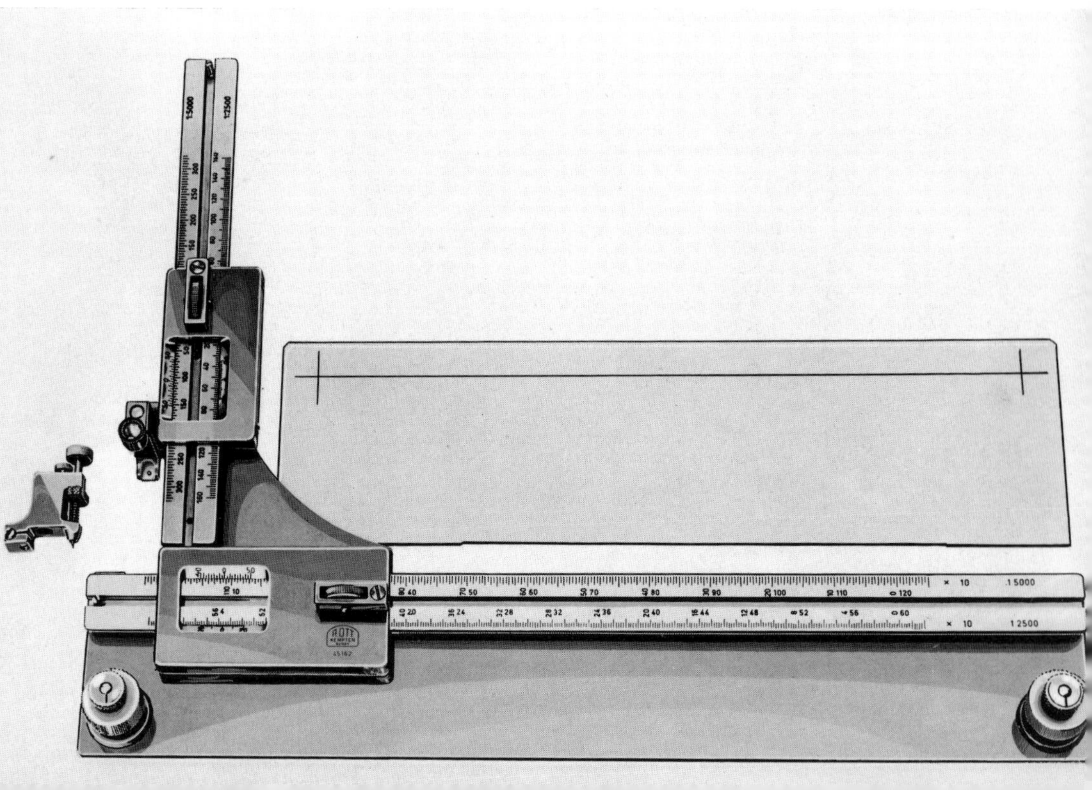

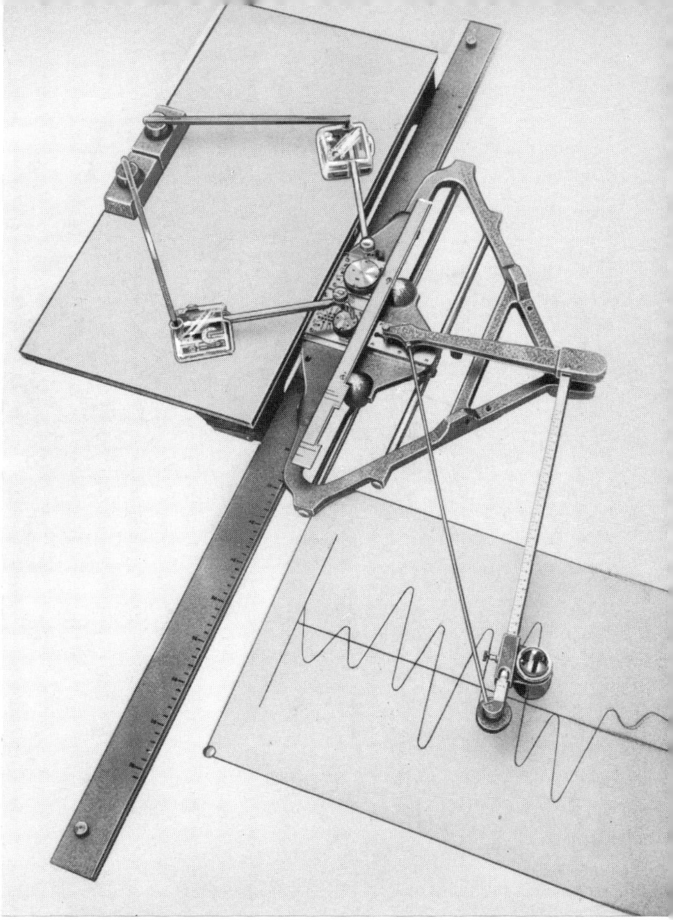

Harmonic analyser, to determine the Fourier expansion of a periodic function

Instrument to determine the tangent or normal to a curve whose graph is given

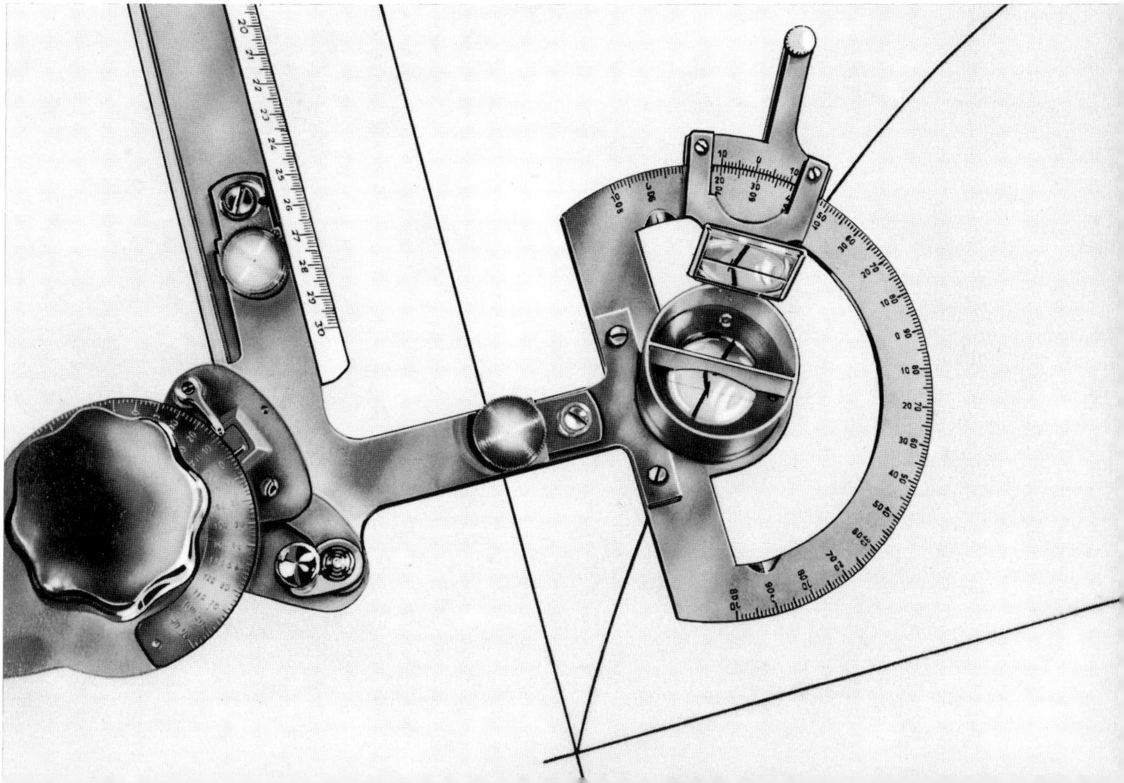

1

2

3

4

5

6

7

48 Famous mathematicians of the 19th/20th century I

1 George Stokes (1819—1903)
2 Richard Dedekind (1831—1916)
3 Georg Frobenius (1849—1917)
4 Georg Cantor (1845—1918)
5 Henri Poincaré (1854—1912)
6 Felix Klein (1849—1925)
7 Emmy Noether (1882—1935)

1

2

3

4

5

6

7

48 Famous mathematicians of the 19th/20th century II

1 David Hilbert (1862—1943)
2 Élie Cartan (1869—1951)
3 Henri Léon Lebesgue (1875—1941)
4 John von Neumann (1903—1957)
5 Hermann Weyl (1885—1955)
6 Jacques Salomon Hadamard (1865—1963)
7 Stefan Banach (1892—1945)

top and *top right*
Signal for the observation of trigonometric nets

50 Surveying

Trigonometric point (TP)

51 Mathematical education I
Practical work of school children

Work on a wall board

Determination of an angle with a hand-made apparatus

Giant slide rule for instructional purposes

Geometrical constructions on the blackboard

52 Mathematical education II

Computations on part of an exhaust system

Application of Pythagoras' theorem

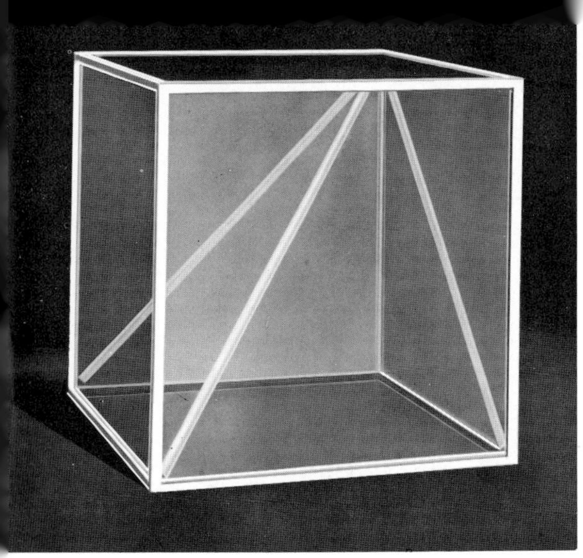

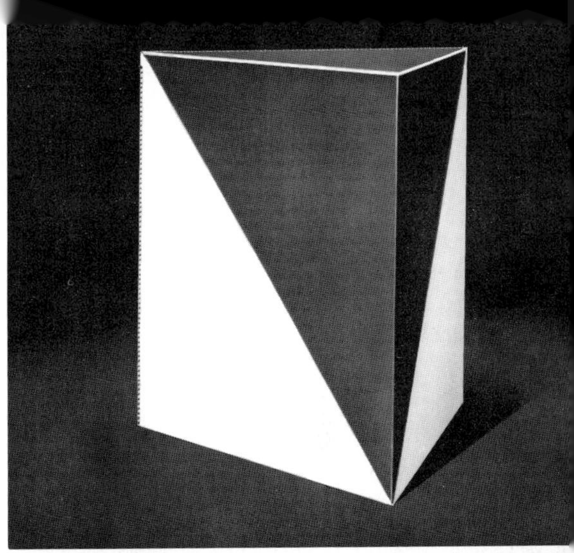

53 Mathematical education III
Models for pupils

1 Cube with surface and space diagonals
2 Prism decomposable into three pyramids of equal volume
3 Cylinder with sections
4 Sphere with plane sections
5 Sections of a right circular cone

(All models are made of plastic)

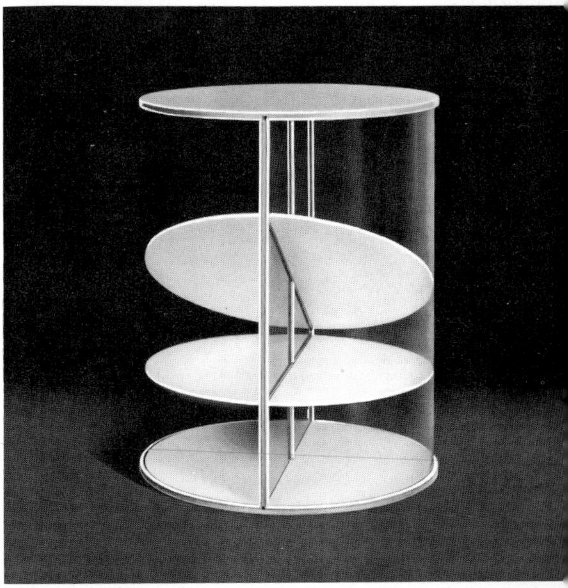

Negative and positive of a photograph

54 Mirror images

Reflection in water (gas holder)

Ship's Diesel engine in a left- and right-hand version

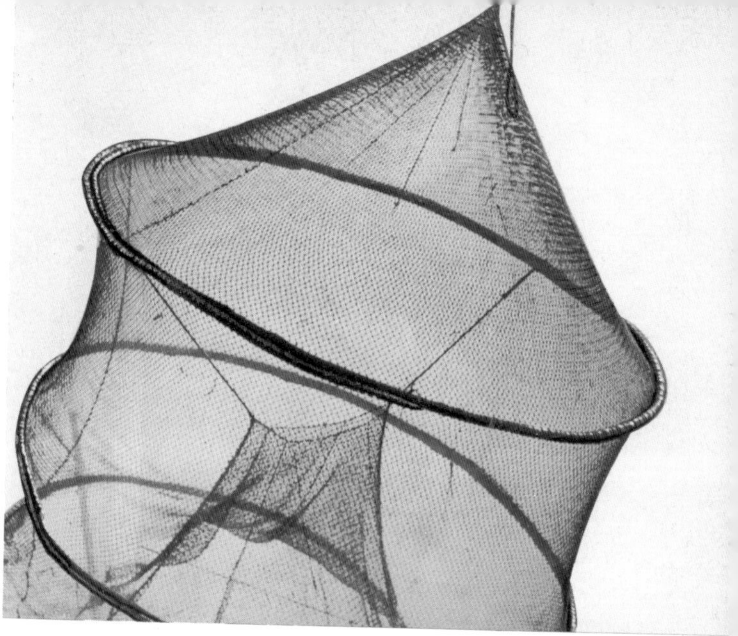

Formation of a minimal surface in a lobster pot

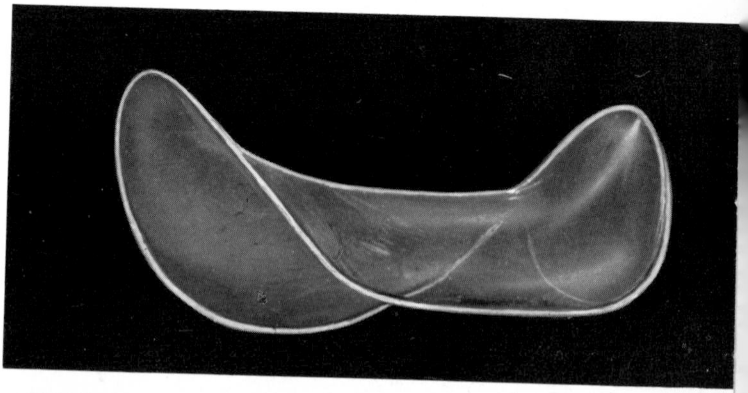

Formation of a minimal surface by a soap film

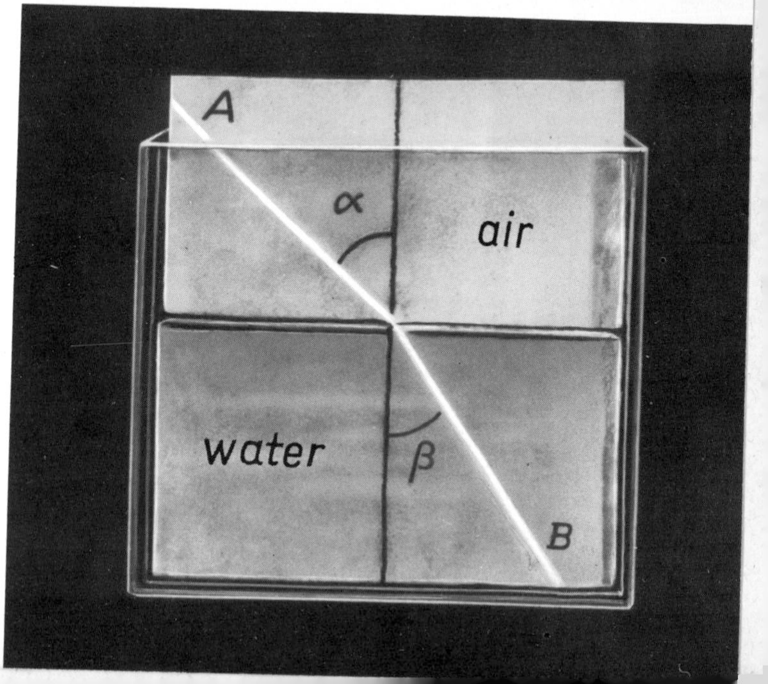

The path of the light ray from A to B is the solution of a minimal problem

Mathematical symbols

Symbol	Explanation
Geometry	
$\sim$	similar
$\cong$	congruent
$\triangle$	triangle; e.g. $\triangle\ ABC$
$\parallel$	parallel
$\nparallel$	not parallel
$\perp$	perpendicular, at right angles
$\sphericalangle$	angle; e.g. $\sphericalangle\ ABC$
$^\circ$	degree; e.g. 90° is a right angle
$'$	minute, $60' = 1^\circ$
$''$	second, $60'' = 1'$
gon	Gon; e.g. 100 gon is a right angle
$\overset{\frown}{AB}$;	arc AB
$\hat{\alpha}$	arc α
AB	segment AB
$\overrightarrow{AB}$	directed segment, from A to B
sin	sine
cos	cosine
tan	tangent
cot	cotangent
arcsin = $\sin^{-1}$ etc.	inverse sine etc.
sinh etc.	hyperbolic sine etc.
Arithmetic, algebra	
$=$	equal
$\equiv$	identical
$\triangleq$	corresponds to; e.g. $100^g \triangleq 90^\circ$
$\neq$	unequal
$<$	smaller than; e.g. $a < b$
$>$	greater than; e.g. $b > a$
$\leqslant$	smaller than or equal to; e.g. $a \leqslant 0$, not positive
$\geqslant$	greater than or equal to; e.g. $a \geqslant 0$, not negative
$+$	plus (sign, operational symbol)
$-$	minus (sign, operational symbol)
$\cdot$, $\times$	times; e.g. $3 \cdot 4$, 3×4
:, /, –	divided by; e.g. 3 : 4, 2/3, $\frac{3}{5}$
$a \mid b$	a divides b, e.g. $3 \mid 12$
$\sum_{i=1}^{n} a_i$	sum; e.g. $\sum_{i=1}^{3} a_i = a_1 + a_2 + a_3$
$\prod_{i=1}^{n} a_i$	product; e.g. $\prod_{i=1}^{3} a_i = a_1 \cdot a_2 \cdot a_3$
a^n	a to the nth; e.g. $a^3 = a \cdot a \cdot a$
$\sqrt{a}$	square root of a
$\sqrt[n]{a}$	nth root of a
$n!$	n factorial; e.g. $3! = 1 \cdot 2 \cdot 3$
$\binom{n}{k}$	n over k; e.g. $\binom{6}{3} = \frac{6 \cdot 5 \cdot 4}{1 \cdot 2 \cdot 3}$
$\lvert a \rvert$	modulus or absolute value of a; e.g. $\lvert -7 \rvert = 7$
$\log_b$	logarithm to the basis b

Symbol	Explanation
lg	decadic or common logarithm, basis 10
ln	natural logarithm, basis e
e	2.718 ...
i	imaginary unit, $i^2 = -1$
$\boldsymbol{a} \cdot \boldsymbol{b}$, $(\boldsymbol{a}, \boldsymbol{b})$	scalar product of two vectors
$\boldsymbol{a} \times \boldsymbol{b}$, $[\boldsymbol{a}, \boldsymbol{b}]$	vector product of two vectors
$(a_{ik}) = \boldsymbol{A}$	matrix with the elements a_{ik}
$\lvert a_{ik} \rvert = \det \boldsymbol{A}$	determinant of a square matrix $\boldsymbol{A}$
$\equiv \pmod{m}$	congruent mod m, e.g. $12 \equiv 7 \pmod{5}$
Analysis	
(a, b)	open interval $a < x < b$
$[a, b]$	closed interval $a \leqslant x \leqslant b$
∞	infinity
π	pi (= 3.14159...)
$\rightarrow$	tends to, converges to
lim	limit
$\approx$	approximately equal
d	symbol of differentiation
$\frac{dy}{dx}$, y'	dy by dx; differential quotient, derivative
$\frac{d^n y}{dx^n}$, $y^{(n)}$	nth derivative
∂	symbol of partial differentiation
∇	nabla operator
$\triangle$	Laplace operator
δf	delta f, variation of f
$\int f(x)\,dx$	indefinite integral
$\int_a^b f(x)\,dx$	definite integral
Set theory	
$\in$	element of, member of; e.g. $a \in \{a, b\}$
$\notin$	not element of; e.g. $c \notin \{a, b\}$
$\subseteq$	subset of, contained in
$\subset$	proper subset, e.g. $\{a\} \subset \{a, b\}$
$\cup$	union, e.g. $\{a\} \cup \{b\} = \{a, b\}$
$\cap$	intersection, e.g. $\{a, b\} \cap \{b, c\} = \{b\}$
$\emptyset$	empty set
Logic	
$\sim$	non (negation)
$\wedge$	and (conjunction)
$\vee$	or (disjunction)
$\rightarrow$	if—then (implication)
$\leftrightarrow$	if and only if (equivalence)
$\exists$	there exists (existential quantifier)
$\forall$	for all (universal quantifier)